Bird's Comprehensive Engineering Mathematics

Why is knowledge of mathematics important in engineering?

A career in any engineering or scientific field will require both basic and advanced mathematics. Without mathematics to determine principles, calculate dimensions and limits, explore variations, prove concepts, and so on, there would be no mobile telephones, televisions, stereo systems, video games, microwave ovens, computers, or virtually anything electronic. There would be no bridges, tunnels, roads, skyscrapers, automobiles, ships, planes, rockets or most things mechanical. There would be no metals beyond the common ones, such as iron and copper, no plastics, no synthetics. In fact, society would most certainly be less advanced without the use of mathematics throughout the centuries and into the future.

Electrical engineers require mathematics to design, develop, test or supervise the manufacturing and installation of electrical equipment, components, or systems for commercial, industrial, military or scientific use.

Mechanical engineers require mathematics to perform engineering duties in planning and designing tools, engines, machines and other mechanically functioning equipment; they oversee installation, operation, maintenance and repair of such equipment as centralised heat, gas, water and steam systems.

Aerospace engineers require mathematics to perform a variety of engineering work in designing, constructing and testing aircraft, missiles and spacecraft; they conduct basic and applied research to evaluate adaptability of materials and equipment to aircraft design and manufacture and recommend improvements in testing equipment and techniques.

Nuclear engineers require mathematics to conduct research on nuclear engineering problems or apply principles and theory of nuclear science to problems concerned with release, control and utilisation of nuclear energy and nuclear waste disposal.

Petroleum engineers require mathematics to devise methods to improve oil and gas well production and determine the need for new or modified tool designs; they oversee drilling and offer technical advice to achieve economical and satisfactory progress.

Industrial engineers require mathematics to design, develop, test and evaluate integrated systems for managing industrial production processes, including human work factors, quality control, inventory control, logistics and material flow, cost analysis and production coordination.

Environmental engineers require mathematics to design, plan or perform engineering duties in the prevention, control and remediation of environmental health hazards, using various engineering disciplines; their work may include waste treatment, site remediation or pollution control technology.

Civil engineers require mathematics in all levels in civil engineering – structural engineering, hydraulics and geotechnical engineering are all fields that employ mathematical tools such as differential equations, tensor analysis, field theory, numerical methods and operations research.

Knowledge of mathematics is therefore needed by each of the engineering disciplines listed above.

It is intended that this text – *Bird's Comprehensive Engineering Mathematics* – will provide a step-by-step approach to learning all the fundamental mathematics needed for your engineering studies.

To Sue

Bird's Comprehensive Engineering Mathematics

Second edition

John Bird, BSc (Hons), CEng, CMath, CSci, FIMA, FIET, FCollT

Routledge
Taylor & Francis Group

LONDON AND NEW YORK

Second edition published 2018
by Routledge
2 Park Square, Milton Park, Abingdon, Oxon OX14 4RN

and by Routledge
711 Third Avenue, New York, NY 10017

Routledge is an imprint of the Taylor & Francis Group, an informa business

© 2018 John Bird

First edition published as *Understanding Engineering Mathematics* by Routledge 2014

British Library Cataloguing-in-Publication Data
A catalogue record for this book is available from the British Library

Library of Congress Cataloging-in-Publication Data
Names: Bird, J. O., author.
Title: Bird's comprehensive engineering mathematics / John Bird.
Other titles: Comprehensive engineering mathematics
Description: Second edition. | New York, NY : Routledge, 2018. | Includes bibliographical references and index.
Identifiers: LCCN 2017052872 (print) | LCCN 2017055894 (ebook) | ISBN 9781351232869 (Adobe PDF) | ISBN 9781351232852 (ePub3) | ISBN 9781351232845 (Mobipocket) | ISBN 9780815378143 (pbk.) | ISBN 9780815378150 (hardback) | ISBN 9781351232876 (ebook)
Subjects: LCSH: Engineering mathematics.
Classification: LCC TA330 (ebook) | LCC TA330 .B514 2018 (print) | DDC 510.24/62–dc23
LC record available at https://lccn.loc.gov/2017052872

ISBN: 978-0-8153-7815-0 (hbk)
ISBN: 978-0-8153-7814-3 (pbk)
ISBN: 978-1-351-23287-6 (ebk)

Typeset in Times
by Servis Filmsetting Ltd, Stockport, Cheshire

Visit the companion website: www.routledge.com/cw/bird

Contents

Section K   Differential equations 827

Section M Laplace transforms   1041

Section N Fourier series   1081

Preface

Studying engineering, whether it is mechanical, electrical, aeronautical, communications, civil, construction or systems engineering, relies heavily on an understanding of mathematics. In fact, it is not possible to study any engineering discipline without a sound knowledge of mathematics. What happens then when a student realises he/she is very weak at mathematics – an increasingly common scenario? The answer may hopefully be found in this textbook *Bird's Comprehensive Engineering Mathematics*, which explains as simply as possible the steps needed to become better at mathematics and hence gain real confidence and understanding in their chosen engineering subject.

Bird's Comprehensive Engineering Mathematics is an amalgam of three books – *Basic Engineering Mathematics, Engineering Mathematics* and *Higher Engineering Mathematics*, all currently published by Routledge. The point about *Bird's Comprehensive Engineering Mathematics* is that it is all-encompassing. We do not have to think 'what course does this book apply to?'. The answer is that it encompasses all courses that include some engineering content in their syllabus, from beginning courses up to degree level.

The primary aim of the material in this text is to provide the fundamental analytical and underpinning knowledge and techniques needed to successfully complete scientific and engineering principles modules covering a wide range of programmes. The material has been designed to enable students to use techniques learned for the analysis, modelling and solution of realistic engineering problems. It also aims to provide some of the more advanced knowledge required for those wishing to pursue careers in a range of engineering disciplines. In addition, the text will be suitable as a valuable reference aid to practising engineers.

In *Bird's Comprehensive Engineering Mathematics*, theory is introduced in each chapter by a full outline of essential definitions, formulae, laws, procedures, etc. The theory is explained simply, but is kept to a minimum, for problem solving is extensively used to establish and exemplify the theory. It is intended that readers will gain real understanding through seeing problems solved and then through solving similar problems themselves.

The material has been ordered into the following **fifteen convenient categories**: number and algebra, further number and algebra, areas and volumes, graphs, geometry and trigonometry, complex numbers, matrices and determinants, vector geometry, differential calculus, integral calculus, differential equations, statistics and probability, Laplace transforms, Fourier series and Z-transforms. Each topic considered in the text is presented in a way that assumes in the reader very little previous knowledge.

With a plethora of engineering courses worldwide it is not possible to have a definitive ordering of material; it is assumed that both students and instructors/lecturers alike will 'dip in' to the text according to the needs of their particular course structure.

The text contains some **1600 worked problems**, **3600 further problems** (with answers), arranged within **384 Practice Exercises**, **255 multiple choice questions arranged into 9 tests**, **35 Revision Tests**, **768 line diagrams** and **15 lists of formulae/revision hints**.

Worked solutions to all 3600 further problems have been prepared and can be **accessed free from the website www.routledge.com/cw/bird** (see page xviii).

At intervals throughout the text are some **35 Revision Tests** to check understanding. For example, Revision Test 1 covers the material in Chapters 1 and 2, Revision Test 2 covers the material in Chapters 3 to 5, Revision Test 3 covers the material in Chapters 6 to 8, and so on. Full solutions to all 400 questions in the Revision Tests are available for lecturers/instructors on the website.

On the **front cover of the paperback version of this text is HMS *Queen Elizabeth* aircraft carrier** sailing for the first time into its home base of Portsmouth on 16 August 2017 following its construction at Rosyth Naval Base in Scotland. The 65,000 tonne carrier is 280 m long and 70 m wide, and has a top speed of 25 knots. When operational, it will carry 36 F-35B fighter jets, 4 Merlin helicopters, with up to 1600 crew.

The *Queen Elizabeth* was constructed at six UK sites and then fitted together with great precision at Rosyth; the planning, design, construction and operation required a considerable understanding of modern engineering and technology.

It is difficult to grasp engineering concepts without a great deal of knowledge of mathematics – and this text will aid potential engineers with much of the essential mathematical principles needed.

'Learning by example' is at the heart of *Bird's Comprehensive Engineering Mathematics*.

JOHN BIRD
Royal Naval Defence College of Marine and Air Engineering, HMS Sultan, formerly University of Portsmouth and Highbury College, Portsmouth

John Bird is the former Head of Applied Electronics in the Faculty of Technology at Highbury College, Portsmouth, UK. More recently, he has combined freelance lecturing at the University of Portsmouth with Examiner responsibilities for Advanced Mathematics with City and Guilds, and examining for International Baccalaureate. He has some 45 years' experience of successfully teaching, lecturing, instructing, educating and planning study programmes for trainee engineers. He is the author of over 130 textbooks on engineering and mathematical subjects with worldwide sales of over one million copies. He is a chartered engineer, a chartered mathematician, a chartered scientist and a Fellow of three professional institutions, and is currently lecturing at the Defence College of Marine and Air Engineering in the Defence College of Technical Training at HMS Sultan, Gosport, Hampshire, UK.

Free Web downloads at www.routledge.com/cw/bird
For students

1. **Full solutions** to the 3600 questions contained in the 384 Practice Exercises

2. Download **Multiple choice questions**

3. **Lists of Essential Formulae**

4. **Famous Engineers/Scientists** – From time to time in the text, 42 famous mathematicians/engineers are referred to with their image and emphasised with an asterisk*. Background information on each of these is available via the website. Mathematicians/Engineers involved are: **Argand, Bessel, Boole, Boyle, Cauchy, Celsius, Charles, Cramer, de Moivre, de Morgan, Descartes, Euler, Faraday, Fourier, Frobenius, Gauss, Henry, Hertz, Hooke, Karnaugh, Kirchhoff, Kutta, Laplace, Legendre, Leibniz, L'Hopital, Maclaurin, Morland, Napier, Newton, Ohm, Pappus, Pascal, Poisson, Pythagoras, Raphson, Rodrigues, Runge, Simpson, Taylor, Wallis and Young**.

For instructors/lecturers

1. **Full solutions** to the 3600 questions contained in the 384 Practice Exercises

2. Full solutions and marking scheme to each of the **35 Revision Tests**

3. **Revision Tests** – available to run off to be given to students

4. Download **Multiple choice questions**

5. **Lists of Essential Formulae**

6. **Illustrations** – all 768 available on PowerPoint

7. **Famous Engineers/Scientists** – 42 are mentioned in the text, as listed above

Section A

Number and algebra

Chapter 1

Basic arithmetic

Why it is important to understand: **Basic arithmetic**

Being numerate, i.e. having an ability to add, subtract, multiply and divide whole numbers with some confidence, goes a long way towards helping you become competent at mathematics. Of course electronic calculators are a marvellous aid to the quite complicated calculations often required in engineering; however, having a feel for numbers 'in our head' can be invaluable when estimating. Do not spend too much time on this chapter because we deal with the calculator later; however, try to have some idea how to do quick calculations in the absence of a calculator. You will feel more confident in dealing with numbers and calculations if you can do this.

At the end of this chapter, you should be able to:

- understand positive and negative integers
- add and subtract whole numbers
- multiply and divide two integers
- multiply numbers up to 12×12 by rote
- determine the highest common factor from a set of numbers
- determine the lowest common multiple from a set of numbers
- appreciate the order of precedence when evaluating expressions
- understand the use of brackets in expressions
- evaluate expressions containing $+$, $-$, $\times$, $\div$ and brackets

1.1 Introduction

Whole Numbers
Whole Numbers are simply the numbers **0, 1, 2, 3, 4, 5, ...**

Counting Numbers
Counting Numbers are whole numbers, but **without the zero**, i.e. **1, 2, 3, 4, 5, ...**

Natural Numbers
Natural Numbers can mean either counting numbers or whole numbers.

Integers
Integers are like whole numbers, but they **also include negative numbers**.

Examples of integers include $\ldots -5, -4, -3, -2, -1, 0, 1, 2, 3, 4, 5, \ldots$

Arithmetic operators
The four basic arithmetic operators are add ($+$), subtract ($-$), multiply ($\times$) and divide ($\div$).
It is assumed that adding, subtracting, multiplying and dividing reasonably small numbers can be achieved without a calculator. However, if revision of this area is needed then some worked problems are included in the following sections.
When **unlike signs** occur together in a calculation, the overall sign is **negative**. For example,

$$5 + (-2) = 5 + -2 = 5 - 2 = \mathbf{3}$$
$$3 + (-4) = 3 + -4 = 3 - 4 = \mathbf{-1}$$

and

$$(+5) \times (-2) = -10$$

Like signs together give an overall **positive sign**. For example,

$$3 - (-4) = 3 - -4 = 3 + 4 = 7$$

and

$$(-6) \times (-4) = +24$$

Prime Numbers

A prime number can be divided, without a remainder, only by itself and by 1. For example, 17 can be divided only by 17 and by 1. Other examples of prime numbers are $2, 3, 5, 7, 11, 13, 19$ and 23.

1.2 Revision of addition and subtraction

You can probably already add two or more numbers together and subtract one number from another. However, if you need a revision then the following worked problems should be helpful.

Problem 1.   Determine $735 + 167$

```
  H T U
  7 3 5
+ 1 6 7
  ─────
  9 0 2
  ─────
  1 1
```

(i) $5 + 7 = 12$. Place 2 in units (U) column. Carry 1 in the tens (T) column.

(ii) $3 + 6 + 1$ (carried) $= 10$. Place the 0 in the tens column. Carry the 1 in the hundreds (H) column.

(iii) $7 + 1 + 1$ (carried) $= 9$. Place the 9 in the hundreds column.

Hence, $735 + 167 = 902$

Problem 2.   Determine $632 - 369$

```
  H T U
  6 3 2
- 3 6 9
  ─────
  2 6 3
  ─────
```

(i) $2 - 9$ is not possible; therefore change one ten into ten units (leaving 2 in the tens column). In the units column, this gives us $12 - 9 = 3$

(ii)   Place 3 in the units column.

(iii) $2 - 6$ is not possible; therefore change one hundred into ten tens (leaving 5 in the hundreds column). In the tens column, this gives us $12 - 6 = 6$

(iv)   Place the 6 in the tens column.

(v) $5 - 3 = 2$

(vi)   Place the 2 in the hundreds column.

Hence, $632 - 369 = 263$

Problem 3.   Add $27, -74, 81$ and -19

This problem is written as $27 - 74 + 81 - 19$

Adding the positive integers:	27
	81
Sum of positive integers is	108
Adding the negative integers:	74
	19
Sum of negative integers is	93

$108 + -93 = 108 - 93$ and taking the sum of the negative integers from the sum of the positive integers gives

	108
	−93
	15

Thus, $27 - 74 + 81 - 19 = 15$

Problem 4.   Subtract -74 from 377

This problem is written as $377 - -74$. Like signs together give an overall positive sign, hence

$$377 - -74 = 377 + 74$$

```
  3 7 7
+   7 4
  ─────
  4 5 1
```

Thus, $377 - -74 = 451$

Problem 5.   Subtract 243 from 126

The problem is $126 - 243$. When the second number is larger than the first, take the smaller number from the larger and make the result negative. Thus,

$$126 - 243 = -(243 - 126) \qquad \begin{array}{r} 2\,4\,3 \\ -\ 1\,2\,6 \\ \hline 1\,1\,7 \end{array}$$

Thus, **$126 - 243 = -117$**

Problem 6.   Subtract 318 from -269

The problem is $-269 - 318$. The sum of the negative integers is

$$\begin{array}{r} 2\,6\,9 \\ +\,3\,1\,8 \\ \hline 5\,8\,7 \end{array}$$

Thus, **$-269 - 318 = -587$**

Now try the following Practice Exercise

Practice Exercise 1   Further problems on addition and subtraction (answers on page 1148)

In Problems 1 to 15, determine the values of the expressions given, without using a calculator.

1. $67\,\text{kg} - 82\,\text{kg} + 34\,\text{kg}$

2. $73\,\text{m} - 57\,\text{m}$

3. $851\,\text{mm} - 372\,\text{mm}$

4. $124 - 273 + 481 - 398$

5. £927 − £114 + £182 − £183 − £247

6. $647 - 872$

7. $2417 - 487 + 2424 - 1778 - 4712$

8. $-38419 - 2177 + 2440 - 799 + 2834$

9. £2715 − £18250 + £11471 − £1509 + £113274

10. $47 + (-74) - (-23)$

11. $813 - (-674)$

12. $3151 - (-2763)$

13. $4872\,\text{g} - 4683\,\text{g}$

14. $-23148 - 47724$

15. $\$53774 - \$38441$

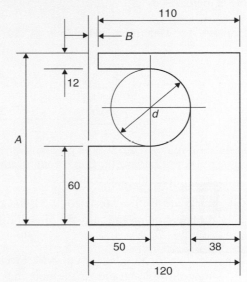

Figure 1.1

16. Figure 1.1 shows the dimensions of a template in millimetres. Calculate the diameter d and dimensions A and B for the template.

1.3 Revision of multiplication and division

You can probably already multiply two numbers together and divide one number by another. However, if you need a revision then the following worked problems should be helpful.

Problem 7.   Determine 86×7

$$\begin{array}{r} \text{H T U} \\ 8\ 6 \\ \times \quad 7 \\ \hline 6\ 0\ 2 \\ \hline 4 \end{array}$$

(i) $7 \times 6 = 42$. Place the 2 in the units (U) column and 'carry' the 4 into the tens (T) column.

(ii) $7 \times 8 = 56$; $56 + 4$ (carried) $= 60$. Place the 0 in the tens column and the 6 in the hundreds (H) column.

Hence, **86 × 7 = 602**

A good grasp of **multiplication tables** is needed when multiplying such numbers; a reminder of the multiplication table up to 12×12 is shown below. Confidence with handling numbers will be greatly improved if this table is memorised.

> **Problem 8.** Determine 764×38

$$\begin{array}{r} 764 \\ \times\ \ 38 \\ \hline 6112 \\ 22920 \\ \hline 29032 \end{array}$$

(i) $8 \times 4 = 32$. Place the 2 in the units column and carry 3 into the tens column.

(ii) $8 \times 6 = 48$; $48 + 3$ (carried) $= 51$. Place the 1 in the tens column and carry the 5 into the hundreds column.

(iii) $8 \times 7 = 56$; $56 + 5$ (carried) $= 61$. Place 1 in the hundreds column and 6 in the thousands column.

(iv)   Place 0 in the units column under the 2. This is done because we are multiplying by tens.

(v) $3 \times 4 = 12$. Place the 2 in the tens column and carry 1 into the hundreds column.

(vi) $3 \times 6 = 18$; $18 + 1$ (carried) $= 19$. Place the 9 in the hundreds column and carry the 1 into the thousands column.

(vii) $3 \times 7 = 21$; $21 + 1$ (carried) $= 22$. Place 2 in the thousands column and 2 in the ten thousands column.

(viii) $6112 + 22920 = 29032$

Hence, **764 × 38 = 29032**

Again, knowing multiplication tables is rather important when multiplying such numbers.

It is appreciated, of course, that such a multiplication can, and probably will, be performed using a **calculator**. However, there are times when a calculator may not be available and it is then useful to be able to calculate the 'long way'.

> **Problem 9.** Multiply 178 by -46

When the numbers have different signs, the result will be negative. (With this in mind, the problem can now be solved by multiplying 178 by 46.) Following the procedure of Problem 8 gives

$$\begin{array}{r} 178 \\ \times\ \ 46 \\ \hline 1068 \\ 7120 \\ \hline 8188 \end{array}$$

Thus, $178 \times 46 = 8188$ and **178 × (−46) = −8188**

Multiplication table

×	2	3	4	5	6	7	8	9	10	11	12
2	4	6	8	10	12	14	16	18	20	22	24
3	6	9	12	15	18	21	24	27	30	33	36
4	8	12	16	20	24	28	32	36	40	44	48
5	10	15	20	25	30	35	40	45	50	55	60
6	12	18	24	30	36	42	48	54	60	66	72
7	14	21	28	35	42	49	56	63	70	77	84
8	16	24	32	40	48	56	64	72	80	88	96
9	18	27	36	45	54	63	72	81	90	99	108
10	20	30	40	50	60	70	80	90	100	110	120
11	22	33	44	55	66	77	88	99	110	121	132
12	24	36	48	60	72	84	96	108	120	132	144

Problem 10. Determine $1834 \div 7$

$$\frac{262}{7 \overline{)1834}}$$

(i) 7 into 18 goes 2, remainder 4. Place the 2 above the 8 of 1834 and carry the 4 remainder to the next digit on the right, making it 43

(ii) 7 into 43 goes 6, remainder 1. Place the 6 above the 3 of 1834 and carry the 1 remainder to the next digit on the right, making it 14

(iii) 7 into 14 goes 2, remainder 0. Place 2 above the 4 of 1834

Hence, $1834 \div 7 = 1834/7 = \dfrac{\mathbf{1834}}{\mathbf{7}} = \mathbf{262}$

The method shown is called **short division**.

Problem 11. Determine $5796 \div 12$

$$
\begin{array}{r}
483 \\
12\overline{)5796} \\
48 \\
\hline
99 \\
-96 \\
\hline
36 \\
-36 \\
\hline
00
\end{array}
$$

(i) 12 into 5 won't go. 12 into 57 goes 4; place 4 above the 7 of 5796

(ii) $4 \times 12 = 48$; place the 48 below the 57 of 5796

(iii) $57 - 48 = 9$

(iv) Bring down the 9 of 5796 to give 99

(v) 12 into 99 goes 8; place 8 above the 9 of 5796

(vi) $8 \times 12 = 96$; place 96 below the 99

(vii) $99 - 96 = 3$

(viii) Bring down the 6 of 5796 to give 36

(ix) 12 into 36 goes 3 exactly

(x) Place the 3 above the final 6

(xi) $3 \times 12 = 36$; Place the 36 below the 36

(xii) $36 - 36 = 0$

Hence, $5796 \div 12 = 5796/12 = \dfrac{\mathbf{5796}}{\mathbf{12}} = \mathbf{483}$

The method shown is called **long division**.

Now try the following Practice Exercise

Practice Exercise 2   Further problems on multiplication and division (answers on page 1148)

Determine the values of the expressions given in Problems 1 to 9, without using a calculator.

1. (a) 78×6 (b) 124×7

2. (a) £261×7 (b) £462×9

3. (a) $783\,\text{kg} \times 11$ (b) $73\,\text{kg} \times 8$

4. (a) $27\,\text{mm} \times 13$ (b) $77\,\text{mm} \times 12$

5. (a) 448×23 (b) $143 \times (-31)$

6. (a) $288\,\text{m} \div 6$ (b) $979\,\text{m} \div 11$

7. (a) $\dfrac{1813}{7}$ (b) $\dfrac{896}{16}$

8. (a) $\dfrac{21424}{13}$ (b) $15900 \div -15$

9. (a) $\dfrac{88737}{11}$ (b) $46858 \div 14$

10. A screw has a mass of 15 grams. Calculate, in kilograms, the mass of 1200 such screws ($1\,\text{kg} = 1000\,\text{g}$).

11. Holes are drilled 35.7 mm apart in a metal plate. If a row of 26 holes is drilled, determine the distance, in centimetres, between the centres of the first and last holes.

12. A builder needs to clear a site of bricks and top soil. The total weight to be removed is 696 tonnes. Trucks can carry a maximum load of 24 tonnes. Determine the number of truck loads needed to clear the site.

1.4 Highest common factors and lowest common multiples

When two or more numbers are multiplied together, the individual numbers are called **factors**. Thus, a factor is a number which divides into another number exactly. The **highest common factor (HCF)** is the largest number which divides into two or more numbers exactly.

For example, consider the numbers 12 and 15

The factors of 12 are 1, 2, 3, 4, 6 and 12 (i.e. all the numbers that divide into 12)

The factors of 15 are 1, 3, 5 and 15 (i.e. all the numbers that divide into 15)

1 and 3 are the only **common factors**; i.e. numbers which are factors of **both** 12 and 15.

Hence, **the HCF of 12 and 15 is 3** since 3 is the highest number which divides into **both** 12 and 15.

A **multiple** is a number which contains another number an exact number of times. The smallest number which is exactly divisible by each of two or more numbers is called the **lowest common multiple (LCM)**.

For example, the multiples of 12 are 12, 24, 36, 48, 60, 72, … and the multiples of 15 are 15, 30, 45, 60, 75, …

60 is a common multiple (i.e. a multiple of **both** 12 and 15) and there are no lower common multiples.

Hence, **the LCM of 12 and 15 is 60** since 60 is the lowest number that both 12 and 15 divide into.

Here are some further problems involving the determination of HCFs and LCMs.

Problem 12. Determine the HCF of the numbers 12, 30 and 42

Probably the simplest way of determining an HCF is to express each number in terms of its lowest factors. This is achieved by repeatedly dividing by the prime numbers 2, 3, 5, 7, 11, 13, … (where possible) in turn. Thus,

$$12 = \boxed{2} \times 2 \times \boxed{3}$$
$$30 = \boxed{2} \quad\quad \times \boxed{3} \times 5$$
$$42 = \boxed{2} \quad\quad \times \boxed{3} \times 7$$

The factors which are common to each of the numbers are 2 in column 1 and 3 in column 3, shown by the broken lines. Hence, **the HCF is 2 × 3**; i.e. **6**. That is, 6 is the largest number which will divide into 12, 30 and 42.

Problem 13. Determine the HCF of the numbers 30, 105, 210 and 1155

Using the method shown in Problem 12:

$$30 = 2 \times \boxed{3} \times \boxed{5}$$
$$105 = \quad\quad \boxed{3} \times \boxed{5} \times 7$$
$$210 = 2 \times \boxed{3} \times \boxed{5} \times 7$$
$$1155 = \quad\quad \boxed{3} \times \boxed{5} \times 7 \times 11$$

The factors which are common to each of the numbers are 3 in column 2 and 5 in column 3. Hence, **the HCF is 3 × 5 = 15**

Problem 14. Determine the LCM of the numbers 12, 42 and 90

The LCM is obtained by finding the lowest factors of each of the numbers, as shown in Problems 12 and 13 above, and then selecting the largest group of any of the factors present. Thus,

$$12 = \boxed{2 \times 2} \times 3$$
$$42 = 2 \quad\quad \times 3 \quad\quad\quad \times \boxed{7}$$
$$90 = 2 \quad\quad \times \boxed{3 \times 3} \times \boxed{5}$$

The largest groups of any of the factors present are shown by the broken lines and are 2 × 2 in 12, 3 × 3 in 90, 5 in 90 and 7 in 42

Hence, **the LCM is 2 × 2 × 3 × 3 × 5 × 7 = 1260** and is the smallest number which 12, 42 and 90 will all divide into exactly.

Problem 15. Determine the LCM of the numbers 150, 210, 735 and 1365

Using the method shown in Problem 14 above:

$$150 = \boxed{2} \times \boxed{3} \times \boxed{5 \times 5}$$
$$210 = 2 \times 3 \times 5 \quad\quad \times 7$$
$$735 = \quad\quad 3 \times 5 \quad\quad \times \boxed{7 \times 7}$$
$$1365 = \quad\quad 3 \times 5 \quad\quad \times 7 \quad\quad \times \boxed{13}$$

Hence, **the LCM is** $\mathbf{2 \times 3 \times 5 \times 5 \times 7 \times 7 \times 13}$
$$\mathbf{= 95550}$$

Now try the following Practice Exercise

Practice Exercise 3   Further problems on highest common factors and lowest common multiples (answers on page 1148)

Find (a) the HCF and (b) the LCM of the following groups of numbers.

1. 8, 12   2. 60, 72

3. 50, 70   4. 270, 900

5. 6, 10, 14   6. 12, 30, 45

7. 10, 15, 70, 105   8. 90, 105, 300

9. 196, 210, 462, 910   10. 196, 350, 770

1.5   Order of operation and brackets

Order of operation

Sometimes addition, subtraction, multiplication, division, powers and brackets may all be involved in a calculation. For example,

$$5 - 3 \times 4 + 24 \div (3 + 5) - 3^2$$

This is an extreme example but will demonstrate the order that is necessary when evaluating.

When we read, we read from left to right. However, with mathematics there is a definite order of precedence which we need to adhere to. The order is as follows:

Brackets
Order (or p**O**wer)
Division
Multiplication
Addition
Subtraction

Notice that the first letters of each word spell **BODMAS**, a handy aide-memoire. **O**rder means p**O**wer. For example, $4^2 = 4 \times 4 = 16$.
$5 - 3 \times 4 + 24 \div (3 + 5) - 3^2$ is   evaluated   as follows:

$5 - 3 \times 4 + 24 \div (3 + 5) - 3^2$

$= 5 - 3 \times 4 + 24 \div 8 - 3^2$ (**B**racket is removed and $3 + 5$ replaced with 8)

$= 5 - 3 \times 4 + 24 \div 8 - 9$ (**O**rder means p**O**wer; in this case, $3^2 = 3 \times 3 = 9$)

$= 5 - 3 \times 4 + 3 - 9$ (**D**ivision: $24 \div 8 = 3$)

$= 5 - 12 + 3 - 9$ (**M**ultiplication: $-3 \times 4 = -12$)

$= 8 - 12 - 9$ (**A**ddition: $5 + 3 = 8$)

$= -13$ (**S**ubtraction: $8 - 12 - 9 = -13$)

In practice, **it does not matter if multiplication is performed before division or if subtraction is performed before addition**. What is important is that **the process of multiplication and division must be completed before addition and subtraction**.

Brackets and operators

The basic laws governing the **use of brackets and operators** are shown by the following examples.

(a) $2 + 3 = 3 + 2$; i.e. the order of numbers when adding does not matter.

(b) $2 \times 3 = 3 \times 2$; i.e. the order of numbers when multiplying does not matter.

(c) $2 + (3 + 4) = (2 + 3) + 4$; i.e. the use of brackets when adding does not affect the result.

(d) $2 \times (3 \times 4) = (2 \times 3) \times 4$; i.e. the use of brackets when multiplying does not affect the result.

(e) $2 \times (3 + 4) = 2(3 + 4) = 2 \times 3 + 2 \times 4$;   i.e.   a number placed outside of a bracket indicates that the whole contents of the bracket must be multiplied by that number.

(f) $(2 + 3)(4 + 5) = (5)(9) = 5 \times 9 = 45$; i.e. adjacent brackets indicate multiplication.

(g) $2[3 + (4 \times 5)] = 2[3 + 20] = 2 \times 23 = 46$;   i.e. when an expression contains inner and outer brackets, **the inner brackets are removed first**.

Here are some further problems in which BODMAS needs to be used.

Problem 16.   Find the value of $6 + 4 \div (5 - 3)$

The order of precedence of operations is remembered by the word BODMAS. Thus,

$$6 + 4 \div (5 - 3) = 6 + 4 \div 2 \qquad \text{(\textbf{Brackets})}$$
$$= 6 + 2 \qquad \text{(\textbf{Division})}$$
$$= 8 \qquad \text{(\textbf{Addition})}$$

Problem 17.   Determine the value of
$$13 - 2 \times 3 + 14 \div (2 + 5)$$

$13 - 2 \times 3 + 14 \div (2 + 5) = 13 - 2 \times 3 + 14 \div 7$ (**B**)
$= 13 - 2 \times 3 + 2$ (**D**)
$= 13 - 6 + 2$ (**M**)
$= 15 - 6$ (**A**)
$= 9$ (**S**)

Problem 18.   Evaluate
$$16 \div (2 + 6) + 18[3 + (4 \times 6) - 21]$$

$16 \div (2+6) + 18[3 + (4 \times 6) - 21]$

$\quad = 16 \div (2+6) + 18[3 + 24 - 21]$ (B: inner bracket is determined first)

$\quad = 16 \div 8 + 18 \times 6$ (B)

$\quad = 2 + 18 \times 6$ (D)

$\quad = 2 + 108$ (M)

$\quad = \mathbf{110}$ (A)

Note that a number outside of a bracket multiplies all that is inside the brackets. In this case,

$18[3 + 24 - 21] = 18[6]$, which means $18 \times 6 = 108$

Problem 19. Find the value of

$$23 - 4(2 \times 7) + \frac{(144 \div 4)}{(14 - 8)}$$

$23 - 4(2 \times 7) + \dfrac{(144 \div 4)}{(14 - 8)} = 23 - 4 \times 14 + \dfrac{36}{6}$ (B)

$\qquad\qquad\qquad\qquad = 23 - 56 + \dfrac{36}{6}$ (M)

$\qquad\qquad\qquad\qquad = 23 - 56 + 6$ (D)

$\qquad\qquad\qquad\qquad = 29 - 56$ (A)

$\qquad\qquad\qquad\qquad = \mathbf{-27}$ (S)

Problem 20. Evaluate

$$\frac{3 + \sqrt{(5^2 - 3^2)} + 2^3}{1 + (4 \times 6) \div (3 \times 4)} + \frac{15 \div 3 + 2 \times 7 - 1}{3 \times \sqrt{4} + 8 - 3^2 + 1}$$

$\dfrac{3 + \sqrt{(5^2 - 3^2)} + 2^3}{1 + (4 \times 6) \div (3 \times 4)} + \dfrac{15 \div 3 + 2 \times 7 - 1}{3 \times \sqrt{4} + 8 - 3^2 + 1}$

$\quad = \dfrac{3 + 4 + 8}{1 + 24 \div 12} + \dfrac{15 \div 3 + 2 \times 7 - 1}{3 \times 2 + 8 - 9 + 1}$

$\quad = \dfrac{3 + 4 + 8}{1 + 2} + \dfrac{5 + 2 \times 7 - 1}{3 \times 2 + 8 - 9 + 1}$

$\quad = \dfrac{15}{3} + \dfrac{5 + 14 - 1}{6 + 8 - 9 + 1}$

$\quad = 5 + \dfrac{18}{6}$

$\quad = 5 + 3 = \mathbf{8}$

Now try the following Practice Exercise

Practice Exercise 4   Further problems on order of precedence and brackets (answers on page 1148)

Evaluate the following expressions.

1. $14 + 3 \times 15$

2. $17 - 12 \div 4$

3. $86 + 24 \div (14 - 2)$

4. $7(23 - 18) \div (12 - 5)$

5. $63 - 8(14 \div 2) + 26$

6. $\dfrac{40}{5} - 42 \div 6 + (3 \times 7)$

7. $\dfrac{(50 - 14)}{3} + 7(16 - 7) - 7$

8. $\dfrac{(7 - 3)(1 - 6)}{4(11 - 6) \div (3 - 8)}$

9. $\dfrac{(3 + 9 \times 6) \div 3 - 2 \div 2}{3 \times 6 + (4 - 9) - 3^2 + 5}$

10. $\dfrac{(4 \times 3^2 + 24) \div 5 + 9 \times 3}{2 \times 3^2 - 15 \div 3} +$

$\qquad\qquad \dfrac{2 + 27 \div 3 + 12 \div 2 - 3^2}{5 + (13 - 2 \times 5) - 4}$

11. $\dfrac{1 + \sqrt{25} + 3 \times 2 - 8 \div 2}{3 \times 4 - \sqrt{(3^2 + 4^2)} + 1} -$

$\qquad\qquad \dfrac{(4 \times 2 + 7 \times 2) \div 11}{\sqrt{9} + 12 \div 2 - 2^3}$

For fully worked solutions to each of the problems in Practice Exercises 1 to 4 in this chapter, go to the website:
www.routledge.com/cw/bird

Chapter 2

Fractions

Why it is important to understand: Fractions

Engineers use fractions all the time. Examples including stress to strain ratios in mechanical engineering, chemical concentration ratios and reaction rates, and ratios in electrical equations to solve for current and voltage. Fractions are also used everywhere in science, from radioactive decay rates to statistical analysis. Calculators are able to handle calculations with fractions. However, there will be times when a quick calculation involving addition, subtraction, multiplication and division of fractions is needed. Again, do not spend too much time on this chapter because we deal with the calculator later; however, try to have some idea how to do quick calculations in the absence of a calculator. You will feel more confident to deal with fractions and calculations if you can do this.

At the end of this chapter, you should be able to:

- understand the terminology numerator, denominator, proper and improper fractions and mixed numbers
- add and subtract fractions
- multiply and divide two fractions
- appreciate the order of precedence when evaluating expressions involving fractions

2.1 Introduction

A mark of 9 out of 14 in an examination may be written as $\dfrac{9}{14}$ or 9/14. $\dfrac{9}{14}$ is an example of a fraction. The number above the line, i.e. 9, is called the **numerator**. The number below the line, i.e. 14, is called the **denominator**.

When the value of the numerator is less than the value of the denominator, the fraction is called a **proper fraction**. $\dfrac{9}{14}$ is an example of a proper fraction.

When the value of the numerator is greater than the value of the denominator, the fraction is called an **improper fraction**. $\dfrac{5}{2}$ is an example of an improper fraction.

A **mixed number** is a combination of a whole number and a fraction. $2\dfrac{1}{2}$ is an example of a mixed number. In fact, $\dfrac{5}{2} = 2\dfrac{1}{2}$

There are a number of everyday examples in which fractions are readily referred to. For example, three people equally sharing a bar of chocolate would have $\dfrac{1}{3}$ each. A supermarket advertises $\dfrac{1}{5}$ off a six-pack of beer; if the beer normally costs £2 then it will now cost £1.60. $\dfrac{3}{4}$ of the employees of a company are women; if the company has 48 employees, then 36 are women.

Calculators are able to handle calculations with fractions. However, to understand a little more about fractions we will in this chapter show how to add, subtract, multiply and divide with fractions without the use of a calculator.

Problem 1. Change the following improper fractions into mixed numbers:

$$\text{(a) } \frac{9}{2} \quad \text{(b) } \frac{13}{4} \quad \text{(c) } \frac{28}{5}$$

(a) $\frac{9}{2}$ means 9 halves and $\frac{9}{2} = 9 \div 2$, and $9 \div 2 = 4$ and 1 half, i.e.

$$\frac{9}{2} = 4\frac{1}{2}$$

(b) $\frac{13}{4}$ means 13 quarters and $\frac{13}{4} = 13 \div 4$, and $13 \div 4 = 3$ and 1 quarter, i.e.

$$\frac{13}{4} = 3\frac{1}{4}$$

(c) $\frac{28}{5}$ means 28 fifths and $\frac{28}{5} = 28 \div 5$, and $28 \div 5 = 5$ and 3 fifths, i.e.

$$\frac{28}{5} = 5\frac{3}{5}$$

Problem 2. Change the following mixed numbers into improper fractions:

$$\text{(a) } 5\frac{3}{4} \quad \text{(b) } 1\frac{7}{9} \quad \text{(c) } 2\frac{3}{7}$$

(a) $5\frac{3}{4}$ means $5 + \frac{3}{4}$. 5 contains $5 \times 4 = 20$ quarters. Thus, $5\frac{3}{4}$ contains $20 + 3 = 23$ quarters, i.e.

$$5\frac{3}{4} = \frac{23}{4}$$

The quick way to change $5\frac{3}{4}$ into an improper fraction is $\frac{4 \times 5 + 3}{4} = \frac{23}{4}$

(b) $1\frac{7}{9} = \frac{9 \times 1 + 7}{9} = \frac{16}{9}$

(c) $2\frac{3}{7} = \frac{7 \times 2 + 3}{7} = \frac{17}{7}$

Problem 3. In a school there are 180 students of which 72 are girls. Express this as a fraction in its simplest form

---

The fraction of girls is $\frac{72}{180}$.

Dividing both the numerator and denominator by the lowest prime number, i.e. 2, gives

$$\frac{72}{180} = \frac{36}{90}$$

Dividing both the numerator and denominator again by 2 gives

$$\frac{72}{180} = \frac{36}{90} = \frac{18}{45}$$

2 will not divide into both 18 and 45, so dividing both the numerator and denominator by the next prime number, i.e. 3, gives

$$\frac{72}{180} = \frac{36}{90} = \frac{18}{45} = \frac{6}{15}$$

Dividing both the numerator and denominator again by 3 gives

$$\frac{72}{180} = \frac{36}{90} = \frac{18}{45} = \frac{6}{15} = \frac{2}{5}$$

So $\frac{72}{180} = \frac{2}{5}$ **in its simplest form.**

Thus, $\frac{2}{5}$ **of the students are girls.**

2.2 Adding and subtracting fractions

When the denominators of two (or more) fractions to be added are the same, the fractions can be added 'on sight'.

For example, $\frac{2}{9} + \frac{5}{9} = \frac{7}{9}$ and $\frac{3}{8} + \frac{1}{8} = \frac{4}{8}$

In the latter example, dividing both the 4 and the 8 by 4 gives $\frac{4}{8} = \frac{1}{2}$, which is the simplified answer. This is called **cancelling**.

Addition and subtraction of fractions is demonstrated in the following worked examples.

Problem 4. Simplify $\frac{1}{3} + \frac{1}{2}$

(i) Make the denominators the same for each fraction. The lowest number that both denominators divide into is called the **lowest common multiple** or **LCM** (see Chapter 1, page 8). In this example, the LCM of 3 and 2 is 6

(ii) 3 divides into 6 twice. Multiplying both numerator and denominator of $\frac{1}{3}$ by 2 gives

$$\frac{1}{3} = \frac{2}{6}$$

(iii) 2 divides into 6, 3 times. Multiplying both numerator and denominator of $\frac{1}{2}$ by 3 gives

$$\frac{1}{2} = \frac{3}{6}$$

(iv) Hence,

$$\frac{1}{3} + \frac{1}{2} = \frac{2}{6} + \frac{3}{6} = \frac{\mathbf{5}}{\mathbf{6}}$$

Problem 5. Simplify $\frac{3}{4} - \frac{7}{16}$

(i) Make the denominators the same for each fraction. The lowest common multiple (LCM) of 4 and 16 is 16.

(ii) 4 divides into 16, 4 times. Multiplying both numerator and denominator of $\frac{3}{4}$ by 4 gives

$$\frac{3}{4} = \frac{12}{16}$$

(iii) $\frac{7}{16}$ already has a denominator of 16.

(iv) Hence,

$$\frac{3}{4} - \frac{7}{16} = \frac{12}{16} - \frac{7}{16} = \frac{\mathbf{5}}{\mathbf{16}}$$

Problem 6. Simplify $4\frac{2}{3} - 1\frac{1}{6}$

$4\frac{2}{3} - 1\frac{1}{6}$ is the same as $\left(4\frac{2}{3}\right) - \left(1\frac{1}{6}\right)$ which is the same as $\left(4 + \frac{2}{3}\right) - \left(1 + \frac{1}{6}\right)$ which is the same as

$4 + \frac{2}{3} - 1 - \frac{1}{6}$ which is the same as $3 + \frac{2}{3} - \frac{1}{6}$ which is the same as $3 + \frac{4}{6} - \frac{1}{6} = 3 + \frac{3}{6} = 3 + \frac{1}{2}$

Thus, $4\frac{2}{3} - 1\frac{1}{6} = \mathbf{3\frac{1}{2}}$

Problem 7. Evaluate $7\frac{1}{8} - 5\frac{3}{7}$

$$7\frac{1}{8} - 5\frac{3}{7} = \left(7 + \frac{1}{8}\right) - \left(5 + \frac{3}{7}\right) = 7 + \frac{1}{8} - 5 - \frac{3}{7}$$

$$= 2 + \frac{1}{8} - \frac{3}{7} = 2 + \frac{7 \times 1 - 8 \times 3}{56}$$

$$= 2 + \frac{7 - 24}{56} = 2 + \frac{-17}{56} = 2 - \frac{17}{56}$$

$$= \frac{112}{56} - \frac{17}{56} = \frac{112 - 17}{56} = \frac{95}{56} = \mathbf{1\frac{39}{56}}$$

Problem 8. Determine the value of

$$4\frac{5}{8} - 3\frac{1}{4} + 1\frac{2}{5}$$

$$4\frac{5}{8} - 3\frac{1}{4} + 1\frac{2}{5} = (4 - 3 + 1) + \left(\frac{5}{8} - \frac{1}{4} + \frac{2}{5}\right)$$

$$= 2 + \frac{5 \times 5 - 10 \times 1 + 8 \times 2}{40}$$

$$= 2 + \frac{25 - 10 + 16}{40}$$

$$= 2 + \frac{31}{40} = \mathbf{2\frac{31}{40}}$$

Now try the following Practice Exercise

Practice Exercise 5   Introduction to fractions (answers on page 1148)

1. Change the improper fraction $\frac{15}{7}$ into a mixed number.

2. Change the improper fraction $\frac{37}{5}$ into a mixed number.

3. Change the mixed number $2\frac{4}{9}$ into an improper fraction.

4. Change the mixed number $8\frac{7}{8}$ into an improper fraction.

5. A box contains 165 paper clips. 60 clips are removed from the box. Express this as a fraction in its simplest form.

6. Order the following fractions from the smallest to the largest.

$$\frac{4}{9}, \frac{5}{8}, \frac{3}{7}, \frac{1}{2}, \frac{3}{5}$$

7. A training college has 375 students of which 120 are girls. Express this as a fraction in its simplest form.

Evaluate, in fraction form, the expressions given in Problems 8 to 20.

8. $\frac{1}{3} + \frac{2}{5}$

9. $\frac{5}{6} - \frac{4}{15}$

10. $\frac{1}{2} + \frac{2}{5}$

11. $\frac{7}{16} - \frac{1}{4}$

12. $\frac{2}{7} + \frac{3}{11}$

13. $\frac{2}{9} - \frac{1}{7} + \frac{2}{3}$

14. $3\frac{2}{5} - 2\frac{1}{3}$

15. $\frac{7}{27} - \frac{2}{3} + \frac{5}{9}$

16. $5\frac{3}{13} + 3\frac{3}{4}$

17. $4\frac{5}{8} - 3\frac{2}{5}$

18. $10\frac{3}{7} - 8\frac{2}{3}$

19. $3\frac{1}{4} - 4\frac{4}{5} + 1\frac{5}{6}$

20. $5\frac{3}{4} - 1\frac{2}{5} - 3\frac{1}{2}$

2.3 Multiplication and division of fractions

Multiplication

To multiply two or more fractions together, the numerators are first multiplied to give a single number and this becomes the new numerator of the combined fraction. The denominators are then multiplied together to give the new denominator of the combined fraction.

For example, $\quad \frac{2}{3} \times \frac{4}{7} = \frac{2 \times 4}{3 \times 7} = \frac{8}{21}$

Problem 9. Simplify $7 \times \frac{2}{5}$

$$7 \times \frac{2}{5} = \frac{7}{1} \times \frac{2}{5} = \frac{7 \times 2}{1 \times 5} = \frac{14}{5} = 2\frac{4}{5}$$

Problem 10. Find the value of $\frac{3}{7} \times \frac{14}{15}$

Dividing numerator and denominator by 3 gives

$$\frac{3}{7} \times \frac{14}{15} = \frac{1}{7} \times \frac{14}{5} = \frac{1 \times 14}{7 \times 5}$$

Dividing numerator and denominator by 7 gives

$$\frac{1 \times 14}{7 \times 5} = \frac{1 \times 2}{1 \times 5} = \frac{2}{5}$$

This process of dividing both the numerator and denominator of a fraction by the same factor(s) is called **cancelling**.

Problem 11. Simplify $\frac{3}{5} \times \frac{4}{9}$

$$\frac{3}{5} \times \frac{4}{9} = \frac{1}{5} \times \frac{4}{3} \text{ by cancelling}$$

$$= \frac{4}{15}$$

Problem 12. Evaluate $1\frac{3}{5} \times 2\frac{1}{3} \times 3\frac{3}{7}$

Mixed numbers **must** be expressed as improper fractions before multiplication can be performed. Thus,

$$1\frac{3}{5} \times 2\frac{1}{3} \times 3\frac{3}{7} = \left(\frac{5}{5} + \frac{3}{5}\right) \times \left(\frac{6}{3} + \frac{1}{3}\right) \times \left(\frac{21}{7} + \frac{3}{7}\right)$$

$$= \frac{8}{5} \times \frac{7}{3} \times \frac{24}{7} = \frac{8 \times 1 \times 8}{5 \times 1 \times 1} = \frac{64}{5}$$

$$= 12\frac{4}{5}$$

Problem 13. Simplify $3\frac{1}{5} \times 1\frac{2}{3} \times 2\frac{3}{4}$

The mixed numbers need to be changed to improper fractions before multiplication can be performed.

$$3\frac{1}{5} \times 1\frac{2}{3} \times 2\frac{3}{4} = \frac{16}{5} \times \frac{5}{3} \times \frac{11}{4}$$

$$= \frac{4}{1} \times \frac{1}{3} \times \frac{11}{1} \text{ by cancelling}$$

$$= \frac{4 \times 1 \times 11}{1 \times 3 \times 1} = \frac{44}{3} = \mathbf{14\frac{2}{3}}$$

Division

The simple rule for division is **change the division sign into a multiplication sign and invert the second fraction**.

For example, $\dfrac{2}{3} \div \dfrac{3}{4} = \dfrac{2}{3} \times \dfrac{4}{3} = \mathbf{\dfrac{8}{9}}$

Problem 14.   Simplify $\dfrac{3}{7} \div \dfrac{8}{21}$

$$\frac{3}{7} \div \frac{8}{21} = \frac{3}{7} \times \frac{21}{8} = \frac{3}{1} \times \frac{3}{8} \text{ by cancelling}$$

$$= \frac{3 \times 3}{1 \times 8} = \frac{9}{8} = \mathbf{1\frac{1}{8}}$$

Problem 15.   Find the value of $5\dfrac{3}{5} \div 7\dfrac{1}{3}$

The mixed numbers must be expressed as improper fractions. Thus,

$$5\frac{3}{5} \div 7\frac{1}{3} = \frac{28}{5} \div \frac{22}{3} = \frac{28}{5} \times \frac{3}{22} = \frac{14}{5} \times \frac{3}{11} = \mathbf{\frac{42}{55}}$$

Problem 16.   Simplify $3\dfrac{2}{3} \times 1\dfrac{3}{4} \div 2\dfrac{3}{4}$

Mixed numbers must be expressed as improper fractions before multiplication and division can be performed:

$$3\frac{2}{3} \times 1\frac{3}{4} \div 2\frac{3}{4} = \frac{11}{3} \times \frac{7}{4} \div \frac{11}{4} = \frac{11}{3} \times \frac{7}{4} \times \frac{4}{11}$$

$$= \frac{1 \times 7 \times 1}{3 \times 1 \times 1} \text{ by cancelling}$$

$$= \frac{7}{3} = \mathbf{2\frac{1}{3}}$$

Now try the following Practice Exercise

Practice Exercise 6   Multiplying and dividing fractions (answers on page 1148)

Evaluate the following.

1. $\dfrac{2}{5} \times \dfrac{4}{7}$

2. $5 \times \dfrac{4}{9}$

3. $\dfrac{3}{4} \times \dfrac{8}{11}$

4. $\dfrac{3}{4} \times \dfrac{5}{9}$

5. $\dfrac{17}{35} \times \dfrac{15}{68}$

6. $\dfrac{3}{5} \times \dfrac{7}{9} \times 1\dfrac{2}{7}$

7. $\dfrac{13}{17} \times 4\dfrac{7}{11} \times 3\dfrac{4}{39}$

8. $\dfrac{1}{4} \times \dfrac{3}{11} \times 1\dfrac{5}{39}$

9. $\dfrac{2}{9} \div \dfrac{4}{27}$

10. $\dfrac{3}{8} \div \dfrac{45}{64}$

11. $\dfrac{3}{8} \div \dfrac{5}{32}$

12. $\dfrac{3}{4} \div 1\dfrac{4}{5}$

13. $2\dfrac{1}{4} \times 1\dfrac{2}{3}$

14. $1\dfrac{1}{3} \div 2\dfrac{5}{9}$

15. $2\dfrac{4}{5} \div \dfrac{7}{10}$

16. $2\dfrac{3}{4} \div 3\dfrac{2}{3}$

17. $\dfrac{1}{9} \times \dfrac{3}{4} \times 1\dfrac{1}{3}$

18. $3\dfrac{1}{4} \times 1\dfrac{3}{5} \div \dfrac{2}{5}$

19. A ship's crew numbers 105, of which $\dfrac{1}{7}$ are women. Of the men, $\dfrac{1}{6}$ are officers. How many male officers are on board?

20. If a storage tank is holding 450 litres when it is three-quarters full, how much will it contain when it is two-thirds full?

21. Three people, P, Q and R, contribute to a fund. P provides 3/5 of the total, Q provides 2/3 of the remainder and R provides £8. Determine (a) the total of the fund and (b) the contributions of P and Q.

22. A tank contains 24 000 litres of oil. Initially, $\dfrac{7}{10}$ of the contents are removed, then $\dfrac{3}{5}$ of the remainder is removed. How much oil is left in the tank?

2.4 Order of operation with fractions

As stated in Chapter 1, sometimes addition, subtraction, multiplication, division, powers and brackets can all be involved in a calculation. A definite order of operation must be adhered to. The order is:

Brackets

Order (or p**O**wer)

Division

Multiplication

Addition

Subtraction

It should be noted that cancelling cannot happen until the final answer has been calculated.
This is demonstrated in the following worked problems.

Problem 17.   Simplify $\dfrac{7}{20} - \dfrac{3}{8} \times \dfrac{4}{5}$

$$\dfrac{7}{20} - \dfrac{3}{8} \times \dfrac{4}{5} = \dfrac{7}{20} - \dfrac{3 \times 1}{2 \times 5} \quad \text{by cancelling} \qquad \text{(M)}$$

$$= \dfrac{7}{20} - \dfrac{3}{10} \qquad \text{(M)}$$

$$= \dfrac{7}{20} - \dfrac{6}{20}$$

$$= \boldsymbol{\dfrac{1}{20}} \qquad \text{(S)}$$

Problem 18.   Simplify $\dfrac{1}{4} - 2\dfrac{1}{5} \times \dfrac{5}{8} + \dfrac{9}{10}$

$$\dfrac{1}{4} - 2\dfrac{1}{5} \times \dfrac{5}{8} + \dfrac{9}{10} = \dfrac{1}{4} - \dfrac{11}{5} \times \dfrac{5}{8} + \dfrac{9}{10}$$

$$= \dfrac{1}{4} - \dfrac{11}{1} \times \dfrac{1}{8} + \dfrac{9}{10} \quad \text{by cancelling}$$

$$= \dfrac{1}{4} - \dfrac{11}{8} + \dfrac{9}{10} \qquad \text{(M)}$$

$$= \dfrac{1 \times 10}{4 \times 10} - \dfrac{11 \times 5}{8 \times 5} + \dfrac{9 \times 4}{10 \times 4}$$

(since the LCM of 4, 8 and 10 is 40)

$$= \dfrac{10}{40} - \dfrac{55}{40} + \dfrac{36}{40}$$

$$= \dfrac{10 - 55 + 36}{40} \qquad \text{(A/S)}$$

$$= -\boldsymbol{\dfrac{9}{40}}$$

Problem 19.   Simplify

$$2\dfrac{1}{2} - \left(\dfrac{2}{5} + \dfrac{3}{4}\right) \div \left(\dfrac{5}{8} \times \dfrac{2}{3}\right)$$

$$2\dfrac{1}{2} - \left(\dfrac{2}{5} + \dfrac{3}{4}\right) \div \left(\dfrac{5}{8} \times \dfrac{2}{3}\right)$$

$$= \dfrac{5}{2} - \left(\dfrac{2 \times 4}{5 \times 4} + \dfrac{3 \times 5}{4 \times 5}\right) \div \left(\dfrac{5}{8} \times \dfrac{2}{3}\right) \qquad \text{(B)}$$

$$= \dfrac{5}{2} - \left(\dfrac{8}{20} + \dfrac{15}{20}\right) \div \left(\dfrac{5}{8} \times \dfrac{2}{3}\right) \qquad \text{(B)}$$

$$= \dfrac{5}{2} - \dfrac{23}{20} \div \left(\dfrac{5}{4} \times \dfrac{1}{3}\right) \quad \text{by cancelling} \qquad \text{(B)}$$

$$= \dfrac{5}{2} - \dfrac{23}{20} \div \dfrac{5}{12} \qquad \text{(B)}$$

$$= \dfrac{5}{2} - \dfrac{23}{20} \times \dfrac{12}{5} \qquad \text{(D)}$$

$$= \dfrac{5}{2} - \dfrac{23}{5} \times \dfrac{3}{5} \quad \text{by cancelling}$$

$$= \dfrac{5}{2} - \dfrac{69}{25} \qquad \text{(M)}$$

$$= \dfrac{5 \times 25}{2 \times 25} - \dfrac{69 \times 2}{25 \times 2} \qquad \text{(S)}$$

$$= \dfrac{125}{50} - \dfrac{138}{50} \qquad \text{(S)}$$

$$= -\boldsymbol{\dfrac{13}{50}}$$

Problem 20.   Evaluate

$$\dfrac{1}{3} \text{ of } \left(5\dfrac{1}{2} - 3\dfrac{3}{4}\right) + 3\dfrac{1}{5} \div \dfrac{4}{5} - \dfrac{1}{2}$$

$$\dfrac{1}{3} \text{ of } \left(5\dfrac{1}{2} - 3\dfrac{3}{4}\right) + 3\dfrac{1}{5} \div \dfrac{4}{5} - \dfrac{1}{2}$$

$$= \dfrac{1}{3} \text{ of } 1\dfrac{3}{4} + 3\dfrac{1}{5} \div \dfrac{4}{5} - \dfrac{1}{2} \qquad \text{(B)}$$

$$= \dfrac{1}{3} \times \dfrac{7}{4} + \dfrac{16}{5} \div \dfrac{4}{5} - \dfrac{1}{2} \qquad \text{(O)}$$

(Note that the 'of ' is replaced with a multiplication sign)

$$= \dfrac{1}{3} \times \dfrac{7}{4} + \dfrac{16}{5} \times \dfrac{5}{4} - \dfrac{1}{2} \qquad \text{(D)}$$

$$= \dfrac{1}{3} \times \dfrac{7}{4} + \dfrac{4}{1} \times \dfrac{1}{1} - \dfrac{1}{2} \quad \text{by cancelling}$$

$$= \frac{7}{12} + \frac{4}{1} - \frac{1}{2} \qquad \text{(M)}$$

$$= \frac{7}{12} + \frac{48}{12} - \frac{6}{12} \qquad \text{(A/S)}$$

$$= \frac{49}{12}$$

$$= \mathbf{4\frac{1}{12}}$$

Now try the following Practice Exercise

Practice Exercise 7   Order of precedence with fractions (answers on page 1148)

Evaluate the following.

1. $2\frac{1}{2} - \frac{3}{5} \times \frac{20}{27}$

2. $\frac{1}{3} - \frac{3}{4} \times \frac{16}{27}$

3. $\frac{1}{2} + \frac{3}{5} \div \frac{9}{15} - \frac{1}{3}$

4. $\frac{1}{5} + 2\frac{2}{3} \div \frac{5}{9} - \frac{1}{4}$

5. $\frac{4}{5} \times \frac{1}{2} - \frac{1}{6} \div \frac{2}{5} + \frac{2}{3}$

6. $\frac{3}{5} - \left(\frac{2}{3} - \frac{1}{2}\right) \div \left(\frac{5}{6} \times \frac{3}{2}\right)$

7. $\frac{1}{2}$ of $\left(4\frac{2}{5} - 3\frac{7}{10}\right) + \left(3\frac{1}{3} \div \frac{2}{3}\right) - \frac{2}{5}$

8. $\dfrac{6\frac{2}{3} \times 1\frac{2}{5} - \frac{1}{3}}{6\frac{3}{4} \div 1\frac{1}{2}}$

9. $1\frac{1}{3} \times 2\frac{1}{5} \div \frac{2}{5}$

10. $\frac{1}{4} \times \frac{2}{5} - \frac{1}{5} \div \frac{2}{3} + \frac{4}{15}$

11. $\dfrac{\frac{2}{3} + 3\frac{1}{5} \times 2\frac{1}{2} + 1\frac{1}{3}}{8\frac{1}{3} \div 3\frac{1}{3}}$

12. $\frac{1}{13}$ of $\left(2\frac{9}{10} - 1\frac{3}{5}\right) + \left(2\frac{1}{3} \div \frac{2}{3}\right) - \frac{3}{4}$

For fully worked solutions to each of the problems in Practice Exercises 5 to 7 in this chapter, go to the website:
www.routledge.com/cw/bird

Revision Test 1 Basic arithmetic and fractions

This assignment covers the material contained in Chapters 1 and 2. *The marks available are shown in brackets at the end of each question.*

1. Evaluate
   $1009 \, \text{cm} - 356 \, \text{cm} - 742 \, \text{cm} + 94 \, \text{cm}.$ (3)

2. Determine £284 × 9   (3)

3. Evaluate
   (a) $-11239 - (-4732) + 9639$
   (b) -164×-12
   (c) 367×-19 (6)

4. Calculate (a) $\$153 \div 9$ (b) $1397 \, \text{g} \div 11$ (4)

5. A small component has a mass of 27 grams. Calculate the mass, in kilograms, of 750 such components.   (3)

6. Find (a) the highest common factor and (b) the lowest common multiple of the following numbers: 15, 40, 75, 120   (7)

Evaluate the expressions in questions 7 to 12.

7. $7 + 20 \div (9 - 5)$ (3)

8. $147 - 21(24 \div 3) + 31$ (3)

9. $40 \div (1 + 4) + 7[8 + (3 \times 8) - 27]$ (5)

10. $\dfrac{(7-3)(2-5)}{3(9-5) \div (2-6)}$ (3)

11. $\dfrac{(7 + 4 \times 5) \div 3 + 6 \div 2}{2 \times 4 + (5 - 8) - 2^2 + 3}$ (5)

12. $\dfrac{(4^2 \times 5 - 8) \div 3 + 9 \times 8}{4 \times 3^2 - 20 \div 5}$ (5)

13. Simplify
    (a) $\dfrac{3}{4} - \dfrac{7}{15}$
    (b) $1\dfrac{5}{8} - 2\dfrac{1}{3} + 3\dfrac{5}{6}$ (8)

14. A training college has 480 students of which 150 are girls. Express this as a fraction in its simplest form.   (2)

15. A tank contains 18 000 litres of oil. Initially, $\dfrac{7}{10}$ of the contents are removed, then $\dfrac{2}{5}$ of the remainder are removed. How much oil is left in the tank? (4)

16. Evaluate
    (a) $1\dfrac{7}{9} \times \dfrac{3}{8} \times 3\dfrac{3}{5}$
    (b) $6\dfrac{2}{3} \div 1\dfrac{1}{3}$
    (c) $1\dfrac{1}{3} \times 2\dfrac{1}{5} \div \dfrac{2}{5}$ (10)

17. Calculate
    (a) $\dfrac{1}{4} \times \dfrac{2}{5} - \dfrac{1}{5} \div \dfrac{2}{3} + \dfrac{4}{15}$
    (b) $\dfrac{\dfrac{2}{3} + 3\dfrac{1}{5} \times 2\dfrac{1}{2} + 1\dfrac{1}{3}}{8\dfrac{1}{3} \div 3\dfrac{1}{3}}$ (8)

18. Simplify $\left\{ \dfrac{1}{13} \text{ of } \left(2\dfrac{9}{10} - 1\dfrac{3}{5} \right) \right\} + \left(2\dfrac{1}{3} \div \dfrac{2}{3} \right) - \dfrac{3}{4}$ (8)

For lecturers/instructors/teachers, fully worked solutions to each of the problems in Revision Test 1, together with a full marking scheme, are available at the website:
www.routledge.com/cw/bird

Chapter 3

Decimals

Why it is important to understand: Decimals

Engineers and scientists use decimal numbers all the time in calculations. Calculators are able to handle calculations with decimals; however, there will be times when a quick calculation involving addition, subtraction, multiplication and division of decimals is needed. Again, do not spend too much time on this chapter because we deal with the calculator later; however, try to have some idea how to do quick calculations involving decimal numbers in the absence of a calculator. You will feel more confident to deal with decimal numbers in calculations if you can do this.

At the end of this chapter, you should be able to:

- convert a decimal number to a fraction and vice versa
- understand and use significant figures and decimal places in calculations
- add and subtract decimal numbers
- multiply and divide decimal numbers

3.1 Introduction

The decimal system of numbers is based on the digits 0 to 9.

There are a number of everyday occurrences in which we use decimal numbers. For example, a radio is, say, tuned to 107.5 MHz FM; 107.5 is an example of a decimal number.

In a shop, a pair of trainers cost, say, £57.95; 57.95 is another example of a decimal number. 57.95 is a decimal fraction, where a decimal point separates the integer, i.e. 57, from the fractional part, i.e. 0.95

57.95 actually means $(5 \times 10) + (7 \times 1)$

$$+ \left(9 \times \frac{1}{10} \right) + \left(5 \times \frac{1}{100} \right)$$

3.2 Converting decimals to fractions and vice versa

Converting decimals to fractions and vice versa is demonstrated below with worked examples.

Problem 1. Convert 0.375 to a proper fraction in its simplest form

(i) 0.375 may be written as $\dfrac{0.375 \times 1000}{1000}$ i.e.

$0.375 = \dfrac{375}{1000}$

(ii) Dividing both numerator and denominator by 5 gives $\dfrac{375}{1000} = \dfrac{75}{200}$

(iii) Dividing both numerator and denominator by 5 again gives $\dfrac{75}{200} = \dfrac{15}{40}$

(iv) Dividing both numerator and denominator by 5 again gives $\dfrac{15}{40} = \dfrac{3}{8}$

Since both 3 and 8 are only divisible by 1, we cannot 'cancel' any further, so $\dfrac{3}{8}$ is the 'simplest form' of the fraction.

Hence, **the decimal fraction** $0.375 = \dfrac{3}{8}$ **as a proper fraction**.

Problem 2. Convert 3.4375 to a mixed number

Initially, the whole number, 3, is ignored.

(i) 0.4375 may be written as $\dfrac{0.4375 \times 10000}{10000}$ i.e. $0.4375 = \dfrac{4375}{10000}$

(ii) Dividing both numerator and denominator by 25 gives $\dfrac{4375}{10000} = \dfrac{175}{400}$

(iii) Dividing both numerator and denominator by 5 gives $\dfrac{175}{400} = \dfrac{35}{80}$

(iv) Dividing both numerator and denominator by 5 again gives $\dfrac{35}{80} = \dfrac{7}{16}$

Since both 5 and 16 are only divisible by 1, we cannot 'cancel' any further, so $\dfrac{7}{16}$ is the 'lowest form' of the fraction.

(v) Hence, $0.4375 = \dfrac{7}{16}$

Thus, **the decimal fraction** $3.4375 = 3\dfrac{7}{16}$ **as a mixed number**.

Problem 3. Express $\dfrac{7}{8}$ as a decimal fraction

To convert a proper fraction to a decimal fraction, the numerator is divided by the denominator.

$$8\overline{)7.000}\quad 0.8\,7\,5$$

(i) 8 into 7 will not go. Place the 0 above the 7

(ii) Place the decimal point above the decimal point of 7.000

(iii) 8 into 70 goes 8, remainder 6. Place the 8 above the first zero after the decimal point and carry the 6 remainder to the next digit on the right, making it 60

(iv) 8 into 60 goes 7, remainder 4. Place the 7 above the next zero and carry the 4 remainder to the next digit on the right, making it 40

(v) 8 into 40 goes 5, remainder 0. Place 5 above the next zero.

Hence, **the proper fraction** $\dfrac{7}{8} = 0.875$ **as a decimal fraction**.

Problem 4. Express $5\dfrac{13}{16}$ as a decimal fraction

For mixed numbers it is only necessary to convert the proper fraction part of the mixed number to a decimal fraction.

$$16\overline{)13.0\,0\,0\,0}\quad 0.8\,1\,2\,5$$

(i) 16 into 13 will not go. Place the 0 above the 3

(ii) Place the decimal point above the decimal point of 13.0000

(iii) 16 into 130 goes 8, remainder 2. Place the 8 above the first zero after the decimal point and carry the 2 remainder to the next digit on the right, making it 20

(iv) 16 into 20 goes 1, remainder 4. Place the 1 above the next zero and carry the 4 remainder to the next digit on the right, making it 40

(v) 16 into 40 goes 2, remainder 8. Place the 2 above the next zero and carry the 8 remainder to the next digit on the right, making it 80

(vi) 16 into 80 goes 5, remainder 0. Place the 5 above the next zero.

(vii) Hence, $\dfrac{13}{16} = 0.8125$

Thus, **the mixed number** $5\dfrac{13}{16} = 5.8125$ **as a decimal fraction**.

Now try the following Practice Exercise

> **Practice Exercise 8   Converting decimals to fractions and vice versa (answers on page 1149)**
>
> 1. Convert 0.65 to a proper fraction.
> 2. Convert 0.036 to a proper fraction.
> 3. Convert 0.175 to a proper fraction.
> 4. Convert 0.048 to a proper fraction.
> 5. Convert the following to proper fractions.
>   (a) 0.66   (b) 0.84   (c) 0.0125
>   (d) 0.282   (e) 0.024
> 6. Convert 4.525 to a mixed number.
> 7. Convert 23.44 to a mixed number.
> 8. Convert 10.015 to a mixed number.
> 9. Convert 6.4375 to a mixed number.
> 10. Convert the following to mixed numbers.
>   (a) 1.82   (b) 4.275   (c) 14.125
>   (d) 15.35   (e) 16.2125
> 11. Express $\dfrac{5}{8}$ as a decimal fraction.
> 12. Express $6\dfrac{11}{16}$ as a decimal fraction.
> 13. Express $\dfrac{7}{32}$ as a decimal fraction.
> 14. Express $11\dfrac{3}{16}$ as a decimal fraction.
> 15. Express $\dfrac{9}{32}$ as a decimal fraction.

3.3   Significant figures and decimal places

A number which can be expressed exactly as a decimal fraction is called a **terminating decimal**. For example,

$$3\frac{3}{16} = 3.1825 \text{ is a terminating decimal.}$$

A number which cannot be expressed exactly as a decimal fraction is called a **non-terminating decimal**. For example,

$$1\frac{5}{7} = 1.7142857\ldots \text{ is a non-terminating decimal.}$$

The answer to a non-terminating decimal may be expressed in two ways, depending on the accuracy required:

(a)   correct to a number of **significant figures**, or

(b)   correct to a number of **decimal places**, i.e. the number of figures after the decimal point.

The last digit in the answer is unaltered if the next digit on the right is in the group of numbers 0, 1, 2, 3 or 4 For example,

$$1.714285\ldots = \mathbf{1.714} \text{ correct to 4 significant figures}$$
$$= \mathbf{1.714} \text{ correct to 3 decimal places}$$

since the next digit on the right in this example is 2. The last digit in the answer is increased by 1 if the next digit on the right is in the group of numbers 5, 6, 7, 8 or 9. For example,

$$1.7142857\ldots = \mathbf{1.7143} \text{ correct to 5 significant figures}$$
$$= \mathbf{1.7143} \text{ correct to 4 decimal places}$$

since the next digit on the right in this example is 8

> **Problem 5.**   Express 15.36815 correct to (a) 2 decimal places,   (b) 3 significant figures, (c) 3 decimal places,   (d) 6 significant figures

(a)   15.36815 = **15.37** correct to 2 decimal places.

(b)   15.36815 = **15.4** correct to 3 significant figures.

(c)   15.36815 = **15.368** correct to 3 decimal places.

(d)   15.36815 = **15.3682** correct to 6 significant figures.

> **Problem 6.**   Express 0.004369 correct to (a) 4 decimal places,   (b) 3 significant figures

(a)   0.004369 = **0.0044** correct to 4 decimal places.

(b)   0.004369 = **0.00437** correct to 3 significant figures.

Note that the zeros to the right of the decimal point do not count as significant figures.

Now try the following Practice Exercise

Practice Exercise 9   Significant figures and decimal places (answers on page 1149)

1. Express 14.1794 correct to 2 decimal places.

2. Express 2.7846 correct to 4 significant figures.

3. Express 65.3792 correct to 2 decimal places.

4. Express 43.2746 correct to 4 significant figures.

5. Express 1.2973 correct to 3 decimal places.

6. Express 0.0005279 correct to 3 significant figures.

3.4 Adding and subtracting decimal numbers

When adding or subtracting decimal numbers, care needs to be taken to ensure that the decimal points are beneath each other. This is demonstrated in the following worked examples.

Problem 7.   Evaluate $46.8 + 3.06 + 2.4 + 0.09$ and give the answer correct to 3 significant figures

The decimal points are placed under each other as shown. Each column is added, starting from the right.

$$
\begin{array}{r}
46.8 \\
3.06 \\
2.4 \\
+\,0.09 \\
\hline
52.35 \\
\hline
11\ \ 1
\end{array}
$$

(i) $6 + 9 = 15$. Place the 5 in the hundredths column. Carry the 1 in the tenths column.

(ii) $8 + 0 + 4 + 0 + 1$ (carried) $= 13$. Place the 3 in the tenths column. Carry the 1 into the units column.

(iii) $6 + 3 + 2 + 0 + 1$ (carried) $= 12$. Place the 2 in the units column. Carry the 1 into the tens column.

(iv) $4 + 1$ (carried) $= 5$. Place the 5 in the hundreds column.

Hence,

$$46.8 + 3.06 + 2.4 + 0.09 = 52.35$$
$$= \mathbf{52.4, \ correct \ to \ 3} $$
$$\mathbf{significant \ figures}$$

Problem 8.   Evaluate $64.46 - 28.77$ and give the answer correct to 1 decimal place

As with addition, the decimal points are placed under each other as shown.

$$
\begin{array}{r}
64.46 \\
-\,28.77 \\
\hline
35.69
\end{array}
$$

(i) $6 - 7$ is not possible; therefore 'borrow' 1 from the tenths column. This gives $16 - 7 = 9$. Place the 9 in the hundredths column.

(ii) $3 - 7$ is not possible; therefore 'borrow' 1 from the units column. This gives $13 - 7 = 6$. Place the 6 in the tenths column.

(iii) $3 - 8$ is not possible; therefore 'borrow' 1 from the hundreds column. This gives $13 - 8 = 5$. Place the 5 in the units column.

(iv) $5 - 2 = 3$. Place the 3 in the hundreds column.

Hence,

$$64.46 - 28.77 = 35.69$$
$$= \mathbf{35.7 \ correct \ to \ 1 \ decimal \ place}$$

Problem 9.   Evaluate $312.64 - 59.826 - 79.66 + 38.5$ and give the answer correct to 4 significant figures

The sum of the positive decimal fractions $=$ $312.64 + 38.5 = 351.14$
The sum of the negative decimal fractions $= 59.826 + 79.66 = 139.486$
Taking the sum of the negative decimal fractions from the sum of the positive decimal fractions gives

$$
\begin{array}{r}
351.140 \\
-\,139.486 \\
\hline
211.654
\end{array}
$$

Hence, $351.140 - 139.486 = \mathbf{211.654} = \mathbf{211.7, \ correct \ to \ 4 \ significant \ figures}$.

Now try the following Practice Exercise

Determine the following without using a calculator.

1. Evaluate $37.69 + 42.6$, correct to 3 significant figures.

2. Evaluate $378.1 - 48.85$, correct to 1 decimal place.

3. Evaluate $68.92 + 34.84 - 31.223$, correct to 4 significant figures.

4. Evaluate $67.841 - 249.55 + 56.883$, correct to 2 decimal places.

5. Evaluate $483.24 - 120.44 - 67.49$, correct to 4 significant figures.

6. Evaluate $738.22 - 349.38 - 427.336 + 56.779$, correct to 1 decimal place.

7. Determine the dimension marked x in the length of the shaft shown in Figure 3.1. The dimensions are in millimetres.

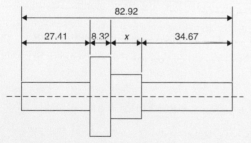

Figure 3.1

3.5 Multiplying and dividing decimal numbers

When multiplying decimal fractions:

(a) the numbers are multiplied as if they were integers, and

(b) the position of the decimal point in the answer is such that there are as many digits to the right of it as the sum of the digits to the right of the decimal points of the two numbers being multiplied together.

This is demonstrated in the following worked examples.

Problem 10.   Evaluate 37.6×5.4

$$
\begin{array}{r}
376 \\
\times 54 \\
\hline
1504 \\
18800 \\
\hline
20304 \\
\hline
\end{array}
$$

(i) $376 \times 54 = 20304$

(ii) As there are $1 + 1 = 2$ digits to the right of the decimal points of the two numbers being multiplied together, $37.\underline{6} \times 5.\underline{4}$, then

$$37.6 \times 5.4 = 203.04$$

Problem 11.   Evaluate $44.25 \div 1.2$, correct to (a) 3 significant figures, (b) 2 decimal places

$$44.25 \div 1.2 = \frac{44.25}{1.2}$$

The denominator is multiplied by 10 to change it into an integer. The numerator is also multiplied by 10 to keep the fraction the same. Thus,

$$\frac{44.25}{1.2} = \frac{44.25 \times 10}{1.2 \times 10} = \frac{442.5}{12}$$

The long division is similar to the long division of integers, and the steps are as shown.

$$
\begin{array}{r}
36.875 \\
12\overline{)442.500} \\
\underline{36} \\
82 \\
\underline{72} \\
105 \\
\underline{96} \\
90 \\
\underline{84} \\
60 \\
\underline{60} \\
0 \\
\end{array}
$$

(i) 12 into 44 goes 3; place the 3 above the second 4 of 442.500

(ii) $3 \times 12 = 36$; place the 36 below the 44 of 442.500

(iii) $44 - 36 = 8$

(iv) Bring down the 2 to give 82

(v) 12 into 82 goes 6; place the 6 above the 2 of 442.500

(vi) $6 \times 12 = 72$; place the 72 below the 82

(vii) $82 - 72 = 10$

(viii) Bring down the 5 to give 105

(ix) 12 into 105 goes 8; place the 8 above the 5 of 442.500

(x) $8 \times 12 = 96$; place the 96 below the 105

(xi) $105 - 96 = 9$

(xii) Bring down the 0 to give 90

(xiii) 12 into 90 goes 7; place the 7 above the first zero of 442.500

(xiv) $7 \times 12 = 84$; place the 84 below the 90

(xv) $90 - 84 = 6$

(xvi) Bring down the 0 to give 60

(xvii) 12 into 60 gives 5 exactly; place the 5 above the second zero of 442.500

(xviii) Hence, $44.25 \div 1.2 = \dfrac{442.5}{12} = \mathbf{36.875}$

So,

(a) $44.25 \div 1.2 = \mathbf{36.9}$, **correct to 3 significant figures**.

(b) $44.25 \div 1.2 = \mathbf{36.88}$, **correct to 2 decimal places**.

Problem 12. Express $7\dfrac{2}{3}$ as a decimal fraction, correct to 4 significant figures

Dividing 2 by 3 gives $\dfrac{2}{3} = 0.666666\ldots$

and $7\dfrac{2}{3} = 7.666666\ldots$

Hence, $7\dfrac{2}{3} = \mathbf{7.667}$, **correct to 4 significant figures**.

Note that 7.6666... is called **7.6 recurring** and is written as $7.\dot{6}$

Now try the following Practice Exercise

Practice Exercise 11   Multiplying and dividing decimal numbers (answers on page 1149)

In Problems 1 to 8, evaluate without using a calculator.

1. Evaluate 3.57×1.4

2. Evaluate 67.92×0.7

3. Evaluate 167.4×2.3

4. Evaluate 342.6×1.7

5. Evaluate $548.28 \div 1.2$

6. Evaluate $478.3 \div 1.1$, correct to 5 significant figures.

7. Evaluate $563.48 \div 0.9$, correct to 4 significant figures.

8. Evaluate $2387.4 \div 1.5$

In Problems 9 to 14, express as decimal fractions to the accuracy stated.

9. $\dfrac{4}{9}$, correct to 3 significant figures.

10. $\dfrac{17}{27}$, correct to 5 decimal places.

11. $1\dfrac{9}{16}$, correct to 4 significant figures.

12. $53\dfrac{5}{11}$, correct to 3 decimal places.

13. $13\dfrac{31}{37}$, correct to 2 decimal places.

14. $8\dfrac{9}{13}$, correct to 3 significant figures.

15. Evaluate $421.8 \div 17$, (a) correct to 4 significant figures and (b) correct to 3 decimal places.

16. Evaluate $\dfrac{0.0147}{2.3}$, (a) correct to 5 decimal places and (b) correct to 2 significant figures.

17. Evaluate (a) $\dfrac{12.\dot{6}}{1.5}$ (b) $5.\dot{2} \times 12$

18. A tank contains 1800 litres of oil. How many tins containing 0.75 litres can be filled from this tank?

For fully worked solutions to each of the problems in Practice Exercises 8 to 11 in this chapter, go to the website:
www.routledge.com/cw/bird

Chapter 4

Using a calculator

Why it is important to understand: Using a calculator

The availability of electronic pocket calculators, at prices which all can afford, has had a considerable impact on engineering education. Engineers and student engineers now use calculators all the time since calculators are able to handle a very wide range of calculations. You will feel more confident to deal with all aspects of engineering studies if you are able to correctly use a calculator accurately.

At the end of this chapter, you should be able to:

- use a calculator to add, subtract, multiply and divide decimal numbers
- use a calculator to evaluate square, cube, reciprocal, power, root and $\times 10^x$ functions
- use a calculator to evaluate expressions containing fractions and trigonometric functions
- use a calculator to evaluate expressions containing π and e^x functions
- appreciate errors and approximations in calculations
- evaluate formulae, given values

4.1 Introduction

In engineering, calculations often need to be performed. For simple numbers it is useful to be able to use mental arithmetic. However, when numbers are larger an electronic calculator needs to be used.

There are several calculators on the market, many of which will be satisfactory for our needs. It is essential to have a **scientific notation calculator** which will have all the necessary functions needed and more.

This chapter assumes you have a **CASIO fx-991ES PLUS calculator**, or similar, as shown in Figure 4.1, on page 26.

Besides straightforward addition, subtraction, multiplication and division, which you will already be able to do, we will check that you can use squares, cubes, powers, reciprocals, roots, fractions and trigonometric functions (the latter in preparation for Chapter 38). There are several other functions on the calculator which we do not need to concern ourselves with at this level.

4.2 Adding, subtracting, multiplying and dividing

Initially, after switching on, press **Mode**.

Of the three possibilities, use **Comp**, which is achieved by pressing **1**.

Next, press **Shift** followed by **Setup** and, of the eight possibilities, use **Mth IO**, which is achieved by pressing **1**.

By all means experiment with the other menu options – refer to your 'User's guide'.

Figure 4.1 A CASIO fx-991ES PLUS Calculator

All calculators have **+, −, × and ÷ functions** and these functions will, no doubt, already have been used in calculations.

Problem 1. Evaluate $364.7 \div 57.5$ correct to 3 decimal places

(i) Type in 364.7

(ii) Press ÷.

(iii) Type in 57.5

(iv) Press = and the fraction $\dfrac{3647}{575}$ appears.

(v) Press the $S \Leftrightarrow D$ function and the decimal answer 6.34260869… appears.

Alternatively, after step (iii) press Shift and = and the decimal will appear.
Hence, $364.7 \div 57.5 = 6.343$, **correct to 3 decimal places**.

Problem 2. Evaluate $\dfrac{12.47 \times 31.59}{70.45 \times 0.052}$ correct to 4 significant figures

(i) Type in 12.47

(ii) Press ×.

(iii) Type in 31.59

(iv) Press ÷.

(v) The denominator must have brackets; i.e. press (.

(vi) Type in 70.45×0.052 and complete the bracket; i.e. press)

(vii) Press = and the answer 107.530518… appears.

Hence, $\dfrac{\mathbf{12.47 \times 31.59}}{\mathbf{70.45 \times 0.052}} = \mathbf{107.5}$, **correct to 4 significant figures**.

Now try the following Practice Exercise

Practice Exercise 12   Addition, subtraction, multiplication and division using a calculator (answers on page 1149)

1. Evaluate $378.37 - 298.651 + 45.64 - 94.562$

2. Evaluate 25.63×465.34 correct to 5 significant figures.

3. Evaluate $562.6 \div 41.3$ correct to 2 decimal places.

4. Evaluate $\dfrac{17.35 \times 34.27}{41.53 \div 3.76}$ correct to 3 decimal places.

5. Evaluate $27.48 + 13.72 \times 4.15$ correct to 4 significant figures.

6. Evaluate $\dfrac{(4.527 + 3.63)}{(452.51 \div 34.75)} + 0.468$ correct to 5 significant figures.

7. Evaluate $52.34 - \dfrac{(912.5 \div 41.46)}{(24.6 - 13.652)}$ correct to 3 decimal places.

8. Evaluate $\dfrac{52.14 \times 0.347 \times 11.23}{19.73 \div 3.54}$ correct to 4 significant figures.

9. Evaluate $\dfrac{451.2}{24.57} - \dfrac{363.8}{46.79}$ correct to 4 significant figures.

10. Evaluate $\dfrac{45.6 - 7.35 \times 3.61}{4.672 - 3.125}$ correct to 3 decimal places.

4.3   Further calculator functions

Square and cube functions

Locate the x^2 and x^3 functions on your calculator and then check the following worked examples.

Problem 3.   Evaluate 2.4^2

(i)   Type in 2.4

(ii)   Press x^2 and 2.4^2 appears on the screen.

(iii)   Press $=$ and the answer $\dfrac{144}{25}$ appears.

(iv)   Press the $S \Leftrightarrow D$ function and the fraction changes to a decimal: 5.76

   Alternatively, after step (ii) press Shift and $=$. Thus, $\mathbf{2.4^2 = 5.76}$

Problem 4.   Evaluate 0.17^2 in engineering form

(i)   Type in 0.17

(ii)   Press x^2 and 0.17^2 appears on the screen.

(iii)   Press Shift and $=$ and the answer 0.0289 appears.

(iv)   Press the ENG function and the answer changes to 28.9×10^{-3}, which is **engineering form**.

Hence, $\mathbf{0.17^2 = 28.9 \times 10^{-3}}$ **in engineering form**. The ENG function is extremely important in engineering calculations.

Problem 5.   Change 348620 into engineering form

(i)   Type in 348620

(ii)   Press $=$ then ENG.

Hence, $\mathbf{348620 = 348.62 \times 10^3}$ **in engineering form**.

Problem 6.   Change 0.0000538 into engineering form

(i)   Type in 0.0000538

(ii)   Press $=$ then ENG.

Hence, $\mathbf{0.0000538 = 53.8 \times 10^{-6}}$ **in engineering form**.

Problem 7.   Evaluate 1.4^3

(i)   Type in 1.4

(ii)   Press x^3 and 1.4^3 appears on the screen.

(iii)   Press $=$ and the answer $\dfrac{343}{125}$ appears.

(iv)   Press the $S \Leftrightarrow D$ function and the fraction changes to a decimal: 2.744

Thus, $\mathbf{1.4^3 = 2.744}$

Now try the following Practice Exercise

Practice Exercise 13   Square and cube functions (answers on page 1149)

1. Evaluate 3.5^2

2. Evaluate 0.19^2

3. Evaluate 6.85^2 correct to 3 decimal places.

4. Evaluate $(0.036)^2$ in engineering form.

5. Evaluate 1.563^2 correct to 5 significant figures.

6. Evaluate 1.3^3

7. Evaluate 3.14^3 correct to 4 significant figures.

8. Evaluate $(0.38)^3$ correct to 4 decimal places.

9. Evaluate $(6.03)^3$ correct to 2 decimal places.

10. Evaluate $(0.018)^3$ in engineering form.

Reciprocal and power functions

The reciprocal of 2 is $\dfrac{1}{2}$, the reciprocal of 9 is $\dfrac{1}{9}$ and the reciprocal of x is $\dfrac{1}{x}$, which from indices may be written as x^{-1}. Locate the reciprocal, i.e. x^{-1}, on the calculator. Also, locate the power function, i.e. $x^{\square}$, on your calculator and then check the following worked examples.

Problem 8. Evaluate $\dfrac{1}{3.2}$

(i) Type in 3.2

(ii) Press x^{-1} and 3.2^{-1} appears on the screen.

(iii) Press $=$ and the answer $\dfrac{5}{16}$ appears.

(iv) Press the $S \Leftrightarrow D$ function and the fraction changes to a decimal: 0.3125

Thus, $\dfrac{1}{3.2} = \mathbf{0.3125}$

Problem 9. Evaluate 1.5^5 correct to 4 significant figures

(i) Type in 1.5

(ii) Press $x^{\square}$ and $1.5^{\square}$ appears on the screen.

(iii) Press 5 and 1.5^5 appears on the screen.

(iv) Press Shift and $=$ and the answer 7.59375 appears.

Thus, $\mathbf{1.5^5 = 7.594}$, **correct to 4 significant figures**.

Problem 10. Evaluate $2.4^6 - 1.9^4$ correct to 3 decimal places

---

(i) Type in 2.4

(ii) Press $x^{\square}$ and $2.4^{\square}$ appears on the screen.

(iii) Press 6 and 2.4^6 appears on the screen.

(iv) The cursor now needs to be moved; this is achieved by using the cursor key (the large blue circular function in the top centre of the calculator). Press $\rightarrow$

(v) Press $-$

(vi) Type in 1.9, press $x^{\square}$, then press 4

(vii) Press $=$ and the answer 178.07087... appears.

Thus, $\mathbf{2.4^6 - 1.9^4 = 178.071}$, **correct to 3 decimal places**.

Now try the following Practice Exercise

Practice Exercise 14 Reciprocal and power functions (answers on page 1149)

1. Evaluate $\dfrac{1}{1.75}$ correct to 3 decimal places.

2. Evaluate $\dfrac{1}{0.0250}$

3. Evaluate $\dfrac{1}{7.43}$ correct to 5 significant figures.

4. Evaluate $\dfrac{1}{0.00725}$ correct to 1 decimal place.

5. Evaluate $\dfrac{1}{0.065} - \dfrac{1}{2.341}$ correct to 4 significant figures.

6. Evaluate 2.1^4

7. Evaluate $(0.22)^5$ correct to 5 significant figures in engineering form.

8. Evaluate $(1.012)^7$ correct to 4 decimal places.

9. Evaluate $(0.05)^6$ in engineering form.

10. Evaluate $1.1^3 + 2.9^4 - 4.4^2$ correct to 4 significant figures.

Root and $\times 10^x$ functions

Locate the square root function $\sqrt{\square}$ and the $\sqrt[\square]{\square}$ function (which is a Shift function located above the $x^{\square}$ function) on your calculator. Also, locate the $\times 10^x$ function and then check the following worked examples.

Problem 11. Evaluate $\sqrt{361}$

(i)   Press the $\sqrt{\square}$ function.

(ii) Type in 361 and $\sqrt{361}$ appears on the screen.

(iii) Press = and the answer 19 appears.

Thus, $\sqrt{361} = \mathbf{19}$

Problem 12. Evaluate $\sqrt[4]{81}$

(i)   Press the $\sqrt[\square]{\square}$ function.

(ii) Type in 4 and $\sqrt[4]{\square}$ appears on the screen.

(iii) Press $\rightarrow$ to move the cursor and then type in 81 and $\sqrt[4]{81}$ appears on the screen.

(iv) Press = and the answer 3 appears.

Thus, $\sqrt[4]{81} = \mathbf{3}$

Problem 13. Evaluate $6 \times 10^5 \times 2 \times 10^{-7}$

(i)   Type in 6

(ii)   Press the $\times 10^x$ function (note, you do not have to use $\times$).

(iii) Type in 5

(iv)   Press $\times$

(v)   Type in 2

(vi)   Press the $\times 10^x$ function.

(vii) Type in -7

(viii) Press = and the answer $\dfrac{3}{25}$ appears.

(ix)   Press the $S \Leftrightarrow D$ function and the fraction changes to a decimal: 0.12

Thus, $6 \times 10^5 \times 2 \times 10^{-7} = \mathbf{0.12}$

Now try the following Practice Exercise

Practice Exercise 15   Root and $\times 10^x$ functions (answers on page 1149)

1. Evaluate $\sqrt{4.76}$ correct to 3 decimal places.

2. Evaluate $\sqrt{123.7}$ correct to 5 significant figures.

3. Evaluate $\sqrt{34528}$ correct to 2 decimal places.

4. Evaluate $\sqrt{0.69}$ correct to 4 significant figures.

5. Evaluate $\sqrt{0.025}$ correct to 4 decimal places.

6. Evaluate $\sqrt[3]{17}$ correct to 3 decimal places.

7. Evaluate $\sqrt[4]{773}$ correct to 4 significant figures.

8. Evaluate $\sqrt[5]{3.12}$ correct to 4 decimal places.

9. Evaluate $\sqrt[3]{0.028}$ correct to 5 significant figures.

10. Evaluate $\sqrt[6]{2451} - \sqrt[4]{46}$ correct to 3 decimal places.

Express the answers to Problems 11 to 15 in engineering form.

11. Evaluate $5 \times 10^{-3} \times 7 \times 10^8$

12. Evaluate $\dfrac{3 \times 10^{-4}}{8 \times 10^{-9}}$

13. Evaluate $\dfrac{6 \times 10^3 \times 14 \times 10^{-4}}{2 \times 10^6}$

14. Evaluate $\dfrac{56.43 \times 10^{-3} \times 3 \times 10^4}{8.349 \times 10^3}$ correct to 3 decimal places.

15. Evaluate $\dfrac{99 \times 10^5 \times 6.7 \times 10^{\ 3}}{36.2 \times 10^{-4}}$ correct to 4 significant figures.

Fractions

Locate the $\dfrac{\square}{\square}$ and $\square\dfrac{\square}{\square}$ functions on your calculator (the latter function is a Shift function found above the $\dfrac{\square}{\square}$ function) and then check the following worked examples.

Problem 14. Evaluate $\dfrac{1}{4} + \dfrac{2}{3}$

(i)   Press the $\dfrac{\square}{\square}$ function.

(ii) Type in 1

(iii) Press $\downarrow$ on the cursor key and type in 4

(iv) $\dfrac{1}{4}$ appears on the screen.

(v) Press → on the cursor key and type in +

(vi) Press the $\dfrac{\square}{\square}$ function.

(vii) Type in 2

(viii) Press ↓ on the cursor key and type in 3

(ix) Press → on the cursor key.

(x) Press = and the answer $\dfrac{11}{12}$ appears.

(xi) Press the $S \Leftrightarrow D$ function and the fraction changes to a decimal: 0.9166666…

Thus, $\dfrac{1}{4} + \dfrac{2}{3} = \dfrac{11}{12} = \mathbf{0.9167}$ as a decimal, correct to 4 decimal places.

It is also possible to deal with **mixed numbers** on the calculator. Press Shift then the $\dfrac{\square}{\square}$ function and $\square\dfrac{\square}{\square}$ appears.

> **Problem 15.** Evaluate $5\dfrac{1}{5} - 3\dfrac{3}{4}$

(i) Press Shift then the $\dfrac{\square}{\square}$ function and $\square\dfrac{\square}{\square}$ appears on the screen.

(ii) Type in 5 then → on the cursor key.

(iii) Type in 1 and ↓ on the cursor key.

(iv) Type in 5 and $5\dfrac{1}{5}$ appears on the screen.

(v) Press → on the cursor key.

(vi) Type in − and then press Shift then the $\dfrac{\square}{\square}$ function and $5\dfrac{1}{5} - \square\dfrac{\square}{\square}$ appears on the screen.

(vii) Type in 3 then → on the cursor key.

(viii) Type in 3 and ↓ on the cursor key.

(ix) Type in 4 and $5\dfrac{1}{5} - 3\dfrac{3}{4}$ appears on the screen.

(x) Press = and the answer $\dfrac{29}{20}$ appears.

(xi) Press $S \Leftrightarrow D$ function and the fraction changes to a decimal: 1.45

Thus, $5\dfrac{1}{5} - 3\dfrac{3}{4} = \dfrac{29}{20} = 1\dfrac{9}{20} = \mathbf{1.45}$ as a decimal.

Now try the following Practice Exercise

> **Practice Exercise 16 Fractions (answers on page 1149)**
>
> 1. Evaluate $\dfrac{4}{5} - \dfrac{1}{3}$ as a decimal, correct to 4 decimal places.
>
> 2. Evaluate $\dfrac{2}{3} - \dfrac{1}{6} + \dfrac{3}{7}$ as a fraction.
>
> 3. Evaluate $2\dfrac{5}{6} + 1\dfrac{5}{8}$ as a decimal, correct to 4 significant figures.
>
> 4. Evaluate $5\dfrac{6}{7} - 3\dfrac{1}{8}$ as a decimal, correct to 4 significant figures.
>
> 5. Evaluate $\dfrac{1}{3} - \dfrac{3}{4} \times \dfrac{8}{21}$ as a fraction.
>
> 6. Evaluate $\dfrac{3}{8} + \dfrac{5}{6} - \dfrac{1}{2}$ as a decimal, correct to 4 decimal places.
>
> 7. Evaluate $\dfrac{3}{4} \times \dfrac{4}{5} - \dfrac{2}{3} \div \dfrac{4}{9}$ as a fraction.
>
> 8. Evaluate $8\dfrac{8}{9} \div 2\dfrac{2}{3}$ as a mixed number.
>
> 9. Evaluate $3\dfrac{1}{5} \times 1\dfrac{1}{3} - 1\dfrac{7}{10}$ as a decimal, correct to 3 decimal places.
>
> 10. Evaluate $\dfrac{\left(4\dfrac{1}{5} - 1\dfrac{2}{3}\right)}{\left(3\dfrac{1}{4} \times 2\dfrac{3}{5}\right)} - \dfrac{2}{9}$ as a decimal, correct to 3 significant figures.

Trigonometric functions

Trigonometric ratios will be covered in Chapter 38. However, very briefly, there are three functions on your calculator that are involved with trigonometry. They are:

sin which is an abbreviation of **sine**

cos which is an abbreviation of **cosine**, and

tan which is an abbreviation of **tangent**

Exactly what these mean will be explained in Chapter 38.

There are two main ways that angles are measured, i.e. in **degrees** or in **radians**. Pressing Shift, Setup and 3 shows degrees, and Shift, Setup and 4 shows radians.

Press 3 and your calculator will be in **degrees mode**, indicated by a small D appearing at the top of the screen.

Press 4 and your calculator will be in **radian mode**, indicated by a small R appearing at the top of the screen.

Locate the sin, cos and tan functions on your calculator and then check the following worked examples.

Problem 16. Evaluate $\sin 38°$

(i) Make sure your calculator is in degrees mode.

(ii) Press sin function and sin(appears on the screen.

(iii) Type in 38 and close the bracket with) and sin (38) appears on the screen.

(iv) Press = and the answer 0.615661475... appears.

Thus, $\sin 38° = 0.6157$, **correct to 4 decimal places**.

Problem 17. Evaluate $5.3 \tan (2.23 \text{ rad})$

(i) Make sure your calculator is in radian mode by pressing Shift then Setup then 4 (a small R appears at the top of the screen).

(ii) Type in 5.3 then press tan function and 5.3 tan(appears on the screen.

(iii) Type in 2.23 and close the bracket with) and 5.3 tan (2.23) appears on the screen.

(iv) Press = and the answer −6.84021262... appears.

Thus, $5.3 \tan (2.23 \text{ rad}) = -6.8402$, **correct to 4 decimal places**.

Now try the following Practice Exercise

Practice Exercise 17   Trigonometric functions (answers on page 1149)

Evaluate the following, each correct to 4 decimal places.

1. Evaluate $\sin 67°$

2. Evaluate $\cos 43°$

3. Evaluate $\tan 71°$

4. Evaluate $\sin 15.78°$

5. Evaluate $\cos 63.74°$

6. Evaluate $\tan 39.55° - \sin 52.53°$

7. Evaluate $\sin(0.437 \text{ rad})$

8. Evaluate $\cos(1.42 \text{ rad})$

9. Evaluate $\tan(5.673 \text{ rad})$

10. Evaluate $\dfrac{(\sin 42.6°)(\tan 83.2°)}{\cos 13.8°}$

π and e^x functions

Press Shift and then press the $\times 10^x$ function key and π appears on the screen. Either press Shift and = (or = and $S \Leftrightarrow D$) and the value of π appears in decimal form as 3.14159265...

Press Shift and then press the ln function key and $e^\square$ appears on the screen. Enter 1 and then press = and $e^1 = e = 2.71828182...$

Now check the following worked examples involving π and e^x functions.

Problem 18. Evaluate 3.57π

(i) Enter 3.57

(ii) Press Shift and the $\times 10^x$ key and 3.57π appears on the screen.

(iii) Either press Shift and = (or = and $S \Leftrightarrow D$) and the value of 3.57π appears in decimal form as 11.2154857...

Hence, $3.57\pi = 11.22$, **correct to 4 significant figures**.

Problem 19. Evaluate $e^{2.37}$

(i) Press Shift and then press the ln function key and $e^\square$ appears on the screen.

(ii) Enter 2.37 and $e^{2.37}$ appears on the screen.

(iii) Press Shift and = (or = and $S \Leftrightarrow D$) and the value of $e^{2.37}$ appears in decimal form as 10.6973922...

Hence, $e^{2.37} = 10.70$, **correct to 4 significant figures**.

Now try the following Practice Exercise

Practice Exercise 18 π and e^x functions (answers on page 1149)

Evaluate the following, each correct to 4 significant figures.

1. 1.59π

2. $2.7(\pi - 1)$

3. $\pi^2 \left(\sqrt{13} - 1 \right)$ 4. $3e^{\pi}$

5. $8.5e^{-2.5}$ 6. $3e^{2.9} - 1.6$

7. $3e^{(2\pi - 1)}$ 8. $2\pi e^{\frac{\pi}{3}}$

9. $\sqrt{\left[\dfrac{5.52\pi}{2e^{-2} \times \sqrt{26.73}} \right]}$ 10. $\sqrt{\left[\dfrac{e^{\left(2 - \sqrt{3}\right)}}{\pi \times \sqrt{8.57}} \right]}$

4.4 Errors and approximations

In all problems in which the measurement of distance, time, mass or other quantities occurs, an exact answer cannot be given; only an answer that is correct to a stated degree of accuracy can be given. To take account of this an **error due to measurement** is said to exist.

To take account of measurement errors it is usual to limit answers so that the result given is **not more than one significant figure greater than the least accurate number given in the data**.

Rounding-off errors can exist with decimal fractions. For example, to state that $\pi = 3.142$ is not strictly correct, but '$\pi = 3.142$ correct to 4 significant figures' is a true statement. (Actually, $\pi = 3.141592653589\ldots$)

It is possible, through an incorrect procedure, to obtain the wrong answer to a calculation. This type of error is simply known as a **blunder**.

An **order of magnitude error** is said to exist if incorrect positioning of the decimal point occurs after a calculation has been completed.

Blunders and order of magnitude errors can be reduced by determining **approximate values of calculations**. Answers which do not seem feasible should be checked and the calculation repeated as necessary.

Obviously, an electronic calculator is an invaluable aid to an engineer. However, an engineer will often need to make a quick mental approximation for a calculation when a calculator is not available. For example, $\dfrac{49.1 \times 18.4 \times 122.1}{61.2 \times 38.1}$ may be approximated to $\dfrac{50 \times 20 \times 120}{60 \times 40}$ and then, by cancelling, $\dfrac{50 \times 20 \times 120}{60 \times 40} = 50$. An accurate answer somewhere between 45 and 55 could therefore be expected. Certainly an answer around 500 or 5 would not be expected. Actually, by calculator $\dfrac{49.1 \times 18.4 \times 122.1}{61.2 \times 38.1} = 47.31$, correct to 4 significant figures.

Problem 20. The area A of a triangle is given by $A = \dfrac{1}{2} \times b \times h$. The base b when measured is found to be 3.26 cm, and the perpendicular height h is 7.5 cm. Determine the area of the triangle.

Area of triangle $= \dfrac{1}{2} \times b \times h = \dfrac{1}{2} \times 3.26 \times 7.5 = 12.225$ cm^2 (by calculator).

The approximate value is $\dfrac{1}{2} \times 3 \times 8 = 12$ cm^2, so there are no obvious blunder or magnitude errors. However, it is not usual in a measurement-type problem to state the answer to accuracy greater than 1 significant figure more than the least accurate number in the data; this is 7.5 cm, so the result should not have more than 3 significant figures.

Thus, **area of triangle = 12.2 cm^2**

Problem 21. State which type of error has been made in the following statements:

(a) $72 \times 31.429 = 2262.9$

(b) $16 \times 0.08 \times 7 = 89.6$

(c) $11.714 \times 0.0088 = 0.3247$, correct to 4 decimal places

(d) $\dfrac{29.74 \times 0.0512}{11.89} = 0.12$, correct to 2 significant figures

(a) $72 \times 31.429 = 2262.888$ (by calculator), hence a **rounding-off error** has occurred. The answer should have stated: $72 \times 31.429 = 2262.9$, correct to 5 significant figures or 2262.9, correct to 1 decimal place.

(b) $16 \times 0.08 \times 7 = 16 \times \dfrac{8}{100} \times 7 = \dfrac{32 \times 7}{25} = \dfrac{224}{25} = 8\dfrac{24}{25} = 8.96$

Hence an **order of magnitude** error has occurred.

(c) 11.714×0.0088 is approximately equal to $12 \times \dfrac{9}{1000}$, i.e. about $\dfrac{108}{1000}$ or 0.108

Thus a **blunder** has been made.

(d) $\dfrac{29.74 \times 0.0512}{11.89} \approx \dfrac{30 \times 5/100}{12} = \dfrac{150/100}{12} = \dfrac{1.5}{12} = \dfrac{15}{120} = \dfrac{1}{8}$ or 0.125, hence no order of magnitude error has occurred. However, by calculator, $\dfrac{29.74 \times 0.0512}{11.89} = 0.128$, correct to 3 significant figures, which equals 0.13 correct to 2 significant figures. Hence a **rounding-off error** has occurred.

Problem 22. Without using a calculator, determine an approximate value of:

(a) $\dfrac{11.7 \times 19.1}{9.3 \times 5.7}$ (b) $\dfrac{2.19 \times 203.6 \times 17.91}{12.1 \times 8.76}$

(a) $\dfrac{11.7 \times 19.1}{9.3 \times 5.7}$ is approximately equal to $\dfrac{10 \times 20}{10 \times 5}$
   i.e. about **4**
   (By calculator, $\dfrac{11.7 \times 19.1}{9.3 \times 5.7} = \mathbf{4.22}$, correct to 3 significant figures.)

(b) $\dfrac{2.19 \times 203.6 \times 17.91}{12.1 \times 8.76} \approx \dfrac{2 \times 200 \times 20}{10 \times 10}$
   $= 2 \times 2 \times 20$ after cancelling,
   i.e. $\dfrac{2.19 \times 203.6 \times 17.91}{12.1 \times 8.76} \approx \mathbf{80}$
   (By calculator, $\dfrac{2.19 \times 203.6 \times 17.91}{12.1 \times 8.76} = \mathbf{75.3}$, correct to 3 significant figures.)

Now try the following Practice Exercise

Practice Exercise 19   Errors and approximations (answers on page 1149)

In Problems 1 to 5, state which type of error, or errors, have been made:

1. $25 \times 0.06 \times 1.4 = 0.21$

2. $137 \times 6.842 = 937.4$

3. $\dfrac{24 \times 0.008}{12.6} = 10.42$

4. For a gas $pV = c$. When pressure p and volume V are measured as $p = 103400$ Pa and $V = 0.54$ m^3 then $c = 55836$ Pa m^3

5. $\dfrac{4.6 \times 0.07}{52.3 \times 0.274} = 0.225$

In Problems 6 to 8, evaluate the expressions approximately, without using a calculator.

6. 4.7×6.3

7. $\dfrac{2.87 \times 4.07}{6.12 \times 0.96}$

8. $\dfrac{72.1 \times 1.96 \times 48.6}{139.3 \times 5.2}$

4.5 Rational and irrational numbers

Rational numbers

A number that can be expressed as a fraction (or ratio) is called a **rational number**. For example, $\dfrac{3}{4}$ is a rational number; so also is **7** $\left(\text{i.e. } \dfrac{7}{1}\right)$, **0.3** $\left(\text{i.e. } \dfrac{3}{10}\right)$ and $\sqrt{9}$ (since it can be simplified to 3, i.e. $\dfrac{3}{1}$)

Irrational numbers

An **irrational number** is one that cannot be expressed as a fraction (or ratio) and has a decimal form consisting of an infinite string of numerals that does not give a repeating pattern. Examples of irrational numbers include π ($= 3.14159265\ldots$), $\sqrt{2}$ ($= 1.414213\ldots$) and e ($= 2.7182818\ldots$)

Now try the following Practice Exercise

Practice Exercise 20   Rational and irrational numbers (answers on page 1149)

State whether the following numbers are rational (R) or irrational (I):

1. 1.5   2. $\sqrt{3}$ 3. $\dfrac{5}{8}$ 4. 0.002   5. π^2 6. 11

7. $\dfrac{6}{0}$ 8. $\sqrt{16}$ 9. 0.11   10. $\left(\sqrt{2}\right)^2$

4.6 Evaluation of formulae

The statement $y = mx + c$ is called a **formula** for y in terms of m, x and c.
y, m, x and c are called **symbols** or **variables**.
When given values of m, x and c we can evaluate y.
There are a large number of formulae used in engineering and in this section we will insert numbers in place of symbols to evaluate engineering quantities.
Just four examples of important formulae are:

1. A straight line graph is of the form $y = mx + c$ (see Chapter 31)

2. Ohm's law states that $V = I \times R$

3. Velocity is expressed as $v = u + at$

4. Force is expressed as $F = m \times a$

Here are some practical examples. Check with your calculator that you agree with the working and answers.

Problem 23. In an electrical circuit the voltage V is given by Ohm's law, i.e. $V = IR$. Find, correct to 4 significant figures, the voltage when $I = 5.36$ A and $R = 14.76\,\Omega$

$$V = IR = (5.36)(14.76)$$

Hence, **voltage $V = 79.11$ V, correct to 4 significant figures**.

Problem 24. The surface area A of a hollow cone is given by $A = \pi r l$. Determine, correct to 1 decimal place, the surface area when $r = 3.0$ cm and $l = 8.5$ cm

$$A = \pi r l = \pi (3.0)(8.5)\,\text{cm}^2$$

Hence, **surface area $A = 80.1\,\text{cm}^2$, correct to 1 decimal place**.

Problem 25. Velocity v is given by $v = u + at$. If $u = 9.54$ m/s, $a = 3.67$ m/s^2 and $t = 7.82$ s, find v, correct to 3 significant figures

$$v = u + at = 9.54 + 3.67 \times 7.82$$
$$= 9.54 + 28.6994 = 38.2394$$

Hence, **velocity $v = 38.2$ m/s, correct to 3 significant figures**.

Problem 26. The area A of a circle is given by $A = \pi r^2$. Determine the area correct to 2 decimal places, given radius $r = 5.23$ m

$$A = \pi r^2 = \pi (5.23)^2 = \pi (27.3529)$$

Hence, **area $A = 85.93\,\text{m}^2$, correct to 2 decimal places**.

Problem 27. Density $= \dfrac{\text{mass}}{\text{volume}}$. Find the density when the mass is 6.45 kg and the volume is $300 \times 10^{-6}\,\text{m}^3$

$$\text{Density} = \frac{\text{mass}}{\text{volume}} = \frac{6.45\,\text{kg}}{300 \times 10^{-6}\,\text{m}^3} = \mathbf{21500\,kg/m^3}$$

Problem 28. The power, P watts, dissipated in an electrical circuit is given by the formula $P = \dfrac{V^2}{R}$. Evaluate the power, correct to 4 significant figures, given that $V = 230$ V and $R = 35.63\,\Omega$

$$P = \frac{V^2}{R} = \frac{(230)^2}{35.63} = \frac{52900}{35.63} = 1484.70390\ldots$$

Press ENG and $1.48470390\ldots \times 10^3$ appears on the screen.

Hence, **power $P = 1485$ W or 1.485 kW correct to 4 significant figures**.

Now try the following Practice Exercise

Practice Exercise 21 Evaluation of formulae (answers on page 1150)

1. The area A of a rectangle is given by the formula $A = lb$. Evaluate the area when $l = 12.4$ cm and $b = 5.37$ cm.

2. The circumference C of a circle is given by the formula $C = 2\pi r$. Determine the circumference given $r = 8.40$ mm.

3. A formula used in connection with gases is $R = \dfrac{PV}{T}$. Evaluate R when $P = 1500$, $V = 5$ and $T = 200$

4. The velocity of a body is given by $v = u + at$. The initial velocity u is measured when time t is 15 seconds and found to be 12 m/s. If the acceleration a is 9.81 m/s^2, calculate the final velocity v.

5. Calculate the current I in an electrical circuit, where $I = V/R$ amperes when the voltage V is measured and found to be 7.2 V and the resistance R is 17.7 Ω.

6. Find the distance s, given that $s = \dfrac{1}{2}gt^2$ when time $t = 0.032$ seconds and acceleration due to gravity $g = 9.81$ m/s^2. Give the answer in millimetres.

7. The energy stored in a capacitor is given by $E = \dfrac{1}{2}CV^2$ joules. Determine the energy when capacitance $C = 5 \times 10^{-6}$ farads and voltage $V = 240$ V.

8. Find the area A of a triangle, given $A = \frac{1}{2}bh$, when the base length b is 23.42 m and the height h is 53.7 m.

9. Resistance R_2 is given by $R_2 = R_1(1 + \alpha t)$. Find R_2, correct to 4 significant figures, when $R_1 = 220$, $\alpha = 0.00027$ and $t = 75.6$

10. Density $= \dfrac{\text{mass}}{\text{volume}}$. Find the density when the mass is 2.462 kg and the volume is 173 cm^3. Give the answer in units of kg/m^3 given 1cm$^3 = 10^{-6}$m^3.

11. Velocity = frequency × wavelength. Find the velocity when the frequency is 1825 Hz and the wavelength is 0.154 m.

12. Evaluate resistance R_T, given

$\dfrac{1}{R_T} = \dfrac{1}{R_1} + \dfrac{1}{R_2} + \dfrac{1}{R_3}$ when $R_1 = 5.5\,\Omega$,

$R_2 = 7.42\,\Omega$ and $R_3 = 12.6\,\Omega$.

Here are some further practical examples. Again, check with your calculator that you agree with the working and answers.

Problem 29. The volume V cm^3 of a right circular cone is given by $V = \frac{1}{3}\pi r^2 h$. Given that radius $r = 2.45$ cm and height $h = 18.7$ cm, find the volume, correct to 4 significant figures

$$V = \frac{1}{3}\pi r^2 h = \frac{1}{3}\pi(2.45)^2(18.7)$$

$$= \frac{1}{3} \times \pi \times 2.45^2 \times 18.7$$

$$= 117.544521\ldots$$

Hence, **volume $V = 117.5$ cm^3, correct to 4 significant figures**.

Problem 30. Force F newtons is given by the formula $F = \dfrac{Gm_1m_2}{d^2}$, where m_1 and m_2 are masses, d their distance apart and G is a constant. Find the value of the force given that $G = 6.67 \times 10^{-11}$, $m_1 = 7.36$, $m_2 = 15.5$ and

$d = 22.6$. Express the answer in standard form, correct to 3 significant figures

$$F = \frac{Gm_1m_2}{d^2} = \frac{(6.67 \times 10^{-11})(7.36)(15.5)}{(22.6)^2}$$

$$= \frac{(6.67)(7.36)(15.5)}{(10^{11})(510.76)} = \frac{1.490}{10^{11}}$$

Hence, **force $F = 1.49 \times 10^{-11}$ newtons, correct to 3 significant figures**.

Problem 31. The time of swing, t seconds, of a simple pendulum is given by $t = 2\pi\sqrt{\dfrac{l}{g}}$

Determine the time, correct to 3 decimal places, given that $l = 12.9$ and $g = 9.81$

$$t = 2\pi\sqrt{\frac{l}{g}} = (2\pi)\sqrt{\frac{12.9}{9.81}} = 7.20510343\ldots$$

Hence, **time $t = 7.205$ seconds, correct to 3 decimal places**.

Problem 32. Resistance, $R\,\Omega$, varies with temperature according to the formula $R = R_0(1 + \alpha t)$. Evaluate R, correct to 3 significant figures, given $R_0 = 14.59$, $\alpha = 0.0043$ and $t = 80$

$$R = R_0(1 + \alpha t) = 14.59[1 + (0.0043)(80)]$$

$$= 14.59(1 + 0.344)$$

$$= 14.59(1.344)$$

Hence, **resistance $R = 19.6\,\Omega$, correct to 3 significant figures**.

Problem 33. The current, I amperes, in an a.c. circuit is given by $I = \dfrac{V}{\sqrt{(R^2 + X^2)}}$. Evaluate the current, correct to 2 decimal places, when $V = 250$ V, $R = 25.0\,\Omega$ and $X = 18.0\,\Omega$.

$$I = \frac{V}{\sqrt{(R^2 + X^2)}} = \frac{250}{\sqrt{(25.0^2 + 18.0^2)}} = 8.11534341\ldots$$

Hence, **current $I = 8.12$ A, correct to 2 decimal places**.

Now try the following Practice Exercise

Practice Exercise 22   Evaluation of formulae (answers on page 1150)

1. Find the total cost of 37 calculators costing £12.65 each and 19 drawing sets costing £6.38 each.

2. Power $= \dfrac{\text{force} \times \text{distance}}{\text{time}}$. Find the power when a force of 3760 N raises an object a distance of 4.73 m in 35 s.

3. The potential difference, V volts, available at battery terminals is given by $V = E - Ir$. Evaluate V when $E = 5.62$, $I = 0.70$ and $R = 4.30$

4. Given force $F = \dfrac{1}{2}m(v^2 - u^2)$, find F when $m = 18.3$, $v = 12.7$ and $u = 8.24$

5. The current, I amperes, flowing in a number of cells is given by $I = \dfrac{nE}{R + nr}$. Evaluate the current when $n = 36$, $E = 2.20$, $R = 2.80$ and $r = 0.50$

6. The time, t seconds, of oscillation for a simple pendulum is given by $t = 2\pi\sqrt{\dfrac{l}{g}}$. Determine the time when $l = 54.32$ and $g = 9.81$

7. Energy, E joules, is given by the formula $E = \dfrac{1}{2}LI^2$. Evaluate the energy when $L = 5.5$ and $I = 1.2$

8. The current, I amperes, in an a.c. circuit is given by $I = \dfrac{V}{\sqrt{(R^2 + X^2)}}$. Evaluate the current when $V = 250$, $R = 11.0$ and $X = 16.2$

9. Distance, s metres, is given by the formula $s = ut + \dfrac{1}{2}at^2$. If $u = 9.50$, $t = 4.60$ and $a = -2.50$, evaluate the distance.

10. The area A of any triangle is given by $A = \sqrt{[s(s-a)(s-b)(s-c)]}$ where $s = \dfrac{a+b+c}{2}$. Evaluate the area, given $a = 3.60$ cm, $b = 4.00$ cm and $c = 5.20$ cm.

11. Given that $a = 0.290$, $b = 14.86$, $c = 0.042$, $d = 31.8$ and $e = 0.650$, evaluate v given that $v = \sqrt{\left(\dfrac{ab}{c} - \dfrac{d}{e}\right)}$

12. Deduce the following information from the train timetable shown in Table 4.1.

    (a) At what time should a man catch a train at Fratton to enable him to be in London Waterloo by 14.23 h?

    (b) A girl leaves Cosham at 12.39 h and travels to Woking. How long does the journey take? And, if the distance between Cosham and Woking is 55 miles, calculate the average speed of the train.

    (c) A man living at Havant has a meeting in London at 15.30 h. It takes around 25 minutes on the underground to reach his destination from London Waterloo. What train should he catch from Havant to comfortably make the meeting?

    (d) Nine trains leave Portsmouth harbour between 12.18 h and 13.15 h. Which train should be taken for the shortest journey time?

For fully worked solutions to each of the problems in Practice Exercises 12 to 22 in this chapter, go to the website:
www.routledge.com/cw/bird

Table 4.1 Train timetable from Portsmouth Harbour to London Waterloo

Portsmouth Harbour – London Waterloo

Saturdays

	OUTWARD	Time	Time	Time	Time	Time	Time	Time	Time	Time
Train Alterations		S04	S03	S08	S02	S03	S04	S04	S01	S02
Portsmouth Harbour	dep	12:18SW	12:22GW	12:22GW	12:45SW	12:45SW	12:45SW	12:54SW	13:12SN	13:15SW
Portsmouth & Southsea	arr	12:21	12:25	12:25	12:48	12:48	12:48	12:57	13:15	13:18
Portsmouth & Southsea	dep	12:24	12:27	12:27	12:50	12:50	12:50	12:59	13:16	13:20
Fratton	arr	12:27	12:30	12:30	12:53	12:53	12:53	13:02	13:19	13:23
Fratton	dep	12:28	12:31	12:31	12:54	12:54	12:54	13:03	13:20	13:24
Hilsea	arr	12:32						13:07		
Hilsea	dep	12:32						13:07		
Cosham	arr		12:38	12:38				13:12		
Cosham	dep		12:39	12:39				13:12		
Bedhampton	arr	12:37								
Bedhampton	dep	12:37								
Havant	arr	12:39			13:03	13:03	13:02		13:29	13:33
Havant	dep	12:40			13:04	13:04	13:04		13:30	13:34
Rowlands Castle	arr	12:46								
Rowlands Castle	dep	12:46								
Chichester	arr								13:40	
Chichester	dep								13:41	
Barnham	arr								13:48	
Barnham	dep								13:49	
Horsham	arr								14:16	
Horsham	dep								14:20	
Crawley	arr								14:28	
Crawley	dep								14:29	
Three Bridges	arr								14:32	
Three Bridges	dep								14:33	
Gatwick Airport	arr								14:37	
Gatwick Airport	dep								14:38	
Horley	arr								14:41	
Horley	dep								14:41	
Redhill	arr								14:47	
Redhill	dep								14:48	
East Croydon	arr								15:00	
East Croydon	dep								15:00	
Petersfield	arr	12:50			13:17	13:17	13:17			13:47
Petersfield	dep	12:57			13:18	13:18	13:18			13:48
Liss	arr	13:02								
Liss	dep	13:02								
Liphook	arr	13:09								
Liphook	dep	13:09								
Haslemere	arr	13:14C			13:31	13:31	13:30C			14:01
Haslemere	dep	13:20SW R			13:32	13:32	13:36SW n			14:02
Guildford	arr	13:55C			13:45	13:45	14:11C			14:15
Guildford	dep	14:02SW			13:47	13:47	14:17SW n			14:17
Portchester	arr							13:17		
Portchester	dep							13:17		
Fareham	arr		12:46	12:46				13:22		
Fareham	dep		12:47	12:47				13:23		
Southampton Central	arr			13:08C						
Southampton Central	dep			13:30SW						
Botley	arr							13:30		
Botley	dep							13:30		
Hedge End	arr							13:34		
Hedge End	dep							13:35		
Eastleigh	arr		13:00C					13:41		
Eastleigh	dep		13:09SW					13:42		
Southampton Airport Parkway	arr			13:37						
Southampton Airport Parkway	dep			13:38						
Winchester	arr		13:17	13:47				13:53		
Winchester	dep		13:18	13:48				13:54		
Micheldever	arr							14:02		
Micheldever	dep							14:02		
Basingstoke	arr		13:34					14:15		
Basingstoke	dep		13:36					14:17		
Farnborough	arr							14:30		
Farnborough	dep							14:31		
Woking	arr	14:11		14:19	13:57	13:57	14:25	14:40		14:25
Woking	dep	14:12		14:21	13:59	13:59	14:26	14:41		14:26
Clapham Junction	arr	14:31	14:12					15:01	15:11C	
Clapham Junction	dep	14:32	14:13						15:21SW	
Vauxhall	arr								15:26	
Vauxhall	dep								15:26	
London Waterloo	arr	14:40	14:24	14:49	14:23	14:27	14:51	15:13	15:31	14:51

Chapter 5

Percentages

Why it is important to understand: **Percentages**

Engineers and scientists use percentages all the time in calculations; calculators are able to handle calculations with percentages. For example, percentage change is commonly used in engineering, statistics, physics, finance, chemistry and economics. When you feel able to do calculations with basic arithmetic, fractions, decimals and percentages, all with the aid of a calculator, then suddenly mathematics doesn't seem quite so difficult.

At the end of this chapter, you should be able to:

- understand the term 'percentage'
- convert decimals to percentages and vice versa
- calculate the percentage of a quantity
- express one quantity as a percentage of another quantity
- calculate percentage error and percentage change

5.1 Introduction

Percentages are used to give a common standard. The use of percentages is very common in many aspects of commercial life, as well as in engineering. Interest rates, sale reductions, pay rises, exams and VAT are all examples of situations in which percentages are used. For this chapter you will need to know about decimals and fractions and be able to use a calculator.

We are familiar with the symbol for percentage, i.e. %. Here are some examples.

- Interest rates indicate the cost at which we can borrow money. If you borrow £8000 at a **6.5% interest rate** for a year, it will cost you 6.5% of the amount borrowed to do so, which will need to be repaid along with the original money you borrowed. If you repay the loan in 1 year, how much interest will you have paid?

- A pair of trainers in a shop cost £60. They are advertised in a sale as **20% off**. How much will you pay?

- If you earn £20 000 p.a. and you receive a **2.5% pay rise**, how much extra will you have to spend the following year?

- A book costing £18 can be purchased on the internet for 30% less. What will be its cost?

When we have completed this chapter on percentages you will be able to understand how to perform the above calculations.

Percentages are fractions having 100 as their denominator. For example, the fraction $\dfrac{40}{100}$ is written as 40% and is read as 'forty per cent'.

The easiest way to understand percentages is to go through some worked examples.

5.2   Percentage calculations

To convert a decimal to a percentage

A decimal number is converted to a percentage by multiplying by 100.

Problem 1.   Express 0.015 as a percentage

To express a decimal number as a percentage, merely multiply by 100, i.e.

$$0.015 = 0.015 \times 100\%$$
$$= \mathbf{1.5\%}$$

Multiplying a decimal number by 100 means moving the decimal point 2 places **to the right**.

Problem 2.   Express 0.275 as a percentage

$$0.275 = 0.275 \times 100\%$$
$$= \mathbf{27.5\%}$$

To convert a percentage to a decimal

A percentage is converted to a decimal number by dividing by 100.

Problem 3.   Express 6.5% as a decimal number

$$6.5\% = \frac{6.5}{100} = \mathbf{0.065}$$

Dividing by 100 means moving the decimal point 2 places **to the left**.

Problem 4.   Express 17.5% as a decimal number

$$17.5\% = \frac{17.5}{100}$$
$$= \mathbf{0.175}$$

To convert a fraction to a percentage

A fraction is converted to a percentage by multiplying by 100.

Problem 5.   Express $\frac{5}{8}$ as a percentage

$$\frac{5}{8} = \frac{5}{8} \times 100\% = \frac{500}{8}\%$$
$$= \mathbf{62.5\%}$$

Problem 6.   Express $\frac{5}{19}$ as a percentage, correct to 2 decimal places

$$\frac{5}{19} = \frac{5}{19} \times 100\%$$
$$= \frac{500}{19}\%$$
$$= 26.3157889\ldots\text{by calculator}$$
$$= \mathbf{26.32\%, \; correct \; to \; 2 \; decimal \; places}$$

Problem 7.   In two successive tests a student gains marks of 57/79 and 49/67. Is the second mark better or worse than the first?

$$57/79 = \frac{57}{79} = \frac{57}{79} \times 100\% = \frac{5700}{79}\%$$
$$= \mathbf{72.15\%}, \; \text{correct to 2 decimal places}$$
$$49/67 = \frac{49}{67} = \frac{49}{67} \times 100\% = \frac{4900}{67}\%$$
$$= \mathbf{73.13\%}, \; \text{correct to 2 decimal places}$$

Hence, **the second test is marginally better than the first test**. This question demonstrates how much easier it is to compare two fractions when they are expressed as percentages.

To convert a percentage to a fraction

A percentage is converted to a fraction by dividing by 100 and then, by cancelling, reducing it to its simplest form.

Problem 8.   Express 75% as a fraction

$$75\% = \frac{75}{100}$$
$$= \mathbf{\frac{3}{4}}$$

The fraction $\dfrac{75}{100}$ is reduced to its simplest form by cancelling, i.e. dividing both numerator and denominator by 25

Problem 9. Express 37.5% as a fraction

$$37.5\% = \frac{37.5}{100}$$

$$= \frac{375}{1000} \quad \text{by multiplying both numerator and denominator by 10}$$

$$= \frac{15}{40} \quad \text{by dividing both numerator and denominator by 25}$$

$$= \frac{3}{8} \quad \text{by dividing both numerator and denominator by 5}$$

Now try the following Practice Exercise

Practice Exercise 23   Percentages (answers on page 1150)

In Problems 1 to 5, express the given numbers as percentages.

1. 0.0032   2. 1.734

3. 0.057   4. 0.374

5. 1.285

6. Express 20% as a decimal number.

7. Express 1.25% as a decimal number.

8. Express $\dfrac{11}{16}$ as a percentage.

9. Express $\dfrac{5}{13}$ as a percentage, correct to 3 decimal places.

10. Express as percentages, correct to 3 significant figures,

    (a) $\dfrac{7}{33}$ (b) $\dfrac{19}{24}$ (c) $1\dfrac{11}{16}$

11. Place the following in order of size, the smallest first, expressing each as a percentage correct to 1 decimal place.

    (a) $\dfrac{12}{21}$ (b) $\dfrac{9}{17}$ (c) $\dfrac{5}{9}$ (d) $\dfrac{6}{11}$

12. Express 65% as a fraction in its simplest form.

13. Express 31.25% as a fraction in its simplest form.

14. Express 56.25% as a fraction in its simplest form.

15. Evaluate A to J in the following table.

Decimal number	Fraction	Percentage
0.5	A	B
C	$\dfrac{1}{4}$	D
E	F	30
G	$\dfrac{3}{5}$	H
I	J	85

16. A resistor has a value of 820 Ω ± 5%. Determine the range of resistance values expected.

5.3 Further percentage calculations

Finding a percentage of a quantity

To find a percentage of a quantity, convert the percentage to a fraction (by dividing by 100) and remember that 'of' means multiply.

Problem 10. Find 27% of £65

$$27\% \text{ of } £65 = \frac{27}{100} \times 65$$

$$= \textbf{£17.55} \text{ by calculator}$$

Problem 11. In a machine shop, it takes 32 minutes to machine a certain part. Using a new tool, the time can be reduced by 12.5%. Calculate the new time taken

$$12.5\% \text{ of } 32 \text{ minutes} = \frac{12.5}{100} \times 32$$

$$= 4 \text{ minutes}$$

Hence, **new time taken** $= 32 - 4 = \textbf{28 minutes}$.
Alternatively, if the time is reduced by 12.5%, it now takes $100\% - 12.5\% = 87.5\%$ of the original time, i.e.

$$87.5\% \text{ of } 32 \text{ minutes} = \frac{87.5}{100} \times 32$$

$$= \textbf{28 minutes}$$

Problem 12. A 160 GB iPod is advertised as costing £190 excluding VAT. If VAT is added at 17.5%, what will be the total cost of the iPod?

$$\text{VAT} = 17.5\% \text{ of } £190 = \frac{17.5}{100} \times 190 = £33.25$$

Total cost of iPod $= £190 + £33.25 = \textbf{£223.25}$

A quicker method to determine the total cost is: $1.175 \times £190 = \textbf{£223.25}$

Expressing one quantity as a percentage of another quantity

To express one quantity as a percentage of another quantity, divide the first quantity by the second then multiply by 100

Problem 13. Express 23 cm as a percentage of 72 cm, correct to the nearest 1%

$$23 \text{ cm as a percentage of } 72 \text{ cm} = \frac{23}{72} \times 100\%$$

$$= 31.94444\ldots\%$$

$$= \textbf{32\%} \text{ correct to the nearest } 1\%$$

Problem 14. Express 47 minutes as a percentage of 2 hours, correct to 1 decimal place

Note that it is essential that the two quantities are in the **same units**.

Working in minute units, 2 hours $= 2 \times 60$

$$= 120 \text{ minutes}$$

$$47 \text{ minutes as a percentage of } 120 \text{ min} = \frac{47}{120} \times 100\%$$

$$= \textbf{39.2\%}, \text{ correct to } 1 \text{ decimal place}$$

Percentage change

Percentage change is given by

$$\frac{\text{new value} - \text{original value}}{\text{original value}} \times 100\%$$

Problem 15. A box of resistors increases in price from £45 to £52. Calculate the percentage change in cost, correct to 3 significant figures

$$\% \text{ change} = \frac{\text{new value} - \text{original value}}{\text{original value}} \times 100\%$$

$$= \frac{52 - 45}{45} \times 100\% = \frac{7}{45} \times 100$$

$$= \textbf{15.6\%} = \textbf{percentage change in cost}$$

Problem 16. A drilling speed should be set to 400 rev/min. The nearest speed available on the machine is 412 rev/min. Calculate the percentage overspeed

$$\% \text{ overspeed} = \frac{\text{available speed} - \text{correct speed}}{\text{correct speed}} \times 100\%$$

$$= \frac{412 - 400}{400} \times 100\% = \frac{12}{400} \times 100\%$$

$$= \textbf{3\%}$$

Now try the following Practice Exercise

Practice Exercise 24   Further percentages (answers on page 1150)

1. Calculate 43.6% of 50 kg.

2. Determine 36% of 27 m.

3. Calculate, correct to 4 significant figures,
   (a) 18% of 2758 tonnes
   (b) 47% of 18.42 grams
   (c) 147% of 14.1 seconds.

4. When 1600 bolts are manufactured, 36 are unsatisfactory. Determine the percentage that is unsatisfactory.

5. Express
   (a) 140 kg as a percentage of 1 t (given 1 t = 1000 kg)
   (b) 47 s as a percentage of 5 min
   (c) 13.4 cm as a percentage of 2.5 m

6. A block of Monel alloy consists of 70% nickel and 30% copper. If it contains 88.2 g of nickel, determine the mass of copper in the block.

7. An athlete runs 5000 m in 15 minutes 20 seconds. With intense training, he is able to reduce this time by 2.5%. Calculate his new time.

Section A

8. A copper alloy comprises 89% copper, 1.5% iron and the remainder aluminium. Find the amount of aluminium, in grams, in a 0.8 kg mass of the alloy.

9. A computer is advertised on the internet at £520, exclusive of VAT. If VAT is payable at 20%, what is the total cost of the computer?

10. Express 325 mm as a percentage of 867 mm, correct to 2 decimal places.

11. A child sleeps on average 9 hours 25 minutes per day. Express this as a percentage of the whole day, correct to 1 decimal place.

12. Express 408 g as a percentage of 2.40 kg.

13. When signing a new contract, a Premiership footballer's pay increases from £15 500 to £21 500 per week. Calculate the percentage pay increase, correct to 3 significant figures.

14. A metal rod 1.80 m long is heated and its length expands by 48.6 mm. Calculate the percentage increase in length.

15. 12.5% of a length of wood is 70 cm. What is the full length?

16. A metal rod 1.20 m long is heated and its length expands by 42 mm. Calculate the percentage increase in length.

17. For each of the following resistors, determine the (i) minimum value, (ii) maximum value:

    (a) 820 $\Omega \pm 20\%$

    (b) 47 k$\Omega \pm 5\%$

18. An engine speed is 2400 rev/min. The speed is increased by 8%. Calculate the new speed.

5.4 More percentage calculations

Percentage error

$$\text{Percentage error} = \frac{\text{error}}{\text{correct value}} \times 100\%$$

Problem 17. The length of a component is measured incorrectly as 64.5 mm. The actual length is 63 mm. What is the percentage error in the measurement?

$$\% \text{ error} = \frac{\text{error}}{\text{correct value}} \times 100\%$$
$$= \frac{64.5 - 63}{63} \times 100\%$$
$$= \frac{1.5}{63} \times 100\% = \frac{150}{63}\%$$
$$= \mathbf{2.38\%}$$

The percentage measurement error is **2.38% too high**, which is sometimes written as **+ 2.38% error**.

Problem 18. The voltage across a component in an electrical circuit is calculated as 50 V using Ohm's law. When measured, the actual voltage is 50.4 V. Calculate, correct to 2 decimal places, the percentage error in the calculation

$$\% \text{ error} = \frac{\text{error}}{\text{correct value}} \times 100\%$$
$$= \frac{50.4 - 50}{50.4} \times 100\%$$
$$= \frac{0.4}{50.4} \times 100\% = \frac{40}{50.4}\%$$
$$= \mathbf{0.79\%}$$

The percentage error in the calculation is **0.79% too low**, which is sometimes written as **−0.79% error**.

Original value

$$\text{Original value} = \frac{\text{new value}}{100 \pm \% \text{ change}} \times 100\%$$

Problem 19. A man pays £149.50 in a sale for a DVD player which is labelled '35% off'. What was the original price of the DVD player?

In this case, it is a 35% reduction in price, so we use $\dfrac{\text{new value}}{100 - \% \text{ change}} \times 100$, i.e. a minus sign in the denominator.

$$\text{Original price} = \frac{\text{new value}}{100 - \% \text{ change}} \times 100$$
$$= \frac{149.5}{100 - 35} \times 100$$
$$= \frac{149.5}{65} \times 100 = \frac{14 950}{65}$$
$$= \mathbf{£230}$$

Problem 20. A couple buy a flat and make an 18% profit by selling it 3 years later for £153 400. Calculate the original cost of the flat

In this case, it is an 18% increase in price, so we use $\dfrac{\text{new value}}{100 + \% \text{ change}} \times 100$, i.e. a plus sign in the denominator.

$$\text{Original cost} = \frac{\text{new value}}{100 + \% \text{ change}} \times 100$$

$$= \frac{153\,400}{100 + 18} \times 100$$

$$= \frac{153\,400}{118} \times 100 = \frac{15\,340\,000}{118}$$

$$= \mathbf{£130\,000}$$

Problem 21. An electrical store makes 40% profit on each widescreen television it sells. If the selling price of a 32 inch HD television is £630, what was the cost to the dealer?

In this case, it is a 40% mark-up in price, so we use $\dfrac{\text{new value}}{100 + \% \text{ change}} \times 100$, i.e. a plus sign in the denominator.

$$\text{Dealer cost} = \frac{\text{new value}}{100 + \% \text{ change}} \times 100$$

$$= \frac{630}{100 + 40} \times 100$$

$$= \frac{630}{140} \times 100 = \frac{63\,000}{140}$$

$$= \mathbf{£450}$$

The dealer buys from the manufacturer for £450 and sells to his customers for £630.

Percentage increase/decrease and interest

$$\text{New value} = \frac{100 + \% \text{ increase}}{100} \times \text{original value}$$

Problem 22. £3600 is placed in an ISA account which pays 3.25% interest per annum. How much is the investment worth after 1 year?

$$\text{Value after 1 year} = \frac{100 + 3.25}{100} \times £3600$$

$$= \frac{103.25}{100} \times £3600$$

$$= 1.0325 \times £3600$$

$$= \mathbf{£3717}$$

Problem 23. The price of a fully installed combination condensing boiler is increased by 6.5%. It originally cost £2400. What is the new price?

$$\text{New price} = \frac{100 + 6.5}{100} \times £2400$$

$$= \frac{106.5}{100} \times £2400 = 1.065 \times £2400$$

$$= \mathbf{£2556}$$

Now try the following Practice Exercise

Practice Exercise 25   Further percentages (answers on page 1150)

1. A machine part has a length of 36 mm. The length is incorrectly measured as 36.9 mm. Determine the percentage error in the measurement.

2. When a resistor is removed from an electrical circuit the current flowing increases from 450 μA to 531 μA. Determine the percentage increase in the current.

3. In a shoe shop sale, everything is advertised as '40% off'. If a lady pays £186 for a pair of Jimmy Choo shoes, what was their original price?

4. Over a four year period a family home increases in value by 22.5% to £214 375. What was the value of the house four years ago?

5. An electrical retailer makes a 35% profit on all its products. What price does the retailer pay for a dishwasher which is sold for £351?

6. The cost of a sports car is £23 500 inclusive of VAT at 17.5%. What is the cost of the car without the VAT added?

7. £8000 is invested in bonds at a building society which is offering a rate of 2.75% per annum. Calculate the value of the investment after 2 years.

8. An electrical contractor earning £36 000 per annum receives a pay rise of 2.5%. He pays 22% of his income as tax and 11% on National Insurance contributions. Calculate the increase he will actually receive per month.

9. Five mates enjoy a meal out. With drinks, the total bill comes to £176. They add a 12.5% tip and divide the amount equally between them. How much does each pay?

10. In December a shop raises the cost of a 40 inch LCD TV costing £920 by 5%. It does not sell and in its January sale it reduces the TV by 5%. What is the sale price of the TV?

11. A man buys a business and makes a 20% profit when he sells it three years later for £222 000. What did he pay originally for the business?

12. A drilling machine should be set to 250 rev/min. The nearest speed available on the machine is 268 rev/min. Calculate the percentage overspeed.

13. Two kilograms of a compound contain 30% of element A, 45% of element B and 25% of element C. Determine the masses of the three elements present.

14. A concrete mixture contains seven parts by volume of ballast, four parts by volume of sand and two parts by volume of cement. Determine the percentage of each of these three constituents correct to the nearest 1% and the mass of cement in a two tonne dry mix, correct to 1 significant figure.

15. In a sample of iron ore, 18% is iron. How much ore is needed to produce 3600 kg of iron?

16. A screw's length is $12.5 \pm 8\%$ mm. Calculate the maximum and minimum possible length of the screw.

17. The output power of an engine is 450 kW. If the efficiency of the engine is 75%, determine the power input.

For fully worked solutions to each of the problems in Practice Exercises 23 to 25 in this chapter, go to the website:
www.routledge.com/cw/bird

Revision Test 2 Decimals, calculations and percentages

This assignment covers the material contained in Chapters 3 to 5. *The marks available are shown in brackets at the end of each question.*

1. Convert 0.048 to a proper fraction. (2)

2. Convert 6.4375 to a mixed number. (3)

3. Express $\dfrac{9}{32}$ as a decimal fraction. (2)

4. Express 0.0784 correct to 2 decimal places. (2)

5. Express 0.0572953 correct to 4 significant figures. (2)

6. Evaluate
   (a) $46.7 + 2.085 + 6.4 + 0.07$
   (b) $68.51 - 136.34$ (4)

7. Determine 2.37×1.2 (3)

8. Evaluate $250.46 \div 1.1$ correct to 1 decimal place. (3)

9. Evaluate $5.\dot{2} \times 12$ (2)

10. Evaluate the following, correct to 4 significant figures: $3.3^2 - 2.7^3 + 1.8^4$ (3)

11. Evaluate $\sqrt{6.72} - \sqrt[3]{2.54}$, correct to 3 decimal places. (3)

12. Evaluate $\dfrac{1}{0.0071} - \dfrac{1}{0.065}$, correct to 4 significant figures. (2)

13. The potential difference, V volts, available at battery terminals is given by $V = E - Ir$. Evaluate V when $E = 7.23$, $I = 1.37$ and $r = 3.60$ (3)

14. Evaluate $\dfrac{4}{9} + \dfrac{1}{5} - \dfrac{3}{8}$ as a decimal, correct to 3 significant figures. (3)

15. Evaluate $\dfrac{16 \times 10^{-6} \times 5 \times 10^9}{2 \times 10^7}$ in engineering form. (2)

16. Evaluate resistance R given $\dfrac{1}{R} = \dfrac{1}{R_1} + \dfrac{1}{R_2} + \dfrac{1}{R_3}$ when $R_1 = 3.6\,\text{k}\Omega$, $R_2 = 7.2\,\text{k}\Omega$ and $R_3 = 13.6\,\text{k}\Omega$. (3)

17. Evaluate $6\dfrac{2}{7} - 4\dfrac{5}{9}$ as a mixed number and as a decimal, correct to 3 decimal places. (3)

18. Evaluate, correct to 3 decimal places:
   $$\sqrt{\left[\dfrac{2e^{1.7} \times 3.67^3}{4.61 \times \sqrt{3\pi}}\right]}$$ (3)

19. If $a = 0.270$, $b = 15.85$, $c = 0.038$, $d = 28.7$ and $e = 0.680$, evaluate v correct to 3 significant figures, given that $v = \sqrt{\left(\dfrac{ab}{c} - \dfrac{d}{e}\right)}$ (4)

20. Evaluate the following, each correct to 2 decimal places.
   (a) $\left(\dfrac{36.2^2 \times 0.561}{27.8 \times 12.83}\right)^3$
   (b) $\sqrt{\left(\dfrac{14.69^2}{\sqrt{17.42} \times 37.98}\right)}$ (4)

21. If $1.6\,\text{km} = 1\,\text{mile}$, determine the speed of 45 miles/hour in kilometres per hour. (2)

22. The area A of a circle is given by $A = \pi r^2$. Find the area of a circle of radius $r = 3.73\,\text{cm}$, correct to 2 decimal places. (3)

23. Evaluate B, correct to 3 significant figures, when $W = 7.20$, $v = 10.0$ and $g = 9.81$, given that $B = \dfrac{Wv^2}{2g}$ (3)

24. Express 56.25% as a fraction in its simplest form. (3)

25. 12.5% of a length of wood is 70 cm. What is the full length? (3)

26. A metal rod 1.20 m long is heated and its length expands by 42 mm. Calculate the percentage increase in length. (2)

27. A man buys a house and makes a 20% profit when he sells it three years later for £312 000. What did he pay for it originally? (3)

For lecturers/instructors/teachers, fully worked solutions to each of the problems in Revision Test 2, together with a full marking scheme, are available at the website:

www.routledge.com/cw/bird

Chapter 6

Ratio and proportion

Why it is important to understand: Ratio and proportion

Why it is important to understand: **Ratio and proportion**

Real-life applications of ratio and proportion are numerous. When you prepare recipes, paint your house or repair gears in a large machine or in a car transmission, you use ratios and proportions.

Builders use ratios all the time; a simple tool may be referred to as a 5/8 or 3/16 wrench. Trusses must have the correct ratio of pitch to support the weight of roof and snow, cement must be the correct mixture to be sturdy, and doctors are always calculating ratios as they determine medications. Almost every job uses ratios in one way or another; ratios are used in building and construction, model making, art and crafts, land surveying, die and tool making, food and cooking, chemical mixing, in automobile manufacturing and in aeroplane and parts making. Engineers use ratios to test structural and mechanical systems for capacity and safety issues. Millwrights use ratio to solve pulley rotation and gear problems. Operating engineers apply ratios to ensure the correct equipment is used to safely move heavy materials such as steel on worksites. It is therefore important that we have some working understanding of ratio and proportion.

At the end of this chapter, you should be able to:

- define ratio
- perform calculations with ratios
- define direct proportion
- perform calculations with direct proportion, including Hooke's law, Charles's law and Ohm's law
- define inverse proportion
- perform calculations with inverse proportion, including Boyle's law

6.1 Introduction

Ratio is a way of comparing amounts of something; it shows how much bigger one thing is than the other. Some practical examples include mixing paint, sand and cement, or screen wash. Gears, map scales, food recipes, scale drawings and metal alloy constituents all use ratios.

Two quantities are in **direct proportion** when they increase or decrease in the **same ratio**. There are several practical engineering laws which rely on direct proportion. Also, calculating currency exchange rates and converting imperial to metric units rely on direct proportion.

Sometimes, as one quantity increases at a particular rate, another quantity decreases at the same rate; this is called **inverse proportion**. For example, the time taken to do a job is inversely proportional to the number of people in a team: double the people, half the time.

When we have completed this chapter on ratio and proportion you will be able to understand, and confidently perform, calculations on the above topics.

For this chapter you will need to know about decimals and fractions and to be able to use a calculator.

6.2   Ratios

Ratios are generally shown as numbers separated by a colon (:), so the ratio of 2 and 7 is written as 2:7 and we read it as a ratio of 'two to seven.'

Some practical examples which are familiar include:

- Mixing 1 measure of screen wash to 6 measures of water; i.e. the ratio of screen wash to water is 1:6

- Mixing 1 shovel of cement to 4 shovels of sand; i.e. the ratio of cement to sand is 1:4

- Mixing 3 parts of red paint to 1 part white, i.e. the ratio of red to white paint is 3:1

Ratio is the number of parts to a mix. The paint mix is 4 parts total, with 3 parts red and 1 part white. 3 parts red paint to 1 part white paint means there is

$$\frac{3}{4} \text{ red paint to } \frac{1}{4} \text{ white paint}$$

Here are some worked examples to help us understand more about ratios.

Problem 1.   In a class, the ratio of female to male students is 6:27. Reduce the ratio to its simplest form

(i)   Both 6 and 27 can be divided by 3

(ii)   Thus, 6:27 is the same as **2:9**

6:27 and 2:9 are called **equivalent ratios**.
It is normal to express ratios in their lowest, or simplest, form. In this example, the simplest form is **2:9** which means for every 2 females in the class there are 9 male students.

Problem 2.   A gear wheel having 128 teeth is in mesh with a 48-tooth gear. What is the gear ratio?

$$\text{Gear ratio} = 128:48$$

A ratio can be simplified by finding common factors.

(i)   128 and 48 can both be divided by 2, i.e. 128:48 is the same as 64:24

(ii)   64 and 24 can both be divided by 8, i.e. 64:24 is the same as 8:3

(iii)   There is no number that divides completely into both 8 and 3 so 8:3 is the simplest ratio, i.e. **the gear ratio is 8:3**

Thus, 128:48 is equivalent to 64:24 which is equivalent to 8:3 and **8:3 is the simplest form**.

Problem 3.   A wooden pole is 2.08 m long. Divide it in the ratio of 7 to 19

(i)   Since the ratio is 7:19, the total number of parts is 7 + 19 = 26 parts.

(ii)   26 parts corresponds to 2.08 m = 208 cm, hence 1 part corresponds to $\frac{208}{26} = 8$

(iii)   Thus, 7 parts corresponds to $7 \times 8 =$ **56 cm** and 19 parts corresponds to $19 \times 8 =$ **152 cm**.

Hence, **2.08 m divides in the ratio of 7:19 as 56 cm to 152 cm**.
(Check: 56 + 152 must add up to 208, otherwise an error has been made.)

Problem 4.   In a competition, prize money of £828 is to be shared among the first three winners in the ratio 5:3:1

(i)   Since the ratio is 5:3:1 the total number of parts is 5 + 3 + 1 = 9 parts.

(ii)   9 parts corresponds to £828

(iii)   1 part corresponds to $\frac{828}{9} =$ **£92**, 3 parts corresponds to $3 \times £92 =$ **£276** and 5 parts corresponds to $5 \times £92 =$ **£460**

Hence, **£828 divides in the ratio of 5:3:1 as £460 to £276 to £92**. (Check: 460 + 276 + 92 must add up to 828, otherwise an error has been made.)

Problem 5.   A map scale is 1:30 000. On the map the distance between two schools is 6 cm.
Determine the actual distance between the schools, giving the answer in kilometres

Actual distance between schools

$$= 6 \times 30\,000\,\text{cm} = 180\,000\,\text{cm}$$

$$= \frac{180\,000}{100}\,\text{m} = 1800\,\text{m}$$

$$= \frac{1800}{1000}\,\text{m} = \mathbf{1.80\,km}$$

(1 mile $\approx$ 1.6 km, hence the schools are just over 1 mile apart.)

Now try the following Practice Exercise

Practice Exercise 26   Ratios (answers on page 1150)

1. In a box of 333 paper clips, 9 are defective. Express the number of non-defective paper clips as a ratio of the number of defective paper clips, in its simplest form.

2. A gear wheel having 84 teeth is in mesh with a 24-tooth gear. Determine the gear ratio in its simplest form.

3. In a box of 2000 nails, 120 are defective. Express the number of non-defective nails as a ratio of the number of defective ones, in its simplest form.

4. A metal pipe 3.36 m long is to be cut into two in the ratio 6 to 15. Calculate the length of each piece.

5. The instructions for cooking a turkey say that it needs to be cooked 45 minutes for every kilogram. How long will it take to cook a 7 kg turkey?

6. In a will, £6440 is to be divided among three beneficiaries in the ratio $4:2:1$. Calculate the amount each receives.

7. A local map has a scale of $1:22\,500$. The distance between two motorways is 2.7 km. How far are they apart on the map?

8. Prize money in a lottery totals £3801 and is shared among three winners in the ratio $4:2:1$. How much does the first prize winner receive?

Here are some further worked examples on ratios.

Problem 6.   Express 45 p as a ratio of £7.65 in its simplest form

(i)   Changing both quantities to the same units, i.e. to pence, gives a ratio of $45:765$

(ii)   Dividing both quantities by 5 gives $45:765 \equiv 9:153$

(iii)   Dividing both quantities by 3 gives $9:153 \equiv 3:51$

(iv)   Dividing both quantities by 3 again gives $3:51 \equiv 1:17$

Thus, **45 p as a ratio of £7.65 is $1:17$**
$45:765, 9:153, 3:51$ and $1:17$ are **equivalent ratios** and $\mathbf{1:17}$ **is the simplest ratio**.

Problem 7.   A glass contains 30 ml of whisky which is 40% alcohol. If 45 ml of water is added and the mixture stirred, what is now the alcohol content?

(i)   The 30 ml of whisky contains 40% alcohol $= \dfrac{40}{100} \times 30 = 12\,\text{ml}$.

(ii)   After 45 ml of water is added we have $30 + 45 = 75\,\text{ml}$ of fluid, of which alcohol is 12 ml.

(iii)   Fraction of alcohol present $= \dfrac{12}{75}$

(iv)   Percentage of alcohol present $= \dfrac{12}{75} \times 100\%$

$$= \mathbf{16\%}$$

Problem 8.   20 tonnes of a mixture of sand and gravel is 30% sand. How many tonnes of sand must be added to produce a mixture which is 40% gravel?

(i)   Amount of sand in 20 tonnes $= 30\%$ of 20 t $= \dfrac{30}{100} \times 20 = 6\,\text{t}$.

(ii)   If the mixture has 6 t of sand then amount of gravel $= 20 - 6 = 14\,\text{t}$.

(iii)   We want this 14 t of gravel to be 40% of the new mixture. 1% would be $\dfrac{14}{40}$ t and 100% of the mixture would be $\dfrac{14}{40} \times 100\,\text{t} = 35\,\text{t}$.

(iv)   If there is 14 t of gravel then amount of sand $= 35 - 14 = 21\,\text{t}$.

(v) We already have 6 t of sand, so **amount of sand to be added to produce a mixture with 40% gravel** $= 21 - 6 = \mathbf{15\,t}$.

(Note 1 tonne $= 1000$ kg.)

Now try the following Practice Exercise

Practice Exercise 27   Further ratios (answers on page 1150)

1. Express 130 g as a ratio of 1.95 kg.

2. In a laboratory, acid and water are mixed in the ratio 2:5. How much acid is needed to make 266 ml of the mixture?

3. A glass contains 30 ml of gin which is 40% alcohol. If 18 ml of water is added and the mixture stirred, determine the new percentage alcoholic content.

4. A wooden beam 4 m long weighs 84 kg. Determine the mass of a similar beam that is 60 cm long.

5. An alloy is made up of metals P and Q in the ratio 3.25:1 by mass. How much of P has to be added to 4.4 kg of Q to make the alloy?

6. 15 000 kg of a mixture of sand and gravel is 20% sand. Determine the amount of sand that must be added to produce a mixture with 30% gravel.

6.3   Direct proportion

Two quantities are in **direct proportion** when they increase or decrease in the **same ratio**. For example, if 12 cans of lager have a mass of 4 kg, then 24 cans of lager will have a mass of 8 kg; i.e. if the quantity of cans doubles then so does the mass. This is direct proportion.

In the previous section we had an example of mixing 1 shovel of cement to 4 shovels of sand; i.e. the ratio of cement to sand was 1:4. So, if we have a mix of 10 shovels of cement and 40 shovels of sand and we wanted to double the amount of the mix then we would need to double both the cement and sand, i.e. 20 shovels of cement and 80 shovels of sand. This is another example of direct proportion.

Here are three laws in engineering which involve direct proportion:

(a) **Hooke's law*** states that, within the elastic limit of a material, the strain ε produced is directly proportional to the stress σ producing it, i.e. $\varepsilon \propto \sigma$ (note than '$\propto$' means 'is proportional to').

(b) **Charles's law*** states that, for a given mass of gas at constant pressure, the volume V is directly proportional to its thermodynamic temperature T, i.e. $V \propto T$.

ROBERT HOOKE
1635 - 1703

* **Who was Hooke?** **Robert Hooke FRS** (28 July 1635–3 March 1703) was an English natural philosopher, architect and polymath who, amongst other things, discovered the law of elasticity. To find out more go to **www.routledge.com/cw/bird**

* **Who was Charles?** **Jacques Alexandre César Charles** (12 November 1746–7 April 1823) was a French inventor, scientist, mathematician and balloonist. Charles's law describes how gases expand when heated. To find out more go to **www.routledge.com/cw/bird**

(c) **Ohm's law*** states that the current I flowing through a fixed resistance is directly proportional to the applied voltage V, i.e. $I \propto V$.

Here are some worked examples to help us understand more about direct proportion.

Problem 9. 3 energy-saving light bulbs cost £7.80. Determine the cost of 7 such light bulbs

(i) 3 light bulbs cost £7.80

(ii) Therefore, 1 light bulb costs $\dfrac{7.80}{3} = £2.60$

Hence, **7 light bulbs cost** $7 \times £2.60 = £18.20$

Problem 10. If 56 litres of petrol costs £59.92, calculate the cost of 32 litres

(i) 56 litres of petrol costs £59.92

* **Who was Ohm? Georg Simon Ohm** (16 March 1789–6 July 1854) was a Bavarian physicist and mathematician who discovered what came to be known as Ohm's law – the direct proportionality between voltage and the resultant electric current. To find out more go to **www.routledge.com/cw/bird**

(ii) Therefore, 1 litre of petrol costs $\dfrac{59.92}{56} = £1.07$

Hence, **32 litres cost** $32 \times 1.07 = £34.24$

Problem 11. Hooke's law states that stress, σ, is directly proportional to strain, ε, within the elastic limit of a material. When, for mild steel, the stress is 63 MPa, the strain is 0.0003. Determine (a) the value of strain when the stress is 42 MPa, (b) the value of stress when the strain is 0.00072

(a) Stress is directly proportional to strain.
  (i) When the stress is 63 MPa, the strain is 0.0003
  (ii) Hence, a stress of 1 MPa corresponds to a strain of $\dfrac{0.0003}{63}$
  (iii) Thus, **the value of strain when the stress is 42 MPa** $= \dfrac{0.0003}{63} \times 42 = \mathbf{0.0002}$

(b) Strain is proportional to stress.
  (i) When the strain is 0.0003, the stress is 63 MPa.
  (ii) Hence, a strain of 0.0001 corresponds to $\dfrac{63}{3}$ MPa.
  (iii) Thus, **the value of stress when the strain is 0.00072** $= \dfrac{63}{3} \times 7.2 = \mathbf{151.2\ MPa}$.

Problem 12. Charles's law states that for a given mass of gas at constant pressure, the volume is directly proportional to its thermodynamic temperature. A gas occupies a volume of 2.4 litres at 600 K. Determine (a) the temperature when the volume is 3.2 litres, (b) the volume at 540 K

(a) Volume is directly proportional to temperature.
  (i) When the volume is 2.4 litres, the temperature is 600 K.
  (ii) Hence, a volume of 1 litre corresponds to a temperature of $\dfrac{600}{2.4}$ K.
  (iii) Thus, **the temperature when the volume is 3.2 litres** $= \dfrac{600}{2.4} \times 3.2 = \mathbf{800\ K}$.

(b) Temperature is proportional to volume.

  (i) When the temperature is 600 K, the volume is 2.4 litres.

  (ii) Hence, a temperature of 1 K corresponds to a volume of $\dfrac{2.4}{600}$ litres.

  (iii) Thus, **the volume at a temperature of 540 K** $= \dfrac{2.4}{600} \times 540 = \textbf{2.16 litres}$.

Now try the following Practice Exercise

Practice Exercise 28   Direct proportion (answers on page 1150)

1. 3 engine parts cost £208.50. Calculate the cost of 8 such parts.

2. If 9 litres of gloss white paint costs £24.75, calculate the cost of 24 litres of the same paint.

3. The total mass of 120 household bricks is 57.6 kg. Determine the mass of 550 such bricks.

4. A simple machine has an effort : load ratio of 3 : 37. Determine the effort, in newtons, to lift a load of 5.55 kN.

5. If 16 cans of lager weighs 8.32 kg, what will 28 cans weigh?

6. Hooke's law states that stress is directly proportional to strain within the elastic limit of a material. When, for copper, the stress is 60 MPa, the strain is 0.000625. Determine (a) the strain when the stress is 24 MPa and (b) the stress when the strain is 0.0005

7. Charles's law states that volume is directly proportional to thermodynamic temperature for a given mass of gas at constant pressure. A gas occupies a volume of 4.8 litres at 330 K. Determine (a) the temperature when the volume is 6.4 litres and (b) the volume when the temperature is 396 K.

8. A machine produces 320 bolts in a day. Calculate the number of bolts produced by 4 machines in 7 days.

Here are some further worked examples on direct proportion.

> **Problem 13.** Some guttering on a house has to decline by 3 mm for every 70 cm to allow rainwater to drain. The gutter spans 8.4 m. How much lower should the low end be?

(i) The guttering has to decline in the ratio 3 : 700 or $\dfrac{3}{700}$

(ii) If d is the vertical drop in 8.4 m or 8400 mm, then the decline must be in the ratio d : 8400 or $\dfrac{d}{8400}$

(iii) Now $\dfrac{d}{8400} = \dfrac{3}{700}$

(iv) Cross-multiplying gives $700 \times d = 8400 \times 3$ from which, $d = \dfrac{8400 \times 3}{700}$

i.e. **$d = 36$ mm, which is how much lower the low end should be to allow rainwater to drain**.

> **Problem 14.** Ohm's law states that the current flowing in a fixed resistance is directly proportional to the applied voltage. When 90 mV is applied across a resistor the current flowing is 3 A. Determine (a) the current when the voltage is 60 mV and (b) the voltage when the current is 4.2 A

(a) Current is directly proportional to the voltage.

  (i) When voltage is 90 mV, the current is 3 A

  (ii) Hence, a voltage of 1 mV corresponds to a current of $\dfrac{3}{90}$ A

  (iii) Thus, **when the voltage is 60 mV, the current** $= 60 \times \dfrac{3}{90} = \textbf{2A}$.

(b) Voltage is directly proportional to the current.

  (i) When current is 3 A, the voltage is 90 mV.

  (ii) Hence, a current of 1 A corresponds to a voltage of $\dfrac{90}{3}$ mV $= 30$ mV.

  (iii) Thus, **when the current is 4.2 A, the voltage** $= 30 \times 4.2 = \textbf{126 mV}$.

Problem 15. Some approximate imperial to metric conversions are shown in Table 6.1. Use the table to determine

(a)   the number of millimetres in 12.5 inches

(b)   a speed of 50 miles per hour in kilometres per hour

(c)   the number of miles in 300 km

(d)   the number of kilograms in 20 pounds weight

(e)   the number of pounds and ounces in 56 kilograms (correct to the nearest ounce)

(f)   the number of litres in 24 gallons

(g)   the number of gallons in 60 litres

Table 6.1

length	1 inch = 2.54 cm
	1 mile = 1.6 km
weight	2.2 lb = 1 kg
	(1 lb = 16 oz)
capacity	1.76 pints = 1 litre
	(8 pints = 1 gallon)

(a)   12.5 inches = 12.5 × 2.54 cm = 31.75 cm

31.73 cm = 31.75 × 10 mm = **317.5 mm**

(b)   50 m.p.h. = 50 × 1.6 km/h = **80 km/h**

(c) $300 \text{ km} = \dfrac{300}{1.6} \text{ miles} = \textbf{187.5 miles}$

(d) $20 \text{ lb} = \dfrac{20}{2.2} \text{ kg} = \textbf{9.09 kg}$

(e)   56 kg = 56 × 2.2 lb = 123.2 lb

0.2 lb = 0.2 × 16 oz = 3.2 oz = 3 oz,   correct   to the nearest ounce.

Thus, 56 kg = **123 lb 3 oz**, correct to the nearest ounce.

(f)   24 gallons = 24 × 8 pints = 192 pints

$192 \text{ pints} = \dfrac{192}{1.76} \text{ litres} = \textbf{109.1 litres}$

(g)   60 litres = 60 × 1.76 pints = 105.6 pints

$105.6 \text{ pints} = \dfrac{105.6}{8} \text{ gallons} = \textbf{13.2 gallons}$

Problem 16. Currency exchange rates for five countries are shown in Table 6.2. Calculate

(a)   how many euros £55 will buy

(b)   the number of Japanese yen which can be bought for £23

(c)   the number of pounds sterling which can be exchanged for 6405 kronor

(d)   the number of American dollars which can be purchased for £92.50

(e)   the number of pounds sterling which can be exchanged for 2926 Swiss francs

Table 6.2

France	£1 = 1.25 euros
Japan	£1 = 150 yen
Norway	£1 = 10.50 kronor
Switzerland	£1 = 1.40 francs
USA	£1 = 1.30 dollars

(a)   £1 = 1.25 euros, hence £55 = 55 × 1.25 euros
= **68.75 euros**

(b)   £1 = 150 yen, hence £23 = 23 × 150 yen
= **3450 yen**

(c) $£1 = 10.50 \text{ kronor, hence } 6405 \text{ kronor} = £\dfrac{6405}{10.50}$
$= \textbf{£610}$

(d)   £1 = 1.30 dollars, hence
£92.50 = 92.50 × 1.30 dollars = **$120.25**

(e)   £1 = 1.40 Swiss francs, hence
$2926 \text{ Swiss francs} = £\dfrac{2926}{1.40} = \textbf{£2090}$

Now try the following Practice Exercise

Practice Exercise 29   Further direct proportion (answers on page 1150)

1.   Ohm's law states that current is proportional to p.d. in an electrical circuit. When a p.d. of 60 mV is applied across a circuit a current of 24 μA flows. Determine (a) the current flowing

when the p.d. is 5 V and (b) the p.d. when the current is 10 mA.

2. The tourist rate for the Swiss franc is quoted in a newspaper as £1 = 1.40 fr. How many francs can be purchased for £310?

3. If 1 inch = 2.54 cm, find the number of millimetres in 27 inches.

4. If 2.2 lb = 1 kg and 1 lb = 16 oz, determine the number of pounds and ounces in 38 kg (correct to the nearest ounce).

5. If 1 litre = 1.76 pints and 8 pints = 1 gallon, determine (a) the number of litres in 35 gallons and (b) the number of gallons in 75 litres.

6. Hooke's law states that stress is directly proportional to strain within the elastic limit of a material. When, for brass, the stress is 21 MPa, the strain is 0.00025. Determine the stress when the strain is 0.00035

7. If 12 inches = 30.48 cm, find the number of millimetres in 23 inches.

8. The tourist rate for the Canadian dollar is quoted in a newspaper as £1 = $1.62. How many Canadian dollars can be purchased for £550?

6.4 Inverse proportion

Two variables, x and y, are in inverse proportion to one another if y is proportional to $\frac{1}{x}$, i.e. $y \propto \frac{1}{x}$ or $y = \frac{k}{x}$ or $k = xy$ where k is a constant, called the **coefficient of proportionality**.

Inverse proportion means that as the value of one variable increases the value of another decreases, and that their product is always the same.

For example, the time for a journey is inversely proportional to the speed of travel. So, if at 30 m.p.h. a journey is completed in 20 minutes, then at 60 m.p.h. the journey would be completed in 10 minutes. Double the speed, half the journey time. (Note that $30 \times 20 = 60 \times 10$)

In another example, the time needed to dig a hole is inversely proportional to the number of people digging. So, if 4 men take 3 hours to dig a hole, then 2 men (working at the same rate) would take 6 hours. Half the men, twice the time. (Note that $4 \times 3 = 2 \times 6$)

Here are some worked examples on inverse proportion.

Problem 17. It is estimated that a team of four designers would take a year to develop an engineering process. How long would three designers take?

If 4 designers take 1 year, then 1 designer would take 4 years to develop the process. Hence, 3 designers would take $\frac{4}{3}$ years, i.e. **1 year 4 months**.

Problem 18. A team of five people can deliver leaflets to every house in a particular area in four hours. How long will it take a team of three people?

If 5 people take 4 hours to deliver the leaflets, then 1 person would take $5 \times 4 = 20$ hours. Hence, 3 people would take $\frac{20}{3}$ hours, i.e. $6\frac{2}{3}$ hours, i.e. **6 hours 40 minutes**.

Problem 19. The electrical resistance R of a piece of wire is inversely proportional to the cross-sectional area A. When $A = 5\,\text{mm}^2$, $R = 7.2$ ohms. Determine (a) the coefficient of proportionality and (b) the cross-sectional area when the resistance is 4 ohms

(a) $R \propto \frac{1}{A}$, i.e. $R = \frac{k}{A}$ or $k = RA$. Hence, when $R = 7.2$ and $A = 5$, the

coefficient of proportionality, $k = (7.2)(5) = \mathbf{36}$

(b) Since $k = RA$ then $A = \frac{k}{R}$. Hence, when $R = 4$,

the cross-sectional area, $A = \frac{36}{4} = \mathbf{9\,mm^2}$

Problem 20. Boyle's law* states that, at constant temperature, the volume V of a fixed mass of gas is inversely proportional to its absolute pressure p. If a gas occupies a volume of $0.08\,\text{m}^3$ at a pressure of 1.5×10^6 pascals, determine (a) the coefficient of proportionality and (b) the volume if the pressure is changed to 4×10^6 pascals

(a) $V \propto \dfrac{1}{p}$ i.e. $V = \dfrac{k}{p}$ or $k = pV$. Hence, the

coefficient of proportionality, k

$$= (1.5 \times 10^6)(0.08) = \mathbf{0.12 \times 10^6}$$

(b) **Volume, $V = \dfrac{k}{p} = \dfrac{0.12 \times 10^6}{4 \times 10^6} = \mathbf{0.03\,m^3}$**

* **Who was Boyle? Robert Boyle** (25 January 1627–31 December 1691) was a natural philosopher, chemist, physicist and inventor. Regarded today as the first modern chemist, he is best known for Boyle's law, which describes the inversely proportional relationship between the absolute pressure and volume of a gas, providing the temperature is kept constant within a closed system. To find out more go to **www.routledge.com/cw/bird**

Now try the following Practice Exercise

Practice Exercise 30   Further inverse proportion (answers on page 1150)

1. A 10 kg bag of potatoes lasts for a week with a family of 7 people. Assuming all eat the same amount, how long will the potatoes last if there are only two in the family?

2. If 8 men take 5 days to build a wall, how long would it take 2 men?

3. If y is inversely proportional to x and $y = 15.3$ when $x = 0.6$, determine (a) the coefficient of proportionality, (b) the value of y when x is 1.5 and (c) the value of x when y is 27.2

4. A car travelling at 50 km/h makes a journey in 70 minutes. How long will the journey take at 70 km/h?

5. Boyle's law states that, for a gas at constant temperature, the volume of a fixed mass of gas is inversely proportional to its absolute pressure. If a gas occupies a volume of 1.5 m^3 at a pressure of 200×10^3 pascals, determine (a) the constant of proportionality, (b) the volume when the pressure is 800×10^3 pascals and (c) the pressure when the volume is 1.25 m^3.

6. The energy received by a surface from a source of heat is inversely proportional to the square of the distance between the heat source and the surface. A surface 1 m from the heat source receives 200 J of energy. Calculate (a) the energy received when the distance is changed to 2.5 m, (b) the distance required if the surface is to receive 800 J of energy.

For fully worked solutions to each of the problems in Practice Exercises 26 to 30 in this chapter, go to the website:
www.routledge.com/cw/bird

Chapter 7

Powers, roots and laws of indices

Why it is important to understand: Powers, roots and laws of indices

Powers and roots are used extensively in mathematics and engineering, so it is important to get a good grasp of what they are and how, and why, they are used. Being able to multiply powers together by adding their indices is particularly useful for disciplines like engineering and electronics, where quantities are often expressed as a value multiplied by some power of ten. In the field of electrical engineering, for example, the relationship between electric current, voltage and resistance in an electrical system is critically important, and yet the typical unit values for these properties can differ by several orders of magnitude. Studying, or working, in an engineering discipline, you very quickly become familiar with powers and roots and laws of indices. This chapter provides an important lead into Chapter 8, which deals with units, prefixes and engineering notation.

At the end of this chapter, you should be able to:

- understand the terms base, index and power
- understand square roots
- perform calculations with powers and roots
- state the laws of indices
- perform calculations using the laws of indices

7.1 Introduction

The manipulation of powers and roots is a crucial underlying skill needed in algebra. In this chapter, powers and roots of numbers are explained, together with the laws of indices.

Many worked examples are included to help understanding.

7.2 Powers and roots

Indices

The number 16 is the same as $2 \times 2 \times 2 \times 2$, and $2 \times 2 \times 2 \times 2$ can be abbreviated to 2^4. When written as 2^4, 2 is called the **base** and the 4 is called the **index** or **power**. 2^4 is read as '**two to the power of four**'.

Similarly, 3^5 is read as '**three to the power of 5**'.

When the indices are 2 and 3 they are given special names; i.e. 2 is called 'squared' and 3 is called 'cubed'. Thus,

4^2 is called '**four squared**' rather than '4 to the power of 2' and 5^3 is called '**five cubed**' rather than '5 to the power of 3'.

When no index is shown, the power is 1. For example, 2 means 2^1

Problem 1. Evaluate (a) 2^6 (b) 3^4

(a) 2^6 means $2 \times 2 \times 2 \times 2 \times 2 \times 2$ (i.e. 2 multiplied by itself 6 times), and $2 \times 2 \times 2 \times 2 \times 2 \times 2 = \textbf{64}$

i.e. $\qquad \textbf{2}^\textbf{6} = \textbf{64}$

(b) 3^4 means $3 \times 3 \times 3 \times 3$ (i.e. 3 multiplied by itself 4 times), and $3 \times 3 \times 3 \times 3 = \textbf{81}$

i.e. $\qquad \textbf{3}^\textbf{4} = \textbf{81}$

Problem 2. Change the following to index form: (a) 32 (b) 625

(a) (i) To express 32 in its lowest factors, 32 is initially divided by the lowest prime number, i.e. 2

(ii) $32 \div 2 = 16$, hence $32 = 2 \times 16$

(iii) 16 is also divisible by 2, i.e. $16 = 2 \times 8$. Thus, $32 = 2 \times 2 \times 8$

(iv) 8 is also divisible by 2, i.e. $8 = 2 \times 4$. Thus, $32 = 2 \times 2 \times 2 \times 4$

(v) 4 is also divisible by 2, i.e. $4 = 2 \times 2$. Thus, $32 = 2 \times 2 \times 2 \times 2 \times 2$

(vi) Thus, $32 = \textbf{2}^\textbf{5}$

(b) (i) 625 is not divisible by the lowest prime number, i.e. 2. The next prime number is 3 and 625 is not divisible by 3 either. The next prime number is 5

(ii) $625 \div 5 = 125$, i.e. $625 = 5 \times 125$

(iii) 125 is also divisible by 5, i.e. $125 = 5 \times 25$. Thus, $625 = 5 \times 5 \times 25$

(iv) 25 is also divisible by 5, i.e. $25 = 5 \times 5$. Thus, $625 = 5 \times 5 \times 5 \times 5$

(v) Thus, $\textbf{625} = \textbf{5}^\textbf{4}$

Problem 3. Evaluate $3^3 \times 2^2$

$$3^3 \times 2^2 = 3 \times 3 \times 3 \times 2 \times 2$$
$$= 27 \times 4$$
$$= \textbf{108}$$

Square roots

When a number is multiplied by itself the product is called a square.

For example, the square of 3 is $3 \times 3 = 3^2 = 9$

A square root is the reverse process; i.e. the value of the base which when multiplied by itself gives the number; i.e. the square root of 9 is 3

The symbol $\sqrt{}$ is used to denote a square root. Thus, $\sqrt{9} = 3$. Similarly, $\sqrt{4} = 2$ and $\sqrt{25} = 5$

Because $-3 \times -3 = 9$, $\sqrt{9}$ also equals -3. Thus, $\sqrt{9} = +3$ or -3 which is usually written as $\sqrt{9} = \pm 3$. Similarly, $\sqrt{16} = \pm 4$ and $\sqrt{36} = \pm 6$

The square root of, say, 9 may also be written in index form as $9^{\frac{1}{2}}$

$$9^{\frac{1}{2}} \equiv \sqrt{9} = \pm 3$$

Problem 4. Evaluate $\dfrac{3^2 \times 2^3 \times \sqrt{36}}{\sqrt{16} \times 4}$ taking only positive square roots

$$\frac{3^2 \times 2^3 \times \sqrt{36}}{\sqrt{16} \times 4} = \frac{3 \times 3 \times 2 \times 2 \times 2 \times 6}{4 \times 4}$$
$$= \frac{9 \times 8 \times 6}{16} = \frac{9 \times 1 \times 6}{2}$$
$$= \frac{9 \times 1 \times 3}{1} \qquad \text{by cancelling}$$
$$= \textbf{27}$$

Problem 5. Evaluate $\dfrac{10^4 \times \sqrt{100}}{10^3}$ taking the positive square root only

$$\frac{10^4 \times \sqrt{100}}{10^3} = \frac{10 \times 10 \times 10 \times 10 \times 10}{10 \times 10 \times 10}$$
$$= \frac{1 \times 1 \times 1 \times 10 \times 10}{1 \times 1 \times 1} \qquad \text{by cancelling}$$
$$= \frac{100}{1}$$
$$= \textbf{100}$$

Now try the following Practice Exercise

Practice Exercise 31   Powers and roots (answers on page 1150)

Evaluate the following without the aid of a calculator.

1. 3^3

2. 2^7

3. 10^5

4. $2^4 \times 3^2 \times 2 \div 3$

5. Change 16 to index form.

6. $25^{\frac{1}{2}}$

7. $64^{\frac{1}{2}}$

8. $\dfrac{10^5}{10^3}$

9. $\dfrac{10^2 \times 10^3}{10^5}$

10. $\dfrac{2^5 \times 64^{\frac{1}{2}} \times 3^2}{\sqrt{144} \times 3}$
    taking positive square roots only.

7.3   Laws of indices

There are six laws of indices.

(1)   From earlier, $2^2 \times 2^3 = (2 \times 2) \times (2 \times 2 \times 2)$

$$= 32$$
$$= 2^5$$

Hence, $2^2 \times 2^3 = 2^5$

or $2^2 \times 2^3 = 2^{2+3}$

This is the first law of indices, which demonstrates that **when multiplying two or more numbers having the same base, the indices are added**.

(2) $\dfrac{2^5}{2^3} = \dfrac{2 \times 2 \times 2 \times 2 \times 2}{2 \times 2 \times 2} = \dfrac{1 \times 1 \times 1 \times 2 \times 2}{1 \times 1 \times 1}$

$$= \dfrac{2 \times 2}{1} = 4 = 2^2$$

Hence, $\dfrac{2^5}{2^3} = 2^2$ or $\dfrac{2^5}{2^3} = 2^{5-3}$

This is the second law of indices, which demonstrates that **when dividing two numbers having the same base, the index in the denominator is subtracted from the index in the numerator**.

(3) $(3^5)^2 = 3^{5 \times 2} = 3^{10}$ and $(2^2)^3 = 2^{2 \times 3} = 2^6$

This is the third law of indices, which demonstrates that **when a number which is raised to a power is raised to a further power, the indices are multiplied**.

(4) $3^0 = 1$ and $17^0 = 1$

This is the fourth law of indices, which states that **when a number has an index of 0, its value is 1**.

(5) $3^{-4} = \dfrac{1}{3^4}$ and $\dfrac{1}{2^{-3}} = 2^3$

This is the fifth law of indices, which demonstrates that **a number raised to a negative power is the reciprocal of that number raised to a positive power**.

(6) $8^{\frac{2}{3}} = \sqrt[3]{8^2} = (2)^2 = 4$ and

$25^{\frac{1}{2}} = \sqrt[2]{25^1} = \sqrt{25^1} = \pm 5$

(Note that $\sqrt{\ } \equiv \sqrt[2]{\ }$)

This is the sixth law of indices, which demonstrates that **when a number is raised to a fractional power the denominator of the fraction is the root of the number and the numerator is the power**.

Here are some worked examples using the laws of indices.

Problem 6.   Evaluate in index form $5^3 \times 5 \times 5^2$

$5^3 \times 5 \times 5^2 = 5^3 \times 5^1 \times 5^2$ (Note that 5 means 5^1)

$$= 5^{3+1+2}$$ from law (1)

$$= \mathbf{5^6}$$

Problem 7.   Evaluate $\dfrac{3^5}{3^4}$

$$\dfrac{3^5}{3^4} = 3^{5-4}$$ from law (2)

$$= 3^1$$

$$= \mathbf{3}$$

Problem 8.   Evaluate $\dfrac{2^4}{2^4}$

$$\dfrac{2^4}{2^4} = 2^{4-4}$$ from law (2)

$$= 2^0$$

But $\dfrac{2^4}{2^4} = \dfrac{2 \times 2 \times 2 \times 2}{2 \times 2 \times 2 \times 2} = \dfrac{16}{16} = 1$

Hence, $\mathbf{2^0 = 1}$ from law (4)

Any number raised to the power of zero equals 1. For example, $6^0 = 1$, $128^0 = 1$, $13742^0 = 1$ and so on.

Problem 9. Evaluate $\dfrac{3 \times 3^2}{3^4}$

$$\frac{3 \times 3^2}{3^4} = \frac{3^1 \times 3^2}{3^4} = \frac{3^{1+2}}{3^4} = \frac{3^3}{3^4} = 3^{3-4} = 3^{-1}$$

from laws (1) and (2)

But $\dfrac{3^3}{3^4} = \dfrac{3 \times 3 \times 3}{3 \times 3 \times 3 \times 3} = \dfrac{1 \times 1 \times 1}{1 \times 1 \times 1 \times 3}$

(by cancelling)

$$= \frac{1}{3}$$

Hence, $\dfrac{3 \times 3^2}{3^4} = 3^{-1} = \dfrac{1}{3}$ from law (5)

Similarly, $2^{-1} = \dfrac{1}{2}, 2^{-5} = \dfrac{1}{2^5}, \dfrac{1}{5^4} = 5^{-4}$ and so on.

Problem 10. Evaluate $\dfrac{10^3 \times 10^2}{10^8}$

$$\frac{10^3 \times 10^2}{10^8} = \frac{10^{3+2}}{10^8} = \frac{10^5}{10^8} \qquad \text{from law (1)}$$

$$= 10^{5-8} = 10^{-3} \qquad \text{from law (2)}$$

$$= \frac{1}{10^{+3}} = \frac{1}{1000} \qquad \text{from law (5)}$$

Hence, $\dfrac{10^3 \times 10^2}{10^8} = 10^{-3} = \dfrac{1}{1000} = \mathbf{0.001}$

Understanding powers of ten is important, especially when dealing with prefixes in Chapter 8. For example,

$$10^2 = 100, 10^3 = 1000, 10^4 = 10\,000,$$

$$10^5 = 100\,000, 10^6 = 1\,000\,000$$

$$10^{-1} = \frac{1}{10} = 0.1, 10^{-2} = \frac{1}{10^2} = \frac{1}{100} = 0.01$$

and so on.

Problem 11. Evaluate (a) $5^2 \times 5^3 \div 5^4$ (b) $(3 \times 3^5) \div (3^2 \times 3^3)$

From laws (1) and (2):

(a) $5^2 \times 5^3 \div 5^4 = \dfrac{5^2 \times 5^3}{5^4} = \dfrac{5^{(2+3)}}{5^4}$

$$= \frac{5^5}{5^4} = 5^{(5-4)} = 5^1 = \mathbf{5}$$

(b) $(3 \times 3^5) \div (3^2 \times 3^3) = \dfrac{3 \times 3^5}{3^2 \times 3^3} = \dfrac{3^{(1+5)}}{3^{(2+3)}}$

$$= \frac{3^6}{3^5} = 3^{6-5} = 3^1 = \mathbf{3}$$

Problem 12. Simplify (a) $(2^3)^4$ (b) $(3^2)^5$, expressing the answers in index form

From law (3):

(a) $(2^3)^4 = 2^{3 \times 4} = \mathbf{2^{12}}$

(b) $(3^2)^5 = 3^{2 \times 5} = \mathbf{3^{10}}$

Problem 13. Evaluate: $\dfrac{(10^2)^3}{10^4 \times 10^2}$

From laws (1) to (4):

$$\frac{(10^2)^3}{10^4 \times 10^2} = \frac{10^{(2 \times 3)}}{10^{(4+2)}} = \frac{10^6}{10^6} = 10^{6-6} = 10^0 = \mathbf{1}$$

Problem 14. Find the value of (a) $\dfrac{2^3 \times 2^4}{2^7 \times 2^5}$ (b) $\dfrac{(3^2)^3}{3 \times 3^9}$

From the laws of indices:

(a) $\dfrac{2^3 \times 2^4}{2^7 \times 2^5} = \dfrac{2^{(3+4)}}{2^{(7+5)}} = \dfrac{2^7}{2^{12}} = 2^{7-12}$

$$= 2^{-5} = \frac{1}{2^5} = \frac{1}{\mathbf{32}}$$

(b) $\dfrac{(3^2)^3}{3 \times 3^9} = \dfrac{3^{2 \times 3}}{3^{1+9}} = \dfrac{3^6}{3^{10}} = 3^{6-10}$

$$= 3^{-4} = \frac{1}{3^4} = \frac{1}{\mathbf{81}}$$

Problem 15. Evaluate (a) $4^{1/2}$ (b) $16^{3/4}$ (c) $27^{2/3}$ (d) $9^{-1/2}$

(a) $4^{1/2} = \sqrt{4} = \mathbf{\pm 2}$

(b) $16^{3/4} = \sqrt[4]{16^3} = (2)^3 = \mathbf{8}$

(Note that it does not matter whether the 4th root of 16 is found first or whether 16 cubed is found first – the same answer will result.)

(c) $27^{2/3} = \sqrt[3]{27^2} = (3)^2 = \mathbf{9}$

(d) $9^{-1/2} = \dfrac{1}{9^{1/2}} = \dfrac{1}{\sqrt{9}} = \dfrac{1}{\pm 3} = \mathbf{\pm \dfrac{1}{3}}$

Now try the following Practice Exercise

Evaluate the following without the aid of a
calculator.

1. $2^2 \times 2 \times 2^4$

2. $3^5 \times 3^3 \times 3$ in index form

3. $\dfrac{2^7}{2^3}$

4. $\dfrac{3^3}{3^5}$

5. 7^0

6. $\dfrac{2^3 \times 2 \times 2^6}{2^7}$

7. $\dfrac{10 \times 10^6}{10^5}$

8. $10^4 \div 10$

9. $\dfrac{10^3 \times 10^4}{10^9}$

10. $5^6 \times 5^2 \div 5^7$

11. $(7^2)^3$ in index form

12. $(3^3)^2$

13. $\dfrac{3^7 \times 3^4}{3^5}$ in index form

14. $\dfrac{(9 \times 3^2)^3}{(3 \times 27)^2}$ in index form

15. $\dfrac{(16 \times 4)^2}{(2 \times 8)^3}$

16. $\dfrac{5^{-2}}{5^{-4}}$

17. $\dfrac{3^2 \times 3^{-4}}{3^3}$

18. $\dfrac{7^2 \times 7^{-3}}{7 \times 7^{-4}}$

19. $\dfrac{2^3 \times 2^{-4} \times 2^5}{2 \times 2^{-2} \times 2^6}$

20. $\dfrac{5^{-7} \times 5^2}{5^{-8} \times 5^3}$

Here are some further worked examples using the laws
of indices.

Problem 16.   Evaluate $\dfrac{3^3 \times 5^7}{5^3 \times 3^4}$

The laws of indices only apply to terms **having the
same base**. Grouping terms having the same base and
then applying the laws of indices to each of the groups
independently gives

$$\frac{3^3 \times 5^7}{5^3 \times 3^4} = \frac{3^3}{3^4} \times \frac{5^7}{5^3} = 3^{(3-4)} \times 5^{(7-3)}$$

$$= 3^{-1} \times 5^4 = \frac{5^4}{3^1} = \frac{625}{3} = \mathbf{208\frac{1}{3}}$$

Problem 17.   Find the value of $\dfrac{2^3 \times 3^5 \times (7^2)^2}{7^4 \times 2^4 \times 3^3}$

$$\frac{2^3 \times 3^5 \times (7^2)^2}{7^4 \times 2^4 \times 3^3} = 2^{3-4} \times 3^{5-3} \times 7^{2 \times 2 - 4}$$

$$= 2^{-1} \times 3^2 \times 7^0$$

$$= \frac{1}{2} \times 3^2 \times 1 = \frac{9}{2} = \mathbf{4\frac{1}{2}}$$

Problem 18.   Evaluate $\dfrac{4^{1.5} \times 8^{1/3}}{2^2 \times 32^{-2/5}}$

$$4^{1.5} = 4^{3/2} = \sqrt{4^3} = 2^3 = 8,$$
$$8^{1/3} = \sqrt[3]{8} = 2,$$
$$2^2 = 4,$$
$$32^{-2/5} = \frac{1}{32^{2/5}} = \frac{1}{\sqrt[5]{32^2}} = \frac{1}{2^2} = \frac{1}{4}$$

Hence, $\dfrac{4^{1.5} \times 8^{1/3}}{2^2 \times 32^{-2/5}} = \dfrac{8 \times 2}{4 \times \frac{1}{4}} = \dfrac{16}{1} = \mathbf{16}$

Alternatively,

$$\frac{4^{1.5} \times 8^{1/3}}{2^2 \times 32^{-2/5}} = \frac{[(2)^2]^{3/2} \times (2^3)^{1/3}}{2^2 \times (2^5)^{-2/5}}$$

$$= \frac{2^3 \times 2^1}{2^2 \times 2^{-2}} = 2^{3+1-2-(-2)} = 2^4 = \mathbf{16}$$

Problem 19.   Evaluate $\dfrac{3^2 \times 5^5 + 3^3 \times 5^3}{3^4 \times 5^4}$

Dividing each term by the HCF (highest common factor)
of the three terms, i.e. $3^2 \times 5^3$, gives

$$\frac{3^2 \times 5^5 + 3^3 \times 5^3}{3^4 \times 5^4} = \frac{\dfrac{3^2 \times 5^5}{3^2 \times 5^3} + \dfrac{3^3 \times 5^3}{3^2 \times 5^3}}{\dfrac{3^4 \times 5^4}{3^2 \times 5^3}}$$

$$= \frac{3^{(2-2)} \times 5^{(5-3)} + 3^{(3-2)} \times 5^0}{3^{(4-2)} \times 5^{(4-3)}}$$

$$= \frac{3^0 \times 5^2 + 3^1 \times 5^0}{3^2 \times 5^1}$$

$$= \frac{1 \times 25 + 3 \times 1}{9 \times 5} = \frac{\mathbf{28}}{\mathbf{45}}$$

Problem 20. Find the value of $\dfrac{3^2 \times 5^5}{3^4 \times 5^4 + 3^3 \times 5^3}$

To simplify the arithmetic, each term is divided by the HCF of all the terms, i.e. $3^2 \times 5^3$. Thus,

$$\frac{3^2 \times 5^5}{3^4 \times 5^4 + 3^3 \times 5^3} = \frac{\dfrac{3^2 \times 5^5}{3^2 \times 5^3}}{\dfrac{3^4 \times 5^4}{3^2 \times 5^3} + \dfrac{3^3 \times 5^3}{3^2 \times 5^3}}$$

$$= \frac{3^{(2-2)} \times 5^{(5-3)}}{3^{(4-2)} \times 5^{(4-3)} + 3^{(3-2)} \times 5^{(3-3)}}$$

$$= \frac{3^0 \times 5^2}{3^2 \times 5^1 + 3^1 \times 5^0}$$

$$= \frac{1 \times 5^2}{3^2 \times 5 + 3 \times 1} = \frac{25}{45 + 3} = \frac{\mathbf{25}}{\mathbf{48}}$$

Problem 21. Simplify $\dfrac{7^{-3} \times 3^4}{3^{-2} \times 7^5 \times 5^{-2}}$ expressing the answer in index form with positive indices

Since $7^{-3} = \dfrac{1}{7^3}$, $\dfrac{1}{3^{-2}} = 3^2$ and $\dfrac{1}{5^{-2}} = 5^2$, then

$$\frac{7^{-3} \times 3^4}{3^{-2} \times 7^5 \times 5^{-2}} = \frac{3^4 \times 3^2 \times 5^2}{7^3 \times 7^5}$$

$$= \frac{3^{(4+2)} \times 5^2}{7^{(3+5)}} = \frac{\mathbf{3^6 \times 5^2}}{\mathbf{7^8}}$$

Problem 22. Simplify $\dfrac{16^2 \times 9^{-2}}{4 \times 3^3 - 2^{-3} \times 8^2}$ expressing the answer in index form with positive indices

Expressing the numbers in terms of their lowest prime numbers gives

$$\frac{16^2 \times 9^{-2}}{4 \times 3^3 - 2^{-3} \times 8^2} = \frac{(2^4)^2 \times (3^2)^{-2}}{2^2 \times 3^3 - 2^{-3} \times (2^3)^2}$$

$$= \frac{2^8 \times 3^{-4}}{2^2 \times 3^3 - 2^{-3} \times 2^6}$$

$$= \frac{2^8 \times 3^{-4}}{2^2 \times 3^3 - 2^3}$$

Dividing each term by the HCF (i.e. 2^2) gives

$$\frac{2^8 \times 3^{-4}}{2^2 \times 3^3 - 2^3} = \frac{2^6 \times 3^{-4}}{3^3 - 2} = \frac{\mathbf{2^6}}{\mathbf{3^4(3^3 - 2)}}$$

Problem 23. Simplify $\dfrac{\left(\dfrac{4}{3}\right)^3 \times \left(\dfrac{3}{5}\right)^{-2}}{\left(\dfrac{2}{5}\right)^{-3}}$ giving the answer with positive indices

Raising a fraction to a power means that both the numerator and the denominator of the fraction are raised to that power, i.e. $\left(\dfrac{4}{3}\right)^3 = \dfrac{4^3}{3^3}$

A fraction raised to a negative power has the same value as the inverse of the fraction raised to a positive power.

Thus, $\left(\dfrac{3}{5}\right)^{-2} = \dfrac{1}{\left(\dfrac{3}{5}\right)^2} = \dfrac{1}{\dfrac{3^2}{5^2}} = 1 \times \dfrac{5^2}{3^2} = \dfrac{5^2}{3^2}$

Similarly, $\left(\dfrac{2}{5}\right)^{-3} = \left(\dfrac{5}{2}\right)^3 = \dfrac{5^3}{2^3}$

Thus, $\dfrac{\left(\dfrac{4}{3}\right)^3 \times \left(\dfrac{3}{5}\right)^{-2}}{\left(\dfrac{2}{5}\right)^{-3}} = \dfrac{\dfrac{4^3}{3^3} \times \dfrac{5^2}{3^2}}{\dfrac{5^3}{2^3}}$

$$= \frac{4^3}{3^3} \times \frac{5^2}{3^2} \times \frac{2^3}{5^3} = \frac{(2^2)^3 \times 2^3}{3^{(3+2)} \times 5^{(3-2)}}$$

$$= \frac{\mathbf{2^9}}{\mathbf{3^5 \times 5}}$$

Now try the following Practice Exercise

Practice Exercise 33   Further problems on indices (answers on page 1151)

In Problems 1 to 4, simplify the expressions given, expressing the answers in index form and with positive indices.

1. $\dfrac{3^3 \times 5^2}{5^4 \times 3^4}$

2. $\dfrac{7^{-2} \times 3^{-2}}{3^5 \times 7^4 \times 7^{-3}}$

3. $\dfrac{4^2 \times 9^3}{8^3 \times 3^4}$

4. $\dfrac{8^{-2} \times 5^2 \times 3^{-4}}{25^2 \times 2^4 \times 9^{-2}}$

In Problems 5 to 15, evaluate the expressions given.

5. $\left(\dfrac{1}{3^2}\right)^{-1}$

6. $81^{0.25}$

7. $16^{-\frac{1}{4}}$

8. $\left(\dfrac{4}{9}\right)^{1/2}$

9. $\dfrac{9^2 \times 7^4}{3^4 \times 7^4 + 3^3 \times 7^2}$

10. $\dfrac{3^3 \times 5^2}{2^3 \times 3^2 - 8^2 \times 9}$

11. $\dfrac{3^3 \times 7^2 - 5^2 \times 7^3}{3^2 \times 5 \times 7^2}$

12. $\dfrac{(2^4)^2 - 3^{-2} \times 4^4}{2^3 \times 16^2}$

13. $\dfrac{\left(\dfrac{1}{2}\right)^3 - \left(\dfrac{2}{3}\right)^{-2}}{\left(\dfrac{3}{2}\right)^2}$

14. $\dfrac{\left(\dfrac{4}{3}\right)^4}{\left(\dfrac{2}{9}\right)^2}$

15. $\dfrac{(3^2)^{3/2} \times (8^{1/3})^2}{(3)^2 \times (4^3)^{1/2} \times (9)^{-1/2}}$

For fully worked solutions to each of the problems in Practice Exercises 31 to 33 in this chapter, go to the website:
www.routledge.com/cw/bird

Chapter 8

Units, prefixes and engineering notation

> **Why it is important to understand:** Units, prefixes and engineering notation
>
> In engineering there are many different quantities to get used to, and hence many units to become familiar with. For example, force is measured in newtons, electric current is measured in amperes and pressure is measured in pascals. Sometimes the units of these quantities are either very large or very small and hence prefixes are used. For example, 1000 pascals may be written as 10^3 Pa, which is written as 1 kPa in prefix form, the k being accepted as a symbol to represent 1000 or 10^3. Studying, or working, in an engineering discipline, you very quickly become familiar with the standard units of measurement, the prefixes used and engineering notation. An electronic calculator is extremely helpful with engineering notation.

At the end of this chapter, you should be able to:

- state the seven SI units
- understand derived units
- recognise common engineering units
- understand common prefixes used in engineering
- express decimal numbers in standard form
- use engineering notation and prefix form with engineering units
- understand and use metric conversions for length, area and volume
- understand and use metric to US/Imperial conversions

8.1 Introduction

Of considerable importance in engineering is a knowledge of units of engineering quantities, the prefixes used with units, and engineering notation.

We need to know, for example, that

$$80\,kV = 80 \times 10^3\,V, \text{ which means } 80\,000 \text{ volts}$$

and $\quad 25\,mA = 25 \times 10^{-3}\,A$, which means 0.025

amperes

and $\quad 50\,nF = 50 \times 10^{-9}\,F$, which means 0.000000050

farads

This is explained in this chapter.

8.2 SI units

The system of units used in engineering and science is the Système Internationale d'Unités (**International System of Units**), usually abbreviated to SI units, and is based on the metric system. This was introduced in 1960 and has now been adopted by the majority of countries as the official system of measurement.

The basic seven units used in the SI system are listed in Table 8.1 with their symbols.

There are, of course, many units other than these seven. These other units are called **derived units** and are defined in terms of the standard units listed in the table.

Table 8.1 Basic SI units

Quantity	Unit	Symbol
Length	metre	m (1 m = 100 cm = 1000 mm)
Mass	kilogram	kg (1 kg = 1000 g)
Time	second	s
Electric current	ampere	A
Thermodynamic temperature	kelvin	K (K = °C + 273)
Luminous intensity	candela	cd
Amount of substance	mole	mol

For example, speed is measured in metres per second, therefore using two of the standard units, i.e. length and time.

Some derived units are given **special names**. For example, force = mass × acceleration has units of kilogram metre per second squared, which uses three of the base units, i.e. kilograms, metres and seconds. The unit of kg m/s^2 is given the special name of a **newton***.

Table 8.2 on page 64 contains a list of some quantities and their units that are common in engineering.

8.3　Common prefixes

SI units may be made larger or smaller by using prefixes which denote multiplication or division by a particular amount.

The most common multiples are listed in Table 8.3 on page 65. A knowledge of indices is needed since all of the prefixes are powers of 10 with indices that are a multiple of 3.

Here are some examples of prefixes used with engineering units.

A **frequency of 15 GHz** means 15×10^9 Hz, which is 15 000 000 000 hertz*,

i.e. 15 gigahertz is written as 15 GHz and is equal to 15 thousand million hertz.

* **Who was Newton?** **Sir Isaac Newton PRS MP** (25 December 1642–20 March 1727) was an English polymath. Newton showed that the motions of objects are governed by the same set of natural laws, by demonstrating the consistency between Kepler's laws of planetary motion and his theory of gravitation. To find out more go to **www.routledge.com/cw/bird**

* **Who was Hertz?** **Heinrich Rudolf Hertz** (22 February 1857–1 January 1894) was the first person to conclusively prove the existence of electromagnetic waves. The scientific unit of frequency – cycles per second – was named the '**hertz**' in his honour. To find out more go to **www.routledge.com/cw/bird**

Table 8.2 Some quantities and their units that are common in engineering

Quantity	Unit	Symbol
Length	metre	m
Area	square metre	m^2
Volume	cubic metre	m^3
Mass	kilogram	kg
Time	second	s
Electric current	ampere	A
Speed, velocity	metre per second	m/s
Acceleration	metre per second squared	m/s^2
Density	kilogram per cubic metre	kg/m^3
Temperature	kelvin or Celsius	K or °C
Angle	radian or degree	rad or °
Angular velocity	radian per second	rad/s
Frequency	hertz	Hz
Force	newton	N
Pressure	pascal	Pa
Energy, work	joule	J
Power	watt	W
Charge, quantity of electricity	coulomb	C
Electric potential	volt	V
Capacitance	farad	F
Electrical resistance	ohm	Ω
Inductance	henry	H
Moment of force	newton metre	N m

Table 8.3 Common SI multiples

Prefix	Name	Meaning	
G	giga	multiply by 10^9	i.e. $\times 1\,000\,000\,000$
M	mega	multiply by 10^6	i.e. $\times 1\,000\,000$
k	kilo	multiply by 10^3	i.e. $\times 1000$
m	milli	multiply by 10^{-3}	i.e. $\times \dfrac{1}{10^3} = \dfrac{1}{1000} = 0.001$
μ	micro	multiply by 10^{-6}	i.e. $\times \dfrac{1}{10^6} = \dfrac{1}{1\,000\,000} = 0.000001$
n	nano	multiply by 10^{-9}	i.e. $\times \dfrac{1}{10^9} = \dfrac{1}{1\,000\,000\,000} = 0.000\,000\,001$
p	pico	multiply by 10^{-12}	i.e. $\times \dfrac{1}{10^{12}} = \dfrac{1}{1\,000\,000\,000\,000} = 0.000\,000\,000\,001$

(Instead of writing $15\,000\,000\,000$ hertz, it is much neater, takes up less space and prevents errors caused by having so many zeros, to write the frequency as 15 GHz.)

A **voltage of 40 MV** means 40×10^6 V, which is $40\,000\,000$ volts,

i.e. 40 megavolts is written as 40 MV and is equal to 40 million volts.

An **inductance of 12 mH** means 12×10^{-3} H or $\dfrac{12}{10^3}$ H or $\dfrac{12}{1000}$ H, which is 0.012 H,

i.e. 12 millihenrys is written as 12 mH and is equal to 12 thousandths of a henry*.

A **time of 150 ns** means 150×10^{-9} s or $\dfrac{150}{10^9}$ s, which is $0.000\,000\,150$ s,

i.e. 150 nanoseconds is written as 150 ns and is equal to 150 thousand millionths of a second.

A **force of 20 kN** means 20×10^3 N, which is 20 000 newtons,

i.e. 20 kilonewtons is written as 20 kN and is equal to 20 thousand newtons.

A **charge of 30 μC** means 30×10^{-6} C or $\dfrac{30}{10^6}$ C, which is $0.000\,030$ C,

i.e. 30 microcoulombs is written as 30 μC and is equal to 30 millionths of a coulomb.

A **capacitance of 45 pF** means 45×10^{-12} F or $\dfrac{45}{10^{12}}$ F, which is $0.000\,000\,000\,045$ F,

* **Who was Henry**? **Joseph Henry** (17 December 1797–13 May 1878) was an American scientist who discovered the electromagnetic phenomenon of self-inductance. He also discovered mutual inductance independently of Michael Faraday, though Faraday was the first to publish his results. Henry was the inventor of a precursor to the electric doorbell and electric relay. The SI unit of inductance, the **henry**, is named in his honour. To find out more go to **www.routledge.com/cw/bird**

i.e. 45 picofarads is written as 45 pF and is equal to 45 million millionths of a farad (named after Michael Faraday*).

In engineering it is important to understand what such quantities as 15 GHz, 40 MV, 12 mH, 150 ns, 20 kN, 30 μC and 45 pF mean.

Now try the following Practice Exercise

Practice Exercise 34   SI units and common prefixes (answers on page 1151)

1. State the SI unit of volume.

2. State the SI unit of capacitance.

3. State the SI unit of area.

4. State the SI unit of velocity.

5. State the SI unit of density.

6. State the SI unit of energy.

* **Who was Faraday? Michael Faraday, FRS** (22 September 1791–25 August 1867) was an English scientist who contributed to the fields of electromagnetism and electrochemistry. His main discoveries include those of electromagnetic induction, diamagnetism and electrolysis. To find out more go to **www.routledge.com/cw/bird**

7. State the SI unit of charge.

8. State the SI unit of power.

9. State the SI unit of angle.

10. State the SI unit of electric potential.

11. State which quantity has the unit kg.

12. State which quantity has the unit symbol Ω.

13. State which quantity has the unit Hz.

14. State which quantity has the unit m/s^2.

15. State which quantity has the unit symbol A.

16. State which quantity has the unit symbol H.

17. State which quantity has the unit symbol m.

18. State which quantity has the unit symbol K.

19. State which quantity has the unit Pa.

20. State which quantity has the unit rad/s.

21. What does the prefix G mean?

22. What is the symbol and meaning of the prefix milli?

23. What does the prefix p mean?

24. What is the symbol and meaning of the prefix mega?

8.4   Standard form

A number written with one digit to the left of the decimal point and multiplied by 10 raised to some power is said to be written in **standard form**.

For example, $43\,645 = 4.3645 \times 10^4$ in standard form

and $0.0534 = 5.34 \times 10^{-2}$ in standard form

Problem 1.   Express in standard form (a) 38.71 (b) 3746 (c) 0.0124

For a number to be in standard form, it is expressed with only one digit to the left of the decimal point. Thus,

(a)   38.71 must be divided by 10 to achieve one digit to the left of the decimal point and it must also be multiplied by 10 to maintain the equality, i.e.

$$38.71 = \frac{38.71}{10} \times 10 = \mathbf{3.871 \times 10} \text{ in standard form}$$

(b) $3746 = \dfrac{3746}{1000} \times 1000 = \mathbf{3.746 \times 10^3}$ in standard form.

(c) $0.0124 = 0.0124 \times \dfrac{100}{100} = \dfrac{1.24}{100} = \mathbf{1.24 \times 10^{-2}}$ in standard form.

Problem 2. Express the following numbers, which are in standard form, as decimal numbers:

(a) 1.725×10^{-2} (b) 5.491×10^4 (c) 9.84×10^0

(a) $1.725 \times 10^{-2} = \dfrac{1.725}{100} = \mathbf{0.01725}$ (i.e. move the decimal point 2 places to the left).

(b) $5.491 \times 10^4 = 5.491 \times 10000 = \mathbf{54910}$ (i.e. move the decimal point 4 places to the right).

(c) $9.84 \times 10^0 = 9.84 \times 1 = \mathbf{9.84}$ (since $10^0 = 1$).

Problem 3. Express in standard form, correct to 3 significant figures, (a) $\dfrac{3}{8}$ (b) $19\dfrac{2}{3}$ (c) $741\dfrac{9}{16}$

(a) $\dfrac{3}{8} = 0.375$, and expressing it in standard form gives

$$0.375 = \mathbf{3.75 \times 10^{-1}}$$

(b) $19\dfrac{2}{3} = 19.\dot{6} = \mathbf{1.97 \times 10}$ in standard form, correct to 3 significant figures.

(c) $741\dfrac{9}{16} = 741.5625 = \mathbf{7.42 \times 10^2}$ in standard form, correct to 3 significant figures.

Problem 4. Express the following numbers, given in standard form, as fractions or mixed numbers, (a) 2.5×10^{-1} (b) 6.25×10^{-2} (c) 1.354×10^2

(a) $2.5 \times 10^{-1} = \dfrac{2.5}{10} = \dfrac{25}{100} = \mathbf{\dfrac{1}{4}}$

(b) $6.25 \times 10^{-2} = \dfrac{6.25}{100} = \dfrac{625}{10000} = \mathbf{\dfrac{1}{16}}$

(c) $1.354 \times 10^2 = 135.4 = 135\dfrac{4}{10} = \mathbf{135\dfrac{2}{5}}$

Problem 5. Evaluate (a) $(3.75 \times 10^3)(6 \times 10^4)$ (b) $\dfrac{3.5 \times 10^5}{7 \times 10^2}$, expressing the answers in standard form

(a) $(3.75 \times 10^3)(6 \times 10^4) = (3.75 \times 6)(10^{3+4})$
$= 22.50 \times 10^7$
$= \mathbf{2.25 \times 10^8}$

(b) $\dfrac{3.5 \times 10^5}{7 \times 10^2} = \dfrac{3.5}{7} \times 10^{5-2} = 0.5 \times 10^3 = \mathbf{5 \times 10^2}$

Now try the following Practice Exercise

Practice Exercise 35 Standard form (answers on page 1151)

In Problems 1 to 5, express in standard form.

1. (a) 73.9 (b) 28.4 (c) 197.62
2. (a) 2748 (b) 33170 (c) 274218
3. (a) 0.2401 (b) 0.0174 (c) 0.00923
4. (a) 1702.3 (b) 10.04 (c) 0.0109
5. (a) $\dfrac{1}{2}$ (b) $11\dfrac{7}{8}$ (c) $130\dfrac{3}{5}$ (d) $\dfrac{1}{32}$

In Problems 6 and 7, express the numbers given as integers or decimal fractions.

6. (a) 1.01×10^3 (b) 9.327×10^2 (c) 5.41×10^4 (d) 7×10^0
7. (a) 3.89×10^{-2} (b) 6.741×10^{-1} (c) 8×10^{-3}

In Problems 8 and 9, evaluate the given expressions, stating the answers in standard form.

8. (a) $(4.5 \times 10^{-2})(3 \times 10^3)$ (b) $2 \times (5.5 \times 10^4)$
9. (a) $\dfrac{6 \times 10^{-3}}{3 \times 10^{-5}}$ (b) $\dfrac{(2.4 \times 10^3)(3 \times 10^{-2})}{(4.8 \times 10^4)}$
10. Write the following statements in standard form.
(a) The density of aluminium is $2710 \, \text{kg m}^{-3}$.

(b)   Poisson's ratio for gold is 0.44

(c)   The impedance of free space is 376.73 Ω.

(d)   The electron rest energy is 0.511 MeV.

(e)   Proton charge–mass ratio is 95 789 700 $C kg^{-1}$.

(f)   The normal volume of a perfect gas is 0.02241 $m^3 mol^{-1}$.

8.5   Engineering notation

In engineering, standard form is not as important as engineering notation. **Engineering notation** is similar to standard form except that the power of 10 **is always a multiple of 3**.

For example, $43 645 = 43.645 \times 10^3$

in engineering notation

and $0.0534 = 53.4 \times 10^{-3}$

in engineering notation

From the list of engineering prefixes on page 65 it is apparent that all prefixes involve powers of 10 that are multiples of 3.
For example, a force of 43 645 N can be rewritten as 43.645×10^3 N and from the list of prefixes can then be expressed as 43.645 kN.

Thus,   **43 645 N ≡ 43.645 kN**

To help further, on your calculator is an 'ENG' button. Enter the number 43 645 into your calculator and then press =. Now press the ENG button and the answer is 43.645×10^3. We then have to appreciate that 10^3 is the prefix 'kilo', giving **43 645 N ≡ 43.645 kN**.
In another example, let a current be 0.0745 A. Enter 0.0745 into your calculator. Press =. Now press ENG and the answer is 74.5×10^{-3}. We then have to appreciate that 10^{-3} is the prefix 'milli', giving **0.0745 A ≡ 74.5 mA**.

Problem 6.   Express the following in engineering notation and in prefix form: (a) 300 000 W (b) 0.000068 H

(a)   Enter 300 000 into the calculator. Press =

Now press ENG and the answer is 300×10^3

From the table of prefixes on page 65, 10^3 corresponds to kilo.

Hence, 300 000 W = **300 × 10^3 W** in engineering notation

= **300 kW** in prefix form.

(b)   Enter 0.000068 into the calculator. Press =

Now press ENG and the answer is 68×10^{-6}

From the table of prefixes on page 65, 10^{-6} corresponds to micro.

Hence, 0.000068 H = **68 × 10^{-6} H** in engineering notation

= **68 μH** in prefix form.

Problem 7.   Express the following in engineering notation and in prefix form:

(a) $42 \times 10^5 \Omega$ (b) 47×10^{-10} F

(a)   Enter 42×10^5 into the calculator. Press =

Now press ENG and the answer is 4.2×10^6

From the table of prefixes on page 65, 10^6 corresponds to mega.

Hence, $42 \times 10^5 \Omega$ = **4.2 × 10^6 Ω** in engineering notation

= **4.2 MΩ** in prefix form.

(b)   Enter $47 \times 10^{-10} = \dfrac{47}{10 000 000 000}$ into the calculator. Press =

Now press ENG and the answer is 4.7×10^{-9}

From the table of prefixes on page 65, 10^{-9} corresponds to nano.

Hence, $47 \div 10^{10}$ F = **4.7 × 10^{-9} F** in engineering notation

= **4.7 nF** in prefix form.

Problem 8.   Rewrite (a) 0.056 mA in μA (b) 16 700 kHz as MHz

(a)   Enter 0.056 ÷ 1000 into the calculator (since milli means ÷1000). Press =

Now press ENG and the answer is 56×10^{-6}

From the table of prefixes on page 65, 10^{-6} corresponds to micro.

Hence, $0.056 \, \text{mA} = \dfrac{0.056}{1000} \, \text{A} = 56 \times 10^{-6} \, \text{A}$

$$= \mathbf{56 \, \mu A}$$

(b) Enter $16\,700 \times 1000$ into the calculator (since kilo means $\times 1000$). Press $=$

Now press ENG and the answer is 16.7×10^6

From the table of prefixes on page 65, 10^6 corresponds to mega.

Hence, $16\,700 \, \text{kHz} = 16\,700 \times 1000 \, \text{Hz}$

$$= 16.7 \times 10^6 \, \text{Hz}$$

$$= \mathbf{16.7 \, MHz}$$

Problem 9. Rewrite (a) $63 \times 10^4 \, \text{V}$ in kV (b) $3100 \, \text{pF}$ in nF

(a) Enter 63×10^4 into the calculator. Press $=$

Now press ENG and the answer is 630×10^3

From the table of prefixes on page 65, 10^3 corresponds to kilo.

Hence, $63 \times 10^4 \, \text{V} = 630 \times 10^3 \, \text{V} = \mathbf{630 \, kV}$

(b) Enter 3100×10^{-12} into the calculator. Press $=$

Now press ENG and the answer is 3.1×10^{-9}

From the table of prefixes on page 65, 10^{-9} corresponds to nano.

Hence, $3100 \, \text{pF} = 3100 \times 10^{-12} \, \text{F} = 3.1 \times 10^{-9} \, \text{F}$

$$= \mathbf{3.1 \, nF}$$

Problem 10. Rewrite (a) $14\,700 \, \text{mm}$ in metres (b) $276 \, \text{cm}$ in metres (c) $3.375 \, \text{kg}$ in grams

(a) $1 \, \text{m} = 1000 \, \text{mm}$, hence

$1 \, \text{mm} = \dfrac{1}{1000} = \dfrac{1}{10^3} = 10^{-3} \, \text{m}$.

Hence, $14\,700 \, \text{mm} = 14\,700 \times 10^{-3} \, \text{m} = \mathbf{14.7 \, m}$

(b) $1 \, \text{m} = 100 \, \text{cm}$, hence $1 \, \text{cm} = \dfrac{1}{100} = \dfrac{1}{10^2} = 10^{-2} \, \text{m}$

Hence, $276 \, \text{cm} = 276 \times 10^{-2} \, \text{m} = \mathbf{2.76 \, m}$

(c) $1 \, \text{kg} = 1000 \, \text{g} = 10^3 \, \text{g}$

Hence, $3.375 \, \text{kg} = 3.375 \times 10^3 \, \text{g} = \mathbf{3375 \, g}$

Now try the following Practice Exercise

Practice Exercise 36   Engineering notation (answers on page 1151)

In Problems 1 to 12, express in engineering notation in prefix form.

1. $60\,000 \, \text{Pa}$

2. $0.00015 \, \text{W}$

3. $5 \times 10^7 \, \text{V}$

4. $5.5 \times 10^{-8} \, \text{F}$

5. $100\,000 \, \text{W}$

6. $0.00054 \, \text{A}$

7. $15 \times 10^5 \, \Omega$

8. $225 \times 10^{-4} \, \text{V}$

9. $35\,000\,000\,000 \, \text{Hz}$

10. $1.5 \times 10^{-11} \, \text{F}$

11. $0.000017 \, \text{A}$

12. $46\,200 \, \Omega$

13. Rewrite $0.003 \, \text{mA}$ in μA

14. Rewrite $2025 \, \text{kHz}$ as MHz

15. Rewrite $5 \times 10^4 \, \text{N}$ in kN

16. Rewrite $300 \, \text{pF}$ in nF

17. Rewrite $6250 \, \text{cm}$ in metres

18. Rewrite $34.6 \, \text{g}$ in kg

In Problems 19 and 20, use a calculator to evaluate in engineering notation.

19. $4.5 \times 10^{-7} \times 3 \times 10^4$

20. $\dfrac{\left(1.6 \times 10^{-5}\right)\left(25 \times 10^3\right)}{\left(100 \times 10^{-6}\right)}$

21. The distance from Earth to the moon is around $3.8 \times 10^8 \, \text{m}$. State the distance in kilometres.

22. The radius of a hydrogen atom is $0.53 \times 10^{-10} \, \text{m}$. State the radius in nanometres.

23. The tensile stress acting on a rod is $5\,600\,000$ Pa. Write this value in engineering notation.

24. The expansion of a rod is $0.0043 \, \text{m}$. Write this in engineering notation.

8.6 Metric conversions

Length in metric units

$$1\text{ m} = 100\text{ cm} = 1000\text{ mm}$$

$$1\text{ cm} = \frac{1}{100}\text{ m} = \frac{1}{10^2}\text{ m} = 10^{-2}\text{ m}$$

$$1\text{ mm} = \frac{1}{1000}\text{ m} = \frac{1}{10^3}\text{ m} = 10^{-3}\text{ m}$$

Problem 11. Rewrite 14700 mm in metres

$1\text{ m} = 1000\text{ mm}$ hence, $1\text{ mm} = \dfrac{1}{1000} = \dfrac{1}{10^3} = 10^{-3}\text{ m}$

Hence, $14700\text{ mm} = 14,700 \times 10^{-3}\text{m} = \mathbf{14.7\ m}$

Problem 12. Rewrite 276 cm in metres

$1\text{ m} = 100\text{ cm}$ hence, $1\text{ cm} = \dfrac{1}{100} = \dfrac{1}{10^2} = 10^{-2}\text{ m}$

Hence, $276\text{ cm} = 276 \times 10^{-2}\text{m} = \mathbf{2.76\ m}$

Now try the following Practice Exercise

Practice Exercise 37   Length in metric units (answers on page 1151)

1. State 2.45 m in millimetres

2. State 1.675 m in centimetres

3. State the number of millimetres in 65.8 cm

4. Rewrite 25400 mm in metres

5. Rewrite 5632 cm in metres

6. State the number of millimetres in 4.356 m

7. How many centimetres are there in 0.875 m?

8. State a length of 465 cm in (a) mm (b) m

9. State a length of 5040 mm in (a) cm (b) m

10. A machine part is measured as 15.0 cm ± 1%. Between what two values would the measurement be? Give the answer in millimetres.

Areas in metric units

Area is a measure of the size or extent of a plane surface. Area is measured in **square units** such as mm^2, cm^2 and m^2.

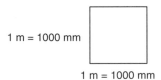

The area of the above square is 1 m^2

$$1\text{ m}^2 = 100\text{ cm} \times 100\text{ cm} = 10000\text{ cm}^2 = 10^4\text{cm}^2$$

i.e. to change from square metres to square centimetres, multiply by 10^4

Hence, $\mathbf{2.5\ m^2 = 2.5 \times 10^4\ cm^2}$

and $\mathbf{0.75\ m^2 = 0.75 \times 10^4\ cm^2}$

Since $\mathbf{1\ m^2 = 10^4\ cm^2}$ then $\mathbf{1cm^2 = \dfrac{1}{10^4}m^2 = 10^{-4}m^2}$

i.e. to change from square centimetres to square metres, multiply by 10^{-4}

Hence, $\mathbf{52\ cm^2 = 52 \times 10^{-4}\ m^2}$

and $\mathbf{643\ cm^2 = 643 \times 10^{-4}\ m^2}$

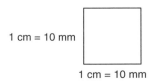

The area of the above square is 1m^2

$$\mathbf{1\,m^2 = 1000\,mm \times 1000\,mm = 1000000\,mm^2 = 10^6\,mm^2}$$

i.e. to change from square metres to square millimetres, multiply by 10^6

Hence, $\mathbf{7.5\,m^2 = 7.5 \times 10^6\,mm^2}$

and $\mathbf{0.63\,m^2 = 0.63 \times 10^6\,mm^2}$

Since $\mathbf{1\,m^2 = 10^6\,mm^2}$

then $\mathbf{1\,mm^2 = \dfrac{1}{10^6}m^2 = 10^{-6}\,m^2}$

i.e. to change from square millimetres to square metres, multiply by 10^{-6}

Hence, $\mathbf{235\,mm^2 = 235 \times 10^{-6}\,m^2}$

and $\mathbf{47\,mm^2 = 47 \times 10^{-6}\,m^2}$

The area of the above square is 1 cm^2

$$1\,cm^2 = 10\,mm \times 10\,mm = 100\,mm^2 = 10^2 mm^2$$

i.e. to change from square centimetres to square millimetres, multiply by 100 or 10^2

Hence, $3.5\,cm^2 = 3.5 \times 10^2 mm^2 = 350\,mm^2$

and $\quad 0.75\,cm^2 = 0.75 \times 10^2 mm^2 = 75\,mm^2$

Since $\quad 1\,cm^2 = 10^2 mm^2$

then $\quad 1\,mm^2 = \dfrac{1}{10^2}cm^2 = 10^{-2}\,cm^2$

i.e. to change from square millimetres to square centimetres, multiply by 10^{-2}

Hence, $250\,mm^2 = 250 \times 10^{-2}cm^2 = 2.5\,cm^2$

and $\quad 85\,mm^2 = 85 \times 10^{-2}cm^2 = 0.85\,cm^2$

Problem 13. Rewrite $12\,m^2$ in square centimetres

$1\,m^2 = 10^4 cm^2$ hence, $\mathbf{12\,m^2 = 12 \times 10^4 cm^2}$

Problem 14. Rewrite $50\,cm^2$ in square metres

$1\,cm^2 = 10^{-4}m^2$ hence, $\mathbf{50\,cm^2 = 50 \times 10^{-4}m^2}$

Problem 15. Rewrite $2.4\,m^2$ in square millimetres

$1\,m^2 = 10^6 mm^2$ hence, $\mathbf{2.4\,m^2 = 2.4 \times 10^6 mm^2}$

Problem 16. Rewrite $147\,mm^2$ in square metres

$1\,mm^2 = 10^{-6}m^2$ hence, $\mathbf{147\,mm^2 = 147 \times 10^{-6}m^2}$

Problem 17. Rewrite $34.5\,cm^2$ in square millimetres

$1\,cm^2 = 10^2 mm^2$

hence, $\mathbf{34.5\,cm^2 = 34.5 \times 10^2 mm^2 = 3450\,mm^2}$

Problem 18. Rewrite $400\,mm^2$ in square centimetres

$1\,mm^2 = 10^{-2}cm^2$

hence, $\mathbf{400\,mm^2 = 400 \times 10^{-2}cm^2 = 4\,cm^2}$

Problem 19. The top of a small rectangular table is 800 mm long and 500 mm wide. Determine its area in (a) mm^2 (b) cm^2 (c) m^2

(a) **Area of rectangular table top** $= l \times b = 800 \times 500$
$$= \mathbf{400{,}000\ mm^2}$$

(b) Since 1 cm = 10 mm then 1 cm^2= 1 cm × 1 cm = 10 mm × 10 mm = 100 mm^2

or $\qquad 1\,mm^2 = \dfrac{1}{100} = 0.01\ cm^2$

Hence, $\mathbf{400{,}000\ mm^2} = 400{,}000 \times 0.01\ cm^2$
$$= \mathbf{4000\ cm^2}$$

(c) $1\,cm^2 = 10^{-4}m^2$
hence, $\mathbf{4000\ cm^2} = 4000 \times 10^{-4}m^2 = \mathbf{0.4\ m^2}$

Now try the following Practice Exercise

Practice Exercise 38   Areas in metric units (answers on page 1151)

1. Rewrite $8\,m^2$ in square centimetres

2. Rewrite $240\,cm^2$ in square metres

3. Rewrite $3.6\,m^2$ in square millimetres

4. Rewrite $350\,mm^2$ in square metres

5. Rewrite $50\,cm^2$ in square millimetres

6. Rewrite $250\,mm^2$ in square centimetres

7. A rectangular piece of metal is 720 mm long and 400 mm wide. Determine its area in

   (a) mm^2 (b) cm^2 (c) m^2

Volumes in metric units

The **volume** of any solid is a measure of the space occupied by the solid.

Volume is measured in **cubic units** such as mm^3, cm^3 and m^3.

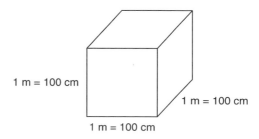

The volume of the cube shown is 1 m^3

$$1\,m^3 = 100\,cm \times 100\,cm \times 100\,cm$$

$$= 1000000\,cm^2 = 10^6\,cm^2$$

$$1\,litre = 1000\,cm^3$$

i.e. to change from cubic metres to cubic centimetres, multiply by 10^6

Hence, $\mathbf{3.2\,m^3 = 3.2 \times 10^6\,cm^3}$

and $\mathbf{0.43\,m^3 = 0.43 \times 10^6\,cm^3}$

Since $1\,m^3 = 10^6\,cm^3$ then $1\,cm^3 = \dfrac{1}{10^6}\,m^3 = 10^{-6}\,m^3$

i.e. to change from cubic centimetres to cubic metres, multiply by 10^{-6}

Hence, $\mathbf{140\,cm^3 = 140 \times 10^{-6}\,m^3}$

and $\mathbf{2500\,cm^3 = 2500 \times 10^{-6}\,m^3}$

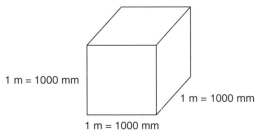

The volume of the cube shown is 1 m^3

$$1\,m^3 = 1000\,mm \times 1000\,mm \times 1000\,mm$$

$$= 1000000000\,mm^3 = 10^9\,mm^3$$

i.e. to change from cubic metres to cubic millimetres, multiply by 10^9

Hence, $\mathbf{4.5\,m^3 = 4.5 \times 10^9\,mm^3}$

and $\mathbf{0.25\,m^3 = 0.25 \times 10^9\,mm^3}$

Since $1\,m^3 = 10^9\,mm^2$

then $1\,mm^3 = \dfrac{1}{10^9}\,m^3 = 10^{-9}\,m^3$

i.e. to change from cubic millimetres to cubic metres, multiply by 10^{-9}

Hence, $\mathbf{500\,mm^3 = 500 \times 10^{-9}\,m^3}$

and $\mathbf{4675\,mm^3 = 4675 \times 10^{-9}\,m^3}$ or $\mathbf{4.675 \times 10^{-6}\,m^3}$

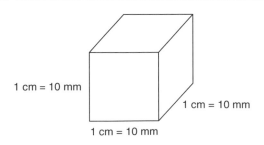

The volume of the cube shown is 1 cm^3

$$1\,cm^3 = 10\,mm \times 10\,mm \times 10\,mm$$

$$= 1000\,mm^3 = 10^3\,mm^3$$

i.e. to change from cubic centimetres to cubic millimetres, multiply by 1000 or 10^3

Hence, $\mathbf{5\,cm^3 = 5 \times 10^3\,mm^3 = 5000\,mm^3}$

and $\mathbf{0.35\,cm^3 = 0.35 \times 10^3\,mm^3 = 350\,mm^3}$

Since $1\,cm^3 = 10^3\,mm^3$

then $1\,mm^3 = \dfrac{1}{10^3}\,cm^3 = 10^{-3}\,cm^3$

i.e. to change from cubic millimetres to cubic centimetres, multiply by 10^{-3}

Hence, $\mathbf{650\,mm^3 = 650 \times 10^{-3}\,cm^3 = 0.65\,cm^3}$

and $\mathbf{75\,mm^3 = 75 \times 10^{-3}\,cm^3 = 0.075\,cm^3}$

Problem 20. Rewrite $1.5\,m^3$ in cubic centimetres

$1\,m^3 = 10^6\,cm^3$ hence, $\mathbf{1.5\,m^3 = 1.5 \times 10^6\,cm^3}$

Problem 21. Rewrite $300\,cm^3$ in cubic metres

$1\,cm^3 = 10^{-6}\,m^3$ hence, $\mathbf{300\,cm^3 = 300 \times 10^{-6}\,m^3}$

Problem 22. Rewrite $0.56\,m^3$ in cubic millimetres

$1\,m^3 = 10^9\,mm^3$

hence, $\mathbf{0.56\,m^3 = 0.56 \times 10^9\,mm^3}$ or $\mathbf{560 \times 10^6\,mm^3}$

Problem 23. Rewrite $1250\,mm^3$ in cubic metres

$1\,mm^3 = 10^{-9}\,m^3$

hence, $\mathbf{1250\,mm^3 = 1250 \times 10^{-9}\,m^3}$ or $\mathbf{1.25 \times 10^{-6}\,m^3}$

Problem 24. Rewrite $8\,cm^3$ in cubic millimetres

$1\,cm^3 = 10^3\,mm^3$

hence, $\mathbf{8\,cm^3 = 8 \times 10^3\,mm^3 = 8000\,mm^3}$

Problem 25. Rewrite $600\,mm^3$ in cubic centimetres

$1\,mm^3 = 10^{-3}\,cm^3$

hence, $\mathbf{600\,mm^3 = 600 \times 10^{-3}\,cm^3 = 0.6\,cm^3}$

Problem 26. A water tank is in the shape of a rectangular prism having length 1.2 m, breadth

50 cm and height 250 mm. Determine the capacity of the tank in (a) m^3 (b) cm^3 (c) litres

Capacity means volume. When dealing with liquids, the word capacity is usually used.

(a) Capacity of water tank $= l \times b \times h$
where $l = 1.2$ m, $b = 50$ cm and $h = 250$ mm.
To use this formula, all dimensions **must** be in the same units.
Thus, $l = 1.2$ m, $b = 0.50$ m and $h = 0.25$ m (since 1 m = 100 cm = 1000 mm)
Hence, **capacity of tank** $= 1.2 \times 0.50 \times 0.25$
$$= \mathbf{0.15\ m^3}$$

(b) $1 m^3 = 10^6 cm^3$
Hence, **capacity** $= 0.15\ m^3 = 0.15 \times 10^6\ cm^3$
$$= \mathbf{150000\ cm^3}$$

(c) 1 litre = 1000 cm^3
Hence, **150000 cm^3** $= \dfrac{150,000}{1000} = \mathbf{150\ litres}$

Now try the following Practice Exercise

Practice Exercise 39   Volumes in metric units (answers on page 1151)

1. Rewrite 2.5 m^3 in cubic centimetres

2. Rewrite 400 cm^3 in cubic metres

3. Rewrite 0.87 m^3 in cubic millimetres

4. Change a volume of 2400000 cm^3 to cubic metres.

5. Rewrite 1500 mm^3 in cubic metres

6. Rewrite 400 mm^3 in cubic centimetres

7. Rewrite 6.4 cm^3 in cubic millimetres

8. Change a volume of 7500 mm^3 to cubic centimetres.

9. An oil tank is in the shape of a rectangular prism having length 1.5 m, breadth 60 cm and height 200 mm. Determine the capacity of the tank in (a) m^3 (b) cm^3 (c) litres

---

8.7 Metric – US/Imperial conversions

The Imperial System (which uses yards, feet, inches, etc. to measure length) was developed over hundreds of years in the UK, then the French developed the Metric System (metres) in 1670, which soon spread through Europe, even to the UK itself in 1960. But the USA and a few other countries still prefer feet and inches.

When converting from metric to imperial units, or vice versa, one of the following tables (8.4 to 8.11) should help.

Table 8.4 Metric to imperial length

Metric	US or Imperial
1 millimetre, mm	0.03937 inch
1 centimetre, cm = 10 mm	0.3937 inch
1 metre, m = 100 cm	1.0936 yard
1 kilometre, km = 1000 m	0.6214 mile

Problem 27. Calculate the number of inches in 350 mm, correct to 2 decimal places

350 mm = 350 × 0.03937 inches = **13.78 inches** from Table 8.4

Problem 28. Calculate the number of inches in 52 cm, correct to 4 significant figures

52 cm = 52 × 0.3937 inches = **20.47 inches** from Table 8.4

Problem 29. Calculate the number of yards in 74 m, correct to 2 decimal places

74 m = 74 × 1.0936 yards = **80.93 yds** from Table 8.4

Problem 30. Calculate the number of miles in 12.5 km, correct to 3 significant figures

12.5 km = 12.5 × 0.6214 miles = **7.77 miles** from Table 8.4

Table 8.5 Imperial to metric length

US or Imperial	Metric
1 inch, in	2.54 cm
1 foot, ft = 12 in	0.3048 m
1 yard, yd = 3 ft	0.9144 m
1 mile = 1760 yd	1.6093 km
1 nautical mile = 2025.4 yd	1.853 km

Problem 31. Calculate the number of centimetres in 35 inches, correct to 1 decimal place

35 inches = 35 × 2.54 cm = **88.9 cm** from Table 8.5

Problem 32. Calculate the number of metres in 66 inches, correct to 2 decimal places

66 inches = $\dfrac{66}{12}$ feet = $\dfrac{66}{12}$ × 0.3048 m = **1.68 m** from Table 8.5

Problem 33. Calculate the number of metres in 50 yards, correct to 2 decimal places

50 yards = 50 × 0.9144 m = **45.72 m** from Table 8.5

Problem 34. Calculate the number of kilometres in 7.2 miles, correct to 2 decimal places

7.2 miles = 7.2 × 1.6093 km = **11.59 km** from Table 8.5

Problem 35. Calculate the number of (a) yards (b) kilometres in 5.2 nautical miles

(a) 5.2 nautical miles = 5.2 × 2025.4 yards
$\qquad\qquad\qquad\qquad$ = **10532 yards** from Table 8.5

(b) 5.2 nautical miles = 5.2 × 1.853 km
$\qquad\qquad\qquad\qquad$ = **9.636 km** from Table 8.5

Table 8.6 Metric to imperial area

Metric	US or Imperial
$1\,cm^2 = 100\,mm^2$	$0.1550\,in^2$
$1\,m^2 = 10000\,cm^2$	$1.1960\,yd^2$
1 hectare, ha = $10000\,m^2$	2.4711 acres
$1\,km^2 = 100$ ha	$0.3861\,mile^2$

Problem 36. Calculate the number of square inches in 47 cm^2, correct to 4 significant figures

47 cm^2 = 47 × 0.1550 in^2 = **7.285 in^2** from Table 8.6

Problem 37. Calculate the number of square yards in 20 m^2, correct to 2 decimal places

20 m^2 = 20 × 1.1960 yd^2 = **23.92 yd^2** from Table 8.6

Problem 38. Calculate the number of acres in 23 hectares of land, correct to 2 decimal places

23 hectares = 23 × 2.4711 acres = **56.84 acres** from Table 8.6

Problem 39. Calculate the number of square miles in a field of 15 km^2 area, correct to 2 decimal places

15 km^2 = 15 × 0.3861 $mile^2$ = **5.79 $mile^2$** from Table 8.6

Table 8.7 Imperial to metric area

US or Imperial	Metric
$1\,in^2$	$6.4516\,cm^2$
$1\,ft^2 = 144\,in^2$	$0.0929\,m^2$
$1\,yd^2 = 9\,ft^2$	$0.8361\,m^2$
1 acre = $4840\,yd^2$	$4046.9\,m^2$
$1\,mile^2 = 640$ acres	$2.59\,km^2$

Problem 40. Calculate the number of square centimetres in 17.5 in^2, correct to the nearest square centimetre

17.5 in^2 = 17.5 × 6.4516 cm^2 = **113 cm^2** from Table 8.7

Problem 41. Calculate the number of square metres in 205 ft^2, correct to 2 decimal places

205 ft^2 = 205 × 0.0929 m^2 = **19.04 m^2** from Table 8.7

Problem 42. Calculate the number of square metres in 11.2 acres, correct to the nearest square metre

11.2 acres = 11.2 × 4046.9 m^2 = **45325 m^2** from Table 8.7

Problem 43.   Calculate the number of square kilometres in 12.6 mile^2, correct to 2 decimal places

12.6 mile^2 = 12.6 × 2.59 km^2 = **32.63 km^2** from Table 8.7

Table 8.8 Metric to imperial volume/capacity

Metric	US or Imperial
1 cm^3	0.0610 in^3
1 dm^3 = 1000 cm^3	0.0353 ft^3
1 m^3 = 1000 dm^3	1.3080 yd^3
1 litre = 1 dm^3 = 1000 cm^3	2.113 fluid pt = 1.7598 pt

Problem 44.   Calculate the number of cubic inches in 123.5 cm^3, correct to 2 decimal places

123.5 cm^3 = 123.5 × 0.0610 in^3 = **7.53 in^3** from Table 8.8

Problem 45.   Calculate the number of cubic feet in 144 dm^3, correct to 3 decimal places

144 dm^3 = 144 × 0.0353 ft^3 = **5.083 ft^3** from Table 8.8

Problem 46.   Calculate the number of cubic yards in 5.75 m^3, correct to 4 significant figures

5.75 m^3 = 5.75 × 1.3080 yd^3 = **7.521 yd^3** from Table 8.8

Problem 47.   Calculate the number of US fluid pints in 6.34 litres of oil, correct to 1 decimal place

6.34 litre = 6.34 × 2.113 US fluid pints = **13.4 US fluid pints** from Table 8.8

Problem 48.   Calculate the number of cubic centimetres in 3.75 in^3, correct to 2 decimal places

3.75 in^3 = 3.75 × 16.387 cm^3 = **61.45 cm^3** from Table 8.9

Table 8.9 Imperial to metric volume/capacity

US or Imperial	Metric
1 in^3	16.387 cm^3
1 ft^3	0.02832 m^3
1 US fl oz = 1.0408 UK fl oz	0.0296 litres
1 US pint (16 fl oz) = 0.8327 UK pt	0.4732 litres
1 US gal (231 in^3) = 0.8327 UK gal	3.7854 litres

Problem 49.   Calculate the number of cubic metres in 210 ft^3, correct to 3 significant figures

210 ft^3 = 210 × 0.02832 m^3 = **5.95 m^3** from Table 8.9

Problem 50.   Calculate the number of litres in 4.32 US pints, correct to 3 decimal places

4.32 US pints = 4.32 × 0.4732 litres = **2.044 litres** from Table 8.9

Problem 51.   Calculate the number of litres in 8.62 US gallons, correct to 2 decimal places

8.62 US gallons = 8.62 × 3.7854 litres = **32.63 litres** from Table 8.9

Table 8.10 Metric to imperial mass

Metric	US or Imperial
1 g = 1000 mg	0.0353 oz
1 kg = 1000 g	2.2046 lb
1 tonne, t = 1000 kg	1.1023 short ton
1 tonne, t = 1000 kg	0.9842 long ton

The British ton is the long ton, which is 2240 pounds, and the US ton is the short ton which is 2000 pounds.

Problem 52.   Calculate the number of ounces in a mass of 1346 g, correct to 2 decimal places

1346 g = 1346 × 0.0353 oz = **47.51 oz** from Table 8.10

Problem 53. Calculate the mass, in pounds, in a 210.4 kg mass, correct to 4 significant figures

210.4 kg = 210.4 × 2.2046 lb = **463.8 lb** from Table 8.10

Problem 54. Calculate the number of short tons in 5000 kg, correct to 2 decimal places

5000 kg = 5 t = 5 × 1.1023 short tons = **5.51 short tons** from Table 8.10

Table 8.11 Imperial to metric mass

US or Imperial	Metric
1 oz = 437.5 grain	28.35 g
1 lb = 16 oz	0.4536 kg
1 stone = 14 lb	6.3503 kg
1 hundredweight, cwt = 112 lb	50.802 kg
1 short ton	0.9072 tonne
1 long ton	1.0160 tonne

Problem 55. Calculate the number of grams in 5.63 oz, correct to 4 significant figures

5.63 oz = 5.63 × 28.35 g = **159.6 g** from Table 8.11

Problem 56. Calculate the number of kilograms in 75 oz, correct to 3 decimal places

$75 \text{ oz} = \dfrac{75}{16}\text{lb} = \dfrac{75}{16} \times 0.4536 \text{ kg} = \textbf{2.126 kg}$ from Table 8.11

Problem 57. Convert 3.25 cwt into (a) pounds (b) kilograms

(a)   3.25 cwt = 3.25 × 112 lb = **364 lb** from Table 8.11

(b)   3.25 cwt = 3.25 × 50.802 kg = **165.1 kg** from Table 8.11

Temperature

To convert from Celsius to Fahrenheit, first multiply by 9/5, then add 32.

To convert from Fahrenheit to Celsius, first subtract 32, then multiply by 5/9

Problem 58. Convert 35° C to degrees Fahrenheit

$F = \dfrac{9}{5}C + 32$ hence, $35° C = \dfrac{9}{5}(35) + 32 = 63 + 32$

$$= \textbf{95° F}$$

Problem 59. Convert 113° F to degrees Celsius

$C = \dfrac{5}{9}(F - 32)$ hence, $113° F = \dfrac{5}{9}(113 - 32) = \dfrac{5}{9}(81)$

$$= \textbf{45° C}$$

Now try the following Practice Exercise

Practice Exercise 40   Metric/Imperial conversions (answers on page 1151)

In the following Problems, use the metric/imperial conversions in Tables 8.4 to 8.11

1.   Calculate the number of inches in 476 mm, correct to 2 decimal places.

2.   Calculate the number of inches in 209 cm, correct to 4 significant figures.

3.   Calculate the number of yards in 34.7 m, correct to 2 decimal places.

4.   Calculate the number of miles in 29.55 km, correct to 2 decimal places.

5.   Calculate the number of centimetres in 16.4 inches, correct to 2 decimal places.

6.   Calculate the number of metres in 78 inches, correct to 2 decimal places.

7.   Calculate the number of metres in 15.7 yards, correct to 2 decimal places.

8.   Calculate the number of kilometres in 3.67 miles, correct to 2 decimal places.

9.   Calculate the number of (a) yards (b) kilometres in 11.23 nautical miles.

10. Calculate the number of square inches in 62.5 cm^2, correct to 4 significant figures.

11. Calculate the number of square yards in 15.2 m^2, correct to 2 decimal places.

12. Calculate the number of acres in 12.5 hectares, correct to 2 decimal places.

13. Calculate the number of square miles in 56.7 km^2, correct to 2 decimal places.

14. Calculate the number of square centimetres in 6.37 in^2, correct to the nearest square centimetre.

15. Calculate the number of square metres in 308.6 ft^2, correct to 2 decimal places.

16. Calculate the number of square metres in 2.5 acres, correct to the nearest square metre.

17. Calculate the number of square kilometres in 21.3 mile^2, correct to 2 decimal places.

18. Calculate the number of cubic inches in 200.7 cm^3, correct to 2 decimal places.

19. Calculate the number of cubic feet in 214.5 dm^3, correct to 3 decimal places.

20. Calculate the number of cubic yards in 13.45 m^3, correct to 4 significant figures.

21. Calculate the number of US fluid pints in 15 litres, correct to 1 decimal place.

22. Calculate the number of cubic centimetres in 2.15 in^3, correct to 2 decimal places.

23. Calculate the number of cubic metres in 175 ft^3, correct to 4 significant figures.

24. Calculate the number of litres in 7.75 US pints, correct to 3 decimal places.

25. Calculate the number of litres in 12.5 US gallons, correct to 2 decimal places.

26. Calculate the number of ounces in 980 g, correct to 2 decimal places.

27. Calculate the mass, in pounds, in 55 kg, correct to 4 significant figures.

28. Calculate the number of short tons in 4000 kg, correct to 3 decimal places.

29. Calculate the number of grams in 7.78 oz, correct to 4 significant figures.

30. Calculate the number of kilograms in 57.5 oz, correct to 3 decimal places.

31. Convert 2.5 cwt into (a) pounds (b) kilograms.

32. Convert 55ºC to degrees Fahrenheit.

33. Convert 167ºF to degrees Celsius.

For fully worked solutions to each of the problems in Practice Exercises 34 to 40 in this chapter, go to the website:
www.routledge.com/cw/bird

Revision Test 3 Ratio, proportion, powers, roots, indices and units

This assignment covers the material contained in Chapters 6 to 8. *The marks available are shown in brackets at the end of each question.*

1. In a box of 1500 nails, 125 are defective. Express the non-defective nails as a ratio of the defective ones, in its simplest form. (3)

2. Prize money in a lottery totals £4500 and is shared among three winners in the ratio 5:3:1. How much does the first prize winner receive? (3)

3. A simple machine has an effort:load ratio of 3:41. Determine the effort, in newtons, to lift a load of 6.15 kN. (3)

4. If 15 cans of lager weigh 7.8 kg, what will 24 cans weigh? (3)

5. Hooke's law states that stress is directly proportional to strain within the elastic limit of a material. When, for brass, the stress is 21 MPa, the strain is 250×10^{-6}. Determine the stress when the strain is 350×10^{-6}. (3)

6. If 12 inches $= 30.48$ cm, find the number of millimetres in 17 inches. (3)

7. If x is inversely proportional to y and $x = 12$ when $y = 0.4$, determine
   (a) the value of x when $y = 3$
   (b) the value of y when $x = 2$ (3)

8. Evaluate
   (a) $3 \times 2^3 \times 2^2$
   (b) $49^{\frac{1}{2}}$ (4)

9. Evaluate $\dfrac{3^2 \times \sqrt{36} \times 2^2}{3 \times 81^{\frac{1}{2}}}$ taking positive square roots only. (3)

10. Evaluate $6^4 \times 6 \times 6^2$ in index form. (3)

11. Evaluate
    (a) $\dfrac{2^7}{2^2}$ (b) $\dfrac{10^4 \times 10 \times 10^5}{10^6 \times 10^2}$ (4)

12. Evaluate
    (a) $\dfrac{2^3 \times 2 \times 2^2}{2^4}$
    (b) $\dfrac{\left(2^3 \times 16\right)^2}{(8 \times 2)^3}$
    (c) $\left(\dfrac{1}{4^2}\right)^{-1}$ (7)

13. Evaluate
    (a) $(27)^{-1/3}$ (b) $\dfrac{\left(\dfrac{3}{2}\right)^{-2} - \dfrac{2}{9}}{\left(\dfrac{2}{3}\right)^2}$ (5)

14. State the SI unit of (a) capacitance (b) electrical potential (c) work (3)

15. State the quantity that has an SI unit of (a) kilograms (b) henrys (c) hertz (d) m^3 (4)

16. Express the following in engineering notation in prefix form.
    (a) 250 000 J (b) 0.05 H
    (c) 2×10^8 W (d) 750×10^{-8} F (4)

17. Rewrite (a) 0.0067 mΛ in μΛ (b) 40×10^4 kV as MV. (2)

18. Rewrite 32 cm^2 in square millimetres. (1)

19. A rectangular tabletop is 1500 mm long and 800 mm wide. Determine its area in
    (a) mm^2 (b) cm^2 (c) m^2 (3)

20. Rewrite 0.065 m^3 in cubic millimetres. (1)

21. Rewrite 20000 mm^3 in cubic metres. (1)

22. Rewrite 8.3 cm^3 in cubic millimetres. (1)

23. A petrol tank is in the shape of a rectangular prism having length 1.0 m, breadth 75 cm and height 120 mm. Determine the capacity of the tank in
    (a) m^3 (b) cm^3 (c) litres (3)

For lecturers/instructors/teachers, fully worked solutions to each of the problems in Revision Test 3, together with a full marking scheme, are available at the website:
www.routledge.com/cw/bird

Chapter 9

Basic algebra

Why it is important to understand: **Basic algebra**

Algebra is one of the most fundamental tools for engineers because it allows them to determine the value of something (length, material constant, temperature, mass, and so on) given values that they do know (possibly other length, material properties, mass). Although the types of problems that mechanical, chemical, civil, environmental and electrical engineers deal with vary, all engineers use algebra to solve problems. An example where algebra is frequently used is in simple electrical circuits, where the resistance is proportional to voltage. Using Ohm's law, or $V = IR$, an engineer simply multiplies the current in a circuit by the resistance to determine the voltage across the circuit. Engineers and scientists use algebra in many ways, and so frequently that they don't even stop to think about it. Depending on what type of engineer you choose to be, you will use varying degrees of algebra, but in all instances algebra lays the foundation for the mathematics you will need to become an engineer.

At the end of this chapter, you should be able to:

- understand basic operations in algebra
- add, subtract, multiply and divide using letters instead of numbers
- state the laws of indices in letters instead of numbers
- simplify algebraic expressions using the laws of indices

9.1 Introduction

We are already familiar with evaluating formulae using a calculator from Chapter 4.

For example, if the length of a football pitch is L and its width is b, then the formula for the area A is given by

$$A = L \times b$$

This is an **algebraic equation**.

If $L = 120\,\text{m}$ and $b = 60\,\text{m}$, then the area

$A = 120 \times 60 = 7200\,\text{m}^2$

The total resistance, R_T, of resistors R_1, R_2 and R_3 connected in series is given by

$$R_T = R_1 + R_2 + R_3$$

This is an **algebraic equation**.

If $R_1 = 6.3\,\text{k}\Omega$, $R_2 = 2.4\,\text{k}\Omega$ and $R_3 = 8.5\,\text{k}\Omega$, then

$$R_T = 6.3 + 2.4 + 8.5 = 17.2\,\text{k}\Omega$$

The temperature in Fahrenheit, F, is given by

$$F = \frac{9}{5}C + 32$$

where C is the temperature in Celsius*. This is an **algebraic equation**.

If $C = 100°C$, then $F = \dfrac{9}{5} \times 100 + 32$

$$= 180 + 32 = 212°F.$$

If you can cope with evaluating formulae then you will be able to cope with algebra.

9.2 Basic operations

Algebra merely uses letters to represent numbers.
If, say, a, b, c and d represent any four numbers, then in algebra:

(a) **$a + a + a + a = 4a$**. For example, if $a = 2$, then $2 + 2 + 2 + 2 = 4 \times 2 = 8$.

(b) **$5b$** means **$5 \times b$**. For example, if $b = 4$, then $5b = 5 \times 4 = 20$.

(c) **$2a + 3b + a - 2b = 2a + a + 3b - 2b = 3a + b$**

Only similar terms can be combined in algebra. The $2a$ and the $+a$ can be combined to give $3a$

and the $3b$ and $-2b$ can be combined to give $1b$, which is written as b.

In addition, with terms separated by $+$ and $-$ signs, the order in which they are written does not matter. In this example, $2a + 3b + a - 2b$ is the same as $2a + a + 3b - 2b$, which is the same as $3b + a + 2a - 2b$, and so on. (Note that the first term, i.e. $2a$, means $+2a$)

(d) **$4abcd = 4 \times a \times b \times c \times d$**

For example, if $a = 3, b = -2, c = 1$ and $d = -5$, then $4abcd = 4 \times 3 \times -2 \times 1 \times -5 = 120$.
(Note that $- \times - = +$)

(e) **$(a)(c)(d)$ means $a \times c \times d$**

Brackets are often used instead of multiplication signs. For example, $(2)(5)(3)$ means $2 \times 5 \times 3 = 30$

(f) **$ab = ba$**
If $a = 2$ and $b = 3$ then 2×3 is exactly the same as 3×2, i.e. 6

(g) **$b^2 = b \times b$** For example, if $b = 3$, then $3^2 = 3 \times 3 = 9$

(h) **$a^3 = a \times a \times a$** For example, if $a = 2$, then $2^3 = 2 \times 2 \times 2 = 8$

Here are some worked examples to help get a feel for basic operations in this introduction to algebra.

Addition and subtraction

Problem 1. Find the sum of $4x, 3x, -2x$ and $-x$

$$4x + 3x + -2x + -x = 4x + 3x - 2x - x$$

$$\text{(Note that} + \times - = -)$$

$$= 4x$$

Problem 2. Find the sum of $5x, 3y, z, -3x, -4y$ and $6z$

$$5x + 3y + z + -3x + -4y + 6z$$

$$= 5x + 3y + z - 3x - 4y + 6z$$

$$= 5x - 3x + 3y - 4y + z + 6z$$

$$= 2x - y + 7z$$

* **Who was Celsius? Anders Celsius** (27 November 1701–25 April 1744) was the Swedish astronomer, who in 1742, proposed the temperature scale which takes his name. To find out more go to **www.routledge.com/cw/bird**

Note that the order can be changed when terms are separated by $+$ and $-$ signs. Only similar terms can be combined.

Problem 3.   Simplify $4x^2 - x - 2y + 5x + 3y$

$$4x^2 - x - 2y + 5x + 3y = 4x^2 + 5x - x + 3y - 2y$$
$$= \mathbf{4x^2 + 4x + y}$$

Problem 4.   Simplify $3xy - 7x + 4xy + 2x$

$$3xy - 7x + 4xy + 2x = 3xy + 4xy + 2x - 7x$$
$$= \mathbf{7xy - 5x}$$

Now try the following Practice Exercise

Practice Exercise 41   Addition and subtraction in algebra (answers on page 1152)

1. Find the sum of $4a, -2a, 3a$ and $-8a$
2. Find the sum of $2a, 5b, -3c, -a, -3b$ and $7c$
3. Simplify $2x - 3x^2 - 7y + x + 4y - 2y^2$
4. Simplify $5ab - 4a + ab + a$
5. Simplify $2x - 3y + 5z - x - 2y + 3z + 5x$
6. Simplify $3 + x + 5x - 2 - 4x$
7. Add $x - 2y + 3$ to $3x + 4y - 1$
8. Subtract $a - 2b$ from $4a + 3b$
9. From $a + b - 2c$ take $3a + 2b - 4c$
10. From $x^2 + xy - y^2$ take $xy - 2x^2$

Multiplication and division

Problem 5.   Simplify $bc \times abc$

$$bc \times abc = a \times b \times b \times c \times c$$
$$= a \times b^2 \times c^2$$
$$= \mathbf{ab^2c^2}$$

Problem 6.   Simplify $-2p \times -3p$

$$- \times - = + \text{ hence, } -2p \times -3p = \mathbf{6p^2}$$

Problem 7.   Simplify $ab \times b^2c \times a$

$$ab \times b^2c \times a = a \times a \times b \times b \times b \times c$$
$$= a^2 \times b^3 \times c$$
$$= \mathbf{a^2b^3c}$$

Problem 8.   Evaluate $3ab + 4bc - abc$ when $a = 3, b = 2$ and $c = 5$

$$3ab + 4bc - abc = 3 \times a \times b + 4 \times b \times c - a \times b \times c$$
$$= 3 \times 3 \times 2 + 4 \times 2 \times 5 - 3 \times 2 \times 5$$
$$= 18 + 40 - 30$$
$$= \mathbf{28}$$

Problem 9.   Determine the value of $5pq^2r^3$, given that $p = 2, q = \dfrac{2}{5}$ and $r = 2\dfrac{1}{2}$

$$5pq^2r^3 = 5 \times p \times q^2 \times r^3$$
$$= 5 \times 2 \times \left(\frac{2}{5}\right)^2 \times \left(2\frac{1}{2}\right)^3$$
$$= 5 \times 2 \times \left(\frac{2}{5}\right)^2 \times \left(\frac{5}{2}\right)^3 \qquad \text{since} \quad 2\frac{1}{2} = \frac{5}{2}$$
$$= \frac{5}{1} \times \frac{2}{1} \times \frac{2}{5} \times \frac{2}{5} \times \frac{5}{2} \times \frac{5}{2} \times \frac{5}{2}$$
$$= \frac{1}{1} \times \frac{1}{1} \times \frac{1}{1} \times \frac{1}{1} \times \frac{1}{1} \times \frac{5}{1} \times \frac{5}{1} \qquad \text{by cancelling}$$
$$= 5 \times 5$$
$$= \mathbf{25}$$

Problem 10.   Multiply $2a + 3b$ by $a + b$

Each term in the first expression is multiplied by a, then each term in the first expression is multiplied by b and the two results are added. The usual layout is shown below.

$$2a + 3b$$
$$a + b$$

Multiplying by a gives $\quad 2a^2 + 3ab$

Multiplying by b gives $\qquad\quad 2ab + 3b^2$

Adding gives $\qquad\qquad 2a^2 + 5ab + 3b^2$

Thus, $(2a + 3b)(a + b) = 2a^2 + 5ab + 3b^2$

Problem 11. Multiply $3x - 2y^2 + 4xy$ by $2x - 5y$

$$3x \;-\; 2y^2 \;+\; 4xy$$
$$2x \;-\; 5y$$

Multiplying by $2x \rightarrow$ $\quad 6x^2 - 4xy^2 + 8x^2y$

Multiplying by $-5y \rightarrow$ $\qquad -20xy^2 \qquad -15xy + 10y^3$

Adding gives $\quad 6x^2 - 24xy^2 + 8x^2y - 15xy + 10y^3$

Thus, $(3x - 2y^2 + 4xy)(2x - 5y)$

$\qquad = 6x^2 - 24xy^2 + 8x^2y - 15xy + 10y^3$

Problem 12. Simplify $2x \div 8xy$

$2x \div 8xy$ means $\dfrac{2x}{8xy}$

$$\frac{2x}{8xy} = \frac{2 \times x}{8 \times x \times y}$$

$$= \frac{1 \times 1}{4 \times 1 \times y} \qquad \text{by cancelling}$$

$$= \frac{1}{4y}$$

Problem 13. Simplify $\dfrac{9a^2bc}{3ac}$

$$\frac{9a^2bc}{3ac} = \frac{9 \times a \times a \times b \times c}{3 \times a \times c}$$

$$= 3 \times a \times b$$

$$= 3ab$$

Problem 14. Divide $2x^2 + x - 3$ by $x - 1$

(i) $2x^2 + x - 3$ is called the **dividend** and $x - 1$ the **divisor**. The usual layout is shown below with the

dividend and divisor both arranged in descending powers of the symbols.

$$
\begin{array}{r}
2x + 3 \\
x - 1 \overline{)\,2x^2 + x - 3} \\
2x^2 - 2x \\
\hline
3x - 3 \\
3x - 3 \\
\hline
\cdot \quad \cdot \\
\hline
\end{array}
$$

(ii) Dividing the first term of the dividend by the first term of the divisor, i.e. $\dfrac{2x^2}{x}$ gives $2x$, which is put above the first term of the dividend as shown.

(iii) The divisor is then multiplied by $2x$, i.e. $2x(x - 1) = 2x^2 - 2x$, which is placed under the dividend as shown. Subtracting gives $3x - 3$

(iv) The process is then repeated, i.e. the first term of the divisor, x, is divided into $3x$, giving $+3$, which is placed above the dividend as shown.

(v) Then $3(x - 1) = 3x - 3$, which is placed under the $3x - 3$. The remainder, on subtraction, is zero, which completes the process.

Thus, $(2x^2 + x - 3) \div (x - 1) = (2x + 3)$

(A check can be made on this answer by multiplying $(2x + 3)$ by $(x - 1)$, which equals $2x^2 + x - 3$)

Problem 15. Simplify $\dfrac{x^3 + y^3}{x + y}$

(i) (iv) (vii)

$$
\begin{array}{r}
x^2 - xy + y^2 \\
x + y \overline{)\,x^3 + 0 \quad + 0 \quad + y^3} \\
x^3 + x^2y \\
\hline
-x^2y \qquad + y^3 \\
-x^2y - xy^2 \\
\hline
xy^2 + y^3 \\
xy^2 + y^3 \\
\hline
\cdot \qquad \cdot \\
\hline
\end{array}
$$

(i) x into x^3 goes x^2. Put x^2 above x^3.

(ii) $x^2(x + y) = x^3 + x^2y$

(iii) Subtract.

(iv) x into $-x^2y$ goes $-xy$. Put $-xy$ above the dividend.

(v) $-xy(x+y) = -x^2y - xy^2$

(vi) Subtract.

(vii) x into xy^2 goes y^2. Put y^2 above the dividend.

(viii) $y^2(x+y) = xy^2 + y^3$

(ix) Subtract.

Thus, $\dfrac{x^3+y^3}{x+y} = x^2 - xy + y^2$

The zeros shown in the dividend are not normally shown, but are included to clarify the subtraction process and to keep similar terms in their respective columns.

Problem 16. Divide $4a^3 - 6a^2b + 5b^3$ by $2a - b$

$$
\begin{array}{r}
2a^2 - 2ab - b^2 \\
2a-b\overline{\smash{)}\,4a^3 - 6a^2b \quad\quad + 5b^3} \\
\underline{4a^3 - 2a^2b} \\
-4a^2b \quad\quad + 5b^3 \\
\underline{-4a^2b + 2ab^2} \\
-2ab^2 + 5b^3 \\
\underline{-2ab^2 + b^3} \\
4b^3
\end{array}
$$

Thus, $\dfrac{4a^3 - 6a^2b + 5b^3}{2a - b} = 2a^2 - 2ab - b^2$, **remainder** $4b^3$

Alternatively, the answer may be expressed as

$$\frac{4a^3 - 6a^2b + 5b^3}{2a - b} = 2a^2 - 2ab - b^2 + \frac{4b^3}{2a - b}$$

Now try the following Practice Exercise

Practice Exercise 42   Basic operations in algebra (answers on page 1152)

1. Simplify $pq \times pq^2r$

2. Simplify $-4a \times -2a$

3. Simplify $3 \times -2q \times -q$

4. Evaluate $3pq - 5qr - pqr$ when $p = 3$, $q = -2$ and $r = 4$

5. Determine the value of $3x^2yz^3$, given that $x = 2$, $y = 1\frac{1}{2}$ and $z = \frac{2}{3}$

6. If $x = 5$ and $y = 6$, evaluate $\dfrac{23(x - y)}{y + xy + 2x}$

7. If $a = 4, b = 3, c = 5$ and $d = 6$, evaluate $\dfrac{3a + 2b}{3c - 2d}$

8. Simplify $2x \div 14xy$

9. Simplify $\dfrac{25x^2yz^3}{5xyz}$

10. Multiply $3a - b$ by $a + b$

11. Multiply $2a - 5b + c$ by $3a + b$

12. Simplify $3a \div 9ab$

13. Simplify $4a^2b \div 2a$

14. Divide $6x^2y$ by $2xy$

15. Divide $2x^2 + xy - y^2$ by $x + y$

16. Divide $3p^2 - pq - 2q^2$ by $p - q$

17. Simplify $(a + b)^2 + (a - b)^2$

9.3   Laws of indices

The laws of indices with numbers were covered in Chapter 7; the laws of indices in algebraic terms are as follows:

(1) $a^m \times a^n = a^{m+n}$

For example, $a^3 \times a^4 = a^{3+4} = a^7$

(2) $\dfrac{a^m}{a^n} = a^{m-n}$ For example, $\dfrac{c^5}{c^2} = c^{5-2} = c^3$

(3) $(a^m)^n = a^{mn}$ For example, $(d^2)^3 = d^{2\times3} = d^6$

(4) $a^{\frac{m}{n}} = \sqrt[n]{a^m}$ For example, $x^{\frac{4}{3}} = \sqrt[3]{x^4}$

(5) $a^{-n} = \dfrac{1}{a^n}$ For example, $3^{-2} = \dfrac{1}{3^2} = \dfrac{1}{9}$

(6) $a^0 = 1$ For example, $17^0 = 1$

Here are some worked examples to demonstrate these laws of indices.

Problem 17. Simplify $a^2b^3c \times ab^2c^5$

$$a^2b^3c \times ab^2c^5 = a^2 \times b^3 \times c \times a \times b^2 \times c^5$$
$$= a^2 \times b^3 \times c^1 \times a^1 \times b^2 \times c^5$$

Grouping together like terms gives

$$a^2 \times a^1 \times b^3 \times b^2 \times c^1 \times c^5$$

Using law (1) of indices gives

$$a^{2+1} \times b^{3+2} \times c^{1+5} = a^3 \times b^5 \times c^6$$

i.e. $\qquad a^2b^3c \times ab^2c^5 = a^3b^5c^6$

Problem 18. Simplify $a^{\frac{1}{3}}b^{\frac{3}{2}}c^{-2} \times a^{\frac{1}{6}}b^{\frac{1}{2}}c$

Using law (1) of indices,

$$a^{\frac{1}{3}}b^{\frac{3}{2}}c^{-2} \times a^{\frac{1}{6}}b^{\frac{1}{2}}c = a^{\frac{1}{3}+\frac{1}{6}} \times b^{\frac{3}{2}+\frac{1}{2}} \times c^{-2+1}$$

$$= a^{\frac{1}{2}}b^2c^{-1}$$

Problem 19. Simplify $\dfrac{x^5y^2z}{x^2yz^3}$

$$\frac{x^5y^2z}{x^2yz^3} = \frac{x^5 \times y^2 \times z}{x^2 \times y \times z^3}$$

$$= \frac{x^5}{x^2} \times \frac{y^2}{y^1} \times \frac{z}{z^3}$$

$$= x^{5-2} \times y^{2-1} \times z^{1-3} \quad \text{by law (2) of indices}$$

$$= x^3 \times y^1 \times z^{-2}$$

$$= x^3yz^{-2} \quad \text{or} \quad \frac{x^3y}{z^2}$$

Problem 20. Simplify $\dfrac{a^3b^2c^4}{abc^{-2}}$ and evaluate when $a = 3, b = \dfrac{1}{4}$ and $c = 2$

Using law (2) of indices,

$$\frac{a^3}{a} = a^{3-1} = a^2, \frac{b^2}{b} = b^{2-1} = b \text{ and}$$

$$\frac{c^4}{c^{-2}} = c^{4--2} = c^6$$

Thus, $\dfrac{a^3b^2c^4}{abc^{-2}} = a^2bc^6$

When $a = 3, b = \dfrac{1}{4}$ and $c = 2$,

$$a^2bc^6 = (3)^2 \left(\frac{1}{4}\right)(2)^6 = (9)\left(\frac{1}{4}\right)(64) = \mathbf{144}$$

Problem 21. Simplify $(p^3)^2(q^2)^4$

Using law (3) of indices gives

$$(p^3)^2(q^2)^4 = p^{3\times2} \times q^{2\times4}$$

$$= p^6q^8$$

Problem 22. Simplify $\dfrac{(mn^2)^3}{(m^{1/2}n^{1/4})^4}$

The brackets indicate that each letter in the bracket must be raised to the power outside. Using law (3) of indices gives

$$\frac{(mn^2)^3}{(m^{1/2}n^{1/4})^4} = \frac{m^{1\times3}n^{2\times3}}{m^{(1/2)\times4}n^{(1/4)\times4}} = \frac{m^3n^6}{m^2n^1}$$

Using law (2) of indices gives

$$\frac{m^3n^6}{m^2n^1} = m^{3-2}n^{6-1} = mn^5$$

Problem 23. Simplify $(a^3b)(a^{-4}b^{-2})$, expressing the answer with positive indices only

Using law (1) of indices gives $a^{3+-4}b^{1+-2} = a^{-1}b^{-1}$

Using law (5) of indices gives $a^{-1}b^{-1} = \dfrac{1}{a^{+1}b^{+1}} = \dfrac{1}{ab}$

Problem 24. Simplify $\dfrac{d^2e^2f^{1/2}}{(d^{3/2}ef^{5/2})^2}$ expressing the answer with positive indices only

Using law (3) of indices gives

$$\frac{d^2e^2f^{1/2}}{(d^{3/2}\ ef^{5/2})^2} = \frac{d^2e^2f^{1/2}}{d^3e^2f^5}$$

Using law (2) of indices gives

$$d^{2-3}e^{2-2}f^{\frac{1}{2}-5} = d^{-1}e^0 f^{-\frac{9}{2}}$$

$$= d^{-1}f^{-\frac{9}{2}} \quad \text{since } e^0 = 1 \text{ from law}$$

$$\text{(6) of indices}$$

$$= \frac{1}{df^{9/2}} \quad \text{from law (5) of indices}$$

Now try the following Practice Exercise

Practice Exercise 43   Laws of indices (answers on page 1152)

In Problems 1 to 18, simplify the following, giving each answer as a power.

1. $z^2 \times z^6$

2. $a \times a^2 \times a^5$

3. $n^8 \times n^{-5}$

4. $b^4 \times b^7$

5. $b^2 \div b^5$

6. $c^5 \times c^3 \div c^4$

7. $\dfrac{m^5 \times m^6}{m^4 \times m^3}$

8. $\dfrac{(x^2)(x)}{x^6}$

9. $\left(x^3\right)^4$

10. $\left(y^2\right)^{-3}$

11. $\left(t \times t^3\right)^2$

12. $\left(c^{-7}\right)^{-2}$

13. $\left(\dfrac{a^2}{a^5}\right)^3$

14. $\left(\dfrac{1}{b^3}\right)^4$

15. $\left(\dfrac{b^2}{b^7}\right)^{-2}$

16. $\dfrac{1}{\left(s^3\right)^3}$

17. $p^3qr^2 \times p^2q^5r \times pqr^2$

18. $\dfrac{x^3y^2z}{x^5yz^3}$

19. Simplify $(x^2y^3z)(x^3yz^2)$ and evaluate when $x = \dfrac{1}{2}, y = 2$ and $z = 3$

20. Simplify $\dfrac{a^5bc^3}{a^2b^3c^2}$ and evaluate when $a = \dfrac{3}{2}, b = \dfrac{1}{2}$ and $c = \dfrac{2}{3}$

Here are some further worked examples on the laws of indices.

Problem 25. Simplify $\dfrac{p^{1/2}q^2r^{2/3}}{p^{1/4}q^{1/2}r^{1/6}}$ and evaluate when $p = 16, q = 9$ and $r = 4$, taking positive roots only

Using law (2) of indices gives $p^{\frac{1}{2}-\frac{1}{4}}q^{2-\frac{1}{2}}r^{\frac{2}{3}-\frac{1}{6}}$

$$p^{\frac{1}{2}-\frac{1}{4}}q^{2-\frac{1}{2}}r^{\frac{2}{3}-\frac{1}{6}} = p^{\frac{1}{4}}q^{\frac{3}{2}}r^{\frac{1}{2}}$$

When $p = 16, q = 9$ and $r = 4$,

$$p^{\frac{1}{4}}q^{\frac{3}{2}}r^{\frac{1}{2}} = 16^{\frac{1}{4}}9^{\frac{3}{2}}4^{\frac{1}{2}}$$

$$= (\sqrt[4]{16})(\sqrt{9^3})(\sqrt{4}) \text{ from law (4) of indices}$$

$$= (2)(3^3)(2) = \mathbf{108}$$

Problem 26. Simplify $\dfrac{x^2y^3 + xy^2}{xy}$

Algebraic expressions of the form $\dfrac{a+b}{c}$ can be split into $\dfrac{a}{c} + \dfrac{b}{c}$. Thus,

$$\frac{x^2y^3 + xy^2}{xy} = \frac{x^2y^3}{xy} + \frac{xy^2}{xy} = x^{2-1}y^{3-1} + x^{1-1}y^{2-1}$$

$$= \mathbf{xy^2 + y}$$

$$\text{(since } x^0 = 1, \text{ from law (6) of indices)}$$

Problem 27. Simplify $\dfrac{x^2y}{xy^2 - xy}$

The highest common factor (HCF) of each of the three terms comprising the numerator and denominator is xy. Dividing each term by xy gives

$$\frac{x^2y}{xy^2 - xy} = \frac{\dfrac{x^2y}{xy}}{\dfrac{xy^2}{xy} - \dfrac{xy}{xy}} = \frac{\mathbf{x}}{\mathbf{y-1}}$$

Problem 28. Simplify $\dfrac{a^2b}{ab^2 - a^{1/2}b^3}$

The HCF of each of the three terms is $a^{1/2}b$. Dividing each term by $a^{1/2}b$ gives

$$\frac{a^2b}{ab^2 - a^{1/2}b^3} = \frac{\dfrac{a^2b}{a^{1/2}b}}{\dfrac{ab^2}{a^{1/2}b} - \dfrac{a^{1/2}b^3}{a^{1/2}b}} = \frac{\mathbf{a^{3/2}}}{\mathbf{a^{1/2}b - b^2}}$$

Problem 29. Simplify $(a^3\sqrt{b}\sqrt{c^5})(\sqrt{a}\sqrt[3]{b^2}c^3)$ and evaluate when $a=\dfrac{1}{4}, b=64$ and $c=1$

Using law (4) of indices, the expression can be written as

$$(a^3\sqrt{b}\sqrt{c^5})(\sqrt{a}\sqrt[3]{b^2}c^3) = \left(a^3 b^{\frac{1}{2}} c^{\frac{5}{2}}\right)\left(a^{\frac{1}{2}} b^{\frac{2}{3}} c^3\right)$$

Using law (1) of indices gives

$$\left(a^3 b^{\frac{1}{2}} c^{\frac{5}{2}}\right)\left(a^{\frac{1}{2}} b^{\frac{2}{3}} c^3\right) = a^{3+\frac{1}{2}} b^{\frac{1}{2}+\frac{2}{3}} c^{\frac{5}{2}+3}$$

$$= a^{\frac{7}{2}} b^{\frac{7}{6}} c^{\frac{11}{2}}$$

It is usual to express the answer in the same form as the question. Hence,

$$a^{\frac{7}{2}} b^{\frac{7}{6}} c^{\frac{11}{2}} = \sqrt{a^7}\sqrt[6]{b^7}\sqrt{c^{11}}$$

When $a=\dfrac{1}{4}, b=64$ and $c=1$,

$$\sqrt{a^7}\sqrt[6]{b^7}\sqrt{c^{11}} = \sqrt{\left(\frac{1}{4}\right)^7\left(\sqrt[6]{64^7}\right)\left(\sqrt{1^{11}}\right)}$$

$$= \left(\frac{1}{2}\right)^7 (2)^7(1) = \mathbf{1}$$

Problem 30. Simplify $\dfrac{(x^2 y^{1/2})(\sqrt{x}\sqrt[3]{y^2})}{(x^5 y^3)^{1/2}}$

Using laws (3) and (4) of indices gives

$$\frac{\left(x^2 y^{1/2}\right)\left(\sqrt{x}\sqrt[3]{y^2}\right)}{(x^5 y^3)^{1/2}} = \frac{\left(x^2 y^{1/2}\right)\left(x^{1/2} y^{2/3}\right)}{x^{5/2} y^{3/2}}$$

Using laws (1) and (2) of indices gives

$$x^{2+\frac{1}{2}-\frac{5}{2}}\, y^{\frac{1}{2}+\frac{2}{3}-\frac{3}{2}} = x^0 y^{-\frac{1}{3}} = \mathbf{y^{-\frac{1}{3}}}\ \text{or}\ \mathbf{\frac{1}{y^{1/3}}}\ \text{or}\ \mathbf{\frac{1}{\sqrt[3]{y}}}$$

from laws (5) and (6) of indices.

Now try the following Practice Exercise

Practice Exercise 44   Laws of indices (answers on page 1152)

1. Simplify $\left(a^{3/2}bc^{-3}\right)\left(a^{1/2}b^{-1/2}c\right)$ and evaluate when $a=3, b=4$ and $c=2$

In Problems 2 to 8, simplify the given expressions.

2. $\dfrac{a^2 b + a^3 b}{a^2 b^2}$

3. $(a^2)^{1/2}(b^2)^3\left(c^{1/2}\right)^3$

4. $\dfrac{(abc)^2}{(a^2 b^{-1} c^{-3})^3}$

5. $\dfrac{p^3 q^2}{pq^2 - p^2 q}$

6. $(\sqrt{x}\sqrt{y^3}\sqrt[3]{z^2})(\sqrt{x}\sqrt{y^3}\sqrt{z^3})$

7. $(e^2 f^3)(e^{-3}f^{-5})$, expressing the answer with positive indices only.

8. $\dfrac{(a^3 b^{1/2} c^{-1/2})(ab)^{1/3}}{(\sqrt{a^3}\sqrt{b}\,c)}$

For fully worked solutions to each of the problems in Practice Exercises 41 to 44 in this chapter, go to the website:
www.routledge.com/cw/bird

Chapter 10

Further algebra

Why it is important to understand: Further algebra

Algebra is a form of mathematics that allows you to work with unknowns. If you do not know what a number is, arithmetic does not allow you to use it in calculations. Algebra has variables. Variables are labels for numbers and measurements you do not yet know. Algebra lets you use these variables in equations and formulae. A basic form of mathematics, algebra is nevertheless among the most commonly used forms of mathematics in the workforce. Although relatively simple, algebra possesses a powerful problem-solving tool used in many fields of engineering. For example, in designing a rocket to go to the moon, an engineer must use algebra to solve for flight trajectory, how long to burn each thruster and at what intensity, and at what angle to lift off. An engineer uses mathematics all the time – and in particular, algebra. Becoming familiar with algebra will make all engineering mathematics studies so much easier.

At the end of this chapter, you should be able to:

- use brackets with basic operations in algebra
- understand factorisation
- factorise simple algebraic expressions
- use the laws of precedence to simplify algebraic expressions

10.1 Introduction

In this chapter the use of brackets and factorisation with algebra is explained, together with further practise with the laws of precedence. Understanding of these topics is often necessary when solving and transposing equations.

10.2 Brackets

With algebra

(a) $2(a+b) = 2a+2b$

(b) $(a+b)(c+d) = a(c+d)+b(c+d)$
$$= ac+ad+bc+bd$$

Here are some worked examples to help understanding of brackets with algebra.

Problem 1. Determine $2b(a-5b)$

$2b(a-5b) = 2b \times a + 2b \times -5b$
$$= 2ba - 10b^2$$
$$= \mathbf{2ab - 10b^2} \qquad \text{(note that } 2ba \text{ is the same as } 2ab\text{)}$$

Problem 2. Determine $(3x + 4y)(x - y)$

$$(3x + 4y)(x - y) = 3x(x - y) + 4y(x - y)$$
$$= 3x^2 - 3xy + 4yx - 4y^2$$
$$= 3x^2 - 3xy + 4xy - 4y^2$$

(note that $4yx$ is the same as $4xy$)

$$= \mathbf{3x^2 + xy - 4y^2}$$

Problem 3. Simplify $3(2x - 3y) - (3x - y)$

$$3(2x - 3y) - (3x - y) = 3 \times 2x - 3 \times 3y - 3x - -y$$

(Note that $-(3x - y) = -1(3x - y)$ and the
-1 multiplies **both** terms in the bracket)

$$= 6x - 9y - 3x + y$$

(Note: $- \times - = +$)

$$= 6x - 3x + y - 9y$$
$$= \mathbf{3x - 8y}$$

Problem 4. Remove the brackets and simplify the expression $(a - 2b) + 5(b - c) - 3(c + 2d)$

$$(a - 2b) + 5(b - c) - 3(c + 2d)$$
$$= a - 2b + 5 \times b + 5 \times -c - 3 \times c - 3 \times 2d$$
$$= a - 2b + 5b - 5c - 3c - 6d$$
$$= \mathbf{a + 3b - 8c - 6d}$$

Problem 5. Simplify $(p + q)(p - q)$

$$(p + q)(p - q) = p(p - q) + q(p - q)$$
$$= p^2 - pq + qp - q^2$$
$$= \mathbf{p^2 - q^2}$$

Problem 6. Simplify $(2x - 3y)^2$

$$(2x - 3y)^2 = (2x - 3y)(2x - 3y)$$
$$= 2x(2x - 3y) - 3y(2x - 3y)$$
$$= 2x \times 2x + 2x \times -3y - 3y \times 2x$$
$$\qquad\qquad -3y \times -3y$$
$$= 4x^2 - 6xy - 6xy + 9y^2$$

(Note: $+ \times - = -$ and $- \times - = +$)

$$= \mathbf{4x^2 - 12xy + 9y^2}$$

Problem 7. Remove the brackets from the expression and simplify $2[x^2 - 3x(y + x) + 4xy]$

$$2[x^2 - 3x(y + x) + 4xy] = 2[x^2 - 3xy - 3x^2 + 4xy]$$

(Whenever more than one type of bracket is involved, always **start with the inner brackets**)

$$= 2[-2x^2 + xy]$$
$$= -4x^2 + 2xy$$
$$= \mathbf{2xy - 4x^2}$$

Problem 8. Remove the brackets and simplify the expression $2a - [3\{2(4a - b) - 5(a + 2b)\} + 4a]$

(i) Removing the innermost brackets gives

$$2a - [3\{8a - 2b - 5a - 10b\} + 4a]$$

(ii) Collecting together similar terms gives

$$2a - [3\{3a - 12b\} + 4a]$$

(iii) Removing the 'curly' brackets gives

$$2a - [9a - 36b + 4a]$$

(iv) Collecting together similar terms gives

$$2a - [13a - 36b]$$

(v) Removing the outer brackets gives

$$2a - 13a + 36b$$

(vi) i.e. $\mathbf{-11a + 36b}$ or $\mathbf{36b - 11a}$

Now try the following Practice Exercise

Practice Exercise 45   Brackets (answers on page 1152)

Expand the brackets in Problems 1 to 28.

1. $(x + 2)(x + 3)$ 2. $(x + 4)(2x + 1)$

3. $(2x + 3)^2$ 4. $(2j - 4)(j + 3)$

5. $(2x + 6)(2x + 5)$ 6. $(pq + r)(r + pq)$

7. $(a + b)(a + b)$ 8. $(x + 6)^2$

9. $(a - c)^2$ 10. $(5x + 3)^2$

11. $(2x - 6)^2$ 12. $(2x - 3)(2x + 3)$

13. $(8x + 4)^2$

14. $(rs + t)^2$

15. $3a(b - 2a)$

16. $2x(x - y)$

17. $(2a - 5b)(a + b)$

18. $3(3p - 2q) - (q - 4p)$

19. $(3x - 4y) + 3(y - z) - (z - 4x)$

20. $(2a + 5b)(2a - 5b)$

21. $(x - 2y)^2$

22. $(3a - b)^2$

23. $2x + [y - (2x + y)]$

24. $3a + 2[a - (3a - 2)]$

25. $4[a^2 - 3a(2b + a) + 7ab]$

26. $3[x^2 - 2x(y + 3x) + 3xy(1 + x)]$

27. $2 - 5[a(a - 2b) - (a - b)^2]$

28. $24p - [2\{3(5p - q) - 2(p + 2q)\} + 3q]$

10.3 Factorisation

The **factors** of 8 are 1, 2, 4 and 8 because 8 divides by 1, 2, 4 and 8

The factors of 24 are 1, 2, 3, 4, 6, 8, 12 and 24 because 24 divides by 1, 2, 3, 4, 6, 8, 12 and 24

The **common factors** of 8 and 24 are 1, 2, 4 and 8 since 1, 2, 4 and 8 are factors of both 8 and 24

The **highest common factor (HCF)** is the largest number that divides into two or more terms.

Hence, the HCF of 8 and 24 is 8, as explained in Chapter 1.

When two or more terms in an algebraic expression contain a common factor, then this factor can be shown outside of a bracket. For example,

$$df + dg = d(f + g)$$

which is just the reverse of

$$d(f + g) = df + dg$$

This process is called **factorisation**.

Here are some worked examples to help understanding of factorising in algebra.

Problem 9.   Factorise $ab - 5ac$

a is common to both terms ab and $-5ac$. a is therefore taken outside of the bracket. What goes inside the bracket?

(i)   What multiplies a to make ab? Answer: b

(ii)   What multiplies a to make $-5ac$? Answer: $-5c$

Hence, $b - 5c$ appears in the bracket. Thus,

$$ab - 5ac = a(b - 5c)$$

Problem 10.   Factorise $2x^2 + 14xy^3$

For the numbers 2 and 14, the highest common factor (HCF) is 2 (i.e. 2 is the largest number that divides into both 2 and 14).

For the x terms, x^2 and x, the HCF is x

Thus, the HCF of $2x^2$ and $14xy^3$ is $2x$

$2x$ is therefore taken outside of the bracket. What goes inside the bracket?

(i)   What multiplies $2x$ to make $2x^2$? Answer: x

(ii)   What multiplies $2x$ to make $14xy^3$? Answer: $7y^3$

Hence, $x + 7y^3$ appears inside the bracket. Thus,

$$2x^2 + 14xy^3 = 2x(x + 7y^3)$$

Problem 11.   Factorise $3x^3y - 12xy^2 + 15xy$

For the numbers 3, 12 and 15, the highest common factor is 3 (i.e. 3 is the largest number that divides into 3, 12 and 15)

For the x terms, x^3, x and x, the HCF is x

For the y terms, y, y^2 and y, the HCF is y

Thus, the HCF of $3x^3y$ and $12xy^2$ and $15xy$ is $3xy$

$3xy$ is therefore taken outside of the bracket. What goes inside the bracket?

(i)   What multiplies $3xy$ to make $3x^3y$? Answer: x^2

(ii)   What multiplies $3xy$ to make $-12xy^2$? Answer: $-4y$

(iii)   What multiplies $3xy$ to make $15xy$? Answer: 5

Hence, $x^2 - 4y + 5$ appears inside the bracket. Thus,

$$3x^3y - 12xy^2 + 15xy = 3xy(x^2 - 4y + 5)$$

Problem 12.   Factorise $25a^2b^5 - 5a^3b^2$

For the numbers 25 and 5, the highest common factor is 5 (i.e. 5 is the largest number that divides into 25 and 5)

For the a terms, a^2 and a^3, the HCF is a^2

For the b terms, b^5 and b^2, the HCF is b^2

Thus, the HCF of $25a^2b^5$ and $5a^3b^2$ is $5a^2b^2$

$5a^2b^2$ is therefore taken outside of the bracket. What goes inside the bracket?

(i) What multiplies $5a^2b^2$ to make $25a^2b^5$? Answer: $5b^3$

(ii) What multiplies $5a^2b^2$ to make $-5a^3b^2$? Answer: $-a$

Hence, $5b^3 - a$ appears in the bracket. Thus,

$$25a^2b^5 - 5a^3b^2 = 5a^2b^2(5b^3 - a)$$

Problem 13. Factorise $ax - ay + bx - by$

The first two terms have a common factor of a and the last two terms a common factor of b. Thus,

$$ax - ay + bx - by = a(x - y) + b(x - y)$$

The two newly formed terms have a common factor of $(x - y)$. Thus,

$$a(x - y) + b(x - y) = (x - y)(a + b)$$

Problem 14. Factorise $2ax - 3ay + 2bx - 3by$

a is a common factor of the first two terms and b a common factor of the last two terms. Thus,

$$2ax - 3ay + 2bx - 3by = a(2x - 3y) + b(2x - 3y)$$

$(2x - 3y)$ is now a common factor. Thus,

$$a(2x - 3y) + b(2x - 3y) = (2x - 3y)(a + b)$$

Alternatively, $2x$ is a common factor of the original first and third terms and $-3y$ is a common factor of the second and fourth terms. Thus,

$$2ax - 3ay + 2bx - 3by = 2x(a + b) - 3y(a + b)$$

$(a + b)$ is now a common factor. Thus,

$$2x(a + b) - 3y(a + b) = (a + b)(2x - 3y)$$

as before.

Problem 15. Factorise $x^3 + 3x^2 - x - 3$

x^2 is a common factor of the first two terms. Thus,

$$x^3 + 3x^2 - x - 3 = x^2(x + 3) - x - 3$$

-1 is a common factor of the last two terms. Thus,

$$x^2(x + 3) - x - 3 = x^2(x + 3) - 1(x + 3)$$

$(x + 3)$ is now a common factor. Thus,

$$x^2(x + 3) - 1(x + 3) = (x + 3)(x^2 - 1)$$

Now try the following Practice Exercise

Practice Exercise 46   Factorisation (answers on page 1152)

Factorise and simplify the following.

1. $2x + 4$ 2. $2xy - 8xz$

3. $pb + 2pc$ 4. $2x + 4xy$

5. $4d^2 - 12df^5$ 6. $4x + 8x^2$

7. $2q^2 + 8qn$ 8. $rs + rp + rt$

9. $x + 3x^2 + 5x^3$ 10. $abc + b^3c$

11. $3x^2y^4 - 15xy^2 + 18xy$

12. $4p^3q^2 - 10pq^3$ 13. $21a^2b^2 - 28ab$

14. $2xy^2 + 6x^2y + 8x^3y$

15. $2x^2y - 4xy^3 + 8x^3y^4$

16. $28y + 7y^2 + 14xy$

17. $\dfrac{3x^2 + 6x - 3xy}{xy + 2y - y^2}$ 18. $\dfrac{abc + 2ab}{2c + 4} - \dfrac{abc}{2c}$

19. $\dfrac{5rs + 15r^3t + 20r}{6r^2t^2 + 8t + 2ts} - \dfrac{r}{2t}$

20. $ay + by + a + b$ 21. $px + qx + py + qy$

22. $ax - ay + bx - by$

23. $2ax + 3ay - 4bx - 6by$

24. $\dfrac{A^3}{p^2g^3} - \dfrac{A^2}{pg^2} + \dfrac{A^5}{pg}$

10.4 Laws of precedence

Sometimes addition, subtraction, multiplication, division, powers and brackets can all be involved in an algebraic expression. With mathematics there is a definite order of precedence (first met in Chapter 1) which we need to adhere to.

With the **laws of precedence** the order is

Brackets

Order (or p**O**wer)

Division

Multiplication

Addition

Subtraction

The first letter of each word spells **BODMAS**.
Here are some examples to help understanding of
BODMAS with algebra.

Problem 16.   Simplify $2x + 3x \times 4x - x$

$$
\begin{aligned}
2x + 3x \times 4x - x &= 2x + 12x^2 - x &\text{(M)}\\
&= 2x - x + 12x^2 \\
&= \mathbf{x + 12x^2} &\text{(S)}\\
&\text{or } \mathbf{x(1 + 12x)} &\text{by factorising}
\end{aligned}
$$

Problem 17.   Simplify $(y + 4y) \times 3y - 5y$

$$
\begin{aligned}
(y + 4y) \times 3y - 5y &= 5y \times 3y - 5y &\text{(B)}\\
&= \mathbf{15y^2 - 5y} &\text{(M)}\\
&\text{or } \mathbf{5y(3y - 1)} &\text{by factorising}
\end{aligned}
$$

Problem 18.   Simplify $p + 2p \times (4p - 7p)$

$$
\begin{aligned}
p + 2p \times (4p - 7p) &= p + 2p \times -3p &\text{(B)}\\
&= \mathbf{p - 6p^2} &\text{(M)}\\
&\text{or } \mathbf{p(1 - 6p)} &\text{by factorising}
\end{aligned}
$$

Problem 19.   Simplify $t \div 2t + 3t - 5t$

$$
\begin{aligned}
t \div 2t + 3t - 5t &= \frac{t}{2t} + 3t - 5t &\text{(D)}\\
&= \frac{1}{2} + 3t - 5t &\text{by cancelling}\\
&= \mathbf{\frac{1}{2} - 2t} &\text{(S)}
\end{aligned}
$$

Problem 20.   Simplify $x \div (4x + x) - 3x$

$$
\begin{aligned}
x \div (4x + x) - 3x &= x \div 5x - 3x &\text{(B)}\\
&= \frac{x}{5x} - 3x &\text{(D)}\\
&= \mathbf{\frac{1}{5} - 3x} &\text{by cancelling}
\end{aligned}
$$

Problem 21.   Simplify $2y \div (6y + 3y - 5y)$

$$
\begin{aligned}
2y \div (6y + 3y - 5y) &= 2y \div 4y &\text{(B)}\\
&= \frac{2y}{4y} &\text{(D)}\\
&= \mathbf{\frac{1}{2}} &\text{by cancelling}
\end{aligned}
$$

Problem 22.   Simplify
$$5a + 3a \times 2a + a \div 2a - 7a$$

$$
\begin{aligned}
5a + 3a \times 2a &+ a \div 2a - 7a \\
&= 5a + 3a \times 2a + \frac{a}{2a} - 7a &\text{(D)}\\
&= 5a + 3a \times 2a + \frac{1}{2} - 7a &\text{by cancelling}\\
&= 5a + 6a^2 + \frac{1}{2} - 7a &\text{(M)}\\
&= -2a + 6a^2 + \frac{1}{2} &\text{(S)}\\
&= \mathbf{6a^2 - 2a + \frac{1}{2}}
\end{aligned}
$$

Problem 23.   Simplify
$$(4y + 3y)2y + y \div 4y - 6y$$

$$
\begin{aligned}
(4y + 3y)2y &+ y \div 4y - 6y \\
&= 7y \times 2y + y \div 4y - 6y &\text{(B)}\\
&= 7y \times 2y + \frac{y}{4y} - 6y &\text{(D)}\\
&= 7y \times 2y + \frac{1}{4} - 6y &\text{by cancelling}\\
&= \mathbf{14y^2 + \frac{1}{4} - 6y} &\text{(M)}
\end{aligned}
$$

Problem 24.   Simplify
$$5b + 2b \times 3b + b \div (4b - 7b)$$

$$
\begin{aligned}
5b + 2b \times 3b &+ b \div (4b - 7b) \\
&= 5b + 2b \times 3b + b \div -3b &\text{(B)}\\
&= 5b + 2b \times 3b + \frac{b}{-3b} &\text{(D)}\\
&= 5b + 2b \times 3b + \frac{1}{-3} &\text{by cancelling}\\
&= 5b + 2b \times 3b - \frac{1}{3} \\
&= \mathbf{5b + 6b^2 - \frac{1}{3}} &\text{(M)}
\end{aligned}
$$

Problem 25.   Simplify
$$(5p + p)(2p + 3p) \div (4p - 5p)$$

$$
\begin{aligned}
(5p + p)(2p + 3p) &\div (4p - 5p) \\
&= (6p)(5p) \div (-p) &\text{(B)}\\
&= 6p \times 5p \div -p
\end{aligned}
$$

$$= 6p \times \frac{5p}{-p} \qquad \text{(D)}$$

$$= 6p \times \frac{5}{-1} \qquad \text{by cancelling}$$

$$= 6p \times -5$$

$$= \mathbf{-30p}$$

Now try the following Practice Exercise

Practice Exercise 47   Laws of precedence (answers on page 1153)

Simplify the following.

1. $3x + 2x \times 4x - x$

2. $(2y + y) \times 4y - 3y$

3. $4b + 3b \times (b - 6b)$

4. $8a \div 2a + 6a - 3a$

5. $6x \div (3x + x) - 4x$

6. $4t \div (5t - 3t + 2t)$

7. $3y + 2y \times 5y + 2y \div 8y - 6y$

8. $(x + 2x)3x + 2x \div 6x - 4x$

9. $5a + 2a \times 3a + a \div (2a - 9a)$

10. $(3t + 2t)(5t + t) \div (t - 3t)$

11. $x \div 5x - x + (2x - 3x)x$

12. $3a + 2a \times 5a + 4a \div 2a - 6a$

For fully worked solutions to each of the problems in Practice Exercises 45 to 47 in this chapter, go to the website:
www.routledge.com/cw/bird

Multiple choice questions Test 1
Number and algebra
This test covers the material in Chapters 1 to 10

Section A

All questions have only one correct answer (answers on page 1200).

1. $1\dfrac{1}{3} + 1\dfrac{2}{3} \div 2\dfrac{2}{3} - \dfrac{1}{3}$ is equal to:

   (a) $1\dfrac{5}{8}$ (b) $\dfrac{19}{24}$ (c) $2\dfrac{1}{21}$ (d) $1\dfrac{2}{7}$

2. The value of $\dfrac{2^{-3}}{2^{-4}} - 1$ is equal to:

   (a) $-\dfrac{1}{2}$ (b) 2 (c) 1 (d) $\dfrac{1}{2}$

3. Four engineers can complete a task in 5 hours. Assuming the rate of work remains constant, six engineers will complete the task in:

   (a) 126 h   (b) 4 h 48 min
   (c) 3 h 20 min   (d) 7 h 30 min

4. In an engineering equation $\dfrac{3^4}{3^r} = \dfrac{1}{9}$. The value of r is:

   (a) -6 (b) 2 (c) 6 (d) -2

5. When $p = 3$, $q = -\dfrac{1}{2}$ and $r = -2$, the engineering expression $2p^2q^3r^4$ is equal to:

   (a) 1296   (b) -36 (c) 36   (d) 18

6. $\dfrac{3}{4} \div 1\dfrac{3}{4}$ is equal to:

   (a) $\dfrac{3}{7}$ (b) $1\dfrac{9}{16}$ (c) $1\dfrac{5}{16}$ (d) $2\dfrac{1}{2}$

7. $(2e - 3f)(e + f)$ is equal to:

   (a) $2e^2 - 3f^2$ (b) $2e^2 - 5ef - 3f^2$
   (c) $2e^2 + 3f^2$ (d) $2e^2 - ef - 3f^2$

8. $16^{-\frac{3}{4}}$ is equal to:

   (a) 8 (b) $-\dfrac{1}{2^3}$ (c) 4 (d) $\dfrac{1}{8}$

9. If $x = \dfrac{57.06 \times 0.0711}{\sqrt{0.0635}}$ cm, which of the following statements is correct?

   (a) $x = 16$ cm, correct to 2 significant figures

   (b) $x = 16.09$ cm, correct to 4 significant figures

   (c) $x = 1.61 \times 10^1$ cm, correct to 3 decimal places

   (d) $x = 16.099$ cm, correct to 3 decimal places

10. $(5.5 \times 10^2)(2 \times 10^3)$ cm in standard form is equal to:

    (a) 11×10^6 cm   (b) 1.1×10^6 cm
    (c) 11×10^5 cm   (d) 1.1×10^5 cm

11. The engineering expression $\dfrac{(16 \times 4)^2}{(8 \times 2)^4}$ is equal to:

    (a) 4 (b) 2^{-4} (c) $\dfrac{1}{2^2}$ (d) 1

12. $\left(16^{-\frac{1}{4}} - 27^{-\frac{2}{3}}\right)$ is equal to:

    (a) -7 (b) $\dfrac{7}{18}$ (c) $1\dfrac{8}{9}$ (d) $-8\dfrac{1}{2}$

13. The value of $\dfrac{2}{5}$ of $\left(4\dfrac{1}{2} - 3\dfrac{1}{4}\right) + 5 \div \dfrac{5}{16} - \dfrac{1}{4}$ is:

    (a) $17\dfrac{7}{20}$ (b) $80\dfrac{1}{2}$ (c) $16\dfrac{1}{4}$ (d) 88

14. $(2x - y)^2$ is equal to:

    (a) $4x^2 + y^2$ (b) $2x^2 - 2xy + y^2$
    (c) $4x^2 - y^2$ (d) $4x^2 - 4xy + y^2$

15. $\left(\sqrt{x}\right)\left(y^{3/2}\right)\left(x^2y\right)$ is equal to:

    (a) $\sqrt{(xy)^5}$ (b) $x^{\sqrt{2}}y^{5/2}$

    (c) $xy^{5/2}$ (d) $x\sqrt{y^3}$

16. $p \times \left(\dfrac{1}{p^{-2}}\right)^{-3}$ is equivalent to:

    (a) $\dfrac{1}{p^{\frac{3}{2}}}$ (b) p^{-5} (c) p^5 (d) $\dfrac{1}{p^{-5}}$

17. $45 + 30 \div (21 - 6) - 2 \times 5 + 1$ is equal to:

    (a) -4 (b) 35 (c) -7 (d) 38

18. Factorising $2xy^2 + 6x^3y - 8x^3y^2$ gives:

    (a) $2x\left(y^2 + 3x^2 - 4x^2y\right)$

    (b) $2xy\left(y + 3x^2y - 4x^2y^2\right)$

    (c) $96x^7y^5$

    (d) $2xy\left(y + 3x^2 - 4x^2y\right)$

19. $\dfrac{62.91}{0.12} + \sqrt{\left(\dfrac{6.36\pi}{2e^{-3} \times \sqrt{73.81}}\right)}$ is equal to:

    (a) 529.08 correct to 5 significant figures

    (b) 529.082 correct to 3 decimal places

    (c) 5.29×10^2

    (d) 529.0 correct to 1 decimal place

20. Expanding $(m - n)^2$ gives:

    (a) $m^2 + n^2$ (b) $m^2 - 2mn - n^2$

    (c) $m^2 - 2mn + n^2$ (d) $m^2 - n^2$

21. $2x^2 - (x - xy) - x(2y - x)$ simplifies to:

    (a) $x(3x - 1 - y)$ (b) $x^2 - 3xy - xy$

    (c) $x(xy - y - 1)$ (d) $3x^2 - x + xy$

22. 11 mm expressed as a percentage of 41 mm is:

    (a) 2.68, correct to 3 significant figures

    (b) 2.6, correct to 2 significant figures

    (c) 26.83, correct to 2 decimal places

    (d) 0.2682, correct to 4 decimal places

23. The current in a component in an electrical circuit is calculated as 25 mA using Ohm's law. When measured, the actual current is 25.2 mA. Correct to 2 decimal places, the percentage error in the calculation is:

    (a) 0.80%   (b) 1.25%   (c) 0.79%   (d) 1.26%

24. $\left(\dfrac{a^2}{a^{-3} \times a}\right)^{-2}$ is equivalent to:

    (a) a^{-8} (b) $\dfrac{1}{a^{-10}}$ (c) $\dfrac{1}{a^{-8}}$ (d) a^{-10}

25. Resistance R is given by the formula $R = \dfrac{\rho\ell}{A}$. When resistivity $\rho = 0.017$ μΩm, length $\ell = 5$ km and cross-sectional area $A = 25$ mm^2, the resistance is:

    (a) 3.4 mΩ   (b) 0.034 Ω   (c) 3.4 kΩ   (d) 3.4 Ω

Chapter 11

Solving simple equations

Why it is important to understand: Solving simple equations

In mathematics, engineering and science, formulae are used to relate physical quantities to each other. They provide rules so that if we know the values of certain quantities, we can calculate the values of others. Equations occur in all branches of engineering. Simple equations always involve one unknown quantity which we try to find when we solve the equation. In reality, we all solve simple equations in our heads all the time without even noticing it. If, for example, you have bought two CDs for the same price, and a DVD, and know that you spent £25 in total and that the DVD was £11, then you actually solve the linear equation $2x + 11 = 25$ to find out that the price of each CD was £7. It is probably true to say that there is no branch of engineering, physics, economics, chemistry and computer science which does not require the solution of simple equations. The ability to solve simple equations is another stepping stone on the way to having confidence to handle engineering mathematics.

At the end of this chapter, you should be able to:

- distinguish between an algebraic expression and an algebraic equation
- maintain the equality of a given equation whilst applying arithmetic operations
- solve linear equations with one unknown including those involving brackets and fractions
- form and solve linear equations involved with practical situations
- evaluate formulae by substitution of data

11.1 Introduction

$3x - 4$ is an example of an **algebraic expression**.
$3x - 4 = 2$ is an example of an **algebraic equation** (i.e. it contains an '=' sign).
An equation is simply a statement that two expressions are equal.

Hence, $A = \pi r^2$ (where A is the area of a circle of radius r)

$\quad F = \dfrac{9}{5}C + 32$ (which relates Fahrenheit and Celsius temperatures)

and $\quad y = 3x + 2$ (which is the equation of a straight line graph)

are all examples of equations.

11.2 Solving equations

To '**solve an equation**' means '**to find the value of the unknown**'. For example, solving $3x - 4 = 2$ means that the value of x is required.

In this example, $x = 2$. How did we arrive at $x = 2$? This is the purpose of this chapter – to show how to solve such equations.

Many equations occur in engineering and it is essential that we can solve them when needed.

Here are some examples to demonstrate how simple equations are solved.

Problem 1. Solve the equation $4x = 20$

Dividing each side of the equation by 4 gives

$$\frac{4x}{4} = \frac{20}{4}$$

i.e. $x = 5$ by cancelling, which is the solution to the equation $4x = 20$.

The same operation **must** be applied to both sides of an equation so that the equality is maintained.

We can do anything we like to an equation, **as long as we do the same to both sides.** This is, in fact, the only rule to remember when solving simple equations (and also when transposing formulae, which we do in Chapter 12).

Problem 2. Solve the equation $\dfrac{2x}{5} = 6$

Multiplying both sides by 5 gives $\quad 5\left(\dfrac{2x}{5}\right) = 5(6)$

Cancelling and removing brackets gives $\quad 2x = 30$

Dividing both sides of the equation by 2 gives

$$\frac{2x}{2} = \frac{30}{2}$$

Cancelling gives $\qquad\qquad\qquad x = 15$

which is the solution of the equation $\dfrac{2x}{5} = 6$

Problem 3. Solve the equation $a - 5 = 8$

Adding 5 to both sides of the equation gives

$$a - 5 + 5 = 8 + 5$$

i.e. $\qquad\qquad\qquad a = 8 + 5$

i.e. $\qquad\qquad\qquad \mathbf{a = 13}$

which is the solution of the equation $a - 5 = 8$

Note that adding 5 to both sides of the above equation results in the -5 moving from the LHS to the RHS, but the sign is changed to $+$.

Problem 4. Solve the equation $x + 3 = 7$

Subtracting 3 from both sides gives $\quad x + 3 - 3 = 7 - 3$

i.e. $\qquad\qquad\qquad\qquad x = 7 - 3$

i.e. $\qquad\qquad\qquad\qquad \mathbf{x = 4}$

which is the solution of the equation $x + 3 = 7$

Note that subtracting 3 from both sides of the above equation results in the $+3$ moving from the LHS to the RHS, but the sign is changed to $-$. So, we can move straight from $x + 3 = 7$ to $x = 7 - 3$

Thus, a term can be moved from one side of an equation to the other **as long as a change in sign is made**.

Problem 5. Solve the equation $6x + 1 = 2x + 9$

In such equations the terms containing x are grouped on one side of the equation and the remaining terms grouped on the other side of the equation. As in Problems 3 and 4, changing from one side of an equation to the other must be accompanied by a change of sign.

Since $\qquad\qquad\qquad 6x + 1 = 2x + 9$

then $\qquad\qquad\qquad 6x - 2x = 9 - 1$

i.e. $\qquad\qquad\qquad 4x = 8$

Dividing both sides by 4 gives $\quad \dfrac{4x}{4} = \dfrac{8}{4}$

Cancelling gives $\qquad\qquad\qquad \mathbf{x = 2}$

which is the solution of the equation $6x + 1 = 2x + 9$.
In the above examples, the solutions can be checked. Thus, in Problem 5, where $6x + 1 = 2x + 9$, if $x = 2$, then

$$\text{LHS of equation} = 6(2) + 1 = 13$$

$$\text{RHS of equation} = 2(2) + 9 = 13$$

Since the left-hand side (LHS) equals the right-hand side (RHS) then $x = 2$ must be the correct solution of the equation.

When solving simple equations, always check your answers by substituting your solution back into the original equation.

Problem 6. Solve the equation $4 - 3p = 2p - 11$

In order to keep the p term positive the terms in p are moved to the RHS and the constant terms to the LHS. Similar to Problem 5, if $4 - 3p = 2p - 11$

then $\qquad\qquad\qquad 4 + 11 = 2p + 3p$

i.e. $\qquad\qquad\qquad 15 = 5p$

Dividing both sides by 5 gives $\quad \dfrac{15}{5} = \dfrac{5p}{5}$

Cancelling gives $\qquad\qquad\qquad \mathbf{3 = p} \quad \text{or} \quad \mathbf{p = 3}$

which is the solution of the equation $4 - 3p = 2p - 11$
By substituting $p = 3$ into the original equation, the solution may be checked.

$$\text{LHS} = 4 - 3(3) = 4 - 9 = -5$$

$$\text{RHS} = 2(3) - 11 = 6 - 11 = -5$$

Since LHS = RHS, the solution $p = 3$ must be correct.

If, in this example, the unknown quantities had been grouped initially on the LHS instead of the RHS, then $-3p - 2p = -11 - 4$

i.e. $-5p = -15$

from which, $\dfrac{-5p}{-5} = \dfrac{-15}{-5}$

and $\boldsymbol{p = 3}$

as before.
It is often easier, however, to work with positive values where possible.

Problem 7.   Solve the equation $3(x - 2) = 9$

Removing the bracket gives $3x - 6 = 9$

Rearranging gives $3x = 9 + 6$

i.e. $3x = 15$

Dividing both sides by 3 gives $\boldsymbol{x = 5}$

which is the solution of the equation $3(x - 2) = 9$
The equation may be checked by substituting $x = 5$ back into the original equation.

Problem 8.   Solve the equation
$4(2r - 3) - 2(r - 4) = 3(r - 3) - 1$

Removing brackets gives

$$8r - 12 - 2r + 8 = 3r - 9 - 1$$

Rearranging gives $8r - 2r - 3r = -9 - 1 + 12 - 8$

i.e. $3r = -6$

Dividing both sides by 3 gives $r = \dfrac{-6}{3} = \boldsymbol{-2}$

which is the solution of the equation

$4(2r - 3) - 2(r - 4) = 3(r - 3) - 1$.

The solution may be checked by substituting $r = -2$ back into the original equation.

$\text{LHS} = 4(-4 - 3) - 2(-2 - 4) = -28 + 12 = -16$

$\text{RHS} = 3(-2 - 3) - 1 = -15 - 1 = -16$

Since LHS = RHS then $r = -2$ is the correct solution.

Now try the following Practice Exercise

Practice Exercise 48   Solving simple equations (answers on page 1153)

Solve the following equations.

1. $2x + 5 = 7$

2. $8 - 3t = 2$

3. $\dfrac{2}{3}c - 1 = 3$

4. $2x - 1 = 5x + 11$

5. $7 - 4p = 2p - 5$

6. $2.6x - 1.3 = 0.9x + 0.4$

7. $2a + 6 - 5a = 0$

8. $3x - 2 - 5x = 2x - 4$

9. $20d - 3 + 3d = 11d + 5 - 8$

10. $2(x - 1) = 4$

11. $16 = 4(t + 2)$

12. $5(f - 2) - 3(2f + 5) + 15 = 0$

13. $2x = 4(x - 3)$

14. $6(2 - 3y) - 42 = -2(y - 1)$

15. $2(3g - 5) - 5 = 0$

16. $4(3x + 1) = 7(x + 4) - 2(x + 5)$

17. $11 + 3(r - 7) = 16 - (r + 2)$

18. $8 + 4(x - 1) - 5(x - 3) = 2(5 - 2x)$

Here are some further worked examples on solving simple equations.

Problem 9.   Solve the equation $\dfrac{4}{x} = \dfrac{2}{5}$

The lowest common multiple (LCM) of the denominators, i.e. the lowest algebraic expression that both x and 5 will divide into, is $5x$.

Multiplying both sides by $5x$ gives

$$5x\left(\dfrac{4}{x}\right) = 5x\left(\dfrac{2}{5}\right)$$

Cancelling gives $5(4) = x(2)$

i.e. $20 = 2x$ (1)

Dividing both sides by 2 gives $\dfrac{20}{2} = \dfrac{2x}{2}$

Cancelling gives   **$10 = x$ or $x = 10$**

which is the solution of the equation $\dfrac{4}{x} = \dfrac{2}{5}$

When there is just one fraction on each side of the equation as in this example, there is a quick way to arrive at equation (1) without needing to find the LCM of the denominators.

We can move from $\dfrac{4}{x} = \dfrac{2}{5}$ to $4 \times 5 = 2 \times x$ by what is called '**cross-multiplication**'.

In general, if $\dfrac{a}{b} = \dfrac{c}{d}$ then $ad = bc$

We can use cross-multiplication when there is one fraction only on each side of the equation.

> **Problem 10.** Solve the equation $\dfrac{3}{t-2} = \dfrac{4}{3t+4}$

Cross-multiplication gives $3(3t+4) - 4(t \quad 2)$

Removing brackets gives $9t + 12 = 4t - 8$

Rearranging gives $9t - 4t = -8 - 12$

i.e. $5t = -20$

Dividing both sides by 5 gives $t = \dfrac{-20}{5} = -4$

which is the solution of the equation $\dfrac{3}{t-2} = \dfrac{4}{3t+4}$

> **Problem 11.** Solve the equation
> $$\dfrac{2y}{5} + \dfrac{3}{4} + 5 = \dfrac{1}{20} - \dfrac{3y}{2}$$

The lowest common multiple (LCM) of the denominators is 20; i.e. the lowest number that 4, 5, 20 and 2 will divide into.
Multiplying each term by 20 gives

$$20\left(\dfrac{2y}{5}\right) + 20\left(\dfrac{3}{4}\right) + 20(5) = 20\left(\dfrac{1}{20}\right) - 20\left(\dfrac{3y}{2}\right)$$

Cancelling gives $4(2y) + 5(3) + 100 = 1 - 10(3y)$

i.e. $8y + 15 + 100 = 1 - 30y$

Rearranging gives $8y + 30y = 1 - 15 - 100$

i.e. $38y = -114$

Dividing both sides by 38 gives $\dfrac{38y}{38} = \dfrac{-114}{38}$

Cancelling gives   **$y = -3$**

which is the solution of the equation

$$\dfrac{2y}{5} + \dfrac{3}{4} + 5 = \dfrac{1}{20} - \dfrac{3y}{2}$$

> **Problem 12.** Solve the equation $\sqrt{x} = 2$

Whenever square root signs are involved in an equation, both sides of the equation must be squared.

Squaring both sides gives $\left(\sqrt{x}\right)^2 = (2)^2$

i.e.   **$x = 4$**

which is the solution of the equation $\sqrt{x} = 2$

> **Problem 13.** Solve the equation $2\sqrt{d} = 8$

Whenever square roots are involved in an equation, the square root term needs to be isolated on its own before squaring both sides.

Cross-multiplying gives $\sqrt{d} = \dfrac{8}{2}$

Cancelling gives $\sqrt{d} = 4$

Squaring both sides gives $\left(\sqrt{d}\right)^2 = (4)^2$

i.e.   **$d = 16$**

which is the solution of the equation $2\sqrt{d} = 8$

> **Problem 14.** Solve the equation $\left(\dfrac{\sqrt{b}+3}{\sqrt{b}}\right) = 2$

Cross-multiplying gives $\sqrt{b} + 3 = 2\sqrt{b}$
Rearranging gives $3 = 2\sqrt{b} - \sqrt{b}$
i.e. $3 = \sqrt{b}$
Squaring both sides gives   **$9 = b$**

which is the solution of the equation $\left(\dfrac{\sqrt{b}+3}{\sqrt{b}}\right) = 2$

> **Problem 15.** Solve the equation $x^2 = 25$

Whenever a square term is involved, the square root of both sides of the equation must be taken.

Taking the square root of both sides gives $\sqrt{x^2} = \sqrt{25}$

i.e. $x = \pm 5$

which is the solution of the equation $x^2 = 25$

Problem 16.   Solve the equation $\dfrac{15}{4t^2} = \dfrac{2}{3}$

We need to rearrange the equation to get the t^2 term on its own.

Cross-multiplying gives $15(3) = 2(4t^2)$

i.e. $45 = 8t^2$

Dividing both sides by 8 gives $\dfrac{45}{8} = \dfrac{8t^2}{8}$

By cancelling $5.625 = t^2$

or $t^2 = 5.625$

Taking the square root of both sides gives

$$\sqrt{t^2} = \sqrt{5.625}$$

i.e. $t = \pm 2.372$

correct to 4 significant figures, which is the solution of the equation $\dfrac{15}{4t^2} = \dfrac{2}{3}$

Now try the following Practice Exercise

Practice Exercise 49   Solving simple equations (answers on page 1153)

Solve the following equations.

1. $\dfrac{1}{5}d + 3 = 4$

2. $2 + \dfrac{3}{4}y = 1 + \dfrac{2}{3}y + \dfrac{5}{6}$

3. $\dfrac{1}{4}(2x - 1) + 3 = \dfrac{1}{2}$

4. $\dfrac{1}{5}(2f - 3) + \dfrac{1}{6}(f - 4) + \dfrac{2}{15} = 0$

5. $\dfrac{1}{3}(3m - 6) - \dfrac{1}{4}(5m + 4) + \dfrac{1}{5}(2m - 9) = -3$

6. $\dfrac{x}{3} - \dfrac{x}{5} = 2$

7. $1 - \dfrac{y}{3} = 3 + \dfrac{y}{3} - \dfrac{y}{6}$

8. $\dfrac{2}{a} = \dfrac{3}{8}$

9. $\dfrac{1}{3n} + \dfrac{1}{4n} = \dfrac{7}{24}$

10. $\dfrac{x+3}{4} = \dfrac{x-3}{5} + 2$

11. $\dfrac{3t}{20} = \dfrac{6-t}{12} + \dfrac{2t}{15} - \dfrac{3}{2}$

12. $\dfrac{y}{5} + \dfrac{7}{20} = \dfrac{5-y}{4}$

13. $\dfrac{v-2}{2v-3} = \dfrac{1}{3}$

14. $\dfrac{2}{a-3} = \dfrac{3}{2a+1}$

15. $\dfrac{x}{4} - \dfrac{x+6}{5} = \dfrac{x+3}{2}$

16. $3\sqrt{t} = 9$

17. $2\sqrt{y} = 5$

18. $4 = \sqrt{\left(\dfrac{3}{a}\right) + 3}$

19. $\dfrac{3\sqrt{x}}{1 - \sqrt{x}} = -6$

20. $10 = 5\sqrt{\left(\dfrac{x}{2} - 1\right)}$

21. $16 = \dfrac{t^2}{9}$

22. $\sqrt{\left(\dfrac{y+2}{y-2}\right)} = \dfrac{1}{2}$

23. $\dfrac{6}{a} = \dfrac{2a}{3}$

24. $\dfrac{11}{2} = 5 + \dfrac{8}{x^2}$

11.3   Practical problems involving simple equations

There are many practical situations in engineering in which solving equations is needed. Here are some worked examples to demonstrate typical practical situations.

Problem 17.   Applying the principle of moments to a beam results in the equation

$$F \times 3 = (7.5 - F) \times 2$$

where F is the force in newtons. Determine the value of F

Removing brackets gives $3F = 15 - 2F$

Rearranging gives $3F + 2F = 15$

i.e. $5F = 15$

Dividing both sides by 5 gives $\dfrac{5F}{5} = \dfrac{15}{5}$

from which force $F = 3N$

Problem 18.   A copper wire has a length L of 1.5 km, a resistance R of 5Ω and a resistivity of $17.2 \times 10^{-6}\,\Omega\,$mm. Find the cross-sectional area a of the wire, given that $R = \dfrac{\rho L}{a}$

Since $R = \dfrac{\rho L}{a}$ then

$$5\Omega = \frac{(17.2 \times 10^{-6}\,\Omega\,\text{mm})(1500 \times 10^{3}\,\text{mm})}{a}$$

From the units given, a is measured in mm^2.

Thus, $5a = 17.2 \times 10^{-6} \times 1500 \times 10^{3}$

and $a = \dfrac{17.2 \times 10^{-6} \times 1500 \times 10^{3}}{5}$

$$= \frac{17.2 \times 1500 \times 10^{3}}{10^{6} \times 5} = \frac{17.2 \times 15}{10 \times 5} = 5.16$$

Hence, the cross-sectional area of the wire is **5.16 mm^2**

Problem 19. $PV = mRT$ is the characteristic gas equation. Find the value of gas constant R when pressure $P = 3 \times 10^{6}\,$Pa, volume $V = 0.90\,\text{m}^3$, mass $m = 2.81\,$kg and temperature $T = 231\,$K

Dividing both sides of $PV = mRT$ by mT gives

$$\frac{PV}{mT} = \frac{mRT}{mT}$$

Cancelling gives $\dfrac{PV}{mT} = R$

Substituting values gives $R = \dfrac{\left(3 \times 10^{6}\right)(0.90)}{(2.81)(231)}$

Using a calculator, **gas constant $R = 4160\,$J/(kg K)**, correct to 4 significant figures.

Problem 20.   A rectangular box with square ends has its length 15 cm greater than its breadth and the total length of its edges is 2.04 m. Find the width of the box and its volume

Let x cm = width = height of box. Then the length of the box is $(x + 15)$ cm, as shown in Figure 11.1.

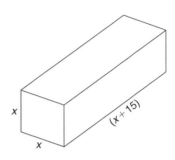

Figure 11.1

The length of the edges of the box is $2(4x) + 4(x + 15)$ cm, which equals 2.04 m or 204 cm.

Hence, $204 = 2(4x) + 4(x + 15)$

$$204 = 8x + 4x + 60$$

$$204 - 60 = 12x$$

i.e. $144 = 12x$

and $x = 12\,$cm

Hence, **the width of the box is 12 cm**.

Volume of box = length × width × height

$$= (x + 15)(x)(x) = (12 + 15)(12)(12)$$

$$= (27)(12)(12)$$

$$= \textbf{3888}\,\textbf{cm}^3$$

Problem 21.   The temperature coefficient of resistance α may be calculated from the formula $R_t = R_0(1 + \alpha t)$. Find α, given $R_t = 0.928$, $R_0 = 0.80$ and $t = 40$

Since $R_t = R_0(1 + \alpha t)$, then

$$0.928 = 0.80[1 + \alpha(40)]$$

$$0.928 = 0.80 + (0.8)(\alpha)(40)$$

$$0.928 - 0.80 = 32\alpha$$

$$0.128 = 32\alpha$$

Hence,
$$\alpha = \frac{0.128}{32} = 0.004$$

Now try the following Practice Exercise

Problem 22. The distance s metres travelled in time t seconds is given by the formula $s = ut + \frac{1}{2}at^2$, where u is the initial velocity in m/s and a is the acceleration in m/s^2. Find the acceleration of the body if it travels 168 m in 6 s, with an initial velocity of 10 m/s

$s = ut + \frac{1}{2}at^2$, and $s = 168, u = 10$ and $t = 6$

Hence,
$$168 = (10)(6) + \frac{1}{2}a(6)^2$$
$$168 = 60 + 18a$$
$$168 - 60 = 18a$$
$$108 = 18a$$
$$a = \frac{108}{18} = 6$$

Hence, **the acceleration of the body is 6 m/s^2**

Problem 23. When three resistors in an electrical circuit are connected in parallel the total resistance R_T is given by $\frac{1}{R_T} = \frac{1}{R_1} + \frac{1}{R_2} + \frac{1}{R_3}$. Find the total resistance when $R_1 = 5\,\Omega, R_2 = 10\,\Omega$ and $R_3 = 30\,\Omega$

$$\frac{1}{R_T} = \frac{1}{5} + \frac{1}{10} + \frac{1}{30} = \frac{6+3+1}{30} = \frac{10}{30} = \frac{1}{3}$$

Taking the reciprocal of both sides gives $\boldsymbol{R_T = 3\,\Omega}$

Alternatively, if $\frac{1}{R_T} = \frac{1}{5} + \frac{1}{10} + \frac{1}{30}$, the LCM of the denominators is $30R_T$

Hence, $30R_T\left(\frac{1}{R_T}\right) = 30R_T\left(\frac{1}{5}\right)$

$$+ 30R_T\left(\frac{1}{10}\right) + 30R_T\left(\frac{1}{30}\right)$$

Cancelling gives $\quad 30 = 6R_T + 3R_T + R_T$

i.e. $\quad 30 = 10R_T$

and $\quad \boldsymbol{R_T = \frac{30}{10} = 3\,\Omega}$, as above.

Now try the following Practice Exercise

Practice Exercise 50   Practical problems involving simple equations (answers on page 1153)

1. A formula used for calculating resistance of a cable is $R = \frac{\rho L}{a}$. Given $R = 1.25, L = 2500$ and $a = 2 \times 10^{-4}$, find the value of ρ.

2. Force F newtons is given by $F = ma$, where m is the mass in kilograms and a is the acceleration in metres per second squared. Find the acceleration when a force of 4 kN is applied to a mass of 500 kg.

3. $PV = mRT$ is the characteristic gas equation. Find the value of m when $P = 100 \times 10^3, V = 3.00, R = 288$ and $T = 300$

4. When three resistors R_1, R_2 and R_3 are connected in parallel, the total resistance R_T is determined from $\frac{1}{R_T} = \frac{1}{R_1} + \frac{1}{R_2} + \frac{1}{R_3}$
    (a) Find the total resistance when $R_1 = 3\,\Omega, R_2 = 6\,\Omega$ and $R_3 = 18\,\Omega$.
    (b) Find the value of R_3 given that $R_T = 3\,\Omega, R_1 = 5\,\Omega$ and $R_2 = 10\,\Omega$.

5. Six digital camera batteries and 3 camcorder batteries cost £96. If a camcorder battery costs £5 more than a digital camera battery, find the cost of each.

6. Ohm's law may be represented by $I = V/R$, where I is the current in amperes, V is the voltage in volts and R is the resistance in ohms. A soldering iron takes a current of 0.30 A from a 240 V supply. Find the resistance of the element.

7. The distance s travelled in time t seconds is given by the formula $s = ut + \frac{1}{2}at^2$ where u is the initial velocity in m/s and a is the acceleration in m/s^2. Calculate the acceleration of the body if it travels 165 m in 3 s, with an initial velocity of 10 m/s.

8. The stress, σ pascals, acting on the reinforcing rod in a concrete column is given in the following equation:
    $$500 \times 10^{-6}\sigma + 2.67 \times 10^5 = 3.55 \times 10^5$$
    Find the value of the stress in MPa.

Here are some further worked examples on solving simple equations in practical situations.

> **Problem 24.** The extension x m of an aluminium tie bar of length l m and cross-sectional area A m^2 when carrying a load of F newtons is given by the modulus of elasticity $E = Fl/Ax$. Find the extension of the tie bar (in mm) if $E = 70 \times 10^9$ N/m^2, $F = 20 \times 10^6$ N, $A = 0.1$m^2 and $l = 1.4$ m

$E = Fl/Ax$, hence

$$70 \times 10^9 \frac{N}{m^2} = \frac{(20 \times 10^6 \text{N})(1.4\,\text{m})}{(0.1\text{m}^2)(x)}$$

(the unit of x is thus metres)

$$70 \times 10^9 \times 0.1 \times x = 20 \times 10^6 \times 1.4$$

$$x = \frac{20 \times 10^6 \times 1.4}{70 \times 10^9 \times 0.1}$$

Cancelling gives $\quad x = \frac{2 \times 1.4}{7 \times 100}$ m

$$= \frac{2 \times 1.4}{7 \times 100} \times 1000 \,\text{mm}$$

$$= 4 \,\text{mm}$$

Hence, **the extension of the tie bar, $x = 4$ mm**.

> **Problem 25.** Power in a d.c. circuit is given by $P = \dfrac{V^2}{R}$ where V is the supply voltage and R is the circuit resistance. Find the supply voltage if the circuit resistance is $1.25\,\Omega$ and the power measured is 320 W

Since $P = \dfrac{V^2}{R}$, then $320 = \dfrac{V^2}{1.25}$

$$(320)(1.25) = V^2$$

i.e. $\qquad\qquad\qquad V^2 = 400$

Supply voltage, $\qquad \mathbf{V = \sqrt{400} = \pm 20\,V}$

> **Problem 26.** A painter is paid £6.30 per hour for a basic 36 hour week and overtime is paid at one and a third times this rate. Determine how many hours the painter has to work in a week to earn £319.20

Basic rate per hour $=$ £6.30 and overtime rate per hour

$$= 1\frac{1}{3} \times £6.30 = £8.40$$

Let the number of overtime hours worked $= x$
Then,

$$(36)(6.30) + (x)(8.40) = 319.20$$

$$226.80 + 8.40x = 319.20$$

$$8.40x = 319.20 - 226.80 = 92.40$$

$$x = \frac{92.40}{8.40} = 11$$

Thus, 11 hours overtime would have to be worked to earn £319.20 per week. Hence, **the total number of hours worked** is $36 + 11$, i.e. **47 hours**.

> **Problem 27.** A formula relating initial and final states of pressures, P_1 and P_2, volumes, V_1 and V_2, and absolute temperatures, T_1 and T_2, of an ideal gas is $\dfrac{P_1 V_1}{T_1} = \dfrac{P_2 V_2}{T_2}$. Find the value of P_2 given $P_1 = 100 \times 10^3$, $V_1 = 1.0$, $V_2 = 0.266$, $T_1 = 423$ and $T_2 = 293$

Since $\dfrac{P_1 V_1}{T_1} = \dfrac{P_2 V_2}{T_2}$

then $\dfrac{(100 \times 10^3)(1.0)}{423} = \dfrac{P_2(0.266)}{293}$

Cross-multiplying gives

$$(100 \times 10^3)(1.0)(293) = P_2(0.266)(423)$$

$$P_2 = \frac{(100 \times 10^3)(1.0)(293)}{(0.266)(423)}$$

Hence, $\qquad \mathbf{P_2 = 260 \times 10^3}$ or $\mathbf{2.6 \times 10^5}$

> **Problem 28.** The stress f in a material of a thick cylinder can be obtained from $\dfrac{D}{d} = \sqrt{\left(\dfrac{f+p}{f-p}\right)}$ Calculate the stress, given that $D = 21.5$, $d = 10.75$ and $p = 1800$

Since $\dfrac{D}{d} = \sqrt{\left(\dfrac{f+p}{f-p}\right)}$ then $\dfrac{21.5}{10.75} = \sqrt{\left(\dfrac{f+1800}{f-1800}\right)}$

i.e.

$$2 = \sqrt{\left(\frac{f + 1800}{f - 1800}\right)}$$

Squaring both sides gives

$$4 = \frac{f + 1800}{f - 1800}$$

Cross-multiplying gives

$$4(f - 1800) = f + 1800$$

$$4f - 7200 = f + 1800$$

$$4f - f = 1800 + 7200$$

$$3f = 9000$$

$$f = \frac{9000}{3} = 3000$$

Hence,　　　　　**stress, $f = 3000$**

Problem 29.　12 workmen employed on a building site earn between them a total of £4035 per week. Labourers are paid £275 per week and craftsmen are paid £380 per week. How many craftsmen and how many labourers are employed?

Let the number of craftsmen be c. The number of labourers is therefore $(12 - c)$.

The wage bill equation is

$$380c + 275(12 - c) = 4035$$

$$380c + 3300 - 275c = 4035$$

$$380c - 275c = 4035 - 3300$$

$$105c = 735$$

$$c = \frac{735}{105} = 7$$

Hence, there are **7 craftsmen** and $(12 - 7)$, i.e. **5 labourers** on the site.

Now try the following Practice Exercise

Practice Exercise 51　Practical problems involving simple equations (answers on page 1153)

1. A rectangle has a length of 20 cm and a width b cm. When its width is reduced by 4 cm its area becomes 160 cm^2. Find the original width and area of the rectangle.

2. Given $R_2 = R_1(1 + \alpha t)$, find α given $R_1 = 5.0$, $R_2 = 6.03$ and $t = 51.5$

3. If $v^2 = u^2 + 2as$, find u given $v = 24$, $a = -40$ and $s = 4.05$

4. The relationship between the temperature on a Fahrenheit scale and that on a Celsius scale is given by $F = \frac{9}{5}C + 32$. Express 113°F in degrees Celsius.

5. If $t = 2\pi\sqrt{\frac{w}{Sg}}$, find the value of S given $w = 1.219$, $g = 9.81$ and $t = 0.3132$

6. Two joiners and five mates earn £1824 between them for a particular job. If a joiner earns £72 more than a mate, calculate the earnings for a joiner and for a mate.

7. An alloy contains 60% by weight of copper, the remainder being zinc. How much copper must be mixed with 50 kg of this alloy to give an alloy containing 75% copper?

8. A rectangular laboratory has a length equal to one and a half times its width and a perimeter of 40 m. Find its length and width.

9. Applying the principle of moments to a beam results in the following equation:

$$F \times 3 = (5 - F) \times 7$$

where F is the force in newtons. Determine the value of F.

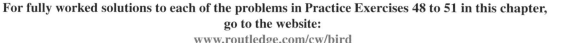

For fully worked solutions to each of the problems in Practice Exercises 48 to 51 in this chapter, go to the website:
www.routledge.com/cw/bird

Section A

Revision Test 4   Algebra and simple equations

This assignment covers the material contained in Chapters 9 to 11. *The marks available are shown in brackets at the end of each question.*

1. Evaluate $3pqr^3 - 2p^2qr + pqr$ when $p = \frac{1}{2}$, $q = -2$ and $r = 1$ (3)

In Problems 2 to 7, simplify the expressions.

2. $\dfrac{9p^2qr^3}{3pq^2r}$ (3)

3. $2(3x - 2y) - (4y - 3x)$ (3)

4. $(x - 2y)(2x + y)$ (3)

5. $p^2q^{-3}r^4 \times pq^2r^{-3}$ (3)

6. $(3a - 2b)^2$ (3)

7. $\dfrac{a^4b^2c}{ab^3c^2}$ (3)

8. Factorise
    (a) $2x^2y^3 - 10xy^2$
    (b) $21ab^2c^3 - 7a^2bc^2 + 28a^3bc^4$ (5)

9. Factorise and simplify
    $\dfrac{2x^2y + 6xy^2}{x + 3y} - \dfrac{x^3y^2}{x^2y}$ (5)

10. Remove the brackets and simplify
    $10a - [3(2a - b) - 4(b - a) + 5b]$ (4)

11. Simplify $x \div 5x - x + (2x - 3x)x$ (4)

12. Simplify $3a + 2a \times 5a + 4a \div 2a - 6a$ (4)

13. Solve the equations
    (a) $3a = 39$
    (b) $2x - 4 = 9$ (3)

14. Solve the equations
    (a) $\dfrac{4}{9}y = 8$
    (b) $6x - 1 = 4x + 5$ (4)

15. Solve the equation
    $5(t - 2) - 3(4 - t) = 2(t + 3) - 40$ (4)

16. Solve the equations:
    (a) $\dfrac{3}{2x + 1} = \dfrac{1}{4x - 3}$
    (b) $2x^2 = 162$ (7)

17. Kinetic energy is given by the formula $E_k = \frac{1}{2}mv^2$ joules, where m is the mass in kilograms and v is the velocity in metres per second. Evaluate the velocity when $E_k = 576 \times 10^{-3}$ J and the mass is 5 kg. (4)

18. An approximate relationship between the number of teeth T on a milling cutter, the diameter of the cutter D and the depth of cut d is given by $T = \dfrac{12.5D}{D + 4d}$. Evaluate d when $T = 10$ and $D = 32$ (5)

19. The modulus of elasticity E is given by the formula $E = \dfrac{FL}{xA}$ where F is force in newtons, L is the length in metres, x is the extension in metres and A the cross-sectional area in square metres. Evaluate A, in square centimetres, when $E = 80 \times 10^9$ N/m^2, $x = 2$ mm, $F = 100 \times 10^3$ N and $L = 2.0$ m. (5)

Chapter 12

Transposing formulae

Why it is important to understand: Transposing formulae

As was mentioned in the last chapter, formulae are used frequently in almost all aspects of engineering in order to relate a physical quantity to one or more others. Many well known physical laws are described using formulae – for example, Ohm's law, $V = I \times R$, or Newton's second law of motion, $F = m \times a$. In an everyday situation, imagine you buy 5 identical items for £20. How much did each item cost? If you divide £20 by 5 to get an answer of £4, you are actually applying transposition of a formula. Transposing formulae is a basic skill required in all aspects of engineering. The ability to transpose formulae is yet another stepping stone on the way to having confidence to handle engineering mathematics.

At the end of this chapter, you should be able to:

- define 'subject of the formula'
- transpose equations whose terms are connected by plus and/or minus signs
- transpose equations that involve fractions
- transpose equations that contain a root or power
- transpose equations in which the subject appears in more than one term

12.1 Introduction

In the formula $I = \dfrac{V}{R}$, I is called the **subject of the formula**.

Similarly, in the formula $y = mx + c$, y is the subject of the formula.

When a symbol other than the subject is required to be the subject, the formula needs to be rearranged to make a new subject. This rearranging process is called **transposing the formula** or **transposition**.

For example, in the above formulae,

$$\text{if } I = \frac{V}{R} \text{ then } V = IR$$

$$\text{and if } y = mx + c \text{ then } x = \frac{y - c}{m}$$

How did we arrive at these transpositions? This is the purpose of this chapter – to show how to transpose formulae. A great many equations occur in engineering and it is essential that we can transpose them when needed.

12.2 Transposing formulae

There are no new rules for transposing formulae. The same rules as were used for simple equations in Chapter 11 are used; i.e. **the balance of an equation must be maintained**: whatever is done to one side of an equation must be done to the other.

It is best that you cover simple equations before trying this chapter.

Here are some worked examples to help understanding of transposing formulae.

> **Problem 1.** Transpose $p = q + r + s$ to make r the subject

The object is to obtain r on its own on the LHS of the equation. Changing the equation around so that r is on the LHS gives

$$q + r + s = p \qquad (1)$$

From Chapter 11 on simple equations, a term can be moved from one side of an equation to the other side as long as the sign is changed.
Rearranging gives $r = p - q - s$
Mathematically, we have subtracted $q + s$ from both sides of equation (1).

> **Problem 2.** If $a + b = w - x + y$, express x as the subject

As stated in Problem 1, a term can be moved from one side of an equation to the other side but with a change of sign.
Hence, rearranging gives $x = w + y - a - b$

> **Problem 3.** Transpose $v = f\lambda$ to make λ the subject

$v = f\lambda$ relates velocity v, frequency f and wavelength λ.

Rearranging gives $\qquad\qquad f\lambda = v$

Dividing both sides by f gives $\qquad \dfrac{f\lambda}{f} = \dfrac{v}{f}$

Cancelling gives $\qquad\qquad \lambda = \dfrac{v}{f}$

> **Problem 4.** When a body falls freely through a height h, the velocity v is given by $v^2 = 2gh$. Express this formula with h as the subject

Rearranging gives $\qquad\qquad 2gh = v^2$

Dividing both sides by $2g$ gives $\qquad \dfrac{2gh}{2g} = \dfrac{v^2}{2g}$

Cancelling gives $\qquad\qquad h = \dfrac{v^2}{2g}$

> **Problem 5.** If $I = \dfrac{V}{R}$, rearrange to make V the subject

$I = \dfrac{V}{R}$ is Ohm's law, where I is the current, V is the voltage and R is the resistance.

Rearranging gives $\qquad\qquad \dfrac{V}{R} = I$

Multiplying both sides by R gives $\quad R\left(\dfrac{V}{R}\right) = R(I)$

Cancelling gives $\qquad\qquad \boldsymbol{V = IR}$

> **Problem 6.** Transpose $a = \dfrac{F}{m}$ for m

$a = \dfrac{F}{m}$ relates acceleration a, force F and mass m.

Rearranging gives $\qquad\qquad \dfrac{F}{m} = a$

Multiplying both sides by m gives $\quad m\left(\dfrac{F}{m}\right) = m(a)$

Cancelling gives $\qquad\qquad F = ma$

Rearranging gives $\qquad\qquad ma = F$

Dividing both sides by a gives $\qquad \dfrac{ma}{a} = \dfrac{F}{a}$

i.e. $\qquad\qquad \boldsymbol{m = \dfrac{F}{a}}$

> **Problem 7.** Rearrange the formula $R = \dfrac{\rho L}{A}$ to make (a) A the subject and (b) L the subject

$R = \dfrac{\rho L}{A}$ relates resistance R of a conductor, resistivity ρ, conductor length L and conductor cross-sectional area A.

(a) Rearranging gives $\qquad\qquad \dfrac{\rho L}{A} = R$

Multiplying both sides by A gives

$$A\left(\dfrac{\rho L}{A}\right) = A(R)$$

Cancelling gives $\qquad\qquad \rho L = AR$

Rearranging gives $\qquad\qquad AR = \rho L$

Dividing both sides by R gives $\qquad \dfrac{AR}{R} = \dfrac{\rho L}{R}$

Cancelling gives $\qquad\qquad A = \dfrac{\rho L}{R}$

(b) Multiplying both sides of $\dfrac{\rho L}{A} = R$ by A gives

$$\rho L = AR$$

Dividing both sides by ρ gives $\qquad \dfrac{\rho L}{\rho} = \dfrac{AR}{\rho}$

Cancelling gives $\qquad\qquad\qquad \boldsymbol{L = \dfrac{AR}{\rho}}$

Problem 8. Transpose $y = mx + c$ to make m the subject

$y = mx + c$ is the equation of a straight line graph, where y is the vertical axis variable, x is the horizontal axis variable, m is the gradient of the graph and c is the y-axis intercept.

Subtracting c from both sides gives $\qquad y - c = mx$

or $\qquad\qquad\qquad\qquad\qquad\qquad mx = y - c$

Dividing both sides by x gives $\qquad \boldsymbol{m = \dfrac{y - c}{x}}$

Now try the following Practice Exercise

Practice Exercise 52   Transposing formulae (answers on page 1153)

Make the symbol indicated the subject of each of the formulae shown and express each in its simplest form.

1. $a + b = c - d - e$ $\qquad (d)$
2. $y = 7x$ $\qquad (x)$
3. $pv = c$ $\qquad (v)$
4. $v = u + at$ $\qquad (a)$
5. $V = IR$ $\qquad (R)$
6. $x + 3y = t$ $\qquad (y)$
7. $c = 2\pi r$ $\qquad (r)$
8. $y = mx + c$ $\qquad (x)$
9. $I = PRT$ $\qquad (T)$
10. $Q = mc\Delta T$ $\qquad (c)$
11. $X_L = 2\pi fL$ $\qquad (L)$
12. $I = \dfrac{E}{R}$ $\qquad (R)$
13. $y = \dfrac{x}{a} + 3$ $\qquad (x)$
14. $F = \dfrac{9}{5}C + 32$ $\qquad (C)$
15. $X_C = \dfrac{1}{2\pi fC}$ $\qquad (f)$
16. $pV = mRT$ $\qquad (R)$

12.3 Further transposing of formulae

Here are some more transposition examples to help us further understand how more difficult formulae are transposed.

Problem 9. Transpose the formula $v = u + \dfrac{Ft}{m}$ to make F the subject

$v = u + \dfrac{Ft}{m}$ relates final velocity v, initial velocity u, force F, mass m and time t. $\left(\dfrac{F}{m} \text{ is acceleration } a. \right)$

Rearranging gives $\qquad\qquad u + \dfrac{Ft}{m} = v$

and $\qquad\qquad\qquad\qquad \dfrac{Ft}{m} = v - u$

Multiplying each side by m gives

$$m\left(\dfrac{Ft}{m} \right) = m(v - u)$$

Cancelling gives $\qquad\qquad Ft = m(v - u)$

Dividing both sides by t gives $\qquad \dfrac{Ft}{t} = \dfrac{m(v - u)}{t}$

Cancelling gives $\boldsymbol{F = \dfrac{m(v - u)}{t}}$ or $\boldsymbol{F = \dfrac{m}{t}(v - u)}$

This shows two ways of expressing the answer. There is often more than one way of expressing a transposed answer. In this case, these equations for F are equivalent; neither one is more correct than the other.

Problem 10. The final length L_2 of a piece of wire heated through $\theta°C$ is given by the formula $L_2 = L_1(1 + \alpha\theta)$ where L_1 is the original length. Make the coefficient of expansion α the subject

Rearranging gives $\qquad\qquad L_1(1 + \alpha\theta) = L_2$

Removing the bracket gives $\qquad L_1 + L_1\alpha\theta = L_2$

Rearranging gives $\qquad L_1\alpha\theta = L_2 - L_1$

Dividing both sides by $L_1\theta$ gives $\dfrac{L_1\alpha\theta}{L_1\theta} = \dfrac{L_2 - L_1}{L_1\theta}$

Cancelling gives $\qquad \boldsymbol{\alpha = \dfrac{L_2 - L_1}{L_1\theta}}$

An alternative method of transposing $L_2 = L_1\,(1 + \alpha\theta)$ for α is:

Dividing both sides by L_1 gives $\qquad \dfrac{L_2}{L_1} = 1 + \alpha\theta$

Subtracting 1 from both sides gives $\dfrac{L_2}{L_1} - 1 = \alpha\theta$

or $\qquad \alpha\theta = \dfrac{L_2}{L_1} - 1$

Dividing both sides by θ gives $\qquad \boldsymbol{\alpha = \dfrac{\dfrac{L_2}{L_1} - 1}{\theta}}$

The two answers $\alpha = \dfrac{L_2 - L_1}{L_1\theta}$ and $\alpha = \dfrac{\dfrac{L_2}{L_1} - 1}{\theta}$ look quite different. They are, however, equivalent. The first answer looks tidier but is no more correct than the second answer.

Problem 11. A formula for the distance s moved by a body is given by $s = \dfrac{1}{2}(v + u)t$. Rearrange the formula to make u the subject

Rearranging gives $\qquad \dfrac{1}{2}(v + u)t = s$

Multiplying both sides by 2 gives $\qquad (v + u)t = 2s$

Dividing both sides by t gives $\qquad \dfrac{(v + u)t}{t} = \dfrac{2s}{t}$

Cancelling gives $\qquad v + u = \dfrac{2s}{t}$

Rearranging gives $\qquad \boldsymbol{u = \dfrac{2s}{t} - v}$ or $\boldsymbol{u = \dfrac{2s - vt}{t}}$

Problem 12. A formula for kinetic energy is $k = \dfrac{1}{2}mv^2$. Transpose the formula to make v the subject

Rearranging gives $\qquad \dfrac{1}{2}mv^2 = k$

Whenever the prospective new subject is a squared term, that term is isolated on the LHS and then the square root of both sides of the equation is taken.

Multiplying both sides by 2 gives $\qquad mv^2 = 2k$

Dividing both sides by m gives $\qquad \dfrac{mv^2}{m} = \dfrac{2k}{m}$

Cancelling gives $\qquad v^2 = \dfrac{2k}{m}$

Taking the square root of both sides gives

$$\sqrt{v^2} = \sqrt{\left(\dfrac{2k}{m}\right)}$$

i.e. $\qquad \boldsymbol{v = \sqrt{\left(\dfrac{2k}{m}\right)}}$

Problem 13. In a right-angled triangle having sides x, y and hypotenuse z, Pythagoras' theorem states $z^2 = x^2 + y^2$. Transpose the formula to find x

Rearranging gives $\qquad x^2 + y^2 = z^2$

and $\qquad x^2 = z^2 - y^2$

Taking the square root of both sides gives

$$\boldsymbol{x = \sqrt{z^2 - y^2}}$$

Problem 14. Transpose $y = \dfrac{ML^2}{8EI}$ to make L the subject

Multiplying both sides by $8EI$ gives $\quad 8EIy = ML^2$

Dividing both sides by M gives $\qquad \dfrac{8EIy}{M} = L^2$

or $\qquad L^2 = \dfrac{8EIy}{M}$

Taking the square root of both sides gives

$$\sqrt{L^2} = \sqrt{\dfrac{8EIy}{M}}$$

i.e. $\qquad \boldsymbol{L = \sqrt{\dfrac{8EIy}{M}}}$

Problem 15. Given $t = 2\pi\sqrt{\dfrac{l}{g}}$, find g in terms of t, l and π

Whenever the prospective new subject is within a square root sign, it is best to isolate that term on the LHS and then to square both sides of the equation.

Rearranging gives $\qquad 2\pi\sqrt{\dfrac{l}{g}} = t$

Dividing both sides by 2π gives $\qquad \sqrt{\dfrac{l}{g}} = \dfrac{t}{2\pi}$

Squaring both sides gives $\qquad \dfrac{l}{g} = \left(\dfrac{t}{2\pi}\right)^2 = \dfrac{t^2}{4\pi^2}$

Cross-multiplying (i.e. multiplying each term by $4\pi^2 g$), gives $\qquad 4\pi^2 l = gt^2$

or $\qquad gt^2 = 4\pi^2 l$

Dividing both sides by t^2 gives $\qquad \dfrac{gt^2}{t^2} = \dfrac{4\pi^2 l}{t^2}$

Cancelling gives $\qquad g = \dfrac{4\pi^2 l}{t^2}$

Problem 16. The impedance Z of an a.c. circuit is given by $Z = \sqrt{R^2 + X^2}$ where R is the resistance. Make the reactance X the subject

Rearranging gives $\qquad \sqrt{R^2 + X^2} = Z$

Squaring both sides gives $\qquad R^2 + X^2 = Z^2$

Rearranging gives $\qquad X^2 = Z^2 - R^2$

Taking the square root of both sides gives

$$X = \sqrt{Z^2 - R^2}$$

Problem 17. The volume V of a hemisphere of radius r is given by $V = \dfrac{2}{3}\pi r^3$. (a) Find r in terms of V. (b) Evaluate the radius when $V = 32\,\text{cm}^3$

(a) Rearranging gives $\qquad \dfrac{2}{3}\pi r^3 = V$

Multiplying both sides by 3 gives $\quad 2\pi r^3 = 3V$

Dividing both sides by 2π gives $\qquad \dfrac{2\pi r^3}{2\pi} = \dfrac{3V}{2\pi}$

Cancelling gives $\qquad r^3 = \dfrac{3V}{2\pi}$

Taking the cube root of both sides gives

$$\sqrt[3]{r^3} = \sqrt[3]{\left(\dfrac{3V}{2\pi}\right)}$$

i.e. $\qquad r = \sqrt[3]{\left(\dfrac{3V}{2\pi}\right)}$

(b) When $V = 32\,\text{cm}^3$,

radius $r = \sqrt[3]{\left(\dfrac{3V}{2\pi}\right)} = \sqrt[3]{\left(\dfrac{3 \times 32}{2\pi}\right)} = 2.48\,\text{cm}$.

Now try the following Practice Exercise

Practice Exercise 53   Further transposing formulae (answers on page 1153)

Make the symbol indicated the subject of each of the formulae shown in Problems 1 to 15 and express each in its simplest form.

1. $S = \dfrac{a}{1-r}$ $\qquad$ (r)

2. $y = \dfrac{\lambda(x-d)}{d}$ $\qquad$ (x)

3. $A = \dfrac{3(F-f)}{L}$ $\qquad$ (f)

4. $y = \dfrac{AB^2}{5CD}$ $\qquad$ (D)

5. $R = R_0(1 + \alpha t)$ $\qquad$ (t)

6. $\dfrac{1}{R} = \dfrac{1}{R_1} + \dfrac{1}{R_2}$ $\qquad$ (R_2)

7. $I = \dfrac{E-e}{R+r}$ $\qquad$ (R)

8. $\dfrac{P_1 V_1}{T_1} = \dfrac{P_2 V_2}{T_2}$ $\qquad$ (V_2)

9. $y = 4ab^2 c^2$ $\qquad$ (b)

10. $\dfrac{a^2}{x^2} + \dfrac{b^2}{y^2} = 1$ $\qquad$ (x)

11. $t = 2\pi\sqrt{\dfrac{L}{g}}$ $\qquad$ (L)

12. $v^2 = u^2 + 2as$ $\qquad$ (u)

13. $A = \dfrac{\pi R^2 \theta}{360}$ $\qquad$ (R)

14. $\dfrac{P_1 V_1}{T_1} = \dfrac{P_2 V_2}{T_2}$ $\qquad$ (T_2)

15. $N = \sqrt{\left(\dfrac{a+x}{y}\right)}$ $\qquad$ (a)

16. Transpose $Z = \sqrt{R^2 + (2\pi f L)^2}$ for L and evaluate L when $Z = 27.82$, $R = 11.76$ and $f = 50$

17. The lift force L on an aircraft is given by: $L = \frac{1}{2}\rho v^2 a c$ where ρ is the density, v is the velocity, a is the area and c is the lift coefficient. Transpose the equation to make the velocity the subject.

18. The angular deflection θ of a beam of electrons due to a magnetic field is given by:

$\theta = k\left(\dfrac{HL}{V^{\frac{1}{2}}}\right)$. Transpose the equation for V.

12.4 More difficult transposing of formulae

Here are some more transposition examples to help us further understand how more difficult formulae are transposed.

Problem 18. (a) Transpose $S = \sqrt{\dfrac{3d(L-d)}{8}}$ to make L the subject. (b) Evaluate L when $d = 1.65$ and $S = 0.82$

The formula $S = \sqrt{\dfrac{3d(L-d)}{8}}$ represents the sag S at the centre of a wire.

(a) Squaring both sides gives $\qquad S^2 = \dfrac{3d(L-d)}{8}$

Multiplying both sides by 8 gives

$$8S^2 = 3d(L-d)$$

Dividing both sides by $3d$ gives $\dfrac{8S^2}{3d} = L - d$

Rearranging gives $\qquad\qquad L = d + \dfrac{8S^2}{3d}$

(b) When $d = 1.65$ and $S = 0.82$,

$$L = d + \frac{8S^2}{3d} = 1.65 + \frac{8 \times 0.82^2}{3 \times 1.65} = \mathbf{2.737}$$

Problem 19. Transpose the formula $p = \dfrac{a^2x + a^2y}{r}$ to make a the subject

Rearranging gives $\qquad \dfrac{a^2x + a^2y}{r} = p$

Multiplying both sides by r gives

$$a^2x + a^2y = rp$$

Factorising the LHS gives $\quad a^2(x + y) = rp$

Dividing both sides by $(x + y)$ gives

$$\frac{a^2(x+y)}{(x+y)} = \frac{rp}{(x+y)}$$

Cancelling gives $\qquad\qquad a^2 = \dfrac{rp}{(x+y)}$

Taking the square root of both sides gives

$$a = \sqrt{\left(\frac{rp}{x+y}\right)}$$

Whenever the letter required as the new subject occurs more than once in the original formula, after rearranging, **factorising** will always be needed.

Problem 20. Make b the subject of the formula $a = \dfrac{x - y}{\sqrt{bd + be}}$

Rearranging gives $\qquad \dfrac{x - y}{\sqrt{bd + be}} = a$

Multiplying both sides by $\sqrt{bd + be}$ gives

$$x - y = a\sqrt{bd + be}$$

or $\qquad\qquad a\sqrt{bd + be} = x - y$

Dividing both sides by a gives $\sqrt{bd + be} = \dfrac{x - y}{a}$

Squaring both sides gives $\qquad bd + be = \left(\dfrac{x-y}{a}\right)^2$

Factorising the LHS gives $\qquad b(d + e) = \left(\dfrac{x-y}{a}\right)^2$

Dividing both sides by $(d + e)$ gives

$$b = \frac{\left(\dfrac{x-y}{a}\right)^2}{(d+e)} \quad \text{or} \quad b = \frac{(x-y)^2}{a^2(d+e)}$$

Problem 21. If $a = \dfrac{b}{1+b}$, make b the subject of the formula

Rearranging gives $\qquad \dfrac{b}{1+b} = a$

Multiplying both sides by $(1 + b)$ gives

$$b = a(1 + b)$$

Removing the bracket gives $\qquad b = a + ab$

Rearranging to obtain terms in b on the LHS gives

$$b - ab = a$$

Factorising the LHS gives $\qquad b(1 - a) = a$

Dividing both sides by $(1 - a)$ gives $\boldsymbol{b = \dfrac{a}{1 - a}}$

> **Problem 22.** Transpose the formula $V = \dfrac{Er}{R + r}$
> to make r the subject

Rearranging gives $\qquad \dfrac{Er}{R + r} = V$

Multiplying both sides by $(R + r)$ gives

$$Er = V(R + r)$$

Removing the bracket gives $\qquad Er = VR + Vr$

Rearranging to obtain terms in r on the LHS gives

$$Er - Vr = VR$$

Factorising gives $\qquad r(E - V) = VR$

Dividing both sides by $(E - V)$ gives $\boldsymbol{r = \dfrac{VR}{E - V}}$

> **Problem 23.** Transpose the formula
> $y = \dfrac{pq^2}{r + q^2} - t$ to make q the subject

Rearranging gives $\qquad \dfrac{pq^2}{r + q^2} - t = y$

and $\qquad \dfrac{pq^2}{r + q^2} = y + t$

Multiplying both sides by $(r + q^2)$ gives

$$pq^2 = (r + q^2)(y + t)$$

Removing brackets gives $\quad pq^2 = ry + rt + q^2 y + q^2 t$

Rearranging to obtain terms in q on the LHS gives

$$pq^2 - q^2 y - q^2 t = ry + rt$$

Factorising gives $q^2(p - y - t) = r(y + t)$

Dividing both sides by $(p - y - t)$ gives

$$q^2 = \frac{r(y + t)}{(p - y - t)}$$

Taking the square root of both sides gives

$$q = \sqrt{\left(\frac{r(y + t)}{p - y - t} \right)}$$

> **Problem 24.** Given that $\dfrac{D}{d} = \sqrt{\left(\dfrac{f + p}{f - p} \right)}$
> express p in terms of D, d and f

Rearranging gives $\qquad \sqrt{\left(\dfrac{f + p}{f - p} \right)} = \dfrac{D}{d}$

Squaring both sides gives $\qquad \left(\dfrac{f + p}{f - p} \right) = \dfrac{D^2}{d^2}$

Cross-multiplying, i.e. multiplying each term
by $d^2(f - p)$, gives $\qquad d^2(f + p) = D^2(f - p)$

Removing brackets gives $\quad d^2 f + d^2 p = D^2 f - D^2 p$

Rearranging to obtain terms in p on the LHS gives

$$d^2 p + D^2 p = D^2 f - d^2 f$$

Factorising gives $\qquad p(d^2 + D^2) = f(D^2 - d^2)$

Dividing both sides by $(d^2 + D^2)$ gives

$$p = \frac{f(D^2 - d^2)}{(d^2 + D^2)}$$

Now try the following Practice Exercise

> **Practice Exercise 54   Further transposing
> formulae (answers on page 1154)**
>
> Make the symbol indicated the subject of each of
> the formulae shown in Problems 1 to 7 and express
> each in its simplest form.
>
> 1. $y = \dfrac{a^2 m - a^2 n}{x}$ $\qquad (a)$
>
> 2. $M = \pi(R^4 - r^4)$ $\qquad (R)$
>
> 3. $x + y = \dfrac{r}{3 + r}$ $\qquad (r)$
>
> 4. $m = \dfrac{\mu L}{L + rCR}$ $\qquad (L)$
>
> 5. $a^2 = \dfrac{b^2 - c^2}{b^2}$ $\qquad (b)$

6. $\dfrac{x}{y} = \dfrac{1+r^2}{1-r^2}$ (r)

7. $\dfrac{p}{q} = \sqrt{\left(\dfrac{a+2b}{a-2b}\right)}$ (b)

8. A formula for the focal length f of a convex lens is $\dfrac{1}{f} = \dfrac{1}{u} + \dfrac{1}{v}$. Transpose the formula to make v the subject and evaluate v when $f = 5$ and $u = 6$

9. The quantity of heat Q is given by the formula $Q = mc(t_2 - t_1)$. Make t_2 the subject of the formula and evaluate t_2 when $m = 10, t_1 = 15$, $c = 4$ and $Q = 1600$

10. The velocity v of water in a pipe appears in the formula $h = \dfrac{0.03Lv^2}{2dg}$. Express v as the subject of the formula and evaluate v when $h = 0.712, L = 150, d = 0.30$ and $g = 9.81$

11. The sag S at the centre of a wire is given by the formula $S = \sqrt{\left(\dfrac{3d(l-d)}{8}\right)}$. Make l the subject of the formula and evaluate l when $d = 1.75$ and $S = 0.80$

12. In an electrical alternating current circuit the impedance Z is given by $Z = \sqrt{\left\{R^2 + \left(\omega L - \dfrac{1}{\omega C}\right)^2\right\}}$. Transpose the formula to make C the subject and hence evaluate C when $Z = 130, R = 120, \omega = 314$ and $L = 0.32$

13. An approximate relationship between the number of teeth T on a milling cutter, the diameter of cutter D and the depth of cut d is given by $T = \dfrac{12.5\,D}{D + 4d}$. Determine the value of D when $T = 10$ and $d = 4$ mm.

14. Make λ, the wavelength of X-rays, the subject of the following formula: $\dfrac{\mu}{\rho} = \dfrac{CZ^4\sqrt{\lambda^5}\,n}{a}$

15. A simply supported beam of length L has a centrally applied load F and a uniformly distributed load of w per metre length of beam. The reaction at the beam support is given by:
$$R = \frac{1}{2}(F + wL)$$
Rearrange the equation to make w the subject. Hence determine the value of w when $L = 4$ m, $F = 8$ kN and $R = 10$ kN.

16. The rate of heat conduction through a slab of material Q is given by the formula $Q = \dfrac{kA(t_1 - t_2)}{d}$ where t_1 and t_2 are the temperatures of each side of the material, A is the area of the slab, d is the thickness of the slab, and k is the thermal conductivity of the material.
Rearrange the formula to obtain an expression for t_2

17. The slip s of a vehicle is given by: $s = \left(1 - \dfrac{r\omega}{v}\right) \times 100\%$ where r is the tyre radius, ω is the angular velocity and v the velocity. Transpose to make r the subject of the formula.

18. The critical load, F newtons, of a steel column may be determined from the formula $L\sqrt{\dfrac{F}{EI}} = n\pi$ where L is the length, EI is the flexural rigidity and n is a positive integer. Transpose for F and hence determine the value of F when $n = 1$, $E = 0.25 \times 10^{12}$ N/m^2, $I = 6.92 \times 10^{-6}$ m^4 and $L = 1.12$ m.

19. The flow of slurry along a pipe on a coal processing plant is given by: $V = \dfrac{\pi p r^4}{8\eta\ell}$ Transpose the equation for r.

20. The deflection head H of a metal structure is given by: $H = \sqrt{\dfrac{I\rho^4 D^2 \ell^{\frac{3}{2}}}{20g}}$ Transpose the formula for length ℓ.

For fully worked solutions to each of the problems in Practice Exercises 52 to 54 in this chapter, go to the website:
www.routledge.com/cw/bird

Chapter 13

Solving simultaneous equations

Why it is important to understand: Solving simultaneous equations

Simultaneous equations arise a great deal in engineering and science, some applications including theory of structures, data analysis, electrical circuit analysis and air traffic control. Systems that consist of a small number of equations can be solved analytically using standard methods from algebra (as explained in this chapter). Systems of large numbers of equations require the use of numerical methods and computers. Matrices are generally used to solve simultaneous equations (as explained in Chapter 48). Solving simultaneous equations is an important skill required in all aspects of engineering.

At the end of this chapter, you should be able to:

- solve simultaneous equations in two unknowns by substitution
- solve simultaneous equations in two unknowns by elimination
- solve simultaneous equations involving practical situations
- solve simultaneous equations in three unknowns

13.1 Introduction

Only one equation is necessary when finding the value of a **single unknown quantity** (as with simple equations in Chapter 11). However, when an equation contains **two unknown quantities** it has an infinite number of solutions. When two equations are available connecting the same two unknown values then a unique solution is possible. Similarly, for three unknown quantities it is necessary to have three equations in order to solve for a particular value of each of the unknown quantities, and so on.

Equations which have to be solved together to find the unique values of the unknown quantities, which are true for each of the equations, are called **simultaneous equations**.

Two methods of solving simultaneous equations analytically are:

(a) by **substitution**, and

(b) by **elimination**.

(A graphical solution of simultaneous equations is shown in Chapter 35)

13.2 Solving simultaneous equations in two unknowns

The method of solving simultaneous equations is demonstrated in the following worked problems.

Problem 1. Solve the following equations for x and y, (a) by substitution and (b) by elimination

$$x + 2y = -1 \qquad (1)$$
$$4x - 3y = 18 \qquad (2)$$

(a) By substitution
From equation (1): $x = -1 - 2y$
Substituting this expression for x into equation (2) gives

$$4(-1 - 2y) - 3y = 18$$

This is now a simple equation in y.
Removing the bracket gives

$$-4 - 8y - 3y = 18$$
$$-11y = 18 + 4 = 22$$
$$y = \frac{22}{-11} = -2$$

Substituting $y = -2$ into equation (1) gives

$$x + 2(-2) = -1$$
$$x - 4 = -1$$
$$x = -1 + 4 = 3$$

Thus, $x = 3$ and $y = -2$ is the solution to the simultaneous equations.
Check: in equation (2), since $x = 3$ and $y = -2$,

$$\text{LHS} = 4(3) - 3(-2) = 12 + 6 = 18 = \text{RHS}$$

(b) By elimination

$$x + 2y = -1 \qquad (1)$$
$$4x - 3y = 18 \qquad (2)$$

If equation (1) is multiplied throughout by 4, the coefficient of x will be the same as in equation (2), giving

$$4x + 8y = -4 \qquad (3)$$

Subtracting equation (3) from equation (2) gives

$$4x - 3y = 18 \qquad (2)$$
$$\underline{4x + 8y = -4} \qquad (3)$$
$$0 - 11y = 22$$

Hence, $y = \dfrac{22}{-11} = -2$

(Note: in the above subtraction,

$$18 - -4 = 18 + 4 = 22)$$

Substituting $y = -2$ into either equation (1) or equation (2) will give $x = 3$, as in method (a). The solution $x = 3$, $y = -2$ is the only pair of values that satisfies both of the original equations.

Problem 2. Solve, by a substitution method, the simultaneous equations

$$3x - 2y = 12 \qquad (1)$$
$$x + 3y = -7 \qquad (2)$$

From equation (2), $x = -7 - 3y$

Substituting for x in equation (1) gives

$$3(-7 - 3y) - 2y = 12$$

i.e. $\qquad -21 - 9y - 2y = 12$
$$-11y = 12 + 21 = 33$$

Hence, $y = \dfrac{33}{-11} = -3$

Substituting $y = -3$ in equation (2) gives

$$x + 3(-3) = -7$$

i.e. $\qquad x - 9 = -7$

Hence $\qquad x = -7 + 9 = 2$

Thus, $x = 2$, $y = -3$ is the solution of the simultaneous equations. (Such solutions should always be checked by substituting values into each of the original two equations.)

Problem 3. Use an elimination method to solve the following simultaneous equations

$$3x + 4y = 5 \qquad (1)$$
$$2x - 5y = -12 \qquad (2)$$

If equation (1) is multiplied throughout by 2 and equation (2) by 3, the coefficient of x will be the same in the newly formed equations. Thus,

$2 \times$ equation (1) gives $\qquad 6x + 8y = 10 \qquad$ (3)

$3 \times$ equation (2) gives $\qquad 6x - 15y = -36 \qquad$ (4)

Equation (3) – equation (4) gives

$$0 + 23y = 46$$

i.e. $\qquad\qquad y = \dfrac{46}{23} = 2$

(Note $+8y - -15y = 8y + 15y = 23y$ and $10 - -36 = 10 + 36 = 46$)

Substituting $y = 2$ in equation (1) gives

$$3x + 4(2) = 5$$

from which $\qquad 3x = 5 - 8 = -3$

and $\qquad\qquad\qquad x = -1$

Checking, by substituting $x = -1$ and $y = 2$ in equation (2), gives

$$\text{LHS} = 2(-1) - 5(2) = -2 - 10 = -12 = \text{RHS}$$

Hence, $x = -1$ and $y = 2$ is the solution of the simultaneous equations.

The elimination method is the most common method of solving simultaneous equations.

Problem 4. Solve

$$7x - 2y = 26 \qquad (1)$$
$$6x + 5y = 29 \qquad (2)$$

When equation (1) is multiplied by 5 and equation (2) by 2, the coefficients of y in each equation are numerically the same, i.e. 10, but are of opposite sign.

$5 \times$ equation (1) gives $\qquad 35x - 10y = 130 \qquad$ (3)

$2 \times$ equation (2) gives $\qquad 12x + 10y = 58 \qquad$ (4)

Adding equations (3) and (4) gives $\qquad 47x + 0 = 188$

Hence, $\qquad\qquad x = \dfrac{188}{47} = 4$

Note that when the signs of common coefficients are **different** the two equations are **added** and when the signs of common coefficients are the **same** the two equations are **subtracted** (as in Problems 1 and 3).

Substituting $x = 4$ in equation (1) gives

$$7(4) - 2y = 26$$
$$28 - 2y = 26$$
$$28 - 26 = 2y$$
$$2 = 2y$$

Hence, $\qquad\qquad\qquad y = 1$

Checking, by substituting $x = 4$ and $y = 1$ in equation (2), gives
$$\text{LHS} = 6(4) + 5(1) = 24 + 5 = 29 = \text{RHS}$$
Thus, the solution is $x = 4, y = 1$

Now try the following Practice Exercise

Practice Exercise 55   Solving simultaneous equations (answers on page 1154)

Solve the following simultaneous equations and verify the results.

1. $2x - y = 6$
   $x + y = 6$

2. $2x - y = 2$
   $x - 3y = -9$

3. $x - 4y = -4$
   $5x - 2y = 7$

4. $3x - 2y = 10$
   $5x + y = 21$

5. $5p + 4q = 6$
   $2p - 3q = 7$

6. $7x + 2y = 11$
   $3x - 5y = -7$

7. $2x - 7y = -8$
   $3x + 4y = 17$

8. $a + 2b = 8$
   $b - 3a = -3$

9. $a + b = 7$
   $a - b = 3$

10. $2x + 5y = 7$
    $x + 3y = 4$

11. $3s + 2t = 12$
    $4s - t = 5$

12. $3x - 2y = 13$
    $2x + 5y = -4$

13. $5m - 3n = 11$
    $3m + n = 8$

14. $8a - 3b = 51$
    $3a + 4b = 14$

15. $5x = 2y$
    $3x + 7y = 41$

16. $5c = 1 - 3d$
    $2d + c + 4 = 0$

13.3 Further solving of simultaneous equations

Here are some further worked problems on solving simultaneous equations.

Problem 5. Solve

$$3p = 2q \qquad (1)$$

$$4p + q + 11 = 0 \qquad (2)$$

Rearranging gives

$$3p - 2q = 0 \qquad (3)$$

$$4p + q = -11 \qquad (4)$$

Multiplying equation (4) by 2 gives

$$8p + 2q = -22 \qquad (5)$$

Adding equations (3) and (5) gives

$$11p + 0 = -22$$

$$p = \frac{-22}{11} = -2$$

Substituting $p = -2$ into equation (1) gives

$$3(-2) = 2q$$

$$-6 = 2q$$

$$q = \frac{-6}{2} = -3$$

Checking, by substituting $p = -2$ and $q = -3$ into equation (2), gives

$$\text{LHS} = 4(-2) + (-3) + 11 = -8 - 3 + 11 = 0 = \text{RHS}$$

Hence, the solution is $\boldsymbol{p = -2, q = -3}$

Problem 6. Solve

$$\frac{x}{8} + \frac{5}{2} = y \qquad (1)$$

$$13 - \frac{y}{3} = 3x \qquad (2)$$

Whenever fractions are involved in simultaneous equations it is often easier to first remove them. Thus, multiplying equation (1) by 8 gives

$$8\left(\frac{x}{8}\right) + 8\left(\frac{5}{2}\right) = 8y$$

i.e. $\qquad\qquad x + 20 = 8y \qquad (3)$

Multiplying equation (2) by 3 gives

$$39 - y = 9x \qquad (4)$$

Rearranging equations (3) and (4) gives

$$x - 8y = -20 \qquad (5)$$

$$9x + y = 39 \qquad (6)$$

Multiplying equation (6) by 8 gives

$$72x + 8y = 312 \qquad (7)$$

Adding equations (5) and (7) gives

$$73x + 0 = 292$$

$$x = \frac{292}{73} = \boldsymbol{4}$$

Substituting $x = 4$ into equation (5) gives

$$4 - 8y = -20$$

$$4 + 20 = 8y$$

$$24 = 8y$$

$$y = \frac{24}{8} = \boldsymbol{3}$$

Checking, substituting $x = 4$ and $y = 3$ in the original equations gives

(1): $\quad \text{LHS} = \dfrac{4}{8} + \dfrac{5}{2} = \dfrac{1}{2} + 2\dfrac{1}{2} = 3 = y = \text{RHS}$

(2): $\quad \text{LHS} = 13 - \dfrac{3}{3} = 13 - 1 = 12$

$\qquad \text{RHS} = 3x = 3(4) = 12$

Hence, the solution is $\boldsymbol{x = 4, y = 3}$

Problem 7. Solve

$$2.5x + 0.75 - 3y = 0$$

$$1.6x = 1.08 - 1.2y$$

It is often easier to remove decimal fractions. Thus, multiplying equations (1) and (2) by 100 gives

$$250x + 75 - 300y = 0 \qquad (1)$$

$$160x = 108 - 120y \qquad (2)$$

Rearranging gives

$$250x - 300y = -75 \qquad (3)$$

$$160x + 120y = 108 \qquad (4)$$

Multiplying equation (3) by 2 gives

$$500x - 600y = -150 \qquad (5)$$

Multiplying equation (4) by 5 gives

$$800x + 600y = 540 \qquad (6)$$

Adding equations (5) and (6) gives

$$1300x + 0 = 390$$

$$x = \frac{390}{1300} = \frac{39}{130} = \frac{3}{10} = \mathbf{0.3}$$

Substituting $x = 0.3$ into equation (1) gives

$$250(0.3) + 75 - 300y = 0$$

$$75 + 75 = 300y$$

$$150 = 300y$$

$$y = \frac{150}{300} = \mathbf{0.5}$$

Checking, by substituting $x = 0.3$ and $y = 0.5$ in equation (2), gives

$$\text{LHS} = 160(0.3) = 48$$

$$\text{RHS} = 108 - 120(0.5) = 108 - 60 = 48$$

Hence, the solution is $x = \mathbf{0.3}, y = \mathbf{0.5}$

Now try the following Practice Exercise

Practice Exercise 56   Solving simultaneous equations (answers on page 1154)

Solve the following simultaneous equations and verify the results.

1. $7p + 11 + 2q = 0$

    $-1 = 3q - 5p$

2. $\dfrac{x}{2} + \dfrac{y}{3} = 4$

    $\dfrac{x}{6} - \dfrac{y}{9} = 0$

3. $\dfrac{a}{2} - 7 = -2b$

    $12 = 5a + \dfrac{2}{3}b$

4. $\dfrac{3}{2}s - 2t = 8$

    $\dfrac{s}{4} + 3t = -2$

5. $\dfrac{x}{5} + \dfrac{2y}{3} = \dfrac{49}{15}$

    $\dfrac{3x}{7} - \dfrac{y}{2} + \dfrac{5}{7} = 0$

6. $v - 1 = \dfrac{u}{12}$

    $u + \dfrac{v}{4} - \dfrac{25}{2} = 0$

7. $1.5x - 2.2y = -18$

    $2.4x + 0.6y = 33$

8. $3b - 2.5a = 0.45$

    $1.6a + 0.8b = 0.8$

13.4 Solving more difficult simultaneous equations

Here are some further worked problems on solving more difficult simultaneous equations.

Problem 8. Solve

$$\frac{2}{x} + \frac{3}{y} = 7 \qquad (1)$$

$$\frac{1}{x} - \frac{4}{y} = -2 \qquad (2)$$

In this type of equation the solution is easier if a substitution is initially made. Let $\dfrac{1}{x} = a$ and $\dfrac{1}{y} = b$

Thus equation (1) becomes $2a + 3b = 7 \qquad (3)$

and equation (2) becomes $a - 4b = -2 \qquad (4)$

Multiplying equation (4) by 2 gives

$$2a - 8b = -4 \qquad (5)$$

Subtracting equation (5) from equation (3) gives

$$0 + 11b = 11$$

i.e. $\mathbf{b = 1}$

Substituting $b = 1$ in equation (3) gives

$$2a + 3 = 7$$

$$2a = 7 - 3 = 4$$

i.e. $\mathbf{a = 2}$

Checking, substituting $a = 2$ and $b = 1$ in equation (4), gives

$$\text{LHS} = 2 - 4(1) = 2 - 4 = -2 = \text{RHS}$$

Hence, $\mathbf{a = 2}$ and $\mathbf{b = 1}$

However, since $\dfrac{1}{x} = a,$ $x = \dfrac{1}{a} = \dfrac{1}{2}$ or 0.5

and since $\dfrac{1}{y} = b,$ $y = \dfrac{1}{b} = \dfrac{1}{1} = 1$

Hence, the solution is $x = \mathbf{0.5}, y = \mathbf{1}$

Problem 9. Solve

$$\frac{1}{2a} + \frac{3}{5b} = 4 \tag{1}$$

$$\frac{4}{a} + \frac{1}{2b} = 10.5 \tag{2}$$

Let $\frac{1}{a} = x$ and $\frac{1}{b} = y$

then

$$\frac{x}{2} + \frac{3}{5}y = 4 \tag{3}$$

$$4x + \frac{1}{2}y = 10.5 \tag{4}$$

To remove fractions, equation (3) is multiplied by 10, giving

$$10\left(\frac{x}{2}\right) + 10\left(\frac{3}{5}y\right) = 10(4)$$

i.e. $\qquad\qquad 5x + 6y = 40 \tag{5}$

Multiplying equation (4) by 2 gives

$$8x + y = 21 \tag{6}$$

Multiplying equation (6) by 6 gives

$$48x + 6y = 126 \tag{7}$$

Subtracting equation (5) from equation (7) gives

$$43x + 0 = 86$$

$$x = \frac{86}{43} = \mathbf{2}$$

Substituting $x = 2$ into equation (3) gives

$$\frac{2}{2} + \frac{3}{5}y = 4$$

$$\frac{3}{5}y = 4 - 1 = 3$$

$$y = \frac{5}{3}(3) = \mathbf{5}$$

Since $\quad \frac{1}{a} = x, \quad a = \frac{1}{x} = \frac{1}{2}$ or 0.5

and since $\quad \frac{1}{b} = y, \quad b = \frac{1}{y} = \frac{1}{5}$ or 0.2

Hence, the solution is $\boldsymbol{a = 0.5}, \boldsymbol{b = 0.2}$, which may be checked in the original equations.

Problem 10. Solve

$$\frac{1}{x + y} = \frac{4}{27} \tag{1}$$

$$\frac{1}{2x - y} = \frac{4}{33} \tag{2}$$

To eliminate fractions, both sides of equation (1) are multiplied by $27(x + y)$, giving

$$27(x + y)\left(\frac{1}{x + y}\right) = 27(x + y)\left(\frac{4}{27}\right)$$

i.e. $\qquad\qquad 27(1) = 4(x + y)$

$$27 = 4x + 4y \tag{3}$$

Similarly, in equation (2) $\quad 33 = 4(2x - y)$

i.e. $\qquad\qquad 33 = 8x - 4y \tag{4}$

Equation (3) + equation (4) gives

$$60 = 12x \text{ and } \boldsymbol{x} = \frac{60}{12} = \mathbf{5}$$

Substituting $x = 5$ in equation (3) gives

$$27 = 4(5) + 4y$$

from which $\qquad 4y = 27 - 20 = 7$

and $\qquad\qquad y = \frac{7}{4} = \mathbf{1}\frac{\mathbf{3}}{\mathbf{4}} \text{ or } \mathbf{1.75}$

Hence, $\boldsymbol{x = 5}, \boldsymbol{y = 1.75}$ is the required solution, which may be checked in the original equations.

Problem 11. Solve

$$\frac{x - 1}{3} + \frac{y + 2}{5} = \frac{2}{15} \tag{1}$$

$$\frac{1 - x}{6} + \frac{5 + y}{2} = \frac{5}{6} \tag{2}$$

Before equations (1) and (2) can be simultaneously solved, the fractions need to be removed and the equations rearranged.

Multiplying equation (1) by 15 gives

$$15\left(\frac{x - 1}{3}\right) + 15\left(\frac{y + 2}{5}\right) = 15\left(\frac{2}{15}\right)$$

i.e. $\qquad 5(x - 1) + 3(y + 2) = 2$

$$5x - 5 + 3y + 6 = 2$$

$$5x + 3y = 2 + 5 - 6$$

Hence, $\qquad\qquad 5x + 3y = 1 \tag{3}$

Multiplying equation (2) by 6 gives

$$6\left(\frac{1-x}{6}\right)+6\left(\frac{5+y}{2}\right)=6\left(\frac{5}{6}\right)$$

i.e. $(1-x)+3(5+y)=5$

$$1-x+15+3y=5$$

$$-x+3y=5-1-15$$

Hence, $-x+3y=-11$ (4)

Thus the initial problem containing fractions can be expressed as

$$5x+3y=1 \qquad (3)$$

$$-x+3y=-11 \qquad (4)$$

Subtracting equation (4) from equation (3) gives

$$6x+0=12$$

$$x=\frac{12}{6}=2$$

Substituting $x=2$ into equation (3) gives

$$5(2)+3y=1$$

$$10+3y=1$$

$$3y=1-10=-9$$

$$y=\frac{-9}{3}=-3$$

Checking, substituting $x=2$, $y=-3$ in equation (4) gives

$$\text{LHS}=-2+3(-3)=-2-9=-11=\text{RHS}$$

Hence, the solution is $x=2, y=-3$

Now try the following Practice Exercise

Practice Exercise 57   Solving more difficult simultaneous equations (answers on page 1154)

In Problems 1 to 7, solve the simultaneous equations and verify the results.

1. $\dfrac{3}{x}+\dfrac{2}{y}=14$

  $\dfrac{5}{x}-\dfrac{3}{y}=-2$

2. $\dfrac{4}{a}-\dfrac{3}{b}=18$

  $\dfrac{2}{a}+\dfrac{5}{b}=-4$

3. $\dfrac{1}{2p}+\dfrac{3}{5q}=5$

  $\dfrac{5}{p}-\dfrac{1}{2q}=\dfrac{35}{2}$

4. $\dfrac{5}{x}+\dfrac{3}{y}=1.1$

  $\dfrac{3}{x}-\dfrac{7}{y}=-1.1$

5. $\dfrac{c+1}{4}-\dfrac{d+2}{3}+1=0$

  $\dfrac{1-c}{5}+\dfrac{3-d}{4}+\dfrac{13}{20}=0$

6. $\dfrac{3r+2}{5}-\dfrac{2s-1}{4}=\dfrac{11}{5}$

  $\dfrac{3+2r}{4}+\dfrac{5-s}{3}=\dfrac{15}{4}$

7. $\dfrac{5}{x+y}=\dfrac{20}{27}$

  $\dfrac{4}{2x-y}=\dfrac{16}{33}$

8. If $5x-\dfrac{3}{y}=1$ and $x+\dfrac{4}{y}=\dfrac{5}{2}$, find the value

  of $\dfrac{xy+1}{y}$

13.5   Practical problems involving simultaneous equations

There are a number of situations in engineering and science in which the solution of simultaneous equations is required. Some are demonstrated in the following worked problems.

Problem 12.   The law connecting friction F and load L for an experiment is of the form $F=aL+b$ where a and b are constants. When $F=5.6\,\text{N}$, $L=8.0\,\text{N}$ and when $F=4.4\,\text{N}$, $L=2.0\,\text{N}$. Find the values of a and b and the value of F when $L=6.5\,\text{N}$

Substituting $F=5.6$ and $L=8.0$ into $F=aL+b$ gives

$$5.6=8.0a+b \qquad (1)$$

Substituting $F=4.4$ and $L=2.0$ into $F=aL+b$ gives

$$4.4=2.0a+b \qquad (2)$$

Subtracting equation (2) from equation (1) gives

$$1.2=6.0a$$

$$a=\frac{1.2}{6.0}=\frac{1}{5}\text{ or }\textbf{0.2}$$

Substituting $a = \dfrac{1}{5}$ into equation (1) gives

$$5.6 = 8.0\left(\frac{1}{5}\right) + b$$

$$5.6 = 1.6 + b$$

$$5.6 - 1.6 = b$$

i.e. $\qquad\qquad b = 4$

Checking, substituting $a = \dfrac{1}{5}$ and $b = 4$ in equation (2) gives

$$\text{RHS} = 2.0\left(\frac{1}{5}\right) + 4 = 0.4 + 4 = 4.4 = \text{LHS}$$

Hence, $a = \dfrac{1}{5}$ and $b = 4$

When $L = 6.5$, $F = aL + b = \dfrac{1}{5}(6.5) + 4 = 1.3 + 4$, i.e. $F = 5.30\,\text{N}$

Problem 13. The equation of a straight line of gradient m and intercept on the y-axis c is $y = mx + c$. If a straight line passes through the point where $x = 1$ and $y = -2$, and also through the point where $x = 3.5$ and $y = 10.5$, find the values of the gradient and the y-axis intercept

Substituting $x = 1$ and $y = -2$ into $y = mx + c$ gives

$$-2 = m + c \qquad\qquad (1)$$

Substituting $x = 3.5$ and $y = 10.5$ into $y = mx + c$ gives

$$10.5 = 3.5m + c \qquad\qquad (2)$$

Subtracting equation (1) from equation (2) gives

$$12.5 = 2.5m, \quad \text{from which,} \quad m = \frac{12.5}{2.5} = 5$$

Substituting $m = 5$ into equation (1) gives

$$-2 = 5 + c$$

$$c = -2 - 5 = -7$$

Checking, substituting $m = 5$ and $c = -7$ in equation (2) gives

$$\text{RHS} = (3.5)(5) + (-7) = 17.5 - 7 = 10.5 = \text{LHS}$$

Hence, the **gradient** $m = 5$ and the y-**axis intercept** $c = -7$

Problem 14. When Kirchhoff's laws* are applied to the electrical circuit shown in Figure 13.1, the currents I_1 and I_2 are connected by the equations

$$27 = 1.5I_1 + 8(I_1 - I_2) \qquad (1)$$

$$-26 = 2I_2 - 8(I_1 - I_2) \qquad (2)$$

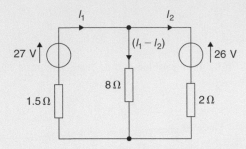

Figure 13.1

Solve the equations to find the values of currents I_1 and I_2

* **Who was Kirchhoff**? **Gustav Robert Kirchhoff** (12 March 1824–17 October 1887) was a German physicist. Concepts in circuit theory and thermal emission are named 'Kirchhoff's laws' after him, as well as a law of thermochemistry. To find out more go to **www.routledge.com/cw/bird**

Removing the brackets from equation (1) gives

$$27 = 1.5I_1 + 8I_1 - 8I_2$$

Rearranging gives

$$9.5I_1 - 8I_2 = 27 \qquad (3)$$

Removing the brackets from equation (2) gives

$$-26 = 2I_2 - 8I_1 + 8I_2$$

Rearranging gives

$$-8I_1 + 10I_2 = -26 \qquad (4)$$

Multiplying equation (3) by 5 gives

$$47.5I_1 - 40I_2 = 135 \qquad (5)$$

Multiplying equation (4) by 4 gives

$$-32I_1 + 40I_2 = -104 \qquad (6)$$

Adding equations (5) and (6) gives

$$15.5I_1 + 0 = 31$$

$$I_1 = \frac{31}{15.5} = 2$$

Substituting $I_1 = 2$ into equation (3) gives

$$9.5(2) - 8I_1 = 27$$

$$19 - 8I_2 = 27$$

$$19 - 27 = 8I_2$$

$$-8 = 8I_2$$

and $\qquad\qquad\qquad I_2 = -1$

Hence, the solution is $I_1 = 2$ and $I_2 = -1$ (which may be checked in the original equations).

> **Problem 15.** The distance s metres from a fixed point of a vehicle travelling in a straight line with constant acceleration, a m/s^2, is given by $s = ut + \frac{1}{2}at^2$, where u is the initial velocity in m/s and t the time in seconds. Determine the initial velocity and the acceleration given that $s = 42$ m when $t = 2$ s, and $s = 144$ m when $t = 4$ s. Also find the distance travelled after 3 s

Substituting $s = 42$ and $t = 2$ into $s = ut + \frac{1}{2}at^2$ gives

$$42 = 2u + \frac{1}{2}a(2)^2$$

i.e. $\qquad\qquad 42 = 2u + 2a \qquad (1)$

Substituting $s = 144$ and $t = 4$ into $s = ut + \frac{1}{2}at^2$ gives

$$144 = 4u + \frac{1}{2}a(4)^2$$

i.e. $\qquad\qquad 144 = 4u + 8a \qquad (2)$

Multiplying equation (1) by 2 gives

$$84 = 4u + 4a \qquad (3)$$

Subtracting equation (3) from equation (2) gives

$$60 = 0 + 4a$$

and $\qquad\qquad a = \frac{60}{4} = 15$

Substituting $a = 15$ into equation (1) gives

$$42 = 2u + 2(15)$$

$$42 - 30 = 2u$$

$$u = \frac{12}{2} = 6$$

Substituting $a = 15$ and $u = 6$ in equation (2) gives

$$\text{RHS} = 4(6) + 8(15) = 24 + 120 = 144 = \text{LHS}$$

Hence, **the initial velocity $u = 6$ m/s and the acceleration $a = 15$ m/s^2**

Distance travelled after 3 s is given by $s = ut + \frac{1}{2}at^2$ where $t = 3, u = 6$ and $a = 15$

Hence, $s = (6)(3) + \frac{1}{2}(15)(3)^2 = 18 + 67.5$

i.e. **distance travelled after 3 s = 85.5 m**

> **Problem 16.** The resistance $R\,\Omega$ of a length of wire at $t\,°\text{C}$ is given by $R = R_0(1 + \alpha t)$, where R_0 is the resistance at $0°\text{C}$ and α is the temperature coefficient of resistance in $/°\text{C}$. Find the values of α and R_0 if $R = 30\,\Omega$ at $50°\text{C}$ and $R = 35\,\Omega$ at $100°\text{C}$

Substituting $R = 30$ and $t = 50$ into $R = R_0(1 + \alpha t)$ gives

$$30 = R_0(1 + 50\alpha) \qquad (1)$$

Substituting $R = 35$ and $t = 100$ into $R = R_0(1 + \alpha t)$ gives

$$35 = R_0(1 + 100\alpha) \qquad (2)$$

Although these equations may be solved by the conventional substitution method, an easier way is to eliminate

R_0 by division. Thus, dividing equation (1) by equation (2) gives

$$\frac{30}{35} = \frac{R_0(1+50\alpha)}{R_0(1+100\alpha)} = \frac{1+50\alpha}{1+100\alpha}$$

Cross-multiplying gives

$$30(1+100\alpha) = 35(1+50\alpha)$$
$$30 + 3000\alpha = 35 + 1750\alpha$$
$$3000\alpha - 1750\alpha = 35 - 30$$
$$1250\alpha = 5$$

i.e. $\qquad \alpha = \frac{5}{1250} = \frac{1}{250} \quad \text{or} \quad \mathbf{0.004}$

Substituting $\alpha = \frac{1}{250}$ into equation (1) gives

$$30 = R_0\left\{1 + (50)\left(\frac{1}{250}\right)\right\}$$
$$30 = R_0(1.2)$$
$$R_0 = \frac{30}{1.2} = \mathbf{25}$$

Checking, substituting $\alpha = \frac{1}{250}$ and $R_0 = 25$ in equation (2) gives

$$\text{RHS} = 25\left\{1 + (100)\left(\frac{1}{250}\right)\right\}$$
$$= 25(1.4) = 35 = \text{LHS}$$

Thus, the solution is $\boldsymbol{\alpha = 0.004\,/°C}$ and $\boldsymbol{R_0 = 25\,\Omega}$.

Problem 17. The molar heat capacity of a solid compound is given by the equation $c = a + bT$, where a and b are constants. When $c = 52, T = 100$ and when $c = 172, T = 400$. Determine the values of a and b

When $c = 52, \quad T = 100$, hence

$$52 = a + 100b \qquad (1)$$

When $c = 172, \quad T = 400$, hence

$$172 = a + 400b \qquad (2)$$

Equation (2) – equation (1) gives

$$120 = 300b$$

from which, $\qquad b = \frac{120}{300} = \mathbf{0.4}$

Substituting $b = 0.4$ in equation (1) gives

$$52 = a + 100(0.4)$$
$$a = 52 - 40 = \mathbf{12}$$

Hence, $\boldsymbol{a = 12}$ and $\boldsymbol{b = 0.4}$

Now try the following Practice Exercise

Practice Exercise 58   Practical problems involving simultaneous equations (answers on page 1154)

1. In a system of pulleys, the effort P required to raise a load W is given by $P = aW + b$, where a and b are constants. If $W = 40$ when $P = 12$ and $W = 90$ when $P = 22$, find the values of a and b.

2. Applying Kirchhoff's laws to an electrical circuit produces the following equations:
$$5 = 0.2I_1 + 2(I_1 - I_2)$$
$$12 = 3I_2 + 0.4I_2 - 2(I_1 - I_2)$$
Determine the values of currents I_1 and I_2.

3. Velocity v is given by the formula $v = u + at$. If $v = 20$ when $t = 2$ and $v = 40$ when $t = 7$, find the values of u and a. Then, find the velocity when $t = 3.5$

4. Three new cars and 4 new vans supplied to a dealer together cost £97 700 and 5 new cars and 2 new vans of the same models cost £103 100. Find the respective costs of a car and a van.

5. $y = mx + c$ is the equation of a straight line of slope m and y-axis intercept c. If the line passes through the point where $x = 2$ and $y = 2$, and also through the point where $x = 5$ and $y = 0.5$, find the slope and y-axis intercept of the straight line.

6. The resistance R ohms of copper wire at $t°C$ is given by $R = R_0(1 + \alpha t)$, where R_0 is the resistance at $0°C$ and α is the temperature coefficient of resistance. If $R = 25.44\,\Omega$ at $30°C$ and $R = 32.17\,\Omega$ at $100°C$, find α and R_0

7. The molar heat capacity of a solid compound is given by the equation $c = a + bT$.

When $c = 70, T = 110$ and when $c = 160$, $T = 290$. Find the values of a and b.

8. In an engineering process, two variables p and q are related by $q = ap + b/p$, where a and b are constants. Evaluate a and b if $q = 13$ when $p = 2$ and $q = 22$ when $p = 5$

9. In a system of forces, the relationship between two forces F_1 and F_2 is given by

$$5F_1 + 3F_2 + 6 = 0$$

$$3F_1 + 5F_2 + 18 = 0$$

Solve for F_1 and F_2

10. For a balanced beam, the equilibrium of forces is given by: $R_1 + R_2 = 12.0$ kN
As a result of taking moments: $0.2R_1 + 7 \times 0.3 + 3 \times 0.6 = 0.8R_2$
Determine the values of the reaction forces R_1 and R_2.

13.6 Solving simultaneous equations in three unknowns

Equations containing three unknowns may be solved using exactly the same procedures as those used with two equations and two unknowns, providing that there are three equations to work with. The method is demonstrated in the following worked problem.

Problem 18. Solve the simultaneous equations.

$$x + y + z = 4 \tag{1}$$

$$2x - 3y + 4z = 33 \tag{2}$$

$$3x - 2y - 2z = 2 \tag{3}$$

There are a number of ways of solving these equations. One method is shown below.
The initial object is to produce two equations with two unknowns. For example, multiplying equation (1) by 4 and then subtracting this new equation from equation (2) will produce an equation with only x and y involved.
Multiplying equation (1) by 4 gives

$$4x + 4y + 4z = 16 \tag{4}$$

Equation (2) – equation (4) gives

$$-2x - 7y = 17 \tag{5}$$

Similarly, multiplying equation (3) by 2 and then adding this new equation to equation (2) will produce another equation with only x and y involved.

Multiplying equation (3) by 2 gives

$$6x - 4y - 4z = 4 \tag{6}$$

Equation (2) + equation (6) gives

$$8x - 7y = 37 \tag{7}$$

Rewriting equation (5) gives

$$-2x - 7y = 17 \tag{5}$$

Now we can use the previous method for solving simultaneous equations in two unknowns.

Equation (7) – equation (5) gives $\qquad 10x = 20$

from which, $\qquad\qquad\qquad\qquad\qquad x = 2$

(Note that $8x - -2x = 8x + 2x = 10x$.)

Substituting $x = 2$ into equation (5) gives

$$-4 - 7y = 17$$

from which, $\qquad -7y = 17 + 4 = 21$

and $\qquad\qquad\qquad\qquad y = -3$

Substituting $x = 2$ and $y = -3$ into equation (1) gives

$$2 - 3 + z = 4$$

from which, $\qquad\qquad\qquad\qquad z = 5$

Hence, the solution of the simultaneous equations is $x = 2, y = -3$ and $z = 5$

Now try the following Practice Exercise

Practice Exercise 59   Simultaneous equations in three unknowns (answers on page 1154)

In Problems 1 to 9, solve the simultaneous equations in three unknowns.

1. $x + 2y + 4z = 16$
   $2x - y + 5z = 18$
   $3x + 2y + 2z = 14$

2. $2x + y - z = 0$
   $3x + 2y + z = 4$
   $5x + 3y + 2z = 8$

3. $3x + 5y + 2z = 6$
   $x - y + 3z = 0$
   $2x + 7y + 3z = -3$

4. $2x + 4y + 5z = 23$
   $3x - y - 2z = 6$
   $4x + 2y + 5z = 31$

5. $2x + 3y + 4z = 36$
$3x + 2y + 3z = 29$
$x + y + z = 11$

6. $4x + y + 3z = 31$
$2x - y + 2z = 10$
$3x + 3y - 2z = 7$

7. $5x + 5y - 4z = 37$
$2x - 2y + 9z = 20$
$-4x + y + z = -14$

8. $6x + 7y + 8z = 13$
$3x + y - z = -11$
$2x - 2y - 2z = -18$

9. $3x + 2y + z = 14$
$7x + 3y + z = 22.5$
$4x - 4y - z = -8.5$

10. Kirchhoff's laws are used to determine the current equations in an electrical network and result in the following:

$i_1 + 8i_2 + 3i_3 = -31$

$3i_1 - 2i_2 + i_3 = -5$

$2i_1 - 3i_2 + 2i_3 = 6$

Determine the values of i_1, i_2 and i_3

11. The forces in three members of a framework are F_1, F_2 and F_3. They are related by the following simultaneous equations.

$1.4F_1 + 2.8F_2 + 2.8F_3 = 5.6$

$4.2F_1 - 1.4F_2 + 5.6F_3 = 35.0$

$4.2F_1 + 2.8F_2 - 1.4F_3 = -5.6$

Find the values of F_1, F_2 and F_3

For fully worked solutions to each of the problems in Practice Exercises 55 to 59 in this chapter, go to the website:
www.routledge.com/cw/bird

This assignment covers the material contained in Chapters 12 and 13. *The marks available are shown in brackets at the end of each question.*

1. Transpose $p - q + r = a - b$ for b (2)

2. Make π the subject of the formula $r = \dfrac{c}{2\pi}$ (2)

3. Transpose $V = \dfrac{1}{3}\pi r^2 h$ for h (2)

4. Transpose $I = \dfrac{E - e}{R + r}$ for E (3)

5. Transpose $k = \dfrac{b}{ad - 1}$ for d (4)

6. Make g the subject of the formula $t = 2\pi\sqrt{\dfrac{L}{g}}$ (3)

7. Transpose $A = \dfrac{\pi R^2 \theta}{360}$ for R (2)

8. Make r the subject of the formula $x + y = \dfrac{r}{3 + r}$ (5)

9. Make L the subject of the formula $m = \dfrac{\mu L}{L + rCR}$ (5)

10. The surface area A of a rectangular prism is given by the formula $A = 2(bh + hl + lb)$. Evaluate b when $A = 11\,750\,\text{mm}^2, h = 25\,\text{mm}$ and $l = 75\,\text{mm}$. (4)

11. The velocity v of water in a pipe appears in the formula $h = \dfrac{0.03\,Lv^2}{2\,dg}$. Evaluate v when $h = 0.384, d = 0.20, L = 80$ and $g = 10$ (5)

12. A formula for the focal length f of a convex lens is $\dfrac{1}{f} = \dfrac{1}{u} + \dfrac{1}{v}$. Evaluate v when $f = 4$ and $u = 20$ (4)

13. Impedance in an a.c. circuit is given by:

$$Z = \sqrt{\left[R^2 + \left(2\pi nL - \dfrac{1}{2\pi nC}\right)^2\right]} \quad \text{where } R \text{ is}$$

the resistance in ohms, L is the inductance in henries, C is the capacitance in farads and n is the frequency of oscillations per second. Given $n = 50, R = 20, L = 0.40$ and $Z = 25$, determine the value of capacitance. (9)

In Questions 14 and 15, solve the simultaneous equations.

14. (a) $2x + y = 6$ (b) $4x - 3y = 11$
      $5x - y = 22$ $3x + 5y = 30$ (9)

15. (a) $3a - 8 + \dfrac{b}{8} = 0$

      $b + \dfrac{a}{2} = \dfrac{21}{4}$

  (b) $\dfrac{2p + 1}{5} - \dfrac{1 - 4q}{2} = \dfrac{5}{2}$

      $\dfrac{1 - 3p}{7} + \dfrac{2q - 3}{5} + \dfrac{32}{35} = 0$ (18)

16. In an engineering process two variables x and y are related by the equation $y = ax + \dfrac{b}{x}$, where a and b are constants. Evaluate a and b if $y = 15$ when $x = 1$ and $y = 13$ when $x = 3$ (5)

17. Kirchhoff's laws are used to determine the current equations in an electrical network and result in the following:

$$i_1 + 8i_2 + 3i_3 = -31$$
$$3i_1 - 2i_2 + i_3 = -5$$
$$2i_1 - 3i_2 + 2i_3 = 6$$

Determine the values of i_1, i_2 and i_3 (10)

18. The forces acting on a beam are given by:
$R_1 + R_2 = 3.3\,\text{kN}$ and
$22 \times 2.7 + 61 \times 0.4 - 12R_1 = 46R_2$
Calculate the reaction forces R_1 and R_2 (8)

For lecturers/instructors/teachers, fully worked solutions to each of the problems in Revision Test 5, together with a full marking scheme, are available at the website:
www.routledge.com/cw/bird

Chapter 14

Solving quadratic equations

Why it is important to understand: Solving quadratic equations

Quadratic equations have many applications in engineering and science; they are used in describing the trajectory of a ball, determining the height of a throw, and in the concept of acceleration, velocity, ballistics and stopping power. In addition, the quadratic equation has been found to be widely evident in a number of natural processes; some of these include the processes by which light is reflected off a lens, water flows down a rocky stream, or even the manner in which fur, spots, or stripes develop on wild animals. When police arrive at the scene of a road accident, they measure the length of the skid marks and assess the road conditions. They can then use a quadratic equation to calculate the speed of the vehicles and hence reconstruct exactly what happened. The U-shape of a parabola can describe the trajectories of water jets in a fountain and a bouncing ball, or be incorporated into structures like the parabolic reflectors that form the base of satellite dishes and car headlights. Quadratic functions can help plot the course of moving objects and assist in determining minimum and maximum values. Most of the objects we use every day, from cars to clocks, would not exist if someone somewhere hadn't applied quadratic functions to their design. Solving quadratic equations is an important skill required in all aspects of engineering.

At the end of this chapter, you should be able to:

- define a quadratic equation
- solve quadratic equations by factorisation
- solve quadratic equations by 'completing the square'
- solve quadratic equations involving practical situations
- solve linear and quadratic equations simultaneously

14.1 Introduction

As stated in Chapter 11, an **equation** is a statement that two quantities are equal and to '**solve an equation**' means 'to find the value of the unknown'. The value of the unknown is called the **root** of the equation.

A **quadratic equation** is one in which the highest power of the unknown quantity is 2. For example, $x^2 - 3x + 1 = 0$ is a quadratic equation.

There are four methods of **solving quadratic equations**. These are:

(a) by factorisation (where possible),

(b) by 'completing the square',

(c) by using the 'quadratic formula' or

(d) graphically (see Chapter 35).

14.2 Solution of quadratic equations by factorisation

Multiplying out $(x + 1)(x - 3)$ gives $x^2 - 3x + x - 3$ i.e. $x^2 - 2x - 3$. The reverse process of moving from $x^2 - 2x - 3$ to $(x + 1)(x - 3)$ is called **factorising**.

If the quadratic expression can be factorised this provides the simplest method of solving a quadratic equation.

For example, if $x^2 - 2x - 3 = 0$, then, by factorising,
$$(x + 1)(x - 3) = 0$$

Hence, either $\quad (x + 1) = 0, \quad$ i.e. $\quad x = -1$

or $\qquad\qquad (x - 3) = 0, \quad$ i.e. $\quad x = 3$

Hence, $x = -1$ and $x = 3$ are the roots of the quadratic equation $x^2 - 2x - 3 = 0$

The technique of factorising is often one of trial and error.

Problem 1. Solve the equation $x^2 + x - 6 = 0$ by factorisation

The factors of x^2 are x and x. These are placed in brackets: $\quad (x \quad)(x \quad)$

The factors of -6 are $+6$ and -1, or -6 and $+1$, or $+3$ and -2, or -3 and $+2$

The only combination to give a middle term of $+x$ is $+3$ and -2,

i.e. $\qquad x^2 + x - 6 = (x + 3)(x - 2)$

The quadratic equation $x^2 + x - 6 = 0$ thus becomes
$$(x + 3)(x - 2) = 0$$

Since the only way that this can be true is for either the first or the second or both factors to be zero, then

either $\qquad (x + 3) = 0, \quad$ i.e. $\quad x = -3$

or $\qquad\quad (x - 2) = 0, \quad$ i.e. $\quad x = 2$

Hence, **the roots of $x^2 + x - 6 = 0$ are $x = -3$** and **$x = 2$**

Problem 2. Solve the equation $x^2 + 2x - 8 = 0$ by factorisation

The factors of x^2 are x and x. These are placed in brackets: $\quad (x \quad)(x \quad)$

The factors of -8 are $+8$ and -1, or -8 and $+1$, or $+4$ and -2, or -4 and $+2$

The only combination to give a middle term of $+2x$ is $+4$ and -2,

i.e. $\qquad x^2 + 2x - 8 = (x + 4)(x - 2)$

(Note that the product of the two inner terms $(4x)$ added to the product of the two outer terms $(-2x)$ must equal the middle term, $+2x$ in this case.)

The quadratic equation $x^2 + 2x - 8 = 0$ thus becomes
$$(x + 4)(x - 2) = 0$$

Since the only way that this can be true is for either the first or the second or both factors to be zero,

either $\qquad (x + 4) = 0, \quad$ i.e. $\quad x = -4$

or $\qquad\quad (x - 2) = 0, \quad$ i.e. $\quad x = 2$

Hence, **the roots of $x^2 + 2x - 8 = 0$ are $x = -4$** and **$x = 2$**

Problem 3. Determine the roots of $x^2 - 6x + 9 = 0$ by factorisation

$x^2 - 6x + 9 = (x - 3)(x - 3), \quad$ i.e. $(x - 3)^2 = 0$

The LHS is known as a **perfect square**.

Hence, $x = 3$ is the only root of the equation $x^2 - 6x + 9 = 0$

Problem 4. Solve the equation $x^2 - 4x = 0$

Factorising gives $\quad x(x - 4) = 0$

If $\qquad\qquad\qquad x(x - 4) = 0,$

either $\qquad\qquad\qquad x = 0 \quad$ or $\quad x - 4 = 0$

i.e. $\qquad\qquad\qquad\quad \boldsymbol{x = 0} \quad$ or $\quad \boldsymbol{x = 4}$

These are the two roots of the given equation. Answers can always be checked by substitution into the original equation.

Problem 5. Solve the equation $x^2 + 3x - 4 = 0$

Factorising gives $\qquad (x - 1)(x + 4) = 0$

Hence, either $\qquad x - 1 = 0 \quad$ or $\quad x + 4 = 0$

i.e. $\qquad\qquad\qquad \boldsymbol{x = 1} \quad$ or $\quad \boldsymbol{x = -4}$

Problem 6. Determine the roots of $4x^2 - 25 = 0$ by factorisation

The LHS of $4x^2 - 25 = 0$ is **the difference of two squares**, $(2x)^2$ and $(5)^2$

By factorising, $\boldsymbol{4x^2 - 25 = (2x + 5)(2x - 5)}$,

i.e. $\qquad (2x + 5)(2x - 5) = 0$

Hence, either $\quad (2x + 5) = 0, \quad$ i.e. $\quad x = -\dfrac{5}{2} = -2.5$

or $\qquad\qquad (2x - 5) = 0, \quad$ i.e. $\quad x = \dfrac{5}{2} = 2.5$

Problem 7.   Solve the equation $x^2 - 5x + 6 = 0$

Factorising gives $(x - 3)(x - 2) = 0$

Hence, either $x - 3 = 0$ or $x - 2 = 0$

i.e. $x = 3$ or $x = 2$

Problem 8.   Solve the equation $x^2 = 15 - 2x$

Rearranging gives $x^2 + 2x - 15 = 0$

Factorising gives $(x + 5)(x - 3) = 0$

Hence, either $x + 5 = 0$ or $x - 3 = 0$

i.e. $x = -5$ or $x = 3$

Problem 9.   Solve the equation $3x^2 - 11x - 4 = 0$ by factorisation

The factors of $3x^2$ are $3x$ and x. These are placed in brackets: $(3x \quad)(x \quad)$
The factors of -4 are -4 and $+1$, or $+4$ and -1, or -2 and $+2$
Remembering that the product of the two inner terms added to the product of the two outer terms must equal $-11x$, the only combination to give this is $+1$ and -4,

i.e. $3x^2 - 11x - 4 = (3x + 1)(x - 4)$

The quadratic equation $3x^2 - 11x - 4 = 0$ thus becomes $(3x + 1)(x - 4) = 0$

Hence, either $(3x + 1) = 0$,   i.e. $x = -\dfrac{1}{3}$

or $(x - 4) = 0$,   i.e. $x = 4$

and both solutions may be checked in the original equation.

Problem 10.   Solve the quadratic equation $4x^2 + 8x + 3 = 0$ by factorising

The factors of $4x^2$ are $4x$ and x or $2x$ and $2x$
The factors of 3 are 3 and 1, or -3 and -1
Remembering that the product of the inner terms added to the product of the two outer terms must equal $+8x$, the only combination that is true (by trial and error) is

$$4x^2 + 8x + 3 = (2x + 3)(2x + 1)$$

Hence, $(2x + 3)(2x + 1) = 0$,   from   which   either $(2x + 3) = 0$ or $(2x + 1) = 0$

Thus, $2x = -3$, from which $x = -\dfrac{3}{2}$ or -1.5

or $2x = -1$, from which $x = -\dfrac{1}{2}$ or -0.5

which may be checked in the original equation.

Problem 11.   Solve the quadratic equation $15x^2 + 2x - 8 = 0$ by factorising

The factors of $15x^2$ are $15x$ and x or $5x$ and $3x$
The factors of -8 are -4 and $+2$, or $+4$ and -2, or -8 and $+1$, or $+8$ and -1
By trial and error the only combination that works is

$$15x^2 + 2x - 8 = (5x + 4)(3x - 2)$$

Hence, $(5x + 4)(3x - 2) = 0$,   from   which   either $5x + 4 = 0$ or $3x - 2 = 0$

Hence, $x = -\dfrac{4}{5}$ or $x = \dfrac{2}{3}$

which may be checked in the original equation.

Problem 12.   The roots of a quadratic equation are $\dfrac{1}{3}$ and -2. Determine the equation in x

If the roots of a quadratic equation are, say, α and β, then $(x - \alpha)(x - \beta) = 0$.

Hence, if $\alpha = \dfrac{1}{3}$ and $\beta = -2$,

$$\left(x - \dfrac{1}{3}\right)(x - (-2)) = 0$$

$$\left(x - \dfrac{1}{3}\right)(x + 2) = 0$$

$$x^2 - \dfrac{1}{3}x + 2x - \dfrac{2}{3} = 0$$

$$x^2 + \dfrac{5}{3}x - \dfrac{2}{3} = 0$$

or $3x^2 + 5x - 2 = 0$

Problem 13.   Find the equation in x whose roots are 5 and -5

If 5 and -5 are the roots of a quadratic equation then

$$(x - 5)(x + 5) = 0$$

i.e. $x^2 - 5x + 5x - 25 = 0$

i.e. $x^2 - 25 = 0$

Problem 14.   Find the equation in x whose roots are 1.2 and -0.4

If 1.2 and -0.4 are the roots of a quadratic equation then

$$(x - 1.2)(x + 0.4) = 0$$

i.e. $x^2 - 1.2x + 0.4x - 0.48 = 0$

i.e. $x^2 - 0.8x - 0.48 = 0$

Now try the following Practice Exercise

Practice Exercise 60   Solving quadratic equations by factorisation (answers on page 1154)

In Problems 1 to 30, solve the given equations by factorisation.

1. $x^2 - 16 = 0$
2. $x^2 + 4x - 32 = 0$
3. $(x + 2)^2 = 16$
4. $4x^2 - 9 = 0$
5. $3x^2 + 4x = 0$
6. $8x^2 - 32 = 0$
7. $x^2 - 8x + 16 = 0$
8. $x^2 + 10x + 25 = 0$
9. $x^2 - 2x + 1 = 0$
10. $x^2 + 5x + 6 = 0$
11. $x^2 + 10x + 21 = 0$
12. $x^2 - x - 2 = 0$
13. $y^2 - y - 12 = 0$
14. $y^2 - 9y + 14 = 0$
15. $x^2 + 8x + 16 = 0$
16. $x^2 - 4x + 4 = 0$
17. $x^2 + 6x + 9 = 0$
18. $x^2 - 9 = 0$
19. $3x^2 + 8x + 4 = 0$
20. $4x^2 + 12x + 9 = 0$
21. $4z^2 - \frac{1}{16} = 0$
22. $x^2 + 3x - 28 = 0$
23. $2x^2 - x - 3 = 0$
24. $6x^2 - 5x + 1 = 0$
25. $10x^2 + 3x - 4 = 0$
26. $21x^2 - 25x = 4$
27. $8x^2 + 13x - 6 = 0$
28. $5x^2 + 13x - 6 = 0$
29. $6x^2 - 5x - 4 = 0$
30. $8x^2 + 2x - 15 = 0$

In Problems 31 to 36, determine the quadratic equations in x whose roots are

31. 3 and 1
32. 2 and -5
33. -1 and -4
34. 2.5 and -0.5
35. 6 and -6
36. 2.4 and -0.7

14.3 Solution of quadratic equations by 'completing the square'

An expression such as x^2 or $(x + 2)^2$ or $(x - 3)^2$ is called a **perfect square**.

If $x^2 = 3$ then $x = \pm\sqrt{3}$

If $(x + 2)^2 = 5$ then $x + 2 = \pm\sqrt{5}$ and $x = -2 \pm \sqrt{5}$

If $(x - 3)^2 = 8$ then $x - 3 = \pm\sqrt{8}$ and $x = 3 \pm \sqrt{8}$

Hence, if a quadratic equation can be rearranged so that one side of the equation is a perfect square and the other side of the equation is a number, then the solution of the equation is readily obtained by taking the square roots of each side as in the above examples. The process of rearranging one side of a quadratic equation into a perfect square before solving is called '**completing the square**'.

$$(x + a)^2 = x^2 + 2ax + a^2$$

Thus, in order to make the quadratic expression $x^2 + 2ax$ into a perfect square, it is necessary to add (half the coefficient of x)^2, i.e. $\left(\frac{2a}{2}\right)^2$ or a^2

For example, $x^2 + 3x$ becomes a perfect square by adding $\left(\frac{3}{2}\right)^2$, i.e.

$$x^2 + 3x + \left(\frac{3}{2}\right)^2 = \left(x + \frac{3}{2}\right)^2$$

The method of completing the square is demonstrated in the following worked problems.

Problem 15.   Solve $2x^2 + 5x = 3$ by completing the square

The procedure is as follows.

(i) Rearrange the equation so that all terms are on the same side of the equals sign (and the coefficient of the x^2 term is positive). Hence,

$$2x^2 + 5x - 3 = 0$$

(ii) Make the coefficient of the x^2 term unity. In this case this is achieved by dividing throughout by 2. Hence,

$$\frac{2x^2}{2} + \frac{5x}{2} - \frac{3}{2} = 0$$

i.e. $x^2 + \frac{5}{2}x - \frac{3}{2} = 0$

(iii) Rearrange the equations so that the x^2 and x terms are on one side of the equals sign and the constant is on the other side. Hence,

$$x^2 + \frac{5}{2}x = \frac{3}{2}$$

(iv) Add to both sides of the equation (half the coefficient of x)^2. In this case the coefficient of x is $\frac{5}{2}$

Half the coefficient squared is therefore $\left(\frac{5}{4}\right)^2$

Thus,

$$x^2 + \frac{5}{2}x + \left(\frac{5}{4}\right)^2 = \frac{3}{2} + \left(\frac{5}{4}\right)^2$$

The LHS is now a perfect square, i.e.

$$\left(x + \frac{5}{4}\right)^2 = \frac{3}{2} + \left(\frac{5}{4}\right)^2$$

(v) Evaluate the RHS. Thus,

$$\left(x + \frac{5}{4}\right)^2 = \frac{3}{2} + \frac{25}{16} = \frac{24 + 25}{16} = \frac{49}{16}$$

(vi) Take the square root of both sides of the equation (remembering that the square root of a number gives a $\pm$ answer). Thus,

$$\sqrt{\left(x + \frac{5}{4}\right)^2} = \sqrt{\left(\frac{49}{16}\right)}$$

i.e. $\quad\quad x + \frac{5}{4} = \pm\frac{7}{4}$

(vii) Solve the simple equation. Thus,

$$x = -\frac{5}{4} \pm \frac{7}{4}$$

i.e. $\quad x = -\frac{5}{4} + \frac{7}{4} = \frac{2}{4} = \frac{1}{2}$ or 0.5

and $\quad x = -\frac{5}{4} - \frac{7}{4} = -\frac{12}{4} = -3$

Hence, $x = 0.5$ or $x = -3$; i.e. **the roots of the equation $2x^2 + 5x = 3$ are 0.5 and -3**

Problem 16. Solve $2x^2 + 9x + 8 = 0$, correct to 3 significant figures, by completing the square

Making the coefficient of x^2 unity gives

$$x^2 + \frac{9}{2}x + 4 = 0$$

Rearranging gives $\quad x^2 + \frac{9}{2}x = -4$

Adding to both sides (half the coefficient of x)^2 gives

$$x^2 + \frac{9}{2}x + \left(\frac{9}{4}\right)^2 = \left(\frac{9}{4}\right)^2 - 4$$

The LHS is now a perfect square. Thus,

$$\left(x + \frac{9}{4}\right)^2 = \frac{81}{16} - 4 = \frac{81}{16} - \frac{64}{16} = \frac{17}{16}$$

Taking the square root of both sides gives

$$x + \frac{9}{4} = \sqrt{\left(\frac{17}{16}\right)} = \pm 1.031$$

Hence, $\quad\quad x = -\frac{9}{4} \pm 1.031$

i.e. $\boldsymbol{x = -1.22}$ or $\boldsymbol{-3.28}$, correct to 3 significant figures.

Problem 17. By completing the square, solve the quadratic equation $4.6y^2 + 3.5y - 1.75 = 0$, correct to 3 decimal places

$$4.6y^2 + 3.5y - 1.75 = 0$$

Making the coefficient of y^2 unity gives

$$y^2 + \frac{3.5}{4.6}y - \frac{1.75}{4.6} = 0$$

and rearranging gives $\quad y^2 + \frac{3.5}{4.6}y = \frac{1.75}{4.6}$

Adding to both sides (half the coefficient of y)^2 gives

$$y^2 + \frac{3.5}{4.6}y + \left(\frac{3.5}{9.2}\right)^2 = \frac{1.75}{4.6} + \left(\frac{3.5}{9.2}\right)^2$$

The LHS is now a perfect square. Thus,

$$\left(y + \frac{3.5}{9.2}\right)^2 = 0.5251654$$

Taking the square root of both sides gives

$$y + \frac{3.5}{9.2} = \sqrt{0.5251654} = \pm 0.7246830$$

Hence, $\quad\quad y = -\frac{3.5}{9.2} \pm 0.7246830$

i.e. $\quad\quad\quad \boldsymbol{y = 0.344}$ or $\boldsymbol{-1.105}$

Now try the following Practice Exercise

Practice Exercise 61   Solving quadratic equations by completing the square (answers on page 1155)

Solve the following equations correct to 3 decimal places by completing the square.

1. $x^2 + 4x + 1 = 0$ 2. $2x^2 + 5x - 4 = 0$
3. $3x^2 - x - 5 = 0$ 4. $5x^2 - 8x + 2 = 0$
5. $4x^2 - 11x + 3 = 0$ 6. $2x^2 + 5x = 2$

14.4 Solution of quadratic equations by formula

Let the general form of a quadratic equation be given by $ax^2 + bx + c = 0$, where a, b and c are constants. Dividing $ax^2 + bx + c = 0$ by a gives

$$x^2 + \frac{b}{a}x + \frac{c}{a} = 0$$

Rearranging gives $x^2 + \frac{b}{a}x = -\frac{c}{a}$

Adding to each side of the equation the square of half the coefficient of the term in x to make the LHS a perfect square gives

$$x^2 + \frac{b}{a}x + \left(\frac{b}{2a}\right)^2 = \left(\frac{b}{2a}\right)^2 - \frac{c}{a}$$

Rearranging gives $\left(x + \frac{b}{a}\right)^2 = \frac{b^2}{4a^2} - \frac{c}{a} = \frac{b^2 - 4ac}{4a^2}$

Taking the square root of both sides gives

$$x + \frac{b}{2a} = \sqrt{\left(\frac{b^2 - 4ac}{4a^2}\right)} = \frac{\pm\sqrt{b^2 - 4ac}}{2a}$$

Hence, $x = -\frac{b}{2a} \pm \frac{\sqrt{b^2 - 4ac}}{2a}$

i.e. the quadratic formula is $x = \dfrac{-b \pm \sqrt{b^2 - 4ac}}{2a}$

(This method of obtaining the formula is completing the square – as shown in the previous section.)
In summary,

if $ax^2 + bx + c = 0$ then $x = \dfrac{-b \pm \sqrt{b^2 - 4ac}}{2a}$

This is known as the **quadratic formula**.

Problem 18.   Solve $x^2 + 2x - 8 = 0$ by using the quadratic formula

Comparing $x^2 + 2x - 8 = 0$ with $ax^2 + bx + c = 0$ gives $a = 1, b = 2$ and $c = -8$
Substituting these values into the quadratic formula

$$x = \frac{-b \pm \sqrt{b^2 - 4ac}}{2a}$$

gives

$$x = \frac{-2 \pm \sqrt{2^2 - 4(1)(-8)}}{2(1)} = \frac{-2 \pm \sqrt{4 + 32}}{2}$$

$$= \frac{-2 \pm \sqrt{36}}{2} = \frac{-2 \pm 6}{2}$$

$$= \frac{-2 + 6}{2} \text{ or } \frac{-2 - 6}{2}$$

Hence, $x = \dfrac{4}{2}$ or $\dfrac{-8}{2}$, i.e. $x = 2$ or $x = -4$

Problem 19.   Solve $3x^2 - 11x - 4 = 0$ by using the quadratic formula

Comparing $3x^2 - 11x - 4 = 0$ with $ax^2 + bx + c = 0$ gives $a = 3, b = -11$ and $c = -4$. Hence,

$$x = \frac{-(-11) \pm \sqrt{(-11)^2 - 4(3)(-4)}}{2(3)}$$

$$= \frac{+11 \pm \sqrt{121 + 48}}{6} = \frac{11 \pm \sqrt{169}}{6}$$

$$= \frac{11 \pm 13}{6} = \frac{11 + 13}{6} \text{ or } \frac{11 - 13}{6}$$

Hence, $x = \dfrac{24}{6}$ or $\dfrac{-2}{6}$, i.e. $x = 4$ or $x = -\dfrac{1}{3}$

Problem 20.   Solve $4x^2 + 7x + 2 = 0$ giving the roots correct to 2 decimal places

Comparing $4x^2 + 7x + 2 = 0$ with $ax^2 + bx + c$ gives $a = 4, b = 7$ and $c = 2$. Hence,

$$x = \frac{-7 \pm \sqrt{7^2 - 4(4)(2)}}{2(4)}$$

$$= \frac{-7 \pm \sqrt{17}}{8} = \frac{-7 \pm 4.123}{8}$$

$$= \frac{-7 + 4.123}{8} \text{ or } \frac{-7 - 4.123}{8}$$

Hence, $x = -0.36$ or -1.39, correct to 2 decimal places.

Problem 21. Use the quadratic formula to solve $\dfrac{x+2}{4} + \dfrac{3}{x-1} = 7$ correct to 4 significant figures

---

Multiplying throughout by $4(x-1)$ gives

$$4(x-1)\frac{(x+2)}{4} + 4(x-1)\frac{3}{(x-1)} = 4(x-1)(7)$$

Cancelling gives $\quad (x-1)(x+2) + (4)(3) = 28(x-1)$

$$x^2 + x - 2 + 12 = 28x - 28$$

Hence, $\qquad\qquad x^2 - 27x + 38 = 0$

Using the quadratic formula,

$$x = \frac{-(-27) \pm \sqrt{(-27)^2 - 4(1)(38)}}{2}$$

$$= \frac{27 \pm \sqrt{577}}{2} = \frac{27 \pm 24.0208}{2}$$

Hence, $\qquad x = \dfrac{27 + 24.0208}{2} = 25.5104$

or $\qquad x = \dfrac{27 - 24.0208}{2} = 1.4896$

Hence, $x = \mathbf{25.51}$ or $\mathbf{1.490}$, correct to 4 significant figures.

Now try the following Practice Exercise

Practice Exercise 62   Solving quadratic equations by formula (answers on page 1155)

Solve the following equations by using the quadratic formula, correct to 3 decimal places.

1. $2x^2 + 5x - 4 = 0$

2. $5.76x^2 + 2.86x - 1.35 = 0$

3. $2x^2 - 7x + 4 = 0$

4. $4x + 5 = \dfrac{3}{x}$

5. $(2x+1) = \dfrac{5}{x-3}$

6. $3x^2 - 5x + 1 = 0$

7. $4x^2 + 6x - 8 = 0$

8. $5.6x^2 - 11.2x - 1 = 0$

9. $3x(x+2) + 2x(x-4) = 8$

10. $4x^2 - x(2x+5) = 14$

11. $\dfrac{5}{x-3} + \dfrac{2}{x-2} = 6$

12. $\dfrac{3}{x-7} + 2x = 7 + 4x$

13. $\dfrac{x+1}{x-1} = x - 3$

14.5 Practical problems involving quadratic equations

There are many **practical problems** in which a quadratic equation has first to be obtained, from given information, before it is solved.

Problem 22. The area of a rectangle is $23.6\,\text{cm}^2$ and its width is $3.10\,\text{cm}$ shorter than its length. Determine the dimensions of the rectangle, correct to 3 significant figures

---

Let the length of the rectangle be x cm. Then the width is $(x - 3.10)$ cm.

$$\text{Area} = \text{length} \times \text{width} = x(x - 3.10) = 23.6$$

i.e. $\qquad\qquad x^2 - 3.10x - 23.6 = 0$

Using the quadratic formula,

$$x = \frac{-(-3.10) \pm \sqrt{(-3.10)^2 - 4(1)(-23.6)}}{2(1)}$$

$$= \frac{3.10 \pm \sqrt{9.61 + 94.4}}{2} = \frac{3.10 \pm 10.20}{2}$$

$$= \frac{13.30}{2} \text{ or } \frac{-7.10}{2}$$

Hence, $x = 6.65\,\text{cm}$ or $-3.55\,\text{cm}$. The latter solution is neglected since length cannot be negative.
Thus, length $x = 6.65\,\text{cm}$ and width $= x - 3.10 = 6.65 - 3.10 = 3.55\,\text{cm}$, i.e. **the dimensions of the rectangle are 6.65 cm by 3.55 cm.**
(Check: Area $= 6.65 \times 3.55 = 23.6\,\text{cm}^2$, correct to 3 significant figures.)

Problem 23. Calculate the diameter of a solid cylinder which has a height of 82.0 cm and a total surface area of 2.0 m²

Total surface area of a cylinder

= curved surface area + 2 circular ends

= $2\pi rh + 2\pi r^2$ (where r = radius and h = height)

Since the total surface area = 2.0 m² and the height h = 82 cm or 0.82 m,

$$2.0 = 2\pi r(0.82) + 2\pi r^2$$

i.e. $$2\pi r^2 + 2\pi r(0.82) - 2.0 = 0$$

Dividing throughout by 2π gives $r^2 + 0.82r - \dfrac{1}{\pi} = 0$

Using the quadratic formula,

$$r = \frac{-0.82 \pm \sqrt{(0.82)^2 - 4(1)\left(-\dfrac{1}{\pi}\right)}}{2(1)}$$

$$= \frac{-0.82 \pm \sqrt{1.94564}}{2} = \frac{-0.82 \pm 1.39486}{2}$$

$$= 0.2874 \text{ or } -1.1074$$

Thus, the radius r of the cylinder is 0.2874 m (the negative solution being neglected).

Hence, the diameter of the cylinder

$$= 2 \times 0.2874$$

$$= \textbf{0.5748 m or 57.5 cm,}$$

correct to 3 significant figures.

Problem 24. The height s metres of a mass projected vertically upwards at time t seconds is $s = ut - \dfrac{1}{2}gt^2$. Determine how long the mass will take after being projected to reach a height of 16 m (a) on the ascent and (b) on the descent, when $u = 30$ m/s and $g = 9.81$ m/s²

When height $s = 16$ m, $16 = 30t - \dfrac{1}{2}(9.81)t^2$

i.e. $$4.905t^2 - 30t + 16 = 0$$

Using the quadratic formula,

$$t = \frac{-(-30) \pm \sqrt{(-30)^2 - 4(4.905)(16)}}{2(4.905)}$$

$$= \frac{30 \pm \sqrt{586.1}}{9.81} = \frac{30 \pm 24.21}{9.81} = 5.53 \text{ or } 0.59$$

Hence, the mass will reach a height of 16 m after 0.59 s on the ascent and after 5.53 s on the descent.

Problem 25. A shed is 4.0 m long and 2.0 m wide. A concrete path of constant width is laid all the way around the shed. If the area of the path is 9.50 m², calculate its width to the nearest centimetre

Figure 14.1 shows a plan view of the shed with its surrounding path of width t metres.

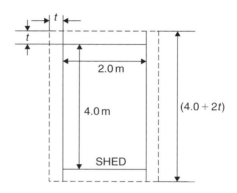

Figure 14.1

$$\text{Area of path} = 2(2.0 \times t) + 2t(4.0 + 2t)$$

i.e. $$9.50 = 4.0t + 8.0t + 4t^2$$

or $$4t^2 + 12.0t - 9.50 = 0$$

Hence,

$$t = \frac{-(12.0) \pm \sqrt{(12.0)^2 - 4(4)(-9.50)}}{2(4)}$$

$$= \frac{-12.0 \pm \sqrt{296.0}}{8} = \frac{-12.0 \pm 17.20465}{8}$$

i.e. $t = 0.6506$ m or -3.65058 m.

Neglecting the negative result, which is meaningless, the width of the path $\textbf{\textit{t}} = \textbf{0.651 m}$ or **65 cm** correct to the nearest centimetre.

Problem 26. If the total surface area of a solid cone is 486.2 cm² and its slant height is 15.3 cm, determine its base diameter.

From Chapter 29, page 309, the total surface area A of a solid cone is given by $A = \pi rl + \pi r^2$, where l is the slant height and r the base radius.

If $A = 482.2$ and $l = 15.3$, then
$$482.2 = \pi r (15.3) + \pi r^2$$

i.e. $\qquad \pi r^2 + 15.3 \pi r - 482.2 = 0$

or $\qquad r^2 + 15.3r - \dfrac{482.2}{\pi} = 0$

Using the quadratic formula,

$$r = \dfrac{-15.3 \pm \sqrt{\left[(15.3)^2 - 4 \left(\dfrac{-482.2}{\pi} \right) \right]}}{2}$$

$$= \dfrac{-15.3 \pm \sqrt{848.0461}}{2} = \dfrac{-15.3 \pm 29.12123}{2}$$

Hence, radius $r = 6.9106$ cm (or -22.21 cm, which is meaningless and is thus ignored).

Thus, the **diameter of the base** $= 2r = 2(6.9106)$
$$= \mathbf{13.82\,cm.}$$

Now try the following Practice Exercise

Practice Exercise 63   Practical problems involving quadratic equations (answers on page 1155)

1. The angle a rotating shaft turns through in t seconds is given by $\theta = \omega t + \dfrac{1}{2} \alpha t^2$. Determine the time taken to complete 4 radians if ω is 3.0 rad/s and α is 0.60 rad/s^2.

2. The power P developed in an electrical circuit is given by $P = 10I - 8I^2$, where I is the current in amperes. Determine the current necessary to produce a power of 2.5 watts in the circuit.

3. The area of a triangle is 47.6 cm^2 and its perpendicular height is 4.3 cm more than its base length. Determine the length of the base correct to 3 significant figures.

4. The sag l in metres in a cable stretched between two supports distance x m apart is given by $l = \dfrac{12}{x} + x$. Determine the distance between the supports when the sag is 20 m.

5. The acid dissociation constant K_a of ethanoic acid is 1.8×10^{-5} mol dm^{-3} for a particular solution. Using the Ostwald dilution law, $K_a = \dfrac{x^2}{v(1-x)}$, determine x, the degree of ionisation, given that $v = 10$ dm^3.

6. A rectangular building is 15 m long by 11 m wide. A concrete path of constant width is laid all the way around the building. If the area of the path is 60.0 m^2, calculate its width correct to the nearest millimetre.

7. The total surface area of a closed cylindrical container is 20.0 m^2. Calculate the radius of the cylinder if its height is 2.80 m.

8. The bending moment M at a point in a beam is given by $M = \dfrac{3x(20 - x)}{2}$, where x metres is the distance from the point of support. Determine the value of x when the bending moment is 50 N m.

9. A tennis court measures 24 m by 11 m. In the layout of a number of courts an area of ground must be allowed for at the ends and at the sides of each court. If a border of constant width is allowed around each court and the total area of the court and its border is 950 m^2, find the width of the border.

10. Two resistors, when connected in series, have a total resistance of 40 ohms. When connected in parallel their total resistance is 8.4 ohms. If one of the resistors has a resistance of R_x ohms,
   (a) show that $R_x^2 - 40R_x + 336 = 0$ and
   (b) calculate the resistance of each.

11. When a ball is thrown vertically upwards its height h varies with time t according to the equation $h = 25t - 4t^2$. Determine the times, correct to 3 significant figures, when the height is 12 m.

12. In an RLC electrical circuit, reactance X is given by $X = \omega L - \dfrac{1}{\omega C}$
   $X = 220\ \Omega$, inductance $L = 800$ mH and capacitance $C = 25\ \mu$F. The angular velocity ω is measured in radians per second. Calculate the value of ω.

14.6   Solution of linear and quadratic equations simultaneously

Sometimes a linear equation and a quadratic equation need to be solved simultaneously. An algebraic method of solution is shown in Problem 27; a graphical solution is shown in Chapter 35, page 389.

Section A

Problem 27. Determine the values of x and y which simultaneously satisfy the equations $y = 5x - 4 - 2x^2$ and $y = 6x - 7$

For a simultaneous solution the values of y must be equal, hence the RHS of each equation is equated.

Thus, $\qquad 5x - 4 - 2x^2 = 6x - 7$

Rearranging gives $\quad 5x - 4 - 2x^2 - 6x + 7 = 0$

i.e. $\qquad\qquad\qquad -x + 3 - 2x^2 = 0$

or $\qquad\qquad\qquad 2x^2 + x - 3 = 0$

Factorising gives $\qquad (2x + 3)(x - 1) = 0$

i.e. $\qquad\qquad\qquad x = -\dfrac{3}{2}$ or $x = 1$

In the equation $y = 6x - 7$,

when $x = -\dfrac{3}{2},\qquad y = 6\left(-\dfrac{3}{2}\right) - 7 = -16$

and when $x = 1,\quad y = 6 - 7 = -1$

(Checking the result in $y = 5x - 4 - 2x^2$:

when $x = -\dfrac{3}{2},\qquad y = 5\left(-\dfrac{3}{2}\right) - 4 - 2\left(-\dfrac{3}{2}\right)^2$

$$= -\frac{15}{2} - 4 - \frac{9}{2} = -16, \text{ as above,}$$

and when $x = 1,\quad y = 5 - 4 - 2 = -1$, as above.)

Hence, the simultaneous solutions occur when

$$x = -\frac{3}{2}, y = -16 \text{ and when } x = 1, y = -1$$

Now try the following Practice Exercise

Practice Exercise 64   Solving linear and quadratic equations simultaneously (answers on page 1155)

Determine the solutions of the following simultaneous equations.

1. $y = x^2 + x + 1$
   $y = 4 - x$

2. $y = 15x^2 + 21x - 11$
   $y = 2x - 1$

3. $2x^2 + y = 4 + 5x$
   $x + y = 4$

For fully worked solutions to each of the problems in Practice Exercises 60 to 64 in this chapter, go to the website:
www.routledge.com/cw/bird

Multiple choice questions Test 2
Equations and transposition
This test covers the material in Chapters 11 to 14

All questions have only one correct answer (answers on page 1200).

1. The relationship between the temperature in degrees Fahrenheit (F) and the temperature in degrees Celsius (C) is given by: $F = \dfrac{9}{5}C + 32$. $135°F$ is equivalent to:

   (a) $43°C$ (b) $57.2°C$ (c) $185.4°C$ (d) $184°C$

2. Transposing $I = \dfrac{V}{R}$ for resistance R gives:

   (a) $I - V$ (b) $\dfrac{V}{I}$ (c) $\dfrac{I}{V}$ (d) VI

3. When two resistors R_1 and R_2 are connected in parallel, the formula $\dfrac{1}{R_T} = \dfrac{1}{R_1} + \dfrac{1}{R_2}$ is used to determine the total resistance R_T. If $R_1 = 470\,\Omega$ and $R_2 = 2.7\,k\Omega$, R_T (correct to 3 significant figures) is equal to:

   (a) $2.68\,\Omega$ (b) $400\,\Omega$ (c) $473\,\Omega$ (d) $3170\,\Omega$

4. Transposing $v = f\lambda$ to make wavelength λ the subject gives:

   (a) $\dfrac{v}{f}$ (b) $v + f$ (c) $f - v$ (d) $\dfrac{f}{v}$

5. Transposing the formula $R = R_0(1 + \alpha t)$ for t gives:

   (a) $\dfrac{R - R_0}{(1 + \alpha)}$ (b) $\dfrac{R - R_0 - 1}{\alpha}$

   (c) $\dfrac{R - R_0}{\alpha R_0}$ (d) $\dfrac{R}{R_0 \alpha}$

6. The current I in an a.c. circuit is given by: $I = \dfrac{V}{\sqrt{R^2 + X^2}}$ When $R = 4.8$, $X = 10.5$ and $I = 15$, the value of voltage V is:

   (a) 173.18 (b) 1.30 (c) 0.98 (d) 229.50

7. The height s of a mass projected vertically upwards at time t is given by: $s = ut - \dfrac{1}{2}gt^2$. When $g = 10$, $t = 1.5$ and $s = 3.75$, the value of u is:

   (a) 10 (b) -5 (c) $+5$ (d) -10

8. The quantity of heat Q is given by the formula $Q = mc(t_2 - t_1)$. When $m = 5$, $t_1 = 20$, $c = 8$ and $Q = 1200$, the value of t_2 is:

   (a) 10 (b) 1.5 (c) 21.5 (d) 50

9. Electrical resistance $R = \dfrac{\rho \ell}{a}$; transposing this equation for ℓ gives:

   (a) $\dfrac{\rho a}{R}$ (b) $\dfrac{R}{a \rho}$ (c) $\dfrac{a}{R \rho}$ (d) $\dfrac{Ra}{\rho}$

10. The solution of the simultaneous equations $3x - 2y = 13$ and $2x + 5y = -4$ is:

    (a) $x = -2$, $y = 3$ (b) $x = 1$, $y = -5$

    (c) $x = 3$, $y = -2$ (d) $x = -7$, $y = 2$

11. A formula for the focal length f of a convex lens is $\dfrac{1}{f} = \dfrac{1}{u} + \dfrac{1}{v}$ When $f = 4$ and $u = 6$, v is:

    (a) -2 (b) 12 (c) $\dfrac{1}{12}$ (d) $-\dfrac{1}{2}$

12. $\text{Volume} = \dfrac{\text{mass}}{\text{density}}$. The density (in kg/m³) when the mass is $2.532\,kg$ and the volume is $162\,cm^3$ is:

    (a) $0.01563\,kg/m^3$ (b) $410.2\,kg/m^3$

    (c) $15\,630\,kg/m^3$ (d) $64.0\,kg/m^3$

13. $PV = mRT$ is the characteristic gas equation. When $P = 100 \times 10^3$, $V = 4.0$, $R = 288$ and $T = 300$, the value of m is:

    (a) 4.630 (b) $313\,600$ (c) 0.216 (d) $100\,592$

14. The quadratic equation in x whose roots are -2 and $+5$ is:

    (a) $x^2 - 3x - 10 = 0$ (b) $x^2 + 7x + 10 = 0$

    (c) $x^2 + 3x - 10 = 0$ (d) $x^2 - 7x - 10 = 0$

15. The area A of a triangular piece of land of sides a, b and c may be calculated using $A = \sqrt{[s(s-a)(s-b)(s-c)]}$ where $s = \dfrac{a+b+c}{2}$.

When $a = 15$ m, $b = 11$ m and $c = 8$ m, the area, correct to the nearest square metre, is:

(a) $1836\,m^2$ (b) $648\,m^2$ (c) $445\,m^2$ (d) $43\,m^2$

16. In a system of pulleys, the effort P required to raise a load W is given by $P = aW + b$, where a and b are constants. If $W = 40$ when $P = 12$ and $W = 90$ when $P = 22$, the values of a and b are:

(a) $a = 5$, $b = \dfrac{1}{4}$

(b) $a = 1$, $b = -28$

(c) $a = \dfrac{1}{3}$, $b = -8$

(d) $a = \dfrac{1}{5}$, $b = 4$

17. Resistance R ohms varies with temperature t according to the formula $R = R_0(1 + \alpha t)$. Given $R = 21\ \Omega$, $\alpha = 0.004$ and $t = 100$, R_0 has a value of:

(a) $21.4\ \Omega$ (b) $29.4\ \Omega$ (c) $15\ \Omega$ (d) $0.067\ \Omega$

18. $8x^2 + 13x - 6 = (x + p)(qx - 3)$. The values of p and q are:

(a) $p = -2$, $q = 4$

(b) $p = 3$, $q = 2$

(c) $p = 2$, $q = 8$

(d) $p = 1$, $q = 8$

19. The height S metres of a mass thrown vertically upwards at time t seconds is given by $S = 80t - 16t^2$. To reach a height of 50 metres on the descent will take the mass:

(a) 0.73 s   (b) 5.56 s   (c) 4.27 s   (d) 81.77 s

20. The final length l_2 of a piece of wire heated through $\theta°C$ is given by the formula $l_2 = l_1(1 + \alpha\theta)$. Transposing, the coefficient of expansion α is given by:

(a) $\dfrac{l_2}{l_1} - \dfrac{1}{\theta}$ (b) $\dfrac{l_2 - l_1}{l_1\theta}$

(c) $l_2 - l_1 - l_1\theta$ (d) $\dfrac{l_1 - l_2}{l_1\theta}$

21. The roots of the quadratic equation $8x^2 + 10x - 3 = 0$ are:

(a) $-\dfrac{1}{4}$ and $\dfrac{3}{2}$ (b) 4 and $\dfrac{2}{3}$

(c) $-\dfrac{3}{2}$ and $\dfrac{1}{4}$ (d) $\dfrac{2}{3}$ and -4

22. The volume V_2 of a material when the temperature is increased is given by $V_2 = V_1\left[1 + \gamma(t_2 - t_1)\right]$. The value of t_2 when $V_2 = 61.5\,cm^3$, $V_1 = 60\,cm^3$, $\gamma = 54 \times 10^{-6}$ and $t_1 = 250$ is:

(a) 213 (b) 463 (c) 713 (d) $28\,028$

23. Current I in an electrical circuit is given by $I = \dfrac{E - e}{R + r}$. Transposing for R gives:

(a) $\dfrac{E - e - Ir}{I}$ (b) $\dfrac{E - e}{I + r}$

(c) $(E - e)(I + r)$ (d) $\dfrac{E - e}{Ir}$

24. The roots of the quadratic equation $2x^2 - 5x + 1 = 0$, correct to 2 decimal places, are:

(a) -0.22 and -2.28

(b) 2.69 and -0.19

(c) 0.19 and -2.69

(d) 2.28 and 0.22

25. Transposing $t = 2\pi\sqrt{\dfrac{l}{g}}$ for g gives:

(a) $\dfrac{(t - 2\pi)^2}{l}$ (b) $\left(\dfrac{2\pi}{t}\right)l^2$

(c) $\dfrac{\sqrt{\dfrac{t}{2\pi}}}{l}$ (d) $\dfrac{4\pi^2 l}{t^2}$

For a copy of this multiple choice test, go to:
www.routledge.com/cw/bird

Section A

Chapter 15

Logarithms

Why it is important to understand: Logarithms

All types of engineers use natural and common logarithms. Chemical engineers use them to measure radioactive decay and pH solutions, both of which are measured on a logarithmic scale. The Richter scale which measures earthquake intensity is a logarithmic scale. Biomedical engineers use logarithms to measure cell decay and growth, and also to measure light intensity for bone mineral density measurements. In electrical engineering, a dB (decibel) scale is very useful for expressing attenuations in radio propagation and circuit gains, and logarithms are used for implementing arithmetic operations in digital circuits. Logarithms are especially useful when dealing with the graphical analysis of non-linear relationships and logarithmic scales are used to linearise data to make data analysis simpler. Understanding and using logarithms is clearly important in all branches of engineering.

At the end of this chapter, you should be able to:

- define base, power, exponent and index
- define a logarithm
- distinguish between common and Napierian (i.e. hyperbolic or natural) logarithms
- evaluate logarithms to any base
- state the laws of logarithms
- simplify logarithmic expressions
- solve equations involving logarithms
- solve indicial equations
- sketch graphs of $\log_{10} x$ and $\log_e x$

15.1 Introduction to logarithms

With the use of calculators firmly established, logarithmic tables are now rarely used for calculation. However, the theory of logarithms is important, for there are several scientific and engineering laws that involve the rules of logarithms.

From Chapter 7, we know that $\qquad 16 = 2^4$

The number 4 is called the **power** or the **exponent** or the **index**. In the expression 2^4, the number 2 is called the **base**.

In another example, we know that $\qquad 64 = 8^2$

In this example, 2 is the power, or exponent, or index. The number 8 is the base.

What is a logarithm?

Consider the expression $16 = 2^4$

An alternative, yet equivalent, way of writing this expression is $\log_2 16 = 4$

This is stated as 'log to the base 2 of 16 equals 4'

We see that the logarithm is the same as the power or index in the original expression. It is the base in the original expression that becomes the base of the logarithm.

The two statements $16 = 2^4$ and

$\log_2 16 = 4$ are equivalent

If we write either of them, we are automatically implying the other.

In general, if a number y can be written in the form a^x, then the index x is called the 'logarithm of y to the base of a', i.e.

if $y = a^x$ then $x = \log_a y$

In another example, if we write down that $64 = 8^2$ then the equivalent statement using logarithms is $\log_8 64 = 2$. In another example, if we write down that $\log_3 27 = 3$ then the equivalent statement using powers is $3^3 = 27$. So the two sets of statements, one involving powers and one involving logarithms, are equivalent.

Common logarithms

From the above, if we write down that $1000 = 10^3$, then $3 = \log_{10} 1000$. This may be checked using the 'log' button on your calculator.

Logarithms having a base of 10 are called **common logarithms** and $\log_{10}$ is usually abbreviated to lg. The following values may be checked using a calculator.

$$\lg 27.5 = 1.4393\ldots$$

$$\lg 378.1 = 2.5776\ldots$$

$$\lg 0.0204 = -1.6903\ldots$$

Napierian logarithms

Logarithms having a base of e (where e is a mathematical constant approximately equal to 2.7183) are called **hyperbolic**, **Napierian** or **natural logarithms**, and $\log_e$ is usually abbreviated to ln. The following values may be checked using a calculator.

$$\ln 3.65 = 1.2947\ldots$$

$$\ln 417.3 = 6.0338\ldots$$

$$\ln 0.182 = -1.7037\ldots$$

Napierian logarithms are explained further in Chapter 16.

Here are some worked problems to help understanding of logarithms.

Problem 1. Evaluate $\log_3 9$

Let $x = \log_3 9$ then $3^x = 9$ from the definition of a logarithm,

i.e. $3^x = 3^2$, from which $x = 2$

Hence, $\log_3 9 = 2$

Problem 2. Evaluate $\log_{10} 10$

Let $x = \log_{10} 10$ then $10^x = 10$ from the definition of a logarithm,

i.e. $10^x = 10^1$, from which $x = 1$

Hence, $\log_{10} 10 = 1$ (which may be checked using a calculator)

Problem 3. Evaluate $\log_{16} 8$

Let $x = \log_{16} 8$ then $16^x = 8$ from the definition of a logarithm,

i.e. $(2^4)^x = 2^3$ i.e. $2^{4x} = 2^3$ from the laws of indices,

from which, $4x = 3$ and $x = \dfrac{3}{4}$

Hence, $\log_{16} 8 = \dfrac{3}{4}$

Problem 4. Evaluate $\lg 0.001$

Let $x = \lg 0.001 = \log_{10} 0.001$
then $10^x = 0.001$
i.e. $10^x = 10^{-3}$ from which, $x = -3$

Hence, $\lg 0.001 = -3$ (which may be checked using a calculator)

Problem 5. Evaluate $\ln e$

Let $x = \ln e = \log_e e$ then $e^x = e$

i.e. $e^x = e^1$,
from which $x = 1$

Hence, $\ln e = 1$ (which may be checked by a calculator)

Problem 6. Evaluate $\log_3 \dfrac{1}{81}$

Let $x = \log_3 \dfrac{1}{81}$ then $3^x = \dfrac{1}{81} = \dfrac{1}{3^4} = 3^{-4}$

from which $\qquad\qquad x = -4$

Hence, $\qquad\qquad \log_3 \dfrac{1}{81} = -4$

Problem 7. Solve the equation $\lg x = 3$

If $\lg x = 3$ then $\log_{10} x = 3$

and $\qquad\qquad x = 10^3$ i.e. $x = 1000$

Problem 8. Solve the equation $\log_2 x = 5$

If $\log_2 x = 5$ then $x = 2^5 = 32$

Problem 9. Solve the equation $\log_5 x = -2$

If $\log_5 x = -2$ then $x = 5^{-2} = \dfrac{1}{5^2} = \dfrac{1}{25}$

Now try the following Practice Exercise

Practice Exercise 65   Laws of logarithms (answers on page 1155)

In Problems 1 to 11, evaluate the given expressions.

1. $\log_{10} 10000$ 2. $\log_2 16$ 3. $\log_5 125$

4. $\log_2 \dfrac{1}{8}$ 5. $\log_8 2$ 6. $\log_7 343$

7. $\lg 100$ 8. $\lg 0.01$ 9. $\log_4 8$

10. $\log_{27} 3$ 11. $\ln e^2$

In Problems 12 to 18, solve the equations.

12. $\log_{10} x = 4$ 13. $\lg x = 5$

14. $\log_3 x = 2$ 15. $\log_4 x = -2\dfrac{1}{2}$

16. $\lg x = -2$ 17. $\log_8 x = -\dfrac{4}{3}$

18. $\ln x = 3$

15.2   Laws of logarithms

There are three laws of logarithms, which apply to any base:

(1) To multiply two numbers:

$$\log (A \times B) = \log A + \log B$$

The following may be checked by using a calculator.

$\qquad$ $\lg 10 = 1$

Also, $\quad \lg 5 + \lg 2 = 0.69897\ldots + 0.301029\ldots = 1$

Hence, $\lg (5 \times 2) = \lg 10 = \lg 5 + \lg 2$

(2) To divide two numbers:

$$\log \left(\frac{A}{B}\right) = \log A - \log B$$

The following may be checked using a calculator.

$$\ln \left(\frac{5}{2}\right) = \ln 2.5 = 0.91629\ldots$$

Also, $\quad \ln 5 - \ln 2 = 1.60943\ldots - 0.69314\ldots$

$$= 0.91629\ldots$$

Hence, $\quad \ln \left(\dfrac{5}{2}\right) = \ln 5 - \ln 2$

(3) To raise a number to a power:

$$\log A^n = n \log A$$

The following may be checked using a calculator.

$$\lg 5^2 = \lg 25 = 1.39794\ldots$$

Also, $\quad 2 \lg 5 = 2 \times 0.69897\ldots = 1.39794\ldots$

Hence, $\quad \lg 5^2 = 2 \lg 5$

Here are some worked problems to help understanding of the laws of logarithms.

Problem 10. Write $\log 4 + \log 7$ as the logarithm of a single number

$\log 4 + \log 7 = \log(7 \times 4)$ $\qquad$ by the first law of logarithms

$\qquad\qquad\qquad = \log 28$

Problem 11. Write $\log 16 - \log 2$ as the logarithm of a single number

$\log 16 - \log 2 = \log \left(\dfrac{16}{2} \right)$ by the second law of logarithms

$\qquad\qquad\qquad = \mathbf{log\,8}$

Problem 12. Write $2 \log 3$ as the logarithm of a single number

$2 \log 3 = \log 3^2$ by the third law of logarithms

$\qquad\quad = \mathbf{log\,9}$

Problem 13. Write $\dfrac{1}{2} \log 25$ as the logarithm of a single number

$\dfrac{1}{2} \log 25 = \log 25^{\frac{1}{2}}$ by the third law of logarithms

$\qquad\qquad = \log \sqrt{25} = \mathbf{log\,5}$

Problem 14. Simplify $\log 64 - \log 128 + \log 32$

$64 = 2^6, 128 = 2^7$ and $32 = 2^5$

Hence, $\log 64 - \log 128 + \log 32$

$\qquad\qquad = \log 2^6 - \log 2^7 + \log 2^5$

$\qquad\qquad = 6 \log 2 - 7 \log 2 + 5 \log 2$

$\qquad\qquad\qquad$ by the third law of logarithms

$\qquad\qquad = \mathbf{4 \log 2}$

Problem 15. Write $\dfrac{1}{2} \log 16 + \dfrac{1}{3} \log 27 - 2 \log 5$ as the logarithm of a single number

$\dfrac{1}{2} \log 16 + \dfrac{1}{3} \log 27 - 2 \log 5$

$\qquad = \log 16^{\frac{1}{2}} + \log 27^{\frac{1}{3}} - \log 5^2$

$\qquad\qquad$ by the third law of logarithms

$\qquad = \log \sqrt{16} + \log \sqrt[3]{27} - \log 25$

$\qquad\qquad$ by the laws of indices

$\qquad = \log 4 + \log 3 - \log 25$

$= \log \left(\dfrac{4 \times 3}{25} \right)$ by the first and second laws of logarithms

$= \log \left(\dfrac{12}{25} \right) = \mathbf{log\,0.48}$

Problem 16. Write (a) $\log 30$ (b) $\log 450$ in terms of $\log 2$, $\log 3$ and $\log 5$ to any base

(a) $\log 30 = \log(2 \times 15) = \log(2 \times 3 \times 5)$

$\qquad\qquad = \mathbf{log\,2 + log\,3 + log\,5}$

$\qquad\qquad\qquad$ by the first law of logarithms

(b) $\log 450 = \log(2 \times 225) = \log(2 \times 3 \times 75)$

$\qquad\qquad = \log(2 \times 3 \times 3 \times 25)$

$\qquad\qquad = \log(2 \times 3^2 \times 5^2)$

$\qquad\qquad = \log 2 + \log 3^2 + \log 5^2$

$\qquad\qquad\qquad$ by the first law of logarithms

i.e. $\log 450 = \mathbf{log\,2 + 2\,log\,3 + 2\,log\,5}$

$\qquad\qquad\qquad$ by the third law of logarithms

Problem 17. Write $\log \left(\dfrac{8 \times \sqrt[4]{5}}{81} \right)$ in terms of $\log 2$, $\log 3$ and $\log 5$ to any base

$\log \left(\dfrac{8 \times \sqrt[4]{5}}{81} \right) = \log 8 + \log \sqrt[4]{5} - \log 81$

$\qquad\qquad\qquad$ by the first and second laws of logarithms

$\qquad\qquad = \log 2^3 + \log 5^{\frac{1}{4}} - \log 3^4$

$\qquad\qquad\qquad$ by the laws of indices

i.e. $\log \left(\dfrac{8 \times \sqrt[4]{5}}{81} \right) = \mathbf{3\,log\,2 + \dfrac{1}{4} log\,5 - 4\,log\,3}$

$\qquad\qquad\qquad$ by the third law of logarithms

Problem 18. Evaluate

$$\frac{\log 25 - \log 125 + \dfrac{1}{2} \log 625}{3 \log 5}$$

$$\frac{\log 25 - \log 125 + \frac{1}{2}\log 625}{3\log 5}$$

$$= \frac{\log 5^2 - \log 5^3 + \frac{1}{2}\log 5^4}{3\log 5}$$

$$= \frac{2\log 5 - 3\log 5 + \frac{4}{2}\log 5}{3\log 5}$$

$$= \frac{1\log 5}{3\log 5} = \frac{1}{3}$$

Problem 19. Solve the equation
$$\log(x-1) + \log(x+8) = 2\log(x+2)$$

$$\text{LHS} = \log(x-1) + \log(x+8) = \log(x-1)(x+8)$$
$$\text{from the first law of logarithms}$$

$$= \log(x^2 + 7x - 8)$$

$$\text{RHS} = 2\log(x+2) = \log(x+2)^2$$
$$\text{from the first law of logarithms}$$

$$= \log(x^2 + 4x + 4)$$

Hence, $\qquad \log(x^2 + 7x - 8) = \log(x^2 + 4x + 4)$

from which, $\qquad x^2 + 7x - 8 = x^2 + 4x + 4$

i.e. $\qquad\qquad 7x - 8 = 4x + 4$

i.e. $\qquad\qquad 3x = 12$

and $\qquad\qquad x = 4$

Problem 20. Solve the equation $\frac{1}{2}\log 4 = \log x$

$$\frac{1}{2}\log 4 = \log 4^{\frac{1}{2}} \quad \text{from the third law of logarithms}$$

$$= \log\sqrt{4} \quad \text{from the laws of indices}$$

Hence, $\qquad \frac{1}{2}\log 4 = \log x$

becomes $\qquad \log\sqrt{4} = \log x$

i.e. $\qquad\qquad \log 2 = \log x$

from which, $\qquad 2 = x$

i.e. the solution of the equation is $x = 2$

Problem 21. Solve the equation
$$\log(x^2 - 3) - \log x = \log 2$$

$$\log(x^2 - 3) - \log x = \log\left(\frac{x^2 - 3}{x}\right) \text{ from the second}$$
$$\text{law of logarithms}$$

Hence, $\qquad \log\left(\frac{x^2 - 3}{x}\right) = \log 2$

from which, $\qquad \frac{x^2 - 3}{x} = 2$

Rearranging gives $\qquad x^2 - 3 = 2x$

and $\qquad\qquad x^2 - 2x - 3 = 0$

Factorising gives $\quad (x-3)(x+1) = 0$

from which, $\qquad\qquad x = 3 \text{ or } x = -1$

$x = -1$ is not a valid solution since the logarithm of a negative number has no real root.

Hence, the solution of the equation is $x = 3$

Now try the following Practice Exercise

**Practice Exercise 66 Laws of logarithms
(answers on page 1155)**

In Problems 1 to 11, write as the logarithm of a single number.

1. $\log 2 + \log 3$ $\qquad\qquad$ 2. $\log 3 + \log 5$

3. $\log 3 + \log 4 - \log 6$

4. $\log 7 + \log 21 - \log 49$

5. $2\log 2 + \log 3$ $\qquad\qquad$ 6. $2\log 2 + 3\log 5$

7. $2\log 5 - \frac{1}{2}\log 81 + \log 36$

8. $\frac{1}{3}\log 8 - \frac{1}{2}\log 81 + \log 27$

9. $\frac{1}{2}\log 4 - 2\log 3 + \log 45$

10. $\frac{1}{4}\log 16 + 2\log 3 - \log 18$

11. $2\log 2 + \log 5 - \log 10$

Simplify the expressions given in Problems 12 to 14.

12. $\log 27 - \log 9 + \log 81$

13. $\log 64 + \log 32 - \log 128$

14. $\log 8 - \log 4 + \log 32$

Evaluate the expressions given in Problems 15 and 16.

15. $\dfrac{\dfrac{1}{2}\log 16 - \dfrac{1}{3}\log 8}{\log 4}$

16. $\dfrac{\log 9 - \log 3 + \dfrac{1}{2}\log 81}{2\log 3}$

Solve the equations given in Problems 17 to 22.

17. $\log x^4 - \log x^3 = \log 5x - \log 2x$

18. $\log 2t^3 - \log t = \log 16 + \log t$

19. $2\log b^2 - 3\log b = \log 8b - \log 4b$

20. $\log(x+1) + \log(x-1) = \log 3$

21. $\dfrac{1}{3}\log 27 = \log(0.5a)$

22. $\log(x^2 - 5) - \log x = \log 4$

15.3   Indicial equations

The laws of logarithms may be used to solve certain equations involving powers, called **indicial equations**.

For example, to solve, say, $3^x = 27$, logarithms to a base of 10 are taken of both sides,

i.e. $\log_{10} 3^x = \log_{10} 27$

and $x\log_{10} 3 = \log_{10} 27$

  by the third law of logarithms

Rearranging gives $x = \dfrac{\log_{10} 27}{\log_{10} 3} = \dfrac{1.43136\ldots}{0.47712\ldots}$

  $= 3$ which may be readily checked.

$\left(\text{Note, } \dfrac{\log 27}{\log 3} \text{ is } \textbf{not} \text{ equal to } \log\dfrac{27}{3}\right)$

Problem 22.   Solve the equation $2^x = 5$, correct to 4 significant figures

Taking logarithms to base 10 of both sides of $2^x = 5$ gives

$$\log_{10} 2^x = \log_{10} 5$$

i.e. $x\log_{10} 2 = \log_{10} 5$

  by the third law of logarithms

Rearranging gives $x = \dfrac{\log_{10} 5}{\log_{10} 2} = \dfrac{0.6989700\ldots}{0.3010299\ldots}$

  $= \textbf{2.322}$, correct to 4 significant figures.

Problem 23.   Solve the equation $2^{x+1} = 3^{2x-5}$ correct to 2 decimal places

Taking logarithms to base 10 of both sides gives

$$\log_{10} 2^{x+1} = \log_{10} 3^{2x-5}$$

i.e. $(x+1)\log_{10} 2 = (2x-5)\log_{10} 3$

$x\log_{10} 2 + \log_{10} 2 = 2x\log_{10} 3 - 5\log_{10} 3$

$x(0.3010) + (0.3010) = 2x(0.4771) - 5(0.4771)$

i.e. $0.3010x + 0.3010 = 0.9542x - 2.3855$

Hence, $2.3855 + 0.3010 = 0.9542x - 0.3010x$

$2.6865 = 0.6532x$

from which $x = \dfrac{2.6865}{0.6532} = \textbf{4.11}$,

  correct to 2 decimal places.

Problem 24.   Solve the equation $x^{2.7} = 34.68$, correct to 4 significant figures

Taking logarithms to base 10 of both sides gives

$$\log_{10} x^{2.7} = \log_{10} 34.68$$

$2.7\log_{10} x = \log_{10} 34.68$

Hence, $\log_{10} x = \dfrac{\log_{10} 34.68}{2.7} = 0.57040$

Thus, $\quad x = \text{antilog } 0.57040 = 10^{0.57040}$

$\qquad\qquad = \textbf{3.719}$, correct to 4 significant figures.

Problem 25. A gas follows the polytropic law $PV^{1.25} = C$. Determine the new volume of the gas, given that its original pressure and volume are 101 kPa and 0.35 m³, respectively, and its final pressure is 1.18 MPa.

If $PV^{1.25} = C$ then $P_1 V_1^{1.25} = P_2 V_2^{1.25}$

$P_1 = 101$ kPa, $P_2 = 1.18$ MPa and $V_1 = 0.35$ m³

$$P_1 V_1^{1.25} = P_2 V_2^{1.25}$$

i.e. $\quad (101 \times 10^3)(0.35)^{1.25} = (1.18 \times 10^6) V_2^{1.25}$

from which, $V_2^{1.25} = \dfrac{(101 \times 10^3)(0.35)^{1.25}}{(1.18 \times 10^6)} = 0.02304$

Taking logarithms of both sides of the equation gives:

$$\log_{10} V_2^{1.25} = \log_{10} 0.02304$$

i.e. $\quad 1.25 \log_{10} V_2 = \log_{10} 0.02304$ from the third law of logarithms

and $\qquad \log_{10} V_2 = \dfrac{\log_{10} 0.02304}{1.25} = -1.3100$

from which, $\quad$ **volume, $V_2 = 10^{-1.3100} = \textbf{0.049 m}^3$**

Now try the following Practice Exercise

Practice Exercise 67 $\quad$ **Indicial equations (answers on page 1155)**

In Problems 1 to 8, solve the indicial equations for x, each correct to 4 significant figures.

1. $3^x = 6.4$ $\qquad$ 2. $2^x = 9$

3. $2^{x-1} = 3^{2x-1}$ $\qquad$ 4. $x^{1.5} = 14.91$

5. $25.28 = 4.2^x$ $\qquad$ 6. $4^{2x-1} = 5^{x+2}$

7. $x^{-0.25} = 0.792$ $\qquad$ 8. $0.027^x = 3.26$

9. The decibel gain n of an amplifier is given by $n = 10 \log_{10}\left(\dfrac{P_2}{P_1}\right)$, where P_1 is the power input and P_2 is the power output. Find the power gain $\dfrac{P_2}{P_1}$ when $n = 25$ decibels.

10. A gas follows the polytropic law $PV^{1.26} = C$. Determine the new volume of the gas, given that its original pressure and volume are 101 kPa and 0.42 m³, respectively, and its final pressure is 1.25 MPa.

15.4 Graphs of logarithmic functions

A graph of $y = \log_{10} x$ is shown in Figure 15.1 and a graph of $y = \log_e x$ is shown in Figure 15.2. Both can be seen to be of similar shape; in fact, the same general shape occurs for a logarithm to any base.

In general, with a logarithm to any base, a, it is noted that

(a) $\quad \textbf{log}_a\,\textbf{1} = \textbf{0}$

Let $\log_a = x$ then $a^x = 1$ from the definition of the logarithm.

If $a^x = 1$ then $x = 0$ from the laws of logarithms.

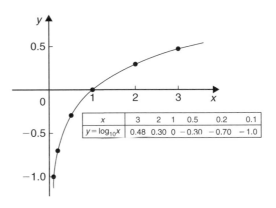

x	3	2	1	0.5	0.2	0.1
$y = \log_{10}x$	0.48	0.30	0	-0.30	-0.70	-1.0

Figure 15.1

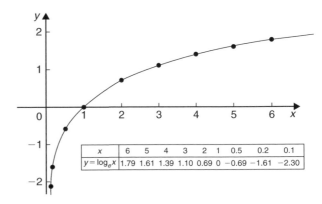

x	6	5	4	3	2	1	0.5	0.2	0.1
$y = \log_e x$	1.79	1.61	1.39	1.10	0.69	0	-0.69	-1.61	-2.30

Figure 15.2

Hence, $\log_a 1 = 0$. In the graphs it is seen that $\log_{10} 1 = 0$ and $\log_e 1 = 0$

(b) $\log_a a = 1$

Let $\log_a a = x$ then $a^x = a$ from the definition of a logarithm.

If $a^x = a$ then $x = 1$

Hence, $\log_a a = 1$. (Check with a calculator that $\log_{10} 10 = 1$ and $\log_e e = 1$)

(c) $\log_a 0 \to -\infty$

Let $\log_a 0 = x$ then $a^x = 0$ from the definition of a logarithm.

If $a^x = 0$, and a is a positive real number, then x must approach minus infinity. (For example, check with a calculator, $2^{-2} = 0.25, 2^{-20} = 9.54 \times 10^{-7}, 2^{-200} = 6.22 \times 10^{-61}$, and so on.)

Hence, $\log_a 0 \to -\infty$

For fully worked solutions to each of the problems in Practice Exercises 65 to 67 in this chapter, go to the website:
www.routledge.com/cw/bird

Chapter 16

Exponential functions

Why it is important to understand: **Exponential functions**

Exponential functions are used in engineering, physics, biology and economics. There are many quantities that grow exponentially; some examples are population, compound interest and charge in a capacitor. With exponential growth, the rate of growth increases as time increases. We also have exponential decay; some examples are radioactive decay, atmospheric pressure, Newton's law of cooling and linear expansion. Understanding and using exponential functions is important in many branches of engineering.

At the end of this chapter, you should be able to:

- evaluate exponential functions using a calculator
- state the exponential series for e^x
- plot graphs of exponential functions
- evaluate Napierian logarithms using a calculator
- solve equations involving Napierian logarithms
- appreciate the many examples of laws of growth and decay in engineering and science
- perform calculations involving the laws of growth and decay

---

16.1 Introduction to exponential functions

An exponential function is one which contains e^x, e being a constant called the exponent and having an approximate value of 2.7183. The exponent arises from the natural laws of growth and decay and is used as a base for natural or Napierian logarithms.

The most common method of evaluating an exponential function is by using a scientific notation **calculator**. Use your calculator to check the following values.

$e^1 = 2.7182818$, correct to 8 significant figures,

$e^{-1.618} = 0.1982949$, correct to 7 significant figures,

$e^{0.12} = 1.1275$, correct to 5 significant figures,

$e^{-1.47} = 0.22993$, correct to 5 decimal places,

$e^{-0.431} = 0.6499$, correct to 4 decimal places,

$e^{9.32} = 11159$, correct to 5 significant figures,

$e^{-2.785} = 0.0617291$, correct to 7 decimal places.

Problem 1. Evaluate the following correct to 4 decimal places using a calculator:

$$0.0256(e^{5.21} - e^{2.49})$$

$0.0256(e^{5.21} - e^{2.49})$

$= 0.0256(183.094058\ldots - 12.0612761\ldots)$

$= \mathbf{4.3784}$, correct to 4 decimal places.

Problem 2. Evaluate the following correct to 4 decimal places using a calculator:

$$5\left(\frac{e^{0.25} - e^{-0.25}}{e^{0.25} + e^{-0.25}}\right)$$

$$5\left(\frac{e^{0.25} - e^{-0.25}}{e^{0.25} + e^{-0.25}}\right)$$

$$= 5\left(\frac{1.28402541\ldots - 0.77880078\ldots}{1.28402541\ldots + 0.77880078\ldots}\right)$$

$$= 5\left(\frac{0.5052246\ldots}{2.0628262\ldots}\right)$$

$$= \mathbf{1.2246}, \text{ correct to 4 decimal places.}$$

Problem 3. The instantaneous voltage v in a capacitive circuit is related to time t by the equation $v = Ve^{-t/CR}$ where V, C and R are constants. Determine v, correct to 4 significant figures, when $t = 50$ ms, $C = 10\,\mu$F, $R = 47\,$kΩ and $V = 300$ volts

$$v = Ve^{-t/CR} = 300e^{(-50\times 10^{-3})/(10\times 10^{-6}\times 47\times 10^{3})}$$

Using a calculator, $v = 300e^{-0.1063829\ldots}$

$$= 300(0.89908025\ldots)$$

$$= \mathbf{269.7\,volts.}$$

Now try the following Practice Exercise

Practice Exercise 68   Evaluating exponential functions (answers on page 1155)

1. Evaluate the following, correct to 4 significant figures.
    (a) $e^{-1.8}$ (b) $e^{-0.78}$ (c) e^{10}

2. Evaluate the following, correct to 5 significant figures.
    (a) $e^{1.629}$ (b) $e^{-2.7483}$ (c) $0.62e^{4.178}$

In Problems 3 and 4, evaluate correct to 5 decimal places.

3. (a) $\dfrac{1}{7}e^{3.4629}$ (b) $8.52e^{-1.2651}$

    (c) $\dfrac{5e^{2.6921}}{3e^{1.1171}}$

4. (a) $\dfrac{5.6823}{e^{-2.1347}}$ (b) $\dfrac{e^{2.1127} - e^{-2.1127}}{2}$

    (c) $\dfrac{4\left(e^{-1.7295} - 1\right)}{e^{3.6817}}$

5. The length of a bar l at a temperature θ is given by $l = l_0 e^{\alpha\theta}$, where l_0 and α are constants. Evaluate l, correct to 4 significant figures, where $l_0 = 2.587$, $\theta = 321.7$ and $\alpha = 1.771 \times 10^{-4}$

6. When a chain of length $2L$ is suspended from two points, $2D$ metres apart on the same horizontal level, $D = k\left\{\ln\left(\dfrac{L + \sqrt{L^2 + k^2}}{k}\right)\right\}$

    Evaluate D when $k = 75$ m and $L = 180$ m.

16.2   The power series for e^x

The value of e^x can be calculated to any required degree of accuracy since it is defined in terms of the following **power series**:

$$e^x = 1 + x + \frac{x^2}{2!} + \frac{x^3}{3!} + \frac{x^4}{4!} + \cdots \qquad (1)$$

(where $3! = 3 \times 2 \times 1$ and is called '**factorial** 3').
The series is valid for all values of x.
The series is said to **converge**; i.e. if all the terms are added, an actual value for e^x (where x is a real number) is obtained. The more terms that are taken, the closer will be the value of e^x to its actual value. The value of the exponent e, correct to say 4 decimal places, may be determined by substituting $x = 1$ in the power series of equation (1). Thus,

$$e^1 = 1 + 1 + \frac{(1)^2}{2!} + \frac{(1)^3}{3!} + \frac{(1)^4}{4!} + \frac{(1)^5}{5!} + \frac{(1)^6}{6!}$$

$$+ \frac{(1)^7}{7!} + \frac{(1)^8}{8!} + \cdots$$

$$= 1 + 1 + 0.5 + 0.16667 + 0.04167 + 0.00833$$

$$+ 0.00139 + 0.00020 + 0.00002 + \cdots$$

$$= 2.71828$$

i.e. $\mathbf{e = 2.7183}$, correct to 4 decimal places.

The value of $e^{0.05}$, correct to say 8 significant figures, is found by substituting $x = 0.05$ in the power series for e^x. Thus,

$$e^{0.05} = 1 + 0.05 + \frac{(0.05)^2}{2!} + \frac{(0.05)^3}{3!} + \frac{(0.05)^4}{4!}$$
$$+ \frac{(0.05)^5}{5!} + \cdots$$
$$= 1 + 0.05 + 0.00125 + 0.000020833$$
$$+ 0.000000260 + 0.0000000026$$

i.e. $\mathbf{e^{0.05} = 1.0512711}$, correct to 8 significant figures.

In this example, successive terms in the series grow smaller very rapidly and it is relatively easy to determine the value of $e^{0.05}$ to a high degree of accuracy. However, when x is nearer to unity or larger than unity, a very large number of terms are required for an accurate result.

If, in the series of equation (1), x is replaced by $-x$, then

$$e^{-x} = 1 + (-x) + \frac{(-x)^2}{2!} + \frac{(-x)^3}{3!} + \cdots$$

i.e. $$\mathbf{e^{-x} = 1 - x + \frac{x^2}{2!} - \frac{x^3}{3!} + \cdots}$$

In a similar manner the power series for e^x may be used to evaluate any exponential function of the form ae^{kx}, where a and k are constants.

In the series of equation (1), let x be replaced by kx. Then

$$ae^{kx} = a\left\{1 + (kx) + \frac{(kx)^2}{2!} + \frac{(kx)^3}{3!} + \cdots\right\}$$

Thus, $$5e^{2x} = 5\left\{1 + (2x) + \frac{(2x)^2}{2!} + \frac{(2x)^3}{3!} + \cdots\right\}$$
$$= 5\left\{1 + 2x + \frac{4x^2}{2} + \frac{8x^3}{6} + \cdots\right\}$$

i.e. $$5e^{2x} = 5\left\{1 + 2x + 2x^2 + \frac{4}{3}x^3 + \cdots\right\}$$

Problem 4.   Determine the value of $5e^{0.5}$, correct to 5 significant figures, by using the power series for e^x

From equation (1),

$$e^x = 1 + x + \frac{x^2}{2!} + \frac{x^3}{3!} + \frac{x^4}{4!} + \cdots$$

Hence, $$e^{0.5} = 1 + 0.5 + \frac{(0.5)^2}{(2)(1)} + \frac{(0.5)^3}{(3)(2)(1)}$$
$$+ \frac{(0.5)^4}{(4)(3)(2)(1)} + \frac{(0.5)^5}{(5)(4)(3)(2)(1)}$$
$$+ \frac{(0.5)^6}{(6)(5)(4)(3)(2)(1)}$$
$$= 1 + 0.5 + 0.125 + 0.020833$$
$$+ 0.0026042 + 0.0002604$$
$$+ 0.0000217$$

i.e. $e^{0.5} = 1.64872$, correct to 6 significant figures.

Hence, $\mathbf{5e^{0.5}} = 5(1.64872) = \mathbf{8.2436}$, correct to 5 significant figures.

Problem 5.   Determine the value of $3e^{-1}$, correct to 4 decimal places, using the power series for e^x

Substituting $x = -1$ in the power series

$$e^x = 1 + x + \frac{x^2}{2!} + \frac{x^3}{3!} + \frac{x^4}{4!} + \cdots$$

gives $$e^{-1} = 1 + (-1) + \frac{(-1)^2}{2!} + \frac{(-1)^3}{3!}$$
$$+ \frac{(-1)^4}{4!} + \cdots$$
$$= 1 - 1 + 0.5 - 0.166667 + 0.041667$$
$$- 0.008333 + 0.001389$$
$$- 0.000198 + \cdots$$
$$= 0.367858 \text{ correct to 6 decimal places.}$$

Hence, $\mathbf{3e^{-1}} = (3)(0.367858) = \mathbf{1.1036}$, correct to 4 decimal places.

Problem 6.   Expand $e^x(x^2 - 1)$ as far as the term in x^5

The power series for e^x is

$$e^x = 1 + x + \frac{x^2}{2!} + \frac{x^3}{3!} + \frac{x^4}{4!} + \frac{x^5}{5!} + \cdots$$

Hence,

$$e^x(x^2 - 1)$$

$$= \left(1 + x + \frac{x^2}{2!} + \frac{x^3}{3!} + \frac{x^4}{4!} + \frac{x^5}{5!} + \cdots\right)(x^2 - 1)$$

$$= \left(x^2 + x^3 + \frac{x^4}{2!} + \frac{x^5}{3!} + \cdots\right)$$

$$\quad - \left(1 + x + \frac{x^2}{2!} + \frac{x^3}{3!} + \frac{x^4}{4!} + \frac{x^5}{5!} + \cdots\right)$$

Grouping like terms gives

$$e^x(x^2 - 1)$$

$$= -1 - x + \left(x^2 - \frac{x^2}{2!}\right) + \left(x^3 - \frac{x^3}{3!}\right)$$

$$\quad + \left(\frac{x^4}{2!} - \frac{x^4}{4!}\right) + \left(\frac{x^5}{3!} - \frac{x^5}{5!}\right) + \cdots$$

$$= \mathbf{-1 - x + \frac{1}{2}x^2 + \frac{5}{6}x^3 + \frac{11}{24}x^4 + \frac{19}{120}x^5}$$

when expanded as far as the term in x^5

Now try the following Practice Exercise

**Practice Exercise 69   Power series for e^x
(answers on page 1155)**

1. Evaluate $5.6e^{-1}$, correct to 4 decimal places, using the power series for e^x.

2. Use the power series for e^x to determine, correct to 4 significant figures, (a) e^2 (b) $e^{-0.3}$ and check your results using a calculator.

3. Expand $(1 - 2x)e^{2x}$ as far as the term in x^4.

4. Expand $(2e^{x^2})(x^{1/2})$ to six terms.

16.3 Graphs of exponential functions

Values of e^x and e^{-x} obtained from a calculator, correct to 2 decimal places, over a range $x = -3$ to $x = 3$, are shown in Table 16.1.

Figure 16.1 shows graphs of $y = e^x$ and $y = e^{-x}$.

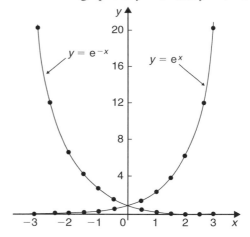

Figure 16.1

Problem 7. Plot a graph of $y = 2e^{0.3x}$ over a range of $x = -3$ to $x = 3$. Then determine the value of y when $x = 2.2$ and the value of x when $y = 1.6$

A table of values is drawn up as shown below.

x	-3	-2	-1	0	1	2	3
$2e^{0.3x}$	0.81	1.10	1.48	2.00	2.70	3.64	4.92

A graph of $y = 2e^{0.3x}$ is shown plotted in Figure 16.2. From the graph, **when $x = 2.2$, $y = 3.87$** and **when $y = 1.6$, $x = -0.74$**

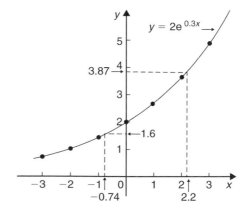

Figure 16.2

Table 16.1

x	-3.0	-2.5	-2.0	-1.5	-1.0	-0.5	0	0.5	1.0	1.5	2.0	2.5	3.0
e^x	0.05	0.08	0.14	0.22	0.37	0.61	1.00	1.65	2.72	4.48	7.39	12.18	20.09
e^{-x}	20.09	12.18	7.39	4.48	2.72	1.65	1.00	0.61	0.37	0.22	0.14	0.08	0.05

Problem 8. Plot a graph of $y = \frac{1}{3}e^{-2x}$ over the range $x = -1.5$ to $x = 1.5$. Determine from the graph the value of y when $x = -1.2$ and the value of x when $y = 1.4$

A table of values is drawn up as shown below.

x	−1.5	−1.0	−0.5	0	0.5	1.0	1.5
$\frac{1}{3}e^{-2x}$	6.70	2.46	0.91	0.33	0.12	0.05	0.02

A graph of $\frac{1}{3}e^{-2x}$ is shown in Figure 16.3.
From the graph, **when $x = -1.2$, $y = 3.67$** and **when $y = 1.4$, $x = -0.72$**

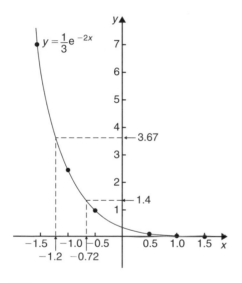

Figure 16.3

Problem 9. The decay of voltage v volts across a capacitor at time t seconds is given by $v = 250e^{-t/3}$. Draw a graph showing the natural decay curve over the first 6 seconds. From the graph, find (a) the voltage after 3.4 s and (b) the time when the voltage is 150 V

A table of values is drawn up as shown below.

t	0	1	2	3
$e^{-t/3}$	1.00	0.7165	0.5134	0.3679
$v = 250e^{-t/3}$	250.0	179.1	128.4	91.97

t	4	5	6
$e^{-t/3}$	0.2636	0.1889	0.1353
$v = 250e^{-t/3}$	65.90	47.22	33.83

The natural decay curve of $v = 250e^{-t/3}$ is shown in Figure 16.4.

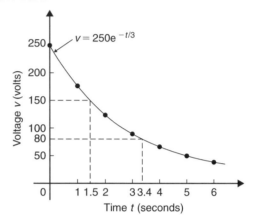

Figure 16.4

From the graph,

(a) when time $t = 3.4$ s, **voltage $v = 80$ V**, and

(b) when **voltage $v = 150$ V, time $t = 1.5$ s**

Now try the following Practice Exercise

Practice Exercise 70   Exponential graphs (answers on page 1155)

1. Plot a graph of $y = 3e^{0.2x}$ over the range $x = -3$ to $x = 3$. Hence determine the value of y when $x = 1.4$ and the value of x when $y = 4.5$

2. Plot a graph of $y = \frac{1}{2}e^{-1.5x}$ over a range $x = -1.5$ to $x = 1.5$ and then determine the value of y when $x = -0.8$ and the value of x when $y = 3.5$

3. In a chemical reaction the amount of starting material C cm^3 left after t minutes is given by $C = 40e^{-0.006t}$. Plot a graph of C against t and determine

    (a) the concentration C after 1 hour.

    (b) the time taken for the concentration to decrease by half.

4. The rate at which a body cools is given by $\theta = 250e^{-0.05t}$ where the excess of temperature of a body above its surroundings at time t minutes is $\theta\,°C$. Plot a graph showing the natural decay curve for the first hour of cooling. Then determine

(a) the temperature after 25 minutes.

(b) the time when the temperature is $195\,°C$.

16.4 Napierian logarithms

Logarithms having a base of e are called **hyperbolic**, **Napierian** or **natural logarithms** and the Napierian logarithm of x is written as $\log_e x$ or, more commonly, as $\ln x$. Logarithms were invented by **John Napier***, a Scotsman (1550–1617).

The most common method of evaluating a Napierian logarithm is by a scientific notation **calculator**. Use your calculator to check the following values:

$\ln 4.328 = 1.46510554\ldots = 1.4651$, correct to 4 decimal places

* **Who was Napier? John Napier of Merchiston** (1550–4 April 1617) is best known as the discoverer of logarithms. The inventor of the so-called 'Napier's bones', Napier also made common the use of the decimal point in arithmetic and mathematics. To find out more go to **www.routledge.com/cw/bird**

$\ln 1.812 = 0.59443$, correct to 5 significant figures

$\ln 1 = 0$

$\ln 527 = 6.2672$, correct to 5 significant figures

$\ln 0.17 = -1.772$, correct to 4 significant figures

$\ln 0.00042 = -7.77526$, correct to 6 significant figures

$\ln e^3 = 3$

$\ln e^1 = 1$

From the last two examples we can conclude that

$$\log_e e^x = x$$

This is useful when **solving equations involving exponential functions**. For example, to solve $e^{3x} = 7$, take Napierian logarithms of both sides, which gives

$$\ln e^{3x} = \ln 7$$

i.e. $\qquad 3x = \ln 7$

from which $\qquad x = \dfrac{1}{3}\ln 7 = \mathbf{0.6486}$,

correct to 4 decimal places.

Problem 10. Evaluate the following, each correct to 5 significant figures:

(a) $\dfrac{1}{2}\ln 4.7291$ (b) $\dfrac{\ln 7.8693}{7.8693}$ (c) $\dfrac{3.17\ln 24.07}{e^{-0.1762}}$

(a) $\dfrac{1}{2}\ln 4.7291 = \dfrac{1}{2}(1.5537349\ldots) = \mathbf{0.77687}$, correct to 5 significant figures.

(b) $\dfrac{\ln 7.8693}{7.8693} = \dfrac{2.06296911\ldots}{7.8693} = \mathbf{0.26215}$, correct to 5 significant figures.

(c) $\dfrac{3.17\ln 24.07}{e^{-0.1762}} = \dfrac{3.17(3.18096625\ldots)}{0.83845027\ldots} = \mathbf{12.027}$, correct to 5 significant figures.

Problem 11. Evaluate the following: (a) $\dfrac{\ln e^{2.5}}{\lg 10^{0.5}}$ (b) $\dfrac{5e^{2.23}\lg 2.23}{\ln 2.23}$ correct to 3 decimal places

(a) $\dfrac{\ln e^{2.5}}{\lg 10^{0.5}} = \dfrac{2.5}{0.5} = \mathbf{5}$

(b) $\dfrac{5e^{2.23}\lg 2.23}{\ln 2.23}$

$= \dfrac{5(9.29986607\ldots)(0.34830486\ldots)}{(0.80200158\ldots)}$

$= \mathbf{20.194}$, correct to 3 decimal places.

Problem 12. Solve the equation $9 = 4e^{-3x}$ to find x, correct to 4 significant figures

Rearranging $9 = 4e^{-3x}$ gives $\dfrac{9}{4} = e^{-3x}$

Taking the reciprocal of both sides gives

$$\frac{4}{9} = \frac{1}{e^{-3x}} = e^{3x}$$

Taking Napierian logarithms of both sides gives

$$\ln\left(\frac{4}{9}\right) = \ln(e^{3x})$$

Since $\log_e e^\alpha = \alpha$, then $\ln\left(\dfrac{4}{9}\right) = 3x$

Hence, $x = \dfrac{1}{3}\ln\left(\dfrac{4}{9}\right) = \dfrac{1}{3}(-0.81093) = \mathbf{-0.2703}$,

correct to 4 significant figures.

Problem 13. Given $32 = 70\left(1 - e^{-\frac{t}{2}}\right)$, determine the value of t, correct to 3 significant figures

Rearranging $32 = 70\left(1 - e^{-\frac{t}{2}}\right)$ gives

$$\frac{32}{70} = 1 - e^{-\frac{t}{2}}$$

and

$$e^{-\frac{t}{2}} = 1 - \frac{32}{70} = \frac{38}{70}$$

Taking the reciprocal of both sides gives

$$e^{\frac{t}{2}} = \frac{70}{38}$$

Taking Napierian logarithms of both sides gives

$$\ln e^{\frac{t}{2}} = \ln\left(\frac{70}{38}\right)$$

i.e. $\dfrac{t}{2} = \ln\left(\dfrac{70}{38}\right)$

from which, $t = 2\ln\left(\dfrac{70}{38}\right) = \mathbf{1.22}$, correct to 3 significant figures.

Problem 14. Solve the equation $2.68 = \ln\left(\dfrac{4.87}{x}\right)$ to find x

From the definition of a logarithm, since $2.68 = \ln\left(\dfrac{4.87}{x}\right)$ then $e^{2.68} = \dfrac{4.87}{x}$

Rearranging gives $x = \dfrac{4.87}{e^{2.68}} = 4.87e^{-2.68}$

i.e. $\qquad x = \mathbf{0.3339}$,

correct to 4 significant figures.

Problem 15. Solve $\dfrac{7}{4} = e^{3x}$ correct to 4 significant figures

Taking natural logs of both sides gives

$$\ln\frac{7}{4} = \ln e^{3x}$$

$$\ln\frac{7}{4} = 3x \ln e$$

Since $\ln e = 1$, $\ln\dfrac{7}{4} = 3x$

i.e. $\qquad 0.55962 = 3x$

i.e. $\qquad x = \mathbf{0.1865}$,

correct to 4 significant figures.

Problem 16. Solve $e^{x-1} = 2e^{3x-4}$ correct to 4 significant figures

Taking natural logarithms of both sides gives

$$\ln\left(e^{x-1}\right) = \ln\left(2e^{3x-4}\right)$$

and by the first law of logarithms,

$$\ln\left(e^{x-1}\right) = \ln 2 + \ln\left(e^{3x-4}\right)$$

i.e. $\qquad x - 1 = \ln 2 + 3x - 4$

Rearranging gives

$$4 - 1 - \ln 2 = 3x - x$$

i.e. $\qquad 3 - \ln 2 = 2x$

from which, $\qquad x = \dfrac{3 - \ln 2}{2} = \mathbf{1.153}$

Problem 17. Solve, correct to 4 significant figures, $\ln(x-2)^2 = \ln(x-2) - \ln(x+3) + 1.6$

Rearranging gives

$$\ln(x-2)^2 - \ln(x-2) + \ln(x+3) = 1.6$$

and by the laws of logarithms,

$$\ln\left\{\frac{(x-2)^2(x+3)}{(x-2)}\right\} = 1.6$$

Cancelling gives

$$\ln\{(x-2)(x+3)\} = 1.6$$

and $$(x-2)(x+3) = e^{1.6}$$

i.e. $$x^2 + x - 6 = e^{1.6}$$

or $$x^2 + x - 6 - e^{1.6} = 0$$

i.e. $$x^2 + x - 10.953 = 0$$

Using the quadratic formula,

$$x = \frac{-1 \pm \sqrt{1^2 - 4(1)(-10.953)}}{2}$$

$$= \frac{-1 \pm \sqrt{44.812}}{2}$$

$$= \frac{-1 \pm 6.6942}{2}$$

i.e. $$x = 2.847 \text{ or } -3.8471$$

$x = -3.8471$ is not valid since the logarithm of a negative number has no real root.

Hence, **the solution of the equation is $x = 2.847$**

Now try the following Practice Exercise

Practice Exercise 71   Evaluating Napierian logarithms (answers on page 1155)

In Problems 1 and 2, evaluate correct to 5 significant figures.

1. (a) $\dfrac{1}{3}\ln 5.2932$ (b) $\dfrac{\ln 82.473}{4.829}$

   (c) $\dfrac{5.62 \ln 321.62}{e^{1.2942}}$

2. (a) $\dfrac{1.786 \ln e^{1.76}}{\lg 10^{1.41}}$ (b) $\dfrac{5e^{-0.1629}}{2\ln 0.00165}$

   (c) $\dfrac{\ln 4.8629 - \ln 2.4711}{5.173}$

In Problems 3 to 16, solve the given equations, each correct to 4 significant figures.

3. $1.5 = 4e^{2t}$

4. $7.83 = 2.91e^{-1.7x}$

5. $16 = 24\left(1 - e^{-\frac{t}{2}}\right)$

6. $5.17 = \ln\left(\dfrac{x}{4.64}\right)$

7. $3.72\ln\left(\dfrac{1.59}{x}\right) = 2.43$

8. $\ln x = 2.40$

9. $24 + e^{2x} = 45$

10. $5 = e^{x+1} - 7$

11. $5 = 8\left(1 - e^{\frac{-x}{2}}\right)$

12. $\ln(x+3) - \ln x = \ln(x-1)$

13. $\ln(x-1)^2 - \ln 3 = \ln(x-1)$

14. $\ln(x+3) + 2 = 12 - \ln(x-2)$

15. $e^{(x+1)} = 3e^{(2x-5)}$

16. $\ln(x+1)^2 = 1.5 - \ln(x-2) + \ln(x+1)$

17. Transpose $b = \ln t - a\ln D$ to make t the subject.

18. If $\dfrac{P}{Q} = 10\log_{10}\left(\dfrac{R_1}{R_2}\right)$, find the value of R_1 when $P = 160$, $Q = 8$ and $R_2 = 5$

19. If $U_2 = U_1 e^{\left(\frac{W}{PV}\right)}$, make W the subject of the formula.

20. The velocity v_2 of a rocket is given by: $v_2 = v_1 + C\ln\left(\dfrac{m_1}{m_2}\right)$ where v_1 is the initial rocket velocity, C is the velocity of the jet exhaust gases, m_1 is the mass of the rocket before the jet engine is fired, and m_2 is the mass of the rocket after the jet engine is switched off. Calculate the velocity of the rocket given $v_1 = 600$ m/s, $C = 3500$ m/s, $m_1 = 8.50 \times 10^4$ kg and $m_2 = 7.60 \times 10^4$ kg.

21. The work done in an isothermal expansion of a gas from pressure p_1 to p_2 is given by:

$$w = w_0 \ln \left(\frac{p_1}{p_2} \right)$$

If the initial pressure $p_1 = 7.0$ kPa, calculate the final pressure p_2 if $w = 3\,w_0$

16.5   Laws of growth and decay

Laws of exponential growth and decay are of the form $y = Ae^{-kx}$ and $y = A(1 - e^{-kx})$, where A and k are constants. When plotted, the form of these equations is as shown in Figure 16.5.

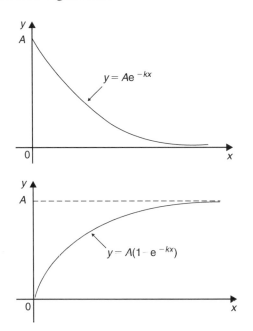

Figure 16.5

The laws occur frequently in engineering and science and examples of quantities related by a natural law include:

(a)   Linear expansion $\qquad\qquad l = l_0 e^{\alpha\theta}$

(b)   Change in electrical resistance
with temperature $\qquad\qquad R_\theta = R_0 e^{\alpha\theta}$

(c)   Tension in belts $\qquad\qquad T_1 = T_0 e^{\mu\theta}$

(d)   Newton's law of cooling $\qquad \theta = \theta_0 e^{-kt}$

(e)   Biological growth $\qquad\qquad y = y_0 e^{kt}$

(f)   Discharge of a capacitor $\qquad q = Q e^{-t/CR}$

(g)   Atmospheric pressure $\qquad p = p_0 e^{-h/c}$

(h)   Radioactive decay $\qquad\qquad N = N_0 e^{-\lambda t}$

(i)   Decay of current in an
inductive circuit $\qquad\qquad i = I e^{-Rt/L}$

(j)   Growth of current in a
capacitive circuit $\qquad\qquad i = I(1 - e^{-t/CR})$

Here are some worked problems to demonstrate the laws of growth and decay.

Problem 18. The resistance R of an electrical conductor at temperature $\theta°C$ is given by $R = R_0 e^{\alpha\theta}$, where α is a constant and $R_0 = 5\,k\Omega$. Determine the value of α correct to 4 significant figures when $R = 6\,k\Omega$ and $\theta = 1500°C$. Also, find the temperature, correct to the nearest degree, when the resistance R is $5.4\,k\Omega$

Transposing $R = R_0 e^{\alpha\theta}$ gives $\dfrac{R}{R_0} = e^{\alpha\theta}$

Taking Napierian logarithms of both sides gives

$$\ln \frac{R}{R_0} = \ln e^{\alpha\theta} = \alpha\theta$$

Hence, $\alpha = \dfrac{1}{\theta} \ln \dfrac{R}{R_0} = \dfrac{1}{1500} \ln \left(\dfrac{6 \times 10^3}{5 \times 10^3} \right)$

$$= \frac{1}{1500}(0.1823215\ldots)$$

$$= 1.215477\ldots \times 10^{-4}$$

Hence, $\boldsymbol{\alpha = 1.215 \times 10^{-4}}$ correct to 4 significant figures.

From above, $\ln \dfrac{R}{R_0} = \alpha\theta$ hence $\theta = \dfrac{1}{\alpha} \ln \dfrac{R}{R_0}$

When $R = 5.4 \times 10^3$, $\alpha = 1.215477\ldots \times 10^{-4}$ and $R_0 = 5 \times 10^3$

$$\theta = \frac{1}{1.215477\ldots \times 10^{-4}} \ln \left(\frac{5.4 \times 10^3}{5 \times 10^3} \right)$$

$$= \frac{10^4}{1.215477\ldots}(7.696104\ldots \times 10^{-2})$$

$$= \boldsymbol{633°C}, \text{ correct to the nearest degree.}$$

Problem 19. In an experiment involving Newton's law of cooling*, the temperature $\theta\,(°C)$ is given by $\theta = \theta_0 e^{-kt}$. Find the value of constant k when $\theta_0 = 56.6°C$, $\theta = 16.5°C$ and $t = 79.0$ seconds

---

Transposing $\theta = \theta_0 e^{-kt}$ gives $\dfrac{\theta}{\theta_0} = e^{-kt}$, from which

$$\frac{\theta_0}{\theta} = \frac{1}{e^{-kt}} = e^{kt}$$

Taking Napierian logarithms of both sides gives

$$\ln\frac{\theta_0}{\theta} = kt$$

from which,

$$k = \frac{1}{t}\ln\frac{\theta_0}{\theta} = \frac{1}{79.0}\ln\left(\frac{56.6}{16.5}\right)$$

$$= \frac{1}{79.0}(1.2326486\ldots)$$

Hence, $k = 0.01560$ or 15.60×10^{-3}

Problem 20.   The current i amperes flowing in a capacitor at time t seconds is given by $i = 8.0(1 - e^{-\frac{t}{CR}})$, where the circuit resistance R is $25\,k\Omega$ and capacitance C is $16\,\mu F$. Determine (a) the current i after 0.5 seconds and (b) the time, to the nearest millisecond, for the current to reach $6.0\,A$. Sketch the graph of current against time

---

(a)   Current $i = 8.0\left(1 - e^{-\frac{t}{CR}}\right)$

$$= 8.0[1 - e^{-0.5/(16\times10^{-6})(25\times10^{3})}]$$

$$= 8.0(1 - e^{-1.25})$$

$$= 8.0(1 - 0.2865047\ldots)$$

$$= 8.0(0.7134952\ldots)$$

$$= \mathbf{5.71\,amperes}$$

(b)   Transposing $i = 8.0\left(1 - e^{-\frac{t}{CR}}\right)$ gives

$$\frac{i}{8.0} = 1 - e^{-\frac{t}{CR}}$$

from which, $e^{-\frac{t}{CR}} = 1 - \dfrac{i}{8.0} = \dfrac{8.0 - i}{8.0}$

Taking the reciprocal of both sides gives

$$e^{\frac{t}{CR}} = \frac{8.0}{8.0 - i}$$

Taking Napierian logarithms of both sides gives

$$\frac{t}{CR} = \ln\left(\frac{8.0}{8.0 - i}\right)$$

Hence,

$$t = CR\ln\left(\frac{8.0}{8.0 - i}\right)$$

When $i = 6.0\,A$,

$$t = (16 \times 10^{-6})(25 \times 10^{3})\ln\left(\frac{8.0}{8.0 - 6.0}\right)$$

i.e. $t = \dfrac{400}{10^{3}}\ln\left(\dfrac{8.0}{2.0}\right) = 0.4\ln 4.0$

$$= 0.4(1.3862943\ldots)$$

$$= 0.5545\,s$$

$$= \mathbf{555\,ms},\ \text{correct to the nearest ms.}$$

A graph of current against time is shown in Figure 16.6.

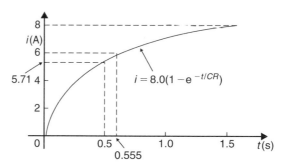

Figure 16.6

Problem 21.   The temperature θ_2 of a winding which is being heated electrically at time t is given by $\theta_2 = \theta_1(1 - e^{-\frac{t}{\tau}})$, where θ_1 is the temperature (in degrees Celsius*) at time $t = 0$ and τ is a constant. Calculate

(a) θ_1, correct to the nearest degree, when θ_2 is $50°C$, t is $30\,s$ and τ is $60\,s$

(b)   the time t, correct to 1 decimal place, for θ_2 to be half the value of θ_1

(a)   Transposing the formula to make θ_1 the subject gives

$$\theta_1 = \frac{\theta_2}{\left(1 - e^{-t/\tau}\right)} = \frac{50}{1 - e^{-30/60}}$$

$$= \frac{50}{1 - e^{-0.5}} = \frac{50}{0.393469\ldots}$$

i.e. $\boldsymbol{\theta_1 = 127°C}$, correct to the nearest degree.

---

* **Who is** Celsius? See page 80. To find out more go to www.routledge.com/cw/bird

(b) Transposing to make t the subject of the formula gives

$$\frac{\theta_2}{\theta_1} = 1 - e^{-\frac{t}{\tau}}$$

from which, $e^{-\frac{t}{\tau}} = 1 - \frac{\theta_2}{\theta_1}$

Hence, $-\frac{t}{\tau} = \ln\left(1 - \frac{\theta_2}{\theta_1}\right)$

i.e. $t = -\tau \ln\left(1 - \frac{\theta_2}{\theta_1}\right)$

Since $\theta_2 = \frac{1}{2}\theta_1$

$$t = -60 \ln\left(1 - \frac{1}{2}\right) = -60 \ln 0.5$$

$$= 41.59 \, s$$

Hence, **the time for the temperature θ_2 to be one half of the value of θ_1 is 41.6 s, correct to 1 decimal place**.

Now try the following Practice Exercise

Practice Exercise 72   Laws of growth and decay (answers on page 1156)

1. The temperature $T\,°C$ of a cooling object varies with time t minutes according to the equation $T = 150 e^{-0.04t}$. Determine the temperature when (a) $t = 0$, (b) $t = 10$ minutes.

2. The pressure p pascals at height h metres above ground level is given by $p = p_0 e^{-h/C}$, where p_0 is the pressure at ground level and C is a constant. Find pressure p when $p_0 = 1.012 \times 10^5$ Pa, height $h = 1420$ m and $C = 71500$

3. The voltage drop v volts across an inductor L henrys at time t seconds is given by $v = 200 e^{-\frac{Rt}{L}}$, where $R = 150\,\Omega$ and $L = 12.5 \times 10^{-3}$ H. Determine (a) the voltage when $t = 160 \times 10^{-6}$ s and (b) the time for the voltage to reach 85 V.

4. The length l metres of a metal bar at temperature $t\,°C$ is given by $l = l_0 e^{\alpha t}$, where l_0 and α are constants. Determine (a) the value of l when $l_0 = 1.894$, $\alpha = 2.038 \times 10^{-4}$ and $t = 250\,°C$ and (b) the value of l_0 when $l = 2.416, t = 310\,°C$ and $\alpha = 1.682 \times 10^{-4}$

5. The temperature $\theta_2\,°C$ of an electrical conductor at time t seconds is given by $\theta_2 = \theta_1(1 - e^{-t/T})$, where θ_1 is the initial temperature and T seconds is a constant. Determine (a) θ_2 when $\theta_1 = 159.9\,°C$, $t = 30$ s and $T = 80$ s and (b) the time t for θ_2 to fall to half the value of θ_1 if T remains at 80 s.

6. A belt is in contact with a pulley for a sector of $\theta = 1.12$ radians and the coefficient of friction between these two surfaces is $\mu = 0.26$. Determine the tension on the taut side of the belt, T newtons, when tension on the slack side is given by $T_0 = 22.7$ newtons, given that these quantities are related by the law $T = T_0 e^{\mu\theta}$

7. The instantaneous current i at time t is given by $i = 10 e^{-t/CR}$ when a capacitor is being charged. The capacitance C is 7×10^{-6} farads and the resistance R is 0.3×10^6 ohms. Determine (a) the instantaneous current when t is 2.5 seconds and (b) the time for the instantaneous current to fall to 5 amperes. Sketch a curve of current against time from $t = 0$ to $t = 6$ seconds.

8. The amount of product x (in mol/cm^3) found in a chemical reaction starting with 2.5 mol/cm^3 of reactant is given by $x = 2.5(1 - e^{-4t})$ where t is the time, in minutes, to form product x. Plot a graph at 30 second intervals up to 2.5 minutes and determine x after 1 minute.

9. The current i flowing in a capacitor at time t is given by $i = 12.5(1 - e^{-t/CR})$, where resistance R is 30 kΩ and the capacitance C is 20 μF. Determine (a) the current flowing after 0.5 seconds and (b) the time for the current to reach 10 amperes.

10. The amount A after n years of a sum invested P is given by the compound interest law $A = P e^{rn/100}$, when the per unit interest rate r is added continuously. Determine, correct to the nearest pound, the amount after eight years for a sum of £1500 invested if the interest rate is 6% per annum.

11. The percentage concentration C of the starting material in a chemical reaction varies with time t according to the equation $C = 100\,e^{-0.004t}$. Determine the concentration when (a) $t = 0$, (b) $t = 100\,\text{s}$, (c) $t = 1000\,\text{s}$

12. The current i flowing through a diode at room temperature is given by: $i = i_S\left(e^{40V} - 1\right)$ amperes. Calculate the current flowing in a silicon diode when the reverse saturation current $i_S = 50\,\text{nA}$ and the forward voltage $V = 0.27\,\text{V}$

13. A formula for chemical decomposition is given by: $C = A\left(1 - e^{-\frac{t}{10}}\right)$ where t is the time in seconds. Calculate the time in milliseconds for a compound to decompose to a value of $C = 0.12$ given $A = 8.5$

14. The mass m of pollutant in a water reservoir decreases according to the law $m = m_0\,e^{-0.1t}$ where t is the time in days and m_0 is the initial mass. Calculate the percentage decrease in the mass after 60 days, correct to 3 decimal places.

15. A metal bar is cooled with water. Its temperature in °C is given by: $\theta = 15 + 1300\,e^{-0.2t}$ where t is the time in minutes. Calculate how long it will take for the temperature θ to decrease to 36°C correct to the nearest second.

For fully worked solutions to each of the problems in Practice Exercises 68 to 72 in this chapter, go to the website:
www.routledge.com/cw/bird

Chapter 17

Inequalities

Why it is important to understand: **Inequalities**

In mathematics, an inequality is a relation that holds between two values when they are different. A working knowledge of inequalities can be beneficial to the practising engineer, and inequalities are central to the definitions of all limiting processes, including differentiation and integration. When exact solutions are unavailable, inconvenient or unnecessary, inequalities can be used to obtain error bounds for numerical approximation. Understanding and using inequalities is important in many branches of engineering.

At the end of this chapter, you should be able to:

- define an inequality
- state simple rules for inequalities
- solve simple inequalities
- solve inequalities involving a modulus
- solve inequalities involving quotients
- solve inequalities involving square functions
- solve quadratic inequalities

17.1 Introduction to inequalities

An **inequality** is any expression involving one of the symbols $<$, $>$, $\leq$ or $\geq$

$p < q$ means p is less than q

$p > q$ means p is greater than q

$p \leq q$ means p is less than or equal to q

$p \geq q$ means p is greater than or equal to q

Some simple rules

(i) When a quantity is **added or subtracted** to both sides of an inequality, the inequality still remains.

For example, if $p < 3$

then $p + 2 < 3 + 2$ (adding 2 to both sides)

and $p - 2 < 3 - 2$ (subtracting 2 from both sides)

(ii) When **multiplying or dividing** both sides of an inequality by a **positive** quantity, say 5, the inequality **remains the same**. For example,

$$\text{if } p > 4 \quad \text{then } 5p > 20 \quad \text{and} \quad \frac{p}{5} > \frac{4}{5}$$

(iii) When **multiplying or dividing** both sides of an inequality by a **negative** quantity, say -3, **the inequality is reversed**. For example,

$$\text{if } p > 1 \quad \text{then } -3p < -3 \quad \text{and} \quad \frac{p}{-3} < \frac{1}{-3}$$

(Note $>$ has changed to $<$ in each example.)

To **solve an inequality** means finding all the values of the variable for which the inequality is true. Knowledge of simple equations and quadratic equations are needed in this chapter.

17.2 Simple inequalities

The solution of some simple inequalities, using only the rules given in Section 17.1, is demonstrated in the following worked problems.

Problem 1. Solve the following inequalities:
(a) $3+x>7$ (b) $3t<6$
(c) $z-2\geq 5$ (d) $\dfrac{p}{3}\leq 2$

(a) Subtracting 3 from both sides of the inequality $3+x>7$ gives:

$$3+x-3>7-3, \text{ i.e. } \boldsymbol{x>4}$$

Hence, all values of x greater than 4 satisfy the inequality.

(b) Dividing both sides of the inequality $3t<6$ by 3 gives:

$$\frac{3t}{3}<\frac{6}{3}, \text{ i.e. } \boldsymbol{t<2}$$

Hence, all values of t less than 2 satisfy the inequality.

(c) Adding 2 to both sides of the inequality $z-2\geq 5$ gives:

$$z-2+2\geq 5+2, \text{ i.e. } \boldsymbol{z\geq 7}$$

Hence, all values of z greater than or equal to 7 satisfy the inequality.

(d) Multiplying both sides of the inequality $\dfrac{p}{3}\leq 2$ by 3 gives:

$$(3)\frac{p}{3}\leq (3)2, \text{ i.e. } \boldsymbol{p\leq 6}$$

Hence, all values of p less than or equal to 6 satisfy the inequality.

Problem 2. Solve the inequality: $4x+1>x+5$

Subtracting 1 from both sides of the inequality $4x+1>x+5$ gives:

$$4x>x+4$$

Subtracting x from both sides of the inequality $4x>x+4$ gives:

$$3x>4$$

Dividing both sides of the inequality $3x>4$ by 3 gives:

$$x>\frac{4}{3}$$

Hence all values of x greater than $\dfrac{4}{3}$ satisfy the inequality:

$$4x+1>x+5$$

Problem 3. Solve the inequality: $3-4t\leq 8+t$

Subtracting 3 from both sides of the inequality $3-4t\leq 8+t$ gives:

$$-4t\leq 5+t$$

Subtracting t from both sides of the inequality $-4t\leq 5+t$ gives:

$$-5t\leq 5$$

Dividing both sides of the inequality $-5t\leq 5$ by -5 gives:

$$\boldsymbol{t\geq -1} \text{ (remembering to reverse the inequality)}$$

Hence, all values of t greater than or equal to -1 satisfy the inequality.

Now try the following Practice Exercise

Practice Exercise 73   Further problems on simple inequalities (answers on page 1156)

Solve the following inequalities:

1. (a) $3t>6$ (b) $2x<10$
2. (a) $\dfrac{x}{2}>1.5$ (b) $x+2\geq 5$
3. (a) $4t-1\leq 3$ (b) $5-x\geq -1$
4. (a) $\dfrac{7-2k}{4}\leq 1$ (b) $3z+2>z+3$
5. (a) $5-2y\leq 9+y$ (b) $1-6x\leq 5+2x$

17.3 Inequalities involving a modulus

The **modulus** of a number is the size of the number, regardless of sign. Vertical lines enclosing the number denote a modulus.

For example, $|4|=4$ and $|-4|=4$ (the modulus of a number is never negative).

The inequality: $|t|<1$ means that all numbers whose actual size, regardless of sign, is less than 1, i.e. any value between -1 and $+1$

Thus $|t| < 1$ means $-1 < t < 1$

Similarly, $|x| > 3$ means all numbers whose actual size, regardless of sign, is greater than 3, i.e. any value greater than 3 and any value less than -3

Thus $|x| > 3$ means $x > 3$ and $x < -3$

Inequalities involving a modulus are demonstrated in the following worked problems.

Problem 4. Solve the following inequality:

$$|3x+1|<4$$

Since $|3x+1|<4$ then $-4<3x+1<4$

Now $-4<3x+1$ becomes $-5<3x$, i.e. $-\dfrac{5}{3}<x$

and $3x+1<4$ becomes $3x<3$, i.e. $x<1$

Hence, these two results together become $-\dfrac{5}{3}<x<1$ and mean that the inequality $|3x+1|<4$ is satisfied for any value of x greater than $-\dfrac{5}{3}$ but less than 1

Problem 5. Solve the inequality: $|1+2t|\le5$

Since $|1+2t|\le5$ then $-5\le1+2t\le5$

Now $-5\le1+2t$ becomes $-6\le2t$, i.e. $-3\le t$

and $1+2t\le5$ becomes $2t\le4$ i.e. $t\le2$

Hence, these two results together become: $-3\le t\le2$

Problem 6. Solve the inequality: $|3z-4|>2$

$|3z-4|>2$ means $3z-4>2$ and $3z-4<-2$,

i.e. $3z>6$ and $3z<2$,

i.e. the inequality: $|3z-4|>2$ is satisfied when

$$z>2 \text{ and } z<\dfrac{2}{3}$$

Now try the following Practice Exercise

Practice Exercise 74   Further problems on inequalities involving a modulus (answers on page 1156)

Solve the following inequalities:

1. $|t+1|<4$

2. $|y+3|\le2$

3. $|2x-1|<4$

4. $|3t-5|>4$

5. $|1-k|\ge3$

17.4 Inequalities involving quotients

If $\dfrac{p}{q}>0$ then $\dfrac{p}{q}$ must be a **positive** value.

For $\dfrac{p}{q}$ to be positive, **either** p is positive **and** q is positive **or** p is negative **and** q is negative.

i.e. $\dfrac{+}{+}=+$ and $\dfrac{-}{-}=+$

If $\dfrac{p}{q}<0$ then $\dfrac{p}{q}$ must be a **negative** value.

For $\dfrac{p}{q}$ to be negative, **either** p is positive **and** q is negative **or** p is negative **and** q is positive.

i.e. $\dfrac{+}{-}=-$ and $\dfrac{-}{+}=-$

This reasoning is used when solving inequalities involving quotients as demonstrated in the following worked problems.

Problem 7. Solve the inequality: $\dfrac{t+1}{3t-6}>0$

Since $\dfrac{t+1}{3t-6}>0$ then $\dfrac{t+1}{3t-6}$ must be **positive**.

For $\dfrac{t+1}{3t-6}$ to be positive,

either (i) $t+1>0$ **and** $3t-6>0$

or (ii) $t+1<0$ **and** $3t-6<0$

(i) If $t+1>0$ then $t>-1$ and if $3t-6>0$ then $3t>6$ and $t>2$

Both of the inequalities $t>-1$ **and** $t>2$ are only true when $t>2$,

i.e. the fraction $\dfrac{t+1}{3t-6}$ is positive when $t>2$

(ii) If $t+1<0$ then $t<-1$ and if $3t-6<0$ then $3t<6$ and $t<2$

Both of the inequalities $t<-1$ **and** $t<2$ are only true when $t<-1$,

i.e. the fraction $\dfrac{t+1}{3t-6}$ is positive when $t<-1$

Summarising, $\dfrac{t+1}{3t-6}>0$ when $t>2$ or $t<-1$

Problem 8. Solve the inequality: $\dfrac{2x+3}{x+2}\le 1$

Since $\dfrac{2x+3}{x+2}\le 1$ then $\dfrac{2x+3}{x+2}-1\le 0$

i.e. $\dfrac{2x+3}{x+2}-\dfrac{x+2}{x+2}\le 0$,

i.e. $\dfrac{2x+3-(x+2)}{x+2}\le 0$ or $\dfrac{x+1}{x+2}\le 0$

For $\dfrac{x+1}{x+2}$ to be negative or zero,

   either (i) $x+1\le 0$ **and** $x+2>0$

   or (ii) $x+1\ge 0$ **and** $x+2<0$

(i) If $x+1\le 0$ then $x\le -1$ and if $x+2>0$ then $x>-2$. (Note that $>$ is used for the denominator, not $\ge$; a zero denominator gives a value for the fraction which is impossible to evaluate.)

Hence, the inequality $\dfrac{x+1}{x+2}\le 0$ is true when x is greater than -2 and less than or equal to -1, which may be written as $-2<x\le -1$

(ii) If $x+1\ge 0$ then $x\ge -1$ and if $x+2<0$ then $x<-2$. It is not possible to satisfy both $x\ge -1$ and $x<-2$ thus no values of x satisfies (ii).

Summarising, $\dfrac{2x+3}{x+2}\le 1$ when $-2<x\le -1$

Now try the following Practice Exercise

Practice Exercise 75 Further problems on inequalities involving quotients (answers on page 1156)

Solve the following inequalitites:

1. $\dfrac{x+4}{6-2x}\ge 0$

2. $\dfrac{2t+4}{t-5}>1$

3. $\dfrac{3z-4}{z+5}\le 2$

4. $\dfrac{2-x}{x+3}\ge 4$

17.5 Inequalities involving square functions

The following two general rules apply when inequalities involve square functions:

(i) if $x^2>k$ then $x>\sqrt{k}$ or $x<-\sqrt{k}$ (1)

(ii) if $x^2<k$ then $-\sqrt{k}<x<\sqrt{k}$ (2)

These rules are demonstrated in the following worked problems.

Problem 9. Solve the inequality: $t^2>9$

Since $t^2>9$ then $t^2-9>0$, i.e. $(t+3)(t-3)>0$ by factorising.
For $(t+3)(t-3)$ to be positive,

   either (i) $(t+3)>0$ **and** $(t-3)>0$

   or (ii) $(t+3)<0$ **and** $(t-3)<0$

(i) If $(t+3)>0$ then $t>-3$ and if $(t-3)>0$ then $t>3$
     Both of these are true only when $t>3$

(ii) If $(t+3)<0$ then $t<-3$ and if $(t-3)<0$ then $t<3$
     Both of these are true only when $t<-3$

Summarising, $t^2>9$ when $t>3$ or $t<-3$

This demonstrates the general rule:

   if $x^2>k$ then $x>\sqrt{k}$ or $x<-\sqrt{k}$ (1)

Problem 10.   Solve the inequality: $x^2 > 4$

From the general rule stated above in equation (1):
if $x^2 > 4$ then $x > \sqrt{4}$ or $x < -\sqrt{4}$

i.e. the inequality: $x^2 > 4$ is satisfied when **$x > 2$ or $x < -2$**

Problem 11.   Solve the inequality: $(2z+1)^2 > 9$

From equation (1), if $(2z+1)^2 > 9$ then

$$2z + 1 > \sqrt{9} \quad \text{or} \quad 2z + 1 < -\sqrt{9}$$

i.e. $2z + 1 > 3$ or $2z + 1 < -3$

i.e. $2z > 2$ or $2z < -4$

i.e.   **$z > 1$ or $z < -2$**

Problem 12.   Solve the inequality: $t^2 < 9$

Since $t^2 < 9$ then $t^2 - 9 < 0$, i.e. $(t+3)(t-3) < 0$ by factorising. For $(t+3)(t-3)$ to be negative,

either   (i) $(t+3) > 0$ **and** $(t-3) < 0$

or   (ii) $(t+3) < 0$ **and** $(t-3) > 0$

(i)   If $(t+3) > 0$ then $t > -3$ and if $(t-3) < 0$ then $t < 3$
Hence (i) is satisfied when $t > -3$ and $t < 3$ which may be written as: **$-3 < t < 3$**

(ii)   If $(t+3) < 0$ then $t < -3$ and if $(t-3) > 0$ then $t > 3$
It is not possible to satisfy both $t < -3$ and $t > 3$, thus no values of t satisfies (ii).

Summarising, $t^2 < 9$ when **$-3 < t < 3$** which means that all values of t between -3 and $+3$ will satisfy the inequality.

This demonstrates the general rule:

$$\text{if } x^2 < k \text{ then} -\sqrt{k} < x < \sqrt{k} \tag{2}$$

Problem 13.   Solve the inequality: $x^2 < 4$

From the general rule stated above in equation (2): if $x^2 < 4$ then $-\sqrt{4} < x < \sqrt{4}$
i.e. the inequality: $x^2 < 4$ is satisfied when:
$$-2 < x < 2$$

Problem 14.   Solve the inequality: $(y-3)^2 \leq 16$

From equation (2), $-\sqrt{16} \leq (y-3) \leq \sqrt{16}$

i.e. $-4 \leq (y-3) \leq 4$

from which, $3 - 4 \leq y \leq 4 + 3$

i.e.   **$-1 \leq y \leq 7$**

Now try the following Practice Exercise

Practice Exercise 76   Further problems on inequalities involving square functions (answers on page 1156)

Solve the following inequalities:

1. $z^2 > 16$

2. $z^2 < 16$

3. $2x^2 \geq 6$

4. $3k^2 - 2 \leq 10$

5. $(t-1)^2 \leq 36$

6. $(t-1)^2 \geq 36$

7. $7 - 3y^2 \leq -5$

8. $(4k+5)^2 > 9$

17.6   Quadratic inequalities

Inequalities involving quadratic expressions are solved using either **factorisation** or **'completing the square'**. For example,

$$x^2 - 2x - 3 \text{ is factorised as } (x+1)(x-3)$$

and $6x^2 + 7x - 5$ is factorised as $(2x-1)(3x+5)$

If a quadratic expression does not factorise, then the technique of 'completing the square' is used. In general, the procedure for $x^2 + bx + c$ is:

$$x^2 + bx + c \equiv \left(x + \frac{b}{2}\right)^2 + c - \left(\frac{b}{2}\right)^2$$

For example, $x^2 + 4x - 7$ does not factorise; completing the square gives:

$$x^2 + 4x - 7 \equiv (x+2)^2 - 7 - 2^2 \equiv (x+2)^2 - 11$$

Similarly,

$$x^2 + 6x - 5 \equiv (x+3)^2 - 5 - 3^2 \equiv (x-3)^2 - 14$$

Solving quadratic inequalities is demonstrated in the following worked problems.

Problem 15. Solve the inequality:
$$x^2 + 2x - 3 > 0$$

Since $x^2 + 2x - 3 > 0$ then $(x-1)(x+3) > 0$ by factorising. For the product $(x-1)(x+3)$ to be positive,

either   (i) $(x-1) > 0$ **and** $(x+3) > 0$

or   (ii) $(x-1) < 0$ **and** $(x+3) < 0$

(i)   Since $(x-1) > 0$ then $x > 1$ and since $(x+3) > 0$ then $x > -3$
Both of these inequalities are satisfied only when **$x > 1$**

(ii)   Since $(x-1) < 0$ then $x < 1$ and since $(x+3) < 0$ then $x < -3$
Both of these inequalities are satisfied only when **$x < -3$**

Summarising, $x^2 + 2x - 3 > 0$ is satisfied when either **$x > 1$ or $x < -3$**

Problem 16. Solve the inequality: $t^2 - 2t - 8 < 0$

Since $t^2 - 2t - 8 < 0$ then $(t-4)(t+2) < 0$ by factorising.
For the product $(t-4)(t+2)$ to be negative,

either   (i) $(t-4) > 0$ **and** $(t+2) < 0$

or   (ii) $(t-4) < 0$ **and** $(t+2) > 0$

(i)   Since $(t-4) > 0$ then $t > 4$ and since $(t+2) < 0$ then $t < -2$
It is not possible to satisfy both $t > 4$ and $t < -2$, thus no values of t satisfies the inequality (i)

(ii)   Since $(t-4) < 0$ then $t < 4$ and since $(t+2) > 0$ then $t > -2$
Hence, (ii) is satisfied when $-2 < t < 4$

Summarising, $t^2 - 2t - 8 < 0$ is satisfied when **$-2 < t < 4$**

Problem 17. Solve the inequality:
$$x^2 + 6x + 3 < 0$$

$x^2 + 6x + 3$ does not factorise; completing the square gives:

$$x^2 + 6x + 3 \equiv (x+3)^2 + 3 - 3^2$$
$$\equiv (x+3)^2 - 6$$

The inequality thus becomes: $(x+3)^2 - 6 < 0$ or $(x+3)^2 < 6$

From equation (2), $-\sqrt{6} < (x+3) < \sqrt{6}$

from which, $(-\sqrt{6} - 3) < x < (\sqrt{6} - 3)$

Hence, $x^2 + 6x + 3 < 0$ is satisfied when **$-5.45 < x < -0.55$**, correct to 2 decimal places.

Problem 18. Solve the inequality:
$$y^2 - 8y - 10 \geq 0$$

$y^2 - 8y - 10$ does not factorise; completing the square gives:

$$y^2 - 8y - 10 \equiv (y-4)^2 - 10 - 4^2$$
$$\equiv (y-4)^2 - 26$$

The inequality thus becomes: $(y-4)^2 - 26 \geq 0$ or $(y-4)^2 \geq 26$
From equation (1), $(y-4) \geq \sqrt{26}$ or $(y-4) \leq -\sqrt{26}$

from which, **$y \geq 4 + \sqrt{26}$ or $y \leq 4 - \sqrt{26}$**

Hence, $y^2 - 8y - 10 \geq 0$ is satisfied when **$y \geq 9.10$ or $y \leq -1.10$**, correct to 2 decimal places.

Now try the following Practice Exercise

Practice Exercise 77   Further problems on quadratic inequalities (answers on page 1156)

Solve the following inequalities:

1. $x^2 - x - 6 > 0$

2. $t^2 + 2t - 8 \leq 0$

3. $2x^2 + 3x - 2 < 0$

4. $y^2 - y - 20 \geq 0$

5. $z^2 + 4z + 4 \leq 4$

6. $x^2 + 6x - 6 \leq 0$

7. $t^2 - 4t - 7 \geq 0$

8. $k^2 - k - 3 \geq 0$

For fully worked solutions to each of the problems in Practice Exercises 73 to 77 in this chapter, go to the website:
www.routledge.com/cw/bird

This assignment covers the material contained in Chapters 14 to 17. *The marks available are shown in brackets at the end of each question.*

1. Solve the following equations by factorisation.
   (a) $x^2 - 9 = 0$ (b) $x^2 + 12x + 36 = 0$
   (c) $x^2 + 3x - 4 = 0$ (d) $3z^2 - z - 4 = 0$
   (9)

2. Solve the following equations, correct to 3 decimal places.
   (a) $5x^2 + 7x - 3 = 0$ (b) $3a^2 + 4a - 5 = 0$
   (8)

3. Solve the equation $3x^2 - x - 4 = 0$ by completing the square.
   (6)

4. Determine the quadratic equation in x whose roots are 1 and -3
   (3)

5. The bending moment M at a point in a beam is given by $M = \dfrac{3x(20 - x)}{2}$ where x metres is the distance from the point of support. Determine the value of x when the bending moment is $50\,\mathrm{N\,m}$.
   (5)

6. The current i flowing through an electronic device is given by $i = (0.005v^2 + 0.014v)$ amperes where v is the voltage. Calculate the values of v when $i = 3 \times 10^{-3}$
   (6)

7. Evaluate the following, correct to 4 significant figures.
   (a) $3.2 \ln 4.92 - 5 \lg 17.9$ (b) $\dfrac{5(1 - e^{-2.65})}{e^{1.73}}$ (4)

8. Solve the following equations.
   (a) $\lg x = 4$ (b) $\ln x = 2$
   (c) $\log_2 x = 5$ (d) $5^x = 2$
   (e) $3^{2t-1} = 7^{t+2}$ (f) $3e^{2x} = 4.2$ (18)

9. Evaluate $\log_{16}\left(\dfrac{1}{8}\right)$
   (4)

10. Write the following as the logarithm of a single number.

(a) $3\log 2 + 2\log 5 - \dfrac{1}{2}\log 16$

(b) $3\log 3 + \dfrac{1}{4}\log 16 - \dfrac{1}{3}\log 27$ (9)

11. Solve the equation
    $$\log(x^2 + 8) - \log(2x) = \log 3$$ (6)

12. Evaluate the following, each correct to 3 decimal places.

(a) $\ln 462.9$

(b) $\ln 0.0753$

(c) $\dfrac{\ln 3.68 - \ln 2.91}{4.63}$ (3)

13. Expand $x e^{3x}$ to six terms.   (5)

14. Evaluate v given that $v = E\left(1 - e^{-\frac{t}{CR}}\right)$ volts when $E = 100\,\mathrm{V}, C = 15\,\mu\mathrm{F}, \quad R = 50\,\mathrm{k}\Omega$ and $t = 1.5\,\mathrm{s}$. Also, determine the time when the voltage is $60\,\mathrm{V}$.   (8)

15. Plot a graph of $y = \dfrac{1}{2}e^{-1.2x}$ over the range $x = -2$ to $x = +1$ and hence determine, correct to 1 decimal place,
    (a) the value of y when $x = -0.75$, and
    (b) the value of x when $y = 4.0$ (8)

16. Solve the following inequalities:

(a) $2 - 5x \leq 9 + 2x$ (b) $|3 + 2t| \leq 6$

(c) $\dfrac{x - 1}{3x + 5} > 0$ (d) $(3t + 2)^2 > 16$

(e) $2x^2 - x - 3 < 0$ (18)

Formulae/revision hints for Section A
Number and algebra

Basic arithmetic

Order of operations is best remembered by **BODMAS** (i.e. brackets, order, division, multiplication, addition, subtraction). Multiplication and division must be completed before addition and subtraction.

A **bracket** means multiplication. For example, $3(2+5) = 3 \times 7 = 21$

Fractions

Here's a quick reminder of some fractions: $\frac{11}{3} = 3\frac{2}{3}$,

$$2\frac{2}{5} = \frac{12}{5}, \quad \frac{1}{5} + \frac{2}{3} = \frac{3+10}{15} = \frac{13}{15}, \quad \frac{2}{3} \times \frac{5}{7} = \frac{2 \times 5}{3 \times 7} = \frac{10}{21},$$

$$\frac{3}{5} \div \frac{7}{10} = \frac{3}{5} \times \frac{10}{7} = \frac{3 \times 2}{1 \times 7} = \frac{6}{7},$$

$$2\frac{3}{5} \times \frac{3}{4} \div \frac{1}{5} = \frac{13}{5} \times \frac{3}{4} \times \frac{5}{1} = \frac{13 \times 3 \times 1}{1 \times 4 \times 1} = \frac{39}{4} = 9\frac{3}{4}$$

SI units

The basic seven units used in the SI system are listed below with their symbols.

Quantity	Unit	Symbol	
Length	metre	m	(1 m = 100 cm = 1000 mm)
Mass	kilogram	kg	(1 kg = 1000 g)
Time	second	s	
Electric current	ampere	A	
Thermodynamic temperature	kelvin	K	(K = °C + 273)
Luminous intensity	candela	cd	
Amount of substance	mole	mol	

Prefixes and their meaning

Prefix	Name	Meaning	
G	giga	multiply by 10^9	i.e. × 1 000 000 000
M	mega	multiply by 10^6	i.e. × 1 000 000
k	kilo	multiply by 10^3	i.e. × 1000
m	milli	multiply by 10^{-3}	i.e. $\times \frac{1}{10^3} = \frac{1}{1000} = 0.001$
μ	micro	multiply by 10^{-6}	i.e. $\times \frac{1}{10^6} = \frac{1}{1\,000\,000}$ $= 0.000001$
n	nano	multiply by 10^{-9}	i.e. $\times \frac{1}{10^9} = \frac{1}{1\,000\,000\,000}$ $= 0.000000001$
p	pico	multiply by 10^{-12}	i.e. $\times \frac{1}{10^{12}} = \frac{1}{1\,000\,000\,000\,000}$ $= 0.000000000001$

Basic operations in algebra

$a + a + a + a = 4a$ For example, if $a = 3$, then $3 + 3 + 3 + 3 = 4 \times 3 = 12$

$5b$ **means** $5 \times b$ For example, if $b = 2$, then $5b = 5 \times 2 = 10$

$3a + 4b + a - 7b = 3a + a + 4b - 7b = 4a - 3b$

$5abcd = 5 \times a \times b \times c \times d$

For example, if $a = 3$, $b = -2$, $c = 1$ and $d = -4$, then $5\,abcd = 5 \times 3 \times -2 \times 1 \times -4 = 120$

$(a)(c)(d)$ **means** $a \times c \times d$ For example, $(2)(5)(4)$ means $2 \times 5 \times 4 = 40$

$ab = ba$ If $a = 2$ and $b = 3$ then 2×3 is exactly the same as 3×2, i.e. 6

$b^2 = b \times b$ For example, if $b = 2$, then $2^2 = 2 \times 2 = 4$

$a^3 = a \times a \times a$ For example, if $a = -2$, then $(-2)^3 = -2 \times -2 \times -2 = -8$

Decimals

$14.786 = 14.79$, correct to 4 significant figures
$ = 14.8$, correct to 1 decimal place

Laws of indices

$$a^m \times a^n = a^{m+n} \qquad \frac{a^m}{a^n} = a^{m-n} \qquad (a^m)^n = a^{m \times n}$$

$$a^{\frac{m}{n}} = \sqrt[n]{a^m} \qquad a^{-n} = \frac{1}{a^n} \qquad a^0 = 1$$

Brackets

$3a(b - 2a) = 3ab - 6a^2$ and $(2x - y)(x + 3y)$
$ = 2x^2 + 6xy - xy - 3y^2 = 2x^2 + 5xy - 3y^2$

Factorisation

$$2ab^2 - 6a^2b + 10a^3b^4 = 2ab(b - 3a + 5a^2b^3)$$

Quadratic formula

If $ax^2 + bx + c = 0$ then $x = \dfrac{-b \pm \sqrt{b^2 - 4ac}}{2a}$

Definition of a logarithm

If $y = a^x$ then $x = \log_a y$

Laws of logarithms

$$\log(A \times B) = \log A + \log B$$

$$\log\left(\frac{A}{B}\right) = \log A - \log B \qquad \log A^n = n \times \log A$$

If $2^x = 11$, $\log_{10}(2^x) = \log_{10}11$, $x\log_{10}2 = \log_{10}11$

and $x = \dfrac{\log_{10}11}{\log_{10}2} = \mathbf{3.459}$

If $5 = 3e^{-2t}$, $\dfrac{5}{3} = e^{-2t}$, $\ln\left(\dfrac{5}{3}\right) = \ln\left(e^{-2t}\right)$,

$$\ln\left(\frac{5}{3}\right) = -2t \text{ and } t = -\frac{1}{2}\ln\left(\frac{5}{3}\right) = \mathbf{-0.2554}$$

Exponential series

$$e^x = 1 + x + \frac{x^2}{2!} + \frac{x^3}{3!} + \ldots. \quad \text{(valid for all values of } x\text{)}$$

For a copy of these formulae/revision hints, go to:
www.routledge.com/cw/bird

Further number and algebra

Chapter 18

Polynomial division and the factor and remainder theorems

Why it is important to understand: *Polynomial division and the factor and remainder theorems*

The study of algebra revolves around using and manipulating polynomials. Polynomials are used in engineering, computer programming, software engineering, in management and in business. Mathematicians, statisticians and engineers of all sciences employ the use of polynomials to solve problems; among them are aerospace engineers, chemical engineers, civil engineers, electrical engineers, environmental engineers, industrial engineers, materials engineers, mechanical engineers and nuclear engineers. The factor and remainder theorems are also employed in engineering software and electronic mathematical applications, through which polynomials of higher degrees and longer arithmetic structures are divided without any complexity. The study of polynomial division and the factor and remainder theorems is therefore of some importance in engineering.

At the end of this chapter, you should be able to:

- divide algebraic expressions using polynomial division
- factorise expressions using the factor theorem
- use the remainder theorem to factorise algebraic expressions

---

18.1 Polynomial division

Before looking at long division in algebra let us revise long division with numbers (we may have forgotten, since calculators do the job for us!).

For example, $\dfrac{208}{16}$ is achieved as follows:

$$
\begin{array}{r}
13 \\
16\overline{)208} \\
\underline{16} \\
48 \\
\underline{48} \\
\cdot\,\cdot
\end{array}
$$

(1)   16 divided into 2 won't go

(2)   16 divided into 20 goes 1

(3)   Put 1 above the zero

(4)   Multiply 16 by 1 giving 16

(5)   Subtract 16 from 20 giving 4

(6)   Bring down the 8

(7)   16 divided into 48 goes 3 times

(8)   Put the 3 above the 8

(9) $3 \times 16 = 48$

(10) $48 - 48 = 0$

Hence $\dfrac{208}{16} = \mathbf{13}$ exactly

Similarly, $\dfrac{172}{15}$ is laid out as follows:

$$
\begin{array}{r}
11 \\
15 \overline{\smash{)}\, 172} \\
\underline{15} \\
22 \\
\underline{15} \\
7 \\
\end{array}
$$

Hence $\dfrac{172}{15} = 11$ remainder 7 or $11 + \dfrac{7}{15} = \mathbf{11\dfrac{7}{15}}$

Below are some examples of division in algebra, which in some respects is similar to long division with numbers.

(Note that a **polynomial** is an expression of the form

$$f(x) = a + bx + cx^2 + dx^3 + \cdots$$

and **polynomial division** is sometimes required when resolving into partial fractions – see Chapter 19)

> **Problem 1.**   Divide $2x^2 + x - 3$ by $x - 1$

$2x^2 + x - 3$ is called the **dividend** and $x - 1$ the **divisor**. The usual layout is shown below with the dividend and divisor both arranged in descending powers of the symbols.

$$
\begin{array}{r}
2x + 3 \\
x - 1 \overline{\smash{)}\, 2x^2 + x - 3} \\
\underline{2x^2 - 2x } \\
3x - 3 \\
\underline{3x - 3} \\
\cdot \quad \cdot \\
\end{array}
$$

Dividing the first term of the dividend by the first term of the divisor, i.e. $\dfrac{2x^2}{x}$, gives $2x$, which is put above the first term of the dividend as shown. The divisor is then multiplied by $2x$, i.e. $2x(x - 1) = 2x^2 - 2x$, which is placed under the dividend as shown. Subtracting gives $3x - 3$. The process is then repeated, i.e. the first term of the divisor, x, is divided into $3x$, giving $+3$, which is placed above the dividend as shown. Then $3(x - 1) = 3x - 3$ which is placed under the $3x - 3$. The remainder, on subtraction, is zero, which completes the process.

The answer, $(2x + 3)$, is called the **quotient**. Thus $\mathbf{(2x^2 + x - 3) \div (x - 1) = (2x + 3)}$

(A check can be made on this answer by multiplying $(2x + 3)$ by $(x - 1)$ which equals $2x^2 + x - 3$)

> **Problem 2.**   Divide $3x^3 + x^2 + 3x + 5$ by $x + 1$

$$
\begin{array}{r}
(1)(4)(7) \\
3x^2 - 2x + 5 \\
x + 1 \overline{\smash{)}\, 3x^3 + x^2 + 3x + 5} \\
\underline{3x^3 + 3x^2 } \\
-2x^2 + 3x + 5 \\
\underline{-2x^2 - 2x } \\
5x + 5 \\
\underline{5x + 5} \\
\cdot \quad \cdot \\
\end{array}
$$

(1) x into $3x^3$ goes $3x^2$. Put $3x^2$ above $3x^3$

(2) $3x^2(x + 1) = 3x^3 + 3x^2$

(3)   Subtract

(4) x into $-2x^2$ goes $-2x$. Put $-2x$ above the dividend

(5) $-2x(x + 1) = -2x^2 - 2x$

(6)   Subtract

(7) x into $5x$ goes 5. Put 5 above the dividend

(8) $5(x + 1) = 5x + 5$

(9)   Subtract

Thus $\dfrac{3x^3 + x^2 + 3x + 5}{x + 1} = \mathbf{3x^2 - 2x + 5}$

Problem 3.   Simplify $\dfrac{x^3 + y^3}{x + y}$

$$
\begin{array}{r}
\ \ (1)\quad(4)\ \ (7)\\
x^2 -\ xy\ +y^2\\
x + y \overline{)\, x^3 +\ 0\ +\ 0\ +y^3\,}\\
\underline{x^3 + x^2 y}\\
-x^2 y + y^3\\
\underline{-x^2 y - x y^2}\\
xy^2 + y^3\\
\underline{xy^2 + y^3}\\
\cdot\ \ \cdot\\
\end{array}
$$

(1) x into x^3 goes x^2. Put x^2 above x^3 of dividend

(2) $x^2(x + y) = x^3 + x^2 y$

(3)   Subtract

(4) x into $-x^2 y$ goes $-xy$. Put $-xy$ above dividend

(5) $-xy(x + y) = -x^2 y - xy^2$

(6)   Subtract

(7) x into xy^2 goes y^2. Put y^2 above dividend

(8) $y^2(x + y) = xy^2 + y^3$

(9)   Subtract

Thus

$$\frac{x^3 + y^3}{x + y} = x^2 - xy + y^2$$

The zeroes shown in the dividend are not normally shown, but are included to clarify the subtraction process and to keep similar terms in their respective columns.

Problem 4.   Divide $(x^2 + 3x - 2)$ by $(x - 2)$

$$
\begin{array}{r}
x\ +5\\
x - 2 \overline{)\, x^2 + 3x -\ 2\,}\\
\underline{x^2 - 2x}\\
5x -\ 2\\
\underline{5x - 10}\\
8\\
\end{array}
$$

Hence

$$\frac{x^2 + 3x - 2}{x - 2} = x + 5 + \frac{8}{x - 2}$$

Problem 5.   Divide $4a^3 - 6a^2 b + 5b^3$ by $2a - b$

$$
\begin{array}{r}
2a^2 - 2ab -\ \ b^2\\
2a - b \overline{)\, 4a^3 - 6a^2 b + 5b^3\,}\\
\underline{4a^3 - 2a^2 b}\\
-4a^2 b + 5b^3\\
\underline{-4a^2 b + 2ab^2}\\
-2ab^2 + 5b^3\\
\underline{-2ab^2 +\ \ b^3}\\
4b^3\\
\end{array}
$$

Thus

$$\frac{4a^3 - 6a^2 b + 5b^3}{2a - b}$$

$$= 2a^2 - 2ab - b^2 + \frac{4b^3}{2a - b}$$

Now try the following Practice Exercise

Practice Exercise 78   Further problems on polynomial division (answers on page 1156)

1.   Divide $(2x^2 + xy - y^2)$ by $(x + y)$

2.   Divide $(3x^2 + 5x - 2)$ by $(x + 2)$

3.   Determine $(10x^2 + 11x - 6) \div (2x + 3)$

4.   Find $\dfrac{14x^2 - 19x - 3}{2x - 3}$

5.   Divide $(x^3 + 3x^2 y + 3xy^2 + y^3)$ by $(x + y)$

6.   Find $(5x^2 - x + 4) \div (x - 1)$

7.   Divide $(3x^3 + 2x^2 - 5x + 4)$ by $(x + 2)$

8.   Determine $(5x^4 + 3x^3 - 2x + 1)/(x - 3)$

18.2   The factor theorem

There is a simple relationship between the factors of a quadratic expression and the roots of the equation obtained by equating the expression to zero.

Section B

For example, consider the quadratic equation
$x^2 + 2x - 8 = 0$
To solve this we may factorise the quadratic expression
$x^2 + 2x - 8$ giving $(x - 2)(x + 4)$
Hence $(x - 2)(x + 4) = 0$
Then, if the product of two numbers is zero, one or both of those numbers must equal zero. Therefore,

either $(x - 2) = 0$, from which $x = 2$
or $(x + 4) = 0$, from which $x = -4$

It is clear then that a factor of $(x - 2)$ indicates a root of $+2$, while a factor of $(x + 4)$ indicates a root of -4
In general, we can therefore say that:

$$\textbf{a factor of } (x - a) \textbf{ corresponds to a}$$
$$\textbf{root of } x = a$$

In practice, we always deduce the roots of a simple quadratic equation from the factors of the quadratic expression, as in the above example. However, we could reverse this process. If, by trial and error, we could determine that $x = 2$ is a root of the equation $x^2 + 2x - 8 = 0$ we could deduce at once that $(x - 2)$ is a factor of the expression $x^2 + 2x - 8$. We wouldn't normally solve quadratic equations this way – but suppose we have to factorise a cubic expression (i.e. one in which the highest power of the variable is 3). A cubic equation might have three simple linear factors and the difficulty of discovering all these factors by trial and error would be considerable. It is to deal with this kind of case that we use the **factor theorem**. This is just a generalised version of what we established above for the quadratic expression. The factor theorem provides a method of factorising any polynomial, $f(x)$, which has simple factors.
A statement of the **factor theorem** says:

$$\textbf{'if } x = a \textbf{ is a root of the equation}$$
$$f(x) = 0, \textbf{ then } (x - a) \textbf{ is a factor of } f(x)\textbf{'}$$

The following worked problems show the use of the factor theorem.

Problem 6. Factorise $x^3 - 7x - 6$ and use it to solve the cubic equation $x^3 - 7x - 6 = 0$

Let $f(x) = x^3 - 7x - 6$

If $x = 1$, then $f(1) = 1^3 - 7(1) - 6 = -12$

If $x = 2$, then $f(2) = 2^3 - 7(2) - 6 = -12$

If $x = 3$, then $f(3) = 3^3 - 7(3) - 6 = 0$

If $f(3) = 0$, then $(x - 3)$ is a factor – from the factor theorem.
We have a choice now. We can divide $x^3 - 7x - 6$ by $(x - 3)$ or we could continue our 'trial and error' by substituting further values for x in the given expression – and hope to arrive at $f(x) = 0$
Let us do both ways. First, dividing out gives:

$$
\begin{array}{r}
x^2 + 3x + 2 \\
x - 3 \overline{)\ x^3 - 0\ \ -7x\ -6} \\
\underline{x^3 - 3x^2} \\
3x^2 - 7x - 6 \\
\underline{3x^2 - 9x} \\
2x - 6 \\
2x - 6 \\
\underline{\cdot\quad\cdot} \\
\end{array}
$$

Hence $\dfrac{x^3 - 7x - 6}{x - 3} = x^2 + 3x + 2$

i.e. $x^3 - 7x - 6 = (x - 3)(x^2 + 3x + 2)$
$x^2 + 3x + 2$ factorises 'on sight' as $(x + 1)(x + 2)$
Therefore

$$x^3 - 7x - 6 = (x - 3)(x + 1)(x + 2)$$

A second method is to continue to substitute values of x into $f(x)$.
Our expression for $f(3)$ was $3^3 - 7(3) - 6$. We can see that if we continue with positive values of x the first term will predominate such that $f(x)$ will not be zero. Therefore, let us try some negative values for x. Therefore $f(-1) = (-1)^3 - 7(-1) - 6 = 0$; hence $(x + 1)$ is a factor (as shown above). Also $f(-2) = (-2)^3 - 7(-2) - 6 = 0$; hence $(x + 2)$ is a factor (also as shown above).
To solve $x^3 - 7x - 6 = 0$, we substitute the factors, i.e.

$$(x - 3)(x + 1)(x + 2) = 0$$

from which, $x = 3$, $x = -1$ and $x = -2$
Note that the values of x, i.e. 3, -1 and -2, are all factors of the constant term, i.e. the 6. This can give us a clue as to what values of x we should consider.

Problem 7. Solve the cubic equation $x^3 - 2x^2 - 5x + 6 = 0$ by using the factor theorem

---

Let $f(x) = x^3 - 2x^2 - 5x + 6$ and let us substitute simple values of x like 1, 2, 3, −1, −2, and so on.

$$f(1) = 1^3 - 2(1)^2 - 5(1) + 6 = 0,$$
$$\text{hence } (x - 1) \text{ is a factor}$$

$$f(2) = 2^3 - 2(2)^2 - 5(2) + 6 \neq 0$$

$$f(3) = 3^3 - 2(3)^2 - 5(3) + 6 = 0,$$
$$\text{hence } (x - 3) \text{ is a factor}$$

$$f(-1) = (-1)^3 - 2(-1)^2 - 5(-1) + 6 \neq 0$$

$$f(-2) = (-2)^3 - 2(-2)^2 - 5(-2) + 6 = 0,$$
$$\text{hence } (x + 2) \text{ is a factor}$$

Hence $x^3 - 2x^2 - 5x + 6 = (x - 1)(x - 3)(x + 2)$
Therefore if $x^3 - 2x^2 - 5x + 6 = 0$
then $(x - 1)(x - 3)(x + 2) = 0$
from which, $x = 1, x = 3 \text{ and } x = -2$
Alternatively, having obtained one factor, i.e. $(x - 1)$ we could divide this into $(x^3 - 2x^2 - 5x + 6)$ as follows:

$$
\begin{array}{r}
x^2 - x - 6 \\
x - 1 \overline{)\ x^3 - 2x^2 - 5x + 6} \\
\underline{x^3 - x^2} \\
-x^2 - 5x + 6 \\
\underline{-x^2 + x} \\
-6x + 6 \\
\underline{-6x + 6} \\
\cdot\ \cdot
\end{array}
$$

Hence $x^3 - 2x^2 - 5x + 6$

$$= (x - 1)(x^2 - x - 6)$$

$$= \mathbf{(x - 1)(x - 3)(x + 2)}$$

Summarising, the factor theorem provides us with a method of factorising simple expressions, and an alternative, in certain circumstances, to polynomial division.

Now try the following Practice Exercise

Practice Exercise 79   Further problems on the factor theorem (answers on page 1156)

Use the factor theorem to factorise the expressions given in Problems 1 to 4.

1. $x^2 + 2x - 3$

2. $x^3 + x^2 - 4x - 4$

3. $2x^3 + 5x^2 - 4x - 7$

4. $2x^3 - x^2 - 16x + 15$

5. Use the factor theorem to factorise $x^3 + 4x^2 + x - 6$ and hence solve the cubic equation $x^3 + 4x^2 + x - 6 = 0$

6. Solve the equation $x^3 - 2x^2 - x + 2 = 0$

18.3   The remainder theorem

Dividing a general quadratic expression $(ax^2 + bx + c)$ by $(x - p)$, where p is any whole number, by long division (see Section 18.1) gives:

$$
\begin{array}{r}
ax + (b + ap) \\
x - p \overline{)\ ax^2 + bx \qquad + c} \\
\underline{ax^2 - apx} \\
(b + ap)x + c \\
\underline{(b + ap)x - (b + ap)p} \\
c + (b + ap)p
\end{array}
$$

The remainder, $c + (b + ap)p = c + bp + ap^2$ or $ap^2 + bp + c$. This is, in fact, what the **remainder theorem** states, i.e.

'if $(ax^2 + bx + c)$ is divided by $(x - p)$, the remainder will be $ap^2 + bp + c$'

If, in the dividend $(ax^2 + bx + c)$, we substitute p for x we get the remainder $ap^2 + bp + c$.
For example, when $(3x^2 - 4x + 5)$ is divided by $(x - 2)$ the remainder is $ap^2 + bp + c$ (where $a = 3$, $b = -4$, $c = 5$ and $p = 2$), i.e. the remainder is

$$3(2)^2 + (-4)(2) + 5 = 12 - 8 + 5 = \mathbf{9}$$

We can check this by dividing $(3x^2 - 4x + 5)$ by $(x - 2)$ by long division:

$$
\begin{array}{r}
3x + 2 \\
x - 2 \overline{)\ 3x^2 - 4x + 5} \\
\underline{3x^2 - 6x} \\
2x + 5 \\
\underline{2x - 4} \\
9
\end{array}
$$

Section B

Similarly, when $(4x^2 - 7x + 9)$ is divided by $(x+3)$, the remainder is $ap^2 + bp + c$ (where $a = 4$, $b = -7$, $c = 9$ and $p = -3$), i.e. the remainder is
$4(-3)^2 + (-7)(-3) + 9 = 36 + 21 + 9 = \mathbf{66}$
Also, when $(x^2 + 3x - 2)$ is divided by $(x - 1)$, the remainder is $1(1)^2 + 3(1) - 2 = \mathbf{2}$
It is not particularly useful, on its own, to know the remainder of an algebraic division. However, if the remainder should be zero then $(x - p)$ is a factor. This is very useful therefore when factorising expressions. For example, when $(2x^2 + x - 3)$ is divided by $(x - 1)$, the remainder is $2(1)^2 + 1(1) - 3 = 0$, which means that $(x - 1)$ is a factor of $(2x^2 + x - 3)$.
In this case the other factor is $(2x + 3)$, i.e.

$$(2x^2 + x - 3) = (x - 1)(2x + 3)$$

The **remainder theorem** may also be stated for a **cubic equation** as:

'if $(ax^3 + bx^2 + cx + d)$ is divided by $(x - p)$, the remainder will be $ap^3 + bp^2 + cp + d$'

As before, the remainder may be obtained by substituting p for x in the dividend.
For example, when $(3x^3 + 2x^2 - x + 4)$ is divided by $(x - 1)$, the remainder is $ap^3 + bp^2 + cp + d$ (where $a = 3$, $b = 2$, $c = -1$, $d = 4$ and $p = 1$), i.e. the remainder is
$3(1)^3 + 2(1)^2 + (-1)(1) + 4 = 3 + 2 - 1 + 4 = \mathbf{8}$
Similarly, when $(x^3 - 7x - 6)$ is divided by $(x - 3)$, the remainder is $1(3)^3 + 0(3)^2 - 7(3) - 6 = 0$, which means that $(x - 3)$ is a factor of $(x^3 - 7x - 6)$.
Here are some more examples on the remainder theorem.

> **Problem 8.** Without dividing out, find the remainder when $2x^2 - 3x + 4$ is divided by $(x - 2)$

By the remainder theorem, the remainder is given by $ap^2 + bp + c$, where $a = 2$, $b = -3$, $c = 4$ and $p = 2$
Hence **the remainder is:**

$$2(2)^2 + (-3)(2) + 4 = 8 - 6 + 4 = \mathbf{6}$$

> **Problem 9.** Use the remainder theorem to determine the remainder when $(3x^3 - 2x^2 + x - 5)$ is divided by $(x + 2)$

By the remainder theorem, the remainder is given by $ap^3 + bp^2 + cp + d$, where $a = 3$, $b = -2$, $c = 1$, $d = -5$ and $p = -2$

Hence **the remainder is:**
$$3(-2)^3 + (-2)(-2)^2 + (1)(-2) + (-5)$$
$$= -24 - 8 - 2 - 5$$
$$= \mathbf{-39}$$

> **Problem 10.** Determine the remainder when $(x^3 - 2x^2 - 5x + 6)$ is divided by (a) $(x - 1)$ and (b) $(x + 2)$. Hence factorise the cubic expression

(a) When $(x^3 - 2x^2 - 5x + 6)$ is divided by $(x - 1)$, the remainder is given by $ap^3 + bp^2 + cp + d$, where $a = 1, b = -2, c = -5, d = 6$ and $p = 1$, i.e.

the remainder $= (1)(1)^3 + (-2)(1)^2$
$$+ (-5)(1) + 6$$
$$= 1 - 2 - 5 + 6 = \mathbf{0}$$

Hence $(x - 1)$ is a factor of $(x^3 - 2x^2 - 5x + 6)$.

(b) When $(x^3 - 2x^2 - 5x + 6)$ is divided by $(x + 2)$, **the remainder is** given by

$$(1)(-2)^3 + (-2)(-2)^2 + (-5)(-2) + 6$$
$$= -8 - 8 + 10 + 6 = \mathbf{0}$$

Hence $(x + 2)$ is also a factor of $(x^3 - 2x^2 - 5x + 6)$. Therefore $(x - 1)(x + 2)(x \quad) = x^3 - 2x^2 - 5x + 6$. To determine the third factor (shown blank) we could

   (i) divide $(x^3 - 2x^2 - 5x + 6)$ by $(x - 1)(x + 2)$

or (ii) use the factor theorem where $f(x) = x^3 - 2x^2 - 5x + 6$ and hope to choose a value of x which makes $f(x) = 0$

or (iii) use the remainder theorem, again hoping to choose a factor $(x - p)$ which makes the remainder zero.

   (i) Dividing $(x^3 - 2x^2 - 5x + 6)$ by $(x^2 + x - 2)$ gives:

$$
\begin{array}{r}
x - 3 \\
x^2 + x - 2 \overline{\smash{\big)}\, x^3 - 2x^2 - 5x + 6} \\
\underline{x^3 + x^2 - 2x} \\
-3x^2 - 3x + 6 \\
\underline{-3x^2 - 3x + 6} \\
\cdot\quad\cdot\quad\cdot
\end{array}
$$

Thus $(x^3 - 2x^2 - 5x + 6)$
$$= (x - 1)(x + 2)(x - 3)$$

(ii) Using the factor theorem, we let

$$f(x) = x^3 - 2x^2 - 5x + 6$$

Then $f(3) = 3^3 - 2(3)^2 - 5(3) + 6$

$$= 27 - 18 - 15 + 6 = 0$$

Hence $(x-3)$ is a factor.

(iii) Using the remainder theorem, when $(x^3 - 2x^2 - 5x + 6)$ is divided by $(x-3)$, the remainder is given by $ap^3 + bp^2 + cp + d$, where $a=1$, $b=-2$, $c=-5$, $d=6$ and $p=3$

Hence the remainder is:

$$1(3)^3 + (-2)(3)^2 + (-5)(3) + 6$$
$$= 27 - 18 - 15 + 6 = 0$$

Hence $(x-3)$ is a factor.

Thus $(x^3 - 2x^2 - 5x + 6)$

$$= (x-1)(x+2)(x-3)$$

Now try the following Practice Exercise

Practice Exercise 80   Further problems on the remainder theorem (answers on page 1156)

1. Find the remainder when $3x^2 - 4x + 2$ is divided by

   (a) $(x-2)$ (b) $(x+1)$

2. Determine the remainder when $x^3 - 6x^2 + x - 5$ is divided by

   (a) $(x+2)$ (b) $(x-3)$

3. Use the remainder theorem to find the factors of $x^3 - 6x^2 + 11x - 6$

4. Determine the factors of $x^3 + 7x^2 + 14x + 8$ and hence solve the cubic equation $x^3 + 7x^2 + 14x + 8 = 0$

5. Determine the value of 'a' if $(x+2)$ is a factor of $(x^3 - ax^2 + 7x + 10)$

6. Using the remainder theorem, solve the equation $2x^3 - x^2 - 7x + 6 = 0$

For fully worked solutions to each of the problems in Practice Exercises 78 to 80 in this chapter, go to the website:
www.routledge.com/cw/bird

Chapter 19

Partial fractions

Why it is important to understand: **Partial fractions**

The algebraic technique of resolving a complicated fraction into partial fractions is often needed by electrical and mechanical engineers for not only determining certain integrals in calculus, but for determining inverse Laplace transforms, and for analysing linear differential equations such as resonant circuits and feedback control systems.

At the end of this chapter, you should be able to:

- understand the term 'partial fraction'
- appreciate the conditions needed to resolve a fraction into partial fractions
- resolve into partial fractions a fraction containing linear factors in the denominator
- resolve into partial fractions a fraction containing repeated linear factors in the denominator
- resolve into partial fractions a fraction containing quadratic factors in the denominator

---

19.1 Introduction to partial fractions

By algebraic addition,

$$\frac{1}{x-2} + \frac{3}{x+1} = \frac{(x+1) + 3(x-2)}{(x-2)(x+1)}$$

$$= \frac{4x-5}{x^2-x-2}$$

The reverse process of moving from $\dfrac{4x-5}{x^2-x-2}$ to $\dfrac{1}{x-2} + \dfrac{3}{x+1}$ is called resolving into **partial fractions**.

In order to resolve an algebraic expression into partial fractions:

(i) the denominator must factorise (in the above example, x^2-x-2 factorises as $(x-2)(x+1)$), and

(ii) the numerator must be at least one degree less than the denominator (in the above example $(4x-5)$ is of degree 1 since the highest powered x term is x^1 and (x^2-x-2) is of degree 2).

When the degree of the numerator is equal to or higher than the degree of the denominator, the numerator must be divided by the denominator (see Problems 3 and 4). There are basically three types of partial fraction and the form of partial fraction used is summarised in Table 19.1, where $f(x)$ is assumed to be of less degree than the relevant denominator and A, B and C are constants to be determined.

(In the latter type in Table 19.1, ax^2+bx+c is a quadratic expression which does not factorise without containing surds or imaginary terms.)

Resolving an algebraic expression into partial fractions is used as a preliminary to integrating certain functions (see Chapter 66).

Table 19.1

Type	Denominator containing	Expression	Form of partial fraction
1	Linear factors (see Problems 1 to 4)	$\dfrac{f(x)}{(x+a)(x-b)(x+c)}$	$\dfrac{A}{(x+a)}+\dfrac{B}{(x-b)}+\dfrac{C}{(x+c)}$
2	Repeated linear factors (see Problems 5 to 7)	$\dfrac{f(x)}{(x+a)^3}$	$\dfrac{A}{(x+a)}+\dfrac{B}{(x+a)^2}+\dfrac{C}{(x+a)^3}$
3	Quadratic factors (see Problems 8 and 9)	$\dfrac{f(x)}{(ax^2+bx+c)(x+d)}$	$\dfrac{Ax+B}{(ax^2+bx+c)}+\dfrac{C}{(x+d)}$

19.2 Worked problems on partial fractions with linear factors

Problem 1. Resolve $\dfrac{11-3x}{x^2+2x-3}$ into partial fractions

The denominator factorises as $(x-1)(x+3)$ and the numerator is of less degree than the denominator. Thus $\dfrac{11-3x}{x^2+2x-3}$ may be resolved into partial fractions.
Let

$$\frac{11-3x}{x^2+2x-3}\equiv\frac{11-3x}{(x-1)(x+3)}\equiv\frac{A}{(x-1)}+\frac{B}{(x+3)},$$

where A and B are constants to be determined,

i.e. $\quad\dfrac{11-3x}{(x-1)(x+3)}\equiv\dfrac{A(x+3)+B(x-1)}{(x-1)(x+3)}$

by algebraic addition.

Since the denominators are the same on each side of the identity then the numerators are equal to each other.

Thus, $11-3x\equiv A(x+3)+B(x-1)$

To determine constants A and B, values of x are chosen to make the term in A or B equal to zero.

When $x=1$, then $11-3(1)\equiv A(1+3)+B(0)$

i.e. $\qquad\qquad 8=4A$

i.e. $\qquad\qquad$ **A = 2**

When $x=-3$, then $11-3(-3)\equiv A(0)+B(-3-1)$

i.e. $\qquad\qquad 20=-4B$

i.e. $\qquad\qquad$ **B = −5**

Thus $\dfrac{\mathbf{11-3x}}{\mathbf{x^2+2x-3}}\equiv\dfrac{2}{(x-1)}+\dfrac{-5}{(x+3)}$

$$\equiv\frac{2}{(x-1)}-\frac{5}{(x+3)}$$

$$\left[\text{Check:}\quad\frac{2}{(x-1)}-\frac{5}{(x+3)}\right.$$

$$=\frac{2(x+3)-5(x-1)}{(x-1)(x+3)}$$

$$\left.=\frac{11-3x}{x^2+2x-3}\right]$$

Problem 2. Convert $\dfrac{2x^2-9x-35}{(x+1)(x-2)(x+3)}$ into the sum of three partial fractions

Let $\quad\dfrac{2x^2-9x-35}{(x+1)(x-2)(x+3)}$

$$\equiv\frac{A}{(x+1)}+\frac{B}{(x-2)}+\frac{C}{(x+3)}$$

$$\equiv\frac{[A(x-2)(x+3)+B(x+1)(x+3)}{(x+1)(x-2)(x+3)}$$

by algebraic addition

Equating the numerators gives:

$$2x^2-9x-35\equiv A(x-2)(x+3)+B(x+1)(x+3)$$
$$+C(x+1)(x-2)$$

Let $x = -1$. Then

$$2(-1)^2 - 9(-1) - 35 \equiv A(-3)(2) + B(0)(2)$$
$$+ C(0)(-3)$$

i.e. $\qquad -24 = -6A$

i.e. $\qquad A = \dfrac{-24}{-6} = \mathbf{4}$

Let $x = 2$. Then

$$2(2)^2 - 9(2) - 35 \equiv A(0)(5) + B(3)(5)$$
$$+ C(3)(0)$$

i.e. $\qquad -45 = 15B$

i.e. $\qquad B = \dfrac{-45}{15} = \mathbf{-3}$

Let $x = -3$. Then

$$2(-3)^2 - 9(-3) - 35 \equiv A(-5)(0) + B(-2)(0)$$
$$+ C(-2)(-5)$$

i.e. $\qquad 10 = 10C$

i.e. $\qquad \mathbf{C = 1}$

Thus $\qquad \dfrac{2x^2 - 9x - 35}{(x+1)(x-2)(x+3)}$

$$\equiv \dfrac{4}{(x+1)} - \dfrac{3}{(x-2)} + \dfrac{1}{(x+3)}$$

Problem 3. Resolve $\dfrac{x^2 + 1}{x^2 - 3x + 2}$ into partial fractions

The denominator is of the same degree as the numerator. Thus dividing out gives:

$$\begin{array}{r} 1 \\ x^2 - 3x + 2 \overline{) x^2 + 1} \\ \underline{x^2 - 3x + 2} \\ 3x - 1 \end{array}$$

For more on polynomial division, see Section 18.1, page 171.

Hence $\qquad \dfrac{x^2 + 1}{x^2 - 3x + 2} \equiv 1 + \dfrac{3x - 1}{x^2 - 3x + 2}$

$$\equiv 1 + \dfrac{3x - 1}{(x-1)(x-2)}$$

Let $\qquad \dfrac{3x - 1}{(x-1)(x-2)} \equiv \dfrac{A}{(x-1)} + \dfrac{B}{(x-2)}$

$$\equiv \dfrac{A(x-2) + B(x-1)}{(x-1)(x-2)}$$

Equating numerators gives:

$$3x - 1 \equiv A(x-2) + B(x-1)$$

Let $x = 1$. $\quad$ Then $\quad 2 = -A$

i.e. $\qquad \mathbf{A = -2}$

Let $x = 2$. $\quad$ Then $\quad \mathbf{5 = B}$

Hence $\quad \dfrac{3x - 1}{(x-1)(x-2)} \equiv \dfrac{-2}{(x-1)} + \dfrac{5}{(x-2)}$

Thus $\quad \dfrac{x^2 + 1}{x^2 - 3x + 2} \equiv 1 - \dfrac{2}{(x-1)} + \dfrac{5}{(x-2)}$

Problem 4. Express $\dfrac{x^3 - 2x^2 - 4x - 4}{x^2 + x - 2}$ in partial fractions

The numerator is of higher degree than the denominator. Thus dividing out gives:

$$\begin{array}{r} x - 3 \\ x^2 + x - 2 \overline{) x^3 - 2x^2 - 4x - 4} \\ \underline{x^3 + x^2 - 2x} \\ -3x^2 - 2x - 4 \\ \underline{-3x^2 - 3x + 6} \\ x - 10 \end{array}$$

Thus $\quad \dfrac{x^3 - 2x^2 - 4x - 4}{x^2 + x - 2} \equiv x - 3 + \dfrac{x - 10}{x^2 + x - 2}$

$$\equiv x - 3 + \dfrac{x - 10}{(x+2)(x-1)}$$

Let $\quad \dfrac{x - 10}{(x+2)(x-1)} \equiv \dfrac{A}{(x+2)} + \dfrac{B}{(x-1)}$

$$\equiv \dfrac{A(x-1) + B(x+2)}{(x+2)(x-1)}$$

Equating the numerators gives:

$$x - 10 \equiv A(x-1) + B(x+2)$$

Let $x = -2$. $\quad$ Then $\quad -12 = -3A$

i.e. $\qquad \mathbf{A = 4}$

Let $x = 1$. Then $-9 = 3B$

i.e. $\mathbf{B = -3}$

Hence $\dfrac{x-10}{(x+2)(x-1)} \equiv \dfrac{4}{(x+2)} - \dfrac{3}{(x-1)}$

Thus $\dfrac{x^3 - 2x^2 - 4x - 4}{x^2 + x - 2}$

$\equiv x - 3 + \dfrac{4}{(x+2)} - \dfrac{3}{(x-1)}$

Now try the following Practice Exercise

Practice Exercise 81   Further problems on partial fractions with linear factors (answers on page 1156)

Resolve the following into partial fractions:

1. $\dfrac{12}{x^2 - 9}$

2. $\dfrac{4(x-4)}{x^2 - 2x - 3}$

3. $\dfrac{x^2 - 3x + 6}{x(x-2)(x-1)}$

4. $\dfrac{3(2x^2 - 8x - 1)}{(x+4)(x+1)(2x-1)}$

5. $\dfrac{x^2 + 9x + 8}{x^2 + x - 6}$

6. $\dfrac{x^2 - x - 14}{x^2 - 2x - 3}$

7. $\dfrac{3x^3 - 2x^2 - 16x + 20}{(x-2)(x+2)}$

19.3 Worked problems on partial fractions with repeated linear factors

Problem 5.   Resolve $\dfrac{2x+3}{(x-2)^2}$ into partial fractions

The denominator contains a repeated linear factor, $(x-2)^2$

Let $\dfrac{2x+3}{(x-2)^2} \equiv \dfrac{A}{(x-2)} + \dfrac{B}{(x-2)^2}$

$\equiv \dfrac{A(x-2) + B}{(x-2)^2}$

Equating the numerators gives:

$$2x + 3 \equiv A(x-2) + B$$

Let $x = 2$.   Then $7 = A(0) + B$

i.e. $\mathbf{B = 7}$

$$2x + 3 \equiv A(x-2) + B$$

$$\equiv Ax - 2A + B$$

Since an identity is true for all values of the unknown, the coefficients of similar terms may be equated.

Hence, equating the coefficients of x gives: $\mathbf{2 = A}$
[Also, as a check, equating the constant terms gives: $3 = -2A + B$. When $A = 2$ and $B = 7$, RHS $= -2(2) + 7 = 3 =$ LHS]

Hence $\dfrac{\mathbf{2x+3}}{\mathbf{(x-2)^2}} \equiv \dfrac{\mathbf{2}}{\mathbf{(x-2)}} + \dfrac{\mathbf{7}}{\mathbf{(x-2)^2}}$

Problem 6.   Express $\dfrac{5x^2 - 2x - 19}{(x+3)(x-1)^2}$ as the sum of three partial fractions

The denominator is a combination of a linear factor and a repeated linear factor.

Let $\dfrac{5x^2 - 2x - 19}{(x+3)(x-1)^2}$

$\equiv \dfrac{A}{(x+3)} + \dfrac{B}{(x-1)} + \dfrac{C}{(x-1)^2}$

$\equiv \dfrac{A(x-1)^2 + B(x+3)(x-1) + C(x+3)}{(x+3)(x-1)^2}$

by algebraic addition

Equating the numerators gives:

$$5x^2 - 2x - 19 \equiv A(x-1)^2 + B(x+3)(x-1)$$
$$+ C(x+3) \quad (1)$$

Let $x = -3$. Then

$$5(-3)^2 - 2(-3) - 19 \equiv A(-4)^2 + B(0)(-4) + C(0)$$

i.e. $32 = 16A$

i.e. $\mathbf{A = 2}$

Let $x = 1$. Then

$$5(1)^2 - 2(1) - 19 \equiv A(0)^2 + B(4)(0) + C(4)$$

i.e. $$-16 = 4C$$

i.e. $$\mathbf{C = -4}$$

Without expanding the RHS of equation (1) it can be seen that equating the coefficients of x^2 gives: $5 = A + B$, and since $A = 2$, $\mathbf{B = 3}$

[Check: identity (1) may be expressed as:

$$5x^2 - 2x - 19 \equiv A(x^2 - 2x + 1)$$
$$+ B(x^2 + 2x - 3) + C(x + 3)$$

i.e. $$5x^2 - 2x - 19 \equiv Ax^2 - 2Ax + A + Bx^2$$
$$+ 2Bx - 3B + Cx + 3C$$

Equating the x term coefficients gives:

$$-2 \equiv -2A + 2B + C$$

When $A = 2$, $B = 3$ and $C = -4$ then
$-2A + 2B + C = -2(2) + 2(3) - 4 = -2 = $ LHS
Equating the constant term gives:

$$-19 \equiv A - 3B + 3C$$

$$\text{RHS} = 2 - 3(3) + 3(-4) = 2 - 9 - 12$$

$$= -19 = \text{LHS}]$$

Hence $\dfrac{5x^2 - 2x - 19}{(x+3)(x-1)^2}$

$$\equiv \frac{2}{(x+3)} + \frac{3}{(x-1)} - \frac{4}{(x-1)^2}$$

Problem 7. Resolve $\dfrac{3x^2 + 16x + 15}{(x+3)^3}$ into partial fractions

Let

$$\frac{3x^2 + 16x + 15}{(x+3)^3} \equiv \frac{A}{(x+3)} + \frac{B}{(x+3)^2} + \frac{C}{(x+3)^3}$$

$$\equiv \frac{A(x+3)^2 + B(x+3) + C}{(x+3)^3}$$

Equating the numerators gives:

$$3x^2 + 16x + 15 \equiv A(x+3)^2 + B(x+3) + C \quad (1)$$

Let $x = -3$. Then

$$3(-3)^2 + 16(-3) + 15 \equiv A(0)^2 + B(0) + C$$

i.e. $$\mathbf{-6 = C}$$

Identity (1) may be expanded as:

$$3x^2 + 16x + 15 \equiv A(x^2 + 6x + 9) + B(x+3) + C$$

i.e. $$3x^2 + 16x + 15 \equiv Ax^2 + 6Ax + 9A$$
$$+ Bx + 3B + C$$

Equating the coefficients of x^2 terms gives:

$$\mathbf{3 = A}$$

Equating the coefficients of x terms gives:

$$16 = 6A + B$$

Since $A = 3$, $\mathbf{B = -2}$

[Check: equating the constant terms gives:

$$15 = 9A + 3B + C$$

When $A = 3$, $B = -2$ and $C = -6$,

$$9A + 3B + C = 9(3) + 3(-2) + (-6)$$
$$= 27 - 6 - 6 = 15 = \text{LHS}]$$

Thus $\dfrac{\mathbf{3x^2 + 16x + 15}}{\mathbf{(x+3)^3}}$

$$\equiv \frac{3}{(x+3)} - \frac{2}{(x+3)^2} - \frac{6}{(x+3)^3}$$

Now try the following Practice Exercise

Practice Exercise 82   Further problems on partial fractions with repeated linear factors (answers on page 1157)

Resolve the following:

1. $\dfrac{4x - 3}{(x+1)^2}$

2. $\dfrac{x^2 + 7x + 3}{x^2(x+3)}$

3. $\dfrac{5x^2 - 30x + 44}{(x-2)^3}$

4. $\dfrac{18 + 21x - x^2}{(x-5)(x+2)^2}$

19.4 Worked problems on partial fractions with quadratic factors

Problem 8. Express $\dfrac{7x^2+5x+13}{(x^2+2)(x+1)}$ in partial fractions

The denominator is a combination of a quadratic factor, (x^2+2), which does not factorise without introducing imaginary surd terms, and a linear factor, $(x+1)$. Let

$$\frac{7x^2+5x+13}{(x^2+2)(x+1)} \equiv \frac{Ax+B}{(x^2+2)} + \frac{C}{(x+1)}$$

$$\equiv \frac{(Ax+B)(x+1)+C(x^2+2)}{(x^2+2)(x+1)}$$

Equating numerators gives:

$$7x^2+5x+13 \equiv (Ax+B)(x+1)+C(x^2+2) \quad (1)$$

Let $x=-1$. Then

$$7(-1)^2+5(-1)+13 \equiv (Ax+B)(0)+C(1+2)$$

i.e. $\qquad\qquad 15 = 3C$

i.e. $\qquad\qquad \mathbf{C=5}$

Identity (1) may be expanded as:

$$7x^2+5x+13 \equiv Ax^2+Ax+Bx+B+Cx^2+2C$$

Equating the coefficients of x^2 terms gives:

$$7 = A+C, \text{and since } C = 5, \mathbf{A=2}$$

Equating the coefficients of x terms gives:

$$5 = A+B, \text{and since } A = 2, \mathbf{B=3}$$

[Check: equating the constant terms gives:

$$13 = B+2C$$

When $B=3$ and $C=5$, $B+2C=3+10=13=$ LHS]

Hence $\dfrac{7x^2+5x+13}{(x^2+2)(x+1)} \equiv \dfrac{2x+3}{(x^2+2)} + \dfrac{5}{(x+1)}$

Problem 9. Resolve $\dfrac{3+6x+4x^2-2x^3}{x^2(x^2+3)}$ into partial fractions

Terms such as x^2 may be treated as $(x+0)^2$, i.e. they are repeated linear factors

Let $\qquad \dfrac{3+6x+4x^2-2x^3}{x^2(x^2+3)}$

$$\equiv \frac{A}{x} + \frac{B}{x^2} + \frac{Cx+D}{(x^2+3)}$$

$$\equiv \frac{Ax(x^2+3)+B(x^2+3)+(Cx+D)x^2}{x^2(x^2+3)}$$

Equating the numerators gives:

$$3+6x+4x^2-2x^3 \equiv Ax(x^2+3)$$
$$+ B(x^2+3)+(Cx+D)x^2$$
$$\equiv Ax^3+3Ax+Bx^2+3B$$
$$+ Cx^3+Dx^2$$

Let $x=0$. Then $3=3B$

i.e. $\qquad\qquad \mathbf{B=1}$

Equating the coefficients of x^3 terms gives:

$$-2 = A+C \qquad\qquad (1)$$

Equating the coefficients of x^2 terms gives:

$$4 = B+D$$

Since $B=1$, $\mathbf{D=3}$

Equating the coefficients of x terms gives:

$$6 = 3A$$

i.e. $\mathbf{A=2}$

From equation (1), since $A=2$, $\mathbf{C=-4}$

Hence $\qquad \dfrac{3+6x+4x^2-2x^3}{x^2(x^2+3)}$

$$\equiv \frac{2}{x} + \frac{1}{x^2} + \frac{-4x+3}{(x^2+3)}$$

$$\equiv \frac{2}{x} + \frac{1}{x^2} + \frac{3-4x}{(x^2+3)}$$

Now try the following Practice Exercise

Practice Exercise 83   Further problems on partial fractions with quadratic factors (answers on page 1157)

Resolve the following:

1. $\dfrac{x^2-x-13}{(x^2+7)(x-2)}$

2. $\dfrac{6x-5}{(x-4)(x^2+3)}$

3. $\dfrac{15+5x+5x^2-4x^3}{x^2(x^2+5)}$

4. $\dfrac{x^3+4x^2+20x-7}{(x-1)^2(x^2+8)}$

5. When solving the differential equation $\dfrac{d^2\theta}{dt^2}-6\dfrac{d\theta}{dt}-10\theta=20-e^{2t}$ by Laplace transforms, for given boundary conditions, the following expression for $\mathcal{L}\{\theta\}$ results:

$$\mathcal{L}\{\theta\}=\dfrac{4s^3-\dfrac{39}{2}s^2+42s-40}{s(s-2)(s^2-6s+10)}$$

Show that the expression can be resolved into partial fractions to give:

$$\mathcal{L}\{\theta\}=\dfrac{2}{s}-\dfrac{1}{2(s-2)}+\dfrac{5s-3}{2(s^2-6s+10)}$$

For fully worked solutions to each of the problems in Practice Exercises 81 to 83 in this chapter, go to the website:
www.routledge.com/cw/bird

Chapter 20

Number sequences

Why it is important to understand: Number sequences

Number sequences are widely used in engineering applications, including computer data structure and sorting algorithms, financial engineering, audio compression and architectural engineering. Thanks to engineers, robots have migrated from factory shop floors (as industrial manipulators) to outer space (as interplanetary explorers), hospitals (as minimally invasive surgical assistants), homes (as vacuum cleaners and lawn mowers) and battlefields (as unmanned air, underwater and ground vehicles). Arithmetic progressions (APs) are used in simulation engineering and in the reproductive cycle of bacteria. Some uses of APs in daily life include uniform increase in speed at regular intervals, completing patterns of objects, calculating simple interest, speed of an aircraft, increase or decrease in the costs of goods, sales and production, and so on. Geometric progressions (GPs) are used in compound interest and the range of speeds on a drilling machine. In fact, GPs are used throughout mathematics, and they have many important applications in physics, engineering, biology, economics, computer science, queuing theory and finance.

At the end of this chapter, you should be able to:

- define and determine simple number sequences
- calculate the nth term of a series
- calculate the nth term of an AP
- calculate the sum of n terms of an AP
- calculate the nth term of a GP
- calculate the sum of n terms of a GP
- calculate the sum to infinity of a GP

20.1 Simple sequences

A set of numbers which are connected by a definite law is called a **series** or a **sequence of numbers**. Each of the numbers in the series is called a **term** of the series.

For example, $1, 3, 5, 7, \ldots$ is a series obtained by adding 2 to the previous term, and $2, 8, 32, 128, \ldots$ is a sequence obtained by multiplying the previous term by 4

Problem 1. Determine the next two terms in the series: $3, 6, 9, 12, \ldots$

We notice that the sequence $3, 6, 9, 12, \ldots$ progressively increases by 3, thus the next two terms will be **15 and 18**

Problem 2. Find the next three terms in the series: $9, 5, 1, \ldots$

We notice that each term in the series $9, 5, 1, \ldots$ progressively decreases by 4, thus the next two terms will be $1 - 4$, i.e. **−3** and $-3 - 4$, i.e. **−7**

Problem 3. Determine the next two terms in the series: $2, 6, 18, 54, \ldots$

We notice that the second term, 6, is three times the first term, the third term, 18, is three times the second term, and that the fourth term, 54, is three times the third term. Hence the fifth term will be $3 \times 54 = $ **162**, and the sixth term will be $3 \times 162 = $ **486**

Now try the following Practice Exercise

Practice Exercise 84   Simple sequences (answers on page 1157)

Determine the next two terms in each of the following series:

1. $5, 9, 13, 17, \ldots$ 2. $3, 6, 12, 24, \ldots$

3. $112, 56, 28, \ldots$ 4. $12, 7, 2, \ldots$

5. $2, 5, 10, 17, 26, 37, \ldots$ 6. $1, 0.1, 0.01, \ldots$

7. $4, 9, 19, 34, \ldots$

20.2   The nth term of a series

If a series is represented by a general expression, say, $2n + 1$, where n is an integer (i.e. a whole number), then by substituting $n = 1, 2, 3, \ldots$ the terms of the series can be determined; in this example, the first three terms will be:

$2(1) + 1, \quad 2(2) + 1, \quad 2(3) + 1, \ldots, \quad$ i.e. $3, 5, 7, \ldots$

What is the nth term of the sequence $1, 3, 5, 7, \ldots$? First, we notice that the gap between each term is 2, hence the law relating the numbers is: '$2n +$ something'.
The second term, $3 = 2n +$ something, hence when $n = 2$ (i.e. the second term of the series), then $3 = 4 +$ something, and the 'something' must be -1

Thus, **the nth term of $1, 3, 5, 7, \ldots$ is $2n - 1$**

Hence the fifth term is given by $2(5) - 1 = 9$, and the twentieth term is $2(20) - 1 = 39$, and so on.

Problem 4. The nth term of a sequence is given by $3n + 1$. Write down the first four terms

The first four terms of the series $3n + 1$ will be:

$3(1) + 1, \quad 3(2) + 1, \quad 3(3) + 1 \quad$ and $\quad 3(4) + 1$

i.e.   **4, 7, 10 and 13**

Problem 5. The nth term of a series is given by $4n - 1$. Write down the first four terms

The first four terms of the series $4n - 1$ will be:

$4(1) - 1, \quad 4(2) - 1, \quad 4(3) - 1 \quad$ and $\quad 4(4) - 1$

i.e.   **3, 7, 11 and 15**

Problem 6. Find the nth term of the series: $1, 4, 7, \ldots$

We notice that the gap between each of the given three terms is 3, hence the law relating the numbers is: '$3n +$ something'.

The second term, $\qquad 4 = 3n +$ something,

so when $n = 2$, then $\quad 4 = 6 +$ something,

so the 'something' must be -2 (from simple equations)

Thus, **the nth term of the series $1, 4, 7, \ldots$ is: $3n - 2$**

Problem 7. Find the nth term of the sequence: $3, 9, 15, 21, \ldots$ Hence determine the 15th term of the series

We notice that the gap between each of the given four terms is 6, hence the law relating the numbers is: '$6n +$ something'.

The second term, $\qquad 9 = 6n +$ something,

so when $n = 2$, then $\quad 9 = 12 +$ something,

so the 'something' must be -3

Thus, **the nth term of the series $3, 9, 15, 21, \ldots$ is: $6n - 3$**
The 15th term of the series is given by $6n - 3$ when $n = 15$
Hence, **the 15th term of the series $3, 9, 15, 21, \ldots$ is: $6(15) - 3 = $ 87**

Problem 8. Find the nth term of the series: $1, 4, 9, 16, 25, \ldots$

This is a special series and does not follow the pattern of the previous examples. Each of the terms in the given series are square numbers,

i.e. $1, 4, 9, 16, 25, \ldots \equiv 1^2, 2^2, 3^2, 4^2, 5^2, \ldots$

Hence the nth term is: n^2

Now try the following Practice Exercise

Practice Exercise 85   The nth term of a series (answers on page 1157)

1. The nth term of a sequence is given by $2n - 1$. Write down the first four terms.

2. The nth term of a sequence is given by $3n + 4$. Write down the first five terms.

3. Write down the first four terms of the sequence given by $5n + 1$.

In Problems 4 to 8, find the nth term in the series:

4. $5, 10, 15, 20, \ldots$ 5. $4, 10, 16, 22, \ldots$

6. $3, 5, 7, 9, \ldots$ 7. $2, 6, 10, 14, \ldots$

8. $9, 12, 15, 18, \ldots$

9. Write down the next two terms of the series: $1, 8, 27, 64, 125, \ldots$

20.3   Arithmetic progressions

When a sequence has a constant difference between successive terms it is called an **arithmetic progression** (often abbreviated to AP).

Examples include:

(i) $1, 4, 7, 10, 13, \ldots$ where the **common difference** is 3

and (ii) $a, a + d, a + 2d, a + 3d, \ldots$ where the common difference is d.

General expression for the nth term of an AP

If the first term of an AP is 'a' and the common difference is 'd' then:

the nth term is: $a + (n - 1)d$

In example (i) above, the 7th term is given by $1 + (7 - 1)3 = 19$, which may be readily checked.

Sum of n terms of an AP

The sum S of an AP can be obtained by multiplying the average of all the terms by the number of terms.

The average of all the terms $= \dfrac{a + l}{2}$, where 'a' is the first term and 'l' is the last term, i.e. $l = a + (n - 1)d$, for n terms.

Hence, the sum of n terms, $S_n = n\left(\dfrac{a + l}{2}\right)$

$$= \frac{n}{2}\{a + [a + (n - 1)d]\}$$

i.e. $$S_n = \frac{n}{2}[2a + (n - 1)d]$$

For example, the sum of the first 7 terms of the series $1, 4, 7, 10, 13, \ldots$ is given by:

$$S_7 = \frac{7}{2}[2(1) + (7 - 1)3] \quad \text{since } a = 1 \text{ and } d = 3$$

$$= \frac{7}{2}[2 + 18] = \frac{7}{2}[20] = \mathbf{70}$$

Here are some worked problems to help understanding of arithmetic progressions.

Problem 9.   Determine (a) the 9th, and (b) the 16th term of the series $2, 7, 12, 17, \ldots$

$2, 7, 12, 17, \ldots$ is an arithmetic progression with a common difference, d, of 5

(a) The nth term of an AP is given by $a + (n - 1)d$
Since the first term $a = 2, d = 5$ and $n = 9$ then the 9th term is: $2 + (9 - 1)5 = 2 + (8)(5)$
$$= 2 + 40 = \mathbf{42}$$

(b) The 16th term is: $2 + (16 - 1)5 = 2 + (15)(5)$
$$= 2 + 75 = \mathbf{77}$$

Problem 10.   The 6th term of an AP is 17 and the 13th term is 38. Determine the 19th term

The nth term of an AP is: $a + (n - 1)d$

The 6th term is: $a + 5d = 17$ (1)

The 13th term is: $a + 12d = 38$ (2)

Equation (2) − equation (1) gives: $7d = 21$ from which,
$$d = \frac{21}{7} = 3$$

Substituting in equation (1) gives: $a + 15 = 17$ from which, $a = 2$

Hence, the 19th term is: $a + (n-1)d = 2 + (19-1)3$

$$= 2 + (18)(3)$$

$$= 2 + 54 = \mathbf{56}$$

Problem 11.   Determine the number of the term whose value is 22 in the series $2.5, 4, 5.5, 7, \ldots$

$2.5, 4, 5.5, 7, \ldots$ is an AP where $a = 2.5$ and $d = 1.5$

Hence, if the nth term is 22 then:

$$a + (n-1)d = 22$$

i.e. $2.5 + (n-1)(1.5) = 22$

i.e. $(n-1)(1.5) = 22 - 2.5 = 19.5$

and $n - 1 = \dfrac{19.5}{1.5} = 13$

from which, $n = 13 + 1 = \mathbf{14}$

i.e.   **the 14th term of the AP is 22**

Problem 12.   Find the sum of the first 12 terms of the series $5, 9, 13, 17, \ldots$

$5, 9, 13, 17, \ldots$ is an AP where $a = 5$ and $d = 4$
The sum of n terms of an AP, $S_n = \dfrac{n}{2}[2a + (n-1)d]$

Hence, the sum of the first 12 terms,

$$S_{12} = \frac{12}{2}[2(5) + (12-1)4]$$

$$= 6[10 + 44] = 6(54) = \mathbf{324}$$

Problem 13.   Find the sum of the first 21 terms of the series $3.5, 4.1, 4.7, 5.3, \ldots$

$3.5, 4.1, 4.7, 5.3, \ldots$ is an AP where $a = 3.5$ and $d = 0.6$

The sum of the first 21 terms,

$$S_{21} = \frac{21}{2}[2a + (n-1)d]$$

$$= \frac{21}{2}[2(3.5) + (21-1)0.6]$$

$$= \frac{21}{2}[7 + 12]$$

$$= \frac{21}{2}(19) = \frac{399}{2} = \mathbf{199.5}$$

Now try the following Practice Exercise

Practice Exercise 86   Arithmetic progressions (answers on page 1157)

1. Find the 11th term of the series $8, 14, 20, 26, \ldots$

2. Find the 17th term of the series $11, 10.7, 10.4, 10.1, \ldots$

3. The 7th term of a series is 29 and the 11th term is 54. Determine the 16th term.

4. Find the 15th term of an arithmetic progression of which the 1st term is 2.5 and the 10th term is 16.

5. Determine the number of the term which is 29 in the series $7, 9.2, 11.4, 13.6, \ldots$

6. Find the sum of the first 11 terms of the series $4, 7, 10, 13, \ldots$

7. Determine the sum of the series $6.5, 8.0, 9.5, 11.0, \ldots, 32$

Here are some further worked problems on arithmetic progressions.

Problem 14.   The sum of seven terms of an AP is 35 and the common difference is 1.2. Determine the first term of the series

$n = 7, d = 1.2$ and $S_7 = 35$. Since the sum of n terms of an AP is given by:

$$S_n = \frac{n}{2}[2a + (n-1)d]$$

Then $35 = \dfrac{7}{2}[2a + (7-1)1.2] = \dfrac{7}{2}[2a + 7.2]$

Hence, $\dfrac{35 \times 2}{7} = 2a + 7.2$

i.e. $10 = 2a + 7.2$

Thus, $2a = 10 - 7.2 = 2.8$

from which, $a = \dfrac{2.8}{2} = 1.4$

i.e.   **the first term, $a = 1.4$**

Problem 15. Three numbers are in arithmetic progression. Their sum is 15 and their product is 80. Determine the three numbers

Let the three numbers be $(a-d)$, a and $(a+d)$

Then, $(a-d) + a + (a+d) = 15$

i.e. $3a = 15$

from which, $a = 5$

Also, $a(a-d)(a+d) = 80$

i.e. $a(a^2 - d^2) = 80$

Since $a = 5$, $5(5^2 - d^2) = 80$

i.e. $125 - 5d^2 = 80$

and $125 - 80 = 5d^2$

i.e. $45 = 5d^2$

from which, $d^2 = \dfrac{45}{5} = 9$

Hence, $d = \sqrt{9} = \pm 3$

The three numbers are thus: $(5-3)$, 5 and $(5+3)$, i.e. **2, 5 and 8**

Problem 16. Find the sum of all the numbers between 0 and 207 which are exactly divisible by 3

The series $3, 6, 9, 12, \dots 207$ is an AP whose first term $a = 3$ and common difference $d = 3$

The last term is: $a + (n-1)d = 207$

i.e. $3 + (n-1)3 = 207$

from which, $(n-1) = \dfrac{207-3}{3} = 68$

Hence, $n = 68 + 1 = 69$

The sum of all 69 terms is given by:

$$S_{69} = \frac{n}{2}[2a + (n-1)d]$$

$$= \frac{69}{2}[2(3) + (69-1)3]$$

$$= \frac{69}{2}[6 + 204] = \frac{69}{2}(210) = \mathbf{7245}$$

Problem 17. The 1st, 12th and last term of an arithmetic progression are: $4, 31.5$, and 376.5, respectively. Determine (a) the number of terms in the series, (b) the sum of all the terms and (c) the 80th term.

(a) Let the AP be $a, a+d, a+2d, \dots, a+(n-1)d$, where $a = 4$

The 12th term is: $a + (12-1)d = 31.5$

i.e. $4 + 11d = 31.5$

from which, $11d = 31.5 - 4 = 27.5$

Hence, $d = \dfrac{27.5}{11} = 2.5$

The last term is: $a + (n-1)d$

i.e. $4 + (n-1)(2.5) = 376.5$

i.e. $(n-1) = \dfrac{376.5 - 4}{2.5} = \dfrac{372.5}{2.5} = 149$

Hence, the number of terms in the series, $n = 149 + 1 = 150$

(b) **Sum of all the terms,**

$$S_{150} = \frac{n}{2}[2a + (n-1)d]$$

$$= \frac{150}{2}[2(4) + (150-1)(2.5)]$$

$$= 75[8 + (149)(2.5)] = 75[8 + 372.5]$$

$$= 75(380.5) = \mathbf{28537.5}$$

(c) **The 80th term is**: $a + (n-1)d = 4 + (80-1)(2.5)$

$$= 4 + (79)(2.5)$$

$$= 4 + 197.5$$

$$= \mathbf{201.5}$$

Now try the following Practice Exercise

Practice Exercise 87   Arithmetic progressions (answers on page 1157)

1. The sum of 15 terms of an arithmetic progression is 202.5 and the common difference is 2. Find the first term of the series.

2. Three numbers are in arithmetic progression. Their sum is 9 and their product is 20.25. Determine the three numbers.

3. Find the sum of all the numbers between 5 and 250 which are exactly divisible by 4

4. Find the number of terms of the series $5, 8, 11, \dots$ of which the sum is 1025

5. Insert four terms between 5 and 22.5 to form an arithmetic progression.

6. The 1st, 10th and last terms of an arithmetic progression are 9, 40.5 and 425.5, respectively. Find (a) the number of terms, (b) the sum of all the terms and (c) the 70th term.

7. On commencing employment a man is paid a salary of £16 000 per annum and receives annual increments of £480. Determine his salary in the 9th year and calculate the total he will have received in the first 12 years.

8. An oil company bores a hole 80 m deep. Estimate the cost of boring if the cost is £30 for drilling the first metre with an increase in cost of £2 per metre for each succeeding metre.

20.4 Geometric progressions

When a sequence has a constant ratio between successive terms it is called a **geometric progression** (often abbreviated to GP). The constant is called the **common ratio, r**.

Examples include

(i) $1, 2, 4, 8, \ldots$ where the common ratio is 2

and (ii) $a, ar, ar^2, ar^3, \ldots$ where the common ratio is r

General expression for the nth term of a GP

If the first term of a GP is 'a' and the common ratio is 'r', then

the nth term is: ar^{n-1}

which can be readily checked from the above examples. For example, the 8th term of the GP $1, 2, 4, 8, \ldots$ is $(1)(2)^7 = \mathbf{128}$, since 'a' $= 1$ and 'r' $= 2$

Sum of n terms of a GP

Let a GP be $a, ar, ar^2, ar^3, \ldots, ar^{n-1}$

then the sum of n terms,

$$S_n = a + ar + ar^2 + ar^3 + \cdots + ar^{n-1} \cdots \quad (1)$$

Multiplying throughout by r gives:

$$rS_n = ar + ar^2 + ar^3 + ar^4 + \cdots ar^{n-1} + ar^n \cdots \quad (2)$$

Subtracting equation (2) from equation (1) gives:

$$S_n - rS_n = a - ar^n$$

i.e. $\qquad S_n(1 - r) = a(1 - r^n)$

Thus, **the sum of n terms, $S_n = \dfrac{a(1 - r^n)}{(1 - r)}$** which is valid when $r < 1$

Subtracting equation (1) from equation (2) gives:

$$S_n = \frac{a(r^n - 1)}{(r - 1)} \qquad \text{which is valid when } r > 1$$

For example, the sum of the first 8 terms of the GP $1, 2, 4, 8, 16, \ldots$ is given by:

$$S_8 = \frac{1(2^8 - 1)}{(2 - 1)} \qquad \text{since '}a\text{'} = 1 \quad \text{and} \quad r = 2$$

i.e. $\quad \mathbf{S_8 = \dfrac{1(256 - 1)}{1} = 255}$

Sum to infinity of a GP

When the common ratio r of a GP is less than unity, the sum of n terms,

$$S_n = \frac{a(1 - r^n)}{(1 - r)}, \text{ which may be written as:}$$

$$S_n = \frac{a}{(1 - r)} - \frac{ar^n}{(1 - r)}$$

Since $r < 1, r^n$ becomes less as n increases, i.e. $r^n \to 0$ as $n \to \infty$

Hence, $\qquad \dfrac{ar^n}{(1 - r)} \to 0$ as $n \to \infty$

Thus, $\qquad S_n \to \dfrac{a}{(1 - r)}$ as $n \to \infty$

The quantity $\dfrac{a}{(1 - r)}$ is called the **sum to infinity, S_∞**, and is the limiting value of the sum of an infinite number of terms,

i.e. $\quad \mathbf{S_\infty = \dfrac{a}{(1 - r)}}$ which is valid when $-1 < r < 1$

Convergence means that the values of the terms must get progressively smaller and the sum of the terms must reach a limiting value.

For example, the function $y = \dfrac{1}{x}$ converges to zero as x increases.

Similarly, the series $1 + \dfrac{1}{2} + \dfrac{1}{4} + \dfrac{1}{8} + \cdots$ is convergent since the value of the terms is getting smaller and the sum of the terms is approaching a limiting value of 2,

i.e. the sum to infinity, $S_\infty = \dfrac{a}{1 - r} = \dfrac{1}{1 - \frac{1}{2}} = \mathbf{2}$

Here are some worked problems to help understanding of geometric progressions.

Problem 18. Determine the 10th term of the series $3, 6, 12, 24, \ldots$

$3, 6, 12, 24, \ldots$ is a geometric progression with a common ratio r of 2

The nth term of a GP is ar^{n-1}, where 'a' is the first term.

Hence, **the 10th term is:**

$(3)(2)^{10-1} = (3)(2)^9 = 3(512) = \mathbf{1536}$

Problem 19. Find the sum of the first seven terms of the series, $0.5, 1.5, 4.5, 13.5, \ldots$

$0.5, 1.5, 4.5, 13.5, \ldots$ is a GP with a common ratio $r = 3$

The sum of n terms, $S_n = \dfrac{a(r^n - 1)}{(r - 1)}$

Hence, **the sum of the first seven terms,**

$$S_7 = \frac{0.5(3^7 - 1)}{(3 - 1)} = \frac{0.5(2187 - 1)}{2} = \mathbf{546.5}$$

Problem 20. The 1st term of a geometric progression is 12 and the 5th term is 55. Determine the 8th term and the 11th term

The 5th term is given by: $ar^4 = 55$, where the first term $a = 12$

Hence, $\quad r^4 = \dfrac{55}{a} = \dfrac{55}{12} \quad$ and $\quad r = \sqrt[4]{\left(\dfrac{55}{12}\right)}$

$$= 1.4631719\ldots$$

The 8th term is: $\quad ar^7 = (12)(1.4631719\ldots)^7$

$$= \mathbf{172.3}$$

The 11th term is: $\quad ar^{10} = (12)(1.4631719\ldots)^{10}$

$$= \mathbf{539.7}$$

Problem 21. Which term of the series $2187, 729, 243, \ldots$ is $\dfrac{1}{9}$?

$2187, 729, 243, \ldots$ is a GP with a common ratio $r = \dfrac{1}{3}$ and first term $a = 2187$

The nth term of a GP is given by: $\quad ar^{n-1}$

Hence, $\quad \dfrac{1}{9} = (2187)\left(\dfrac{1}{3}\right)^{n-1} \quad$ from which,

$$\left(\frac{1}{3}\right)^{n-1} = \frac{1}{(9)(2187)} = \frac{1}{3^2 3^7} = \frac{1}{3^9} = \left(\frac{1}{3}\right)^9$$

Thus, $\quad (n - 1) = 9 \quad$ from which, $\quad \mathbf{n = 9 + 1 = 10}$

i.e. $\qquad \dfrac{1}{9}$ **is the 10th term of the GP**

Problem 22. Find the sum of the first nine terms of the series $72.0, 57.6, 46.08, \ldots$

The common ratio, $\qquad r = \dfrac{ar}{a} = \dfrac{57.6}{72.0} = 0.8$

$$\left(\text{also } \frac{ar^2}{ar} = \frac{46.08}{57.6} = 0.8\right)$$

The sum of nine terms, $\qquad S_9 = \dfrac{a(1 - r^n)}{(1 - r)}$

$$= \frac{72.0(1 - 0.8^9)}{(1 - 0.8)}$$

$$= \frac{72.0(1 - 0.1342)}{0.2}$$

$$= \mathbf{311.7}$$

Problem 23. Find the sum to infinity of the series $3, 1, \dfrac{1}{3}, \ldots$

$3, 1, \dfrac{1}{3}, \ldots$ is a GP of common ratio, $r = \dfrac{1}{3}$

The sum to infinity, $S_\infty = \dfrac{a}{1 - r} = \dfrac{3}{1 - \frac{1}{3}} = \dfrac{3}{\frac{2}{3}}$

$$= \frac{9}{2} = \mathbf{4.5}$$

Now try the following Practice Exercise

Practice Exercise 88   Geometric progressions (answers on page 1157)

1. Find the 10th term of the series $5, 10, 20, 40, \ldots$

2. Determine the sum of the first seven terms of the series $0.25, 0.75, 2.25, 6.75, \ldots$

3. The 1st term of a geometric progression is 4 and the 6th term is 128. Determine the 8th and 11th terms.

4. Find the sum of the first seven terms of the series $2, 5, 12.5, \ldots$ (correct to 4 significant figures).

Section B

5. Determine the sum to infinity of the series $4, 2, 1, \ldots$

6. Find the sum to infinity of the series $2\frac{1}{2}, -1\frac{1}{4}, \frac{5}{8}, \ldots$

Here are some further worked problems on geometric progressions.

Problem 24. In a geometric progression the 6th term is 8 times the 3rd term and the sum of the 7th and 8th terms is 192. Determine (a) the common ratio, (b) the 1st term, and (c) the sum of the 5th to 11th terms, inclusive

(a) Let the GP be $a, ar, ar^2, ar^3, \ldots, ar^{n-1}$

The 3rd term $= ar^2$ and the 6th term $= ar^5$

The 6th term is 8 times the 3rd

Hence, $ar^5 = 8ar^2$ from which,

$r^3 = 8$ and $r = \sqrt[3]{8}$

i.e. **the common ratio $r = 2$**

(b) The sum of the 7th and 8th terms is 192. Hence $ar^6 + ar^7 = 192$
Since $r = 2$, then $64a + 128a = 192$

$192a = 192$ from which, **the 1st term, $a = 1$**

(c) **The sum of the 5th to 11th terms** (inclusive) **is given by**:

$$S_{11} - S_4 = \frac{a(r^{11} - 1)}{(r - 1)} - \frac{a(r^4 - 1)}{(r - 1)}$$

$$= \frac{1(2^{11} - 1)}{(2 - 1)} - \frac{1(2^4 - 1)}{(2 - 1)}$$

$$= (2^{11} - 1) - (2^4 - 1) = 2^{11} - 2^4$$

$$= 2048 - 16 = \mathbf{2032}$$

Problem 25. A tool hire firm finds that their net return from hiring tools is decreasing by 10% per annum. If their net gain on a certain tool this year is £400, find the possible total of all future profits from this tool (assuming the tool lasts forever)

The net gain forms a series: $£400 + £400 \times 0.9 + £400 \times 0.9^2 + \cdots$, which is a GP with $a = 400$ and $r = 0.9$

The sum to infinity, $S_\infty = \dfrac{a}{(1 - r)} = \dfrac{400}{(1 - 0.9)}$

$$= \mathbf{£4000}$$

$$= \textbf{total future profits}$$

Problem 26. If £100 is invested at compound interest of 8% per annum, determine (a) the value after 10 years and (b) the time, correct to the nearest year, it takes to reach more than £300

(a) Let the GP be $a, ar, ar^2, \ldots ar^n$

The 1st term, $a = £100$ and the common ratio, $r = 1.08$

Hence, the 2nd term is: $ar = (100)(1.08) = £108$, which is the value after 1 year, the 3rd term is: $ar^2 = (100)(1.08)^2 = £116.64$, which is the value after 2 years, and so on.

Thus, **the value after 10 years**

$$= ar^{10} = (100)(1.08)^{10} = \mathbf{£215.89}$$

(b) When £300 has been reached, $300 = ar^n$

i.e. $\qquad\qquad 300 = 100(1.08)^n$

and $\qquad\qquad 3 = (1.08)^n$

Taking logarithms to base 10 of both sides gives:

$$\lg 3 = \lg(1.08)^n = n \lg(1.08)$$

$$\text{by the laws of logarithms}$$

from which, $\quad n = \dfrac{\lg 3}{\lg 1.08} = 14.3$

Hence, it will take 15 years to reach more than £300

Problem 27. A drilling machine is to have 6 speeds ranging from 50 rev/min to 750 rev/min. If the speeds form a geometric progression determine their values, each correct to the nearest whole number

Let the GP of n terms be given by: $a, ar, ar^2, \ldots, ar^{n-1}$

The 1st term $a = 50$ rev/min

The 6th term is given by ar^{6-1}, which is 750 rev/min, i.e. $\qquad ar^5 = 750$

from which, $\quad r^5 = \dfrac{750}{a} = \dfrac{750}{50} = 15$

Thus, the common ratio, $\quad r = \sqrt[5]{15} = 1.7188$

The 1st term is: $a = 50\,\text{rev/min}$

The 2nd term is: $ar = (50)(1.7188) = 85.94,$

The 3rd term is: $ar^2 = (50)(1.7188)^2 = 147.71,$

The 4th term is: $ar^3 = (50)(1.7188)^3 = 253.89,$

The 5th term is: $ar^4 = (50)(1.7188)^4 = 436.39,$

The 6th term is: $ar^5 = (50)(1.7188)^5 = 750.06$

Hence, correct to the nearest whole number, **the 6 speeds of the drilling machine are**:

50, 86, 148, 254, 436 and 750 rev/min

Now try the following Practice Exercise

Practice Exercise 89   Geometric progressions (answers on page 1157)

1. In a geometric progression the 5th term is 9 times the 3rd term and the sum of the 6th and 7th terms is 1944. Determine (a) the common ratio, (b) the first term and (c) the sum of the 4th to 10th terms inclusive.

2. Which term of the series $3, 9, 27, \ldots$ is $59\,049$?

3. The value of a lathe originally valued at £3000 depreciates 15% per annum. Calculate its value after 4 years. The machine is sold when its value is less than £550. After how many years is the lathe sold?

4. If the population of Great Britain is 64 million and is decreasing at 2.4% per annum, what will be the population in 5 years' time?

5. 100 g of a radioactive substance disintegrates at a rate of 3% per annum. How much of the substance is left after 11 years?

6. If £250 is invested at compound interest of 6% per annum, determine (a) the value after 15 years and (b) the time, correct to the nearest year, it takes to reach £750.

7. A drilling machine is to have 8 speeds ranging from 100 rev/min to 1000 rev/min. If the speeds form a geometric progression, determine their values, each correct to the nearest whole number.

20.5   Combinations and permutations

A **combination** is the number of selections of r different items from n distinguishable items when order of selection is ignored. A combination is denoted by ^nC_r or $\binom{n}{r}$

where
$$^nC_r = \frac{n!}{r!\,(n-r)!}$$

where, for example, 4! denotes $4 \times 3 \times 2 \times 1$ and is termed 'factorial 4'.

Thus,
$$^5C_3 = \frac{5!}{3!\,(5-3)!} = \frac{5 \times 4 \times 3 \times 2 \times 1}{(3 \times 2 \times 1)(2 \times 1)}$$
$$= \frac{120}{6 \times 2} = 10$$

For example, the five letters A, B, C, D, E can be arranged in groups of three as follows: ABC, ABD, ABE, ACD, ACE, ADE, BCD, BCE, BDE, CDE, i.e. there are ten groups. The above calculation $^5C_3$ produces the answer of 10 combinations without having to list all of them.

A **permutation** is the number of ways of selecting $r \leq n$ objects from n distinguishable objects when order of selection is important. A permutation is denoted by ^nP_r or $_nP_r$

where $^nP_r = n(n-1)(n-2)\ldots(n-r+1)$

or $^nP_r = \frac{n!}{(n-r)!}$

Thus, $^4P_2 = \frac{4!}{(4-2)!} = \frac{4!}{2!}$

$$= \frac{4 \times 3 \times 2}{2} = 12$$

Problem 28.   Evaluate: (a) $^7C_4$ (b) $^{10}C_6$

(a) $^7C_4 = \frac{7!}{4!\,(7-4)!} = \frac{7!}{4!\,3!}$

$$= \frac{7 \times 6 \times 5 \times 4 \times 3 \times 2}{(4 \times 3 \times 2)(3 \times 2)} = \mathbf{35}$$

(b) $^{10}C_6 = \frac{10!}{6!\,(10-6)!} = \frac{10!}{6!\,4!} = \mathbf{210}$

Section B

Problem 29. Evaluate: (a) $^6P_2$ (b) $^9P_5$

(a) $^6P_2 = \dfrac{6!}{(6-2)!} = \dfrac{6!}{4!}$

$\qquad = \dfrac{6 \times 5 \times 4 \times 3 \times 2}{4 \times 3 \times 2} = \textbf{30}$

(b) $^9P_5 = \dfrac{9!}{(9-5)!} = \dfrac{9!}{4!}$

$\qquad = \dfrac{9 \times 8 \times 7 \times 6 \times 5 \times 4!}{4!} = \textbf{15 120}$

Now try the following Practice Exercise

Practice Exercise 90   Further problems on permutations and combinations (answers on page 1157)

Evaluate the following:

1. (a) $^9C_6$ (b) $^3C_1$

2. (a) $^6C_2$ (b) $^8C_5$

3. (a) $^4P_2$ (b) $^7P_4$

4. (a) $^{10}P_3$ (b) $^8P_5$

For fully worked solutions to each of the problems in Practice Exercises 84 to 90 in this chapter, go to the website:
www.routledge.com/cw/bird

This assignment covers the material contained in Chapters 18 to 20. *The marks available are shown in brackets at the end of each question.*

1. Factorise $x^3 + 4x^2 + x - 6$ using the factor theorem. Hence solve the equation

$$x^3 + 4x^2 + x - 6 = 0 \qquad (6)$$

2. Use the remainder theorem to find the remainder when $2x^3 + x^2 - 7x - 6$ is divided by

   (a) $(x-2)$ (b) $(x+1)$

   Hence factorise the cubic expression. (7)

3. Resolve into partial fractions:

   $$\frac{x-11}{x^2-x-2} \qquad (6)$$

4. Resolve into partial fractions:

   $$\frac{3-x}{(x^2+3)(x+3)} \qquad (8)$$

5. Resolve into partial fractions:

   $$\frac{x^3-6x+9}{x^2+x-2} \qquad (11)$$

6. Determine the 20th term of the series 15.6, 15, 14.4, 13.8, ... (3)

7. The sum of 13 terms of an arithmetic progression is 286 and the common difference is 3. Determine the first term of the series. (5)

8. An engineer earns £21 000 per annum and receives annual increments of £600. Determine the salary in the 9th year and calculate the total earnings in the first 11 years. (5)

9. Determine the 11th term of the series 1.5, 3, 6, 12, ... (2)

10. Find the sum of the first eight terms of the series $1, 2\frac{1}{2}, 6\frac{1}{4}, \ldots$, correct to 1 decimal place. (4)

11. Determine the sum to infinity of the series 5, $1, \frac{1}{5}, \ldots$ (3)

12. A machine is to have seven speeds ranging from 25 rev/min to 500 rev/min. If the speeds form a geometric progression, determine their value, each correct to the nearest whole number. (10)

Section B

Chapter 21

The binomial series

Why it is important to understand: **The binomial series**

There are many applications of the binomial theorem in every part of algebra, and in general with permutations, combinations and probability. It is also used in atomic physics where it is used to count s, p, d and f orbitals. There are applications of the binomial series in financial mathematics to determine the number of stock price paths that leads to a particular stock price at maturity.

At the end of this chapter, you should be able to:

- define a binomial expression
- use Pascal's triangle to expand a binomial expression
- state the general binomial expansion of $(a + x)^n$ and $(1 + x)^n$
- use the binomial series to expand expressions of the form $(a + x)^n$ for positive, negative and fractional values of n
- determine the rth term of a binomial expansion
- use the binomial expansion with practical applications

21.1 Pascal's triangle

A **binomial expression** is one that contains two terms connected by a plus or minus sign. Thus $(p + q)$, $(a + x)^2$, $(2x + y)^3$ are examples of binomial expression. Expanding $(a + x)^n$ for integer values of n from 0 to 6 gives the results shown at the top of page 197. From the results the following patterns emerge:

(i) 'a' decreases in power moving from left to right.

(ii) 'x' increases in power moving from left to right.

(iii) The coefficients of each term of the expansions are symmetrical about the middle coefficient when n is even and symmetrical about the two middle coefficients when n is odd.

(iv) The coefficients are shown separately in Table 21.1 and this arrangement is known as

$$(a+x)^0 = \qquad\qquad 1$$
$$(a+x)^1 = \qquad\qquad a+x$$
$$(a+x)^2 = (a+x)(a+x) = \qquad a^2+2ax+x^2$$
$$(a+x)^3 = (a+x)^2(a+x) = \qquad a^3+3a^2x+3ax^2+x^3$$
$$(a+x)^4 = (a+x)^3(a+x) = \qquad a^4+4a^3x+6a^2x^2+4ax^3+x^4$$
$$(a+x)^5 = (a+x)^4(a+x) = \qquad a^5+5a^4x+10a^3x^2+10a^2x^3+5ax^4+x^5$$
$$(a+x)^6 = (a+x)^5(a+x) = \qquad a^6+6a^5x+15a^4x^2+20a^3x^3+15a^2x^4+6ax^5+x^6$$

Pascal's triangle*. A coefficient of a term may be obtained by adding the two adjacent coefficients immediately above in the previous row. This is shown by the triangles in Table 21.1, where, for example, $1+3=4$, $10+5=15$, and so on.

(v) Pascal's triangle method is used for expansions of the form $(a+x)^n$ for integer values of n less than about 8.

Problem 1. Use the Pascal's triangle method to determine the expansion of $(a+x)^7$

From Table 21.1, the row the Pascal's triangle corresponding to $(a+x)^6$ is as shown in (1) below. Adding

* **Who was Pascal?** Blaise Pascal (19 June 1623–19 August 1662) was a French polymath. A child prodigy, he wrote a significant treatise on the subject of projective geometry at the age of sixteen, and later corresponded with Pierre de Fermat on probability theory, strongly influencing the development of modern economics and social science. To find out more go to **www.routledge.com/cw/bird**

Table 21.1

$(a+x)^0$				1			
$(a+x)^1$			1		1		
$(a+x)^2$			1	2	1		
$(a+x)^3$		1	3	3	1		
$(a+x)^4$	1	4	6	4	1		
$(a+x)^5$	1	5	10	10	5	1	
$(a+x)^6$	1	6	15	20	15	6	1

adjacent coefficients gives the coefficients of $(a+x)^7$ as shown in (2) below.

$$1 \quad 6 \quad 15 \quad 20 \quad 15 \quad 6 \quad 1 \qquad (1)$$
$$1 \quad 7 \quad 21 \quad 35 \quad 35 \quad 21 \quad 7 \quad 1 \qquad (2)$$

The first and last terms of the expansion of $(a+x)^7$ and a^7 and x^7, respectively. The powers of 'a' decrease and the powers of 'x' increase moving from left to right. Hence,

$$(a+x)^7 = a^7+7a^6x+21a^5x^2+35a^4x^3+35a^3x^4$$
$$+21a^2x^5+7ax^6+x^7$$

Problem 2. Determine, using Pascal's triangle method, the expansion of $(2p-3q)^5$

Comparing $(2p-3q)^5$ with $(a+x)^5$ shows that $a=2p$ and $x=-3q$

Using Pascal's triangle method:
$$(a+x)^5 = a^5+5a^4x+10a^3x^2+10a^2x^3+\cdots$$

Hence

$$(2p-3q)^5 = (2p)^5+5(2p)^4(-3q)$$
$$+10(2p)^3(-3q)^2$$
$$+10(2p)^2(-3q)^3$$
$$+5(2p)(-3q)^4+(-3q)^5$$

i.e. $(2p - 3q)^5 = 32p^5 - 240p^4q + 720p^3q^2$
$$- 1080p^2q^3 + 810pq^4 - 243q^5$$

Now try the following Practice Exercise

Practice Exercise 91   Further problems on Pascal's triangle (answers on page 1157)

1. Use Pascal's triangle to expand $(x - y)^7$

2. Expand $(2a + 3b)^5$ using Pascal's triangle.

21.2   The binomial series

The **binomial series** or **binomial theorem** is a formula for raising a binomial expression to any power without lengthy multiplication. The general binomial expansion of $(a + x)^n$ is given by:

$$(a + x)^n = a^n + na^{n-1}x + \frac{n(n-1)}{2!}a^{n-2}x^2$$
$$+ \frac{n(n-1)(n-2)}{3!}a^{n-3}x^3$$
$$+ \cdots + x^n \quad (1)$$

where, for example, 3! denotes $3 \times 2 \times 1$ and is termed '**factorial** 3'.

With the binomial theorem n may be a fraction, a decimal fraction or a positive or negative integer.

In the general expansion of $(a + x)^n$ it is noted that the 4th term is:

$$\frac{n(n-1)(n-2)}{3!}a^{n-3}x^3$$

The number 3 is very evident in this expression.

For any term in a binomial expansion, say the rth term, $(r - 1)$ is very evident. It may therefore be reasoned that **the rth term of the expansion $(a + x)^n$ is:**

$$\frac{n(n-1)(n-2) \ldots \text{to } (r-1) \text{ terms}}{(r-1)!}a^{n-(r-1)}x^{r-1}$$

If $a = 1$ in the binomial expansion of $(a + x)^n$ then:

$$(1 + x)^n = 1 + nx + \frac{n(n-1)}{2!}x^2$$
$$+ \frac{n(n-1)(n-2)}{3!}x^3 + \cdots$$

which is valid for $-1 < x < 1$

When x is small compared with 1 then:

$$(1 + x)^n \approx 1 + nx$$

21.3   Worked problems on the binomial series

Problem 3. Use the binomial series to determine the expansion of $(2 + x)^7$

The binomial expansion is given by:

$$(a + x)^n = a^n + na^{n-1}x + \frac{n(n-1)}{2!}a^{n-2}x^2$$
$$+ \frac{n(n-1)(n-2)}{3!}a^{n-3}x^3 + \cdots$$

When $a = 2$ and $n = 7$:

$$(2 + x)^7 = 2^7 + 7(2)^6 + \frac{(7)(6)}{(2)(1)}(2)^5x^2$$
$$+ \frac{(7)(6)(5)}{(3)(2)(1)}(2)^4x^3 + \frac{(7)(6)(5)(4)}{(4)(3)(2)(1)}(2)^3x^4$$
$$+ \frac{(7)(6)(5)(4)(3)}{(5)(4)(3)(2)(1)}(2)^2x^5$$
$$+ \frac{(7)(6)(5)(4)(3)(2)}{(6)(5)(4)(3)(2)(1)}(2)x^6$$
$$+ \frac{(7)(6)(5)(4)(3)(2)(1)}{(7)(6)(5)(4)(3)(2)(1)}x^7$$

i.e. $(2 + x)^7 = 128 + 448x + 672x^2 + 560x^3$
$$+ 280x^4 + 84x^5 + 14x^6 + x^7$$

Problem 4. Use the binomial series to determine the expansion of $(2a - 3b)^5$

From equation (1), the binomial expansion is given by:

$$(a + x)^n = a^n + na^{n-1}x + \frac{n(n-1)}{2!}a^{n-2}x^2$$
$$+ \frac{n(n-1)(n-2)}{3!}a^{n-3}x^3 + \cdots$$

When $a = 2a$, $x = -3b$ and $n = 5$:
$$(2a - 3b)^5 = (2a)^5 + 5(2a)^4(-3b)$$
$$+ \frac{(5)(4)}{(2)(1)}(2a)^3(-3b)^2$$
$$+ \frac{(5)(4)(3)}{(3)(2)(1)}(2a)^2(-3b)^3$$

$$+ \frac{(5)(4)(3)(2)}{(4)(3)(2)(1)}(2a)(-3b)^4$$

$$+ \frac{(5)(4)(3)(2)(1)}{(5)(4)(3)(2)(1)}(-3b)^5$$

i.e. $(2a - 3b)^5 = 32a^5 - 240a^4b + 720a^3b^2$
$$- 1080a^2b^3 + 810ab^4 - 243b^5$$

Problem 5. Expand $\left(c - \dfrac{1}{c}\right)^5$ using the binomial series

$$\left(c - \frac{1}{c}\right)^5 = c^5 + 5c^4\left(-\frac{1}{c}\right) + \frac{(5)(4)}{(2)(1)}c^3\left(-\frac{1}{c}\right)^2$$

$$+ \frac{(5)(4)(3)}{(3)(2)(1)}c^2\left(-\frac{1}{c}\right)^3$$

$$+ \frac{(5)(4)(3)(2)}{(4)(3)(2)(1)}c\left(-\frac{1}{c}\right)^4$$

$$+ \frac{(5)(4)(3)(2)(1)}{(5)(4)(3)(2)(1)}\left(-\frac{1}{c}\right)^5$$

i.e. $\left(c - \dfrac{1}{c}\right)^5 = c^5 - 5c^3 + 10c - \dfrac{10}{c} + \dfrac{5}{c^3} - \dfrac{1}{c^5}$

Problem 6. Without fully expanding $(3 + x)^7$, determine the 5th term

The rth term of the expansion $(a + x)^n$ is given by:

$$\frac{n(n - 1)(n - 2)\dots\text{to } (r - 1) \text{ terms}}{(r - 1)!}a^{n-(r-1)}x^{r-1}$$

Substituting $n = 7$, $a = 3$ and $r - 1 = 5 - 1 = 4$ gives:

$$\frac{(7)(6)(5)(4)}{(4)(3)(2)(1)}(3)^{7-4}x^4$$

i.e. the 5th term of $(3 + x)^7 = 35(3)^3x^4 = \mathbf{945x^4}$

Problem 7. Find the middle term of $\left(2p - \dfrac{1}{2q}\right)^{10}$

In the expansion of $(a + x)^{10}$ there are $10 + 1$, i.e. 11 terms. Hence the middle term is the 6th. Using the general expression for the rth term where $a = 2p$,

$x = -\dfrac{1}{2q}$, $n = 10$ and $r - 1 = 5$ gives:

$$\frac{(10)(9)(8)(7)(6)}{(5)(4)(3)(2)(1)}(2p)^{10-5}\left(-\frac{1}{2q}\right)^5$$

$$= 252(32p^5)\left(-\frac{1}{32q^5}\right)$$

Hence the middle term of $\left(2q - \dfrac{1}{2q}\right)^{10}$ is $-252\dfrac{p^5}{q^5}$

Problem 8. Evaluate $(1.002)^9$ using the binomial theorem correct to (a) 3 decimal places and (b) 7 significant figures

$$(1 + x)^n = 1 + nx + \frac{n(n - 1)}{2!}x^2$$

$$+ \frac{n(n - 1)(n - 2)}{3!}x^3 + \cdots$$

$$(1.002)^9 = (1 + 0.002)^9$$

Substituting $x = 0.002$ and $n = 9$ in the general expansion for $(1 + x)^n$ gives:

$$(1 + 0.002)^9 = 1 + 9(0.002) + \frac{(9)(8)}{(2)(1)}(0.002)^2$$

$$+ \frac{(9)(8)(7)}{(3)(2)(1)}(0.002)^3 + \cdots$$

$$= 1 + 0.018 + 0.000144$$

$$+ 0.000000672 + \cdots$$

$$= 1.018144672\dots$$

Hence, $(1.002)^9 = \mathbf{1.018, \text{ correct to 3 decimal places}}$

$$= \mathbf{1.018145, \text{ correct to}}$$

$$\mathbf{7 \text{ significant figures}}$$

Problem 9. Determine the value of $(3.039)^4$, correct to 6 significant figures using the binomial theorem

$(3.039)^4$ may be written in the form $(1 + x)^n$ as:

$$(3.039)^4 = (3 + 0.039)^4$$

$$= \left[3\left(1 + \frac{0.039}{3}\right)\right]^4$$

$$= 3^4(1 + 0.013)^4$$

$$(1+0.013)^4 = 1 + 4(0.013) + \frac{(4)(3)}{(2)(1)}(0.013)^2$$

$$+ \frac{(4)(3)(2)}{(3)(2)(1)}(0.013)^3 + \cdots$$

$$= 1 + 0.052 + 0.001014$$

$$+ 0.000008788 + \cdots$$

$$= 1.0530228,$$

correct to 8 significant figures

Hence $(3.039)^4 = 3^4(1.0530228)$

$$= \mathbf{85.2948, \ correct \ to}$$

$$\mathbf{6 \ significant \ figures}$$

Now try the following Practice Exercise

Practice Exercise 92   Further problems on the binomial series (answers on page 1157)

1. Use the binomial theorem to expand $(a+2x)^4$

2. Use the binomial theorem to expand $(2-x)^6$

3. Expand $(2x-3y)^4$

4. Determine the expansion of $\left(2x + \frac{2}{x}\right)^5$

5. Expand $(p+2q)^{11}$ as far as the 5th term

6. Determine the 6th term of $\left(3p + \frac{q}{3}\right)^{13}$

7. Determine the middle term of $(2a-5b)^8$

8. Use the binomial theorem to determine, correct to 4 decimal places:
    (a) $(1.003)^8$ (b) $(0.98)^7$

9. Evaluate $(4.044)^6$ correct to 2 decimal places.

21.4 Further worked problems on the binomial series

Problem 10.

(a) Expand $\dfrac{1}{(1+2x)^3}$ in ascending powers of x as far as the term in x^3, using the binomial series

(b) State the limits of x for which the expansion is valid

(a) Using the binomial expansion of $(1+x)^n$, where $n = -3$ and x is replaced by $2x$ gives:

$$\frac{1}{(1+2x)^3} = (1+2x)^{-3}$$

$$= 1 + (-3)(2x) + \frac{(-3)(-4)}{2!}(2x)^2$$

$$+ \frac{(-3)(-4)(-5)}{3!}(2x)^3 + \cdots$$

$$= 1 - 6x + 24x^2 - 80x^3 +$$

(b) The expansion is valid provided $|2x| < 1$,

i.e. $|x| < \dfrac{1}{2}$ or $-\dfrac{1}{2} < x < \dfrac{1}{2}$

Problem 11.

(a) Expand $\dfrac{1}{(4-x)^2}$ in ascending powers of x as far as the term in x^3, using the binomial theorem

(b) What are the limits of x for which the expansion in (a) is true?

(a) $\dfrac{1}{(4-x)^2} = \dfrac{1}{\left[4\left(1-\frac{x}{4}\right)\right]^2} = \dfrac{1}{4^2\left(1-\frac{x}{4}\right)^2}$

$$= \frac{1}{16}\left(1 - \frac{x}{4}\right)^{-2}$$

Using the expansion of $(1+x)^n$

$$\frac{1}{(4-x)^2} = \frac{1}{16}\left(1 - \frac{x}{4}\right)^{-2}$$

$$= \frac{1}{16}\left[1 + (-2)\left(-\frac{x}{4}\right)\right.$$

$$+ \frac{(-2)(-3)}{2!}\left(-\frac{x}{4}\right)^2$$

$$\left. + \frac{(-2)(-3)(-4)}{3!}\left(-\frac{x}{4}\right)^3 + \ldots\right]$$

$$= \frac{1}{16}\left(1 + \frac{x}{2} + \frac{3x^2}{16} + \frac{x^3}{16} + \cdots\right)$$

(b) The expansion in (a) is true provided $\left|\dfrac{x}{4}\right| < 1$,

i.e. $|x| < 4$ or $-4 < x < 4$

Problem 12.   Use the binomial theorem to expand $\sqrt{4+x}$ in ascending powers of x to four terms. Give the limits of x for which the expansion is valid

$$\sqrt{4+x} = \sqrt{4\left(1+\frac{x}{4}\right)}$$

$$= \sqrt{4}\sqrt{1+\frac{x}{4}}$$

$$= 2\left(1+\frac{x}{4}\right)^{\frac{1}{2}}$$

Using the expansion of $(1+x)^n$,

$$2\left(1+\frac{x}{4}\right)^{\frac{1}{2}}$$

$$= 2\left[1+\left(\frac{1}{2}\right)\left(\frac{x}{4}\right)+\frac{(1/2)(-1/2)}{2!}\left(\frac{x}{4}\right)^2\right.$$

$$\left.+\frac{(1/2)(-1/2)(-3/2)}{3!}\left(\frac{x}{4}\right)^3+\cdots\right]$$

$$= 2\left(1+\frac{x}{8}-\frac{x^2}{128}+\frac{x^3}{1024}-\cdots\right)$$

$$= \mathbf{2}+\frac{\mathbf{x}}{\mathbf{4}}-\frac{\mathbf{x^2}}{\mathbf{64}}+\frac{\mathbf{x^3}}{\mathbf{512}}-\cdots$$

This is valid when $\left|\dfrac{x}{4}\right| < 1$,

i.e. $\left|\dfrac{x}{4}\right| < \mathbf{4}$ or $-\mathbf{4} < \mathbf{x} < \mathbf{4}$

Problem 13. Expand $\dfrac{1}{\sqrt{1-2t}}$ in ascending powers of t as far as the term in t^3. State the limits of t for which the expression is valid

$$\frac{1}{\sqrt{1-2t}}$$

$$= (1-2t)^{-\frac{1}{2}}$$

$$= 1+\left(-\frac{1}{2}\right)(-2t)+\frac{(-1/2)(-3/2)}{2!}(-2t)^2$$

$$+\frac{(-1/2)(-3/2)(-5/2)}{3!}(-2t)^3+\cdots$$

$$\text{using the expansion for} (1+x)^n$$

$$= \mathbf{1}+\mathbf{t}+\frac{\mathbf{3}}{\mathbf{2}}\mathbf{t^2}+\frac{\mathbf{5}}{\mathbf{2}}\mathbf{t^3}+\cdots$$

The expression is valid when $|2t| < 1$,

i.e. $|\mathbf{t}| < \dfrac{\mathbf{1}}{\mathbf{2}}$ or $-\dfrac{\mathbf{1}}{\mathbf{2}} < \mathbf{t} < \dfrac{\mathbf{1}}{\mathbf{2}}$

Problem 14. Simplify $\dfrac{\sqrt[3]{1-3x}\sqrt{1+x}}{\left(1+\dfrac{x}{2}\right)^3}$ given that powers of x above the first may be neglected

$$\frac{\sqrt[3]{1-3x}\sqrt{1+x}}{\left(1+\dfrac{x}{2}\right)^3}$$

$$= (1-3x)^{\frac{1}{3}}(1+x)^{\frac{1}{2}}\left(1+\frac{x}{2}\right)^{-3}$$

$$\approx \left[1+\left(\frac{1}{3}\right)(-3x)\right]\left[1+\left(\frac{1}{2}\right)(x)\right]\left[1+(-3)\left(\frac{x}{2}\right)\right]$$

when expanded by the binomial theorem as far as the x term only,

$$= (1-x)\left(1+\frac{x}{2}\right)\left(1-\frac{3x}{2}\right)$$

$$= \left(1-x+\frac{x}{2}-\frac{3x}{2}\right) \quad \text{when powers of } x \text{ higher than unity are neglected}$$

$$= (\mathbf{1-2x})$$

Problem 15. Express $\dfrac{\sqrt{1+2x}}{\sqrt[3]{1-3x}}$ as a power series as far as the term in x^2. State the range of values of x for which the series is convergent

$$\frac{\sqrt{1+2x}}{\sqrt[3]{1-3x}} = (1+2x)^{\frac{1}{2}}(1-3x)^{-\frac{1}{3}}$$

$$(1+2x)^{\frac{1}{2}} = 1+\left(\frac{1}{2}\right)(2x)$$

$$+\frac{(1/2)(-1/2)}{2!}(2x)^2+\cdots$$

$$= 1+x-\frac{x^2}{2}+\cdots \text{ which is valid for}$$

$$|2x| < 1, \text{ i.e. } |x| < \frac{1}{2}$$

$$(1-3x)^{-\frac{1}{3}} = 1+(-1/3)(-3x)$$

$$+\frac{(-1/3)(-4/3)}{2!}(-3x)^2+\cdots$$

$$= 1+x+2x^2+\cdots \text{ which is valid for}$$

$$|3x| < 1, \text{ i.e. } |x| < \frac{1}{3}$$

Section B

Hence $\dfrac{\sqrt{1+2x}}{\sqrt[3]{1-3x}}$

$= (1+2x)^{\frac{1}{2}}(1-3x)^{\frac{1}{3}}$

$= \left(1+x-\dfrac{x^2}{2}+\cdots\right)(1+x+2x^2+\cdots)$

$= 1+x+2x^2+x+x^2-\dfrac{x^2}{2}$

neglecting terms of higher power than 2

$= \mathbf{1+2x+\dfrac{5}{2}x^2}$

The series is convergent if $-\dfrac{1}{3}<x<\dfrac{1}{3}$

Now try the following Practice Exercise

Practice Exercise 93   Further problems on the binomial series (answers on page 1158)

In Problems 1 to 5 expand in ascending powers of x as far as the term in x^3, using the binomial theorem. State in each case the limits of x for which the series is valid.

1. $\dfrac{1}{(1-x)}$

2. $\dfrac{1}{(1+x)^2}$

3. $\dfrac{1}{(2+x)^3}$

4. $\sqrt{2+x}$

5. $\dfrac{1}{\sqrt{1+3x}}$

6. Expand $(2+3x)^{-6}$ to three terms. For what values of x is the expansion valid?

7. When x is very small show that:

(a) $\dfrac{1}{(1-x)^2\sqrt{1-x}} \approx 1+\dfrac{5}{2}x$

(b) $\dfrac{(1-2x)}{(1-3x)^4} \approx 1+10x$

(c) $\dfrac{\sqrt{1+5x}}{\sqrt[3]{1-2x}} \approx 1+\dfrac{19}{6}x$

8. If x is very small such that x^2 and higher powers may be neglected, determine the power series for $\dfrac{\sqrt{x+4}\ \sqrt[3]{8-x}}{\sqrt[5]{(1+x)^3}}$

9. Express the following as power series in ascending powers of x as far as the term in x^2. State in each case the range of x for which the series is valid.

(a) $\sqrt{\dfrac{1-x}{1+x}}$ (b) $\dfrac{(1+x)\ \sqrt[3]{(1-3x)^2}}{\sqrt{1+x^2}}$

21.5 Practical problems involving the binomial theorem

Binomial expansions may be used for numerical approximations, for calculations with small variations and in probability theory.

Problem 16. The radius of a cylinder is reduced by 4% and its height is increased by 2%. Determine the approximate percentage change in (a) its volume and (b) its curved surface area (neglecting the products of small quantities)

Volume of cylinder $=\pi r^2 h$

Let r and h be the original values of radius and height. The new values are $0.96r$ or $(1-0.04)r$ and $1.02h$ or $(1+0.02)h$

(a) New volume $=\pi[(1-0.04)r]^2[(1+0.02)h]$
$$= \pi r^2 h(1-0.04)^2(1+0.02)$$

Now $(1-0.04)^2 = 1-2(0.04)+(0.04)^2$
$$= (1-0.08),\ \text{neglecting powers of small terms}$$

Hence new volume

$\approx \pi r^2 h(1-0.08)(1+0.02)$

$\approx \pi r^2 h(1-0.08+0.02)$, neglecting products of small terms

$\approx \pi r^2 h(1-0.06)$ or $0.94\pi r^2 h$, i.e. 94% of the original volume

Hence the volume is reduced by approximately 6%

(b) Curved surface area of cylinder $= 2\pi rh$

New surface area
$$= 2\pi[(1-0.04)r][(1+0.02)h]$$

$$= 2\pi rh(1-0.04)(1+0.02)$$

$$\approx 2\pi rh(1-0.04+0.02), \text{ neglecting products of small terms}$$

$$\approx 2\pi rh(1-0.02) \text{ or } 0.98(2\pi rh), \text{ i.e. 98\% of the original surface area}$$

Hence the curved surface area is reduced by approximately 2%

Problem 17. The second moment of area of a rectangle through its centroid is given by $\dfrac{bl^3}{12}$. Determine the approximate change in the second moment of area if b is increased by 3.5% and l is reduced by 2.5%

New values of b and l are $(1+0.035)b$ and $(1-0.025)l$, respectively.
New second moment of area
$$= \frac{1}{12}[(1+0.035)b][(1-0.025)l]^3$$

$$= \frac{bl^3}{12}(1+0.035)(1-0.025)^3$$

$$\approx \frac{bl^3}{12}(1+0.035)(1-0.075), \text{ neglecting powers of small terms}$$

$$\approx \frac{bl^3}{12}(1+0.035-0.075), \text{ neglecting products of small terms}$$

$$\approx \frac{bl^3}{12}(1-0.040) \text{ or } (0.96)\frac{bl^3}{12}, \text{ i.e. 96\% of the original second moment of area}$$

Hence the second moment of area is reduced by approximately 4%

Problem 18. The resonant frequency of a vibrating shaft is given by: $f = \dfrac{1}{2\pi}\sqrt{\dfrac{k}{I}}$ where k is the stiffness and I is the inertia of the shaft. Use the binomial theorem to determine the approximate percentage error in determining the frequency using the measured values of k and I when the measured value of k is 4% too large and the measured value of I is 2% too small

---

Let f, k and I be the true values of frequency, stiffness and inertia, respectively. Since the measured value of stiffness, k_1, is 4% too large, then

$$k_1 = \frac{104}{100}k = (1+0.04)k$$

The measured value of inertia, I_1, is 2% too small, hence

$$I_1 = \frac{98}{100}I = (1-0.02)I$$

The measured value of frequency,

$$f_1 = \frac{1}{2\pi}\sqrt{\frac{k_1}{I_1}} = \frac{1}{2\pi}k_1^{\frac{1}{2}}I_1^{-\frac{1}{2}}$$

$$= \frac{1}{2\pi}[(1+0.04)k]^{\frac{1}{2}}[(1-0.02)I]^{-\frac{1}{2}}$$

$$= \frac{1}{2\pi}(1+0.04)^{\frac{1}{2}}k^{\frac{1}{2}}(1-0.02)^{-\frac{1}{2}}I^{-\frac{1}{2}}$$

$$= \frac{1}{2\pi}k^{\frac{1}{2}}I^{-\frac{1}{2}}(1+0.04)^{\frac{1}{2}}(1-0.02)^{-\frac{1}{2}}$$

i.e. $f_1 = f(1+0.04)^{\frac{1}{2}}(1-0.02)^{-\frac{1}{2}}$

$$\approx f\left[1+\left(\frac{1}{2}\right)(0.04)\right]\left[1+\left(-\frac{1}{2}\right)(-0.02)\right]$$

$$\approx f(1+0.02)(1+0.01)$$

Neglecting the products of small terms,

$$f_1 \approx (1+0.02+0.01)f \approx 1.03f$$

Thus the percentage error in f based on the measured values of k and I is approximately $[(1.03)(100)-100]$, i.e. **3% too large**

Now try the following Practice Exercise

Practice Exercise 94 Further practical problems involving the binomial theorem (answers on page 1158)

1. Pressure p and volume v are related by $pv^3 = c$, where c is a constant. Determine the approximate percentage change in c when p is increased by 3% and v decreased by 1.2%

2. Kinetic energy is given by $\frac{1}{2}mv^2$. Determine the approximate change in the kinetic energy

when mass m is increased by 2.5% and the velocity v is reduced by 3%

3. An error of $+1.5\%$ was made when measuring the radius of a sphere. Ignoring the products of small quantities, determine the approximate error in calculating (a) the volume, and (b) the surface area.

4. The power developed by an engine is given by $I = k$ PLAN, where k is a constant. Determine the approximate percentage change in the power when P and A are each increased by 2.5% and L and N are each decreased by 1.4%

5. The radius of a cone is increased by 2.7% and its height reduced by 0.9%. Determine the approximate percentage change in its volume, neglecting the products of small terms.

6. The electric field strength H due to a magnet of length $2l$ and moment M at a point on its axis distance x from the centre is given by:

$$H = \frac{M}{2l}\left\{\frac{1}{(x-l)^2} - \frac{1}{(x+l)^2}\right\}$$

Show that if l is very small compared with x, then $H \approx \dfrac{2M}{x^3}$

7. The shear stress τ in a shaft of diameter D under a torque T is given by: $\tau = \dfrac{kT}{\pi D^3}$. Determine the approximate percentage error in calculating τ if T is measured 3% too small and D 1.5% too large.

8. The energy W stored in a flywheel is given by: $W = kr^5 N^2$, where k is a constant, r is the radius and N the number of revolutions. Determine the approximate percentage change in W when r is increased by 1.3% and N is decreased by 2%

9. In a series electrical circuit containing inductance L and capacitance C the resonant frequency is given by: $f_r = \dfrac{1}{2\pi\sqrt{LC}}$. If the values of L and C used in the calculation are 2.6% too large and 0.8% too small, respectively, determine the approximate percentage error in the frequency.

10. The viscosity η of a liquid is given by: $\eta = \dfrac{kr^4}{vl}$, where k is a constant. If there is an error in r of $+2\%$, in v of $+4\%$ and l of -3%, what is the resultant error in η?

11. A magnetic pole, distance x from the plane of a coil of radius r, and on the axis of the coil, is subject to a force F when a current flows in the coil. The force is given by: $F = \dfrac{kx}{\sqrt{(r^2+x^2)^5}}$, where k is a constant. Use the binomial theorem to show that when x is small compared to r, then $F \approx \dfrac{kx}{r^5} - \dfrac{5kx^3}{2r^7}$

12. The flow of water through a pipe is given by: $G = \sqrt{\dfrac{(3d)^5 H}{L}}$. If d decreases by 2% and H by 1%, use the binomial theorem to estimate the decrease in G.

For fully worked solutions to each of the problems in Practice Exercises 91 to 94 in this chapter, go to the website:
www.routledge.com/cw/bird

Chapter 22

Maclaurin's series

Why it is important to understand: Maclaurin's series

One of the simplest kinds of function to deal with, in either algebra or calculus, is a polynomial (i.e. an expression of the form $a + bx + cx^2 + dx^3 + \ldots$). Polynomials are easy to substitute numerical values into, and they are easy to differentiate. One useful application of Maclaurin's series is to approximate, to a polynomial, functions which are not already in polynomial form. In the simple theory of flexure of beams, the slope, bending moment, shearing force, load and other quantities are functions of a derivative of y with respect to x. The elastic curve of a transversely loaded beam can be represented by the Maclaurin series. Substitution of the values of the derivatives gives a direct solution of beam problems. Another application of Maclaurin's series is in relating inter-atomic potential functions. At this stage, not all of the above applications will have been met or understood; however, sufficient to say that Maclaurin's series has a number of applications in engineering and science.

At the end of this chapter, you should be able to:

- determine simple derivatives
- derive Maclaurin's theorem
- appreciate the conditions of Maclaurin's series
- use Maclaurin's series to determine the power series for simple trigonometric, logarithmic and exponential functions
- evaluate a definite integral using Maclaurin's series
- state L'Hôpital's rule
- determine limiting values of functions

22.1 Introduction

Some mathematical functions may be represented as power series, containing terms in ascending powers of the variable. For example,

$$e^x = 1 + x + \frac{x^2}{2!} + \frac{x^3}{3!} + \cdots$$

$$\sin x = x - \frac{x^3}{3!} + \frac{x^5}{5!} - \frac{x^7}{7!} + \cdots$$

$$\text{and} \cosh x = 1 + \frac{x^2}{2!} + \frac{x^4}{4!} + \cdots$$

(as introduced in Chapter 24).

Using a series, called **Maclaurin's* series**, mixed functions containing, say, algebraic, trigonometric and exponential functions, may be expressed solely as algebraic

Page Sculp.
MACLAURIN,
London Pub.ᵈ as the Act directs Nov.ᵗ 1ˢᵗ 1813, by G.Jones.

* **Who was** Maclaurin? **Colin Maclaurin** (February 1698– 14 June 1746) was a Scottish mathematician who made important contributions to geometry and algebra. **The Maclaurin series** are named after him. To find out more go to **www.routledge.com/cw/bird**

functions, and differentiation and integration can often be more readily performed.

y or $f(x)$	$\frac{dy}{dx}$ or $f'(x)$
ax^n	anx^{n-1}
$\sin ax$	$a\cos ax$
$\cos ax$	$-a\sin ax$
e^{ax}	ae^{ax}
$\ln ax$	$\frac{1}{x}$
$\sinh ax$	$a\cosh ax$
$\cosh ax$	$a\sinh ax$

To expand a function using Maclaurin's theorem, some knowledge of differentiation is needed. (More on differentiation is given in Chapters 52 and 53.) Above is a revision of derivatives of the main functions needed in this chapter.

Given a general function $f(x)$, then $f'(x)$ is the first derivative, $f''(x)$ is the second derivative, and so on. Also, $f(0)$ means the value of the function when $x = 0$, $f'(0)$ means the value of the first derivative when $x = 0$, and so on.

22.2 Derivation of Maclaurin's theorem

Let the power series for $f(x)$ be

$$f(x) = a_0 + a_1 x + a_2 x^2 + a_3 x^3 + a_4 x^4 + a_5 x^5 + \cdots \tag{1}$$

where $a_0, a_1, a_2, \ldots$ are constants.

When $x = 0, f(0) = a_0$

Differentiating equation (1) with respect to x gives:

$$f'(x) = a_1 + 2a_2 x + 3a_3 x^2 + 4a_4 x^3 + 5a_5 x^4 + \cdots \tag{2}$$

When $x = 0, f'(0) = a_1$

Differentiating equation (2) with respect to x gives:

$$f''(x) = 2a_2 + (3)(2)a_3 x + (4)(3)a_4 x^2 + (5)(4)a_5 x^3 + \cdots \tag{3}$$

When $x = 0$, $f''(0) = 2a_2 = 2!\, a_2$, i.e. $a_2 = \dfrac{f''(0)}{2!}$

Differentiating equation (3) with respect to x gives:

$$f'''(x) = (3)(2)a_3 + (4)(3)(2)a_4 x$$

$$+ (5)(4)(3)a_5 x^2 + \cdots \quad (4)$$

When $x = 0$, $f'''(0) = (3)(2)a_3 = 3!a_3$, i.e. $a_3 = \dfrac{f'''(0)}{3!}$

Continuing the same procedure gives $a_4 = \dfrac{f^{iv}(0)}{4!}$,

$a_5 = \dfrac{f^v(0)}{5!}$, and so on.

Substituting for $a_0, a_1, a_2, \ldots$ in equation (1) gives:

$$f(x) = f(0) + f'(0)x + \frac{f''(0)}{2!}x^2$$

$$+ \frac{f'''(0)}{3!}x^3 + \cdots$$

i.e. $\quad f(x) = f(0) + xf'(0) + \dfrac{x^2}{2!}f''(0)$

$$+ \frac{x^3}{3!}f'''(0) + \cdots \quad (5)$$

Equation (5) is a mathematical statement called **Maclaurin's theorem** or **Maclaurin's series**.

22.3 Conditions of Maclaurin's series

Maclaurin's series may be used to represent any function, say $f(x)$, as a power series provided that at $x = 0$ the following three conditions are met:

(a) $\boldsymbol{f(0) \neq \infty}$

For example, for the function $f(x) = \cos x$, $f(0) = \cos 0 = 1$, thus $\cos x$ meets the condition. However, if $f(x) = \ln x$, $f(0) = \ln 0 = -\infty$, thus $\ln x$ does not meet this condition.

(b) $\boldsymbol{f'(0), f''(0), f'''(0), \ldots \neq \infty}$

For example, for the function $f(x) = \cos x$, $f'(0) = -\sin 0 = 0$, $f''(0) = -\cos 0 = -1$, and so on; thus $\cos x$ meets this condition. However, if $f(x) = \ln x$, $f'(0) = \frac{1}{0} = \infty$, thus $\ln x$ does not meet this condition.

(c) **The resultant Maclaurin's series must be convergent**

In general, this means that the values of the terms, or groups of terms, must get progressively smaller and the sum of the terms must reach a limiting value.

For example, the series $1 + \frac{1}{2} + \frac{1}{4} + \frac{1}{8} + \cdots$ is convergent since the value of the terms is getting smaller and the sum of the terms is approaching a limiting value of 2.

22.4 Worked problems on Maclaurin's series

Problem 1. Determine the first four terms of the power series for $\cos x$

The values of $f(0)$, $f'(0)$, $f''(0)$, $\ldots$ in the Maclaurin's series are obtained as follows:

$$f(x) = \cos x \qquad f(0) = \cos 0 = 1$$
$$f'(x) = -\sin x \qquad f'(0) = -\sin 0 = 0$$
$$f''(x) = -\cos x \qquad f''(0) = -\cos 0 = -1$$
$$f'''(x) = \sin x \qquad f'''(0) = \sin 0 = 0$$
$$f^{iv}(x) = \cos x \qquad f^{iv}(0) = \cos 0 = 1$$
$$f^v(x) = -\sin x \qquad f^v(0) = -\sin 0 = 0$$
$$f^{vi}(x) = -\cos x \qquad f^{vi}(0) = -\cos 0 = -1$$

Substituting these values into equation (5) gives:

$$f(x) = \cos x = 1 + x(0) + \frac{x^2}{2!}(-1) + \frac{x^3}{3!}(0)$$

$$+ \frac{x^4}{4!}(1) + \frac{x^5}{5!}(0) + \frac{x^6}{6!}(-1) + \cdots$$

i.e. $\quad \boldsymbol{\cos x = 1 - \dfrac{x^2}{2!} + \dfrac{x^4}{4!} - \dfrac{x^6}{6!} + \cdots}$

Problem 2. Determine the power series for $\cos 2\theta$

Replacing x with 2θ in the series obtained in Problem 1 gives:

$$\cos 2\theta = 1 - \frac{(2\theta)^2}{2!} + \frac{(2\theta)^4}{4!} - \frac{(2\theta)^6}{6!} + \cdots$$

$$= 1 - \frac{4\theta^2}{2} + \frac{16\theta^4}{24} - \frac{64\theta^6}{720} + \cdots$$

i.e. $\boldsymbol{\cos 2\theta = 1 - 2\theta^2 + \dfrac{2}{3}\theta^4 - \dfrac{4}{45}\theta^6 + \cdots}$

Problem 3. Using Maclaurin's series, find the first four (non-zero) terms for the function $f(x) = \sin x$

$$f(x) = \sin x \qquad f(0) = \sin 0 = 0$$
$$f'(x) = \cos x \qquad f'(0) = \cos 0 = 1$$
$$f''(x) = -\sin x \qquad f''(0) = -\sin 0 = 0$$
$$f'''(x) = -\cos x \qquad f'''(0) = -\cos 0 = -1$$
$$f^{iv}(x) = \sin x \qquad f^{iv}(0) = \sin 0 = 0$$
$$f^{v}(x) = \cos x \qquad f^{v}(0) = \cos 0 = 1$$
$$f^{vi}(x) = -\sin x \qquad f^{vi}(0) = -\sin 0 = 0$$
$$f^{vii}(x) = -\cos x \qquad f^{vii}(0) = -\cos 0 = -1$$

Substituting the above values into Maclaurin's series of equation (5) gives:

$$\sin x = 0 + x(1) + \frac{x^2}{2!}(0) + \frac{x^3}{3!}(-1) + \frac{x^4}{4!}(0)$$

$$+ \frac{x^5}{5!}(1) + \frac{x^6}{6!}(0) + \frac{x^7}{7!}(-1) + \cdots\cdots$$

i.e. $\mathbf{\sin x = x - \dfrac{x^3}{3!} + \dfrac{x^5}{5!} - \dfrac{x^7}{7!} + \cdots}$

Problem 4. Using Maclaurin's series, find the first five terms for the expansion of the function $f(x) = e^{3x}$

$$f(x) = e^{3x} \qquad f(0) = e^0 = 1$$
$$f'(x) = 3e^{3x} \qquad f'(0) = 3e^0 = 3$$
$$f''(x) = 9e^{3x} \qquad f''(0) = 9e^0 = 9$$
$$f'''(x) = 27e^{3x} \qquad f'''(0) = 27e^0 = 27$$
$$f^{iv}(x) = 81e^{3x} \qquad f^{iv}(0) = 81e^0 = 81$$

Substituting the above values into Maclaurin's series of equation (5) gives:

$$e^{3x} = 1 + x(3) + \frac{x^2}{2!}(9) + \frac{x^3}{3!}(27)$$

$$+ \frac{x^4}{4!}(81) + \cdots\cdots$$

$$e^{3x} = 1 + 3x + \frac{9x^2}{2!} + \frac{27x^3}{3!} + \frac{81x^4}{4!} + \cdots$$

i.e. $\mathbf{e^{3x} = 1 + 3x + \dfrac{9x^2}{2} + \dfrac{9x^3}{2} + \dfrac{27x^4}{8} + \cdots}$

Problem 5. Determine the power series for $\tan x$ as far as the term in x^3

$$f(x) = \tan x$$
$$f(0) = \tan 0 = 0$$
$$f'(x) = \sec^2 x$$
$$f'(0) = \sec^2 0 = \frac{1}{\cos^2 0} = 1$$
$$f''(x) = (2\sec x)(\sec x \tan x)$$
$$= 2\sec^2 x \tan x$$
$$f''(0) = 2\sec^2 0 \tan 0 = 0$$
$$f'''(x) = (2\sec^2 x)(\sec^2 x)$$
$$+ (\tan x)(4\sec x \sec x \tan x), \text{ by the}$$
$$\text{product rule,}$$
$$= 2\sec^4 x + 4\sec^2 x \tan^2 x$$
$$f'''(0) = 2\sec^4 0 + 4\sec^2 0 \tan^2 0 = 2$$

Substituting these values into equation (5) gives:

$$f(x) = \tan x = 0 + (x)(1) + \frac{x^2}{2!}(0) + \frac{x^3}{3!}(2)$$

i.e. $\mathbf{\tan x = x + \dfrac{1}{3}x^3}$

Problem 6. Expand $\ln(1+x)$ to five terms

$$f(x) = \ln(1+x) \qquad f(0) = \ln(1+0) = 0$$

$$f'(x) = \frac{1}{(1+x)} \qquad f'(0) = \frac{1}{1+0} = 1$$

$$f''(x) = \frac{-1}{(1+x)^2} \qquad f''(0) = \frac{-1}{(1+0)^2} = -1$$

$$f'''(x) = \frac{2}{(1+x)^3} \qquad f'''(0) = \frac{2}{(1+0)^3} = 2$$

$$f^{iv}(x) = \frac{-6}{(1+x)^4} \qquad f^{iv}(0) = \frac{-6}{(1+0)^4} = -6$$

$$f^{v}(x) = \frac{24}{(1+x)^5} \qquad f^{v}(0) = \frac{24}{(1+0)^5} = 24$$

Substituting these values into equation (5) gives:

$$f(x) = \ln(1+x) = 0 + x(1) + \frac{x^2}{2!}(-1)$$

$$+ \frac{x^3}{3!}(2) + \frac{x^4}{4!}(-6) + \frac{x^5}{5!}(24)$$

i.e. $\ln(1+x) = x - \dfrac{x^2}{2} + \dfrac{x^3}{3} - \dfrac{x^4}{4} + \dfrac{x^5}{5} - \cdots$

Problem 7.   Expand $\ln(1-x)$ to five terms

Replacing x by $-x$ in the series for $\ln(1+x)$ in Problem 6 gives:

$$\ln(1-x) = (-x) - \frac{(-x)^2}{2} + \frac{(-x)^3}{3}$$

$$- \frac{(-x)^4}{4} + \frac{(-x)^5}{5} - \cdots$$

i.e. $\ln(1-x) = -x - \dfrac{x^2}{2} - \dfrac{x^3}{3} - \dfrac{x^4}{4} - \dfrac{x^5}{5} - \cdots$

Problem 8.   Determine the power series for $\ln\left(\dfrac{1+x}{1-x}\right)$

$\ln\left(\dfrac{1+x}{1-x}\right) = \ln(1+x) - \ln(1-x)$ by the laws of logarithms, and from Problems 6 and 7,

$$\ln\left(\frac{1+x}{1-x}\right) = \left(x - \frac{x^2}{2} + \frac{x^3}{3} - \frac{x^4}{4} + \frac{x^5}{5} - \cdots\right)$$

$$- \left(-x - \frac{x^2}{2} - \frac{x^3}{3} - \frac{x^4}{4} - \frac{x^5}{5} - \cdots\right)$$

$$= 2x + \frac{2}{3}x^3 + \frac{2}{5}x^5 + \cdots$$

i.e. $\ln\left(\dfrac{1+x}{1-x}\right) = 2\left(x + \dfrac{x^3}{3} + \dfrac{x^5}{5} + \cdots\right)$

Problem 9.   Use Maclaurin's series to find the expansion of $(2+x)^4$

$$f(x) = (2+x)^4 \qquad f(0) = 2^4 = 16$$
$$f'(x) = 4(2+x)^3 \qquad f'(0) = 4(2)^3 = 32$$

$$f''(x) = 12(2+x)^2 \qquad f''(0) = 12(2)^2 = 48$$
$$f'''(x) = 24(2+x)^1 \qquad f'''(0) = 24(2) = 48$$
$$f^{iv}(x) = 24 \qquad\qquad f^{iv}(0) = 24$$

Substituting in equation (5) gives:

$$(2+x)^4$$

$$= f(0) + xf'(0) + \frac{x^2}{2!}f''(0) + \frac{x^3}{3!}f'''(0) + \frac{x^4}{4!}f^{iv}(0)$$

$$= 16 + (x)(32) + \frac{x^2}{2!}(48) + \frac{x^3}{3!}(48) + \frac{x^4}{4!}(24)$$

$$= 16 + 32x + 24x^2 + 8x^3 + x^4$$

(This expression could have been obtained by applying the binomial theorem.)

Problem 10.   Expand $e^{\frac{x}{2}}$ as far as the term in x^4

$$f(x) = e^{\frac{x}{2}} \qquad\qquad f(0) = e^0 = 1$$

$$f'(x) = \frac{1}{2}e^{\frac{x}{2}} \qquad f'(0) = \frac{1}{2}e^0 = \frac{1}{2}$$

$$f''(x) = \frac{1}{4}e^{\frac{x}{2}} \qquad f''(0) = \frac{1}{4}e^0 = \frac{1}{4}$$

$$f'''(x) = \frac{1}{8}e^{\frac{x}{2}} \qquad f'''(0) = \frac{1}{8}e^0 = \frac{1}{8}$$

$$f^{iv}(x) = \frac{1}{16}e^{\frac{x}{2}} \qquad f^{iv}(0) = \frac{1}{16}e^0 = \frac{1}{16}$$

Substituting in equation (5) gives:

$$e^{\frac{x}{2}} = f(0) + xf'(0) + \frac{x^2}{2!}f''(0)$$

$$+ \frac{x^3}{3!}f'''(0) + \frac{x^4}{4!}f^{iv}(0) + \cdots$$

$$= 1 + (x)\left(\frac{1}{2}\right) + \frac{x^2}{2!}\left(\frac{1}{4}\right) + \frac{x^3}{3!}\left(\frac{1}{8}\right)$$

$$+ \frac{x^4}{4!}\left(\frac{1}{16}\right) + \cdots$$

i.e. $e^{\frac{x}{2}} = 1 + \dfrac{1}{2}x + \dfrac{1}{8}x^2 + \dfrac{1}{48}x^3 + \dfrac{1}{384}x^4 + \cdots$

Section B

Problem 11. Develop a series for $\sinh x$ using Maclaurin's series

$$f(x) = \sinh x \qquad f(0) = \sinh 0 = \frac{e^0 - e^{-0}}{2} = 0$$

$$f'(x) = \cosh x \qquad f'(0) = \cosh 0 = \frac{e^0 + e^{-0}}{2} = 1$$

$$f''(x) = \sinh x \qquad f''(0) = \sinh 0 = 0$$

$$f'''(x) = \cosh x \qquad f'''(0) = \cosh 0 = 1$$

$$f^{\text{iv}}(x) = \sinh x \qquad f^{\text{iv}}(0) = \sinh 0 = 0$$

$$f^{\text{v}}(x) = \cosh x \qquad f^{\text{v}}(0) = \cosh 0 = 1$$

Substituting in equation (5) gives:

$$\sinh x = f(0) + x f'(0) + \frac{x^2}{2!} f''(0) + \frac{x^3}{3!} f'''(0)$$

$$+ \frac{x^4}{4!} f^{\text{iv}}(0) + \frac{x^5}{5!} f^{\text{v}}(0) + \cdots$$

$$= 0 + (x)(1) + \frac{x^2}{2!}(0) + \frac{x^3}{3!}(1) + \frac{x^4}{4!}(0)$$

$$+ \frac{x^5}{5!}(1) + \cdots$$

i.e. $\mathbf{\sinh x = x + \dfrac{x^3}{3!} + \dfrac{x^5}{5!} + \cdots}$

(as obtained in Section 24.5, page 235)

Problem 12. Produce a power series for $\cos^2 2x$ as far as the term in x^6

From double angle formulae, $\cos 2A = 2\cos^2 A - 1$ (see Chapter 44).

from which, $\quad \cos^2 A = \dfrac{1}{2}(1 + \cos 2A)$

and $\quad \cos^2 2x = \dfrac{1}{2}(1 + \cos 4x)$

From Problem 1,

$$\cos x = 1 - \frac{x^2}{2!} + \frac{x^4}{4!} - \frac{x^6}{6!} + \cdots$$

hence $\quad \cos 4x = 1 - \dfrac{(4x)^2}{2!} + \dfrac{(4x)^4}{4!} - \dfrac{(4x)^6}{6!} + \cdots$

$$= 1 - 8x^2 + \frac{32}{3}x^4 - \frac{256}{45}x^6 + \cdots$$

Thus $\quad \cos^2 2x = \dfrac{1}{2}(1 + \cos 4x)$

$$= \frac{1}{2}\left(1 + 1 - 8x^2 + \frac{32}{3}x^4 - \frac{256}{45}x^6 + \cdots\right)$$

i.e. $\mathbf{\cos^2 2x = 1 - 4x^2 + \dfrac{16}{3}x^4 - \dfrac{128}{45}x^6 + \cdots}$

Now try the following Practice Exercise

Practice Exercise 95 Further problems on Maclaurin's series (answers on page 1158)

1. Determine the first four terms of the power series for $\sin 2x$ using Maclaurin's series.

2. Use Maclaurin's series to produce a power series for $\cosh 3x$ as far as the term in x^6.

3. Use Maclaurin's theorem to determine the first three terms of the power series for $\ln(1 + e^x)$.

4. Determine the power series for $\cos 4t$ as far as the term in t^6.

5. Expand $e^{\frac{3}{2}x}$ in a power series as far as the term in x^3.

6. Develop, as far as the term in x^4, the power series for $\sec 2x$.

7. Expand $e^{2\theta}\cos 3\theta$ as far as the term in θ^2 using Maclaurin's series.

8. Determine the first three terms of the series for $\sin^2 x$ by applying Maclaurin's theorem.

9. Use Maclaurin's series to determine the expansion of $(3 + 2t)^4$.

22.5 Numerical integration using Maclaurin's series

The value of many integrals cannot be determined using the various analytical methods. In Chapter 71, the trapezoidal, mid-ordinate and Simpson's rules are used to numerically evaluate such integrals. Another method of finding the approximate value of a definite integral is to express the function as a power series using Maclaurin's series, and then integrating each algebraic term in turn. This is demonstrated in the following worked problems.

As a reminder, the general solution of integrals of the form $\int ax^n dx$, where a and n are constants, is given by:

$$\int ax^n dx = \frac{ax^{n+1}}{n+1} + c$$

Problem 13. Evaluate $\int_{0.1}^{0.4} 2e^{\sin\theta}\,d\theta$, correct to 3 significant figures

A power series for $e^{\sin\theta}$ is first obtained using Maclaurin's series.

$$f(\theta) = e^{\sin\theta} \qquad f(0) = e^{\sin 0} = e^0 = 1$$

$$f'(\theta) = \cos\theta\,e^{\sin\theta} \qquad f'(0) = \cos 0\,e^{\sin 0} = (1)e^0 = 1$$

$$f''(\theta) = (\cos\theta)(\cos\theta\,e^{\sin\theta}) + (e^{\sin\theta})(-\sin\theta),$$

$$\text{by the product rule,}$$

$$= e^{\sin\theta}(\cos^2\theta - \sin\theta);$$

$$f''(0) = e^0(\cos^2 0 - \sin 0) = 1$$

$$f'''(\theta) = (e^{\sin\theta})[(2\cos\theta(-\sin\theta) - \cos\theta)$$

$$+ (\cos^2\theta - \sin\theta)(\cos\theta\,e^{\sin\theta})]$$

$$= e^{\sin\theta}\cos\theta[-2\sin\theta - 1 + \cos^2\theta - \sin\theta]$$

$$f'''(0) = e^0\cos 0[(0 - 1 + 1 - 0)] = 0$$

Hence, from equation (5):

$$e^{\sin\theta} = f(0) + \theta f'(0) + \frac{\theta^2}{2!}f''(0) + \frac{\theta^3}{3!}f'''(0) + \cdots$$

$$= 1 + \theta + \frac{\theta^2}{2} + 0$$

Thus $\int_{0.1}^{0.4} 2e^{\sin\theta}\,d\theta = \int_{0.1}^{0.4} 2\left(1 + \theta + \frac{\theta^2}{2}\right)d\theta$

$$= \int_{0.1}^{0.4}(2 + 2\theta + \theta^2)\,d\theta$$

$$= \left[2\theta + \frac{2\theta^2}{2} + \frac{\theta^3}{3}\right]_{0.1}^{0.4}$$

$$= \left(0.8 + (0.4)^2 + \frac{(0.4)^3}{3}\right)$$

$$- \left(0.2 + (0.1)^2 + \frac{(0.1)^3}{3}\right)$$

$$= 0.98133 - 0.21033$$

$$= \mathbf{0.771}, \text{correct to 3 significant figures.}$$

Problem 14. Evaluate $\int_0^1 \frac{\sin\theta}{\theta}\,d\theta$ using Maclaurin's series, correct to 3 significant figures

Let $f(\theta) = \sin\theta \qquad f(0) = 0$

$$f'(\theta) = \cos\theta \qquad f'(0) = 1$$

$$f''(\theta) = -\sin\theta \qquad f''(0) = 0$$

$$f'''(\theta) = -\cos\theta \qquad f'''(0) = -1$$

$$f^{iv}(\theta) = \sin\theta \qquad f^{iv}(0) = 0$$

$$f^{v}(\theta) = \cos\theta \qquad f^{v}(0) = 1$$

Hence from equation (5):

$$\sin\theta = f(0) + \theta f'(0) + \frac{\theta^2}{2!}f''(0) + \frac{\theta^3}{3!}f'''(0)$$

$$+ \frac{\theta^4}{4!}f^{iv}(0) + \frac{\theta^5}{5!}f^{v}(0) + \cdots$$

$$= 0 + \theta(1) + \frac{\theta^2}{2!}(0) + \frac{\theta^3}{3!}(-1)$$

$$+ \frac{\theta^4}{4!}(0) + \frac{\theta^5}{5!}(1) + \cdots$$

i.e. $\sin\theta = \theta - \dfrac{\theta^3}{3!} + \dfrac{\theta^5}{5!} - \cdots$

Hence

$$\int_0^1 \frac{\sin\theta}{\theta}\,d\theta$$

$$= \int_0^1 \frac{\left(\theta - \dfrac{\theta^3}{3!} + \dfrac{\theta^5}{5!} - \dfrac{\theta^7}{7!} + \cdots\right)}{\theta}\,d\theta$$

$$= \int_0^1 \left(1 - \frac{\theta^2}{6} + \frac{\theta^4}{120} - \frac{\theta^6}{5040} + \cdots\right)d\theta$$

$$= \left[\theta - \frac{\theta^3}{18} + \frac{\theta^5}{600} - \frac{\theta^7}{7(5040)} + \cdots\right]_0^1$$

$$= 1 - \frac{1}{18} + \frac{1}{600} - \frac{1}{7(5040)} + \cdots$$

$$= \mathbf{0.946}, \text{correct to 3 significant figures.}$$

Problem 15. Evaluate $\int_0^{0.4} x \ln(1+x)\,dx$ using Maclaurin's theorem, correct to 3 decimal places

From Problem 6,

$$\ln(1+x) = x - \frac{x^2}{2} + \frac{x^3}{3} - \frac{x^4}{4} + \frac{x^5}{5} - \cdots$$

Hence $\displaystyle\int_0^{0.4} x \ln(1+x)\,dx$

$$= \int_0^{0.4} x\left(x - \frac{x^2}{2} + \frac{x^3}{3} - \frac{x^4}{4} + \frac{x^5}{5} - \cdots\right) dx$$

$$= \int_0^{0.4} \left(x^2 - \frac{x^3}{2} + \frac{x^4}{3} - \frac{x^5}{4} + \frac{x^6}{5} - \cdots\right) dx$$

$$= \left[\frac{x^3}{3} - \frac{x^4}{8} + \frac{x^5}{15} - \frac{x^6}{24} + \frac{x^7}{35} - \cdots\right]_0^{0.4}$$

$$= \left(\frac{(0.4)^3}{3} - \frac{(0.4)^4}{8} + \frac{(0.4)^5}{15} - \frac{(0.4)^6}{24}\right.$$

$$\left. + \frac{(0.4)^7}{35} - \cdots\right) - (0)$$

$$= 0.02133 - 0.0032 + 0.0006827 - \cdots$$

$$= \mathbf{0.019}, \text{ correct to 3 decimal places.}$$

Now try the following Practice Exercise

Practice Exercise 96   Further problems on numerical integration using Maclaurin's series (answers on page 1158)

1. Evaluate $\int_{0.2}^{0.6} 3e^{\sin\theta}\,d\theta$, correct to 3 decimal places, using Maclaurin's series.

2. Use Maclaurin's theorem to expand $\cos 2\theta$ and hence evaluate, correct to 2 decimal places,
$$\int_0^1 \frac{\cos 2\theta}{\theta^{\frac{1}{3}}}\,d\theta$$

3. Determine the value of $\int_0^1 \sqrt{\theta}\cos\theta\,d\theta$, correct to 2 significant figures, using Maclaurin's series.

4. Use Maclaurin's theorem to expand $\sqrt{x}\ln(x+1)$ as a power series. Hence evaluate, correct to 3 decimal places, $\int_0^{0.5} \sqrt{x}\ln(x+1)\,dx$.

22.6   Limiting values

It is sometimes necessary to find limits of the form $\displaystyle\lim_{x\to a}\left\{\frac{f(x)}{g(x)}\right\}$, where $f(a)=0$ and $g(a)=0$

For example,

$$\lim_{x\to 1}\left\{\frac{x^2+3x-4}{x^2-7x+6}\right\} = \frac{1+3-4}{1-7+6} = \frac{0}{0}$$

and $\frac{0}{0}$ is generally referred to as indeterminate.
For certain limits a knowledge of series can sometimes help.

For example,

$$\lim_{x\to 0}\left\{\frac{\tan x - x}{x^3}\right\}$$

$$\equiv \lim_{x\to 0}\left\{\frac{x + \frac{1}{3}x^3 + \cdots - x}{x^3}\right\} \text{ from Problem 5}$$

$$= \lim_{x\to 0}\left\{\frac{\frac{1}{3}x^3 + \cdots}{x^3}\right\} = \lim_{x\to 0}\left\{\frac{1}{3}\right\} = \frac{1}{3}$$

Similarly,

$$\lim_{x\to 0}\left\{\frac{\sinh x}{x}\right\}$$

$$\equiv \lim_{x\to 0}\left\{\frac{x + \frac{x^3}{3!} + \frac{x^5}{5!} +}{x}\right\} \text{ from Problem 11}$$

$$= \lim_{x\to 0}\left\{1 + \frac{x^2}{3!} + \frac{x^4}{5!} + \cdots\right\} = \mathbf{1}$$

However, a knowledge of series does not help with examples such as $\displaystyle\lim_{x\to 1}\left\{\frac{x^2+3x-4}{x^2-7x+6}\right\}$

L'Hôpital's* rule will enable us to determine such limits when the differential coefficients of the numerator and denominator can be found.

L'Hôpital's rule states:

$$\lim_{x \to a}\left\{\frac{f(x)}{g(x)}\right\} = \lim_{x \to a}\left\{\frac{f'(x)}{g'(x)}\right\} \quad \text{provided} \quad g'(a) \neq 0$$

It can happen that $\lim\limits_{x \to a}\left\{\dfrac{f'(x)}{g'(x)}\right\}$ is still $\frac{0}{0}$; if so, the numerator and denominator are differentiated again (and again) until a non-zero value is obtained for the denominator.

The following worked problems demonstrate how L'Hôpital's rule is used. Refer to Chapters 52 and 53 for methods of differentiation.

Problem 16. Determine $\lim\limits_{x \to 1}\left\{\dfrac{x^2 + 3x - 4}{x^2 - 7x + 6}\right\}$

The first step is to substitute $x = 1$ into both numerator and denominator. In this case we obtain $\frac{0}{0}$. It is

* **Who was** L'Hôpital? **Guillaume François Antoine, Marquis de l'Hôpital** (1661–2 February 1704, Paris) was a French mathematician. His name is firmly associated with l'Hôpital's rule for calculating limits involving indeterminate forms 0/0 and ∞/∞. To find out more go to **www.routledge.com/cw/bird**

only when we obtain such a result that we then use L'Hôpital's rule. Hence applying L'Hôpital's rule,

$$\lim_{x \to 1}\left\{\frac{x^2 + 3x - 4}{x^2 - 7x + 6}\right\} = \lim_{x \to 1}\left\{\frac{2x + 3}{2x - 7}\right\}$$

i.e. both numerator and denominator have been differentiated

$$= \frac{5}{-5} = -\mathbf{1}$$

Problem 17. Determine $\lim\limits_{x \to 0}\left\{\dfrac{\sin x - x}{x^2}\right\}$

Substituting $x = 0$ gives

$$\lim_{x \to 0}\left\{\frac{\sin x - x}{x^2}\right\} = \frac{\sin 0 - 0}{0} = \frac{0}{0}$$

Applying L'Hôpital's rule gives

$$\lim_{x \to 0}\left\{\frac{\sin x - x}{x^2}\right\} = \lim_{x \to 0}\left\{\frac{\cos x - 1}{2x}\right\}$$

Substituting $x = 0$ gives

$$\frac{\cos 0 - 1}{0} = \frac{1 - 1}{0} = \frac{0}{0} \quad \text{again}$$

Applying L'Hôpital's rule again gives

$$\lim_{x \to 0}\left\{\frac{\cos x - 1}{2x}\right\} = \lim_{x \to 0}\left\{\frac{-\sin x}{2}\right\} = \mathbf{0}$$

Problem 18. Determine $\lim\limits_{x \to 0}\left\{\dfrac{x - \sin x}{x - \tan x}\right\}$

Substituting $x = 0$ gives

$$\lim_{x \to 0}\left\{\frac{x - \sin x}{x - \tan x}\right\} = \frac{0 - \sin 0}{0 - \tan 0} = \frac{0}{0}$$

Applying L'Hôpital's rule gives

$$\lim_{x \to 0}\left\{\frac{x - \sin x}{x - \tan x}\right\} = \lim_{x \to 0}\left\{\frac{1 - \cos x}{1 - \sec^2 x}\right\}$$

Substituting $x = 0$ gives

$$\lim_{x \to 0}\left\{\frac{1 - \cos x}{1 - \sec^2 x}\right\} = \frac{1 - \cos 0}{1 - \sec^2 0} = \frac{1 - 1}{1 - 1} = \frac{0}{0} \quad \text{again}$$

Applying L'Hôpital's rule gives

$$\lim_{x \to 0}\left\{\frac{1-\cos x}{1-\sec^2 x}\right\} = \lim_{x \to 0}\left\{\frac{\sin x}{(-2\sec x)(\sec x \tan x)}\right\}$$

$$= \lim_{x \to 0}\left\{\frac{\sin x}{-2\sec^2 x \tan x}\right\}$$

Substituting $x = 0$ gives

$$\frac{\sin 0}{-2\sec^2 0 \tan 0} = \frac{0}{0} \quad \text{again}$$

Applying L'Hôpital's rule gives

$$\lim_{x \to 0}\left\{\frac{\sin x}{-2\sec^2 x \tan x}\right\}$$

$$= \lim_{x \to 0}\left\{\frac{\cos x}{\begin{array}{c}(-2\sec^2 x)(\sec^2 x)\\ + (\tan x)(-4\sec^2 x \tan x)\end{array}}\right\}$$

using the product rule

Substituting $x = 0$ gives

$$\frac{\cos 0}{-2\sec^4 0 - 4\sec^2 0 \tan^2 0} = \frac{1}{-2-0}$$

$$= -\frac{1}{2}$$

$$\text{Hence } \lim_{x \to 0}\left\{\frac{x - \sin x}{x - \tan x}\right\} = -\frac{\mathbf{1}}{\mathbf{2}}$$

Now try the following Practice Exercise

Determine the following limiting values

1. $\displaystyle\lim_{x \to 1}\left\{\frac{x^3 - 2x + 1}{2x^3 + 3x - 5}\right\}$

2. $\displaystyle\lim_{x \to 0}\left\{\frac{\sin x}{x}\right\}$

3. $\displaystyle\lim_{x \to 0}\left\{\frac{\ln(1+x)}{x}\right\}$

4. $\displaystyle\lim_{x \to 0}\left\{\frac{x^2 - \sin 3x}{3x + x^2}\right\}$

5. $\displaystyle\lim_{\theta \to 0}\left\{\frac{\sin\theta - \theta\cos\theta}{\theta^3}\right\}$

6. $\displaystyle\lim_{t \to 1}\left\{\frac{\ln t}{t^2 - 1}\right\}$

7. $\displaystyle\lim_{x \to 0}\left\{\frac{\sinh x - \sin x}{x^3}\right\}$

8. $\displaystyle\lim_{\theta \to \frac{\pi}{2}}\left\{\frac{\sin\theta - 1}{\ln\sin\theta}\right\}$

9. $\displaystyle\lim_{t \to 0}\left\{\frac{\sec t - 1}{t \sin t}\right\}$

For fully worked solutions to each of the problems in Practice Exercises 95 to 97 in this chapter, go to the website:
www.routledge.com/cw/bird

This assignment covers the material contained in Chapters 21 and 22. *The marks available are shown in brackets at the end of each question.*

1. Use the binomial series to expand
   $(2a - 3b)^6$ (7)

2. Expand the following in ascending powers of t as far as the term in t^3

   (a) $\dfrac{1}{1+t}$ (b) $\dfrac{1}{\sqrt{1-3t}}$

   For each case, state the limits for which the expansion is valid.   (9)

3. The modulus of rigidity G is given by $G = \dfrac{R^4\theta}{L}$
   where R is the radius, θ the angle of twist and L the length. Find the approximate percentage error in G when R is measured 1.5% too large, θ is measured 3% too small and L is measured 1% too small.   (8)

4. Use Maclaurin's series to determine a power series for $e^{2x} \cos 3x$ as far as the term in x^2 (10)

5. Show, using Maclaurin's series, that the first four terms of the power series for $\cosh 2x$ are given by:

   $$\cosh 2x = 1 + 2x^2 + \frac{2}{3}x^4 + \frac{4}{45}x^6$$ (11)

6. Expand the function $x^2 \ln(1 + \sin x)$ using Maclaurin's series and hence evaluate

   $$\int_0^{\frac{1}{2}} x^2 \ln(1 + \sin x)\, dx$$

   correct to 2 significant figures.   (15)

Section B

Chapter 23

Solving equations by iterative methods

Why it is important to understand: Solving equations by iterative methods

There are many, many different types of equations used in every branch of engineering and science. There are straightforward methods for solving simple, quadratic and simultaneous equations. However, there are many other types of equations than these three. Great progress has been made in the engineering and scientific disciplines regarding the use of iterative methods for linear systems. In engineering it is important that we can solve any equation; iterative methods help us do that.

At the end of this chapter, you should be able to:

- define iterative methods
- use the method of bisection to solve equations
- use an algebraic method of successive approximations to solve equations
- state the Newton–Raphson formula
- use Newton's method to solve equations

23.1 Introduction to iterative methods

Many equations can only be solved graphically or by methods of successive approximations to the roots, called **iterative methods**. Three methods of successive approximations are (i) bisection method, introduced in Section 23.2, (ii) an algebraic method, introduction in Section 23.3, and (iii) by using the Newton–Raphson formula, given in Section 23.4.

Each successive approximation method relies on a reasonably good first estimate of the value of a root being made. One way of determining this is to sketch a graph of the function, say $y = f(x)$, and determine the approximate values of roots from the points where the graph cuts the x-axis. Another way is by using a functional notation method. This method uses the property that the value of the graph of $f(x) = 0$ changes sign for values of x just before and just after the value of a root. For example, one root of the equation $x^2 - x - 6 = 0$ is $x = 3$. Using functional notation:

$$f(x) = x^2 - x - 6$$
$$f(2) = 2^2 - 2 - 6 = -4$$
$$f(4) = 4^2 - 4 - 6 = +6$$

It can be seen from these results that the value of $f(x)$ changes from -4 at $f(2)$ to $+6$ at $f(4)$, indicating that a root lies between 2 and 4. This is shown more clearly in Figure 23.1.

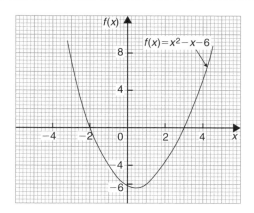

Figure 23.1

23.2   The bisection method

As shown above, by using functional notation it is possible to determine the vicinity of a root of an equation by the occurrence of a change of sign, i.e. if x_1 and x_2 are such that $f(x_1)$ and $f(x_2)$ have opposite signs, there is at least one root of the equation $f(x)=0$ in the interval between x_1 and x_2 (provided $f(x)$ is a continuous function). In the **method of bisection** the mid-point of the interval, i.e. $x_3 = \dfrac{x_1 + x_2}{2}$, is taken, and from the sign of $f(x_3)$ it can be deduced whether a root lies in the half interval to the left or right of x_3. Whichever half interval is indicated, its mid-point is then taken and the procedure repeated. The method often requires many iterations and is therefore slow, but never fails to eventually produce the root. The procedure stops when two successive values of x are equal – to the required degree of accuracy.

The method of bisection is demonstrated in Problems 1 to 3.

> **Problem 1.**   Use the method of bisection to find the positive root of the equation $5x^2 + 11x - 17 = 0$, correct to 3 significant figures

Let $f(x) = 5x^2 + 11x - 17$

Then, using functional notation:

$$f(0) = -17$$

$$f(1) = 5(1)^2 + 11(1) - 17 = -1$$

$$f(2) = 5(2)^2 + 11(2) - 17 = +25$$

Since there is a change of sign from negative to positive there must be a root of the equation between $x = 1$ and $x = 2$. This is shown graphically in Figure 23.2.

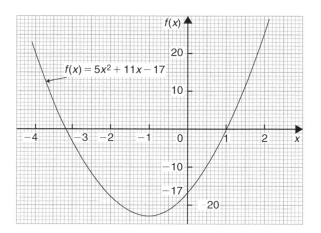

Figure 23.2

The method of bisection suggests that the root is at $\dfrac{1+2}{2} = 1.5$, i.e. the interval between 1 and 2 has been bisected.

Hence

$$f(1.5) = 5(1.5)^2 + 11(1.5) - 17$$

$$= +10.75$$

Since $f(1)$ is negative, $f(1.5)$ is positive, and $f(2)$ is also positive, a root of the equation must lie between $x = 1$ and $x = 1.5$, since a **sign change** has occurred between $f(1)$ and $f(1.5)$.

Bisecting this interval gives $\dfrac{1+1.5}{2}$ i.e. 1.25 as the next root.

Hence

$$f(1.25) = 5(1.25)^2 + 11x - 17$$

$$= +4.5625$$

Since $f(1)$ is negative and $f(1.25)$ is positive, a root lies between $x = 1$ and $x = 1.25$.

Bisecting this interval gives $\dfrac{1+1.25}{2}$ i.e. 1.125

Hence

$$f(1.125) = 5(1.125)^2 + 11(1.125) - 17$$

$$= +1.703125$$

Section B

Since $f(1)$ is negative and $f(1.125)$ is positive, a root lies between $x = 1$ and $x = 1.125$

Bisecting this interval gives $\dfrac{1 + 1.125}{2}$ i.e. 1.0625

Hence

$$f(1.0625) = 5(1.0625)^2 + 11(1.0625) - 17$$

$$= +0.33203125$$

Since $f(1)$ is negative and $f(1.0625)$ is positive, a root lies between $x = 1$ and $x = 1.0625$

Bisecting this interval gives $\dfrac{1 + 1.0625}{2}$ i.e. 1.03125

Hence

$$f(1.03125) = 5(1.03125)^2 + 11(1.03125) - 17$$

$$= -0.338867\ldots$$

Since $f(1.03125)$ is negative and $f(1.0625)$ is positive, a root lies between $x = 1.03125$ and $x = 1.0625$

Bisecting this interval gives

$$\frac{1.03125 + 1.0625}{2} \text{ i.e. } 1.046875$$

Hence

$$f(1.046875) = 5(1.046875)^2 + 11(1.046875) - 17$$

$$= -0.0046386\ldots$$

Since $f(1.046875)$ is negative and $f(1.0625)$ is positive, a root lies between $x = 1.046875$ and $x = 1.0625$

Bisecting this interval gives

$$\frac{1.046875 + 1.0625}{2} \text{ i.e. } \mathbf{1.0546875}$$

The last three values obtained for the root are 1.03125, 1.046875 and 1.0546875. The last two values are both 1.05, correct to 3 significant figures. We therefore stop the iterations here.

Thus, correct to 3 significant figures, the positive root of $5x^2 + 11x - 17 = 0$ is 1.05

> **Problem 2.** Use the bisection method to determine the positive root of the equation $x + 3 = e^x$, correct to 3 decimal places

Let $f(x) = x + 3 - e^x$

Then, using functional notation:

$$f(0) = 0 + 3 - e^0 = +2$$
$$f(1) = 1 + 3 - e^1 = +1.2817\ldots$$
$$f(2) = 2 + 3 - e^2 = -2.3890\ldots$$

Since $f(1)$ is positive and $f(2)$ is negative, a root lies between $x = 1$ and $x = 2$. A sketch of $f(x) = x + 3 - e^x$, i.e. $x + 3 = e^x$ is shown in Figure 23.3.

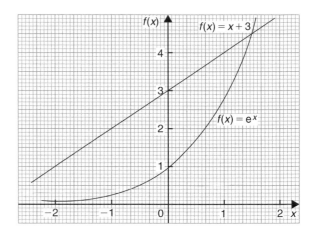

Figure 23.3

Bisecting the interval between $x = 1$ and $x = 2$ gives $\dfrac{1 + 2}{2}$ i.e. 1.5

Hence

$$f(1.5) = 1.5 + 3 - e^{1.5}$$

$$= +0.01831\ldots$$

Since $f(1.5)$ is positive and $f(2)$ is negative, a root lies between $x = 1.5$ and $x = 2$

Bisecting this interval gives $\dfrac{1.5 + 2}{2}$ i.e. 1.75

Hence

$$f(1.75) = 1.75 + 3 - e^{1.75}$$

$$= -1.00460\ldots$$

Since $f(1.75)$ is negative and $f(1.5)$ is positive, a root lies between $x = 1.75$ and $x = 1.5$

Bisecting this interval gives $\dfrac{1.75 + 1.5}{2}$ i.e. 1.625

Hence

$$f(1.625) = 1.625 + 3 - e^{1.625}$$

$$= -0.45341\ldots$$

Since $f(1.625)$ is negative and $f(1.5)$ is positive, a root lies between $x = 1.625$ and $x = 1.5$

Bisecting this interval gives $\dfrac{1.625 + 1.5}{2}$ i.e. 1.5625
Hence

$$f(1.5625) = 1.5625 + 3 - e^{1.5625}$$

$$= -0.20823\ldots$$

Since $f(1.5625)$ is negative and $f(1.5)$ is positive, a root lies between $x = 1.5625$ and $x = 1.5$

Bisecting this interval gives

$$\frac{1.5625 + 1.5}{2} \text{ i.e. } 1.53125$$

Hence

$$f(1.53125) = 1.53125 + 3 - e^{1.53125}$$

$$= -0.09270\ldots$$

Since $f(1.53125)$ is negative and $f(1.5)$ is positive, a root lies between $x = 1.53125$ and $x = 1.5$

Bisecting this interval gives

$$\frac{1.53125 + 1.5}{2} \text{ i.e. } 1.515625$$

Hence

$$f(1.515625) = 1.515625 + 3 - e^{1.515625}$$

$$= -0.03664\ldots$$

Since $f(1.515625)$ is negative and $f(1.5)$ is positive, a root lies between $x = 1.515625$ and $x = 1.5$

Bisecting this interval gives

$$\frac{1.515625 + 1.5}{2} \text{ i.e. } 1.5078125$$

Hence

$$f(1.5078125) = 1.5078125 + 3 - e^{1.5078125}$$

$$= -0.009026\ldots$$

Since $f(1.5078125)$ is negative and $f(1.5)$ is positive, a root lies between $x = 1.5078125$ and $x = 1.5$

Bisecting this interval gives

$$\frac{1.5078125 + 1.5}{2} \text{ i.e. } 1.50390625$$

Hence

$$f(1.50390625) = 1.50390625 + 3 - e^{1.50390625}$$

$$= +0.004676\ldots$$

Since $f(1.50390625)$ is positive and $f(1.5078125)$ is negative, a root lies between $x = 1.50390625$ and $x = 1.5078125$
Bisecting this interval gives

$$\frac{1.50390625 + 1.5078125}{2} \text{ i.e. } 1.505859375$$

Hence

$$f(1.505859375) = 1.505859375 + 3 - e^{1.505859375}$$

$$= -0.0021666\ldots$$

Since $f(1.505859375)$ is negative and $f(1.50390625)$ is positive, a root lies between $x = 1.505859375$ and $x = 1.50390625$

Bisecting this interval gives

$$\frac{1.505859375 + 1.50390625}{2} \text{ i.e. } 1.504882813$$

Hence

$$f(1.504882813) = 1.504882813 + 3 - e^{1.504882813}$$

$$= +0.001256\ldots$$

Since $f(1.504882813)$ is positive and $f(1.505859375)$ is negative,

a root lies between $x = 1.504882813$ and $x = 1.505859375$

Bisecting this interval gives

$$\frac{1.504882813 + 1.505859375}{2} \text{ i.e. } \mathbf{1.505371094}$$

The last two values of x are 1.504882813 and 1.505371094, i.e. both are equal to 1.505, correct to 3 decimal places.

Hence the root of $x + 3 = e^x$ is $x = $ 1.505, correct to 3 decimal places.

The above is a lengthy procedure and it is probably easier to present the data in a table as shown below.

x_1	x_2	$x_3 = \dfrac{x_1 + x_2}{2}$	$f(x_3)$
		0	+2
		1	+1.2817...
		2	−2.3890...
1	2	1.5	+0.0183...
1.5	2	1.75	−1.0046...
1.5	1.75	1.625	−0.4534...
1.5	1.625	1.5625	−0.2082...
1.5	1.5625	1.53125	−0.0927...
1.5	1.53125	1.515625	−0.0366...
1.5	1.515625	1.5078125	−0.0090...
1.5	1.5078125	1.50390625	+0.0046...
1.50390625	1.5078125	1.505859375	−0.0021...
1.50390625	1.505859375	**1.504882813**	+0.0012...
1.504882813	1.505859375	**1.505388282**	

Problem 3. Solve, correct to 2 decimal places, the equation $2\ln x + x = 2$ using the method of bisection

Let $f(x) = 2\ln x + x - 2$

$$f(0.1) = 2\ln(0.1) + 0.1 - 2 = -6.5051\ldots$$

(Note that $\ln 0$ is infinite, which is why $x = 0$ was not chosen.)

$$f(1) = 2\ln 1 + 1 - 2 = -1$$

$$f(2) = 2\ln 2 + 2 - 2 = +1.3862\ldots$$

A change of sign indicates a root lies between $x = 1$ and $x = 2$

Since $2\ln x + x = 2$ then $2\ln x = -x + 2$; sketches of $2\ln x$ and $-x + 2$ are shown in Figure 23.4.

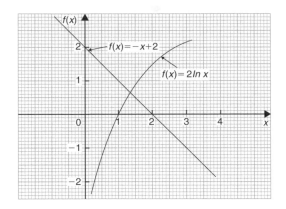

Figure 23.4

As in Problem 2, a table of values is produced to reduce space.

x_1	x_2	$x_3 = \dfrac{x_1 + x_2}{2}$	$f(x_3)$
		0.1	−6.5051...
		1	−1
		2	+1.3862...
1	2	1.5	+0.3109...
1	1.5	1.25	−0.3037...
1.25	1.5	1.375	+0.0119...
1.25	1.375	1.3125	−0.1436...
1.3125	1.375	1.34375	−0.0653...
1.34375	1.375	1.359375	−0.0265...
1.359375	1.375	**1.3671875**	−0.0073...
1.3671875	1.375	**1.37109375**	+0.0023...

The last two values of x_3 are both equal to 1.37 when expressed to 2 decimal places. We therefore stop the iterations.

Hence, the solution of $2\ln x + x = 2$ is $x = 1.37$, correct to 2 decimal places.

Now try the following Practice Exercise

Practice Exercise 98   Further problems on the bisection method (answers on page 1158)

Use the method of bisection to solve the following equations to the accuracy stated.

1. Find the positive root of the equation $x^2 + 3x - 5 = 0$, correct to 3 significant figures, using the method of bisection.

2. Using the bisection method solve $e^x - x = 2$, correct to 4 significant figures.

3. Determine the positive root of $x^2 = 4\cos x$, correct to 2 decimal places, using the method of bisection. (Have your calculator set on 'radians', not 'degrees'.)

4. Solve $x - 2 - \ln x = 0$ for the root near to 3, correct to 3 decimal places, using the bisection method.

5. Solve, correct to 4 significant figures, $x - 2\sin^2 x = 0$ using the bisection method. (Have your calculator set on 'radians', not 'degrees'.)

23.3 An algebraic method of successive approximations

This method can be used to solve equations of the form:

$$a + bx + cx^2 + dx^3 + \cdots = 0,$$

where $a, b, c, d, \ldots$ are constants.
Procedure:

First approximation

(a) Using a graphical or the functional notation method (see Section 23.1) determine an approximate value of the root required, say x_1

Second approximation

(b) Let the true value of the root be $(x_1 + \delta_1)$

(c) Determine x_2, the approximate value of $(x_1 + \delta_1)$ by determining the value of $f(x_1 + \delta_1) = 0$, but neglecting terms containing products of δ_1

Third approximation

(d) Let the true value of the root be $(x_2 + \delta_2)$

(e) Determine x_3, the approximate value of $(x_2 + \delta_2)$, by determining the value of $f(x_2 + \delta_2) = 0$, but neglecting terms containing products of δ_2

(f) The fourth and higher approximations are obtained in a similar way.

Using the techniques given in paragraphs (b) to (f), it is possible to continue getting values nearer and nearer to the required root. The procedure is repeated until the value of the required root does not change on two consecutive approximations, when expressed to the required degree of accuracy.

Problem 4. Use an algebraic method of successive approximations to determine the value of the negative root of the quadratic equation: $4x^2 - 6x - 7 = 0$, correct to 3 significant figures. Check the value of the root by using the quadratic formula

A first estimate of the values of the roots is made by using the functional notation method

$$f(x) = 4x^2 - 6x - 7$$
$$f(0) = 4(0)^2 - 6(0) - 7 = -7$$
$$f(-1) = 4(-1)^2 - 6(-1) - 7 = 3$$

These results show that the negative root lies between 0 and -1, since the value of $f(x)$ changes sign between $f(0)$ and $f(-1)$ (see Section 23.1). The procedure given above for the root lying between 0 and -1 is followed.

First approximation

(a) Let a first approximation be such that it divides the interval 0 to -1 in the ratio of -7 to 3, i.e. let $x_1 = -0.7$

[-7 to $+3$ is an interval of 10; hence, a first approximation is $\dfrac{7}{10}$ from zero towards -1, i.e. -0.7, or $\dfrac{3}{10}$ back from -1, i.e. -0.7]

Second approximation

(b) Let the true value of the root, x_2, be $(x_1 + \delta_1)$.

(c) Let $f(x_1 + \delta_1) = 0$, then, since $x_1 = -0.7$,

$$4(-0.7 + \delta_1)^2 - 6(-0.7 + \delta_1) - 7 = 0$$
$$\text{Hence, } 4[(-0.7)^2 + (2)(-0.7)(\delta_1) + \delta_1^2]$$
$$- (6)(-0.7) - 6\delta_1 - 7 = 0$$

Neglecting terms containing products of δ_1 gives:

$$1.96 - 5.6\delta_1 + 4.2 - 6\delta_1 - 7 \approx 0$$

i.e. $-5.6\delta_1 - 6\delta_1 = -1.96 - 4.2 + 7$

i.e. $\delta_1 \approx \dfrac{-1.96 - 4.2 + 7}{-5.6 - 6}$

$$\approx \dfrac{0.84}{-11.6}$$

$$\approx -0.0724$$

Thus, x_2, a second approximation to the root, is $[-0.7+(-0.0724)]$

i.e. $x_2 = -0.7724$, correct to 4 significant figures. (Since the question asked for 3 significant-figure accuracy, it is usual to work to one figure greater than this.)

The procedure given in (b) and (c) is now repeated for $x_2 = -0.7724$

Third approximation

(d) Let the true value of the root, x_3, be $(x_2 + \delta_2)$

(e) Let $f(x_2 + \delta_2) = 0$, then, since $x_2 = -0.7724$,

$$4(-0.7724 + \delta_2)^2 - 6(-0.7724 + \delta_2) - 7 = 0$$

$$4[(-0.7724)^2 + (2)(-0.7724)(\delta_2) + \delta_2^2]$$

$$- (6)(-0.7724) - 6\delta_2 - 7 = 0$$

Neglecting terms containing products of δ_2 gives:

$$2.3864 - 6.1792\,\delta_2 + 4.6344 - 6\delta_2 - 7 \approx 0$$

i.e. $\delta_2 \approx \dfrac{-2.3864 - 4.6344 + 7}{-6.1792 - 6}$

$$\approx \dfrac{-0.0208}{-12.1792}$$

$$\approx +0.001708$$

Thus, x_3, the third approximation to the root, is $(-0.7724 + 0.001708)$,

i.e. $x_3 = -0.7707$, correct to 4 significant figures (or -0.771 correct to 3 significant figures).

Fourth approximation

(f) The procedure given for the second and third approximations is now repeated for

$$x_3 = -0.7707$$

Let the true value of the root, x_4, be $(x_3 + \delta_3)$

Let $f(x_3 + \delta_3) = 0$, then, since $x_3 = -0.7707$,

$$4(-0.7707 + \delta_3)^2 - 6(-0.7707$$

$$+ \delta_3) - 7 = 0$$

$$4[(-0.7707)^2 + (2)(-0.7707)\,\delta_3 + \delta_3^2]$$

$$- 6(-0.7707) - 6\delta_3 - 7 = 0$$

Neglecting terms containing products of δ_3 gives:

$$2.3759 - 6.1656\,\delta_3 + 4.6242 - 6\delta_3 - 7 \approx 0$$

i.e. $\delta_3 \approx \dfrac{-2.3759 - 4.6242 + 7}{-6.1656 - 6}$

$$\approx \dfrac{-0.0001}{-12.156}$$

$$\approx +0.00000822$$

Thus, x_4, the fourth approximation to the root, is $(-0.7707 + 0.00000822)$,

i.e. $x_4 = -0.7707$, correct to 4 significant figures, and -0.771, correct to 3 significant figures.

Since the values of the roots are the same on two consecutive approximations, when stated to the required degree of accuracy, then the negative root of $4x^2 - 6x - 7 = 0$ is **−0.771, correct to 3 significant figures**.

[Checking, using the quadratic formula:

$$x = \dfrac{-(-6) \pm \sqrt{[(-6)^2 - (4)(4)(-7)]}}{(2)(4)}$$

$$= \dfrac{6 \pm 12.166}{8} = \textbf{−0.771 and 2.27},$$

correct to 3 significant figures]

(**Note on accuracy and errors.** Depending on the accuracy of evaluating the $f(x + \delta)$ terms, one or two iterations (i.e. successive approximations) might be saved. However, it is not usual to work to more than about 4 significant-figure accuracy in this type of calculation. If a small error is made in calculations, the only likely effect is to increase the number of iterations.)

> **Problem 5.** Determine the value of the smallest positive root of the equation $3x^3 - 10x^2 + 4x + 7 = 0$, correct to 3 significant figures, using an algebraic method of successive approximations

The functional notation method is used to find the value of the first approximation.

$$f(x) = 3x^3 - 10x^2 + 4x + 7$$

$$f(0) = 3(0)^3 - 10(0)^2 + 4(0) + 7 = 7$$

$$f(1) = 3(1)^3 - 10(1)^2 + 4(1) + 7 = 4$$

$$f(2) = 3(2)^3 - 10(2)^2 + 4(2) + 7 = -1$$

Following the above procedure:

First approximation

(a) Let the first approximation be such that it divides the interval 1 to 2 in the ratio of 4 to -1, i.e. let x_1 be 1.8

[4 to -1 is an interval of 5; hence, a first approximation is $\dfrac{4}{5}$ from 1 towards 2, i.e. 1.8, or $\dfrac{1}{5}$ back from 2, i.e. 1.8]

Second approximation

(b) Let the true value of the root, x_2, be $(x_1 + \delta_1)$

(c) Let $f(x_1 + \delta_1) = 0$, then since $x_1 = 1.8$,

$$3(1.8 + \delta_1)^3 - 10(1.8 + \delta_1)^2$$
$$+ 4(1.8 + \delta_1) + 7 = 0$$

Neglecting terms containing products of δ_1 and using the binomial series gives:

$$3[1.8^3 + 3(1.8)^2 \delta_1] - 10[1.8^2 + (2)(1.8)\delta_1]$$
$$+ 4(1.8 + \delta_1) + 7 \approx 0$$

$$3(5.832 + 9.720\,\delta_1) - 32.4 - 36\,\delta_1$$
$$+ 7.2 + 4\delta_1 + 7 \approx 0$$

$$17.496 + 29.16\,\delta_1 - 32.4 - 36\,\delta_1$$
$$+ 7.2 + 4\delta_1 + 7 \approx 0$$

$$\delta_1 \approx \frac{-17.496 + 32.4 - 7.2 - 7}{29.16 - 36 + 4}$$

$$\approx -\frac{0.704}{2.84} \approx -0.2479$$

Thus $x_2 \approx 1.8 - 0.2479 = 1.5521$

Third approximation

(d) Let the true value of the root, x_3, be $(x_2 + \delta_2)$

(e) Let $f(x_2 + \delta_2) = 0$, then since $x_2 = 1.5521$,

$$3(1.5521 + \delta_2)^3 - 10(1.5521 + \delta_2)^2$$
$$+ 4(1.5521 + \delta_2) + 7 = 0$$

Neglecting terms containing products of δ_2 gives:

$$11.217 + 21.681\,\delta_2 - 24.090 - 31.042\,\delta_2$$
$$+ 6.2084 + 4\delta_2 + 7 \approx 0$$

$$\delta_2 \approx \frac{-11.217 + 24.090 - 6.2084 - 7}{21.681 - 31.042 + 4}$$

$$\approx \frac{-0.3354}{-5.361}$$

$$\approx 0.06256$$

Thus $x_3 \approx 1.5521 + 0.06256 \approx 1.6147$

(f) Values of x_4 and x_5 are found in a similar way.

$$f(x_3 + \delta_3) = 3(1.6147 + \delta_3)^3 - 10(1.6147$$
$$+ \delta_3)^2 + 4(1.6147 + \delta_3) + 7 = 0$$

giving $\delta_3 \approx 0.003175$ and $x_4 \approx 1.618$, i.e. 1.62 correct to 3 significant figures.

$$f(x_4 + \delta_4) = 3(1.618 + \delta_4)^3 - 10(1.618$$
$$+ \delta_4)^2 + 4(1.618 + \delta_4) + 7 = 0$$

giving $\delta_4 \approx 0.0000417$, and $x_5 \approx 1.62$, correct to 3 significant figures.
Since x_4 and x_5 are the same when expressed to the required degree of accuracy, then the required root is **1.62**, correct to 3 significant figures.

Now try the following Practice Exercise

Practice Exercise 99   Further problems on solving equations by an algebraic method of successive approximations (answers on page 1158)

Use an algebraic method of successive approximation to solve the following equations to the accuracy stated.

1. $3x^2 + 5x - 17 = 0$, correct to 3 significant figures.

2. $x^3 - 2x + 14 = 0$, correct to 3 decimal places.

3. $x^4 - 3x^3 + 7x - 5.5 = 0$, correct to 3 significant figures.

4. $x^4 + 12x^3 - 13 = 0$, correct to 4 significant figures.

Section B

23.4   The Newton–Raphson method

The **Newton–Raphson formula**,* often just referred to as **Newton's method**, may be stated as follows:

If r_1 is the approximate value of a real root of the equation $f(x) = 0$, then a closer approximation to the root r_2 is given by:

$$r_2 = r_1 - \frac{f(r_1)}{f'(r_1)}$$

The advantages of Newton's method over the algebraic method of successive approximations is that it can be used for any type of mathematical equation (i.e. ones containing trigonometric, exponential, logarithmic, hyperbolic and algebraic functions), and it is usually easier to apply than the algebraic method.

Note that in the formula for Newton's method, $f'(r_1)$ means the differential coefficient of $f(r_1)$; see Chapter 52 for some help on differentiation.

> **Problem 6.** Use Newton's method to determine the positive root of the quadratic equation $5x^2 + 11x - 17 = 0$, correct to 3 significant figures. Check the value of the root by using the quadratic formula

The functional notation method is used to determine the first approximation to the root.

$$f(x) = 5x^2 + 11x - 17$$

$$f(0) = 5(0)^2 + 11(0) - 17 = -17$$

$$f(1) = 5(1)^2 + 11(1) - 17 = -1$$

$$f(2) = 5(2)^2 + 11(2) - 17 = 25$$

This shows that the value of the root is close to $x = 1$

Let the first approximation to the root, r_1, be 1

Newton's formula states that a closer approximation,

$$r_2 = r_1 - \frac{f(r_1)}{f'(r_1)}$$

$$f(x) = 5x^2 + 11x - 17,$$

thus, $f(r_1) = 5(r_1)^2 + 11(r_1) - 17$

$$= 5(1)^2 + 11(1) - 17 = -1$$

---

* **Who was** Newton? See page 63, then go to **www.routledge. com/cw/bird**
*Joseph Raphson** was an English mathematician known best for the Newton–Raphson method for approximating the roots of an equation. To find out more go to **www.routledge.com/cw/ bird**

$f'(x)$ is the differential coefficient of $f(x)$,

i.e. $f'(x) = 10x + 11$

Thus $f'(r_1) = 10(r_1) + 11$

$$= 10(1) + 11 = 21$$

By Newton's formula, a better approximation to the root is:

$$r_2 = 1 - \frac{-1}{21} = 1 - (-0.048) = 1.05,$$

correct to 3 significant figures.

A still better approximation to the root, r_3, is given by:

$$r_3 = r_2 - \frac{f(r_2)}{f'(r_2)}$$

$$= 1.05 - \frac{[5(1.05)^2 + 11(1.05) - 17]}{[10(1.05) + 11]}$$

$$= 1.05 - \frac{0.0625}{21.5}$$

$$= 1.05 - 0.003 = 1.047,$$

i.e. 1.05, correct to 3 significant figures.

Since the values of r_2 and r_3 are the same when expressed to the required degree of accuracy, the required root is **1.05**, correct to 3 significant figures. Checking, using the quadratic equation formula,

$$x = \frac{-11 \pm \sqrt{[121 - 4(5)(-17)]}}{(2)(5)}$$

$$= \frac{-11 \pm 21.47}{10}$$

The positive root is 1.047, i.e. **1.05**, correct to 3 significant figures. (This root was determined in Problem 1 using the bisection method; Newton's method is clearly quicker.)

> **Problem 7.** Taking the first approximation as 2, determine the root of the equation $x^2 - 3\sin x + 2\ln(x+1) = 3.5$, correct to 3 significant figures, by using Newton's method

Newton's formula states that $r_2 = r_1 - \dfrac{f(r_1)}{f'(r_1)}$, where r_1 is a first approximation to the root and r_2 is a better approximation to the root.

Since $f(x) = x^2 - 3\sin x + 2\ln(x+1) - 3.5$

$$f(r_1) = f(2) = 2^2 - 3\sin 2 + 2\ln 3 - 3.5,$$

where sin2 means the sine of 2 radians

$$= 4 - 2.7279 + 2.1972 - 3.5$$

$$= -0.0307$$

$$f'(x) = 2x - 3\cos x + \frac{2}{x+1}$$

$$f'(r_1) = f'(2) = 2(2) - 3\cos 2 + \frac{2}{3}$$

$$= 4 + 1.2484 + 0.6667$$

$$= 5.9151$$

Hence, $\quad r_2 = r_1 - \dfrac{f(r_1)}{f'(r_1)}$

$$= 2 - \frac{-0.0307}{5.9151}$$

$$= 2.005 \text{ or } 2.01, \text{ correct to 3 significant}$$
$$\text{figures.}$$

A still better approximation to the root, r_3, is given by:

$$r_3 = r_2 - \frac{f(r_2)}{f'(r_2)}$$

$$= 2.005 - \frac{[(2.005)^2 - 3\sin 2.005 + 2\ln 3.005 - 3.5]}{\left[2(2.005) - 3\cos 2.005 + \dfrac{2}{2.005 + 1}\right]}$$

$$= 2.005 - \frac{(-0.00104)}{5.9376} = 2.005 + 0.000175$$

i.e. $r_3 = 2.01$, correct to 3 significant figures.

Since the values of r_2 and r_3 are the same when expressed to the required degree of accuracy, then the required root is **2.01**, correct to 3 significant figures.

Problem 8. Use Newton's method to find the positive root of:

$$(x+4)^3 - e^{1.92x} + 5\cos\frac{x}{3} = 9,$$

correct to 3 significant figures

The functional notational method is used to determine the approximate value of the root.

$$f(x) = (x+4)^3 - e^{1.92x} + 5\cos\frac{x}{3} - 9$$

$$f(0) = (0+4)^3 - e^0 + 5\cos 0 - 9 = 59$$

$$f(1) = 5^3 - e^{1.92} + 5\cos\frac{1}{3} - 9 \approx 114$$

$$f(2) = 6^3 - e^{3.84} + 5\cos\frac{2}{3} - 9 \approx 164$$

$$f(3) = 7^3 - e^{5.76} + 5\cos 1 - 9 \approx 19$$

$$f(4) = 8^3 - e^{7.68} + 5\cos\frac{4}{3} - 9 \approx -1660$$

From these results, let a first approximation to the root be $r_1 = 3$.

Newton's formula states that a better approximation to the root,

$$r_2 = r_1 - \frac{f(r_1)}{f'(r_1)}$$

$$f(r_1) = f(3) = 7^3 - e^{5.76} + 5\cos 1 - 9$$

$$= 19.35$$

$$f'(x) = 3(x+4)^2 - 1.92e^{1.92x} - \frac{5}{3}\sin\frac{x}{3}$$

$$f'(r_1) = f'(3) = 3(7)^2 - 1.92e^{5.76} - \frac{5}{3}\sin 1$$

$$= -463.7$$

Thus, $\quad r_2 = 3 - \dfrac{19.35}{-463.7} = 3 + 0.042$

$$= 3.042 = 3.04,$$
$$\text{correct to 3 significant figures.}$$

Similarly, $\quad r_3 = 3.042 - \dfrac{f(3.042)}{f'(3.042)}$

$$= 3.042 - \frac{(-1.146)}{(-513.1)}$$

$$= 3.042 - 0.0022 = 3.0398 = 3.04,$$
$$\text{correct to 3 significant figures.}$$

Since r_2 and r_3 are the same when expressed to the required degree of accuracy, then the required root is **3.04**, correct to 3 significant figures.

Now try the following Practice Exercise

Practice Exercise 100   Further problems on Newton's method (answers on page 1158)

In Problems 1 to 7, use **Newton's method** to solve the equations given to the accuracy stated.

1. $x^2 - 2x - 13 = 0$, correct to 3 decimal places.

2. $3x^3 - 10x = 14$, correct to 4 significant figures.

3. $x^4 - 3x^3 + 7x = 12$, correct to 3 decimal places.

4. $3x^4 - 4x^3 + 7x - 12 = 0$, correct to 3 decimal places.

5. $3 \ln x + 4x = 5$, correct to 3 decimal places.

6. $x^3 = 5 \cos 2x$, correct to 3 significant figures.

7. $300e^{-2\theta} + \dfrac{\theta}{2} = 6$, correct to 3 significant figures.

8. Solve the equations in Problems 1 to 5, Exercise 98, page 220 and Problems 1 to 4, Exercise 99, page 223 using Newton's method.

9. A Fourier analysis of the instantaneous value of a waveform can be represented by:

$$y = \left(t + \frac{\pi}{4} \right) + \sin t + \frac{1}{8} \sin 3t$$

Use Newton's method to determine the value of t near to 0.04, correct to 4 decimal places, when the amplitude, y, is 0.880

10. A damped oscillation of a system is given by the equation:

$$y = -7.4e^{0.5t} \sin 3t$$

Determine the value of t near to 4.2, correct to 3 significant figures, when the magnitude y of the oscillation is zero.

11. The critical speeds of oscillation, λ, of a loaded beam are given by the equation:

$$\lambda^3 - 3.250\lambda^2 + \lambda - 0.063 = 0$$

Determine the value of λ which is approximately equal to 3.0 by Newton's method, correct to 4 decimal places.

For fully worked solutions to each of the problems in Practice Exercises 98 to 100 in this chapter, go to the website:
www.routledge.com/cw/bird

Chapter 24

Hyperbolic functions

Why it is important to understand: **Hyperbolic functions**

There are two combinations of e^x and e^{-x} which are used so often in engineering that they are given their own name. They are the hyperbolic sine, sinh, and the hyperbolic cosine, cosh. They are of interest because they have many properties analogous to those of trigonometric functions and because they arise in the study of falling bodies, hanging cables, ocean waves, and many other phenomena in science and engineering. The shape of a chain hanging under gravity is well described by cosh x and the deformation of uniform beams can be expressed in terms of hyperbolic tangents. Other applications of hyperbolic functions are found in fluid dynamics, optics, heat, mechanical engineering, and in astronomy when dealing with the curvature of light in the presence of black holes.

At the end of this chapter, you should be able to:

- define a hyperbolic function
- state practical applications of hyperbolic functions
- define sinh x, cosh x, tanh x, cosech x, sech x and coth x
- evaluate hyperbolic functions
- sketch graphs of hyperbolic functions
- state Osborne's rule
- prove simple hyperbolic identities
- solve equations involving hyperbolic functions
- derive the series expansions for cosh x and sinh x

24.1 Introduction to hyperbolic functions

Functions which are associated with the geometry of the conic section called a hyperbola are called **hyperbolic functions**. Applications of hyperbolic functions include transmission line theory and catenary problems. By definition:

(i) Hyperbolic sine of x,

$$\sinh x = \frac{e^x - e^{-x}}{2} \qquad (1)$$

'sinh x' is often abbreviated to 'sh x' and is pronounced as 'shine x'

(ii) Hyperbolic cosine of x,

$$\cosh x = \frac{e^x + e^{-x}}{2} \qquad (2)$$

'cosh x' is often abbreviated to 'ch x' and is pronounced as 'kosh x'

(iii) Hyperbolic tangent of x,

$$\tanh x = \frac{\sinh x}{\cosh x} = \frac{e^x - e^{-x}}{e^x + e^{-x}} \qquad (3)$$

'tanh x' is often abbreviated to 'th x' and is pronounced as 'than x'

(iv)   Hyperbolic cosecant of x,

$$\mathbf{cosech}\ x = \frac{1}{\mathbf{sinh}\,x} = \frac{2}{\mathbf{e}^x - \mathbf{e}^{-x}} \qquad (4)$$

'cosech x' is pronounced as 'coshec x'

(v)   Hyperbolic secant of x,

$$\mathbf{sech}\ x = \frac{1}{\mathbf{cosh}\,x} = \frac{2}{\mathbf{e}^x + \mathbf{e}^{-x}} \qquad (5)$$

'sech x' is pronounced as 'shec x'

(vi)   Hyperbolic cotangent of x,

$$\mathbf{coth}\,x = \frac{1}{\mathbf{tanh}\,x} = \frac{\mathbf{e}^x + \mathbf{e}^{-x}}{\mathbf{e}^x - \mathbf{e}^{-x}} \qquad (6)$$

'coth x' is pronounced as 'koth x'

Some properties of hyperbolic functions

Replacing x by 0 in equation (1) gives:

$$\sinh 0 = \frac{e^0 - e^{-0}}{2} = \frac{1 - 1}{2} = 0$$

Replacing x by 0 in equation (2) gives:

$$\cosh 0 = \frac{e^0 + e^{-0}}{2} = \frac{1 + 1}{2} = 1$$

If a function of x, $f(-x) = -f(x)$, then $f(x)$ is called an **odd function** of x. Replacing x by $-x$ in equation (1) gives:

$$\sinh(-x) = \frac{e^{-x} - e^{-(-x)}}{2} = \frac{e^{-x} - e^x}{2}$$

$$= -\left(\frac{e^x - e^{-x}}{2}\right) = -\sinh x$$

Replacing x by $-x$ in equation (3) gives:

$$\tanh(-x) = \frac{e^{-x} - e^{-(-x)}}{e^{-x} + e^{-(-x)}} = \frac{e^{-x} - e^x}{e^{-x} + e^x}$$

$$= -\left(\frac{e^x - e^{-x}}{e^x + e^{-x}}\right) = -\tanh x$$

Hence **sinh x and tanh$x$ are both odd functions** (see Section 24.2), as also are cosech $x\left(= \dfrac{1}{\sinh x}\right)$ and coth $x\left(= \dfrac{1}{\tanh x}\right)$

If a function of x, $f(-x) = f(x)$, then $f(x)$ is called an **even function** of x. Replacing x by $-x$ in equation (2) gives:

$$\cosh(-x) = \frac{e^{-x} + e^{-(-x)}}{2} = \frac{e^{-x} + e^x}{2}$$

$$= \cosh x$$

Hence **cosh x is an even function** (see Section 24.2), as also is sech $x\left(= \dfrac{1}{\cosh x}\right)$

Hyperbolic functions may be evaluated easiest using a calculator. Many scientific notation calculators actually possess sinh and cosh functions; however, if a calculator does not contain these functions, then the definitions given above may be used.

Problem 1. Evaluate sinh 5.4, correct to 4 significant figures

Using a calculator,

(i)   press hyp

(ii)   press 1 and sinh(appears

(iii)   type in 5.4

(iv)   press) to close the brackets

(v)   press = and 110.7009498 appears

Hence, **sinh 5.4 = 110.7**, correct to 4 significant figures.

Alternatively, $\sinh 5.4 = \dfrac{1}{2}(e^{5.4} - e^{-5.4})$

$$= \frac{1}{2}(221.406416\ldots - 0.00451658\ldots)$$

$$= \frac{1}{2}(221.401899\ldots)$$

$$= \mathbf{110.7},\ \text{correct to 4 significant figures.}$$

Problem 2. Evaluate cosh 1.86, correct to 3 decimal places

Using a calculator with the procedure similar to that used in Problem 1,

$$\mathbf{cosh\,1.86 = 3.290},\ \text{correct to 3 decimal places.}$$

Problem 3. Evaluate th 0.52, correct to 4 significant figures

Using a calculator with the procedure similar to that used in Problem 1,

th 0.52 = 0.4777, correct to 4 significant figures.

Problem 4. Evaluate cosech 1.4, correct to 4 significant figures

$$\text{cosech } 1.4 = \frac{1}{\sinh 1.4}$$

Using a calculator,

(i)   press hyp

(ii)   press 1 and sinh(appears

(iii) type in 1.4

(iv)   press) to close the brackets

(v)   press = and 1.904301501 appears

(vi)   press x^{-1}

(vii) press = and 0.5251269293 appears

Hence, **cosech 1.4 = 0.5251**, correct to 4 significant figures.

Problem 5. Evaluate sech 0.86, correct to 4 significant figures

$$\text{sech } 0.86 = \frac{1}{\cosh 0.86}$$

Using a calculator with the procedure similar to that used in Problem 4,

sech 0.86 = 0.7178, correct to 4 significant figures.

Problem 6. Evaluate coth 0.38, correct to 3 decimal places

$$\text{coth } 0.38 = \frac{1}{\tanh 0.38}$$

Using a calculator with the procedure similar to that used in Problem 4,

coth 0.38 = 2.757, correct to 3 decimal places.

Now try the following Practice Exercise

Practice Exercise 101   Further problems on evaluating hyperbolic functions (answers on page 1158)

In Problems 1 to 6, evaluate correct to 4 significant figures.

1. (a) sh 0.64   (b) sh 2.182

2. (a) ch 0.72   (b) ch 2.4625

3. (a) th 0.65   (b) th 1.81

4. (a) cosech 0.543   (b) cosech 3.12

5. (a) sech 0.39   (b) sech 2.367

6. (a) coth 0.444   (b) coth 1.843

7. A telegraph wire hangs so that its shape is described by $y = 50 \,\text{ch}\, \dfrac{x}{50}$. Evaluate, correct to 4 significant figures, the value of y when $x = 25$

8. The length l of a heavy cable hanging under gravity is given by $l = 2c \,\text{sh}\,(L/2c)$. Find the value of l when $c = 40$ and $L = 30$

9. $V^2 = 0.55 L \tanh(6.3 d/L)$ is a formula for velocity V of waves over the bottom of shallow water, where d is the depth and L is the wavelength. If $d = 8.0$ and $L = 96$, calculate the value of V

24.2   Graphs of hyperbolic functions

A graph of **$y = \sinh x$** may be plotted using calculator values of hyperbolic functions. The curve is shown in Figure 24.1. Since the graph is symmetrical about the origin, $\sinh x$ is an **odd function** (as stated in Section 24.1).

A graph of **$y = \cosh x$** may be plotted using calculator values of hyperbolic functions. The curve is shown in Figure 24.2. Since the graph is symmetrical about the y-axis, $\cosh x$ is an **even function** (as stated in Section 24.1). The shape of $y = \cosh x$ is that of a heavy rope or chain hanging freely under gravity and is called a **catenary**. Examples include transmission lines, a telegraph wire or a fisherman's line. It is used in the design of roofs and arches. Graphs of $y = \tanh x$, $y = \text{cosech}\, x$, $y = \text{sech}\, x$ and $y = \coth x$ are deduced in Problems 7 and 8.

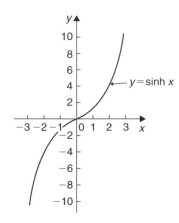

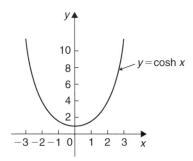

Figure 24.1

Figure 24.2

(a)   A graph of $y = \tanh x$ is shown in Figure 24.3(a)

(b)   A graph of $y = \coth x$ is shown in Figure 24.3(b)

Both graphs are symmetrical about the origin, thus $\tanh x$ and $\coth x$ are odd functions.

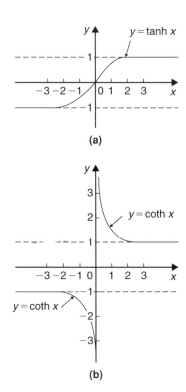

Figure 24.3

Problem 7.   Sketch graphs of (a) $y = \tanh x$ and (b) $y = \coth x$ for values of x between -3 and 3

A table of values is drawn up as shown below.

x	-3	-2	-1
$\operatorname{sh} x$	-10.02	-3.63	-1.18
$\operatorname{ch} x$	10.07	3.76	1.54
$y = \operatorname{th} x = \dfrac{\operatorname{sh} x}{\operatorname{ch} x}$	-0.995	-0.97	-0.77
$y = \coth x = \dfrac{\operatorname{ch} x}{\operatorname{sh} x}$	-1.005	-1.04	-1.31

x	0	1	2	3
$\operatorname{sh} x$	0	1.18	3.63	10.02
$\operatorname{ch} x$	1	1.54	3.76	10.07
$y = \operatorname{th} x = \dfrac{\operatorname{sh} x}{\operatorname{ch} x}$	0	0.77	0.97	0.995
$y = \coth x = \dfrac{\operatorname{ch} x}{\operatorname{sh} x}$	$\pm\infty$	1.31	1.04	1.005

Problem 8.   Sketch graphs of (a) $y = \operatorname{cosech} x$ and (b) $y = \operatorname{sech} x$ from $x = -4$ to $x = 4$, and, from the graphs, determine whether they are odd or even functions

A table of values is drawn up as shown below.

x	-4	-3	-2	-1
$\operatorname{sh} x$	-27.29	-10.02	-3.63	-1.18
$\operatorname{cosech} x = \dfrac{1}{\operatorname{sh} x}$	-0.04	-0.10	-0.28	-0.85
$\operatorname{ch} x$	27.31	10.07	3.76	1.54
$\operatorname{sech} x = \dfrac{1}{\operatorname{ch} x}$	0.04	0.10	0.27	0.65

x		0	1	2	3	4
sh x		0	1.18	3.63	10.02	27.29
cosech $x = \dfrac{1}{\text{sh}\,x}$		$\pm\infty$	0.85	0.28	0.10	0.04
ch x		1	1.54	3.76	10.07	27.31
sech $x = \dfrac{1}{\text{ch}\,x}$		1	0.65	0.27	0.10	0.04

(a) A graph of $y = \text{cosech}\,x$ is shown in Figure 24.4(a). The graph is symmetrical about the origin and is thus an **odd function**.

(b) A graph of $y = \text{sech}\,x$ is shown in Figure 24.4(b). The graph is symmetrical about the y-axis and is thus an **even function**.

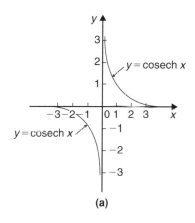

(a)

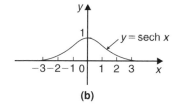

(b)

Figure 24.4

24.3 Hyperbolic identities

For every trigonometric identity there is a corresponding hyperbolic identity. **Hyperbolic identities** may be proved by either

(i) replacing sh x by $\dfrac{e^x - e^{-x}}{2}$ and ch x by $\dfrac{e^x + e^{-x}}{2}$, or

(ii) by using **Osborne's rule**, which states: '*the six trigonometric ratios used in trigonometrical identities relating general angles may be replaced by their corresponding hyperbolic functions, but the sign of any direct or implied product of two sines must be changed*'.

For example, since $\cos^2 x + \sin^2 x = 1$ then, by Osborne's rule, $\text{ch}^2 x - \text{sh}^2 x = 1$, i.e. the trigonometric functions have been changed to their corresponding hyperbolic functions and since $\sin^2 x$ is a product of two sines the sign is changed from $+$ to $-$. Table 24.1 on page 232 shows some trigonometric identities and their corresponding hyperbolic identities.

> **Problem 9.** Prove the hyperbolic identities
> (a) $\text{ch}^2 x - \text{sh}^2 x = 1$ (b) $1 - \text{th}^2 x = \text{sech}^2 x$
> (c) $\coth^2 x - 1 = \text{cosech}^2 x$

(a) $\text{ch}\,x + \text{sh}\,x = \left(\dfrac{e^x + e^{-x}}{2}\right) + \left(\dfrac{e^x - e^{-x}}{2}\right) = e^x$

$\text{ch}\,x - \text{sh}\,x = \left(\dfrac{e^x + e^{-x}}{2}\right) - \left(\dfrac{e^x - e^{-x}}{2}\right)$

$= e^{+-x}$

$(\text{ch}\,x + \text{sh}\,x)(\text{ch}\,x - \text{sh}\,x) = (e^x)(e^{-x}) = e^0 = 1$

i.e. $\mathbf{ch^2 x - sh^2 x = 1}$ \hfill (1)

(b) Dividing each term in equation (1) by $\text{ch}^2 x$ gives:

$$\frac{\text{ch}^2 x}{\text{ch}^2 x} - \frac{\text{sh}^2 x}{\text{ch}^2 x} = \frac{1}{\text{ch}^2 x}$$

i.e. $\mathbf{1 - th^2 x = sech^2 x}$

(c) Dividing each term in equation (1) by $\text{sh}^2 x$ gives:

$$\frac{\text{ch}^2 x}{\text{sh}^2 x} - \frac{\text{sh}^2 x}{\text{sh}^2 x} = \frac{1}{\text{sh}^2 x}$$

i.e. $\mathbf{coth^2 x - 1 = cosech^2 x}$

> **Problem 10.** Prove, using Osborne's rule
> (a) $\text{ch}\,2A = \text{ch}^2 A + \text{sh}^2 A$
> (b) $1 - \text{th}^2 x = \text{sech}^2 x$

Section B

Table 24.1

Trigonometric identity	Corresponding hyperbolic identity
$\cos^2 x + \sin^2 x = 1$	$\text{ch}^2 x - \text{sh}^2 x = 1$
$1 + \tan^2 x = \sec^2 x$	$1 - \text{th}^2 x = \text{sech}^2 x$
$\cot^2 x + 1 = \text{cosec}^2 x$	$\coth^2 x - 1 = \text{cosech}^2 x$
Compound angle formulae	
$\sin(A \pm B) = \sin A \cos B \pm \cos A \sin B$	$\text{sh}(A \pm B) = \text{sh} A \, \text{ch} B \pm \text{ch} A \, \text{sh} B$
$\cos(A \pm B) = \cos A \cos B \mp \sin A \sin B$	$\text{ch}(A \pm B) = \text{ch} A \, \text{ch} B \pm \text{sh} A \, \text{sh} B$
$\tan(A \pm B) = \dfrac{\tan A \pm \tan B}{1 \mp \tan A \tan B}$	$\text{th}(A \pm B) = \dfrac{\text{th} A \pm \text{th} B}{1 \pm \text{th} A \, \text{th} B}$
Double angles	
$\sin 2x = 2 \sin x \cos x$	$\text{sh} 2x = 2 \, \text{sh} x \, \text{ch} x$
$\cos 2x = \cos^2 x - \sin^2 x$	$\text{ch} 2x = \text{ch}^2 x + \text{sh}^2 x$
$\quad = 2 \cos^2 x - 1$	$\quad = 2 \text{ch}^2 x - 1$
$\quad = 1 - 2 \sin^2 x$	$\quad = 1 + 2 \text{sh}^2 x$
$\tan 2x = \dfrac{2 \tan x}{1 - \tan^2 x}$	$\text{th} 2x = \dfrac{2 \, \text{th} x}{1 + \text{th}^2 x}$

(a) From trigonometric ratios,

$$\cos 2A = \cos^2 A - \sin^2 A \qquad (1)$$

Osborne's rule states that trigonometric ratios may be replaced by their corresponding hyperbolic functions but the sign of any product of two sines has to be changed. In this case, $\sin^2 A = (\sin A)(\sin A)$, i.e. a product of two sines, thus the sign of the corresponding hyperbolic function, $\text{sh}^2 A$, is changed from $+$ to $-$. Hence, from (1), $\mathbf{ch\,2A = ch^2 A + sh^2 A}$

(b) From trigonometric ratios,

$$1 + \tan^2 x = \sec^2 x \qquad (2)$$

and $\tan^2 x = \dfrac{\sin^2 x}{\cos^2 x} = \dfrac{(\sin x)(\sin x)}{\cos^2 x}$

i.e. a product of two sines.

Hence, in equation (2), the trigonometric ratios are changed to their equivalent hyperbolic function and the sign of $\text{th}^2 x$ changed from $+$ to $-$, i.e. $\mathbf{1 - th^2 x = sech^2 x}$

Problem 11. Prove that $1 + 2 \text{sh}^2 x = \text{ch} 2x$

Left-hand side (LHS)

$$= 1 + 2 \text{sh}^2 x = 1 + 2 \left(\frac{e^x - e^{-x}}{2} \right)^2$$

$$= 1 + 2 \left(\frac{e^{2x} - 2 e^x e^{-x} + e^{-2x}}{4} \right)$$

$$= 1 + \frac{e^{2x} - 2 + e^{-2x}}{2}$$

$$= 1 + \left(\frac{e^{2x} + e^{-2x}}{2} \right) - \frac{2}{2}$$

$$= \frac{e^{2x} + e^{-2x}}{2} = \text{ch} 2x = \text{RHS}$$

Problem 12. Show that $\text{th}^2 x + \text{sech}^2 x = 1$

$$\text{LHS} = \text{th}^2 x + \text{sech}^2 x = \frac{\text{sh}^2 x}{\text{ch}^2 x} + \frac{1}{\text{ch}^2 x}$$

$$= \frac{\text{sh}^2 x + 1}{\text{ch}^2 x}$$

Since $\text{ch}^2 x - \text{sh}^2 x = 1$ then $1 + \text{sh}^2 x = \text{ch}^2 x$

Thus $\dfrac{\text{sh}^2 x + 1}{\text{ch}^2 x} = \dfrac{\text{ch}^2 x}{\text{ch}^2 x} = 1 = \text{RHS}$

Problem 13. Given $Ae^x + Be^{-x} \equiv 4\,\text{ch}\,x - 5\,\text{sh}\,x$, determine the values of A and B

$$Ae^x + Be^{-x} \equiv 4\,\text{ch}\,x - 5\,\text{sh}\,x$$

$$= 4\left(\frac{e^x + e^{-x}}{2}\right) - 5\left(\frac{e^x - e^{-x}}{2}\right)$$

$$= 2e^x + 2e^{-x} - \frac{5}{2}e^x + \frac{5}{2}e^{-x}$$

$$= -\frac{1}{2}e^x + \frac{9}{2}e^{-x}$$

Equating coefficients gives: $A = -\dfrac{1}{2}$ and $B = 4\dfrac{1}{2}$

Problem 14. If $4e^x - 3e^{-x} \equiv P\,\text{sh}\,x + Q\,\text{ch}\,x$, determine the values of P and Q

$$4e^x - 3e^{-x} \equiv P\,\text{sh}\,x + Q\,\text{ch}\,x$$

$$= P\left(\frac{e^x - e^{-x}}{2}\right) + Q\left(\frac{e^x + e^{-x}}{2}\right)$$

$$= \frac{P}{2}e^x - \frac{P}{2}e^{-x} + \frac{Q}{2}e^x + \frac{Q}{2}e^{-x}$$

$$= \left(\frac{P + Q}{2}\right)e^x + \left(\frac{Q - P}{2}\right)e^{-x}$$

Equating coefficients gives:

$$4 = \frac{P + Q}{2} \quad \text{and} \quad -3 = \frac{Q - P}{2}$$

i.e. $P + Q = 8$ \hfill (1)

$-P + Q = -6$ \hfill (2)

Adding equations (1) and (2) gives: $2Q = 2$, i.e. $\boldsymbol{Q = 1}$

Substituting in equation (1) gives: $\boldsymbol{P = 7}$

Now try the following Practice Exercise

Practice Exercise 102　Further problems on hyperbolic identities (answers on page 1158)

In Problems 1 to 4, prove the given identities.

1.　(a) $\text{ch}(P - Q) \equiv \text{ch}\,P\,\text{ch}\,Q - \text{sh}\,P\,\text{sh}\,Q$

　　(b) $\text{ch}\,2x \equiv \text{ch}^2 x + \text{sh}^2 x$

2.　(a) $\text{coth}\,x \equiv 2\,\text{cosech}\,2x + \text{th}\,x$

　　(b) $\text{ch}\,2\theta - 1 \equiv 2\,\text{sh}^2 \theta$

3.　(a) $\text{th}(A - B) \equiv \dfrac{\text{th}\,A - \text{th}\,B}{1 - \text{th}\,A\,\text{th}\,B}$

　　(b) $\text{sh}\,2A \equiv 2\,\text{sh}\,A\,\text{ch}\,A$

4.　(a) $\text{sh}(A + B) \equiv \text{sh}\,A\,\text{ch}\,B + \text{ch}\,A\,\text{sh}\,B$

　　(b) $\dfrac{\text{sh}^2 x + \text{ch}^2 x - 1}{2\,\text{ch}^2 x\,\text{coth}^2 x} \equiv \tanh^4 x$

5.　Given $Pe^x - Qe^{-x} \equiv 6\,\text{ch}\,x - 2\,\text{sh}\,x$, find P and Q.

6.　If $5e^x - 4e^{-x} \equiv A\,\text{sh}\,x + B\,\text{ch}\,x$, find A and B.

24.4　Solving equations involving hyperbolic functions

Equations such as $\sinh x = 3.25$ or $\coth x = 3.478$ may be determined using a calculator. This is demonstrated in Problems 15 to 21.

Problem 15. Solve the equation $\text{sh}\,x = 3$, correct to 4 significant figures

If $\sinh x = 3$, then $x = \sinh^{-1} 3$
This can be determined by calculator.

(i)　Press hyp

(ii)　Choose 4, which is $\sinh^{-1}$

(iii)　Type in 3

(iv)　Close bracket)

(v)　Press $=$ and the answer is 1.818446459

i.e. the solution of $\text{sh}\,x = 3$ is: $\boldsymbol{x = 1.818}$, correct to 4 significant figures.

Problem 16. Solve the equation $\text{ch}\,x = 1.52$, correct to 3 decimal places

Using a calculator with a similar procedure as in Problem 15, check that:

$$x = 0.980, \text{ correct to 3 decimal places.}$$

With reference to Figure 24.2, it can be seen that there will be two values corresponding to $y = \cosh x = 1.52$. Hence, $x = \pm 0.980$

Problem 17. Solve the equation $\tanh \theta = 0.256$, correct to 4 significant figures

Using a calculator with a similar procedure as in Problem 15, check that

$$\theta = 0.2618, \text{ correct to 4 significant figures.}$$

Problem 18. Solve the equation $\operatorname{sech} x = 0.4562$, correct to 3 decimal places

If $\operatorname{sech} x = 0.4562$, then $x = \operatorname{sech}^{-1} 0.4562 = \cosh^{-1}\left(\dfrac{1}{0.4562}\right)$ since $\cosh = \dfrac{1}{\operatorname{sech}}$

i.e. $x = 1.421$, correct to 3 decimal places.

With reference to the graph of $y = \operatorname{sech} x$ in Figure 24.4, it can be seen that there will be two values corresponding to $y = \operatorname{sech} x = 0.4562$
Hence, $x = \pm 1.421$

Problem 19. Solve the equation $\operatorname{cosech} y = -0.4458$, correct to 4 significant figures

If $\operatorname{cosech} y = -0.4458$, then $y = \operatorname{cosech}^{-1}(-0.4458)$
$= \sinh^{-1}\left(\dfrac{1}{-0.4458}\right)$ since $\sinh = \dfrac{1}{\operatorname{cosech}}$
i.e. $y = -1.547$, correct to 4 significant figures.

Problem 20. Solve the equation $\coth A = 2.431$, correct to 3 decimal places

If $\coth A = 2.431$, then $A = \coth^{-1} 2.431 = \tanh^{-1}\left(\dfrac{1}{2.431}\right)$ since $\tanh = \dfrac{1}{\coth}$
i.e. $A = 0.437$, correct to 3 decimal places.

Problem 21. A chain hangs in the form given by $y = 40 \operatorname{ch} \dfrac{x}{40}$. Determine, correct to 4 significant figures, (a) the value of y when x is 25, and (b) the value of x when $y = 54.30$

(a) $y = 40 \operatorname{ch} \dfrac{x}{40}$, and when $x = 25$,

$$y = 40 \operatorname{ch} \dfrac{25}{40} = 40 \operatorname{ch} 0.625$$

$$= 40(1.2017536\ldots) = 48.07$$

(b) When $y = 54.30$, $54.30 = 40 \operatorname{ch} \dfrac{x}{40}$, from which

$$\operatorname{ch} \dfrac{x}{40} = \dfrac{54.30}{40} = 1.3575$$

Hence, $\dfrac{x}{40} = \cosh^{-1} 1.3575 = \pm 0.822219\ldots$
(see Figure 24.2 for the reason as to why the answer is $\pm$) from which, $x = 40(\pm 0.822219\ldots) = \pm 32.89$

Equations of the form $a \operatorname{ch} x + b \operatorname{sh} x = c$, where a, b and c are constants may be solved either by:

(a) plotting graphs of $y = a \operatorname{ch} x + b \operatorname{sh} x$ and $y = c$ and noting the points of intersection, or more accurately,

(b) by adopting the following procedure:

  (i) Change $\operatorname{sh} x$ to $\left(\dfrac{e^x - e^{-x}}{2}\right)$ and $\operatorname{ch} x$ to $\left(\dfrac{e^x + e^{-x}}{2}\right)$

  (ii) Rearrange the equation into the form $pe^x + qe^{-x} + r = 0$, where p, q and r are constants.

  (iii) Multiply each term by e^x, which produces an equation of the form $p(e^x)^2 + re^x + q = 0$ (since $(e^{-x})(e^x) = e^0 = 1$)

  (iv) Solve the quadratic equation $p(e^x)^2 + re^x + q = 0$ for e^x by factorising or by using the quadratic formula.

  (v) Given $e^x = a$ constant (obtained by solving the equation in (iv)), take Napierian logarithms of both sides to give $x = \ln(\text{constant})$

This procedure is demonstrated in Problem 22.

Problem 22. Solve the equation $2.6 \operatorname{ch} x + 5.1 \operatorname{sh} x = 8.73$, correct to 4 decimal places

Following the above procedure:

(i) $2.6\,\text{ch}\,x + 5.1\,\text{sh}\,x = 8.73$

i.e. $2.6\left(\dfrac{e^x + e^{-x}}{2}\right) + 5.1\left(\dfrac{e^x - e^{-x}}{2}\right) = 8.73$

(ii) $1.3e^x + 1.3e^{-x} + 2.55e^x - 2.55e^{-x} = 8.73$

i.e. $3.85e^x - 1.25e^{-x} - 8.73 = 0$

(iii) $3.85(e^x)^2 - 8.73e^x - 1.25 = 0$

(iv) e^x

$$= \frac{-(-8.73) \pm \sqrt{[(-8.73)^2 - 4(3.85)(-1.25)]}}{2(3.85)}$$

$$= \frac{8.73 \pm \sqrt{95.463}}{7.70} = \frac{8.73 \pm 9.7705}{7.70}$$

Hence $e^x = 2.4027$ or $e^x = -0.1351$

(v) $x = \ln 2.4027$ or $x = \ln(-0.1351)$ which has no real solution.

Hence $x = \mathbf{0.8766}$, correct to 4 decimal places.

Now try the following Practice Exercise

Practice Exercise 103   Further problems on hyperbolic equations (answers on page 1159)

In Problems 1 to 9, solve the given equations correct to 4 decimal places.

1.   (a) $\sinh x = 1$ (b) $\text{sh}\,A = -2.43$

2.   (a) $\cosh B = 1.87$ (b) $2\,\text{ch}\,x = 3$

3.   (a) $\tanh y = -0.76$ (b) $3\,\text{th}\,x = 2.4$

4.   (a) $\text{sech}\,B = 0.235$ (b) $\text{sech}\,Z = 0.889$

5.   (a) $\text{cosech}\,\theta = 1.45$ (b) $5\,\text{cosech}\,x = 4.35$

6.   (a) $\coth x = 2.54$ (b) $2\coth y = -3.64$

7. $3.5\,\text{sh}\,x + 2.5\,\text{ch}\,x = 0$

8. $2\,\text{sh}\,x + 3\,\text{ch}\,x = 5$

9. $4\,\text{th}\,x - 1 = 0$

10.   A chain hangs so that its shape is of the form $y = 56\cosh\left(\dfrac{x}{56}\right)$. Determine, correct to 4 significant figures, (a) the value of y when x is 35, and (b) the value of x when y is 62.35

24.5 Series expansions for cosh x and sinh x

By definition,

$$e^x = 1 + x + \frac{x^2}{2!} + \frac{x^3}{3!} + \frac{x^4}{4!} + \frac{x^5}{5!} + \cdots$$

from Chapter 16.
Replacing x by $-x$ gives:

$$e^{-x} = 1 - x + \frac{x^2}{2!} - \frac{x^3}{3!} + \frac{x^4}{4!} - \frac{x^5}{5!} + \cdots.$$

$$\cosh x = \frac{1}{2}(e^x + e^{-x})$$

$$= \frac{1}{2}\left[\left(1 + x + \frac{x^2}{2!} + \frac{x^3}{3!} + \frac{x^4}{4!} + \frac{x^5}{5!} + \cdots\right) \right.$$

$$\left. + \left(1 - x + \frac{x^2}{2!} - \frac{x^3}{3!} + \frac{x^4}{4!} - \frac{x^5}{5!} + \cdots\right)\right]$$

$$= \frac{1}{2}\left[\left(2 + \frac{2x^2}{2!} + \frac{2x^4}{4!} + \cdots\right)\right]$$

i.e. $\cosh x = 1 + \dfrac{x^2}{2!} + \dfrac{x^4}{4!} + \cdots$ (which is valid for all values of x). $\cosh x$ is an even function and contains only even powers of x in its expansion.

$$\sinh x = \frac{1}{2}(e^x - e^{-x})$$

$$= \frac{1}{2}\left[\left(1 + x + \frac{x^2}{2!} + \frac{x^3}{3!} + \frac{x^4}{4!} + \frac{x^5}{5!} + \cdots\right) \right.$$

$$\left. - \left(1 - x + \frac{x^2}{2!} - \frac{x^3}{3!} + \frac{x^4}{4!} - \frac{x^5}{5!} + \cdots\right)\right]$$

$$= \frac{1}{2}\left[2x + \frac{2x^3}{3!} + \frac{2x^5}{5!} + \cdots\right]$$

i.e. $\sinh x = x + \dfrac{x^3}{3!} + \dfrac{x^5}{5!} + \cdots$ (which is valid for all values of x). $\sinh x$ is an odd function and contains only odd powers of x in its series expansion.

Problem 23.   Using the series expansion for ch x evaluate ch 1 correct to 4 decimal places

Section B

$\mathrm{ch}\,x = 1 + \dfrac{x^2}{2!} + \dfrac{x^4}{4!} + \cdots$ from above

Let $x = 1$,

then $\mathrm{ch}\,1 = 1 + \dfrac{1^2}{2 \times 1} + \dfrac{1^4}{4 \times 3 \times 2 \times 1}$

$\qquad\qquad + \dfrac{1^6}{6 \times 5 \times 4 \times 3 \times 2 \times 1} + \cdots$

$\qquad\quad = 1 + 0.5 + 0.04167 + 0.001389 + \cdots$

i.e. **ch 1 = 1.5431**, correct to 4 decimal places, which may be checked by using a calculator.

Problem 24. Determine, correct to 3 decimal places, the value of sh 3 using the series expansion for $\mathrm{sh}\,x$

$\mathrm{sh}\,x = x + \dfrac{x^3}{3!} + \dfrac{x^5}{5!} + \cdots$ from above

Let $x = 3$, then

$\mathrm{sh}\,3 = 3 + \dfrac{3^3}{3!} + \dfrac{3^5}{5!} + \dfrac{3^7}{7!} + \dfrac{3^9}{9!} + \dfrac{3^{11}}{11!} + \cdots$

$\qquad\;\; = 3 + 4.5 + 2.025 + 0.43393 + 0.05424$

$\qquad\qquad + 0.00444 + \cdots$

i.e. **sh 3 = 10.018**, correct to 3 decimal places.

Problem 25. Determine the power series for $2\,\mathrm{ch}\left(\dfrac{\theta}{2}\right) - \mathrm{sh}\,2\theta$ as far as the term in θ^5

In the series expansion for $\mathrm{ch}\,x$, let $x = \dfrac{\theta}{2}$ then:

$2\,\mathrm{ch}\left(\dfrac{\theta}{2}\right) = 2\left[1 + \dfrac{(\theta/2)^2}{2!} + \dfrac{(\theta/2)^4}{4!} + \cdots\right]$

$\qquad\qquad = 2 + \dfrac{\theta^2}{4} + \dfrac{\theta^4}{192} + \cdots$

In the series expansion for $\mathrm{sh}\,x$, let $x = 2\theta$, then:

$\mathrm{sh}\,2\theta = 2\theta + \dfrac{(2\theta)^3}{3!} + \dfrac{(2\theta)^5}{5!} + \cdots$

$\qquad\quad = 2\theta + \dfrac{4}{3}\theta^3 + \dfrac{4}{15}\theta^5 + \cdots$

Hence

$2\,\mathrm{ch}\left(\dfrac{\theta}{2}\right) - \mathrm{sh}\,2\theta = \left(2 + \dfrac{\theta^2}{4} + \dfrac{\theta^4}{192} + \cdots\right)$

$\qquad\qquad\qquad - \left(2\theta + \dfrac{4}{3}\theta^3 + \dfrac{4}{15}\theta^5 + \cdots\right)$

$\qquad\quad = 2 - 2\theta + \dfrac{\theta^2}{4} - \dfrac{4}{3}\theta^3 + \dfrac{\theta^4}{192}$

$\qquad\qquad - \dfrac{4}{15}\theta^5 + \cdots$ as far the term in $\boldsymbol{\theta^5}$

Now try the following Practice Exercise

Practice Exercise 104   Further problems on series expansions for cosh x and sinh x (answers on page 1159)

1. Use the series expansion for $\mathrm{ch}\,x$ to evaluate, correct to 4 decimal places: (a) ch 1.5 (b) ch 0.8

2. Use the series expansion for $\mathrm{sh}\,x$ to evaluate, correct to 4 decimal places: (a) sh 0.5 (b) sh 2

3. Expand the following as a power series as far as the term in x^5: (a) sh 3x (b) ch 2x

In Problems 4 and 5, prove the given identities, the series being taken as far as the term in θ^5 only.

4. $\mathrm{sh}\,2\theta - \mathrm{sh}\,\theta \equiv \theta + \dfrac{7}{6}\theta^3 + \dfrac{31}{120}\theta^5$

5. $2\,\mathrm{sh}\dfrac{\theta}{2} - \mathrm{ch}\dfrac{\theta}{2} \equiv -1 + \theta - \dfrac{\theta^2}{8} + \dfrac{\theta^3}{24} - \dfrac{\theta^4}{384}$

$\qquad\qquad\qquad\qquad\qquad + \dfrac{\theta^5}{1920}$

For fully worked solutions to each of the problems in Practice Exercises 101 to 104 in this chapter, go to the website:
www.routledge.com/cw/bird

This assignment covers the material contained in Chapters 23 and 24. *The marks available are shown in brackets at the end of each question.*

1. Use the method of bisection to evaluate the root of the equation: $x^3 + 5x = 11$ in the range $x = 1$ to $x = 2$, correct to 3 significant figures. (10)

2. Repeat Question 1 using an algebraic method of successive approximations. (16)

3. The solution to a differential equation associated with the path taken by a projectile for which the resistance to motion is proportional to the velocity is given by:
$$y = 2.5(e^x - e^{-x}) + x - 25$$
Use Newton's method to determine the value of x, correct to 2 decimal places, for which the value of y is zero. (10)

4. Evaluate correct to 4 significant figures:

   (a) $\sinh 2.47$ (b) $\tanh 0.6439$

   (c) $\operatorname{sech} 1.385$ (d) $\operatorname{cosech} 0.874$ (6)

5. The increase in resistance of strip conductors due to eddy currents at power frequencies is given by:
$$\lambda = \frac{\alpha t}{2}\left[\frac{\sinh \alpha t + \sin \alpha t}{\cosh \alpha t - \cos \alpha t}\right]$$

Calculate λ, correct to 5 significant figures, when $\alpha = 1.08$ and $t = 1$ (5)

6. If $A \operatorname{ch} x - B \operatorname{sh} x \equiv 4e^x - 3e^{-x}$ determine the values of A and B. (6)

7. Solve the following equation:
$$3.52 \operatorname{ch} x + 8.42 \operatorname{sh} x = 5.32$$
correct to 4 decimal places. (7)

Section B

Chapter 25

Binary, octal and hexadecimal numbers

Why it is important to understand: **Binary, octal and hexadecimal numbers**

There are infinite ways to represent a number. The four commonly associated with modern computers and digital electronics are decimal, binary, octal and hexadecimal. All four number systems are equally capable of representing any number. Furthermore, a number can be perfectly converted between the various number systems without any loss of numeric value. At a first look, it seems like using any number system other than decimal is complicated and unnecessary. However, since the job of electrical and software engineers is to work with digital circuits, engineers require number systems that can best transfer information between the human world and the digital circuit world. Thus the way in which a number is represented can make it easier for the engineer to perceive the meaning of the number as it applies to a digital circuit, i.e. the appropriate number system can actually make things less complicated.

At the end of this chapter, you should be able to:

- recognise a binary number
- convert binary to decimal and vice versa
- add binary numbers
- recognise an octal number
- convert decimal to binary via octal and vice versa
- recognise a hexadecimal number
- convert from hexadecimal to decimal and vice versa
- convert from binary to hexadecimal and vice versa

25.1 Introduction

Man's earliest number or counting system was probably developed to help determine how many possessions a person had. As daily activities became more complex, numbers became more important in trade, time, distance, and all other phases of human life. Ever since people discovered that it was necessary to count objects, they have been looking for easier ways to do so. The **abacus**, developed by the Chinese, is one of the earliest known calculators; it is still in use in some parts of the world.

Blaise Pascal* invented the first adding machine in 1642. Twenty years later, an Englishman, **Sir Samuel Morland***, developed a more compact device that could multiply, add and subtract. About 1672, **Gottfried Wilhelm von Leibniz*** perfected a machine that could perform all the basic operations (add, subtract, multiply, divide), as well as extract the square root. Modern electronic digital computers still use von Leibniz's principles.

Computers are now employed wherever repeated calculations or the processing of huge amounts of data are needed. The greatest applications are found in the military, scientific and commercial fields. They have applications that range from mail sorting and engineering design, to the identification and destruction of enemy targets. The advantages of digital computers include speed, accuracy and man-power savings. Often computers are able to take over routine jobs and release personnel for more important work that cannot be handled by a computer. People and computers do not normally speak the same language. Methods of translating information into forms that are understandable and usable to both are necessary. Humans generally speak in words and numbers expressed in the decimal number system, while computers only understand coded electronic pulses that represent digital information.

All data in modern computers is stored as a series of **bits**, a bit being a <u>bi</u>nary digi<u>t</u>, and can have one of two values, the numbers 0 and 1. The most basic form of representing computer data is to represent a piece of data as a string of 1s and 0s, one for each bit. This is called a **binary** or base-2 number.

Because binary notation requires so many bits to represent relatively small numbers, two further compact notations are often used, called **octal** and **hexadecimal**. Computer programmers who design sequences of number codes instructing a computer what to do, would have a very difficult task if they were forced to work with nothing but long strings of 1s and 0s, the 'native language' of any digital circuit.

Octal notation represents data as base-8 numbers, with each digit in an octal number representing three bits. Similarly, hexadecimal notation uses base-16 numbers, representing four bits with each digit. Octal numbers use only the digits 0–7, while hexadecimal numbers use all ten base-10 digits (0–9) and the letters A–F (representing the numbers 10–15).

This chapter explains how to convert between the decimal, binary, octal and hexadecimal systems.

* **Who was Leibniz?** **Gottfried Wilhelm Leibniz** (sometimes von Leibniz theorem) (1 July 1646–14 November 1716) was a German mathematician and philosopher. Leibniz developed infinitesimal calculus and invented the Leibniz wheel. To find out more go to **www.routledge.com/cw/bird**

* **Who was Morland?** **Sir Samuel Morland, 1st Baronet** (1625–1695), was an English academic, diplomat, spy, inventor and mathematician of the seventeenth century, a polymath credited with early developments in relation to computing, hydraulics and steam power. To find out more go to **www.routledge.com/cw/bird**

* **Who was Pascal?** For image and resume of Pascal, see page 197. To find out more go to **www.routledge.com/cw/bird**

25.2   Binary numbers

The system of numbers in everyday use is the **denary** or **decimal** system of numbers, using the digits 0 to 9. It has ten different digits (0, 1, 2, 3, 4, 5, 6, 7, 8 and 9) and is said to have a **radix** or **base** of 10.

The **binary** system of numbers has a radix of 2 and uses only the digits 0 and 1.

(a) Conversion of binary to decimal

The decimal number 234.5 is equivalent to

$$2 \times 10^2 + 3 \times 10^1 + 4 \times 10^0 + 5 \times 10^{-1}$$

In the binary system of numbers, the base is 2, so 1101.1 is equivalent to:

$$1 \times 2^3 + 1 \times 2^2 + 0 \times 2^1 + 1 \times 2^0 + 1 \times 2^{-1}$$

Section B

Thus, the decimal number equivalent to the binary number 1101.1 is:

$$8 + 4 + 0 + 1 + \frac{1}{2} \text{ that is } 13.5$$

i.e. $1101.1_2 = 13.5_{10}$, the suffixes 2 and 10 denoting binary and decimal systems of numbers, respectively.

Problem 1. Convert 1010_2 to a decimal number

From above:

$$1010_2 = 1 \times 2^3 + 0 \times 2^2 + 1 \times 2^1 + 0 \times 2^0$$
$$= 8 + 0 + 2 + 0$$
$$= 10_{10}$$

Problem 2. Convert 11011_2 to a decimal number

$$11011_2 = 1 \times 2^4 + 1 \times 2^3 + 0 \times 2^2 + 1 \times 2^1 + 1 \times 2^0$$
$$- 16 + 8 + 0 + 2 + 1$$
$$= 27_{10}$$

Problem 3. Convert 0.1011_2 to a decimal fraction

$$0.1011_2 = 1 \times 2^{-1} + 0 \times 2^{-2} + 1 \times 2^{-3} + 1 \times 2^{-4}$$
$$= 1 \times \frac{1}{2} + 0 \times \frac{1}{2^2} + 1 \times \frac{1}{2^3} + 1 \times \frac{1}{2^4}$$
$$= \frac{1}{2} + \frac{1}{8} + \frac{1}{16}$$
$$= 0.5 + 0.125 + 0.0625$$
$$= 0.6875_{10}$$

Problem 4. Convert 101.0101_2 to a decimal number

$$101.0101_2 = 1 \times 2^2 + 0 \times 2^1 + 1 \times 2^0 + 0 \times 2^{-1}$$
$$+ 1 \times 2^{-2} + 0 \times 2^{-3} + 1 \times 2^{-4}$$
$$= 4 + 0 + 1 + 0 + 0.25 + 0 + 0.0625$$
$$= 5.3125_{10}$$

Now try the following Practice Exercise

(b) Conversion of decimal to binary

An integer decimal number can be converted to a corresponding binary number by repeatedly dividing by 2 and noting the remainder at each stage, as shown below for 39_{10}.

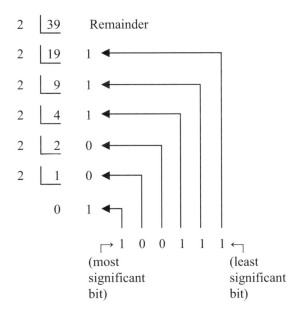

The result is obtained by writing the top digit of the remainder as the least significant bit (the least significant bit is the one on the right). The bottom bit of the remainder is the most significant bit, i.e. the bit on the left.

Thus, $39_{10} = 100111_2$

The fractional part of a decimal number can be converted to a binary number by repeatedly multiplying by 2, as shown below for the fraction 0.625

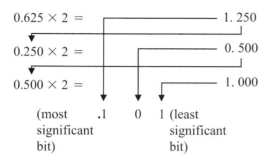

$$0.625 \times 2 = \quad 1.\,250$$
$$0.250 \times 2 = \quad 0.\,500$$
$$0.500 \times 2 = \quad 1.\,000$$

(most significant bit)   .1   0   1   (least significant bit)

For fractions, the most significant bit of the result is the top bit obtained from the integer part of multiplication by 2. The least significant bit of the result is the bottom bit obtained from the integer part of multiplication by 2.

Thus, $\mathbf{0.625_{10} = 0.101_2}$

Problem 5.   Convert 49_{10} to a binary number

From above, repeatedly dividing by 2 and noting the remainder gives:

```
2  | 49      Remainder
2  | 24    1
2  | 12    0
2  |  6    0
2  |  3    0
2  |  1    1
   |  0    1
```

→1   1   0   0   0   1←
(most significant bit)   (least significant bit)

Thus, $\mathbf{49_{10} = 110001_2}$

Problem 6.   Convert 56_{10} to a binary number

The integer part is repeatedly divided by 2, giving:

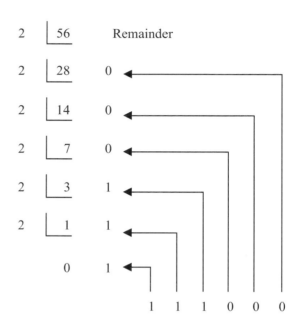

```
2  | 56      Remainder
2  | 28    0
2  | 14    0
2  |  7    0
2  |  3    1
2  |  1    1
   |  0    1
```

1   1   1   0   0   0

Thus, $\mathbf{56_{10} = 111000_2}$

Problem 7.   Convert 0.40625_{10} to a binary number

From above, repeatedly multiplying by 2 gives:

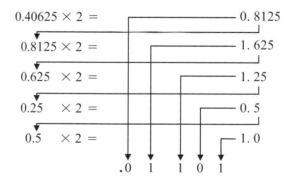

$$0.40625 \times 2 = \quad 0.\,8125$$
$$0.8125 \times 2 = \quad 1.\,625$$
$$0.625 \ \times 2 = \quad 1.\,25$$
$$0.25 \ \ \times 2 = \quad 0.\,5$$
$$0.5 \ \ \ \times 2 = \quad 1.\,0$$

.0   1   1   0   1

i.e. $\mathbf{0.40625_{10} = 0.01101_2}$

Problem 8.   Convert 58.3125_{10} to a binary number

Section B

The integer part is repeatedly divided by 2, giving:

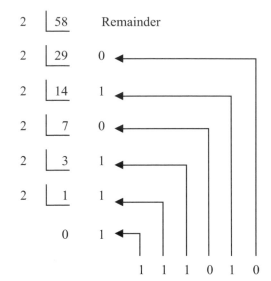

The fractional part is repeatedly multiplied by 2, giving:

$$0.3125 \times 2 = 0.625$$
$$0.625 \times 2 = 1.25$$
$$0.25 \times 2 = 0.5$$
$$0.5 \times 2 = 1.0$$

.0 1 0 1

Thus, **$58.3125_{10} = 111010.0101_2$**

Now try the following Practice Exercise

Practice Exercise 106   Conversion of decimal to binary numbers (answers on page 1159)

In Problems 1 to 5, convert the decimal numbers given to binary numbers.

1. (a) 5   (b) 15   (c) 19   (d) 29

2. (a) 31   (b) 42   (c) 57   (d) 63

3. (a) 47   (b) 60   (c) 73   (d) 84

4. (a) 0.25   (b) 0.21875
    (c) 0.28125   (d) 0.59375

5. (a) 47.40625   (b) 30.8125
    (c) 53.90625   (d) 61.65625

(c) Binary addition

Binary addition of two/three bits is achieved according to the following rules:

	sum	carry		sum	carry
$0+0=0$		0	$0+0+0=0$		0
$0+1=1$		0	$0+0+1=1$		0
$1+0=1$		0	$0+1+0=1$		0
$1+1=0$		1	$0+1+1=0$		1
			$1+0+0=1$		0
			$1+0+1=0$		1
			$1+1+0=0$		1
			$1+1+1=1$		1

These rules are demonstrated in the following worked problems.

Problem 9.   Perform the binary addition:
$1001 + 10110$

$$\begin{array}{r} 1001 \\ + \ 10110 \\ \hline \mathbf{11111} \end{array}$$

Problem 10.   Perform the binary addition:
$11111 + 10101$

$$\begin{array}{r} 11111 \\ + \ 10101 \\ \hline \text{sum } \mathbf{110100} \end{array}$$
carry 11111

Problem 11.   Perform the binary addition:
$1101001 + 1110101$

$$\begin{array}{r} 1101001 \\ + \ 1110101 \\ \hline \text{sum } \mathbf{11011110} \end{array}$$
carry 11   1

Problem 12.   Perform the binary addition:
$1011101 + 1100001 + 110101$

$$
\begin{array}{r}
1011101 \\
1100001 \\
+ \underline{110101} \\
\text{sum } \mathbf{11110011} \\
\text{carry } 111111\ 1
\end{array}
$$

Now try the following Practice Exercise

Practice Exercise 107   Binary addition (answers on page 1159)

Perform the following binary additions:

1. $10 + 11$

2. $101 + 110$

3. $1101 + 111$

4. $1111 + 11101$

5. $110111 + 10001$

6. $10000101 + 10000101$

7. $11101100 + 111001011$

8. $110011010 + 11100011$

9. $10110 + 1011 + 11011$

10. $111 + 10101 + 11011$

11. $1101 + 1001 + 11101$

12. $100011 + 11101 + 101110$

13. $101 + 111 + 1011 + 1101$

25.3   Octal numbers

For decimal integers containing several digits, repeatedly dividing by 2 can be a lengthy process. In this case, it is usually easier to convert a decimal number to a binary number via the octal system of numbers. This system has a radix of 8, using the digits 0, 1, 2, 3, 4, 5, 6 and 7. The decimal number equivalent to the octal number $\mathbf{4317_8}$ is:

$$4 \times 8^3 + 3 \times 8^2 + 1 \times 8^1 + 7 \times 8^0$$

i.e. $4 \times 512 + 3 \times 64 + 1 \times 8 + 7 \times 1 = \mathbf{2255_{10}}$

An integer decimal number can be converted to a corresponding octal number by repeatedly dividing by 8 and noting the remainder at each stage, as shown below for 493_{10}

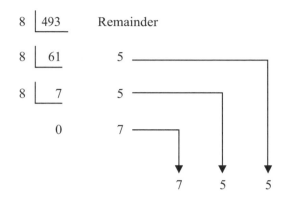

Thus, $\mathbf{493_{10} = 755_8}$

The fractional part of a decimal number can be converted to an octal number by repeatedly multiplying by 8, as shown below for the fraction 0.4375_{10}

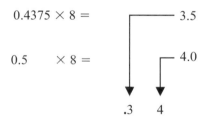

For fractions, the most significant bit is the top integer obtained by multiplication of the decimal fraction by 8, thus,

$$\mathbf{0.4375_{10} = 0.34_8}$$

The natural binary code for digits 0 to 7 is shown in Table 25.1, and an octal number can be converted to a binary number by writing down the three bits corresponding to the octal digit.

Thus, $\mathbf{437_8 = 100\ 011\ 111_2}$

and $\mathbf{26.35_8 = 010\ 110.011\ 101_2}$

The '0' on the extreme left does not signify anything, thus

$$\mathbf{26.35_8 = 10\ 110.011\ 101_2}$$

Conversion of decimal to binary via octal is demonstrated in the following worked problems.

Table 25.1

Octal digit	Natural binary number
0	000
1	001
2	010
3	011
4	100
5	101
6	110
7	111

Problem 13. Convert 3714_{10} to a binary number via octal

Dividing repeatedly by 8 and noting the remainder, gives:

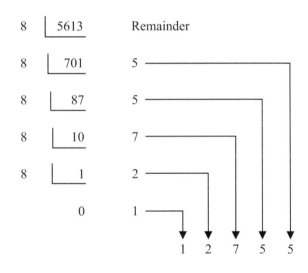

```
8 | 3714    Remainder
8 |  464    2 ─────────────────────┐
8 |   58    0 ──────────────┐       │
8 |    7    2 ───────┐       │       │
     0     7 ─┐       │       │       │
             ▼       ▼       ▼       ▼
             7       2       0       2
```

From Table 25.1, $7202_8 = 111\ 010\ 000\ 010_2$

i.e. $3714_{10} = \mathbf{111\ 010\ 000\ 010_2}$

Problem 14. Convert 0.59375_{10} to a binary number via octal

Multiplying repeatedly by 8 and noting the integer values, gives:

$0.59375 \times 8 = \quad 4.75$

$0.75 \quad \times 8 = \quad 6.00$

$\qquad\qquad\qquad .4 \quad 6$

Thus, $0.59375_{10} = 0.46_8$

From Table 25.1, $0.46_8 = 0.100\ 110_2$

i.e. $\mathbf{0.59375_{10} = 0.100\ 11_2}$

Problem 15. Convert 5613.90625_{10} to a binary number via octal

The integer part is repeatedly divided by 8, noting the remainder, giving:

```
8 | 5613    Remainder
8 |  701    5 ──────────────────────────┐
8 |   87    5 ───────────────────┐        │
8 |   10    7 ────────────┐        │        │
8 |    1    2 ─────┐        │        │        │
     0     1 ─┐     │        │        │        │
             ▼     ▼        ▼        ▼        ▼
             1     2        7        5        5
```

This octal number is converted to a binary number (see Table 25.1)

$$12755_8 = 001\ 010\ 111\ 101\ 101_2$$

i.e. $5613_{10} = 1\ 010\ 111\ 101\ 101_2$

The fractional part is repeatedly multiplied by 8 and noting the integer part, giving:

$$0.90625 \times 8 = \quad \quad 7.25$$

$$0.25 \quad \times 8 = \quad \quad 2.00$$

$$.7 \quad 2$$

This octal fraction is converted to a binary number (see Table 25.1)

$$0.72_8 = 0.111\,010_2$$

i.e. $0.90625_{10} = 0.111\,01_2$

Thus, $\mathbf{5613.90625_{10} = 1\,010\,111\,101\,101.111\,01_2}$

Problem 16.   Convert $11\,110\,011.100\,01_2$ to a decimal number via octal.

Grouping the binary number in threes from the binary point gives:

$$011\,110\,011.100\,010_2$$

Using Table 25.1 to convert this binary number to an octal number gives:

$$363.42_8$$

and $363.42_8 = 3 \times 8^2 + 6 \times 8^1 + 3 \times 8^0$

$$+ 4 \times 8^{-1} + 2 \times 8^{-2}$$

$$= 192 + 48 + 3 + 0.5 + 0.03125_{10}$$

$$= 243.53125_{10}$$

i.e. $\mathbf{11\,110\,011.100\,01_2 = 363.42_8 = 243.53125_{10}}$

Now try the following Practice Exercise

Practice Exercise 108   Conversion between decimal and binary numbers via octal (answers on page 1159)

In Problems 1 to 3, convert the decimal numbers given to binary numbers via octal.

1.   (a) 343   (b) 572   (c) 1265

2.   (a) 0.46875   (b) 0.6875   (c) 0.71875

3.   (a) 247.09375 (b) 514.4375 (c) 1716.78125

4.   Convert the following binary numbers to decimal numbers via octal:

   (a) 111.011 1   (b) 101 001.01
   (c) 1 110 011 011 010.001 1

25.4   Hexadecimal numbers

The hexadecimal system is particularly important in computer programming, since four bits (each consisting of a one or zero) can be succinctly expressed using a single hexadecimal digit. Two hexadecimal digits represent numbers from 0 to 255, a common range used, for example, to specify colours. Thus, in the HTML language of the web, colours are specified using three pairs of hexadecimal digits RRGGBB, where RR is the amount of red, GG the amount of green, and BB the amount of blue. A **hexadecimal numbering system** has a radix of 16 and uses the following 16 distinct digits:

$$0, 1, 2, 3, 4, 5, 6, 7, 8, 9, A, B, C, D, E \text{ and } F$$

'A' corresponds to 10 in the denary system, B to 11, C to 12, and so on.

(a) Converting from hexadecimal to decimal

For example, $1A_{16} = 1 \times 16^1 + A \times 16^0$

$$= 1 \times 16^1 + 10 \times 1$$

$$= 16 + 10 = 26$$

i.e. $\mathbf{1A_{16} = 26_{10}}$

Similarly, $\mathbf{2E_{16}} = 2 \times 16^1 + E \times 16^0$

$$= 2 \times 16^1 + 14 \times 16^0$$

$$= 32 + 14 = \mathbf{46_{10}}$$

and $\mathbf{1BF_{16}} = 1 \times 16^2 + B \times 16^1 + F \times 16^0$

$$= 1 \times 16^2 + 11 \times 16^1 + 15 \times 16^0$$

$$= 256 + 176 + 15 = \mathbf{447_{10}}$$

Table 25.2 compares decimal, binary, octal and hexadecimal numbers and shows, for example, that

$$23_{10} = 10111_2 = 27_8 = 17_{16}$$

Problem 17.   Convert the following hexadecimal numbers into their decimal equivalents:
(a) $7A_{16}$ (b) $3F_{16}$

(a) $7A_{16} = 7 \times 16^1 + A \times 16^0 = 7 \times 16 + 10 \times 1$
$$= 112 + 10 = 122$$
   Thus, $\mathbf{7A_{16} = 122_{10}}$

(b) $3F_{16} = 3 \times 16^1 + F \times 16^0 = 3 \times 16 + 15 \times 1$
$$= 48 + 15 = 63$$
   Thus, $\mathbf{3F_{16} = 63_{10}}$

Problem 18.   Convert the following hexadecimal numbers into their decimal equivalents:
(a) $C9_{16}$ (b) BD_{16}

Section B

Table 25.2

Decimal	Binary	Octal	Hexadecimal
0	0000	0	0
1	0001	1	1
2	0010	2	2
3	0011	3	3
4	0100	4	4
5	0101	5	5
6	0110	6	6
7	0111	7	7
8	1000	10	8
9	1001	11	9
10	1010	12	A
11	1011	13	B
12	1100	14	C
13	1101	15	D
14	1110	16	E
15	1111	17	F
16	10000	20	10
17	10001	21	11
18	10010	22	12
19	10011	23	13
20	10100	24	14
21	10101	25	15
22	10110	26	16
23	10111	27	17
24	11000	30	18
25	11001	31	19
26	11010	32	1A
27	11011	33	1B
28	11100	34	1C
29	11101	35	1D
30	11110	36	1E
31	11111	37	1F
32	100000	40	20

(a) $C9_{16} = C \times 16^1 + 9 \times 16^0 = 12 \times 16 + 9 \times 1$
$$= 192 + 9 = 201$$
Thus, $\qquad$ **$C9_{16} = 201_{10}$**

(b) $BD_{16} = B \times 16^1 + D \times 16^0 = 11 \times 16 + 13 \times 1$
$$= 176 + 13 = 189$$
Thus, $\qquad$ **$BD_{16} = 189_{10}$**

Problem 19. Convert $1A4E_{16}$ into a decimal number

$1A4E_{16} = 1 \times 16^3 + A \times 16^2 + 4 \times 16^1 + E \times 16^0$
$$= 1 \times 16^3 + 10 \times 16^2 + 4 \times 16^1 + 14 \times 16^0$$
$$= 1 \times 4096 + 10 \times 256 + 4 \times 16 + 14 \times 1$$
$$= 4096 + 2560 + 64 + 14 = 6734$$

Thus, **$1A4E_{16} = 6734_{10}$**

(b) Converting from decimal to hexadecimal

This is achieved by repeatedly dividing by 16 and noting the remainder at each stage, as shown below for 26_{10}

Hence, $\qquad$ **$26_{10} = 1A_{16}$**

Similarly, for 446_{10}

Thus, $\qquad$ **$446_{10} = 1BE_{16}$**

Problem 20. Convert the following decimal numbers into their hexadecimal equivalents:
(a) 39_{10} (b) 107_{10}

(a)   16 | 39   Remainder

     16 | 2 $7 \equiv 7_{16}$

          0     $2 \equiv 2_{16}$

                        2   7

              (most   (least
              significant   significant
              bit)   bit)

Hence, $39_{10} = 27_{16}$

(b)   16 | 107   Remainder

     16 | 6 $11 \equiv B_{16}$

          0     $6 \equiv 6_{16}$

                        6   B

Hence, $107_{10} = 6B_{16}$

Problem 21. Convert the following decimal numbers into their hexadecimal equivalents:
(a) 164_{10} (b) 239_{10}

(a)   16 | 164   Remainder

     16 | 10 $4 \equiv 4_{16}$

          0     $10 \equiv A_{16}$

                        A   4

Hence, $164_{10} = A4_{16}$

(b)   16 | 239   Remainder

     16 | 14 $15 \equiv F_{16}$

          0     $14 \equiv E_{16}$

                        E   F

Hence, $239_{10} = EF_{16}$

Now try the following Practice Exercise

Practice Exercise 109   Hexadecimal numbers (answers on page 1159)

In Problems 1 to 4, convert the given hexadecimal numbers into their decimal equivalents.

1. $E7_{16}$ 2. $2C_{16}$ 3. 98_{16} 4. $2F1_{16}$

In Problems 5 to 8, convert the given decimal numbers into their hexadecimal equivalents.

5. 54_{10} 6. 200_{10} 7. 91_{10} 8. 238_{10}

(c) Converting from binary to hexadecimal

The binary bits are arranged in groups of four, starting from right to left, and a hexadecimal symbol is assigned to each group. For example, the binary number 1110011110101001 is initially grouped in fours as:
1110   0111   1010   1001

and a hexadecimal symbol assigned to each group as:
E   7   A   9

from Table 25.2

Hence, $1110011110101001_2 = E7A9_{16}$

Problem 22. Convert the following binary numbers into their hexadecimal equivalents:
(a) 11010110_2 (b) 1100111_2

(a)   Grouping bits in fours from the right gives:
     1101   0110

     and assigning hexadecimal symbols to each group gives:

     D   6

     from Table 25.2

     Thus, $11010110_2 = D6_{16}$

Section B

(b) Grouping bits in fours from the right gives:

0110   0111

and assigning hexadecimal symbols to each group gives:

6   7

from Table 25.2

Thus, $1100111_2 = 67_{16}$

Problem 23. Convert the following binary numbers into their hexadecimal equivalents:
(a) 11001111_2 (b) 110011110_2

(a) Grouping bits in fours from the right gives:

1100   1111

and assigning hexadecimal symbols to each group gives:

C   F

from Table 25.2

Thus, $11001111_2 = CF_{16}$

(b) Grouping bits in fours from the right gives:

0001   1001   1110

and assigning hexadecimal symbols to each group gives:

1   9   E

from Table 25.2

Thus, $110011110_2 = 19E_{16}$

(d) Converting from hexadecimal to binary

The above procedure is reversed, thus, for example,

$6CF3_{16} = 0110\ 1100\ 1111\ 0011$ from Table 25.2

i.e. $6CF3_{16} = 110110011110011_2$

Problem 24. Convert the following hexadecimal numbers into their binary equivalents:
(a) $3F_{16}$ (b) $A6_{16}$

(a) Spacing out hexadecimal digits gives:

3   F

and converting each into binary gives:

0011   1111

from Table 25.2

Thus, $3F_{16} = 111111_2$

(b) Spacing out hexadecimal digits gives:

A   6

and converting each into binary gives:

1010   0110

from Table 25.2

Thus, $A6_{16} = 10100110_2$

Problem 25. Convert the following hexadecimal numbers into their binary equivalents:
(a) $7B_{16}$ (b) $17D_{16}$

(a) Spacing out hexadecimal digits gives:

7   B

and converting each into binary gives:

0111   1011

from Table 25.2

Thus, $7B_{16} = 1111011_2$

(b) Spacing out hexadecimal digits gives:

1   7   D

and converting each into binary gives:

0001   0111   1101

from Table 25.2

Thus, $17D_{16} = 101111101_2$

Now try the following Practice Exercise

Practice Exercise 110   Hexadecimal numbers (answers on page 1159)

In Problems 1 to 4, convert the given binary numbers into their hexadecimal equivalents.

1. 11010111_2 2. 11101010_2

3. 10001011_2 4. 10100101_2

In Problems 5 to 8, convert the given hexadecimal numbers into their binary equivalents.

5. 37_{16} 6. ED_{16}

7. $9F_{16}$ 8. $A21_{16}$

For fully worked solutions to each of the problems in Practice Exercises 105 to 110 in this chapter, go to the website:
www.routledge.com/cw/bird

Chapter 26

Boolean algebra and logic circuits

Why it is important to understand: Boolean algebra and logic circuits

Logic circuits are the basis for modern digital computer systems; to appreciate how computer systems operate an understanding of digital logic and Boolean algebra is needed. Boolean algebra (named after its developer, George Boole), is the algebra of digital logic circuits all computers use. Boolean algebra is the algebra of binary systems. A logic gate is a physical device implementing a Boolean function, performing a logical operation on one or more logic inputs, and produces a single logic output. Logic gates are implemented using diodes or transistors acting as electronic switches, but can also be constructed using electromagnetic relays, fluidic relays, pneumatic relays, optics, molecules or even mechanical elements. Learning Boolean algebra for logic analysis, learning about gates that process logic signals and learning how to design some smaller logic circuits is clearly of importance to computer engineers.

At the end of this chapter, you should be able to:

- draw a switching circuit and truth table for a two-input and three-input or-function and state its Boolean expression
- draw a switching circuit and truth table for a two-input and three-input and-function and state its Boolean expression
- produce the truth table for a two-input not-function and state its Boolean expression
- simplify Boolean expressions using the laws and rules of Boolean algebra
- simplify Boolean expressions using de Morgan's laws
- simplify Boolean expressions using Karnaugh maps
- draw a circuit diagram symbol and truth table for a three-input and-gate and state its Boolean expression
- draw a circuit diagram symbol and truth table for a three-input or-gate and state its Boolean expression
- draw a circuit diagram symbol and truth table for a three-input invert (or not)-gate and state its Boolean expression
- draw a circuit diagram symbol and truth table for a three-input nand-gate and state its Boolean expression
- draw a circuit diagram symbol and truth table for a three-input nor-gate and state its Boolean expression
- devise logic systems for particular Boolean expressions
- use universal gates to devise logic circuits for particular Boolean expressions

26.1 Boolean algebra and switching circuits

A **two-state device** is one whose basic elements can only have one of two conditions. Thus, two-way switches, which can either be on or off, and the binary numbering system, having the digits 0 and 1 only, are two-state devices. In **Boolean algebra** (named after **George Boole***), if A represents one state, then $\overline{A}$, called 'not-A', represents the second state.

The or-function

In Boolean algebra, the **or**-function for two elements A and B is written as $A + B$, and is defined as 'A, or B, or both A and B'. The equivalent electrical circuit for a two-input **or**-function is given by two switches connected in parallel. With reference to Figure 26.1(a), the lamp will be on when A is on, when B is on, or when both A and B are on. In the table shown in Figure 26.1(b), all

the possible switch combinations are shown in columns 1 and 2, in which a 0 represents a switch being off and a 1 represents the switch being on, these columns being called the inputs. Column 3 is called the output and a 0 represents the lamp being off and a 1 represents the lamp being on. Such a table is called a **truth table**.

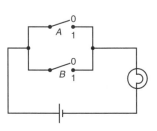

1	2	3
\multicolumn{2}{c}{Input (switches)}	Output (lamp)
A	B	$Z = A + B$
0	0	0
0	1	1
1	0	1
1	1	1

(a) Switching circuit for or-function (b) Truth table for or-function

Figure 26.1

The and-function

In Boolean algebra, the **and**-function for two elements A and B is written as $A \cdot B$ and is defined as 'both A and B'. The equivalent electrical circuit for a two-input **and**-function is given by two switches connected in series. With reference to Figure 26.2(a) the lamp will be on only when both A and B are on. The truth table for a two-input **and**-function is shown in Figure 26.2(b).

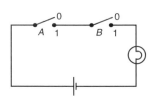

\multicolumn{2}{c}{Input (switches)}	Output (lamp)
A	B	$Z = A \cdot B$
0	0	0
0	1	0
1	0	0
1	1	1

(a) Switching circuit for and-function (b) Truth table for and-function

Figure 26.2

The not-function

In Boolean algebra, the **not**-function for element A is written as $\overline{A}$, and is defined as 'the opposite to A'. Thus if A means switch A is on, $\overline{A}$ means that switch A is off. The truth table for the **not**-function is shown in Table 26.1.

In the above, the Boolean expressions, equivalent switching circuits and truth tables for the three functions used in Boolean algebra are given for a two-input

Table 26.1

Input A	Output $Z = \overline{A}$
0	1
1	0

system. A system may have more than two inputs and the Boolean expression for a three-input **or**-function having elements A, B and C is $A + B + C$. Similarly, a three-input **and**-function is written as $A \cdot B \cdot C$. The equivalent electrical circuits and truth tables for three-input **or**- and **and**-functions are shown in Figures 26.3(a) and (b) respectively.

To achieve a given output, it is often necessary to use combinations of switches connected both in series and in parallel. If the output from a switching circuit is given by the Boolean expression $Z = A \cdot B + \overline{A} \cdot \overline{B}$, the truth table is as shown in Figure 26.4(a). In this table, columns 1 and 2 give all the possible combinations of A and B. Column 3 corresponds to $A \cdot B$ and column 4 to $\overline{A} \cdot \overline{B}$, i.e. a 1 output is obtained when $A=0$ and when $B=0$. Column 5 is the **or**-function applied to columns 3 and 4 giving an output of $Z = A \cdot B + \overline{A} \cdot \overline{B}$. The corresponding switching circuit is shown in Figure 26.4(b) in which A and B are connected in series to give $A \cdot B$, $\overline{A}$ and $\overline{B}$ are connected in series to give $\overline{A} \cdot \overline{B}$, and $A \cdot B$ and $\overline{A} \cdot \overline{B}$ are connected in parallel to give $A \cdot B + \overline{A} \cdot \overline{B}$. The circuit symbols used are such that A means the switch

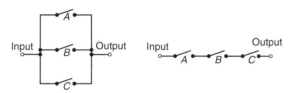

Input $A\ B\ C$	Output $Z = A + B + C$
0 0 0	0
0 0 1	1
0 1 0	1
0 1 1	1
1 0 0	1
1 0 1	1
1 1 0	1
1 1 1	1

(a) The or-function electrical circuit and truth table

Input $A\ B\ C$	Output $Z = A \cdot B \cdot C$
0 0 0	0
0 0 1	0
0 1 0	0
0 1 1	0
1 0 0	0
1 0 1	0
1 1 0	0
1 1 1	1

(b) The and-function electrical circuit and truth table

Figure 26.3

1 A	2 B	3 $A \cdot B$	4 $\overline{A} \cdot \overline{B}$	5 $Z = A \cdot B + \overline{A} \cdot \overline{B}$
0	0	0	1	1
0	1	0	0	0
1	0	0	0	0
1	1	1	0	1

(a) Truth table for $Z = A \cdot B + \overline{A} \cdot \overline{B}$

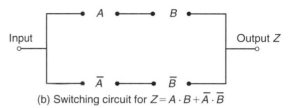

(b) Switching circuit for $Z = A \cdot B + \overline{A} \cdot \overline{B}$

Figure 26.4

is on when A is 1, $\overline{A}$ means the switch is on when A is 0, and so on.

Problem 1. Derive the Boolean expression and construct a truth table for the switching circuit shown in Figure 26.5

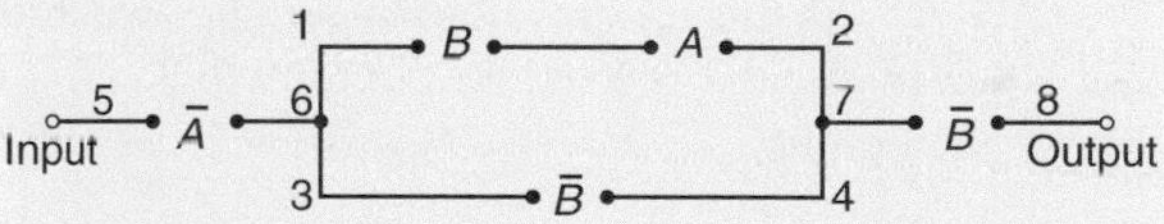

Figure 26.5

The switches between 1 and 2 in Figure 26.5 are in series and have a Boolean expression of $B \cdot A$. The parallel circuit 1 to 2 and 3 to 4 have a Boolean expression of $(B \cdot A + \overline{B})$. The parallel circuit can be treated as a single switching unit, giving the equivalent of switches 5 to 6, 6 to 7 and 7 to 8 in series. Thus the output is given by:

$$Z = \overline{A} \cdot (B \cdot A + \overline{B}) \cdot \overline{B}$$

The truth table is as shown in Table 26.2. Columns 1 and 2 give all the possible combinations of switches A and B. Column 3 is the **and**-function applied to columns 1 and 2, giving $B \cdot A$. Column 4 is $\overline{B}$, i.e. the opposite to column 2. Column 5 is the **or**-function applied to columns 3 and 4. Column 6 is $\overline{A}$, i.e. the opposite to column 1. The output is column 7 and is obtained by applying the **and**-function to columns 4, 5 and 6.

Table 26.2

1	2	3	4	5	6	7
A	B	$B \cdot A$	$\overline{B}$	$B \cdot A + \overline{B}$	$\overline{A}$	$Z = \overline{A} \cdot (B \cdot A + \overline{B}) \cdot \overline{B}$
0	0	0	1	1	1	1
0	1	0	0	0	1	0
1	0	0	1	1	0	0
1	1	1	0	1	0	0

Problem 2. Derive the Boolean expression and construct a truth table for the switching circuit shown in Figure 26.6

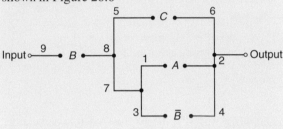

Figure 26.6

The parallel circuit 1 to 2 and 3 to 4 gives $(A + \overline{B})$ and this is equivalent to a single switching unit between 7 and 2. The parallel circuit 5 to 6 and 7 to 2 gives $C + (A + \overline{B})$ and this is equivalent to a single switching unit between 8 and 2. The series circuit 9 to 8 and 8 to 2 gives the output

$$Z = B \cdot [C + (A + \overline{B})]$$

The truth table is shown in Table 26.3. Columns 1, 2 and 3 give all the possible combinations of A, B and C. Column 4 is $\overline{B}$ and is the opposite to column 2. Column 5 is the **or**-function applied to columns 1 and 4, giving $(A + \overline{B})$. Column 6 is the **or**-function applied to columns 3 and 5 giving $C + (A + \overline{B})$. The output is given in column 7 and is obtained by applying the **and**-function to columns 2 and 6, giving $Z = B \cdot [C + (A + \overline{B})]$.

Problem 3. Construct a switching circuit to meet the requirements of the Boolean expression:

$$Z = A \cdot \overline{C} + \overline{A} \cdot B + \overline{A} \cdot B \cdot \overline{C}$$

Construct the truth table for this circuit

The three terms joined by **or**-functions, $(+)$, indicate three parallel branches.

Table 26.3

1	2	3	4	5	6	7
A	B	C	$\overline{B}$	$A + \overline{B}$	$C + (A + \overline{B})$	$Z = B \cdot [C + (A + \overline{B})]$
0	0	0	1	1	1	0
0	0	1	1	1	1	0
0	1	0	0	0	0	0
0	1	1	0	0	1	1
1	0	0	1	1	1	0
1	0	1	1	1	1	0
1	1	0	0	1	1	1
1	1	1	0	1	1	1

having: branch 1 A **and** $\overline{C}$ in series

           branch 2 $\overline{A}$ **and** B in series

and   branch 3 $\overline{A}$ **and** B **and** $\overline{C}$ in series

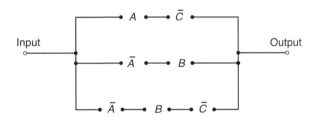

Figure 26.7

Hence the required switching circuit is as shown in Figure 26.7. The corresponding truth table is shown in Table 26.4.

Table 26.4

1	2	3	4	5	6	7	8	9
A	B	C	$\overline{C}$	$A \cdot \overline{C}$	$\overline{A}$	$\overline{A} \cdot B$	$\overline{A} \cdot B \cdot \overline{C}$	$Z = A \cdot \overline{C} + \overline{A} \cdot B + \overline{A} \cdot B \cdot \overline{C}$
0	0	0	1	0	1	0	0	0
0	0	1	0	0	1	0	0	0
0	1	0	1	0	1	1	1	1
0	1	1	0	0	1	1	0	1
1	0	0	1	1	0	0	0	1
1	0	1	0	0	0	0	0	0
1	1	0	1	1	0	0	0	1
1	1	1	0	0	0	0	0	0

Column 4 is $\overline{C}$, i.e. the opposite to column 3.

Column 5 is $A \cdot \overline{C}$, obtained by applying the **and**-function to columns 1 and 4.

Column 6 is $\overline{A}$, the opposite to column 1.

Column 7 is $\overline{A} \cdot B$, obtained by applying the **and**-function to columns 2 and 6.

Column 8 is $\overline{A} \cdot B \cdot \overline{C}$, obtained by applying the **and**-function to columns 4 and 7.

Column 9 is the output, obtained by applying the **or**-function to columns 5, 7 and 8.

Problem 4. Derive the Boolean expression and construct the switching circuit for the truth table given in Table 26.5

Table 26.5

	A	B	C	Z
1	0	0	0	1
2	0	0	1	0
3	0	1	0	1
4	0	1	1	1
5	1	0	0	0
6	1	0	1	1
7	1	1	0	0
8	1	1	1	0

Examination of the truth table shown in Table 26.5 shows that there is a 1 output in the Z-column in rows 1, 3, 4 and 6. Thus, the Boolean expression and switching circuit should be such that a 1 output is obtained for row 1 **or** row 3 **or** row 4 **or** row 6. In row 1, A is 0 **and** B is 0 **and** C is 0 and this corresponds to the Boolean expression $\overline{A} \cdot \overline{B} \cdot \overline{C}$. In row 3, A is 0 **and** B is 1 **and** C is 0, i.e. the Boolean expression is $\overline{A} \cdot B \cdot \overline{C}$. Similarly in rows 4 and 6, the Boolean expressions are $\overline{A} \cdot B \cdot C$ and $A \cdot \overline{B} \cdot C$ respectively. Hence the Boolean expression is:

$$Z = \overline{A} \cdot \overline{B} \cdot \overline{C} + \overline{A} \cdot B \cdot \overline{C} + \overline{A} \cdot B \cdot C + A \cdot \overline{B} \cdot C$$

The corresponding switching circuit is shown in Figure 26.8. The four terms are joined by **or**-functions, (+), and are represented by four parallel circuits. Each

term has three elements joined by an **and**-function, and is represented by three elements connected in series.

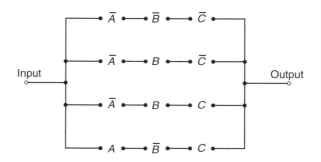

Figure 26.8

Now try the following Practice Exercise

Practice Exercise 111   Further problems on Boolean algebra and switching circuits (answers on page 1159)

In Problems 1 to 4, determine the Boolean expressions and construct truth tables for the switching circuits given.

1.   The circuit shown in Figure 26.9

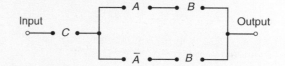

Figure 26.9

2.   The circuit shown in Figure 26.10

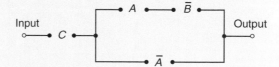

Figure 26.10

3.   The circuit shown in Figure 26.11

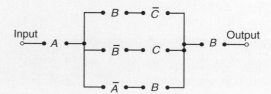

Figure 26.11

4. The circuit shown in Figure 26.12

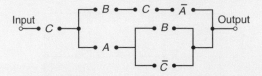

Figure 26.12

In Problems 5 to 7, construct switching circuits to meet the requirements of the Boolean expressions given.

5. $A \cdot C + A \cdot \overline{B} \cdot C + A \cdot B$

6. $A \cdot B \cdot C \cdot (A + B + C)$

7. $A \cdot (A \cdot \overline{B} \cdot C + B \cdot (A + \overline{C}))$

In Problems 8 to 10, derive the Boolean expressions and construct the switching circuits for the truth table stated.

8. Table 26.6, column 4

Table 26.6

1 A	2 B	3 C	4	5	6
0	0	0	0	1	1
0	0	1	1	0	0
0	1	0	0	0	1
0	1	1	0	1	0
1	0	0	0	1	1
1	0	1	0	0	1
1	1	0	1	0	0
1	1	1	0	0	0

9. Table 26.6, column 5

10. Table 26.6, column 6

26.2 Simplifying Boolean expressions

A Boolean expression may be used to describe a complex switching circuit or logic system. If the Boolean expression can be simplified, then the number of switches or logic elements can be reduced, resulting in

a saving in cost. Three principal ways of simplifying Boolean expressions are:

(a) by using the laws and rules of Boolean algebra (see Section 26.3),

(b) by applying de Morgan's laws (see Section 26.4), and

(c) by using Karnaugh maps (see Section 26.5).

26.3 Laws and rules of Boolean algebra

A summary of the principal laws and rules of Boolean algebra are given in Table 26.7. The way in which these laws and rules may be used to simplify Boolean expressions is shown in Problems 5 to 10.

Table 26.7

Ref.	Name	Rule or law
1	Commutative laws	$A + B = B + A$
2		$A \cdot B = B \cdot A$
3	Associative laws	$(A + B) + C = A + (B + C)$
4		$(A \cdot B) \cdot C = A \cdot (B \cdot C)$
5	Distributive laws	$A \cdot (B + C) = A \cdot B + A \cdot C$
6		$A + (B \cdot C)$ $= (A + B) \cdot (A + C)$
7	Sum rules	$A + 0 = A$
8		$A + 1 = 1$
9		$A + A = A$
10		$A + \overline{A} = 1$
11	Product rules	$A \cdot 0 = 0$
12		$A \cdot 1 = A$
13		$A \cdot A = A$
14		$A \cdot \overline{A} = 0$
15	Absorption rules	$A + A \cdot B = A$
16		$A \cdot (A + B) = A$
17		$A + \overline{A} \cdot B = A + B$

Problem 5. Simplify the Boolean expression:
$$\overline{P} \cdot \overline{Q} + \overline{P} \cdot Q + P \cdot \overline{Q}$$

With reference to Table 26.7: *Reference*

$\overline{P} \cdot \overline{Q} + \overline{P} \cdot Q + P \cdot \overline{Q}$

$= \overline{P} \cdot (\overline{Q} + Q) + P \cdot \overline{Q}$ 5

$= \overline{P} \cdot 1 + P \cdot \overline{Q}$ 10

$= \boldsymbol{\overline{P} + P \cdot \overline{Q}}$ 12

Problem 6. Simplify:
$$(P + \overline{P} \cdot Q) \cdot (Q + \overline{Q} \cdot P)$$

With reference to Table 26.7: *Reference*

$(P + \overline{P} \cdot Q) \cdot (Q + \overline{Q} \cdot P)$

$= P \cdot (Q + \overline{Q} \cdot P)$

$\qquad + \overline{P} \cdot Q \cdot (Q + \overline{Q} \cdot P)$ 5

$= P \cdot Q + P \cdot \overline{Q} \cdot P + \overline{P} \cdot Q \cdot Q$

$\qquad + \overline{P} \cdot Q \cdot \overline{Q} \cdot P$ 5

$= P \cdot Q + P \cdot \overline{Q} + \overline{P} \cdot Q$

$\qquad + \overline{P} \cdot Q \cdot \overline{Q} \cdot P$ 13

$= P \cdot Q + P \cdot \overline{Q} + \overline{P} \cdot Q + 0$ 14

$= P \cdot Q + P \cdot \overline{Q} + \overline{P} \cdot Q$ 7

$= P \cdot (Q + \overline{Q}) + \overline{P} \cdot Q$ 5

$= P \cdot 1 + \overline{P} \cdot Q$ 10

$= \boldsymbol{P + \overline{P} \cdot Q}$ 12

Problem 7. Simplify:
$$F \cdot G \cdot \overline{H} + F \cdot G \cdot H + \overline{F} \cdot G \cdot H$$

With reference to Table 26.7: *Reference*

$F \cdot G \cdot \overline{H} + F \cdot G \cdot H + \overline{F} \cdot G \cdot H$

$= F \cdot G \cdot (\overline{H} + H) + \overline{F} \cdot G \cdot H$ 5

$= F \cdot G \cdot 1 + \overline{F} \cdot G \cdot H$ 10

$= F \cdot G + \overline{F} \cdot G \cdot H$ 12

$= \boldsymbol{G \cdot (F + \overline{F} \cdot H)}$ 5

Problem 8. Simplify:
$$\overline{F} \cdot \overline{G} \cdot H + \overline{F} \cdot G \cdot H + F \cdot \overline{G} \cdot H + F \cdot G \cdot H$$

With reference to Table 26.7: *Reference*

$\overline{F} \cdot \overline{G} \cdot H + \overline{F} \cdot G \cdot H + F \cdot \overline{G} \cdot H + F \cdot G \cdot H$

$= \overline{G} \cdot H \cdot (\overline{F} + F) + G \cdot H \cdot (\overline{F} + F)$ 5

$= \overline{G} \cdot H \cdot 1 + G \cdot H \cdot 1$ 10

$= \overline{G} \cdot H + G \cdot H$ 12

$= H \cdot (\overline{G} + G)$ 5

$= H \cdot 1 = \boldsymbol{H}$ 10 and 12

Problem 9. Simplify:
$$A \cdot \overline{C} + \overline{A} \cdot (B + C) + A \cdot B \cdot (C + \overline{B})$$

using the rules of Boolean algebra

With reference to Table 26.7: *Reference*

$A \cdot \overline{C} + \overline{A} \cdot (B + C) + A \cdot B \cdot (C + \overline{B})$

$= A \cdot \overline{C} + \overline{A} \cdot B + \overline{A} \cdot C + A \cdot B \cdot C$

$\qquad\qquad + A \cdot B \cdot \overline{B}$ 5

$= A \cdot \overline{C} + \overline{A} \cdot B + \overline{A} \cdot C + A \cdot B \cdot C$

$\qquad\qquad + A \cdot 0$ 14

$= A \cdot \overline{C} + \overline{A} \cdot B + \overline{A} \cdot C + A \cdot B \cdot C$ 11 and 7

$= A \cdot (\overline{C} + B \cdot C) + \overline{A} \cdot B + \overline{A} \cdot C$ 5

$= A \cdot (\overline{C} + B) + \overline{A} \cdot B + \overline{A} \cdot C$ 17

$= A \cdot \overline{C} + A \cdot B + \overline{A} \cdot B + \overline{A} \cdot C$ 5

$= A \cdot \overline{C} + B \cdot (A + \overline{A}) + \overline{A} \cdot C$ 5

$= A \cdot \overline{C} + B \cdot 1 + \overline{A} \cdot C$ 10

$= \boldsymbol{A \cdot \overline{C} + B + \overline{A} \cdot C}$ 12

Problem 10. Simplify the expression
$P \cdot \overline{Q} \cdot R + P \cdot Q \cdot (\overline{P} + R) + Q \cdot R \cdot (\overline{Q} + P)$
using the rules of Boolean algebra

With reference to Table 26.7: *Reference*

$P \cdot \overline{Q} \cdot R + P \cdot Q \cdot (\overline{P} + R) + Q \cdot R \cdot (\overline{Q} + P)$

$= P \cdot \overline{Q} \cdot R + P \cdot Q \cdot \overline{P} + P \cdot Q \cdot R$

$\qquad\qquad + Q \cdot R \cdot \overline{Q} + Q \cdot R \cdot P$ 5

$= P \cdot \overline{Q} \cdot R + 0 \cdot Q + P \cdot Q \cdot R + 0 \cdot R$

$\qquad\qquad\qquad + P \cdot Q \cdot R$ 14

$= P \cdot \overline{Q} \cdot R + P \cdot Q \cdot R + P \cdot Q \cdot R$ 7 and 11

$= P \cdot \overline{Q} \cdot R + P \cdot Q \cdot R$ 9

$= P \cdot R \cdot (Q + \overline{Q})$ 5

$= P \cdot R \cdot 1$ 10

$= \boldsymbol{P \cdot R}$ 12

Now try the following Practice Exercise

Practice Exercise 112   Further problems on the laws the rules of Boolean algebra (answers on page 1160)

Use the laws and rules of Boolean algebra given in Table 26.7 to simplify the following expressions:

1. $\overline{P} \cdot \overline{Q} + \overline{P} \cdot Q$

2. $\overline{P} \cdot Q + P \cdot Q + \overline{P} \cdot \overline{Q}$

3. $\overline{F} \cdot \overline{G} + F \cdot \overline{G} + \overline{G} \cdot (F + \overline{F})$

4. $F \cdot \overline{G} + F \cdot (G + \overline{G}) + F \cdot G$

5. $(P + P \cdot Q) \cdot (Q + Q \cdot P)$

6. $\overline{F} \cdot \overline{G} \cdot H + \overline{F} \cdot G \cdot H + F \cdot \overline{G} \cdot H$

7. $F \cdot \overline{G} \cdot \overline{H} + F \cdot G \cdot H + \overline{F} \cdot G \cdot H$

8. $\overline{P} \cdot \overline{Q} \cdot \overline{R} + \overline{P} \cdot Q \cdot R + P \cdot \overline{Q} \cdot \overline{R}$

9. $\overline{F} \cdot \overline{G} \cdot \overline{H} + \overline{F} \cdot \overline{G} \cdot H + F \cdot \overline{G} \cdot \overline{H}$
   $+ F \cdot \overline{G} \cdot H$

10. $F \cdot \overline{G} \cdot H + F \cdot G \cdot H + F \cdot G \cdot \overline{H}$
    $+ \overline{F} \cdot G \cdot \overline{H}$

11. $R \cdot (\overline{P} \cdot Q + P \cdot \overline{Q}) + \overline{R} \cdot (\overline{P} \cdot \overline{Q} + \overline{P} \cdot Q)$

12. $\overline{R} \cdot (\overline{P} \cdot \overline{Q} + P \cdot Q + P \cdot \overline{Q})$
    $+ P \cdot (Q \cdot R + \overline{Q} \cdot R)$

26.4   De Morgan's laws

De Morgan's[*] **laws** may be used to simplify **not-**functions having two or more elements. The laws state that:

$$\overline{A + B} = \overline{A} \cdot \overline{B} \quad \text{and} \quad \overline{A \cdot B} = \overline{A} + \overline{B}$$

and may be verified by using a truth table (see Problem 11). The application of de Morgan's laws in simplifying Boolean expressions is shown in Problems 12 and 13.

Problem 11. Verify that $\overline{A + B} = \overline{A} \cdot \overline{B}$

* **Who was** de Morgan? **Augustus de Morgan** (27 June 1806–18 March 1871) was a British mathematician and logician. He formulated **de Morgan's laws** and introduced the term mathematical induction. To find out more go to **www.routledge.com/cw/bird**

A Boolean expression may be verified by using a truth table. In Table 26.8, columns 1 and 2 give all the possible arrangements of the inputs A and B. Column 3 is the **or**-function applied to columns 1 and 2 and column 4 is the **not**-function applied to column 3. Columns 5 and 6 are the **not**-function applied to columns 1 and 2 respectively and column 7 is the **and**-function applied to columns 5 and 6.

Table 26.8

1	2	3	4	5	6	7
A	B	$A+B$	$\overline{A+B}$	$\overline{A}$	$\overline{B}$	$\overline{A}\cdot\overline{B}$
0	0	0	1	1	1	1
0	1	1	0	1	0	0
1	0	1	0	0	1	0
1	1	1	0	0	0	0

Since columns 4 and 7 have the same pattern of 0s and 1s this verifies that $\overline{A+B}=\overline{A}\cdot\overline{B}$

Problem 12. Simplify the Boolean expression $(\overline{A\cdot B})+(\overline{\overline{A}+B})$ by using de Morgan's laws and the rules of Boolean algebra

Applying de Morgan's law to the first term gives:

$$\overline{A\cdot B}=\overline{A}+\overline{B}=A+\overline{B}\quad\text{since }\overline{\overline{A}}=A$$

Applying de Morgan's law to the second term gives:

$$\overline{\overline{A}+B}=\overline{\overline{A}}\cdot\overline{B}=A\cdot\overline{B}$$

Thus, $(\overline{A\cdot B})+(\overline{\overline{A}+B})=(A+\overline{B})+A\cdot\overline{B}$

Removing the bracket and reordering gives:
$A+A\cdot\overline{B}+\overline{B}$
But, by rule 15, Table 26.7, $A+A\cdot B=A$. It follows that: $A+A\cdot\overline{B}=A$

Thus: $(\overline{A\cdot B})+(\overline{\overline{A}+B})=A+\overline{B}$

Problem 13. Simplify the Boolean expression $(A\cdot\overline{B}+C)\cdot(\overline{A}+B\cdot\overline{C})$ by using de Morgan's laws and the rules of Boolean algebra

Applying de Morgan's laws to the first term gives:

$$\overline{A\cdot\overline{B}+C}=\overline{A\cdot\overline{B}}\cdot\overline{C}=(\overline{A}+\overline{\overline{B}})\cdot\overline{C}$$
$$=(\overline{A}+B)\cdot\overline{C}=\overline{A}\cdot\overline{C}+B\cdot\overline{C}$$

Applying de Morgan's law to the second term gives:

$$\overline{A+\overline{B\cdot\overline{C}}}=\overline{A}+(\overline{B}+\overline{\overline{C}})=\overline{A}+(\overline{B}+C)$$

Thus $(\overline{A\cdot\overline{B}+C})\cdot(\overline{A}+\overline{B\cdot\overline{C}})$

$$=(\overline{A}\cdot\overline{C}+B\cdot\overline{C})\cdot(\overline{A}+\overline{B}+C)$$
$$=\overline{A}\cdot\overline{A}\cdot\overline{C}+\overline{A}\cdot\overline{B}\cdot\overline{C}+\overline{A}\cdot\overline{C}\cdot C$$
$$\quad+\overline{A}\cdot B\cdot\overline{C}+B\cdot\overline{B}\cdot\overline{C}+B\cdot\overline{C}\cdot C$$

But from Table 26.7, $\overline{A}\cdot\overline{A}=\overline{A}$ and $\overline{C}\cdot C=B\cdot\overline{B}=0$ Hence the Boolean expression becomes:

$$\overline{A}\cdot\overline{C}+\overline{A}\cdot\overline{B}\cdot\overline{C}+\overline{A}\cdot B\cdot\overline{C}$$
$$=\overline{A}\cdot\overline{C}\cdot(1+\overline{B}+B)$$
$$=\overline{A}\cdot\overline{C}\cdot(1+B)$$
$$=\overline{A}\cdot\overline{C}$$

Thus: $(\overline{A\cdot\overline{B}+C})\cdot(\overline{A}+\overline{B\cdot\overline{C}})=\overline{A}\cdot\overline{C}$

Now try the following Practice Exercise

Practice Exercise 113   Further problems on simplifying Boolean expressions using de Morgan's laws (answers on page 1161)

Use de Morgan's laws and the rules of Boolean algebra given in Table 26.7 to simplify the following expressions.

1. $(\overline{A\cdot B})\cdot(\overline{\overline{A}\cdot B})$

2. $\overline{(A+B\cdot\overline{C})}+(\overline{A\cdot B}+C)$

3. $(\overline{A\cdot B+B\cdot\overline{C}})\cdot\overline{A\cdot\overline{B}}$

4. $(\overline{A\cdot\overline{B}+B\cdot\overline{C}})+(\overline{A\cdot B})$

5. $\overline{(P\cdot\overline{Q}+\overline{P}\cdot R)}\cdot(\overline{P\cdot\overline{Q}\cdot R})$

Section B

26.5 Karnaugh maps

Two-variable Karnaugh maps

A truth table for a two-variable expression is shown in Table 26.9(a), the '1' in the third row output showing that $Z = A \cdot \overline{B}$. Each of the four possible Boolean expressions associated with a two-variable function can be depicted as shown in Table 26.9(b) in which one cell is allocated to each row of the truth table. A matrix similar to that shown in Table 26.9(b) can be used to depict $Z = A \cdot \overline{B}$, by putting a 1 in the cell corresponding to $A \cdot \overline{B}$ and 0s in the remaining cells. This method of depicting a Boolean expression is called a two-variable **Karnaugh* map**, and is shown in Table 26.9(c).

To simplify a two-variable Boolean expression, the Boolean expression is depicted on a Karnaugh map, as outlined above. Any cells on the map having either a common vertical side or a common horizontal side are grouped together to form a **couple**. (This is a coupling

* **Who is Karnaugh?** **Maurice Karnaugh** (4 October 1924 in New York City) is an American physicist, famous for the Karnaugh map used in Boolean algebra. To find out more go to **www.routledge.com/cw/bird**

Table 26.9

| Inputs | | Output | Boolean |
A	B	Z	expression
0	0	0	$\overline{A} \cdot \overline{B}$
0	1	0	$\overline{A} \cdot B$
1	0	1	$A \cdot \overline{B}$
1	1	0	$A \cdot B$

(a)

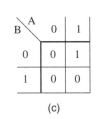

(b)   (c)

together of cells, not just combining two together.) The simplified Boolean expression for a couple is given by those variables common to all cells in the couple. See Problem 14.

Three-variable Karnaugh maps

A truth table for a three-variable expression is shown in Table 26.10(a), the 1s in the output column showing that:

$$Z = \overline{A} \cdot \overline{B} \cdot C + \overline{A} \cdot B \cdot C + A \cdot B \cdot \overline{C}$$

Each of the eight possible Boolean expressions associated with a three-variable function can be depicted as shown in Table 26.10(b) in which one cell is allocated to each row of the truth table. A matrix similar to that shown in Table 26.10(b) can be used to depict $Z = \overline{A} \cdot \overline{B} \cdot C + \overline{A} \cdot B \cdot C + A \cdot B \cdot \overline{C}$, by putting 1s in the cells corresponding to the Boolean terms on the right of the Boolean equation and 0s in the remaining cells. This method of depicting a three-variable Boolean expression is called a three-variable Karnaugh map, and is shown in Table 26.10(c).

To simplify a three-variable Boolean expression, the Boolean expression is depicted on a Karnaugh map as outlined above. Any cells on the map having common edges either vertically or horizontally are grouped together to form couples of four cells or two cells. During coupling the horizontal lines at the top and bottom of the cells are taken as a common edge, as are the vertical lines on the left and right of the cells. The simplified Boolean expression for a couple is given by those variables common to all cells in the couple. See Problems 15 and 17.

Table 26.10

Inputs			Output	Boolean
A	B	C	Z	expression
0	0	0	0	$\overline{A}\cdot\overline{B}\cdot\overline{C}$
0	0	1	1	$\overline{A}\cdot\overline{B}\cdot C$
0	1	0	0	$\overline{A}\cdot B\cdot\overline{C}$
0	1	1	1	$\overline{A}\cdot B\cdot C$
1	0	0	0	$A\cdot\overline{B}\cdot\overline{C}$
1	0	1	0	$A\cdot\overline{B}\cdot C$
1	1	0	1	$A\cdot B\cdot\overline{C}$
1	1	1	0	$A\cdot B\cdot C$

(a)

A·B C	0.0 $(\overline{A}\cdot\overline{B})$	0.1 $(\overline{A}\cdot B)$	1.1 $(A\cdot B)$	1.0 $(A\cdot\overline{B})$
$0(\overline{C})$	$\overline{A}\cdot\overline{B}\cdot\overline{C}$	$\overline{A}\cdot B\cdot\overline{C}$	$A\cdot B\cdot\overline{C}$	$A\cdot\overline{B}\cdot\overline{C}$
$1(C)$	$\overline{A}\cdot\overline{B}\cdot C$	$\overline{A}\cdot B\cdot C$	$A\cdot B\cdot C$	$A\cdot\overline{B}\cdot C$

(b)

A.B C	0.0	0.1	1.1	1.0
0	0	0	1	0
1	1	1	0	0

(c)

Four-variable Karnaugh maps

A truth table for a four-variable expression is shown in Table 26.11(a), the 1s in the output column showing that:

$$Z = \overline{A}\cdot\overline{B}\cdot C\cdot\overline{D} + \overline{A}\cdot B\cdot C\cdot\overline{D}$$
$$+ A\cdot\overline{B}\cdot C\cdot\overline{D} + A\cdot B\cdot C\cdot\overline{D}$$

Each of the 16 possible Boolean expressions associated with a four-variable function can be depicted as shown in Table 26.11(b), in which one cell is allocated to each row of the truth table. A matrix similar to that shown in Table 26.11(b) can be used to depict

$$Z = \overline{A}\cdot\overline{B}\cdot C\cdot\overline{D} + \overline{A}\cdot B\cdot C\cdot\overline{D}$$
$$+ A\cdot\overline{B}\cdot C\cdot\overline{D} + A\cdot B\cdot C\cdot\overline{D}$$

by putting 1s in the cells corresponding to the Boolean terms on the right of the Boolean equation and 0s in the remaining cells. This method of depicting a four-variable expression is called a four-variable Karnaugh map, and is shown in Table 26.11(c).

To simplify a four-variable Boolean expression, the Boolean expression is depicted on a Karnaugh map as outlined above. Any cells on the map having common edges either vertically or horizontally are grouped together to form couples of eight cells, four cells or two cells. During coupling, the horizontal lines at the top and bottom of the cells may be considered to be common edges, as are the vertical lines on the left and the right of the cells. The simplified Boolean expression for a couple is given by those variables common to all cells in the couple. See Problems 18 and 19.

Summary of procedure when simplifying a Boolean expression using a Karnaugh map

(a) Draw a four, eight or sixteen-cell matrix, depending on whether there are two, three or four variables.

(b) Mark in the Boolean expression by putting 1s in the appropriate cells.

(c) Form couples of 8, 4 or 2 cells having common edges, forming the largest groups of cells possible. (Note that a cell containing a 1 may be used more than once when forming a couple. Also note that each cell containing a 1 must be used at least once.)

(d) The Boolean expression for the couple is given by the variables which are common to all cells in the couple.

Problem 14. Use Karnaugh map techniques to simplify the expression $\overline{P}\cdot\overline{Q} + \overline{P}\cdot Q$

Using the above procedure:

(a) The two-variable matrix is drawn and is shown in Table 26.12.

(b) The term $\overline{P}\cdot\overline{Q}$ is marked with a 1 in the top left-hand cell, corresponding to $P=0$ and $Q=0$; $\overline{P}\cdot Q$ is marked with a 1 in the bottom left-hand cell corresponding to $P=0$ and $Q=1$

(c) The two cells containing 1s have a common horizontal edge and thus a vertical couple can be formed.

Table 26.11

| Inputs | | | | Output | Boolean |
A	B	C	D	Z	expression
0	0	0	0	0	$\overline{A}\cdot\overline{B}\cdot\overline{C}\cdot\overline{D}$
0	0	0	1	0	$\overline{A}\cdot\overline{B}\cdot\overline{C}\cdot D$
0	0	1	0	1	$\overline{A}\cdot\overline{B}\cdot C\cdot\overline{D}$
0	0	1	1	0	$\overline{A}\cdot\overline{B}\cdot C\cdot D$
0	1	0	0	0	$\overline{A}\cdot B\cdot\overline{C}\cdot\overline{D}$
0	1	0	1	0	$\overline{A}\cdot B\cdot\overline{C}\cdot D$
0	1	1	0	1	$\overline{A}\cdot B\cdot C\cdot\overline{D}$
0	1	1	1	0	$\overline{A}\cdot B\cdot C\cdot D$
1	0	0	0	0	$A\cdot\overline{B}\cdot\overline{C}\cdot\overline{D}$
1	0	0	1	0	$A\cdot\overline{B}\cdot\overline{C}\cdot D$
1	0	1	0	1	$A\cdot\overline{B}\cdot C\cdot\overline{D}$
1	0	1	1	0	$A\cdot\overline{B}\cdot C\cdot D$
1	1	0	0	0	$A\cdot B\cdot\overline{C}\cdot\overline{D}$
1	1	0	1	0	$A\cdot B\cdot\overline{C}\cdot D$
1	1	1	0	1	$A\cdot B\cdot C\cdot\overline{D}$
1	1	1	1	0	$A\cdot B\cdot C\cdot D$

(a)

A·B C·D	0.0 $(\overline{A}\cdot\overline{B})$	0.1 $(\overline{A}\cdot B)$	1.1 $(A\cdot B)$	1.0 $(A\cdot\overline{B})$
0.0 $(\overline{C}\cdot\overline{D})$	$\overline{A}\cdot\overline{B}\cdot\overline{C}\cdot\overline{D}$	$\overline{A}\cdot B\cdot\overline{C}\cdot\overline{D}$	$A\cdot B\cdot\overline{C}\cdot\overline{D}$	$A\cdot\overline{B}\cdot\overline{C}\cdot\overline{D}$
0.1 $(\overline{C}\cdot D)$	$\overline{A}\cdot\overline{B}\cdot\overline{C}\cdot D$	$\overline{A}\cdot B\cdot\overline{C}\cdot D$	$A\cdot B\cdot\overline{C}\cdot D$	$A\cdot\overline{B}\cdot\overline{C}\cdot D$
1.1 $(C\cdot D)$	$\overline{A}\cdot\overline{B}\cdot C\cdot D$	$\overline{A}\cdot B\cdot C\cdot D$	$A\cdot B\cdot C\cdot D$	$A\cdot\overline{B}\cdot C\cdot D$
1.0 $(C\cdot\overline{D})$	$\overline{A}\cdot\overline{B}\cdot C\cdot\overline{D}$	$\overline{A}\cdot B\cdot C\cdot\overline{D}$	$A\cdot B\cdot C\cdot\overline{D}$	$A\cdot\overline{B}\cdot C\cdot\overline{D}$

(b)

A·B C·D	0.0	0.1	1.1	1.0
0.0	0	0	0	0
0.1	0	0	0	0
1.1	0	0	0	0
1.0	1	1	1	1

(c)

Table 26.12

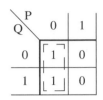

(d) The variable common to both cells in the couple is $P=0$, i.e. $\overline{P}$ thus

$$\overline{P}\cdot\overline{Q}+\overline{P}\cdot Q=\overline{P}$$

Problem 15. Simplify the expression $\overline{X}\cdot Y\cdot\overline{Z}+\overline{X}\cdot\overline{Y}\cdot Z+X\cdot Y\cdot\overline{Z}+X\cdot\overline{Y}\cdot Z$ by using Karnaugh map techniques

Using the above procedure:

(a) A three-variable matrix is drawn and is shown in Table 26.13.

Table 26.13

Z \ X·Y	0.0	0.1	1.1	1.0
0	0	1	1	0
1	1	0	0	1

(b) The 1s on the matrix correspond to the expression given, i.e. for $\overline{X}\cdot Y\cdot\overline{Z}$, $X=0$, $Y=1$ and $Z=0$ and hence corresponds to the cell in the two rows and second column, and so on.

(c) Two couples can be formed as shown. The couple in the bottom row may be formed since the vertical lines on the left and right of the cells are taken as a common edge.

(d) The variables common to the couple in the top row are $Y=1$ and $Z=0$, that is, $Y\cdot\overline{Z}$ and the variables common to the couple in the bottom row are $Y=0$, $Z=1$, that is, $\overline{Y}\cdot Z$. Hence:

$$\overline{X}\cdot Y\cdot\overline{Z}+\overline{X}\cdot\overline{Y}\cdot Z+X\cdot Y\cdot\overline{Z}$$

$$+X\cdot\overline{Y}\cdot Z=Y\cdot\overline{Z}+\overline{Y}\cdot Z$$

Problem 16. Use a Karnaugh map technique to simplify the expression $(\overline{A}\cdot B)\cdot(\overline{A}+B)$

Using the procedure, a two-variable matrix is drawn and is shown in Table 26.14.

$\overline{A}\cdot B$ corresponds to the bottom left-hand cell and $(\overline{A}\cdot B)$ must therefore be all cells except this one,

Table 26.14

B \ A	0	1
0	1	1 2
1		1

marked with a 1 in Table 26.14. $\overline{(\overline{A}+B)}$ corresponds to all the cells except the top right-hand cell marked with a 2 in Table 26.14. Hence $(\overline{A}+B)$ must correspond to the cell marked with a 2. The expression $(\overline{A \cdot B}) \cdot (\overline{\overline{A}+B})$ corresponds to the cell having both 1 and 2 in it, i.e.

$$(\overline{A \cdot B}) \cdot (\overline{\overline{A}+B}) = A \cdot \overline{B}$$

Problem 17. Simplify $\overline{(P+\overline{Q} \cdot R)} + \overline{(P \cdot Q + \overline{R})}$ using a Karnaugh map technique

The term $(P+\overline{Q} \cdot R)$ corresponds to the cells marked 1 on the matrix in Table 26.15(a), hence $\overline{(P+\overline{Q} \cdot R)}$ corresponds to the cells marked 2. Similarly, $(P \cdot Q + \overline{R})$ corresponds to the cells marked 3 in Table 26.15(a), hence $\overline{(P \cdot Q + \overline{R})}$ corresponds to the cells marked 4. The expression $\overline{(P+\overline{Q} \cdot R)} + \overline{(P \cdot Q + \overline{R})}$ corresponds to cells marked with either a 2 or with a 4 and is shown in Table 26.15(b) by Xs. These cells may be coupled as shown. The variables common to the group of four cells is $P=0$, i.e. $\overline{P}$, and those common to the group of two cells are $Q=0$, $R=1$, i.e. $\overline{Q} \cdot R$

Thus: $\overline{(P+\overline{Q} \cdot R)} + \overline{(P \cdot Q + \overline{R})} = \overline{P} + \overline{Q} \cdot R$

Table 26.15

R \ P·Q	0.0	0.1	1.1	1.0
0	3 2	3 2	3 1	3 1
1	4 1	4 2	3 1	4 1

(a)

R \ P·Q	0.0	0.1	1.1	1.0
0	X	X		
1	X	X		X

(b)

Problem 18. Use Karnaugh map techniques to simplify the expression:

$$A \cdot B \cdot \overline{C} \cdot \overline{D} + A \cdot B \cdot C \cdot D + \overline{A} \cdot B \cdot C \cdot D$$
$$+ A \cdot B \cdot C \cdot \overline{D} + \overline{A} \cdot B \cdot C \cdot \overline{D}$$

Using the procedure, a four-variable matrix is drawn and is shown in Table 26.16. The 1s marked on the matrix correspond to the expression given. Two couples can be formed as shown. The four-cell couple has $B=1$, $C=1$,

i.e. $B \cdot C$ as the common variables to all four cells and the two-cell couple has $A \cdot B \cdot \overline{D}$ as the common variables to both cells. Hence, the expression simplifies to:

$$B \cdot C + A \cdot B \cdot \overline{D} \quad \text{i.e.} \quad B \cdot (C + A \cdot \overline{D})$$

Table 26.16

C·D \ A·B	0.0	0.1	1.1	1.0
0.0			1	
0.1				
1.1			1	1
1.0			1	1

Problem 19. Simplify the expression $\overline{A} \cdot \overline{B} \cdot \overline{C} \cdot \overline{D} + A \cdot \overline{B} \cdot \overline{C} \cdot \overline{D} + \overline{A} \cdot \overline{B} \cdot C \cdot \overline{D} + A \cdot \overline{B} \cdot C \cdot \overline{D} + A \cdot B \cdot C \cdot D$ by using Karnaugh map techniques

The Karnaugh map for the expression is shown in Table 26.17. Since the top and bottom horizontal lines are common edges and the vertical lines on the left and right of the cells are common, then the four corner cells form a couple, $\overline{B} \cdot \overline{D}$ (the cells can be considered as if they are stretched to completely cover a sphere, as far as common edges are concerned). The cell $A \cdot B \cdot C \cdot D$ cannot be coupled with any other. Hence the expression simplifies to

$$\overline{B} \cdot \overline{D} + A \cdot B \cdot C \cdot D$$

Table 26.17

C·D \ A·B	0.0	0.1	1.1	1.0
0.0	1			1
0.1				
1.1			1	
1.0	1			1

Now try the following Practice Exercise

Practice Exercise 114   Further problems on simplifying Boolean expressions using Karnaugh maps (answers on page 1161)

In Problems 1 to 11 use Karnaugh map techniques to simplify the expressions given.

1. $\overline{X}\cdot Y+X\cdot Y$

2. $\overline{X}\cdot\overline{Y}+\overline{X}\cdot Y+X\cdot Y$

3. $(\overline{P}\cdot\overline{Q})\cdot(\overline{\overline{P}\cdot Q})$

4. $A\cdot\overline{C}+\overline{A}\cdot(B+C)+A\cdot B\cdot(C+\overline{B})$

5. $\overline{P}\cdot\overline{Q}\cdot\overline{R}+\overline{P}\cdot Q\cdot\overline{R}+P\cdot Q\cdot\overline{R}$

6. $\overline{P}\cdot\overline{Q}\cdot\overline{R}+P\cdot Q\cdot\overline{R}+P\cdot Q\cdot R+P\cdot\overline{Q}\cdot R$

7. $\overline{A}\cdot\overline{B}\cdot\overline{C}\cdot\overline{D}+\overline{A}\cdot B\cdot\overline{C}\cdot\overline{D}+\overline{A}\cdot B\cdot\overline{C}\cdot D$

8. $\overline{A}\cdot\overline{B}\cdot C\cdot D+\overline{A}\cdot\overline{B}\cdot C\cdot\overline{D}+A\cdot\overline{B}\cdot C\cdot\overline{D}$

9. $\overline{A}\cdot B\cdot\overline{C}\cdot D+A\cdot B\cdot\overline{C}\cdot D+A\cdot B\cdot C\cdot D$
    $+A\cdot\overline{B}\cdot\overline{C}\cdot D+A\cdot\overline{B}\cdot C\cdot D$

10. $\overline{A}\cdot\overline{B}\cdot\overline{C}\cdot D+A\cdot B\cdot\overline{C}\cdot\overline{D}+A\cdot\overline{B}\cdot\overline{C}\cdot\overline{D}$
    $+A\cdot B\cdot C\cdot\overline{D}+A\cdot\overline{B}\cdot C\cdot\overline{D}$

11. $A\cdot B\cdot\overline{C}\cdot\overline{D}+\overline{A}\cdot\overline{B}\cdot\overline{C}\cdot\overline{D}+\overline{A}\cdot B\cdot C\cdot D$
    $+\overline{A}\cdot\overline{B}\cdot C\cdot D+A\cdot\overline{B}\cdot\overline{C}\cdot D+\overline{A}\cdot B\cdot C\cdot\overline{D}$
    $+\overline{A}\cdot B\cdot C\cdot\overline{D}$

26.6 Logic circuits

In practice, logic gates are used to perform the **and**, **or** and **not**-functions introduced in Section 26.1. Logic gates can be made from switches, magnetic devices or fluidic devices, but most logic gates in use are electronic devices. Various logic gates are available. For example, the Boolean expression $(A\cdot B\cdot C)$ can be produced using a three-input **and**-gate and $(C+D)$ by using a two-input **or**-gate. The principal gates in common use are introduced below. The term 'gate' is used in the same sense as a normal gate, the open state being indicated by a binary '1' and the closed state by a binary '0'. A gate will only open when the requirements of the gate are met and, for example, there will only be a '1' output on a two-input **and**-gate when both the inputs to the gate are at a '1' state.

The and-gate

The different symbols used for a three-input **and**-gate are shown in Figure 26.13(a) and the truth table is shown

BRITISH   AMERICAN

(a)

INPUTS			OUTPUT
A	B	C	$Z=A\cdot B\cdot C$
0	0	0	0
0	0	1	0
0	1	0	0
0	1	1	0
1	0	0	0
1	0	1	0
1	1	0	0
1	1	1	1

(b)

Figure 26.13

in Figure 26.13(b). This shows that there will only be a '1' output when A is 1 and B is 1 and C is 1, written as:

$$Z = A\cdot B\cdot C$$

The or-gate

The different symbols used for a three-input **or**-gate are shown in Figure 26.14(a) and the truth table is shown in Figure 26.14(b). This shows that there will be a '1' output when A is 1, or B is 1, or C is 1, or any combination of A, B or C is 1, written as:

$$Z = A+B+C$$

The invert-gate or not-gate

The different symbols used for an **invert**-gate are shown in Figure 26.15(a) and the truth table is shown in Figure 26.15(b). This shows that a '0' input gives a '1' output and vice versa, i.e. it is an 'opposite to' function. The invert of A is written $\overline{A}$ and is called 'not-A'.

The nand-gate

The different symbols used for a **nand**-gate are shown in Figure 26.16(a) and the truth table is shown in

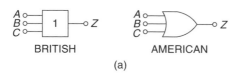

BRITISH   AMERICAN

(a)

| INPUTS | | | OUTPUT |
A	B	C	$Z = A + B + C$
0	0	0	0
0	0	1	1
0	1	0	1
0	1	1	1
1	0	0	1
1	0	1	1
1	1	0	1
1	1	1	1

(b)

Figure 26.14

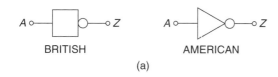

BRITISH   AMERICAN

(a)

| INPUT | OUTPUT |
A	$Z = \bar{A}$
0	1
1	0

(b)

Figure 26.15

Figure 26.16(b). This gate is equivalent to an **and**-gate and an **invert**-gate in series (not-and=nand) and the output is written as:

$$Z = \overline{A \cdot B \cdot C}$$

The nor-gate

The different symbols used for a **nor**-gate are shown in Figure 26.17(a) and the truth table is shown in Figure 26.17(b). This gate is equivalent to an **or**-gate and an **invert**-gate in series (not-or=nor), and the output is written as:

$$Z = \overline{A + B + C}$$

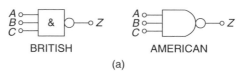

BRITISH   AMERICAN

(a)

| INPUTS | | | | OUTPUT |
A	B	C	$A.B.C.$	$Z = \overline{A.B.C.}$
0	0	0	0	1
0	0	1	0	1
0	1	0	0	1
0	1	1	0	1
1	0	0	0	1
1	0	1	0	1
1	1	0	0	1
1	1	1	1	0

(b)

Figure 26.16

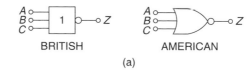

BRITISH   AMERICAN

(a)

| INPUTS | | | | OUTPUT |
A	B	C	$A + B + C$	$Z = \overline{A + B + C}$
0	0	0	0	1
0	0	1	1	0
0	1	0	1	0
0	1	1	1	0
1	0	0	1	0
1	0	1	1	0
1	1	0	1	0
1	1	1	1	0

(b)

Figure 26.17

Combinational logic networks

In most logic circuits, more than one gate is needed to give the required output. Except for the **invert**-gate, logic gates generally have two, three or four inputs and are confined to one function only. Thus, for example, a two-input **or**-gate or a four-input **and**-gate can be used when designing a logic circuit. The way in which logic gates are used to generate a given output is shown in Problems 20 to 23.

Problem 20. Devise a logic system to meet the requirements of: $Z = A \cdot \overline{B} + C$

With reference to Figure 26.18 an **invert**-gate, shown as (1), gives $\overline{B}$. The **and**-gate, shown as (2), has inputs of A and $\overline{B}$, giving $A \cdot \overline{B}$. The **or**-gate, shown as (3), has inputs of $A \cdot \overline{B}$ and C, giving:

$$Z = A \cdot \overline{B} + C$$

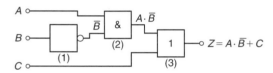

Figure 26.18

Problem 21. Devise a logic system to meet the requirements of $(P + \overline{Q}) \cdot (\overline{R} + S)$

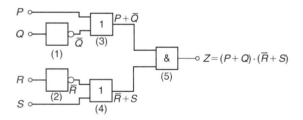

Figure 26.19

The logic system is shown in Figure 26.19. The given expression shows that two **invert**-functions are needed to give $\overline{Q}$ and $\overline{R}$ and these are shown as gates (1) and (2). Two **or**-gates, shown as (3) and (4), give $(P + \overline{Q})$ and $(\overline{R} + S)$ respectively. Finally, an **and**-gate, shown as (5), gives the required output.

$$Z = (P + \overline{Q}) \cdot (\overline{R} + S)$$

Problem 22. Devise a logic circuit to meet the requirements of the output given in Table 26.18, using as few gates as possible.

Table 26.18

Inputs			Output
A	B	C	Z
0	0	0	0
0	0	1	0
0	1	0	0
0	1	1	0
1	0	0	0
1	0	1	1
1	1	0	1
1	1	1	1

The '1' outputs in rows 6, 7 and 8 of Table 26.18 show that the Boolean expression is:

$$Z = A \cdot \overline{B} \cdot C + A \cdot B \cdot \overline{C} + A \cdot B \cdot C$$

The logic circuit for this expression can be built using three 3-input **and**-gates and one 3-input **or**-gate, together with two **invert**-gates. However, the number of gates required can be reduced by using the techniques introduced in Sections 26.3 to 26.5, resulting in the cost of the circuit being reduced. Any of the techniques can be used, and in this case, the rules of Boolean algebra (see Table 26.7) are used.

$$Z = A \cdot \overline{B} \cdot C + A \cdot B \cdot \overline{C} + A \cdot B \cdot C$$
$$= A \cdot [\overline{B} \cdot C + B \cdot \overline{C} + B \cdot C]$$
$$= A \cdot [\overline{B} \cdot C + B(\overline{C} + C)] = A \cdot [\overline{B} \cdot C + B]$$
$$= A \cdot [B + \overline{B} \cdot C] = A \cdot [B + C]$$

The logic circuit to give this simplified expression is shown in Figure 26.20.

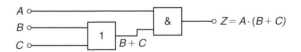

Figure 26.20

Problem 23. Simplify the expression:

$$Z = \overline{P} \cdot \overline{Q} \cdot \overline{R} \cdot \overline{S} + \overline{P} \cdot \overline{Q} \cdot \overline{R} \cdot S + \overline{P} \cdot Q \cdot \overline{R} \cdot \overline{S}$$
$$+ \overline{P} \cdot Q \cdot \overline{R} \cdot S + P \cdot \overline{Q} \cdot \overline{R} \cdot \overline{S}$$

and devise a logic circuit to give this output

The given expression is simplified using the Karnaugh map techniques introduced in Section 26.5. Two couples are formed, as shown in Figure 26.21(a), and the simplified expression becomes:

$$Z = \overline{Q} \cdot \overline{R} \cdot \overline{S} + \overline{P} \cdot \overline{R}$$

i.e. $Z = \overline{R} \cdot (\overline{P} + \overline{Q} \cdot \overline{S})$

The logic circuit to produce this expression is shown in Figure 26.21(b).

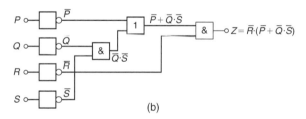

(a)

(b)

Figure 26.21

Now try the following Practice Exercise

Practice Exercise 115   Further problems on logic circuits (answers on page 1161)

In Problems 1 to 4, devise logic systems to meet the requirements of the Boolean expressions given.

1. $Z = \overline{A} + B \cdot C$

2. $Z = A \cdot \overline{B} + B \cdot \overline{C}$

3. $Z = A \cdot B \cdot \overline{C} + \overline{A} \cdot \overline{B} \cdot C$

4. $Z = (\overline{A} + B) \cdot (\overline{C} + D)$

In Problems 5 to 7, simplify the expression given in the truth table and devise a logic circuit to meet the requirements stated.

5. Column 4 of Table 26.19

6. Column 5 of Table 26.19

Table 26.19

1	2	3	4	5	6
A	B	C	Z_1	Z_2	Z_3
0	0	0	0	0	0
0	0	1	1	0	0
0	1	0	0	0	1
0	1	1	1	1	1
1	0	0	0	1	0
1	0	1	1	1	1
1	1	0	1	0	1
1	1	1	1	1	1

7. Column 6 of Table 26.19

In Problems 8 to 12, simplify the Boolean expressions given and devise logic circuits to give the requirements of the simplified expressions.

8. $\overline{P} \cdot \overline{Q} + \overline{P} \cdot Q + P \cdot Q$

9. $\overline{P} \cdot \overline{Q} \cdot \overline{R} + P \cdot Q \cdot \overline{R} + P \cdot \overline{Q} \cdot \overline{R}$

10. $P \cdot \overline{Q} \cdot R + P \cdot \overline{Q} \cdot \overline{R} + \overline{P} \cdot \overline{Q} \cdot \overline{R}$

11. $\overline{A} \cdot \overline{B} \cdot \overline{C} \cdot \overline{D} + A \cdot \overline{B} \cdot \overline{C} \cdot \overline{D} + \overline{A} \cdot \overline{B} \cdot C \cdot \overline{D}$
     $+ \overline{A} \cdot B \cdot C \cdot \overline{D} + A \cdot \overline{B} \cdot C \cdot \overline{D}$

12. $\overline{(\overline{P} \cdot Q \cdot R)} \cdot \overline{(P + Q \cdot R)}$

26.7   Universal logic gates

The function of any of the five logic gates in common use can be obtained by using either **nand**-gates or **nor**-gates and when used in this manner, the gate selected is called a **universal gate**. The way in which a universal **nand**-gate is used to produce the **invert, and, or** and **nor**-functions is shown in Problem 24. The way in which a universal **nor**-gate is used to produce the **invert, or, and** and **nand**-function is shown in Problem 25.

Problem 24. Show how **invert, and, or** and **nor**-functions can be produced using nand-gates only

A single input to a **nand**-gate gives the **invert**-function, as shown in Figure 26.22(a). When two **nand**-gates are connected, as shown in Figure 26.22(b), the output from the first gate is $\overline{A \cdot B \cdot C}$ and this is inverted by the second gate, giving $Z = \overline{\overline{A \cdot B \cdot C}} = A \cdot B \cdot C$ i.e. the **and**-function is produced. When $\overline{A}$, $\overline{B}$ and $\overline{C}$ are the inputs to a **nand**-gate, the output is $\overline{\overline{A} \cdot \overline{B} \cdot \overline{C}}$

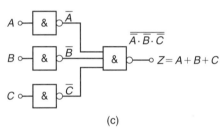

$A \circ\!\!-\!\!\boxed{\&}\!\!-\!\!\circ Z = \overline{A}$

(a)

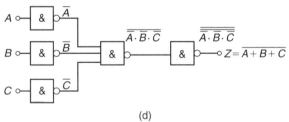

$\overline{A \cdot B \cdot C}$ $\overline{\overline{A \cdot B \cdot C}}$

$\begin{matrix} A \circ \\ B \circ \\ C \circ \end{matrix}\!\!\boxed{\&}\!\!-\!\!\boxed{\&}\!\!-\!\!\circ Z = A \cdot B \cdot C$

(b)

$A \circ\!\!-\!\!\boxed{\&}\!\!-\!\!\overline{A}$

$B \circ\!\!-\!\!\boxed{\&}\!\!-\!\!\overline{B}$ $\overline{\overline{A} \cdot \overline{B} \cdot \overline{C}}$
$\qquad\qquad\boxed{\&}\!\!-\!\!\circ Z = A + B + C$

$C \circ\!\!-\!\!\boxed{\&}\!\!-\!\!\overline{C}$

(c)

$A \circ\!\!-\!\!\boxed{\&}\!\!-\!\!\overline{A}$

$B \circ\!\!-\!\!\boxed{\&}\!\!-\!\!\overline{B}$ $\overline{\overline{A} \cdot \overline{B} \cdot \overline{C}}$ $\overline{\overline{\overline{A} \cdot \overline{B} \cdot \overline{C}}}$
$\qquad\qquad\boxed{\&}\!\!-\!\!\boxed{\&}\!\!-\!\!\circ Z = \overline{A + B + C}$

$C \circ\!\!-\!\!\boxed{\&}\!\!-\!\!\overline{C}$

(d)

Figure 26.22

By de Morgan's law, $\overline{A \cdot B \cdot C} = \overline{A} + \overline{B} + \overline{C} = A + B + C$, i.e. a **nand**-gate is used to produce the **or**-function. The logic circuit is shown in Figure 26.22(c). If the output from the logic circuit in Figure 26.22(c) is inverted by adding an additional **nand**-gate, the output becomes the invert of an **or**-function, i.e. the **nor**-function, as shown in Figure 26.22(d).

Problem 25. Show how **invert, or, and** and **nand**-functions can be produced by using **nor**-gates only

A single input to a **nor**-gate gives the **invert**-function, as shown in Figure 26.23(a). When two **nor**-gates are connected, as shown in Figure 26.23(b), the output from the first gate is $\overline{A + B + C}$ and this is inverted by the

second gate, giving $Z = \overline{\overline{A + B + C}} = A + B + C$, i.e. the **or**-function is produced. Inputs of $\overline{A}$, $\overline{B}$ and $\overline{C}$ to a **nor**-gate give an output of $\overline{\overline{A} + \overline{B} + \overline{C}}$

By de Morgan's law, $\overline{\overline{A} + \overline{B} + \overline{C}} = \overline{\overline{A}} \cdot \overline{\overline{B}} \cdot \overline{\overline{C}} = A \cdot B \cdot C$, i.e. the **nor**-gate can be used to produce the **and**-function. The logic circuit is shown in Figure 26.23(c). When the output of the logic circuit, shown in Figure 26.23(c), is inverted by adding an additional **nor**-gate, the output then becomes the invert of an **or**-function, i.e. the **nor**-function as shown in Figure 26.23(d).

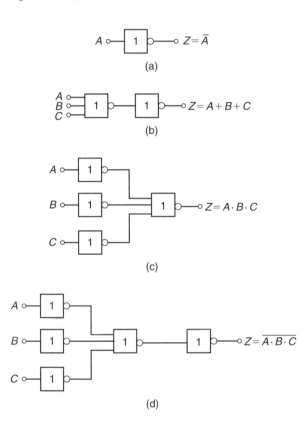

$A \circ\!\!-\!\!\boxed{1}\!\!-\!\!\circ Z = \overline{A}$

(a)

$\begin{matrix} A \circ \\ B \circ \\ C \circ \end{matrix}\!\!\boxed{1}\!\!-\!\!\boxed{1}\!\!-\!\!\circ Z = A + B + C$

(b)

$A \circ\!\!-\!\!\boxed{1}$

$B \circ\!\!-\!\!\boxed{1}\qquad\boxed{1}\!\!-\!\!\circ Z = A \cdot B \cdot C$

$C \circ\!\!-\!\!\boxed{1}$

(c)

$A \circ\!\!-\!\!\boxed{1}$

$B \circ\!\!-\!\!\boxed{1}\qquad\boxed{1}\!\!-\!\!\boxed{1}\!\!-\!\!\circ Z = \overline{A \cdot B \cdot C}$

$C \circ\!\!-\!\!\boxed{1}$

(d)

Figure 26.23

Problem 26. Design a logic circuit, using **nand**-gates having not more than three inputs, to meet the requirements of the Boolean expression:

$$Z = \overline{A} + \overline{B} + C + \overline{D}$$

When designing logic circuits, it is often easier to start at the output of the circuit. The given expression shows there are four variables joined by **or**-functions. From the principles introduced in Problem 24, if a four-input **nand**-gate is used to give the expression given, the inputs are $\overline{\overline{A}}$, $\overline{\overline{B}}$, $\overline{C}$ and $\overline{\overline{D}}$ that is A, B, $\overline{C}$ and D.

However, the problem states that three inputs are not to be exceeded so two of the variables are joined, i.e. the inputs to the three input **nand**-gate, shown as gate (1) in Figure 26.24, are A, B, $\overline{C}$ and D. From Problem 24, the **and**-function is generated by using two **nand**-gates connected in series, as shown by gates (2) and (3) in Figure 26.24. The logic circuit required to produce the given expression is as shown in Figure 26.24.

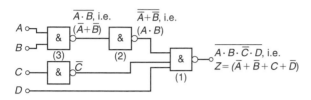

Figure 26.24

Problem 27.   Use **nor**-gates only to design a logic circuit to meet the requirements of the expressions:

$$Z = \overline{D} \cdot (\overline{A} + B + \overline{C})$$

It is usual in logic circuit design to start the design at the output. From Problem 25, the **and**-function between $\overline{D}$ and the terms in the bracket can be produced by using inputs of $\overline{D}$ and $\overline{A} + B + \overline{C}$ to a **nor**-gate, i.e. by de Morgan's law, inputs of D and $A \cdot \overline{B} \cdot C$. Again, with reference to Problem 25, inputs of $\overline{A} \cdot B$ and $\overline{C}$ to a **nor**-gate give an output of $\overline{\overline{A} + B + \overline{C}}$, which by de Morgan's law is $A \cdot \overline{B} \cdot C$. The logic circuit to produce the required expression is as shown in Figure 26.25.

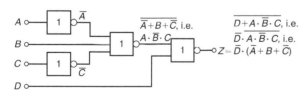

Figure 26.25

Problem 28.   An alarm indicator in a grinding mill complex should be activated if (a) the power supply to all mills is off and (b) the hopper feeding the mills is less than 10% full, and (c) if less than two of the three grinding mills are in action. Devise a logic system to meet these requirements

Let variable A represent the power supply on to all the mills, then $\overline{A}$ represents the power supply off. Let B represent the hopper feeding the mills being more than 10% full, then $\overline{B}$ represents the hopper being less than 10%

full. Let C, D and E represent the three mills respectively being in action, then $\overline{C}$, $\overline{D}$ and $\overline{E}$ represent the three mills respectively not being in action. The required expression to activate the alarm is:

$$Z = \overline{A} \cdot \overline{B} \cdot (\overline{C} + \overline{D} + \overline{E})$$

There are three variables joined by **and**-functions in the output, indicating that a three-input **and**-gate is required, having inputs of $\overline{A}$, $\overline{B}$ and $(\overline{C} + \overline{D} + \overline{E})$. The term $(\overline{C} + \overline{D} + \overline{E})$ is produced by a three-input **nand**-gate. When variables C, D and E are the inputs to a **nand**-gate, the output is $\overline{C \cdot D \cdot E}$ which, by de Morgan's law, is $\overline{C} + \overline{D} + \overline{E}$. Hence the required logic circuit is as shown in Figure 26.26.

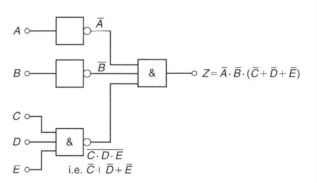

Figure 26.26

Now try the following Practice Exercise

Practice Exercise 116   Further problems on universal logic gates (answers on page 1161)

In Problems 1 to 3, use **nand**-gates only to devise the logic systems stated.

1. $Z = A + B \cdot C$

2. $Z = A \cdot \overline{B} + B \cdot \overline{C}$

3. $Z = A \cdot B \cdot \overline{C} + \overline{A} \cdot \overline{B} \cdot C$

In Problems 4 to 6, use **nor**-gates only to devise the logic systems stated.

4. $Z = (\overline{A} + B) \cdot (\overline{C} + D)$

5. $Z = A \cdot \overline{B} + B \cdot \overline{C} + C \cdot \overline{D}$

6. $Z = \overline{P} \cdot Q + P \cdot (Q + R)$

7. In a chemical process, three of the transducers used are P, Q and R, giving output signals of either 0 or 1. Devise a logic system to give a 1 output when:

(a) P and Q and R all have 0 outputs, or when:

(b) P is 0 and (Q is 1 or R is 0)

8. Lift doors should close (Z) if:

(a) the master switch (A) is on and either

(b) a call (B) is received from any other floor, or

(c) the doors (C) have been open for more than 10 seconds, or

(d) the selector push within the lift (D) is pressed for another floor.

Devise a logic circuit to meet these requirements.

9. A water tank feeds three separate processes. When any two of the processes are in operation at the same time, a signal is required to start a pump to maintain the head of water in the tank. Devise a logic circuit using **nor**-gates only to give the required signal.

10. A logic signal is required to give an indication when:

(a) the supply to an oven is on, and

(b) the temperature of the oven exceeds 210°C, or

(c) the temperature of the oven is less than 190°C.

Devise a logic circuit using **nand**-gates only to meet these requirements.

For fully worked solutions to each of the problems in Practice Exercises 111 to 116 in this chapter, go to the website:
www.routledge.com/cw/bird

This assignment covers the material contained in Chapters 25 and 26. *The marks available are shown in brackets at the end of each question.*

1. Convert the following to decimal numbers:
   (a) 11010_2 (b) 101110_2 (6)

2. Convert the following decimal numbers into binary numbers:
   (a) 53   (b) 29   (8)

3. Determine the binary addition:
   $1011 + 11011$ (3)

4. Convert the hexadecimal number 3B into its binary equivalent.   (3)

5. Convert 173_{10} into hexadecimal.   (4)

6. Convert 1011011_2 into hexadecimal.   (3)

7. Convert DF_{16} into its binary equivalent.   (3)

8. Use the laws and rules of Boolean algebra to simplify the following expressions:
   (a) $B \cdot (A + \overline{B}) + A \cdot \overline{B}$
   (b) $\overline{A} \cdot \overline{B} \cdot \overline{C} + \overline{A} \cdot B \cdot \overline{C} + \overline{A} \cdot B \cdot C + \overline{A} \cdot \overline{B} \cdot C$ (9)

9. Simplify the Boolean expression:
   $\overline{A \cdot \overline{B}} + A \cdot B \cdot \overline{C}$ using de Morgan's laws.   (5)

10. Use a Karnaugh map to simplify the Boolean expression:
    $$\overline{A} \cdot \overline{B} \cdot \overline{C} + \overline{A} \cdot B \cdot \overline{C} + \overline{A} \cdot B \cdot C + A \cdot \overline{B} \cdot C$$ (8)

11. A clean room has two entrances, each having two doors, as shown in Figure RT10.1. A warning bell must sound if both doors A and B or doors C and D are open at the same time. Write down the Boolean expression depicting this occurrence, and devise a logic network to operate the bell using nand-gates only.   (8)

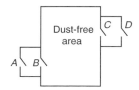

Figure RT10.1

Multiple choice questions Test 3
Logarithms, further number and algebra
This test covers the material in Chapters 15 to 26

Section B

All questions have only one correct answer (answers on page 1200).

1. $\log_{16} 8$ is equal to:

   (a) $\dfrac{1}{2}$ (b) 144   (c) $\dfrac{3}{4}$ (d) 2

2. $(p+x)^4 = p^4 + 4p^3x + 6p^2x^2 + 4px^3 + x^4$. Using Pascal's triangle, the third term of $(p+x)^5$ is:

   (a) $10p^2x^3$ (b) $5p^4x$ (c) $5p^3x^2$ (d) $10p^3x^2$

3. The value of $\dfrac{\ln 2}{e^2 \lg 2}$, correct to 3 significant figures, is:

   (a) 0.0588   (b) 0.312   (c) 17.0   (d) 3.209

4. In decimal, the binary number 1101101 is equal to:

   (a) 218   (b) 31   (c) 127   (d) 109

5. If $\log_2 x = 3$ then:

   (a) $x = 8$ (b) $x = \dfrac{3}{2}$ (c) $x = 9$ (d) $x = \dfrac{2}{3}$

6. The pressure p pascals at height h metres above ground level is given by $p = p_0 e^{-h/k}$, where p_0 is the pressure at ground level and k is a constant. When p_0 is 1.01×10^5 Pa and the pressure at a height of 1500 m is 9.90×10^4 Pa, the value of k, correct to 3 significant figures is:

   (a) 1.33×10^{-5} (b) 75000   (c) 173000   (d) 197

7. The 5th term of an arithmetic progression is 18 and the 12th term is 46. The 18th term is:

   (a) 72   (b) 74   (c) 68   (d) 70

8. The value of $\dfrac{3.67 \ln 21.28}{e^{-0.189}}$, correct to 4 significant figures, is:

   (a) 9.289   (b) 13.56
   (c) 13.5566   (d) -3.844×10^9

9. The Boolean expression $A + \overline{A}.B$ is equivalent to:

   (a) A (b) B (c) $A + B$ (d) $A + \overline{A}$

10. $\log_3 9$ is equal to:

    (a) 3   (b) 27   (c) $\dfrac{1}{3}$ (d) 2

11. The second moment of area of a rectangle through its centroid is given by $\dfrac{bl^3}{12}$. Using the binomial theorem, the approximate percentage change in the second moment of area if b is increased by 3% and l is reduced by 2% is:

    (a) -3% (b) $+1\%$ (c) $+3\%$ (d) -6%

12. The motion of a particle in an electrostatic field is described by the equation $y = x^3 + 3x^2 + 5x - 28$. When $x = 2$, y is approximately zero. Using one iteration of the Newton–Raphson method, a better approximation (correct to 2 decimal places) is:

    (a) 1.89   (b) 2.07   (c) 2.11   (d) 1.93

13. In hexadecimal, the decimal number 123 is:

    (a) 7B   (b) 123   (c) 173   (d) 1111011

14. $6x^2 - 5x - 6$ divided by $2x - 3$ gives:

    (a) $2x - 1$ (b) $3x + 2$ (c) $3x - 2$ (d) $6x + 1$

15. The 1st term of a geometric progression is 9 and the 4th term is 45. The 8th term is:

    (a) 225   (b) 150.5   (c) 384.7   (d) 657.9

16. The solution of the inequality $\dfrac{3t + 2}{t + 1} \le 1$ is:

    (a) $t \ge -2\dfrac{1}{2}$ (b) $-1 < t \le \dfrac{1}{2}$

    (c) $t < -1$ (d) $-\dfrac{1}{2} < t \le 1$

17. The Boolean expression $\overline{P}.\overline{Q} + \overline{P}.Q$ is equivalent to:

    (a) $\overline{P}$ (b) $\overline{Q}$ (c) P (d) Q

18. In the equation $5.0 = 3.0 \ln\left(\dfrac{2.9}{x}\right)$, x has a value correct to 3 significant figures of:

    (a) 1.59   (b) 0.392   (c) 0.548   (d) 0.0625

19. The solution of the inequality $x^2 - x - 2 < 0$ is:

    (a) $1 < x < -2$ (b) $x > 2$

    (c) $-1 < x < 2$ (d) $x < -1$

20. The current i amperes flowing in a capacitor at time t seconds is given by $i = 10(1 - e^{-t/CR})$, where resistance R is 25×10^3 ohms and capacitance C is 16×10^{-6} farads. When current i reaches 7 amperes, the time t is:

    (a) -0.48 s   (b) 0.14 s   (c) 0.21 s   (d) 0.48 s

21. The Boolean expression $\overline{F}.\overline{G}.\overline{H} + \overline{F}.\overline{G}.H$ is equivalent to:

    (a) $F.G$ (b) $F.\overline{G}$ (c) $\overline{F}.H$ (d) $\overline{F}.\overline{G}$

22. $(x^3 - x^2 - x + 1)$ divided by $(x - 1)$ gives:

    (a) $x^2 - x - 1$ (b) $x^2 + 1$

    (c) $x^2 - 1$ (d) $x^2 + x - 1$

23. In hexadecimal, the decimal number 237 is equivalent to:

    (a) 355   (b) ED   (c) 1010010011   (d) DE

24. When $(2x - 1)^5$ is expanded by the binomial series, the fourth term is:

    (a) $-40x^2$ (b) $10x^2$ (c) $40x^2$ (d) $-10x^2$

25. $(1/4)\log 16 - 2\log 5 + (1/3)\log 27$ is equivalent to:

    (a) $-\log 20$ (b) $\log 0.24$ (c) $-\log 5$ (d) $\log 3$

Section B

COMPANION @ WEBSITE

Formulae/revision hints for Section B
Further number and algebra

Factor theorem

If $x = a$ is a root of the equation $f(x) = 0$,

then $(x - a)$ is a factor of $f(x)$

Remainder theorem

If $(ax^2 + bx + c)$ is divided by $(x - p)$, the

remainder will be: $ap^2 + bp + c$

or if $(ax^3 + bx^2 + cx + d)$ is divided by $(x - p)$, the

remainder will be: $ap^3 + bp^2 + cp + d$

Arithmetic progression

If $a =$ first term and $d =$ common difference, then the arithmetic progression is:

$$a, a + d, a + 2d, \ldots.$$

The nth term is: $a + (n - 1)d$

Sum of n terms, $S_n = \dfrac{n}{2}[2a + (n - 1)d]$

Geometric progression

If $a =$ first term and $r =$ common ratio, then the geometric progression is:

$$a, ar, ar^2, \ldots.$$

The nth term is: ar^{n-1}

Sum of n terms, $S_n = \dfrac{a(1 - r^n)}{(1 - r)}$ or $\dfrac{a(r^n - 1)}{(r - 1)}$

If $-1 < r < 1$, $S_\infty = \dfrac{a}{(1 - r)}$

Partial fractions

Provided that the numerator $f(x)$ is of less degree than the relevant denominator, the following identities are typical examples of the form of partial fractions used:

$$\frac{f(x)}{(x+a)(x+b)(x+c)} \equiv \frac{A}{(x+a)} + \frac{B}{(x+b)} + \frac{C}{(x+c)}$$

$$\frac{f(x)}{(x+a)^3(x+b)} \equiv \frac{A}{(x+a)} + \frac{B}{(x+a)^2} + \frac{C}{(x+a)^3} + \frac{D}{(x+b)}$$

$$\frac{f(x)}{(ax^2+bx+c)(x+d)} \equiv \frac{Ax+B}{(ax^2+bx+c)} + \frac{C}{(x+d)}$$

Binomial series

$$(a+b)^n = a^n + na^{n-1}b + \frac{n(n-1)}{2!}a^{n-2}b^2$$
$$+ \frac{n(n-1)(n-2)}{3!}a^{n-3}b^3 + \ldots.$$

$$(1+x)^n = 1 + nx + \frac{n(n-1)}{2!}x^2$$
$$+ \frac{n(n-1)(n-2)}{3!}x^3 + \ldots.$$

Maclaurin's series

$$f(x) = f(0) + x f'(0) + \frac{x^2}{2!} f''(0) + \frac{x^3}{3!} f'''(0) + \ldots.$$

Hyperbolic functions

$$\sinh x = \frac{e^x - e^{-x}}{2} \qquad \operatorname{cosech} x = \frac{1}{\sinh x} = \frac{2}{e^x - e^{-x}}$$

$$\cosh x = \frac{e^x + e^{-x}}{2} \qquad \operatorname{sech} x = \frac{1}{\cosh x} = \frac{2}{e^x + e^{-x}}$$

$$\tanh x = \frac{e^x - e^{-x}}{e^x + e^{-x}} \qquad \coth x = \frac{1}{\tanh x} = \frac{e^x + e^{-x}}{e^x - e^{-x}}$$

$$\cosh^2 x - \sinh^2 x = 1 \qquad 1 - \tanh^2 x = \operatorname{sech}^2 x$$

$$\coth^2 x - 1 = \operatorname{cosech}^2 x$$

Newton–Raphson iterative method

If r_1 is the approximate value for a real root of the equation $f(x) = 0$, then a closer approximation to the root, r_2, is given by:

$$r_2 = r_1 - \frac{f(r_1)}{f'(r_1)}$$

Boolean algebra

Laws and rules of Boolean algebra

Commutative Laws: $A + B = B + A$

$A.B = B.A$

Associative Laws: $A + B + C = (A + B) + C$

$A.B.C = (A.B).C$

Distributive Laws: $A.(B + C) = A.B + A.C$

$A + (B.C) = (A + B).(A + C)$

Sum rules:

$A + \overline{A} = 1$

$A + 1 = 1$

$A + 0 = A$

$A + A = A$

Product rules: $A.\overline{A} = 0$

$A.0 = 0$

$A.1 = A$

$A.A = A$

Absorption rules: $A + A.B = A$

$A.(A + B) = A$

$A + \overline{A}.B = A + B$

De Morgan's Laws: $\overline{A + B} = \overline{A}.\overline{B}$

$\overline{A.B} = \overline{A} + \overline{B}$

For a copy of these formulae/revision hints, go to:
www.routledge.com/cw/bird

Section B

Section C

Areas and volumes

Chapter 27

Areas of common shapes

Why it is important to understand: Areas of common shapes

To paint, wallpaper or panel a wall, you must know the total area of the wall so you can buy the appropriate amount of finish. When designing a new building, or seeking planning permission, it is often necessary to specify the total floor area of the building. In construction, calculating the area of a gable end of a building is important when determining the number of bricks and mortar to order. When using a bolt, the most important thing is that it is long enough for your particular application and it may also be necessary to calculate the shear area of the bolt connection. Ridge vents allow a home to properly vent, while disallowing rain or other forms of precipitation to leak into the attic or crawlspace underneath the roof. Equal amounts of cool air and warm air flowing through the vents is paramount for proper heat exchange. Calculating how much surface area is available on the roof aids in determining how long the ridge vent should run. A race track is an oval shape, and it is sometimes necessary to find the perimeter of the inside of a race track. Arches are everywhere, from sculptures and monuments to pieces of architecture and strings on musical instruments; finding the height of an arch or its cross-sectional area is often required. Determining the cross-sectional areas of beam structures is vitally important in design engineering. There are thus a large number of situations in engineering where determining area is important.

At the end of this chapter, you should be able to:

- state the SI unit of area
- identify common polygons – triangle, quadrilateral, pentagon, hexagon, heptagon and octagon
- identify common quadrilaterals – rectangle, square, parallelogram, rhombus and trapezium
- calculate areas of quadrilaterals and circles
- appreciate that areas of similar shapes are proportional to the squares of the corresponding linear dimensions

27.1 Introduction

Area is a measure of the size or extent of a plane surface. Area is measured in **square units** such as mm^2, cm^2 and m^2. This chapter deals with finding the areas of common shapes.

In engineering it is often important to be able to calculate simple areas of various shapes. In everyday life its important to be able to measure area to, say, lay a carpet, order sufficient paint for a decorating job or order sufficient bricks for a new wall.

On completing this chapter you will be able to recognise common shapes and be able to find the areas of rectangles, squares, parallelograms, triangles, trapeziums and circles.

27.2 Common shapes

Polygons

A polygon is a closed plane figure bounded by straight lines. A polygon which has

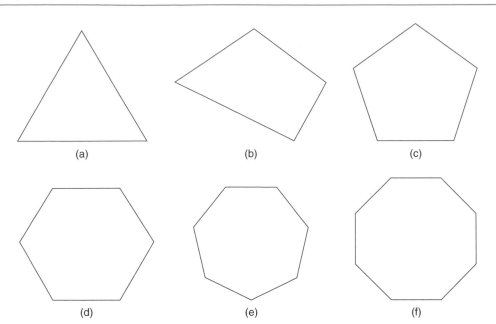

Figure 27.1

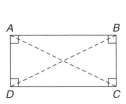

Figure 27.2

Figure 27.4

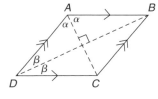

Figure 27.5

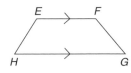

Figure 27.6

3 sides is called a **triangle** – see Figure 27.1(a)
4 sides is called a **quadrilateral** – see Figure 27.1(b)
5 sides is called a **pentagon** – see Figure 27.1(c)
6 sides is called a **hexagon** – see Figure 27.1(d)
7 sides is called a **heptagon** – see Figure 27.1(e)
8 sides is called an **octagon** – see Figure 27.1(f)

Quadrilaterals

There are five types of quadrilateral, these being rectangle, square, parallelogram, rhombus and trapezium. If the opposite corners of any quadrilateral are joined by a straight line, two triangles are produced. Since the sum of the angles of a triangle is 180°, the sum of the angles of a quadrilateral is 360°.

Rectangle

In the rectangle *ABCD* shown in Figure 27.2,

(a) all four angles are right angles,

(b) the opposite sides are parallel and equal in length, and

(c) diagonals *AC* and *BD* are equal in length and bisect one another.

Square

In the square *PQRS* shown in Figure 27.3,

(a) all four angles are right angles,

(b) the opposite sides are parallel,

(c) all four sides are equal in length, and

(d) diagonals *PR* and *QS* are equal in length and bisect one another at right angles.

Parallelogram

In the parallelogram *WXYZ* shown in Figure 27.4,

(a) opposite angles are equal,

(b) opposite sides are parallel and equal in length, and

(c) diagonals *WY* and *XZ* bisect one another.

Rhombus

In the rhombus *ABCD* shown in Figure 27.5,

(a) opposite angles are equal,

(b) opposite angles are bisected by a diagonal,

(c) opposite sides are parallel,

(d) all four sides are equal in length, and

(e) diagonals *AC* and *BD* bisect one another at right angles.

Trapezium

In the trapezium *EFGH* shown in Figure 27.6,

(a) only one pair of sides is parallel.

> **Problem 1.**   State the types of quadrilateral shown in Figure 27.7 and determine the angles marked *a* to *l*

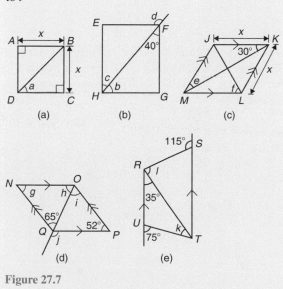

Figure 27.7

(a) **ABCD is a square**
The diagonals of a square bisect each of the right angles, hence

$$a = \frac{90°}{2} = \mathbf{45°}$$

(b) **EFGH is a rectangle**
In triangle *FGH*, $40° + 90° + b = 180°$, since the angles in a triangle add up to $180°$, from which $b = \mathbf{50°}$. Also, $c = \mathbf{40°}$ (alternate angles between

parallel lines *EF* and *HG*). (Alternatively, *b* and *c* are complementary; i.e. add up to $90°$)
$d = 90° + c$ (external angle of a triangle equals the sum of the interior opposite angles), hence $\mathbf{d} = 90° + 40° = \mathbf{130°}$ (or $\angle EFH = 50°$ and $d = 180° - 50° = 130°$)

(c) **JKLM is a rhombus**
The diagonals of a rhombus bisect the interior angles and the opposite internal angles are equal. Thus, $\angle JKM = \angle MKL = \angle JMK = \angle LMK = 30°$, hence, $\mathbf{e = 30°}$
In triangle *KLM*, $30° + \angle KLM + 30° = 180°$ (the angles in a triangle add up to $180°$), hence, $\angle KLM = 120°$. The diagonal *JL* bisects $\angle KLM$, hence, $\mathbf{f} = \dfrac{120°}{2} = \mathbf{60°}$

(d) **NOPQ is a parallelogram**
$\mathbf{g = 52°}$ since the opposite interior angles of a parallelogram are equal.
In triangle *NOQ*, $g + h + 65° = 180°$ (the angles in a triangle add up to $180°$), from which $\mathbf{h} = 180° - 65° - 52° = \mathbf{63°}$
$\mathbf{i = 65°}$ (alternate angles between parallel lines *NQ* and *OP*).
$\mathbf{j} = 52° + i = 52° + 65° = \mathbf{117°}$ (the external angle of a triangle equals the sum of the interior opposite angles). (Alternatively, $\angle PQO = h = 63°$; hence, $j = 180° - 63° = 117°$)

(e) **RSTU is a trapezium**
$35° + k = 75°$ (external angle of a triangle equals the sum of the interior opposite angles), hence, $\mathbf{k = 40°}$
$\angle STR = 35°$ (alternate angles between parallel lines *RU* and *ST*). $l + 35° = 115°$ (external angle of a triangle equals the sum of the interior opposite angles), hence, $\mathbf{l} = 115° - 35° = \mathbf{80°}$

Now try the following Practice Exercise

> **Practice Exercise 117   Common shapes (answers on page 1162)**
>
> 1. Find the angles *p* and *q* in Figure 27.8(a).
>
> 2. Find the angles *r* and *s* in Figure 27.8(b).
>
> 3. Find the angle *t* in Figure 27.8(c).

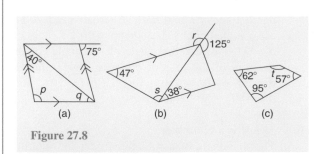

Figure 27.8

27.3 Areas of common shapes

The formulae for the areas of common shapes are shown in Table 27.1.
Here are some worked problems to demonstrate how the formulae are used to determine the area of common shapes.

Problem 2. Calculate the area and length of the perimeter of the square shown in Figure 27.9

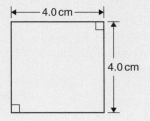

Figure 27.9

$$\textbf{Area of square} = x^2 = (4.0)^2 = 4.0\,\text{cm} \times 4.0\,\text{cm}$$
$$= \textbf{16.0}\,\textbf{cm}^2$$

(Note the unit of area is $\text{cm} \times \text{cm} = \text{cm}^2$; i.e. square centimetres or centimetres squared.)

$$\textbf{Perimeter of square} = 4.0\,\text{cm} + 4.0\,\text{cm} + 4.0\,\text{cm}$$
$$+ 4.0\,\text{cm} = \textbf{16.0}\,\textbf{cm}$$

Problem 3. Calculate the area and length of the perimeter of the rectangle shown in Figure 27.10

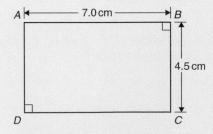

Figure 27.10

$$\textbf{Area of rectangle} = l \times b = 7.0 \times 4.5$$
$$= \textbf{31.5}\,\textbf{cm}^2$$

$$\textbf{Perimeter of rectangle} = 7.0\,\text{cm} + 4.5\,\text{cm}$$
$$+ 7.0\,\text{cm} + 4.5\,\text{cm}$$
$$= \textbf{23.0}\,\textbf{cm}$$

Problem 4. Calculate the area of the parallelogram shown in Figure 27.11

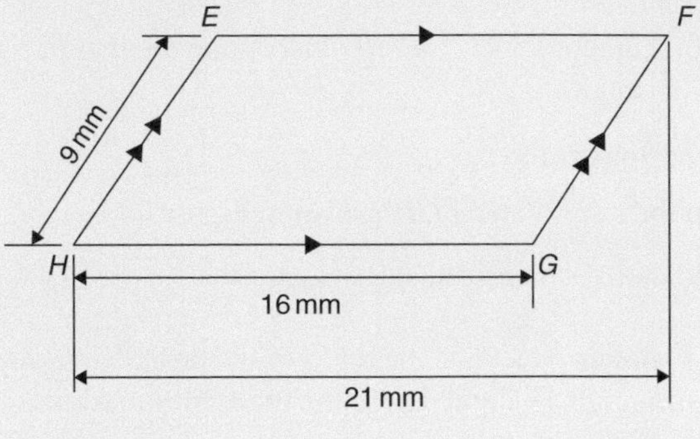

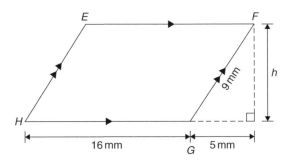

Figure 27.11

Area of a parallelogram = base × perpendicular height
The perpendicular height h is not shown in Figure 27.11 but may be found using Pythagoras' theorem (see Chapter 38).
From Figure 27.12, $9^2 = 5^2 + h^2$, from which $h^2 = 9^2 - 5^2 = 81 - 25 = 56$
Hence, perpendicular height,

$$h = \sqrt{56} = 7.48\,\text{mm}.$$

Figure 27.12

Hence, **area of parallelogram** *EFGH*

$$= 16\,\text{mm} \times 7.48\,\text{mm}$$
$$= \textbf{120}\,\textbf{mm}^2$$

Table 27.1 Formulae for the areas of common shapes

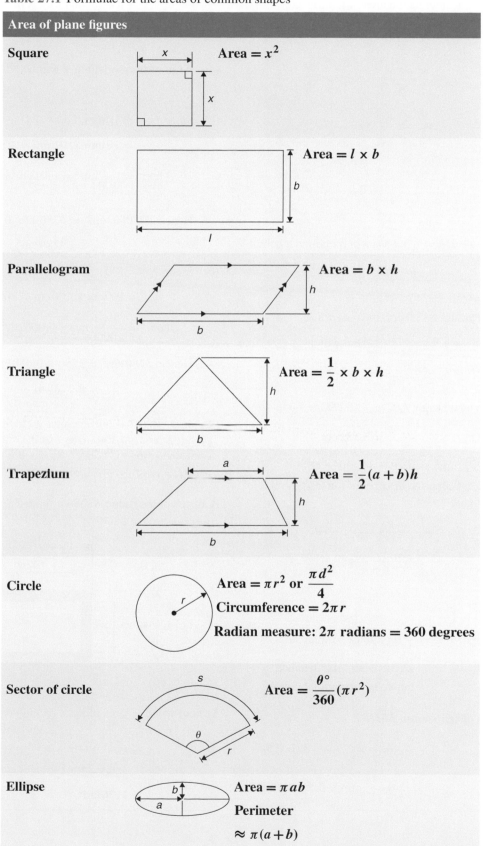

Area of plane figures		
Square		$\text{Area} = x^2$
Rectangle		$\text{Area} = l \times b$
Parallelogram		$\text{Area} = b \times h$
Triangle		$\text{Area} = \dfrac{1}{2} \times b \times h$
Trapezium		$\text{Area} = \dfrac{1}{2}(a+b)h$
Circle		$\text{Area} = \pi r^2 \text{ or } \dfrac{\pi d^2}{4}$ $\text{Circumference} = 2\pi r$ Radian measure: 2π radians $= 360$ degrees
Sector of circle		$\text{Area} = \dfrac{\theta^\circ}{360}(\pi r^2)$
Ellipse		$\text{Area} = \pi a b$ Perimeter $\approx \pi(a+b)$

Problem 5. Calculate the area of the triangle shown in Figure 27.13

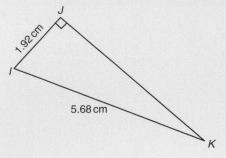

Figure 27.13

Area of triangle $IJK = \dfrac{1}{2} \times$ base $\times$ perpendicular height

$$= \dfrac{1}{2} \times IJ \times JK$$

To find JK, Pythagoras' theorem is used; i.e.

$$5.68^2 = 1.92^2 + JK^2, \text{ from which}$$

$$JK = \sqrt{5.68^2 - 1.92^2} = 5.346 \text{cm}$$

Hence, **area of triangle IJK** $= \dfrac{1}{2} \times 1.92 \times 5.346$

$$= \mathbf{5.132 cm^2}$$

Problem 6. Calculate the area of the trapezium shown in Figure 27.14

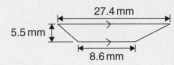

Figure 27.14

Area of a trapezium $= \dfrac{1}{2} \times$ (sum of parallel sides)

$$\times \text{ (perpendicular distance}$$
$$\text{between the parallel sides)}$$

Hence, **area of trapezium $LMNO$**

$$= \dfrac{1}{2} \times (27.4 + 8.6) \times 5.5$$

$$= \dfrac{1}{2} \times 36 \times 5.5 = \mathbf{99 mm^2}$$

Problem 7. A rectangular tray is 820 mm long and 400 mm wide. Find its area in (a) mm^2 (b) cm^2 (c) m^2

(a) **Area of tray** $=$ length $\times$ width $= 820 \times 400$

$$= \mathbf{328000 mm^2}$$

(b) Since $1 \text{cm} = 10 \text{mm}$, $1 \text{cm}^2 = 1 \text{cm} \times 1 \text{cm}$

$$= 10 \text{mm} \times 10 \text{mm} = 100 \text{mm}^2, \text{ or}$$

$$1 \text{mm}^2 = \dfrac{1}{100} \text{cm}^2 = 0.01 \text{cm}^2$$

Hence, $\mathbf{328000 mm^2} = 328000 \times 0.01 \text{cm}^2$

$$= \mathbf{3280 cm^2}$$

(c) Since $1 \text{m} = 100 \text{cm}$, $1 \text{m}^2 = 1 \text{m} \times 1 \text{m}$

$$= 100 \text{cm} \times 100 \text{cm} = 10000 \text{cm}^2, \text{ or}$$

$$1 \text{cm}^2 = \dfrac{1}{10000} \text{m}^2 = 0.0001 \text{m}^2$$

Hence, $\mathbf{3280 cm^2} = 3280 \times 0.0001 \text{m}^2$

$$= \mathbf{0.3280 m^2}$$

Problem 8. The outside measurements of a picture frame are 100 cm by 50 cm. If the frame is 4 cm wide, find the area of the wood used to make the frame

A sketch of the frame is shown shaded in Figure 27.15.

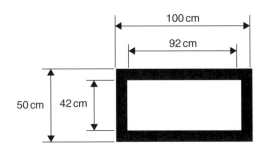

Figure 27.15

Area of wood $=$ area of large rectangle $-$ area of small rectangle

$$= (100 \times 50) - (92 \times 42)$$

$$= 5000 - 3864$$

$$= \mathbf{1136 cm^2}$$

Problem 9.   Find the cross-sectional area of the girder shown in Figure 27.16

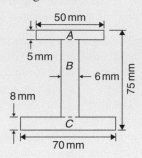

Figure 27.16

The girder may be divided into three separate rectangles, as shown.

$$\text{Area of rectangle } A = 50 \times 5 = 250 \, \text{mm}^2$$

$$\text{Area of rectangle } B = (75 - 8 - 5) \times 6$$

$$= 62 \times 6 = 372 \, \text{mm}^2$$

$$\text{Area of rectangle } C = 70 \times 8 = 560 \, \text{mm}^2$$

Total area of girder $= 250 + 372 + 560$

$$= \textbf{1182} \, \textbf{mm}^2 \text{ or } \textbf{11.82} \, \textbf{cm}^2$$

Problem 10.   Figure 27.17 shows the gable end of a building. Determine the area of brickwork in the gable end

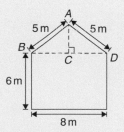

Figure 27.17

The shape is that of a rectangle and a triangle.

$$\text{Area of rectangle } = 6 \times 8 = 48 \, \text{m}^2$$

$$\text{Area of triangle } = \frac{1}{2} \times \text{base} \times \text{height}$$

$CD = 4 \, \text{m}$ and $AD = 5 \, \text{m}$, hence $AC = 3 \, \text{m}$ (since it is a 3, 4, 5 triangle – or by Pythagoras).

$$\text{Hence, area of triangle } ABD = \frac{1}{2} \times 8 \times 3 = 12 \, \text{m}^2$$

Total area of brickwork $= 48 + 12$

$$= \textbf{60} \, \textbf{m}^2$$

Now try the following Practice Exercise

Practice Exercise 118   Areas of common shapes (answers on page 1162)

1. Name the types of quadrilateral shown in Figure 27.18(i) to (iv) and determine for each (a) the area and (b) the perimeter.

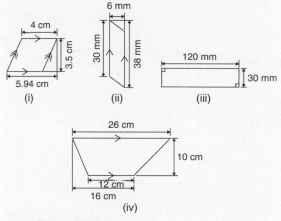

Figure 27.18

2. A rectangular plate is 85 mm long and 42 mm wide. Find its area in square centimetres.

3. A rectangular field has an area of 1.2 hectares and a length of 150 m. If 1 hectare $= 10000 \, \text{m}^2$, find (a) the field's width and (b) the length of a diagonal.

4. Find the area of a triangle whose base is 8.5 cm and perpendicular height is 6.4 cm.

5. A square has an area of 162 cm^2. Determine the length of a diagonal.

6. A rectangular picture has an area of 0.96 m^2. If one of the sides has a length of 800 mm, calculate, in millimetres, the length of the other side.

7. Determine the area of each of the angle iron sections shown in Figure 27.19.

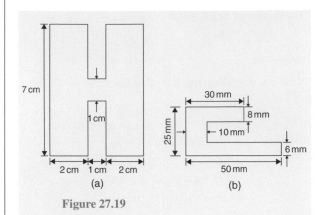

Figure 27.19

8. Figure 27.20 shows a 4 m wide path around the outside of a 41 m by 37 m garden. Calculate the area of the path.

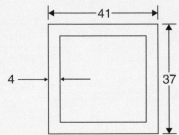

Figure 27.20

9. The area of a trapezium is 13.5 cm² and the perpendicular distance between its parallel sides is 3 cm. If the length of one of the parallel sides is 5.6 cm, find the length of the other parallel side.

10. Calculate the area of the steel plate shown in Figure 27.21.

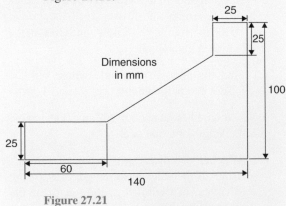

Figure 27.21

11. Determine the area of an equilateral triangle of side 10.0 cm.

12. If paving slabs are produced in 250 mm by 250 mm squares, determine the number of slabs required to cover an area of 2 m².

13. Figure 27.22 shows a plan view of an office block to be built. The walls will have a height of 8 m, and it is necessary to make an estimate of the number of bricks required to build the walls. Assuming that any doors and windows are ignored in the calculation and that 48 bricks are required to build 1 m² of wall, calculate the number of external bricks required.

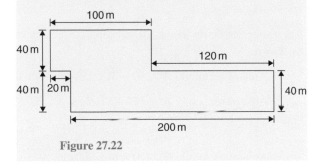

Figure 27.22

Here are some further worked problems on finding the areas of common shapes using the formulae in Table 27.1, page 281.

Problem 11. Find the area of a circle having a radius of 5 cm

$$\textbf{Area of circle} = \pi r^2 = \pi (5)^2 = 25\pi = \textbf{78.54 cm}^2$$

Problem 12. Find the area of a circle having a diameter of 15 mm

$$\textbf{Area of circle} = \frac{\pi d^2}{4} = \frac{\pi (15)^2}{4} = \frac{225\pi}{4} = \textbf{176.7 mm}^2$$

Problem 13. Find the area of a circle having a circumference of 70 mm

Circumference, $c = 2\pi r$, hence

$$\text{radius}, r = \frac{c}{2\pi} = \frac{70}{2\pi} = \frac{35}{\pi} \text{ mm}$$

$$\textbf{Area of circle} = \pi r^2 = \pi \left(\frac{35}{\pi}\right)^2 = \frac{35^2}{\pi}$$

$$= \textbf{389.9 mm}^2 \text{ or } \textbf{3.899 cm}^2$$

Problem 14. Calculate the area of the sector of a circle having radius 6 cm with angle subtended at centre 50°

$$\textbf{Area of sector} = \frac{\theta°}{360}(\pi r^2) = \frac{50}{360}(\pi 6^2)$$

$$= \frac{50 \times \pi \times 36}{360} = \textbf{15.71 cm}^2$$

Problem 15. Calculate the area of the sector of a circle having diameter 80 mm with angle subtended at centre 107°42′

If diameter = 80 mm then radius $r = 40$ mm, and

$$\text{area of sector} = \frac{107°42′}{360}(\pi 40^2) = \frac{107\frac{42}{60}}{360}(\pi 40^2)$$

$$= \frac{107.7}{360}(\pi 40^2)$$

$$= \textbf{1504 mm}^2 \text{ or } \textbf{15.04 cm}^2$$

Problem 16. A hollow shaft has an outside diameter of 5.45 cm and an inside diameter of 2.25 cm. Calculate the cross-sectional area of the shaft

The cross-sectional area of the shaft is shown by the shaded part in Figure 27.23 (often called an **annulus**).

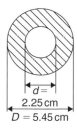

Figure 27.23

Area of shaded part = area of large circle – area of small circle

$$= \frac{\pi D^2}{4} - \frac{\pi d^2}{4} = \frac{\pi}{4}(D^2 - d^2)$$

$$= \frac{\pi}{4}(5.45^2 - 2.25^2)$$

$$= \textbf{19.35 cm}^2$$

Problem 17. The major axis of an ellipse is 15.0 cm and the minor axis is 9.0 cm. Find its area and approximate perimeter

If the major axis = 15.0 cm, then the semi-major axis = 7.5 cm. If the minor axis = 9.0 cm, then the semi-minor axis = 4.5 cm. Hence, from Table 27.1,

$$\textbf{area} = \pi ab = \pi(7.5)(4.5) = \textbf{106.0 cm}^2$$

and $\textbf{perimeter} \approx \pi(a + b) = \pi(7.5 + 4.5)$
$$= 12.0\pi = \textbf{37.7 cm}$$

Now try the following Practice Exercise

Practice Exercise 119   Areas of common shapes (answers on page 1162)

1. A rectangular garden measures 40 m by 15 m. A 1 m flower border is made round the two shorter sides and one long side. A circular swimming pool of diameter 8 m is constructed in the middle of the garden. Find, correct to the nearest square metre, the area remaining.

2. Determine the area of circles having (a) a radius of 4 cm (b) a diameter of 30 mm (c) a circumference of 200 mm.

3. An annulus has an outside diameter of 60 mm and an inside diameter of 20 mm. Determine its area.

4. If the area of a circle is 320 mm², find (a) its diameter and (b) its circumference.

5. Calculate the areas of the following sectors of circles.
   (a) radius 9 cm, angle subtended at centre 75°.
   (b) diameter 35 mm, angle subtended at centre 48°37′.

6. Determine the shaded area of the template shown in Figure 27.24.

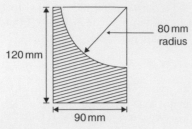

Figure 27.24

7. An archway consists of a rectangular opening topped by a semi-circular arch, as shown in Figure 27.25. Determine the area of the opening if the width is 1 m and the greatest height is 2 m.

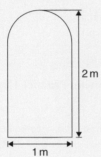

Figure 27.25

8. The major axis of an ellipse is 200 mm and the minor axis 100 mm. Determine the area and approximate perimeter of the ellipse.

9. If fencing costs £15 per metre, find the cost (to the nearest pound) of enclosing an elliptical plot of land which has major and minor diameter lengths of 120 m and 80 m.

10. A cycling track is in the form of an ellipse, the axes being 250 m and 150 m respectively for the inner boundary, and 270 m and 170 m for the outer boundary. Calculate the area of the track.

Here are some further worked problems of common shapes.

Problem 18. Calculate the area of a regular octagon if each side is 5 cm and the width across the flats is 12 cm

An octagon is an eight-sided polygon. If radii are drawn from the centre of the polygon to the vertices then eight equal triangles are produced, as shown in Figure 27.26.

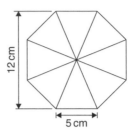

Figure 27.26

$$\text{Area of one triangle} = \frac{1}{2} \times \text{base} \times \text{height}$$

$$= \frac{1}{2} \times 5 \times \frac{12}{2} = 15 \text{cm}^2$$

Area of octagon $= 8 \times 15 = \mathbf{120\,cm^2}$

Problem 19. Determine the area of a regular hexagon which has sides 8 cm long

A hexagon is a six-sided polygon which may be divided into six equal triangles as shown in Figure 27.27. The angle subtended at the centre of each triangle is $360° \div 6 = 60°$. The other two angles in the triangle add up to $120°$ and are equal to each other. Hence, each of the triangles is equilateral with each angle $60°$ and each side 8 cm.

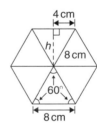

Figure 27.27

$$\text{Area of one triangle} = \frac{1}{2} \times \text{base} \times \text{height}$$

$$= \frac{1}{2} \times 8 \times h$$

h is calculated using Pythagoras' theorem:

$$8^2 = h^2 + 4^2$$

from which $h = \sqrt{8^2 - 4^2} = 6.928 \text{cm}$

Hence,

$$\text{Area of one triangle} = \frac{1}{2} \times 8 \times 6.928 = 27.71 \text{cm}^2$$

Area of hexagon $= 6 \times 27.71$

$$= \mathbf{166.3\,cm^2}$$

Problem 20. Figure 27.28 shows a plan of a floor of a building which is to be carpeted. Calculate the area of the floor in square metres. Calculate the cost, correct to the nearest pound, of carpeting the floor with carpet costing £16.80 per m², assuming 30% extra carpet is required due to wastage in fitting

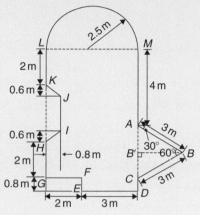

Figure 27.28

Area of floor plan

= area of triangle ABC + area of semicircle

+ area of rectangle $CGLM$

+ area of rectangle $CDEF$

− area of trapezium $HIJK$

Triangle ABC is equilateral since $AB = BC = 3$ m and, hence, angle $B'CB = 60°$.

$$\sin B'CB = BB'/3$$

i.e. $$BB' = 3\sin 60° = 2.598\,\text{m}.$$

$$\text{Area of triangle } ABC = \frac{1}{2}(AC)(BB')$$

$$= \frac{1}{2}(3)(2.598) = 3.897\,\text{m}^2$$

$$\text{Area of semicircle } = \frac{1}{2}\pi r^2 = \frac{1}{2}\pi(2.5)^2$$

$$= 9.817\,\text{m}^2$$

$$\text{Area of } CGLM = 5 \times 7 = 35\,\text{m}^2$$

$$\text{Area of } CDEF = 0.8 \times 3 = 2.4\,\text{m}^2$$

$$\text{Area of } HIJK = \frac{1}{2}(KH + IJ)(0.8)$$

Since $MC = 7$ m then $LG = 7$ m, hence $JI = 7 - 5.2 = 1.8$ m. Hence,

$$\text{Area of } HIJK = \frac{1}{2}(3 + 1.8)(0.8) = 1.92\,\text{m}^2$$

Total floor area $= 3.897 + 9.817 + 35 + 2.4 - 1.92$

$$= 49.194\,\text{m}^2$$

To allow for 30% wastage, amount of carpet required $= 1.3 \times 49.194 = 63.95\,\text{m}^2$

Cost of carpet at £16.80 per m²
$= 63.95 \times 16.80 = $ **£1074**, correct to the nearest pound.

Now try the following Practice Exercise

Practice Exercise 120   Areas of common shapes (answers on page 1162)

1. Calculate the area of a regular octagon if each side is 20 mm and the width across the flats is 48.3 mm.

2. Determine the area of a regular hexagon which has sides 25 mm.

3. A plot of land is in the shape shown in Figure 27.29. Determine

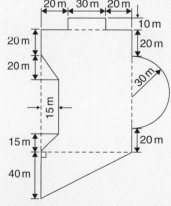

Figure 27.29

(a) its area in hectares (1 ha $= 10^4\,\text{m}^2$),

(b) the length of fencing required, to the nearest metre, to completely enclose the plot of land.

Section C

27.4 Areas of similar shapes

Figure 27.30 shows two squares, one of which has sides three times longer than the other.

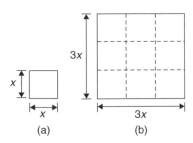

Figure 27.30

$$\text{Area of Figure } 27.30(a) = (x)(x) = x^2$$

$$\text{Area of Figure } 27.30(b) = (3x)(3x) = 9x^2$$

Hence, Figure 27.30(b) has an area $(3)^2$; i.e. 9 times the area of Figure 27.30(a).

In summary, **the areas of similar shapes are proportional to the squares of corresponding linear dimensions**.

> **Problem 21.** A rectangular garage is shown on a building plan having dimensions 10 mm by 20 mm. If the plan is drawn to a scale of 1 to 250, determine the true area of the garage in square metres

$$\text{Area of garage on the plan} = 10\,\text{mm} \times 20\,\text{mm}$$

$$= 200\,\text{mm}^2$$

Since the areas of similar shapes are proportional to the squares of corresponding dimensions,

$$\text{True area of garage} = 200 \times (250)^2$$

$$= 12.5 \times 10^6 \,\text{mm}^2$$

$$= \frac{12.5 \times 10^6}{10^6}\,\text{m}^2$$

$$\text{since } 1\,\text{m}^2 = 10^6 \,\text{mm}^2$$

$$= \mathbf{12.5\,m^2}$$

Now try the following Practice Exercise

> **Practice Exercise 121 Areas of similar shapes (answers on page 1162)**
>
> 1. The area of a park on a map is 500 mm². If the scale of the map is 1 to 40 000, determine the true area of the park in hectares (1 hectare = 10^4 m²).
>
> 2. A model of a boiler is made having an overall height of 75 mm corresponding to an overall height of the actual boiler of 6 m. If the area of metal required for the model is 12 500 mm², determine, in square metres, the area of metal required for the actual boiler.
>
> 3. The scale of an Ordnance Survey map is 1 : 2500. A circular sports field has a diameter of 8 cm on the map. Calculate its area in hectares, giving your answer correct to 3 significant figures (1 hectare = 10^4 m²).

For fully worked solutions to each of the problems in Practice Exercises 117 to 121 in this chapter, go to the website:
www.routledge.com/cw/bird

Chapter 28

The circle and its properties

Why it is important to understand: **The circle and its properties**

A circle is one of the fundamental shapes of geometry; it consists of all the points that are equidistant from a central point. Knowledge of calculations involving circles is needed with crank mechanisms, with determinations of latitude and longitude, with pendulums, and even in the design of paper clips. The floodlit area at a football ground, the area an automatic garden sprayer sprays and the angle of lap of a belt drive all rely on calculations involving the arc of a circle. The ability to handle calculations involving circles and their properties is clearly essential in several branches of engineering design.

At the end of this chapter, you should be able to:

- define a circle
- state some properties of a circle — including radius, circumference, diameter, semicircle, quadrant, tangent, sector, chord, segment and arc
- appreciate the angle in a semicircle is a right angle
- define a radian, and change radians to degrees, and vice versa
- determine arc length, area of a circle and area of a sector of a circle
- state the equation of a circle
- sketch a circle given its equation
- understand linear and angular velocity, performing simple calculations
- appreciate centripetal force and acceleration, performing simple calculations

28.1 Introduction

A **circle** is a plane figure enclosed by a curved line, every point on which is equidistant from a point within, called the **centre**.

In Chapter 27, worked problems on the areas of circles and sectors were demonstrated. In this chapter, properties of circles are listed and arc lengths are calculated, together with more practical examples on the areas of sectors of circles. In addition, the equation of a circle is explained and application to linear and angular velocity and centripetal force and acceleration are explored.

28.2 Properties of circles

(a) The distance from the centre to the curve is called the **radius**, r, of the circle (see OP in Figure 28.1).

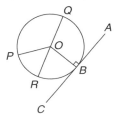

Figure 28.1

(b) The boundary of a circle is called the **circumference**, c.

(c) Any straight line passing through the centre and touching the circumference at each end is called the **diameter**, d (see QR in Figure 28.1). Thus, $d = 2r$

(d) The ratio $\dfrac{\text{circumference}}{\text{diameter}}$ is a constant for any circle. This constant is denoted by the Greek letter π (pronounced 'pie'), where $\pi = 3.14159$, correct to 5 decimal places (check with your calculator). Hence, $\dfrac{c}{d} = \pi$ or $c = \pi d$ or $c = 2\pi r$

(e) A **semicircle** is one half of a whole circle.

(f) A **quadrant** is one quarter of a whole circle.

(g) A **tangent** to a circle is a straight line which meets the circle at one point only and does not cut the circle when produced. AC in Figure 28.1 is a tangent to the circle since it touches the curve at point B only. If radius OB is drawn, **angle ABO is a right angle**.

(h) The **sector** of a circle is the part of a circle between radii (for example, the portion OXY of Figure 28.2 is a sector). If a sector is less than a semicircle it is called a **minor sector**; if greater than a semicircle it is called a **major sector**.

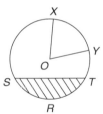

Figure 28.2

(i) The **chord** of a circle is any straight line which divides the circle into two parts and is terminated at each end by the circumference. ST, in Figure 28.2, is a chord.

(j) **Segment** is the name given to the parts into which a circle is divided by a chord. If the segment is less than a semicircle it is called a **minor segment** (see shaded area in Figure 28.2). If the segment is greater than a semicircle it is called a **major segment** (see the unshaded area in Figure 28.2).

(k) An **arc** is a portion of the circumference of a circle. The distance SRT in Figure 28.2 is called a **minor arc** and the distance $SXYT$ is called a **major arc**.

(l) The angle at the centre of a circle, subtended by an arc, is double the angle at the circumference

Figure 28.3

subtended by the same arc. With reference to Figure 28.3,

Angle $AOC = 2 \times$ angle ABC

(m) The angle in a semicircle is a right angle (see angle BQP in Figure 28.3).

Problem 1. Find the circumference of a circle of radius 12.0 cm

Circumference, $c = 2 \times \pi \times \text{radius} = 2\pi r = 2\pi(12.0)$
$$= \textbf{75.40 cm}$$

Problem 2. If the diameter of a circle is 75 mm, find its circumference

Circumference, $c = \pi \times \text{diameter} = \pi d = \pi(75)$
$$= \textbf{235.6 mm}$$

Problem 3. Determine the radius of a circular pond if its perimeter is 112 m

Perimeter $=$ circumference, $c = 2\pi r$
Hence, **radius of pond**, $r = \dfrac{c}{2\pi} = \dfrac{112}{2\pi} = \textbf{17.83 cm}$

Problem 4. In Figure 28.4, AB is a tangent to the circle at B. If the circle radius is 40 mm and $AB = 150$ mm, calculate the length AO

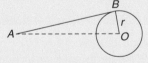

Figure 28.4

A tangent to a circle is at right angles to a radius drawn from the point of contact; i.e. $\textbf{ABO} = \textbf{90}°$. Hence, using Pythagoras' theorem,
$$AO^2 = AB^2 + OB^2$$
from which, $\quad AO = \sqrt{AB^2 + OB^2}$
$$= \sqrt{150^2 + 40^2} = \textbf{155.2 mm}$$

Now try the following Practice Exercise

Practice Exercise 122   Properties of a circle (answers on page 1162)

1. Calculate the length of the circumference of a circle of radius 7.2 cm.

2. If the diameter of a circle is 82.6 mm, calculate the circumference of the circle.

3. Determine the radius of a circle whose circumference is 16.52 cm.

4. Find the diameter of a circle whose perimeter is 149.8 cm.

5. A crank mechanism is shown in Figure 28.5, where XY is a tangent to the circle at point X. If the circle radius OX is 10 cm and length OY is 40 cm, determine the length of the connecting rod XY.

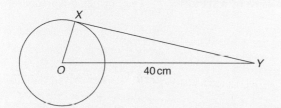

Figure 28.5

6. If the circumference of the earth is 40 000 km at the equator, calculate its diameter.

7. Calculate the length of wire in the paper clip shown in Figure 28.6. The dimensions are in millimetres.

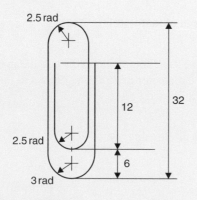

Figure 28.6

28.3 Radians and degrees

One **radian** is defined as the angle subtended at the centre of a circle by an arc equal in length to the radius. With reference to Figure 28.7, for arc length s,

$$\theta \text{ radians} = \frac{s}{r}$$

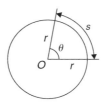

Figure 28.7

When s = whole circumference ($= 2\pi r$) then

$$\theta = \frac{s}{r} = \frac{2\pi r}{r} = 2\pi$$

i.e. **2π radians $= 360°$ or π radians $= 180°$**

Thus, **$1\,\text{rad} = \dfrac{180°}{\pi} = 57.30°$**, correct to 2 decimal places.

Since π rad $= 180°$, then $\dfrac{\pi}{2} = 90°$, $\dfrac{\pi}{3} = 60°$, $\dfrac{\pi}{4} = 45°$, and so on.

Problem 5. Convert to radians: (a) $125°$ (b) $69°47'$

(a) Since $180° = \pi$ rad, $1° = \dfrac{\pi}{180}$ rad, therefore

$$125° = 125\left(\frac{\pi}{180}\right)\text{rad} = \textbf{2.182 radians}$$

(b) $69°47' = 69\dfrac{47°}{60} = 69.783°$ (or, with your calculator, enter $69°47'$ using $°\,'\,''$ function, press $=$ and press $°\,'\,''$ again),
    and $69.783° = 69.783\left(\dfrac{\pi}{180}\right)$ rad

$$= \textbf{1.218 radians}$$

Problem 6. Convert to degrees and minutes: (a) 0.749 radians (b) $3\pi/4$ radians

(a) Since π rad $= 180°$, $1\,\text{rad} = \dfrac{180°}{\pi}$

   therefore $0.749\,\text{rad} = 0.749\left(\dfrac{180}{\pi}\right)^{°} = 42.915°$

$0.915° = (0.915 \times 60)' = 55'$, correct to the nearest minute,

Hence, **0.749 radians = 42°55′**

(b) Since $1 \text{ rad} = \left(\dfrac{180}{\pi}\right)°$ then

$$\frac{3\pi}{4} \text{ rad} = \frac{3\pi}{4}\left(\frac{180}{\pi}\right)° = \frac{3}{4}(180)° = \mathbf{135°}$$

Problem 7. Express in radians, in terms of π, (a) $150°$ (b) $270°$ (c) $37.5°$

Since $180° = \pi \text{ rad}$, $1° = \dfrac{\pi}{180} \text{ rad}$

(a) $150° = 150\left(\dfrac{\pi}{180}\right) \text{rad} = \dfrac{\mathbf{5\pi}}{\mathbf{6}} \textbf{ rad}$

(b) $270° = 270\left(\dfrac{\pi}{180}\right) \text{rad} = \dfrac{\mathbf{3\pi}}{\mathbf{2}} \textbf{ rad}$

(c) $37.5° = 37.5\left(\dfrac{\pi}{180}\right) \text{rad} = \dfrac{75\pi}{360} \text{rad} = \dfrac{\mathbf{5\pi}}{\mathbf{24}} \textbf{ rad}$

Now try the following Practice Exercise

Practice Exercise 123   Radians and degrees (answers on page 1162)

1. Convert to radians in terms of π:

    (a) $30°$ (b) $75°$ (c) $225°$

2. Convert to radians, correct to 3 decimal places:

    (a) $48°$ (b) $84°51'$ (c) $232°15'$

3. Convert to degrees:

    (a) $\dfrac{7\pi}{6} \text{rad}$ (b) $\dfrac{4\pi}{9} \text{rad}$ (c) $\dfrac{7\pi}{12} \text{rad}$

4. Convert to degrees and minutes:

    (a) 0.0125 rad (b) 2.69 rad
    (c) 7.241 rad

5. A car engine speed is 1000 rev/min. Convert this speed into rad/s.

28.4 Arc length and area of circles and sectors

Arc length

From the definition of the radian in the previous section and Figure 28.7,

arc length, $s = r\theta$ where θ is in radians

Area of a circle

From Chapter 27, for any circle, area $= \pi \times (\text{radius})^2$

i.e. $\mathbf{area = \pi r^2}$

Since $r = \dfrac{d}{2}$, $\mathbf{area = \pi r^2}$ or $\dfrac{\boldsymbol{\pi d^2}}{\mathbf{4}}$

Area of a sector

$$\mathbf{Area\ of\ a\ sector = \frac{\theta}{360}(\pi r^2)} \text{ when } \theta \text{ is in degrees}$$

$$= \frac{\theta}{2\pi}(\pi r^2)$$

$$= \frac{1}{2}r^2\theta \quad \text{when } \theta \text{ is in radians}$$

Problem 8. A hockey pitch has a semicircle of radius 14.63 m around each goal net. Find the area enclosed by the semicircle, correct to the nearest square metre

---

Area of a semicircle $= \dfrac{1}{2}\pi r^2$

When $r = 14.63\,\text{m}$, area $= \dfrac{1}{2}\pi (14.63)^2$

i.e. **area of semicircle $= 336\,\text{m}^2$**

Problem 9. Find the area of a circular metal plate having a diameter of 35.0 mm, correct to the nearest square millimetre

---

Area of a circle $= \pi r^2 = \dfrac{\pi d^2}{4}$

When $d = 35.0\,\text{mm}$, area $= \dfrac{\pi (35.0)^2}{4}$

i.e. **area of circular plate $= 962\,\text{mm}^2$**

Problem 10. Find the area of a circle having a circumference of 60.0 mm

Circumference, $c = 2\pi r$

from which radius, $r = \dfrac{c}{2\pi} = \dfrac{60.0}{2\pi} = \dfrac{30.0}{\pi}$

Area of a circle $= \pi r^2$

i.e. **area** $= \pi \left(\dfrac{30.0}{\pi}\right)^2 = \mathbf{286.5\,mm^2}$

Problem 11. Find the length of the arc of a circle of radius 5.5 cm when the angle subtended at the centre is 1.20 radians

Length of arc, $s = r\theta$, where θ is in radians.
Hence, **arc length,** $s = (5.5)(1.20) = \mathbf{6.60\,cm}$

Problem 12. Determine the diameter and circumference of a circle if an arc of length 4.75 cm subtends an angle of 0.91 radians

Since arc length, $s = r\theta$ then

radius, $r = \dfrac{s}{\theta} = \dfrac{4.75}{0.91} = 5.22$ cm

Diameter $= 2 \times$ radius $= 2 \times 5.22 = \mathbf{10.44\,cm}$

Circumference, $c = \pi d = \pi(10.44) = \mathbf{32.80\,cm}$

Problem 13. If an angle of $125°$ is subtended by an arc of a circle of radius 8.4 cm, find the length of (a) the minor arc and (b) the major arc, correct to 3 significant figures

Since $180° = \pi$ rad then $1° = \left(\dfrac{\pi}{180}\right)$ rad and

$125° = 125\left(\dfrac{\pi}{180}\right)$ rad

(a) Length of minor arc,

$s = r\theta = (8.4)(125)\left(\dfrac{\pi}{180}\right) = \mathbf{18.3\,cm}$,

correct to 3 significant figures

(b) Length of major arc $=$ (circumference $-$ minor arc) $= 2\pi(8.4) - 18.3 = \mathbf{34.5\,cm}$, correct to 3 significant figures.
(Alternatively, major arc $= r\theta$
$= 8.4(360 - 125)\left(\dfrac{\pi}{180}\right) = \mathbf{34.5\,cm}$)

Problem 14. Determine the angle, in degrees and minutes, subtended at the centre of a circle of diameter 42 mm by an arc of length 36 mm. Calculate also the area of the minor sector formed

Since length of arc, $s = r\theta$ then $\theta = \dfrac{s}{r}$

Radius, $r = \dfrac{\text{diameter}}{2} = \dfrac{42}{2} = 21$ mm,

hence $\theta = \dfrac{s}{r} = \dfrac{36}{21} = 1.7143$ radians.

$1.7143\,\text{rad} = 1.7143 \times \left(\dfrac{180}{\pi}\right)^{\circ} = 98.22° = \mathbf{98°\,13'} =$ angle subtended at centre of circle.

From equation (2),

$$\textbf{area of sector} = \frac{1}{2}r^2\theta = \frac{1}{2}(21)^2(1.7143)$$
$$= \mathbf{378\,mm^2}$$

Problem 15. A football stadium floodlight can spread its illumination over an angle of $45°$ to a distance of 55 m. Determine the maximum area that is floodlit.

$\textbf{Floodlit area} = \text{area of sector} = \dfrac{1}{2}r^2\theta$

$= \dfrac{1}{2}(55)^2\left(45 \times \dfrac{\pi}{180}\right)$

$= \mathbf{1188\,m^2}$

Problem 16. An automatic garden sprayer produces spray to a distance of 1.8 m and revolves through an angle α which may be varied. If the desired spray catchment area is to be 2.5 m², to what should angle α be set, correct to the nearest degree?

Area of sector $= \dfrac{1}{2}r^2\theta$, hence $2.5 = \dfrac{1}{2}(1.8)^2\alpha$

from which, $\alpha = \dfrac{2.5 \times 2}{1.8^2} = 1.5432$ radians

$1.5432\,\text{rad} = \left(1.5432 \times \dfrac{180}{\pi}\right)^{\circ} = 88.42°$

Hence, **angle** $\alpha = \mathbf{88°}$, correct to the nearest degree.

Problem 17. The angle of a tapered groove is checked using a 20 mm diameter roller as shown in

Section C

Figure 28.8. If the roller lies 2.12 mm below the top of the groove, determine the value of angle θ

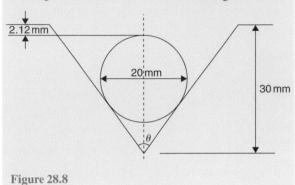

Figure 28.8

In Figure 28.9, triangle *ABC* is right-angled at *C* (see property (g) in Section 28.2).

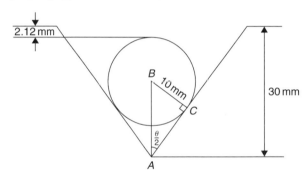

Figure 28.9

Length $BC = 10$ mm (i.e. the radius of the circle), and $AB = 30 - 10 - 2.12 = 17.88$ mm, from Figure 28.9.

Hence, $\sin\dfrac{\theta}{2} = \dfrac{10}{17.88}$ and $\dfrac{\theta}{2} = \sin^{-1}\left(\dfrac{10}{17.88}\right) = 34°$

and **angle $\theta = 68°$**

Now try the following Practice Exercise

Practice Exercise 124   Arc length and area of circles and sectors (answers on page 1162)

1. Calculate the area of a circle of radius 6.0 cm, correct to the nearest square centimetre.

2. The diameter of a circle is 55.0 mm. Determine its area, correct to the nearest square millimetre.

3. The perimeter of a circle is 150 mm. Find its area, correct to the nearest square millimetre.

4. Find the area of the sector, correct to the nearest square millimetre, of a circle having a radius of 35 mm with angle subtended at centre of 75°.

5. An annulus has an outside diameter of 49.0 mm and an inside diameter of 15.0 mm. Find its area correct to 4 significant figures.

6. Find the area, correct to the nearest square metre, of a 2 m wide path surrounding a circular plot of land 200 m in diameter.

7. A rectangular park measures 50 m by 40 m. A 3 m flower bed is made round the two longer sides and one short side. A circular fish pond of diameter 8.0 m is constructed in the centre of the park. It is planned to grass the remaining area. Find, correct to the nearest square metre, the area of grass.

8. With reference to Figure 28.10, determine (a) the perimeter and (b) the area.

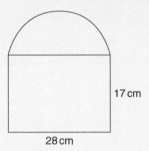

Figure 28.10

9. Find the area of the shaded portion of Figure 28.11.

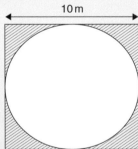

Figure 28.11

10. Find the length of an arc of a circle of radius 8.32 cm when the angle subtended at the centre is 2.14 radians. Calculate also the area of the minor sector formed.

11. If the angle subtended at the centre of a circle of diameter 82 mm is 1.46 rad, find the lengths of the (a) minor arc and (b) major arc.

12. A pendulum of length 1.5 m swings through an angle of 10° in a single swing. Find, in centimetres, the length of the arc traced by the pendulum bob.

13. Determine the shaded area of the section shown in Figure 28.12.

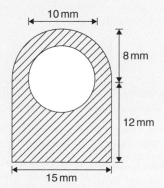

Figure 28.12

14. Determine the length of the radius and circumference of a circle if an arc length of 32.6 cm subtends an angle of 3.76 radians.

15. Determine the angle of lap, in degrees and minutes, if 180 mm of a belt drive are in contact with a pulley of diameter 250 mm.

16. Determine the number of complete revolutions a motorcycle wheel will make in travelling 2 km if the wheel's diameter is 85.1 cm.

17. A floodlight at a sports ground spreads its illumination over an angle of 40° to a distance of 48 m. Determine (a) the angle in radians and (b) the maximum area that is floodlit.

18. Find the area swept out in 50 minutes by the minute hand of a large floral clock if the hand is 2 m long.

19. Determine (a) the shaded area in Figure 28.13 and (b) the percentage of the whole sector that the shaded area represents.

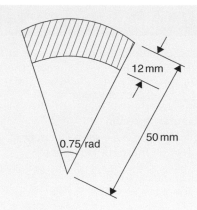

Figure 28.13

20. Determine the length of steel strip required to make the clip shown in Figure 28.14.

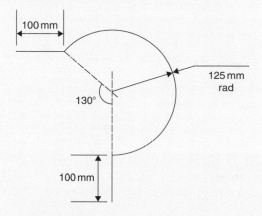

Figure 28.14

21. A 50° tapered hole is checked with a 40 mm diameter ball as shown in Figure 28.15. Determine the length shown as x.

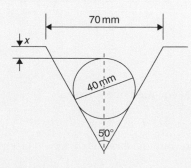

Figure 28.15

28.5 The equation of a circle

The simplest equation of a circle, centre at the origin and radius r, is given by

$$x^2 + y^2 = r^2$$

from Pythagoras' theorem
For example, Figure 28.16 shows a circle $x^2 + y^2 = 9$
i.e. $x^2 + y^2 = 3^2$ from Pythagoras

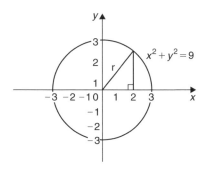

Figure 28.16

More generally, the equation of a circle, centre (a, b) and radius r, is given by

$$(x - a)^2 + (y - b)^2 = r^2 \qquad (1)$$

Figure 28.17 shows a circle $(x - 2)^2 + (y - 3)^2 = 4$

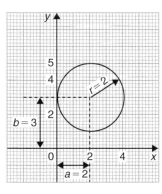

Figure 28.17

The general equation of a circle is

$$x^2 + y^2 + 2ex + 2fy + c = 0 \qquad (2)$$

Multiplying out the bracketed terms in equation (1) gives

$$x^2 - 2ax + a^2 + y^2 - 2by + b^2 = r^2$$

Comparing this with equation (2) gives

$$2e = -2a, \text{ i.e. } a = -\frac{2e}{2}$$

and $$2f = -2b, \text{ i.e. } b = -\frac{2f}{2}$$

and $$c = a^2 + b^2 - r^2, \text{ i.e. } r = \sqrt{a^2 + b^2 - c}$$

Thus, for example, the equation

$$x^2 + y^2 - 4x - 6y + 9 = 0$$

represents a circle with centre,

$$a = -\left(\frac{-4}{2}\right), b = -\left(\frac{-6}{2}\right) \text{ i.e. at } (2, 3) \text{ and}$$

radius, $$r = \sqrt{2^2 + 3^2 - 9} = 2$$

Hence, $x^2 + y^2 - 4x - 6y + 9 = 0$ is the circle shown in Figure 28.17 (which may be checked by multiplying out the brackets in the equation $(x - 2)^2 + (y - 3)^2 = 4$).

Problem 18. Determine (a) the radius and (b) the co-ordinates of the centre of the circle given by the equation $x^2 + y^2 + 8x - 2y + 8 = 0$

$x^2 + y^2 + 8x - 2y + 8 = 0$ is of the form shown in equation (2),

where $$a = -\left(\frac{8}{2}\right) = -4, b = -\left(\frac{-2}{2}\right) = 1$$

and $$r = \sqrt{(-4)^2 + 1^2 - 8} = \sqrt{9} = 3$$

Hence, $x^2 + y^2 + 8x - 2y + 8 = 0$ represents a circle **centre $(-4, 1)$** and **radius 3**, as shown in Figure 28.18.

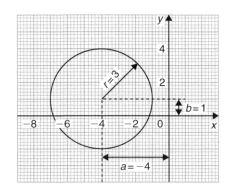

Figure 28.18

Alternatively, $x^2 + y^2 + 8x - 2y + 8 = 0$ may be rearranged as

$$(x + 4)^2 + (y - 1)^2 - 9 = 0$$

i.e. $\qquad (x + 4)^2 + (y - 1)^2 = 3^2$

which represents a circle, **centre (−4, 1)** and **radius 3**, as stated above.

> **Problem 19.** Sketch the circle given by the equation $x^2 + y^2 - 4x + 6y - 3 = 0$

The equation of a circle, centre (a, b), radius r is given by

$$(x - a)^2 + (y - b)^2 = r^2$$

The general equation of a circle is
$x^2 + y^2 + 2ex + 2fy + c = 0$

From above $a = -\dfrac{2e}{2}, b = -\dfrac{2f}{2}$ and $r = \sqrt{a^2 + b^2 - c}$

Hence, if $x^2 + y^2 - 4x + 6y - 3 = 0$

then $\qquad a = -\left(\dfrac{-4}{2}\right) = 2, b = -\left(\dfrac{6}{2}\right) = -3 \qquad$ and

$r = \sqrt{2^2 + (-3)^2 - (-3)} = \sqrt{16} = 4$

Thus, the circle has **centre (2, −3)** and **radius 4**, as shown in Figure 28.19.

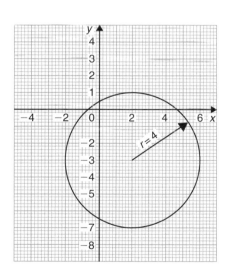

Figure 28.19

Alternatively, $x^2 + y^2 - 4x + 6y - 3 = 0$ may be rearranged as

$$(x - 2)^2 + (y + 3)^2 - 3 - 13 = 0$$

i.e. $\qquad (x - 2)^2 + (y + 3)^2 = 4^2$

which represents a circle, **centre (2, −3)** and **radius 4**, as stated above.

Now try the following Practice Exercise

> **Practice Exercise 125   The equation of a circle (answers on page 1163)**
>
> 1. Determine (a) the radius and (b) the co-ordinates of the centre of the circle given by the equation $x^2 + y^2 - 6x + 8y + 21 = 0$
>
> 2. Sketch the circle given by the equation $x^2 + y^2 - 6x + 4y - 3 = 0$
>
> 3. Sketch the curve $x^2 + (y - 1)^2 - 25 = 0$
>
> 4. Sketch the curve $x = 6\sqrt{\left[1 - \left(\dfrac{y}{6}\right)^2\right]}$

28.6   Linear and angular velocity

Linear velocity

Linear velocity v is defined as the rate of change of linear displacement s with respect to time t. For motion in a straight line:

$$\text{linear velocity} = \frac{\text{change of displacement}}{\text{change of time}}$$

i.e. $\qquad v = \dfrac{s}{t} \qquad (1)$

The unit of linear velocity is metres per second (m/s).

Angular velocity

The speed of revolution of a wheel or a shaft is usually measured in revolutions per minute or revolutions per second but these units do not form part of a coherent system of units. The basis in SI units is the angle turned through in one second.

Angular velocity is defined as the rate of change of angular displacement θ, with respect to time t. For an object rotating about a fixed axis at a constant speed:

$$\text{angular velocity} = \frac{\text{angle turned through}}{\text{time taken}}$$

i.e. $\qquad \omega = \dfrac{\theta}{t} \qquad (2)$

The unit of angular velocity is radians per second (rad/s). An object rotating at a constant speed of n revolutions per second subtends an angle of $2\pi n$

Section C

radians in one second, i.e. its angular velocity ω is given by:

$$\omega = 2\pi n \text{ rad/s} \qquad (3)$$

From page 292, $s = r\theta$ and from equation (2) above, $\theta = \omega t$

hence
$$s = r(\omega t)$$

from which
$$\frac{s}{t} = \omega r$$

However, from equation (1) $v = \dfrac{s}{t}$

hence
$$v = \omega r \qquad (4)$$

Equation (4) gives the relationship between linear velocity v and angular velocity ω.

Problem 20. A wheel of diameter 540 mm is rotating at $\dfrac{1500}{\pi}$ rev/min. Calculate the angular velocity of the wheel and the linear velocity of a point on the rim of the wheel

---

From equation (3), angular velocity $\omega = 2\pi n$ where n is the speed of revolution in rev/s. Since in this case

$$n = \frac{1500}{\pi} \text{ rev/min} = \frac{1500}{60\pi} = \text{rev/s, then}$$

$$\textbf{angular velocity } \omega = 2\pi\left(\frac{1500}{60\pi}\right) = \textbf{50 rad/s}$$

The linear velocity of a point on the rim, $v = \omega r$, where r is the radius of the wheel, i.e.

$$\frac{540}{2} \text{ mm} = \frac{0.54}{2} \text{ m} = 0.27 \text{ m}$$

Thus **linear velocity** $v = \omega r = (50)(0.27)$

$$= \textbf{13.5 m/s}$$

Problem 21. A car is travelling at 64.8 km/h and has wheels of diameter 600 mm.

(a) Find the angular velocity of the wheels in both rad/s and rev/min

(b) If the speed remains constant for 1.44 km, determine the number of revolutions made by the wheel, assuming no slipping occurs

---

(a) Linear velocity $v = 64.8$ km/h

$$= 64.8 \frac{\text{km}}{\text{h}} \times 1000 \frac{\text{m}}{\text{km}} \times \frac{1}{3600} \frac{\text{h}}{\text{s}} = 18 \text{ m/s}$$

The radius of a wheel $= \dfrac{600}{2} = 300$ mm

$$= 0.3 \text{ m}$$

From equation (4), $v = \omega r$, from which,

$$\textbf{angular velocity } \omega = \frac{v}{r} = \frac{18}{0.3}$$

$$= \textbf{60 rad/s}$$

From equation (3), angular velocity, $\omega = 2\pi n$, where n is in rev/s.

Hence angular speed $n = \dfrac{\omega}{2\pi} = \dfrac{60}{2\pi}$ rev/s

$$= 60 \times \frac{60}{2\pi} \text{ rev/min}$$

$$= \textbf{573 rev/min}$$

(b) From equation (1), since $v = s/t$ then the time taken to travel 1.44 km, i.e. 1440 m at a constant speed of 18 m/s is given by:

$$\text{time } t = \frac{s}{v} = \frac{1440 \text{ m}}{18 \text{ m/s}} = 80 \text{ s}$$

Since a wheel is rotating at 573 rev/min, then in 80/60 minutes it makes

$$573 \text{ rev/min} \times \frac{80}{60} \text{ min} = \textbf{764 revolutions}$$

Now try the following Practice Exercise

Practice Exercise 126   Further problems on linear and angular velocity (answers on page 1163)

1. A pulley driving a belt has a diameter of 300 mm and is turning at $2700/\pi$ revolutions per minute. Find the angular velocity of the pulley and the linear velocity of the belt assuming that no slip occurs.

2. A bicycle is travelling at 36 km/h and the diameter of the wheels of the bicycle is 500 mm. Determine the linear velocity of a point on the rim of one of the wheels of the bicycle, and the angular velocity of the wheels.

3. A train is travelling at 108 km/h and has wheels of diameter 800 mm.

(a) Determine the angular velocity of the wheels in both rad/s and rev/min.

(b) If the speed remains constant for 2.70 km, determine the number of revolutions made by a wheel, assuming no slipping occurs.

28.7 Centripetal force

When an object moves in a circular path at constant speed, its direction of motion is continually changing and hence its velocity (which depends on both magnitude and direction) is also continually changing. Since acceleration is the (change in velocity)/(time taken), the object has an acceleration. Let the object be moving with a constant angular velocity of ω and a tangential velocity of magnitude v and let the change of velocity for a small change of angle of θ ($=\omega t$) be V in Figure 28.20. Then $v_2 - v_1 = V$. The vector diagram is shown in Figure 28.20(b) and since the magnitudes of v_1 and v_2 are the same, i.e. v, the vector diagram is an isosceles triangle.

Bisecting the angle between v_2 and v_1 gives:

$$\sin\frac{\theta}{2} = \frac{V/2}{v_2} = \frac{V}{2v}$$

i.e. $V = 2v\sin\dfrac{\theta}{2}$ (1)

Since $\theta = \omega t$ then

$$t = \frac{\theta}{\omega} \qquad (2)$$

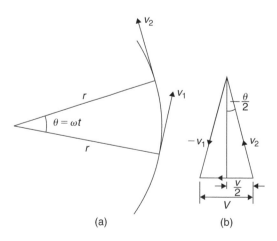

(a)   (b)

Figure 28.20

Dividing equation (1) by equation (2) gives:

$$\frac{V}{t} = \frac{2v\sin(\theta/2)}{(\theta/\omega)} = \frac{v\omega\sin(\theta/2)}{(\theta/2)}$$

For small angles $\dfrac{\sin(\theta/2)}{(\theta/2)} \approx 1$, (see page 455, Section 38.9)

hence $\dfrac{V}{t} = \dfrac{\text{change of velocity}}{\text{change of time}}$

$$= \text{acceleration } a = v\omega$$

However, $\omega = \dfrac{v}{r}$ (from Section 28.6)

thus $v\omega = v \cdot \dfrac{v}{r} = \dfrac{v^2}{r}$

i.e. the acceleration a is $\dfrac{v^2}{r}$ and is towards the centre of the circle of motion (along V). It is called the **centripetal acceleration**. If the mass of the rotating object is m, then by Newton's second law, the **centripetal force is** $\dfrac{mv^2}{r}$ and its direction is towards the centre of the circle of motion.

Problem 22. A vehicle of mass 750 kg travels around a bend of radius 150 m, at 50.4 km/h. Determine the centripetal force acting on the vehicle

The centripetal force is given by $\dfrac{mv^2}{r}$ and its direction is towards the centre of the circle.

Mass $m = 750$ kg, $v = 50.4$ km/h

$$= \frac{50.4 \times 1000}{60 \times 60} \text{ m/s}$$

$$= 14 \text{ m/s}$$

and radius $r = 150$ m,

thus **centripetal force** $= \dfrac{750(14)^2}{150} = \mathbf{980\,N}$

Problem 23. An object is suspended by a thread 250 mm long and both object and thread move in a horizontal circle with a constant angular velocity of 2.0 rad/s. If the tension in the thread is 12.5 N, determine the mass of the object

Centripetal force (i.e. tension in thread),

$$F = \frac{mv^2}{r} = 12.5\,\text{N}$$

Angular velocity $\omega = 2.0$ rad/s and radius $r = 250$ mm $= 0.25$ m

Since linear velocity $v = \omega r$, $v = (2.0)(0.25) = 0.5$ m/s

Since $F = \dfrac{mv^2}{r}$, then mass $m = \dfrac{Fr}{v^2}$,

i.e. **mass of object,** $m = \dfrac{(12.5)(0.25)}{0.5^2} = \mathbf{12.5\ kg}$

Problem 24. An aircraft is turning at constant altitude, the turn following the arc of a circle of radius 1.5 km. If the maximum allowable acceleration of the aircraft is 2.5 g, determine the maximum speed of the turn in km/h. Take g as 9.8 m/s^2

The acceleration of an object turning in a circle is $\dfrac{v^2}{r}$. Thus, to determine the maximum speed of turn,

$\dfrac{v^2}{r} = 2.5\,g$, from which,

$$\textbf{velocity, } v = \sqrt{(2.5\,gr)} = \sqrt{(2.5)(9.8)(1500)}$$

$$= \sqrt{36750} = 191.7\,\text{m/s}$$

and $191.7\,\text{m/s} = 191.7 \times \dfrac{60 \times 60}{1000}\,\text{km/h} = \mathbf{690\,km/h}$

Now try the following Practice Exercise

Practice Exercise 127   Further problems on centripetal force (answers on page 1163)

1. Calculate the tension in a string when it is used to whirl a stone of mass 200 g round in a horizontal circle of radius 90 cm with a constant speed of 3 m/s.

2. Calculate the centripetal force acting on a vehicle of mass 1 tonne when travelling around a bend of radius 125 m at 40 km/h. If this force should not exceed 750 N, determine the reduction in speed of the vehicle to meet this requirement.

3. A speed-boat negotiates an S-bend consisting of two circular arcs of radii 100 m and 150 m. If the speed of the boat is constant at 34 km/h, determine the change in acceleration when leaving one arc and entering the other.

**For fully worked solutions to each of the problems in Practice Exercises 122 to 127 in this chapter,
go to the website:**
www.routledge.com/cw/bird

This assignment covers the material contained in Chapters 27 and 28. *The marks available are shown in brackets at the end of each question.*

1. A rectangular metal plate has an area of 9600 cm². If the length of the plate is 1.2 m, calculate the width, in centimetres.   (3)

2. Calculate the cross-sectional area of the angle iron section shown in Figure RT11.1, the dimensions being in millimetres.   (4)

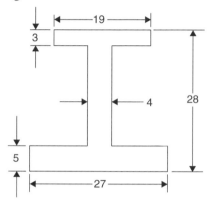

Figure RT11.1

3. Find the area of the trapezium *MNOP* shown in Figure RT11.2 when $a = 6.3$ cm, $b = 11.7$ cm and $h = 5.5$ cm.   (3)

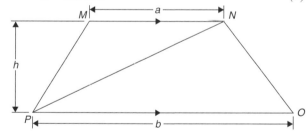

Figure RT11.2

4. Find the area of the triangle *DEF* in Figure RT11.3, correct to 2 decimal places.   (4)

5. A rectangular park measures 150 m by 70 m. A 2 m flower border is constructed round the two longer sides and one short side. A circular fish pond of diameter 15 m is in the centre of the park and the remainder of the park is grass. Calculate, correct to the nearest square metre, the area of (a) the fish pond, (b) the flower borders and (c) the grass.   (6)

6. A swimming pool is 55 m long and 10 m wide. The perpendicular depth at the deep end is 5 m and at

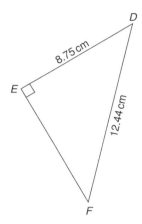

Figure RT11.3

the shallow end is 1.5 m, the slope from one end to the other being uniform. The inside of the pool needs two coats of a protective paint before it is filled with water. Determine how many litres of paint will be needed if 1 litre covers 10 m².   (7)

7. Find the area of an equilateral triangle of side 20.0 cm.   (4)

8. A steel template is of the shape shown in Figure RT11.4, the circular area being removed. Determine the area of the template, in square centimetres, correct to 1 decimal place.   (8)

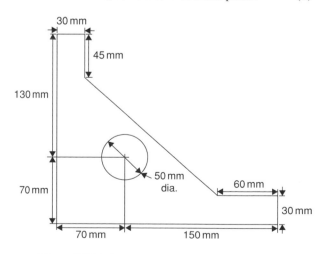

Figure RT11.4

9. The area of a plot of land on a map is 400 mm². If the scale of the map is 1 to 50 000,

Section C

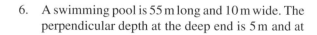

determine the true area of the land in hectares (1 hectare = 10^4m^2). (4)

10. Determine the shaded area in Figure RT11.5, correct to the nearest square centimetre. (3)

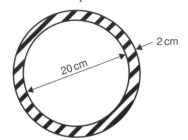

Figure RT11.5

11. Determine the diameter of a circle, correct to the nearest millimetre, whose circumference is 178.4 cm. (2)

12. Calculate the area of a circle of radius 6.84 cm, correct to 1 decimal place. (2)

13. The circumference of a circle is 250 mm. Find its area, correct to the nearest square millimetre. (4)

14. Find the area of the sector of a circle having a radius of 50.0 mm, with angle subtended at centre of 120°. (3)

15. Determine the total area of the shape shown in Figure RT11.6, correct to 1 decimal place. (7)

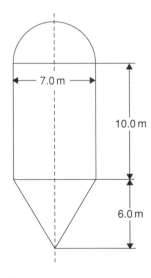

Figure RT11.6

16. The radius of a circular cricket ground is 75 m. The boundary is painted with white paint and 1 tin of paint will paint a line 22.5 m long. How many tins of paint are needed? (3)

17. Find the area of a 1.5 m wide path surrounding a circular plot of land 100 m in diameter. (3)

18. A cyclometer shows 2530 revolutions in a distance of 3.7 km. Find the diameter of the wheel in centimetres, correct to 2 decimal places. (4)

19. The minute hand of a wall clock is 10.5 cm long. How far does the tip travel in the course of 24 hours? (4)

20. Convert
    (a) 125°47′ to radians.
    (b) 1.724 radians to degrees and minutes. (4)

21. Calculate the length of metal strip needed to make the clip shown in Figure RT11.7. (7)

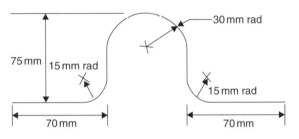

Figure RT11.7

22. A lorry has wheels of radius 50 cm. Calculate the number of complete revolutions a wheel makes (correct to the nearest revolution) when travelling 3 miles (assume 1 mile = 1.6 km). (4)

23. The equation of a circle is
    $x^2 + y^2 + 12x - 4y + 4 = 0$. Determine
    (a) the diameter of the circle.
    (b) the co-ordinates of the centre of the circle. (7)

For lecturers/instructors/teachers, fully worked solutions to each of the problems in Revision Test 11, together with a full marking scheme, are available at the website:
www.routledge.com/cw/bird

Section C

Chapter 29

Volumes and surface areas of common solids

Why it is important to understand: Volumes of common solids

There are many practical applications where volumes and surface areas of common solids are required. Examples include determining capacities of oil, water, petrol and fish tanks, ventilation shafts and cooling towers, determining volumes of blocks of metal, ball-bearings, boilers and buoys, and calculating the cubic metres of concrete needed for a path. Finding the surface areas of loudspeaker diaphragms and lampshades provide further practical examples. Understanding these calculations is essential for the many practical applications in engineering, construction, architecture and science.

At the end of this chapter, you should be able to:

- state the SI unit of volume
- calculate the volumes and surface areas of cuboids, cylinders, prisms, pyramids, cones and spheres
- calculate volumes and surface areas of frusta of pyramids and cones
- calculate the frustum and zone of a sphere
- calculate volumes of regular solids using the prismoidal rule
- appreciate that volumes of similar bodies are proportional to the cubes of the corresponding linear dimensions

29.1 Introduction

The **volume** of any solid is a measure of the space occupied by the solid. Volume is measured in **cubic units** such as mm^3, cm^3 and m^3.

This chapter deals with finding volumes of common solids; in engineering it is often important to be able to calculate volume or capacity to estimate, say, the amount of liquid, such as water, oil or petrol, in different shaped containers.

A **prism** is a solid with a constant cross-section and with two ends parallel. The shape of the end is used to describe the prism. For example, there are rectangular prisms (called cuboids), triangular prisms and circular prisms (called cylinders).

On completing this chapter you will be able to calculate the volumes and surface areas of rectangular and other prisms, cylinders, pyramids, cones and spheres, together with frusta of pyramids and cones. Volumes of similar shapes are also considered.

29.2 Volumes and surface areas of common shapes

Cuboids or rectangular prisms

A cuboid is a solid figure bounded by six rectangular faces; all angles are right angles and opposite faces are equal. A typical cuboid is shown in Figure 29.1 with length l, breadth b and height h.

$$\textbf{Volume of cuboid} = l \times b \times h$$

and

$$\textbf{surface area} = 2bh + 2hl + 2lb = 2(bh + hl + lb)$$

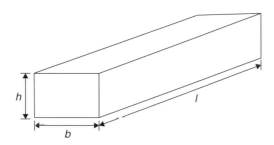

Figure 29.1

A **cube** is a square prism. If all the sides of a cube are x then

$$\textbf{Volume} = x^3 \textbf{ and surface area} = 6x^2$$

Problem 1. A cuboid has dimensions of 12 cm by 4 cm by 3 cm. Determine (a) its volume and (b) its total surface area

The cuboid is similar to that in Figure 29.1, with $l = 12$ cm, $b = 4$ cm and $h = 3$ cm.

(a) **Volume of cuboid** $= l \times b \times h = 12 \times 4 \times 3$
$$= \textbf{144 cm}^3$$

(b) **Surface area** $= 2(bh + hl + lb)$
$$= 2(4 \times 3 + 3 \times 12 + 12 \times 4)$$
$$= 2(12 + 36 + 48)$$
$$= 2 \times 96 = \textbf{192 cm}^2$$

Problem 2. An oil tank is the shape of a cube, each edge being of length 1.5 m. Determine (a) the maximum capacity of the tank in m^3 and litres and (b) its total surface area ignoring input and output orifices

(a) **Volume of oil tank** = volume of cube
$$= 1.5 \text{ m} \times 1.5 \text{ m} \times 1.5 \text{ m}$$
$$= 1.5^3 \text{ m}^3 = \textbf{3.375 m}^3$$

$1 \text{ m}^3 = 100 \text{ cm} \times 100 \text{ cm} \times 100 \text{ cm} = 10^6 \text{ cm}^3$. Hence,
$$\text{volume of tank} = 3.375 \times 10^6 \text{ cm}^3$$

1 litre $= 1000 \text{ cm}^3$, hence **oil tank capacity**
$$= \frac{3.375 \times 10^6}{1000} \text{ litres} = \textbf{3375 litres}$$

(b) Surface area of one side $= 1.5 \text{ m} \times 1.5 \text{ m}$
$$= 2.25 \text{ m}^2$$

A cube has six identical sides, hence

$$\textbf{total surface area of oil tank} = \textbf{6} \times \textbf{2.25}$$
$$= \textbf{13.5 m}^2$$

Problem 3. A water tank is the shape of a rectangular prism having length 2 m, breadth 75 cm and height 500 mm. Determine the capacity of the tank in (a) m^3 (b) cm^3 (c) litres

Capacity means volume; when dealing with liquids, the word capacity is usually used.

The water tank is similar in shape to that in Figure 29.1, with $l = 2$ m, $b = 75$ cm and $h = 500$ mm.

(a) Capacity of water tank $= l \times b \times h$. To use this formula, all dimensions **must** be in the same units. Thus, $l = 2$ m, $b = 0.75$ m and $h = 0.5$ m (since 1 m = 100 cm = 1000 mm). Hence,

$$\textbf{capacity of tank} = 2 \times 0.75 \times 0.5 = \textbf{0.75 m}^3$$

(b) $1 \text{ m}^3 = 1 \text{ m} \times 1 \text{ m} \times 1 \text{ m}$
$$= 100 \text{ cm} \times 100 \text{ cm} \times 100 \text{ cm}$$

i.e. $\textbf{1 m}^3 = \textbf{1 000 000} = \textbf{10}^6 \textbf{ cm}^3$. Hence,
$$\textbf{capacity} = 0.75 \text{ m}^3 = 0.75 \times 10^6 \text{ cm}^3$$
$$= \textbf{750 000 cm}^3$$

(c) 1 litre $= 1000 \text{ cm}^3$. Hence,
$$\textbf{750 000 cm}^3 = \frac{750 000}{1000} = \textbf{750 litres}$$

Cylinders

A cylinder is a circular prism. A cylinder of radius r and height h is shown in Figure 29.2.

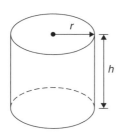

Figure 29.2

$$\textbf{Volume} = \pi r^2 h$$

$$\textbf{Curved surface area} = 2\pi r h$$

$$\textbf{Total surface area} = 2\pi r h + 2\pi r^2$$

Total surface area means the curved surface area plus the area of the two circular ends.

> **Problem 4.**   A solid cylinder has a base diameter of 12 cm and a perpendicular height of 20 cm. Calculate (a) the volume and (b) the total surface area

(a) $\textbf{Volume} = \pi r^2 h = \pi \times \left(\dfrac{12}{2}\right)^2 \times 20$

$$= 720\pi = \textbf{2262 cm}^3$$

(b)   **Total surface area**

$$= 2\pi r h + 2\pi r^2$$

$$= (2 \times \pi \times 6 \times 20) + (2 \times \pi \times 6^2)$$

$$= 240\pi + 72\pi = 312\pi = \textbf{980 cm}^2$$

> **Problem 5.**   A copper pipe has the dimensions shown in Figure 29.3. Calculate the volume of copper in the pipe, in cubic metres

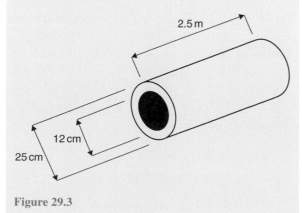

Figure 29.3

Outer diameter, $D = 25\,\text{cm} = 0.25\,\text{m}$ and inner diameter, $d = 12\,\text{cm} = 0.12\,\text{m}$.

Area of cross-section of copper

$$= \frac{\pi D^2}{4} - \frac{\pi d^2}{4} = \frac{\pi (0.25)^2}{4} - \frac{\pi (0.12)^2}{4}$$

$$= 0.0491 - 0.0113 = 0.0378\,\text{m}^2$$

Hence, **volume of copper**

$$= (\text{cross-sectional area}) \times \text{length of pipe}$$

$$= 0.0378 \times 2.5 = \textbf{0.0945 m}^3$$

More prisms

A right-angled triangular prism is shown in Figure 29.4 with dimensions b, h and l.

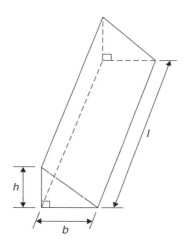

Figure 29.4

$$\textbf{Volume} = \frac{1}{2}bhl$$

and

$$\textbf{surface area} = \textbf{area of each end}$$

$$\textbf{+ area of three sides}$$

Notice that the volume is given by the area of the end (i.e. area of triangle $= \frac{1}{2}bh$) multiplied by the length l. In fact, the volume of any shaped prism is given by the area of an end multiplied by the length.

> **Problem 6.**   Determine the volume (in cm^3) of the shape shown in Figure 29.5

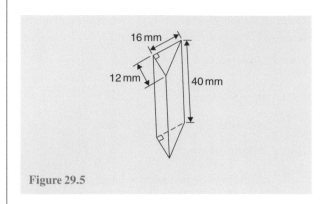

Figure 29.5

The solid shown in Figure 29.5 is a triangular prism. The volume V of any prism is given by $V = Ah$, where A is the cross-sectional area and h is the perpendicular height. Hence,

$$\textbf{volume} = \frac{1}{2} \times 16 \times 12 \times 40 = 3840\,\text{mm}^3$$

$$= \textbf{3.840}\,\textbf{cm}^3$$

$$(\text{since } 1\,\text{cm}^3 = 1000\,\text{mm}^3)$$

Problem 7. Calculate the volume of the right-angled triangular prism shown in Figure 29.6. Also, determine its total surface area

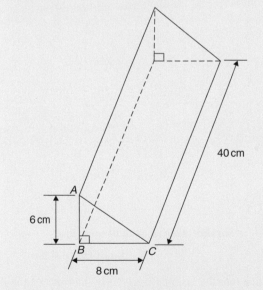

Figure 29.6

Volume of right-angled triangular prism

$$= \frac{1}{2}bhl = \frac{1}{2} \times 8 \times 6 \times 40$$

i.e.   **volume = 960 cm³**

Total surface area = area of each end + area of three sides.

In triangle ABC, $AC^2 = AB^2 + BC^2$

from which, $AC = \sqrt{AB^2 + BC^2} = \sqrt{6^2 + 8^2}$

$$= 10\,\text{cm}$$

Hence, total surface area

$$= 2\left(\frac{1}{2}bh\right) + (AC \times 40) + (BC \times 40) + (AB \times 40)$$

$$= (8 \times 6) + (10 \times 40) + (8 \times 40) + (6 \times 40)$$

$$= 48 + 400 + 320 + 240$$

i.e.   **total surface area = 1008 cm²**

Problem 8. Calculate the volume and total surface area of the solid prism shown in Figure 29.7

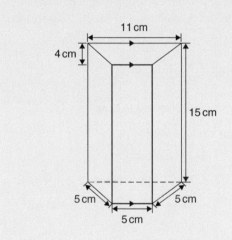

Figure 29.7

The solid shown in Figure 29.7 is a **trapezoidal prism**.

Volume of prism = cross-sectional area × height

$$= \frac{1}{2}(11 + 5)4 \times 15 = 32 \times 15$$

$$= \textbf{480}\,\textbf{cm}^3$$

Surface area of prism

$$= \text{sum of two trapeziums} + 4 \text{ rectangles}$$

$$= (2 \times 32) + (5 \times 15) + (11 \times 15) + 2(5 \times 15)$$

$$= 64 + 75 + 165 + 150 = \textbf{454}\,\textbf{cm}^2$$

Now try the following Practice Exercise

Practice Exercise 128   Volumes and surface areas of common shapes (answers on page 1163)

1. Change a volume of $1\,200\,000\,cm^3$ to cubic metres.

2. Change a volume of $5000\,mm^3$ to cubic centimetres.

3. A metal cube has a surface area of $24\,cm^2$. Determine its volume.

4. A rectangular block of wood has dimensions of 40 mm by 12 mm by 8 mm. Determine
   (a) its volume, in cubic millimetres
   (b) its total surface area in square millimetres.

5. Determine the capacity, in litres, of a fish tank measuring 90 cm by 60 cm by 1.8 m, given $1\,litre = 1000\,cm^3$.

6. A rectangular block of metal has dimensions of 40 mm by 25 mm by 15 mm. Determine its volume in cm^3. Find also its mass if the metal has a density of $9\,g/cm^3$.

7. Determine the maximum capacity, in litres, of a fish tank measuring 50 cm by 40 cm by 2.5 m $(1\,litre = 1000\,cm^3)$.

8. Determine how many cubic metres of concrete are required for a 120 m long path, 150 mm wide and 80 mm deep.

9. A cylinder has a diameter 30 mm and height 50 mm. Calculate
   (a) its volume in cubic centimetres, correct to 1 decimal place
   (b) the total surface area in square centimetres, correct to 1 decimal place.

10. Find (a) the volume and (b) the total surface area of a right-angled triangular prism of length 80 cm and whose triangular end has a base of 12 cm and perpendicular height 5 cm.

11. A steel ingot whose volume is $2\,m^3$ is rolled out into a plate which is 30 mm thick and 1.80 m wide. Calculate the length of the plate in metres.

12. The volume of a cylinder is $75\,cm^3$. If its height is 9.0 cm, find its radius.

13. Calculate the volume of a metal tube whose outside diameter is 8 cm and whose inside diameter is 6 cm, if the length of the tube is 4 m.

14. The volume of a cylinder is $400\,cm^3$. If its radius is 5.20 cm, find its height. Also determine its curved surface area.

15. A cylinder is cast from a rectangular piece of alloy 5 cm by 7 cm by 12 cm. If the length of the cylinder is to be 60 cm, find its diameter.

16. Find the volume and the total surface area of a regular hexagonal bar of metal of length 3 m if each side of the hexagon is 6 cm.

17. A block of lead 1.5 m by 90 cm by 750 mm is hammered out to make a square sheet 15 mm thick. Determine the dimensions of the square sheet, correct to the nearest centimetre.

18. How long will it take a tap dripping at a rate of $800\,mm^3/s$ to fill a 3-litre can?

19. A cylinder is cast from a rectangular piece of alloy 5.20 cm by 6.50 cm by 19.33 cm. If the height of the cylinder is to be 52.0 cm, determine its diameter, correct to the nearest centimetre.

20. How much concrete is required for the construction of the path shown in Figure 29.8, if the path is 12 cm thick?

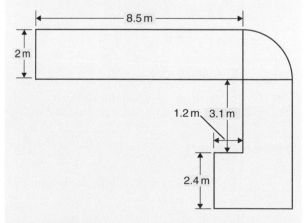

Figure 29.8

Pyramids

Volume of any pyramid

$$= \frac{1}{3} \times \text{area of base} \times \text{perpendicular height}$$

A square-based pyramid is shown in Figure 29.9 with base dimension x by x and the perpendicular height of the pyramid h. For the square-base pyramid shown,

$$\text{volume} = \frac{1}{3}x^2 h$$

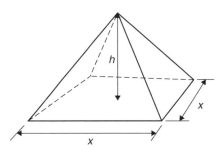

Figure 29.9

Problem 9.   A square pyramid has a perpendicular height of 16 cm. If a side of the base is 6 cm, determine the volume of a pyramid

Volume of pyramid

$$= \frac{1}{3} \times \text{area of base} \times \text{perpendicular height}$$

$$= \frac{1}{3} \times (6 \times 6) \times 16$$

$$= \mathbf{192\,cm^3}$$

Problem 10.   Determine the volume and the total surface area of the square pyramid shown in Figure 29.10 if its perpendicular height is 12 cm

Volume of pyramid

$$= \frac{1}{3}(\text{area of base}) \times \text{perpendicular height}$$

$$= \frac{1}{3}(5 \times 5) \times 12$$

$$= \mathbf{100\,cm^3}$$

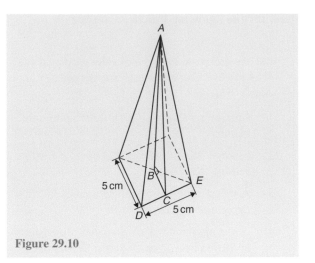

Figure 29.10

The total surface area consists of a square base and four equal triangles.

Area of triangle ADE

$$= \frac{1}{2} \times \text{base} \times \text{perpendicular height}$$

$$= \frac{1}{2} \times 5 \times AC$$

The length AC may be calculated using Pythagoras' theorem on triangle ABC, where $AB = 12\,\text{cm}$ and $BC = \frac{1}{2} \times 5 = 2.5\,\text{cm}$.

$$AC = \sqrt{AB^2 + BC^2} = \sqrt{12^2 + 2.5^2} = 12.26\,\text{cm}$$

Hence,

$$\text{area of triangle } ADE = \frac{1}{2} \times 5 \times 12.26 = 30.65\,\text{cm}^2$$

Total surface area of pyramid $= (5 \times 5) + 4(30.65)$

$$= \mathbf{147.6\,cm^2}$$

Problem 11.   A rectangular prism of metal having dimensions of 5 cm by 6 cm by 18 cm is melted down and recast into a pyramid having a rectangular base measuring 6 cm by 10 cm. Calculate the perpendicular height of the pyramid, assuming no waste of metal

Volume of rectangular prism $= 5 \times 6 \times 18 = 540\,\text{cm}^3$

Volume of pyramid

$$= \frac{1}{3} \times \text{area of base} \times \text{perpendicular height}$$

Hence, $540 = \frac{1}{3} \times (6 \times 10) \times h$

from which, $h = \dfrac{3 \times 540}{6 \times 10} = 27\,\text{cm}$

i.e. **perpendicular height of pyramid = 27 cm**

Cones

A cone is a circular-based pyramid. A cone of base radius r and perpendicular height h is shown in Figure 29.11.

$$\text{Volume} = \frac{1}{3} \times \text{area of base} \times \text{perpendicular height}$$

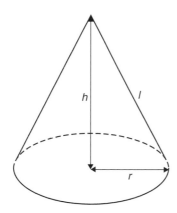

Figure 29.11

i.e.
$$\textbf{Volume} = \frac{1}{3}\boldsymbol{\pi r^2 h}$$

$$\textbf{Curved surface area} = \boldsymbol{\pi r l}$$

$$\textbf{Total surface area} = \boldsymbol{\pi r l + \pi r^2}$$

Problem 12. Calculate the volume, in cubic centimetres, of a cone of radius 30 mm and perpendicular height 80 mm

$\text{Volume of cone} = \dfrac{1}{3}\pi r^2 h = \dfrac{1}{3} \times \pi \times 30^2 \times 80$

$\qquad\qquad\qquad = 75398.2236\ldots\,\text{mm}^3$

$1\,\text{cm} = 10\,\text{mm}$ and
$1\,\text{cm}^3 = 10\,\text{mm} \times 10\,\text{mm} \times 10\,\text{mm} = 10^3\,\text{mm}^3$, or

$$\textbf{1\,mm}^3 = \textbf{10}^{-3}\,\textbf{cm}^3$$

Hence, $75398.2236\ldots\,\text{mm}^3$
$\qquad = 75398.2236\ldots \times 10^{-3}\,\text{cm}^3$
i.e.

$$\textbf{volume} = \textbf{75.40\,cm}^3$$

Alternatively, from the question, $r = 30\,\text{mm} = 3\,\text{cm}$ and $h = 80\,\text{mm} = 8\,\text{cm}$. Hence,

$$\textbf{volume} = \frac{1}{3}\pi r^2 h = \frac{1}{3} \times \pi \times 3^2 \times 8 = \textbf{75.40\,cm}^3$$

Problem 13.   Determine the volume and total surface area of a cone of radius 5 cm and perpendicular height 12 cm

The cone is shown in Figure 29.12.

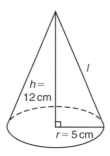

Figure 29.12

$$\textbf{Volume of cone} = \frac{1}{3}\pi r^2 h = \frac{1}{3} \times \pi \times 5^2 \times 12$$

$$= \textbf{314.2\,cm}^3$$

$\text{Total surface area} = \text{curved surface area} + \text{area of base}$
$$= \pi r l + \pi r^2$$

From Figure 29.12, slant height l may be calculated using Pythagoras' theorem:

$$l = \sqrt{12^2 + 5^2} = 13\,\text{cm}$$

Hence, **total surface area** $= (\pi \times 5 \times 13) + (\pi \times 5^2)$

$$= \textbf{282.7\,cm}^2$$

Spheres

For the sphere shown in Figure 29.13:

$$\textbf{Volume} = \frac{4}{3}\boldsymbol{\pi r^3} \quad \text{and} \quad \textbf{surface area} = \textbf{4}\boldsymbol{\pi r^2}$$

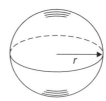

Figure 29.13

Problem 14. Find the volume and surface area of a sphere of diameter 10 cm

Since diameter $= 10$ cm, radius, $r = 5$ cm.

$$\textbf{Volume of sphere} = \frac{4}{3}\pi r^3 = \frac{4}{3} \times \pi \times 5^3$$

$$= \textbf{523.6 cm}^3$$

$$\textbf{Surface area of sphere} = 4\pi r^2 = 4 \times \pi \times 5^2$$

$$= \textbf{314.2 cm}^2$$

Problem 15. The surface area of a sphere is 201.1 cm². Find the diameter of the sphere and hence its volume

Surface area of sphere $= 4\pi r^2$.

Hence, $201.1 \text{ cm}^2 = 4 \times \pi \times r^2$,

from which $\qquad r^2 = \dfrac{201.1}{4 \times \pi} = 16.0$

and $\qquad$ radius, $r = \sqrt{16.0} = 4.0$ cm

from which, **diameter** $= 2 \times r = 2 \times 4.0 = \textbf{8.0 cm}$

$$\textbf{Volume of sphere} = \frac{4}{3}\pi r^3 = \frac{4}{3} \times \pi \times (4.0)^3$$

$$= \textbf{268.1 cm}^3$$

Now try the following Practice Exercise

Practice Exercise 129   Volumes and surface areas of common shapes (answers on page 1163)

1. If a cone has a diameter of 80 mm and a perpendicular height of 120 mm, calculate its volume in cm³ and its curved surface area.

2. A square pyramid has a perpendicular height of 4 cm. If a side of the base is 2.4 cm long, find the volume and total surface area of the pyramid.

3. A sphere has a diameter of 6 cm. Determine its volume and surface area.

4. If the volume of a sphere is 566 cm³, find its radius.

5. A pyramid having a square base has a perpendicular height of 25 cm and a volume of 75 cm³. Determine, in centimetres, the length of each side of the base.

6. A cone has a base diameter of 16 mm and a perpendicular height of 40 mm. Find its volume correct to the nearest cubic millimetre.

7. Determine (a) the volume and (b) the surface area of a sphere of radius 40 mm.

8. The volume of a sphere is 325 cm³. Determine its diameter.

9. Given the radius of the Earth is 6380 km, calculate, in engineering notation

   (a) its surface area in km²

   (b) its volume in km³

10. An ingot whose volume is 1.5 m³ is to be made into ball bearings whose radii are 8.0 cm. How many bearings will be produced from the ingot, assuming 5% wastage?

11. A spherical chemical storage tank has an internal diameter of 5.6 m. Calculate the storage capacity of the tank, correct to the nearest cubic metre. If 1 litre = 1000 cm³, determine the tank capacity in litres.

29.3 Summary of volumes and surface areas of common solids

A summary of volumes and surface areas of regular solids is shown in Table 29.1.

Table 29.1 Volumes and surface areas of regular solids

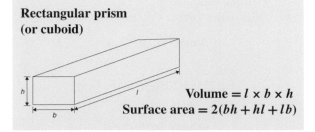

Rectangular prism (or cuboid)

$$\textbf{Volume} = l \times b \times h$$

$$\textbf{Surface area} = 2(bh + hl + lb)$$

Cylinder

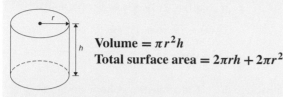

$$\text{Volume} = \pi r^2 h$$
$$\text{Total surface area} = 2\pi rh + 2\pi r^2$$

Triangular prism

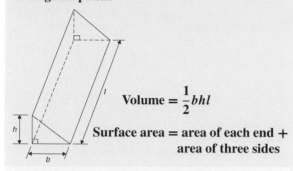

$$\text{Volume} = \frac{1}{2}bhl$$

$$\text{Surface area} = \text{area of each end} +$$
$$\text{area of three sides}$$

Pyramid

$$\text{Volume} = \frac{1}{3} \times A \times h$$

$$\text{Total surface area} =$$
$$\text{sum of areas of triangles}$$
$$\text{forming sides} + \text{area of base}$$

Cone

$$\text{Volume} = \frac{1}{3}\pi r^2 h$$

$$\text{Curved surface area} = \pi rl$$
$$\text{Total surface area} = \pi rl + \pi r^2$$

Sphere

$$\text{Volume} = \frac{4}{3}\pi r^3$$
$$\text{Surface area} = 4\pi r^2$$

29.4 More complex volumes and surface areas

Here are some worked problems involving more complex and composite solids.

Problem 16. A wooden section is shown in Figure 29.14. Find (a) its volume in m³ and (b) its total surface area

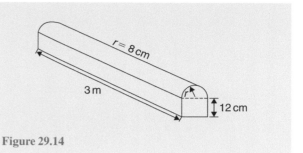

Figure 29.14

(a) The section of wood is a prism whose end comprises a rectangle and a semicircle. Since the radius of the semicircle is 8 cm, the diameter is 16 cm. Hence, the rectangle has dimensions 12 cm by 16 cm.

$$\text{Area of end} = (12 \times 16) + \frac{1}{2}\pi 8^2 = 292.5\,\text{cm}^2$$

Volume of wooden section

$$= \text{area of end} \times \text{perpendicular height}$$
$$= 292.5 \times 300 = \mathbf{87\,750\,cm^3}$$
$$= \frac{87750}{10^6}\,\text{m}^3, \text{ since } 1\,\text{m}^3 = 10^6\,\text{cm}^3$$
$$= \mathbf{0.08775\,m^3}$$

(b) The total surface area comprises the two ends (each of area 292.5 cm²), three rectangles and a curved surface (which is half a cylinder). Hence,

total surface area

$$= (2 \times 292.5) + 2(12 \times 300)$$
$$+ (16 \times 300) + \frac{1}{2}(2\pi \times 8 \times 300)$$
$$= 585 + 7200 + 4800 + 2400\pi$$
$$= \mathbf{20\,125\,cm^2} \quad \text{or} \quad \mathbf{2.0125\,m^2}$$

Problem 17. A pyramid has a rectangular base 3.60 cm by 5.40 cm. Determine the volume and total surface area of the pyramid if each of its sloping edges is 15.0 cm

The pyramid is shown in Figure 29.15. To calculate the volume of the pyramid, the perpendicular height EF is required. Diagonal BD is calculated using Pythagoras' theorem,

i.e. $$BD = \sqrt{[3.60^2 + 5.40^2]} = 6.490\,\text{cm}$$

Hence, $$EB = \frac{1}{2}BD = \frac{6.490}{2} = 3.245\,\text{cm}$$

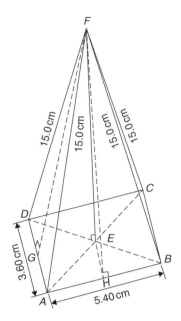

Figure 29.15

Using Pythagoras' theorem on triangle BEF gives

$$BF^2 = EB^2 + EF^2$$

from which $EF = \sqrt{(BF^2 - EB^2)}$

$$= \sqrt{15.0^2 - 3.245^2} = 14.64 \text{cm}$$

Volume of pyramid

$$= \frac{1}{3}(\text{area of base})(\text{perpendicular height})$$

$$= \frac{1}{3}(3.60 \times 5.40)(14.64) = \textbf{94.87 cm}^3$$

Area of triangle ADF (which equals triangle BCF) $= \frac{1}{2}(AD)(FG)$, where G is the mid-point of AD. Using Pythagoras' theorem on triangle FGA gives

$$FG = \sqrt{[15.0^2 - 1.80^2]} = 14.89 \text{cm}$$

Hence, area of triangle $ADF = \frac{1}{2}(3.60)(14.89)$

$$= 26.80 \text{cm}^2$$

Similarly, if H is the mid-point of AB,

$$FH = \sqrt{15.0^2 - 2.70^2} = 14.75 \text{cm}$$

Hence, area of triangle ABF (which equals triangle CDF) $= \frac{1}{2}(5.40)(14.75) = 39.83 \text{cm}^2$

Total surface area of pyramid

$$= 2(26.80) + 2(39.83) + (3.60)(5.40)$$

$$= 53.60 + 79.66 + 19.44$$

$$= \textbf{152.7 cm}^2$$

Problem 18. Calculate the volume and total surface area of a hemisphere of diameter 5.0 cm

Volume of hemisphere $= \frac{1}{2}(\text{volume of sphere})$

$$= \frac{2}{3}\pi r^3 = \frac{2}{3}\pi\left(\frac{5.0}{2}\right)^3$$

$$= \textbf{32.7 cm}^3$$

Total surface area

$$= \text{curved surface area} + \text{area of circle}$$

$$= \frac{1}{2}(\text{surface area of sphere}) + \pi r^2$$

$$= \frac{1}{2}(4\pi r^2) + \pi r^2$$

$$= 2\pi r^2 + \pi r^2 = 3\pi r^2 = 3\pi\left(\frac{5.0}{2}\right)^2$$

$$= \textbf{58.9 cm}^2$$

Problem 19. A rectangular piece of metal having dimensions 4 cm by 3 cm by 12 cm is melted down and recast into a pyramid having a rectangular base measuring 2.5 cm by 5 cm. Calculate the perpendicular height of the pyramid

Volume of rectangular prism of metal $= 4 \times 3 \times 12$

$$= 144 \text{cm}^3$$

Volume of pyramid

$$= \frac{1}{3}(\text{area of base})(\text{perpendicular height})$$

Assuming no waste of metal,

$$144 = \frac{1}{3}(2.5 \times 5)(\text{height})$$

i.e. **perpendicular height of pyramid** $= \dfrac{144 \times 3}{2.5 \times 5}$

$$= \textbf{34.56 cm}$$

Problem 20. A rivet consists of a cylindrical head of diameter 1 cm and depth 2 mm, and a shaft of diameter 2 mm and length 1.5 cm. Determine the volume of metal in 2000 such rivets

Radius of cylindrical head $= \dfrac{1}{2}$ cm $= 0.5$ cm and

height of cylindrical head $= 2$ mm $= 0.2$ cm

Hence, volume of cylindrical head

$$= \pi r^2 h = \pi (0.5)^2 (0.2) = 0.1571 \, \text{cm}^3$$

Volume of cylindrical shaft

$$= \pi r^2 h = \pi \left(\frac{0.2}{2}\right)^2 (1.5) = 0.0471 \, \text{cm}^3$$

Total volume of 1 rivet $= 0.1571 + 0.0471$

$$= 0.2042 \, \text{cm}^3$$

Volume of metal in 2000 such rivets

$$= 2000 \times 0.2042 = \mathbf{408.4 \, cm^3}$$

Problem 21. A solid metal cylinder of radius 6 cm and height 15 cm is melted down and recast into a shape comprising a hemisphere surmounted by a cone. Assuming that 8% of the metal is wasted in the process, determine the height of the conical portion if its diameter is to be 12 cm

Volume of cylinder $= \pi r^2 h = \pi \times 6^2 \times 15$

$$= 540\pi \, \text{cm}^3$$

If 8% of metal is lost then 92% of 540π gives the volume of the new shape, shown in Figure 29.16.

Figure 29.16

Hence, the volume of (hemisphere + cone)

$$= 0.92 \times 540\pi \, \text{cm}^3$$

i.e. $\dfrac{1}{2}\left(\dfrac{4}{3}\pi r^3\right) + \dfrac{1}{3}\pi r^2 h = 0.92 \times 540\pi$

Dividing throughout by π gives

$$\frac{2}{3}r^3 + \frac{1}{3}r^2 h = 0.92 \times 540$$

Since the diameter of the new shape is to be 12 cm, radius $r = 6$ cm,

hence $\dfrac{2}{3}(6)^3 + \dfrac{1}{3}(6)^2 h = 0.92 \times 540$

$$144 + 12h = 496.8$$

i.e. **height of conical portion,**

$$h = \frac{496.8 - 144}{12} = \mathbf{29.4 \, cm}$$

Problem 22. A block of copper having a mass of 50 kg is drawn out to make 500 m of wire of uniform cross-section. Given that the density of copper is 8.91 g/cm^3, calculate (a) the volume of copper, (b) the cross-sectional area of the wire and (c) the diameter of the cross-section of the wire

(a) A density of 8.91 g/cm^3 means that 8.91 g of copper has a volume of 1 cm^3, or 1 g of copper has a volume of $(1 \div 8.91)$ cm^3.

$$\text{Density} = \frac{\text{mass}}{\text{volume}}$$

from which $\quad \text{volume} = \dfrac{\text{mass}}{\text{density}}$

Hence, 50 kg, i.e. 50 000 g, has a

$$\mathbf{volume} = \frac{\text{mass}}{\text{density}} = \frac{50000}{8.91} \, \text{cm}^3 = \mathbf{5612 \, cm^3}$$

(b) Volume of wire $=$ area of circular cross-section

$$\times \text{ length of wire.}$$

Hence, $5612 \, \text{cm}^3 = \text{area} \times (500 \times 100 \, \text{cm})$

from which, **area** $= \dfrac{5612}{500 \times 100} \, \text{cm}^2$

$$= \mathbf{0.1122 \, cm^2}$$

(c)   Area of circle $= \pi r^2$ or $\dfrac{\pi d^2}{4}$

hence, $0.1122 = \dfrac{\pi d^2}{4}$

from which, $d = \sqrt{\left(\dfrac{4 \times 0.1122}{\pi}\right)} = 0.3780\,\text{cm}$

i.e. **diameter of cross-section is 3.780 mm**

Problem 23.   A boiler consists of a cylindrical section of length 8 m and diameter 6 m, on one end of which is surmounted a hemispherical section of diameter 6 m and on the other end a conical section of height 4 m and base diameter 6 m. Calculate the volume of the boiler and the total surface area

---

The boiler is shown in Figure 29.17.

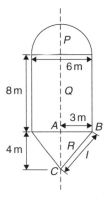

Figure 29.17

Volume of hemisphere, $P = \dfrac{2}{3}\pi r^3$

$= \dfrac{2}{3} \times \pi \times 3^3 = 18\pi\,\text{m}^3$

Volume of cylinder, $Q = \pi r^2 h = \pi \times 3^2 \times 8$

$= 72\pi\,\text{m}^3$

Volume of cone, $R = \dfrac{1}{3}\pi r^2 h = \dfrac{1}{3} \times \pi \times 3^2 \times 4$

$= 12\pi\,\text{m}^3$

Total volume of boiler $= 18\pi + 72\pi + 12\pi$

$= 102\pi = \mathbf{320.4\,m^3}$

Surface area of hemisphere, $P = \dfrac{1}{2}(4\pi r^2)$

$= 2 \times \pi \times 3^2 = 18\pi\,\text{m}^2$

Curved surface area of cylinder, $Q = 2\pi rh$

$= 2 \times \pi \times 3 \times 8$

$= 48\pi\,\text{m}^2$

The slant height of the cone, l, is obtained by Pythagoras' theorem on triangle ABC, i.e.

$l = \sqrt{(4^2 + 3^2)} = 5$

Curved surface area of cone,

$R = \pi rl = \pi \times 3 \times 5 = 15\pi\,\text{m}^2$

Total surface area of boiler $= 18\pi + 48\pi + 15\pi$

$= 81\pi = \mathbf{254.5\,m^2}$

Now try the following Practice Exercise

Practice Exercise 130   More complex volumes and surface areas (answers on page 1163)

1.   Find the total surface area of a hemisphere of diameter 50 mm.

2.   Find (a) the volume and (b) the total surface area of a hemisphere of diameter 6 cm.

3.   Determine the mass of a hemispherical copper container whose external and internal radii are 12 cm and 10 cm, assuming that 1 cm^3 of copper weighs 8.9 g.

4.   A metal plumb bob comprises a hemisphere surmounted by a cone. If the diameter of the hemisphere and cone are each 4 cm and the total length is 5 cm, find its total volume.

5.   A marquee is in the form of a cylinder surmounted by a cone. The total height is 6 m and the cylindrical portion has a height of 3.5 m with a diameter of 15 m. Calculate the surface area of material needed to make the marquee, assuming 12% of the material is wasted in the process.

6.   Determine (a) the volume and (b) the total surface area of the following solids.

   (i)   A cone of radius 8.0 cm and perpendicular height 10 cm.

   (ii)   A sphere of diameter 7.0 cm.

   (iii)   A hemisphere of radius 3.0 cm.

(iv)   A 2.5 cm by 2.5 cm square pyramid of perpendicular height 5.0 cm.

(v)   A 4.0 cm by 6.0 cm rectangular pyramid of perpendicular height 12.0 cm.

(vi)   A 4.2 cm by 4.2 cm square pyramid whose sloping edges are each 15.0 cm.

(vii)   A pyramid having an octagonal base of side 5.0 cm and perpendicular height 20 cm.

7.   A metal sphere weighing 24 kg is melted down and recast into a solid cone of base radius 8.0 cm. If the density of the metal is 8000 kg/m^3 determine

(a)   the diameter of the metal sphere.

(b)   the perpendicular height of the cone, assuming that 15% of the metal is lost in the process.

8.   Find the volume of a regular hexagonal pyramid if the perpendicular height is 16.0 cm and the side of the base is 3.0 cm.

9.   A buoy consists of a hemisphere surmounted by a cone. The diameter of the cone and hemisphere is 2.5 m and the slant height of the cone is 4.0 m. Determine the volume and surface area of the buoy.

10.   A petrol container is in the form of a central cylindrical portion 5.0 m long with a hemispherical section surmounted on each end. If the diameters of the hemisphere and cylinder are both 1.2 m, determine the capacity of the tank in litres (1 litre = 1000 cm^3).

11.   Figure 29.18 shows a metal rod section. Determine its volume and total surface area.

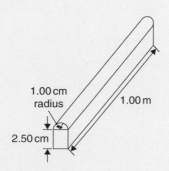

Figure 29.18

12.   Find the volume (in cm^3) of the die-casting shown in Figure 29.19. The dimensions are in millimetres.

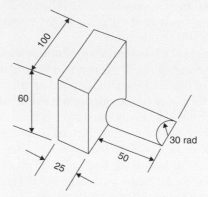

Figure 29.19

13.   The cross-section of part of a circular ventilation shaft is shown in Figure 29.20, ends *AB* and *CD* being open. Calculate

(a)   the volume of the air, correct to the nearest litre, contained in the part of the system shown, neglecting the sheet metal thickness (given 1 litre = 1000 cm^3).

(b)   the cross-sectional area of the sheet metal used to make the system, in square metres.

(c)   the cost of the sheet metal if the material costs £11.50 per square metre, assuming that 25% extra metal is required due to wastage.

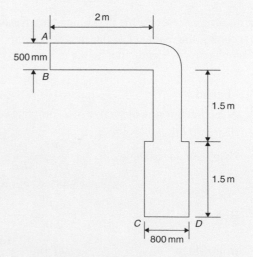

Figure 29.20

29.5 Volumes and surface areas of frusta of pyramids and cones

The **frustum** of a pyramid or cone is the portion remaining when a part containing the vertex is cut off by a plane parallel to the base.

The **volume of a frustum of a pyramid or cone** is given by the volume of the whole pyramid or cone minus the volume of the small pyramid or cone cut off.

The **surface area of the sides of a frustum of a pyramid or cone** is given by the surface area of the whole pyramid or cone minus the surface area of the small pyramid or cone cut off. This gives the lateral surface area of the frustum. If the total surface area of the frustum is required then the surface area of the two parallel ends are added to the lateral surface area.

There is an alternative method for finding the volume and surface area of a **frustum of a cone**. With reference to Figure 29.21,

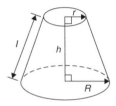

Figure 29.21

$$\text{Volume} = \frac{1}{3}\pi h(R^2 + Rr + r^2)$$

$$\text{Curved surface area} = \pi l(R + r)$$

$$\text{Total surface area} = \pi l(R + r) + \pi r^2 + \pi R^2$$

Problem 24. Determine the volume of a frustum of a cone if the diameter of the ends are 6.0 cm and 4.0 cm and its perpendicular height is 3.6 cm

(i) Method 1

A section through the vertex of a complete cone is shown in Figure 29.22.

Using similar triangles, $\dfrac{AP}{DP} = \dfrac{DR}{BR}$

(see page 435, Section 37.5)

Hence, $\dfrac{AP}{2.0} = \dfrac{3.6}{1.0}$

from which $AP = \dfrac{(2.0)(3.6)}{1.0}$

$= 7.2 \text{ cm}$

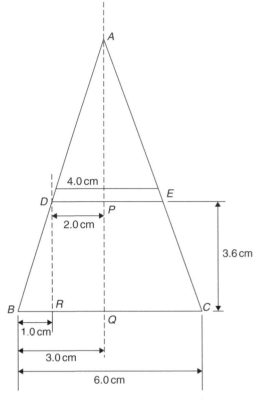

Figure 29.22

The height of the large cone $= 3.6 + 7.2$

$= 10.8 \text{ cm}$

Volume of frustum of cone

$= $ volume of large cone

$\quad -$ volume of small cone cut off

$= \dfrac{1}{3}\pi(3.0)^2(10.8) - \dfrac{1}{3}\pi(2.0)^2(7.2)$

$= 101.79 - 30.16 = \mathbf{71.6\,cm^3}$

(ii) Method 2

From above, volume of the frustum of a cone

$= \dfrac{1}{3}\pi h(R^2 + Rr + r^2)$

where $\quad R = 3.0 \text{ cm}$,

$\quad r = 2.0 \text{ cm} \quad \text{and} \quad h = 3.6 \text{ cm}$

Hence, **volume of frustum**

$= \dfrac{1}{3}\pi(3.6)\left[(3.0)^2 + (3.0)(2.0) + (2.0)^2\right]$

$= \dfrac{1}{3}\pi(3.6)(19.0) = \mathbf{71.6\,cm^3}$

Problem 25. Find the total surface area of the frustum of the cone in Problem 24

(i) Method 1

Curved surface area of frustum = curved surface area of large cone − curved surface area of small cone cut off.

From Figure 29.22, using Pythagoras' theorem,

$$AB^2 = AQ^2 + BQ^2$$

from which $AB = \sqrt{[10.8^2 + 3.0^2]} = 11.21\,cm$

and $AD^2 = AP^2 + DP^2$

from which $AD = \sqrt{[7.2^2 + 2.0^2]} = 7.47\,cm$

Curved surface area of large cone $= \pi rl$

$$= \pi(BQ)(AB) = \pi(3.0)(11.21)$$

$$= 105.65\,cm^2$$

and curved surface area of small cone

$$= \pi(DP)(AD) = \pi(2.0)(7.47) = 46.94\,cm^2$$

Hence, curved surface area of frustum

$$= 105.65 - 46.94$$

$$= 58.71\,cm^2$$

Total surface area of frustum

$$= \text{curved surface area}$$
$$\quad + \text{area of two circular ends}$$

$$= 58.71 + \pi(2.0)^2 + \pi(3.0)^2$$

$$= 58.71 + 12.57 + 28.27 = \mathbf{99.6\,cm^2}$$

(ii) Method 2

From page 316, total surface area of frustum

$$= \pi l(R + r) + \pi r^2 + \pi R^2$$

where $l = BD = 11.21 - 7.47 = 3.74\,cm$, $R = 3.0\,cm$ and $r = 2.0\,cm$. Hence,

total surface area of frustum

$$= \pi(3.74)(3.0 + 2.0) + \pi(2.0)^2 + \pi(3.0)^2$$

$$= \mathbf{99.6\,cm^2}$$

Problem 26. A storage hopper is in the shape of a frustum of a pyramid. Determine its volume if the ends of the frustum are squares of sides 8.0 m and 4.6 m, respectively, and the perpendicular height between its ends is 3.6 m

The frustum is shown shaded in Figure 29.23(a) as part of a complete pyramid. A section perpendicular to the base through the vertex is shown in Figure 29.23(b).

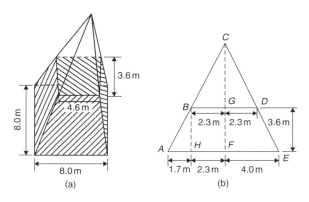

Figure 29.23

By similar triangles $\dfrac{CG}{BG} = \dfrac{BH}{AH}$

From which, height

$$CG = BG\left(\frac{BH}{AH}\right) = \frac{(2.3)(3.6)}{1.7} = 4.87\,m$$

Height of complete pyramid $= 3.6 + 4.87 = 8.47\,m$

Volume of large pyramid $= \dfrac{1}{3}(8.0)^2(8.47)$

$$= 180.69\,m^3$$

Volume of small pyramid cut off $= \dfrac{1}{3}(4.6)^2(4.87)$

$$= 34.35\,m^3$$

Hence, **volume of storage hopper** $= 180.69 - 34.35$

$$= \mathbf{146.3\,m^3}$$

Problem 27. Determine the lateral surface area of the storage hopper in Problem 26

The lateral surface area of the storage hopper consists of four equal trapeziums. From Figure 29.24,

$$\text{Area of trapezium } PRSU = \frac{1}{2}(PR + SU)(QT)$$

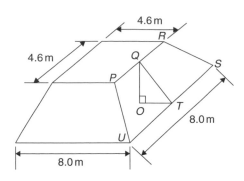

Figure 29.24

$OT = 1.7\,\text{m}$ (same as AH in Figure 29.23(b) and $OQ = 3.6\,\text{m}$. By Pythagoras' theorem,

$$QT = \sqrt{(OQ^2 + OT^2)} = \sqrt{[3.6^2 + 1.7^2]} = 3.98\,\text{m}$$

Area of trapezium $PRSU$

$$= \frac{1}{2}(4.6 + 8.0)(3.98) = 25.07\,\text{m}^2$$

Lateral surface area of hopper $= 4(25.07)$

$$= \mathbf{100.3\,m^2}$$

Problem 28. A lampshade is in the shape of a frustum of a cone. The vertical height of the shade is 25.0 cm and the diameters of the ends are 20.0 cm and 10.0 cm, respectively. Determine the area of the material needed to form the lampshade, correct to 3 significant figures

The curved surface area of a frustum of a cone $= \pi l(R + r)$ from page 316. Since the diameters of the ends of the frustum are 20.0 cm and 10.0 cm, from Figure 29.25,

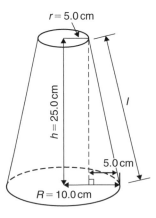

Figure 29.25

$r = 5.0\,\text{cm}, R = 10.0\,\text{cm}$

and $\quad l = \sqrt{[25.0^2 + 5.0^2]} = 25.50\,\text{cm}$

from Pythagoras' theorem.
Hence, curved surface area

$$= \pi(25.50)(10.0 + 5.0) = 1201.7\,\text{cm}^2$$

i.e. the area of material needed to form the lampshade is **1200 cm²**, correct to 3 significant figures.

Problem 29. A cooling tower is in the form of a cylinder surmounted by a frustum of a cone, as shown in Figure 29.26. Determine the volume of air space in the tower if 40% of the space is used for pipes and other structures

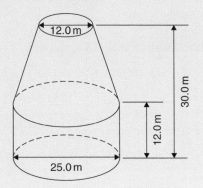

Figure 29.26

Volume of cylindrical portion $= \pi r^2 h$

$$= \pi \left(\frac{25.0}{2}\right)^2 (12.0)$$

$$= 5890\,\text{m}^3$$

Volume of frustum of cone $= \frac{1}{3}\pi h(R^2 + Rr + r^2)$

where $\qquad h = 30.0 - 12.0 = 18.0\,\text{m}$,

$$R = 25.0 \div 2 = 12.5\,\text{m}$$

and $\qquad r = 12.0 \div 2 = 6.0\,\text{m}$

Hence, volume of frustum of cone

$$= \frac{1}{3}\pi(18.0)\left[(12.5)^2 + (12.5)(6.0) + (6.0)^2\right]$$

$$= 5038\,\text{m}^3$$

Total volume of cooling tower $= 5890 + 5038$

$$= 10\,928\,\text{m}^3$$

If 40% of space is occupied then

volume of air space $= 0.6 \times 10928 = \mathbf{6557\,m^3}$

Now try the following Practice Exercise

Practice Exercise 131   Volumes and surface areas of frusta of pyramids and cones (answers on page 1163)

1. The radii of the faces of a frustum of a cone are 2.0 cm and 4.0 cm and the thickness of the frustum is 5.0 cm. Determine its volume and total surface area.

2. A frustum of a pyramid has square ends, the squares having sides 9.0 cm and 5.0 cm, respectively. Calculate the volume and total surface area of the frustum if the perpendicular distance between its ends is 8.0 cm.

3. A cooling tower is in the form of a frustum of a cone. The base has a diameter of 32.0 m, the top has a diameter of 14.0 m and the vertical height is 24.0 m. Calculate the volume of the tower and the curved surface area.

4. A loudspeaker diaphragm is in the form of a frustum of a cone. If the end diameters are 28.0 cm and 6.00 cm and the vertical distance between the ends is 30.0 cm, find the area of material needed to cover the curved surface of the speaker.

5. A rectangular prism of metal having dimensions 4.3 cm by 7.2 cm by 12.4 cm is melted down and recast into a frustum of a square pyramid, 10% of the metal being lost in the process. If the ends of the frustum are squares of sides 3 cm and 8 cm, respectively, find the thickness of the frustum.

6. Determine the volume and total surface area of a bucket consisting of an inverted frustum of a cone, of slant height 36.0 cm and end diameters 55.0 cm and 35.0 cm.

7. A cylindrical tank of diameter 2.0 m and perpendicular height 3.0 m is to be replaced by a tank of the same capacity but in the form of a frustum of a cone. If the diameters of the ends of the frustum are 1.0 m and 2.0 m, respectively, determine the vertical height required.

29.6 The frustum and zone of a sphere

Volume of sphere $= \frac{4}{3}\pi r^3$ and the surface area of sphere $= 4\pi r^2$

A **frustum of a sphere** is the portion contained between two parallel planes. In Figure 29.27, *PQRS* is a frustum of the sphere. A **zone of a sphere** is the curved surface of a frustum. With reference to Figure 29.27:

> **Surface area of a zone of a sphere** $= 2\pi rh$
> **Volume of frustum of sphere**
> $$= \frac{\pi h}{6}\left(h^2 + 3r_1^2 + 3r_2^2\right)$$

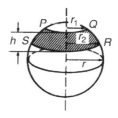

Figure 29.27

Problem 30. Determine the volume of a frustum of a sphere of diameter 49.74 cm if the diameter of the ends of the frustum are 24.0 cm and 40.0 cm, and the height of the frustum is 7.00 cm

From above, volume of frustum of a sphere

$$= \frac{\pi h}{6}(h^2 + 3r_1^2 + 3r_2^2)$$

where $h = 7.00\,\text{cm}$, $r_1 = 24.0/2 = 12.0\,\text{cm}$ and $r_2 = 40.0/2 = 20.0\,\text{cm}$.

Hence volume of frustum

$$= \frac{\pi(7.00)}{6}[(7.00)^2 + 3(12.0)^2 + 3(20.0)^2]$$

$$= \mathbf{6161\,cm^3}$$

Problem 31. Determine for the frustum of Problem 30 the curved surface area of the frustum

The curved surface area of the frustum $=$ surface area of zone $= 2\pi rh$ (from above), where $r =$ radius of sphere $= 49.74/2 = 24.87\,\text{cm}$ and $h = 7.00\,\text{cm}$. Hence, surface area of zone $= 2\pi(24.87)(7.00) = \mathbf{1094\,cm^2}$

Problem 32. The diameters of the ends of the frustum of a sphere are 14.0 cm and 26.0 cm, respectively, and the thickness of the frustum is 5.0 cm. Determine, correct to 3 significant figures (a) the volume of the frustum of the sphere, (b) the radius of the sphere and (c) the area of the zone formed

The frustum is shown shaded in the cross-section of Figure 29.28.

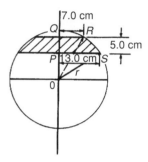

Figure 29.28

(a) Volume of frustum of sphere

$$= \frac{\pi h}{6}(h^2 + 3r_1^2 + 3r_2^2)$$

from above, where $h = 5.0$ cm, $r_1 = 14.0/2 = 7.0$ cm and $r_2 = 26.0/2 = 13.0$ cm. Hence volume of frustum of sphere

$$= \frac{\pi(5.0)}{6}[(5.0)^2 + 3(7.0)^2 + 3(13.0)^2]$$

$$= \frac{\pi(5.0)}{6}[25.0 + 147.0 + 507.0]$$

$$= \mathbf{1780\,cm^3} \text{ correct to 3 significant figures.}$$

(b) The radius, r, of the sphere may be calculated using Figure 29.28. Using Pythagoras' theorem:

$$OS^2 = PS^2 + OP^2$$

i.e. $\quad r^2 = (13.0)^2 + OP^2 \qquad (1)$

$$OR^2 = QR^2 + OQ^2$$

i.e. $\quad r^2 = (7.0)^2 + OQ^2$

However, $OQ = QP + OP = 5.0 + OP$, therefore

$$r^2 = (7.0)^2 + (5.0 + OP)^2 \qquad (2)$$

Equating equations (1) and (2) gives:

$$(13.0)^2 + OP^2 = (7.0)^2 + (5.0 + OP)^2$$

$$169.0 + OP^2 = 49.0 + 25.0$$

$$+ 10.0(OP) + OP^2$$

$$169.0 = 74.0 + 10.0(OP)$$

Hence

$$OP = \frac{169.0 - 74.0}{10.0} = 9.50 \text{ cm}$$

Substituting $OP = 9.50$ cm into equation (1) gives:

$$r^2 = (13.0)^2 + (9.50)^2$$

from which $r = \sqrt{13.0^2 + 9.50^2}$

i.e. **radius of sphere, $r = 16.1$ cm**

(c) Area of zone of sphere

$$= 2\pi rh = 2\pi(16.1)(5.0)$$

$$= \mathbf{506\,cm^2}, \text{ correct to 3 significant figures.}$$

Problem 33. A frustum of a sphere of diameter 12.0 cm is formed by two parallel planes, one through the diameter and the other distance h from the diameter. The curved surface area of the frustum is required to be $\frac{1}{4}$ of the total surface area of the sphere. Determine (a) the volume and surface area of the sphere, (b) the thickness h of the frustum, (c) the volume of the frustum and (d) the volume of the frustum expressed as a percentage of the sphere

(a) Volume of sphere,

$$V = \frac{4}{3}\pi r^3 = \frac{4}{3}\pi\left(\frac{12.0}{2}\right)^3$$

$$= \mathbf{904.8\,cm^3}$$

Surface area of sphere

$$= 4\pi r^2 = 4\pi\left(\frac{12.0}{2}\right)^2$$

$$= \mathbf{452.4\,cm^2}$$

(b) Curved surface area of frustum

$$= \tfrac{1}{4} \times \text{surface area of sphere}$$

$$= \tfrac{1}{4} \times 452.4 = 113.1 \text{ cm}^2$$

From above,

$$113.1 = 2\pi rh = 2\pi\left(\frac{12.0}{2}\right)h$$

Hence thickness of frustum

$$h = \frac{113.1}{2\pi(6.0)} = \mathbf{3.0\,cm}$$

(c) Volume of frustum,

$$V = \frac{\pi h}{6}(h^2 + 3r_1^2 + 3r_2^2)$$

where $h = 3.0\,cm$, $r_2 = 6.0\,cm$ and

$$r_1 = \sqrt{OQ^2 - OP^2}$$

from Figure 29.29,

i.e. $r_1 = \sqrt{6.0^2 - 3.0^2} = 5.196\,cm$

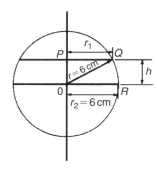

Figure 29.29

Hence volume of frustum

$$= \frac{\pi(3.0)}{6}[(3.0)^2 + 3(5.196)^2 + 3(6.0)^2]$$

$$= \frac{\pi}{2}[9.0 + 81 + 108.0] = \mathbf{311.0\,cm^3}$$

(d) $\dfrac{\text{Volume of frustum}}{\text{Volume of sphere}} = \dfrac{311.0}{904.8} \times 100\%$

$$= \mathbf{34.37\%}$$

> **Problem 34.** A spherical storage tank is filled with liquid to a depth of 20 cm. If the internal diameter of the vessel is 30 cm, determine the number of litres of liquid in the container (1 litre = 1000 cm^3)

The liquid is represented by the shaded area in the section shown in Figure 29.30. The volume of liquid comprises a hemisphere and a frustum of thickness 5 cm. Hence volume of liquid

$$= \frac{2}{3}\pi r^3 + \frac{\pi h}{6}[h^2 + 3r_1^2 + 3r_2^2]$$

where $r_2 = 30/2 = 15\,cm$ and

$$r_1 = \sqrt{15^2 - 5^2} = 14.14\,cm$$

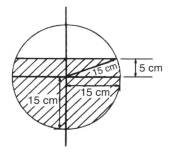

Figure 29.30

Volume of liquid

$$= \frac{2}{3}\pi(15)^3 + \frac{\pi(5)}{6}[5^2 + 3(14.14)^2 + 3(15)^2]$$

$$= 7069 + 3403 = 10\,470\,cm^3$$

Since 1 litre = 1000 cm^3, the number of litres of liquid

$$= \frac{10\,470}{1000} = \mathbf{10.47\,litres}$$

Now try the following Practice Exercise

> **Practice Exercise 132   Further problems on frustums and zones of spheres (answers on page 1163)**
>
> 1. Determine the volume and surface area of a frustum of a sphere of diameter 47.85 cm, if the radii of the ends of the frustum are 14.0 cm and 22.0 cm and the height of the frustum is 10.0 cm.
>
> 2. Determine the volume (in cm^3) and the surface area (in cm^2) of a frustum of a sphere if the diameters of the ends are 80.0 mm and 120.0 mm and the thickness is 30.0 mm.
>
> 3. A sphere has a radius of 6.50 cm. Determine its volume and surface area. A frustum of the sphere is formed by two parallel planes, one through the diameter and the other at a distance h from the diameter. If the curved surface area of the frustum is to be $\frac{1}{5}$ of the surface area of the sphere, find the height h and the volume of the frustum.
>
> 4. A sphere has a diameter of 32.0 mm. Calculate the volume (in cm^3) of the frustum of the sphere contained between two parallel

planes distances 12.0 mm and 10.00 mm from the centre and on opposite sides of it.

5. A spherical storage tank is filled with liquid to a depth of 30.0 cm. If the inner diameter of the vessel is 45.0 cm determine the number of litres of liquid in the container (1 litre = 1000 cm³).

29.7 Prismoidal rule

The prismoidal rule applies to a solid of length x divided by only three equidistant plane areas, A_1, A_2 and A_3, as shown in Figure 29.31, and is merely an extension of Simpson's rule (see Chapter 30) – but for volumes.

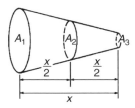

Figure 29.31

With reference to Figure 29.31

$$\text{Volume, } V = \frac{x}{6}[A_1 + 4A_2 + A_3]$$

The prismoidal rule gives precise values of volume for regular solids such as pyramids, cones, spheres and prismoids.

Problem 35. A container is in the shape of a frustum of a cone. Its diameter at the bottom is 18 cm and at the top 30 cm. If the depth is 24 cm determine the capacity of the container, correct to the nearest litre, by the prismoidal rule (1 litre = 1000 cm³)

The container is shown in Figure 29.32. At the mid-point, i.e. at a distance of 12 cm from one end, the radius r_2 is $(9+15)/2 = 12$ cm, since the sloping side changes uniformly.
    Volume of container by the prismoidal rule

$$= \frac{x}{6}[A_1 + 4A_2 + A_3]$$

from above, where $x = 24$ cm, $A_1 = \pi(15)^2$ cm², $A_2 = \pi(12)^2$ cm² and $A_3 = \pi(9)^2$ cm²

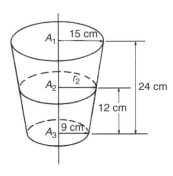

Figure 29.32

Hence volume of container

$$= \frac{24}{6}[\pi(15)^2 + 4\pi(12)^2 + \pi(9)^2]$$

$$= 4[706.86 + 1809.56 + 254.47]$$

$$= 11\,080 \text{ cm}^3 = \frac{11\,080}{1000} \text{ litres}$$

= 11 litres, correct to the nearest litre

(Check: Volume of frustum of cone

$$= \tfrac{1}{3}\pi h[R^2 + Rr + r^2] \quad \text{from Section 29.5}$$

$$= \tfrac{1}{3}\pi(24)[(15)^2 + (15)(9) + (9)^2]$$

$$= 11\,080 \text{ cm}^3(\text{as shown above})$$

Problem 36. A frustum of a sphere of radius 13 cm is formed by two parallel planes on opposite sides of the centre, each at a distance of 5 cm from the centre. Determine the volume of the frustum (a) by using the prismoidal rule, and (b) by using the formula for the volume of a frustum of a sphere

The frustum of the sphere is shown by the section in Figure 29.33.
Radius $r_1 = r_2 = PQ = \sqrt{13^2 - 5^2} = 12$ cm, by Pythagoras' theorem.

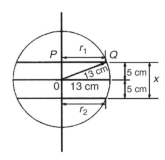

Figure 29.33

(a) Using the prismoidal rule, volume of frustum,

$$V = \frac{x}{6}[A_1 + 4A_2 + A_3]$$

$$= \frac{10}{6}[\pi(12)^2 + 4\pi(13)^2 + \pi(12)^2]$$

$$= \frac{10\pi}{6}[144 + 676 + 144] = \textbf{5047 cm}^3$$

(b) Using the formula for the volume of a frustum of a sphere:

Volume $$V = \frac{\pi h}{6}(h^2 + 3r_1^2 + 3r_2^2)$$

$$= \frac{\pi(10)}{6}[10^2 + 3(12)^2 + 3(12)^2]$$

$$= \frac{10\pi}{6}(100 + 432 + 432)$$

$$= \textbf{5047 cm}^3$$

Problem 37. A hole is to be excavated in the form of a prismoid. The bottom is to be a rectangle 16 m long by 12 m wide; the top is also a rectangle, 26 m long by 20 m wide. Find the volume of earth to be removed, correct to 3 significant figures, if the depth of the hole is 6.0 m

---

The prismoid is shown in Figure 29.34. Let A_1 represent the area of the top of the hole, i.e. $A_1 = 20 \times 26 = 520\,\text{m}^2$. Let A_3 represent the area of the bottom of the hole, i.e. $A_3 = 16 \times 12 = 192\,\text{m}^2$. Let A_2 represent the rectangular area through the middle of the hole parallel to areas A_1 and A_3. The length of this rectangle is $(26 + 16)/2 = 21$ m and the width is $(20 + 12)/2 = 16$ m, assuming the sloping edges are uniform. Thus area $A_2 = 21 \times 16 = 336\,\text{m}^2$.

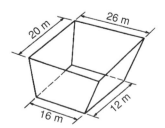

Figure 29.34

Using the prismoidal rule,

$$\text{volume of hole} = \frac{x}{6}[A_1 + 4A_2 + A_3]$$

$$= \frac{6}{6}[520 + 4(336) + 192]$$

$$= 2056\,\text{m}^3 = \textbf{2060 m}^3,$$

correct to 3 significant figures.

Problem 38. The roof of a building is in the form of a frustum of a pyramid with a square base of side 5.0 m. The flat top is a square of side 1.0 m and all the sloping sides are pitched at the same angle. The vertical height of the flat top above the level of the eaves is 4.0 m. Calculate, using the prismoidal rule, the volume enclosed by the roof

---

Let area of top of frustum be $A_1 = (1.0)^2 = 1.0\,\text{m}^2$
Let area of bottom of frustum be $A_3 = (5.0)^2 = 25.0\,\text{m}^2$
Let area of section through the middle of the frustum parallel to A_1 and A_3 be A_2. The length of the side of the square forming A_2 is the average of the sides forming A_1 and A_3, i.e. $(1.0 + 5.0)/2 = 3.0$ m. Hence $A_2 = (3.0)^2 = 9.0\,\text{m}^2$.
Using the prismoidal rule,

$$\text{volume of frustum} = \frac{x}{6}[A_1 + 4A_2 + A_3]$$

$$= \frac{4.0}{6}[1.0 + 4(9.0) + 25.0]$$

Hence, **volume enclosed by roof = 41.3 m³**

Now try the following Practice Exercise

Practice Exercise 133   Further problems on the prismoidal rule (answers on page 1163)

1. Use the prismoidal rule to find the volume of a frustum of a sphere contained between two parallel planes on opposite sides of the centre each of radius 7.0 cm and each 4.0 cm from the centre.

2. Determine the volume of a cone of perpendicular height 16.0 cm and base diameter 10.0 cm by using the prismoidal rule.

3. A bucket is in the form of a frustum of a cone. The diameter of the base is 28.0 cm and the diameter of the top is 42.0 cm. If the height is 32.0 cm, determine the capacity of the bucket (in litres) using the prismoidal rule (1 litre = 1000 cm³).

4. Determine the capacity of a water reservoir, in litres, the top being a 30.0 m by 12.0 m rectangle, the bottom being a 20.0 m by 8.0 m rectangle and the depth being 5.0 m (1 litre = 1000 cm³).

29.8   Volumes of similar shapes

Figure 29.35 shows two cubes, one of which has sides three times longer than those of the other.

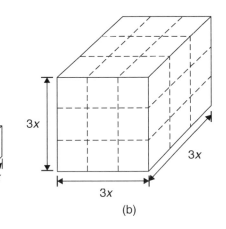

(a)   (b)

Figure 29.35

Volume of Figure 29.35(a) = $(x)(x)(x) = x^3$

Volume of Figure 29.35(b) = $(3x)(3x)(3x) = 27x^3$

Hence, Figure 29.35(b) has a volume $(3)^3$, i.e. 27 times the volume of Figure 29.35(a).

Summarising, **the volumes of similar bodies are proportional to the cubes of corresponding linear dimensions.**

> **Problem 39.** A car has a mass of 1000 kg. A model of the car is made to a scale of 1 to 50. Determine the mass of the model if the car and its model are made of the same material

$$\frac{\text{Volume of model}}{\text{Volume of car}} = \left(\frac{1}{50}\right)^3$$

since the volume of similar bodies are proportional to the cube of corresponding dimensions.
Mass = density × volume and, since both car and model are made of the same material,

$$\frac{\text{Mass of model}}{\text{Mass of car}} = \left(\frac{1}{50}\right)^3$$

Hence, mass of model

$$= (\text{mass of car})\left(\frac{1}{50}\right)^3 = \frac{1000}{50^3} = \mathbf{0.008\,kg}\ \text{or}\ \mathbf{8\,g}$$

Now try the following Practice Exercise

Practice Exercise 134   Volumes of similar shapes (answers on page 1163)

1. The diameters of two spherical bearings are in the ratio 2:5. What is the ratio of their volumes?

2. An engineering component has a mass of 400 g. If each of its dimensions are reduced by 30%, determine its new mass.

For fully worked solutions to each of the problems in Practice Exercises 128 to 134 in this chapter, go to the website:
www.routledge.com/cw/bird

Chapter 30

Irregular areas and volumes, and mean values

Why it is important to understand: Irregular areas and volumes and mean values of waveforms

Surveyors, farmers and landscapers often need to determine the area of irregularly shaped pieces of land to work with the land properly. There are many applications in business, economics and the sciences, including all aspects of engineering, where finding the areas of irregular shapes, the volumes of solids, and the lengths of irregular shaped curves are important applications. Typical earthworks include roads, railway beds, causeways, dams and canals. Other common earthworks are land grading to reconfigure the topography of a site, or to stabilise slopes. Engineers need to concern themselves with issues of geotechnical engineering (such as soil density and strength) and with quantity estimation to ensure that soil volumes in the cuts match those of the fills, while minimising the distance of movement. Simpson's rule is a staple of scientific data analysis and engineering; it is widely used, for example, by naval architects to numerically determine hull offsets and cross-sectional areas to determine volumes and centroids of ships or lifeboats. There are therefore plenty of examples where irregular areas and volumes need to be determined by engineers.

At the end of this chapter, you should be able to:

- use the trapezoidal rule to determine irregular areas
- use the mid-ordinate rule to determine irregular areas
- use Simpson's rule to determine irregular areas
- estimate the volume of irregular solids
- determine the mean values of waveforms

30.1 Areas of irregular figures

Areas of irregular plane surfaces may be approximately determined by using

(a) a planimeter,

(b) the trapezoidal rule,

(c) the mid-ordinate rule or

(d) Simpson's rule.

Such methods may be used by, for example, engineers estimating areas of indicator diagrams of steam engines, surveyors estimating areas of plots of land or naval architects estimating areas of water planes or transverse sections of ships.

(a) **A planimeter** is an instrument for directly measuring small areas bounded by an irregular curve. There are many different kinds of planimeters but all operate in a similar way. A pointer on the planimeter is used to trace around the boundary of the shape. This induces a movement in another part of the instrument and a reading of this is used to establish the area of the shape.

(b) **Trapezoidal rule**

To determine the area $PQRS$ in Figure 30.1,

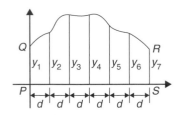

Figure 30.1

(i) Divide base PS into any number of equal intervals, each of width d (the greater the number of intervals, the greater the accuracy).

(ii) Accurately measure ordinates y_1, y_2, y_3, etc.

(iii) Area $PQRS$

$$\approx d\left[\frac{y_1 + y_7}{2} + y_2 + y_3 + y_4 + y_5 + y_6\right]$$

In general, the trapezoidal rule states

$$\textbf{Area} \approx \binom{\textbf{width of}}{\textbf{interval}}\left[\frac{1}{2}\binom{\textbf{first} + \textbf{last}}{\textbf{ordinate}}\right.$$
$$\left. + \binom{\textbf{sum of}}{\substack{\textbf{remaining} \\ \textbf{ordinates}}}\right]$$

(c) **Mid-ordinate rule**

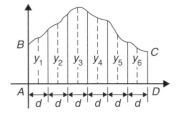

Figure 30.2

To determine the area $ABCD$ of Figure 30.2,

(i) Divide base AD into any number of equal intervals, each of width d (the greater the number of intervals, the greater the accuracy).

(ii) Erect ordinates in the middle of each interval (shown by broken lines in Figure 30.2).

(iii) Accurately measure ordinates y_1, y_2, y_3, etc.

(iv) Area $ABCD$

$$\approx d(y_1 + y_2 + y_3 + y_4 + y_5 + y_6)$$

In general, the mid-ordinate rule states

Area $\approx$ (width of interval)(sum of mid-ordinates)

(d) **Simpson's rule***

To determine the area $PQRS$ of Figure 30.1,

(i) Divide base PS into an **even** number of intervals, each of width d (the greater the number of intervals, the greater the accuracy).

(ii) Accurately measure ordinates y_1, y_2, y_3, etc.

(iii) Area $PQRS \approx \dfrac{d}{3}[(y_1 + y_7) + 4(y_2 + y_4 + y_6) + 2(y_3 + y_5)]$

In general, Simpson's rule states

$$\textbf{Area} \approx \frac{1}{3}\binom{\textbf{width of}}{\textbf{interval}}\left[\binom{\textbf{first} + \textbf{last}}{\textbf{ordinate}}\right.$$
$$\left. + 4\binom{\textbf{sum of even}}{\textbf{ordinates}} + 2\binom{\textbf{sum of remaining}}{\textbf{odd ordinates}}\right]$$

* **Who was Simpson? Thomas Simpson FRS** (20 August 1710–14 May 1761) was the British mathematician who invented Simpson's rule to approximate definite integrals. To find out more go to **www.routledge.com/cw/bird**

Section C

Problem 1.   A car starts from rest and its speed is measured every second for 6 s.

Time t (s)	0	1	2	3	4	5	6
Speed v (m/s)	0	2.5	5.5	8.75	12.5	17.5	24.0

Determine the distance travelled in 6 seconds (i.e. the area under the v/t graph), using (a) the trapezoidal rule (b) the mid-ordinate rule (c) Simpson's rule

A graph of speed/time is shown in Figure 30.3.

(a)   **Trapezoidal rule** (see (b) above)
The time base is divided into six strips, each of width 1 s, and the length of the ordinates measured. Thus,

$$\text{area} \approx (1)\left[\left(\frac{0+24.0}{2}\right) + 2.5 + 5.5\right.$$

$$\left. + 8.75 + 12.5 + 17.5\right] = \mathbf{58.75\,m}$$

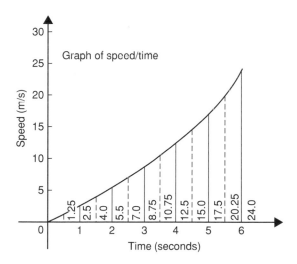

Figure 30.3

(b)   **Mid-ordinate rule** (see (c) above)
The time base is divided into six strips each of width 1 s. Mid-ordinates are erected as shown in Figure 30.3 by the broken lines. The length of each mid-ordinate is measured. Thus,

$$\text{area} \approx (1)[1.25 + 4.0 + 7.0 + 10.75 + 15.0$$

$$+ 20.25] = \mathbf{58.25\,m}$$

(c)   **Simpson's rule** (see (d) above)
The time base is divided into six strips each of width 1 s and the length of the ordinates measured. Thus,

$$\text{area} \approx \frac{1}{3}(1)[(0+24.0) + 4(2.5 + 8.75$$

$$+ 17.5) + 2(5.5 + 12.5)] = \mathbf{58.33\,m}$$

Problem 2.   A river is 15 m wide. Soundings of the depth are made at equal intervals of 3 m across the river and are as shown below.

Depth (m)	0	2.2	3.3	4.5	4.2	2.4	0

Calculate the cross-sectional area of the flow of water at this point using Simpson's rule

From (d) above,

$$\text{Area} \approx \frac{1}{3}(3)[(0+0) + 4(2.2+4.5+2.4) + 2(3.3+4.2)]$$

$$= (1)[0 + 36.4 + 15] = \mathbf{51.4\,m^2}$$

Now try the following Practice Exercise

Practice Exercise 135   Areas of irregular figures (answers on page 1163)

1.   Plot a graph of $y = 3x - x^2$ by completing a table of values of y from $x = 0$ to $x = 3$. Determine the area enclosed by the curve, the x-axis and ordinates $x = 0$ and $x = 3$ by (a) the trapezoidal rule (b) the mid-ordinate rule (c) Simpson's rule.

2.   Plot the graph of $y = 2x^2 + 3$ between $x = 0$ and $x = 4$. Estimate the area enclosed by the curve, the ordinates $x = 0$ and $x = 4$ and the x-axis by an approximate method.

3.   The velocity of a car at one-second intervals is given in the following table.

Time t (s)	0	1	2	3	4	5	6
Velocity v (m/s)	0	2.0	4.5	8.0	14.0	21.0	29.0

Determine the distance travelled in six seconds (i.e. the area under the v/t graph) using Simpson's rule.

Section C

4. The shape of a piece of land is shown in Figure 30.4. To estimate the area of the land, a surveyor takes measurements at intervals of 50 m, perpendicular to the straight portion, with the results shown (the dimensions being in metres). Estimate the area of the land in hectares (1 ha = 10^4 m^2).

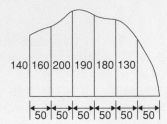

140 | 160 | 200 | 190 | 180 | 130

|← 50 →|← 50 →|← 50 →|← 50 →|← 50 →|← 50 →|

Figure 30.4

5. The deck of a ship is 35 m long. At equal intervals of 5 m the width is given by the following table.

| Width (m) | 0 | 2.8 | 5.2 | 6.5 | 5.8 | 4.1 | 3.0 | 2.3 |

Estimate the area of the deck.

30.2 Volumes of irregular solids

If the cross-sectional areas $A_1, A_2, A_3, \ldots$ of an irregular solid bounded by two parallel planes are known at equal intervals of width d (as shown in Figure 30.5), by Simpson's rule

$$\textbf{Volume, } V = \frac{d}{3}[(A_1 + A_7) + 4(A_2 + A_4 + A_6)$$

$$+ 2(A_3 + A_5)]$$

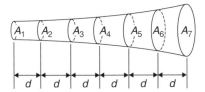

Figure 30.5

Problem 3. A tree trunk is 12 m in length and has a varying cross-section. The cross-sectional areas at intervals of 2 m measured from one end are 0.52, 0.55, 0.59, 0.63, 0.72, 0.84 and 0.97 m^2. Estimate the volume of the tree trunk

A sketch of the tree trunk is similar to that shown in Figure 30.5 above, where $d = 2$ m, $A_1 = 0.52$ m^2, $A_2 = 0.55$ m^2, and so on.

Using Simpson's rule for volumes gives

$$\textbf{Volume} = \frac{2}{3}[(0.52 + 0.97) + 4(0.55 + 0.63 + 0.84)$$

$$+ 2(0.59 + 0.72)]$$

$$= \frac{2}{3}[1.49 + 8.08 + 2.62] = \textbf{8.13 m}^3$$

Problem 4. The areas of seven horizontal cross-sections of a water reservoir at intervals of 10 m are 210, 250, 320, 350, 290, 230 and 170 m^2. Calculate the capacity of the reservoir in litres

Using Simpson's rule for volumes gives

$$\text{Volume} = \frac{10}{3}[(210 + 170) + 4(250 + 350 + 230)$$

$$+ 2(320 + 290)]$$

$$= \frac{10}{3}[380 + 3320 + 1220] = \textbf{16 400 m}^3$$

$16\,400$ m$^3 = 16\,400 \times 10^6$ cm^3
Since 1 litre $= 1000$ cm^3

$$\text{capacity of reservoir} = \frac{16400 \times 10^6}{1000} \text{litres}$$

$$= 16\,400\,000 = \textbf{16.4} \times \textbf{10}^6 \textbf{ litres}$$

Now try the following Practice Exercise

Practice Exercise 136   Volumes of irregular solids (answers on page 1163)

1. The areas of equidistantly spaced sections of the underwater form of a small boat are as follows:
1.76, 2.78, 3.10, 3.12, 2.61, 1.24 and 0.85 m^2. Determine the underwater volume if the sections are 3 m apart.

2. To estimate the amount of earth to be removed when constructing a cutting, the cross-sectional areas at intervals of 8 m were estimated as follows:
0, 2.8, 3.7, 4.5, 4.1, 2.6 and 0 m^2
Estimate the volume of earth to be excavated.

3. The circumference of a 12 m long log of timber of varying circular cross-section is measured at intervals of 2 m along its length and the results are as follows. Estimate the volume of the timber in cubic metres.

Distance from one end (m)	0	2	4	6
Circumference (m)	2.80	3.25	3.94	4.32

Distance from one end (m)	8	10	12
Circumference (m)	5.16	5.82	6.36

30.3 Mean or average values of waveforms

The mean or average value, y, of the waveform shown in Figure 30.6 is given by

$$y = \frac{\text{area under curve}}{\text{length of base, } b}$$

If the mid-ordinate rule is used to find the area under the curve, then

$$y \approx \frac{\text{sum of mid-ordinates}}{\text{number of mid-ordinates}}$$

$$\left(\approx \frac{y_1 + y_2 + y_3 + y_4 + y_5 + y_6 + y_7}{7} \text{ for Figure 30.6} \right)$$

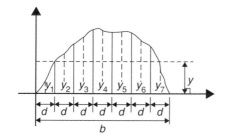

Figure 30.6

For a **sine wave**, the mean or average value

(a) over one complete cycle is **zero** (see Figure 30.7(a)),

(b) over half a cycle is **0.637 × maximum value** or $\frac{2}{\pi}$ **× maximum value**,

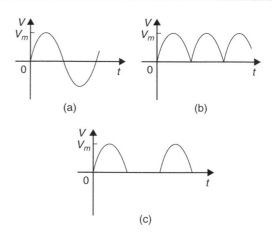

Figure 30.7

(c) of a full-wave rectified waveform (see Figure 30.7(b)) is **0.637 × maximum value**,

(d) of a half-wave rectified waveform (see Figure 30.7(c)) is **0.318 × maximum value** or $\frac{1}{\pi}$ **× maximum value**.

Problem 5. Determine the average values over half a cycle of the periodic waveforms shown in Figure 30.8

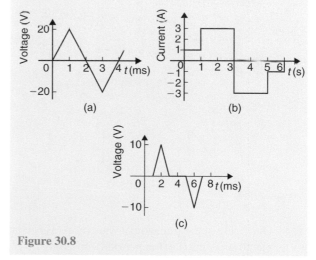

Figure 30.8

(a) Area under triangular waveform (a) for a half cycle is given by

$$\text{Area} = \frac{1}{2}(\text{base})(\text{perpendicular height})$$

$$= \frac{1}{2}(2 \times 10^{-3})(20) = 20 \times 10^{-3} \text{ Vs}$$

Average value of waveform

$$= \frac{\text{area under curve}}{\text{length of base}}$$

$$= \frac{20 \times 10^{-3}\,\text{Vs}}{2 \times 10^{-3}\,\text{s}} = \mathbf{10\,V}$$

(b) Area under waveform (b) for a half cycle
$$= (1 \times 1) + (3 \times 2) = 7\,\text{As}$$

Average value of waveform $= \dfrac{\text{area under curve}}{\text{length of base}}$

$$= \frac{7\,\text{As}}{3\,\text{s}} = \mathbf{2.33\,A}$$

(c) A half cycle of the voltage waveform (c) is completed in 4 ms.

$$\text{Area under curve} = \frac{1}{2}\{(3-1)10^{-3}\}(10)$$

$$= 10 \times 10^{-3}\,\text{Vs}$$

Average value of waveform $= \dfrac{\text{area under curve}}{\text{length of base}}$

$$= \frac{10 \times 10^{-3}\,\text{Vs}}{4 \times 10^{-3}\,\text{s}} = \mathbf{2.5\,V}$$

Problem 6. Determine the mean value of current over one complete cycle of the periodic waveforms shown in Figure 30.9

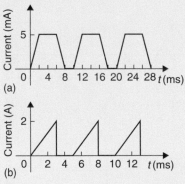

(a)

(b)

Figure 30.9

(a) One cycle of the trapezoidal waveform (a) is completed in 10 ms (i.e. the periodic time is 10 ms).

Area under curve = area of trapezium

$$= \frac{1}{2}(\text{sum of parallel sides})(\text{perpendicular}$$
$$\text{distance between parallel sides})$$

$$= \frac{1}{2}\{(4+8) \times 10^{-3}\}(5 \times 10^{-3})$$

$$= 30 \times 10^{-6}\,\text{As}$$

Mean value over one cycle $= \dfrac{\text{area under curve}}{\text{length of base}}$

$$= \frac{30 \times 10^{-6}\,\text{As}}{10 \times 10^{-3}\,\text{s}}$$

$$= \mathbf{3\,mA}$$

(b) One cycle of the saw-tooth waveform (b) is completed in 5 ms.

$$\text{Area under curve} = \frac{1}{2}(3 \times 10^{-3})(2)$$

$$= 3 \times 10^{-3}\,\text{As}$$

Mean value over one cycle $= \dfrac{\text{area under curve}}{\text{length of base}}$

$$= \frac{3 \times 10^{-3}\,\text{As}}{5 \times 10^{-3}\,\text{s}}$$

$$= \mathbf{0.6\,A}$$

Problem 7. The power used in a manufacturing process during a 6 hour period is recorded at intervals of 1 hour as shown below.

Time (h)	0	1	2	3	4	5	6
Power (kW)	0	14	29	51	45	23	0

Plot a graph of power against time and, by using the mid-ordinate rule, determine (a) the area under the curve and (b) the average value of the power

The graph of power/time is shown in Figure 30.10.

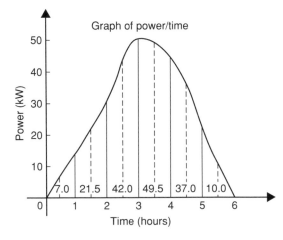

Figure 30.10

(a) The time base is divided into 6 equal intervals, each of width 1 hour. Mid-ordinates are erected (shown by broken lines in Figure 30.10) and measured. The values are shown in Figure 30.10.

Area under curve

$$\approx \text{(width of interval)(sum of mid-ordinates)}$$

$$= (1)[7.0 + 21.5 + 42.0 + 49.5 + 37.0 + 10.0]$$

$$= \mathbf{167\,kWh} \text{ (i.e. a measure of electrical energy)}$$

(b) **Average value of waveform** $\approx \dfrac{\text{area under curve}}{\text{length of base}}$

$$= \frac{167\,\text{kWh}}{6\,\text{h}}$$

$$= \mathbf{27.83\,kW}$$

Alternatively, average value

$$\approx \frac{\text{sum of mid-ordinates}}{\text{number of mid-ordinates}}$$

Problem 8. Figure 30.11 shows a sinusoidal output voltage of a full-wave rectifier. Determine, using the mid-ordinate rule with 6 intervals, the mean output voltage

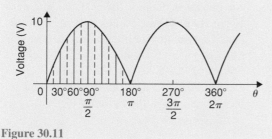

Figure 30.11

One cycle of the output voltage is completed in π radians or 180°. The base is divided into 6 intervals, each of width 30°. The mid-ordinate of each interval will lie at 15°, 45°, 75°, etc.

At 15° the height of the mid-ordinate is $10\sin 15° = 2.588\,\text{V}$

At 45° the height of the mid-ordinate is $10\sin 45° = 7.071\,\text{V}$, and so on.

The results are tabulated below.

Mid-ordinate	Height of mid-ordinate
15°	$10\sin 15° = 2.588\,\text{V}$
45°	$10\sin 45° = 7.071\,\text{V}$
75°	$10\sin 75° = 9.659\,\text{V}$
105°	$10\sin 105° = 9.659\,\text{V}$
135°	$10\sin 135° = 7.071\,\text{V}$
165°	$10\sin 165° = 2.588\,\text{V}$
Sum of mid-ordinates = 38.636 V	

Mean or average value of output voltage

$$\approx \frac{\text{sum of mid ordinates}}{\text{number of mid-ordinates}} = \frac{38.636}{6} = \mathbf{6.439\,V}$$

(With a larger number of intervals a more accurate answer may be obtained.)

For a sine wave the actual mean value is $0.637 \times$ maximum value, which in this problem gives 6.37 V

Problem 9. An indicator diagram for a steam engine is shown in Figure 30.12. The base line has been divided into 6 equally spaced intervals and the lengths of the 7 ordinates measured with the results shown in centimetres. Determine

(a) the area of the indicator diagram using Simpson's rule

(b) the mean pressure in the cylinder given that 1 cm represents 100 kPa

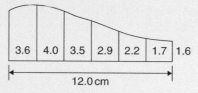

3.6 | 4.0 | 3.5 | 2.9 | 2.2 | 1.7 | 1.6

12.0 cm

Figure 30.12

(a) The width of each interval is $\dfrac{12.0}{6}$ cm. Using Simpson's rule,

$$\text{area} \approx \frac{1}{3}(2.0)[(3.6 + 1.6) + 4(4.0 + 2.9 + 1.7)$$
$$+ 2(3.5 + 2.2)]$$

$$= \frac{2}{3}[5.2 + 34.4 + 11.4] = \mathbf{34\,cm^2}$$

(b) Mean height of ordinates $= \dfrac{\text{area of diagram}}{\text{length of base}}$

$$= \frac{34}{12} = 2.83\,\text{cm}$$

Since 1 cm represents 100 kPa,

mean pressure in the cylinder
$= 2.83\,\text{cm} \times 100\,\text{kPa/cm} = \mathbf{283\,kPa}$

Now try the following Practice Exercise

Practice Exercise 137   Mean or average values of waveforms (answers on page 1163)

1. Determine the mean value of the periodic waveforms shown in Figure 30.13 over a half cycle.

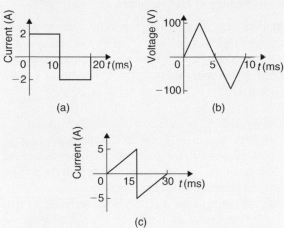

(a)

(b)

(c)

Figure 30.13

2. Find the average value of the periodic waveforms shown in Figure 30.14 over one complete cycle.

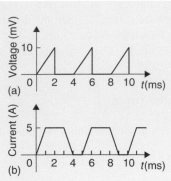

(a)

(b)

Figure 30.14

3. An alternating current has the following values at equal intervals of 5 ms:

Time (ms)	0	5	10	15	20	25	30
Current (A)	0	0.9	2.6	4.9	5.8	3.5	0

Plot a graph of current against time and estimate the area under the curve over the 30 ms period, using the mid-ordinate rule, and determine its mean value.

4. Determine, using an approximate method, the average value of a sine wave of maximum value 50 V for (a) a half cycle (b) a complete cycle.

5. An indicator diagram of a steam engine is 12 cm long. Seven evenly spaced ordinates, including the end ordinates, are measured as follows:
5.90, 5.52, 4.22, 3.63, 3.32, 3.24 and 3.16 cm. Determine the area of the diagram and the mean pressure in the cylinder if 1 cm represents 90 kPa.

For fully worked solutions to each of the problems in Practice Exercises 135 to 137 in this chapter, go to the website:
www.routledge.com/cw/bird

Section C

This assignment covers the material contained in Chapters 29 and 30. *The marks available are shown in brackets at the end of each question.*

1. A rectangular block of alloy has dimensions of 60 mm by 30 mm by 12 mm. Calculate the volume of the alloy in cubic centimetres. (3)

2. Determine how many cubic metres of concrete are required for a 120 m long path, 400 mm wide and 10 cm deep. (3)

3. Find the volume of a cylinder of radius 5.6 cm and height 15.5 cm. Give the answer correct to the nearest cubic centimetre. (3)

4. A garden roller is 0.35 m wide and has a diameter of 0.20 m. What area will it roll in making 40 revolutions? (4)

5. Find the volume of a cone of height 12.5 cm and base diameter 6.0 cm, correct to 1 decimal place. (3)

6. Find (a) the volume and (b) the total surface area of the right-angled triangular prism shown in Figure RT12.1. (9)

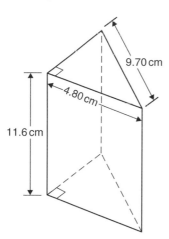

11.6 cm

Figure RT12.1

7. A pyramid having a square base has a volume of 86.4 cm^3. If the perpendicular height is 20 cm, determine the length of each side of the base. (4)

8. A copper pipe is 80 m long. It has a bore of 80 mm and an outside diameter of 100 mm. Calculate, in cubic metres, the volume of copper in the pipe. (4)

9. Find (a) the volume and (b) the surface area of a sphere of diameter 25 mm. (4)

10. A piece of alloy with dimensions 25 mm by 60 mm by 1.60 m is melted down and recast into a cylinder whose diameter is 150 mm. Assuming no wastage, calculate the height of the cylinder in centimetres, correct to 1 decimal place. (4)

11. Determine the volume (in cubic metres) and the total surface area (in square metres) of a solid metal cone of base radius 0.5 m and perpendicular height 1.20 m. Give answers correct to 2 decimal places. (6)

12. A rectangular storage container has dimensions 3.2 m by 90 cm by 60 cm. Determine its volume in (a) m^3 (b) cm^3. (4)

13. Calculate (a) the volume and (b) the total surface area of a 10 cm by 15 cm rectangular pyramid of height 20 cm. (8)

14. A water container is of the form of a central cylindrical part 3.0 m long and diameter 1.0 m, with a hemispherical section surmounted at each end as shown in Figure RT12.2. Determine the maximum capacity of the container, correct to the nearest litre (1 litre = 1000 cm^3).

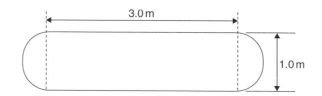

Figure RT12.2

(5)

15. Find the total surface area of a bucket consisting of an inverted frustum of a cone of slant height 35.0 cm and end diameters 60.0 cm and 40.0 cm. (4)

16. A boat has a mass of 20 000 kg. A model of the boat is made to a scale of 1 to 80. If the model is made of the same material as the boat, determine the mass of the model (in grams). (3)

17. Plot a graph of $y = 3x^2 + 5$ from $x = 1$ to $x = 4$. Estimate, correct to 2 decimal places, using 6 intervals, the area enclosed by the curve, the ordinates $x = 1$ and $x = 4$, and the x-axis by
    (a) the trapezoidal rule
    (b) the mid-ordinate rule
    (c) Simpson's rule. (16)

18. A circular cooling tower is 20 m high. The inside diameter of the tower at different heights is given in the following table.

Height (m)	0	5.0	10.0	15.0	20.0
Diameter (m)	16.0	13.3	10.7	8.6	8.0

Determine the area corresponding to each diameter and hence estimate the capacity of the tower in cubic metres. (7)

19. A vehicle starts from rest and its velocity is measured every second for 6 seconds, with the following results.

Time t (s)	0	1	2	3	4	5	6
Velocity v (m/s)	0	1.2	2.4	3.7	5.2	6.0	9.2

Using Simpson's rule, calculate
(a) the distance travelled in 6 s (i.e. the area under the v/t graph),
(b) the average speed over this period. (6)

Formulae/revision hints for Section C
Areas and volumes

Formulae for the areas of common shapes

Area of plane figures

Square

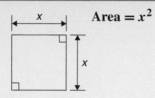

$$\text{Area} = x^2$$

Rectangle

$$\text{Area} = l \times b$$

Parallelogram

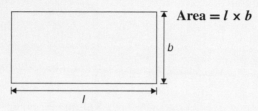

$$\text{Area} = b \times h$$

Triangle

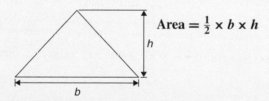

$$\text{Area} = \tfrac{1}{2} \times b \times h$$

Trapezium

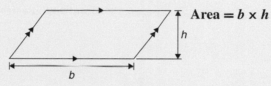

$$\text{Area} = \tfrac{1}{2}(a + b)h$$

Circle

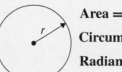

$$\text{Area} = \pi r^2 \text{ or } \frac{\pi d^2}{4}$$

$$\text{Circumference} = 2\pi r$$

Radian measure: 2π radians $= 360$ degrees

Sector of circle

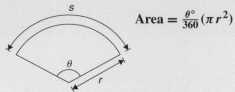

$$\text{Area} = \frac{\theta^\circ}{360}(\pi r^2)$$

Ellipse

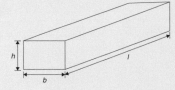

$$\text{Area} = \pi ab$$
$$\text{Perimeter} \approx \pi(a + b)$$

Volumes and surface areas of regular solids

Rectangular prism (or cuboid)

$$\text{Volume} = l \times b \times h$$
$$\text{Surface area} = 2(bh + hl + lb)$$

Cylinder

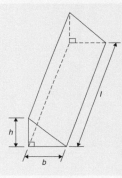

$$\text{Volume} = \pi r^2 h$$
$$\text{Total surface area} = 2\pi rh + 2\pi r^2$$

Triangular prism

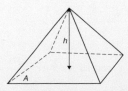

$$\text{Volume} = \frac{1}{2}bhl$$

$$\text{Surface area} = \text{area of each end} +$$
$$\text{area of three sides}$$

Pyramid

$$\text{Volume} = \frac{1}{3} \times A \times h$$

$$\text{Total surface area} =$$
$$\text{sum of areas of triangles}$$
$$\text{forming sides} + \text{area of base}$$

Cone

$$\text{Volume} = \frac{1}{3}\pi r^2 h$$

$$\text{Curved surface area} = \pi rl$$
$$\text{Total surface area} = \pi rl + \pi r^2$$

Sphere

$$\text{Volume} = \tfrac{4}{3}\pi r^3$$
$$\text{Surface area} = 4\pi r^2$$

Frustum of cone

$$\text{Volume} = \tfrac{1}{3}\pi h(R^2 + Rr + r^2)$$
$$\text{Curved surface area} = \pi l(R + r)$$
$$\text{Total surface area} = \pi l(R + r) + \pi r^2 + \pi R^2$$

Frustum of a sphere

$$\text{Surface area of a zone of a sphere} = 2\pi rh$$
$$\text{Volume of frustum of sphere} = \frac{\pi h}{6}\left(h^2 + 3r_1^2 + 3r_2^2\right)$$

Prismoidal rule

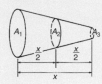

$$\text{Volume, } V = \tfrac{x}{6}[A_1 + 4A_2 + A_3]$$

Trapezoidal rule

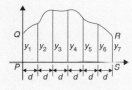

$$\text{Area} = \left(\begin{array}{c}\text{width of} \\ \text{interval}\end{array}\right)\left[\frac{1}{2}\left(\begin{array}{c}\text{first} + \text{last} \\ \text{ordinate}\end{array}\right) + \left(\begin{array}{c}\text{sum of} \\ \text{remaining} \\ \text{ordinates}\end{array}\right)\right]$$

Mid-ordinate rule

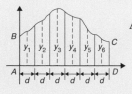

$$\text{Area} = (\text{width of interval})(\text{sum of mid-ordinates})$$

Simpson's rule

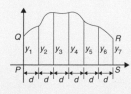

$$\text{Area} = \frac{1}{3}\left(\begin{array}{c}\text{width of} \\ \text{interval}\end{array}\right)\left[\left(\begin{array}{c}\text{first} + \text{last} \\ \text{ordinate}\end{array}\right) + 4\left(\begin{array}{c}\text{sum of even} \\ \text{ordinates}\end{array}\right) + 2\left(\begin{array}{c}\text{sum of remaining} \\ \text{odd ordinates}\end{array}\right)\right]$$

Volumes of irregular solids

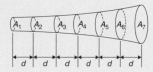

$$\text{Volume, } V = \tfrac{d}{3}[(A_1 + A_7) + 4(A_2 + A_4 + A_6) + 2(A_3 + A_5)]$$

Section C

For a sine wave

over half a cycle, mean value $= \dfrac{2}{\pi} \times$ maximum value

of a full-wave rectified waveform, mean value $= \dfrac{2}{\pi} \times$ maximum value

of a half-wave rectified waveform, mean value $= \dfrac{1}{\pi} \times$ maximum value

Graphs

Chapter 31

Straight line graphs

Why it is important to understand: **Straight line graphs**

Graphs have a wide range of applications in engineering and in physical sciences because of their inherent simplicity. A graph can be used to represent almost any physical situation involving discrete objects and the relationship among them. If two quantities are directly proportional and one is plotted against the other, a straight line is produced. Examples include an applied force on the end of a spring plotted against spring extension, the speed of a flywheel plotted against time and strain in a wire plotted against stress (Hooke's law). In engineering, the straight line graph is the most basic graph to draw and evaluate.

At the end of this chapter, you should be able to:

- understand rectangular axes, scales and co-ordinates
- plot given co-ordinates and draw the best straight line graph
- determine the gradient of a straight line graph
- estimate the vertical-axis intercept
- state the equation of a straight line graph
- plot straight line graphs involving practical engineering examples

31.1 Introduction to graphs

A graph is a visual representation of information, showing how one quantity varies with another related quantity.

We often see graphs in newspapers or in business reports, in travel brochures and government publications. For example, a graph of the share price (in pence) over a six month period for a drinks company, Fizzy Pops, is shown in Figure 31.1.

Generally, we see that the share price increases to a high of 400 p in June, but dips down to around 280 p in August before recovering slightly in September.

A graph should convey information more quickly to the reader than if the same information was explained in words.

When this chapter is completed you should be able to draw up a table of values, plot co-ordinates, determine the gradient and state the equation of a straight line graph. Some typical practical examples are included in which straight lines are used.

31.2 Axes, scales and co-ordinates

We are probably all familiar with reading a map to locate a town, or a local map to locate a particular street. For example, a street map of central Portsmouth is shown in Figure 31.2. Notice the squares drawn horizontally and vertically on the map; this is called a **grid** and enables us to locate a place of interest or a particular road. Most maps contain such a grid.

We locate places of interest on the map by stating a letter and a number – this is called the **grid reference**.

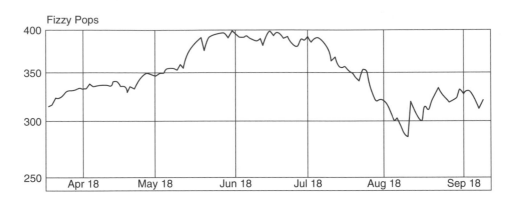

Figure 31.1

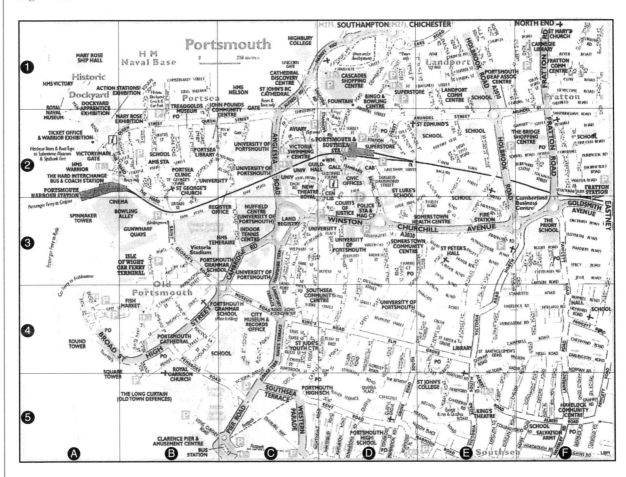

Figure 31.2 Reprinted with permission from AA Media Ltd.

For example, on the map, the Portsmouth & Southsea station is in square D2, King's Theatre is in square E5, HMS Warrior is in square A2, Gunwharf Quays is in square B3 and High Street is in square B4.

Portsmouth & Southsea station is located by moving horizontally along the bottom of the map until the square labelled D is reached and then moving vertically upwards until square 2 is met.

The letter/number, D2, is referred to as **co-ordinates**; i.e. co-ordinates are used to locate the position of a point on a map. If you are familiar with using a map in this way then you should have no difficulties with graphs, because similar co-ordinates are used with graphs.

As stated earlier, a **graph** is a visual representation of information, showing how one quantity varies with another related quantity. The most common method of showing a relationship between two sets of data is to use a pair of reference axes – these are two lines drawn at right angles to each other, often called **Cartesian** (named after **Descartes***) or **rectangular axes**, as shown in Figure 31.3.

The horizontal axis is labelled the x-axis and the vertical axis is labelled the y-axis. The point where x is 0 and y is 0 is called the **origin**.

x values have **scales** that are positive to the right of the origin and negative to the left. y values have scales that are positive up from the origin and negative down from the origin.

Co-ordinates are written with brackets and a comma in between two numbers. For example, point A is shown with co-ordinates (3, 2) and is located by starting at the

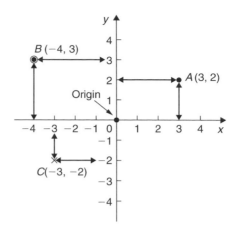

Figure 31.3

origin and moving 3 units in the positive x direction (i.e. to the right) and then 2 units in the positive y direction (i.e. up).

When co-ordinates are stated the first number is always the x value and the second number is always the y value. In Figure 31.3, point B has co-ordinates $(-4, 3)$ and point C has co-ordinates $(-3, -2)$

31.3 Straight line graphs

The distances travelled by a car in certain periods of time are shown in the following table of values.

Time (s)	10	20	30	40	50	60
Distance travelled (m)	50	100	150	200	250	300

We will plot time on the horizontal (or x) axis with a scale of 1 cm = 10 s.

We will plot distance on the vertical (or y) axis with a scale of 1 cm = 50 m.

(When choosing scales it is better to choose ones such as 1 cm = 1 unit, 1 cm = 2 units or 1 cm = 10 units because doing so makes reading values between these values easier.)

With the above data, the (x, y) co-ordinates become (time, distance) co-ordinates; i.e. the co-ordinates are (10, 50), (20, 100), (30, 150), and so on.

The co-ordinates are shown plotted in Figure 31.4 using crosses. (Alternatively, a dot or a dot and circle may be used, as shown in Figure 31.3.)

A straight line is drawn through the plotted co-ordinates as shown in Figure 31.4.

* **Who was Descartes?** **René Descartes** (31 March 1596–11 February 1650) was a French philosopher, mathematician and writer. He wrote many influential texts including *Meditations on First Philosophy*. Descartes is best known for the philosophical statement '*Cogito ergo sum*' (I think, therefore I am), found in part IV of *Discourse on the Method*. To find out more go to **www.routledge.com/cw/bird**

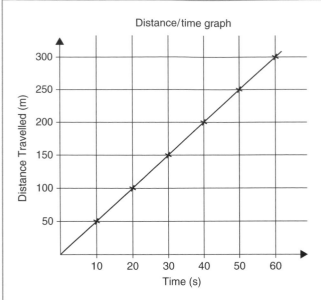

Figure 31.4

Student task

The following table gives the force F newtons which, when applied to a lifting machine, overcomes a corresponding load of L newtons.

F (newtons)	19	35	50	93	125	147
L (newtons)	40	120	230	410	540	680

1. Plot L horizontally and F vertically.

2. Scales are normally chosen such that the graph occupies as much space as possible on the graph paper. So in this case, the following scales are chosen.

   > Horizontal axis (i.e. L): 1 cm = 50 N

   Vertical axis (i.e. F): 1 cm = 10 N

3. Draw the axes and label them L (newtons) for the horizontal axis and F (newtons) for the vertical axis.

4. Label the origin as 0.

5. Write on the horizontal scaling at 100, 200, 300, and so on, every 2 cm.

6. Write on the vertical scaling at 10, 20, 30, and so on, every 1 cm.

7. Plot on the graph the co-ordinates (40, 19), (120, 35), (230, 50), (410, 93), (540, 125) and (680, 147), marking each with a cross or a dot.

8. Using a ruler, draw the best straight line through the points. You will notice that not all of the points lie exactly on a straight line. This is quite normal with experimental values. In a practical situation it would be surprising if all of the points lay exactly on a straight line.

9. Extend the straight line at each end.

10. From the graph, determine the force applied when the load is 325 N. It should be close to 75 N. This process of finding an equivalent value within the given data is called **interpolation**. Similarly, determine the load that a force of 45 N will overcome. It should be close to 170 N.

11. From the graph, determine the force needed to overcome a 750 N load. It should be close to 161 N. This process of finding an equivalent value outside the given data is called **extrapolation**. To extrapolate we need to have extended the straight line drawn. Similarly, determine the force applied when the load is zero. It should be close to 11 N. The point where the straight line crosses the vertical axis is called the **vertical-axis intercept**. So, in this case, the vertical-axis intercept = 11 N at co-ordinates (0, 11).

The graph you have drawn should look something like Figure 31.5 shown below.

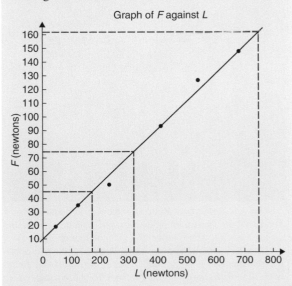

Figure 31.5

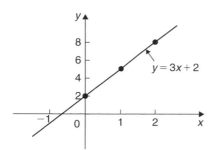

Figure 31.6 Graph of y/x

In another example, let the relationship between two variables x and y be $y = 3x + 2$

When $x = 0$, $y = 0 + 2 = 2$

When $x = 1$, $y = 3 + 2 = 5$

When $x = 2$, $y = 6 + 2 = 8$, and so on.

The co-ordinates $(0, 2)$, $(1, 5)$ and $(2, 8)$ have been produced and are plotted, with others, as shown in Figure 31.6.

When the points are joined together **a straight line graph results**, i.e. $y = 3x + 2$ is a straight line graph.

Summary of general rules to be applied when drawing graphs

(a) Give the graph a title clearly explaining what is being illustrated.

(b) Choose scales such that the graph occupies as much space as possible on the graph paper being used.

(c) Choose scales so that interpolation is made as easy as possible. Usually scales such as $1\,\text{cm} = 1\,\text{unit}$, $1\,\text{cm} = 2\,\text{units}$ or $1\,\text{cm} = 10\,\text{units}$ are used. Awkward scales such as $1\,\text{cm} = 3\,\text{units}$ or $1\,\text{cm} = 7$ units should not be used.

(d) The scales need not start at zero, particularly when starting at zero produces an accumulation of points within a small area of the graph paper.

(e) The co-ordinates, or points, should be clearly marked. This is achieved by a cross, or a dot and circle, or just by a dot (see Figure 31.3).

(f) A statement should be made next to each axis explaining the numbers represented with their appropriate units.

(g) Sufficient numbers should be written next to each axis without cramping.

> **Problem 1.** Plot the graph $y = 4x + 3$ in the range $x = -3$ to $x = +4$. From the graph, find (a) the value of y when $x = 2.2$ and (b) the value of x when $y = -3$

Whenever an equation is given and a graph is required, a table giving corresponding values of the variable is necessary. The table is achieved as follows:

When $x = -3$, $y = 4x + 3 = 4(-3) + 3$

$$= -12 + 3 = -9$$

When $x = -2$, $y = 4(-2) + 3$

$$= -8 + 3 = -5, \text{ and so on.}$$

Such a table is shown below.

x	-3	-2	-1	0	1	2	3	4
y	-9	-5	-1	3	7	11	15	19

The co-ordinates $(-3, -9)$, $(-2, -5)$, $(-1, -1)$, and so on, are plotted and joined together to produce the straight line shown in Figure 31.7. (Note that the scales used on the x and y axes do not have to be the same.) From the graph:

(a) when $x = 2.2$, $\mathbf{y = 11.8}$, and

(b) when $y = -3$, $\mathbf{x = -1.5}$

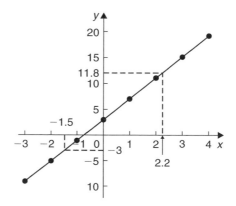

Figure 31.7

Section D

Now try the following Practice Exercise

Practice Exercise 138   Straight line graphs (answers on page 1163)

1. Assuming graph paper measuring 20 cm by 20 cm is available, suggest suitable scales for the following ranges of values.
   (a) Horizontal axis: 3 V to 55 V; vertical axis: 10 Ω to 180 Ω.

   (b) Horizontal axis: 7 m to 86 m; vertical axis: 0.3 V to 1.69 V.

   (c) Horizontal axis: 5 N to 150 N; vertical axis: 0.6 mm to 3.4 mm.

2. Corresponding values obtained experimentally for two quantities are

x	−5	−3	−1	0	2	4
y	−13	−9	−5	−3	1	5

   Plot a graph of y (vertically) against x (horizontally) to scales of 2 cm = 1 for the horizontal x-axis and 1 cm = 1 for the vertical y-axis. (This graph will need the whole of the graph paper, with the origin somewhere in the centre of the paper.)
   From the graph, find

   (a) the value of y when $x = 1$

   (b) the value of y when $x = -2.5$

   (c) the value of x when $y = -6$

   (d) the value of x when $y = 5$

3. Corresponding values obtained experimentally for two quantities are

x	−2.0	−0.5	0	1.0	2.5	3.0	5.0
y	−13.0	−5.5	−3.0	2.0	9.5	12.0	22.0

   Use a horizontal scale for x of 1 cm = $\frac{1}{2}$ unit and a vertical scale for y of 1 cm = 2 units and draw a graph of x against y. Label the graph and each of its axes. By interpolation, find from the graph the value of y when x is 3.5

4. Draw a graph of $y - 3x + 5 = 0$ over a range of $x = -3$ to $x = 4$. Hence determine
   (a) the value of y when $x = 1.3$

   (b) the value of x when $y = -9.2$

5. The speed n rev/min of a motor changes when the voltage V across the armature is varied. The results are shown in the following table.

n (rev/min)	560	720	900	1010	1240	1410
V (volts)		80	100	120	140	160	180

   It is suspected that one of the readings taken of the speed is inaccurate. Plot a graph of speed (horizontally) against voltage (vertically) and find this value. Find also
   (a) the speed at a voltage of 132 V

   (b) the voltage at a speed of 1300 rev/min.

31.4 Gradients, intercepts and equations of graphs

Gradients

The **gradient or slope** of a straight line is the ratio of the change in the value of y to the change in the value of x between any two points on the line. If, as x increases ($\rightarrow$), y also increases ($\uparrow$), then the gradient is positive. In Figure 31.8(a) a straight line graph $y = 2x + 1$ is shown. To find the gradient of this straight line, choose two points on the straight line graph, such as A and C.

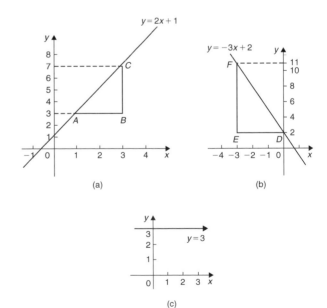

(a)

(b)

(c)

Figure 31.8

Then construct a right-angled triangle, such as ABC, where BC is vertical and AB is horizontal.

Then, **gradient of AC** $= \dfrac{\text{change in } y}{\text{change in } x} = \dfrac{CB}{BA}$

$$= \dfrac{7-3}{3-1} = \dfrac{4}{2} = \mathbf{2}$$

In Figure 31.8(b) a straight line graph $y = -3x + 2$ is shown. To find the gradient of this straight line, choose two points on the straight line graph, such as D and F. Then construct a right-angled triangle, such as DEF, where EF is vertical and DE is horizontal.

Then, **gradient of DF** $= \dfrac{\text{change in } y}{\text{change in } x} = \dfrac{FE}{ED}$

$$= \dfrac{11-2}{-3-0} = \dfrac{9}{-3} = \mathbf{-3}$$

Figure 31.8(c) shows a straight line graph $y = 3$. **Since the straight line is horizontal the gradient is zero.**

The y-axis intercept

The value of y when $x = 0$ is called the **y-axis intercept**. In Figure 31.8(a) the y-axis intercept is 1 and in Figure 31.8(b) the y-axis intercept is 2

The equation of a straight line graph

The general equation of a straight line graph is

$$y = mx + c$$

where m is the gradient and c is the y-axis intercept. Thus, as we have found in Figure 31.8(a), $y = 2x + 1$ represents a straight line of gradient 2 and y-axis intercept 1. So, given the equation $y = 2x + 1$, we are able to state, on sight, that the gradient $= 2$ and the y-axis intercept $= 1$, without the need for any analysis.
Similarly, in Figure 31.8(b) $y = -3x + 2$ represents a straight line of gradient -3 and y-axis intercept 2
In Figure 31.8(c), $y = 3$ may be rewritten as $y = 0x + 3$ and therefore represents a straight line of gradient 0 and y-axis intercept 3
Here are some worked problems to help understanding of gradients, intercepts and equations of graphs.

Problem 2. Plot the following graphs on the same axes in the range $x = -4$ to $x = +4$ and determine the gradient of each.

(a) $y = x$ (b) $y = x + 2$

(c) $y = x + 5$ (d) $y = x - 3$

A table of co-ordinates is produced for each graph.

(a) $y = x$

x	−4	−3	−2	−1	0	1	2	3	4
y	−4	−3	−2	−1	0	1	2	3	4

(b) $y = x + 2$

x	−4	−3	−2	−1	0	1	2	3	4
y	−2	−1	0	1	2	3	4	5	6

(c) $y = x + 5$

x	−4	−3	−2	−1	0	1	2	3	4
y	1	2	3	4	5	6	7	8	9

(d) $y = x - 3$

x	−4	−3	−2	−1	0	1	2	3	4
y	−7	−6	−5	−4	−3	−2	−1	0	1

The co-ordinates are plotted and joined for each graph. The results are shown in Figure 31.9. Each of the straight lines produced is parallel to the others; i.e. the slope or gradient is the same for each.
To find the gradient of any straight line, say, $y = x - 3$, a horizontal and vertical component needs to be constructed. In Figure 31.9, AB is constructed vertically at $x = 4$ and BC is constructed horizontally at $y = -3$

The gradient of $AC = \dfrac{AB}{BC} = \dfrac{1 - (-3)}{4 - 0} = \dfrac{4}{4} = 1$

i.e. the gradient of the straight line $y = x - 3$ is 1, which could have been deduced 'on sight' since $y = 1x - 3$ represents a straight line graph with gradient 1 and y-axis intercept of -3
The actual positioning of AB and BC is unimportant because the gradient is also given by

$$\dfrac{DE}{EF} = \dfrac{-1 - (-2)}{2 - 1} = \dfrac{1}{1} = 1$$

The slope or gradient of each of the straight lines in Figure 31.9 is thus 1 since they are parallel to each other.

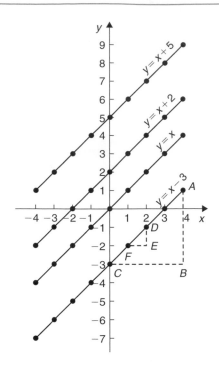

Figure 31.9

Problem 3. Plot the following graphs on the same axes between the values $x = -3$ to $x = +3$ and determine the gradient and y-axis intercept of each.

(a) $y = 3x$ (b) $y = 3x + 7$

(c) $y = -4x + 4$ (d) $y = -4x - 5$

A table of co-ordinates is drawn up for each equation.

(a) $y = 3x$

x	-3	-2	-1	0	1	2	3
y	-9	-6	-3	0	3	6	9

(b) $y = 3x + 7$

x	-3	-2	-1	0	1	2	3
y	-2	1	4	7	10	13	16

(c) $y = -4x + 4$

x	-3	-2	-1	0	1	2	3
y	16	12	8	4	0	-4	-8

(d) $y = -4x - 5$

x	-3	-2	-1	0	1	2	3
y	7	3	-1	-5	-9	-13	-17

Each of the graphs is plotted as shown in Figure 31.10 and each is a straight line. $y = 3x$ and $y = 3x + 7$ are parallel to each other and thus have the same gradient. The gradient of AC is given by

$$\frac{CB}{BA} = \frac{16 - 7}{3 - 0} = \frac{9}{3} = 3$$

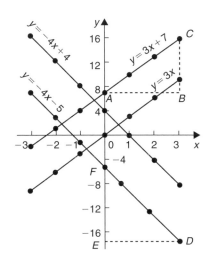

Figure 31.10

Hence, **the gradients of both $y = 3x$ and $y = 3x + 7$ are 3**, which could have been deduced 'on sight'.
$y = -4x + 4$ and $y = -4x - 5$ are parallel to each other and thus have the same gradient. The gradient of DF is given by

$$\frac{FE}{ED} = \frac{-5 - (-17)}{0 - 3} = \frac{12}{-3} = -4$$

Hence, **the gradient of both $y = -4x + 4$ and $y = -4x - 5$ is -4**, which, again, could have been deduced 'on sight'.

The y-axis intercept means the value of y where the straight line cuts the y-axis. From Figure 31.10,

    $y = 3x$ cuts the y-axis at $y = 0$
    $y = 3x + 7$ cuts the y-axis at $y = +7$
    $y = -4x + 4$ cuts the y-axis at $y = +4$
    $y = -4x - 5$ cuts the y-axis at $y = -5$

Some general conclusions can be drawn from the graphs shown in Figures 31.9 and 31.10. When an equation is of the form $y = mx + c$, where m and c are constants, then

(a)   a graph of y against x produces a straight line,

(b) m represents the slope or gradient of the line, and

(c) c represents the y-axis intercept.

Thus, given an equation such as $y = 3x + 7$, it may be deduced 'on sight' that its gradient is $+3$ and its y-axis intercept is $+7$, as shown in Figure 31.10. Similarly, if $y = -4x - 5$, the gradient is -4 and the y-axis intercept is -5, as shown in Figure 31.10.

When plotting a graph of the form $y = mx + c$, only two co-ordinates need be determined. When the co-ordinates are plotted a straight line is drawn between the two points. Normally, three co-ordinates are determined, the third one acting as a check.

> **Problem 4.** Plot the graph $3x + y + 1 = 0$ and $2y - 5 = x$ on the same axes and find their point of intersection

Rearranging $3x + y + 1 = 0$ gives $y = -3x - 1$

Rearranging $2y - 5 = x$ gives $2y = x + 5$ and

$$y = \frac{1}{2}x + 2\frac{1}{2}$$

Since both equations are of the form $y = mx + c$, both are straight lines. Knowing an equation is a straight line means that only two co-ordinates need to be plotted and a straight line drawn through them. A third co-ordinate is usually determined to act as a check. A table of values is produced for each equation as shown below.

x	1	0	−1
$-3x - 1$	−4	−1	2

x	2	0	−3
$\frac{1}{2}x + 2\frac{1}{2}$	$3\frac{1}{2}$	$2\frac{1}{2}$	1

The graphs are plotted as shown in Figure 31.11.
The **two straight lines are seen to intersect at $(-1, 2)$**

> **Problem 5.** If graphs of y against x were to be plotted for each of the following, state (i) the gradient and (ii) the y-axis intercept.
>
> (a) $y = 9x + 2$ (b) $y = -4x + 7$
> (c) $y = 3x$ (d) $y = -5x - 3$
> (e) $y = 6$ (f) $y = x$

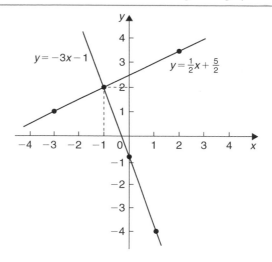

Figure 31.11

If $y = mx + c$ then $m =$ **gradient** and $c = y$**-axis intercept**.

(a)   If $y = 9x + 2$, then   (i) **gradient = 9**

                                           (ii) **y-axis intercept = 2**

(b)   If $y = -4x + 7$, then   (i) **gradient = −4**

                                           (ii) **y-axis intercept = 7**

(c)   If $y = 3x$ i.e. $y = 3x + 0$,

     then   (i) **gradient = 3**

                                           (ii) **y-axis intercept = 0**

    i.e. the straight line passes through the origin.

(d)   If $y = -5x - 3$, then   (i) **gradient = −5**

                                           (ii) **y-axis intercept = −3**

(e)   If $y = 6$ i.e. $y = 0x + 6$,

     then   (i) **gradient = 0**

                                           (ii) **y-axis intercept = 6**

    i.e. $y = 6$ is a straight horizontal line.

(f)   If $y = x$ i.e. $y = 1x + 0$,

     then   (i) **gradient = 1**

                                           (ii) **y-axis intercept = 0**

Since $y = x$, as x increases, y increases by the same amount; i.e. **y is directly proportional to x**.

> **Problem 6.** Without drawing graphs, determine the gradient and y-axis intercept for each of the following equations.
>
> (a) $y + 4 = 3x$ (b) $2y + 8x = 6$ (c) $3x = 4y + 7$

If $y = mx + c$ then m = **gradient** and c = **y-axis intercept**.

(a) Transposing $y + 4 = 3x$ gives $y = 3x - 4$

Hence, **gradient = 3** and **y-axis intercept = -4**

(b) Transposing $2y + 8x = 6$ gives $2y = -8x + 6$

Dividing both sides by 2 gives $y = -4x + 3$

Hence, **gradient = -4** and **y-axis intercept = 3**

(c) Transposing $3x = 4y + 7$ gives $3x - 7 = 4y$
or $\qquad\qquad 4y = 3x - 7$

Dividing both sides by 4 gives $y = \dfrac{3}{4}x - \dfrac{7}{4}$
or $\qquad\qquad y = 0.75x - 1.75$

Hence, **gradient = 0.75** and **y-axis intercept = -1.75**

Problem 7. Without plotting graphs, determine the gradient and y-axis intercept values of the following equations.
(a) $y = 7x - 3$ $\qquad$ (b) $3y = -6x + 2$

(c) $y - 2 = 4x + 9$ $\qquad$ (d) $\dfrac{y}{3} = \dfrac{x}{2} - \dfrac{1}{5}$

(e) $2x + 9y + 1 = 0$

(a) $y = 7x - 3$ is of the form $y = mx + c$

Hence, **gradient, $m = 7$** and **y-axis intercept, $c = -3$**

(b) Rearranging $3y = -6x + 2$ gives $y = -\dfrac{6x}{3} + \dfrac{2}{3}$,

i.e. $\qquad\qquad y = -2x + \dfrac{2}{3}$
which is of the form $y = mx + c$

Hence, **gradient, $m = -2$** and **y-axis intercept, $c = \dfrac{2}{3}$**

(c) Rearranging $y - 2 = 4x + 9$ gives $y = 4x + 11$
Hence, **gradient = 4** and **y-axis intercept = 11**

(d) Rearranging $\dfrac{y}{3} = \dfrac{x}{2} - \dfrac{1}{5}$ gives $y = 3\left(\dfrac{x}{2} - \dfrac{1}{5}\right)$
$\qquad\qquad\qquad = \dfrac{3}{2}x - \dfrac{3}{5}$

Hence, **gradient = $\dfrac{3}{2}$** and **y-axis intercept = $-\dfrac{3}{5}$**

(e) Rearranging $2x + 9y + 1 = 0$ gives $9y = -2x - 1$,
i.e. $y = -\dfrac{2}{9}x - \dfrac{1}{9}$

Hence, **gradient = $-\dfrac{2}{9}$** and

y-axis intercept = $-\dfrac{1}{9}$

Problem 8. Determine for the straight line shown in Figure 31.12 (a) the gradient and (b) the equation of the graph

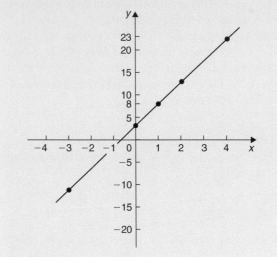

Figure 31.12

(a) A right-angled triangle ABC is constructed on the graph as shown in Figure 31.13.

$$\textbf{Gradient} = \frac{AC}{CB} = \frac{23 - 8}{4 - 1} = \frac{15}{3} = \mathbf{5}$$

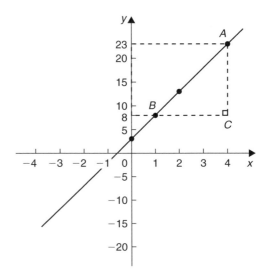

Figure 31.13

(b) The y-axis intercept at $x = 0$ is seen to be at $y = 3$

$y = mx + c$ is a straight line graph where $m = $ gradient and $c = y$-axis intercept.

From above, $m = 5$ and $c = 3$

Hence, the equation of the graph is $\boldsymbol{y = 5x + 3}$

Problem 9. Determine the equation of the straight line shown in Figure 31.14

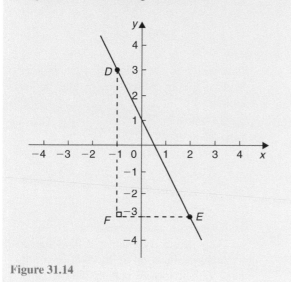

Figure 31.14

The triangle DEF is shown constructed in Figure 31.14.

$$\text{Gradient of } DE = \frac{DF}{FE} = \frac{3 - (-3)}{-1 - 2} = \frac{6}{-3} = -2$$

and the y-axis intercept $= 1$
Hence, **the equation of the straight line is $y = mx + c$**
i.e. $\boldsymbol{y = -2x + 1}$

Problem 10. The velocity of a body was measured at various times and the results obtained were as follows:

Velocity v (m/s)	8	10.5	13	15.5	18	20.5	23
Time t (s)	1	2	3	4	5	6	7

Plot a graph of velocity (vertically) against time (horizontally) and determine the equation of the graph

Suitable scales are chosen and the co-ordinates (1, 8), (2, 10.5), (3, 13), and so on, are plotted as shown in Figure 31.15.

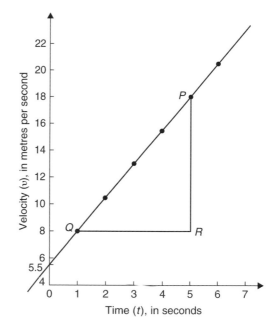

Figure 31.15

The right-angled triangle PRQ is constructed on the graph as shown in Figure 31.15

$$\textbf{Gradient of } \boldsymbol{PQ} = \frac{PR}{RQ} = \frac{18 - 8}{5 - 1} = \frac{10}{4} = \textbf{2.5}$$

The **vertical axis intercept** is at $v = \textbf{5.5 m/s}$.
The equation of a straight line graph is $y = mx + c$. In this case, t corresponds to x and v corresponds to y. Hence, the equation of the graph shown in Figure 31.15 is $v = mt + c$. But, from above, gradient, $m = 2.5$ and v-axis intercept, $c = 5.5$
Hence, **the equation of the graph is $v = 2.5t + 5.5$**

Problem 11. Determine the gradient of the straight line graph passing through the co-ordinates (a) $(-2, 5)$ and $(3, 4)$, and (b) $(-2, -3)$ and $(-1, 3)$

From Figure 31.16, a straight line graph passing through co-ordinates (x_1, y_1) and (x_2, y_2) has a gradient given by

$$m = \frac{y_2 - y_1}{x_2 - x_1}$$

(a) A straight line passes through $(-2, 5)$ and $(3, 4)$, hence $x_1 = -2$, $y_1 = 5$, $x_2 = 3$ and $y_2 = 4$, hence,

$$\textbf{gradient, } \boldsymbol{m} = \frac{y_2 - y_1}{x_2 - x_1} = \frac{4 - 5}{3 - (-2)} = -\frac{1}{5}$$

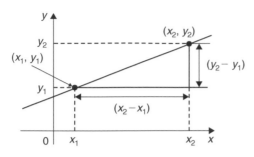

Figure 31.16

(b) A straight line passes through $(-2, -3)$ and $(-1, 3)$, hence $x_1 = -2$, $y_1 = -3$, $x_2 = -1$ and $y_2 = 3$, hence, **gradient,**

$$m = \frac{y_2 - y_1}{x_2 - x_1} = \frac{3 - (-3)}{-1 - (-2)} = \frac{3 + 3}{-1 + 2} = \frac{6}{1} = 6$$

Now try the following Practice Exercise

Practice Exercise 139   Gradients, intercepts and equations of graphs (answers on page 1164)

1. The equation of a line is $4y = 2x + 5$. A table of corresponding values is produced and is shown below. Complete the table and plot a graph of y against x. Find the gradient of the graph.

x	-4	-3	-2	-1	0	1	2	3	4
y		-0.25			1.25			3.25	

2. Determine the gradient and intercept on the y-axis for each of the following equations.
    (a) $y = 4x - 2$ (b) $y = -x$
    (c) $y = -3x - 4$ (d) $y = 4$

3. Find the gradient and intercept on the y-axis for each of the following equations.
    (a) $2y - 1 = 4x$ (b) $6x - 2y = 5$
    (c) $3(2y - 1) = \dfrac{x}{4}$

Determine the gradient and y-axis intercept for each of the equations in Problems 4 and 5 and sketch the graphs.

4. (a) $y = 6x - 3$ (b) $y = -2x + 4$
    (c) $y = 3x$ (d) $y = 7$

5. (a) $2y + 1 = 4x$ (b) $2x + 3y + 5 = 0$
    (c) $3(2y - 4) = \dfrac{x}{3}$ (d) $5x - \dfrac{y}{2} - \dfrac{7}{3} = 0$

6. Determine the gradient of the straight line graphs passing through the co-ordinates:
    (a) $(2, 7)$ and $(-3, 4)$
    (b) $(-4, -1)$ and $(-5, 3)$
    (c) $\left(\dfrac{1}{4}, -\dfrac{3}{4}\right)$ and $\left(-\dfrac{1}{2}, \dfrac{5}{8}\right)$

7. State which of the following equations will produce graphs which are parallel to one another.
    (a) $y - 4 = 2x$ (b) $4x = -(y + 1)$
    (c) $x = \dfrac{1}{2}(y + 5)$ (d) $1 + \dfrac{1}{2}y = \dfrac{3}{2}x$
    (e) $2x = \dfrac{1}{2}(7 - y)$

8. Draw on the same axes the graphs of $y = 3x - 5$ and $3y + 2x = 7$. Find the co-ordinates of the point of intersection. Check the result obtained by solving the two simultaneous equations algebraically.

9. Plot the graphs $y = 2x + 3$ and $2y = 15 - 2x$ on the same axes and determine their point of intersection.

10. Draw on the same axes the graphs of $y = 3x - 1$ and $y + 2x = 4$. Find the co-ordinates of the point of intersection.

11. A piece of elastic is tied to a support so that it hangs vertically and a pan, on which weights can be placed, is attached to the free end. The length of the elastic is measured as various weights are added to the pan and the results obtained are as follows:

Load, W (N)	5	10	15	20	25
Length, l (cm)	60	72	84	96	108

Plot a graph of load (horizontally) against length (vertically) and determine

(a) the value of length when the load is 17 N

(b) the value of load when the length is 74 cm

(c) its gradient

(d) the equation of the graph

12. The following table gives the effort P to lift a load W with a small lifting machine.

W (N)	10	20	30	40	50	60
P (N)	5.1	6.4	8.1	9.6	10.9	12.4

Plot W horizontally against P vertically and show that the values lie approximately on a straight line. Determine the probable relationship connecting P and W in the form $P = aW + b$.

13. In an experiment the speeds N rpm of a flywheel slowly coming to rest were recorded against the time t in minutes. Plot the results and show that N and t are connected by an equation of the form $N = at + b$. Find probable values of a and b.

t (min)	2	4	6	8	10	12	14
N (rev/min)	372	333	292	252	210	177	132

31.5 Practical problems involving straight line graphs

When a set of co-ordinate values are given or are obtained experimentally and it is believed that they follow a law of the form $y = mx + c$, if a straight line can be drawn reasonably close to most of the co-ordinate values when plotted, this verifies that a law of the form $y = mx + c$ exists. From the graph, constants m (i.e. gradient) and c (i.e. y-axis intercept) can be determined. Here are some worked problems in which practical situations are featured.

Problem 12. The temperature in degrees Celsius* and the corresponding values in degrees Fahrenheit

* **Who was** Celsius? See page 80, then go to **www.routledge. com/cw/bird**

are shown in the table below. Construct rectangular axes, choose suitable scales and plot a graph of degrees Celsius (on the horizontal axis) against degrees Fahrenheit (on the vertical scale).

°C	10	20	40	60	80	100
°F	50	68	104	140	176	212

From the graph find (a) the temperature in degrees Fahrenheit at 55°C, (b) the temperature in degrees Celsius at 167°F, (c) the Fahrenheit temperature at 0°C and (d) the Celsius temperature at 230°F

The co-ordinates (10, 50), (20, 68), (40, 104), and so on are plotted as shown in Figure 31.17. When the co-ordinates are joined, a straight line is produced. Since a straight line results, there is a linear relationship between degrees Celsius and degrees Fahrenheit.

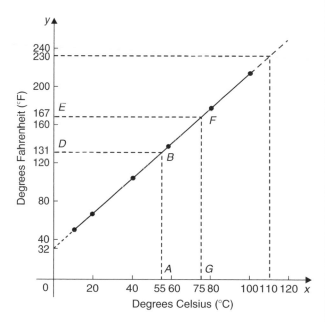

Figure 31.17

(a) To find the Fahrenheit temperature at 55°C, a vertical line AB is constructed from the horizontal axis to meet the straight line at B. The point where the horizontal line BD meets the vertical axis indicates the equivalent Fahrenheit temperature.

Hence, **55°C is equivalent to 131°F**.

This process of finding an equivalent value in between the given information in the above table is called **interpolation**.

(b) To find the Celsius temperature at 167°F, a horizontal line *EF* is constructed as shown in Figure 31.17. The point where the vertical line *FG* cuts the horizontal axis indicates the equivalent Celsius temperature.

Hence, 167°F is equivalent to 75°C.

(c) If the graph is assumed to be linear even outside of the given data, the graph may be extended at both ends (shown by broken lines in Figure 31.17).

From Figure 31.17, **0°C corresponds to 32°F**.

(d) **230°F is seen to correspond to 110°C.**

The process of finding equivalent values outside of the given range is called **extrapolation**.

Problem 13. In an experiment on Charles's law,* the value of the volume of gas, V m^3, was measured for various temperatures, T°C. The results are shown below.

V m^3	25.0	25.8	26.6	27.4	28.2	29.0
T°C	60	65	70	75	80	85

Plot a graph of volume (vertical) against temperature (horizontal) and from it find (a) the temperature when the volume is 28.6 m^3 and (b) the volume when the temperature is 67°C

If a graph is plotted with both the scales starting at zero then the result is as shown in Figure 31.18. All of the points lie in the top right-hand corner of the graph, making interpolation difficult. A more accurate graph is obtained if the temperature axis starts at 55°C and the volume axis starts at 24.5 m^3. The axes corresponding to these values are shown by the broken lines in Figure 31.18 and are called **false axes**, since the origin is not now at zero. A magnified version of this relevant part of the graph is shown in Figure 31.19. From the graph,

(a) When the volume is 28.6 m^3, the equivalent temperature is **82.5°C.**

---

* **Who was Charles?** See page 49, then go to **www.routledge. com/cw/bird**

(b) When the temperature is 67°C, the equivalent volume is **26.1 m^3**.

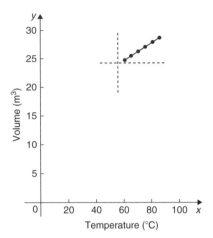

Figure 31.18

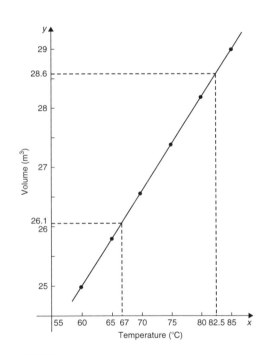

Figure 31.19

Problem 14. In an experiment demonstrating Hooke's law,* the strain in an aluminium wire was measured for various stresses. The results were:

---

* **Who was Hooke?** See page 49, then go to **www.routledge. com/cw/bird**

Stress (N/mm^2)	4.9	8.7	15.0
Strain	0.00007	0.00013	0.00021

Stress (N/mm^2)	18.4	24.2	27.3
Strain	0.00027	0.00034	0.00039

Plot a graph of stress (vertically) against strain (horizontally). Find (a) Young's modulus of elasticity* for aluminium, which is given by the gradient of the graph, (b) the value of the strain at a stress of 20 N/mm^2 and (c) the value of the stress when the strain is 0.00020

The co-ordinates (0.00007, 4.9), (0.00013, 8.7), and so on, are plotted as shown in Figure 31.20. The graph produced is the best straight line which can be drawn corresponding to these points. (With experimental results it is unlikely that all the points will lie exactly

* **Who was Young?** **Thomas Young** (13 June 1773–10 May 1829) was an English polymath. He is famous for having partly deciphered Egyptian hieroglyphics (specifically the Rosetta Stone). Young made notable scientific contributions to the fields of vision, light, solid mechanics, energy, physiology, language, musical harmony and Egyptology. To find out more go to **www.routledge.com/cw/bird**

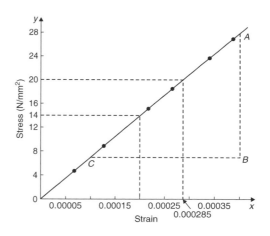

Figure 31.20

on a straight line.) The graph, and each of its axes, are labelled. Since the straight line passes through the origin, stress is directly proportional to strain for the given range of values.

(a)   The gradient of the straight line AC is given by

$$\frac{AB}{BC} = \frac{28 - 7}{0.00040 - 0.00010} = \frac{21}{0.00030}$$

$$= \frac{21}{3 \times 10^{-4}} = \frac{7}{10^{-4}} = 7 \times 10^4$$

$$= 70\,000\,\text{N/mm}^2$$

Thus, **Young's modulus of elasticity for aluminium is 70 000 N/mm^2**. Since $1\,\text{m}^2 = 10^6\,\text{mm}^2$, 70 000 N/mm^2 is equivalent to $70\,000 \times 10^6\,\text{N/m}^2$, i.e. **70 × 10^9 N/m^2 (or pascals)**.

From Figure 31.20,

(b)   The value of the strain at a stress of 20 N/mm^2 is **0.000285**

(c)   The value of the stress when the strain is 0.00020 is **14 N/mm^2**.

Problem 15.   The following values of resistance R ohms and corresponding voltage V volts are obtained from a test on a filament lamp.

R ohms	30	48.5	73	107	128
V volts	16	29	52	76	94

Choose suitable scales and plot a graph with R representing the vertical axis and V the horizontal axis. Determine (a) the gradient of the graph, (b) the

R-axis intercept value, (c) the equation of the graph, (d) the value of resistance when the voltage is 60 V and (e) the value of the voltage when the resistance is 40 ohms. (f) If the graph were to continue in the same manner, what value of resistance would be obtained at 110 V?

The co-ordinates (16, 30), (29, 48.5), and so on are shown plotted in Figure 31.21, where the best straight line is drawn through the points.

(a) The slope or gradient of the straight line *AC* is given by

$$\frac{AB}{BC} = \frac{135 - 10}{100 - 0} = \frac{125}{100} = \mathbf{1.25}$$

(Note that the vertical line *AB* and the horizontal line *BC* may be constructed anywhere along the length of the straight line. However, calculations are made easier if the horizontal line *BC* is carefully chosen; in this case, 100)

(b) The *R*-axis intercept is at $R = \mathbf{10\,ohms}$ (by extrapolation).

(c) The equation of a straight line is $y = mx + c$, where *y* is plotted on the vertical axis and *x* on the horizontal axis. *m* represents the gradient and *c* the *y*-axis intercept. In this case, *R* corresponds to *y*,

V corresponds to *x*, $m = 1.25$ and $c = 10$. Hence, the equation of the graph is $R = \mathbf{(1.25\,V + 10)\,\Omega}$

From Figure 31.21,

(d) When the voltage is 60 V, the resistance is **85 Ω**

(e) When the resistance is 40 ohms, the voltage is **24 V**

(f) By extrapolation, when the voltage is 110 V, the resistance is **147 Ω**

Problem 16. Experimental tests to determine the breaking stress σ of rolled copper at various temperatures *t* gave the following results.

Stress σ (N/cm^2)	8.46	8.04	7.78
Temperature $t(^\circ\text{C})$	70	200	280

Stress σ (N/cm^2)	7.37	7.08	6.63
Temperature $t(^\circ\text{C})$	410	500	640

Show that the values obey the law $\sigma = at + b$, where *a* and *b* are constants, and determine approximate values for *a* and *b*. Use the law to determine the stress at 250°C and the temperature when the stress is 7.54 N/cm^2

The co-ordinates (70, 8.46), (200, 8.04), and so on, are plotted as shown in Figure 31.22. Since the graph is a straight line then the values obey the law $\sigma = at + b$,

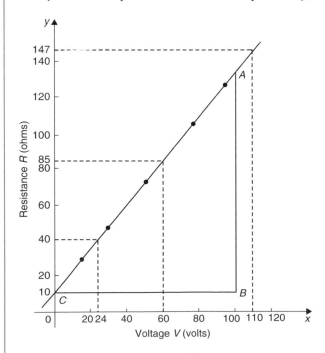

Figure 31.21

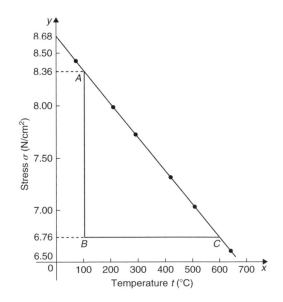

Figure 31.22

and the gradient of the straight line is

$$a = \frac{AB}{BC} = \frac{8.36 - 6.76}{100 - 600} = \frac{1.60}{-500} = -0.0032$$

Vertical axis intercept, $b = 8.68$

Hence, the law of the graph is $\sigma = -0.0032t + 8.68$

When the temperature is 250°C, stress σ is given by

$$\sigma = -0.0032(250) + 8.68 = 7.88\,\text{N/cm}^2$$

Rearranging $\sigma = -0.0032t + 8.68$ gives

$$0.0032t = 8.68 - \sigma, \quad \text{i.e.} \quad t = \frac{8.68 - \sigma}{0.0032}$$

Hence, when the stress, $\sigma = 7.54\,\text{N/cm}^2$,

$$\textbf{temperature,}\ t = \frac{8.68 - 7.54}{0.0032} = \textbf{356.3°C}$$

Now try the following Practice Exercise

Practice Exercise 140   Practical problems involving straight line graphs (answers on page 1164)

1. The resistance R ohms of a copper winding is measured at various temperatures $t\,°C$ and the results are as follows:

R (ohms)	112	120	126	131	134
$t\,°C$	20	36	48	58	64

Plot a graph of R (vertically) against t (horizontally) and find from it (a) the temperature when the resistance is $122\,\Omega$ and (b) the resistance when the temperature is 52°C.

2. The speed of a motor varies with armature voltage as shown by the following experimental results.

n (rev/min)	285	517	615
V (volts)	60	95	110

n (rev/min)	750	917	1050
V (volts)	130	155	175

Plot a graph of speed (horizontally) against voltage (vertically) and draw the best straight line through the points. Find from the graph (a) the speed at a voltage of 145 V and (b) the voltage at a speed of 400 rev/min.

3. The following table gives the force F newtons which, when applied to a lifting machine, overcomes a corresponding load of L newtons.

Force F (newtons)	25	47	64
Load L (newtons)	50	140	210

Force F (newtons)	120	149	187
Load L (newtons)	430	550	700

Choose suitable scales and plot a graph of F (vertically) against L (horizontally). Draw the best straight line through the points. Determine from the graph

(a) the gradient,

(b) the F-axis intercept,

(c) the equation of the graph,

(d) the force applied when the load is 310 N, and

(e) the load that a force of 160 N will overcome.

(f) If the graph were to continue in the same manner, what value of force will be needed to overcome a 800 N load?

4. The following table gives the results of tests carried out to determine the breaking stress σ of rolled copper at various temperatures, t.

Stress σ (N/cm^2)	8.51	8.07	7.80
Temperature $t(°C)$	75	220	310

Stress σ (N/cm^2)	7.47	7.23	6.78
Temperature $t(°C)$	420	500	650

Plot a graph of stress (vertically) against temperature (horizontally). Draw the best straight line through the plotted co-ordinates. Determine the slope of the graph and the vertical axis intercept.

5. The velocity v of a body after varying time intervals t was measured as follows:

t (seconds)	2	5	8
v (m/s)	16.9	19.0	21.1

t (seconds)	11	15	18
v (m/s)	23.2	26.0	28.1

Plot v vertically and t horizontally and draw a graph of velocity against time. Determine from the graph (a) the velocity after $10\,s$, (b) the time at $20\,m/s$ and (c) the equation of the graph.

6. The mass m of a steel joist varies with length L as follows:

mass, m (kg)	80	100	120	140	160
length, L (m)	3.00	3.74	4.48	5.23	5.97

Plot a graph of mass (vertically) against length (horizontally). Determine the equation of the graph.

7. The crushing strength of mortar varies with the percentage of water used in its preparation, as shown below.

Crushing strength, F (tonnes)	1.67	1.40	1.13
% of water used, $w\%$	6	9	12

Crushing strength, F (tonnes)	0.86	0.59	0.32
% of water used, $w\%$	15	18	21

Plot a graph of F (vertically) against w (horizontally).

(a) Interpolate and determine the crushing strength when 10% water is used.

(b) Assuming the graph continues in the same manner, extrapolate and determine the percentage of water used when the crushing strength is 0.15 tonnes.

(c) What is the equation of the graph?

8. In an experiment demonstrating Hooke's law, the strain in a copper wire was measured for various stresses. The results were

Stress (pascals)	10.6×10^6	18.2×10^6	24.0×10^6
Strain	0.00011	0.00019	0.00025

Stress (pascals)	30.7×10^6	39.4×10^6
Strain	0.00032	0.00041

Plot a graph of stress (vertically) against strain (horizontally). Determine

(a) Young's modulus of elasticity for copper, which is given by the gradient of the graph,

(b) the value of strain at a stress of 21×10^6 Pa,

(c) the value of stress when the strain is 0.00030

9. An experiment with a set of pulley blocks gave the following results.

Effort, E (newtons)	9.0	11.0	13.6
Load, L (newtons)	15	25	38

Effort, E (newtons)	17.4	20.8	23.6
Load, L (newtons)	57	74	88

Plot a graph of effort (vertically) against load (horizontally). Determine
(a) the gradient,
(b) the vertical axis intercept,
(c) the law of the graph,
(d) the effort when the load is 30 N,
(e) the load when the effort is 19 N.

10. The variation of pressure p in a vessel with temperature T is believed to follow a law of the form $p = aT + b$, where a and b are constants. Verify this law for the results given below and determine the approximate values of a and b. Hence, determine the pressures at temperatures of 285 K and 310 K and the temperature at a pressure of 250 kPa.

Pressure, p (kPa)	244	247	252
Temperature, T (K)	273	277	282

Pressure, p (kPa)	258	262	267
Temperature, T (K)	289	294	300

Section D

For fully worked solutions to each of the problems in Practice Exercises 138 to 140 in this chapter, go to the website:
www.routledge.com/cw/bird

Chapter 32

Graphs reducing non-linear laws to linear form

Why it is important to understand: **Graphs reducing non-linear laws to linear form**

Graphs are important tools for analysing and displaying data between two experimental quantities. Many times situations occur in which the relationship between the variables is not linear. By manipulation, a straight line graph may be plotted to produce a law relating the two variables. Sometimes this involves using the laws of logarithms. The relationship between the resistance of wire and its diameter is not a linear one. Similarly, the periodic time of oscillations of a pendulum does not have a linear relationship with its length, and the head of pressure and the flow velocity are not linearly related. There are thus plenty of examples in engineering where determination of law is needed.

At the end of this chapter, you should be able to:

- understand what is meant by determination of law
- prepare co-ordinates for a non-linear relationship between two variables
- plot prepared co-ordinates and draw a straight line graph
- determine the gradient and vertical-axis intercept of a straight line graph
- state the equation of a straight line graph
- plot straight line graphs involving practical engineering examples
- determine straight line laws involving logarithms: $y = ax^n$, $y = ab^x$ and $y = ae^{bx}$
- plot straight line graphs involving logarithms

32.1   Introduction

In Chapter 31 we discovered that the equation of a straight line graph is of the form $y = mx + c$, where m is the gradient and c is the y-axis intercept. This chapter explains how the law of a graph can still be determined even when it is not of the linear form $y = mx + c$. The method used is called **determination of law** and is explained in the following sections.

32.2   Determination of law

Frequently, the relationship between two variables, say x and y, is not a linear one; i.e. when x is plotted against y a curve results. In such cases the non-linear equation may be modified to the linear form, $y = mx + c$, so that the constants, and thus the law relating the variables, can be determined. This technique is called '**determination of law**'.

Some examples of the reduction of equations to linear form include

(i) $y = ax^2 + b$ compares with $Y = mX + c$, where $m = a$, $c = b$ and $X = x^2$.
Hence, y is plotted vertically against x^2 horizontally to produce a straight line graph of gradient a and y-axis intercept b.

(ii) $y = \dfrac{a}{x} + b$, i.e. $y = a\left(\dfrac{1}{x}\right) + b$

y is plotted vertically against $\dfrac{1}{x}$ horizontally to produce a straight line graph of gradient a and y-axis intercept b.

(iii) $y = ax^2 + bx$
Dividing both sides by x gives $\dfrac{y}{x} = ax + b$.

Comparing with $Y = mX + c$ shows that $\dfrac{y}{x}$ is plotted vertically against x horizontally to produce a straight line graph of gradient a and $\dfrac{y}{x}$ axis intercept b.

Here are some worked problems to demonstrate determination of law.

Problem 1.   Experimental values of x and y, shown below, are believed to be related by the law $y = ax^2 + b$. By plotting a suitable graph, verify this law and determine approximate values of a and b

x	1	2	3	4	5
y	9.8	15.2	24.2	36.5	53.0

If y is plotted against x, a curve results and it is not possible to determine the values of constants a and b from the curve.
Comparing $y = ax^2 + b$ with $Y = mX + c$ shows that y is to be plotted vertically against x^2 horizontally. A table of values is drawn up as shown below.

x	1	2	3	4	5
x^2	1	4	9	16	25
y	9.8	15.2	24.2	36.5	53.0

A graph of y against x^2 is shown in Figure 32.1, with the best straight line drawn through the points. **Since a straight line graph results, the law is verified**.

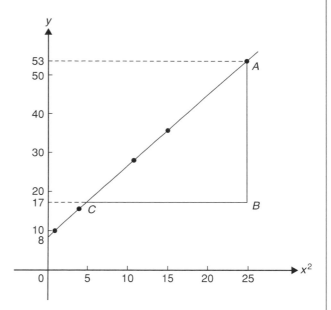

Figure 32.1

From the graph,   **gradient,** $a = \dfrac{AB}{BC} = \dfrac{53 - 17}{25 - 5}$

$$= \dfrac{36}{20} = \mathbf{1.8}$$

and the **y-axis intercept, $b = 8.0$**

Hence, the law of the graph is $y = \mathbf{1.8}x^2 + \mathbf{8.0}$

Problem 2.   Values of load L newtons and distance d metres obtained experimentally are shown in the following table.

Load, L(N)	32.3	29.6	27.0	23.2
Distance, d(m)	0.75	0.37	0.24	0.17

Load, L(N)	18.3	12.8	10.0	6.4
Distance, d(m)	0.12	0.09	0.08	0.07

(a)   Verify that load and distance are related by a law of the form $L = \dfrac{a}{d} + b$ and determine approximate values of a and b.

(b)   Hence, calculate the load when the distance is 0.20 m and the distance when the load is 20 N.

(a)   Comparing $L = \dfrac{a}{d} + b$ i.e. $L = a\left(\dfrac{1}{d}\right) + b$ with $Y = mX + c$ shows that L is to be plotted vertically against $\dfrac{1}{d}$ horizontally. Another table of values is drawn up as shown below.

L	32.3	29.6	27.0	23.2	18.3	12.8	10.0	6.4
d	0.75	0.37	0.24	0.17	0.12	0.09	0.08	0.07
$\dfrac{1}{d}$	1.33	2.70	4.17	5.88	8.33	11.11	12.50	14.29

A graph of L against $\dfrac{1}{d}$ is shown in Figure 32.2. **A straight line can be drawn through the points, which verifies that load and distance are related by a law of the form $L = \dfrac{a}{d} + b$.**

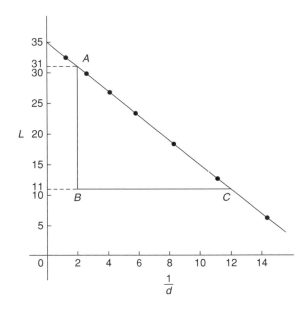

Figure 32.2

Gradient of straight line, $a = \dfrac{AB}{BC} = \dfrac{31 - 11}{2 - 12}$

$$= \dfrac{20}{-10} = -2$$

L-axis intercept, $b = 35$

Hence, the law of the graph is $L = -\dfrac{2}{d} + 35$

(b)   When the distance $d = 0.20\,\text{m}$,

load, $L = \dfrac{-2}{0.20} + 35 = 25.0\,\text{N}$

Rearranging $L = -\dfrac{2}{d} + 35$ gives

$\dfrac{2}{d} = 35 - L$ and $d = \dfrac{2}{35 - L}$

Hence,   when   the   load $L = 20\,\text{N}$,   **distance,**

$d = \dfrac{2}{35 - 20} = \dfrac{2}{15} = \mathbf{0.13\,m}$

Problem 3.   The solubility s of potassium chlorate is shown by the following table.

$t\,°C$	10	20	30	40	50	60	80	100
s	4.9	7.6	11.1	15.4	20.4	26.4	40.6	58.0

The relationship between s and t is thought to be of the form $s = 3 + at + bt^2$. Plot a graph to test the supposition and use the graph to find approximate values of a and b. Hence, calculate the solubility of potassium chlorate at $70\,°C$

Rearranging $s = 3 + at + bt^2$ gives

$$s - 3 = at + bt^2$$

and $\dfrac{s - 3}{t} = a + bt$

or $\dfrac{s - 3}{t} = bt + a$

which is of the form $Y = mX + c$

This shows that $\dfrac{s - 3}{t}$ is to be plotted vertically and t horizontally, with gradient b and vertical axis intercept a.

Another table of values is drawn up as shown below.

t	10	20	30	40	50	60	80	100
s	4.9	7.6	11.1	15.4	20.4	26.4	40.6	58.0
$\dfrac{s - 3}{t}$	0.19	0.23	0.27	0.31	0.35	0.39	0.47	0.55

A graph of $\dfrac{s - 3}{t}$ against t is shown plotted in Figure 32.3. **A straight line fits the points, which shows that s and t are related by $s = 3 + at + bt^2$.**

Gradient of straight line, $b = \dfrac{AB}{BC} = \dfrac{0.39 - 0.19}{60 - 10}$

$$= \dfrac{0.20}{50} = \mathbf{0.004}$$

Vertical axis intercept, $a = \mathbf{0.15}$
Hence, the law of the graph is $s = 3 + 0.15t + 0.004t^2$.
The solubility of potassium chlorate at $70\,°C$ is given by

$$s = 3 + 0.15(70) + 0.004(70)^2$$

$$= 3 + 10.5 + 19.6 = \mathbf{33.1}$$

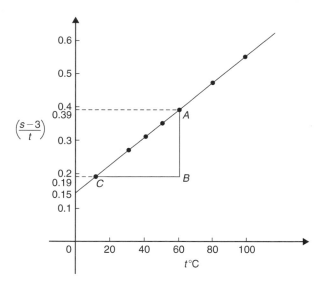

Figure 32.3

Now try the following Practice Exercise

Practice Exercise 141   Graphs reducing non-linear laws to linear form (answers on page 1164)

In Problems 1 to 5, x and y are two related variables and all other letters denote constants. For the stated laws to be verified it is necessary to plot graphs of the variables in a modified form. State for each (a) what should be plotted on the vertical axis, (b) what should be plotted on the horizontal axis, (c) the gradient and (d) the vertical axis intercept.

1. $y = d + cx^2$ 2. $y - a = b\sqrt{x}$

3. $y - e = \dfrac{f}{x}$ 4. $y - cx = bx^2$

5. $y = \dfrac{a}{x} + bx$

6. In an experiment the resistance of wire is measured for wires of different diameters with the following results.

R (ohms)	1.64	1.14	0.89	0.76	0.63
d (mm)	1.10	1.42	1.75	2.04	2.56

It is thought that R is related to d by the law $R = \dfrac{a}{d^2} + b$, where a and b are constants. Verify this and find the approximate values for

a and b. Determine the cross-sectional area needed for a resistance reading of 0.50 ohms.

7. Corresponding experimental values of two quantities x and y are given below.

x	1.5	3.0	4.5	6.0	7.5	9.0
y	11.5	25.0	47.5	79.0	119.5	169.0

By plotting a suitable graph, verify that y and x are connected by a law of the form $y = kx^2 + c$, where k and c are constants. Determine the law of the graph and hence find the value of x when y is 60.0

8. Experimental results of the safe load L kN, applied to girders of varying spans, d m, are shown below.

Span, d (m)	2.0	2.8	3.6	4.2	4.8
Load, L (kN)	475	339	264	226	198

It is believed that the relationship between load and span is $L = c/d$, where c is a constant. Determine (a) the value of constant c and (b) the safe load for a span of 3.0 m.

9. The following results give corresponding values of two quantities x and y which are believed to be related by a law of the form $y = ax^2 + bx$, where a and b are constants.

x	3.4	5.2	6.5	7.3	9.1	12.4
y	33.86	55.54	72.80	84.10	111.4	168.1

Verify the law and determine approximate values of a and b. Hence, determine (a) the value of y when x is 8.0 and (b) the value of x when y is 146.5

32.3 Revision of laws of logarithms

The laws of logarithms were stated in Chapter 15 as follows:

$$\log(A \times B) = \log A + \log B \tag{1}$$

$$\log \left(\frac{A}{B} \right) = \log A - \log B \tag{2}$$

$$\log A^n = n \times \log A \tag{3}$$

Also,　　$\ln e = 1$ and if, say, $\lg x = 1.5$,
　　　　then $x = 10^{1.5} = 31.62$

Further, if $3^x = 7$ then $\lg 3^x = \lg 7$ and $x \lg 3 = \lg 7$,
from which $x = \dfrac{\lg 7}{\lg 3} = 1.771$

These laws and techniques are used whenever non-linear laws of the form $y = ax^n$, $y = ab^x$ and $y = ae^{bx}$ are reduced to linear form with the values of a and b needing to be calculated. This is demonstrated in the following section.

32.4 Determination of laws involving logarithms

Examples of the reduction of equations to linear form involving logarithms include

(a)　**$y = ax^n$**

Taking logarithms to **a base of 10** of both sides gives

$$\lg y = \lg(ax^n)$$

$$= \lg a + \lg x^n \quad \text{by law (1)}$$

i.e.　　　$\lg y = n \lg x + \lg a$　by law (3)

which compares with $Y = mX + c$

and shows that **lg y is plotted vertically against lg x horizontally** to produce a straight line **graph of gradient n and lg y-axis intercept lg a**.
See Problems 4 and 5 to demonstrate how this law is determined.

(b)　**$y = ab^x$**

Taking logarithms to **a base of 10** of both sides gives

$$\lg y = \lg(ab^x)$$

i.e.　　$\lg y = \lg a + \lg b^x$　by law (1)

$$\lg y = \lg a + x \lg b \quad \text{by law (3)}$$

i.e.　　$\lg y = x \lg b + \lg a$

or　　　$\lg y = (\lg b)x + \lg a$

which compares with

$$Y = mX + c$$

and shows that **lg y is plotted vertically against x horizontally** to produce **a straight line graph of gradient lg b and lg y-axis intercept lg a**.

See Problem 6 to demonstrate how this law is determined.

(c)　**$y = ae^{bx}$**

Taking logarithms to **a base of e** of both sides gives

$$\ln y = \ln(ae^{bx})$$

i.e.　　$\ln y = \ln a + \ln e^{bx}$　by law (1)

i.e.　　$\ln y = \ln a + bx \ln e$　by law (3)

i.e.　　$\ln y = bx + \ln a$　　since $\ln e = 1$

which compares with

$$Y = mX + c$$

and shows that **ln y is plotted vertically against x horizontally** to produce **a straight line graph of gradient b and ln y-axis intercept ln a**.
See Problem 7 to demonstrate how this law is determined.

Problem 4. The current flowing in, and the power dissipated by, a resistor are measured experimentally for various values and the results are as shown below.

Current, I (amperes)	2.2	3.6	4.1	5.6	6.8
Power, P (watts)	116	311	403	753	1110

Show that the law relating current and power is of the form $P = RI^n$, where R and n are constants, and determine the law

---

Taking logarithms to a base of 10 of both sides of $P = RI^n$ gives

$$\lg P = \lg(RI^n) = \lg R + \lg I^n = \lg R + n \lg I$$
$$\text{by the laws of logarithms}$$

i.e.　　$\lg P = n \lg I + \lg R$

which is of the form $Y = mX + c$,
showing that $\lg P$ is to be plotted vertically against $\lg I$ horizontally.
A table of values for $\lg I$ and $\lg P$ is drawn up as shown below.

I	2.2	3.6	4.1	5.6	6.8
$\lg I$	0.342	0.556	0.613	0.748	0.833
P	116	311	403	753	1110
$\lg P$	2.064	2.493	2.605	2.877	3.045

A graph of $\lg P$ against $\lg I$ is shown in Figure 32.4 and, **since a straight line results, the law $P = RI^n$ is verified**.

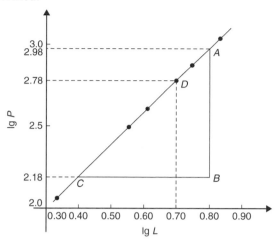

Figure 32.4

Gradient of straight line, $n = \dfrac{AB}{BC} = \dfrac{2.98 - 2.18}{0.8 - 0.4}$

$$= \dfrac{0.80}{0.4} = 2$$

It is not possible to determine the vertical axis intercept on sight since the horizontal axis scale does not start at zero. Selecting any point from the graph, say point D, where $\lg I = 0.70$ and $\lg P = 2.78$ and substituting values into

$$\lg P = n \lg I + \lg R$$

gives $\qquad 2.78 = (2)(0.70) + \lg R$

from which, $\qquad \lg R = 2.78 - 1.40 = 1.38$

Hence, $\qquad\qquad R = \text{antilog } 1.38 = 10^{1.38} = \mathbf{24.0}$

Hence, **the law of the graph is $P = 24.0I^2$**

Problem 5. The periodic time, T, of oscillation of a pendulum is believed to be related to its length, L, by a law of the form $T = kL^n$, where k and n are constants. Values of T were measured for various lengths of the pendulum and the results are as shown below.

Periodic time, T(s)	1.0	1.3	1.5	1.8	2.0	2.3
Length, L(m)	0.25	0.42	0.56	0.81	1.0	1.32

Show that the law is true and determine the approximate values of k and n. Hence find the periodic time when the length of the pendulum is 0.75 m

From para (a), page 364, if $\qquad T = kL^n$

then $\qquad\qquad\qquad \lg T = n \lg L + \lg k$

and comparing with $\qquad Y = mX + c$

shows that $\lg T$ is plotted vertically against $\lg L$ horizontally, with gradient n and vertical-axis intercept $\lg k$.

A table of values for $\lg T$ and $\lg L$ is drawn up as shown below.

T	1.0	1.3	1.5	1.8	2.0	2.3
$\lg T$	0	0.114	0.176	0.255	0.301	0.362
L	0.25	0.42	0.56	0.81	1.0	1.32
$\lg L$	−0.602	−0.377	−0.252	−0.092	0	0.121

A graph of $\lg T$ against $\lg L$ is shown in Figure 32.5 and the **law $T = kL^n$ is true since a straight line results**.

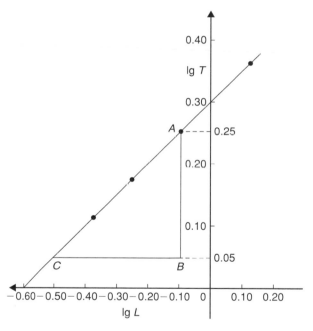

Figure 32.5

From the graph, gradient of straight line,

$$\boldsymbol{n} = \dfrac{AB}{BC} = \dfrac{0.25 - 0.05}{-0.10 - (-0.50)} = \dfrac{0.20}{0.40} = \dfrac{1}{2}$$

Vertical axis intercept, $\lg k = 0.30$. Hence, $k = \text{antilog } 0.30 = 10^{0.30} = \mathbf{2.0}$

Hence, **the law of the graph is $T = 2.0L^{1/2}$ or $T = 2.0\sqrt{L}$**

When length $L = 0.75$ m, $\mathbf{T} = 2.0\sqrt{0.75} = \mathbf{1.73\,s}$

Problem 6. Quantities x and y are believed to be related by a law of the form $y = ab^x$, where a and b are constants. The values of x and corresponding values of y are

x	0	0.6	1.2	1.8	2.4	3.0
y	5.0	9.67	18.7	36.1	69.8	135.0

Verify the law and determine the approximate values of a and b. Hence determine (a) the value of y when x is 2.1 and (b) the value of x when y is 100

From para (b), page 364, if $\quad y = ab^x$

then $\qquad\qquad\qquad \lg y = (\lg b)x + \lg a$

and comparing with $\qquad Y = mX + c$

shows that $\lg y$ is plotted vertically and x horizontally, with gradient $\lg b$ and vertical-axis intercept $\lg a$. Another table is drawn up as shown below.

x	0	0.6	1.2	1.8	2.4	3.0
y	5.0	9.67	18.7	36.1	69.8	135.0
$\lg y$	0.70	0.99	1.27	1.56	1.84	2.13

A graph of $\lg y$ against x is shown in Figure 32.6 and, **since a straight line results, the law $y = ab^x$ is verified**.

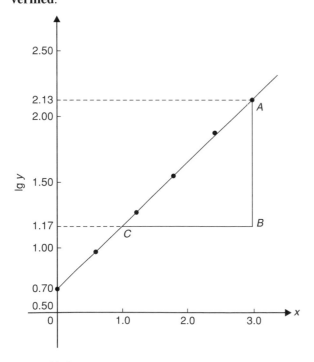

Figure 32.6

Gradient of straight line,

$$\lg b = \frac{AB}{BC} = \frac{2.13 - 1.17}{3.0 - 1.0} = \frac{0.96}{2.0} = 0.48$$

Hence, $b = $ antilog $0.48 = 10^{0.48} = \mathbf{3.0}$, correct to 2 significant figures.

Vertical axis intercept, $\lg a = 0.70$, from which

$$a = \text{antilog } 0.70$$
$$= 10^{0.70} = \mathbf{5.0}, \text{ correct to}$$
$$\text{2 significant figures.}$$

Hence, **the law of the graph is $y = 5.0(3.0)^x$**

(a) When $x = 2.1$, $\mathbf{y = 5.0(3.0)^{2.1} = 50.2}$

(b) When $y = 100$, $100 = 5.0(3.0)^x$, from which
$$100/5.0 = (3.0)^x$$

i.e. $\quad 20 = (3.0)^x$

Taking logarithms of both sides gives
$$\lg 20 = \lg(3.0)^x = x \lg 3.0$$

Hence, $x = \dfrac{\lg 20}{\lg 3.0} = \dfrac{1.3010}{0.4771} = \mathbf{2.73}$

Problem 7. The current i mA flowing in a capacitor which is being discharged varies with time t ms, as shown below.

i (mA)	203	61.14	22.49	6.13	2.49	0.615
t (ms)	100	160	210	275	320	390

Show that these results are related by a law of the form $i = Ie^{t/T}$, where I and T are constants. Determine the approximate values of I and T

Taking Napierian logarithms of both sides of
$$i = Ie^{t/T}$$

gives $\qquad \ln i = \ln(Ie^{t/T}) = \ln I + \ln e^{t/T}$
$$= \ln I + \frac{t}{T}\ln e$$

i.e. $\qquad \ln i = \ln I + \frac{t}{T}$ since $\ln e = 1$

or $\qquad \ln i = \left(\frac{1}{T}\right)t + \ln I$

which compares with $\quad y = mx + c$

showing that $\ln i$ is plotted vertically against t horizontally, with gradient $\frac{1}{T}$ and vertical-axis intercept $\ln I$.

Another table of values is drawn up as shown below.

t	100	160	210	275	320	390
i	203	61.14	22.49	6.13	2.49	0.615
$\ln i$	5.31	4.11	3.11	1.81	0.91	−0.49

A graph of $\ln i$ against t is shown in Figure 32.7 and, since a straight line results, **the law $i = Ie^{t/T}$ is verified**.

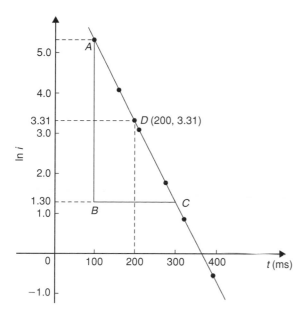

Figure 32.7

Gradient of straight line, $\dfrac{1}{T} = \dfrac{AB}{BC} = \dfrac{5.30 - 1.30}{100 - 300}$

$$= \dfrac{4.0}{-200} = -0.02$$

Hence, $$T = \dfrac{1}{-0.02} = \mathbf{-50}$$

Selecting any point on the graph, say point D, where $t = 200$ and $\ln i = 3.31$, and substituting

into $$\ln i = \left(\dfrac{1}{T}\right)t + \ln I$$

gives $$3.31 = -\dfrac{1}{50}(200) + \ln I$$

from which, $\ln I = 3.31 + 4.0 = 7.31$

and $I = $ antilog $7.31 = e^{7.31} = 1495$ or **1500** correct to 3 significant figures.

Hence, **the law of the graph is $i = 1500e^{-t/50}$**

Now try the following Practice Exercise

Practice Exercise 142   Determination of law involving logarithms (answers on page 1164)

In Problems 1 to 3, x and y are two related variables and all other letters denote constants. For the stated laws to be verified it is necessary to plot graphs of the variables in a modified form. State for each (a) what should be plotted on the vertical axis, (b) what should be plotted on the horizontal axis, (c) the gradient and (d) the vertical axis intercept.

1. $y = ba^x$ 2. $y = kx^L$ 3. $\dfrac{y}{m} = e^{nx}$

4. The luminosity I of a lamp varies with the applied voltage V and the relationship between I and V is thought to be $I = kV^n$. Experimental results obtained are

I (candelas)	1.92	4.32	9.72
V (volts)	40	60	90

I (candelas)	15.87	23.52	30.72
V (volts)	115	140	160

Verify that the law is true and determine the law of the graph. Also determine the luminosity when 75 V is applied across the lamp.

5. The head of pressure h and the flow velocity v are measured and are believed to be connected by the law $v = ah^b$, where a and b are constants. The results are as shown below.

h	10.6	13.4	17.2	24.6	29.3
v	9.77	11.0	12.44	14.88	16.24

Verify that the law is true and determine values of a and b.

6. Experimental values of x and y are measured as follows.

x	0.4	0.9	1.2	2.3	3.8
y	8.35	13.47	17.94	51.32	215.20

The law relating x and y is believed to be of the form $y = ab^x$, where a and b are constants.

Determine the approximate values of a and b. Hence, find the value of y when x is 2.0 and the value of x when y is 100

7. The activity of a mixture of radioactive isotopes is believed to vary according to the law $R = R_0 t^{-c}$, where R_0 and c are constants. Experimental results are shown below.

R	9.72	2.65	1.15	0.47	0.32	0.23
t	2	5	9	17	22	28

Verify that the law is true and determine approximate values of R_0 and c.

8. Determine the law of the form $y = ae^{kx}$ which relates the following values.

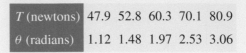

y	0.0306	0.285	0.841	5.21	173.2	1181
x	−4.0	5.3	9.8	17.4	32.0	40.0

9. The tension T in a belt passing round a pulley wheel and in contact with the pulley over an angle of θ radians is given by $T = T_0 e^{\mu\theta}$, where T_0 and μ are constants. Experimental results obtained are

T (newtons)	47.9	52.8	60.3	70.1	80.9
θ (radians)	1.12	1.48	1.97	2.53	3.06

Determine approximate values of T_0 and μ. Hence, find the tension when θ is 2.25 radians and the value of θ when the tension is 50.0 newtons.

For fully worked solutions to each of the problems in Practice Exercises 141 and 142 in this chapter, go to the website:
www.routledge.com/cw/bird

Chapter 33

Graphs with logarithmic scales

Why it is important to understand: Graphs with logarithmic scales

As mentioned in previous chapters, graphs are important tools for analysing and displaying data between two experimental quantities and many times situations occur in which the relationship between the variables is not linear. By manipulation, a straight line graph may be plotted to produce a law relating the two variables. Knowledge of logarithms may be used to simplify plotting the relation between one variable and another. In particular, we consider those situations in which one of the variables requires scaling because the range of its data values is very large in comparison to the range of the other variable. Log-log and log-linear graph paper is available to make the plotting process easier.

At the end of this chapter, you should be able to:

- understand logarithmic scales
- understand log-log and log-linear graph paper
- plot a graph of the form $y = ax^n$ using log-log graph paper and determine constants 'a' and 'n'
- plot a graph of the form $y = a\,b^x$ using log-linear graph paper and determine constants 'a' and 'b'
- plot a graph of the form $y = ae^{kx}$ using log-linear graph paper and determine constants 'a' and 'k'

33.1 Logarithmic scales and logarithmic graph paper

Graph paper is available where the scale markings along the horizontal and vertical axes are proportional to the logarithms of the numbers. Such graph paper is called **log-log graph paper**.

A **logarithmic scale** is shown in Figure 33.1 where the distance between, say, 1 and 2, is proportional to $\lg 2 - \lg 1$, i.e. 0.3010 of the total distance from 1 to 10. Similarly, the distance between 7 and 8 is proportional to $\lg 8 - \lg 7$, i.e. 0.05799 of the total distance from 1 to

10. Thus the distance between markings progressively decreases as the numbers increase from 1 to 10.

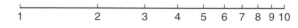

Figure 33.1

With log-log graph paper the scale markings are from 1 to 9, and this pattern can be repeated several times. The number of times the pattern of markings is repeated on an axis signifies the number of **cycles**. When the vertical axis has, say, 3 sets of values from 1 to 9, and the horizontal axis has, say, 2 sets of values from 1 to 9, then this

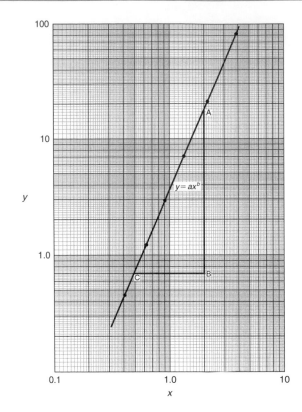

Figure 33.2

log-log graph paper is called 'log 3 cycle × 2 cycle' (see Figure 33.2). Many different arrangements are available ranging from 'log 1 cycle × 1 cycle' through to 'log 5 cycle × 5 cycle'.

To depict a set of values, say, from 0.4 to 161, on an axis of log-log graph paper, 4 cycles are required, from 0.1 to 1, 1 to 10, 10 to 100 and 100 to 1000

33.2 Graphs of the form $y = ax^n$

Taking logarithms to a base of 10 of both sides of $y = ax^n$ gives:

$$\lg y = \lg (ax^n) = \lg a + \lg x^n$$

i.e.
$$\lg y = n\lg x + \lg a$$

which compares with
$$Y = mX + c$$

Thus, by plotting $\lg y$ vertically against $\lg x$ horizontally, a straight line results, i.e. the equation $y = ax^n$ is reduced to linear form. With log-log graph paper available x and y may be plotted directly, without having first to determine their logarithms, as shown in Chapter 32.

Problem 1. Experimental values of two related quantities x and y are shown below:

x	0.41	0.63	0.92	1.36	2.17	3.95
y	0.45	1.21	2.89	7.10	20.79	82.46

The law relating x and y is believed to be $y = ax^b$, where 'a' and 'b' are constants. Verify that this law is true and determine the approximate values of 'a' and 'b'

If $y = ax^b$ then $\lg y = b \lg x + \lg a$, from above, which is of the form $Y = mX + c$, showing that to produce a straight line graph $\lg y$ is plotted vertically against $\lg x$ horizontally. x and y may be plotted directly onto log-log graph paper as shown in Figure 33.2. The values of y range from 0.45 to 82.46 and 3 cycles are needed (i.e. 0.1 to 1, 1 to 10 and 10 to 100). The values of x range from 0.41 to 3.95 and 2 cycles are needed (i.e. 0.1 to 1 and 1 to 10). Hence 'log 3 cycle × 2 cycle' is used as shown in Figure 33.2 where the axes are marked and the points plotted. **Since the points lie on a straight line the law $y = ax^b$ is verified**.

To evaluate constants 'a' and 'b':

Method 1. Any two points on the straight line, say points A and C, are selected, and AB and BC are measured (say in centimetres). Then,

$$\text{gradient, } b = \frac{AB}{BC} = \frac{11.5\,\text{units}}{5\,\text{units}} = 2.3$$

Since $\lg y = b \lg x + \lg a$, when $x = 1$, $\lg x = 0$ and $\lg y = \lg a$.

The straight line crosses the ordinate $x = 1.0$ at $y = 3.5$.

Hence, $\lg a = \lg 3.5$, i.e. $a = 3.5$

Method 2. Any two points on the straight line, say points A and C, are selected. A has co-ordinates (2, 17.25) and C has co-ordinates (0.5, 0.7).

Since $y = ax^b$ then $17.25 = a(2)^b$ (1)

and $0.7 = a(0.5)^b$ (2)

i.e. two simultaneous equations are produced and may be solved for 'a' and 'b'.

Dividing equation (1) by equation (2) to eliminate 'a' gives:

$$\frac{17.25}{0.7} = \frac{(2)^b}{(0.5)^b} = \left(\frac{2}{0.5}\right)^b$$

i.e. $24.643 = (4)^b$

Taking logarithms of both sides gives:

$\lg 24.643 = b \lg 4$,

i.e. $b = \dfrac{\lg 24.643}{\lg 4}$

    $= 2.3$, correct to 2 significant figures.

Substituting $b = 2.3$ in equation (1) gives:
$17.25 = a(2)^{2.3}$

i.e. $a = \dfrac{17.25}{(2)^{2.3}} = \dfrac{17.25}{4.925}$

    $= 3.5$, correct to 2 significant figures.

Hence, the law of the graph is: $y = 3.5x^{2.3}$

> **Problem 2.**   The power dissipated by a resistor
> was measured for varying values of current flowing
> in the resistor and the results are as shown:
>
> | Current, I (amperes) | 1.4 | 4.7 | 6.8 | 9.1 | 11.2 | 13.1 |
> |---|---|---|---|---|---|---|
> | Power, P (watts) | 49 | 552 | 1156 | 2070 | 3136 | 4290 |
>
> Prove that the law relating current and power is of
> the form $P = RI^n$, where R and n are constants,
> and determine the law. Hence calculate the power
> when the current is 12 A and the current when the
> power is 1000 W.

Since $P = RI^n$ then $\lg P = n \lg I + \lg R$, which is of
the form $Y = mX + c$, showing that to produce a straight
line graph $\lg P$ is plotted vertically against $\lg I$ horizontally. Power values range from 49 to 4290, hence 3
cycles of log-log graph paper are needed (10 to 100, 100
to 1000 and 1000 to 10000). Current values range from
1.4 to 13.1, hence 2 cycles of log-log graph paper are
needed (1 to 10 and 10 to 100). Thus 'log 3 cycles $\times$
2 cycles' is used as shown in Figure 33.3 (or, if not
available, graph paper having a larger number of cycles
per axis can be used).

The co-ordinates are plotted and **a straight line results
which proves that the law relating current and power
is of the form $P = RI^n$.**

Gradient of straight line, $n = \dfrac{AB}{BC} = \dfrac{14 \text{ units}}{7 \text{ units}} = 2$

At point $C, I = 2$ and $P = 100$. Substituting these
values into $P = RI^n$ gives:

$$100 = R(2)^2$$

Hence, $R = \dfrac{100}{2^2} = 25$

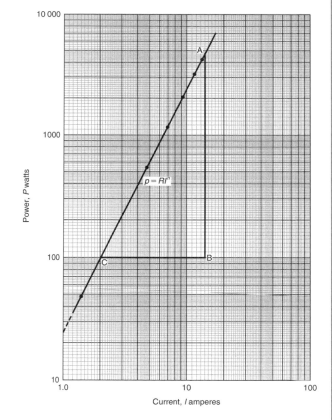

Figure 33.3

which may have been found from the intercept on the
$I = 1.0$ axis in Figure 33.3.

Hence, the law of the graph is: $P = 25I^2$

When current $I = 12$,
power $P = 25(12)^2 = 3600$ watts (which may be read
from the graph).

When power $P = 1000, 1000 = 25I^2$

Hence $I^2 = \dfrac{1000}{25} = 40$

from which,   **current $I = \sqrt{40} = 6.32$A**

> **Problem 3.**   The pressure p and volume v of a gas
> are believed to be related by a law of the form
> $p = cv^n$, where c and n are constants. Experimental
> values of p and corresponding values of v obtained
> in a laboratory are:
>
> | p pascals | 2.28×10^5 | 8.04×10^5 | 2.03×10^6 |
> |---|---|---|---|
> | v m^3 | 3.2×10^{-2} | 1.3×10^{-2} | 6.7×10^{-3} |

p (pascals)	5.05×10^6	1.82×10^7
v (m^3)	3.5×10^{-3}	1.4×10^{-3}

Verify that the law is true and determine approximate values of c and n.

Since $p = cv^n$, then $\lg p = n \lg v + \lg c$, which is of the form $Y = mX + c$, showing that to produce a straight line graph $\lg p$ is plotted vertically against $\lg v$ horizontally. The co-ordinates are plotted on 'log 3 cycle $\times$ 2 cycle' graph paper as shown in Figure 33.4. With the data expressed in standard form, the axes are marked in standard form also. Since a straight line results the law $p = cv^n$ is verified.

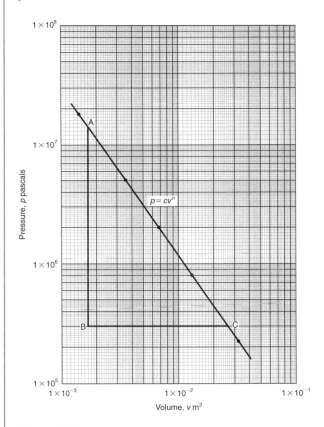

Figure 33.4

The straight line has a negative gradient and the value of the gradient is given by:

$$\frac{AB}{BC} = \frac{14\,\text{units}}{10\,\text{units}} = 1.4, \text{ hence } n = \mathbf{-1.4}$$

Selecting any point on the straight line, say point C, having co-ordinates $(2.63 \times 10^{-2}, 3 \times 10^5)$, and substituting these values in $p = cv^n$ gives

$$3 \times 10^5 = c(2.63 \times 10^{-2})^{-1.4}$$

Hence, $\quad c = \dfrac{3 \times 10^5}{(2.63 \times 10^{-2})^{-1.4}} = \dfrac{3 \times 10^5}{(0.0263)^{-1.4}}$

$$= \frac{3 \times 10^5}{1.63 \times 10^2}$$

$$= \mathbf{1840}, \text{ correct to 3}$$
$$\text{significant figures.}$$

Hence, the law of the graph is: $p = 1840\,v^{-1.4}$ or $pv^{1.4} = 1840$

Now try the following Practice Exercise

Practice Exercise 143   Logarithmic graphs of the form $y = ax^n$ (answers on page 1164)

1. Quantities x and y are believed to be related by a law of the form $y = ax^n$, where a and n are constants. Experimental values of x and corresponding values of y are:

x	0.8	2.3	5.4	11.5	21.6	42.9
y	8	54	250	974	3028	10410

    Show that the law is true and determine the values of 'a' and 'n'. Hence determine the value of y when x is 7.5 and the value of x when y is 5000

2. Show from the following results of voltage V and admittance Y of an electrical circuit that the law connecting the quantities is of the form $V = kY^n$, and determine the values of k and n.

Voltage, V (volts)	2.88	2.05	1.60	1.22	0.96
Admittance, Y (siemens)	0.52	0.73	0.94	1.23	1.57

3. Quantities x and y are believed to be related by a law of the form $y = mx^n$. The values of x and corresponding values of y are:

x	0.5	1.0	1.5	2.0	2.5	3.0
y	0.53	3.0	8.27	16.97	29.69	46.77

    Verify the law and find the values of m and n.

33.3   Graphs of the form $y = ab^x$

Taking logarithms to a base of 10 of both sides of $y = ab^x$ gives:

$$\lg y = \lg (ab^x) = \lg a + \lg b^x$$

$$= \lg a + x\lg b$$

i.e.
$$\lg y = (\lg b)x + \lg a$$

which compares with $Y = mX + c$

Thus, by plotting $\lg y$ vertically against x horizontally a straight line results, i.e. the graph $y = ab^x$ is reduced to linear form. In this case, graph paper having a linear horizontal scale and a logarithmic vertical scale may be used. This type of graph paper is called **log-linear graph paper**, and is specified by the number of cycles on the logarithmic scale. For example, graph paper having 3 cycles on the logarithmic scale is called 'log 3 cycle $\times$ linear' graph paper.

> **Problem 4.** Experimental values of quantities x and y are believed to be related by a law of the form $y = ab^x$, where 'a' and 'b' are constants. The values of x and corresponding values of y are:
>
> | x | 0.7 | 1.4 | 2.1 | 2.9 | 3.7 | 4.3 |
> |---|---|---|---|---|---|---|
> | y | 18.4 | 45.1 | 111 | 308 | 858 | 1850 |
>
> Verify the law and determine the approximate values of 'a' and 'b'. Hence evaluate (i) the value of y when x is 2.5, and (ii) the value of x when y is 1200

Since $y = ab^x$ then $\lg y = (\lg b)x + \lg a$ (from above), which is of the form $Y = mX + c$, showing that to produce a straight line graph $\lg y$ is plotted vertically against x horizontally. Using log-linear graph paper, values of x are marked on the horizontal scale to cover the range 0.7 to 4.3. Values of y range from 18.4 to 1850 and 3 cycles are needed (i.e. 10 to 100, 100 to 1000 and 1000 to 10000). Thus, using 'log 3 cycles $\times$ linear' graph paper the points are plotted as shown in Figure 33.5.
A straight line is drawn through the co-ordinates, hence the law $y = ab^x$ is verified.
Gradient of straight line, $\lg b = AB/BC$. Direct measurement (say in centimetres) is not made with log-linear graph paper since the vertical scale is logarithmic and the horizontal scale is linear.

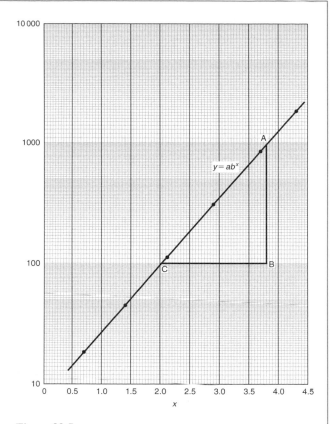

Figure 33.5

Hence,
$$\lg b = \frac{AB}{BC} = \frac{\lg 1000 - \lg 100}{3.82 - 2.02}$$

$$= \frac{3 - 2}{1.80} = \frac{1}{1.80} = 0.5556$$

Hence,
$$b = \text{antilog } 0.5556$$

$$= 10^{0.5556} = \mathbf{3.6}, \text{ correct to 2 significant figures.}$$

Point A has co-ordinates (3.82, 1000).
Substituting these values into $y = ab^x$

gives:
$$1000 = a(3.6)^{3.82}$$

i.e.
$$a = \frac{1000}{(3.6)^{3.82}} = \mathbf{7.5}, \text{ correct to 2 significant figures.}$$

Hence, the law of the graph is: $y = 7.5(3.6)^x$

(i)   When $x = 2.5$, $y = 7.5(3.6)^{2.5} = \mathbf{184}$

(ii)   When $y = 1200$, $1200 = 7.5(3.6)^x$,

$$\text{hence } (3.6)^x = \frac{1200}{7.5} = 160$$

Taking logarithms gives:

$$x \lg 3.6 = \lg 160$$

i.e.

$$x = \frac{\lg 160}{\lg 3.6} = \frac{2.2041}{0.5563} = 3.96$$

Now try the following Practice Exercise

Practice Exercise 144   Logarithmic graphs of the form $y = ab^x$ (answers on page 1164)

1. Experimental values of p and corresponding values of q are shown below.

p	−13.2	−27.9	−62.2	−383.2	−1581	−2931
q	0.30	0.75	1.23	2.32	3.17	3.54

   Show that the law relating p and q is $p = ab^q$, where 'a' and 'b' are constants. Determine (i) values of 'a' and 'b', and state the law, (ii) the value of p when q is 2.0, and (iii) the value of q when p is −2000

33.4   Graphs of the form $y = ae^{kx}$

Taking logarithms to a base of e of both sides of $y = ae^{kx}$ gives:

$$\ln y = \ln(ae^{kx}) = \ln a + \ln e^{kx}$$

$$= \ln a + kx \ln e$$

i.e. $\ln y = kx + \ln a$ since $\ln e = 1$

which compares with $Y = mX + c$

Thus, by plotting $\ln y$ vertically against x horizontally, a straight line results, i.e. the equation $y = ae^{kx}$ is reduced to linear form. In this case, graph paper having a linear horizontal scale and a logarithmic vertical scale may be used.

Problem 5.   The data given below is believed to be related by a law of the form $y = ae^{kx}$, where 'a' and 'b' are constants. Verify that the law is true and determine approximate values of 'a' and 'b'. Also, determine the value of y when x is 3.8 and the value of x when y is 85

x	−1.2	0.38	1.2	2.5	3.4	4.2	5.3
y	9.3	22.2	34.8	71.2	117	181	332

Since $y = ae^{kx}$ then $\ln y = kx + \ln a$ (from above), which is of the form $Y = mX + c$, showing that to produce a straight line graph $\ln y$ is plotted vertically against x horizontally. The value of y ranges from 9.3 to 332 hence 'log 3 cycle × linear' graph paper is used. The plotted co-ordinates are shown in Figure 33.6 and **since a straight line passes through the points the law $y = ae^{kx}$ is verified**.

Gradient of straight line,

$$k = \frac{AB}{BC} = \frac{\ln 100 - \ln 10}{3.12 - (-1.08)} = \frac{2.3026}{4.20}$$

$$= 0.55, \text{ correct to 2 significant figures.}$$

Since $\ln y = kx + \ln a$, when $x = 0$, $\ln y = \ln a$ i.e. $y = a$
The vertical axis intercept value at $x = 0$ is 18, hence, $a = 18$

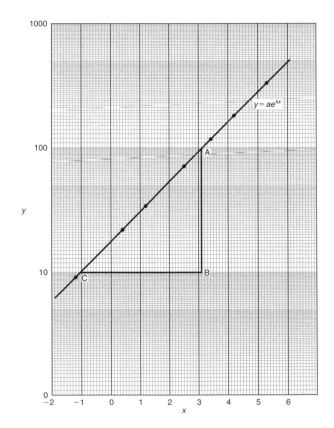

Figure 33.6

The law of the graph is thus: $y = 18e^{0.55x}$

When x is 3.8,

$$y = 18e^{0.55(3.8)} = 18e^{2.09} = 18(8.0849) = \mathbf{146}$$

When y is 85, $85 = 18e^{0.55x}$

Hence, $e^{0.55x} = \dfrac{85}{18} = 4.7222$

and $0.55x = \ln 4.7222 = 1.5523$

Hence, $x = \dfrac{1.5523}{0.55} = \mathbf{2.82}$

Problem 6.   The voltage, v volts, across an inductor is believed to be related to time, t ms, by the law $v = Ve^{t/T}$, where V and T are constants. Experimental results obtained arc:

v (volts)	883	347	90	55.5	18.6	5.2
t (ms)	10.4	21.6	37.8	43.6	56.7	72.0

Show that the law relating voltage and time is as stated, and determine the approximate values of V and T. Find also the value of voltage after 25 ms and the time when the voltage is 30.0 V

Since $v = Ve^{t/T}$ then: $\ln v = \ln(Ve^{t/T})$

$$= \ln V + \ln e^{t/T} = \ln V + \frac{t}{T} \ln e = \frac{t}{T} + \ln V$$

i.e. $\ln v = \dfrac{1}{T} t + \ln V$ which is of the form $Y = mX + c$.

Using 'log 3 cycle × linear' graph paper, the points are plotted as shown in Figure 33.7.
Since the points are joined by a straight line the law $v = Ve^{t/T}$ is verified.

Gradient of straight line,

$$\frac{1}{T} = \frac{AB}{BC} = \frac{\ln 100 - \ln 10}{36.5 - 64.2} = \frac{2.3026}{-27.7}$$

Hence, $T = \dfrac{-27.7}{2.3026} = \mathbf{-12.0}$, correct to 3 significant figures.

Since the straight line does not cross the vertical axis at $t = 0$ in Figure 33.7, the value of V is determined by selecting any point, say A, having co-ordinates (36.5, 100) and substituting these values into $v = Ve^{t/T}$.

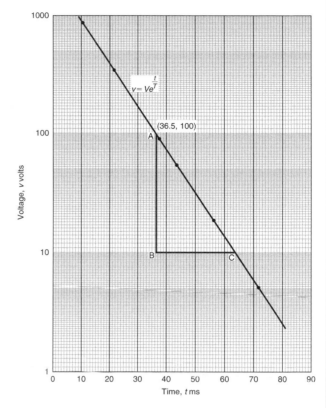

Figure 33.7

Thus, $100 = Ve^{36.5/-12.0}$

i.e. $V = \dfrac{100}{e^{-36.5/12.0}}$

$= \mathbf{2090\ volts}$, correct to 3 significant figures.

Hence, the law of the graph is: $v = 2090e^{-t/12.0}$

When time $t = 25$ ms, voltage $v = 2090e^{-25/12.0}$

$$= \mathbf{260\ V}$$

When the voltage is 30.0 volts, $30.0 = 2090e^{-t/12.0}$

hence $e^{-t/12.0} = \dfrac{30.0}{2090}$ and $e^{t/12.0} = \dfrac{2090}{30.0} = 69.67$

Taking Napierian logarithms gives:

$$\frac{t}{12.0} = \ln 69.67 = 4.2438$$

from which,   time $t = (12.0)(4.2438) = \mathbf{50.9\ ms}$

Section D

Now try the following Practice Exercise

Practice Exercise 145   Logarithmic graphs of the form $y = ae^{kx}$ (answers on page 1164)

1. Atmospheric pressure p is measured at varying altitudes h and the results are as shown below:

Altitude, h (m)	500	1500	3000	5000	8000
pressure, p (cm)	73.39	68.42	61.60	53.56	43.41

Show that the quantities are related by the law $p = ae^{kh}$, where 'a' and 'k' are constants. Determine the values of 'a' and 'k' and state the law. Find also the atmospheric pressure at 10 000 m.

2. At particular times, t minutes, measurements are made of the temperature, $\theta°$C, of a cooling liquid and the following results are obtained:

Temperature, θ (°C)	92.2	55.9	33.9	20.6	12.5
Time, t (minutes)	10	20	30	40	50

Prove that the quantities follow a law of the form $\theta = \theta_0 e^{kt}$, where θ_0 and k are constants, and determine the approximate values of θ_0 and k.

For fully worked solutions to each of the problems in Practice Exercises 143 to 145 in this chapter, go to the website:
www.routledge.com/cw/bird

This assignment covers the material contained in Chapters 31 to 33. *The marks available are shown in brackets at the end of each question.*

1. Determine the value of P in the following table of values.

x	0	1	4
$y = 3x - 5$	-5	-2	P

(2)

2. Assuming graph paper measuring 20cm by 20cm is available, suggest suitable scales for the following ranges of values.

Horizontal axis: 5N to 70N; vertical axis: 20mm to 190mm. (2)

3. Corresponding values obtained experimentally for two quantities are:

x	-5	-3	-1	0	2	4
y	-17	-11	-5	-2	4	10

Plot a graph of y (vertically) against x (horizontally) to scales of 1 cm = 1 for the horizontal x-axis and 1 cm = 2 for the vertical y-axis. From the graph, find
(a) the value of y when $x = 3$
(b) the value of y when $x = -4$
(c) the value of x when $y = 1$
(d) the value of x when $y = -20$ (8)

4. If graphs of y against x were to be plotted for each of the following, state (i) the gradient, and (ii) the y-axis intercept.
(a) $y = -5x + 3$ (b) $y = 7x$
(c) $2y + 4 = 5x$ (d) $5x + 2y = 6$
(e) $2x - \dfrac{y}{3} = \dfrac{7}{6}$ (10)

5. Plot the graphs $y = 3x + 2$ and $\dfrac{y}{2} + x = 6$ on the same axes and determine the co-ordinates of their point of intersection. (7)

6. The resistance R ohms of a copper winding is measured at various temperatures $t°C$ and the results are as follows.

R (Ω)	38	47	55	62	72
t (°C)	16	34	50	64	84

Plot a graph of R (vertically) against t (horizontally) and find from it
(a) the temperature when the resistance is 50Ω,
(b) the resistance when the temperature is 72°C,
(c) the gradient,
(d) the equation of the graph. (10)

7. x and y are two related variables and all other letters denote constants. For the stated laws to be verified it is necessary to plot graphs of the variables in a modified form. State for each
(a) what should be plotted on the vertical axis,
(b) what should be plotted on the horizontal axis,
(c) the gradient,
(d) the vertical axis intercept.
(i) $y = p + rx^2$ (ii) $y = \dfrac{a}{x} + bx$ (4)

8. The following results give corresponding values of two quantities x and y which are believed to be related by a law of the form $y = ax^2 + bx$ where a and b are constants.

y	33.9	55.5	72.8	84.1	111.4	168.1
x	3.4	5.2	6.5	7.3	9.1	12.4

Verify the law and determine approximate values of a and b.
Hence determine (i) the value of y when x is 8.0 and (ii) the value of x when y is 146.5 (18)

9. By taking logarithms of both sides of $y = kx^n$, show that $\lg y$ needs to be plotted vertically and $\lg x$ needs to be plotted horizontally to produce a straight line graph. Also, state the gradient and vertical-axis intercept. (6)

10. By taking logarithms of both sides of $y = ae^{kx}$ show that $\ln y$ needs to be plotted vertically and x needs to be plotted horizontally to produce a straight line graph. Also, state the gradient and vertical-axis intercept. (6)

11. State the minimum number of cycles on logarithmic graph paper needed to plot a set of values ranging from 0.073 to 490 (2)

12. Show from the following results of voltage V and admittance Y of an electrical circuit that the law connecting the quantities is of the form $V = kY^n$ and determine the values of k and n.

Voltage, V(volts)	2.88	2.05	1.60	1.22	0.96
Admittance, Y(siemens)	0.52	0.73	0.94	1.23	1.57

(11)

13. The current i flowing in a discharging capacitor varies with time t as shown.

i (mA)	50.0	17.0	5.8	1.7	0.58	0.24
t (ms)	200	255	310	375	425	475

Show that these results are connected by the law of the form $i = Ie^{\frac{t}{T}}$ where I is the initial current flowing and T is a constant. Determine approximate values of constants I and T. (14)

For lecturers/instructors/teachers, fully worked solutions to each of the problems in Revision Test 13, together with a full marking scheme, are available at the website:
www.routledge.com/cw/bird

Chapter 34

Polar curves

Why it is important to understand: Polar curves

Lighting engineers study photometrics, which is the process for measuring the intensity of light. Such intensity varies according to how close or how far you are from LEDs and other bulbs. Engineers use polar curves to plot the level of intensity at any given point, i.e. a polar curve tells an engineer how LEDs spread light across a room or surface. This distribution may be narrow, wide, indirect or direct. By using polar curves, an engineer can arrange the right number and type of LEDs for a lighting project. For example, a designer may want specific lighting effects in a room. Creating polar curves before installation gives the designer and lighting engineer a chance to check light distribution and intensity. This saves time and money, and ensures success.

At the end of this chapter, you should be able to:

- appreciate the use of polar curves
- plot polar curves of the form $r = a \sin \theta$, $r = a \sin n\theta$, $r = a \sin^2 \theta$, $r = a \cos^2 \theta$ and $r = a\theta$
- plot polar graphs of the form $r = a(1 + \cos \theta)$ and $r = a + b \cos \theta$

34.1 Introduction to polar curves

With Cartesian co-ordinates the equation of a curve is expressed as a general relationship between x and y, i.e. $y = f(x)$. Similarly, with **polar co-ordinates**, the equation of a curve is expressed in the form $r = \mathbf{f}(\boldsymbol{\theta})$. When a graph of $r = f(\theta)$ is required, a table of values needs to be drawn up and the co-ordinates (r, θ) plotted. Polar curves are demonstrated in the following worked problems.

34.2 Worked problems on polar curves

Problem 1. Plot the polar graph of $r = 5 \sin \theta$ between $\theta = 0°$ and $\theta = 360°$ using increments of $30°$

A table of values at $30°$ intervals is produced as shown below.

θ	0	30°	60°	90°	120°	150°	180°
$r = 5\sin\theta$	0	2.50	4.33	5.00	4.33	2.50	0

θ	210°	240°	270°	300°	330°	360°
$r = 5\sin\theta$	−2.50	−4.33	−5.00	−4.33	−2.50	0

The graph is plotted as shown in Figure 34.1.
Initially the zero line 0A is constructed and then the broken lines in Figure 34.1 at $30°$ intervals are produced. The maximum value of r is 5.00 hence 0A is scaled and circles drawn as shown with the largest at a radius of 5 units. The polar co-ordinates $(0, 0°)$, $(2.50, 30°)$, $(4.33, 60°)$, $(5.00, 90°)$ ... are plotted and shown as points 0, B, C, D, ... in Figure 34.1. When polar co-ordinate $(0, 180°)$ is plotted and the points joined with a smooth curve a complete circle is seen to have been produced. When plotting the next point, $(-2.50, 210°)$, since r is negative it is plotted in the opposite direction to $210°$, i.e.

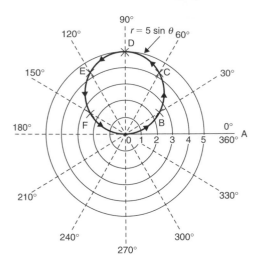

Figure 34.1

2.50 units long on the 30° axis. Hence the point $(-2.50, 210°)$ is equivalent to the point $(2.50, 30°)$. Similarly, $(-4.33, 240°)$ is the same point as $(4.33, 60°)$.

When all the co-ordinates are plotted the graph $r = 5 \sin \theta$ appears as a single circle; it is, in fact, two circles, one on top of the other.

In general, a polar curve $r = a \sin \theta$ is as shown in Figure 34.2.

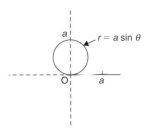

Figure 34.2

In a similar manner to that explained above, it may be shown that the polar curve $r = a \cos \theta$ is as sketched in Figure 34.3.

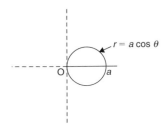

Figure 34.3

Problem 2. Plot the polar graph of $r = 4 \sin^2 \theta$ between $\theta = 0$ and $\theta = 2\pi$ radians using intervals of $\dfrac{\pi}{6}$

A table of values is produced as shown below.

θ	0	$\dfrac{\pi}{6}$	$\dfrac{\pi}{3}$	$\dfrac{\pi}{2}$	$\dfrac{2\pi}{3}$	$\dfrac{5\pi}{6}$	π
$r = 4\sin^2\theta$	0	1	3	4	3	1	0

θ	$\dfrac{7\pi}{6}$	$\dfrac{4\pi}{3}$	$\dfrac{3\pi}{2}$	$\dfrac{5\pi}{3}$	$\dfrac{11\pi}{6}$	2π
$r = 4\sin^2\theta$	1	3	4	3	1	0

The zero line 0A is first constructed and then the broken lines at intervals of $\frac{\pi}{6}$ rad (or 30°) are produced. The maximum value of r is 4, hence 0A is scaled and circles produced with the largest at a radius of 4 units, as shown in Figure 34.4.

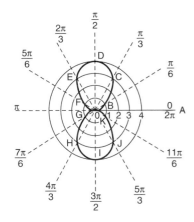

Figure 34.4

The polar co-ordinates $(0, 0)$, $\left(1, \dfrac{\pi}{6}\right)$, $\left(3, \dfrac{\pi}{3}\right)$, ... $(0, \pi)$ are plotted and shown as points 0, B, C, D, E, F, 0, respectively. Then $\left(1, \dfrac{7\pi}{6}\right)$, $\left(3, \dfrac{4\pi}{3}\right)$, ... $(0, 0)$ are plotted as shown by points G, H, I, J, K, 0 respectively. Thus two distinct loops are produced as shown in Figure 34.4.

In general, a polar curve $r = a \sin^2 \theta$ is as shown in Figure 34.5.

In a similar manner it may be shown that the polar curve $r = a \cos^2 \theta$ is as sketched in Figure 34.6.

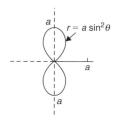

Figure 34.5

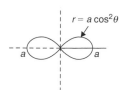

Figure 34.6

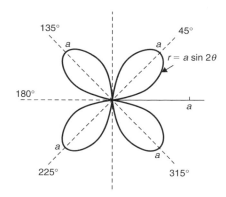

Figure 34.8

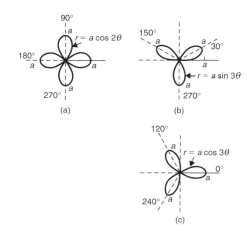

Figure 34.9

> **Problem 3.** Plot the polar graph of $r = 3 \sin 2\theta$ between $\theta = 0°$ and $\theta = 360°$, using $15°$ intervals

As in previous examples a table of values may be produced.

The polar graph $r = 3 \sin 2\theta$ is plotted as shown in Figure 34.7 and is seen to contain four similar shaped loops displaced at $90°$ from each other.

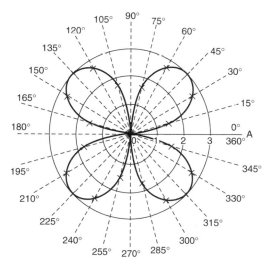

Figure 34.7

In general, a polar curve $r = a \sin 2\theta$ is as shown in Figure 34.8.

In a similar manner it may be shown that polar curves of $r = a \cos 2\theta$, $r = a \sin 3\theta$ and $r = a \cos 3\theta$ are as sketched in Figure 34.9.

> **Problem 4.** Sketch the polar curve $r = 2\theta$ between $\theta = 0$ and $\theta = \dfrac{5\pi}{2}$ rad at intervals of $\dfrac{\pi}{6}$

A table of values may be produced and the polar graph of $r = 2\theta$ is shown in Figure 34.10 and is seen to be an ever-increasing spiral.

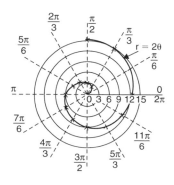

Figure 34.10

Problem 5. Plot the polar curve $r = 5(1 + \cos\theta)$ from $\theta = 0°$ to $\theta = 360°$, using $30°$ intervals

A table of values may be produced and the polar curve $r = 5(1 + \cos\theta)$ is shown in Figure 34.11.

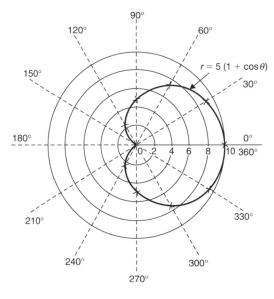

Figure 34.11

In general, a polar curve $r = a(1 + \cos\theta)$ is as shown in Figure 34.12 and the shape is called a **cardioid**.

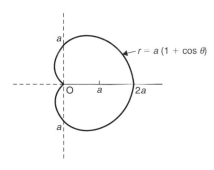

Figure 34.12

In a similar manner it may be shown that the polar curve $r = a + b\cos\theta$ varies in shape according to the relative values of a and b. When $a = b$ the polar curve shown in Figure 34.12 results.
When $a < b$ the general shape shown in Figure 34.13(a) results and when $a > b$ the general shape shown in Figure 34.13(b) results.

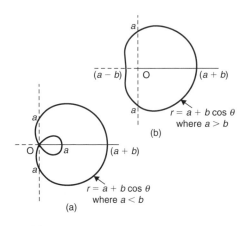

Figure 34.13

Now try the following Practice Exercise

Practice Exercise 146   Polar curves (answers on page 1164)

In Problems 1 to 6, sketch the given polar curves.

1. (a) $r = 3\sin\theta$ (b) $r = 4\cos\theta$

2. (a) $r = 2\sin^2\theta$ (b) $r = 7\cos^2\theta$

3. (a) $r = 5\sin 2\theta$ (b) $r = 4\cos 2\theta$

4. (a) $r = 3\cos 3\theta$ (b) $r = 9\sin 3\theta$

5. (a) $r = 5\theta$ (b) $r = 2(1 + \cos\theta)$

6. (a) $r = 1 + 2\cos\theta$ (b) $r = 3 + \cos\theta$

For fully worked solutions to each of the problems in Practice Exercise 146 in this chapter, go to the website:
www.routledge.com/cw/bird

Chapter 35

Graphical solution of equations

Why it is important to understand: Graphical solution of equations

It has been established in previous chapters that the solution of linear, quadratic, simultaneous and cubic equations occur often in engineering and science and may be solved using algebraic means. Being able to solve equations graphically provides another method to aid understanding and interpretation of equations. Engineers, including architects, surveyors and a variety of engineers in fields such as biomedical, chemical, electrical, mechanical and nuclear, all use equations which need solving by one means or another.

At the end of this chapter, you should be able to:

- solve two simultaneous equations graphically
- solve a quadratic equation graphically
- solve a linear and a quadratic equation simultaneously by graphical means
- solve a cubic equation graphically

35.1 Graphical solution of simultaneous equations

Linear simultaneous equations in two unknowns may be solved graphically by

(a) plotting the two straight lines on the same axes, and

(b) noting their point of intersection.

The co-ordinates of the point of intersection give the required solution.
Here are some worked problems to demonstrate the graphical solution of simultaneous equations.

Problem 1. Solve graphically the simultaneous equations

$$2x - y = 4$$
$$x + y = 5$$

Rearranging each equation into $y = mx + c$ form gives

$$y = 2x - 4$$
$$y = -x + 5$$

Only three co-ordinates need be calculated for each graph since both are straight lines.

x	0	1	2
$y = 2x - 4$	-4	-2	0

x	0	1	2
$y = -x + 5$	5	4	3

Each of the graphs is plotted as shown in Figure 35.1. The point of intersection is at (3, 2) and since this is the only point which lies simultaneously on both lines then $x = 3$, $y = 2$ is the solution of the simultaneous equations.

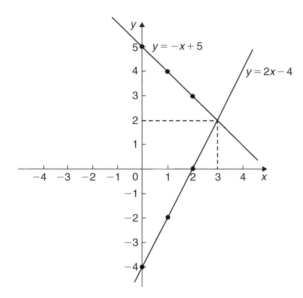

Figure 35.1

> **Problem 2.** Solve graphically the equations
> $$1.20x + y = 1.80$$
> $$x - 5.0y = 8.50$$

Rearranging each equation into $y = mx + c$ form gives

$$y = -1.20x + 1.80 \qquad (1)$$

$$y = \frac{x}{5.0} - \frac{8.5}{5.0}$$

i.e. $\qquad y = 0.20x - 1.70 \qquad (2)$

Three co-ordinates are calculated for each equation as shown below.

x	0	1	2
$y = -1.20x + 1.80$	1.80	0.60	-0.60

x	0	1	2
$y = 0.20x - 1.70$	-1.70	-1.50	-1.30

The two sets of co-ordinates are plotted as shown in Figure 35.2. The point of intersection is (2.50, −1.20). Hence, the solution of the simultaneous equations is $x = 2.50$, $y = -1.20$

(It is sometimes useful to initially sketch the two straight lines to determine the region where the point of intersection is. Then, for greater accuracy, a graph having a smaller range of values can be drawn to 'magnify' the point of intersection.)

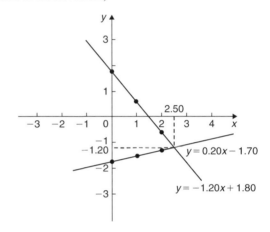

Figure 35.2

Now try the following Practice Exercise

> **Practice Exercise 147   Graphical solution of simultaneous equations (answers on page 1165)**
>
> In Problems 1 to 6, solve the simultaneous equations graphically.
>
> 1. $y = 3x - 2$ 2. $x + y = 2$
> $y = -x + 6$ $3y - 2x = 1$
>
> 3. $y = 5 - x$ 4. $3x + 4y = 5$
> $x - y = 2$ $2x - 5y + 12 = 0$
>
> 5. $1.4x - 7.06 = 3.2y$ 6. $3x - 2y = 0$
> $2.1x - 6.7y = 12.87$ $4x + y + 11 = 0$
>
> 7. The friction force F newtons and load L newtons are connected by a law of the form $F = aL + b$, where a and b are constants. When $F = 4\,\text{N}$, $L = 6\,\text{N}$ and when $F = 2.4\,\text{N}$, $L = 2\,\text{N}$. Determine graphically the values of a and b.

35.2 Graphical solution of quadratic equations

A general **quadratic equation** is of the form $y = ax^2 + bx + c$, where a, b and c are constants and a is not equal to zero.

A graph of a quadratic equation always produces a shape called a **parabola**.

The gradients of the curves between 0 and A and between B and C in Figure 35.3 are positive, whilst the gradient between A and B is negative. Points such as A and B are called **turning points**. At A the gradient is zero and, as x increases, the gradient of the curve changes from positive just before A to negative just after. Such a point is called a **maximum value**. At B the gradient is also zero and, as x increases, the gradient of the curve changes from negative just before B to positive just after. Such a point is called a **minimum value**.

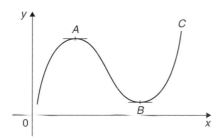

Figure 35.3

Following are three examples of solutions using **quadratic graphs**.

(a) $y = ax^2$

Graphs of $y = x^2$, $y = 3x^2$ and $y = \dfrac{1}{2}x^2$ are shown in Figure 35.4. All have minimum values at the origin (0, 0).

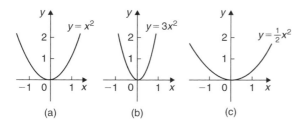

Figure 35.4

Graphs of $y = -x^2$, $y = -3x^2$ and $y = -\dfrac{1}{2}x^2$ are shown in Figure 35.5. All have maximum values at the origin (0, 0).

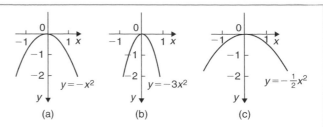

Figure 35.5

When $y = ax^2$,

   (i) curves are symmetrical about the y-axis,

  (ii) the magnitude of a affects the gradient of the curve, and

 (iii) the sign of a determines whether it has a maximum or minimum value.

(b) $y = ax^2 + c$

Graphs of $y = x^2 + 3$, $y = x^2 - 2$, $y = -x^2 + 2$ and $y = -2x^2 - 1$ are shown in Figure 35.6.

When $y = ax^2 + c$,

   (i) curves are symmetrical about the y-axis,

  (ii) the magnitude of a affects the gradient of the curve, and

 (iii) the constant c is the y-axis intercept.

(c) $y = ax^2 + bx + c$

Whenever b has a value other than zero the curve is displaced to the right or left of the y-axis.

When b/a is positive, the curve is displaced $b/2a$ to the left of the y-axis, as shown in Figure 35.7(a).

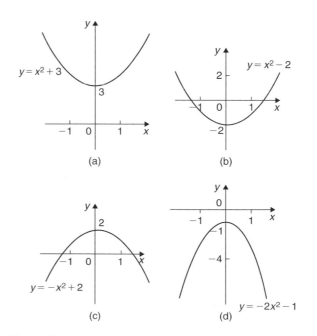

Figure 35.6

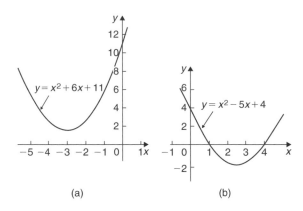

Figure 35.7

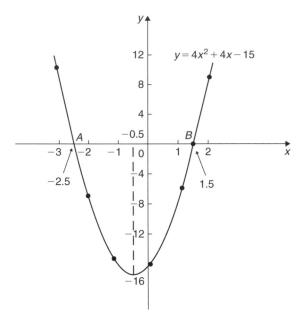

Figure 35.8

When b/a is negative, the curve is displaced $b/2a$ to the right of the y-axis, as shown in Figure 35.7(b).

Quadratic equations of the form $ax^2 + bx + c = 0$ may be solved graphically by

(a) plotting the graph $y = ax^2 + bx + c$, and

(b) noting the points of intersection on the x-axis (i.e. where $y = 0$).

The x values of the points of intersection give the required solutions since at these points both $y = 0$ and $ax^2 + bx + c = 0$

The number of solutions, or roots, of a quadratic equation depends on how many times the curve cuts the x-axis. There can be no real roots, as in Figure 35.7(a), one root, as in Figures 35.4 and 35.5, or two roots, as in Figure 35.7(b).

Here are some worked problems to demonstrate the graphical solution of quadratic equations.

Problem 3. Solve the quadratic equation $4x^2 + 4x - 15 = 0$ graphically, given that the solutions lie in the range $x = -3$ to $x = 2$. Determine also the co-ordinates and nature of the turning point of the curve

---

Let $y = 4x^2 + 4x - 15$. A table of values is drawn up as shown below.

x		-3	-2	-1	0	1	2
$y = 4x^2 + 4x - 15$		9	-7	-15	-15	-7	9

A graph of $y = 4x^2 + 4x - 15$ is shown in Figure 35.8. The only points where $y = 4x^2 + 4x - 15$ and $y = 0$ are the points marked A and B. This occurs at $x = -2.5$

and $x = 1.5$ and these are the solutions of the quadratic equation $4x^2 + 4x - 15 = 0$
By substituting $x = -2.5$ and $x = 1.5$ into the original equation the solutions may be checked.
The curve has a turning point at $(-0.5, -16)$ and the nature of the point is a **minimum**.
An alternative graphical method of solving $4x^2 + 4x - 15 = 0$ is to rearrange the equation as $4x^2 = -4x + 15$ and then plot two separate graphs – in this case, $y = 4x^2$ and $y = -4x + 15$. Their points of intersection give the roots of the equation $4x^2 = -4x + 15$, i.e. $4x^2 + 4x - 15 = 0$. This is shown in Figure 35.9, where the roots are $x = -2.5$ and $x = 1.5$, as before.

Problem 4. Solve graphically the quadratic equation $-5x^2 + 9x + 7.2 = 0$ given that the solutions lie between $x = -1$ and $x = 3$. Determine also the co-ordinates of the turning point and state its nature

---

Let $y = -5x^2 + 9x + 7.2$. A table of values is drawn up as shown below.

x		-1	-0.5	0	1
$y = -5x^2 + 9x + 7.2$		-6.8	1.45	7.2	11.2

x		2	2.5	3
$y = -5x^2 + 9x + 7.2$		5.2	-1.55	-10.8

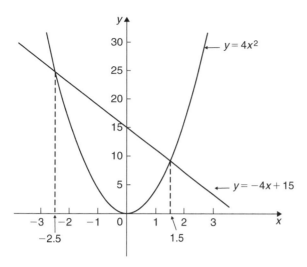

Figure 35.9

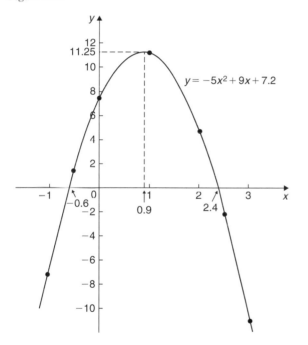

Figure 35.10

A graph of $y = -5x^2 + 9x + 7.2$ is shown plotted in Figure 35.10. The graph crosses the x-axis (i.e. where $y = 0$) at $x = -0.6$ and $x = 2.4$ and these are the solutions of the quadratic equation $-5x^2 + 9x + 7.2 = 0$. The turning point is a **maximum**, having co-ordinates **(0.9, 11.25)**

> **Problem 5.** Plot a graph of $y = 2x^2$ and hence solve the equations
> (a) $2x^2 - 8 = 0$ (b) $2x^2 - x - 3 = 0$

A graph of $y = 2x^2$ is shown in Figure 35.11.

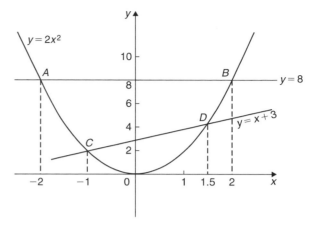

Figure 35.11

(a) Rearranging $2x^2 - 8 = 0$ gives $2x^2 = 8$ and the solution of this equation is obtained from the points of intersection of $y = 2x^2$ and $y = 8$; i.e. at co-ordinates $(-2, 8)$ and $(2, 8)$, shown as A and B, respectively, in Figure 35.11. Hence, the solutions of $2x^2 - 8 = 0$ are $x = -2$ and $x = +2$

(b) Rearranging $2x^2 - x - 3 = 0$ gives $2x^2 = x + 3$ and the solution of this equation is obtained from the points of intersection of $y = 2x^2$ and $y = x + 3$; i.e. at C and D in Figure 35.11. Hence, the solutions of $2x^2 - x - 3 = 0$ are $x = -1$ and $x = 1.5$

> **Problem 6.** Plot the graph of $y = -2x^2 + 3x + 6$ for values of x from $x = -2$ to $x = 4$. Use the graph to find the roots of the following equations.
> (a) $-2x^2 + 3x + 6 = 0$ (b) $-2x^2 + 3x + 2 = 0$
> (c) $-2x^2 + 3x + 9 = 0$ (d) $-2x^2 + x + 5 = 0$

A table of values for $y = -2x^2 + 3x + 6$ is drawn up as shown below.

x	-2	-1	0	1	2	3	4
y	-8	1	6	7	4	-3	-14

A graph of $y = -2x^2 + 3x + 6$ is shown in Figure 35.12.

(a) The parabola $y = -2x^2 + 3x + 6$ and the straight line $y = 0$ intersect at A and B, where $x = -1.13$ and $x = 2.63$, and these are the roots of the equation $-2x^2 + 3x + 6 = 0$

(b) Comparing $y = -2x^2 + 3x + 6$ (1)

with $0 = -2x^2 + 3x + 2$ (2)

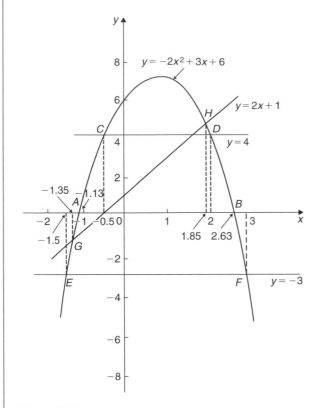

Figure 35.12

shows that, if 4 is added to both sides of equation (2), the RHS of both equations will be the same. Hence, $4 = -2x^2 + 3x + 6$. The solution of this equation is found from the points of intersection of the line $y = 4$ and the parabola $y = -2x^2 + 3x + 6$; i.e. points C and D in Figure 35.12. Hence, the roots of $-2x^2 + 3x + 2 = 0$ are $x = -0.5$ and $x = 2$

(c) $-2x^2 + 3x + 9 = 0$ may be rearranged as $-2x^2 + 3x + 6 = -3$ and the solution of this equation is obtained from the points of intersection of the line $y = -3$ and the parabola $y = -2x^2 + 3x + 6$; i.e. at points E and F in Figure 35.12. Hence, the roots of $-2x^2 + 3x + 9 = 0$ are $x = -1.5$ and $x = 3$

(d) Comparing $y = -2x^2 + 3x + 6$ (3)

with $0 = -2x^2 + x + 5$ (4)

shows that, if $2x + 1$ is added to both sides of equation (4), the RHS of both equations will be the same. Hence, equation (4) may be written as $2x + 1 = -2x^2 + 3x + 6$. The solution of this equation is found from the points of intersection of

the line $y = 2x + 1$ and the parabola $y = -2x^2 + 3x + 6$; i.e. points G and H in Figure 35.12. Hence, the roots of $-2x^2 + x + 5 = 0$ are $x = -1.35$ and $x = 1.85$

Now try the following **Practice Exercise**

Practice Exercise 148   Solving quadratic equations graphically (answers on page 1165)

1. Sketch the following graphs and state the nature and co-ordinates of their respective turning points.
   (a) $y = 4x^2$ (b) $y = 2x^2 - 1$
   (c) $y = -x^2 + 3$ (d) $y = -\frac{1}{2}x^2 - 1$

Solve graphically the quadratic equations in Problems 2 to 5 by plotting the curves between the given limits. Give answers correct to 1 decimal place.

2. $4x^2 - x - 1 = 0$; $x = -1$ to $x = 1$

3. $x^2 - 3x = 27$; $x = -5$ to $x = 8$

4. $2x^2 - 6x - 9 = 0$; $x = -2$ to $x = 5$

5. $2x(5x - 2) = 39.6$; $x = -2$ to $x = 3$

6. Solve the quadratic equation $2x^2 + 7x + 6 = 0$ graphically, given that the solutions lie in the range $x = -3$ to $x = 1$. Determine also the nature and co-ordinates of its turning point.

7. Solve graphically the quadratic equation $10x^2 - 9x - 11.2 = 0$, given that the roots lie between $x = -1$ and $x = 2$

8. Plot a graph of $y = 3x^2$ and hence solve the following equations.
   (a) $3x^2 - 8 = 0$ (b) $3x^2 - 2x - 1 = 0$

9. Plot the graphs $y = 2x^2$ and $y = 3 - 4x$ on the same axes and find the co-ordinates of the points of intersection. Hence, determine the roots of the equation $2x^2 + 4x - 3 = 0$

10. Plot a graph of $y = 10x^2 - 13x - 30$ for values of x between $x = -2$ and $x = 3$. Solve the equation $10x^2 - 13x - 30 = 0$ and from the graph determine
    (a) the value of y when x is 1.3,
    (b) the value of x when y is 10,
    (c) the roots of the equation $10x^2 - 15x - 18 = 0$

35.3 Graphical solution of linear and quadratic equations simultaneously

The solution of **linear and quadratic equations simultaneously** may be achieved graphically by

(a) plotting the straight line and parabola on the same axes, and

(b) noting the points of intersection.

The co-ordinates of the points of intersection give the required solutions.
Here is a worked problem to demonstrate the simultaneous solution of a linear and quadratic equation.

Problem 7. Determine graphically the values of x and y which simultaneously satisfy the equations $y = 2x^2 - 3x - 4$ and $y = 2 - 4x$

$y = 2x^2 - 3x - 4$ is a parabola and a table of values is drawn up as shown below.

x	-2	-1	0	1	2	3
y	10	1	-4	-5	-2	5

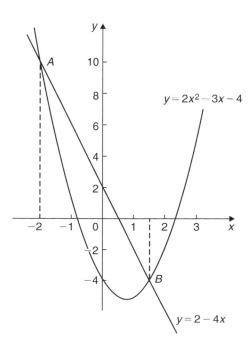

Figure 35.13

$y = 2 - 4x$ is a straight line and only three co-ordinates need be calculated:

x	0	1	2
y	2	-2	-6

The two graphs are plotted in Figure 35.13 and the points of intersection, shown as A and B, are at co-ordinates $(-2, 10)$ and $(1.5, -4)$. Hence, the simultaneous solutions occur when $x = -2$, $y = 10$ and when $x = 1.5$, $y = -4$

These solutions may be checked by substituting into each of the original equations.

Now try the following Practice Exercise

Practice Exercise 149   Solving linear and quadratic equations simultaneously (answers on page 1165)

1. Determine graphically the values of x and y which simultaneously satisfy the equations $y = 2(x^2 - 2x - 4)$ and $y + 4 = 3x$

2. Plot the graph of $y = 4x^2 - 8x - 21$ for values of x from -2 to $+4$. Use the graph to find the roots of the following equations.

    (a) $4x^2 - 8x - 21 = 0$

    (b) $4x^2 - 8x - 16 = 0$

    (c) $4x^2 - 6x - 18 = 0$

35.4 Graphical solution of cubic equations

A **cubic equation** of the form $ax^3 + bx^2 + cx + d = 0$ may be solved graphically by

(a) plotting the graph $y = ax^3 + bx^2 + cx + d$, and

(b) noting the points of intersection on the x-axis (i.e. where $y = 0$)

The x-values of the points of intersection give the required solution since at these points both $y = 0$ and $ax^3 + bx^2 + cx + d = 0$

The number of solutions, or roots, of a cubic equation depends on how many times the curve cuts the x-axis and there can be one, two or three possible roots, as shown in Figure 35.14.

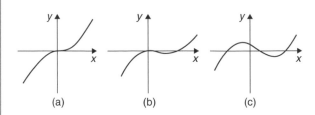

Figure 35.14

Here are some worked problems to demonstrate the graphical solution of cubic equations.

Problem 8.   Solve graphically the cubic equation $4x^3 - 8x^2 - 15x + 9 = 0$, given that the roots lie between $x = -2$ and $x = 3$. Determine also the co-ordinates of the turning points and distinguish between them

Let $y = 4x^3 - 8x^2 - 15x + 9$. A table of values is drawn up as shown below.

x	-2	-1	0	1	2	3
y	-25	12	9	-10	-21	0

A graph of $y = 4x^3 - 8x^2 - 15x + 9$ is shown in Figure 35.15.

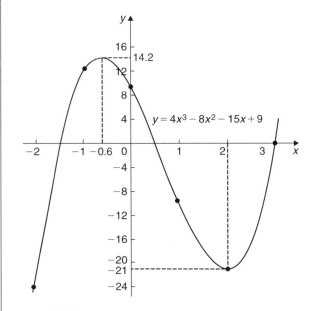

Figure 35.15

The graph crosses the x-axis (where $y = 0$) at $x = -1.5$, $x = 0.5$ and $x = 3$ and these are the solutions to the cubic equation $4x^3 - 8x^2 - 15x + 9 = 0$
The turning points occur at $(-0.6, 14.2)$, which is a **maximum**, and $(2, -21)$, which is a **minimum**.

Problem 9.   Plot the graph of $y = 2x^3 - 7x^2 + 4x + 4$ for values of x between $x = -1$ and $x = 3$. Hence, determine the roots of the equation $2x^3 - 7x^2 + 4x + 4 = 0$

A table of values is drawn up as shown below.

x	-1	0	1	2	3
y	-9	4	3	0	7

A graph of $y = 2x^3 - 7x^2 + 4x + 4$ is shown in Figure 35.16. The graph crosses the x-axis at $x = -0.5$ and touches the x-axis at $x = 2$

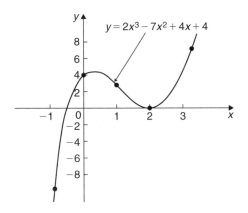

Figure 35.16

Hence the solutions of the equation $2x^3 - 7x^2 + 4x + 4 = 0$ are $x = -0.5$ and $x = 2$

Now try the following Practice Exercise

Practice Exercise 150   Solving cubic equations (answers on page 1165)

1. Plot the graph $y = 4x^3 + 4x^2 - 11x - 6$ between $x = -3$ and $x = 2$ and use the graph to solve the cubic equation $4x^3 + 4x^2 - 11x - 6 = 0$

2. By plotting a graph of $y = x^3 - 2x^2 - 5x + 6$ between $x = -3$ and $x = 4$, solve the equation $x^3 - 2x^2 - 5x + 6 = 0$. Determine also the co-ordinates of the turning points and distinguish between them.

In Problems 3 to 6, solve graphically the cubic equations given, each correct to 2 significant figures.

3. $x^3 - 1 = 0$

4. $x^3 - x^2 - 5x + 2 = 0$

5. $x^3 - 2x^2 = 2x - 2$

6. $2x^3 - x^2 - 9.08x + 8.28 = 0$

7. Show that the cubic equation $8x^3 + 36x^2 + 54x + 27 = 0$ has only one real root and determine its value.

For fully worked solutions to each of the problems in Practice Exercises 147 to 150 in this chapter, go to the website:
www.routledge.com/cw/bird

Chapter 36

Functions and their curves

Why it is important to understand: **Functions and their curves**

Graphs and diagrams provide a simple and powerful approach to a variety of problems that are typical to computer science in general, and software engineering in particular; graphical transformations have many applications in software engineering problems. Periodic functions are used throughout engineering and science to describe oscillations, waves and other phenomena that exhibit periodicity. Engineers use many basic mathematical functions to represent, say, the input/output of systems – linear, quadratic, exponential, sinusoidal, and so on, and knowledge of these are needed to determine how these are used to generate some of the more unusual input/output signals such as the square wave, saw-tooth wave and fully-rectified sine wave. Understanding of continuous and discontinuous functions, odd and even functions, inverse functions and asymptotes are helpful in this – it's all part of the 'language of engineering'.

At the end of this chapter, you should be able to:

- recognise standard curves and their equations – straight line, quadratic, cubic, trigonometric, circle, ellipse, hyperbola, rectangular hyperbola, logarithmic function, exponential function and polar curves
- perform simple graphical transformations
- define a periodic function
- define continuous and discontinuous functions
- define odd and even functions
- define inverse functions
- determine asymptotes
- sketch curves

36.1 Definition of a function

In mathematics, a **function** is a relation between a set of inputs and a set of permissible outputs, with the property that each input is related to exactly one output. An example is the function that relates each real number x to its square x^2. The output of a function f corresponding to an input x is denoted by $f(x)$ (read as f of x). In this example, if the input is 2, then the output is 4, i.e. if $f(x) = x^2$ then $f(2) = 2^2 = 4$
The input variable(s) are sometimes referred to as the argument(s) of the function.

Similarly, if $f(x) = 2x + 3$, then $f(5) = 2(5) + 3 = 13$, and $f(2) = 2(2) + 3 = 7$

36.2 Standard curves

When a mathematical equation is known, co-ordinates may be calculated for a limited range of values, and the equation may be represented pictorially as a graph, within this range of calculated values. Sometimes it is useful to show all the characteristic features of an equation, and in this case a sketch depicting the equation can be drawn, in which all the important features are shown, but the accurate plotting of points is less important. This technique is called 'curve sketching' and can involve the use of differential calculus, with, for example, calculations involving turning points.

If, say, y depends on, say, x, then y is said to be a function of x and the relationship is expressed as $y = f(x)$; x is called the independent variable and y is the dependent variable.

In engineering and science, corresponding values are obtained as a result of tests or experiments.

Here is a brief resumé of standard curves, some of which have been met earlier in this text.

(i) Straight line

The general equation of a straight line is $y = mx + c$, where m is the gradient and c is the y-axis intercept. Two examples are shown in Figure 36.1

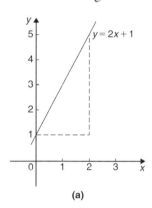

(a)

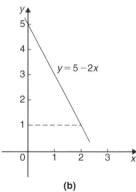

(b)

Figure 36.1

(ii) Quadratic graphs

The general equation of a quadratic graph is $y = ax^2 + bx + c$, and its shape is that of a parabola. The simplest example of a quadratic graph, $y = x^2$, is shown in Figure 36.2.

(iii) Cubic equations

The general equation of a cubic graph is $y = ax^3 + bx^2 + cx + d$

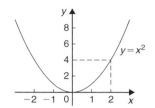

Figure 36.2

The simplest example of a cubic graph, $y = x^3$, is shown in Figure 36.3.

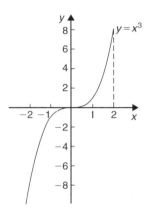

Figure 36.3

(iv) Trigonometric functions (see Chapter 39, page 459)

Graphs of $y = \sin\theta$, $y = \cos\theta$ and $y = \tan\theta$ are shown in Figure 36.4.

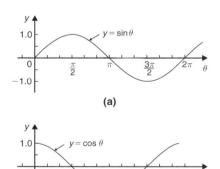

(a)

(b)

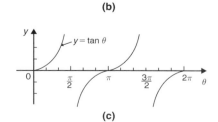

(c)

Figure 36.4

(v) Circle (see Chapter 28, page 296)

The simplest equation of a circle is $x^2 + y^2 = r^2$, with centre at the origin and radius r, as shown in Figure 36.5.

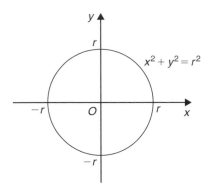

Figure 36.5

More generally, the equation of a circle, centre (a, b), radius r, is given by:

$$(x - a)^2 + (y - b)^2 = r^2$$

Figure 36.6 shows a circle

$$(x - 2)^2 + (y - 3)^2 = 4$$

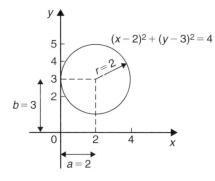

Figure 36.6

(vi) Ellipse

The equation of an ellipse is

$$\frac{x^2}{a^2} + \frac{y^2}{b^2} = 1$$

and the general shape is as shown in Figure 36.7.
The length AB is called the **major axis** and CD the **minor axis**.
In the above equation, 'a' is the semi-major axis and 'b' is the semi-minor axis.

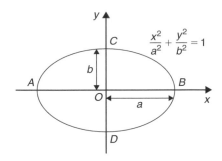

Figure 36.7

(Note that if $b = a$, the equation becomes $\frac{x^2}{a^2} + \frac{y^2}{a^2} = 1$, i.e. $x^2 + y^2 = a^2$, which is a circle of radius a.)

(vii) Hyperbola

The equation of a hyperbola is

$$\frac{x^2}{a^2} - \frac{y^2}{b^2} = 1$$

and the general shape is shown in Figure 36.8. The curve is seen to be symmetrical about both the x- and y-axes. The distance AB in Figure 36.8 is given by $2a$

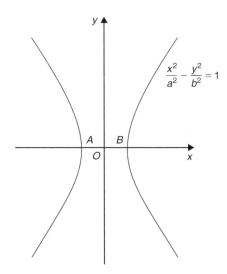

Figure 36.8

(viii) Rectangular Hyperbola

The equation of a rectangular hyperbola is $xy = c$ or $y = \frac{c}{x}$ (where c is a constant) and the general shape is shown in Figure 36.9.

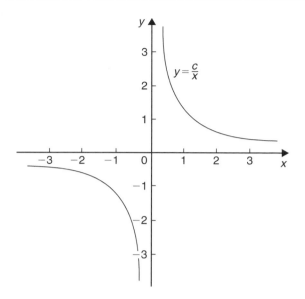

Figure 36.9

(ix) Logarithmic function (see Chapter 15, page 144)

$y = \ln x$ and $y = \lg x$ are both of the general shape shown in Figure 36.10

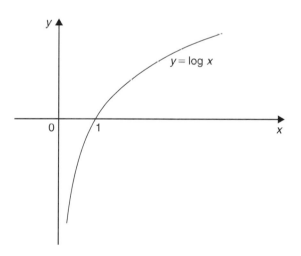

Figure 36.10

(x) Exponential functions (see Chapter 16, page 149)

$y = e^x$ is of the general shape shown in Figure 36.11.

(xi) Polar curves (see Chapter 34)

The equation of a polar curve is of the form $r = f(\theta)$. An example of a polar curve, $r = a \sin \theta$, is shown in Figure 36.12.

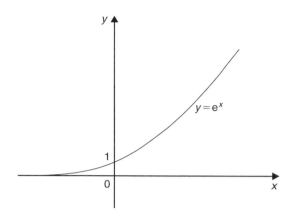

Figure 36.11

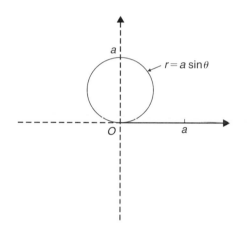

Figure 36.12

36.3   Simple transformations

From the graph of $y = f(x)$ it is possible to deduce the graphs of other functions which are transformations of $y = f(x)$. For example, knowing the graph of $y = f(x)$ can help us draw the graphs of $y = af(x)$, $y = f(x) + a$, $y = f(x + a)$, $y = f(ax)$, $y = -f(x)$ and $y = f(-x)$.

(i) $y = af(x)$

For each point (x_1, y_1) on the graph of $y = f(x)$ there exists a point (x_1, ay_1) on the graph of $y = af(x)$. Thus the graph of $y = af(x)$ can be obtained by stretching $y = f(x)$ parallel to the y-axis by a scale factor 'a'

Graphs of $y = x + 1$ and $y = 3(x + 1)$ are shown in Figure 36.13(a) and graphs of $y = \sin \theta$ and $y = 2 \sin \theta$ are shown in Figure 36.13(b).

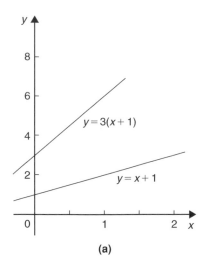

(a)

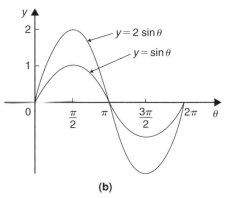

(b)

Figure 36.13

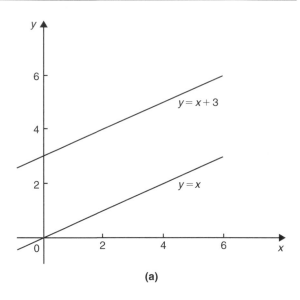

(a)

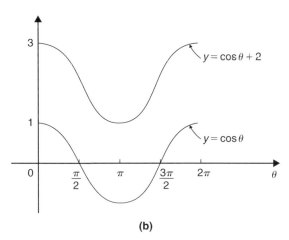

(b)

(ii) $y = f(x) + a$

The graph of $y = f(x)$ is translated by 'a' units parallel to the y-axis to obtain $y = f(x) + a$. For example, if $f(x) = x$, $y = f(x) + 3$ becomes $y = x + 3$, as shown in Figure 36.14(a). Similarly, if $f(\theta) = \cos\theta$, then $y = f(\theta) + 2$ becomes $y = \cos\theta + 2$, as shown in Figure 36.14(b). Also, if $f(x) = x^2$, then $y = f(x) + 3$ becomes $y = x^2 + 3$, as shown in Figure 36.14(c).

(iii) $y = f(x + a)$

The graph of $y = f(x)$ is translated by 'a' units parallel to the x-axis to obtain $y = f(x + a)$. If 'a' > 0 it moves $y = f(x)$ in the negative direction on the x-axis (i.e. to the left), and if 'a' < 0 it moves $y = f(x)$ in the positive direction on the x-axis (i.e. to the right). For example, if $f(x) = \sin x$, $y = f\left(x - \dfrac{\pi}{3}\right)$ becomes $y = \sin\left(x - \dfrac{\pi}{3}\right)$ as shown in Figure 36.15(a) and $y = \sin\left(x + \dfrac{\pi}{4}\right)$ is shown in Figure 36.15(b).

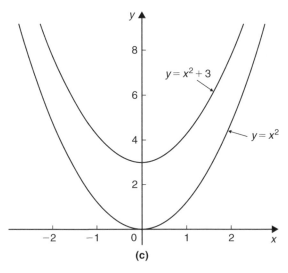

(c)

Figure 36.14

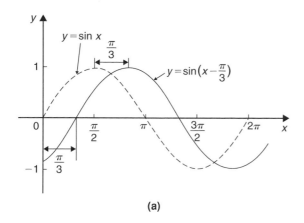

(a)

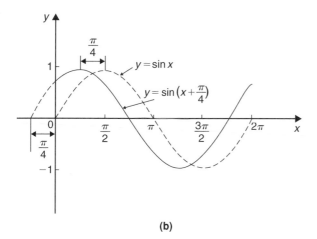

(b)

Figure 36.15

Similarly graphs of $y = x^2$, $y = (x-1)^2$ and $y = (x+2)^2$ are shown in Figure 36.16.

(iv) $y = f(ax)$

For each point (x_1, y_1) on the graph of $y = f(x)$, there exists a point $\left(\dfrac{x_1}{a}, y_1\right)$ on the graph of $y = f(ax)$. Thus the graph of $y = f(ax)$ can be obtained by stretching $y = f(x)$ parallel to the x-axis by a scale factor $\dfrac{1}{a}$

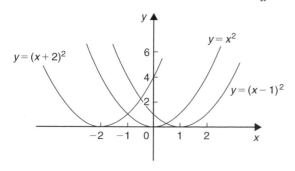

Figure 36.16

For example, if $f(x) = (x-1)^2$, and $a = \dfrac{1}{2}$, then

$$f(ax) = \left(\frac{x}{2} - 1\right)^2$$

Both of these curves are shown in Figure 36.17(a).

Similarly, $y = \cos x$ and $y = \cos 2x$ are shown in Figure 36.17(b).

(v) $y = -f(x)$

The graph of $y = -f(x)$ is obtained by reflecting $y = f(x)$ in the x-axis. For example, graphs of $y = e^x$ and $y = -e^x$ are shown in Figure 36.18(a) and graphs of $y = x^2 + 2$ and $y = -(x^2+2)$ are shown in Figure 36.18(b).

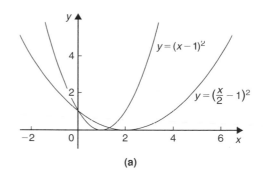

(a)

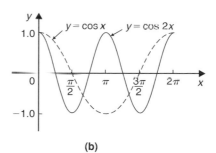

(b)

Figure 36.17

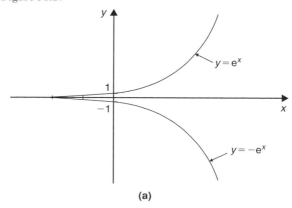

(a)

Figure 36.18

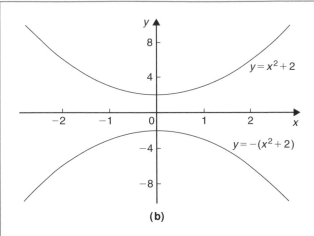

(b)

Figure 36.18 *(Continued)*

(vi) $y = f(-x)$

The graph of $y = f(-x)$ is obtained by reflecting $y = f(x)$ in the y-axis. For example, graphs of $y = x^3$ and $y = (-x)^3 = -x^3$ are shown in Figure 36.19(a)

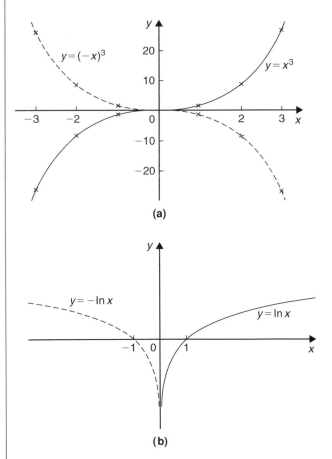

Figure 36.19

and graphs of $y = \ln x$ and $y = -\ln x$ are shown in Figure 36.19(b).

> **Problem 1.** Sketch the following graphs, showing relevant points:
>
> (a) $y = (x-4)^2$ (b) $y = x^3 - 8$

(a) In Figure 36.20 a graph of $y = x^2$ is shown by the broken line. The graph of $y = (x-4)^2$ is of the form $y = f(x+a)$. Since $a = -4$, then $y = (x-4)^2$ is translated 4 units to the right of $y = x^2$, parallel to the x-axis. (See Section (iii) above.)

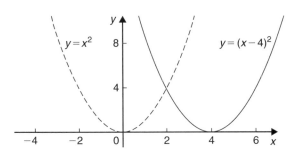

Figure 36.20

(b) In Figure 36.21 a graph of $y = x^3$ is shown by the broken line. The graph of $y = x^3 - 8$ is of the form $y = f(x) + a$. Since $a = -8$, then $y = x^3 - 8$ is translated 8 units down from $y = x^3$, parallel to the y-axis. (See Section (ii) above.)

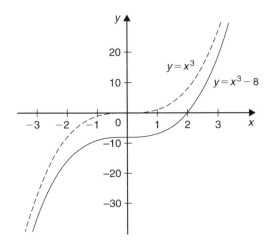

Figure 36.21

Problem 2.   Sketch the following graphs, showing relevant points:

(a) $y = 5 - (x+2)^3$ (b) $y = 1 + 3\sin 2x$

(a)   Figure 36.22(a) shows a graph of $y = x^3$. Figure 36.22(b) shows a graph of $y = (x+2)^3$ (see $f(x+a)$, Section (iii) above).

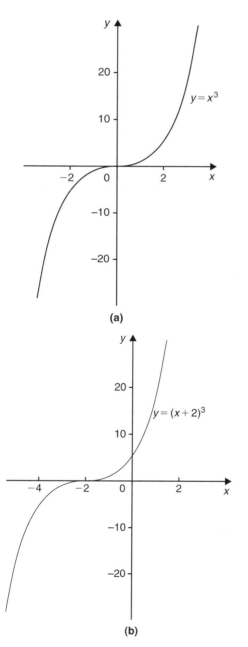

(a)

(b)

Figure 36.22

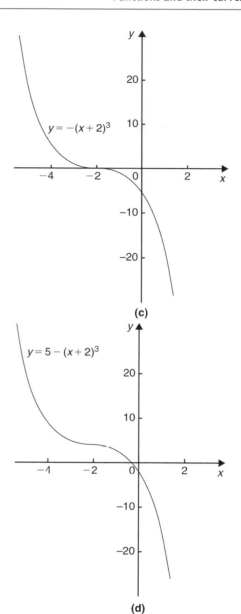

(c)

(d)

Figure 36.22 *(Continued)*

Figure 36.22(c) shows a graph of $y = -(x+2)^3$ (see $-f(x)$, Section (v) above). Figure 36.22(d) shows the graph of $y = 5 - (x+2)^3$ (see $f(x) + a$, Section (ii) above).

(b)   Figure 36.23(a) shows a graph of $y = \sin x$. Figure 36.23(b) shows a graph of $y = \sin 2x$ (see $f(ax)$, Section (iv) above).

Figure 36.23(c) shows a graph of $y = 3\sin 2x$ (see $a\,f(x)$, Section (i) above). Figure 36.23(d) shows a graph of $y = 1 + 3\sin 2x$ (see $f(x) + a$, Section (ii) above).

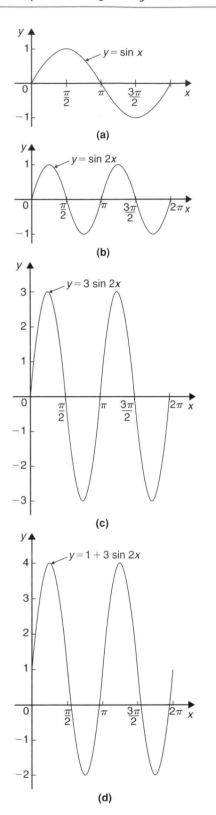

Figure 36.23

Now try the following Practice Exercise

Practice Exercise 151   Further problems on simple transformations with curve sketching (answers on page 1165)

Sketch the following graphs, showing relevant points:

1. $y = 3x - 5$

2. $y = -3x + 4$

3. $y = x^2 + 3$

4. $y = (x - 3)^2$

5. $y = (x - 4)^2 + 2$

6. $y = x - x^2$

7. $y = x^3 + 2$

8. $y = 1 + 2\cos 3x$

9. $y = 3 - 2\sin\left(x + \dfrac{\pi}{4}\right)$

10. $y = 2\ln x$

36.4   Periodic functions

A function $f(x)$ is said to be **periodic** if $f(x + T) = f(x)$ for all values of x, where T is some positive number. T is the interval between two successive repetitions and is called the period of the function $f(x)$. For example, $y = \sin x$ is periodic in x with period 2π since $\sin x = \sin(x + 2\pi) = \sin(x + 4\pi)$, and so on. Similarly, $y = \cos x$ is a periodic function with period 2π since $\cos x = \cos(x + 2\pi) = \cos(x + 4\pi)$, and so on. In general, if $y = \sin \omega t$ or $y = \cos \omega t$ then the period of the waveform is $2\pi / \omega$. The function shown in Figure 36.24

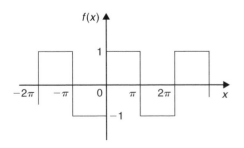

Figure 36.24

is also periodic of period 2π and is defined by:

$$f(x) = \begin{cases} -1, & \text{when } -\pi \le x \le 0 \\ 1, & \text{when } 0 \le x \le \pi \end{cases}$$

36.5 Continuous and discontinuous functions

If a graph of a function has no sudden jumps or breaks it is called a **continuous function**, examples being the graphs of sine and cosine functions. However, other graphs make finite jumps at a point or points in the interval. The square wave shown in Figure 36.24 has **finite discontinuities** as $x = \pi, 2\pi, 3\pi$, and so on, and is therefore a discontinuous function. $y = \tan x$ is another example of a discontinuous function.

36.6 Even and odd functions

Even functions

A function $y = f(x)$ is said to be even if $f(-x) = f(x)$ for all values of x. Graphs of even functions are always symmetrical about the y-axis (i.e. is a mirror image). Two examples of even functions are $y = x^2$ and $y = \cos x$ as shown in Figure 36.25.

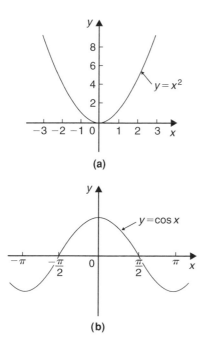

(a)

(b)

Figure 36.25

Odd functions

A function $y = f(x)$ is said to be odd if $f(-x) = -f(x)$ for all values of x. Graphs of odd functions are always symmetrical about the origin. Two examples of odd functions are $y = x^3$ and $y = \sin x$ as shown in Figure 36.26.

Many functions are neither even nor odd, two such examples being shown in Figure 36.27.

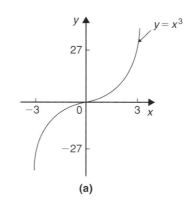

(a)

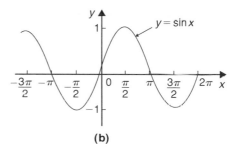

(b)

Figure 36.26

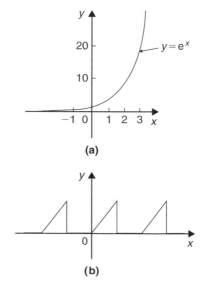

(a)

(b)

Figure 36.27

Problem 3. Sketch the following functions and state whether they are even or odd functions:

(a) $y = \tan x$

(b) $f(x) = \begin{cases} 2, & \text{when } 0 \leq x \leq \dfrac{\pi}{2} \\[2mm] -2, & \text{when } \dfrac{\pi}{2} \leq x \leq \dfrac{3\pi}{2} \\[2mm] 2, & \text{when } \dfrac{3\pi}{2} \leq x \leq 2\pi \end{cases}$

and is periodic of period 2π

(a) A graph of $y = \tan x$ is shown in Figure 36.28(a) and is symmetrical about the origin and is thus an **odd function** (i.e. $\tan(-x) = -\tan x$).

(b) A graph of $f(x)$ is shown in Figure 36.28(b) and is symmetrical about the $f(x)$ axis hence the function is an **even** one ($f(-x) = f(x)$).

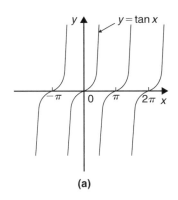

(a)

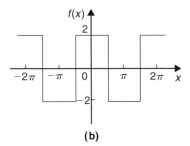

(b)

Figure 36.28

Problem 4. Sketch the following graphs and state whether the functions are even, odd or neither even nor odd:

(a) $y = \ln x$

(b) $f(x) = x$ in the range $-\pi$ to π and is periodic of period 2π

(a) A graph of $y = \ln x$ is shown in Figure 36.29(a) and the curve is neither symmetrical about the y-axis nor symmetrical about the origin and is thus **neither even nor odd**.

(b) A graph of $y = x$ in the range $-\pi$ to π is shown in Figure 36.29(b) and is symmetrical about the origin and is thus an **odd function**.

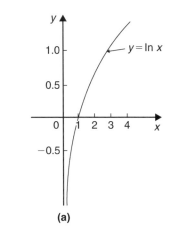

(a)

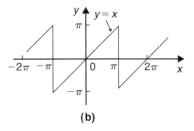

(b)

Figure 36.29

Now try the following Practice Exercise

Practice Exercise 152 Further problems on even and odd functions (answers on page 1168)

In Problems 1 and 2, determine whether the given functions are even, odd or neither even nor odd.

1. (a) x^4 (b) $\tan 3x$ (c) $2e^{3t}$ (d) $\sin^2 x$

2. (a) $5t^3$ (b) $e^x + e^{-x}$ (c) $\dfrac{\cos \theta}{\theta}$ (d) e^x

3. State whether the following functions, which are periodic of period 2π, are even or odd:

(a) $f(\theta) = \begin{cases} \theta, & \text{when } -\pi \le \theta \le 0 \\ -\theta, & \text{when } 0 \le \theta \le \pi \end{cases}$

(b) $f(x) = \begin{cases} x, & \text{when } -\dfrac{\pi}{2} \le x \le \dfrac{\pi}{2} \\ 0, & \text{when } \dfrac{\pi}{2} \le x \le \dfrac{3\pi}{2} \end{cases}$

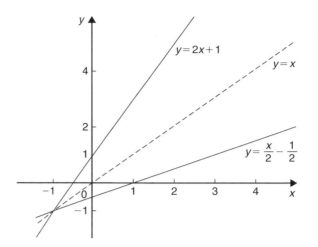

Figure 36.30

36.7 Inverse functions

If y is a function of x, the graph of y against x can be used to find x when any value of y is given. Thus the graph also expresses that x is a function of y. Two such functions are called **inverse functions**.

In general, given a function $y = f(x)$, its inverse may be obtained by interchanging the roles of x and y and then transposing for y. The inverse function is denoted by $y = f^{-1}(x)$.

For example, if $y = 2x + 1$, the inverse is obtained by

(i) transposing for x, i.e. $x = \dfrac{y-1}{2} = \dfrac{y}{2} - \dfrac{1}{2}$ and

(ii) interchanging x and y, giving the inverse as $y = \dfrac{x}{2} - \dfrac{1}{2}$

Thus if $f(x) = 2x + 1$, then $f^{-1}(x) = \dfrac{x}{2} - \dfrac{1}{2}$

A graph of $f(x) = 2x + 1$ and its inverse $f^{-1}(x) = \dfrac{x}{2} - \dfrac{1}{2}$ is shown in Figure 36.30 and $f^{-1}(x)$ is seen to be a reflection of $f(x)$ in the line $y = x$

Similarly, if $y = x^2$, the inverse is obtained by

(i) transposing for x, i.e. $x = \pm\sqrt{y}$ and

(ii) interchanging x and y, giving the inverse $y = \pm\sqrt{x}$

Hence the inverse has two values for every value of x. Thus $f(x) = x^2$ does not have a single inverse. In such a case the domain of the original function may be restricted to $y = x^2$ for $x > 0$. Thus the inverse is then $y = +\sqrt{x}$. A graph of $f(x) = x^2$ and its inverse $f^{-1}(x) = \sqrt{x}$ for $x > 0$ is shown in Figure 36.31 and, again, $f^{-1}(x)$ is seen to be a reflection of $f(x)$ in the line $y = x$

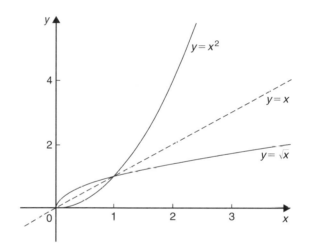

Figure 36.31

It is noted from the latter example that not all functions have an inverse. An inverse, however, can be determined if the range is restricted.

Problem 5. Determine the inverse for each of the following functions:
(a) $f(x) = x - 1$ (b) $f(x) = x^2 - 4$ $(x > 0)$
(c) $f(x) = x^2 + 1$

(a) If $y = f(x)$, then $y = x - 1$
Transposing for x gives $x = y + 1$
Interchanging x and y gives $y = x + 1$
Hence if $f(x) = x - 1$, then $\boldsymbol{f^{-1}(x) = x + 1}$

(b)   If $y = f(x)$, then $y = x^2 - 4 \quad (x > 0)$
Transposing for x gives $x = \sqrt{y + 4}$
Interchanging x and y gives $y = \sqrt{x + 4}$
Hence if $f(x) = x^2 - 4 \quad (x > 0)$ then
$f^{-1}(x) = \sqrt{x + 4}$ **if $x > -4$**

(c)   If $y = f(x)$, then $y = x^2 + 1$
Transposing for x gives $x = \sqrt{y - 1}$
Interchanging x and y gives $y = \sqrt{x - 1}$, which has two values.
Hence there is no inverse of $f(x) = x^2 + 1$, since the domain of $f(x)$ is not restricted.

Inverse trigonometric functions

If $y = \sin x$, then x is the angle whose sine is y. Inverse trigonometrical functions are denoted by prefixing the function with 'arc' or, more commonly, $^{-1}$. Hence transposing $y = \sin x$ for x gives $x = \sin^{-1} y$. Interchanging x and y gives the inverse $y = \sin^{-1} x$.
Similarly, $y = \cos^{-1} x$, $y = \tan^{-1} x$, $y = \sec^{-1} x$, $y = \csc^{-1} x$ and $y = \cot^{-1} x$ are all inverse trigonometric functions. The angle is always expressed in radians.

Inverse trigonometric functions are periodic so it is necessary to specify the smallest or principal value of the angle. For $\sin^{-1} x$, $\tan^{-1} x$, $\csc^{-1} x$ and $\cot^{-1} x$, the principal value is in the range $-\dfrac{\pi}{2} < y < \dfrac{\pi}{2}$. For $\cos^{-1} x$ and $\sec^{-1} x$ the principal value is in the range $0 < y < \pi$. Graphs of the six inverse trigonometric functions are shown in Figure 59.1, page 678.

Problem 6.   Determine the principal values of

(a) arcsin 0.5   (b) arctan(-1)

(c) $\arccos\left(-\dfrac{\sqrt{3}}{2}\right)$ (d) $\operatorname{arccosec}(\sqrt{2})$

Using a calculator,

(a) $\arcsin 0.5 \equiv \sin^{-1} 0.5 = 30°$

$$= \frac{\pi}{6} \text{ rad or } \mathbf{0.5236 \, rad}$$

(b) $\arctan(-1) \equiv \tan^{-1}(-1) = -45°$

$$= -\frac{\pi}{4} \text{ rad or } \mathbf{-0.7854 \, rad}$$

(c) $\arccos\left(-\dfrac{\sqrt{3}}{2}\right) \equiv \cos^{-1}\left(-\dfrac{\sqrt{3}}{2}\right) = 150°$

$$= \frac{5\pi}{6} \text{ rad or } \mathbf{2.6180 \, rad}$$

(d) $\operatorname{arccosec}(\sqrt{2}) = \arcsin\left(\dfrac{1}{\sqrt{2}}\right)$

$$\equiv \sin^{-1}\left(\frac{1}{\sqrt{2}}\right) = 45°$$

$$= \frac{\pi}{4} \text{ rad or } \mathbf{0.7854 \, rad}$$

Problem 7.   Evaluate (in radians), correct to 3 decimal places: $\sin^{-1} 0.30 + \cos^{-1} 0.65$

$\sin^{-1} 0.30 = 17.4576° = 0.3047 \, \text{rad}$

$\cos^{-1} 0.65 = 49.4584° = 0.8632 \, \text{rad}$

Hence $\sin^{-1} 0.30 + \cos^{-1} 0.65$
$= 0.3047 + 0.8632 = \mathbf{1.168 \, rad}$, correct to 3 decimal places.

Now try the following Practice Exercise

Practice Exercise 153   Further problems on inverse functions (answers on page 1168)

Determine the inverse of the functions given in Problems 1 to 4.

1. $f(x) = x + 1$

2. $f(x) = 5x - 1$

3. $f(x) = x^3 + 1$

4. $f(x) = \dfrac{1}{x} + 2$

Determine the principal value of the inverse functions in Problems 5 to 11.

5. $\sin^{-1}(-1)$

6. $\cos^{-1} 0.5$

7. $\tan^{-1} 1$

8. $\cot^{-1} 2$

9. $\operatorname{cosec}^{-1} 2.5$

10. $\sec^{-1} 1.5$

11. $\sin^{-1}\left(\dfrac{1}{\sqrt{2}}\right)$

12. Evaluate x, correct to 3 decimal places:

$$x = \sin^{-1}\frac{1}{3} + \cos^{-1}\frac{4}{5} - \tan^{-1}\frac{8}{9}$$

13. Evaluate y, correct to 4 significant figures:

$$y = 3\sec^{-1}\sqrt{2} - 4\operatorname{cosec}^{-1}\sqrt{2}$$
$$+ 5\cot^{-1} 2$$

36.8   Asymptotes

If a table of values for the function $y = \dfrac{x+2}{x+1}$ is drawn up for various values of x and then y plotted against x, the graph would be as shown in Figure 36.32. The straight lines AB, i.e. $x = -1$, and CD, i.e. $y = 1$, are known as **asymptotes**.

An asymptote to a curve is defined as a straight line to which the curve approaches as the distance from the origin increases. Alternatively, an asymptote can be considered as a tangent to the curve at infinity.

Asymptotes parallel to the x- and y-axes

There is a simple rule which enables asymptotes parallel to the x- and y-axes to be determined. For a curve $y = f(x)$:

(i) the asymptotes parallel to the x-axis are found by equating the coefficient of the highest power of x to zero.

(ii) the asymptotes parallel to the y-axis are found by equating the coefficient of the highest power of y to zero.

With the above example $y = \dfrac{x+2}{x+1}$, rearranging gives:

$$y(x+1) = x+2$$

i.e. $yx + y - x - 2 = 0$ (1)

and $x(y-1) + y - 2 = 0$

The coefficient of the highest power of x (in this case x^1) is $(y-1)$. Equating to zero gives: $y - 1 = 0$

From which, $y = 1$, which is an asymptote of $y = \dfrac{x+2}{x+1}$ as shown in Figure 36.32

Returning to equation (1): $yx + y - x - 2 = 0$

from which, $y(x+1) - x - 2 = 0$

The coefficient of the highest power of y (in this case y^1) is $(x+1)$. Equating to zero gives: $x + 1 = 0$ from which, $x = -1$, which is another asymptote of $y = \dfrac{x+2}{x+1}$ as shown in Figure 36.32

> **Problem 8.**   Determine the asymptotes for the function $y = \dfrac{x-3}{2x+1}$ and hence sketch the curve

Rearranging $y = \dfrac{x-3}{2x+1}$ gives: $y(2x+1) = x-3$

i.e. $2xy + y = x - 3$

or $2xy + y - x + 3 = 0$

and $x(2y - 1) + y + 3 = 0$

Equating the coefficient of the highest power of x to zero gives: $2y - 1 = 0$ from which, $y = \frac{1}{2}$ which is an asymptote.

Since $y(2x+1) = x - 3$ then equating the coefficient of the highest power of y to zero gives: $2x + 1 = 0$ from which, $x = -\frac{1}{2}$ which is also an asymptote.

When $x = 0$, $y = \dfrac{x-3}{2x+1} = \dfrac{-3}{1} = -3$ and when $y = 0$,

$0 = \dfrac{x-3}{2x+1}$ from which, $x - 3 = 0$ and $x = 3$

A sketch of $y = \dfrac{x-3}{2x+1}$ is shown in Figure 36.33

> **Problem 9.**   Determine the asymptotes parallel to the x- and y-axes for the function $x^2 y^2 = 9(x^2 + y^2)$

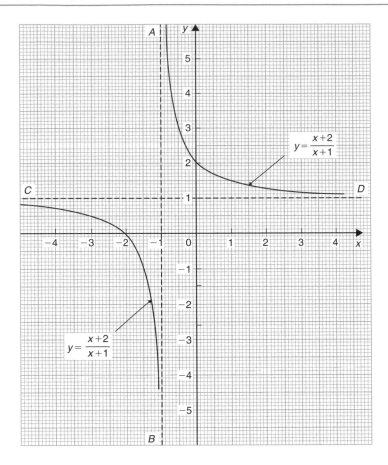

Figure 36.32

Asymptotes parallel to the x-axis:

Rearranging $x^2 y^2 = 9(x^2 + y^2)$ gives

$$x^2 y^2 - 9x^2 - 9y^2 = 0$$

hence $x^2(y^2 - 9) - 9y^2 = 0$

Equating the coefficient of the highest power of x to zero gives $y^2 - 9 = 0$ from which, $y^2 = 9$ and $y = \pm 3$

Asymptotes parallel to the y-axis:

Since $x^2 y^2 - 9x^2 - 9y^2 = 0$

then $y^2(x^2 - 9) - 9x^2 = 0$

Equating the coefficient of the highest power of y to zero gives $x^2 - 9 = 0$ from which, $x^2 = 9$ and $x = \pm 3$

Hence asymptotes occur at $y = \pm 3$ and $x = \pm 3$

Other asymptotes

To determine asymptotes other than those parallel to the x- and y-axes a simple procedure is:

(i) substitute $y = mx + c$ in the given equation

(ii) simplify the expression

(iii) equate the coefficients of the two highest powers of x to zero and determine the values of m and c. $y = mx + c$ gives the asymptote.

Problem 10. Determine the asymptotes for the function: $y(x+1) = (x-3)(x+2)$ and sketch the curve

Following the above procedure:

(i) Substituting $y = mx + c$ into $y(x+1) = (x-3)(x+2)$ gives:

$$(mx + c)(x + 1) = (x - 3)(x + 2)$$

(ii) Simplifying gives

$$mx^2 + mx + cx + c = x^2 - x - 6$$

and $(m-1)x^2 + (m+c+1)x + c + 6 = 0$

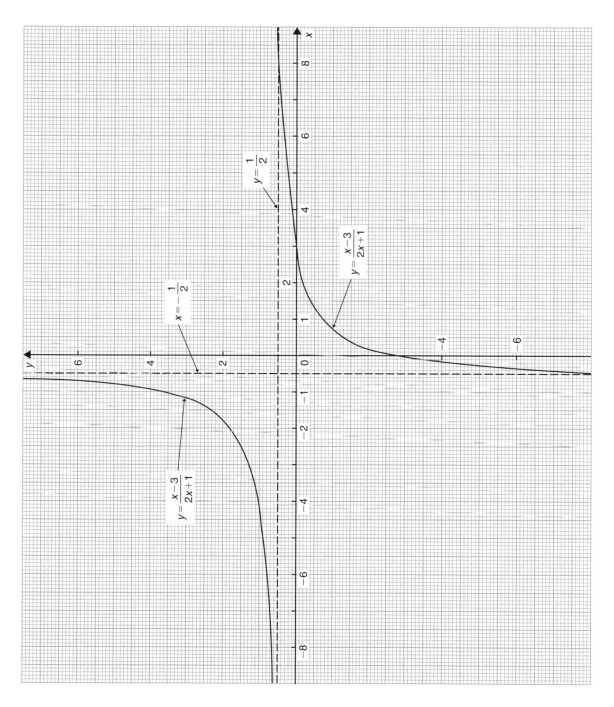

Figure 36.33

(iii) Equating the coefficient of the highest power of x to zero gives $m - 1 = 0$ from which, **$m = 1$**

Equating the coefficient of the next highest power of x to zero gives $m + c + 1 = 0$
and since $m = 1$, $1 + c + 1 = 0$ from which, **$c = -2$**
Hence $y = mx + c = 1x - 2$

i.e. **$y = x - 2$ is an asymptote**.

To determine any asymptotes parallel to the x-axis:

Rearranging $\quad y(x+1) = (x-3)(x+2)$

gives $\qquad yx + y = x^2 - x - 6$

The coefficient of the highest power of x (i.e. x^2) is 1. Equating this to zero gives $1 = 0$, which is not an equation of a line. Hence there is no asymptote parallel to the x-axis.
To determine any asymptotes parallel to the y-axis:
Since $y(x+1) = (x-3)(x+2)$ the coefficient of the highest power of y is $x+1$. Equating this to zero gives $x + 1 = 0$, from which, $x = -1$. Hence **$x = -1$** is an asymptote.
When **$x = 0$**, $y(1) = (-3)(2)$, i.e. **$y = -6$**
When **$y = 0$**, $0 = (x-3)(x+2)$, i.e. **$x = 3$ and $x = -2$**
A sketch of the function $y(x+1) = (x-3)(x+2)$ is shown in Figure 36.34.

> **Problem 11.** Determine the asymptotes for the function $x^3 - xy^2 + 2x - 9 = 0$

Following the procedure:

(i) Substituting $y = mx + c$ gives
$x^3 - x(mx+c)^2 + 2x - 9 = 0$

(ii) Simplifying gives

$$x^3 - x[m^2x^2 + 2mcx + c^2] + 2x - 9 = 0$$

i.e. $\quad x^3 - m^2x^3 - 2mcx^2 - c^2x + 2x - 9 = 0$

and $\quad x^3(1 - m^2) - 2mcx^2 - c^2x + 2x - 9 = 0$

(iii) Equating the coefficient of the highest power of x (i.e. x^3 in this case) to zero gives $1 - m^2 = 0$, from which, $m = \pm 1$

Equating the coefficient of the next highest power of x (i.e. x^2 in this case) to zero gives $-2mc = 0$, from which, $c = 0$

Hence $y = mx + c = \pm 1x + 0$, i.e. **$y = x$ and $y = -x$ are asymptotes**.

To determine any asymptotes parallel to the x- and y-axes for the function $x^3 - xy^2 + 2x - 9 = 0$:

Equating the coefficient of the highest power of x term to zero gives $1 = 0$, which is not an equation of a line. Hence there is no asymptote parallel with the x-axis.

Equating the coefficient of the highest power of y term to zero gives $-x = 0$, from which $x = 0$

Hence $x = 0$, $y = x$ and $y = -x$ are asymptotes for the function $x^3 - xy^2 + 2x - 9 = 0$

> **Problem 12.** Find the asymptotes for the function $y = \dfrac{x^2 + 1}{x}$ and sketch a graph of the function

Rearranging $y = \dfrac{x^2+1}{x}$ gives $yx = x^2 + 1$
Equating the coefficient of the highest power x term to zero gives $1 = 0$, hence there is no asymptote parallel to the x-axis.

Equating the coefficient of the highest power y term to zero gives $x = 0$

Hence there is an asymptote at $x = 0$ (i.e. the y-axis).

To determine any other asymptotes we substitute $y = mx + c$ into $yx = x^2 + 1$, which gives

$$(mx + c)x = x^2 + 1$$

i.e. $\qquad mx^2 + cx = x^2 + 1$

and $\quad (m - 1)x^2 + cx - 1 = 0$

Equating the coefficient of the highest power x term to zero gives $m - 1 = 0$, from which $m = 1$
Equating the coefficient of the next highest power x term to zero gives $c = 0$. Hence $y = mx + c = 1x + 0$, i.e. **$y = x$ is an asymptote**.

A sketch of $y = \dfrac{x^2 + 1}{x}$ is shown in Figure 36.35.
It is possible to determine maximum/minimum points on the graph (see Chapter 54).

Since $\qquad y = \dfrac{x^2 + 1}{x} = \dfrac{x^2}{x} + \dfrac{1}{x} = x + x^{-1}$

then $\qquad \dfrac{dy}{dx} = 1 - x^{-2} = 1 - \dfrac{1}{x^2} = 0$

for a turning point.

Section D

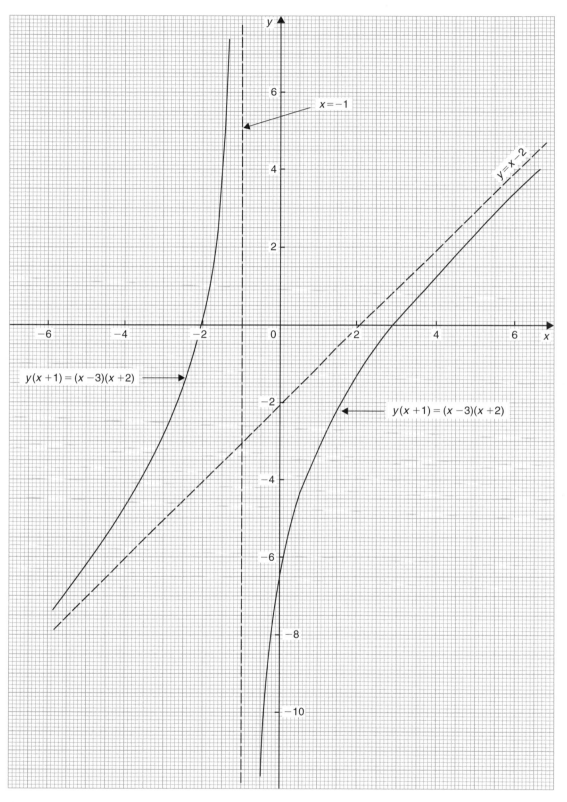

Figure 36.34

Hence $1 = \dfrac{1}{x^2}$ and $x^2 = 1$, from which, $x = \pm 1$

When $x = 1$,

$$y = \frac{x^2+1}{x} = \frac{1+1}{1} = 2$$

and when $x = -1$,

$$y = \frac{(-1)^2+1}{-1} = -2$$

i.e. $(1, 2)$ and $(-1, -2)$ are the co-ordinates of the turning points. $\dfrac{d^2 y}{dx^2} = 2x^{-3} = \dfrac{2}{x^3}$; when $x = 1$, $\dfrac{d^2 y}{dx^2}$ is positive,

which indicates a minimum point and when $x = -1$, $\dfrac{d^2 y}{dx^2}$ is negative, which indicates a maximum point, as shown in Figure 36.35.

Now try the following Practice Exercise

Practice Exercise 154   Further problems on asymptotes (answers on page 1168)

In Problems 1 to 3, determine the asymptotes parallel to the x- and y-axes.

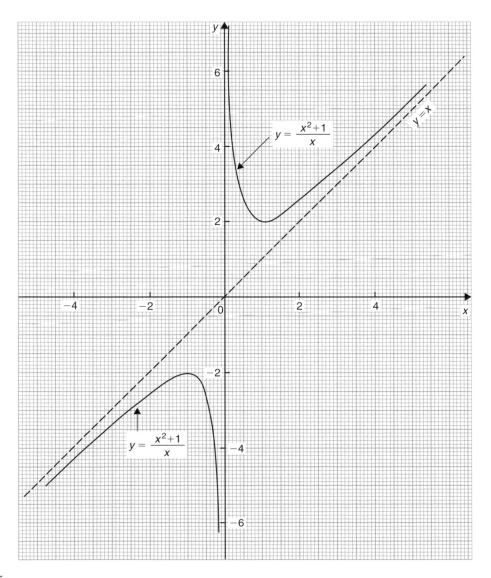

Figure 36.35

1. $y = \dfrac{x-2}{x+1}$

2. $y^2 = \dfrac{x}{x-3}$

3. $y = \dfrac{x(x+3)}{(x+2)(x+1)}$

In Problems 4 and 5, determine all the asymptotes.

4. $8x - 10 + x^3 - xy^2 = 0$

5. $x^2(y^2 - 16) = y$

In Problems 6 and 7, determine the asymptotes and sketch the curves.

6. $y = \dfrac{x^2 - x - 4}{x+1}$

7. $xy^2 - x^2y + 2x - y = 5$

36.9   Brief guide to curve sketching

The following steps will give information from which the graphs of many types of functions $y = f(x)$ can be sketched.

(i)   Use calculus to determine the location and nature of maximum and minimum points (see Chapter 54).

(ii)   Determine where the curve cuts the x- and y-axes.

(iii)   Inspect the equation for symmetry.

(a)   If the equation is unchanged when $-x$ is substituted for x, the graph will be symmetrical about the y-axis (i.e. it is an **even function**).

(b)   If the equation is unchanged when $-y$ is substituted for y, the graph will be symmetrical about the x-axis.

(c)   If $f(-x) = -f(x)$, the graph is symmetrical about the origin (i.e. it is an **odd function**).

(iv)   Check for any asymptotes.

36.10   Worked problems on curve sketching

Problem 13.   Sketch the graphs of

(a) $y = 2x^2 + 12x + 20$

(b) $y = -3x^2 + 12x - 15$

(a) $y = 2x^2 + 12x + 20$ is a parabola since the equation is a quadratic. To determine the turning point:

$\text{Gradient} = \dfrac{dy}{dx} = 4x + 12 = 0$ for a turning point.

Hence $4x = -12$ and $x = -3$

When $x = -3$, $y = 2(-3)^2 + 12(-3) + 20 = 2$

Hence $(-3, 2)$ are the co-ordinates of the turning point

$\dfrac{d^2 y}{dx^2} = 4$, which is positive, hence $(-3, 2)$ is a minimum point.

When $x = 0$, $y = 20$, hence the curve cuts the y-axis at $y = 20$

Thus knowing the curve passes through $(-3, 2)$ and $(0, 20)$ and appreciating the general shape of a parabola results in the sketch given in Figure 36.36

(b) $y = -3x^2 + 12x - 15$ is also a parabola (but 'upside down' due to the minus sign in front of the x^2 term).

$\text{Gradient} = \dfrac{dy}{dx} = -6x + 12 = 0$ for a turning point.

Hence $6x = 12$ and $x = 2$

When $x = 2$, $y = -3(2)^2 + 12(2) - 15 = -3$

Hence $(2, -3)$ are the co-ordinates of the turning point

$\dfrac{d^2 y}{dx^2} = -6$, which is negative, hence $(2, -3)$ is a maximum point.

When $x = 0$, $y = -15$, hence the curve cuts the axis at $y = -15$

The curve is shown sketched in Figure 36.36

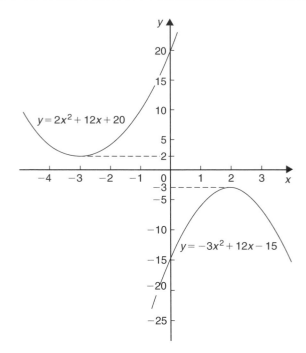

Figure 36.36

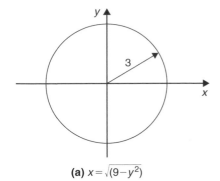

(a) $x = \sqrt{(9 - y^2)}$

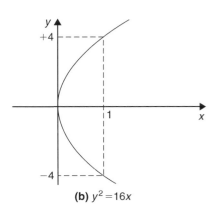

(b) $y^2 = 16x$

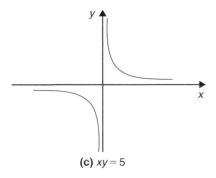

(c) $xy = 5$

Figure 36.37

Problem 14. Sketch the curves depicting the following equations:

(a) $x = \sqrt{9 - y^2}$ (b) $y^2 = 16x$

(c) $xy = 5$

(a)   Squaring both sides of the equation and transposing gives $x^2 + y^2 = 9$. Comparing this with the standard equation of a circle, centre origin and radius a, i.e. $x^2 + y^2 = a^2$, shows that $x^2 + y^2 = 9$ represents a circle, centre origin and radius 3. A sketch of this circle is shown in Figure 36.37(a).

(b)   The equation $y^2 = 16x$ is symmetrical about the x-axis and having its vertex at the origin $(0, 0)$. Also, when $x = 1$, $y = \pm 4$. A sketch of this parabola is shown in Figure 36.37(b).

(c)   The equation $y = \dfrac{a}{x}$ represents a rectangular hyperbola lying entirely within the first and third quadrants. Transposing $xy = 5$ gives $y = \dfrac{5}{x}$, and therefore represents the rectangular hyperbola shown in Figure 36.37(c).

Problem 15. Sketch the curves depicting the following equations:

(a) $4x^2 = 36 - 9y^2$ (b) $3y^2 + 15 = 5x^2$

(a)   By dividing throughout by 36 and transposing, the equation $4x^2 = 36 - 9y^2$ can be written as $\dfrac{x^2}{9} + \dfrac{y^2}{4} = 1$. The equation of an ellipse is of the form $\dfrac{x^2}{a^2} + \dfrac{y^2}{b^2} = 1$, where $2a$ and $2b$ represent the length of the axes of the ellipse. Thus $\dfrac{x^2}{3^2} + \dfrac{y^2}{2^2} = 1$ represents an ellipse, having its axes coinciding with the x- and y-axes of a rectangular co-ordinate system, the major axis being 2(3), i.e. 6 units long, and the minor axis 2(2), i.e. 4 units long, as shown in Figure 36.38(a).

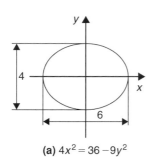

(a) $4x^2 = 36 - 9y^2$

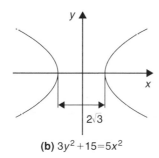

(b) $3y^2 + 15 = 5x^2$

Figure 36.38

(b) Dividing $3y^2 + 15 = 5x^2$ throughout by 15 and transposing gives $\dfrac{x^2}{3} - \dfrac{y^2}{5} = 1$. The equation $\dfrac{x^2}{a^2} - \dfrac{y^2}{b^2} = 1$ represents a hyperbola which is symmetrical about both the x- and y-axes, the distance between the vertices being given by $2a$.

Thus a sketch of $\dfrac{x^2}{3} - \dfrac{y^2}{5} = 1$ is as shown in Figure 36.38(b), having a distance of $2\sqrt{3}$ between its vertices.

Problem 16. Describe the shape of the curves represented by the following equations:

(a) $x = 2\sqrt{\left[1 - \left(\dfrac{y}{2}\right)^2\right]}$ (b) $\dfrac{y^2}{8} = 2x$

(c) $y = 6\left(1 - \dfrac{x^2}{16}\right)^{1/2}$

(a) Squaring the equation gives $x^2 = 4\left[1 - \left(\dfrac{y}{2}\right)^2\right]$ and transposing gives $x^2 = 4 - y^2$, i.e.

$x^2 + y^2 = 4$. Comparing this equation with $x^2 + y^2 = a^2$ shows that $x^2 + y^2 = 4$ is the equation of a **circle** having centre at the origin $(0, 0)$ and radius of 2 units.

(b) Transposing $\dfrac{y^2}{8} = 2x$ gives $y = 4\sqrt{x}$. Thus $\dfrac{y^2}{8} = 2x$ is the equation of a **parabola** having its axis of symmetry coinciding with the x-axis and its vertex at the origin of a rectangular co-ordinate system.

(c) $y = 6\left(1 - \dfrac{x^2}{16}\right)^{1/2}$ can be transposed to $\dfrac{y}{6} = \left(1 - \dfrac{x^2}{16}\right)^{1/2}$ and squaring both sides gives $\dfrac{y^2}{36} = 1 - \dfrac{x^2}{16}$, i.e. $\dfrac{x^2}{16} + \dfrac{y^2}{36} = 1$

This is the equation of an **ellipse**, centre at the origin of a rectangular co-ordinate system, the major axis coinciding with the y-axis and being $2\sqrt{36}$, i.e. 12 units long. The minor axis coincides with the x-axis and is $2\sqrt{16}$, i.e. 8 units long.

Problem 17. Describe the shape of the curves represented by the following equations:

(a) $\dfrac{x}{5} = \sqrt{\left[1 + \left(\dfrac{y}{2}\right)^2\right]}$ (b) $\dfrac{y}{4} = \dfrac{15}{2x}$

(a) Since $\dfrac{x}{5} = \sqrt{\left[1 + \left(\dfrac{y}{2}\right)^2\right]}$

$\dfrac{x^2}{25} = 1 + \left(\dfrac{y}{2}\right)^2$

i.e. $\dfrac{x^2}{25} - \dfrac{y^2}{4} = 1$

This is a **hyperbola** which is symmetrical about both the x- and y-axes, the vertices being $2\sqrt{25}$, i.e. 10 units apart.

(With reference to Section 36.2 (vii), a is equal to ± 5)

(b) The equation $\dfrac{y}{4} = \dfrac{15}{2x}$ is of the form $y = \dfrac{a}{x}$, $a = \dfrac{60}{2} = 30$

This represents a **rectangular hyperbola**, symmetrical about both the x- and y-axes, and lying entirely in the first and third quadrants, similar in shape to the curves shown in Figure 36.9, page 395.

Now try the following Practice Exercise

Practice Exercise 155   Further problems on curve sketching (answers on page 1168)

1. Sketch the graphs of (a) $y = 3x^2 + 9x + \dfrac{7}{4}$

    (b) $y = -5x^2 + 20x + 50$

In Problems 2 to 8, sketch the curves depicting the equations given.

2. $x = 4\sqrt{\left[1 - \left(\dfrac{y}{4}\right)^2\right]}$

3. $\sqrt{x} = \dfrac{y}{9}$

4. $y^2 = \dfrac{x^2 - 16}{4}$

5. $\dfrac{y^2}{5} = 5 - \dfrac{x^2}{2}$

6. $x = 3\sqrt{1 + y^2}$

7. $x^2 y^2 = 9$

8. $x = \dfrac{1}{3}\sqrt{(36 - 18y^2)}$

9. Sketch the circle given by the equation $x^2 + y^2 - 4x + 10y + 25 = 0$

In Problems 10 to 15, describe the shape of the curves represented by the equations given.

10. $y = \sqrt{[3(1 - x^2)]}$

11. $y = \sqrt{[3(x^2 - 1)]}$

12. $y = \sqrt{9 - x^2}$

13. $y = 7x^{-1}$

14. $y = (3x)^{1/2}$

15. $y^2 - 8 = -2x^2$

For fully worked solutions to each of the problems in Practice Exercises 151 to 155 in this chapter, go to the website:
www.routledge.com/cw/bird

This assignment covers the material contained in Chapters 34 to 36. *The marks available are shown in brackets at the end of each question.*

1. Sketch the following polar curves:

    (a) $r = 3 \sin \theta$ (b) $r = 5 \cos^2 2\theta$
    (c) $r = 2 \sin 3\theta$ (d) $r = 4(1 + \cos \theta)$ (16)

2. Plot a graph of $y = 2x^2$ from $x = -3$ to $x = +3$ and hence solve the equations:

    (a) $2x^2 - 8 = 0$ (b) $2x^2 - 4x - 6 = 0$ (10)

3. Plot the graph of $y = x^3 + 4x^2 + x - 6$ for values of x between $x = -4$ and $x = 2$. Hence determine the roots of the equation $x^3 + 4x^2 + x - 6 = 0$ (7)

4. Sketch the following graphs, showing the relevant points:

    (a) $y = (x - 2)^2$ (b) $y = 3 - \cos 2x$

    (c) $f(x) = \begin{cases} -1 & -\pi \leq x \leq -\dfrac{\pi}{2} \\ x & -\dfrac{\pi}{2} \leq x \leq \dfrac{\pi}{2} \\ 1 & \dfrac{\pi}{2} \leq x \leq \pi \end{cases}$ (10)

5. Determine the inverse of $f(x) = 3x + 1$ (3)

6. Evaluate, correct to 3 decimal places:

    $2 \tan^{-1} 1.64 + \sec^{-1} 2.43 - 3 \csc^{-1} 3.85$ (4)

For lecturers/instructors/teachers, fully worked solutions to each of the problems in Revision Test 14, together with a full marking scheme, are available at the website:
www.routledge.com/cw/bird

Multiple choice questions Test 4
Areas, volumes and graphs
This test covers the material in Chapters 27 to 36

All questions have only one correct answer (answers on page 1200).

1. A hollow shaft has an outside diameter of 6.0 cm and an inside diameter of 4.0 cm. The cross-sectional area of the shaft is:

   (a) 6283 mm²   (b) 1257 mm²

   (c) 1571 mm²   (d) 628 mm²

2. The speed of a car at 1 second intervals is given in the following table:

Time t (s)	0	1	2	3	4	5	6
Speed v (m/s)	0	2.5	5.0	9.0	15.0	22.0	30.0

   The distance travelled in 6 s (i.e. the area under the v/t graph) using the trapezoidal rule is:

   (a) 83.5 m   (b) 68 m   (c) 68.5 m   (d) 204 m

3. An arc of a circle of length 5.0 cm subtends an angle of 2 radians. The circumference of the circle is:

   (a) 2.5 cm   (b) 10.0 cm   (c) 5.0 cm   (d) 15.7 cm

4. A graph of resistance against voltage for an electrical circuit is shown in Figure M4.1. The equation relating resistance R and voltage V is:

   (a) $R = 1.45\,V + 40$

   (b) $R = 0.8\,V + 20$

   (c) $R = 1.45\,V + 20$

   (d) $R = 1.25\,V + 20$

5. The mean value of a sine wave over half a cycle is:

   (a) $0.318 \times$ maximum value

   (b) $0.707 \times$ maximum value

   (c)   the peak value

   (d) $0.637 \times$ maximum value

6. The area of the path shown shaded in Figure M4.2 is:

   (a) 300 m²   (b) 234 m²

   (c) 124 m²   (d) 66 m²

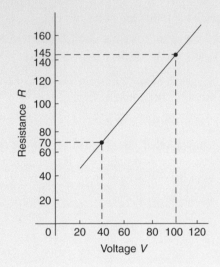

Figure M4.1

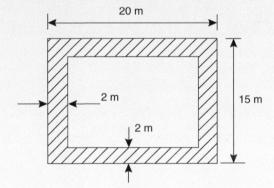

Figure M4.2

7. A vehicle has a mass of 2000 kg. A model of the vehicle is made to a scale of 1 to 100. If the vehicle and model are made of the same material, the mass of the model is:

   (a) 2 g   (b) 20 kg   (c) 200 g   (d) 20 g

8. A graph of y against x, two engineering quantities, produces a straight line.
   A table of values is shown below:

x	2	−1	p
y	9	3	5

   The value of p is:

   (a) $-\dfrac{1}{2}$ (b) -2 (c) 3   (d) 0

9. If the circumference of a circle is 100 mm, its area is:

   (a) $314.2\,\text{cm}^2$ (b) $7.96\,\text{cm}^2$

   (c) $31.83\,\text{mm}^2$ (d) $78.54\,\text{cm}^2$

10. An indicator diagram for a steam engine is as shown in Figure M4.3. The base has been divided into 6 equally spaced intervals and the lengths of the 7 ordinates measured, with the results shown in centimetres. Using Simpson's rule the area of the indicator diagram is:

   (a) $32\,\text{cm}^2$ (b) $17.9\,\text{cm}^2$ (c) $16\,\text{cm}^2$ (d) $96\,\text{cm}^2$

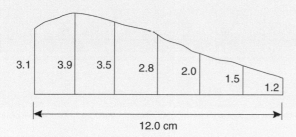

Figure M4.3

11. The equation of a circle is
    $x^2 + y^2 - 2x + 4y - 4 = 0.$

    Which of the following statements is correct?

    (a)   The circle has centre $(1, -2)$ and radius 4

    (b)   The circle has centre $(-1, 2)$ and radius 2

    (c)   The circle has centre $(-1, -2)$ and radius 4

    (d)   The circle has centre $(1, -2)$ and radius 3

12. The surface area of a sphere of diameter 40 mm is:
    (a) $50.27\,\text{cm}^2$ (b) $33.51\,\text{cm}^2$
    (c) $268.08\,\text{cm}^2$ (d) $201.06\,\text{cm}^2$

Questions 13 to 16 relate to the following information. x and y are two related engineering variables and p and q are constants.

For the law $y - p = \dfrac{q}{x}$ to be verified it is necessary to plot a graph of the variables.

13. On the vertical axis is plotted:
    (a) y (b) p (c) q (d) x

14. On the horizontal axis is plotted:

    (a) x (b) $\dfrac{q}{x}$ (c) $\dfrac{1}{x}$ (d) p

15. The gradient of the graph is:

    (a) y (b) p (c) q (d) x

16. The vertical axis intercept is:

    (a) y (b) p (c) q (d) x

17. A water tank is in the shape of a rectangular prism having length 1.5 m, breadth 60 cm and height 300 mm. If 1 litre $= 1000\ \text{cm}^3$, the capacity of the tank is:

    (a) 27 litre   (b) 2.7 litre

    (c) 2700 litre   (d) 270 litre

18. A pendulum of length 1.2 m swings through an angle of $12°$ in a single swing. The length of arc traced by the pendulum bob is:

    (a) 14.40 cm   (b) 25.13 cm

    (c) 10.00 cm   (d) 45.24 cm

19. Here are four equations in x and y. When x is plotted against y, in each case a straight line results.
    (i) $y + 3 = 3x$ (ii) $y + 3x = 3$ (iii) $\dfrac{y}{2} - \dfrac{3}{2} = x$
    (iv) $\dfrac{y}{3} = x + \dfrac{2}{3}$

    Which of these equations are parallel to each other?

    (a) (i) and (ii)   (b) (i) and (iv)

    (c) (ii) and (iii)   (d) (ii) and (iv)

20. A wheel on a car has a diameter of 800 mm. If the car travels 5 miles, the number of complete revolutions the wheel makes (given $1\,\text{km} = \frac{5}{8}$ mile) is:

    (a) 1989   (b) 1591   (c) 3183   (d) 10000

21. A rectangular building is shown on a building plan having dimensions 20 mm by 10 mm. If the plan is drawn to a scale of 1 to 300, the true area of the building in m^2 is:

    (a) $60000\,\text{m}^2$ (b) $18\,\text{m}^2$

    (c) $0.06\,\text{m}^2$ (d) $1800\,\text{m}^2$

Questions 22 to 25 relate to the following information. A straight line graph is plotted for the equation $y = ax^n$, where y and x are the variables and a and n are constants.

22. On the vertical axis is plotted:

    (a) y (b) x (c) $\lg y$ (d) a

23. On the horizontal axis is plotted:

    (a) $\lg x$ (b) x (c) x^n (d) a

24. The gradient of the graph is given by:

    (a) y (b) a (c) x (d) n

25. The vertical axis intercept is given by:

    (a) n (b) $\lg a$ (c) x (d) $\lg y$

26. The total surface area of a cylinder of length 20 cm and diameter 6 cm is:

    (a) $56.55\,\text{cm}^2$ (b) $433.54\,\text{cm}^2$

    (c) $980.18\,\text{cm}^2$ (d) $226.19\,\text{cm}^2$

27. The equation of the graph shown in Figure M4.4 is:

    (a) $x(x+1) = \dfrac{15}{4}$

    (b) $4x^2 - 4x - 15 = 0$

    (c) $x^2 - 4x - 5 = 0$

    (d) $4x^2 + 4x - 15 = 0$

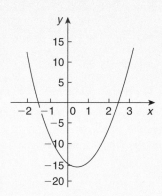

Figure M4.4

28. The total surface area of a solid hemisphere of diameter 6.0 cm is:

    (a) $84.82\,\text{cm}^2$ (b) $339.3\,\text{cm}^2$

    (c) $226.2\,\text{cm}^2$ (d) $56.55\,\text{cm}^2$

29. Which of the straight lines shown in Figure M4.5 has the equation $y + 4 = 2x$?

    (a) (iv)   (b) (ii)   (c) (iii)   (d) (i)

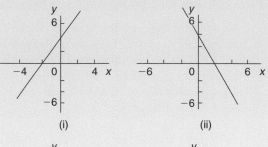

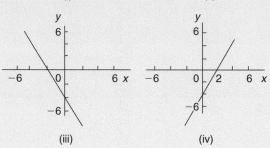

Figure M4.5

30. A graph relating effort E (plotted vertically) against load L (plotted horizontally) for a set of pulleys is given by $L + 30 = 6E$. The gradient of the graph is:

    (a) $\dfrac{1}{6}$ (b) 5   (c) 6   (d) $\dfrac{1}{5}$

31. The relationship between two related engineering variables x and y is $y - cx = bx^2$ where b and c are constants. To produce a straight line graph it is necessary to plot:

    (a) x vertically against y horizontally

    (b) y vertically against x^2 horizontally

    (c) $\dfrac{y}{x}$ vertically against x horizontally

    (d) y vertically against x horizontally

32. In an experiment demonstrating Hooke's law, the strain in a copper wire was measured for various stresses. The results included

Stress (megapascals)	18.24	24.00	39.36
Strain	0.00019	0.00025	0.00041

When stress is plotted vertically against strain horizontally a straight line graph results. Young's modulus of elasticity for copper, which is given by the gradient of the graph, is:

(a) 96000 Pa   (b) 1.04×10^{-11} Pa

(c) 96 Pa   (d) 96×10^9 Pa

33.   To depict a set of values from 0.05 to 275, the minimum number of cycles required on logarithmic graph paper is:

(a) 2   (b) 3   (c) 4   (d) 5

34.   The graph of $y = 2\tan 3\theta$ is:

(a)   a continuous, periodic, even function

(b)   a discontinuous, non-periodic, odd function

(c)   a discontinuous, periodic, odd function

(d)   a continuous, non-periodic, even function

Questions 35 to 38 relate to the following information. A straight line graph is plotted for the equation $y = ae^{bx}$, where y and x are the variables and a and b are constants.

35.   On the vertical axis is plotted:

(a) y (b) x (c) $\ln y$ (d) a

36.   On the horizontal axis is plotted:

(a) $\ln x$ (b) x (c) e^x (d) a

37.   The gradient of the graph is given by:

(a) y (b) a (c) x (d) b

38.   The vertical axis intercept is given by:

(a) b (b) $\ln a$ (c) x (d) $\ln y$

39.   The outside measurements of a picture frame are 80 cm by 30 cm. If the frame is 3 cm wide, the area of the metal used to make the frame is:

(a) 624 cm^2 (b) 2079 cm^2

(c) 660 mm^2 (d) 588 cm^2

40.   A cylindrical, copper pipe, 1.8 m long, has an outside diameter of 300 mm and an inside diameter of 180 mm. The volume of copper in the pipe, in cubic metres is:

(a) 0.3257 m^3 (b) 0.0814 m^3

(c) 8.143 m^3 (d) 814.3 m^3

Formulae/revision hints for Section D
Graphs

Equations of functions

Equation of a straight line:
$y = mx + c$ where m is the gradient and c is the y-axis intercept

If $y = ax^n$ then $\lg y = n \lg x + \lg a$

If $y = ab^x$ then $\lg y = (\lg b)x + \lg a$

If $y = ae^{kx}$ then $\ln y = kx + \ln a$

Equation of a parabola: $y = ax^2 + bx + c$

Circle, centre (a, b), radius r:
$(x - a)^2 + (y - b)^2 = r^2$

Equation of an ellipse, centre at origin, semi-axes a and b:
$$\frac{x^2}{a^2} + \frac{y^2}{b^2} = 1$$

Equation of a hyperbola: $\dfrac{x^2}{a^2} - \dfrac{y^2}{b^2} = 1$

Equation of a rectangular hyperbola: $xy = c$

Odd and even functions

A function $y = f(x)$ is **odd** if $f(-x) = -f(x)$ for all values of x
Graphs of odd functions are always symmetrical about the origin.

A function $y = f(x)$ is **even** if $f(-x) = f(x)$ for all values of x
Graphs of even functions are always symmetrical about the y-axis.

Geometry and trigonometry

Chapter 37

Angles and triangles

Why it is important to understand: Angles and triangles

Knowledge of angles and triangles is very important in engineering. Trigonometry is needed in surveying and architecture, for building structures/systems, designing bridges and solving scientific problems. Trigonometry is also used in electrical engineering; the functions that relate angles and side lengths in right-angled triangles are useful in expressing how a.c. electric current varies with time. Engineers use triangles to determine how much force it will take to move along an incline, GPS satellite receivers use triangles to determine exactly where they are in relation to satellites orbiting hundreds of miles away. Whether you want to build a skateboard ramp, a stairway, or a bridge, you can't escape trigonometry.

At the end of this chapter, you should be able to:

- define an angle – acute, right, obtuse, reflex, complementary, supplementary
- define parallel lines, transversal, vertically opposite angles, corresponding angles, alternate angles, interior angles
- define degrees, minutes, seconds, radians
- add and subtract angles
- state types of triangles – acute, right, obtuse, equilateral, isosceles, scalene
- define hypotenuse, adjacent and opposite sides with reference to an angle in a right-angled triangle
- recognise congruent triangles
- recognise similar triangles
- construct triangles, given certain sides and angles

37.1 Introduction

Trigonometry is a subject that involves the measurement of sides and angles of triangles and their relationship to each other. This chapter involves the measurement of angles and introduces types of triangle.

37.2 Angular measurement

An **angle** is the amount of rotation between two straight lines. Angles may be measured either in **degrees** or in **radians**.

If a circle is divided into 360 equal parts, then each part is called **1 degree** and is written as **1°**

i.e. **1 revolution = 360°**

or **1 degree is $\dfrac{1}{360}$th of a revolution**

Some angles are given **special names**.

- Any angle between 0° and 90° is called an **acute angle**.
- An angle equal to 90° is called a **right angle**.

- Any angle between 90° and 180° is called an **obtuse angle**.
- Any angle greater than 180° and less than 360° is called a **reflex angle**.
- An angle of 180° lies on a **straight line**.
- If two angles add up to 90° they are called **complementary angles**.
- If two angles add up to 180° they are called **supplementary angles**.
- **Parallel lines** are straight lines which are in the same plane and never meet. Such lines are denoted by arrows, as in Figure 37.1.
- A straight line which crosses two parallel lines is called a **transversal** (see *MN* in Figure 37.1).

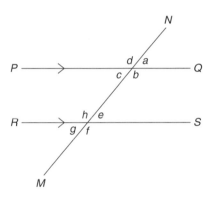

Figure 37.1

With reference to Figure 37.1,

(a) $a = c$, $b = d$, $e = g$ and $f = h$. Such pairs of angles are called **vertically opposite angles**.

(b) $a = e$, $b = f$, $c = g$ and $d = h$. Such pairs of angles are called **corresponding angles**.

(c) $c = e$ and $b = h$. Such pairs of angles are called **alternate angles**.

(d) $b + e = 180°$ and $c + h = 180°$. Such pairs of angles are called **interior angles**.

Minutes and seconds

One degree may be sub-divided into 60 parts, called **minutes**.

i.e.   **1 degree = 60 minutes**

which is written as   **1° = 60′**

41 degrees and 29 minutes is written as 41°29′. 41°29′ is equivalent to $41\dfrac{29°}{60} = 41.483°$ as a decimal, correct to 3 decimal places by calculator.

1 minute further subdivides into 60 seconds,

i.e.   **1 minute = 60 seconds**

which is written as   **1′ = 60″**

(Notice that for minutes, 1 dash is used and for seconds, 2 dashes are used.)

For example, 56 degrees, 36 minutes and 13 seconds is written as 56°36′13″

Radians and degrees

One radian is defined as the angle subtended at the centre of a circle by an arc equal in length to the radius. (For more on circles, see Chapter 28.)

With reference to Figure 37.2, for arc length s,

θ radians $= \dfrac{s}{r}$

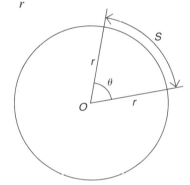

Figure 37.2

When s is the whole circumference, i.e. when $s = 2\pi r$,

$$\theta = \frac{s}{r} = \frac{2\pi r}{r} = 2\pi$$

In one revolution, $\theta = 360°$. Hence, the relationship between **degrees and radians** is

$$360° = 2\pi \text{ radians or } \mathbf{180° = \pi \ rad}$$

i.e. $\mathbf{1 \ rad = \dfrac{180°}{\pi} \approx 57.30°}$

Since π rad $= 180°$, then $\dfrac{\pi}{2}$rad $= 90°$, $\dfrac{\pi}{4}$rad $= 45°$, $\dfrac{\pi}{3}$rad $= 60°$ and $\dfrac{\pi}{6}$rad $= 30°$

Here are some worked examples on angular measurement.

Problem 1.   Evaluate $43°29' + 27°43'$

$$\begin{array}{r} 43°\ 29' \\ +\ \ 27°\ 43' \\ \hline 71°\ 12' \\ 1° \end{array}$$

(i) $29' + 43' = 72'$

(ii)   Since $60' = 1°, 72' = 1°12'$

(iii) The $12'$ is placed in the minutes column and $1°$ is carried in the degrees column.

(iv) $43° + 27° + 1°$ (carried) $= 71°$. Place $71°$ in the degrees column.

This answer can be obtained using the **calculator** as follows.

1. Enter 43   2. Press $°\ ' \ '' $ 3. Enter 29
4. Press $°\ ' \ '' $ 5. Press $+$ 6. Enter 27
7. Press $°\ ' \ '' $ 8. Enter 43   9. Press $°\ ' \ '' $
10. Press $=$ Answer $= \mathbf{71°12'}$

Thus, $\mathbf{43°29' + 27°43' = 71°12'}$

Problem 2.   Evaluate $84°13' - 56°39'$

$$\begin{array}{r} 84°\ 13' \\ -\ \ 56°\ 39' \\ \hline 27°\ 34' \end{array}$$

(i) $13' - 39'$ cannot be done.

(ii) $1°$ or $60'$ is 'borrowed' from the degrees column, which leaves $83°$ in that column.

(iii) $(60' + 13') - 39' = 34'$, which is placed in the minutes column.

(iv) $83° - 56° = 27°$, which is placed in the degrees column.

This answer can be obtained using the **calculator** as follows.

1. Enter 84   2. Press $°\ ' \ '' $ 3. Enter 13
4. Press $°\ ' \ '' $ 5. Press $-$ 6. Enter 56
7. Press $°\ ' \ '' $ 8. Enter 39   9. Press $°\ ' \ '' $
10. Press $=$ Answer $= \mathbf{27°34'}$

Thus, $\mathbf{84°13' - 56°39' = 27°34'}$

Problem 3.   Evaluate $19°\ 51'47'' + 63°27'34''$

$$\begin{array}{r} 19°\ 51'\ 47'' \\ +\ \ 63°\ 27'\ 34'' \\ \hline 83°\ 19'\ 21'' \\ 1°\ \ 1' \end{array}$$

(i) $47'' + 34'' = 81''$

(ii)   Since $60'' = 1', 81'' = 1'21''$

(iii) The $21''$ is placed in the seconds column and $1'$ is carried in the minutes column.

(iv) $51' + 27' + 1' = 79'$

(v)   Since $60' = 1°, 79' = 1°19'$

(vi)   The $19'$ is placed in the minutes column and $1°$ is carried in the degrees column.

(vii) $19° + 63° + 1°$ (carried) $= 83°$. Place $83°$ in the degrees column.

This answer can be obtained using the **calculator** as follows.

1. Enter 19   2. Press $°\ ' \ '' $ 3. Enter 51
4. Press $°\ ' \ '' $ 5. Enter 47   6. Press $°\ ' \ '' $
7. Press $+$ 8. Enter 63   9. Press $°\ ' \ '' $
10. Enter 27   11. Press $°\ ' \ '' $
12. Enter 34   13. Press $°\ ' \ '' $
14. Press $=$ Answer $= \mathbf{83°19'21''}$

Thus, $\mathbf{19°51'47'' + 63°27'34'' = 83°19'21''}$

Problem 4.   Convert $39°27'$ to degrees in decimal form

$$39°27' = 39\frac{27°}{60}$$

$$\frac{27°}{60} = 0.45°\ \text{ by calculator}$$

Hence, $\mathbf{39°27' = 39\dfrac{27°}{60} = 39.45°}$

This answer can be obtained using the **calculator** as follows.

1. Enter 39   2. Press $°\ ' \ '' $ 3. Enter 27
4. Press $°\ ' \ '' $ 5. Press $=$ 6. Press $°\ ' \ '' $
Answer $= \mathbf{39.45°}$

Problem 5.   Convert $63°26'51''$ to degrees in decimal form, correct to 3 decimal places

$$63°26'51'' = 63°26\frac{51'}{60} = 63°26.85'$$

$$63°26.85' = 63\frac{26.85°}{60} = 63.4475°$$

Hence, $\mathbf{63°26'51'' = 63.448°}$ correct to 3 decimal places.

This answer can be obtained using the **calculator** as follows.

1. Enter 63   2. Press ° ' '' '''   3. Enter 26
4. Press ° ' '' '''   5. Enter 51   6. Press ° ' '' '''
7. Press =   8. Press ° ' '' '''   Answer = **63.4475°**

Problem 6. Convert 53.753° to degrees, minutes and seconds

$$0.753° = 0.753 \times 60' = 45.18'$$

$$0.18' = 0.18 \times 60'' = 11''$$
to the nearest second

Hence, **53.753° = 53°45′11″**
This answer can be obtained using the **calculator** as follows.
1. Enter 53.753   2. Press =
3. Press ° ' '' '''   Answer = **53°45′10.8″**

Now try the following Practice Exercise

Practice Exercise 156   Angular measurement (answers on page 1168)

1. Evaluate $52°39' + 29°48'$

2. Evaluate $76°31' - 48°37'$

3. Evaluate $77°22' + 41°36' - 67°47'$

4. Evaluate $41°37'16'' + 58°29'36''$

5. Evaluate $54°37'42'' - 38°53'25''$

6. Evaluate $79°26'19'' - 45°58'56'' + 53°21'38''$

7. Convert 72°33′ to degrees in decimal form.

8. Convert 27°45′15″ to degrees correct to 3 decimal places.

9. Convert 37.952° to degrees and minutes.

10. Convert 58.381° to degrees, minutes and seconds.

Here are some further worked examples on angular measurement.

Problem 7. State the general name given to the following angles: (a) 157° (b) 49° (c) 90° (d) 245°

(a) Any angle between 90° and 180° is called an obtuse angle.

Thus, **157° is an obtuse angle**.

(b) Any angle between 0° and 90° is called an acute angle.

Thus, **49° is an acute angle**.

(c) An angle equal to 90° is called a **right angle**.

(d) Any angle greater than 180° and less than 360° is called a reflex angle.

Thus, **245° is a reflex angle**.

Problem 8. Find the angle complementary to 48° 39′

If two angles add up to 90° they are called **complementary angles**. Hence, **the angle complementary to 48°39′ is**

$$90° - 48°39' = \mathbf{41°21'}$$

Problem 9. Find the angle supplementary to 74°25′

If two angles add up to 180° they are called **supplementary angles**. Hence, **the angle supplementary to 74°25′ is**

$$180° - 74°25' = \mathbf{105°\,35'}$$

Problem 10. Evaluate angle θ in each of the diagrams shown in Figure 37.3

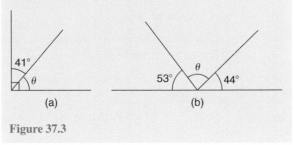

Figure 37.3

(a) The symbol shown in Figure 37.4 is called a **right angle** and it equals 90°.

Figure 37.4

Hence, from Figure 37.3(a),
$$\theta + 41° = 90°$$
from which, $$\boldsymbol{\theta} = 90° - 41° = \mathbf{49°}$$

(b) An angle of 180° lies on a straight line. Hence, from Figure 37.3(b),

$$180° = 53° + \theta + 44°$$

from which, $\boldsymbol{\theta} = 180° - 53° - 44° = \mathbf{83°}$

Problem 11. Evaluate angle θ in the diagram shown in Figure 37.5

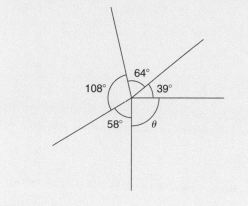

Figure 37.5

There are 360° in a complete revolution of a circle. Thus, $360° = 58° + 108° + 64° + 39° + \theta$
from which, $\theta = 360° - 58° - 108° - 64° - 39° = \mathbf{91°}$

Problem 12. Two straight lines AB and CD intersect at 0. If $\angle AOC$ is 43°, find $\angle AOD$, $\angle DOB$ and $\angle BOC$

From Figure 37.6, $\angle AOD$ is supplementary to $\angle AOC$. Hence, $\angle AOD = 180° - 43° = \mathbf{137°}$
When two straight lines intersect, the vertically opposite angles are equal.
Hence, $\angle DOB = \mathbf{43°}$ and $\angle BOC\ \mathbf{137°}$

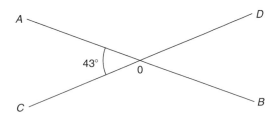

Figure 37.6

Problem 13. Determine angle β in Figure 37.7

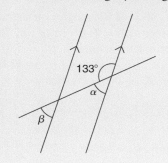

Figure 37.7

$\alpha = 180° - 133° = 47°$ (i.e. supplementary angles).
$\alpha = \beta = \mathbf{47°}$ (corresponding angles between parallel lines).

Problem 14. Determine the value of angle θ in Figure 37.8

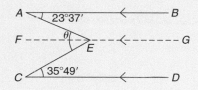

Figure 37.8

Let a straight line FG be drawn through E such that FG is parallel to AB and CD.
$\angle BAE = \angle AEF$ (alternate angles between parallel lines AB and FG), hence $\angle AEF = 23°37'$
$\angle ECD = \angle FEC$ (alternate angles between parallel lines FG and CD), hence $\angle FEC = 35°49'$

$$\text{Angle } \boldsymbol{\theta} = \angle AEF + \angle FEC = 23°37' + 35°49'$$
$$= \mathbf{59°26'}$$

Problem 15. Determine angles c and d in Figure 37.9

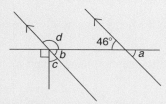

Figure 37.9

$a = b = 46°$ (corresponding angles between parallel lines).

Also, $b + c + 90° = 180°$ (angles on a straight line).

Hence, $46° + c + 90° = 180°$, from which, $c = 44°$

b and d are supplementary, hence $d = 180° - 46°$

$$= 134°$$

Alternatively, $90° + c = d$ (vertically opposite angles).

Problem 16. Convert the following angles to radians, correct to 3 decimal places.
(a) $73°$ (b) $25°37'$

Although we may be more familiar with degrees, radians is the SI unit of angular measurement in engineering (1 radian $\approx 57.3°$).

(a) Since $180° = \pi$ rad then $1° = \dfrac{\pi}{180}$ rad

Hence, $\mathbf{73°} = 73 \times \dfrac{\pi}{180}$ rad $= \mathbf{1.274\ rad}$

(b) $25°37' = 25\dfrac{37°}{60} = 25.616666\ldots$

Hence, $\mathbf{25°37'} = 25.616666\ldots°$

$$= 25.616666\ldots \times \dfrac{\pi}{180}\text{rad}$$

$$= \mathbf{0.447\ rad}$$

Problem 17. Convert 0.743 rad to degrees and minutes

Since $180° = \pi$ rad then 1 rad $= \dfrac{180°}{\pi}$

Hence, $\mathbf{0.743\ rad} = 0.743 \times \dfrac{180°}{\pi} = 42.57076\ldots°$

$$= \mathbf{42°34'}$$

Now try the following Practice Exercise

Practice Exercise 157   Further angular measurement (answers on page 1168)

1. State the general name given to an angle of $197°$.

2. State the general name given to an angle of $136°$.

3. State the general name given to an angle of $49°$.

4. State the general name given to an angle of $90°$.

5. Determine the angles complementary to the following.
   (a) $69°$ (b) $27°37'$ (c) $41°3'43''$

6. Determine the angles supplementary to
   (a) $78°$ (b) $15°$ (c) $169°41'11''$

7. Find the values of angle θ in diagrams (a) to (i) of Figure 37.10.

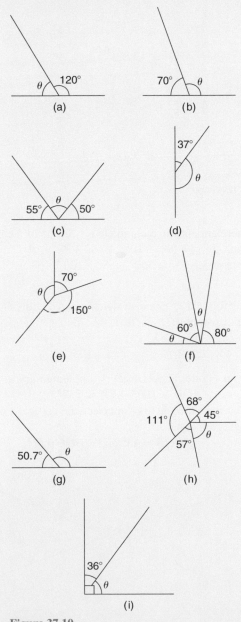

Figure 37.10

8. With reference to Figure 37.11, what is the name given to the line *XY*? Give examples of each of the following:

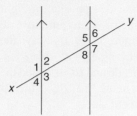

Figure 37.11

(a) vertically opposite angles
(b) supplementary angles
(c) corresponding angles
(d) alternate angles.

9. In Figure 37.12, find angle α.

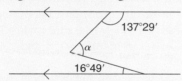

Figure 37.12

10. In Figure 37.13, find angles *a*, *b* and *c*.

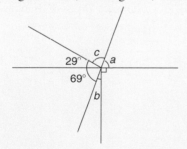

Figure 37.13

11. Find angle β in Figure 37.14.

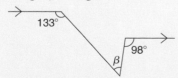

Figure 37.14

12. Convert $76°$ to radians, correct to 3 decimal places.

13. Convert $34°40'$ to radians, correct to 3 decimal places.

14. Convert 0.714 rad to degrees and minutes.

37.3   Triangles

A **triangle** is a figure enclosed by three straight lines. **The sum of the three angles of a triangle is equal to 180°.**

Types of triangle

An **acute-angled triangle** is one in which all the angles are acute; i.e. all the angles are less than 90°. An example is shown in triangle *ABC* in Figure 37.15(a).

A **right-angled triangle** is one which contains a right angle; i.e. one in which one of the angles is 90°. An example is shown in triangle *DEF* in Figure 37.15(b).

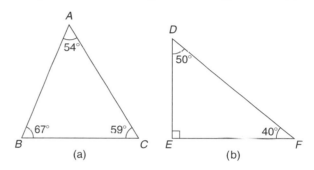

Figure 37.15

An **obtuse-angled triangle** is one which contains an obtuse angle; i.e. one angle which lies between 90° and 180°. An example is shown in triangle *PQR* in Figure 37.16(a).

An **equilateral triangle** is one in which all the sides and all the angles are equal; i.e. each is 60°. An example is shown in triangle *ABC* in Figure 37.16(b).

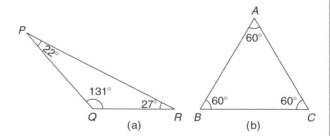

Figure 37.16

An **isosceles triangle** is one in which two angles and two sides are equal. An example is shown in triangle *EFG* in Figure 37.17(a).

A **scalene triangle** is one with unequal angles and therefore unequal sides. An example of an acute angled scalene triangle is shown in triangle *ABC* in Figure 37.17(b).

Section E

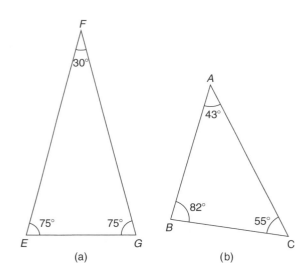

Figure 37.17

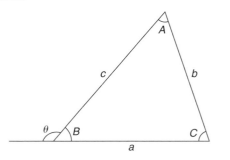

Figure 37.18

With reference to Figure 37.18,

(a) Angles A, B and C are called **interior angles** of the triangle.

(b) Angle θ is called an **exterior angle** of the triangle and is equal to the sum of the two opposite interior angles; i.e. $\theta = A + C$

(c) $a + b + c$ is called the **perimeter** of the triangle.

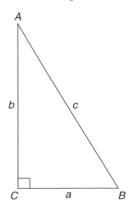

Figure 37.19

A right-angled triangle ABC is shown in Figure 37.19. The point of intersection of two lines is called a vertex (plural **vertices**); the three vertices of the triangle are labelled as A, B and C, respectively. The right angle is angle C. The side opposite the right angle is given the special name of the **hypotenuse**. The hypotenuse, length AB in Figure 37.19, is always the longest side of a right-angled triangle. With reference to angle B, AC is the **opposite** side and BC is called the **adjacent** side. With reference to angle A, BC is the **opposite** side and AC is the **adjacent** side.

Often sides of a triangle are labelled with lower case letters, a being the side opposite angle A, b being the side opposite angle B and c being the side opposite angle C. So, in the triangle ABC, length $AB = c$, length $BC = a$ and length $AC = b$. Thus, c is the hypotenuse in the triangle ABC.

$\angle$ is the symbol used for 'angle'. For example, in the triangle shown, $\angle C = 90°$. Another way of indicating an angle is to use all three letters. For example, $\angle ABC$ actually means $\angle B$; i.e. we take the middle letter as the angle. Similarly, $\angle BAC$ means $\angle A$ and $\angle ACB$ means $\angle C$.

Here are some worked examples to help us understand more about triangles.

Problem 18. Name the types of triangle shown in Figure 37.20

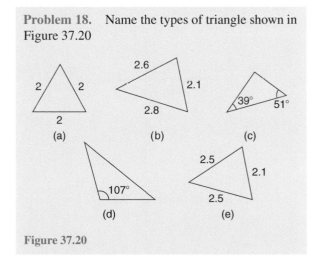

Figure 37.20

(a) **Equilateral triangle** (since all three sides are equal).

(b) **Acute-angled scalene triangle** (since all the angles are less than $90°$).

(c) **Right-angled triangle** ($39° + 51° = 90°$; hence, the third angle must be $90°$, since there are $180°$ in a triangle).

(d) **Obtuse-angled scalene triangle** (since one of the angles lies between 90° and 180°).

(e) **Isosceles triangle** (since two sides are equal).

Problem 19. In the triangle *ABC* shown in Figure 37.21, with reference to angle θ, which side is the adjacent?

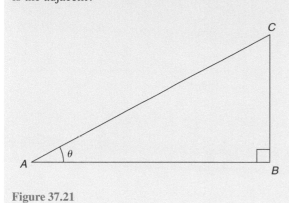

Figure 37.21

The triangle is right-angled; thus, side *AC* is the hypotenuse. With reference to angle θ, the opposite side is *BC*. The remaining side, *AB*, **is the adjacent side**.

Problem 20. In the triangle shown in Figure 37.22, determine angle θ

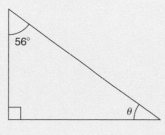

Figure 37.22

The sum of the three angles of a triangle is equal to 180°.
The triangle is right-angled. Hence,
$$90° + 56° + \angle\theta = 180°$$

from which, $\angle\theta = 180° - 90° - 56° = \mathbf{34°}$

Problem 21. Determine the value of θ and α in Figure 37.23

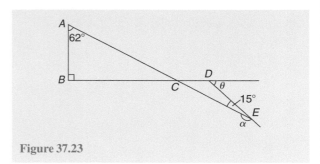

Figure 37.23

In triangle *ABC*, $\angle A + \angle B + \angle C = 180°$ (the angles in a triangle add up to 180°).
Hence, $\angle C = 180° - 90° - 62° = 28°$. Thus, $\angle DCE = 28°$ (vertically opposite angles).
$\theta = \angle DCE + \angle DEC$ (the exterior angle of a triangle is equal to the sum of the two opposite interior angles).
Hence, $\angle\theta = 28° + 15° = \mathbf{43°}$
$\angle\alpha$ and $\angle DEC$ are supplementary; thus,
$\alpha = 180° - 15° = \mathbf{165°}$

Problem 22. *ABC* is an isosceles triangle in which the unequal angle *BAC* is 56°. *AB* is extended to *D* as shown in Figure 37.24. Find, for the triangle, $\angle ABC$ and $\angle ACB$. Also, calculate $\angle DBC$

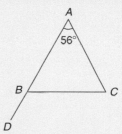

Figure 37.24

Since triangle *ABC* is isosceles, two sides – i.e. *AB* and *AC* – are equal and two angles – i.e. $\angle ABC$ and $\angle ACB$ – are equal.
The sum of the three angles of a triangle is equal to 180°.
Hence, $\angle ABC + \angle ACB = 180° - 56° = 124°$
Since $\angle ABC = \angle ACB$ then
$$\angle\mathbf{ABC} = \angle\mathbf{ACB} = \frac{124°}{2} = \mathbf{62°}$$

An angle of 180° lies on a straight line; hence,
$\angle ABC + \angle DBC = 180°$ from which,
$\angle\mathbf{DBC} = 180° - \angle ABC = 180° - 62° = \mathbf{118°}$
Alternatively, $\angle DBC = \angle A + \angle C$ (exterior angle equals sum of two interior opposite angles),
i.e. $\angle\mathbf{DBC} = 56° + 62° = \mathbf{118°}$

Problem 23. Find angles a, b, c, d and e in Figure 37.25

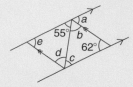

Figure 37.25

$a = 62°$ and $c = 55°$ (alternate angles between parallel lines).
$55° + b + 62° = 180°$ (angles in a triangle add up to $180°$); hence, $b = 180° - 55° - 62° = 63°$
$b = d = 63°$ (alternate angles between parallel lines).
$e + 55° + 63° = 180°$ (angles in a triangle add up to $180°$); hence, $e = 180° - 55° - 63° = 62°$
Check: $e = a = 62°$ (corresponding angles between parallel lines).

Now try the following Practice Exercise

Practice Exercise 158   Triangles (answers on page 1168)

1. Name the types of triangle shown in diagrams (a) to (f) in Figure 37.26.

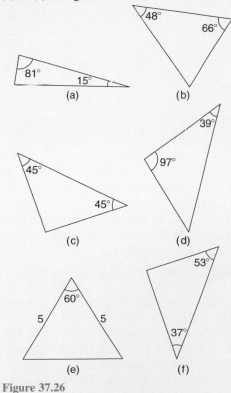

Figure 37.26

2. Find the angles a to f in Figure 37.27.

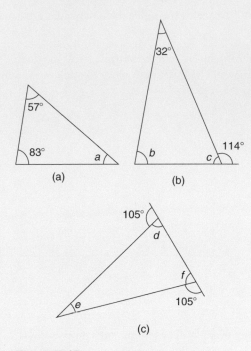

Figure 37.27

3. In the triangle *DEF* of Figure 37.28, which side is the hypotenuse? With reference to angle *D*, which side is the adjacent?

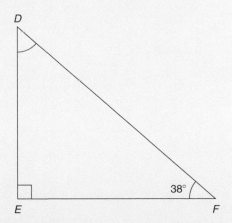

Figure 37.28

4. In triangle *DEF* of Figure 37.28, determine angle *D*.

5. *MNO* is an isosceles triangle in which the unequal angle is $65°$ as shown in Figure 37.29. Calculate angle θ.

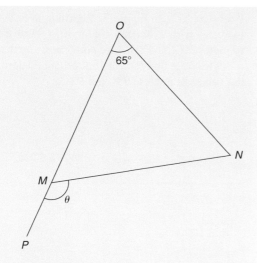

Figure 37.29

6. Determine $\angle\phi$ and $\angle x$ in Figure 37.30.

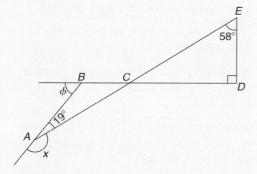

Figure 37.30

7. In Figure 37.31(a) and (b), find angles w, x, y and z. What is the name given to the types of triangle shown in (a) and (b)?

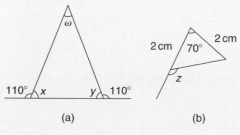

(a)   (b)

Figure 37.31

8. Find the values of angles a to g in Figure 37.32(a) and (b).

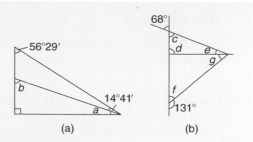

(a)   (b)

Figure 37.32

9. Find the unknown angles a to k in Figure 37.33.

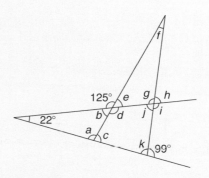

Figure 37.33

10. Triangle ABC has a right angle at B and $\angle BAC$ is $34°$. BC is produced to D. If the bisectors of $\angle ABC$ and $\angle ACD$ meet at E, determine $\angle BEC$.

11. If in Figure 37.34 triangle BCD is equilateral, find the interior angles of triangle ABE.

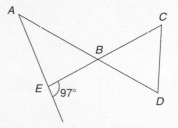

Figure 37.34

37.4 Congruent triangles

Two triangles are said to be **congruent** if they are equal in all respects; i.e. three angles and three sides in one triangle are equal to three angles and three sides in the other triangle. Two triangles are congruent if

Section E

(a) the three sides of one are equal to the three sides of the other (SSS),

(b) two sides of one are equal to two sides of the other and the angles included by these sides are equal (SAS),

(c) two angles of the one are equal to two angles of the other and any side of the first is equal to the corresponding side of the other (ASA), or

(d) their hypotenuses are equal and one other side of one is equal to the corresponding side of the other (RHS).

Problem 24. State which of the pairs of triangles shown in Figure 37.35 are congruent and name their sequence

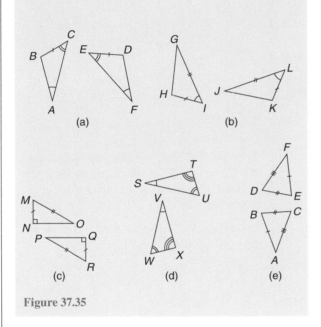

Figure 37.35

(a) Congruent *ABC*, *FDE* (angle, side, angle; i.e. ASA).

(b) Congruent *GIH*, *JLK* (side, angle, side; i.e. SAS).

(c) Congruent *MNO*, *RQP* (right angle, hypotenuse, side; i.e. RHS).

(d) Not necessarily congruent. It is not indicated that any side coincides.

(e) Congruent *ABC*, *FED* (side, side, side; i.e. SSS).

Problem 25. In Figure 37.36, triangle *PQR* is isosceles with *Z*, the mid-point of *PQ*. Prove that triangles *PXZ* and *QYZ* are congruent and that triangles *RXZ* and *RYZ* are congruent. Determine the values of angles *RPZ* and *RXZ*

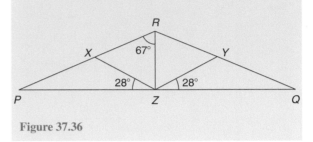

Figure 37.36

Since triangle *PQR* is isosceles, $PR = RQ$ and thus $\angle QPR = \angle RQP$.

$\angle RXZ = \angle QPR + 28°$ and $\angle RYZ = \angle RQP + 28°$ (exterior angles of a triangle equal the sum of the two interior opposite angles). Hence, $\angle RXZ = \angle RYZ$.

$\angle PXZ = 180° - \angle RXZ$ and $\angle QYZ = 180° - \angle RYZ$. Thus, $\angle PXZ = \angle QYZ$.

Triangles *PXZ* and *QYZ* are congruent since $\angle XPZ = \angle YQZ$, $PZ = ZQ$ and $\angle XZP = \angle YZQ$ (ASA). Hence, $XZ = YZ$.

Triangles *PRZ* and *QRZ* are congruent since $PR = RQ$, $\angle RPZ = \angle RQZ$ and $PZ = ZQ$ (SAS). Hence, $\angle RZX = \angle RZY$.

Triangles *RXZ* and *RYZ* are congruent since $\angle RXZ = \angle RYZ$, $XZ = YZ$ and $\angle RZX = \angle RZY$ (ASA).

$\angle QRZ = 67°$ and thus $\angle PRQ = 67° + 67° = 134°$

Hence, $\angle \boldsymbol{RPZ} = \angle RQZ = \dfrac{180° - 134°}{2} = \boldsymbol{23°}$

$\angle \boldsymbol{RXZ} = 23° + 28° = \boldsymbol{51°}$ (external angle of a triangle equals the sum of the two interior opposite angles).

Now try the following Practice Exercise

Practice Exercise 159   Congruent triangles (answers on page 1169)

1. State which of the pairs of triangles in Figure 37.37 are congruent and name their sequences.

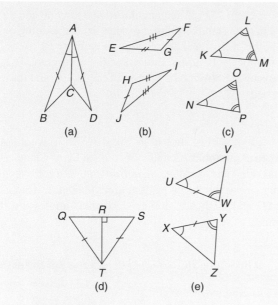

Figure 37.37

2. In a triangle ABC, $AB = BC$ and D and E are points on AB and BC, respectively, such that $AD = CE$. Show that triangles AEB and CDB are congruent.

37.5   Similar triangles

Two triangles are said to be **similar** if the angles of one triangle are equal to the angles of the other triangle. With reference to Figure 37.38, triangles ABC and PQR are similar and the corresponding sides are in proportion to each other,

i.e.
$$\frac{p}{a} = \frac{q}{b} = \frac{r}{c}$$

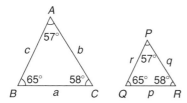

Figure 37.38

Problem 26.   In Figure 37.39, find the length of side a

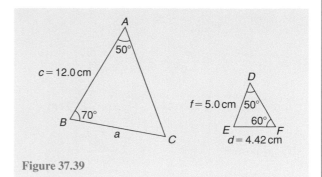

Figure 37.39

In triangle ABC, $50° + 70° + \angle C = 180°$, from which $\angle C = 60°$

In triangle DEF, $\angle E = 180° - 50° - 60° = 70°$

Hence, triangles ABC and DEF are similar, since their angles are the same. Since corresponding sides are in proportion to each other,

$$\frac{a}{d} = \frac{c}{f} \quad \text{i.e.} \quad \frac{a}{4.42} = \frac{12.0}{5.0}$$

Hence, **side $a = \dfrac{12.0}{5.0}(4.42) = $ 10.61 cm**.

Problem 27.   In Figure 37.40, find the dimensions marked r and p

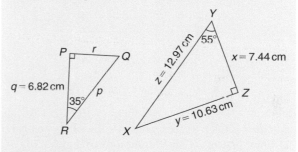

Figure 37.40

In triangle PQR, $\angle Q = 180° - 90° - 35° = 55°$
In triangle XYZ, $\angle X = 180° - 90° - 55° = 35°$
Hence, triangles PQR and ZYX are similar since their angles are the same. The triangles may he redrawn as shown in Figure 37.41.

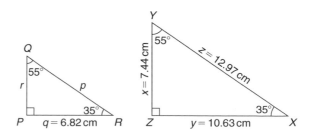

Figure 37.41

By proportion: $\dfrac{p}{z} = \dfrac{r}{x} = \dfrac{q}{y}$

i.e. $\dfrac{p}{12.97} = \dfrac{r}{7.44} = \dfrac{6.82}{10.63}$

from which, $r = 7.44\left(\dfrac{6.82}{10.63}\right) = \mathbf{4.77\,cm}$

By proportion: $\dfrac{p}{z} = \dfrac{q}{y}$ i.e. $\dfrac{p}{12.97} = \dfrac{6.82}{10.63}$

Hence, $p = 12.97\left(\dfrac{6.82}{10.63}\right) = \mathbf{8.32\,cm}$

Problem 28. In Figure 37.42, show that triangles *CBD* and *CAE* are similar and hence find the length of *CD* and *BD*. The lengths shown are in centimetres.

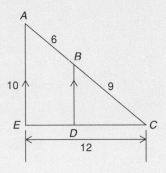

Figure 37.42

Since *BD* is parallel to *AE* then $\angle CBD = \angle CAE$ and $\angle CDB = \angle CEA$ (corresponding angles between parallel lines). Also, $\angle C$ is common to triangles *CBD* and *CAE*.

Since the angles in triangle *CBD* are the same as in triangle *CAE*, the triangles are similar. Hence,

by proportion: $\dfrac{CB}{CA} = \dfrac{CD}{CE}\left(= \dfrac{BD}{AE}\right)$

i.e. $\dfrac{9}{6+9} = \dfrac{CD}{12}$, from which

$$CD = 12\left(\dfrac{9}{15}\right) = \mathbf{7.2\,cm}$$

Also, $\dfrac{9}{15} = \dfrac{BD}{10}$, from which

$$BD = 10\left(\dfrac{9}{15}\right) = \mathbf{6\,cm}$$

Problem 29. A rectangular shed 2 m wide and 3 m high stands against a perpendicular building of height 5.5 m. A ladder is used to gain access to the roof of the building. Determine the minimum distance between the bottom of the ladder and the shed

A side view is shown in Figure 37.43, where *AF* is the minimum length of the ladder. Since *BD* and *CF* are parallel, $\angle ADB = \angle DFE$ (corresponding angles between parallel lines). Hence, triangles *BAD* and *EDF* are similar since their angles are the same.

$$AB = AC - BC = AC - DE = 5.5 - 3 = 2.5\,m$$

By proportion: $\dfrac{AB}{DE} = \dfrac{BD}{EF}$ i.e. $\dfrac{2.5}{3} = \dfrac{2}{EF}$

Hence, $EF = 2\left(\dfrac{3}{2.5}\right) = \mathbf{2.4\,m} = \mathbf{minimum\ distance}$ **from bottom of ladder to the shed**.

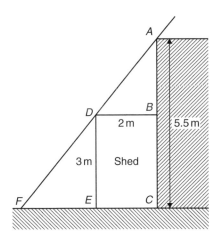

Figure 37.43

Now try the following Practice Exercise

Practice Exercise 160   Similar triangles (answers on page 1169)

1. In Figure 37.44, find the lengths x and y.

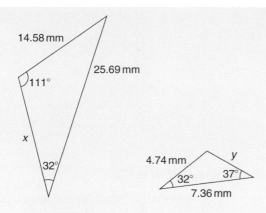

Figure 37.44

2. *PQR* is an equilateral triangle of side 4 cm. When *PQ* and *PR* are produced to *S* and *T*, respectively, *ST* is found to be parallel with *QR*. If *PS* is 9 cm, find the length of *ST*. *X* is a point on *ST* between *S* and *T* such that the line *PX* is the bisector of ∠*SPT*. Find the length of *PX*.

3. In Figure 37.45, find
   (a) the length of *BC* when *AB* = 6 cm, *DE* = 8 cm and *DC* = 3 cm,
   (b) the length of *DE* when *EC* = 2 cm, *AC* = 5 cm and *AB* = 10 cm.

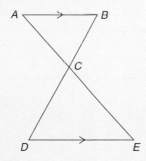

Figure 37.45

4. In Figure 37.46, *AF* = 8 m, *AB* = 5 m and *BC* = 3 m. Find the length of *BD*.

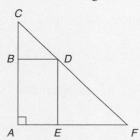

Figure 37.46

37.6 Construction of triangles

To construct any triangle, the following drawing instruments are needed:

(a)   ruler and/or straight edge

(b)   compass

(c)   protractor

(d)   pencil.

Here are some worked problems to demonstrate triangle construction.

Problem 30. Construct a triangle whose sides are 6 cm, 5 cm and 3 cm

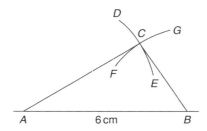

Figure 37.47

With reference to Figure 37.47:

(i)   Draw a straight line of any length and, with a pair of compasses, mark out 6 cm length and label it *AB*.

(ii)   Set compass to 5 cm and with centre at *A* describe arc *DE*.

(iii)   Set compass to 3 cm and with centre at *B* describe arc *FG*.

(iv)   The intersection of the two curves at *C* is the vertex of the required triangle. Join *AC* and *BC* by straight lines.

It may be proved by measurement that the ratio of the angles of a triangle is not equal to the ratio of the sides (i.e. in this problem, the angle opposite the 3 cm side is not equal to half the angle opposite the 6 cm side).

Problem 31. Construct a triangle *ABC* such that *a* = 6 cm, *b* = 3 cm and ∠*C* = 60°

With reference to Figure 37.48:

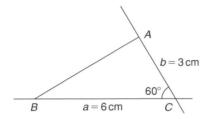

Figure 37.48

(i)   Draw a line BC, 6 cm long.

(ii) Using a protractor centred at C, make an angle of $60°$ to BC.

(iii) From C measure a length of 3 cm and label A.

(iv) Join B to A by a straight line.

Problem 32.   Construct a triangle PQR given that $QR = 5$ cm, $\angle Q = 70°$ and $\angle R = 44°$

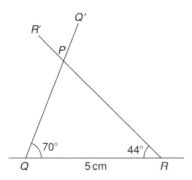

Figure 37.49

With reference to Figure 37.49:

(i)   Draw a straight line 5 cm long and label it QR.

(ii) Use a protractor centred at Q and make an angle of $70°$. Draw QQ'.

(iii) Use a protractor centred at R and make an angle of $44°$. Draw RR'.

(iv) The intersection of QQ' and RR' forms the vertex P of the triangle.

Problem 33.   Construct a triangle XYZ given that $XY = 5$ cm, the hypotenuse $YZ = 6.5$ cm and $\angle X = 90°$

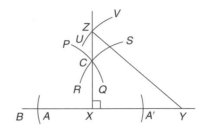

Figure 37.50

With reference to Figure 37.50:

(i)   Draw a straight line 5 cm long and label it XY.

(ii) Produce XY any distance to B. With compass centred at X make an arc at A and A'. (The length XA and XA' is arbitrary.) With compass centred at A draw the arc PQ. With the same compass setting and centred at A', draw the arc RS. Join the intersection of the arcs, C to X, and a right angle to XY is produced at X. (Alternatively, a protractor can be used to construct a $90°$ angle.)

(iii) The hypotenuse is always opposite the right angle. Thus, YZ is opposite $\angle X$. Using a compass centred at Y and set to 6.5 cm, describe the arc UV.

(iv) The intersection of the arc UV with XC produced, forms the vertex Z of the required triangle. Join YZ with a straight line.

Now try the following Practice Exercise

Practice Exercise 161   Construction of triangles (answers on page 1169)

In the following, construct the triangles ABC for the given sides/angles.

1. $a = 8$ cm, $b = 6$ cm and $c = 5$ cm.

2. $a = 40$ mm, $b = 60$ mm and $C = 60°$

3. $a = 6$ cm, $C = 45°$ and $B = 75°$

4. $c = 4$ cm, $A = 130°$ and $C = 15°$

5. $a = 90$ mm, $B = 90°$, hypotenuse $= 105$ mm.

For fully worked solutions to each of the problems in Practice Exercises 156 to 161 in this chapter, go to the website:
www.routledge.com/cw/bird

Section E

Chapter 38

Introduction to trigonometry

Why it is important to understand: Introduction to trigonometry

There are an enormous number of uses of trigonometry and trigonometric functions. Fields that use trigonometry or trigonometric functions include astronomy (especially for locating apparent positions of celestial objects, in which spherical trigonometry is essential) and hence navigation (on the oceans, in aircraft, and in space), music theory, acoustics, optics, analysis of financial markets, electronics, probability theory, statistics, biology, medical imaging (CAT scans and ultrasound), pharmacy, chemistry, number theory (and hence cryptology), seismology, meteorology, oceanography, many physical sciences, land surveying and geodesy (a branch of earth sciences), architecture, phonetics, economics, electrical engineering, mechanical engineering, civil engineering, computer graphics, cartography, crystallography and game development. It is clear that a good knowledge of trigonometry is essential in many fields of engineering.

At the end of this chapter, you should be able to:

- state the theorem of Pythagoras and use it to find the unknown side of a right-angled triangle
- define sine, cosine and tangent of an angle in a right-angled triangle
- evaluate trigonometric ratios of angles
- define reciprocal trigonometric ratios secant, cosecant and cotangent
- determine the fractional and surd forms of common trigonometric ratios
- solve right-angled triangles
- understand angles of elevation and depression
- appreciate trigonometric approximations for small angles

38.1 Introduction

Trigonometry is a subject that involves the measurement of sides and angles of triangles and their relationship to each other.

The theorem of Pythagoras and trigonometric ratios are used with right-angled triangles only. However, there are many practical examples in engineering where knowledge of right-angled triangles is very important.

In this chapter, three trigonometric ratios – i.e. sine, cosine and tangent – are defined and then evaluated using a calculator. Finally, solving right-angled triangle problems using Pythagoras and trigonometric ratios is demonstrated, together with some practical examples involving angles of elevation and depression.

38.2 The theorem of Pythagoras

The **theorem of Pythagoras*** states:
In any right-angled triangle, the square of the hypotenuse is equal to the sum of the squares of the other two sides.
In the right-angled triangle ABC shown in Figure 38.1, this means

$$b^2 = a^2 + c^2 \qquad (1)$$

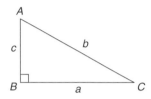

Figure 38.1

If the lengths of any two sides of a right-angled triangle are known, the length of the third side may be calculated by Pythagoras' theorem.

From equation (1): $\quad b = \sqrt{a^2 + c^2}$

* **Who was Pythagoras?** **Pythagoras of Samos** (born about 570 BC and died about 495 BC) was an Ionian Greek philosopher and mathematician. He is best known for the Pythagorean theorem, which states that in a right-angled triangle $a^2 + b^2 = c^2$. To find out more go to **www.routledge.com/cw/bird**

Transposing equation (1) for a gives $a^2 = b^2 - c^2$, from which $a = \sqrt{b^2 - c^2}$
Transposing equation (1) for c gives $c^2 = b^2 - a^2$, from which $c = \sqrt{b^2 - a^2}$
Here are some worked problems to demonstrate the theorem of Pythagoras.

Problem 1. In Figure 38.2, find the length of BC

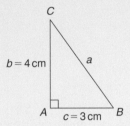

Figure 38.2

From Pythagoras, $\quad a^2 = b^2 + c^2$

i.e. $\qquad a^2 = 4^2 + 3^2$

$\qquad\qquad = 16 + 9 = 25$

Hence, $\qquad a = \sqrt{25} = 5\,\text{cm}$

$\sqrt{25} = \pm 5$ but in a practical example like this an answer of $a = -5\,\text{cm}$ has no meaning, so we take only the positive answer.

Thus $\qquad a = BC = 5\,\text{cm}$

ABC **is a 3, 4, 5 triangle**. There are not many right-angled triangles which have integer values (i.e. whole numbers) for all three sides.

Problem 2. In Figure 38.3, find the length of EF

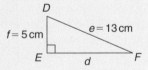

Figure 38.3

By Pythagoras' theorem, $\qquad e^2 = d^2 + f^2$

Hence, $\qquad\qquad 13^2 = d^2 + 5^2$

$\qquad\qquad\qquad 169 = d^2 + 25$

$\qquad\qquad\qquad d^2 = 169 - 25 = 144$

Thus, $\qquad\qquad d = \sqrt{144} = 12\,\text{cm}$

i.e. $\qquad\qquad d = EF = 12\,\text{cm}$

DEF **is a 5, 12, 13 triangle**, another right-angled triangle which has integer values for all three sides.

Problem 3. Two aircraft leave an airfield at the same time. One travels due north at an average speed of 300 km/h and the other due west at an average speed of 220 km/h. Calculate their distance apart after 4 hours

After 4 hours, the first aircraft has travelled

$$4 \times 300 = 1200 \text{ km due north}$$

and the second aircraft has travelled

$$4 \times 220 = 880 \text{ km due west,}$$

as shown in Figure 38.4. The distance apart after 4 hours = BC.

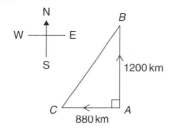

Figure 38.4

From Pythagoras' theorem,

$$BC^2 = 1200^2 + 880^2$$
$$= 1440000 + 774400 = 2214400$$

and $\qquad BC = \sqrt{2214400} = 1488 \text{ km.}$

Hence, distance apart after 4 hours = 1488 km.

Now try the following Practice Exercise

Practice Exercise 162   Theorem of Pythagoras (answers on page 1169)

1. Find the length of side x in Figure 38.5.

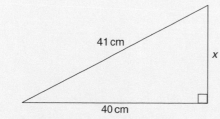

Figure 38.5

2. Find the length of side x in Figure 38.6(a).

3. Find the length of side x in Figure 38.6(b), correct to 3 significant figures.

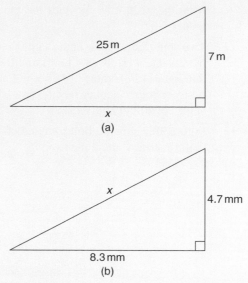

Figure 38.6

4. In a triangle ABC, $AB = 17$ cm, $BC = 12$ cm and $\angle ABC = 90°$. Determine the length of AC, correct to 2 decimal places.

5. A tent peg is 4.0 m away from a 6.0 m high tent. What length of rope, correct to the nearest centimetre, runs from the top of the tent to the peg?

6. In a triangle ABC, $\angle B$ is a right angle, $AB = 6.92$ cm and $BC = 8.78$ cm. Find the length of the hypotenuse.

7. In a triangle CDE, $D = 90°$, $CD = 14.83$ mm and $CE = 28.31$ mm. Determine the length of DE.

8. Show that if a triangle has sides of 8, 15 and 17 cm it is right-angled.

9. Triangle PQR is isosceles, Q being a right angle. If the hypotenuse is 38.46 cm find (a) the lengths of sides PQ and QR and (b) the value of $\angle QPR$.

10. A man cycles 24 km due south and then 20 km due east. Another man, starting at the same time and position as the first man, cycles 32 km due east and then 7 km due south. Find the distance between the two men.

11. A ladder 3.5 m long is placed against a perpendicular wall with its foot 1.0 m from the wall. How far up the wall (to the nearest centimetre) does the ladder reach? If the foot of the ladder is now moved 30 cm further away

Section E

from the wall, how far does the top of the ladder fall?

12. Two ships leave a port at the same time. One travels due west at 18.4 knots and the other due south at 27.6 knots. If 1 knot = 1 nautical mile per hour, calculate how far apart the two ships are after 4 hours.

13. Figure 38.7 shows a bolt rounded off at one end. Determine the dimension h.

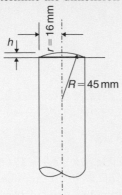

Figure 38.7

14. Figure 38.8 shows a cross-section of a component that is to be made from a round bar. If the diameter of the bar is 74 mm, calculate the dimension x.

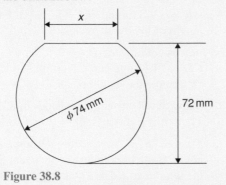

Figure 38.8

38.3 Sines, cosines and tangents

With reference to angle θ in the right-angled triangle ABC shown in Figure 38.9,

$$\text{sine}\,\theta = \frac{\textbf{opposite side}}{\textbf{hypotenuse}}$$

'Sine' is abbreviated to 'sin', thus $\sin\theta = \dfrac{BC}{AC}$

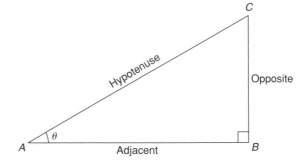

Figure 38.9

Also, $\textbf{cosine}\,\theta = \dfrac{\textbf{adjacent side}}{\textbf{hypotenuse}}$

'Cosine' is abbreviated to 'cos', thus $\cos\theta = \dfrac{AB}{AC}$

Finally, $\textbf{tangent}\,\theta = \dfrac{\textbf{opposite side}}{\textbf{adjacent side}}$

'Tangent' is abbreviated to 'tan', thus $\tan\theta = \dfrac{BC}{AB}$

These three trigonometric ratios only apply to right-angled triangles. Remembering these three equations is very important and the mnemonic '**SOH CAH TOA**' is one way of remembering them.

  SOH indicates <u>s</u>in = <u>o</u>pposite ÷ <u>h</u>ypotenuse
  CAH indicates <u>c</u>os = <u>a</u>djacent ÷ <u>h</u>ypotenuse
  TOA indicates <u>t</u>an = <u>o</u>pposite ÷ <u>a</u>djacent
Here are some worked problems to help familiarise ourselves with trigonometric ratios.

Problem 4. In triangle PQR shown in Figure 38.10, determine $\sin\theta$, $\cos\theta$ and $\tan\theta$

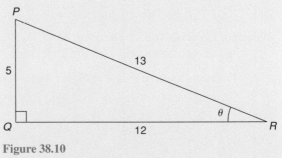

Figure 38.10

$$\sin\theta = \frac{\text{opposite side}}{\text{hypotenuse}} = \frac{PQ}{PR} = \frac{5}{13} = \textbf{0.3846}$$

$$\cos\theta = \frac{\text{adjacent side}}{\text{hypotenuse}} = \frac{QR}{PR} = \frac{12}{13} = \textbf{0.9231}$$

$$\tan\theta = \frac{\text{opposite side}}{\text{adjacent side}} = \frac{PQ}{QR} = \frac{5}{12} = \textbf{0.4167}$$

Problem 5. In triangle ABC of Figure 38.11, determine length AC, $\sin C$, $\cos C$, $\tan C$, $\sin A$, $\cos A$ and $\tan A$

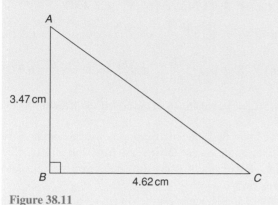

Figure 38.11

By Pythagoras, $AC^2 = AB^2 + BC^2$

i.e. $AC^2 = 3.47^2 + 4.62^2$

from which $AC = \sqrt{3.47^2 + 4.62^2} = \textbf{5.778\,cm}$

$$\sin C = \frac{\text{opposite side}}{\text{hypotenuse}} = \frac{AB}{AC} = \frac{3.47}{5.778} = \textbf{0.6006}$$

$$\cos C = \frac{\text{adjacent side}}{\text{hypotenuse}} = \frac{BC}{AC} = \frac{4.62}{5.778} = \textbf{0.7996}$$

$$\tan C = \frac{\text{opposite side}}{\text{adjacent side}} = \frac{AB}{BC} = \frac{3.47}{4.62} = \textbf{0.7511}$$

$$\sin A = \frac{\text{opposite side}}{\text{hypotenuse}} = \frac{BC}{AC} = \frac{4.62}{5.778} = \textbf{0.7996}$$

$$\cos A = \frac{\text{adjacent side}}{\text{hypotenuse}} = \frac{AB}{AC} = \frac{3.47}{5.778} = \textbf{0.6006}$$

$$\tan A = \frac{\text{opposite side}}{\text{adjacent side}} = \frac{BC}{AB} = \frac{4.62}{3.47} = \textbf{1.3314}$$

Problem 6. If $\tan B = \dfrac{8}{15}$, determine the value of $\sin B$, $\cos B$, $\sin A$ and $\tan A$

A right-angled triangle ABC is shown in Figure 38.12. If $\tan B = \dfrac{8}{15}$, then $AC = 8$ and $BC = 15$

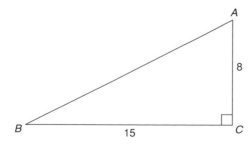

Figure 38.12

By Pythagoras, $AB^2 = AC^2 + BC^2$

i.e. $AB^2 = 8^2 + 15^2$

from which $AB = \sqrt{8^2 + 15^2} = 17$

$$\sin B = \frac{AC}{AB} = \frac{8}{17} \text{ or } \textbf{0.4706}$$

$$\cos B = \frac{BC}{AB} = \frac{15}{17} \text{ or } \textbf{0.8824}$$

$$\sin A = \frac{BC}{AB} = \frac{15}{17} \text{ or } \textbf{0.8824}$$

$$\tan A = \frac{BC}{AC} = \frac{15}{8} \text{ or } \textbf{1.8750}$$

Problem 7. Point A lies at co-ordinate $(2, 3)$ and point B at $(8, 7)$. Determine (a) the distance AB and (b) the gradient of the straight line AB

(a) Points A and B are shown in Figure 38.13(a).

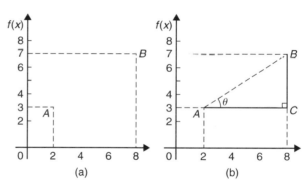

Figure 38.13

In Figure 38.13(b), the horizontal and vertical lines AC and BC are constructed. Since ABC is a right-angled triangle, and $AC = (8-2) = 6$ and $BC = (7-3) = 4$, by Pythagoras' theorem,

$$AB^2 = AC^2 + BC^2 = 6^2 + 4^2$$

and $AB = \sqrt{6^2 + 4^2} = \sqrt{52}$

$$= \textbf{7.211} \text{ correct to 3 decimal places.}$$

(b) The gradient of AB is given by $\tan\theta$, i.e.

$$\textbf{gradient} = \tan\theta = \frac{BC}{AC} = \frac{4}{6} = \frac{\textbf{2}}{\textbf{3}}$$

Now try the following Practice Exercise

Practice Exercise 163   Trigonometric ratios (answers on page 1169)

1. Sketch a triangle XYZ such that $\angle Y = 90°$, $XY = 9$ cm and $YZ = 40$ cm. Determine $\sin Z$, $\cos Z$, $\tan X$ and $\cos X$.

2. In triangle ABC shown in Figure 38.14, find $\sin A$, $\cos A$, $\tan A$, $\sin B$, $\cos B$ and $\tan B$.

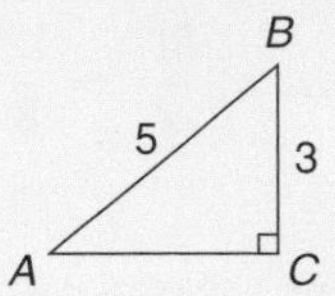

Figure 38.14

3. If $\cos A = \dfrac{15}{17}$, find $\sin A$ and $\tan A$, in fraction form.

4. If $\tan X = \dfrac{15}{112}$, find $\sin X$ and $\cos X$, in fraction form.

5. For the right-angled triangle shown in Figure 38.15, find (a) $\sin\alpha$ (b) $\cos\theta$ (c) $\tan\theta$.

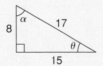

Figure 38.15

6. If $\tan\theta = \dfrac{7}{24}$, find (a) $\sin\theta$ and (b) $\cos\theta$ in fraction form.

7. Point P lies at co-ordinate $(-3, 1)$ and point Q at $(5, -4)$. Determine
   (a) the distance PQ
   (b) the gradient of the straight line PQ

38.4 Evaluating trigonometric ratios of acute angles

The easiest way to evaluate trigonometric ratios of any angle is to use a calculator. Use a calculator to check the following (each correct to 4 decimal places).

$\sin 29° = \mathbf{0.4848}$ $\sin 53.62° = \mathbf{0.8051}$

$\cos 67° = \mathbf{0.3907}$ $\cos 83.57° = \mathbf{0.1120}$

$\tan 34° = \mathbf{0.6745}$ $\tan 67.83° = \mathbf{2.4541}$

$\sin 67°43' = \sin 67\dfrac{43°}{60} = \sin 67.7166666\ldots° = \mathbf{0.9253}$

$\cos 13°28' = \cos 13\dfrac{28°}{60} = \cos 13.466666\ldots° = \mathbf{0.9725}$

$\tan 56°54' = \tan 56\dfrac{54°}{60} = \tan 56.90° = \mathbf{1.5340}$

If we know the value of a trigonometric ratio and need to find the angle we use the **inverse function** on our calculators. For example, using shift and sin on our calculator gives $\sin^{-1}($

If, for example, we know the sine of an angle is 0.5 then the value of the angle is given by

$$\sin^{-1} 0.5 = \mathbf{30°} \text{ (Check that } \sin 30° = 0.5)$$

Similarly, if

$$\cos\theta = 0.4371 \text{ then } \theta = \cos^{-1} 0.4371 = \mathbf{64.08°}$$

and if

$$\tan A = 3.5984 \text{ then } A = \tan^{-1} 3.5984 = \mathbf{74.47°}$$

each correct to 2 decimal places.

Use your calculator to check the following worked problems.

Problem 8. Determine, correct to 4 decimal places, $\sin 43°39'$

$$\sin 43°39' = \sin 43\dfrac{39°}{60} = \sin 43.65° = \mathbf{0.6903}$$

This answer can be obtained using the **calculator** as follows:

1. Press sin   2. Enter 43   3. Press° '''
4. Enter 39   5. Press° '''   6. Press)
7. Press =   Answer = **0.6902512...**

Problem 9. Determine, correct to 3 decimal places, $6\cos 62°12'$

$$6\cos 62°12' = 6\cos 62\dfrac{12°}{60} = 6\cos 62.20° = \mathbf{2.798}$$

This answer can be obtained using the **calculator** as follows:

1. Enter 6   2. Press cos   3. Enter 62

4. Press° '''   5. Enter 12   6. Press° '''

7. Press)   8. Press =   Answer = **2.798319**...

> **Problem 10.** Evaluate sin 1.481, correct to 4 significant figures

sin 1.481 means the sine of 1.481 **radians**. (If there is no degrees sign, i.e. °, then radians are assumed.) Therefore a calculator needs to be on the radian function. Hence, sin 1.481 = **0.9960**

> **Problem 11.** Evaluate cos(3π/5), correct to 4 significant figures

As in Problem 10, 3π/5 is in radians.
Hence, **cos(3π/5)** = cos 1.884955... = **−0.3090**
Since, from page 424, π radians = 180°,
$3\pi/5 \text{ rad} = \frac{3}{5} \times 180° = 108°$.
i.e. 3π/5 rad = 108°. Check with your calculator that
cos 108° = −0.3090

> **Problem 12.** Evaluate tan 2.93, correct to 4 significant figures

Again, since there is no degrees sign, 2.93 means 2.93 radians.
Hence, **tan 2.93 = −0.2148**
It is important to know when to have your calculator on either degrees mode or radian mode. A lot of mistakes can arise from this if we are not careful.

> **Problem 13.** Find the acute angle $\sin^{-1} 0.4128$ in degrees, correct to 2 decimal places

$\sin^{-1} 0.4128$ means 'the angle whose sine is 0.4128'. Using a calculator,

1. Press shift   2. Press sin   3. Enter 0.4128

4. Press)   5. Press =

The answer 24.380848... is displayed.
Hence, $\sin^{-1} 0.4128 = 24.38°$ correct to 2 decimal places.

> **Problem 14.** Find the acute angle $\cos^{-1} 0.2437$ in degrees and minutes

$\cos^{-1} 0.2437$ means 'the angle whose cosine is 0.2437'. Using a calculator,

1. Press shift   2. Press cos   3. Enter 0.2437

4. Press)   5. Press =

The answer 75.894979... is displayed.

6. Press ° ''' and 75° 53' 41.93'' is displayed.

Hence, $\cos^{-1} 0.2437 = 75.89° = 77°54'$ correct to the nearest minute.

> **Problem 15.** Find the acute angle $\tan^{-1} 7.4523$ in degrees and minutes

$\tan^{-1} 7.4523$ means 'the angle whose tangent is 7.4523'. Using a calculator,

1. Press shift   2. Press tan   3. Enter 7.4523

4. Press)   5. Press =

The answer 82.357318... is displayed.

6. Press ° ''' and 82° 21' 26.35'' is displayed.

Hence, $\tan^{-1} 7.4523 = 82.36° = 82°21'$ correct to the nearest minute.

> **Problem 16.** In triangle *EFG* in Figure 38.16, calculate angle *G*

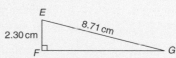

Figure 38.16

With reference to ∠G, the two sides of the triangle given are the opposite side *EF* and the hypotenuse *EG*; hence, sine is used,

i.e. $\sin G = \dfrac{2.30}{8.71} = 0.26406429\ldots$

from which, $G = \sin^{-1} 0.26406429\ldots$

i.e. $G = 15.311360°$

Hence,   **∠G = 15.31°** or **15°19'**

> **Problem 17.** Evaluate the following expression, correct to 3 significant figures

$$\frac{4.2 \tan 49°26' - 3.7 \sin 66°1'}{7.1 \cos 29°34'}$$

By calculator:

$$\tan 49°26' = \tan\left(49\frac{26}{60}\right)° = 1.1681,$$

$$\sin 66°1' = 0.9137 \text{ and } \cos 29°34' = 0.8698$$

Hence, $\dfrac{4.2 \tan 49°26' - 3.7 \sin 66°1'}{7.1 \cos 29°34'}$

$$= \frac{(4.2 \times 1.1681) - (3.7 \times 0.9137)}{(7.1 \times 0.8698)}$$

$$= \frac{4.9060 - 3.3807}{6.1756} = \frac{1.5253}{6.1756}$$

$$= 0.2470 = \mathbf{0.247},$$

correct to 3 significant figures.

Now try the following Practice Exercise

Practice Exercise 164   Evaluating trigonometric ratios (answers on page 1169)

1. Determine, correct to 4 decimal places, $3 \sin 66°41'$

2. Determine, correct to 3 decimal places, $5 \cos 14°15'$

3. Determine, correct to 4 significant figures, $7 \tan 79°9'$

4. Determine

    (a) sine $\dfrac{2\pi}{3}$ (b) $\cos 1.681$ (c) $\tan 3.672$

5. Find the acute angle $\sin^{-1} 0.6734$ in degrees, correct to 2 decimal places.

6. Find the acute angle $\cos^{-1} 0.9648$ in degrees, correct to 2 decimal places.

7. Find the acute angle $\tan^{-1} 3.4385$ in degrees, correct to 2 decimal places.

8. Find the acute angle $\sin^{-1} 0.1381$ in degrees and minutes.

9. Find the acute angle $\cos^{-1} 0.8539$ in degrees and minutes.

10. Find the acute angle $\tan^{-1} 0.8971$ in degrees and minutes.

11. In the triangle shown in Figure 38.17, determine angle θ, correct to 2 decimal places.

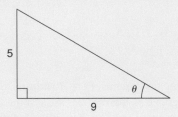

Figure 38.17

12. In the triangle shown in Figure 38.18, determine angle θ in degrees and minutes.

Figure 38.18

13. Evaluate, correct to 4 decimal places, $\dfrac{4.5 \cos 67°34' - \sin 90°}{2 \tan 45°}$

14. Evaluate, correct to 4 significant figures, $\dfrac{(3 \sin 37.83°)(2.5 \tan 57.48°)}{4.1 \cos 12.56°}$

15. For the supported beam AB shown in Figure 38.19, determine (a) the angle the supporting stay CD makes with the beam, i.e. θ, correct to the nearest degree, and (b) the length of the stay, CD, correct to the nearest centimetre.

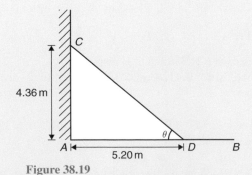

Figure 38.19

38.5   Reciprocal ratios

With reference to angle θ in the right-angled triangle ABC shown in Figure 38.20:

$$\textbf{secant } \theta = \frac{\textbf{hypotenuse}}{\textbf{adjacent side}}$$

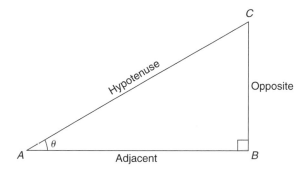

Figure 38.20

'Secant' is abbreviated to 'sec', thus $\textbf{sec } \theta = \dfrac{AC}{AB}$

Also, $\textbf{cosecant } \theta = \dfrac{\textbf{hypotenuse}}{\textbf{opposite side}}$

'Cosecant' is abbreviated to 'cosec', thus $\textbf{cosec } \theta = \dfrac{AC}{BC}$

Finally, $\textbf{cotangent } \theta = \dfrac{\textbf{adjacent side}}{\textbf{opposite side}}$

'Cotangent' is abbreviated to 'cot', thus $\textbf{cot } \theta = \dfrac{AB}{BC}$

From above,

(i) $\dfrac{\sin\theta}{\cos\theta} = \dfrac{\frac{BC}{AC}}{\frac{AB}{AC}} = \dfrac{BC}{AB} = \tan\theta$ i.e. $\textbf{tan } \theta = \dfrac{\sin\theta}{\cos\theta}$

(ii) $\dfrac{\cos\theta}{\sin\theta} = \dfrac{\frac{AB}{AC}}{\frac{BC}{AC}} = \dfrac{AB}{BC} = \cot\theta$ i.e. $\textbf{cot } \theta = \dfrac{\cos\theta}{\sin\theta}$

(iii) $\textbf{sec } \theta = \dfrac{1}{\cos\theta}$

(iv) $\textbf{cosec } \theta = \dfrac{1}{\sin\theta}$ (Note 's' and 'c' go together)

(v) $\textbf{cot } \theta = \dfrac{1}{\tan\theta}$

Secants, cosecants and cotangents are called the **reciprocal ratios**.

Problem 18.   If $\cos X = \dfrac{9}{41}$ determine the value of the other five trigonometry ratios.

Figure 38.21 shows a right-angled triangle XYZ

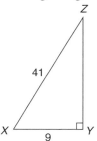

Figure 38.21

Since $\cos X = \dfrac{9}{41}$, then $XY = 9$ units and $XZ = 41$ units.

Using Pythagoras' theorem: $41^2 = 9^2 + YZ^2$ from which $YZ = \sqrt{41^2 - 9^2} = 40$ units.

Thus, $\sin X = \dfrac{40}{41}$, $\tan X = \dfrac{40}{9} = 4\dfrac{4}{9}$,

$\text{cosec } X = \dfrac{41}{40} = 1\dfrac{1}{40}$, $\sec X = \dfrac{41}{9} = 4\dfrac{5}{9}$

and $\cot X = \dfrac{9}{40}$

Problem 19.   If $\sin\theta = 0.625$ and $\cos\theta = 0.500$ determine the values of $\text{cosec } \theta$, $\sec\theta$, $\tan\theta$ and $\cot\theta$

$\text{cosec } \theta = \dfrac{1}{\sin\theta} = \dfrac{1}{0.625} = \textbf{1.60}$

$\sec\theta = \dfrac{1}{\cos\theta} = \dfrac{1}{0.500} = \textbf{2.00}$

$\tan\theta = \dfrac{\sin\theta}{\cos\theta} = \dfrac{0.625}{0.500} = \textbf{1.25}$

$\cot\theta = \dfrac{\cos\theta}{\sin\theta} = \dfrac{0.500}{0.625} = \textbf{0.80}$

Problem 20.   Evaluate, correct to 4 decimal places: (a) secant $161°$ (b) secant $302°29'$

(a) $\sec 161° = \dfrac{1}{\cos 161°} = \textbf{-1.0576}$

(b) $\sec 302°29' = \dfrac{1}{\cos 302°29'} = \dfrac{1}{\cos 302\frac{29°}{60}}$

$= \textbf{1.8620}$

Problem 21. Evaluate, correct to 4 significant figures: (a) cosecant $279.16°$ (b) cosecant $49°7'$

(a) $\operatorname{cosec} 279.16° = \dfrac{1}{\sin 279.16°} = \mathbf{-1.013}$

(b) $\operatorname{cosec} 49°7' = \dfrac{1}{\sin 49°7'} = \dfrac{1}{\sin 49\dfrac{7°}{60}} = \mathbf{1.323}$

Problem 22. Evaluate, correct to 4 decimal places: (a) cotangent $17.49°$ (b) cotangent $163°52'$

(a) $\cot 17.49° = \dfrac{1}{\tan 17.49°} = \mathbf{3.1735}$

(b) $\cot 163°52' = \dfrac{1}{\tan 163°52'} = \dfrac{1}{\tan 163\dfrac{52°}{60}}$

$\qquad = \mathbf{-3.4570}$

Problem 23. Evaluate, correct to 4 decimal places:

(a) secant 5.37 $\qquad$ (b) cosecant $\pi/4$

(c) cotangent $5\pi/24$

(a) With no degrees sign, it is assumed that 5.37 means 5.37 radians.
$\qquad$ Hence $\sec 5.37 = \dfrac{1}{\cos 5.37} = \mathbf{1.6361}$

(b) $\operatorname{cosec}(\pi/4) = \dfrac{1}{\sin(\pi/4)} = \dfrac{1}{\sin 0.785398\ldots}$

$\qquad\qquad\qquad\qquad = \mathbf{1.4142}$

(c) $\cot(5\pi/24) = \dfrac{1}{\tan(5\pi/24)} = \dfrac{1}{\tan 0.654498\ldots}$

$\qquad\qquad\qquad\qquad = \mathbf{1.3032}$

Problem 24. Determine the acute angles:

(a) $\sec^{-1} 2.3164$ $\qquad$ (b) $\operatorname{cosec}^{-1} 1.1784$

(c) $\cot^{-1} 2.1273$

(a) $\sec^{-1} 2.3164 = \cos^{-1}\left(\dfrac{1}{2.3164}\right) = \cos^{-1} 0.4317\ldots$

$\qquad\qquad = \mathbf{64.42°}$ or $\mathbf{64°25'}$ or $\mathbf{1.124\ radians}$

(b) $\operatorname{cosec}^{-1} 1.1784 = \sin^{-1}\left(\dfrac{1}{1.1784}\right) =$

$\sin^{-1} 0.8486\ldots = \mathbf{58.06°}$ or $\mathbf{58°4'}$ or $\mathbf{1.013\ radians}$

(c) $\cot^{-1} 2.1273 = \tan^{-1}\left(\dfrac{1}{2.1273}\right) = \tan^{-1} 0.4700\ldots$

$\qquad\qquad = \mathbf{25.18°}$ or $\mathbf{25°11'}$ or $\mathbf{0.439\ radians}$

Problem 25. Evaluate the following expression, correct to 4 significant figures:

$$\frac{4 \sec 32°10' - 2 \cot 15°19'}{3 \operatorname{cosec} 63°8' \tan 14°57'}$$

By calculator:
$\qquad \sec 32°10' = 1.1813, \quad \cot 15°19' = 3.6512$

$\qquad \operatorname{cosec} 63°8' = 1.1210, \quad \tan 14°57' = 0.2670$

Hence, $\dfrac{4 \sec 32°10' - 2 \cot 15°19'}{3 \operatorname{cosec} 63°8' \tan 14°57'}$

$\qquad = \dfrac{4(1.1813) - 2(3.6512)}{3(1.1210)(0.2670)} = \dfrac{4.7252 - 7.3024}{0.8979}$

$\qquad = \dfrac{-2.5772}{0.8979} = \mathbf{-2.870},$

$\qquad\qquad\qquad\qquad$ correct to 4 significant figures.

Now try the following Practice Exercise

Practice Exercise 165 $\qquad$ Reciprocal ratios (answers on page 1169)

1. If $\cos X = \dfrac{7}{25}$ determine the value of the other five trigonometric ratios.

2. If $\sin\theta = 0.40$ and $\cos\theta = 0.50$ determine the values of $\operatorname{cosec}\theta$, $\sec\theta$, $\tan\theta$ and $\cot\theta$

In Problems 3 to 6, evaluate correct to 4 decimal places.

3. (a) secant $73°$ (b) secant $286.45°$
    (c) secant $155°41'$

4. (a) cosecant $213°$ (b) cosecant $15.62°$
    (c) cosecant $311°50'$

5. (a) cotangent $71°$ (b) cotangent $151.62°$
    (c) cotangent $321°23'$

6. (a) $\sec \dfrac{\pi}{8}$ (b) cosec 2.961 (c) cot 2.612

In Problems 7 to 9, determine the acute angle in degrees (correct to 2 decimal places), degrees and minutes, and in radians (correct to 3 decimal places).

7. $\sec^{-1} 1.6214$

8. $\operatorname{cosec}^{-1} 2.4891$

9. $\cot^{-1} 1.9614$

10. Evaluate $\dfrac{6.4 \operatorname{cosec} 29°5' - \sec 81°}{2 \cot 12°}$ correct to 4 significant figures.

11. If $\tan x = 1.5276$, determine $\sec x$, $\operatorname{cosec} x$, and $\cot x$. (Assume x is an acute angle.)

In Problems 12 and 13, evaluate correct to 4 significant figures.

12. $3 \cot 14°15' \sec 23°9'$

13. $\dfrac{\operatorname{cosec} 27°19' + \sec 45°29'}{1 - \operatorname{cosec} 27°19' \sec 45°29'}$

38.6 Fractional and surd forms of trigonometric ratios

In Figure 38.22, ABC is an equilateral triangle of side 2 units. AD bisects angle A and bisects the side BC. Using Pythagoras' theorem on triangle ABD gives:

$$AD = \sqrt{2^2 - 1^2} = \sqrt{3}$$

Hence, $\quad \sin 30° = \dfrac{BD}{AB} = \dfrac{1}{2}, \ \cos 30° = \dfrac{AD}{AB} = \dfrac{\sqrt{3}}{2}$

and $\quad \tan 30° = \dfrac{BD}{AD} = \dfrac{1}{\sqrt{3}}$

$$\sin 60° = \dfrac{AD}{AB} = \dfrac{\sqrt{3}}{2}, \ \cos 60° = \dfrac{BD}{AB} = \dfrac{1}{2}$$

and $\quad \tan 60° = \dfrac{AD}{BD} = \sqrt{3}$

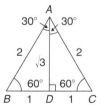

Figure 38.22

Figure 38.23

In Figure 38.23, PQR is an isosceles triangle with $PQ = QR = 1$ unit. By Pythagoras' theorem, $PR = \sqrt{1^2 + 1^2} = \sqrt{2}$

Hence,

$$\sin 45° = \dfrac{1}{\sqrt{2}}, \quad \cos 45° = \dfrac{1}{\sqrt{2}} \text{ and } \tan 45° = 1$$

A quantity that is not exactly expressible as a rational number is called a **surd**. For example, $\sqrt{2}$ and $\sqrt{3}$ are called surds because they cannot be expressed as a fraction and the decimal part may be continued indefinitely. For example,

$$\sqrt{2} = 1.4142135\ldots, \text{ and } \sqrt{3} = 1.7320508\ldots$$

From above,

$$\sin 30° = \cos 60°, \sin 45° = \cos 45° \text{ and}$$
$$\sin 60° = \cos 30°$$

In general,

$$\sin \theta = \cos(90° - \theta) \quad \text{and} \quad \cos \theta = \sin(90° - \theta)$$

For example, it may be checked by calculator that $\sin 25° = \cos 65°$, $\sin 42° = \cos 48°$ and $\cos 84°\ 10' = \sin 5°\ 50'$, and so on.

Problem 26. Using surd forms, evaluate:

$$\dfrac{3 \tan 60° - 2 \cos 30°}{\tan 30°}$$

From above, $\tan 60° = \sqrt{3}$, $\cos 30° = \dfrac{\sqrt{3}}{2}$ and $\tan 30° = \dfrac{1}{\sqrt{3}}$, hence

$$\dfrac{3 \tan 60° - 2 \cos 30°}{\tan 30°} = \dfrac{3\left(\sqrt{3}\right) - 2\left(\dfrac{\sqrt{3}}{2}\right)}{\dfrac{1}{\sqrt{3}}}$$

$$= \dfrac{3\sqrt{3} - \sqrt{3}}{\dfrac{1}{\sqrt{3}}} = \dfrac{2\sqrt{3}}{\dfrac{1}{\sqrt{3}}}$$

$$= 2\sqrt{3}\left(\dfrac{\sqrt{3}}{1}\right) = 2(3) = \mathbf{6}$$

Now try the following Practice Exercise

Evaluate the following, without using a calculator, leaving where necessary in surd form:

1. $3 \sin 30° - 2 \cos 60°$

2. $5 \tan 60° - 3 \sin 60°$

3. $\dfrac{\tan 60°}{3 \tan 30°}$

4. $(\tan 45°)(4 \cos 60° - 2 \sin 60°)$

5. $\dfrac{\tan 60° - \tan 30°}{1 + \tan 30° \tan 60°}$

38.7   Solving right-angled triangles

'Solving a right-angled triangle' means 'finding the unknown sides and angles'. This is achieved using

(a)   the theorem of Pythagoras and/or

(b)   trigonometric ratios.

Six pieces of information describe a triangle completely; i.e. three sides and three angles. As long as at least three facts are known, the other three can usually be calculated.

Here are some worked problems to demonstrate the solution of right-angled triangles.

Problem 27.   In the triangle ABC shown in Figure 38.24, find the lengths AC and AB

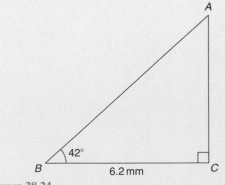

Figure 38.24

There is usually more than one way to solve such a triangle.

In triangle ABC,

$$\tan 42° = \frac{AC}{BC} = \frac{AC}{6.2}$$

(Remember SOH CAH TOA)

Transposing gives

$$AC = 6.2 \tan 42° = \mathbf{5.583\,mm}$$

$$\cos 42° = \frac{BC}{AB} = \frac{6.2}{AB}, \text{ from which}$$

$$AB = \frac{6.2}{\cos 42°} = \mathbf{8.343\,mm}$$

Alternatively,   by   Pythagoras, $AB^2 = AC^2 + BC^2$ from   which $\boldsymbol{AB} = \sqrt{AC^2 + BC^2} = \sqrt{5.583^2 + 6.2^2}$

$$= \sqrt{69.609889} = \mathbf{8.343\,mm}.$$

Problem 28.   Sketch a right-angled triangle ABC such that $B = 90°$, $AB = 5$ cm and $BC = 12$ cm. Determine the length of AC and hence evaluate $\sin A, \cos C$ and $\tan A$

Triangle ABC is shown in Figure 38.25.

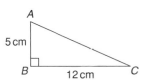

Figure 38.25

By Pythagoras' theorem, $AC = \sqrt{5^2 + 12^2} = 13$

By definition: $\sin A = \dfrac{\text{opposite side}}{\text{hypotenuse}} = \dfrac{12}{13}$ or $\mathbf{0.9231}$

(Remember SOH CAH TOA)

$$\cos C = \frac{\text{adjacent side}}{\text{hypotenuse}} = \frac{12}{13} \text{ or } \mathbf{0.9231}$$

and $\tan A = \dfrac{\text{opposite side}}{\text{adjacent side}} = \dfrac{12}{5}$ or $\mathbf{2.400}$

Problem 29.   In triangle PQR shown in Figure 38.26, find the lengths of PQ and PR

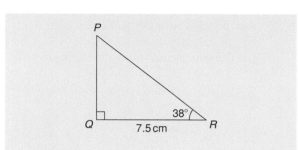

Figure 38.26

$$\tan 38° = \frac{PQ}{QR} = \frac{PQ}{7.5}, \text{ hence}$$

$$PQ = 7.5 \tan 38° = 7.5(0.7813) = \textbf{5.860 cm}$$

$$\cos 38° = \frac{QR}{PR} = \frac{7.5}{PR}, \text{ hence}$$

$$PR = \frac{7.5}{\cos 38°} = \frac{7.5}{0.7880} = \textbf{9.518 cm}$$

Check: using Pythagoras' theorem,
$(7.5)^2 + (5.860)^2 = 90.59 = (9.518)^2$

Problem 30. Solve the triangle ABC shown in Figure 38.27

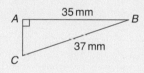

Figure 38.27

To 'solve the triangle ABC' means 'to find the length AC and angles B and C'.

$$\sin C = \frac{35}{37} = 0.94595, \text{ hence}$$

$$C = \sin^{-1} 0.94595 = \textbf{71.08°}$$

$B = 180° - 90° - 71.08° = \textbf{18.92°}$ (since the angles in a triangle add up to $180°$)

$$\sin B = \frac{AC}{37}, \text{ hence}$$

$$AC = 37 \sin 18.92° = 37(0.3242) = \textbf{12.0 mm}$$

or, using Pythagoras' theorem, $37^2 = 35^2 + AC^2$, from which $AC = \sqrt{(37^2 - 35^2)} = \textbf{12.0 mm}$.

Problem 31. Solve triangle XYZ given $\angle X = 90°$, $\angle Y = 23°17'$ and $YZ = 20.0$ mm

It is always advisable to make a reasonably accurate sketch so as to visualise the expected magnitudes of unknown sides and angles. Such a sketch is shown in Figure 38.28.

$$\angle Z = 180° - 90° - 23°17' = \textbf{66°43'}$$

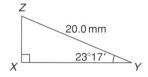

Figure 38.28

$$\sin 23°17' = \frac{XZ}{20.0}, \text{ hence } \textbf{XZ} = 20.0 \sin 23°17'$$

$$= 20.0(0.3953) = \textbf{7.906 mm}$$

$$\cos 23°17' = \frac{XY}{20.0}, \text{ hence } \textbf{XY} = 20.0 \cos 23°17'$$

$$= 20.0(0.9186) = \textbf{18.37 mm}$$

Check: using Pythagoras' theorem,
$(18.37)^2 + (7.906)^2 = 400.0 = (20.0)^2$

Now try the following Practice Exercise

Practice Exercise 167   Solving right-angled triangles (answers on page 1169)

1. Calculate the dimensions shown as x in Figures 38.29(a) to (f), each correct to 4 significant figures.

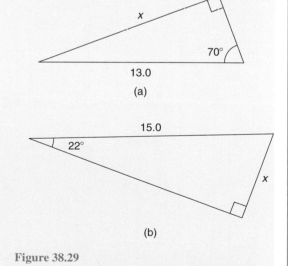

Figure 38.29

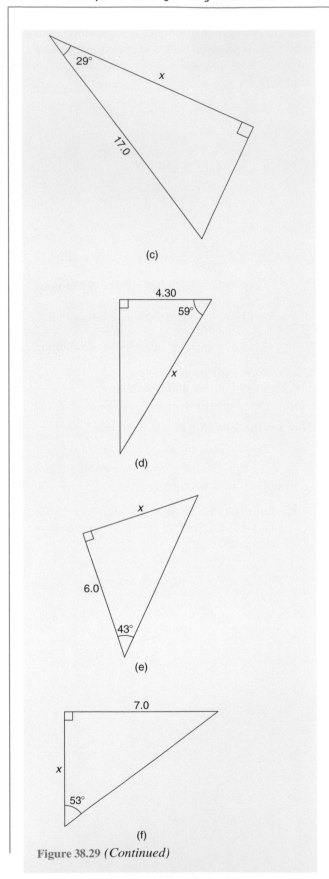

(c)

(d)

(e)

(f)

Figure 38.29 *(Continued)*

2. Find the unknown sides and angles in the right-angled triangles shown in Figure 38.30(a) to (f). The dimensions shown are in centimetres.

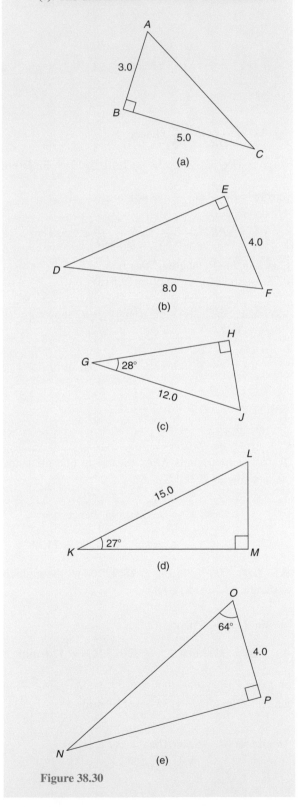

(a)

(b)

(c)

(d)

(e)

Figure 38.30

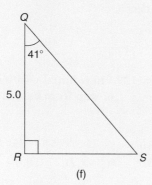

(f)

Figure 38.30 *(Continued)*

3. A ladder rests against the top of the perpendicular wall of a building and makes an angle of 73° with the ground. If the foot of the ladder is 2 m from the wall, calculate the height of the building.

4. Determine the length x in Figure 38.31.

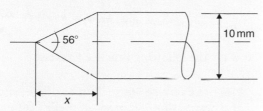

Figure 38.31

5. A symmetrical part of a bridge lattice is shown in Figure 38.32. If $AB = 6$ m, angle $BAD = 56°$ and E is the mid-point of $ABCD$, determine the height h, correct to the nearest centimetre.

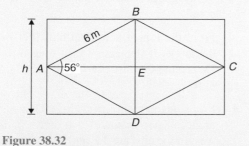

Figure 38.32

38.8 Angles of elevation and depression

If, in Figure 38.33, BC represents horizontal ground and AB a vertical flagpole, the **angle of elevation** of the top of the flagpole, A, from the point C is the angle that the

imaginary straight line AC must be raised (or elevated) from the horizontal CB; i.e. angle θ.

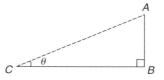

Figure 38.33

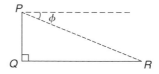

Figure 38.34

If, in Figure 38.34, PQ represents a vertical cliff and R a ship at sea, the **angle of depression** of the ship from point P is the angle through which the imaginary straight line PR must be lowered (or depressed) from the horizontal to the ship; i.e. angle ϕ. (Note, $\angle PRQ$ is also ϕ − **alternate angles** between parallel lines – see page 424.)

> **Problem 32.** An electricity pylon stands on horizontal ground. At a point 80 m from the base of the pylon, the angle of elevation of the top of the pylon is 23°. Calculate the height of the pylon to the nearest metre

Figure 38.35 shows the pylon AB and the angle of elevation of A from point C is 23°.

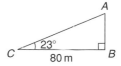

Figure 38.35

$$\tan 23° = \frac{AB}{BC} = \frac{AB}{80}$$

Hence, height of pylon $AB = 80 \tan 23°$
$$= 80(0.4245) = 33.96\,\text{m}$$
$$= \textbf{34\,m to the nearest metre}.$$

> **Problem 33.** A surveyor measures the angle of elevation of the top of a perpendicular building as 19°. He moves 120 m nearer to the building and finds the angle of elevation is now 47°. Determine the height of the building

The building PQ and the angles of elevation are shown in Figure 38.36.

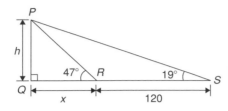

Figure 38.36

In triangle PQS, $\tan 19° = \dfrac{h}{x + 120}$

Hence, $h = \tan 19°(x + 120)$

i.e. $h = 0.3443(x + 120)$ (1)

In triangle PQR, $\tan 47° = \dfrac{h}{x}$

Hence, $h = \tan 47°(x)$

i.e. $h = 1.0724x$ (2)

Equating equations (1) and (2) gives

$$0.3443(x + 120) = 1.0724x$$
$$0.3443x + (0.3443)(120) = 1.0724x$$
$$(0.3443)(120) = (1.0724 - 0.3443)x$$
$$41.316 = 0.7281x$$
$$x = \frac{41.316}{0.7281} = 56.74\,\text{m}$$

From equation (2), **height of building**, $h = 1.0724x = 1.0724(56.74) = \mathbf{60.85\,m}$.

> **Problem 34.** The angle of depression of a ship viewed at a particular instant from the top of a 75 m vertical cliff is 30°. Find the distance of the ship from the base of the cliff at this instant. The ship is sailing away from the cliff at constant speed and 1 minute later its angle of depression from the top of the cliff is 20°. Determine the speed of the ship in km/h

Figure 38.37 shows the cliff AB, the initial position of the ship at C and the final position at D. Since the angle of depression is initially 30°, $\angle ACB = 30°$ (alternate angles between parallel lines).

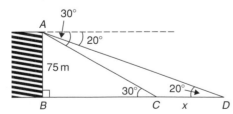

Figure 38.37

$\tan 30° = \dfrac{AB}{BC} = \dfrac{75}{BC}$ hence,

$$BC = \frac{75}{\tan 30°}$$

$$= \mathbf{129.9\,m} = \textbf{initial position}$$
$$\textbf{of ship from base of cliff}$$

In triangle ABD,

$$\tan 20° = \frac{AB}{BD} = \frac{75}{BC + CD} = \frac{75}{129.9 + x}$$

Hence, $129.9 + x = \dfrac{75}{\tan 20°} = 206.06\,\text{m}$

from which $x = 206.06 - 129.9 = 76.16\,\text{m}$

Thus, the ship sails 76.16 m in 1 minute; i.e. 60 s,

Hence, **speed of ship** $= \dfrac{\text{distance}}{\text{time}} = \dfrac{76.16}{60}$ m/s

$$= \frac{76.16 \times 60 \times 60}{60 \times 1000}\,\text{km/h} = \mathbf{4.57\,km/h}$$

Now try the following Practice Exercise

Practice Exercise 168   Angles of elevation and depression (answers on page 1169)

1. A vertical tower stands on level ground. At a point 105 m from the foot of the tower the angle of elevation of the top is 19°. Find the height of the tower.

2. If the angle of elevation of the top of a vertical 30 m high aerial is 32°, how far is it to the aerial?

3. From the top of a vertical cliff 90.0 m high the angle of depression of a boat is 19°50′. Determine the distance of the boat from the cliff.

4. From the top of a vertical cliff 80.0 m high the angles of depression of two buoys lying due west of the cliff are 23° and 15°, respectively. How far apart are the buoys?

5. From a point on horizontal ground a surveyor measures the angle of elevation of the top of a flagpole as 18°40′. He moves 50 m nearer to the flagpole and measures the angle of elevation as 26°22′. Determine the height of the flagpole.

6. A flagpole stands on the edge of the top of a building. At a point 200 m from the building the angles of elevation of the top and bottom of the pole are 32° and 30°, respectively. Calculate the height of the flagpole.

7. From a ship at sea, the angles of elevation of the top and bottom of a vertical lighthouse standing on the edge of a vertical cliff are 31° and 26°, respectively. If the lighthouse is 25.0 m high, calculate the height of the cliff.

8. From a window 4.2 m above horizontal ground the angle of depression of the foot of a building across the road is 24° and the angle of elevation of the top of the building is 34°. Determine, correct to the nearest centimetre, the width of the road and the height of the building.

9. The elevation of a tower from two points, one due west of the tower and the other due east of it, are 20° and 24°, respectively, and the two points of observation are 300 m apart. Find the height of the tower to the nearest metre.

38.9 Trigonometric approximations for small angles

If angle x is a small angle (i.e. less than about 5°) and is expressed in radians, then the following trigonometric approximations may be shown to be true:

(i) $\sin x \approx x$

(ii) $\tan x \approx x$

(iii) $\cos x \approx 1 - \dfrac{x^2}{2}$

For example, let $x = 1°$, i.e. $1 \times \dfrac{\pi}{180} = 0.01745$ radians, correct to 5 decimal places. By calculator, $\sin 1° = 0.01745$ and $\tan 1° = 0.01746$, showing that: $\sin x = x \approx \tan x$ when $x = 0.01745$ radians. Also, $\cos 1° = 0.99985$; when $x = 1°$, i.e. 0.01745 radians,

$$1 - \frac{x^2}{2} = 1 - \frac{0.01745^2}{2} = 0.99985,$$

correct to 5 decimal places, showing that

$$\cos x = 1 - \frac{x^2}{2} \text{ when } x = 0.01745 \text{ radians.}$$

Similarly, let $x = 5°$, i.e. $5 \times \dfrac{\pi}{180} = 0.08727$ radians, correct to 5 decimal places.

By calculator, $\sin 5° = 0.08716$, thus $\sin x \approx x$,

                   $\tan 5° = 0.08749$, thus $\tan x \approx x$,

and $\cos 5° = 0.99619$;

since $x = 0.08727$ radians,

$$1 - \frac{x^2}{2} = 1 - \frac{0.08727^2}{2} = 0.99619 \text{ showing that:}$$

$$\cos x = 1 - \frac{x^2}{2} \text{ when } x = 0.0827 \text{ radians.}$$

If $\sin x \approx x$ for small angles, then $\dfrac{\sin x}{x} \approx 1$, and this relationship can occur in engineering considerations.

For fully worked solutions to each of the problems in Practice Exercises 162 to 168 in this chapter, go to the website:
www.routledge.com/cw/bird

Revision Test 15   Angles, triangles and trigonometry

This assignment covers the material contained in Chapters 37 and 38. *The marks available are shown in brackets at the end of each question.*

1. Determine $38°48' + 17°23'$. (2)

2. Determine $47°43'12'' - 58°35'53'' + 26°17'29''$. (3)

3. Change $42.683°$ to degrees and minutes. (2)

4. Convert $77°42'34''$ to degrees, correct to 3 decimal places. (3)

5. Determine angle θ in Figure RT15.1. (3)

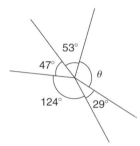

Figure RT15.1

6. Determine angle θ in the triangle in Figure RT15.2. (2)

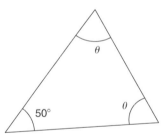

Figure RT15.2

7. Determine angle θ in the triangle in Figure RT15.3. (2)

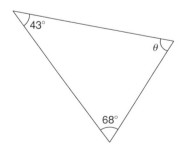

Figure RT15.3

8. In Figure RT15.4, if triangle ABC is equilateral, determine $\angle CDE$. (3)

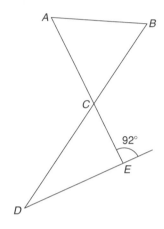

Figure RT15.4

9. Find angle J in Figure RT15.5. (2)

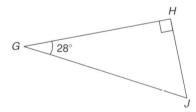

Figure RT15.5

10. Determine angle θ in Figure RT15.6. (3)

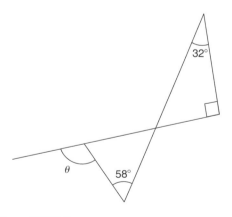

Figure RT15.6

11. State the angle (a) supplementary to $49°$ (b) complementary to $49°$. (2)

12. In Figure RT15.7, determine angles x, y and z. (3)

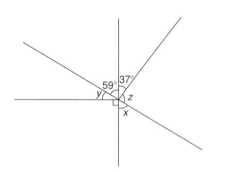

Figure RT15.7

13. In Figure RT15.8, determine angles a to e. (5)

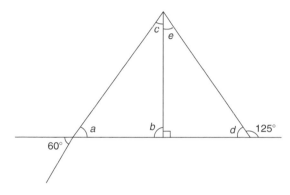

Figure RT15.8

14. In Figure RT15.9, determine the length of AC. (4)

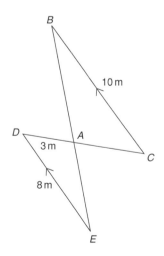

Figure RT15.9

15. In triangle JKL in Figure RT15.10, find
    (a) the length KJ correct to 3 significant figures.
    (b) $\sin L$ and $\tan K$, each correct to 3 decimal places. (4)

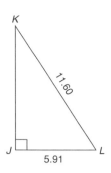

Figure RT15.10

16. Two ships leave a port at the same time. Ship X travels due west at $30\,$km/h and ship Y travels due north. After 4 hours the two ships are $130\,$km apart. Calculate the velocity of ship Y. (4)

17. If $\sin A = \dfrac{12}{37}$, find $\tan A$ in fraction form. (3)

18. Evaluate $5\tan 62°11'$, correct to 3 significant figures. (2)

19. Determine the acute angle $\cos^{-1} 0.3649$ in degrees and minutes. (2)

20. In triangle PQR in Figure RT15.11, find angle P in decimal form, correct to 2 decimal places. (2)

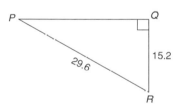

Figure RT15.11

21. Evaluate, correct to 3 significant figures, $3\tan 81.27° - 5\cos 7.32° - 6\sin 54.81°$. (2)

22. In triangle ABC in Figure RT15.12, find lengths AB and AC, correct to 2 decimal places. (4)

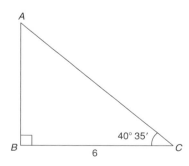

Figure RT15.12

Section E

23. From a point P, the angle of elevation of a 40 m high electricity pylon is $20°$. How far is point P from the base of the pylon, correct to the nearest metre? (3)

24. Figure RT15.13 shows a plan view of a kite design. Calculate the lengths of the dimensions shown as a and b. (4)

25. In Figure RT15.13, evaluate (a) angle θ (b) angle α. (5)

26. Determine the area of the plan view of a kite shown in Figure RT15.13 (4)

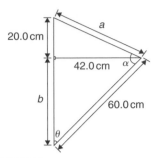

Figure RT15.13

27. Solve the following equations in the range $0°$ to $360°$ (a) $\sin^{-1}(-0.4161)=x$
(b) $\cot^{-1}(2.4198)=\theta$ (7)

For lecturers/instructors/teachers, fully worked solutions to each of the problems in Revision Test 15, together with a full marking scheme, are available at the website:
www.routledge.com/cw/bird

Chapter 39

Trigonometric waveforms

Why it is important to understand: Trigonometric waveforms

Trigonometric graphs are commonly used in all areas of science and engineering for modelling many different natural and mechanical phenomena such as waves, engines, acoustics, electronics, populations, UV intensity, growth of plants and animals, and so on. Periodic trigonometric graphs mean that the shape repeats itself exactly after a certain amount of time. Anything that has a regular cycle, like the tides, temperatures, rotation of the Earth, and so on, can be modelled using a sine or cosine curve. The most common periodic signal waveform that is used in electrical and electronic engineering is the sinusoidal waveform. However, an alternating a.c. waveform may not always take the shape of a smooth shape based around the sine and cosine function; a.c. waveforms can also take the shape of square or triangular waves, i.e. complex waves. In engineering, it is therefore important to have some clear understanding of sine and cosine waveforms.

At the end of this chapter, you should be able to:

- sketch sine, cosine and tangent waveforms
- determine angles of any magnitude
- understand cycle, amplitude, period, periodic time, frequency, lagging/leading angles with reference to sine and cosine waves
- perform calculations involving sinusoidal form $A \sin(\omega t \pm \alpha)$
- define a complex wave and harmonic analysis
- use harmonic synthesis to construct a complex waveform

39.1 Graphs of trigonometric functions

By drawing up tables of values from $0°$ to $360°$, graphs of $y = \sin A$, $y = \cos A$ and $y = \tan A$ may be plotted. Values obtained with a calculator (correct to 3 decimal places – which is more than sufficient for plotting graphs), using $30°$ intervals, are shown below, with the respective graphs shown in Figure 39.1.

(a) $y = \sin A$

A	0	30°	60°	90°	120°	150°	180°
$\sin A$	0	0.500	0.866	1.000	0.866	0.500	0

A	210°	240°	270°	300°	330°	360°
$\sin A$	−0.500	−0.866	−1.000	−0.866	−0.500	0

(b) $y = \cos A$

A	0	30°	60°	90°	120°	150°	180°
cos A	1.000	0.866	0.500	0	−0.500	−0.866	−1.000

A	210°	240°	270°	300°	330°	360°
cos A	−0.866	−0.500	0	0.500	0.866	1.000

(c) $y = \tan A$

A	0	30°	60°	90°	120°	150°	180°
tan A	0	0.577	1.732	∞	−1.732	−0.577	0

A	210°	240°	270°	300°	330°	360°
tan A	0.577	1.732	∞	−1.732	−0.577	0

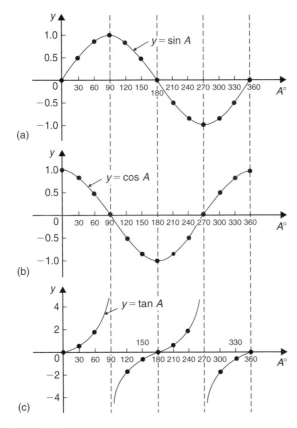

Figure 39.1

From Figure 39.1 it is seen that

(a) Sine and cosine graphs oscillate between peak values of ± 1

(b) The cosine curve is the same shape as the sine curve but displaced by 90°

(c) The sine and cosine curves are continuous and they repeat at intervals of 360° and the tangent curve appears to be discontinuous and repeats at intervals of 180°

39.2 Angles of any magnitude

Figure 39.2 shows rectangular axes XX' and YY' intersecting at origin 0. As with graphical work, measurements made to the right and above 0 are positive, while those to the left and downwards are negative.

Let $0A$ be free to rotate about 0. By convention, when $0A$ moves anticlockwise angular measurement is considered positive, and vice versa.

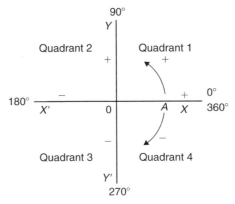

Figure 39.2

Let $0A$ be rotated anticlockwise so that θ_1 is any angle in the first quadrant and let perpendicular AB be constructed to form the right-angled triangle $0AB$ in Figure 39.3. Since all three sides of the triangle are positive, the trigonometric ratios sine, cosine and tangent will all be positive in the first quadrant. (Note: $0A$ is always positive since it is the radius of a circle.)

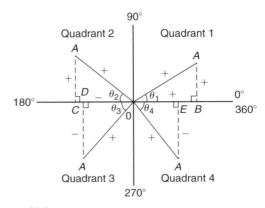

Figure 39.3

Let 0A be further rotated so that θ_2 is any angle in the second quadrant and let AC be constructed to form the right-angled triangle $0AC$. Then,

$$\sin\theta_2 = \frac{+}{+} = + \qquad \cos\theta_2 = \frac{-}{+} = -$$

$$\tan\theta_2 = \frac{+}{-} = -$$

Let 0A be further rotated so that θ_3 is any angle in the third quadrant and let AD be constructed to form the right-angled triangle $0AD$. Then,

$$\sin\theta_3 = \frac{-}{+} = - \qquad \cos\theta_3 = \frac{-}{+} = -$$

$$\tan\theta_3 = \frac{-}{-} = +$$

Let 0A be further rotated so that θ_4 is any angle in the fourth quadrant and let AE be constructed to form the right-angled triangle $0AE$. Then,

$$\sin\theta_4 = \frac{-}{+} = - \qquad \cos\theta_4 = \frac{+}{+} = +$$

$$\tan\theta_4 = \frac{-}{+} = -$$

The above results are summarised in Figure 39.4, in which all three trigonometric ratios are positive in the first quadrant, only sine is positive in the second quadrant, only tangent is positive in the third quadrant and only cosine is positive in the fourth quadrant.

The underlined letters in Figure 39.4 spell the word CAST when starting in the fourth quadrant and moving in an anticlockwise direction.

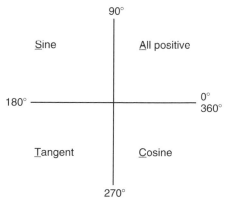

Figure 39.4

It is seen that, in the first quadrant of Figure 39.1, all of the curves have positive values; in the second only sine is positive; in the third only tangent is positive; and in the fourth only cosine is positive – exactly as summarised in Figure 39.4.

A knowledge of angles of any magnitude is needed when finding, for example, all the angles between $0°$ and $360°$ whose sine is, say, 0.3261. If 0.3261 is entered into a calculator and then the inverse sine key pressed (or $\sin^{-1}$ key), the answer $19.03°$ appears. However, there is a second angle between $0°$ and $360°$ which the calculator does not give. Sine is also positive in the second quadrant (either from CAST or from Figure 39.1(a)). The other angle is shown in Figure 39.5 as angle θ, where $\theta = 180° - 19.03° = 160.97°$. Thus, $19.03°$ **and** $160.97°$ are the angles between $0°$ and $360°$ whose sine is 0.3261 (check that $\sin 160.97° = 0.3261$ on your calculator).

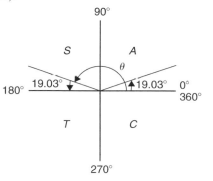

Figure 39.5

Be careful! Your calculator only gives you one of these answers. The second answer needs to be deduced from a knowledge of angles of any magnitude, as shown in the following worked problems.

Problem 1. Determine all of the angles between $0°$ and $360°$ whose sine is -0.4638

The angles whose sine is -0.4638 occur in the third and fourth quadrants since sine is negative in these quadrants – see Figure 39.6.

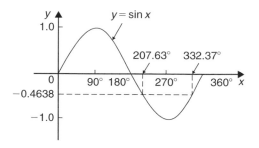

Figure 39.6

From Figure 39.7, $\theta = \sin^{-1} 0.4638 = 27.63°$. Measured from $0°$, the two angles between $0°$ and $360°$

whose sine is -0.4638 are $180° + 27.63°$ i.e. **207.63°** and $360° - 27.63°$, i.e. **332.37°**. (Note that if a calculator is used to determine $\sin^{-1}(-0.4638)$ it only gives one answer: $-27.632588°$)

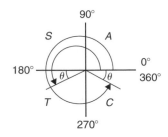

Figure 39.7

Problem 2. Determine all of the angles between $0°$ and $360°$ whose tangent is 1.7629

A tangent is positive in the first and third quadrants – see Figure 39.8.

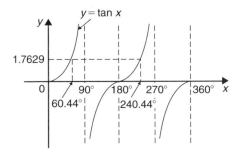

Figure 39.8

From Figure 39.9, $\theta = \tan^{-1} 1.7629 = 60.44°$. Measured from $0°$, the two angles between $0°$ and $360°$ whose tangent is 1.7629 are **60.44°** and $180° + 60.44°$, i.e. **240.44°**

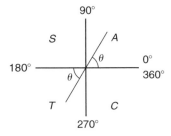

Figure 39.9

Problem 3. Solve the equation $\cos^{-1}(-0.2348) = \alpha$ for angles of α between $0°$ and $360°$

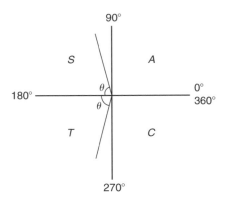

Figure 39.10

Cosine is positive in the first and fourth quadrants and thus negative in the second and third quadrants – see Figure 39.10 or from Figure 39.1(b).

In Figure 39.10, angle $\theta = \cos^{-1}(0.2348) = 76.42°$. Measured from $0°$, the two angles whose cosine is -0.2348 are $\alpha = 180° - 76.42°$, i.e. **103.58°** and $\alpha = 180° + 76.42°$, i.e. **256.42°**

Now try the following Practice Exercise

Practice Exercise 169   Angles of any magnitude (answers on page 1170)

1. Determine all of the angles between $0°$ and $360°$ whose sine is
   (a) 0.6792   (b) -0.1483

2. Solve the following equations for values of x between $0°$ and $360°$

   (a) $x = \cos^{-1} 0.8739$

   (b) $x = \cos^{-1}(-0.5572)$

3. Find the angles between $0°$ to $360°$ whose tangent is
   (a) 0.9728   (b) -2.3420

In Problems 4 to 6, solve the given equations in the range $0°$ to $360°$, giving the answers in degrees and minutes.

4. $\cos^{-1}(-0.5316) = t$

5. $\sin^{-1}(-0.6250) = \alpha$

6. $\tan^{-1} 0.8314 = \theta$

39.3 The production of sine and cosine waves

In Figure 39.11, let OR be a vector 1 unit long and free to rotate anticlockwise about 0. In one revolution a circle is produced and is shown with $15°$ sectors. Each radius arm has a vertical and a horizontal component. For example, at $30°$, the vertical component is TS and the horizontal component is OS.

From triangle OST,

$$\sin 30° = \frac{TS}{TO} = \frac{TS}{1} \quad \text{i.e.} \quad TS = \sin 30°$$

and $\quad \cos 30° = \dfrac{OS}{TO} = \dfrac{OS}{1} \quad \text{i.e.} \quad OS = \cos 30°$

Sine waves

The vertical component TS may be projected across to $T'S'$, which is the corresponding value of $30°$ on the graph of y against angle $x°$. If all such vertical components as TS are projected onto the graph, a **sine wave** is produced as shown in Figure 39.11.

Actually, the easiest method of drawing a sine wave of $y = \sin x$ is to:

(1) use a calculator to produce a table of values for y from $x = 0°$ to $x = 360°$ at $30°$ intervals, as shown below,

$x(°)$	0	30	60	90	120	150	180
$y = \sin x$	0	0.50	0.87	1.00	0.87	0.50	0

$x(°)$	210	240	270	300	330	360
$y = \sin x$	−0.50	−0.87	−1.00	−0.87	−0.50	0

and (2) to plot the co-ordinates $(0,0)$, $(30°, 0.50)$, $(60°, 0.87)$, and so on.

The graph of $y = \sin x$ should look exactly like Figure 39.11.

Cosine waves

If all horizontal components such as OS are projected onto a graph of y against angle $x°$, a **cosine wave** is produced. It is easier to visualise these projections by redrawing the circle with the radius arm OR initially in a vertical position as shown in Figure 39.12.

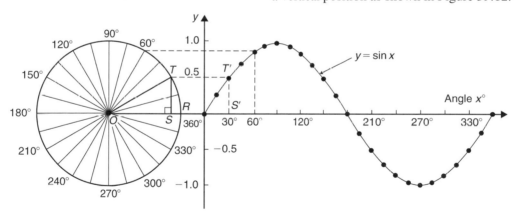

Figure 39.11

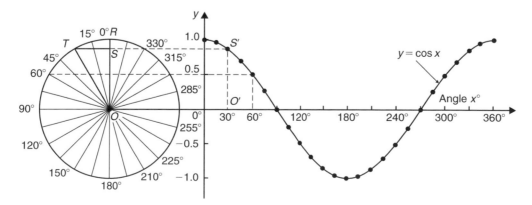

Figure 39.12

As with the sine wave above, the easiest method of drawing a cosine wave of $y = \cos x$ is to:

(1) use a calculator to produce a table of values for y from $x = 0°$ to $x = 360°$ at $30°$ intervals, as shown below,

$x(°)$	0	30	60	90	120	150	180
$y = \cos x$	1.00	0.87	0.50	0	−0.50	−0.87	−1.00

$x(°)$	210	240	270	300	330	360
$y = \cos x$	−0.87	−0.50	0	0.50	0.87	1.00

and (2) to plot the co-ordinates $(0, 1.00)$, $(30°, 0.87)$, $(60°, 0.50)$, and so on.

The graph of $y = \cos x$ should look exactly like Figure 39.12.

It is seen from Figures 39.11 and 39.12 that a cosine curve is of the same form as the sine curve but is displaced by $90°$ (or $\pi/2$ radians). Both sine and cosine waves repeat every $360°$

39.4 Terminology involved with sine and cosine waves

Sine waves are extremely important in engineering, with examples occurring with alternating currents and voltages – the mains supply is a sine wave – and with simple harmonic motion.

Cycle

When a sine wave has passed through a complete series of values, both positive and negative, it is said to have completed one **cycle**. One cycle of a sine wave is shown in Figure 39.1(a) on page 460 and in Figure 39.11.

Amplitude

The amplitude is the maximum value reached in a half cycle by a sine wave. Another name for **amplitude** is **peak value** or **maximum value**.

A sine wave $y = 5 \sin x$ has an amplitude of 5, a sine wave $v = 200 \sin 314t$ has an amplitude of 200 and the sine wave $y = \sin x$ shown in Figure 39.11 has an amplitude of 1

Period

The waveforms $y = \sin x$ and $y = \cos x$ repeat themselves every $360°$. Thus, for each, the **period** is $360°$. A waveform of $y = \tan x$ has a period of $180°$ (from Figure 39.1(c)).

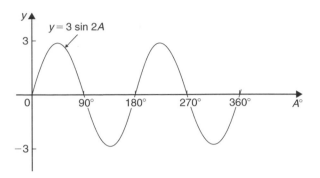

Figure 39.13

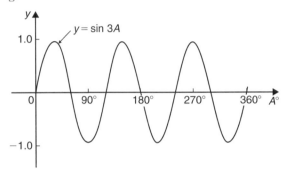

Figure 39.14

A graph of $y = 3 \sin 2A$, as shown in Figure 39.13, has an **amplitude of 3** and **period 180°**

A graph of $y = \sin 3A$, as shown in Figure 39.14, has an **amplitude of 1** and **period of 120°**

A graph of $y = 4 \cos 2x$, as shown in Figure 39.15, has an **amplitude of 4** and a **period of 180°**.

In general, **if $y = A \sin px$ or $y = A \cos px$,**

amplitude $= A$ and period $= \dfrac{360°}{p}$

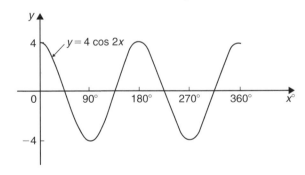

Figure 39.15

Problem 4. Sketch $y = 2 \sin \dfrac{3}{5} A$ over one cycle

Amplitude $= 2$; period $= \dfrac{360°}{\frac{3}{5}} = \dfrac{360° \times 5}{3} = 600°$

A sketch of $y = 2\sin\dfrac{3}{5}A$ is shown in Figure 39.16.

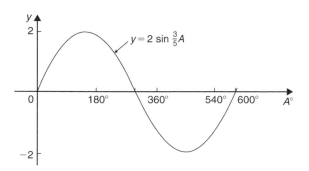

Figure 39.16

Periodic time

In practice, the horizontal axis of a sine wave will be time. The time taken for a sine wave to complete one cycle is called the **periodic time, T**.

In the sine wave of voltage v (volts) against time t (milliseconds) shown in Figure 39.17, the amplitude is 10 V and the periodic time is 20 ms; i.e. $T = 20\,\text{ms}$.

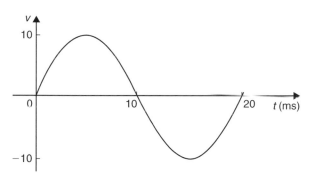

Figure 39.17

Frequency

The number of cycles completed in one second is called the **frequency, f** and is measured in **hertz, Hz**.

$$f = \frac{1}{T} \text{ or } T = \frac{1}{f}$$

Problem 5. Determine the frequency of the sine wave shown in Figure 39.17

In the sine wave shown in Figure 39.17, $T = 20\,\text{ms}$, hence

$$\textbf{frequency, } f = \frac{1}{T} = \frac{1}{20 \times 10^{-3}} = \textbf{50\,Hz}$$

Problem 6. If a waveform has a frequency of 200 kHz, determine the periodic time

If a waveform has a frequency of 200 kHz, the periodic time, T, is given by

$$\textbf{periodic time, } T = \frac{1}{f} = \frac{1}{200 \times 10^3}$$
$$= 5 \times 10^{-6}\text{s} = \textbf{5\,}\boldsymbol{\mu}\textbf{s}$$

Lagging and leading angles

A sine or cosine curve may not always start at $0°$. To show this, a periodic function is represented by $y = A\sin(x \pm \alpha)$ where α is a phase displacement compared with $y = A\sin x$. For example, $y = \sin A$ is shown by the broken line in Figure 39.18 and, on the same axes, $y = \sin(A - 60°)$ is shown. **The graph $y = \sin(A - 60°)$ is said to lag $y = \sin A$ by $60°$.**

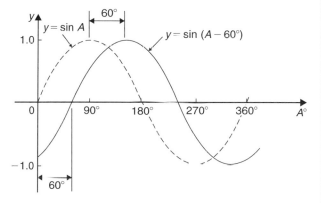

Figure 39.18

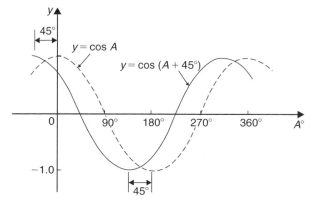

Figure 39.19

In another example, $y = \cos A$ is shown by the broken line in Figure 39.19 and, on the same axes, $y = \cos(A + 45°)$ is shown. **The graph $y = \cos(A + 45°)$ is said to lead $y = \cos A$ by $45°$.**

Problem 7. Sketch $y = 5\sin(A + 30°)$ from $A = 0°$ to $A = 360°$

Amplitude = 5 and **period** $= 360°/1 = $ **360°**

$5\sin(A + 30°)$ **leads** $5\sin A$ by $30°$ (i.e. starts $30°$ earlier).
A sketch of $y = 5\sin(A + 30°)$ is shown in Figure 39.20.

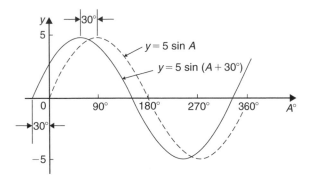

Figure 39.20

Problem 8. Sketch $y = 7\sin(2A - \pi/3)$ in the range $0 \le A \le 360°$

Amplitude = 7 and **period** $= 2\pi/2 = $ **π radians**

In general, $y = \sin(pt - \alpha)$ **lags** $y = \sin pt$ by α/p, hence $7\sin(2A - \pi/3)$ lags $7\sin 2A$ by $(\pi/3)/2$, i.e.

$$\pi/6 \text{ rad or } 30°$$

A sketch of $y = 7\sin(2A - \pi/3)$ is shown in Figure 39.21.

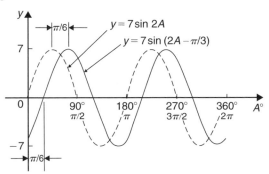

Figure 39.21

Problem 9. Sketch $y = 2\cos(\omega t - 3\pi/10)$ over one cycle

Amplitude = 2 and **period** $= 2\pi/\omega$ **rad**

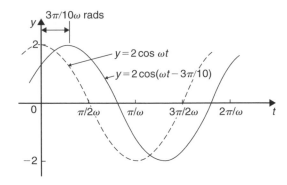

Figure 39.22

$2\cos(\omega t - 3\pi/10)$ **lags** $2\cos\omega t$ by $3\pi/10\omega$ seconds. A sketch of $y = 2\cos(\omega t - 3\pi/10)$ is shown in Figure 39.22.

Now try the following Practice Exercise

Practice Exercise 170   Trigonometric waveforms (answers on page 1170)

1. A sine wave is given by $y = 5\sin 3x$. State its peak value.

2. A sine wave is given by $y = 4\sin 2x$. State its period in degrees.

3. A periodic function is given by $y = 30\cos 5x$. State its maximum value.

4. A periodic function is given by $y = 25\cos 3x$. State its period in degrees.

In Problems 5 to 11, state the amplitude and period of the waveform and sketch the curve between $0°$ and $360°$.

5. $y = \cos 3A$ 6. $y = 2\sin\dfrac{5x}{2}$

7. $y = 3\sin 4t$ 8. $y = 5\cos\dfrac{\theta}{2}$

9. $y = \dfrac{7}{2}\sin\dfrac{3x}{8}$ 10. $y = 6\sin(t - 45°)$

11. $y = 4\cos(2\theta + 30°)$

12. The frequency of a sine wave is $200\,\text{Hz}$. Calculate the periodic time.

13. Calculate the frequency of a sine wave that has a periodic time of $25\,\text{ms}$.

14. Calculate the periodic time for a sine wave having a frequency of $10\,\text{kHz}$.

15. An alternating current completes 15 cycles in $24\,\text{ms}$. Determine its frequency.

16. Graphs of $y_1 = 2\sin x$ and $y_2 = 3\sin(x + 50°)$ are drawn on the same axes. Is y_2 lagging or leading y_1?

17. Graphs of $y_1 = 6\sin x$ and $y_2 = 5\sin(x - 70°)$ are drawn on the same axes. Is y_1 lagging or leading y_2?

39.5   Sinusoidal form: $A\sin(\omega t \pm \alpha)$

If a sine wave is expressed in the form $y = A\sin(\omega t \pm \alpha)$ then

(a) A = amplitude

(b) ω = angular velocity = $2\pi f$ rad/s

(c) frequency, $f = \dfrac{\omega}{2\pi}$ hertz

(d) periodic time, $T = \dfrac{2\pi}{\omega}$ seconds $\left(\text{i.e. } T = \dfrac{1}{f}\right)$

(e) α = angle of lead or lag (compared with $y = A\sin\omega t$)

Here are some worked problems involving the sinusoidal form $A\sin(\omega t \pm \alpha)$

Problem 10. An alternating current is given by $i = 30\sin(100\pi t + 0.35)$ amperes. Find the (a) amplitude, (b) frequency, (c) periodic time and (d) phase angle (in degrees and minutes)

(a) $i = 30\sin(100\pi t + 0.35)A$; hence, **amplitude = 30 A**

(b) Angular velocity, $\omega = 100\pi$ rad/s, hence
$$\textbf{frequency,}\, f = \frac{\omega}{2\pi} = \frac{100\pi}{2\pi} = \textbf{50 Hz}$$

(c) **Periodic time,** $T = \dfrac{1}{f} = \dfrac{1}{50} = \textbf{0.02 s}$ or **20 ms**

(d) 0.35 is the angle in **radians**. The relationship between radians and degrees is
$$360° = 2\pi \text{ radians or } \textbf{180°} = \pi\,\textbf{radians}$$
from which,
$$\textbf{1°} = \frac{\pi}{180}\textbf{rad and 1 rad} = \frac{180°}{\pi}\ (\approx 57.30°)$$
Hence, **phase angle,** $\alpha = 0.35$ rad
$$= \left(0.35 \times \frac{180}{\pi}\right)^{\circ} = \textbf{20.05° or 20°3' leading}$$
$$i = 30\sin(100\pi t)$$

Problem 11. An oscillating mechanism has a maximum displacement of 2.5 m and a frequency of 60 Hz. At time $t = 0$ the displacement is 90 cm. Express the displacement in the general form $A\sin(\omega t \pm \alpha)$

Amplitude = maximum displacement = 2.5 m

Angular velocity, $\omega = 2\pi f = 2\pi(60) = 120\pi$ rad/s

Hence,   **displacement** $= 2.5\sin(120\pi t + \alpha)$ **m**

When $t = 0$, displacement = 90 cm = 0.90 m

hence, $0.90 = 2.5\sin(0 + \alpha)$

i.e. $\sin\alpha = \dfrac{0.90}{2.5} = 0.36$

Hence, $\alpha = \sin^{-1}0.36 = 21.10°$
$$= 21°6' = 0.368\,\text{rad}$$

Thus, **displacement** $= 2.5\sin(120\pi t + 0.368)$ **m**

Problem 12. The instantaneous value of voltage in an a.c. circuit at any time t seconds is given by $v = 340\sin(50\pi t - 0.541)$ volts. Determine the (a) amplitude, frequency, periodic time and phase angle (in degrees), (b) value of the voltage when $t = 0$, (c) value of the voltage when $t = 10$ ms, (d) time when the voltage first reaches 200 V and (e) time when the voltage is a maximum. Also, (f) sketch one cycle of the waveform

(a)   **Amplitude = 340 V**

    Angular velocity, $\omega = 50\pi$

    Frequency, $f = \dfrac{\omega}{2\pi} = \dfrac{50\pi}{2\pi} = \textbf{25 Hz}$

    Periodic time, $T = \dfrac{1}{f} = \dfrac{1}{25} = \textbf{0.04 s}$ or **40 ms**

    Phase angle $= 0.541$ rad $= \left(0.541 \times \dfrac{180}{\pi}\right)^{\circ}$
$$= \textbf{31° lagging } v = 340\sin(50\pi t)$$

(b)   **When** $t = 0$,
$$v = 340\sin(0 - 0.541)$$
$$= 340\sin(-31°) = \textbf{−175.1 V}$$

(c)   **When** $t = 10$ **ms,**
$$v = 340\sin(50\pi \times 10 \times 10^{-3} - 0.541)$$
$$= 340\sin(1.0298)$$
$$= 340\sin 59° = \textbf{291.4 volts}$$

(d) When $v = 200$ volts,

$$200 = 340 \sin(50\pi t - 0.541)$$

$$\frac{200}{340} = \sin(50\pi t - 0.541)$$

Hence, $(50\pi t - 0.541) = \sin^{-1} \dfrac{200}{340}$

$$= 36.03° \text{ or } 0.628875 \text{ rad}$$

$$50\pi t = 0.628875 + 0.541$$

$$= 1.169875$$

Hence, when $v = 200$ V,

time, $t = \dfrac{1.169875}{50\pi} = \textbf{7.448 ms}$

(e) When the voltage is a maximum, $v = 340$ V.

Hence $340 = 340 \sin(50\pi t - 0.541)$

$$1 = \sin(50\pi t - 0.541)$$

$$50\pi t - 0.541 = \sin^{-1} 1 = 90° \text{ or } 1.5708 \text{ rad}$$

$$50\pi t = 1.5708 + 0.541 = 2.1118$$

Hence, **time,** $t = \dfrac{2.1118}{50\pi} = \textbf{13.44 ms}$

(f) A sketch of $v = 340 \sin(50\pi t - 0.541)$ volts is shown in Figure 39.23.

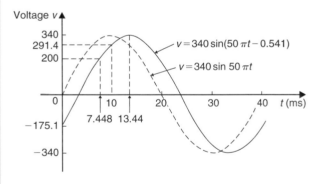

Figure 39.23

Now try the following Practice Exercise

Practice Exercise 171   Sinusoidal form
$A \sin(\omega t \pm \alpha)$ **(answers on page 1170)**

In Problems 1 to 3 find the (a) amplitude, (b) frequency, (c) periodic time and (d) phase angle (stating whether it is leading or lagging $\sin \omega t$) of the alternating quantities given.

1. $i = 40 \sin(50\pi t + 0.29)$ mA

2. $y = 75 \sin(40t - 0.54)$ cm

3. $v = 300 \sin(200\pi t - 0.412)$ V

4. A sinusoidal voltage has a maximum value of 120 V and a frequency of 50 Hz. At time $t = 0$, the voltage is (a) zero and (b) 50 V. Express the instantaneous voltage v in the form $v = A \sin(\omega t \pm \alpha)$.

5. An alternating current has a periodic time of 25 ms and a maximum value of 20 A. When time $= 0$, current $i = -10$ amperes. Express the current i in the form $i = A \sin(\omega t \pm \alpha)$.

6. An oscillating mechanism has a maximum displacement of 3.2 m and a frequency of 50 Hz. At time $t = 0$ the displacement is 150 cm. Express the displacement in the general form $A \sin(\omega t \pm \alpha)$.

7. The current in an a.c. circuit at any time t seconds is given by

$$i = 5 \sin(100\pi t - 0.432) \text{ amperes}$$

Determine the
(a) amplitude, frequency, periodic time and phase angle (in degrees),

(b) value of current at $t = 0$,

(c) value of current at $t = 8$ ms,

(d) time when the current is first a maximum,

(e) time when the current first reaches 3A.

Also,

(f) sketch one cycle of the waveform showing relevant points.

39.6   Complex waveforms

A waveform that is not sinusoidal is called a **complex wave. Harmonic analysis** is the process of resolving a complex periodic waveform into a series of sinusoidal components of ascending order of frequency. Many of the waveforms met in practice can be represented by the following mathematical expression.

$$v = V_{1m} \sin(\omega t + \alpha_1) + V_{2m} \sin(2\omega t + \alpha_2)$$

$$+ \cdots + V_{nm} \sin(n\omega t + \alpha_n)$$

and the magnitude of their harmonic components together with their phase may be calculated using **Fourier series** (see Chapters 101 to 104). **Numerical**

methods are used to analyse waveforms for which simple mathematical expressions cannot be obtained. A numerical method of harmonic analysis is explained in Chapter 105 on page 1109. In a laboratory, waveform analysis may be performed using a **waveform analyser** which produces a direct readout of the component waves present in a complex wave.

By adding the instantaneous values of the fundamental and progressive harmonics of a complex wave for given instants in time, the shape of a complex waveform can be gradually built up. This graphical procedure is known as **harmonic synthesis** (synthesis meaning 'the putting together of parts or elements so as to make up a complex whole'). For more on the addition of alternating waveforms, see Chapter 50.

Some examples of harmonic synthesis are considered in the following worked problems.

> **Problem 13.** Use harmonic synthesis to construct the complex voltage given by:
>
> $$v_1 = 100 \sin \omega t + 30 \sin 3\omega t \text{ volts}$$

The waveform is made up of a fundamental wave of maximum value 100 V and frequency, $f = \omega/2\pi$ hertz and a third harmonic component of maximum value 30 V and frequency $= 3\omega/2\pi (=3f)$, the fundamental and third harmonics being initially in phase with each other.

In Figure 39.24, the fundamental waveform is shown by the broken line plotted over one cycle, the periodic time T being $2\pi/\omega$ seconds. On the same axis is plotted $30 \sin 3\omega t$, shown by the dotted line, having a maximum value of 30 V and for which three cycles are completed in time T seconds. At zero time, $30 \sin 3\omega t$ is in phase with $100 \sin \omega t$.

The fundamental and third harmonics are combined by adding ordinates at intervals to produce the waveform for v_1, as shown. For example, at time $T/12$ seconds, the fundamental has a value of 50 V and the third harmonic a value of 30 V. Adding gives a value of 80 V for waveform v_1 at time $T/12$ seconds. Similarly, at time $T/4$ seconds, the fundamental has a value of 100 V and the third harmonic a value of -30 V. After addition, the resultant waveform v_1 is 70 V at $T/4$. The procedure is continued between $t=0$ and $t=T$ to produce the complex waveform for v_1. The negative half-cycle of waveform v_1 is seen to be identical in shape to the positive half-cycle.

If further odd harmonics of the appropriate amplitude and phase were added to v_1 a good approximation to a **square wave** would result.

> **Problem 14.** Construct the complex voltage given by:
>
> $$v_2 = 100 \sin \omega t + 30 \sin\left(3\omega t + \frac{\pi}{2}\right) \text{ volts}$$

The peak value of the fundamental is 100 volts and the peak value of the third harmonic is 30 V. However, the third harmonic has a phase displacement of $\dfrac{\pi}{2}$ radian

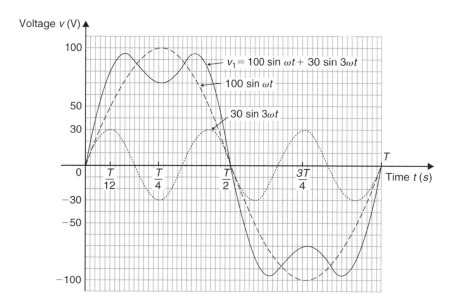

Figure 39.24

leading (i.e. leading $30\sin 3\omega t$ by $\dfrac{\pi}{2}$ radian). Note that, since the periodic time of the fundamental is T seconds, the periodic time of the third harmonic is $T/3$ seconds, and a phase displacement of $\dfrac{\pi}{2}$ radian or $\dfrac{1}{4}$ cycle of the third harmonic represents a time interval of $(T/3) \div 4$, i.e. $T/12$ seconds.

Figure 39.25 shows graphs of $100\sin\omega t$ and $30\sin\left(3\omega t + \dfrac{\pi}{2}\right)$ over the time for one cycle of the fundamental. When ordinates of the two graphs are added at intervals, the resultant waveform v_2 is as shown. If the negative half-cycle in Figure 39.25 is reversed it can be seen that the shape of the positive and negative half-cycles are identical.

Problems 13 and 14 demonstrate that whenever odd harmonics are added to a fundamental waveform, whether initially in phase with each other or not, the positive and negative half-cycles of the resultant complex wave are identical in shape. This is a feature of waveforms containing the fundamental and odd harmonics.

> **Problem 15.** Use harmonic synthesis to construct the complex current given by:
>
> $$i_1 = 10\sin\omega t + 4\sin 2\omega t \text{ amperes}$$

Current i_1 consists of a fundamental component, $10\sin\omega t$, and a second harmonic component, $4\sin 2\omega t$, the components being initially in phase with each other.

The fundamental and second harmonic are shown plotted separately in Figure 39.26. By adding ordinates at intervals, the complex waveform representing i_1 is produced as shown. It is noted that if all the values in the negative half-cycle were reversed then this half-cycle would appear as a mirror image of the positive half-cycle about a vertical line drawn through time, $t = T/2$.

> **Problem 16.** Construct the complex current given by:
>
> $$i_2 = 10\sin\omega t + 4\sin\left(2\omega t + \dfrac{\pi}{2}\right) \text{ amperes}$$

The fundamental component, $10\sin\omega t$, and the second harmonic component, having an amplitude of $4\,$A and a phase displacement of $\dfrac{\pi}{2}$ radian leading (i.e. leading $4\sin 2\omega t$ by $\dfrac{\pi}{2}$ radian or $T/8$ seconds), are shown plotted separately in Figure 39.27. By adding ordinates at intervals, the complex waveform for i_2 is produced as shown. The positive and negative half-cycles of the resultant waveform are seen to be quite dissimilar.

From Problems 15 and 16 it is seen that whenever even harmonics are added to a fundamental component:

(a) if the harmonics are initially in phase, the negative half-cycle, when reversed, is a mirror image of the positive half-cycle about a vertical line drawn through time, $t = T/2$.

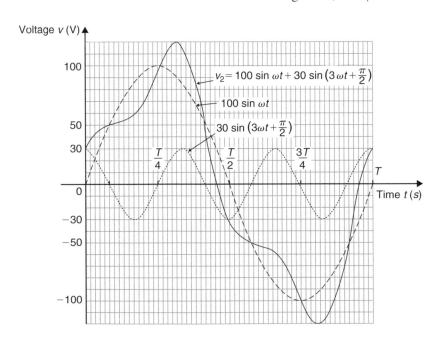

Figure 39.25

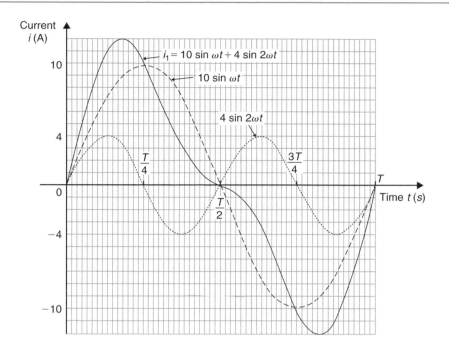

Figure 39.26

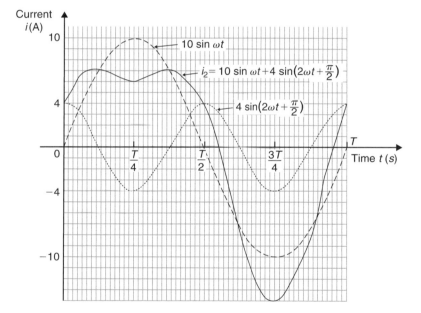

Figure 39.27

(b)   if the harmonics are initially out of phase with each other, the positive and negative half-cycles are dissimilar.

These are features of waveforms containing the fundamental and even harmonics.

Problem 17.   Use harmonic synthesis to construct the complex current expression given by:

$$i = 32 + 50\sin\omega t + 20\sin\left(2\omega t - \frac{\pi}{2}\right) \text{ mA}$$

The current i comprises three components – a 32 mA d.c. component, a fundamental of amplitude 50 mA and a second harmonic of amplitude 20 mA, lagging by $\dfrac{\pi}{2}$ radian. The fundamental and second harmonic are shown separately in Figure 39.28. Adding ordinates at intervals gives the complex waveform

$$50\sin\omega t + 20\sin\left(2\omega t - \frac{\pi}{2}\right)$$

This waveform is then added to the 32 mA d.c. component to produce the waveform i as shown. The effect of the d.c. component is to shift the whole wave 32 mA upward. The waveform approaches that expected from a **half-wave rectifier**.

Problem 18. A complex waveform v comprises a fundamental voltage of 240 V rms and frequency 50 Hz, together with a 20% third harmonic which has a phase angle lagging by $3\pi/4$ rad at time $t = 0$. (a) Write down an expression to represent voltage v. (b) Use harmonic synthesis to sketch the complex waveform representing voltage v over one cycle of the fundamental component

(a) A fundamental voltage having an rms value of 240 V has a maximum value, or amplitude, of $\sqrt{2}\,(240)$, i.e. 339.4 V.

If the fundamental frequency is 50 Hz then angular velocity, $\omega = 2\pi f = 2\pi\,(50) = 100\pi$ rad/s. Hence the fundamental voltage is represented by $339.4\sin 100\pi t$ volts. Since the fundamental frequency is 50 Hz, the time for one cycle of the fundamental is given by $T = 1/f = 1/50$ s or 20 ms.

The third harmonic has an amplitude equal to 20% of 339.4 V, i.e. 67.9 V. The frequency of the third harmonic component is $3 \times 50 = 150$ Hz, thus the angular velocity is $2\pi\,(150)$, i.e. 300π rad/s. Hence the third harmonic voltage is represented by $67.9\sin(300\pi t - 3\pi/4)$ volts. Thus

voltage, $\boldsymbol{v = 339.4\sin 100\pi t}$

$$\boldsymbol{+\,67.9\sin\,(300\pi t - 3\pi/4)\,volts}$$

(b) One cycle of the fundamental, $339.4\sin 100\pi t$, is shown sketched in Figure 39.29, together with three cycles of the third harmonic component, $67.9\sin(300\pi t - 3\pi/4)$, initially lagging

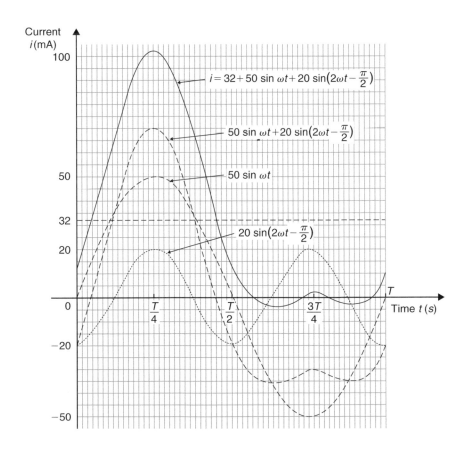

Figure 39.28

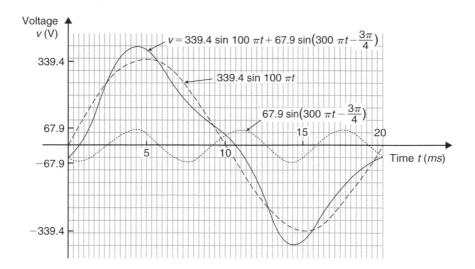

Figure 39.29

by $3\pi/4$ rad. By adding ordinates at intervals, the complex waveform representing voltage is produced as shown.

Now try the following Practice Exercise

Practice Exercises 172   Further problems on harmonic synthesis with complex waveforms (answers on page 1170)

1. A complex current waveform i comprises a fundamental current of 50 A rms and frequency 100 Hz, together with a 24% third harmonic, both being in phase with each other at zero time. (a) Write down an expression to represent current i. (b) Sketch the complex waveform of current using harmonic synthesis over one cycle of the fundamental.

2. A complex voltage waveform v is comprised of a 212.1 V rms fundamental voltage at a frequency of 50 Hz, a 30% second harmonic component lagging by $\pi/2$ rad, and a 10% fourth harmonic component leading by $\pi/3$ rad. (a) Write down an expression to represent voltage v. (b) Sketch the complex voltage waveform using harmonic synthesis over one cycle of the fundamental waveform.

3. A voltage waveform is represented by:

    $$v = 20 + 50\sin\omega t$$
    $$+ 20\sin(2\omega t - \pi/2) \text{ volts}$$

    Draw the complex waveform over one cycle of the fundamental by using harmonic synthesis.

4. Write down an expression representing a current i having a fundamental component of amplitude 16 A and frequency 1 kHz, together with its third and fifth harmonics being respectively one-fifth and one-tenth the amplitude of the fundamental, all components being in phase at zero time. Sketch the complex current waveform for one cycle of the fundamental using harmonic synthesis.

5. A voltage waveform is described by

    $$v = 200\sin 377t + 80\sin\left(1131t + \frac{\pi}{4}\right)$$
    $$+ 20\sin\left(1885t - \frac{\pi}{3}\right) \text{ volts}$$

    Determine (a) the fundamental and harmonic frequencies of the waveform, (b) the percentage third harmonic and (c) the percentage fifth harmonic. Sketch the voltage waveform using harmonic synthesis over one cycle of the fundamental.

For fully worked solutions to each of the problems in Practice Exercises 169 to 172 in this chapter, go to the website:
www.routledge.com/cw/bird

Section E

Chapter 40

Cartesian and polar co-ordinates

Why it is important to understand: **Cartesian and polar co-ordinates**

Applications where polar co-ordinates would be used include terrestrial navigation with sonar-like devices, and those in engineering and science involving energy radiation patterns. Applications where Cartesian co-ordinates would be used include any navigation on a grid and anything involving raster graphics (i.e. bitmap – a dot matrix data structure representing a generally rectangular grid of pixels). The ability to change from Cartesian to polar co-ordinates is vitally important when using complex numbers with a.c. electrical circuit theory and with vector geometry.

At the end of this chapter, you should be able to:

- change from Cartesian to polar co-ordinates
- change from polar to Cartesian co-ordinates
- use a scientific notation calculator to change from Cartesian to polar co-ordinates and vice versa

40.1 Introduction

There are two ways in which the position of a point in a plane can be represented. These are

(a) Cartesian co-ordinates (named after Descartes*), i.e. (x, y).

(b) Polar co-ordinates, i.e. (r, θ), where r is a radius from a fixed point and θ is an angle from a fixed point.

---

* **Who was Descartes**? See page 343. To find out more go to www.routledge.com/cw/bird

40.2 Changing from Cartesian to polar co-ordinates

In Figure 40.1, if lengths x and y are known then the length of r can be obtained from Pythagoras' theorem (see Chapter 38) since OPQ is a right-angled triangle.

Hence, $r^2 = (x^2 + y^2)$, from which $\quad r = \sqrt{x^2 + y^2}$

From trigonometric ratios (see Chapter 38), $\tan\theta = \dfrac{y}{x}$

from which $\quad \theta = \tan^{-1}\dfrac{y}{x}$

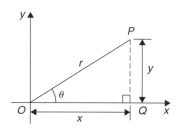

Figure 40.1

$r = \sqrt{x^2 + y^2}$ and $\theta = \tan^{-1}\dfrac{y}{x}$ are the two formulae we need to change from Cartesian to polar co-ordinates. The angle θ, which may be expressed in degrees or radians, must **always** be measured from the positive x-axis; i.e. measured from the line OQ in Figure 40.1. It is suggested that when changing from Cartesian to polar co-ordinates a diagram should always be sketched.

Problem 1. Change the Cartesian co-ordinates (3, 4) into polar co-ordinates

A diagram representing the point (3, 4) is shown in Figure 40.2.

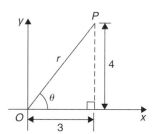

Figure 40.2

From Pythagoras' theorem, $r = \sqrt{3^2 + 4^2} = 5$ (note that -5 has no meaning in this context).

By trigonometric ratios, $\theta = \tan^{-1}\dfrac{4}{3} = 53.13°$ or 0.927 rad.

Note that $53.13° = 53.13 \times \dfrac{\pi}{180}$ rad $= 0.927$ rad.

Hence, **(3, 4) in Cartesian co-ordinates corresponds to (5, 53.13°) or (5, 0.927 rad) in polar co-ordinates**.

Problem 2. Express in polar co-ordinates the position $(-4, 3)$

A diagram representing the point using the Cartesian co-ordinates $(-4, 3)$ is shown in Figure 40.3.

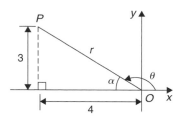

Figure 40.3

From Pythagoras' theorem, $r = \sqrt{4^2 + 3^2} = 5$

By trigonometric ratios, $\alpha = \tan^{-1}\dfrac{3}{4} = 36.87°$ or 0.644 rad

Hence, $\theta = 180° - 36.87° = 143.13°$

or $\theta = \pi - 0.644 = 2.498$ rad.

Hence, **the position of point P in polar co-ordinate form is (5, 143.13°) or (5, 2.498 rad)**.

Problem 3. Express $(-5, -12)$ in polar co-ordinates

A sketch showing the position $(-5, -12)$ is shown in Figure 40.4.

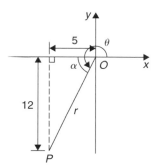

Figure 40.4

$r = \sqrt{5^2 + 12^2} = 13$ and $\alpha = \tan^{-1}\dfrac{12}{5} = 67.38°$ or 1.176 rad

Hence, $\theta = 180° + 67.38° = 247.38°$

or $\theta = \pi + 1.176 = 4.318$ rad.

Thus, **$(-5, -12)$ in Cartesian co-ordinates corresponds to (13, 247.38°) or (13, 4.318 rad) in polar co-ordinates**.

Problem 4. Express $(2, -5)$ in polar co-ordinates

A sketch showing the position $(2, -5)$ is shown in Figure 40.5.

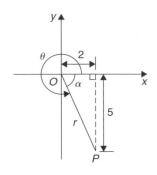

Figure 40.5

$r = \sqrt{2^2 + 5^2} = \sqrt{29} = 5.385$, correct to 3 decimal places.

$\alpha = \tan^{-1}\dfrac{5}{2} = 68.20°$ or 1.190 rad

Hence, $\theta = 360° - 68.20° = 291.80°$
or $\theta = 2\pi - 1.190 = 5.093$ rad.

Thus, **(2, −5) in Cartesian co-ordinates corresponds to (5.385, 291.80°) or (5.385, 5.093 rad) in polar co-ordinates**.

Now try the following Practice Exercise

Practice Exercise 173   Changing from Cartesian to polar co-ordinates (answers on page 1170)

In Problems 1 to 8, express the given Cartesian co-ordinates as polar co-ordinates, correct to 2 decimal places, in both degrees and radians.

1. $(3, 5)$ 2. $(6.18, 2.35)$

3. $(-2, 4)$ 4. $(-5.4, 3.7)$

5. $(-7, -3)$ 6. $(-2.4, -3.6)$

7. $(5, -3)$ 8. $(9.6, -12.4)$

40.3 Changing from polar to Cartesian co-ordinates

From the right-angled triangle OPQ in Figure 40.6,

$$\cos\theta = \frac{x}{r} \text{ and } \sin\theta = \frac{y}{r} \text{ from trigonometric ratios}$$

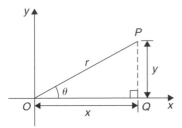

Figure 40.6

Hence, $x = r\cos\theta$ and $y = r\sin\theta$.

If lengths r and angle θ are known then $x = r\cos\theta$ and $y = r\sin\theta$ are the two formulae we need to change from polar to Cartesian co-ordinates.

Problem 5.   Change $(4, 32°)$ into Cartesian co-ordinates

A sketch showing the position $(4, 32°)$ is shown in Figure 40.7.

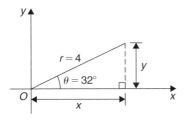

Figure 40.7

Now $x = r\cos\theta = 4\cos 32° = 3.39$

and $y = r\sin\theta = 4\sin 32° = 2.12$

Hence, **(4, 32°) in polar co-ordinates corresponds to (3.39, 2.12) in Cartesian co-ordinates**.

Problem 6.   Express $(6, 137°)$ in Cartesian co-ordinates

A sketch showing the position $(6, 137°)$ is shown in Figure 40.8.

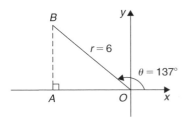

Figure 40.8

$$x = r\cos\theta = 6\cos 137^\circ = -4.388$$

which corresponds to length OA in Figure 40.8.

$$y = r\sin\theta = 6\sin 137^\circ = 4.092$$

which corresponds to length AB in Figure 40.8.

Thus, **(6, 137°) in polar co-ordinates corresponds to (−4.388, 4.092) in Cartesian co-ordinates**.

(Note that when changing from polar to Cartesian co-ordinates it is not quite so essential to draw a sketch. Use of $x = r\cos\theta$ and $y = r\sin\theta$ automatically produces the correct values and signs.)

Problem 7.   Express (4.5, 5.16 rad) in Cartesian co-ordinates

A sketch showing the position (4.5, 5.16 rad) is shown in Figure 40.9.

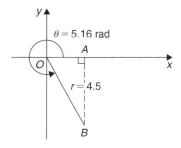

Figure 40.9

$$x = r\cos\theta = 4.5\cos 5.16 = 1.948$$

which corresponds to length OA in Figure 40.9.

$$y = r\sin\theta = 4.5\sin 5.16 = -4.057$$

which corresponds to length AB in Figure 40.9.

Thus, **(1.948, −4.057) in Cartesian co-ordinates corresponds to (4.5, 5.16 rad) in polar co-ordinates**.

Now try the following Practice Exercise

Practice Exercise 174   Changing polar to Cartesian co-ordinates (answers on page 1170)

In Problems 1 to 8, express the given polar co-ordinates as Cartesian co-ordinates, correct to 3 decimal places.

1. (5, 75°)   2. (4.4, 1.12 rad)

3. (7, 140°)   4. (3.6, 2.5 rad)

5. (10.8, 210°)   6. (4, 4 rad)

7. (1.5, 300°)   8. (6, 5.5 rad)

9. Figure 40.10 shows five equally spaced holes on an 80 mm pitch circle diameter. Calculate their co-ordinates relative to axes Ox and Oy in (a) polar form, (b) Cartesian form.

10. In Figure 40.10, calculate the shortest distance between the centres of two adjacent holes.

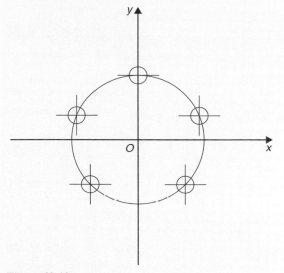

Figure 40.10

40.4   Use of Pol/Rec functions on calculators

Another name for Cartesian co-ordinates is **rectangular** co-ordinates. Many scientific notation calculators have **Pol** and **Rec** functions. 'Rec' is an abbreviation of 'rectangular' (i.e. Cartesian) and 'Pol' is an abbreviation of 'polar'. Check the operation manual for your particular calculator to determine how to use these two functions. They make changing from Cartesian to polar co-ordinates, and vice versa, so much quicker and easier. For example, with the Casio fx-991ES calculator, or similar, to change the Cartesian number (3, 4) into polar form, the following procedure is adopted.

1. Press 'shift'   2. Press 'Pol'   3. Enter 3

4. Enter 'comma' (obtained by 'shift' then))

5. Enter 4   6. Press)   7. Press =

The answer is $r = 5, \theta = 53.13°$

Hence, **(3, 4) in Cartesian form is the same as (5, 53.13°) in polar form**.

If the angle is required in **radians**, then before repeating the above procedure press 'shift', 'mode' and then 4 to change your calculator to radian mode.

Similarly, to change the polar form number (7, 126°) into Cartesian or rectangular form, adopt the following procedure.

1. Press 'shift'   2. Press 'Rec'

3. Enter 7   4. Enter 'comma'

5. Enter 126 (assuming your calculator is in degrees mode)

6. Press)   7. Press =

The answer is $X = -4.11$ and, scrolling across, $Y = 5.66$, correct to 2 decimal places.

Hence, **(7, 126°) in polar form is the same as (−4.11, 5.66) in rectangular or Cartesian form**.

Now return to Practice Exercises 173 and 174 in this chapter and use your calculator to determine the answers, and see how much more quickly they may be obtained.

For fully worked solutions to each of the problems in Practice Exercises 173 and 174 in this chapter, go to the website:
www.routledge.com/cw/bird

Chapter 41

Non-right-angled triangles and some practical applications

Why it is important to understand: Non-right-angled triangles and some practical applications

As was mentioned earlier, fields that use trigonometry include astronomy, navigation, music theory, acoustics, optics, electronics, probability theory, statistics, biology, medical imaging (CAT scans and ultrasound), pharmacy, chemistry, seismology, meteorology, oceanography, many physical sciences, land surveying, architecture, economics, electrical engineering, mechanical engineering, civil engineering, computer graphics, cartography and crystallography. There are so many examples where triangles are involved in engineering, and the ability to solve such triangles is of great importance.

At the end of this chapter, you should be able to:

- state and use the sine rule
- state and use the cosine rule
- use various formulae to determine the area of any triangle
- apply the sine and cosine rules to solving practical trigonometric problems

41.1 The sine and cosine rules

To 'solve a triangle' means 'to find the values of unknown sides and angles'. If a triangle is **right-angled**, trigonometric ratios and the theorem of Pythagoras* may be used for its solution, as shown in Chapter 38. However, for a **non-right-angled triangle**,

trigonometric ratios and Pythagoras' theorem cannot be used. Instead, two rules, called the **sine rule** and the **cosine rule**, are used.

The sine rule

With reference to triangle ABC of Figure 41.1, the **sine rule** states

$$\frac{a}{\sin A} = \frac{b}{\sin B} = \frac{c}{\sin C}$$

---

* **Who was Pythagoras?** See page 440. To find out more go to www.routledge.com/cw/bird

Figure 41.1

The rule may be used only when

(a)   1 side and any 2 angles are initially given, or

(b)   2 sides and an angle (not the included angle) are initially given.

The cosine rule

With reference to triangle ABC of Figure 41.1, the **cosine rule** states

$$a^2 = b^2 + c^2 - 2bc \cos A$$

or $$b^2 = a^2 + c^2 - 2ac \cos B$$

or $$c^2 = a^2 + b^2 - 2ab \cos C$$

The rule may be used only when

(a)   2 sides and the included angle are initially given, or

(b)   3 sides are initially given.

41.2   Area of any triangle

The **area of any triangle** such as ABC of Figure 41.1 is given by

(a) $\dfrac{1}{2}$ **× base × perpendicular height**

or (b) $\dfrac{1}{2}ab \sin C$ or $\dfrac{1}{2}ac \sin B$ or $\dfrac{1}{2}bc \sin A$

or (c) $\sqrt{[s(s-a)(s-b)(s-c)]}$ where $s = \dfrac{a+b+c}{2}$

This latter formula is called **Heron's formula** (or sometimes **Hero's formula**).

41.3   Worked problems on the solution of triangles and their areas

Problem 1.   In a triangle XYZ, $\angle X = 51°$, $\angle Y = 67°$ and $YZ = 15.2$ cm. Solve the triangle and find its area

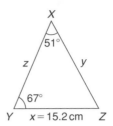

Figure 41.2

The triangle XYZ is shown in Figure 41.2. Solving the triangle means finding $\angle Z$ and sides XZ and XY. Since the angles in a triangle add up to $180°$, $Z = 180° - 51° - 67° = \mathbf{62°}$

Applying the sine rule, $\dfrac{15.2}{\sin 51°} = \dfrac{y}{\sin 67°}$

$$= \dfrac{z}{\sin 62°}$$

Using $\dfrac{15.2}{\sin 51°} = \dfrac{y}{\sin 67°}$

and transposing gives $y = \dfrac{15.2 \sin 67°}{\sin 51°}$

$$= \mathbf{18.00\,cm} = XZ$$

Using $\dfrac{15.2}{\sin 51°} = \dfrac{z}{\sin 62°}$

and transposing gives $z = \dfrac{15.2 \sin 62°}{\sin 51°}$

$$= \mathbf{17.27\,cm} = XY$$

Area of triangle $XYZ = \dfrac{1}{2}xy \sin Z$

$$= \dfrac{1}{2}(15.2)(18.00) \sin 62° = \mathbf{120.8\,cm^2}$$

(or area $= \dfrac{1}{2}xz \sin Y = \dfrac{1}{2}(15.2)(17.27) \sin 67°$

$$= \mathbf{120.8\,cm^2})$$

It is always worth checking with triangle problems that the longest side is opposite the largest angle and vice versa. In this problem, Y is the largest angle and XZ is the longest of the three sides.

Problem 2.   Solve the triangle ABC given $B = 78°51'$, $AC = 22.31$ mm and $AB = 17.92$ mm. Also find its area

Triangle ABC is shown in Figure 41.3. Solving the triangle means finding angles A and C and side BC.

Section E

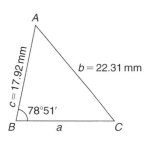

Figure 41.3

Applying the sine rule, $\dfrac{22.31}{\sin 78°51'} = \dfrac{17.92}{\sin C}$

from which $\sin C = \dfrac{17.92 \sin 78°51'}{22.31} = 0.7881$

Hence, $C = \sin^{-1} 0.7881 = 52°0'$ or $128°0'$

Since $B = 78°51'$, C cannot be $128°0'$, since $128°0' + 78°51'$ is greater than $180°$.
Thus, only $\boldsymbol{C = 52°0'}$ is valid.
Angle $A = 180° - 78°51' - 52°0' = \mathbf{49°9'}$

Applying the sine rule, $\dfrac{a}{\sin 49°9'} = \dfrac{22.31}{\sin 78°51'}$

from which $a = \dfrac{22.31 \sin 49°9'}{\sin 78°51'} = \mathbf{17.20\,mm}$

Hence, $\boldsymbol{A = 49°9', C = 52°0'}$ and $\boldsymbol{BC = 17.20\,mm}$.

Area of triangle *ABC* $= \dfrac{1}{2} ac \sin B$

$= \dfrac{1}{2}(17.20)(17.92)\sin 78°51'$

$= \mathbf{151.2\,mm^2}$

Problem 3. Solve the triangle *PQR* and find its area given that $QR = 36.5$ mm, $PR = 29.6$ mm and $\angle Q = 36°$

Triangle *PQR* is shown in Figure 41.4.

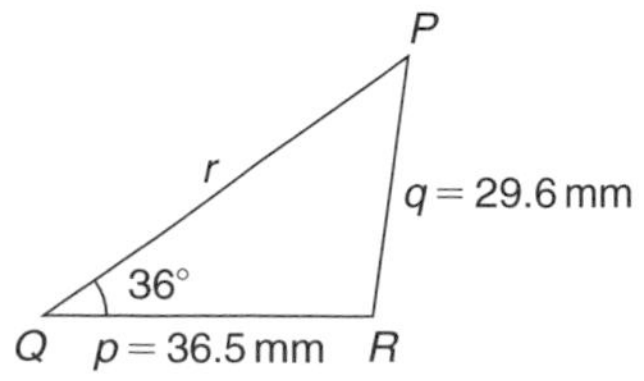

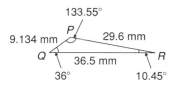

Figure 41.4

Applying the sine rule, $\dfrac{29.6}{\sin 36°} = \dfrac{36.5}{\sin P}$

from which $\sin P = \dfrac{36.5 \sin 36°}{29.6} = 0.7248$

Hence, $P = \sin^{-1} 0.7248 = 46.45°$ or $133.55°$

When $P = 46.45°$ and $Q = 36°$ then
$R = 180° - 46.45° - 36° = 97.55°$
When $P = 133.55°$ and $Q = 36°$ then
$R = 180° - 133.55° - 36° = 10.45°$

Thus, in this problem, there are **two** separate sets of results and both are feasible solutions. Such a situation is called the **ambiguous case**.

Case 1. $P = 46.45°$, $Q = 36°$, $R = 97.55°$, $p = 36.5$ mm and $q = 29.6$ mm

From the sine rule, $\dfrac{r}{\sin 97.55°} = \dfrac{29.6}{\sin 36°}$

from which $r = \dfrac{29.6 \sin 97.55°}{\sin 36°} = \mathbf{49.92\,mm} = \boldsymbol{PQ}$

Area of *PQR* $= \dfrac{1}{2} pq \sin R = \dfrac{1}{2}(36.5)(29.6)\sin 97.55°$

$= \mathbf{535.5\,mm^2}$

Case 2. $P = 133.55°$, $Q = 36°$, $R = 10.45°$, $p = 36.5$ mm and $q = 29.6$ mm

From the sine rule, $\dfrac{r}{\sin 10.45°} = \dfrac{29.6}{\sin 36°}$

from which $r = \dfrac{29.6 \sin 10.45°}{\sin 36°} = \mathbf{9.134\,mm} = \boldsymbol{PQ}$

Area of *PQR* $= \dfrac{1}{2} pq \sin R = \dfrac{1}{2}(36.5)(29.6)\sin 10.45°$

$= \mathbf{97.98\,mm^2}$

The triangle *PQR* for case 2 is shown in Figure 41.5.

Figure 41.5

Now try the following Practice Exercise

Practice Exercise 175 Solution of triangles and their areas (answers on page 1170)

In Problems 1 and 2, use the sine rule to solve the triangles *ABC* and find their areas.

1. $A = 29°, B = 68°, b = 27$ mm
2. $B = 71°26', C = 56°32', b = 8.60$ cm

In Problems 3 and 4, use the sine rule to solve the triangles *DEF* and find their areas.

3. $d = 17\,\text{cm}, f = 22\,\text{cm}, F = 26°$

4. $d = 32.6\,\text{mm}, e = 25.4\,\text{mm}, D = 104°22'$

In Problems 5 and 6, use the sine rule to solve the triangles *JKL* and find their areas.

5. $j = 3.85\,\text{cm}, k = 3.23\,\text{cm}, K = 36°$

6. $k = 46\,\text{mm}, l = 36\,\text{mm}, L = 35°$

41.4 Further worked problems on the solution of triangles and their areas

Problem 4. Solve triangle *DEF* and find its area given that $EF = 35.0\,\text{mm}, DE = 25.0\,\text{mm}$ and $\angle E = 64°$

Triangle *DEF* is shown in Figure 41.6. Solving the triangle means finding angles *D* and *F* and side *DF*. Since two sides and the angle in between the two sides are given, the cosine rule needs to be used.

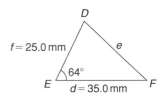

Figure 41.6

Applying the cosine rule, $\quad e^2 = d^2 + f^2 - 2df\cos E$

i.e. $\quad e^2 = (35.0)^2 + (25.0)^2 - [2(35.0)(25.0)\cos 64°]$

$$= 1225 + 625 - 767.15$$

$$= 1082.85$$

from which $\quad e = \sqrt{1082.85}$

$$= \mathbf{32.91\,mm = DF}$$

Applying the sine rule, $\quad \dfrac{32.91}{\sin 64°} = \dfrac{25.0}{\sin F}$

from which $\quad \sin F = \dfrac{25.0\sin 64°}{32.91} = 0.6828$

Thus, $\quad \angle F = \sin^{-1} 0.6828 = 43°4'$ or $136°56'$

$F = 136°56'$ is not possible in this case since $136°56' + 64°$ is greater than $180°$. Thus, only $\mathbf{F = 43°4'}$ is valid. Then $\angle D = 180° - 64° - 43°4' = \mathbf{72°56'}$

Area of triangle *DEF* $= \dfrac{1}{2}df\sin E$

$$= \dfrac{1}{2}(35.0)(25.0)\sin 64° = \mathbf{393.2\,mm^2}$$

Problem 5. A triangle *ABC* has sides $a = 9.0\,\text{cm}, b = 7.5\,\text{cm}$ and $c = 6.5\,\text{cm}$. Determine its three angles and its area

Triangle *ABC* is shown in Figure 41.7. It is usual first to calculate the largest angle to determine whether the triangle is acute or obtuse. In this case the largest angle is *A* (i.e. opposite the longest side).

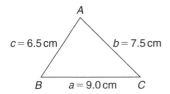

Figure 41.7

Applying the cosine rule, $\quad a^2 = b^2 + c^2 - 2bc\cos A$

from which $\quad 2bc\cos A = b^2 + c^2 - a^2$

and $\quad \cos A = \dfrac{b^2 + c^2 - a^2}{2bc} = \dfrac{7.5^2 + 6.5^2 - 9.0^2}{2(7.5)(6.5)}$

$$= 0.1795$$

Hence, $\quad A = \cos^{-1} 0.1795 = \mathbf{79.67°}$
(or $280.33°$, which is clearly impossible)

The triangle is thus acute angled since cos *A* is positive. (If cos *A* had been negative, angle *A* would be obtuse; i.e. would lie between $90°$ and $180°$)

Applying the sine rule, $\quad \dfrac{9.0}{\sin 79.67°} = \dfrac{7.5}{\sin B}$

from which $\quad \sin B = \dfrac{7.5\sin 79.67°}{9.0} = 0.8198$

Hence, $\quad B = \sin^{-1} 0.8198 = \mathbf{55.07°}$

and $\quad C = 180° - 79.67° - 55.07° = \mathbf{45.26°}$

Area $= \sqrt{[s(s-a)(s-b)(s-c)]}$, where

$$s = \dfrac{a+b+c}{2} = \dfrac{9.0+7.5+6.5}{2} = 11.5\,\text{cm}$$

Hence,

$$\mathbf{area} = \sqrt{[11.5(11.5-9.0)(11.5-7.5)(11.5-6.5)]}$$

$$= \sqrt{[11.5(2.5)(4.0)(5.0)]} = \mathbf{23.98\,cm^2}$$

Alternatively, **area** $= \dfrac{1}{2} ac \sin B$

$$= \dfrac{1}{2}(9.0)(6.5) \sin 55.07° = \mathbf{23.98\,cm^2}$$

Problem 6. Solve triangle XYZ, shown in Figure 41.8, and find its area given that $Y = 128°$, $XY = 7.2\,cm$ and $YZ = 4.5\,cm$

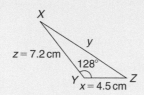

Figure 41.8

Applying the cosine rule,

$$y^2 = x^2 + z^2 - 2xz \cos Y$$

$$= 4.5^2 + 7.2^2 - [2(4.5)(7.2) \cos 128°]$$

$$= 20.25 + 51.84 - [-39.89]$$

$$= 20.25 + 51.84 + 39.89 = 112.0$$

$$y = \sqrt{112.0} = \mathbf{10.58\,cm = XZ}$$

Applying the sine rule, $\dfrac{10.58}{\sin 128°} = \dfrac{7.2}{\sin Z}$

from which $\sin Z = \dfrac{7.2 \sin 128°}{10.58} = 0.5363$

Hence, $Z = \sin^{-1} 0.5363 = \mathbf{32.43°}$
(or $147.57°$, which is not possible)

Thus, $X = 180° - 128° - 32.43° = \mathbf{19.57°}$

Area of XYZ $= \dfrac{1}{2} xz \sin Y = \dfrac{1}{2}(4.5)(7.2) \sin 128°$

$$= \mathbf{12.77\,cm^2}$$

Now try the following Practice Exercise

Practice Exercise 176   Solution of triangles and their areas (answers on page 1170)

In Problems 1 and 2, use the cosine and sine rules to solve the triangles PQR and find their areas.

1. $q = 12\,cm, r = 16\,cm, P = 54°$

2. $q = 3.25\,m, r = 4.42\,m, P = 105°$

In Problems 3 and 4, use the cosine and sine rules to solve the triangles XYZ and find their areas.

3. $x = 10.0\,cm, y = 8.0\,cm, z = 7.0\,cm$

4. $x = 21\,mm, y = 34\,mm, z = 42\,mm$

41.5 Practical situations involving trigonometry

There are a number of **practical situations** in which the use of trigonometry is needed to find unknown sides and angles of triangles. This is demonstrated in the following worked problems.

Problem 7. A room 8.0 m wide has a span roof which slopes at $33°$ on one side and $40°$ on the other. Find the length of the roof slopes, correct to the nearest centimetre

A section of the roof is shown in Figure 41.9.

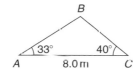

Figure 41.9

Angle at ridge, $B = 180° - 33° - 40° = 107°$

From the sine rule, $\dfrac{8.0}{\sin 107°} = \dfrac{a}{\sin 33°}$

from which $a = \dfrac{8.0 \sin 33°}{\sin 107°} = \mathbf{4.556\,m = BC}$

Also from the sine rule, $\dfrac{8.0}{\sin 107°} = \dfrac{c}{\sin 40°}$

from which $c = \dfrac{8.0 \sin 40°}{\sin 107°} = \mathbf{5.377\,m = AB}$

Hence, **the roof slopes are 4.56 m and 5.38 m**, correct to the nearest centimetre.

Problem 8. A man leaves a point walking at 6.5 km/h in a direction E $20°$ N (i.e. a bearing of $70°$). A cyclist leaves the same point at the same time in a direction E $40°$ S (i.e. a bearing of $130°$) travelling at a constant speed. Find the average speed of the cyclist if the walker and cyclist are 80 km apart after 5 hours

After 5 hours the walker has travelled $5 \times 6.5 = 32.5$ km (shown as AB in Figure 41.10). If AC is the distance the cyclist travels in 5 hours then $BC = 80$ km.

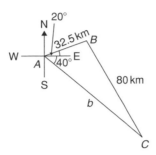

Figure 41.10

Applying the sine rule, $\dfrac{80}{\sin 60°} = \dfrac{32.5}{\sin C}$

from which $\sin C = \dfrac{32.5 \sin 60°}{80} = 0.3518$

Hence, $C = \sin^{-1} 0.3518 = 20.60°$

(or 159.40°, which is not possible)

and $B = 180° - 60° - 20.60° = 99.40°$

Applying the sine rule again, $\dfrac{80}{\sin 60°} = \dfrac{b}{\sin 99.40°}$

from which $b = \dfrac{80 \sin 99.40°}{\sin 60°} = 91.14$ km

Since the cyclist travels 91.14 km in 5 hours,

average speed $= \dfrac{\text{distance}}{\text{time}} = \dfrac{91.14}{5} = \textbf{18.23 km/h}$

Problem 9. Two voltage phasors are shown in Figure 41.11. If $V_1 = 40$ V and $V_2 = 100$ V, determine the value of their resultant (i.e. length OA) and the angle the resultant makes with V_1

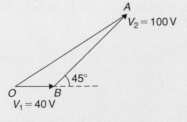

Figure 41.11

Angle $OBA = 180° - 45° = 135°$

Applying the cosine rule,

$$OA^2 = V_1^2 + V_2^2 - 2V_1 V_2 \cos OBA$$
$$= 40^2 + 100^2 - \{2(40)(100)\cos 135°\}$$
$$= 1600 + 10000 - \{-5657\}$$
$$= 1600 + 10000 + 5657 = 17257$$

Thus, **resultant**, $OA = \sqrt{17257} = \textbf{131.4 V}$

Applying the sine rule, $\dfrac{131.4}{\sin 135°} = \dfrac{100}{\sin AOB}$

from which $\sin AOB = \dfrac{100 \sin 135°}{131.4} = 0.5381$

Hence, angle $AOB = \sin^{-1} 0.5381 = 32.55°$ (or 147.45°, which is not possible)

Hence, **the resultant voltage is 131.4 volts at 32.55° to V_1**

Problem 10. In Figure 41.12, PR represents the inclined jib of a crane and is 10.0 m long. PQ is 4.0 m long. Determine the inclination of the jib to the vertical and the length of tie QR.

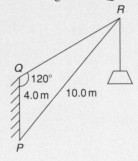

Figure 41.12

Applying the sine rule, $\dfrac{PR}{\sin 120°} = \dfrac{PQ}{\sin R}$

from which $\sin R = \dfrac{PQ \sin 120°}{PR} = \dfrac{(4.0)\sin 120°}{10.0}$
$$= 0.3464$$

Hence, $\angle R = \sin^{-1} 0.3464 = 20.27°$ (or 159.73°, which is not possible)

$\angle P = 180° - 120° - 20.27° = \textbf{39.73°}$, **which is the inclination of the jib to the vertical.**

Applying the sine rule, $\dfrac{10.0}{\sin 120°} = \dfrac{QR}{\sin 39.73°}$

from which **length of tie**, $QR = \dfrac{10.0 \sin 39.73°}{\sin 120°}$
$$= \textbf{7.38 m}$$

Now try the following Practice Exercise

Practice Exercise 177   Practical situations involving trigonometry (answers on page 1171)

1. A ship P sails at a steady speed of 45 km/h in a direction of W 32° N (i.e. a bearing of 302°) from a port. At the same time another ship Q leaves the port at a steady speed of 35 km/h in a direction N 15° E (i.e. a bearing of 015°). Determine their distance apart after 4 hours.

2. Two sides of a triangular plot of land are 52.0 m and 34.0 m, respectively. If the area of the plot is 620 m², find (a) the length of fencing required to enclose the plot and (b) the angles of the triangular plot.

3. A jib crane is shown in Figure 41.13. If the tie rod PR is 8.0 m long and PQ is 4.5 m long, determine (a) the length of jib RQ and (b) the angle between the jib and the tie rod.

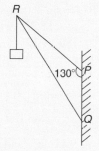

Figure 41.13

4. A building site is in the form of a quadrilateral, as shown in Figure 41.14, and its area is 1510 m². Determine the length of the perimeter of the site.

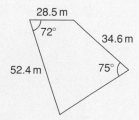

Figure 41.14

5. Determine the length of members BF and EB in the roof truss shown in Figure 41.15.

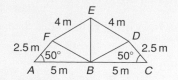

Figure 41.15

6. A laboratory 9.0 m wide has a span roof which slopes at 36° on one side and 44° on the other. Determine the lengths of the roof slopes.

7. PQ and QR are the phasors representing the alternating currents in two branches of a circuit. Phasor PQ is 20.0 A and is horizontal. Phasor QR (which is joined to the end of PQ to form triangle PQR) is 14.0 A and is at an angle of 35° to the horizontal. Determine the resultant phasor PR and the angle it makes with phasor PQ.

41.6   Further practical situations involving trigonometry

Problem 11. A vertical aerial stands on horizontal ground. A surveyor positioned due east of the aerial measures the elevation of the top as 48°. He moves due south 30.0 m and measures the elevation as 44°. Determine the height of the aerial

In Figure 41.16, DC represents the aerial, A is the initial position of the surveyor and B his final position.

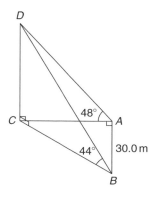

Figure 41.16

From triangle ACD, $\tan 48° = \dfrac{DC}{AC}$ from which

$$AC = \frac{DC}{\tan 48°}$$

Similarly, from triangle BCD, $BC = \dfrac{DC}{\tan 44°}$

For triangle ABC, using Pythagoras' theorem,

$$BC^2 = AB^2 + AC^2$$

$$\left(\frac{DC}{\tan 44°}\right)^2 = (30.0)^2 + \left(\frac{DC}{\tan 48°}\right)^2$$

$$DC^2\left(\frac{1}{\tan^2 44°} - \frac{1}{\tan^2 48°}\right) = 30.0^2$$

$$DC^2(1.072323 - 0.810727) = 30.0^2$$

$$DC^2 = \frac{30.0^2}{0.261596} = 3440.4$$

Hence, **height of aerial, $DC = \sqrt{3340.4} = 58.65\,\text{m}$**

Problem 12. A crank mechanism of a petrol engine is shown in Figure 41.17. Arm OA is 10.0 cm long and rotates clockwise about O. The connecting rod AB is 30.0 cm long and end B is constrained to move horizontally.

(a) For the position shown in Figure 41.17, determine the angle between the connecting rod AB and the horizontal, and the length of OB.

(b) How far does B move when angle AOB changes from 50° to 120°?

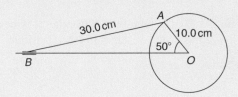

Figure 41.17

(a) Applying the sine rule, $\dfrac{AB}{\sin 50°} = \dfrac{AO}{\sin B}$

from which $\sin B = \dfrac{AO \sin 50°}{AB} = \dfrac{10.0 \sin 50°}{30.0}$

$$= 0.2553$$

Hence, $B = \sin^{-1} 0.2553 = 14.79°$ (or 165.21°, which is not possible)

Hence, **the connecting rod AB makes an angle of 14.79° with the horizontal**.

Angle $OAB = 180° - 50° - 14.79° = 115.21°$

Applying the sine rule, $\dfrac{30.0}{\sin 50°} = \dfrac{OB}{\sin 115.21°}$

from which $OB = \dfrac{30.0 \sin 115.21°}{\sin 50°}$

$$= \mathbf{35.43\,cm}$$

(b) Figure 41.18 shows the initial and final positions of the crank mechanism.

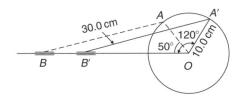

Figure 41.18

In triangle $OA'B'$, applying the sine rule,

$$\frac{30.0}{\sin 120°} = \frac{10.0}{\sin A'B'O}$$

from which $\sin A'B'O = \dfrac{10.0 \sin 120°}{30.0} = 0.28868$

Hence, $\quad A'B'O = \sin^{-1} 0.28868 = 16.78°$ (or 163.22°, which is not possible)

Angle $OA'B' = 180° - 120° - 16.78° = 43.22°$

Applying the sine rule, $\dfrac{30.0}{\sin 120°} = \dfrac{OB'}{\sin 43.22°}$

from which $OB' = \dfrac{30.0 \sin 43.22°}{\sin 120°} = 23.72\,\text{cm}$

Since $OB = 35.43\,\text{cm}$ and $OB' = 23.72\,\text{cm}$, $BB' = 35.43 - 23.72 = 11.71\,\text{cm}$

Hence, **B moves 11.71 cm when angle AOB changes from 50° to 120°**

Problem 13. The area of a field is in the form of a quadrilateral $ABCD$ as shown in Figure 41.19. Determine its area

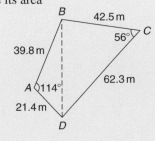

Figure 41.19

A diagonal drawn from B to D divides the quadrilateral into two triangles.

Area of quadrilateral *ABCD*

$$= \text{area of triangle } ABD$$
$$+ \text{area of triangle } BCD$$

$$= \frac{1}{2}(39.8)(21.4)\sin 114°$$

$$+ \frac{1}{2}(42.5)(62.3)\sin 56°$$

$$= 389.04 + 1097.5$$

$$= \mathbf{1487\,m^2}$$

Now try the following Practice Exercise

Practice Exercise 178   More practical situations involving trigonometry (answers on page 1171)

1. Three forces acting on a fixed point are represented by the sides of a triangle of dimensions 7.2 cm, 9.6 cm and 11.0 cm. Determine the angles between the lines of action of the three forces.

2. A vertical aerial *AB*, 9.60 m high, stands on ground which is inclined 12° to the horizontal. A stay connects the top of the aerial *A* to a point *C* on the ground 10.0 m downhill from *B*, the foot of the aerial. Determine (a) the length of the stay and (b) the angle the stay makes with the ground.

3. A reciprocating engine mechanism is shown in Figure 41.20. The crank *AB* is 12.0 cm long and the connecting rod *BC* is 32.0 cm long. For the position shown determine the length of *AC* and the angle between the crank and the connecting rod.

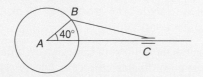

Figure 41.20

4. From Figure 41.20, determine how far *C* moves, correct to the nearest millimetre, when angle *CAB* changes from 40° to 160°, *B* moving in an anticlockwise direction.

5. A surveyor standing W 25° S of a tower measures the angle of elevation of the top of the tower as 46°30′. From a position E 23° S from the tower the elevation of the top is 37°15′. Determine the height of the tower if the distance between the two observations is 75 m.

6. Calculate, correct to 3 significant figures, the co-ordinates *x* and *y* to locate the hole centre at *P* shown in Figure 41.21.

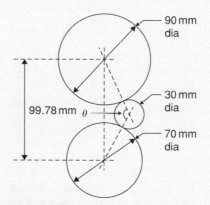

Figure 41.21

7. An idler gear, 30 mm in diameter, has to be fitted between a 70 mm diameter driving gear and a 90 mm diameter driven gear, as shown in Figure 41.22. Determine the value of angle θ between the centre lines.

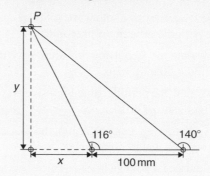

Figure 41.22

8. 16 holes are equally spaced on a pitch circle of 70 mm diameter. Determine the length of the chord joining the centres of two adjacent holes.

For fully worked solutions to each of the problems in Practice Exercises 175 to 178 in this chapter, go to the website:
www.routledge.com/cw/bird

This assignment covers the material contained in Chapters 39 to 41. *The marks available are shown in brackets at the end of each question.*

1. A sine wave is given by $y = 8\sin 4x$. State its peak value and its period, in degrees. (2)

2. A periodic function is given by $y = 15\tan 2x$. State its period in degrees. (2)

3. The frequency of a sine wave is 800 Hz. Calculate the periodic time in milliseconds. (2)

4. Calculate the frequency of a sine wave that has a periodic time of 40 µs. (2)

5. Calculate the periodic time for a sine wave having a frequency of 20 kHz. (2)

6. An alternating current completes 12 cycles in 16 ms. What is its frequency? (3)

7. A sinusoidal voltage is given by $e = 150\sin(500\pi t - 0.25)$ volts. Determine the
    (a) amplitude,
    (b) frequency,
    (c) periodic time,
    (d) phase angle (stating whether it is leading or lagging $150\sin 500\pi t$). (4)

8. Determine the acute angles in degrees, degrees and minutes, and radians.
    (a) $\sin^{-1} 0.4721$ (b) $\cos^{-1} 0.8457$
    (c) $\tan^{-1} 1.3472$ (9)

9. Sketch the following curves, labelling relevant points.
    (a) $y = 4\cos(\theta + 45°)$ (b) $y = 5\sin(2t - 60°)$ (8)

10. The current in an alternating current circuit at any time t seconds is given by $i = 120\sin(100\pi t + 0.274)$ amperes. Determine
    (a) the amplitude, frequency, periodic time and phase angle (with reference to $120\sin 100\pi t$),
    (b) the value of current when $t = 0$,
    (c) the value of current when $t = 6$ ms.
    Sketch one cycle of the oscillation. (16)

11. A triangular plot of land ABC is shown in Figure RT16.1. Solve the triangle and determine its area. (10)

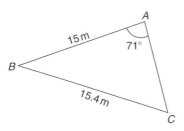

Figure RT16.1

12. A car is travelling 20 m above sea level. It then travels 500 m up a steady slope of 17°. Determine, correct to the nearest metre, how high the car is now above sea level. (3)

13. Figure RT16.2 shows a roof truss PQR with rafter $PQ = 3$ m. Calculate the length of
    (a) the roof rise PP',
    (b) rafter PR,
    (c) the roof span QR.
    Find also (d) the cross-sectional area of the roof truss. (11)

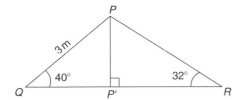

Figure RT16.2

14. Solve triangle ABC given $b = 10$ cm, $c = 15$ cm and $\angle A = 60°$. (10)

15. Change the following Cartesian co-ordinates into polar co-ordinates, correct to 2 decimal places, in both degrees and in radians.
    (a) $(-2.3, 5.4)$ (b) $(7.6, -9.2)$ (10)

16. Change the following polar co-ordinates into Cartesian co-ordinates, correct to 3 decimal places.
    (a) $(6.5, 132°)$ (b) $(3, 3\,\text{rad})$ (6)

Chapter 42

Trigonometric identities and equations

Why it is important to understand: Trigonometric identities and equations

In engineering, trigonometric identities occur often, examples being in the more advanced areas of calculus to generate derivatives and integrals, with tensors/vectors, and with differential equations and partial differential equations. One of the skills required for more advanced work in mathematics, especially in calculus, is the ability to use identities to write expressions in alternative forms. In software engineering, working, say, on the next *Ratatouille* or *Toy Story*, trigonometric identities are needed for computer graphics; an RF engineer working on the next-generation mobile phone will also need trigonometric identities. In addition, identities are needed in electrical engineering when dealing with a.c. power, and wave addition/subtraction and the solutions of trigonometric equations often require knowledge of trigonometric identities.

At the end of this chapter, you should be able to:

- state simple trigonometric identities
- prove simple identities
- solve equations of the form $b \sin A + c = 0$
- solve equations of the form $a \sin^2 A + c = 0$
- solve equations of the form $a \sin^2 A + b \sin A + c = 0$
- solve equations requiring trigonometric identities

42.1 Trigonometric identities

A trigonometric identity is a relationship that is true for all values of the unknown variable.

$$\tan\theta = \frac{\sin\theta}{\cos\theta} \quad \cot\theta = \frac{\cos\theta}{\sin\theta} \quad \sec\theta = \frac{1}{\cos\theta}$$

$$\csc\theta = \frac{1}{\sin\theta} \quad \text{and} \quad \cot\theta = \frac{1}{\tan\theta}$$

are examples of trigonometric identities from Chapter 38.

Applying Pythagoras' theorem to the right-angled triangle shown in Figure 42.1 gives:

$$a^2 + b^2 = c^2 \tag{1}$$

Figure 42.1

Dividing each term of equation (1) by c^2 gives:

$$\frac{a^2}{c^2} + \frac{b^2}{c^2} = \frac{c^2}{c^2}$$

i.e.

$$\left(\frac{a}{c}\right)^2 + \left(\frac{b}{c}\right)^2 = 1$$

$$(\cos\theta)^2 + (\sin\theta)^2 = 1$$

Hence $\quad \cos^2\theta + \sin^2\theta = 1 \qquad (2)$

Dividing each term of equation (1) by a^2 gives:

$$\frac{a^2}{a^2} + \frac{b^2}{a^2} = \frac{c^2}{a^2}$$

i.e.

$$1 + \left(\frac{b}{a}\right)^2 = \left(\frac{c}{a}\right)^2$$

Hence $\quad 1 + \tan^2\theta = \sec^2\theta \qquad (3)$

Dividing each term of equation (1) by b^2 gives:

$$\frac{a^2}{b^2} + \frac{b^2}{b^2} = \frac{c^2}{b^2} \qquad (4)$$

i.e.

$$\left(\frac{a}{b}\right)^2 + 1 = \left(\frac{c}{b}\right)^2$$

Hence $\quad \cot^2\theta + 1 = \operatorname{cosec}^2\theta \qquad (5)$

Equations (2), (3) and (4) are three further examples of trigonometric identities.

42.2 Worked problems on trigonometric identities

Problem 1. Prove the identity
$$\sin^2\theta \cot\theta \sec\theta = \sin\theta$$

With trigonometric identities it is necessary to start with the left-hand side (LHS) and attempt to make it equal to the right-hand side (RHS) or vice versa. It is often useful to change all of the trigonometric ratios into sines and cosines where possible. Thus

$$\text{LHS} = \sin^2\theta \cot\theta \sec\theta$$

$$= \sin^2\theta \left(\frac{\cos\theta}{\sin\theta}\right)\left(\frac{1}{\cos\theta}\right)$$

$$= \sin\theta \text{ (by cancelling)} = \text{RHS}$$

Problem 2. Prove that:
$$\frac{\tan x + \sec x}{\sec x \left(1 + \dfrac{\tan x}{\sec x}\right)} = 1$$

$$\text{LHS} = \frac{\tan x + \sec x}{\sec x \left(1 + \dfrac{\tan x}{\sec x}\right)}$$

$$= \frac{\dfrac{\sin x}{\cos x} + \dfrac{1}{\cos x}}{\left(\dfrac{1}{\cos x}\right)\left(1 + \dfrac{\dfrac{\sin x}{\cos x}}{\dfrac{1}{\cos x}}\right)}$$

$$= \frac{\dfrac{\sin x + 1}{\cos x}}{\left(\dfrac{1}{\cos x}\right)\left[1 + \left(\dfrac{\sin x}{\cos x}\right)\left(\dfrac{\cos x}{1}\right)\right]}$$

$$= \frac{\dfrac{\sin x + 1}{\cos x}}{\left(\dfrac{1}{\cos x}\right)[1 + \sin x]}$$

$$= \left(\frac{\sin x + 1}{\cos x}\right)\left(\frac{\cos x}{1 + \sin x}\right)$$

$$= 1 \text{ (by cancelling)} = \text{RHS}$$

Problem 3. Prove that: $\dfrac{1 + \cot\theta}{1 + \tan\theta} = \cot\theta$

$$\text{LHS} = \frac{1 + \cot\theta}{1 + \tan\theta} = \frac{1 + \dfrac{\cos\theta}{\sin\theta}}{1 + \dfrac{\sin\theta}{\cos\theta}} = \frac{\dfrac{\sin\theta + \cos\theta}{\sin\theta}}{\dfrac{\cos\theta + \sin\theta}{\cos\theta}}$$

$$= \left(\frac{\sin\theta + \cos\theta}{\sin\theta}\right)\left(\frac{\cos\theta}{\cos\theta + \sin\theta}\right)$$

$$= \frac{\cos\theta}{\sin\theta} = \cot\theta = \text{RHS}$$

Problem 4. Show that:
$$\cos^2\theta - \sin^2\theta = 1 - 2\sin^2\theta$$

From equation (2), $\cos^2\theta + \sin^2\theta = 1$, from which, $\cos^2\theta = 1 - \sin^2\theta$

Hence, $\quad \text{LHS} = \cos^2\theta - \sin^2\theta$

$$= (1 - \sin^2\theta) - \sin^2\theta$$

$$= 1 - \sin^2\theta - \sin^2\theta$$

$$= 1 - 2\sin^2\theta = \text{RHS}$$

Problem 5. Prove that:
$$\sqrt{\frac{1 - \sin x}{1 + \sin x}} = \sec x - \tan x$$

$$\text{LHS} = \sqrt{\frac{1 - \sin x}{1 + \sin x}}$$

$$= \sqrt{\frac{(1 - \sin x)(1 - \sin x)}{(1 + \sin x)(1 - \sin x)}}$$

$$= \sqrt{\frac{(1 - \sin x)^2}{(1 - \sin^2 x)}}$$

Since $\cos^2 x + \sin^2 x = 1$ then $1 - \sin^2 x = \cos^2 x$

$$\text{LHS} = \sqrt{\frac{(1 - \sin x)^2}{(1 - \sin^2 x)}} = \sqrt{\frac{(1 - \sin x)^2}{\cos^2 x}}$$

$$= \frac{1 - \sin x}{\cos x} = \frac{1}{\cos x} - \frac{\sin x}{\cos x}$$

$$= \sec x - \tan x = \text{RHS}$$

Now try the following Practice Exercise

Practice Exercise 179   Further problems on trigonometric identities (answers on page 1171)

Prove the following trigonometric identities:

1. $\sin x \cot x = \cos x$

2. $\dfrac{1}{\sqrt{1 - \cos^2 \theta}} = \text{cosec}\,\theta$

3. $2\cos^2 A - 1 = \cos^2 A - \sin^2 A$

4. $\dfrac{\cos x - \cos^3 x}{\sin x} = \sin x \cos x$

5. $(1 + \cot \theta)^2 + (1 - \cot \theta)^2 = 2\,\text{cosec}^2 \theta$

6. $\dfrac{\sin^2 x (\sec x + \text{cosec}\,x)}{\cos x \tan x} = 1 + \tan x$

42.3   Trigonometric equations

Equations which contain trigonometric ratios are called **trigonometric equations**. There are usually an infinite number of solutions to such equations; however, solutions are often restricted to those between 0° and 360°. A knowledge of angles of any magnitude is essential in the solution of trigonometric equations and calculators cannot be relied upon to give all the solutions (as shown in Chapter 39). Figure 42.2 shows a summary for angles of any magnitude.

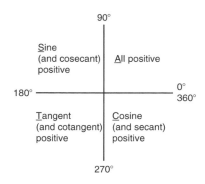

Figure 42.2

Equations of the type $a \sin^2 A + b \sin A + c = 0$

(i)   When $a = 0$, $b \sin A + c = 0$, hence $\sin A = -\dfrac{c}{b}$ and $A = \sin^{-1}\left(-\dfrac{c}{b}\right)$

There are two values of A between 0° and 360° that satisfy such an equation, provided $-1 \le \dfrac{c}{b} \le 1$ (see Problems 6 to 9).

(ii)   When $b = 0$, $a \sin^2 A + c = 0$, hence $\sin^2 A = -\dfrac{c}{a}$, $\sin A = \sqrt{-\dfrac{c}{a}}$ and $A = \sin^{-1}\sqrt{-\dfrac{c}{a}}$

If either a or c is a negative number, then the value within the square root sign is positive. Since when a square root is taken there is a positive and negative answer there are four values of A between 0° and 360° which satisfy such an equation, provided $-1 \le \dfrac{c}{b} \le 1$ (see Problems 10 and 11).

(iii)   When a, b and c are all non-zero:
$a \sin^2 A + b \sin A + c = 0$ is a quadratic equation in which the unknown is $\sin A$. The solution of a quadratic equation is obtained either by factorising (if possible) or by using the quadratic formula:

$$\sin A = \frac{-b \pm \sqrt{b^2 - 4ac}}{2a}$$

(see Problems 12 and 13).

(iv)   Often the trigonometric identities $\cos^2 A + \sin^2 A = 1$, $1 + \tan^2 A = \sec^2 A$ and $\cot^2 A + 1 = \text{cosec}^2 A$ need to be used to reduce equations to one of the above forms (see Problems 14 to 16).

42.4 Worked problems (i) on trigonometric equations

Problem 6. Solve the trigonometric equation: $5\sin\theta + 3 = 0$ for values of θ from $0°$ to $360°$

$5\sin\theta + 3 = 0$, from which
$\sin\theta = -3/5 = -0.6000$

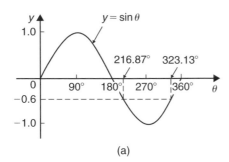

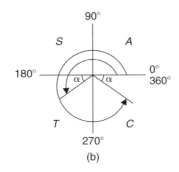

Figure 42.3

Hence $\theta = \sin^{-1}(-0.6000)$. Sine is negative in the third and fourth quadrants (see Figure 42.3). The acute angle $\sin^{-1}(0.6000) = 36.87°$ (shown as α in Figure 42.3(b)).
Hence $\theta = 180° + 36.87°$, i.e. **216.87°**
or $\theta = 360° - 36.87°$, i.e. **323.13°**

Problem 7. Solve: $1.5\tan x - 1.8 = 0$ for $0° \le x \le 360°$

$1.5\tan x - 1.8 = 0$, from which

$\tan x = \dfrac{1.8}{1.5} = 1.2000$

Hence $x = \tan^{-1} 1.2000$

Tangent is positive in the first and third quadrants (see Figure 42.4).

The acute angle $\tan^{-1} 1.2000 = 50.19°$

Hence, $x = \mathbf{50.19°}$ or $180° + 50.19° = \mathbf{230.19°}$

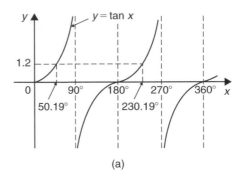

(a)

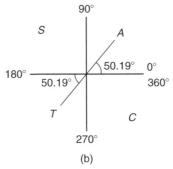

(b)

Figure 42.4

Problem 8. Solve for θ in the range $0° \le \theta \le 360°$ for $2\sin\theta = \cos\theta$

Dividing both sides by $\cos\theta$ gives: $\dfrac{2\sin\theta}{\cos\theta} = 1$

From Section 42.1, $\tan\theta = \dfrac{\sin\theta}{\cos\theta}$

hence $2\tan\theta = 1$

Dividing by 2 gives: $\tan\theta = \dfrac{1}{2}$

from which, $\theta = \tan^{-1}\dfrac{1}{2}$

Since tangent is positive in the first and third quadrants,
$\theta = \mathbf{26.57°}$ **and** $\mathbf{206.57°}$

Problem 9. Solve: $4\sec t = 5$ for values of t between $0°$ and $360°$

$4\sec t = 5$, from which $\sec t = \frac{5}{4} = 1.2500$

Hence $t = \sec^{-1} 1.2500$
Secant $(= 1/\text{cosine})$ is positive in the first and fourth quadrants (see Figure 42.5). The acute angle $\sec^{-1} 1.2500 = 36.87°$. Hence

$t = \mathbf{36.87°}$ or $360° - 36.87° = \mathbf{323.13°}$

Section E

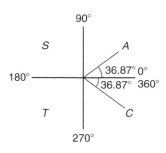

Figure 42.5

Now try the following Practice Exercise

Practice Exercise 180   Further problems on trigonometric equations (answers on page 1171)

Solve the following equations for angles between $0°$ and $360°$

1. $4 - 7\sin\theta = 0$

2. $3\,\text{cosec}\,A + 5.5 = 0$

3. $4(2.32 - 5.4\cot t) = 0$

In Problems 4 to 6, solve for θ in the range $0° \le \theta \le 360°$

4. $\sec\theta = 2$

5. $\cot\theta = 0.6$

6. $\text{cosec}\,\theta = 1.5$

In Problems 7 to 9, solve for x in the range $-180° \le x \le 180°$

7. $\sec x = -1.5$

8. $\cot x = 1.2$

9. $\text{cosec}\,x = -2$

In Problems 10 and 11, solve for θ in the range $0° \le \theta \le 360°$

10. $3\sin\theta = 2\cos\theta$

11. $5\cos\theta = -\sin\theta$

42.5 Worked problems (ii) on trigonometric equations

Problem 10.   Solve: $2 - 4\cos^2 A = 0$ for values of A in the range $0° < A < 360°$

$2 - 4\cos^2 A = 0$, from which $\cos^2 A = \dfrac{2}{4} = 0.5000$

Hence $\cos A = \sqrt{0.5000} = \pm 0.7071$ and
$A = \cos^{-1}(\pm 0.7071)$

Cosine is positive in quadrant one and four and negative in quadrants two and three. Thus in this case there are four solutions, one in each quadrant (see Figure 42.6). The acute angle $\cos^{-1} 0.7071 = 45°$

Hence $\boldsymbol{A = 45°, 135°, 225°\ \text{or}\ 315°}$

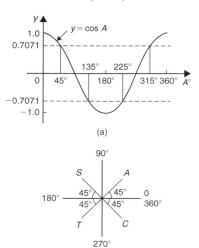

Figure 42.6

Problem 11.   Solve: $\frac{1}{2}\cot^2 y = 1.3$ for $0° < y < 360°$

$\frac{1}{2}\cot^2 y = 1.3$, from which, $\cot^2 y = 2(1.3) = 2.6$

Hence $\cot y = \sqrt{2.6} = \pm 1.6125$, and
$y = \cot^{-1}(\pm 1.6125)$. There are four solutions, one in each quadrant. The acute angle $\cot^{-1} 1.6125 = 31.81°$

Hence $\boldsymbol{y = 31.81°, 148.19°, 211.81°\ \text{or}\ 328.19°}$

Now try the following Practice Exercise

Practice Exercise 181   Further problems on trigonometric equations (answers on page 1171)

Solve the following equations for angles between $0°$ and $360°$

1. $5\sin^2 y = 3$

2. $\cos^2\theta = 0.25$

3. $\tan^2 x = 3$

4. $5 + 3\csc^2 D = 8$

5. $2\cot^2\theta = 5$

42.6 Worked problems (iii) on trigonometric equations

Problem 12. Solve the equation:
$8\sin^2\theta + 2\sin\theta - 1 = 0$, for all values of θ between $0°$ and $360°$

Factorising $8\sin^2\theta + 2\sin\theta - 1 = 0$ gives
$(4\sin\theta - 1)(2\sin\theta + 1) = 0$

Hence $4\sin\theta - 1 = 0$, from which,

$\sin\theta = \frac{1}{4} = 0.2500$, or $2\sin\theta + 1 = 0$, from which,
$\sin\theta = -\frac{1}{2} = -0.5000$
(Instead of factorising, the quadratic formula can, of course, be used.) $\theta = \sin^{-1} 0.2500 = 14.48°$ or $165.52°$, since sine is positive in the first and second quadrants, or $\theta = \sin^{-1}(-0.5000) = 210°$ or $330°$, since sine is negative in the third and fourth quadrants. Hence

$$\theta = 14.48°, 165.52°, 210° \text{ or } 330°$$

Problem 13. Solve:
$6\cos^2\theta + 5\cos\theta - 6 = 0$ for values of θ from $0°$ to $360°$

Factorising $6\cos^2\theta + 5\cos\theta - 6 = 0$ gives
$(3\cos\theta - 2)(2\cos\theta + 3) = 0$

Hence $3\cos\theta - 2 = 0$, from which,

$\cos\theta = \frac{2}{3} = 0.6667$, or $2\cos\theta + 3 = 0$, from which,
$\cos = -\frac{3}{2} = -1.5000$

The minimum value of a cosine is -1, hence the latter expression has no solution and is thus neglected. Hence

$$\theta = \cos^{-1} 0.6667 = 48.18° \text{ or } 311.82°$$

since cosine is positive in the first and fourth quadrants.

Now try the following Practice Exercise

Practice Exercise 182   Further problems on trigonometric equations (answers on page 1171)

Solve the following equations for angles between $0°$ and $360°$

1. $15\sin^2 A + \sin A - 2 = 0$

2. $8\tan^2\theta + 2\tan\theta = 15$

3. $2\csc^2 t - 5\csc t = 12$

4. $2\cos^2\theta + 9\cos\theta - 5 = 0$

42.7 Worked problems (iv) on trigonometric equations

Problem 14. Solve: $5\cos^2 t + 3\sin t - 3 = 0$ for values of t from $0°$ to $360°$

Since $\cos^2 t + \sin^2 t = 1$, $\cos^2 t = 1 - \sin^2 t$.
Substituting for $\cos^2 t$ in $5\cos^2 t + 3\sin t - 3 = 0$ gives

$$5(1 - \sin^2 t) + 3\sin t - 3 = 0$$
$$5 - 5\sin^2 t + 3\sin t - 3 = 0$$
$$-5\sin^2 t + 3\sin t + 2 = 0$$
$$5\sin^2 t - 3\sin t - 2 = 0$$

Factorising gives $(5\sin t + 2)(\sin t - 1) = 0$. Hence $5\sin t + 2 = 0$, from which, $\sin t = -\frac{2}{5} = -0.4000$, or $\sin t - 1 = 0$, from which, $\sin t = 1$
$t = \sin^{-1}(-0.4000) = 203.58°$ or $336.42°$, since sine is negative in the third and fourth quadrants, or $t = \sin^{-1} 1 = 90°$. Hence

$$t = 90°, 203.58° \text{ or } 336.42°$$

as shown in Figure 42.7.

Problem 15. Solve: $18\sec^2 A - 3\tan A = 21$ for values of A between $0°$ and $360°$

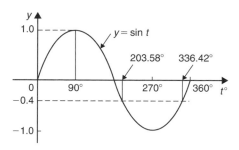

Figure 42.7

$1 + \tan^2 A = \sec^2 A$. Substituting for $\sec^2 A$ in $18 \sec^2 A - 3 \tan A = 21$ gives

$$18(1 + \tan^2 A) - 3 \tan A = 21$$

i.e. $18 + 18 \tan^2 A - 3 \tan A - 21 = 0$

$$18 \tan^2 A - 3 \tan A - 3 = 0$$

Factorising gives $(6 \tan A - 3)(3 \tan A + 1) = 0$

Hence $6 \tan A - 3 = 0$, from which, $\tan A = \frac{3}{6} = 0.5000$ or $3 \tan A + 1 = 0$, from which, $\tan A = -\frac{1}{3} = -0.3333$ Thus $A = \tan^{-1}(0.5000) = 26.57°$ or $206.57°$, since tangent is positive in the first and third quadrants, or $A = \tan^{-1}(-0.3333) = 161.57°$ or $341.57°$, since tangent is negative in the second and fourth quadrants. Hence

$$A = 26.57°, \, 161.57°, \, 206.57° \text{ or } 341.57°$$

Problem 16.   Solve: $3 \operatorname{cosec}^2\theta - 5 = 4 \cot\theta$ in the range $0 < \theta < 360°$

$\cot^2\theta + 1 = \operatorname{cosec}^2\theta$. Substituting for $\operatorname{cosec}^2\theta$ in $3 \operatorname{cosec}^2\theta - 5 = 4 \cot\theta$ gives

$$3(\cot^2\theta + 1) - 5 = 4 \cot\theta$$

$$3 \cot^2\theta + 3 - 5 = 4 \cot\theta$$

$$3 \cot^2\theta - 4 \cot\theta - 2 = 0$$

Since the left-hand side does not factorise, the quadratic formula is used.

Thus, $\cot\theta = \dfrac{-(-4) \pm \sqrt{(-4)^2 - 4(3)(-2)}}{2(3)}$

$$= \frac{4 \pm \sqrt{16 + 24}}{6} = \frac{4 \pm \sqrt{40}}{6}$$

$$= \frac{10.3246}{6} \text{ or } -\frac{2.3246}{6}$$

Hence $\cot\theta = 1.7208$ or -0.3874
$\theta = \cot^{-1} 1.7208 = 30.17°$ or $210.17°$, since cotangent is positive in the first and third quadrants, or $\theta = \cot^{-1}(-0.3874) = 111.18°$ or $291.18°$, since cotangent is negative in the second and fourth quadrants.

Hence, $\theta = \mathbf{30.17°, \, 111.18°, \, 210.17° \text{ or } 291.18°}$

Now try the following Practice Exercise

Practice Exercise 183   Further problems on trigonometric equations (answers on page 1171)

Solve the following equations for angles between $0°$ and $360°$

1. $2 \cos^2\theta + \sin\theta = 1$

2. $4 \cos^2 t + 5 \sin t = 3$

3. $2 \cos\theta - 4 \sin^2\theta = 0$

4. $3 \cos\theta + 2 \sin^2\theta = 3$

5. $12 \sin^2\theta - 6 = \cos\theta$

6. $16 \sec x - 2 = 14 \tan^2 x$

7. $4 \cot^2 A - 6 \operatorname{cosec} A + 6 = 0$

8. $5 \sec t + 2 \tan^2 t = 3$

9. $2.9 \cos^2 a - 7 \sin a + 1 = 0$

10. $3 \operatorname{cosec}^2\beta = 8 - 7 \cot\beta$

11. $\cot\theta = \sin\theta$

12. $\tan\theta + 3 \cot\theta = 5 \sec\theta$

For fully worked solutions to each of the problems in Practice Exercises 179 to 183 in this chapter, go to the website:
www.routledge.com/cw/bird

Section E

Chapter 43

The relationship between trigonometric and hyperbolic functions

> **Why it is important to understand: The relationship between trigonometric and hyperbolic functions**
>
> There are similarities between notations used for hyperbolic and trigonometric functions. Both trigonometric and hyperbolic functions have applications in many areas of engineering. For example, the shape of a chain hanging under gravity (known as a catenary curve) is described by the hyperbolic cosine, cosh x, and the deformation of uniform beams can be expressed in terms of hyperbolic tangents. Hyperbolic functions are also used in electrical engineering applications such as transmission line theory. Einstein's special theory of relativity used hyperbolic functions and both trigonometric and hyperbolic functions are needed for certain areas of more advanced integral calculus. There arc many identities showing relationships between hyperbolic and trigonometric functions; these may be used to evaluate trigonometric and hyperbolic functions of complex numbers.

At the end of this chapter, you should be able to:

- identify the relationship between trigonometric and hyperbolic functions
- state hyperbolic identities

43.1 The relationship between trigonometric and hyperbolic functions

In Chapter 46, it is shown that

$$\cos\theta + j\sin\theta = e^{j\theta} \qquad (1)$$
$$\text{and} \quad \cos\theta - j\sin\theta = e^{-j\theta} \qquad (2)$$

Adding equations (1) and (2) gives:

$$\cos\theta = \frac{1}{2}(e^{j\theta} + e^{-j\theta}) \qquad (3)$$

Subtracting equation (2) from equation (1) gives:

$$\sin\theta = \frac{1}{2j}(e^{j\theta} - e^{-j\theta}) \qquad (4)$$

Substituting $j\theta$ for θ in equations (3) and (4) gives:

$$\cos j\theta = \frac{1}{2}(e^{j(j\theta)} + e^{-j(j\theta)})$$

and $\sin j\theta = \dfrac{1}{2j}(e^{j(j\theta)} - e^{-j(j\theta)})$

Since $j^2 = -1$, $\cos j\theta = \frac{1}{2}(e^{-\theta} + e^{\theta}) = \frac{1}{2}(e^{\theta} + e^{-\theta})$

Hence from Chapter 24, $\cos j\theta = \cosh\theta$ (5)

Similarly, $\sin j\theta = \dfrac{1}{2j}(e^{-\theta} - e^{\theta}) = -\dfrac{1}{2j}(e^{\theta} - e^{-\theta})$

$$= \dfrac{-1}{j}\left[\dfrac{1}{2}(e^{\theta} - e^{-\theta})\right]$$

$$= -\dfrac{1}{j}\sinh\theta \quad \text{(see Chapter 24)}$$

But $-\dfrac{1}{j} = -\dfrac{1}{j} \times \dfrac{j}{j} = -\dfrac{j}{j^2} = j$

hence $\sin j\theta = j\sinh\theta$ (6)

Equations (5) and (6) may be used to verify that in all standard trigonometric identities, $j\theta$ may be written for θ and the identity still remains true.

Problem 1. Verify that $\cos^2 j\theta + \sin^2 j\theta = 1$

From equation (5), $\cos j\theta = \cosh\theta$, and from equation (6), $\sin j\theta = j\sinh\theta$.

Thus, $\cos^2 j\theta + \sin^2 j\theta = \cosh^2\theta + j^2\sinh^2\theta$,   and since $j^2 = -1$,

$$\cos^2 j\theta + \sin^2 j\theta = \cosh^2\theta - \sinh^2\theta$$

But from Chapter 24, Problem 9 (a), page 231

$$\cosh^2\theta - \sinh^2\theta = 1,$$

hence $\cos^2 j\theta + \sin^2 j\theta = 1$

Problem 2. Verify that $\sin j2A = 2\sin jA\cos jA$

From equation (6), writing $2A$ for θ, $\sin j2A = j\sinh 2A$, and from Chapter 24, Table 24.1, page 232, $\sinh 2A = 2\sinh A\cosh A$

Hence, $\sin j2A = j(2\sinh A\cosh A)$

But, $\sinh A = \frac{1}{2}(e^A - e^{-A})$ and $\cosh A = \frac{1}{2}(e^A + e^{-A})$

Hence, $\sin j2A = j2\left(\dfrac{e^A - e^{-A}}{2}\right)\left(\dfrac{e^A + e^{-A}}{2}\right)$

$$= -\dfrac{2}{j}\left(\dfrac{e^A - e^{-A}}{2}\right)\left(\dfrac{e^A + e^{-A}}{2}\right)$$

$$= -\dfrac{2}{j}\left(\dfrac{\sin jA}{j}\right)(\cos jA)$$

$$= 2\sin jA\cos jA \text{ since } j^2 = -1$$

i.e. $\sin j2A = 2\sin jA\cos jA$

Now try the following Practice Exercise

Practice Exercise 184   Further problems on the relationship between trigonometric and hyperbolic functions (answers on page 1171)

Verify the following identities by expressing in exponential form.

1. $\sin j(A + B) = \sin jA\cos jB + \cos jA\sin jB$

2. $\cos j(A - B) = \cos jA\cos jB + \sin jA\sin jB$

3. $\cos j2A = 1 - 2\sin^2 jA$

4. $\sin jA\cos jB = \frac{1}{2}[\sin j(A + B) + \sin j(A - B)]$

5. $\sin jA - \sin jB$

$$-2\cos j\left(\dfrac{A + B}{2}\right)\sin j\left(\dfrac{A - B}{2}\right)$$

43.2 Hyperbolic identities

From Chapter 24, $\cosh\theta = \frac{1}{2}(e^{\theta} + e^{-\theta})$

Substituting $j\theta$ for θ gives:

$\cosh j\theta = \frac{1}{2}(e^{j\theta} + e^{-j\theta}) = \cos\theta$, from equation (3),

i.e. $\cosh j\theta = \cos\theta$ (7)

Similarly, from Chapter 24,

$$\sinh\theta = \frac{1}{2}(e^{\theta} - e^{-\theta})$$

Substituting $j\theta$ for θ gives:

$\sinh j\theta = \frac{1}{2}(e^{j\theta} - e^{-j\theta}) = j\sin\theta$, from equation (4).

Hence $\quad \sinh j\theta = j\sin\theta$ $\qquad$ (8)

$$\tan j\theta = \frac{\sin j\theta}{\cosh j\theta}$$

From equations (5) and (6),

$$\frac{\sin j\theta}{\cos j\theta} = \frac{j\sinh\theta}{\cosh\theta} = j\tanh\theta$$

Hence $\quad \tan j\theta = j\tanh\theta$ $\qquad$ (9)

Similarly, $\quad \tanh j\theta = \dfrac{\sinh j\theta}{\cosh j\theta}$

From equations (7) and (8),

$$\frac{\sinh j\theta}{\cosh j\theta} = \frac{j\sin\theta}{\cos\theta} = j\tan\theta$$

Hence $\quad \tanh j\theta = j\tan\theta$ $\qquad$ (10)

Two methods are commonly used to verify hyperbolic identities. These are (a) by substituting $j\theta$ (and $j\phi$) in the corresponding trigonometric identity and using the relationships given in equations (5) to (10) (see Problems 3 to 5) and (b) by applying Osborne's rule given in Chapter 24, page 231.

Problem 3. By writing jA for θ in $\cot^2\theta + 1 = \operatorname{cosec}^2\theta$, determine the corresponding hyperbolic identity

Substituting jA for θ gives:

$$\cot^2 jA + 1 = \operatorname{cosec}^2 jA,$$

i.e. $\quad \dfrac{\cos^2 jA}{\sin^2 jA} + 1 = \dfrac{1}{\sin^2 jA}$

But from equation (5), $\cos jA = \cosh A$

and from equation (6), $\sin jA = j\sinh A$

Hence $\quad \dfrac{\cosh^2 A}{j^2\sinh^2 A} + 1 = \dfrac{1}{j^2\sinh^2 A}$

and since $\quad j^2 = -1, -\dfrac{\cosh^2 A}{\sinh^2 A} + 1 = -\dfrac{1}{\sinh^2 A}$

Multiplying throughout by -1, gives:

$$\frac{\cosh^2 A}{\sinh^2 A} - 1 = \frac{1}{\sinh^2 A}$$

i.e. $\quad \mathbf{\coth^2 A - 1 = \operatorname{cosech}^2 A}$

Problem 4. By substituting jA and jB for θ and ϕ respectively in the trigonometric identity for $\cos\theta - \cos\phi$, show that

$$\cosh A - \cosh B = 2\sinh\left(\frac{A+B}{2}\right)\sinh\left(\frac{A-B}{2}\right)$$

$$\cos\theta - \cos\phi = -2\sin\left(\frac{\theta+\phi}{2}\right)\sin\left(\frac{\theta-\phi}{2}\right)$$

(see Chapter 44, page 509)

thus $\cos jA - \cos jB$

$$= -2\sin j\left(\frac{A+B}{2}\right)\sin j\left(\frac{A-B}{2}\right)$$

But from equation (5), $\cos jA = \cosh A$

and from equation (6), $\sin jA = j\sinh A$

Hence, $\cosh A - \cosh B$

$$= -2\, j\sinh\left(\frac{A+B}{2}\right)\, j\sinh\left(\frac{A-B}{2}\right)$$

$$= -2\, j^2\sinh\left(\frac{A+B}{2}\right)\sinh\left(\frac{A-B}{2}\right)$$

But $\quad j^2 = -1$, hence

$$\mathbf{\cosh A - \cosh B = 2\sinh\left(\frac{A+B}{2}\right)\sinh\left(\frac{A-B}{2}\right)}$$

Problem 5. Develop the hyperbolic identity corresponding to $\sin 3\theta = 3\sin\theta - 4\sin^3\theta$ by writing jA for θ

Substituting jA for θ gives:

$$\sin 3\,jA = 3\sin jA - 4\sin^3 jA$$

and since from equation (6),

$$\sin jA = j\sinh A,$$

$$j\sinh 3A = 3j\sinh A - 4\,j^3\sinh^3 A$$

Dividing throughout by j gives:

$$\sinh 3A = 3\sinh A - j^2 4\sinh^3 A$$

But $\quad j^2 = -1$, hence

$$\mathbf{\sinh 3A = 3\sinh A + 4\sinh^3 A}$$

(An examination of Problems 3 to 5 shows that whenever the trigonometric identity contains a term which is the product of two sines, or the implied product of two sines [e.g. $\tan^2 \theta = \sin^2 \theta / \cos^2 \theta$, thus $\tan^2 \theta$ is the implied product of two sines], the sign of the corresponding term in the hyperbolic function changes. This relationship between trigonometric and hyperbolic functions is known as Osborne's rule, as discussed in Chapter 24, page 231.)

Now try the following Practice Exercise

Practice Exercise 185   Further problems on hyperbolic identities (answers on page 1171)

In Problems 1 to 7, use the substitution $A = j\theta$ (and $B = j\phi$) to obtain the hyperbolic identities corresponding to the trigonometric identities given.

1. $1 + \tan^2 A = \sec^2 A$

2. $\cos(A + B) = \cos A \cos B - \sin A \sin B$

3. $\sin(A - B) = \sin A \cos B - \cos A \sin B$

4. $\tan 2A = \dfrac{2 \tan A}{1 - \tan^2 A}$

5. $\cos A \sin B = \dfrac{1}{2}[\sin(A + B) - \sin(A - B)]$

6. $\sin^3 A = \dfrac{3}{4} \sin A - \dfrac{1}{4} \sin 3A$

7. $\cot^2 A(\sec^2 A - 1) = 1$

For fully worked solutions to each of the problems in Practice Exercises 184 and 185 in this chapter, go to the website:
www.routledge.com/cw/bird

Chapter 44

Compound angles

Why it is important to understand: Compound angles

It is often necessary to rewrite expressions involving sines, cosines and tangents in alternative forms. To do this, formulae known as trigonometric identities are used, as explained previously. Compound angle (or sum and difference) formulae and double angles are further commonly used identities. Compound angles are required, for example, in the analysis of acoustics (where a beat is an interference between two sounds of slightly different frequencies), and with phase detectors (which is a frequency mixer, analogue multiplier or logic circuit that generates a voltage signal which represents the difference in phase between two signal inputs). Many rational functions of sine and cosine are difficult to integrate without compound angle formulae.

At the end of this chapter, you should be able to:

- state compound angle formulae for $\sin(A \pm B), \cos(A \pm B)$ and $\tan(A \pm B)$
- convert a $\sin \omega t + b \cos \omega t$ into $R \sin(\omega t + \alpha)$
- derive double angle formulae
- change products of sines and cosines into sums or differences
- change sums or differences of sines and cosines into products
- develop expressions for power in a.c. circuits – purely resistive, inductive and capacitive circuits, R–L and R–C circuits

44.1   Compound angle formulae

An electric current i may be expressed as $i = 5\sin(\omega t - 0.33)$ amperes. Similarly, the displacement x of a body from a fixed point can be expressed as $x = 10\sin(2t + 0.67)$ metres. The angles $(\omega t - 0.33)$ and $(2t + 0.67)$ are called **compound angles** because they are the sum or difference of two angles. The **compound angle formulae** for sines and cosines of the sum and difference of two angles A and B are:

$$\sin(A + B) = \sin A \cos B + \cos A \sin B$$

$$\sin(A - B) = \sin A \cos B - \cos A \sin B$$

$$\cos(A + B) = \cos A \cos B - \sin A \sin B$$

$$\cos(A - B) = \cos A \cos B + \sin A \sin B$$

(Note, $\sin(A + B)$ is **not** equal to $(\sin A + \sin B)$, and so on.)

The formulae stated above may be used to derive two further compound angle formulae:

$$\tan(A + B) = \frac{\tan A + \tan B}{1 - \tan A \tan B}$$

$$\tan(A - B) = \frac{\tan A - \tan B}{1 + \tan A \tan B}$$

The compound angle formulae are true for all values of A and B, and by substituting values of A and B into the formulae they may be shown to be true.

Problem 1. Expand and simplify the following expressions:
(a) $\sin(\pi + \alpha)$ (b) $-\cos(90° + \beta)$
(c) $\sin(A - B) - \sin(A + B)$

(a) $\sin(\pi + \alpha) = \sin\pi\cos\alpha + \cos\pi\sin\alpha$ (from the formula for $\sin(A + B)$)
$$= (0)(\cos\alpha) + (-1)(\sin\alpha) = -\sin\alpha$$

(b) $-\cos(90° + \beta)$

$$= -[\cos 90°\cos\beta - \sin 90°\sin\beta]$$
$$= -[(0)(\cos\beta) - (1)(\sin\beta)] = \sin\beta$$

(c) $\sin(A - B) - \sin(A + B)$

$$= [\sin A\cos B - \cos A\sin B]$$
$$- [\sin A\cos B + \cos A\sin B]$$
$$= -2\cos A\sin B$$

Problem 2. Prove that
$$\cos(y - \pi) + \sin\left(y + \frac{\pi}{2}\right) = 0$$

$\cos(y - \pi) = \cos y\cos\pi + \sin y\sin\pi$
$$= (\cos y)(-1) + (\sin y)(0)$$
$$= -\cos y$$
$\sin\left(y + \frac{\pi}{2}\right) = \sin y\cos\frac{\pi}{2} + \cos y\sin\frac{\pi}{2}$
$$= (\sin y)(0) + (\cos y)(1) = \cos y$$

Hence $\cos(y - \pi) + \sin\left(y + \frac{\pi}{2}\right)$

$$= (-\cos y) + (\cos y) = 0$$

Problem 3. Show that
$$\tan\left(x + \frac{\pi}{4}\right)\tan\left(x - \frac{\pi}{4}\right) = -1$$

$$\tan\left(x + \frac{\pi}{4}\right) = \frac{\tan x + \tan\frac{\pi}{4}}{1 - \tan x\tan\frac{\pi}{4}}$$

from the formula for $\tan(A + B)$

$$= \frac{\tan x + 1}{1 - (\tan x)(1)} = \left(\frac{1 + \tan x}{1 - \tan x}\right)$$

since $\tan\frac{\pi}{4} = 1$

$$\tan\left(x - \frac{\pi}{4}\right) = \frac{\tan x - \tan\frac{\pi}{4}}{1 + \tan x\tan\frac{\pi}{4}} = \left(\frac{\tan x - 1}{1 + \tan x}\right)$$

Hence $\tan\left(x + \frac{\pi}{4}\right)\tan\left(x - \frac{\pi}{4}\right)$

$$= \left(\frac{1 + \tan x}{1 - \tan x}\right)\left(\frac{\tan x - 1}{1 + \tan x}\right)$$

$$= \frac{\tan x - 1}{1 - \tan x} = \frac{-(1 - \tan x)}{1 - \tan x} = -1$$

Problem 4. If $\sin P = 0.8142$ and $\cos Q = 0.4432$ evaluate, correct to 3 decimal places:
(a) $\sin(P - Q)$, (b) $\cos(P + Q)$ and
(c) $\tan(P + Q)$,
using the compound angle formulae.

Since $\sin P = 0.8142$ then
$P = \sin^{-1} 0.8142 = 54.51°$
Thus $\cos P = \cos 54.51° = 0.5806$ and
$\tan P = \tan 54.51° = 1.4025$

Since $\cos Q = 0.4432$, $Q = \cos^{-1} 0.4432 = 63.69°$
Thus $\sin Q = \sin 63.69° = 0.8964$ and
$\tan Q = \tan 63.69° = 2.0225$

(a) $\sin(P - Q)$

$$= \sin P\cos Q - \cos P\sin Q$$
$$= (0.8142)(0.4432) - (0.5806)(0.8964)$$
$$= 0.3609 - 0.5204 = -0.160$$

(b) $\cos(P + Q)$

$$= \cos P\cos Q - \sin P\sin Q$$
$$= (0.5806)(0.4432) - (0.8142)(0.8964)$$
$$= 0.2573 - 0.7298 = -0.473$$

(c) $\tan(P + Q)$

$$= \frac{\tan P + \tan Q}{1 - \tan P\tan Q} = \frac{(1.4025) + (2.0225)}{1 - (1.4025)(2.0225)}$$

$$= \frac{3.4250}{-1.8366} = -1.865$$

Section E

Problem 5. Solve the equation

$$4\sin(x - 20°) = 5\cos x$$

for values of x between $0°$ and $90°$

$4\sin(x - 20°) = 4[\sin x \cos 20° - \cos x \sin 20°],$
$\qquad\qquad$ from the formula for $\sin(A - B)$

$\qquad = 4[\sin x (0.9397) - \cos x (0.3420)]$

$\qquad = 3.7588\sin x - 1.3680\cos x$

Since $4\sin(x - 20°) = 5\cos x$ then
$3.7588\sin x - 1.3680\cos x = 5\cos x$
Rearranging gives:

$3.7588\sin x = 5\cos x + 1.3680\cos x$

$\qquad\qquad = 6.3680\cos x$

and $\qquad \dfrac{\sin x}{\cos x} = \dfrac{6.3680}{3.7588} = 1.6942$

i.e. $\tan x = 1.6942$, and $x = \tan^{-1} 1.6942 = 59.449°$ or
59°27′

[Check: LHS $= 4\sin(59.449° - 20°)$

$\qquad\qquad = 4\sin 39.449° = 2.542$

$\qquad$ RHS $= 5\cos x = 5\cos 59.449° = 2.542$]

Now try the following Practice Exercise

Practice Exercise 186   Further problems on compound angle formulae (answers on page 1171)

1. Reduce the following to the sine of one angle:

    (a) $\sin 37° \cos 21° + \cos 37° \sin 21°$
    (b) $\sin 7t \cos 3t - \cos 7t \sin 3t$

2. Reduce the following to the cosine of one angle:

    (a) $\cos 71° \cos 33° - \sin 71° \sin 33°$
    (b) $\cos\dfrac{\pi}{3}\cos\dfrac{\pi}{4} + \sin\dfrac{\pi}{3}\sin\dfrac{\pi}{4}$

3. Show that:
    (a) $\sin\left(x + \dfrac{\pi}{3}\right) + \sin\left(x + \dfrac{2\pi}{3}\right) = \sqrt{3}\cos x$

and

    (b) $-\sin\left(\dfrac{3\pi}{2} - \phi\right) = \cos\phi$

4. Prove that:
    (a) $\sin\left(\theta + \dfrac{\pi}{4}\right) - \sin\left(\theta - \dfrac{3\pi}{4}\right)$

    $\qquad\qquad = \sqrt{2}(\sin\theta + \cos\theta)$

    (b) $\dfrac{\cos(270° + \theta)}{\cos(360° - \theta)} = \tan\theta$

5. Given $\cos A = 0.42$ and $\sin B = 0.73$ evaluate
    (a) $\sin(A - B)$, (b) $\cos(A - B)$, and
    (c) $\tan(A + B)$, correct to 4 decimal places.

In Problems 6 and 7, solve the equations for values of θ between $0°$ and $360°$.

6. $3\sin(\theta + 30°) = 7\cos\theta$

7. $4\sin(\theta - 40°) = 2\sin\theta$

44.2   Conversion of $a \sin \omega t + b \cos \omega t$ into $R \sin(\omega t + \alpha)$

(i) $R\sin(\omega t + \alpha)$ represents a sine wave of maximum value R, periodic time $2\pi/\omega$, frequency $\omega/2\pi$ and leading $R\sin\omega t$ by angle α (see Chapter 39).

(ii) $R\sin(\omega t + \alpha)$ may be expanded using the compound-angle formula for $\sin(A + B)$, where $A = \omega t$ and $B = \alpha$. Hence,

$R\sin(\omega t + \alpha)$

$\qquad = R[\sin\omega t \cos\alpha + \cos\omega t \sin\alpha]$

$\qquad = R\sin\omega t \cos\alpha + R\cos\omega t \sin\alpha$

$\qquad = (R\cos\alpha)\sin\omega t + (R\sin\alpha)\cos\omega t$

(iii) If $a = R\cos\alpha$ and $b = R\sin\alpha$, where a and b are constants, then $R\sin(\omega t + \alpha) = a\sin\omega t + b\cos\omega t$, i.e. a sine and cosine function of the same frequency when added produce a sine wave of the same frequency (which is further demonstrated in Chapter 50).

(iv) Since $a = R\cos\alpha$, then $\cos\alpha = a/R$, and since $b = R\sin\alpha$, then $\sin\alpha = b/R$

Section E

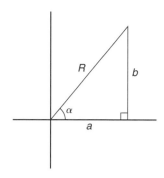

Figure 44.1

If the values of a and b are known then the values of R and α may be calculated. The relationship between constants a, b, R and α is shown in Figure 44.1.

From Figure 44.1, by Pythagoras' theorem:

$$R = \sqrt{a^2 + b^2}$$

and from trigonometric ratios:

$$\alpha = \tan^{-1} b/a$$

Problem 6. Find an expression for $3\sin\omega t + 4\cos\omega t$ in the form $R\sin(\omega t + \alpha)$ and sketch graphs of $3\sin\omega t$, $4\cos\omega t$ and $R\sin(\omega t + \alpha)$ on the same axes

Let $3\sin\omega t + 4\cos\omega t = R\sin(\omega t + \alpha)$

then $3\sin\omega t + 4\cos\omega t$

$$= R[\sin\omega t \cos\alpha + \cos\omega t \sin\alpha]$$
$$= (R\cos\alpha)\sin\omega t + (R\sin\alpha)\cos\omega t$$

Equating coefficients of $\sin\omega t$ gives:

$$3 = R\cos\alpha, \text{ from which, } \cos\alpha = \frac{3}{R}$$

Equating coefficients of $\cos\omega t$ gives:

$$4 = R\sin\alpha, \text{ from which, } \sin\alpha = \frac{4}{R}$$

There is only one quadrant where both $\sin\alpha$ **and** $\cos\alpha$ are positive, and this is the first, as shown in Figure 44.2. From Figure 44.2, by Pythagoras' theorem:

$$R = \sqrt{(3^2 + 4^2)} = 5$$

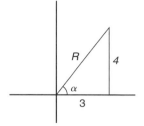

Figure 44.2

From trigonometric ratios: $\alpha = \tan^{-1}\frac{4}{3} = 53.13°$ or 0.927 radians.

Hence $3\sin\omega t + 4\cos\omega t = 5\sin(\omega t + 0.927)$

A sketch of $3\sin\omega t$, $4\cos\omega t$ and $5\sin(\omega t + 0.927)$ is shown in Figure 44.3.

Two periodic functions of the same frequency may be combined by,

(a) plotting the functions graphically and combining ordinates at intervals, or

(b) resolution of phasors by drawing or calculation.

Problem 6, together with Problems 7 and 8 following, demonstrate a third method of combining waveforms. See also Chapter 50 for further methods of combining waveforms.

Problem 7. Express $4.6\sin\omega t - 7.3\cos\omega t$ in the form $R\sin(\omega t + \alpha)$

Let $4.6\sin\omega t - 7.3\cos\omega t = R\sin(\omega t + \alpha)$

then $4.6\sin\omega t - 7.3\cos\omega t$

$$= R[\sin\omega t \cos\alpha + \cos\omega t \sin\alpha]$$
$$= (R\cos\alpha)\sin\omega t + (R\sin\alpha)\cos\omega t$$

Equating coefficients of $\sin\omega t$ gives:

$$4.6 = R\cos\alpha, \text{ from which, } \cos\alpha = \frac{4.6}{R}$$

Equating coefficients of $\cos\omega t$ gives:

$$-7.3 = R\sin\alpha, \text{ from which, } \sin\alpha = \frac{-7.3}{R}$$

There is only one quadrant where cosine is positive **and** sine is negative, i.e. the fourth quadrant, as shown in Figure 44.4. By Pythagoras' theorem:

$$R = \sqrt{[(4.6)^2 + (-7.3)^2]} = 8.628$$

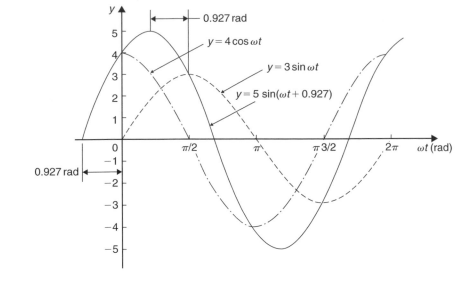

Figure 44.3

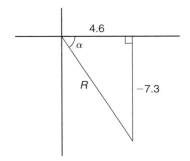

Figure 44.4

By trigonometric ratios:

$$\alpha = \tan^{-1}\left(\frac{-7.3}{4.6}\right)$$

$$= -57.78° \text{ or } -1.008 \text{ radians.}$$

Hence

$$\mathbf{4.6\sin\omega t - 7.3\cos\omega t = 8.628\sin(\omega t - 1.008)}$$

> **Problem 8.** Express $-2.7\sin\omega t - 4.1\cos\omega t$ in the form $R\sin(\omega t + \alpha)$

Let $-2.7\sin\omega t - 4.1\cos\omega t = R\sin(\omega t + \alpha)$

$$= R[\sin\omega t\cos\alpha + \cos\omega t\sin\alpha]$$

$$= (R\cos\alpha)\sin\omega t + (R\sin\alpha)\cos\omega t$$

Equating coefficients gives:

$$-2.7 = R\cos\alpha, \text{ from which, } \cos\alpha = \frac{-2.7}{R}$$

and $\quad -4.1 = R\sin\alpha$, from which, $\sin\alpha = \dfrac{-4.1}{R}$

There is only one quadrant in which both cosine **and** sine are negative, i.e. the third quadrant, as shown in Figure 44.5. From Figure 44.5,

$$R = \sqrt{[(-2.7)^2 + (-4.1)^2]} = 4.909$$

and $\quad \theta = \tan^{-1}\dfrac{4.1}{2.7} = 56.63°$

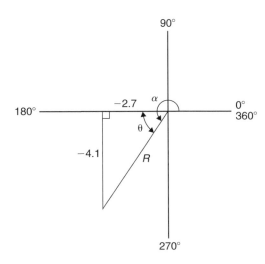

Figure 44.5

Hence $\alpha = 180° + 56.63° = 236.63°$ or 4.130 radians. **Thus,**

$$-2.7\sin\omega t - 4.1\cos\omega t = 4.909\sin(\omega t + 4.130)$$

An angle of $236.63°$ is the same as $-123.37°$ or -2.153 radians.

Hence $-2.7\sin\omega t - 4.1\cos\omega t$ may be expressed also as $\mathbf{4.909\sin(\omega t - 2.153)}$, which is preferred since it is the **principal value** (i.e. $-\pi \leq \alpha \leq \pi$).

> **Problem 9.** Express $3\sin\theta + 5\cos\theta$ in the form $R\sin(\theta + \alpha)$, and hence solve the equation $3\sin\theta + 5\cos\theta = 4$, for values of θ between $0°$ and $360°$

Let $3\sin\theta + 5\cos\theta = R\sin(\theta + \alpha)$

$$= R[\sin\theta\cos\alpha + \cos\theta\sin\alpha]$$

$$= (R\cos\alpha)\sin\theta + (R\sin\alpha)\cos\theta$$

Equating coefficients gives:

$$3 = R\cos\alpha, \text{from which}, \cos\alpha = \frac{3}{R}$$

and $5 = R\sin\alpha$, from which, $\sin\alpha = \dfrac{5}{R}$

Since both $\sin\alpha$ and $\cos\alpha$ are positive, R lies in the first quadrant, as shown in Figure 44.6.

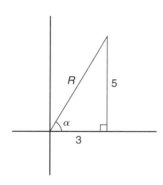

Figure 44.6

From Figure 44.6, $R = \sqrt{(3^2 + 5^2)} = 5.831$ and $\alpha = \tan^{-1}\frac{5}{3} = 59.04°$

Hence $3\sin\theta + 5\cos\theta = 5.831\sin(\theta + 59.04°)$

However, $3\sin\theta + 5\cos\theta = 4$

Thus $5.831\sin(\theta + 59.04°) = 4$, from which

$$(\theta + 59.04°) = \sin^{-1}\left(\frac{4}{5.831}\right)$$

i.e. $\theta + 59.04° = 43.32°$ or $136.68°$

Hence $\theta = 43.32° - 59.04° = -15.72°$

or $\theta = 136.68° - 59.04° = 77.64°$

Since $-15.72°$ is the same as $-15.72° + 360°$, i.e. $344.28°$, then the solutions are $\theta = 77.64°$ or $344.28°$, which may be checked by substituting into the original equation.

> **Problem 10.** Solve the equation $3.5\cos A - 5.8\sin A = 6.5$ for $0° \leq A \leq 360°$

Let $3.5\cos A - 5.8\sin A = R\sin(A + \alpha)$

$$= R[\sin A\cos\alpha + \cos A\sin\alpha]$$

$$= (R\cos\alpha)\sin A + (R\sin\alpha)\cos A$$

Equating coefficients gives:

$$3.5 = R\sin\alpha, \text{from which}, \sin\alpha = \frac{3.5}{R}$$

and $-5.8 = R\cos\alpha$, from which, $\cos\alpha = \dfrac{-5.8}{R}$

There is only one quadrant in which both sine is positive **and** cosine is negative, i.e. the second, as shown in Figure 44.7.

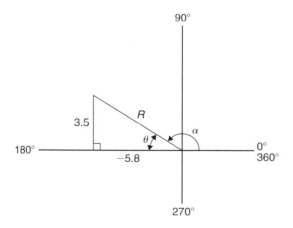

Figure 44.7

From Figure 44.7, $R = \sqrt{[(3.5)^2 + (-5.8)^2]} = 6.774$ and $\theta = \tan^{-1}\dfrac{3.5}{5.8} = 31.11°$

Hence $\alpha = 180° - 31.11° = 148.89°$

Thus

$$3.5\cos A - 5.8\sin A = 6.774\sin(A + 144.89°) = 6.5$$

Hence $\sin(A + 148.89°) = \dfrac{6.5}{6.774}$, from which,

$$(A + 148.89°) = \sin^{-1}\dfrac{6.5}{6.774}$$

$$= 73.65° \text{ or } 106.35°$$

Thus $A = 73.65° - 148.89° = -75.24°$

$$\equiv (-75.24° + 360°) = 284.76°$$

or $A = 106.35° - 148.89° = -42.54°$

$$\equiv (-42.54° + 360°) = 317.46°$$

The solutions are thus **A = 284.76° or 317.46°**, which may be checked in the original equation.

Now try the following Practice Exercise

Practice Exercise 187   Further problems on the conversion of $a\sin\omega t + b\cos\omega t$ into $R\sin(\omega t + \alpha)$ (answers on page 1171)

In Problems 1 to 4, change the functions into the form $R\sin(\omega t \pm \alpha)$.

1. $5\sin\omega t + 8\cos\omega t$

2. $4\sin\omega t - 3\cos\omega t$

3. $-7\sin\omega t + 4\cos\omega t$

4. $-3\sin\omega t - 6\cos\omega t$

5. Solve the following equations for values of θ between $0°$ and $360°$: (a) $2\sin\theta + 4\cos\theta = 3$ (b) $12\sin\theta - 9\cos\theta = 7$

6. Solve the following equations for $0° < A < 360°$: (a) $3\cos A + 2\sin A = 2.8$ (b) $12\cos A - 4\sin A = 11$

7. Solve the following equations for values of θ between $0°$ and $360°$: (a) $3\sin\theta + 4\cos\theta = 3$ (b) $2\cos\theta + \sin\theta = 2$

8. Solve the following equations for values of θ between $0°$ and $360°$: (a) $6\cos\theta + \sin\theta = \sqrt{3}$ (b) $2\sin 3\theta + 8\cos 3\theta = 1$

9. The third harmonic of a wave motion is given by $4.3\cos 3\theta - 6.9\sin 3\theta$. Express this in the form $R\sin(3\theta \pm \alpha)$

10. The displacement x metres of a mass from a fixed point about which it is oscillating is given by $x = 2.4\sin\omega t + 3.2\cos\omega t$, where t

is the time in seconds. Express x in the form $R\sin(\omega t + \alpha)$

11. Two voltages, $v_1 = 5\cos\omega t$ and $v_2 = -8\sin\omega t$ are inputs to an analogue circuit. Determine an expression for the output voltage if this is given by $(v_1 + v_2)$

12. The motion of a piston moving in a cylinder can be described by:

    $$x = (5\cos 2t + 5\sin 2t)\text{ cm}$$

    Express x in the form $R\sin(\omega t + \alpha)$

44.3   Double angles

(i) If, in the compound angle formula for $\sin(A + B)$, we let $B = A$ then

$$\sin 2A = 2\sin A\cos A$$

Also, for example,

$$\sin 4A = 2\sin 2A\cos 2A$$

and $\sin 8A = 2\sin 4A\cos 4A$, and so on.

(ii) If, in the compound angle formula for $\cos(A + B)$, we let $B = A$ then

$$\cos 2A = \cos^2 A - \sin^2 A$$

Since $\cos^2 A + \sin^2 A = 1$, then $\cos^2 A = 1 - \sin^2 A$, and $\sin^2 A = 1 - \cos^2 A$, and two further formulae for $\cos 2A$ can be produced.

Thus $\cos 2A = \cos^2 A - \sin^2 A$

$$= (1 - \sin^2 A) - \sin^2 A$$

i.e. $\cos 2A = 1 - 2\sin^2 A$

and $\cos 2A = \cos^2 A - \sin^2 A$

$$= \cos^2 A - (1 - \cos^2 A)$$

i.e. $\cos 2A = 2\cos^2 A - 1$

Also, for example,

$$\cos 4A = \cos^2 2A - \sin^2 2A \text{ or}$$

$$1 - 2\sin^2 2A \text{ or}$$

$$2\cos^2 2A - 1$$

and $\cos 6A = \cos^2 3A - \sin^2 3A \text{ or}$

$$1 - 2\sin^2 3A \text{ or}$$

$$2\cos^2 3A - 1,$$

and so on.

(iii) If, in the compound angle formula for $\tan(A+B)$, we let $B=A$ then

$$\tan 2A = \frac{2\tan A}{1-\tan^2 A}$$

Also, for example,

$$\tan 4A = \frac{2\tan 2A}{1-\tan^2 2A}$$

and $\tan 5A = \dfrac{2\tan\frac{5}{2}A}{1-\tan^2\frac{5}{2}A}$ and so on.

Problem 11. $I_3\sin 3\theta$ is the third harmonic of a waveform. Express the third harmonic in terms of the first harmonic $\sin\theta$, when $I_3=1$

When $I_3=1$,

$I_3\sin 3\theta = \sin 3\theta = \sin(2\theta+\theta)$

$\qquad = \sin 2\theta\cos\theta + \cos 2\theta\sin\theta,$

$\qquad\qquad\qquad$ from the $\sin(A+B)$ formula

$\qquad = (2\sin\theta\cos\theta)\cos\theta + (1-2\sin^2\theta)\sin\theta,$

$\qquad\qquad\qquad$ from the double angle expansions

$\qquad = 2\sin\theta\cos^2\theta + \sin\theta - 2\sin^3\theta$

$\qquad = 2\sin\theta(1-\sin^2\theta) + \sin\theta - 2\sin^3\theta,$

$\qquad\qquad\qquad$ (since $\cos^2\theta = 1-\sin^2\theta$)

$\qquad = 2\sin\theta - 2\sin^3\theta + \sin\theta - 2\sin^3\theta$

i.e. $\boldsymbol{\sin 3\theta = 3\sin\theta - 4\sin^3\theta}$

Problem 12. Prove that $\dfrac{1-\cos 2\theta}{\sin 2\theta} = \tan\theta$

$\text{LHS} = \dfrac{1-\cos 2\theta}{\sin 2\theta} = \dfrac{1-(1-2\sin^2\theta)}{2\sin\theta\cos\theta}$

$\qquad = \dfrac{2\sin^2\theta}{2\sin\theta\cos\theta} = \dfrac{\sin\theta}{\cos\theta}$

$\qquad = \tan\theta = \text{RHS}$

Problem 13. Prove that

$\qquad \cot 2x + \operatorname{cosec} 2x = \cot x$

$\text{LHS} = \cot 2x + \operatorname{cosec} 2x = \dfrac{\cos 2x}{\sin 2x} + \dfrac{1}{\sin 2x}$

$\qquad = \dfrac{\cos 2x+1}{\sin 2x}$

$\qquad = \dfrac{(2\cos^2 x - 1)+1}{\sin 2x}$

$\qquad = \dfrac{2\cos^2 x}{\sin 2x} = \dfrac{2\cos^2 x}{2\sin x\cos x}$

$\qquad = \dfrac{\cos x}{\sin x} = \cot x = \text{RHS}$

Problem 14. Solve the equation
$\cos 2\theta + 3\sin\theta = 2$ for θ in the range $0° \le \theta \le 360°$

Replacing the double angle term with the relationship $\cos 2\theta = 1-2\sin^2\theta$ gives:

$$1 - 2\sin^2\theta + 3\sin\theta = 2$$

Rearranging gives: $\qquad -2\sin^2\theta + 3\sin\theta - 1 = 0$
or $\qquad\qquad\qquad\quad 2\sin^2\theta - 3\sin\theta + 1 = 0$
which is a quadratic in $\sin\theta$
Using the quadratic formula or by factorising gives:

$$(2\sin\theta - 1)(\sin\theta - 1) = 0$$

from which, $\quad 2\sin\theta - 1 = 0 \quad$ or $\quad \sin\theta - 1 = 0$
and $\qquad\qquad\qquad \sin\theta = \frac{1}{2} \quad$ or $\quad \sin\theta = 1$
from which, $\qquad\quad \boldsymbol{\theta = 30° \text{ or } 150° \text{ or } 90°}$

Now try the following Practice Exercise

Practice Exercise 188   Further problems on double angles (answers on page 1171)

1. The power p in an electrical circuit is given by $p = \dfrac{v^2}{R}$. Determine the power in terms of V, R and $\cos 2t$ when $v = V\cos t$.

2. Prove the following identities:
   (a) $1 - \dfrac{\cos 2\phi}{\cos^2\phi} = \tan^2\phi$

   (b) $\dfrac{1+\cos 2t}{\sin^2 t} = 2\cot^2 t$

   (c) $\dfrac{(\tan 2x)(1+\tan x)}{\tan x} = \dfrac{2}{1-\tan x}$

   (d) $2\operatorname{cosec} 2\theta\cos 2\theta = \cot\theta - \tan\theta$

Section E

3. If the third harmonic of a waveform is given by $V_3 \cos 3\theta$, express the third harmonic in terms of the first harmonic $\cos\theta$, when $V_3 = 1$

In Problems 4 to 8, solve for θ in the range $-180° \le \theta \le 180°$

4. $\cos 2\theta = \sin\theta$

5. $3\sin 2\theta + 2\cos\theta = 0$

6. $\sin 2\theta + \cos\theta = 0$

7. $\cos 2\theta + 2\sin\theta = -3$

8. $\tan\theta + \cot\theta = 2$

44.4 Changing products of sines and cosines into sums or differences

(i) $\sin(A+B) + \sin(A-B) = 2\sin A \cos B$ (from the formulae in Section 44.1)

i.e. $\sin A \cos B$

$$= \tfrac{1}{2}[\sin(A+B) + \sin(A-B)] \qquad (1)$$

(ii) $\sin(A+B) - \sin(A-B) = 2\cos A \sin B$

i.e. $\cos A \sin B$

$$= \tfrac{1}{2}[\sin(A+B) - \sin(A-B)] \qquad (2)$$

(iii) $\cos(A+B) + \cos(A-B) = 2\cos A \cos B$

i.e. $\cos A \cos B$

$$= \tfrac{1}{2}[\cos(A+B) + \cos(A-B)] \qquad (3)$$

(iv) $\cos(A+B) - \cos(A-B) = -2\sin A \sin B$

i.e. $\sin A \sin B$

$$= -\tfrac{1}{2}[\cos(A+B) - \cos(A-B)] \qquad (4)$$

Problem 15.   Express $\sin 4x \cos 3x$ as a sum or difference of sines and cosines

From equation (1),

$\sin 4x \cos 3x = \tfrac{1}{2}[\sin(4x+3x) + \sin(4x-3x)]$

$$= \tfrac{1}{2}(\sin 7x + \sin x)$$

Problem 16.   Express $2\cos 5\theta \sin 2\theta$ as a sum or difference of sines or cosines

From equation (2),

$2\cos 5\theta \sin 2\theta = 2\left\{\dfrac{1}{2}[\sin(5\theta + 2\theta) - \sin(5\theta - 2\theta)]\right\}$

$$= \sin 7\theta - \sin 3\theta$$

Problem 17.   Express $3\cos 4t \cos t$ as a sum or difference of sines or cosines

From equation (3),

$3\cos 4t \cos t = 3\left\{\dfrac{1}{2}[\cos(4t+t) + \cos(4t-t)]\right\}$

$$= \dfrac{3}{2}(\cos 5t + \cos 3t)$$

Thus, if the integral $\int 3\cos 4t \cos t\, dt$ was required (for integration, see Chapter 63), then

$$\int 3\cos 4t \cos t\, dt = \int \dfrac{3}{2}(\cos 5t + \cos 3t)\, dt$$

$$= \dfrac{3}{2}\left[\dfrac{\sin 5t}{5} + \dfrac{\sin 3t}{3}\right] + c$$

Problem 18.   In an alternating current circuit, voltage $v = 5\sin\omega t$ and current $i = 10\sin(\omega t - \pi/6)$. Find an expression for the instantaneous power p at time t given that $p = vi$, expressing the answer as a sum or difference of sines and cosines

$p = vi = (5\sin\omega t)[10\sin(\omega t - \pi/6)]$

$$= 50\sin\omega t \sin(\omega t - \pi/6)$$

From equation (4),

$50\sin\omega t \sin(\omega t - \pi/6)$

$$= (50)\left[-\tfrac{1}{2}\left\{\cos(\omega t + \omega t - \pi/6)\right.\right.$$

$$\left.\left. - \cos\left[\omega t - (\omega t - \pi/6)\right]\right\}\right]$$

$$= -25\{\cos(2\omega t - \pi/6) - \cos\pi/6\}$$

i.e. instantaneous power,

$$p = 25[\cos\pi/6 - \cos(2\omega t - \pi/6)]$$

Now try the following Practice Exercise

> **Practice Exercise 189** **Further problems on changing products of sines and cosines into sums or differences (answers on page 1172)**
>
> In Problems 1 to 5, express as sums or differences:
>
> 1. $\sin 7t \cos 2t$
>
> 2. $\cos 8x \sin 2x$
>
> 3. $2 \sin 7t \sin 3t$
>
> 4. $4 \cos 3\theta \cos \theta$
>
> 5. $3 \sin \dfrac{\pi}{3} \cos \dfrac{\pi}{6}$
>
> 6. Determine $\int 2 \sin 3t \cos t \, dt$
>
> 7. Evaluate $\displaystyle\int_0^{\frac{\pi}{2}} 4 \cos 5x \cos 2x \, dx$
>
> 8. Solve the equation: $2 \sin 2\phi \sin \phi = \cos \phi$ in the range $\phi = 0$ to $\phi = 180°$

44.5 Changing sums or differences of sines and cosines into products

In the compound angle formula let,

$$(A + B) = X$$

and

$$(A - B) = Y$$

Solving the simultaneous equations gives:

$$A = \frac{X + Y}{2} \text{ and } B = \frac{X - Y}{2}$$

Thus $\sin(A + B) + \sin(A - B) = 2 \sin A \cos B$ becomes,

$$\sin X + \sin Y = 2 \sin\left(\frac{X + Y}{2}\right) \cos\left(\frac{X - Y}{2}\right) \quad (5)$$

Similarly,

$$\sin X - \sin Y = 2 \cos\left(\frac{X + Y}{2}\right) \sin\left(\frac{X - Y}{2}\right) \quad (6)$$

$$\cos X + \cos Y = 2 \cos\left(\frac{X + Y}{2}\right) \cos\left(\frac{X - Y}{2}\right) \quad (7)$$

$$\cos X - \cos Y = -2 \sin\left(\frac{X + Y}{2}\right) \sin\left(\frac{X - Y}{2}\right) \quad (8)$$

Problem 19. Express $\sin 5\theta + \sin 3\theta$ as a product

From equation (5),

$$\sin 5\theta + \sin 3\theta = 2 \sin\left(\frac{5\theta + 3\theta}{2}\right) \cos\left(\frac{5\theta - 3\theta}{2}\right)$$

$$= \mathbf{2 \sin 4\theta \cos \theta}$$

Problem 20. Express $\sin 7x - \sin x$ as a product

From equation (6),

$$\sin 7x - \sin x = 2 \cos\left(\frac{7x + x}{2}\right) \sin\left(\frac{7x - x}{2}\right)$$

$$= \mathbf{2 \cos 4x \sin 3x}$$

Problem 21. Express $\cos 2t - \cos 5t$ as a product

From equation (8),

$$\cos 2t - \cos 5t = -2 \sin\left(\frac{2t + 5t}{2}\right) \sin\left(\frac{2t - 5t}{2}\right)$$

$$= -2 \sin \frac{7}{2}t \sin\left(-\frac{3}{2}t\right) = \mathbf{2 \sin \frac{7}{2}t \sin \frac{3}{2}t}$$

$$\left(\text{since } \sin\left(-\frac{3}{2}t\right) = -\sin \frac{3}{2}t \right)$$

Problem 22. Show that $\dfrac{\cos 6x + \cos 2x}{\sin 6x + \sin 2x} = \cot 4x$

From equation (7),

$$\cos 6x + \cos 2x = 2 \cos 4x \cos 2x$$

From equation (5),

$$\sin 6x + \sin 2x = 2 \sin 4x \cos 2x$$

Hence

$$\frac{\cos 6x + \cos 2x}{\sin 6x + \sin 2x} = \frac{2 \cos 4x \cos 2x}{2 \sin 4x \cos 2x}$$

$$= \frac{\cos 4x}{\sin 4x} = \mathbf{\cot 4x}$$

Problem 23.   Solve the equation $\cos 4\theta + \cos 2\theta = 0$ for θ in the range $0° \leq \theta \leq 360°$

From equation (7),

$$\cos 4\theta + \cos 2\theta = 2 \cos \left(\frac{4\theta + 2\theta}{2} \right) \cos \left(\frac{4\theta - 2\theta}{2} \right)$$

Hence, $2 \cos 3\theta \cos \theta = 0$

Dividing by 2 gives: $\cos 3\theta \cos \theta = 0$

Hence, either $\cos 3\theta = 0$ or $\cos \theta = 0$

Thus, $3\theta = \cos^{-1} 0$ or $\theta = \cos^{-1} 0$

from which, $3\theta = 90°$ or $270°$ or $450°$ or $630°$ or $810°$ or $990°$

and $\boldsymbol{\theta = 30°, 90°, 150°, 210°, 270°\ \text{or}\ 330°}$

Now try the following Practice Exercise

Practice Exercise 190   Further problems on changing sums or differences of sines and cosines into products (answers on page 1172)

In Problems 1 to 5, express as products:

1. $\sin 3x + \sin x$

2. $\frac{1}{2}(\sin 9\theta - \sin 7\theta)$

3. $\cos 5t + \cos 3t$

4. $\frac{1}{8}(\cos 5t - \cos t)$

5. $\frac{1}{2}\left(\cos \frac{\pi}{3} + \cos \frac{\pi}{4} \right)$

6.   Show that:

   (a) $\dfrac{\sin 4x - \sin 2x}{\cos 4x + \cos 2x} = \tan x$

   (b) $\frac{1}{2}\{\sin(5x - \alpha) - \sin(x + \alpha)\}$
   $= \cos 3x \sin(2x - \alpha)$

In Problems 7 and 8, solve for θ in the range $0° \leq \theta \leq 180°$.

7. $\cos 6\theta + \cos 2\theta = 0$

8. $\sin 3\theta - \sin \theta = 0$

In Problems 9 and 10, solve in the range $0°$ to $360°$.

9. $\cos 2x = 2 \sin x$

10. $\sin 4t + \sin 2t = 0$

44.6   Power waveforms in a.c. circuits

(a) Purely resistive a.c. circuits

Let a voltage $v = V_m \sin \omega t$ be applied to a circuit comprising resistance only. The resulting current is $i = I_m \sin \omega t$, and the corresponding instantaneous power, p, is given by:

$$p = vi = (V_m \sin \omega t)(I_m \sin \omega t)$$

i.e. $p = V_m I_m \sin^2 \omega t$

From double angle formulae of Section 44.3,

$$\cos 2A = 1 - 2\sin^2 A, \text{ from which,}$$

$$\sin^2 A = \tfrac{1}{2}(1 - \cos 2A) \text{ thus}$$

$$\sin^2 \omega t = \tfrac{1}{2}(1 - \cos 2\omega t)$$

Then power $p = V_m I_m \left[\tfrac{1}{2}(1 - \cos 2\omega t) \right]$

i.e. $\boldsymbol{p = \tfrac{1}{2} V_m I_m (1 - \cos 2\omega t)}$

The waveforms of v, i and p are shown in Figure 44.8. The waveform of power repeats itself after π/ω seconds and hence the power has a frequency twice that of voltage and current. The power is always positive, having a maximum value of $V_m I_m$. The average or mean value of the power is $\frac{1}{2} V_m I_m$

The rms value of voltage $V = 0.707 V_m$, i.e. $V = \dfrac{V_m}{\sqrt{2}}$

from which, $V_m = \sqrt{2}\,V$

Similarly, the rms value of current, $I = \dfrac{I_m}{\sqrt{2}}$, from which, $I_m = \sqrt{2}\,I$. Hence the average power, P, developed in a purely resistive a.c. circuit is given by $P = \frac{1}{2} V_m I_m = \frac{1}{2}(\sqrt{2}V)(\sqrt{2}I) = V I$ watts.

Also, power $P = I^2 R$ or V^2/R as for a d.c. circuit, since $V = I R$.

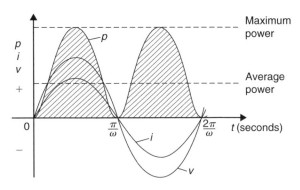

Figure 44.8

Summarising, the average power P in a purely resistive a.c. circuit is given by

$$P = VI = I^2 R = \frac{V^2}{R}$$

where V and I are rms values.

(b) Purely inductive a.c. circuits

Let a voltage $v = V_m \sin \omega t$ be applied to a circuit containing pure inductance (theoretical case). The resulting current is $i = I_m \sin\left(\omega t - \frac{\pi}{2}\right)$ since current lags voltage by $\frac{\pi}{2}$ radians or $90°$ in a purely inductive circuit, and the corresponding instantaneous power, p, is given by:

$$p = vi = (V_m \sin \omega t) I_m \sin\left(\omega t - \frac{\pi}{2}\right)$$

i.e. $\quad p = V_m I_m \sin \omega t \sin\left(\omega t - \frac{\pi}{2}\right)$

However,

$$\sin\left(\omega t - \frac{\pi}{2}\right) = -\cos \omega t \text{ thus}$$
$$p = -V_m I_m \sin \omega t \cos \omega t$$

Rearranging gives:

$$p = -\tfrac{1}{2} V_m I_m (2 \sin \omega t \cos \omega t)$$

However, from double angle formulae,

$$2 \sin \omega t \cos \omega t = \sin 2\omega t$$

Thus   **power**, $p = -\tfrac{1}{2} V_m I_m \sin 2\omega t$

The waveforms of v, i and p are shown in Figure 44.9. The frequency of power is twice that of voltage and current. For the power curve shown in Figure 44.9, the area above the horizontal axis is equal to the area below, thus over a complete cycle the average power P is zero. It is noted that when v and i are both positive, power p is positive and energy is delivered from the source to the inductance; when v and i have opposite signs, power p is negative and energy is returned from the inductance to the source.

In general, when the current through an inductance is increasing, energy is transferred from the circuit to the magnetic field, but this energy is returned when the current is decreasing.

Summarising, **the average power P in a purely inductive a.c. circuit is zero**.

(c) Purely capacitive a.c. circuits

Let a voltage $v = V_m \sin \omega t$ be applied to a circuit containing pure capacitance. The resulting current is $i = I_m \sin\left(\omega t + \frac{\pi}{2}\right)$, since current leads voltage by $90°$ in a purely capacitive circuit, and the corresponding instantaneous power, p, is given by:

$$p = vi = (V_m \sin \omega t) I_m \sin\left(\omega t + \frac{\pi}{2}\right)$$

i.e. $\quad p = V_m I_m \sin \omega t \sin\left(\omega t + \frac{\pi}{2}\right)$

However, $\quad \sin\left(\omega t + \frac{\pi}{2}\right) = \cos \omega t$

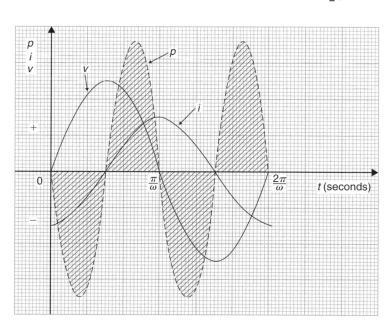

Figure 44.9

Section E

thus $\qquad p = V_m I_m \sin\omega t \cos\omega t$

Rearranging gives $p = \frac{1}{2} V_m I_m (2\sin\omega t \cos\omega t)$

Thus **power, $p = \frac{1}{2} V_m I_m \sin 2\omega t$**

The waveforms of v, i and p are shown in Figure 44.10. Over a complete cycle the average power P is zero. When the voltage across a capacitor is increasing, energy is transferred from the circuit to the electric field, but this energy is returned when the voltage is decreasing.

Summarising, **the average power P in a purely capacitive a.c. circuit is zero**.

(d) R–L or R–C a.c. circuits

Let a voltage $v = V_m \sin\omega t$ be applied to a circuit containing resistance and inductance or resistance and capacitance. Let the resulting current be $i = I_m \sin(\omega t + \phi)$, where phase angle ϕ will be positive for an R–C circuit and negative for an R–L circuit. The corresponding instantaneous power, p, is given by:

$$p = vi = (V_m \sin\omega t) I_m \sin(\omega t + \phi)$$

i.e. $p = V_m I_m \sin\omega t \sin(\omega t + \phi)$

Products of sine functions may be changed into differences of cosine functions as shown in Section 44.4,
i.e. $\sin A \sin B = -\frac{1}{2}[\cos(A + B) - \cos(A - B)]$

Substituting $\omega t = A$ and $(\omega t + \phi) = B$ gives:

power, $p = V_m I_m \{ -\frac{1}{2}[\cos(\omega t + \omega t + \phi)$
$\qquad\qquad\qquad - \cos(\omega t - (\omega t + \phi))]\}$

i.e. $p = \frac{1}{2} V_m I_m [\cos(-\phi) - \cos(2\omega t + \phi)]$

However, $\cos(-\phi) = \cos\phi$
Thus $p = \frac{1}{2} V_m I_m [\cos\phi - \cos(2\omega t + \phi)]$
The instantaneous power p thus consists of

(i) a sinusoidal term, $-\frac{1}{2} V_m I_m \cos(2\omega t + \phi)$ which has a mean value over a cycle of zero, and

(ii) a constant term, $\frac{1}{2} V_m I_m \cos\phi$ (since ϕ is constant for a particular circuit).

Thus the average value of power, $P = \frac{1}{2} V_m I_m \cos\phi$. Since $V_m = \sqrt{2}V$ and $I_m = \sqrt{2}I$, average power,

$P = \frac{1}{2}(\sqrt{2}V)(\sqrt{2}I)\cos\phi$

i.e. $P = VI\cos\phi$

The waveforms of v, i and p are shown in Figure 44.11 for an R–L circuit. The waveform of power is seen to pulsate at twice the supply frequency. The areas of the power curve (shown shaded) above the horizontal time axis represent power supplied to the load; the small areas below the axis represent power being returned to the supply from the inductance as the magnetic field collapses.

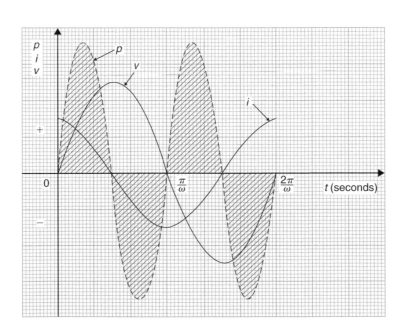

Figure 44.10

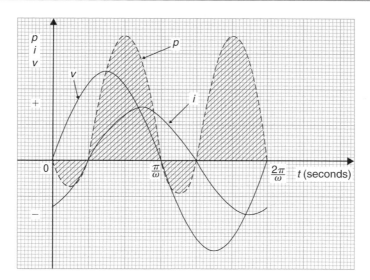

Figure 44.11

A similar shape of power curve is obtained for an R–C circuit, the small areas below the horizontal axis representing power being returned to the supply from the charged capacitor. The difference between the areas above and below the horizontal axis represents the heat loss due to the circuit resistance. Since power is dissipated only in a pure resistance, the alternative equations for power, $P = I_R^2 R$, may be used, where I_R is the rms current flowing through the resistance.

Summarising, **the average power P in a circuit containing resistance and inductance and/or capacitance, whether in series or in parallel, is given by** $P = VI \cos\phi$ or $P = I_R^2 R$ (V, I and I_R being rms values).

For fully worked solutions to each of the problems in Practice Exercises 186 to 190 in this chapter, go to the website:
www.routledge.com/cw/bird

Section E

This assignment covers the material contained in Chapters 42 to 44. *The marks available are shown in brackets at the end of each question.*

1. Prove the following identities:

   (a) $\sqrt{\left[\dfrac{1-\cos^2\theta}{\cos^2\theta}\right]}=\tan\theta$

   (b) $\cos\left(\dfrac{3\pi}{2}+\phi\right)=\sin\phi$

   (c) $\dfrac{\sin^2 x}{1+\cos 2x}=\frac{1}{2}\tan^2 x$ (9)

2. Solve the following trigonometric equations in the range $0°\le x\le 360°$:

   (a) $4\cos x+1=0$

   (b) $3.25\,\mathrm{cosec}\,x=5.25$

   (c) $5\sin^2 x+3\sin x=4$

   (d) $2\sec^2\theta+5\tan\theta=3$ (18)

3. Solve the equation $5\sin(\theta-\pi/6)=8\cos\theta$ for values $0\le\theta\le 2\pi$ (8)

4. Express $5.3\cos t-7.2\sin t$ in the form $R\sin(t+\alpha)$. Hence solve the equation $5.3\cos t-7.2\sin t=4.5$ in the range $0\le t\le 2\pi$ (12)

5. Determine $\int 2\cos 3t\sin t\,dt$ (3)

Multiple choice questions Test 5
Trigonometry
This test covers the material in Chapters 37 to 44

All questions have only one correct answer (answers on page 1200).

1. In the right-angled triangle ABC shown in Figure M5.1, sine A is given by:

   (a) b/a (b) c/b (c) b/c (d) a/b

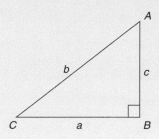

   Figure M5.1

2. In the right-angled triangle ABC shown in Figure M5.1, cosine C is given by:

   (a) a/b (b) c/b (c) a/c (d) b/a

3. In the right-angled triangle shown in Figure M5.1, tangent A is given by:

   (a) b/c (b) a/c (c) a/b (d) c/a

4. $\dfrac{3\pi}{4}$ radians is equivalent to:

   (a) $135°$ (b) $270°$ (c) $45°$ (d) $67.5°$

5. In the triangular template ABC shown in Figure M5.2, the length AC is:

   (a) 6.17 cm   (b) 11.17 cm

   (c) 9.22 cm   (d) 12.40 cm

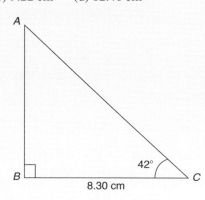

   Figure M5.2

6. $(-4, 3)$ in polar co-ordinates is:

   (a) $(5, 2.498$ rad$)$ (b) $(7, \ 36.87°)$

   (c) $(5, 36.87°)$ (d) $(5, 323.13°)$

7. Correct to 3 decimal places, $\sin(-2.6$ rad$)$ is:

   (a) 0.516 (b) -0.045 (c) -0.516 (d) 0.045

8. For the right-angled triangle PQR shown in Figure M5.3, angle R is equal to:

   (a) $41.41°$ (b) $48.59°$ (c) $36.87°$ (d) $53.13°$

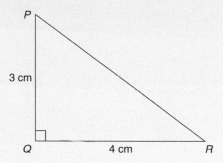

   Figure M5.3

9. If $\cos A = \dfrac{12}{13}$, then $\sin A$ is equal to:

   (a) $\dfrac{5}{13}$ (b) $\dfrac{13}{12}$ (c) $\dfrac{5}{12}$ (d) $\dfrac{12}{5}$

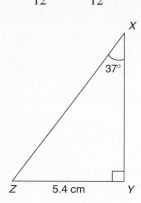

   Figure M5.4

10. The area of triangle XYZ in Figure M5.4 is:

    (a) 24.22 cm^2 (b) 19.35 cm^2

    (c) 38.72 cm^2 (d) 32.16 cm^2

Section E

11. The value, correct to 3 decimal places, of $\cos\left(\dfrac{-3\pi}{4}\right)$ is:

    (a) 0.999 (b) 0.707 (c) −0.999 (d) −0.707

12. A triangle has sides $a = 9.0$ cm, $b = 8.0$ cm and $c = 6.0$ cm. Angle A is equal to:

    (a) 82.42° (b) 56.49° (c) 78.58° (d) 79.87°

13. In the right-angled triangle ABC shown in Figure M5.5, secant C is given by:

    (a) $\dfrac{a}{b}$ (b) $\dfrac{a}{c}$ (c) $\dfrac{b}{c}$ (d) $\dfrac{b}{a}$

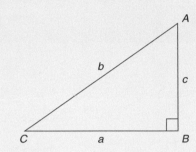

Figure M5.5

14. In the right-angled triangle ABC shown in Figure M5.5, cotangent C is given by:

    (a) $\dfrac{a}{b}$ (b) $\dfrac{b}{c}$ (c) $\dfrac{c}{b}$ (d) $\dfrac{a}{c}$

15. In the right-angled triangle ABC shown in Figure M5.5, cosecant A is given by:

    (a) $\dfrac{c}{a}$ (b) $\dfrac{b}{a}$ (c) $\dfrac{a}{b}$ (d) $\dfrac{b}{c}$

16. Tan 60° is equivalent to:

    (a) $\dfrac{1}{\sqrt{3}}$ (b) $\dfrac{\sqrt{3}}{2}$ (c) $\dfrac{1}{2}$ (d) $\sqrt{3}$

17. An alternating current is given by: $i = 15\sin(100\pi t - 0.25)$ amperes. When time $t = 5$ ms, the current i has a value of:

    (a) 0.35 A (b) 14.53 A (c) 15 A (d) 0.41 A

18. Correct to 4 significant figures, the value of sec 161° is:

    (a) −1.058 (b) 0.3256 (c) 3.072 (d) −0.9455

19. Which of the following trigonometric identities is true?

    (a) $\operatorname{cosec}\theta = \dfrac{1}{\cos\theta}$

    (b) $\cot\theta = \dfrac{1}{\sin\theta}$

    (c) $\dfrac{\sin\theta}{\cos\theta} = \tan\theta$

    (d) $\sec\theta = \dfrac{1}{\sin\theta}$

20. The displacement x metres of a mass from a fixed point about which it is oscillating is given by $x = 3\cos\omega t - 4\sin\omega t$, where t is the time in seconds. x may be expressed as:

    (a) $5\sin(\omega t + 2.50)$ metres

    (b) $7\sin(\omega t - 36.87°)$ metres

    (c) $5\sin\omega t$ metres

    (d) $-\sin(\omega t - 2.50)$ metres

21. The solutions of the equation $2\tan x - 7 = 0$ for $0° \le x \le 360°$ are:

    (a) 105.95° and 254.05°

    (b) 74.05° and 254.05°

    (c) 74.05° and 285.95°

    (d) 254.05° and 285.95°

22. A sinusoidal current is given by: $i = R\sin(\omega t + \alpha)$. Which of the following statements is incorrect?

    (a) R is the average value of the current

    (b) frequency $= \dfrac{\omega}{2\pi}$ Hz

    (c) $\omega = $ angular velocity

    (d) periodic time $= \dfrac{2\pi}{\omega}$ s

23. The trigonometric expression $\cos^2\theta - \sin^2\theta$ is equivalent to:

    (a) $2\sin^2\theta - 1$

    (b) $1 + 2\sin^2\theta$

    (c) $2\sin^2\theta + 1$

    (d) $1 - 2\sin^2\theta$

24. A vertical tower stands on level ground. At a point 100 m from the foot of the tower the angle of elevation of the top is 20°. The height of the tower is:

    (a) 274.7 m   (b) 36.4 m   (c) 34.3 m   (d) 94.0 m

25. (7, 141°) in Cartesian co-ordinates is:

    (a) (5.44, −4.41)   (b) (−5.44, −4.41)

    (c) (5.44, 4.41)   (d) (−5.44, 4.41)

26. If $\tan A = 1.4276$, $\sec A$ is equal to:

    (a) 0.8190 (b) 0.5737 (c) 0.7005 (d) 1.743

27. The acute angle $\cot^{-1} 2.562$ is equal to:

    (a) 67.03° (b) 21.32° (c) 22.97° (d) 68.68°

28. Cos 30° is equivalent to:

    (a) $\dfrac{1}{2}$ (b) $\dfrac{2}{\sqrt{3}}$ (c) $\dfrac{\sqrt{3}}{2}$ (d) $\dfrac{1}{\sqrt{3}}$

29. The angles between 0° and 360° whose tangent is −1.7624 are:

    (a) 60.43° and 240.43°

    (b) 119.57° and 299.57°

    (c) 119.57° and 240.43°

    (d) 150.43° and 299.57°

30. In the triangular template *DEF* shown in Figure M5.6, angle *F* is equal to:

    (a) 43.5°   (b) 28.6°   (c) 116.4°   (d) 101.5°

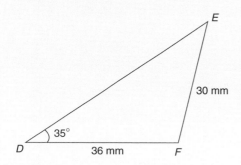

Figure M5.6

31. The area of the triangular template *DEF* shown in Figure M5.6 is:

    (a) 529.2 mm²   (b) 258.5 mm²
    (c) 483.7 mm²   (d) 371.7 mm²

32. In the range $0° \leq \theta \leq 360°$ the solutions of the trigonometric equation $9\tan^2\theta - 12\tan\theta + 4 = 0$ are:

    (a) 33.69°, 146.31°, 213.69° and 326.31°

    (b) 33.69° and 213.69°

    (c) 146.31° and 213.69°

    (d) 146.69° and 326.31°

33. An alternating voltage v is given by $v = 100 \sin\left(100\pi t + \dfrac{\pi}{4}\right)$ volts. When $v = 50$ volts, the time t is equal to:

    (a) 0.093 s   (b) −0.908 ms

    (c) −0.833 ms   (d) −0.162 s

34. The acute angle $\operatorname{cosec}^{-1} 1.429$ is equal to:

    (a) 55.02° (b) 45.59° (c) 44.41° (d) 34.98°

35. The area of triangle *PQR* is given by:

    (a) $\dfrac{1}{2}pr \cos Q$

    (b) $\sqrt{[(s-p)(s-q)(s-r)]}$ where $s = \dfrac{p+q+r}{2}$

    (c) $\dfrac{1}{2}rq \sin P$

    (d) $\dfrac{1}{2}pq \sin Q$

36. The values of θ that are true for the equation $5\sin\theta + 2 = 0$ in the range $\theta = 0°$ to $\theta = 360°$ are:

    (a) 23.58° and 336.42°

    (b) 23.58° and 203.58°

    (c) 156.42° and 336.42°

    (d) 203.58° and 336.42°

37. In triangle *ABC* in Figure M5.7, length *AC* is:

    (a) 14.90 cm   (b) 18.15 cm

    (c) 13.16 cm   (d) 14.04 cm

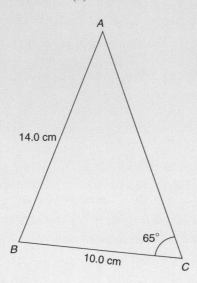

Figure M5.7

38. The acute angle $\sec^{-1} 2.4178$ is equal to:

    (a) 24.43°   (b) 22.47°   (c) 0.426 rad   (d) 65.57°

39. The solutions of the equation $3 - 5\cos^2 A = 0$ for values of *A* in the range $0° \leq A \leq 360°$ are:

    (a)   39.23° and 320.77°

    (b)   39.23°, 140.77°, 219.23° and 320.77°

    (c)   140.77° and 219.23°

    (d)   53.13°, 126.87°, 233.13° and 306.87°

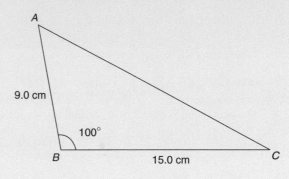

Figure M5.8

40. In triangle *ABC* in Figure M5.8, the length *AC* is:

    (a) 18.79 cm   (b) 70.89 cm

    (c) 22.89 cm   (d) 16.10 cm

Theorem of Pythagoras: $b^2 = a^2 + c^2$

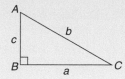

$$\sin C = \frac{c}{b} \quad \cos C = \frac{a}{b} \quad \tan C = \frac{c}{a} \quad \sec C = \frac{b}{a}$$

$$\operatorname{cosec} C = \frac{b}{c} \quad \cot C = \frac{a}{c}$$

Trigonometric ratios for angles of any magnitude

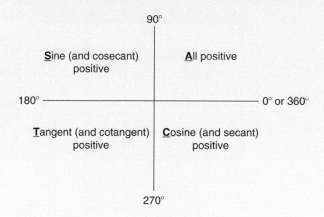

For a **general sinusoidal function** $y = A\sin(\omega t \pm \alpha)$, then

A = amplitude ω = angular velocity = $2\pi f$ rad/s

$\dfrac{2\pi}{\omega}$ = periodic time, T seconds $\dfrac{\omega}{2\pi}$ = frequency, f hertz

α = angle of lead or lag (compared with $y = A\sin\omega t$)

$$180^\circ = \pi \text{ rad} \qquad 1 \text{ rad} = \frac{180^\circ}{\pi}$$

Cartesian and polar co-ordinates

If co-ordinate $(x, y) = (r, \theta)$ then
$r = \sqrt{x^2 + y^2}$ and $\theta = \tan^{-1}\dfrac{y}{x}$

If co-ordinate $(r, \theta) = (x, y)$ then
$x = r\cos\theta$ and $y = r\sin\theta$

Triangle formulae

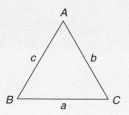

Sine rule $\dfrac{a}{\sin A} = \dfrac{b}{\sin B} = \dfrac{c}{\sin C}$

Cosine rule $a^2 = b^2 + c^2 - 2bc\cos A$

Area of any triangle

(i) $\dfrac{1}{2} \times$ base $\times$ perpendicular height

(ii) $\dfrac{1}{2}ab\sin C$ or $\dfrac{1}{2}ac\sin B$ or $\dfrac{1}{2}bc\sin A$

(iii) $\sqrt{[s(s-a)(s-b)(s-c)]}$ where $s = \dfrac{a+b+c}{2}$

Identities

$$\sec\theta = \frac{1}{\cos\theta} \qquad \operatorname{cosec}\theta = \frac{1}{\sin\theta} \qquad \cot\theta = \frac{1}{\tan\theta}$$

$$\tan\theta = \frac{\sin\theta}{\cos\theta}$$

$$\cos^2\theta + \sin^2\theta = 1 \qquad 1 + \tan^2\theta = \sec^2\theta$$

$$\cot^2\theta + 1 = \operatorname{cosec}^2\theta$$

Compound angle formulae

$$\sin(A \pm B) = \sin A \cos B \pm \cos A \sin B$$

$$\cos(A \pm B) = \cos A \cos B \mp \sin A \sin B$$

$$\tan(A \pm B) = \frac{\tan A \pm \tan B}{1 \mp \tan A \tan B}$$

If $R\sin(\omega t + \alpha) = a \sin \omega t + b \cos \omega t$,
then $a = R\cos\alpha$, $b = R\sin\alpha$, $R = \sqrt{(a^2 + b^2)}$
and $\alpha = \tan^{-1}\dfrac{b}{a}$

Double angles

$$\sin 2A = 2\sin A \cos A$$

$$\cos 2A = \cos^2 A - \sin^2 A = 2\cos^2 A - 1 = 1 - 2\sin^2 A$$

$$\tan 2A = \frac{2\tan A}{1 - \tan^2 A}$$

Products of sines and cosines into sums or differences

$$\sin A \cos B = \frac{1}{2}[\sin(A + B) + \sin(A - B)]$$

$$\cos A \sin B = \frac{1}{2}[\sin(A + B) - \sin(A - B)]$$

$$\cos A \cos B = \frac{1}{2}[\cos(A + B) + \cos(A - B)]$$

$$\sin A \sin B = -\frac{1}{2}[\cos(A + B) - \cos(A - B)]$$

Sums or differences of sines and cosines into products

$$\sin x + \sin y = 2\sin\left(\frac{x+y}{2}\right)\cos\left(\frac{x-y}{2}\right)$$

$$\sin x - \sin y = 2\cos\left(\frac{x+y}{2}\right)\sin\left(\frac{x-y}{2}\right)$$

$$\cos x + \cos y = 2\cos\left(\frac{x+y}{2}\right)\cos\left(\frac{x-y}{2}\right)$$

$$\cos x - \cos y = -2\sin\left(\frac{x+y}{2}\right)\sin\left(\frac{x-y}{2}\right)$$

For a copy of these formulae/revision hints, go to:
www.routledge.com/cw/bird

Complex numbers

Chapter 45

Complex numbers

Why it is important to understand: Complex numbers

Complex numbers are used in many scientific fields, including engineering, electromagnetism, quantum physics and applied mathematics, such as chaos theory. Any physical motion which is periodic, such as an oscillating beam, string, wire, pendulum, electronic signal or electromagnetic wave can be represented by a complex number function. This can make calculations with the various components simpler than with real numbers and sines and cosines. In control theory, systems are often transformed from the time domain to the frequency domain using the Laplace transform. In fluid dynamics, complex functions are used to describe potential flow in two dimensions. In electrical engineering, the Fourier transform is used to analyse varying voltages and currents. Complex numbers are used in signal analysis and other fields for a convenient description for periodically varying signals. This use is also extended into digital signal processing and digital image processing, which utilise digital versions of Fourier analysis (and wavelet analysis) to transmit, compress, restore and otherwise process digital audio signals, still images and video signals. Knowledge of complex numbers is clearly absolutely essential for further studies in so many engineering disciplines.

At the end of this chapter, you should be able to:

- define a complex number
- solve quadratic equations with imaginary roots
- use an Argand diagram to represent a complex number pictorially
- add, subtract, multiply and divide Cartesian complex numbers
- solve complex equations
- convert a Cartesian complex number into polar form, and vice versa
- multiply and divide polar form complex numbers
- apply complex numbers to practical applications

45.1 Cartesian complex numbers

There are several applications of complex numbers in science and engineering, in particular in electrical alternating current theory and in mechanical vector analysis.

There are two main forms of complex number – **Cartesian form** (named after Descartes*) **and polar form** – and both are explained in this chapter.

If we can add, subtract, multiply and divide complex numbers in both forms and represent the numbers on an Argand diagram, then a.c. theory and vector analysis become considerably easier.

---

* **Who was** Descartes? See page 343. To find out more go to **www.routledge.com/cw/bird**

(i) If the quadratic equation $x^2 + 2x + 5 = 0$ is solved using the quadratic formula then,

$$x = \frac{-2 \pm \sqrt{[(2)^2 - (4)(1)(5)]}}{2(1)}$$

$$= \frac{-2 \pm \sqrt{[-16]}}{2} = \frac{-2 \pm \sqrt{[(16)(-1)]}}{2}$$

$$= \frac{-2 \pm \sqrt{16}\sqrt{-1}}{2} = \frac{-2 \pm 4\sqrt{-1}}{2}$$

$$= -1 \pm 2\sqrt{-1}$$

It is not possible to evaluate $\sqrt{-1}$ in real terms. However, if an operator j is defined as $j = \sqrt{-1}$ then the solution may be expressed as $x = -1 \pm j2$

(ii) $-1 + j2$ and $-1 - j2$ are known as **complex numbers**. Both solutions are of the form $a + jb$, 'a' being termed the **real part** and jb the **imaginary part**. A complex number of the form $a + jb$ is called a **Cartesian complex number**.

(iii) In pure mathematics the symbol i is used to indicate $\sqrt{-1}$ (i being the first letter of the word imaginary). However, i is the symbol of electric current in engineering, and to avoid possible confusion the next letter in the alphabet, j, is used to represent $\sqrt{-1}$

Problem 1. Solve the quadratic equation $x^2 + 4 = 0$

Since $x^2 + 4 = 0$ then $x^2 = -4$ and $x = \sqrt{-4}$

i.e. $x = \sqrt{[(-1)(4)]} = \sqrt{(-1)}\sqrt{4} = j(\pm 2)$

$= \pm j2$, (since $j = \sqrt{-1}$)

(Note that $\pm j2$ may also be written $\pm 2j$)

Problem 2. Solve the quadratic equation $2x^2 + 3x + 5 = 0$

Using the quadratic formula,

$$x = \frac{-3 \pm \sqrt{[(3)^2 - 4(2)(5)]}}{2(2)}$$

$$= \frac{-3 \pm \sqrt{-31}}{4} = \frac{-3 \pm \sqrt{(-1)}\sqrt{31}}{4}$$

$$= \frac{-3 \pm j\sqrt{31}}{4}$$

Hence $x = -\dfrac{3}{4} \pm j\dfrac{\sqrt{31}}{4}$ or $-0.750 \pm j1.392$,

correct to 3 decimal places.

(Note, a graph of $y = 2x^2 + 3x + 5$ does not cross the x-axis and hence $2x^2 + 3x + 5 = 0$ has no real roots.)

Problem 3. Evaluate

(a) j^3 (b) j^4 (c) j^{23} (d) $\dfrac{-4}{j^9}$

(a) $j^3 = j^2 \times j = (-1) \times j = -j$, since $j^2 = -1$

(b) $j^4 = j^2 \times j^2 = (-1) \times (-1) = 1$

(c) $j^{23} = j \times j^{22} = j \times (j^2)^{11} = j \times (-1)^{11}$

$= j \times (-1) = -j$

(d) $j^9 = j \times j^8 = j \times (j^2)^4 = j \times (-1)^4$

$= j \times 1 = j$

Hence $\dfrac{-4}{j^9} = \dfrac{-4}{j} = \dfrac{-4}{j} \times \dfrac{-j}{-j} = \dfrac{4j}{-j^2}$

$= \dfrac{4j}{-(-1)} = 4j$ or $j4$

Now try the following Practice Exercise

Practice Exercise 191 Further problems on the introduction to Cartesian complex numbers (answers on page 1172)

In Problems 1 to 9, solve the quadratic equations.

1. $x^2 + 25 = 0$

2. $x^2 - 2x + 2 = 0$

3. $x^2 - 4x + 5 = 0$

4. $x^2 - 6x + 10 = 0$

5. $2x^2 - 2x + 1 = 0$

6. $x^2 - 4x + 8 = 0$

7. $25x^2 - 10x + 2 = 0$

8. $2x^2 + 3x + 4 = 0$

9. $4t^2 - 5t + 7 = 0$

10. Evaluate (a) j^8 (b) $-\dfrac{1}{j^7}$ (c) $\dfrac{4}{2j^{13}}$

45.2 The Argand diagram

A complex number may be represented pictorially on rectangular or Cartesian axes. The horizontal (or x) axis is used to represent the real axis and the vertical (or y) axis is used to represent the imaginary axis. Such a diagram is called an **Argand diagram***. In Figure 45.1, the point A represents the complex number $(3 + j2)$ and is obtained by plotting the co-ordinates $(3, \ j2)$ as in graphical work. Figure 45.1 also shows the Argand points B, C and D representing the complex numbers $(-2 + j4)$, $(-3 - j5)$ and $(1 - j3)$, respectively.

45.3 Addition and subtraction of complex numbers

Two complex numbers are added/subtracted by adding/subtracting separately the two real parts and the two imaginary parts.

For example, if $Z_1 = a + jb$ and $Z_2 = c + jd$,

then $\quad Z_1 + Z_2 = (a + jb) + (c + jd)$
$$= (a + c) + j(b + d)$$

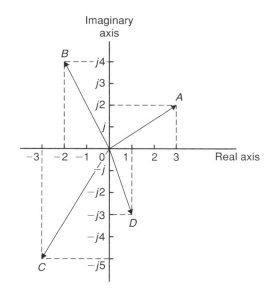

Figure 45.1

and $\quad Z_1 - Z_2 = (a + jb) - (c + jd)$
$$= (a - c) + j(b - d)$$

Thus, for example,

$$(2 + j3) + (3 - j4) = 2 + j3 + 3 - j4$$
$$= \mathbf{5 - j1}$$

and $\quad (2 + j3) - (3 - j4) = 2 + j3 - 3 + j4$
$$= \mathbf{-1 + j\,7}$$

The addition and subtraction of complex numbers may be achieved graphically as shown in the Argand diagram of Figure 45.2. $(2 + j3)$ is represented by vector $\mathbf{OP}$ and $(3 - j4)$ by vector $\mathbf{OQ}$. In Figure 45.2(a) by vector addition (i.e. the diagonal of the parallelogram – see page 579) $\mathbf{OP + OQ = OR}$. R is the point $(5, -j1)$

Hence $(2 + j3) + (3 - j4) = \mathbf{5 - j1}$

In Figure 45.2(b), vector $\mathbf{OQ}$ is reversed (shown as OQ') since it is being subtracted. (Note $\mathbf{OQ} = 3 - j4$ and $\mathbf{OQ'} = -(3 - j4) = -3 + j4$)

$\mathbf{OP - OQ = OP + OQ' = OS}$ is found to be the Argand point $(-1, j7)$

Hence $(2 + j3) - (3 - j4) = \mathbf{-1 + j7}$

* **Who was Argand?** **Jean-Robert Argand** (18 July 1768–13 August 1822) was a highly influential mathematician. He privately published a landmark essay on the representation of imaginary quantities which became known as the **Argand diagram**. To find out more go to **www.routledge.com/cw/bird**

Problem 4. Given $Z_1 = 2 + j4$ and $Z_2 = 3 - j$ determine (a) $Z_1 + Z_2$, (b) $Z_1 - Z_2$, (c) $Z_2 - Z_1$ and show the results on an Argand diagram

Section F

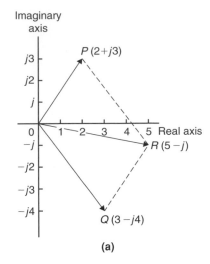

(a)

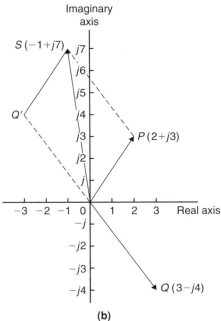

(b)

Figure 45.2

(a) $Z_1 + Z_2 = (2 + j4) + (3 - j)$

$$= (2 + 3) + j(4 - 1) = \mathbf{5 + j3}$$

(b) $Z_1 - Z_2 = (2 + j4) - (3 - j)$

$$= (2 - 3) + j(4 - (-1)) = \mathbf{-1 + j5}$$

(c) $Z_2 - Z_1 = (3 - j) - (2 + j4)$

$$= (3 - 2) + j(-1 - 4) = \mathbf{1 - j5}$$

Each result is shown in the Argand diagram of Figure 45.3.

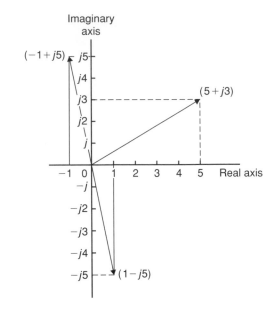

Figure 45.3

45.4 Multiplication and division of complex numbers

(i) **Multiplication of complex numbers** is achieved by assuming all quantities involved are real and then using $j^2 = -1$ to simplify.

Hence $(a + jb)(c + jd)$

$$= ac + a(jd) + (jb)c + (jb)(jd)$$
$$= ac + jad + jbc + j^2bd$$

$$= (ac - bd) + j(ad + bc),$$
$$\text{since } j^2 = -1$$

Thus $(3 + j2)(4 - j5)$

$$= 12 - j15 + j8 - j^2 10$$
$$= (12 - (-10)) + j(-15 + 8)$$
$$= \mathbf{22 - j7}$$

(ii) The **complex conjugate** of a complex number is obtained by changing the sign of the imaginary part. Hence the complex conjugate of $a + jb$ is $a - jb$. The product of a complex number and its complex conjugate is always a real number.

For example,

$$(3 + j4)(3 - j4) = 9 - j12 + j12 - j^2 16$$
$$= 9 + 16 = 25$$

$[(a + jb)(a - jb)$ may be evaluated 'on sight' as $a^2 + b^2]$

(iii) **Division of complex numbers** is achieved by multiplying both numerator and denominator by the complex conjugate of the denominator.

For example,

$$\frac{2-j5}{3+j4}=\frac{2-j5}{3+j4}\times\frac{(3-j4)}{(3-j4)}$$

$$=\frac{6-j8-j15+j^220}{3^2+4^2}$$

$$=\frac{-14-j23}{25}=\frac{-14}{25}-j\frac{23}{25}$$

or $-0.56-j0.92$

Problem 5. If $Z_1=1-j3$, $Z_2=-2+j5$ and $Z_3=-3-j4$, determine in $a+jb$ form:

(a) Z_1Z_2 (b) $\dfrac{Z_1}{Z_3}$

(c) $\dfrac{Z_1Z_2}{Z_1+Z_2}$ (d) $Z_1Z_2Z_3$

(a) $Z_1Z_2=(1-j3)(-2+j5)$

$$=-2+j5+j6-j^215$$

$$=(-2+15)+j(5+6),\text{ since } j^2=-1,$$

$$=13+j11$$

(b) $\dfrac{Z_1}{Z_3}=\dfrac{1-j3}{-3-j4}=\dfrac{1-j3}{-3-j4}\times\dfrac{-3+j4}{-3+j4}$

$$=\frac{-3+j4+j9-j^212}{3^2+4^2}$$

$$=\frac{9+j13}{25}=\frac{9}{25}+j\frac{13}{25}$$

or $0.36+j0.52$

(c) $\dfrac{Z_1Z_2}{Z_1+Z_2}=\dfrac{(1-j3)(-2+j5)}{(1-j3)+(-2+j5)}$

$$=\frac{13+j11}{-1+j2},\text{ from part (a),}$$

$$=\frac{13+j11}{-1+j2}\times\frac{-1-j2}{-1-j2}$$

$$=\frac{-13-j26-j11-j^222}{1^2+2^2}$$

$$=\frac{9-j37}{5}=\frac{9}{5}-j\frac{37}{5}\text{ or } 1.8-j7.4$$

(d) $Z_1Z_2Z_3=(13+j11)(-3-j4)$, since

$$Z_1Z_2=13+j11,\text{ from part (a)}$$

$$=-39-j52-j33-j^244$$

$$=(-39+44)-j(52+33)$$

$$=5-j85$$

Problem 6. Evaluate:

(a) $\dfrac{2}{(1+j)^4}$ (b) $j\left(\dfrac{1+j3}{1-j2}\right)^2$

(a) $(1+j)^2=(1+j)(1+j)=1+j+j+j^2$

$$=1+j+j-1=j2$$

$(1+j)^4=[(1+j)^2]^2=(j2)^2=j^24=-4$

Hence $\dfrac{2}{(1+j)^4}=\dfrac{2}{-4}=-\dfrac{1}{2}$

(b) $\dfrac{1+j3}{1-j2}=\dfrac{1+j3}{1-j2}\times\dfrac{1+j2}{1+j2}$

$$=\frac{1+j2+j3+j^26}{1^2+2^2}=\frac{-5+j5}{5}$$

$$=-1+j1=-1+j$$

$\left(\dfrac{1+j3}{1-j2}\right)^2=(-1+j)^2=(-1+j)(-1+j)$

$$=1-j-j+j^2=-j2$$

Hence $j\left(\dfrac{1+j3}{1-j2}\right)^2=j(-j2)=-j^22=2$,

since $j^2=-1$

Now try the following Practice Exercise

Practice Exercise 192 Further problems on operations involving Cartesian complex numbers (answers on page 1172)

1. Evaluate (a) $(3+j2)+(5-j)$ and (b) $(-2+j6)-(3-j2)$ and show the results on an Argand diagram.

2. Write down the complex conjugates of (a) $3+j4$, (b) $2-j$

3. If $z=2+j$ and $w=3-j$ evaluate
   (a) $z+w$ (b) $w-z$ (c) $3z-2w$
   (d) $5z+2w$ (e) $j(2w-3z)$ (f) $2jw-jz$

In Problems 4 to 8, evaluate in $a + jb$ form given $Z_1 = 1 + j2$, $Z_2 = 4 - j3$, $Z_3 = -2 + j3$ and $Z_4 = -5 - j$

4. (a) $Z_1 + Z_2 - Z_3$ (b) $Z_2 - Z_1 + Z_4$

5. (a) $Z_1 Z_2$ (b) $Z_3 Z_4$

6. (a) $Z_1 Z_3 + Z_4$ (b) $Z_1 Z_2 Z_3$

7. (a) $\dfrac{Z_1}{Z_2}$ (b) $\dfrac{Z_1 + Z_3}{Z_2 - Z_4}$

8. (a) $\dfrac{Z_1 Z_3}{Z_1 + Z_3}$ (b) $Z_2 + \dfrac{Z_1}{Z_4} + Z_3$

9. Evaluate (a) $\dfrac{1-j}{1+j}$ (b) $\dfrac{1}{1+j}$

10. Show that $\dfrac{-25}{2}\left(\dfrac{1+j2}{3+j4} - \dfrac{2-j5}{-j}\right)$
$= 57 + j24$

45.5 Complex equations

If two complex numbers are equal, then their real parts are equal and their imaginary parts are equal. Hence if $a + jb = c + jd$, then $a = c$ and $b = d$.

Problem 7. Solve the complex equations:
(a) $2(x + jy) = 6 - j3$

(b) $(1 + j2)(-2 - j3) = a + jb$

(a) $2(x + jy) = 6 - j3$ hence $2x + j2y = 6 - j3$
Equating the real parts gives:
$2x = 6$, i.e. $x = 3$

Equating the imaginary parts gives:
$2y = -3$, i.e. $y = -\frac{3}{2}$

(b) $(1 + j2)(-2 - j3) = a + jb$
$-2 - j3 - j4 - j^2 6 = a + jb$
Hence $4 - j7 = a + jb$
Equating real and imaginary terms gives:
$a = 4$ and $b = -7$

Problem 8. Solve the equations:
(a) $(2 - j3) = \sqrt{(a + jb)}$

(b) $(x - j2y) + (y - j3x) = 2 + j3$

(a) $(2 - j3) = \sqrt{(a + jb)}$
Hence $(2 - j3)^2 = a + jb$,
i.e. $(2 - j3)(2 - j3) = a + jb$
Hence $4 - j6 - j6 + j^2 9 = a + jb$
and $-5 - j12 = a + jb$
Thus $a = -5$ and $b = -12$

(b) $(x - j2y) + (y - j3x) = 2 + j3$
Hence $(x + y) + j(-2y - 3x) = 2 + j3$
Equating real and imaginary parts gives:
$$x + y = 2 \tag{1}$$
$$\text{and } -3x - 2y = 3 \tag{2}$$
i.e. two simultaneous equations to solve.
Multiplying equation (1) by 2 gives:
$$2x + 2y = 4 \tag{3}$$
Adding equations (2) and (3) gives:
$-x = 7$, i.e. $x = -7$

From equation (1), $y = 9$, which may be checked in equation (2).

Now try the following Practice Exercise

Practice Exercise 193 Further problems on complex equations (answers on page 1172)

In Problems 1 to 4 solve the complex equations.

1. $(2 + j)(3 - j2) = a + jb$

2. $\dfrac{2 + j}{1 - j} = j(x + jy)$

3. $(2 - j3) = \sqrt{(a + jb)}$

4. $(x - j2y) - (y - jx) = 2 + j$

5. If $Z = R + j\omega L + 1/j\omega C$, express Z in $(a + jb)$ form when $R = 10$, $L = 5$, $C = 0.04$ and $\omega = 4$

45.6 The polar form of a complex number

(i) Let a complex number z be $x + jy$ as shown in the Argand diagram of Figure 45.4. Let distance OZ be r and the angle OZ makes with the positive real axis be θ.

From trigonometry, $x = r\cos\theta$ and
$$y = r\sin\theta$$
Hence $Z = x + jy$ $= r\cos\theta + jr\sin\theta$
$$= r(\cos\theta + j\sin\theta)$$
$Z = r(\cos\theta + j\sin\theta)$ is usually abbreviated to
$Z = r\angle\theta$ which is known as the **polar form** of
a complex number.

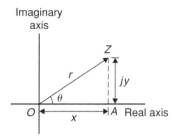

Figure 45.4

(ii) r is called the **modulus** (or magnitude) of Z and
is written as mod Z or $|Z|$
r is determined using Pythagoras' theorem on
triangle OAZ in Figure 45.4,

i.e. $r = \sqrt{(x^2 + y^2)}$

(iii) θ is called the **argument** (or amplitude) of Z and
is written as arg Z.
By trigonometry on triangle OAZ,

$$\arg Z = \theta = \tan^{-1}\frac{y}{x}$$

(iv) Whenever changing from Cartesian form to polar
form, or vice versa, a sketch is invaluable for
determining the quadrant in which the complex
number occurs.

Problem 9. Determine the modulus and argument
of the complex number $Z = 2 + j3$, and express Z
in polar form

$Z = 2 + j3$ lies in the first quadrant as shown in
Figure 45.5.

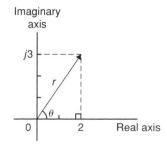

Figure 45.5

Modulus, $|Z| = r = \sqrt{(2^2 + 3^2)} = \sqrt{13}$ or **3.606**, correct
to 3 decimal places.

Argument, arg $Z = \theta = \tan^{-1}\frac{3}{2}$

$$= \mathbf{56.31°} \text{ or } \mathbf{56°19'}$$

In polar form, $2 + j3$ is written as **3.606∠56.31°**

Problem 10. Express the following complex
numbers in polar form:

(a) $3 + j4$ (b) $-3 + j4$

(c) $-3 - j4$ (d) $3 - j4$

(a) $3 + j4$ is shown in Figure 45.6 and lies in the first
quadrant.

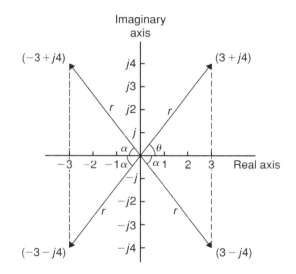

Figure 45.6

Modulus, $r = \sqrt{(3^2 + 4^2)} = 5$ and   argument
$\theta = \tan^{-1}\frac{4}{3} = 53.13°$

Hence **3 + j4 = 5∠53.13°**

(b) $-3 + j4$ is shown in Figure 45.6 and lies in the
second quadrant.

Modulus, $r = 5$ and   angle $\alpha = 53.13°$, from
part (a).

Argument $= 180° - 53.13° = 126.87°$ (i.e.   the
argument must be measured from the positive real
axis).

Hence **−3 + j4 = 5∠126.87°**

(c) $-3 - j4$ is shown in Figure 45.6 and lies in the
third quadrant.

Modulus, $r = 5$ and $\alpha = 53.13°$, as above.

Hence the argument $= 180° + 53.13° = 233.13°$, which is the same as $-126.87°$

Hence $(-3 - j4) = 5\angle 233.13°$ or $5\angle -126.87°$

(By convention the **principal value** is normally used, i.e. the numerically least value, such that $-\pi < \theta < \pi$)

(d) $3 - j4$ is shown in Figure 45.6 and lies in the fourth quadrant.

Modulus, $r = 5$ and angle $\alpha = 53.13°$, as above.

Hence $(3 - j4) = 5\angle -53.13°$

> **Problem 11.** Convert (a) $4\angle 30°$ (b) $7\angle -145°$ into $a + jb$ form, correct to 4 significant figures

(a) $4\angle 30°$ is shown in Figure 45.7(a) and lies in the first quadrant.

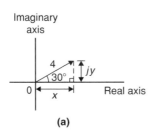

(a)

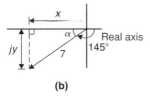

(b)

Figure 45.7

Using trigonometric ratios, $x = 4\cos 30° = 3.464$ and $y = 4\sin 30° = 2.000$

Hence $4\angle 30° = 3.464 + j2.000$

(b) $7\angle -145°$ is shown in Figure 45.7(b) and lies in the third quadrant.

Angle $\alpha = 180° - 145° = 35°$

Hence $x = 7\cos 35° = 5.734$

and $y = 7\sin 35° = 4.015$

Hence $7\angle -145° = -5.734 - j4.015$

Alternatively

$7\angle -145° = 7\cos(-145°) + j7\sin(-145°)$

$$= -5.734 - j4.015$$

Calculator

Using the '**Pol**' and '**Rec**' functions on a calculator enables changing from Cartesian to polar and vice versa to be achieved more quickly.

Since complex numbers are used with vectors and with electrical engineering a.c. theory, it is essential that the calculator can be used quickly and accurately.

45.7 Multiplication and division in polar form

If $Z_1 = r_1 \angle \theta_1$ and $Z_2 = r_2 \angle \theta_2$ then:

(i) $Z_1 Z_2 = r_1 r_2 \angle (\theta_1 + \theta_2)$ and

(ii) $\dfrac{Z_1}{Z_2} = \dfrac{r_1}{r_2} \angle (\theta_1 - \theta_2)$

> **Problem 12.** Determine, in polar form:
>
> (a) $8\angle 25° \times 4\angle 60°$
>
> (b) $3\angle 16° \times 5\angle -44° \times 2\angle 80°$

(a) $8\angle 25° \times 4\angle 60° = (8 \times 4)\angle(25° + 60°) = 32\angle 85°$

(b) $3\angle 16° \times 5\angle -44° \times 2\angle 80°$

$$= (3 \times 5 \times 2)\angle[16° + (-44°) + 80°] = 30\angle 52°$$

> **Problem 13.** Evaluate, in polar form:
>
> (a) $\dfrac{16\angle 75°}{2\angle 15°}$ (b) $\dfrac{10\angle \dfrac{\pi}{4} \times 12\angle \dfrac{\pi}{2}}{6\angle -\dfrac{\pi}{3}}$

(a) $\dfrac{16\angle 75°}{2\angle 15°} = \dfrac{16}{2}\angle(75° - 15°) = 8\angle 60°$

(b) $\dfrac{10\angle \dfrac{\pi}{4} \times 12\angle \dfrac{\pi}{2}}{6\angle -\dfrac{\pi}{3}} = \dfrac{10 \times 12}{6}\angle\left(\dfrac{\pi}{4} + \dfrac{\pi}{2} - \left(-\dfrac{\pi}{3}\right)\right)$

$$= 20\angle \dfrac{13\pi}{12} \text{ or } 20\angle -\dfrac{11\pi}{12} \text{ or}$$

$$20\angle 195° \text{ or } 20\angle -165°$$

Problem 14. Evaluate, in polar form:
$2\angle 30° + 5\angle -45° - 4\angle 120°$

Addition and subtraction in polar form is not possible directly. Each complex number has to be converted into Cartesian form first.

$$2\angle 30° = 2(\cos 30° + j \sin 30°)$$

$$= 2\cos 30° + j2\sin 30° = 1.732 + j1.000$$

$$5\angle -45° = 5(\cos(-45°) + j \sin(-45°))$$

$$= 5\cos(-45°) + j5\sin(-45°)$$

$$= 3.536 - j3.536$$

$$4\angle 120° = 4(\cos 120° + j \sin 120°)$$

$$= 4\cos 120° + j4\sin 120°$$

$$= -2.000 + j3.464$$

Hence $2\angle 30° + 5\angle -45° - 4\angle 120°$

$$= (1.732 + j1.000) + (3.536 - j3.536)$$

$$- (-2.000 + j3.464)$$

$$= 7.268 - j6.000, \text{ which lies in the fourth quadrant}$$

$$= \sqrt{[(7.268)^2 + (6.000)^2]}\angle \tan^{-1}\left(\frac{-6.000}{7.268}\right)$$

$$= \mathbf{9.425\angle -39.54°}$$

Now try the following Practice Exercise

Practice Exercise 194   Further problems on polar form (answers on page 1172)

1. Determine the modulus and argument of
    (a) $2 + j4$ (b) $-5 - j2$ (c) $j(2 - j)$

In Problems 2 and 3, express the given Cartesian complex numbers in polar form, leaving answers in surd form.

2. (a) $2 + j3$ (b) -4 (c) $-6 + j$

3. (a) $-j3$ (b) $(-2 + j)^3$ (c) $j^3(1 - j)$

In Problems 4 and 5, convert the given polar complex numbers into $(a + jb)$ form, giving answers correct to 4 significant figures.

4. (a) $5\angle 30°$ (b) $3\angle 60°$ (c) $7\angle 45°$

5. (a) $6\angle 125°$ (b) $4\angle \pi$ (c) $3.5\angle -120°$

In Problems 6 to 8, evaluate in polar form.

6. (a) $3\angle 20° \times 15\angle 45°$
    (b) $2.4\angle 65° \times 4.4\angle -21°$

7. (a) $6.4\angle 27° \div 2\angle -15°$
    (b) $5\angle 30° \times 4\angle 80° \div 10\angle -40°$

8. (a) $4\angle \dfrac{\pi}{6} + 3\angle \dfrac{\pi}{8}$
    (b) $2\angle 120° + 5.2\angle 58° - 1.6\angle -40°$

45.8 Applications of complex numbers

There are several applications of complex numbers in science and engineering, in particular in electrical alternating current theory and in mechanical vector analysis.

The effect of multiplying a phasor by j is to rotate it in a positive direction (i.e. anticlockwise) on an Argand diagram through 90° without altering its length. Similarly, multiplying a phasor by $-j$ rotates the phasor through $-90°$. These facts are used in a.c. theory since certain quantities in the phasor diagrams lie at 90° to each other. For example, in the $R-L$ series circuit shown in Figure 45.8(a), V_L leads I by 90° (i.e. I lags V_L by 90°) and may be written as jV_L, the vertical axis being regarded as the imaginary axis of an Argand diagram. Thus $V_R + jV_L = V$ and since $V_R = IR$, $V = IX_L$ (where X_L is the inductive reactance, $2\pi fL$ ohms) and $V = IZ$ (where Z is the impedance) then $R + jX_L = Z$

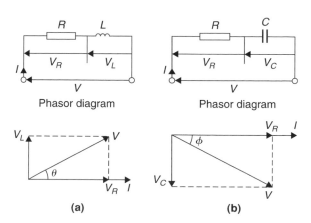

Figure 45.8

Similarly, for the $R-C$ circuit shown in Figure 45.8(b), V_C lags I by $90°$ (i.e. I leads V_C by $90°$) and $V_R - jV_C = V$, from which $R - jX_C = Z$ (where X_C is the capacitive reactance $\dfrac{1}{2\pi fC}$ ohms).

Problem 15. Determine the resistance and series inductance (or capacitance) for each of the following impedances, assuming a frequency of 50 Hz:

(a) $(4.0 + j7.0)\ \Omega$ (b) $-j20\ \Omega$

(c) $15\angle{-60°}\ \Omega$

(a) Impedance, $Z = (4.0 + j7.0)\ \Omega$ hence, **resistance = 4.0 Ω** and reactance = 7.00 Ω.

Since the imaginary part is positive, the reactance is inductive,

i.e. $X_L = 7.0\ \Omega$

Since $X_L = 2\pi fL$ then **inductance**,

$$L = \frac{X_L}{2\pi f} = \frac{7.0}{2\pi(50)} = 0.0223\,\text{H or } 22.3\,\text{mH}$$

(b) Impedance, $Z = -j20$, i.e. $Z = (0 - j20)\ \Omega$ hence **resistance = 0** and reactance = 20 Ω. Since the imaginary part is negative, the reactance is capacitive, i.e. $X_C = 20\ \Omega$ and since $X_C = \dfrac{1}{2\pi fC}$ then:

$$\text{capacitance, } C = \frac{1}{2\pi fX_C} = \frac{1}{2\pi(50)(20)}\,\text{F}$$

$$= \frac{10^6}{2\pi(50)(20)}\,\mu\text{F} = 159.2\,\mu\text{F}$$

(c) Impedance, Z

$$= 15\angle{-60°} = 15[\cos(-60°) + j\sin(-60°)]$$

$$= 7.50 - j12.99\ \Omega$$

Hence **resistance = 7.50 Ω** and capacitive reactance, $X_C = 12.99\ \Omega$

Since $X_C = \dfrac{1}{2\pi fC}$ then **capacitance**,

$$C = \frac{1}{2\pi fX_C} = \frac{10^6}{2\pi(50)(12.99)}\,\mu\text{F}$$

$$= 245\,\mu\text{F}$$

Problem 16. An alternating voltage of 240 V, 50 Hz is connected across an impedance of $(60 - j100)\ \Omega$. Determine (a) the resistance, (b) the capacitance, (c) the magnitude of the impedance and its phase angle and (d) the current flowing

(a) Impedance $Z = (60 - j100)\ \Omega$.

Hence **resistance = 60 Ω**

(b) Capacitive reactance $X_C = 100\ \Omega$ and since $X_C = \dfrac{1}{2\pi fC}$ then

$$\text{capacitance, } C = \frac{1}{2\pi fX_C} = \frac{1}{2\pi(50)(100)}$$

$$= \frac{10^6}{2\pi(50)(100)}\,\mu\text{F}$$

$$= 31.83\,\mu\text{F}$$

(c) Magnitude of impedance,

$$|Z| = \sqrt{[(60)^2 + (-100)^2]} = 116.6\ \Omega$$

$$\text{Phase angle, arg } Z = \tan^{-1}\left(\frac{-100}{60}\right) = -59.04°$$

(d) Current flowing, $I = \dfrac{V}{Z} = \dfrac{240\angle{0°}}{116.6\angle{-59.04°}}$

$$= 2.058\angle{59.04°}\,\text{A}$$

The circuit and phasor diagrams are as shown in Figure 45.8(b).

Problem 17. For the parallel circuit shown in Figure 45.9, determine the value of current I and its phase relative to the 240 V supply, using complex numbers

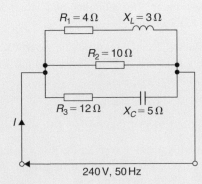

Figure 45.9

Current $I = \dfrac{V}{Z}$. Impedance Z for the three-branch parallel circuit is given by:

$$\frac{1}{Z} = \frac{1}{Z_1} + \frac{1}{Z_2} + \frac{1}{Z_3}$$

where $Z_1 = 4 + j3$, $Z_2 = 10$ and $Z_3 = 12 - j5$

Admittance, $Y_1 = \dfrac{1}{Z_1} = \dfrac{1}{4 + j3}$

$$= \frac{1}{4 + j3} \times \frac{4 - j3}{4 - j3} = \frac{4 - j3}{4^2 + 3^2}$$

$$= 0.160 - j0.120 \text{ siemens}$$

Admittance, $Y_2 = \dfrac{1}{Z_2} = \dfrac{1}{10} = 0.10 \text{ siemens}$

Admittance, $Y_3 = \dfrac{1}{Z_3} = \dfrac{1}{12 - j5}$

$$= \frac{1}{12 - j5} \times \frac{12 + j5}{12 + j5} = \frac{12 + j5}{12^2 + 5^2}$$

$$= 0.0710 + j0.0296 \text{ siemens}$$

Total admittance, $Y = Y_1 + Y_2 + Y_3$

$$= (0.160 - j0.120) + (0.10)$$

$$+ (0.0710 + j0.0296)$$

$$= 0.331 - j0.0904$$

$$= 0.343\angle -15.28° \text{ siemens}$$

Current $I = \dfrac{V}{Z} = VY$

$$= (240\angle 0°)(0.343\angle -15.28°)$$

$$= \mathbf{82.32 \angle -15.28° \, A}$$

Problem 18. Determine the magnitude and direction of the resultant of the three coplanar forces given below, when they act at a point.

Force A, 10 N acting at 45° from the positive horizontal axis

Force B, 8 N acting at 120° from the positive horizontal axis

Force C, 15 N acting at 210° from the positive horizontal axis

The space diagram is shown in Figure 45.10. The forces may be written as complex numbers.
Thus force A, $f_A = 10\angle 45°$, force B, $f_B = 8\angle 120°$ and force C, $f_C = 15\angle 210°$

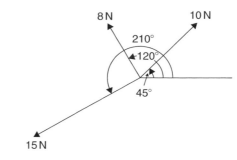

Figure 45.10

The resultant force

$$\begin{aligned}
&= f_A + f_B + f_C \\
&= 10\angle 45° + 8\angle 120° + 15\angle 210° \\
&= 10(\cos 45° + j\sin 45°) + 8(\cos 120° \\
&\quad + j\sin 120°) + 15(\cos 210° + j\sin 210°) \\
&= (7.071 + j7.071) + (-4.00 + j6.928) \\
&\qquad\qquad\qquad\qquad + (-12.99 - j7.50) \\
&= -9.919 + j6.499
\end{aligned}$$

Magnitude of resultant force

$$= \sqrt{[(-9.919)^2 + (6.499)^2]} = \mathbf{11.86\,N}$$

Direction of resultant force

$$= \tan^{-1}\left(\frac{6.499}{-9.919}\right) = \mathbf{146.77°}$$

(since $-9.919 + j6.499$ lies in the second quadrant).

Now try the following Practice Exercise

Practice Exercise 195   Further problems on applications of complex numbers (answers on page 1172)

1. Determine the resistance R and series inductance L (or capacitance C) for each of the following impedances assuming the frequency to be 50 Hz.

    (a) $(3 + j8)\,\Omega$ (b) $(2 - j3)\,\Omega$
    (c) $j14\,\Omega$ (d) $8\angle -60°\,\Omega$

2. Two impedances, $Z_1 = (3 + j6)\,\Omega$ and $Z_2 = (4 - j3)\,\Omega$ are connected in series to a supply voltage of 120 V. Determine the magnitude of the current and its phase angle relative to the voltage.

3. If the two impedances in Problem 2 are connected in parallel determine the current flowing and its phase relative to the 120 V supply voltage.

4. A series circuit consists of a 12 Ω resistor, a coil of inductance 0.10 H and a capacitance of 160 μF. Calculate the current flowing and its phase relative to the supply voltage of 240 V, 50 Hz. Determine also the power factor of the circuit (Note that power factor $= \cos\theta$).

5. For the circuit shown in Figure 45.11, determine the current I flowing and its phase relative to the applied voltage.

6. Determine, using complex numbers, the magnitude and direction of the resultant of the coplanar forces given below, which are acting at a point. Force A, 5 N acting horizontally, force B, 9 N acting at an angle of 135° to force A, force C, 12 N acting at an angle of 240° to force A.

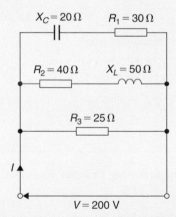

Figure 45.11

7. A delta-connected impedance Z_A is given by:

$$Z_A = \frac{Z_1 Z_2 + Z_2 Z_3 + Z_3 Z_1}{Z_2}$$

Determine Z_A in both Cartesian and polar form given $Z_1 = (10 + j0)\,\Omega$, $Z_2 = (0 - j10)\,\Omega$ and $Z_3 = (10 + j10)\,\Omega$.

8. In the hydrogen atom, the angular momentum, p, of the de Broglie wave is given by: $p\psi = -\left(\dfrac{jh}{2\pi}\right)(\pm jm\psi)$. Determine an expression for p.

9. An aircraft P flying at a constant height has a velocity of $(400 + j300)$ km/h. Another aircraft Q at the same height has a velocity of $(200 - j600)$ km/h. Determine (a) the velocity of P relative to Q, and (b) the velocity of Q relative to P. Express the answers in polar form, correct to the nearest km/h.

10. Three vectors are represented by P, $2\angle 30°$, Q, $3\angle 90°$ and R, $4\angle -60°$. Determine in polar form the vectors represented by (a) $P + Q + R$, (b) $P - Q - R$

11. In a Schering bridge circuit,
$$Z_X = (R_X - jX_{C_X}), \quad Z_2 = -jX_{C_2},$$
$$Z_3 = \frac{(R_3)(-jX_{C_3})}{(R_3 - jX_{C_3})} \quad \text{and} \quad Z_4 = R_4$$
where $X_C = \dfrac{1}{2\pi fC}$
At balance: $(Z_X)(Z_3) = (Z_2)(Z_4)$
Show that at balance $R_X = \dfrac{C_3 R_4}{C_2}$ and
$$C_X = \frac{C_2 R_3}{R_4}$$

12. An amplifier has a transfer function, T, given by: $T = \dfrac{500}{1 + j\omega(5 \times 10^{-4})}$ where ω is the angular frequency. The gain of the amplifier is given by the modulus of T and the phase is given by the argument of T. If $\omega = 2000$ rad/s, determine the gain and the phase (in degrees).

13. The sending end current of a transmission line is given by: $I_S = \dfrac{V_S}{Z_0} \tanh PL$. Calculate the value of the sending current, in polar form, given $V_S = 200$V, $Z_0 = 560 + j420\,\Omega$, $P = 0.20$ and $L = 10$

45.9 Using a calculator with complex numbers

With a calculator such as the CASIO 991ES PLUS it is possible, using the *complex mode*, to achieve many of the calculations in this chapter much more quickly.

Here is just one example to demonstrate its usefulness. Note that not all calculators possess the 'complex mode'.

To determine, in polar form, the following calculation:

$$\frac{(15.6 - j9.7)(7.3\angle 21.9°)}{(8.8 + j17.3)}$$

using the fx-991ES PLUS calculator:

1. Make sure the calculator is on 'degrees' – i.e. shift mode 3

2. Press 'Mode 2' and CMPLX should appear at the top of the screen

3. Press the fraction function $\frac{\square}{\square}$

4. Type in $(15.6 - 9.7i)$ Note that i is obtained by 'Shift, Eng' and must be placed after 9.7

5. Type in $(7.3\angle 21.9°)$ Note that $\angle$ is obtained by 'Shift, $(-)$'

6. Press ↓ on the cursor key and enter $(8.8 + 17.3i)$

7. Press '=' and an answer in Cartesian form results: $2.0185\ldots - 6.607\ldots i$

8. If an answer in polar form is needed, then press 'Shift, 2' and then press '3', and then press '=' which gives: $6.91\angle - 73.01°$, correct to 2 decimal places.

There are aspects of engineering, such as with calculations involving a.c. networks, where use of the calculator complex mode becomes so helpful.

For fully worked solutions to each of the problems in Practice Exercises 191 to 195 in this chapter, go to the website:
www.routledge.com/cw/bird

Chapter 46

De Moivre's theorem

Why it is important to understand: De Moivre's theorem

There are many, many examples of the use of complex numbers in engineering and science. De Moivre's theorem has several uses, including finding powers and roots of complex numbers, solving polynomial equations, calculating trigonometric identities, and for evaluating the sums of trigonometric series. The theorem is also used to calculate exponential and logarithmic functions of complex numbers. De Moivre's theorem has applications in electrical engineering and physics.

At the end of this chapter, you should be able to:

- state De Moivre's theorem
- calculate powers of complex numbers
- calculate roots of complex numbers
- state the exponential form of a complex number
- convert Cartesian/polar form into exponential form and vice versa
- determine loci in the complex plane

46.1 Introduction

From multiplication of complex numbers in polar form,

$$(r\angle\theta) \times (r\angle\theta) = r^2\angle 2\theta$$

Similarly, $(r\angle\theta) \times (r\angle\theta) \times (r\angle\theta) = r^3\angle 3\theta$, and so on. In general, **De Moivre's theorem*** states:

$$[r\angle\theta]^n = r^n\angle n\theta$$

The theorem is true for all positive, negative and fractional values of n. The theorem is used to determine powers and roots of complex numbers.

46.2 Powers of complex numbers

For example $[3\angle 20°]^4 = 3^4\angle(4 \times 20°) = 81\angle 80°$ by De Moivre's theorem.

Problem 1. Determine, in polar form:
(a) $[2\angle 35°]^5$ (b) $(-2+j3)^6$

* **Who was de Moivre?** **Abraham de Moivre** (26 May 1667–27 November 1754) was a French mathematician famous for **de Moivre's formula**, which links complex numbers and trigonometry, and for his work on the normal distribution and probability theory. To find out more go to **www.routledge.com/cw/bird**

(a) $[2\angle 35°]^5 = 2^5\angle(5 \times 35°)$,
$$\text{from De Moivre's theorem}$$
$$= \mathbf{32\angle 175°}$$

(b) $(-2+j3) = \sqrt{[(-2)^2+(3)^2]}\angle\tan^{-1}\dfrac{3}{-2}$
$$= \sqrt{13}\angle 123.69°, \text{ since } -2+j3$$
$$\text{lies in the second quadrant}$$

$(-2+j3)^6 = [\sqrt{13}\angle 123.69°]^6$
$$= (\sqrt{13})^6\angle(6 \times 123.69°),$$
$$\text{by De Moivre's theorem}$$
$$= 2197\angle 742.14°$$
$$= 2197\angle 382.14°(\text{since } 742.14$$
$$\equiv 742.14° - 360° = 382.14°)$$
$$= \mathbf{2197\angle 22.14°}(\text{since } 382.14°$$
$$\equiv 382.14° - 360° = 22.14°)$$
$$\text{or } \mathbf{2197\angle 22°8'}$$

Problem 2. Determine the value of $(-7+j5)^4$, expressing the result in polar and rectangular forms

---

$(-7+j5) = \sqrt{[(-7)^2+5^2]}\angle\tan^{-1}\dfrac{5}{7}$
$$= \sqrt{74}\angle 144.46°$$

(Note, by considering the Argand diagram, $-7+j5$ must represent an angle in the second quadrant and **not** in the fourth quadrant.)

Applying De Moivre's theorem:

$(-7+j5)^4 = [\sqrt{74}\angle 144.46°]^4$
$$= \sqrt{74}^4\angle 4 \times 144.46°$$
$$= 5476\angle 577.84°$$
$$= \mathbf{5476\angle 217.84°}$$
$$\text{or } \mathbf{5476\angle 217°50'} \text{ in polar form}$$

Since $r\angle\theta = r\cos\theta + jr\sin\theta$,

$5476\angle 217.84° = 5476\cos 217.84°$
$$+ j5476\sin 217.84°$$
$$= -4325 - j3359$$

i.e. $\quad \mathbf{(-7+j5)^4 = -4325 - j3359}$

$$\text{in rectangular form}$$

Now try the following Practice Exercise

Practice Exercise 196   Further problems on powers of complex numbers (answers on page 1172)

1. Determine in polar form (a) $[1.5\angle 15°]^5$ (b) $(1 + j2)^6$

2. Determine in polar and Cartesian forms (a) $[3\angle 41°]^4$ (b) $(-2 - j)^5$

3. Convert $(3 - j)$ into polar form and hence evaluate $(3 - j)^7$, giving the answer in polar form.

In Problems 4 to 7, express in both polar and rectangular forms.

4. $(6 + j5)^3$

5. $(3 - j8)^5$

6. $(-2 + j7)^4$

7. $(-16 - j9)^6$

46.3 Roots of complex numbers

The **square root** of a complex number is determined by letting $n = 1/2$ in De Moivre's theorem,

i.e. $\sqrt{[r\angle\theta]} = [r\angle 0]^{\frac{1}{2}} = r^{\frac{1}{2}}\angle\frac{1}{2}\theta = \sqrt{r}\angle\frac{\theta}{2}$

There are two square roots of a real number, equal in size but opposite in sign.

Problem 3. Determine the two square roots of the complex number $(5 + j12)$ in polar and Cartesian forms and show the roots on an Argand diagram

$(5 + j12) = \sqrt{[5^2 + 12^2]}\angle\tan^{-1}\left(\frac{12}{5}\right)$

$= 13\angle 67.38°$

When determining square roots two solutions result. To obtain the second solution one way is to express $13\angle 67.38°$ also as $13\angle(67.38° + 360°)$, i.e. $13\angle 427.38°$. When the angle is divided by 2 an angle less than 360° is obtained.

Hence

$\sqrt{(5 + j12)} = \sqrt{[13\angle 67.38°]}$ and $\sqrt{[13\angle 427.38°]}$

$= [13\angle 67.38°]^{\frac{1}{2}}$ and $[13\angle 427.38°]^{\frac{1}{2}}$

$= 13^{\frac{1}{2}}\angle\left(\frac{1}{2} \times 67.38°\right)$ and

$13^{\frac{1}{2}}\angle\left(\frac{1}{2} \times 427.38°\right)$

$= \sqrt{13}\angle 33.69°$ and $\sqrt{13}\angle 213.69°$

$= 3.61\angle 33.69°$ and $3.61\angle 213.69°$

Thus, in polar form, the two roots are $3.61\angle 33.69°$ and $3.61\angle -146.31°$

$\sqrt{13}\angle 33.69° = \sqrt{13}(\cos 33.69° + j\sin 33.69°)$

$= 3.0 + j2.0$

$\sqrt{13}\angle 213.69° = \sqrt{13}(\cos 213.69° + j\sin 213.69°)$

$= -3.0 - j2.0$

Thus, in Cartesian form the two roots are $\pm(3.0 + j2.0)$

From the Argand diagram shown in Figure 46.1 the two roots are seen to be 180° apart, which is always true when finding square roots of complex numbers.

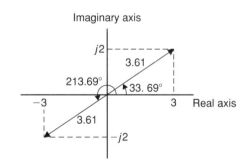

Figure 46.1

In general, **when finding the nth root of a complex number, there are n solutions**. For example, there are three solutions to a cube root, five solutions to a fifth root, and so on. In the solutions to the roots of a complex number, the modulus, r, is always the same, but the arguments, θ, are different. It is shown in Problem 3 that arguments are symmetrically spaced on an Argand diagram and are $(360/n)^\circ$ apart, where n is the number of the roots required. Thus if one of the solutions to the cube root of a complex number is, say, $5\angle 20^\circ$, the other two roots are symmetrically spaced $(360/3)^\circ$, i.e. 120° from this root and the three roots are $5\angle 20^\circ$, $5\angle 140^\circ$ and $5\angle 260^\circ$.

Problem 4. Find the roots of $[(5+j3)]^{\frac{1}{2}}$ in rectangular form, correct to 4 significant figures

$$(5+j3) = \sqrt{34}\angle 30.96^\circ$$

Applying De Moivre's theorem:

$$(5+j3)^{\frac{1}{2}} = \sqrt{34}^{\frac{1}{2}}\angle \tfrac{1}{2} \times 30.96^\circ$$

$$= 2.415\angle 15.48^\circ \text{ or } 2.415\angle 15^\circ 29'$$

The second root may be obtained as shown above, i.e. having the same modulus but displaced $(360/2)^\circ$ from the first root.

Thus, $(5+j3)^{\frac{1}{2}} = 2.415\angle(15.48^\circ + 180^\circ)$

$$= 2.415\angle 195.48^\circ$$

In rectangular form:

$$2.415\angle 15.48^\circ = 2.415\cos 15.48^\circ$$
$$+ j2.415\sin 15.48^\circ$$
$$= 2.327 + j0.6446$$

and $\quad 2.415\angle 195.48^\circ = 2.415\cos 195.48^\circ$
$$+ j2.415\sin 195.48^\circ$$
$$= -2.327 - j0.6446$$

Hence $\quad [(5+j3)]^{\frac{1}{2}} = \mathbf{2.415\angle 15.48^\circ}$ and

$$\mathbf{2.415\angle 195.48^\circ} \text{ or}$$

$$\mathbf{\pm(2.327 + j0.6446)}$$

Problem 5. Express the roots of $(-14+j3)^{\frac{-2}{5}}$ in polar form

$$(-14+j3) = \sqrt{205}\angle 167.905^\circ$$

$$(-14+j3)^{\frac{-2}{5}} = \sqrt{205}^{\frac{-2}{5}}\angle\left[\left(-\frac{2}{5}\right) \times 167.905^\circ\right]$$

$$= 0.3449\angle -67.162^\circ$$

$$\text{or } 0.3449\angle -67^\circ 10'$$

There are five roots to this complex number,

$$\left(x^{\frac{-2}{5}} = \frac{1}{x^{\frac{2}{5}}} = \frac{1}{\sqrt[5]{x^2}}\right)$$

The roots are symmetrically displaced from one another $(360/5)^\circ$, i.e. 72° apart round an Argand diagram.

Thus the required roots are $\mathbf{0.3449\angle -67^\circ 10'}$, $\mathbf{0.3449\angle 4^\circ 50'}$, $\mathbf{0.3449\angle 76^\circ 50'}$, $\mathbf{0.3449\angle 148^\circ 50'}$ and $\mathbf{0.3449\angle 220^\circ 50'}$

Now try the following Practice Exercise

Practice Exercise 197 Further problems on the roots of complex numbers (answers on page 1173)

In Problems 1 to 3, determine the two square roots of the given complex numbers in Cartesian form and show the results on an Argand diagram.

1. (a) $1+j$ (b) j

2. (a) $3-j4$ (b) $-1-j2$

3. (a) $7\angle 60^\circ$ (b) $12\angle\dfrac{3\pi}{2}$

In Problems 4 to 7, determine the moduli and arguments of the complex roots.

4. $(3+j4)^{\frac{1}{3}}$

5. $(-2+j)^{\frac{1}{4}}$

6. $(-6-j5)^{\frac{1}{2}}$

7. $(4-j3)^{\frac{-2}{3}}$

8. For a transmission line, the characteristic impedance Z_0 and the propagation coefficient

γ are given by:

$$Z_0 = \sqrt{\left(\frac{R + j\omega L}{G + j\omega C}\right)} \quad \text{and}$$

$$\gamma = \sqrt{[(R + j\omega L)(G + j\omega C)]}$$

Given $R = 25\,\Omega, L = 5 \times 10^{-3}\,\text{H}$, $G = 80 \times 10^{-6}$ siemens, $C = 0.04 \times 10^{-6}\,\text{F}$ and $\omega = 2000\,\pi$ rad/s, determine, in polar form, Z_0 and γ

46.4 The exponential form of a complex number

Certain mathematical functions may be expressed as power series (for example, by Maclaurin's series – see Chapter 22), three examples being:

(i) $\quad e^x = 1 + x + \dfrac{x^2}{2!} + \dfrac{x^3}{3!} + \dfrac{x^4}{4!} + \dfrac{x^5}{5!} + \cdots$ (1)

(ii) $\quad \sin x = x - \dfrac{x^3}{3!} + \dfrac{x^5}{5!} - \dfrac{x^7}{7!} + \cdots$ (2)

(iii) $\quad \cos x = 1 - \dfrac{x^2}{2!} + \dfrac{x^4}{4!} - \dfrac{x^6}{6!} + \cdots$ (3)

Replacing x in equation (1) by the imaginary number $j\theta$ gives:

$$e^{j\theta} = 1 + j\theta + \frac{(j\theta)^2}{2!} + \frac{(j\theta)^3}{3!} + \frac{(j\theta)^4}{4!} + \frac{(j\theta)^5}{5!} + \cdots$$

$$= 1 + j\theta + \frac{j^2\theta^2}{2!} + \frac{j^3\theta^3}{3!} + \frac{j^4\theta^4}{4!} + \frac{j^5\theta^5}{5!} + \cdots$$

By definition, $j = \sqrt{(-1)}$, hence $j^2 = -1$, $j^3 = -j$, $j^4 = 1$, $j^5 = j$, and so on.

Thus $e^{j\theta} = 1 + j\theta - \dfrac{\theta^2}{2!} - j\dfrac{\theta^3}{3!} + \dfrac{\theta^4}{4!} + j\dfrac{\theta^5}{5!} - \cdots$

Grouping real and imaginary terms gives:

$$e^{j\theta} = \left(1 - \frac{\theta^2}{2!} + \frac{\theta^4}{4!} - \cdots\right)$$

$$+ j\left(\theta - \frac{\theta^3}{3!} + \frac{\theta^5}{5!} - \cdots\right)$$

However, from equations (2) and (3):

$$\left(1 - \frac{\theta^2}{2!} + \frac{\theta^4}{4!} - \cdots\right) = \cos\theta$$

and $\left(\theta - \dfrac{\theta^3}{3!} + \dfrac{\theta^5}{5!} - \cdots\right) = \sin\theta$

Thus $\qquad e^{j\theta} = \cos\theta + j\sin\theta$ (4)

Writing $-\theta$ for θ in equation (4), gives:

$$e^{j(-\theta)} = \cos(-\theta) + j\sin(-\theta)$$

However, $\cos(-\theta) = \cos\theta$ and $\sin(-\theta) = -\sin\theta$

Thus $\qquad e^{-j\theta} = \cos\theta - j\sin\theta$ (5)

The polar form of a complex number z is: $z = r(\cos\theta + j\sin\theta)$. But, from equation (4), $\cos\theta + j\sin\theta = e^{j\theta}$.

Therefore $\qquad z = re^{j\theta}$

When a complex number is written in this way, it is said to be expressed in **exponential form**.

There are therefore three ways of expressing a complex number:

1. $z = (a + jb)$, called **Cartesian** or **rectangular form**,

2. $z = r(\cos\theta + j\sin\theta)$ or $r\angle\theta$, called **polar form**, and

3. $z = re^{j\theta}$ called **exponential form**.

The exponential form is obtained from the polar form. For example, $4\angle 30°$ becomes $4e^{j\frac{\pi}{6}}$ in exponential form. (Note that in $re^{j\theta}$, θ must be in radians.)

Problem 6. Change $(3 - j4)$ into (a) polar form, (b) exponential form

(a) $(3 - j4) = \mathbf{5\angle -53.13°}$ or $\mathbf{5\angle -0.927}$

in polar form

(b) $(3 - j4) = 5\angle -0.927 = \mathbf{5e^{-j0.927}}$

in exponential form

Problem 7. Convert $7.2e^{j1.5}$ into rectangular form

$7.2e^{j1.5} = 7.2\angle 1.5$ rad $(= 7.2\angle 85.94°)$ in polar form

$= 7.2\cos 1.5 + j7.2\sin 1.5$

$= \mathbf{(0.509 + j7.182)}$ in rectangular form

Problem 8. Express $z=2e^{1+j\frac{\pi}{3}}$ in Cartesian form

$z = (2e^1)\left(e^{j\frac{\pi}{3}}\right)$ by the laws of indices

$= (2e^1)\angle\frac{\pi}{3}$ (or $2e\angle60°$) in polar form

$= 2e\left(\cos\frac{\pi}{3} + j\sin\frac{\pi}{3}\right)$

$= (2.718 + j4.708)$ in Cartesian form

Problem 9. Change $6e^{2-j3}$ into $(a+jb)$ form

$6e^{2-j3} = (6e^2)(e^{-j3})$ by the laws of indices

$= 6e^2\angle-3$ rad (or $6e^2\angle-179.89°$) in polar form

$= 6e^2[\cos(-3) + j\sin(-3)]$

$= (-43.89 - j6.26)$ in $(a+jb)$ form

Problem 10. If $z=4e^{j1.3}$, determine $\ln z$ (a) in Cartesian form, and (b) in polar form

If $z = re^{j\theta}$ then $\ln z = \ln(re^{j\theta})$

$= \ln r + \ln e^{j\theta}$

i.e. $\ln z = \ln r + j\theta$,

by the laws of logarithms

(a) Thus if $z=4e^{j1.3}$ then $\ln z = \ln(4e^{j1.3})$

$= \ln 4 + j1.3$

(or $1.386 + j1.300$) in Cartesian form.

(b) $(1.386 + j1.300) = 1.90\angle43.17°$ or $1.90\angle0.753$ in polar form.

Problem 11. Given $z=3e^{1-j}$, find $\ln z$ in polar form

If $z = 3e^{1-j}$, then

$\ln z = \ln(3e^{1-j})$

$= \ln 3 + \ln e^{1-j}$

$= \ln 3 + 1 - j$

$= (1 + \ln 3) - j$

$= 2.0986 - j1.0000$

$= 2.325\angle-25.48°$ or $2.325\angle-0.445$

Problem 12. Determine, in polar form, $\ln(3+j4)$

$\ln(3+j4) = \ln[5\angle0.927] = \ln[5e^{j0.927}]$

$= \ln 5 + \ln(e^{j0.927})$

$= \ln 5 + j0.927$

$= 1.609 + j0.927$

$= 1.857\angle29.95°$ or $1.857\angle0.523$

Now try the following Practice Exercise

Practice Exercise 198 Further problems on the exponential form of complex numbers (answers on page 1173)

1. Change $(5+j3)$ into exponential form.

2. Convert $(-2.5+j4.2)$ into exponential form.

3. Change $3.6e^{j2}$ into Cartesian form.

4. Express $2e^{3+j\frac{\pi}{6}}$ in $(a+jb)$ form.

5. Convert $1.7e^{1.2-j2.5}$ into rectangular form.

6. If $z=7e^{j2.1}$, determine $\ln z$ (a) in Cartesian form, and (b) in polar form.

7. Given $z=4e^{1.5-j2}$, determine $\ln z$ in polar form.

8. Determine in polar form (a) $\ln(2+j5)$ (b) $\ln(-4-j3)$

9. When displaced electrons oscillate about an equilibrium position the displacement x is given by the equation:

$$x = Ae^{\left\{-\frac{ht}{2m}+j\frac{\sqrt{(4mf-h^2)}}{2m-a}t\right\}}$$

Determine the real part of x in terms of t, assuming $(4mf - h^2)$ is positive.

46.5 Introduction to locus problems

The locus is a set of points that satisfy a certain condition. For example, the locus of points that are, say, 2 units from point C, refers to the set of all points that are

2 units from C; this would be a circle with centre C as shown in Figure 46.2.

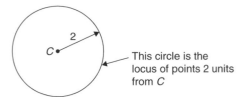

Figure 46.2

It is sometimes needed to find the locus of a point which moves in the Argand diagram according to some stated condition. Loci (the plural of locus) are illustrated by the following worked problems.

Problem 13. Determine the locus defined by $|z| = 4$, given that $z = x + jy$

If $z = x + jy$, then on an Argand diagram as shown in Figure 46.3, the **modulus z**,

$$|z| = \sqrt{x^2 + y^2}$$

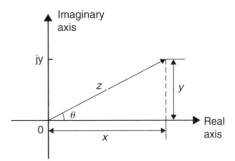

Figure 46.3

In this case, $\sqrt{x^2 + y^2} = 4$ from which, $x^2 + y^2 = 4^2$
From Chapter 28, $x^2 + y^2 = 4^2$ **is a circle, with centre at the origin and with radius 4**
The locus (or path) of $|z| = 4$ is shown in Figure 46.4

Problem 14. Determine the locus defined by arg $z = \frac{\pi}{4}$, given that $z = x + jy$

In Figure 46.3 above, $\theta = \tan^{-1}\left(\frac{y}{x}\right)$
where θ is called the **argument** and is written as
$\textbf{arg } z = \textbf{tan}^{-1}\left(\frac{y}{x}\right)$

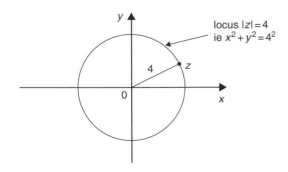

Figure 46.4

Hence, in this example, $\tan^{-1}\left(\frac{y}{x}\right) = \frac{\pi}{4}$ i.e. $\frac{y}{x} = \tan\frac{\pi}{4} = \tan 45° = 1$

Thus, if $\frac{y}{x} = 1$, then $y = x$

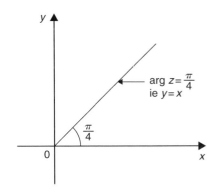

Figure 46.5

Hence, the locus (or path) of arg $z = \frac{\pi}{4}$ **is a straight line $y = x$ (with y > 0) as shown in Figure 46.5.**

Problem 15. If $z = x + jy$, determine the locus defined by arg $(z - 1) = \frac{\pi}{6}$

If arg $(z - 1) = \frac{\pi}{6}$, then arg $(x + jy - 1) = \frac{\pi}{6}$

i.e. arg $[(x - 1) + jy] = \frac{\pi}{6}$

In Figure 46.3 above, $\theta = \tan^{-1}\left(\frac{y}{x}\right)$
i.e. arg $z = \tan^{-1}\left(\frac{y}{x}\right)$

Hence, in this example, $\tan^{-1}\left(\frac{y}{x - 1}\right) = \frac{\pi}{6}$

i.e. $\frac{y}{x - 1} = \tan\frac{\pi}{6} = \tan 30° = \frac{1}{\sqrt{3}}$

Thus, if $\dfrac{y}{x-1} = \dfrac{1}{\sqrt{3}}$, then $y = \dfrac{1}{\sqrt{3}}(x-1)$

Hence, the locus of $\arg(z-1) = \dfrac{\pi}{6}$ is a straight line

$$y = \dfrac{1}{\sqrt{3}}x - \dfrac{1}{\sqrt{3}}$$

Problem 16. Determine the locus defined by $|z-2| = 3$, given that $z = x + jy$

If $z = x + jy$, then
$|z-2| = |x + jy - 2| = |(x-2) + jy| = 3$
On the Argand diagram shown in Figure 46.3,
$|z| = \sqrt{x^2 + y^2}$
Hence, in this case, $|z-2| = \sqrt{(x-2)^2 + y^2} = 3$,
from which $(x-2)^2 + y^2 = 3^2$
From Chapter 28, $(x-a)^2 + (y-b)^2 = r^2$ is a circle, with centre (a,b) and radius r.
Hence, $(x-2)^2 + y^2 = 3^2$ is a circle, with centre $(2,0)$ and radius 3
The locus of $|z-2| = 3$ is shown in Figure 46.6

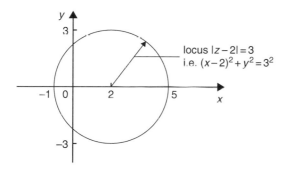

Figure 46.6

Problem 17. If $z = x + jy$, determine the locus defined by $\left|\dfrac{z-1}{z+1}\right| = 3$

$z - 1 = x + jy - 1 = (x-1) + jy$

$z + 1 = x + jy + 1 = (x+1) + jy$

Hence, $\left|\dfrac{z-1}{z+1}\right| = \left|\dfrac{(x-1)+jy}{(x+1)+jy}\right| = \dfrac{\sqrt{(x-1)^2 + y^2}}{\sqrt{(x+1)^2 + y^2}}$
$= 3$

and squaring both sides gives:

$$\dfrac{(x-1)^2 + y^2}{(x+1)^2 + y^2} = 9$$

from which,

$$(x-1)^2 + y^2 = 9\left[(x+1)^2 + y^2\right]$$

$$x^2 - 2x + 1 + y^2 = 9\left[x^2 + 2x + 1 + y^2\right]$$

$$x^2 - 2x + 1 + y^2 = 9x^2 + 18x + 9 + 9y^2$$

$$0 = 8x^2 + 20x + 8 + 8y^2$$

i.e.

$$8x^2 + 20x + 8 + 8y^2 = 0$$

and dividing by 4 gives:

$2x^2 + 5x + 2 + 2y^2 = 0$ which is the equation of the locus

Rearranging gives: $x^2 + \dfrac{5}{2}x + y^2 = -1$

Completing the square (see Chapter 14) gives:

$$\left(x + \dfrac{5}{4}\right)^2 - \dfrac{25}{16} + y^2 = -1$$

i.e. $\left(x + \dfrac{5}{4}\right)^2 + y^2 = -1 + \dfrac{25}{16}$

i.e. $\left(x + \dfrac{5}{4}\right)^2 + y^2 = \dfrac{9}{16}$

i.e. $\left(x + \dfrac{5}{4}\right)^2 + y^2 = \left(\dfrac{3}{4}\right)^2$ which is the equation of a circle.

Hence the locus defined by $\left|\dfrac{z-1}{z+1}\right| = 3$ is a circle of centre $\left(-\dfrac{5}{4}, 0\right)$ and radius $\dfrac{3}{4}$

Problem 18. If $z = x + jy$, determine the locus defined by $\arg\left(\dfrac{z+1}{z}\right) = \dfrac{\pi}{4}$

$$\left(\dfrac{z+1}{z}\right) = \left(\dfrac{(x+1)+jy}{x+jy}\right) = \dfrac{[(x+1)+jy](x-jy)}{(x+jy)(x-jy)}$$

$$= \dfrac{x(x+1) - j(x+1)y + jxy + y^2}{x^2 + y^2}$$

$$= \dfrac{x^2 + x - jxy - jy + jxy + y^2}{x^2 + y^2}$$

$$= \dfrac{x^2 + x - jy + y^2}{x^2 + y^2}$$

$$= \frac{(x^2 + x + y^2) - jy}{x^2 + y^2}$$

$$= \frac{x^2 + x + y^2}{x^2 + y^2} - \frac{jy}{x^2 + y^2}$$

Since $\arg\left(\dfrac{z+1}{z}\right) = \dfrac{\pi}{4}$ then

$$\tan^{-1}\left(\frac{\dfrac{-y}{x^2 + y^2}}{\dfrac{x^2 + x + y^2}{x^2 + y^2}}\right) = \frac{\pi}{4}$$

i.e. $\qquad \tan^{-1}\left(\dfrac{-y}{x^2 + x + y^2}\right) = \dfrac{\pi}{4}$

from which,

$$\frac{-y}{x^2 + x + y^2} = \tan\frac{\pi}{4} = 1$$

Hence,

$$-y = x^2 + x + y^2$$

Hence, **the locus defined by** $\arg\left(\dfrac{z+1}{z}\right) = \dfrac{\pi}{4}$ **is:**

$$x^2 + x + y + y^2 = 0$$

Completing the square gives:

$$\left(x + \frac{1}{2}\right)^2 + \left(y + \frac{1}{2}\right)^2 = \frac{1}{2}$$

which is a circle centre $\left(-\dfrac{1}{2}, -\dfrac{1}{2}\right)$ **and radius** $\dfrac{1}{\sqrt{2}}$

Problem 19. Determine the locus defined by $|z - j| = |z - 3|$, given that $z = x + jy$

Since $\qquad |z - j| = |z - 3|$

then $\qquad |x + j(y-1)| = |(x-3) + jy|$

and $\qquad \sqrt{x^2 + (y-1)^2} = \sqrt{(x-3)^2 + y^2}$

Squaring both sides gives: $x^2 + (y-1)^2 = (x-3)^2 + y^2$

i.e. $\qquad x^2 + y^2 - 2y + 1 = x^2 - 6x + 9 + y^2$

from which, $\quad -2y + 1 = -6x + 9$

i.e. $\qquad 6x - 8 = 2y$

or $\qquad y = 3x - 4$

Hence, the locus defined by $|z-j| = |z-3|$ **is a straight line:** $y = 3x - 4$

Now try the following Practice Exercise

Practice Exercise 199 Locus problems (answers on page 1173)

For each of the following, if $z = x + jy$, (a) determine the equation of the locus, (b) sketch the locus

1. $|z| = 2$ $\qquad$ 2. $|z| = 5$
3. $\arg(z - 2) = \dfrac{\pi}{3}$ $\quad$ 4. $\arg(z + 1) = \dfrac{\pi}{6}$
5. $|z - 2| = 4$ $\qquad$ 6. $|z + 3| = 5$
7. $\left|\dfrac{z+1}{z-1}\right| = 3$ $\quad$ 8. $\left|\dfrac{z-1}{z}\right| = \sqrt{2}$
9. $\arg\left|\dfrac{z-1}{z}\right| = \dfrac{\pi}{4}$ $\quad$ 10. $\arg\left|\dfrac{z+2}{z}\right| = \dfrac{\pi}{4}$
11. $|z + j| = |z + 2|$ $\quad$ 12. $|z - 4| = |z - 2j|$
13. $|z - 1| = |z|$

This assignment covers the material contained in Chapters 45 and 46. *The marks available are shown in brackets at the end of each question.*

1. Solve the quadratic equation $x^2 - 2x + 5 = 0$ and show the roots on an Argand diagram. (9)

2. If $Z_1 = 2 + j5$, $Z_2 = 1 - j3$ and $Z_3 = 4 - j$ determine, in both Cartesian and polar forms, the value of $\dfrac{Z_1 Z_2}{Z_1 + Z_2} + Z_3$, correct to 2 decimal places. (9)

3. Three vectors are represented by A, $4.2\angle 45°$, B, $5.5\angle -32°$ and $C, 2.8\angle 75°$. Determine in polar form the resultant D, where $D = B + C - A$ (8)

4. Two impedances, $Z_1 = (2 + j7)$ ohms and $Z_2 = (3 - j4)$ ohms, are connected in series to a supply voltage V of $150\angle 0°$ V. Determine the magnitude of the current I and its phase angle relative to the voltage. (6)

5. Determine in both polar and rectangular forms:

   (a) $[2.37\angle 35°]^4$ (b) $[3.2 - j4.8]^5$

   (c) $\sqrt{[-1 - j3]}$ (18)

Section F

Complex numbers

$z = a + jb = r(\cos\theta + j\sin\theta) = r\angle\theta = re^{j\theta}$

where $j^2 = -1$

Modulus $r = |z| = \sqrt{(a^2 + b^2)}$

Argument $\theta = \arg z = \tan^{-1}\left(\dfrac{b}{a}\right)$

Addition: $(a + jb) + (c + jd) = (a + c) + j(b + d)$

Subtraction: $(a + jb) - (c + jd) = (a - c) + j(b - d)$

Complex equations: If $m + jn = p + jq$
then $m = p$ and $n = q$

Multiplication: If $z_1 = r_1\angle\theta_1$ and $z_2 = r_2\angle\theta_2$

$$z_1 z_2 = r_1 r_2 \angle(\theta_1 + \theta_2)$$

Division: If $z_1 = r_1\angle\theta_1$ and $z_2 = r_2\angle\theta_2$

$$\frac{z_1}{z_2} = \frac{r_1}{r_2}\angle(\theta_1 - \theta_2)$$

De Moivre's theorem:
$$[r\angle\theta]^n = r^n \angle n\theta = r^n(\cos n\theta + j\sin n\theta) = re^{j\theta}$$

Section F

Matrices and determinants

The theory of matrices and determinants

At the end of this chapter, you should be able to:

- understand matrix notation
- add, subtract and multiply 2 by 2 and 3 by 3 matrices
- recognise the unit matrix
- calculate the determinant of a 2 by 2 matrix
- determine the inverse (or reciprocal) of a 2 by 2 matrix
- calculate the determinant of a 3 by 3 matrix
- determine the inverse (or reciprocal) of a 3 by 3 matrix

47.1 Matrix notation

Matrices and determinants are mainly used for the solution of linear simultaneous equations. The theory of matrices and determinants is dealt with in this chapter and this theory is then used in Chapter 48 to solve simultaneous equations.

The coefficients of the variables for linear simultaneous equations may be shown in matrix form. The coefficients of x and y in the simultaneous equations

$$x + 2y = 3$$
$$4x - 5y = 6$$

become $\begin{pmatrix} 1 & 2 \\ 4 & -5 \end{pmatrix}$ in matrix notation.

Similarly, the coefficients of p, q and r in the equations

$$1.3p - 2.0q + r = 7$$
$$3.7p + 4.8q - 7r = 3$$
$$4.1p + 3.8q + 12r = -6$$

become $\begin{pmatrix} 1.3 & -2.0 & 1 \\ 3.7 & 4.8 & -7 \\ 4.1 & 3.8 & 12 \end{pmatrix}$ in matrix form.

The numbers within a matrix are called an **array** and the coefficients forming the array are called the **elements** of the matrix. The number of rows in a matrix is usually specified by m and the number of columns by n and a matrix referred to as an 'm by n' matrix. Thus, $\begin{pmatrix} 2 & 3 & 6 \\ 4 & 5 & 7 \end{pmatrix}$ is a '2 by 3' matrix. Matrices cannot be expressed as a single numerical value, but they can often be simplified or combined, and unknown element values can be determined by comparison methods. Just as there are rules for addition, subtraction, multiplication and division of numbers in arithmetic, rules for these operations can be applied to matrices and the rules of matrices are such that they obey most of those governing the algebra of numbers.

47.2 Addition, subtraction and multiplication of matrices

(i) Addition of matrices

Corresponding elements in two matrices may be added to form a single matrix.

Problem 1. Add the matrices

(a) $\begin{pmatrix} 2 & -1 \\ -7 & 4 \end{pmatrix}$ and $\begin{pmatrix} -3 & 0 \\ 7 & -4 \end{pmatrix}$ and

(b) $\begin{pmatrix} 3 & 1 & -4 \\ 4 & 3 & 1 \\ 1 & 4 & -3 \end{pmatrix}$ and $\begin{pmatrix} 2 & 7 & -5 \\ -2 & 1 & 0 \\ 6 & 3 & 4 \end{pmatrix}$

(a) Adding the corresponding elements gives:

$$\begin{pmatrix} 2 & -1 \\ -7 & 4 \end{pmatrix} + \begin{pmatrix} -3 & 0 \\ 7 & -4 \end{pmatrix}$$

$$= \begin{pmatrix} 2+(-3) & -1+0 \\ -7+7 & 4+(-4) \end{pmatrix}$$

$$= \begin{pmatrix} -1 & -1 \\ 0 & 0 \end{pmatrix}$$

(b) Adding the corresponding elements gives:

$$\begin{pmatrix} 3 & 1 & -4 \\ 4 & 3 & 1 \\ 1 & 4 & -3 \end{pmatrix} + \begin{pmatrix} 2 & 7 & -5 \\ -2 & 1 & 0 \\ 6 & 3 & 4 \end{pmatrix}$$

$$= \begin{pmatrix} 3+2 & 1+7 & -4+(-5) \\ 4+(-2) & 3+1 & 1+0 \\ 1+6 & 4+3 & -3+4 \end{pmatrix}$$

$$= \begin{pmatrix} 5 & 8 & -9 \\ 2 & 4 & 1 \\ 7 & 7 & 1 \end{pmatrix}$$

(ii) Subtraction of matrices

If A is a matrix and B is another matrix, then $(A - B)$ is a single matrix formed by subtracting the elements of B from the corresponding elements of A.

Problem 2. Subtract

(a) $\begin{pmatrix} -3 & 0 \\ 7 & -4 \end{pmatrix}$ from $\begin{pmatrix} 2 & -1 \\ -7 & 4 \end{pmatrix}$ and

(b) $\begin{pmatrix} 2 & 7 & -5 \\ -2 & 1 & 0 \\ 6 & 3 & 4 \end{pmatrix}$ from $\begin{pmatrix} 3 & 1 & -4 \\ 4 & 3 & 1 \\ 1 & 4 & -3 \end{pmatrix}$

To find matrix A minus matrix B, the elements of B are taken from the corresponding elements of A. Thus:

(a) $\begin{pmatrix} 2 & -1 \\ -7 & 4 \end{pmatrix} - \begin{pmatrix} -3 & 0 \\ 7 & -4 \end{pmatrix}$

$$= \begin{pmatrix} 2-(-3) & -1-0 \\ -7-7 & 4-(-4) \end{pmatrix}$$

$$= \begin{pmatrix} 5 & -1 \\ -14 & 8 \end{pmatrix}$$

(b) $\begin{pmatrix} 3 & 1 & -4 \\ 4 & 3 & 1 \\ 1 & 4 & -3 \end{pmatrix} - \begin{pmatrix} 2 & 7 & -5 \\ -2 & 1 & 0 \\ 6 & 3 & 4 \end{pmatrix}$

$$= \begin{pmatrix} 3-2 & 1-7 & -4-(-5) \\ 4-(-2) & 3-1 & 1-0 \\ 1-6 & 4-3 & -3-4 \end{pmatrix}$$

$$= \begin{pmatrix} 1 & -6 & 1 \\ 6 & 2 & 1 \\ -5 & 1 & -7 \end{pmatrix}$$

Problem 3. If

$A = \begin{pmatrix} -3 & 0 \\ 7 & -4 \end{pmatrix}$, $B = \begin{pmatrix} 2 & -1 \\ -7 & 4 \end{pmatrix}$ and

$C = \begin{pmatrix} 1 & 0 \\ -2 & -4 \end{pmatrix}$ find $A + B - C$

$$A + B = \begin{pmatrix} -1 & -1 \\ 0 & 0 \end{pmatrix}$$

(from Problem 1)

Hence, $A + B - C = \begin{pmatrix} -1 & -1 \\ 0 & 0 \end{pmatrix} - \begin{pmatrix} 1 & 0 \\ -2 & -4 \end{pmatrix}$

$$= \begin{pmatrix} -1-1 & -1-0 \\ 0-(-2) & 0-(-4) \end{pmatrix}$$

$$= \begin{pmatrix} -2 & -1 \\ 2 & 4 \end{pmatrix}$$

Alternatively $A + B - C$

$$= \begin{pmatrix} -3 & 0 \\ 7 & -4 \end{pmatrix} + \begin{pmatrix} 2 & -1 \\ -7 & 4 \end{pmatrix} - \begin{pmatrix} 1 & 0 \\ -2 & -4 \end{pmatrix}$$

$$= \begin{pmatrix} -3+2-1 & 0+(-1)-0 \\ 7+(-7)-(-2) & -4+4-(-4) \end{pmatrix}$$

$$= \begin{pmatrix} -2 & -1 \\ 2 & 4 \end{pmatrix} \text{ as obtained previously}$$

(iii) Multiplication

When a matrix is multiplied by a number, called **scalar multiplication**, a single matrix results in which each element of the original matrix has been multiplied by the number.

Problem 4. If $A = \begin{pmatrix} -3 & 0 \\ 7 & -4 \end{pmatrix}$,

$B = \begin{pmatrix} 2 & -1 \\ -7 & 4 \end{pmatrix}$ and $C = \begin{pmatrix} 1 & 0 \\ -2 & -4 \end{pmatrix}$ find

$2A - 3B + 4C$

For scalar multiplication, each element is multiplied by the scalar quantity, hence

$$2A = 2\begin{pmatrix} -3 & 0 \\ 7 & -4 \end{pmatrix} = \begin{pmatrix} -6 & 0 \\ 14 & -8 \end{pmatrix}$$

$$3B = 3\begin{pmatrix} 2 & -1 \\ -7 & 4 \end{pmatrix} = \begin{pmatrix} 6 & -3 \\ -21 & 12 \end{pmatrix}$$

and $4C = 4\begin{pmatrix} 1 & 0 \\ -2 & -4 \end{pmatrix} = \begin{pmatrix} 4 & 0 \\ -8 & -16 \end{pmatrix}$

Hence $2A - 3B + 4C$

$$= \begin{pmatrix} -6 & 0 \\ 14 & -8 \end{pmatrix} - \begin{pmatrix} 6 & -3 \\ -21 & 12 \end{pmatrix} + \begin{pmatrix} 4 & 0 \\ -8 & -16 \end{pmatrix}$$

$$= \begin{pmatrix} -6-6+4 & 0-(-3)+0 \\ 14-(-21)+(-8) & -8-12+(-16) \end{pmatrix}$$

$$= \begin{pmatrix} -8 & 3 \\ 27 & -36 \end{pmatrix}$$

When a matrix A is multiplied by another matrix B, a single matrix results in which elements are obtained from the sum of the products of the corresponding rows of A and the corresponding columns of B.

Two matrices A and B may be multiplied together, provided the number of elements in the rows of matrix A is equal to the number of elements in the columns of matrix B. In general terms, when multiplying a matrix of dimensions (m by n) by a matrix of dimensions (n by r), the resulting matrix has dimensions (m by r). Thus a 2 by 3 matrix multiplied by a 3 by 1 matrix gives a matrix of dimensions 2 by 1

Problem 5. If $A = \begin{pmatrix} 2 & 3 \\ 1 & -4 \end{pmatrix}$ and $B = \begin{pmatrix} -5 & 7 \\ -3 & 4 \end{pmatrix}$ find $A \times B$

Let $A \times B = C$ where $C = \begin{pmatrix} C_{11} & C_{12} \\ C_{21} & C_{22} \end{pmatrix}$

C_{11} is the sum of the products of the first row elements of A and the first column elements of B, taken one at a time,

i.e. $C_{11} = (2 \times (-5)) + (3 \times (-3)) = -19$

C_{12} is the sum of the products of the first row elements of A and the second column elements of B, taken one at a time,

i.e. $C_{12} = (2 \times 7) + (3 \times 4) = 26$

C_{21} is the sum of the products of the second row elements of A and the first column elements of B, taken one at a time,

i.e. $C_{21} = (1 \times (-5)) + (-4 \times (-3)) = 7$

Finally, C_{22} is the sum of the products of the second row elements of A and the second column elements of B, taken one at a time,

i.e. $C_{22} = (1 \times 7) + ((-4) \times 4) = -9$

Thus, $A \times B = \begin{pmatrix} -19 & 26 \\ 7 & -9 \end{pmatrix}$

Problem 6. Simplify

$$\begin{pmatrix} 3 & 4 & 0 \\ -2 & 6 & -3 \\ 7 & -4 & 1 \end{pmatrix} \times \begin{pmatrix} 2 \\ 5 \\ -1 \end{pmatrix}$$

The sum of the products of the elements of each row of the first matrix and the elements of the second matrix (called a **column matrix**), are taken one at a time. Thus:

$$\begin{pmatrix} 3 & 4 & 0 \\ -2 & 6 & -3 \\ 7 & -4 & 1 \end{pmatrix} \times \begin{pmatrix} 2 \\ 5 \\ -1 \end{pmatrix}$$

$$= \begin{pmatrix} (3 \times 2) & + (4 \times 5) & + (0 \times (-1)) \\ (-2 \times 2) & + (6 \times 5) & + (-3 \times (-1)) \\ (7 \times 2) & + (-4 \times 5) & + (1 \times (-1)) \end{pmatrix}$$

$$= \begin{pmatrix} 26 \\ 29 \\ -7 \end{pmatrix}$$

Problem 7. If $A = \begin{pmatrix} 3 & 4 & 0 \\ -2 & 6 & -3 \\ 7 & -4 & 1 \end{pmatrix}$ and

$B = \begin{pmatrix} 2 & -5 \\ 5 & -6 \\ -1 & -7 \end{pmatrix}$, find $A \times B$

The sum of the products of the elements of each row of the first matrix and the elements of each column of the second matrix are taken one at a time. Thus:

$$\begin{pmatrix} 3 & 4 & 0 \\ -2 & 6 & -3 \\ 7 & -4 & 1 \end{pmatrix} \times \begin{pmatrix} 2 & -5 \\ 5 & -6 \\ -1 & -7 \end{pmatrix}$$

$$= \begin{pmatrix} [(3 \times 2) & [(3 \times (-5)) \\ +(4 \times 5) & +(4 \times (-6)) \\ +(0 \times (-1))] & +(0 \times (-7))] \\ [(-2 \times 2) & [(-2 \times (-5)) \\ +(6 \times 5) & +(6 \times (-6)) \\ +(-3 \times (-1))] & +(-3 \times (-7))] \\ [(7 \times 2) & [(7 \times (-5)) \\ +(-4 \times 5) & +(-4 \times (-6)) \\ +(1 \times (-1))] & +(1 \times (-7))] \end{pmatrix}$$

$$= \begin{pmatrix} 26 & -39 \\ 29 & -5 \\ -7 & -18 \end{pmatrix}$$

Problem 8. Determine

$$\begin{pmatrix} 1 & 0 & 3 \\ 2 & 1 & 2 \\ 1 & 3 & 1 \end{pmatrix} \times \begin{pmatrix} 2 & 2 & 0 \\ 1 & 3 & 2 \\ 3 & 2 & 0 \end{pmatrix}$$

The sum of the products of the elements of each row of the first matrix and the elements of each column of the second matrix are taken one at a time. Thus:

$$\begin{pmatrix} 1 & 0 & 3 \\ 2 & 1 & 2 \\ 1 & 3 & 1 \end{pmatrix} \times \begin{pmatrix} 2 & 2 & 0 \\ 1 & 3 & 2 \\ 3 & 2 & 0 \end{pmatrix}$$

$$= \begin{pmatrix} [(1 \times 2) & [(1 \times 2) & [(1 \times 0) \\ +(0 \times 1) & +(0 \times 3) & +(0 \times 2) \\ +(3 \times 3)] & +(3 \times 2)] & +(3 \times 0)] \\ [(2 \times 2) & [(2 \times 2) & [(2 \times 0) \\ +(1 \times 1) & +(1 \times 3) & +(1 \times 2) \\ +(2 \times 3)] & +(2 \times 2)] & +(2 \times 0)] \\ [(1 \times 2) & [(1 \times 2) & [(1 \times 0) \\ +(3 \times 1) & +(3 \times 3) & +(3 \times 2) \\ +(1 \times 3)] & +(1 \times 2)] & +(1 \times 0)] \end{pmatrix}$$

$$= \begin{pmatrix} 11 & 8 & 0 \\ 11 & 11 & 2 \\ 8 & 13 & 6 \end{pmatrix}$$

In algebra, the commutative law of multiplication states that $a \times b = b \times a$. For matrices, this law is only true in a few special cases, and in general $A \times B$ is **not** equal to $B \times A$

Problem 9. If $A = \begin{pmatrix} 2 & 3 \\ 1 & 0 \end{pmatrix}$ and

$B = \begin{pmatrix} 2 & 3 \\ 0 & 1 \end{pmatrix}$ show that $A \times B \neq B \times A$

$$A \times B = \begin{pmatrix} 2 & 3 \\ 1 & 0 \end{pmatrix} \times \begin{pmatrix} 2 & 3 \\ 0 & 1 \end{pmatrix}$$

$$= \begin{pmatrix} [(2 \times 2) + (3 \times 0)] & [(2 \times 3) + (3 \times 1)] \\ [(1 \times 2) + (0 \times 0)] & [(1 \times 3) + (0 \times 1)] \end{pmatrix}$$

$$= \begin{pmatrix} 4 & 9 \\ 2 & 3 \end{pmatrix}$$

$$B \times A = \begin{pmatrix} 2 & 3 \\ 0 & 1 \end{pmatrix} \times \begin{pmatrix} 2 & 3 \\ 1 & 0 \end{pmatrix}$$

$$= \begin{pmatrix} [(2 \times 2) + (3 \times 1)] & [(2 \times 3) + (3 \times 0)] \\ [(0 \times 2) + (1 \times 1)] & [(0 \times 3) + (1 \times 0)] \end{pmatrix}$$

$$= \begin{pmatrix} 7 & 6 \\ 1 & 0 \end{pmatrix}$$

Since $\begin{pmatrix} 4 & 9 \\ 2 & 3 \end{pmatrix} \neq \begin{pmatrix} 7 & 6 \\ 1 & 0 \end{pmatrix}$, then $A \times B \neq B \times A$

Now try the following Practice Exercise

Practice Exercise 200   Further problems on addition, subtraction and multiplication of matrices (answers on page 1173)

In Problems 1 to 13, the matrices A to K are:

$$A = \begin{pmatrix} 3 & -1 \\ -4 & 7 \end{pmatrix} \quad B = \begin{pmatrix} 5 & 2 \\ -1 & 6 \end{pmatrix}$$

$$C = \begin{pmatrix} -1.3 & 7.4 \\ 2.5 & -3.9 \end{pmatrix}$$

$$D = \begin{pmatrix} 4 & -7 & 6 \\ -2 & 4 & 0 \\ 5 & 7 & -4 \end{pmatrix}$$

$$E = \begin{pmatrix} 3 & 6 & 2 \\ 5 & -3 & 7 \\ -1 & 0 & 2 \end{pmatrix}$$

$$F = \begin{pmatrix} 3.1 & 2.4 & 6.4 \\ -1.6 & 3.8 & -1.9 \\ 5.3 & 3.4 & -4.8 \end{pmatrix} \quad G = \begin{pmatrix} 6 \\ -2 \end{pmatrix}$$

$$H = \begin{pmatrix} -2 \\ 5 \end{pmatrix} \quad J = \begin{pmatrix} 4 \\ -11 \\ 7 \end{pmatrix} \quad K = \begin{pmatrix} 1 & 0 \\ 0 & 1 \\ 1 & 0 \end{pmatrix}$$

In Problems 1 to 12, perform the matrix operation stated.

1. $A + B$

2. $D + E$

3. $A - B$

4. $A + B - C$

5. $5A + 6B$

6. $2D + 3E - 4F$

7. $A \times H$

8. $A \times B$

9. $A \times C$

10. $D \times J$

11. $E \times K$

12. $D \times F$

13. Show that $A \times C \neq C \times A$

47.3 The unit matrix

A **unit matrix, I**, is one in which all elements of the leading diagonal (\) have a value of 1 and all other elements have a value of 0. Multiplication of a matrix by I is the equivalent of multiplying by 1 in arithmetic.

47.4 The determinant of a 2 by 2 matrix

The **determinant** of a 2 by 2 matrix, $\begin{pmatrix} a & b \\ c & d \end{pmatrix}$ is defined as $(ad - bc)$.

The elements of the determinant of a matrix are written between vertical lines. Thus, the determinant of $\begin{pmatrix} 3 & -4 \\ 1 & 6 \end{pmatrix}$ is written as $\begin{vmatrix} 3 & -4 \\ 1 & 6 \end{vmatrix}$ and is equal to $(3 \times 6) - (-4 \times 1)$, i.e. $18 - (-4)$ or 22. Hence the determinant of a matrix can be expressed as a single numerical value, i.e. $\begin{vmatrix} 3 & -4 \\ 1 & 6 \end{vmatrix} = 22$

Problem 10. Determine the value of

$$\begin{vmatrix} 3 & -2 \\ 7 & 4 \end{vmatrix}$$

$$\begin{vmatrix} 3 & -2 \\ 7 & 4 \end{vmatrix} = (3 \times 4) - (-2 \times 7)$$

$$= 12 - (-14) = \mathbf{26}$$

Problem 11. Evaluate $\begin{vmatrix} (1+j) & j2 \\ -j3 & (1-j4) \end{vmatrix}$

$$\begin{vmatrix} (1+j) & j2 \\ -j3 & (1-j4) \end{vmatrix} = (1+j)(1-j4) - (j2)(-j3)$$

$$= 1 - j4 + j - j^2 4 + j^2 6$$

$$= 1 - j4 + j - (-4) + (-6)$$

since from Chapter 45, $j^2 = -1$

$$= 1 - j4 + j + 4 - 6$$

$$= \mathbf{-1 - j\,3}$$

Problem 12. Evaluate $\begin{vmatrix} 5\angle 30° & 2\angle -60° \\ 3\angle 60° & 4\angle -90° \end{vmatrix}$

$$\begin{vmatrix} 5\angle 30° & 2\angle -60° \\ 3\angle 60° & 4\angle -90° \end{vmatrix} = (5\angle 30°)(4\angle -90°)$$

$$- (2\angle -60°)(3\angle 60°)$$

$$= (20\angle -60°) - (6\angle 0°)$$

from page 530

$$= (10 - j17.32) - (6 + j0)$$

$$= \mathbf{(4 - j\,17.32)} \text{ or } \mathbf{17.78\angle -77°}$$

Now try the following Practice Exercise

Practice Exercise 201   Further problems on 2 by 2 determinants (answers on page 1173)

1. Calculate the determinant of $\begin{pmatrix} 3 & -1 \\ -4 & 7 \end{pmatrix}$

2. Calculate the determinant of $\begin{pmatrix} -2 & 5 \\ 3 & -6 \end{pmatrix}$

3. Calculate the determinant of $\begin{pmatrix} -1.3 & 7.4 \\ 2.5 & -3.9 \end{pmatrix}$

4. Evaluate $\begin{vmatrix} j2 & -j3 \\ (1+j) & j \end{vmatrix}$

5. Evaluate $\begin{vmatrix} 2\angle 40° & 5\angle -20° \\ 7\angle -32° & 4\angle -117° \end{vmatrix}$

6. Given matrix $A = \begin{pmatrix} (x-2) & 6 \\ 2 & (x-3) \end{pmatrix}$ determine values of x for which $|A| = 0$

47.5 The inverse or reciprocal of a 2 by 2 matrix

The inverse of matrix A is A^{-1} such that $A \times A^{-1} = I$, the unit matrix.
Let matrix A be $\begin{pmatrix} 1 & 2 \\ 3 & 4 \end{pmatrix}$ and let the inverse matrix, A^{-1} be $\begin{pmatrix} a & b \\ c & d \end{pmatrix}$.

Then, since $A \times A^{-1} = I$,

$$\begin{pmatrix} 1 & 2 \\ 3 & 4 \end{pmatrix} \times \begin{pmatrix} a & b \\ c & d \end{pmatrix} = \begin{pmatrix} 1 & 0 \\ 0 & 1 \end{pmatrix}$$

Multiplying the matrices on the left-hand side, gives

$$\begin{pmatrix} a+2c & b+2d \\ 3a+4c & 3b+4d \end{pmatrix} = \begin{pmatrix} 1 & 0 \\ 0 & 1 \end{pmatrix}$$

Equating corresponding elements gives:

$$b+2d = 0, \text{i.e. } b = -2d$$

and $3a+4c = 0, \text{i.e. } a = -\frac{4}{3}c$

Substituting for a and b gives:

$$\begin{pmatrix} -\frac{4}{3}c+2c & -2d+2d \\ 3\left(-\frac{4}{3}c\right)+4c & 3(-2d)+4d \end{pmatrix} = \begin{pmatrix} 1 & 0 \\ 0 & 1 \end{pmatrix}$$

i.e.

$$\begin{pmatrix} \frac{2}{3}c & 0 \\ 0 & -2d \end{pmatrix} = \begin{pmatrix} 1 & 0 \\ 0 & 1 \end{pmatrix}$$

showing that $\frac{2}{3}c = 1$, i.e. $c = \frac{3}{2}$ and $-2d = 1$, i.e. $d = -\frac{1}{2}$

Since $b = -2d$, $b = 1$ and since $a = -\frac{4}{3}c$, $a = -2$

Thus the inverse of matrix $\begin{pmatrix} 1 & 2 \\ 3 & 4 \end{pmatrix}$ is $\begin{pmatrix} a & b \\ c & d \end{pmatrix}$ that is,

$$\begin{pmatrix} -2 & 1 \\ \frac{3}{2} & -\frac{1}{2} \end{pmatrix}$$

There is, however, **a quicker method of obtaining the inverse** of a 2 by 2 matrix.

For any matrix $\begin{pmatrix} p & q \\ r & s \end{pmatrix}$ the inverse may be obtained by:

(i)   interchanging the positions of p and s,

(ii) changing the signs of q and r, and

(iii) multiplying this new matrix by the reciprocal of the determinant of $\begin{pmatrix} p & q \\ r & s \end{pmatrix}$

Thus the inverse of matrix $\begin{pmatrix} 1 & 2 \\ 3 & 4 \end{pmatrix}$ is

$$\frac{1}{4-6} \begin{pmatrix} 4 & -2 \\ -3 & 1 \end{pmatrix} = \begin{pmatrix} -2 & 1 \\ \frac{3}{2} & -\frac{1}{2} \end{pmatrix}$$

as obtained previously.

Problem 13. Determine the inverse of $\begin{pmatrix} 3 & -2 \\ 7 & 4 \end{pmatrix}$

The inverse of matrix $\begin{pmatrix} p & q \\ r & s \end{pmatrix}$ is obtained by interchanging the positions of p and s, changing the signs of q and r and multiplying by the reciprocal of the determinant $\begin{vmatrix} p & q \\ r & s \end{vmatrix}$. Thus, the inverse of

$$\begin{pmatrix} 3 & -2 \\ 7 & 4 \end{pmatrix} = \frac{1}{(3 \times 4) - (-2 \times 7)} \begin{pmatrix} 4 & 2 \\ -7 & 3 \end{pmatrix}$$

$$= \frac{1}{26} \begin{pmatrix} 4 & 2 \\ -7 & 3 \end{pmatrix} = \begin{pmatrix} \frac{2}{13} & \frac{1}{13} \\ \frac{-7}{26} & \frac{3}{26} \end{pmatrix}$$

Now try the following Practice Exercise

Practice Exercise 202 Further problems on the inverse of 2 by 2 matrices (answers on page 1174)

1. Determine the inverse of $\begin{pmatrix} 3 & -1 \\ -4 & 7 \end{pmatrix}$

2. Determine the inverse of $\begin{pmatrix} \dfrac{1}{2} & \dfrac{2}{3} \\ -\dfrac{1}{3} & -\dfrac{3}{5} \end{pmatrix}$

3. Determine the inverse of $\begin{pmatrix} -1.3 & 7.4 \\ 2.5 & -3.9 \end{pmatrix}$

47.6 The determinant of a 3 by 3 matrix

(i) The **minor** of an element of a 3 by 3 matrix is the value of the 2 by 2 determinant obtained by covering up the row and column containing that element.

Thus for the matrix $\begin{pmatrix} 1 & 2 & 3 \\ 4 & 5 & 6 \\ 7 & 8 & 9 \end{pmatrix}$ the minor of element 4 is obtained by covering the row (4 5 6) and the column $\begin{pmatrix} 1 \\ 4 \\ 7 \end{pmatrix}$, leaving the 2 by 2 determinant $\begin{vmatrix} 2 & 3 \\ 8 & 9 \end{vmatrix}$, i.e. the minor of element 4 is $(2 \times 9) - (3 \times 8) = -6$

(ii) The sign of a minor depends on its position within the matrix, the sign pattern being $\begin{pmatrix} + & - & + \\ - & + & - \\ + & - & + \end{pmatrix}$

Thus the signed-minor of element 4 in the matrix $\begin{pmatrix} 1 & 2 & 3 \\ 4 & 5 & 6 \\ 7 & 8 & 9 \end{pmatrix}$ is $-\begin{vmatrix} 2 & 3 \\ 8 & 9 \end{vmatrix} = -(-6) = 6$

The signed-minor of an element is called the **cofactor** of the element.

(iii) **The value of a 3 by 3 determinant is the sum of the products of the elements and their cofactors of any row or any column of the corresponding 3 by 3 matrix.**

There are thus six different ways of evaluating a 3×3 determinant – and all should give the same value.

Problem 14. Find the value of
$$\begin{vmatrix} 3 & 4 & -1 \\ 2 & 0 & 7 \\ 1 & -3 & -2 \end{vmatrix}$$

The value of this determinant is the sum of the products of the elements and their cofactors, of any row or of any column. If the second row or second column is selected, the element 0 will make the product of the element and its cofactor zero and reduce the amount of arithmetic to be done to a minimum.

Supposing a second row expansion is selected.

The minor of 2 is the value of the determinant remaining when the row and column containing the 2 (i.e. the second row and the first column) are covered up.

Thus the cofactor of element 2 is $\begin{vmatrix} 4 & -1 \\ -3 & -2 \end{vmatrix}$ i.e. -11

The sign of element 2 is minus (see (ii) above), hence the cofactor of element 2 (the signed-minor) is $+11$

Similarly, the minor of element 7 is $\begin{vmatrix} 3 & 4 \\ 1 & -3 \end{vmatrix}$ i.e. -13, and its cofactor is $+13$. Hence the value of the sum of the products of the elements and their cofactors is $2 \times 11 + 7 \times 13$, i.e.

$$\begin{vmatrix} 3 & 4 & -1 \\ 2 & 0 & 7 \\ 1 & -3 & -2 \end{vmatrix} = 2(11) + 0 + 7(13) = \mathbf{113}$$

The same result will be obtained whichever row or column is selected. For example, the third column expansion is

$$(-1)\begin{vmatrix} 2 & 0 \\ 1 & -3 \end{vmatrix} - 7\begin{vmatrix} 3 & 4 \\ 1 & -3 \end{vmatrix} + (-2)\begin{vmatrix} 3 & 4 \\ 2 & 0 \end{vmatrix}$$

$$= 6 + 91 + 16 = \mathbf{113}, \text{ as obtained previously.}$$

Problem 15. Evaluate $\begin{vmatrix} 1 & 4 & -3 \\ -5 & 2 & 6 \\ -1 & -4 & 2 \end{vmatrix}$

Using the first row: $\begin{vmatrix} 1 & 4 & -3 \\ -5 & 2 & 6 \\ -1 & -4 & 2 \end{vmatrix}$

$$= 1\begin{vmatrix} 2 & 6 \\ -4 & 2 \end{vmatrix} - 4\begin{vmatrix} -5 & 6 \\ -1 & 2 \end{vmatrix} + (-3)\begin{vmatrix} -5 & 2 \\ -1 & -4 \end{vmatrix}$$

$$= (4 + 24) - 4(-10 + 6) - 3(20 + 2)$$

$$= 28 + 16 - 66 = \mathbf{-22}$$

Using the second column: $\begin{vmatrix} 1 & 4 & -3 \\ -5 & 2 & 6 \\ -1 & -4 & 2 \end{vmatrix}$

$$= -4 \begin{vmatrix} -5 & 6 \\ -1 & 2 \end{vmatrix} + 2 \begin{vmatrix} 1 & -3 \\ -1 & 2 \end{vmatrix} - (-4) \begin{vmatrix} 1 & -3 \\ -5 & 6 \end{vmatrix}$$

$$= -4(-10+6) + 2(2-3) + 4(6-15)$$

$$= 16 - 2 - 36 = \mathbf{-22}$$

Problem 16. Determine the value of

$$\begin{vmatrix} j2 & (1+j) & 3 \\ (1-j) & 1 & j \\ 0 & j4 & 5 \end{vmatrix}$$

Using the first column, the value of the determinant is:

$$(j2) \begin{vmatrix} 1 & j \\ j4 & 5 \end{vmatrix} - (1-j) \begin{vmatrix} (1+j) & 3 \\ j4 & 5 \end{vmatrix}$$

$$+ (0) \begin{vmatrix} (1+j) & 3 \\ 1 & j \end{vmatrix}$$

$$= j2(5 - j^2 4) - (1-j)(5 + j5 - j12) + 0$$

$$= j2(9) - (1-j)(5 - j7)$$

$$= j18 - [5 - j7 - j5 + j^2 7]$$

$$= j18 - [-2 - j12]$$

$$= j18 + 2 + j12 = \mathbf{2 + j\,30} \text{ or } \mathbf{30.07\angle 86.19°}$$

Now try the following Practice Exercise

Practice Exercise 203   Further problems on 3 by 3 determinants (answers on page 1174)

1. Find the matrix of minors of

$$\begin{pmatrix} 4 & -7 & 6 \\ -2 & 4 & 0 \\ 5 & 7 & -4 \end{pmatrix}$$

2. Find the matrix of cofactors of

$$\begin{pmatrix} 4 & -7 & 6 \\ -2 & 4 & 0 \\ 5 & 7 & -4 \end{pmatrix}$$

3. Calculate the determinant of

$$\begin{pmatrix} 4 & -7 & 6 \\ -2 & 4 & 0 \\ 5 & 7 & -4 \end{pmatrix}$$

4. Evaluate $\begin{vmatrix} 8 & -2 & -10 \\ 2 & -3 & -2 \\ 6 & 3 & 8 \end{vmatrix}$

5. Calculate the determinant of

$$\begin{pmatrix} 3.1 & 2.4 & 6.4 \\ -1.6 & 3.8 & -1.9 \\ 5.3 & 3.4 & -4.8 \end{pmatrix}$$

6. Evaluate $\begin{vmatrix} j2 & 2 & j \\ (1+j) & 1 & -3 \\ 5 & -j4 & 0 \end{vmatrix}$

7. Evaluate $\begin{vmatrix} 3\angle 60° & j2 & 1 \\ 0 & (1+j) & 2\angle 30° \\ 0 & 2 & j5 \end{vmatrix}$

8. Find the eigenvalues λ that satisfy the following equations:

    (a) $\begin{vmatrix} (2-\lambda) & 2 \\ -1 & (5-\lambda) \end{vmatrix} = 0$

    (b) $\begin{vmatrix} (5-\lambda) & 7 & -5 \\ 0 & (4-\lambda) & -1 \\ 2 & 8 & (-3-\lambda) \end{vmatrix} = 0$

(You may need to refer to Chapter 18, pages 173–177, for the solution of cubic equations.)

47.7 The inverse or reciprocal of a 3 by 3 matrix

The **adjoint** of a matrix A is obtained by:

(i) forming a matrix B of the cofactors of A, and

(ii) **transposing** matrix B to give B^T, where B^T is the matrix obtained by writing the rows of B as the columns of B^T. Then **adj** $A = B^T$

The **inverse of matrix** A, A^{-1} is given by

$$A^{-1} = \frac{\text{adj } A}{|A|}$$

where adj A is the adjoint of matrix A and $|A|$ is the determinant of matrix A.

Problem 17. Determine the inverse of the matrix
$$\begin{pmatrix} 3 & 4 & -1 \\ 2 & 0 & 7 \\ 1 & -3 & -2 \end{pmatrix}$$

The inverse of matrix A, $A^{-1} = \dfrac{\text{adj } A}{|A|}$

The adjoint of A is found by:

(i) obtaining the matrix of the cofactors of the elements, and

(ii) transposing this matrix.

The cofactor of element 3 is $+\begin{vmatrix} 0 & 7 \\ -3 & -2 \end{vmatrix} = 21$

The cofactor of element 4 is $-\begin{vmatrix} 2 & 7 \\ 1 & -2 \end{vmatrix} = 11$, and so on.

The matrix of cofactors is $\begin{pmatrix} 21 & 11 & -6 \\ 11 & -5 & 13 \\ 28 & -23 & -8 \end{pmatrix}$

The transpose of the matrix of cofactors, i.e. the adjoint of the matrix, is obtained by writing the rows as columns, and is $\begin{pmatrix} 21 & 11 & 28 \\ 11 & -5 & -23 \\ -6 & 13 & -8 \end{pmatrix}$

From Problem 14, the determinant of $\begin{vmatrix} 3 & 4 & -1 \\ 2 & 0 & 7 \\ 1 & -3 & -2 \end{vmatrix}$ is 113

Hence the inverse of $\begin{pmatrix} 3 & 4 & -1 \\ 2 & 0 & 7 \\ 1 & -3 & -2 \end{pmatrix}$ is

$\dfrac{\begin{pmatrix} 21 & 11 & 28 \\ 11 & -5 & -23 \\ -6 & 13 & -8 \end{pmatrix}}{113}$ or $\dfrac{1}{113}\begin{pmatrix} 21 & 11 & 28 \\ 11 & -5 & -23 \\ -6 & 13 & -8 \end{pmatrix}$

Problem 18. Find the inverse of
$$\begin{pmatrix} 1 & 5 & -2 \\ 3 & -1 & 4 \\ -3 & 6 & -7 \end{pmatrix}$$

$\text{Inverse} = \dfrac{\text{adjoint}}{\text{determinant}}$

The matrix of cofactors is $\begin{pmatrix} -17 & 9 & 15 \\ 23 & -13 & -21 \\ 18 & -10 & -16 \end{pmatrix}$

The transpose of the matrix of cofactors (i.e. the adjoint) is $\begin{pmatrix} -17 & 23 & 18 \\ 9 & -13 & -10 \\ 15 & -21 & -16 \end{pmatrix}$

The determinant of $\begin{pmatrix} 1 & 5 & -2 \\ 3 & -1 & 4 \\ -3 & 6 & -7 \end{pmatrix}$

$= 1(7 - 24) - 5(-21 + 12) - 2(18 - 3)$

$= -17 + 45 - 30 = -2$

Hence the inverse of $\begin{pmatrix} 1 & 5 & -2 \\ 3 & -1 & 4 \\ -3 & 6 & -7 \end{pmatrix}$

$= \dfrac{\begin{pmatrix} -17 & 23 & 18 \\ 9 & -13 & -10 \\ 15 & -21 & -16 \end{pmatrix}}{-2}$

$= \begin{pmatrix} \mathbf{8.5} & \mathbf{-11.5} & \mathbf{-9} \\ \mathbf{-4.5} & \mathbf{6.5} & \mathbf{5} \\ \mathbf{-7.5} & \mathbf{10.5} & \mathbf{8} \end{pmatrix}$

Now try the following Practice Exercise

Practice Exercise 204   Further problems on the inverse of a 3 by 3 matrix (answers on page 1174)

1. Write down the transpose of
$$\begin{pmatrix} 4 & -7 & 6 \\ -2 & 4 & 0 \\ 5 & 7 & -4 \end{pmatrix}$$

2. Write down the transpose of
$$\begin{pmatrix} 3 & 6 & \frac{1}{2} \\ 5 & -\frac{2}{3} & 7 \\ -1 & 0 & \frac{3}{5} \end{pmatrix}$$

3. Determine the adjoint of

$$\begin{pmatrix} 4 & -7 & 6 \\ -2 & 4 & 0 \\ 5 & 7 & -4 \end{pmatrix}$$

4. Determine the adjoint of

$$\begin{pmatrix} 3 & 6 & \frac{1}{2} \\ 5 & -\frac{2}{3} & 7 \\ -1 & 0 & \frac{3}{5} \end{pmatrix}$$

5. Find the inverse of

$$\begin{pmatrix} 4 & -7 & 6 \\ -2 & 4 & 0 \\ 5 & 7 & -4 \end{pmatrix}$$

6. Find the inverse of $\begin{pmatrix} 3 & 6 & \frac{1}{2} \\ 5 & -\frac{2}{3} & 7 \\ -1 & 0 & \frac{3}{5} \end{pmatrix}$

For fully worked solutions to each of the problems in Practice Exercises 200 to 204 in this chapter, go to the website:
www.routledge.com/cw/bird

Chapter 48

Applications of matrices and determinants

> **Why it is important to understand:** Applications of matrices and determinants
>
> As mentioned previously, matrices are used to solve problems, for example, in electrical circuits, optics, quantum mechanics, statics, robotics, genetics and much more, and for calculating forces, vectors, tensions, masses, loads and a lot of other factors that must be accounted for in engineering. In the main, matrices and determinants are used to solve a system of simultaneous linear equations. The simultaneous solution of multiple equations finds its way into many common engineering problems. In fact, modern structural engineering analysis techniques are all about solving systems of equations simultaneously. Eigenvalues and eigenvectors, which are based on matrix theory, are very important in engineering and science. For example, car designers analyse eigenvalues in order to damp out the noise in a car, eigenvalue analysis is used in the design of car stereo systems, eigenvalues can be used to test for cracks and deformities in a solid, and oil companies use eigenvalues analysis to explore land for oil.

At the end of this chapter, you should be able to:

- solve simultaneous equations in two and three unknowns using matrices
- solve simultaneous equations in two and three unknowns using determinants
- solve simultaneous equations using Cramer's rule
- solve simultaneous equations using Gaussian elimination
- determine the eigenvalues of a 2 by 2 and 3 by 3 matrix
- determine the eigenvectors of a 2 by 2 and 3 by 3 matrix

48.1 Solution of simultaneous equations by matrices

The procedure for solving linear simultaneous equations in **two unknowns using matrices** is:

(i) write the equations in the form

$$a_1 x + b_1 y = c_1$$
$$a_2 x + b_2 y = c_2$$

(ii) write the matrix equation corresponding to these equations, i.e. $\begin{pmatrix} a_1 & b_1 \\ a_2 & b_2 \end{pmatrix} \times \begin{pmatrix} x \\ y \end{pmatrix} = \begin{pmatrix} c_1 \\ c_2 \end{pmatrix}$

(iii) determine the inverse matrix of $\begin{pmatrix} a_1 & b_1 \\ a_2 & b_2 \end{pmatrix}$

i.e. $\dfrac{1}{a_1 b_2 - b_1 a_2} \begin{pmatrix} b_2 & -b_1 \\ -a_2 & a_1 \end{pmatrix}$

(from Chapter 47),

(iv) multiply each side of (ii) by the inverse matrix, and

(v) solve for x and y by equating corresponding elements.

Problem 1. Use matrices to solve the simultaneous equations:
$$3x + 5y - 7 = 0 \quad (1)$$
$$4x - 3y - 19 = 0 \quad (2)$$

(i) Writing the equations in the $a_1 x + b_1 y = c$ form gives:
$$3x + 5y = 7$$
$$4x - 3y = 19$$

(ii) The matrix equation is
$$\begin{pmatrix} 3 & 5 \\ 4 & -3 \end{pmatrix} \times \begin{pmatrix} x \\ y \end{pmatrix} = \begin{pmatrix} 7 \\ 19 \end{pmatrix}$$

(iii) The inverse of matrix $\begin{pmatrix} 3 & 5 \\ 4 & -3 \end{pmatrix}$ is
$$\dfrac{1}{3 \times (-3) - 5 \times 4} \begin{pmatrix} -3 & -5 \\ -4 & 3 \end{pmatrix}$$

i.e. $\begin{pmatrix} \dfrac{3}{29} & \dfrac{5}{29} \\ \dfrac{4}{29} & \dfrac{-3}{29} \end{pmatrix}$

(iv) Multiplying each side of (ii) by (iii) and remembering that $A \times A^{-1} = I$, the unit matrix, gives:
$$\begin{pmatrix} 1 & 0 \\ 0 & 1 \end{pmatrix} \begin{pmatrix} x \\ y \end{pmatrix} = \begin{pmatrix} \dfrac{3}{29} & \dfrac{5}{29} \\ \dfrac{4}{29} & \dfrac{-3}{29} \end{pmatrix} \times \begin{pmatrix} 7 \\ 19 \end{pmatrix}$$

Thus $\begin{pmatrix} x \\ y \end{pmatrix} = \begin{pmatrix} \dfrac{21}{29} + \dfrac{95}{29} \\ \dfrac{28}{29} - \dfrac{57}{29} \end{pmatrix}$

i.e. $\begin{pmatrix} x \\ y \end{pmatrix} = \begin{pmatrix} 4 \\ -1 \end{pmatrix}$

(v) By comparing corresponding elements:
$$x = 4 \quad \text{and} \quad y = -1$$
Checking:

equation (1),
$$3 \times 4 + 5 \times (-1) - 7 = 0 = \text{RHS}$$
equation (2),
$$4 \times 4 - 3 \times (-1) - 19 = 0 = \text{RHS}$$

The procedure for solving linear simultaneous equations in **three unknowns using matrices** is:

(i) write the equations in the form
$$a_1 x + b_1 y + c_1 z = d_1$$
$$a_2 x + b_2 y + c_2 z = d_2$$
$$a_3 x + b_3 y + c_3 z = d_3$$

(ii) write the matrix equation corresponding to these equations, i.e.
$$\begin{pmatrix} a_1 & b_1 & c_1 \\ a_2 & b_2 & c_2 \\ a_3 & b_3 & c_3 \end{pmatrix} \times \begin{pmatrix} x \\ y \\ z \end{pmatrix} = \begin{pmatrix} d_1 \\ d_2 \\ d_3 \end{pmatrix}$$

(iii) determine the inverse matrix of
$$\begin{pmatrix} a_1 & b_1 & c_1 \\ a_2 & b_2 & c_2 \\ a_3 & b_3 & c_3 \end{pmatrix} \text{ (see Chapter 47)}$$

(iv) multiply each side of (ii) by the inverse matrix, and

(v) solve for x, y and z by equating the corresponding elements.

Problem 2. Use matrices to solve the simultaneous equations:
$$x + y + z - 4 = 0 \quad (1)$$
$$2x - 3y + 4z - 33 = 0 \quad (2)$$
$$3x - 2y - 2z - 2 = 0 \quad (3)$$

(i) Writing the equations in the $a_1 x + b_1 y + c_1 z = d_1$ form gives:
$$x + y + z = 4$$
$$2x - 3y + 4z = 33$$
$$3x - 2y - 2z = 2$$

(ii) The matrix equation is

$$\begin{pmatrix} 1 & 1 & 1 \\ 2 & -3 & 4 \\ 3 & -2 & -2 \end{pmatrix} \times \begin{pmatrix} x \\ y \\ z \end{pmatrix} = \begin{pmatrix} 4 \\ 33 \\ 2 \end{pmatrix}$$

(iii) The inverse matrix of

$$A = \begin{pmatrix} 1 & 1 & 1 \\ 2 & -3 & 4 \\ 3 & -2 & -2 \end{pmatrix}$$

is given by

$$A^{-1} = \frac{\text{adj } A}{|A|}$$

The adjoint of A is the transpose of the matrix of the cofactors of the elements (see Chapter 47). The matrix of cofactors is

$$\begin{pmatrix} 14 & 16 & 5 \\ 0 & -5 & 5 \\ 7 & -2 & -5 \end{pmatrix}$$

and the transpose of this matrix gives

$$\text{adj } A = \begin{pmatrix} 14 & 0 & 7 \\ 16 & -5 & -2 \\ 5 & 5 & -5 \end{pmatrix}$$

The determinant of A, i.e. the sum of the products of elements and their cofactors, using a first row expansion is

$$1 \begin{vmatrix} -3 & 4 \\ -2 & -2 \end{vmatrix} - 1 \begin{vmatrix} 2 & 4 \\ 3 & -2 \end{vmatrix} + 1 \begin{vmatrix} 2 & -3 \\ 3 & -2 \end{vmatrix}$$

$$= (1 \times 14) - (1 \times (-16)) + (1 \times 5) = 35$$

Hence the inverse of A,

$$A^{-1} = \frac{1}{35} \begin{pmatrix} 14 & 0 & 7 \\ 16 & -5 & -2 \\ 5 & 5 & -5 \end{pmatrix}$$

(iv) Multiplying each side of (ii) by (iii), and remembering that $A \times A^{-1} = I$, the unit matrix, gives

$$\begin{pmatrix} 1 & 0 & 0 \\ 0 & 1 & 0 \\ 0 & 0 & 1 \end{pmatrix} \times \begin{pmatrix} x \\ y \\ z \end{pmatrix} = \frac{1}{35} \begin{pmatrix} 14 & 0 & 7 \\ 16 & -5 & -2 \\ 5 & 5 & -5 \end{pmatrix} \times \begin{pmatrix} 4 \\ 33 \\ 2 \end{pmatrix}$$

$$\begin{pmatrix} x \\ y \\ z \end{pmatrix} = \frac{1}{35} \begin{pmatrix} (14 \times 4) + (0 \times 33) + (7 \times 2) \\ (16 \times 4) + ((-5) \times 33) + ((-2) \times 2) \\ (5 \times 4) + (5 \times 33) + ((-5) \times 2) \end{pmatrix}$$

$$= \frac{1}{35} \begin{pmatrix} 70 \\ -105 \\ 175 \end{pmatrix}$$

$$= \begin{pmatrix} 2 \\ -3 \\ 5 \end{pmatrix}$$

(v) By comparing corresponding elements, $x = 2$, $y = -3$, $z = 5$, which can be checked in the original equations.

Now try the following Practice Exercise

Practice Exercise 205 Further problems on solving simultaneous equations using matrices (answers on page 1174)

In Problems 1 to 5, use **matrices** to solve the simultaneous equations given.

1. $3x + 4y = 0$
   $2x + 5y + 7 = 0$

2. $2p + 5q + 14.6 = 0$
   $3.1p + 1.7q + 2.06 = 0$

3. $x + 2y + 3z = 5$
   $2x - 3y - z = 3$
   $-3x + 4y + 5z = 3$

4. $3a + 4b - 3c = 2$
   $-2a + 2b + 2c = 15$
   $7a - 5b + 4c = 26$

5. $p + 2q + 3r + 7.8 = 0$
   $2p + 5q - r - 1.4 = 0$
   $5p - q + 7r - 3.5 = 0$

6. In two closed loops of an electrical circuit, the currents flowing are given by the simultaneous equations:

   $I_1 + 2I_2 + 4 = 0$
   $5I_1 + 3I_2 - 1 = 0$

   Use matrices to solve for I_1 and I_2

7. The relationship between the displacement, s, velocity, v, and acceleration, a, of a piston is given by the equations:

$s + 2v + 2a = 4$
$3s - v + 4a = 25$
$3s + 2v - a = -4$

Use matrices to determine the values of s, v and a

8. In a mechanical system, acceleration $\ddot{x}$, velocity $\dot{x}$ and distance x are related by the simultaneous equations:

$3.4\ddot{x} + 7.0\dot{x} - 13.2x = -11.39$
$-6.0\ddot{x} + 4.0\dot{x} + 3.5x = 4.98$
$2.7\ddot{x} + 6.0\dot{x} + 7.1x = 15.91$

Use matrices to find the values of $\ddot{x}$, $\dot{x}$ and x

48.2 Solution of simultaneous equations by determinants

When solving linear simultaneous equations in **two unknowns using determinants:**

(i) write the equations in the form

$a_1 x + b_1 y + c_1 = 0$

$a_2 x + b_2 y + c_2 = 0$

and then

(ii) the solution is given by

$$\frac{x}{D_x} = \frac{-y}{D_y} = \frac{1}{D}$$

where $D_x = \begin{vmatrix} b_1 & c_1 \\ b_2 & c_2 \end{vmatrix}$

i.e. the determinant of the coefficients left when the x-column is covered up,

$$D_y = \begin{vmatrix} a_1 & c_1 \\ a_2 & c_2 \end{vmatrix}$$

i.e. the determinant of the coefficients left when the y-column is covered up,

and $D = \begin{vmatrix} a_1 & b_1 \\ a_2 & b_2 \end{vmatrix}$

i.e. the determinant of the coefficients left when the constants-column is covered up.

Problem 3. Solve the following simultaneous equations using determinants:

$$3x - 4y = 12$$

$$7x + 5y = 6.5$$

Following the above procedure:

(i) $3x - 4y - 12 = 0$
$7x + 5y - 6.5 = 0$

(ii) $\dfrac{x}{\begin{vmatrix} -4 & -12 \\ 5 & -6.5 \end{vmatrix}} = \dfrac{-y}{\begin{vmatrix} 3 & -12 \\ 7 & -6.5 \end{vmatrix}} = \dfrac{1}{\begin{vmatrix} 3 & -4 \\ 7 & 5 \end{vmatrix}}$

i.e. $\dfrac{x}{(-4)(-6.5) - (-12)(5)}$

$= \dfrac{-y}{(3)(-6.5) - (-12)(7)}$

$= \dfrac{1}{(3)(5) - (-4)(7)}$

i.e. $\dfrac{x}{26 + 60} = \dfrac{-y}{-19.5 + 84} = \dfrac{1}{15 + 28}$

i.e. $\dfrac{x}{86} = \dfrac{-y}{64.5} = \dfrac{1}{43}$

Since $\dfrac{x}{86} = \dfrac{1}{43}$ then $x = \dfrac{86}{43} = 2$

and since

$\dfrac{-y}{64.5} = \dfrac{1}{43}$ then $y = -\dfrac{64.5}{43} = -1.5$

Problem 4. The velocity of a car, accelerating at uniform acceleration a between two points, is given by $v = u + at$, where u is its velocity when passing the first point and t is the time taken to pass between the two points. If $v = 21$ m/s when $t = 3.5$ s and $v = 33$ m/s when $t = 6.1$ s, use determinants to find the values of u and a, each correct to 4 significant figures

Substituting the given values in $v = u + at$ gives:

$21 = u + 3.5a$ (1)

$33 = u + 6.1a$ (2)

(i) The equations are written in the form
$$a_1 x + b_1 y + c_1 = 0,$$
i.e. $\quad u + 3.5a - 21 = 0$
and $\quad u + 6.1a - 33 = 0$

(ii) The solution is given by
$$\frac{u}{D_u} = \frac{-a}{D_a} = \frac{1}{D}$$

where D_u is the determinant of coefficients left when the u column is covered up,

i.e. $\qquad D_u = \begin{vmatrix} 3.5 & -21 \\ 6.1 & -33 \end{vmatrix}$

$$= (3.5)(-33) - (-21)(6.1)$$
$$= 12.6$$

Similarly, $\quad D_a = \begin{vmatrix} 1 & -21 \\ 1 & -33 \end{vmatrix}$

$$= (1)(-33) - (-21)(1)$$
$$= -12$$

and $\qquad D = \begin{vmatrix} 1 & 3.5 \\ 1 & 6.1 \end{vmatrix}$

$$= (1)(6.1) - (3.5)(1) = 2.6$$

Thus $\quad \dfrac{u}{12.6} = \dfrac{-a}{-12} = \dfrac{1}{2.6}$

i.e. $\qquad u = \dfrac{12.6}{2.6} = \mathbf{4.846\,m/s}$

and $\qquad a = \dfrac{12}{2.6} = \mathbf{4.615\,m/s^2},$

each correct to 4 significant figures.

Problem 5. Applying Kirchhoff's laws to an electric circuit results in the following equations:

$$(9 + j12)I_1 - (6 + j8)I_2 = 5$$
$$-(6 + j8)I_1 + (8 + j3)I_2 = (2 + j4)$$

Solve the equations for I_1 and I_2

Following the procedure:

(i) $(9 + j12)I_1 - (6 + j8)I_2 - 5 = 0$

$-(6 + j8)I_1 + (8 + j3)I_2 - (2 + j4) = 0$

(ii) $\dfrac{I_1}{\begin{vmatrix} -(6+j8) & -5 \\ (8+j3) & -(2+j4) \end{vmatrix}}$

$$= \frac{-I_2}{\begin{vmatrix} (9+j12) & -5 \\ -(6+j8) & -(2+j4) \end{vmatrix}}$$

$$= \frac{1}{\begin{vmatrix} (9+j12) & -(6+j8) \\ -(6+j8) & (8+j3) \end{vmatrix}}$$

$$\frac{I_1}{(-20+j40) + (40+j15)}$$
$$= \frac{-I_2}{(30-j60) - (30+j40)}$$
$$= \frac{1}{(36+j123) - (-28+j96)}$$

from the theory of complex numbers in Chapter 45

$$\frac{I_1}{20 + j55} = \frac{-I_2}{-j100}$$
$$= \frac{1}{64 + j27}$$

Hence $I_1 = \dfrac{20 + j55}{64 + j27}$

$$= \frac{58.52\angle 70.02°}{69.46\angle 22.87°} = \mathbf{0.84\angle 47.15°\,A}$$

and $\quad I_2 = \dfrac{100\angle 90°}{69.46\angle 22.87°}$

$$= \mathbf{1.44\angle 67.13°\,A}$$

When solving simultaneous equations in **three unknowns using determinants:**

(i) Write the equations in the form

$$a_1 x + b_1 y + c_1 z + d_1 = 0$$
$$a_2 x + b_2 y + c_2 z + d_2 = 0$$
$$a_3 x + b_3 y + c_3 z + d_3 = 0$$

and then

(ii) the solution is given by

$$\frac{x}{D_x} = \frac{-y}{D_y} = \frac{z}{D_z} = \frac{-1}{D}$$

where D_x is $\begin{vmatrix} b_1 & c_1 & d_1 \\ b_2 & c_2 & d_2 \\ b_3 & c_3 & d_3 \end{vmatrix}$

i.e. the determinant of the coefficients obtained by covering up the x-column,

D_y is $\begin{vmatrix} a_1 & c_1 & d_1 \\ a_2 & c_2 & d_2 \\ a_3 & c_3 & d_3 \end{vmatrix}$

i.e. the determinant of the coefficients obtained by covering up the y-column,

$$D_z \text{ is } \begin{vmatrix} a_1 & b_1 & d_1 \\ a_2 & b_2 & d_2 \\ a_3 & b_3 & d_3 \end{vmatrix}$$

i.e. the determinant of the coefficients obtained by covering up the z-column,

$$\text{and } D \text{ is } \begin{vmatrix} a_1 & b_1 & c_1 \\ a_2 & b_2 & c_2 \\ a_3 & b_3 & c_3 \end{vmatrix}$$

i.e. the determinant of the coefficients obtained by covering up the constants-column.

> **Problem 6.** A d.c. circuit comprises three closed loops. Applying Kirchhoff's laws to the closed loops gives the following equations for current flow in milliamperes:
>
> $$2I_1 + 3I_2 - 4I_3 = 26$$
> $$I_1 - 5I_2 - 3I_3 = -87$$
> $$-7I_1 + 2I_2 + 6I_3 = 12$$
>
> Use determinants to solve for I_1, I_2 and I_3

(i) Writing the equations in the $a_1 x + b_1 y + c_1 z + d_1 = 0$ form gives:

$$2I_1 + 3I_2 - 4I_3 - 26 = 0$$
$$I_1 - 5I_2 - 3I_3 + 87 = 0$$
$$-7I_1 + 2I_2 + 6I_3 - 12 = 0$$

(ii) the solution is given by

$$\frac{I_1}{D_{I_1}} = \frac{-I_2}{D_{I_2}} = \frac{I_3}{D_{I_3}} = \frac{-1}{D}$$

where D_{I_1} is the determinant of coefficients obtained by covering up the I_1 column, i.e.

$$D_{I_1} = \begin{vmatrix} 3 & -4 & -26 \\ -5 & -3 & 87 \\ 2 & 6 & -12 \end{vmatrix}$$

$$= (3) \begin{vmatrix} -3 & 87 \\ 6 & -12 \end{vmatrix} - (-4) \begin{vmatrix} -5 & 87 \\ 2 & -12 \end{vmatrix}$$

$$+ (-26) \begin{vmatrix} -5 & -3 \\ 2 & 6 \end{vmatrix}$$

$$= 3(-486) + 4(-114) - 26(-24)$$

$$= -1290$$

$$D_{I_2} = \begin{vmatrix} 2 & -4 & -26 \\ 1 & -3 & 87 \\ -7 & 6 & -12 \end{vmatrix}$$

$$= (2)(36 - 522) - (-4)(-12 + 609)$$
$$+ (-26)(6 - 21)$$

$$= -972 + 2388 + 390$$

$$= \mathbf{1806}$$

$$D_{I_3} = \begin{vmatrix} 2 & 3 & -26 \\ 1 & -5 & 87 \\ -7 & 2 & -12 \end{vmatrix}$$

$$= (2)(60 - 174) - (3)(-12 + 609)$$
$$+ (-26)(2 - 35)$$

$$= -228 - 1791 + 858 = \mathbf{-1161}$$

$$\text{and } D = \begin{vmatrix} 2 & 3 & -4 \\ 1 & -5 & -3 \\ -7 & 2 & 6 \end{vmatrix}$$

$$= (2)(-30 + 6) - (3)(6 - 21)$$
$$+ (-4)(2 - 35)$$

$$= -48 + 45 + 132 = \mathbf{129}$$

Thus

$$\frac{I_1}{-1290} = \frac{-I_2}{1806} = \frac{I_3}{-1161} = \frac{-1}{129}$$

giving

$$I_1 = \frac{-1290}{-129} = \mathbf{10\,mA}$$

$$I_2 = \frac{1806}{129} = \mathbf{14\,mA}$$

$$I_3 = \frac{1161}{129} = \mathbf{9\,mA}$$

Now try the following Practice Exercise

> **Practice Exercise 206   Further problems on solving simultaneous equations using determinants (answers on page 1174)**
>
> In Problems 1 to 5, use **determinants** to solve the simultaneous equations given.
>
> 1. $3x - 5y = -17.6$
> $7y - 2x - 22 = 0$

2. $2.3m - 4.4n = 6.84$
    $8.5n - 6.7m = 1.23$

3. $3x + 4y + z = 10$
    $2x - 3y + 5z + 9 = 0$
    $x + 2y - z = 6$

4. $1.2p - 2.3q - 3.1r + 10.1 = 0$
    $4.7p + 3.8q - 5.3r - 21.5 = 0$
    $3.7p - 8.3q + 7.4r + 28.1 = 0$

5. $\dfrac{x}{2} - \dfrac{y}{3} + \dfrac{2z}{5} = -\dfrac{1}{20}$
    $\dfrac{x}{4} + \dfrac{2y}{3} - \dfrac{z}{2} = \dfrac{19}{40}$
    $x + y - z = \dfrac{59}{60}$

6. In a system of forces, the relationship between two forces F_1 and F_2 is given by:
    $5F_1 + 3F_2 + 6 = 0$
    $3F_1 + 5F_2 + 18 = 0$
    Use determinants to solve for F_1 and F_2

7. Applying mesh-current analysis to an a.c. circuit results in the following equations:
    $(5 - j4)I_1 - (-j4)I_2 = 100\angle 0°$
    $(4 + j3 - j4)I_2 - (-j4)I_1 = 0$
    Solve the equations for I_1 and I_2

8. Kirchhoff's laws are used to determine the current equations in an electrical network and show that
    $i_1 + 8i_2 + 3i_3 = -31$
    $3i_1 - 2i_2 + i_3 = -5$
    $2i_1 - 3i_2 + 2i_3 = 6$
    Use determinants to find the values of i_1, i_2 and i_3

9. The forces in three members of a framework are F_1, F_2 and F_3. They are related by the simultaneous equations shown below.
    $1.4F_1 + 2.8F_2 + 2.8F_3 = 5.6$
    $4.2F_1 - 1.4F_2 + 5.6F_3 = 35.0$
    $4.2F_1 + 2.8F_2 - 1.4F_3 = -5.6$
    Find the values of F_1, F_2 and F_3 using determinants.

10. Mesh-current analysis produces the following three equations:
    $20\angle 0° = (5 + 3 - j4)I_1 - (3 - j4)I_2$
    $10\angle 90° = (3 - j4 + 2)I_2 - (3 - j4)I_1 - 2I_3$
    $-15\angle 0° - 10\angle 90° = (12 + 2)I_3 - 2I_2$
    Solve the equations for the loop currents I_1, I_2 and I_3

48.3 Solution of simultaneous equations using Cramer's rule

Cramer's[*] rule states that if

$$a_{11}x + a_{12}y + a_{13}z = b_1$$

$$a_{21}x + a_{22}y + a_{23}z = b_2$$

$$a_{31}x + a_{32}y + a_{33}z = b_3$$

then $x = \dfrac{D_x}{D}$, $y = \dfrac{D_y}{D}$ and $z = \dfrac{D_z}{D}$

where $D = \begin{vmatrix} a_{11} & a_{12} & a_{13} \\ a_{21} & a_{22} & a_{23} \\ a_{31} & a_{32} & a_{33} \end{vmatrix}$

* **Who was Cramer? Gabriel Cramer** (31 July 1704–4 January 1752) was a Swiss mathematician. His articles cover a wide range of subjects including the study of geometric problems, the history of mathematics, philosophy, and the date of Easter. Cramer's most famous book is a work which Cramer modelled on Newton's memoir on cubic curves. To find out more go to **www.routledge.com/cw/bird**

$$D_x = \begin{vmatrix} b_1 & a_{12} & a_{13} \\ b_2 & a_{22} & a_{23} \\ b_3 & a_{32} & a_{33} \end{vmatrix}$$

i.e. the x-column has been replaced by the RHS b column,

$$D_y = \begin{vmatrix} a_{11} & b_1 & a_{13} \\ a_{21} & b_2 & a_{23} \\ a_{31} & b_3 & a_{33} \end{vmatrix}$$

i.e. the y-column has been replaced by the RHS b column,

$$D_z = \begin{vmatrix} a_{11} & a_{12} & b_1 \\ a_{21} & a_{22} & b_2 \\ a_{31} & a_{32} & b_3 \end{vmatrix}$$

i.e. the z-column has been replaced by the RHS b column.

Problem 7. Solve the following simultaneous equations using Cramer's rule.

$$x + y + z = 4$$
$$2x - 3y + 4z = 33$$
$$3x - 2y - 2z = 2$$

(This is the same as Problem 2 and a comparison of methods may be made.) Following the above method:

$$D = \begin{vmatrix} 1 & 1 & 1 \\ 2 & -3 & 4 \\ 3 & -2 & -2 \end{vmatrix}$$
$$= 1(6 - (-8)) - 1((-4) - 12)$$
$$\quad + 1((-4) - (-9)) = 14 + 16 + 5 = \mathbf{35}$$

$$D_x = \begin{vmatrix} 4 & 1 & 1 \\ 33 & -3 & 4 \\ 2 & -2 & -2 \end{vmatrix}$$
$$= 4(6 - (-8)) - 1((-66) - 8)$$
$$\quad + 1((-66) - (-6)) = 56 + 74 - 60 = \mathbf{70}$$

$$D_y = \begin{vmatrix} 1 & 4 & 1 \\ 2 & 33 & 4 \\ 3 & 2 & -2 \end{vmatrix}$$
$$= 1((-66) - 8) - 4((-4) - 12) + 1(4 - 99)$$
$$= -74 + 64 - 95 = \mathbf{-105}$$

$$D_z = \begin{vmatrix} 1 & 1 & 4 \\ 2 & -3 & 33 \\ 3 & -2 & 2 \end{vmatrix}$$

$$= 1((-6) - (-66)) - 1(4 - 99)$$
$$\quad + 4((-4) - (-9)) = 60 + 95 + 20 = \mathbf{175}$$

Hence

$$x = \frac{D_x}{D} = \frac{70}{35} = \mathbf{2}, \quad y = \frac{D_y}{D} = \frac{-105}{35} = \mathbf{-3}$$

and $z = \dfrac{D_z}{D} = \dfrac{175}{35} = \mathbf{5}$

Now try the following Practice Exercise

Practice Exercise 207   Further problems on solving simultaneous equations using Cramer's rule (answers on page 1174)

1. Repeat Problems 3, 4, 5, 7 and 8 of Exercise 205 on page 561, using Cramer's rule.

2. Repeat Problems 3, 4, 8 and 9 of Exercise 206 on page 564, using Cramer's rule.

48.4   Solution of simultaneous equations using the Gaussian elimination method

Consider the following simultaneous equations:

$$x + y + z = 4 \tag{1}$$
$$2x - 3y + 4z = 33 \tag{2}$$
$$3x - 2y - 2z = 2 \tag{3}$$

Leaving equation (1) as it is gives:

$$x + y + z = 4 \tag{1}$$

Equation (2) $- 2 \times$ equation (1) gives:

$$0 - 5y + 2z = 25 \tag{2$'$}$$

and equation (3) $- 3 \times$ equation (1) gives:

$$0 - 5y - 5z = -10 \tag{3$'$}$$

Leaving equations (1) and (2$'$) as they are gives:

$$x + y + z = 4 \tag{1}$$
$$0 - 5y + 2z = 25 \tag{2$'$}$$

Equation (3$'$) $-$ equation (2$'$) gives:

$$0 + 0 - 7z = -35 \tag{3$''$}$$

By appropriately manipulating the three original equations we have deliberately obtained zeros in the positions shown in equations (2$'$) and (3$''$).

Working backwards, from equation (3″),

$$z = \frac{-35}{-7} = 5,$$

from equation (2′),

$$-5y + 2(5) = 25,$$

from which,

$$y = \frac{25 - 10}{-5} = -3$$

and from equation (1),

$$x + (-3) + 5 = 4,$$

from which,

$$x = 4 + 3 - 5 = 2$$

(This is the same example as Problems 2 and 7, and a comparison of methods can be made.) The above method is known as the **Gaussian* elimination method**.

* **Who was Gauss?** **Johann Carl Friedrich Gauss** (30 April 1777–23 February 1855) was a German mathematician and physical scientist who contributed significantly to many fields, including number theory, statistics, electrostatics, astronomy and optics. To find out more go to **www.routledge.com/cw/bird**

We conclude from the above example that if

$$a_{11}x + a_{12}y + a_{13}z = b_1$$

$$a_{21}x + a_{22}y + a_{23}z = b_2$$

$$a_{31}x + a_{32}y + a_{33}z = b_3$$

the three-step **procedure** to solve simultaneous equations in three unknowns using the **Gaussian elimination method** is:

1. Equation $(2) - \dfrac{a_{21}}{a_{11}} \times$ equation (1) to form equation (2′) and equation $(3) - \dfrac{a_{31}}{a_{11}} \times$ equation (1) to form equation (3′).

2. Equation $(3') - \dfrac{a_{32}}{a_{22}} \times$ equation (2′) to form equation (3″).

3. Determine z from equation (3″), then y from equation (2′) and finally, x from equation (1).

Problem 8. A d.c. circuit comprises three closed loops. Applying Kirchhoff's laws to the closed loops gives the following equations for current flow in milliamperes:

$$2I_1 + 3I_2 - 4I_3 = 26 \tag{1}$$

$$I_1 - 5I_2 - 3I_3 = -87 \tag{2}$$

$$-7I_1 + 2I_2 + 6I_3 = 12 \tag{3}$$

Use the Gaussian elimination method to solve for I_1, I_2 and I_3

(This is the same example as Problem 6 on page 564, and a comparison of methods may be made.)

Following the above procedure:

1. $2I_1 + 3I_2 - 4I_3 = 26 \tag{1}$
   Equation $(2) - \dfrac{1}{2} \times$ equation (1) gives:
   $$0 - 6.5I_2 - I_3 = -100 \tag{2'}$$

   Equation $(3) - \dfrac{-7}{2} \times$ equation (1) gives:
   $$0 + 12.5I_2 - 8I_3 = 103 \tag{3'}$$

2. $2I_1 + 3I_2 - 4I_3 = 26 \tag{1}$
   $$0 - 6.5I_2 - I_3 = -100 \tag{2'}$$
   Equation $(3') - \dfrac{12.5}{-6.5} \times$ equation (2′) gives:
   $$0 + 0 - 9.923I_3 = -89.308 \tag{3''}$$

3. From equation (3''),

$$I_3 = \frac{-89.308}{-9.923} = 9\,\text{mA},$$

from equation (2'), $-6.5I_2 - 9 = -100,$

from which, $I_2 = \frac{-100 + 9}{-6.5} = 14\,\text{mA}$

and from equation (1), $2I_1 + 3(14) - 4(9) = 26,$

from which, $I_1 = \frac{26 - 42 + 36}{2} = \frac{20}{2}$

$$= 10\,\text{mA}$$

Now try the following Practice Exercise

Practice Exercise 208   Further problems on solving simultaneous equations using Gaussian elimination (answers on page 1174)

1. In a mass–spring–damper system, the acceleration $\ddot{x}$ m/s^2, velocity $\dot{x}$ m/s and displacement x m are related by the following simultaneous equations:

$$6.2\ddot{x} + 7.9\dot{x} + 12.6x = 18.0$$

$$7.5\ddot{x} + 4.8\dot{x} + 4.8x = 6.39$$

$$13.0\ddot{x} + 3.5\dot{x} - 13.0x = -17.4$$

By using Gaussian elimination, determine the acceleration, velocity and displacement for the system, correct to 2 decimal places.

2. The tensions, T_1, T_2 and T_3 in a simple framework are given by the equations:

$$5T_1 + 5T_2 + 5T_3 = 7.0$$

$$T_1 + 2T_2 + 4T_3 = 2.4$$

$$4T_1 + 2T_2 \qquad = 4.0$$

Determine T_1, T_2 and T_3 using Gaussian elimination.

3. Repeat Problems 3, 4, 5, 7 and 8 of Exercise 205 on page 561, using the Gaussian elimination method.

4. Repeat Problems 3, 4, 8 and 9 of Exercise 206 on page 564, using the Gaussian elimination method.

48.5   Eigenvalues and eigenvectors

In practical applications, such as coupled oscillations and vibrations, equations of the form:

$$A\,x = \lambda x$$

occur, where A is a square matrix and λ (Greek lambda) is a number. Whenever $x \neq 0$, the values of λ are called the **eigenvalues** of the matrix A; the corresponding solutions of the equation $A\,x = \lambda x$ are called the **eigenvectors** of A.

Sometimes, instead of the term *eigenvalues*, **characteristic values** or **latent roots** are used. Also, instead of the term *eigenvectors*, **characteristic vectors** is used.

From above, if $A\,x = \lambda x$ then $A\,x - \lambda x = 0$ i.e. $(A - \lambda I)x = 0$ where I is the unit matrix. If $x \neq 0$ then $|A - \lambda I| = 0$

$|A - \lambda I|$ is called the **characteristic determinant** of A and $|A - \lambda I| = 0$ is called the **characteristic equation**. Solving the characteristic equation will give the value(s) of the eigenvalues, as demonstrated in the following worked problems.

Problem 9. Determine the eigenvalues of the matrix $A = \begin{pmatrix} 3 & 4 \\ 2 & 1 \end{pmatrix}$

The eigenvalue is determined by solving the characteristic equation $|A - \lambda I| = 0$

i.e. $\left| \begin{pmatrix} 3 & 4 \\ 2 & 1 \end{pmatrix} - \lambda \begin{pmatrix} 1 & 0 \\ 0 & 1 \end{pmatrix} \right| = 0$

i.e. $\left| \begin{pmatrix} 3 & 4 \\ 2 & 1 \end{pmatrix} - \begin{pmatrix} \lambda & 0 \\ 0 & \lambda \end{pmatrix} \right| = 0$

i.e $\left| \begin{matrix} 3 - \lambda & 4 \\ 2 & 1 - \lambda \end{matrix} \right| = 0$

(Given a square matrix, we can get used to going straight to this characteristic equation.)

Hence, $(3 - \lambda)(1 - \lambda) - (2)(4) = 0$

i.e $3 - 3\lambda - \lambda + \lambda^2 - 8 = 0$

and $\lambda^2 - 4\lambda - 5 = 0$

and $(\lambda - 5)(\lambda + 1) = 0$

from which, $\lambda - 5 = 0$ i.e. $\boldsymbol{\lambda = 5}$ or $\lambda + 1 = 0$ i.e. $\boldsymbol{\lambda = -1}$

(Instead of factorising, the quadratic formula could be used; even electronic calculators can solve quadratic equations.)

Hence, the eigenvalues of the matrix $\begin{pmatrix} 3 & 4 \\ 2 & 1 \end{pmatrix}$ are 5 and -1

Problem 10.   Determine the eigenvectors of the matrix $A = \begin{pmatrix} 3 & 4 \\ 2 & 1 \end{pmatrix}$

From Problem 9, the eigenvalues of $\begin{pmatrix} 3 & 4 \\ 2 & 1 \end{pmatrix}$ are $\lambda_1 = 5$ and $\lambda_2 = -1$
Using the equation $(A - \lambda I)x = 0$ for $\lambda_1 = 5$

then $\quad \begin{pmatrix} 3-5 & 4 \\ 2 & 1-5 \end{pmatrix} \begin{pmatrix} x_1 \\ x_2 \end{pmatrix} = \begin{pmatrix} 0 \\ 0 \end{pmatrix}$

i.e. $\quad \begin{pmatrix} -2 & 4 \\ 2 & -4 \end{pmatrix} \begin{pmatrix} x_1 \\ x_2 \end{pmatrix} = \begin{pmatrix} 0 \\ 0 \end{pmatrix}$

from which,

$$-2x_1 + 4x_2 = 0$$

and

$$2x_1 - 4x_2 = 0$$

From either of these two equations, $x_1 = 2x_2$
Hence, whatever value x_2 is, the value of x_1 will be two times greater. Hence the simplest eigenvector is:

$$x_1 = \begin{pmatrix} 2 \\ 1 \end{pmatrix}$$

Using the equation $(A - \lambda I)x = 0$ for $\lambda_2 = -1$

then $\quad \begin{pmatrix} 3--1 & 4 \\ 2 & 1--1 \end{pmatrix} \begin{pmatrix} x_1 \\ x_2 \end{pmatrix} = \begin{pmatrix} 0 \\ 0 \end{pmatrix}$

i.e. $\quad \begin{pmatrix} 4 & 4 \\ 2 & 2 \end{pmatrix} \begin{pmatrix} x_1 \\ x_2 \end{pmatrix} = \begin{pmatrix} 0 \\ 0 \end{pmatrix}$

from which,

$$4x_1 + 4x_2 = 0$$

and

$$2x_1 + 2x_2 = 0$$

From either of these two equations, $x_1 = -x_2$ or $x_2 = -x_1$
Hence, whatever value x_1 is, the value of x_2 will be -1 times greater. Hence the simplest eigenvector is:

$$x_2 = \begin{pmatrix} 1 \\ -1 \end{pmatrix}$$

Summarising, $x_1 = \begin{pmatrix} 2 \\ 1 \end{pmatrix}$ is an eigenvector corresponding to $\lambda_1 = 5$

and $x_2 = \begin{pmatrix} 1 \\ -1 \end{pmatrix}$ is an eigenvector corresponding to $\lambda_2 = -1$

Problem 11.   Determine the eigenvalues of the matrix $A = \begin{pmatrix} 5 & -2 \\ -9 & 2 \end{pmatrix}$

The eigenvalue is determined by solving the characteristic equation $|A - \lambda I| = 0$

i.e. $\quad \begin{vmatrix} 5-\lambda & -2 \\ -9 & 2-\lambda \end{vmatrix} = 0$

Hence, $(5 - \lambda)(2 - \lambda) - (-9)(-2) = 0$

i.e. $10 - 5\lambda - 2\lambda + \lambda^2 - 18 = 0$

and $\lambda^2 - 7\lambda - 8 = 0$

i.e. $(\lambda - 8)(\lambda + 1) = 0$

from which, $\lambda - 8 = 0$ i.e. $\lambda = 8$ or $\lambda + 1 = 0$ i.e. $\lambda = -1$ (or use the quadratic formula). **Hence, the eigenvalues of the matrix $\begin{pmatrix} 5 & -2 \\ -9 & 2 \end{pmatrix}$ are 8 and -1**

Problem 12.   Determine the eigenvectors of the matrix $A = \begin{pmatrix} 5 & -2 \\ -9 & 2 \end{pmatrix}$

From Problem 11, the eigenvalues of $\begin{pmatrix} 5 & -2 \\ -9 & 2 \end{pmatrix}$ are $\lambda_1 = 8$ and $\lambda_2 = -1$
Using the equation $(A - \lambda I)x = 0$ for $\lambda_1 = 8$

then $\quad \begin{pmatrix} 5-8 & -2 \\ -9 & 2-8 \end{pmatrix} \begin{pmatrix} x_1 \\ x_2 \end{pmatrix} = \begin{pmatrix} 0 \\ 0 \end{pmatrix}$

i.e. $\quad \begin{pmatrix} -3 & -2 \\ -9 & -6 \end{pmatrix} \begin{pmatrix} x_1 \\ x_2 \end{pmatrix} = \begin{pmatrix} 0 \\ 0 \end{pmatrix}$

from which,

$$-3x_1 - 2x_2 = 0$$

and

$$-9x_1 - 6x_2 = 0$$

From either of these two equations, $3x_1 = -2x_2$ or $x_1 = -\frac{2}{3}x_2$

Hence, if $x_2 = 3$, $x_1 = -2$. Hence the simplest eigenvector is: $x_1 = \begin{pmatrix} -2 \\ 3 \end{pmatrix}$

Using the equation $(A - \lambda I)x = 0$ for $\lambda_2 = -1$

then
$$\begin{pmatrix} 5--1 & -2 \\ -9 & 2--1 \end{pmatrix}\begin{pmatrix} x_1 \\ x_2 \end{pmatrix} = \begin{pmatrix} 0 \\ 0 \end{pmatrix}$$

i.e.
$$\begin{pmatrix} 6 & -2 \\ -9 & 3 \end{pmatrix}\begin{pmatrix} x_1 \\ x_2 \end{pmatrix} = \begin{pmatrix} 0 \\ 0 \end{pmatrix}$$

from which,

$$6x_1 - 2x_2 = 0$$

and

$$-9x_1 + 3x_2 = 0$$

From either of these two equations, $x_2 = 3x_1$

Hence, if $x_1 = 1, x_2 = 3$. Hence the simplest eigenvector is: $x_2 = \begin{pmatrix} 1 \\ 3 \end{pmatrix}$

Summarising, $x_1 = \begin{pmatrix} -2 \\ 3 \end{pmatrix}$ is an eigenvector corresponding to $\lambda_1 = 8$ and $x_2 = \begin{pmatrix} 1 \\ 3 \end{pmatrix}$ is an eigenvector corresponding to $\lambda_2 = -1$

Problem 13. Determine the eigenvalues of the matrix $A = \begin{pmatrix} 1 & 2 & 1 \\ 6 & -1 & 0 \\ -1 & -2 & -1 \end{pmatrix}$

The eigenvalue is determined by solving the characteristic equation $|A - \lambda I| = 0$

i.e.
$$\begin{vmatrix} 1-\lambda & 2 & 1 \\ 6 & -1-\lambda & 0 \\ -1 & -2 & -1-\lambda \end{vmatrix} = 0$$

Hence, using the top row:

$$(1-\lambda)[(-1-\lambda)(-1-\lambda) - (-2)(0)] - 2[6(-1-\lambda) - (-1)(0)] + 1[(6)(-2) - (-1)(-1-\lambda)] = 0$$

i.e. $(1-\lambda)[1+\lambda+\lambda+\lambda^2] - 2[-6-6\lambda] + 1[-12-1-\lambda] = 0$

i.e. $(1-\lambda)[\lambda^2 + 2\lambda + 1] + 12 + 12\lambda - 13 - \lambda = 0$

and $\lambda^2 + 2\lambda + 1 - \lambda^3 - 2\lambda^2 - \lambda + 12$
$$+12\lambda - 13 - \lambda = 0$$

i.e. $-\lambda^3 - \lambda^2 + 12\lambda = 0$

or $\lambda^3 + \lambda^2 - 12\lambda = 0$

i.e. $\lambda(\lambda^2 + \lambda - 12) = 0$

i.e. $\lambda(\lambda - 3)(\lambda + 4) = 0$ by factorising from which, $\lambda = 0, \lambda = 3$ or $\lambda = -4$

Hence, the eigenvalues of the matrix $\begin{pmatrix} 1 & 2 & 1 \\ 6 & -1 & 0 \\ -1 & -2 & -1 \end{pmatrix}$ are 0, 3 and -4

Problem 14. Determine the eigenvectors of the matrix $A = \begin{pmatrix} 1 & 2 & 1 \\ 6 & -1 & 0 \\ -1 & -2 & -1 \end{pmatrix}$

From Problem 13, the eigenvalues of $\begin{pmatrix} 1 & 2 & 1 \\ 6 & -1 & 0 \\ -1 & -2 & -1 \end{pmatrix}$ are $\lambda_1 = 0$, $\lambda_2 = 3$ and $\lambda_3 = -4$

Using the equation $(A - \lambda I)x = 0$ for $\lambda_1 = 0$

then
$$\begin{pmatrix} 1-0 & 2 & 1 \\ 6 & -1-0 & 0 \\ -1 & -2 & -1-0 \end{pmatrix}\begin{pmatrix} x_1 \\ x_2 \\ x_3 \end{pmatrix} = \begin{pmatrix} 0 \\ 0 \\ 0 \end{pmatrix}$$

i.e.
$$\begin{pmatrix} 1 & 2 & 1 \\ 6 & -1 & 0 \\ -1 & -2 & -1 \end{pmatrix}\begin{pmatrix} x_1 \\ x_2 \\ x_3 \end{pmatrix} = \begin{pmatrix} 0 \\ 0 \\ 0 \end{pmatrix}$$

from which,

$$x_1 + 2x_2 + x_3 = 0$$

$$6x_1 - x_2 = 0$$

$$-x_1 - 2x_2 - x_3 = 0$$

From the second equation,

$$6x_1 = x_2$$

Substituting in the first equation,

$$x_1 + 12x_1 + x_3 = 0 \quad \text{i.e.} \quad -13x_1 = x_3$$

Hence, when $x_1 = 1$, $x_2 = 6$ and $x_3 = -13$
Hence the simplest eigenvector corresponding to $\lambda_1 = 0$ is: $x_1 = \begin{pmatrix} 1 \\ 6 \\ -13 \end{pmatrix}$

Using the equation $(A - \lambda I)x = 0$ for $\lambda_2 = 3$

then
$$\begin{pmatrix} 1-3 & 2 & 1 \\ 6 & -1-3 & 0 \\ -1 & -2 & -1-3 \end{pmatrix}\begin{pmatrix} x_1 \\ x_2 \\ x_3 \end{pmatrix} = \begin{pmatrix} 0 \\ 0 \\ 0 \end{pmatrix}$$

i.e.
$$\begin{pmatrix} -2 & 2 & 1 \\ 6 & -4 & 0 \\ -1 & -2 & -4 \end{pmatrix}\begin{pmatrix} x_1 \\ x_2 \\ x_3 \end{pmatrix} = \begin{pmatrix} 0 \\ 0 \\ 0 \end{pmatrix}$$

from which,

$$-2x_1 + 2x_2 + x_3 = 0$$

$$6x_1 - 4x_2 = 0$$

$$-x_1 - 2x_2 - 4x_3 = 0$$

From the second equation,

$$3x_1 = 2x_2$$

Substituting in the first equation,

$$-2x_1 + 3x_1 + x_3 = 0 \quad \text{i.e.} \quad x_3 = -x_1$$

Hence, if $x_2 = 3$, then $x_1 = 2$ and $x_3 = -2$

Hence the simplest eigenvector corresponding to $\lambda_2 = 3$

is: $x_2 = \begin{pmatrix} 2 \\ 3 \\ -2 \end{pmatrix}$

Using the equation $(A \quad \lambda I)x = 0$ for $\lambda_2 = -4$

then $\begin{pmatrix} 1--4 & 2 & 1 \\ 6 & -1--4 & 0 \\ -1 & -2 & -1--4 \end{pmatrix} \begin{pmatrix} x_1 \\ x_2 \\ x_3 \end{pmatrix}$

$$= \begin{pmatrix} 0 \\ 0 \\ 0 \end{pmatrix}$$

i.e.

$$\begin{pmatrix} 5 & 2 & 1 \\ 6 & 3 & 0 \\ -1 & -2 & 3 \end{pmatrix} \begin{pmatrix} x_1 \\ x_2 \\ x_3 \end{pmatrix} = \begin{pmatrix} 0 \\ 0 \\ 0 \end{pmatrix}$$

from which,

$$5x_1 + 2x_2 + x_3 = 0$$

$$6x_1 + 3x_2 = 0$$

$$-x_1 - 2x_2 + 3x_3 = 0$$

From the second equation,

$$x_2 = -2x_1$$

Substituting in the first equation,

$$5x_1 - 4x_1 + x_3 = 0 \quad \text{i.e.} \quad x_3 = -x_1$$

Hence, if $x_1 = -1$, then $x_2 = 2$ and $x_3 = 1$

Hence the simplest eigenvector corresponding to

$\lambda_2 = -4$ is: $x_3 = \begin{pmatrix} -1 \\ 2 \\ 1 \end{pmatrix}$

Problem 15. Determine the eigenvalues of the

matrix $A = \begin{pmatrix} 1 & -4 & -2 \\ 0 & 3 & 1 \\ 1 & 2 & 4 \end{pmatrix}$

The eigenvalue is determined by solving the characteristic equation $|A - \lambda I| = 0$

i.e. $\begin{vmatrix} 1-\lambda & -4 & -2 \\ 0 & 3-\lambda & 1 \\ 1 & 2 & 4-\lambda \end{vmatrix} = 0$

Hence, using the top row:

$$(1-\lambda)[(3-\lambda)(4-\lambda) - (2)(1)] - (-4)[0 - (1)(1)]$$
$$- 2[0 - 1(3-\lambda)] = 0$$

i.e. $(1-\lambda)[12 - 3\lambda - 4\lambda + \lambda^2 - 2] + 4[-1]$
$$- 2[-3 + \lambda] = 0$$

i.e. $(1-\lambda)[\lambda^2 - 7\lambda + 10] - 4 + 6 - 2\lambda = 0$

and $\lambda^2 - 7\lambda + 10 - \lambda^3 + 7\lambda^2 - 10\lambda - 4$
$$+ 6 - 2\lambda = 0$$

i.e. $-\lambda^3 + 8\lambda^2 - 19\lambda + 12 = 0$

or $\lambda^3 - 8\lambda^2 + 19\lambda - 12 = 0$

To solve a cubic equation, the **factor theorem** (see Chapter 18) may be used. (Alternatively, electronic calculators can solve such cubic equations.)

Let $f(\lambda) = \lambda^3 - 8\lambda^2 + 19\lambda - 12$

If $\lambda = 1$, then $f(1) = 1^3 - 8(1)^2 + 19(1) - 12 = 0$

Since $f(1) = 0$ then $(\lambda - 1)$ is a factor.

If $\lambda = 2$, then $f(2) = 2^3 - 8(2)^2 + 19(2) - 12 \neq 0$

If $\lambda = 3$, then $f(3) = 3^3 - 8(3)^2 + 19(3) - 12 = 0$

Since $f(3) = 0$ then $(\lambda - 3)$ is a factor.

If $\lambda = 4$, then $f(4) = 4^3 - 8(4)^2 + 19(4) - 12 = 0$

Since $f(4) = 0$ then $(\lambda - 4)$ is a factor.

Thus, since $\lambda^3 - 8\lambda^2 + 19\lambda - 12 = 0$

then $(\lambda - 1)(\lambda - 3)(\lambda - 4) = 0$

from which, $\lambda = 1$ or $\lambda = 3$ or $\lambda = 4$

Hence, the eigenvalues of the matrix $\begin{pmatrix} 1 & -4 & -2 \\ 0 & 3 & 1 \\ 1 & 2 & 4 \end{pmatrix}$ **are 1, 3 and 4**

Problem 16. Determine the eigenvectors of the

matrix $A = \begin{pmatrix} 1 & -4 & -2 \\ 0 & 3 & 1 \\ 1 & 2 & 4 \end{pmatrix}$

From Problem 15, the eigenvalues of $\begin{pmatrix} 1 & -4 & -2 \\ 0 & 3 & 1 \\ 1 & 2 & 4 \end{pmatrix}$

are $\lambda_1 = 1$, $\lambda_2 = 3$ and $\lambda_3 = 4$

Using the equation $(A - \lambda I)x = 0$ for $\lambda_1 = 1$

then $\begin{pmatrix} 1-1 & -4 & -2 \\ 0 & 3-1 & 1 \\ 1 & 2 & 4-1 \end{pmatrix} \begin{pmatrix} x_1 \\ x_2 \\ x_3 \end{pmatrix} = \begin{pmatrix} 0 \\ 0 \\ 0 \end{pmatrix}$

i.e. $\begin{pmatrix} 0 & -4 & -2 \\ 0 & 2 & 1 \\ 1 & 2 & 3 \end{pmatrix} \begin{pmatrix} x_1 \\ x_2 \\ x_3 \end{pmatrix} = \begin{pmatrix} 0 \\ 0 \\ 0 \end{pmatrix}$

from which,

$$-4x_2 - 2x_3 = 0$$
$$2x_2 + x_3 = 0$$
$$x_1 + 2x_2 + 3x_3 = 0$$

From the first two equations,

$$x_3 = -2x_2 \quad \text{(i.e. if } x_2 = 1, x_3 = -2)$$

From the last equation,
$$x_1 = -2x_2 - 3x_3 \quad \text{i.e.} \quad x_1 = -2x_2 - 3(-2x_2)$$

i.e.
$$x_1 = 4x_2 \quad \text{(i.e. if } x_2 = 1, x_1 = 4)$$

Hence the simplest eigenvector corresponding to $\lambda_1 = 1$

is: $x_1 = \begin{pmatrix} 4 \\ 1 \\ -2 \end{pmatrix}$

Using the equation $(A - \lambda I)x = 0$ for $\lambda_2 = 3$

then $\begin{pmatrix} 1-3 & -4 & -2 \\ 0 & 3-3 & 1 \\ 1 & 2 & 4-3 \end{pmatrix} \begin{pmatrix} x_1 \\ x_2 \\ x_3 \end{pmatrix} = \begin{pmatrix} 0 \\ 0 \\ 0 \end{pmatrix}$

i.e. $\begin{pmatrix} -2 & -4 & -2 \\ 0 & 0 & 1 \\ 1 & 2 & 1 \end{pmatrix} \begin{pmatrix} x_1 \\ x_2 \\ x_3 \end{pmatrix} = \begin{pmatrix} 0 \\ 0 \\ 0 \end{pmatrix}$

from which,

$$-2x_1 - 4x_2 - 2x_3 = 0$$
$$x_3 = 0$$
$$x_1 + 2x_2 + x_3 = 0$$

Since $x_3 = 0$, $x_1 = -2x_2$ (i.e. if $x_2 = 1$, $x_1 = -2$)
Hence the simplest eigenvector corresponding to $\lambda_2 = 3$

is: $x_2 = \begin{pmatrix} -2 \\ 1 \\ 0 \end{pmatrix}$

Using the equation $(A - \lambda I)x = 0$ for $\lambda_3 = 4$

then $\begin{pmatrix} 1-4 & -4 & -2 \\ 0 & 3-4 & 1 \\ 1 & 2 & 4-4 \end{pmatrix} \begin{pmatrix} x_1 \\ x_2 \\ x_3 \end{pmatrix} = \begin{pmatrix} 0 \\ 0 \\ 0 \end{pmatrix}$

i.e. $\begin{pmatrix} -3 & -4 & -2 \\ 0 & -1 & 1 \\ 1 & 2 & 0 \end{pmatrix} \begin{pmatrix} x_1 \\ x_2 \\ x_3 \end{pmatrix} = \begin{pmatrix} 0 \\ 0 \\ 0 \end{pmatrix}$

from which,

$$-3x_1 - 4x_2 - 2x_3 = 0$$
$$-x_2 + x_3 = 0$$
$$x_1 + 2x_2 = 0$$

from which, $x_3 = x_2$ and $x_1 = -2x_2$ (i.e. if $x_2 = 1$, $x_1 = -2$ and $x_3 = 1$)
Hence the simplest eigenvector corresponding to $\lambda_3 = 4$

is: $x_3 = \begin{pmatrix} -2 \\ 1 \\ 1 \end{pmatrix}$

Now try the following Practice Exercise

Practice Exercise 209   Eigenvalues and eigenvectors (answers on page 1174)

For each of the following matrices, determine their (a) eigenvalues (b) eigenvectors

1. $\begin{pmatrix} 2 & -4 \\ -1 & -1 \end{pmatrix}$ 2. $\begin{pmatrix} 3 & 6 \\ 1 & 4 \end{pmatrix}$

3. $\begin{pmatrix} 3 & 1 \\ -2 & 0 \end{pmatrix}$ 4. $\begin{pmatrix} -1 & -1 & 1 \\ -4 & 2 & 4 \\ -1 & 1 & 5 \end{pmatrix}$

5. $\begin{pmatrix} 1 & -1 & 0 \\ -1 & 2 & -1 \\ 0 & -1 & 1 \end{pmatrix}$ 6. $\begin{pmatrix} 2 & 2 & -2 \\ 1 & 3 & 1 \\ 1 & 2 & 2 \end{pmatrix}$

7. $\begin{pmatrix} 1 & 1 & 2 \\ 0 & 2 & 2 \\ -1 & 1 & 3 \end{pmatrix}$

For fully worked solutions to each of the problems in Practice Exercises 205 to 209 in this chapter, go to the website:
www.routledge.com/cw/bird

Revision Test 19 Matrices and determinants

This assignment covers the material contained in Chapters 47 and 48. *The marks available are shown in brackets at the end of each question.*

In Questions 1 to 5, the matrices stated are:

$$A = \begin{pmatrix} -5 & 2 \\ 7 & -8 \end{pmatrix} \quad B = \begin{pmatrix} 1 & 6 \\ -3 & -4 \end{pmatrix}$$

$$C = \begin{pmatrix} j3 & (1+j2) \\ (-1-j4) & -j2 \end{pmatrix}$$

$$D = \begin{pmatrix} 2 & -1 & 3 \\ -5 & 1 & 0 \\ 4 & -6 & 2 \end{pmatrix} \quad E = \begin{pmatrix} -1 & 3 & 0 \\ 4 & -9 & 2 \\ -5 & 7 & 1 \end{pmatrix}$$

1. Determine $A \times B$ (4)

2. Calculate the determinant of matrix C (4)

3. Determine the inverse of matrix A (4)

4. Determine $E \times D$ (9)

5. Calculate the determinant of matrix D (6)

6. Solve the following simultaneous equations:

$$4x - 3y = 17$$
$$x + y + 1 = 0$$

   using matrices. (7)

7. Use determinants to solve the following simultaneous equations:

$$4x + 9y + 2z = 21$$
$$-8x + 6y - 3z = 41$$
$$3x + y - 5z = -73 \quad (11)$$

8. The simultaneous equations representing the currents flowing in an unbalanced, three-phase, star-connected, electrical network are as follows:

$$2.4I_1 + 3.6I_2 + 4.8I_3 = 1.2$$
$$-3.9I_1 + 1.3I_2 - 6.5I_3 = 2.6$$
$$1.7I_1 + 11.9I_2 + 8.5I_3 = 0$$

   Using matrices, solve the equations for I_1, I_2 and I_3 (11)

9. For the matrix $\begin{pmatrix} 2 & -4 \\ -1 & -1 \end{pmatrix}$ find the
   (a) eigenvalues (b) eigenvectors. (14)

Matrices:

If $A = \begin{pmatrix} a & b \\ c & d \end{pmatrix}$ and $B = \begin{pmatrix} e & f \\ g & h \end{pmatrix}$ then

$$A + B = \begin{pmatrix} a+e & b+f \\ c+g & d+h \end{pmatrix}$$

$$A - B = \begin{pmatrix} a-e & b-f \\ c-g & d-h \end{pmatrix}$$

$$A \times B = \begin{pmatrix} ae+bg & af+bh \\ ce+dg & cf+dh \end{pmatrix}$$

$$A^{-1} = \frac{1}{ad-bc} \begin{pmatrix} d & -b \\ -c & a \end{pmatrix}$$

If $A = \begin{pmatrix} a_1 & b_1 & c_1 \\ a_2 & b_2 & c_2 \\ a_3 & b_3 & c_3 \end{pmatrix}$ then $A^{-1} = \dfrac{B^T}{|A|}$

where $B^T =$ transpose of cofactors of matrix A

Determinants:

$$\begin{vmatrix} a & b \\ c & d \end{vmatrix} = ad - bc$$

$$\begin{vmatrix} a_1 & b_1 & c_1 \\ a_2 & b_2 & c_2 \\ a_3 & b_3 & c_3 \end{vmatrix} = a_1 \begin{vmatrix} b_2 & c_2 \\ b_3 & c_3 \end{vmatrix} - b_1 \begin{vmatrix} a_2 & c_2 \\ a_3 & c_3 \end{vmatrix}$$
$$+ c_1 \begin{vmatrix} a_2 & b_2 \\ a_3 & b_3 \end{vmatrix}$$

Eigenvalues: For a matrix, A, the eigenvalues are determined by solving the characteristic equation $|A - \lambda I| = 0$

Eigenvectors: For a matrix, A, eigenvectors are determined by solving the equation $(A - \lambda I)x = 0$ for eigenvalue λ

Section H

Vector geometry

Chapter 49

Vectors

Why it is important to understand: Vectors

Vectors are an important part of the language of science, mathematics and engineering. They are used to discuss multivariable calculus, electrical circuits with oscillating currents, stress and strain in structures and materials, and flows of atmospheres and fluids, and they have many other applications. Resolving a vector into components is a precursor to computing things with or about a vector quantity. Because position, velocity, acceleration, force, momentum, and angular momentum are all vector quantities, resolving vectors into components is a most important skill required in any engineering studies.

At the end of this chapter, you should be able to:

- distinguish between scalars and vectors
- recognise how vectors are represented
- add vectors using the nose-to-tail method
- add vectors using the parallelogram method
- resolve vectors into their horizontal and vertical components
- add vectors by calculation – horizontal and vertical components, complex numbers
- perform vector subtraction
- understand relative velocity
- understand i, j, k notation

49.1 Introduction

This chapter initially explains the difference between scalar and vector quantities and shows how a vector is drawn and represented.

Any object that is acted upon by an external force will respond to that force by moving in the line of the force. However, if two or more forces act simultaneously, the result is more difficult to predict; the ability to add two or more vectors then becomes important.

This chapter thus shows how vectors are added and subtracted, both by drawing and by calculation, and finding the resultant of two or more vectors has many uses in engineering. (Resultant means the single vector which would have the same effect as the individual vectors.)

Relative velocities and vector i, j, k notation are also briefly explained.

49.2 Scalars and vectors

The time taken to fill a water tank may be measured as, say, 50 s. Similarly, the temperature in a room may be measured as, say, 16°C, or the mass of a bearing may be measured as, say, 3 kg.

Quantities such as time, temperature and mass are entirely defined by a numerical value and are called **scalars** or **scalar quantities**.

Not all quantities are like this. Some are defined by more than just size; some also have direction. For example, the velocity of a car is 90 km/h due west, or a force of

---

20 N acts vertically downwards, or an acceleration of $10\,m/s^2$ acts at $50°$ to the horizontal.

Quantities such as velocity, force and acceleration, which **have both a magnitude and a direction**, are called **vectors**.

Now try the following Practice Exercise

Practice Exercise 210   Further problems on scalar and vector quantities (answers on page 1175)

1. State the difference between scalar and vector quantities.

In Problems 2 to 9, state whether the quantities given are scalar (S) or vector (V)

2. A temperature of $70°C$

3. $5\,m^3$ volume

4. A downward force of $20\,N$

5. $500\,J$ of work

6. $30\,cm^2$ area

7. A south-westerly wind of $10\,knots$

8. $50\,m$ distance

9. An acceleration of $15\,m/s^2$ at $60°$ to the horizontal

49.3   Drawing a vector

A vector quantity can be represented graphically by a line, drawn so that:

(a)   the **length** of the line denotes the magnitude of the quantity, and

(b)   the **direction** of the line denotes the direction in which the vector quantity acts.

An arrow is used to denote the sense, or direction, of the vector.

The arrow end of a vector is called the 'nose' and the other end the 'tail'.

For example, a force of 9 N acting at $45°$ to the horizontal is shown in Figure 49.1.

Note that an angle of $+45°$ is drawn from the horizontal and moves **anticlockwise**.

A velocity of $20\,m/s$ at $-60°$ is shown in Figure 49.2.

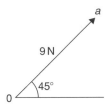

Figure 49.1

Note that an angle of $-60°$ is drawn from the horizontal and moves **clockwise**.

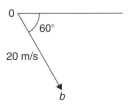

Figure 49.2

Representing a vector

There are a number of ways of representing vector quantities. These include:

1. Using **bold print**

2. $\overrightarrow{AB}$ where an arrow above two capital letters denotes the sense of direction, where A is the starting point and B the end point of the vector

3. $\overline{AB}$ or $\overline{a}$ i.e. a line over the top of letters

4. $\underline{a}$ i.e. an underlined letter.

The force of 9 N at $45°$ shown in Figure 49.1 may be represented as:

$$0a \quad \text{or} \quad \overrightarrow{0a} \quad \text{or} \quad \overline{0a}$$

The magnitude of the force is $0a$

Similarly, the velocity of $20\,m/s$ at $-60°$ shown in Figure 49.2 may be represented as:

$$0b \quad \text{or} \quad \overrightarrow{0b} \quad \text{or} \quad \overline{0b}$$

The magnitude of the velocity is $0b$

In this chapter a vector quantity is denoted by **bold print**.

49.4   Addition of vectors by drawing

Adding two or more vectors by drawing assumes that a ruler, pencil and protractor are available. Results obtained by drawing are naturally not as accurate as those obtained by calculation.

(a) Nose-to-tail method

Two force vectors, F_1 and F_2, are shown in Figure 49.3. When an object is subjected to more than one force, the resultant of the forces is found by the addition of vectors.

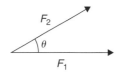

Figure 49.3

To add forces F_1 and F_2:

(i) Force F_1 is drawn to scale horizontally, shown as **0a** in Figure 49.4

(ii) From the nose of F_1, force F_2 is drawn at angle θ to the horizontal, shown as **ab**

(iii) The resultant force is given by length **0b**, which may be measured.

This procedure is called the '**nose-to-tail**' or '**triangle**' method.

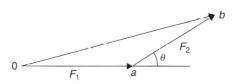

Figure 49.4

(b) Parallelogram method

To add the two force vectors, F_1 and F_2, of Figure 49.3:

(i) A line cb is constructed which is parallel to and equal in length to **0a** (see Figure 49.5)

(ii) A line ab is constructed which is parallel to and equal in length to **0c**

(iii) The resultant force is given by the diagonal of the parallelogram, i.e. length **0b**

This procedure is called the '**parallelogram**' method.

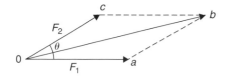

Figure 49.5

Problem 1. A force of 5 N is inclined at an angle of 45° to a second force of 8 N, both forces acting at a point. Find the magnitude of the resultant of these two forces and the direction of the resultant with respect to the 8 N force by (a) the 'nose-to-tail' method, and (b) the 'parallelogram' method

The two forces are shown in Figure 49.6. (Although the 8 N force is shown horizontal, it could have been drawn in any direction.)

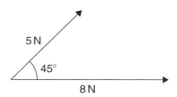

Figure 49.6

(a) 'Nose-to tail' method

(i) The 8 N force is drawn horizontally 8 units long, shown as **0a** in Figure 49.7

(ii) From the nose of the 8 N force, the 5 N force is drawn 5 units long at an angle of 45° to the horizontal, shown as **ab**

(iii) The resultant force is given by length **0b** and is measured as **12 N** and angle θ is measured as **17°**

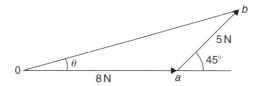

Figure 49.7

(b) 'Parallelogram' method

(i) In Figure 49.8, a line is constructed which is parallel to and equal in length to the 8 N force

(ii) A line is constructed which is parallel to and equal in length to the 5 N force

(iii) The resultant force is given by the diagonal of the parallelogram, i.e. length **0b**, and is measured as **12 N** and angle θ is measured as **17°**

Thus, **the resultant of the two force vectors in Figure 49.6 is 12 N at 17° to the 8 N force**.

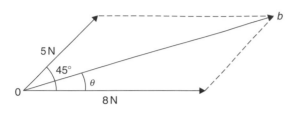

Figure 49.8

Problem 2. Forces of 15 N and 10 N are at an angle of 90° to each other, as shown in Figure 49.9. Find, by drawing, the magnitude of the resultant of these two forces and the direction of the resultant with respect to the 15 N force

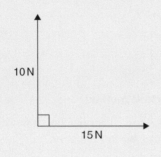

Figure 49.9

Using the 'nose-to-tail' method:

(i) The 15 N force is drawn horizontally 15 units long as shown in Figure 49.10

(ii) From the nose of the 15 N force, the 10 N force is drawn 10 units long at an angle of 90° to the horizontal as shown

(iii) The resultant force is shown as R and is measured as **18 N** and angle θ is measured as **34°**

Thus, **the resultant of the two force vectors is 18 N at 34° to the 15 N force**.

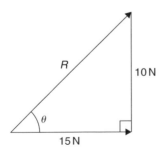

Figure 49.10

Problem 3. Velocities of 10 m/s, 20 m/s and 15 m/s act as shown in Figure 49.11. Determine, by

drawing, the magnitude of the resultant velocity and its direction relative to the horizontal

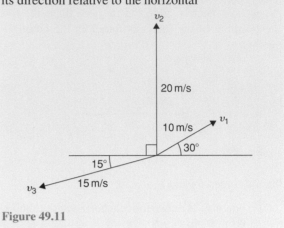

Figure 49.11

When more than two vectors are being added the 'nose-to-tail' method is used.

The order in which the vectors are added does not matter. In this case the order taken is v_1, then v_2, then v_3. However, if a different order is taken the same result will occur.

(i) v_1 is drawn 10 units long at an angle of 30° to the horizontal, shown as **0a** in Figure 49.12

(ii) From the nose of v_1, v_2 is drawn 20 units long at an angle of 90° to the horizontal, shown as **ab**

(iii) From the nose of v_2, v_3 is drawn 15 units long at an angle of 195° to the horizontal, shown as **br**

(iv) The resultant velocity is given by length **0r** and is measured as **22 m/s** and the angle measured to the horizontal is **105°**

Thus, **the resultant of the three velocities is 22 m/s at 105° to the horizontal**.

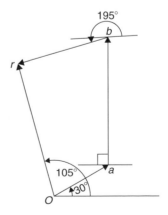

Figure 49.12

Problems 1 to 3 have demonstrated how vectors are added to determine their resultant and their direction. However, drawing to scale is time-consuming and not highly accurate. The following sections demonstrate how to determine resultant vectors by calculation using horizontal and vertical components and, where possible, by Pythagoras' theorem.

49.5 Resolving vectors into horizontal and vertical components

A force vector F is shown in Figure 49.13 at angle θ to the horizontal. Such a vector can be resolved into two components such that the vector addition of the components is equal to the original vector.

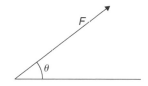

Figure 49.13

The two components usually taken are a **horizontal component** and a **vertical component**.

If a right-angled triangle is constructed as shown in Figure 49.14, then $0a$ is called the horizontal component of F and ab is called the vertical component of F.

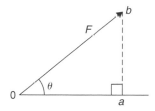

Figure 49.14

From trigonometry (see Chapter 38),

$$\cos\theta = \frac{0a}{0b} \quad \text{from which,} \quad 0a = 0b\cos\theta$$
$$= F\cos\theta$$

i.e. **the horizontal component of $F = F\cos\theta$**

and $\sin\theta = \dfrac{ab}{0b}$ from which, $ab = 0b\sin\theta$
$$= F\sin\theta$$

i.e. **the vertical component of $F = F\sin\theta$**

Problem 4. Resolve the force vector of 50 N at an angle of 35° to the horizontal into its horizontal and vertical components

The **horizontal component** of the 50 N force, $0a = 50\cos35° = \mathbf{40.96\,N}$
The **vertical component** of the 50 N force, $ab = 50\sin35° = \mathbf{28.68\,N}$
The horizontal and vertical components are shown in Figure 49.15.

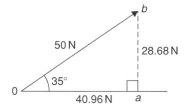

Figure 49.15

(Checking: by Pythagoras, $0b = \sqrt{40.96^2 + 28.68^2}$
$$= 50\,\text{N}$$

and $\theta = \tan^{-1}\left(\dfrac{28.68}{40.96}\right) = 35°$

Thus, the vector addition of components 40.96 N and 28.68 N is 50 N at 35°)

Problem 5. Resolve the velocity vector of 20 m/s at an angle of −30° to the horizontal into horizontal and vertical components

The **horizontal component** of the 20 m/s velocity, $0a = 20\cos(-30°) = \mathbf{17.32\,m/s}$
The **vertical component** of the 20 m/s velocity, $ab = 20\sin(-30°) = \mathbf{-10\,m/s}$
The horizontal and vertical components are shown in Figure 49.16.

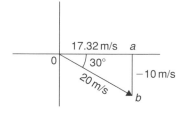

Figure 49.16

Problem 6. Resolve the displacement vector of 40 m at an angle of 120° into horizontal and vertical components

The **horizontal component** of the 40 m displacement,
$0a = 40\cos 120° = \mathbf{-20.0\,m}$
The **vertical component** of the 40 m displacement,
$ab = 40\sin 120° = \mathbf{34.64\,m}$
The horizontal and vertical components are shown in Figure 49.17.

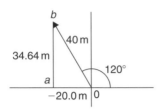

Figure 49.17

49.6 Addition of vectors by calculation

Two force vectors, F_1 and F_2, are shown in Figure 49.18, F_1 being at an angle of θ_1 and F_2 being at an angle of θ_2.

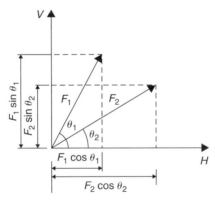

Figure 49.18

A method of adding two vectors together is to use horizontal and vertical components.
The horizontal component of force F_1 is $F_1\cos\theta_1$ and the horizontal component of force F_2 is $F_2\cos\theta_2$
The total horizontal component of the two forces,
$\mathbf{H = F_1\cos\theta_1 + F_2\cos\theta_2}$
The vertical component of force F_1 is $F_1\sin\theta_1$ and the vertical component of force F_2 is $F_2\sin\theta_2$
The total vertical component of the two forces,
$\mathbf{V = F_1\sin\theta_1 + F_2\sin\theta_2}$
Since we have H and V, the resultant of F_1 and F_2 is obtained by using the theorem of Pythagoras. From Figure 49.19, $\quad 0b^2 = H^2 + V^2$
i.e. $\qquad \mathbf{resultant = \sqrt{H^2 + V^2}} \qquad$ at an angle
given by $\boldsymbol{\theta = \tan^{-1}\left(\dfrac{V}{H}\right)}$

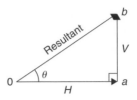

Figure 49.19

Problem 7. A force of 5 N is inclined at an angle of 45° to a second force of 8 N, both forces acting at a point. Calculate the magnitude of the resultant of these two forces and the direction of the resultant with respect to the 8 N force

The two forces are shown in Figure 49.20.

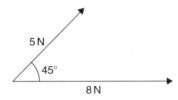

Figure 49.20

The horizontal component of the 8 N force is $8\cos 0°$ and the horizontal component of the 5 N force is $5\cos 45°$
The total horizontal component of the two forces,

$$H = 8\cos 0° + 5\cos 45° = 8 + 3.5355$$

$$= \mathbf{11.5355}$$

The vertical component of the 8 N force is $8\sin 0°$ and the vertical component of the 5 N force is $5\sin 45°$
The total vertical component of the two forces,

$$V = 8\sin 0° + 5\sin 45° = 0 + 3.5355$$

$$= \mathbf{3.5355}$$

From Figure 49.21, magnitude of resultant vector

$$= \sqrt{H^2 + V^2}$$

$$= \sqrt{11.5355^2 + 3.5355^2} = \mathbf{12.07\,N}$$

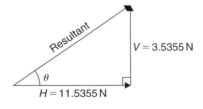

Figure 49.21

The direction of the resultant vector,

$$\theta = \tan^{-1}\left(\frac{V}{H}\right) = \tan^{-1}\left(\frac{3.5355}{11.5355}\right)$$

$$= \tan^{-1} 0.30648866\ldots = \mathbf{17.04°}$$

Thus, **the resultant of the two forces is a single vector of 12.07 N at 17.04° to the 8 N vector**.

Perhaps an easier and quicker method of calculating the magnitude and direction of the resultant is to use **complex numbers** (see Chapter 45).

In this example, the **resultant**

$$= 8\angle 0° + 5\angle 45°$$

$$= (8\cos 0° + j8\sin 0°) + (5\cos 45° + j5\sin 45°)$$

$$= (8 + j0) + (3.536 + j3.536)$$

$$= (11.536 + j3.536)\,\text{N} \quad \text{or} \quad \mathbf{12.07\angle 17.04°\,N}$$

as obtained above using horizontal and vertical components.

Problem 8. Forces of 15 N and 10 N are at an angle of 90° to each other, as shown in Figure 49.22. Calculate the magnitude of the resultant of these two forces and its direction with respect to the 15 N force

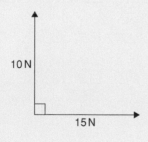

Figure 49.22

The horizontal component of the 15 N force is $15\cos 0°$ and the horizontal component of the 10 N force is $10\cos 90°$

The total horizontal component of the two forces,

$$H = 15\cos 0° + 10\cos 90° = 15 + 0 = \mathbf{15}$$

The vertical component of the 15 N force is $15\sin 0°$ and the vertical component of the 10 N force is $10\sin 90°$

The total vertical component of the two forces,

$$V = 15\sin 0° + 10\sin 90° = 0 + 10 = \mathbf{10}$$

Magnitude of resultant vector

$$= \sqrt{H^2 + V^2} = \sqrt{15^2 + 10^2} = \mathbf{18.03\,N}$$

The direction of the resultant vector,

$$\theta = \tan^{-1}\left(\frac{V}{H}\right) = \tan^{-1}\left(\frac{10}{15}\right) = \mathbf{33.69°}$$

Thus, **the resultant of the two forces is a single vector of 18.03 N at 33.69° to the 15 N vector**.

There is an alternative method of calculating the resultant vector in this case.
If we used the triangle method, then the diagram would be as shown in Figure 49.23.

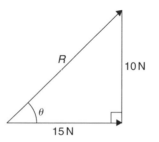

Figure 49.23

Since a right-angled triangle results, then we could use **Pythagoras' theorem** without needing to go through the procedure for horizontal and vertical components. In fact, the horizontal and vertical components are 15 N and 10 N respectively.
This is, of course, a special case. Pythagoras **can only be used when there is an angle of 90° between vectors**. This is demonstrated in the next Problem.

Problem 9. Calculate the magnitude and direction of the resultant of the two acceleration vectors shown in Figure 49.24

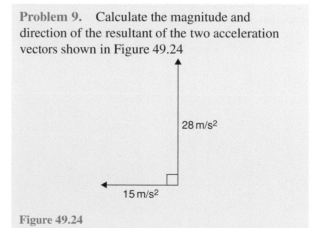

Figure 49.24

The 15 m/s² acceleration is drawn horizontally, shown as **0a** in Figure 49.25.

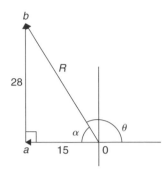

Figure 49.25

From the nose of the $15\,\text{m/s}^2$ acceleration, the $28\,\text{m/s}^2$ acceleration is drawn at an angle of $90°$ to the horizontal, shown as **ab**.
The resultant acceleration, R, is given by length **0b**.
Since a right-angled triangle results, the theorem of Pythagoras may be used.

$$0b = \sqrt{15^2 + 28^2} = \mathbf{31.76\,m/s^2}$$

and $$\alpha = \tan^{-1}\left(\frac{28}{15}\right) = \mathbf{61.82°}$$

Measuring from the horizontal,
$\boldsymbol{\theta = 180° - 61.82° = 118.18°}$
Thus, **the resultant of the two accelerations is a single vector of 31.76 m/s² at 118.18° to the horizontal**.

Problem 10. Velocities of 10 m/s, 20 m/s and 15 m/s act as shown in Figure 49.26. Calculate the magnitude of the resultant velocity and its direction relative to the horizontal.

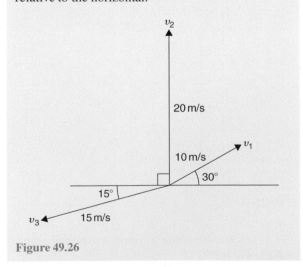

Figure 49.26

The horizontal component of the 10 m/s velocity is $10\cos 30° = 8.660\,\text{m/s}$,

the horizontal component of the 20 m/s velocity is $20\cos 90° = 0\,\text{m/s}$,

and the horizontal component of the 15 m/s velocity is $15\cos 195° = -14.489\,\text{m/s}$.

The total horizontal component of the three velocities,

$$H = 8.660 + 0 - 14.489 = \mathbf{-5.829\,m/s}$$

The vertical component of the 10 m/s velocity is $10\sin 30° = 5\,\text{m/s}$,

the vertical component of the 20 m/s velocity is $20\sin 90° = 20\,\text{m/s}$,

and the vertical component of the 15 m/s velocity is $15\sin 195° = -3.882\,\text{m/s}$.

The total vertical component of the three velocities,

$$V = 5 + 20 - 3.882 = \mathbf{21.118\,m/s}$$

From Figure 49.27, magnitude of resultant vector,

$$R = \sqrt{H^2 + V^2} = \sqrt{5.829^2 + 21.118^2} = \mathbf{21.91\,m/s}$$

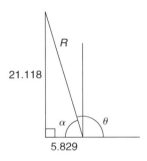

Figure 49.27

The direction of the resultant vector,

$$\alpha = \tan^{-1}\left(\frac{V}{H}\right) = \tan^{-1}\left(\frac{21.118}{5.829}\right) = \mathbf{74.57°}$$

Measuring from the horizontal,

$$\theta = 180° - 74.57° = \mathbf{105.43°}$$

Thus, **the resultant of the three velocities is a single vector of 21.91 m/s at 105.43° to the horizontal**.

Using **complex numbers**, from Figure 49.26,

$$\text{resultant} = 10\angle 30° + 20\angle 90° + 15\angle 195°$$

$$= (10\cos 30° + j10\sin 30°)$$

$$+ (20\cos 90° + j20\sin 90°)$$

$$+ (15\cos 195° + j15\sin 195°)$$

$$= (8.660 + j5.000) + (0 + j20.000)$$

$$+ (-14.489 - j3.882)$$

$$= (-5.829 + j21.118)\,\text{N} \quad \text{or}$$

$$\mathbf{21.91 \angle 105.43° \, N}$$

as obtained above using horizontal and vertical components.

The method used to add vectors by calculation will not be specified – the choice is yours, but probably the quickest and easiest method is by using complex numbers.

Now try the following Practice Exercise

Practice Exercise 211   Further problems on addition of vectors by calculation (answers on page 1175)

1. A force of 7 N is inclined at an angle of 50° to a second force of 12 N, both forces acting at a point. Calculate the magnitude of the resultant of the two forces, and the direction of the resultant with respect to the 12 N force.

2. Velocities of 5 m/s and 12 m/s act at a point at 90° to each other. Calculate the resultant velocity and its direction relative to the 12 m/s velocity.

3. Calculate the magnitude and direction of the resultant of the two force vectors shown in Figure 49.28.

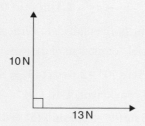

Figure 49.28

4. Calculate the magnitude and direction of the resultant of the two force vectors shown in Figure 49.29.

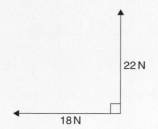

Figure 49.29

5. A displacement vector s_1 is 30 m at 0°. A second displacement vector s_2 is 12 m at 90°. Calculate magnitude and direction of the resultant vector $s_1 + s_2$

6. Three forces of 5 N, 8 N and 13 N act as shown in Figure 49.30. Calculate the magnitude and direction of the resultant force.

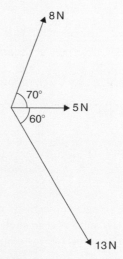

Figure 49.30

7. If velocity $v_1 = 25$ m/s at 60° and $v_2 = 15$ m/s at $-30°$, calculate the magnitude and direction of $v_1 + v_2$

8. Calculate the magnitude and direction of the resultant vector of the force system shown in Figure 49.31.

Section H

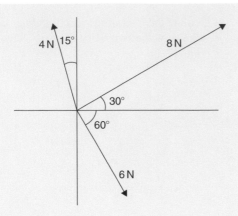

Figure 49.31

9. Calculate the magnitude and direction of the resultant vector of the system shown in Figure 49.32.

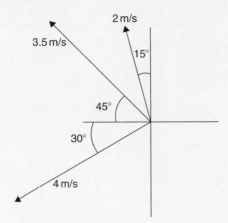

Figure 49.32

10. An object is acted upon by two forces of magnitude 10 N and 8 N at an angle of $60°$ to each other. Determine the resultant force on the object.

11. A ship heads in a direction of E 20° S at a speed of 20 knots while the current is 4 knots in a direction of N 30° E. Determine the speed and actual direction of the ship.

49.7 Vector subtraction

In Figure 49.33, a force vector F is represented by oa. The vector $(-oa)$ can be obtained by drawing a vector

from o in the opposite sense to oa but having the same magnitude, shown as ob in Figure 49.33, i.e. $ob = (-oa)$

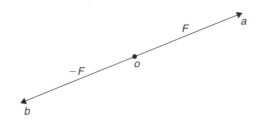

Figure 49.33

For two vectors acting at a point, as shown in Figure 49.34(a), the resultant of vector addition is: $os = oa + ob$
Figure 49.34(b) shows vectors $ob + (-oa)$, that is, $ob - oa$ and the vector equation is $ob - oa = od$. Comparing od in Figure 49.34(b) with the broken line ab in Figure 49.34(a) shows that the second diagonal of the 'parallelogram' method of vector addition gives the magnitude and direction of vector subtraction of oa from ob

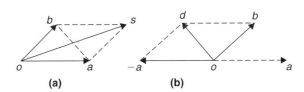

(a) **(b)**

Figure 49.34

Problem 11. Accelerations of $a_1 = 1.5 \, \text{m/s}^2$ at $90°$ and $a_2 = 2.6 \, \text{m/s}^2$ at $145°$ act at a point. Find $a_1 + a_2$ and $a_1 - a_2$ (i) by drawing a scale vector diagram, and (ii) by calculation

(i) The scale vector diagram is shown in Figure 49.35. By measurement,

$$a_1 + a_2 = 3.7 \, \text{m/s}^2 \text{ at } 126°$$

$$a_1 - a_2 = 2.1 \, \text{m/s}^2 \text{ at } 0°$$

(ii) Resolving horizontally and vertically gives:
Horizontal component of $a_1 + a_2$,

$$H = 1.5 \cos 90° + 2.6 \cos 145° = -2.13$$

Vertical component of $a_1 + a_2$,

$$V = 1.5 \sin 90° + 2.6 \sin 145° = 2.99$$

From Figure 49.36, magnitude of $a_1 + a_2$,

$$R = \sqrt{(-2.13)^2 + 2.99^2} = 3.67 \, \text{m/s}^2$$

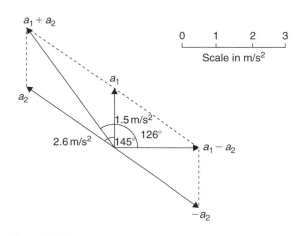

Figure 49.35

In Figure 49.36, $\alpha = \tan^{-1}\left(\dfrac{2.99}{2.13}\right) = 54.53°$ and
$$\theta = 180° - 54.53° = \mathbf{125.47°}$$

Thus, $\quad a_1 + a_2 = \mathbf{3.67\,m/s^2\ at\ 125.47°}$

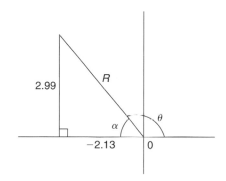

Figure 49.36

Horizontal component of $a_1 - a_2$
$$= 1.5\cos 90° - 2.6\cos 145° = \mathbf{2.13}$$

Vertical component of $a_1 - a_2$
$$= 1.5\sin 90° - 2.6\sin 145° = 0$$
Magnitude of $a_1 - a_2 = \sqrt{2.13^2 + 0^2}$
$$= \mathbf{2.13\,m/s^2}$$

Direction of $a_1 - a_2 = \tan^{-1}\left(\dfrac{0}{2.13}\right) = \mathbf{0°}$

Thus, $\quad a_1 - a_2 = \mathbf{2.13\,m/s^2\ at\ 0°}$

Problem 12. Calculate the resultant of
(i) $v_1 - v_2 + v_3$ and (ii) $v_2 - v_1 - v_3$ when
$v_1 = 22$ units at $140°$, $v_2 = 40$ units at $190°$ and
$v_3 = 15$ units at $290°$

(i) The vectors are shown in Figure 49.37.

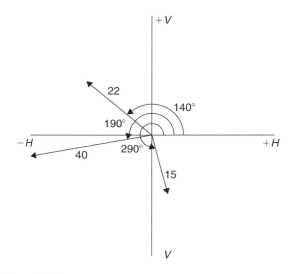

Figure 49.37

The horizontal component of
$$v_1 - v_2 + v_3 = (22\cos 140°) - (40\cos 190°)$$
$$+ (15\cos 290°)$$
$$= (-16.85) - (-39.39) + (5.13)$$
$$= \mathbf{27.67\,units}$$

The vertical component of
$$v_1 - v_2 + v_3 = (22\sin 140°) - (40\sin 190°)$$
$$+ (15\sin 290°)$$
$$= (14.14) - (-6.95) + (-14.10)$$
$$= \mathbf{6.99\,units}$$

The magnitude of the resultant,
$$\mathbf{R = \sqrt{27.67^2 + 6.99^2} = 28.54\,units}$$

The direction of the resultant $R = \tan^{-1}\left(\dfrac{6.99}{27.67}\right)$
$$= 14.18°$$

Thus, $v_1 - v_2 + v_3 = \mathbf{28.54\ units\ at\ 14.18°}$

Using **complex numbers**,
$$v_1 - v_2 + v_3 = 22\angle140° - 40\angle190° + 15\angle290°$$
$$= (-16.853 + j14.141)$$
$$- (-39.392 - j6.946)$$
$$+ (5.130 - j14.095)$$
$$= 27.669 + j6.992 = \mathbf{28.54\angle14.18°}$$

(ii) The horizontal component of

$$v_2 - v_1 - v_3 = (40\cos 190°) - (22\cos 140°)$$
$$- (15\cos 290°)$$
$$= (-39.39) - (-16.85) - (5.13)$$
$$= \mathbf{-27.67 \ units}$$

The vertical component of

$$v_2 - v_1 - v_3 = (40\sin 190°) - (22\sin 140°)$$
$$- (15\sin 290°)$$
$$= (-6.95) - (14.14) - (-14.10)$$
$$= \mathbf{-6.99 \ units}$$

From Figure 49.38 the magnitude of the resultant,

$$R = \sqrt{(-27.67)^2 + (-6.99)^2} = \mathbf{28.54 \ units}$$

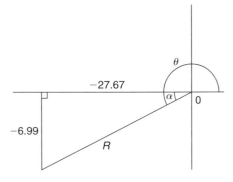

Figure 49.38

and $\alpha = \tan^{-1}\left(\dfrac{6.99}{27.67}\right) = 14.18°$, from which,

$$\theta = 180° + 14.18° = 194.18°$$

Thus, $v_2 - v_1 - v_3 = \mathbf{28.54 \ units \ at \ 194.18°}$

This result is as expected, since $v_2 - v_1 - v_3 = -(v_1 - v_2 + v_3)$ and the vector 28.54 units at 194.18° is minus times (i.e. is 180° out of phase with) the vector 28.54 units at 14.18°

Using **complex numbers**,

$$v_2 - v_1 - v_3 = 40\angle 190° - 22\angle 140° - 15\angle 290°$$
$$= (-39.392 - j6.946)$$
$$- (-16.853 + j14.141)$$
$$- (5.130 - j14.095)$$
$$= -27.669 - j6.992$$
$$= \mathbf{28.54\angle -165.82°} \ \text{or}$$
$$\mathbf{28.54\angle 194.18°}$$

Now try the following Practice Exercise

Practice Exercise 212   Further problems on vector subtraction (answers on page 1175)

1. Forces of $F_1 = 40$N at 45° and $F_2 = 30$N at 125° act at a point. Determine by drawing and by calculation: (a) $F_1 + F_2$ (b) $F_1 - F_2$

2. Calculate the resultant of (a) $v_1 + v_2 - v_3$ and (b) $v_3 - v_2 + v_1$ when $v_1 = 15$m/s at 85°, $v_2 = 25$m/s at 175° and $v_3 = 12$m/s at 235°

49.8   Relative velocity

For relative velocity problems, some fixed datum point needs to be selected. This is often a fixed point on the earth's surface. In any vector equation, only the start and finish points affect the resultant vector of a system. Two different systems are shown in Figure 49.39, but in each of the systems the resultant vector is ad.

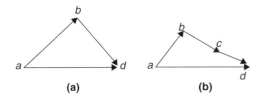

Figure 49.39

The vector equation of the system shown in Figure 49.39(a) is:

$$ad = ab + bd$$

and that for the system shown in Figure 49.39(b) is:

$$ad = ab + bc + cd$$

Thus in vector equations of this form, only the first and last letters, 'a' and 'd', respectively, fix the magnitude and direction of the resultant vector. This principle is used in relative velocity problems.

Problem 13. Two cars, P and Q, are travelling towards the junction of two roads which are at right angles to one another. Car P has a velocity of 45 km/h due east and car Q a velocity of 55 km/h due south. Calculate (i) the velocity of car P relative to car Q, and (ii) the velocity of car Q relative to car P

(i) The directions of the cars are shown in Figure 49.40(a), called a **space diagram**. The velocity diagram is shown in Figure 49.40(b), in which **pe** is taken as the velocity of car P relative to point e on the earth's surface. The velocity of P relative to Q is vector **pq** and the vector equation is **pq** = **pe** + **eq**. Hence the vector directions are as shown, **eq** being in the opposite direction to **qe**.

From the geometry of the vector triangle, the magnitude of **pq** = $\sqrt{45^2 + 55^2}$ = 71.06 km/h and the direction of **pq** = $\tan^{-1}\left(\dfrac{55}{45}\right)$ = 50.71°

i.e. **the velocity of car P relative to car Q is 71.06 km/h at 50.71°**

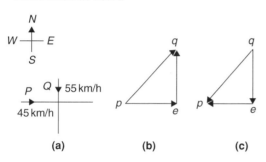

(a)   **(b)**   **(c)**

Figure 49.40

(ii) The velocity of car Q relative to car P is given by the vector equation **qp** = **qe** + **ep** and the vector diagram is as shown in Figure 49.40(c), having **ep** opposite in direction to **pe**.

From the geometry of this vector triangle, the magnitude of **qp** = $\sqrt{45^2 + 55^2}$ = 71.06 km/h and the direction of **qp** = $\tan^{-1}\left(\dfrac{55}{45}\right)$ = 50.71° but must lie in the third quadrant, i.e. the required angle is: 180° + 50.71° = 230.71°

i.e. **the velocity of car Q relative to car P is 71.06 km/h at 230.71°**

Now try the following Practice Exercise

Practice Exercise 213   Further problems on relative velocity (answers on page 1175)

1. A car is moving along a straight horizontal road at 79.2 km/h and rain is falling vertically downwards at 26.4 km/h. Find the velocity of the rain relative to the driver of the car.

2. Calculate the time needed to swim across a river 142 m wide when the swimmer can swim at 2 km/h in still water and the river is flowing at 1 km/h. At what angle to the bank should the swimmer swim?

3. A ship is heading in a direction N 60° E at a speed which in still water would be 20 km/h. It is carried off course by a current of 8 km/h in a direction of E 50° S. Calculate the ship's actual speed and direction.

49.9 i, j and k notation

A method of completely specifying the direction of a vector in space relative to some reference point is to use three unit vectors, i, j and k, mutually at right angles to each other, as shown in Figure 49.41.

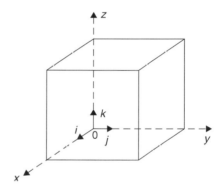

Figure 49.41

Calculations involving vectors given in i, j, k notation are carried out in exactly the same way as standard algebraic calculations, as shown in the problem below.

Problem 14.   Determine:
$(3i + 2j + 2k) - (4i - 3j + 2k)$

$(3i + 2j + 2k) - (4i - 3j + 2k) = 3i + 2j + 2k$
$$- 4i + 3j - 2k$$
$$= -i + 5j$$

Problem 15.   Given $p = 3i + 2k$, $q = 4i - 2j + 3k$ and $r = -3i + 5j - 4k$ determine:

(a) $-r$ (b) $3p$ (c) $2p + 3q$ (d) $-p + 2r$
(e) $0.2p + 0.6q - 3.2r$

(a) $-r = -(-3i + 5j - 4k) = \mathbf{+3i - 5j + 4k}$

(b) $3p = 3(3i + 2k) = \mathbf{9i + 6k}$

(c) $2p + 3q = 2(3i + 2k) + 3(4i - 2j + 3k)$

$= 6i + 4k + 12i - 6j + 9k$

$= \mathbf{18i - 6j + 13k}$

(d) $-p + 2r = -(3i + 2k) + 2(-3i + 5j - 4k)$

$= -3i - 2k + (-6i + 10j - 8k)$

$= -3i - 2k - 6i + 10j - 8k$

$= \mathbf{-9i + 10j - 10k}$

(e) $0.2p + 0.6q - 3.2r = 0.2(3i + 2k)$

$+ 0.6(4i - 2j + 3k) - 3.2(-3i + 5j - 4k)$

$= 0.6i + 0.4k + 2.4i - 1.2j + 1.8k$

$+ 9.6i - 16j + 12.8k$

$= \mathbf{12.6i - 17.2j + 15k}$

Now try the following Practice Exercise

Practice Exercise 214   Further problems on i, j, k notation (answers on page 1175)

Given that $p = 2i + 0.5j - 3k$, $q = -i + j + 4k$ and $r = 6j - 5k$, evaluate and simplify the following vectors in i, j, k form:

1. $-q$
2. $2p$
3. $q + r$
4. $-q + 2p$
5. $3q + 4r$
6. $q - 2p$
7. $p + q + r$
8. $p + 2q + 3r$
9. $2p + 0.4q + 0.5r$
10. $7r - 2q$

For fully worked solutions to each of the problems in Practice Exercises 210 to 214 in this chapter, go to the website:
www.routledge.com/cw/bird

Chapter 50

Methods of adding alternating waveforms

Why it is important to understand: Methods of adding alternating waveforms

In electrical engineering, a phasor is a rotating vector representing a quantity such as an alternating current or voltage that varies sinusoidally. Sometimes it is necessary when studying sinusoidal quantities to add together two alternating waveforms, for example in an a.c. series circuit that are not in-phase with each other. Electrical engineers, electronics engineers, electronic engineering technicians and aircraft engineers all use phasor diagrams to visualise complex constants and variables. So, given oscillations to add and subtract, the required rotating vectors are constructed, called a phasor diagram, and graphically the resulting sum and/or difference oscillations are added or calculated. Phasors may be used to analyse the behaviour of electrical and mechanical systems that have reached a kind of equilibrium called sinusoidal steady state. Hence, discovering different methods of combining sinusoidal waveforms is of some importance in certain areas of engineering.

At the end of this chapter, you should be able to:

- determine the resultant of two phasors by graph plotting
- determine the resultant of two or more phasors by drawing
- determine the resultant of two phasors by the sine and cosine rules
- determine the resultant of two or more phasors by horizontal and vertical components
- determine the resultant of two or more phasors by complex numbers

50.1 Combination of two periodic functions

There are a number of instances in engineering and science where waveforms have to be combined and where it is required to determine the single phasor (called the resultant) that could replace two or more separate phasors. Uses are found in electrical alternating current theory, in mechanical vibrations, in the addition of forces and with sound waves.

There are a number of methods of determining the resultant waveform. These include:

(a) by drawing the waveforms and adding graphically
(b) by drawing the phasors and measuring the resultant
(c) by using the cosine and sine rules
(d) by using horizontal and vertical components
(e) by using complex numbers.

50.2 Plotting periodic functions

This may be achieved by sketching the separate functions on the same axes and then adding (or subtracting) ordinates at regular intervals. This is demonstrated in the following Problems.

Problem 1. Plot the graph of $y_1 = 3\sin A$ from $A = 0°$ to $A = 360°$. On the same axes plot $y_2 = 2\cos A$. By adding ordinates, plot $y_R = 3\sin A + 2\cos A$ and obtain a sinusoidal expression for this resultant waveform

$y_1 = 3\sin A$ and $y_2 = 2\cos A$ are shown plotted in Figure 50.1. Ordinates may be added at, say, 15° intervals. For example,

at 0°, $y_1 + y_2 = 0 + 2 = 2$
at 15°, $y_1 + y_2 = 0.78 + 1.93 = 2.71$
at 120°, $y_1 + y_2 = 2.60 + -1 = 1.6$
at 210°, $y_1 + y_2 = -1.50 - 1.73 = -3.23$, and so on.

The resultant waveform, shown by the broken line, has the same period, i.e. 360°, and thus the same frequency as the single phasors. The maximum value, or amplitude, of the resultant is 3.6. The resultant waveform **leads** $y_1 = 3\sin A$ by 34° or $34 \times \dfrac{\pi}{180}$ rad $= 0.593$ rad.

The sinusoidal expression for the resultant waveform is:

$$y_R = 3.6\sin(A + 34°) \quad \text{or}$$

$$y_R = 3.6\sin(A + 0.593)$$

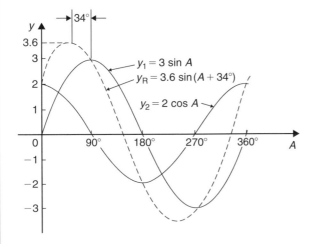

Figure 50.1

Problem 2. Plot the graphs of $y_1 = 4\sin\omega t$ and $y_2 = 3\sin(\omega t - \pi/3)$ on the same axes, over one cycle. By adding ordinates at intervals, plot $y_R = y_1 + y_2$ and obtain a sinusoidal expression for the resultant waveform

$y_1 = 4\sin\omega t$ and $y_2 = 3\sin(\omega t - \pi/3)$ are shown plotted in Figure 50.2.

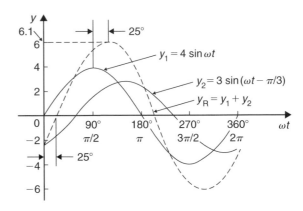

Figure 50.2

Ordinates are added at 15° intervals and the resultant is shown by the broken line. The amplitude of the resultant is 6.1 and it **lags** y_1 by 25° or 0.436 rad.

Hence, the sinusoidal expression for the resultant waveform is:

$$y_R = 6.1\sin(\omega t - 0.436)$$

Problem 3. Determine a sinusoidal expression for $y_1 - y_2$ when $y_1 = 4\sin\omega t$ and $y_2 = 3\sin(\omega t - \pi/3)$

y_1 and y_2 are shown plotted in Figure 50.3. At 15° intervals y_2 is subtracted from y_1. For example:

at 0°, $y_1 - y_2 = 0 - (-2.6) = +2.6$

at 30°, $y_1 - y_2 = 2 - (-1.5) = +3.5$

at 150°, $y_1 - y_2 = 2 - 3 = -1$, and so on.

The amplitude, or peak value of the resultant (shown by the broken line) is 3.6 and it leads y_1 by 45° or 0.79 rad. Hence,

$$y_1 - y_2 = 3.6\sin(\omega t + 0.79)$$

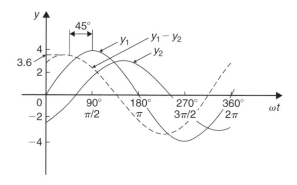

Figure 50.3

> **Problem 4.** Two alternating currents are given by:
> $i_1 = 20 \sin \omega t$ amperes and
> $i_2 = 10 \sin \left(\omega t + \dfrac{\pi}{3} \right)$ amperes.
> By drawing the waveforms on the same axes and adding, determine the sinusoidal expression for the resultant $i_1 + i_2$

i_1 and i_2 are shown plotted in Figure 50.4. The resultant waveform for $i_1 + i_2$ is shown by the broken line. It has the same period, and hence frequency, as i_1 and i_2

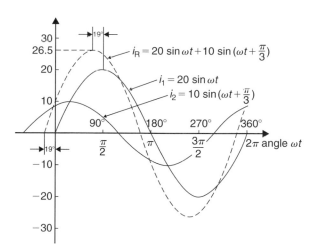

Figure 50.4

The amplitude or peak value is 26.5 A.
The resultant waveform leads the waveform of $i_1 = 20 \sin \omega t$ by 19° or 0.33 rad.
Hence, the sinusoidal expression for the resultant $i_1 + i_2$ is given by:

$$i_R = i_1 + i_2 = 26.5 \sin (\omega t + 0.33) \text{ A}$$

Now try the following Practice Exercise

> **Practice Exercise 215 Further problems on plotting periodic functions (answers on page 1175)**
>
> 1. Plot the graph of $y = 2 \sin A$ from $A = 0°$ to $A = 360°$. On the same axes plot $y = 4 \cos A$. By adding ordinates at intervals plot $y = 2 \sin A + 4 \cos A$ and obtain a sinusoidal expression for the waveform.
>
> 2. Two alternating voltages are given by $v_1 = 10 \sin \omega t$ volts and $v_2 = 14 \sin(\omega t + \pi/3)$ volts. By plotting v_1 and v_2 on the same axes over one cycle obtain a sinusoidal expression for (a) $v_1 + v_2$ (b) $v_1 - v_2$
>
> 3. Express $12 \sin \omega t + 5 \cos \omega t$ in the form $A \sin(\omega t \pm \alpha)$ by drawing and measurement.

50.3 Determining resultant phasors by drawing

The resultant of two periodic functions may be found from their relative positions when the time is zero. For example, if $y_1 = 4 \sin \omega t$ and $y_2 = 3 \sin(\omega t - \pi/3)$, then each may be represented as phasors as shown in Figure 50.5, y_1 being 4 units long and drawn horizontally and y_2 being 3 units long, lagging y_1 by $\pi/3$ radians or 60°. To determine the resultant of $y_1 + y_2$, y_1 is drawn horizontally as shown in Figure 50.6 and y_2 is joined to the end of y_1 at 60° to the horizontal. The resultant is given by y_R. This is the same as the diagonal of a parallelogram that is shown completed in Figure 50.7. (See page 579 for more on the nose-to-tail and the parallelogram methods of adding phasors.)
Resultant y_R, in Figures 50.6 and 50.7, may be determined by drawing the phasors and their directions to scale and measuring using a ruler and protractor.

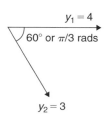

Figure 50.5

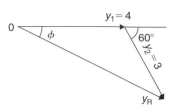

Figure 50.6

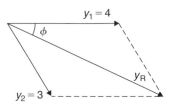

Figure 50.7

In this example, y_R is measured as 6 units long and angle ϕ is measured as $25°$.

$$25° = 25 \times \frac{\pi}{180} \text{ radians} = 0.44 \text{ rad}$$

Hence, summarising, by drawing:

$$\boldsymbol{y_R = y_1 + y_2} = 4\sin\omega t + 3\sin(\omega t - \pi/3)$$
$$= \boldsymbol{6\sin(\omega t - 0.44)}$$

If the resultant phasor $y_R = y_1 - y_2$ is required, then y_2 is still 3 units long but is drawn in the opposite direction, as shown in Figure 50.8. By measurement, $y_1 - y_2 = 3.6\sin(\omega t + 0.80)$

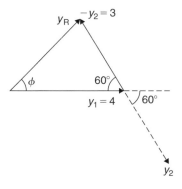

Figure 50.8

> **Problem 5.** Two alternating currents are given by: $i_1 = 20\sin\omega t$ amperes and $i_2 = 10\sin\left(\omega t + \dfrac{\pi}{3}\right)$ amperes. Determine $i_1 + i_2$ by drawing phasors

The relative positions of i_1 and i_2 at time $t = 0$ are shown as phasors in Figure 50.9, where $\dfrac{\pi}{3}\text{rad} = 60°$

The phasor diagram in Figure 50.10 is drawn to scale with a ruler and protractor.

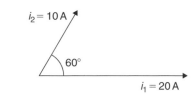

Figure 50.9

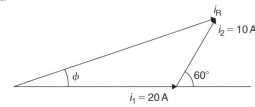

Figure 50.10

The resultant i_R is shown and is measured as 26 A and angle ϕ as $19°$ or 0.33 rad leading i_1. Hence, by drawing and measuring:

$$i_R = i_1 + i_2 = 26\sin(\omega t + 0.33)\text{A}$$

> **Problem 6.** For the currents in Problem 5, determine $i_1 - i_2$ by drawing phasors

At time $t = 0$, current i_1 is drawn 20 units long horizontally as shown by $0a$ in Figure 50.11. Current i_2 is shown, drawn 10 units long and leading by $60°$. The current $-i_2$ is drawn in the opposite direction to the line of i_2, shown as ab in Figure 50.11. The resultant i_R is given by $0b$ lagging by angle ϕ.

By measurement, $i_R = 17\text{A}$ and $\phi = 30°$ or 0.52 rad

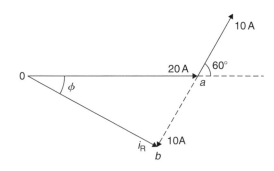

Figure 50.11

Hence, by drawing phasors:

$$i_R = i_1 - i_2 = 17\sin(\omega t - 0.52)\text{A}$$

Now try the following Practice Exercise

Practice Exercise 216   Further problems on determining resultant phasors by drawing (answers on page 1175)

1. Determine a sinusoidal expression for $2\sin\theta + 4\cos\theta$ by drawing phasors.

2. If $v_1 = 10\sin\omega t$ volts and $v_2 = 14\sin(\omega t + \pi/3)$ volts, determine by drawing phasors sinusoidal expressions for (a) $v_1 + v_2$ (b) $v_1 - v_2$

3. Express $12\sin\omega t + 5\cos\omega t$ in the form $R\sin(\omega t \pm \alpha)$ by drawing phasors.

50.4   Determining resultant phasors by the sine and cosine rules

As stated earlier, the resultant of two periodic functions may be found from their relative positions when the time is zero. For example, if $y_1 = 5\sin\omega t$ and $y_2 = 4\sin(\omega t - \pi/6)$, then each may be represented by phasors as shown in Figure 50.12, y_1 being 5 units long and drawn horizontally and y_2 being 4 units long, lagging y_1 by $\pi/6$ radians or 30°. To determine the resultant of $y_1 + y_2$, y_1 is drawn horizontally as shown in Figure 50.13 and y_2 is joined to the end of y_1 at $\pi/6$ radians, i.e. 30° to the horizontal. The resultant is given by y_R

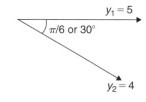

Figure 50.12

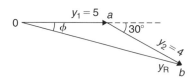

Figure 50.13

From Chapter 41, using the cosine rule on triangle $0ab$ of Figure 50.13 gives:

$$y_R^2 = 5^2 + 4^2 - [2(5)(4)\cos 150°]$$
$$= 25 + 16 - (-34.641)$$
$$= 75.641$$

from which, $y_R = \sqrt{75.641} = 8.697$

Using the sine rule, $\dfrac{8.697}{\sin 150°} = \dfrac{4}{\sin\phi}$

from which, $\sin\phi = \dfrac{4\sin 150°}{8.697}$
$$= 0.22996$$

and $\phi = \sin^{-1} 0.22996$
$$= 13.29° \text{ or } 0.232 \text{ rad}$$

Hence, $y_R = y_1 + y_2 = 5\sin\omega t + 4\sin(\omega t - \pi/6)$
$$= 8.697\sin(\omega t - 0.232)$$

Problem 7.   Given $y_1 = 2\sin\omega t$ and $y_2 = 3\sin(\omega t + \pi/4)$, obtain an expression, by calculation, for the resultant, $y_R = y_1 + y_2$

When time $t = 0$, the position of phasors y_1 and y_2 are as shown in Figure 50.14(a). To obtain the resultant, y_1 is drawn horizontally, 2 units long, and y_2 is drawn 3 units long at an angle of $\pi/4$ rads or 45° and joined to the end of y_1 as shown in Figure 50.14(b).

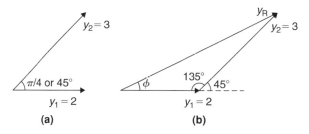

(a)   (b)

Figure 50.14

From Figure 50.14(b), and using the cosine rule:

$$y_R^2 = 2^2 + 3^2 - [2(2)(3)\cos 135°]$$
$$= 4 + 9 - [-8.485] = 21.49$$

Hence, $y_R = \sqrt{21.49} = 4.6357$

Using the sine rule: $\dfrac{3}{\sin\phi} = \dfrac{4.6357}{\sin 135°}$

from which, $\sin\phi = \dfrac{3\sin 135°}{4.6357} = 0.45761$

Hence, $$\phi = \sin^{-1} 0.45761$$

$$= 27.23° \text{ or } 0.475 \text{ rad}$$

Thus, by calculation, $\quad y_R = \mathbf{4.635\sin(\omega t + 0.475)}$

Problem 8. Determine

$20\sin\omega t + 10\sin\left(\omega t + \dfrac{\pi}{3}\right) A$ using the cosine and sine rules

---

From the phasor diagram of Figure 50.15, and using the cosine rule:

$$i_R^2 = 20^2 + 10^2 - [2(20)(10)\cos 120°]$$

$$= 700$$

Hence, $i_R = \sqrt{700} = \mathbf{26.46 \ A}$

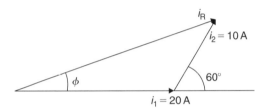

Figure 50.15

Using the sine rule gives : $\dfrac{10}{\sin\phi} = \dfrac{26.46}{\sin 120°}$

from which, $\quad \sin\phi = \dfrac{10\sin 120°}{26.46}$

$$= 0.327296$$

and $\quad \phi = \sin^{-1} 0.327296 = \mathbf{19.10°}$

$$= 19.10 \times \dfrac{\pi}{180} = \mathbf{0.333 \ rad}$$

Hence, by cosine and sine rules,

$$i_R = i_1 + i_2 = \mathbf{26.46\sin(\omega t + 0.333) \ A}$$

Now try the following Practice Exercise

Practice Exercise 217 Resultant phasors by the sine and cosine rules (answers on page 1175)

1. Determine, using the cosine and sine rules, a sinusoidal expression for:
   $y = 2\sin A + 4\cos A$

2. Given $v_1 = 10\sin\omega t$ volts and $v_2 = 14\sin(\omega t + \pi/3)$ volts, use the cosine and sine rules to determine sinusoidal expressions for (a) $v_1 + v_2$ (b) $v_1 - v_2$

In Problems 3 to 5, express the given expressions in the form $A\sin(\omega t \pm \alpha)$ by using the cosine and sine rules.

3. $12\sin\omega t + 5\cos\omega t$

4. $7\sin\omega t + 5\sin\left(\omega t + \dfrac{\pi}{4}\right)$

5. $6\sin\omega t + 3\sin\left(\omega t - \dfrac{\pi}{6}\right)$

6. The sinusoidal currents in two parallel branches of an electrical network are $400\sin\omega t$ and $750\sin(\omega t - \pi/3)$, both measured in milliamperes. Determine the total current flowing into the parallel arrangement. Give the answer in sinusoidal form and in amperes.

50.5 Determining resultant phasors by horizontal and vertical components

If a right-angled triangle is constructed as shown in Figure 50.16, then $0a$ is called the horizontal component of F and ab is called the vertical component of F

From trigonometry (see Chapter 38),

$$\cos\theta = \dfrac{0a}{0b} \quad \text{from which,}$$

$$0a = 0b\cos\theta = F\cos\theta$$

i.e. **the horizontal component of F, $H = F\cos\theta$**

and $\sin\theta = \dfrac{ab}{0b}$ from which $ab = 0b\sin\theta$

$$= F\sin\theta$$

i.e. **the vertical component of F, $V = F\sin\theta$**

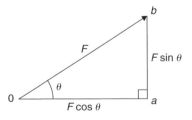

Figure 50.16

Determining resultant phasors by horizontal and vertical components is demonstrated in the following Problems.

> **Problem 9.** Two alternating voltages are given by $v_1 = 15\sin\omega t$ volts and $v_2 = 25\sin(\omega t - \pi/6)$ volts. Determine a sinusoidal expression for the resultant $v_R = v_1 + v_2$ by finding horizontal and vertical components

The relative positions of v_1 and v_2 at time $t=0$ are shown in Figure 50.17(a) and the phasor diagram is shown in Figure 50.17(b).

The horizontal component of v_R,
$$H = 15\cos 0° + 25\cos(-30°) = 0a + ab = \textbf{36.65 V}$$

The vertical component of v_R,
$$V = 15\sin 0° + 25\sin(-30°) = bc = \textbf{-12.50 V}$$

Hence, $v_R = 0c = \sqrt{36.65^2 + (-12.50)^2}$

by Pythagoras' theorem

$$= \textbf{38.72 volts}$$

$$\tan\phi = \frac{V}{H} = \frac{-12.50}{36.65} = -0.3411$$

from which, $\phi = \tan^{-1}(-0.3411) = -18.83°$

or -0.329 radians

Hence, $v_R = v_1 + v_2 = \textbf{38.72}\sin(\omega t - \textbf{0.329})\textbf{V}$

> **Problem 10.** For the voltages in Problem 9, determine the resultant $v_R = v_1 - v_2$ using horizontal and vertical components

The horizontal component of v_R,
$$H = 15\cos 0° - 25\cos(-30°) = \textbf{-6.65V}$$

The vertical component of v_R,
$$V = 15\sin 0° - 25\sin(-30°) = \textbf{12.50V}$$

Hence, $v_R = \sqrt{(-6.65)^2 + (12.50)^2}$

by Pythagoras' theorem

$$= \textbf{14.16 volts}$$

$$\tan\phi = \frac{V}{H} = \frac{12.50}{-6.65} = -1.8797$$

from which, $\phi = \tan^{-1}(-1.8797) = 118.01°$

or 2.06 radians

(See Chapter 39, page 460)

Hence,
$$v_R = v_1 - v_2 = \textbf{14.16}\sin(\omega t + \textbf{2.06})\textbf{V}$$

The phasor diagram is shown in Figure 50.18.

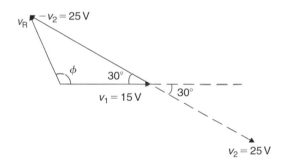

Figure 50.18

> **Problem 11.** Determine
> $$20\sin\omega t + 10\sin\left(\omega t + \frac{\pi}{3}\right) \text{A using horizontal and}$$
> vertical components

From the phasors shown in Figure 50.19:
Total horizontal component,
$$H = 20\cos 0° + 10\cos 60° = \textbf{25.0}$$
Total vertical component,
$$V = 20\sin 0° + 10\sin 60° = \textbf{8.66}$$
By Pythagoras, the resultant, $i_R = \sqrt{[25.0^2 + 8.66^2]}$
$$= \textbf{26.46 A}$$

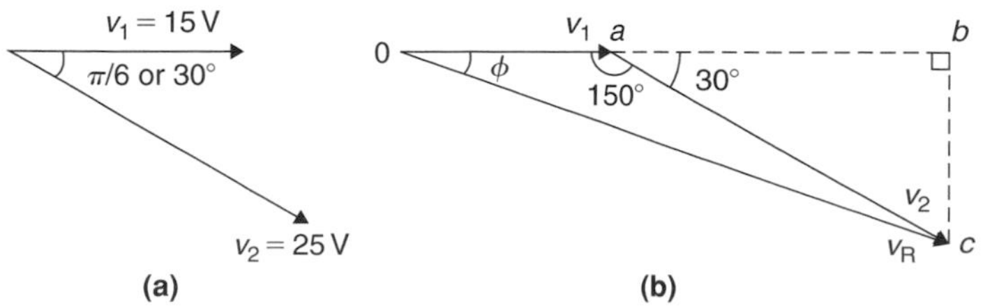

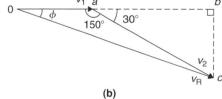

(a)   (b)

Figure 50.17

Section H

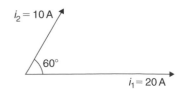

Figure 50.19

Phase angle, $\phi = \tan^{-1}\left(\dfrac{8.66}{25.0}\right) = \mathbf{19.11°}$

or **0.333 rad**

Hence, by using horizontal and vertical components,

$$20\sin\omega t + 10\sin\left(\omega t + \frac{\pi}{3}\right) = \mathbf{26.46\sin(\omega t + 0.333)}$$

Now try the following Practice Exercise

Practice Exercise 218   Further problems on resultant phasors by horizontal and vertical components (answers on page 1175)

In Problems 1 to 4, express the combination of periodic functions in the form $A\sin(\omega t \pm \alpha)$ by horizontal and vertical components:

1. $7\sin\omega t + 5\sin\left(\omega t + \dfrac{\pi}{4}\right)$

2. $6\sin\omega t + 3\sin\left(\omega t - \dfrac{\pi}{6}\right)$

3. $i = 25\sin\omega t - 15\sin\left(\omega t + \dfrac{\pi}{3}\right)$

4. $x = 9\sin\left(\omega t + \dfrac{\pi}{3}\right) - 7\sin\left(\omega t - \dfrac{3\pi}{8}\right)$

5. The voltage drops across two components when connected in series across an a.c. supply are: $v_1 = 200\sin 314.2t$ and $v_2 = 120\sin(314.2t - \pi/5)$ volts, respectively. Determine the:

    (a) voltage of the supply (given by $v_1 + v_2$) in the form $A\sin(\omega t \pm \alpha)$

    (b) frequency of the supply

6. If the supply to a circuit is $v = 20\sin 628.3t$ volts and the voltage drop across one of the components is $v_1 = 15\sin(628.3t - 0.52)$ volts, calculate the:

    (a) voltage drop across the remainder of the circuit, given by $v - v_1$, in the form $A\sin(\omega t \pm \alpha)$

    (b) supply frequency

    (c) periodic time of the supply

7. The voltages across three components in a series circuit when connected across an a.c. supply are:

    $$v_1 = 25\sin\left(300\pi t + \frac{\pi}{6}\right)\text{ volts,}$$

    $$v_2 = 40\sin\left(300\pi t - \frac{\pi}{4}\right)\text{ volts, and}$$

    $$v_3 = 50\sin\left(300\pi t + \frac{\pi}{3}\right)\text{ volts.}$$

    Calculate the:

    (a) supply voltage, in sinusoidal form, in the form $A\sin(\omega t \pm \alpha)$

    (b) frequency of the supply

    (c) periodic time

8. In an electrical circuit, two components are connected in series. The voltage across the first component is given by $80\sin(\omega t + \pi/3)$ volts, and the voltage across the second component is given by $150\sin(\omega t - \pi/4)$ volts. Determine the total supply voltage to the two components. Give the answer in sinusoidal form.

50.6   Determining resultant phasors by complex numbers

As stated earlier, the resultant of two periodic functions may be found from their relative positions when the time is zero. For example, if $y_1 = 5\sin\omega t$ and $y_2 = 4\sin(\omega t - \pi/6)$, then each may be represented by phasors as shown in Figure 50.20, y_1 being 5 units long and drawn horizontally and y_2 being 4 units long, lagging y_1 by $\pi/6$ radians or 30°. To determine the resultant of $y_1 + y_2$, y_1 is drawn horizontally, as shown in Figure 50.21, and y_2 is joined to the end of y_1 at $\pi/6$ radians, i.e. 30° to the horizontal. The resultant is given by y_R

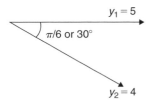

Figure 50.20

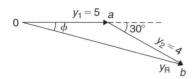

Figure 50.21

From Chapter 45, in polar form

$$y_R = 5\angle 0 + 4\angle - \frac{\pi}{6}$$

$$= 5\angle 0° + 4\angle - 30°$$

$$= (5 + j0) + (3.464 - j2.0)$$

$$= 8.464 - j2.0 = 8.70\angle - 13.29°$$

$$= 8.70\angle -0.23 \text{rad}$$

Hence, by using complex numbers, the resultant in sinusoidal form is:

$$y_1 + y_2 = 5\sin\omega t + 4\sin(\omega t - \pi/6)$$

$$= \mathbf{8.70\sin(\omega t - 0.23)}$$

Problem 12. Two alternating voltages are given by $v_1 = 15\sin\omega t$ volts and $v_2 = 25\sin(\omega t - \pi/6)$ volts. Determine a sinusoidal expression for the resultant $v_R = v_1 + v_2$ by using complex numbers

The relative positions of v_1 and v_2 at time $t = 0$ are shown in Figure 50.22(a) and the phasor diagram is shown in Figure 50.22(b).

In polar form, $v_R = v_1 + v_2 = 15\angle 0 + 25\angle - \frac{\pi}{6}$

$$= 15\angle 0° + 25\angle - 30°$$

$$= (15 + j0) + (21.65 - j12.5)$$

$$= 36.65 - j12.5 = 38.72\angle - 18.83°$$

$$= 38.72\angle - 0.329 \text{ rad}$$

Hence, by using complex numbers, the resultant in sinusoidal form is:

$$v_R = v_1 + v_2 = 15\sin\omega t + 25\sin(\omega t - \pi/6)$$

$$= \mathbf{38.72\sin(\omega t - 0.329) \text{ volts}}$$

Problem 13. For the voltages in Problem 12, determine the resultant $v_R = v_1 - v_2$ using complex numbers

In polar form, $y_R = v_1 - v_2 = 15\angle 0 - 25\angle - \frac{\pi}{6}$

$$= 15\angle 0° - 25\angle - 30°$$

$$= (15 + j0) - (21.65 - j12.5)$$

$$= -6.65 + j12.5 = 14.16\angle 118.01°$$

$$= 14.16\angle 2.06 \text{ rad}$$

Hence, by using complex numbers, the resultant in sinusoidal form is:

$$y_1 - y_2 = 15\sin\omega t - 25\sin(\omega t - \pi/6)$$

$$= \mathbf{14.16\sin(\omega t - 2.06) \text{ volts}}$$

Problem 14. Determine
$20\sin\omega t + 10\sin\left(\omega t + \frac{\pi}{3}\right)$ A using complex numbers

From the phasors shown in Figure 50.23, the resultant may be expressed in polar form as:

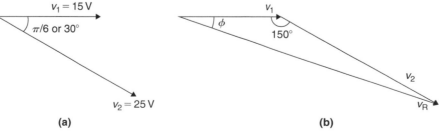

Figure 50.23

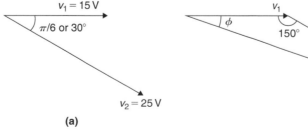

(a)

(b)

Figure 50.22

$$i_R = 20\angle 0° + 10\angle 60°$$

i.e. $i_R = (20 + j0) + (5 + j8.66)$

$$= (25 + j8.66) = \mathbf{26.46\angle 19.11°A} \quad \text{or}$$

$$\mathbf{26.46\angle 0.333\,rad\,A}$$

Hence, by using complex numbers, the resultant in sinusoidal form is:
$$i_R = i_1 + i_2 = \mathbf{26.46\sin(\omega t + 0.333)A}$$

Problem 15. If the supply to a circuit is $v = 30\sin 100\pi t$ volts and the voltage drop across one of the components is $v_1 = 20\sin(100\pi t - 0.59)$ volts, calculate the:

(a)   voltage drop across the remainder of the circuit, given by $v - v_1$, in the form $A\sin(\omega t \pm \alpha)$

(b)   supply frequency

(c)   periodic time of the supply

(d)   rms value of the supply voltage

(a)   Supply voltage, $v = v_1 + v_2$ where v_2 is the voltage across the remainder of the circuit.

Hence, $v_2 = v - v_1 = 30\sin 100\pi t$

$$- 20\sin(100\pi t - 0.59)$$

$$= 30\angle 0 - 20\angle - 0.59\,\text{rad}$$

$$= (30 + j0) - (16.619 - j11.127)$$

$$= 13.381 + j11.127$$

$$= 17.40\angle 0.694\,\text{rad}$$

Hence, by using complex numbers, the resultant in sinusoidal form is:

$$v - v_1 = 30\sin 100\pi t - 20\sin(100\pi t - 0.59)$$

$$= \mathbf{17.40\sin(100\pi t + 0.694)\,volts}$$

(b)   **Supply frequency,** $f = \dfrac{\omega}{2\pi} = \dfrac{100\pi}{2\pi} = \mathbf{50\,Hz}$

(c)   **Periodic time,** $T = \dfrac{1}{f} = \dfrac{1}{50} = \mathbf{0.02\,s}$ or **20 ms**

(d)   From Chapter 73, page 794, Problem 6, **rms value of supply voltage,** $= \dfrac{1}{\sqrt{2}} \times 30$
$$= \mathbf{21.21\,volts}$$

Now try the following Practice Exercise

Practice Exercise 219   Further problems on resultant phasors by complex numbers (answers on page 1175)

In Problems 1 to 4, express the combination of periodic functions in the form $A\sin(\omega t \pm \alpha)$ by using complex numbers:

1. $8\sin\omega t + 5\sin\left(\omega t + \dfrac{\pi}{4}\right)$

2. $6\sin\omega t + 9\sin\left(\omega t - \dfrac{\pi}{6}\right)$

3. $v = 12\sin\omega t - 5\sin\left(\omega t - \dfrac{\pi}{4}\right)$

4. $x = 10\sin\left(\omega t + \dfrac{\pi}{3}\right) - 8\sin\left(\omega t - \dfrac{3\pi}{8}\right)$

5.   The voltage drops across two components when connected in series across an a.c. supply are: $v_1 = 240\sin 314.2t$ and $v_2 = 150\sin(314.2t - \pi/5)$ volts, respectively. Determine the:

(a)   voltage of the supply (given by $v_1 + v_2$) in the form $A\sin(\omega t \pm \alpha)$

(b)   frequency of the supply

6.   If the supply to a circuit is $v = 25\sin 200\pi t$ volts and the voltage drop across one of the components is $v_1 = 18\sin(200\pi t - 0.43)$ volts, calculate the:

(a)   voltage drop across the remainder of the circuit, given by $v - v_1$, in the form $A\sin(\omega t \pm \alpha)$

(b)   supply frequency

(c)   periodic time of the supply

7.   The voltages across three components in a series circuit when connected across an a.c. supply are:

$$v_1 = 20\sin\left(300\pi t - \dfrac{\pi}{6}\right)\text{ volts,}$$

$$v_2 = 30\sin\left(300\pi t + \dfrac{\pi}{4}\right)\text{ volts, and}$$

$$v_3 = 60\sin\left(300\pi t - \dfrac{\pi}{3}\right)\text{ volts.}$$

Calculate the:

(a) supply voltage, in sinusoidal form, in the form $A\sin(\omega t \pm \alpha)$

(b) frequency of the supply

(c) periodic time

(d) rms value of the supply voltage

8. Measurements made at a substation at peak demand of the current in the red, yellow and blue phases of a transmission system are: $I_{red} = 1248\angle - 15°$A, $I_{yellow} = 1120\angle - 135°$A and $I_{blue} = 1310\angle 95°$A. Determine the current in the neutral cable if the sum of the currents flows through it.

For fully worked solutions to each of the problems in Practice Exercises 215 to 219 in this chapter, go to the website:
www.routledge.com/cw/bird

Chapter 51

Scalar and vector products

Why it is important to understand: Scalar and vector products

Common applications of the scalar product in engineering and physics is to test whether two vectors are perpendicular or to find the angle between two vectors when they are expressed in Cartesian form or to find the component of one vector in the direction of another. In mechanical engineering the scalar and vector product is used with forces, displacement, moments, velocities and the determination of work done. In electrical engineering, scalar and vector product calculations are important in electromagnetic calculations; most electromagnetic equations use vector calculus, which is based on the scalar and vector product. Such applications include electric motors and generators and power transformers, and so on. Knowledge of scalar and vector products thus has important engineering applications.

At the end of this chapter, you should be able to:

- define a unit triad
- determine the scalar (or dot) product of two vectors
- calculate the angle between two vectors
- determine the direction cosines of a vector
- apply scalar products to practical situations
- determine the vector (or cross) product of two vectors
- apply vector products to practical situations
- determine the vector equation of a line

51.1 The unit triad

When a vector x of magnitude x units and direction $\theta°$ is divided by the magnitude of the vector, the result is a vector of unit length at angle $\theta°$. The unit vector for a velocity of 10 m/s at 50° is $\dfrac{10\,\text{m/s at }50°}{10\,\text{m/s}}$, i.e. 1 at 50°.

In general, the unit vector for oa is $\dfrac{oa}{|oa|}$, the oa being a vector and having both magnitude and direction and $|oa|$ being the magnitude of the vector only.

As explained on page 589, one method of completely specifying the direction of a vector in space relative to some reference point is to use three unit vectors, mutually at right angles to each other, as shown in Figure 51.1. Such a system is called a **unit triad**.

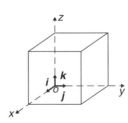

Figure 51.1

In Figure 51.2, one way to get from o to r is to move x units along $\boldsymbol{i}$ to point a, then y units in direction $\boldsymbol{j}$ to get to b and finally z units in direction $\boldsymbol{k}$ to get to r. The vector $\boldsymbol{or}$ is specified as

$$\boldsymbol{or} = x\boldsymbol{i} + y\boldsymbol{j} + z\boldsymbol{k}$$

Problem 1. With reference to three axes drawn mutually at right angles, depict the vectors (i) $\boldsymbol{op} = 4\boldsymbol{i} + 3\boldsymbol{j} - 2\boldsymbol{k}$ and (ii) $\boldsymbol{or} = 5\boldsymbol{i} - 2\boldsymbol{j} + 2\boldsymbol{k}$

The required vectors are depicted in Figure 51.3, $\boldsymbol{op}$ being shown in Figure 51.3(a) and $\boldsymbol{or}$ in Figure 51.3(b).

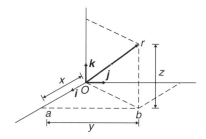

Figure 51.2

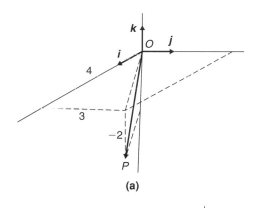

(a)

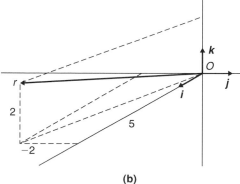

(b)

Figure 51.3

51.2 The scalar product of two vectors

When vector $\boldsymbol{oa}$ is multiplied by a scalar quantity, say k, the magnitude of the resultant vector will be k times the magnitude of $\boldsymbol{oa}$ and its direction will remain the same. Thus $2 \times (5\,\text{N at } 20°)$ results in a vector of magnitude $10\,\text{N at } 20°$.

One of the products of two vector quantities is called the **scalar** or **dot product** of two vectors and is defined as the product of their magnitudes multiplied by the cosine of the angle between them. The scalar product of $\boldsymbol{oa}$ and $\boldsymbol{ob}$ is shown as $\boldsymbol{oa} \cdot \boldsymbol{ob}$. For vectors $\boldsymbol{oa} = oa$ at θ_1, and $\boldsymbol{ob} = ob$ at θ_2 where $\theta_2 > \theta_1$, **the scalar product is**:

$$\boldsymbol{oa} \cdot \boldsymbol{ob} = oa\, ob\, \cos(\theta_2 - \theta_1)$$

For vectors $\boldsymbol{v_1}$ and $\boldsymbol{v_2}$ shown in Figure 51.4, the scalar product is:

$$\boldsymbol{v_1} \cdot \boldsymbol{v_2} = v_1 v_2 \cos\theta$$

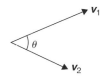

Figure 51.4

The commutative law of algebra, $a \times b = b \times a$ applies to scalar products. This is demonstrated in Figure 51.5. Let $\boldsymbol{oa}$ represent vector $\boldsymbol{v_1}$ and $\boldsymbol{ob}$ represent vector $\boldsymbol{v_2}$. Then:

$$\boldsymbol{oa} \cdot \boldsymbol{ob} = v_1 v_2 \cos\theta \text{ (by definition of a scalar product)}$$

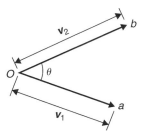

Figure 51.5

Similarly, $\boldsymbol{ob} \cdot \boldsymbol{oa} = v_2 v_1 \cos\theta = v_1 v_2 \cos\theta$ by the commutative law of algebra. Thus $\boldsymbol{oa} \cdot \boldsymbol{ob} = \boldsymbol{ob} \cdot \boldsymbol{oa}$

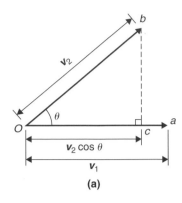

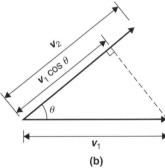

Figure 51.6

The projection of ob on oa is shown in Figure 51.6(a) and by the geometry of triangle obc, it can be seen that the projection is $v_2 \cos\theta$. Since, by definition

$$oa \cdot ob = v_1(v_2 \cos\theta),$$

it follows that

$$oa \cdot ob = v_1 \text{ (the projection of } v_2 \text{ on } v_1)$$

Similarly the projection of oa on ob is shown in Figure 51.6(b) and is $v_1 \cos\theta$. Since by definition

$$ob \cdot oa = v_2(v_1 \cos\theta),$$

it follows that

$$ob \cdot oa = v_2 \text{(the projection of } v_1 \text{ on } v_2)$$

This shows that the scalar product of two vectors is the product of the magnitude of one vector and the magnitude of the projection of the other vector on it. The **angle between two vectors** can be expressed in terms of the vector constants as follows:
Because $a \cdot b = a\,b \cos\theta$,

then

$$\cos\theta = \frac{a \cdot b}{ab} \qquad (1)$$

Let $\quad a = a_1 i + a_2 j + a_3 k$

and $\quad b = b_1 i + b_2 j + b_3 k$

$$a \cdot b = (a_1 i + a_2 j + a_3 k) \cdot (b_1 i + b_2 j + b_3 k)$$

Multiplying out the brackets gives:

$$a \cdot b = a_1 b_1 i \cdot i + a_1 b_2 i \cdot j + a_1 b_3 i \cdot k$$
$$+ a_2 b_1 j \cdot i + a_2 b_2 j \cdot j + a_2 b_3 j \cdot k$$
$$+ a_3 b_1 k \cdot i + a_3 b_2 k \cdot j + a_3 b_3 k \cdot k$$

However, the unit vectors i, j and k all have a magnitude of 1 and $i \cdot i = (1)(1) \cos 0° = 1, i \cdot j = (1)(1) \cos 90° = 0,$ $i \cdot k = (1)(1) \cos 90° = 0$ and similarly $j \cdot j = 1, j \cdot k = 0$ and $k \cdot k = 1$. Thus, only terms containing $i \cdot i, \; j \cdot j$ or $k \cdot k$ in the expansion above will not be zero.
Thus, the scalar product

$$a \cdot b = a_1 b_1 + a_2 b_2 + a_3 b_3 \qquad (2)$$

Both a and b in equation (1) can be expressed in terms of a_1, b_1, a_2, b_2, a_3 and b_3

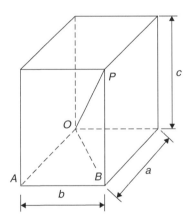

Figure 51.7

From the geometry of Figure 51.7, the length of diagonal OP in terms of side lengths a, b and c can be obtained from Pythagoras' theorem as follows:

$$OP^2 = OB^2 + BP^2 \text{ and}$$
$$OB^2 = OA^2 + AB^2$$

Thus, $\quad OP^2 = OA^2 + AB^2 + BP^2$
$$= a^2 + b^2 + c^2$$
in terms of side lengths

Thus, the **length** or **modulus** or **magnitude** or **norm of vector _OP_** is given by:

$$OP = \sqrt{(a^2 + b^2 + c^2)} \qquad (3)$$

Relating this result to the two vectors $a_1 i + a_2 j + a_3 k$ and $b_1 i + b_2 j + b_3 k$, gives:

$$a = \sqrt{(a_1^2 + a_2^2 + a_3^2)}$$

and $b = \sqrt{(b_1^2 + b_2^2 + b_3^2)}$

That is, from equation (1),

$$\cos\theta = \frac{a_1 b_1 + a_2 b_2 + a_3 b_3}{\sqrt{(a_1^2 + a_2^2 + a_3^2)}\sqrt{(b_1^2 + b_2^2 + b_3^2)}} \qquad (4)$$

> **Problem 2.** Find vector _a_ joining points _P_ and _Q_ where point _P_ has co-ordinates (4, −1, 3) and point _Q_ has co-ordinates (2, 5, 0). Also, find |_a_|, the magnitude or norm of _a_

Let O be the origin, i.e. its co-ordinates are (0, 0, 0). The position vectors of _P_ and _Q_ are given by:

$$OP = 4i - j + 3k \text{ and } OQ = 2i + 5j$$

By the addition law of vectors $OP + PQ = OQ$

Hence $a = PQ = OQ - OP$

i.e. $a = PQ = (2i + 5j) - (4i - j + 3k)$

$$= -2i + 6j - 3k$$

From equation (3), the magnitude or norm of _a_,

$$|a| = \sqrt{(a^2 + b^2 + c^2)}$$

$$= \sqrt{[(-2)^2 + 6^2 + (-3)^2]} = \sqrt{49} = 7$$

> **Problem 3.** If $p = 2i + j - k$ and $q = i - 3j + 2k$ determine:
>
> (i) $p \cdot q$ (ii) $p + q$
>
> (iii) $|p + q|$ (iv) $|p| + |q|$

(i) From equation (2),

if $\qquad p = a_1 i + a_2 j + a_3 k$

and $\qquad q = b_1 i + b_2 j + b_3 k$

then $\qquad p \cdot q = a_1 b_1 + a_2 b_2 + a_3 b_3$

When $\qquad p = 2i + j - k,$

$$a_1 = 2, \ a_2 = 1 \text{ and } a_3 = -1$$

and when $q = i - 3j + 2k,$

$$b_1 = 1, \ b_2 = -3 \text{ and } b_3 = 2$$

Hence $p \cdot q = (2)(1) + (1)(-3) + (-1)(2)$

i.e. $\qquad \boldsymbol{p \cdot q = -3}$

(ii) $p + q = (2i + j - k) + (i - 3j + 2k)$

$$= 3i - 2j + k$$

(iii) $|p + q| = |3i - 2j + k|$

From equation (3),

$$|p + q| = \sqrt{[3^2 + (-2)^2 + 1^2]} = \sqrt{14}$$

(iv) From equation (3),

$$|p| = |2i + j - k|$$
$$= \sqrt{[2^2 + 1^2 + (-1)^2]} = \sqrt{6}$$

Similarly,

$$|q| = |i - 3j + 2k|$$
$$= \sqrt{[1^2 + (-3)^2 + 2^2]} = \sqrt{14}$$

Hence $|p| + |q| = \sqrt{6} + \sqrt{14} = \boldsymbol{6.191}$, correct to 3 decimal places.

> **Problem 4.** Determine the angle between vectors _oa_ and _ob_ when
>
> $$oa = i + 2j - 3k$$
>
> and $ob = 2i - j + 4k$

An equation for $\cos\theta$ is given in equation (4)

$$\cos\theta = \frac{a_1 b_1 + a_2 b_2 + a_3 b_3}{\sqrt{(a_1^2 + a_2^2 + a_3^2)}\sqrt{(b_1^2 + b_2^2 + b_3^2)}}$$

Since $oa = i + 2j - 3k$,

$$a_1 = 1, a_2 = 2 \text{ and } a_3 = -3$$

Since $ob = 2i - j + 4k$,

$$b_1 = 2, b_2 = -1 \text{ and } b_3 = 4$$

Thus,

$$\cos\theta = \frac{(1 \times 2) + (2 \times -1) + (-3 \times 4)}{\sqrt{(1^2 + 2^2 + (-3)^2)}\sqrt{(2^2 + (-1)^2 + 4^2)}}$$

$$= \frac{-12}{\sqrt{14}\sqrt{21}} = -0.6999$$

i.e. $\theta = 134.4°$ or $225.6°$

By sketching the position of the two vectors as shown in Problem 1, it will be seen that $225.6°$ is not an acceptable answer.

Thus the angle between the vectors oa and ob, $\theta = 134.4°$

Direction cosines

From Figure 51.2, $or = xi + yj + zk$ and from equation (3), $|or| = \sqrt{x^2 + y^2 + z^2}$

If or makes angles of α, β and γ with the co-ordinate axes i, j and k respectively, then:

The direction cosines are:

$$\cos\alpha = \frac{x}{\sqrt{x^2 + y^2 + z^2}}$$

$$\cos\beta = \frac{y}{\sqrt{x^2 + y^2 + z^2}}$$

and $\cos\gamma = \dfrac{z}{\sqrt{x^2 + y^2 + z^2}}$

such that $\cos^2\alpha + \cos^2\beta + \cos^2\gamma = 1$

The values of $\cos\alpha$, $\cos\beta$ and $\cos\gamma$ are called the **direction cosines** of or

> **Problem 5.** Find the direction cosines of $3i + 2j + k$

$$\sqrt{x^2 + y^2 + z^2} = \sqrt{3^2 + 2^2 + 1^2} = \sqrt{14}$$

The direction cosines are:

$$\cos\alpha = \frac{x}{\sqrt{x^2 + y^2 + z^2}} = \frac{3}{\sqrt{14}} = \mathbf{0.802}$$

$$\cos\beta = \frac{y}{\sqrt{x^2 + y^2 + z^2}} = \frac{2}{\sqrt{14}} = \mathbf{0.535}$$

and $\cos\gamma = \dfrac{z}{\sqrt{x^2 + y^2 + z^2}} = \dfrac{1}{\sqrt{14}} = \mathbf{0.267}$

(and hence $\alpha = \cos^{-1} 0.802 = 36.7°$, $\beta = \cos^{-1} 0.535 = 57.7°$ and $\gamma = \cos^{-1} 0.267 = 74.5°$)
Note that $\cos^2\alpha + \cos^2\beta + \cos^2\gamma$
$$= 0.802^2 + 0.535^2 + 0.267^2 = 1$$

Practical application of scalar product

> **Problem 6.** A constant force of $F = 10i + 2j - k$ newtons displaces an object from $A = i + j + k$ to $B = 2i - j + 3k$ (in metres). Find the work done in newton metres

One of the applications of scalar products is to calculate the work done by a constant force when moving a body. The work done is the product of the applied force and the distance moved in the direction of the force.

i.e. **work done** $= F \cdot d$

The principles developed in Problem 13, page 588, apply equally to this problem when determining the displacement. From the sketch shown in Figure 51.8,

$$AB = AO + OB = OB - OA$$

that is $AB = (2i - j + 3k) - (i + j + k)$

$$= i - 2j + 2k$$

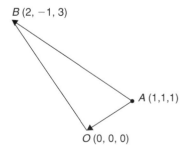

$B (2, -1, 3)$

$A (1,1,1)$

$O (0, 0, 0)$

Figure 51.8

The work done is $F \cdot d$, that is $F \cdot AB$ in this case

i.e. **work done** $= (10i + 2j - k) \cdot (i - 2j + 2k)$

But from equation (2),

$$a \cdot b = a_1 b_1 + a_2 b_2 + a_3 b_3$$

Hence **work done** $=$
$(10 \times 1) + (2 \times (-2)) + ((-1) \times 2) = \mathbf{4\,Nm}$.
(Theoretically, it is quite possible to get a negative answer to a 'work done' problem. This indicates that the force must be in the opposite sense to that given, in order to give the displacement stated.)

Now try the following Practice Exercise

Practice Exercise 220   Further problems on scalar products (answers on page 1175)

1. Find the scalar product $a \cdot b$ when
    (i) $a = i + 2j - k$ and $b = 2i + 3j + k$

    (ii) $a = i - 3j + k$ and $b = 2i + j + k$

Given $p = 2i - 3j$, $q = 4j - k$ and
$r = i + 2j - 3k$, determine the quantities stated in Problems 2 to 8.

2. (a) $p \cdot q$ (b) $p \cdot r$

3. (a) $q \cdot r$ (b) $r \cdot q$

4. (a) $|p|$ (b) $|r|$

5. (a) $p \cdot (q + r)$ (b) $2r \cdot (q - 2p)$

6. (a) $|p + r|$ (b) $|p| + |r|$

7. Find the angle between (a) p and q (b) q and r

8. Determine the direction cosines of (a) p (b) q (c) r

9. Determine the angle between the forces:

    $$F_1 = 3i + 4j + 5k \text{ and}$$

    $$F_2 = i + j + k$$

10. Find the angle between the velocity vectors $v_1 = 5i + 2j + 7k$ and $v_2 = 4i + j - k$

11. Calculate the work done by a force $F = (-5i + j + 7k)\,N$ when its point of application moves from point $(-2i - 6j + k)\,m$ to the point $(i - j + 10k)\,m$.

51.3   Vector products

A second product of two vectors is called the **vector** or **cross product** and is defined in terms of its modulus and the magnitudes of the two vectors and the sine of the angle between them. The vector product of vectors oa and ob is written as $oa \times ob$ and is defined by:

$$|oa \times ob| = oa\,ob\sin\theta$$

where θ is the angle between the two vectors.
The direction of $oa \times ob$ is perpendicular to both oa and ob, as shown in Figure 51.9.

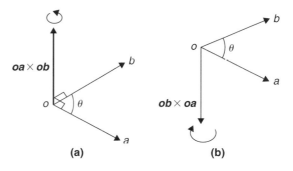

(a)   **(b)**

Figure 51.9

The direction is obtained by considering that a right-handed screw is screwed along $oa \times ob$ with its head at the origin and if the direction of $oa \times ob$ is correct, the head should rotate from oa to ob, as shown in Figure 51.9(a). It follows that the direction of $ob \times oa$ is as shown in Figure 51.9(b). Thus $oa \times ob$ is not equal to $ob \times oa$. The magnitudes of $oa\,ob\sin\theta$ are the same but their directions are $180°$ displaced, i.e.

$$oa \times ob = -ob \times oa$$

The vector product of two vectors may be expressed in terms of the unit vectors. Let two vectors, a and b, be such that:

$$a = a_1 i + a_2 j + a_3 k \text{ and}$$
$$b = b_1 i + b_2 j + b_3 k$$

Then,

$$a \times b = (a_1 i + a_2 j + a_3 k) \times (b_1 i + b_2 j + b_3 k)$$

$$= a_1 b_1 i \times i + a_1 b_2 i \times j$$

$$+ a_1 b_3 i \times k + a_2 b_1 j \times i + a_2 b_2 j \times j$$

$$+ a_2 b_3 j \times k + a_3 b_1 k \times i + a_3 b_2 k \times j$$

$$+ a_3 b_3 k \times k$$

But by the definition of a vector product,

$$i \times j = k, \ j \times k = i \text{ and } k \times i = j$$

Also $i \times i = j \times j = k \times k = (1)(1) \sin 0° = 0$

Remembering that $a \times b = -b \times a$ gives:

$$a \times b = a_1 b_2 k - a_1 b_3 j - a_2 b_1 k + a_2 b_3 i$$
$$+ a_3 b_1 j - a_3 b_2 i$$

Grouping the i, j and k terms together gives:

$$a \times b = (a_2 b_3 - a_3 b_2)i + (a_3 b_1 - a_1 b_3)j$$
$$+ (a_1 b_2 - a_2 b_1)k$$

The vector product can be written in determinant form as:

$$a \times b = \begin{vmatrix} i & j & k \\ a_1 & a_2 & a_3 \\ b_1 & b_2 & b_3 \end{vmatrix} \qquad (5)$$

The 3×3 determinant $\begin{vmatrix} i & j & k \\ a_1 & a_2 & a_3 \\ b_1 & b_2 & b_3 \end{vmatrix}$ is evaluated as:

$$i \begin{vmatrix} a_2 & a_3 \\ b_2 & b_3 \end{vmatrix} - j \begin{vmatrix} a_1 & a_3 \\ b_1 & b_3 \end{vmatrix} + k \begin{vmatrix} a_1 & a_2 \\ b_1 & b_2 \end{vmatrix}$$

where

$$\begin{vmatrix} a_2 & a_3 \\ b_2 & b_3 \end{vmatrix} = a_2 b_3 - a_3 b_2,$$

$$\begin{vmatrix} a_1 & a_3 \\ b_1 & b_3 \end{vmatrix} = a_1 b_3 - a_3 b_1 \text{ and}$$

$$\begin{vmatrix} a_1 & a_2 \\ b_1 & b_2 \end{vmatrix} = a_1 b_2 - a_2 b_1$$

The magnitude of the vector product of two vectors can be found by expressing it in scalar product form and then using the relationship

$$a \cdot b = a_1 b_1 + a_2 b_2 + a_3 b_3 \quad \text{from equation (2)}$$

Squaring both sides of a vector product equation gives:

$$(|a \times b|)^2 = a^2 b^2 \sin^2 \theta = a^2 b^2 (1 - \cos^2 \theta)$$
$$\text{from page 490, equation (2)}$$
$$= a^2 b^2 - a^2 b^2 \cos^2 \theta \qquad (6)$$

It is stated in Section 51.2 that $a \cdot b = ab \cos \theta$, hence
$$a \cdot a = a^2 \cos \theta$$

But $\theta = 0°$, thus $a \cdot a = a^2$

Also, $\cos \theta = \dfrac{a \cdot b}{ab}$

Multiplying both sides of this equation by $a^2 b^2$ and squaring gives:

$$a^2 b^2 \cos^2 \theta = \frac{a^2 b^2 (a \cdot b)^2}{a^2 b^2} = (a \cdot b)^2$$

Substituting in equation (6) above for $a^2 = a \cdot a, b^2 = b \cdot b$ and $a^2 b^2 \cos^2 \theta = (a \cdot b)^2$ gives:

$$(|a \times b|)^2 = (a \cdot a)(b \cdot b) - (a \cdot b)^2$$

That is,

$$|a \times b| = \sqrt{[(a \cdot a)(b \cdot b) - (a \cdot b)^2]} \qquad (7)$$

Problem 7. For the vectors $a = i + 4j - 2k$ and $b = 2i - j + 3k$ find (i) $a \times b$ and (ii) $|a \times b|$

(i) From equation (5),

$$a \times b = \begin{vmatrix} i & j & k \\ a_1 & a_2 & a_3 \\ b_1 & b_2 & b_3 \end{vmatrix}$$

$$= i \begin{vmatrix} a_2 & a_3 \\ b_2 & b_3 \end{vmatrix} - j \begin{vmatrix} a_1 & a_3 \\ b_1 & b_3 \end{vmatrix} + k \begin{vmatrix} a_1 & a_2 \\ b_1 & b_2 \end{vmatrix}$$

Hence

$$a \times b = \begin{vmatrix} i & j & k \\ 1 & 4 & -2 \\ 2 & -1 & 3 \end{vmatrix}$$

$$= i \begin{vmatrix} 4 & -2 \\ -1 & 3 \end{vmatrix} - j \begin{vmatrix} 1 & -2 \\ 2 & 3 \end{vmatrix} + k \begin{vmatrix} 1 & 4 \\ 2 & -1 \end{vmatrix}$$

$$= i(12 - 2) - j(3 + 4) + k(-1 - 8)$$

$$= 10i - 7j - 9k$$

(ii) From equation (7)

$$|a \times b| = \sqrt{[(a \cdot a)(b \cdot b) - (a \cdot b)^2]}$$

Now $a \cdot a = (1)(1) + (4 \times 4) + (-2)(-2)$
$$= 21$$
$$b \cdot b = (2)(2) + (-1)(-1) + (3)(3)$$
$$= 14$$

and $\quad \boldsymbol{a \cdot b} = (1)(2) + (4)(-1) + (-2)(3)$

$$= -8$$

Thus $\quad |\boldsymbol{a \times b}| = \sqrt{(21 \times 14 - 64)}$

$$= \sqrt{230} = \mathbf{15.17}$$

Problem 8. If $p = 4i + j - 2k$, $q = 3i - 2j + k$ and $r = i - 2k$ find (a) $(p - 2q) \times r$ (b) $p \times (2r \times 3q)$

(a) $(p - 2q) \times r = [4i + j - 2k$

$$- 2(3i - 2j + k)] \times (i - 2k)$$

$$= (-2i + 5j - 4k) \times (i - 2k)$$

$$= \begin{vmatrix} i & j & k \\ -2 & 5 & -4 \\ 1 & 0 & -2 \end{vmatrix}$$

from equation (5)

$$= i \begin{vmatrix} 5 & -4 \\ 0 & -2 \end{vmatrix} - j \begin{vmatrix} -2 & -4 \\ 1 & -2 \end{vmatrix}$$

$$+ k \begin{vmatrix} -2 & 5 \\ 1 & 0 \end{vmatrix}$$

$$= i(-10 - 0) - j(4 + 4)$$

$$+ k(0 - 5), \text{ i.e.}$$

$$(p - 2q) \times r = -10i - 8j - 5k$$

(b) $(2r \times 3q) = (2i - 4k) \times (9i - 6j + 3k)$

$$= \begin{vmatrix} i & j & k \\ 2 & 0 & -4 \\ 9 & -6 & 3 \end{vmatrix}$$

$$= i(0 - 24) - j(6 + 36)$$

$$+ k(-12 - 0)$$

$$= -24i - 42j - 12k$$

Hence

$$p \times (2r \times 3q) = (4i + j - 2k)$$

$$\times (-24i - 42j - 12k)$$

$$= \begin{vmatrix} i & j & k \\ 4 & 1 & -2 \\ -24 & -42 & -12 \end{vmatrix}$$

$$= i(-12 - 84) - j(-48 - 48)$$

$$+ k(-168 + 24)$$

$$= \mathbf{-96i + 96j - 144k}$$

$$\text{or} \ \mathbf{-48(2i - 2j + 3k)}$$

Practical applications of vector products

Problem 9. Find the moment and the magnitude of the moment of a force of $(i + 2j - 3k)$ newtons about point B having co-ordinates $(0, 1, 1)$, when the force acts on a line through A whose co-ordinates are $(1, 3, 4)$

The moment M about point B of a force vector F which has a position vector of r from A is given by:

$$M = r \times F$$

r is the vector from B to A, i.e. $r = BA$
But $\ BA = BO + OA = OA - OB$ (see Problem 13, page 588), that is:

$$r = (i + 3j + 4k) - (j + k)$$

$$= i + 2j + 3k$$

Moment,

$$M = r \times F = (i + 2j + 3k) \times (i + 2j - 3k)$$

$$= \begin{vmatrix} i & j & k \\ 1 & 2 & 3 \\ 1 & 2 & -3 \end{vmatrix}$$

$$= i(-6 - 6) - j(-3 - 3)$$

$$+ k(2 - 2)$$

$$= \mathbf{-12i + 6j} \ \text{Nm}$$

The magnitude of M,

$$|M| = |r \times F|$$

$$= \sqrt{[(r \cdot r)(F \cdot F) - (r \cdot F)^2]}$$

$$r \cdot r = (1)(1) + (2)(2) + (3)(3) = 14$$

$$F \cdot F = (1)(1) + (2)(2) + (-3)(-3) = 14$$

$$r \cdot F = (1)(1) + (2)(2) + (3)(-3) = -4$$

$$|M| = \sqrt{[14 \times 14 - (-4)^2]}$$

$$= \sqrt{180} \ \text{Nm} = \mathbf{13.42 \ Nm}$$

Problem 10. The axis of a circular cylinder coincides with the z-axis and it rotates with an angular velocity of $(2i - 5j + 7k)$ rad/s. Determine the tangential velocity at a point P on the cylinder, whose co-ordinates are $(j + 3k)$ metres, and also determine the magnitude of the tangential velocity

The velocity v of point P on a body rotating with angular velocity ω about a fixed axis is given by:

$$v = \omega \times r$$

where r is the point on vector P

Thus $v = (2i - 5j + 7k) \times (j + 3k)$

$$= \begin{vmatrix} i & j & k \\ 2 & -5 & 7 \\ 0 & 1 & 3 \end{vmatrix}$$

$$= i(-15 - 7) - j(6 - 0) + k(2 - 0)$$

$$= (-22i - 6j + 2k)\,\text{m/s}$$

The magnitude of v,

$$|v| = \sqrt{[(\omega \cdot \omega)(r \cdot r) - (\omega \cdot r)^2]}$$

$$\omega \cdot \omega = (2)(2) + (-5)(-5) + (7)(7) = 78$$

$$r \cdot r = (0)(0) + (1)(1) + (3)(3) = 10$$

$$\omega \cdot r = (2)(0) + (-5)(1) + (7)(3) = 16$$

Hence,

$$|v| = \sqrt{(78 \times 10 - 16^2)}$$

$$= \sqrt{524}\,\text{m/s} = \mathbf{22.89\,m/s}$$

Now try the following Practice Exercise

Practice Exercise 221   Further problems on vector products (answers on page 1175)

In Problems 1 to 4, determine the quantities stated when

$p = 3i + 2k, q = i - 2j + 3k$ and

$r = -4i + 3j - k$

1.   (a) $p \times q$ (b) $q \times p$

2.   (a) $|p \times r|$ (b) $|r \times q|$

3.   (a) $2p \times 3r$ (b) $(p + r) \times q$

4.   (a) $p \times (r \times q)$ (b) $(3p \times 2r) \times q$

5.   For vectors $p = 4i - j + 2k$ and $q = -2i + 3j - 2k$ determine: (i) $p \cdot q$ (ii) $p \times q$ (iii) $|p \times q|$ (iv) $q \times p$ and (v) the angle between the vectors.

6.   For vectors $a = -7i + 4j + \frac{1}{2}k$ and $b = 6i - 5j - k$ find (i) $a \cdot b$ (ii) $a \times b$ (iii) $|a \times b|$ (iv) $b \times a$ and (v) the angle between the vectors.

7.   Forces of $(i + 3j), (-2i - j), (-i - 2j)$ newtons act at three points having position vectors of $(2i + 5j), 4j$ and $(-i + j)$ metres, respectively. Calculate the magnitude of the moment.

8.   A force of $(2i - j + k)$ newtons acts on a line through point P having co-ordinates (0, 3, 1) metres. Determine the moment vector and its magnitude about point Q having co-ordinates (4, 0, −1) metres.

9.   A sphere is rotating with angular velocity ω about the z-axis of a system, the axis coinciding with the axis of the sphere. Determine the velocity vector and its magnitude at position $(-5i + 2j - 7k)$ m, when the angular velocity is $(i + 2j)$ rad/s.

10.   Calculate the velocity vector and its magnitude for a particle rotating about the z-axis at an angular velocity of $(3i - j + 2k)$ rad/s when the position vector of the particle is at $(i - 5j + 4k)$ m.

51.4   Vector equation of a line

The equation of a straight line may be determined, given that it passes through the point A with position vector a relative to O, and is parallel to vector b. Let r be the position vector of a point P on the line, as shown in Figure 51.10.

By vector addition, $OP = OA + AP$

i.e. $r = a + AP$

However, as the straight line through A is parallel to the free vector b (**free vector** means one that has the same

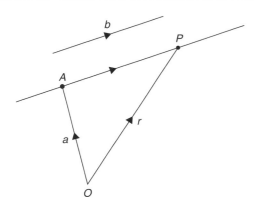

Figure 51.10

magnitude, direction and sense), then $AP = \lambda b$, where λ is a scalar quantity. Hence, from above,

$$r = a + \lambda b \qquad (8)$$

If, say, $r = xi + yj + zk$, $a = a_1 i + a_2 j + a_3 k$ and $b = b_1 i + b_2 j + b_3 k$, then from equation (8),

$$xi + yj + zk = (a_1 i + a_2 j + a_3 k)$$
$$+ \lambda(b_1 i + b_2 j + b_3 k)$$

Hence $x = a_1 + \lambda b_1$, $y = a_2 + \lambda b_2$ and $z = a_3 + \lambda b_3$. Solving for λ gives:

$$\frac{x - a_1}{b_1} = \frac{y - a_2}{b_2} = \frac{z - a_3}{b_3} = \lambda \qquad (9)$$

Equation (9) is the standard Cartesian form for the vector equation of a straight line.

Problem 11. (a) Determine the vector equation of the line through the point with position vector $2i + 3j - k$ which is parallel to the vector $i - 2j + 3k$

(b) Find the point on the line corresponding to $\lambda = 3$ in the resulting equation of part (a)

(c) Express the vector equation of the line in standard Cartesian form

(a) From equation (8),

$$r = a + \lambda b$$

i.e. $r = (2i + 3j - k) + \lambda(i - 2j + 3k)$

or $r = (2 + \lambda)i + (3 - 2\lambda)j + (3\lambda - 1)k$

which is the vector equation of the line.

(b) When $\lambda = 3$, $r = 5i - 3j + 8k$

(c) From equation (9),

$$\frac{x - a_1}{b_1} = \frac{y - a_2}{b_2} = \frac{z - a_3}{b_3} = \lambda$$

Since $a = 2i + 3j - k$, then $a_1 = 2$,

$$a_2 = 3 \text{ and } a_3 = -1 \text{ and}$$

$$b = i - 2j + 3k, \text{ then}$$

$$b_1 = 1, b_2 = -2 \text{ and } b_3 = 3$$

Hence, the Cartesian equations are:

$$\frac{x - 2}{1} = \frac{y - 3}{-2} = \frac{z - (-1)}{3} = \lambda$$

i.e. $x - 2 = \dfrac{3 - y}{2} = \dfrac{z + 1}{3} = \lambda$

Problem 12. The equation

$$\frac{2x - 1}{3} = \frac{y + 4}{3} = \frac{-z + 5}{2}$$

represents a straight line. Express this in vector form

Comparing the given equation with equation (9) shows that the coefficients of x, y and z need to be equal to unity.

Thus $\dfrac{2x - 1}{3} = \dfrac{y + 4}{3} = \dfrac{-z + 5}{2}$ becomes:

$$\frac{x - \frac{1}{2}}{\frac{3}{2}} = \frac{y + 4}{3} = \frac{z - 5}{-2}$$

Again, comparing with equation (9) shows that

$$a_1 = \frac{1}{2}, a_2 = -4 \text{ and } a_3 = 5 \text{ and}$$

$$b_1 = \frac{3}{2}, b_2 = 3 \text{ and } b_3 = -2$$

In vector form the equation is:

$$r = (a_1 + \lambda b_1)i + (a_2 + \lambda b_2)j + (a_3 + \lambda b_3)k,$$
$$\text{from equation (8)}$$

i.e. $r = \left(\dfrac{1}{2} + \dfrac{3}{2}\lambda\right)i + (-4 + 3\lambda)j + (5 - 2\lambda)k$

or $r = \dfrac{1}{2}(1 + 3\lambda)i + (3\lambda - 4)j + (5 - 2\lambda)k$

Section H

Now try the following Practice Exercise

Practice Exercise 222   Further problems on the vector equation of a line (answers on page 1176)

1. Find the vector equation of the line through the point with position vector $5i - 2j + 3k$ which is parallel to the vector $2i + 7j - 4k$. Determine the point on the line corresponding to $\lambda = 2$ in the resulting equation.

2. Express the vector equation of the line in Problem 1 in standard Cartesian form.

In Problems 3 and 4, express the given straight line equations in vector form.

3. $\dfrac{3x - 1}{4} = \dfrac{5y + 1}{2} = \dfrac{4 - z}{3}$

4. $2x + 1 = \dfrac{1 - 4y}{5} = \dfrac{3z - 1}{4}$

For fully worked solutions to each of the problems in Practice Exercises 220 to 222 in this chapter, go to the website:
www.routledge.com/cw/bird

This assignment covers the material contained in Chapters 49 to 51. *The marks available are shown in brackets at the end of each question.*

1. State whether the following are scalar or vector quantities:
   (a) A temperature of 50°C
   (b) A downward force of 80 N
   (c) 300 J of work
   (d) A south-westerly wind of 15 knots
   (e) 70 m distance
   (f) An acceleration of 25 m/s² at 30° to the horizontal (6)

2. Calculate the resultant and direction of the force vectors shown in Figure RT20.1, correct to 2 decimal places. (7)

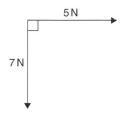

Figure RT20.1

3. Four coplanar forces act at a point A as shown in Figure RT20.2. Determine the value and direction of the resultant force by (a) drawing and (b) calculation using horizontal and vertical components. (10)

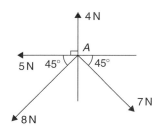

Figure RT20.2

4. The instantaneous values of two alternating voltages are given by:

   $$v_1 = 150\sin(\omega t + \pi/3) \text{ volts and}$$

   $$v_2 = 90\sin(\omega t - \pi/6) \text{ volts}$$

   Plot the two voltages on the same axes to scales of 1 cm = 50 volts and 1 cm = $\dfrac{\pi}{6}$ rad.

   Obtain a sinusoidal expression for the resultant $v_1 + v_2$ in the form $R\sin(\omega t + \alpha)$: (a) by adding ordinates at intervals and (b) by calculation. (13)

5. If velocity $v_1 = 26$ m/s at $52°$ and $v_2 = 17$ m/s at $-28°$, calculate the magnitude and direction of $v_1 + v_2$, correct to 2 decimal places, using complex numbers. (10)

6. Given $a = -3i + 3j + 5k$, $b = 2i - 5j + 7k$ and $c = 3i + 6j - 4k$, determine the following:
   (i) $-4b$ (ii) $a + b - c$ (iii) $5b - 3c$ (8)

7. If $a = 2i + 4j - 5k$ and $b = 3i - 2j + 6k$ determine:
   (i) $a \cdot b$ (ii) $|a + b|$ (iii) $a \times b$ (iv) the angle between a and b (14)

8. Determine the work done by a force of F newtons acting at a point A on a body, when A is displaced to point B, the co-ordinates of A and B being $(2, 5, -3)$ and $(1, -3, 0)$ metres respectively, and when $F = 2i - 5j + 4k$ newtons. (4)

9. A force of $F = 3i - 4j + k$ newtons acts on a line passing through a point P. Determine moment M and its magnitude of the force F about a point Q when P has co-ordinates $(4, -1, 5)$ metres and Q has co-ordinates $(4, 0, -3)$ metres. (8)

Section H

For lecturers/instructors/teachers, fully worked solutions to each of the problems in Revision Test 20, together with a full marking scheme, are available at the website:
www.routledge.com/cw/bird

Multiple choice questions Test 6
Complex numbers, matrices and vectors
This test covers the material in Chapters 45 to 51

All questions have only one correct answer (answers on page 1200).

Section H

1. $\dfrac{5}{j^6}$ is equivalent to:

   (a) $j5$ (b) -5 (c) $-j5$ (d) 5

2. Two voltage phasors are shown in Figure M6.1. If $V_1 = 40$ volts and $V_2 = 100$ volts, the resultant (i.e. length OA) is:

   (a)   131.4 volts at $32.55°$ to V_1

   (b)   105.0 volts at $32.55°$ to V_1

   (c)   131.4 volts at $68.30°$ to V_1

   (d)   105.0 volts at $42.31°$ to V_1

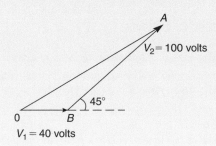

Figure M6.1

3. A force of 4 N is inclined at an angle of $45°$ to a second force of 7 N, both forces acting at a point, as shown in Figure M6.2. The magnitude of the resultant of these two forces and the direction of the resultant with respect to the 7 N force is:

   (a) 3 N at $45°$ (b) 5 N at $146°$

   (c) 11 N at $135°$ (d) 10.2 N at $16°$

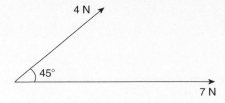

Figure M6.2

4. The matrix product $\begin{pmatrix} 2 & 3 \\ -1 & 4 \end{pmatrix}\begin{pmatrix} 1 & -5 \\ -2 & 6 \end{pmatrix}$ is equal to:

   (a) $\begin{pmatrix} -13 \\ 26 \end{pmatrix}$ (b) $\begin{pmatrix} 3 & -2 \\ -3 & 10 \end{pmatrix}$

   (c) $\begin{pmatrix} -4 & 8 \\ -9 & 29 \end{pmatrix}$ (d) $\begin{pmatrix} 1 & -2 \\ -3 & -2 \end{pmatrix}$

5. $\dfrac{(1-j)^2}{(-j)^3}$ is equal to:

   (a) -2 (b) $-j2$ (c) $2-j2$ (d) $j2$

Questions 6 and 7 relate to the following information. Two voltage phasors V_1 and V_2 are shown in Figure M6.3.

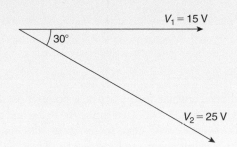

Figure M6.3

6. The resultant $V_1 + V_2$ is given by:

   (a)   14.16 V at $62°$ to V_1

   (b)   38.72 V at $-19°$ to V_1

   (c)   38.72 V at $161°$ to V_1

   (d)   14.16 V at $118°$ to V_1

7. The resultant $V_1 - V_2$ is given by:

   (a)   38.72 V at $-19°$ to V_1

   (b)   14.16 V at $62°$ to V_1

   (c)   38.72 V at $161°$ to V_1

   (d)   14.16 V at $118°$ to V_1

8. The inverse of the matrix $\begin{pmatrix} 5 & -3 \\ -2 & 1 \end{pmatrix}$ is:

   (a) $\begin{pmatrix} -5 & -3 \\ 2 & -1 \end{pmatrix}$ (b) $\begin{pmatrix} -1 & -3 \\ -2 & -5 \end{pmatrix}$

   (c) $\begin{pmatrix} -1 & 3 \\ 2 & -5 \end{pmatrix}$ (d) $\begin{pmatrix} 1 & 3 \\ 2 & 5 \end{pmatrix}$

9. $(-4 - j3)$ in polar form is:

   (a) $5\angle -143.13°$ (b) $5\angle 126.87°$

   (c) $5\angle 143.13°$ (d) $5\angle -126.87°$

10. The magnitude of the resultant of velocities of 3 m/s at 20° and 7 m/s at 120° when acting simultaneously at a point is:

    (a) 21 m/s (b) 10 m/s (c) 7.12 m/s (d) 4 m/s

11. The inverse of the matrix $\begin{pmatrix} 2 & -3 \\ 1 & -4 \end{pmatrix}$ is:

    (a) $\begin{pmatrix} 0.8 & 0.6 \\ -0.2 & -0.4 \end{pmatrix}$

    (b) $\begin{pmatrix} -4 & 3 \\ -1 & 2 \end{pmatrix}$

    (c) $\begin{pmatrix} 0.8 & -0.6 \\ 0.2 & -0.4 \end{pmatrix}$

    (d) $\begin{pmatrix} -0.4 & -0.6 \\ 0.2 & 0.8 \end{pmatrix}$

12. $(1 + j)^4$ is equivalent to:

    (a) 4   (b) $-j4$ (c) $j4$ (d) -4

13. For the following simultaneous equations:

    $$3x - 4y + 10 = 0$$

    $$5y - 2x = 9$$

    using matrices, the value of x is:

    (a) -2 (b) 1   (c) 2   (d) -1

14. Three forces of 2 N, 3 N and 4 N act as shown in Figure M6.4. The magnitude of the resultant force is:

    (a) 8.08 N   (b) 9 N   (c) 7.17 N   (d) 1 N

15. $2\angle\dfrac{\pi}{3} + 3\angle\dfrac{\pi}{6}$ in polar form is:

    (a) $5\angle\dfrac{\pi}{2}$ (b) $4.84\angle0.84$

    (c) $6\angle0.55$ (d) $4.84\angle0.73$

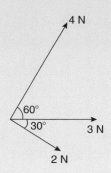

Figure M6.4

16. The value of the determinant $\begin{vmatrix} 2 & -1 & 4 \\ 0 & 1 & 5 \\ 6 & 0 & -1 \end{vmatrix}$ is:

    (a) 4   (b) 52   (c) -56 (d) 8

17. The value(s) of λ given $\begin{vmatrix} (2-\lambda) & -1 \\ -4 & (-1-\lambda) \end{vmatrix} = 0$ is:

    (a) -2 (b) $+2$ and -3 (c) -2 and $+3$ (d) 3

Questions 18 and 19 relate to the following information. Two alternating voltages are given by:

$v_1 = 2\sin\omega t$ and $v_2 = 3\sin\left(\omega t + \dfrac{\pi}{4}\right)$ volts.

18. Which of the phasor diagrams shown in Figure M6.5 represents $v_R = v_1 + v_2$?

    (a) (i)   (b) (ii)   (c) (iii)   (d) (iv)

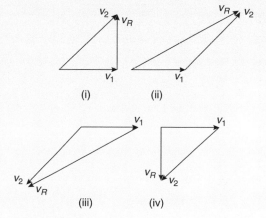

Figure M6.5

19. Which of the phasor diagrams shown represents $v_R = v_1 - v_2$?

    (a) (i)   (b) (ii)   (c) (iii)   (d) (iv)

20. The two square roots of $(-3 + j4)$ are:

(a) $\pm(1 + j2)$ (b) $\pm(0.71 + j2.12)$

(c) $\pm(1 - j2)$ (d) $\pm(0.71 - j2.12)$

21. The value of $\begin{vmatrix} j2 & -(1+j) \\ (1-j) & 1 \end{vmatrix}$ is:

(a) $-j2$ (b) 2 (c) $2(1+j)$ (d) $-2 + j2$

22. $[2\angle 30°]^4$ in Cartesian form is:

(a) $(0.50 + j0.06)$

(b) $(-8 + j13.86)$

(c) $(-4 + j6.93)$

(d) $(13.86 + j8)$

23. For the following simultaneous equations:

$$3x - 4y + 2z = 11$$
$$5x + 3y - 2z = -17$$
$$2x - 3y + 5z = 19$$

using Cramer's rule, the value of y is:

(a) -2 (b) 3 (c) 1 (d) -1

24. The argument of $j(1-j)$ is:

(a) $\sqrt{2}$ (b) $-45°$ (c) 0 (d) $\dfrac{\pi}{4}$rad

25. The modulus of $j(1-j)$ is:

(a) 1 (b) $\sqrt{2}$ (c) 2 (d) $45°$

Section H

Let $\boldsymbol{a} = a_1\boldsymbol{i} + a_2\boldsymbol{j} + a_3\boldsymbol{k}$ and $\boldsymbol{b} = b_1\boldsymbol{i} + b_2\boldsymbol{j} + b_3\boldsymbol{k}$

Scalar or dot product

$\boldsymbol{a} \cdot \boldsymbol{b} = a_1b_1 + a_2b_2 + a_3b_3$

$|\boldsymbol{a}| = \sqrt{a_1^2 + a_2^2 + a_3^2}$

$\cos\theta = \dfrac{a \cdot b}{|a|\,|b|}$

Vector or cross product

$$\boldsymbol{a} \times \boldsymbol{b} = \begin{vmatrix} i & j & k \\ a_1 & a_2 & a_3 \\ b_1 & b_2 & b_3 \end{vmatrix}$$

$|a \times b| = \sqrt{[(a \cdot a)(b \cdot b) - (a \cdot b)^2]}$

Section H

For a copy of these formulae/revision hints, go to:
www.routledge.com/cw/bird

Section I

Differential calculus

Chapter 52

Introduction to differentiation

Why it is important to understand: **Introduction to differentiation**

There are many practical situations engineers have to analyse which involve quantities that are varying. Typical examples include the stress in a loaded beam, the temperature of an industrial chemical, the rate at which the speed of a vehicle is increasing or decreasing, the current in an electrical circuit or the torque on a turbine blade. Differential calculus, or differentiation, is a mathematical technique for analysing the way in which functions change. There are many methods and rules of differentiation which are individually covered in the following chapters. A good knowledge of algebra, in particular laws of indices, is essential. This chapter explains how to differentiate the five most common functions, providing an important base for future chapters.

At the end of this chapter, you should be able to:

- state that calculus comprises two parts – differential and integral calculus
- understand functional notation
- describe the gradient of a curve and limiting value
- differentiate simple functions from first principles
- differentiate $y = ax^n$ by the general rule
- differentiate sine and cosine functions
- differentiate exponential and logarithmic functions

52.1 Introduction to calculus

Calculus is a branch of mathematics involving or leading to calculations dealing with continuously varying functions.

Calculus is a subject that falls into two parts:

(i) **differential calculus** (or **differentiation**) and

(ii) **integral calculus** (or **integration**).

Differentiation is used in calculations involving velocity and acceleration, rates of change and maximum and minimum values of curves.

52.2 Functional notation

A 'function' is defined in Section 36.1, page 392.
In an equation such as $y = 3x^2 + 2x - 5$, y is said to be a function of x and may be written as $y = f(x)$.

An equation written in the form $f(x) = 3x^2 + 2x - 5$ is termed **functional notation**. The value of $f(x)$ when $x = 0$ is denoted by $f(0)$, and the value of $f(x)$ when $x = 2$ is denoted by $f(2)$ and so on. Thus, when $f(x) = 3x^2 + 2x - 5$, then

$$f(0) = 3(0)^2 + 2(0) - 5 = -5$$

and $f(2) = 3(2)^2 + 2(2) - 5 = 11$ and so on.

Problem 1. If $f(x) = 4x^2 - 3x + 2$ find: $f(0), f(3), f(-1)$ and $f(3) - f(-1)$

$$f(x) = 4x^2 - 3x + 2$$

$$f(0) = 4(0)^2 - 3(0) + 2 = \mathbf{2}$$

$$f(3) = 4(3)^2 - 3(3) + 2$$

$$= 36 - 9 + 2 = \mathbf{29}$$

$$f(-1) = 4(-1)^2 - 3(-1) + 2$$

$$= 4 + 3 + 2 = \mathbf{9}$$

$$f(3) - f(-1) = 29 - 9 = \mathbf{20}$$

Problem 2. Given that $f(x) = 5x^2 + x - 7$ determine:
(i) $f(2) \div f(1)$ (iii) $f(3+a) - f(3)$
(ii) $f(3+a)$ (iv) $\dfrac{f(3+a) - f(3)}{a}$

$$f(x) = 5x^2 + x - 7$$

(i) $f(2) = 5(2)^2 + 2 - 7 = 15$

$f(1) = 5(1)^2 + 1 - 7 = -1$

$f(2) \div f(1) = \dfrac{15}{-1} = \mathbf{-15}$

(ii) $f(3+a) = 5(3+a)^2 + (3+a) - 7$

$$= 5(9 + 6a + a^2) + (3+a) - 7$$

$$= 45 + 30a + 5a^2 + 3 + a - 7$$

$$= \mathbf{41 + 31a + 5a^2}$$

(iii) $f(3) = 5(3)^2 + 3 - 7 = 41$

$$f(3+a) - f(3) = (41 + 31a + 5a^2) - (41)$$

$$= \mathbf{31a + 5a^2}$$

(iv) $\dfrac{f(3+a) - f(3)}{a} = \dfrac{31a + 5a^2}{a} = \mathbf{31 + 5a}$

Now try the following Practice Exercise

Practice Exercise 223   Further problems on functional notation (answers on page 1176)

1. If $f(x) = 6x^2 - 2x + 1$ find $f(0)$, $f(1)$, $f(2), f(-1)$ and $f(-3)$

2. If $f(x) = 2x^2 + 5x - 7$, find $f(1), f(2), f(-1)$, $f(2) - f(-1)$

3. Given $f(x) = 3x^3 + 2x^2 - 3x + 2$, prove that $f(1) = \frac{1}{7}f(2)$

4. If $f(x) = -x^2 + 3x + 6$, find $f(2), f(2+a)$, $f(2+a) - f(2)$ and $\dfrac{f(2+a) - f(2)}{a}$

52.3   The gradient of a curve

(a) For the curve shown in Figure 52.1, let the points A and B have co-ordinates (x_1, y_1) and (x_2, y_2), respectively. In functional notation, $y_1 = f(x_1)$ and $y_2 = f(x_2)$ as shown.
The gradient of the chord AB (i.e. the average rate of change from x_1 to x_2) is given by

$$\frac{BC}{AC} = \frac{BD - CD}{ED}$$

$$= \frac{f(x_2) - f(x_1)}{(x_2 - x_1)}$$

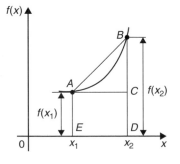

Figure 52.1

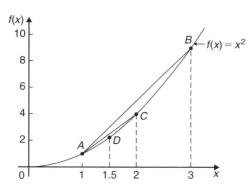

Figure 52.2

(b) For the curve $f(x) = x^2$ shown in Figure 52.2:

(i) the gradient of chord AB is given by

$$\frac{f(3) - f(1)}{3 - 1} = \frac{9 - 1}{2} = 4$$

(ii) the gradient of chord AC is given by

$$\frac{f(2) - f(1)}{2 - 1} = \frac{4 - 1}{1} = 3$$

(iii) the gradient of chord AD is given by

$$\frac{f(1.5) - f(1)}{1.5 - 1} = \frac{2.25 - 1}{0.5} = 2.5$$

(iv) if E is the point on the curve $(1.1, f(1.1))$, then the gradient of chord AE is given by

$$\frac{f(1.1) - f(1)}{1.1 - 1}$$
$$= \frac{1.21 - 1}{0.1} = 2.1$$

(v) if F is the point on the curve $(1.01, f(1.01))$, then the gradient of chord AF is given by

$$\frac{f(1.01) - f(1)}{1.01 - 1}$$
$$= \frac{1.0201 - 1}{0.01} = 2.01$$

Thus as point B moves closer and closer to point A the gradient of the chord approaches nearer and nearer to the value 2. This is called the **limiting value** of the gradient of the chord AB and when B coincides with A the chord becomes the tangent to the curve.

(c) If a tangent is drawn at a point P on a curve, then the gradient of this tangent is said to be the **gradient of the curve** at P. In Figure 52.3, the gradient of the curve at P is equal to the gradient of the tangent PQ.

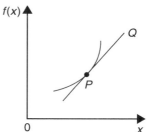

Figure 52.3

Now try the following Practice Exercise

Practice Exercise 224   A further problem on the gradient of a curve (answers on page 1176)

1. Plot the curve $f(x) = 4x^2 - 1$ for values of x from $x = -1$ to $x = +4$. Label the co-ordinates $(3, f(3))$ and $(1, f(1))$ as J and K, respectively. Join points J and K to form the chord JK. Determine the gradient of chord JK. By moving J nearer and nearer to K determine the gradient of the tangent of the curve at K.

52.4 Differentiation from first principles

(i) In Figure 52.4, A and B are two points very close together on a curve, δx (delta x) and δy (delta y) representing small increments in the x and y directions, respectively.

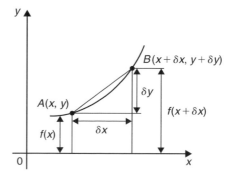

Figure 52.4

Gradient of chord $\quad AB = \dfrac{\delta y}{\delta x}$

However, $\qquad \delta y = f(x + \delta x) - f(x)$

Hence $\qquad \dfrac{\delta y}{\delta x} = \dfrac{f(x + \delta x) - f(x)}{\delta x}$

As δx approaches zero, $\dfrac{\delta y}{\delta x}$ approaches a limiting value and the gradient of the chord approaches the gradient of the tangent at A.

(ii) When determining the gradient of a tangent to a curve there are two notations used. The gradient of the curve at A in Figure 52.4 can either be written as:

$$\lim_{\delta x \to 0} \frac{\delta y}{\delta x} \text{ or } \lim_{\delta x \to 0} \left\{ \frac{f(x + \delta x) - f(x)}{\delta x} \right\}$$

In **Leibniz*** notation, $\dfrac{dy}{dx} = \lim\limits_{\delta x \to 0} \dfrac{\delta y}{\delta x}$

* **Who was Leibniz?** **Gottfried Wilhelm Leibniz** (sometimes von Leibniz) (1 July, 1646–14 November, 1716) was a German mathematician and philosopher. Leibniz developed infinitesimal calculus and invented the Leibniz wheel. To find out more go to **www.routledge.com/cw/bird**

In **functional notation**,

$$f'(x) = \lim_{\delta x \to 0} \left\{ \frac{f(x + \delta x) - f(x)}{\delta x} \right\}$$

(iii) $\dfrac{dy}{dx}$ is the same as $f'(x)$ and is called the **differential coefficient**, or more commonly, the **derivative**. The process of finding the differential coefficient is called **differentiation**.

Summarising, the differential coefficient,

$$\frac{dy}{dx} = f'(x) = \lim_{\delta x \to 0} \frac{\delta y}{\delta x}$$

$$= \lim_{\delta x \to 0} \left\{ \frac{f(x + \delta x) - f(x)}{\delta x} \right\}$$

Problem 3. Differentiate from first principles $f(x) = x^2$ and determine the value of the gradient of the curve at $x = 2$

To 'differentiate from first principles' means 'to find $f'(x)$' by using the expression

$$f'(x) = \lim_{\delta x \to 0} \left\{ \frac{f(x + \delta x) - f(x)}{\delta x} \right\}$$

$$f(x) = x^2$$

Substituting $(x + \delta x)$ for x gives
$f(x + \delta x) = (x + \delta x)^2 = x^2 + 2x\delta x + \delta x^2$, hence

$$f'(x) = \lim_{\delta x \to 0} \left\{ \frac{(x^2 + 2x\delta x + \delta x^2) - (x^2)}{\delta x} \right\}$$

$$= \lim_{\delta x \to 0} \left\{ \frac{2x\delta x + \delta x^2}{\delta x} \right\} = \lim_{\delta x \to 0} \{2x + \delta x\}$$

As $\delta x \to 0$, $[2x + \delta x] \to [2x + 0]$. Thus $f'(x) = 2x$, i.e. the differential coefficient of x^2 is $2x$. At $x = 2$, the gradient of the curve, $f'(x) = 2(2) = \mathbf{4}$

Problem 4. Find the differential coefficient of $y = 5x$

By definition, $\dfrac{dy}{dx} = f'(x)$

$$= \lim_{\delta x \to 0} \left\{ \frac{f(x + \delta x) - f(x)}{\delta x} \right\}$$

The function being differentiated is $y = f(x) = 5x$. Substituting $(x + \delta x)$ for x gives:

$f(x+\delta x) = 5(x+\delta x) = 5x + 5\delta x$. Hence

$$\frac{dy}{dx} = f'(x) = \lim_{\delta x \to 0}\left\{\frac{(5x+5\delta x)-(5x)}{\delta x}\right\}$$

$$= \lim_{\delta x \to 0}\left\{\frac{5\delta x}{\delta x}\right\} = \lim_{\delta x \to 0}\{5\}$$

Since the term δx does not appear in [5] the limiting value as $\delta x \to 0$ of [5] is 5. Thus $\dfrac{dy}{dx} = \mathbf{5}$, i.e. the differential coefficient of $5x$ is 5. The equation $y = 5x$ represents a straight line of gradient 5 (see Chapter 31). The 'differential coefficient' (i.e. $\dfrac{dy}{dx}$ or $f'(x)$) means 'the gradient of the curve', and since the slope of the line $y = 5x$ is 5 this result can be obtained by inspection. Hence, in general, if $y = kx$ (where k is a constant), then the gradient of the line is k and $\dfrac{dy}{dx}$ or $f'(x) = k$

Problem 5. Find the derivative of $y = 8$

$y = f(x) = 8$. Since there are no x-values in the original equation, substituting $(x+\delta x)$ for x still gives $f(x+\delta x) = 8$. Hence

$$\frac{dy}{dx} = f'(x) = \lim_{\delta x \to 0}\left\{\frac{f(x+\delta x)-f(x)}{\delta x}\right\}$$

$$= \lim_{\delta x \to 0}\left\{\frac{8-8}{\delta x}\right\} = 0$$

Thus, when $y = 8$, $\dfrac{dy}{dx} = \mathbf{0}$

The equation $y = 8$ represents a straight horizontal line and the gradient of a horizontal line is zero, hence the result could have been determined by inspection. 'Finding the derivative' means 'finding the gradient', hence, in general, for any horizontal line if $y = k$ (where k is a constant) then $\dfrac{dy}{dx} = 0$

Problem 6. Differentiate from first principles $f(x) = 2x^3$

Substituting $(x+\delta x)$ for x gives

$$f(x+\delta x) = 2(x+\delta x)^3$$

$$= 2(x+\delta x)(x^2 + 2x\delta x + \delta x^2)$$

$$= 2(x^3 + 3x^2\delta x + 3x\delta x^2 + \delta x^3)$$

$$= 2x^3 + 6x^2\delta x + 6x\delta x^2 + 2\delta x^3$$

$$\frac{dy}{dx} = f'(x) = \lim_{\delta x \to 0}\left\{\frac{f(x+\delta x)-f(x)}{\delta x}\right\}$$

$$= \lim_{\delta x \to 0}\left\{\frac{(2x^3+6x^2\delta x+6x\delta x^2+2\delta x^3)-(2x^3)}{\delta x}\right\}$$

$$= \lim_{\delta x \to 0}\left\{\frac{6x^2\delta x+6x\delta x^2+2\delta x^3}{\delta x}\right\}$$

$$= \lim_{\delta x \to 0}\{6x^2 + 6x\delta x + 2\delta x^2\}$$

Hence $f'(x) = \mathbf{6x^2}$, i.e. the differential coefficient of $2x^3$ is $6x^2$

Problem 7. Find the differential coefficient of $y = 4x^2 + 5x - 3$ and determine the gradient of the curve at $x = -3$

$$y = f(x) = 4x^2 + 5x - 3$$

$$f(x+\delta x) = 4(x+\delta x)^2 + 5(x+\delta x) - 3$$

$$= 4(x^2 + 2x\delta x + \delta x^2) + 5x + 5\delta x - 3$$

$$= 4x^2 + 8x\delta x + 4\delta x^2 + 5x + 5\delta x - 3$$

$$\frac{dy}{dx} = f'(x) = \lim_{\delta x \to 0}\left\{\frac{f(x+\delta x)-f(x)}{\delta x}\right\}$$

$$= \lim_{\delta x \to 0}\left\{\frac{\begin{array}{c}(4x^2+8x\delta x+4\delta x^2+5x+5\delta x-3)\\-(4x^2+5x-3)\end{array}}{\delta x}\right\}$$

$$= \lim_{\delta x \to 0}\left\{\frac{8x\delta x+4\delta x^2+5\delta x}{\delta x}\right\}$$

$$= \lim_{\delta x \to 0}\{8x + 4\delta x + 5\}$$

i.e. $\dfrac{dy}{dx} = f'(x) = \mathbf{8x+5}$

At $x = -3$, the gradient of the curve

$$= \frac{dy}{dx} = f'(x) = 8(-3) + 5 = \mathbf{-19}$$

Differentiation from first principles can be a lengthy process and it would not be convenient to go through this procedure every time we want to differentiate a function. In reality we do not have to, because a set of general rules have evolved from the above procedure, which we consider in the following section.

Now try the following Practice Exercise

Practice Exercise 225 Further problems on differentiation from first principles (answers on page 1176)

In Problems 1 to 12, differentiate from first principles.

1. $y = x$

2. $y = 7x$

3. $y = 4x^2$

4. $y = 5x^3$

5. $y = -2x^2 + 3x - 12$

6. $y = 23$

7. $f(x) = 9x$

8. $f(x) = \dfrac{2x}{3}$

9. $f(x) = 9x^2$

10. $f(x) = -7x^3$

11. $f(x) = x^2 + 15x - 4$

12. $f(x) = 4$

13. Determine $\dfrac{d}{dx}(4x^3)$ from first principles.

14. Find $\dfrac{d}{dx}(3x^2 + 5)$ from first principles.

52.5 Differentiation of $y = ax^n$ by the general rule

From differentiation by first principles, a general rule for differentiating ax^n emerges where a and n are any constants. This rule is:

$$\text{if } y = ax^n \text{ then } \frac{dy}{dx} = anx^{n-1}$$

$$\text{or, if } f(x) = ax^n \text{ then } f'(x) = anx^{n-1}$$

(Each of the results obtained in Problems 3 to 7 may be deduced by using this general rule.)
When differentiating, results can be expressed in a number of ways.

For example:

(i)   if $y = 3x^2$ then $\dfrac{dy}{dx} = 6x$,

(ii)   if $f(x) = 3x^2$ then $f'(x) = 6x$,

(iii)   the differential coefficient of $3x^2$ is $6x$,

(iv)   the derivative of $3x^2$ is $6x$, and

(v) $\dfrac{d}{dx}(3x^2) = 6x$

Addition rule for differentiation

If $y = f(x) + g(x)$ then $\dfrac{dy}{dx} = \dfrac{d}{dx}[f(x)] + \dfrac{d}{dx}[g(x)]$

For example, if $y = 3x^2 + 2x$ then

$$\frac{dy}{dx} = \frac{d}{dx}[3x^2] + \frac{d}{dx}[2x] = 6x + 2$$

Problem 8. Using the general rule, differentiate the following with respect to x:

(a) $y = 5x^7$ (b) $y = 3\sqrt{x}$ (c) $y = \dfrac{4}{x^2}$

(a) Comparing $y = 5x^7$ with $y = ax^n$ shows that $a = 5$ and $n = 7$. Using the general rule,

$$\frac{dy}{dx} = anx^{n-1} = (5)(7)x^{7-1} = \mathbf{35x^6}$$

(b) $y = 3\sqrt{x} = 3x^{\frac{1}{2}}$. Hence $a = 3$ and $n = \dfrac{1}{2}$

$$\frac{dy}{dx} = anx^{n-1} = (3)\frac{1}{2}x^{\frac{1}{2}-1}$$

$$= \frac{3}{2}x^{-\frac{1}{2}} = \frac{3}{2x^{\frac{1}{2}}} = \frac{3}{2\sqrt{x}}$$

(c) $y = \dfrac{4}{x^2} = 4x^{-2}$. Hence $a = 4$ and $n = -2$

$$\frac{dy}{dx} = anx^{n-1} = (4)(-2)x^{-2-1}$$

$$= -8x^{-3} = -\frac{8}{x^3}$$

Problem 9. Find the differential coefficient of $y = \dfrac{2}{5}x^3 - \dfrac{4}{x^3} + 4\sqrt{x^5} + 7$

$$y = \frac{2}{5}x^3 - \frac{4}{x^3} + 4\sqrt{x^5} + 7$$

i.e. $$y = \frac{2}{5}x^3 - 4x^{-3} + 4x^{5/2} + 7$$

$$\frac{dy}{dx} = \left(\frac{2}{5}\right)(3)x^{3-1} - (4)(-3)x^{-3-1}$$

$$+ (4)\left(\frac{5}{2}\right)x^{(5/2)-1} + 0$$

$$= \frac{6}{5}x^2 + 12x^{-4} + 10x^{3/2}$$

i.e. $$\frac{dy}{dx} = \frac{6}{5}x^2 + \frac{12}{x^4} + 10\sqrt{x^3}$$

Problem 10. If $f(t) = 5t + \dfrac{1}{\sqrt{t^3}}$ find $f'(t)$

$$f(t) = 5t + \frac{1}{\sqrt{t^3}} = 5t + \frac{1}{t^{\frac{3}{2}}} = 5t^1 + t^{-\frac{3}{2}}$$

Hence $$f'(t) = (5)(1)t^{1-1} + \left(-\frac{3}{2}\right)t^{-\frac{3}{2}-1}$$

$$= 5t^0 - \frac{3}{2}t^{-\frac{5}{2}}$$

i.e. $$f'(t) = 5 - \frac{3}{2t^{\frac{5}{2}}} = \mathbf{5} - \frac{\mathbf{3}}{\mathbf{2}\sqrt{t^5}}$$

Problem 11. Differentiate $y = \dfrac{(x+2)^2}{x}$ with respect to x

$$y = \frac{(x+2)^2}{x} = \frac{x^2 + 4x + 4}{x}$$

$$= \frac{x^2}{x} + \frac{4x}{x} + \frac{4}{x}$$

i.e. $$y = x + 4 + 4x^{-1}$$

Hence $$\frac{dy}{dx} = 1 + 0 + (4)(-1)x^{-1-1}$$

$$= 1 - 4x^{-2} = \mathbf{1} - \frac{\mathbf{4}}{x^2}$$

Now try the following Practice Exercise

Practice Exercise 226   Further problems on differentiation of $y = ax^n$ by the general rule (answers on page 1176)

In Problems 1 to 8, determine the differential coefficients with respect to the variable.

1. $y = 7x^4$

2. $y = \sqrt{x}$

3. $y = \sqrt{t^3}$

4. $y = 6 + \dfrac{1}{x^3}$

5. $y = 3x - \dfrac{1}{\sqrt{x}} + \dfrac{1}{x}$

6. $y = \dfrac{5}{x^2} - \dfrac{1}{\sqrt{x^7}} + 2$

7. $y = 3(t-2)^2$

8. $y = (x+1)^3$

9.   Using the general rule for ax^n, check the results of Problems 1 to 12 of Exercise 225, page 626.

10. Differentiate $f(x) = 6x^2 - 3x + 5$ and find the gradient of the curve at (a) $x = -1$, and (b) $x = 2$

11. Find the differential coefficient of $y = 2x^3 + 3x^2 - 4x - 1$ and determine the gradient of the curve at $x = 2$

12. Determine the derivative of $y = -2x^3 + 4x + 7$ and determine the gradient of the curve at $x = -1.5$

52.6 Differentiation of sine and cosine functions

Figure 52.5(a) shows a graph of $y = \sin\theta$. The gradient is continually changing as the curve moves from 0 to A to B to C to D. The gradient, given by $\dfrac{dy}{d\theta}$, may be plotted in a corresponding position below $y = \sin\theta$, as shown in Figure 52.5(b).

(i)   At 0, the gradient is positive and is at its steepest. Hence $0'$ is the maximum positive value.

(ii) Between 0 and A the gradient is positive but is decreasing in value until at A the gradient is zero, shown as A'.

(iii) Between A and B the gradient is negative but is increasing in value until at B the gradient is at its steepest. Hence B' is a maximum negative value.

(iv) If the gradient of $y = \sin\theta$ is further investigated between B and C and C and D then the resulting graph of $\dfrac{dy}{d\theta}$ is seen to be a cosine wave.

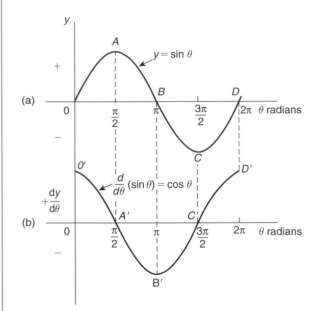

Figure 52.5

Hence the rate of change of $\sin\theta$ is $\cos\theta$, i.e.

if $\qquad y = \sin\theta$ then $\dfrac{dy}{d\theta} = \cos\theta$

It may also be shown that:

if $\qquad y = \sin a\theta$, $\dfrac{dy}{d\theta} = a\cos a\theta$

$\qquad\qquad$ (where a is a constant)

and if $\qquad y = \sin(a\theta + \alpha)$, $\dfrac{dy}{d\theta} = a\cos(a\theta + \alpha)$

$\qquad\qquad$ (where a and α are constants).

If a similar exercise is followed for $y = \cos\theta$ then the graphs of Figure 52.6 result, showing $\dfrac{dy}{d\theta}$ to be a graph of $\sin\theta$, but displaced by π radians. If each point on the curve $y = \sin\theta$ (as shown in Figure 52.5(a)) were to be made negative, (i.e. $+\dfrac{\pi}{2}$ is made $-\dfrac{\pi}{2}$, $-\dfrac{3\pi}{2}$ is made $+\dfrac{3\pi}{2}$, and so on) then the graph shown in Figure 52.6(b) would result. This latter graph therefore represents the curve of $-\sin\theta$.

Thus, if $y = \cos\theta$, $\dfrac{dy}{d\theta} = -\sin\theta$

It may also be shown that:

if $\qquad y = \cos a\theta$, $\dfrac{dy}{d\theta} = -a\sin a\theta$

$\qquad\qquad$ (where a is a constant)

and if $\qquad y = \cos(a\theta + \alpha)$, $\dfrac{dy}{d\theta} = -a\sin(a\theta + \alpha)$

$\qquad\qquad$ (where a and α are constants).

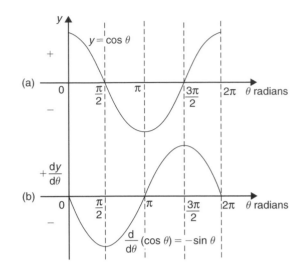

Figure 52.6

Problem 12. Differentiate the following with respect to the variable: (a) $y = 2\sin 5\theta$
(b) $f(t) = 3\cos 2t$

(a) $y = 2\sin 5\theta$

$\dfrac{dy}{d\theta} = (2)(5)\cos 5\theta = \mathbf{10\cos 5\theta}$

(b) $f(t) = 3\cos 2t$

$f'(t) = (3)(-2)\sin 2t = \mathbf{-6\sin 2t}$

Problem 13. Find the differential coefficient of $y = 7\sin 2x - 3\cos 4x$

$y = 7\sin 2x - 3\cos 4x$

$\dfrac{dy}{dx} = (7)(2)\cos 2x - (3)(-4)\sin 4x$

$\qquad = \mathbf{14\cos 2x + 12\sin 4x}$

Problem 14. Differentiate the following with respect to the variable:

(a) $f(\theta) = 5\sin(100\pi\theta - 0.40)$
(b) $f(t) = 2\cos(5t + 0.20)$

(a) If $f(\theta) = 5\sin(100\pi\theta - 0.40)$

$\quad f'(\theta) = 5[100\pi\cos(100\pi\theta - 0.40)]$

$\qquad = \mathbf{500\pi\cos(100\pi\theta - 0.40)}$

(b) If $f(t) = 2\cos(5t + 0.20)$

$\quad f'(t) = 2[-5\sin(5t + 0.20)]$

$\qquad = \mathbf{-10\sin(5t + 0.20)}$

Problem 15. An alternating voltage is given by: $v = 100 \sin 200t$ volts, where t is the time in seconds. Calculate the rate of change of voltage when (a) $t = 0.005$ s and (b) $t = 0.01$ s

$v = 100 \sin 200t$ volts. The rate of change of v is given by $\dfrac{dv}{dt}$

$$\frac{dv}{dt} = (100)(200)\cos 200t = 20\,000\cos 200t$$

(a) When $t = 0.005$ s,

$$\frac{dv}{dt} = 20\,000\cos(200)(0.005) = 20\,000\cos 1$$

cos 1 means 'the cosine of 1 radian' (make sure your calculator is on radians – not degrees).

Hence $\dfrac{dv}{dt} =$ **10 806 volts per second**

(b) When $t = 0.01$ s,

$$\frac{dv}{dt} = 20\,000\cos(200)(0.01) = 20\,000\cos 2$$

Hence $\dfrac{dv}{dt} =$ **−8323 volts per second**

Now try the following Practice Exercise

Practice Exercise 227   Further problems on the differentiation of sine and cosine functions (answers on page 1176)

1. Differentiate with respect to x: (a) $y = 4\sin 3x$ (b) $y = 2\cos 6x$

2. Given $f(\theta) = 2\sin 3\theta - 5\cos 2\theta$, find $f'(\theta)$

3. An alternating current is given by $i = 5\sin 100t$ amperes, where t is the time in seconds. Determine the rate of change of current when $t = 0.01$ seconds

4. $v = 50\sin 40t$ volts represents an alternating voltage where t is the time in seconds. At a time of 20×10^{-3} seconds, find the rate of change of voltage

5. If $f(t) = 3\sin(4t + 0.12) - 2\cos(3t - 0.72)$ determine $f'(t)$

52.7   Differentiation of e^{ax} and ln ax

A graph of $y = e^x$ is shown in Figure 52.7(a) from page 149, Figure 16.1. The gradient of the curve at any point

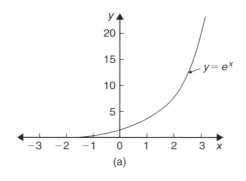

(a)

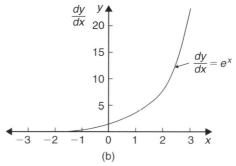

(b)

Figure 52.7

is given by $\dfrac{dy}{dx}$ and is continually changing. By drawing tangents to the curve at many points on the curve and measuring the gradient of the tangents, values of $\dfrac{dy}{dx}$ for corresponding values of x may be obtained. These values are shown graphically in Figure 52.7(b). The graph of $\dfrac{dy}{dx}$ against x is identical to the original graph of $y = e^x$. It follows that:

$$\text{if } y = e^x, \text{ then } \frac{dy}{dx} = e^x$$

It may also be shown that

$$\text{if } y = e^{ax}, \text{ then } \frac{dy}{dx} = ae^{ax}$$

Therefore if $y = 2e^{6x}$, then $\dfrac{dy}{dx} = (2)(6e^{6x}) = \mathbf{12e^{6x}}$

A graph of $y = \ln x$ is shown in Figure 52.8(a) from page 144, Figure 15.2. The gradient of the curve at any point is given by $\dfrac{dy}{dx}$ and is continually changing. By drawing tangents to the curve at many points on the curve and measuring the gradient of the tangents, values of $\dfrac{dy}{dx}$ for corresponding values of x may be obtained. These values are shown graphically in Figure 52.8(b). The graph of $\dfrac{dy}{dx}$ against x is the graph of $\dfrac{dy}{dx} = \dfrac{1}{x}$

Section I

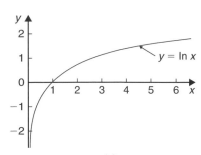

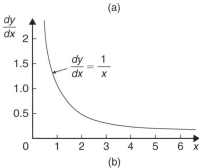

Figure 52.8

It follows that: if $y = \ln x$, then $\dfrac{dy}{dx} = \dfrac{1}{x}$
It may also be shown that

$$\text{if } y = \ln ax, \text{ then } \frac{dy}{dx} = \frac{1}{x}$$

(Note that in the latter expression 'a' does not appear in the $\dfrac{dy}{dx}$ term.)

Thus if $y = \ln 4x$, then $\dfrac{dy}{dx} = \dfrac{1}{x}$

Problem 16. Differentiate the following with respect to the variable: (a) $y = 3e^{2x}$
(b) $f(t) = \dfrac{4}{3e^{5t}}$

(a) If $y = 3e^{2x}$ then $\dfrac{dy}{dx} = (3)(2e^{2x}) = \mathbf{6e^{2x}}$

(b) If $f(t) = \dfrac{4}{3e^{5t}} = \dfrac{4}{3}e^{-5t}$, then

$$f'(t) = \frac{4}{3}(-5e^{-5t}) = -\frac{20}{3}e^{-5t} = -\frac{\mathbf{20}}{\mathbf{3e^{5t}}}$$

Problem 17. Differentiate $y = 5 \ln 3x$

If $y = 5 \ln 3x$, then $\dfrac{dy}{dx} = (5)\left(\dfrac{1}{x}\right) = \dfrac{\mathbf{5}}{\mathbf{x}}$

Now try the following Practice Exercise

Practice Exercise 228 Further problems on the differentiation of e^{ax} and ln ax (answers on page 1176)

1. Differentiate with respect to x:

    (a) $y = 5e^{3x}$ (b) $y = \dfrac{2}{7e^{2x}}$

2. Given $f(\theta) = 5 \ln 2\theta - 4 \ln 3\theta$, determine $f'(\theta)$

3. If $f(t) = 4 \ln t + 2$, evaluate $f'(t)$ when $t = 0.25$

4. Evaluate $\dfrac{dy}{dx}$ when $x = 1$, given

    $y = 3e^{4x} - \dfrac{5}{2e^{3x}} + 8 \ln 5x$. Give the answer correct to 3 significant figures.

Section I

For fully worked solutions to each of the problems in Practice Exercises 223 to 228 in this chapter, go to the website:
www.routledge.com/cw/bird

Methods of differentiation

At the end of this chapter, you should be able to:

- differentiate common functions
- differentiate a product using the product rule
- differentiate a quotient using the quotient rule
- differentiate a function of a function
- differentiate successively

53.1 Differentiation of common functions

The **standard derivatives** summarised below were derived in Chapter 52 and are true for all real values of x.

y or $f(x)$	$\dfrac{dy}{dx}$ of $f'(x)$
ax^n	anx^{n-1}
$\sin ax$	$a\cos ax$
$\cos ax$	$-a\sin ax$
e^{ax}	ae^{ax}
$\ln ax$	$\dfrac{1}{x}$

The **differential coefficient of a sum or difference** is the sum or difference of the differential coefficients of the separate terms.

Thus, if $f(x) = p(x) + q(x) - r(x)$, (where f, p, q and r are functions), then $f'(x) = p'(x) + q'(x) - r'(x)$

Differentiation of common functions is demonstrated in the following Problems.

Problem 1. Find the differential coefficients of:

(a) $y = 12x^3$ (b) $y = \dfrac{12}{x^3}$

If $y = ax^n$ then $\dfrac{dy}{dx} = anx^{n-1}$

(a) Since $y = 12x^3$, $a = 12$ and $n = 3$, thus

$$\dfrac{dy}{dx} = (12)(3)x^{3-1} = \mathbf{36x^2}$$

(b) $y = \dfrac{12}{x^3}$ is rewritten in the standard ax^n form as

$y = 12x^{-3}$ and in the general rule $a = 12$ and $n = -3$

Thus $\dfrac{dy}{dx} = (12)(-3)x^{-3-1}$

$= -36x^{-4} = -\dfrac{36}{x^4}$

Problem 2. Differentiate: (a) $y = 6$ (b) $y = 6x$

(a) $y = 6$ may be written as $y = 6x^0$, i.e. in the general rule $a = 6$ and $n = 0$

Hence $\dfrac{dy}{dx} = (6)(0)x^{0-1} = 0$

In general, **the differential coefficient of a constant is always zero**.

(b) Since $y = 6x$, in the general rule $a = 6$ and $n = 1$

Hence $\dfrac{dy}{dx} = (6)(1)x^{1-1} = 6x^0 = 6$

In general, the differential coefficient of kx, where k is a constant, is always k.

Problem 3. Find the derivatives of:

(a) $y = 3\sqrt{x}$ (b) $y = \dfrac{5}{\sqrt[3]{x^4}}$

(a) $y = 3\sqrt{x}$ is rewritten in the standard differential form as $y = 3x^{1/2}$

In the general rule, $a = 3$ and $n = \dfrac{1}{2}$

Thus $\dfrac{dy}{dx} = (3)\left(\dfrac{1}{2}\right)x^{\frac{1}{2}-1} = \dfrac{3}{2}x^{-\frac{1}{2}}$

$= \dfrac{3}{2x^{1/2}} = \dfrac{3}{2\sqrt{x}}$

(b) $y = \dfrac{5}{\sqrt[3]{x^4}} = \dfrac{5}{x^{4/3}} = 5x^{-4/3}$

In the general rule, $a = 5$ and $n = -\dfrac{4}{3}$

Thus $\dfrac{dy}{dx} = (5)\left(-\dfrac{4}{3}\right)x^{(-4/3)-1}$

$= \dfrac{-20}{3}x^{-7/3} = \dfrac{-20}{3x^{7/3}} = \dfrac{-20}{3\sqrt[3]{x^7}}$

Problem 4. Differentiate:

$y = 5x^4 + 4x - \dfrac{1}{2x^2} + \dfrac{1}{\sqrt{x}} - 3$ with respect to x

$y = 5x^4 + 4x - \dfrac{1}{2x^2} + \dfrac{1}{\sqrt{x}} - 3$ is rewritten as

$y = 5x^4 + 4x - \dfrac{1}{2}x^{-2} + x^{-1/2} - 3$

When differentiating a sum, each term is differentiated in turn.

Thus $\dfrac{dy}{dx} = (5)(4)x^{4-1} + (4)(1)x^{1-1} - \dfrac{1}{2}(-2)x^{-2-1}$

$+ (1)\left(-\dfrac{1}{2}\right)x^{(-1/2)-1} - 0$

$= 20x^3 + 4 + x^{-3} - \dfrac{1}{2}x^{-3/2}$

i.e. $\dfrac{dy}{dx} = 20x^3 + 4 + \dfrac{1}{x^3} - \dfrac{1}{2\sqrt{x^3}}$

Problem 5. Find the differential coefficients of:
(a) $y = 3\sin 4x$ (b) $f(t) = 2\cos 3t$ with respect to the variable

(a) When $y = 3\sin 4x$ then $\dfrac{dy}{dx} = (3)(4\cos 4x)$

$= \mathbf{12\cos 4x}$

(b) When $f(t) = 2\cos 3t$ then
$f'(t) = (2)(-3\sin 3t) = -6\sin 3t$

Problem 6. Determine the derivatives of:

(a) $y = 3e^{5x}$ (b) $f(\theta) = \dfrac{2}{e^{3\theta}}$ (c) $y = 6\ln 2x$

(a) When $y = 3e^{5x}$ then $\dfrac{dy}{dx} = (3)(5)e^{5x} = 15e^{5x}$

(b) $f(\theta) = \dfrac{2}{e^{3\theta}} = 2e^{-3\theta}$, thus

$f'(\theta) = (2)(-3)e^{-3\theta} = -6e^{-3\theta} = \dfrac{-6}{e^{3\theta}}$

(c) When $y = 6\ln 2x$ then $\dfrac{dy}{dx} = 6\left(\dfrac{1}{x}\right) = \dfrac{6}{x}$

Problem 7. Find the gradient of the curve $y = 3x^4 - 2x^2 + 5x - 2$ at the points $(0, -2)$ and $(1, 4)$

The gradient of a curve at a given point is given by the corresponding value of the derivative. Thus, since $y = 3x^4 - 2x^2 + 5x - 2$ then

the gradient $= \dfrac{dy}{dx} = 12x^3 - 4x + 5$

At the point $(0, -2)$, $x = 0$

Thus the gradient $= 12(0)^3 - 4(0) + 5 = \mathbf{5}$

At the point $(1, 4)$, $x = 1$

Thus the gradient $= 12(1)^3 - 4(1) + 5 = \mathbf{13}$

Problem 8. Determine the co-ordinates of the point on the graph $y = 3x^2 - 7x + 2$ where the gradient is -1

The gradient of the curve is given by the derivative.

When $y = 3x^2 - 7x + 2$ then $\dfrac{dy}{dx} = 6x - 7$

Since the gradient is -1 then $6x - 7 = -1$, from which, $x = 1$

When $x = 1$, $y = 3(1)^2 - 7(1) + 2 = -2$

Hence the gradient is -1 at the point $(1, -2)$

Now try the following Practice Exercise

Practice Exercise 229   Further problems on differentiating common functions (answers on page 1176)

In Problems 1 to 6, find the differential coefficients of the given functions with respect to the variable.

1. (a) $5x^5$ (b) $2.4x^{3.5}$ (c) $\dfrac{1}{x}$

2. (a) $\dfrac{-4}{x^2}$ (b) 6 (c) $2x$

3. (a) $2\sqrt{x}$ (b) $3\sqrt[3]{x^5}$ (c) $\dfrac{4}{\sqrt{x}}$

4. (a) $\dfrac{-3}{\sqrt[3]{x}}$ (b) $(x - 1)^2$ (c) $2\sin 3x$

5. (a) $-4\cos 2x$ (b) $2e^{6x}$ (c) $\dfrac{3}{e^{5x}}$

6. (a) $4\ln 9x$ (b) $\dfrac{e^x - e^{-x}}{2}$ (c) $\dfrac{1 - \sqrt{x}}{x}$

7. Find the gradient of the curve $y = 2t^4 + 3t^3 - t + 4$ at the points $(0, 4)$ and $(1, 8)$

8. Find the co-ordinates of the point on graph $y = 5x^2 - 3x + 1$ where the gradient is 2

9. (a) Differentiate
    $$y = \frac{2}{\theta^2} + 2\ln 2\theta - 2(\cos 5\theta + 3\sin 2\theta) - \frac{2}{e^{3\theta}}$$
    (b) Evaluate $\dfrac{dy}{d\theta}$ when $\theta = \dfrac{\pi}{2}$, correct to 4 significant figures.

10. Evaluate $\dfrac{ds}{dt}$, correct to 3 significant figures, when $t = \dfrac{\pi}{6}$ given $s = 3\sin t - 3 + \sqrt{t}$

11. A mass, m, is held by a spring with a stiffness constant k. The potential energy, p, of the system is given by: $p = \dfrac{1}{2}kx^2 - mgx$ where x is the displacement and g is acceleration due to gravity. The system is in equilibrium if $\dfrac{dp}{dx} = 0$. Determine the expression for x for system equilibrium.

12. The current i flowing in an inductor of inductance 100 mH is given by: $i = 5\sin 100t$ amperes, where t is the time t in seconds. The voltage v across the inductor is given by: $v = L\dfrac{di}{dt}$ volts. Determine the voltage when $t = 10$ ms.

53.2 Differentiation of a product

When $y = uv$, and u and v are both functions of x,

then $$\frac{dy}{dx} = u\frac{dv}{dx} + v\frac{du}{dx}$$

This is known as the **product rule**.

Problem 9. Find the differential coefficient of: $y = 3x^2 \sin 2x$

$3x^2 \sin 2x$ is a product of two terms $3x^2$ and $\sin 2x$. Let $u = 3x^2$ and $v = \sin 2x$

Using the product rule:

$$\frac{dy}{dx} = \underset{\downarrow}{u} \quad \underset{\downarrow}{\frac{dv}{dx}} \quad + \quad \underset{\downarrow}{v} \quad \underset{\downarrow}{\frac{du}{dx}}$$

gives: $\dfrac{dy}{dx} = (3x^2)(2\cos 2x) + (\sin 2x)(6x)$

i.e. $\dfrac{dy}{dx} = 6x^2 \cos 2x + 6x \sin 2x$

$= \mathbf{6x(x\cos 2x + \sin 2x)}$

Section I

Note that the differential coefficient of a product is **not** obtained by merely differentiating each term and multiplying the two answers together. The product rule formula **must** be used when differentiating products.

Problem 10. Find the rate of change of y with respect to x given $y = 3\sqrt{x}\ln 2x$

The rate of change of y with respect to x is given by $\dfrac{dy}{dx}$

$y = 3\sqrt{x}\ln 2x = 3x^{1/2}\ln 2x$, which is a product.

Let $u = 3x^{1/2}$ and $v = \ln 2x$

Then $\dfrac{dy}{dx} = \underset{\downarrow}{u}\ \underset{\downarrow}{\dfrac{dv}{dx}} + \underset{\downarrow}{v}\ \underset{\downarrow}{\dfrac{du}{dx}}$

$= (3x^{1/2})\left(\dfrac{1}{x}\right) + (\ln 2x)\left[3\left(\dfrac{1}{2}\right)x^{(1/2)-1}\right]$

$= 3x^{(1/2)-1} + (\ln 2x)\left(\dfrac{3}{2}\right)x^{-1/2}$

$= 3x^{-1/2}\left(1 + \dfrac{1}{2}\ln 2x\right)$

i.e. $\dfrac{dy}{dx} = \dfrac{3}{\sqrt{x}}\left(1 + \dfrac{1}{2}\ln 2x\right)$

Problem 11. Differentiate: $y = x^3\cos 3x\ln x$

Let $u = x^3\cos 3x$ (i.e. a product) and $v = \ln x$

Then $\dfrac{dy}{dx} = u\dfrac{dv}{dx} + v\dfrac{du}{dx}$

where $\dfrac{du}{dx} = (x^3)(-3\sin 3x) + (\cos 3x)(3x^2)$

and $\dfrac{dv}{dx} = \dfrac{1}{x}$

Hence $\dfrac{dy}{dx} = (x^3\cos 3x)\left(\dfrac{1}{x}\right)$

$\qquad\qquad + (\ln x)[-3x^3\sin 3x + 3x^2\cos 3x]$

$= x^2\cos 3x + 3x^2\ln x(\cos 3x - x\sin 3x)$

i.e. $\dfrac{dy}{dx} = x^2\{\cos 3x + 3\ln x(\cos 3x - x\sin 3x)\}$

Problem 12. Determine the rate of change of voltage, given $v = 5t\sin 2t$ volts, when $t = 0.2\,\text{s}$

Rate of change of voltage

$= \dfrac{dv}{dt} = (5t)(2\cos 2t) + (\sin 2t)(5)$

$= 10t\cos 2t + 5\sin 2t$

When $t = 0.2$,

$\dfrac{dv}{dt} = 10(0.2)\cos 2(0.2) + 5\sin 2(0.2)$

$= 2\cos 0.4 + 5\sin 0.4$

(where cos 0.4 means the cosine of 0.4 radians $= 0.92106$)

Hence $\dfrac{dv}{dt} = 2(0.92106) + 5(0.38942)$

$= 1.8421 + 1.9471 = 3.7892$

i.e. **the rate of change of voltage when $t = 0.2\,\text{s}$ is 3.79 volts/s, correct to 3 significant figures.**

Now try the following Practice Exercise

Practice Exercise 230   Further problems on differentiating products (answers on page 1176)

In Problems 1 to 8, differentiate the given products with respect to the variable.

1. $x\sin x$

2. $x^2 e^{2x}$

3. $x^2\ln x$

4. $2x^3\cos 3x$

5. $\sqrt{x^3}\ln 3x$

6. $e^{3t}\sin 4t$

7. $e^{4\theta}\ln 3\theta$

8. $e^t\ln t\cos t$

9. Evaluate $\dfrac{di}{dt}$, correct to 4 significant figures, when $t = 0.1$, and $i = 15t\sin 3t$

10. Evaluate $\dfrac{dz}{dt}$, correct to 4 significant figures, when $t = 0.5$, given that $z = 2e^{3t}\sin 2t$

53.3 Differentiation of a quotient

When $y = \dfrac{u}{v}$, and u and v are both functions of x

then
$$\frac{dy}{dx} = \frac{v\dfrac{du}{dx} - u\dfrac{dv}{dx}}{v^2}$$

This is known as the **quotient rule**.

Problem 13. Find the differential coefficient of
$$y = \frac{4\sin 5x}{5x^4}$$

$\dfrac{4\sin 5x}{5x^4}$ is a quotient. Let $u = 4\sin 5x$ and $v = 5x^4$
(Note that v is **always** the denominator and u the numerator.)

$$\frac{dy}{dx} = \frac{v\dfrac{du}{dx} - u\dfrac{dv}{dx}}{v^2}$$

where $\dfrac{du}{dx} = (4)(5)\cos 5x = 20\cos 5x$

and $\dfrac{dv}{dx} = (5)(4)x^3 = 20x^3$

Hence $\dfrac{dy}{dx} = \dfrac{(5x^4)(20\cos 5x) - (4\sin 5x)(20x^3)}{(5x^4)^2}$

$$= \frac{100x^4\cos 5x - 80x^3\sin 5x}{25x^8}$$

$$= \frac{20x^3[5x\cos 5x - 4\sin 5x]}{25x^8}$$

i.e. $\dfrac{dy}{dx} = \dfrac{4}{5x^5}(5x\cos 5x - 4\sin 5x)$

Note that the differential coefficient is **not** obtained by merely differentiating each term in turn and then dividing the numerator by the denominator. The quotient formula **must** be used when differentiating quotients.

Problem 14. Determine the differential coefficient of $y = \tan ax$

$y = \tan ax = \dfrac{\sin ax}{\cos ax}$ (from Chapter 42). Differentiation of $\tan ax$ is thus treated as a quotient with $u = \sin ax$ and $v = \cos ax$

$$\frac{dy}{dx} = \frac{v\dfrac{du}{dx} - u\dfrac{dv}{dx}}{v^2}$$

$$= \frac{(\cos ax)(a\cos ax) - (\sin ax)(-a\sin ax)}{(\cos ax)^2}$$

$$= \frac{a\cos^2 ax + a\sin^2 ax}{(\cos ax)^2}$$

$$= \frac{a(\cos^2 ax + \sin^2 ax)}{\cos^2 ax}$$

$$= \frac{a}{\cos^2 ax} \text{ since } \cos^2 ax + \sin^2 ax = 1$$

(see Chapter 42)

Hence $\dfrac{dy}{dx} = a\sec^2 ax$ since $\sec^2 ax = \dfrac{1}{\cos^2 ax}$
(see Chapter 38)

Problem 15. Find the derivative of $y = \sec ax$

$y = \sec ax = \dfrac{1}{\cos ax}$ (i.e. a quotient). Let $u = 1$ and $v = \cos ax$

$$\frac{dy}{dx} = \frac{v\dfrac{du}{dx} - u\dfrac{dv}{dx}}{v^2}$$

$$= \frac{(\cos ax)(0) - (1)(-a\sin ax)}{(\cos ax)^2}$$

$$= \frac{a\sin ax}{\cos^2 ax} = a\left(\frac{1}{\cos ax}\right)\left(\frac{\sin ax}{\cos ax}\right)$$

i.e. $\dfrac{dy}{dx} = a\sec ax\tan ax$

Problem 16. Differentiate $y = \dfrac{te^{2t}}{2\cos t}$

The function $\dfrac{te^{2t}}{2\cos t}$ is a quotient, whose numerator is a product.
Let $u = te^{2t}$ and $v = 2\cos t$

then $\dfrac{du}{dt} = (t)(2e^{2t}) + (e^{2t})(1)$ (by the product rule)

and $\dfrac{dv}{dt} = -2\sin t$

Hence $\dfrac{dy}{dt} = \dfrac{v\dfrac{du}{dt} - u\dfrac{dv}{dt}}{v^2}$

$$= \frac{(2\cos t)[2te^{2t} + e^{2t}] - (te^{2t})(-2\sin t)}{(2\cos t)^2}$$

$$= \frac{4te^{2t}\cos t + 2e^{2t}\cos t + 2te^{2t}\sin t}{4\cos^2 t}$$

$$= \frac{2e^{2t}[2t\cos t + \cos t + t\sin t]}{4\cos^2 t}$$

i.e. $\dfrac{dy}{dt} = \dfrac{e^{2t}}{2\cos^2 t}(2t\cos t + \cos t + t\sin t)$

Problem 17. Determine the gradient of the curve $y = \dfrac{5x}{2x^2 + 4}$ at the point $\left(\sqrt{3}, \dfrac{\sqrt{3}}{2}\right)$

Let $y = 5x$ and $v = 2x^2 + 4$

$$\frac{dy}{dx} = \frac{v\dfrac{du}{dx} - u\dfrac{dv}{dx}}{v^2} = \frac{(2x^2 + 4)(5) - (5x)(4x)}{(2x^2 + 4)^2}$$

$$= \frac{10x^2 + 20 - 20x^2}{(2x^2 + 4)^2} = \frac{20 - 10x^2}{(2x^2 + 4)^2}$$

At the point $\left(\sqrt{3}, \dfrac{\sqrt{3}}{2}\right)$, $x = \sqrt{3}$,

hence the gradient $= \dfrac{dy}{dx} = \dfrac{20 - 10(\sqrt{3})^2}{[2(\sqrt{3})^2 + 4]^2}$

$$= \frac{20 - 30}{100} = -\frac{1}{10}$$

Now try the following Practice Exercise

Practice Exercise 231   Further problems on differentiating quotients (answers on page 1177)

In Problems 1 to 7, differentiate the quotients with respect to the variable.

1. $\dfrac{\sin x}{x}$

2. $\dfrac{2\cos 3x}{x^3}$

3. $\dfrac{2x}{x^2 + 1}$

4. $\dfrac{\sqrt{x}}{\cos x}$

5. $\dfrac{3\sqrt{\theta^3}}{2\sin 2\theta}$

6. $\dfrac{\ln 2t}{\sqrt{t}}$

7. $\dfrac{2xe^{4x}}{\sin x}$

8. Find the gradient of the curve $y = \dfrac{2x}{x^2 - 5}$ at the point $(2, -4)$

9. Evaluate $\dfrac{dy}{dx}$ at $x = 2.5$, correct to 3 significant figures, given $y = \dfrac{2x^2 + 3}{\ln 2x}$

53.4 Function of a function

It is often easier to make a substitution before differentiating.

If y is a function of x then $\dfrac{dy}{dx} = \dfrac{dy}{du} \times \dfrac{du}{dx}$

This is known as the **'function of a function'** rule (or sometimes the **chain rule**).

For example, if $y = (3x - 1)^9$ then, by making the substitution $u = (3x - 1)$, $y = u^9$, which is of the 'standard' from.

Hence $\dfrac{dy}{du} = 9u^8$ and $\dfrac{du}{dx} = 3$

Then $\dfrac{dy}{dx} = \dfrac{dy}{du} \times \dfrac{du}{dx} = (9u^8)(3) = 27u^8$

Rewriting u as $(3x - 1)$ gives: $\dfrac{dy}{dx} = 27(3x - 1)^8$

Since y is a function of u, and u is a function of x, then y is a function of a function of x.

Problem 18. Differentiate $y = 3\cos(5x^2 + 2)$

Let $u = 5x^2 + 2$ then $y = 3\cos u$

Hence $\dfrac{du}{dx} = 10x$ and $\dfrac{dy}{du} = -3\sin u$

Using the function of a function rule,

$$\frac{dy}{dx} = \frac{dy}{du} \times \frac{du}{dx} = (-3\sin u)(10x) = -30x\sin u$$

Rewriting u as $5x^2 + 2$ gives:

$$\frac{dy}{dx} = -30x\sin(5x^2 + 2)$$

Problem 19. Find the derivative of $y = (4t^3 - 3t)^6$

Let $u = 4t^3 - 3t$, then $y = u^6$

Hence $\dfrac{du}{dt} = 12t^2 - 3$ and $\dfrac{dy}{dt} = 6u^5$

Using the function of a function rule,

$$\frac{dy}{dt} = \frac{dy}{du} \times \frac{du}{dt} = (6u^5)(12t^2 - 3)$$

Rewriting u as $(4t^3 - 3t)$ gives:

$$\frac{dy}{dt} = 6(4t^3 - 3t)^5(12t^2 - 3)$$

$$= 18(4t^2 - 1)(4t^3 - 3t)^5$$

Problem 20. Determine the differential coefficient of: $y = \sqrt{3x^2 + 4x - 1}$

$y = \sqrt{3x^2 + 4x - 1} = (3x^2 + 4x - 1)^{1/2}$

Let $u = 3x^2 + 4x - 1$ then $y = u^{1/2}$

Hence $\dfrac{du}{dx} = 6x + 4$ and $\dfrac{dy}{du} = \dfrac{1}{2}u^{-1/2} = \dfrac{1}{2\sqrt{u}}$

Using the function of a function rule,

$$\frac{dy}{dx} = \frac{dy}{du} \times \frac{du}{dx} = \left(\frac{1}{2\sqrt{u}}\right)(6x + 4) = \frac{3x + 2}{\sqrt{u}}$$

i.e. $\quad \dfrac{dy}{dx} = \dfrac{3x + 2}{\sqrt{3x^2 + 4x - 1}}$

Problem 21. Differentiate $y = 3\tan^4 3x$

Let $u = \tan 3x$ then $y = 3u^4$

Hence $\quad \dfrac{du}{dx} = 3\sec^2 3x$ (from Problem 14),

and $\quad \dfrac{dy}{du} = 12u^3$

Then $\quad \dfrac{dy}{dx} = \dfrac{dy}{du} \times \dfrac{du}{dx} = (12u^3)(3\sec^2 3x)$

$$= 12(\tan 3x)^3(3\sec^2 3x)$$

i.e. $\quad \dfrac{dy}{dx} = 36\tan^3 3x\sec^2 3x$

Problem 22. Find the differential coefficient of
$$y = \frac{2}{(2t^3 - 5)^4}$$

$y = \dfrac{2}{(2t^3 - 5)^4} = 2(2t^3 - 5)^{-4}$. Let $u = (2t^3 - 5)$,
then $y = 2u^{-4}$

Hence $\dfrac{du}{dt} = 6t^2$ and $\dfrac{dy}{du} = -8u^{-5} = \dfrac{-8}{u^5}$

Then $\dfrac{dy}{dt} = \dfrac{dy}{du} \times \dfrac{du}{dt} = \left(\dfrac{-8}{u^5}\right)(6t^2) = \dfrac{-48t^2}{(2t^3 - 5)^5}$

Now try the following Practice Exercise

Practice Exercise 232   Further problems on the function of a function (answers on page 1177)

In Problems 1 to 9, find the differential coefficients with respect to the variable.

1. $(2x - 1)^6$

2. $(2x^3 - 5x)^5$

3. $2\sin(3\theta - 2)$

4. $2\cos^5 \alpha$

5. $\dfrac{1}{(x^3 - 2x + 1)^5}$

6. $5e^{2t+1}$

7. $2\cot(5t^2 + 3)$

8. $6\tan(3y + 1)$

9. $2e^{\tan\theta}$

10. Differentiate $\theta\sin\left(\theta - \dfrac{\pi}{3}\right)$ with respect to θ, and evaluate, correct to 3 significant figures, when $\theta = \dfrac{\pi}{2}$

11. The extension, x metres, of an un-damped vibrating spring after t seconds is given by:

$$x = 0.54\cos(0.3t - 0.15) + 3.2$$

Calculate the speed of the spring, given by $\dfrac{dx}{dt}$, when (a) $t = 0$, (b) $t = 2$ s

---

53.5   Successive differentiation

When a function $y = f(x)$ is differentiated with respect to x the differential coefficient is written as $\dfrac{dy}{dx}$ or $f'(x)$. If the expression is differentiated again, the second differential coefficient is obtained and is written as $\dfrac{d^2y}{dx^2}$ (pronounced dee two y by dee x squared) or $f''(x)$

(pronounced f double-dash x). By successive differentiation further higher derivatives such as $\dfrac{d^3 y}{dx^3}$ and $\dfrac{d^4 y}{dx^4}$ may be obtained.

Thus if $y = 3x^4$,

$$\frac{dy}{dx} = 12x^3, \quad \frac{d^2 y}{dx^2} = 36x^2,$$

$$\frac{d^3 y}{dx^3} = 72x, \quad \frac{d^4 y}{dx^4} = 72 \text{ and } \frac{d^5 y}{dx^5} = 0$$

Problem 23. If $f(x) = 2x^5 - 4x^3 + 3x - 5$, find $f''(x)$

$$f(x) = 2x^5 - 4x^3 + 3x - 5$$

$$f'(x) = 10x^4 - 12x^2 + 3$$

$$f''(x) = 40x^3 - 24x = \mathbf{4x(10x^2 - 6)}$$

Problem 24. If $y = \cos x - \sin x$, evaluate x, in the range $0 \le x \le \dfrac{\pi}{2}$, when $\dfrac{d^2 y}{dx^2}$ is zero

Since $y = \cos x - \sin x$, $\dfrac{dy}{dx} = -\sin x - \cos x$ and

$$\frac{d^2 y}{dx^2} = -\cos x + \sin x$$

When $\dfrac{d^2 y}{dx^2}$ is zero, $-\cos x + \sin x = 0$,

i.e. $\sin x = \cos x$ or $\dfrac{\sin x}{\cos x} = 1$

Hence $\tan x = 1$ and $x = \tan^{-1} 1 = \mathbf{45°}$ **or** $\dfrac{\pi}{4}$ **rads**
in the range $0 \le x \le \dfrac{\pi}{2}$

Problem 25. Given $y = 2xe^{-3x}$ show that
$$\frac{d^2 y}{dx^2} + 6\frac{dy}{dx} + 9y = 0$$

$y = 2xe^{-3x}$ (i.e. a product)

Hence $\dfrac{dy}{dx} = (2x)(-3e^{-3x}) + (e^{-3x})(2)$

$$= -6xe^{-3x} + 2e^{-3x}$$

$$\frac{d^2 y}{dx^2} = [(-6x)(-3e^{-3x}) + (e^{-3x})(-6)]$$

$$+ (-6e^{-3x})$$

$$= 18xe^{-3x} - 6e^{-3x} - 6e^{-3x}$$

i.e. $\dfrac{d^2 y}{dx^2} = 18xe^{-3x} - 12e^{-3x}$

Substituting values into $\dfrac{d^2 y}{dx^2} + 6\dfrac{dy}{dx} + 9y$ gives:

$$(18xe^{-3x} - 12e^{-3x}) + 6(-6xe^{-3x} + 2e^{-3x})$$

$$= 18xe^{-3x} - 12e^{-3x} - 36xe^{-3x} \quad +9(2xe^{-3x})$$

$$+ 12e^{-3x} + 18xe^{-3x} = 0$$

Thus when $y = 2xe^{-3x}$, $\dfrac{d^2 y}{dx^2} + 6\dfrac{dy}{dx} + 9y = 0$

Problem 26. Evaluate $\dfrac{d^2 y}{d\theta^2}$ when $\theta = 0$ given: $y = 4 \sec 2\theta$

Since $y = 4 \sec 2\theta$, then

$$\frac{dy}{d\theta} = (4)(2) \sec 2\theta \tan 2\theta \text{ (from Problem 15)}$$

$$= 8 \sec 2\theta \tan 2\theta \text{ (i.e. a product)}$$

$$\frac{d^2 y}{d\theta^2} = (8 \sec 2\theta)(2 \sec^2 2\theta)$$

$$+ (\tan 2\theta)[(8)(2) \sec 2\theta \tan 2\theta]$$

$$= 16 \sec^3 2\theta + 16 \sec 2\theta \tan^2 2\theta$$

When $\theta = 0$,

$$\frac{d^2 y}{d\theta^2} = 16 \sec^3 0 + 16 \sec 0 \tan^2 0$$

$$= 16(1) + 16(1)(0) = \mathbf{16}$$

Now try the following Practice Exercise

Practice Exercise 233 Further problems on successive differentiation (answers on page 1177)

1. If $y = 3x^4 + 2x^3 - 3x + 2$ find
   (a) $\dfrac{d^2 y}{dx^2}$ (b) $\dfrac{d^3 y}{dx^3}$

2. (a) Given $f(t) = \dfrac{2}{5}t^2 - \dfrac{1}{t^3} + \dfrac{3}{t} - \sqrt{t} + 1$
      determine $f''(t)$
   (b) Evaluate $f''(t)$ when $t = 1$

3. The charge q on the plates of a capacitor is given by $q = CVe^{-\frac{t}{CR}}$, where t is the time, C

is the capacitance and R the resistance. Determine (a) the rate of change of charge, which is given by $\dfrac{dq}{dt}$, (b) the rate of change of current, which is given by $\dfrac{d^2q}{dt^2}$

In Problems 4 and 5, find the second differential coefficient with respect to the variable.

4. (a) $3\sin 2t + \cos t$ (b) $2\ln 4\theta$

5. (a) $2\cos^2 x$ (b) $(2x-3)^4$

6. Evaluate $f''(\theta)$ when $\theta = 0$ given $f(\theta) = 2\sec 3\theta$

7. Show that the differential equation $\dfrac{d^2y}{dx^2} - 4\dfrac{dy}{dx} + 4y = 0$ is satisfied when $y = xe^{2x}$

8. Show that, if P and Q are constants and $y = P\cos(\ln t) + Q\sin(\ln t)$, then

$$t^2\frac{d^2y}{dt^2} + t\frac{dy}{dt} + y = 0$$

9. The displacement, s, of a mass in a vibrating system is given by: $s = (1+t)\,e^{-\omega t}$ where ω is the natural frequency of vibration. Show that: $\dfrac{d^2s}{dt^2} + 2\omega\dfrac{ds}{dt} + \omega^2 s = 0$

For fully worked solutions to each of the problems in Practice Exercises 229 to 233 in this chapter, go to the website:
www.routledge.com/cw/bird

Chapter 54

Some applications of differentiation

Why it is important to understand: Some applications of differentiation

In the previous two chapters some basic differentiation techniques were explored, sufficient to allow us to look at some applications of differential calculus. Some practical rates of change problems are initially explained, followed by some practical velocity and acceleration problems. Determining maximum and minimum points and points of inflexion on curves, together with some practical maximum and minimum problems follow. Tangents and normals to curves and errors and approximations complete this initial look at some practical applications of differentiation. In general, with these applications, the differentiation tends to be straightforward.

At the end of this chapter, you should be able to:

- determine rates of change using differentiation
- solve velocity and acceleration problems
- understand turning points
- determine the turning points on a curve and determine their nature
- solve practical problems involving maximum and minimum values
- determine points of inflexion on a curve
- determine tangents and normals to a curve
- determine small changes in functions

54.1 Rates of change

If a quantity y depends on and varies with a quantity x then the rate of change of y with respect to x is $\dfrac{dy}{dx}$

Thus, for example, the rate of change of pressure p with height h is $\dfrac{dp}{dh}$

A rate of change with respect to time is usually just called 'the rate of change', the 'with respect to time' being assumed. Thus, for example, a rate of change of current, i, is $\dfrac{di}{dt}$ and a rate of change of temperature, θ, is $\dfrac{d\theta}{dt}$, and so on.

Problem 1. The length l metres of a certain metal rod at temperature $\theta°C$ is given by $l = 1 + 0.00005\theta + 0.0000004\theta^2$. Determine the rate of change of length, in mm/°C, when the temperature is (a) 100°C and (b) 400°C

The rate of change of length means $\dfrac{dl}{d\theta}$

Since length $l = 1 + 0.00005\theta + 0.0000004\theta^2$,

then $\quad \dfrac{dl}{d\theta} = 0.00005 + 0.0000008\theta$

(a) When $\theta = 100°C$,

$$\dfrac{dl}{d\theta} = 0.00005 + (0.0000008)(100)$$

$$= 0.00013\,\text{m/}°C = \mathbf{0.13\,mm/}°\mathbf{C}$$

(b) When $\theta = 400°C$,

$$\dfrac{dl}{d\theta} = 0.00005 + (0.0000008)(400)$$

$$= 0.00037\,\text{m/}°C = \mathbf{0.37\,mm/}°\mathbf{C}$$

Problem 2. The luminous intensity I candelas of a lamp at varying voltage V is given by: $I = 4 \times 10^{-4}\,V^2$. Determine the voltage at which the light is increasing at a rate of 0.6 candelas per volt

The rate of change of light with respect to voltage is given by $\dfrac{dI}{dV}$

Since $I = 4 \times 10^{-4}\,V^2$, $\dfrac{dI}{dV} = (4 \times 10^{-4})(2)V$

$$= 8 \times 10^{-4}\,V$$

When the light is increasing at 0.6 candelas per volt then $+0.6 = 8 \times 10^{-4}\,V$, from which, voltage

$$V = \dfrac{0.6}{8 \times 10^{-4}} = 0.075 \times 10^{+4} = \mathbf{750\ volts}$$

Problem 3. Newton's law of cooling is given by: $\theta = \theta_0 e^{-kt}$, where the excess of temperature at zero time is $\theta_0\,°C$ and at time t seconds is $\theta\,°C$. Determine the rate of change of temperature after 40 s, given that $\theta_0 = 16°C$ and $k = 0.03$

The rate of change of temperture is $\dfrac{d\theta}{dt}$

Since $\theta = \theta_0 e^{-kt}$ then $\dfrac{d\theta}{dt} = (\theta_0)(-k)e^{-kt}$

$$= -k\theta_0 e^{-kt}$$

When $\theta_0 = 16$, $k = 0.03$ and $t = 40$ then

$$\dfrac{d\theta}{dt} = -(0.03)(16)e^{-(0.03)(40)}$$

$$= -0.48e^{-1.2} = \mathbf{-0.145°C/s}$$

Problem 4. The displacement s cm of the end of a stiff spring at time t seconds is given by: $s = ae^{-kt} \sin 2\pi ft$. Determine the velocity of the end of the spring after 1 s, if $a = 2$, $k = 0.9$ and $f = 5$

Velocity $v = \dfrac{ds}{dt}$ where $s = ae^{-kt} \sin 2\pi ft$ (i.e. a product)

Using the product rule,

$$\dfrac{ds}{dt} = (ae^{-kt})(2\pi f \cos 2\pi ft)$$

$$+ (\sin 2\pi ft)(-ake^{-kt})$$

When $a = 2$, $k = 0.9$, $f = 5$ and $t = 1$,

velocity, v $= (2e^{-0.9})(2\pi 5 \cos 2\pi 5)$

$$+ (\sin 2\pi 5)(-2)(0.9)e^{-0.9}$$

$$= 25.5455 \cos 10\pi - 0.7318 \sin 10\pi$$

$$= 25.5455(1) - 0.7318(0)$$

$$= \mathbf{25.55\ cm/s}$$

(Note that $\cos 10\pi$ means 'the cosine of 10π radians', *not* degrees, and $\cos 10\pi = \cos 2\pi = 1$)

Now try the following Practice Exercise

Practice Exercise 234   Further problems on rates of change (answers on page 1177)

1. An alternating current, i amperes, is given by $i = 10 \sin 2\pi ft$, where f is the frequency in hertz and t the time in seconds. Determine the rate of change of current when $t = 20\,\text{ms}$, given that $f = 150\,\text{Hz}$.

2. The luminous intensity, I candelas, of a lamp is given by $I = 6 \times 10^{-4}\,V^2$, where V is the voltage. Find (a) the rate of change of luminous intensity with voltage when $V = 200$ volts, and (b) the voltage at which the light is increasing at a rate of 0.3 candelas per volt.

3. The voltage across the plates of a capacitor at any time t seconds is given by $v = Ve^{-t/CR}$, where V, C and R are constants. Given $V = 300$ volts, $C = 0.12 \times 10^{-6}$ farads and $R = 4 \times 10^6$ ohms, find (a) the initial rate of

change of voltage, and (b) the rate of change of voltage after 0.5 s

4. The pressure p of the atmosphere at height h above ground level is given by $p = p_0 e^{-h/c}$, where p_0 is the pressure at ground level and c is a constant. Determine the rate of change of pressure with height when $p_0 = 1.013 \times 10^5$ pascals and $c = 6.05 \times 10^4$ at 1450 metres.

5. The volume, v cubic metres, of water in a reservoir varies with time t, in minutes. When a valve is opened the relationship between v and t is given by: $v = 2 \times 10^4 - 20t^2 - 10t^3$. Calculate the rate of change of water volume at the time when $t = 3$ minutes.

54.2 Velocity and acceleration

When a car moves a distance x metres in a time t seconds along a straight road, if the **velocity** v is constant then $v = \dfrac{x}{t}$ m/s, i.e. the gradient of the distance/time graph shown in Figure 54.1 is constant.

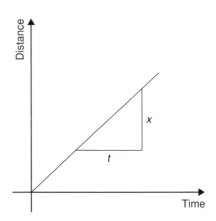

Figure 54.1

If, however, the velocity of the car is not constant then the distance/time graph will not be a straight line. It may be as shown in Figure 54.2.

The average velocity over a small time δt and distance δx is given by the gradient of the chord AB, i.e. the average velocity over time δt is $\dfrac{\delta x}{\delta t}$. As $\delta t \to 0$, the chord AB becomes a tangent, such that at point A, the velocity is given by: $v = \dfrac{dx}{dt}$

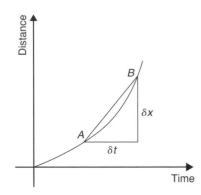

Figure 54.2

Hence the velocity of the car at any instant is given by the gradient of the distance/time graph. If an expression for the distance x is known in terms of time t then the velocity is obtained by differentiating the expression.

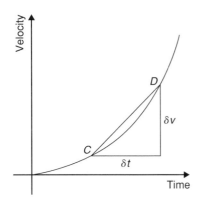

Figure 54.3

The **acceleration** a of the car is defined as the rate of change of velocity. A velocity/time graph is shown in Figure 54.3. If δv is the change in v and δt the corresponding change in time, then $a = \dfrac{\delta v}{\delta t}$. As $\delta t \to 0$, the chord CD becomes a tangent, such that at point C, the acceleration is given by: $a = \dfrac{dv}{dt}$

Hence the acceleration of the car at any instant is given by the gradient of the velocity/time graph. If an expression for velocity is known in terms of time t then the acceleration is obtained by differentiating the expression.

Acceleration $\quad a = \dfrac{dv}{dt}$

However, $\qquad v = \dfrac{dx}{dt}$

Hence $\qquad a = \dfrac{d}{dt}\left(\dfrac{dx}{dt}\right) = \dfrac{d^2 x}{dt^2}$

The acceleration is given by the second differential coefficient of distance x with respect to time t.

Summarising, if a body moves a distance x metres in a time t seconds then:

(i)   distance $x = f(t)$

(ii)   velocity $v = f'(t)$ or $\dfrac{dx}{dt}$, which is the gradient of the distance/time graph

(iii)   acceleration $a = \dfrac{dv}{dt} = f''(t)$ or $\dfrac{d^2x}{dt^2}$, which is the gradient of the velocity/time graph.

Another notation is **Newton's notation** for differentiation (also called the dot notation for differentiation), which requires placing a dot over the dependent variable and is often used for time derivatives such as velocity and acceleration, i.e. if distance, $x = f(t)$, then velocity, $v = \dot{x} = f'(t)$ and acceleration, $a = \ddot{x} = f''(t)$

Problem 5.   The distance x metres moved by a car in a time t seconds is given by $x = 3t^3 - 2t^2 + 4t - 1$. Determine the velocity and acceleration when (a) $t = 0$, and (b) $t = 1.5$ s

Distance $x = 3t^3 - 2t^2 + 4t - 1$ m

Velocity $v = \dfrac{dx}{dt} = 9t^2 - 4t + 4$ m/s

Acceleration $a = \dfrac{d^2x}{dt^2} = 18t - 4$ m/s^2

(a)   When time $t = 0$,

velocity $v = 9(0)^2 - 4(0) + 4 = \mathbf{4\,m/s}$

and acceleration $a = 18(0) - 4 = \mathbf{-4\,m/s^2}$

(i.e. a deceleration)

(b)   When time $t = 1.5$ s,

velocity $v = 9(1.5)^2 - 4(1.5) + 4 = \mathbf{18.25\,m/s}$

and acceleration $a = 18(1.5) - 4 = \mathbf{23\,m/s^2}$

Problem 6.   Supplies are dropped from a helicopter and the distance fallen in a time t seconds is given by: $x = \frac{1}{2}gt^2$, where $g = 9.8$ m/s^2. Determine the velocity and acceleration of the supplies after it has fallen for 2 seconds

Distance $x = \dfrac{1}{2}gt^2 = \dfrac{1}{2}(9.8)t^2 = 4.9t^2$ m

Velocity $v = \dfrac{dv}{dt} = 9.8t$ m/s

and acceleration $a = \dfrac{d^2x}{dt^2} = 9.8$ m/s^2

When time $t = 2$ s,
velocity $v = (9.8)(2) = \mathbf{19.6\,m/s}$
and **acceleration $a = 9.8$ m/s^2** (which is acceleration due to gravity).

Problem 7.   The distance x metres travelled by a vehicle in time t seconds after the brakes are applied is given by: $x = 20t - \frac{5}{3}t^2$. Determine (a) the speed of the vehicle (in km/h) at the instant the brakes are applied, and (b) the distance the car travels before it stops

(a)   Distance, $x = 20t - \dfrac{5}{3}t^2$

Hence velocity $v = \dfrac{dx}{dt} = 20 - \dfrac{10}{3}t$

At the instant the brakes are applied, time $= 0$

Hence

$$\textbf{velocity } v = 20\,\text{m/s} = \frac{20 \times 60 \times 60}{1000}\,\text{km/h}$$

$$= \mathbf{72\,km/h}$$

(Note: changing from m/s to km/h merely involves multiplying by 3.6)

(b)   When the car finally stops, the velocity is zero, i.e. $v = 20 - \dfrac{10}{3}t = 0$, from which, $20 = \dfrac{10}{3}t$, giving $t = 6$ s. Hence the distance travelled before the car stops is given by:

$$x = 20t - \frac{5}{3}t^2 = 20(6) - \frac{5}{3}(6)^2$$

$$= 120 - 60 = \mathbf{60\,m}$$

Problem 8.   The angular displacement θ radians of a flywheel varies with time t seconds and follows the equation: $\theta = 9t^2 - 2t^3$. Determine (a) the angular velocity and acceleration of the flywheel when time, $t = 1$ s, and (b) the time when the angular acceleration is zero

(a)   Angular displacement $\theta = 9t^2 - 2t^3$ rad

Angular velocity $\omega = \dfrac{d\theta}{dt} = 18t - 6t^2$ rad/s

When time $t = 1$ s,

$\omega = 18(1) - 6(1)^2 = \mathbf{12\,rad/s}$

Angular acceleration $\alpha = \dfrac{d^2\theta}{dt^2} = 18 - 12t$ rad/s^2

When time $t = 1$ s, $\alpha = 18 - 12(1)$

$$= \mathbf{6\,rad/s^2}$$

(b) When the angular acceleration is zero, $18 - 12t = 0$, from which, $18 = 12t$, giving time, $t = \mathbf{1.5\,s}$

Problem 9. The displacement x cm of the slide valve of an engine is given by: $x = 2.2 \cos 5\pi t + 3.6 \sin 5\pi t$. Evaluate the velocity (in m/s) when time $t = 30$ ms

---

Displacement $x = 2.2 \cos 5\pi t + 3.6 \sin 5\pi t$

Velocity $v = \dfrac{dx}{dt} = (2.2)(-5\pi) \sin 5\pi t$

$$+ (3.6)(5\pi) \cos 5\pi t$$

$$= -11\pi \sin 5\pi t + 18\pi \cos 5\pi t \text{ cm/s}$$

When time $t = 30$ ms,

velocity $= -11\pi \sin(5\pi \times 30 \times 10^{-3})$

$$+ 18\pi \cos(5\pi \times 30 \times 10^{-3})$$

$$= -11\pi \sin 0.4712 + 18\pi \cos 0.4712$$

$$= -11\pi \sin 27° + 18\pi \cos 27°$$

$$= -15.69 + 50.39$$

$$= 34.7 \text{ cm/s} = \mathbf{0.347\,m/s}$$

Now try the following Practice Exercise

Practice Exercise 235   Further problems on velocity and acceleration (answers on page 1177)

1. A missile fired from ground level rises x metres vertically upwards in t seconds and $x = 100t - \dfrac{25}{2}t^2$. Find (a) the initial velocity of the missile, (b) the time when the height of the missile is a maximum, (c) the maximum height reached and (d) the velocity with which the missile strikes the ground.

2. The distance s metres travelled by a car in t seconds after the brakes are applied is given by $s = 25t - 2.5t^2$. Find (a) the speed of the car (in km/h) when the brakes are applied and (b) the distance the car travels before it stops.

3. The equation $\theta = 10\pi + 24t - 3t^2$ gives the angle θ, in radians, through which a wheel turns in t seconds. Determine (a) the time the wheel takes to come to rest and (b) the angle turned through in the last second of movement.

4. At any time t seconds the distance x metres of a particle moving in a straight line from a fixed point is given by: $x = 4t + \ln(1 - t)$. Determine (a) the initial velocity and acceleration, (b) the velocity and acceleration after 1.5 s and (c) the time when the velocity is zero.

5. The angular displacement θ of a rotating disc is given by: $\theta = 6 \sin \dfrac{t}{4}$, where t is the time in seconds. Determine (a) the angular velocity of the disc when t is 1.5 s, (b) the angular acceleration when t is 5.5 s and (c) the first time when the angular velocity is zero.

6. $x = \dfrac{20t^3}{3} - \dfrac{23t^2}{2} + 6t + 5$ represents the distance, x metres, moved by a body in t seconds. Determine (a) the velocity and acceleration at the start, (b) the velocity and acceleration when $t = 3$ s, (c) the values of t when the body is at rest, (d) the value of t when the acceleration is 37 m/s² and (e) the distance travelled in the third second.

7. A particle has a displacement s given by $s = 30t + 27t^2 - 3t^3$ metres, where time t is in seconds. Determine the time at which the acceleration will be zero.

54.3 Turning points

In Figure 54.4, the gradient (or rate of change) of the curve changes from positive between O and P to negative between P and Q, and then positive again between Q and R. At point P, the gradient is zero and, as x

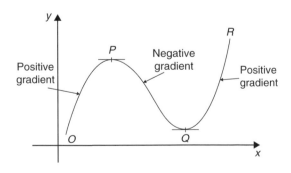

Figure 54.4

increases, the gradient of the curve changes from positive just before P to negative just after. Such a point is called a **maximum point** and appears as the 'crest of a wave'. At point Q, the gradient is also zero and, as x increases, the gradient of the curve changes from negative just before Q to positive just after. Such a point is called a **minimum point**, and appears as the 'bottom of a valley'. Points such as P and Q are given the general name of **turning points**.

It is possible to have a turning point, the gradient on either side of which is the same. Such a point is given the special name of a **point of inflexion**, and examples are shown in Figure 54.5.

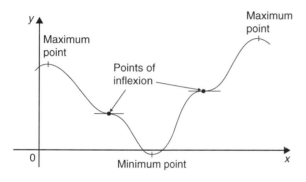

Figure 54.5

Maximum and minimum points and points of inflexion are given the general term of **stationary points**.

Procedure for finding and distinguishing between stationary points.

(i)   Given $y = f(x)$, determine $\dfrac{dy}{dx}$ (i.e. $f'(x)$)

(ii)   Let $\dfrac{dy}{dx} = 0$ and solve for the values of x

(iii)   Substitute the values of x into the original equation, $y = f(x)$, to find the corresponding y-ordinate values. This establishes the co-ordinates of the stationary points.

To determine the nature of the stationary points:
Either

(iv)   Find $\dfrac{d^2y}{dx^2}$ and substitute into it the values of x found in (ii).

   If the result is: (a)   positive – the point is a minimum one,
   (b)   negative – the point is a maximum one,
   (c)   zero – the point is a point of inflexion (see Section 54.5)

or

(v)   Determine the sign of the gradient of the curve just before and just after the stationary points. If the sign change for the gradient of the curve is:
   (a)   positive to negative – the point is a maximum one,
   (b)   negative to positive – the point is a minimum one,
   (c)   positive to positive or negative to negative – the point is a point of inflexion (see Section 54.5).

Problem 10.   Locate the turning point on the curve $y = 3x^2 - 6x$ and determine its nature by examining the sign of the gradient on either side

Following the above procedure:

(i)   Since $y = 3x^2 - 6x$, $\dfrac{dy}{dx} = 6x - 6$

(ii)   At a turning point, $\dfrac{dy}{dx} = 0$, hence $6x - 6 = 0$, from which, $x = 1$.

(iii)   When $x = 1$, $y = 3(1)^2 - 6(1) = -3$

   Hence the co-ordinates of the turning point are (1, −3)

(v)   If x is slightly less than 1, say, 0.9, then $\dfrac{dy}{dx} = 6(0.9) - 6 = -0.6$, i.e. negative.

   If x is slightly greater than 1, say, 1.1, then $\dfrac{dy}{dx} = 6(1.1) - 6 = 0.6$, i.e. positive.

   Since the gradient of the curve is negative just before the turning point and positive just after (i.e. $-\cup+$), **(1, −3) is a minimum point.**

Problem 11.   Find the maximum and minimum values of the curve $y = x^3 - 3x + 5$ by (a) examining the gradient on either side of the turning points, and (b) determining the sign of the second derivative

Since $y = x^3 - 3x + 5$ then $\dfrac{dy}{dx} = 3x^2 - 3$

For a maximum or minimum value $\dfrac{dy}{dx} = 0$

Hence $3x^2 - 3 = 0$

from which, $3x^2 = 3$

and $x = \pm 1$

When $x = 1$, $y = (1)^3 - 3(1) + 5 = 3$

When $x = -1$, $y = (-1)^3 - 3(-1) + 5 = 7$

Hence (1, 3) and (−1, 7) are the co-ordinates of the turning points.

(a) Considering the point (1, 3):

If x is slightly less than 1, say 0.9, then $\dfrac{dy}{dx} = 3(0.9)^2 - 3$, which is negative.

If x is slightly more than 1, say 1.1, then $\dfrac{dy}{dx} = 3(1.1)^2 - 3$, which is positive.

Since the gradient changes from negative to positive, **the point (1, 3) is a minimum point.**

Considering the point (−1, 7):

If x is slightly less than −1, say −1.1, then $\dfrac{dy}{dx} = 3(-1.1)^2 - 3$, which is positive.

If x is slightly more than −1, say −0.9, then $\dfrac{dy}{dx} = 3(-0.9)^2 - 3$, which is negative.

Since the gradient changes from positive to negative, **the point (−1, 7) is a maximum point.**

(b) Since $\dfrac{dy}{dx} = 3x^2 - 3$, then $\dfrac{d^2y}{dx^2} = 6x$

When $x = 1$, $\dfrac{d^2y}{dx^2}$ is positive, hence (1, 3) is a **minimum value.**

When $x = -1$, $\dfrac{d^2y}{dx^2}$ is negative, hence (−1, 7) is a **maximum value.**

Thus the maximum value is 7 and the minimum value is 3.

It can be seen that the second differential method of determining the nature of the turning points is, in this case, quicker than investigating the gradient.

Problem 12. Locate the turning point on the following curve and determine whether it is a maximum or minimum point: $y = 4\theta + e^{-\theta}$

Since $y = 4\theta + e^{-\theta}$ then $\dfrac{dy}{d\theta} = 4 - e^{-\theta} = 0$ for a maximum or minimum value.

Hence $4 = e^{-\theta}$ and $\dfrac{1}{4} = e^{\theta}$

giving $\theta = \ln \dfrac{1}{4} = -1.3863$

When $\theta = -1.3863$,

$y = 4(-1.3863) + e^{-(-1.3863)} = 5.5452 + 4.0000$

$= -1.5452$

Thus (−1.3863, −1.5452) are the co-ordinates of the turning point.

$$\frac{d^2y}{d\theta^2} = e^{-\theta}$$

When $\theta = -1.3863$, $\dfrac{d^2y}{d\theta^2} = e^{+1.3863} = 4.0$, which is positive, hence **(−1.3863, −1.5452) is a minimum point.**

Problem 13. Determine the co-ordinates of the maximum and minimum values of the graph $y = \dfrac{x^3}{3} - \dfrac{x^2}{2} - 6x + \dfrac{5}{3}$ and distinguish between them. Sketch the graph

Following the given procedure:

(i) Since $y = \dfrac{x^3}{3} - \dfrac{x^2}{2} - 6x + \dfrac{5}{3}$ then

$$\frac{dy}{dx} = x^2 - x - 6$$

(ii) At a turning point, $\dfrac{dy}{dx} = 0$

Hence $\qquad x^2 - x - 6 = 0$

i.e. $\qquad (x + 2)(x - 3) = 0$

from which $\qquad x = -2 \text{ or } x = 3$

(iii) When $x = -2$

$$y = \frac{(-2)^3}{3} - \frac{(-2)^2}{2} - 6(-2) + \frac{5}{3} = 9$$

When $x = 3$, $y = \dfrac{(3)^3}{3} - \dfrac{(3)^2}{2} - 6(3) + \dfrac{5}{3}$

$$= -11\frac{5}{6}$$

Thus the co-ordinates of the turning points are (−2, 9) and $\left(3, -11\dfrac{5}{6}\right)$

(iv) Since $\dfrac{dy}{dx} = x^2 - x - 6$ then $\dfrac{d^2y}{dx^2} = 2x - 1$

When $x=-2$, $\dfrac{d^2y}{dx^2}=2(-2)-1=-5$, which is negative.

Hence $(-2, 9)$ is a maximum point.

When $x=3$, $\dfrac{d^2y}{dx^2}=2(3)-1=5$, which is positive.

Hence $\left(3,-11\dfrac{5}{6}\right)$ is a minimum point.

Knowing $(-2, 9)$ is a maximum point (i.e. crest of a wave), and $\left(3,-11\dfrac{5}{6}\right)$ is a minimum point (i.e. bottom of a valley) and that when $x=0$, $y=\dfrac{5}{3}$, a sketch may be drawn as shown in Figure 54.6.

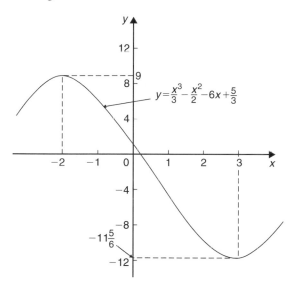

Figure 54.6

Problem 14. Determine the turning points on the curve $y=4\sin x-3\cos x$ in the range $x=0$ to $x=2\pi$ radians, and distinguish between them. Sketch the curve over one cycle

Since $y=4\sin x-3\cos x$ then

$\dfrac{dy}{dx}=4\cos x+3\sin x=0$, for a turning point,

from which, $4\cos x=-3\sin x$ and $\dfrac{-4}{3}=\dfrac{\sin x}{\cos x}$

$\qquad\qquad\qquad = \tan x$

Hence $x=\tan^{-1}\left(\dfrac{-4}{3}\right)=126.87°$ or $306.87°$, since tangent is negative in the second and fourth quadrants.

When $x=126.87°$,

$\qquad y=4\sin 126.87°-3\cos 126.87°=5$

When $x=306.87°$

$\qquad y=4\sin 306.87°-3\cos 306.87°=-5$

$\qquad 126.87°=\left(125.87°\times\dfrac{\pi}{180}\right)$ radians

$\qquad\qquad =2.214\,\text{rad}$

$\qquad 306.87°=\left(306.87°\times\dfrac{\pi}{180}\right)$ radians

$\qquad\qquad =5.356\,\text{rad}$

Hence $(2.214, 5)$ and $(5.356, -5)$ are the co-ordinates of the turning points.

$$\dfrac{d^2y}{dx^2}=-4\sin x+3\cos x$$

When $x=2.214\,\text{rad}$,

$\dfrac{d^2y}{dx^2}=-4\sin 2.214+3\cos 2.214$, which is negative.

Hence $(2.214, 5)$ is a maximum point.

When $x=5.356\,\text{rad}$,

$\dfrac{d^2y}{dx^2}=-4\sin 5.356+3\cos 5.356$, which is positive.

Hence $(5.356, -5)$ is a minimum point.

A sketch of $y=4\sin x-3\cos x$ is shown in Figure 54.7.

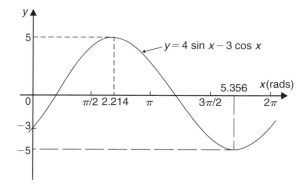

Figure 54.7

Now try the following Practice Exercise

Practice Exercise 236   Further problems on turning points (answers on page 1177)

In Problems 1 to 11, find the turning points and distinguish between them.

1. $y = x^2 - 6x$

2. $y = 8 + 2x - x^2$

3. $y = x^2 - 4x + 3$

4. $y = 3 + 3x^2 - x^3$

5. $y = 3x^2 - 4x + 2$

6. $x = \theta(6 - \theta)$

7. $y = 4x^3 + 3x^2 - 60x - 12$

8. $y = 5x - 2\ln x$

9. $y = 2x - e^x$

10. $y = t^3 - \dfrac{t^2}{2} - 2t + 4$

11. $x = 8t + \dfrac{1}{2t^2}$

12. Determine the maximum and minimum values on the graph $y = 12\cos\theta - 5\sin\theta$ in the range $\theta = 0$ to $\theta = 360°$. Sketch the graph over one cycle showing relevant points.

13. Show that the curve $y = \dfrac{2}{3}(t-1)^3 + 2t(t-2)$ has a maximum value of $\dfrac{2}{3}$ and a minimum value of -2

54.4 Practical problems involving maximum and minimum values

There are many **practical problems** involving maximum and minimum values which occur in science and engineering. Usually, an equation has to be determined from given data, and rearranged where necessary, so that it contains only one variable. Some examples are demonstrated in Problems 15 to 20.

Problem 15.   A rectangular area is formed having a perimeter of 40 cm. Determine the length and breadth of the rectangle if it is to enclose the maximum possible area

Let the dimensions of the rectangle be x and y. Then the perimeter of the rectangle is $(2x + 2y)$. Hence

$$2x + 2y = 40, \quad \text{or} \quad x + y = 20 \qquad (1)$$

Since the rectangle is to enclose the maximum possible area, a formula for area A must be obtained in terms of one variable only.

Area $A = xy$. From equation (1), $x = 20 - y$

Hence, area $A = (20 - y)y = 20y - y^2$

$\dfrac{dA}{dy} = 20 - 2y = 0$ for a turning point, from which, $y = 10$ cm.

$\dfrac{d^2 A}{dy^2} = -2$, which is negative, giving a maximum point.

When $y = 10$ cm, $x = 10$ cm, from equation (1).

Hence the length and breadth of the rectangle are each 10 cm, i.e. a square gives the maximum possible area. When the perimeter of a rectangle is 40 cm, the maximum possible area is $10 \times 10 = \mathbf{100\,cm^2}$

Problem 16.   A rectangular sheet of metal having dimensions 20 cm by 12 cm has squares removed from each of the four corners and the sides bent upwards to form an open box. Determine the maximum possible volume of the box

The squares to be removed from each corner are shown in Figure 54.8, having sides x cm. When the sides are bent upwards the dimensions of the box will be: length $(20 - 2x)$ cm, breadth $(12 - 2x)$ cm and height x cm.

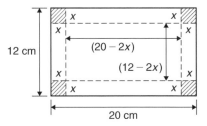

Figure 54.8

Volume of box, $V = (20 - 2x)(12 - 2x)(x)$

$$= 240x - 64x^2 + 4x^3$$

$\dfrac{dV}{dx} = 240 - 128x + 12x^2 = 0$ for a turning point.

Hence $4(60 - 32x + 3x^2) = 0$, i.e. $3x^2 - 32x + 60 = 0$

Using the quadratic formula,

$$x = \frac{32 \pm \sqrt{(-32)^2 - 4(3)(60)}}{2(3)}$$

$$= 8.239 \, \text{cm or } 2.427 \, \text{cm.}$$

Since the breadth is $(12 - 2x)$ cm then $x = 8.239$ cm is not possible and is neglected.
Hence $x = 2.427$ cm.

$$\frac{d^2 V}{dx^2} = -128 + 24x$$

When $x = 2.427$, $\frac{d^2 V}{dx^2}$ is negative, giving a maximum value.

The dimensions of the box are:
length $= 20 - 2(2.427) = 15.146$ cm,
breadth $= 12 - 2(2.427) = 7.146$ cm and
height $= 2.427$ cm.

$$\textbf{Maximum volume} = (15.146)(7.146)(2.427)$$

$$= \textbf{262.7 cm}^3$$

Problem 17.   Determine the height and radius of a cylinder of volume $200 \, \text{cm}^3$ which has the least surface area

Let the cylinder have radius r and perpendicular height h.

From Chapter 29, volume of cylinder,

$$V = \pi r^2 h = 200 \tag{1}$$

Surface area of cylinder, $A = 2\pi r h + 2\pi r^2$

Least surface area means minimum surface area and a formula for the surface area in terms of one variable only is required.

From equation (1), $h = \dfrac{200}{\pi r^2}$ \hfill (2)

Hence surface area,

$$A = 2\pi r \left(\frac{200}{\pi r^2}\right) + 2\pi r^2$$

$$= \frac{400}{r} + 2\pi r^2 = 400r^{-1} + 2\pi r^2$$

$$\frac{dA}{dr} = \frac{-400}{r^2} + 4\pi r = 0, \text{ for a turning point.}$$

Hence $4\pi r = \dfrac{400}{r^2}$

and $r^3 = \dfrac{400}{4\pi}$

from which, $r = \sqrt[3]{\dfrac{100}{\pi}} = 3.169 \, \text{cm}$

$$\frac{d^2 A}{dr^2} = \frac{800}{r^3} + 4\pi$$

When $r = 3.169$ cm, $\dfrac{d^2 A}{dr^2}$ is positive, giving a minimum value.
From equation (2), when $r = 3.169$ cm,

$$h = \frac{200}{\pi (3.169)^2} = 6.339 \, \text{cm}$$

Hence for the least surface area, a cylinder of volume $200 \, \text{cm}^3$ has a radius of $3.169 \, \text{cm}$ and height of $6.339 \, \text{cm}$.

Problem 18.   Determine the area of the largest piece of rectangular ground that can be enclosed by 100 m of fencing, if part of an existing straight wall is used as one side

Let the dimensions of the rectangle be x and y as shown in Figure 54.9, where PQ represents the straight wall.

From Figure 54.9, $x + 2y = 100$ \hfill (1)

Area of rectangle, $A = xy$ \hfill (2)

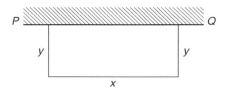

Figure 54.9

Since the maximum area is required, a formula for area A is needed in terms of one variable only.
From equation (1), $x = 100 - 2y$
Hence, area $A = xy = (100 - 2y)y = 100y - 2y^2$
$\dfrac{dA}{dy} = 100 - 4y = 0$, for a turning point, from which,
$y = 25$ m.
$\dfrac{d^2 A}{dy^2} = -4$, which is negative, giving a maximum value.
When $y = 25$ m, $x = 50$ m, from equation (1).

Hence the **maximum possible area**
$$= xy = (50)(25) = \textbf{1250 m}^2$$

Problem 19. An open rectangular box with square ends is fitted with an overlapping lid which covers the top and the front face. Determine the maximum volume of the box if $6\,\text{m}^2$ of metal are used in its construction

A rectangular box having square ends of side x and length y is shown in Figure 54.10.

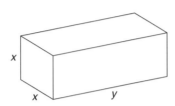

Figure 54.10

Surface area of box, A, consists of two ends and five faces (since the lid also covers the front face).

Hence $\quad A = 2x^2 + 5xy = 6$ $\qquad\qquad$ (1)

Since it is the maximum volume required, a formula for the volume in terms of one variable only is needed. Volume of box, $V = x^2 y$
From equation (1),

$$y = \frac{6 - 2x^2}{5x} = \frac{6}{5x} - \frac{2x}{5} \qquad\qquad (2)$$

Hence volume $V = x^2 y = x^2 \left(\frac{6}{5x} - \frac{2x}{5} \right)$

$$= \frac{6x}{5} - \frac{2x^3}{5}$$

$\dfrac{\mathrm{d}V}{\mathrm{d}x} = \dfrac{6}{5} - \dfrac{6x^2}{5} = 0$ for a maximum or minimum value.
Hence $6 = 6x^2$, giving $x = 1\,\text{m}$ ($x = -1$ is not possible, and is thus neglected).

$$\frac{\mathrm{d}^2 V}{\mathrm{d}x^2} = \frac{-12x}{5}$$

When $x = 1$, $\dfrac{\mathrm{d}^2 V}{\mathrm{d}x^2}$ is negative, giving a maximum value.

From equation (2), when $x = 1$, $y = \dfrac{6}{5(1)} - \dfrac{2(1)}{5} = \dfrac{4}{5}$

Hence the maximum volume of the box is given by
$$V = x^2 y = (1)^2 \left(\frac{4}{5} \right) = \frac{4}{5}\,\textbf{m}^3$$

Problem 20. Find the diameter and height of a cylinder of maximum volume which can be cut from a sphere of radius 12 cm

A cylinder of radius r and height h is shown enclosed in a sphere of radius $R = 12\,\text{cm}$ in Figure 54.11.

Volume of cylinder, $\quad V = \pi r^2 h$ $\qquad\qquad$ (1)

Using the right-angled triangle OPQ shown in Figure 54.11,

$$r^2 + \left(\frac{h}{2} \right)^2 = R^2 \text{ by Pythagoras' theorem,}$$

i.e. $\qquad r^2 + \dfrac{h^2}{4} = 144$ $\qquad\qquad$ (2)

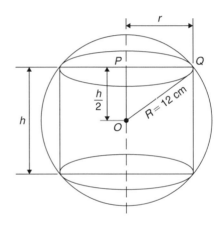

Figure 54.11

Since the maximum volume is required, a formula for the volume V is needed in terms of one variable only.

From equation (2), $r^2 = 144 - \dfrac{h^2}{4}$

Substituting into equation (1) gives:

$$V = \pi \left(144 - \frac{h^2}{4} \right) h = 144\pi h - \frac{\pi h^3}{4}$$

$\dfrac{\mathrm{d}V}{\mathrm{d}h} = 144\pi - \dfrac{3\pi h^2}{4} = 0$, for a maximum or minimum value.

Hence $144\pi = \dfrac{3\pi h^2}{4}$, from which,

$$h = \sqrt{\frac{(144)(4)}{3}} = 13.86\,\text{cm}.$$

$$\frac{\mathrm{d}^2 V}{\mathrm{d}h^2} = \frac{-6\pi h}{4}$$

When $h = 13.86$, $\dfrac{d^2V}{dh^2}$ is negative, giving a maximum value.

From equation (2),

$r^2 = 144 - \dfrac{h^2}{4} = 144 - \dfrac{13.86^2}{4}$, from which, radius $r = 9.80\,\text{cm}$

Diameter of cylinder $= 2r = 2(9.80) = 19.60\,\text{cm}$.

Hence the cylinder having the maximum volume that can be cut from a sphere of radius 12 cm is one in which the diameter is 19.60 cm and the height is 13.86 cm.

Now try the following Practice Exercise

Practice Exercise 237   Further problems on practical maximum and minimum problems (answers on page 1177)

1. The speed, v, of a car (in m/s) is related to time t s by the equation $v = 3 + 12t - 3t^2$. Determine the maximum speed of the car in km/h.

2. Determine the maximum area of a rectangular piece of land that can be enclosed by 1200 m of fencing.

3. A shell is fired vertically upwards and its vertical height, x metres, is given by: $x = 24t - 3t^2$, where t is the time in seconds. Determine the maximum height reached.

4. A lidless box with square ends is to be made from a thin sheet of metal. Determine the least area of the metal for which the volume of the box is 3.5 m^3.

5. A closed cylindrical container has a surface area of 400 cm^2. Determine the dimensions for maximum volume.

6. Calculate the height of a cylinder of maximum volume that can be cut from a cone of height 20 cm and base radius 80 cm.

7. The power developed in a resistor R by a battery of emf E and internal resistance r is given by $P = \dfrac{E^2 R}{(R + r)^2}$. Differentiate P with respect to R and show that the power is a maximum when $R = r$.

8. Find the height and radius of a closed cylinder of volume 125 cm^3 which has the least surface area.

9. Resistance to motion, F, of a moving vehicle, is given by: $F = \dfrac{5}{x} + 100x$. Determine the minimum value of resistance.

10. An electrical voltage E is given by: $E = (15 \sin 50\pi t + 40 \cos 50\pi t)$ volts, where t is the time in seconds. Determine the maximum value of voltage.

11. The fuel economy E of a car, in miles per gallon, is given by:

$$E = 21 + 2.10 \times 10^{-2} v^2 - 3.80 \times 10^{-6} v^4$$

where v is the speed of the car in miles per hour.

Determine, correct to 3 significant figures, the most economical fuel consumption, and the speed at which it is achieved.

12. The horizontal range of a projectile, x, launched with velocity u at an angle θ to the horizontal is given by: $x = \dfrac{2u^2 \sin\theta \cos\theta}{g}$

To achieve maximum horizontal range, determine the angle the projectile should be launched at.

13. The signalling range, x, of a submarine cable is given by the formula $x = r^2 \ln\left(\dfrac{1}{r}\right)$, where r is the ratio of the radii of the conductor and cable. Determine the value of r for maximum range.

54.5   Points of inflexion

As mentioned earlier in the chapter, it is possible to have a turning point, the gradient on either side of which is the same. This is called a **point of inflexion**.

Procedure to determine points of inflexion:

(i) Given $y = f(x)$, determine $\dfrac{dy}{dx}$ and $\dfrac{d^2y}{dx^2}$

(ii) Solve the equation $\dfrac{d^2y}{dx^2} = 0$

(iii) Test whether there is a change of sign occurring in $\dfrac{d^2y}{dx^2}$. This is achieved by substituting into the expression for $\dfrac{d^2y}{dx^2}$ firstly a value of x just less than the solution and then a value just greater than the solution.

Section I

(iv)   A point of inflexion has been found if $\dfrac{d^2y}{dx^2} = 0$ **and** there is a change of sign.

This procedure is demonstrated in the following Problems.

> **Problem 21.**   Determine the point(s) of inflexion (if any) on the graph of the function
>
> $$y = x^3 - 6x^2 + 9x + 5$$
>
> Find also any other turning points. Hence, sketch the graph

Using the above procedure:

(i)   Given $y = x^3 - 6x^2 + 9x + 5$,

$\dfrac{dy}{dx} = 3x^2 - 12x + 9$ and $\dfrac{d^2y}{dx^2} = 6x - 12$

(ii)   Solving the equation $\dfrac{d^2y}{dx^2} = 0$ gives: $6x - 12 = 0$

from which, $6x = 12$ and $x = 2$

Hence, if there is a point of inflexion, it occurs at $x = 2$

(iii)   Taking a value just less than 2, say, 1.9:

$\dfrac{d^2y}{dx^2} = 6x - 12 = 6(1.9) - 12$, which is negative.

Taking a value just greater than 2, say, 2.1:

$\dfrac{d^2y}{dx^2} = 6x - 12 = 6(2.1) - 12$, which is positive.

(iv)   Since a change of sign has occurred a point of inflexion exists at $x = 2$

When $x = 2$, $y = 2^3 - 6(2)^2 + 9(2) + 5 = 7$

i.e. **a point of inflexion occurs at the co-ordinates (2, 7)**

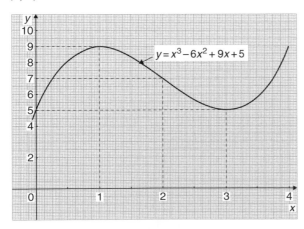

Figure 54.12

From above, $\dfrac{dy}{dx} = 3x^2 - 12x + 9 = 0$ for a turning point

i.e. $x^2 - 4x + 3 = 0$

Using the quadratic formula or factorising (or by calculator), the solution is: $x = 1$ or $x = 3$

Since $y = x^3 - 6x^2 + 9x + 5$, then

when $x = 1$, $y = 1^3 - 6(1)^2 + 9(1) + 5 = 9$

and when $x = 3$, $y = 3^3 - 6(3)^2 + 9(3) + 5 = 5$

Hence, there are turning points at $(1, 9)$ and at $(3, 5)$

Since $\dfrac{d^2y}{dx^2} = 6x - 12$, when $x = 1$, $\dfrac{d^2y}{dx^2} = 6(1) - 12$ which is negative – hence a maximum point

and when $x = 3$, $\dfrac{d^2y}{dx^2} = 6(3) - 12$ which is positive – hence a minimum point

Thus, **(1, 9) is a maximum point and (3, 5) is a minimum point**.

A sketch of the graph $y = x^3 - 6x^2 + 9x + 5$ is shown in Figure 54.12

> **Problem 22.**   Determine the point(s) of inflexion (if any) on the graph of the function
>
> $$y = x^4 - 24x^2 + 5x + 60$$

Using the above procedure:

(i)   Given $y = x^4 - 24x^2 + 5x + 60$,

$\dfrac{dy}{dx} = 4x^3 - 48x + 5$ and $\dfrac{d^2y}{dx^2} = 12x^2 - 48$

(ii)   Solving the equation $\dfrac{d^2y}{dx^2} = 0$ gives:

$12x^2 - 48 = 0$ from which, $12x^2 = 48$ and $x^2 = 4$ from which, $x = \sqrt{4} = \pm 2$

Hence, if there are points of inflexion, they occur at $x = 2$ and at $x = -2$

(iii)   Taking a value just less than 2, say, 1.9:

$\dfrac{d^2y}{dx^2} = 12x^2 - 48 = 12(1.9)^2 - 48$, which is negative.

Taking a value just greater than 2, say, 2.1:

$\dfrac{d^2y}{dx^2} = 12x^2 - 48 = 12(2.1)^2 - 48$, which is positive.

Taking a value just less than -2, say, -2.1:

$\dfrac{d^2y}{dx^2} = 12x^2 - 48 = 12(-2.1)^2 - 48$, which is positive.

Taking a value just greater than -2, say, -1.9:

$$\frac{d^2y}{dx^2} = 12x^2 - 48 = 12(-1.9)^2 - 48, \text{ which is}$$

negative.

(iv)   Since changes of signs have occurred, points of inflexion exist at $x = 2$ and $x = -2$

When $x = 2$,
$y = 2^4 - 24(2)^2 + 5(2) + 60 = -10$
When $x = -2$,
$y = (-2)^4 - 24(-2)^2 + 5(-2) + 60 = -30$
i.e. **points of inflexion occur at the co-ordinates (2, -10) and at (-2, -30)**

Now try the following Practice Exercise

Practice Exercise 238   Further problems on points of inflexion (answers on page 1177)

1. Find the points of inflexion (if any) on the graph of the function
    $$y = \frac{1}{3}x^3 - \frac{1}{2}x^2 - 2x + \frac{1}{12}$$

2. Find the points of inflexion (if any) on the graph of the function
    $$y = 4x^3 + 3x^2 - 18x - \frac{5}{8}$$

3. Find the point(s) of inflexion on the graph of the function $y = x + \sin x$ for $0 \langle x \langle 2\pi$

4. Find the point(s) of inflexion on the graph of the function $y = 3x^3 - 27x^2 + 15x + 17$

5. Find the point(s) of inflexion on the graph of the function $y = 2xe^{-x}$

6. The displacement, s, of a particle is given by: $s = 3t^3 - 9t^2 + 10$. Determine the maximum, minimum and point of inflexion of s.

54.6   Tangents and normals

Tangents

The equation of the tangent to a curve $y = f(x)$ at the point (x_1, y_1) is given by:

$$y - y_1 = m(x - x_1)$$

where $m = \dfrac{dy}{dx} = $ gradient of the curve at (x_1, y_1)

Problem 23.   Find the equation of the tangent to the curve $y = x^2 - x - 2$ at the point $(1, -2)$

Gradient, $m = \dfrac{dy}{dx} = 2x - 1$

At the point $(1, -2)$, $x = 1$ and $m = 2(1) - 1 = 1$
Hence the equation of the tangent is:

$$y - y_1 = m(x - x_1)$$

i.e. $y - -2 = 1(x - 1)$

i.e. $y + 2 = x - 1$

or $\boldsymbol{y = x - 3}$

The graph of $y = x^2 - x - 2$ is shown in Figure 54.13. The line AB is the tangent to the curve at the point C, i.e. $(1, -2)$, and the equation of this line is $y = x - 3$

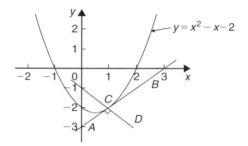

Figure 54.13

Normals

The normal at any point on a curve is the line that passes through the point and is at right angles to the tangent. Hence, in Figure 54.13, the line CD is the normal.
It may be shown that if two lines are at right angles then the product of their gradients is -1. Thus if m is the gradient of the tangent, then the gradient of the normal is $-\dfrac{1}{m}$
Hence the equation of the normal at the point (x_1, y_1) is given by:

$$y - y_1 = -\frac{1}{m}(x - x_1)$$

Problem 24.   Find the equation of the normal to the curve $y = x^2 - x - 2$ at the point $(1, -2)$

$m = 1$ from Problem 23, hence the equation of the normal is $y - y_1 = -\dfrac{1}{m}(x - x_1)$

i.e. $y - -2 = -\dfrac{1}{1}(x - 1)$

i.e. $y + 2 = -x + 1$ or $\boldsymbol{y = -x - 1}$

Thus the line *CD* in Figure 54.13 has the equation $y = -x - 1$

Problem 25. Determine the equations of the tangent and normal to the curve $y = \dfrac{x^3}{5}$ at the point $\left(-1, -\dfrac{1}{5}\right)$

Gradient *m* of curve $y = \dfrac{x^3}{5}$ is given by

$$m = \frac{dy}{dx} = \frac{3x^2}{5}$$

At the point $\left(-1, -\dfrac{1}{5}\right)$, $x = -1$ and

$$m = \frac{3(-1)^2}{5} = \frac{3}{5}$$

Equation of the tangent is:

$$y - y_1 = m(x - x_1)$$

i.e. $y - -\dfrac{1}{5} = \dfrac{3}{5}(x - -1)$

i.e. $y + \dfrac{1}{5} = \dfrac{3}{5}(x + 1)$

or $5y + 1 = 3x + 3$

or   **$5y - 3x = 2$**

Equation of the normal is:

$$y - y_1 = -\frac{1}{m}(x - x_1)$$

i.e. $y - -\dfrac{1}{5} = \dfrac{-1}{\left(\dfrac{3}{5}\right)}(x - -1)$

i.e. $y + \dfrac{1}{5} = -\dfrac{5}{3}(x + 1)$

i.e. $y + \dfrac{1}{5} = -\dfrac{5}{3}x - \dfrac{5}{3}$

Multiplying each term by 15 gives:

$$15y + 3 = -25x - 25$$

Hence **equation of the normal** is:

$$\mathbf{15y + 25x + 28 = 0}$$

Now try the following Practice Exercise

Practice Exercise 239 Further problems on tangents and normals (answers on page 1177)

For the following curves, at the points given, find (a) the equation of the tangent and (b) the equation of the normal.

1. $y = 2x^2$ at the point $(1, 2)$

2. $y = 3x^2 - 2x$ at the point $(2, 8)$

3. $y = \dfrac{x^3}{2}$ at the point $\left(-1, -\dfrac{1}{2}\right)$

4. $y = 1 + x - x^2$ at the point $(-2, -5)$

5. $\theta = \dfrac{1}{t}$ at the point $\left(3, \dfrac{1}{3}\right)$

54.7 Small changes

If y is a function of x, i.e. $y = f(x)$, and the approximate change in y corresponding to a small change δx in x is required, then:

$$\frac{\delta y}{\delta x} \approx \frac{dy}{dx}$$

and $\delta y \approx \dfrac{dy}{dx} \cdot \delta x$ or $\delta y \approx f'(x) \cdot \delta x$

Problem 26. Given $y = 4x^2 - x$, determine the approximate change in y if x changes from 1 to 1.02

Since $y = 4x^2 - x$, then $\dfrac{dy}{dx} = 8x - 1$

Approximate change in y,

$$\delta y \approx \frac{dy}{dx} \cdot \delta x \approx (8x - 1)\delta x$$

When $x = 1$ and $\delta x = 0.02$, $\delta y \approx [8(1) - 1](0.02)$

$$\approx \mathbf{0.14}$$

(Obviously, in this case, the exact value of δy may be obtained by evaluating y when $x = 1.02$, i.e. $y = 4(1.02)^2 - 1.02 = 3.1416$ and then subtracting from it the value of y when $x = 1$, i.e. $y = 4(1)^2 - 1 = 3$, giving $\delta y = 3.1416 - 3 = \mathbf{0.1416}$. Using $\delta y = \dfrac{dy}{dx} \cdot \delta x$ above

Section I

gave 0.14, which shows that the formula gives the approximate change in y for a small change in x)

Problem 27. The time of swing T of a pendulum is given by $T = k\sqrt{l}$, where k is a constant. Determine the percentage change in the time of swing if the length of the pendulum l changes from 32.1 cm to 32.0 cm

If $\qquad T = k\sqrt{l} = kl^{1/2}$

then $\quad \dfrac{dT}{dl} = k\left(\dfrac{1}{2}l^{-1/2}\right) = \dfrac{k}{2\sqrt{l}}$

Approximate change in T,

$$\delta T \approx \dfrac{dT}{dl}\delta l \approx \left(\dfrac{k}{2\sqrt{l}}\right)\delta l \approx \left(\dfrac{k}{2\sqrt{l}}\right)(-0.1)$$

(negative since l decreases)

Percentage change

$$= \left(\dfrac{\text{approximate change in } T}{\text{original value of } T}\right)100\%$$

$$= \dfrac{\left(\dfrac{k}{2\sqrt{l}}\right)(-0.1)}{k\sqrt{l}} \times 100\%$$

$$= \left(\dfrac{-0.1}{2l}\right)100\% = \left(\dfrac{-0.1}{2(32.1)}\right)100\%$$

$$= -0.156\%$$

Hence the percentage change in the time of swing is a decrease of 0.156%

Problem 28. A circular template has a radius of 10 cm (± 0.02). Determine the possible error in calculating the area of the template. Find also the percentage error

Area of circular template, $A = \pi r^2$, hence

$$\dfrac{dA}{dr} = 2\pi r$$

Approximate change in area,

$$\delta A \approx \dfrac{dA}{dr} \cdot \delta r \approx (2\pi r)\delta r$$

When $r = 10$ cm and $\delta r = 0.02$,

$$\delta A = (2\pi 10)(0.02) \approx 0.4\pi \text{ cm}^2$$

i.e. the possible error in calculating the template area is approximately 1.257 cm^2

$$\textbf{Percentage error} \approx \left(\dfrac{0.4\pi}{\pi(10)^2}\right)100\% = \textbf{0.40\%}$$

Now try the following Practice Exercise

Practice Exercise 240 Further problems on small changes (answers on page 1178)

1. Determine the change in y if x changes from 2.50 to 2.51 when (a) $y = 2x - x^2$ (b) $y = \dfrac{5}{x}$

2. The pressure p and volume v of a mass of gas are related by the equation $pv = 50$. If the pressure increases from 25.0 to 25.4, determine the approximate change in the volume of the gas. Find also the percentage change in the volume of the gas

3. Determine the approximate increase in (a) the volume, and (b) the surface area of a cube of side x cm if x increases from 20.0 cm to 20.05 cm

4. The radius of a sphere decreases from 6.0 cm to 5.96 cm. Determine the approximate change in (a) the surface area, and (b) the volume

5. The rate of flow of a liquid through a tube is given by Poiseuilles' equation as: $Q = \dfrac{p\pi r^4}{8\eta L}$ where Q is the rate of flow, p is the pressure difference between the ends of the tube, r is the radius of the tube, L is the length of the tube and η is the coefficient of viscosity of the liquid. η is obtained by measuring Q, p, r and L. If Q can be measured accurate to $\pm 0.5\%$, p accurate to $\pm 3\%$, r accurate to $\pm 2\%$ and L accurate to $\pm 1\%$, calculate the maximum possible percentage error in the value of η

Section I

For fully worked solutions to each of the problems in Practice Exercises 234 to 240 in this chapter, go to the website:
www.routledge.com/cw/bird

This assignment covers the material contained in Chapters 52 to 54. *The marks available are shown in brackets at the end of each question.*

1. Differentiate the following with respect to the variable:

   (a) $y = 5 + 2\sqrt{x^3} - \dfrac{1}{x^2}$ (b) $s = 4e^{2\theta} \sin 3\theta$

   (c) $y = \dfrac{3 \ln 5t}{\cos 2t}$ (d) $x = \dfrac{2}{\sqrt{t^2 - 3t + 5}}$

   (15)

2. If $f(x) = 2.5x^2 - 6x + 2$ find the co-ordinates at the point at which the gradient is -1 (5)

3. The displacement s cm of the end of a stiff spring at time t seconds is given by: $s = ae^{-kt} \sin 2\pi f t$. Determine the velocity and acceleration of the end of the spring after 2 seconds if $a = 3$, $k = 0.75$ and $f = 20$ (10)

4. Find the co-ordinates of the turning points on the curve $y = 3x^3 + 6x^2 + 3x - 1$ and distinguish between them. Find also the point(s) of inflexion.   (14)

5. The heat capacity C of a gas varies with absolute temperature θ as shown:

   $C = 26.50 + 7.20 \times 10^{-3}\theta - 1.20 \times 10^{-6}\theta^2$

   Determine the maximum value of C and the temperature at which it occurs.   (7)

6. Determine for the curve $y = 2x^2 - 3x$ at the point $(2, 2)$: (a) the equation of the tangent (b) the equation of the normal.   (7)

7. A rectangular block of metal with a square cross-section has a total surface area of $250 \, \text{cm}^2$. Find the maximum volume of the block of metal.   (7)

Chapter 55

Differentiation of parametric equations

Why it is important to understand: Differentiation of parametric equations

Rather than using a single equation to define two variables with respect to one another, parametric equations exist as a set that relates the two variables to one another with respect to a third variable. Some curves are easier to describe using a pair of parametric equations. The co-ordinates x and y of the curve are given using a third variable t, such as $x = f(t)$ and $y = g(t)$, where t is referred to as the parameter. Hence, for a given value of t, a point (x, y) is determined. For example, let t be the time while x and y are the positions of a particle; the parametric equations then describe the path of the particle at different times. Parametric equations are useful in defining three-dimensional curves and surfaces, such as determining the velocity or acceleration of a particle following a three-dimensional path. CAD systems use parametric versions of equations. Sometimes in engineering differentiation of parametric equations is necessary; for example, when determining the radius of curvature of part of the surface when finding the surface tension of a liquid. Knowledge of standard differentials and the function of a function rule from previous chapters is all that is needed.

At the end of this chapter, you should be able to:

- recognise parametric equations – ellipse, parabola, hyperbola, rectangular hyperbola, cardioids, asteroid and cycloid
- differentiate parametric equations

55.1 Introduction to parametric equations

Certain mathematical functions can be expressed more simply by expressing, say, x and y separately in terms of a third variable. For example, $y = r \sin\theta$, $x = r \cos\theta$. Then, any value given to θ will produce a pair of values for x and y, which may be plotted to provide a curve of $y = f(x)$

The third variable, θ, is called a **parameter** and the two expressions for y and x are called **parametric equations**.

The above example of $y = r \sin\theta$ and $x = r \cos\theta$ are the parametric equations for a circle. The equation of any point on a circle, centre at the origin and of radius r is given by: $x^2 + y^2 = r^2$, as shown in Chapter 28.

To show that $y = r \sin\theta$ and $x = r \cos\theta$ are suitable parametric equations for such a circle:

Left-hand side of equation
$$= x^2 + y^2$$
$$= (r\cos\theta)^2 + (r\sin\theta)^2$$
$$= r^2\cos^2\theta + r^2\sin^2\theta$$
$$= r^2(\cos^2\theta + \sin^2\theta)$$
$$= r^2 = \text{right-hand side}$$
(since $\cos^2\theta + \sin^2\theta = 1$, as shown in Chapter 42)

55.2 Some common parametric equations

The following are some of the most common parametric equations, and Figure 55.1 shows typical shapes of these curves.

(a) Ellipse   (b) Parabola

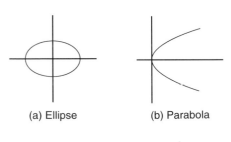

(c) Hyperbola   (d) Rectangular hyperbola

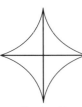

(e) Cardioid   (f) Astroid

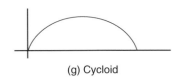

(g) Cycloid

Figure 55.1

(a)   Ellipse $x = a\cos\theta$, $y = b\sin\theta$
(b)   Parabola $x = at^2$, $y = 2at$
(c)   Hyperbola $x = a\sec\theta$, $y = b\tan\theta$

(d)   Rectangular hyperbola $x = ct$, $y = \dfrac{c}{t}$
(e)   Cardioid $x = a(2\cos\theta - \cos 2\theta)$, $y = a(2\sin\theta - \sin 2\theta)$
(f)   Astroid $x = a\cos^3\theta$, $y = a\sin^3\theta$
(g)   Cycloid $x = a(\theta - \sin\theta)$, $y = a(1 - \cos\theta)$

55.3 Differentiation in parameters

When x and y are given in terms of a parameter, say θ, then by the function of a function rule of differentiation (from Chapter 53):

$$\frac{dy}{dx} = \frac{dy}{d\theta} \times \frac{d\theta}{dx}$$

It may be shown that this can be written as:

$$\frac{dy}{dx} = \frac{\dfrac{dy}{d\theta}}{\dfrac{dx}{d\theta}} \qquad (1)$$

For the second differential,

$$\frac{d^2y}{dx^2} = \frac{d}{dx}\left(\frac{dy}{dx}\right) = \frac{d}{d\theta}\left(\frac{dy}{dx}\right)\cdot\frac{d\theta}{dx}$$

or

$$\frac{d^2y}{dx^2} = \frac{\dfrac{d}{d\theta}\left(\dfrac{dy}{dx}\right)}{\dfrac{dx}{d\theta}} \qquad (2)$$

Problem 1. Given $x = 5\theta - 1$ and $y = 2\theta(\theta - 1)$, determine $\dfrac{dy}{dx}$ in terms of θ

$x = 5\theta - 1$, hence $\dfrac{dx}{d\theta} = 5$

$y = 2\theta(\theta - 1) = 2\theta^2 - 2\theta$,

hence $\dfrac{dy}{d\theta} = 4\theta - 2 = 2(2\theta - 1)$

From equation (1),

$$\frac{dy}{dx} = \frac{\dfrac{dy}{d\theta}}{\dfrac{dx}{d\theta}} = \frac{2(2\theta - 1)}{5} \text{ or } \frac{2}{5}(2\theta - 1)$$

Problem 2. The parametric equations of a function are given by $y = 3\cos 2t$, $x = 2\sin t$. Determine expressions for (a) $\dfrac{dy}{dx}$ (b) $\dfrac{d^2y}{dx^2}$

(a) $y = 3 \cos 2t$, hence $\dfrac{dy}{dt} = -6 \sin 2t$

$x = 2 \sin t$, hence $\dfrac{dx}{dt} = 2 \cos t$

From equation (1),

$$\frac{dy}{dx} = \frac{\dfrac{dy}{dt}}{\dfrac{dx}{dt}} = \frac{-6 \sin 2t}{2 \cos t} = \frac{-6(2 \sin t \cos t)}{2 \cos t}$$

from double angles, Chapter 44

i.e. $\dfrac{\mathbf{d}y}{\mathbf{d}x} = \mathbf{-6 \sin} t$

(b)   From equation (2),

$$\frac{d^2 y}{dx^2} = \frac{\dfrac{d}{dt}\left(\dfrac{dy}{dx}\right)}{\dfrac{dx}{dt}} = \frac{\dfrac{d}{dt}(-6 \sin t)}{2 \cos t} = \frac{-6 \cos t}{2 \cos t}$$

i.e. $\dfrac{\mathbf{d}^2 y}{\mathbf{d}x^2} = \mathbf{-3}$

Problem 3.   The equation of a tangent drawn to a curve at point (x_1, y_1) is given by:

$$y - y_1 = \frac{dy_1}{dx_1}(x - x_1)$$

Determine the equation of the tangent drawn to the parabola $x = 2t^2$, $y = 4t$ at the point t

---

At point t, $x_1 = 2t^2$, hence $\dfrac{dx_1}{dt} = 4t$

and $y_1 = 4t$, hence $\dfrac{dy_1}{dt} = 4$

From equation (1),

$$\frac{dy}{dx} = \frac{\dfrac{dy}{dt}}{\dfrac{dx}{dt}} = \frac{4}{4t} = \frac{1}{t}$$

Hence, the equation of the tangent is:

$$y - 4t = \frac{1}{t}(x - 2t^2)$$

Problem 4.   The parametric equations of a cycloid are $x = 4(\theta - \sin \theta)$, $y = 4(1 - \cos \theta)$

Determine (a) $\dfrac{dy}{dx}$ (b) $\dfrac{d^2 y}{dx^2}$

(a) $x = 4(\theta - \sin \theta)$

hence $\dfrac{dx}{d\theta} = 4 - 4 \cos \theta = 4(1 - \cos \theta)$

$y = 4(1 - \cos \theta)$, hence $\dfrac{dy}{d\theta} = 4 \sin \theta$

From equation (1),

$$\frac{dy}{dx} = \frac{\dfrac{dy}{d\theta}}{\dfrac{dx}{d\theta}} = \frac{4 \sin \theta}{4(1 - \cos \theta)} = \frac{\mathbf{\sin \theta}}{\mathbf{(1 - \cos \theta)}}$$

(b)   From equation (2),

$$\frac{d^2 y}{dx^2} - \frac{\dfrac{d}{d\theta}\left(\dfrac{dy}{dx}\right)}{\dfrac{dx}{d\theta}} = \frac{\dfrac{d}{d\theta}\left(\dfrac{\sin \theta}{1 - \cos \theta}\right)}{4(1 - \cos \theta)}$$

$$- \frac{\dfrac{(1 - \cos \theta)(\cos \theta) - (\sin \theta)(\sin \theta)}{(1 - \cos \theta)^2}}{4(1 - \cos \theta)}$$

$$= \frac{\cos \theta - \cos^2 \theta - \sin^2 \theta}{4(1 - \cos \theta)^3}$$

$$= \frac{\cos \theta - (\cos^2 \theta + \sin^2 \theta)}{4(1 - \cos \theta)^3}$$

$$= \frac{\cos \theta - 1}{4(1 - \cos \theta)^3}$$

$$= \frac{-(1 - \cos \theta)}{4(1 - \cos \theta)^3} = \frac{\mathbf{-1}}{\mathbf{4(1 - \cos \theta)^2}}$$

Now try the following Practice Exercise

Practice Exercise 241   Further problems on differentiation of parametric equations (answers on page 1178)

1.   Given $x = 3t - 1$ and $y = t(t - 1)$, determine $\dfrac{dy}{dx}$ in terms of t

2.   A parabola has parametric equations: $x = t^2$, $y = 2t$. Evaluate $\dfrac{dy}{dx}$ when $t = 0.5$

Section I

3. The parametric equations for an ellipse are $x = 4\cos\theta$, $y = \sin\theta$. Determine (a) $\dfrac{dy}{dx}$ (b) $\dfrac{d^2 y}{dx^2}$

4. Evaluate $\dfrac{dy}{dx}$ at $\theta = \dfrac{\pi}{6}$ radians for the hyperbola whose parametric equations are $x = 3\sec\theta$, $y = 6\tan\theta$

5. The parametric equations for a rectangular hyperbola are $x = 2t$, $y = \dfrac{2}{t}$. Evaluate $\dfrac{dy}{dx}$ when $t = 0.40$

The equation of a tangent drawn to a curve at point (x_1, y_1) is given by:

$$y - y_1 = \frac{dy_1}{dx_1}(x - x_1)$$

Use this in Problems 6 and 7.

6. Determine the equation of the tangent drawn to the ellipse $x = 3\cos\theta$, $y = 2\sin\theta$ at $\theta = \dfrac{\pi}{6}$

7. Determine the equation of the tangent drawn to the rectangular hyperbola $x = 5t$, $y = \dfrac{5}{t}$ at $t = 2$

55.4 Further worked problems on differentiation of parametric equations

Problem 5. The equation of the normal drawn to a curve at point (x_1, y_1) is given by:

$$y - y_1 = -\frac{1}{\frac{dy_1}{dx_1}}(x - x_1)$$

Determine the equation of the normal drawn to the astroid $x = 2\cos^3\theta$, $y = 2\sin^3\theta$ at the point $\theta = \dfrac{\pi}{4}$

$x = 2\cos^3\theta$, hence $\dfrac{dx}{d\theta} = -6\cos^2\theta\sin\theta$

$y = 2\sin^3\theta$, hence $\dfrac{dy}{d\theta} = 6\sin^2\theta\cos\theta$

From equation (1),

$$\frac{dy}{dx} = \frac{\frac{dy}{d\theta}}{\frac{dx}{d\theta}} = \frac{6\sin^2\theta\cos\theta}{-6\cos^2\theta\sin\theta} = -\frac{\sin\theta}{\cos\theta} = -\tan\theta$$

When $\theta = \dfrac{\pi}{4}$, $\dfrac{dy}{dx} = -\tan\dfrac{\pi}{4} = -1$

$x_1 = 2\cos^3\dfrac{\pi}{4} = 0.7071$ and $y_1 = 2\sin^3\dfrac{\pi}{4} = 0.7071$

Hence, **the equation of the normal is**:

$$y - 0.7071 = -\frac{1}{-1}(x - 0.7071)$$

i.e. $y - 0.7071 = x - 0.7071$

i.e. $\quad\quad\quad \boldsymbol{y = x}$

Problem 6. The parametric equations for a hyperbola are $x = 2\sec\theta$, $y = 4\tan\theta$. Evaluate (a) $\dfrac{dy}{dx}$ (b) $\dfrac{d^2 y}{dx^2}$, correct to 4 significant figures, when $\theta = 1$ radian

(a) $x = 2\sec\theta$, hence $\dfrac{dx}{d\theta} = 2\sec\theta\tan\theta$

$y = 4\tan\theta$, hence $\dfrac{dy}{d\theta} = 4\sec^2\theta$

From equation (1),

$$\frac{dy}{dx} = \frac{\frac{dy}{d\theta}}{\frac{dx}{d\theta}} = \frac{4\sec^2\theta}{2\sec\theta\tan\theta} = \frac{2\sec\theta}{\tan\theta}$$

$$= \frac{2\left(\dfrac{1}{\cos\theta}\right)}{\left(\dfrac{\sin\theta}{\cos\theta}\right)} = \frac{2}{\sin\theta} \text{ or } 2\operatorname{cosec}\theta$$

When $\theta = 1$ rad, $\dfrac{dy}{dx} = \dfrac{2}{\sin 1} = \mathbf{2.377}$, correct to 4 significant figures.

(b) From equation (2),

$$\frac{d^2 y}{dx^2} = \frac{\frac{d}{d\theta}\left(\frac{dy}{dx}\right)}{\frac{dx}{d\theta}} = \frac{\frac{d}{d\theta}(2\operatorname{cosec}\theta)}{2\sec\theta\tan\theta}$$

$$= \frac{-2\operatorname{cosec}\theta\cot\theta}{2\sec\theta\tan\theta}$$

$$= \frac{-\left(\dfrac{1}{\sin\theta}\right)\left(\dfrac{\cos\theta}{\sin\theta}\right)}{\left(\dfrac{1}{\cos\theta}\right)\left(\dfrac{\sin\theta}{\cos\theta}\right)}$$

$$= -\left(\dfrac{\cos\theta}{\sin^2\theta}\right)\left(\dfrac{\cos^2\theta}{\sin\theta}\right)$$

$$= -\dfrac{\cos^3\theta}{\sin^3\theta} = -\cot^3\theta$$

When $\theta = 1$ rad, $\dfrac{d^2y}{dx^2} = -\cot^3 1 = -\dfrac{1}{(\tan 1)^3} =$ **−0.2647**, correct to 4 significant figures.

Problem 7. When determining the surface tension of a liquid, the radius of curvature, ρ, of part of the surface is given by:

$$\rho = \frac{\sqrt{\left[1+\left(\dfrac{dy}{dx}\right)^2\right]^3}}{\dfrac{d^2y}{dx^2}}$$

Find the radius of curvature of the part of the surface having the parametric equations $x = 3t^2$, $y = 6t$ at the point $t = 2$

$x = 3t^2$, hence $\dfrac{dx}{dt} = 6t$

$y = 6t$, hence $\dfrac{dy}{dt} = 6$

From equation (1), $\dfrac{dy}{dx} = \dfrac{\dfrac{dy}{dt}}{\dfrac{dx}{dt}} = \dfrac{6}{6t} = \dfrac{1}{t}$

From equation (2),

$$\frac{d^2y}{dx^2} = \frac{\dfrac{d}{dt}\left(\dfrac{dy}{dx}\right)}{\dfrac{dx}{dt}} = \frac{\dfrac{d}{dt}\left(\dfrac{1}{t}\right)}{6t} = \frac{-\dfrac{1}{t^2}}{6t} = -\frac{1}{6t^3}$$

Hence, radius of curvature, $\rho = \dfrac{\sqrt{\left[1+\left(\dfrac{dy}{dx}\right)^2\right]^3}}{\dfrac{d^2y}{dx^2}}$

$$= \frac{\sqrt{\left[1+\left(\dfrac{1}{t}\right)^2\right]^3}}{-\dfrac{1}{6t^3}}$$

When $t = 2$, $\rho = \dfrac{\sqrt{\left[1+\left(\dfrac{1}{2}\right)^2\right]^3}}{-\dfrac{1}{6(2)^3}} = \dfrac{\sqrt{(1.25)^3}}{-\dfrac{1}{48}}$

$$= -48\sqrt{(1.25)^3} = \mathbf{-67.08}$$

Now try the following Practice Exercise

Practice Exercise 242   Further problems on differentiation of parametric equations (answers on page 1178)

1. A cycloid has parametric equations $x = 2(\theta - \sin\theta)$, $y = 2(1 - \cos\theta)$. Evaluate, at $\theta = 0.62$ rad, correct to 4 significant figures, (a) $\dfrac{dy}{dx}$ (b) $\dfrac{d^2y}{dx^2}$

The equation of the normal drawn to a curve at point (x_1, y_1) is given by:

$$y - y_1 = -\frac{1}{\dfrac{dy_1}{dx_1}}(x - x_1)$$

Use this in Problems 2 and 3.

2. Determine the equation of the normal drawn to the parabola $x = \dfrac{1}{4}t^2$, $y = \dfrac{1}{2}t$ at $t = 2$

3. Find the equation of the normal drawn to the cycloid $x = 2(\theta - \sin\theta)$, $y = 2(1 - \cos\theta)$ at $\theta = \dfrac{\pi}{2}$ rad.

4. Determine the value of $\dfrac{d^2y}{dx^2}$, correct to 4 significant figures, at $\theta = \dfrac{\pi}{6}$ rad for the cardioid $x = 5(2\theta - \cos 2\theta)$, $y = 5(2\sin\theta - \sin 2\theta)$

5. The radius of curvature, ρ, of part of a surface when determining the surface tension of a liquid is given by:

$$\rho = \frac{\left[1 + \left(\dfrac{dy}{dx} \right)^2 \right]^{3/2}}{\dfrac{d^2 y}{dx^2}}$$

Find the radius of curvature (correct to 4 significant figures) of the part of the surface having parametric equations

(a) $x = 3t$, $y = \dfrac{3}{t}$ at the point $t = \dfrac{1}{2}$

(b) $x = 4 \cos^3 t$, $y = 4 \sin^3 t$ at $t = \dfrac{\pi}{6}$ rad

For fully worked solutions to each of the problems in Practice Exercises 241 and 242 in this chapter, go to the website:
www.routledge.com/cw/bird

Chapter 56

Differentiation of implicit functions

Why it is important to understand: Differentiation of implicit functions

Differentiation of implicit functions is another special technique, but it occurs often enough to be important. It is needed for more complicated problems involving different rates of change. Up to this chapter we have been finding derivatives of functions of the form $y = f(x)$; unfortunately not all functions fall into this form. However, implicit differentiation is nothing more than a special case of the function of a function (or chain rule) for derivatives. Engineering applications where implicit differentiation is needed are found in optics, electronics, control, and even some thermodynamics.

At the end of this chapter, you should be able to:

- recognise implicit functions
- differentiate simple implicit functions
- differentiate implicit functions containing products and quotients

56.1 Implicit functions

When an equation can be written in the form $y = f(x)$ it is said to be an **explicit function** of x. Examples of explicit functions include

$$y = 2x^3 - 3x + 4, \quad y = 2x \ln x$$

and $\quad y = \dfrac{3e^x}{\cos x}$

In these examples y may be differentiated with respect to x by using standard derivatives, the product rule and the quotient rule of differentiation, respectively.

Sometimes with equations involving, say, y and x, it is impossible to make y the subject of the formula. The equation is then called an **implicit function** and examples of such functions include $y^3 + 2x^2 = y^2 - x$ and $\sin y = x^2 + 2xy$

56.2 Differentiating implicit functions

It is possible to **differentiate an implicit function** by using the **function of a function rule**, which may be stated as

$$\frac{du}{dx} = \frac{du}{dy} \times \frac{dy}{dx}$$

Thus, to differentiate y^3 with respect to x, the substitution $u = y^3$ is made, from which, $\dfrac{du}{dy} = 3y^2$

Hence, $\dfrac{d}{dx}(y^3) = (3y^2) \times \dfrac{dy}{dx}$, by the function of a function rule.

A simple rule for differentiating an implicit function is summarised as:

$$\frac{d}{dx}[f(y)] = \frac{d}{dy}[f(y)] \times \frac{dy}{dx} \qquad (1)$$

Problem 1.   Differentiate the following functions with respect to x: (a) $2y^4$ (b) $\sin 3t$

(a)   Let $u = 2y^4$, then, by the function of a function rule:

$$\frac{du}{dx} = \frac{du}{dy} \times \frac{dy}{dx} = \frac{d}{dy}(2y^4) \times \frac{dy}{dx}$$

$$= 8y^3 \frac{dy}{dx}$$

(b)   Let $u = \sin 3t$, then, by the function of a function rule:

$$\frac{du}{dx} = \frac{du}{dt} \times \frac{dt}{dx} = \frac{d}{dt}(\sin 3t) \times \frac{dt}{dx}$$

$$= 3 \cos 3t \frac{dt}{dx}$$

Problem 2.   Differentiate the following functions with respect to x:

$$\text{(a) } 4 \ln 5y \quad \text{(b) } \frac{1}{5}e^{3\theta - 2}$$

(a)   Let $u = 4 \ln 5y$, then, by the function of a function rule:

$$\frac{du}{dx} = \frac{du}{dy} \times \frac{dy}{dx} = \frac{d}{dy}(4 \ln 5y) \times \frac{dy}{dx}$$

$$= \frac{4}{y} \frac{dy}{dx}$$

(b)   Let $u = \dfrac{1}{5}e^{3\theta - 2}$, then, by the function of a function rule:

$$\frac{du}{dx} = \frac{du}{d\theta} \times \frac{d\theta}{dx} = \frac{d}{d\theta}\left(\frac{1}{5}e^{3\theta - 2}\right) \times \frac{d\theta}{dx}$$

$$= \frac{3}{5}e^{3\theta - 2} \frac{d\theta}{dx}$$

Now try the following Practice Exercise

Practice Exercise 243   Further problems on differentiating implicit functions (answers on page 1178)

In Problems 1 and 2, differentiate the given functions with respect to x.

1.   (a) $3y^5$ (b) $2 \cos 4\theta$ (c) $\sqrt{k}$

2.   (a) $\dfrac{5}{2} \ln 3t$ (b) $\dfrac{3}{4}e^{2y+1}$ (c) $2 \tan 3y$

3.   Differentiate the following with respect to y:

(a) $3 \sin 2\theta$ (b) $4\sqrt{x^3}$ (c) $\dfrac{2}{e^t}$

4.   Differentiate the following with respect to u:

(a) $\dfrac{2}{(3x+1)}$ (b) $3 \sec 2\theta$ (c) $\dfrac{2}{\sqrt{y}}$

56.3   Differentiating implicit functions containing products and quotients

The product and quotient rules of differentiation must be applied when differentiating functions containing products and quotients of two variables.

For example, $\dfrac{d}{dx}(x^2 y) = (x^2)\dfrac{d}{dx}(y) + (y)\dfrac{d}{dx}(x^2),$

$$\text{by the product rule}$$

$$= (x^2)\left(1\frac{dy}{dx}\right) + y(2x),$$

$$\text{by using equation (1)}$$

$$= x^2 \frac{dy}{dx} + 2xy$$

Problem 3.   Determine $\dfrac{d}{dx}(2x^3 y^2)$

In the product rule of differentiation let $u = 2x^3$ and $v = y^2$.

Thus $\dfrac{d}{dx}(2x^3 y^2) = (2x^3)\dfrac{d}{dx}(y^2) + (y^2)\dfrac{d}{dx}(2x^3)$

$$= (2x^3)\left(2y\frac{dy}{dx}\right) + (y^2)(6x^2)$$

$$= 4x^3 y\frac{dy}{dx} + 6x^2 y^2$$

$$= 2x^2 y\left(2x\frac{dy}{dx} + 3y\right)$$

Problem 4. Find $\dfrac{d}{dx}\left(\dfrac{3y}{2x}\right)$

In the quotient rule of differentiation let $u=3y$ and $v=2x$.

Thus $\dfrac{d}{dx}\left(\dfrac{3y}{2x}\right)=\dfrac{(2x)\dfrac{d}{dx}(3y)-(3y)\dfrac{d}{dx}(2x)}{(2x)^2}$

$$=\dfrac{(2x)\left(3\dfrac{dy}{dx}\right)-(3y)(2)}{4x^2}$$

$$=\dfrac{6x\dfrac{dy}{dx}-6y}{4x^2}=\dfrac{3}{2x^2}\left(x\dfrac{dy}{dx}-y\right)$$

Problem 5. Differentiate $z=x^2+3x\cos 3y$ with respect to y

$$\dfrac{dz}{dy}=\dfrac{d}{dy}(x^2)+\dfrac{d}{dy}(3x\cos 3y)$$

$$=2x\dfrac{dx}{dy}+\left[(3x)(-3\sin 3y)+(\cos 3y)\left(3\dfrac{dx}{dy}\right)\right]$$

$$=2x\dfrac{dx}{dy}-9x\sin 3y+3\cos 3y\dfrac{dx}{dy}$$

Now try the following Practice Exercise

Practice Exercise 244   Further problems on differentiating implicit functions involving products and quotients (answers on page 1178)

1. Determine $\dfrac{d}{dx}(3x^2y^3)$

2. Find $\dfrac{d}{dx}\left(\dfrac{2y}{5x}\right)$

3. Determine $\dfrac{d}{du}\left(\dfrac{3u}{4v}\right)$

4. Given $z=3\sqrt{y}\cos 3x$ find $\dfrac{dz}{dx}$

5. Determine $\dfrac{dz}{dy}$ given $z=2x^3\ln y$

56.4   Further implicit differentiation

An implicit function such as $3x^2+y^2-5x+y=2$ may be differentiated term by term with respect to x. This gives:

$$\dfrac{d}{dx}(3x^2)+\dfrac{d}{dx}(y^2)-\dfrac{d}{dx}(5x)+\dfrac{d}{dx}(y)=\dfrac{d}{dx}(2)$$

i.e. $\qquad 6x+2y\dfrac{dy}{dx}-5+1\dfrac{dy}{dx}=0,$

using equation (1) and standard derivatives.

An expression for the derivative $\dfrac{dy}{dx}$ in terms of x and y may be obtained by rearranging this latter equation. Thus:

$$(2y+1)\dfrac{dy}{dx}=5-6x$$

from which, $\qquad \dfrac{dy}{dx}=\dfrac{5-6x}{2y+1}$

Problem 6. Given $2y^2-5x^4-2-7y^3=0$, determine $\dfrac{dy}{dx}$

Each term in turn is differentiated with respect to x:

Hence $\qquad \dfrac{d}{dx}(2y^2)-\dfrac{d}{dx}(5x^4)-\dfrac{d}{dx}(2)-\dfrac{d}{dx}(7y^3)$

$$=\dfrac{d}{dx}(0)$$

i.e. $\qquad 4y\dfrac{dy}{dx}-20x^3-0-21y^2\dfrac{dy}{dx}=0$

Rearranging gives:

$$(4y-21y^2)\dfrac{dy}{dx}=20x^3$$

i.e. $\qquad \dfrac{dy}{dx}=\dfrac{20x^3}{(4y-21y^2)}$

Problem 7. Determine the values of $\dfrac{dy}{dx}$ when $x=4$ given that $x^2+y^2=25$

Differentiating each term in turn with respect to x gives:

$$\dfrac{d}{dx}(x^2)+\dfrac{d}{dx}(y^2)=\dfrac{d}{dx}(25)$$

i.e. $\qquad 2x+2y\dfrac{dy}{dx}=0$

Section I

Hence
$$\frac{dy}{dx} = -\frac{2x}{2y} = -\frac{x}{y}$$

Since $x^2 + y^2 = 25$, when $x = 4$, $y = \sqrt{(25 - 4^2)} = \pm 3$

Thus, when $x = 4$ and $y = \pm 3$, $\dfrac{dy}{dx} = -\dfrac{4}{\pm 3} = \pm\dfrac{4}{3}$

$x^2 + y^2 = 25$ is the equation of a circle, centre at the origin and radius 5, as shown in Figure 56.1. At $x = 4$, the two gradients are shown.

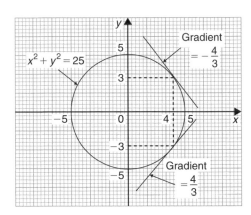

Figure 56.1

Above, $x^2 + y^2 = 25$ was differentiated implicitly; actually, the equation could be transposed to $y = \sqrt{(25 - x^2)}$ and differentiated using the function of a function rule. This gives

$$\frac{dy}{dx} = \frac{1}{2}(25 - x^2)^{-\frac{1}{2}}(-2x) = -\frac{x}{\sqrt{(25 - x^2)}}$$

and when $x = 4$, $\dfrac{dy}{dx} = -\dfrac{4}{\sqrt{(25 - 4^2)}} = \pm\dfrac{4}{3}$ as obtained above.

Problem 8.

(a) Find $\dfrac{dy}{dx}$ in terms of x and y given

$$4x^2 + 2xy^3 - 5y^2 = 0$$

(b) Evalate $\dfrac{dy}{dx}$ when $x = 1$ and $y = 2$

(a) Differentiating each term in turn with respect to x gives:

$$\frac{d}{dx}(4x^2) + \frac{d}{dx}(2xy^3) - \frac{d}{dx}(5y^2) = \frac{d}{dx}(0)$$

i.e. $8x + \left[(2x)\left(3y^2\dfrac{dy}{dx}\right) + (y^3)(2)\right]$

$$- 10y\frac{dy}{dx} = 0$$

i.e. $8x + 6xy^2\dfrac{dy}{dx} + 2y^3 - 10y\dfrac{dy}{dx} = 0$

Rearranging gives:

$$8x + 2y^3 = (10y - 6xy^2)\frac{dy}{dx}$$

and $\dfrac{dy}{dx} = \dfrac{8x + 2y^3}{10y - 6xy^2} = \dfrac{4x + y^3}{y(5 - 3xy)}$

(b) When $x = 1$ and $y = 2$,

$$\frac{dy}{dx} = \frac{4(1) + (2)^3}{2[5 - (3)(1)(2)]} = \frac{12}{-2} = -6$$

Problem 9.   Find the gradients of the tangents drawn to the circle $x^2 + y^2 - 2x - 2y = 3$ at $x = 2$

The gradient of the tangent is given by $\dfrac{dy}{dx}$

Differentiating each term in turn with respect to x gives:

$$\frac{d}{dx}(x^2) + \frac{d}{dx}(y^2) - \frac{d}{dx}(2x) - \frac{d}{dx}(2y) = \frac{d}{dx}(3)$$

i.e. $2x + 2y\dfrac{dy}{dx} - 2 - 2\dfrac{dy}{dx} = 0$

Hence $(2y - 2)\dfrac{dy}{dx} = 2 - 2x$

from which $\dfrac{dy}{dx} = \dfrac{2 - 2x}{2y - 2} = \dfrac{1 - x}{y - 1}$

The value of y when $x = 2$ is determined from the original equation

Hence $(2)^2 + y^2 - 2(2) - 2y = 3$

i.e. $4 + y^2 - 4 - 2y = 3$

or $y^2 - 2y - 3 = 0$

Factorising gives: $(y + 1)(y - 3) = 0$, from which $y = -1$ or $y = 3$
When $x = 2$ and $y = -1$,

$$\frac{dy}{dx} = \frac{1 - x}{y - 1} = \frac{1 - 2}{-1 - 1} = \frac{-1}{-2} = \frac{1}{2}$$

When $x=2$ and $y=3$,

$$\frac{dy}{dx} = \frac{1-2}{3-1} = \frac{-1}{2}$$

Hence the gradients of the tangents are $\pm\dfrac{1}{2}$

The circle having the given equation has its centre at $(1, 1)$ and radius $\sqrt{5}$ (see Chapter 28) and is shown in Figure 56.2 with the two gradients of the tangents.

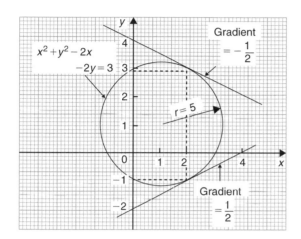

Figure 56.2

Problem 10. Pressure p and volume v of a gas are related by the law $pv^{\gamma}=k$, where γ and k are constants. Show that the rate of change of pressure $\dfrac{dp}{dt} = -\gamma\dfrac{p}{v}\dfrac{dv}{dt}$

Since $pv^{\gamma}=k$, then $p=\dfrac{k}{v^{\gamma}}=kv^{-\gamma}$

$$\frac{dp}{dt} = \frac{dp}{dv} \times \frac{dv}{dt}$$

by the function of a function rule

$$\frac{dp}{dv} = \frac{d}{dv}(kv^{-\gamma})$$

$$= -\gamma kv^{-\gamma-1} = \frac{-\gamma k}{v^{\gamma+1}}$$

$$\frac{dp}{dt} = \frac{-\gamma k}{v^{\gamma+1}} \times \frac{dv}{dt}$$

Since $\qquad k = pv^{\gamma}$

$$\frac{dp}{dt} = \frac{-\gamma(pv^{\gamma})}{v^{\gamma+1}}\frac{dv}{dt} = \frac{-\gamma pv^{\gamma}}{v^{\gamma}v^{1}}\frac{dv}{dt}$$

i.e. $\qquad \dfrac{dp}{dt} = -\gamma\dfrac{p}{v}\dfrac{dv}{dt}$

Now try the following Practice Exercise

Practice Exercise 245   Further problems on implicit differentiation (answers on page 1178)

In Problems 1 and 2 determine $\dfrac{dy}{dx}$

1. $x^2+y^2+4x-3y+1=0$

2. $2y^3-y+3x-2=0$

3. Given $x^2+y^2=9$ evaluate $\dfrac{dy}{dx}$ when $x=\sqrt{5}$ and $y=2$

In Problems 4 to 7, determine $\dfrac{dy}{dx}$

4. $x^2+2x\sin 4y=0$

5. $3y^2+2xy-4x^2=0$

6. $2x^2y+3x^3=\sin y$

7. $3y+2x\ln y=y^4+x$

8. If $3x^2+2x^2y^3-\dfrac{5}{4}y^2=0$ evaluate $\dfrac{dy}{dx}$ when $x=\dfrac{1}{2}$ and $y=1$

9. Determine the gradients of the tangents drawn to the circle $x^2+y^2=16$ at the point where $x=2$. Give the answer correct to 4 significant figures.

10. Find the gradients of the tangents drawn to the ellipse $\dfrac{x^2}{4}+\dfrac{y^2}{9}=2$ at the point where $x=2$

11. Determine the gradient of the curve $3xy+y^2=-2$ at the point $(1, -2)$

For fully worked solutions to each of the problems in Practice Exercises 243 to 245 in this chapter, go to the website:
www.routledge.com/cw/bird

Section I

Chapter 57

Logarithmic differentiation

Why it is important to understand: Logarithmic differentiation

Logarithmic differentiation is a means of differentiating algebraically complicated functions, or functions for which the ordinary rules of differentiation do not apply. The technique is performed in cases where it is easier to differentiate the logarithm of a function rather than the function itself. Logarithmic differentiation relies on the function of a function rule (i.e. chain rule) as well as properties of logarithms (in particular, the natural logarithm, or logarithm to the base e) to transform products into sums and divisions into subtractions, and can also be applied to functions raised to the power of variables of functions. Logarithmic differentiation occurs often enough in engineering calculations to make it an important technique.

At the end of this chapter, you should be able to:

- state the laws of logarithms
- differentiate simple logarithmic functions
- differentiate an implicit function involving logarithms
- differentiate more difficult logarithmic functions involving products and quotients
- differentiate functions of the form $y = [f(x)]^x$

57.1 Introduction to logarithmic differentiation

With certain functions containing more complicated products and quotients, differentiation is often made easier if the logarithm of the function is taken before differentiating. This technique, called **'logarithmic differentiation'** is achieved with a knowledge of (i) the laws of logarithms, (ii) the differential coefficients of logarithmic functions, and (iii) the differentiation of implicit functions.

57.2 Laws of logarithms

Three laws of logarithms may be expressed as:

(i) $\log (A \times B) = \log A + \log B$

(ii) $\log \left(\dfrac{A}{B} \right) = \log A - \log B$

(iii) $\log A^n = n \log A$

In calculus, Napierian logarithms (i.e. logarithms to a base of 'e') are invariably used. Thus for two functions $f(x)$ and $g(x)$ the laws of logarithms may be expressed as:

(i) $\ln[f(x) \cdot g(x)] = \ln f(x) + \ln g(x)$

(ii) $\ln \left(\dfrac{f(x)}{g(x)} \right) = \ln f(x) - \ln g(x)$

(iii) $\ln[f(x)]^n = n \ln f(x)$

Taking Napierian logarithms of both sides of the equation $y = \dfrac{f(x) \cdot g(x)}{h(x)}$ gives:

$$\ln y = \ln \left(\frac{f(x) \cdot g(x)}{h(x)} \right)$$

which may be simplified using the above laws of logarithms, giving:

$$\ln y = \ln f(x) + \ln g(x) - \ln h(x)$$

This latter form of the equation is often easier to differentiate.

57.3 Differentiation of logarithmic functions

The differential coefficient of the logarithmic function $\ln x$ is given by:

$$\frac{d}{dx}(\ln x) = \frac{1}{x}$$

More generally, it may be shown that:

$$\frac{d}{dx}[\ln f(x)] = \frac{f'(x)}{f(x)} \tag{1}$$

For example, if $y = \ln(3x^2 + 2x - 1)$ then,

$$\frac{dy}{dx} = \frac{6x + 2}{3x^2 + 2x - 1}$$

Similarly, if $y = \ln(\sin 3x)$ then

$$\frac{dy}{dx} = \frac{3\cos 3x}{\sin 3x} = 3\cot 3x$$

Now try the following Practice Exercise

Practice Exercise 246 Differentiating logarithmic functions (answers on page 1178)

Differentiate the following using the laws for logarithms.

1. $\ln(4x - 10)$
2. $\ln(\cos 3x)$
3. $\ln(3x^3 + x)$
4. $\ln(5x^2 + 10x - 7)$
5. $\ln 8x$
6. $\ln(x^2 - 1)$
7. $3\ln 4x$
8. $2\ln(\sin x)$
9. $\ln(4x^3 - 6x^2 + 3x)$

57.4 Differentiation of further logarithmic functions

As explained in Chapter 56, by using the function of a function rule:

$$\frac{d}{dx}(\ln y) = \left(\frac{1}{y}\right)\frac{dy}{dx} \tag{2}$$

Differentiation of an expression such as

$y = \dfrac{(1+x)^2\sqrt{(x-1)}}{x\sqrt{(x+2)}}$ may be achieved by using the product and quotient rules of differentiation; however, the working would be rather complicated. With logarithmic differentiation the following procedure is adopted:

(i) Take Napierian logarithms of both sides of the equation.

$$\text{Thus } \ln y = \ln\left\{\frac{(1+x)^2\sqrt{(x-1)}}{x\sqrt{(x+2)}}\right\}$$

$$= \ln\left\{\frac{(1+x)^2(x-1)^{\frac{1}{2}}}{x(x+2)^{\frac{1}{2}}}\right\}$$

(ii) Apply the laws of logarithms.

$$\text{Thus } \ln y = \ln(1+x)^2 + \ln(x-1)^{\frac{1}{2}}$$
$$- \ln x - \ln(x+2)^{\frac{1}{2}}, \text{ by laws (i)}$$
$$\text{and (ii) of Section 57.2}$$

i.e. $\ln y = 2\ln(1+x) + \dfrac{1}{2}\ln(x-1)$
$$- \ln x - \frac{1}{2}\ln(x+2), \text{ by law (iii)}$$
$$\text{of Section 57.2}$$

(iii) Differentiate each term in turn with respect to x using equations (1) and (2).

$$\text{Thus } \frac{1}{y}\frac{dy}{dx} = \frac{2}{(1+x)} + \frac{\frac{1}{2}}{(x-1)} - \frac{1}{x} - \frac{\frac{1}{2}}{(x+2)}$$

(iv) Rearrange the equation to make $\dfrac{dy}{dx}$ the subject.

$$\text{Thus } \frac{dy}{dx} = y\left\{\frac{2}{(1+x)} + \frac{1}{2(x-1)} - \frac{1}{x} - \frac{1}{2(x+2)}\right\}$$

(v) Substitute for y in terms of x.

$$\text{Thus } \frac{dy}{dx} = \frac{(1+x)^2\sqrt{(x-1)}}{x\sqrt{(x+2)}}\left\{\frac{2}{(1+x)} + \frac{1}{2(x-1)} - \frac{1}{x} - \frac{1}{2(x+2)}\right\}$$

Problem 1. Use logarithmic differentiation to differentiate $y = \dfrac{(x+1)(x-2)^3}{(x-3)}$

Following the above procedure:

(i) Since $y = \dfrac{(x+1)(x-2)^3}{(x-3)}$

then $\ln y = \ln\left\{\dfrac{(x+1)(x-2)^3}{(x-3)}\right\}$

(ii) $\ln y = \ln(x+1) + \ln(x-2)^3 - \ln(x-3)$,

by laws (i) and (ii) of Section 57.2,

i.e. $\ln y = \ln(x+1) + 3\ln(x-2) - \ln(x-3)$,

by law (iii) of Section 57.2

(iii) Differentiating with respect to x gives:

$$\frac{1}{y}\frac{dy}{dx} = \frac{1}{(x+1)} + \frac{3}{(x-2)} - \frac{1}{(x-3)}$$

by using equations (1) and (2).

(iv) Rearranging gives:

$$\frac{dy}{dx} = y\left\{\frac{1}{(x+1)} + \frac{3}{(x-2)} - \frac{1}{(x-3)}\right\}$$

(v) Substituting for y gives:

$$\frac{dy}{dx} = \frac{(x+1)(x-2)^3}{(x-3)}\left\{\frac{1}{(x+1)} + \frac{3}{(x-2)} - \frac{1}{(x-3)}\right\}$$

Problem 2. Differentiate $y = \dfrac{\sqrt{(x-2)^3}}{(x+1)^2(2x-1)}$ with respect to x and evaluate $\dfrac{dy}{dx}$ when $x=3$

Using logarithmic differentiation and following the above procedure:

(i) Since $y = \dfrac{\sqrt{(x-2)^3}}{(x+1)^2(2x-1)}$

then $\ln y = \ln\left\{\dfrac{\sqrt{(x-2)^3}}{(x+1)^2(2x-1)}\right\}$

$= \ln\left\{\dfrac{(x-2)^{\frac{3}{2}}}{(x+1)^2(2x-1)}\right\}$

(ii) $\ln y = \ln(x-2)^{\frac{3}{2}} - \ln(x+1)^2 - \ln(2x-1)$

i.e. $\ln y = \frac{3}{2}\ln(x-2) - 2\ln(x+1) - \ln(2x-1)$

(iii) $\dfrac{1}{y}\dfrac{dy}{dx} = \dfrac{\frac{3}{2}}{(x-2)} - \dfrac{2}{(x+1)} - \dfrac{2}{(2x-1)}$

(iv) $\dfrac{dy}{dx} = y\left\{\dfrac{3}{2(x-2)} - \dfrac{2}{(x+1)} - \dfrac{2}{(2x-1)}\right\}$

(v) $\dfrac{dy}{dx} = \dfrac{\sqrt{(x-2)^3}}{(x+1)^2(2x-1)}\left\{\dfrac{3}{2(x-2)}\right.$

$\left. -\dfrac{2}{(x+1)} - \dfrac{2}{(2x-1)}\right\}$

When $x = 3$, $\dfrac{dy}{dx} = \dfrac{\sqrt{(1)^3}}{(4)^2(5)}\left(\dfrac{3}{2} - \dfrac{2}{4} - \dfrac{2}{5}\right)$

$= \pm\dfrac{1}{80}\left(\dfrac{3}{5}\right) = \pm\dfrac{3}{400}$ or ± 0.0075

Problem 3. Given $y = \dfrac{3e^{2\theta}\sec 2\theta}{\sqrt{(\theta-2)}}$ determine $\dfrac{dy}{d\theta}$

Using logarithmic differentiation and following the procedure:

(i) Since $y = \dfrac{3e^{2\theta}\sec 2\theta}{\sqrt{(\theta-2)}}$

then $\ln y = \ln\left\{\dfrac{3e^{2\theta}\sec 2\theta}{\sqrt{(\theta-2)}}\right\}$

$= \ln\left\{\dfrac{3e^{2\theta}\sec 2\theta}{(\theta-2)^{\frac{1}{2}}}\right\}$

(ii) $\ln y = \ln 3e^{2\theta} + \ln\sec 2\theta - \ln(\theta-2)^{\frac{1}{2}}$

i.e. $\ln y = \ln 3 + \ln e^{2\theta} + \ln\sec 2\theta$

$-\frac{1}{2}\ln(\theta-2)$

i.e. $\ln y = \ln 3 + 2\theta + \ln\sec 2\theta - \frac{1}{2}\ln(\theta-2)$

(iii) Differentiating with respect to θ gives:

$$\frac{1}{y}\frac{dy}{d\theta} = 0 + 2 + \frac{2\sec 2\theta\tan 2\theta}{\sec 2\theta} - \frac{\frac{1}{2}}{(\theta-2)}$$

from equations (1) and (2)

(iv) Rearranging gives:

$$\frac{dy}{d\theta} = y\left\{2 + 2\tan 2\theta - \frac{1}{2(\theta-2)}\right\}$$

(v)   Substituting for y gives:

$$\frac{dy}{d\theta} = \frac{3e^{2\theta}\sec 2\theta}{\sqrt{(\theta-2)}}\left\{2+2\tan 2\theta - \frac{1}{2(\theta-2)}\right\}$$

Problem 4. Differentiate $y=\dfrac{x^3\ln 2x}{e^x\sin x}$ with respect to x

Using logarithmic differentiation and following the procedure gives:

(i) $\ln y = \ln\left\{\dfrac{x^3\ln 2x}{e^x\sin x}\right\}$

(ii) $\ln y = \ln x^3 + \ln(\ln 2x) - \ln(e^x) - \ln(\sin x)$
      i.e. $\ln y = 3\ln x + \ln(\ln 2x) - x - \ln(\sin x)$

(iii) $\dfrac{1}{y}\dfrac{dy}{dx} = \dfrac{3}{x} + \dfrac{\frac{1}{x}}{\ln 2x} - 1 - \dfrac{\cos x}{\sin x}$

(iv) $\dfrac{dy}{dx} = y\left\{\dfrac{3}{x} + \dfrac{1}{x\ln 2x} - 1 - \cot x\right\}$

(v) $\dfrac{dy}{dx} = \dfrac{x^3\ln 2x}{e^x\sin x}\left\{\dfrac{3}{x} + \dfrac{1}{x\ln 2x} - 1 - \cot x\right\}$

Now try the following Practice Exercise

Practice Exercise 247   Further problems on differentiating logarithmic functions (answers on page 1178)

In Problems 1 to 6, use logarithmic differentiation to differentiate the given functions with respect to the variable.

1. $y = \dfrac{(x-2)(x+1)}{(x-1)(x+3)}$

2. $y = \dfrac{(x+1)(2x+1)^3}{(x-3)^2(x+2)^4}$

3. $y = \dfrac{(2x-1)\sqrt{(x+2)}}{(x-3)\sqrt{(x+1)^3}}$

4. $y = \dfrac{e^{2x}\cos 3x}{\sqrt{(x-4)}}$

5. $y = 3\theta\sin\theta\cos\theta$

6. $y = \dfrac{2x^4\tan x}{e^{2x}\ln 2x}$

7. Evaluate $\dfrac{dy}{dx}$ when $x=1$ given
    $$y = \frac{(x+1)^2\sqrt{2x-1}}{\sqrt{(x+3)^3}}$$

8. Evaluate $\dfrac{dy}{d\theta}$, correct to 3 significant figures, when $\theta=\dfrac{\pi}{4}$ given $y=\dfrac{2e^\theta\sin\theta}{\sqrt{\theta^5}}$

57.5   Differentiation of $[f(x)]^x$

Whenever an expression to be differentiated contains a term raised to a power which is itself a function of the variable, then logarithmic differentiation must be used. For example, the differentiation of expressions such as x^x, $(x+2)^x$, $\sqrt[x]{(x-1)}$ and x^{3x+2} can only be achieved using logarithmic differentiation.

Problem 5. Determine $\dfrac{dy}{dx}$ given $y=x^x$

Taking Napierian logarithms of both sides of $y=x^x$ gives:
$\ln y = \ln x^x = x\ln x$, by law (iii) of Section 57.2
Differentiating both sides with respect to x gives:
$\dfrac{1}{y}\dfrac{dy}{dx} = (x)\left(\dfrac{1}{x}\right) + (\ln x)(1)$, using the product rule
i.e. $\dfrac{1}{y}\dfrac{dy}{dx} = 1 + \ln x$
from which, $\dfrac{dy}{dx} = y(1+\ln x)$
i.e. $\dfrac{dy}{dx} = x^x(1+\ln x)$

Problem 6. Evaluate $\dfrac{dy}{dx}$ when $x=-1$ given $y=(x+2)^x$

Taking Napierian logarithms of both sides of $y=(x+2)^x$ gives:
$\ln y = \ln(x+2)^x = x\ln(x+2)$, by law (iii) of Section 57.2
Differentiating both sides with respect to x gives:
$$\frac{1}{y}\frac{dy}{dx} = (x)\left(\frac{1}{x+2}\right) + [\ln(x+2)](1),$$
by the product rule.

Hence $\dfrac{dy}{dx} = y\left(\dfrac{x}{x+2} + \ln(x+2)\right)$

$= (x+2)^x \left\{\dfrac{x}{x+2} + \ln(x+2)\right\}$

When $x = -1$, $\dfrac{dy}{dx} = (1)^{-1}\left(\dfrac{-1}{1} + \ln 1\right)$

$= (+1)(-1) = -1$

Problem 7.   Determine (a) the differential coefficient of $y = \sqrt[x]{(x-1)}$ and (b) evaluate $\dfrac{dy}{dx}$ when $x = 2$

(a) $y = \sqrt[x]{(x-1)} = (x-1)^{\frac{1}{x}}$, since by the laws of indices $\sqrt[n]{a^m} = a^{\frac{m}{n}}$

Taking Napierian logarithms of both sides gives:

$\ln y = \ln(x-1)^{\frac{1}{x}} = \dfrac{1}{x}\ln(x-1)$,

by law (iii) of Section 57.2

Differentiating each side with respect to x gives:

$\dfrac{1}{y}\dfrac{dy}{dx} = \left(\dfrac{1}{x}\right)\left(\dfrac{1}{x-1}\right) + [\ln(x-1)]\left(\dfrac{-1}{x^2}\right)$

by the product rule.

Hence $\dfrac{dy}{dx} = y\left\{\dfrac{1}{x(x-1)} - \dfrac{\ln(x-1)}{x^2}\right\}$

i.e. $\dfrac{dy}{dx} = \sqrt[x]{(x-1)}\left\{\dfrac{1}{x(x-1)} - \dfrac{\ln(x-1)}{x^2}\right\}$

(b)   When $x = 2$, $\dfrac{dy}{dx} = \sqrt[2]{(1)}\left\{\dfrac{1}{2(1)} - \dfrac{\ln(1)}{4}\right\}$

$= \pm 1\left\{\dfrac{1}{2} - 0\right\} = \pm\dfrac{1}{2}$

Problem 8.   Differentiate x^{3x+2} with respect to x

Let $y = x^{3x+2}$
Taking Napierian logarithms of both sides gives:
$\ln y = \ln x^{3x+2}$

i.e. $\ln y = (3x+2)\ln x$, by law (iii) of Section 57.2
Differentiating each term with respect to x gives:

$\dfrac{1}{y}\dfrac{dy}{dx} = (3x+2)\left(\dfrac{1}{x}\right) + (\ln x)(3)$,

by the product rule.

Hence $\dfrac{dy}{dx} = y\left\{\dfrac{3x+2}{x} + 3\ln x\right\}$

$= x^{3x+2}\left\{\dfrac{3x+2}{x} + 3\ln x\right\}$

$= x^{3x+2}\left\{3 + \dfrac{2}{x} + 3\ln x\right\}$

Now try the following Practice Exercise

Practice Exercise 248   Further problems on differentiating $[f(x)]^x$ type functions (answers on page 1179)

In Problems 1 to 4, differentiate with respect to x

1. $y = x^{2x}$

2. $y = (2x-1)^x$

3. $y = \sqrt[x]{(x+3)}$

4. $y = 3x^{4x+1}$

5.   Show that when $y = 2x^x$ and $x = 1$, $\dfrac{dy}{dx} = 2$

6.   Evaluate $\dfrac{d}{dx}\left\{\sqrt[x]{(x-2)}\right\}$ when $x = 3$

7.   Show that if $y = \theta^\theta$ and $\theta = 2$, $\dfrac{dy}{d\theta} = 6.77$, correct to 3 significant figures.

Section I

This assignment covers the material contained in Chapters 55 to 57. *The marks available are shown in brackets at the end of each question.*

1. A cycloid has parametric equations given by: $x = 5(\theta - \sin\theta)$ and $y = 5(1 - \cos\theta)$. Evaluate (a) $\dfrac{dy}{dx}$ (b) $\dfrac{d^2 y}{dx^2}$ when $\theta = 1.5$ radians. Give answers correct to 3 decimal places. (8)

2. Determine the equation of (a) the tangent and (b) the normal, drawn to an ellipse $x = 4\cos\theta$, $y = \sin\theta$ at $\theta = \dfrac{\pi}{3}$ (8)

3. Determine expressions for $\dfrac{dz}{dy}$ for each of the following functions:
   (a) $z = 5y^2 \cos x$ (b) $z = x^2 + 4xy - y^2$ (5)

4. If $x^2 + y^2 + 6x + 8y + 1 = 0$, find $\dfrac{dy}{dx}$ in terms of x and y (4)

5. Determine the gradient of the tangents drawn to the hyperbola $x^2 - y^2 = 8$ at $x = 3$ (4)

6. Use logarithmic differentiation to differentiate $y = \dfrac{(x+1)^2 \sqrt{(x-2)}}{(2x-1)\sqrt[3]{(x-3)^4}}$ with respect to x (6)

7. Differentiate $y = \dfrac{3e^\theta \sin 2\theta}{\sqrt{\theta^5}}$ and hence evaluate $\dfrac{dy}{d\theta}$, correct to 2 decimal places, when $\theta = \dfrac{\pi}{3}$ (9)

8. Evaluate $\dfrac{d}{dt}[\sqrt[3]{(2t+1)}]$ when $t = 2$, correct to 4 significant figures. (6)

Section I

Chapter 58

Differentiation of hyperbolic functions

Why it is important to understand: Differentiation of hyperbolic functions

Hyperbolic functions have applications in many areas of engineering. For example, the shape formed by a wire freely hanging between two points (known as a catenary curve) is described by the hyperbolic cosine. Hyperbolic functions are also used in electrical engineering applications and for solving differential equations. Differentiation of hyperbolic functions is quite straightforward.

At the end of this chapter, you should be able to:

- derive the differential coefficients of hyperbolic functions
- differentiate hyperbolic functions

58.1 Standard differential coefficients of hyperbolic functions

From Chapter 24,

$$\frac{d}{dx}(\sinh x) = \frac{d}{dx}\left(\frac{e^x - e^{-x}}{2}\right) = \left[\frac{e^x - (-e^{-x})}{2}\right]$$

$$= \left(\frac{e^x + e^{-x}}{2}\right) = \cosh x$$

If $y = \sinh ax$, where 'a' is a constant, then
$$\frac{dy}{dx} = a\cosh ax$$

$$\frac{d}{dx}(\cosh x) = \frac{d}{dx}\left(\frac{e^x + e^{-x}}{2}\right) = \left[\frac{e^x + (-e^{-x})}{2}\right]$$

$$= \left(\frac{e^x - e^{-x}}{2}\right) = \sinh x$$

If $y = \cosh ax$, where 'a' is a constant, then
$$\frac{dy}{dx} = a\sinh ax$$

Using the quotient rule of differentiation the derivatives of $\tanh x$, $\operatorname{sech} x$, $\operatorname{cosech} x$ and $\coth x$ may be determined using the above results.

Problem 1. Determine the differential coefficient of: (a) $\operatorname{th} x$ (b) $\operatorname{sech} x$

(a) $$\frac{d}{dx}(\operatorname{th} x) = \frac{d}{dx}\left(\frac{\operatorname{sh} x}{\operatorname{ch} x}\right)$$

$$= \frac{(\operatorname{ch} x)(\operatorname{ch} x) - (\operatorname{sh} x)(\operatorname{sh} x)}{\operatorname{ch}^2 x}$$

using the quotient rule

$$= \frac{\operatorname{ch}^2 x - \operatorname{sh}^2 x}{\operatorname{ch}^2 x} = \frac{1}{\operatorname{ch}^2 x} = \operatorname{sech}^2 x$$

(b) $\dfrac{d}{dx}(\operatorname{sech} x) = \dfrac{d}{dx}\left(\dfrac{1}{\operatorname{ch} x}\right)$

$\qquad = \dfrac{(\operatorname{ch} x)(0) - (1)(\operatorname{sh} x)}{\operatorname{ch}^2 x}$

$\qquad = \dfrac{-\operatorname{sh} x}{\operatorname{ch}^2 x} = -\left(\dfrac{1}{\operatorname{ch} x}\right)\left(\dfrac{\operatorname{sh} x}{\operatorname{ch} x}\right)$

$\qquad = -\operatorname{sech} x\,\operatorname{th} x$

Problem 2.   Determine $\dfrac{dy}{d\theta}$ given
(a) $y = \operatorname{cosech}\theta$ (b) $y = \operatorname{coth}\theta$

(a) $\dfrac{d}{d\theta}(\operatorname{cosech}\theta) = \dfrac{d}{d\theta}\left(\dfrac{1}{\operatorname{sh}\theta}\right)$

$\qquad = \dfrac{(\operatorname{sh}\theta)(0) - (1)(\operatorname{ch}\theta)}{\operatorname{sh}^2\theta}$

$\qquad = \dfrac{-\operatorname{ch}\theta}{\operatorname{sh}^2\theta} = -\left(\dfrac{1}{\operatorname{sh}\theta}\right)\left(\dfrac{\operatorname{ch}\theta}{\operatorname{sh}\theta}\right)$

$\qquad = -\operatorname{cosech}\theta\,\operatorname{coth}\theta$

(b) $\dfrac{d}{d\theta}(\operatorname{coth}\theta) = \dfrac{d}{d\theta}\left(\dfrac{\operatorname{ch}\theta}{\operatorname{sh}\theta}\right)$

$\qquad = \dfrac{(\operatorname{sh}\theta)(\operatorname{sh}\theta) - (\operatorname{ch}\theta)(\operatorname{ch}\theta)}{\operatorname{sh}^2\theta}$

$\qquad = \dfrac{\operatorname{sh}^2\theta - \operatorname{ch}^2\theta}{\operatorname{sh}^2\theta} = \dfrac{-(\operatorname{ch}^2\theta - \operatorname{sh}^2\theta)}{\operatorname{sh}^2\theta}$

$\qquad = \dfrac{-1}{\operatorname{sh}^2\theta} = -\operatorname{cosech}^2\theta$

Summary of differential coefficients

y or $f(x)$	$\dfrac{dy}{dx}$ or $f'(x)$
$\sinh ax$	$a\cosh ax$
$\cosh ax$	$a\sinh ax$
$\tanh ax$	$a\operatorname{sech}^2 ax$
$\operatorname{sech} ax$	$-a\operatorname{sech} ax\,\tanh ax$
$\operatorname{cosech} ax$	$-a\operatorname{cosech} ax\,\coth ax$
$\coth ax$	$-a\operatorname{cosech}^2 ax$

58.2   Further worked problems on differentiation of hyperbolic functions

Problem 3.   Differentiate the following with respect to x:

(a) $y = 4\operatorname{sh} 2x - \dfrac{3}{7}\operatorname{ch} 3x$

(b) $y = 5\operatorname{th}\dfrac{x}{2} - 2\coth 4x$

(a) $y = 4\operatorname{sh} 2x - \dfrac{3}{7}\operatorname{ch} 3x$

$\dfrac{dy}{dx} = 4(2\cosh 2x) - \dfrac{3}{7}(3\sinh 3x)$

$\qquad = 8\cosh 2x - \dfrac{9}{7}\sinh 3x$

(b) $y = 5\operatorname{th}\dfrac{x}{2} - 2\coth 4x$

$\dfrac{dy}{dx} = 5\left(\dfrac{1}{2}\operatorname{sech}^2\dfrac{x}{2}\right) - 2(-4\operatorname{cosech}^2 4x)$

$\qquad = \dfrac{5}{2}\operatorname{sech}^2\dfrac{x}{2} + 8\operatorname{cosech}^2 4x$

Problem 4.   Differentiate the following with respect to the variable: (a) $y = 4\sin 3t\,\operatorname{ch} 4t$
(b) $y = \ln(\operatorname{sh} 3\theta) - 4\operatorname{ch}^2 3\theta$

(a) $y = 4\sin 3t\,\operatorname{ch} 4t$ (i.e. a product)

$\dfrac{dy}{dt} = (4\sin 3t)(4\operatorname{sh} 4t) + (\operatorname{ch} 4t)(4)(3\cos 3t)$

$\qquad = 16\sin 3t\,\operatorname{sh} 4t + 12\operatorname{ch} 4t\cos 3t$

$\qquad = 4(4\sin 3t\,\operatorname{sh} 4t + 3\cos 3t\,\operatorname{ch} 4t)$

(b) $y = \ln(\operatorname{sh} 3\theta) - 4\operatorname{ch}^2 3\theta$

$\qquad$ (i.e. a function of a function)

$\dfrac{dy}{d\theta} = \left(\dfrac{1}{\operatorname{sh} 3\theta}\right)(3\operatorname{ch} 3\theta) - (4)(2\operatorname{ch} 3\theta)(3\operatorname{sh} 3\theta)$

$\qquad = 3\coth 3\theta - 24\operatorname{ch} 3\theta\,\operatorname{sh} 3\theta$

$\qquad = 3(\coth 3\theta - 8\operatorname{ch} 3\theta\,\operatorname{sh} 3\theta)$

Section I

Problem 5. Show that the differential coefficient of

$$y = \frac{3x^2}{\text{ch}\,4x} \quad \text{is:} \quad 6x\,\text{sech}\,4x\,(1 - 2x\,\text{th}\,4x)$$

$$y = \frac{3x^2}{\text{ch}\,4x} \quad \text{(i.e. a quotient)}$$

$$\frac{dy}{dx} = \frac{(\text{ch}\,4x)(6x) - (3x^2)(4\,\text{sh}\,4x)}{(\text{ch}\,4x)^2}$$

$$= \frac{6x(\text{ch}\,4x - 2x\,\text{sh}\,4x)}{\text{ch}^2\,4x}$$

$$= 6x\left[\frac{\text{ch}\,4x}{\text{ch}^2\,4x} - \frac{2x\,\text{sh}\,4x}{\text{ch}^2\,4x}\right]$$

$$= 6x\left[\frac{1}{\text{ch}\,4x} - 2x\left(\frac{\text{sh}\,4x}{\text{ch}\,4x}\right)\left(\frac{1}{\text{ch}\,4x}\right)\right]$$

$$= 6x[\text{sech}\,4x - 2x\,\text{th}\,4x\,\text{sech}\,4x]$$

$$= \mathbf{6x\,sech\,4x\,(1 - 2x\,th\,4x)}$$

Now try the following Practice Exercise

In Problems 1 to 5, differentiate the given functions with respect to the variable:

1. (a) $3\,\text{sh}\,2x$ (b) $2\,\text{ch}\,5\theta$ (c) $4\,\text{th}\,9t$

2. (a) $\frac{2}{3}\text{sech}\,5x$ (b) $\frac{5}{8}\text{cosech}\,\frac{t}{2}$ (c) $2\,\text{coth}\,7\theta$

3. (a) $2\ln(\text{sh}\,x)$ (b) $\frac{3}{4}\ln\left(\text{th}\left(\frac{\theta}{2}\right)\right)$

4. (a) $\text{sh}\,2x\,\text{ch}\,2x$ (b) $3e^{2x}\,\text{th}\,2x$

5. (a) $\frac{3\,\text{sh}\,4x}{2x^3}$ (b) $\frac{\text{ch}\,2t}{\cos 2t}$

For fully worked solutions to each of the problems in Practice Exercise 249 in this chapter, go to the website:
www.routledge.com/cw/bird

Chapter 59

Differentiation of inverse trigonometric and hyperbolic functions

As has been mentioned earlier, hyperbolic functions have applications in many areas of engineering. For example, the hyperbolic sine arises in the gravitational potential of a cylinder, the hyperbolic cosine function is the shape of a hanging cable, the hyperbolic tangent arises in the calculation of and rapidity of special relativity, the hyperbolic secant arises in the profile of a laminar jet and the hyperbolic cotangent arises in the Langevin function for magnetic polarisation. So there are plenty of applications for inverse functions in engineering, and this chapter explains how to differentiate inverse trigonometric and hyperbolic functions.

At the end of this chapter, you should be able to:

- understand inverse functions
- differentiate inverse trigonometric functions
- evaluate inverse hyperbolic functions using logarithmic forms
- differentiate inverse hyperbolic functions

59.1 Inverse functions

If $y = 3x - 2$, then by transposition, $x = \dfrac{y+2}{3}$. Interchanging x and y gives $y = \dfrac{x+2}{3}$. The function $x = \dfrac{y+2}{3}$ is called the **inverse function** of $y = 3x - 2$ (see page 403).

Inverse trigonometric functions are denoted by prefixing the function with 'arc' or, more commonly, by using the $^{-1}$ notation. For example, if $y = \sin x$, then $x = \arcsin y$ or $x = \sin^{-1} y$. Similarly, if $y = \cos x$, then $x = \arccos y$ or $x = \cos^{-1} y$, and so on. In this chapter the $^{-1}$ notation will be used. A sketch of each of the inverse trigonometric functions is shown in Figure 59.1.

Inverse hyperbolic functions are denoted by prefixing the function with 'ar' or, more commonly, by using the $^{-1}$ notation. For example, if $y = \sinh x$, then $x = \operatorname{arsinh} y$ or $x = \sinh^{-1} y$. Similarly, if $y = \operatorname{sech} x$, then $x = \operatorname{arsech} y$ or $x = \operatorname{sech}^{-1} y$, and so on. In this chapter the $^{-1}$ notation will be used. A sketch of each of the inverse hyperbolic functions is shown in Figure 59.2.

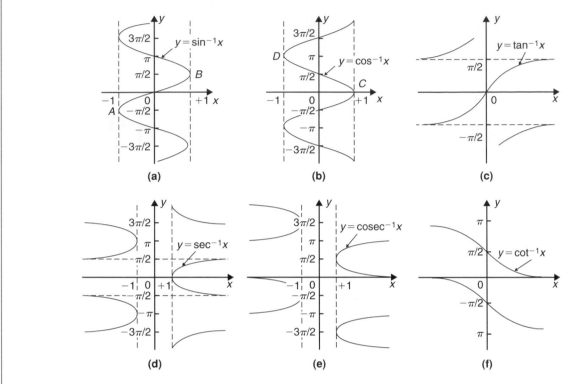

Figure 59.1

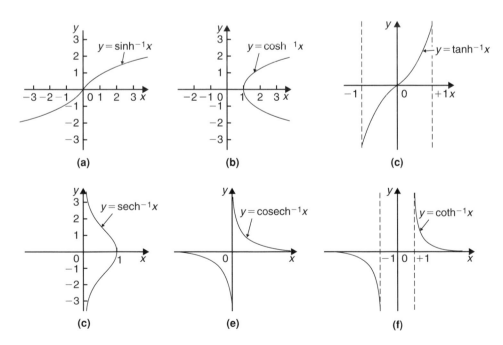

Figure 59.2

59.2 Differentiation of inverse trigonometric functions

(i) If $y = \sin^{-1} x$, then $x = \sin y$

Differentiating both sides with respect to y gives:

$$\frac{dx}{dy} = \cos y = \sqrt{1 - \sin^2 y}$$

since $\cos^2 y + \sin^2 y = 1$, i.e. $\dfrac{dx}{dy} = \sqrt{1 - x^2}$

However, $\dfrac{dy}{dx} = \dfrac{1}{\dfrac{dx}{dy}}$

Hence, when $y = \sin^{-1} x$ then

$$\frac{dy}{dx} = \frac{1}{\sqrt{1 - x^2}}$$

(ii) A sketch of part of the curve of $y = \sin^{-1} x$ is shown in Figure 59.1(a). The principal value of $\sin^{-1} x$ is defined as the value lying between $-\pi/2$ and $\pi/2$. The gradient of the curve between points A and B is positive for all values of x and thus only the positive value is taken when evaluating $\dfrac{1}{\sqrt{1 - x^2}}$

(iii) Given $y = \sin^{-1} \dfrac{x}{a}$ then $\dfrac{x}{a} = \sin y$ and $x = a \sin y$

Hence $\dfrac{dx}{dy} = a \cos y = a\sqrt{1 - \sin^2 y}$

$$= a\sqrt{\left[1 - \left(\frac{x}{a}\right)^2\right]} = a\sqrt{\left(\frac{a^2 - x^2}{a^2}\right)}$$

$$= \frac{a\sqrt{a^2 - x^2}}{a} = \sqrt{a^2 - x^2}$$

Thus $\dfrac{dy}{dx} = \dfrac{1}{\dfrac{dx}{dy}} = \dfrac{1}{\sqrt{a^2 - x^2}}$

i.e. **when $y = \sin^{-1} \dfrac{x}{a}$ then $\dfrac{dy}{dx} = \dfrac{1}{\sqrt{a^2 - x^2}}$**

Since integration is the reverse process of differentiation then:

$$\int \frac{1}{\sqrt{a^2 - x^2}} \, dx = \sin^{-1} \frac{x}{a} + c$$

(iv) Given $y = \sin^{-1} f(x)$ the function of a function rule may be used to find $\dfrac{dy}{dx}$

Let $u = f(x)$ then $y = \sin^{-1} u$

Then $\dfrac{du}{dx} = f'(x)$ and $\dfrac{dy}{du} = \dfrac{1}{\sqrt{1 - u^2}}$

(see para. (i))

Thus $\dfrac{dy}{dx} = \dfrac{dy}{du} \times \dfrac{du}{dx} = \dfrac{1}{\sqrt{1 - u^2}} \, f'(x)$

$$= \frac{f'(x)}{\sqrt{1 - [f(x)]^2}}$$

(v) The differential coefficients of the remaining inverse trigonometric functions are obtained in a similar manner to that shown above and a summary of the results is shown in Table 59.1.

Table 59.1 Differential coefficients of inverse trigonometric functions

	y or $f(x)$	$\dfrac{dy}{dx}$ or $f'(x)$
(i)	$\sin^{-1} \dfrac{x}{a}$	$\dfrac{1}{\sqrt{a^2 - x^2}}$
	$\sin^{-1} f(x)$	$\dfrac{f'(x)}{\sqrt{1 - [f(x)]^2}}$
(ii)	$\cos^{-1} \dfrac{x}{a}$	$\dfrac{-1}{\sqrt{a^2 - x^2}}$
	$\cos^{-1} f(x)$	$\dfrac{-f'(x)}{\sqrt{1 - [f(x)]^2}}$
(iii)	$\tan^{-1} \dfrac{x}{a}$	$\dfrac{a}{a^2 + x^2}$
	$\tan^{-1} f(x)$	$\dfrac{f'(x)}{1 + [f(x)]^2}$
(iv)	$\sec^{-1} \dfrac{x}{a}$	$\dfrac{a}{x\sqrt{x^2 - a^2}}$
	$\sec^{-1} f(x)$	$\dfrac{f'(x)}{f(x)\sqrt{[f(x)]^2 - 1}}$
(v)	$\operatorname{cosec}^{-1} \dfrac{x}{a}$	$\dfrac{-a}{x\sqrt{x^2 - a^2}}$
	$\operatorname{cosec}^{-1} f(x)$	$\dfrac{-f'(x)}{f(x)\sqrt{[f(x)]^2 - 1}}$
(vi)	$\cot^{-1} \dfrac{x}{a}$	$\dfrac{-a}{a^2 + x^2}$
	$\cot^{-1} f(x)$	$\dfrac{-f'(x)}{1 + [f(x)]^2}$

Problem 1. Find $\dfrac{dy}{dx}$ given $y = \sin^{-1} 5x^2$

From Table 59.1(i), if

$$y = \sin^{-1} f(x) \text{ then } \frac{dy}{dx} = \frac{f'(x)}{\sqrt{1 - [f(x)]^2}}$$

Hence, if $y = \sin^{-1} 5x^2$ then $f(x) = 5x^2$ and $f'(x) = 10x$

Thus $\dfrac{dy}{dx} = \dfrac{10x}{\sqrt{1 - (5x^2)^2}} = \dfrac{\mathbf{10x}}{\sqrt{\mathbf{1 - 25x^4}}}$

Problem 2.

(a) Show that if $y = \cos^{-1} x$ then
$$\frac{dy}{dx} = \frac{1}{\sqrt{1 - x^2}}$$

(b) Hence obtain the differential coefficient of $y = \cos^{-1}(1 - 2x^2)$

(a) If $y = \cos^{-1} x$ then $x = \cos y$

Differentiating with respect to y gives:

$$\frac{dx}{dy} = -\sin y = -\sqrt{1 - \cos^2 y}$$
$$= -\sqrt{1 - x^2}$$

Hence $\quad \dfrac{dy}{dx} = \dfrac{1}{\dfrac{dx}{dy}} = -\dfrac{1}{\sqrt{1 - x^2}}$

The principal value of $y = \cos^{-1} x$ is defined as the angle lying between 0 and π, i.e. between points C and D shown in Figure 59.1(b). The gradient of the curve is negative between C and D and thus the differential coefficient $\dfrac{dy}{dx}$ is negative as shown above.

(b) If $y = \cos^{-1} f(x)$ then by letting $u = f(x)$, $y = \cos^{-1} u$

Then $\quad \dfrac{dy}{du} = -\dfrac{1}{\sqrt{1 - u^2}}$ (from part (a))

and $\quad \dfrac{du}{dx} = f'(x)$

From the function of a function rule,

$$\frac{dy}{dx} = \frac{dy}{du} \cdot \frac{du}{dx} = -\frac{1}{\sqrt{1 - u^2}} f'(x)$$
$$= \frac{-f'(x)}{\sqrt{1 - [f(x)]^2}}$$

Hence, when $y = \cos^{-1}(1 - 2x^2)$

then $\quad \dfrac{dy}{dx} = \dfrac{-(-4x)}{\sqrt{1 - [1 - 2x^2]^2}}$

$$= \frac{4x}{\sqrt{1 - (1 - 4x^2 + 4x^4)}} = \frac{4x}{\sqrt{(4x^2 - 4x^4)}}$$

$$= \frac{4x}{\sqrt{[4x^2(1 - x^2)]}} = \frac{4x}{2x\sqrt{1 - x^2}} = \frac{\mathbf{2}}{\sqrt{\mathbf{1 - x^2}}}$$

Problem 3. Determine the differential coefficient of $y = \tan^{-1} \dfrac{x}{a}$ and show that the differential coefficient of $\tan^{-1} \dfrac{2x}{3}$ is $\dfrac{6}{9 + 4x^2}$

If $y = \tan^{-1} \dfrac{x}{a}$ then $\dfrac{x}{a} = \tan y$ and $x = a \tan y$

$\dfrac{dx}{dy} = a \sec^2 y = a(1 + \tan^2 y)$ since
$$\sec^2 y = 1 + \tan^2 y$$

$$= a\left[1 + \left(\frac{x}{a}\right)^2\right] = a\left(\frac{a^2 + x^2}{a^2}\right)$$

$$= \frac{a^2 + x^2}{a}$$

Hence $\quad \dfrac{dy}{dx} = \dfrac{1}{\dfrac{dx}{dy}} = \dfrac{a}{a^2 + x^2}$

The principal value of $y = \tan^{-1} x$ is defined as the angle lying between $-\dfrac{\pi}{2}$ and $\dfrac{\pi}{2}$ and the gradient $\left(\text{i.e. } \dfrac{dy}{dx}\right)$ between these two values is always positive (see Figure 59.1(c)).

Comparing $\tan^{-1} \dfrac{2x}{3}$ with $\tan^{-1} \dfrac{x}{a}$ shows that $a = \dfrac{3}{2}$

Hence if $y = \tan^{-1} \dfrac{2x}{3}$ then

$$\frac{dy}{dx} = \frac{\dfrac{3}{2}}{\left(\dfrac{3}{2}\right)^2 + x^2} = \frac{\dfrac{3}{2}}{\dfrac{9}{4} + x^2} = \frac{\dfrac{3}{2}}{\dfrac{9 + 4x^2}{4}}$$

$$= \frac{\dfrac{3}{2}(4)}{9 + 4x^2} = \frac{\mathbf{6}}{\mathbf{9 + 4x^2}}$$

Problem 4.   Find the differential coefficient of $y = \ln(\cos^{-1} 3x)$

Let $u = \cos^{-1} 3x$ then $y = \ln u$

By the function of a function rule,

$$\frac{dy}{dx} = \frac{dy}{du} \cdot \frac{du}{dx} = \frac{1}{u} \times \frac{d}{dx}(\cos^{-1} 3x)$$

$$= \frac{1}{\cos^{-1} 3x} \left\{ \frac{-3}{\sqrt{1-(3x)^2}} \right\}$$

i.e. $\dfrac{d}{dx}[\ln(\cos^{-1} 3x)] = \dfrac{-3}{\sqrt{1-9x^2}\,\cos^{-1} 3x}$

Problem 5.   If $y = \tan^{-1} \dfrac{3}{t^2}$ find $\dfrac{dy}{dt}$

Using the general form from Table 59.1(iii),

$$f(t) = \frac{3}{t^2} = 3t^{-2}$$

from which $f'(t) = \dfrac{-6}{t^3}$

Hence $\dfrac{d}{dt}\left(\tan^{-1}\dfrac{3}{t^2}\right) = \dfrac{f'(t)}{1+[f(t)]^2}$

$$= \frac{-\dfrac{6}{t^3}}{\left\{1+\left(\dfrac{3}{t^2}\right)^2\right\}} = \frac{-\dfrac{6}{t^3}}{\dfrac{t^4+9}{t^4}}$$

$$= \left(-\frac{6}{t^3}\right)\left(\frac{t^4}{t^4+9}\right) = -\frac{6t}{t^4+9}$$

Problem 6.   Differentiate $y = \dfrac{\cot^{-1} 2x}{1+4x^2}$

Using the quotient rule:

$$\frac{dy}{dx} = \frac{(1+4x^2)\left(\dfrac{-2}{1+(2x)^2}\right) - (\cot^{-1} 2x)(8x)}{(1+4x^2)^2}$$

from Table 59.1(vi)

$$= \frac{-2(1+4x\cot^{-1} 2x)}{(1+4x^2)^2}$$

Problem 7.   Differentiate $y = x\,\mathrm{cosec}^{-1} x$

Using the product rule:

$$\frac{dy}{dx} = (x)\left[\frac{-1}{x\sqrt{x^2-1}}\right] + (\mathrm{cosec}^{-1} x)\,(1)$$

from Table 59.1(v)

$$= \frac{-1}{\sqrt{x^2-1}} + \mathrm{cosec}^{-1} x$$

Problem 8.   Show that if

$$y = \tan^{-1}\left(\frac{\sin t}{\cos t - 1}\right) \text{ then } \frac{dy}{dt} = \frac{1}{2}$$

If $f(t) = \left(\dfrac{\sin t}{\cos t - 1}\right)$

then $f'(t) = \dfrac{(\cos t - 1)(\cos t) - (\sin t)(-\sin t)}{(\cos t - 1)^2}$

$$= \frac{\cos^2 t - \cos t + \sin^2 t}{(\cos t - 1)^2} = \frac{1-\cos t}{(\cos t - 1)^2}$$

since $\sin^2 t + \cos^2 t = 1$

$$= \frac{-(\cos t - 1)}{(\cos t - 1)^2} = \frac{-1}{\cos t - 1}$$

Using Table 59.1(iii), when

$$y = \tan^{-1}\left(\frac{\sin t}{\cos t - 1}\right)$$

then $\dfrac{dy}{dt} = \dfrac{\dfrac{-1}{\cos t - 1}}{1+\left(\dfrac{\sin t}{\cos t - 1}\right)^2}$

$$= \frac{\dfrac{-1}{\cos t - 1}}{\dfrac{(\cos t - 1)^2 + (\sin t)^2}{(\cos t - 1)^2}}$$

$$= \left(\frac{-1}{\cos t - 1}\right)\left(\frac{(\cos t - 1)^2}{\cos^2 t - 2\cos t + 1 + \sin^2 t}\right)$$

$$= \frac{-(\cos t - 1)}{2 - 2\cos t} = \frac{1-\cos t}{2(1-\cos t)} = \frac{1}{2}$$

Now try the following Practice Exercise

Practice Exercise 250   Further problems on differentiating inverse trigonometric functions (answers on page 1179)

In Problems 1 to 6, differentiate with respect to the variable.

1. (a) $\sin^{-1} 4x$ (b) $\sin^{-1} \dfrac{x}{2}$

2. (a) $\cos^{-1} 3x$ (b) $\dfrac{2}{3} \cos^{-1} \dfrac{x}{3}$

3. (a) $3 \tan^{-1} 2x$ (b) $\dfrac{1}{2} \tan^{-1} \sqrt{x}$

4. (a) $2 \sec^{-1} 2t$ (b) $\sec^{-1} \dfrac{3}{4} x$

5. (a) $\dfrac{5}{2} \operatorname{cosec}^{-1} \dfrac{\theta}{2}$ (b) $\operatorname{cosec}^{-1} x^2$

6. (a) $3 \cot^{-1} 2t$ (b) $\cot^{-1} \sqrt{\theta^2 - 1}$

7. Show that the differential coefficient of $\tan^{-1}\left(\dfrac{x}{1-x^2}\right)$ is $\dfrac{1+x^2}{1-x^2+x^4}$

In Problems 8 to 11, differentiate with respect to the variable.

8. (a) $2x \sin^{-1} 3x$ (b) $t^2 \sec^{-1} 2t$

9. (a) $\theta^2 \cos^{-1} (\theta^2 - 1)$ (b) $(1 - x^2) \tan^{-1} x$

10. (a) $2\sqrt{t} \cot^{-1} t$ (b) $x \operatorname{cosec}^{-1} \sqrt{x}$

11. (a) $\dfrac{\sin^{-1} 3x}{x^2}$ (b) $\dfrac{\cos^{-1} x}{\sqrt{1-x^2}}$

59.3 Logarithmic forms of inverse hyperbolic functions

Inverse hyperbolic functions may be evaluated most conveniently when expressed in a **logarithmic form**.

For example, if $y = \sinh^{-1} \dfrac{x}{a}$ then $\dfrac{x}{a} = \sinh y$.

From Chapter 24, $e^y = \cosh y + \sinh y$ and $\cosh^2 y - \sinh^2 y = 1$, from which,

$\cosh y = \sqrt{1 + \sinh^2 y}$ which is positive since $\cosh y$ is always positive (see Figure 24.2, page 230).

Hence $e^y = \sqrt{1 + \sinh^2 y} + \sinh y$

$$= \sqrt{\left[1 + \left(\frac{x}{a}\right)^2\right]} + \frac{x}{a} = \sqrt{\left(\frac{a^2 + x^2}{a^2}\right)} + \frac{x}{a}$$

$$= \frac{\sqrt{a^2 + x^2}}{a} + \frac{x}{a} \quad \text{or} \quad \frac{x + \sqrt{a^2 + x^2}}{a}$$

Taking Napierian logarithms of both sides gives:

$$y = \ln\left\{\frac{x + \sqrt{a^2 + x^2}}{a}\right\}$$

Hence, $\sinh^{-1} \dfrac{x}{a} = \ln\left\{\dfrac{x + \sqrt{a^2 + x^2}}{a}\right\}$ (1)

Thus to evaluate $\sinh^{-1} \dfrac{3}{4}$, let $x = 3$ and $a = 4$ in equation (1).

Then $\sinh^{-1} \dfrac{3}{4} = \ln\left\{\dfrac{3 + \sqrt{4^2 + 3^2}}{4}\right\}$

$$= \ln\left(\frac{3+5}{4}\right) = \ln 2 = 0.6931$$

By similar reasoning to the above it may be shown that:

$$\cosh^{-1} \frac{x}{a} = \ln\left\{\frac{x + \sqrt{x^2 - a^2}}{a}\right\}$$

and $\tanh^{-1} \dfrac{x}{a} = \dfrac{1}{2}\ln\left(\dfrac{a+x}{a-x}\right)$

Problem 9.   Evaluate, correct to 4 decimal places, $\sinh^{-1} 2$

From above, $\sinh^{-1} \dfrac{x}{a} = \ln\left\{\dfrac{x + \sqrt{a^2 + x^2}}{a}\right\}$

With $x = 2$ and $a = 1$,

$\sinh^{-1} 2 = \ln\left\{\dfrac{2 + \sqrt{1^2 + 2^2}}{1}\right\}$

$$= \ln(2 + \sqrt{5}) = \ln 4.2361$$

$$= \mathbf{1.4436, \ correct \ to \ 4 \ decimal \ places}.$$

Using a calculator,

(i) press hyp

(ii) press 4 and $\sinh^{-1}($ appears

(iii) type in 2

(iv)   press) to close the brackets

(v)   press = and 1.443635475 appears

Hence, **sinh^{-1}2 = 1.4436**, correct to 4 decimal places.

Problem 10.   Show that

$\tanh^{-1}\dfrac{x}{a}=\dfrac{1}{2}\ln\left(\dfrac{a+x}{a-x}\right)$ and evaluate, correct

to 4 decimal places, $\tanh^{-1}\dfrac{3}{5}$

If $y=\tanh^{-1}\dfrac{x}{a}$ then $\dfrac{x}{a}=\tanh y$.

From Chapter 24,

$$\tanh y=\frac{\sinh y}{\cosh y}=\frac{\frac{1}{2}(e^y-e^{-y})}{\frac{1}{2}(e^y+e^{-y})}=\frac{e^{2y}-1}{e^{2y}+1}$$

by dividing each term by e^{-y}

Thus, $\dfrac{x}{a}=\dfrac{e^{2y}-1}{e^{2y}+1}$

from which, $x(e^{2y}+1)=a(e^{2y}-1)$

Hence $x+a=ae^{2y}-xe^{2y}=e^{2y}(a-x)$

from which $e^{2y}=\left(\dfrac{a+x}{a-x}\right)$

Taking Napierian logarithms of both sides gives:

$$2y=\ln\left(\frac{a+x}{a-x}\right)$$

and $y=\dfrac{1}{2}\ln\left(\dfrac{a+x}{a-x}\right)$

Hence, **$\tanh^{-1}\dfrac{x}{a}=\dfrac{1}{2}\ln\left(\dfrac{a+x}{a-x}\right)$**

Substituting $x=3$ and $a=5$ gives:

$$\tanh^{-1}\frac{3}{5}=\frac{1}{2}\ln\left(\frac{5+3}{5-3}\right)=\frac{1}{2}\ln 4$$

$$=\mathbf{0.6931},\text{ correct to 4 decimal places.}$$

Problem 11.   Prove that

$\cosh^{-1}\dfrac{x}{a}=\ln\left\{\dfrac{x+\sqrt{x^2-a^2}}{a}\right\}$

and hence evaluate $\cosh^{-1}1.4$ correct to
4 decimal places

If $y=\cosh^{-1}\dfrac{x}{a}$ then $\dfrac{x}{a}=\cosh y$

$$e^y=\cosh y+\sinh y=\cosh y\pm\sqrt{\cosh^2 y-1}$$

$$=\frac{x}{a}\pm\sqrt{\left[\left(\frac{x}{a}\right)^2-1\right]}=\frac{x}{a}\pm\frac{\sqrt{x^2-a^2}}{a}$$

$$=\frac{x\pm\sqrt{x^2-a^2}}{a}$$

Taking Napierian logarithms of both sides gives:

$$y=\ln\left\{\frac{x\pm\sqrt{x^2-a^2}}{a}\right\}$$

Thus, assuming the principal value,

$$\mathbf{\cosh^{-1}\frac{x}{a}=\ln\left\{\frac{x+\sqrt{x^2-a^2}}{a}\right\}}$$

$\cosh^{-1}1.4=\cosh^{-1}\dfrac{14}{10}=\cosh^{-1}\dfrac{7}{5}$

In the equation for $\cosh^{-1}\dfrac{x}{a}$, let $x=7$ and $a=5$

Then $\cosh^{-1}\dfrac{7}{5}=\ln\left\{\dfrac{7+\sqrt{7^2-5^2}}{5}\right\}$

$$=\ln 2.3798=\mathbf{0.8670},$$

correct to 4 decimal places.

Now try the following Practice Exercise

Practice Exercise 251   Further problems on logarithmic forms of the inverse hyperbolic functions (answers on page 1179)

In Problems 1 to 3, use logarithmic equivalents of inverse hyperbolic functions to evaluate correct to 4 decimal places.

1.   (a) $\sinh^{-1}\dfrac{1}{2}$ (b) $\sinh^{-1}4$ (c) $\sinh^{-1}0.9$

2.   (a) $\cosh^{-1}\dfrac{5}{4}$ (b) $\cosh^{-1}3$ (c) $\cosh^{-1}4.3$

3.   (a) $\tanh^{-1}\dfrac{1}{4}$ (b) $\tanh^{-1}\dfrac{5}{8}$ (c) $\tanh^{-1}0.7$

59.4 Differentiation of inverse hyperbolic functions

If $y = \sinh^{-1}\dfrac{x}{a}$ then $\dfrac{x}{a} = \sinh y$ and $x = a \sinh y$

$\dfrac{dx}{dy} = a \cosh y$ (from Chapter 58).

Also $\cosh^2 y - \sinh^2 y = 1$, from which,

$$\cosh y = \sqrt{1 + \sinh^2 y} = \sqrt{\left[1 + \left(\frac{x}{a}\right)^2\right]}$$

$$= \frac{\sqrt{a^2 + x^2}}{a}$$

Hence $\dfrac{dx}{dy} = a \cosh y = \dfrac{a\sqrt{a^2 + x^2}}{a} = \sqrt{a^2 + x^2}$

Then $\dfrac{dy}{dx} = \dfrac{1}{\dfrac{dx}{dy}} = \dfrac{1}{\sqrt{a^2 + x^2}}$

(An alternative method of differentiating $\sinh^{-1}\dfrac{x}{a}$ is to differentiate the logarithmic form

$\ln\left\{\dfrac{x + \sqrt{a^2 + x^2}}{a}\right\}$ with respect to x)

From the sketch of $y = \sinh^{-1} x$ shown in Figure 59.2(a) it is seen that the gradient $\left(\text{i.e. } \dfrac{dy}{dx}\right)$ is always positive.

It may be shown that

$$\frac{d}{dx}(\sinh^{-1} x) = \frac{1}{\sqrt{x^2 + 1}}$$

or more generally

$$\frac{d}{dx}[\sinh^{-1} f(x)] = \frac{f'(x)}{\sqrt{[f(x)]^2 + 1}}$$

by using the function of a function rule as in Section 59.2(iv).

The remaining inverse hyperbolic functions are differentiated in a similar manner to that shown above and the results are summarised in Table 59.2.

Table 59.2 Differential coefficients of inverse hyperbolic functions

	y or $f(x)$	$\dfrac{dy}{dx}$ or $f'(x)$
(i)	$\sinh^{-1}\dfrac{x}{a}$	$\dfrac{1}{\sqrt{x^2 + a^2}}$
	$\sinh^{-1} f(x)$	$\dfrac{f'(x)}{\sqrt{[f(x)]^2 + 1}}$
(ii)	$\cosh^{-1}\dfrac{x}{a}$	$\dfrac{1}{\sqrt{x^2 - a^2}}$
	$\cosh^{-1} f(x)$	$\dfrac{f'(x)}{\sqrt{[f(x)]^2 - 1}}$
(iii)	$\tanh^{-1}\dfrac{x}{a}$	$\dfrac{a}{a^2 - x^2}$
	$\tanh^{-1} f(x)$	$\dfrac{f'(x)}{1 - [f(x)]^2}$
(iv)	$\text{sech}^{-1}\dfrac{x}{a}$	$\dfrac{-a}{x\sqrt{a^2 - x^2}}$
	$\text{sech}^{-1} f(x)$	$\dfrac{-f'(x)}{f(x)\sqrt{1 - [f(x)]^2}}$
(v)	$\text{cosech}^{-1}\dfrac{x}{a}$	$\dfrac{-a}{x\sqrt{x^2 + a^2}}$
	$\text{cosech}^{-1} f(x)$	$\dfrac{-f'(x)}{f(x)\sqrt{[f(x)]^2 + 1}}$
(vi)	$\coth^{-1}\dfrac{x}{a}$	$\dfrac{a}{a^2 - x^2}$
	$\coth^{-1} f(x)$	$\dfrac{f'(x)}{1 - [f(x)]^2}$

Problem 12. Find the differential coefficient of $y = \sinh^{-1} 2x$

From Table 59.2(i),

$$\frac{d}{dx}[\sinh^{-1} f(x)] = \frac{f'(x)}{\sqrt{[f(x)]^2 + 1}}$$

Hence $\dfrac{d}{dx}(\sinh^{-1} 2x) = \dfrac{2}{\sqrt{[(2x)^2 + 1]}}$

$$= \frac{2}{\sqrt{[4x^2 + 1]}}$$

Problem 13.   Determine
$$\frac{d}{dx}\left[\cosh^{-1}\sqrt{(x^2+1)}\right]$$

If $y=\cosh^{-1} f(x)$, $\dfrac{dy}{dx}=\dfrac{f'(x)}{\sqrt{[f(x)]^2-1}}$

If $y=\cosh^{-1}\sqrt{(x^2+1)}$, then $f(x)=\sqrt{(x^2+1)}$ and
$$f'(x)=\frac{1}{2}(x+1)^{-1/2}(2x)=\frac{x}{\sqrt{(x^2+1)}}$$

Hence, $\dfrac{d}{dx}\left[\cosh^{-1}\sqrt{(x^2+1)}\right]$

$$=\frac{\dfrac{x}{\sqrt{(x^2+1)}}}{\sqrt{\left[\left(\sqrt{(x^2+1)}\right)^2-1\right]}}=\frac{\dfrac{x}{\sqrt{(x^2+1)}}}{\sqrt{(x^2+1-1)}}$$

$$=\frac{\dfrac{x}{\sqrt{(x^2+1)}}}{x}=\frac{1}{\sqrt{(x^2+1)}}$$

Problem 14.   Show that $\dfrac{d}{dx}\left[\tanh^{-1}\dfrac{x}{a}\right]=\dfrac{a}{a^2-x^2}$
and hence determine the differential coefficient of $\tanh^{-1}\dfrac{4x}{3}$

If $y=\tanh^{-1}\dfrac{x}{a}$ then $\dfrac{x}{a}=\tanh y$ and $x=a\tanh y$

$\dfrac{dx}{dy}=a\operatorname{sech}^2 y=a(1-\tanh^2 y)$, since

$$1-\operatorname{sech}^2 y=\tanh^2 y$$

$$=a\left[1-\left(\frac{x}{a}\right)^2\right]=a\left(\frac{a^2-x^2}{a^2}\right)=\frac{a^2-x^2}{a}$$

Hence $\dfrac{dy}{dx}=\dfrac{1}{\dfrac{dx}{dy}}=\dfrac{a}{a^2-x^2}$

Comparing $\tanh^{-1}\dfrac{4x}{3}$ with $\tanh^{-1}\dfrac{x}{a}$ shows that $a=\dfrac{3}{4}$

Hence $\dfrac{d}{dx}\left[\tanh^{-1}\dfrac{4x}{3}\right]=\dfrac{\dfrac{3}{4}}{\left(\dfrac{3}{4}\right)^2-x^2}=\dfrac{\dfrac{3}{4}}{\dfrac{9}{16}-x^2}$

$$=\frac{\dfrac{3}{4}}{\dfrac{9-16x^2}{16}}=\frac{3}{4}\cdot\frac{16}{(9-16x^2)}=\frac{\mathbf{12}}{\mathbf{9-16x^2}}$$

Problem 15.   Differentiate $\operatorname{cosech}^{-1}(\sinh\theta)$

From Table 59.2(v),

$$\frac{d}{dx}[\operatorname{cosech}^{-1} f(x)]=\frac{-f'(x)}{f(x)\sqrt{[f(x)]^2+1}}$$

Hence $\dfrac{d}{d\theta}[\operatorname{cosech}^{-1}(\sinh\theta)]$

$$=\frac{-\cosh\theta}{\sinh\theta\sqrt{[\sinh^2\theta+1]}}$$

$$=\frac{-\cosh\theta}{\sinh\theta\sqrt{\cosh^2\theta}}\quad\text{since }\cosh^2\theta-\sinh^2\theta=1$$

$$=\frac{-\cosh\theta}{\sinh\theta\cosh\theta}=\frac{-1}{\sinh\theta}=\mathbf{-\operatorname{cosech}\theta}$$

Problem 16.   Find the differential coefficient of $y=\operatorname{sech}^{-1}(2x-1)$

From Table 59.2(iv),

$$\frac{d}{dx}[\operatorname{sech}^{-1} f(x)]=\frac{-f'(x)}{f(x)\sqrt{1-[f(x)]^2}}$$

Hence, $\dfrac{d}{dx}[\operatorname{sech}^{-1}(2x-1)]$

$$=\frac{-2}{(2x-1)\sqrt{[1-(2x-1)^2]}}$$

$$=\frac{-2}{(2x-1)\sqrt{[1-(4x^2-4x+1)]}}$$

$$=\frac{-2}{(2x-1)\sqrt{(4x-4x^2)}}=\frac{-2}{(2x-1)\sqrt{[4x(1-x)]}}$$

$$=\frac{-2}{(2x-1)2\sqrt{[x(1-x)]}}=\frac{\mathbf{-1}}{\mathbf{(2x-1)\sqrt{[x(1-x)]}}}$$

Problem 17.   Show that
$$\frac{d}{dx}[\coth^{-1}(\sin x)]=\sec x$$

From Table 59.2(vi),

$$\frac{d}{dx}[\coth^{-1} f(x)]=\frac{f'(x)}{1-[f(x)]^2}$$

Section I

Hence $\dfrac{d}{dx}[\coth^{-1}(\sin x)] = \dfrac{\cos x}{[1 - (\sin x)^2]}$

$= \dfrac{\cos x}{\cos^2 x}$ since $\cos^2 x + \sin^2 x = 1$

$= \dfrac{1}{\cos x} = \sec x$

Problem 18. Differentiate
$y = (x^2 - 1)\tanh^{-1} x$

Using the product rule,

$\dfrac{dy}{dx} = (x^2 - 1)\left(\dfrac{1}{1 - x^2}\right) + (\tanh^{-1} x)(2x)$

$= \dfrac{-(1 - x^2)}{(1 - x^2)} + 2x \, \tanh^{-1} x = \mathbf{2x \, \tanh^{-1} x - 1}$

Problem 19. Determine $\displaystyle\int \dfrac{dx}{\sqrt{(x^2 + 4)}}$

Since $\dfrac{d}{dx}\left(\sinh^{-1}\dfrac{x}{a}\right) = \dfrac{1}{\sqrt{(x^2 + a^2)}}$

then $\displaystyle\int \dfrac{dx}{\sqrt{(x^2 + a^2)}} = \sinh^{-1}\dfrac{x}{a} + c$

Hence $\displaystyle\int \dfrac{1}{\sqrt{(x^2 + 4)}}\,dx = \int \dfrac{1}{\sqrt{(x^2 + 2^2)}}\,dx$

$= \sinh^{-1}\dfrac{x}{2} + c$

Problem 20. Determine $\displaystyle\int \dfrac{4}{\sqrt{(x^2 - 3)}}\,dx$

Since $\dfrac{d}{dx}\left(\cosh^{-1}\dfrac{x}{a}\right) = \dfrac{1}{\sqrt{(x^2 - a^2)}}$

then $\displaystyle\int \dfrac{1}{\sqrt{(x^2 - a^2)}}\,dx = \cosh^{-1}\dfrac{x}{a} + c$

Hence $\displaystyle\int \dfrac{4}{\sqrt{(x^2 - 3)}}\,dx = 4\int \dfrac{1}{\sqrt{[x^2 - (\sqrt{3})^2]}}\,dx$

$= \mathbf{4\cosh^{-1}\dfrac{x}{\sqrt{3}} + c}$

Problem 21. Find $\displaystyle\int \dfrac{2}{(9 - 4x^2)}\,dx$

Since $\dfrac{d}{dx}\tanh^{-1}\dfrac{x}{a} = \dfrac{a}{a^2 - x^2}$

then $\displaystyle\int \dfrac{a}{a^2 - x^2}\,dx = \tanh^{-1}\dfrac{x}{a} + c$

i.e. $\displaystyle\int \dfrac{1}{a^2 - x^2}\,dx = \dfrac{1}{a}\tanh^{-1}\dfrac{x}{a} + c$

Hence $\displaystyle\int \dfrac{2}{(9 - 4x^2)}\,dx = 2\int \dfrac{1}{4\left(\frac{9}{4} - x^2\right)}\,dx$

$= \dfrac{1}{2}\int \dfrac{1}{\left[\left(\frac{3}{2}\right)^2 - x^2\right]}\,dx$

$= \dfrac{1}{2}\left[\dfrac{1}{\left(\frac{3}{2}\right)}\tanh^{-1}\dfrac{x}{\left(\frac{3}{2}\right)} + c\right]$

i.e. $\displaystyle\int \dfrac{2}{(9 - 4x^2)}\,dx = \mathbf{\dfrac{1}{3}\tanh^{-1}\dfrac{2x}{3} + c}$

Now try the following Practice Exercise

Practice Exercise 252   Further problems on differentiation of inverse hyperbolic functions (answers on page 1179)

In Problems 1 to 11, differentiate with respect to the variable.

1. (a) $\sinh^{-1}\dfrac{x}{3}$ (b) $\sinh^{-1} 4x$

2. (a) $2\cosh^{-1}\dfrac{t}{3}$ (b) $\dfrac{1}{2}\cosh^{-1} 2\theta$

3. (a) $\tanh^{-1}\dfrac{2x}{5}$ (b) $3\tanh^{-1} 3x$

4. (a) $\mathrm{sech}^{-1}\dfrac{3x}{4}$ (b) $-\dfrac{1}{2}\mathrm{sech}^{-1} 2x$

5. (a) $\mathrm{cosech}^{-1}\dfrac{x}{4}$ (b) $\dfrac{1}{2}\mathrm{cosech}^{-1} 4x$

6. (a) $\coth^{-1}\dfrac{2x}{7}$ (b) $\dfrac{1}{4}\coth^{-1} 3t$

7. (a) $2\sinh^{-1}\sqrt{(x^2 - 1)}$

    (b) $\dfrac{1}{2}\cosh^{-1}\sqrt{(x^2 + 1)}$

8. (a) $\mathrm{sech}^{-1}(x - 1)$ (b) $\tanh^{-1}(\tanh x)$

9. (a) $\cosh^{-1}\left(\dfrac{t}{t-1}\right)$

    (b) $\coth^{-1}(\cos x)$

10. (a) $\theta\ \sinh^{-1}\theta$ (b) $\sqrt{x}\ \cosh^{-1}\ x$

11. (a) $\dfrac{2\ \mathrm{sech}^{-1}\sqrt{t}}{t^2}$ (b) $\dfrac{\tanh^{-1}x}{(1-x^2)}$

12. Show that $\dfrac{\mathrm{d}}{\mathrm{d}x}[x\cosh^{-1}(\cosh x)]=2x$

In Problems 13 to 15, determine the given integrals.

13. (a) $\displaystyle\int \dfrac{1}{\sqrt{(x^2+9)}}\ \mathrm{d}x$

    (b) $\displaystyle\int \dfrac{3}{\sqrt{(4x^2+25)}}\ \mathrm{d}x$

14. (a) $\displaystyle\int \dfrac{1}{\sqrt{(x^2-16)}}\ \mathrm{d}x$

    (b) $\displaystyle\int \dfrac{1}{\sqrt{(t^2-5)}}\ \mathrm{d}t$

15. (a) $\displaystyle\int \dfrac{\mathrm{d}\theta}{\sqrt{(36+\theta^2)}}$ (b) $\displaystyle\int \dfrac{3}{(16-2x^2)}\ \mathrm{d}x$

For fully worked solutions to each of the problems in Practice Exercises 250 to 252 in this chapter, go to the website:
www.routledge.com/cw/bird

Section I

Chapter 60

Partial differentiation

Why it is important to understand: **Partial differentiation**

First-order partial derivatives can be used for numerous applications, from determining the volume of different shapes to analysing anything from water to heat flow. Second-order partial derivatives are used in many fields of engineering. One of its applications is used in solving problems related to dynamics of rigid bodies and in determination of forces and strength of materials. Partial differentiation is used to estimate errors in calculated quantities that depend on more than one uncertain experimental measurement. Thermodynamic energy functions (enthalpy, Gibbs free energy, Helmholtz free energy) are functions of two or more variables. Most thermodynamic quantities (temperature, entropy, heat capacity) can be expressed as derivatives of these functions. Many laws of nature are best expressed as relations between the partial derivatives of one or more quantities. Partial differentiation is hence important in many branches of engineering.

At the end of this chapter, you should be able to:

- determine first-order partial derivatives
- determine second-order partial derivatives

60.1 Introduction to partial derivatives

In engineering, it sometimes happens that the variation of one quantity depends on changes taking place in two, or more, other quantities. For example, the volume V of a cylinder is given by $V = \pi r^2 h$. The volume will change if either radius r or height h is changed. The formula for volume may be stated mathematically as $V = f(r, h)$, which means 'V is some function of r and h'. Some other practical examples include:

(i) time of oscillation, $t = 2\pi \sqrt{\dfrac{l}{g}}$ i.e. $t = f(l, g)$

(ii) torque $T = I\alpha$, i.e. $T = f(I, \alpha)$

(iii) pressure of an ideal gas $p = \dfrac{mRT}{V}$
i.e. $p = f(T, V)$

(iv) resonant frequency $f_r = \dfrac{1}{2\pi \sqrt{LC}}$
i.e. $f_r = f(L, C)$, and so on.

When differentiating a function having two variables, **one variable is kept constant and the differential coefficient of the other variable is found with respect to that variable**. The differential coefficient obtained is called a **partial derivative** of the function.

60.2 First-order partial derivatives

A 'curly dee', ∂, is used to denote a differential coefficient in an expression containing more than one variable.

Hence if $V = \pi r^2 h$ then $\dfrac{\partial V}{\partial r}$ means 'the partial derivative of V with respect to r, with h remaining constant'. Thus,

$$\frac{\partial V}{\partial r} = (\pi h)\frac{d}{dr}(r^2) = (\pi h)(2r) = 2\pi rh$$

Similarly, $\frac{\partial V}{\partial h}$ means 'the partial derivative of V with respect to h, with r remaining constant'. Thus,

$$\frac{\partial V}{\partial h} = (\pi r^2)\frac{d}{dh}(h) = (\pi r^2)(1) = \pi r^2$$

$\frac{\partial V}{\partial r}$ and $\frac{\partial V}{\partial h}$ are examples of **first order partial derivatives**, since $n=1$ when written in the form $\frac{\partial^n V}{\partial r^n}$

First order partial derivatives are used when finding the total differential, rates of change and errors for functions of two or more variables (see Chapter 61), when finding maxima, minima and saddle points for functions of two variables (see Chapter 62), and with partial differential equations (see Chapter 84).

Problem 1. If $z = 5x^4 + 2x^3y^2 - 3y$ find
(a) $\frac{\partial z}{\partial x}$ and (b) $\frac{\partial z}{\partial y}$

(a) To find $\frac{\partial z}{\partial x}$, y is kept constant.

Since $z = 5x^4 + (2y^2)x^3 - (3y)$

then,

$$\frac{\partial z}{\partial x} = \frac{d}{dx}(5x^4) + (2y^2)\frac{d}{dx}(x^3) - (3y)\frac{d}{dx}(1)$$

$$= 20x^3 + (2y^2)(3x^2) - 0$$

Hence $\frac{\partial z}{\partial x} = 20x^3 + 6x^2y^2$

(b) To find $\frac{\partial z}{\partial y}$, x is kept constant.

Since $z = (5x^4) + (2x^3)y^2 - 3y$

then,

$$\frac{\partial z}{\partial y} = (5x^4)\frac{d}{dy}(1) + (2x^3)\frac{d}{dy}(y^2) - 3\frac{d}{dy}(y)$$

$$= 0 + (2x^3)(2y) - 3$$

Hence $\frac{\partial z}{\partial y} = 4x^3y - 3$

Problem 2. Given $y = 4\sin 3x \cos 2t$, find $\frac{\partial y}{\partial x}$ and $\frac{\partial y}{\partial t}$

To find $\frac{\partial y}{\partial x}$, t is kept constant.

Hence $\quad \frac{\partial y}{\partial x} = (4\cos 2t)\frac{d}{dx}(\sin 3x)$

$$= (4\cos 2t)(3\cos 3x)$$

i.e. $\quad \dfrac{\partial y}{\partial x} = 12\cos 3x \cos 2t$

To find $\frac{\partial y}{\partial t}$, x is kept constant.

Hence $\quad \frac{\partial y}{\partial t} = (4\sin 3x)\frac{d}{dt}(\cos 2t)$

$$= (4\sin 3x)(-2\sin 2t)$$

i.e. $\quad \dfrac{\partial y}{\partial t} = -8\sin 3x \sin 2t$

Problem 3. If $z = \sin xy$ show that
$$\frac{1}{y}\frac{\partial z}{\partial x} = \frac{1}{x}\frac{\partial z}{\partial y}$$

$\dfrac{\partial z}{\partial x} = y\cos xy$, since y is kept constant.

$\dfrac{\partial z}{\partial y} = x\cos xy$, since x is kept constant.

$$\frac{1}{y}\frac{\partial z}{\partial x} = \left(\frac{1}{y}\right)(y\cos xy) = \cos xy$$

and $\quad \dfrac{1}{x}\dfrac{\partial z}{\partial y} = \left(\dfrac{1}{x}\right)(x\cos xy) = \cos xy$

Hence $\quad \dfrac{1}{y}\dfrac{\partial z}{\partial x} = \dfrac{1}{x}\dfrac{\partial z}{\partial y}$

Problem 4. Determine $\dfrac{\partial z}{\partial x}$ and $\dfrac{\partial z}{\partial y}$ when
$$z = \frac{1}{\sqrt{(x^2 + y^2)}}$$

$$z = \frac{1}{\sqrt{(x^2+y^2)}} = (x^2+y^2)^{\frac{-1}{2}}$$

$$\frac{\partial z}{\partial x} = -\frac{1}{2}(x^2+y^2)^{\frac{-3}{2}}(2x), \text{ by the function of a}$$

function of a function rule (keeping y constant)

$$= \frac{-x}{(x^2+y^2)^{\frac{3}{2}}} = \frac{-x}{\sqrt{(x^2+y^2)^3}}$$

$$\frac{\partial z}{\partial y} = -\frac{1}{2}(x^2+y^2)^{\frac{-3}{2}}(2y) \text{ (keeping } x \text{ constant)}$$

$$= \frac{-y}{\sqrt{(x^2+y^2)^3}}$$

Problem 5. Pressure p of a mass of gas is given by $pV = mRT$, where m and R are constants, V is the volume and T the temperature. Find expressions for $\dfrac{\partial p}{\partial T}$ and $\dfrac{\partial p}{\partial V}$

Since $pV = mRT$ then $p = \dfrac{mRT}{V}$

To find $\dfrac{\partial p}{\partial T}$, V is kept constant.

Hence $\dfrac{\partial p}{\partial T} = \left(\dfrac{mR}{V}\right)\dfrac{\mathrm{d}}{\mathrm{d}T}(T) = \dfrac{mR}{V}$

To find $\dfrac{\partial p}{\partial V}$, T is kept constant.

Hence $\dfrac{\partial p}{\partial V} = (mRT)\dfrac{\mathrm{d}}{\mathrm{d}V}\left(\dfrac{1}{V}\right)$

$$= (mRT)(-V^{-2}) = \dfrac{-mRT}{V^2}$$

Problem 6. The time of oscillation, t, of a pendulum is given by $t = 2\pi\sqrt{\dfrac{l}{g}}$ where l is the length of the pendulum and g the free fall acceleration due to gravity. Determine $\dfrac{\partial t}{\partial l}$ and $\dfrac{\partial t}{\partial g}$

To find $\dfrac{\partial t}{\partial l}$, g is kept constant.

$$t = 2\pi\sqrt{\dfrac{l}{g}} = \left(\dfrac{2\pi}{\sqrt{g}}\right)\sqrt{l} = \left(\dfrac{2\pi}{\sqrt{g}}\right)l^{\frac{1}{2}}$$

Hence $\dfrac{\partial t}{\partial l} = \left(\dfrac{2\pi}{\sqrt{g}}\right)\dfrac{\mathrm{d}}{\mathrm{d}l}(l^{\frac{1}{2}}) = \left(\dfrac{2\pi}{\sqrt{g}}\right)\left(\dfrac{1}{2}l^{\frac{-1}{2}}\right)$

$$= \left(\dfrac{2\pi}{\sqrt{g}}\right)\left(\dfrac{1}{2\sqrt{l}}\right) = \dfrac{\pi}{\sqrt{lg}}$$

To find $\dfrac{\partial t}{\partial g}$, l is kept constant.

$$t = 2\pi\sqrt{\dfrac{l}{g}} = (2\pi\sqrt{l})\left(\dfrac{1}{\sqrt{g}}\right)$$

$$= (2\pi\sqrt{l})g^{\frac{-1}{2}}$$

Hence $\dfrac{\partial t}{\partial g} = (2\pi\sqrt{l})\left(-\dfrac{1}{2}g^{\frac{-3}{2}}\right)$

$$= (2\pi\sqrt{l})\left(\dfrac{-1}{2\sqrt{g^3}}\right)$$

$$= \dfrac{-\pi\sqrt{l}}{\sqrt{g^3}} = -\pi\sqrt{\dfrac{l}{g^3}}$$

Now try the following Practice Exercise

Practice Exercise 253   Further problems on first-order partial derivatives (answers on page 1180)

In Problems 1 to 6, find $\dfrac{\partial z}{\partial x}$ and $\dfrac{\partial z}{\partial y}$

1. $z = 2xy$

2. $z = x^3 - 2xy + y^2$

3. $z = \dfrac{x}{y}$

4. $z = \sin(4x + 3y)$

5. $z = x^3y^2 - \dfrac{y}{x^2} + \dfrac{1}{y}$

6. $z = \cos 3x \sin 4y$

7. The volume of a cone of height h and base radius r is given by $V = \frac{1}{3}\pi r^2 h$. Determine $\dfrac{\partial V}{\partial h}$ and $\dfrac{\partial V}{\partial r}$

8. The resonant frequency f_r in a series electrical circuit is given by $f_r = \dfrac{1}{2\pi\sqrt{LC}}$. Show that $\dfrac{\partial f_r}{\partial L} = \dfrac{-1}{4\pi\sqrt{CL^3}}$

9. An equation resulting from plucking a string is:

$$y = \sin\left(\frac{n\pi}{L}\right) x \left\{k \cos\left(\frac{n\pi b}{L}\right) t + c \sin\left(\frac{n\pi b}{L}\right) t\right\}$$

Determine $\frac{\partial y}{\partial t}$ and $\frac{\partial y}{\partial x}$

10. In a thermodynamic system, $k = Ae^{\frac{T\Delta S - \Delta H}{RT}}$, where R, k and A are constants.

Find (a) $\frac{\partial k}{\partial T}$ (b) $\frac{\partial A}{\partial T}$ (c) $\frac{\partial(\Delta S)}{\partial T}$ (d) $\frac{\partial(\Delta H)}{\partial T}$

60.3 Second-order partial derivatives

As with ordinary differentiation, where a differential coefficient may be differentiated again, a partial derivative may be differentiated partially again to give higher order partial derivatives.

(i) Differentiating $\frac{\partial V}{\partial r}$ of Section 60.2 with respect to r, keeping h constant, gives $\frac{\partial}{\partial r}\left(\frac{\partial V}{\partial r}\right)$ which is written as $\frac{\partial^2 V}{\partial r^2}$

Thus if $V = \pi r^2 h$,

then $\frac{\partial^2 V}{\partial r^2} = \frac{\partial}{\partial r}(2\pi rh) = 2\pi h$

(ii) Differentiating $\frac{\partial V}{\partial h}$ with respect to h, keeping r constant, gives $\frac{\partial}{\partial h}\left(\frac{\partial V}{\partial h}\right)$ which is written as $\frac{\partial^2 V}{\partial h^2}$

Thus $\frac{\partial^2 V}{\partial h^2} = \frac{\partial}{\partial h}(\pi r^2) = 0$

(iii) Differentiating $\frac{\partial V}{\partial h}$ with respect to r, keeping h constant, gives $\frac{\partial}{\partial r}\left(\frac{\partial V}{\partial h}\right)$ which is written as $\frac{\partial^2 V}{\partial r \partial h}$. Thus,

$$\frac{\partial^2 V}{\partial r \partial h} = \frac{\partial}{\partial r}\left(\frac{\partial V}{\partial h}\right) = \frac{\partial}{\partial r}(\pi r^2) = 2\pi r$$

(iv) Differentiating $\frac{\partial V}{\partial r}$ with respect to h, keeping r constant, gives $\frac{\partial}{\partial h}\left(\frac{\partial V}{\partial r}\right)$, which is written as $\frac{\partial^2 V}{\partial h \partial r}$. Thus,

$$\frac{\partial^2 V}{\partial h \partial r} = \frac{\partial}{\partial h}\left(\frac{\partial V}{\partial r}\right) = \frac{\partial}{\partial h}(2\pi rh) = 2\pi r$$

(v) $\frac{\partial^2 V}{\partial r^2}$, $\frac{\partial^2 V}{\partial h^2}$, $\frac{\partial^2 V}{\partial r \partial h}$ and $\frac{\partial^2 V}{\partial h \partial r}$ are examples of **second-order partial derivatives**.

(vi) It is seen from (iii) and (iv) that $\frac{\partial^2 V}{\partial r \partial h} = \frac{\partial^2 V}{\partial h \partial r}$ and such a result is always true for continuous functions (i.e. a graph of the function which has no sudden jumps or breaks).

Second-order partial derivatives are used in the solution of partial differential equations, in waveguide theory, in such areas of thermodynamics covering entropy and the continuity theorem, and when finding maxima, minima and saddle points for functions of two variables (see Chapter 62).

Problem 7. Given $z = 4x^2y^3 - 2x^3 + 7y^2$, find (a) $\frac{\partial^2 z}{\partial x^2}$ (b) $\frac{\partial^2 z}{\partial y^2}$ (c) $\frac{\partial^2 z}{\partial x \partial y}$ (d) $\frac{\partial^2 z}{\partial y \partial x}$

(a) $\frac{\partial z}{\partial x} = 8xy^3 - 6x^2$

$$\frac{\partial^2 z}{\partial x^2} = \frac{\partial}{\partial x}\left(\frac{\partial z}{\partial x}\right) = \frac{\partial}{\partial x}(8xy^3 - 6x^2)$$
$$= 8y^3 - 12x$$

(b) $\frac{\partial z}{\partial y} = 12x^2y^2 + 14y$

$$\frac{\partial^2 z}{\partial y^2} = \frac{\partial}{\partial y}\left(\frac{\partial z}{\partial y}\right) = \frac{\partial}{\partial y}(12x^2y^2 + 14y)$$
$$= 24x^2y + 14$$

(c) $\frac{\partial^2 z}{\partial x \partial y} = \frac{\partial}{\partial x}\left(\frac{\partial z}{\partial y}\right) = \frac{\partial}{\partial x}(12x^2y^2 + 14y) = 24xy^2$

(d) $\frac{\partial^2 z}{\partial y \partial x} = \frac{\partial}{\partial y}\left(\frac{\partial z}{\partial x}\right) = \frac{\partial}{\partial y}(8xy^3 - 6x^2) = 24xy^2$

$\left(\text{It is noted that } \frac{\partial^2 z}{\partial x \partial y} = \frac{\partial^2 z}{\partial y \partial x}\right)$

Problem 8. Show that when $z = e^{-t} \sin\theta$,

(a) $\dfrac{\partial^2 z}{\partial t^2} = -\dfrac{\partial^2 z}{\partial \theta^2}$, and (b) $\dfrac{\partial^2 z}{\partial t \partial\theta} = \dfrac{\partial^2 z}{\partial\theta\partial t}$

(a) $\dfrac{\partial z}{\partial t} = -e^{-t}\sin\theta$ and $\dfrac{\partial^2 z}{\partial t^2} = e^{-t}\sin\theta$

$\dfrac{\partial z}{\partial\theta} = e^{-t}\cos\theta$ and $\dfrac{\partial^2 z}{\partial\theta^2} = -e^{-t}\sin\theta$

Hence $\dfrac{\partial^2 z}{\partial t^2} = -\dfrac{\partial^2 z}{\partial\theta^2}$

(b) $\dfrac{\partial^2 z}{\partial t\partial\theta} = \dfrac{\partial}{\partial t}\left(\dfrac{\partial z}{\partial\theta}\right) = \dfrac{\partial}{\partial t}(e^{-t}\cos\theta)$

$\qquad = -e^{-t}\cos\theta$

$\dfrac{\partial^2 z}{\partial\theta\partial t} = \dfrac{\partial}{\partial\theta}\left(\dfrac{\partial z}{\partial t}\right) = \dfrac{\partial}{\partial\theta}(-e^{-t}\sin\theta)$

$\qquad = -e^{-t}\cos\theta$

Hence $\dfrac{\partial^2 z}{\partial t\partial\theta} = \dfrac{\partial^2 z}{\partial\theta\,\partial t}$

Problem 9. Show that if $z = \dfrac{x}{y}\ln y$, then

(a) $\dfrac{\partial z}{\partial y} = x\dfrac{\partial^2 z}{\partial y\partial x}$ and (b) evaluate $\dfrac{\partial^2 z}{\partial y^2}$ when $x = -3$ and $y = 1$

(a) To find $\dfrac{\partial z}{\partial x}$, y is kept constant.

Hencc $\dfrac{\partial z}{\partial x} = \left(\dfrac{1}{y}\ln y\right)\dfrac{\mathrm{d}}{\mathrm{d}x}(x) = \dfrac{1}{y}\ln y$

To find $\dfrac{\partial z}{\partial y}$, x is kept constant.

Hence

$\dfrac{\partial z}{\partial y} = (x)\dfrac{\mathrm{d}}{\mathrm{d}y}\left(\dfrac{\ln y}{y}\right)$

$= (x)\left\{\dfrac{(y)\left(\dfrac{1}{y}\right) - (\ln y)(1)}{y^2}\right\}$

using the quotient rule

$= x\left(\dfrac{1 - \ln y}{y^2}\right) = \dfrac{x}{y^2}(1 - \ln y)$

$\dfrac{\partial^2 z}{\partial y\partial x} = \dfrac{\partial}{\partial y}\left(\dfrac{\partial z}{\partial x}\right) = \dfrac{\partial}{\partial y}\left(\dfrac{\ln y}{y}\right)$

$= \dfrac{(y)\left(\dfrac{1}{y}\right) - (\ln y)(1)}{y^2}$

using the quotient rule

$= \dfrac{1}{y^2}(1 - \ln y)$

Hence $x\dfrac{\partial^2 z}{\partial y\partial x} = \dfrac{x}{y^2}(1 - \ln y) = \dfrac{\partial z}{\partial y}$

(b) $\dfrac{\partial^2 z}{\partial y^2} = \dfrac{\partial}{\partial y}\left(\dfrac{\partial z}{\partial y}\right) = \dfrac{\partial}{\partial y}\left\{\dfrac{x}{y^2}(1 - \ln y)\right\}$

$= (x)\dfrac{\mathrm{d}}{\mathrm{d}y}\left(\dfrac{1 - \ln y}{y^2}\right)$

$= (x)\left\{\dfrac{(y^2)\left(-\dfrac{1}{y}\right) - (1 - \ln y)(2y)}{y^4}\right\}$

using the quotient rule

$= \dfrac{x}{y^4}[-y - 2y + 2y\ln y]$

$= \dfrac{xy}{y^4}[-3 + 2\ln y] = \dfrac{x}{y^3}(2\ln y - 3)$

When $x = -3$ and $y = 1$,

$\dfrac{\partial^2 z}{\partial y^2} = \dfrac{(-3)}{(1)^3}(2\ln 1 - 3) = (-3)(-3) = \mathbf{9}$

Now try the following Practice Exercise

Practice Exercise 254   Further problems on second-order partial derivatives (answers on page 1180)

In Problems 1 to 4, find (a) $\dfrac{\partial^2 z}{\partial x^2}$ (b) $\dfrac{\partial^2 z}{\partial y^2}$ (c) $\dfrac{\partial^2 z}{\partial x\partial y}$ (d) $\dfrac{\partial^2 z}{\partial y\partial x}$

1. $z = (2x - 3y)^2$

2. $z = 2\ln xy$

3. $z = \dfrac{(x - y)}{(x + y)}$

4. $z = \sinh x\cosh 2y$

5. Given $z = x^2 \sin(x - 2y)$, find (a) $\dfrac{\partial^2 z}{\partial x^2}$ and

    (b) $\dfrac{\partial^2 z}{\partial y^2}$

    Show also that $\dfrac{\partial^2 z}{\partial x \partial y} = \dfrac{\partial^2 z}{\partial y \partial x}$
    $$= 2x^2 \sin(x - 2y) - 4x \cos(x - 2y)$$

6. Find $\dfrac{\partial^2 z}{\partial x^2}, \dfrac{\partial^2 z}{\partial y^2}$ and show that $\dfrac{\partial^2 z}{\partial x \partial y} = \dfrac{\partial^2 z}{\partial y \partial x}$

    when $z = \cos^{-1} \dfrac{x}{y}$

7. Given $z = \sqrt{\left(\dfrac{3x}{y} \right)}$, show that

    $\dfrac{\partial^2 z}{\partial x \partial y} = \dfrac{\partial^2 z}{\partial y \partial x}$ and evaluate $\dfrac{\partial^2 z}{\partial x^2}$ when

    $x = \dfrac{1}{2}$ and $y = 3$

8. An equation used in thermodynamics is the Benedict–Webb–Rubine equation of state for the expansion of a gas. The equation is:

    $$p = \frac{RT}{V} + \left(B_0 RT - A_0 - \frac{C_0}{T^2} \right) \frac{1}{V^2}$$

    $$+ (bRT - a) \frac{1}{V^3} + \frac{A\alpha}{V^6}$$

    $$+ \frac{C \left(1 + \frac{\gamma}{V^2} \right)}{T^2} \left(\frac{1}{V^3} \right) e^{-\frac{\gamma}{V^2}}$$

    Show that $\dfrac{\partial^2 p}{\partial T^2}$

    $$= \frac{6}{V^2 T^4} \left\{ \frac{C}{V} \left(1 + \frac{\gamma}{V^2} \right) e^{-\frac{\gamma}{V^2}} - C_0 \right\}$$

For fully worked solutions to each of the problems in Practice Exercises 253 and 254 in this chapter, go to the website:
www.routledge.com/cw/bird

Section I

Chapter 61

Total differential, rates of change and small changes

Why it is important to understand: Total differential, rates of change and small changes

This chapter looks at some applications of partial differentiation, i.e. the total differential for variables which may all be changing at the same time, rates of change, where different quantities have different rates of change, and small changes, where approximate errors may be calculated in situations where small changes in the variables associated with the quantity occur.

At the end of this chapter, you should be able to:

- determine the total differential of a function having more than one variable
- determine the rates of change of functions having more than one variable
- determine an approximate error in a function having more than one variable

61.1 Total differential

In Chapter 60, partial differentiation is introduced for the case where only one variable changes at a time, the other variables being kept constant. In practice, variables may all be changing at the same time.

If $z = f(u, v, w, \ldots)$, then the **total differential, dz**, is given by the sum of the separate partial differentials of z,

i.e. $dz = \dfrac{\partial z}{\partial u}\, du + \dfrac{\partial z}{\partial v}\, dv + \dfrac{\partial z}{\partial w}\, dw + \cdots$ (1)

Problem 1. If $z = f(x, y)$ and $z = x^2 y^3 + \dfrac{2x}{y} + 1$, determine the total differential, dz.

The total differential is the sum of the partial differentials,

i.e. $dz = \dfrac{\partial z}{\partial x}\, dx + \dfrac{\partial z}{\partial y}\, dy$

$\dfrac{\partial z}{\partial x} = 2xy^3 + \dfrac{2}{y}$ (i.e. y is kept constant)

$\dfrac{\partial z}{\partial y} = 3x^2 y^2 - \dfrac{2x}{y^2}$ (i.e. x is kept constant)

Hence $dz = \left(2xy^3 + \dfrac{2}{y}\right) dx + \left(3x^2 y^2 - \dfrac{2x}{y^2}\right) dy$

Problem 2. If $z = f(u, v, w)$ and $z = 3u^2 - 2v + 4w^3 v^2$ find the total differential, dz

The total differential

$$dz = \dfrac{\partial z}{\partial u}\, du + \dfrac{\partial z}{\partial v}\, dv + \dfrac{\partial z}{\partial w}\, dw$$

$\dfrac{\partial z}{\partial u} = 6u$ (i.e. v and w are kept constant)

$$\frac{\partial z}{\partial v} = -2 + 8w^3 v$$

(i.e. u and w are kept constant)

$$\frac{\partial z}{\partial w} = 12w^2 v^2 \text{ (i.e. } u \text{ and } v \text{ are kept constant)}$$

Hence

$$\mathbf{d}z = 6u\,\mathbf{d}u + (8vw^3 - 2)\,\mathbf{d}v + (12v^2w^2)\,\mathbf{d}w$$

Problem 3. The pressure p, volume V and temperature T of a gas are related by $pV = kT$, where k is a constant. Determine the total differentials (a) dp and (b) dT in terms of p, V and T

(a) Total differential $dp = \dfrac{\partial p}{\partial T}\,dT + \dfrac{\partial p}{\partial V}\,dV$

Since $pV = kT$ then $p = \dfrac{kT}{V}$

hence $\dfrac{\partial p}{\partial T} = \dfrac{k}{V}$ and $\dfrac{\partial p}{\partial V} = -\dfrac{kT}{V^2}$

Thus $dp = \dfrac{k}{V}\,dT - \dfrac{kT}{V^2}\,dV$

Since $pV = kT, k = \dfrac{pV}{T}$

Hence $dp = \dfrac{\left(\dfrac{pV}{T}\right)}{V}\,dT - \dfrac{\left(\dfrac{pV}{T}\right)T}{V^2}\,dV$

i.e. $\mathbf{d}p = \dfrac{p}{T}\,\mathbf{d}T - \dfrac{p}{V}\,\mathbf{d}V$

(b) Total differential $dT = \dfrac{\partial T}{\partial p}\,dp + \dfrac{\partial T}{\partial V}\,dV$

Since $pV = kT, T = \dfrac{pV}{k}$

hence $\dfrac{\partial T}{\partial p} = \dfrac{V}{k}$ and $\dfrac{\partial T}{\partial V} = \dfrac{p}{k}$

Thus $dT = \dfrac{V}{k}\,dp + \dfrac{p}{k}\,dV$ and substituting $k = \dfrac{pV}{T}$ gives:

$$dT = \frac{V}{\left(\dfrac{pV}{T}\right)}\,dp + \frac{p}{\left(\dfrac{pV}{T}\right)}\,dV$$

i.e. $\mathbf{d}T = \dfrac{T}{p}\,\mathbf{d}p + \dfrac{T}{V}\,\mathbf{d}V$

Now try the following Practice Exercise

Practice Exercise 255   Further problems on the total differential (answers on page 1180)

In Problems 1 to 5, find the total differential dz

1. $z = x^3 + y^2$

2. $z = 2xy - \cos x$

3. $z = \dfrac{x - y}{x + y}$

4. $z = x \ln y$

5. $z = xy + \dfrac{\sqrt{x}}{y} - 4$

6. If $z = f(a, b, c)$ and $z = 2ab - 3b^2 c + abc$, find the total differential, dz

7. Given $u = \ln \sin(xy)$ show that $du = \cot(xy)(y\,dx + x\,dy)$

61.2 Rates of change

Sometimes it is necessary to solve problems in which different quantities have different rates of change. From equation (1), the rate of change of z, $\dfrac{dz}{dt}$ is given by:

$$\frac{dz}{dt} = \frac{\partial z}{\partial u}\frac{du}{dt} + \frac{\partial z}{\partial v}\frac{dv}{dt} + \frac{\partial z}{\partial w}\frac{dw}{dt} + \cdots \qquad (2)$$

Problem 4. If $z = f(x, y)$ and $z = 2x^3 \sin 2y$, find the rate of change of z, correct to 4 significant figures, when x is 2 units and y is $\pi/6$ radians and when x is increasing at 4 units/s and y is decreasing at 0.5 units/s

Using equation (2), the rate of change of z,

$$\frac{dz}{dt} = \frac{\partial z}{\partial x}\frac{dx}{dt} + \frac{\partial z}{\partial y}\frac{dy}{dt}$$

Since $z = 2x^3 \sin 2y$, then

$$\frac{\partial z}{\partial x} = 6x^2 \sin 2y \text{ and } \frac{\partial z}{\partial y} = 4x^3 \cos 2y$$

Section I

Since x is increasing at 4 units/s, $\dfrac{\mathrm{d}x}{\mathrm{d}t} = +4$

and since y is decreasing at 0.5 units/s, $\dfrac{\mathrm{d}y}{\mathrm{d}t} = -0.5$

Hence $\dfrac{\mathrm{d}z}{\mathrm{d}t} = (6x^2 \sin 2y)(+4) + (4x^3 \cos 2y)(-0.5)$

$\qquad = 24x^2 \sin 2y - 2x^3 \cos 2y$

When $x = 2$ units and $y = \dfrac{\pi}{6}$ radians, then

$\dfrac{\mathrm{d}z}{\mathrm{d}t} = 24(2)^2 \sin[2(\pi/6)] - 2(2)^3 \cos[2(\pi/6)]$

$\qquad = 83.138 - 8.0$

Hence the rate of change of z, $\dfrac{\mathrm{d}z}{\mathrm{d}t} = \mathbf{75.14\,units/s}$, correct to 4 significant figures.

Problem 5. The height of a right circular cone is increasing at 3 mm/s and its radius is decreasing at 2 mm/s. Determine, correct to 3 significant figures, the rate at which the volume is changing (in cm^3/s) when the height is 3.2 cm and the radius is 1.5 cm

Volume of a right circular cone, $V = \dfrac{1}{3}\pi r^2 h$

Using equation (2), the rate of change of volume,

$$\frac{\mathrm{d}V}{\mathrm{d}t} = \frac{\partial V}{\partial r}\frac{\mathrm{d}r}{\mathrm{d}t} + \frac{\partial V}{\partial h}\frac{\mathrm{d}h}{\mathrm{d}t}$$

$$\frac{\partial V}{\partial r} = \frac{2}{3}\pi r h \quad \text{and} \quad \frac{\partial V}{\partial h} = \frac{1}{3}\pi r^2$$

Since the height is increasing at 3 mm/s,

i.e. 0.3 cm/s, then $\dfrac{\mathrm{d}h}{\mathrm{d}t} = +0.3$

and since the radius is decreasing at 2 mm/s,

i.e. 0.2 cm/s, then $\dfrac{\mathrm{d}r}{\mathrm{d}t} = -0.2$

Hence $\dfrac{\mathrm{d}V}{\mathrm{d}t} = \left(\dfrac{2}{3}\pi r h\right)(-0.2) + \left(\dfrac{1}{3}\pi r^2\right)(+0.3)$

$\qquad = \dfrac{-0.4}{3}\pi r h + 0.1\pi r^2$

However, $h = 3.2$ cm and $r = 1.5$ cm

Hence $\dfrac{\mathrm{d}V}{\mathrm{d}t} = \dfrac{-0.4}{3}\pi(1.5)(3.2) + (0.1)\pi(1.5)^2$

$\qquad = -2.011 + 0.707 = -1.304\,\text{cm}^3/\text{s}$

Thus the rate of change of volume is 1.30 cm^3/s decreasing.

Problem 6. The area A of a triangle is given by $A = \frac{1}{2}ac \sin B$, where B is the angle between sides a and c. If a is increasing at 0.4 units/s, c is decreasing at 0.8 units/s and B is increasing at 0.2 units/s, find the rate of change of the area of the triangle, correct to 3 significant figures, when a is 3 units, c is 4 units and B is $\pi/6$ radians

Using equation (2), the rate of change of area,

$$\frac{\mathrm{d}A}{\mathrm{d}t} = \frac{\partial A}{\partial a}\frac{\mathrm{d}a}{\mathrm{d}t} + \frac{\partial A}{\partial c}\frac{\mathrm{d}c}{\mathrm{d}t} + \frac{\partial A}{\partial B}\frac{\mathrm{d}B}{\mathrm{d}t}$$

Since $A = \dfrac{1}{2}ac \sin B$, $\dfrac{\partial A}{\partial a} = \dfrac{1}{2}c \sin B$,

$\dfrac{\partial A}{\partial c} = \dfrac{1}{2}a \sin B$ and $\dfrac{\partial A}{\partial B} = \dfrac{1}{2}ac \cos B$

$\dfrac{\mathrm{d}a}{\mathrm{d}t} = 0.4$ units/s, $\dfrac{\mathrm{d}c}{\mathrm{d}t} = -0.8$ units/s

and $\dfrac{\mathrm{d}B}{\mathrm{d}t} = 0.2$ units/s

Hence $\dfrac{\mathrm{d}A}{\mathrm{d}t} = \left(\dfrac{1}{2}c \sin B\right)(0.4) + \left(\dfrac{1}{2}a \sin B\right)(-0.8)$

$\qquad\qquad + \left(\dfrac{1}{2}ac \cos B\right)(0.2)$

When $a = 3$, $c = 4$ and $B = \dfrac{\pi}{6}$ then:

$\dfrac{\mathbf{d}A}{\mathbf{d}t} = \left(\dfrac{1}{2}(4)\sin\dfrac{\pi}{6}\right)(0.4) + \left(\dfrac{1}{2}(3)\sin\dfrac{\pi}{6}\right)(-0.8)$

$\qquad\qquad + \left(\dfrac{1}{2}(3)(4)\cos\dfrac{\pi}{6}\right)(0.2)$

$\qquad = 0.4 - 0.6 + 1.039 = \mathbf{0.839\,units^2/s}$, correct to 3 significant figures.

Problem 7. Determine the rate of increase of diagonal AC of the rectangular solid, shown in Figure 61.1, correct to 2 significant figures, if the sides x, y and z increase at 6 mm/s, 5 mm/s and 4 mm/s when these three sides are 5 cm, 4 cm and 3 cm, respectively

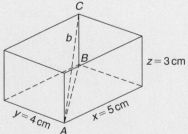

Figure 61.1

Diagonal $AB = \sqrt{(x^2 + y^2)}$

Diagonal $AC = \sqrt{(BC^2 + AB^2)}$

$$= \sqrt{\left[z^2 + \left\{\sqrt{(x^2 + y^2)}\right\}^2\right]}$$

$$= \sqrt{(z^2 + x^2 + y^2)}$$

Let $AC = b$, then $b = \sqrt{(x^2 + y^2 + z^2)}$

Using equation (2), the rate of change of diagonal b is given by:

$$\frac{db}{dt} = \frac{\partial b}{\partial x}\frac{dx}{dt} + \frac{\partial b}{\partial y}\frac{dy}{dt} + \frac{\partial b}{\partial z}\frac{dz}{dt}$$

Since $b = \sqrt{(x^2 + y^2 + z^2)}$

$$\frac{\partial b}{\partial x} = \frac{1}{2}(x^2 + y^2 + z^2)^{\frac{-1}{2}}(2x) = \frac{x}{\sqrt{(x^2 + y^2 + z^2)}}$$

Similarly, $\dfrac{\partial b}{\partial y} = \dfrac{y}{\sqrt{(x^2 + y^2 + z^2)}}$

and $\dfrac{\partial b}{\partial z} = \dfrac{z}{\sqrt{(x^2 + y^2 + z^2)}}$

$$\frac{dx}{dt} = 6\,\text{mm/s} = 0.6\,\text{cm/s},$$

$$\frac{dy}{dt} = 5\,\text{mm/s} = 0.5\,\text{cm/s},$$

and $\dfrac{dz}{dt} = 4\,\text{mm/s} = 0.4\,\text{cm/s}$

Hence $\dfrac{db}{dt} = \left[\dfrac{x}{\sqrt{(x^2 + y^2 + z^2)}}\right](0.6)$

$$+ \left[\frac{y}{\sqrt{(x^2 + y^2 + z^2)}}\right](0.5)$$

$$+ \left[\frac{z}{\sqrt{(x^2 + y^2 + z^2)}}\right](0.4)$$

When $x = 5\,\text{cm}$, $y = 4\,\text{cm}$ and $z = 3\,\text{cm}$, then:

$$\frac{db}{dt} = \left[\frac{5}{\sqrt{(5^2 + 4^2 + 3^2)}}\right](0.6)$$

$$+ \left[\frac{4}{\sqrt{(5^2 + 4^2 + 3^2)}}\right](0.5)$$

$$+ \left[\frac{3}{\sqrt{(5^2 + 4^2 + 3^2)}}\right](0.4)$$

$$= 0.4243 + 0.2828 + 0.1697 = 0.8768\,\text{cm/s}$$

Hence the rate of increase of diagonal AC is 0.88 cm/s or 8.8 mm/s, correct to 2 significant figures.

Now try the following Practice Exercise

Practice Exercise 256   Further problems on rates of change (answers on page 1180)

1. The radius of a right cylinder is increasing at a rate of 8 mm/s and the height is decreasing at a rate of 15 mm/s. Find the rate at which the volume is changing in cm^3/s when the radius is 40 mm and the height is 150 mm.

2. If $z = f(x, y)$ and $z = 3x^2y^5$, find the rate of change of z when x is 3 units and y is 2 units, when x is decreasing at 5 units/s and y is increasing at 2.5 units/s.

3. Find the rate of change of k, correct to 4 significant figures, given the following data: $k = f(a, b, c)$; $k = 2b\ln a + c^2 e^a$; a is increasing at 2 cm/s; b is decreasing at 3 cm/s; c is decreasing at 1 cm/s; $a = 1.5\,\text{cm}$, $b = 6\,\text{cm}$ and $c = 8\,\text{cm}$.

4. A rectangular box has sides of length x cm, y cm and z cm. Sides x and z are expanding at rates of 3 mm/s and 5 mm/s respectively and side y is contracting at a rate of 2 mm/s. Determine the rate of change of volume when x is 3 cm, y is 1.5 cm and z is 6 cm.

5. Find the rate of change of the total surface area of a right circular cone at the instant when the base radius is 5 cm and the height is 12 cm if the radius is increasing at 5 mm/s and the height is decreasing at 15 mm/s.

61.3   Small changes

It is often useful to find an approximate value for the change (or error) of a quantity caused by small changes (or errors) in the variables associated with the quantity. If $z = f(u, v, w, \ldots)$ and $\delta u, \delta v, \delta w, \ldots$ denote **small changes** in $u, v, w, \ldots$ respectively, then the corresponding approximate change δz in z is obtained from equation (1) by replacing the differentials by the small changes.

Thus $\delta z \approx \dfrac{\partial z}{\partial u} \delta u + \dfrac{\partial z}{\partial v} \delta v + \dfrac{\partial z}{\partial w} \delta w + \cdots$ $\qquad$ (3)

Problem 8. Pressure p and volume V of a gas are connected by the equation $pV^{1.4} = k$. Determine the approximate percentage error in k when the pressure is increased by 4% and the volume is decreased by 1.5%

Using equation (3), the approximate error in k,

$$\delta k \approx \frac{\partial k}{\partial p} \delta p + \frac{\partial k}{\partial V} \delta V$$

Let p, V and k refer to the initial values.

Since $\quad k = pV^{1.4}$ then $\dfrac{\partial k}{\partial p} = V^{1.4}$

and $\quad \dfrac{\partial k}{\partial V} = 1.4pV^{0.4}$

Since the pressure is increased by 4%, the change in pressure $\delta p = \dfrac{4}{100} \times p = 0.04p$

Since the volume is decreased by 1.5%, the change in volume $\delta V = \dfrac{-1.5}{100} \times V = -0.015V$

Hence the approximate error in k,

$$\delta k \approx (V)^{1.4}(0.04p) + (1.4pV^{0.4})(-0.015V)$$

$$\approx pV^{1.4}[0.04 - 1.4(0.015)]$$

$$\approx pV^{1.4}[0.019] \approx \frac{1.9}{100}pV^{1.4} \approx \frac{1.9}{100}k$$

i.e. **the approximate error in k is a 1.9% increase**.

Problem 9. Modulus of rigidity $G = (R^4\theta)/L$, where R is the radius, θ the angle of twist and L the length. Determine the approximate percentage error in G when R is increased by 2%, θ is reduced by 5% and L is increased by 4%

Using $\quad \delta G \approx \dfrac{\partial G}{\partial R} \delta R + \dfrac{\partial G}{\partial \theta} \delta\theta + \dfrac{\partial G}{\partial L} \delta L$

Since $\quad G = \dfrac{R^4\theta}{L}, \dfrac{\partial G}{\partial R} = \dfrac{4R^3\theta}{L}, \dfrac{\partial G}{\partial \theta} = \dfrac{R^4}{L}$

and $\quad \dfrac{\partial G}{\partial L} = \dfrac{-R^4\theta}{L^2}$

Since R is increased by 2%, $\delta R = \dfrac{2}{100}R = 0.02R$

Similarly, $\delta\theta = -0.05\theta$ and $\delta L = 0.04L$

Hence $\delta G \approx \left(\dfrac{4R^3\theta}{L}\right)(0.02R) + \left(\dfrac{R^4}{L}\right)(-0.05\theta)$

$$+ \left(-\dfrac{R^4\theta}{L^2}\right)(0.04L)$$

$$\approx \frac{R^4\theta}{L}[0.08 - 0.05 - 0.04] \approx -0.01\frac{R^4\theta}{L}$$

i.e. $\quad \delta G \approx -\dfrac{1}{100}G$

Hence the approximate percentage error in G is a 1% decrease.

Problem 10. The second moment of area of a rectangle is given by $I = (bl^3)/3$. If b and l are measured as 40 mm and 90 mm respectively and the measurement errors are -5 mm in b and $+8$ mm in l, find the approximate error in the calculated value of I

Using equation (3), the approximate error in I,

$$\delta I \approx \frac{\partial I}{\partial b} \delta b + \frac{\partial I}{\partial l} \delta l$$

$$\frac{\partial I}{\partial b} = \frac{l^3}{3} \text{ and } \frac{\partial I}{\partial l} = \frac{3bl^2}{3} = bl^2$$

$$\delta b = -5 \text{ mm and } \delta l = +8 \text{ mm}$$

Hence $\delta I \approx \left(\dfrac{l^3}{3}\right)(-5) + (bl^2)(+8)$

Since $b = 40$ mm and $l = 90$ mm then

$$\delta I \approx \left(\frac{90^3}{3}\right)(-5) + 40(90)^2(8)$$

$$\approx -1\,215\,000 + 2\,592\,000$$

$$\approx 1\,377\,000 \text{ mm}^4 \approx 137.7 \text{ cm}^4$$

Hence the approximate error in the calculated value of I is a 137.7 cm^4 increase.

Section I

Problem 11. The time of oscillation t of a pendulum is given by $t = 2\pi\sqrt{\dfrac{l}{g}}$. Determine the approximate percentage error in t when l has an error of 0.2% too large and g 0.1% too small

Using equation (3), the approximate change in t,

$$\delta t \approx \frac{\partial t}{\partial l}\delta l + \frac{\partial t}{\partial g}\delta g$$

Since $t = 2\pi\sqrt{\dfrac{l}{g}}$, $\dfrac{\partial t}{\partial l} = \dfrac{\pi}{\sqrt{lg}}$

and $\dfrac{\partial t}{\partial g} = -\pi\sqrt{\dfrac{l}{g^3}}$ (from Problem 6, Chapter 60)

$$\delta l = \frac{0.2}{100}l = 0.002\,l \text{ and } \delta g = -0.001g$$

hence $\delta t \approx \dfrac{\pi}{\sqrt{lg}}(0.002l) + -\pi\sqrt{\dfrac{l}{g^3}}(-0.001\,g)$

$$\approx 0.002\pi\sqrt{\frac{l}{g}} + 0.001\pi\sqrt{\frac{l}{g}}$$

$$\approx (0.001)\left[2\pi\sqrt{\frac{l}{g}}\right] + 0.0005\left[2\pi\sqrt{\frac{l}{g}}\right]$$

$$\approx 0.0015t \approx \frac{0.15}{100}t$$

Hence the approximate error in t is a 0.15% increase.

Now try the following Practice Exercise

Practice Exercise 257   Further problems on small changes (answers on page 1180)

1. The power P consumed in a resistor is given by $P = V^2/R$ watts. Determine the approximate change in power when V increases by 5% and R decreases by 0.5% if the original values of V and R are 50 volts and 12.5 ohms, respectively.

2. An equation for heat generated H is $H = i^2 Rt$. Determine the error in the calculated value of H if the error in measuring current i is +2%, the error in measuring resistance R is −3% and the error in measuring time t is +1%

3. $f_r = \dfrac{1}{2\pi\sqrt{LC}}$ represents the resonant frequency of a series-connected circuit containing inductance L and capacitance C. Determine the approximate percentage change in f_r when L is decreased by 3% and C is increased by 5%

4. The second moment of area of a rectangle about its centroid parallel to side b is given by $I = bd^3/12$. If b and d are measured as 15 cm and 6 cm respectively and the measurement errors are +12 mm in b and −1.5 mm in d, find the error in the calculated value of I

5. The side b of a triangle is calculated using $b^2 = a^2 + c^2 - 2ac\cos B$. If a, c and B are measured as 3 cm, 4 cm and $\pi/4$ radians, respectively, and the measurement errors which occur are +0.8 cm, −0.5 cm and +$\pi/90$ radians, respectively, determine the error in the calculated value of b

6. Q factor in a resonant electrical circuit is given by: $Q = \dfrac{1}{R}\sqrt{\dfrac{L}{C}}$. Find the percentage change in Q when L increases by 4%, R decreases by 3% and C decreases by 2%

7. The rate of flow of gas in a pipe is given by: $v = \dfrac{C\sqrt{d}}{\sqrt[6]{T^5}}$, where C is a constant, d is the diameter of the pipe and T is the thermodynamic temperature of the gas. When determining the rate of flow experimentally, d is measured and subsequently found to be in error by +1.4%, and T has an error of −1.8%. Determine the percentage error in the rate of flow based on the measured values of d and T

For fully worked solutions to each of the problems in Practice Exercises 255 to 257 in this chapter, go to the website:
www.routledge.com/cw/bird

Section I

Chapter 62

Maxima, minima and saddle points for functions of two variables

---

Why it is important to understand: **Maxima, minima and saddle points for functions of two variables**

The problem of finding the maximum and minimum values of functions is encountered in mechanics, physics, geometry and many other fields. Finding maxima, minima and saddle points for functions of two variables requires the application of partial differentiation (which was explained in the previous chapters). This is demonstrated in this chapter.

At the end of this chapter, you should be able to:

- understand functions of two independent variables
- explain a saddle point
- determine the maxima, minima and saddle points for a function of two variables
- sketch a contour map for functions of two variables

62.1 Functions of two independent variables

If a relation between two real variables, x and y, is such that when x is given, y is determined, then y is said to be a function of x and is denoted by $y = f(x)$; x is called the independent variable and y the dependent variable. If $y = f(u, v)$, then y is a function of two independent variables u and v. For example, if, say, $y = f(u, v) = 3u^2 - 2v$ then when $u = 2$ and $v = 1$, $y = 3(2)^2 - 2(1) = 10$. This may be written as $f(2, 1) = 10$. Similarly, if $u = 1$ and $v = 4$, $f(1, 4) = -5$

Consider a function of two variables x and y defined by $z = f(x, y) = 3x^2 - 2y$. If $(x, y) = (0, 0)$, then $f(0, 0) = 0$ and if $(x, y) = (2, 1)$, then $f(2, 1) = 10$. Each pair of numbers, (x, y), may be represented by a point P in the (x, y) plane of a rectangular Cartesian co-ordinate system, as shown in Figure 62.1. The corresponding value of $z = f(x, y)$ may be represented by a line PP' drawn parallel to the z-axis. Thus, if, for example, $z = 3x^2 - 2y$, as above, and P is the co-ordinate $(2, 3)$ then the length of PP' is $3(2)^2 - 2(3) = 6$. Figure 62.2 shows that when a large number of (x, y) co-ordinates are taken for a function $f(x, y)$, and then $f(x, y)$ calculated for each, a large number of lines such as PP' can be constructed, and in

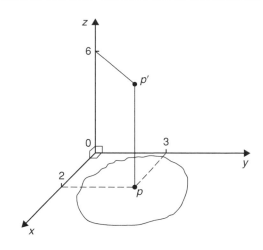

Figure 62.1

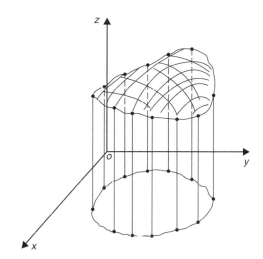

Figure 62.2

the limit when all points in the (x, y) plane are considered, a surface is seen to result as shown in Figure 62.2. Thus the function $z = f(x, y)$ represents a surface and not a curve.

62.2 Maxima, minima and saddle points

Partial differentiation is used when determining stationary points for functions of two variables. A function $f(x, y)$ is said to be a maximum at a point (x, y) if the value of the function there is greater than at all points in the immediate vicinity, and is a minimum if less than at all points in the immediate vicinity. Figure 62.3 shows geometrically a maximum value of a function of two variables and it is seen that the surface $z = f(x, y)$ is

higher at $(x, y) = (a, b)$ than at any point in the immediate vicinity. Figure 62.4 shows a minimum value of a function of two variables and it is seen that the surface $z = f(x, y)$ is lower at $(x, y) = (p, q)$ than at any point in the immediate vicinity.

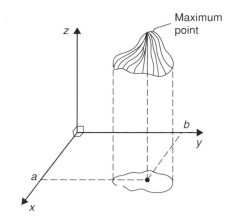

Figure 62.3

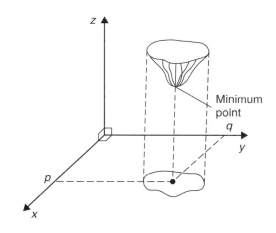

Figure 62.4

If $z = f(x, y)$ and a maximum occurs at (a, b), the curve lying in the two planes $x = a$ and $y = b$ must also have a maximum point (a, b), as shown in Figure 62.5. Consequently, the tangents (shown as t_1 and t_2) to the curves at (a, b) must be parallel to Ox and Oy, respectively. This requires that $\dfrac{\partial z}{\partial x} = 0$ and $\dfrac{\partial z}{\partial y} = 0$ at all maximum and minimum values, and the solution of these equations gives the stationary (or critical) points of z.

With functions of two variables there are three types of stationary points possible, these being a maximum

Section I

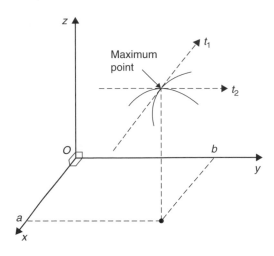

Figure 62.5

point, a minimum point and a **saddle point**. A saddle point Q is shown in Figure 62.6 and is such that a point Q is a maximum for curve 1 and a minimum for curve 2.

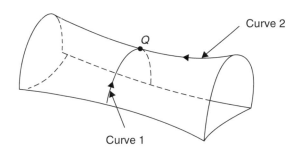

Figure 62.6

62.3 Procedure to determine maxima, minima and saddle points for functions of two variables

Given $z = f(x, y)$:

(i)   determine $\dfrac{\partial z}{\partial x}$ and $\dfrac{\partial z}{\partial y}$

(ii)   for stationary points, $\dfrac{\partial z}{\partial x} = 0$ and $\dfrac{\partial z}{\partial y} = 0,$

(iii)   solve the simultaneous equations $\dfrac{\partial z}{\partial x} = 0$ and $\dfrac{\partial z}{\partial y} = 0$ for x and y, which gives the co-ordinates of the stationary points,

(iv)   determine $\dfrac{\partial^2 z}{\partial x^2}, \dfrac{\partial^2 z}{\partial y^2}$ and $\dfrac{\partial^2 z}{\partial x \partial y}$

(v)   for each of the co-ordinates of the stationary points, substitute values of x and y into $\dfrac{\partial^2 z}{\partial x^2}, \dfrac{\partial^2 z}{\partial y^2}$ and $\dfrac{\partial^2 z}{\partial x \partial y}$ and evaluate each,

(vi)   evaluate $\left(\dfrac{\partial^2 z}{\partial x \partial y}\right)^2$ for each stationary point,

(vii)   substitute the values of $\dfrac{\partial^2 z}{\partial x^2}, \dfrac{\partial^2 z}{\partial y^2}$ and $\dfrac{\partial^2 z}{\partial x \partial y}$ into the equation

$$\Delta = \left(\frac{\partial^2 z}{\partial x \partial y}\right)^2 - \left(\frac{\partial^2 z}{\partial x^2}\right)\left(\frac{\partial^2 z}{\partial y^2}\right)$$

and evaluate,

(viii)   (a)   if $\boldsymbol{\Delta > 0}$ then the stationary point is a **saddle point**

(b)   if $\boldsymbol{\Delta < 0}$ and $\dfrac{\partial^2 z}{\partial x^2} < 0$, then the stationary point is a **maximum point**

and

(c)   if $\boldsymbol{\Delta < 0}$ and $\dfrac{\partial^2 z}{\partial x^2} > 0$, then the stationary point is a **minimum point.**

62.4 Worked problems on maxima, minima and saddle points for functions of two variables

Problem 1.   Show that the function $z = (x-1)^2 + (y-2)^2$ has one stationary point only and determine its nature. Sketch the surface represented by z and produce a contour map in the x–y plane

Following the above procedure:

(i) $\dfrac{\partial z}{\partial x} = 2(x-1)$ and $\dfrac{\partial z}{\partial y} = 2(y-2)$

(ii) $2(x-1) = 0$ \hfill (1)

$2(y-2) = 0$ \hfill (2)

(iii) From equations (1) and (2), $x = 1$ and $y = 2$, thus the only stationary point exists at $(1, 2)$

(iv) Since $\dfrac{\partial z}{\partial x} = 2(x - 1) = 2x - 2$, $\dfrac{\partial^2 z}{\partial x^2} = 2$

and since $\dfrac{\partial z}{\partial y} = 2(y - 2) = 2y - 4$, $\dfrac{\partial^2 z}{\partial y^2} = 2$

and $\dfrac{\partial^2 z}{\partial x \partial y} = \dfrac{\partial}{\partial x}\left(\dfrac{\partial z}{\partial y}\right) = \dfrac{\partial}{\partial x}(2y - 4) = 0$

(v) $\dfrac{\partial^2 z}{\partial x^2} = \dfrac{\partial^2 z}{\partial y^2} = 2$ and $\dfrac{\partial^2 z}{\partial x \partial y} = 0$

(vi) $\left(\dfrac{\partial^2 z}{\partial x \partial y}\right)^2 = 0$

(vii) $\Delta = (0)^2 - (2)(2) = -4$

(viii) Since $\Delta < 0$ and $\dfrac{\partial^2 z}{\partial x^2} > 0$, **the stationary point (1, 2) is a minimum**.

The surface $z = (x - 1)^2 + (y - 2)^2$ is shown in three dimensions in Figure 62.7. Looking down towards the x–y plane from above, it is possible to produce a **contour map**. A contour is a line on a map which gives places having the same vertical height above a datum line (usually the mean sea-level on a geographical map).

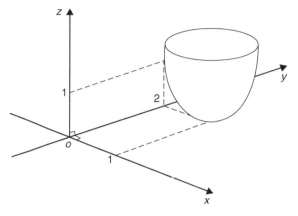

Figure 62.7

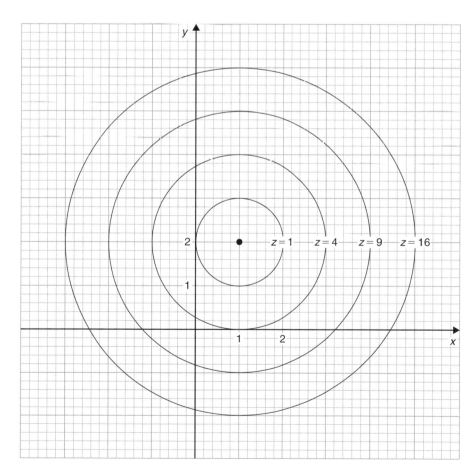

Figure 62.8

A contour map for $z = (x-1)^2 + (y-2)^2$ is shown in Figure 62.8. The values of z are shown on the map and these give an indication of the rise and fall to a stationary point.

> **Problem 2.** Find the stationary points of the surface $f(x, y) = x^3 - 6xy + y^3$ and determine their nature

Let $z = f(x, y) = x^3 - 6xy + y^3$

Following the procedure:

(i) $\dfrac{\partial z}{\partial x} = 3x^2 - 6y$ and $\dfrac{\partial z}{\partial y} = -6x + 3y^2$

(ii) for stationary points, $3x^2 - 6y = 0$ (1)

and $\quad\quad\quad -6x + 3y^2 = 0$ (2)

(iii) from equation (1), $3x^2 = 6y$

and $\quad y = \dfrac{3x^2}{6} = \dfrac{1}{2}x^2$

and substituting in equation (2) gives:

$$-6x + 3\left(\frac{1}{2}x^2\right)^2 = 0$$

$$-6x + \frac{3}{4}x^4 = 0$$

$$3x\left(\frac{x^3}{4} - 2\right) = 0$$

from which, $x = 0$ or $\dfrac{x^3}{4} - 2 = 0$

i.e. $x^3 = 8$ and $x = 2$

When $x = 0$, $y = 0$ and when $x = 2$, $y = 2$ from equations (1) and (2)

Thus stationary points occur at (0, 0) and (2, 2)

(iv) $\dfrac{\partial^2 z}{\partial x^2} = 6x$, $\dfrac{\partial^2 z}{\partial y^2} = 6y$ and $\dfrac{\partial^2 z}{\partial x \partial y} = \dfrac{\partial}{\partial x}\left(\dfrac{\partial z}{\partial y}\right)$

$$= \frac{\partial}{\partial x}(-6x + 3y^2) = -6$$

(v) for (0, 0) $\dfrac{\partial^2 z}{\partial x^2} = 0$, $\dfrac{\partial^2 z}{\partial y^2} = 0$

and $\quad \dfrac{\partial^2 z}{\partial x \partial y} = -6$

for (2, 2), $\dfrac{\partial^2 z}{\partial x^2} = 12$, $\dfrac{\partial^2 z}{\partial y^2} = 12$

and $\quad \dfrac{\partial^2 z}{\partial x \partial y} = -6$

(vi) for (0, 0), $\left(\dfrac{\partial^2 z}{\partial x \partial y}\right)^2 = (-6)^2 = 36$

for (2, 2), $\left(\dfrac{\partial^2 z}{\partial x \partial y}\right)^2 = (-6)^2 = 36$

(vii) $\Delta_{(0, 0)} = \left(\dfrac{\partial^2 z}{\partial x \partial y}\right)^2 - \left(\dfrac{\partial^2 z}{\partial x^2}\right)\left(\dfrac{\partial^2 z}{\partial y^2}\right)$

$$= 36 - (0)(0) = 36$$

$$\Delta_{(2, 2)} = 36 - (12)(12) = -108$$

(viii) Since $\Delta_{(0, 0)} > 0$ then **(0, 0) is a saddle point**.

Since $\Delta_{(2, 2)} < 0$ and $\dfrac{\partial^2 z}{\partial x^2} > 0$, then **(2, 2) is a minimum point**.

Now try the following Practice Exercise

> **Practice Exercise 258   Further problems on maxima, minima and saddle points for functions of two variables (answers on page 1180)**
>
> 1. Find the stationary point of the surface $f(x, y) = x^2 + y^2$ and determine its nature. Sketch the contour map represented by z
>
> 2. Find the maxima, minima and saddle points for the following functions:
>
>   (a) $f(x, y) = x^2 + y^2 - 2x + 4y + 8$
>   (b) $f(x, y) = x^2 - y^2 - 2x + 4y + 8$
>   (c) $f(x, y) = 2x + 2y - 2xy - 2x^2 - y^2 + 4$
>
> 3. Determine the stationary values of the function $f(x, y) = x^3 - 6x^2 - 8y^2$ and distinguish between them.
>
> 4. Locate the stationary point of the function $z = 12x^2 + 6xy + 15y^2$
>
> 5. Find the stationary points of the surface $z = x^3 - xy + y^3$ and distinguish between them.

62.5 Further worked problems on maxima, minima and saddle points for functions of two variables

Problem 3. Find the co-ordinates of the stationary points on the surface

$$z = (x^2 + y^2)^2 - 8(x^2 - y^2)$$

and distinguish between them. Sketch the approximate contour map associated with z

Following the procedure:

(i) $\dfrac{\partial z}{\partial x} = 2(x^2 + y^2)2x - 16x$ and

$\dfrac{\partial z}{\partial y} = 2(x^2 + y^2)2y + 16y$

(ii) for stationary points,

$$2(x^2 + y^2)2x - 16x = 0$$

i.e. $\quad 4x^3 + 4xy^2 - 16x = 0 \qquad (1)$

and $\quad 2(x^2 + y^2)2y + 16y = 0$

i.e. $\qquad 4y(x^2 + y^2 + 4) = 0 \qquad (2)$

(iii) From equation (1), $y^2 = \dfrac{16x - 4x^3}{4x} = 4 - x^2$

Substituting $y^2 = 4 - x^2$ in equation (2) gives

$$4y(x^2 + 4 - x^2 + 4) = 0$$

i.e. $32y = 0$ and $y = 0$

When $y = 0$ in equation (1), $\quad 4x^3 - 16x = 0$

i.e. $\qquad\qquad\qquad\qquad\qquad 4x(x^2 - 4) = 0$

from which, $x = 0$ or $x = \pm 2$

The co-ordinates of the stationary points are (0, 0), (2, 0) and (−2, 0)

(iv) $\dfrac{\partial^2 z}{\partial x^2} = 12x^2 + 4y^2 - 16,$

$\dfrac{\partial^2 z}{\partial y^2} = 4x^2 + 12y^2 + 16$ and $\dfrac{\partial^2 z}{\partial x \partial y} = 8xy$

(v) For the point (0, 0),

$\dfrac{\partial^2 z}{\partial x^2} = -16, \dfrac{\partial^2 z}{\partial y^2} = 16$ and $\dfrac{\partial^2 z}{\partial x \partial y} = 0$

For the point (2, 0),

$\dfrac{\partial^2 z}{\partial x^2} = 32, \dfrac{\partial^2 z}{\partial y^2} = 32$ and $\dfrac{\partial^2 z}{\partial x \partial y} = 0$

For the point (−2, 0),

$\dfrac{\partial^2 z}{\partial x^2} = 32, \dfrac{\partial^2 z}{\partial y^2} = 32$ and $\dfrac{\partial^2 z}{\partial x \partial y} = 0$

(vi) $\left(\dfrac{\partial^2 z}{\partial x \partial y}\right)^2 = 0$ for each stationary point

(vii) $\Delta_{(0, 0)} = (0)^2 - (-16)(16) = 256$

$\Delta_{(2, 0)} = (0)^2 - (32)(32) = -1024$

$\Delta_{(-2, 0)} = (0)^2 - (32)(32) = -1024$

(viii) Since $\Delta_{(0, 0)} > 0$, **the point (0, 0) is a saddle point**.

Since $\Delta_{(0, 0)} < 0$ and $\left(\dfrac{\partial^2 z}{\partial x^2}\right)_{(2,0)} > 0$, **the point (2, 0) is a minimum point**.

Since $\Delta_{(-2, 0)} < 0$ and $\left(\dfrac{\partial^2 z}{\partial x^2}\right)_{(-2,0)} > 0$, **the point (−2, 0) is a minimum point**.

Looking down towards the x–y plane from above, an approximate contour map can be constructed to represent the value of z. Such a map is shown in Figure 62.9. To produce a contour map requires a large number of x–y co-ordinates to be chosen and the values of z at each co-ordinate calculated. Here are a few examples of points used to construct the contour map.

When $z = 0$, $\quad 0 = (x^2 + y^2)^2 - 8(x^2 - y^2)$

In addition, when, say, $y = 0$ (i.e. on the x-axis)

$$0 = x^4 - 8x^2, \text{ i.e. } x^2(x^2 - 8) = 0$$

from which, $\quad x = 0$ or $x = \pm\sqrt{8}$

Hence the contour $z = 0$ crosses the x-axis at 0 and $\pm\sqrt{8}$, i.e. at co-ordinates (0, 0), (2.83, 0) and (−2.83, 0), shown as points, S, a and b, respectively.

When $z = 0$ and $x = 2$ then

$$0 = (4 + y^2)^2 - 8(4 - y^2)$$

i.e. $0 = 16 + 8y^2 + y^4 - 32 + 8y^2$

i.e. $0 = y^4 + 16y^2 - 16$

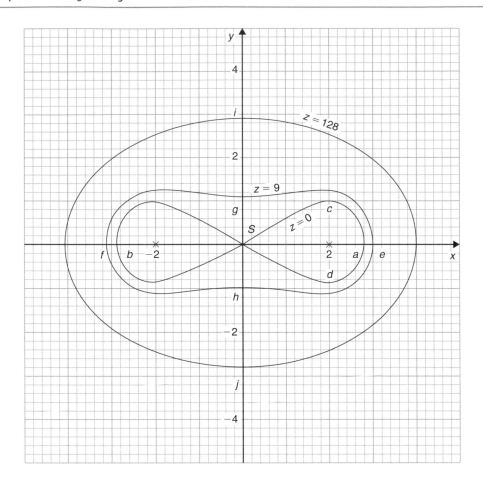

Figure 62.9

Let $y^2 = p$, then $p^2 + 16p - 16 = 0$ and

$$p = \frac{-16 \pm \sqrt{16^2 - 4(1)(-16)}}{2}$$

$$= \frac{-16 \pm 17.89}{2}$$

$$= 0.945 \text{ or } -16.945$$

Hence $y = \sqrt{p} = \sqrt{(0.945)}$ or $\sqrt{(-16.945)}$

$$= \pm 0.97 \text{ or complex roots.}$$

Hence the $z = 0$ contour passes through the co-ordinates $(2, 0.97)$ and $(2, -0.97)$ shown as c and d in Figure 62.9.

Similarly, for the **$z = 9$** contour, when $y = 0$,

$$9 = (x^2 + 0^2)^2 - 8(x^2 - 0^2)$$

i.e. $\quad 9 = x^4 - 8x^2$

i.e. $\quad x^4 - 8x^2 - 9 = 0$

Hence $(x^2 - 9)(x^2 + 1) = 0$

from which, $x = \pm 3$ or complex roots.

Thus the $z = 9$ contour passes through $(3, 0)$ and $(-3, 0)$, shown as e and f in Figure 62.9.

If $z = 9$ and $x = 0$, $\quad 9 = y^4 + 8y^2$

i.e. $\quad\quad y^4 + 8y^2 - 9 \quad = \quad 0$

i.e. $\quad (y^2 + 9)(y^2 - 1) \quad = \quad 0$

from which, $y = \pm 1$ or complex roots.

Thus the $z = 9$ contour also passes through $(0, 1)$ and $(0, -1)$, shown as g and h in Figure 62.9.

When, say, $x = 4$ and $y = 0$,

$$z = (4^2)^2 - 8(4^2) = 128$$

When $z = 128$ and $x = 0$, $128 = y^4 + 8y^2$

i.e. $\quad y^4 + 8y^2 - 128 = 0$

i.e. $\quad (y^2 + 16)(y^2 - 8) = 0$

from which, $y = \pm\sqrt{8}$ or complex roots.

Thus the $z = 128$ contour passes through (0, 2.83) and (0, −2.83), shown as i and j in Figure 62.9.

In a similar manner many other points may be calculated with the resulting approximate contour map shown in Figure 62.9. It is seen that two 'hollows' occur at the minimum points, and a 'cross-over' occurs at the saddle point S, which is typical of such contour maps.

> **Problem 4.** Show that the function
>
> $$f(x, y) = x^3 - 3x^2 - 4y^2 + 2$$
>
> has one saddle point and one maximum point. Determine the maximum value

Let $z = f(x, y) = x^3 - 3x^2 - 4y^2 + 2$

Following the procedure:

(i) $\dfrac{\partial z}{\partial x} = 3x^2 - 6x$ and $\dfrac{\partial z}{\partial y} = -8y$

(ii) for stationary points, $3x^2 - 6x = 0$ (1)

     and $-8y = 0$ (2)

(iii) From equation (1), $3x(x-2) = 0$ from which, $x = 0$ and $x = 2$

     From equation (2), $y = 0$

     Hence the stationary points are (0, 0) and (2, 0)

(iv) $\dfrac{\partial^2 z}{\partial x^2} = 6x - 6$, $\dfrac{\partial^2 z}{\partial y^2} = -8$ and $\dfrac{\partial^2 z}{\partial x \partial y} = 0$

(v) For the point $(0, 0)$

$$\frac{\partial^2 z}{\partial x^2} = -6, \frac{\partial^2 z}{\partial y^2} = -8 \text{ and } \frac{\partial^2 z}{\partial x \partial y} = 0$$

     For the point (2, 0)

$$\frac{\partial^2 z}{\partial x^2} = 6, \frac{\partial^2 z}{\partial y^2} = -8 \text{ and } \frac{\partial^2 z}{\partial x \partial y} = 0$$

(vi) $\left(\dfrac{\partial^2 z}{\partial x \partial y}\right)^2 = (0)^2 = 0$

(vii) $\Delta_{(0, 0)} = 0 - (-6)(-8) = -48$
$\Delta_{(2, 0)} = 0 - (6)(-8) = 48$

(viii) Since $\Delta_{(0, 0)} < 0$ and $\left(\dfrac{\partial^2 z}{\partial x^2}\right)_{(0, 0)} < 0$, **the point (0, 0) is a maximum point** and hence **the maximum value is 0**

Since $\Delta_{(2, 0)} > 0$, **the point (2, 0) is a saddle point**.

The value of z at the saddle point is $2^3 - 3(2)^2 - 4(0)^2 + 2 = -2$

An approximate contour map representing the surface $f(x, y)$ is shown in Figure 62.10 where a 'hollow effect' is seen surrounding the maximum point and a 'cross-over' occurs at the saddle point S.

> **Problem 5.** An open rectangular container is to have a volume of $62.5\,m^3$. Determine the least surface area of material required

Let the dimensions of the container be x, y and z as shown in Figure 62.11

Volume, $V = xyz = 62.5$ (1)

Surface area, $S = xy + 2yz + 2xz$ (2)

From equation (1), $z = \dfrac{62.5}{xy}$

Substituting in equation (2) gives:

$$S = xy + 2y\left(\frac{62.5}{xy}\right) + 2x\left(\frac{62.5}{xy}\right)$$

i.e. $S = xy + \dfrac{125}{x} + \dfrac{125}{y}$

which is a function of two variables

$$\frac{\partial S}{\partial x} = y - \frac{125}{x^2} = 0 \text{ for a stationary point,}$$

hence $x^2 y = 125$ (3)

$$\frac{\partial S}{\partial y} = x - \frac{125}{y^2} = 0 \text{ for a stationary point,}$$

hence $xy^2 = 125$ (4)

Dividing equation (3) by (4) gives:

$$\frac{x^2 y}{xy^2} = 1, \text{ i.e. } \frac{x}{y} = 1, \text{ i.e. } x = y$$

Substituting $y = x$ in equation (3) gives $x^3 = 125$, from which, $x = 5\,m$.

Hence $y = 5\,m$ also.

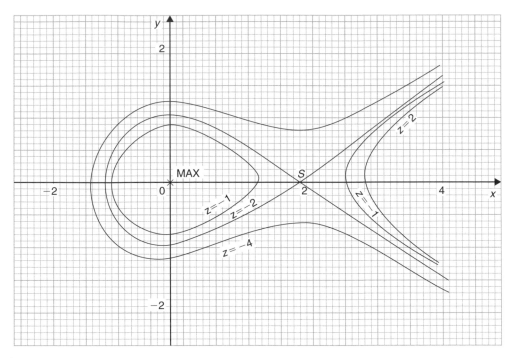

Figure 62.10

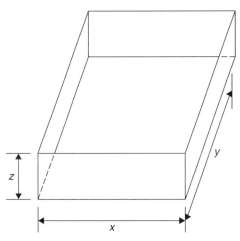

Figure 62.11

From equation (1), $(5)(5)z = 62.5$

from which, $z = \dfrac{62.5}{25} = 2.5\,\text{m}$

$$\frac{\partial^2 S}{\partial x^2} = \frac{250}{x^3},\ \frac{\partial^2 S}{\partial y^2} = \frac{250}{y^3}\ \text{and}\ \frac{\partial^2 S}{\partial x\,\partial y} = 1$$

When $x = y = 5$, $\dfrac{\partial^2 S}{\partial x^2} = 2$, $\dfrac{\partial^2 S}{\partial y^2} = 2$ and $\dfrac{\partial^2 S}{\partial x\,\partial y} = 1$

$$\Delta = (1)^2 - (2)(2) = -3$$

Since $\Delta < 0$ and $\dfrac{\partial^2 S}{\partial x^2} > 0$, then the surface area S is a **minimum**.

Hence the minimum dimensions of the container to have a volume of $62.5\,\text{m}^3$ are **5 m by 5 m by 2.5 m**.

From equation (2), **minimum surface area**, S
$$= (5)(5) + 2(5)(2.5) + 2(5)(2.5)$$

$$= 75\,\text{m}^2$$

Now try the following Practice Exercise

Practice Exercise 259   Further problems on maxima, minima and saddle points for functions of two variables (answers on page 1181)

1. The function $z = x^2 + y^2 + xy + 4x - 4y + 3$ has one stationary value. Determine its co-ordinates and its nature.

2. An open rectangular container is to have a volume of $32\,\text{m}^3$. Determine the dimensions and the total surface area such that the total surface area is a minimum.

3. Determine the stationary values of the function

$$f(x, y) = x^4 + 4x^2y^2 - 2x^2 + 2y^2 - 1$$

and distinguish between them.

4. Determine the stationary points of the surface
    $f(x, y) = x^3 - 6x^2 - y^2$

5. Locate the stationary points on the surface

$$f(x, y) = 2x^3 + 2y^3 - 6x - 24y + 16$$

and determine their nature.

6. A large marquee is to be made in the form of a rectangular box-like shape with canvas covering on the top, back and sides. Determine the minimum surface area of canvas necessary if the volume of the marquee is $250\,m^3$.

For fully worked solutions to each of the problems in Practice Exercises 258 and 259 in this chapter, go to the website:
www.routledge.com/cw/bird

This assignment covers the material contained in Chapters 58 to 62. *The marks available are shown in brackets at the end of each question.*

1. Differentiate the following functions with respect to x:

    (a) $5 \ln (\text{sh} x)$ (b) $3 \text{ch}^3 2x$

    (c) $e^{2x} \text{sech} 2x$ $\qquad$ (7)

2. Differentiate the following functions with respect to the variable:

    (a) $y = \dfrac{1}{5} \cos^{-1} \dfrac{x}{2}$

    (b) $y = 3e^{\sin^{-1} t}$

    (c) $y = \dfrac{2 \sec^{-1} 5x}{x}$

    (d) $y = 3 \sinh^{-1} \sqrt{(2x^2 - 1)}$ $\qquad$ (14)

3. Evaluate the following, each correct to 3 decimal places:

    (a) $\sinh^{-1} 3$ (b) $\cosh^{-1} 2.5$ (c) $\tanh^{-1} 0.8$ $\qquad$ (6)

4. If $z = f(x, y)$ and $z = x \cos(x + y)$ determine

    $$\frac{\partial z}{\partial x}, \frac{\partial z}{\partial y}, \frac{\partial^2 z}{\partial x^2}, \frac{\partial^2 z}{\partial y^2}, \frac{\partial^2 z}{\partial x \partial y} \text{ and } \frac{\partial^2 z}{\partial y \partial x} \qquad (12)$$

5. The magnetic field vector H due to a steady current I flowing around a circular wire of radius r and at a distance x from its centre is given by

    $$H = \pm \frac{I}{2} \frac{\partial}{\partial x} \left(\frac{x}{\sqrt{r^2 + x^2}} \right)$$

    Show that $\quad H = \pm \dfrac{r^2 I}{2 \sqrt{(r^2 + x^2)^3}}$ $\qquad$ (7)

6. If $x y z = c$, where c is constant, show that

    $$dz = -z \left(\frac{dx}{x} + \frac{dy}{y} \right) \qquad (6)$$

7. An engineering function $z = f(x, y)$ and $z = e^{\frac{y}{2}} \ln(2x + 3y)$. Determine the rate of increase of z, correct to 4 significant figures, when $x = 2$ cm, $y = 3$ cm, x is increasing at 5 cm/s and y is increasing at 4 cm/s. (8)

8. The volume V of a liquid of viscosity coefficient η delivered after time t when passed through a tube of length L and diameter d by a pressure p is given by $V = \dfrac{p d^4 t}{128 \eta L}$. If the errors in V, p and L are 1%, 2% and 3% respectively, determine the error in η (8)

9. Determine and distinguish between the stationary values of the function

    $$f(x, y) = x^3 - 6x^2 - 8y^2$$

    and sketch an approximate contour map to represent the surface $f(x, y)$ (20)

10. An open, rectangular fish tank is to have a volume of 13.5 m^3. Determine the least surface area of glass required. (12)

For lecturers/instructors/teachers, fully worked solutions to each of the problems in Revision Test 23, together with a full marking scheme, are available at the website:
www.routledge.com/cw/bird

Multiple choice questions Test 7
Differentiation and its applications
This test covers the material in Chapters 52 to 62

All questions have only one correct answer (answers on page 1200).

1. Differentiating $y = 4x^5$ gives:

   (a) $\dfrac{dy}{dx} = \dfrac{2}{3}x^6$

   (b) $\dfrac{dy}{dx} = 20x^4$

   (c) $\dfrac{dy}{dx} = 4x^6$

   (d) $\dfrac{dy}{dx} = 5x^4$

2. The gradient of the curve $y = -2x^3 + 3x + 5$ at $x = 2$ is:

   (a) -21 (b) 27 (c) -16 (d) -5

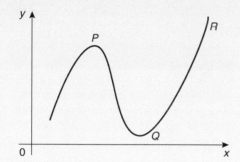

Figure M7.1

3. For the curve shown in Figure M7.1, which of the following statements is incorrect?

   (a) P is a turning point

   (b) Q is a minimum point

   (c) R is a maximum value

   (d) Q is a stationary value

4. If $y = 5\sqrt{x^3} - 2$, $\dfrac{dy}{dx}$ is equal to:

   (a) $\dfrac{15}{2}\sqrt{x}$

   (b) $2\sqrt{x^5} - 2x + c$

   (c) $\dfrac{5}{2}\sqrt{x} - 2$

   (d) $5\sqrt{x} - 2x$

5. An alternating current is given by $i = 4\sin 150t$ amperes, where t is the time in seconds. The rate of change of current at $t = 0.025$ s is:

   (a) 3.99 A/s   (b) -492.3 A/s

   (c) -3.28 A/s   (d) 598.7 A/s

6. The resistance to motion F of a moving vehicle is given by $F = \dfrac{5}{x} + 100x$. The minimum value of resistance is:

   (a) -44.72 (b) 0.2236 (c) 44.72 (d) -0.2236

7. Differentiating $i = 3\sin 2t - 2\cos 3t$ with respect to t gives:

   (a) $3\cos 2t + 2\sin 3t$

   (b) $6(\sin 2t - \cos 3t)$

   (c) $\dfrac{3}{2}\cos 2t + \dfrac{2}{3}\sin 3t$

   (d) $6(\cos 2t + \sin 3t)$

8. Given $y = 3e^x + 2\ln 3x$, $\dfrac{dy}{dx}$ is equal to:

   (a) $6e^x + \dfrac{2}{3x}$

   (b) $3e^x + \dfrac{2}{x}$

   (c) $6e^x + \dfrac{2}{x}$

   (d) $3e^x + \dfrac{2}{3}$

9. The vertical displacement, s, of a prototype model in a tank is given by $s = 40\sin(0.1t)$ mm, where t is the time in seconds. The vertical velocity of the model, in mm/s, is:

   (a) $-\cos 0.1t$ (b) $400\cos 0.1t$

   (c) $-400\cos 0.1t$ (d) $4\cos 0.1t$

10. The equation of a curve is $y = 2x^3 - 6x + 1$. The maximum value of the curve is:

    (a) -3 (b) 1 (c) 5 (d) -6

11. Given $f(t) = 3t^4 - 2$, $f'(t)$ is equal to:

    (a) $12t^3 - 2$

    (b) $\frac{3}{4}t^5 - 2t + c$

    (c) $12t^3$

    (d) $3t^5 - 2$

12. The current i in a circuit at time t seconds is given by $i = 0.20(1 - e^{-20t})$ A. When time $t = 0.1\,s$, the rate of change of current is:

    (a) -1.022 A/s (b) 0.541 A/s

    (c) 0.173 A/s (d) 0.373 A/s

13. The gradient of the curve $y = 4x^2 - 7x + 3$ at the point $(1, 0)$ is

    (a) 3 (b) 1 (c) 0 (d) -7

14. The velocity of a car (in m/s) is related to time t seconds by the equation $v = 4.5 + 18t - 4.5t^2$. The maximum speed of the car, in km/h, is:

    (a) 81 (b) 6.25 (c) 22.5 (d) 77

15. An alternating voltage is given by $v = 10\sin 300t$ volts, where t is the time in seconds. The rate of change of voltage when $t = 0.01\,s$ is:

    (a) -2996 V/s (b) 157 V/s

    (c) -2970 V/s (d) 0.523 V/s

16. If $f(t) = 5t - \dfrac{1}{\sqrt{t}}$ $f'(t)$ is equal to:

    (a) $5 + \dfrac{1}{2\sqrt{t^3}}$

    (b) $5 - 2\sqrt{t}$

    (c) $\dfrac{5t^2}{2} - 2\sqrt{t} + c$

    (d) $5 + \dfrac{1}{\sqrt{t^3}}$

17. The equation of a curve is $y = 2x^3 - 6x + 1$. The minimum value of the curve is:

    (a) -6 (b) 1 (c) 5 (d) -3

18. The length ℓ metres of a certain metal rod at temperature $t°C$ is given by $\ell = 1 + 4 \times 10^{-5}t + 4 \times 10^{-7}t^2$. The rate of change of length, in mm/°C, when the temperature is $400°C$, is:

    (a) 3.6×10^{-4} (b) 1.00036

    (c) 0.36 (d) 3.2×10^{-4}

19. If $y = 3x^2 - \ln 5x$ then $\dfrac{d^2 y}{dx^2}$ is equal to:

    (a) $6 + \dfrac{1}{5x^2}$ (b) $6x - \dfrac{1}{x}$ (c) $6 - \dfrac{1}{5x}$ (d) $6 + \dfrac{1}{x^2}$

20. The turning point on the curve $y = x^2 - 4x$ is at:

    (a) $(2, 0)$ (b) $(0, 4)$ (c) $(-2, 12)$ (d) $(2, -4)$

21. An alternating current, i amperes, is given by $i = 100\sin 2\pi ft$ amperes, where f is the frequency in hertz and t is the time in seconds. The rate of change of current when $t = 12$ ms and $f = 50$ Hz is:

    (a) $31\,348$ A/s (b) -58.78 A/s

    (c) 627.0 A/s (d) -25416 A/s

22. If $f(t) = e^{2t}\ln 2t$, $f'(t)$ is equal to:

    (a) $\dfrac{2e^{2t}}{t}$ (b) $e^{2t}\left(\dfrac{1}{t} + 2\ln 2t\right)$

    (c) $\dfrac{e^{2t}}{2t}$ (d) $\dfrac{e^{2t}}{2t} + 2e^{2t}\ln 2t$

23. Given $x = 3t - 1$ and $y = 3t(t - 1)$ then:

    (a) $\dfrac{dy}{dx} = 2t - 1$

    (b) $\dfrac{d^2 y}{dx^2} = 2$

    (c) $\dfrac{d^2 y}{dx^2} = \dfrac{1}{2}$

    (d) $\dfrac{dy}{dx} = \dfrac{1}{2t - 1}$

24. $\dfrac{d}{dx}\left(3x^2 y^5\right)$ is equal to:

    (a) $6xy^5$

    (b) $3xy^4\left(2y\dfrac{dx}{dy} + 5x\right)$

    (c) $15x^2 y^4 + 6xy^5$

    (d) $3xy^4\left(5x\dfrac{dy}{dx} + 2y\right)$

25. If $y = 3x^{2x}$ then $\dfrac{dy}{dx}$ is equal to:

    (a) $6x^{2x}(1 + \ln x)$

    (b) $\ln 3 + 2x\ln x$

    (c) $2x(3x)^x$

    (d) $3x^{2x}(2x\ln 3x)$

Formulae/revision hints for Section I
Differential calculus

Standard derivatives

y or $f(x)$	$\dfrac{dy}{dx}$ or $f'(x)$
ax^n	anx^{n-1}
$\sin ax$	$a \cos ax$
$\cos ax$	$-a \sin ax$
$\tan ax$	$a \sec^2 ax$
$\sec ax$	$a \sec ax \tan ax$
$\operatorname{cosec} ax$	$-a \operatorname{cosec} ax \cot ax$
$\cot ax$	$-a \operatorname{cosec}^2 ax$
e^{ax}	ae^{ax}
$\ln ax$	$\dfrac{1}{x}$
$\sinh ax$	$a \cosh ax$
$\cosh ax$	$a \sinh ax$
$\tanh ax$	$a \operatorname{sech}^2 ax$
$\operatorname{sech} ax$	$-a \operatorname{sech} ax \tanh ax$
$\operatorname{cosech} ax$	$-a \operatorname{cosech} ax \coth ax$
$\coth ax$	$-a \operatorname{cosech}^2 ax$
$\sin^{-1} \dfrac{x}{a}$	$\dfrac{1}{\sqrt{a^2 - x^2}}$
$\sin^{-1} f(x)$	$\dfrac{f'(x)}{\sqrt{1 - [f(x)]^2}}$
$\cos^{-1} \dfrac{x}{a}$	$\dfrac{-1}{\sqrt{a^2 - x^2}}$
$\cos^{-1} f(x)$	$\dfrac{-f'(x)}{\sqrt{1 - [f(x)]^2}}$
$\tan^{-1} \dfrac{x}{a}$	$\dfrac{a}{a^2 + x^2}$
$\tan^{-1} f(x)$	$\dfrac{f'(x)}{1 + [f(x)]^2}$
$\sec^{-1} \dfrac{x}{a}$	$\dfrac{a}{x\sqrt{x^2 - a^2}}$
$\sec^{-1} f(x)$	$\dfrac{f'(x)}{f(x)\sqrt{[f(x)]^2 - 1}}$
$\operatorname{cosec}^{-1} \dfrac{x}{a}$	$\dfrac{-a}{x\sqrt{x^2 - a^2}}$
$\operatorname{cosec}^{-1} f(x)$	$\dfrac{-f'(x)}{f(x)\sqrt{[f(x)]^2 - 1}}$
$\cot^{-1} \dfrac{x}{a}$	$\dfrac{-a}{a^2 + x^2}$
$\cot^{-1} f(x)$	$\dfrac{-f'(x)}{1 + [f(x)]^2}$
$\sinh^{-1} \dfrac{x}{a}$	$\dfrac{1}{\sqrt{x^2 + a^2}}$
$\sinh^{-1} f(x)$	$\dfrac{f'(x)}{\sqrt{[f(x)]^2 + 1}}$
$\cosh^{-1} \dfrac{x}{a}$	$\dfrac{1}{\sqrt{x^2 - a^2}}$
$\cosh^{-1} f(x)$	$\dfrac{f'(x)}{\sqrt{[f(x)]^2 - 1}}$
$\tanh^{-1} \dfrac{x}{a}$	$\dfrac{a}{a^2 - x^2}$
$\tanh^{-1} f(x)$	$\dfrac{f'(x)}{1 - [f(x)]^2}$
$\operatorname{sech}^{-1} \dfrac{x}{a}$	$\dfrac{-a}{x\sqrt{a^2 - x^2}}$
$\operatorname{sech}^{-1} f(x)$	$\dfrac{-f'(x)}{f(x)\sqrt{1 - [f(x)]^2}}$
$\operatorname{cosech}^{-1} \dfrac{x}{a}$	$\dfrac{-a}{x\sqrt{x^2 + a^2}}$
$\operatorname{cosech}^{-1} f(x)$	$\dfrac{-f'(x)}{f(x)\sqrt{[f(x)]^2 + 1}}$
$\coth^{-1} \dfrac{x}{a}$	$\dfrac{a}{a^2 - x^2}$
$\coth^{-1} f(x)$	$\dfrac{f'(x)}{1 - [f(x)]^2}$

Product rule: When $y = uv$ and u and v are functions of x then:

$$\frac{dy}{dx} = u\frac{dv}{dx} + v\frac{du}{dx}$$

Quotient rule: When $y = \dfrac{u}{v}$ and u and v are functions of x then:

$$\frac{dy}{dx} = \frac{v\dfrac{du}{dx} - u\dfrac{dv}{dx}}{v^2}$$

Function of a function

If u is a function of x then: $\dfrac{dy}{dx} = \dfrac{dy}{du} \times \dfrac{du}{dx}$

Parametric differentiation

If x and y are both functions of θ, then:

$$\frac{dy}{dx} = \frac{\dfrac{dy}{d\theta}}{\dfrac{dx}{d\theta}} \quad \text{and} \quad \frac{d^2y}{dx^2} = \frac{\dfrac{d}{d\theta}\left(\dfrac{dy}{dx}\right)}{\dfrac{dx}{d\theta}}$$

Implicit function

$$\frac{d}{dx}[f(y)] = \frac{d}{dy}[f(y)] \times \frac{dy}{dx}$$

Maximum and minimum values

If $y = f(x)$ then $\dfrac{dy}{dx} = 0$ for stationary points.

Let a solution of $\dfrac{dy}{dx} = 0$ be $x = a$; if the value of $\dfrac{d^2y}{dx^2}$ when $x = a$ is:

positive, the point is a *minimum*; *negative*, the point is a *maximum*.

Points of inflexion

(i) Given $y = f(x)$, determine $\dfrac{dy}{dx}$ and $\dfrac{d^2y}{dx^2}$

(ii) Solve the equation $\dfrac{d^2y}{dx^2} = 0$

(iii) Test whether there is a change of sign occurring in $\dfrac{d^2y}{dx^2}$. This is achieved by substituting into the expression for $\dfrac{d^2y}{dx^2}$ firstly a value of x just less than the solution and then a value just greater than the solution.

(iv) A point of inflexion has been found if $\dfrac{d^2y}{dx^2} = 0$ **and** there is a change of sign.

Velocity and acceleration: If distance $x = f(t)$, then velocity $v = f'(t)$ or $\dfrac{dx}{dt}$ and acceleration $a = f''(t)$ or $\dfrac{d^2x}{dt^2}$

Tangents and normals

Equation of tangent to curve $y = f(x)$ at the point (x_1, y_1) is:

$$y - y_1 = m(x - x_1) \quad \text{where } m = \text{gradient of curve at } (x_1, y_1)$$

Equation of normal to curve $y = f(x)$ at the point (x_1, y_1) is:

$$y - y_1 = -\frac{1}{m}(x - x_1)$$

Partial differentiation

Total differential: If $z = f(u, v, ..)$, then the total differential,

$$dz = \frac{\partial z}{\partial u}du + \frac{\partial z}{\partial v}dv + \dots.$$

Rate of change: If $z = f(u, v, ..)$ and $\dfrac{du}{dt}, \dfrac{dv}{dt}, \dots$ denote the rate of change of u, v, $\dots$ respectively, then the rate of change of z,

$$\frac{dz}{dt} = \frac{\partial z}{\partial u}\cdot\frac{du}{dt} + \frac{\partial z}{\partial v}\cdot\frac{dv}{dt} + \dots$$

Small changes: If $z = f(u, v, \dots)$ and δx, δy, $\dots$ denote small changes in x, y, $\dots$ respectively, then the corresponding change,

$$\delta z \approx \frac{\partial z}{\partial x}\delta x + \frac{\partial z}{\partial y}\delta y + \dots$$

To determine maxima, minima and saddle points for functions of two variables

Given $z = f(x, y)$,

(i) determine $\dfrac{\partial z}{\partial x}$ and $\dfrac{\partial z}{\partial y}$

(ii) for stationary points, $\dfrac{\partial z}{\partial x} = 0$ and $\dfrac{\partial z}{\partial y} = 0$

Section I

(iii)   solve the simultaneous equations $\dfrac{\partial z}{\partial x} = 0$ and $\dfrac{\partial z}{\partial y} = 0$ for x and y, which gives the co-ordinates of the stationary points

(iv)   determine $\dfrac{\partial^2 z}{\partial x^2}$, $\dfrac{\partial^2 z}{\partial y^2}$ and $\dfrac{\partial^2 z}{\partial x \partial y}$

(v)   for each of the co-ordinates of the stationary points, substitute values of x and y into $\dfrac{\partial^2 z}{\partial x^2}$, $\dfrac{\partial^2 z}{\partial y^2}$ and $\dfrac{\partial^2 z}{\partial x \partial y}$ and evaluate each

(vi)   evaluate $\left(\dfrac{\partial^2 z}{\partial x \partial y}\right)^2$ for each stationary point

(vii)   substitute the values of $\dfrac{\partial^2 z}{\partial x^2}$, $\dfrac{\partial^2 z}{\partial y^2}$ and $\dfrac{\partial^2 z}{\partial x \partial y}$ into the equation $\Delta = \left(\dfrac{\partial^2 z}{\partial x \partial y}\right)^2 - \left(\dfrac{\partial^2 z}{\partial x^2}\right)\left(\dfrac{\partial^2 z}{\partial y^2}\right)$ and evaluate

(viii)   (a) if $\Delta > \mathbf{0}$ then the stationary point is a **saddle point**

(b) if $\Delta < \mathbf{0}$ and $\dfrac{\partial^2 z}{\partial x^2} < \mathbf{0}$, then the stationary point is a **maximum point**

(c) if $\Delta < \mathbf{0}$ and $\dfrac{\partial^2 z}{\partial x^2} > \mathbf{0}$, then the stationary point is a **minimum point**.

For a copy of these formulae/revision hints, go to:
www.routledge.com/cw/bird

Section I

Integral calculus

Chapter 63

Standard integration

Why it is important to understand: Standard integration

Engineering is all about problem solving and many problems in engineering can be solved using calculus. Physicists, chemists, engineers, and many other scientific and technical specialists use calculus in their everyday work; it is a technique of fundamental importance. Both integration and differentiation have numerous applications in engineering and science and some typical examples include determining areas, mean and rms values, volumes of solids of revolution, centroids, second moments of area, differential equations and Fourier series. Besides the standard integrals covered in this chapter, there are a number of other methods of integration covered in future chapters. For any further studies in engineering, differential and integral calculus are unavoidable.

At the end of this chapter, you should be able to:

- understand that integration is the reverse process of differentiation
- determine integrals of the form ax^n where n is fractional, zero or a positive or negative integer
- integrate standard functions – $\cos ax$, $\sin ax$, $\sec^2 ax$, $\operatorname{cosec}^2 ax$, $\operatorname{cosec} ax$, $\cot ax$, $\sec ax$, $\tan ax$, e^{ax}, $\frac{1}{x}$
- evaluate definite integrals

63.1 The process of integration

The process of integration reverses the process of differentiation. In differentiation, if $f(x) = 2x^2$ then $f'(x) = 4x$. Thus the integral of $4x$ is $2x^2$, i.e. integration is the process of moving from $f'(x)$ to $f(x)$. By similar reasoning, the integral of $2t$ is t^2

Integration is a process of summation or adding parts together and an elongated S, shown as $\int$, is used to replace the words 'the integral of'. Hence, from above, $\int 4x = 2x^2$ and $\int 2t$ is t^2

In differentiation, the differential coefficient $\dfrac{dy}{dx}$ indicates that a function of x is being differentiated with respect to x, the dx indicating that it is 'with respect to x'. In integration the variable of integration is shown by adding d (the variable) after the function to be integrated.

Thus $\displaystyle\int 4x\,dx$ means 'the integral of $4x$ with respect to x'

and $\displaystyle\int 2t\,dt$ means 'the integral of $2t$ with respect to t'

As stated above, the differential coefficient of $2x^2$ is $4x$, hence $\int 4x\,dx = 2x^2$. However, the differential coefficient of $2x^2 + 7$ is also $4x$. Hence $\int 4x\,dx$ is also equal to $2x^2 + 7$. To allow for the possible presence of a constant, whenever the process of integration is performed, a constant 'c' is added to the result.

Thus $\displaystyle\int 4x\,dx = 2x^2 + c$ and $\displaystyle\int 2t\,dt = t^2 + c$

'c' is called the **arbitrary constant of integration**.

63.2 The general solution of integrals of the form ax^n

The general solution of integrals of the form $\int ax^n\,dx$, where a and n are constants, is given by:

$$\int ax^n\,dx = \frac{ax^{n+1}}{n+1} + c$$

This rule is true when n is fractional, zero or a positive or negative integer, with the exception of $n=-1$

Using this rule gives:

(i) $\int 3x^4\,dx = \frac{3x^{4+1}}{4+1} + c = \frac{3}{5}x^5 + c$

(ii) $\int \frac{2}{x^2}\,dx = \int 2x^{-2}\,dx = \frac{2x^{-2+1}}{-2+1} + c$

$= \frac{2x^{-1}}{-1} + c = \frac{-2}{x} + c$, and

(iii) $\int \sqrt{x}\,dx = \int x^{\frac{1}{2}}\,dx = \frac{x^{\frac{1}{2}+1}}{\frac{1}{2}+1} + c = \frac{x^{\frac{3}{2}}}{\frac{3}{2}} + c$

$= \frac{2}{3}\sqrt{x^3} + c$

Each of these three results may be checked by differentiation.

(a) The integral of a constant k is $kx+c$. For example,

$$\int 8\,dx = 8x + c$$

(b) When a sum of several terms is integrated the result is the sum of the integrals of the separate terms. For example,

$$\int (3x + 2x^2 - 5)\,dx$$

$$= \int 3x\,dx + \int 2x^2\,dx - \int 5\,dx$$

$$= \frac{3x^2}{2} + \frac{2x^3}{3} - 5x + c$$

63.3 Standard integrals

Since integration is the reverse process of differentiation the **standard integrals** listed in Table 63.1 may be deduced and readily checked by differentiation.

Table 63.1 Standard integrals

(i) $\int ax^n\,dx = \frac{ax^{n+1}}{n+1} + c$

  (except when $n=-1$)

(ii) $\int \cos ax\,dx = \frac{1}{a}\sin ax + c$

(iii) $\int \sin ax\,dx = -\frac{1}{a}\cos ax + c$

(iv) $\int \sec^2 ax\,dx = \frac{1}{a}\tan ax + c$

(v) $\int \operatorname{cosec}^2 ax\,dx = -\frac{1}{a}\cot ax + c$

(vi) $\int \operatorname{cosec} ax\cot ax\,dx = -\frac{1}{a}\operatorname{cosec} ax + c$

(vii) $\int \sec ax\tan ax\,dx = \frac{1}{a}\sec ax + c$

(viii) $\int e^{ax}\,dx = \frac{1}{a}e^{ax} + c$

(ix) $\int \frac{1}{x}\,dx = \ln x + c$

Problem 1. Determine (a) $\int 5x^2\,dx$ (b) $\int 2t^3\,dt$

The standard integral, $\int ax^n\,dx = \frac{ax^{n+1}}{n+1} + c$

(a) When $a=5$ and $n=2$, then

$$\int 5x^2\,dx = \frac{5x^{2+1}}{2+1} + c = \frac{5x^3}{3} + c$$

(b) When $a=2$ and $n=3$, then

$$\int 2t^3\,dt = \frac{2t^{3+1}}{3+1} + c = \frac{2t^4}{4} + c = \frac{1}{2}t^4 + c$$

Each of these results may be checked by differentiating them.

Problem 2. Determine

$$\int \left(4 + \frac{3}{7}x - 6x^2\right) dx$$

$\int(4 + \frac{3}{7}x - 6x^2)\,dx$ may be written as $\int 4\,dx + \int \frac{3}{7}x\,dx - \int 6x^2\,dx$, i.e. each term is integrated separately. (This splitting up of terms only applies, however, for addition and subtraction.)

Hence $\displaystyle\int \left(4 + \frac{3}{7}x - 6x^2\right)dx$

$$= 4x + \left(\frac{3}{7}\right)\frac{x^{1+1}}{1+1} - (6)\frac{x^{2+1}}{2+1} + c$$

$$= 4x + \left(\frac{3}{7}\right)\frac{x^2}{2} - (6)\frac{x^3}{3} + c$$

$$= 4x + \frac{3}{14}x^2 - 2x^3 + c$$

Note that when an integral contains more than one term there is no need to have an arbitrary constant for each; just a single constant at the end is sufficient.

Problem 3. Determine

(a) $\displaystyle\int \frac{2x^3 - 3x}{4x}\,dx$ (b) $\displaystyle\int (1-t)^2\,dt$

(a) Rearranging into standard integral form gives:

$$\int \frac{2x^3 - 3x}{4x}\,dx$$

$$= \int \frac{2x^3}{4x} - \frac{3x}{4x}\,dx = \int \frac{x^2}{2} - \frac{3}{4}\,dx$$

$$= \left(\frac{1}{2}\right)\frac{x^{2+1}}{2+1} - \frac{3}{4}x + c$$

$$= \left(\frac{1}{2}\right)\frac{x^3}{3} - \frac{3}{4}x + c = \frac{1}{6}x^3 - \frac{3}{4}x + c$$

(b) Rearranging $\displaystyle\int (1-t)^2\,dt$ gives:

$$\int (1 - 2t + t^2)\,dt = t - \frac{2t^{1+1}}{1+1} + \frac{t^{2+1}}{2+1} + c$$

$$= t - \frac{2t^2}{2} + \frac{t^3}{3} + c$$

$$= t - t^2 + \frac{1}{3}t^3 + c$$

This problem shows that functions often have to be rearranged into the standard form of $\int ax^n\,dx$ before it is possible to integrate them.

Problem 4. Determine $\displaystyle\int \frac{3}{x^2}\,dx$

$\displaystyle\int \frac{3}{x^2}\,dx = \int 3x^{-2}\,dx$. Using the standard integral, $\int ax^n\,dx$ when $a = 3$ and $n = -2$ gives:

$$\int 3x^{-2}\,dx = \frac{3x^{-2+1}}{-2+1} + c = \frac{3x^{-1}}{-1} + c$$

$$= -3x^{-1} + c = \frac{-3}{x} + c$$

Problem 5. Determine $\int 3\sqrt{x}\,dx$

For fractional powers it is necessary to appreciate $\sqrt[n]{a^m} = a^{\frac{m}{n}}$

$$\int 3\sqrt{x}\,dx = \int 3x^{\frac{1}{2}}\,dx = \frac{3x^{\frac{1}{2}+1}}{\frac{1}{2}+1} + c$$

$$= \frac{3x^{\frac{3}{2}}}{\frac{3}{2}} + c = 2x^{\frac{3}{2}} + c = 2\sqrt{x^3} + c$$

Problem 6. Determine $\displaystyle\int \frac{-5}{9\sqrt[4]{t^3}}\,dt$

$$\int \frac{-5}{9\sqrt[4]{t^3}}\,dt = \int \frac{-5}{9t^{\frac{3}{4}}}\,dt = \int \left(-\frac{5}{9}\right)t^{-\frac{3}{4}}\,dt$$

$$= \left(-\frac{5}{9}\right)\frac{t^{-\frac{3}{4}+1}}{-\frac{3}{4}+1} + c$$

$$= \left(-\frac{5}{9}\right)\frac{t^{\frac{1}{4}}}{\frac{1}{4}} + c = \left(-\frac{5}{9}\right)\left(\frac{4}{1}\right)t^{\frac{1}{4}} + c$$

$$= -\frac{20}{9}\sqrt[4]{t} + c$$

Problem 7. Determine $\int \dfrac{(1+\theta)^2}{\sqrt{\theta}}\,d\theta$

$\int \dfrac{(1+\theta)^2}{\sqrt{\theta}}\,d\theta = \int \dfrac{(1+2\theta+\theta^2)}{\sqrt{\theta}}\,d\theta$

$= \int \left(\dfrac{1}{\theta^{\frac{1}{2}}} + \dfrac{2\theta}{\theta^{\frac{1}{2}}} + \dfrac{\theta^2}{\theta^{\frac{1}{2}}} \right) d\theta$

$= \int \left(\theta^{\frac{-1}{2}} + 2\theta^{1-\left(\frac{1}{2}\right)} + \theta^{2-\left(\frac{1}{2}\right)} \right) d\theta$

$= \int \left(\theta^{\frac{-1}{2}} + 2\theta^{\frac{1}{2}} + \theta^{\frac{3}{2}} \right) d\theta$

$= \dfrac{\theta^{\left(\frac{-1}{2}\right)+1}}{-\frac{1}{2}+1} + \dfrac{2\theta^{\left(\frac{1}{2}\right)+1}}{\frac{1}{2}+1} + \dfrac{\theta^{\left(\frac{3}{2}\right)+1}}{\frac{3}{2}+1} + c$

$= \dfrac{\theta^{\frac{1}{2}}}{\frac{1}{2}} + \dfrac{2\theta^{\frac{3}{2}}}{\frac{3}{2}} + \dfrac{\theta^{\frac{5}{2}}}{\frac{5}{2}} + c$

$= 2\theta^{\frac{1}{2}} + \dfrac{4}{3}\theta^{\frac{3}{2}} + \dfrac{2}{5}\theta^{\frac{5}{2}} + c$

$= \mathbf{2\sqrt{\theta} + \dfrac{4}{3}\sqrt{\theta^3} + \dfrac{2}{5}\sqrt{\theta^5} + c}$

Problem 8. Determine
(a) $\int 4\cos 3x\,dx$ (b) $\int 5\sin 2\theta\,d\theta$

(a) From Table 63.1(ii),

$\int 4\cos 3x\,dx = (4)\left(\dfrac{1}{3}\right)\sin 3x + c$

$= \dfrac{4}{3}\sin 3x + c$

(b) From Table 63.1(iii),

$\int 5\sin 2\theta\,d\theta = (5)\left(-\dfrac{1}{2}\right)\cos 2\theta + c$

$= -\dfrac{5}{2}\cos 2\theta + c$

Problem 9. Determine
(a) $\int 7\sec^2 4t\,dt$ (b) $3\int \operatorname{cosec}^2 2\theta\,d\theta$

(a) From Table 63.1(iv),

$\int 7\sec^2 4t\,dt = (7)\left(\dfrac{1}{4}\right)\tan 4t + c$

$= \dfrac{7}{4}\tan 4t + c$

(b) From Table 63.1(v),

$3\int \operatorname{cosec}^2 2\theta\,d\theta = (3)\left(-\dfrac{1}{2}\right)\cot 2\theta + c$

$= -\dfrac{3}{2}\cot 2\theta + c$

Problem 10. Determine
(a) $\int 5e^{3x}\,dx$ (b) $\int \dfrac{2}{3e^{4t}}\,dt$

(a) From Table 63.1(viii),

$\int 5e^{3x}\,dx = (5)\left(\dfrac{1}{3}\right)e^{3x} + c = \dfrac{5}{3}e^{3x} + c$

(b) $\int \dfrac{2}{3e^{4t}}\,dt = \int \dfrac{2}{3}e^{-4t}\,dt = \left(\dfrac{2}{3}\right)\left(-\dfrac{1}{4}\right)e^{-4t} + c$

$= -\dfrac{1}{6}e^{-4t} + c = -\dfrac{1}{6e^{4t}} + c$

Problem 11. Determine
(a) $\int \dfrac{3}{5x}\,dx$ (b) $\int \left(\dfrac{2m^2+1}{m}\right)dm$

(a) $\int \dfrac{3}{5x}\,dx = \int \left(\dfrac{3}{5}\right)\left(\dfrac{1}{x}\right)dx = \dfrac{3}{5}\ln x + c$

(from Table 63.1(ix))

(b) $\int \left(\dfrac{2m^2+1}{m}\right)dm = \int \left(\dfrac{2m^2}{m} + \dfrac{1}{m}\right)dm$

$= \int \left(2m + \dfrac{1}{m}\right)dm$

$= \dfrac{2m^2}{2} + \ln m + c$

$= m^2 + \ln m + c$

Now try the following Practice Exercise

Practice Exercise 260   Further problems on standard integrals (answers on page 1181)

In Problems 1 to 12, determine the indefinite integrals.

1. (a) $\int 4\,dx$ (b) $\int 7x\,dx$

2. (a) $\int \frac{2}{5}x^2\,dx$ (b) $\int \frac{5}{6}x^3\,dx$

3. (a) $\int \left(\frac{3x^2-5x}{x}\right)dx$ (b) $\int (2+\theta)^2\,d\theta$

4. (a) $\int \frac{4}{3x^2}\,dx$ (b) $\int \frac{3}{4x^4}\,dx$

5. (a) $2\int \sqrt{x^3}\,dx$ (b) $\int \frac{1}{4}\sqrt[4]{x^5}\,dx$

6. (a) $\int \frac{-5}{\sqrt{t^3}}\,dt$ (b) $\int \frac{3}{7\sqrt[5]{x^4}}\,dx$

7. (a) $\int 3\cos 2x\,dx$ (b) $\int 7\sin 3\theta\,d\theta$

8. (a) $\int \frac{3}{4}\sec^2 3x\,dx$ (b) $\int 2\,\text{cosec}^2\,4\theta\,d\theta$

9. (a) $5\int \cot 2t\,\text{cosec}\,2t\,dt$

    (b) $\int \frac{4}{3}\sec 4t\tan 4t\,dt$

10. (a) $\int \frac{3}{4}e^{2x}\,dx$ (b) $\frac{2}{3}\int \frac{dx}{e^{5x}}$

11. (a) $\int \frac{2}{3x}\,dx$ (b) $\int \left(\frac{u^2-1}{u}\right)du$

12. (a) $\int \frac{(2+3x)^2}{\sqrt{x}}\,dx$ (b) $\int \left(\frac{1}{t}+2t\right)^2\,dt$

63.4   Definite integrals

Integrals containing an arbitrary constant c in their results are called **indefinite integrals** since their precise value cannot be determined without further information. **Definite integrals** are those in which limits are applied. If an expression is written as $[x]_a^b$, 'b' is called the **upper**

limit and 'a' the **lower limit**. The operation of applying the limits is defined as $[x]_a^b = (b) - (a)$

The increase in the value of the integral x^2 as x increases from 1 to 3 is written as $\int_1^3 x^2\,dx$

Applying the limits gives:

$$\int_1^3 x^2\,dx = \left[\frac{x^3}{3}+c\right]_1^3 = \left(\frac{3^3}{3}+c\right) - \left(\frac{1^3}{3}+c\right)$$

$$= (9+c) - \left(\frac{1}{3}+c\right) = 8\frac{2}{3}$$

Note that the 'c' term always cancels out when limits are applied and it need not be shown with definite integrals.

Problem 12.   Evaluate

(a) $\int_1^2 3x\,dx$ (b) $\int_{-2}^3 (4-x^2)\,dx$

(a) $\int_1^2 3x\,dx = \left[\frac{3x^2}{2}\right]_1^2 = \left\{\frac{3}{2}(2)^2\right\} - \left\{\frac{3}{2}(1)^2\right\}$

$$= 6 - 1\frac{1}{2} = 4\frac{1}{2}$$

(b) $\int_{-2}^3 (4-x^2)\,dx = \left[4x - \frac{x^3}{3}\right]_{-2}^3$

$$= \left\{4(3) - \frac{(3)^3}{3}\right\} - \left\{4(-2) - \frac{(-2)^3}{3}\right\}$$

$$= \{12-9\} - \left\{-8 - \frac{-8}{3}\right\}$$

$$= \{3\} - \left\{-5\frac{1}{3}\right\} = 8\frac{1}{3}$$

Problem 13.   Evaluate $\int_1^4 \left(\frac{\theta+2}{\sqrt{\theta}}\right)d\theta$, **taking positive square roots only**

$$\int_1^4 \left(\frac{\theta+2}{\sqrt{\theta}}\right)d\theta = \int_1^4 \left(\frac{\theta}{\theta^{\frac{1}{2}}} + \frac{2}{\theta^{\frac{1}{2}}}\right)d\theta$$

$$= \int_1^4 \left(\theta^{\frac{1}{2}} + 2\theta^{\frac{-1}{2}}\right)d\theta$$

$$= \left[\frac{\theta^{\left(\frac{1}{2}\right)+1}}{\frac{1}{2}+1} + \frac{2\theta^{\left(\frac{-1}{2}\right)+1}}{-\frac{1}{2}+1}\right]_1^4$$

$$= \left[\frac{\theta^{\frac{3}{2}}}{\frac{3}{2}} + \frac{2\theta^{\frac{1}{2}}}{\frac{1}{2}} \right]_1^4 = \left[\frac{2}{3}\sqrt{\theta^3} + 4\sqrt{\theta} \right]_1^4$$

$$= \left\{ \frac{2}{3}\sqrt{(4)^3} + 4\sqrt{4} \right\} - \left\{ \frac{2}{3}\sqrt{(1)^3} + 4\sqrt{(1)} \right\}$$

$$= \left\{ \frac{16}{3} + 8 \right\} - \left\{ \frac{2}{3} + 4 \right\}$$

$$= 5\frac{1}{3} + 8 - \frac{2}{3} - 4 = 8\frac{2}{3}$$

Problem 14. Evaluate $\int_0^{\frac{\pi}{2}} 3\sin 2x \, dx$

$$\int_0^{\frac{\pi}{2}} 3\sin 2x \, dx$$

$$= \left[(3)\left(-\frac{1}{2}\right)\cos 2x \right]_0^{\frac{\pi}{2}} = \left[-\frac{3}{2}\cos 2x \right]_0^{\frac{\pi}{2}}$$

$$= \left\{ -\frac{3}{2}\cos 2\left(\frac{\pi}{2}\right) \right\} - \left\{ -\frac{3}{2}\cos 2(0) \right\}$$

$$= \left\{ -\frac{3}{2}\cos \pi \right\} - \left\{ -\frac{3}{2}\cos 0 \right\}$$

$$= \left\{ -\frac{3}{2}(-1) \right\} - \left\{ -\frac{3}{2}(1) \right\} = \frac{3}{2} + \frac{3}{2} = 3$$

Problem 15. Evaluate $\int_1^2 4\cos 3t \, dt$

$$\int_1^2 4\cos 3t \, dt = \left[(4)\left(\frac{1}{3}\right)\sin 3t \right]_1^2 = \left[\frac{4}{3}\sin 3t \right]_1^2$$

$$= \left\{ \frac{4}{3}\sin 6 \right\} - \left\{ \frac{4}{3}\sin 3 \right\}$$

Note that limits of trigonometric functions are always expressed in radians – thus, for example, $\sin 6$ means the sine of 6 radians $= -0.279415\ldots$

Hence $\int_1^2 4\cos 3t \, dt$

$$= \left\{ \frac{4}{3}(-0.279415\ldots) \right\} - \left\{ \frac{4}{3}(0.141120\ldots) \right\}$$

$$= (-0.37255) - (0.18816) = -0.5607$$

Problem 16. Evaluate

(a) $\int_1^2 4e^{2x} \, dx$ (b) $\int_1^4 \frac{3}{4u} \, du$,

each correct to 4 significant figures

(a) $\int_1^2 4e^{2x} \, dx = \left[\frac{4}{2}e^{2x} \right]_1^2 = 2[e^{2x}]_1^2 = 2[e^4 - e^2]$

$$= 2[54.5982 - 7.3891] = 94.42$$

(b) $\int_1^4 \frac{3}{4u} \, du = \left[\frac{3}{4}\ln u \right]_1^4 = \frac{3}{4}[\ln 4 - \ln 1]$

$$= \frac{3}{4}[1.3863 - 0] = 1.040$$

Now try the following Practice Exercise

Practice Exercise 261   Further problems on definite integrals (answers on page 1181)

In Problems 1 to 8, evaluate the definite integrals (where necessary, correct to 4 significant figures).

1. (a) $\int_1^4 5x^2 \, dx$ (b) $\int_{-1}^1 -\frac{3}{4}t^2 \, dt$

2. (a) $\int_{-1}^2 (3 - x^2) \, dx$ (b) $\int_1^3 (x^2 - 4x + 3) \, dx$

3. (a) $\int_0^\pi \frac{3}{2}\cos\theta \, d\theta$ (b) $\int_0^{\frac{\pi}{2}} 4\cos\theta \, d\theta$

4. (a) $\int_{\frac{\pi}{6}}^{\frac{\pi}{3}} 2\sin 2\theta \, d\theta$ (b) $\int_0^2 3\sin t \, dt$

5. (a) $\int_0^1 5\cos 3x \, dx$ (b) $\int_0^{\frac{\pi}{6}} 3\sec^2 2x \, dx$

6. (a) $\int_1^2 \text{cosec}^2 4t \, dt$

    (b) $\int_{\frac{\pi}{4}}^{\frac{\pi}{2}} (3\sin 2x - 2\cos 3x) \, dx$

7. (a) $\int_0^1 3e^{3t} \, dt$ (b) $\int_{-1}^2 \frac{2}{3e^{2x}} \, dx$

8. (a) $\displaystyle\int_2^3 \frac{2}{3x}\,dx$ (b) $\displaystyle\int_1^3 \frac{2x^2+1}{x}\,dx$

9. The entropy change ΔS for an ideal gas is given by:

$$\Delta S = \int_{T_1}^{T_2} C_v \frac{dT}{T} - R\int_{V_1}^{V_2} \frac{dV}{V}\,\text{J/K}$$

where T is the thermodynamic temperature, V is the volume and $R=8.314$. Determine the entropy change when a gas expands from 1 litre to 3 litres for a temperature rise from 100 K to 400 K given that:

$$C_v = 45 + 6\times 10^{-3}T + 8\times 10^{-6}T^2$$

10. The p.d. between boundaries a and b of an electric field is given by: $V = \displaystyle\int_a^b \frac{Q}{2\pi r \varepsilon_0 \varepsilon_r}\,dr$

If $a=10$, $b=20$, $Q=2\times 10^{-6}$ coulombs, $\varepsilon_0 = 8.85 \times 10^{-12}$ and $\varepsilon_r = 2.77$, show that $V = 9\,\text{kV}$

11. The average value of a complex voltage waveform is given by:

$$V_{AV} = \frac{1}{\pi}\int_0^\pi (10\sin\omega t + 3\sin 3\omega t$$
$$+\,2\sin 5\omega t)\,d(\omega t)$$

Evaluate V_{AV} correct to 2 decimal places.

12. The volume of liquid in a tank is given by: $v = \int_{t_1}^{t_2} q\,dt$. Determine the volume of a chemical, given $q = (5 - 0.05t + 0.003t^2)\text{m}^3/\text{s}$, $t_1 = 0$ and $t_2 = 16\,\text{s}$

For some **applications of integration**, i.e. areas under curves, mean and rms values, volumes of solids of revolution, centroids and second moments of area, go to Chapters 72 to 76.

For fully worked solutions to each of the problems in Practice Exercises 260 and 261 in this chapter, go to the website:
www.routledge.com/cw/bird

Chapter 64

Integration using algebraic substitutions

Why it is important to understand: **Integration using algebraic substitutions**

As intimated in the previous chapter, most complex engineering problems cannot be solved without calculus. Calculus has widespread applications in science, economics and engineering and can solve many problems for which algebra alone is insufficient. For example, calculus is needed to calculate the force exerted on a particle a specific distance from an electrically charged wire, and is needed for computations involving arc length, centre of mass, work and pressure. Sometimes the integral is not a standard one; in these cases it may be possible to replace the variable of integration with a function of a new variable. A change in variable can reduce an integral to a standard form, and this is demonstrated in this chapter.

At the end of this chapter, you should be able to:

- appreciate when an algebraic substitution is required to determine an integral
- integrate functions which require an algebraic substitution
- determine definite integrals where an algebraic substitution is required
- appreciate that it is possible to change the limits when determining a definite integral

64.1 Introduction

Functions which require integrating are not always in the 'standard form' shown in Chapter 63. However, it is often possible to change a function into a form which can be integrated by using either:

(i) an algebraic substitution (see Section 64.2),

(ii) a trigonometric or hyperbolic substitution (see Chapter 65),

(iii) partial fractions (see Chapter 66),

(iv) the $t = \tan \theta/2$ substitution (see Chapter 67),

(v) integration by parts (see Chapter 68), or

(vi) reduction formulae (see Chapter 69).

64.2 Algebraic substitutions

With **algebraic substitutions**, the substitution usually made is to let u be equal to $f(x)$ such that $f(u)\,du$ is a standard integral. It is found that integrals of the forms,

$$k \int [f(x)]^n f'(x)\,dx \text{ and } k \int \frac{f'(x)}{[f(x)]^n}\,dx$$

(where k and n are constants) can both be integrated by substituting u for $f(x)$.

64.3   Worked problems on integration using algebraic substitutions

---

Problem 1.   Determine $\int \cos(3x+7)\,dx$

$\int \cos(3x+7)\,dx$ is not a standard integral of the form shown in Table 63.1, page 720, thus an algebraic substitution is made.

Let $u = 3x+7$ then $\dfrac{du}{dx} = 3$ and rearranging gives $dx = \dfrac{du}{3}$. Hence,

$$\int \cos(3x+7)\,dx = \int (\cos u)\frac{du}{3} = \int \frac{1}{3}\cos u\,du,$$

$$\text{which is a standard integral}$$

$$= \frac{1}{3}\sin u + c$$

Rewriting u as $(3x+7)$ gives:

$$\int \cos(3x+7)\,dx = \frac{1}{3}\,\sin(3x+7)+c,$$

which may be checked by differentiating it.

Problem 2.   Find $\int (2x-5)^7\,dx$

$(2x-5)$ may be multiplied by itself 7 times and then each term of the result integrated. However, this would be a lengthy process, and thus an algebraic substitution is made.

Let $u = (2x-5)$ then $\dfrac{du}{dx} = 2$ and $dx = \dfrac{du}{2}$

Hence

$$\int (2x-5)^7\,dx = \int u^7\frac{du}{2} = \frac{1}{2}\int u^7\,du$$

$$= \frac{1}{2}\left(\frac{u^8}{8}\right)+c = \frac{1}{16}u^8 + c$$

Rewriting u as $(2x-5)$ gives:

$$\int (2x-5)^7\,dx = \frac{1}{16}(2x-5)^8 + c$$

Problem 3.   Find $\int \dfrac{4}{(5x-3)}\,dx$

Let $u = (5x-3)$ then $\dfrac{du}{dx} = 5$ and $dx = \dfrac{du}{5}$

Hence

$$\int \frac{4}{(5x-3)}\,dx = \int \frac{4}{u}\frac{du}{5} = \frac{4}{5}\int \frac{1}{u}\,du$$

$$= \frac{4}{5}\ln u + c = \frac{4}{5}\ln(5x-3)+c$$

Problem 4.   Evaluate $\int_0^1 2e^{6x-1}\,dx$, correct to 4 significant figures

Let $u = 6x-1$ then $\dfrac{du}{dx} = 6$ and $dx = \dfrac{du}{6}$

Hence

$$\int 2e^{6x-1}\,dx = \int 2e^u\frac{du}{6} = \frac{1}{3}\int e^u\,du$$

$$= \frac{1}{3}e^u + c = \frac{1}{3}e^{6x-1}+c$$

Thus

$$\int_0^1 2e^{6x-1}\,dx = \frac{1}{3}[e^{6x-1}]_0^1 = \frac{1}{3}[e^5 - e^{-1}] = \mathbf{49.35},$$

$$\text{correct to 4 significant figures.}$$

Problem 5.   Determine $\int 3x(4x^2+3)^5\,dx$

Let $u = (4x^2+3)$ then $\dfrac{du}{dx} = 8x$ and $dx = \dfrac{du}{8x}$

Hence

$$\int 3x(4x^2+3)^5\,dx = \int 3x(u)^5\frac{du}{8x}$$

$$= \frac{3}{8}\int u^5\,du,\text{ by cancelling.}$$

The original variable 'x' has been completely removed and the integral is now only in terms of u and is a standard integral.

Hence $\dfrac{3}{8}\displaystyle\int u^5\,du = \dfrac{3}{8}\left(\dfrac{u^6}{6}\right)+c$

$$= \frac{1}{16}u^6 + c = \frac{1}{16}(4x^2+3)^6 + c$$

Problem 6.   Evaluate $\int_0^{\frac{\pi}{6}} 24 \sin^5 \theta \cos \theta \, d\theta$

Let $u = \sin \theta$ then $\dfrac{du}{d\theta} = \cos \theta$ and $d\theta = \dfrac{du}{\cos \theta}$

Hence $\int 24 \sin^5 \theta \cos \theta \, d\theta = \int 24 u^5 \cos \theta \dfrac{du}{\cos \theta}$

$$= 24 \int u^5 \, du, \text{ by cancelling}$$

$$= 24 \dfrac{u^6}{6} + c = 4u^6 + c = 4(\sin \theta)^6 + c$$

$$= 4 \sin^6 \theta + c$$

Thus $\int_0^{\frac{\pi}{6}} 24 \sin^5 \theta \cos \theta \, d\theta = [4 \sin^6 \theta]_0^{\frac{\pi}{6}}$

$$= 4 \left[\left(\sin \dfrac{\pi}{6} \right)^6 - (\sin 0)^6 \right]$$

$$= 4 \left[\left(\dfrac{1}{2} \right)^6 - 0 \right] = \dfrac{1}{16} \text{ or } \mathbf{0.0625}$$

Now try the following Practice Exercise

Practice Exercise 262   Further problems on integration using algebraic substitutions (answers on page 1181)

In Problems 1 to 6, integrate with respect to the variable.

1. $2 \sin(4x + 9)$

2. $3 \cos(2\theta - 5)$

3. $4 \sec^2(3t + 1)$

4. $\dfrac{1}{2}(5x - 3)^6$

5. $\dfrac{-3}{(2x - 1)}$

6. $3e^{3\theta + 5}$

In Problems 7 to 10, evaluate the definite integrals correct to 4 significant figures.

7. $\int_0^1 (3x + 1)^5 \, dx$

8. $\int_0^2 x \sqrt{(2x^2 + 1)} \, dx$

9. $\int_0^{\frac{\pi}{3}} 2 \sin \left(3t + \dfrac{\pi}{4} \right) dt$

10. $\int_0^1 3 \cos(4x - 3) \, dx$

11.   The mean time to failure, M years, for a set of components is given by:

$$M = \int_0^4 (1 - 0.25t)^{1.5} \, dt$$

Determine the mean time to failure.

64.4   Further worked problems on integration using algebraic substitutions

Problem 7.   Find $\int \dfrac{x}{2 + 3x^2} \, dx$

Let $u = 2 + 3x^2$ then $\dfrac{du}{dx} = 6x$ and $dx = \dfrac{du}{6x}$
Hence

$$\int \dfrac{x}{2 + 3x^2} \, dx = \int \dfrac{x}{u} \dfrac{du}{6x} = \dfrac{1}{6} \int \dfrac{1}{u} \, du,$$

$$\text{by cancelling}$$

$$= \dfrac{1}{6} \ln u + c = \dfrac{1}{6} \ln(2 + 3x^2) + c$$

Problem 8.   Determine $\int \dfrac{2x}{\sqrt{(4x^2 - 1)}} \, dx$

Let $u = 4x^2 - 1$ then $\dfrac{du}{dx} = 8x$ and $dx = \dfrac{du}{8x}$

Hence $\int \dfrac{2x}{\sqrt{(4x^2 - 1)}} \, dx = \int \dfrac{2x}{\sqrt{u}} \dfrac{du}{8x}$

$$= \dfrac{1}{4} \int \dfrac{1}{\sqrt{u}} \, du, \text{ by cancelling}$$

$$= \dfrac{1}{4} \int u^{\frac{-1}{2}} \, du = \dfrac{1}{4} \left[\dfrac{u^{\left(\frac{-1}{2} \right) + 1}}{-\dfrac{1}{2} + 1} \right] + c$$

$$= \frac{1}{4} \left[\frac{u^{\frac{1}{2}}}{\frac{1}{2}} \right] + c = \frac{1}{2}\sqrt{u} + c$$

$$= \frac{1}{2}\sqrt{(4x^2 - 1)} + c$$

Problem 9. Show that

$$\int \tan\theta \, d\theta = \ln(\sec\theta) + c$$

$\int \tan\theta \, d\theta = \int \frac{\sin\theta}{\cos\theta} \, d\theta$. Let $u = \cos\theta$

then $\dfrac{du}{d\theta} = -\sin\theta$ and $d\theta = \dfrac{-du}{\sin\theta}$

Hence

$$\int \frac{\sin\theta}{\cos\theta} \, d\theta = \int \frac{\sin\theta}{u} \left(\frac{-du}{\sin\theta} \right)$$

$$= -\int \frac{1}{u} \, du = -\ln u + c$$

$$= -\ln(\cos\theta) + c = \ln(\cos\theta)^{-1} + c,$$

by the laws of logarithms.

Hence $\displaystyle\int \tan\theta \, d\theta = \ln(\sec\theta) + c$

since $(\cos\theta)^{-1} = \dfrac{1}{\cos\theta} = \sec\theta$

64.5 Change of limits

When evaluating definite integrals involving substitutions it is sometimes more convenient to **change the limits** of the integral, as shown in Problems 10 and 11.

Problem 10. Evaluate $\int_1^3 5x\sqrt{(2x^2 + 7)}\,dx$, taking positive values of square roots only

Let $u = 2x^2 + 7$, then $\dfrac{du}{dx} = 4x$ and $dx = \dfrac{du}{4x}$
It is possible in this case to change the limits of integration. Thus when $x = 3$, $u = 2(3)^2 + 7 = 25$ and when $x = 1$, $u = 2(1)^2 + 7 = 9$

Hence

$$\int_{x=1}^{x=3} 5x\sqrt{(2x^2 + 7)}\,dx = \int_{u=9}^{u=25} 5x\sqrt{u}\frac{du}{4x}$$

$$= \frac{5}{4} \int_9^{25} \sqrt{u}\,du$$

$$= \frac{5}{4} \int_9^{25} u^{\frac{1}{2}}\,du$$

Thus the limits have been changed, and it is unnecessary to change the integral back in terms of x.

Thus $\displaystyle\int_{x=1}^{x=3} 5x\sqrt{(2x^2 + 7)}\,dx = \frac{5}{4} \left[\frac{u^{\frac{3}{2}}}{3/2} \right]_9^{25}$

$$= \frac{5}{6} \left[\sqrt{u^3} \right]_9^{25} = \frac{5}{6} \left[\sqrt{25^3} - \sqrt{9^3} \right]$$

$$= \frac{5}{6}(125 - 27) = \mathbf{81\frac{2}{3}}$$

Problem 11. Evaluate $\displaystyle\int_0^2 \frac{3x}{\sqrt{(2x^2 + 1)}}\,dx$, taking positive values of square roots only

Let $u = 2x^2 + 1$ then $\dfrac{du}{dx} = 4x$ and $dx = \dfrac{du}{4x}$

Hence $\displaystyle\int_0^2 \frac{3x}{\sqrt{(2x^2 + 1)}}\,dx = \int_{x=0}^{x=2} \frac{3x}{\sqrt{u}}\frac{du}{4x}$

$$= \frac{3}{4} \int_{x=0}^{x=2} u^{\frac{-1}{2}}\,du$$

Since $u = 2x^2 + 1$, when $x = 2$, $u = 9$ and when $x = 0$, $u = 1$

Thus $\dfrac{3}{4} \displaystyle\int_{x=0}^{x=2} u^{\frac{-1}{2}}\,du = \frac{3}{4} \int_{u=1}^{u=9} u^{\frac{-1}{2}}\,du$

i.e. the limits have been changed

$$= \frac{3}{4} \left[\frac{u^{\frac{1}{2}}}{\frac{1}{2}} \right]_1^9 = \frac{3}{2} \left[\sqrt{9} - \sqrt{1} \right] = \mathbf{3}$$

taking positive values of square roots only.

Now try the following Practice Exercise

Practice Exercise 263   Further problems on integration using algebraic substitutions (answers on page 1181)

In Problems 1 to 7, integrate with respect to the variable.

1. $2x(2x^2 - 3)^5$

2. $5\cos^5 t \sin t$

3. $3\sec^2 3x \tan 3x$

4. $2t\sqrt{(3t^2 - 1)}$

5. $\dfrac{\ln\theta}{\theta}$

6. $3\tan 2t$

7. $\dfrac{2e^t}{\sqrt{(e^t + 4)}}$

In Problems 8 to 10, evaluate the definite integrals correct to 4 significant figures.

8. $\displaystyle\int_0^1 3x\,e^{(2x^2-1)}\,dx$

9. $\displaystyle\int_0^{\frac{\pi}{2}} 3\sin^4\theta\cos\theta\,d\theta$

10. $\displaystyle\int_0^1 \dfrac{3x}{(4x^2-1)^5}\,dx$

11. The electrostatic potential on all parts of a conducting circular disc of radius r is given by the equation:

$$V = 2\pi\sigma\int_0^9 \frac{R}{\sqrt{R^2 + r^2}}\,dR$$

Solve the equation by determining the integral.

12. In the study of a rigid rotor the following integration occurs:

$$Z_r = \int_0^\infty (2J+1)e^{\frac{-J(J+1)h^2}{8\pi^2 IkT}}\,dJ$$

Determine Z_r for constant temperature T assuming h, I and k are constants.

13. In electrostatics,

$$E = \int_0^\pi \left\{ \frac{a^2\sigma\sin\theta}{2\varepsilon\sqrt{(a^2 - x^2 - 2ax\cos\theta)}}d\theta \right\}$$

where a, σ and ε are constants, x is greater than a, and x is independent of θ. Show that $E = \dfrac{a^2\sigma}{\varepsilon x}$

14. The time taken, t hours, for a vehicle to reach a velocity of 130 km/h with an initial speed of 60 km/h is given by:

$$t = \int_{60}^{130} \frac{dv}{650 - 3v}$$

where v is the velocity in km/h. Determine t, correct to the nearest second.

For fully worked solutions to each of the problems in Practice Exercises 262 and 263 in this chapter, go to the website:
www.routledge.com/cw/bird

Chapter 65

Integration using trigonometric and hyperbolic substitutions

Why it is important to understand: Integration using trigonometric and hyperbolic substitutions

Calculus is the most powerful branch of mathematics. It is capable of computing many quantities accurately, which cannot be calculated using any other branch of mathematics. Many integrals are not 'standard' ones that we can determine from a list of results. Some need substitutions to rearrange them into a standard form. There are a number of trigonometric and hyperbolic substitutions that may be used for certain integrals to change them into a form that can be integrated. These are explained in this chapter, which provides another piece of the integral calculus jigsaw.

At the end of this chapter, you should be able to:

- integrate functions of the form $\sin^2 x$, $\cos^2 x$, $\tan^2 x$ and $\cot^2 x$
- integrate functions having powers of sines and cosines
- integrate functions that are products of sines and cosines
- integrate using the $\sin \theta$ substitution
- integrate using the $\tan \theta$ substitution
- integrate using the $\sinh \theta$ substitution
- integrate using the $\cosh \theta$ substitution

65.1 Introduction

Table 65.1 gives a summary of the integrals that require the use of **trigonometric and hyperbolic substitutions** and their application is demonstrated in Problems 1 to 27.

65.2 Worked problems on integration of $\sin^2 x$, $\cos^2 x$, $\tan^2 x$ and $\cot^2 x$

Problem 1. Evaluate $\displaystyle\int_0^{\frac{\pi}{4}} 2\cos^2 4t \, dt$

Since $\cos 2t = 2\cos^2 t - 1$ (from Chapter 44),

then $\cos^2 t = \dfrac{1}{2}(1 + \cos 2t)$ and

$$\cos^2 4t = \dfrac{1}{2}(1 + \cos 8t)$$

Table 65.1 Integrals using trigonometric and hyperbolic substitutions

$f(x)$	$\int f(x)\mathrm{d}x$	Method	See problem
1. $\cos^2 x$	$\frac{1}{2}\left(x + \frac{\sin 2x}{2}\right) + c$	Use $\cos 2x = 2\cos^2 x - 1$	1
2. $\sin^2 x$	$\frac{1}{2}\left(x - \frac{\sin 2x}{2}\right) + c$	Use $\cos 2x = 1 - 2\sin^2 x$	2
3. $\tan^2 x$	$\tan x - x + c$	Use $1 + \tan^2 x = \sec^2 x$	3
4. $\cot^2 x$	$-\cot x - x + c$	Use $\cot^2 x + 1 = \operatorname{cosec}^2 x$	4
5. $\cos^m x \sin^n x$	(a) If either m or n is odd (but not both), use $\cos^2 x + \sin^2 x = 1$		5, 6
	(b) If both m and n are even, use either $\cos 2x = 2\cos^2 x - 1$ or $\cos 2x = 1 - 2\sin^2 x$		7, 8
6. $\sin A \cos B$		Use $\frac{1}{2}[\sin(A+B) + \sin(A-B)]$	9
7. $\cos A \sin B$		Use $\frac{1}{2}[\sin(A+B) - \sin(A-B)]$	10
8. $\cos A \cos B$		Use $\frac{1}{2}[\cos(A+B) + \cos(A-B)]$	11
9. $\sin A \sin B$		Use $-\frac{1}{2}[\cos(A+B) - \cos(A-B)]$	12
10. $\dfrac{1}{\sqrt{(a^2 - x^2)}}$	$\sin^{-1}\dfrac{x}{a} + c$	Use $x = a\sin\theta$ substitution	13, 14
11. $\sqrt{(a^2 - x^2)}$	$\dfrac{a^2}{2}\sin^{-1}\dfrac{x}{a} + \dfrac{x}{2}\sqrt{(a^2 - x^2)} + c$	Use $x = a\sin\theta$ substitution	15, 16
12. $\dfrac{1}{a^2 + x^2}$	$\dfrac{1}{a}\tan^{-1}\dfrac{x}{a} + c$	Use $x = a\tan\theta$ substitution	17–19
13. $\dfrac{1}{\sqrt{(x^2 + a^2)}}$	$\sinh^{-1}\dfrac{x}{a} + c$ or $\ln\left\{\dfrac{x + \sqrt{(x^2 + a^2)}}{a}\right\} + c$	Use $x = a\sinh\theta$ substitution	20–22
14. $\sqrt{(x^2 + a^2)}$	$\dfrac{a^2}{2}\sinh^{-1}\dfrac{x}{a} + \dfrac{x}{2}\sqrt{(x^2 + a^2)} + c$	Use $x = a\sinh\theta$ substitution	23
15. $\dfrac{1}{\sqrt{(x^2 - a^2)}}$	$\cosh^{-1}\dfrac{x}{a} + c$ or $\ln\left\{\dfrac{x + \sqrt{(x^2 - a^2)}}{a}\right\} + c$	Use $x = a\cosh\theta$ substitution	24, 25
16. $\sqrt{(x^2 - a^2)}$	$\dfrac{x}{2}\sqrt{(x^2 - a^2)} - \dfrac{a^2}{2}\cosh^{-1}\dfrac{x}{a} + c$	Use $x = a\cosh\theta$ substitution	26, 27

Hence $\displaystyle\int_0^{\frac{\pi}{4}} 2\cos^2 4t\,dt$

$$= 2\int_0^{\frac{\pi}{4}} \frac{1}{2}(1+\cos 8t)\,dt$$

$$= \left[t + \frac{\sin 8t}{8}\right]_0^{\frac{\pi}{4}}$$

$$= \left[\frac{\pi}{4} + \frac{\sin 8\left(\frac{\pi}{4}\right)}{8}\right] - \left[0 + \frac{\sin 0}{8}\right]$$

$$= \frac{\pi}{4} \text{ or } \mathbf{0.7854}$$

Problem 2.   Determine $\int \sin^2 3x\,dx$

Since $\cos 2x = 1 - 2\sin^2 x$ (from Chapter 44),

then $\sin^2 x = \dfrac{1}{2}(1 - \cos 2x)$ and

$$\sin^2 3x = \frac{1}{2}(1 - \cos 6x)$$

Hence $\displaystyle\int \sin^2 3x\,dx = \int \frac{1}{2}(1 - \cos 6x)\,dx$

$$= \frac{1}{2}\left(x - \frac{\sin 6x}{6}\right) + c$$

Problem 3.   Find $3\int \tan^2 4x\,dx$

Since $1 + \tan^2 x = \sec^2 x$, then $\tan^2 x = \sec^2 x - 1$ and $\tan^2 4x = \sec^2 4x - 1$

Hence $\displaystyle 3\int \tan^2 4x\,dx = 3\int (\sec^2 4x - 1)\,dx$

$$= 3\left(\frac{\tan 4x}{4} - x\right) + c$$

Problem 4.   Evaluate $\displaystyle\int_{\frac{\pi}{6}}^{\frac{\pi}{3}} \frac{1}{2}\cot^2 2\theta\,d\theta$

Since $\cot^2\theta + 1 = \operatorname{cosec}^2\theta$, then $\cot^2\theta = \operatorname{cosec}^2\theta - 1$ and $\cot^2 2\theta = \operatorname{cosec}^2 2\theta - 1$

Hence $\displaystyle\int_{\frac{\pi}{6}}^{\frac{\pi}{3}} \frac{1}{2}\cot^2 2\theta\,d\theta$

$$= \frac{1}{2}\int_{\frac{\pi}{6}}^{\frac{\pi}{3}} (\operatorname{cosec}^2 2\theta - 1)\,d\theta = \frac{1}{2}\left[\frac{-\cot 2\theta}{2} - \theta\right]_{\frac{\pi}{6}}^{\frac{\pi}{3}}$$

$$= \frac{1}{2}\left[\left(\frac{-\cot 2\left(\frac{\pi}{3}\right)}{2} - \frac{\pi}{3}\right) - \left(\frac{-\cot 2\left(\frac{\pi}{6}\right)}{2} - \frac{\pi}{6}\right)\right]$$

$$= \frac{1}{2}[(0.2887 - 1.0472) - (-0.2887 - 0.5236)]$$

$$= \mathbf{0.0269}$$

Now try the following Practice Exercise

Practice Exercise 264   Further problems on integration of $\sin^2 x$, $\cos^2 x$, $\tan^2 x$ and $\cot^2 x$ (answers on page 1181)

In Problems 1 to 4, integrate with respect to the variable.

1. $\sin^2 2x$

2. $3\cos^2 t$

3. $5\tan^2 3\theta$

4. $2\cot^2 2t$

In Problems 5 to 8, evaluate the definite integrals, correct to 4 significant figures.

5. $\displaystyle\int_0^{\frac{\pi}{3}} 3\sin^2 3x\,dx$

6. $\displaystyle\int_0^{\frac{\pi}{4}} \cos^2 4x\,dx$

7. $\displaystyle\int_0^1 2\tan^2 2t\,dt$

8. $\displaystyle\int_{\frac{\pi}{6}}^{\frac{\pi}{3}} \cot^2\theta\,d\theta$

65.3   Worked problems on integration of powers of sines and cosines

Problem 5.   Determine $\int \sin^5\theta\,d\theta$

Since $\cos^2\theta + \sin^2\theta = 1$ then $\sin^2\theta = (1 - \cos^2\theta)$

Hence $\displaystyle\int \sin^5\theta\,d\theta$

$$= \int \sin\theta(\sin^2\theta)^2\,d\theta = \int \sin\theta(1 - \cos^2\theta)^2\,d\theta$$

$$= \int \sin\theta(1 - 2\cos^2\theta + \cos^4\theta)\,d\theta$$

$$= \int (\sin\theta - 2\sin\theta\cos^2\theta + \sin\theta\cos^4\theta)\,d\theta$$

$$= -\cos\theta + \frac{2\cos^3\theta}{3} - \frac{\cos^5\theta}{5} + c$$

Whenever a power of a cosine is multiplied by a sine of power 1, or vice versa, the integral may be determined by inspection, as shown.

In general, $\displaystyle\int \cos^n\theta\,\sin\theta\,d\theta = \frac{-\cos^{n+1}\theta}{(n+1)} + c$

and $\displaystyle\int \sin^n\theta\,\cos\theta\,d\theta = \frac{\sin^{n+1}\theta}{(n+1)} + c$

Problem 6. Evaluate $\displaystyle\int_0^{\frac{\pi}{2}} \sin^2 x\,\cos^3 x\,dx$

$$\int_0^{\frac{\pi}{2}} \sin^2 x\,\cos^3 x\,dx = \int_0^{\frac{\pi}{2}} \sin^2 x\,\cos^2 x\,\cos x\,dx$$

$$= \int_0^{\frac{\pi}{2}} (\sin^2 x)(1 - \sin^2 x)(\cos x)\,dx$$

$$= \int_0^{\frac{\pi}{2}} (\sin^2 x\,\cos x - \sin^4 x\,\cos x)\,dx$$

$$= \left[\frac{\sin^3 x}{3} - \frac{\sin^5 x}{5} \right]_0^{\frac{\pi}{2}}$$

$$= \left[\frac{\left(\sin\frac{\pi}{2}\right)^3}{3} - \frac{\left(\sin\frac{\pi}{2}\right)^5}{5} \right] - [0 - 0]$$

$$= \frac{1}{3} - \frac{1}{5} = \frac{2}{15} \text{ or } \mathbf{0.1333}$$

Problem 7. Evaluate $\displaystyle\int_0^{\frac{\pi}{4}} 4\cos^4\theta\,d\theta$, correct to 4 significant figures

$$\int_0^{\frac{\pi}{4}} 4\cos^4\theta\,d\theta = 4\int_0^{\frac{\pi}{4}} (\cos^2\theta)^2\,d\theta$$

$$= 4\int_0^{\frac{\pi}{4}} \left[\frac{1}{2}(1 + \cos 2\theta) \right]^2\,d\theta$$

$$= \int_0^{\frac{\pi}{4}} (1 + 2\cos 2\theta + \cos^2 2\theta)\,d\theta$$

$$= \int_0^{\frac{\pi}{4}} \left[1 + 2\cos 2\theta + \frac{1}{2}(1 + \cos 4\theta) \right]\,d\theta$$

$$= \int_0^{\frac{\pi}{4}} \left(\frac{3}{2} + 2\cos 2\theta + \frac{1}{2}\cos 4\theta \right)\,d\theta$$

$$= \left[\frac{3\theta}{2} + \sin 2\theta + \frac{\sin 4\theta}{8} \right]_0^{\frac{\pi}{4}}$$

$$= \left[\frac{3}{2}\left(\frac{\pi}{4}\right) + \sin\frac{2\pi}{4} + \frac{\sin 4(\pi/4)}{8} \right] - [0]$$

$$= \frac{3\pi}{8} + 1 = \mathbf{2.178},$$

correct to 4 significant figures.

Problem 8. Find $\int \sin^2 t\,\cos^4 t\,dt$

$$\int \sin^2 t\,\cos^4 t\,dt = \int \sin^2 t\,(\cos^2 t)^2\,dt$$

$$= \int \left(\frac{1 - \cos 2t}{2} \right)\left(\frac{1 + \cos 2t}{2} \right)^2\,dt$$

$$= \frac{1}{8}\int (1 - \cos 2t)(1 + 2\cos 2t + \cos^2 2t)\,dt$$

$$= \frac{1}{8}\int (1 + 2\cos 2t + \cos^2 2t - \cos 2t$$
$$\qquad\qquad - 2\cos^2 2t - \cos^3 2t)\,dt$$

$$= \frac{1}{8}\int (1 + \cos 2t - \cos^2 2t - \cos^3 2t)\,dt$$

$$= \frac{1}{8}\int \left[1 + \cos 2t - \left(\frac{1 + \cos 4t}{2} \right) \right.$$
$$\qquad\qquad \left. - \cos 2t(1 - \sin^2 2t) \right]\,dt$$

$$= \frac{1}{8}\int \left(\frac{1}{2} - \frac{\cos 4t}{2} + \cos 2t\,\sin^2 2t \right)\,dt$$

$$= \frac{1}{8}\left(\frac{t}{2} - \frac{\sin 4t}{8} + \frac{\sin^3 2t}{6} \right) + c$$

Now try the following Practice Exercise

In Problems 1 to 6, integrate with respect to the variable.

1. $\sin^3\theta$

2. $2\cos^3 2x$

3. $2\sin^3 t \cos^2 t$

4. $\sin^3 x \cos^4 x$

5. $2\sin^4 2\theta$

6. $\sin^2 t \cos^2 t$

65.4 Worked problems on integration of products of sines and cosines

Problem 9. Determine $\int \sin 3t \cos 2t \, dt$

$$\int \sin 3t \cos 2t \, dt$$

$$= \int \frac{1}{2}[\sin(3t + 2t) + \sin(3t - 2t)]dt,$$

from 6 of Table 65.1 (page 732), which follows from Section 44.4, page 508,

$$= \frac{1}{2}\int (\sin 5t + \sin t)\, dt$$

$$= \frac{1}{2}\left(\frac{-\cos 5t}{5} - \cos t\right) + c$$

Problem 10. Find $\int \frac{1}{3}\cos 5x \sin 2x \, dx$

$$\int \frac{1}{3}\cos 5x \sin 2x \, dx$$

$$= \frac{1}{3}\int \frac{1}{2}[\sin(5x + 2x) - \sin(5x - 2x)]dx,$$

from 7 of Table 65.1

$$= \frac{1}{6}\int (\sin 7x - \sin 3x)\, dx$$

$$= \frac{1}{6}\left(\frac{-\cos 7x}{7} + \frac{\cos 3x}{3}\right) + c$$

Problem 11. Evaluate $\int_0^1 2\cos 6\theta \cos \theta \, d\theta$, correct to 4 decimal places

$$\int_0^1 2\cos 6\theta \cos \theta \, d\theta$$

$$= 2\int_0^1 \frac{1}{2}[\cos(6\theta + \theta) + \cos(6\theta - \theta)]d\theta,$$

from 8 of Table 65.1

$$= \int_0^1 (\cos 7\theta + \cos 5\theta)\, d\theta = \left[\frac{\sin 7\theta}{7} + \frac{\sin 5\theta}{5}\right]_0^1$$

$$= \left(\frac{\sin 7}{7} + \frac{\sin 5}{5}\right) - \left(\frac{\sin 0}{7} + \frac{\sin 0}{5}\right)$$

'sin 7' means 'the sine of 7 radians' ($\equiv 401°4'$) and $\sin 5 \equiv 286°29'$.

Hence $\int_0^1 2\cos 6\theta \cos \theta \, d\theta$

$$= (0.09386 + (-0.19178)) - (0)$$

$$= \mathbf{-0.0979}, \text{ correct to 4 decimal places.}$$

Problem 12. Find $3\int \sin 5x \sin 3x \, dx$

$$3\int \sin 5x \sin 3x \, dx$$

$$= 3\int -\frac{1}{2}[\cos(5x + 3x) - \cos(5x - 3x)]dx,$$

from 9 of Table 65.1

$$= -\frac{3}{2}\int (\cos 8x - \cos 2x)\, dx$$

$$= -\frac{3}{2}\left(\frac{\sin 8x}{8} - \frac{\sin 2x}{2}\right) + c \quad \text{or}$$

$$\frac{3}{16}(4\sin 2x - \sin 8x) + c$$

Now try the following Practice Exercise

Practice Exercise 266   Further problems on integration of products of sines and cosines (answers on page 1182)

In Problems 1 to 4, integrate with respect to the variable.

1. $\sin 5t \cos 2t$

2. $2\sin 3x \sin x$

3. $3\cos 6x \cos x$

4. $\frac{1}{2}\cos 4\theta \sin 2\theta$

In Problems 5 to 8, evaluate the definite integrals.

5. $\displaystyle\int_0^{\frac{\pi}{2}} \cos 4x \cos 3x \, dx$

6. $\displaystyle\int_0^1 2 \sin 7t \cos 3t \, dt$

7. $-4 \displaystyle\int_0^{\frac{\pi}{3}} \sin 5\theta \sin 2\theta \, d\theta$

8. $\displaystyle\int_1^2 3 \cos 8t \sin 3t \, dt$

65.5 Worked problems on integration using the $\sin\theta$ substitution

Problem 13. Determine $\displaystyle\int \frac{1}{\sqrt{(a^2 - x^2)}} \, dx$

Let $x = a \sin\theta$, then $\dfrac{dx}{d\theta} = a \cos\theta$ and $dx = a \cos\theta \, d\theta$

Hence $\displaystyle\int \frac{1}{\sqrt{(a^2 - x^2)}} \, dx$

$= \displaystyle\int \frac{1}{\sqrt{(a^2 - a^2 \sin^2\theta)}} a \cos\theta \, d\theta$

$= \displaystyle\int \frac{a \cos\theta \, d\theta}{\sqrt{[a^2(1 - \sin^2\theta)]}}$

$= \displaystyle\int \frac{a \cos\theta \, d\theta}{\sqrt{(a^2 \cos^2\theta)}}$, since $\sin^2\theta + \cos^2\theta = 1$

$= \displaystyle\int \frac{a \cos\theta \, d\theta}{a \cos\theta} = \int d\theta = \theta + c$

Since $x = a \sin\theta$, then $\sin\theta = \dfrac{x}{a}$ and $\theta = \sin^{-1} \dfrac{x}{a}$

Hence $\displaystyle\int \frac{1}{\sqrt{(a^2 - x^2)}} \, dx = \mathbf{\sin^{-1} \dfrac{x}{a} + c}$

Problem 14. Evaluate $\displaystyle\int_0^3 \frac{1}{\sqrt{(9 - x^2)}} \, dx$

From Problem 13, $\displaystyle\int_0^3 \frac{1}{\sqrt{(9 - x^2)}} \, dx$

$= \left[\sin^{-1} \dfrac{x}{3}\right]_0^3$, since $a = 3$

$= (\sin^{-1} 1 - \sin^{-1} 0) = \dfrac{\pi}{2}$ or **1.5708**

Problem 15. Find $\displaystyle\int \sqrt{(a^2 - x^2)} \, dx$

Let $x = a \sin\theta$ then $\dfrac{dx}{d\theta} = a \cos\theta$ and $dx = a \cos\theta \, d\theta$

Hence $\displaystyle\int \sqrt{(a^2 - x^2)} \, dx$

$= \displaystyle\int \sqrt{(a^2 - a^2 \sin^2\theta)} \, (a \cos\theta \, d\theta)$

$= \displaystyle\int \sqrt{[a^2(1 - \sin^2\theta)]} \, (a \cos\theta \, d\theta)$

$= \displaystyle\int \sqrt{(a^2 \cos^2\theta)} \, (a \cos\theta \, d\theta)$

$= \displaystyle\int (a \cos\theta)(a \cos\theta \, d\theta)$

$= a^2 \displaystyle\int \cos^2\theta \, d\theta = a^2 \int \left(\frac{1 + \cos 2\theta}{2}\right) d\theta$

$\qquad\qquad$ (since $\cos 2\theta = 2 \cos^2\theta - 1$)

$= \dfrac{a^2}{2}\left(\theta + \dfrac{\sin 2\theta}{2}\right) + c$

$= \dfrac{a^2}{2}\left(\theta + \dfrac{2 \sin\theta \cos\theta}{2}\right) + c$

$\qquad\qquad$ since from Chapter 44, $\sin 2\theta = 2 \sin\theta \cos\theta$

$= \dfrac{a^2}{2}[\theta + \sin\theta \cos\theta] + c$

Since $x = a \sin\theta$, then $\sin\theta = \dfrac{x}{a}$ and $\theta = \sin^{-1} \dfrac{x}{a}$

Also, $\cos^2\theta + \sin^2\theta = 1$, from which,

$\cos\theta = \sqrt{(1 - \sin^2\theta)} = \sqrt{\left[1 - \left(\dfrac{x}{a}\right)^2\right]}$

$= \sqrt{\left(\dfrac{a^2 - x^2}{a^2}\right)} = \dfrac{\sqrt{(a^2 - x^2)}}{a}$

Thus $\displaystyle\int \sqrt{(a^2 - x^2)} \, dx = \dfrac{a^2}{2}[\theta + \sin\theta \cos\theta] + c$

$= \dfrac{a^2}{2}\left[\sin^{-1} \dfrac{x}{a} + \left(\dfrac{x}{a}\right) \dfrac{\sqrt{(a^2 - x^2)}}{a}\right] + c$

$= \dfrac{a^2}{2} \sin^{-1} \dfrac{x}{a} + \dfrac{x}{2}\sqrt{(a^2 - x^2)} + c$

Problem 16. Evaluate $\int_0^4 \sqrt{(16-x^2)}\,dx$

From Problem 15, $\int_0^4 \sqrt{(16-x^2)}\,dx$

$= \left[\frac{16}{2}\sin^{-1}\frac{x}{4} + \frac{x}{2}\sqrt{(16-x^2)}\right]_0^4$

$= \left[8\sin^{-1}1 + 2\sqrt{(0)}\right] - [8\sin^{-1}0 + 0]$

$= 8\sin^{-1}1 = 8\left(\frac{\pi}{2}\right) = 4\pi$ or **12.57**

Now try the following Practice Exercise

Practice Exercise 267   Further problems on integration using the $\sin\theta$ substitution (answers on page 1182)

1. Determine $\int \dfrac{5}{\sqrt{(4-t^2)}}\,dt$

2. Determine $\int \dfrac{3}{\sqrt{(9-x^2)}}\,dx$

3. Determine $\int \sqrt{(4-x^2)}\,dx$

4. Determine $\int \sqrt{(16-9t^2)}\,dt$

5. Evaluate $\int_0^4 \dfrac{1}{\sqrt{(16-x^2)}}\,dx$

6. Evaluate $\int_0^1 \sqrt{(9-4x^2)}\,dx$

65.6 Worked problems on integration using the $\tan\theta$ substitution

Problem 17. Determine $\int \dfrac{1}{(a^2+x^2)}\,dx$

Let $x = a\tan\theta$ then $\dfrac{dx}{d\theta} = a\sec^2\theta$ and $dx = a\sec^2\theta\,d\theta$

Hence $\int \dfrac{1}{(a^2+x^2)}\,dx$

$= \int \dfrac{1}{(a^2 + a^2\tan^2\theta)}(a\sec^2\theta\,d\theta)$

$= \int \dfrac{a\sec^2\theta\,d\theta}{a^2(1+\tan^2\theta)}$

$= \int \dfrac{a\sec^2\theta\,d\theta}{a^2\sec^2\theta}$, since $1+\tan^2\theta = \sec^2\theta$

$= \int \dfrac{1}{a}\,d\theta = \dfrac{1}{a}(\theta) + c$

Since $x = a\tan\theta$, $\theta = \tan^{-1}\dfrac{x}{a}$

Hence $\int \dfrac{1}{(a^2+x^2)}\,dx = \dfrac{1}{a}\tan^{-1}\dfrac{x}{a} + c$

Problem 18. Evaluate $\int_0^2 \dfrac{1}{(4+x^2)}\,dx$

From Problem 17, $\int_0^2 \dfrac{1}{(4+x^2)}\,dx$

$= \dfrac{1}{2}\left[\tan^{-1}\dfrac{x}{2}\right]_0^2$ since $a = 2$

$= \dfrac{1}{2}(\tan^{-1}1 - \tan^{-1}0) = \dfrac{1}{2}\left(\dfrac{\pi}{4} - 0\right)$

$= \dfrac{\pi}{8}$ or **0.3927**

Problem 19. Evaluate $\int_0^1 \dfrac{5}{(3+2x^2)}\,dx$, correct to 4 decimal places

$\int_0^1 \dfrac{5}{(3+2x^2)}\,dx = \int_0^1 \dfrac{5}{2[(3/2)+x^2]}\,dx$

$= \dfrac{5}{2}\int_0^1 \dfrac{1}{[\sqrt{(3/2)}]^2 + x^2}\,dx$

$= \dfrac{5}{2}\left[\dfrac{1}{\sqrt{(3/2)}}\tan^{-1}\dfrac{x}{\sqrt{(3/2)}}\right]_0^1$

$= \dfrac{5}{2}\sqrt{\left(\dfrac{2}{3}\right)}\left[\tan^{-1}\sqrt{\left(\dfrac{2}{3}\right)} - \tan^{-1}0\right]$

$= (2.0412)[0.6847 - 0]$

$= \mathbf{1.3976}$, correct to 4 decimal places.

Now try the following Practice Exercise

> **Practice Exercise 268   Further problems on integration using the $\tan\theta$ substitution (answers on page 1182)**
>
> 1. Determine $\displaystyle\int \frac{3}{4+t^2}\,dt$
>
> 2. Determine $\displaystyle\int \frac{5}{16+9\theta^2}\,d\theta$
>
> 3. Evaluate $\displaystyle\int_0^1 \frac{3}{1+t^2}\,dt$
>
> 4. Evaluate $\displaystyle\int_0^3 \frac{5}{4+x^2}\,dx$

65.7 Worked problems on integration using the $\sinh\theta$ substitution

> **Problem 20. Determine $\displaystyle\int \frac{1}{\sqrt{(x^2+a^2)}}\,dx$**

Let $x = a\sinh\theta$, then $\dfrac{dx}{d\theta} = a\cosh\theta$ and $dx = a\cosh\theta\,d\theta$

Hence $\displaystyle\int \frac{1}{\sqrt{(x^2+a^2)}}\,dx$

$= \displaystyle\int \frac{1}{\sqrt{(a^2\sinh^2\theta + a^2)}}(a\cosh\theta\,d\theta)$

$= \displaystyle\int \frac{a\cosh\theta\,d\theta}{\sqrt{(a^2\cosh^2\theta)}}$

$\qquad$ since $\cosh^2\theta - \sinh^2\theta = 1$

$= \displaystyle\int \frac{a\cosh\theta}{a\cosh\theta}\,d\theta = \int d\theta = \theta + c$

$= \mathbf{sinh^{-1}}\dfrac{x}{a} + c$, since $x = a\sinh\theta$

It is shown on page 682 that

$$\sinh^{-1}\frac{x}{a} = \ln\left\{\frac{x+\sqrt{(x^2+a^2)}}{a}\right\}$$

which provides an alternative solution to

$$\int \frac{1}{\sqrt{(x^2+a^2)}}\,dx$$

> **Problem 21. Evaluate $\displaystyle\int_0^2 \frac{1}{\sqrt{(x^2+4)}}\,dx$, correct to 4 decimal places**

$$\int_0^2 \frac{1}{\sqrt{(x^2+4)}}\,dx = \left[\sinh^{-1}\frac{x}{2}\right]_0^2 \text{ or}$$

$$\left[\ln\left\{\frac{x+\sqrt{(x^2+4)}}{2}\right\}\right]_0^2$$

from Problem 20, where $a = 2$

Using the logarithmic form,

$$\int_0^2 \frac{1}{\sqrt{(x^2+4)}}\,dx$$

$$= \left[\ln\left(\frac{2+\sqrt{8}}{2}\right) - \ln\left(\frac{0+\sqrt{4}}{2}\right)\right]$$

$$= \ln 2.4142 - \ln 1 = \mathbf{0.8814},$$

$$\text{correct to 4 decimal places.}$$

> **Problem 22. Evaluate $\displaystyle\int_1^2 \frac{2}{x^2\sqrt{(1+x^2)}}\,dx$, correct to 3 significant figures**

Since the integral contains a term of the form $\sqrt{(a^2+x^2)}$, then let $x = \sinh\theta$, from which $\dfrac{dx}{d\theta} = \cosh\theta$ and $dx = \cosh\theta\,d\theta$

Hence $\displaystyle\int \frac{2}{x^2\sqrt{(1+x^2)}}\,dx$

$= \displaystyle\int \frac{2(\cosh\theta\,d\theta)}{\sinh^2\theta\sqrt{(1+\sinh^2\theta)}}$

$= 2\displaystyle\int \frac{\cosh\theta\,d\theta}{\sinh^2\theta\cosh\theta}$

$\qquad$ since $\cosh^2\theta - \sinh^2\theta = 1$

$= 2\displaystyle\int \frac{d\theta}{\sinh^2\theta} = 2\int \text{cosech}^2\theta\,d\theta$

$= -2\coth\theta + c$

$\coth\theta = \dfrac{\cosh\theta}{\sinh\theta} = \dfrac{\sqrt{(1+\sinh^2\theta)}}{\sinh\theta} = \dfrac{\sqrt{(1+x^2)}}{x}$

Hence $\displaystyle\int_1^2 \frac{2}{x^2\sqrt{(1+x^2)}}\,\mathrm{d}x$

$= -[2\coth\theta]_1^2 = -2\left[\dfrac{\sqrt{(1+x^2)}}{x}\right]_1^2$

$= -2\left[\dfrac{\sqrt{5}}{2} - \dfrac{\sqrt{2}}{1}\right] = \mathbf{0.592},$

correct to 3 significant figures.

Problem 23. Find $\displaystyle\int \sqrt{(x^2+a^2)}\,\mathrm{d}x$

Let $x = a\sinh\theta$ then $\dfrac{\mathrm{d}x}{\mathrm{d}\theta} = a\cosh\theta$ and
$\mathrm{d}x = a\cosh\theta\,\mathrm{d}\theta$

Hence $\displaystyle\int \sqrt{(x^2+a^2)}\,\mathrm{d}x$

$= \displaystyle\int \sqrt{(a^2\sinh^2\theta + a^2)}\,(a\cosh\theta\,\mathrm{d}\theta)$

$= \displaystyle\int \sqrt{[a^2(\sinh^2\theta + 1)]}\,(a\cosh\theta\,\mathrm{d}\theta)$

$= \displaystyle\int \sqrt{(a^2\cosh^2\theta)}\,(a\cosh\theta\,\mathrm{d}\theta),$

$\qquad\qquad$ since $\cosh^2\theta - \sinh^2\theta = 1$

$= \displaystyle\int (a\cosh\theta)(a\cosh\theta)\,\mathrm{d}\theta = a^2\int \cosh^2\theta\,\mathrm{d}\theta$

$= a^2\displaystyle\int \left(\frac{1 + \cosh 2\theta}{2}\right)\mathrm{d}\theta$

$= \dfrac{a^2}{2}\left(\theta + \dfrac{\sinh 2\theta}{2}\right) + c$

$= \dfrac{a^2}{2}[\theta + \sinh\theta\cosh\theta] + c$

$\qquad\qquad$ since $\sinh 2\theta = 2\sinh\theta\cosh\theta$

Since $x = a\sinh\theta$, then $\sinh\theta = \dfrac{x}{a}$ and $\theta = \sinh^{-1}\dfrac{x}{a}$

Also since $\cosh^2\theta - \sinh^2\theta = 1$

then $\cosh\theta = \sqrt{(1 + \sinh^2\theta)}$

$= \sqrt{\left[1 + \left(\dfrac{x}{a}\right)^2\right]} = \sqrt{\left(\dfrac{a^2 + x^2}{a^2}\right)}$

$= \dfrac{\sqrt{(a^2 + x^2)}}{a}$

Hence $\displaystyle\int \sqrt{(x^2+a^2)}\,\mathrm{d}x$

$= \dfrac{a^2}{2}\left[\sinh^{-1}\dfrac{x}{a} + \left(\dfrac{x}{a}\right)\dfrac{\sqrt{(x^2+a^2)}}{a}\right] + c$

$= \dfrac{a^2}{2}\sinh^{-1}\dfrac{x}{a} + \dfrac{x}{2}\sqrt{(x^2+a^2)} + c$

Now try the following Practice Exercise

Practice Exercise 269   Further problems on integration using the $\sinh\theta$ substitution (answers on page 1182)

1. Find $\displaystyle\int \frac{2}{\sqrt{(x^2+16)}}\,\mathrm{d}x$

2. Find $\displaystyle\int \frac{3}{\sqrt{(9+5x^2)}}\,\mathrm{d}x$

3. Find $\displaystyle\int \sqrt{(x^2+9)}\,\mathrm{d}x$

4. Find $\displaystyle\int \sqrt{(4t^2+25)}\,\mathrm{d}t$

5. Evaluate $\displaystyle\int_0^3 \frac{4}{\sqrt{(t^2+9)}}\,\mathrm{d}t$

6. Evaluate $\displaystyle\int_0^1 \sqrt{(16+9\theta^2)}\,\mathrm{d}\theta$

65.8 Worked problems on integration using the $\cosh\theta$ substitution

Problem 24. Determine $\displaystyle\int \frac{1}{\sqrt{(x^2-a^2)}}\,\mathrm{d}x$

Let $x = a\cosh\theta$ then $\dfrac{\mathrm{d}x}{\mathrm{d}\theta} = a\sinh\theta$ and
$\mathrm{d}x = a\sinh\theta\,\mathrm{d}\theta$

Hence $\displaystyle\int \frac{1}{\sqrt{(x^2-a^2)}}\,\mathrm{d}x$

$= \displaystyle\int \frac{1}{\sqrt{(a^2\cosh^2\theta - a^2)}}\,(a\sinh\theta\,\mathrm{d}\theta)$

$= \displaystyle\int \frac{a\sinh\theta\,\mathrm{d}\theta}{\sqrt{[a^2(\cosh^2\theta - 1)]}}$

$$= \int \frac{a\sinh\theta\,d\theta}{\sqrt{(a^2\sinh^2\theta)}}$$

$$\text{since } \cosh^2\theta - \sinh^2\theta = 1$$

$$= \int \frac{a\sinh\theta\,d\theta}{a\sinh\theta} = \int d\theta = \theta + c$$

$$= \cosh^{-1}\frac{x}{a} + c \quad \text{since } x = a\cosh\theta$$

It is shown on page 682 that

$$\cosh^{-1}\frac{x}{a} = \ln\left\{\frac{x + \sqrt{(x^2 - a^2)}}{a}\right\}$$

which provides an alternative solution to

$$\int \frac{1}{\sqrt{(x^2 - a^2)}}\,dx$$

Problem 25. Determine $\int \frac{2x - 3}{\sqrt{(x^2 - 9)}}\,dx$

$$\int \frac{2x - 3}{\sqrt{(x^2 - 9)}}\,dx = \int \frac{2x}{\sqrt{(x^2 - 9)}}\,dx$$

$$-\int \frac{3}{\sqrt{(x^2 - 9)}}\,dx$$

The first integral is determined using the algebraic substitution $u = (x^2 - 9)$, and the second integral is of the form $\int \frac{1}{\sqrt{(x^2 - a^2)}}\,dx$ (see Problem 24)

Hence $\int \frac{2x}{\sqrt{(x^2 - 9)}}\,dx - \int \frac{3}{\sqrt{(x^2 - 9)}}\,dx$

$$= 2\sqrt{(x^2 - 9)} - 3\cosh^{-1}\frac{x}{3} + c$$

Problem 26. $\int \sqrt{(x^2 - a^2)}\,dx$

Let $x = a\cosh\theta$ then $\frac{dx}{d\theta} = a\sinh\theta$ and $dx = a\sinh\theta\,d\theta$

Hence $\int \sqrt{(x^2 - a^2)}\,dx$

$$= \int \sqrt{(a^2\cosh^2\theta - a^2)}\,(a\sinh\theta\,d\theta)$$

$$= \int \sqrt{[a^2(\cosh^2\theta - 1)]}\,(a\sinh\theta\,d\theta)$$

$$= \int \sqrt{(a^2\sinh^2\theta)}\,(a\sinh\theta\,d\theta)$$

$$= a^2\int \sinh^2\theta\,d\theta = a^2\int \left(\frac{\cosh 2\theta - 1}{2}\right)d\theta$$

$$\text{since } \cosh 2\theta = 1 + 2\sinh^2\theta$$

from Table 24.1, page 232,

$$= \frac{a^2}{2}\left[\frac{\sinh 2\theta}{2} - \theta\right] + c$$

$$= \frac{a^2}{2}[\sinh\theta\cosh\theta - \theta] + c$$

$$\text{since } \sinh 2\theta = 2\sinh\theta\cosh\theta$$

Since $x = a\cosh\theta$ then $\cosh\theta = \frac{x}{a}$ and

$$\theta = \cosh^{-1}\frac{x}{a}$$

Also, since $\cosh^2\theta - \sinh^2\theta = 1$, then

$$\sinh\theta = \sqrt{(\cosh^2\theta - 1)}$$

$$= \sqrt{\left[\left(\frac{x}{a}\right)^2 - 1\right]} = \frac{\sqrt{(x^2 - a^2)}}{a}$$

Hence $\int \sqrt{(x^2 - a^2)}\,dx$

$$= \frac{a^2}{2}\left[\frac{\sqrt{(x^2 - a^2)}}{a}\left(\frac{x}{a}\right) - \cosh^{-1}\frac{x}{a}\right] + c$$

$$= \frac{x}{2}\sqrt{(x^2 - a^2)} - \frac{a^2}{2}\cosh^{-1}\frac{x}{a} + c$$

Problem 27. Evaluate $\int_2^3 \sqrt{(x^2 - 4)}\,dx$

$$\int_2^3 \sqrt{(x^2 - 4)}\,dx = \left[\frac{x}{2}\sqrt{(x^2 - 4)} - \frac{4}{2}\cosh^{-1}\frac{x}{2}\right]_2^3$$

from Problem 26, when $a = 2$

$$= \left(\frac{3}{5} \sqrt{5} - 2 \cosh^{-1} \frac{3}{2} \right)$$

$$- (0 - 2 \cosh^{-1} 1)$$

Since $\cosh^{-1} \dfrac{x}{a} = \ln \left\{ \dfrac{x + \sqrt{(x^2 - a^2)}}{a} \right\}$ then

$$\cosh^{-1} \frac{3}{2} = \ln \left\{ \frac{3 + \sqrt{(3^2 - 2^2)}}{2} \right\}$$

$$= \ln 2.6180 = 0.9624$$

Similarly, $\cosh^{-1} 1 = 0$

Hence $\displaystyle\int_2^3 \sqrt{(x^2 - 4)} \, dx$

$$= \left[\frac{3}{2} \sqrt{5} - 2(0.9624) \right] - [0]$$

$$= \mathbf{1.429}, \text{ correct to 4 significant figures.}$$

Now try the following Practice Exercise

Practice Exercise 270   Further problems on integration using the $\cosh \theta$ substitution (answers on page 1182)

1. Find $\displaystyle\int \frac{1}{\sqrt{(t^2 - 16)}} \, dt$

2. Find $\displaystyle\int \frac{3}{\sqrt{(4x^2 - 9)}} \, dx$

3. Find $\displaystyle\int \sqrt{(\theta^2 - 9)} \, d\theta$

4. Find $\displaystyle\int \sqrt{(4\theta^2 - 25)} \, d\theta$

5. Evaluate $\displaystyle\int_1^2 \frac{2}{\sqrt{(x^2 - 1)}} \, dx$

6. Evaluate $\displaystyle\int_2^3 \sqrt{(t^2 - 4)} \, dt$

For fully worked solutions to each of the problems in Practice Exercises 264 to 270 in this chapter, go to the website:
www.routledge.com/cw/bird

Revision Test 24   Standard integration and some substitution methods

This assignment covers the material contained in Chapters 63 to 65. *The marks available are shown in brackets at the end of each question.*

1. Determine: (a) $\displaystyle\int 3\sqrt{t^5}\,dt$ (b) $\displaystyle\int \frac{2}{\sqrt[3]{x^2}}\,dx$

   (c) $\displaystyle\int (2+\theta)^2\,d\theta$ (9)

2. Evaluate the following integrals, each correct to 4 significant figures:

   (a) $\displaystyle\int_0^{\frac{\pi}{3}} 3\sin 2t\,dt$ (b) $\displaystyle\int_1^2 \left(\frac{2}{x^2}+\frac{1}{x}+\frac{3}{4}\right)dx$

   (c) $\displaystyle\int_0^1 \frac{3}{e^{2t}}\,dt$ (15)

3. Determine the following integrals:

   (a) $\displaystyle\int 5(6t+5)^7\,dt$ (b) $\displaystyle\int \frac{3\ln x}{x}\,dx$

   (c) $\displaystyle\int \frac{2}{\sqrt{(2\theta-1)}}\,d\theta$ (12)

4. Evaluate the following definite integrals:

   (a) $\displaystyle\int_0^{\frac{\pi}{2}} 2\sin\left(2t+\frac{\pi}{3}\right)dt$ (b) $\displaystyle\int_0^1 3x\,e^{4x^2-3}\,dx$ (10)

5. Determine the following integrals:

   (a) $\displaystyle\int \cos^3 x\sin^2 x\,dx$ (b) $\displaystyle\int \frac{2}{\sqrt{(9-4x^2)}}\,dx$

   (c) $\displaystyle\int \frac{2}{\sqrt{(4x^2-9)}}\,dx$ (14)

6. Evaluate the following definite integrals, correct to 4 significant figures:

   (a) $\displaystyle\int_0^{\frac{\pi}{2}} 3\sin^2 t\,dt$ (b) $\displaystyle\int_0^{\frac{\pi}{3}} 3\cos 5\theta\sin 3\theta\,d\theta$

   (c) $\displaystyle\int_0^2 \frac{5}{4+x^2}\,dx$ (15)

For lecturers/instructors/teachers, fully worked solutions to each of the problems in Revision Test 24, together with a full marking scheme, are available at the website:
www.routledge.com/cw/bird

Chapter 66

Integration using partial fractions

Why it is important to understand: Integration using partial fractions

Sometimes expressions which at first sight look impossible to integrate using standard techniques may in fact be integrated by first expressing them as simpler partial fractions and then using earlier learned techniques. As explained in Chapter 19, the algebraic technique of resolving a complicated fraction into partial fractions is often needed by electrical and mechanical engineers for not only determining certain integrals in calculus, but for determining inverse Laplace transforms, and for analysing linear differential equations like resonant circuits and feedback control systems.

At the end of this chapter, you should be able to:

- integrate functions using partial fractions with linear factors
- integrate functions using partial fractions with repeated linear factors
- integrate functions using partial fractions with quadratic factors

---

66.1 Introduction

The process of expressing a fraction in terms of simpler fractions – called **partial fractions** – is discussed in Chapter 19, with the forms of partial fractions used being summarised in Table 19.1, page 179.

Certain functions have to be resolved into partial fractions before they can be integrated, as demonstrated in the following Problems.

66.2 Worked problems on integration using partial fractions with linear factors

Problem 1. Determine $\int \dfrac{11 - 3x}{x^2 + 2x - 3} \, dx$

As shown in Problem 1, page 179:

$$\frac{11 - 3x}{x^2 + 2x - 3} \equiv \frac{2}{(x-1)} - \frac{5}{(x+3)}$$

Hence $\int \dfrac{11 - 3x}{x^2 + 2x - 3}\, dx$

$$= \int \left\{ \frac{2}{(x-1)} - \frac{5}{(x+3)} \right\} dx$$

$$= 2\ln(x-1) - 5\ln(x+3) + c$$

(by algebraic substitutions – see Chapter 64)

or $\ln \left\{ \dfrac{(x-1)^2}{(x+3)^5} \right\} + c$ by the laws of logarithms.

Problem 2. Find

$$\int \frac{2x^2 - 9x - 35}{(x+1)(x-2)(x+3)}\, dx$$

It was shown in Problem 2, page 179:

$$\frac{2x^2 - 9x - 35}{(x+1)(x-2)(x+3)} \equiv \frac{4}{(x+1)} - \frac{3}{(x-2)} + \frac{1}{(x+3)}$$

Hence $\int \dfrac{2x^2 - 9x - 35}{(x+1)(x-2)(x+3)}\, dx$

$$\equiv \int \left\{ \frac{4}{(x+1)} - \frac{3}{(x-2)} + \frac{1}{(x+3)} \right\} dx$$

$$= 4\ln(x+1) - 3\ln(x-2) + \ln(x+3) + c$$

or $\ln \left\{ \dfrac{(x+1)^4(x+3)}{(x-2)^3} \right\} + c$

Problem 3. Determine $\displaystyle\int \frac{x^2 + 1}{x^2 - 3x + 2}\, dx$

By dividing out (since the numerator and denominator are of the same degree) and resolving into partial fractions, it was shown in Problem 3, page 180:

$$\frac{x^2 + 1}{x^2 - 3x + 2} \equiv 1 - \frac{2}{(x-1)} + \frac{5}{(x-2)}$$

Hence $\int \dfrac{x^2 + 1}{x^2 - 3x + 2}\, dx$

$$\equiv \int \left\{ 1 - \frac{2}{(x-1)} + \frac{5}{(x-2)} \right\} dx$$

$$= (x-2)\ln(x-1) + 5\ln(x-2) + c$$

or $x + \ln \left\{ \dfrac{(x-2)^5}{(x-1)^2} \right\} + c$

Problem 4. Evaluate

$$\int_2^3 \frac{x^3 - 2x^2 - 4x - 4}{x^2 + x - 2}\, dx$$

correct to 4 significant figures

By dividing out and resolving into partial fractions it was shown in Problem 4, page 180:

$$\frac{x^3 - 2x^2 - 4x - 4}{x^2 + x - 2} \equiv x - 3 + \frac{4}{(x+2)} - \frac{3}{(x-1)}$$

Hence $\int_2^3 \dfrac{x^3 - 2x^2 - 4x - 4}{x^2 + x - 2}\, dx$

$$\equiv \int_2^3 \left\{ x - 3 + \frac{4}{(x+2)} - \frac{3}{(x-1)} \right\} dx$$

$$= \left[\frac{x^2}{2} - 3x + 4\ln(x+2) - 3\ln(x-1) \right]_2^3$$

$$= \left(\frac{9}{2} - 9 + 4\ln 5 - 3\ln 2 \right)$$

$$\qquad\qquad - (2 - 6 + 4\ln 4 - 3\ln 1)$$

$$= -1.687, \text{ correct to 4 significant figures.}$$

Now try the following Practice Exercise

Practice Exercise 271 Further problems on integration using partial fractions with linear factors (answers on page 1182)

In Problems 1 to 5, integrate with respect to x.

1. $\displaystyle\int \frac{12}{(x^2 - 9)}\, dx$

2. $\displaystyle\int \frac{4(x-4)}{(x^2 - 2x - 3)}\, dx$

3. $\int \dfrac{3(2x^2 - 8x - 1)}{(x+4)(x+1)(2x-1)}\,dx$

4. $\int \dfrac{x^2 + 9x + 8}{x^2 + x - 6}\,dx$

5. $\int \dfrac{3x^3 - 2x^2 - 16x + 20}{(x-2)(x+2)}\,dx$

In Problems 6 and 7, evaluate the definite integrals correct to 4 significant figures.

6. $\int_3^4 \dfrac{x^2 - 3x + 6}{x(x-2)(x-1)}\,dx$

7. $\int_4^6 \dfrac{x^2 - x - 14}{x^2 - 2x - 3}\,dx$

8. Determine the value of k, given that:
$$\int_0^1 \dfrac{(x-k)}{(3x+1)(x+1)}\,dx = 0$$

9. The velocity constant k of a given chemical reaction is given by:
$$kt = \int \left(\dfrac{1}{(3-0.4x)(2-0.6x)} \right) dx$$
where $x = 0$ when $t = 0$. Show that:
$$kt = \ln \left\{ \dfrac{2(3-0.4x)}{3(2-0.6x)} \right\}$$

10. The velocity, v, of an object in a medium at time t seconds is given by: $t = \int_{20}^{80} \dfrac{dv}{v(2v-1)}$
Evaluate t, in milliseconds, correct to 2 decimal places.

66.3 Worked problems on integration using partial fractions with repeated linear factors

Problem 5. Determine $\int \dfrac{2x+3}{(x-2)^2}\,dx$

It was shown in Problem 5, page 181:
$$\dfrac{2x+3}{(x-2)^2} \equiv \dfrac{2}{(x-2)} + \dfrac{7}{(x-2)^2}$$

Thus $\int \dfrac{2x+3}{(x-2)^2}\,dx \equiv \int \left\{ \dfrac{2}{(x-2)} + \dfrac{7}{(x-2)^2} \right\} dx$
$$= 2\ln(x-2) - \dfrac{7}{(x-2)} + c$$

$\left[\int \dfrac{7}{(x-2)^2}\,dx \text{ is determined using the algebraic substitution } u = (x-2) - \text{see Chapter 64.} \right]$

Problem 6. Find $\int \dfrac{5x^2 - 2x - 19}{(x+3)(x-1)^2}\,dx$

It was shown in Problem 6, page 181:
$$\dfrac{5x^2 - 2x - 19}{(x+3)(x-1)^2} \equiv \dfrac{2}{(x+3)} + \dfrac{3}{(x-1)} - \dfrac{4}{(x-1)^2}$$

Hence $\int \dfrac{5x^2 - 2x - 19}{(x+3)(x-1)^2}\,dx$
$$\equiv \int \left\{ \dfrac{2}{(x+3)} + \dfrac{3}{(x-1)} - \dfrac{4}{(x-1)^2} \right\} dx$$
$$= 2\ln(x+3) + 3\ln(x-1) + \dfrac{4}{(x-1)} + c$$
or $\ln\left\{ (x+3)^2(x-1)^3 \right\} + \dfrac{4}{(x-1)} + c$

Problem 7. Evaluate
$$\int_{-2}^1 \dfrac{3x^2 + 16x + 15}{(x+3)^3}\,dx$$
correct to 4 significant figures

It was shown in Problem 7, page 182:
$$\dfrac{3x^2 + 16x + 15}{(x+3)^3} \equiv \dfrac{3}{(x+3)} - \dfrac{2}{(x+3)^2} - \dfrac{6}{(x+3)^3}$$

Hence $\int \dfrac{3x^2 + 16x + 15}{(x+3)^3}\,dx$

$$\equiv \int_{-2}^{1} \left\{ \frac{3}{(x+3)} - \frac{2}{(x+3)^2} - \frac{6}{(x+3)^3} \right\} dx$$

$$= \left[3\ln(x+3) + \frac{2}{(x+3)} + \frac{3}{(x+3)^2} \right]_{-2}^{1}$$

$$= \left(3\ln 4 + \frac{2}{4} + \frac{3}{16} \right) - \left(3\ln 1 + \frac{2}{1} + \frac{3}{1} \right)$$

$$= -0.1536, \text{ correct to 4 significant figures.}$$

Now try the following Practice Exercise

Practice Exercise 272   Further problems on integration using partial fractions with repeated linear factors (answers on page 1182)

In Problems 1 and 2, integrate with respect to x.

1. $\displaystyle\int \frac{4x-3}{(x+1)^2} dx$

2. $\displaystyle\int \frac{5x^2 - 30x + 44}{(x-2)^3} dx$

In Problems 3 and 4, evaluate the definite integrals correct to 4 significant figures.

3. $\displaystyle\int_{1}^{2} \frac{x^2 + 7x + 3}{x^2(x+3)}$

4. $\displaystyle\int_{6}^{7} \frac{18 + 21x - x^2}{(x-5)(x+2)^2} dx$

5. Show that $\displaystyle\int_{0}^{1} \left(\frac{4t^2 + 9t + 8}{(t+2)(t+1)^2} \right) dt = 2.546$,

   correct to 4 significant figures.

66.4 Worked problems on integration using partial fractions with quadratic factors

Problem 8.   Find $\displaystyle\int \frac{3 + 6x + 4x^2 - 2x^3}{x^2(x^2+3)} dx$

It was shown in Problem 9, page 183:

$$\frac{3 + 6x + 4x^2 - 2x^3}{x^2(x^2+3)} \equiv \frac{2}{x} + \frac{1}{x^2} + \frac{3 - 4x}{(x^2+3)}$$

Thus $\displaystyle\int \frac{3 + 6x + 4x^2 - 2x^3}{x^2(x^2+3)} dx$

$$\equiv \int \left(\frac{2}{x} + \frac{1}{x^2} + \frac{(3-4x)}{(x^2+3)} \right) dx$$

$$= \int \left\{ \frac{2}{x} + \frac{1}{x^2} + \frac{3}{(x^2+3)} - \frac{4x}{(x^2+3)} \right\} dx$$

$$\int \frac{3}{(x^2+3)} dx = 3 \int \frac{1}{x^2 + (\sqrt{3})^2} dx$$

$$= \frac{3}{\sqrt{3}} \tan^{-1} \frac{x}{\sqrt{3}}, \text{ from 12, Table 65.1, page 732.}$$

$\displaystyle\int \frac{4x}{x^2+3} dx$ is determined using the algebraic substitution $u = (x^2+3)$

Hence $\displaystyle\int \left\{ \frac{2}{x} + \frac{1}{x^2} + \frac{3}{(x^2+3)} - \frac{4x}{(x^2+3)} \right\} dx$

$$= 2\ln x - \frac{1}{x} + \frac{3}{\sqrt{3}} \tan^{-1} \frac{x}{\sqrt{3}} - 2\ln(x^2+3) + c$$

$$= \ln \left(\frac{x}{x^2+3} \right)^2 - \frac{1}{x} + \sqrt{3} \tan^{-1} \frac{x}{\sqrt{3}} + c$$

Problem 9.   Determine $\displaystyle\int \frac{1}{(x^2 - a^2)} dx$

Let $\dfrac{1}{(x^2 - a^2)} \equiv \dfrac{A}{(x-a)} + \dfrac{B}{(x+a)}$

$$\equiv \frac{A(x+a) + B(x-a)}{(x+a)(x-a)}$$

Equating the numerators gives:

$$1 \equiv A(x+a) + B(x-a)$$

Let $x = a$, then $A = \dfrac{1}{2a}$, and let $x = -a$, then

$B = -\dfrac{1}{2a}$

Hence $\int \dfrac{1}{(x^2 - a^2)}\, dx$

$\equiv \int \dfrac{1}{2a}\left[\dfrac{1}{(x-a)} - \dfrac{1}{(x+a)}\right] dx$

$= \dfrac{1}{2a}[\ln(x-a) - \ln(x+a)] + c$

$= \dfrac{1}{2a}\ln\left(\dfrac{x-a}{x+a}\right) + c$

Problem 10.   Evaluate

$$\int_3^4 \dfrac{3}{(x^2 - 4)}\, dx$$

correct to 3 significant figures

From Problem 9,

$\int_3^4 \dfrac{3}{(x^2 - 4)}\, dx = 3\left[\dfrac{1}{2(2)}\ln\left(\dfrac{x-2}{x+2}\right)\right]_3^4$

$= \dfrac{3}{4}\left[\ln\dfrac{2}{6} - \ln\dfrac{1}{5}\right]$

$= \dfrac{3}{4}\ln\dfrac{5}{3} = \mathbf{0.383}$, correct to 3
significant figures.

Problem 11.   Determine $\int \dfrac{1}{(a^2 - x^2)}\, dx$

Using partial fractions, let

$\dfrac{1}{(a^2 - x^2)} \equiv \dfrac{1}{(a-x)(a+x)} \equiv \dfrac{A}{(a-x)} + \dfrac{B}{(a+x)}$

$\equiv \dfrac{A(a+x) + B(a-x)}{(a-x)(a+x)}$

Then $1 \equiv A(a+x) + B(a-x)$

Let $x = a$ then $A = \dfrac{1}{2a}$. Let $x = -a$ then $B = \dfrac{1}{2a}$

Hence $\int \dfrac{1}{(a^2 - x^2)}\, dx$

$= \int \dfrac{1}{2a}\left[\dfrac{1}{(a-x)} + \dfrac{1}{(a+x)}\right] dx$

$= \dfrac{1}{2a}[-\ln(a-x) + \ln(a+x)] + c$

$= \dfrac{1}{2a}\ln\left(\dfrac{a+x}{a-x}\right) + c$

Problem 12.   Evaluate

$$\int_0^2 \dfrac{5}{(9 - x^2)}\, dx$$

correct to 4 decimal places

From Problem 11,

$\int_0^2 \dfrac{5}{(9 - x^2)}\, dx = 5\left[\dfrac{1}{2(3)}\ln\left(\dfrac{3+x}{3-x}\right)\right]_0^2$

$= \dfrac{5}{6}\left[\ln\dfrac{5}{1} - \ln 1\right]$

$= \mathbf{1.3412}$, correct to 4 decimal places.

Now try the following **Practice Exercise**

**Practice Exercise 273   Further problems on
integration using partial fractions with
quadratic factors (answers on page 1182)**

1.   Determine $\int \dfrac{x^2 - x - 13}{(x^2 + 7)(x - 2)}\, dx$

In Problems 2 to 4, evaluate the definite integrals
correct to 4 significant figures.

2. $\displaystyle\int_5^6 \dfrac{6x - 5}{(x - 4)(x^2 + 3)}\, dx$

3. $\displaystyle\int_1^2 \dfrac{4}{(16 - x^2)}\, dx$

4. $\displaystyle\int_4^5 \dfrac{2}{(x^2 - 9)}\, dx$

5.   Show   that $\displaystyle\int_1^2 \left(\dfrac{2 + \theta + 6\theta^2 - 2\theta^3}{\theta^2(\theta^2 + 1)}\right) d\theta$

$= 1.606$, correct to 4 significant figures.

**For fully worked solutions to each of the problems in Practice Exercises 271 to 273 in this chapter,
go to the website:**
www.routledge.com/cw/bird

The $t = \tan\dfrac{\theta}{2}$ substitution

At the end of this chapter, you should be able to:

- develop formulae for $\sin\theta$, $\cos\theta$ and $d\theta$ in terms of t, where $t = \tan\theta/2$
- integrate functions using $t = \tan\theta/2$ substitution

67.1 Introduction

Integrals of the form $\displaystyle\int \dfrac{1}{a\cos\theta + b\sin\theta + c}\,d\theta$, where a, b and c are constants, may be determined by using the substitution $t = \tan\dfrac{\theta}{2}$. The reason is explained below.

If angle A in the right-angled triangle ABC shown in Figure 67.1 is made equal to $\dfrac{\theta}{2}$ then, since tangent $= \dfrac{\text{opposite}}{\text{adjacent}}$, if $BC = t$ and $AB = 1$, then $\tan\dfrac{\theta}{2} = t$

By Pythagoras' theorem, $AC = \sqrt{1 + t^2}$

Therefore $\sin\dfrac{\theta}{2} = \dfrac{t}{\sqrt{1 + t^2}}$ and $\cos\dfrac{\theta}{2} = \dfrac{1}{\sqrt{1 + t^2}}$ Since $\sin 2x = 2\sin x \cos x$ (from double angle formulae, Chapter 44), then

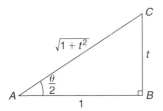

Figure 67.1

$$\sin\theta = 2\sin\dfrac{\theta}{2}\cos\dfrac{\theta}{2}$$

$$= 2\left(\dfrac{t}{\sqrt{1 + t^2}}\right)\left(\dfrac{t}{\sqrt{1 + t^2}}\right)$$

i.e. $\qquad \sin\theta = \dfrac{2t}{(1 + t^2)} \qquad (1)$

Since $\cos 2x = \cos^2\frac{\theta}{2} - \sin^2\frac{\theta}{2}$

$$= \left(\frac{1}{\sqrt{1+t^2}}\right)^2 - \left(\frac{t}{\sqrt{1+t^2}}\right)^2$$

i.e. $\qquad\qquad \cos\theta = \dfrac{1-t^2}{1+t^2} \qquad\qquad (2)$

Also, since $t = \tan\dfrac{\theta}{2}$,

$\dfrac{dt}{d\theta} = \dfrac{1}{2}\sec^2\dfrac{\theta}{2} = \dfrac{1}{2}\left(1 + \tan^2\dfrac{\theta}{2}\right)$ from trigonometric identities,

i.e. $\qquad\qquad \dfrac{dt}{d\theta} = \dfrac{1}{2}(1+t^2)$

from which, $\qquad\qquad d\theta = \dfrac{2\,dt}{1+t^2} \qquad (3)$

Equations (1), (2) and (3) are used to determine integrals of the form $\displaystyle\int \dfrac{1}{a\cos\theta + b\sin\theta + c}\,d\theta$ where a, b or c may be zero.

67.2 Worked problems on the $t = \tan\dfrac{\theta}{2}$ substitution

Problem 1. Determine $\displaystyle\int \dfrac{d\theta}{\sin\theta}$

If $t = \tan\dfrac{\theta}{2}$ then $\sin\theta = \dfrac{2t}{1+t^2}$ and $d\theta = \dfrac{2\,dt}{1+t^2}$ from equations (1) and (3).

Thus $\displaystyle\int \dfrac{d\theta}{\sin\theta} = \int \dfrac{1}{\sin\theta}\,d\theta$

$$= \int \dfrac{1}{\dfrac{2t}{1+t^2}}\left(\dfrac{2\,dt}{1+t^2}\right)$$

$$= \int \dfrac{1}{t}\,dt = \ln t + c$$

Hence $\displaystyle\int \dfrac{d\theta}{\sin\theta} = \ln\left(\tan\dfrac{\theta}{2}\right) + c$

Problem 2. Determine $\displaystyle\int \dfrac{dx}{\cos x}$

If $t = \tan\dfrac{x}{2}$ then $\cos x = \dfrac{1-t^2}{1+t^2}$ and $dx = \dfrac{2\,dt}{1+t^2}$ from equations (2) and (3).

Thus $\displaystyle\int \dfrac{dx}{\cos x} = \int \dfrac{1}{\dfrac{1-t^2}{1+t^2}}\left(\dfrac{2\,dt}{1+t^2}\right)$

$$= \int \dfrac{2}{1-t^2}\,dt$$

$\dfrac{2}{1-t^2}$ may be resolved into partial fractions (see Chapter 19).

Let $\qquad \dfrac{2}{1-t^2} = \dfrac{2}{(1-t)(1+t)}$

$$= \dfrac{A}{(1-t)} + \dfrac{B}{(1+t)}$$

$$= \dfrac{A(1+t) + B(1-t)}{(1-t)(1+t)}$$

Hence $\qquad 2 = A(1+t) + B(1-t)$

When $\qquad t = 1, 2 = 2A$, from which, $A = 1$

When $\qquad t = -1, 2 = 2B$, from which, $B = 1$

Hence $\displaystyle\int \dfrac{2\,dt}{1-t^2} = \int \dfrac{1}{(1-t)} + \dfrac{1}{(1+t)}\,dt$

$$= -\ln(1-t) + \ln(1+t) + c$$

$$= \ln\left\{\dfrac{(1+t)}{(1-t)}\right\} + c$$

Thus $\displaystyle\int \dfrac{dx}{\cos x} = \ln\left\{\dfrac{1+\tan\dfrac{x}{2}}{1-\tan\dfrac{x}{2}}\right\} + c$

Note that since $\tan\dfrac{\pi}{4} = 1$, the above result may be written as:

$$\int \dfrac{dx}{\cos x} = \ln\left\{\dfrac{\tan\dfrac{\pi}{4} + \tan\dfrac{x}{2}}{1 - \tan\dfrac{\pi}{4}\tan\dfrac{x}{2}}\right\} + c$$

$$= \ln\left\{\tan\left(\dfrac{\pi}{4} + \dfrac{x}{2}\right)\right\} + c$$

from compound angles, Chapter 44.

Section J

Problem 3. Determine $\displaystyle\int \frac{dx}{1+\cos x}$

If $t = \tan \dfrac{x}{2}$ then $\cos x = \dfrac{1-t^2}{1+t^2}$ and $dx = \dfrac{2\,dt}{1+t^2}$ from equations (2) and (3).

Thus $\displaystyle\int \frac{dx}{1+\cos x} = \int \frac{1}{1+\cos x}\,dx$

$$= \int \frac{1}{1+\dfrac{1-t^2}{1+t^2}}\left(\frac{2\,dt}{1+t^2}\right)$$

$$= \int \frac{1}{\dfrac{(1+t^2)+(1-t^2)}{1+t^2}}\left(\frac{2\,dt}{1+t^2}\right)$$

$$= \int dt$$

Hence $\displaystyle\int \frac{dx}{1+\cos x} = t + c = \tan \frac{x}{2} + c$

Problem 4. Determine $\displaystyle\int \frac{d\theta}{5+4\cos\theta}$

If $t = \tan \dfrac{\theta}{2}$ then $\cos\theta = \dfrac{1-t^2}{1+t^2}$ and $dx = \dfrac{2\,dt}{1+t^2}$ from equations (2) and (3).

Thus $\displaystyle\int \frac{d\theta}{5+4\cos\theta} = \int \frac{\left(\dfrac{2\,dt}{1+t^2}\right)}{5+4\left(\dfrac{1-t^2}{1+t^2}\right)}$

$$= \int \frac{\left(\dfrac{2\,dt}{1+t^2}\right)}{\dfrac{5(1+t^2)+4(1-t^2)}{(1+t^2)}}$$

$$= 2\int \frac{dt}{t^2+9} = 2\int \frac{dt}{t^2+3^2}$$

$$= 2\left(\frac{1}{3}\tan^{-1}\frac{t}{3}\right) + c$$

from 12 of Table 65.1, page 732. Hence

$$\int \frac{d\theta}{5+4\cos\theta} = \frac{2}{3}\tan^{-1}\left(\frac{1}{3}\tan\frac{\theta}{2}\right) + c$$

Now try the following Practice Exercise

Practice Exercise 274 **Further problems on the $t = \tan \dfrac{\theta}{2}$ substitution (answers on page 1182)**

Integrate the following with respect to the variable:

1. $\displaystyle\int \frac{d\theta}{1+\sin\theta}$

2. $\displaystyle\int \frac{dx}{1-\cos x + \sin x}$

3. $\displaystyle\int \frac{d\alpha}{3+2\cos\alpha}$

4. $\displaystyle\int \frac{dx}{3\sin x - 4\cos x}$

67.3 Further worked problems on the $t = \tan \dfrac{\theta}{2}$ substitution

Problem 5. Determine $\displaystyle\int \frac{dx}{\sin x + \cos x}$

If $t = \tan \dfrac{x}{2}$ then $\sin x = \dfrac{2t}{1+t^2}$, $\cos x = \dfrac{1-t^2}{1+t^2}$ and $dx = \dfrac{2\,dt}{1+t^2}$ from equations (1), (2) and (3).

Thus

$$\int \frac{dx}{\sin x + \cos x} = \int \frac{\dfrac{2\,dt}{1+t^2}}{\left(\dfrac{2t}{1+t^2}\right)+\left(\dfrac{1-t^2}{1+t^2}\right)}$$

$$= \int \frac{\dfrac{2\,dt}{1+t^2}}{\dfrac{2t+1-t^2}{1+t^2}} = \int \frac{2\,dt}{1+2t-t^2}$$

$$= \int \frac{-2\,dt}{t^2-2t-1} = \int \frac{-2\,dt}{(t-1)^2-2}$$

$$= \int \frac{2\,dt}{(\sqrt{2})^2-(t-1)^2}$$

$$= 2\left[\frac{1}{2\sqrt{2}}\ln\left\{\frac{\sqrt{2}+(t-1)}{\sqrt{2}-(t-1)}\right\}\right] + c$$

(see Problem 11, Chapter 66, page 747),

i.e. $\displaystyle\int \frac{dx}{\sin x + \cos x}$

$$= \frac{1}{\sqrt{2}}\ln\left\{\frac{\sqrt{2}-1+\tan\dfrac{x}{2}}{\sqrt{2}+1-\tan\dfrac{x}{2}}\right\} + c$$

Problem 6. Determine
$$\int \frac{dx}{7 - 3\sin x + 6\cos x}$$

From equations (1) and (3),

$$\int \frac{dx}{7 - 3\sin x + 6\cos x}$$

$$= \int \frac{\dfrac{2\,dt}{1+t^2}}{7 - 3\left(\dfrac{2t}{1+t^2}\right) + 6\left(\dfrac{1-t^2}{1+t^2}\right)}$$

$$= \int \frac{\dfrac{2\,dt}{1+t^2}}{\dfrac{7(1+t^2)-3(2t)+6(1-t^2)}{1+t^2}}$$

$$= \int \frac{2\,dt}{7 + 7t^2 - 6t + 6 - 6t^2}$$

$$= \int \frac{2\,dt}{t^2 - 6t + 13} = \int \frac{2\,dt}{(t-3)^2 + 2^2}$$

$$= 2\left[\frac{1}{2}\tan^{-1}\left(\frac{t-3}{2}\right)\right] + c$$

from 12, Table 65.1, page 732. Hence

$$\int \frac{dx}{7 - 3\sin x + 6\cos x}$$

$$= \tan^{-1}\left(\frac{\tan\dfrac{x}{2}-3}{2}\right) + c$$

Problem 7. Determine $\displaystyle\int \frac{d\theta}{4\cos\theta + 3\sin\theta}$

From equations (1) to (3),

$$\int \frac{d\theta}{4\cos\theta + 3\sin\theta}$$

$$= \int \frac{\dfrac{2\,dt}{1+t^2}}{4\left(\dfrac{1-t^2}{1+t^2}\right) + 3\left(\dfrac{2t}{1+t^2}\right)}$$

$$= \int \frac{2\,dt}{4 - 4t^2 + 6t} = \int \frac{dt}{2 + 3t - 2t^2}$$

$$= -\frac{1}{2}\int \frac{dt}{t^2 - \dfrac{3}{2}t - 1}$$

$$= -\frac{1}{2}\int \frac{dt}{\left(t - \dfrac{3}{4}\right)^2 - \dfrac{25}{16}}$$

$$= \frac{1}{2}\int \frac{dt}{\left(\dfrac{5}{4}\right)^2 - \left(t - \dfrac{3}{4}\right)^2}$$

$$= \frac{1}{2}\left[\frac{1}{2\left(\dfrac{5}{4}\right)}\ln\left\{\frac{\dfrac{5}{4}+\left(t-\dfrac{3}{4}\right)}{\dfrac{5}{4}-\left(t-\dfrac{3}{4}\right)}\right\}\right] + c$$

from Problem 11, Chapter 66, page 747

$$= \frac{1}{5}\ln\left\{\frac{\dfrac{1}{2}+t}{2-t}\right\} + c$$

Hence $\displaystyle\int \frac{d\theta}{4\cos\theta + 3\sin\theta}$

$$= \frac{1}{5}\ln\left\{\frac{\dfrac{1}{2}+\tan\dfrac{\theta}{2}}{2-\tan\dfrac{\theta}{2}}\right\} + c$$

or $\displaystyle \frac{1}{5}\ln\left\{\frac{1+2\tan\dfrac{\theta}{2}}{4-2\tan\dfrac{\theta}{2}}\right\} + c$

Now try the following Practice Exercise

Practice Exercise 275 Further problems on the $t = \tan\dfrac{\theta}{2}$ substitution (answers on page 1183)

In Problems 1 to 4, integrate with respect to the variable.

1. $\displaystyle\int \frac{\mathrm{d}\theta}{5 + 4\sin\theta}$

2. $\displaystyle\int \frac{\mathrm{d}x}{1 + 2\sin x}$

3. $\displaystyle\int \frac{\mathrm{d}p}{3 - 4\sin p + 2\cos p}$

4. $\displaystyle\int \frac{\mathrm{d}\theta}{3 - 4\sin\theta}$

5. Show that
$$\int \frac{\mathrm{d}t}{1 + 3\cos t} = \frac{1}{2\sqrt{2}}\ln\left\{\frac{\sqrt{2} + \tan\dfrac{t}{2}}{\sqrt{2} - \tan\dfrac{t}{2}}\right\} + c$$

6. Show that $\displaystyle\int_0^{\pi/3} \frac{3\,\mathrm{d}\theta}{\cos\theta} = 3.95$, correct to 3 significant figures.

7. Show that
$$\int_0^{\pi/2} \frac{\mathrm{d}\theta}{2 + \cos\theta} = \frac{\pi}{3\sqrt{3}}$$

For fully worked solutions to each of the problems in Practice Exercises 274 and 275 in this chapter, go to the website:
www.routledge.com/cw/bird

Chapter 68

Integration by parts

> *Why it is important to understand:* **Integration by parts**
>
> **Integration by parts is a very important technique that is used often in engineering and science. It is frequently used to change the integral of a product of functions into an ideally simpler integral. It is the foundation for the theory of differential equations and is used with Fourier series. We have looked at standard integrals followed by various techniques to change integrals into standard ones; integration by parts is a particularly important technique for integrating a product of two functions.**

At the end of this chapter, you should be able to:

- appreciate when integration by parts is required
- integrate functions using integration by parts
- evaluate definite integrals using integration by parts

68.1 Introduction

From the product rule of differentiation:

$$\frac{d}{dx}(uv) = v\frac{du}{dx} + u\frac{dv}{dx}$$

where u and v are both functions of x.

Rearranging gives: $u\dfrac{dv}{dx} = \dfrac{d}{dx}(uv) - v\dfrac{du}{dx}$

Integrating both sides with respect to x gives:

$$\int u\frac{dv}{dx}\,dx = \int \frac{d}{dx}(uv)\,dx - \int v\frac{du}{dx}\,dx$$

i.e.
$$\int u\frac{dv}{dx}\,dx = uv - \int v\frac{du}{dx}\,dx$$

or
$$\int u\,dv = uv - \int v\,du$$

This is known as the **integration by parts formula** and provides a method of integrating such products of simple functions as $\int x e^x\,dx$, $\int t\sin t\,dt$, $\int e^\theta \cos\theta\,d\theta$ and $\int x\ln x\,dx$

Given a product of two terms to integrate, the initial choice is: 'which part to make equal to u' and 'which part to make equal to v'. The choice must be such that the 'u part' becomes a constant after successive differentiation and the 'dv part' can be integrated from standard integrals. Invariably, the following rule holds: if a product to be integrated contains an algebraic term (such as x, t^2 or 3θ) then this term is chosen as the u part. The one exception to this rule is when a '$\ln x$' term is involved; in this case $\ln x$ is chosen as the 'u part'.

68.2 Worked problems on integration by parts

Problem 1. Determine $\int x\cos x\,dx$

From the integration by parts formula,

$$\int u\,dv = uv - \int v\,du$$

Let $u = x$, from which $\dfrac{du}{dx} = 1$, i.e. $du = dx$ and let $dv = \cos x\,dx$, from which $v = \int \cos x\,dx = \sin x$. Expressions for u, du and v are now substituted into the 'by parts' formula as shown below.

$$\int \boxed{u}\,\boxed{dv} = \boxed{u}\,\boxed{v} - \int \boxed{v}\,\boxed{du}$$

$$\int \boxed{x}\,\boxed{\cos x\,dx} = \boxed{(x)}\,\boxed{(\sin x)} - \int \boxed{(\sin x)}\,\boxed{(dx)}$$

i.e. $\displaystyle\int x \cos x\,dx = x \sin x - (-\cos x) + c$

$$= x\sin x + \cos x + c$$

(This result may be checked by differentiating the right-hand side,

i.e. $\dfrac{d}{dx}(x \sin x + \cos x + c)$

$\quad = [(x)(\cos x) + (\sin x)(1)] - \sin x + 0$

$\qquad$ using the product rule

$\quad = x \cos x$, which is the function being

$\qquad$ integrated)

Problem 2. Find $\int 3te^{2t}\,dt$

Let $u = 3t$, from which, $\dfrac{du}{dt} = 3$, i.e. $du = 3\,dt$ and

let $dv = e^{2t}\,dt$, from which, $v = \int e^{2t}\,dt = \dfrac{1}{2}e^{2t}$

Substituting into $\int u\,dv = uv - \int v\,du$ gives:

$$\int 3te^{2t}\,dt = (3t)\left(\frac{1}{2}e^{2t}\right) - \int \left(\frac{1}{2}e^{2t}\right)(3\,dt)$$

$$= \frac{3}{2}te^{2t} - \frac{3}{2}\int e^{2t}\,dt$$

$$= \frac{3}{2}te^{2t} - \frac{3}{2}\left(\frac{e^{2t}}{2}\right) + c$$

Hence

$$\int 3te^{2t}\,dt = \tfrac{3}{2}e^{2t}\left(t - \tfrac{1}{2}\right) + c,$$

which may be checked by differentiating.

Problem 3. Evaluate $\displaystyle\int_0^{\frac{\pi}{2}} 2\theta \sin\theta\,d\theta$

Let $u = 2\theta$, from which, $\dfrac{du}{d\theta} = 2$, i.e. $du = 2\,d\theta$ and let $dv = \sin\theta\,d\theta$, from which,

$$v = \int \sin\theta\,d\theta = -\cos\theta$$

Substituting into $\int u\,dv = uv - \int v\,du$ gives:

$$\int 2\theta \sin\theta\,d\theta = (2\theta)(-\cos\theta) - \int(-\cos\theta)(2\,d\theta)$$

$$= -2\theta\cos\theta + 2\int \cos\theta\,d\theta$$

$$= -2\theta\cos\theta + 2\sin\theta + c$$

Hence $\displaystyle\int_0^{\frac{\pi}{2}} 2\theta \sin\theta\,d\theta$

$$= \left[-2\theta\cos\theta + 2\sin\theta\right]_0^{\frac{\pi}{2}}$$

$$= \left[-2\left(\frac{\pi}{2}\right)\cos\frac{\pi}{2} + 2\sin\frac{\pi}{2}\right] - [0 + 2\sin 0]$$

$$= (-0 + 2) - (0 + 0) = \mathbf{2}$$

$$\text{since } \cos\frac{\pi}{2} = 0 \text{ and } \sin\frac{\pi}{2} = 1$$

Problem 4. Evaluate $\displaystyle\int_0^1 5xe^{4x}\,dx$, correct to 3 significant figures

Let $u = 5x$, from which $\dfrac{du}{dx} = 5$, i.e. $du = 5\,dx$ and

let $dv = e^{4x}\,dx$, from which, $v = \int e^{4x}\,dx = \frac{1}{4}e^{4x}$

Substituting into $\int u\,dv = uv - \int v\,du$ gives:

$$\int 5xe^{4x}\,dx = (5x)\left(\frac{e^{4x}}{4}\right) - \int \left(\frac{e^{4x}}{4}\right)(5\,dx)$$

$$= \frac{5}{4}xe^{4x} - \frac{5}{4}\int e^{4x}\,dx$$

$$= \frac{5}{4}xe^{4x} - \frac{5}{4}\left(\frac{e^{4x}}{4}\right) + c$$

$$= \frac{5}{4}e^{4x}\left(x - \frac{1}{4}\right) + c$$

Hence $\displaystyle\int_0^1 5xe^{4x}\,dx$

$$= \left[\frac{5}{4}e^{4x}\left(x - \frac{1}{4}\right)\right]_0^1$$

$$= \left[\frac{5}{4}e^4\left(1 - \frac{1}{4}\right)\right] - \left[\frac{5}{4}e^0\left(0 - \frac{1}{4}\right)\right]$$

$$= \left(\frac{15}{16}e^4\right) - \left(-\frac{5}{16}\right)$$

$$= 51.186 + 0.313 = 51.499 = \mathbf{51.5},$$

correct to 3 significant figures.

Problem 5. Determine $\int x^2 \sin x\,dx$

Let $u = x^2$, from which, $\dfrac{du}{dx} = 2x$, i.e. $du = 2x\,dx$, and let $dv = \sin x\,dx$, from which,

$$v = \int \sin x\,dx = -\cos x$$

Substituting into $\int u\,dv = uv - \int v\,du$ gives:

$$\int x^2 \sin x\,dx = (x^2)(-\cos x) - \int (-\cos x)(2x\,dx)$$

$$= -x^2 \cos x + 2\left[\int x \cos x\,dx\right]$$

The integral, $\int x \cos x\,dx$, is not a 'standard integral' and it can only be determined by using the integration by parts formula again.
From Problem 1, $\int x \cos x\,dx = x \sin x + \cos x$

Hence $\displaystyle\int x^2 \sin x\,dx$

$$= -x^2 \cos x + 2\{x \sin x + \cos x\} + c$$

$$= -x^2 \cos x + 2x \sin x + 2 \cos x + c$$

$$= \mathbf{(2 - x^2)\cos x + 2x \sin x + c}$$

In general, if the algebraic term of a product is of power n, then the integration by parts formula is applied n times.

Now try the following Practice Exercise

Practice Exercise 276   Further problems on integration by parts (answers on page 1183)

Determine the integrals in Problems 1 to 5 using integration by parts.

1. $\displaystyle\int xe^{2x}\,dx$

2. $\displaystyle\int \frac{4x}{e^{3x}}\,dx$

3. $\displaystyle\int x \sin x\,dx$

4. $\displaystyle\int 5\theta \cos 2\theta\,d\theta$

5. $\displaystyle\int 3t^2 e^{2t}\,dt$

Evaluate the integrals in Problems 6 to 9, correct to 4 significant figures.

6. $\displaystyle\int_0^2 2xe^x\,dx$

7. $\displaystyle\int_0^{\frac{\pi}{4}} x \sin 2x\,dx$

8. $\displaystyle\int_0^{\frac{\pi}{2}} t^2 \cos t\,dt$

9. $\displaystyle\int_1^2 3x^2 e^{\frac{x}{2}}\,dx$

68.3 Further worked problems on integration by parts

Problem 6. Find $\int x \ln x\,dx$

The logarithmic function is chosen as the 'u part'.

Thus when $u = \ln x$, then $\dfrac{du}{dx} = \dfrac{1}{x}$, i.e. $du = \dfrac{dx}{x}$

Letting $dv = x\,dx$ gives $v = \int x\,dx = \dfrac{x^2}{2}$
Substituting into $\int u\,dv = uv - \int v\,du$ gives:

$$\int x \ln x\,dx = (\ln x)\left(\frac{x^2}{2}\right) - \int \left(\frac{x^2}{2}\right)\frac{dx}{x}$$

$$= \frac{x^2}{2}\ln x - \frac{1}{2}\int x\,dx$$

$$= \frac{x^2}{2}\ln x - \frac{1}{2}\left(\frac{x^2}{2}\right) + c$$

Hence $\int x \ln x \, dx = \dfrac{x^2}{2}\left(\ln x - \dfrac{1}{2}\right) + c$ or

$$\dfrac{x^2}{4}(2\ln x - 1) + c$$

Problem 7. Determine $\int \ln x \, dx$

$\int \ln x \, dx$ is the same as $\int (1) \ln x \, dx$

Let $u = \ln x$, from which, $\dfrac{du}{dx} = \dfrac{1}{x}$, i.e. $du = \dfrac{dx}{x}$
and let $dv = 1 \, dx$, from which, $v = \int 1 \, dx = x$
Substituting into $\int u \, dv = uv - \int v \, du$ gives:

$$\int \ln x \, dx = (\ln x)(x) - \int x \, \dfrac{dx}{x}$$

$$= x \ln x - \int dx = x \ln x - x + c$$

Hence $\int \ln x \, dx = x(\ln x - 1) + c$

Problem 8. Evaluate $\int_1^9 \sqrt{x} \ln x \, dx$, correct to 3 significant figures

Let $u = \ln x$, from which $du = \dfrac{dx}{x}$
and let $dv = \sqrt{x} \, dx = x^{\frac{1}{2}} \, dx$, from which,

$$v = \int x^{\frac{1}{2}} \, dx = \dfrac{2}{3} x^{\frac{3}{2}}$$

Substituting into $\int u \, dv = uv - \int v \, du$ gives:

$$\int \sqrt{x} \ln x \, dx = (\ln x)\left(\dfrac{2}{3} x^{\frac{3}{2}}\right) - \int \left(\dfrac{2}{3} x^{\frac{3}{2}}\right)\left(\dfrac{dx}{x}\right)$$

$$= \dfrac{2}{3}\sqrt{x^3} \ln x - \dfrac{2}{3}\int x^{\frac{1}{2}} \, dx$$

$$= \dfrac{2}{3}\sqrt{x^3} \ln x - \dfrac{2}{3}\left(\dfrac{2}{3} x^{\frac{3}{2}}\right) + c$$

$$= \dfrac{2}{3}\sqrt{x^3}\left[\ln x - \dfrac{2}{3}\right] + c$$

Hence $\int_1^9 \sqrt{x} \ln x \, dx$

$$= \left[\dfrac{2}{3}\sqrt{x^3}\left(\ln x - \dfrac{2}{3}\right)\right]_1^9$$

$$= \left[\dfrac{2}{3}\sqrt{9^3}\left(\ln 9 - \dfrac{2}{3}\right)\right] - \left[\dfrac{2}{3}\sqrt{1^3}\left(\ln 1 - \dfrac{2}{3}\right)\right]$$

$$= \left[18\left(\ln 9 - \dfrac{2}{3}\right)\right] - \left[\dfrac{2}{3}\left(0 - \dfrac{2}{3}\right)\right]$$

$$= 27.550 + 0.444 = 27.994 = \mathbf{28.0},$$

correct to 3 significant figures.

Problem 9. Find $\int e^{ax}\cos bx \, dx$

When integrating a product of an exponential and a sine or cosine function it is immaterial which part is made equal to 'u'.

Let $u = e^{ax}$, from which $\dfrac{du}{dx} = ae^{ax}$,

i.e. $du = ae^{ax} \, dx$ and let $dv = \cos bx \, dx$, from which,

$$v = \int \cos bx \, dx = \dfrac{1}{b}\sin bx$$

Substituting into $\int u \, dv = uv - \int v \, du$ gives:

$$\int e^{ax}\cos bx \, dx$$

$$= (e^{ax})\left(\dfrac{1}{b}\sin bx\right) - \int \left(\dfrac{1}{b}\sin bx\right)(ae^{ax} \, dx)$$

$$= \dfrac{1}{b}e^{ax}\sin bx - \dfrac{a}{b}\left[\int e^{ax}\sin bx \, dx\right] \qquad (1)$$

$\int e^{ax}\sin bx \, dx$ is now determined separately using integration by parts again.

Let $u = e^{ax}$ then $du = ae^{ax} \, dx$, and let $dv = \sin bx \, dx$, from which

$$v = \int \sin bx \, dx = -\dfrac{1}{b}\cos bx$$

Substituting into the integration by parts formula gives:

$$\int e^{ax}\sin bx \, dx = (e^{ax})\left(-\dfrac{1}{b}\cos bx\right)$$

$$- \int \left(-\dfrac{1}{b}\cos bx\right)(ae^{ax} \, dx)$$

$$= -\dfrac{1}{b}e^{ax}\cos bx + \dfrac{a}{b}\int e^{ax}\cos bx \, dx$$

Substituting this result into equation (1) gives:

$$\int e^{ax}\cos bx \, dx = \dfrac{1}{b}e^{ax}\sin bx - \dfrac{a}{b}\left[-\dfrac{1}{b}e^{ax}\cos bx\right.$$

$$\left. + \dfrac{a}{b}\int e^{ax}\cos bx \, dx\right]$$

$$= \frac{1}{b}e^{ax}\sin bx + \frac{a}{b^2}e^{ax}\cos bx$$

$$- \frac{a^2}{b^2}\int e^{ax}\cos bx\,dx$$

The integral on the far right of this equation is the same as the integral on the left-hand side and thus they may be combined.

$$\int e^{ax}\cos bx\,dx + \frac{a^2}{b^2}\int e^{ax}\cos bx\,dx$$

$$= \frac{1}{b}e^{ax}\sin bx + \frac{a}{b^2}e^{ax}\cos bx$$

i.e. $\left(1 + \dfrac{a^2}{b^2}\right)\displaystyle\int e^{ax}\cos bx\,dx$

$$= \frac{1}{b}e^{ax}\sin bx + \frac{a}{b^2}e^{ax}\cos bx$$

i.e. $\left(\dfrac{b^2+a^2}{b^2}\right)\displaystyle\int e^{ax}\cos bx\,dx$

$$= \frac{e^{ax}}{b^2}(b\sin bx + a\cos bx)$$

Hence $\displaystyle\int e^{ax}\cos bx\,dx$

$$= \left(\frac{b^2}{b^2+a^2}\right)\left(\frac{e^{ax}}{b^2}\right)(b\sin bx + a\cos bx)$$

$$= \frac{e^{ax}}{a^2+b^2}(b\sin bx + a\cos bx) + c$$

Using a similar method to above, that is, integrating by parts twice, the following result may be proved:

$$\int e^{ax}\sin bx\,dx$$

$$= \frac{e^{ax}}{a^2+b^2}(a\sin bx - b\cos bx) + c \qquad (2)$$

Problem 10. Evaluate $\displaystyle\int_0^{\frac{\pi}{4}} e^t \sin 2t\,dt$, correct to 4 decimal places

Comparing $\int e^t \sin 2t\,dt$ with $\int e^{ax}\sin bx\,dx$ shows that $x=t$, $a=1$ and $b=2$

Hence, substituting into equation (2) gives:

$$\int_0^{\frac{\pi}{4}} e^t \sin 2t\,dt$$

$$= \left[\frac{e^t}{1^2+2^2}(1\sin 2t - 2\cos 2t)\right]_0^{\frac{\pi}{4}}$$

$$= \left[\frac{e^{\frac{\pi}{4}}}{5}\left(\sin 2\left(\frac{\pi}{4}\right) - 2\cos 2\left(\frac{\pi}{4}\right)\right)\right]$$

$$- \left[\frac{e^0}{5}(\sin 0 - 2\cos 0)\right]$$

$$= \left[\frac{e^{\frac{\pi}{4}}}{5}(1-0)\right] - \left[\frac{1}{5}(0-2)\right] = \frac{e^{\frac{\pi}{4}}}{5} + \frac{2}{5}$$

$$= \mathbf{0.8387},\ \text{correct to 4 decimal places.}$$

Now try the following Practice Exercise

Practice Exercise 277 Further problems on integration by parts (answers on page 1183)

Determine the integrals in Problems 1 to 5 using integration by parts.

1. $\displaystyle\int 2x^2 \ln x\,dx$

2. $\displaystyle\int 2\ln 3x\,dx$

3. $\displaystyle\int x^2 \sin 3x\,dx$

4. $\displaystyle\int 2e^{5x}\cos 2x\,dx$

5. $\displaystyle\int 2\theta \sec^2\theta\,d\theta$

Evaluate the integrals in Problems 6 to 9, correct to 4 significant figures.

6. $\displaystyle\int_1^2 x\ln x\,dx$

7. $\displaystyle\int_0^1 2e^{3x}\sin 2x\,dx$

8. $\displaystyle\int_0^{\frac{\pi}{2}} e^t \cos 3t \, dt$

9. $\displaystyle\int_1^4 \sqrt{x^3} \ln x \, dx$

10. In determining a Fourier series to represent $f(x) = x$ in the range $-\pi$ to π, Fourier coefficients are given by:

$$a_n = \frac{1}{\pi} \int_{-\pi}^{\pi} x \cos nx \, dx$$

and $\quad b_n = \dfrac{1}{\pi} \displaystyle\int_{-\pi}^{\pi} x \sin nx \, dx$

where n is a positive integer. Show by using integration by parts that $a_n = 0$ and

$$b_n = -\frac{2}{n} \cos n\pi$$

11. The equation $C = \displaystyle\int_0^1 e^{-0.4\theta} \cos 1.2\theta \, d\theta$

and $\qquad S = \displaystyle\int_0^1 e^{-0.4\theta} \sin 1.2\theta \, d\theta$

are involved in the study of damped oscillations. Determine the values of C and S.

For fully worked solutions to each of the problems in Practice Exercises 276 and 277 in this chapter, go to the website:
www.routledge.com/cw/bird

Revision Test 25   *Integration using partial fractions, tan θ/2 and 'by parts'*

This assignment covers the material contained in Chapters 66 to 68. *The marks available are shown in brackets at the end of each question.*

1. Determine:

   (a) $\displaystyle\int \frac{x-11}{x^2-x-2}\,dx$

   (b) $\displaystyle\int \frac{3-x}{(x^2+3)(x+3)}\,dx$ (21)

2. Evaluate $\displaystyle\int_1^2 \frac{3}{x^2(x+2)}\,dx$ correct to 4 significant figures.   (12)

3. Determine: $\displaystyle\int \frac{dx}{2\sin x + \cos x}$ (8)

4. Evaluate $\displaystyle\int_{\frac{\pi}{3}}^{\frac{\pi}{2}} \frac{dx}{3-2\sin x}$ correct to 3 decimal places.   (10)

5. Determine the following integrals:

   (a) $\displaystyle\int 5x\,e^{2x}\,dx$ (b) $\displaystyle\int t^2 \sin 2t\,dt$ (14)

6. Evaluate correct to 3 decimal places:

   $\displaystyle\int_1^4 \sqrt{x}\ln x\,dx$ (10)

Chapter 69

Reduction formulae

Why it is important to understand: **Reduction formulae**

When an integral contains a power of n, then it may sometimes be rewritten, using integration by parts, in terms of a similar integral containing $(n-1)$ or $(n-2)$, and so on. The result is a recurrence relation, i.e. an equation that recursively defines a sequence, once one or more initial terms are given; each further term of the sequence is defined as a function of the preceding terms. It may sound difficult but it's actually quite straightforward and is just one more technique for integrating certain functions.

At the end of this chapter, you should be able to:

- appreciate when reduction formulae might be used
- integrate functions of the form $\int x^n e^x \, dx$ using reduction formulae
- integrate functions of the form $\int x^n \cos x \, dx$ and $\int x^n \sin x \, dx$ using reduction formulae
- integrate functions of the form $\int \sin^n x \, dx$ and $\int \cos^n x \, dx$ using reduction formulae
- integrate further functions using reduction formulae

69.1 Introduction

When using integration by parts in Chapter 68, an integral such as $\int x^2 e^x \, dx$ requires integration by parts twice. Similarly, $\int x^3 e^x \, dx$ requires integration by parts three times. Thus, integrals such as $\int x^5 e^x \, dx$, $\int x^6 \cos x \, dx$ and $\int x^8 \sin 2x \, dx$, for example, would take a long time to determine using integration by parts. **Reduction formulae** provide a quicker method for determining such integrals and the method is demonstrated in the following sections.

69.2 Using reduction formulae for integrals of the form $\int x^n e^x \, dx$

To determine $\int x^n e^x \, dx$ using integration by parts,

let $\quad u = x^n$ from which,

$$\frac{du}{dx} = nx^{n-1} \text{ and } du = nx^{n-1} \, dx$$

and $\quad dv = e^x \, dx$ from which,

$$v = \int e^x \, dx = e^x$$

Thus, $\quad \displaystyle\int x^n e^x \, dx = x^n e^x - \int e^x nx^{n-1} \, dx$

using the integration by parts formula

$$= x^n e^x - n \int x^{n-1} e^x \, dx$$

The integral on the far right is seen to be of the same form as the integral on the left-hand side, except that n has been replaced by $n-1$

Thus, if we let,

$$\int x^n e^x \, dx = I_n$$

then $\displaystyle\int x^{n-1}\mathrm{e}^x\,\mathrm{d}x = I_{n-1}$

Hence $\displaystyle\int x^n\mathrm{e}^x\,\mathrm{d}x = x^n\mathrm{e}^x - n\int x^{n-1}\mathrm{e}^x\,\mathrm{d}x$

can be written as:

$$I_n = x^n\mathrm{e}^x - nI_{n-1} \qquad (1)$$

Equation (1) is an example of a reduction formula since it expresses an integral in n in terms of the same integral in $n-1$

Problem 1.   Determine $\int x^2\mathrm{e}^x\,\mathrm{d}x$ using a reduction formula

Using equation (1) with $n=2$ gives:

$$\int x^2\mathrm{e}^x\,\mathrm{d}x = I_2 = x^2\mathrm{e}^x - 2I_1$$

and $\qquad I_1 = x^1\mathrm{e}^x - 1I_0$

$$I_0 = \int x^0\mathrm{e}^x\,\mathrm{d}x = \int \mathrm{e}^x\,\mathrm{d}x = \mathrm{e}^x + c_1$$

Hence $\qquad I_2 = x^2\mathrm{e}^x - 2[x\mathrm{e}^x - 1I_0]$

$$= x^2\mathrm{e}^x - 2[x\mathrm{e}^x - 1(\mathrm{e}^x + c_1)]$$

i.e. $\displaystyle\int x^2\mathrm{e}^x\,\mathrm{d}x = x^2\mathrm{e}^x - 2x\mathrm{e}^x + 2\mathrm{e}^x + 2c_1$

$$= \mathrm{e}^x(x^2 - 2x + 2) + c$$

$$\text{(where } c = 2c_1)$$

As with integration by parts, in the following examples the constant of integration will be added at the last step with indefinite integrals.

Problem 2.   Use a reduction formula to determine $\int x^3\mathrm{e}^x\,\mathrm{d}x$

From equation (1), $I_n = x^n\mathrm{e}^x - nI_{n-1}$

Hence $\displaystyle\int x^3\mathrm{e}^x\,\mathrm{d}x = I_3 = x^3\mathrm{e}^x - 3I_2$

$$I_2 = x^2\mathrm{e}^x - 2I_1$$
$$I_1 = x^1\mathrm{e}^x - 1I_0$$

and $\qquad I_0 = \displaystyle\int x^0\mathrm{e}^x\,\mathrm{d}x = \int \mathrm{e}^x\,\mathrm{d}x = \mathrm{e}^x$

Thus $\displaystyle\int x^3\mathrm{e}^x\,\mathrm{d}x = x^3\mathrm{e}^x - 3[x^2\mathrm{e}^x - 2I_1]$

$$= x^3\mathrm{e}^x - 3[x^2\mathrm{e}^x - 2(x\mathrm{e}^x - I_0)]$$

$$= x^3\mathrm{e}^x - 3[x^2\mathrm{e}^x - 2(x\mathrm{e}^x - \mathrm{e}^x)]$$

$$= x^3\mathrm{e}^x - 3x^2\mathrm{e}^x + 6(x\mathrm{e}^x - \mathrm{e}^x)$$

$$= x^3\mathrm{e}^x - 3x^2\mathrm{e}^x + 6x\mathrm{e}^x - 6\mathrm{e}^x$$

i.e. $\displaystyle\int x^3\mathrm{e}^x\,\mathrm{d}x = \mathrm{e}^x(x^3 - 3x^2 + 6x - 6) + c$

Now try the following Practice Exercise

Practice Exercise 278   Further problems on using reduction formulae for integrals of the form $\int x^n\mathrm{e}^x\,\mathrm{d}x$ (answers on page 1183)

1. Use a reduction formula to determine $\int x^4\mathrm{e}^x\,\mathrm{d}x$

2. Determine $\int t^3\mathrm{e}^{2t}\,\mathrm{d}t$ using a reduction formula.

3. Use the result of Problem 2 to evaluate $\int_0^1 5t^3\mathrm{e}^{2t}\,\mathrm{d}t$, correct to 3 decimal places.

69.3   Using reduction formulae for integrals of the form $\int x^n\cos x\,\mathrm{d}x$ and $\int x^n\sin x\,\mathrm{d}x$

(a) $\int x^n\cos x\,\mathrm{d}x$

Let $I_n = \int x^n\cos x\,\mathrm{d}x$ then, using integration by parts:

if $\qquad u = x^n$ then $\dfrac{\mathrm{d}u}{\mathrm{d}x} = nx^{n-1}$

and if $\quad \mathrm{d}v = \cos x\,\mathrm{d}x$ then

$$v = \int \cos x\,\mathrm{d}x = \sin x$$

Hence $\quad I_n = x^n\sin x - \displaystyle\int (\sin x)nx^{n-1}\,\mathrm{d}x$

$$= x^n\sin x - n\int x^{n-1}\sin x\,\mathrm{d}x$$

Using integration by parts again, this time with $u = x^{n-1}$:

$$\frac{\mathrm{d}u}{\mathrm{d}x} = (n-1)x^{n-2}, \text{ and } \mathrm{d}v = \sin x\,\mathrm{d}x$$

from which

$$v = \int \sin x\,\mathrm{d}x = -\cos x$$

Hence $I_n = x^n \sin x - n\left[x^{n-1}(-\cos x) \right.$

$$\left. - \int(-\cos x)(n-1)x^{n-2}\,dx \right]$$

$$= x^n \sin x + nx^{n-1}\cos x$$

$$- n(n-1)\int x^{n-2}\cos x\,dx$$

i.e. $I_n = x^n \sin x + nx^{n-1}\cos x$

$$- n(n-1)I_{n-2} \tag{2}$$

Problem 3.   Use a reduction formula to determine $\int x^2 \cos x\,dx$

Using the reduction formula of equation (2):

$$\int x^2 \cos x\,dx = I_2$$

$$= x^2 \sin x + 2x^1 \cos x - 2(1)I_0$$

and $\qquad I_0 = \int x^0 \cos x\,dx$

$$= \int \cos x\,dx = \sin x$$

Hence

$$\int x^2 \cos x\,dx = x^2 \sin x + 2x\cos x - 2\sin x + c$$

Problem 4.   Evaluate $\int_1^2 4t^3 \cos t\,dt$, correct to 4 significant figures

Let us firstly find a reduction formula for $\int t^3 \cos t\,dt$.

From equation (2),

$$\int t^3 \cos t\,dt = I_3 = t^3 \sin t + 3t^2 \cos t - 3(2)I_1$$

and

$$I_1 = t^1 \sin t + 1t^0 \cos t - 1(0)I_{n-2}$$

$$= t\sin t + \cos t$$

Hence

$$\int t^3 \cos t\,dt = t^3 \sin t + 3t^2 \cos t$$

$$- 3(2)[t\sin t + \cos t]$$

$$= t^3 \sin t + 3t^2 \cos t - 6t\sin t - 6\cos t$$

Thus

$$\int_1^2 4t^3 \cos t\,dt$$

$$= [4(t^3 \sin t + 3t^2 \cos t - 6t\sin t - 6\cos t)]_1^2$$

$$= [4(8\sin 2 + 12\cos 2 - 12\sin 2 - 6\cos 2)]$$

$$- [4(\sin 1 + 3\cos 1 - 6\sin 1 - 6\cos 1)]$$

$$= (-24.53628) - (-23.31305)$$

$$= -1.223$$

Problem 5.   Determine a reduction formula for $\int_0^\pi x^n \cos x\,dx$ and hence evaluate $\int_0^\pi x^4 \cos x\,dx$, correct to 2 decimal places

From equation (2),

$$I_n = x^n \sin x + nx^{n-1}\cos x - n(n-1)I_{n-2}$$

Hence $\int_0^\pi x^n \cos x\,dx = [x^n \sin x + nx^{n-1}\cos x]_0^\pi$

$$- n(n-1)I_{n-2}$$

$$= [(\pi^n \sin \pi + n\pi^{n-1}\cos \pi)$$

$$- (0+0)] - n(n-1)I_{n-2}$$

$$= -n\pi^{n-1} - n(n-1)I_{n-2}$$

Hence

$$\int_0^\pi x^4 \cos x\,dx = I_4$$

$$= -4\pi^3 - 4(3)I_2 \text{ since } n = 4$$

When $n = 2$,

$$\int_0^\pi x^2 \cos x\,dx = I_2 = -2\pi^1 - 2(1)I_0$$

and $\qquad I_0 = \int_0^\pi x^0 \cos x\,dx$

$$= \int_0^\pi \cos x\,dx$$

$$= [\sin x]_0^\pi = 0$$

Hence

$$\int_0^\pi x^4 \cos x\,dx = -4\pi^3 - 4(3)[-2\pi - 2(1)(0)]$$

$$= -4\pi^3 + 24\pi \text{ or } -48.63,$$

correct to 2 decimal places.

(b) $\int x^n \sin x \, dx$

Let $I_n = \int x^n \sin x \, dx$

Using integration by parts, if $u = x^n$ then
$\dfrac{du}{dx} = nx^{n-1}$ and if $dv = \sin x \, dx$ then
$v = \int \sin x \, dx = -\cos x$. Hence

$$\int x^n \sin x \, dx$$

$$= I_n = x^n(-\cos x) - \int (-\cos x) nx^{n-1} \, dx$$

$$= -x^n \cos x + n \int x^{n-1} \cos x \, dx$$

Using integration by parts again, with $u = x^{n-1}$, from
which, $\dfrac{du}{dx} = (n-1)x^{n-2}$ and $dv = \cos x$, from which,
$v = \int \cos x \, dx = \sin x$. Hence

$$I_n = -x^n \cos x + n \left[x^{n-1}(\sin x) \right.$$

$$\left. - \int (\sin x)(n-1)x^{n-2} \, dx \right]$$

$$= -x^n \cos x + nx^{n-1}(\sin x)$$

$$- n(n-1) \int x^{n-2} \sin x \, dx$$

i.e. $\quad I_n = -x^n \cos x + nx^{n-1} \sin x - n(n-1)I_{n-2}$ (3)

Problem 6.   Use a reduction formula to determine
$\int x^3 \sin x \, dx$

Using equation (3),

$$\int x^3 \sin x \, dx = I_3$$

$$= -x^3 \cos x + 3x^2 \sin x - 3(2)I_1$$

and $\qquad I_1 = -x^1 \cos x + 1x^0 \sin x$

$$= -x \cos x + \sin x$$

Hence

$$\int x^3 \sin x \, dx = -x^3 \cos x + 3x^2 \sin x$$

$$- 6[-x \cos x + \sin x]$$

$$= -x^3 \cos x + 3x^2 \sin x$$

$$+ 6x \cos x - 6 \sin x + c$$

Problem 7.   Evaluate $\displaystyle\int_0^{\frac{\pi}{2}} 3\theta^4 \sin \theta \, d\theta$, correct to 2
decimal places

From equation (3),

$$I_n = [-x^n \cos x + nx^{n-1}(\sin x)]_0^{\frac{\pi}{2}} - n(n-1)I_{n-2}$$

$$= \left[\left(-\left(\frac{\pi}{2}\right)^n \cos \frac{\pi}{2} + n\left(\frac{\pi}{2}\right)^{n-1} \sin \frac{\pi}{2} \right) - (0) \right]$$

$$- n(n-1)I_{n-2}$$

$$= n\left(\frac{\pi}{2}\right)^{n-1} - n(n-1)I_{n-2}$$

Hence

$$\int_0^{\frac{\pi}{2}} 3\theta^4 \sin \theta \, d\theta = 3 \int_0^{\frac{\pi}{2}} \theta^4 \sin \theta \, d\theta$$

$$= 3I_4$$

$$= 3\left[4\left(\frac{\pi}{2}\right)^3 - 4(3)I_2 \right]$$

$$I_2 = 2\left(\frac{\pi}{2}\right)^1 - 2(1)I_0 \text{ and}$$

$$I_0 = \int_0^{\frac{\pi}{2}} \theta^0 \sin \theta \, d\theta = [-\cos x]_0^{\frac{\pi}{2}}$$

$$= [-0 - (-1)] = 1$$

Hence

$$3 \int_0^{\frac{\pi}{2}} \theta^4 \sin \theta \, d\theta$$

$$= 3I_4$$

$$= 3\left[4\left(\frac{\pi}{2}\right)^3 - 4(3)\left\{ 2\left(\frac{\pi}{2}\right)^1 - 2(1)I_0 \right\} \right]$$

$$= 3\left[4\left(\frac{\pi}{2}\right)^3 - 4(3)\left\{ 2\left(\frac{\pi}{2}\right)^1 - 2(1)(1) \right\} \right]$$

$$= 3\left[4\left(\frac{\pi}{2}\right)^3 - 24\left(\frac{\pi}{2}\right)^1 + 24 \right]$$

$$= 3(15.503 - 37.699 + 24)$$

$$= 3(1.8039) = \mathbf{5.41}$$

Now try the following Practice Exercise

> **Practice Exercise 279   Further problems on reduction formulae for integrals of the form $\int x^n \cos x \, dx$ and $\int x^n \sin x \, dx$ (answers on page 1183)**
>
> 1. Use a reduction formula to determine $\int x^5 \cos x \, dx$
>
> 2. Evaluate $\int_0^\pi x^5 \cos x \, dx$, correct to 2 decimal places.
>
> 3. Use a reduction formula to determine $\int x^5 \sin x \, dx$
>
> 4. Evaluate $\int_0^\pi x^5 \sin x \, dx$, correct to 2 decimal places.

69.4 Using reduction formulae for integrals of the form $\int \sin^n x \, dx$ and $\int \cos^n x \, dx$

(a) $\int \sin^n x \, dx$

Let $I_n = \int \sin^n x \, dx \equiv \int \sin^{n-1} x \sin x \, dx$ from laws of indices.

Using integration by parts, let $u = \sin^{n-1} x$, from which,

$$\frac{du}{dx} = (n-1)\sin^{n-2} x \cos x \quad \text{and}$$

$$du = (n-1)\sin^{n-2} x \cos x \, dx$$

and let $dv = \sin x \, dx$, from which,
$v = \int \sin x \, dx = -\cos x$. Hence,

$$I_n = \int \sin^{n-1} x \sin x \, dx$$

$$= (\sin^{n-1} x)(-\cos x)$$

$$\quad - \int (-\cos x)(n-1)\sin^{n-2} x \cos x \, dx$$

$$= -\sin^{n-1} x \cos x$$

$$\quad + (n-1)\int \cos^2 x \sin^{n-2} x \, dx$$

$$= -\sin^{n-1} x \cos x$$

$$\quad + (n-1)\int (1 - \sin^2 x)\sin^{n-2} x \, dx$$

$$= -\sin^{n-1} x \cos x$$

$$\quad + (n-1)\left\{ \int \sin^{n-2} x \, dx - \int \sin^n x \, dx \right\}$$

i.e. $I_n = -\sin^{n-1} x \cos x$

$$+ (n-1)I_{n-2} - (n-1)I_n$$

i.e. $I_n + (n-1)I_n$

$$= -\sin^{n-1} x \cos x + (n-1)I_{n-2}$$

and $nI_n = -\sin^{n-1} x \cos x + (n-1)I_{n-2}$

from which,

$$\int \sin^n x \, dx =$$

$$I_n = -\frac{1}{n}\sin^{n-1} x \cos x + \frac{n-1}{n}I_{n-2} \qquad (4)$$

> **Problem 8.**   Use a reduction formula to determine $\int \sin^4 x \, dx$

Using equation (4),

$$\int \sin^4 x \, dx = I_4 = -\frac{1}{4}\sin^3 x \cos x + \frac{3}{4}I_2$$

$$I_2 = -\frac{1}{2}\sin^1 x \cos x + \frac{1}{2}I_0$$

and $I_0 = \int \sin^0 x \, dx = \int 1 \, dx = x$

Hence

$$\int \sin^4 x \, dx = I_4 = -\frac{1}{4}\sin^3 x \cos x$$

$$+ \frac{3}{4}\left[-\frac{1}{2}\sin x \cos x + \frac{1}{2}(x) \right]$$

$$= -\frac{1}{4}\sin^3 x \cos x - \frac{3}{8}\sin x \cos x$$

$$+ \frac{3}{8}x + c$$

> **Problem 9.**   Evaluate $\int_0^1 4\sin^5 t \, dt$, correct to 3 significant figures

Using equation (4),

$$\int \sin^5 t \, dt = I_5 = -\frac{1}{5}\sin^4 t \cos t + \frac{4}{5}I_3$$

$$I_3 = -\frac{1}{3}\sin^2 t \cos t + \frac{2}{3}I_1$$

and $I_1 = -\frac{1}{1}\sin^0 t \cos t + 0 = -\cos t$

Hence

$$\int \sin^5 t \, dt = -\frac{1}{5} \sin^4 t \cos t$$

$$+ \frac{4}{5} \left[-\frac{1}{3} \sin^2 t \cos t + \frac{2}{3}(-\cos t) \right]$$

$$= -\frac{1}{5} \sin^4 t \cos t - \frac{4}{15} \sin^2 t \cos t$$

$$- \frac{8}{15} \cos t + c$$

and $\displaystyle\int_0^1 4 \sin^5 t \, dt$

$$= 4 \left[-\frac{1}{5} \sin^4 t \cos t - \frac{4}{15} \sin^2 t \cos t - \frac{8}{15} \cos t \right]_0^1$$

$$= 4 \left[\left(-\frac{1}{5} \sin^4 1 \cos 1 - \frac{4}{15} \sin^2 1 \cos 1 \right. \right.$$

$$\left. \left. - \frac{8}{15} \cos 1 \right) - \left(-0 - 0 - \frac{8}{15} \right) \right]$$

$$= 4[(-0.054178 - 0.1020196$$

$$- 0.2881612) - (-0.533333)]$$

$$= 4(0.0889745) = \mathbf{0.356}$$

Problem 10. Determine a reduction formula for $\displaystyle\int_0^{\frac{\pi}{2}} \sin^n x \, dx$ and hence evaluate $\displaystyle\int_0^{\frac{\pi}{2}} \sin^6 x \, dx$

From equation (4),

$$\int \sin^n x \, dx$$

$$= I_n = -\frac{1}{n} \sin^{n-1} x \cos x + \frac{n-1}{n} I_{n-2}$$

Hence

$$\int_0^{\frac{\pi}{2}} \sin^n x \, dx = \left[-\frac{1}{n} \sin^{n-1} x \cos x \right]_0^{\frac{\pi}{2}} + \frac{n-1}{n} I_{n-2}$$

$$= [0 - 0] + \frac{n-1}{n} I_{n-2}$$

i.e. $\qquad I_n = \dfrac{n-1}{n} I_{n-2}$

Hence

$$\int_0^{\frac{\pi}{2}} \sin^6 x \, dx = I_6 = \frac{5}{6} I_4$$

$$I_4 = \frac{3}{4} I_2, \quad I_2 = \frac{1}{2} I_0$$

and $\qquad I_0 = \displaystyle\int_0^{\frac{\pi}{2}} \sin^0 x \, dx = \int_0^{\frac{\pi}{2}} 1 \, dx = \frac{\pi}{2}$

Thus

$$\int_0^{\frac{\pi}{2}} \sin^6 x \, dx = I_6 = \frac{5}{6} I_4 = \frac{5}{6} \left[\frac{3}{4} I_2 \right]$$

$$= \frac{5}{6} \left[\frac{3}{4} \left\{ \frac{1}{2} I_0 \right\} \right]$$

$$= \frac{5}{6} \left[\frac{3}{4} \left\{ \frac{1}{2} \left[\frac{\pi}{2} \right] \right\} \right] = \frac{15}{96} \pi$$

(b) $\int \cos^n x \, dx$

Let $I_n = \int \cos^n x \, dx \equiv \int \cos^{n-1} x \cos x \, dx$ from laws of indices.

Using integration by parts, let $u = \cos^{n-1} x$ from which,

$$\frac{du}{dx} = (n-1) \cos^{n-2} x (-\sin x)$$

and $\qquad du = (n-1) \cos^{n-2} x (-\sin x) \, dx$

and let $\quad dv = \cos x \, dx$

from which, $v = \displaystyle\int \cos x \, dx = \sin x$

Then

$$I_n = (\cos^{n-1} x)(\sin x)$$

$$- \int (\sin x)(n-1) \cos^{n-2} x (-\sin x) \, dx$$

$$= (\cos^{n-1} x)(\sin x)$$

$$+ (n-1) \int \sin^2 x \cos^{n-2} x \, dx$$

$$= (\cos^{n-1} x)(\sin x)$$

$$+ (n-1) \int (1 - \cos^2 x) \cos^{n-2} x \, dx$$

$$= (\cos^{n-1} x)(\sin x)$$

$$+ (n-1) \left\{ \int \cos^{n-2} x \, dx - \int \cos^n x \, dx \right\}$$

i.e. $I_n = (\cos^{n-1} x)(\sin x) + (n-1)I_{n-2} - (n-1)I_n$

i.e. $I_n + (n-1)I_n = (\cos^{n-1} x)(\sin x) + (n-1)I_{n-2}$

i.e. $nI_n = (\cos^{n-1} x)(\sin x) + (n-1)I_{n-2}$

Thus $\quad I_n = \dfrac{1}{n}\cos^{n-1} x \sin x + \dfrac{n-1}{n}I_{n-2}$ (5)

> **Problem 11.** Use a reduction formula to determine $\int \cos^4 x \, dx$

Using equation (5),

$$\int \cos^4 x \, dx = I_4 = \frac{1}{4}\cos^3 x \sin x + \frac{3}{4}I_2$$

and $\quad I_2 = \dfrac{1}{2}\cos x \sin x + \dfrac{1}{2}I_0$

and $\quad I_0 = \displaystyle\int \cos^0 x \, dx$

$$= \int 1 \, dx = x$$

Hence $\displaystyle\int \cos^4 x \, dx$

$$= \frac{1}{4}\cos^3 x \sin x + \frac{3}{4}\left(\frac{1}{2}\cos x \sin x + \frac{1}{2}x\right)$$

$$= \frac{1}{4}\cos^3 x \sin x + \frac{3}{8}\cos x \sin x + \frac{3}{8}x + c$$

> **Problem 12.** Determine a reduction formula for $\displaystyle\int_0^{\frac{\pi}{2}} \cos^n x \, dx$ and hence evaluate $\displaystyle\int_0^{\frac{\pi}{2}} \cos^5 x \, dx$

From equation (5),

$$\int \cos^n x \, dx = \frac{1}{n}\cos^{n-1} x \sin x + \frac{n-1}{n}I_{n-2}$$

and hence

$$\int_0^{\frac{\pi}{2}} \cos^n x \, dx = \left[\frac{1}{n}\cos^{n-1} x \sin x\right]_0^{\frac{\pi}{2}}$$

$$+ \frac{n-1}{n}I_{n-2}$$

$$= [0-0] + \frac{n-1}{n}I_{n-2}$$

i.e. $\displaystyle\int_0^{\frac{\pi}{2}} \cos^n x \, dx = I_n = \frac{n-1}{n}I_{n-2}$ (6)

(Note that this is the same reduction formula as for $\displaystyle\int_0^{\frac{\pi}{2}} \sin^n x \, dx$ (in Problem 10) and the result is usually known as **Wallis's* formula**.)

Thus, from equation (6),

$$\int_0^{\frac{\pi}{2}} \cos^5 x \, dx = \frac{4}{5}I_3, \qquad I_3 = \frac{2}{3}I_1$$

and $\quad I_1 = \displaystyle\int_0^{\frac{\pi}{2}} \cos^1 x \, dx$

$$= [\sin x]_0^{\frac{\pi}{2}} = (1-0) = 1$$

Hence $\displaystyle\int_0^{\frac{\pi}{2}} \cos^5 x \, dx = \frac{4}{5}I_3 = \frac{4}{5}\left[\frac{2}{3}I_1\right]$

$$= \frac{4}{5}\left[\frac{2}{3}(1)\right] = \frac{8}{15}$$

* **Who was Wallis?** **John Wallis** (23 November 1616–28 October 1703) was an English mathematician partially credited for the development of infinitesimal calculus, and is also credited with introducing the symbol ∞ for infinity. To find out more go to **www.routledge.com/cw/bird**

Now try the following Practice Exercise

> **Practice Exercise 280   Further problems on reduction formulae for integrals of the form $\int \sin^n x\, dx$ and $\int \cos^n x\, dx$ (answers on page 1183)**
>
> 1. Use a reduction formula to determine $\int \sin^7 x\, dx$
>
> 2. Evaluate $\int_0^\pi 3\sin^3 x\, dx$ using a reduction formula.
>
> 3. Evaluate $\int_0^{\frac{\pi}{2}} \sin^5 x\, dx$ using a reduction formula.
>
> 4. Determine, using a reduction formula, $\int \cos^6 x\, dx$
>
> 5. Evaluate $\int_0^{\frac{\pi}{2}} \cos^7 x\, dx$

69.5   Further reduction formulae

The following Problems demonstrate further examples where integrals can be determined using reduction formulae.

> **Problem 13.**   Determine a reduction formula for $\int \tan^n x\, dx$ and hence find $\int \tan^7 x\, dx$

Let $I_n = \int \tan^n x\, dx \equiv \int \tan^{n-2} x \tan^2 x\, dx$

by the laws of indices

$$= \int \tan^{n-2} x\, (\sec^2 x - 1)\, dx$$

since $1 + \tan^2 x = \sec^2 x$

$$= \int \tan^{n-2} x \sec^2 x\, dx - \int \tan^{n-2} x\, dx$$

$$= \int \tan^{n-2} x \sec^2 x\, dx - I_{n-2}$$

i.e. $I_n = \dfrac{\tan^{n-1} x}{n-1} - I_{n-2}$

When $n = 7$,

$$I_7 = \int \tan^7 x\, dx = \frac{\tan^6 x}{6} - I_5$$

$$I_5 = \frac{\tan^4 x}{4} - I_3 \quad \text{and} \quad I_3 = \frac{\tan^2 x}{2} - I_1$$

$$I_1 = \int \tan x\, dx = \ln(\sec x)$$

from Problem 9, Chapter 64, page 729.
Thus

$$\int \tan^7 x\, dx = \frac{\tan^6 x}{6} - \left[\frac{\tan^4 x}{4} \right.$$
$$\left. - \left(\frac{\tan^2 x}{2} - \ln(\sec x) \right) \right]$$

Hence $\int \tan^7 x\, dx$

$$= \frac{1}{6}\tan^6 x - \frac{1}{4}\tan^4 x + \frac{1}{2}\tan^2 x$$

$$- \ln(\sec x) + c$$

> **Problem 14.**   Evaluate, using a reduction formula, $\int_0^{\frac{\pi}{2}} \sin^2 t \cos^6 t\, dt$

$$\int_0^{\frac{\pi}{2}} \sin^2 t \cos^6 t\, dt = \int_0^{\frac{\pi}{2}} (1 - \cos^2 t) \cos^6 t\, dt$$

$$= \int_0^{\frac{\pi}{2}} \cos^6 t\, dt - \int_0^{\frac{\pi}{2}} \cos^8 t\, dt$$

If $\qquad I_n = \int_0^{\frac{\pi}{2}} \cos^n t\, dt$

then

$$\int_0^{\frac{\pi}{2}} \sin^2 t \cos^6 t\, dt = I_6 - I_8$$

and from equation (6),

$$I_6 = \frac{5}{6} I_4 = \frac{5}{6}\left[\frac{3}{4} I_2 \right]$$

$$= \frac{5}{6}\left[\frac{3}{4} \left(\frac{1}{2} I_0 \right) \right]$$

and $\quad I_0 = \int_0^{\frac{\pi}{2}} \cos^0 t\, dt$

$$= \int_0^{\frac{\pi}{2}} 1\, dt = [x]_0^{\frac{\pi}{2}} = \frac{\pi}{2}$$

Hence $I_6 = \dfrac{5}{6} \cdot \dfrac{3}{4} \cdot \dfrac{1}{2} \cdot \dfrac{\pi}{2}$

$$= \dfrac{15\pi}{96} \text{ or } \dfrac{5\pi}{32}$$

Similarly, $I_8 = \dfrac{7}{8} I_6 = \dfrac{7}{8} \cdot \dfrac{5\pi}{32}$

Thus

$$\int_0^{\frac{\pi}{2}} \sin^2 t \cos^6 t \, dt = I_6 - I_8$$

$$= \dfrac{5\pi}{32} - \dfrac{7}{8} \cdot \dfrac{5\pi}{32}$$

$$= \dfrac{1}{8} \cdot \dfrac{5\pi}{32} = \dfrac{5\pi}{256}$$

Problem 15. Use integration by parts to determine a reduction formula for $\int (\ln x)^n \, dx$. Hence determine $\int (\ln x)^3 \, dx$

Let $I_n = \int (\ln x)^n \, dx$.
Using integration by parts, let $u = (\ln x)^n$, from which,

$$\dfrac{du}{dx} = n(\ln x)^{n-1} \left(\dfrac{1}{x} \right)$$

and $\quad du = n(\ln x)^{n-1} \left(\dfrac{1}{x} \right) dx$

and let $dv = dx$, from which, $v = \int dx = x$

Then $\quad I_n = \int (\ln x)^n \, dx$

$$= (\ln x)^n (x) - \int (x) n (\ln x)^{n-1} \left(\dfrac{1}{x} \right) dx$$

$$= x(\ln x)^n - n \int (\ln x)^{n-1} \, dx$$

i.e. $\boldsymbol{I_n = x(\ln x)^n - nI_{n-1}}$

When $n = 3$,

$$\int (\ln x)^3 \, dx = I_3 = x(\ln x)^3 - 3I_2$$

$I_2 = x(\ln x)^2 - 2I_1$ and $I_1 = \int \ln x \, dx = x(\ln x - 1)$ from Problem 7, page 756.

Hence

$$\int (\ln x)^3 \, dx = x(\ln x)^3 - 3[x(\ln x)^2 - 2I_1] + c$$

$$= x(\ln x)^3 - 3[x(\ln x)^2$$
$$- 2[x(\ln x - 1)]] + c$$

$$= x(\ln x)^3 - 3[x(\ln x)^2$$
$$- 2x \ln x + 2x] + c$$

$$= x(\ln x)^3 - 3x(\ln x)^2$$
$$+ 6x \ln x - 6x + c$$

$$= x[(\ln x)^3 - 3(\ln x)^2$$

$$+ 6 \ln x - 6] + c$$

Now try the following Practice Exercise

For fully worked solutions to each of the problems in Practice Exercises 278 to 281 in this chapter, go to the website:
www.routledge.com/cw/bird

Chapter 70

Double and triple integrals

Why it is important to understand: **Double and triple integrals**

Double and triple integrals have engineering applications in finding areas, masses and forces of two-dimensional regions, and in determining volumes, average values of functions, centres of mass, moments of inertia and surface areas. A multiple integral is a type of definite integral extended to functions of more than one real variable. This chapter explains how to evaluate double and triple integrals and completes the many techniques of integral calculus explained in the preceding chapters.

At the end of this chapter, you should be able to:

- evaluate a double integral
- evaluate a triple integral

70.1 Double integrals

The procedure to determine a double integral of the form: $\int_{y_1}^{y_2} \int_{x_1}^{x_2} f(x,y)\mathbf{d}x\,\mathbf{d}y$ is:

(i) integrate $f(x,y)$ with respect to x between the limits of $x = x_1$ and $x = x_2$ (where y is regarded as being a constant), and

(ii) integrate the result in (i) with respect to y between the limits of $y = y_1$ and $y = y_2$

It is seen from this procedure that to determine a double integral we start with the innermost integral and then work outwards.

Double integrals may be used to determine areas under curves, second moments of area, centroids and moments of inertia.

(Sometimes $\int_{y_1}^{y_2} \int_{x_1}^{x_2} f(x,y)\mathrm{d}x\,\mathrm{d}y$ is written

as: $\int_{y_1}^{y_2} \mathrm{d}y \int_{x_1}^{x_2} f(x,y)\mathrm{d}x$. All this means is that the right-hand side integral is determined first.)

Determining double integrals is demonstrated in the following worked problems.

Problem 1. Evaluate $\int_{1}^{3} \int_{2}^{5} (2x - 3y)\mathrm{d}x\,\mathrm{d}y$

Following the above procedure:

(i) $(2x - 3y)$ is integrated with respect to x between $x = 2$ and $x = 5$, with y regarded as a constant

i.e. $\int_{2}^{5} (2x - 3y)\mathrm{d}x = \left[\frac{2x^2}{2} - (3y)x\right]_{2}^{5}$

$= \left[x^2 - 3xy\right]_{2}^{5}$

$= \left[\left(5^2 - 3(5)y\right) - \left(2^2 - 3(2)y\right)\right]$

$= (25 - 15y) - (4 - 6y)$

$= 25 - 15y - 4 + 6y$

$= 21 - 9y$

(ii) $\int_{1}^{3} \int_{2}^{5} (2x - 3y)\mathrm{d}x\,\mathrm{d}y = \int_{1}^{3} (21 - 9y)\mathrm{d}y$

$= \left[21y - \frac{9y^2}{2}\right]_{1}^{3}$

$$= \left[\left(21(3) - \frac{9(3)^2}{2} \right) - \left(21(1) - \frac{9(1)^2}{2} \right) \right]$$

$$= (63 - 40.5) - (21 - 4.5)$$

$$= 63 - 40.5 - 21 + 4.5 = 6$$

Hence, $\displaystyle \int_1^3 \int_2^5 (2x - 3y)dx\,dy = 6$

Problem 2. Evaluate $\displaystyle \int_0^4 \int_1^2 (3x^2 - 2)dx\,dy$

Following the above procedure:

(i) $(3x^2 - 2)$ is integrated with respect to x between $x = 1$ and $x = 2$,

i.e. $\displaystyle \int_1^2 (3x^2 - 2)dx = \left[\frac{3x^3}{3} - 2x \right]_1^2$

$$= \left[\left(2^3 - 2(2) \right) - \left(1^3 - 2(1) \right) \right]$$

$$= (8 - 4) - (1 - 2)$$

$$= 8 - 4 - 1 + 2 = 5$$

(ii) $\displaystyle \int_0^4 \int_1^2 (3x^2 - 2)dx\,dy = \int_0^4 (5)dy$

$$= [5y]_0^4$$

$$= [(5(4)) - (5(0))]$$

$$= 20 - 0 = 20$$

Hence, $\displaystyle \int_0^4 \int_1^2 (3x^2 - 2)dx\,dy = 20$

Problem 3. Evaluate $\displaystyle \int_1^3 \int_0^2 (2x^2 y)dx\,dy$

Following the above procedure:

(i) $(2x^2 y)$ is integrated with respect to x between $x = 0$ and $x = 2$,

i.e. $\displaystyle \int_0^2 (2x^2 y)dx = \left[\frac{2x^3 y}{3} \right]_0^2$

$$= \left[\left(\frac{2(2)^3 y}{3} \right) - (0) \right] = \frac{16}{3} y$$

(ii) $\displaystyle \int_1^3 \int_0^2 (2x^2 y)dx\,dy = \int_1^3 \left(\frac{16}{3} y \right) dy = \left[\frac{16 y^2}{6} \right]_1^3$

$$= \left[\left(\frac{16(3)^2}{6} \right) - \left(\frac{16(1)^2}{6} \right) \right]$$

$$= 24 - 2.67 = 21.33$$

Hence, $\displaystyle \int_1^3 \int_0^2 (2x^2 y)dx\,dy = \mathbf{21.33}$

Problem 4. Evaluate $\displaystyle \int_1^3 dy \int_0^2 (2x^2 y)dx$

With this configuration:

(i) $(2x^2 y)$ is integrated with respect to x between $x = 0$ and $x = 2$,

i.e. $\displaystyle \int_0^2 (2x^2 y)dx = \left[\frac{2x^3 y}{3} \right]_0^2$

$$= \left[\left(\frac{2(2)^3 y}{3} \right) - (0) \right]$$

$$= \frac{16}{3} y$$

(ii) $\displaystyle \int_1^3 dy \int_0^2 (2x^2 y)dx = \int_1^3 dy \left(\frac{16}{3} y \right)$

$$= \int_1^3 \left(\frac{16}{3} y \right) dy = \left[\frac{16 y^2}{6} \right]_1^3$$

$$= \left[\left(\frac{16(3)^2}{6} \right) - \left(\frac{16(1)^2}{6} \right) \right]$$

$$= 24 - 2.67 = 21.33$$

Hence, $\displaystyle \int_1^3 dy \int_0^2 (2x^2 y)dx = \mathbf{21.33}$

The last two worked problems show that

$\displaystyle \int_1^3 \int_0^2 (2x^2 y)dx\,dy$ gives the same answer as

$\displaystyle \int_1^3 dy \int_0^2 2x^2 y\,dx$

Problem 5. Evaluate $\displaystyle \int_1^4 \int_0^\pi (2 + \sin 2\theta)d\theta\,dr$

Following the above procedure:

(i) $(2 + \sin 2\theta)$ is integrated with respect to θ between $\theta = 0$ and $\theta = \pi$,

i.e. $\displaystyle \int_0^\pi (2 + \sin 2\theta)dx = \left[2\theta - \frac{1}{2} \cos 2\theta \right]_0^\pi$

$$= \left[\left(2\pi - \frac{1}{2}\cos 2\pi \right) - \left(0 - \frac{1}{2}\cos 0 \right) \right]$$

$$= (2\pi - 0.5) - (0 - 0.5) = 2\pi$$

(ii) $\displaystyle\int_1^4 \int_0^\pi (2 + \sin 2\theta) \mathrm{d}\theta \mathrm{d}r = \int_1^4 (2\pi) \mathrm{d}r$

$$= [2\pi r]_1^4 = [(2\pi(4)) - (2\pi(1))]$$

$$= 8\pi - 2\pi = \mathbf{6\pi \ or \ 18.85}$$

Hence, $\displaystyle\int_1^4 \int_0^\pi (2 + \sin 2\theta) \mathrm{d}\theta \mathrm{d}r = \mathbf{18.85}$

Now try the following Practice Exercise

**Practice Exercise 282   Double integrals
(answers on page 1183)**

Evaluate the double integrals in Problems 1 to 8.

1. $\displaystyle\int_0^3 \int_2^4 2\mathrm{d}x \, \mathrm{d}y$

2. $\displaystyle\int_1^2 \int_2^3 (2x - y)\mathrm{d}x \, \mathrm{d}y$

3. $\displaystyle\int_1^2 \int_2^3 (2x - y)\mathrm{d}y \, \mathrm{d}x$

4. $\displaystyle\int_1^5 \int_{-1}^2 (x - 5y)\mathrm{d}x \, \mathrm{d}y$

5. $\displaystyle\int_1^6 \int_2^5 (x^2 + 4y)\mathrm{d}x \, \mathrm{d}y$

6. $\displaystyle\int_1^4 \int_{-3}^2 (3xy^2)\mathrm{d}x \, \mathrm{d}y$

7. $\displaystyle\int_{-1}^3 \int_0^\pi (3 + \sin 2\theta)\mathrm{d}\theta \mathrm{d}r$

8. $\displaystyle\int_1^3 \mathrm{d}x \int_2^4 (40 - 2xy)\mathrm{d}y$

9. The volume of a solid, V, bounded by the curve $4 - x - y$ between the limits $x = 0$ to $x = 1$ and $y = 0$ to $y = 2$ is given by:
$$V = \int_0^2 \int_0^1 (4 - x - y)\mathrm{d}x \, \mathrm{d}y$$
Evaluate V

10. The second moment of area, I, of a 5 cm by 3 cm rectangle about an axis through one corner perpendicular to the plane of the figure is given by:
$$I = \int_0^5 \int_0^3 (x^2 + y^2)\mathrm{d}y \, \mathrm{d}x$$
Evaluate I

70.2   Triple integrals

The procedure to determine a triple integral of the form:
$$\int_{z_1}^{z_2} \int_{y_1}^{y_2} \int_{x_1}^{x_2} f(x,y,z)\mathrm{d}x \, \mathrm{d}y \, \mathrm{d}z \ \text{is:}$$

(i)   integrate $f(x, y, z)$ with respect to x between the limits of $x = x_1$ and $x = x_2$ (where y and z are regarded as being constants),

(ii) integrate the result in (i) with respect to y between the limits of $y = y_1$ and $y = y_2$ and

(iii) integrate the result in (ii) with respect to z between the limits of $z = z_1$ and $z = z_2$

It is seen from this procedure that to determine a triple integral we start with the innermost integral and then work outwards.

Determining triple integrals is demonstrated in the following Problems.

Problem 6.   Evaluate
$$\int_1^2 \int_{-1}^3 \int_0^2 (x - 3y + z)\mathrm{d}x \, \mathrm{d}y \, \mathrm{d}z$$

Following the above procedure:

(i) $(x - 3y + z)$ is integrated with respect to x between $x = 0$ and $x = 2$, with y and z regarded as constants,

i.e. $\displaystyle\int_0^2 (x - 3y + z)\mathrm{d}x = \left[\frac{x^2}{2} - (3y)x + (z)x \right]_0^2$

$$= \left[\left(\frac{2^2}{2} - (3y)(2) + (z)(2) \right) - (0) \right]$$

$$= 2 - 6y + 2z$$

(ii) $(2 - 6y + 2z)$ is integrated with respect to y between $y = -1$ and $y = 3$, with z regarded as a constant, i.e.

$\displaystyle\int_{-1}^3 (2 - 6y + 2z)\mathrm{d}y = \left[2y - \frac{6y^2}{2} + (2z)y \right]_{-1}^3$

$$= \left[\left(2(3) - \frac{6(3)^2}{2} + (2z)(3) \right) \right.$$

$$\left. - \left(2(-1) - \frac{6(-1)^2}{2} + (2z)(-1) \right) \right]$$

$$= [(6 - 27 + 6z) - (-2 - 3 - 2z)]$$

$$= 6 - 27 + 6z + 2 + 3 + 2z = 8z - 16$$

(iii) $(8z - 16)$ is integrated with respect to z between $z = 1$ and $z = 2$

i.e. $\displaystyle\int_1^2 (8z - 16)\mathrm{d}z = \left[\frac{8z^2}{2} - 16z\right]_1^2$

$= \left[\left(\frac{8(2)^2}{2} - 16(2)\right) - \left(\frac{8(1)^2}{2} - 16(1)\right)\right]$

$= [(16 - 32) - (4 - 16)]$

$= 16 - 32 - 4 + 16 = -4$

Hence, $\displaystyle\int_1^2 \int_{-1}^3 \int_0^2 (x - 3y + z)\mathrm{d}x\,\mathrm{d}y\,\mathrm{d}z = -4$

Problem 7.   Evaluate
$\displaystyle\int_1^3 \int_0^2 \int_0^1 (2a^2 - b^2 + 3c^2)\mathrm{d}a\,\mathrm{d}b\,\mathrm{d}c$

Following the above procedure:

(i) $(2a^2 - b^2 + 3c^2)$ is integrated with respect to 'a' between $a = 0$ and $a = 1$, with 'b' and 'c' regarded as constants,

i.e. $\displaystyle\int_0^1 (2a^2 - b^2 + 3c^2)\mathrm{d}a$

$= \left[\frac{2a^3}{3} - (b^2)a + (3c^2)a\right]_0^1$

$= \left[\left(\frac{2}{3} - (b^2) + (3c^2)\right) - (0)\right]$

$= \frac{2}{3} - b^2 + 3c^2$

(ii) $\left(\dfrac{2}{3} - b^2 + 3c^2\right)$ is integrated with respect to 'b' between $b = 0$ and $b = 2$, with c regarded as a constant, i.e.

$\displaystyle\int_0^2 \left(\frac{2}{3} - b^2 + 3c^2\right)\mathrm{d}b = \left[\frac{2}{3}b - \frac{b^3}{3} + (3c^2)b\right]_0^2$

$= \left[\left(\frac{2}{3}(2) - \frac{(2)^3}{3} + (3c^2)(2)\right) - (0)\right]$

$= \left(\frac{4}{3} - \frac{8}{3} + 6c^2\right) - (0) = 6c^2 - \frac{4}{3}$

(iii) $\left(6c^2 - \dfrac{4}{3}\right)$ is integrated with respect to 'c' between $c = 1$ and $c = 3$

i.e. $\displaystyle\int_1^3 \left(6c^2 - \frac{4}{3}\right)\mathrm{d}c = \left[\frac{6c^3}{3} - \frac{4}{3}c\right]_1^3$

$= \left[(54 - 4) - \left(2 - \frac{4}{3}\right)\right]$

$= [(50) - (0.67)] = \mathbf{49.33}$

Hence, $\displaystyle\int_1^3 \int_0^2 \int_0^1 (2a^2 - b^2 + 3c^2)\mathrm{d}a\,\mathrm{d}b\,\mathrm{d}c = \mathbf{49.33}$

Now try the following Practice Exercise

Practice Exercise 283   Triple integrals (answers on page 1183)

Evaluate the triple integrals in Problems 1 to 7.

1. $\displaystyle\int_1^2 \int_2^3 \int_0^1 (8xyz)\mathrm{d}z\,\mathrm{d}x\,\mathrm{d}y$

2. $\displaystyle\int_1^2 \int_2^3 \int_0^1 (8xyz)\mathrm{d}x\,\mathrm{d}y\,\mathrm{d}z$

3. $\displaystyle\int_0^2 \int_{-1}^2 \int_1^3 \left(x + y^2 + z^3\right)\mathrm{d}x\,\mathrm{d}y\,\mathrm{d}z$

4. $\displaystyle\int_2^3 \int_{-1}^1 \int_0^2 (x^2 + 5y^2 - 2z)\mathrm{d}x\,\mathrm{d}y\,\mathrm{d}z$

5. $\displaystyle\int_0^\pi \int_0^\pi \int_0^\pi (xy\sin z)\mathrm{d}x\,\mathrm{d}y\,\mathrm{d}z$

6. $\displaystyle\int_0^4 \int_{-2}^{-1} \int_1^2 (xy)\mathrm{d}x\,\mathrm{d}y\,\mathrm{d}z$

7. $\displaystyle\int_1^3 \int_0^2 \int_{-1}^1 (xz + y)\mathrm{d}x\,\mathrm{d}y\,\mathrm{d}z$

8.   A box shape X is described by the triple integral: $X = \displaystyle\int_0^3 \int_0^2 \int_0^1 (x + y + z)\mathrm{d}z\,\mathrm{d}y\,\mathrm{d}x$
     Evaluate X

For fully worked solutions to each of the problems in Practice Exercises 282 and 283 in this chapter, go to the website:
www.routledge.com/cw/bird

Chapter 71

Numerical integration

Why it is important to understand: Numerical integration

There are two main reasons for why there is a need to do numerical integration – analytical integration may be impossible or infeasible, or it may be necessary to integrate tabulated data rather than known functions. As has been mentioned before, there are many applications for integration. For example, Maxwell's equations can be written in integral form. Numerical solutions of Maxwell's equations can be directly used for a huge number of engineering applications. Integration is involved in practically every physical theory in some way – vibration, distortion under weight, or one of many types of fluid flow – be it heat flow, air flow (over a wing), or water flow (over a ship's hull, through a pipe or perhaps even groundwater flow regarding a contaminant), and so on; all these things can be either directly solved by integration (for simple systems), or some type of numerical integration (for complex systems). Numerical integration is also essential for the evaluation of integrals of functions available only at discrete points; such functions often arise in the numerical solution of differential equations or from experimental data taken at discrete intervals. Engineers therefore often require numerical integration and this chapter explains the procedures available.

At the end of this chapter, you should be able to:

- appreciate the need for numerical integration
- evaluate integrals using the trapezoidal rule
- evaluate integrals using the mid-ordinate rule
- evaluate integrals using Simpson's rule
- apply numerical integration to practical situations

71.1 Introduction

Even with advanced methods of integration there are many mathematical functions which cannot be integrated by analytical methods and thus approximate methods have then to be used. In many cases, such as in modelling airflow around a car, an exact answer may not even be strictly necessary. Approximate methods of definite integrals may be determined by what is termed **numerical integration**.

It may be shown that determining the value of a definite integral is, in fact, finding the area between a curve, the horizontal axis and the specified ordinates. Three methods of finding approximate areas under curves are the trapezoidal rule, the mid-ordinate rule and Simpson's rule, and these rules are used as a basis for numerical integration.

71.2 The trapezoidal rule

Let a required definite integral be denoted by $\int_a^b y \, dx$ and be represented by the area under the graph of $y = f(x)$ between the limits $x = a$ and $x = b$ as shown in Figure 71.1.

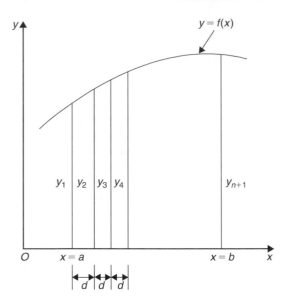

Figure 71.1

Let the range of integration be divided into n equal intervals each of width d, such that $nd = b - a$, i.e.

$$d = \frac{b-a}{n}$$

The ordinates are labelled $y_1, y_2, y_3, \ldots, y_{n+1}$ as shown. An approximation to the area under the curve may be determined by joining the tops of the ordinates by straight lines. Each interval is thus a trapezium, and since the area of a trapezium is given by:

$$\text{area} = \frac{1}{2}(\text{sum of parallel sides}) (\text{perpendicular}$$

$$\text{distance between them})$$

then

$$\int_a^b y \, dx \approx \frac{1}{2}(y_1 + y_2)d + \frac{1}{2}(y_2 + y_3)d$$

$$+ \frac{1}{2}(y_3 + y_4)d + \cdots \frac{1}{2}(y_n + y_{n+1})d$$

$$\approx d\left[\frac{1}{2}y_1 + y_2 + y_3 + y_4 + \cdots + y_n\right.$$

$$\left. + \frac{1}{2}y_{n+1}\right]$$

i.e. **the trapezoidal rule states:**

$$\int_a^b y \, dx \approx \left(\begin{array}{c}\textbf{width of}\\ \textbf{interval}\end{array}\right)\left\{\frac{1}{2}\left(\begin{array}{c}\textbf{first}+\textbf{last}\\ \textbf{ordinate}\end{array}\right)\right.$$

$$\left. + \left(\begin{array}{c}\textbf{sum of remaining}\\ \textbf{ordinates}\end{array}\right)\right\} \quad (1)$$

Problem 1. (a) Use integration to evaluate, correct to 3 decimal places, $\int_1^3 \frac{2}{\sqrt{x}} \, dx$ (b) Use the trapezoidal rule with 4 intervals to evaluate the integral in part (a), correct to 3 decimal places

(a) $\int_1^3 \frac{2}{\sqrt{x}} \, dx = \int_1^3 2x^{-\frac{1}{2}} \, dx$

$$= \left[\frac{2x^{\left(\frac{-1}{2}\right)+1}}{-\frac{1}{2}+1}\right]_1^3 = \left[4x^{\frac{1}{2}}\right]_1^3$$

$$= 4\left[\sqrt{x}\right]_1^3 = 4\left[\sqrt{3} - \sqrt{1}\right]$$

$$= \textbf{2.928}, \text{correct to 3 decimal places.}$$

(b) The range of integration is the difference between the upper and lower limits, i.e. $3 - 1 = 2$. Using the trapezoidal rule with 4 intervals gives an interval width $d = \frac{3-1}{4} = 0.5$ and ordinates situated at 1.0, 1.5, 2.0, 2.5 and 3.0. Corresponding values of $\frac{2}{\sqrt{x}}$ are shown in the table below, each correct to 4 decimal places (which is one more decimal place than required in the problem).

x	$\frac{2}{\sqrt{x}}$
1.0	2.0000
1.5	1.6330
2.0	1.4142
2.5	1.2649
3.0	1.1547

From equation (1):

$$\int_1^3 \frac{2}{\sqrt{x}} \, dx \approx (0.5)\left\{\frac{1}{2}(2.0000 + 1.1547)\right.$$

$$\left. + 1.6330 + 1.4142 + 1.2649\right\}$$

$$= \textbf{2.945}, \text{correct to 3 decimal places.}$$

This problem demonstrates that even with just 4 intervals a close approximation to the true value of 2.928 (correct to 3 decimal places) is obtained using the trapezoidal rule.

Problem 2. Use the trapezoidal rule with 8 intervals to evaluate $\int_1^3 \frac{2}{\sqrt{x}}\,dx$, correct to 3 decimal places

With 8 intervals, the width of each is $\frac{3-1}{8}$ i.e. 0.25, giving ordinates at 1.00, 1.25, 1.50, 1.75, 2.00, 2.25, 2.50, 2.75 and 3.00. Corresponding values of $\frac{2}{\sqrt{x}}$ are shown in the table below.

x	$\dfrac{2}{\sqrt{x}}$
1.00	2.0000
1.25	1.7889
1.50	1.6330
1.75	1.5119
2.00	1.4142
2.25	1.3333
2.50	1.2649
2.75	1.2060
3.00	1.1547

From equation (1):

$$\int_1^3 \frac{2}{\sqrt{x}}\,dx \approx (0.25)\left\{\frac{1}{2}(2.000+1.1547)+1.7889\right.$$
$$+1.6330+1.5119+1.4142$$
$$\left.+1.3333+1.2649+1.2060\right\}$$

$$= \mathbf{2.932}, \text{correct to 3 decimal places.}$$

This problem demonstrates that the greater the number of intervals chosen (i.e. the smaller the interval width) the more accurate will be the value of the definite integral. The exact value is found when the number of intervals is infinite, which is, of course, what the process of integration is based upon.

Problem 3. Use the trapezoidal rule to evaluate $\int_0^{\frac{\pi}{2}} \frac{1}{1+\sin x}\,dx$ using 6 intervals. Give the answer correct to 4 significant figures

With 6 intervals, each will have a width of $\dfrac{\frac{\pi}{2}-0}{6}$ i.e. $\dfrac{\pi}{12}$ rad (or $15°$) and the ordinates occur at $0, \dfrac{\pi}{12}, \dfrac{\pi}{6}, \dfrac{\pi}{4}, \dfrac{\pi}{3}, \dfrac{5\pi}{12}$ and $\dfrac{\pi}{2}$

Corresponding values of $\dfrac{1}{1+\sin x}$ are shown in the table below.

x	$\dfrac{1}{1+\sin x}$
0	1.0000
$\dfrac{\pi}{12}$ (or $15°$)	0.79440
$\dfrac{\pi}{6}$ (or $30°$)	0.66667
$\dfrac{\pi}{4}$ (or $45°$)	0.58579
$\dfrac{\pi}{3}$ (or $60°$)	0.53590
$\dfrac{5\pi}{12}$ (or $75°$)	0.50867
$\dfrac{\pi}{2}$ (or $90°$)	0.50000

From equation (1):

$$\int_0^{\frac{\pi}{2}} \frac{1}{1+\sin x}\,dx \approx \left(\frac{\pi}{12}\right)\left\{\frac{1}{2}(1.00000+0.50000)\right.$$
$$+0.79440+0.66667$$
$$+0.58579+0.53590$$
$$\left.+0.50867\right\}$$

$$= \mathbf{1.006}, \text{correct to 4}$$
$$\text{significant figures.}$$

Now try the following Practice Exercise

Practice Exercise 284   Further problems on the trapezoidal rule (answers on page 1183)

In Problems 1 to 4, evaluate the definite integrals using the **trapezoidal rule**, giving the answers correct to 3 decimal places.

1. $\int_0^1 \dfrac{2}{1+x^2}\,dx$ (use 8 intervals).

2. $\int_1^3 2\ln 3x\,dx$ (use 8 intervals).

3. $\int_0^{\frac{\pi}{3}} \sqrt{(\sin\theta)}\,d\theta$ (use 6 intervals).

4. $\int_0^{1.4} e^{-x^2}\,dx$ (use 7 intervals).

71.3   The mid-ordinate rule

Let a required definite integral be denoted again by $\int_a^b y\,dx$ and represented by the area under the graph of $y = f(x)$ between the limits $x = a$ and $x = b$, as shown in Figure 71.2.

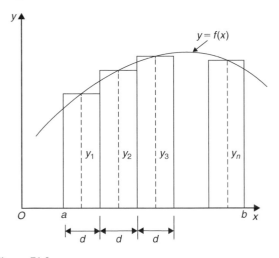

Figure 71.2

With the mid-ordinate rule each interval of width d is assumed to be replaced by a rectangle of height equal to the ordinate at the middle point of each interval, shown as $y_1, y_2, y_3, \ldots, y_n$ in Figure 71.2.

Thus $\int_a^b y\,dx \approx dy_1 + dy_2 + dy_3 + \cdots + dy_n$

$\approx d(y_1 + y_2 + y_3 + \cdots + y_n)$

i.e. **the mid-ordinate rule states:**

$$\boxed{\int_a^b y\,dx \approx \textbf{(width of interval) (sum of mid-ordinates)}}$$ (2)

Problem 4.   Use the mid-ordinate rule with (a) 4 intervals, (b) 8 intervals, to evaluate $\int_1^3 \dfrac{2}{\sqrt{x}}\,dx$, correct to 3 decimal places

(a)   With 4 intervals, each will have a width of $\dfrac{3-1}{4}$, i.e. 0.5 and the ordinates will occur at 1.0, 1.5, 2.0, 2.5 and 3.0. Hence the mid-ordinates y_1, y_2, y_3 and y_4 occur at 1.25, 1.75, 2.25 and 2.75. Corresponding values of $\dfrac{2}{\sqrt{x}}$ are shown in the following table.

x	$\dfrac{2}{\sqrt{x}}$
1.25	1.7889
1.75	1.5119
2.25	1.3333
2.75	1.2060

From equation (2):

$$\int_1^3 \dfrac{2}{\sqrt{x}}\,dx \approx (0.5)[1.7889 + 1.5119$$
$$+ 1.3333 + 1.2060]$$
$$= \textbf{2.920}, \text{correct to 3 decimal places.}$$

(b)   With 8 intervals, each will have a width of 0.25 and the ordinates will occur at 1.00, 1.25, 1.50, 1.75, … and thus mid-ordinates at 1.125, 1.375, 1.625, 1.875, …
Corresponding values of $\dfrac{2}{\sqrt{x}}$ are shown in the following table.

x	$\dfrac{2}{\sqrt{x}}$
1.125	1.8856
1.375	1.7056
1.625	1.5689
1.875	1.4606
2.125	1.3720
2.375	1.2978
2.625	1.2344
2.875	1.1795

From equation (2):

$$\int_1^3 \frac{2}{\sqrt{x}}\,dx \approx (0.25)[1.8856 + 1.7056$$

$$+ 1.5689 + 1.4606 + 1.3720$$

$$+ 1.2978 + 1.2344 + 1.1795]$$

$$= \mathbf{2.926}, \text{ correct to 3 decimal places.}$$

As previously, the greater the number of intervals the nearer the result is to the true value (of 2.928, correct to 3 decimal places).

Problem 5. Evaluate $\int_0^{2.4} e^{\frac{-x^2}{3}}\,dx$, correct to 4 significant figures, using the mid-ordinate rule with 6 intervals

With 6 intervals, each will have a width of $\dfrac{2.4 - 0}{6}$, i.e. 0.40, and the ordinates will occur at 0, 0.40, 0.80, 1.20, 1.60, 2.00 and 2.40 and thus mid-ordinates at 0.20, 0.60, 1.00, 1.40, 1.80 and 2.20. Corresponding values of $e^{\frac{-x^2}{3}}$ are shown in the following table.

x	$e^{\frac{-x^2}{3}}$
0.20	0.98676
0.60	0.88692
1.00	0.71653
1.40	0.52031
1.80	0.33960
2.20	0.19922

From equation (2):

$$\int_0^{2.4} e^{\frac{-x^2}{3}}\,dx \approx (0.40)[0.98676 + 0.88692$$

$$+ 0.71653 + 0.52031$$

$$+ 0.33960 + 0.19922]$$

$$= \mathbf{1.460}, \text{ correct to 4 significant figures.}$$

Now try the following Practice Exercise

In Problems 1 to 4, evaluate the definite integrals using the **mid-ordinate rule**, giving the answers correct to 3 decimal places.

1. $\displaystyle\int_0^2 \frac{3}{1+t^2}\,dt$ (use 8 intervals).

2. $\displaystyle\int_0^{\frac{\pi}{2}} \frac{1}{1+\sin\theta}\,d\theta$ (use 6 intervals).

3. $\displaystyle\int_1^3 \frac{\ln x}{x}\,dx$ (use 10 intervals).

4. $\displaystyle\int_0^{\frac{\pi}{3}} \sqrt{(\cos^3 x)}\,dx$ (use 6 intervals).

71.4 Simpson's rule

The approximation made with the trapezoidal rule is to join the top of two successive ordinates by a straight line, i.e. by using a linear approximation of the form $a + bx$. With **Simpson's*** rule, the approximation made is to join the tops of three successive ordinates by a parabola, i.e. by using a quadratic approximation of the form $a + bx + cx^2$.

Figure 71.3 shows a parabola $y = a + bx + cx^2$ with ordinates y_1, y_2 and y_3 at $x = -d$, $x = 0$ and $x = d$ respectively.

Thus the width of each of the two intervals is d. The area enclosed by the parabola, the x-axis and ordinates $x = -d$ and $x = d$ is given by:

$$\int_{-d}^{d} (a + bx + cx^2)\,dx = \left[ax + \frac{bx^2}{2} + \frac{cx^3}{3} \right]_{-d}^{d}$$

$$= \left(ad + \frac{bd^2}{2} + \frac{cd^3}{3} \right) - \left(-ad + \frac{bd^2}{2} - \frac{cd^3}{3} \right)$$

$$= 2ad + \frac{2}{3}cd^3 \text{ or } \frac{1}{3}d(6a + 2cd^2) \tag{3}$$

---

* **Who was Simpson?** **Thomas Simpson FRS** (20 August 1710–14 May 1761) was the British mathematician who invented Simpson's rule to approximate definite integrals. For an image of Simpson, see page 326. To find out more go to **www.routledge.com/cw/bird**

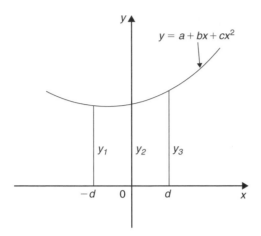

Figure 71.3

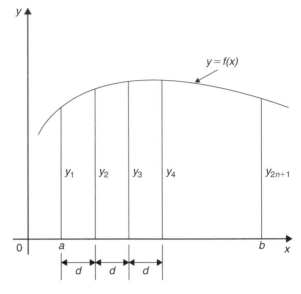

Figure 71.4

Since $y = a + bx + cx^2$,

at $\quad x = -d, y_1 = a - bd + cd^2$

at $\quad x = 0, \ y_2 = a$

and at $\quad x = d, \ y_3 = a + bd + cd^2$

Hence $\quad y_1 + y_3 = 2a + 2cd^2$

And $\quad y_1 + 4y_2 + y_3 = 6a + 2cd^2 \qquad (4)$

Thus the area under the parabola between $x = -d$ and $x = d$ in Figure 71.3 may be expressed as $\frac{1}{3}d(y_1 + 4y_2 + y_3)$, from equations (3) and (4), and the result is seen to be independent of the position of the origin.

Let a definite integral be denoted by $\int_a^b y\,\mathrm{d}x$ and represented by the area under the graph of $y = f(x)$ between the limits $x = a$ and $x = b$, as shown in Figure 71.4. The range of integration, $b - a$, is divided into an **even** number of intervals, say $2n$, each of width d. Since an even number of intervals is specified, an odd number of ordinates, $2n + 1$, exists. Let an approximation to the curve over the first two intervals be a parabola of the form $y = a + bx + cx^2$ which passes through the tops of the three ordinates y_1, y_2 and y_3. Similarly, let an approximation to the curve over the next two intervals be the parabola which passes through the tops of the ordinates y_3, y_4 and y_5, and so on.

Then $\displaystyle\int_a^b y\,\mathrm{d}x$

$\approx \dfrac{1}{3}d(y_1 + 4y_2 + y_3) + \dfrac{1}{3}d(y_3 + 4y_4 + y_5)$

$\qquad + \dfrac{1}{3}d(y_{2n-1} + 4y_{2n} + y_{2n+1})$

$\approx \dfrac{1}{3}d[(y_1 + y_{2n+1}) + 4(y_2 + y_4 + \cdots + y_{2n})$

$\qquad\qquad + 2(y_3 + y_5 + \cdots + y_{2n-1})]$

i.e. **Simpson's rule states:**

$$\int_a^b y\,\mathrm{d}x \approx \frac{1}{3}\begin{pmatrix}\textbf{width of}\\\textbf{interval}\end{pmatrix}\left\{\begin{pmatrix}\textbf{first + last}\\\textbf{ordinate}\end{pmatrix}\right.$$
$$+ 4\begin{pmatrix}\textbf{sum of even}\\\textbf{ordinates}\end{pmatrix}$$
$$\left.+ 2\begin{pmatrix}\textbf{sum of remaining}\\\textbf{odd ordinates}\end{pmatrix}\right\} \qquad (5)$$

Note that Simpson's rule can only be applied when an even number of intervals is chosen, i.e. an odd number of ordinates.

Problem 6. Use Simpson's rule with (a) 4 intervals, (b) 8 intervals, to evaluate $\displaystyle\int_1^3 \frac{2}{\sqrt{x}}\,\mathrm{d}x$, correct to 3 decimal places

---

(a) With 4 intervals, each will have a width of $\dfrac{3-1}{4}$, i.e. 0.5, and the ordinates will occur at 1.0, 1.5, 2.0, 2.5 and 3.0. The values of the ordinates are as shown in the table of Problem 1(b), page 774.

Thus, from equation (5):

$$\int_1^3 \frac{2}{\sqrt{x}}\,dx \approx \frac{1}{3}(0.5)\,[(2.0000 + 1.1547)$$

$$+\ 4(1.6330 + 1.2649) + 2(1.4142)]$$

$$=\frac{1}{3}(0.5)[3.1547 + 11.5916$$

$$+\ 2.8284]$$

$$= \textbf{2.929}, \text{ correct to 3 decimal places.}$$

(b) With 8 intervals, each will have a width of $\dfrac{3-1}{8}$, i.e. 0.25, and the ordinates occur at 1.00, 1.25, 1.50, 1.75, ..., 3.0. The values of the ordinates are as shown in the table in Problem 2, page 775.
Thus, from equation (5):

$$\int_1^3 \frac{2}{\sqrt{x}}\,dx \approx \frac{1}{3}(0.25)\,[(2.0000 + 1.1547)$$

$$+\ 4(1.7889 + 1.5119 + 1.3333$$

$$+\ 1.2060) + 2(1.6330 + 1.4142$$

$$+\ 1.2649)]$$

$$=\frac{1}{3}(0.25)[3.1547 + 23.3604$$

$$+\ 8.6242]$$

$$= \textbf{2.928}, \text{ correct to 3 decimal places.}$$

It is noted that the latter answer is exactly the same as that obtained by integration. In general, Simpson's rule is regarded as the most accurate of the three approximate methods used in numerical integration.

Problem 7. Evaluate

$$\int_0^{\frac{\pi}{3}} \sqrt{\left(1 - \frac{1}{3}\sin^2\theta\right)}\,d\theta$$

correct to 3 decimal places, using Simpson's rule with 6 intervals

With 6 intervals, each will have a width of $\dfrac{\dfrac{\pi}{3} - 0}{6}$ i.e. $\dfrac{\pi}{18}$ rad (or 10°), and the ordinates will occur at

$$0, \frac{\pi}{18}, \frac{\pi}{9}, \frac{\pi}{6}, \frac{2\pi}{9}, \frac{5\pi}{18} \text{ and } \frac{\pi}{3}$$

Corresponding values of $\sqrt{\left(1 - \dfrac{1}{3}\sin^2\theta\right)}$ are shown in the table below.

θ	0	$\dfrac{\pi}{18}$ (or 10°)	$\dfrac{\pi}{9}$ (or 20°)	$\dfrac{\pi}{6}$ (or 30°)
$\sqrt{\left(1 - \dfrac{1}{3}\sin^2\theta\right)}$	1.0000	0.9950	0.9803	0.9574

θ	$\dfrac{2\pi}{9}$ (or 40°)	$\dfrac{5\pi}{18}$ (or 50°)	$\dfrac{\pi}{3}$ (or 60°)
$\sqrt{\left(1 - \dfrac{1}{3}\sin^2\theta\right)}$	0.9286	0.8969	0.8660

From equation (5)

$$\int_0^{\frac{\pi}{3}} \sqrt{\left(1 - \frac{1}{3}\sin^2\theta\right)}\,d\theta$$

$$\approx \frac{1}{3}\left(\frac{\pi}{18}\right)[(1.0000 + 0.8660) + 4(0.9950$$

$$+\ 0.9574 + 0.8969)$$

$$+\ 2(0.9803 + 0.9286)]$$

$$=\frac{1}{3}\left(\frac{\pi}{18}\right)[1.8660 + 11.3972 + 3.8178]$$

$$= \textbf{0.994}, \text{ correct to 3 decimal places.}$$

Problem 8. An alternating current i has the following values at equal intervals of 2.0 milliseconds:

Time (ms)	Current i (A)
0	0
2.0	3.5
4.0	8.2
6.0	10.0

8.0	7.3
10.0	2.0
12.0	0

Charge, q, in millicoulombs, is given by

$$q = \int_0^{12.0} i \, dt$$

Use Simpson's rule to determine the approximate charge in the 12 millisecond period

From equation (5):

$$\text{Charge, } q = \int_0^{12.0} i \, dt \approx \frac{1}{3}(2.0)\left[(0+0)+4(3.5\right.$$

$$\left.+10.0+2.0)+2(8.2+7.3)\right]$$

$$= 62 \, \text{mC}$$

Now try the following Practice Exercise

Practice Exercise 286   Further problems on Simpson's rule (answers on page 1184)

In Problems 1 to 5, evaluate the definite integrals using **Simpson's rule**, giving the answers correct to 3 decimal places.

1. $\int_0^{\frac{\pi}{2}} \sqrt{(\sin x)} \, dx$ (use 6 intervals).

2. $\int_0^{1.6} \frac{1}{1+\theta^4} \, d\theta$ (use 8 intervals).

3. $\int_{0.2}^{1.0} \frac{\sin\theta}{\theta} \, d\theta$ (use 8 intervals).

4. $\int_0^{\frac{\pi}{2}} x \cos x \, dx$ (use 6 intervals).

5. $\int_0^{\frac{\pi}{3}} e^{x^2} \sin 2x \, dx$ (use 10 intervals).

In Problems 6 and 7, evaluate the definite integrals using (a) integration, (b) the trapezoidal rule, (c) the mid-ordinate rule, (d) Simpson's rule. Give answers correct to 3 decimal places.

6. $\int_1^4 \frac{4}{x^3} \, dx$ (use 6 intervals).

7. $\int_2^6 \frac{1}{\sqrt{(2x-1)}} \, dx$ (use 8 intervals).

In Problems 8 and 9, evaluate the definite integrals using (a) the trapezoidal rule, (b) the mid-ordinate rule, (c) Simpson's rule. Use 6 intervals in each case and give answers correct to 3 decimal places.

8. $\int_0^3 \sqrt{(1+x^4)} \, dx$

9. $\int_{0.1}^{0.7} \frac{1}{\sqrt{(1-y^2)}} \, dy$

10. A vehicle starts from rest and its velocity is measured every second for 8 s, with values as follows:

time t (s)	velocity v (ms^{-1})
0	0
1.0	0.4
2.0	1.0
3.0	1.7
4.0	2.9
5.0	4.1
6.0	6.2
7.0	8.0
8.0	9.4

The distance travelled in 8.0 s is given by

$$\int_0^{8.0} v \, dt$$

Estimate this distance using Simpson's rule, giving the answer correct to 3 significant figures.

11. A pin moves along a straight guide so that its velocity v (m/s) when it is a distance x(m) from the beginning of the guide at time t(s) is given in the table below.

t (s)	v (m/s)
0	0
0.5	0.052
1.0	0.082
1.5	0.125
2.0	0.162
2.5	0.175
3.0	0.186
3.5	0.160
4.0	0

Use Simpson's rule with 8 intervals to determine the approximate total distance travelled by the pin in the 4.0 s period.

71.5   Accuracy of numerical integration

For a function with an increasing gradient, the trapezoidal rule will tend to over-estimate and the mid-ordinate rule will tend to under-estimate (but by half as much). The appropriate combination of the two in Simpson's rule eliminates this error term, giving a rule which will perfectly model anything up to a cubic, and have a proportionately lower error for any function of greater complexity.

In general, for a given number of strips, Simpson's rule is considered the most accurate of the three numerical methods.

For fully worked solutions to each of the problems in Practice Exercises 284 to 286 in this chapter, go to the website:
www.routledge.com/cw/bird

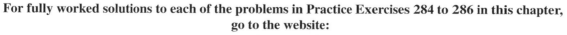

Revision Test 26   *Reduction formulae, double and triple integrals, and numerical integration*

This assignment covers the material contained in Chapters 69 to 71. *The marks available are shown in brackets at the end of each question.*

1. Use reduction formulae to determine:

   (a) $\displaystyle\int x^3 e^{3x}\,dx$ (b) $\displaystyle\int t^4 \sin t\,dt$ (14)

2. Evaluate $\displaystyle\int_0^{\frac{\pi}{2}} \cos^6 x\,dx$ using a reduction formula.   (6)

3. Evaluate $\displaystyle\int_1^3 \int_0^{\pi} (1 + \sin 3\theta)\,d\theta\,dr$ correct to 4 significant figures.   (7)

4. Evaluate $\displaystyle\int_0^3 \int_{-1}^1 \int_1^2 \left(2x + y^2 + 4z^3\right) dx\,dy\,dz$ (9)

5. Evaluate $\displaystyle\int_1^3 \frac{5}{x^2}\,dx$ using (a) integration, (b) the trapezoidal rule, (c) the mid-ordinate rule,

   (d) Simpson's rule. In each of the approximate methods use 8 intervals and give the answers correct to 3 decimal places.   (19)

6. An alternating current i has the following values at equal intervals of 5 ms:

Time t(ms)	0	5	10	15	20	25	30
Current i(A)	0	4.8	9.1	12.7	8.8	3.5	0

   Charge q, in coulombs, is given by

   $$q = \int_0^{30 \times 10^{-3}} i\,dt$$

   Use Simpson's rule to determine the approximate charge in the 30 ms period.   (5)

For lecturers/instructors/teachers, fully worked solutions to each of the problems in Revision Test 26, together with a full marking scheme, are available at the website:
www.routledge.com/cw/bird

Chapter 72

Areas under and between curves

Why it is important to understand: Areas under and between curves

One of the important applications of integration is to find the area bounded by a curve. Often such an area can have a physical significance, like the work done by a motor, or the distance travelled by a vehicle. Other examples where finding the area under a curve is important can involve position, velocity, force, charge density, resistivity and current density. Hence, finding the area under and between curves is of some importance in engineering and science.

At the end of this chapter, you should be able to:

- appreciate that the area under a curve of a known function is a definite integral
- state practical examples where areas under a curve need to be accurately determined
- calculate areas under curves using integration
- calculate areas between curves using integration

72.1 Area under a curve

The area shown shaded in Figure 72.1 may be determined using approximate methods (such as the trapezoidal rule, the mid-ordinate rule or Simpson's rule) or, more precisely, by using integration.

(i) Let A be the area shown shaded in Figure 72.1 and let this area be divided into a number of strips each of width δx. One such strip is shown and let the area of this strip be δA.

Then: $\delta A \approx y\delta x$ (1)

The accuracy of statement (1) increases when the width of each strip is reduced, i.e. area A is divided into a greater number of strips.

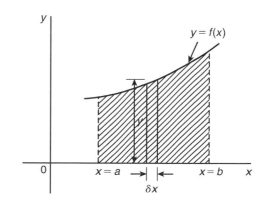

Figure 72.1

(ii) Area A is equal to the sum of all the strips from $x=a$ to $x=b$, i.e. $A = \underset{\delta x \to 0}{\text{limit}} \sum_{x=a}^{x=b} y \, \delta x$ (2)

(iii) From statement (1), $\dfrac{\delta A}{\delta x} \approx y$ (3)

In the limit, as δx approaches zero, $\dfrac{\delta A}{\delta x}$ becomes the differential coefficient $\dfrac{dA}{dx}$

Hence $\underset{\delta x \to 0}{\text{limit}} \left(\dfrac{\delta A}{\delta x} \right) = \dfrac{dA}{dx} = y$, from statement (3).

By integration,

$$\int \dfrac{dA}{dx} \, dx = \int y \, dx \quad \text{i.e.} \quad A = \int y \, dx$$

The ordinates $x = a$ and $x = b$ limit the area and such ordinate values are shown as limits. Hence

$$A = \int_a^b y \, dx \quad (4)$$

(iv) Equating statements (2) and (4) gives:

$$\textbf{Area } A = \underset{\delta x \to 0}{\text{limit}} \sum_{x=a}^{x=b} y \, \delta x = \int_a^b y \, dx$$

$$= \int_a^b f(x) \, dx$$

(v) If the area between a curve $x = f(y)$, the y-axis and ordinates $y = p$ and $y = q$ is required, then

$$\text{area} = \int_p^q x \, dy$$

Thus, determining the area under a curve by integration merely involves evaluating a definite integral.
There are several instances in engineering and science where the area beneath a curve needs to be accurately determined. For example, **the area between limits of a:**

**velocity/time graph gives distance travelled,
force/distance graph gives work done,
voltage/current graph gives power, and so on.**

Should a curve drop below the x-axis, then $y \; (= f(x))$ becomes negative and $f(x) \, dx$ is negative. When determining such areas by integration, a negative sign is placed before the integral. For the curve shown in Figure 72.2, the total shaded area is given by (area $E +$ area $F +$ area G).
By integration, **total shaded area**

$$= \int_a^b f(x) \, dx - \int_b^c f(x) \, dx + \int_c^d f(x) \, dx$$

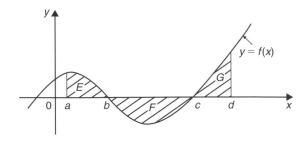

Figure 72.2

(Note that this is **not** the same as $\displaystyle\int_a^d f(x) \, dx$)
It is usually necessary to sketch a curve in order to check whether it crosses the x-axis.

72.2 Worked problems on the area under a curve

Problem 1. Determine the area enclosed by $y = 2x + 3$, the x-axis and ordinates $x = 1$ and $x = 4$

$y = 2x + 3$ is a straight line graph as shown in Figure 72.3, where the required area is shown shaded. By integration,

$$\text{shaded area} = \int_1^4 y \, dx$$

$$= \int_1^4 (2x + 3) \, dx$$

$$- \left[\frac{2x^2}{2} + 3x \right]_1^4$$

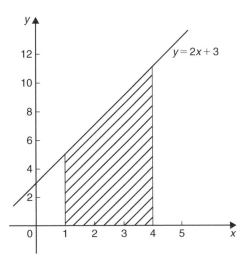

Figure 72.3

$$= [(16+12) - (1+3)]$$
$$= 24 \text{ square units}$$

(This answer may be checked since the shaded area is a trapezium.

Area of trapezium

$$= \frac{1}{2} \begin{pmatrix} \text{sum of parallel} \\ \text{sides} \end{pmatrix} \begin{pmatrix} \text{perpendicular distance} \\ \text{between parallel sides} \end{pmatrix}$$

$$= \frac{1}{2}(5+11)(3)$$

$$= 24 \text{ square units})$$

Problem 2. The velocity v of a body t seconds after a certain instant is: $(2t^2 + 5)$ m/s. Find by integration how far it moves in the interval from $t=0$ to $t=4$ s

Since $2t^2 + 5$ is a quadratic expression, the curve $v = 2t^2 + 5$ is a parabola cutting the v-axis at $v=5$, as shown in Figure 72.4.

The distance travelled is given by the area under the v/t curve (shown shaded in Figure 72.4).

By integration,

$$\text{shaded area} = \int_0^4 v \, dt$$

$$= \int_0^4 (2t^2 + 5) \, dt$$

$$= \left[\frac{2t^3}{3} + 5t \right]_0^4$$

$$= \left(\frac{2(4^3)}{3} + 5(4) \right) - (0)$$

i.e. **distance travelled $= 62.67$ m**

Problem 3. Sketch the graph $y = x^3 + 2x^2 - 5x - 6$ between $x = -3$ and $x = 2$ and determine the area enclosed by the curve and the x-axis

A table of values is produced and the graph sketched as shown in Figure 72.5, where the area enclosed by the curve and the x-axis is shown shaded.

x	-3	-2	-1	0	1	2
x^3	-27	-8	-1	0	1	8
$2x^2$	18	8	2	0	2	8
$-5x$	15	10	5	0	-5	-10
-6	-6	-6	-6	-6	-6	-6
y	0	4	0	-6	-8	0

Shaded area $= \displaystyle\int_{-3}^{-1} y \, dx - \int_{-1}^{2} y \, dx$, the minus sign before the second integral being necessary since the enclosed area is below the x-axis.

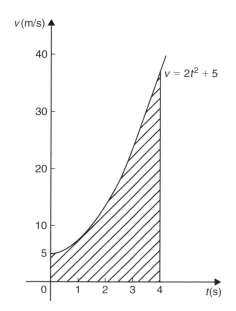

Figure 72.4

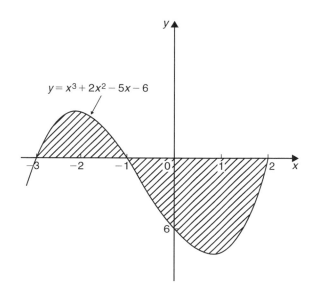

Figure 72.5

Hence shaded area

$$= \int_{-3}^{-1} (x^3 + 2x^2 - 5x - 6)\, dx$$

$$- \int_{-1}^{2} (x^3 + 2x^2 - 5x - 6)\, dx$$

$$= \left[\frac{x^4}{4} + \frac{2x^3}{3} - \frac{5x^2}{2} - 6x \right]_{-3}^{-1}$$

$$- \left[\frac{x^4}{4} + \frac{2x^3}{3} - \frac{5x^2}{2} - 6x \right]_{-1}^{2}$$

$$= \left[\left\{ \frac{1}{4} - \frac{2}{3} - \frac{5}{2} + 6 \right\} \right.$$

$$- \left\{ \frac{81}{4} - 18 - \frac{45}{2} + 18 \right\} \right]$$

$$- \left[\left\{ 4 + \frac{16}{3} - 10 - 12 \right\} \right.$$

$$- \left\{ \frac{1}{4} - \frac{2}{3} - \frac{5}{2} + 6 \right\} \right]$$

$$= \left[\left\{ 3\frac{1}{12} \right\} - \left\{ -2\frac{1}{4} \right\} \right]$$

$$- \left[\left\{ -12\frac{2}{3} \right\} - \left\{ 3\frac{1}{12} \right\} \right]$$

$$= \left[5\frac{1}{3} \right] - \left[-15\frac{3}{4} \right]$$

$$= 21\frac{1}{12} \quad \text{or} \quad \textbf{21.08 square units}$$

> **Problem 4.** Determine the area enclosed by the curve $y = 3x^2 + 4$, the x-axis and ordinates $x = 1$ and $x = 4$ by (a) the trapezoidal rule, (b) the mid-ordinate rule, (c) Simpson's rule, and (d) integration

The curve $y = 3x^2 + 4$ is shown plotted in Figure 72.6.

(a) **By the trapezoidal rule,**

$$\text{Area} = \begin{pmatrix} \text{width of} \\ \text{interval} \end{pmatrix} \left[\frac{1}{2} \begin{pmatrix} \text{first} + \text{last} \\ \text{ordinate} \end{pmatrix} + \begin{pmatrix} \text{sum of} \\ \text{remaining} \\ \text{ordinates} \end{pmatrix} \right]$$

Selecting 6 intervals each of width 0.5 gives:

$$\text{Area} = (0.5) \left[\frac{1}{2}(7 + 52) + 10.75 + 16 \right.$$

$$\left. + 22.75 + 31 + 40.75 \right]$$

$$= \textbf{75.375 square units}$$

x	0	1.0	1.5	2.0	2.5	3.0	3.5	4.0
y	4	7	10.75	16	22.75	31	40.75	52

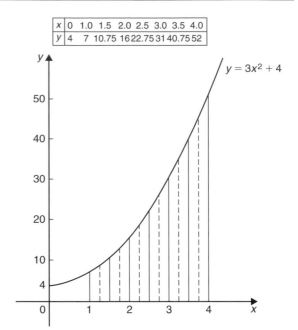

Figure 72.6

(b) **By the mid-ordinate rule,**
area = (width of interval) (sum of mid-ordinates). Selecting 6 intervals, each of width 0.5, gives the mid-ordinates as shown by the broken lines in Figure 72.6.

Thus, area $= (0.5)(8.5 + 13 + 19 + 26.5$

$$+ 35.5 + 46)$$

$$= \textbf{74.25 square units}$$

(c) **By Simpson's rule,**

$$\text{area} = \frac{1}{3} \begin{pmatrix} \text{width of} \\ \text{interval} \end{pmatrix} \left[\begin{pmatrix} \text{first} + \text{last} \\ \text{ordinates} \end{pmatrix} \right.$$

$$+ 4 \begin{pmatrix} \text{sum of even} \\ \text{ordinates} \end{pmatrix}$$

$$\left. + 2 \begin{pmatrix} \text{sum of remaining} \\ \text{odd ordinates} \end{pmatrix} \right]$$

Selecting 6 intervals, each of width 0.5, gives:

$$\text{area} = \frac{1}{3}(0.5)[(7 + 52) + 4(10.75 + 22.75$$

$$+ 40.75) + 2(16 + 31)]$$

$$= \textbf{75 square units}$$

(d) **By integration**, shaded area

$$= \int_1^4 y\, dx$$

$$= \int_1^4 (3x^2 + 4)\, dx$$

$$= \left[x^3 + 4x\right]_1^4$$

$$= \mathbf{75\ square\ units}$$

Integration gives the precise value for the area under a curve. In this case Simpson's rule is seen to be the most accurate of the three approximate methods.

Problem 5. Find the area enclosed by the curve $y = \sin 2x$, the x-axis and the ordinates $x = 0$ and $x = \pi/3$

A sketch of $y = \sin 2x$ is shown in Figure 72.7.

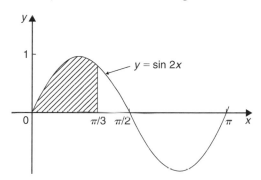

Figure 72.7

(Note that $y = \sin 2x$ has a period of $\dfrac{2\pi}{2}$, i.e. π radians.)

$$\text{Shaded area} = \int_0^{\pi/3} y\, dx$$

$$= \int_0^{\pi/3} \sin 2x\, dx$$

$$= \left[-\frac{1}{2}\cos 2x\right]_0^{\pi/3}$$

$$= \left\{-\frac{1}{2}\cos\frac{2\pi}{3}\right\} - \left\{-\frac{1}{2}\cos 0\right\}$$

$$= \left\{-\frac{1}{2}\left(-\frac{1}{2}\right)\right\} - \left\{-\frac{1}{2}(1)\right\}$$

$$= \frac{1}{4} + \frac{1}{2} = \frac{3}{4}\ \mathbf{square\ units}$$

Now try the following Practice Exercise

Unless otherwise stated all answers are in square units.

1. Show by integration that the area of the triangle formed by the line $y = 2x$, the ordinates $x = 0$ and $x = 4$ and the x-axis is 16 square units.

2. Sketch the curve $y = 3x^2 + 1$ between $x = -2$ and $x = 4$. Determine by integration the area enclosed by the curve, the x-axis and ordinates $x = -1$ and $x = 3$. Use an approximate method to find the area and compare your result with that obtained by integration.

In Problems 3 to 8, find the area enclosed between the given curves, the horizontal axis and the given ordinates.

3. $y = 5x;\quad x = 1,\, x = 4$

4. $y = 2x^2 - x + 1;\quad x = -1,\, x = 2$

5. $y = 2\sin 2\theta;\quad \theta = 0,\, \theta = \dfrac{\pi}{4}$

6. $\theta = t + e^t;\quad t = 0,\, t = 2$

7. $y = 5\cos 3t;\quad t = 0,\, t = \dfrac{\pi}{6}$

8. $y = (x - 1)(x - 3);\quad x = 0,\, x = 3$

72.3 Further worked problems on the area under a curve

Problem 6. A gas expands according to the law $pv = \text{constant}$. When the volume is $3\,\text{m}^3$ the pressure is $150\,\text{kPa}$. Given that

$$\text{work done} = \int_{v_1}^{v_2} p\, dv,$$ determine the work

done as the gas expands from $2\,\text{m}^3$ to a volume of $6\,\text{m}^3$

$pv = \text{constant}$. When $v = 3\,\text{m}^3$ and $p = 150\,\text{kPa}$ the constant is given by $(3 \times 150) = 450\,\text{kPa}\,\text{m}^3$ or $450\,\text{kJ}$.

Hence $pv = 450$, or $p = \dfrac{450}{v}$

Work done $= \int_2^6 \frac{450}{v} dv$

$= \Big[450 \ln v\Big]_2^6 = 450[\ln 6 - \ln 2]$

$= 450 \ln \frac{6}{2} = 450 \ln 3 = \mathbf{494.4\,kJ}$

Problem 7.   Determine the area enclosed by the curve $y = 4\cos\left(\dfrac{\theta}{2}\right)$, the θ-axis and ordinates $\theta = 0$ and $\theta = \dfrac{\pi}{2}$

The curve $y = 4\cos(\theta/2)$ is shown in Figure 72.8.

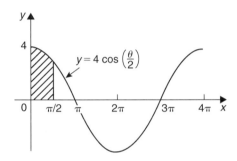

Figure 72.8

(Note that $y = 4\cos\left(\dfrac{\theta}{2}\right)$ has a maximum value of 4 and period $2\pi/(1/2)$, i.e. 4π rads.)

Shaded area $= \int_0^{\pi/2} y\, d\theta = \int_0^{\pi/2} 4\cos\dfrac{\theta}{2} d\theta$

$= \left[4\left(\dfrac{1}{\frac{1}{2}}\right)\sin\dfrac{\theta}{2}\right]_0^{\pi/2}$

$= \left(8\sin\dfrac{\pi}{4}\right) - (8\sin 0)$

$= \mathbf{5.657\,square\ units}$

Problem 8.   Determine the area bounded by the curve $y = 3e^{t/4}$, the t-axis and ordinates $t = -1$ and $t = 4$, correct to 4 significant figures

---

A table of values is produced as shown.

t	-1	0	1	2	3	4
$y = 3e^{t/4}$	2.34	3.0	3.85	4.95	6.35	8.15

Since all the values of y are positive the area required is wholly above the t-axis.

Hence area $= \int_1^4 y\, dt$

$= \int_1^4 3e^{t/4} dt = \left[\dfrac{3}{\left(\frac{1}{4}\right)} e^{t/4}\right]_{-1}^4$

$= 12\left[e^{t/4}\right]_{-1}^4 = 12(e^1 - e^{-1/4})$

$= 12(2.7183 - 0.7788)$

$= 12(1.9395) = \mathbf{23.27\ square\ units}$

Problem 9.   Sketch the curve $y = x^2 + 5$ between $x = -1$ and $x = 4$. Find the area enclosed by the curve, the x-axis and the ordinates $x = 0$ and $x = 3$. Determine also, by integration, the area enclosed by the curve and the y-axis, between the same limits

---

A table of values is produced and the curve $y = x^2 + 5$ plotted as shown in Figure 72.9.

x	-1	0	1	2	3
y	6	5	6	9	14

Shaded area $= \int_0^3 y\, dx = \int_0^3 (x^2 + 5)\, dx$

$= \left[\dfrac{x^3}{5} + 5x\right]_0^3$

$= \mathbf{24\ square\ units}$

When $x = 3$, $y = 3^2 + 5 = 14$, and when $x = 0$, $y = 5$
Since $y = x^2 + 5$ then $x^2 = y - 5$ and $x = \sqrt{y - 5}$

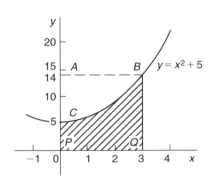

Figure 72.9

The area enclosed by the curve $y = x^2 + 5$ (i.e. $x = \sqrt{y-5}$), the y-axis and the ordinates $y = 5$ and $y = 14$ (i.e. area ABC of Fig. 72.9) is given by:

$$\text{Area} = \int_{y=5}^{y=14} x \, dy = \int_{5}^{14} \sqrt{y-5} \, dy$$

$$= \int_{5}^{14} (y-5)^{1/2} \, dy$$

Let $u = y - 5$, then $\dfrac{du}{dy} = 1$ and $dy = du$

Hence $\displaystyle\int (y-5)^{1/2} dy = \int u^{1/2} du = \dfrac{2}{3} u^{3/2}$

(for algebraic substitutions, see Chapter 64)

Since $u = y - 5$ then

$$\int_{5}^{14} \sqrt{y-5} \, dy = \frac{2}{3} \Big[(y-5)^{3/2} \Big]_{5}^{14}$$

$$= \frac{2}{3} [\sqrt{9^3} - 0]$$

$$= \mathbf{18 \text{ square units}}$$

(Check: From Figure 72.9, area $BCPQ + $ area $ABC = 24 + 18 = 42$ square units, which is the area of rectangle $ABQP$)

Problem 10. Determine the area between the curve $y = x^3 - 2x^2 - 8x$ and the x-axis

$$y = x^3 - 2x^2 - 8x = x(x^2 - 2x - 8)$$
$$= x(x+2)(x-4)$$

When $y = 0$, then $x = 0$ or $(x+2) = 0$ or $(x-4) = 0$, i.e. when $y = 0$, $x = 0$ or -2 or 4, which means that the curve crosses the x-axis at 0, -2 and 4. Since the curve is a continuous function, only one other co-ordinate value needs to be calculated before a sketch of the curve can be produced. When $x = 1$, $y = -9$, showing that the part of the curve between $x = 0$ and $x = 4$ is negative. A sketch of $y = x^3 - 2x^2 - 8x$ is shown in Figure 72.10. (Another method of sketching Figure 72.10 would have been to draw up a table of values.)

$$\text{Shaded area} = \int_{-2}^{0} (x^3 - 2x^2 - 8x) \, dx$$

$$- \int_{0}^{4} (x^3 - 2x^2 - 8x) \, dx$$

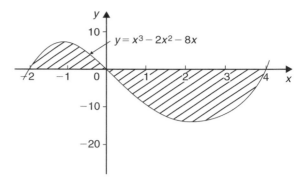

Figure 72.10

$$= \left[\frac{x^4}{4} - \frac{2x^3}{3} - \frac{8x^2}{2} \right]_{-2}^{0}$$

$$- \left[\frac{x^4}{4} - \frac{2x^3}{3} - \frac{8x^2}{2} \right]_{0}^{4}$$

$$= \left(6\frac{2}{3} \right) - \left(-42\frac{2}{3} \right)$$

$$= \mathbf{49\frac{1}{3} \text{ square units}}$$

Now try the following Practice Exercise

Practice Exercise 288   Further problems on areas under curves (answers on page 1184)

In Problems 1 and 2, find the area enclosed between the given curves, the horizontal axis and the given ordinates.

1. $y = 2x^3$; $x = -2, x = 2$

2. $xy = 4$; $x = 1, x = 4$

3. The force F newtons acting on a body at a distance x metres from a fixed point is given by: $F = 3x + 2x^2$. If work done $= \displaystyle\int_{x_1}^{x_2} F \, dx$, determine the work done when the body moves from the position where $x = 1$ m to that where $x = 3$ m

4. Find the area between the curve $y = 4x - x^2$ and the x-axis.

5. Determine the area enclosed by the curve $y = 5x^2 + 2$, the x-axis and the ordinates $x = 0$ and $x = 3$. Find also the area enclosed by the curve and the y-axis between the same limits.

6. Calculate the area enclosed between $y = x^3 - 4x^2 - 5x$ and the x-axis.

7. The velocity v of a vehicle t seconds after a certain instant is given by: $v = (3t^2 + 4)$ m/s. Determine how far it moves in the interval from $t = 1$ s to $t = 5$ s

8. A gas expands according to the law $pv =$ constant. When the volume is $2\,\text{m}^3$ the pressure is $250\,\text{kPa}$. Find the work done as the gas expands from $1\,\text{m}^3$ to a volume of $4\,\text{m}^3$, given that work done $= \int_{v_1}^{v_2} p\,dv$

72.4 The area between curves

The area enclosed between curves $y = f_1(x)$ and $y = f_2(x)$ (shown shaded in Figure 72.11) is given by:

$$\text{shaded area} = \int_a^b f_2(x)\,dx - \int_a^b f_1(x)\,dx$$

$$= \int_a^b [f_2(x) - f_1(x)]\,dx$$

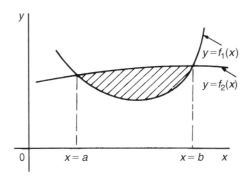

Figure 72.11

Problem 11. Determine the area enclosed between the curves $y = x^2 + 1$ and $y = 7 - x$

At the points of intersection, the curves are equal. Thus, equating the y-values of each curve gives: $x^2 + 1 = 7 - x$, from which $x^2 + x - 6 = 0$. Factorising gives $(x - 2)(x + 3) = 0$, from which, $x = 2$ and $x = -3$. By firstly determining the points of intersection the range of x-values has been found. Tables of values are produced as shown below.

x	-3	-2	-1	0	1	2
$y = x^2 + 1$	10	5	2	1	2	5

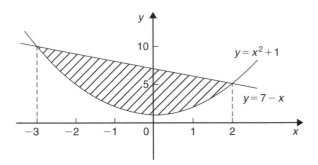

x	-3	0	2
$y = 7 - x$	10	7	5

A sketch of the two curves is shown in Figure 72.12.

Figure 72.12

$$\text{Shaded area} = \int_{-3}^{2} (7 - x)\,dx - \int_{-3}^{2} (x^2 + 1)\,dx$$

$$= \int_{-3}^{2} [(7 - x) - (x^2 + 1)]\,dx$$

$$= \int_{-3}^{2} (6 - x - x^2)\,dx$$

$$= \left[6x - \frac{x^2}{2} - \frac{x^3}{3} \right]_{-3}^{2}$$

$$= \left(12 - 2 - \frac{8}{3} \right) - \left(-18 - \frac{9}{2} + 9 \right)$$

$$= \left(7\frac{1}{3} \right) - \left(-13\frac{1}{2} \right)$$

$$= 20\frac{5}{6} \text{ square units}$$

Problem 12. (a) Determine the co-ordinates of the points of intersection of the curves $y = x^2$ and $y^2 = 8x$. (b) Sketch the curves $y = x^2$ and $y^2 = 8x$ on the same axes. (c) Calculate the area enclosed by the two curves

(a) At the points of intersection the co-ordinates of the curves are equal. When $y = x^2$ then $y^2 = x^4$

Hence at the points of intersection $x^4 = 8x$, by equating the y^2 values.

Thus $x^4 - 8x = 0$, from which $x(x^3 - 8) = 0$, i.e. $x = 0$ or $(x^3 - 8) = 0$

Hence at the points of intersection $x = 0$ or $x = 2$

When $x = 0$, $y = 0$ and when $x = 2$, $y = 2^2 = 4$

Hence the points of intersection of the curves $y=x^2$ and $y^2=8x$ are (0, 0) and (2, 4)

(b) A sketch of $y=x^2$ and $y^2=8x$ is shown in Figure 72.13.

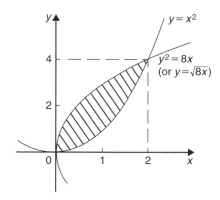

Figure 72.13

(c) **Shaded area** $= \displaystyle\int_0^2 \{\sqrt{8x} - x^2\}\,dx$

$$= \int_0^2 \{(\sqrt{8})x^{1/2} - x^2\}\,dx$$

$$= \left[(\sqrt{8})\frac{x^{3/2}}{\left(\frac{3}{2}\right)} - \frac{x^3}{3}\right]_0^2$$

$$= \left\{\frac{\sqrt{8}\sqrt{8}}{\left(\frac{3}{2}\right)} - \frac{8}{3}\right\} - \{0\}$$

$$= \frac{16}{3} - \frac{8}{3} = \frac{8}{3}$$

$$= 2\frac{2}{3} \textbf{ square units}$$

Problem 13. Determine by integration the area bounded by the three straight lines $y=4-x$, $y=3x$ and $3y=x$

Each of the straight lines is shown sketched in Figure 72.14.

$$\text{Shaded area} = \int_0^1 \left(3x - \frac{x}{3}\right)dx + \int_1^3 \left[(4-x) - \frac{x}{3}\right]dx$$

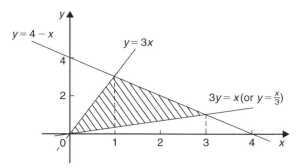

Figure 72.14

$$= \left[\frac{3x^2}{2} - \frac{x^2}{6}\right]_0^1 + \left[4x - \frac{x^2}{2} - \frac{x^2}{6}\right]_1^3$$

$$= \left[\left(\frac{3}{2} - \frac{1}{6}\right) - (0)\right]$$

$$+ \left[\left(12 - \frac{9}{2} - \frac{9}{6}\right) - \left(4 - \frac{1}{2} - \frac{1}{6}\right)\right]$$

$$= \left(1\frac{1}{3}\right) + \left(6 - 3\frac{1}{3}\right)$$

$$= \textbf{4 square units}$$

Now try the following Practice Exercise

Practice Exercise 289   Further problems on areas between curves (answers on page 1184)

1. Determine the co-ordinates of the points of intersection and the area enclosed between the parabolas $y^2=3x$ and $x^2=3y$

2. Sketch the curves $y=x^2+3$ and $y=7-3x$ and determine the area enclosed by them

3. Determine the area enclosed by the curves $y=\sin x$ and $y=\cos x$ and the y-axis

4. Determine the area enclosed by the three straight lines $y=3x$, $2y=x$ and $y+2x=5$

For fully worked solutions to each of the problems in Practice Exercises 287 to 289 in this chapter, go to the website:
www.routledge.com/cw/bird

Chapter 73

Mean and root mean square values

> **Why it is important to understand:** Mean and root mean square values
>
> Electrical currents and voltages often vary with time, and engineers may wish to know the average or mean value of such a current or voltage over some particular time interval. The mean value of a time-varying function is defined in terms of an integral. An associated quantity is the root mean square (rms) value of a current which is used, for example, in the calculation of the power dissipated by a resistor. Mean and rms values are required with alternating currents and voltages, pressure of sound waves, and much more.

At the end of this chapter, you should be able to:

- determine the mean or average value of a function over a given range using integration
- define an rms value
- determine the rms value of a function over a given range using integration

73.1 Mean or average values

(i) The mean or average value of the curve shown in Figure 73.1, between $x = a$ and $x = b$, is given by:
mean or average value,

$$\bar{y} = \frac{\text{area under curve}}{\text{length of base}}$$

(ii) When the area under a curve may be obtained by integration then:
mean or average value,

$$\bar{y} = \frac{\displaystyle\int_a^b y\,dx}{b - a}$$

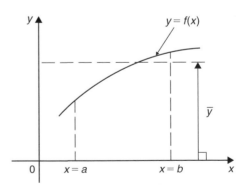

Figure 73.1

i.e. $\quad \bar{y} = \dfrac{1}{b - a}\displaystyle\int_a^b f(x)\,dx$

(iii) For a periodic function, such as a sine wave, the mean value is assumed to be 'the mean value over half a cycle', since the mean value over a complete cycle is zero.

Problem 1. Determine, using integration, the mean value of $y = 5x^2$ between $x = 1$ and $x = 4$

Mean value,

$$\bar{y} = \frac{1}{4-1} \int_1^4 y \, dx = \frac{1}{3} \int_1^4 5x^2 \, dx$$

$$= \frac{1}{3} \left[\frac{5x^3}{3} \right]_1^4 = \frac{5}{9} [x^3]_1^4 = \frac{5}{9}(64 - 1) = \mathbf{35}$$

Problem 2. A sinusoidal voltage is given by $v = 100 \sin \omega t$ volts. Determine the mean value of the voltage over half a cycle using integration

Half a cycle means the limits are 0 to π radians. Mean value,

$$\bar{v} = \frac{1}{\pi - 0} \int_0^\pi v \, d(\omega t)$$

$$= \frac{1}{\pi} \int_0^\pi 100 \sin \omega t \, d(\omega t) = \frac{100}{\pi} \left[-\cos \omega t \right]_0^\pi$$

$$= \frac{100}{\pi} [(-\cos \pi) - (-\cos 0)]$$

$$= \frac{100}{\pi} [(+1) - (-1)] = \frac{200}{\pi}$$

$$= \mathbf{63.66 \ volts}$$

(Note that for a sine wave,

$$\mathbf{mean \ value = \frac{2}{\pi} \times maximum \ value}$$

In this case, mean value $= \dfrac{2}{\pi} \times 100 = 63.66$ V)

Problem 3. Calculate the mean value of $y = 3x^2 + 2$ in the range $x = 0$ to $x = 3$ by (a) the mid-ordinate rule and (b) integration

(a) A graph of $y = 3x^2 + 2$ over the required range is shown in Figure 73.2 using the following table:

x	0	0.5	1.0	1.5	2.0	2.5	3.0
y	2.0	2.75	5.0	8.75	14.0	20.75	29.0

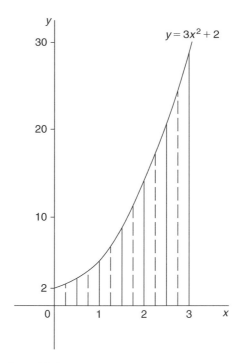

Figure 73.2

Using the mid-ordinate rule, mean value

$$= \frac{\text{area under curve}}{\text{length of base}}$$

$$= \frac{\text{sum of mid-ordinates}}{\text{number of mid-ordinates}}$$

Selecting 6 intervals, each of width 0.5, the mid-ordinates are erected as shown by the broken lines in Figure 73.2.

$$\text{Mean value} = \frac{\begin{array}{c} 2.2 + 3.7 + 6.7 + 11.2 \\ + 17.2 + 24.7 \end{array}}{6}$$

$$= \frac{65.7}{6} = \mathbf{10.95}$$

(b) By integration, mean value

$$= \frac{1}{3 - 0} \int_0^3 y \, dx = \frac{1}{3} \int_0^3 (3x^2 + 2) \, dx$$

$$= \frac{1}{3} [x^3 + 2x]_0^3 = \frac{1}{3} [(27 + 6) - (0)]$$

$$= \mathbf{11}$$

The answer obtained by integration is exact; greater accuracy may be obtained by the mid-ordinate rule if a larger number of intervals are selected.

Problem 4. The number of atoms, N, remaining in a mass of material during radioactive decay after time t seconds is given by: $N = N_0 e^{-\lambda t}$, where N_0 and λ are constants. Determine the mean number of atoms in the mass of material for the time period $t = 0$ and $t = \dfrac{1}{\lambda}$

Mean number of atoms

$$= \frac{1}{\frac{1}{\lambda} - 0} \int_0^{1/\lambda} N \, dt = \frac{1}{\frac{1}{\lambda}} \int_0^{1/\lambda} N_0 e^{-\lambda t} \, dt$$

$$= \lambda N_0 \int_0^{1/\lambda} e^{-\lambda t} \, dt = \lambda N_0 \left[\frac{e^{-\lambda t}}{-\lambda} \right]_0^{1/\lambda}$$

$$= -N_0 [e^{-\lambda(1/\lambda)} - e^0] = -N_0 [e^{-1} - e^0]$$

$$= +N_0 [e^0 - e^{-1}] = N_0 [1 - e^{-1}] = \mathbf{0.632 N_0}$$

Now try the following Practice Exercise

Practice Exercise 290   Further problems on mean or average values (answers on page 1184)

1. Determine the mean value of (a) $y = 3\sqrt{x}$ from $x = 0$ to $x = 4$ (b) $y = \sin 2\theta$ from $\theta = 0$ to $\theta = \dfrac{\pi}{4}$ (c) $y = 4e^t$ from $t = 1$ to $t = 4$

2. Calculate the mean value of $y = 2x^2 + 5$ in the range $x = 1$ to $x = 4$ by (a) the mid-ordinate rule, and (b) integration

3. The speed v of a vehicle is given by: $v = (4t + 3)$ m/s, where t is the time in seconds. Determine the average value of the speed from $t = 0$ to $t = 3$ s

4. Find the mean value of the curve $y = 6 + x - x^2$ which lies above the x-axis by using an approximate method. Check the result using integration

5. The vertical height h km of a missile varies with the horizontal distance d km, and is given by $h = 4d - d^2$. Determine the mean height of the missile from $d = 0$ to $d = 4$ km

6. The velocity v of a piston moving with simple harmonic motion at any time t is given by: $v = c \sin \omega t$, where c is a constant. Determine the mean velocity between $t = 0$ and $t = \dfrac{\pi}{\omega}$

73.2   Root mean square values

The **root mean square value** of a quantity is 'the square root of the mean value of the squared values of the quantity' taken over an interval. With reference to Figure 73.1, the rms value of $y = f(x)$ over the range $x = a$ to $x = b$ is given by:

$$\text{rms value} = \sqrt{\frac{1}{b - a} \int_a^b y^2 \, dx}$$

One of the principal applications of rms values is with alternating currents and voltages. The rms value of an alternating current is defined as that current which will give the same heating effect as the equivalent direct current.

Problem 5. Determine the rms value of $y = 2x^2$ between $x = 1$ and $x = 4$

Rms value

$$= \sqrt{\frac{1}{4 - 1} \int_1^4 y^2 \, dx} = \sqrt{\frac{1}{3} \int_1^4 (2x^2)^2 \, dx}$$

$$= \sqrt{\frac{1}{3} \int_1^4 4x^4 \, dx} = \sqrt{\frac{4}{3} \left[\frac{x^5}{5} \right]_1^4}$$

$$= \sqrt{\frac{4}{15}(1024 - 1)} = \sqrt{272.8} = \mathbf{16.5}$$

Problem 6. A sinusoidal voltage has a maximum value of 100 V. Calculate its rms value

A sinusoidal voltage v having a maximum value of 100 V may be written as: $v = 100 \sin \theta$. Over the range $\theta = 0$ to $\theta = \pi$,
rms value

$$= \sqrt{\frac{1}{\pi - 0} \int_0^\pi v^2 \, d\theta}$$

$$= \sqrt{\frac{1}{\pi} \int_0^\pi (100 \sin \theta)^2 \, d\theta}$$

$$= \sqrt{\frac{10\,000}{\pi} \int_0^\pi \sin^2 \theta \, d\theta}$$

which is not a 'standard' integral. It is shown in Chapter 44 that $\cos 2A = 1 - 2\sin^2 A$ and this formula is used whenever $\sin^2 A$ needs to be integrated.
Rearranging $\cos 2A = 1 - 2\sin^2 A$ gives $\sin^2 A = \frac{1}{2}(1 - \cos 2A)$

Hence $\sqrt{\dfrac{10\,000}{\pi} \displaystyle\int_0^{\pi} \sin^2 \theta \, d\theta}$

$= \sqrt{\dfrac{10\,000}{\pi} \displaystyle\int_0^{\pi} \frac{1}{2}(1 - \cos 2\theta) \, d\theta}$

$= \sqrt{\dfrac{10\,000}{\pi} \frac{1}{2}\left[\theta - \dfrac{\sin 2\theta}{2}\right]_0^{\pi}}$

$= \sqrt{\dfrac{10\,000}{\pi} \frac{1}{2}\left[\left(\pi - \dfrac{\sin 2\pi}{2}\right) - \left(0 - \dfrac{\sin 0}{2}\right)\right]}$

$= \sqrt{\dfrac{10\,000}{\pi} \frac{1}{2}[\pi]} = \sqrt{\dfrac{10\,000}{2}}$

$= \dfrac{100}{\sqrt{2}} = \textbf{70.71 volts}$

(Note that for a sine wave,

$$\textbf{rms value} = \dfrac{1}{\sqrt{2}} \times \textbf{maximum value.}$$

In this case, rms value $= \dfrac{1}{\sqrt{2}} \times 100 = 70.71 \text{ V})$

Problem 7. In a frequency distribution the average distance from the mean, y, is related to the variable, x, by the equation $y = 2x^2 - 1$. Determine, correct to 3 significant figures, the rms deviation from the mean for values of x from -1 to $+4$

Rms deviation

$= \sqrt{\dfrac{1}{4 - -1} \displaystyle\int_{-1}^{4} y^2 \, dx}$

$= \sqrt{\dfrac{1}{5} \displaystyle\int_{-1}^{4} (2x^2 - 1)^2 \, dx}$

$= \sqrt{\dfrac{1}{5} \displaystyle\int_{-1}^{4} (4x^4 - 4x^2 + 1) \, dx}$

$= \sqrt{\dfrac{1}{5}\left[\dfrac{4x^5}{5} - \dfrac{4x^3}{3} + x\right]_{-1}^{4}}$

$= \sqrt{\dfrac{1}{5}\left[\left(\dfrac{4}{5}(4)^5 - \dfrac{4}{3}(4)^3 + 4\right) - \left(\dfrac{4}{5}(-1)^5 - \dfrac{4}{3}(-1)^3 + (-1)\right)\right]}$

$= \sqrt{\dfrac{1}{5}[(737.87) - (-0.467)]}$

$= \sqrt{\dfrac{1}{5}[738.34]}$

$= \sqrt{147.67} = 12.152 = \textbf{12.2,}$

correct to 3 significant figures.

Now try the following Practice Exercise

Practice Exercise 291 Further problems on root mean square values (answers on page 1184)

1. Determine the rms values of:
   (a) $y = 3x$ from $x = 0$ to $x = 4$
   (b) $y = t^2$ from $t = 1$ to $t = 3$
   (c) $y = 25\sin\theta$ from $\theta = 0$ to $\theta = 2\pi$

2. Calculate the rms values of:
   (a) $y = \sin 2\theta$ from $\theta = 0$ to $\theta = \dfrac{\pi}{4}$
   (b) $y = 1 + \sin t$ from $t = 0$ to $t = 2\pi$
   (c) $y = 3\cos 2x$ from $x = 0$ to $x = \pi$
   (Note that $\cos^2 t = \dfrac{1}{2}(1 + \cos 2t)$, from Chapter 44)

3. The distance, p, of points from the mean value of a frequency distribution is related to the variable, q, by the equation $p = \dfrac{1}{q} + q$. Determine the standard deviation (i.e. the rms value), correct to 3 significant figures, for values from $q = 1$ to $q = 3$

4. A current, $i = 30\sin 100\pi t$ amperes, is applied across an electric circuit. Determine its mean and rms values, each correct to 4 significant figures, over the range $t = 0$ to $t = 10$ ms

5. A sinusoidal voltage has a peak value of 340 V. Calculate its mean and rms values, correct to 3 significant figures

6. Determine the form factor, correct to 3 significant figures, of a sinusoidal voltage of maximum value 100 volts, given that form

$$\text{factor} = \frac{\text{rms value}}{\text{average value}}$$

7. A wave is defined by the equation:

$$v = E_1 \sin \omega t + E_3 \sin 3\omega t$$

where, E_1, E_3 and ω are constants.

Determine the rms value of v over the interval $0 \le t \le \dfrac{\pi}{\omega}$

For fully worked solutions to each of the problems in Practice Exercises 290 and 291 in this chapter, go to the website:
www.routledge.com/cw/bird

Chapter 74

Volumes of solids of revolution

Why it is important to understand: **Volumes of solids of revolution**

Revolving a plane figure about an axis generates a volume. The solid generated by rotating a plane area about an axis in its plane is called a solid of revolution, and integration may be used to calculate such a volume. There are many applications of volumes of solids of revolution in engineering, and particularly in manufacturing.

At the end of this chapter, you should be able to:

- understand the term volume of a solid of revolution
- determine the volume of a solid of revolution for functions between given limits using integration

---

74.1 Introduction

If the area under the curve $y = f(x)$ (shown in Figure 74.1(a)) between $x = a$ and $x = b$ is rotated $360°$ about the x-axis, then a volume known as a **solid of revolution** is produced as shown in Figure 74.1(b). The volume of such a solid may be determined precisely using integration.

(i) Let the area shown in Figure 74.1(a) be divided into a number of strips each of width δx. One such strip is shown shaded.

(ii) When the area is rotated $360°$ about the x-axis, each strip produces a solid of revolution approximating to a circular disc of radius y and thickness δx. Volume of disc = (circular cross-sectional area) (thickness) $= (\pi y^2)(\delta x)$

(iii) Total volume, V, between ordinates $x = a$ and $x = b$ is given by:

$$\textbf{Volume } V = \lim_{\delta x \to 0} \sum_{x=a}^{x=b} \pi y^2 \delta x = \int_a^b \pi y^2 \, \mathrm{d}x$$

If a curve $x = f(y)$ is rotated about the y-axis $360°$ between the limits $y = c$ and $y = d$, as shown in Figure 74.2, then the volume generated is given by:

$$\textbf{Volume } V = \lim_{\delta y \to 0} \sum_{y=c}^{y=d} \pi x^2 \delta y = \int_c^d \pi x^2 \, \mathrm{d}y$$

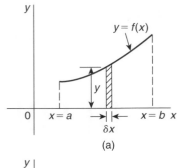

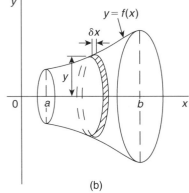

Figure 74.1

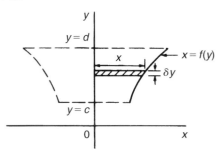

Figure 74.2

74.2 Worked problems on volumes of solids of revolution

Problem 1. Determine the volume of the solid of revolution formed when the curve $y = 2$ is rotated $360°$ about the x-axis between the limits $x = 0$ and $x = 3$

When $y = 2$ is rotated $360°$ about the x-axis between $x = 0$ and $x = 3$ (see Fig. 74.3):
volume generated

$$= \int_0^3 \pi y^2 \, \mathrm{d}x = \int_0^3 \pi (2)^2 \, \mathrm{d}x$$

$$= \int_0^3 4\pi \, \mathrm{d}x = 4\pi [x]_0^3 = \mathbf{12\pi \text{ cubic units}}$$

(Check: The volume generated is a cylinder of radius 2 and height 3.
Volume of cylinder $= \pi r^2 h = \pi (2)^2 (3) = \mathbf{12\pi \text{ cubic units}}$.)

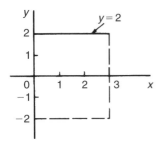

Figure 74.3

Problem 2. Find the volume of the solid of revolution when the curve $y = 2x$ is rotated one revolution about the x-axis between the limits $x = 0$ and $x = 5$

When $y = 2x$ is revolved one revolution about the x-axis between $x = 0$ and $x = 5$ (see Fig. 74.4) then:

volume generated

$$= \int_0^5 \pi y^2 \mathrm{d}x = \int_0^5 \pi (2x)^2 \mathrm{d}x$$

$$= \int_0^5 4\pi x^2 \mathrm{d}x = 4\pi \left[\frac{x^3}{3} \right]_0^5$$

$$= \frac{500\pi}{3} = \mathbf{166\frac{2}{3}\pi \text{ cubic units}}$$

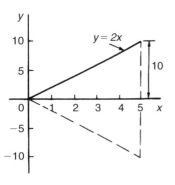

Figure 74.4

(Check: The volume generated is a cone of radius 10 and height 5. Volume of cone

$$= \frac{1}{3}\pi r^2 h = \frac{1}{3}\pi(10)^2 5 = \frac{500\pi}{3}$$

$$= 166\frac{2}{3}\pi \text{ cubic units.})$$

Problem 3. The curve $y = x^2 + 4$ is rotated one revolution about the x-axis between the limits $x = 1$ and $x = 4$. Determine the volume of the solid of revolution produced

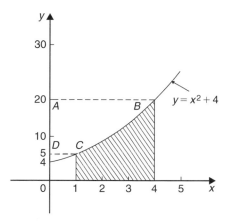

Figure 74.5

Revolving the shaded area shown in Figure 74.5 about the x-axis $360°$ produces a solid of revolution given by:

$$\text{Volume} = \int_1^4 \pi y^2 \, dx = \int_1^4 \pi(x^2 + 4)^2 dx$$

$$= \int_1^4 \pi(x^4 + 8x^2 + 16) \, dx$$

$$= \pi\left[\frac{x^5}{5} + \frac{8x^3}{3} + 16x\right]_1^4$$

$$= \pi[(204.8 + 170.67 + 64)$$

$$- (0.2 + 2.67 + 16)]$$

$$= \mathbf{420.6\pi \text{ cubic units}}$$

Problem 4. If the curve in Problem 3 is revolved about the y-axis between the same limits, determine the volume of the solid of revolution produced

The volume produced when the curve $y = x^2 + 4$ is rotated about the y-axis between $y = 5$ (when $x = 1$)

and $y = 20$ (when $x = 4$), i.e. rotating area $ABCD$ of Figure 74.5 about the y-axis is given by:

$$\text{volume} = \int_5^{20} \pi x^2 \, dy$$

Since $y = x^2 + 4$, then $x^2 = y - 4$

$$\text{Hence volume} = \int_5^{20} \pi(y - 4) \, dy = \pi\left[\frac{y^2}{2} - 4y\right]_5^{20}$$

$$= \pi[(120) - (-7.5)]$$

$$= \mathbf{127.5\pi \text{ cubic units}}$$

Now try the following Practice Exercise

Practice Exercise 292   Further problems on volumes of solids of revolution (answers on page 1184)

(Answers are in cubic units and in terms of π)

In Problems 1 to 5, determine the volume of the solid of revolution formed by revolving the areas enclosed by the given curve, the x-axis and the given ordinates through one revolution about the x-axis.

1. $y = 5x$; $x = 1, x = 4$

2. $y = x^2$; $x = -2, x = 3$

3. $y = 2x^2 + 3$; $x = 0, x = 2$

4. $\dfrac{y^2}{4} = x$; $x = 1, x = 5$

5. $xy = 3$; $x = 2, x = 3$

In Problems 6 to 8, determine the volume of the solid of revolution formed by revolving the areas enclosed by the given curves, the y-axis and the given ordinates through one revolution about the y-axis.

6. $y = x^2$; $y = 1, y = 3$

7. $y = 3x^2 - 1$; $y = 2, y = 4$

8. $y = \dfrac{2}{x}$; $y = 1, y = 3$

9. The curve $y = 2x^2 + 3$ is rotated about (a) the x-axis between the limits $x = 0$ and $x = 3$, and (b) the y-axis, between the same limits. Determine the volume generated in each case.

74.3 Further worked problems on volumes of solids of revolution

Problem 5. The area enclosed by the curve $y = 3e^{\frac{x}{3}}$, the x-axis and ordinates $x = -1$ and $x = 3$ is rotated $360°$ about the x-axis. Determine the volume generated

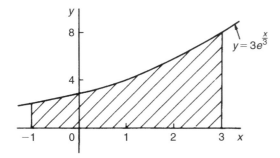

Figure 74.6

A sketch of $y = 3e^{\frac{x}{3}}$ is shown in Figure 74.6. When the shaded area is rotated $360°$ about the x-axis then:

$$\text{volume generated} = \int_{-1}^{3} \pi y^2 \, dx$$

$$= \int_{-1}^{3} \pi \left(3e^{\frac{x}{3}}\right)^2 dx$$

$$= 9\pi \int_{-1}^{3} e^{\frac{2x}{3}} \, dx$$

$$= 9\pi \left[\frac{e^{\frac{2x}{3}}}{\frac{2}{3}}\right]_{-1}^{3}$$

$$= \frac{27\pi}{2}\left(e^2 - e^{-\frac{2}{3}}\right)$$

$$= \mathbf{92.82\pi \text{ cubic units}}$$

Problem 6. Determine the volume generated when the area above the x-axis bounded by the curve $x^2 + y^2 = 9$ and the ordinates $x = 3$ and $x = -3$ is rotated one revolution about the x-axis

Figure 74.7 shows the part of the curve $x^2 + y^2 = 9$ lying above the x-axis, Since, in general, $x^2 + y^2 = r^2$ represents a circle, centre 0 and radius r, then $x^2 + y^2 = 9$ represents a circle, centre 0 and radius 3.

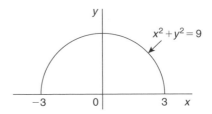

Figure 74.7

When the semi-circular area of Figure 74.7 is rotated one revolution about the x-axis then:

$$\text{volume generated} = \int_{-3}^{3} \pi y^2 dx$$

$$= \int_{-3}^{3} \pi (9 - x^2) \, dx$$

$$= \pi \left[9x - \frac{x^3}{3}\right]_{-3}^{3}$$

$$= \pi[(18) - (-18)]$$

$$= \mathbf{36\pi \text{ cubic units}}$$

(Check: The volume generated is a sphere of radius 3. Volume of sphere $= \frac{4}{3}\pi r^3 = \frac{4}{3}\pi(3)^3 = \mathbf{36\pi \text{ cubic units}}$.)

Problem 7. Calculate the volume of a frustum of a sphere of radius 4 cm that lies between two parallel planes at 1 cm and 3 cm from the centre and on the same side of it

The volume of a frustum of a sphere may be determined by integration by rotating the curve $x^2 + y^2 = 4^2$ (i.e. a circle, centre 0, radius 4) one revolution about the x-axis, between the limits $x = 1$ and $x = 3$ (i.e. rotating the shaded area of Figure 74.8).

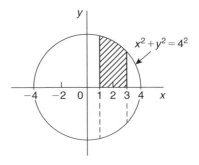

Figure 74.8

Volume of frustum $= \int_1^3 \pi y^2 \, dx$

$$= \int_1^3 \pi (4^2 - x^2) \, dx$$

$$= \pi \left[16x - \frac{x^3}{3} \right]_1^3$$

$$= \pi \left[(39) - \left(15\frac{2}{3} \right) \right]$$

$$= 23\frac{1}{3} \pi \text{ cubic units}$$

Problem 8.   The area enclosed between the two parabolas $y = x^2$ and $y^2 = 8x$ of Problem 12, Chapter 72, page 790, is rotated $360°$ about the x-axis. Determine the volume of the solid produced

The area enclosed by the two curves is shown in Figure 72.13, page 791. The volume produced by revolving the shaded area about the x-axis is given by: [(volume produced by revolving $y^2 = 8x$) − (volume produced by revolving $y = x^2$)]

i.e.   **volume** $= \int_0^2 \pi (8x) \, dx - \int_0^2 \pi (x^4) \, dx$

$$= \pi \int_0^2 (8x - x^4) \, dx = \pi \left[\frac{8x^2}{2} - \frac{x^5}{5} \right]_0^2$$

$$= \pi \left[\left(16 - \frac{32}{5} \right) - (0) \right]$$

$$= \mathbf{9.6\pi \text{ cubic units}}$$

Now try the following Practice Exercise

Practice Exercise 293   Further problems on volumes of solids of revolution (answers on page 1184)

(Answers to volumes are in cubic units and in terms of π)

In Problems 1 and 2, determine the volume of the solid of revolution formed by revolving the areas enclosed by the given curve, the x-axis and the given ordinates through one revolution about the x-axis.

1. $y = 4e^x$; $x = 0$, $x = 2$

2. $y = \sec x$; $x = 0$, $x = \dfrac{\pi}{4}$

In Problems 3 and 4, determine the volume of the solid of revolution formed by revolving the areas enclosed by the given curves, the y-axis and the given ordinates through one revolution about the y-axis.

3. $x^2 + y^2 = 16$; $y = 0$, $y = 4$

4. $x\sqrt{y} = 2$; $y = 2$, $y = 3$

5.   Determine the volume of a plug formed by the frustum of a sphere of radius 6 cm which lies between two parallel planes at 2 cm and 4 cm from the centre and on the same side of it.

 (The equation of a circle, centre 0, radius r is $x^2 + y^2 = r^2$)

6.   The area enclosed between the two curves $x^2 = 3y$ and $y^2 = 3x$ is rotated about the x-axis. Determine the volume of the solid formed.

7.   The portion of the curve $y = x^2 + \dfrac{1}{x}$ lying between $x = 1$ and $x = 3$ is revolved $360°$ about the x-axis. Determine the volume of the solid formed.

8.   Calculate the volume of the frustum of a sphere of radius 5 cm that lies between two parallel planes at 3 cm and 2 cm from the centre and on opposite sides of it.

9.   Sketch the curves $y = x^2 + 2$ and $y - 12 = 3x$ from $x = -3$ to $x = 6$. Determine (a) the co-ordinates of the points of intersection of the two curves, and (b) the area enclosed by the two curves. (c) If the enclosed area is rotated $360°$ about the x-axis, calculate the volume of the solid produced.

For fully worked solutions to each of the problems in Practice Exercises 292 and 293 in this chapter, go to the website:
www.routledge.com/cw/bird

Chapter 75

Centroids of simple shapes

Why it is important to understand: **Centroids of simple shapes**

The centroid of an area is similar to the centre of mass of a body. Calculating the centroid involves only the geometrical shape of the area; the centre of gravity will equal the centroid if the body has constant density. Centroids of basic shapes can be intuitive – such as the centre of a circle. Centroids of more complex shapes can be found using integral calculus – as long as the area, volume or line of an object can be described by a mathematical equation. Centroids are of considerable importance in manufacturing, and in mechanical, civil and structural design engineering.

At the end of this chapter, you should be able to:

- define a centroid
- determine the centroid of an area between a curve and the *x*-axis for functions between given limits using integration
- determine the centroid of an area between a curve and the *y*-axis for functions between given limits using integration
- state the theorem of Pappus
- determine the centroid of an area using the theorem of Pappus

75.1 Centroids

A **lamina** is a thin, flat sheet having uniform thickness. The **centre of gravity** of a lamina is the point where it balances perfectly, i.e. the lamina's **centre of mass**. When dealing with an area (i.e. a lamina of negligible thickness and mass) the term **centre of area** or **centroid** is used for the point where the centre of gravity of a lamina of that shape would lie.

75.2 The first moment of area

The **first moment of area** is defined as the product of the area and the perpendicular distance of its centroid from a given axis in the plane of the area. In Figure 75.1, the first moment of area A about axis XX is given by (Ay) cubic units.

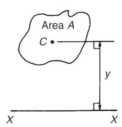

Figure 75.1

75.3 Centroid of area between a curve and the *x*-axis

(i) Figure 75.2 shows an area *PQRS* bounded by the curve $y = f(x)$, the *x*-axis and ordinates $x = a$ and $x = b$. Let this area be divided into a large number of strips, each of width δx. A typical strip is shown shaded drawn at point (x, y) on $f(x)$. The area of the strip is approximately rectangular and is given by $y\delta x$. The centroid, C, has co-ordinates $\left(x, \dfrac{y}{2}\right)$

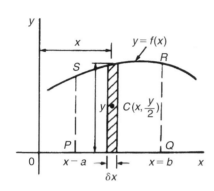

Figure 75.2

(ii) First moment of area of shaded strip about axis $Oy = (y\delta x)(x) = xy\delta x$
Total first moment of area *PQRS* about axis $Oy = \lim_{\delta x \to 0}\sum_{x=a}^{x=b} xy\delta x = \int_a^b xy\,dx$

(iii) First moment of area of shaded strip about axis $Ox = (y\delta x)\left(\dfrac{y}{2}\right) = \dfrac{1}{2}y^2 x$
Total first moment of area *PQRS* about axis $Ox = \lim_{\delta x \to 0}\sum_{x=a}^{x=b}\dfrac{1}{2}y^2\delta x = \dfrac{1}{2}\int_a^b y^2\,dx$

(iv) Area of *PQRS*, $A = \int_a^b y\,dx$ (from Chapter 72)

(v) Let $\bar{x}$ and $\bar{y}$ be the distances of the centroid of area A about Oy and Ox, respectively, then: $(\bar{x})(A) =$ total first moment of area A about axis $Oy = \int_a^b xy\,dx$

from which, $\quad \bar{x} = \dfrac{\displaystyle\int_a^b xy\,dx}{\displaystyle\int_a^b y\,dx}$

and $(\bar{y})(A) =$ total moment of area A about axis $Ox = \dfrac{1}{2}\int_a^b y^2\,dx$

from which, $\quad \bar{y} = \dfrac{\dfrac{1}{2}\displaystyle\int_a^b y^2\,dx}{\displaystyle\int_a^b y\,dx}$

75.4 Centroid of area between a curve and the *y*-axis

If $\bar{x}$ and $\bar{y}$ are the distances of the centroid of area *EFGH* in Figure 75.3 from Oy and Ox, respectively, then, by similar reasoning as above:

$$(\bar{x})(\text{total area}) = \lim_{\delta y \to 0}\sum_{y=c}^{y=d} x\delta y\left(\dfrac{x}{2}\right) = \dfrac{1}{2}\int_c^d x^2\,dy$$

from which, $\quad \bar{x} = \dfrac{\dfrac{1}{2}\displaystyle\int_c^d x^2\,dy}{\displaystyle\int_c^d x\,dy}$

and $(\bar{y})(\text{total area}) = \lim_{\delta y \to 0}\sum_{y=c}^{y=d}(x\delta y)\,y = \int_c^d xy\,dy$

from which, $\quad \bar{y} = \dfrac{\displaystyle\int_c^d xy\,dy}{\displaystyle\int_c^d x\,dy}$

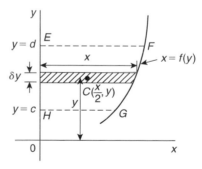

Figure 75.3

75.5 Worked problems on centroids of simple shapes

Problem 1. Show, by integration, that the centroid of a rectangle lies at the intersection of the diagonals

Let a rectangle be formed by the line $y = b$, the *x*-axis and ordinates $x = 0$ and $x = l$, as shown in Figure 75.4. Let the co-ordinates of the centroid C of this area be $(\bar{x}, \bar{y})$

By integration, $\bar{x} = \dfrac{\displaystyle\int_0^l xy\,dx}{\displaystyle\int_0^l y\,dx} = \dfrac{\displaystyle\int_0^l (x)(b)\,dx}{\displaystyle\int_0^l b\,dx}$

$= \dfrac{\left[b\dfrac{x^2}{2}\right]_0^l}{[bx]_0^l} = \dfrac{\dfrac{bl^2}{2}}{bl} = \dfrac{l}{2}$

and $\bar{y} = \dfrac{\dfrac{1}{2}\displaystyle\int_0^l y^2\,dx}{\displaystyle\int_0^l y\,dx} = \dfrac{\dfrac{1}{2}\displaystyle\int_0^l b^2\,dx}{bl}$

$= \dfrac{\dfrac{1}{2}[b^2 x]_0^l}{bl} = \dfrac{\dfrac{b^2 l}{2}}{bl} = \dfrac{b}{2}$

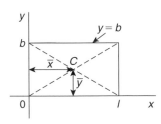

Figure 75.4

i.e. **the centroid lies at** $\left(\dfrac{l}{2}, \dfrac{b}{2}\right)$ **which is at the intersection of the diagonals.**

Problem 2. Find the position of the centroid of the area bounded by the curve $y = 3x^2$, the x-axis and the ordinates $x = 0$ and $x = 2$

If, $(\bar{x}, \bar{y})$ are the co-ordinates of the centroid of the given area then:

$\bar{x} = \dfrac{\displaystyle\int_0^2 xy\,dx}{\displaystyle\int_0^2 y\,dx} = \dfrac{\displaystyle\int_0^2 x(3x^2)\,dx}{\displaystyle\int_0^2 3x^2\,dx}$

$= \dfrac{\displaystyle\int_0^2 3x^3\,dx}{\displaystyle\int_0^2 3x^2\,dx} = \dfrac{\left[\dfrac{3x^4}{4}\right]_0^2}{[x^3]_0^2} = \dfrac{12}{8} = \mathbf{1.5}$

$\bar{y} = \dfrac{\dfrac{1}{2}\displaystyle\int_0^2 y^2\,dx}{\displaystyle\int_0^2 y\,dx} = \dfrac{\dfrac{1}{2}\displaystyle\int_0^2 (3x^2)^2\,dx}{8}$

$= \dfrac{\dfrac{1}{2}\displaystyle\int_0^2 9x^4\,dx}{8} = \dfrac{\dfrac{9}{2}\left[\dfrac{x^5}{5}\right]_0^2}{8} = \dfrac{\dfrac{9}{2}\left(\dfrac{32}{5}\right)}{8}$

$= \dfrac{18}{5} = \mathbf{3.6}$

Hence the centroid lies at (1.5, 3.6)

Problem 3. Determine by integration the position of the centroid of the area enclosed by the line $y = 4x$, the x-axis and ordinates $x = 0$ and $x = 3$

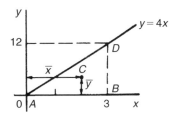

Figure 75.5

Let the co-ordinates of the area be $(\bar{x}, \bar{y})$ as shown in Figure 75.5.

Then $\bar{x} = \dfrac{\displaystyle\int_0^3 xy\,dx}{\displaystyle\int_0^3 y\,dx} = \dfrac{\displaystyle\int_0^3 (x)(4x)\,dx}{\displaystyle\int_0^3 4x\,dx}$

$= \dfrac{\displaystyle\int_0^3 4x^2\,dx}{\displaystyle\int_0^3 4x\,dx} = \dfrac{\left[\dfrac{4x^3}{3}\right]_0^3}{[2x^2]_0^3} = \dfrac{36}{18} = 2$

$\bar{y} = \dfrac{\dfrac{1}{2}\displaystyle\int_0^3 y^2\,dx}{\displaystyle\int_0^3 y\,dx} = \dfrac{\dfrac{1}{2}\displaystyle\int_0^3 (4x)^2\,dx}{18}$

$= \dfrac{\dfrac{1}{2}\displaystyle\int_0^3 16x^2\,dx}{18} = \dfrac{\dfrac{1}{2}\left[\dfrac{16x^3}{3}\right]_0^3}{18} = \dfrac{72}{18} = 4$

Hence the centroid lies at (2, 4)

In Figure 75.5, *ABD* is a right-angled triangle. The centroid lies 4 units from *AB* and 1 unit from *BD*, showing that the centroid of a triangle lies at one-third of the perpendicular height above any side as base.

Now try the following Practice Exercise

Practice Exercise 294   Further problems on centroids of simple shapes (answers on page 1184)

In Problems 1 to 5, find the position of the centroids of the areas bounded by the given curves, the *x*-axis and the given ordinates.

1. $y = 2x$; $x = 0$, $x = 3$

2. $y = 3x + 2$; $x = 0$, $x = 4$

3. $y = 5x^2$; $x = 1$, $x = 4$

4. $y = 2x^3$; $x = 0$, $x = 2$

5. $y = x(3x + 1)$; $x = -1$, $x = 0$

75.6 Further worked problems on centroids of simple shapes

Problem 4. Determine the co-ordinates of the centroid of the area lying between the curve $y = 5x - x^2$ and the *x*-axis

$y = 5x - x^2 = x(5 - x)$. When $y = 0$, $x = 0$ or $x = 5$, Hence the curve cuts the *x*-axis at 0 and 5 as shown in Figure 75.6. Let the co-ordinates of the centroid be $(\bar{x}, \bar{y})$ then, by integration,

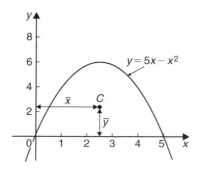

Figure 75.6

$$\bar{x} = \frac{\int_0^5 xy\,dx}{\int_0^5 y\,dx} = \frac{\int_0^5 x(5x - x^2)dx}{\int_0^5 (5x - x^2)dx}$$

$$= \frac{\int_0^5 (5x^2 - x^3)dx}{\int_0^5 (5x - x^2)dx} = \frac{\left[\frac{5x^3}{3} - \frac{x^4}{4}\right]_0^5}{\left[\frac{5x^2}{2} - \frac{x^3}{3}\right]_0^5}$$

$$= \frac{\frac{625}{3} - \frac{625}{4}}{\frac{125}{2} - \frac{125}{3}} = \frac{\frac{625}{12}}{\frac{125}{6}}$$

$$= \left(\frac{625}{12}\right)\left(\frac{6}{125}\right) = \frac{5}{2} = \textbf{2.5}$$

$$\bar{y} = \frac{\frac{1}{2}\int_0^5 y^2\,dx}{\int_0^5 y\,dx} = \frac{\frac{1}{2}\int_0^5 (5x - x^2)^2\,dx}{\int_0^5 (5x - x^2)\,dx}$$

$$= \frac{\frac{1}{2}\int_0^5 (25x^2 - 10x^3 + x^4)\,dx}{\frac{125}{6}}$$

$$= \frac{\frac{1}{2}\left[\frac{25x^3}{3} - \frac{10x^4}{4} + \frac{x^5}{5}\right]_0^5}{\frac{125}{6}}$$

$$= \frac{\frac{1}{2}\left(\frac{25(125)}{3} - \frac{6250}{4} + 625\right)}{\frac{125}{6}} = \textbf{2.5}$$

Hence the centroid of the area lies at (2.5, 2.5).
(Note from Figure 75.6 that the curve is symmetrical about $x = 2.5$ and thus $\bar{x}$ could have been determined 'on sight'.)

Problem 5. Locate the centroid of the area enclosed by the curve $y = 2x^2$, the *y*-axis and ordinates $y = 1$ and $y = 4$, correct to 3 decimal places

From Section 75.4,

$$\bar{x} = \frac{\dfrac{1}{2}\displaystyle\int_1^4 x^2\,dy}{\displaystyle\int_1^4 x\,dy} = \frac{\dfrac{1}{2}\displaystyle\int_1^4 \dfrac{y}{2}\,dy}{\displaystyle\int_1^4 \sqrt{\dfrac{y}{2}}\,dy}$$

$$= \frac{\dfrac{1}{2}\left[\dfrac{y^2}{4}\right]_1^4}{\left[\dfrac{2y^{3/2}}{3\sqrt{2}}\right]_1^4} = \frac{\dfrac{15}{8}}{\dfrac{14}{3\sqrt{2}}} = \mathbf{0.568}$$

and

$$\bar{y} = \frac{\displaystyle\int_1^4 xy\,dy}{\displaystyle\int_1^4 x\,dy} = \frac{\displaystyle\int_1^4 \sqrt{\dfrac{y}{2}}(y)\,dy}{\dfrac{14}{3\sqrt{2}}}$$

$$= \frac{\displaystyle\int_1^4 \dfrac{y^{3/2}}{\sqrt{2}}\,dy}{\dfrac{14}{3\sqrt{2}}} = \frac{\dfrac{1}{\sqrt{2}}\left[\dfrac{y^{5/2}}{\dfrac{5}{2}}\right]_1^4}{\dfrac{14}{3\sqrt{2}}}$$

$$= \frac{\dfrac{2}{5\sqrt{2}}(31)}{\dfrac{14}{3\sqrt{2}}} = \mathbf{2.657}$$

Hence the position of the centroid is at (0.568, 2.657)

> **Problem 6.**   Locate the position of the centroid enclosed by the curves $y = x^2$ and $y^2 = 8x$

Figure 75.7 shows the two curves intersection at (0, 0) and (2, 4). These are the same curves as used in Prob-

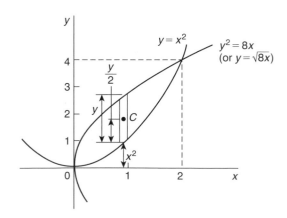

Figure 75.7

lem 12, Chapter 72 where the shaded area was calculated as $2\frac{2}{3}$ square units. Let the co-ordinates of centroid C be $\bar{x}$ and $\bar{y}$.

By integration, $\bar{x} = \dfrac{\displaystyle\int_0^2 xy\,dx}{\displaystyle\int_0^2 y\,dx}$

The value of y is given by the height of the typical strip shown in Figure 75.7, i.e. $y = \sqrt{8x} - x^2$. Hence,

$$\bar{x} = \frac{\displaystyle\int_0^2 x(\sqrt{8x} - x^2)\,dx}{2\frac{2}{3}} = \frac{\displaystyle\int_0^2 (\sqrt{8}x^{3/2} - x^3)}{2\frac{2}{3}}$$

$$= \frac{\left[\sqrt{8}\dfrac{x^{5/2}}{\dfrac{5}{2}} - \dfrac{x^4}{4}\right]_0^2}{2\frac{2}{3}} = \frac{\left(\sqrt{8}\dfrac{\sqrt{25}}{\dfrac{5}{2}} - 4\right)}{2\frac{2}{3}}$$

$$= \frac{2\dfrac{2}{5}}{2\dfrac{2}{3}} = \mathbf{0.9}$$

Care needs to be taken when finding $\bar{y}$ in such examples as this. From Figure 75.7, $y = \sqrt{8x} - x^2$ and $\dfrac{y}{2} = \dfrac{1}{2}(\sqrt{8x} - x^2)$. The perpendicular distance from centroid C of the strip to Ox is $\dfrac{1}{2}(\sqrt{8x} - x^2) + x^2$. Taking moments about Ox gives:

(total area) $(\bar{y}) = \sum_{x=0}^{x=2}$(area of strip) (perpendicular distance of centroid of strip to Ox)

Hence (area) $(\bar{y})$

$$= \int \left[\sqrt{8x} - x^2\right]\left[\frac{1}{2}(\sqrt{8x} - x^2) + x^2\right]dx$$

i.e. $\left(2\dfrac{2}{3}\right)(\bar{y}) = \displaystyle\int_0^2 \left[\sqrt{8x} - x^2\right]\left(\dfrac{\sqrt{8x}}{2} + \dfrac{x^2}{2}\right)dx$

$$= \int_0^2 \left(\frac{8x}{2} - \frac{x^4}{2}\right)dx = \left[\frac{8x^2}{4} - \frac{x^5}{10}\right]_0^2$$

$$= \left(8 - 3\frac{1}{5}\right) - (0) = 4\frac{4}{5}$$

Hence $\bar{y} = \dfrac{4\frac{4}{5}}{2\frac{2}{3}} = \mathbf{1.8}$

Thus the position of the centroid of the enclosed area in Figure 75.7 is at (0.9, 1.8).

Now try the following Practice Exercise

Practice Exercise 295   Further problems on centroids of simple shapes (answers on page 1184)

1. Determine the position of the centroid of a sheet of metal formed by the curve $y = 4x - x^2$ which lies above the x-axis

2. Find the co-ordinates of the centroid of the area that lies between curve $\dfrac{y}{x} = x - 2$ and the x-axis

3. Determine the co-ordinates of the centroid of the area formed between the curve $y = 9 - x^2$ and the x-axis

4. Determine the centroid of the area lying between $y = 4x^2$, the y-axis and the ordinates $y = 0$ and $y = 4$

5. Find the position of the centroid of the area enclosed by the curve $y = \sqrt{5x}$, the x-axis and the ordinate $x = 5$

6. Sketch the curve $y^2 = 9x$ between the limits $x = 0$ and $x = 4$. Determine the position of the centroid of this area

7. Calculate the points of intersection of the curves $x^2 = 4y$ and $\dfrac{y^2}{4} = x$, and determine the position of the centroid of the area enclosed by them

8. Sketch the curves $y = 2x^2 + 5$ and $y - 8 = x(x + 2)$ on the same axes and determine their points of intersection. Calculate the co-ordinates of the centroid of the area enclosed by the two curves

75.7   Theorem of Pappus

A theorem of Pappus* states:
'If a plane area is rotated about an axis in its own plane but not intersecting it, the volume of the solid formed is given by the product of the area and the distance moved by the centroid of the area.'
With reference to Figure 75.8, when the curve $y = f(x)$ is rotated one revolution about the x-axis between the limits $x = a$ and $x = b$, the volume V generated is given by:
volume $V = (A)(2\pi\bar{y})$, from which,

$$\bar{y} = \frac{V}{2\pi A}$$

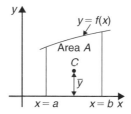

Figure 75.8

Problem 7. Determine the position of the centroid of a semicircle of radius r by using the theorem of Pappus. Check the answer by using integration (given that the equation of a circle, centre 0, radius r is $x^2 + y^2 = r^2$)

A semicircle is shown in Figure 75.9 with its diameter lying on the x-axis and its centre at the origin. Area of semicircle $= \dfrac{\pi r^2}{2}$. When the area is rotated about the x-axis one revolution a sphere is generated of volume $\dfrac{4}{3}\pi r^3$

Let centroid C be at a distance $\bar{y}$ from the origin as shown in Figure 75.9. From the theorem of Pappus, volume generated $=$ area $\times$ distance moved through by

---

* **Who was Pappus?** – **Pappus of Alexandria** (c. 290–c. 350) was one of the last great Greek mathematicians of Antiquity. *Collection*, his best-known work, is a compendium of mathematics in eight volumes. It covers a wide range of topics, including geometry, recreational mathematics, doubling the cube, polygons and polyhedra. To find out more go to **www.routledge.com/cw/bird**

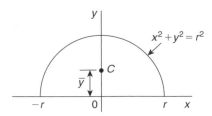

Figure 75.9

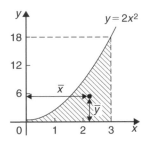

Figure 75.10

centroid i.e.

$$\frac{4}{3}\pi r^3 = \left(\frac{\pi r^2}{2}\right)(2\pi \bar{y})$$

Hence $\qquad \bar{y} = \dfrac{\dfrac{4}{3}\pi r^3}{\pi^2 r^2} = \dfrac{4r}{3\pi}$

By integration,

$$\bar{y} = \frac{\dfrac{1}{2}\displaystyle\int_{-r}^{r} y^2 \, dx}{\text{area}}$$

$$= \frac{\dfrac{1}{2}\displaystyle\int_{-r}^{r}(r^2 - x^2)\,dx}{\dfrac{\pi r^2}{2}} = \frac{\dfrac{1}{2}\left[r^2 x - \dfrac{x^3}{3}\right]_{-r}^{r}}{\dfrac{\pi r^2}{2}}$$

$$= \frac{\dfrac{1}{2}\left[\left(r^3 - \dfrac{r^3}{3}\right) - \left(-r^3 + \dfrac{r^3}{3}\right)\right]}{\dfrac{\pi r^2}{2}} = \frac{4r}{3\pi}$$

Hence the centroid of a semicircle lies on the axis of symmetry, distance $\dfrac{4r}{3\pi}$ (or $0.424\,r$) from its diameter.

Problem 8.   Calculate the area bounded by the curve $y = 2x^2$, the x-axis and ordinates $x = 0$ and $x = 3$. (b) If this area is revolved (i) about the x-axis and (ii) about the y-axis, find the volumes of the solids produced. (c) Locate the position of the centroid using (i) integration, and (ii) the theorem of Pappus

---

(a)   The required area is shown shaded in Figure 75.10.

$$\text{Area} = \int_0^3 y \, dx = \int_0^3 2x^2 \, dx = \left[\frac{2x^3}{3}\right]_0^3$$

$$= \textbf{18 square units}$$

(b)   (i) When the shaded area of Figure 75.10 is revolved $360°$ about the x-axis, the volume generated

$$= \int_0^3 \pi y^2 \, dx = \int_0^3 \pi (2x^2)^2 \, dx$$

$$= \int_0^3 4\pi x^4 \, dx = 4\pi \left[\frac{x^5}{5}\right]_0^3 = 4\pi\left(\frac{243}{5}\right)$$

$$= \textbf{194.4}\pi \textbf{ cubic units}$$

(ii) When the shaded area of Figure 75.10 is revolved $360°$ about the y-axis, the volume generated $=$ (volume generated by $x = 3$) $-$ (volume generated by $y = 2x^2$)

$$= \int_0^{18} \pi (3)^2 \, dy - \int_0^{18} \pi \left(\frac{y}{2}\right) dy$$

$$= \pi \int_0^{18}\left(9 - \frac{y}{2}\right) dy = \pi \left[9y - \frac{y^2}{4}\right]_0^{18}$$

$$= \textbf{81}\pi \textbf{ cubic units}$$

(c)   If the co-ordinates of the centroid of the shaded area in Figure 75.10 are $(\bar{x}, \bar{y})$ then:

(i) by integration,

$$\bar{x} = \frac{\displaystyle\int_0^3 xy \, dx}{\displaystyle\int_0^3 y \, dx} = \frac{\displaystyle\int_0^3 x(2x^2) \, dx}{18}$$

$$= \frac{\displaystyle\int_0^3 2x^3 \, dx}{18} = \frac{\left[\dfrac{2x^4}{4}\right]_0^3}{18} = \frac{81}{36} = \textbf{2.25}$$

$$\bar{y} = \frac{\frac{1}{2}\int_0^3 y^2 dx}{\int_0^3 y \, dx} = \frac{\frac{1}{2}\int_0^3 (2x^2)^2 dx}{18}$$

$$= \frac{\frac{1}{2}\int_0^3 4x^4 dx}{18} = \frac{\frac{1}{2}\left[\frac{4x^5}{5}\right]_0^3}{18} = \mathbf{5.4}$$

(ii) using the theorem of Pappus:
Volume generated when shaded area is revolved about $Oy = (\text{area})(2\pi\bar{x})$

i.e. $\qquad 81\pi = (18)(2\pi\bar{x}),$

from which, $\qquad \bar{x} = \dfrac{81\pi}{36\pi} = \mathbf{2.25}$

Volume generated when shaded area is revolved about $Ox = (\text{area})(2\pi\bar{y})$

i.e. $\qquad 194.4\pi = (18)(2\pi\bar{y})$

from which, $\qquad \bar{y} = \dfrac{194.4\pi}{36\pi} = \mathbf{5.4}$

Hence the centroid of the shaded area in Figure 75.10 is at (2.25, 5.4).

Problem 9. A cylindrical pillar of diameter 400 mm has a groove cut round its circumference. The section of the groove is a semicircle of diameter 50 mm. Determine the volume of material removed, in cubic centimetres, correct to 4 significant figures

---

A part of the pillar showing the groove is shown in Figure 75.11.

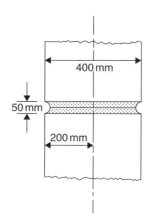

Figure 75.11

The distance of the centroid of the semicircle from its base is $\dfrac{4r}{3\pi}$ (see Problem 7) $= \dfrac{4(25)}{3\pi} = \dfrac{100}{3\pi}$ mm.
The distance of the centroid from the centre of the pillar $= \left(200 - \dfrac{100}{3\pi}\right)$ mm.
The distance moved by the centroid in one revolution

$$= 2\pi\left(200 - \frac{100}{3\pi}\right) = \left(400\pi - \frac{200}{3}\right)\text{mm}.$$

From the theorem of Pappus,
volume $=$ area $\times$ distance moved by centroid

$$= \left(\frac{1}{2}\pi 25^2\right)\left(400\pi - \frac{200}{3}\right) = 1\,168\,250\,\text{mm}^3$$

Hence the volume of material removed is 1168 cm^3, correct to 4 significant figures.

Problem 10. A metal disc has a radius of 5.0 cm and is of thickness 2.0 cm. A semicircular groove of diameter 2.0 cm is machined centrally around the rim to form a pulley. Determine, using Pappus' theorem, the volume and mass of metal removed and the volume and mass of the pulley if the density of the metal is 8000 kg m^{-3}

A side view of the rim of the disc is shown in Figure 75.12.

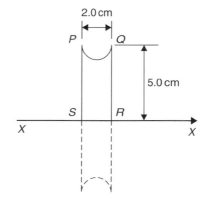

Figure 75.12

When area $PQRS$ is rotated about axis XX the volume generated is that of the pulley. The centroid of the semicircular area removed is at a distance of $\dfrac{4r}{3\pi}$ from its diameter (see Problem 7), i.e. $\dfrac{4(1.0)}{3\pi}$, i.e.

0.424 cm from *PQ*. Thus the distance of the centroid from *XX* is (5.0 − 0.424), i.e. 4.576 cm. The distance moved through in one revolution by the centroid is $2\pi(4.576)$ cm.

Area of semicircle

$$= \frac{\pi r^2}{2} = \frac{\pi(1.0)^2}{2} = \frac{\pi}{2}\ \text{cm}^2$$

By the theorem of Pappus, volume generated

$$= \text{area} \times \text{distance moved by centroid}$$

$$= \left(\frac{\pi}{2}\right)(2\pi)(4.576)$$

i.e. **volume of metal removed = 45.16 cm^3**

Mass of metal removed = density × volume

$$= 8000\,\text{kg m}^{-3} \times \frac{45.16}{10^6}\,\text{m}^3$$

$$= \textbf{0.3613 kg or 361.3 g}$$

Volume of pulley = volume of cylindrical disc
$$\qquad\qquad\quad - \text{volume of metal removed}$$

$$= \pi(5.0)^2(2.0) - 45.16 = \textbf{111.9 cm}^3$$

Mass of pulley = density × volume

$$= 8000\,\text{kg m}^{-3} \times \frac{111.9}{10^6}\,\text{m}^3$$

$$= \textbf{0.8952 kg or 895.2 g}$$

Now try the following Practice Exercise

Practice Exercise 296   Further problems on the theorem of Pappus (answers on page 1184)

1. A right-angled isosceles triangle having a hypotenuse of 8 cm is revolved one revolution about one of its equal sides as axis. Determine the volume of the solid generated using Pappus' theorem

2. A rectangle measuring 10.0 cm by 6.0 cm rotates one revolution about one of its longest sides as axis. Determine the volume of the resulting cylinder by using the theorem of Pappus

3. Using (a) the theorem of Pappus and (b) integration, determine the position of the centroid of a metal template in the form of a quadrant of a circle of radius 4 cm. (The equation of a circle, centre 0, radius *r* is $x^2 + y^2 = r^2$)

4. (a) Determine the area bounded by the curve $y = 5x^2$, the *x*-axis and the ordinates $x = 0$ and $x = 3$
    (b) If this area is revolved 360° about (i) the *x*-axis and (ii) the *y*-axis, find the volumes of the solids of revolution produced in each case
    (c) Determine the co-ordinates of the centroid of the area using (i) integral calculus, and (ii) the theorem of Pappus

5. A metal disc has a radius of 7.0 cm and is of thickness 2.5 cm. A semicircular groove of diameter 2.0 cm is machined centrally around the rim to form a pulley. Determine the volume of metal removed using Pappus' theorem and express this as a percentage of the original volume of the disc. Find also the mass of metal removed if the density of the metal is $7800\,\text{kg m}^{-3}$

For fully worked solutions to each of the problems in Practice Exercises 294 to 296 in this chapter, go to the website:
www.routledge.com/cw/bird

Chapter 76

Second moments of area

Why it is important to understand: Second moments of area

The second moment of area is a property of a cross-section that can be used to predict the resistance of a beam to bending and deflection around an axis that lies in the cross-sectional plane. The stress in, and deflection of, a beam under load depends not only on the load but also on the geometry of the beam's cross-section; larger values of second moment cause smaller values of stress and deflection. This is why beams with larger second moments of area, such as I-beams, are used in building construction in preference to other beams with the same cross-sectional area. The second moment of area has applications in many scientific disciplines, including fluid mechanics, engineering mechanics and biomechanics – for example, to study the structural properties of bone during bending. The static roll stability of a ship depends on the second moment of area of the waterline section – short, fat ships are stable, long, thin ones are not. It is clear that calculations involving the second moment of area are very important in many areas of engineering.

At the end of this chapter, you should be able to:

- define second moment of area and radius of gyration
- derive second moments of area of regular sections – rectangle, triangle, circle and semicircle
- state the parallel and perpendicular axis theorems
- determine second moment of area and radius of gyration of regular sections using a table of derived results
- determine second moment of area and radius of gyration of composite areas using a table of derived results

76.1 Second moments of area and radius of gyration

The **first moment of area** about a fixed axis of a lamina of area A, perpendicular distance y from the centroid of the lamina is defined as Ay cubic units. The **second moment of area** of the same lamina as above is given by Ay^2, i.e. the perpendicular distance from the centroid of the area to the fixed axis is squared. Second moments of areas are usually denoted by I and have limits of mm^4, cm^4, and so on.

Radius of gyration

Several areas, $a_1, a_2, a_3, \ldots$ at distances $y_1, y_2, y_3, \ldots$ from a fixed axis, may be replaced by a single area A, where $A = a_1 + a_2 + a_3 + \cdots$ at distance k from the axis, such that $Ak^2 = \sum ay^2$. k is called the **radius of gyration** of area A about the given axis. Since $Ak^2 = \sum ay^2 = I$ then the radius of gyration, $k = \sqrt{\dfrac{I}{A}}$

The second moment of area is a quantity much used in the theory of bending of beams, in the torsion of shafts, and in calculations involving water planes and centres of pressure.

76.2 Second moment of area of regular sections

The procedure to determine the second moment of area of regular sections about a given axis is (i) to find the second moment of area of a typical element and (ii) to sum all such second moments of area by integrating between appropriate limits.

For example, the second moment of area of the rectangle shown in Figure 76.1 about axis PP is found by initially considering an elemental strip of width δx, parallel to and distance x from axis PP. Area of shaded strip $= b\delta x$. Second moment of area of the shaded strip about $PP = (x^2)(b\delta x)$

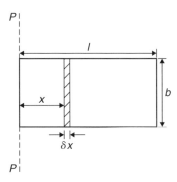

Figure 76.1

The second moment of area of the whole rectangle about PP is obtained by summing all such strips between $x = 0$ and $x = l$, i.e. $\sum_{x=0}^{x=l} x^2 b\delta x$. It is a fundamental theorem of integration that

$$\underset{\delta x \to 0}{\text{limit}} \sum_{x=0}^{x=l} x^2 b\delta x = \int_0^l x^2 b \, dx$$

Thus the second moment of area of the rectangle about PP is given by $\quad b\int_0^l x^2 dx = b\left[\dfrac{x^3}{3}\right]_0^l = \dfrac{bl^3}{3}$

Since the total area of the rectangle, $A = lb$, then

$$I_{pp} = (lb)\left(\frac{l^2}{3}\right) = \frac{Al^2}{3}$$

$I_{pp} = Ak_{pp}^2$ thus $k_{pp}^2 = \dfrac{l^2}{3}$

i.e. the radius of gyration about axes PP,

$$k_{pp} = \sqrt{\frac{l^2}{3}} = \frac{l}{\sqrt{3}}$$

76.3 Parallel axis theorem

In Figure 76.2, axis GG passes through the centroid C of area A. Axes DD and GG are in the same plane, are parallel to each other and distance d apart. The parallel axis theorem states:

$$I_{DD} = I_{GG} + Ad^2$$

Using the parallel axis theorem the second moment of area of a rectangle about an axis through the centroid may be determined. In the rectangle shown in Figure 76.3, $I_{pp} = \dfrac{bl^3}{3}$ (from above). From the parallel axis theorem $I_{pp} = I_{GG} + (bl)\left(\dfrac{l}{2}\right)^2$

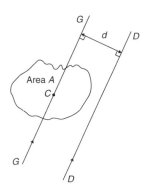

Figure 76.2

i.e. $\qquad \dfrac{bl^3}{3} = I_{GG} + \dfrac{bl^3}{4}$

from which, $\quad \boldsymbol{I_{GG}} = \dfrac{bl^3}{3} - \dfrac{bl^3}{4} = \boldsymbol{\dfrac{bl^3}{12}}$

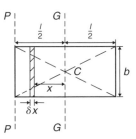

Figure 76.3

76.4 Perpendicular axis theorem

In Figure 76.4, axes OX, OY and OZ are mutually perpendicular. If OX and OY lie in the plane

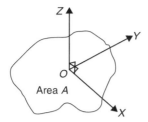

Figure 76.4

of area A then the perpendicular axis theorem states:

$$I_{OZ} = I_{OX} + I_{OY}$$

76.5   Summary of derived results

In Section 76.2 the second moment of area and radius of gyration of a rectangle coinciding with its breadth was found be $\frac{bl^3}{3}$ and $\frac{l}{\sqrt{3}}$ respectively. A summary of standard results for the second moment of area and radius of gyration of regular sections, derived in a similar manner, are listed in Table 76.1. In Section 76.6 are some Problems involving calculations using this table. (It should be noted that there are computer programs, such as Auto-CAD, that can determine second moment of area for any shape.)

Table 76.1 Summary of standard results of the second moments of areas of regular sections

Shape	Position of axis	Second moment of area, I	Radius of gyration, k
Rectangle length l breadth b	(1) Coinciding with b	$\dfrac{bl^3}{3}$	$\dfrac{l}{\sqrt{3}}$
	(2) Coinciding with l	$\dfrac{lb^3}{3}$	$\dfrac{b}{\sqrt{3}}$
	(3) Through centroid, parallel to b	$\dfrac{bl^3}{12}$	$\dfrac{l}{\sqrt{12}}$
	(4) Through centroid, parallel to l	$\dfrac{lb^3}{12}$	$\dfrac{b}{\sqrt{12}}$
Triangle Perpendicular height h base b	(1) Coinciding with b	$\dfrac{bh^3}{12}$	$\dfrac{h}{\sqrt{6}}$
	(2) Through centroid, parallel to base	$\dfrac{bh^3}{36}$	$\dfrac{h}{\sqrt{18}}$
	(3) Through vertex, parallel to base	$\dfrac{bh^3}{4}$	$\dfrac{h}{\sqrt{2}}$
Circle radius r	(1) Through centre, perpendicular to plane (i.e. polar axis)	$\dfrac{\pi r^4}{2}$	$\dfrac{r}{\sqrt{2}}$
	(2) Coinciding with diameter	$\dfrac{\pi r^4}{4}$	$\dfrac{r}{2}$
	(3) About a tangent	$\dfrac{5\pi r^4}{4}$	$\dfrac{\sqrt{5}}{2}r$
Semicircle radius r	Coinciding with diameter	$\dfrac{\pi r^4}{8}$	$\dfrac{r}{2}$

76.6 Worked problems on second moments of area of regular sections

Problem 1. Determine the second moment of area and the radius of gyration about axes AA, BB and CC for the rectangle shown in Figure 76.5

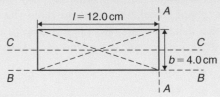

Figure 76.5

From Table 76.1, the second moment of area about axis AA, $I_{AA} = \dfrac{bl^3}{3} = \dfrac{(4.0)(12.0)^3}{3} = \mathbf{2304\,cm^4}$

Radius of gyration,

$$k_{AA} = \frac{l}{\sqrt{3}} = \frac{12.0}{\sqrt{3}} = \mathbf{6.93\,cm}$$

Similarly, $\quad I_{BB} = \dfrac{lb^3}{3} = \dfrac{(12.0)(4.0)^3}{3} = \mathbf{256\,cm^4}$

and $\quad k_{BB} = \dfrac{b}{\sqrt{3}} = \dfrac{4.0}{\sqrt{3}} = \mathbf{2.31\,cm}$

The second moment of area about the centroid of a rectangle is $\dfrac{bl^3}{12}$ when the axis through the centroid is parallel with the breadth, b. In this case, the axis CC is parallel with the length l.

Hence $\quad I_{CC} = \dfrac{lb^3}{12} = \dfrac{(12.0)(4.0)^3}{12} = \mathbf{64\,cm^4}$

and $\quad k_{CC} = \dfrac{b}{\sqrt{12}} = \dfrac{4.0}{\sqrt{12}} = \mathbf{1.15\,cm}$

Problem 2. Find the second moment of area and the radius of gyration about axis PP for the rectangle shown in Figure 76.6

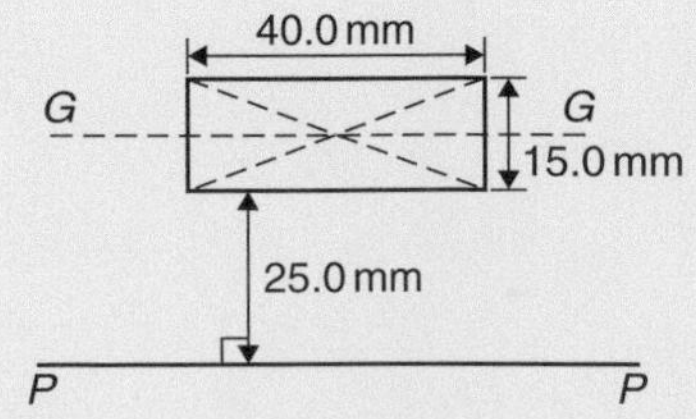

Figure 76.6

$I_{GG} = \dfrac{lb^3}{12}$ where $l = 40.0$ mm and $b = 15.0$ mm

Hence $I_{GG} = \dfrac{(40.0)(15.0)^3}{12} = 11\,250\,mm^4$

From the parallel axis theorem, $I_{PP} = I_{GG} + Ad^2$, where $A = 40.0 \times 15.0 = 600\,mm^2$ and $d = 25.0 + 7.5 = 32.5\,mm$, the perpendicular distance between GG and PP.

Hence, $\quad\quad I_{PP} = 11\,250 + (600)(32.5)^2$

$$= \mathbf{645\,000\,mm^4}$$

$$I_{PP} = Ak_{PP}^2$$

from which, $\quad k_{PP} = \sqrt{\dfrac{I_{PP}}{\text{area}}}$

$$= \sqrt{\dfrac{645\,000}{600}} = \mathbf{32.79\,mm}$$

Problem 3. Determine the second moment of area and radius of gyration about axis QQ of the triangle BCD shown in Figure 76.7

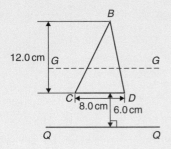

Figure 76.7

Using the parallel axis theorem: $I_{QQ} = I_{GG} + Ad^2$, where I_{GG} is the second moment of area about the centroid of the triangle,

i.e. $\dfrac{bh^3}{36} = \dfrac{(8.0)(12.0)^3}{36} = 384\,cm^4$, A is the area of the triangle $= \frac{1}{2}bh = \frac{1}{2}(8.0)(12.0) = 48\,cm^2$ and d is the distance between axes GG and $QQ = 6.0 + \frac{1}{3}(12.0) = 10\,cm$.

Hence the second moment of area about axis QQ,

$$I_{QQ} = 384 + (48)(10)^2 = \mathbf{5184\,cm^4}$$

Radius of gyration,

$$k_{QQ} = \sqrt{\dfrac{I_{QQ}}{\text{area}}} = \sqrt{\dfrac{5184}{48}} = \mathbf{10.4\,cm}$$

Problem 4.   Determine the second moment of area and radius of gyration of the circle shown in Figure 76.8 about axis YY

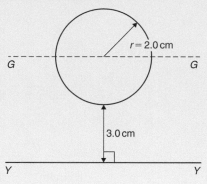

Figure 76.8

In Figure 76.8, $I_{GG} = \dfrac{\pi r^4}{4} = \dfrac{\pi}{4}(2.0)^4 = 4\pi\ \text{cm}^4$. Using the parallel axis theorem, $I_{YY} = I_{GG} + Ad^2$, where $d = 3.0 + 2.0 = 5.0\ \text{cm}$

Hence $I_{YY} = 4\pi + [\pi(2.0)^2](5.0)^2$

$= 4\pi + 100\pi = 104\pi = \mathbf{327\ cm^4}$

Radius of gyration,

$$k_{YY} = \sqrt{\dfrac{I_{YY}}{\text{area}}} = \sqrt{\dfrac{104\pi}{\pi(2.0)^2}} = \sqrt{26} = \mathbf{5.10\ cm}$$

Problem 5.   Determine the second moment of area and radius of gyration for the semicircle shown in Figure 76.9 about axis XX

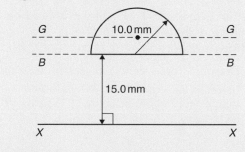

Figure 76.9

The centroid of a semicircle lies at $\dfrac{4r}{3\pi}$ from its diameter.

Using the parallel axis theorem: $I_{BB} = I_{GG} + Ad^2$,

where $I_{BB} = \dfrac{\pi r^4}{8}$ (from Table 76.1)

$= \dfrac{\pi(10.0)^4}{8} = 3927\ \text{mm}^4,$

$$A = \dfrac{\pi r^2}{2} = \dfrac{\pi(10.0)^2}{2} = 157.1\ \text{mm}^2$$

and $d = \dfrac{4r}{3\pi} = \dfrac{4(10.0)}{3\pi} = 4.244\ \text{mm}$

Hence $3927 = I_{GG} + (157.1)(4.244)^2$

i.e. $3927 = I_{GG} + 2830,$

from which, $I_{GG} = 3927 - 2830 = 1097\ \text{mm}^4$

Using the parallel axis theorem again:

$$I_{XX} = I_{GG} + A(15.0 + 4.244)^2$$

i.e. $I_{XX} = 1097 + (157.1)(19.244)^2$

$= 1097 + 58\,179 = 59\,276\ \text{mm}^4$ or $\mathbf{59\,280\ mm^4}$,

correct to 4 significant figures.

Radius of gyration, $k_{XX} = \sqrt{\dfrac{I_{XX}}{\text{area}}} = \sqrt{\dfrac{59\,276}{157.1}}$

$= \mathbf{19.42\ mm}$

Problem 6.   Determine the polar second moment of area of the propeller shaft cross-section shown in Figure 76.10

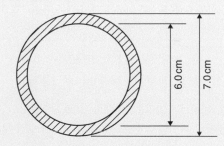

Figure 76.10

The polar second moment of area of a circle $= \dfrac{\pi r^4}{2}$

The polar second moment of area of the shaded area is given by the polar second moment of area of the 7.0 cm diameter circle minus the polar second moment of area of the 6.0 cm diameter circle. Hence the polar second moment of area of the cross-section shown

$$= \dfrac{\pi}{2}\left(\dfrac{7.0}{2}\right)^4 - \dfrac{\pi}{2}\left(\dfrac{6.0}{2}\right)^4$$

$= 235.7 - 127.2 = \mathbf{108.5\ cm^4}$

Problem 7. Determine the second moment of area and radius of gyration of a rectangular lamina of length 40 mm and width 15 mm about an axis through one corner, perpendicular to the plane of the lamina

The lamina is shown in Figure 76.11.

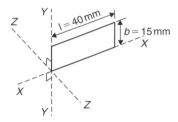

Figure 76.11

From the perpendicular axis theorem:

$$I_{ZZ} = I_{XX} + I_{YY}$$

$$I_{XX} = \frac{lb^3}{3} = \frac{(40)(15)^3}{3} = 45\,000 \,\text{mm}^4$$

and $I_{YY} = \frac{bl^3}{3} = \frac{(15)(40)^3}{3} = 320\,000 \,\text{mm}^4$

Hence $I_{ZZ} = 45\,000 + 320\,000$

$$= \textbf{365\,000 mm}^4 \text{ or } \textbf{36.5 cm}^4$$

Radius of gyration,

$$k_{ZZ} = \sqrt{\frac{I_{ZZ}}{\text{area}}} = \sqrt{\frac{365\,000}{(40)(15)}}$$

$$= \textbf{24.7 mm} \text{ or } \textbf{2.47 cm}$$

Problem 8. A solid circular section bar of diameter 20 mm is subjected to a pure bending moment of 0.3 kN m. If $E = 2 \times 10^{11}$ N/m², determine the resulting radius of curvature of the neutral layer of this beam and the maximum bending stress

In the bending of beams the following formula is used:
$$\frac{\sigma}{y} = \frac{M}{I} = \frac{E}{R} \text{ where}$$

$\sigma = $ the stress due to bending moment M occurring at a distance y from the neutral axis NA,

$I = $ the second moment of area of the beam's cross-section about NA,

$E = $ Young's modulus of elasticity of the beams' material and

$R = $ radius of curvature of the neutral layer of the beam due to the bending moment M.

From Table 76.1, page 813,

$$I = \frac{\pi r^4}{4} = \frac{\pi \times 10^4}{4} = \textbf{7854 mm}^4$$

Now, $\mathbf{M} = 0.3 \text{ kN m} \times 1000 \frac{\text{N}}{\text{kN}} \times 1000 \frac{\text{mm}}{\text{m}}$

$$= \textbf{3} \times \textbf{10}^5 \textbf{N mm}$$

and $\mathbf{E} = 2 \times 10^{11} \frac{\text{N}}{\text{m}^2} \times 1 \frac{\text{m}}{1000 \text{ mm}} \times \frac{\text{m}}{1000 \text{ mm}}$

$$= \textbf{2} \times \textbf{10}^5 \textbf{N/mm}^2$$

Using $\frac{M}{I} = \frac{E}{R}$ radius of curvature,

$$R = \frac{EI}{M} = 2 \times 10^5 \frac{\text{N}}{\text{mm}^2} \times \frac{7854 \text{ mm}^4}{3 \times 10^5 \text{ N mm}}$$

i.e. $\mathbf{R} = \textbf{5236 mm} = \textbf{5.24 m}$

Using $\frac{\sigma}{y} = \frac{M}{I}$ maximum stress due to bending,

$$\hat{\sigma} = \frac{M\hat{y}}{I}$$

where $\hat{y} = $ outermost fibre from the neutral axis
$$= \frac{d}{2} = \frac{20}{2} = 10 \text{ mm}$$

Hence, **maximum bending stress,**

$$\hat{\sigma} = \frac{M\hat{y}}{I} = \frac{3 \times 10^5 \text{N mm} \times 10 \text{ mm}}{7854 \text{ mm}^4}$$

$$= \textbf{382 N/mm}^2 = \textbf{382} \times \textbf{10}^6 \textbf{N/m}^2 = \textbf{382 MPa}$$

Now try the following Practice Exercise

Practice Exercise 297   Further problems on second moments of area of regular sections (answers on page 1184)

1. Determine the second moment of area and radius of gyration for the rectangle shown in Figure 76.12 about (a) axis *AA* (b) axis *BB*, and (c) axis *CC*

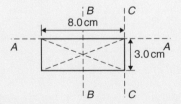

Figure 76.12

2. Determine the second moment of area and radius of gyration for the triangle shown in

Section J

Figure 76.13 about (a) axis *DD* (b) axis *EE*, and (c) an axis through the centroid of the triangle parallel to axis *DD*

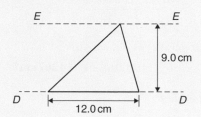

Figure 76.13

3. For the circle shown in Figure 76.14, find the second moment of area and radius of gyration about (a) axis *FF*, and (b) axis *HH*

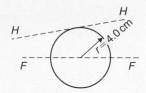

Figure 76.14

4. For the semicircle shown in Figure 76.15, find the second moment of area and radius of gyration about axis *JJ*

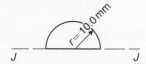

Figure 76.15

5. For each of the areas shown in Figure 76.16 determine the second moment of area and radius of gyration about axis *LL*, by using the parallel axis theorem

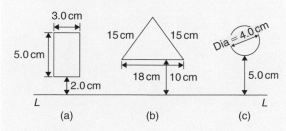

Figure 76.16

6. Calculate the radius of gyration of a rectangular door 2.0 m high by 1.5 m wide about a vertical axis through its hinge

7. A circular door of a boiler is hinged so that it turns about a tangent. If its diameter is 1.0 m, determine its second moment of area and radius of gyration about the hinge

8. A circular cover, centre 0, has a radius of 12.0 cm. A hole of radius 4.0 cm and centre *X*, where $OX = 6.0$ cm, is cut in the cover. Determine the second moment of area and the radius of gyration of the remainder about a diameter through 0 perpendicular to *OX*

76.7 Worked problems on second moments of area of composite areas

In engineering practice, use of sections which are built up of many simple sections is very common. Such sections are called **composite sections** or **built-up sections**. The following problems explain how to obtain the second moment of area and radius of gyration about a given axis for such sections.

Problem 9. Determine correct to 3 significant figures the second moment of area about *XX* for the composite area shown in Figure 76.17

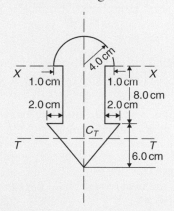

Figure 76.17

For the semicircle, $\quad I_{XX} = \dfrac{\pi r^4}{8} = \dfrac{\pi (4.0)^4}{8}$

$$= 100.5 \, \text{cm}^4$$

For the rectangle, $I_{XX} = \dfrac{bl^3}{3} = \dfrac{(6.0)(8.0)^3}{3}$

$$= 1024 \, \text{cm}^4$$

For the triangle, about axis TT through centroid C_T,

$$I_{TT} = \frac{bh^3}{36} = \frac{(10)(6.0)^3}{36} = 60 \, \text{cm}^4$$

By the parallel axis theorem, the second moment of area of the triangle about axis XX

$$= 60 + \left[\tfrac{1}{2}(10)(6.0)\right]\left[8.0 + \tfrac{1}{3}(6.0)\right]^2 = 3060 \, \text{cm}^4$$

Total second moment of area about XX

$$= 100.5 + 1024 + 3060 = 4184.5 = \mathbf{4180 \, cm^4},$$

correct to 3 significant figures.

Problem 10. Determine the second moment of area and the radius of gyration about axis XX for the I-section shown in Figure 76.18

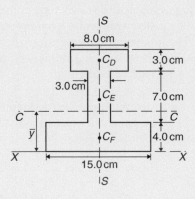

Figure 76.18

The I-section is divided into three rectangles, D, E and F and their centroids denoted by C_D, C_E and C_F respectively.

For rectangle D:
The second moment of area about C_D (an axis through C_D parallel to XX)

$$= \frac{bl^3}{12} = \frac{(8.0)(3.0)^3}{12} = 18 \, \text{cm}^4$$

Using the parallel axis theorem: $I_{XX} = 18 + Ad^2$ where $A = (8.0)(3.0) = 24 \, \text{cm}^2$ and $d = 12.5 \, \text{cm}$

Hence $I_{XX} = 18 + 24(12.5)^2 = \mathbf{3768 \, cm^4}$

For rectangle E:
The second moment of area about C_E (an axis through C_E parallel to XX)

$$= \frac{bl^3}{12} = \frac{(3.0)(7.0)^3}{12} = 85.75 \, \text{cm}^4$$

Using the parallel axis theorem:
$$I_{XX} = 85.75 + (7.0)(3.0)(7.5)^2 = \mathbf{1267 \, cm^4}$$

For rectangle F:

$$I_{XX} = \frac{bl^3}{3} = \frac{(15.0)(4.0)^3}{3} = \mathbf{320 \, cm^4}$$

Total second moment of area for the I-section about axis XX,

$$I_{XX} = 3768 + 1267 + 320 = \mathbf{5355 \, cm^4}$$

Total area of I-section

$$= (8.0)(3.0) + (3.0)(7.0) + (15.0)(4.0) = 105 \, \text{cm}^2$$

Radius of gyration,

$$k_{XX} = \sqrt{\frac{I_{XX}}{\text{area}}} = \sqrt{\frac{5355}{105}} = \mathbf{7.14 \, cm}$$

Now try the following Practice Exercise

Practice Exercise 298 Further problems on second moments of area of composite areas (answers on page 1185)

1. For the sections shown in Figure 76.19, find the second moment of area and the radius of gyration about axis XX.

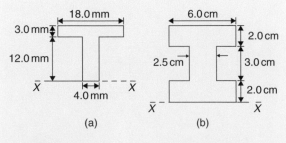

Figure 76.19

2. Determine the second moment of area about the given axes for the shapes shown in Figure 76.20. (In Figure 76.20(b), the circular area is removed.)

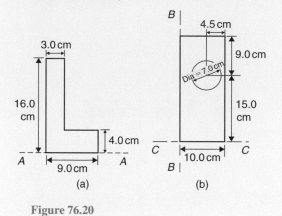

(a)

(b)

Figure 76.20

3. Find the second moment of area and radius of gyration about the axis *XX* for the beam section shown in Figure 76.21.

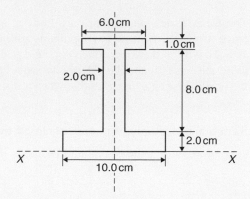

Figure 76.21

For fully worked solutions to each of the problems in Practice Exercises 297 and 298 in this chapter, go to the website:
www.routledge.com/cw/bird

Revision Test 27　Applications of integration

This assignment covers the material contained in Chapters 72 to 76. *The marks available are shown in brackets at the end of each question.*

1. The force F newtons acting on a body at a distance x metres from a fixed point is given by: $F = 2x + 3x^2$. If work done $= \int_{x_1}^{x_2} F \, dx$, determine the work done when the body moves from the position when $x = 1$ m to that when $x = 4$ m
   (4)

2. Sketch and determine the area enclosed by the curve $y = 3 \sin \dfrac{\theta}{2}$, the θ-axis and ordinates $\theta = 0$ and $\theta = \dfrac{2\pi}{3}$
   (4)

3. Calculate the area between the curve $y = x^3 - x^2 - 6x$ and the x-axis
   (10)

4. A voltage $v = 25 \sin 50\pi t$ volts is applied across an electrical circuit. Determine, using integration, its mean and r.m.s. values over the range $t = 0$ to $t = 20$ ms, each correct to 4 significant figures
   (12)

5. Sketch on the same axes the curves $x^2 = 2y$ and $y^2 = 16x$ and determine the co-ordinates of the points of intersection. Determine (a) the area enclosed by the curves and (b) the volume of the solid produced if the area is rotated one revolution about the x-axis
   (13)

6. Calculate the position of the centroid of the sheet of metal formed by the x-axis and the part of the curve $y = 5x - x^2$ which lies above the x-axis
   (9)

7. A cylindrical pillar of diameter 500 mm has a groove cut around its circumference as shown in Figure RT27.1. The section of the groove is a semicircle of diameter 40 mm. Given that the centroid of a semicircle from its base is $\dfrac{4r}{3\pi}$, use the theorem

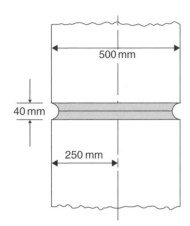

Figure RT27.1

of Pappus to determine the volume of material removed, in cm^3, correct to 3 significant figures
   (8)

8. For each of the areas shown in Figure RT27.2, determine the second moment of area and radius of gyration about axis XX
   (15)

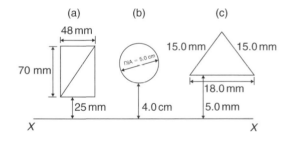

Figure RT27.2

9. A circular door is hinged so that it turns about a tangent. If its diameter is 1.0 m, find its second moment of area and radius of gyration about the hinge.
   (5)

For lecturers/instructors/teachers, fully worked solutions to each of the problems in Revision Test 27, together with a full marking scheme, are available at the website:
www.routledge.com/cw/bird

Multiple choice questions Test 8
Integration and its applications
This test covers the material in Chapters 63 to 76

All questions have only one correct answer (answers on page 1200).

1. $\int (5 - 3t^2)\,dt$ is equal to:

   (a) $5 - t^3 + c$ (b) $-3t^3 + c$

   (c) $-6t + c$ (d) $5t - t^3 + c$

2. $\int \left(\dfrac{5x - 1}{x} \right) dx$ is equal to:

   (a) $5x - \ln x + c$ (b) $\dfrac{5x^2 - x}{\dfrac{x^2}{2}}$

   (c) $\dfrac{5x^2}{2} + \dfrac{1}{x^2} + c$ (d) $5x + \dfrac{1}{x^2} + c$

3. The value of $\displaystyle\int_0^1 (3\sin 2\theta - 4\cos\theta)\,d\theta$, correct to 4 significant figures, is:

   (a) -0.06890 (b) -1.242

   (c) -2.742 (d) -1.569

4. $\int x e^{2x}\,dx$ is:

   (a) $\dfrac{x^2}{4} e^{2x} + c$ (b) $2e^{2x} + c$

   (c) $\dfrac{e^{2x}}{4}(2x - 1) + c$ (d) $2e^{2x}(x - 2) + c$

5. A vehicle has a velocity $v = (2 + 3t)$ m/s after t seconds. The distance travelled is equal to the area under the v/t graph. In the first 3 seconds the vehicle has travelled:

   (a) 11 m   (b) 33 m   (c) 13.5 m   (d) 19.5 m

6. The area, in square units, enclosed by the curve $y = 2x + 3$, the x-axis and ordinates $x = 1$ and $x = 4$ is:

   (a) 28   (b) 2   (c) 24   (d) 39

7. $\int \dfrac{2}{9} t^3\,dt$ is equal to:

   (a) $\dfrac{t^4}{18} + c$ (b) $\dfrac{2}{3} t^2 + c$ (c) $\dfrac{2}{9} t^4 + c$ (d) $\dfrac{2}{9} t^3 + c$

8. Evaluating $\displaystyle\int_0^{\pi/3} 3\sin 3x\,dx$ gives:

   (a) 1.503   (b) 2   (c) -18 (d) 6

9. The mean value of $y = 2x^2$ between $x = 1$ and $x = 3$ is:

   (a) 2   (b) 4   (c) $4\dfrac{1}{3}$ (d) $8\dfrac{2}{3}$

10. $\int \ln x\,dx$ is equal to:

    (a) $x(\ln x - 1) + c$ (b) $\dfrac{1}{x} + c$

    (c) $x \ln x - 1 + c$ (d) $\dfrac{1}{x} + \dfrac{1}{x^2} + c$

11. $\displaystyle\int_2^3 \dfrac{3}{x^2 + x - 2}\,dx$ is equal to:

    (a) $3\ln 2.5$ (b) $\dfrac{1}{3}\lg 1.6$ (c) $\ln 40$ (d) $\ln 1.6$

12. $\int (5\sin 3t - 3\cos 5t)\,dt$ is equal to:

    (a) $-5\cos 3t + 3\sin 5t + c$

    (b) $15(\cos 3t + \sin 3t) + c$

    (c) $-\dfrac{5}{3}\cos 3t - \dfrac{3}{5}\sin 5t + c$

    (d) $\dfrac{3}{5}\cos 3t - \dfrac{5}{3}\sin 5t + c$

13. $\int (\sqrt{x} - 3)\,dx$ is equal to:

    (a) $\dfrac{3}{2}\sqrt{x^3} - 3x + c$ (b) $\dfrac{2}{3}\sqrt{x^3} + c$

    (c) $\dfrac{1}{2\sqrt{x}} + c$ (d) $\dfrac{2}{3}\sqrt{x^3} - 3x + c$

14. The r.m.s. value of $y = x^2$ between $x = 1$ and $x = 3$, correct to 2 decimal places, is:

    (a) 2.08   (b) 4.92   (c) 6.96   (d) 24.2

15. The value of $\displaystyle\int_0^{\pi/6} 2\sin\left(3t + \dfrac{\pi}{2}\right)dt$ is:

    (a) 6   (b) $-\dfrac{2}{3}$ (c) -6 (d) $\dfrac{2}{3}$

16. The volume of the solid of revolution when the curve $y = 2x$ is rotated one revolution about the x-axis between the limits $x = 0$ and $x = 4$ cm is:

(a) $85\frac{1}{3}\pi$ cm^3 (b) 8 cm^3

(c) $85\frac{1}{3}$ cm^3 (d) 64π cm^3

17. The area enclosed by the curve $y = 3\cos 2\theta$, the ordinates $\theta = 0$ and $\theta = \dfrac{\pi}{4}$ and the θ axis is:

(a) -3 (b) 6   (c) 1.5   (d) 3

18. $\displaystyle\int \left(1 + \frac{4}{e^{2x}}\right) dx$ is equal to:

(a) $\dfrac{8}{e^{2x}} + c$ (b) $x - \dfrac{2}{e^{2x}} + c$

(c) $x + \dfrac{4}{e^{2x}}$ (d) $x - \dfrac{8}{e^{2x}} + c$

19. Evaluating $\int_1^2 2e^{3t}\,dt$, correct to 4 significant figures, gives:

(a) 2300   (b) 255.6   (c) 766.7   (d) 282.3

20. A metal template is bounded by the curve $y = x^2$, the x-axis and ordinates $x = 0$ and $x = 2$. The x-co-ordinate of the centroid of the area is:

(a) 1.0   (b) 2.0   (c) 1.5   (d) 2.5

21. The area under a force/distance graph gives the work done. The shaded area shown between p and q in Figure M8.1 is:

(a) $c(\ln p - \ln q)$ (b) $-\dfrac{c}{2}\left(\dfrac{1}{q^2} - \dfrac{1}{p^2}\right)$

(c) $\dfrac{c}{2}(\ln q - \ln p)$ (d) $c\ln\dfrac{q}{p}$

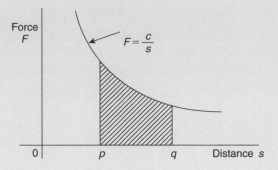

Figure M8.1

22. The value of $\int_0^{\pi/3} 16\cos^4\theta\sin\theta\,d\theta$ is:

(a) -0.1 (b) 3.1   (c) 0.1   (d) -3.1

23. $\int_0^{\pi/2} 2\sin^3 t\,dt$ is equal to:

(a) 1.33   (b) -0.25 (c) -1.33 (d) 0.25

24. Using the theorem of Pappus, the position of the centroid of a semicircle of radius r lies on the axis of symmetry at a distance from the diameter of:

(a) $\dfrac{3\pi}{4r}$ (b) $\dfrac{3r}{4\pi}$ (c) $\dfrac{4r}{3\pi}$ (d) $\dfrac{4\pi}{3r}$

25. An alternating current i has the following values at equal intervals of 2 ms:

Time t (ms)	0	2.0	4.0	6.0	8.0	10.0	12.0
Current I (A)	0	4.6	7.4	10.8	8.5	3.7	0

Charge q (in millicoulombs) is given by $q = \int_0^{12.0} i\,dt$. Using the trapezoidal rule, the approximate charge in the 12 ms period is:

(a) 70 mC   (b) 72.1 mC   (c) 35 mC   (d) 216.4 mC

Formulae/revision hints for Section J
Integral calculus

Standard integrals

y	$\int y\,dx$
ax^n	$a\dfrac{x^{n+1}}{n+1}+c$ (except where $n=-1$)
$\cos ax$	$\dfrac{1}{a}\sin ax+c$
$\sin ax$	$-\dfrac{1}{a}\cos ax+c$
$\sec^2 ax$	$\dfrac{1}{a}\tan ax+c$
$\operatorname{cosec}^2 ax$	$-\dfrac{1}{a}\cot ax+c$
$\operatorname{cosec} ax\cot ax$	$-\dfrac{1}{a}\operatorname{cosec} ax+c$
$\sec ax\tan ax$	$\dfrac{1}{a}\sec ax+c$
e^{ax}	$\dfrac{1}{a}e^{ax}+c$
$\dfrac{1}{x}$	$\ln x+c$
$\tan ax$	$\dfrac{1}{a}\ln(\sec ax)+c$
$\cos^2 x$	$\dfrac{1}{2}\left(x+\dfrac{\sin 2x}{2}\right)+c$
$\sin^2 x$	$\dfrac{1}{2}\left(x-\dfrac{\sin 2x}{2}\right)+c$
$\tan^2 x$	$\tan x-x+c$
$\cot^2 x$	$-\cot x-x+c$
$\dfrac{1}{\sqrt{(a^2-x^2)}}$	$\sin^{-1}\dfrac{x}{a}+c$

$$\sqrt{(a^2-x^2)}\qquad \frac{a^2}{2}\sin^{-1}\frac{x}{a}+\frac{x}{2}\sqrt{(a^2-x^2)}+c$$

$$\frac{1}{a^2+x^2}\qquad \frac{1}{a}\tan^{-1}\frac{x}{a}+c$$

$$\frac{1}{\sqrt{(x^2+a^2)}}\qquad \sinh^{-1}\frac{x}{a}+c \text{ or } \ln\left[\frac{x+\sqrt{(x^2+a^2)}}{a}\right]+c$$

$$\sqrt{(x^2+a^2)}\qquad \frac{a^2}{2}\sinh^{-1}\frac{x}{a}+\frac{x}{2}\sqrt{(x^2+a^2)}+c$$

$$\frac{1}{\sqrt{(x^2-a^2)}}\qquad \cosh^{-1}\frac{x}{a}+c \text{ or } \ln\left[\frac{x+\sqrt{(x^2-a^2)}}{a}\right]+c$$

$$\sqrt{(x^2-a^2)}\qquad \frac{x}{2}\sqrt{(x^2-a^2)}-\frac{a^2}{2}\cosh^{-1}\frac{x}{a}+c$$

$t=\tan\dfrac{\theta}{2}$ substitution

To determine $\displaystyle\int \frac{1}{a\cos\theta+b\sin\theta+c}\,d\theta$

let $\quad \sin\theta=\dfrac{2t}{(1+t^2)}\qquad \cos\theta=\dfrac{1-t^2}{1+t^2}$

and $\quad d\theta=\dfrac{2\,dt}{(1+t^2)}$

Integration by parts

If u and v are both functions of x then:

$$\int u\frac{dv}{dx}\,dx = uv - \int v\frac{du}{dx}\,dx$$

Reduction formulae

$$\int x^n e^x\,dx = I_n = x^n e^x - nI_{n-1}$$

$$\int x^n \cos x\,dx = I_n = x^n \sin x + nx^{n-1}\cos x - n(n-1)I_{n-2}$$

$$\int_0^\pi x^n \cos x\,dx = I_n = -n\pi^{n-1} - n(n-1)I_{n-2}$$

$$\int x^n \sin x \, dx = I_n = -x^n \cos x + n x^{n-1} \sin x$$
$$- n(n-1)I_{n-2}$$

$$\int \sin^n x \, dx = I_n = -\frac{1}{n} \sin^{n-1} x \cos x + \frac{n-1}{n} I_{n-2}$$

$$\int \cos^n x \, dx = I_n = \frac{1}{n} \cos^{n-1} \sin x + \frac{n-1}{n} I_{n-2}$$

$$\int_0^{\pi/2} \sin^n x \, dx = \int_0^{\pi/2} \cos^n x \, dx = I_n = \frac{n-1}{n} I_{n-2}$$

$$\int \tan^n x \, dx = I_n = \frac{\tan^{n-1} x}{n-1} - I_{n-2}$$

$$\int (\ln x)^n dx = I_n = x(\ln x)^n - n I_{n-1}$$

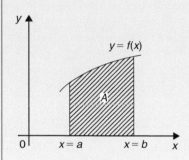

Figure FJ.1

With reference to Figure FJ.1

Area under a curve

$$\text{area } A = \int_a^b y \, dx$$

Mean value

$$\text{mean value} = \frac{1}{b-a} \int_a^b y \, dx$$

R.m.s. value

$$\text{r.m.s. value} = \sqrt{\left\{ \frac{1}{b-a} \int_a^b y^2 \, dx \right\}}$$

Volume of solid of revolution

$$\text{volume} = \int_a^b \pi y^2 \, dx \text{ about the } x\text{-axis}$$

Centroids

With reference to Figure FJ.2:

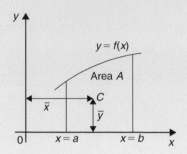

Figure FJ.2

$$\overline{x} = \frac{\int_a^b xy \, dx}{\int_a^b y \, dx} \quad \text{and} \quad \overline{y} = \frac{\frac{1}{2} \int_a^b y^2 \, dx}{\int_a^b y \, dx}$$

Theorem of Pappus

With reference to Figure FJ.2, when the curve is rotated one revolution about the x-axis between the limits $x = a$ and $x = b$, the volume V generated is given by:

$$v = 2\pi A \overline{y}$$

Parallel axis theorem

If C is the centroid of area A in Figure FJ.3 then

$$A k_{BB}^2 = A k_{GG}^2 + A d^2 \quad \text{or} \quad k_{BB}^2 = k_{GG}^2 + d^2$$

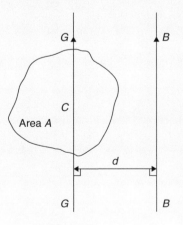

Figure FJ.3

Perpendicular axis theorem

If OX and OY lie in the plane of area A in Figure FJ.4 then

$$A\,k^2_{OZ} = A\,k^2_{OX} + A\,k^2_{OY} \quad \text{or} \quad k^2_{OZ} = k^2_{OX} + k^2_{OY}$$

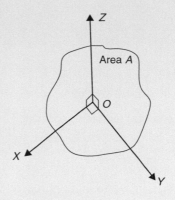

Figure FJ.4

Second moment of area and radius of gyration

Shape	Position of axis	Second moment of area, I	Radius of gyration, k
Rectangle length l breadth b	(1) Coinciding with b	$\dfrac{bl^3}{3}$	$\dfrac{l}{\sqrt{3}}$
	(2) Coinciding with l	$\dfrac{lb^3}{3}$	$\dfrac{b}{\sqrt{3}}$
	(3) Through centroid, parallel to b	$\dfrac{bl^3}{12}$	$\dfrac{l}{\sqrt{12}}$
	(4) Through centroid, parallel to l	$\dfrac{lb^3}{12}$	$\dfrac{b}{\sqrt{12}}$
Triangle Perpendicular height h base b	(1) Coinciding with b	$\dfrac{bh^3}{12}$	$\dfrac{h}{\sqrt{6}}$
	(2) Through centroid, parallel to base	$\dfrac{bh^3}{36}$	$\dfrac{h}{\sqrt{18}}$
	(3) Through vertex, parallel to base	$\dfrac{bh^3}{4}$	$\dfrac{h}{\sqrt{2}}$
Circle radius r	(1) Through centre, perpendicular to plane (i.e. polar axis)	$\dfrac{\pi r^4}{2}$	$\dfrac{r}{\sqrt{2}}$
	(2) Coinciding with diameter	$\dfrac{\pi r^4}{4}$	$\dfrac{r}{2}$
	(3) About a tangent	$\dfrac{5\pi r^4}{4}$	$\dfrac{\sqrt{5}}{2}r$
Semicircle radius r	Coinciding with diameter	$\dfrac{\pi r^4}{8}$	$\dfrac{r}{2}$

Numerical integration

Trapezoidal rule

$$\int y\,\mathrm{d}x \approx \left(\begin{array}{c}\text{width of}\\ \text{interval}\end{array}\right)\left[\frac{1}{2}\left(\begin{array}{c}\text{first}+\text{last}\\ \text{ordinates}\end{array}\right)\right.$$
$$\left.+\ \text{sum of remaining ordinates}\right]$$

Mid-ordinate rule

$$\int y\,\mathrm{d}x \approx \ (\text{width of interval})(\text{sum of mid-ordinates})$$

Simpson's rule

$$\int y\,\mathrm{d}x \approx \frac{1}{3}\left(\begin{array}{c}\text{width of}\\ \text{interval}\end{array}\right)\left[\left(\begin{array}{c}\text{first}+\text{last}\\ \text{ordinate}\end{array}\right)\right.$$
$$\left.+\ 4\left(\begin{array}{c}\text{sum of even}\\ \text{ordinates}\end{array}\right)+2\left(\begin{array}{c}\text{sum of remaining}\\ \text{odd ordinates}\end{array}\right)\right]$$

Section K

Differential equations

Chapter 77

Solution of first-order differential equations by separation of variables

Why it is important to understand: **Solution of first-order differential equations by separation of variables**

Differential equations play an important role in modelling virtually every physical, technical or biological process, from celestial motion, to bridge design, to interactions between neurons. Further applications are found in fluid dynamics with the design of containers and funnels, in heat conduction analysis with the design of heat spreaders in microelectronics, in rigid-body dynamic analysis, with falling objects, and in exponential growth of current in an R–L circuit, to name but a few. This chapter introduces first-order differential equations – the subject is clearly of great importance in many different areas of engineering.

At the end of this chapter, you should be able to:

- sketch a family of curves given a simple derivative
- define a differential equation – first order, second order, general solution, particular solution, boundary conditions
- solve a differential equation of the form $\dfrac{dy}{dx} = f(x)$
- solve a differential equation of the form $\dfrac{dy}{dx} = f(y)$
- solve a differential equation of the form $\dfrac{dy}{dx} = f(x).f(y)$

77.1 Family of curves

Integrating both sides of the derivative $\dfrac{dy}{dx} = 3$ with respect to x gives $y = \int 3\, dx$, i.e. $y = 3x + c$, where c is an arbitrary constant.

$y = 3x + c$ represents a **family of curves**, each of the curves in the family depending on the value of c. Examples include $y = 3x + 8$, $y = 3x + 3$, $y = 3x$ and $y = 3x - 10$, and these are shown in Figure 77.1. Each are straight lines of gradient 3. A particular curve of a family may be determined when a point on the curve

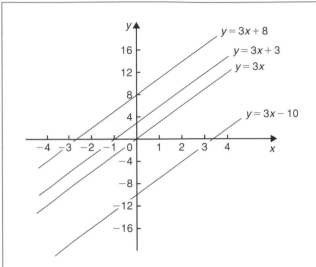

Figure 77.1

is specified. Thus, if $y = 3x + c$ passes through the point $(1, 2)$ then $2 = 3(1) + c$, from which, $c = -1$. The equation of the curve passing through $(1, 2)$ is therefore $y = 3x - 1$

> **Problem 1.** Sketch the family of curves given by the equation $\dfrac{dy}{dx} = 4x$ and determine the equation of one of these curves which passes through the point $(2, 3)$

Integrating both sides of $\dfrac{dy}{dx} = 4x$ with respect to x gives:

$$\int \frac{dy}{dx}\, dx = \int 4x\, dx, \quad \text{i.e. } y = 2x^2 + c$$

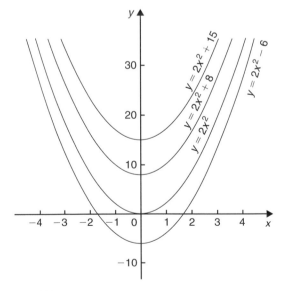

Figure 77.2

Some members of the family of curves having an equation $y = 2x^2 + c$ include $y = 2x^2 + 15$, $y = 2x^2 + 8$, $y = 2x^2$ and $y = 2x^2 - 6$, and these are shown in Figure 77.2. To determine the equation of the curve passing through the point $(2, 3)$, $x = 2$ and $y = 3$ are substituted into the equation $y = 2x^2 + c$

Thus $3 = 2(2)^2 + c$, from which $c = 3 - 8 = -5$

Hence, the equation of the curve passing through the point $(2, 3)$ is $y = 2x^2 - 5$

Now try the following Practice Exercise

> **Practice Exercise 299   Further problems on families of curves (answers on page 1185)**
>
> 1. Sketch a family of curves represented by each of the following differential equations:
>
>   (a) $\dfrac{dy}{dx} = 6$ (b) $\dfrac{dy}{dx} = 3x$ (c) $\dfrac{dy}{dx} = x + 2$
>
> 2. Sketch the family of curves given by the equation $\dfrac{dy}{dx} = 2x + 3$ and determine the equation of one of these curves which passes through the point $(1, 3)$

77.2   Differential equations

A **differential equation** is one that contains differential coefficients.
Examples include

$$\text{(i) } \frac{dy}{dx} = 7x \quad \text{and} \quad \text{(ii) } \frac{d^2y}{dx^2} + 5\frac{dy}{dx} + 2y = 0$$

Differential equations are classified according to the highest derivative which occurs in them. Thus, example (i) above is a **first-order differential equation**, and example (ii) is a **second-order differential equation**. The **degree** of a differential equation is that of the highest power of the highest differential which the equation contains after simplification.

Thus $\left(\dfrac{d^2x}{dt^2}\right)^3 + 2\left(\dfrac{dx}{dt}\right)^5 = 7$ is a second-order differential equation of degree three.
Starting with a differential equation it is possible, by integration and by being given sufficient data to determine unknown constants, to obtain the original function. This process is called **'solving the differential equation'**. A solution to a differential equation

which contains one or more arbitrary constants of integration is called the **general solution** of the differential equation.

When additional information is given so that constants may be calculated, the **particular solution** of the differential equation is obtained. The additional information is called **boundary conditions**. It was shown in Section 77.1 that $y = 3x + c$ is the general solution of the differential equation $\dfrac{dy}{dx} = 3$

Given the boundary conditions $x = 1$ and $y = 2$, this produces the particular solution of $y = 3x - 1$

Equations which can be written in the form

$$\frac{dy}{dx} = f(x), \frac{dy}{dx} = f(y) \text{ and } \frac{dy}{dx} = f(x) \cdot f(y)$$

can all be solved by integration. In each case it is possible to separate the y to one side of the equation and the x to the other. Solving such equations is therefore known as solution by **separation of variables**.

77.3   The solution of equations of the form $\dfrac{dy}{dx} = f(x)$

A differential equation of the form $\dfrac{dy}{dx} = f(x)$ is solved by direct integration,

i.e.
$$y = \int f(x) \, dx$$

Problem 2. Determine the general solution of $x\dfrac{dy}{dx} = 2 - 4x^3$

Rearranging $x\dfrac{dy}{dx} = 2 - 4x^3$ gives:

$$\frac{dy}{dx} = \frac{2 - 4x^3}{x} = \frac{2}{x} - \frac{4x^3}{x} = \frac{2}{x} - 4x^2$$

Integrating both sides gives:

$$y = \int \left(\frac{2}{x} - 4x^2 \right) dx$$

i.e. $\quad y = 2\ln x - \dfrac{4}{3}x^3 + c$

which is the general solution.

Problem 3. Find the particular solution of the differential equation $5\dfrac{dy}{dx} + 2x = 3$, given the boundary conditions $y = 1\dfrac{2}{5}$ when $x = 2$

Since $5\dfrac{dy}{dx} + 2x = 3$ then $\dfrac{dy}{dx} = \dfrac{3 - 2x}{5} = \dfrac{3}{5} - \dfrac{2x}{5}$

Hence $\quad y = \int \left(\dfrac{3}{5} - \dfrac{2x}{5} \right) dx$

i.e. $\quad y = \dfrac{3x}{5} - \dfrac{x^2}{5} + c$

which is the general solution.

Substituting the boundary conditions $y = 1\dfrac{2}{5}$ and $x = 2$ to evaluate c gives:

$1\dfrac{2}{5} = \dfrac{6}{5} - \dfrac{4}{5} + c$, from which, $c = 1$

Hence, the particular solution is $y = \dfrac{3x}{5} - \dfrac{x^2}{5} + 1$

Problem 4. Solve the equation $2t \left(t - \dfrac{d\theta}{dt} \right) = 5$, given $\theta = 2$ when $t = 1$

Rearranging gives:

$$t - \frac{d\theta}{dt} = \frac{5}{2t} \quad \text{and} \quad \frac{d\theta}{dt} = t - \frac{5}{2t}$$

Integrating gives:

$$\theta = \int \left(t - \frac{5}{2t} \right) dt$$

i.e. $\quad \theta = \dfrac{t^2}{2} - \dfrac{5}{2}\ln t + c$

which is the general solution.

When $\theta = 2$, $t = 1$, thus $2 = \dfrac{1}{2} - \dfrac{5}{2}\ln 1 + c$ from which, $c = \dfrac{3}{2}$

Hence the particular solution is:

$$\theta = \frac{t^2}{2} - \frac{5}{2}\ln t + \frac{3}{2}$$

i.e. $\quad \boldsymbol{\theta = \dfrac{1}{2}(t^2 - 5\ln t + 3)}$

Problem 5. The bending moment M of the beam is given by $\dfrac{dM}{dx} = -w(l - x)$, where w and x are constants. Determine M in terms of x given: $M = \dfrac{1}{2}wl^2$ when $x = 0$

$$\frac{dM}{dx} = -w(l - x) = -wl + wx$$

Integrating with respect to x gives:

$$M = -wlx + \frac{wx^2}{2} + c$$

which is the general solution.

Section K

When $M = \frac{1}{2}wl^2, x = 0$

Thus $\frac{1}{2}wl^2 = -wl(0) + \frac{w(0)^2}{2} + c$

from which, $c = \frac{1}{2}wl^2$

Hence the particular solution is:

$$M = -wlx + \frac{w(x)^2}{2} + \frac{1}{2}wl^2$$

i.e. $M = \frac{1}{2}w(l^2 - 2lx + x^2)$

or $M = \frac{1}{2}w(l-x)^2$

Now try the following Practice Exercise

Practice Exercise 300 Further problems on equations of the form $\frac{dy}{dx} = f(x)$ (answers on page 1185)

In Problems 1 to 5, solve the differential equations.

1. $\frac{dy}{dx} = \cos 4x - 2x$

2. $2x\frac{dy}{dx} = 3 - x^3$

3. $\frac{dy}{dx} + x = 3$, given $y = 2$ when $x = 1$

4. $3\frac{dy}{d\theta} + \sin\theta = 0$, given $y = \frac{2}{3}$ when $\theta = \frac{\pi}{3}$

5. $\frac{1}{e^x} + 2 = x - 3\frac{dy}{dx}$, given $y = 1$ when $x = 0$

6. The gradient of a curve is given by:

$$\frac{dy}{dx} + \frac{x^2}{2} = 3x$$

Find the equation of the curve if it passes through the point $\left(1, \frac{1}{3}\right)$

7. The acceleration, a, of a body is equal to its rate of change of velocity, $\frac{dv}{dt}$. Find an equation for v in terms of t, given that when $t = 0$, velocity $v = u$.

8. An object is thrown vertically upwards with an initial velocity, u, of $20\,\text{m/s}$. The motion of the object follows the differential equation $\frac{ds}{dt} = u - gt$, where s is the height of the object in metres at time t seconds and $g = 9.8\,\text{m/s}^2$.

Determine the height of the object after 3 seconds if $s = 0$ when $t = 0$

77.4 The solution of equations of the form $\frac{dy}{dx} = f(y)$

A differential equation of the form $\frac{dy}{dx} = f(y)$ is initially rearranged to give $dx = \frac{dy}{f(y)}$ and then the solution is obtained by direct integration,

i.e. $$\int dx = \int \frac{dy}{f(y)}$$

Problem 6. Find the general solution of $\frac{dy}{dx} = 3 + 2y$

Rearranging $\frac{dy}{dx} = 3 + 2y$ gives:

$$dx = \frac{dy}{3 + 2y}$$

Integrating both sides gives:

$$\int dx = \int \frac{dy}{3 + 2y}$$

Thus, by using the substitution $u = (3 + 2y)$ – see Chapter 64,

$$x = \tfrac{1}{2}\ln(3 + 2y) + c \qquad (1)$$

it is possible to give the general solution of a differential equation in a different form. For example, if $c = \ln k$, where k is a constant, then:

$$x = \tfrac{1}{2}\ln(3 + 2y) + \ln k,$$

i.e. $x = \ln(3 + 2y)^{\frac{1}{2}} + \ln k$

or $x = \ln[k\sqrt{(3 + 2y)}] \qquad (2)$

by the laws of logarithms, from which,

$$e^x = k\sqrt{(3 + 2y)} \qquad (3)$$

Equations (1), (2) and (3) are all acceptable general solutions of the differential equation

$$\frac{dy}{dx} = 3 + 2y$$

Problem 7.   Determine the particular solution of $(y^2-1)\dfrac{dy}{dx}=3y$ given that $y=1$ when $x=2\dfrac{1}{6}$

Rearranging gives:

$$dx=\left(\dfrac{y^2-1}{3y}\right)dy=\left(\dfrac{y}{3}-\dfrac{1}{3y}\right)dy$$

Integrating gives:

$$\int dx=\int\left(\dfrac{y}{3}-\dfrac{1}{3y}\right)dy$$

i.e. $x=\dfrac{y^2}{6}-\dfrac{1}{3}\ln y+c$

which is the general solution.

When $y=1$, $x=2\frac{1}{6}$, thus $2\frac{1}{6}=\frac{1}{6}-\frac{1}{3}\ln 1+c$, from which, $c=2$

Hence the particular solution is:

$$x=\dfrac{y^2}{6}-\dfrac{1}{3}\ln y+2$$

Problem 8.   (a) The variation of resistance, R ohms, of an aluminium conductor with temperature $\theta°C$ is given by $\dfrac{dR}{d\theta}=\alpha R$, where α is the temperature coefficient of resistance of aluminium. If $R=R_0$ when $\theta=0°C$, solve the equation for R. (b) If $\alpha=38\times10^{-4}/°C$, determine the resistance of an aluminium conductor at $50°C$, correct to 3 significant figures, when its resistance at $0°C$ is $24.0\ \Omega$

(a) $\dfrac{dR}{d\theta}=\alpha R$ is of the form $\dfrac{dy}{dx}=f(y)$

Rearranging gives: $d\theta=\dfrac{dR}{\alpha R}$

Integrating both sides gives:

$$\int d\theta=\int\dfrac{dR}{\alpha R}$$

i.e. $\theta=\dfrac{1}{\alpha}\ln R+c$

which is the general solution.
Substituting the boundary conditions $R=R_0$ when $\theta=0$ gives:

$$0=\dfrac{1}{\alpha}\ln R_0+c$$

from which $c=-\dfrac{1}{\alpha}\ln R_0$

Hence the particular solution is

$$\theta=\dfrac{1}{\alpha}\ln R-\dfrac{1}{\alpha}\ln R_0=\dfrac{1}{\alpha}(\ln R-\ln R_0)$$

i.e. $\theta=\dfrac{1}{\alpha}\ln\left(\dfrac{R}{R_0}\right)$ or $\alpha\theta=\ln\left(\dfrac{R}{R_0}\right)$

Hence $e^{\alpha\theta}=\dfrac{R}{R_0}$ from which, $\boldsymbol{R=R_0e^{\alpha\theta}}$

(b)   Substituting $\alpha=38\times10^{-4}$, $R_0=24.0$ and $\theta=50$ into $R=R_0e^{\alpha\theta}$ gives the resistance at $50°C$, i.e.
$R_{50}=24.0\,e^{(38\times10^{-4}\times50)}=\textbf{29.0 ohms}$

Now try the following Practice Exercise

Practice Exercise 301   Further problems on equations of the form $\dfrac{dy}{dx}=f(y)$ (answers on page 1185)

In Problems 1 to 3, solve the differential equations.

1. $\dfrac{dy}{dx}=2+3y$

2. $\dfrac{dy}{dx}=2\cos^2 y$

3. $(y^2+2)\dfrac{dy}{dx}=5y$, given $y=1$ when $x=\dfrac{1}{2}$

4.   The current in an electric circuit is given by the equation

$$Ri+L\dfrac{di}{dt}=0$$

where L and R are constants. Show that $i=Ie^{\frac{-Rt}{L}}$, given that $i=I$ when $t=0$

5.   The velocity of a chemical reaction is given by $\dfrac{dx}{dt}=k(a-x)$, where x is the amount transferred in time t, k is a constant and a is the concentration at time $t=0$ when $x=0$. Solve the equation and determine x in terms of t.

6.   (a)   Charge Q coulombs at time t seconds is given by the differential equation $R\dfrac{dQ}{dt}+\dfrac{Q}{C}=0$, where C is the capacitance in farads and R the resistance in ohms. Solve the equation for Q given that $Q=Q_0$ when $t=0$

(b) A circuit possesses a resistance of $250 \times 10^3 \, \Omega$ and a capacitance of $8.5 \times 10^{-6} \, \text{F}$, and after 0.32 seconds the charge falls to 8.0 C. Determine the initial charge and the charge after 1 second, each correct to 3 significant figures.

7. A differential equation relating the difference in tension T, pulley contact angle θ and coefficient of friction μ is $\dfrac{dT}{d\theta} = \mu T$. When $\theta = 0$, $T = 150 \, \text{N}$ and $\mu = 0.30$ as slipping starts. Determine the tension at the point of slipping when $\theta = 2$ radians. Determine also the value of θ when T is 300 N.

8. The rate of cooling of a body is given by $\dfrac{d\theta}{dt} = k\theta$, where k is a constant. If $\theta = 60^\circ \text{C}$ when $t = 2$ minutes and $\theta = 50^\circ \text{C}$ when $t = 5$ minutes, determine the time taken for θ to fall to 40°C, correct to the nearest second.

77.5 The solution of equations of the form $\dfrac{dy}{dx} = f(x) \cdot f(y)$

A differential equation of the form $\dfrac{dy}{dx} = f(x) \cdot f(y)$, where $f(x)$ is a function of x only and $f(y)$ is a function of y only, may be rearranged as $\dfrac{dy}{f(y)} = f(x) \, dx$, and then the solution is obtained by direct integration, i.e.

$$\int \frac{dy}{f(y)} = \int f(x) \, dx$$

Problem 9. Solve the equation $4xy \dfrac{dy}{dx} = y^2 - 1$

Separating the variables gives:

$$\left(\frac{4y}{y^2 - 1} \right) dy = \frac{1}{x} \, dx$$

Integrating both sides gives:

$$\int \left(\frac{4y}{y^2 - 1} \right) dy = \int \left(\frac{1}{x} \right) dx$$

Using the substitution $u = y^2 - 1$, the general solution is:

$$2\ln(y^2 - 1) = \ln x + c \qquad (1)$$

or $\qquad \ln(y^2 - 1)^2 - \ln x = c$

from which, $\quad \ln \left\{ \dfrac{(y^2 - 1)^2}{x} \right\} = c$

and $\qquad \dfrac{(y^2 - 1)^2}{x} = e^c \qquad (2)$

If in equation (1), $c = \ln A$, where A is a different constant,

then $\quad \ln(y^2 - 1)^2 = \ln x + \ln A$

i.e. $\quad \ln(y^2 - 1)^2 = \ln Ax$

i.e. $\quad (y^2 - 1)^2 = Ax \qquad (3)$

Equations (1) to (3) are thus three valid solutions of the differential equations

$$4xy \frac{dy}{dx} = y^2 - 1$$

Problem 10. Determine the particular solution of $\dfrac{d\theta}{dt} = 2e^{3t - 2\theta}$, given that $t = 0$ when $\theta = 0$

$$\frac{d\theta}{dt} = 2e^{3t - 2\theta} = 2(e^{3t})(e^{-2\theta})$$

by the laws of indices.

Separating the variables gives:

$$\frac{d\theta}{e^{-2\theta}} = 2e^{3t} \, dt$$

i.e. $\quad e^{2\theta} \, d\theta = 2e^{3t} \, dt$

Integrating both sides gives:

$$\int e^{2\theta} \, d\theta = \int 2e^{3t} \, dt$$

Thus the general solution is:

$$\frac{1}{2} e^{2\theta} = \frac{2}{3} e^{3t} + c$$

When $t = 0$, $\theta = 0$, thus:

$$\frac{1}{2} e^0 = \frac{2}{3} e^0 + c$$

from which, $c = \dfrac{1}{2} - \dfrac{2}{3} = -\dfrac{1}{6}$

Hence the particular solution is:

$$\frac{1}{2}e^{2\theta} = \frac{2}{3}e^{3t} - \frac{1}{6}$$

or $3e^{2\theta} = 4e^{3t} - 1$

Problem 11. Find the curve which satisfies the equation $xy = (1+x^2)\dfrac{dy}{dx}$ and passes through the point $(0, 1)$

Separating the variables gives:

$$\frac{x}{(1+x^2)}dx = \frac{dy}{y}$$

Integrating both sides gives:

$$\tfrac{1}{2}\ln(1+x^2) = \ln y + c$$

When $x=0$, $y=1$ thus $\frac{1}{2}\ln 1 = \ln 1 + c$, from which, $c=0$

Hence the particular solution is $\frac{1}{2}\ln(1+x^2) = \ln y$

i.e. $\ln(1+x^2)^{\frac{1}{2}} = \ln y$, from which, $(1+x^2)^{\frac{1}{2}} = y$

Hence the equation of the curve is $y = \sqrt{(1+x^2)}$

Problem 12. The current i in an electric circuit containing resistance R and inductance L in series with a constant voltage source E is given by the differential equation $E - L\left(\dfrac{di}{dt}\right) = Ri$. Solve the equation and find i in terms of time t given that when $t=0$, $i=0$

In the $R-L$ series circuit shown in Figure 77.3, the supply p.d., E, is given by

$$E = V_R + V_L$$

$$V_R = iR \text{ and } V_L = L\frac{di}{dt}$$

Hence $E = iR + L\dfrac{di}{dt}$

from which $E - L\dfrac{di}{dt} = Ri$

Most electrical circuits can be reduced to a differential equation.

Rearranging $E - L\dfrac{di}{dt} = Ri$ gives $\dfrac{di}{dt} = \dfrac{E - Ri}{L}$

and separating the variables gives:

$$\frac{di}{E - Ri} = \frac{dt}{L}$$

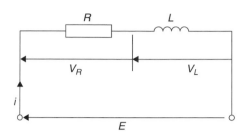

Figure 77.3

Integrating both sides gives:

$$\int \frac{di}{E - Ri} = \int \frac{dt}{L}$$

Hence the general solution is:

$$-\frac{1}{R}\ln(E - Ri) = \frac{t}{L} + c$$

(by making a substitution $u = E - Ri$, see Chapter 64).

When $t=0$, $i=0$, thus $-\dfrac{1}{R}\ln E = c$

Thus the particular solution is:

$$-\frac{1}{R}\ln(E - Ri) = \frac{t}{L} - \frac{1}{R}\ln E$$

Transposing gives:

$$-\frac{1}{R}\ln(E - Ri) + \frac{1}{R}\ln E = \frac{t}{L}$$

$$\frac{1}{R}[\ln E - \ln(E - Ri)] = \frac{t}{L}$$

$$\ln\left(\frac{E}{E - Ri}\right) = \frac{Rt}{L}$$

from which $\dfrac{E}{E - Ri} = e^{\frac{Rt}{L}}$

Hence $\dfrac{E - Ri}{E} = e^{\frac{-Rt}{L}}$ and $E - Ri = Ee^{\frac{-Rt}{L}}$ and

$Ri = E - Ee^{\frac{-Rt}{L}}$

Hence current,

$$i = \frac{E}{R}\left(1 - e^{\frac{-Rt}{L}}\right)$$

which represents the law of growth of current in an inductive circuit as shown in Fig. 77.4.

Section K

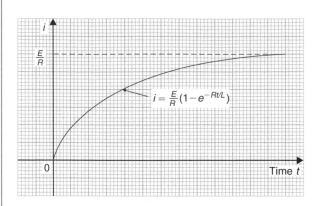

Figure 77.4

Problem 13. For an adiabatic expansion of a gas

$$C_v \frac{dp}{p} + C_p \frac{dV}{V} = 0$$

where C_p and C_v are constants. Given $n = \dfrac{C_p}{C_v}$ show that $pV^n = \text{constant}$

Separating the variables gives:

$$C_v \frac{dp}{p} = -C_p \frac{dV}{V}$$

Integrating both sides gives:

$$C_v \int \frac{dp}{p} = -C_p \int \frac{dV}{V}$$

i.e. $C_v \ln p = -C_p \ln V + k$

Dividing throughout by constant C_v gives:

$$\ln p = -\frac{C_p}{C_v} \ln V + \frac{k}{C_v}$$

Since $\dfrac{C_p}{C_v} = n$, then $\ln p + n \ln V = K$

where $K = \dfrac{k}{C_v}$

i.e. $\ln p + \ln V^n = K$ or $\ln\ pV^n = K$, by the laws of logarithms.

Hence $pV^n = e^K$, i.e. $\boldsymbol{pV^n = \textbf{constant}}$.

Now try the following Practice Exercise

Practice Exercise 302   Further problems on equations of the form $\dfrac{dy}{dx} = f(x) \cdot f(y)$ (answers on page 1185)

In Problems 1 to 4, solve the differential equations.

1. $\dfrac{dy}{dx} = 2y \cos x$

2. $(2y-1)\dfrac{dy}{dx} = (3x^2+1)$, given $x=1$ when $y=2$

3. $\dfrac{dy}{dx} = e^{2x-y}$, given $x=0$ when $y=0$

4. $2y(1-x) + x(1+y)\dfrac{dy}{dx} = 0$, given $x=1$ when $y=1$

5. Show that the solution of the equation $\dfrac{y^2+1}{x^2+1} = \dfrac{y}{x}\dfrac{dy}{dx}$ is of the form

$$\sqrt{\left(\frac{y^2+1}{x^2+1}\right)} = \text{constant}.$$

6. Solve $xy = (1-x^2)\dfrac{dy}{dx}$ for y, given $x=0$ when $y=1$

7. Determine the equation of the curve which satisfies the equation $xy\dfrac{dy}{dx} = x^2 - 1$, and which passes through the point $(1, 2)$

8. The p.d., V, between the plates of a capacitor C charged by a steady voltage E through a resistor R is given by the equation $CR\dfrac{dV}{dt} + V = E$

   (a) Solve the equation for V given that at $t=0$, $V=0$

   (b) Calculate V, correct to 3 significant figures, when $E=25$V, $C=20 \times 10^{-6}$F, $R=200 \times 10^3 \, \Omega$ and $t=3.0$s

9. Determine the value of p, given that $x^3\dfrac{dy}{dx} = p - x$, and that $y=0$ when $x=2$ and when $x=6$

For fully worked solutions to each of the problems in Practice Exercises 299 to 302 in this chapter, go to the website:
www.routledge.com/cw/bird

Chapter 78

Homogeneous first-order differential equations

Why it is important to understand: **Homogeneous first-order differential equations**

As was previously stated, differential equations play a prominent role in engineering, physics, economics, and other disciplines. Applications are many and varied; for example, they are involved in combined heat conduction and convection with the design of heating and cooling chambers, in fluid mechanics analysis, in heat transfer analysis, in kinematic analysis of rigid body dynamics, with exponential decay in radioactive material, Newton's law of cooling and in mechanical oscillations. This chapter explains how to solve a particular type of differential equation – the homogeneous first-order type.

At the end of this chapter, you should be able to:

- recognise a homogeneous differential equation
- solve a differential equation of the form $P\dfrac{\mathrm{d}y}{\mathrm{d}x} = Q$

78.1 Introduction

Certain first-order differential equations are not of the 'variable-separable' type, but can be made separable by changing the variable.

An equation of the form $P\dfrac{\mathrm{d}y}{\mathrm{d}x} = Q$, where P and Q are functions of both x and y of the same degree throughout, is said to be **homogeneous** in y and x. For example, $f(x, y) = x^2 + 3xy + y^2$ is a homogeneous function since each of the three terms are of degree 2. However, $f(x, y) = \dfrac{x^2 - y}{2x^2 + y^2}$ is not homogeneous since the term in y in the numerator is of degree 1 and the other three terms are of degree 2

78.2 Procedure to solve differential equations of the form $P\dfrac{\mathrm{d}y}{\mathrm{d}x} = Q$

(i) Rearrange $P\dfrac{\mathrm{d}y}{\mathrm{d}x} = Q$ into the form $\dfrac{\mathrm{d}y}{\mathrm{d}x} = \dfrac{Q}{P}$

(ii) Make the substitution $y = vx$ (where v is a function of x), from which, $\dfrac{\mathrm{d}y}{\mathrm{d}x} = v(1) + x\dfrac{\mathrm{d}v}{\mathrm{d}x}$, by the product rule.

(iii) Substitute for both y and $\dfrac{\mathrm{d}y}{\mathrm{d}x}$ in the equation $\dfrac{\mathrm{d}y}{\mathrm{d}x} = \dfrac{Q}{P}$. Simplify, by cancelling, and an equation results in which the variables are separable.

(iv) Separate the variables and solve using the method shown in Chapter 77.

(v) Substitute $v = \dfrac{y}{x}$ to solve in terms of the original variables.

78.3 Worked problems on homogeneous first-order differential equations

Problem 1. Solve the differential equation: $y - x = x\dfrac{dy}{dx}$, given $x = 1$ when $y = 2$.

Using the above procedure:

(i) Rearranging $y - x = x\dfrac{dy}{dx}$ gives:

$$\frac{dy}{dx} = \frac{y - x}{x}$$

which is homogeneous in x and y

(ii) Let $y = vx$, then $\dfrac{dy}{dx} = v + x\dfrac{dv}{dx}$

(iii) Substituting for y and $\dfrac{dy}{dx}$ gives:

$$v + x\frac{dv}{dx} = \frac{vx - x}{x} = \frac{x(v-1)}{x} = v - 1$$

(iv) Separating the variables gives:

$$x\frac{dv}{dx} = v - 1 - v = -1, \text{ i.e. } dv = -\frac{1}{x}dx$$

Integrating both sides gives:

$$\int dv = \int -\frac{1}{x}dx$$

Hence, $v = -\ln x + c$

(v) Replacing v by $\dfrac{y}{x}$ gives: $\dfrac{y}{x} = -\ln x + c$, which is the general solution.

When $x = 1, y = 2$, thus: $\dfrac{2}{1} = -\ln 1 + c$ from which, $c = 2$

Thus, the particular solution is: $\dfrac{y}{x} = -\ln x + 2$

or $y = -x(\ln x - 2)$ or $y = x(2 - \ln x)$

Problem 2. Find the particular solution of the equation: $x\dfrac{dy}{dx} = \dfrac{x^2 + y^2}{y}$, given the boundary conditions that $y = 4$ when $x = 1$

Using the procedure of Section 78.2:

(i) Rearranging $x\dfrac{dy}{dx} = \dfrac{x^2 + y^2}{y}$ gives:

$$\frac{dy}{dx} = \frac{x^2 + y^2}{xy}$$ which is homogeneous in x and y since each of the three terms on the right-hand side are of the same degree (i.e. degree 2).

(ii) Let $y = vx$ then $\dfrac{dy}{dx} = v + x\dfrac{dv}{dx}$

(iii) Substituting for y and $\dfrac{dy}{dx}$ in the equation $\dfrac{dy}{dx} = \dfrac{x^2 + y^2}{xy}$ gives:

$$v + x\frac{dv}{dx} = \frac{x^2 + v^2x^2}{x(vx)} = \frac{x^2 + v^2x^2}{vx^2} = \frac{1 + v^2}{v}$$

(iv) Separating the variables gives:

$$x\frac{dv}{dx} = \frac{1 + v^2}{v} - v = \frac{1 + v^2 - v^2}{v} = \frac{1}{v}$$

Hence, $v\,dv = \dfrac{1}{x}dx$

Integrating both sides gives:

$$\int v\,dv = \int \frac{1}{x}dx \text{ i.e. } \frac{v^2}{2} = \ln x + c$$

(v) Replacing v by $\dfrac{y}{x}$ gives: $\dfrac{y^2}{2x^2} = \ln x + c$, which is the general solution.

When $x = 1, y = 4$, thus: $\dfrac{16}{2} = \ln 1 + c$ from which, $c = 8$

Hence, the particular solution is: $\dfrac{y^2}{2x^2} = \ln x + 8$

or $y^2 = 2x^2(8 + \ln x)$

Now try the following Practice Exercise

78.4   Further worked problems on homogeneous first-order differential equations

Problem 3.   Solve the equation:
$7x(x-y)dy = 2(x^2 + 6xy - 5y^2)dx$
given that $x = 1$ when $y = 0$

Using the procedure of Section 78.2:

(i) Rearranging   gives: $\dfrac{dy}{dx} = \dfrac{2x^2 + 12xy - 10y^2}{7x^2 - 7xy}$
which is homogeneous in x and y since each of the terms on the right-hand side is of degree 2

(ii) Let $y = vx$ then $\dfrac{dy}{dx} = v + x\dfrac{dv}{dx}$

(iii) Substituting for y and $\dfrac{dy}{dx}$ gives:

$$v + x\frac{dv}{dx} = \frac{2x^2 + 12x(vx) - 10(vx)^2}{7x^2 - 7x(vx)}$$

$$= \frac{2 + 12v - 10v^2}{7 - 7v}$$

(iv) Separating the variables gives:

$$x\frac{dv}{dx} = \frac{2 + 12v - 10v^2}{7 - 7v} - v$$

$$= \frac{(2 + 12v - 10v^2) - v(7 - 7v)}{7 - 7v}$$

$$= \frac{2 + 5v - 3v^2}{7 - 7v}$$

Hence, $\dfrac{7 - 7v}{2 + 5v - 3v^2}\,dv = \dfrac{dx}{x}$

Integrating both sides gives:

$$\int \left(\frac{7 - 7v}{2 + 5v - 3v^2}\right)dv = \int \frac{1}{x}\,dx$$

Resolving $\dfrac{7 - 7v}{2 + 5v - 3v^2}$ into partial fractions gives: $\dfrac{4}{(1 + 3v)} - \dfrac{1}{(2 - v)}$ (see Chapter 19)

Hence, $\displaystyle\int \left(\frac{4}{(1 + 3v)} - \frac{1}{(2 - v)}\right)dv = \int \frac{1}{x}dx$

i.e. $\dfrac{4}{3}\ln(1 + 3v) + \ln(2 - v) = \ln x + c$

(v) Replacing v by $\dfrac{y}{x}$ gives:

$$\frac{4}{3}\ln\left(1 + \frac{3y}{x}\right) + \ln\left(2 - \frac{y}{x}\right) = \ln + c$$

or $\dfrac{4}{3}\ln\left(\dfrac{x + 3y}{x}\right) + \ln\left(\dfrac{2x - y}{x}\right) = \ln + c$

When $x = 1, y = 0$, thus: $\dfrac{4}{3}\ln 1 + \ln 2 = \ln 1 + c$
from which, $c = \ln 2$

Hence, the particular solution is:

$$\frac{4}{3}\ln\left(\frac{x + 3y}{x}\right) + \ln\left(\frac{2x - y}{x}\right) = \ln + \ln 2$$

i.e. $\ln\left(\dfrac{x + 3y}{x}\right)^{\frac{4}{3}}\left(\dfrac{2x - y}{x}\right) = \ln(2x)$

from the laws of logarithms

i.e. $\left(\dfrac{x + 3y}{x}\right)^{\frac{4}{3}}\left(\dfrac{2x - y}{x}\right) = 2x$

Problem 4. Show that the solution of the differential equation: $x^2 - 3y^2 + 2xy\dfrac{dy}{dx} = 0$ is: $y = x\sqrt{(8x+1)}$, given that $y = 3$ when $x = 1$

Using the procedure of Section 78.2:

(i) Rearranging gives:

$$2xy\frac{dy}{dx} = 3y^2 - x^2 \quad \text{and} \quad \frac{dy}{dx} = \frac{3y^2 - x^2}{2xy}$$

(ii) Let $y = vx$ then $\dfrac{dy}{dx} = v + x\dfrac{dv}{dx}$

(iii) Substituting for y and $\dfrac{dy}{dx}$ gives:

$$v + x\frac{dv}{dx} = \frac{3(vx)^2 - x^2}{2x(vx)} = \frac{3v^2 - 1}{2v}$$

(iv) Separating the variables gives:

$$x\frac{dv}{dx} = \frac{3v^2 - 1}{2v} - v = \frac{3v^2 - 1 - 2v^2}{2v} = \frac{v^2 - 1}{2v}$$

Hence, $\dfrac{2v}{v^2 - 1}\,dv = \dfrac{1}{x}\,dx$

Integrating both sides gives:

$$\int \frac{2v}{v^2 - 1}\,dv = \int \frac{1}{x}\,dx$$

i.e. $\ln(v^2 - 1) = \ln x + c$

(v) Replacing v by $\dfrac{y}{x}$ gives:

$$\ln\left(\frac{y^2}{x^2} - 1\right) = \ln x + c$$

which is the general solution.

When $y = 3, x = 1$, thus: $\ln\left(\dfrac{9}{1} - 1\right) = \ln 1 + c$

from which, $c = \ln 8$

Hence, the particular solution is:

$$\ln\left(\frac{y^2}{x^2} - 1\right) = \ln x + \ln 8 = \ln 8x$$

by the laws of logarithms.

Hence, $\left(\dfrac{y^2}{x^2} - 1\right) = 8x$ i.e. $\dfrac{y^2}{x^2} = 8x + 1$ and $y^2 = x^2(8x + 1)$

i.e. $$\boldsymbol{y = x\sqrt{(8x+1)}}$$

Now try the following Practice Exercise

Practice Exercise 304 Further problems on homogeneous first-order differential equations (answers on page 1185)

1. Solve the differential equation:

$$xy^3\,dy = (x^4 + y^4)dx$$

2. Solve: $(9xy - 11xy)\dfrac{dy}{dx} = 11y^2 - 16xy + 3x^2$

3. Solve the differential equation:

$2x\dfrac{dy}{dx} = x + 3y$, given that when $x = 1$, $y = 1$

4. Show that the solution of the differential equation: $2xy\dfrac{dy}{dx} = x^2 + y^2$ can be expressed as: $x = K(x^2 - y^2)$, where K is a constant.

5. Determine the particular solution of $\dfrac{dy}{dx} = \dfrac{x^3 + y^3}{xy^2}$, given that $x = 1$ when $y = 4$

6. Show that the solution of the differential equation: $\dfrac{dy}{dx} = \dfrac{y^3 - xy^2 - x^2y - 5x^3}{xy^2 - x^2y - 2x^3}$ is of the form:

$$\frac{y^2}{2x^2} + \frac{4y}{x} + 18\ln\left(\frac{y - 5x}{x}\right) = \ln x + 42$$

when $x = 1$ and $y = 6$

For fully worked solutions to each of the problems in Practice Exercises 303 and 304 in this chapter, go to the website:
www.routledge.com/cw/bird

Chapter 79

Linear first-order differential equations

Why it is important to understand: Linear first-order differential equations

As has been stated in previous chapters, differential equations have many applications in engineering and science. For example, first-order differential equations model phenomena of cooling, population growth, radioactive decay, mixture of salt solutions, series circuits, survivability with AIDS, draining a tank, economics and finance, drug distribution, pursuit problem and harvesting of renewable natural resources. This chapter explains how to solve another specific type of equation, the linear first-order differential equation.

At the end of this chapter, you should be able to:

- recognise a linear differential equation
- solve a differential equation of the form $\dfrac{dy}{dx} + Py = Q$ where P and Q are functions of x only

79.1 Introduction

An equation of the form $\dfrac{dy}{dx} + Py = Q$, where P and Q are functions of x only is called a **linear differential equation** since y and its derivatives are of the first degree.

(i) The solution of $\dfrac{dy}{dx} + Py = Q$ is obtained by multiplying throughout by what is termed an **integrating factor**.

(ii) Multiplying $\dfrac{dy}{dx} + Py = Q$ by, say, R, a function of x only, gives:

$$R\frac{dy}{dx} + RPy = RQ \qquad (1)$$

(iii) The differential coefficient of a product Ry is obtained using the product rule,

i.e. $\dfrac{d}{dx}(Ry) = R\dfrac{dy}{dx} + y\dfrac{dR}{dx}$

which is the same as the left-hand side of equation (1), when R is chosen such that

$$RP = \frac{dR}{dx}$$

(iv) If $\dfrac{dR}{dx} = RP$, then separating the variables gives $\dfrac{dR}{R} = P\,dx$

Integrating both sides gives:

$$\int \frac{dR}{R} = \int P\,dx \quad \text{i.e. } \ln R = \int P\,dx + c$$

from which,

$$R = e^{\int P\,dx + c} = e^{\int P\,dx}\,e^{c}$$

i.e. $R = Ae^{\int P\,dx}$, where $A = e^c = $ a constant.

(v) Substituting $R = Ae^{\int P\,dx}$ in equation (1) gives:

$$Ae^{\int P\,dx}\left(\frac{dy}{dx}\right) + Ae^{\int P\,dx}Py = Ae^{\int P\,dx}Q$$

i.e. $e^{\int P\,dx}\left(\frac{dy}{dx}\right) + e^{\int P\,dx}Py = e^{\int P\,dx}Q$ (2)

(vi) The left-hand side of equation (2) is

$$\frac{d}{dx}\left(ye^{\int P\,dx}\right)$$

which may be checked by differentiating $ye^{\int P\,dx}$ with respect to x, using the product rule.

(vii) From equation (2),

$$\frac{d}{dx}\left(ye^{\int P\,dx}\right) = e^{\int P\,dx}Q$$

Integrating both sides gives:

$$ye^{\int P\,dx} = \int e^{\int P\,dx}Q\,dx \qquad (3)$$

(viii) $c^{\int P\,dx}$ is the **integrating factor**.

79.2 Procedure to solve differential equations of the form $\dfrac{dy}{dx} + Py = Q$

(i) Rearrange the differential equation into the form $\frac{dy}{dx} + Py = Q$, where P and Q are functions of x.

(ii) Determine $\int P\,dx$

(iii) Determine the integrating factor $e^{\int P\,dx}$

(iv) Substitute $e^{\int P\,dx}$ into equation (3).

(v) Integrate the right-hand side of equation (3) to give the general solution of the differential equation. Given boundary conditions, the particular solution may be determined.

79.3 Worked problems on linear first-order differential equations

Problem 1. Solve $\dfrac{1}{x}\dfrac{dy}{dx} + 4y = 2$ given the boundary conditions $x = 0$ when $y = 4$

Using the above procedure:

(i) Rearranging gives $\frac{dy}{dx} + 4xy = 2x$, which is of the form $\frac{dy}{dx} + Py = Q$, where $P = 4x$ and $Q = 2x$

(ii) $\int P\,dx = \int 4x\,dx = 2x^2$

(iii) Integrating factor $e^{\int P\,dx} = e^{2x^2}$

(iv) Substituting into equation (3) gives:

$$ye^{2x^2} = \int e^{2x^2}(2x)\,dx$$

(v) Hence the general solution is:

$$ye^{2x^2} = \tfrac{1}{2}e^{2x^2} + c$$

by using the substitution $u = 2x^2$. When $x = 0$, $y = 4$, thus $4e^0 = \tfrac{1}{2}e^0 + c$, from which, $c = \tfrac{7}{2}$

Hence the particular solution is

$$ye^{2x^2} = \tfrac{1}{2}e^{2x^2} + \tfrac{7}{2}$$

or $y = \tfrac{1}{2} + \tfrac{7}{2}e^{-2x^2}$ or $y = \tfrac{1}{2}\left(1 + 7e^{-2x^2}\right)$

Problem 2. Show that the solution of the equation $\dfrac{dy}{dx} + 1 = -\dfrac{y}{x}$ is given by $y = \dfrac{3-x^2}{2x}$, given $x = 1$ when $y = 1$

Using the procedure of Section 79.2:

(i) Rearranging gives: $\frac{dy}{dx} + \left(\frac{1}{x}\right)y = -1$, which is of the form $\frac{dy}{dx} + Py = Q$, where $P = \frac{1}{x}$ and $Q = -1$. (Note that Q can be considered to be $-1x^0$, i.e. a function of x)

(ii) $\int P\,dx = \int \frac{1}{x}\,dx = \ln x$

(iii) Integrating factor $e^{\int P\,dx} = e^{\ln x} = x$ (from the definition of logarithm).

(iv) Substituting into equation (3) gives:

$$yx = \int x(-1)\,dx$$

(v) Hence the general solution is:

$$yx = \frac{-x^2}{2} + c$$

When $x=1$, $y=1$, thus $1=\dfrac{-1}{2}+c$, from

which, $c=\dfrac{3}{2}$

Hence the particular solution is:

$$yx = \dfrac{-x^2}{2}+\dfrac{3}{2}$$

i.e. $2yx = 3 - x^2$ and $y = \dfrac{3-x^2}{2x}$

Problem 3. Determine the particular solution of $\dfrac{dy}{dx} - x + y = 0$, given that $x=0$ when $y=2$

Using the procedure of Section 79.2:

(i) Rearranging gives $\dfrac{dy}{dx}+y=x$, which is of the form $\dfrac{dy}{dx}+Py=Q$, where $P=1$ and $Q=x$. (In this case P can be considered to be $1x^0$, i.e. a function of x)

(ii) $\int P\,dx = \int 1\,dx = x$

(iii) Integrating factor $e^{\int P\,dx} = e^x$

(iv) Substituting in equation (3) gives:

$$ye^x = \int e^x(x)\,dx \qquad (4)$$

(v) $\int e^x(x)\,dx$ is determined using integration by parts (see Chapter 68).

$$\int xe^x\,dx = xe^x - e^x + c$$

Hence from equation (4): $ye^x = xe^x - e^x + c$, which is the general solution.

When $x=0$, $y=2$ thus $2e^0 = 0 - e^0 + c$, from which, $c=3$

Hence the particular solution is:

$$ye^x = xe^x - e^x + 3 \text{ or } y = x - 1 + 3e^{-x}$$

Now try the following Practice Exercise

Practice Exercise 305   Further problems on linear first-order differential equations (answers on page 1185)

Solve the following differential equations.

1. $x\dfrac{dy}{dx}=3-y$

2. $\dfrac{dy}{dx}=x(1-2y)$

3. $t\dfrac{dy}{dt}-5t=-y$

4. $x\left(\dfrac{dy}{dx}+1\right)=x^3-2y$, given $x=1$ when $y=3$

5. $\dfrac{1}{x}\dfrac{dy}{dx}+y=1$

6. $\dfrac{dy}{dx}+x=2y$

79.4   Further worked problems on linear first-order differential equations

---

Problem 4. Solve the differential equation $\dfrac{dy}{d\theta} = \sec\theta + y\tan\theta$ given the boundary conditions $y=1$ when $\theta=0$

Using the procedure of Section 79.2:

(i) Rearranging gives $\dfrac{dy}{d\theta} - (\tan\theta)y = \sec\theta$, which is of the form $\dfrac{dy}{d\theta}+Py=Q$, where $P=-\tan\theta$ and $Q-\sec\theta$

(ii) $\int P\,dx = \int -\tan\theta\,d\theta = -\ln(\sec\theta)$
$$= \ln(\sec\theta)^{-1} = \ln(\cos\theta)$$

(iii) Integrating factor $e^{\int P\,d\theta} = e^{\ln(\cos\theta)} = \cos\theta$ (from the definition of a logarithm).

(iv) Substituting in equation (3) gives:

$$y\cos\theta = \int \cos\theta(\sec\theta)\,d\theta$$

i.e. $y\cos\theta = \int d\theta$

(v) Integrating gives: $y\cos\theta = \theta + c$, which is the general solution. When $\theta=0$, $y=1$, thus $1\cos 0 = 0 + c$, from which, $c=1$
Hence the particular solution is:

$$y\cos\theta = \theta + 1 \text{ or } y = (\theta+1)\sec\theta$$

Problem 5.

(a)   Find the general solution of the equation

$$(x-2)\frac{dy}{dx}+\frac{3(x-1)}{(x+1)}y=1$$

(b)   Given the boundary conditions that $y=5$ when $x=-1$, find the particular solution of the equation given in (a)

(a)   Using the procedure of Section 79.2:

(i)   Rearranging gives:

$$\frac{dy}{dx}+\frac{3(x-1)}{(x+1)(x-2)}y=\frac{1}{(x-2)}$$

which is of the form

$$\frac{dy}{dx}+Py=Q, \text{ where } P=\frac{3(x-1)}{(x+1)(x-2)}$$

and $Q=\dfrac{1}{(x-2)}$

(ii) $\displaystyle\int P\,dx=\int\frac{3(x-1)}{(x+1)(x-2)}\,dx,$ which   is
integrated using partial fractions.

Let $\dfrac{3x-3}{(x+1)(x-2)}$

$$\equiv\frac{A}{(x+1)}+\frac{B}{(x-2)}$$

$$\equiv\frac{A(x-2)+B(x+1)}{(x+1)(x-2)}$$

from which, $3x-3=A(x-2)+B(x+1)$
When $x=-1$,

$$-6=-3A, \text{ from which, } A=2$$

When $x=2$,

$$3=3B, \text{ from which, } B=1$$

Hence $\displaystyle\int\frac{3x-3}{(x+1)(x-2)}\,dx$

$$=\int\left[\frac{2}{x+1}+\frac{1}{x-2}\right]dx$$

$$=2\ln(x+1)+\ln(x-2)$$

$$=\ln[(x+1)^2(x-2)]$$

(iii)   Integrating factor

$$e^{\int P\,dx}=e^{\ln[(x+1)^2(x-2)]}=(x+1)^2(x-2)$$

(iv)   Substituting in equation (3) gives:

$$y(x+1)^2(x-2)$$

$$=\int(x+1)^2(x-2)\frac{1}{x-2}\,dx$$

$$=\int(x+1)^2\,dx$$

(v)   **Hence the general solution is**:

$$y(x+1)^2(x-2)=\tfrac{1}{3}(x+1)^3+c$$

(b)   When $x=-1$, $y=5$, thus $5(0)(-3)=0+c$, from
which, $c=0$

Hence $y(x+1)^2(x-2)=\tfrac{1}{3}(x+1)^3$

i.e. $y=\dfrac{(x+1)^3}{3(x+1)^2(x-2)}$

and hence **the particular solution is**

$$y=\frac{(x+1)}{3(x-2)}$$

Now try the following Practice Exercise

**Practice Exercise 306   Further problems
on linear first-order differential equations
(answers on page 1185)**

In Problems 1 and 2, solve the differential equations.

1. $\cot x\dfrac{dy}{dx}=1-2y$, given $y=1$ when $x=\dfrac{\pi}{4}$

2. $t\dfrac{d\theta}{dt}+\sec t\,(t\sin t+\cos t)\theta=\sec t,$ given
$t=\pi$ when $\theta=1$

3.   Given   the   equation $x\dfrac{dy}{dx}=\dfrac{2}{x+2}-y$ show
that the particular solution is $y=\dfrac{2}{x}\ln(x+2)$,
given the boundary conditions that $x=-1$
when $y=0$

4. Show that the solution of the differential equation

$$\frac{dy}{dx} - 2(x+1)^3 = \frac{4}{(x+1)}y$$

is $y = (x+1)^4 \ln(x+1)^2$, given that $x = 0$ when $y = 0$

5. Show that the solution of the differential equation

$$\frac{dy}{dx} + ky = a \sin bx$$

is given by:

$$y = \left(\frac{a}{k^2 + b^2}\right)(k \sin bx - b \cos bx)$$

$$+ \left(\frac{k^2 + b^2 + ab}{k^2 + b^2}\right) e^{-kx},$$

given $y = 1$ when $x = 0$

6. The equation $\dfrac{dv}{dt} = -(av + bt)$, where a and b are constants, represents an equation of motion when a particle moves in a resisting medium. Solve the equation for v given that $v = u$ when $t = 0$

7. In an alternating current circuit containing resistance R and inductance L the current i is given by: $Ri + L\dfrac{di}{dt} = E_0 \sin \omega t$. Given $i = 0$ when $t = 0$, show that the solution of the equation is given by:

$$i = \left(\frac{E_0}{R^2 + \omega^2 L^2}\right)(R \sin \omega t - \omega L \cos \omega t)$$

$$+ \left(\frac{E_0 \omega L}{R^2 + \omega^2 L^2}\right) e^{-Rt/L}$$

8. The concentration, C, of impurities of an oil purifier varies with time t and is described by the equation

$a\dfrac{dC}{dt} = b + dm - Cm$, where a, b, d and m are constants. Given $C = c_0$ when $t = 0$, solve the equation and show that:

$$C = \left(\frac{b}{m} + d\right)(1 - e^{-mt/a}) + c_0 e^{-mt/a}$$

9. The equation of motion of a train is given by: $m\dfrac{dv}{dt} = mk(1 - e^{-t}) - mcv$, where v is the speed, t is the time and m, k and c are constants. Determine the speed, v, given $v = 0$ at $t = 0$

For fully worked solutions to each of the problems in Practice Exercises 305 and 306 in this chapter, go to the website:
www.routledge.com/cw/bird

Chapter 80

Numerical methods for first-order differential equations

Why it is important to understand: Numerical methods for first-order differential equations

Most physical systems can be described in mathematical terms through differential equations. Specific types of differential equation have been solved in the preceding chapters, i.e. the separable-variable type, the homogeneous type and the linear type. However, differential equations such as those used to solve real-life problems may not necessarily be directly solvable, i.e. do not have closed form solutions. Instead, solutions can be approximated using numerical methods, and in science and engineering a numeric approximation to the solution is often good enough to solve a problem. Various numerical methods are explained in this chapter.

At the end of this chapter, you should be able to:

- state the reason for solving differential equations using numerical methods
- obtain a numerical solution to a first-order differential equation using Euler's method
- obtain a numerical solution to a first-order differential equation using the Euler–Cauchy method
- obtain a numerical solution to a first-order differential equation using the Runge–Kutta method

80.1 Introduction

Not all first-order differential equations may be solved by separating the variables (as in Chapter 77) or by the integrating factor method (as in Chapter 79). A number of other analytical methods of solving differential equations exist. However, the range of differential equations that can be solved by such analytical methods is fairly restricted.

Where a differential equation and known boundary conditions are given, an approximate solution may be obtained by applying a **numerical method**. There are a number of such numerical methods available and the simplest of these is called **Euler's*** **method**.

80.2   Euler's method

From Chapter 22, Maclaurin's series may be stated as:

$$f(x) = f(0) + x\, f'(0) + \frac{x^2}{2!} f''(0) + \cdots$$

Hence at some point $f(h)$ in Figure 80.1:

$$f(h) = f(0) + h\, f'(0) + \frac{h^2}{2!} f''(0) + \cdots$$

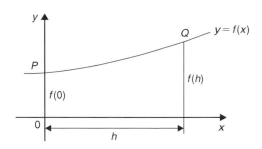

Figure 80.1

If the y-axis and origin are moved a units to the left, as shown in Figure 80.2, the equation of the same curve relative to the new axis becomes $y = f(a + x)$ and the function value at P is $f(a)$.

At point Q in Figure 80.2:

$$f(a+h) = f(a) + h\, f'(a) + \frac{h^2}{2!} f''(a) + \cdots \qquad (1)$$

which is a statement called **Taylor's*** **series**.

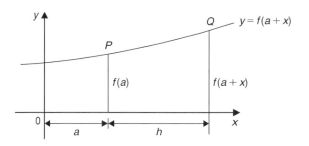

Figure 80.2

* **Who was Euler**? **Leonhard Euler** (15 April 1707–18 September 1783) was a pioneering Swiss mathematician and physicist who made important discoveries in infinitesimal calculus and graph theory. He also introduced much of the modern mathematical terminology and notation. To find out more go to **www.routledge.com/cw/bird**

* **Who was** Taylor? **Brook Taylor** (18 August 1685–29 December 1731) solved the problem of the centre of oscillation of a body. In 1712 Taylor was elected to the Royal Society. Taylor added a new branch to mathematics now called the 'calculus of finite differences', invented integration by parts, and discovered the celebrated series known as Taylor's expansion. To find out more go to **www.routledge.com/cw/bird**

Section K

If h is the interval between two new ordinates y_0 and y_1, as shown in Figure 80.3, and if $f(a) = y_0$ and $y_1 = f(a+h)$, then Euler's method states:

$$f(a+h) = f(a) + hf'(a)$$

i.e. $$y_1 = y_0 + h(y')_0 \qquad (2)$$

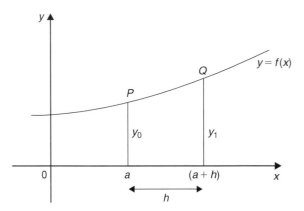

Figure 80.3

The approximation used with Euler's method is to take only the first two terms of Taylor's series shown in equation (1).

Hence if y_0, h and $(y')_0$ are known, y_1, which is an approximate value for the function at Q in Figure 80.3, can be calculated.

Euler's method is demonstrated in the Problems following.

80.3 Worked problems on Euler's method

Problem 1. Obtain a numerical solution of the differential equation

$$\frac{dy}{dx} = 3(1+x) - y$$

given the initial conditions that $x = 1$ when $y = 4$, for the range $x = 1.0$ to $x = 2.0$ with intervals of 0.2

Draw the graph of the solution

$$\frac{dy}{dx} = y' = 3(1+x) - y$$

With $x_0 = 1$ and $y_0 = 4$, $(y')_0 = 3(1+1) - 4 = \mathbf{2}$

By Euler's method:

$$y_1 = y_0 + h(y')_0, \text{ from equation (2)}$$

Hence $y_1 = 4 + (0.2)(2) = \mathbf{4.4}$, since $h = 0.2$

At point Q in Figure 80.4, $x_1 = 1.2$, $y_1 = 4.4$

and $(y')_1 = 3(1+x_1) - y_1$

i.e. $(y')_1 = 3(1+1.2) - 4.4 = \mathbf{2.2}$

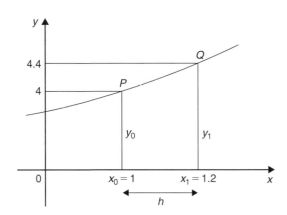

Figure 80.4

If the values of x, y and y' found for point Q are regarded as new starting values of x_0, y_0 and $(y')_0$, the above process can be repeated and values found for the point R shown in Figure 80.5.

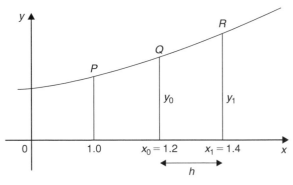

Figure 80.5

Thus at point R,

$$y_1 = y_0 + h(y')_0 \quad \text{from equation (2)}$$

$$= 4.4 + (0.2)(2.2) = \mathbf{4.84}$$

When $x_1 = 1.4$ and $y_1 = 4.84$,
$(y')_1 = 3(1 + 1.4) - 4.84 = \mathbf{2.36}$

This step-by-step Euler's method can be continued and it is easiest to list the results in a table, as shown in Table 80.1. The results for lines 1 to 3 have been produced above.

Table 80.1

	x_0	y_0	$(y')_0$
1.	1	4	2
2.	1.2	4.4	2.2
3.	1.4	4.84	2.36
4.	1.6	5.312	2.488
5.	1.8	5.8096	2.5904
6.	2.0	6.32768	

For line 4, where $x_0 = 1.6$:

$$y_1 = y_0 + h(y')_0$$

$$= 4.84 + (0.2)(2.36) = \mathbf{5.312}$$

and $(y')_0 = 3(1 + 1.6) - 5.312 = \mathbf{2.488}$

For line 5, where $x_0 = 1.8$:

$$y_1 = y_0 + h(y')_0$$

$$= 5.312 + (0.2)(2.488) = \mathbf{5.8096}$$

and $(y')_0 = 3(1 + 1.8) - 5.8096 = \mathbf{2.5904}$

For line 6, where $x_0 = 2.0$:

$$y_1 = y_0 + h(y')_0$$

$$= 5.8096 + (0.2)(2.5904)$$

$$= \mathbf{6.32768}$$

(As the range is 1.0 to 2.0 there is no need to calculate $(y')_0$ in line 6.) The particular solution is given by the value of y against x.

A graph of the solution of $\dfrac{dy}{dx} = 3(1 + x) - y$ with initial conditions $x = 1$ and $y = 4$ is shown in Figure 80.6. In practice it is probably best to plot the graph as each calculation is made, which checks that there is a smooth progression and that no calculation errors have occurred.

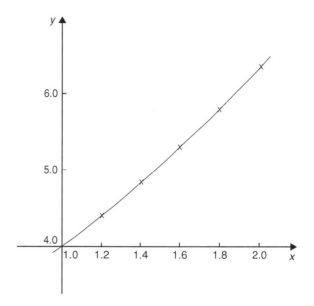

Figure 80.6

Problem 2.   Use Euler's method to obtain a numerical solution of the differential equation $\dfrac{dy}{dx} + y = 2x$, given the initial conditions that at $x = 0$, $y = 1$, for the range $x = 0(0.2)1.0$. Draw the graph of the solution in this range

$x = 0(0.2)1.0$ means that x ranges from 0 to 1.0 in equal intervals of 0.2 (i.e. $h = 0.2$ in Euler's method).

$$\frac{dy}{dx} + y = 2x,$$

hence $\dfrac{dy}{dx} = 2x - y$, i.e. $y' = 2x - y$

If initially $x_0 = 0$ and $y_0 = 1$, then

$$(y')_0 = 2(0) - 1 = -1$$

Hence line 1 in Table 80.2 can be completed with $x = 0$, $y = 1$ and $y'(0) = -1$

Table 80.2

	x_0	y_0	$(y')_0$
1.	0	1	-1
2.	0.2	0.8	-0.4
3.	0.4	0.72	0.08
4.	0.6	0.736	0.464
5.	0.8	0.8288	0.7712
6.	1.0	0.98304	

For line 2, where $x_0 = 0.2$ and $h = 0.2$:

$$y_1 = y_0 + h(y'), \quad \text{from equation (2)}$$

$$= 1 + (0.2)(-1) = 0.8$$

and $(y')_0 = 2x_0 - y_0 = 2(0.2) - 0.8 = -0.4$

For line 3, where $x_0 = 0.4$:

$$y_1 = y_0 + h(y')_0$$

$$= 0.8 + (0.2)(-0.4) = 0.72$$

and $(y')_0 = 2x_0 - y_0 = 2(0.4) - 0.72 = 0.08$

For line 4, where $x_0 = 0.6$:

$$y_1 = y_0 + h(y')_0$$

$$= 0.72 + (0.2)(0.08) = 0.736$$

and $(y')_0 = 2x_0 - y_0 = 2(0.6) - 0.736 = 0.464$

For line 5, where $x_0 = 0.8$:

$$y_1 = y_0 + h(y')_0$$

$$= 0.736 + (0.2)(0.464) = 0.8288$$

and $(y')_0 = 2x_0 - y_0 = 2(0.8) - 0.8288 = 0.7712$

For line 6, where $x_0 = 1.0$:

$$y_1 = y_0 + h(y')_0$$

$$= 0.8288 + (0.2)(0.7712) = 0.98304$$

As the range is 0 to 1.0, $(y')_0$ in line 6 is not needed. A graph of the solution of $\dfrac{dy}{dx} + y = 2x$, with initial conditions $x = 0$ and $y = 1$, is shown in Figure 80.7.

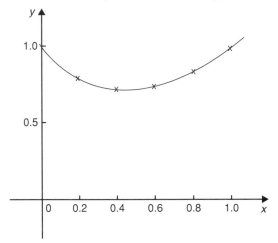

Figure 80.7

Problem 3.

(a)   Obtain a numerical solution, using Euler's method, of the differential equation $\dfrac{dy}{dx} = y - x$, with the initial conditions that at $x = 0$, $y = 2$, for the range $x = 0(0.1)0.5$. Draw the graph of the solution

(b)   By an analytical method (using the integrating factor method of Chapter 79), the solution of the above differential equation is given by $y = x + 1 + e^x$

Determine the percentage error at $x = 0.3$

(a) $\dfrac{dy}{dx} = y' = y - x$

If initially $x_0 = 0$ and $y_0 = 2$, then $(y')_0 = y_0 - x_0 = 2 - 0 = 2$
Hence line 1 of Table 80.3 is completed.

For line 2, where $x_0 = 0.1$:

$$y_1 = y_0 + h(y')_0, \quad \text{from equation (2),}$$

$$= 2 + (0.1)(2) = 2.2$$

Table 80.3

	x_0	y_0	$(y')_0$
1.	0	2	2
2.	0.1	2.2	2.1
3.	0.2	2.41	2.21
4.	0.3	2.631	2.331
5.	0.4	2.8641	2.4641
6.	0.5	3.11051	

and $(y')_0 = y_0 - x_0$

$$= 2.2 - 0.1 = \mathbf{2.1}$$

For line 3, where $x_0 = 0.2$:

$$\mathbf{y_1} = y_0 + h(y')_0$$

$$= 2.2 + (0.1)(2.1) = \mathbf{2.41}$$

and $(\mathbf{y'})_0 = y_0 - x_0 = 2.41 - 0.2 = \mathbf{2.21}$

For line 4, where $x_0 = 0.3$:

$$\mathbf{y_1} = y_0 + h(y')_0$$

$$= 2.41 + (0.1)(2.21) = \mathbf{2.631}$$

and $(\mathbf{y'})_0 = y_0 - x_0$

$$= 2.631 - 0.3 = \mathbf{2.331}$$

For line 5, where $x_0 = 0.4$:

$$\mathbf{y_1} = y_0 + h(y')_0$$

$$= 2.631 + (0.1)(2.331) = \mathbf{2.8641}$$

and $(\mathbf{y'})_0 = y_0 - x_0$

$$= 2.8641 - 0.4 = \mathbf{2.4641}$$

For line 6, where $x_0 = 0.5$:

$$\mathbf{y_1} = y_0 + h(y')_0$$

$$= 2.8641 + (0.1)(2.4641) = \mathbf{3.11051}$$

A graph of the solution of $\dfrac{dy}{dx} = y - x$ with $x = 0$, $y = 2$ is shown in Figure 80.8.

(b) If the solution of the differential equation $\dfrac{dy}{dx} = y - x$ is given by $y = x + 1 + e^x$, then when $x = 0.3$, $y = 0.3 + 1 + e^{0.3} = 2.649859$

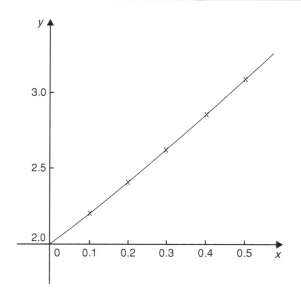

Figure 80.8

By Euler's method, when $x = 0.3$ (i.e. line 4 in Table 80.3), $y = 2.631$

Percentage error

$$= \left(\frac{\text{actual} - \text{estimated}}{\text{actual}} \right) \times 100\%$$

$$= \left(\frac{2.649859 - 2.631}{2.649859} \right) \times 100\%$$

$$= \mathbf{0.712\%}$$

Euler's method of numerical solution of differential equations is simple, but approximate. The method is most useful when the interval h is small.

Now try the following Practice Exercise

Practice Exercise 307   Further problems on Euler's method (answers on page 1186)

1. Use Euler's method to obtain a numerical solution of the differential equation $\dfrac{dy}{dx} = 3 - \dfrac{y}{x}$, with the initial conditions that $x = 1$ when $y = 2$, for the range $x = 1.0$ to $x = 1.5$ with intervals of 0.1. Draw the graph of the solution in this range

2. Obtain a numerical solution of the differential equation $\frac{1}{x}\frac{dy}{dx} + 2y = 1$, given the initial conditions that $x = 0$ when $y = 1$, in the range $x = 0(0.2)1.0$

3. (a) The differential equation $\frac{dy}{dx} + 1 = -\frac{y}{x}$ has the initial conditions that $y = 1$ at $x = 2$. Produce a numerical solution of the differential equation in the range $x = 2.0(0.1)2.5$

   (b) If the solution of the differential equation by an analytical method is given by $y = \frac{4}{x} - \frac{x}{2}$, determine the percentage error at $x = 2.2$

4. Use Euler's method to obtain a numerical solution of the differential equation $\frac{dy}{dx} = x - \frac{2y}{x}$, given the initial conditions that $y = 1$ when $x = 2$, in the range $x = 2.0(0.2)3.0$

   If the solution of the differential equation is given by $y = \frac{x^2}{4}$, determine the percentage error by using Euler's method when $x = 2.8$

80.4 The Euler–Cauchy method

In Euler's method of Section 80.2, the gradient $(y')_0$ at $P_{(x_0, y_0)}$ in Figure 80.9 across the whole interval h is

used to obtain an approximate value of y_1 at point Q. QR in Figure 80.9 is the resulting error in the result.

In an improved Euler method, called the **Euler–Cauchy*** **method**, the gradient at $P_{(x_0, y_0)}$ across half the interval is used and then continues with a line whose gradient approximates to the gradient of the curve at x_1, shown in Figure 80.10.

Let y_{P_1} be the predicted value at point R using Euler's method, i.e. length RZ, where

$$y_{P_1} = y_0 + h(y')_0 \tag{3}$$

The error shown as QT in Figure 80.10 is now less than the error QR used in the basic Euler method and the calculated results will be of greater accuracy. The corrected value, y_{C_1} in the improved Euler method is given by:

$$y_{C_1} = y_0 + \frac{1}{2}h[(y')_0 + f(x_1, y_{P_1})] \tag{4}$$

* **Who was Cauchy?** **Baron Augustin-Louis Cauchy** (21 August 1789–23 May 1857) was a French mathematician who became an early pioneer of analysis. A prolific writer, he wrote approximately 800 research articles. To find out more go to **www.routledge.com/cw/bird**
* **Who was Euler?** For image and resumé of Leonhard Euler, see page 847. To find out more go to **www.routledge.com/cw/bird**

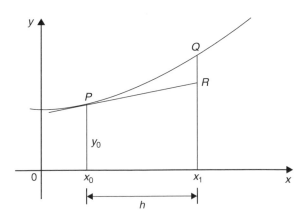

Figure 80.9

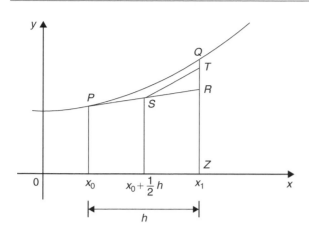

Figure 80.10

Table 80.4

	x	y	y'
1.	0	2	2
2.	0.1	2.205	2.105
3.	0.2	2.421025	2.221025
4.	0.3	2.649232625	2.349232625
5.	0.4	2.89090205	2.49090205
6.	0.5	3.147446765	

The following worked problems demonstrate how equations (3) and (4) are used in the Euler–Cauchy method.

> **Problem 4.** Apply the Euler–Cauchy method to solve the differential equation
>
> $$\frac{dy}{dx} = y - x$$
>
> in the range 0(0.1)0.5, given the initial conditions that at $x=0$, $y=2$

$$\frac{dy}{dx} = y' = y - x$$

Since the initial conditions are $x_0 = 0$ and $y_0 = 2$ then $(y')_0 = 2 - 0 = 2$. Interval $h = 0.1$, hence $x_1 = x_0 + h = 0 + 0.1 = 0.1$
From equation (3),

$$y_{P_1} = y_0 + h(y')_0 = 2 + (0.1)(2) = 2.2$$

From equation (4),

$$y_{C_1} = y_0 + \tfrac{1}{2}h[(y')_0 + f(x_1, y_{P_1})]$$
$$= y_0 + \tfrac{1}{2}h[(y')_0 + (y_{P_1} - x_1)]$$

in this case

$$= 2 + \tfrac{1}{2}(0.1)[2 + (2.2 - 0.1)] = \mathbf{2.205}$$
$$(y')_1 = y_{C_1} - x_1 = 2.205 - 0.1 = 2.105$$

If we produce a table of values, as in Euler's method, we have so far determined lines 1 and 2 of Table 80.4. The results in line 2 are now taken as x_0, y_0 and $(y')_0$ for the next interval and the process is repeated.

For line 3, $x_1 = 0.2$

$$y_{P_1} = y_0 + h(y')_0 = 2.205 + (0.1)(2.105)$$
$$= 2.4155$$
$$y_{C_1} = y_0 + \tfrac{1}{2}h[(y')_0 + f(x_1, y_{P_1})]$$
$$= 2.205 + \tfrac{1}{2}(0.1)[2.105 + (2.4155 - 0.2)]$$
$$= \mathbf{2.421025}$$
$$(y')_0 = y_{C_1} - x_1 = 2.421025 - 0.2 = 2.221025$$

For line 4, $x_1 = 0.3$

$$y_{P_1} = y_0 + h(y')_0$$
$$= 2.421025 + (0.1)(2.221025)$$
$$= 2.6431275$$
$$y_{C_1} = y_0 + \tfrac{1}{2}h[(y')_0 + f(x_1, y_{P_1})]$$
$$= 2.421025 + \tfrac{1}{2}(0.1)[2.221025$$
$$+ (2.6431275 - 0.3)]$$
$$= \mathbf{2.649232625}$$
$$(y')_0 = y_{C_1} - x_1 = 2.649232625 - 0.3$$
$$= 2.349232625$$

For line 5, $x_1 = 0.4$

$$y_{P_1} = y_0 + h(y')_0$$
$$= 2.649232625 + (0.1)(2.349232625)$$
$$= 2.884155887$$

$$y_{C_1} = y_0 + \tfrac{1}{2} h[(y')_0 + f(x_1, y_{P_1})]$$

$$= 2.649232625 + \tfrac{1}{2}(0.1)[2.349232625$$

$$+ (2.884155887 - 0.4)]$$

$$= \mathbf{2.89090205}$$

$$(y')_0 = y_{C_1} - x_1 = 2.89090205 - 0.4$$

$$= 2.49090205$$

For line 6, $x_1 = 0.5$

$$y_{P_1} = y_0 + h(y')_0$$

$$= 2.89090205 + (0.1)(2.49090205)$$

$$= 3.139992255$$

$$y_{C_1} = y_0 + \tfrac{1}{2}h[(y')_0 + f(x_1, y_{P_1})]$$

$$= 2.89090205 + \tfrac{1}{2}(0.1)[2.49090205$$

$$+ (3.139992255 - 0.5)]$$

$$= \mathbf{3.147446765}$$

Problem 4 is the same example as Problem 3 and Table 80.5 shows a comparison of the results, i.e. it compares the results of Tables 80.3 and 80.4. $\dfrac{\mathrm{d}y}{\mathrm{d}x} = y - x$ may be solved analytically by the integrating factor method of Chapter 79 with the solution $y = x + 1 + \mathrm{e}^x$. Substituting values of x of $0, 0.1, 0.2, \ldots$ gives the exact values shown in Table 80.5.

The percentage error for each method for each value of x is shown in Table 80.6. For example, when $x = 0.3$,

Table 80.6

x	Error in Euler method	Error in Euler–Cauchy method
0	0	0
0.1	0.234%	0.00775%
0.2	0.472%	0.0156%
0.3	0.712%	0.0236%
0.4	0.959%	0.0319%
0.5	1.214%	0.0405%

% error with Euler method

$$= \left(\frac{\text{actual} - \text{estimated}}{\text{actual}} \right) \times 100\%$$

$$= \left(\frac{2.649858808 - 2.631}{2.649858808} \right) \times 100\%$$

$$= \mathbf{0.712\%}$$

% error with Euler–Cauchy method

$$= \left(\frac{2.649858808 - 2.649232625}{2.649858808} \right) \times 100\%$$

$$= \mathbf{0.0236\%}$$

This calculation and the others listed in Table 80.6 show the Euler–Cauchy method to be more accurate than the Euler method.

Table 80.5

	x	Euler method y	Euler–Cauchy method y	Exact value $y = x + 1 + \mathrm{e}^x$
1.	0	2	2	2
2.	0.1	2.2	2.205	2.205170918
3.	0.2	2.41	2.421025	2.421402758
4.	0.3	2.631	2.649232625	2.649858808
5.	0.4	2.8641	2.89090205	2.891824698
6.	0.5	3.11051	3.147446765	3.148721271

Problem 5.   Obtain a numerical solution of the differential equation

$$\frac{dy}{dx} = 3(1+x) - y$$

in the range 1.0(0.2)2.0, using the Euler–Cauchy method, given the initial conditions that $x = 1$ when $y = 4$

This is the same as Problem 1 on page 848, and a comparison of values may be made.

$$\frac{dy}{dx} = y' = 3(1+x) - y \quad \text{i.e. } y' = 3 + 3x - y$$

$$x_0 = 1.0, \ y_0 = 4 \text{ and } h = 0.2$$

$$(y')_0 = 3 + 3x_0 - y_0 = 3 + 3(1.0) - 4 = 2$$

$x_1 = 1.2$ and from equation (3),

$$y_{P_1} = y_0 + h(y')_0 = 4 + 0.2(2) = 4.4$$

$$y_{C_1} = y_0 + \tfrac{1}{2}h[(y')_0 + f(x_1, y_{P_1})]$$

$$= y_0 + \tfrac{1}{2}h[(y')_0 + (3 + 3x_1 - y_{P_1})]$$

$$= 4 + \tfrac{1}{2}(0.2)[2 + (3 + 3(1.2) - 4.4)]$$

$$= \mathbf{4.42}$$

$$(y')_1 = 3 + 3x_1 - y_{P_1} = 3 + 3(1.2) - 4.42 = 2.18$$

Thus the first two lines of Table 80.7 have been completed.

Table 80.7

	x_0	y_0	y'_0
1.	1.0	4	2
2.	1.2	4.42	2.18
3.	1.4	4.8724	2.3276
4.	1.6	5.351368	2.448632
5.	1.8	5.85212176	2.54787824
6.	2.0	6.370739843	

For line 3, $x_1 = 1.4$

$$y_{P_1} = y_0 + h(y')_0 = 4.42 + 0.2(2.18) = 4.856$$

$$y_{C_1} = y_0 + \tfrac{1}{2}h[(y')_0 + (3 + 3x_1 - y_{P_1})]$$

$$= 4.42 + \tfrac{1}{2}(0.2)[2.18 + (3 + 3(1.4) - 4.856)]$$

$$= \mathbf{4.8724}$$

$$(y')_1 = 3 + 3x_1 - y_{P_1} = 3 + 3(1.4) - 4.8724$$

$$= 2.3276$$

For line 4, $x_1 = 1.6$

$$y_{P_1} = y_0 + h(y')_0 = 4.8724 + 0.2(2.3276)$$

$$= 5.33792$$

$$y_{C_1} = y_0 + \tfrac{1}{2}h[(y')_0 + (3 + 3x_1 - y_{P_1})]$$

$$= 4.8724 + \tfrac{1}{2}(0.2)[2.3276$$

$$+ (3 + 3(1.6) - 5.33792)]$$

$$= \mathbf{5.351368}$$

$$(y')_1 = 3 + 3x_1 - y_{P_1}$$

$$= 3 + 3(1.6) - 5.351368$$

$$= 2.448632$$

For line 5, $x_1 = 1.8$

$$y_{P_1} = y_0 + h(y')_0 = 5.351368 + 0.2(2.448632)$$

$$= 5.8410944$$

$$y_{C_1} = y_0 + \tfrac{1}{2}h[(y')_0 + (3 + 3x_1 - y_{P_1})]$$

$$= 5.351368 + \tfrac{1}{2}(0.2)[2.448632$$

$$+ (3 + 3(1.8) - 5.8410944)]$$

$$= \mathbf{5.85212176}$$

$$(y')_1 = 3 + 3x_1 - y_{P_1}$$

$$= 3 + 3(1.8) - 5.85212176$$

$$= 2.54787824$$

Section K

For line 6, $x_1 = 2.0$

$$y_{P_1} = y_0 + h(y')_0$$

$$= 5.85212176 + 0.2(2.54787824)$$

$$= 6.361697408$$

$$y_{C_1} = y_0 + \tfrac{1}{2}h[(y')_0 + (3 + 3x_1 - y_{P_1})]$$

$$= 5.85212176 + \tfrac{1}{2}(0.2)[2.54787824$$

$$+ (3 + 3(2.0) - 6.361697408)]$$

$$= \mathbf{6.370739843}$$

Problem 6. Using the integrating factor method the solution of the differential equation $\dfrac{dy}{dx} = 3(1+x) - y$ of Problem 5 is $y = 3x + e^{1-x}$.
When $x = 1.6$, compare the accuracy, correct to 3 decimal places, of the Euler and the Euler–Cauchy methods

When $x = 1.6$, $y = 3x + e^{1-x} = 3(1.6) + e^{1-1.6} = 4.8 + e^{-0.6} = 5.348811636$

From Table 80.1, page 849, by Euler's method, when $x = 1.6$, $y = 5.312$

% error in the Euler method

$$= \left(\frac{5.348811636 - 5.312}{5.348811636} \right) \times 100\%$$

$$= \mathbf{0.688\%}$$

From Table 80.7 of Problem 5, by the Euler–Cauchy method, when $x = 1.6$, $y = 5.351368$

% error in the Euler–Cauchy method

$$= \left(\frac{5.348811636 - 5.351368}{5.348811636} \right) \times 100\%$$

$$= \mathbf{-0.048\%}$$

The Euler–Cauchy method is seen to be more accurate than the Euler method when $x = 1.6$

Now try the following Practice Exercise

Practice Exercise 308 Further problems on an improved Euler method (answers on page 1186)

1. Apply the Euler–Cauchy method to solve the differential equation

    $$\frac{dy}{dx} = 3 - \frac{y}{x}$$

    for the range $1.0(0.1)1.5$, given the initial conditions that $x = 1$ when $y = 2$

2. Solving the differential equation in Problem 1 by the integrating factor method gives $y = \dfrac{3}{2}x + \dfrac{1}{2x}$. Determine the percentage error, correct to 3 significant figures, when $x = 1.3$ using (a) Euler's method, and (b) the Euler–Cauchy method.

3. (a) Apply the Euler–Cauchy method to solve the differential equation

        $$\frac{dy}{dx} - x = y$$

        for the range $x = 0$ to $x = 0.5$ in increments of 0.1, given the initial conditions that when $x = 0$, $y = 1$

    (b) The solution of the differential equation in part (a) is given by $y = 2e^x - x - 1$. Determine the percentage error, correct to 3 decimal places, when $x = 0.4$

4. Obtain a numerical solution of the differential equation

    $$\frac{1}{x}\frac{dy}{dx} + 2y = 1$$

    using the Euler–Cauchy method in the range $x = 0(0.2)1.0$, given the initial conditions that $x = 0$ when $y = 1$

80.5 The Runge–Kutta method

The **Runge–Kutta*** method for solving first order differential equations is widely used and provides a high degree of accuracy. Again, as with the two previous methods, the Runge–Kutta method is a step-by-step process where results are tabulated for a range of values of x. Although several intermediate calculations are needed at each stage, the method is fairly straightforward.

The 7 step **procedure for the Runge–Kutta method**, without proof, is as follows.

To solve the differential equation $\dfrac{dy}{dx} = f(x, y)$ given the initial condition $y = y_0$ at $x = x_0$ for a range of values of $x = x_0(h)x_n$:

1. Identify x_0, y_0 and h, and values of $x_1, x_2, x_3, \ldots$

2. Evaluate $k_1 = f(x_n, y_n)$ starting with $n = 0$

3. Evaluate $k_2 = f\left(x_n + \dfrac{h}{2}, \, y_n + \dfrac{h}{2}k_1\right)$

4. Evaluate $k_3 = f\left(x_n + \dfrac{h}{2}, \, y_n + \dfrac{h}{2}k_2\right)$

5. Evaluate $k_4 = f(x_n + h, \, y_n + hk_3)$

6. Use the values determined from steps 2 to 5 to evaluate:

$$y_{n+1} = y_n + \frac{h}{6}\{k_1 + 2k_2 + 2k_3 + k_4\}$$

7. Repeat steps 2 to 6 for $n = 1, 2, 3, \ldots$

* **Who was Runge**? **Carl David Tolmé Runge** (1856–1927) was a German mathematician, physicist, and spectroscopist. He was co-developer of the Runge–Kutta method in the field of numerical analysis. The Runge crater on the Moon is named after him. To find out more go to **www.routledge.com/cw/bird**

* **Who was Kutta**? **Martin Wilhelm Kutta** (3 November 1867–25 December 1944) wrote a thesis that contains the now famous Runge–Kutta method for solving ordinary differential equations. To find out more go to **www.routledge.com/cw/bird**

Section K

Thus, step 1 is given, and steps 2 to 5 are intermediate steps leading to step 6. It is usually most convenient to construct a table of values.

The Runge–Kutta method is demonstrated in the following Problems.

Problem 7. Use the Runge–Kutta method to solve the differential equation:

$$\frac{dy}{dx} = y - x$$

in the range $0(0.1)0.5$, given the initial conditions that at $x = 0$, $y = 2$

Using the above procedure:

1. $x_0 = 0$, $y_0 = 2$ and since $h = 0.1$, and the range is from $x = 0$ to $x = 0.5$, then $x_1 = 0.1$, $x_2 = 0.2$, $x_3 = 0.3$, $x_4 = 0.4$, and $x_5 = 0.5$

Let $n = 0$ to determine y_1:

2. $k_1 = f(x_0, y_0) = f(0, 2)$;
   since $\dfrac{dy}{dx} = y - x$, $f(0, 2) = 2 - 0 = \mathbf{2}$

3. $k_2 = f\left(x_0 + \dfrac{h}{2}, y_0 + \dfrac{h}{2}k_1\right)$

   $= f\left(0 + \dfrac{0.1}{2}, 2 + \dfrac{0.1}{2}(2)\right)$

   $= f(0.05, 2.1) = 2.1 - 0.05 = \mathbf{2.05}$

4. $k_3 = f\left(x_0 + \dfrac{h}{2}, y_0 + \dfrac{h}{2}k_2\right)$

   $= f\left(0 + \dfrac{0.1}{2}, 2 + \dfrac{0.1}{2}(2.05)\right)$

   $= f(0.05, 2.1025)$

   $= 2.1025 - 0.05 = \mathbf{2.0525}$

5. $k_4 = f(x_0 + h, y_0 + hk_3)$

   $= f(0 + 0.1, 2 + 0.1(2.0525))$

   $= f(0.1, 2.20525)$

   $= 2.20525 - 0.1 = \mathbf{2.10525}$

6. $y_{n+1} = y_n + \dfrac{h}{6}\{k_1 + 2k_2 + 2k_3 + k_4\}$ and when $n = 0$:

$$y_1 = y_0 + \frac{h}{6}\{k_1 + 2k_2 + 2k_3 + k_4\}$$

$$= 2 + \frac{0.1}{6}\{2 + 2(2.05) + 2(2.0525)$$

$$+ 2.10525\}$$

$$= 2 + \frac{0.1}{6}\{12.31025\} = \mathbf{2.205171}$$

A table of values may be constructed as shown in Table 80.8. The working has been shown for the first two rows.

Let $n = 1$ to determine y_2:

2. $k_1 = f(x_1, y_1) = f(0.1, 2.205171)$; since

   $\dfrac{dy}{dx} = y - x$, $f(0.1, 2.205171)$

   $= 2.205171 - 0.1 = \mathbf{2.105171}$

3. $k_2 = f\left(x_1 + \dfrac{h}{2}, y_1 + \dfrac{h}{2}k_1\right)$

   $= f\left(0.1 + \dfrac{0.1}{2}, 2.205171 + \dfrac{0.1}{2}(2.105171)\right)$

   $= f(0.15, 2.31042955)$

   $= 2.31042955 - 0.15 = \mathbf{2.160430}$

4. $k_3 = f\left(x_1 + \dfrac{h}{2}, y_1 + \dfrac{h}{2}k_2\right)$

   $= f\left(0.1 + \dfrac{0.1}{2}, 2.205171 + \dfrac{0.1}{2}(2.160430)\right)$

   $= f(0.15, 2.3131925) = 2.3131925 - 0.15$

   $= \mathbf{2.163193}$

5. $k_4 = f(x_1 + h, y_1 + hk_3)$

   $= f(0.1 + 0.1, 2.205171 + 0.1(2.163193))$

   $= f(0.2, 2.421490)$

   $= 2.421490 - 0.2 = \mathbf{2.221490}$

6. $y_{n+1} = y_n + \dfrac{h}{6}\{k_1 + 2k_2 + 2k_3 + k_4\}$

   and when $n = 1$:

Table 80.8

n	x_n	k_1	k_2	k_3	k_4	y_n
0	**0**					**2**
1	**0.1**	2.0	2.05	2.0525	2.10525	**2.205171**
2	**0.2**	2.105171	2.160430	2.163193	2.221490	**2.421403**
3	**0.3**	2.221403	2.282473	2.285527	2.349956	**2.649859**
4	**0.4**	2.349859	2.417339	2.420726	2.491932	**2.891824**
5	**0.5**	2.491824	2.566415	2.570145	2.648838	**3.148720**

$$y_2 = y_1 + \frac{h}{6}\{k_1 + 2k_2 + 2k_3 + k_4\}$$

$$= 2.205171 + \frac{0.1}{6}\{2.105171 + 2(2.160430)$$

$$+ 2(2.163193) + 2.221490\}$$

$$= 2.205171 + \frac{0.1}{6}\{12.973907\} = \mathbf{2.421403}$$

This completes the third row of Table 80.8. In a similar manner y_3, y_4 and y_5 can be calculated and the results are as shown in Table 80.8. Such a table is best produced by using a **spreadsheet**, such as Microsoft Excel.

This problem is the same as Problem 3, page 850, which used Euler's method, and Problem 4, page 853, which used the improved Euler's method, and a comparison of results can be made.

The differential equation $\frac{dy}{dx} = y - x$ may be solved analytically using the integrating factor method of Chapter 79, with the solution:

$$y = x + 1 + e^x$$

Substituting values of x of 0, 0.1, 0.2, ..., 0.5 will give the exact values. A comparison of the results obtained by Euler's method, the Euler–Cauchy method and the Runge–Kutta method, together with the exact values, is shown in Table 80.9.

It is seen from Table 80.9 that **the Runge–Kutta method is exact, correct to 5 decimal places**.

Problem 8. Obtain a numerical solution of the differential equation: $\frac{dy}{dx} = 3(1+x) - y$ in the range 1.0(0.2)2.0, using the Runge–Kutta method, given the initial conditions that $x = 1.0$ when $y = 4.0$

Using the above procedure:

1. $x_0 = 1.0$, $y_0 = 4.0$ and since $h = 0.2$, and the range is from $x = 1.0$ to $x = 2.0$, then $x_1 = 1.2$, $x_2 = 1.4$, $x_3 = 1.6$, $x_4 = 1.8$, and $x_5 = 2.0$

Let $n = 0$ to determine y_1:

2. $k_1 = f(x_0, y_0) = f(1.0, 4.0)$; since

$$\frac{dy}{dx} = 3(1+x) - y,$$

$$f(1.0, 4.0) = 3(1 + 1.0) - 4.0 = \mathbf{2.0}$$

3. $k_2 = f\left(x_0 + \frac{h}{2}, y_0 + \frac{h}{2}k_1\right)$

$$= f\left(1.0 + \frac{0.2}{2}, 4.0 + \frac{0.2}{2}(2)\right)$$

$$= f(1.1, 4.2) = 3(1 + 1.1) - 4.2 = \mathbf{2.1}$$

4. $k_3 = f\left(x_0 + \frac{h}{2}, y_0 + \frac{h}{2}k_2\right)$

$$= f\left(1.0 + \frac{0.2}{2}, 4.0 + \frac{0.2}{2}(2.1)\right)$$

$$= f(1.1, 4.21)$$

$$= 3(1 + 1.1) - 4.21 = \mathbf{2.09}$$

5. $k_4 = f(x_0 + h, y_0 + hk_3)$

$$= f(1.0 + 0.2, 4.1 + 0.2(2.09))$$

$$= f(1.2, 4.418)$$

$$= 3(1 + 1.2) - 4.418 = \mathbf{2.182}$$

Table 80.9

x	Euler's method y	Euler–Cauchy method y	Runge–Kutta method y	Exact value $y = x + 1 + e^x$
0	2	2	2	2
0.1	2.2	2.205	2.205171	2.205170918
0.2	2.41	2.421025	2.421403	2.421402758
0.3	2.631	2.649232625	2.649859	2.649858808
0.4	2.8641	2.89090205	2.891824	2.891824698
0.5	3.11051	3.147446765	3.148720	3.148721271

6. $y_{n+1} = y_n + \dfrac{h}{6}\{k_1 + 2k_2 + 2k_3 + k_4\}$ and   when $n = 0$:

$$y_1 = y_0 + \frac{h}{6}\{k_1 + 2k_2 + 2k_3 + k_4\}$$

$$= 4.0 + \frac{0.2}{6}\{2.0 + 2(2.1) + 2(2.09) + 2.182\}$$

$$= 4.0 + \frac{0.2}{6}\{12.562\} = \mathbf{4.418733}$$

A table of values is compiled in Table 80.10. The working has been shown for the first two rows.

Let $n = 1$ to determine y_2:

2. $k_1 = f(x_1, y_1) = f(1.2, 4.418733)$; since

$$\frac{\mathrm{d}y}{\mathrm{d}x} = 3(1 + x) - y, \ \ f(1.2, 4.418733)$$

$$= 3(1 + 1.2) - 4.418733 = \mathbf{2.181267}$$

3. $k_2 = f\left(x_1 + \dfrac{h}{2}, y_1 + \dfrac{h}{2}k_1\right)$

$$= f\left(1.2 + \frac{0.2}{2}, 4.418733 + \frac{0.2}{2}(2.181267)\right)$$

$$= f(1.3, 4.636860)$$

$$= 3(1 + 1.3) - 4.636860 = \mathbf{2.263140}$$

4. $k_3 = f\left(x_1 + \dfrac{h}{2}, y_1 + \dfrac{h}{2}k_2\right)$

$$= f\left(1.2 + \frac{0.2}{2}, 4.418733 + \frac{0.2}{2}(2.263140)\right)$$

$$= f(1.3, 4.645047) = 3(1 + 1.3) - 4.645047$$

$$= \mathbf{2.254953}$$

5. $k_4 = f(x_1 + h, y_1 + hk_3)$

$$= f(1.2 + 0.2, 4.418733 + 0.2(2.254953))$$

Table 80.10

n	x_n	k_1	k_2	k_3	k_4	y_n
0	**1.0**					**4.0**
1	**1.2**	2.0	2.1	2.09	2.182	**4.418733**
2	**1.4**	2.181267	2.263140	2.254953	2.330276	**4.870324**
3	**1.6**	2.329676	2.396708	2.390005	2.451675	**5.348817**
4	**1.8**	2.451183	2.506065	2.500577	2.551068	**5.849335**
5	**2.0**	2.550665	2.595599	2.591105	2.632444	**6.367886**

$= f(1.4, 4.869724) = 3(1 + 1.4) - 4.869724$

$= \mathbf{2.330276}$

6. $y_{n+1} = y_n + \dfrac{h}{6}\{k_1 + 2k_2 + 2k_3 + k_4\}$ and   when $n = 1$:

$$\mathbf{y_2} = y_1 + \dfrac{h}{6}\{k_1 + 2k_2 + 2k_3 + k_4\}$$

$$= 4.418733 + \frac{0.2}{6}\{2.181267 + 2(2.263140)$$

$$+\, 2(2.254953) + 2.330276\}$$

$$= 4.418733 + \frac{0.2}{6}\{13.547729\} = \mathbf{4.870324}$$

This completes the third row of Table 80.10. In a similar manner y_3, y_4 and y_5 can be calculated and the results are as shown in Table 80.10. As in the previous problem, such a table is best produced by using a **spreadsheet**.

This problem is the same as Problem 1, page 848, which used Euler's method, and Problem 5, page 855, which used the Euler–Cauchy method, and a comparison of results can be made.

The differential equation $\dfrac{dy}{dx} = 3(1 + x) - y$ may be solved analytically using the integrating factor method of Chapter 79, with the solution:

$$y = 3x + e^{1-x}$$

Substituting values of x of $1.0, 1.2, 1.4, \ldots, 2.0$ will give the exact values. A comparison of the results obtained by Euler's method, the Euler–Cauchy method and the Runge–Kutta method, together with the exact values, is shown in Table 80.11.

It is seen from Table 80.11 that **the Runge–Kutta method is exact, correct to 4 decimal places**.

The percentage error in the Runge–Kutta method when, say, $x = 1.6$ is:

$$\left(\frac{5.348811636 - 5.348817}{5.348811636}\right) \times 100\% = \mathbf{-0.0001\%}$$

From Problem 6, page 856, when $x = 1.6$, the percentage error for the Euler method was 0.688%, and for the Euler–Cauchy method -0.048%. Clearly, the Runge–Kutta method is the most accurate of the three methods.

Now try the following Practice Exercise

Practice Exercise 309   Further problems on the Runge–Kutta method (answers on page 1187)

1. Apply the Runge–Kutta method to solve the differential equation: $\dfrac{dy}{dx} = 3 - \dfrac{y}{x}$ for the range $1.0(0.1)1.5$, given that the initial conditions that $x = 1$ when $y = 2$

2. Obtain a numerical solution of the differential equation: $\dfrac{1}{x}\dfrac{dy}{dx} + 2y = 1$ using the Runge–Kutta method in the range $x = 0(0.2)1.0$, given the initial conditions that $x = 0$ when $y = 1$

Section K

Table 80.11

x	Euler's method y	Euler–Cauchy method y	Runge–Kutta method y	Exact value $y = 3x + e^{1-x}$
1.0	4	4	4	4
1.2	4.4	4.42	4.418733	4.418730753
1.4	4.84	4.8724	4.870324	4.870320046
1.6	5.312	5.351368	5.348817	5.348811636
1.8	5.8096	5.85212176	5.849335	5.849328964
2.0	6.32768	6.370739847	6.367886	6.367879441

3. (a) The differential equation: $\dfrac{dy}{dx} + 1 = -\dfrac{y}{x}$ has the initial conditions that $y = 1$ at $x = 2$. Produce a numerical solution of the differential equation, correct to 6 decimal places, using the Runge–Kutta method in the range $x = 2.0(0.1)2.5$

(b) If the solution of the differential equation by an analytical method is given by: $y = \dfrac{4}{x} - \dfrac{x}{2}$ determine the percentage error at $x = 2.2$

For fully worked solutions to each of the problems in Practice Exercises 307 to 309 in this chapter, go to the website:
www.routledge.com/cw/bird

This assignment covers the material contained in Chapters 77 to 80. *The marks available are shown in brackets at the end of each question.*

1. Solve the differential equation: $x\dfrac{dy}{dx}+x^2=5$ given that $y=2.5$ when $x=1$ (4)

2. Determine the equation of the curve which satisfies the differential equation $2xy\dfrac{dy}{dx}=x^2+1$ and which passes through the point $(1, 2)$ (5)

3. A capacitor C is charged by applying a steady voltage E through a resistance R. The p.d. between the plates, V, is given by the differential equation:

    $$CR\dfrac{dV}{dt}+V=E$$

    (a) Solve the equation for V given that when time $t=0$, $V=0$

    (b) Evaluate voltage V when $E=50\,\text{V}$, $C=10\,\mu\text{F}$, $R=200\,\text{k}\Omega$ and $t=1.2\,\text{s}$ (14)

4. Show that the solution to the differential equation:

    $4x\dfrac{dy}{dx}=\dfrac{x^2+y^2}{y}$ is of the form

    $3y^2=\sqrt{x}\left(\sqrt{x^3}-1\right)$ given that $y=0$ when $x=1$ (12)

5. Show that the solution to the differential equation

    $$x\cos x\dfrac{dy}{dx}+(x\sin x+\cos x)y=1$$

    is given by: $xy=\sin x+k\cos x$ where k is a constant. (11)

6. (a) Use Euler's method to obtain a numerical solution of the differential equation:

    $$\dfrac{dy}{dx}=\dfrac{y}{x}+x^2-2$$

    given the initial conditions that $x=1$ when $y=3$, for the range $x=1.0\,(0.1)\,1.5$

    (b) Apply the Euler–Cauchy method to the differential equation given in part (a) over the same range.

    (c) Apply the integrating factor method to solve the differential equation in part (a) analytically.

    (d) Determine the percentage error, correct to 3 significant figures, in each of the two numerical methods when $x=1.2$ (30)

7. Use the Runge–Kutta method to solve the differential equation: $\dfrac{dy}{dx}=\dfrac{y}{x}+x^2-2$ in the range $1.0(0.1)1.5$, given the initial conditions that at $x=1$, $y=3$. Work to an accuracy of 6 decimal places. (24)

For lecturers/instructors/teachers, fully worked solutions to each of the problems in Revision Test 28, together with a full marking scheme, are available at the website:
www.routledge.com/cw/bird

Section K

Chapter 81

Second-order differential equations of the form

$$a\frac{\mathrm{d}^2 y}{\mathrm{d}x^2} + b\frac{\mathrm{d}y}{\mathrm{d}x} + cy = 0$$

Why it is important to understand: Second-order differential equations of the form $a\dfrac{\mathrm{d}^2 y}{\mathrm{d}x^2} + b\dfrac{\mathrm{d}y}{\mathrm{d}x} + cy = 0$

Second-order differential equations have many engineering applications. These include free-vibration analysis with simple and damped mass-spring systems, resonant and non-resonant vibration analysis, with modal analysis, time-varying mechanical forces or pressure, fluid-induced vibration such as intermittent wind, forced electrical and mechanical oscillations, tidal waves, acoustics, ultrasonic and random movements of support. This chapter explains the procedure to solve second-order differential equations of the form $a\dfrac{\mathrm{d}^2 y}{\mathrm{d}x^2} + b\dfrac{\mathrm{d}y}{\mathrm{d}x} + cy = 0$

At the end of this chapter, you should be able to:

- identify and solve the auxiliary equation of a second-order differential equation

- solve a second-order differential equation of the form $a\dfrac{\mathrm{d}^2 y}{\mathrm{d}x^2} + b\dfrac{\mathrm{d}y}{\mathrm{d}x} + cy = 0$

81.1 Introduction

An equation of the form $a\dfrac{\mathrm{d}^2 y}{\mathrm{d}x^2} + b\dfrac{\mathrm{d}y}{\mathrm{d}x} + cy = 0$, where a, b and c are constants, is called a **linear second-order differential equation with constant coefficients**. When the right-hand side of the differential equation is zero, it is referred to as a **homogeneous differential equation**. When the right-hand side is not equal to zero (as in Chapter 82) it is referred to as a **non-homogeneous differential equation**.

There are numerous engineering examples of second-order differential equations. Three examples are:

(i) $L\dfrac{\mathrm{d}^2 q}{\mathrm{d}t^2} + R\dfrac{\mathrm{d}q}{\mathrm{d}t} + \dfrac{1}{C}q = 0$, representing an equation for charge q in an electrical circuit containing resistance R, inductance L and capacitance C in series.

(ii) $m\dfrac{d^2s}{dt^2} + a\dfrac{ds}{dt} + ks = 0$, defining a mechanical system, where s is the distance from a fixed point after t seconds, m is a mass, a the damping factor and k the spring stiffness.

(iii) $\dfrac{d^2y}{dx^2} + \dfrac{P}{EI}y = 0$, representing an equation for the deflected profile y of a pin-ended uniform strut of length l subjected to a load P. E is Young's modulus and I is the second moment of area.

If D represents $\dfrac{d}{dx}$ and D^2 represents $\dfrac{d^2}{dx^2}$ then the above equation may be stated as $(aD^2 + bD + c)y = 0$. This equation is said to be in **'D-operator'** form.

If $y = Ae^{mx}$ then $\dfrac{dy}{dx} = Ame^{mx}$ and $\dfrac{d^2y}{dx^2} = Am^2e^{mx}$

Substituting these values into $a\dfrac{d^2y}{dx^2} + b\dfrac{dy}{dx} + cy = 0$ gives:

$$a(Am^2e^{mx}) + b(Ame^{mx}) + c(Ae^{mx}) = 0$$

i.e. $$Ae^{mx}(am^2 + bm + c) = 0$$

Thus $y = Ae^{mx}$ is a solution of the given equation provided that $(am^2 + bm + c) = 0$. $am^2 + bm + c = 0$ is called the **auxiliary equation**, and since the equation is a quadratic, m may be obtained either by factorising or by using the quadratic formula. Since, in the auxiliary equation, a, b and c are real values, then the equation may have either

(i)   two different real roots (when $b^2 > 4ac$) or

(ii)   two equal real roots (when $b^2 = 4ac$) or

(iii)   two complex roots (when $b^2 < 4ac$)

81.2   Procedure to solve differential equations of the form $a\dfrac{d^2y}{dx^2} + b\dfrac{dy}{dx} + cy = 0$

(a)   Rewrite the differential equation

$$a\dfrac{d^2y}{dx^2} + b\dfrac{dy}{dx} + cy = 0$$

as $(aD^2 + bD + c)y = 0$

(b)   Substitute m for D and solve the auxiliary equation $am^2 + bm + c = 0$ for m

(c)   If the roots of the auxiliary equation are:

(i)   **real and different**, say $m = \alpha$ and $m = \beta$, then the general solution is

$$y = Ae^{\alpha x} + Be^{\beta x}$$

(ii)   **real and equal**, say $m = \alpha$ twice, then the general solution is

$$y = (Ax + B)e^{\alpha x}$$

(iii)   **complex**, say $m = \alpha \pm j\beta$, then the general solution is

$$y = e^{\alpha x}\{A\cos \beta x + B\sin \beta x\}$$

(d)   Given boundary conditions, constants A and B may be determined and the **particular solution** of the differential equation obtained.

The particular solutions obtained in the Problems of Section 81.3 may each be verified by substituting expressions for y, $\dfrac{dy}{dx}$ and $\dfrac{d^2y}{dx^2}$ into the original equation.

81.3   Worked problems on differential equations of the form $a\dfrac{d^2y}{dx^2} + b\dfrac{dy}{dx} + cy = 0$

Problem 1.   By applying Kirchhoff's voltage law to a circuit the following differential equation is obtained: $2\dfrac{d^2y}{dx^2} + 5\dfrac{dy}{dx} - 3y = 0$. Determine the general solution. Find also the particular solution given that when $x = 0$, $y = 4$ and $\dfrac{dy}{dx} = 9$

Using the above procedure:

(a) $2\dfrac{d^2y}{dx^2} + 5\dfrac{dy}{dx} - 3y = 0$ in D-operator form is $(2D^2 + 5D - 3)y = 0$, where $D \equiv \dfrac{d}{dx}$

(b)   Substituting m for D gives the auxiliary equation

$$2m^2 + 5m - 3 = 0$$

Factorising gives: $(2m - 1)(m + 3) = 0$, from which, $m = \frac{1}{2}$ or $m = -3$

(c)   Since the roots are real and different the **general solution is $y = Ae^{\frac{1}{2}x} + Be^{-3x}$**

(d) When $x=0$, $y=4$,

hence $\quad 4 = A + B \qquad$ (1)

Since $\quad y = Ae^{\frac{1}{2}x} + Be^{-3x}$

then $\quad \dfrac{dy}{dx} = \dfrac{1}{2}Ae^{\frac{1}{2}x} - 3Be^{-3x}$

When $\quad x=0, \dfrac{dy}{dx} = 9$

thus $\quad 9 = \dfrac{1}{2}A - 3B \qquad$ (2)

Solving the simultaneous equations (1) and (2) gives $A=6$ and $B=-2$

Hence the particular solution is

$$y = 6e^{\frac{1}{2}x} - 2e^{-3x}$$

Problem 2. Find the general solution of $9\dfrac{d^2y}{dt^2} - 24\dfrac{dy}{dt} + 16y = 0$ and also the particular solution given the boundary conditions that when $t=0$, $y = \dfrac{dy}{dt} = 3$

Using the procedure of Section 81.2:

(a) $9\dfrac{d^2y}{dt^2} - 24\dfrac{dy}{dt} + 16y = 0$ in D-operator form is

$(9D^2 - 24D + 16)y = 0$ where $D \equiv \dfrac{d}{dt}$

(b) Substituting m for D gives the auxiliary equation $9m^2 - 24m + 16 = 0$

Factorising gives: $(3m-4)(3m-4)=0$, i.e. $m = \dfrac{4}{3}$ twice.

(c) Since the roots are real and equal, **the general solution is $y = (At+B)e^{\frac{4}{3}t}$**

(d) When $t=0$, $y=3$ hence $3=(0+B)e^0$, i.e. $B=3$

Since $y = (At+B)e^{\frac{4}{3}t}$

then $\dfrac{dy}{dt} = (At+B)\left(\dfrac{4}{3}e^{\frac{4}{3}t}\right) + Ae^{\frac{4}{3}t}$, by the product rule.

When $\quad t=0, \dfrac{dy}{dt} = 3$

thus $\quad 3 = (0+B)\dfrac{4}{3}e^0 + Ae^0$

i.e. $3 = \dfrac{4}{3}B + A$ from which, $A=-1$, since $B=3$

Hence the particular solution is

$$y = (-t+3)e^{\frac{4}{3}t} \text{ or } y = (3-t)e^{\frac{4}{3}t}$$

Problem 3. Solve the differential equation $\dfrac{d^2y}{dx^2} + 6\dfrac{dy}{dx} + 13y = 0$, given that when $x=0$, $y=3$ and $\dfrac{dy}{dx} = 7$

Using the procedure of Section 81.2:

(a) $\dfrac{d^2y}{dx^2} + 6\dfrac{dy}{dx} + 13y = 0$ in D-operator form is

$(D^2 + 6D + 13)y = 0$, where $D \equiv \dfrac{d}{dx}$

(b) Substituting m for D gives the auxiliary equation $m^2 + 6m + 13 = 0$

Using the quadratic formula:

$$m = \dfrac{-6 \pm \sqrt{[(6)^2 - 4(1)(13)]}}{2(1)}$$

$$= \dfrac{-6 \pm \sqrt{(-16)}}{2}$$

i.e. $m = \dfrac{-6 \pm j4}{2} = -3 \pm j2$

(c) Since the roots are complex, **the general solution is**

$$y = e^{-3x}(A\cos 2x + B\sin 2x)$$

(d) When $x=0$, $y=3$, hence $3 = e^0(A\cos 0 + B\sin 0)$, i.e. $A=3$

Since $y = e^{-3x}(A\cos 2x + B\sin 2x)$

then $\dfrac{dy}{dx} = e^{-3x}(-2A\sin 2x + 2B\cos 2x)$

$\qquad\qquad - 3e^{-3x}(A\cos 2x + B\sin 2x)$,

by the product rule,

$\qquad = e^{-3x}[(2B - 3A)\cos 2x$

$\qquad\qquad - (2A + 3B)\sin 2x]$

When $x=0$, $\dfrac{dy}{dx} = 7$,

hence $7 = e^0[(2B - 3A)\cos 0 - (2A + 3B)\sin 0]$
i.e. $7 = 2B - 3A$, from which, $B=8$, since $A=3$

Hence the particular solution is

$$y = e^{-3x}(3\cos 2x + 8\sin 2x)$$

Since, from Chapter 44, page 502,

$a\cos\omega t+b\sin\omega t=R\sin(\omega t+\alpha)$, where

$R=\sqrt{(a^2+b^2)}$ and $\alpha=\tan^{-1}\frac{a}{b}$ then

$3\cos 2x+8\sin 2x$

$$=\sqrt{(3^2+8^2)}\sin(2x+\tan^{-1}\tfrac{3}{8})$$

$$=\sqrt{73}\sin(2x+20.56°)$$

$$=\sqrt{73}\sin(2x+0.359)$$

Thus the particular solution may also be expressed as

$$y=\sqrt{73}\,e^{-3x}\sin(2x+0.359)$$

Now try the following Practice Exercise

Practice Exercise 310   Further problems on differential equations of the form

$a\frac{d^2y}{dx^2}+b\frac{dy}{dx}+cy=0$ **(answers on page 1187)**

In Problems 1 to 3, determine the general solution of the given differential equations.

1. $6\frac{d^2y}{dt^2}-\frac{dy}{dt}-2y=0$

2. $4\frac{d^2\theta}{dt^2}+4\frac{d\theta}{dt}+\theta=0$

3. $\frac{d^2y}{dx^2}+2\frac{dy}{dx}+5y=0$

In Problems 4 to 9, find the particular solution of the given differential equations for the stated boundary conditions.

4. $6\frac{d^2y}{dx^2}+5\frac{dy}{dx}-6y=0$; when $x=0$, $y=5$ and $\frac{dy}{dx}=-1$

5. $4\frac{d^2y}{dt^2}-5\frac{dy}{dt}+y=0$; when $t=0$, $y=1$ and $\frac{dy}{dt}=-2$

6. $(9D^2+30D+25)y=0$, where $D\equiv\frac{d}{dx}$; when $x=0$, $y=0$ and $\frac{dy}{dx}=2$

7. $\frac{d^2x}{dt^2}-6\frac{dx}{dt}+9x=0$; when $t=0$, $x=2$ and $\frac{dx}{dt}=0$

8. $\frac{d^2y}{dx^2}+6\frac{dy}{dx}+13y=0$; when $x=0$, $y=4$ and $\frac{dy}{dx}=0$

9. $(4D^2+20D+125)\theta=0$, where $D\equiv\frac{d}{dt}$; when $t=0$, $\theta=3$ and $\frac{d\theta}{dt}=2.5$

81.4   Further worked problems on practical differential equations of the form $a\frac{d^2y}{dx^2}+b\frac{dy}{dx}+cy=0$

Problem 4. The equation of motion of a body oscillating on the end of a spring is

$$\frac{d^2x}{dt^2}+100x=0$$

where x is the displacement in metres of the body from its equilibrium position after time t seconds. Determine x in terms of t given that at time $t=0$, $x=2m$ and $\frac{dx}{dt}=0$

An equation of the form $\frac{d^2x}{dt^2}+m^2x=0$ is a differential equation representing simple harmonic motion (SHM). Using the procedure of Section 81.2:

(a) $\frac{d^2x}{dt^2}+100x=0$ in D-operator form is $(D^2+100)x=0$

(b) The auxiliary equation is $m^2+100=0$, i.e. $m^2=-100$ and $m=\sqrt{(-100)}$, i.e. $m=\pm j10$

(c) Since the roots are complex, the general solution is $x=e^0(A\cos 10t+B\sin 10t)$,

i.e. $x=(A\cos 10t+B\sin 10t)$ **metres**

(d) When $t=0$, $x=2$, thus $2=A$

$$\frac{dx}{dt}=-10A\sin 10t+10B\cos 10t$$

When $t=0$, $\dfrac{dx}{dt}=0$

thus $0=-10A\sin 0+10B\cos 0$, i.e. $B=0$

Hence the particular solution is

$$x=2\cos 10t \text{ metres}$$

Problem 5. Given the differential equation $\dfrac{d^2V}{dt^2}=\omega^2 V$, where ω is a constant, show that its solution may be expressed as:

$$V=7\cosh\omega t+3\sinh\omega t$$

given the boundary conditions that when

$t=0$, $V=7$ and $\dfrac{dV}{dt}=3\omega$

Using the procedure of Section 81.2:

(a) $\dfrac{d^2V}{dt^2}=\omega^2 V$, i.e. $\dfrac{d^2V}{dt^2}-\omega^2 V=0$ in D-operator form is $(\mathrm{D}^2-\omega^2)v=0$, where $\mathrm{D}\equiv\dfrac{d}{dx}$

(b) The auxiliary equation is $m^2-\omega^2=0$, from which, $m^2=\omega^2$ and $m=\pm\omega$

(c) Since the roots are real and different, **the general solution is**

$$V=Ae^{\omega t}+Be^{-\omega t}$$

(d) When $t=0$, $V=7$ hence $7=A+B$ \hfill (1)

$$\dfrac{dV}{dt}=A\omega e^{\omega t}-B\omega e^{-\omega t}$$

When $t=0$, $\dfrac{dV}{dt}=3\omega$

thus $3\omega=A\omega-B\omega$

i.e. $3=A-B$ \hfill (2)

From equations (1) and (2), $A=5$ and $B=2$

Hence the particular solution is

$$V=5e^{\omega t}+2e^{-\omega t}$$

Since $\sinh\omega t=\tfrac{1}{2}(e^{\omega t}-e^{-\omega t})$

and $\cosh\omega t=\tfrac{1}{2}(e^{\omega t}+e^{-\omega t})$

then $\sinh\omega t+\cosh\omega t=e^{\omega t}$

and $\cosh\omega t-\sinh\omega t=e^{-\omega t}$ from Chapter 24.

Hence the particular solution may also be written as

$$V=5(\sinh\omega t+\cosh\omega t)$$
$$+2(\cosh\omega t-\sinh\omega t)$$

i.e. $V=(5+2)\cosh\omega t+(5-2)\sinh\omega t$

i.e. $\boldsymbol{V=7\cosh\omega t+3\sinh\omega t}$

Problem 6. The equation

$$\dfrac{d^2i}{dt^2}+\dfrac{R}{L}\dfrac{di}{dt}+\dfrac{1}{LC}i=0$$

represents a current i flowing in an electrical circuit containing resistance R, inductance L and capacitance C connected in series. If $R=200$ ohms, $L=0.20$ henry and $C=20\times10^{-6}$ farads, solve the equation for i given the boundary conditions that when $t=0$, $i=0$ and $\dfrac{di}{dt}=100$

Using the procedure of Section 81.2:

(a) $\dfrac{d^2i}{dt^2}+\dfrac{R}{L}\dfrac{di}{dt}+\dfrac{1}{LC}i=0$ in D-operator form is

$$\left(\mathrm{D}^2+\dfrac{R}{L}\mathrm{D}+\dfrac{1}{LC}\right)i=0 \text{ where } \mathrm{D}\equiv\dfrac{d}{dt}$$

(b) The auxiliary equation is $m^2+\dfrac{R}{L}m+\dfrac{1}{LC}=0$

Hence $m=\dfrac{-\dfrac{R}{L}\pm\sqrt{\left[\left(\dfrac{R}{L}\right)^2-4(1)\left(\dfrac{1}{LC}\right)\right]}}{2}$

When $R=200$, $L=0.20$ and $C=20\times10^{-6}$, then

$$m=\dfrac{-\dfrac{200}{0.20}\pm\sqrt{\left[\left(\dfrac{200}{0.20}\right)^2-\dfrac{4}{(0.20)(20\times10^{-6})}\right]}}{2}$$

$$=\dfrac{-1000\pm\sqrt{0}}{2}=-500$$

(c)   Since the two roots are real and equal (i.e. -500 twice, since for a second-order differential equation there must be two solutions), **the general solution is $i=(At+B)e^{-500t}$**

(d)   When $t=0$, $i=0$, hence $B=0$

$$\frac{di}{dt}=(At+B)(-500e^{-500t})+(e^{-500t})(A),$$

by the product rule

When $t=0$, $\dfrac{di}{dt}=100$, thus $100=-500B+A$

i.e. $A=100$, since $B=0$

Hence the particular solution is

$$i=100te^{-500t}$$

Problem 7.   The oscillations of a heavily damped pendulum satisfy the differential equation $\dfrac{d^2 x}{dt^2}+6\dfrac{dx}{dt}+8x=0$, where x cm is the displacement of the bob at time t seconds. The initial displacement is equal to $+4$ cm and the initial velocity $\left(\text{i.e. } \dfrac{dx}{dt}\right)$ is 8 cm/s. Solve the equation for x

Using the procedure of Section 81.2:

(a) $\dfrac{d^2 x}{dt^2}+6\dfrac{dx}{dt}+8x=0$ in D-operator form is $(D^2+6D+8)x=0$, where $D\equiv\dfrac{d}{dt}$

(b)   The auxiliary equation is $m^2+6m+8=0$

Factorising gives: $(m+2)(m+4)=0$, from which $m=-2$ or $m=-4$

(c)   Since the roots are real and different, **the general solution is $x=Ae^{-2t}+Be^{-4t}$**

(d)   Initial displacement means that time $t=0$. At this instant, $x=4$

Thus $4=A+B$ \hspace{2cm} (1)

Velocity,

$$\frac{dx}{dt}=-2Ae^{-2t}-4Be^{-4t}$$

$$\frac{dx}{dt}=8\,\text{cm/s when } t=0,$$

thus $8=-2A-4B$ \hspace{2cm} (2)

From equations (1) and (2),

$$A=12 \text{ and } B=-8$$

Hence the particular solution is

$$x=12e^{-2t}-8e^{-4t}$$

i.e. **displacement, $x=4(3e^{-2t}-2e^{-4t})$ cm**

Now try the following Practice Exercise

Practice Exercise 311   Further problems on second-order differential equations of the form $a\dfrac{d^2 y}{dx^2}+b\dfrac{dy}{dx}+cy=0$ (answers on page 1187)

1.   The charge, q, on a capacitor in a certain electrical circuit satisfies the differential equation $\dfrac{d^2 q}{dt^2}+4\dfrac{dq}{dt}+5q=0$. Initially (i.e. when $t=0$), $q=Q$ and $\dfrac{dq}{dt}=0$. Show that the charge in the circuit can be expressed as: $q=\sqrt{5}\,Qe^{-2t}\sin(t+0.464)$

2.   A body moves in a straight line so that its distance s metres from the origin after time t seconds is given by $\dfrac{d^2 s}{dt^2}+a^2 s=0$, where a is a constant. Solve the equation for s given that $s=c$ and $\dfrac{ds}{dt}=0$ when $t=\dfrac{2\pi}{a}$

3.   The motion of the pointer of a galvanometer about its position of equilibrium is represented by the equation

$$I\frac{d^2\theta}{dt^2}+K\frac{d\theta}{dt}+F\theta=0$$

If I, the moment of inertia of the pointer about its pivot, is 5×10^{-3}, K, the resistance due to friction at unit angular velocity, is 2×10^{-2} and F, the force on the spring necessary to produce unit displacement, is 0.20, solve the equation for θ in terms of t given that when $t=0$, $\theta=0.3$ and $\dfrac{d\theta}{dt}=0$

Section K

4. Determine an expression for x for a differential equation $\dfrac{d^2x}{dt^2}+2n\dfrac{dx}{dt}+n^2x=0$ which represents a critically damped oscillator, given that at time $t=0$, $x=s$ and $\dfrac{dx}{dt}=u$

5. $L\dfrac{d^2i}{dt^2}+R\dfrac{di}{dt}+\dfrac{1}{C}i=0$ is an equation representing current i in an electric circuit. If inductance L is 0.25 henry, capacitance C is 29.76×10^{-6} farads and R is 250 ohms, solve the equation for i given the boundary conditions that when $t=0$, $i=0$ and $\dfrac{di}{dt}=34$

6. The displacement s of a body in a damped mechanical system, with no external forces, satisfies the following differential equation:

$$2\dfrac{d^2s}{dt^2}+6\dfrac{ds}{dt}+4.5s=0$$

where t represents time. If initially, when $t=0, s=0$ and $\dfrac{ds}{dt}=4$, solve the differential equation for s in terms of t

For fully worked solutions to each of the problems in Practice Exercises 310 and 311 in this chapter, go to the website:
www.routledge.com/cw/bird

Chapter 82

Second-order differential equations of the form

$$a\frac{d^2y}{dx^2} + b\frac{dy}{dx} + cy = f(x)$$

Why it is important to understand: Second-order differential equations of the form $a\dfrac{d^2y}{dx^2} + b\dfrac{dy}{dx} + cy = f(x)$

As stated in the previous chapter, second-order differential equations have many engineering applications. Differential equations govern the fundamental operation of important areas such as automobile dynamics, tyre dynamics, aerodynamics, acoustics, active control systems, including speed control, engine performance and emissions control, climate control, ABS control systems, airbag deployment systems, structural dynamics of buildings, bridges and dams, for example, earthquake and wind engineering, industrial process control, control and operation of automation (robotic) systems, the operation of the electric power grid, electric power generation, orbital dynamics of satellite systems, heat transfer from electrical equipment (including computer chips), economic systems, biological systems, chemical systems, and so on. This chapter explains the procedure to solve second-order differential equations of the form $a\dfrac{d^2y}{dx^2} + b\dfrac{dy}{dx} + cy = f(x)$ for different functions $f(x)$.

At the end of this chapter, you should be able to:

- identify the complementary function and particular integral of a second order differential equation
- solve a second order differential equation of the form $a\dfrac{d^2y}{dx^2} + b\dfrac{dy}{dx} + cy = f(x)$

82.1 Complementary function and particular integral

If in the differential equation

$$a\frac{d^2y}{dx^2} + b\frac{dy}{dx} + cy = f(x) \tag{1}$$

the substitution $y = u + v$ is made then:

$$a\frac{d^2(u+v)}{dx^2} + b\frac{d(u+v)}{dx} + c(u+v) = f(x)$$

Rearranging gives:

$$\left(a\frac{d^2u}{dx^2}+b\frac{du}{dx}+cu\right)+\left(a\frac{d^2v}{dx^2}+b\frac{dv}{dx}+cv\right)$$
$$= f(x)$$

If we let

$$a\frac{d^2v}{dx^2}+b\frac{dv}{dx}+cv = f(x) \qquad (2)$$

then

$$a\frac{d^2u}{dx^2}+b\frac{du}{dx}+cu = 0 \qquad (3)$$

The general solution, u, of equation (3) will contain two unknown constants, as required for the general solution of equation (1). The method of solution of equation (3) is shown in Chapter 81. The function u is called the **complementary function (C.F.)**.

If the particular solution, v, of equation (2) can be determined without containing any unknown constants then $y = u + v$ will give the general solution of equation (1). The function v is called the **particular integral (P.I.)**. Hence the general solution of equation (1) is given by:

$$y = \textbf{C.F.} + \textbf{P.I.}$$

82.2 Procedure to solve differential equations of the form

$$a\frac{d^2y}{dx^2}+b\frac{dy}{dx}+cy = f(x)$$

(i) Rewrite the given differential equation as $(aD^2+bD+c)y = f(x)$

(ii) Substitute m for D, and solve the auxiliary equation $am^2+bm+c=0$ for m

(iii) Obtain the complementary function, u, which is achieved using the same procedure as in Section 81.2(c), page 865.

(iv) To determine the particular integral, v, firstly assume a particular integral which is suggested by $f(x)$, but which contains undetermined coefficients. Table 82.1 gives some suggested substitutions for different functions $f(x)$.

(v) Substitute the suggested P.I. into the differential equation $(aD^2+bD+c)v = f(x)$ and equate relevant coefficients to find the constants introduced.

(vi) The general solution is given by $y = \text{C.F.} + \text{P.I.}$, i.e. $y = u + v$

(vii) Given boundary conditions, arbitrary constants in the C.F. may be determined and the particular solution of the differential equation obtained.

82.3 Worked problems on differential equations of the form $a\dfrac{d^2y}{dx^2}+b\dfrac{dy}{dx}+cy = f(x)$ where $f(x)$ is a constant or polynomial

Problem 1. Solve the differential equation $\dfrac{d^2y}{dx^2}+\dfrac{dy}{dx}-2y = 4$

Using the procedure of Section 82.2:

(i) $\dfrac{d^2y}{dx^2}+\dfrac{dy}{dx}-2y=4$ in D-operator form is $(D^2+D-2)y=4$

(ii) Substituting m for D gives the auxiliary equation $m^2+m-2=0$. Factorising gives: $(m-1)(m+2)=0$, from which $m=1$ or $m=-2$

(iii) Since the roots are real and different, the C.F., $u = Ae^x+Be^{-2x}$

(iv) Since the term on the right-hand side of the given equation is a constant, i.e. $f(x)=4$, let the P.I. also be a constant, say $v=k$ (see Table 82.1(a)).

(v) Substituting $v=k$ into $(D^2+D-2)v=4$ gives $(D^2+D-2)k=4$. Since $D(k)=0$ and $D^2(k)=0$ then $-2k=4$, from which, $k=-2$ Hence the P.I., $v=-2$

(vi) The general solution is given by $y=u+v$, i.e. $y=Ae^x+Be^{-2x}-2$

Problem 2. Determine the particular solution of the equation $\dfrac{d^2y}{dx^2}-3\dfrac{dy}{dx}=9$, given the boundary conditions that when $x=0$, $y=0$ and $\dfrac{dy}{dx}=0$

Table 82.1 Form of particular integral for different functions

Type	*Straightforward cases* Try as particular integral:	*'Snag' cases* Try as particular integral:	*See problem*
(a) $f(x)=$ a constant	$v=k$	$v=kx$ (used when C.F. contains a constant)	1, 2
(b) $f(x)=$ polynomial (i.e. $f(x)=L+Mx+Nx^2+\cdots$ where any of the coefficients may be zero)	$v=a+bx+cx^2+\cdots$		3
(c) $f(x)=$ an exponential function (i.e. $f(x)=Ae^{ax}$)	$v=ke^{ax}$	(i) $v=kxe^{ax}$ (used when e^{ax} appears in the C.F.)	4, 5
		(ii) $v=kx^2e^{ax}$ (used when e^{ax} **and** xe^{ax} both appear in the C.F.)	6
(d) $f(x)=$ a sine or cosine function (i.e. $f(x)=a\sin px+b\cos px$, where a or b may be zero)	$v=A\sin px+B\cos px$	$v=x(A\sin px+B\cos px)$ (used when $\sin px$ and/or $\cos px$ appears in the C.F.)	7, 8
(e) $f(x)=$ a sum e.g. (i) $f(x)=4x^2-3\sin 2x$ (ii) $f(x)=2-x+e^{3x}$	(i) $v=ax^2+bx+c$ $\quad+d\sin 2x+e\cos 2x$ (ii) $v=ax+b+ce^{3x}$		9
(f) $f(x)=$ a product e.g. $f(x)=2e^x\cos 2x$	$v=e^x(A\sin 2x+B\cos 2x)$		10

Using the procedure of Section 82.2:

(i) $\dfrac{d^2y}{dx^2}-3\dfrac{dy}{dx}=9$ in D-operator form is $(D^2-3D)y=9$

(ii) Substituting m for D gives the auxiliary equation $m^2-3m=0$. Factorising gives: $m(m-3)=0$, from which, $m=0$ or $m=3$

(iii) Since the roots are real and different, the C.F., $u=Ae^0+Be^{3x}$, i.e. $\boldsymbol{u=A+Be^{3x}}$

(iv) Since the C.F. contains a constant (i.e. A) then let the P.I., $v=kx$ (see Table 82.1(a)).

(v) Substituting $v=kx$ into $(D^2-3D)v=9$ gives $(D^2-3D)kx=9$
$D(kx)=k$ and $D^2(kx)=0$
Hence $(D^2-3D)kx=0-3k=9$, from which, $k=-3$
Hence the P.I., $\boldsymbol{v=-3x}$

(vi) The general solution is given by $y=u+v$, i.e.
$\boldsymbol{y=A+Be^{3x}-3x}$

(vii) When $x=0$, $y=0$, thus $0=A+Be^0-0$, i.e.
$0=A+B$ \hfill (1)
$\dfrac{dy}{dx}=3Be^{3x}-3$; $\dfrac{dy}{dx}=0$ when $x=0$, thus
$0=3Be^0-3$ from which, $B=1$. From equation (1), $A=-1$

Hence the particular solution is

$$y=-1+1e^{3x}-3x$$

i.e. $\boldsymbol{y=e^{3x}-3x-1}$

Problem 3. Solve the differential equation
$2\dfrac{d^2y}{dx^2}-11\dfrac{dy}{dx}+12y=3x-2$

Using the procedure of Section 82.2:

(i) $2\dfrac{d^2y}{dx^2} - 11\dfrac{dy}{dx} + 12y = 3x - 2$ in D-operator form is

$$(2D^2 - 11D + 12)y = 3x - 2$$

(ii) Substituting m for D gives the auxiliary equation $2m^2 - 11m + 12 = 0$. Factorising gives: $(2m - 3)(m - 4) = 0$, from which, $m = \frac{3}{2}$ or $m = 4$

(iii) Since the roots are real and different, the C.F.,
$$u = Ae^{\frac{3}{2}x} + Be^{4x}$$

(iv) Since $f(x) = 3x - 2$ is a polynomial, let the P.I., $v = ax + b$ (see Table 82.1(b)).

(v) Substituting $v = ax + b$ into $(2D^2 - 11D + 12)v = 3x - 2$ gives:

$$(2D^2 - 11D + 12)(ax + b) = 3x - 2$$

i.e. $2D^2(ax + b) - 11D(ax + b)$
$$+ 12(ax + b) = 3x - 2$$

i.e. $0 - 11a + 12ax + 12b = 3x - 2$

Equating the coefficients of x gives: $12a = 3$, from which, $a = \frac{1}{4}$

Equating the constant terms gives:
$$-11a + 12b = -2$$

i.e. $-11\left(\frac{1}{4}\right) + 12b = -2$ from which,

$$12b = -2 + \frac{11}{4} = \frac{3}{4} \text{ i.e. } b = \frac{1}{16}$$

Hence the P.I., $v = ax + b = \dfrac{1}{4}x + \dfrac{1}{16}$

(vi) The general solution is given by $y = u + v$, i.e.

$$y = Ae^{\frac{3}{2}x} + Be^{4x} + \frac{1}{4}x + \frac{1}{16}$$

Now try the following Practice Exercise

Practice Exercise 312   Further problems on differential equations of the form

$a\dfrac{d^2y}{dx^2} + b\dfrac{dy}{dx} + cy = f(x)$ where $f(x)$ is a

constant or polynomial (answers on page 1187)

In Problems 1 and 2, find the general solutions of the given differential equations.

1. $2\dfrac{d^2y}{dx^2} + 5\dfrac{dy}{dx} - 3y = 6$

2. $6\dfrac{d^2y}{dx^2} + 4\dfrac{dy}{dx} - 2y = 3x - 2$

In Problems 3 and 4, find the particular solutions of the given differential equations.

3. $3\dfrac{d^2y}{dx^2} + \dfrac{dy}{dx} - 4y = 8$; when $x = 0$, $y = 0$ and $\dfrac{dy}{dx} = 0$

4. $9\dfrac{d^2y}{dx^2} - 12\dfrac{dy}{dx} + 4y = 3x - 1$; when $x = 0$, $y = 0$ and $\dfrac{dy}{dx} = -\dfrac{4}{3}$

5. The charge q in an electric circuit at time t satisfies the equation $L\dfrac{d^2q}{dt^2} + R\dfrac{dq}{dt} + \dfrac{1}{C}q = E$, where L, R, C and E are constants. Solve the equation given $L = 2$H, $C = 200 \times 10^{-6}$ F and $E = 250$ V, when (a) $R = 200\,\Omega$ and (b) R is negligible. Assume that when $t = 0$, $q = 0$ and $\dfrac{dq}{dt} = 0$

6. In a galvanometer the deflection θ satisfies the differential equation $\dfrac{d^2\theta}{dt^2} + 4\dfrac{d\theta}{dt} + 4\theta = 8$. Solve the equation for θ given that when $t = 0$, $\theta = \dfrac{d\theta}{dt} = 2$

82.4   Worked problems on differential equations of the form $a\dfrac{d^2y}{dx^2} + b\dfrac{dy}{dx} + cy = f(x)$ where $f(x)$ is an exponential function

Problem 4.   Solve the equation
$\dfrac{d^2y}{dx^2} - 2\dfrac{dy}{dx} + y = 3e^{4x}$ given the boundary conditions that when $x = 0$, $y = -\dfrac{2}{3}$ and $\dfrac{dy}{dx} = 4\dfrac{1}{3}$

Using the procedure of Section 82.2:

(i) $\dfrac{d^2y}{dx^2} - 2\dfrac{dy}{dx} + y = 3e^{4x}$ in D-operator form is
$(D^2 - 2D + 1)y = 3e^{4x}$

(ii)   Substituting m for D gives the auxiliary equation $m^2 - 2m + 1 = 0$. Factorising gives: $(m-1)(m-1) = 0$, from which, $m = 1$ twice.

(iii)   Since the roots are real and equal the C.F., $u = (Ax + B)e^x$

(iv)   Let the particular integral, $v = ke^{4x}$ (see Table 82.1(c)).

(v)   Substituting $v = ke^{4x}$ into
$(D^2 - 2D + 1)v = 3e^{4x}$ gives:
$$(D^2 - 2D + 1)ke^{4x} = 3e^{4x}$$
i.e. $D^2(ke^{4x}) - 2D(ke^{4x}) + 1(ke^{4x}) = 3e^{4x}$
i.e. $16ke^{4x} - 8ke^{4x} + ke^{4x} = 3e^{4x}$

Hence $9ke^{4x} = 3e^{4x}$, from which, $k = \frac{1}{3}$
Hence the P.I., $v = ke^{4x} = \frac{1}{3}e^{4x}$

(vi)   The general solution is given by $y = u + v$, i.e.
$y = (Ax + B)e^x + \frac{1}{3}e^{4x}$

(vii)   When $x = 0$, $y = -\frac{2}{3}$ thus
$-\frac{2}{3} = (0 + B)e^0 + \frac{1}{3}e^0$, from which, $B = -1$
$\dfrac{dy}{dx} = (Ax + B)e^x + e^x(A) + \frac{4}{3}e^{4x}$
When $x = 0$, $\dfrac{dy}{dx} = 4\frac{1}{3}$, thus $\dfrac{13}{3} = B + A + \frac{4}{3}$

from which, $A = 4$, since $B = -1$
Hence the particular solution is:

$$y = (4x - 1)e^x + \tfrac{1}{3}e^{4x}$$

Problem 5.   Solve the differential equation
$$2\dfrac{d^2y}{dx^2} - \dfrac{dy}{dx} - 3y = 5e^{\frac{3}{2}x}$$

Using the procedure of Section 82.2:

(i) $2\dfrac{d^2y}{dx^2} - \dfrac{dy}{dx} - 3y = 5e^{\frac{3}{2}x}$ in D-operator form is
$(2D^2 - D - 3)y = 5e^{\frac{3}{2}x}$

(ii)   Substituting m for D gives the auxiliary equation $2m^2 - m - 3 = 0$. Factorising gives: $(2m - 3)(m + 1) = 0$, from which $m = \frac{3}{2}$ or $m = -1$. Since the roots are real and different then the C.F., $u = Ae^{\frac{3}{2}x} + Be^{-x}$

(iii)   Since $e^{\frac{3}{2}x}$ appears in the C.F. **and** in the right-hand side of the differential equation, let the P.I., $v = kxe^{\frac{3}{2}x}$ (see Table 82.1(c), snag case (i)).

(iv)   Substituting $v = kxe^{\frac{3}{2}x}$ into $(2D^2 - D - 3)v = 5e^{\frac{3}{2}x}$
gives: $(2D^2 - D - 3)kxe^{\frac{3}{2}x} = 5e^{\frac{3}{2}x}$

$$D\left(kxe^{\frac{3}{2}x}\right) = (kx)\left(\tfrac{3}{2}e^{\frac{3}{2}x}\right) + \left(e^{\frac{3}{2}x}\right)(k),$$

by the product rule,

$$= ke^{\frac{3}{2}x}\left(\tfrac{3}{2}x + 1\right)$$

$$D^2\left(kxe^{\frac{3}{2}x}\right) = D\left[ke^{\frac{3}{2}x}\left(\tfrac{3}{2}x + 1\right)\right]$$

$$= \left(ke^{\frac{3}{2}x}\right)\left(\tfrac{3}{2}\right)$$

$$+ \left(\tfrac{3}{2}x + 1\right)\left(\tfrac{3}{2}ke^{\frac{3}{2}x}\right)$$

$$= ke^{\frac{3}{2}x}\left(\tfrac{9}{4}x + 3\right)$$

Hence $(2D^2 - D - 3)\left(kxe^{\frac{3}{2}x}\right)$

$$= 2\left[ke^{\frac{3}{2}x}\left(\tfrac{9}{4}x + 3\right)\right] - \left[ke^{\frac{3}{2}x}\left(\tfrac{3}{2}x + 1\right)\right]$$

$$- 3\left[kxe^{\frac{3}{2}x}\right] = 5e^{\frac{3}{2}x}$$

i.e. $\frac{9}{2}kxe^{\frac{3}{2}x} + 6ke^{\frac{3}{2}x} - \frac{3}{2}xke^{\frac{3}{2}x} - ke^{\frac{3}{2}x}$

$$- 3kxe^{\frac{3}{2}x} = 5e^{\frac{3}{2}x}$$

Equating coefficients of $e^{\frac{3}{2}x}$ gives: $5k = 5$, from which, $k = 1$
Hence the P.I., $v = kxe^{\frac{3}{2}x} = xe^{\frac{3}{2}x}$

(v)   The general solution is $y = u + v$, i.e.
$y = Ae^{\frac{3}{2}x} + Be^{-x} + xe^{\frac{3}{2}x}$

Problem 6.   Solve $\dfrac{d^2y}{dx^2} - 4\dfrac{dy}{dx} + 4y = 3e^{2x}$

Using the procedure of Section 82.2:

(i) $\dfrac{d^2y}{dx^2} - 4\dfrac{dy}{dx} + 4y = 3e^{2x}$ in D-operator form is
$(D^2 - 4D + 4)y = 3e^{2x}$

Section K

(ii) Substituting m for D gives the auxiliary equation $m^2 - 4m + 4 = 0$. Factorising gives: $(m-2)(m-2) = 0$, from which $m = 2$ twice.

(iii) Since the roots are real and equal, the C.F., $u = (Ax+B)e^{2x}$

(iv) Since e^{2x} **and** xe^{2x} both appear in the C.F., let the P.I., $v = kx^2e^{2x}$ (see Table 82.1(c), snag case (ii)).

(v) Substituting $v = kx^2e^{2x}$ into $(D^2 - 4D + 4)v = 3e^{2x}$ gives: $(D^2 - 4D + 4)(kx^2e^{2x}) = 3e^{2x}$

$$D(kx^2e^{2x}) = (kx^2)(2e^{2x}) + (e^{2x})(2kx)$$
$$= 2ke^{2x}(x^2 + x)$$
$$D^2(kx^2e^{2x}) = D[2ke^{2x}(x^2 + x)]$$
$$= (2ke^{2x})(2x+1) + (x^2+x)(4ke^{2x})$$
$$= 2ke^{2x}(4x + 1 + 2x^2)$$

Hence $(D^2 - 4D + 4)(kx^2e^{2x})$
$$= [2ke^{2x}(4x + 1 + 2x^2)]$$
$$- 4[2ke^{2x}(x^2 + x)] + 4[kx^2e^{2x}]$$
$$= 3e^{2x}$$

from which, $2ke^{2x} = 3e^{2x}$ and $k = \frac{3}{2}$

Hence the P.I., $v = kx^2e^{2x} = \frac{3}{2}x^2e^{2x}$

(vi) The general solution, $y = u + v$, i.e.

$$y = (Ax+B)e^{2x} + \tfrac{3}{2}x^2e^{2x}$$

Now try the following Practice Exercise

Practice Exercise 313   Further problems on differential equations of the form
$a\dfrac{d^2y}{dx^2} + b\dfrac{dy}{dx} + cy = f(x)$ **where** $f(x)$ **is an exponential function (answers on page 1187)**

In Problems 1 to 4, find the general solutions of the given differential equations.

1. $\dfrac{d^2y}{dx^2} - \dfrac{dy}{dx} - 6y = 2e^x$

2. $\dfrac{d^2y}{dx^2} - 3\dfrac{dy}{dx} - 4y = 3e^{-x}$

3. $\dfrac{d^2y}{dx^2} + 9y = 26e^{2x}$

4. $9\dfrac{d^2y}{dt^2} - 6\dfrac{dy}{dt} + y = 12e^{\frac{t}{3}}$

In Problems 5 and 6, find the particular solutions of the given differential equations.

5. $5\dfrac{d^2y}{dx^2} + 9\dfrac{dy}{dx} - 2y = 3e^x$, when $x = 0$, $y = \dfrac{1}{4}$
and $\dfrac{dy}{dx} = 0$

6. $\dfrac{d^2y}{dt^2} - 6\dfrac{dy}{dt} + 9y = 4e^{3t}$, when $t = 0$, $y = 2$
and $\dfrac{dy}{dt} = 0$

82.5 Worked problems on differential equations of the form $a\dfrac{d^2y}{dx^2} + b\dfrac{dy}{dx} + cy = f(x)$ where $f(x)$ is a sine or cosine function

Problem 7. Solve the differential equation
$2\dfrac{d^2y}{dx^2} + 3\dfrac{dy}{dx} - 5y = 6\sin 2x$

Using the procedure of Section 82.2:

(i) $2\dfrac{d^2y}{dx^2} + 3\dfrac{dy}{dx} - 5y = 6\sin 2x$ in D-operator form
is $(2D^2 + 3D - 5)y = 6\sin 2x$

(ii) The auxiliary equation is $2m^2 + 3m - 5 = 0$, from which,

$$(m-1)(2m+5) = 0$$

i.e. $m = 1$ or $m = -\frac{5}{2}$

(iii) Since the roots are real and different, the C.F.,
$u = Ae^x + Be^{-\frac{5}{2}x}$

(iv) Let the P.I., $v = A\sin 2x + B\cos 2x$ (see Table 82.1(d)).

(v) Substituting $v = A\sin 2x + B\cos 2x$ into

$(2D^2 + 3D - 5)v = 6\sin 2x$ gives:
$(2D^2 + 3D - 5)(A\sin 2x + B\cos 2x) = 6\sin 2x$

$D(A\sin 2x + B\cos 2x)$

$$= 2A\cos 2x - 2B\sin 2x$$

$D^2(A\sin 2x + B\cos 2x)$

$$= D(2A\cos 2x - 2B\sin 2x)$$

$$= -4A\sin 2x - 4B\cos 2x$$

Hence $(2D^2 + 3D - 5)(A\sin 2x + B\cos 2x)$

$$= -8A\sin 2x - 8B\cos 2x + 6A\cos 2x$$

$$- 6B\sin 2x - 5A\sin 2x - 5B\cos 2x$$

$$= 6\sin 2x$$

Equating coefficient of $\sin 2x$ gives:

$$-13A - 6B = 6 \qquad (1)$$

Equating coefficients of $\cos 2x$ gives:

$$6A - 13B = 0 \qquad (2)$$

$6 \times (1)$ gives: $-78A - 36B = 36 \qquad (3)$

$13 \times (2)$ gives: $78A - 169B = 0 \qquad (4)$

$(3) + (4)$ gives: $-205B = 36$

from which, $B = \dfrac{-36}{205}$

Substituting $B = \dfrac{-36}{205}$ into equation (1) or (2)

gives $A = \dfrac{-78}{205}$

Hence the P.I., $v = \dfrac{-78}{205}\sin 2x - \dfrac{36}{205}\cos 2x$

(vi) The general solution, $y = u + v$, i.e.

$$y = Ae^x + Be^{-\frac{5}{2}x}$$

$$- \frac{2}{205}(39\sin 2x + 18\cos 2x)$$

Problem 8.   Solve $\dfrac{d^2y}{dx^2} + 16y = 10\cos 4x$ given

$y = 3$ and $\dfrac{dy}{dx} = 4$ when $x = 0$

Using the procedure of Section 82.2:

(i) $\dfrac{d^2y}{dx^2} + 16y = 10\cos 4x$ in D-operator form is

$$(D^2 + 16)y = 10\cos 4x$$

(ii) The auxiliary equation is $m^2 + 16 = 0$, from which $m = \sqrt{-16} = \pm j4$

(iii) Since the roots are complex the C.F.,
$u = e^0(A\cos 4x + B\sin 4x)$

i.e. $\boldsymbol{u = A\cos 4x + B\sin 4x}$

(iv) Since $\sin 4x$ occurs in the C.F. **and** in the right-hand side of the given differential equation, let the P.I., $v = x(C\sin 4x + D\cos 4x)$ (see Table 82.1(d), snag case – constants C and D are used since A and B have already been used in the C.F.).

(v) Substituting $v = x(C\sin 4x + D\cos 4x)$ into $(D^2 + 16)v = 10\cos 4x$ gives:

$$(D^2 + 16)[x(C\sin 4x + D\cos 4x)] = 10\cos 4x$$

$$D[x(C\sin 4x + D\cos 4x)]$$

$$= x(4C\cos 4x - 4D\sin 4x)$$

$$+ (C\sin 4x + D\cos 4x)(1),$$

$$\text{by the product rule}$$

$$D^2[x(C\sin 4x + D\cos 4x)]$$

$$= x(-16C\sin 4x - 16D\cos 4x)$$

$$+ (4C\cos 4x - 4D\sin 4x)$$

$$+ (4C\cos 4x - 4D\sin 4x)$$

Hence $(D^2 + 16)[x(C\sin 4x + D\cos 4x)]$

$$= -16Cx\sin 4x - 16Dx\cos 4x + 4C\cos 4x$$

$$- 4D\sin 4x + 4C\cos 4x - 4D\sin 4x$$

$$+ 16Cx\sin 4x + 16Dx\cos 4x$$

$$= 10\cos 4x$$

i.e. $-8D\sin 4x + 8C\cos 4x = 10\cos 4x$

Equating coefficients of $\cos 4x$ gives:

$8C = 10$, from which $C = \dfrac{10}{8} = \dfrac{5}{4}$

Equating coefficients of $\sin 4x$ gives:
$-8D = 0$, from which $D = 0$

Hence the P.I., $\boldsymbol{v = x\left(\frac{5}{4}\sin 4x\right)}$

(vi) The general solution, $y = u + v$, i.e.

$$\boldsymbol{y = A\cos 4x + B\sin 4x + \tfrac{5}{4}x\sin 4x}$$

(vii)   When $x=0$, $y=3$, thus

$$3=A\cos 0 + B\sin 0 + 0, \text{ i.e. } A=3$$

$$\frac{dy}{dx} = -4A\sin 4x + 4B\cos 4x$$

$$+\tfrac{5}{4}x(4\cos 4x) + \tfrac{5}{4}\sin 4x$$

When $x=0$, $\frac{dy}{dx}=4$, thus

$$4 = -4A\sin 0 + 4B\cos 0 + 0 + \tfrac{5}{4}\sin 0$$

i.e. $4=4B$, from which, $B=1$

Hence the particular solution is

$$y = 3\cos 4x + \sin 4x + \tfrac{5}{4}x\sin 4x$$

Now try the following Practice Exercise

In Problems 1 to 3, find the general solutions of the given differential equations.

1. $2\dfrac{d^2y}{dx^2} - \dfrac{dy}{dx} - 3y = 25\sin 2x$

2. $\dfrac{d^2y}{dx^2} - 4\dfrac{dy}{dx} + 4y = 5\cos x$

3. $\dfrac{d^2y}{dx^2} + y = 4\cos x$

4. Find the particular solution of the differential equation $\dfrac{d^2y}{dx^2} - 3\dfrac{dy}{dx} - 4y = 3\sin x$; when $x=0$, $y=0$ and $\dfrac{dy}{dx}=0$

5. A differential equation representing the motion of a body is $\dfrac{d^2y}{dt^2} + n^2 y = k\sin pt$, where k, n and p are constants. Solve the equation (given $n\neq 0$ and $p^2 \neq n^2$) given that when $t=0$, $y=\dfrac{dy}{dt}=0$

6. The motion of a vibrating mass is given by $\dfrac{d^2y}{dt^2} + 8\dfrac{dy}{dt} + 20y = 300\sin 4t$. Show that the

general solution of the differential equation is given by:

$$y = e^{-4t}(A\cos 2t + B\sin 2t)$$

$$+ \frac{15}{13}(\sin 4t - 8\cos 4t)$$

7. $L\dfrac{d^2q}{dt^2} + R\dfrac{dq}{dt} + \dfrac{1}{C}q = V_0\sin\omega t$ represents the variation of capacitor charge in an electric circuit. Determine an expression for q at time t seconds given that $R=40\,\Omega$, $L=0.02\,\text{H}$, $C=50\times 10^{-6}\,\text{F}$, $V_0=540.8\,\text{V}$ and $\omega=200\,\text{rad/s}$ and given the boundary conditions that when $t=0$, $q=0$ and $\dfrac{dq}{dt}=4.8$

82.6   Worked problems on differential equations of the form $a\dfrac{d^2y}{dx^2}+b\dfrac{dy}{dx}+cy=f(x)$ where $f(x)$ is a sum or a product

Problem 9.   Solve $\dfrac{d^2y}{dx^2} + \dfrac{dy}{dx} - 6y = 12x - 50\sin x$

Using the procedure of Section 82.2:

(i) $\dfrac{d^2y}{dx^2} + \dfrac{dy}{dx} - 6y = 12x - 50\sin x$ in D-operator form is

$$(D^2 + D - 6)y = 12x - 50\sin x$$

(ii)   The auxiliary equation is $(m^2 + m - 6)=0$, from which,

$$(m-2)(m+3)=0,$$

i.e. $m=2$ or $m=-3$

(iii)   Since the roots are real and different, the C.F., $u = Ae^{2x} + Be^{-3x}$

(iv)   Since the right hand side of the given differential equation is the sum of a polynomial and a sine function let the P.I. $v = ax + b + c\sin x + d\cos x$ (see Table 82.1(e)).

(v)   Substituting v into

$(D^2 + D - 6)v = 12x - 50\sin x$ gives:

$(D^2 + D - 6)(ax + b + c\sin x + d\cos x)$

$$= 12x - 50\sin x$$

$D(ax + b + c\sin x + d\cos x)$

$$= a + c\cos x - d\sin x$$

$D^2(ax + b + c\sin x + d\cos x)$

$$= -c\sin x - d\cos x$$

Hence $(D^2 + D - 6)(v)$

$$= (-c\sin x - d\cos x) + (a + c\cos x$$
$$\quad - d\sin x) - 6(ax + b + c\sin x + d\cos x)$$

$$= 12x - 50\sin x$$

Equating constant terms gives:

$$a - 6b = 0 \tag{1}$$

Equating coefficients of x gives: $-6a = 12$, from which, $a = -2$

Hence, from (1), $b = -\frac{1}{3}$

Equating the coefficients of $\cos x$ gives:

$$-d + c - 6d = 0$$

i.e. $c - 7d = 0 \tag{2}$

Equating the coefficients of $\sin x$ gives:

$$-c - d - 6c = -50$$

i.e. $-7c - d = -50 \tag{3}$

Solving equations (2) and (3) gives: $c = 7$ and $d = 1$
Hence the P.I.,

$$v = -2x - \tfrac{1}{3} + 7\sin x + \cos x$$

(vi)   The general solution, $y = u + v$,

i.e. $y = Ae^{2x} + Be^{-3x} - 2x$

$$-\tfrac{1}{3} + 7\sin x + \cos x$$

Problem 10.   Solve the differential equation $\dfrac{d^2y}{dx^2} - 2\dfrac{dy}{dx} + 2y = 3e^x\cos 2x$, given that when $x = 0$, $y = 2$ and $\dfrac{dy}{dx} = 3$

Using the procedure of Section 82.2:

(i) $\dfrac{d^2y}{dx^2} - 2\dfrac{dy}{dx} + 2y = 3e^x\cos 2x$ in   D-operator form is

$$(D^2 - 2D + 2)y = 3e^x\cos 2x$$

(ii)   The auxiliary equation is $m^2 - 2m + 2 = 0$
Using the quadratic formula,

$$m = \frac{2 \pm \sqrt{[4 - 4(1)(2)]}}{2}$$

$$= \frac{2 \pm \sqrt{-4}}{2} = \frac{2 \pm j2}{2} \quad \text{i.e. } m = 1 \pm j1$$

(iii)   Since the roots are complex, the C.F.,
$u = e^x(A\cos x + B\sin x)$

(iv)   Since the right-hand side of the given differential equation is a product of an exponential and a cosine function, let the P.I., $v = e^x(C\sin 2x + D\cos 2x)$ (see Table 82.1(f) – again, constants C and D are used since A and B have already been used for the C.F.).

(v)   Substituting v into $(D^2 - 2D + 2)v = 3e^x\cos 2x$ gives:

$(D^2 - 2D + 2)[e^x(C\sin 2x + D\cos 2x)]$
$$= 3e^x\cos 2x$$

$D(v) = e^x(2C\cos 2x - 2D\sin 2x)$
$$+ e^x(C\sin 2x + D\cos 2x)$$
$$(\equiv e^x\{(2C + D)\cos 2x$$
$$+ (C - 2D)\sin 2x\})$$

$D^2(v) = e^x(-4C\sin 2x - 4D\cos 2x)$
$$+ e^x(2C\cos 2x - 2D\sin 2x)$$
$$+ e^x(2C\cos 2x - 2D\sin 2x)$$
$$+ e^x(C\sin 2x + D\cos 2x)$$
$$\equiv e^x\{(-3C - 4D)\sin 2x$$
$$+ (4C - 3D)\cos 2x\}$$

Hence $(D^2 - 2D + 2)v$

$$= e^x\{(-3C - 4D)\sin 2x$$
$$+ (4C - 3D)\cos 2x\}$$
$$- 2e^x\{(2C + D)\cos 2x$$
$$+ (C - 2D)\sin 2x\}$$
$$+ 2e^x(C\sin 2x + D\cos 2x)$$

$$= 3e^x\cos 2x$$

Equating coefficients of $e^x \sin 2x$ gives:

$$-3C - 4D - 2C + 4D + 2C = 0$$

i.e. $-3C = 0$, from which, $C = 0$

Equating coefficients of $e^x \cos 2x$ gives:

$$4C - 3D - 4C - 2D + 2D = 3$$

i.e. $-3D = 3$, from which, $D = -1$

Hence the P.I., $v = e^x(-\cos 2x)$

(vi) The general solution, $y = u + v$, i.e.

$$y = e^x(A \cos x + B \sin x) - e^x \cos 2x$$

(vii) When $x = 0$, $y = 2$ thus

$$2 = e^0(A \cos 0 + B \sin 0) - e^0 \cos 0$$

i.e. $\quad 2 = A - 1$, from which, $A = 3$

$$\frac{dy}{dx} = e^x(-A \sin x + B \cos x)$$
$$+ e^x(A \cos x + B \sin x)$$
$$- [e^x(-2 \sin 2x) + e^x \cos 2x]$$

When $\quad x = 0$, $\dfrac{dy}{dx} = 3$

thus $\quad 3 = e^0(-A \sin 0 + B \cos 0)$
$$+ e^0(A \cos 0 + B \sin 0)$$
$$- e^0(-2 \sin 0) - e^0 \cos 0$$

i.e. $\quad 3 = B + A - 1$, from which,

$$B = 1, \text{ since } A = 3$$

Hence the particular solution is

$$y = e^x(3 \cos x + \sin x) - e^x \cos 2x$$

Now try the following Practice Exercise

For fully worked solutions to each of the problems in Practice Exercises 312 to 315 in this chapter, go to the website:
www.routledge.com/cw/bird

Chapter 83

Power series methods of solving ordinary differential equations

Why it is important to understand: **Power series methods of solving ordinary differential equations**

The differential equations studied so far have all had closed form solutions, that is, their solutions could be expressed in terms of elementary functions, such as exponential, trigonometric, polynomial, and logarithmic functions, and most such elementary functions have expansions in terms of power series. However, there are a whole class of functions which are not elementary functions and which occur frequently in mathematical physics and engineering. These equations can sometimes be solved by discovering a power series that satisfies the differential equation, but the solution series may not be summable to an elementary function. In this chapter the methods of solution to such equations are explained.

At the end of this chapter, you should be able to:

- appreciate the reason for using power series methods to solve differential equations
- determine higher-order differential coefficients as a series
- use Leibniz's theorem to obtain the nth derivative of a given function
- obtain a power series solution of a differential equation by the Leibniz–Maclaurin method
- obtain a power series solution of a differential equation by the Frobenius method
- determine the general power series solution of Bessel's equation
- express Bessel's equation in terms of gamma functions
- determine the general power series solution of Legendre's equation
- determine Legendre polynomials
- determine Legendre polynomials using Rodrigues' formula

83.1 Introduction

Second-order ordinary differential equations that cannot be solved by analytical methods (as shown in Chapters 81 and 82), i.e. those involving variable coefficients, can often be solved in the form of an infinite series of powers of the variable. This chapter looks at some of the methods that make this possible – by the Leibniz–Maclaurin and Frobinius methods, involving Bessel's and Legendre's equations, Bessel and gamma functions and Legendre's polynomials. Before introducing Leibniz's theorem, some trends with higher differential coefficients are considered. To better understand this chapter it is necessary to be able to:

(i) differentiate standard functions (as explained in Chapters 53 and 58),

(ii) appreciate the binomial theorem (as explained in Chapter 21), and

(iii) use Maclaurin's theorem (as explained in Chapter 22).

83.2 Higher-order differential coefficients as series

The following is an extension of successive differentiation (see page 637), but looking for trends, or series, as the differential coefficient of common functions rises.

(i) If $y = e^{ax}$, then $\dfrac{dy}{dx} = ae^{ax}$, $\dfrac{d^2y}{dx^2} = a^2e^{ax}$, and so on.

If we abbreviate $\dfrac{dy}{dx}$ as y', $\dfrac{d^2y}{dx^2}$ as y'', ... and $\dfrac{d^ny}{dx^n}$ as $y^{(n)}$, then $y' = ae^{ax}$, $y'' = a^2e^{ax}$, and the emerging pattern gives: $\quad y^{(n)} = a^n e^{ax} \quad$ (1)

For example, if $y = 3e^{2x}$, then
$$\dfrac{d^7y}{dx^7} = y^{(7)} = 3(2^7)e^{2x} = 384e^{2x}$$

(ii) If $y = \sin ax$,
$$y' = a\cos ax = a\sin\left(ax + \dfrac{\pi}{2}\right)$$
$$y'' = -a^2\sin ax = a^2\sin(ax + \pi)$$
$$= a^2\sin\left(ax + \dfrac{2\pi}{2}\right)$$

$$y''' = -a^3\cos x$$
$$= a^3\sin\left(ax + \dfrac{3\pi}{2}\right) \text{ and so on.}$$

In general, $\quad y^{(n)} = a^n \sin\left(ax + \dfrac{n\pi}{2}\right) \quad$ (2)

For example, if $y = \sin 3x$,

then $\dfrac{d^5y}{dx^5} = y^{(5)}$
$$= 3^5\sin\left(3x + \dfrac{5\pi}{2}\right) = 3^5\sin\left(3x + \dfrac{\pi}{2}\right)$$
$$= 243\cos 3x$$

(iii) If $y = \cos ax$,
$$y' = -a\sin ax = a\cos\left(ax + \dfrac{\pi}{2}\right)$$
$$y'' = -a^2\cos ax = a^2\cos\left(ax + \dfrac{2\pi}{2}\right)$$
$$y''' = a^3\sin ax = a^3\cos\left(ax + \dfrac{3\pi}{2}\right) \text{ and so on.}$$

In general, $\quad y^{(n)} = a^n \cos\left(ax + \dfrac{n\pi}{2}\right) \quad$ (3)

For example, if $y = 4\cos 2x$,

then $\dfrac{d^6y}{dx^6} = y^{(6)} = 4(2^6)\cos\left(2x + \dfrac{6\pi}{2}\right)$
$$= 4(2^6)\cos(2x + 3\pi)$$
$$= 4(2^6)\cos(2x + \pi)$$
$$= -256\cos 2x$$

(iv) If $y = x^a$, $y' = ax^{a-1}$, $y'' = a(a-1)x^{a-2}$, $y''' = a(a-1)(a-2)x^{a-3}$, and $y^{(n)} = a(a-1)(a-2)\ldots\ldots(a-n+1)x^{a-n}$

or $\quad y^{(n)} = \dfrac{a!}{(a-n)!}x^{a-n} \quad$ (4)

where a is a positive integer.

For example, if $y = 2x^6$, then $\dfrac{d^4y}{dx^4} = y^{(4)}$
$$= (2)\dfrac{6!}{(6-4)!}x^{6-4}$$
$$= (2)\dfrac{6 \times 5 \times 4 \times 3 \times 2 \times 1}{2 \times 1}x^2$$
$$= 720x^2$$

(v) If $y = \sinh ax$, $y' = a \cosh ax$

$$y'' = a^2 \sinh ax$$

$$y''' = a^3 \cosh ax, \text{ and so on}$$

Since $\sinh ax$ is not periodic (see graph on page 230), it is more difficult to find a general statement for $y^{(n)}$. However, this is achieved with the following general series:

$$y^{(n)} = \frac{a^n}{2}\{[1 + (-1)^n]\sinh ax$$

$$+ [1 - (-1)^n]\cosh ax\} \qquad (5)$$

For example, if

$$y = \sinh 2x, \text{ then } \frac{d^5 y}{dx^5} = y^{(5)}$$

$$= \frac{2^5}{2}\{[1 + (-1)^5]\sinh 2x$$

$$+ [1 - (-1)^5]\cosh 2x\}$$

$$= \frac{2^5}{2}\{[0]\sinh 2x + [2]\cosh 2x\}$$

$$= 32 \cosh 2x$$

(vi) If $y = \cosh ax$,

$$y' = a \sinh ax$$

$$y'' = a^2 \cosh ax$$

$$y''' = a^3 \sinh ax, \text{ and so on}$$

Since $\cosh ax$ is not periodic (see graph on page 230), again it is more difficult to find a general statement for $y^{(n)}$. However, this is achieved with the following general series:

$$y^{(n)} = \frac{a^n}{2}\{[1 - (-1)^n]\sinh ax$$

$$+ [1 + (-1)^n]\cosh ax\} \qquad (6)$$

For example, if $y = \frac{1}{9} \cosh 3x$,

then $\dfrac{d^7 y}{dx^7} = y^{(7)} = \left(\dfrac{1}{9}\right)\dfrac{3^7}{2}(2 \sinh 3x)$

$$= 243 \sinh 3x$$

(vii) If $y = \ln ax$, $y' = \dfrac{1}{x}$, $y'' = -\dfrac{1}{x^2}$, $y''' = \dfrac{2}{x^3}$, and so on.

In general, $\qquad y^{(n)} = (-1)^{n-1}\dfrac{(n-1)!}{x^n}$ $\qquad (7)$

For example, if $y = \ln 5x$, then

$$\frac{d^6 y}{dx^6} = y^{(6)} = (-1)^{6-1}\left(\frac{5!}{x^6}\right) = -\frac{120}{x^6}$$

Note that if $y = \ln x$, $y' = \dfrac{1}{x}$; if in equation (7),

$n = 1$ then $y' = (-1)^0 \dfrac{(0)!}{x^1}$

$(-1)^0 = 1$ and if $y' = \dfrac{1}{x}$ then $(\mathbf{0})! = \mathbf{1}$ (check that $(-1)^0 = 1$ and $(0)! = 1$ on a calculator).

Now try the following Practice Exercise

Practice Exercise 316   Further problems on higher-order differential coefficients as series (answers on page 1188)

Determine the following derivatives:

1. (a) $y^{(4)}$ when $y = e^{2x}$ (b) $y^{(5)}$ when $y = 8 e^{\frac{t}{2}}$

2. (a) $y^{(4)}$ when $y = \sin 3t$
    (b) $y^{(7)}$ when $y = \dfrac{1}{50} \sin 5\theta$

3. (a) $y^{(8)}$ when $y = \cos 2x$
    (b) $y^{(9)}$ when $y = 3 \cos \dfrac{2}{3}t$

4. (a) $y^{(7)}$ when $y = 2x^9$ (b) $y^{(6)}$ when $y = \dfrac{t^7}{8}$

5. (a) $y^{(7)}$ when $y = \dfrac{1}{4} \sinh 2x$
    (b) $y^{(6)}$ when $y = 2 \sinh 3x$

6. (a) $y^{(7)}$ when $y = \cosh 2x$
    (b) $y^{(8)}$ when $y = \dfrac{1}{9} \cosh 3x$

7. (a) $y^{(4)}$ when $y = 2\ln 3\theta$
    (b) $y^{(7)}$ when $y = \dfrac{1}{3} \ln 2t$

83.3   Leibniz's theorem

If $y = uv$ (8)

where u and v are each functions of x, then by using the product rule,

$$y' = uv' + vu'$$ (9)

$$y'' = uv'' + v'u' + vu'' + u'v'$$

$$= u''v + 2u'v' + uv''$$ (10)

$$y''' = u''v' + vu''' + 2u'v'' + 2v'u'' + uv''' + v''u'$$

$$= u'''v + 3u''v' + 3u'v'' + uv'''$$ (11)

$$y^{(4)} = u^{(4)}v + 4u^{(3)}v^{(1)} + 6u^{(2)}v^{(2)}$$

$$+ 4u^{(1)}v^{(3)} + uv^{(4)}$$ (12)

From equations (8) to (12) it is seen that

(a)   the nth derivative of u decreases by 1 moving from left to right,

(b)   the nth derivative of v increases by 1 moving from left to right,

(c)   the coefficients $1, 4, 6, 4, 1$ are the normal binomial coefficients (see page 197).

In fact, $(uv)^{(n)}$ may be obtained by expanding $(u + v)^{(n)}$ using the binomial theorem (see page 198), where the 'powers' are interpreted as derivatives. Thus, expanding $(u + v)^{(n)}$ gives:

$$y^{(n)} = (uv)^{(n)} = u^{(n)}v + nu^{(n-1)}v^{(1)}$$

$$+ \frac{n(n-1)}{2!}u^{(n-2)}v^{(2)}$$

$$+ \frac{n(n-1)(n-2)}{3!}u^{(n-3)}v^{(3)} + \cdots$$ (13)

Equation (13) is a statement of **Leibniz's theorem**[*], which can be used to differentiate a product n times. The theorem is demonstrated in the following worked problems.

Problem 1.   Determine $y^{(n)}$ when $y = x^2 e^{3x}$

For a product $y = uv$, the function taken as

(i) u is the one whose nth derivative can readily be determined (from equations (1) to (7)),

---

(ii) v is the one whose derivative reduces to zero after a few stages of differentiation.

Thus, when $y = x^2 e^{3x}$, $v = x^2$, since its third derivative is zero, and $u = e^{3x}$ since the nth derivative is known from equation (1), i.e. $3^n e^{ax}$

Using Leibniz's theorem (equation (13)),

$$y^{(n)} = u^{(n)}v + nu^{(n-1)}v^{(1)} + \frac{n(n-1)}{2!}u^{(n-2)}v^{(2)}$$

$$+ \frac{n(n-1)(n-2)}{3!}u^{(n-3)}v^{(3)} + \cdots$$

where in this case $v = x^2$, $v^{(1)} = 2x$, $v^{(2)} = 2$ and $v^{(3)} = 0$

Hence, $y^{(n)} = (3^n e^{3x})(x^2) + n(3^{n-1}e^{3x})(2x)$

$$+ \frac{n(n-1)}{2!}(3^{n-2}e^{3x})(2)$$

$$+ \frac{n(n-1)(n-2)}{3!}(3^{n-3}e^{3x})(0)$$

$$= 3^{n-2}e^{3x}(3^2x^2 + n(3)(2x)$$

$$+ n(n-1) + 0)$$

i.e. $$y^{(n)} = e^{3x}3^{n-2}(9x^2 + 6nx + n(n-1))$$

Problem 2.   If $x^2y'' + 2xy' + y = 0$ show that:
$$xy^{(n+2)} + 2(n+1)xy^{(n+1)} + (n^2 + n + 1)y^{(n)} = 0$$

Differentiating each term of $x^2y'' + 2xy' + y = 0$
n times, using Leibniz's theorem of equation (13), gives:

$$\left\{ y^{(n+2)}x^2 + ny^{(n+1)}(2x) + \frac{n(n-1)}{2!}y^{(n)}(2) + 0 \right\}$$

$$+ \{ y^{(n+1)}(2x) + ny^{(n)}(2) + 0 \} + \{ y^{(n)} \} = 0$$

i.e. $x^2y^{(n+2)} + 2nxy^{(n+1)} + n(n-1)y^{(n)}$

$$+ 2xy^{(n+1)} + 2ny^{(n)} + y^{(n)} = 0$$

i.e. $x^2y^{(n+2)} + 2(n+1)xy^{(n+1)}$

$$+ (n^2 - n + 2n + 1)y^{(n)} = 0$$

or $$x^2y^{(n+2)} + 2(n+1)xy^{(n+1)}$$

$$+ (n^2 + n + 1)y^{(n)} = 0$$

Problem 3.   Differentiate the following differential equation n times:
$$(1 + x^2)y'' + 2xy' - 3y = 0$$

By Leibniz's equation, equation (13),

$$\left\{y^{(n+2)}(1+x^2)+n\,y^{(n+1)}(2x)+\frac{n(n-1)}{2!}y^{(n)}(2)+0\right\}$$

$$+2\{y^{(n+1)}(x)+n\,y^{(n)}(1)+0\}-3\{y^{(n)}\}=0$$

i.e. $(1+x^2)y^{(n+2)}+2nxy^{(n+1)}+n(n-1)y^{(n)}$

$$+2xy^{(n+1)}+2ny^{(n)}-3y^{(n)}=0$$

or $(1+x^2)y^{(n+2)}+2(n+1)xy^{(n+1)}$

$$+(n^2-n+2n-3)y^{(n)}=0$$

i.e. $(1+x^2)y^{(n+2)}+2(n+1)xy^{(n+1)}$

$$+(n^2+n-3)y^{(n)}=0$$

Problem 4. Find the 5th derivative of $y=x^4\sin x$

If $y=x^4\sin x$, then using Leibniz's equation with $u=\sin x$ and $v=x^4$ gives:

$$y^{(n)}=\left[\sin\left(x+\frac{n\pi}{2}\right)x^4\right]$$

$$+n\left[\sin\left(x+\frac{(n-1)\pi}{2}\right)4x^3\right]$$

$$+\frac{n(n-1)}{2!}\left[\sin\left(x+\frac{(n-2)\pi}{2}\right)12x^2\right]$$

$$+\frac{n(n-1)(n-2)}{3!}\left[\sin\left(x+\frac{(n-3)\pi}{2}\right)24x\right]$$

$$+\frac{n(n-1)(n-2)(n-3)}{4!}\left[\sin\left(x\right.\right.$$

$$\left.\left.+\frac{(n-4)\pi}{2}\right)24\right]$$

and $y^{(5)}=x^4\sin\left(x+\frac{5\pi}{2}\right)+20x^3\sin(x+2\pi)$

$$+\frac{(5)(4)}{2}(12x^2)\sin\left(x+\frac{3\pi}{2}\right)$$

$$+\frac{(5)(4)(3)}{(3)(2)}(24x)\sin(x+\pi)$$

$$+\frac{(5)(4)(3)(2)}{(4)(3)(2)}(24)\sin\left(x+\frac{\pi}{2}\right)$$

Since $\sin\left(x+\frac{5\pi}{2}\right)\equiv\sin\left(x+\frac{\pi}{2}\right)\equiv\cos x,$

$\sin(x+2\pi)\equiv\sin x,\ \sin\left(x+\frac{3\pi}{2}\right)\equiv-\cos x,$

and $\sin(x+\pi)\equiv-\sin x,$

then $y^{(5)}=x^4\cos x+20x^3\sin x+120x^2(-\cos x)$

$$+240x(-\sin x)+120\cos x$$

i.e. $y^{(5)}=(x^4-120x^2+120)\cos x$

$$+(20x^3-240x)\sin x$$

Now try the following Practice Exercise

Practice Exercise 317   Further problems on Leibniz's theorem (answers on page 1188)

Use the theorem of Leibniz in the following problems:

1. Obtain the nth derivative of: x^2y

2. If $y=x^3e^{2x}$ find $y^{(n)}$ and hence $y^{(3)}$

3. Determine the 4th derivative of: $y=2x^3e^{-x}$

4. If $y=x^3\cos x$ determine the 5th derivative.

5. Find an expression for $y^{(4)}$ if $y=e^{-t}\sin t$

6. If $y=x^5\ln 2x$, find $y^{(3)}$

7. Given $2x^2y''+xy'+3y=0$, show that $2x^2y^{(n+2)}+(4n+1)xy^{(n+1)}$
$$+(2n^2-n+3)y^{(n)}=0$$

8. If $y=(x^3+2x^2)e^{2x}$, determine an expansion for $y^{(5)}$

83.4 Power series solution by the Leibniz–Maclaurin method

For second-order differential equations that cannot be solved by algebraic methods, the **Leibniz–Maclaurin*** **method** produces a solution in the form of infinite

---

* **Who is** Maclaurin? For image and resume of Maclaurin, see page 206. For more information, go to **www.routledge.com/cw/bird**

series of powers of the unknown variable. The following simple **5-step procedure** may be used in the Leibniz–Maclaurin method:

(i) Differentiate the given equation n times, using the Leibniz theorem of equation (13),

(ii) rearrange the result to obtain the recurrence relation at $x = 0$

(iii) determine the values of the derivatives at $x = 0$, i.e. find $(y)_0$ and $(y')_0$

(iv) substitute in the Maclaurin expansion for $y = f(x)$ (see page 207, equation (5)),

(v) simplify the result where possible and apply boundary condition (if given).

The Leibniz–Maclaurin method is demonstrated, using the above procedure, in the following worked problems.

Problem 5. Determine the power series solution of the differential equation:
$$\frac{d^2 y}{dx^2} + x\frac{dy}{dx} + 2y = 0 \text{ using Leibniz–Maclaurin's}$$
method, given the boundary conditions that at
$$x = 0, y = 1 \text{ and } \frac{dy}{dx} = 2$$

Following the above procedure:

(i) The differential equation is rewritten as: $y'' + xy' + 2y = 0$ and from the Leibniz theorem of equation (13), each term is differentiated n times, which gives:

$$y^{(n+2)} + \{y^{(n+1)}(x) + n\,y^{(n)}(1) + 0\} + 2\,y^{(n)} = 0$$

i.e. $\qquad y^{(n+2)} + xy^{(n+1)} + (n+2)\,y^{(n)} = 0$

$$(14)$$

(ii) At $x = 0$, equation (14) becomes:

$$y^{(n+2)} + (n+2)\,y^{(n)} = 0$$

from which, $\quad y^{(n+2)} = -(n+2)\,y^{(n)}$

This equation is called a **recurrence relation** or **recurrence formula**, because each recurring term depends on a previous term.

(iii) Substituting $n = 0, 1, 2, 3, \ldots$ will produce a set of relationships between the various coefficients.

For $n = 0$, $\quad (y'')_0 = -2(y)_0$

$n = 1$, $\quad (y''')_0 = -3(y')_0$

$n = 2$, $\quad (y^{(4)})_0 = -4(y'')_0 = -4\{-2(y)_0\}$

$$= 2 \times 4(y)_0$$

$n = 3$, $\quad (y^{(5)})_0 = -5(y''')_0 = -5\{-3(y')_0\}$

$$= 3 \times 5(y')_0$$

$n = 4$, $\quad (y^{(6)})_0 = -6(y^{(4)})_0 = -6\{2 \times 4(y)_0\}$

$$= -2 \times 4 \times 6(y)_0$$

$n = 5$, $\quad (y^{(7)})_0 = -7(y^{(5)})_0 = -7\{3 \times 5(y')_0\}$

$$= -3 \times 5 \times 7(y')_0$$

$n = 6$, $\quad (y^{(8)})_0 = -8(y^{(6)})_0 =$

$$-8\{-2 \times 4 \times 6(y)_0\} = 2 \times 4 \times 6 \times 8(y)_0$$

(iv) Maclaurin's theorem from page 207 may be written as:

$$y = (y)_0 + x(y')_0 + \frac{x^2}{2!}(y'')_0 + \frac{x^3}{3!}(y''')_0$$

$$+ \frac{x^4}{4!}(y^{(4)})_0 + \cdots$$

Substituting the above values into Maclaurin's theorem gives:

$$y = (y)_0 + x(y')_0 + \frac{x^2}{2!}\{-2(y)_0\}$$

$$+ \frac{x^3}{3!}\{-3(y')_0\} + \frac{x^4}{4!}\{2 \times 4(y)_0\}$$

$$+ \frac{x^5}{5!}\{3 \times 5(y')_0\} + \frac{x^6}{6!}\{-2 \times 4 \times 6(y)_0\}$$

$$+ \frac{x^7}{7!}\{-3 \times 5 \times 7(y')_0\}$$

$$+ \frac{x^8}{8!}\{2 \times 4 \times 6 \times 8(y)_0\}$$

(v)   Collecting similar terms together gives:

$$y = (y)_0 \left\{ 1 - \frac{2x^2}{2!} + \frac{2 \times 4x^4}{4!} \right.$$

$$- \frac{2 \times 4 \times 6x^6}{6!} + \frac{2 \times 4 \times 6 \times 8x^8}{8!}$$

$$\left. - \cdots \right\} + (y')_0 \left\{ x - \frac{3x^3}{3!} + \frac{3 \times 5x^5}{5!} \right.$$

$$\left. - \frac{3 \times 5 \times 7x^7}{7!} + \cdots \right\}$$

i.e. $y = (y)_0 \left\{ 1 - \frac{x^2}{1} + \frac{x^4}{1 \times 3} - \frac{x^6}{3 \times 5} \right.$

$$\left. + \frac{x^8}{3 \times 5 \times 7} - \cdots \right\}$$

$$+ (y')_0 \left\{ \frac{x}{1} - \frac{x^3}{1 \times 2} + \frac{x^5}{2 \times 4} \right.$$

$$\left. - \frac{x^7}{2 \times 4 \times 6} + \cdots \right\}$$

The boundary conditions are that at $x = 0$, $y = 1$ and $\frac{dy}{dx} = 2$, i.e. $(y)_0 = 1$ and $(y')_0 = 2$

Hence, the power series solution of the differential equation: $\frac{d^2y}{dx^2} + x\frac{dy}{dx} + 2y = 0$ is:

$$y = \left\{ 1 - \frac{x^2}{1} + \frac{x^4}{1 \times 3} - \frac{x^6}{3 \times 5} \right.$$

$$\left. + \frac{x^8}{3 \times 5 \times 7} - \cdots \right\} + 2 \left\{ \frac{x}{1} - \frac{x^3}{1 \times 2} \right.$$

$$\left. + \frac{x^5}{2 \times 4} - \frac{x^7}{2 \times 4 \times 6} + \cdots \right\}$$

Problem 6.   Determine the power series solution of the differential equation:
$\frac{d^2y}{dx^2} + \frac{dy}{dx} + xy = 0$ given the boundary conditions that at $x = 0$, $y = 0$ and $\frac{dy}{dx} = 1$, using the Leibniz–Maclaurin method

Following the above procedure:

(i)   The differential equation is rewritten as: $y'' + y' + xy = 0$ and from the Leibniz theorem of equation (13), each term is differentiated n times, which gives:

$$y^{(n+2)} + y^{(n+1)} + y^{(n)}(x) + n\, y^{(n-1)}(1) + 0 = 0$$

i.e. $y^{(n+2)} + y^{(n+1)} + xy^{(n)} + n\, y^{(n-1)} = 0$

$$\tag{15}$$

(ii)   At $x = 0$, equation (15) becomes:

$$y^{(n+2)} + y^{(n+1)} + n\, y^{(n-1)} = 0$$

from which, $y^{(n+2)} = -\{ y^{(n+1)} + n\, y^{(n-1)} \}$

This is the **recurrence relation** and applies for $n \geq 1$

(iii)   Substituting $n = 1, 2, 3, \ldots$ will produce a set of relationships between the various coefficients.

For $n = 1$, $(y''')_0 = -\{(y'')_0 + (y)_0\}$

$n = 2$, $(y^{(4)})_0 = -\{(y''')_0 + 2(y')_0\}$

$n = 3$, $(y^{(5)})_0 = -\{(y^{(4)})_0 + 3(y'')_0\}$

$n = 4$, $(y^{(6)})_0 = -\{(y^{(5)})_0 + 4(y''')_0\}$

$n = 5$, $(y^{(7)})_0 = -\{(y^{(6)})_0 + 5(y^{(4)})_0\}$

$n = 6$, $(y^{(8)})_0 = -\{(y^{(7)})_0 + 6(y^{(5)})_0\}$

From the given boundary conditions, at $x = 0$, $y = 0$, thus $(y)_0 = 0$, and at $x = 0$, $\frac{dy}{dx} = 1$, thus $(y')_0 = 1$

From the given differential equation, $y'' + y' + xy = 0$, and, at $x = 0$, $(y'')_0 + (y')_0 + (0)y = 0$ from which, $(y'')_0 = -(y')_0 = -1$

Thus, $(y)_0 = \mathbf{0}$, $(y')_0 = \mathbf{1}$, $(y'')_0 = \mathbf{-1}$,

$$(y''')_0 = -\{(y'')_0 + (y)_0\} = -(-1 + 0) = \mathbf{1}$$

$$(y^{(4)})_0 = -\{(y''')_0 + 2(y')_0\}$$

$$= -[1 + 2(1)] = \mathbf{-3}$$

$$(y^{(5)})_0 = -\{(y^{(4)})_0 + 3(y'')_0\}$$

$$= -[-3 + 3(-1)] = \mathbf{6}$$

$$(y^{(6)})_0 = -\{(y^{(5)})_0 + 4(y''')_0\}$$

$$= -[6 + 4(1)] = \mathbf{-10}$$

$$(y^{(7)})_0 = -\{(y^{(6)})_0 + 5(y^{(4)})_0\}$$

$$= -[-10 + 5(-3)] = \mathbf{25}$$

$$(y^{(8)})_0 = -\{(y^{(7)})_0 + 6(y^{(5)})_0\}$$

$$= -[25 + 6(6)] = \mathbf{-61}$$

(iv)   Maclaurin's theorem states:

$$y = (y)_0 + x(y')_0 + \frac{x^2}{2!}(y'')_0 + \frac{x^3}{3!}(y''')_0$$

$$+ \frac{x^4}{4!}(y^{(4)})_0 + \cdots$$

and substituting the above values into Maclaurin's theorem gives:

$$y = 0 + x(1) + \frac{x^2}{2!}\{-1\} + \frac{x^3}{3!}\{1\} + \frac{x^4}{4!}\{-3\}$$

$$+ \frac{x^5}{5!}\{6\} + \frac{x^6}{6!}\{-10\} + \frac{x^7}{7!}\{25\}$$

$$+ \frac{x^8}{8!}\{-61\} + \cdots$$

(v)   Simplifying, the power series solution of the differential equation: $\dfrac{d^2y}{dx^2} + \dfrac{dy}{dx} + xy = 0$ is given by:

$$y = x - \frac{x^2}{2!} + \frac{x^3}{3!} - \frac{3x^4}{4!} + \frac{6x^5}{5!} - \frac{10x^6}{6!} + \frac{25x^7}{7!} - \frac{61x^8}{8!} + \cdots$$

Now try the following Practice Exercise

Practice Exercise 318   Further problems on power series solutions by the Leibniz–Maclaurin method (answers on page 1188)

1.   Determine the power series solution of the differential equation: $\dfrac{d^2y}{dx^2} + 2x\dfrac{dy}{dx} + y = 0$ using the Leibniz–Maclaurin method, given that at $x=0$, $y=1$ and $\dfrac{dy}{dx} = 2$

2.   Show that the power series solution of the differential equation: $(x+1)\dfrac{d^2y}{dx^2} + (x-1)\dfrac{dy}{dx} - 2y = 0$, using the Leibniz–Maclaurin method, is given by: $y = 1 + x^2 + e^{-x}$ given the boundary conditions that at $x=0$, $y=2$ and $\dfrac{dy}{dx} = -1$

3.   Find the particular solution of the differential equation: $(x^2+1)\dfrac{d^2y}{dx^2} + x\dfrac{dy}{dx} - 4y = 0$ using the Leibniz–Maclaurin method, given

the boundary conditions that at $x=0$, $y=1$ and $\dfrac{dy}{dx} = 1$

4.   Use the Leibniz–Maclaurin method to determine the power series solution for the differential equation: $x\dfrac{d^2y}{dx^2} + \dfrac{dy}{dx} + xy = 1$, given that at $x=0$, $y=1$ and $\dfrac{dy}{dx} = 2$

83.5   Power series solution by the Frobenius method

A differential equation of the form $y'' + Py' + Qy = 0$, where P and Q are both functions of x, such that the equation can be represented by a power series, may be solved by the **Frobenius* method**.

* **Who was** Frobenius? **Ferdinand Georg Frobenius** (26 October 1849–3 August 1917) was a German mathematician best known for his contributions to the theory of elliptic functions, differential equations and to group theory. He is also known for determinantal identities, known as Frobenius–Stickelberger formulae. To find out more go to **www.routledge.com/cw/bird**

The following **4-step procedure** may be used in the Frobenius method:

(i) Assume a trial solution of the form
$$y = x^c \left\{ a_0 + a_1 x + a_2 x^2 + a_3 x^3 + \cdots + a_r x^r + \cdots \right\}$$

(ii) differentiate the trial series,

(iii) substitute the results in the given differential equation,

(iv) equate coefficients of corresponding powers of the variable on each side of the equation; this enables index c and coefficients a_1, a_2, a_3, ... from the trial solution, to be determined.

This introductory treatment of the Frobenius method covering the simplest cases is demonstrated, using the above procedure, in the following Problems.

Problem 7. Determine, using the Frobenius method, the general power series solution of the differential equation: $3x \dfrac{d^2 y}{dx^2} + \dfrac{dy}{dx} - y = 0$

The differential equation may be rewritten as: $3xy'' + y' - y = 0$

(i) Let a trial solution be of the form

$$y = x^c \left\{ a_0 + a_1 x + a_2 x^2 + a_3 x^3 + \cdots \right.$$
$$\left. + a_r x^r + \cdots \right\} \quad (16)$$

where $a_0 \neq 0$,

i.e. $y = a_0 x^c + a_1 x^{c+1} + a_2 x^{c+2} + a_3 x^{c+3}$
$$+ \cdots + a_r x^{c+r} + \cdots \quad (17)$$

(ii) Differentiating equation (17) gives:

$$y' = a_0 c x^{c-1} + a_1 (c+1) x^c$$
$$+ a_2 (c+2) x^{c+1} + \cdots$$
$$+ a_r (c+r) x^{c+r-1} + \cdots$$

and $y'' = a_0 c (c-1) x^{c-2} + a_1 c (c+1) x^{c-1}$
$$+ a_2 (c+1)(c+2) x^c + \cdots$$
$$+ a_r (c+r-1)(c+r) x^{c+r-2} + \cdots$$

(iii) Substituting y, y' and y'' into each term of the given equation $3xy'' + y' - y = 0$ gives:

$$3xy'' = 3a_0 c (c-1) x^{c-1} + 3a_1 c (c+1) x^c$$
$$+ 3a_2 (c+1)(c+2) x^{c+1} + \cdots$$
$$+ 3a_r (c+r-1)(c+r) x^{c+r-1} + \cdots \quad (a)$$

$$y' = a_0 c x^{c-1} + a_1 (c+1) x^c + a_2 (c+2) x^{c+1}$$
$$+ \cdots + a_r (c+r) x^{c+r-1} + \cdots \quad (b)$$

$$-y = -a_0 x^c - a_1 x^{c+1} - a_2 x^{c+2} - a_3 x^{c+3}$$
$$- \cdots - a_r x^{c+r} - \cdots \quad (c)$$

(iv) The sum of these three terms forms the left-hand side of the equation. Since the right-hand side is zero, the coefficients of each power of x can be equated to zero.
For example, the coefficient of x^{c-1} is equated to zero, giving: $3a_0 c (c-1) + a_0 c = 0$

or $a_0 c [3c - 3 + 1] = a_0 c (3c - 2) = 0 \quad (18)$

The coefficient of x^c is equated to zero giving:
$3a_1 c (c+1) + a_1 (c+1) - a_0 = 0$

i.e. $a_1 (3c^2 + 3c + c + 1) - a_0$
$$= a_1 (3c^2 + 4c + 1) - a_0 = 0$$

or $a_1 (3c+1)(c+1) - a_0 = 0 \quad (19)$

In each of series (a), (b) and (c) an x^c term is involved, after which, a general relationship can be obtained for x^{c+r}, where $r \geq 0$
In series (a) and (b), terms in x^{c+r-1} are present; replacing r by $(r+1)$ will give the corresponding terms in x^{c+r}, which occurs in all three equations, i.e.

in series (a), $3a_{r+1}(c+r)(c+r+1) x^{c+r}$

in series (b), $a_{r+1}(c+r+1) x^{c+r}$

in series (c), $-a_r x^{c+r}$

Equating the total coefficients of x^{c+r} to zero gives:

$$3a_{r+1}(c+r)(c+r+1) + a_{r+1}(c+r+1)$$
$$- a_r = 0$$

which simplifies to:

$$a_{r+1}\{(c+r+1)(3c+3r+1)\} - a_r = 0 \qquad (20)$$

Equation (18), which was formed from the coefficients of the lowest power of x, i.e. x^{c-1}, is called the **indicial equation**, from which the value of c is obtained. From equation (18), since $a_0 \neq 0$, then $c = 0$ or $c = \dfrac{2}{3}$

(a) When $c = 0$:

From equation (19), if $c = 0$, $a_1(1 \times 1) - a_0 = 0$, i.e. $a_1 = a_0$

From equation (20), if $c = 0$,
$a_{r+1}(r+1)(3r+1) - a_r = 0$,

i.e. $a_{r+1} = \dfrac{a_r}{(r+1)(3r+1)} \qquad r \geq 0$

Thus, when $r = 1$, $a_2 = \dfrac{a_1}{(2 \times 4)} = \dfrac{a_0}{(2 \times 4)}$

since $a_1 = a_0$

when $r = 2$, $a_3 = \dfrac{a_2}{(3 \times 7)} = \dfrac{a_0}{(2 \times 4)(3 \times 7)}$

or $\dfrac{a_0}{(2 \times 3)(4 \times 7)}$

when $r = 3$, $a_4 = \dfrac{a_3}{(4 \times 10)}$

$= \dfrac{a_0}{(2 \times 3 \times 4)(4 \times 7 \times 10)}$

and so on.

From equation (16), the trial solution was:

$$y = x^c\{a_0 + a_1 x + a_2 x^2 + a_3 x^3 + \cdots + a_r x^r + \cdots\}$$

Substituting $c = 0$ and the above values of $a_1, a_2, a_3, \ldots$ into the trial solution gives:

$$y = x^0 \left\{ a_0 + a_0 x + \left(\frac{a_0}{(2 \times 4)} \right) x^2 \right.$$

$$+ \left(\frac{a_0}{(2 \times 3)(4 \times 7)} \right) x^3$$

$$\left. + \left(\frac{a_0}{(2 \times 3 \times 4)(4 \times 7 \times 10)} \right) x^4 + \cdots \right\}$$

i.e. $y = a_0 \left\{ 1 + x + \dfrac{x^2}{(2 \times 4)} + \dfrac{x^3}{(2 \times 3)(4 \times 7)} \right.$

$$\left. + \frac{x^4}{(2 \times 3 \times 4)(4 \times 7 \times 10)} + \cdots \right\} \qquad (21)$$

(b) When $c = \dfrac{2}{3}$:

From equation (19), if $c = \dfrac{2}{3}$, $a_1(3)\left(\dfrac{5}{3}\right) - a_0 = 0$, i.e.

$$a_1 = \frac{a_0}{5}$$

From equation (20), if $c = \dfrac{2}{3}$

$$a_{r+1}\left(\frac{2}{3} + r + 1\right)(2 + 3r + 1) - a_r = 0,$$

i.e. $a_{r+1}\left(r + \dfrac{5}{3}\right)(3r + 3) - a_r$

$$= a_{r+1}(3r^2 + 8r + 5) - a_r = 0,$$

i.e. $a_{r+1} = \dfrac{a_r}{(r+1)(3r+5)} \qquad r \geq 0$

Thus, when $r = 1$, $a_2 = \dfrac{a_1}{(2 \times 8)} = \dfrac{a_0}{(2 \times 5 \times 8)}$

since $a_1 = \dfrac{a_0}{5}$

when $r = 2$, $a_3 = \dfrac{a_2}{(3 \times 11)}$

$$= \frac{a_0}{(2 \times 3)(5 \times 8 \times 11)}$$

when $r = 3$, $a_4 = \dfrac{a_3}{(4 \times 14)}$

$$= \frac{a_0}{(2 \times 3 \times 4)(5 \times 8 \times 11 \times 14)}$$

and so on.

From equation (16), the trial solution was:

$$y = x^c\{a_0 + a_1 x + a_2 x^2 + a_3 x^3 + \cdots + a_r x^r + \cdots\}$$

Substituting $c = \dfrac{2}{3}$ and the above values of $a_1, a_2, a_3, \ldots$ into the trial solution gives:

$$y = x^{\frac{2}{3}} \left\{ a_0 + \left(\frac{a_0}{5}\right) x + \left(\frac{a_0}{2 \times 5 \times 8}\right) x^2 \right.$$

$$+ \left(\frac{a_0}{(2 \times 3)(5 \times 8 \times 11)} \right) x^3$$

$$\left. + \left(\frac{a_0}{(2 \times 3 \times 4)(5 \times 8 \times 11 \times 14)} \right) x^4 + \cdots \right\}$$

i.e. $y = a_0 x^{\frac{2}{3}} \left\{ 1 + \dfrac{x}{5} + \dfrac{x^2}{(2 \times 5 \times 8)} \right.$

$+ \dfrac{x^3}{(2 \times 3)(5 \times 8 \times 11)}$

$\left. + \dfrac{x^4}{(2 \times 3 \times 4)(5 \times 8 \times 11 \times 14)} + \cdots \right\}$ (22)

Since a_0 is an arbitrary (non-zero) constant in each solution, its value could well be different.
Let $a_0 = A$ in equation (21), and $a_0 = B$ in equation (22). Also, if the first solution is denoted by $u(x)$ and the second by $v(x)$, then the general solution of the given differential equation is $y = u(x) + v(x)$. Hence,

$y = A \left\{ 1 + x + \dfrac{x^2}{(2 \times 4)} + \dfrac{x^3}{(2 \times 3)(4 \times 7)} \right.$

$\left. + \dfrac{x^4}{(2 \times 3 \times 4)(4 \times 7 \times 10)} + \cdots \right\}$

$+ B x^{\frac{2}{3}} \left\{ 1 + \dfrac{x}{5} + \dfrac{x^2}{(2 \times 5 \times 8)} \right.$

$+ \dfrac{x^3}{(2 \times 3)(5 \times 8 \times 11)}$

$\left. + \dfrac{x^4}{(2 \times 3 \times 4)(5 \times 8 \times 11 \times 14)} + \cdots \right\}$

Problem 8. Use the Frobenius method to determine the general power series solution of the differential equation:
$$2x^2 \frac{d^2 y}{dx^2} - x \frac{dy}{dx} + (1 - x)y = 0$$

The differential equation may be rewritten as:
$2x^2 y'' - xy' + (1 - x)y = 0$

(i) Let a trial solution be of the form

$y = x^c \{ a_0 + a_1 x + a_2 x^2 + a_3 x^3 + \cdots$

$+ a_r x^r + \cdots \}$ (23)

where $a_0 \neq 0$,

i.e. $y = a_0 x^c + a_1 x^{c+1} + a_2 x^{c+2} + a_3 x^{c+3}$

$+ \cdots + a_r x^{c+r} + \cdots$ (24)

(ii) Differentiating equation (24) gives:

$y' = a_0 c x^{c-1} + a_1 (c+1) x^c + a_2 (c+2) x^{c+1}$

$+ \cdots + a_r (c+r) x^{c+r-1} + \cdots$

and $y'' = a_0 c(c-1) x^{c-2} + a_1 c(c+1) x^{c-1}$

$+ a_2 (c+1)(c+2) x^c + \cdots$

$+ a_r (c+r-1)(c+r) x^{c+r-2} + \cdots$

(iii) Substituting y, y' and y'' into each term of the given equation $2x^2 y'' - xy' + (1 - x)y = 0$ gives:

$2x^2 y'' = 2a_0 c(c-1) x^c + 2a_1 c(c+1) x^{c+1}$

$+ 2a_2 (c+1)(c+2) x^{c+2} + \cdots$

$+ 2a_r (c+r-1)(c+r) x^{c+r} + \cdots$ (a)

$-xy' = -a_0 c x^c - a_1 (c+1) x^{c+1}$

$- a_2 (c+2) x^{c+2} - \cdots$

$- a_r (c+r) x^{c+r} - \cdots$ (b)

$(1-x)y = (1-x)(a_0 x^c + a_1 x^{c+1} + a_2 x^{c+2}$

$+ a_3 x^{c+3} + \cdots + a_r x^{c+r} + \cdots)$

$= a_0 x^c + a_1 x^{c+1} + a_2 x^{c+2} + a_3 x^{c+3}$

$+ \cdots + a_r x^{c+r} + \cdots$

$- a_0 x^{c+1} - a_1 x^{c+2} - a_2 x^{c+3}$

$- a_3 x^{c+4} - \cdots - a_r x^{c+r+1} - \cdots$ (c)

(iv) The **indicial equation**, which is obtained by equating the coefficient of the lowest power of x to zero, gives the value(s) of c. Equating the total coefficients of x^c (from equations (a) to (c)) to zero gives:

$2a_0 c(c-1) - a_0 c + a_0 = 0$

i.e. $a_0 [2c(c-1) - c + 1] = 0$

i.e. $a_0 [2c^2 - 2c - c + 1] = 0$

i.e. $a_0 [2c^2 - 3c + 1] = 0$

i.e. $a_0 [(2c - 1)(c - 1)] = 0$

from which, $\quad c = 1$ or $c = \dfrac{1}{2}$

The coefficient of the general term, i.e. x^{c+r}, gives (from equations (a) to (c)):

$2a_r (c+r-1)(c+r) - a_r (c+r)$

$+ a_r - a_{r-1} = 0$

from which,

$a_r [2(c+r-1)(c+r) - (c+r) + 1] = a_{r-1}$

Section K

and $a_r = \dfrac{a_{r-1}}{2(c+r-1)(c+r)-(c+r)+1}$ (25)

(a) With $c=1$:

$$a_r = \dfrac{a_{r-1}}{2(r)(1+r)-(1+r)+1}$$

$$= \dfrac{a_{r-1}}{2r+2r^2-1-r+1}$$

$$= \dfrac{a_{r-1}}{2r^2+r} = \dfrac{a_{r-1}}{r(2r+1)}$$

Thus, when $r=1$,

$$a_1 = \dfrac{a_0}{1(2+1)} = \dfrac{a_0}{1\times 3}$$

when $r=2$,

$$a_2 = \dfrac{a_1}{2(4+1)} = \dfrac{a_1}{(2\times 5)}$$

$$= \dfrac{a_0}{(1\times 3)(2\times 5)} \text{ or } \dfrac{a_0}{(1\times 2)\times(3\times 5)}$$

when $r=3$,

$$a_3 = \dfrac{a_2}{3(6+1)} = \dfrac{a_2}{3\times 7}$$

$$= \dfrac{a_0}{(1\times 2\times 3)\times(3\times 5\times 7)}$$

when $r=4$,

$$a_4 = \dfrac{a_3}{4(8+1)} = \dfrac{a_3}{4\times 9}$$

$$= \dfrac{a_0}{(1\times 2\times 3\times 4)\times(3\times 5\times 7\times 9)}$$

and so on.

From equation (23), the trial solution was:

$$y = x^c\left\{a_0 + a_1 x + a_2 x^2 + a_3 x^3 + \cdots\right.$$

$$\left. + a_r x^r + \cdots\right\}$$

Substituting $c=1$ and the above values of $a_1, a_2, a_3, \ldots$ into the trial solution gives:

$$y = x^1\left\{a_0 + \dfrac{a_0}{(1\times 3)}x + \dfrac{a_0}{(1\times 2)\times(3\times 5)}x^2\right.$$

$$+ \dfrac{a_0}{(1\times 2\times 3)\times(3\times 5\times 7)}x^3$$

$$+ \dfrac{a_0}{(1\times 2\times 3\times 4)\times(3\times 5\times 7\times 9)}x^4$$

$$\left. + \cdots\right\}$$

i.e. $y = a_0 x^1\left\{1 + \dfrac{x}{(1\times 3)} + \dfrac{x^2}{(1\times 2)\times(3\times 5)}\right.$

$$+ \dfrac{x^3}{(1\times 2\times 3)\times(3\times 5\times 7)}$$

$$+ \dfrac{x^4}{(1\times 2\times 3\times 4)\times(3\times 5\times 7\times 9)}$$

$$\left. + \cdots\right\} \quad (26)$$

(b) With $c=\dfrac{1}{2}$:

$$a_r = \dfrac{a_{r-1}}{2(c+r-1)(c+r)-(c+r)+1}$$

from equation (25)

i.e. $a_r = \dfrac{a_{r-1}}{2\left(\dfrac{1}{2}+r-1\right)\left(\dfrac{1}{2}+r\right)-\left(\dfrac{1}{2}+r\right)+1}$

$$= \dfrac{a_{r-1}}{2\left(r-\dfrac{1}{2}\right)\left(r+\dfrac{1}{2}\right)-\dfrac{1}{2}-r+1}$$

$$= \dfrac{a_{r-1}}{2\left(r^2-\dfrac{1}{4}\right)-\dfrac{1}{2}-r+1}$$

$$= \dfrac{a_{r-1}}{2r^2-\dfrac{1}{2}-\dfrac{1}{2}-r+1} = \dfrac{a_{r-1}}{2r^2-r}$$

$$= \dfrac{a_{r-1}}{r(2r-1)}$$

Thus, when $r=1$, $a_1 = \dfrac{a_0}{1(2-1)} = \dfrac{a_0}{1\times 1}$

when $r=2$, $a_2 = \dfrac{a_1}{2(4-1)} = \dfrac{a_1}{(2\times 3)}$

$$= \dfrac{a_0}{(2\times 3)}$$

when $r=3$, $a_3 = \dfrac{a_2}{3(6-1)} = \dfrac{a_2}{3\times 5}$

$$= \dfrac{a_0}{(2\times 3)\times(3\times 5)}$$

when $r=4$, $a_4 = \dfrac{a_3}{4(8-1)} = \dfrac{a_3}{4\times 7}$

$$= \dfrac{a_0}{(2\times 3\times 4)\times(3\times 5\times 7)}$$

and so on.

From equation (23), the trial solution was:

$$y = x^c \left\{ a_0 + a_1 x + a_2 x^2 + a_3 x^3 + \cdots \right.$$
$$\left. + a_r x^r + \cdots \right\}$$

Substituting $c = \dfrac{1}{2}$ and the above values of $a_1, a_2, a_3, \ldots$ into the trial solution gives:

$$y = x^{\frac{1}{2}} \left\{ a_0 + a_0 x + \frac{a_0}{(2 \times 3)} x^2 + \frac{a_0}{(2 \times 3) \times (3 \times 5)} x^3 \right.$$
$$\left. + \frac{a_0}{(2 \times 3 \times 4) \times (3 \times 5 \times 7)} x^4 + \cdots \right\}$$

i.e. $y = a_0 x^{\frac{1}{2}} \left\{ 1 + x + \dfrac{x^2}{(2 \times 3)} \right.$

$$+ \frac{x^3}{(2 \times 3) \times (3 \times 5)}$$

$$+ \frac{x^4}{(2 \times 3 \times 4) \times (3 \times 5 \times 7)}$$

$$\left. + \cdots \right\} \qquad (27)$$

Since a_0 is an arbitrary (non-zero) constant in each solution, its value could well be different. Let $a_0 = A$ in equation (26), and $a_0 = B$ in equation (27). Also, if the first solution is denoted by $u(x)$ and the second by $v(x)$, then the general solution of the given differential equation is $y = u(x) + v(x)$,

i.e. $y = Ax \left\{ 1 + \dfrac{x}{(1 \times 3)} + \dfrac{x^2}{(1 \times 2) \times (3 \times 5)} \right.$

$$+ \frac{x^3}{(1 \times 2 \times 3) \times (3 \times 5 \times 7)}$$

$$+ \frac{x^4}{(1 \times 2 \times 3 \times 4) \times (3 \times 5 \times 7 \times 9)}$$

$$\left. + \cdots \right\} + Bx^{\frac{1}{2}} \left\{ 1 + x + \frac{x^2}{(2 \times 3)} \right.$$

$$+ \frac{x^3}{(2 \times 3) \times (3 \times 5)}$$

$$+ \frac{x^4}{(2 \times 3 \times 4) \times (3 \times 5 \times 7)} \left. + \cdots \right\}$$

Section K

Problem 9. Use the Frobenius method to determine the general power series solution of the differential equation: $\dfrac{d^2 y}{dx^2} - 2y = 0$

The differential equation may be rewritten as: $y'' - 2y = 0$

(i) Let a trial solution be of the form

$$y = x^c \left\{ a_0 + a_1 x + a_2 x^2 + a_3 x^3 + \cdots \right.$$
$$\left. + a_r x^r + \cdots \right\} \qquad (28)$$

where $a_0 \neq 0$,

i.e. $y = a_0 x^c + a_1 x^{c+1} + a_2 x^{c+2} + a_3 x^{c+3}$
$$+ \cdots + a_r x^{c+r} + \cdots \qquad (29)$$

(ii) Differentiating equation (29) gives:

$$y' = a_0 c x^{c-1} + a_1 (c+1) x^c + a_2 (c+2) x^{c+1}$$
$$+ \cdots + a_r (c+r) x^{c+r-1} + \cdots$$

and $y'' = a_0 c (c-1) x^{c-2} + a_1 c (c+1) x^{c-1}$
$$+ a_2 (c+1)(c+2) x^c + \cdots$$
$$+ a_r (c+r-1)(c+r) x^{c+r-2} + \cdots$$

(iii) Replacing r by $(r+2)$ in $a_r (c+r-1)(c+r) \, x^{c+r-2}$ gives: $a_{r+2} (c+r+1)(c+r+2) x^{c+r}$
Substituting y and y'' into each term of the given equation $y'' - 2y = 0$ gives:

$$y'' - 2y = a_0 c (c-1) x^{c-2} + a_1 c (c+1) x^{c-1}$$
$$+ [a_2 (c+1)(c+2) - 2a_0] x^c + \cdots$$
$$+ [a_{r+2} (c+r+1)(c+r+2)$$
$$- 2a_r] x^{c+r} + \cdots = 0 \qquad (30)$$

(iv) The **indicial equation** is obtained by equating the coefficient of the lowest power of x to zero.

Hence, $a_0 c (c-1) = 0$ from which, $c = 0$ or $c = 1$ since $a_0 \neq 0$

For the term in x^{c-1}, i.e. $a_1 c (c+1) = 0$

With $c = 1$, $a_1 = 0$; however, when $c = 0$, a_1 **is indeterminate**, since any value of a_1 combined with the zero value of c would make the product zero.

For the term in x^c,

$$a_2(c+1)(c+2) - 2a_0 = 0 \text{ from which,}$$

$$a_2 = \frac{2a_0}{(c+1)(c+2)} \qquad (31)$$

For the term in x^{c+r},

$$a_{r+2}(c+r+1)(c+r+2) - 2a_r = 0$$

from which,

$$a_{r+2} = \frac{2a_r}{(c+r+1)(c+r+2)} \qquad (32)$$

(a) When $c = 0$:

a_1 is indeterminate, and from equation (31)

$$a_2 = \frac{2a_0}{(1 \times 2)} = \frac{2a_0}{2!}$$

In general, $a_{r+2} = \dfrac{2a_r}{(r+1)(r+2)}$ and

when $r = 1$, $a_3 = \dfrac{2a_1}{(2 \times 3)} = \dfrac{2a_1}{(1 \times 2 \times 3)} = \dfrac{2a_1}{3!}$

when $r = 2$, $a_4 = \dfrac{2a_2}{3 \times 4} = \dfrac{4a_0}{4!}$

Hence, $y = x^0 \left\{ a_0 + a_1 x + \dfrac{2a_0}{2!}x^2 + \dfrac{2a_1}{3!}x^3 \right.$

$$\left. + \frac{4a_0}{4!}x^4 + \cdots \right\}$$

from equation (28)

$$= a_0 \left\{ 1 + \frac{2x^2}{2!} + \frac{4x^4}{4!} + \cdots \right\}$$

$$+ a_1 \left\{ x + \frac{2x^3}{3!} + \frac{4x^5}{5!} + \cdots \right\}$$

Since a_0 and a_1 are arbitrary constants depending on boundary conditions, let $a_0 = P$ and $a_1 = Q$, then:

$$y = P \left\{ 1 + \frac{2x^2}{2!} + \frac{4x^4}{4!} + \cdots \right\}$$

$$+ Q \left\{ x + \frac{2x^3}{3!} + \frac{4x^5}{5!} + \cdots \right\} \qquad (33)$$

(b) When $c = 1$:

$a_1 = 0$, and from equation (31),

$$a_2 = \frac{2a_0}{(2 \times 3)} = \frac{2a_0}{3!}$$

Since $c = 1$, $a_{r+2} = \dfrac{2a_r}{(c+r+1)(c+r+2)}$

$$= \frac{2a_r}{(r+2)(r+3)}$$

from equation (32) and when $r = 1$,

$$a_3 = \frac{2a_1}{(3 \times 4)} = 0 \text{ since } a_1 = 0$$

when $r = 2$,

$$a_4 = \frac{2a_2}{(4 \times 5)} = \frac{2}{(4 \times 5)} \times \frac{2a_0}{3!} = \frac{4a_0}{5!}$$

when $r = 3$,

$$a_5 = \frac{2a_3}{(5 \times 6)} = 0$$

Hence, when $c = 1$,

$$y = x^1 \left\{ a_0 + \frac{2a_0}{3!}x^2 + \frac{4a_0}{5!}x^4 + \cdots \right\}$$

from equation (28)

i.e. $y = a_0 \left\{ x + \dfrac{2x^3}{3!} + \dfrac{4x^5}{5!} + \cdots \right\}$

Again, a_0 is an arbitrary constant; let $a_0 = K$,

then $\quad y = K \left\{ x + \dfrac{2x^3}{3!} + \dfrac{4x^5}{5!} + \cdots \right\}$

However, this latter solution is not a separate solution, for it is the same form as the second series in equation (33). Hence, equation (33), with its two arbitrary constants P and Q, gives the general solution. This is always the case when the two values of c differ by an integer (i.e. whole number). From the above three Problems, the following can be deduced, and in future assumed:

(i) if two solutions of the indicial equation differ by a quantity *not* an integer, then two independent solutions $y = u(x) + v(x)$ result, the general solution of which is $y = Au + Bv$ (note: Problem 7 had $c = 0$ and $\dfrac{2}{3}$ and Problem 8 had $c = 1$ and $\dfrac{1}{2}$; in neither case did c differ by an integer)

(ii) if two solutions of the indicial equation *do* differ by an integer, as in Problem 9 where $c = 0$ and 1, and if one coefficient is indeterminate, as with when $c = 0$, then the complete solution is always given by using this value of c. Using the second value of c, i.e. $c = 1$ in Problem 9, always gives a series which is one of the series in the first solution.

Now try the following Practice Exercise

Practice Exercise 319   Further problems on power series solution by the Frobenius method (answers on page 1188)

1. Produce, using Frobenius' method, a power series solution for the differential equation:
    $$2x\frac{d^2y}{dx^2} + \frac{dy}{dx} - y = 0$$

2. Use the Frobenius method to determine the general power series solution of the differential equation $\dfrac{d^2y}{dx^2} + y = 0$

3. Determine the power series solution of the differential equation: $3x\dfrac{d^2y}{dx^2} + 4\dfrac{dy}{dx} - y = 0$ using the Frobenius method.

4. Show, using the Frobenius method, that the power series solution of the differential equation $\dfrac{d^2y}{dx^2} - y = 0$ may be expressed as $y = P\cosh x + Q\sinh x$, where P and Q are constants. (Hint: check the series expansions for $\cosh x$ and $\sinh x$ on page 235.)

83.6 Bessel's equation and Bessel's functions

One of the most important differential equations in applied mathematics is **Bessel's* equation**.
It is of the form:

$$x^2\frac{d^2y}{dx^2} + x\frac{dy}{dx} + (x^2 - v^2)y = 0$$

where v is a real constant. The equation, which has applications in electric fields, vibrations and heat conduction, may be solved using Frobenius' method of the previous section.

Problem 10.   Determine the general power series solution of Bessel's equation

Bessel's equation $x^2\dfrac{d^2y}{dx^2} + x\dfrac{dy}{dx} + (x^2 - v^2)y = 0$ may be rewritten as: $x^2y'' + xy' + (x^2 - v^2)y = 0$

Using the Frobenius method from page 889:

(i)   Let a trial solution be of the form
$$y = x^c\{a_0 + a_1x + a_2x^2 + a_3x^3 + \cdots$$
$$+ a_rx^r + \cdots\} \tag{34}$$
$$\text{where } a_0 \neq 0$$

i.e. $y = a_0x^c + a_1x^{c+1} + a_2x^{c+2} + a_3x^{c+3}$
$$+ \cdots + a_rx^{c+r} + \cdots \tag{35}$$

(ii)   Differentiating equation (35) gives:
$$y' = a_0cx^{c-1} + a_1(c+1)x^c$$
$$+ a_2(c+2)x^{c+1} + \cdots$$
$$+ a_r(c+r)x^{c+r-1} + \cdots$$

* **Who was Bessel? Friedrich Wilhelm Bessel** (22 July 1784–17 March 1846) was a German mathematician, astronomer, and the systematiser of the Bessel functions. Bessel produced a refinement on the orbital calculations for Halley's Comet and produced precise positions for 3222 stars. To find out more go to **www.routledge.com/cw/bird**

Section K

and $y'' = a_0 c(c-1)x^{c-2} + a_1 c(c+1)x^{c-1}$

$$+ a_2(c+1)(c+2)x^c + \cdots$$

$$+ a_r(c+r-1)(c+r)x^{c+r-2} + \cdots$$

(iii) Substituting y, y' and y'' into each term of the given equation $x^2 y'' + xy' + (x^2 - v^2)y = 0$ gives:

$$a_0 c(c-1)x^c + a_1 c(c+1)x^{c+1}$$

$$+ a_2(c+1)(c+2)x^{c+2} + \cdots$$

$$+ a_r(c+r-1)(c+r)x^{c+r} + \cdots + a_0 c x^c$$

$$+ a_1(c+1)x^{c+1} + a_2(c+2)x^{c+2} + \cdots$$

$$+ a_r(c+r)x^{c+r} + \cdots + a_0 x^{c+2} + a_1 x^{c+3}$$

$$+ a_2 x^{c+4} + \cdots + a_r x^{c+r+2} + \cdots - a_0 v^2 x^c$$

$$- a_1 v^2 x^{c+1} - \cdots - a_r v^2 x^{c+r} + \cdots = 0$$

(36)

(iv) The **indicial equation** is obtained by equating the coefficient of the lowest power of x to zero.

Hence, $\quad a_0 c(c-1) + a_0 c - a_0 v^2 = 0$

from which, $\quad a_0[c^2 - c + c - v^2] = 0$

i.e. $\quad a_0[c^2 - v^2] = 0$

from which, $\quad c = +v$ or $c = -v$ since $a_0 \neq 0$

For the term in x^{c+r},

$$a_r(c+r-1)(c+r) + a_r(c+r) + a_{r-2}$$

$$- a_r v^2 = 0$$

$$a_r[(c+r-1)(c+r) + (c+r) - v^2] = -a_{r-2}$$

i.e. $\quad a_r[(c+r)(c+r-1+1) - v^2] = -a_{r-2}$

i.e. $\quad a_r[(c+r)^2 - v^2] = -a_{r-2}$

i.e. the **recurrence relation** is:

$$a_r = \frac{a_{r-2}}{v^2 - (c+r)^2} \quad \text{for} \quad r \geq 2$$

(37)

For the term in x^{c+1},

$$a_1[c(c+1) + (c+1) - v^2] = 0$$

i.e. $\qquad\qquad a_1[(c+1)^2 - v^2] = 0$

but if $c = v$ $\qquad\qquad a_1[(v+1)^2 - v^2] = 0$

i.e. $\qquad\qquad a_1[2v+1] = 0$

Similarly, if $c = -v$ $\qquad\qquad a_1[1-2v] = 0$

The terms $(2v+1)$ *and* $(1-2v)$ cannot both be zero since v is a real constant, hence $a_1 = 0$

Since $\quad a_1 = 0$, then from equation (37) $a_3 = a_5 = a_7 = \ldots = 0$

and

$$a_2 = \frac{a_0}{v^2 - (c+2)^2}$$

$$a_4 = \frac{a_0}{[v^2 - (c+2)^2][v^2 - (c+4)^2]}$$

$$a_6 = \frac{a_0}{[v^2 - (c+2)^2][v^2 - (c+4)^2][v^2 - (c+6)^2]}$$

and so on.

When $c = +v$:

$$a_2 = \frac{a_0}{v^2 - (v+2)^2} = \frac{a_0}{v^2 - v^2 - 4v - 4}$$

$$= \frac{-a_0}{4+4v} = \frac{-a_0}{2^2(v+1)}$$

$$a_4 = \frac{a_0}{\left[v^2 - (v+2)^2\right]\left[v^2 - (v+4)^2\right]}$$

$$= \frac{a_0}{[-2^2(v+1)][-2^3(v+2)]}$$

$$= \frac{a_0}{2^5(v+1)(v+2)}$$

$$= \frac{a_0}{2^4 \times 2(v+1)(v+2)}$$

$$a_6 = \frac{a_0}{[v^2-(v+2)^2][v^2-(v+4)^2][v^2-(v+6)^2]}$$

$$= \frac{a_0}{[2^4 \times 2(v+1)(v+2)][-12(v+3)]}$$

$$= \frac{-a_0}{2^4 \times 2(v+1)(v+2) \times 2^2 \times 3(v+3)}$$

$$= \frac{-a_0}{2^6 \times 3!(v+1)(v+2)(v+3)} \quad \text{and so on.}$$

The resulting solution for $c = +v$ is given by:

$$y = u =$$

$$Ax^v \left\{ 1 - \frac{x^2}{2^2(v+1)} + \frac{x^4}{2^4 \times 2!\,(v+1)(v+2)} \right.$$

$$\left. - \frac{x^6}{2^6 \times 3!\,(v+1)(v+2)(v+3)} + \cdots \right\} \tag{38}$$

which is valid provided v is not a negative integer and where A is an arbitrary constant.

When $c = -v$:

$$a_2 = \frac{a_0}{v^2 - (-v+2)^2} = \frac{a_0}{v^2 - (v^2 - 4v + 4)}$$

$$= \frac{-a_0}{4 - 4v} = \frac{-a_0}{2^2(v-1)}$$

$$a_4 = \frac{a_0}{[2^2(v-1)][v^2 - (-v+4)^2]}$$

$$= \frac{a_0}{[2^2(v-1)][2^3(v-2)]}$$

$$= \frac{a_0}{2^4 \times 2(v-1)(v-2)}$$

Similarly, $a_6 = \dfrac{a_0}{2^6 \times 3!\,(v-1)(v-2)(v-3)}$

Hence,

$$y = w =$$

$$Bx^{-v} \left\{ 1 + \frac{x^2}{2^2(v-1)} + \frac{x^4}{2^4 \times 2!\,(v-1)(v-2)} \right.$$

$$\left. + \frac{x^6}{2^6 \times 3!\,(v-1)(v-2)(v-3)} + \cdots \right\}$$

which is valid provided v is not a positive integer and where B is an arbitrary constant.
The complete solution of Bessel's equation:

$$x^2 \frac{d^2 y}{dx^2} + x\frac{dy}{dx} + \left(x^2 - v^2\right)y = 0 \text{ is:}$$

$$y = u + w =$$

$$Ax^v \left\{ 1 - \frac{x^2}{2^2(v+1)} + \frac{x^4}{2^4 \times 2!\,(v+1)(v+2)} \right.$$

$$\left. - \frac{x^6}{2^6 \times 3!\,(v+1)(v+2)(v+3)} + \cdots \right\}$$

$$+ Bx^{-v} \left\{ 1 + \frac{x^2}{2^2(v-1)} \right.$$

$$+ \frac{x^4}{2^4 \times 2!\,(v-1)(v-2)}$$

$$\left. + \frac{x^6}{2^6 \times 3!\,(v-1)(v-2)(v-3)} + \cdots \right\} \tag{39}$$

The gamma function

The solution of the Bessel equation of Problem 10 may be expressed in terms of **gamma functions**. Γ is the upper case Greek letter gamma, and the gamma function $\Gamma(x)$ is defined by the integral

$$\Gamma(x) = \int_0^\infty t^{x-1}e^{-t}\,dt \tag{40}$$

and is convergent for $x > 0$

From equation (40), $\Gamma(x+1) = \displaystyle\int_0^\infty t^x e^{-t}\,dt$

and by using integration by parts (see page 753):

$$\Gamma(x+1) = \left[(t^x)\left(\frac{e^{-t}}{-1}\right) \right]_0^\infty$$

$$- \int_0^\infty \left(\frac{e^{-t}}{-1}\right) x\,t^{x-1}\,dx$$

$$= (0 - 0) + x \int_0^\infty e^{-t}t^{x-1}\,dt$$

$$= x\Gamma(x) \quad \text{from equation (40)}$$

This is an important recurrence relation for gamma functions.

Thus, since $\Gamma(x+1) = x\Gamma(x)$

then similarly, $\Gamma(x+2) = (x+1)\Gamma(x+1)$

$$= (x+1)x\Gamma(x) \tag{41}$$

and $\Gamma(x+3) = (x+2)\Gamma(x+2)$

$$= (x+2)(x+1)x\Gamma(x)$$

and so on.

These relationships involving gamma functions are used with Bessel functions.

Bessel functions

The power series solution of the Bessel equation may be written in terms of gamma functions, as shown in Problem 11 below.

Problem 11. Show that the power series solution of the Bessel equation of worked Problem 10 may be written in terms of the Bessel functions $J_v(x)$ and $J_{-v}(x)$ as:

$$AJ_v(x) + BJ_{-v}(x)$$

$$= \left(\frac{x}{2}\right)^v \left\{ \frac{1}{\Gamma(v+1)} - \frac{x^2}{2^2(1!)\Gamma(v+2)} \right.$$

$$\left. + \frac{x^4}{2^4(2!)\Gamma(v+4)} - \cdots \right\}$$

$$+ \left(\frac{x}{2}\right)^{-v} \left\{ \frac{1}{\Gamma(1-v)} - \frac{x^2}{2^2(1!)\Gamma(2-v)} \right.$$

$$\left. + \frac{x^4}{2^4(2!)\Gamma(3-v)} - \cdots \right\}$$

From Problem 10 above, **when $c = +v$,**

$$a_2 = \frac{-a_0}{2^2(v+1)}$$

If we let $a_0 = \dfrac{1}{2^v \Gamma(v+1)}$

then

$$a_2 = \frac{-1}{2^2(v+1)\,2^v\Gamma(v+1)} = \frac{-1}{2^{v+2}(v+1)\Gamma(v+1)}$$

$$= \frac{-1}{2^{v+2}\Gamma(v+2)} \quad \text{from equation (41)}$$

Similarly, $a_4 = \dfrac{a_2}{v^2 - (c+4)^2}$ from equation (37)

$$= \frac{a_2}{(v-c-4)(v+c+4)} = \frac{a_2}{-4(2v+4)}$$

$$\qquad\qquad\qquad\qquad\qquad \text{since } c = v$$

$$= \frac{-a_2}{2^3(v+2)} = \frac{-1}{2^3(v+2)}\frac{-1}{2^{v+2}\Gamma(v+2)}$$

$$= \frac{1}{2^{v+4}(2!)\Gamma(v+3)}$$

$$\qquad\qquad \text{since } (v+2)\Gamma(v+2) = \Gamma(v+3)$$

and $a_6 = \dfrac{-1}{2^{v+6}(3!)\Gamma(v+4)}$ and so on.

The **recurrence relation** is:

$$a_r = \frac{(-1)^{r/2}}{2^{v+r}\left(\frac{r}{2}!\right)\Gamma\left(v+\frac{r}{2}+1\right)}$$

And if we let $r = 2k$, then

$$a_{2k} = \frac{(-1)^k}{2^{v+2k}(k!)\Gamma(v+k+1)} \qquad (42)$$

$$\text{for } k = 1, 2, 3, \ldots$$

Hence, it is possible to write the new form for equation (38) as:

$$y = Ax^v \left\{ \frac{1}{2^v\Gamma(v+1)} - \frac{x^2}{2^{v+2}(1!)\Gamma(v+2)} \right.$$

$$\left. + \frac{x^4}{2^{v+4}(2!)\Gamma(v+4)} - \cdots \right\}$$

This is called *the Bessel function of the first order kind, of order v,* and is denoted by $J_v(x)$,

i.e. $J_v(x) = \left(\dfrac{x}{2}\right)^v \left\{ \dfrac{1}{\Gamma(v+1)} - \dfrac{x^2}{2^2(1!)\Gamma(v+2)} \right.$

$$\left. + \frac{x^4}{2^4(2!)\Gamma(v+4)} - \cdots \right\}$$

$$\text{provided } v \text{ is not a negative integer.}$$

For the second solution, **when $c = -v$,** replacing v by $-v$ in equation (42) above gives:

$$a_{2k} = \frac{(-1)^k}{2^{2k-v}(k!)\,\Gamma(k-v+1)}$$

from which, when $k = 0$, $a_0 = \dfrac{(-1)^0}{2^{-v}(0!)\Gamma(1-v)} =$

$\dfrac{1}{2^{-v}\Gamma(1-v)}$ since $0! = 1$ (see page 883)

when $k = 1$, $a_2 = \dfrac{(-1)^1}{2^{2-v}(1!)\,\Gamma(1-v+1)}$

$$= \frac{-1}{2^{2-v}(1!)\Gamma(2-v)}$$

when $k = 2$, $a_4 = \dfrac{(-1)^2}{2^{4-v}(2!)\Gamma(2-v+1)}$

$$= \frac{1}{2^{4-v}(2!)\Gamma(3-v)}$$

when $k = 3$, $a_6 = \dfrac{(-1)^3}{2^{6-v}(3!)\,\Gamma(3-v+1)}$

$$= \frac{1}{2^{6-v}(3!)\Gamma(4-v)} \quad \text{and so on.}$$

Hence, $y = Bx^{-v}\left\{\dfrac{1}{2^{-v}\Gamma(1-v)} - \dfrac{x^2}{2^{2-v}(1!)\Gamma(2-v)}\right.$

$\left. + \dfrac{x^4}{2^{4-v}(2!)\Gamma(3-v)} - \cdots\right\}$

i.e. $J_{-v}(x) = \left(\dfrac{x}{2}\right)^{-v}\left\{\dfrac{1}{\Gamma(1-v)} - \dfrac{x^2}{2^2(1!)\Gamma(2-v)}\right.$

$\left. + \dfrac{x^4}{2^4(2!)\Gamma(3-v)} - \cdots\right\}$

provided v is not a positive integer.

$J_v(x)$ and $J_{-v}(x)$ are two independent solutions of the Bessel equation; the complete solution is:

$y = AJ_v(x) + BJ_{-v}(x)$ where A and B are constants

i.e. $\boldsymbol{y = AJ_v(x) + BJ_{-v}(x)}$

$= A\left(\dfrac{x}{2}\right)^v\left\{\dfrac{1}{\Gamma(v+1)} - \dfrac{x^2}{2^2(1!)\Gamma(v+2)}\right.$

$\left. + \dfrac{x^4}{2^4(2!)\Gamma(v+4)} - \cdots\right\}$

$+ B\left(\dfrac{x}{2}\right)^{-v}\left\{\dfrac{1}{\Gamma(1-v)} - \dfrac{x^2}{2^2(1!)\Gamma(2-v)}\right.$

$\left. + \dfrac{x^4}{2^4(2!)\Gamma(3-v)} - \cdots\right\}$

In general terms: $J_v(x) = \left(\dfrac{x}{2}\right)^v \displaystyle\sum_{k=0}^{\infty} \dfrac{(-1)^k x^{2k}}{2^{2k}(k!)\Gamma(v+k+1)}$

and $J_{-v}(x) = \left(\dfrac{x}{2}\right)^{-v} \displaystyle\sum_{k=0}^{\infty} \dfrac{(-1)^k x^{2k}}{2^{2k}(k!)\Gamma(k-v+1)}$

Another Bessel function

It may be shown that another series for $J_n(x)$ is given by:

$J_n(x) = \left(\dfrac{x}{2}\right)^n\left\{\dfrac{1}{n!} - \dfrac{1}{(n+1)!}\left(\dfrac{x}{2}\right)^2\right.$

$\left. + \dfrac{1}{(2!)(n+2)!}\left(\dfrac{x}{2}\right)^4 - \cdots\right\}$

From this series two commonly used function are derived,

i.e. $J_0(x) = \dfrac{1}{(0!)} - \dfrac{1}{(1!)^2}\left(\dfrac{x}{2}\right)^2 + \dfrac{1}{(2!)^2}\left(\dfrac{x}{2}\right)^4$

$- \dfrac{1}{(3!)^2}\left(\dfrac{x}{2}\right)^6 + \cdots$

$= 1 - \dfrac{x^2}{2^2(1!)^2} + \dfrac{x^4}{2^4(2!)^2} - \dfrac{x^6}{2^6(3!)^2} + \cdots$

and $J_1(x) = \dfrac{x}{2}\left\{\dfrac{1}{(1!)} - \dfrac{1}{(1!)(2!)}\left(\dfrac{x}{2}\right)^2\right.$

$\left. + \dfrac{1}{(2!)(3!)}\left(\dfrac{x}{2}\right)^4 - \cdots\right\}$

$= \dfrac{x}{2} - \dfrac{x^3}{2^3(1!)(2!)} + \dfrac{x^5}{2^5(2!)(3!)}$

$- \dfrac{x^7}{2^7(3!)(4!)} + \cdots$

Tables of Bessel functions are available for a range of values of n and x, and in these, $J_0(x)$ and $J_1(x)$ are most commonly used.

Graphs of $J_0(x)$, which looks similar to a cosine, and $J_1(x)$, which looks similar to a sine, are shown in Figure 83.1.

Now try the following Practice Exercise

Practice Exercise 320   Further problems on Bessel's equation and Bessel's functions (answers on page 1188)

1. Determine the power series solution of Bessel's equation: $x^2\dfrac{d^2y}{dx^2} + x\dfrac{dy}{dx} + (x^2 - v^2)y = 0$ when $v = 2$, up to and including the term in x^4

2. Find the power series solution of the Bessel function: $x^2y'' + xy' + \left(x^2 - v^2\right)y = 0$ in terms of the Bessel function $J_3(x)$ when $v = 3$. Give the answer up to and including the term in x^7

3. Evaluate the Bessel functions $J_0(x)$ and $J_1(x)$ when $x = 1$, correct to 3 decimal places.

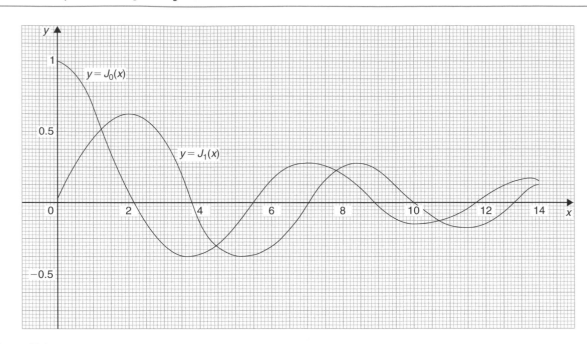

Figure 83.1

83.7 Legendre's equation and Legendre polynomials

Another important differential equation in physics and engineering applications is **Legendre's* equation** of the form: $(1 - x^2)\dfrac{d^2 y}{dx^2} - 2x\dfrac{dy}{dx} + k(k+1)y = 0$ or $(1 - x^2)y'' - 2xy' + k(k+1)y = 0$, where k is a real constant.

> **Problem 12.** Determine the general power series solution of Legendre's equation

To solve Legendre's equation
$(1 - x^2)y'' - 2xy' + k(k+1)y = 0$ using the Frobenius method:

(i) Let a trial solution be of the form

$$y = x^c \left\{ a_0 + a_1 x + a_2 x^2 + a_3 x^3 \right.$$
$$\left. + \cdots + a_r x^r + \cdots \right\} \quad (43)$$

where $a_0 \neq 0$,

i.e. $y = a_0 x^c + a_1 x^{c+1} + a_2 x^{c+2} + a_3 x^{c+3}$
$$+ \cdots + a_r x^{c+r} + \cdots \quad (44)$$

* **Who was Legendre?** **Adrien-Marie Legendre** (18 September 1752–10 January 1833) was a French mathematician. Legendre developed the least-squares method, which is used in linear regression, signal processing, statistics, and curve fitting. To find out more go to **www.routledge.com/cw/bird**

(ii) Differentiating equation (44) gives:

$$y' = a_0 c x^{c-1} + a_1(c+1)x^c$$

$$+ a_2(c+2)x^{c+1} + \cdots$$

$$+ a_r(c+r)x^{c+r-1} + \cdots$$

and $y'' = a_0 c(c-1)x^{c-2} + a_1 c(c+1)x^{c-1}$

$$+ a_2(c+1)(c+2)x^c + \cdots$$

$$+ a_r(c+r-1)(c+r)x^{c+r-2} + \cdots$$

(iii) Substituting y, y' and y'' into each term of the given equation:

$$(1-x^2)\,y'' - 2xy' + k(k+1)y = 0 \text{ gives:}$$

$$a_0 c(c-1)x^{c-2} + a_1 c(c+1)x^{c-1}$$

$$+ a_2(c+1)(c+2)x^c + \cdots$$

$$+ a_r(c+r-1)(c+r)x^{c+r-2} + \cdots$$

$$- a_0 c(c-1)x^c - a_1 c(c+1)x^{c+1}$$

$$- a_2(c+1)(c+2)x^{c+2} - \cdots$$

$$- a_r(c+r-1)(c+r)x^{c+r} - \cdots - 2a_0 c x^c$$

$$- 2a_1(c+1)x^{c+1} - 2a_2(c+2)x^{c+2} - \cdots$$

$$- 2a_r(c+r)x^{c+r} - \cdots + k^2 a_0 x^c$$

$$+ k^2 a_1 x^{c+1} + k^2 a_2 x^{c+2} + \cdots + k^2 a_r x^{c+r}$$

$$+ \cdots + k a_0 x^c + k a_1 x^{c+1} + \cdots$$

$$+ k a_r x^{c+r} + \cdots = 0 \qquad (45)$$

(iv) The **indicial equation** is obtained by equating the coefficient of the lowest power of x (i.e. x^{c-2}) to zero. Hence, $a_0 c(c-1) = 0$, from which $c = 0$ or $c = 1$ since $a_0 \neq 0$

For the term in x^{c-1}, i.e. $a_1 c(c+1) = 0$ With $c = 1$, $a_1 = 0$; however, when $c = 0$, a_1 is indeterminate, since any value of a_1 combined with the zero value of c would make the product zero.

For the term in x^{c+r},

$$a_{r+2}(c+r+1)(c+r+2) - a_r(c+r-1)$$

$$(c+r) - 2a_r(c+r) + k^2 a_r + k a_r = 0$$

from which,

$$a_{r+2} = \frac{a_r[(c+r-1)(c+r) + 2(c+r) - k^2 - k]}{(c+r+1)(c+r+2)}$$

$$= \frac{a_r[(c+r)(c+r+1) - k(k+1)]}{(c+r+1)(c+r+2)} \qquad (46)$$

When $c = 0$:

$$a_{r+2} = \frac{a_r[r(r+1) - k(k+1)]}{(r+1)(r+2)}$$

For $r = 0$,

$$a_2 = \frac{a_0[-k(k+1)]}{(1)(2)}$$

For $r = 1$,

$$a_3 = \frac{a_1[(1)(2) - k(k+1)]}{(2)(3)}$$

$$= \frac{-a_1[k^2 + k - 2]}{3!} = \frac{a_1(k-1)(k+2)}{3!}$$

For $r = 2$,

$$a_4 = \frac{a_2[(2)(3) - k(k+1)]}{(3)(4)} = \frac{-a_2[k^2 + k - 6]}{(3)(4)}$$

$$= \frac{-a_2(k+3)(k-2)}{(3)(4)}$$

$$= \frac{-(k+3)(k-2)}{(3)(4)} \cdot \frac{a_0[-k(k+1)]}{(1)(2)}$$

$$= \frac{a_0 k(k+1)(k+3)(k-2)}{4!}$$

For $r = 3$,

$$a_5 = \frac{a_3[(3)(4) - k(k+1)]}{(4)(5)} = \frac{-a_3[k^2 + k - 12]}{(4)(5)}$$

$$= \frac{-a_3(k+4)(k-3)}{(4)(5)}$$

$$= \frac{-(k+4)(k-3)}{(4)(5)} \cdot \frac{-a_1(k-1)(k+2)}{(2)(3)}$$

$$= \frac{a_1(k-1)(k-3)(k+2)(k+4)}{5!} \text{ and so on.}$$

Substituting values into equation (43) gives:

$$y = x^0 \left\{ a_0 + a_1 x - \frac{a_0 k(k+1)}{2!} x^2 \right.$$

$$- \frac{a_1(k-1)(k+2)}{3!} x^3$$

$$+ \frac{a_0 k(k+1)(k-2)(k+3)}{4!} x^4$$

$$+ \frac{a_1(k-1)(k-3)(k+2)(k+4)}{5!} x^5$$

$$\left. + \cdots \right\}$$

i.e. $y = a_0 \left\{ 1 - \frac{k(k+1)}{2!} x^2 \right.$

$$\left. + \frac{k(k+1)(k-2)(k+3)}{4!} x^4 - \cdots \right\}$$

$$+ a_1 \left\{ x - \frac{(k-1)(k+2)}{3!} x^3 \right.$$

$$\left. + \frac{(k-1)(k-3)(k+2)(k+4)}{5!} x^5 - \cdots \right\} \quad (47)$$

From page 894, it was stated that if two solutions of the indicial equation differ by an integer, as in this case, where $c=0$ and 1, and if one coefficient is indeterminate, as with when $c=0$, then the complete solution is always given by using this value of c. Using the second value of c, i.e. $c=1$ in this problem, will give a series which is one of the series in the first solution. (This may be checked for $c=1$ and where $a_1=0$; the result will be the first part of equation (47) above.)

Legendre's polynomials

(A **polynomial** is an expression of the form: $f(x) = a + bx + cx^2 + dx^3 + \cdots$). When k in equation (47) above is an integer, say, n, one of the solution series terminates after a finite number of terms. For example, if $k=2$, then the first series terminates after the term in x^2. The resulting polynomial in x, denoted by $P_n(x)$, is called a **Legendre polynomial**. Constants a_0 and a_1 are chosen so that $y=1$ when $x=1$. This is demonstrated in the following Problems.

Problem 13. Determine the Legendre polynomial $P_2(x)$

Since in $P_2(x)$, $n=k=2$, then from the first part of equation (47), i.e. the even powers of x:

$$y = a_0 \left\{ 1 - \frac{2(3)}{2!} x^2 + 0 \right\} = a_0 \{ 1 - 3x^2 \}$$

a_0 is chosen to make $y=1$ when $x=1$

i.e. $1 = a_0 \{ 1 - 3(1)^2 \} = -2a_0$, from which, $a_0 = -\dfrac{1}{2}$

Hence, $P_2(x) = -\dfrac{1}{2}(1 - 3x^2) = \dfrac{1}{2}(3x^2 - 1)$

Problem 14. Determine the Legendre polynomial $P_3(x)$

Since in $P_3(x)$, $n=k=3$, then from the second part of equation (47), i.e. the odd powers of x:

$$y = a_1 \left\{ x - \frac{(k-1)(k+2)}{3!} x^3 \right.$$

$$\left. + \frac{(k-1)(k-3)(k+2)(k+4)}{5!} x^5 - \cdots \right\}$$

i.e. $y = a_1 \left\{ x - \dfrac{(2)(5)}{3!} x^3 + \dfrac{(2)(0)(5)(7)}{5!} x^5 \right\}$

$$= a_1 \left\{ x - \frac{5}{3} x^3 + 0 \right\}$$

a_1 is chosen to make $y=1$ when $x=1$

i.e. $1 = a_1 \left\{ 1 - \dfrac{5}{3} \right\} = a_1 \left(-\dfrac{2}{3} \right)$ from which, $a_1 = -\dfrac{3}{2}$

Hence, $P_3(x) = -\dfrac{3}{2} \left(x - \dfrac{5}{3} x^3 \right)$ or $P_3(x) = \dfrac{1}{2}(5x^3 - 3x)$

Rodrigues' formula

An alternative method of determining Legendre polynomials is by using **Rodrigues'** * **formula**, which states:

$$P_n(x) = \frac{1}{2^n n!} \frac{d^n (x^2 - 1)^n}{dx^n} \tag{48}$$

This is demonstrated in the following Problems.

Problem 15. Determine the Legendre polynomial $P_2(x)$ using Rodrigues' formula.

In Rodrigues' formula, $P_n(x) = \dfrac{1}{2^n n!} \dfrac{d^n (x^2 - 1)^n}{dx^n}$ and when $n = 2$,

$$P_2(x) = \frac{1}{2^2 2!} \frac{d^2 (x^2 - 1)^2}{dx^2} = \frac{1}{2^3} \frac{d^2 (x^4 - 2x^2 + 1)}{dx^2}$$

* **Who was Rodrigues? Benjamin Olinde Rodrigues** (1795–1851), was a French banker, mathematician and social reformer. Rodrigues is remembered for three results: Rodrigues' rotation formula for vectors; the Rodrigues formula about series of orthogonal polynomials; and the Euler–Rodrigues parameters. To find out more go to **www.routledge.com/cw/bird**

$$\frac{d}{dx}(x^4 - 2x^2 + 1) = 4x^3 - 4x$$

and $\dfrac{d^2 (x^4 - 2x^2 + 1)}{dx^2} = \dfrac{d(4x^3 - 4x)}{dx} = 12x^2 - 4$

Hence, $P_2(x) = \dfrac{1}{2^3} \dfrac{d^2 (x^4 - 2x^2 + 1)}{dx^2} = \dfrac{1}{8}(12x^2 - 4)$

i.e. $P_2(x) = \dfrac{1}{2}(3x^2 - 1)$, the same as in Problem 13.

Problem 16. Determine the Legendre polynomial $P_3(x)$ using Rodrigues' formula

In Rodrigues' formula, $P_n(x) = \dfrac{1}{2^n n!} \dfrac{d^n (x^2 - 1)^n}{dx^n}$ and when $n = 3$,

$$P_3(x) = \frac{1}{2^3 3!} \frac{d^3 (x^2 - 1)^3}{dx^3}$$

$$= \frac{1}{2^3 (6)} \frac{d^3 (x^2 - 1)(x^4 - 2x^2 + 1)}{dx^3}$$

$$= \frac{1}{(8)(6)} \frac{d^3 (x^6 - 3x^4 + 3x^2 - 1)}{dx^3}$$

$$\frac{d(x^6 - 3x^4 + 3x^2 - 1)}{dx} = 6x^5 - 12x^3 + 6x$$

$$\frac{d(6x^5 - 12x^3 + 6x)}{dx} = 30x^4 - 36x^2 + 6$$

and $\dfrac{d(30x^4 - 36x^2 + 6)}{dx} = 120x^3 - 72x$

Hence, $P_3(x) = \dfrac{1}{(8)(6)} \dfrac{d^3 (x^6 - 3x^4 + 3x^2 - 1)}{dx^3}$

$$= \frac{1}{(8)(6)}(120x^3 - 72x) = \frac{1}{8}(20x^3 - 12x)$$

i.e. $P_3(x) = \dfrac{1}{2}(5x^3 - 3x)$, the same as in Problem 14.

Now try the following Practice Exercise

Practice Exercise 321   Legendre's equation and Legendre polynomials (answers on page 1188)

1. Determine the power series solution of the Legendre equation:

$$\left(1 - x^2\right) y'' - 2xy' + k(k+1)y = 0 \text{ when}$$
(a) $k=0$ (b) $k=2$, up to and including the term in x^5

2. Find the following Legendre polynomials:
(a) $P_1(x)$ (b) $P_4(x)$ (c) $P_5(x)$

For fully worked solutions to each of the problems in Practice Exercises 316 to 321 in this chapter, go to the website:
www.routledge.com/cw/bird

Chapter 84

An introduction to partial differential equations

Why it is important to understand: **An introduction to partial differential equations**

In engineering, physics and economics, quantities are frequently encountered – for example, energy – that depend on many variables, such as position, velocity and temperature. Usually this dependency is expressed through a partial differential equation, and solving these equations is important for understanding these complex relationships. Solving ordinary differential equations involves finding a function (or a set of functions) of one independent variable, but partial differential equations are for functions of two or more variables. Examples of physical models using partial differential equations are the heat equation for the evolution of the temperature distribution in a body, the wave equation for the motion of a wave front, the flow equation for the flow of fluids and Laplace's equation for an electrostatic potential or elastic strain field. In such cases, not only are the initial conditions needed, but also boundary conditions for the region in which the model applies; thus boundary value problems have to be solved. This chapter provides an introduction to the often complex subject of partial differential equation.

At the end of this chapter, you should be able to:

- recognise some important engineering partial differential equations
- solve a partial differential equation by direct partial integration
- solve differential equations by separating the variables
- solve the wave equation $\dfrac{\partial^2 u}{\partial x^2} = \dfrac{1}{c^2}\dfrac{\partial^2 u}{\partial t^2}$
- solve the heat conduction equation $\dfrac{\partial^2 u}{\partial x^2} = \dfrac{1}{c^2}\dfrac{\partial u}{\partial t}$
- solve Laplace's equation $\dfrac{\partial^2 u}{\partial x^2} + \dfrac{\partial^2 u}{\partial y^2} = 0$

84.1 Introduction

A partial differential equation is an equation that contains one or more partial derivatives. Examples include:

(i) $\quad a\dfrac{\partial u}{\partial x}+b\dfrac{\partial u}{\partial y}=c$

(ii) $\quad \dfrac{\partial^2 u}{\partial x^2}=\dfrac{1}{c^2}\dfrac{\partial u}{\partial t}$

(known as the heat conduction equation)

(iii) $\quad \dfrac{\partial^2 u}{\partial x^2}+\dfrac{\partial^2 u}{\partial y^2}=0$

(known as Laplace's equation)

Equation (i) is a **first-order partial differential equation**, and equations (ii) and (iii) are **second-order partial differential equations** since the highest power of the differential is 2.

Partial differential equations occur in many areas of engineering and technology; electrostatics, heat conduction, magnetism, wave motion, hydrodynamics and aerodynamics all use models that involve partial differential equations. Such equations are difficult to solve, but techniques have been developed for the simpler types. In fact, for all but for the simplest cases, there are a number of numerical methods of solutions of partial differential equations available.

To be able to solve simple partial differential equations knowledge of the following is required:

(a) partial integration,

(b) first- and second-order partial differentiation – as explained in Chapter 60, and

(c) the solution of ordinary differential equations – as explained in Chapters 77–82.

It should be appreciated that whole books have been written on partial differential equations and their solutions. This chapter does no more than introduce the topic.

84.2 Partial integration

Integration is the reverse process of differentiation.

Thus, if, for example, $\dfrac{\partial u}{\partial t}=5\cos x\sin t$ is integrated partially with respect to t, then the $5\cos x$ term is considered as a constant,

and $\quad u=\displaystyle\int 5\cos x\sin t\,dt=(5\cos x)\int\sin t\,dt$

$\qquad = (5\cos x)(-\cos t)+c$

$\qquad = \mathbf{-5\cos x\cos t}+f(x)$

Similarly, if $\dfrac{\partial^2 u}{\partial x\partial y}=6x^2\cos 2y$ is integrated partially with respect to y,

then $\quad \dfrac{\partial u}{\partial x}=\displaystyle\int 6x^2\cos 2y\,dy=\left(6x^2\right)\int\cos 2y\,dy$

$\qquad = \left(6x^2\right)\left(\dfrac{1}{2}\sin 2y\right)+f(x)$

$\qquad = 3x^2\sin 2y+f(x)$

and integrating $\dfrac{\partial u}{\partial x}$ partially with respect to x gives:

$u=\displaystyle\int [3x^2\sin 2y+f(x)]\,dx$

$\quad = x^3\sin 2y+(x)f(x)+g(y)$

$f(x)$ and $g(y)$ are functions that may be determined if extra information, called **boundary conditions** or **initial conditions**, are known.

84.3 Solution of partial differential equations by direct partial integration

The simplest form of partial differential equations occurs when a solution can be determined by direct partial integration. This is demonstrated in the following Problems.

Problem 1. Solve the differential equation $\dfrac{\partial^2 u}{\partial x^2}=6x^2(2y-1)$ given the boundary conditions that at $x=0$, $\dfrac{\partial u}{\partial x}=\sin 2y$ and $u=\cos y$

Since $\dfrac{\partial^2 u}{\partial x^2}=6x^2(2y-1)$ then integrating partially with respect to x gives:

$\dfrac{\partial u}{\partial x}=\displaystyle\int 6x^2(2y-1)\,dx=(2y-1)\int 6x^2\,dx$

$\qquad = (2y-1)\dfrac{6x^3}{3}+f(y)$

$\qquad = 2x^3(2y-1)+f(y)$

where $f(y)$ is an arbitrary function.

From the boundary conditions, when $x=0$,

$$\frac{\partial u}{\partial x} = \sin 2y$$

Hence, $\qquad \sin 2y = 2(0)^3(2y-1) + f(y)$

from which, $\qquad f(y) = \sin 2y$

Now $\qquad \dfrac{\partial u}{\partial x} = 2x^3(2y-1) + \sin 2y$

Integrating partially with respect to x gives:

$$u = \int [2x^3(2y-1) + \sin 2y]\,dx$$

$$= \frac{2x^4}{4}(2y-1) + x(\sin 2y) + F(y)$$

From the boundary conditions, when $x=0$, $u = \cos y$, hence

$$\cos y = \frac{(0)^4}{2}(2y-1) + (0)\sin 2y + F(y)$$

from which, $F(y) = \cos y$

Hence, the solution of $\dfrac{\partial^2 u}{\partial x^2} = 6x^2(2y-1)$ for the given boundary conditions is:

$$u = \frac{x^4}{2}(2y-1) + x\sin y + \cos y$$

Problem 2. Solve the differential equation:
$\dfrac{\partial^2 u}{\partial x \partial y} = \cos(x+y)$ given that $\dfrac{\partial u}{\partial x} = 2$ when $y = 0$,
and $u = y^2$ when $x = 0$

Since $\dfrac{\partial^2 u}{\partial x \partial y} = \cos(x+y)$ then integrating partially with respect to y gives:

$$\frac{\partial u}{\partial x} = \int \cos(x+y)\,dy = \sin(x+y) + f(x)$$

From the boundary conditions, $\dfrac{\partial u}{\partial x} = 2$ when $y = 0$, hence

$$2 = \sin x + f(x)$$

from which, $\quad f(x) = 2 - \sin x$

i.e. $\qquad \dfrac{\partial u}{\partial x} = \sin(x+y) + 2 - \sin x$

Integrating partially with respect to x gives:

$$u = \int [\sin(x+y) + 2 - \sin x]\,dx$$

$$= -\cos(x+y) + 2x + \cos x + f(y)$$

From the boundary conditions, $u = y^2$ when $x = 0$, hence

$$y^2 = -\cos y + 0 + \cos 0 + f(y)$$

$$= 1 - \cos y + f(y)$$

from which, $\ f(y) = y^2 - 1 + \cos y$

Hence, the solution of $\dfrac{\partial^2 u}{\partial x \partial y} = \cos(x+y)$ is given by:

$$u = -\cos(x+y) + 2x + \cos x + y^2 - 1 + \cos y$$

Problem 3. Verify that
$\phi(x, y, z) = \dfrac{1}{\sqrt{x^2+y^2+z^2}}$ satisfies the partial
differential equation $\dfrac{\partial^2 \phi}{\partial x^2} + \dfrac{\partial^2 \phi}{\partial y^2} + \dfrac{\partial^2 \phi}{\partial z^2} = 0$

The partial differential equation

$\dfrac{\partial^2 \phi}{\partial x^2} + \dfrac{\partial^2 \phi}{\partial y^2} + \dfrac{\partial^2 \phi}{\partial z^2} = 0$ is called **Laplace's equation**.

If $\ \phi(x, y, z) = \dfrac{1}{\sqrt{x^2+y^2+z^2}} = (x^2 + y^2 + z^2)^{-\frac{1}{2}}$

then differentiating partially with respect to x gives:

$$\frac{\partial \phi}{\partial x} = -\frac{1}{2}(x^2 + y^2 + z^2)^{-\frac{3}{2}}(2x)$$

$$= -x(x^2 + y^2 + z^2)^{-\frac{3}{2}}$$

and $\quad \dfrac{\partial^2 \phi}{\partial x^2} = (-x)\left[-\dfrac{3}{2}(x^2 + y^2 + z^2)^{-\frac{5}{2}}(2x)\right]$

$$+ (x^2 + y^2 + z^2)^{-\frac{3}{2}}(-1)$$
$$\text{by the product rule}$$

$$= \frac{3x^2}{(x^2 + y^2 + z^2)^{\frac{5}{2}}} - \frac{1}{(x^2 + y^2 + z^2)^{\frac{3}{2}}}$$

$$= \frac{(3x^2) - (x^2 + y^2 + z^2)}{(x^2 + y^2 + z^2)^{\frac{5}{2}}}$$

Similarly, it may be shown that

$$\frac{\partial^2 \phi}{\partial y^2} = \frac{(3y^2) - (x^2 + y^2 + z^2)}{(x^2 + y^2 + z^2)^{\frac{5}{2}}}$$

and $\dfrac{\partial^2 \phi}{\partial z^2} = \dfrac{(3z^2) - (x^2 + y^2 + z^2)}{(x^2 + y^2 + z^2)^{\frac{5}{2}}}$

Thus,

$$\frac{\partial^2 \phi}{\partial x^2} + \frac{\partial^2 \phi}{\partial y^2} + \frac{\partial^2 \phi}{\partial z^2} = \frac{(3x^2) - (x^2 + y^2 + z^2)}{(x^2 + y^2 + z^2)^{\frac{5}{2}}}$$

$$+ \frac{(3y^2) - (x^2 + y^2 + z^2)}{(x^2 + y^2 + z^2)^{\frac{5}{2}}}$$

$$+ \frac{(3z^2) - (x^2 + y^2 + z^2)}{(x^2 + y^2 + z^2)^{\frac{5}{2}}}$$

$$= \frac{\begin{pmatrix} 3x^2 - (x^2 + y^2 + z^2) \\ +3y^2 - (x^2 + y^2 + z^2) \\ +3z^2 - (x^2 + y^2 + z^2) \end{pmatrix}}{(x^2 + y^2 + z^2)^{\frac{5}{2}}} = 0$$

Thus, $\dfrac{1}{\sqrt{x^2 + y^2 + z^2}}$ satisfies the Laplace equation

$$\frac{\partial^2 \phi}{\partial x^2} + \frac{\partial^2 \phi}{\partial y^2} + \frac{\partial^2 \phi}{\partial z^2} = 0$$

Now try the following Practice Exercise

Practice Exercise 322   Further problems on the solution of partial differential equations by direct partial integration (answers on page 1189)

1. Determine the general solution of $\dfrac{\partial u}{\partial y} = 4ty$

2. Solve $\dfrac{\partial u}{\partial t} = 2t \cos\theta$ given that $u = 2t$ when $\theta = 0$

3. Verify that $u(\theta, t) = \theta^2 + \theta t$ is a solution of $\dfrac{\partial u}{\partial \theta} - 2\dfrac{\partial u}{\partial t} = t$

4. Verify that $u = e^{-y} \cos x$ is a solution of $\dfrac{\partial^2 u}{\partial x^2} + \dfrac{\partial^2 u}{\partial y^2} = 0$

5. Solve $\dfrac{\partial^2 u}{\partial x \partial y} = 8e^y \sin 2x$ given that at $y = 0$, $\dfrac{\partial u}{\partial x} = \sin x$, and at $x = \dfrac{\pi}{2}, u = 2y^2$

6. Solve $\dfrac{\partial^2 u}{\partial x^2} = y(4x^2 - 1)$ given that at $x = 0$, $u = \sin y$ and $\dfrac{\partial u}{\partial x} = \cos 2y$

7. Solve $\dfrac{\partial^2 u}{\partial x \partial t} = \sin(x + t)$ given that $\dfrac{\partial u}{\partial x} = 1$ when $t = 0$, and when $u = 2t$ when $x = 0$

8. Show that $u(x, y) = xy + \dfrac{x}{y}$ is a solution of
$$2x\frac{\partial^2 u}{\partial x \partial y} + y\frac{\partial^2 u}{\partial y^2} = 2x$$

9. Find the particular solution of the differential equation $\dfrac{\partial^2 u}{\partial x \partial y} = \cos x \cos y$ given the initial conditions that when $y = \pi$, $\dfrac{\partial u}{\partial x} = x$, and when $x = \pi, u = 2\cos y$

10. Verify that $\phi(x, y) = x \cos y + e^x \sin y$ satisfies the differential equation
$$\frac{\partial^2 \phi}{\partial x^2} + \frac{\partial^2 \phi}{\partial y^2} + x \cos y = 0$$

84.4 Some important engineering partial differential equations

There are many types of partial differential equations. Some typically found in engineering and science include:

(a) The **wave equation**, where the equation of motion is given by:

$$\frac{\partial^2 u}{\partial x^2} = \frac{1}{c^2}\frac{\partial^2 u}{\partial t^2}$$

where $c^2 = \dfrac{T}{\rho}$, with T being the tension in a string and ρ being the mass/unit length of the string.

(b) The **heat conduction equation** is of the form:

$$\frac{\partial^2 u}{\partial x^2} = \frac{1}{c^2}\frac{\partial u}{\partial t}$$

where $c^2 = \dfrac{h}{\sigma\rho}$, with h being the thermal conductivity of the material, σ the specific heat of the material, and ρ the mass/unit length of material.

(c) **Laplace's* equation**, used extensively with electrostatic fields, is of the form:

$$\frac{\partial^2 u}{\partial x^2} + \frac{\partial^2 u}{\partial y^2} + \frac{\partial^2 u}{\partial z^2} = 0$$

(d) The **transmission equation**, where the potential u in a transmission cable is of the form:

$$\frac{\partial^2 u}{\partial x^2} = A\frac{\partial^2 u}{\partial t^2} + B\frac{\partial u}{\partial t} + Cu \text{ where } A, B \text{ and } C \text{ are}$$

constants.

Some of these equations are used in the next sections.

* **Who was Laplace? Pierre-Simon, marquis de Laplace** (23 March 1749–5 March 1827) was a French mathematician and astronomer who formulated Laplace's equation, and pioneered the Laplace transform which appears in many branches of mathematical physics. To find out more go to **www.routledge.com/cw/bird**

84.5 Separating the variables

Let $u(x,t) = X(x)T(t)$, where $X(x)$ is a function of x only and $T(t)$ is a function of t only, be a trial solution to the wave equation $\dfrac{\partial^2 u}{\partial x^2} = \dfrac{1}{c^2}\dfrac{\partial^2 u}{\partial t^2}$. If the trial solution is simplified to $u = XT$, then $\dfrac{\partial u}{\partial x} = X'T$ and $\dfrac{\partial^2 u}{\partial x^2} = X''T$

Also $\dfrac{\partial u}{\partial t} = XT'$ and $\dfrac{\partial^2 u}{\partial t^2} = XT''$

Substituting into the partial differential equation $\dfrac{\partial^2 u}{\partial x^2} = \dfrac{1}{c^2}\dfrac{\partial^2 u}{\partial t^2}$ gives:

$$X''T = \frac{1}{c^2}XT''$$

Separating the variables gives:

$$\frac{X''}{X} = \frac{1}{c^2}\frac{T''}{T}$$

Let $\mu = \dfrac{X''}{X} = \dfrac{1}{c^2}\dfrac{T''}{T}$ where μ is a constant.

Thus, since $\mu = \dfrac{X''}{X}$ (a function of x only), it must be independent of t; and, since $\mu = \dfrac{1}{c^2}\dfrac{T''}{T}$ (a function of t only), it must be independent of x.

If μ is independent of x *and* t, it can only be a constant. If $\mu = \dfrac{X''}{X}$ then $X'' = \mu X$ or $X'' - \mu X = 0$ and if $\mu = \dfrac{1}{c^2}\dfrac{T''}{T}$ then $T'' = c^2\mu T$ or $T'' - c^2\mu T = 0$

Such ordinary differential equations are of the form found in Chapter 81, and their solutions will depend on whether $\mu > 0$, $\mu = 0$ or $\mu < 0$

Problem 4 will be a reminder of solving ordinary differential equations of this type.

> **Problem 4.** Find the general solution of the following differential equations:
>
> (a) $X'' - 4X = 0$ (b) $T'' + 4T = 0$

(a) If $X'' - 4X = 0$ then the auxiliary equation (see Chapter 81) is:

$$m^2 - 4 = 0 \text{ i.e. } m^2 = 4 \text{ from which,}$$

$$m = +2 \text{ or } m = -2$$

Thus, the general solution is:

$$X = Ae^{2x} + Be^{-2x}$$

(b) If $T'' + 4T = 0$ then the auxiliary equation is:

$$m^2 + 4 = 0 \text{ i.e. } m^2 = -4 \text{ from which,}$$

$$m = \sqrt{-4} = \pm j2$$

Thus, the general solution is:

$$T = e^0 \{A \cos 2t + B \sin 2t\} = A \cos 2t + B \sin 2t$$

Now try the following Practice Exercise

Practice Exercise 323   Further problems on revising the solution of ordinary differential equations (answers on page 1189)

1. Solve $T'' = c^2 \mu T$ given $c = 3$ and $\mu = 1$

2. Solve $T'' - c^2 \mu T = 0$ given $c = 3$ and $\mu = -1$

3. Solve $X'' = \mu X$ given $\mu = 1$

4. Solve $X'' - \mu X = 0$ given $\mu = -1$

84.6   The wave equation

An **elastic string** is a string with elastic properties, i.e. the string satisfies Hooke's law. Figure 84.1 shows a flexible elastic string stretched between two points at $x = 0$ and $x = L$ with uniform tension T. The string will vibrate if the string is displace slightly from its initial position of rest and released, the end points remaining fixed. The position of any point P on the string depends on its distance from one end, and on the instant in time. Its displacement u at any time t can be expressed as $u = f(x, t)$, where x is its distance from 0.
The equation of motion is as stated in Section 84.4(a),
i.e. $\dfrac{\partial^2 u}{\partial x^2} = \dfrac{1}{c^2} \dfrac{\partial^2 u}{\partial t^2}$

The boundary and initial conditions are:

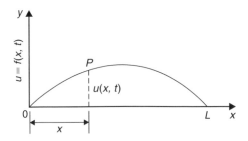

Figure 84.1

(i) The string is fixed at both ends, i.e. $x = 0$ and $x = L$ for all values of time t.

Hence, $u(x, t)$ becomes:

$$\left. \begin{array}{r} u(0, t) = 0 \\ u(L, t) = 0 \end{array} \right\} \text{ for all values of } t \geq 0$$

(ii) If the initial deflection of P at $t = 0$ is denoted by $f(x)$ then $u(x, 0) = f(x)$

(iii) Let the initial velocity of P be $g(x)$, then

$$\left[\frac{\partial u}{\partial t} \right]_{t=0} = g(x)$$

Initially a **trial solution** of the form $u(x, t) = X(x)T(t)$ is assumed, where $X(x)$ is a function of x only and $T(t)$ is a function of t only. The trial solution may be simplified to $u = XT$ and the variables separated as explained in the previous section to give:

$$\frac{X''}{X} = \frac{1}{c^2} \frac{T''}{T}$$

When both sides are equated to a constant μ this results in two ordinary differential equations:

$$T'' - c^2 \mu T = 0 \quad \text{and} \quad X'' - \mu X = 0$$

Three cases are possible, depending on the value of μ.

Case 1: $\mu > 0$

For convenience, let $\mu = p^2$, where p is a real constant. Then the equations

$$X'' - p^2 X = 0 \quad \text{and} \quad T'' - c^2 p^2 T = 0$$

have solutions: $X = A e^{px} + B e^{-px}$ and
$T = C e^{cpt} + D e^{-cpt}$ where A, B, C and D are constants.
But $X = 0$ at $x = 0$, hence $0 = A + B$ i.e. $B = -A$ and
$X = 0$ at $x = L$, hence
$0 = A e^{pL} + B e^{-pL} = A(e^{pL} - e^{-pL})$
Assuming $(e^{pL} - e^{-pL})$ is not zero, then $A = 0$ and since
$B = -A$, then $B = 0$ also.
This corresponds to the string being stationary; since it is non-oscillatory, this solution will be disregarded.

Case 2: $\mu = 0$

In this case, since $\mu = p^2 = 0$, $T'' = 0$ and $X'' = 0$. We will assume that $T(t) \neq 0$. Since $X'' = 0$, $X' = a$ and $X = ax + b$ where a and b are constants. But $X = 0$ at $x = 0$, hence $b = 0$ and $X = ax$ and $X = 0$ at $x = L$, hence $a = 0$. Thus, again, the solution is non-oscillatory and is also disregarded.

Section K

Case 3: $\mu < 0$

For convenience,
let $\mu = -p^2$ then $X'' + p^2 X = 0$ from which,

$$X = A\cos px + B\sin px \qquad (1)$$

and $T'' + c^2 p^2 T = 0$ from which,

$$T = C\cos cpt + D\sin cpt \qquad (2)$$

(see Problem 4 above).

Thus, the suggested solution $u = XT$ now becomes:

$$u = \{A\cos px + B\sin px\}\{C\cos cpt + D\sin cpt\} \qquad (3)$$

Applying the boundary conditions:

(i) $u = 0$ when $x = 0$ for all values of t,

thus $0 = \{A\cos 0 + B\sin 0\}\{C\cos cpt + D\sin cpt\}$

i.e. $\qquad 0 = A\{C\cos cpt + D\sin cpt\}$

from which, $A = 0$, (since $\{C\cos cpt + D\sin cpt\} \neq 0$)

Hence, $\qquad u = \{B\sin px\}\{C\cos cpt + D\sin cpt\} \qquad (4)$

(ii) $u = 0$ when $x = L$ for all values of t

Hence, $\quad 0 = \{B\sin pL\}\{C\cos cpt + D\sin cpt\}$

Now $B \neq 0$ or $u(x, t)$ would be identically zero.
Thus $\sin pL = 0$, i.e. $pL = n\pi$ or $p = \dfrac{n\pi}{L}$ for integer values of n.
Substituting in equation (4) gives:

$$u = \left\{B\sin\frac{n\pi x}{L}\right\}\left\{C\cos\frac{cn\pi t}{L} + D\sin\frac{cn\pi t}{L}\right\}$$

i.e. $u = \sin\dfrac{n\pi x}{L}\left\{A_n\cos\dfrac{cn\pi t}{L} + B_n\sin\dfrac{cn\pi t}{L}\right\}$

(where constant $A_n = BC$ and $B_n = BD$). There will be many solutions, depending on the value of n. Thus, more generally,

$$u_n(x, t) = \sum_{n=1}^{\infty}\left\{\sin\frac{n\pi x}{L}\left(A_n\cos\frac{cn\pi t}{L}\right.\right.$$
$$\left.\left. + B_n\sin\frac{cn\pi t}{L}\right)\right\} \qquad (5)$$

To find A_n and B_n we put in the initial conditions not yet taken into account.

(i) At $t = 0$, $u(x, 0) = f(x)$ for $0 \leq x \leq L$
Hence, from equation (5),

$$u(x, 0) = f(x) = \sum_{n=1}^{\infty}\left\{A_n\sin\frac{n\pi x}{L}\right\} \qquad (6)$$

(ii) Also at $t = 0$, $\left[\dfrac{\partial u}{\partial t}\right]_{t=0} = g(x)$ for $0 \leq x \leq L$

Differentiating equation (5) with respect to t gives:

$$\frac{\partial u}{\partial t} = \sum_{n=1}^{\infty}\left\{\sin\frac{n\pi x}{L}\left(A_n\left(-\frac{cn\pi}{L}\sin\frac{cn\pi t}{L}\right)\right.\right.$$
$$\left.\left. + B_n\left(\frac{cn\pi}{L}\cos\frac{cn\pi t}{L}\right)\right)\right\}$$

and when $t = 0$,

$$g(x) = \sum_{n=1}^{\infty}\left\{\sin\frac{n\pi x}{L}B_n\frac{cn\pi}{L}\right\}$$

i.e. $g(x) = \dfrac{c\pi}{L}\displaystyle\sum_{n=1}^{\infty}\left\{B_n n\sin\dfrac{n\pi x}{L}\right\} \qquad (7)$

From Fourier series (see page 1104) it may be shown that:
A_n is twice the mean value of $f(x)\sin\dfrac{n\pi x}{L}$ between $x = 0$ and $x = L$

i.e. $\quad A_n = \dfrac{2}{L}\displaystyle\int_0^L f(x)\sin\dfrac{n\pi x}{L}dx$

$$\text{for } n = 1, 2, 3, \ldots (8)$$

and $B_n\left(\dfrac{cn\pi}{L}\right)$ is twice the mean value of

$g(x)\sin\dfrac{n\pi x}{L}$ between $x = 0$ and $x = L$

i.e. $\quad B_n = \dfrac{L}{cn\pi}\left(\dfrac{2}{L}\right)\displaystyle\int_0^L g(x)\sin\dfrac{n\pi x}{L}dx$

or $\quad B_n = \dfrac{2}{cn\pi}\displaystyle\int_0^L g(x)\sin\dfrac{n\pi x}{L}dx \qquad (9)$

Summary of solution of the wave equation

The above may seem complicated; however, a practical problem may be solved using the following **8-point procedure**:

1. Identify clearly the initial and boundary conditions.

2. Assume a solution of the form $u = XT$ and express the equations in terms of X and T and their derivatives.

3. Separate the variables by transposing the equation and equate each side to a constant, say, μ; two separate equations are obtained, one in x and the other in t.

4. Let $\mu = -p^2$ to give an oscillatory solution.

5. The two solutions are of the form:

$$X = A\cos px + B\sin px$$

and $T = C\cos cpt + D\sin cpt$

Then $u(x,t) = \{A\cos px + B\sin px\}\{C\cos cpt + D\sin cpt\}$

6. Apply the boundary conditions to determine constants A and B.

7. Determine the general solution as an infinite sum.

8. Apply the remaining initial and boundary conditions and determine the coefficients A_n and B_n from equations (8) and (9), using Fourier series techniques.

Problem 5. Figure 84.2 shows a stretched string of length 50 cm which is set oscillating by displacing its mid-point a distance of 2 cm from its rest position and releasing it with zero velocity. Solve the wave equation: $\dfrac{\partial^2 u}{\partial x^2} = \dfrac{1}{c^2}\dfrac{\partial^2 u}{\partial t^2}$ where $c^2 = 1$, to determine the resulting motion $u(x,t)$

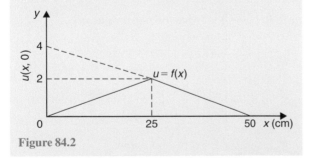

Figure 84.2

Following the above procedure,

1. The boundary and initial conditions given are:

$$\left.\begin{array}{l} u(0,t) = 0 \\ u(50,t) = 0 \end{array}\right\} \text{ i.e. fixed end points}$$

$$u(x,0) = f(x) = \frac{2}{25}x \quad 0 \le x \le 25$$

$$= -\frac{2}{25}x + 4 = \frac{100 - 2x}{25}$$

$$25 \le x \le 50$$

(Note: $y = mx + c$ is a straight line graph, so the gradient, m, between 0 and 25 is 2/25 and the y-axis intercept is zero, thus $y = f(x) = \dfrac{2}{25}x + 0$; between 25 and 50, the gradient $= -2/25$ and the y-axis intercept is at 4, thus $f(x) = -\dfrac{2}{25}x + 4$)

$$\left[\frac{\partial u}{\partial t}\right]_{t=0} = 0 \quad \text{i.e. zero initial velocity}$$

2. Assuming a solution $u = XT$, where X is a function of x only, and T is a function of t only, then $\dfrac{\partial u}{\partial x} = X'T$ and $\dfrac{\partial^2 u}{\partial x^2} = X''T$ and $\dfrac{\partial u}{\partial y} = XT'$ and $\dfrac{\partial^2 u}{\partial y^2} = XT''$. Substituting into the partial differential equation, $\dfrac{\partial^2 u}{\partial x^2} = \dfrac{1}{c^2}\dfrac{\partial^2 u}{\partial t^2}$ gives:

$$X''T = \frac{1}{c^2}XT'' \quad \text{i.e. } X''T = XT'' \text{ since } c^2 = 1$$

3. Separating the variables gives: $\dfrac{X''}{X} = \dfrac{T''}{T}$

Let constant,

$$\mu = \frac{X''}{X} = \frac{T''}{T} \quad \text{then } \mu = \frac{X''}{X} \text{ and } \mu = \frac{T''}{T}$$

from which,

$$X'' - \mu X = 0 \quad \text{and} \quad T'' - \mu T = 0$$

4. Letting $\mu = -p^2$ to give an oscillatory solution gives:

$$X'' + p^2 X = 0 \quad \text{and} \quad T'' + p^2 T = 0$$

The auxiliary equation for each is: $m^2 + p^2 = 0$ from which, $m = \sqrt{-p^2} = \pm jp$

5. Solving each equation gives:
$X = A\cos px + B\sin px$, and
$T = C\cos pt + D\sin pt$
Thus,
$u(x,t) = \{A\cos px + B\sin px\}\{C\cos pt + D\sin pt\}$.

6. Applying the boundary conditions to determine constants A and B gives:

   (i) $u(0,t)=0$, hence $0=A\{C\cos pt+D\sin pt\}$ from which we conclude that $A=0$
   Therefore,

   $$u(x,t)=B\sin px\{C\cos pt+D\sin pt\} \quad \text{(a)}$$

   (ii) $u(50,t)=0$, hence
   $0=B\sin 50p\{C\cos pt+D\sin pt\} \quad B\neq 0$,
   hence $\sin 50p=0$ from which, $50p=n\pi$ and
   $p=\dfrac{n\pi}{50}$

7. Substituting in equation (a) gives:

   $$u(x,t)=B\sin\frac{n\pi x}{50}\left\{C\cos\frac{n\pi t}{50}+D\sin\frac{n\pi t}{50}\right\}$$

   or, more generally,

   $$u_n(x,t)=\sum_{n=1}^{\infty}\sin\frac{n\pi x}{50}\left\{A_n\cos\frac{n\pi t}{50}\right.$$
   $$\left.+B_n\sin\frac{n\pi t}{50}\right\} \quad \text{(b)}$$

   where $A_n=BC$ and $B_n=BD$

8. From equation (8),

   $$A_n=\frac{2}{L}\int_0^L f(x)\sin\frac{n\pi x}{L}\,dx$$
   $$=\frac{2}{50}\left[\int_0^{25}\left(\frac{2}{25}x\right)\sin\frac{n\pi x}{50}\,dx\right.$$
   $$\left.+\int_{25}^{50}\left(\frac{100-2x}{25}\right)\sin\frac{n\pi x}{50}\,dx\right]$$

   Each integral is determined using integration by parts (see Chapter 68, page 753) with the result:

   $$A_n=\frac{16}{n^2\pi^2}\sin\frac{n\pi}{2}$$

   From equation (9),

   $$B_n=\frac{2}{cn\pi}\int_0^L g(x)\sin\frac{n\pi x}{L}\,dx$$
   $$\left[\frac{\partial u}{\partial t}\right]_{t=0}=0=g(x) \text{ thus, } B_n=0$$

Substituting into equation (b) gives:

$$u_n(x,t)=\sum_{n=1}^{\infty}\sin\frac{n\pi x}{50}\left\{A_n\cos\frac{n\pi t}{50}\right.$$
$$\left.+B_n\sin\frac{n\pi t}{50}\right\}$$
$$=\sum_{n=1}^{\infty}\sin\frac{n\pi x}{50}\left\{\frac{16}{n^2\pi^2}\sin\frac{n\pi}{2}\cos\frac{n\pi t}{50}\right.$$
$$\left.+(0)\sin\frac{n\pi t}{50}\right\}$$

Hence,

$$u(x,t)=\frac{16}{\pi^2}\sum_{n=1}^{\infty}\frac{1}{n^2}\sin\frac{n\pi x}{50}\sin\frac{n\pi}{2}\cos\frac{n\pi t}{50}$$

For stretched string problems as in Problem 5 above, the main parts of the procedure are:

1. Determine A_n from equation (8).

   Note that $\dfrac{2}{L}\displaystyle\int_0^L f(x)\sin\dfrac{n\pi x}{L}\,dx$ is **always** equal to $\dfrac{8d}{n^2\pi^2}\sin\dfrac{n\pi}{2}$ (see Figure 84.3)

2. Determine B_n from equation (9)

3. Substitute in equation (5) to determine $u(x,t)$

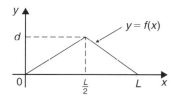

Figure 84.3

Now try the following Practice Exercise

Practice Exercise 324   Further problems on the wave equation (answers on page 1189)

1. An elastic string is stretched between two points 40 cm apart. Its centre point is displaced 1.5 cm from its position of rest at right angles to the original direction of the string and then released with zero velocity. Determine the subsequent motion $u(x,t)$ by applying the wave equation $\dfrac{\partial^2 u}{\partial x^2}=\dfrac{1}{c^2}\dfrac{\partial^2 u}{\partial t^2}$ with $c^2=9$

2. The centre point of an elastic string between two points P and Q, 80 cm apart, is deflected a distance of 1 cm from its position of rest perpendicular to PQ and released initially with zero velocity. Apply the wave equation $\dfrac{\partial^2 u}{\partial x^2} = \dfrac{1}{c^2}\dfrac{\partial^2 u}{\partial t^2}$, where $c = 8$, to determine the motion of a point distance x from P at time t.

84.7 The heat conduction equation

The heat conduction equation $\dfrac{\partial^2 u}{\partial x^2} = \dfrac{1}{c^2}\dfrac{\partial u}{\partial t}$ is solved in a similar manner to that for the wave equation; the equation differs only in that the right-hand side contains a first partial derivative instead of the second.

The conduction of heat in a uniform bar depends on the initial distribution of temperature and on the physical properties of the bar, i.e. the thermal conductivity, h, the specific heat of the material, σ, and the mass per unit length, ρ, of the bar. In the above equation, $c^2 = \dfrac{h}{\sigma\rho}$

With a uniform bar insulated, except at its ends, any heat flow is along the bar and, at any instant, the temperature u at a point P is a function of its distance x from one end, and of the time t. Consider such a bar, shown in Figure 84.4, where the bar extends from $x = 0$ to $x = L$, the temperature of the ends of the bar is maintained at zero, and the initial temperature distribution along the bar is defined by $f(x)$.

Thus, the boundary conditions can be expressed as:

$$\left.\begin{array}{l} u(0, t) = 0 \\ u(L, t) = 0 \end{array}\right\} \text{ for all } t \geq 0$$

and $u(x, 0) = f(x)$ for $0 \leq x \leq L$

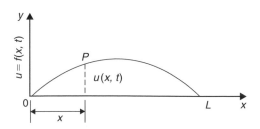

Figure 84.4

As with the wave equation, a solution of the form $u(x, t) = X(x)T(t)$ is assumed, where X is a function of x only and T is a function of t only. If the trial solution is simplified to $u = XT$, then

$$\frac{\partial u}{\partial x} = X'T \quad \frac{\partial^2 u}{\partial x^2} = X''T \text{ and } \frac{\partial u}{\partial t} = XT'$$

Substituting into the partial differential equation, $\dfrac{\partial^2 u}{\partial x^2} = \dfrac{1}{c^2}\dfrac{\partial u}{\partial t}$ gives:

$$X''T = \frac{1}{c^2}XT'$$

Separating the variables gives:

$$\frac{X''}{X} = \frac{1}{c^2}\frac{T'}{T}$$

Let $\;-p^2 = \dfrac{X''}{X} = \dfrac{1}{c^2}\dfrac{T'}{T}$ where $-p^2$ is a constant.

If $-p^2 = \dfrac{X''}{X}$ then $X'' = -p^2 X$ or $X'' + p^2 X = 0$, giving $X = A\cos px + B\sin px$

and if $-p^2 = \dfrac{1}{c^2}\dfrac{T'}{T}$ then $\dfrac{T'}{T} = -p^2 c^2$ and integrating with respect to t gives:

$$\int \frac{T'}{T}\,dt = \int -p^2 c^2\,dt$$

from which, $\ln T = -p^2 c^2 t + c_1$
The left-hand integral is obtained by an algebraic substitution (see Chapter 64).

If $\ln T = -p^2 c^2 t + c_1$ then
$T = e^{-p^2 c^2 t + c_1} = e^{-p^2 c^2 t}e^{c_1}$ i.e. $\boldsymbol{T = k\,e^{-p^2 c^2 t}}$ (where constant $k = e^{c_1}$).
Hence, $u(x, t) = XT = \{A\cos px + B\sin px\}k\,e^{-p^2 c^2 t}$
i.e. $u(x, t) = \{P\cos px + Q\sin px\}e^{-p^2 c^2 t}$ where $P = Ak$ and $Q = Bk$.
Applying the boundary conditions $u(0, t) = 0$ gives:
$0 = \{P\cos 0 + Q\sin 0\}e^{-p^2 c^2 t} = Pe^{-p^2 c^2 t}$ from which, $P = 0$ and $u(x, t) = Q\sin px\,e^{-p^2 c^2 t}$
Also, $u(L, t) = 0$ thus, $0 = Q\sin pL\,e^{-p^2 c^2 t}$ and since $Q \neq 0$ then $\sin pL = 0$ from which, $pL = n\pi$ or $p = \dfrac{n\pi}{L}$
where $n = 1, 2, 3, \ldots$
There are therefore many values of $u(x, t)$.
Thus, in general,

$$u(x, t) = \sum_{n=1}^{\infty}\left\{ Q_n\,e^{-p^2 c^2 t}\sin\frac{n\pi x}{L}\right\}$$

Applying the remaining boundary condition, that when $t=0, u(x,t)=f(x)$ for $0 \le x \le L$, gives:

$$f(x) = \sum_{n=1}^{\infty} \left\{ Q_n \sin \frac{n\pi x}{L} \right\}$$

From Fourier series, $Q_n = 2 \times$ mean value of $f(x) \sin \frac{n\pi x}{L}$ from x to L

Hence, $\quad Q_n = \frac{2}{L} \int_0^L f(x) \sin \frac{n\pi x}{L} \, dx$

Thus, $\quad u(x,t) =$

$$\frac{2}{L} \sum_{n=1}^{\infty} \left\{ \left(\int_0^L f(x) \sin \frac{n\pi x}{L} \, dx \right) e^{-p^2 c^2 t} \sin \frac{n\pi x}{L} \right\}$$

This method of solution is demonstrated in the following Problem.

Problem 6. A metal bar, insulated along its sides, is 1 m long. It is initially at room temperature of 15°C and at time $t=0$, the ends are placed into ice at 0°C. Find an expression for the temperature at a point P at a distance x m from one end at any time t seconds after $t=0$

The temperature u along the length of bar is shown in Figure 84.5.

The heat conduction equation is $\dfrac{\partial^2 u}{\partial x^2} = \dfrac{1}{c^2} \dfrac{\partial u}{\partial t}$ and the given boundary conditions are:

$$u(0,t) = 0, \ u(1,t) = 0 \quad \text{and} \quad u(x,0) = 15$$

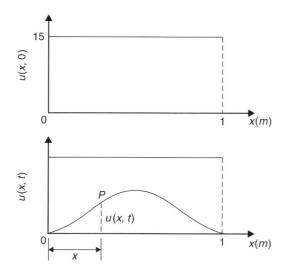

Figure 84.5

Assuming a solution of the form $u = XT$, then, from above,

$$X = A \cos px + B \sin px$$

and $\quad T = k e^{-p^2 c^2 t}$

Thus, the general solution is given by:

$$u(x,t) = \{P \cos px + Q \sin px\} e^{-p^2 c^2 t}$$

$$u(0,t) = 0 \text{ thus } 0 = P e^{-p^2 c^2 t}$$

from which, $P=0$ and $u(x,t) = \{Q \sin px\} e^{-p^2 c^2 t}$

Also, $u(1,t) = 0$ thus $0 = \{Q \sin p\} e^{-p^2 c^2 t}$

Since $Q \neq 0$, $\sin p = 0$ from which, $p = n\pi$ where $n = 1, 2, 3, \ldots$

Hence, $u(x,t) = \sum_{n=1}^{\infty} \left\{ Q_n e^{-p^2 c^2 t} \sin n\pi x \right\}$

The final initial condition given was that at $t=0$, $u = 15$, i.e. $u(x,0) = f(x) = 15$

Hence, $15 = \sum_{n=1}^{\infty} \{Q_n \sin n\pi x\}$ where, from Fourier coefficients, $Q_n = 2 \times$ mean value of $15 \sin n\pi x$ from $x=0$ to $x=1$,

i.e. $\quad Q_n = \frac{2}{1} \int_0^1 15 \sin n\pi x \, dx = 30 \left[-\frac{\cos n\pi x}{n\pi} \right]_0^1$

$$= -\frac{30}{n\pi} [\cos n\pi - \cos 0]$$

$$= \frac{30}{n\pi} (1 - \cos n\pi)$$

$$= 0 \text{ (when } n \text{ is even) and } \frac{60}{n\pi} \text{(when } n \text{ is odd)}$$

Hence, the required solution is:

$$u(x,t) = \sum_{n=1}^{\infty} \left\{ Q_n e^{-p^2 c^2 t} \sin n\pi x \right\}$$

$$= \frac{60}{\pi} \sum_{n(\text{odd})=1}^{\infty} \frac{1}{n} (\sin n\pi x) e^{-n^2 \pi^2 c^2 t}$$

Now try the following Practice Exercise

Practice Exercise 325   Further problems on the heat conduction equation (answers on page 1189)

1. A metal bar, insulated along its sides, is 4 m long. It is initially at a temperature of 10°C

and at time $t=0$, the ends are placed into ice at $0°C$. Find an expression for the temperature at a point P at a distance x m from one end at any time t seconds after $t=0$

2. An insulated uniform metal bar, 8 m long, has the temperature of its ends maintained at $0°C$, and at time $t=0$ the temperature distribution $f(x)$ along the bar is defined by $f(x)=x(8-x)$. If $c^2=1$, solve the heat conduction equation $\dfrac{\partial^2 u}{\partial x^2}=\dfrac{1}{c^2}\dfrac{\partial u}{\partial t}$ to determine the temperature u at any point in the bar at time t.

3. The ends of an insulated rod PQ, 20 units long, are maintained at $0°C$. At time $t=0$, the temperature within the rod rises uniformly from each end, reaching $4°C$ at the mid-point of PQ. Find an expression for the temperature $u(x, t)$ at any point in the rod, distant x from P at any time t after $t=0$. Assume the heat conduction equation to be $\dfrac{\partial^2 u}{\partial x^2}=\dfrac{1}{c^2}\dfrac{\partial u}{\partial t}$ and take $c^2=1$

84.8 Laplace's equation

The distribution of electrical potential, or temperature, over a plane area subject to certain boundary conditions, can be described by **Laplace's**[*] equation. The potential at a point P in a plane (see Figure 84.6) can be indicated by an ordinate axis and is a function of its position, i.e. $z=u(x, y)$, where $u(x, y)$ is the solution of the Laplace two-dimensional equation $\dfrac{\partial^2 u}{\partial x^2}+\dfrac{\partial^2 u}{\partial y^2}=0$

The method of solution of Laplace's equation is similar to the previous examples, as shown below.

Figure 84.7 shows a rectangle $OPQR$ bounded by the lines $x=0, y=0, x=a$ and $y=b$, for which we are required to find a solution of the equation $\dfrac{\partial^2 u}{\partial x^2}+\dfrac{\partial^2 u}{\partial y^2}=0$. The solution $z=(x, y)$ will give, say, the potential at any point within the rectangle $OPQR$. The boundary conditions are:

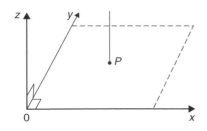

Figure 84.6

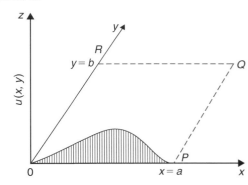

Figure 84.7

$u=0$ when $x=0$ i.e. $u(0, y)=0$ for $0\le y\le b$

$u=0$ when $x=a$ i.e. $u(a, y)=0$ for $0\le y\le b$

$u=0$ when $y=b$ i.e. $u(x, b)=0$ for $0\le x\le a$

$u=f(x)$ when $y=0$ i.e. $u(x, 0)=f(x)$

for $0\le x\le a$

As with previous partial differential equations, a solution of the form $u(x, y)=X(x)Y(y)$ is assumed, where X is a function of x only, and Y is a function of y only. Simplifying to $u=XY$, determining partial derivatives, and substituting into $\dfrac{\partial^2 u}{\partial x^2}+\dfrac{\partial^2 u}{\partial y^2}=0$ gives

$X''Y+XY''=0$

Separating the variables gives: $\dfrac{X''}{X}=-\dfrac{Y''}{Y}$

Letting each side equal a constant, $-p^2$, gives the two equations:

$X''+p^2 X=0$ and $Y''-p^2 Y=0$

from which, $X=A\cos px+B\sin px$ and

$Y=C\,e^{py}+D\,e^{-py}$ or $Y=C\cosh py+D\sinh py$ (see Problem 5, page 868 for this conversion).

This latter form can also be expressed as:

$Y=E\sinh p(y+\phi)$ by using compound angles.

Hence $u(x, y)=XY$

$=\{A\cos px+B\sin px\}\{E\sinh p(y+\phi)\}$

or $u(x, y)$

$$= \{P \cos px + Q \sin px\}\{\sinh p(y + \phi)\}$$

where $P = AE$ and $Q = BE$

The first boundary condition is: $u(0, y) = 0$, hence $0 = P \sinh p(y + \phi)$, from which $P = 0$. Hence, $u(x, y) = Q \sin px \sinh p(y + \phi)$
The second boundary condition is: $u(a, y) = 0$, hence $0 = Q \sin pa \sinh p(y + \phi)$, from which, $\sin pa = 0$, hence, $pa = n\pi$ or $p = \dfrac{n\pi}{a}$ for $n = 1, 2, 3, \ldots$
The third boundary condition is: $u(x, b) = 0$, hence, $0 = Q \sin px \sinh p(b + \phi)$, from which, $\sinh p(b + \phi) = 0$ and $\phi = -b$
Hence, $u(x, y) = Q \sin px \sinh p(y - b)$
$\qquad\qquad = Q_1 \sin px \sinh p(b - y)$ where $Q_1 = -Q$
Since there are many solutions for integer values of n,

$$u(x, y) = \sum_{n=1}^{\infty} Q_n \sin px \sinh p(b - y)$$

$$= \sum_{n=1}^{\infty} Q_n \sin \frac{n\pi x}{a} \sinh \frac{n\pi}{a}(b - y)$$

The fourth boundary condition is: $u(x, 0) = f(x)$,

hence, $f(x) = \displaystyle\sum_{n=1}^{\infty} Q_n \sin \frac{n\pi x}{a} \sinh \frac{n\pi b}{a}$

i.e. $f(x) = \displaystyle\sum_{n=1}^{\infty} \left(Q_n \sinh \frac{n\pi b}{a}\right) \sin \frac{n\pi x}{a}$

From Fourier series coefficients,

$$\left(Q_n \sinh \frac{n\pi b}{a}\right) = 2 \times \text{the mean value of}$$

$$f(x) \sin \frac{n\pi x}{a} \text{ from } x = 0 \text{ to } x = a$$

i.e. $= \displaystyle\int_0^a f(x) \sin \frac{n\pi x}{a} \, dx$ from which

$$Q_n \text{ may be determined.}$$

This is demonstrated in the following problem.

Problem 7. A square plate is bounded by the lines $x = 0, y = 0, x = 1$ and $y = 1$. Apply the Laplace equation $\dfrac{\partial^2 u}{\partial x^2} + \dfrac{\partial^2 u}{\partial y^2} = 0$ to determine the potential distribution $u(x, y)$ over the plate, subject to the following boundary conditions:

$u = 0$ when $x = 0$ $0 \leq y \leq 1$

$u = 0$ when $x = 1$ $0 \leq y \leq 1$

$u = 0$ when $y = 0$ $0 \leq x \leq 1$

$u = 4$ when $y = 1$ $0 \leq x \leq 1$

Initially a solution of the form $u(x, y) = X(x)Y(y)$ is assumed, where X is a function of x only, and Y is a function of y only. Simplifying to $u = XY$, determining partial derivatives, and substituting into $\dfrac{\partial^2 u}{\partial x^2} + \dfrac{\partial^2 u}{\partial y^2} = 0$
gives $X''Y + XY'' = 0$
Separating the variables gives: $\dfrac{X''}{X} = -\dfrac{Y''}{Y}$
Letting each side equal a constant, $-p^2$, gives the two equations:

$$X'' + p^2 X = 0 \quad \text{and} \quad Y'' - p^2 Y = 0$$

from which, $X = A \cos px + B \sin px$
and $Y = Ce^{py} + De^{-py}$
or $Y = C \cosh py + D \sinh py$
or $Y = E \sinh p(y + \phi)$

Hence $u(x, y) = XY$
$\qquad\qquad = \{A \cos px + B \sin px\}\{E \sinh p(y + \phi)\}$
or $u(x, y)$
$\qquad\qquad = \{P \cos px + Q \sin px\}\{\sinh p(y + \phi)\}$
where $P = AE$ and $Q = BE$

The first boundary condition is: $u(0, y) = 0$, hence $0 = P \sinh p(y + \phi)$ from which, $P = 0$
Hence, $u(x, y) = Q \sin px \sinh p(y + \phi)$
The second boundary condition is: $u(1, y) = 0$, hence $0 = Q \sin p(1) \sinh p(y + \phi)$, from which $\sin p = 0$, hence, $p = n\pi$ for $n = 1, 2, 3, \ldots$
The third boundary condition is: $u(x, 0) = 0$, hence, $0 = Q \sin px \sinh p(\phi)$, from which $\sinh p(\phi) = 0$ and $\phi = 0$
Hence, $u(x, y) = Q \sin px \sinh py$
Since there are many solutions for integer values of n,

$$u(x, y) = \sum_{n=1}^{\infty} Q_n \sin px \sinh py$$

$$= \sum_{n=1}^{\infty} Q_n \sin n\pi x \sinh n\pi y \qquad\qquad \text{(a)}$$

The fourth boundary condition is: $u(x, 1) = 4 = f(x)$,

hence, $f(x) = \sum_{n=1}^{\infty} Q_n \sin n\pi x \sinh n\pi$ (1)

From Fourier series coefficients,

$Q_n \sinh n\pi = 2 \times$ the mean value of

$$f(x) \sin n\pi x \text{ from } x = 0 \text{ to } x = 1$$

i.e. $= \dfrac{2}{1} \displaystyle\int_0^1 4 \sin n\pi x \, dx$

$= 8 \left[-\dfrac{\cos n\pi x}{n\pi} \right]_0^1$

$= -\dfrac{8}{n\pi} (\cos n\pi - \cos 0)$

$= \dfrac{8}{n\pi} (1 - \cos n\pi)$

$= 0$ (for even values of n)

$= \dfrac{16}{n\pi}$ (for odd values of n)

Hence, $Q_n = \dfrac{16}{n\pi (\sinh n\pi)} = \dfrac{16}{n\pi} \operatorname{cosech} n\pi$

Hence, from equation (a),

$$\boldsymbol{u(x, y)} = \sum_{n=1}^{\infty} Q_n \sin n\pi x \sinh n\pi y$$

$$= \dfrac{\mathbf{16}}{\boldsymbol{\pi}} \sum_{\boldsymbol{n(\text{odd})=1}}^{\infty} \dfrac{\mathbf{1}}{\boldsymbol{n}} (\operatorname{\mathbf{cosech}} \boldsymbol{n\pi} \, \mathbf{\sin} \boldsymbol{n\pi x} \, \mathbf{\sinh} \boldsymbol{n\pi y})$$

Now try the following Practice Exercise

Practice Exercise 326   Further problems on the Laplace equation (answers on page 1189)

1. A rectangular plate is bounded by the lines $x = 0, y = 0, x = 1$ and $y = 3$. Apply the Laplace equation $\dfrac{\partial^2 u}{\partial x^2} + \dfrac{\partial^2 u}{\partial y^2} = 0$ to determine the potential distribution $u(x, y)$ over the plate, subject to the following boundary conditions:

    $u = 0$ when $x = 0$ $0 \le y \le 2$
    $u = 0$ when $x = 1$ $0 \le y \le 2$
    $u = 0$ when $y = 2$ $0 \le x \le 1$
    $u = 5$ when $y = 3$ $0 \le x \le 1$

2. A rectangular plate is bounded by the lines $x = 0, y = 0, x = 3, y = 2$. Determine the potential distribution $u(x, y)$ over the rectangle using the Laplace equation

    $\dfrac{\partial^2 u}{\partial x^2} + \dfrac{\partial^2 u}{\partial y^2} = 0$, subject to the following boundary conditions:

    $u(0, y) = 0$ $0 \le y \le 2$

    $u(3, y) = 0$ $0 \le y \le 2$

    $u(x, 2) = 0$ $0 \le x \le 3$

    $u(x, 0) = x(3 - x)$ $0 \le x \le 3$

For fully worked solutions to each of the problems in Practice Exercises 322 to 326 in this chapter, go to the website:
www.routledge.com/cw/bird

This assignment covers the material contained in Chapters 81 to 84. *The marks available are shown in brackets at the end of each question.*

1. Find the particular solution of the following differential equations:

   (a) $12\dfrac{d^2 y}{dt^2} - 3y = 0$ given that when $t = 0$, $y = 3$ and $\dfrac{dy}{dt} = \dfrac{1}{2}$

   (b) $\dfrac{d^2 y}{dx^2} + 2\dfrac{dy}{dx} + 2y = 10e^x$ given that when $x = 0$, $y = 0$ and $\dfrac{dy}{dx} = 1$ (20)

2. In a galvanometer the deflection θ satisfies the differential equation:

   $$\frac{d^2 \theta}{dt^2} + 2\frac{d\theta}{dt} + \theta = 4$$

   Solve the equation for θ given that when $t = 0$, $\theta = 0$ and $\dfrac{d\theta}{dt} = 0$ (12)

3. Determine $y^{(n)}$ when $y = 2x^3 e^{4x}$ (10)

4. Determine the power series solution of the differential equation $\dfrac{d^2 y}{dx^2} + 2x\dfrac{dy}{dx} + y = 0$ using the Leibniz–Maclaurin method, given the boundary conditions that at $x = 0$, $y = 2$ and $\dfrac{dy}{dx} = 1$ (20)

5. Use the Frobenius method to determine the general power series solution of the differential equation: $\dfrac{d^2 y}{dx^2} + 4y = 0$ (21)

6. Determine the general power series solution of Bessel's equation:

   $$x^2\frac{d^2 y}{dx^2} + x\frac{dy}{dx} + (x^2 - v^2)y = 0$$

   and hence state the series up to and including the term in x^6 when $v = +3$ (26)

7. Determine the general solution of $\dfrac{\partial u}{\partial x} = 5xy$ (2)

8. Solve the differential equation $\dfrac{\partial^2 u}{\partial x^2} = x^2(y - 3)$ given the boundary conditions that at $x = 0$, $\dfrac{\partial u}{\partial x} = \sin y$ and $u = \cos y$ (6)

9. Figure RT29.1 shows a stretched string of length 40 cm which is set oscillating by displacing its mid-point a distance of 1 cm from its rest position and releasing it with zero velocity. Solve the wave equation: $\dfrac{\partial^2 u}{\partial x^2} = \dfrac{1}{c^2}\dfrac{\partial^2 u}{\partial t^2}$ where $c^2 = 1$, to determine the resulting motion $u(x, t)$ (23)

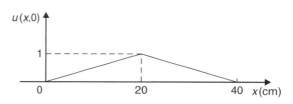

Figure RT29.1

First-order differential equations

Separation of variables

If $\dfrac{dy}{dx} = f(x)$ then $y = \displaystyle\int f(x)\,dx$

If $\dfrac{dy}{dx} = f(y)$ then $\displaystyle\int dx = \int \dfrac{dy}{f(y)}$

If $\dfrac{dy}{dx} = f(x).\,f(y)$ then $\displaystyle\int \dfrac{dy}{f(y)} = \int f(x)\,dx$

Homogeneous equations

If $P\dfrac{dy}{dx} = Q$, where P and Q are functions of both x and y of the same degree throughout (i.e. a homogeneous first-order differential equation), then:

(i) Rearrange $P\dfrac{dy}{dx} = Q$ into the form $\dfrac{dy}{dx} = \dfrac{Q}{P}$

(ii) Make the substitution $y = vx$ (where v is a function of x), from which, by the product rule,

$$\dfrac{dy}{dx} = v(1) + x\dfrac{dv}{dx}$$

(iii) Substitute for both y and $\dfrac{dy}{dx}$ in the equation $\dfrac{dy}{dx} = \dfrac{Q}{P}$

(iv) Simplify, by cancelling, and then separate the variables and solve using the $\dfrac{dy}{dx} = f(x).\,f(y)$ method

(v) Substitute $v = \dfrac{y}{x}$ to solve in terms of the original variables.

Linear first order

If $\dfrac{dy}{dx} + Py = Q$, where P and Q are functions of x only (i.e. a linear first-order differential equation), then

(i) determine the integrating factor, $e^{\int P\,dx}$

(ii) substitute the integrating factor (I.F.) into the equation

$$y(\text{I.F.}) = \int (\text{I.F.})\,Q\,dx$$

(iii) determine the integral $\int (\text{I.F.})\,Q\,dx$

Numerical solutions of first-order differential equations

Euler's method: $y_1 = y_0 + h\left(y'\right)_0$

Euler–Cauchy method: $y_{P_1} = y_0 + h\left(y'\right)_0$

and $y_{C_1} = y_0 + \dfrac{1}{2}h\left[\left(y'\right)_0 + f(x_1, y_{P_1})\right]$

Runge–Kutta method:

To solve the differential equation $\dfrac{dy}{dx} = f(x,\ y)$ given the initial condition $y = y_0$ at $x = x_0$ for a range of values of $x = x_0(h)x_n$:

1. Identify x_0, y_0 and h, and values of $x_1, x_2, x_3, \dots$.

2. Evaluate $k_1 = f\left(x_n, y_n\right)$, starting with $n = 0$

3. Evaluate $k_2 = f\left(x_n + \dfrac{h}{2}, y_n + \dfrac{h}{2}k_1\right)$

4. Evaluate $k_3 = f\left(x_n + \dfrac{h}{2}, y_n + \dfrac{h}{2}k_2\right)$

5. Evaluate $k_4 = f\left(x_n + h, y_n + hk_3\right)$

6. Use the values determined from steps 2 to 5 to evaluate:

$$y_{n+1} = y_n + \dfrac{h}{6}\{k_1 + 2k_2 + 2k_3 + k_4\}$$

7. Repeat steps 2 to 6 for $n = 1, 2, 3, \dots$

Second-order differential equations

If $a\dfrac{d^2 y}{dx^2} + b\dfrac{dy}{dx} + cy = 0$ (where a, b and c are constants) then:

(i) rewrite the differential equation as $(aD^2 + bD + c)y = 0$

(ii) substitute m for D and solve the auxiliary equation $am^2 + bm + c = 0$

(iii) if the roots of the auxiliary equation are:
   (a) **real and different**, say $m = \alpha$ and $m = \beta$, then the general solution is
   $$y = Ae^{\alpha x} + Be^{\beta x}$$

   (b) **real and equal**, say $m = \alpha$ twice, then the general solution is
   $$y = (Ax + B)e^{\alpha x}$$

   (c) **complex**, say $m = \alpha \pm j\beta$, then the general solution is
   $$y = e^{\alpha x}(A \cos \beta x + B \sin \beta x)$$

(iv) given boundary conditions, constants A and B can be determined and the particular solution obtained.

If $a\dfrac{d^2 y}{dx^2} + b\dfrac{dy}{dx} + cy = f(x)$ then:

(i) rewrite the differential equation as $(aD^2 + bD + c)y = 0$

(ii) substitute m for D and solve the auxiliary equation $am^2 + bm + c = 0$

(iii) obtain the complementary function (C.F.), u, as per (iii) above

(iv) to find the particular integral, v, first assume a particular integral which is suggested by $f(x)$, but which contains undetermined coefficients (see Table 82.1, page 873 for guidance)

(v) substitute the suggested particular integral into the original differential equation and equate relevant coefficients to find the constants introduced

(vi) the general solution is given by $y = u + v$

(vii) given boundary conditions, arbitrary constants in the C.F. can be determined and the particular solution obtained

Higher derivatives

y	$y^{(n)}$
e^{ax}	$a^n e^{ax}$
$\sin ax$	$a^n \sin\left(ax + \dfrac{n\pi}{2}\right)$
$\cos ax$	$a^n \cos\left(ax + \dfrac{n\pi}{2}\right)$
x^a	$\dfrac{a!}{(a-n)!}x^{a-n}$
$\sinh ax$	$\dfrac{a^n}{2}\left\{\left[1 + (-1)^n\right]\sinh ax + \left[1 - (-1)^n\right]\cosh ax\right\}$
$\cosh ax$	$\dfrac{a^n}{2}\left\{\left[1 - (-1)^n\right]\sinh ax + \left[1 + (-1)^n\right]\cosh ax\right\}$
$\ln ax$	$(-1)^{n-1}\dfrac{(n-1)!}{x^n}$

Leibniz's theorem

To find the nth derivative of a product $y = uv$:

$$y^{(n)} = (uv)^{(n)} = u^{(n)}v + nu^{(n-1)}v^{(1)}$$
$$+ \frac{n(n-1)}{2!}u^{(n-2)}v^{(2)}$$
$$+ \frac{n(n-1)(n-2)}{3!}u^{(n-3)}v^{(3)} + \dots$$

Power series solutions of second-order differential equations

(a) **Leibniz–Maclaurin method**
   (i) Differentiate the given equation n times, using the Leibniz theorem
   (ii) rearrange the result to obtain the recurrence relation at $x = 0$
   (iii) determine the values of the derivatives at $x = 0$, i.e. find $(y)_0$ and $(y')_0$
   (iv) substitute in the Maclaurin expansion for $y = f(x)$
   (v) simplify the result where possible and apply boundary condition (if given).

(b) **Frobenius method**

    (i) Assume a trial solution of the form:
$$y = x^c\{a_0 + a_1 x + a_2 x^2 + a_3 x^3 + \ldots + a_r x^r + \ldots\} a_0 \neq 0$$

    (ii) differentiate the trial series to find y' and y'',

    (iii) substitute the results in the given differential equation,

    (iv) equate coefficients of corresponding powers of the variable on each side of the equation: this enables index c and coefficients $a_1, a_2, a_3, \ldots$ from the trial solution, to be determined.

Bessel's equation

The solution of $x^2 \dfrac{d^2 y}{dx^2} + x\dfrac{dy}{dx} + (x^2 - v^2)y = 0$ is:

$$y = Ax^v \left\{ 1 - \frac{x^2}{2^2(v+1)} + \frac{x^4}{2^4 \times 2!(v+1)(v+2)} \right.$$
$$\left. - \frac{x^6}{2^6 \times 3!(v+1)(v+2)(v+3)} + \ldots \right\}$$

$$+ Bx^{-v} \left\{ 1 + \frac{x^2}{2^2(v-1)} + \frac{x^4}{2^4 \times 2!(v-1)(v-2)} \right.$$
$$\left. + \frac{x^6}{2^6 \times 3!(v-1)(v-2)(v-3)} + \ldots \right\}$$

or, in terms of **Bessel functions** and **gamma functions**:

$$y = AJ_v(x) + BJ_{-v}(x)$$

$$= A\left(\frac{x}{2}\right)^v \left\{ \frac{1}{\Gamma(v+1)} - \frac{x^2}{2^2(1!)\Gamma(v+2)} \right.$$
$$\left. + \frac{x^4}{2^4(2!)\Gamma(v+4)} - \ldots \right\}$$

$$+ B\left(\frac{x}{2}\right)^{-v} \left\{ \frac{1}{\Gamma(1-v)} - \frac{x^2}{2^2(1!)\Gamma(2-v)} \right.$$
$$\left. + \frac{x^4}{2^4(2!)\Gamma(3-v)} - \ldots \right\}$$

In general terms:

$$J_v(x) = \left(\frac{x}{2}\right)^v \sum_{k=0}^{\infty} \frac{(-1)^k x^{2k}}{2^{2k}(k!)\,\Gamma(v+k+1)}$$

and $J_{-v}(x) = \left(\dfrac{x}{2}\right)^{-v} \sum_{k=0}^{\infty} \dfrac{(-1)^k x^{2k}}{2^{2k}(k!)\,\Gamma(k-v+1)}$

and in particular:

$$J_n(x) = \left(\frac{x}{2}\right)^n \left\{ \frac{1}{n!} - \frac{1}{(n+1)!}\left(\frac{x}{2}\right)^2 \right.$$
$$\left. + \frac{1}{(2!)(n+2)!}\left(\frac{x}{2}\right)^4 - \ldots \right\}$$

$$J_0(x) = 1 - \frac{x^2}{2^2(1!)^2} + \frac{x^4}{2^4(2!)^2} - \frac{x^6}{2^6(3!)^2} + \ldots$$

and $J_1(x) = \dfrac{x}{2} - \dfrac{x^3}{2^3(1!)(2!)}$

$$+ \frac{x^5}{2^5(2!)(3!)} - \frac{x^7}{2^7(3!)(4!)} + \ldots$$

Legendres' equation

The solution of $(1-x^2)\dfrac{d^2 y}{dx^2} - 2x\dfrac{dy}{dx} + k(k+1)y = 0$

is: $y = a_0 \left\{ 1 - \dfrac{k(k+1)}{2!}x^2 + \right.$

$$\left. \frac{k(k+1)(k-2)(k+3)}{4!}x^4 - \ldots \right\}$$

$$+ a_1 \left\{ x - \frac{(k-1)(k+2)}{3!}x^3 + \right.$$

$$\left. \frac{(k-1)(k-3)(k+2)(k+4)}{5!}x^5 \ldots \right\}$$

Rodrigues' formula

$$P_n(x) = \frac{1}{2^n\,n!}\frac{d^n\,(x^2-1)^n}{dx^n}$$

Section L

Statistics and probability

Chapter 85

Presentation of statistical data

Why it is important to understand: **Presentation of statistical data**

Statistics is the study of the collection, organisation, analysis and interpretation of data. It deals with all aspects of this, including the planning of data collection in terms of the design of surveys and experiments. Statistics is applicable to a wide variety of academic disciplines, including natural and social sciences, engineering, government and business. Statistical methods can be used for summarising or describing a collection of data. Engineering statistics combines engineering and statistics. Design of experiments is a methodology for formulating scientific and engineering problems using statistical models. Quality control and process control use statistics as a tool to manage conformance to specifications of manufacturing processes and their products. Time and methods engineering use statistics to study repetitive operations in manufacturing in order to set standards and find optimum manufacturing procedures. Reliability engineering measures the ability of a system to perform for its intended function (and time) and has tools for improving performance. Probabilistic design involves the use of probability in product and system design. System identification uses statistical methods to build mathematical models of dynamical systems from measured data. System identification also includes the optimal design of experiments for efficiently generating informative data for fitting such models. This chapter introduces the presentation of statistical data.

At the end of this chapter, you should be able to:

- distinguish between discrete and continuous data
- present data diagrammatically – pictograms, horizontal and vertical bar charts, percentage component bar charts, pie diagrams
- produce a tally diagram for a set of data
- form a frequency distribution from a tally diagram
- construct a histogram from a frequency distribution
- construct a frequency polygon from a frequency distribution
- produce a cumulative frequency distribution from a set of grouped data
- construct an ogive from a cumulative frequency distribution

85.1 Some statistical terminology

Discrete and continuous data

Data are obtained largely by two methods:

(a) By counting – for example, the number of stamps sold by a post office in equal periods of time.

(b) By measurement – for example, the heights of a group of people.

When data are obtained by counting and only whole numbers are possible, the data are called **discrete**. Measured data can have any value within certain limits and are called **continuous**.

> **Problem 1.** Data are obtained on the topics given below. State whether they are discrete or continuous data.
>
> (a) The number of days on which rain falls in a month for each month of the year.
>
> (b) The mileage travelled by each of a number of salesmen.
>
> (c) The time that each of a batch of similar batteries lasts.
>
> (d) The amount of money spent by each of several families on food.

(a) The number of days on which rain falls in a given month must be an integer value and is obtained by **counting** the number of days. Hence, these data are **discrete**.

(b) A salesman can travel any number of miles (and parts of a mile) between certain limits and these data are **measured**. Hence, the data are **continuous**.

(c) The time that a battery lasts is **measured** and can have any value between certain limits. Hence, these data are **continuous**.

(d) The amount of money spent on food can only be expressed correct to the nearest pence, the amount being **counted**. Hence, these data are **discrete**.

Now try the following Practice Exercise

> **Practice Exercise 327   Discrete and continuous data (answers on page 1189)**
>
> In the following problems, state whether data relating to the topics given are discrete or continuous.

1. (a) The amount of petrol produced daily, for each of 31 days, by a refinery.
   (b) The amount of coal produced daily by each of 15 miners.
   (c) The number of bottles of milk delivered daily by each of 20 milkmen.
   (d) The size of 10 samples of rivets produced by a machine.

2. (a) The number of people visiting an exhibition on each of 5 days.
   (b) The time taken by each of 12 athletes to run 100 metres.
   (c) The value of stamps sold in a day by each of 20 post offices.
   (d) The number of defective items produced in each of 10 one-hour periods by a machine.

Further statistical terminology

A **set** is a group of data and an individual value within the set is called a **member** of the set. Thus, if the masses of five people are measured correct to the nearest 0.1 kilogram and are found to be 53.1 kg, 59.4 kg, 62.1 kg, 77.8 kg and 64.4 kg, then the set of masses in kilograms for these five people is

$$\{53.1, 59.4, 62.1, 77.8, 64.4\}$$

and one of the members of the set is 59.4

A set containing all the members is called a **population**. A **sample** is a set of data collected and/or selected from a statistical population by a defined procedure. Typically, the population is very large, making a census or a complete enumeration of all the values in the population impractical or impossible. The sample usually represents a subset of manageable size. Samples are collected and statistics are calculated from the samples so that inferences or extrapolations can be made from the sample to the population. Thus, all car registration numbers form a population but the registration numbers of, say, 20 cars taken at random throughout the country are a sample drawn from that population.

The number of times that the value of a member occurs in a set is called the **frequency** of that member. Thus, in the set $\{2, 3, 4, 5, 4, 2, 4, 7, 9\}$, member 4 has a frequency of three, member 2 has a frequency of 2 and the other members have a frequency of one.

The **relative frequency** with which any member of a set occurs is given by the ratio

$$\frac{\text{frequency of member}}{\text{total frequency of all members}}$$

For the set $\{2, 3, 5, 4, 7, 5, 6, 2, 8\}$, the relative frequency of member 5 is $\frac{2}{9}$. Often, relative frequency is expressed as a percentage and the **percentage relative frequency** is

$$(\text{relative frequency} \times 100)\%$$

85.2   Presentation of ungrouped data

Ungrouped data can be presented diagrammatically in several ways and these include

(a) **pictograms**, in which pictorial symbols are used to represent quantities (see Problem 2),

(b) **horizontal bar charts**, having data represented by equally spaced horizontal rectangles (see Problem 3) and

(c) **vertical bar charts**, in which data are represented by equally spaced vertical rectangles (see Problem 4).

Trends in ungrouped data over equal periods of time can be presented diagrammatically by a **percentage component bar chart**. In such a chart, equally spaced rectangles of any width, but whose height corresponds to 100%, are constructed. The rectangles are then subdivided into values corresponding to the percentage relative frequencies of the members (see Problem 5).

A **pie diagram** is used to show diagrammatically the parts making up the whole. In a pie diagram, the area of a circle represents the whole and the areas of the sectors of the circle are made proportional to the parts which make up the whole (see Problem 6).

Problem 2.   The number of television sets repaired in a workshop by a technician in six one-month periods is as shown below. Present these data as a pictogram

Month	January	February	March
Number repaired	11	6	15

Month	April	May	June
Number repaired	9	13	8

Each symbol shown in Figure 85.1 represents two television sets repaired. Thus, in January, $5\frac{1}{2}$ symbols are used to represent the 11 sets repaired; in February, 3 symbols are used to represent the 6 sets repaired, and so on.

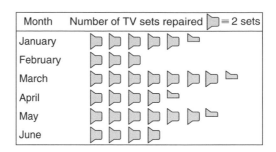

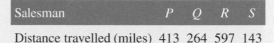

Figure 85.1

Problem 3.   The distances in miles travelled by four salesmen in a week are as shown below.

Salesman	P	Q	R	S
Distance travelled (miles)	413	264	597	143

Use a horizontal bar chart to represent these data diagrammatically

Equally spaced horizontal rectangles of any width, but whose length is proportional to the distance travelled, are used. Thus, the length of the rectangle for salesman P is proportional to 413 miles, and so on. The horizontal bar chart depicting these data is shown in Figure 85.2.

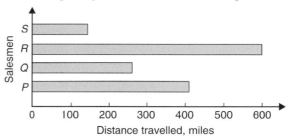

Figure 85.2

Problem 4.   The number of issues of tools or materials from a store in a factory is observed for seven one-hour periods in a day and the results of the survey are as follows.

Period	1	2	3	4	5	6	7
Number of issues	34	17	9	5	27	13	6

Present these data on a vertical bar chart

In a vertical bar chart, equally spaced vertical rectangles of any width, but whose height is proportional to the quantity being represented, are used. Thus, the height of the rectangle for period 1 is proportional to 34 units, and so on. The vertical bar chart depicting these data is shown in Figure 85.3.

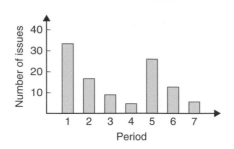

Figure 85.3

Problem 5. The numbers of various types of dwellings sold by a company annually over a three-year period are as shown below. Draw percentage component bar charts to present these data

	Year 1	Year 2	Year 3
4-roomed bungalows	24	17	7
5-roomed bungalows	38	71	118
4-roomed houses	44	50	53
5-roomed houses	64	82	147
6-roomed houses	30	30	25

A table of percentage relative frequency values, correct to the nearest 1%, is the first requirement. Since

$$\text{percentage relative frequency}$$
$$= \frac{\text{frequency of member} \times 100}{\text{total frequency}}$$

then for 4-roomed bungalows in year 1

$$\text{percentage relative frequency}$$
$$= \frac{24 \times 100}{24 + 38 + 44 + 64 + 30} = 12\%$$

The percentage relative frequencies of the other types of dwellings for each of the three years are similarly calculated and the results are as shown in the table below.

	Year 1	Year 2	Year 3
4-roomed bungalows	12%	7%	2%
5-roomed bungalows	19%	28%	34%
4-roomed houses	22%	20%	15%
5-roomed houses	32%	33%	42%
6-roomed houses	15%	12%	7%

The percentage component bar chart is produced by constructing three equally spaced rectangles of any width, corresponding to the three years. The heights of the rectangles correspond to 100% relative frequency and are subdivided into the values in the table of percentages shown above. A key is used (different types of shading or different colour schemes) to indicate corresponding percentage values in the rows of the table of percentages. The percentage component bar chart is shown in Figure 85.4.

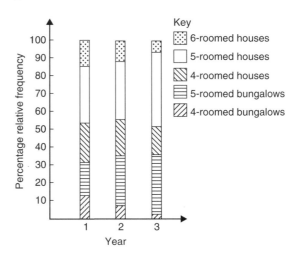

Figure 85.4

Problem 6. The retail price of a product costing £2 is made up as follows: materials 10 p, labour 20 p, research and development 40 p, overheads 70 p, profit 60 p. Present these data on a pie diagram

A circle of any radius is drawn. The area of the circle represents the whole, which in this case is £2. The circle is subdivided into sectors so that the areas of the sectors are proportional to the parts; i.e. the parts which make up the total retail price. For the area of a sector to be proportional to a part, the angle at the centre of the circle must be proportional to that part. The whole, £2 or 200 p, corresponds to 360°. Therefore,

$$10\,\text{p corresponds to } 360 \times \frac{10}{200} \text{ degrees, i.e. } 18°$$

$$20\,\text{p corresponds to } 360 \times \frac{20}{200} \text{ degrees, i.e. } 36°$$

and so on, giving the angles at the centre of the circle for the parts of the retail price as 18°, 36°, 72°, 126° and 108°, respectively.

The pie diagram is shown in Figure 85.5.

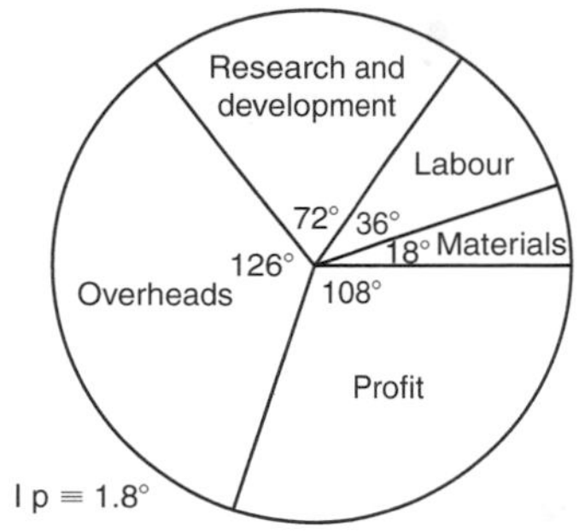

Figure 85.5

Problem 7.

(a) Using the data given in Figure 85.2 only, calculate the amount of money paid to each salesman for travelling expenses if they are paid an allowance of 37 p per mile

(b) Using the data presented in Figure 85.4, comment on the housing trends over the three-year period

(c) Determine the profit made by selling 700 units of the product shown in Figure 85.5

(a) By measuring the length of rectangle P, the mileage covered by salesman P is equivalent to 413 miles. Hence **salesman *P*** receives a travelling allowance of

$$\frac{£413 \times 37}{100} \text{ i.e. } \mathbf{£152.81}$$

Similarly, for **salesman *Q***, the miles travelled are 264 and his allowance is

$$\frac{£264 \times 37}{100} \text{ i.e. } \mathbf{£97.68}$$

Salesman *R* travels 597 miles and he receives

$$\frac{£597 \times 37}{100} \text{ i.e. } \mathbf{£220.89}$$

Finally, **salesman *S*** receives

$$\frac{£143 \times 37}{100} \text{ i.e. } \mathbf{£52.91}$$

(b) An analysis of Figure 85.4 shows that 5-roomed bungalows and 5-roomed houses are becoming more popular, the greatest change in the three years being a 15% increase in the sales of 5-roomed bungalows.

(c) Since 1.8° corresponds to 1 p and the profit occupies 108° of the pie diagram, the profit per unit is

$$\frac{108 \times 1}{1.8} \text{ i.e. } 60\,\text{p}$$

The profit when selling 700 units of the product is

$$£\frac{700 \times 60}{100} \text{ i.e. } \mathbf{£420}$$

Now try the following Practice Exercise

Practice Exercise 328 Presentation of ungrouped data (answers on page 1189)

1. The number of vehicles passing a stationary observer on a road in six 10-minute intervals is as shown. Draw a pictogram to represent these data.

Period of time	1	2	3	4	5	6
Number of vehicles	35	44	62	68	49	41

2. The number of components produced by a factory in a week is as shown below.

Day	Mon	Tues	Wed	Thur	Fri
Number of components	1580	2190	1840	2385	1280

Show these data on a pictogram.

3. For the data given in Problem 1 above, draw a horizontal bar chart.

4. Present the data given in Problem 2 above on a horizontal bar chart.

5. For the data given in Problem 1 above, construct a vertical bar chart.

6. Depict the data given in Problem 2 above on a vertical bar chart.

7. A factory produces three different types of components. The percentages of each of these components produced for three one-month periods are as shown below. Show this information on percentage component bar charts and comment on the changing trend

Section L

in the percentages of the types of component produced.

Month	1	2	3
Component P	20	35	40
Component Q	45	40	35
Component R	35	25	25

8. A company has five distribution centres and the mass of goods in tonnes sent to each centre during four one-week periods is as shown.

Week	1	2	3	4
Centre A	147	160	174	158
Centre B	54	63	77	69
Centre C	283	251	237	211
Centre D	97	104	117	144
Centre E	224	218	203	194

Use a percentage component bar chart to present these data and comment on any trends.

9. The employees in a company can be split into the following categories: managerial 3, supervisory 9, craftsmen 21, semi-skilled 67, others 44. Show these data on a pie diagram.

10. The way in which an apprentice spent his time over a one-month period is as follows: drawing office 44 hours, production 64 hours, training 12 hours, at college 28 hours. Use a pie diagram to depict this information.

11. (a) With reference to Figure 85.5, determine the amount spent on labour and materials to produce 1650 units of the product.
    (b) If in year 2 of Figure 85.4 1% corresponds to 2.5 dwellings, how many bungalows are sold in that year?

12. (a) If the company sells 23 500 units per annum of the product depicted in Figure 85.5, determine the cost of their overheads per annum.
    (b) If 1% of the dwellings represented in year 1 of Figure 85.4 corresponds to 2 dwellings, find the total number of houses sold in that year.

85.3 Presentation of grouped data

When the number of members in a set is small, say ten or fewer, the data can be represented diagrammatically without further analysis, by means of pictograms, bar charts, percentage components bar charts or pie diagrams (as shown in Section 85.2).

For sets having more than ten members, those members having similar values are grouped together in **classes** to form a **frequency distribution**. To assist in accurately counting members in the various classes, a **tally diagram** is used (see Problems 8 and 12).

A **frequency distribution** is merely a table showing classes and their corresponding frequencies (see Problems 8 and 12). The new set of values obtained by forming a frequency distribution is called **grouped data**. The terms used in connection with grouped data are shown in Figure 85.6(a). The size or range of a class is given by the **upper class boundary value** minus the **lower class boundary value** and in Figure 85.6(b) is $7.65 - 7.35$; i.e. 0.30. The **class interval** for the class shown in Figure 85.6(b) is 7.4 to 7.6 and the class mid-point value is given by

$$\frac{(\text{upper class boundary value}) + (\text{lower class boundary value})}{2}$$

and in Figure 85.6(b) is

$$\frac{7.65 + 7.35}{2} \quad \text{i.e. } 7.5$$

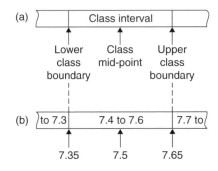

Figure 85.6

One of the principal ways of presenting grouped data diagrammatically is to use a **histogram**, in which the **areas** of vertical, adjacent rectangles are made proportional to frequencies of the classes (see Problem 9). When class intervals are equal, the heights of the rectangles of a histogram are equal to the frequencies of the classes. For histograms having unequal class intervals, the area must be proportional to the frequency. Hence, if the class interval of class A is twice the class interval of class B, then for equal frequencies the height of the rectangle representing *A* is half that of *B* (see Problem 11).

Another method of presenting grouped data diagrammatically is to use a **frequency polygon**, which is the graph produced by plotting frequency against class mid-point values and joining the co-ordinates with straight lines (see Problem 12).

A **cumulative frequency distribution** is a table showing the cumulative frequency for each value of upper class boundary. The cumulative frequency for a particular value of upper class boundary is obtained by adding the frequency of the class to the sum of the previous frequencies. A cumulative frequency distribution is formed in Problem 13.

The curve obtained by joining the co-ordinates of cumulative frequency (vertically) against upper class boundary (horizontally) is called an **ogive** or a **cumulative frequency distribution curve** (see Problem 13).

Problem 8. The data given below refer to the gain of each of a batch of 40 transistors, expressed correct to the nearest whole number. Form a frequency distribution for these data having seven classes

81	83	87	74	76	89	82	84
86	76	77	71	86	85	87	88
84	81	80	81	73	89	82	79
81	79	78	80	85	77	84	78
83	79	80	83	82	79	80	77

The **range** of the data is the value obtained by taking the value of the smallest member from that of the largest member. Inspection of the set of data shows that range $= 89 - 71 = 18$. The size of each class is given approximately by the range divided by the number of classes. Since 7 classes are required, the size of each class is $18 \div 7$; that is, approximately 3. To achieve seven equal classes spanning a range of values from 71 to 89, the class intervals are selected as 70–72, 73–75 and so on.

To assist with accurately determining the number in each class, a **tally diagram** is produced, as shown in Table 85.1(a). This is obtained by listing the classes in the left-hand column and then inspecting each of the 40 members of the set in turn and allocating them to the appropriate classes by putting 1s in the appropriate rows. Every fifth '1' allocated to a particular row is shown as an oblique line crossing the four previous 1s, to help with final counting.

Table 85.1(a)

Class	Tally
70–72	1
73–75	11
76–78	⊥⊥⊥⊤ 11
79–81	⊥⊥⊥⊤ ⊥⊥⊥⊤ 11
82–84	⊥⊥⊥⊤ 1111
85–87	⊥⊥⊥⊤ 1
88–90	111

Table 85.1(b)

Class	Class mid-point	Frequency
70–72	71	1
73–75	74	2
76–78	77	7
79–81	80	12
82–84	83	9
85–87	86	6
88–90	89	3

A **frequency distribution** for the data is shown in Table 85.1(b) and lists classes and their corresponding frequencies, obtained from the tally diagram. (Class mid-point values are also shown in the table, since they are used for constructing the histogram for these data (see Problem 9).)

Problem 9. Construct a histogram for the data given in Table 85.1(b)

The histogram is shown in Figure 85.7. The width of the rectangles corresponds to the upper class boundary

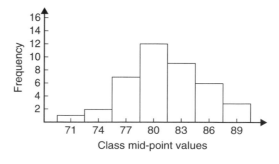

Figure 85.7

values minus the lower class boundary values and the heights of the rectangles correspond to the class frequencies. The easiest way to draw a histogram is to mark the class mid-point values on the horizontal scale and draw the rectangles symmetrically about the appropriate class mid-point values and touching one another.

Problem 10. The amount of money earned weekly by 40 people working part-time in a factory, correct to the nearest £10, is shown below. Form a frequency distribution having 6 classes for these data

80	90	70	110	90	160	110	80
140	30	90	50	100	110	60	100
80	90	110	80	100	90	120	70
130	170	80	120	100	110	40	110
50	100	110	90	100	70	110	80

Inspection of the set given shows that the majority of the members of the set lie between £80 and £110 and that there is a much smaller number of extreme values ranging from £30 to £170. If equal class intervals are selected, the frequency distribution obtained does not give as much information as one with unequal class intervals. Since the majority of the members lie between £80 and £110, the class intervals in this range are selected to be smaller than those outside of this range. There is no unique solution, one possible solution is shown in Table 85.2.

Table 85.2

Class	Frequency
20–40	2
50–70	6
80–90	12
100–110	14
120–140	4
150–170	2

Problem 11. Draw a histogram for the data given in Table 85.2

When dealing with unequal class intervals, the histogram must be drawn so that the areas (and not the heights) of the rectangles are proportional to the frequencies of the classes. The data given are shown in columns 1 and 2 of Table 85.3. Columns 3 and 4 give the upper and lower class boundaries, respectively. In column 5, the class ranges (i.e. upper class boundary minus lower class boundary values) are listed. The heights of the rectangles are proportional to the ratio

Table 85.3

1 Class	2 Frequency	3 Upper class boundary	4 Lower class boundary	5 Class range	6 Height of rectangle
20–40	2	45	15	30	$\frac{2}{30} = \frac{1}{15}$
50–70	6	75	45	30	$\frac{6}{30} = \frac{3}{15}$
80–90	12	95	75	20	$\frac{12}{20} = \frac{9}{15}$
100–110	14	115	95	20	$\frac{14}{20} = \frac{10\frac{1}{2}}{15}$
120–140	4	145	115	30	$\frac{4}{30} = \frac{2}{15}$
150–170	2	175	145	30	$\frac{2}{30} = \frac{1}{15}$

$\dfrac{\text{frequency}}{\text{class range}}$, as shown in column 6. The histogram is shown in Figure 85.8.

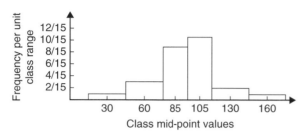

Figure 85.8

Problem 12. The masses of 50 ingots in kilograms are measured correct to the nearest 0.1 kg and the results are as shown below. Produce a frequency distribution having about 7 classes for these data and then present the grouped data as a frequency polygon and a histogram

8.0	8.6	8.2	7.5	8.0	9.1	8.5	7.6	8.2	7.8
8.3	7.1	8.1	8.3	8.7	7.8	8.7	8.5	8.4	8.5
7.7	8.4	7.9	8.8	7.2	8.1	7.8	8.2	7.7	7.5
8.1	7.4	8.8	8.0	8.4	8.5	8.1	7.3	9.0	8.6
7.4	8.2	8.4	7.7	8.3	8.2	7.9	8.5	7.9	8.0

The **range** of the data is the member having the largest value minus the member having the smallest value. Inspection of the set of data shows that **range** $= 9.1 - 7.1 = $ **2.0**

The size of each class is given approximately by

$$\frac{\text{range}}{\text{number of classes}}$$

Since about seven classes are required, the size of each class is $2.0 \div 7$, i.e. approximately 0.3, and thus the **class limits** are selected as 7.1 to 7.3, 7.4 to 7.6, 7.7 to 7.9, and so on. The **class mid-point** for the 7.1 to 7.3 class is

$$\frac{7.35 + 7.05}{2} \text{ i.e. } 7.2$$

the class mid-point for the 7.4 to 7.6 class is

$$\frac{7.65 + 7.35}{2} \text{ i.e. } 7.5$$

and so on.

To assist with accurately determining the number in each class, a **tally diagram** is produced as shown in Table 85.4. This is obtained by listing the classes in the left-hand column and then inspecting each of the 50 members of the set of data in turn and allocating it to the appropriate class by putting a '1' in the appropriate row. Each fifth '1' allocated to a particular row is marked as an oblique line to help with final counting.

Table 85.4

Class	Tally
7.1 to 7.3	111
7.4 to 7.6	̶I̶I̶I̶I̶
7.7 to 7.9	̶I̶I̶I̶I̶ 1111
8.0 to 8.2	̶I̶I̶I̶I̶ ̶I̶I̶I̶I̶ 1111
8.3 to 8.5	̶I̶I̶I̶I̶ ̶I̶I̶I̶I̶ 1
8.6 to 8.8	̶I̶I̶I̶I̶ 1
8.9 to 9.1	11

A **frequency distribution** for the data is shown in Table 85.5 and lists classes and their corresponding frequencies. Class mid-points are also shown in this table since they are used when constructing the frequency polygon and histogram.

Table 85.5

Class	Class mid-point	Frequency
7.1 to 7.3	7.2	3
7.4 to 7.6	7.5	5
7.7 to 7.9	7.8	9
8.0 to 8.2	8.1	14
8.3 to 8.5	8.4	11
8.6 to 8.8	8.7	6
8.9 to 9.1	9.0	2

A **frequency polygon** is shown in Figure 85.9, the co-ordinates corresponding to the class mid-point/frequency values given in Table 85.5. The co-ordinates are joined by straight lines and the polygon is 'anchored-down' at each end by joining to the next class mid-point value and zero frequency.

A **histogram** is shown in Figure 85.10, the width of a rectangle corresponding to (upper class boundary

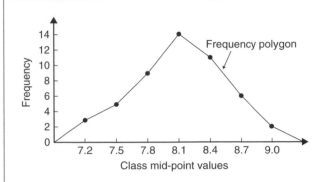

Figure 85.9

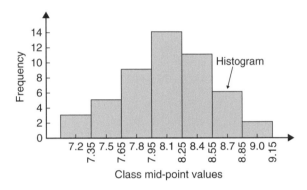

Figure 85.10

value – lower class boundary value) and height corresponding to the class frequency. The easiest way to draw a histogram is to mark class mid-point values on the horizontal scale and to draw the rectangles symmetrically about the appropriate class mid-point values and touching one another. A histogram for the data given in Table 85.5 is shown in Figure 85.10.

Problem 13. The frequency distribution for the masses in kilograms of 50 ingots is

7.1 to 7.3	3
7.4 to 7.6	5
7.7 to 7.9	9
8.0 to 8.2	14
8.3 to 8.5	11
8.6 to 8.8	6
8.9 to 9.1	2

Form a cumulative frequency distribution for these data and draw the corresponding ogive

A **cumulative frequency distribution** is a table giving values of cumulative frequency for the values of upper class boundaries and is shown in Table 85.6. Columns 1 and 2 show the classes and their frequencies. Column 3 lists the upper class boundary values for the classes given in column 1. Column 4 gives the cumulative frequency values for all frequencies less than the upper class boundary values given in column 3. Thus, for example, for the 7.7 to 7.9 class shown in row 3, the cumulative frequency value is the sum of all frequencies having values of less than 7.95, i.e. $3 + 5 + 9 = 17$, and so on.

Table 85.6

1 Class	2 Frequency	3 Upper class boundary less than	4 Cumulative frequency
7.1–7.3	3	7.35	3
7.4–7.6	5	7.65	8
7.7–7.9	9	7.95	17
8.0–8.2	14	8.25	31
8.3–8.5	11	8.55	42
8.6–8.8	6	8.85	48
8.9–9.1	2	9.15	50

The **ogive** for the cumulative frequency distribution given in Table 85.6 is shown in Figure 85.11. The co-ordinates corresponding to each upper class boundary/cumulative frequency value are plotted and

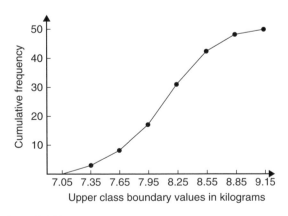

Figure 85.11

the co-ordinates are joined by straight lines (not the best curve drawn through the co-ordinates as in experimental work). The ogive is 'anchored' at its start by adding the co-ordinate (7.05, 0).

Now try the following Practice Exercise

Practice Exercise 329   Presentation of grouped data (answers on page 1189)

1. The masses in kilograms, correct to the nearest one-tenth of a kilogram, of 60 bars of metal are as shown. Form a frequency distribution of about 8 classes for these data.

39.8	40.1	40.3	40.0	40.6	39.7	40.0	40.4	39.6	39.3
39.6	40.7	40.2	39.9	40.3	40.2	40.4	39.9	39.8	40.0
40.2	40.1	40.3	39.7	39.9	40.5	39.9	40.5	40.0	39.9
40.1	40.8	40.0	40.0	40.1	40.2	40.1	40.0	40.2	39.9
39.7	39.8	40.4	39.7	39.9	39.5	40.1	40.1	39.9	40.2
39.5	40.6	40.0	40.1	39.8	39.7	39.5	40.2	39.9	40.3

2. Draw a histogram for the frequency distribution given in the solution of Problem 1.

3. The information given below refers to the value of resistance in ohms of a batch of 48 resistors of similar value. Form a frequency distribution for the data, having about 6 classes, and draw a frequency polygon and histogram to represent these data diagrammatically.

21.0	22.4	22.8	21.5	22.6	21.1	21.6	22.3
22.9	20.5	21.8	22.2	21.0	21.7	22.5	20.7
23.2	22.9	21.7	21.4	22.1	22.2	22.3	21.3
22.1	21.8	22.0	22.7	21.7	21.9	21.1	22.6
21.4	22.4	22.3	20.9	22.8	21.2	22.7	21.6
22.2	21.6	21.3	22.1	21.5	22.0	23.4	21.2

4. The times taken in hours to the failure of 50 specimens of a metal subjected to fatigue failure tests are as shown. Form a frequency distribution, having about 7 classes and unequal class intervals, for these data.

28	22	23	20	12	24	37	28	21	25
21	14	30	23	27	13	23	7	26	19
24	22	26	3	21	24	28	40	27	24
20	25	23	26	47	21	29	26	22	33
27	9	13	35	20	16	20	25	18	22

5. Form a cumulative frequency distribution and hence draw the ogive for the frequency distribution given in the solution to Problem 3.

6. Draw a histogram for the frequency distribution given in the solution to Problem 4.

7. The frequency distribution for a batch of 50 capacitors of similar value, measured in microfarads, is

10.5–10.9	2
11.0–11.4	7
11.5–11.9	10
12.0–12.4	12
12.5–12.9	11
13.0–13.4	8

Form a cumulative frequency distribution for these data.

8. Draw an ogive for the data given in the solution of Problem 7.

9. The diameter in millimetres of a reel of wire is measured in 48 places and the results are as shown.

2.10	2.29	2.32	2.21	2.14	2.22
2.28	2.18	2.17	2.20	2.23	2.13
2.26	2.10	2.21	2.17	2.28	2.15
2.16	2.25	2.23	2.11	2.27	2.34
2.24	2.05	2.29	2.18	2.24	2.16
2.15	2.22	2.14	2.27	2.09	2.21
2.11	2.17	2.22	2.19	2.12	2.20
2.23	2.07	2.13	2.26	2.16	2.12

Section L

(a)   Form a frequency distribution of diameters having about 6 classes.

(b)   Draw a histogram depicting the data.

(c)   Form a cumulative frequency distribution.

(d)   Draw an ogive for the data.

For fully worked solutions to each of the problems in Practice Exercises 327 to 329 in this chapter, go to the website:
www.routledge.com/cw/bird

Chapter 86

Mean, median, mode and standard deviation

Why it is important to understand: Mean, median, mode and standard deviation

Statistics is a field of mathematics that pertains to data analysis. In many real-life situations, it is helpful to describe data by a single number that is most representative of the entire collection of numbers. Such a number is called a measure of central tendency; the most commonly used measures are mean, median, mode and standard deviation, the latter being the average distance between the actual data and the mean. Statistics is important in the field of engineering since it provides tools to analyse collected data. For example, a chemical engineer may wish to analyse temperature measurements from a mixing tank. Statistical methods can be used to determine how reliable and reproducible the temperature measurements are, how much the temperature varies within the data set, what future temperatures of the tank may be, and how confident the engineer can be in the temperature measurements made. When performing statistical analysis on a set of data, the mean, median, mode and standard deviation are all helpful values to calculate; this chapter explains how to determine these measures of central tendency.

At the end of this chapter, you should be able to:

- determine the mean, median and mode for a set of ungrouped data
- determine the mean, median and mode for a set of grouped data
- draw a histogram from a set of grouped data
- determine the mean, median and mode from a histogram
- calculate the standard deviation from a set of ungrouped data
- calculate the standard deviation from a set of grouped data
- determine the quartile values from an ogive
- determine quartile, decile and percentile values from a set of data

---

86.1 Measures of central tendency

A single value, which is representative of a set of values, may be used to give an indication of the general size of the members in a set, the word '**average**' often being used to indicate the single value. The statistical term used for 'average' is the 'arithmetic mean' or just the '**mean**'.

Other measures of central tendency may be used and these include the **median** and the **modal** values.

86.2 Mean, median and mode for discrete data

Mean

The **arithmetic mean value** is found by adding together the values of the members of a set and dividing by the number of members in the set. Thus, the mean of the set of numbers $\{4, 5, 6, 9\}$ is

$$\frac{4+5+6+9}{4} \text{ i.e. } 6$$

In general, the mean of the set $\{x_1, x_2, x_3, \ldots x_n\}$ is

$$\overline{x} = \frac{x_1 + x_2 + x_3 + \cdots + x_n}{n} \text{ written as } \frac{\sum x}{n}$$

where $\sum$ is the Greek letter 'sigma' and means 'the sum of ' and $\overline{x}$ (called x-bar) is used to signify a mean value.

Median

The **median value** often gives a better indication of the general size of a set containing extreme values. The set $\{7, 5, 74, 10\}$ has a mean value of 24, which is not really representative of any of the values of the members of the set. The median value is obtained by

(a) **ranking** the set in ascending order of magnitude, and

(b) selecting the value of the **middle member** for sets containing an odd number of members or finding the value of the mean of the two middle members for sets containing an even number of members.

For example, the set $\{7, 5, 74, 10\}$ is ranked as $\{5, 7, 10, 74\}$ and, since it contains an even number of members (four in this case), the mean of 7 and 10 is taken, giving a median value of 8.5. Similarly, the set $\{3, 81, 15, 7, 14\}$ is ranked as $\{3, 7, 14, 15, 81\}$ and the median value is the value of the middle member, i.e. 14

Mode

The **modal value**, or **mode**, is the most commonly occurring value in a set. If two values occur with the same frequency, the set is '**bi-modal**'. The set

$\{5, 6, 8, 2, 5, 4, 6, 5, 3\}$ has a modal value of 5, since the member having a value of 5 occurs the most, i.e. three times.

Problem 1. Determine the mean, median and mode for the set $\{2, 3, 7, 5, 5, 13, 1, 7, 4, 8, 3, 4, 3\}$

The mean value is obtained by adding together the values of the members of the set and dividing by the number of members in the set. Thus,

mean value, $\overline{x}$

$$= \frac{2+3+7+5+5+13+1+7+4+8+3+4+3}{13}$$

$$= \frac{65}{13} = 5$$

To obtain the median value the set is ranked, that is, placed in ascending order of magnitude, and since the set contains an odd number of members the value of the middle member is the median value. Ranking the set gives $\{1, 2, 3, 3, 3, 4, 4, 5, 5, 7, 7, 8, 13\}$. The middle term is the seventh member; i.e. 4. Thus, the **median value is 4**.

The **modal value** is the value of the most commonly occurring member and is **3**, which occurs three times, all other members only occurring once or twice.

Problem 2. The following set of data refers to the amount of money in £s taken by a news vendor for 6 days. Determine the mean, median and modal values of the set
$\{27.90, 34.70, 54.40, 18.92, 47.60, 39.68\}$

Mean value

$$= \frac{27.90 + 34.70 + 54.40 + 18.92 + 47.60 + 39.68}{6}$$

$$= \textbf{£37.20}$$

The ranked set is
$\{18.92, 27.90, 34.70, 39.68, 47.60, 54.40\}$
Since the set has an even number of members, the mean of the middle two members is taken to give the median value; i.e.

$$\text{median value} = \frac{34.70 + 39.68}{2} = \textbf{£37.19}$$

Since no two members have the same value, this set has **no mode**.

Now try the following Practice Exercise

Practice Exercise 330   Mean, median and mode for discrete data (answers on page 1190)

In Problems 1 to 4, determine the mean, median and modal values for the sets given.

1. {3, 8, 10, 7, 5, 14, 2, 9, 8}

2. {26, 31, 21, 29, 32, 26, 25, 28}

3. {4.72, 4.71, 4.74, 4.73, 4.72, 4.71, 4.73, 4.72}

4. {73.8, 126.4, 40.7, 141.7, 28.5, 237.4, 157.9}

86.3 Mean, median and mode for grouped data

The mean value for a set of grouped data is found by determining the sum of the (frequency × class mid-point values) and dividing by the sum of the frequencies; i.e.

$$\textbf{mean value}, \bar{x} = \frac{f_1 x_1 + f_2 x_2 + \cdots f_n x_n}{f_1 + f_2 + \cdots + f_n} = \frac{\sum(fx)}{\sum f}$$

where f is the frequency of the class having a mid-point value of x, and so on.

Problem 3.   The frequency distribution for the value of resistance in ohms of 48 resistors is as shown. Determine the mean value of resistance

20.5–20.9	3
21.0–21.4	10
21.5–21.9	11
22.0–22.4	13
22.5–22.9	9
23.0–23.4	2

The class mid-point/frequency values are 20.7 3, 21.2 10, 21.7 11, 22.2 13, 22.7 9 and 23.2 2
For grouped data, the mean value is given by

$$\bar{x} = \frac{\sum(fx)}{\sum f}$$

where f is the class frequency and x is the class mid-point value. Hence mean value,

$$\bar{x} = \frac{\begin{array}{l}(3 \times 20.7) + (10 \times 21.2) + (11 \times 21.7)\\ + (13 \times 22.2) + (9 \times 22.7) + (2 \times 23.2)\end{array}}{48}$$

$$= \frac{1052.1}{48} = 21.919\ldots$$

i.e. **the mean value is 21.9 ohms**, correct to 3 significant figures.

Histograms

The mean, median and modal values for grouped data may be determined from a **histogram**. In a histogram, frequency values are represented vertically and variable values horizontally. The mean value is given by the value of the variable corresponding to a vertical line drawn through the centroid of the histogram. The median value is obtained by selecting a variable value such that the area of the histogram to the left of a vertical line drawn through the selected variable value is equal to the area of the histogram on the right of the line. The modal value is the variable value obtained by dividing the width of the highest rectangle in the histogram in proportion to the heights of the adjacent rectangles. The method of determining the mean, median and modal values from a histogram is shown in Problem 4.

Problem 4.   The time taken in minutes to assemble a device is measured 50 times and the results are as shown. Draw a histogram depicting the data and hence determine the mean, median and modal values of the distribution

14.5–15.5	5
16.5–17.5	8
18.5–19.5	16
20.5–21.5	12
22.5–23.5	6
24.5–25.5	3

The histogram is shown in Figure 86.1. The mean value lies at the centroid of the histogram. With reference to

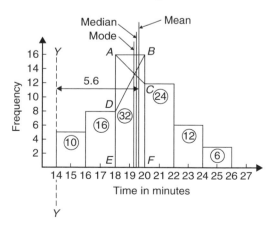

Figure 86.1

any arbitrary axis, say YY shown at a time of 14 minutes, the position of the horizontal value of the centroid can be obtained from the relationship $AM = \sum(am)$, where A is the area of the histogram, M is the horizontal distance of the centroid from the axis YY, a is the area of a rectangle of the histogram and m is the distance of the centroid of the rectangle from YY. The areas of the individual rectangles are shown circled on the histogram, giving a total area of 100 square units. The positions, m, of the centroids of the individual rectangles are $1, 3, 5, \ldots$ units from YY. Thus

$$100M = (10 \times 1) + (16 \times 3) + (32 \times 5) + (24 \times 7)$$
$$+ (12 \times 9) + (6 \times 11)$$

i.e. $M = \dfrac{560}{100} = 5.6$ units from YY

Thus, the position of the **mean** with reference to the time scale is $14 + 5.6$, i.e. **19.6 minutes**.

The median is the value of time corresponding to a vertical line dividing the total area of the histogram into two equal parts. The total area is 100 square units, hence the vertical line must be drawn to give 50 units of area on each side. To achieve this with reference to Figure 86.1, rectangle $ABFE$ must be split so that $50 - (10 + 16)$ units of area lie on one side and $50 - (24 + 12 + 6)$ units of area lie on the other. This shows that the area of $ABFE$ is split so that 24 units of area lie to the left of the line and 8 units of area lie to the right; i.e. the vertical line must pass through 19.5 minutes. Thus, the **median value** of the distribution is **19.5 minutes**.

The mode is obtained by dividing the line AB, which is the height of the highest rectangle, proportionally to the heights of the adjacent rectangles. With reference to Figure 86.1, this is achieved by joining AC and BD and drawing a vertical line through the point of intersection of these two lines. This gives the **mode** of the distribution, which is **19.3 minutes**.

Now try the following Practice Exercise

Practice Exercise 331   Mean, median and mode for grouped data (answers on page 1190)

1.   21 bricks have a mean mass of 24.2 kg and 29 similar bricks have a mass of 23.6 kg. Determine the mean mass of the 50 bricks.

2.   The frequency distribution given below refers to the heights in centimetres of 100 people.

Determine the mean value of the distribution, correct to the nearest millimetre.

150–156	5
157–163	18
164–170	20
171–177	27
178–184	22
185–191	8

3.   The gain of 90 similar transistors is measured and the results are as shown. By drawing a histogram of this frequency distribution, determine the mean, median and modal values of the distribution.

83.5–85.5	6
86.5–88.5	39
89.5–91.5	27
92.5–94.5	15
95.5–97.5	3

4.   The diameters, in centimetres, of 60 holes bored in engine castings are measured and the results are as shown. Draw a histogram depicting these results and hence determine the mean, median and modal values of the distribution.

2.011–2.014	7
2.016–2.019	16
2.021–2.024	23
2.026–2.029	9
2.031–2.034	5

86.4   Standard deviation

Discrete data

The standard deviation of a set of data gives an indication of the amount of dispersion, or the scatter, of members of the set from the measure of central tendency. Its value is the root-mean-square value of the members of the set and for discrete data is obtained as follows.

(i)   Determine the measure of central tendency, usually the mean value (occasionally the median or modal values are specified).

(ii)   Calculate the deviation of each member of the set from the mean, giving

$$(x_1 - \overline{x}), (x_2 - \overline{x}), (x_3 - \overline{x}), \ldots$$

(iii) Determine the squares of these deviations; i.e.
$(x_1 - \overline{x})^2, (x_2 - \overline{x})^2, (x_3 - \overline{x})^2, \ldots$

(iv) Find the sum of the squares of the deviations, i.e.
$(x_1 - \overline{x})^2 + (x_2 - \overline{x})^2 + (x_3 - \overline{x})^2, \ldots$

(v) Divide by the number of members in the set, n, giving
$$\frac{(x_1 - \overline{x})^2 + (x_2 - \overline{x})^2 + \left(x^3 - \overline{x}\right)^2 + \cdots}{n}$$

(vi) Determine the square root of (v).

The standard deviation is indicated by σ (the Greek letter small 'sigma') and is written mathematically as

$$\textbf{standard deviation}, \sigma = \sqrt{\left\{ \frac{\sum (x - \overline{x})^2}{n} \right\}}$$

where x is a member of the set, $\overline{x}$ is the mean value of the set and n is the number of members in the set. The value of standard deviation gives an indication of the distance of the members of a set from the mean value. The set {1, 4, 7, 10, 13} has a mean value of 7 and a standard deviation of about 4.2. The set {5, 6, 7, 8, 9} also has a mean value of 7 but the standard deviation is about 1.4. This shows that the members of the second set are mainly much closer to the mean value than the members of the first set. The method of determining the standard deviation for a set of discrete data is shown in Problem 5. Note that in the above formula for standard deviation, $\dfrac{\sum (x - \bar{x})^2}{n}$ is referred to as the **variance**,

i.e. $\sigma = \sqrt{\text{variance}}$ or variance $= \sigma^2$

Problem 5. Determine the standard deviation from the mean of the set of numbers {5, 6, 8, 4, 10, 3}, correct to 4 significant figures

The arithmetic mean, $\overline{x} = \dfrac{\sum x}{n}$

$$= \frac{5 + 6 + 8 + 4 + 10 + 3}{6} = 6$$

Standard deviation, $\sigma = \sqrt{\left\{ \dfrac{\sum (x - \overline{x})^2}{n} \right\}}$

The $(x - \overline{x})^2$ values are $(5 - 6)^2, (6 - 6)^2, (8 - 6)^2,$ $(4 - 6)^2, (10 - 6)^2$ and $(3 - 6)^2$

The sum of the $(x - \overline{x})^2$ values,

i.e. $\sum (x - \overline{x})^2$, is $1 + 0 + 4 + 4 + 16 + 9 = 34$

and $\dfrac{\sum (x - \overline{x})^2}{n} = \dfrac{34}{6} = 5.\dot{6}$

since there are 6 members in the set.

Hence, **standard deviation**,

$$\sigma = \sqrt{\left\{ \frac{\sum (x - \overline{x})^2}{n} \right\}} = \sqrt{5.\dot{6}} = \textbf{2.380}$$

correct to 4 significant figures.

Grouped data

For grouped data,

$$\textbf{standard deviation}, \sigma = \sqrt{\left\{ \frac{\sum \{ f (x - \overline{x})^2 \}}{\sum f} \right\}}$$

where f is the class frequency value, x is the class mid-point value and $\overline{x}$ is the mean value of the grouped data. The method of determining the standard deviation for a set of grouped data is shown in Problem 6.

Problem 6. The frequency distribution for the values of resistance in ohms of 48 resistors is as shown. Calculate the standard deviation from the mean of the resistors, correct to 3 significant figures

20.5–20.9	3
21.0–21.4	10
21.5–21.9	11
22.0–22.4	13
22.5–22.9	9
23.0–23.4	2

The standard deviation for grouped data is given by

$$\sigma = \sqrt{\left\{ \frac{\sum \{ f (x - \overline{x})^2 \}}{\sum f} \right\}}$$

From Problem 3, the distribution mean value is $\overline{x} = 21.92$, correct to 2 significant figures.

The 'x-values' are the class mid-point values, i.e. $20.7, 21.2, 21.7, \ldots$

Thus, the $(x - \overline{x})^2$ values are $(20.7 - 21.92)^2,$ $(21.2 - 21.92)^2, (21.7 - 21.92)^2, \ldots$

and the $f(x - \overline{x})^2$ values are $3(20.7 - 21.92)^2,$ $10(21.2 - 21.92)^2, 11(21.7 - 21.92)^2, \ldots$

The $\sum f(x - \overline{x})^2$ values are

$$4.4652 + 5.1840 + 0.5324 + 1.0192$$

$$+5.4756 + 3.2768 = 19.9532$$

$$\frac{\sum\{f(x - \overline{x})^2\}}{\sum f} = \frac{19.9532}{48} = 0.41569$$

and **standard deviation**,

$$\sigma = \sqrt{\left\{\frac{\sum\{f(x - \overline{x})^2\}}{\sum f}\right\}} = \sqrt{0.41569}$$

$$= \mathbf{0.645}, \text{ correct to 3 significant figures.}$$

Now try the following Practice Exercise

Practice Exercise 332   Standard deviation (answers on page 1190)

1. Determine the standard deviation from the mean of the set of numbers

$$\{35, 22, 25, 23, 28, 33, 30\}$$

correct to 3 significant figures.

2. The values of capacitances, in microfarads, of ten capacitors selected at random from a large batch of similar capacitors are

   34.3, 25.0, 30.4, 34.6, 29.6, 28.7,

   33.4, 32.7, 29.0 and 31.3

   Determine the standard deviation from the mean for these capacitors, correct to 3 significant figures.

3. The tensile strengths in megapascals for 15 samples of tin were determined and found to be

   34.61, 34.57, 34.40, 34.63, 34.63, 34.51,

   34.49, 34.61, 34.52, 34.55, 34.58, 34.53,

   34.44, 34.48 and 34.40

   Calculate the mean and standard deviation from the mean for these 15 values, correct to 4 significant figures.

4. Calculate the standard deviation from the mean for the mass of the 50 bricks given in Problem 1 of Practice Exercise 331, page 940, correct to 3 significant figures.

5. Determine the standard deviation from the mean, correct to 4 significant figures, for the heights of the 100 people given in Problem 2 of Practice Exercise 331, page 940.

6. Calculate the standard deviation from the mean for the data given in Problem 4 of Practice Exercise 331, page 940, correct to 3 decimal places.

86.5   Quartiles, deciles and percentiles

Other measures of dispersion which are sometimes used are the quartile, decile and percentile values. The **quartile values** of a set of discrete data are obtained by selecting the values of members which divide the set into four equal parts. Thus, for the set {2, 3, 4, 5, 5, 7, 9, 11, 13, 14, 17} there are 11 members and the values of the members dividing the set into four equal parts are 4, 7 and 13. These values are signified by Q_1, Q_2 and Q_3, and called the first, second and third quartile values, respectively. It can be seen that the second quartile value, Q_2, is the value of the middle member and hence is the median value of the set.

For grouped data the ogive may be used to determine the quartile values. In this case, points are selected on the vertical cumulative frequency values of the ogive, such that they divide the total value of cumulative frequency into four equal parts. Horizontal lines are drawn from these values to cut the ogive. The values of the variable corresponding to these cutting points on the ogive give the quartile values (see Problem 7).

When a set contains a large number of members, the set can be split into ten parts, each containing an equal number of members. These ten parts are then called **deciles**. For sets containing a very large number of members, the set may be split into 100 parts, each containing an equal number of members. One of these parts is called a **percentile**.

Problem 7. The frequency distribution given below refers to the overtime worked by a group of craftsmen during each of 48 working weeks in a year. Draw an ogive for these data and hence determine the quartile values.

25–29	5
30–34	4
35–39	7
40–44	11
45–49	12
50–54	8
55–59	1

The cumulative frequency distribution (i.e. upper class boundary/cumulative frequency values) is

29.5   5,   34.5   9,   39.5 16,   44.5 27,
49.5 39,   54.5 47,   59.5   48

The ogive is formed by plotting these values on a graph, as shown in Figure 86.2. The total frequency is divided into four equal parts, each having a range of $48 \div 4$, i.e. 12. This gives cumulative frequency values of 0 to 12 corresponding to the first quartile, 12 to 24 corresponding to the second quartile, 24 to 36 corresponding to the third quartile and 36 to 48 corresponding to the fourth quartile of the distribution; i.e. the distribution is divided into four equal parts. The quartile values are those of the variable corresponding to cumulative frequency values of 12, 24 and 36, marked Q_1, Q_2 and Q_3 in Figure 86.2. These values, correct to the nearest hour, are **37 hours**, **43 hours** and **48 hours**, respectively. The Q_2 value is also equal to the median value of the distribution. One measure of the dispersion of a distribution is called the **semi-interquartile range** and is given by $(Q_3 - Q_1) \div 2$ and is $(48 - 37) \div 2$ in this case; i.e. $5\frac{1}{2}$ **hours**.

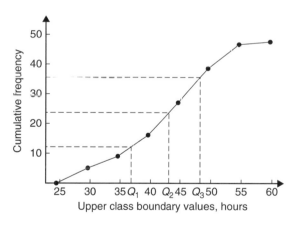

Figure 86.2

Problem 8. Determine the numbers contained in the (a) 41st to 50th percentile group and (b) 8th decile group of the following set of numbers.

14   22   17   21   30   28   37   7   23   32
24   17   20   22   27   19   26   21   15   29

The set is ranked, giving

7   14   15   17   17   19   20   21   21   22
22   23   24   26   27   28   29   30   32   37

(a) There are 20 numbers in the set, hence the first 10% will be the two numbers 7 and 14, the second 10% will be 15 and 17, and so on. Thus, the 41st to 50th percentile group will be the numbers **21 and 22**.

(b) The first decile group is obtained by splitting the ranked set into 10 equal groups and selecting the first group; i.e. the numbers 7 and 14. The second decile group is the numbers 15 and 17, and so on. Thus, the 8th decile group contains the numbers **27 and 28**.

Now try the following Practice Exercise

Practice Exercise 333   **Quartiles, deciles and percentiles (answers on page 1190)**

1. The number of working days lost due to accidents for each of 12 one-monthly periods are as shown. Determine the median and first and third quartile values for this data.

   27   37   40   28   23   30   35   24   30   32   31   28

2. The number of faults occurring on a production line in a nine-week period are as shown below. Determine the median and quartile values for the data.

   30   27   25   24   27   37   31   27   35

3. Determine the quartile values and semi-interquartile range for the frequency distribution given in Problem 2 of Practice Exercise 331, page 940.

4. Determine the numbers contained in the 5th decile group and in the 61st to 70th percentile groups for the following set of numbers.

   40   46   28   32   37   42   50   31   48   45
   32   38   27   33   40   35   25   42   38   41

5. Determine the numbers in the 6th decile group and in the 81st to 90th percentile groups for the following set of numbers.

   43 47 30 25 15 51 17 21 37 33 44 56 40 49 22
   36 44 33 17 35 58 51 35 44 40 31 41 55 50 16

For fully worked solutions to each of the problems in Practice Exercises 330 to 333 in this chapter, go to the website:
www.routledge.com/cw/bird

Section L

Chapter 87

Probability

Why it is important to understand: Probability

Engineers deal with uncertainty in their work, often with precision and analysis, and probability theory is widely used to model systems in engineering and scientific applications. There are a number of examples where probability is used in engineering. For example, with electronic circuits, scaling down the power and energy of such circuits reduces the reliability and predictability of many individual elements, but the circuits must nevertheless be engineered so that the overall circuit is reliable. Centres for disease control need to decide whether to institute massive vaccination or other preventative measures in the face of globally threatening, possibly mutating diseases in humans and animals. System designers must weigh the costs and benefits of measures for reliability and security, such as levels of backups and firewalls, in the face of uncertainty about threats from equipment failures or malicious attackers. Models incorporating probability theory have been developed and are continuously being improved for understanding the brain, gene pools within populations, weather and climate forecasts, microelectronic devices, and imaging systems such as computer aided tomography (CAT) scan and radar. The electric power grid, including power generating stations, transmission lines and consumers, is a complex system with many redundancies; however, breakdowns occur, and guidance for investment comes from modelling the most likely sequences of events that could cause outage. Similar planning and analysis is done for communication networks, transportation networks, water and other infrastructure. Probabilities, permutations and combinations are used daily in many different fields that range from gambling and games, to mechanical or structural failure rates, to rates of detection in medical screening. Uncertainty is clearly all around us, in our daily lives and in many professions. Use of standard deviation is widely used when results of opinion polls are described. The language of probability theory lets people break down complex problems, and to argue about pieces of them with each other, and then aggregate information about subsystems to analyse a whole system. This chapter briefly introduces the important subject of probability.

At the end of this chapter, you should be able to:

- define probability
- define expectation, dependent event, independent event and conditional probability
- state the addition and multiplication laws of probability
- use the laws of probability in simple calculations
- use the laws of probability in practical situations
- determine permutations and combinations
- Understand and use Bayes' theorem

87.1 Introduction to probability

Probability

The **probability** of something happening is the likelihood or chance of it happening. Values of probability lie between 0 and 1, where 0 represents an absolute impossibility and 1 represents an absolute certainty. The probability of an event happening usually lies somewhere between these two extreme values and is expressed as either a proper or decimal fraction. Examples of probability are

that a length of copper wire has zero resistance at 100°C	0
that a fair, six-sided dice will stop with a 3 upwards	$\frac{1}{6}$ or 0.1667
that a fair coin will land with a head upwards	$\frac{1}{2}$ or 0.5
that a length of copper wire has some resistance at 100°C	1

If p is the probability of an event happening and q is the probability of the same event not happening, then the total probability is $p+q$ and is equal to unity, since it is an absolute certainty that the event either will or will not occur; i.e. $p+q=1$

> **Problem 1.** Determine the probabilities of selecting at random (a) a man and (b) a woman from a crowd containing 20 men and 33 women

(a) The probability of selecting at random a man, p, is given by the ratio

$$\frac{\text{number of men}}{\text{number in crowd}}$$

i.e. $p = \dfrac{20}{20+33} = \dfrac{20}{53}$ or **0.3774**

(b) The probability of selecting at random a woman, q, is given by the ratio

$$\frac{\text{number of women}}{\text{number in crowd}}$$

i.e. $q = \dfrac{33}{20+33} = \dfrac{33}{53}$ or **0.6226**

(Check: the total probability should be equal to 1:

$p = \dfrac{20}{53}$ and $q = \dfrac{33}{53}$, thus the total probability,

$$p+q = \frac{20}{53} + \frac{33}{53} = 1$$

hence no obvious error has been made.)

Expectation

The **expectation**, E, of an event happening is defined in general terms as the product of the probability p of an event happening and the number of attempts made, n; i.e. $E = pn$

Thus, since the probability of obtaining a 3 upwards when rolling a fair dice is 1/6, the expectation of getting a 3 upwards on four throws of the dice is

$$\frac{1}{6} \times 4, \text{ i.e. } \frac{2}{3}$$

Thus **expectation is the average occurrence of an event**.

> **Problem 2.** Find the expectation of obtaining a 4 upwards with 3 throws of a fair dice

Expectation is the average occurrence of an event and is defined as the probability times the number of attempts. The probability, p, of obtaining a 4 upwards for one throw of the dice is 1/6

If 3 attempts are made, $n = 3$ and the expectation, E, is pn, i.e.

$$E = \frac{1}{6} \times 3 = \frac{1}{2} \text{ or } \mathbf{0.50}$$

Dependent events

A **dependent event** is one in which the probability of an event happening affects the probability of another ever happening. Let 5 transistors be taken at random from a batch of 100 transistors for test purposes and the probability of there being a defective transistor, p_1, be determined. At some later time, let another 5 transistors be taken at random from the 95 remaining transistors in the batch and the probability of there being a defective transistor, p_2, be determined. The value of p_2 is different from p_1 since the batch size has effectively altered from 100 to 95; i.e. probability p_2 is dependent on probability p_1. Since transistors are drawn and then another 5 transistors are drawn without replacing the first 5,

Section L

the second random selection is said to be **without replacement**.

Independent events

An independent event is one in which the probability of an event happening does not affect the probability of another event happening. If 5 transistors are taken at random from a batch of transistors and the probability of a defective transistor, p_1, is determined and the process is repeated after the original 5 have been replaced in the batch to give p_2, then p_1 is equal to p_2. Since the 5 transistors are replaced between draws, the second selection is said to be **with replacement**.

87.2 Laws of probability

The addition law of probability

The addition law of probability is recognised by the word '**or**' joining the probabilities.
If p_A is the probability of event A happening and p_B is the probability of event B happening, the probability of **event A or event B** happening is given by $p_A + p_B$
Similarly, the probability of events **A or B or C or** $\ldots N$ happening is given by

$$p_A + p_B + p_C + \cdots + p_N$$

The multiplication law of probability

The multiplication law of probability is recognised by the word '**and**' joining the probabilities.
If p_A is the probability of event A happening and p_B is the probability of event B happening, the probability of **event A and event B** happening is given by $p_A \times p_B$
Similarly, the probability of events **A and B and C and** $\ldots N$ happening is given by

$$p_A \times p_B \times p_C \times \ldots \times p_N$$

Here are some Problems to demonstrate probability.

> **Problem 3.** Calculate the probabilities of selecting at random
>
> (a) the winning horse in a race in which 10 horses are running
>
> (b) the winning horses in both the first and second races if there are 10 horses in each race

(a) Since only one of the ten horses can win, the probability of selecting at random the winning horse is

$$\frac{\text{number of winners}}{\text{number of horses}}, \text{ i.e. } \frac{1}{10} \quad \text{or} \quad \mathbf{0.10}$$

(b) The probability of selecting the winning horse in the first race is $\dfrac{1}{10}$
The probability of selecting the winning horse in the second race is $\dfrac{1}{10}$
The probability of selecting the winning horses in the first **and** second races is given by the multiplication law of probability; i.e.

$$\mathbf{probability} = \frac{1}{10} \times \frac{1}{10} = \frac{1}{100} \quad \text{or} \quad \mathbf{0.01}$$

> **Problem 4.** The probability of a component failing in one year due to excessive temperature is 1/20, that due to excessive vibration is 1/25 and that due to excessive humidity is 1/50. Determine the probabilities that during a one-year period a component
>
> (a) fails due to excessive temperature and excessive vibration
>
> (b) fails due to excessive vibration or excessive humidity
>
> (c) will not fail because of both excessive temperature and excessive humidity

Let p_A be the probability of failure due to excessive temperature, then

$$p_A = \frac{1}{20} \quad \text{and} \quad \overline{p_A} = \frac{19}{20}$$

(where $\overline{p_A}$ is the probability of not failing)

Let p_B be the probability of failure due to excessive vibration, then

$$p_B = \frac{1}{25} \quad \text{and} \quad \overline{p_B} = \frac{24}{25}$$

Let p_C be the probability of failure due to excessive humidity, then

$$p_C = \frac{1}{50} \quad \text{and} \quad \overline{p_C} = \frac{49}{50}$$

(a) The probability of a component failing due to excessive temperature **and** excessive vibration is given by

$$p_A \times p_B = \frac{1}{20} \times \frac{1}{25} = \frac{1}{500} \quad \text{or} \quad \mathbf{0.002}$$

(b)   The probability of a component failing due to excessive vibration **or** excessive humidity is

$$p_B + p_C = \frac{1}{25} + \frac{1}{50} = \frac{3}{50} \quad \text{or} \quad \mathbf{0.06}$$

(c)   The probability that a component will not fail due to excessive temperature **and** will not fail due to excess humidity is

$$\overline{p_A} \times \overline{p_C} = \frac{19}{20} \times \frac{49}{50} = \frac{931}{1000} \quad \text{or} \quad \mathbf{0.931}$$

Problem 5.   A batch of 100 capacitors contains 73 which are within the required tolerance values and 17 which are below the required tolerance values, the remainder being above the required tolerance values. Determine the probabilities that, when randomly selecting a capacitor and then a second capacitor,

(a)   both are within the required tolerance values when selecting with replacement

(b)   the first one drawn is below and the second one drawn is above the required tolerance value, when selection is without replacement

(a)   The probability of selecting a capacitor within the required tolerance values is 73/100. The first capacitor drawn is now replaced and a second one is drawn from the batch of 100. The probability of this capacitor being within the required tolerance values is also 73/100. Thus, the probability of selecting a capacitor within the required tolerance values for both the first **and** the second draw is

$$\frac{73}{100} \times \frac{73}{100} = \frac{5329}{10000} \quad \text{or} \quad \mathbf{0.5329}$$

(b)   The probability of obtaining a capacitor below the required tolerance values on the first draw is 17/100. There are now only 99 capacitors left in the batch, since the first capacitor is not replaced. The probability of drawing a capacitor above the required tolerance values on the second draw is 10/99, since there are $(100 - 73 - 17)$, i.e. 10, capacitors above the required tolerance value. Thus, the probability of randomly selecting a capacitor below the required tolerance values and subsequently randomly selecting a capacitor above the tolerance values is

$$\frac{17}{100} \times \frac{10}{99} = \frac{170}{9900} = \frac{17}{990} \quad \text{or} \quad \mathbf{0.0172}$$

Now try the following Practice Exercise

Practice Exercise 334   Laws of probability (answers on page 1190)

1.   In a batch of 45 lamps 10 are faulty. If one lamp is drawn at random, find the probability of it being (a) faulty, (b) satisfactory.

2.   A box of fuses that are all the same shape and size comprises 23 2 A fuses, 47 5 A fuses and 69 13 A fuses. Determine the probability of selecting at random (a) a 2 A fuse, (b) a 5 A fuse, (c) a 13 A fuse.

3.   (a)   Find the probability of having a 2 upwards when throwing a fair 6-sided dice.

   (b)   Find the probability of having a 5 upwards when throwing a fair 6-sided dice.

   (c)   Determine the probability of having a 2 and then a 5 on two successive throws of a fair 6-sided dice.

4.   Determine the probability that the total score is 8 when two like dice are thrown.

5.   The probability of event A happening is $\frac{3}{5}$ and the probability of event B happening is $\frac{2}{3}$. Calculate the probabilities of

   (a)   both A and B happening.

   (b)   only event A happening, i.e. event A happening and event B not happening.

   (c)   only event B happening.

   (d)   either A, or B, or A and B happening.

6.   When testing 1000 soldered joints, 4 failed during a vibration test and 5 failed due to having a high resistance. Determine the probability of a joint failing due to

   (a)   vibration

   (b)   high resistance

   (c)   vibration or high resistance

   (d)   vibration and high resistance

Section L

Here are some Problems on probability.

Problem 6. A batch of 40 components contains 5 which are defective. A component is drawn at random from the batch and tested and then a second component is drawn. Determine the probability that neither of the components is defective when drawn (a) with replacement and (b) without replacement

---

(a) **With replacement**

The probability that the component selected on the first draw is satisfactory is 35/40 i.e. 7/8. The component is now replaced and a second draw is made. The probability that this component is also satisfactory is 7/8. Hence, the probability that both the first component drawn and the second component drawn are satisfactory is

$$\frac{7}{8} \times \frac{7}{8} = \frac{49}{64} \quad \text{or} \quad \mathbf{0.7656}$$

(b) **Without replacement**

The probability that the first component drawn is satisfactory is 7/8. There are now only 34 satisfactory components left in the batch and the batch number is 39. Hence, the probability of drawing a satisfactory component on the second draw is 34/39. Thus, the probability that the first component drawn and the second component drawn are satisfactory, i.e. neither is defective, is

$$\frac{7}{8} \times \frac{34}{39} = \frac{238}{312} \quad \text{or} \quad \mathbf{0.7628}$$

Problem 7. A batch of 40 components contains 5 which are defective. If a component is drawn at random from the batch and tested and then a second component is drawn at random, calculate the probability of having one defective component, both (a) with replacement and (b) without replacement

---

The probability of having one defective component can be achieved in two ways. If p is the probability of drawing a defective component and q is the probability of drawing a satisfactory component, then the probability of having one defective component is given by drawing a satisfactory component and then a defective component **or** by drawing a defective component and then a satisfactory one; i.e. by $q \times p + p \times q$

(a) **With replacement**

$$p = \frac{5}{40} = \frac{1}{8} \quad \text{and} \quad q = \frac{35}{40} = \frac{7}{8}$$

Hence, the probability of having one defective component is

$$\frac{1}{8} \times \frac{7}{8} + \frac{7}{8} \times \frac{1}{8}$$

i.e. $\quad \dfrac{7}{64} + \dfrac{7}{64} = \dfrac{7}{32} \quad$ or $\quad \mathbf{0.2188}$

(b) **Without replacement**

$p_1 = \dfrac{1}{8}$ and $q_1 = \dfrac{7}{8}$ on the first of the two draws

The batch number is now 39 for the second draw, thus,

$$p_2 = \frac{5}{39} \quad \text{and} \quad q_2 = \frac{35}{39}$$

$$p_1 q_2 + q_1 p_2 = \frac{1}{8} \times \frac{35}{39} + \frac{7}{8} \times \frac{5}{39}$$

$$= \frac{35 + 35}{312}$$

$$= \frac{70}{312} \quad \text{or} \quad \mathbf{0.2244}$$

Problem 8. A box contains 74 brass washers, 86 steel washers and 40 aluminium washers. Three washers are drawn at random from the box without replacement. Determine the probability that all three are steel washers

---

Assume, for clarity of explanation, that a washer is drawn at random, then a second, then a third (although this assumption does not affect the results obtained). The total number of washers is

$$74 + 86 + 40, \quad \text{i.e.} \quad 200$$

The probability of randomly selecting a steel washer on the first draw is 86/200. There are now 85 steel washers in a batch of 199. The probability of randomly selecting a steel washer on the second draw is 85/199. There are now 84 steel washers in a batch of 198. The probability of randomly selecting a steel washer on the third draw is 84/198. Hence, the probability of selecting a steel washer on the first draw **and** the second draw **and** the third draw is

$$\frac{86}{200} \times \frac{85}{199} \times \frac{84}{198} = \frac{614040}{7880400} = \mathbf{0.0779}$$

Problem 9. For the box of washers given in Problem 8 above, determine the probability that there are no aluminium washers drawn when three washers are drawn at random from the box without replacement

The probability of not drawing an aluminium washer on the first draw is $1 - \left(\dfrac{40}{200}\right)$ i.e. 160/200. There are now 199 washers in the batch of which 159 are not made of aluminium. Hence, the probability of not drawing an aluminium washer on the second draw is 159/199. Similarly, the probability of not drawing an aluminium washer on the third draw is 158/198. Hence the probability of not drawing an aluminium washer on the first **and** second **and** third draws is

$$\frac{160}{200} \times \frac{159}{199} \times \frac{158}{198} = \frac{4019520}{7880400} = \mathbf{0.5101}$$

Problem 10. For the box of washers in Problem 8 above, find the probability that there are two brass washers and either a steel or an aluminium washer when three are drawn at random, without replacement

Two brass washers (A) and one steel washer (B) can be obtained in any of the following ways.

1st draw	2nd draw	3rd draw
A	A	B
A	B	A
B	A	A

Two brass washers and one aluminium washer (C) can also be obtained in any of the following ways.

1st draw	2nd draw	3rd draw
A	A	C
A	C	A
C	A	A

Thus, there are six possible ways of achieving the combinations specified. If A represents a brass washer, B a steel washer and C an aluminium washer, the combinations and their probabilities are as shown.

Draw			Probability
First	Second	Third	
A	A	B	$\dfrac{74}{200} \times \dfrac{73}{199} \times \dfrac{86}{198} = 0.0590$
A	B	A	$\dfrac{74}{200} \times \dfrac{86}{199} \times \dfrac{73}{198} = 0.0590$
B	A	A	$\dfrac{86}{200} \times \dfrac{74}{199} \times \dfrac{73}{198} = 0.0590$
A	A	C	$\dfrac{74}{200} \times \dfrac{73}{199} \times \dfrac{40}{198} = 0.0274$
A	C	A	$\dfrac{74}{200} \times \dfrac{40}{199} \times \dfrac{73}{198} = 0.0274$
C	A	A	$\dfrac{40}{200} \times \dfrac{74}{199} \times \dfrac{73}{198} = 0.0274$

The probability of having the first combination **or** the second **or** the third, and so on, is given by the sum of the probabilities; i.e. by $3 \times 0.0590 + 3 \times 0.0274$, i.e. **0.2592**

Now try the following Practice Exercise

Practice Exercise 335   Laws of probability (answers on page 1190)

1. The probability that component A will operate satisfactorily for 5 years is 0.8 and that B will operate satisfactorily over that same period of time is 0.75. Find the probabilities that in a 5-year period

   (a) both components will operate satisfactorily.

   (b) only component A will operate satisfactorily.

   (c) only component B will operate satisfactorily.

2. In a particular street, 80% of the houses have landline telephones. If two houses selected at random are visited, calculate the probabilities that

   (a) they both have a telephone.

   (b) one has a telephone but the other does not.

Section L

3.   Veroboard pins are packed in packets of 20 by a machine. In a thousand packets, 40 have less than 20 pins. Find the probability that if 2 packets are chosen at random, one will contain less than 20 pins and the other will contain 20 pins or more.

4.   A batch of 1 kW fire elements contains 16 which are within a power tolerance and 4 which are not. If 3 elements are selected at random from the batch, calculate the probabilities that

   (a)   all three are within the power tolerance.

   (b)   two are within but one is not within the power tolerance.

5.   An amplifier is made up of three transistors, A, B and C. The probabilities of A, B or C being defective are 1/20, 1/25 and 1/50, respectively. Calculate the percentage of amplifiers produced

   (a)   which work satisfactorily.

   (b)   which have just one defective transistor.

6.   A box contains 14 40 W lamps, 28 60 W lamps and 58 25 W lamps, all the lamps being of the same shape and size. Three lamps are drawn at random from the box, first one, then a second, then a third. Determine the probabilities of

   (a)   getting one 25 W, one 40 W and one 60 W lamp with replacement.

   (b)   getting one 25 W, one 40 W and one 60 W lamp without replacement.

   (c)   getting either one 25 W and two 40 W or one 60 W and two 40 W lamps with replacement.

87.3   Permutations and combinations

Permutations

If n different objects are available, they can be arranged in different orders of selection. Each different ordered arrangement is called a **permutation**. For example, permutations of the three letters X, Y and Z taken together are:

$$XYZ, XZY, YXZ, YZX, ZXY \text{ and } ZYX$$

This can be expressed as $^3P_3 = 6$, the upper 3 denoting the number of items from which the arrangements are made, and the lower 3 indicating the number of items used in each arrangement.

If we take the same three letters XYZ two at a time, the permutations

$$XY, YX, XZ, ZX, YZ, ZY$$

can be found, and denoted by $^3P_2 = 6$

(Note that the order of the letters matter in permutations, i.e. YX is a different permutation from XY.) In general,
$^nP_r = n(n-1)(n-2)\ldots(n-r+1)$ or

$$^nP_r = \frac{n!}{(n-r)!} \text{ as stated in Chapter 20}$$

For example, $^5P_4 = 5(4)(3)(2) = 120$ or
$$^5P_4 = \frac{5!}{(5-4)!} = \frac{5!}{1!} = (5)(4)(3)(2) = 120$$

Also, $^3P_3 = 6$ from above; using $^nP_r = \dfrac{n!}{(n-r)!}$ gives

$$^3P_3 = \frac{3!}{(3-3)!} = \frac{6}{0!}. \text{ Since this must equal 6, then } 0! = 1$$

(check this with your calculator).

Combinations

If selections of the three letters X, Y, Z are made without regard to the order of the letters in each group, i.e. XY is now the same as YX, for example, then each group is called a **combination**. The number of possible combinations is denoted by ^nC_r, where n is the total number of items and r is the number in each selection. In general,

$$^nC_r = \frac{n!}{r!(n-r)!}$$

For example,

$$^5C_4 = \frac{5!}{4!(5-4)!} = \frac{5!}{4!}$$

$$= \frac{5 \times 4 \times 3 \times 2 \times 1}{4 \times 3 \times 2 \times 1} = 5$$

> **Problem 11.**   Calculate the number of permutations there are of: (a) 5 distinct objects taken 2 at a time, (b) 4 distinct objects taken 2 at a time

(a) $^5P_2 = \dfrac{5!}{(5-2)!} = \dfrac{5!}{3!} = \dfrac{5 \times 4 \times 3 \times 2}{3 \times 2} = \mathbf{20}$

(b) $^4P_2 = \dfrac{4!}{(4-2)!} = \dfrac{4!}{2!} = \mathbf{12}$

Problem 12. Calculate the number of combinations there are of: (a) 5 distinct objects taken 2 at a time, (b) 4 distinct objects taken 2 at a time

(a) $^5C_2 = \dfrac{5!}{2!\,(5-2)!} = \dfrac{5!}{2!\,3!}$

$= \dfrac{5 \times 4 \times 3 \times 2 \times 1}{(2 \times 1)(3 \times 2 \times 1)} = \mathbf{10}$

(b) $^4C_2 = \dfrac{4!}{2!\,(4-2)!} = \dfrac{4!}{2!\,2!} = \mathbf{6}$

Problem 13. A class has 24 students. Four can represent the class at an exam board. How many combinations are possible when choosing this group?

Number of combinations possible,

$$^nC_r = \dfrac{n!}{r!\,(n-r!)}$$

i.e. $^{24}C_4 = \dfrac{24!}{4!\,(24-4)!} = \dfrac{24!}{4!\,20!} = \mathbf{10\,626}$

Problem 14. In how many ways can a team of eleven be picked from sixteen possible players?

Number of ways $= {}^nC_r = {}^{16}C_{11}$

$$= \dfrac{16!}{11!\,(16-11)!} = \dfrac{16!}{11!\,5!} = \mathbf{4368}$

Now try the following Practice Exercise

Practice Exercise 336   Further problems on permutations and combinations (answers on page 1190)

1. Calculate the number of permutations there are of: (a) 15 distinct objects taken 2 at a time, (b) 9 distinct objects taken 4 at a time

2. Calculate the number of combinations there are of: (a) 12 distinct objects taken 5 at a time, (b) 6 distinct objects taken 4 at a time

3. In how many ways can a team of six be picked from ten possible players?

4. 15 boxes can each hold one object. In how many ways can 10 identical objects be placed in the boxes?

5. Six numbers between 1 and 49 are chosen when playing the National Lottery. Determine the probability of winning the top prize (i.e. 6 correct numbers!) if 10 tickets were purchased and six different numbers were chosen on each ticket.

87.4   Bayes' theorem

Bayes' theorem is one of probability theory (originally stated by the Reverend Thomas Bayes), and may be seen as a way of understanding how the probability that a theory is true is affected by a new piece of evidence. The theorem has been used in a wide variety of contexts, ranging from marine biology to the development of 'Bayesian' spam blockers for email systems; in science, it has been used to try to clarify the relationship between theory and evidence. Insights in the philosophy of science involving confirmation, falsification and other topics can be made more precise, and sometimes extended or corrected, by using Bayes' theorem.

Bayes' theorem may be stated mathematically as:

$$P(A_1|B)$$

$$= \dfrac{P(B|A_1)\,P(A_1)}{P(B|A_1)\,P(A_1) + P(B|A_2)\,P(A_2) + \dots}$$

or $P(A_i|B)$

$$= \dfrac{P(B|A_i)\,P(A_i)}{\displaystyle\sum_{j=1}^{n} P(B|A_j)\,P(A_j)} \quad (i = 1, 2, \dots, n)$$

where $P(A|B)$ is the probability of A *given* B, i.e. *after* B is observed.

$P(A)$ and $P(B)$ are the probabilities of A and B without regard to each other, and $P(B|A)$ is the probability of observing event B given that A is true.

In the Bayes theorem formula, 'A' represents a theory or hypothesis that is to be tested, and 'B' represents a new piece of evidence that seems to confirm or disprove the theory. Bayes' theorem is demonstrated in the following worked problem.

Problem 15. An outdoor degree ceremony is taking place tomorrow, 5 July, in the hot climate of Dubai. In recent years it has rained only 2 days in the four-month period June to September. However, the weather forecaster has predicted rain for tomorrow. When it actually rains, the weatherman correctly forecasts rain 85% of the time. When it doesn't rain, he incorrectly forecasts rain 15% of the time. Determine the probability that it will rain tomorrow.

There are two possible mutually exclusive events occurring here – it either rains or it does not rain. Also, a third event occurs when the weatherman predicts rain.

Let the notation for these events be:

Event A_1 It rains at the ceremony

Event A_2 It does not rain at the ceremony

Event B The weatherman predicts rain

The probability values are:

$$P(A_1) = \frac{2}{30 + 31 + 31 + 30} = \frac{1}{61}$$

(i.e. it rains 2 days in the months June to September)

$$P(A_2) = \frac{120}{30 + 31 + 31 + 30} = \frac{60}{61}$$

(i.e. it does not rain for 120 of the 122 days

in the months June to September)

$$P(B|A_1) = 0.85$$

(i.e. when it rains, the weatherman predicts rain

85% of the time)

$$P(B|A_2) = 0.15$$

(i.e. when it does not rain, the weatherman

predicts rain 15% of the time)

Using Bayes' theorem to determine the probability that it will rain tomorrow, given the forecast of rain by the weatherman:

$$P(A_1|B) = \frac{P(B|A_1)P(A_1)}{P(B|A_1)P(A_1) + P(B|A_2)P(A_2)}$$

$$= \frac{(0.85)\left(\frac{1}{61}\right)}{0.85 \times \frac{1}{61} + 0.15 \times \frac{60}{61}} = \frac{0.0139344}{0.1614754}$$

$$= \mathbf{0.0863} \text{ or } \mathbf{8.63\%}$$

Even when the weatherman predicts rain, it rains only between 8% and 9% of the time. **Hence, there is a good chance it will not rain tomorrow in Dubai for the degree ceremony.**

Now try the following Practice Exercise

Practice Exercise 337   Bayes' theorem (answers on page 1190)

1.   Machines A, B and C produce similar vehicle engine parts. Of the total output, machine A produces 35%, machine B 20% and machine C 45%. The proportions of the output from each machine that do not conform to the specification are 7% for A, 4% for B and 3% for C. Determine the proportion of those parts that do not conform to the specification that are produced by machine A.

2.   A doctor is called to see a sick child. The doctor has prior information that 85% of sick children in that area have the flu, while the other 15% are sick with measles. For simplicity, assume that there are no other maladies in that area. A symptom of measles is a rash. The probability of children developing a rash and having measles is 94% and the probability of children with flu occasionally also developing a rash is 7%. Upon examining the child, the doctor finds a rash. Determine the probability that the child has measles.

3.   In a study, oncologists were asked what the odds of breast cancer would be in a woman

who was initially thought to have a 1% risk of cancer but who ended up with a positive mammogram result (a mammogram accurately classifies about 80% of cancerous tumours and 90% of benign tumours). 95 out of a hundred oncologists estimated the probability of cancer to be about 75%. Use Bayes' theorem to determine the probability of cancer.

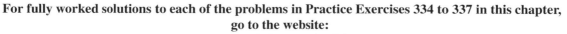

For fully worked solutions to each of the problems in Practice Exercises 334 to 337 in this chapter, go to the website:
www.routledge.com/cw/bird

Section L

This assignment covers the material contained in Chapters 85 to 87. *The marks available are shown in brackets at the end of each question.*

1. A company produces five products in the following proportions:

Product A	24
Product B	6
Product C	15
Product D	9
Product E	18

Draw (a) a horizontal bar chart and (b) a pie diagram to represent these data visually. (9)

2. State whether the data obtained on the following topics are likely to be discrete or continuous.
   (a) the number of books in a library
   (b) the speed of a car
   (c) the time to failure of a light bulb. (3)

3. Draw a histogram, frequency polygon and ogive for the data given below, which refer to the diameter of 50 components produced by a machine.

Class intervals	Frequency
1.30–1.32 mm	4
1.33–1.35 mm	7
1.36–1.38 mm	10
1.39–1.41 mm	12
1.42–1.44 mm	8
1.45–1.47 mm	5
1.48–1.50 mm	4

(16)

4. Determine for the 10 measurements of lengths shown below:
   (a) the arithmetic mean, (b) the median, (c) the mode and (d) the standard deviation:
   28 m, 20 m, 32 m, 44 m, 28 m, 30 m, 30 m, 26 m, 28 m and 34 m (9)

5. The length in millimetres of 100 bolts is as shown below.

50–56	6
57–63	16
64–70	22
71–77	30
78–84	19
85–91	7

Determine for the sample
(a) the mean value
(b) the standard deviation, correct to 4 significant figures. (10)

6. The number of faulty components in a factory in a 12-week period is

   14 12 16 15 10 13 15 11 16 19 17 19

   Determine the median and the first and third quartile values. (7)

7. Determine the probability of winning a prize in a lottery by buying 10 tickets when there are 10 prizes and a total of 5000 tickets sold. (4)

8. The probabilities of an engine failing are given by: p_1, failure due to overheating; p_2, failure due to ignition problems; p_3, failure due to fuel blockage. When $p_1 = \frac{1}{8}$, $p_2 = \frac{1}{5}$ and $p_3 = \frac{2}{7}$, determine the probabilities of:
   (a) all three failures occurring,
   (b) the first and second but not the third failure occurring,
   (c) only the second failure occurring,
   (d) the first or the second failure occurring but not the third. (12)

9. A sample of 50 resistors contains 44 which are within the required tolerance value, 4 which are below and the remainder which are above. Determine the probability of selecting from the sample a resistor which is
   (a) below the required tolerance
   (b) above the required tolerance.

Now two resistors are selected at random from the sample. Determine the probability, correct to 3 decimal places, that neither resistor is defective when drawn

(c) with replacement

(d) without replacement.

(e) If a resistor is drawn at random from the batch and tested and then a second resistor is drawn from those left, calculate the probability of having one defective component when selection is without replacement.   (15)

For lecturers/instructors/teachers, fully worked solutions to each of the problems in Revision Test 30, together with a full marking scheme, are available at the website:
www.routledge.com/cw/bird

Chapter 88

The binomial and Poisson distributions

Why it is important to understand: The binomial and Poisson distributions

The binomial distribution is used only when both of two conditions are met – the test has only two possible outcomes, and the sample must be random. If both of these conditions are met, then this distribution may be used to predict the probability of a desired result. For example, a binomial distribution may be used in determining whether a new drug being tested has or has not contributed to alleviating symptoms of a disease. Common applications of this distribution range from scientific and engineering applications to military and medical ones, in quality assurance, genetics and in experimental design.

A Poisson distribution has several applications, and is essentially a derived limiting case of the binomial distribution. It is most applicably relevant to a situation in which the total number of successes is known, but the number of trials is not. An example of such a situation would be if the mean expected number of cancer cells present per sample is known and it was required to determine the probability of finding 1.5 times that amount of cells in any given sample; this is an example when the Poisson distribution would be used. The Poisson distribution has widespread applications in analysing traffic flow, in fault prediction on electric cables, in the prediction of randomly occurring accidents and in reliability engineering.

At the end of this chapter, you should be able to:

- define the binomial distribution
- use the binomial distribution
- apply the binomial distribution to industrial inspection
- draw a histogram of probabilities
- define the Poisson distribution
- apply the Poisson distribution to practical situations

88.1 The binomial distribution

The binomial distribution deals with two numbers only, these being the probability that an event will happen, p, and the probability that an event will not happen, q. Thus, when a coin is tossed, if p is the probability of the coin landing with a head upwards, q is the probability of the coin landing with a tail upwards. $p + q$ must always be equal to unity. A binomial distribution can be used for finding, say, the probability of getting three heads in seven tosses of the coin, or in industry for determining defect rates as a result of sampling. One way of defining a binomial distribution is as follows:

'If p is the probability that an event will happen and q is the probability that the event will not happen, then the probabilities that the event will happen 0, 1, 2, 3, …, n times in n trials are given by the successive terms of the expansion of (q + p)ⁿ, taken from left to right.'

The binomial expansion of $(q + p)^n$ is:

$$q^n + nq^{n-1}p + \frac{n(n-1)}{2!}q^{n-2}p^2$$
$$+ \frac{n(n-1)(n-2)}{3!}q^{n-3}p^3 + \cdots$$

from Chapter 21.

This concept of a binomial distribution is used in Problems 1 and 2.

> **Problem 1.** Determine the probabilities of having (a) at least 1 girl and (b) at least 1 girl and 1 boy in a family of 4 children, assuming equal probability of male and female birth

The probability of a girl being born, p, is 0.5 and the probability of a girl not being born (male birth), q, is also 0.5. The number in the family, n, is 4. From above, the probabilities of 0, 1, 2, 3, 4 girls in a family of 4 are given by the successive terms of the expansion of $(q + p)^4$ taken from left to right. From the binomial expansion:

$$(q + p)^4 = q^4 + 4q^3p + 6q^2p^2 + 4qp^3 + p^4$$

Hence the probability of no girls is q^4,

i.e. $\qquad\qquad 0.5^4 = 0.0625$

the probability of 1 girl is $4q^3p$,

i.e. $\qquad\qquad 4 \times 0.5^3 \times 0.5 = 0.2500$

the probability of 2 girls is $6q^2p^2$,

i.e. $\qquad\qquad 6 \times 0.5^2 \times 0.5^2 = 0.3750$

the probability of 3 girls is $4qp^3$,

i.e. $\qquad\qquad 4 \times 0.5 \times 0.5^3 = 0.2500$

the probability of 4 girls is p^4,

i.e. $\qquad\qquad 0.5^4 = 0.0625$

$$\text{Total probability, } (q + p)^4 = 1.0000$$

(a) The probability of having at least one girl is the sum of the probabilities of having 1, 2, 3 and 4 girls, i.e.

$$0.2500 + 0.3750 + 0.2500 + 0.0625 = \mathbf{0.9375}$$

(Alternatively, the probability of having at least 1 girl is: 1 − (the probability of having no girls), i.e. 1 − 0.0625, giving **0.9375**, as obtained previously.)

(b) The probability of having at least 1 girl and 1 boy is given by the sum of the probabilities of having: 1 girl and 3 boys, 2 girls and 2 boys and 3 girls and 2 boys, i.e.

$$0.2500 + 0.3750 + 0.2500 = \mathbf{0.8750}$$

(Alternatively, this is also the probability of having 1 − (probability of having no girls + probability of having no boys), i.e. 1 − 2 × 0.0625 = **0.8750**, as obtained previously.)

> **Problem 2.** A dice is rolled 9 times. Find the probabilities of having a 4 upwards (a) 3 times and (b) less than 4 times

Let p be the probability of having a 4 upwards. Then $p = 1/6$, since dice have six sides.

Let q be the probability of not having a 4 upwards. Then $q = 5/6$. The probabilities of having a 4 upwards 0, 1, 2, …, n times are given by the successive terms of the expansion of $(q + p)^n$, taken from left to right. From the binomial expansion:

$$(q + p)^9 = q^9 + 9q^8p + 36q^7p^2 + 84q^6p^3 + \cdots$$

The probability of having a 4 upwards no times is

$$q^9 = (5/6)^9 = 0.1938$$

The probability of having a 4 upwards once is

$$9q^8p = 9(5/6)^8(1/6) = 0.3489$$

The probability of having a 4 upwards twice is

$$36q^7p^2 = 36(5/6)^7(1/6)^2 = 0.2791$$

The probability of having a 4 upwards 3 times is

$$84q^6p^3 = 84(5/6)^6(1/6)^3 = 0.1302$$

(a) The probability of having a 4 upwards 3 times is **0.1302**

(b) The probability of having a 4 upwards less than 4 times is the sum of the probabilities of having a 4 upwards 0, 1, 2, and 3 times, i.e.

$$0.1938 + 0.3489 + 0.2791 + 0.1302 = \mathbf{0.9520}$$

Industrial inspection

In industrial inspection, p is often taken as the probability that a component is defective and q is the probability that the component is satisfactory. In this case, a binomial distribution may be defined as:

> 'The probabilities that 0, 1, 2, 3,..., n *components are defective in a sample of* n *components, drawn at random from a large batch of components, are given by the successive terms of the expansion of* (q+p)^n, *taken from left to right.*'

This definition is used in Problems 3 and 4.

> **Problem 3.** A machine is producing a large number of bolts automatically. In a box of these bolts, 95% are within the allowable tolerance values with respect to diameter, the remainder being outside of the diameter tolerance values. Seven bolts are drawn at random from the box. Determine the probabilities that (a) two and (b) more than two of the seven bolts are outside of the diameter tolerance values

Let p be the probability that a bolt is outside of the allowable tolerance values, i.e. is defective, and let q be the probability that a bolt is within the tolerance values, i.e. is satisfactory. Then $p = 5\%$, i.e. 0.05 per unit and $q = 95\%$, i.e. 0.95 per unit. The sample number is 7 The probabilities of drawing $0, 1, 2, \ldots, n$ defective bolts are given by the successive terms of the expansion of $(q + p)^n$, taken from left to right. In this problem

$$(q + p)^n = (0.95 + 0.05)^7$$

$$= 0.95^7 + 7 \times 0.95^6 \times 0.05$$

$$+ 21 \times 0.95^5 \times 0.05^2 + \cdots$$

Thus the probability of no defective bolts is

$$0.95^7 = 0.6983$$

The probability of 1 defective bolt is

$$7 \times 0.95^6 \times 0.05 = 0.2573$$

The probability of 2 defective bolts is

$$21 \times 0.95^5 \times 0.05^2 = 0.0406, \text{ and so on.}$$

(a) The probability that two bolts are outside of the diameter tolerance values is **0.0406**

(b) To determine the probability that more than two bolts are defective, the sum of the probabilities of 3 bolts, 4 bolts, 5 bolts, 6 bolts and 7 bolts being defective can be determined. An easier way to find this sum is to find $1 - (\text{sum of } 0 \text{ bolts, } 1 \text{ bolt}$ and 2 bolts being defective), since the sum of all the terms is unity. Thus, the probability of there being more than two bolts outside of the tolerance values is:

$$1 - (0.6983 + 0.2573 + 0.0406), \text{ i.e. } \mathbf{0.0038}$$

> **Problem 4.** A package contains 50 similar components and inspection shows that four have been damaged during transit. If six components are drawn at random from the contents of the package, determine the probabilities that in this sample (a) one and (b) fewer than three are damaged

The probability of a component being damaged, p, is 4 in 50, i.e. 0.08 per unit. Thus, the probability of a component not being damaged, q, is $1 - 0.08$, i.e. 0.92. The probability of there being $0, 1, 2, \ldots, 6$ damaged components is given by the successive terms of $(q + p)^6$, taken from left to right.

$$(q + p)^6 = q^6 + 6q^5 p + 15q^4 p^2 + 20q^3 p^3 + \cdots$$

(a) The probability of one damaged component is

$$6q^5 p = 6 \times 0.92^5 \times 0.08 = \mathbf{0.3164}$$

(b) The probability of fewer than three damaged components is given by the sum of the probabilities of 0, 1 and 2 damaged components.

$$q^6 + 6q^5 p + 15q^4 p^2$$

$$= 0.92^6 + 6 \times 0.92^5 \times 0.08$$

$$+ 15 \times 0.92^4 \times 0.08^2$$

$$= 0.6064 + 0.3164 + 0.0688 = \mathbf{0.9916}$$

Histogram of probabilities

The terms of a binomial distribution may be represented pictorially by drawing a histogram, as shown in Problem 5.

> **Problem 5.** The probability of a student successfully completing a course of study in three years is 0.45. Draw a histogram showing the probabilities of $0, 1, 2, \ldots, 10$ students successfully completing the course in three years

Let p be the probability of a student successfully completing a course of study in three years and q be the probability of not doing so. Then $p = 0.45$ and $q = 0.55$. The number of students, n, is 10

The probabilities of $0, 1, 2, \ldots, 10$ students successfully completing the course are given by the successive terms of the expansion of $(q + p)^{10}$, taken from left to right.

$$(q + p)^{10} = q^{10} + 10q^9 p + 45q^8 p^2 + 120q^7 p^3$$
$$+ 210q^6 p^4 + 252q^5 p^5 + 210q^4 p^6$$
$$+ 120q^3 p^7 + 45q^2 p^8 + 10qp^9 + p^{10}$$

Substituting $q = 0.55$ and $p = 0.45$ in this expansion gives the values of the successive terms as: 0.0025, 0.0207, 0.0763, 0.1665, 0.2384, 0.2340, 0.1596, 0.0746, 0.0229, 0.0042 and 0.0003. The histogram depicting these probabilities is shown in Figure 88.1.

Now try the following Practice Exercise

Practice Exercise 338   Further problems on the binomial distribution (answers on page 1190)

1. Concrete blocks are tested and it is found that, on average, 7% fail to meet the required specification. For a batch of nine blocks, determine the probabilities that (a) three blocks and (b) less than four blocks will fail to meet the specification.

2. If the failure rate of the blocks in Problem 1 rises to 15%, find the probabilities that (a) no blocks and (b) more than two blocks will fail to meet the specification in a batch of nine blocks.

3. The average number of employees absent from a firm each day is 4%. An office within the firm has seven employees. Determine the probabilities that (a) no employee and (b) three employees will be absent on a particular day.

4. A manufacturer estimates that 3% of his output of a small item is defective. Find the probabilities that in a sample of ten items (a) less than two and (b) more than two items will be defective.

5. Five coins are tossed simultaneously. Determine the probabilities of having 0, 1, 2, 3, 4 and 5 heads upwards, and draw a histogram depicting the results.

6. If the probability of rain falling during a particular period is 2/5, find the probabilities of having 0, 1, 2, 3, 4, 5, 6 and 7 wet days in a week. Show these results on a histogram.

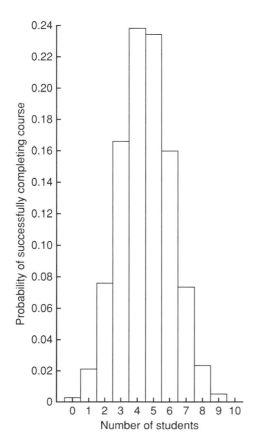

Figure 88.1

7. An automatic machine produces, on average, 10% of its components outside of the tolerance required. In a sample of ten components from this machine, determine the probability of having three components outside of the tolerance required by assuming a binomial distribution.

88.2 The Poisson distribution

When the number of trials, n, in a binomial distribution becomes large (usually taken as larger than 10), the calculations associated with determining the values of the terms becomes laborious. If n is large and p is small, and the product np is less than 5, a very good approximation to a binomial distribution is given by the corresponding **Poisson*** **distribution**, in which calculations are usually simpler.

The Poisson approximation to a binomial distribution may be defined as follows:

* **Who was Poisson? Siméon Denis Poisson** (21 June 1781–25 April 1840), was a French mathematician, geometer, and physicist. His work on the theory of electricity and magnetism virtually created a new branch of mathematical physics, and his study of celestial mechanics discussed the stability of the planetary orbits. To find out more go to **www.routledge.com/cw/bird**

'The probabilities that an event will happen 0, 1, 2, 3, ..., n *times in* n *trials are given by the successive terms of the expression*

$$e^{-\lambda}\left(1 + \lambda + \frac{\lambda^2}{2!} + \frac{\lambda^3}{3!} + \cdots\right)$$

taken from left to right.'

The symbol λ is the expectation of an event happening and is equal to np.

Problem 6. If 3% of the gearwheels produced by a company are defective, determine the probabilities that in a sample of 80 gearwheels (a) two and (b) more than two will be defective

The sample number, n, is large, the probability of a defective gearwheel, p, is small and the product np is 80×0.03, i.e. 2.4, which is less than 5

Hence a Poisson approximation to a binomial distribution may be used. The expectation of a defective gearwheel, $\lambda = np = 2.4$

The probabilities of $0, 1, 2, \ldots$ defective gearwheels are given by the successive terms of the expression

$$e^{-\lambda}\left(1 + \lambda + \frac{\lambda^2}{2!} + \frac{\lambda^3}{3!} + \cdots\right)$$

taken from left to right, i.e. by

$$e^{-\lambda}, \lambda e^{-\lambda}, \frac{\lambda^2 e^{-\lambda}}{2!}, \ldots$$

Thus, probability of no defective gearwheels is

$$e^{-\lambda} = e^{-2.4} = 0.0907$$

probability of 1 defective gearwheel is

$$\lambda e^{-\lambda} = 2.4e^{-2.4} = 0.2177$$

probability of 2 defective gearwheels is

$$\frac{\lambda^2 e^{-\lambda}}{2!} = \frac{2.4^2 e^{-2.4}}{2 \times 1} = 0.2613$$

(a) The probability of having 2 defective gearwheels is **0.2613**

(b) The probability of having more than 2 defective gearwheels is $1 -$ (the sum of the probabilities of having 0, 1, and 2 defective gearwheels), i.e.

$$1 - (0.0907 + 0.2177 + 0.2613)$$

that is, **0.4303**

The principal use of a Poisson distribution is to determine the theoretical probabilities when p, the probability of an event happening, is known, but q, the probability of the event not happening, is unknown. For example, the average number of goals scored per match by a football team can be calculated, but it is not possible to quantify the number of goals which were not scored. In this type of problem, a Poisson distribution may be defined as follows:

'The probabilities of an event occurring 0, 1, 2, 3, ... times are given by the successive terms of the expression

$$e^{-\lambda}\left(1 + \lambda + \frac{\lambda^2}{2!} + \frac{\lambda^3}{3!} + \cdots\right)$$

taken from left to right.'

The symbol λ is the value of the average occurrence of the event.

Problem 7. A production department has 35 similar milling machines. The number of breakdowns on each machine averages 0.06 per week. Determine the probabilities of having (a) one and (b) less than three machines breaking down in any week

Since the average occurrence of a breakdown is known but the number of times when a machine did not break down is unknown, a Poisson distribution must be used. The expectation of a breakdown for 35 machines is 35×0.06, i.e. 2.1 breakdowns per week. The probabilities of a breakdown occurring $0, 1, 2, \ldots$ times are given by the successive terms of the expression

$$e^{-\lambda}\left(1 + \lambda + \frac{\lambda^2}{2!} + \frac{\lambda^3}{3!} + \cdots\right)$$

taken from left to right.

Hence probability of no breakdowns

$$e^{-\lambda} = e^{-2.1} = 0.1225$$

probability of 1 breakdown is

$$\lambda e^{-\lambda} = 2.1 e^{-2.1} = 0.2572$$

probability of 2 breakdowns is

$$\frac{\lambda^2 e^{-\lambda}}{2!} = \frac{2.1^2 e^{-2.1}}{2 \times 1} = 0.2700$$

(a) The probability of 1 breakdown per week is **0.2572**

(b) The probability of less than 3 breakdowns per week is the sum of the probabilities of 0, 1 and 2 breakdowns per week,

i.e. $0.1225 + 0.2572 + 0.2700$, i.e. **0.6497**

Histogram of probabilities

The terms of a Poisson distribution may be represented pictorially by drawing a histogram, as shown in Problem 8.

Problem 8. The probability of a person having an accident in a certain period of time is 0.0003. For a population of 7500 people, draw a histogram showing the probabilities of 0, 1, 2, 3, 4, 5 and 6 people having an accident in this period.

The probabilities of $0, 1, 2, \ldots$ people having an accident are given by the terms of expression

$$e^{-\lambda}\left(1 + \lambda + \frac{\lambda^2}{2!} + \frac{\lambda^3}{3!} + \cdots\right)$$

taken from left to right.
The average occurrence of the event, λ, is 7500×0.0003, i.e. 2.25

The probability of no people having an accident is

$$e^{-\lambda} = e^{-2.25} = 0.1054$$

The probability of 1 person having an accident is

$$\lambda e^{-\lambda} = 2.25 e^{-2.25} = 0.2371$$

The probability of 2 people having an accident is

$$\frac{\lambda^2 e^{-\lambda}}{2!} = \frac{2.25^2 e^{-2.25}}{2!} = 0.2668$$

and so on, giving probabilities of 0.2001, 0.1126, 0.0506 and 0.0190 for 3, 4, 5 and 6, respectively, having an accident. The histogram for these probabilities is shown in Figure 88.2.

Section L

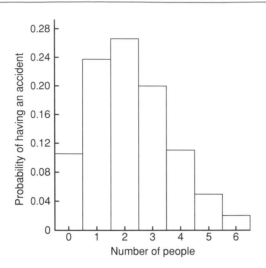

Figure 88.2

Now try the following Practice Exercise

Practice Exercise 339   Further problems on the Poisson distribution (answers on page 1191)

1. In Problem 7 of Exercise 338, page 959, determine the probability of having three components outside of the required tolerance using the Poisson distribution.

2. The probability that an employee will go to hospital in a certain period of time is 0.0015. Use a Poisson distribution to determine the probability of more than two employees going to hospital during this period of time if there are 2000 employees on the payroll.

3. When packaging a product, a manufacturer finds that one packet in twenty is underweight. Determine the probabilities that in a box of 72 packets (a) two and (b) less than four will be underweight.

4. A manufacturer estimates that 0.25% of his output of a component are defective. The components are marketed in packets of 200. Determine the probability of a packet containing less than three defective components.

5. The demand for a particular tool from a store is, on average, five times a day and the demand follows a Poisson distribution. How many of these tools should be kept in the stores so that the probability of there being one available when required is greater than 10%?

6. Failure of a group of particular machine tools follows a Poisson distribution with a mean value of 0.7. Determine the probabilities of 0, 1, 2, 3, 4 and 5 failures in a week and present these results on a histogram.

Chapter 89

The normal distribution

Why it is important to understand: The normal distribution

A normal distribution is a very important statistical data distribution pattern occurring in many natural phenomena, such as height, blood pressure, lengths of objects produced by machines, marks in a test, errors in measurements, and so on. In general, when data are gathered, we expect to see a particular pattern to the data, called a normal distribution. This is a distribution where the data are evenly distributed around the mean in a very regular way, which when plotted as a histogram will result in a bell curve. The normal distribution is the most important of all probability distributions; it is applied directly to many practical problems in every engineering discipline. There are two principal applications of the normal distribution to engineering and reliability. One application deals with the analysis of items which exhibit failure to wear, such as mechanical devices – frequently the wear-out failure distribution is sufficiently close to normal that the use of this distribution for predicting or assessing reliability is valid. Another application is in the analysis of manufactured items and their ability to meet specifications. No two parts made to the same specification are exactly alike; the variability of parts leads to a variability in systems composed of those parts. The design must take this variability into account, otherwise the system may not meet the specification requirement due to the combined effect of part variability.

At the end of this chapter, you should be able to:

- recognise a normal curve
- use the normal distribution in calculations
- test for a normal distribution using probability paper

89.1 Introduction to the normal distribution

When data are obtained, they can frequently be considered to be a sample (i.e. a few members) drawn at random from a large population (i.e. a set having many members). If the sample number is large, it is theoretically possible to choose class intervals which are very small, but which still have a number of members falling within each class. A frequency polygon of this data then has a large number of small line segments and approximates to a continuous curve. Such a curve is called a **frequency or a distribution curve**.

An extremely important symmetrical distribution curve is called the **normal curve** and is as shown in Figure 89.1. This curve can be described by a mathematical equation and is the basis of much of the work done in more advanced statistics. Many natural occurrences such as the heights or weights of a group of people, the sizes of components produced by a particular machine and the life length of certain components approximate to a normal distribution.

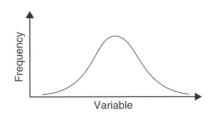

Figure 89.1

Normal distribution curves can differ from one another in the following four ways:

(a) by having different mean values

(b) by having different values of standard deviations

(c) the variables having different values and different units and

(d) by having different areas between the curve and the horizontal axis.

A normal distribution curve is **standardised** as follows:

(a) The mean value of the unstandardised curve is made the origin, thus making the mean value, $\bar{x}$, zero.

(b) The horizontal axis is scaled in standard deviations. This is done by letting $z = \dfrac{x - \bar{x}}{\sigma}$, where z is called the **normal standard variate**, x is the value of the variable, $\bar{x}$ is the mean value of the distribution and σ is the standard deviation of the distribution.

(c) The area between the normal curve and the horizontal axis is made equal to unity.

When a normal distribution curve has been standardised, the normal curve is called a **standardised normal curve** or a **normal probability curve**, and any normally distributed data may be represented by the **same** normal probability curve.

The area under part of a normal probability curve is directly proportional to probability, and the value of the shaded area shown in Figure 89.2 can be determined by evaluating:

$$\int \frac{1}{\sqrt{(2\pi)}} e^{\left(\frac{z^2}{2}\right)} dz, \quad \text{where } z = \frac{x - \bar{x}}{\sigma}$$

To save repeatedly determining the values of this function, tables of partial areas under the standardised normal curve are available in many mathematical formulae books, and such a table is shown in Table 89.1, on page 966.

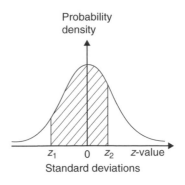

Figure 89.2

> **Problem 1.** The mean height of 500 people is 170 cm and the standard deviation is 9 cm. Assuming the heights are normally distributed, determine the number of people likely to have heights between 150 cm and 195 cm

The mean value, $\bar{x}$, is 170 cm and corresponds to a normal standard variate value, z, of zero on the standardised normal curve. A height of 150 cm has a z-value given by $z = \dfrac{x - \bar{x}}{\sigma}$ standard deviations, i.e. $\dfrac{150 - 170}{9}$ or -2.22 standard deviations. Using a table of partial areas beneath the standardised normal curve (see Table 89.1), a z-value of -2.22 corresponds to an area of 0.4868 between the mean value and the ordinate $z = -2.22$. The negative z-value shows that it lies to the left of the $z = 0$ ordinate.

This area is shown shaded in Figure 89.3(a). Similarly, 195 cm has a z-value of $\dfrac{195 - 170}{9}$ that is 2.78 standard deviations. From Table 89.1, this value of z corresponds to an area of 0.4973, the positive value of z showing that it lies to the right of the $z = 0$ ordinate. This area is shown shaded in Figure 89.3(b). The total area shaded in Figures 89.3(a) and (b) is shown in Figure 89.3(c) and is $0.4868 + 0.4973$, i.e. 0.9841 of the total area beneath the curve.

However, the area is directly proportional to probability. Thus, the probability that a person will have a height of between 150 and 195 cm is 0.9841. For a group of 500 people, 500×0.9841, i.e. **492 people are likely to have heights in this range**. The value of 500×0.9841 is 492.05, but since answers based on a normal probability distribution can only be approximate, results are usually given correct to the nearest whole number.

> **Problem 2.** For the group of people given in Problem 1, find the number of people likely to have heights of less than 165 cm

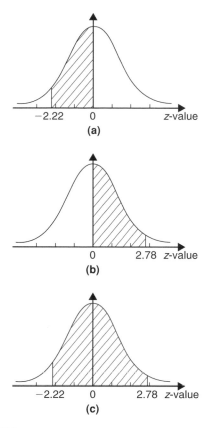

Figure 89.3

A height of 165 cm corresponds to $\dfrac{165-170}{9}$ i.e. -0.56 standard deviations.

The area between $z=0$ and $z=-0.56$ (from Table 89.1) is 0.2123, shown shaded in Figure 89.4(a). The total area under the standardised normal curve is unity and since the curve is symmetrical, it follows that the total area to the left of the $z=0$ ordinate is 0.5000. Thus the area to the left of the $z=-0.56$ ordinate ('left' means 'less than', 'right' means 'more than') is 0.5000 − 0.2123, i.e. 0.2877 of the total area, which is shown shaded in Figure 89.4(b). The area is directly proportional to probability and since the total area beneath the standardised normal curve is unity, the probability of a person's height being less than 165 cm is 0.2877. For a group of 500 people, 500 × 0.2877, i.e. **144 people are likely to have heights of less than 165 cm**.

Problem 3. For the group of people given in Problem 1, find how many people are likely to have heights of more than 194 cm

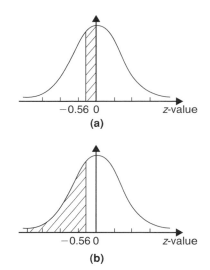

Figure 89.4

194 cm corresponds to a z-value of $\dfrac{194-170}{9}$ that is, 2.67 standard deviations. From Table 89.1, the area between $z=0$, $z=2.67$ and the standardised normal curve is 0.4962, shown shaded in Figure 89.5(a). Since the standardised normal curve is symmetrical, the total area to the right of the $z=0$ ordinate is 0.5000, hence the shaded area shown in Figure 89.5(b) is 0.5000 − 0.4962, i.e. 0.0038. This area represents the probability of a person having a height of more than 194 cm, and for 500 people, the number of people likely to have a height of more than 194 cm is 0.0038 × 500, i.e. **2 people**.

Problem 4. A batch of 1500 lemonade bottles have an average content of 753 ml and the standard deviation of the content is 1.8 ml. If the volumes of the contents are normally distributed, find

(a) the number of bottles likely to contain less than 750 ml

(b) the number of bottles likely to contain between 751 and 754 ml

(c) the number of bottles likely to contain more than 757 ml and

(d) the number of bottles likely to contain between 750 and 751 ml

(a) The z-value corresponding to 750 ml is given by $\dfrac{x-\overline{x}}{\sigma}$ i.e. $\dfrac{750-753}{1.8}=-1.67$ standard deviations. From Table 89.1, the area between $z=0$ and $z=-1.67$ is 0.4525. Thus the area to the

Table 89.1 Partial areas under the standardised normal curve

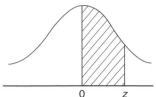

$z = \dfrac{x - \bar{x}}{\sigma}$	0	1	2	3	4	5	6	7	8	9
0.0	0.0000	0.0040	0.0080	0.0120	0.0159	0.0199	0.0239	0.0279	0.0319	0.0359
0.1	0.0398	0.0438	0.0478	0.0517	0.0557	0.0596	0.0636	0.0678	0.0714	0.0753
0.2	0.0793	0.0832	0.0871	0.0910	0.0948	0.0987	0.1026	0.1064	0.1103	0.1141
0.3	0.1179	0.1217	0.1255	0.1293	0.1331	0.1388	0.1406	0.1443	0.1480	0.1517
0.4	0.1554	0.1591	0.1628	0.1664	0.1700	0.1736	0.1772	0.1808	0.1844	0.1879
0.5	0.1915	0.1950	0.1985	0.2019	0.2054	0.2086	0.2123	0.2157	0.2190	0.2224
0.6	0.2257	0.2291	0.2324	0.2357	0.2389	0.2422	0.2454	0.2486	0.2517	0.2549
0.7	0.2580	0.2611	0.2642	0.2673	0.2704	0.2734	0.2760	0.2794	0.2823	0.2852
0.8	0.2881	0.2910	0.2939	0.2967	0.2995	0.3023	0.3051	0.3078	0.3106	0.3133
0.9	0.3159	0.3186	0.3212	0.3238	0.3264	0.3289	0.3315	0.3340	0.3365	0.3389
1.0	0.3413	0.3438	0.3451	0.3485	0.3508	0.3531	0.3554	0.3577	0.3599	0.3621
1.1	0.3643	0.3665	0.3686	0.3708	0.3729	0.3749	0.3770	0.3790	0.3810	0.3830
1.2	0.3849	0.3869	0.3888	0.3907	0.3925	0.3944	0.3962	0.3980	0.3997	0.4015
1.3	0.4032	0.4049	0.4066	0.4082	0.4099	0.4115	0.4131	0.4147	0.4162	0.4177
1.4	0.4192	0.4207	0.4222	0.4236	0.4251	0.4265	0.4279	0.4292	0.4306	0.4319
1.5	0.4332	0.4345	0.4357	0.4370	0.4382	0.4394	0.4406	0.4418	0.4430	0.4441
1.6	0.4452	0.4463	0.4474	0.4484	0.4495	0.4505	0.4515	0.4525	0.4535	0.4545
1.7	0.4554	0.4564	0.4573	0.4582	0.4591	0.4599	0.4608	0.4616	0.4625	0.4633
1.8	0.4641	0.4649	0.4656	0.4664	0.4671	0.4678	0.4686	0.4693	0.4699	0.4706
1.9	0.4713	0.4719	0.4726	0.4732	0.4738	0.4744	0.4750	0.4756	0.4762	0.4767
2.0	0.4772	0.4778	0.4783	0.4785	0.4793	0.4798	0.4803	0.4808	0.4812	0.4817
2.1	0.4821	0.4826	0.4830	0.4834	0.4838	0.4842	0.4846	0.4850	0.4854	0.4857
2.2	0.4861	0.4864	0.4868	0.4871	0.4875	0.4878	0.4881	0.4884	0.4887	0.4890
2.3	0.4893	0.4896	0.4898	0.4901	0.4904	0.4906	0.4909	0.4911	0.4913	0.4916
2.4	0.4918	0.4920	0.4922	0.4925	0.4927	0.4929	0.4931	0.4932	0.4934	0.4936
2.5	0.4938	0.4940	0.4941	0.4943	0.4945	0.4946	0.4948	0.4949	0.4951	0.4952
2.6	0.4953	0.4955	0.4956	0.4957	0.4959	0.4960	0.4961	0.4962	0.4963	0.4964
2.7	0.4965	0.4966	0.4967	0.4968	0.4969	0.4970	0.4971	0.4972	0.4973	0.4974
2.8	0.4974	0.4975	0.4976	0.4977	0.4977	0.4978	0.4979	0.4980	0.4980	0.4981
2.9	0.4981	0.4982	0.4982	0.4983	0.4984	0.4984	0.4985	0.4985	0.4986	0.4986
3.0	0.4987	0.4987	0.4987	0.4988	0.4988	0.4989	0.4989	0.4989	0.4990	0.4990
3.1	0.4990	0.4991	0.4991	0.4991	0.4992	0.4992	0.4992	0.4992	0.4993	0.4993
3.2	0.4993	0.4993	0.4994	0.4994	0.4994	0.4994	0.4994	0.4995	0.4995	0.4995
3.3	0.4995	0.4995	0.4995	0.4996	0.4996	0.4996	0.4996	0.4996	0.4996	0.4997
3.4	0.4997	0.4997	0.4997	0.4997	0.4997	0.4997	0.4997	0.4997	0.4997	0.4998
3.5	0.4998	0.4998	0.4998	0.4998	0.4998	0.4998	0.4998	0.4998	0.4998	0.4998
3.6	0.4998	0.4998	0.4999	0.4999	0.4999	0.4999	0.4999	0.4999	0.4999	0.4999
3.7	0.4999	0.4999	0.4999	0.4999	0.4999	0.4999	0.4999	0.4999	0.4999	0.4999
3.8	0.4999	0.4999	0.4999	0.4999	0.4999	0.4999	0.4999	0.4999	0.4999	0.4999
3.9	0.5000	0.5000	0.5000	0.5000	0.5000	0.5000	0.5000	0.5000	0.5000	0.5000

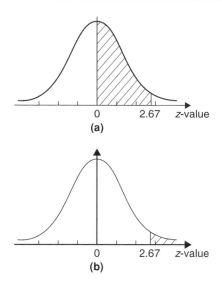

Figure 89.5

left of the $z=-1.67$ ordinate is $0.5000-0.4525$ (see Problem 2), i.e. 0.0475. This is the probability of a bottle containing less than 750 ml. Thus, for a batch of 1500 bottles, it is likely that 1500×0.0475, i.e. **71 bottles will contain less than 750 ml**.

(b) The z-values corresponding to 751 and 754 ml are $\dfrac{751-753}{1.8}$ and $\dfrac{754-753}{1.8}$ i.e. -1.11 and 0.56, respectively. From Table 89.1, the areas corresponding to these values are 0.3665 and 0.2123, respectively. Thus the probability of a bottle containing between 751 and 754 ml is $0.3665+0.2123$ (see Problem 1), i.e. 0.5788. For 1500 bottles, it is likely that 1500×0.5788, i.e. **868 bottles will contain between 751 and 754 ml**.

(c) The z-value corresponding to 757 ml is $\dfrac{757-753}{1.8}$, i.e. 2.22 standard deviations. From Table 89.1, the area corresponding to a z-value of 2.22 is 0.4868. The area to the right of the $z=2.22$ ordinate is $0.5000-0.4868$ (see Problem 3), i.e. 0.0132. Thus, for 1500 bottles, it is likely that 1500×0.0132, i.e. **20 bottles will have contents of more than 757 ml**.

(d) The z-value corresponding to 750 ml is -1.67 (see part (a)), and the z-value corresponding to 751 ml is -1.11 (see part (b)). The areas corresponding to these z-values are 0.4525 and 0.3665, respectively, and both these areas lie on the left of the $z=0$ ordinate. The area between $z=-1.67$ and $z=-1.11$ is $0.4525-0.3665$, i.e. 0.0860 and this is the probability of a bottle having contents between 750 and 751 ml. For 1500 bottles, it is likely that 1500×0.0860, i.e. **129 bottles will be in this range**.

Now try the following Practice Exercise

Practice Exercise 340   Further problems on the introduction to the normal distribution (answers on page 1191)

1. A component is classed as defective if it has a diameter of less than 69 mm. In a batch of 350 components, the mean diameter is 75 mm and the standard deviation is 2.8 mm. Assuming the diameters are normally distributed, determine how many are likely to be classed as defective.

2. The masses of 800 people are normally distributed, having a mean value of 64.7 kg and a standard deviation of 5.4 kg. Find how many people are likely to have masses of less than 54.4 kg.

3. 500 tins of paint have a mean content of 1010 ml and the standard deviation of the contents is 8.7 ml. Assuming the volumes of the contents are normally distributed, calculate the number of tins likely to have contents whose volumes are less than (a) 1025 ml, (b) 1000 ml and (c) 995 ml.

4. For the 350 components in Problem 1, if those having a diameter of more than 81.5 mm are rejected, find, correct to the nearest component, the number likely to be rejected due to being oversized.

5. For the 800 people in Problem 2, determine how many are likely to have masses of more than (a) 70 kg and (b) 62 kg.

6. The mean diameter of holes produced by a drilling machine bit is 4.05 mm and the standard deviation of the diameters is 0.0028 mm. For twenty holes drilled using this machine, determine, correct to the nearest whole number, how many are likely to have diameters of between (a) 4.048 and 4.0553 mm and

(b) 4.052 and 4.056 mm, assuming the diameters are normally distributed.

7. The IQs of 400 children have a mean value of 100 and a standard deviation of 14. Assuming that IQs are normally distributed, determine the number of children likely to have IQs of between (a) 80 and 90, (b) 90 and 110 and (c) 110 and 130

8. The mean mass of active material in tablets produced by a manufacturer is 5.00 g and the standard deviation of the masses is 0.036 g. In a bottle containing 100 tablets, find how many tablets are likely to have masses of (a) between 4.88 and 4.92 g, (b) between 4.92 and 5.04 g and (c) more than 5.04 g

89.2 Testing for a normal distribution

It should never be assumed that because data is continuous it automatically follows that it is normally distributed. One way of checking that data is normally distributed is by using **normal probability paper**, often just called **probability paper**. This is special graph paper which has linear markings on one axis and percentage probability values from 0.01 to 99.99 on the other axis (see Figures 89.6 and 89.7). The divisions on the probability axis are such that a straight line graph results for normally distributed data when percentage cumulative frequency values are plotted against upper class boundary values. If the points do not lie in a reasonably straight line, then the data is not normally distributed. The method used to test the normality of a distribution is shown in Problems 5 and 6. The mean value and standard deviation of normally distributed data may be determined using normal probability paper. For normally distributed data, the area beneath the standardised normal curve and a z-value of unity (i.e. one standard deviation) may be obtained from Table 89.1. For one standard deviation, this area is 0.3413, i.e. 34.13%. An area of ± 1 standard deviation is symmetrically placed on either side of the $z = 0$ value, i.e. is symmetrically placed on either side of the 50% cumulative frequency value. Thus an area corresponding to ± 1 standard deviation extends from percentage cumulative frequency values of $(50 + 34.13)\%$ to $(50 - 34.13)\%$, i.e. from 84.13% to 15.87%. For most purposes, these values are taken as 84% and 16%. Thus, when using normal probability

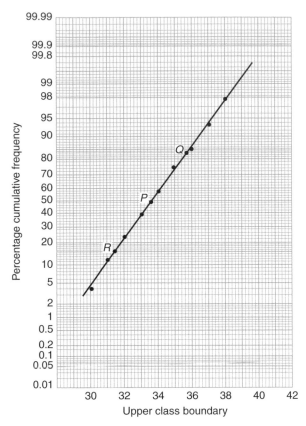

Figure 89.6

paper, the standard deviation of the distribution is given by:

$$\frac{\left(\begin{array}{l} \text{variable value for 84\% cumulative frequency} - \\ \text{variable value for 16\% cumulative frequency} \end{array} \right)}{2}$$

Problem 5. Use normal probability paper to determine whether the data given below, which refers to the masses of 50 copper ingots, is approximately normally distributed. If the data is normally distributed, determine the mean and standard deviation of the data from the graph drawn

Class mid-point value (kg)	Frequency
29.5	2
30.5	4
31.5	6
32.5	8
33.5	9
34.5	8

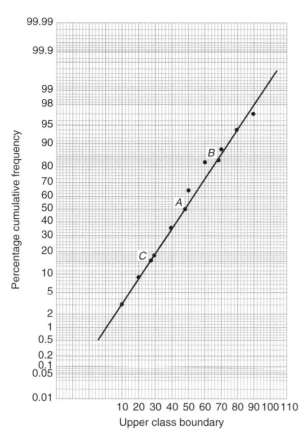

Figure 89.7

Class mid-point value (kg)	Frequency
35.5	6
36.5	4
37.5	2
38.5	1

To test the normality of a distribution, the upper class boundary/percentage cumulative frequency values are plotted on normal probability paper. The upper class boundary values are: 30, 31, 32, ..., 38, 39. The corresponding cumulative frequency values (for 'less than' the upper class boundary values) are: 2, $(4+2)=6$, $(6+4+2)=12$, 20, 29, 37, 43, 47, 49 and 50. The corresponding percentage cumulative frequency values are $\frac{2}{50} \times 100 = 4$, $\frac{6}{50} \times 100 = 12$, 24, 40, 58, 74, 86, 94, 98 and 100%

The co-ordinates of upper class boundary/percentage cumulative frequency values are plotted as shown

in Figure 89.6. When plotting these values, it will always be found that the co-ordinate for the 100% cumulative frequency value cannot be plotted, since the maximum value on the probability scale is 99.99. **Since the points plotted in Figure 89.6 lie very nearly in a straight line, the data is approximately normally distributed**. The mean value and standard deviation can be determined from Figure 89.6. Since a normal curve is symmetrical, the mean value is the value of the variable corresponding to a 50% cumulative frequency value, shown as point P on the graph. This shows that the **mean value is 33.6 kg**. The standard deviation is determined using the 84% and 16% cumulative frequency values, shown as Q and R in Figure 89.6. The variable values for Q and R are 35.7 and 31.4, respectively; thus two standard deviations correspond to $35.7 - 31.4$, i.e. 4.3, showing that the standard deviation of the distribution is approximately $\frac{4.3}{2}$ i.e. **2.15 standard deviations**.

The mean value and standard deviation of the distribution can be calculated using

$$\text{mean, } \bar{x} = \frac{(\sum fx)}{(\sum f)}$$

and standard deviation,

$$\sigma = \sqrt{\left\{ \frac{(\sum [f(x - \bar{x})^2])}{(\sum f)} \right\}}$$

where f is the frequency of a class and x is the class mid-point value. Using these formulae gives a mean value of the distribution of 33.6 (as obtained graphically) and a standard deviation of 2.12, showing that the graphical method of determining the mean and standard deviation give quite realistic results.

Problem 6. Use normal probability paper to determine whether the data given below is normally distributed. Use the graph and assume a normal distribution, whether this is so or not, to find approximate values of the mean and standard deviation of the distribution

Class mid-point values	Frequency
5	1
15	2
25	3
35	6

Class mid-point values	Frequency
45	9
55	6
65	2
75	2
85	1
95	1

To test the normality of a distribution, the upper class boundary/percentage cumulative frequency values are plotted on normal probability paper. The upper class boundary values are: $10, 20, 30, \ldots, 90$ and 100. The corresponding cumulative frequency values are $1, 1+2=3$, $1+2+3=6$, $12, 21, 27, 29, 31, 32$ and 33. The percentage cumulative frequency values are $\frac{1}{33} \times 100 = 3$, $\frac{3}{33} \times 100 = 9, 18, 36, 64, 82, 88, 94, 97$ and 100%

The co-ordinates of upper class boundary values/percentage cumulative frequency values are plotted as shown in Figure 89.7. Although six of the points lie approximately in a straight line, three points corresponding to upper class boundary values of 50, 60 and 70 are not close to the line and indicate that **the distribution is not normally** distributed. However, if a normal distribution is assumed, the mean value corresponds to the variable value at a cumulative frequency of 50% and, from Figure 89.7, point A is **48**. The value of the standard deviation of the distribution can be obtained from the variable values corresponding to the 84% and 16% cumulative frequency values, shown as B and C in Figure 89.7 and give: $2\sigma = 69 - 28$, i.e. the standard deviation $\sigma = \mathbf{20.5}$. The calculated values of the mean and standard deviation of the distribution are 45.9 and 19.4 respectively, showing that errors are introduced if the graphical method of determining these values is used for data which is not normally distributed.

Now try the following Practice Exercise

Practice Exercise 341 Further problems on testing for a normal distribution (answers on page 1191)

1. A frequency distribution of 150 measurements is as shown:

Class mid-point value	Frequency
26.4	5
26.6	12
26.8	24
27.0	36
27.2	36
27.4	25
27.6	12

Use normal probability paper to show that this data approximates to a normal distribution and hence determine the approximate values of the mean and standard deviation of the distribution. Use the formula for mean and standard deviation to verify the results obtained.

2. A frequency distribution of the class mid-point values of the breaking loads for 275 similar fibres is as shown below:

Load (kN)	17	19	21	23	25	27	29	31
Frequency	9	23	55	78	64	28	14	4

Use normal probability paper to show that this distribution is approximately normally distributed and determine the mean and standard deviation of the distribution (a) from the graph and (b) by calculation.

For fully worked solutions to each of the problems in Practice Exercises 340 and 341 in this chapter, go to the website:
www.routledge.com/cw/bird

Chapter 90

Linear correlation

Why it is important to understand: Linear correlation

Correlation coefficients measure the strength of association between two variables. The most common correlation coefficient, called the product-moment correlation coefficient, measures the strength of the linear association between variables. A positive value indicates a positive correlation and the higher the value, the stronger the correlation. Similarly, a negative value indicates a negative correlation and the lower the value the stronger the correlation. This chapter explores linear correlation and the meaning of values obtained calculating the coefficient of correlation.

At the end of this chapter, you should be able to:

- recognise linear correlation
- state the Pearson product-moment formula
- appreciate the significance of a coefficient of correlation
- determine the linear coefficient of correlation between two given variables

90.1 Introduction to linear correlation

Correlation is a measure of the amount of association existing between two variables. For linear correlation, if points are plotted on a graph and all the points lie on a straight line, then **perfect linear correlation** is said to exist. When a straight line having a positive gradient can reasonably be drawn through points on a graph, **positive or direct linear correlation** exists, as shown in Figure 90.1(a). Similarly, when a straight line having a negative gradient can reasonably be drawn through points on a graph, **negative or inverse linear correlation** exists, as shown in Figure 90.1(b). When there is no apparent relationship between co-ordinate values plotted on a graph then **no correlation** exists between the points, as shown in Figure 90.1(c). In statistics, when two variables are being investigated, the location of the co-ordinates on a rectangular co-ordinate system is called a **scatter diagram** – as shown in Figure 90.1.

90.2 The Pearson product-moment formula for determining the linear correlation coefficient

The Pearson product-moment correlation coefficient (or **Pearson correlation coefficient**, for short) is a measure of the strength of a linear association between two variables and is denoted by the symbol r. A Pearson product-moment correlation attempts to draw a line of best fit through the data of two variables, and the Pearson correlation coefficient, r, indicates how far away all these data points are from this line of best fit (how well the data points fit this new model/line of best fit). The **product-moment formula** states:

coefficient of correlation,

$$r = \frac{\sum xy}{\sqrt{\{(\sum x^2)(\sum y^2)\}}} \tag{1}$$

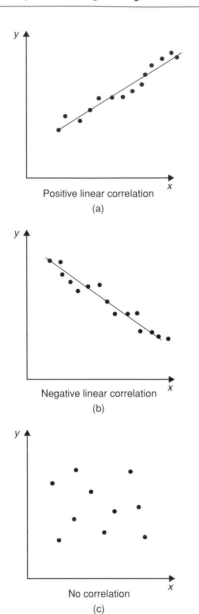

Figure 90.1

where the x-values are the values of the deviations of co-ordinates X from $\overline{X}$, their mean value, and the y-values are the values of the deviations of co-ordinates Y from $\overline{Y}$, their mean value. That is, $x = (X - \overline{X})$ and $y = (Y - \overline{Y})$. The results of this determination give values of r lying between $+1$ and -1, where $+1$ indicates perfect direct correlation, -1 indicates perfect inverse correlation and 0 indicates that no correlation exists. Between these values, the smaller the value of r, the less is the amount of correlation which exists. Generally, values of r in the ranges 0.7 to 1 and -0.7 to -1 show that a fair amount of correlation exists.

90.3 The significance of a coefficient of correlation

When the value of the coefficient of correlation has been obtained from the product-moment formula, some care is needed before coming to conclusions based on this result. Checks should be made to ascertain the following two points:

(a) that a 'cause and effect' relationship exists between the variables; it is relatively easy, mathematically, to show that some correlation exists between, say, the number of ice creams sold in a given period of time and the number of chimneys swept in the same period of time, although there is no relationship between these variables;

(b) that a linear relationship exists between the variables; the product-moment formula given in Section 90.2 is based on linear correlation. Perfect non-linear correlation may exist (for example, the co-ordinates exactly following the curve $y = x^3$), but this gives a low value of coefficient of correlation since the value of r is determined using the product-moment formula, based on a linear relationship.

90.4 Worked problems on linear correlation

Problem 1. In an experiment to determine the relationship between force on a wire and the resulting extension, the following data are obtained:

Force (N)	10	20	30	40	50	60	70
Extension (mm)	0.22	0.40	0.61	0.85	1.20	1.45	1.70

Determine the linear coefficient of correlation for this data.

Let X be the variable force values and Y be the dependent variable extension values. The coefficient of correlation is given by:

$$r = \frac{\sum xy}{\sqrt{\{(\sum x^2)(\sum y^2)\}}}$$

where $x = (X - \overline{X})$ and $y = (Y - \overline{Y})$, $\overline{X}$ and $\overline{Y}$ being the mean values of the X and Y values, respectively. Using a tabular method to determine the quantities of this formula gives:

X	Y	$x=(X-\overline{X})$	$y=(Y-\overline{Y})$
10	0.22	−30	−0.699
20	0.40	−20	−0.519
30	0.61	−10	−0.309
40	0.85	0	−0.069
50	1.20	10	0.281
60	1.45	20	0.531
70	1.70	30	0.781

$$\sum X=280, \quad \overline{X}=\frac{280}{7}=40$$

$$\sum Y=6.43, \quad \overline{Y}=\frac{6.43}{7}=0.919$$

xy	x^2	y^2
20.97	900	0.489
10.38	400	0.269
3.09	100	0.095
0	0	0.005
2.81	100	0.079
10.62	400	0.282
23.43	900	0.610
$\sum xy=71.30$	$\sum x^2=2800$	$\sum y^2=1.829$

Thus $\quad r=\dfrac{71.3}{\sqrt{[2800\times1.829]}}=\mathbf{0.996}$

This shows that a **very good direct correlation exists** between the values of force and extension.

Problem 2. The relationship between expenditure on welfare services and absenteeism for similar periods of time is shown below for a small company.

Expenditure (£'000)	3.5	5.0	7.0	10	12	15	18
Days lost	241	318	174	110	147	122	86

Determine the coefficient of linear correlation for this data.

Let X be the expenditure in thousands of pounds and Y be the days lost.

The coefficient of correlation,

$$r=\frac{\sum xy}{\sqrt{\{(\sum x^2)(\sum y^2)\}}}$$

where $x=(X-\overline{X})$ and $y=(Y-\overline{Y})$, $\overline{X}$ and $\overline{Y}$ being the mean values of X and Y, respectively. Using a tabular approach:

X	Y	$x=(X-\overline{X})$	$y=(Y-\overline{Y})$
3.5	241	−6.57	69.9
5.0	318	−5.07	146.9
7.0	174	−3.07	2.9
10	110	−0.07	−61.1
12	147	1.93	−24.1
15	122	4.93	−49.1
18	86	7.93	−85.1

$$\sum X=70.5, \quad \overline{X}=\frac{70.5}{7}=10.07$$

$$\sum Y=1198, \quad \overline{Y}=\frac{1198}{7}=171.1$$

xy	x^2	y^2
−459.2	43.2	4886
−744.8	25.7	21580
−8.9	9.4	8
4.3	0	3733
−46.5	3.7	581
−242.1	24.3	2411
−674.8	62.9	7242
$\sum xy=-2172$	$\sum x^2=169.2$	$\sum y^2=40441$

Thus

$$r=\frac{-2172}{\sqrt{[169.2\times40441]}}=\mathbf{-0.830}$$

This shows that there is **fairly good inverse correlation** between the expenditure on welfare and days lost due to absenteeism.

Section L

Problem 3. The relationship between monthly car sales and income from the sale of petrol for a garage is as shown:

Cars sold   2 5 3 12 14 7 3 28 14 7 3 13

Income from petrol sales   12 9 13 21 17 22 31 47 17 10 9 11 (£'000)

Determine the linear coefficient of correlation between these quantities.

Let X represent the number of cars sold and Y the income, in thousands of pounds, from petrol sales. Using the tabular approach:

X	Y	$x=(X-\overline{X})$	$y=(Y-\overline{Y})$
2	12	-7.25	-6.25
5	9	-4.25	-9.25
3	13	-6.25	-5.25
12	21	2.75	2.75
14	17	4.75	-1.25
7	22	-2.25	3.75
3	31	-6.25	12.75
28	47	18.75	28.75
14	17	4.75	-1.25
7	10	-2.25	-8.25
3	9	-6.25	-9.25
13	11	3.75	-7.25

$$\sum X=111, \quad \overline{X}=\frac{111}{12}=9.25$$

$$\sum Y=219, \quad \overline{Y}=\frac{219}{12}=18.25$$

xy	x^2	y^2
45.3	52.6	39.1
39.3	18.1	85.6
32.8	39.1	27.6
7.6	7.6	7.6
-5.9	22.6	1.6
-8.4	5.1	14.1
-79.7	39.1	162.6
539.1	351.6	826.6
-5.9	22.6	1.6
18.6	5.1	68.1
57.8	39.1	85.6
-27.2	14.1	52.6
$\sum xy=613.4$	$\sum x^2=616.7$	$\sum y^2=1372.7$

The coefficient of correlation,

$$r = \frac{\sum xy}{\sqrt{\{(\sum x^2)(\sum y^2)\}}}$$

$$= \frac{613.4}{\sqrt{\{(616.7)(1372.7)\}}} = \mathbf{0.667}$$

Thus, there is **no appreciable correlation** between petrol and car sales.

Now try the following Practice Exercise

Practice Exercise 342   Further problems on linear correlation (answers on page 1191)

In Problems 1 to 3, determine the coefficient of correlation for the data given, correct to 3 decimal places.

1. | X | 14 | 18 | 23 | 30 | 50 |
    |---|---|---|---|---|---|
    | Y | 900 | 1200 | 1600 | 2100 | 3800 |

2. | X | 2.7 | 4.3 | 1.2 | 1.4 | 4.9 |
    |---|---|---|---|---|---|
    | Y | 11.9 | 7.10 | 33.8 | 25.0 | 7.50 |

3. | X | 24 | 41 | 9 | 18 | 73 |
    |---|---|---|---|---|---|
    | Y | 39 | 46 | 90 | 30 | 98 |

4. In an experiment to determine the relationship between the current flowing in an electrical circuit and the applied voltage, the results obtained are:

Current (mA)	5	11	15	19	24	28	33
Applied voltage (V)	2	4	6	8	10	12	14

Determine, using the product-moment formula, the coefficient of correlation for these results.

5. A gas is being compressed in a closed cylinder and the values of pressures and corresponding volumes at constant temperature are as shown:

Pressure (kPa)	Volume (m³)
160	0.034
180	0.036
200	0.030
220	0.027
240	0.024
260	0.025
280	0.020
300	0.019

Find the coefficient of correlation for these values.

6. The relationship between the number of miles travelled by a group of engineering salesmen in ten equal time periods and the corresponding value of orders taken is given below. Calculate the coefficient of correlation using the product-moment formula for these values.

Miles travelled	Orders taken (£'000)
1370	23
1050	17
980	19
1770	22
1340	27
1560	23
2110	30
1540	23
1480	25
1670	19

7. The data shown below refer to the number of times machine tools had to be taken out of service, in equal time periods, due to faults occurring and the number of hours worked by maintenance teams. Calculate the coefficient of correlation for this data.

Machines out of service	4	13	2	9	16	8	7
Maintenance hours	400	515	360	440	570	380	415

For fully worked solutions to each of the problems in Practice Exercise 342 in this chapter, go to the website:
www.routledge.com/cw/bird

Chapter 91

Linear regression

Why it is important to understand: Linear regression

The general process of fitting data to a linear combination of basic functions is termed linear regression. Linear least-squares regression is by far the most widely used modelling method; it is what most people mean when they say they have used 'regression', 'linear regression' or 'least-squares' to fit a model to their data. Not only is linear least-squares regression the most widely used modelling method, but it has been adapted to a broad range of situations that are outside its direct scope. It plays a strong underlying role in many other modelling methods. This chapter explains how regression lines are determined.

At the end of this chapter, you should be able to:

* explain linear regression
* understand least-squares regression lines
* determine, for two variables X and Y, the equations of the regression lines of X on Y and Y on X

91.1   Introduction to linear regression

Regression analysis, usually termed **regression**, is used to draw the line of 'best fit' through co-ordinates on a graph. The techniques used enable a mathematical equation of the straight line form $y = mx + c$ to be deduced for a given set of co-ordinate values, the line being such that the sum of the deviations of the co-ordinate values from the line is a minimum, i.e. it is the line of 'best fit'. When a regression analysis is made, it is possible to obtain two lines of best fit, depending on which variable is selected as the dependent variable and which variable is the independent variable. For example, in a resistive electrical circuit, the current flowing is directly proportional to the voltage applied to the circuit. There are two ways of obtaining experimental values relating the current and voltage. Either, certain voltages are applied to the circuit and the current values are measured, in which case the voltage is the independent variable and the current is the dependent

variable; or, the voltage can be adjusted until a desired value of current is flowing and the value of voltage is measured, in which case the current is the independent value and the voltage is the dependent value.

91.2   The least-squares regression lines

For a given set of co-ordinate values, (X_1, Y_1), $(X_2, Y_2), \ldots, (X_n, Y_n)$ let the X values be the independent variables and the Y values be the dependent values. Also let $D_1, \ldots, D_n$ be the vertical distances between the line shown as PQ in Figure 91.1 and the points representing the co-ordinate values. The least-squares regression line, i.e. the line of best fit, is the line which makes the value of $D_1^2 + D_2^2 + \cdots + D_n^2$ a minimum value.
The equation of the least-squares regression line is usually written as $Y = a_0 + a_1 X$, where a_0 is the Y-axis intercept value and a_1 is the gradient of the line (analogous to c and m in the equation $y = mx + c$). The

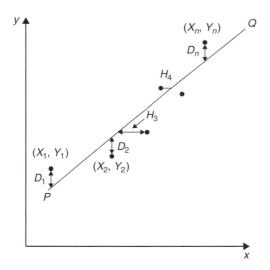

Figure 91.1

values of a_0 and a_1 to make the sum of the 'deviations squared' a minimum can be obtained from the two equations:

$$\sum Y = a_0 N + a_1 \sum X \qquad (1)$$

$$\sum (XY) = a_0 \sum X + a_1 \sum X^2 \qquad (2)$$

where X and Y are the co-ordinate values, N is the number of co-ordinates and a_0 and a_1 are called the **regression coefficients** of Y on X. Equations (1) and (2) are called the **normal equations** of the regression lines of Y on X. The regression line of Y on X is used to estimate values of Y for given values of X. If the Y values (vertical axis) are selected as the independent variables, the horizontal distances between the line shown as PQ in Figure 91.1 and the co-ordinate values (H_3, H_4, etc.) are taken as the deviations. The equation of the regression line is of the form: $X = b_0 + b_1 Y$ and the normal equations become:

$$\sum X = b_0 N + b_1 \sum Y \qquad (3)$$

$$\sum (XY) = b_0 \sum Y + b_1 \sum Y^2 \qquad (4)$$

where X and Y are the co-ordinate values, b_0 and b_1 are the regression coefficients of X on Y and N is the number of co-ordinates. These normal equations are of the regression line of X on Y, which is slightly different to the regression line of Y on X. The regression line of X on Y is used to estimated values of X for given values of Y. The regression line of Y on X is used to determine any value of Y corresponding to a given value of X. If the value of Y lies within the range of Y values of the extreme co-ordinates, the process of finding the corresponding value of X is called **linear interpolation**. If it lies outside of the range of Y values of the extreme co-ordinates than the process is called **linear extrapolation** and the assumption must be made that the line of best fit extends outside of the range of the co-ordinate values given.

By using the regression line of X on Y, values of X corresponding to given values of Y may be found by either interpolation or extrapolation.

91.3 Worked problems on linear regression

Problem 1. In an experiment to determine the relationship between frequency and the inductive reactance of an electrical circuit, the following results were obtained:

Frequency (Hz)	Inductive reactance (ohms)
50	30
100	65
150	90
200	130
250	150
300	190
350	200

Determine the equation of the regression line of inductive reactance on frequency, assuming a linear relationship

Since the regression line of inductive reactance on frequency is required, the frequency is the independent variable, X, and the inductive reactance is the dependent variable, Y. The equation of the regression line of Y on X is:

$$Y = a_0 + a_1 X$$

and the regression coefficients a_0 and a_1 are obtained by using the normal equations

$$\sum Y = a_0 N + a_1 \sum X$$

and $$\sum XY = a_0 \sum X + a_1 \sum X^2$$

(from equations (1) and (2))

A tabular approach is used to determine the summed quantities.

Frequency, X	Inductive reactance, Y	X^2
50	30	2500
100	65	10 000
150	90	22 500
200	130	40 000
250	150	62 500
300	190	90 000
350	200	122 500
$\sum X = 1400$	$\sum Y = 855$	$\sum X^2 = 350\,000$

XY	Y^2
1500	900
6500	4225
13 500	8100
26 000	16 900
37 500	22 500
57 000	36 100
70 000	40 000
$\sum XY = 212\,000$	$\sum Y^2 = 128\,725$

The number of co-ordinate values given, N, is 7. Substituting in the normal equations gives:

$$855 = 7a_0 + 1400a_1 \tag{1}$$

$$212000 = 1400a_0 + 350000a_1 \tag{2}$$

$1400 \times (1)$ gives:

$$1197000 = 9800a_0 + 1960000a_1 \tag{3}$$

$7 \times (2)$ gives:

$$1484000 = 9800a_0 + 2450000a_1 \tag{4}$$

$(4) - (3)$ gives:

$$287000 = 0 + 490000a_1$$

from which, $a_1 = \dfrac{287000}{490000} = 0.586$

Substituting $a_1 = 0.586$ in equation (1) gives:

$$855 = 7a_0 + 1400(0.586)$$

i.e. $a_0 = \dfrac{855 - 820.4}{7} = 4.94$

Thus the equation of the regression line of inductive reactance on frequency is:

$$Y = 4.94 + 0.586X$$

Problem 2. For the data given in Problem 1, determine the equation of the regression line of frequency on inductive reactance, assuming a linear relationship

In this case, the inductive reactance is the independent variable X and the frequency is the dependent variable Y. From equations 3 and 4, the equation of the regression line of X on Y is:

$$X = b_0 + b_1 Y$$

and the normal equations are

$$\sum X = b_0 N + b_1 \sum Y$$

and $\sum XY = b_0 \sum Y + b_1 \sum Y^2$

From the table shown in Problem 1, the simultaneous equations are:

$$1400 = 7b_0 + 855b_1$$

$$212000 = 855b_0 + 128725b_1$$

Solving these equations in a similar way to that in Problem 1 gives:

$$b_0 = -6.15$$

and $b_1 = 1.69$, correct to 3 significant figures.

Thus the equation of the regression line of frequency on inductive reactance is:

$$X = -6.15 + 1.69Y$$

Problem 3. Use the regression equations calculated in Problems 1 and 2 to find (a) the value of inductive reactance when the frequency is 175 Hz and (b) the value of frequency when the inductive reactance is 250 ohms, assuming the line of best fit extends outside of the given co-ordinate values. Draw a graph showing the two regression lines.

(a) From Problem 1, the regression equation of inductive reactance on frequency is
$Y = 4.94 + 0.586X$. When the frequency, X, is 175 Hz, $Y = 4.94 + 0.586(175) = 107.5$, correct to 4 significant figures, i.e. the inductive reactance is **107.5 ohms** when the frequency is 175 Hz.

(b) From Problem 2, the regression equation of frequency on inductive reactance is
$X = -6.15 + 1.69Y$. When the inductive reactance, Y, is 250 ohms,
$X = -6.15 + 1.69(250) = 416.4$ Hz, correct to 4 significant figures, i.e. the frequency is **416.4 Hz** when the inductive reactance is 250 ohms.

The graph depicting the two regression lines is shown in Figure 91.2. To obtain the regression line of inductive reactance on frequency the regression line equation $Y = 4.94 + 0.586X$ is used, and X (frequency) values of 100 and 300 have been selected in order to find the corresponding Y values. These values gave the co-ordinates as (100, 63.5) and (300, 180.7), shown as points A and B in Figure 91.2. Two co-ordinates for the regression line of frequency on inductive reactance are calculated using the equation $X = -6.15 + 1.69Y$, the values of inductive reactance of 50 and 150 being used to obtain the co-ordinate values. These values gave co-ordinates (78.4, 50) and (247.4, 150), shown as points C and D in Figure 91.2.

It can be seen from Figure 91.2 that to the scale drawn, the two regression lines coincide. Although it is not necessary to do so, the co-ordinate values are also shown to indicate that the regression lines do appear to be the lines of best fit. A graph showing co-ordinate values is called a **scatter diagram** in statistics.

Problem 4. The experimental values relating centripetal force and radius, for a mass travelling at constant velocity in a circle, are as shown:

Force (N)	5	10	15	20	25	30	35	40
Radius (cm)	55	30	16	12	11	9	7	5

Determine the equations of (a) the regression line of force on radius and (b) the regression line of radius on force. Hence, calculate the force at a radius of 40 cm and the radius corresponding to a force of 32 newtons

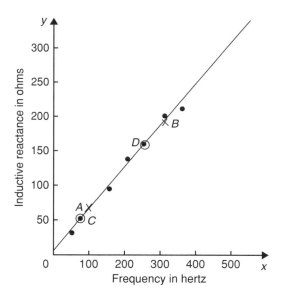

Figure 91.2

Let the radius be the independent variable X, and the force be the dependent variable Y. (This decision is usually based on a 'cause' corresponding to X and an 'effect' corresponding to Y.)

(a) The equation of the regression line of force on radius is of the form $Y = a_0 + a_1X$ and the constants a_0 and a_1 are determined from the normal equations:
$$\sum Y = a_0 N + a_1 \sum X$$
and $\sum XY = a_0 \sum X + a_1 \sum X^2$
$$\text{(from equations (1) and (2))}$$

Using a tabular approach to determine the values of the summations gives:

Radius, X	Force, Y	X^2
55	5	3025
30	10	900
16	15	256
12	20	144
11	25	121
9	30	81
7	35	49
5	40	25
$\sum X = 145$	$\sum Y = 180$	$\sum X^2 = 4601$

Section L

XY	Y^2
275	25
300	100
240	225
240	400
275	625
270	900
245	1225
200	1600
$\sum XY = 2045$	$\sum Y^2 = 5100$

Thus $\quad 180 = 8a_0 + 145a_1$

and $\quad 2045 = 145a_0 + 4601a_1$

Solving these simultaneous equations gives $a_0 = 33.7$ and $a_1 = -0.617$, correct to 3 significant figures. Thus the equation of the regression line of force on radius is:

$$Y = 33.7 - 0.617X$$

(b) The equation of the regression line of radius on force is of the form $X = b_0 + b_1 Y$ and the constants b_0 and b_1 are determined from the normal equations:

$$\sum X = b_0 N + b_1 \sum Y$$
and $\quad \sum XY = b_0 \sum Y + b_1 \sum Y^2$

(from equations (3) and (4))

The values of the summations have been obtained in part (a) giving:

$$145 = 8b_0 + 180b_1$$

and $\quad 2045 = 180b_0 + 5100b_1$

Solving these simultaneous equations gives $b_0 = 44.2$ and $b_1 = -1.16$, correct to 3 significant figures. Thus the equation of the regression line of radius on force is:

$$X = 44.2 - 1.16Y$$

The force, Y, at a radius of 40 cm, is obtained from the regression line of force on radius, i.e. $y = 33.7 - 0.617(40) = 9.02$,

i.e. **the force at a radius of 40 cm is 9.02 N**

The radius, X, when the force is 32 newtons is obtained from the regression line of radius on force, i.e. $X = 44.2 - 1.16(32) = 7.08$

i.e. **the radius when the force is 32 N is 7.08 cm**.

Now try the following Practice Exercise

Practice Exercise 343   Further problems on linear regression (answers on page 1191)

In Problems 1 and 2, determine the equation of the regression line of Y on X, correct to 3 significant figures.

1.
X	14	18	23	30	50
Y	900	1200	1600	2100	3800

2.
X	6	3	9	15	2	14	21	13
Y	1.3	0.7	2.0	3.7	0.5	2.9	4.5	2.7

In Problems 3 and 4, determine the equations of the regression lines of X on Y for the data stated, correct to 3 significant figures.

3. The data given in Problem 1

4. The data given in Problem 2

5. The relationship between the voltage applied to an electrical circuit and the current flowing is as shown:

Current (mA)	Applied voltage (V)
2	5
4	11
6	15
8	19
10	24
12	28
14	33

Assuming a linear relationship, determine the equation of the regression line of applied voltage, Y, on current, X, correct to 4 significant figures.

6. For the data given in Problem 5, determine the equation of the regression line of current on applied voltage, correct to 3 significant figures.

7. Draw the scatter diagram for the data given in Problem 5 and show the regression lines of applied voltage on current and current on applied voltage. Hence determine the values of (a) the applied voltage needed to give a current of 3 mA and (b) the current flowing when the applied voltage is 40 volts, assuming the regression lines are still true outside of the range of values given.

8. In an experiment to determine the relationship between force and momentum, a force, X, is applied to a mass by placing the mass on an inclined plane, and the time, Y, for the velocity to change from u m/s to v m/s is measured. The results obtained are as follows:

Force (N)	Time (s)
11.4	0.56
18.7	0.35
11.7	0.55
12.3	0.52
14.7	0.43
18.8	0.34
19.6	0.31

Determine the equation of the regression line of time on force, assuming a linear relationship between the quantities, correct to 3 significant figures.

9. Find the equation for the regression line of force on time for the data given in Problem 8, correct to 3 decimal places.

10. Draw a scatter diagram for the data given in Problem 8 and show the regression lines of time on force and force on time. Hence find (a) the time corresponding to a force of 16 N, and (b) the force at a time of 0.25 s, assuming the relationship is linear outside of the range of values given.

Section L

For fully worked solutions to each of the problems in Practice Exercise 343 in this chapter, go to the website:
www.routledge.com/cw/bird

This assignment covers the material contained in Chapters 88 to 91. *The marks available are shown in brackets at the end of each question.*

1. A machine produces 15% defective components. In a sample of 5, drawn at random, calculate, using the binomial distribution, the probability that:

    (a)   there will be 4 defective items,

    (b)   there will be not more than 3 defective items,

    (c)   all the items will be non-defective.

    Draw a histogram showing the probabilities of 0, 1, 2,…, 5 defective items.   (20)

2. 2% of the light bulbs produced by a company are defective. Determine, using the Poisson distribution, the probability that in a sample of 80 bulbs: (a) 3 bulbs will be defective, (b) not more than 3 bulbs will be defective, (c) at least 2 bulbs will be defective.   (13)

3. Some engineering components have a mean length of 20 mm and a standard deviation of 0.25 mm. Assume that the data on the lengths of the components are normally distributed.

    In a batch of 500 components, determine the number of components likely to:

    (a)   have a length of less than 19.95 mm,

    (b)   be between 19.95 mm and 20.15 mm,

    (c)   be longer than 20.54 mm.   (15)

4. In a factory, cans are packed with an average of 1.0 kg of a compound and the masses are normally distributed about the average value. The standard deviation of a sample of the contents of the cans is 12 g. Determine the percentage of cans containing (a) less than 985 g, (b) more than 1030 g, (c) between 985 g and 1030 g.   (10)

5. The data given below give the experimental values obtained for the torque output, X, from an electric motor and the current, Y, taken from the supply.

Torque X	Current Y
0	3
1	5
2	6
3	6
4	9
5	11
6	12
7	12
8	14
9	13

Determine the linear coefficient of correlation for the data.   (18)

6. Some results obtained from a tensile test on a steel specimen are shown below:

Tensile force (kN)	Extension (mm)
4.8	3.5
9.3	8.2
12.8	10.1
17.7	15.6
21.6	18.4
26.0	20.8

Assuming a linear relationship:

(a)   determine the equation of the regression line of extension on force,

(b)   determine the equation of the regression line of force on extension,

(c)   estimate (i) the value of extension when the force is 16 kN, and (ii) the value of force when the extension is 17 mm.   (24)

Chapter 92

Sampling and estimation theories

Why it is important to understand: Sampling and estimation theories

Estimation theory is a branch of statistics and signal processing that deals with estimating the values of parameters based on measured/empirical data that has a random component. Estimation theory can be found at the heart of many electronic signal processing systems designed to extract information; these systems include radar, sonar, speech, image, communications, control and seismology. This chapter introduces some of the principles involved with sampling and estimation theories.

At the end of this chapter, you should be able to:

- understand sampling distributions
- determine the standard error of the means
- understand point and interval estimates and confidence intervals
- calculate confidence limits
- estimate the mean and standard deviation of a population from sample data
- estimate the mean of a population based on a small sample data using a Student's *t* distribution

92.1   Introduction

The concepts of elementary sampling theory and estimation theories introduced in this chapter will provide the basis for a more detailed study of inspection, control and quality control techniques used in industry. Such theories can be quite complicated; in this chapter a full treatment of the theories and the derivation of formulae have been omitted for clarity – basic concepts only have been developed.

92.2   Sampling distributions

In statistics, it is not always possible to take into account all the members of a set, and in these circumstances a

sample, or many samples, are drawn from a population. Usually when the word sample is used, it means that a **random sample** is taken. If each member of a population has the same chance of being selected, then a sample taken from that population is called random. A sample which is not random is said to be **biased** and this usually occurs when some influence affects the selection.

When it is necessary to make predictions about a population based on random sampling, often many samples of, say, *N* members are taken, before the predictions are made. If the mean value and standard deviation of each of the samples is calculated, it is found that the results vary from sample to sample, even though the samples are all taken from the same population. In the theories introduced in the following sections, it is important to know whether the differences in the values obtained are due to

---

chance or whether the differences obtained are related in some way. If M samples of N members are drawn at random from a population, the mean values for the M samples together form a set of data. Similarly, the standard deviations of the M samples collectively form a set of data. Sets of data based on many samples drawn from a population are called **sampling distributions**. They are often used to describe the chance fluctuations of mean values and standard deviations based on random sampling.

92.3 The sampling distribution of the means

Suppose that it is required to obtain a sample of two items from a set containing five items. If the set is the five letters A, B, C, D and E, then the different samples which are possible are:

$$AB, AC, AD, AE, BC, BD, BE,$$
$$CD, CE \text{ and } DE,$$

that is, 10 different samples. The number of possible different samples in this case is given by $\dfrac{5 \times 4}{2 \times 1}$ i.e. 10. Similarly, the number of different ways in which a sample of three items can be drawn from a set having ten members can be shown to be $\dfrac{10 \times 9 \times 8}{3 \times 2 \times 1}$ i.e. 120. It follows that when a small sample is drawn from a large population, there are very many different combinations of members possible. With so many different samples possible, quite a large variation can occur in the mean values of various samples taken from the same population.

Usually, the greater the number of members in a sample, the closer will be the mean value of the sample to that of the population. Consider the set of numbers 3, 4, 5, 6, and 7. For a sample of two members, the lowest value of the mean is $\dfrac{3+4}{2}$, i.e. 3.5; the highest is $\dfrac{6+7}{2}$, i.e. 6.5, giving a range of mean values of $6.5 - 3.5 = 3$. For a sample of 3 members, the range is $\dfrac{3+4+5}{3}$ to $\dfrac{5+6+7}{3}$ that is, 2. As the number in the sample increases, the range decreases until, in the limit, if the sample contains all the members of the set, the range of mean values is zero. When many samples are drawn from a population and a sample distribution of the mean values of the sample is formed, the range of the mean values is small provided the number in the sample is

large. Because the range is small it follows that the standard deviation of all the mean values will also be small, since it depends on the distance of the mean values from the distribution mean. The relationship between the standard deviation of the mean values of a sampling distribution and the number in each sample can be expressed as follows:

Theorem 1 'If all possible samples of size N are drawn from a finite population, N_p, without replacement, and the standard deviation of the mean values of the sampling distribution of means is determined then:

$$\sigma_{\bar{x}} = \frac{\sigma}{\sqrt{N}} \sqrt{\left(\frac{N_p - N}{N_p - 1} \right)}$$

where $\sigma_{\bar{x}}$ is the standard deviation of the sampling distribution of means and σ is the standard deviation of the population.'

The standard deviation of a sampling distribution of mean values is called the **standard error of the means**, thus

standard error of the means,

$$\sigma_{\bar{x}} = \frac{\sigma}{\sqrt{N}} \sqrt{\left(\frac{N_p - N}{N_p - 1} \right)} \tag{1}$$

Equation (1) is used for a finite population of size N_p and/or for sampling without replacement. The word 'error' in the 'standard error of the means' does not mean that a mistake has been made but rather that there is a degree of uncertainty in predicting the mean value of a population based on the mean values of the samples. The formula for the standard error of the means is true for all values of the number in the sample, N. When N_p is very large compared with N or when the population is infinite (this can be considered to be the case when sampling is done with replacement), the correction factor $\sqrt{\left(\frac{N_p - N}{N_p - 1} \right)}$ approaches unity and equation (1) becomes

$$\sigma_{\bar{x}} = \frac{\sigma}{\sqrt{N}} \tag{2}$$

Equation (2) is used for an infinite population and/or for sampling with replacement.

Problem 1. Verify Theorem 1 above for the set of numbers $\{3, 4, 5, 6, 7\}$ when the sample size is 2

The only possible different samples of size 2 which can be drawn from this set without replacement are:

$$(3,4), (3,5), (3,6), (3,7), (4,5),$$
$$(4,6), (4,7), (5,6), (5,7) \text{ and } (6,7)$$

The mean values of these samples form the following sampling distribution of means:

$$3.5, \ 4, \ 4.5, \ 5, \ 4.5, \ 5, \ 5.5, \ 5.5, \ 6 \text{ and } 6.5$$

The mean of the sampling distributions of means,

$$\mu_{\bar{x}} = \frac{\left(\begin{array}{c} 3.5 + 4 + 4.5 + 5 + 4.5 + 5 \\ + 5.5 + 5.5 + 6 + 6.5 \end{array}\right)}{10} = \frac{50}{10} = 5$$

The standard deviation of the sampling distribution of means,

$$\sigma_{\bar{x}} = \sqrt{\left[\frac{\begin{array}{c}(3.5-5)^2 + (4-5)^2 + (4.5-5)^2 \\ + (5-5)^2 + \cdots + (6.5-5)^2\end{array}}{10}\right]}$$

$$= \sqrt{\frac{7.5}{10}} = \pm 0.866$$

Thus, **the standard error of the means is 0.866**. The standard deviation of the population,

$$\sigma = \sqrt{\left[\frac{\begin{array}{c}(3-5)^2 + (4-5)^2 + (5-5)^2 \\ + (6-5)^2 + (7-5)\end{array}}{5}\right]}$$

$$= \sqrt{2} = \pm 1.414$$

But from Theorem 1:

$$\sigma_{\bar{x}} = \frac{\sigma}{\sqrt{N}} \sqrt{\left(\frac{N_p - N}{N_p - 1}\right)}$$

and substituting for N_p, N and σ in equation (1) gives:

$$\sigma_{\bar{x}} = \frac{\pm 1.414}{\sqrt{2}} \sqrt{\left(\frac{5-2}{5-1}\right)} = \sqrt{\frac{3}{4}} = \pm 0.866$$

as obtained by considering all samples from the population. Thus Theorem 1 is verified.

In Problem 1 above, it can be seen that the mean of the population,

$$\left(\frac{3+4+5+6+7}{5}\right)$$

is 5 and also that the mean of the sampling distribution of means, $\mu_{\bar{x}}$ is 5. This result is generalised in Theorem 2.

Theorem 2 'If all possible samples of size N are drawn from a population of size N_p and the mean value of the sampling distribution of means $\mu_{\bar{x}}$ is determined then

$$\mu_{\bar{x}} = \mu \qquad (3)$$

where μ is the mean value of the population.'

In practice, all possible samples of size N are not drawn from the population. However, if the sample size is large (usually taken as 30 or more), then the relationship between the mean of the sampling distribution of means and the mean of the population is very near to that shown in equation (3). Similarly, the relationship between the standard error of the means and the standard deviation of the population is very near to that shown in equation (2). Another important property of a sampling distribution is that when the sample size, N, is large, **the sampling distribution of means approximates to a normal distribution**, of mean value $\mu_{\bar{x}}$ and standard deviation $\sigma_{\bar{x}}$. This is true for all normally distributed populations and also for populations which are not normally distributed provided the population size is at least twice as large as the sample size. This property of normality of a sampling distribution is based on a special case of the 'central limit theorem', an important theorem relating to sampling theory. Because the sampling distribution of means and standard deviations is normally distributed, the table of the partial areas under the standardised normal curve (shown in Table 89.1 on page 966) can be used to determine the probabilities of a particular sample lying between, say, ± 1 standard deviation, and so on. This point is expanded in Problem 3.

Problem 2. The heights of 3000 people are normally distributed with a mean of 175 cm and a standard deviation of 8 cm. If random samples are taken of 40 people, predict the standard deviation and the mean of the sampling distribution of means if sampling is done (a) with replacement, and (b) without replacement

For the population: number of members, $N_p = 3000$; standard deviation, $\sigma = 8$ cm; mean, $\mu = 175$ cm.

For the samples: number in each sample, $N = 40$.

(a) When sampling is done **with replacement**, the total number of possible samples (two or more can be the same) is infinite. Hence, from equation (2) the **standard error of the mean (i.e. the standard deviation of the sampling distribution of means)**

$$\sigma_{\bar{x}} = \frac{\sigma}{\sqrt{N}} = \frac{8}{\sqrt{40}} = 1.265\,\text{cm}$$

From equation (3), **the mean of the sampling distribution**

$$\mu_{\bar{x}} = \mu = 175\,\text{cm}$$

(b) When sampling is done **without replacement**, the total number of possible samples is finite and hence equation (1) applies. Thus **the standard error of the means**

$$\sigma_{\bar{x}} = \frac{\sigma}{\sqrt{N}} \sqrt{\left(\frac{N_p - N}{N_p - 1}\right)}$$

$$= \frac{8}{\sqrt{40}} \sqrt{\left(\frac{3000 - 40}{3000 - 1}\right)}$$

$$= (1.265)(0.9935) = 1.257\,\text{cm}$$

As stated, following equation (3), provided the sample size is large, the mean of the sampling distribution of means is the same for both finite and infinite populations. Hence, from equation (3),

$$\mu_{\bar{x}} = 175\,\text{cm}$$

Problem 3. 1500 ingots of a metal have a mean mass of 6.5 kg and a standard deviation of 0.5 kg. Find the probability that a sample of 60 ingots chosen at random from the group, without replacement, will have a combined mass of (a) between 378 and 396 kg, and (b) more than 399 kg

---

For the population: numbers of members, $N_p = 1500$; standard deviation, $\sigma = 0.5\,\text{kg}$; mean $\mu = 6.5\,\text{kg}$.
For the sample: number in sample, $N = 60$
If many samples of 60 ingots had been drawn from the group, then the mean of the sampling distribution of means, $\mu_{\bar{x}}$ would be equal to the mean of the population. Also, the standard error of means is given by

$$\sigma_{\bar{x}} = \frac{\sigma}{\sqrt{N}} \sqrt{\left(\frac{N_p - N}{N_p - 1}\right)}$$

In addition, the sample distribution would have been approximately normal. Assume that the sample given in the problem is one of many samples. For many (theoretical) samples:

the mean of the sampling distribution of means, $\mu_{\bar{x}} = \mu = 6.5\,\text{kg}$.

Also, the standard error of the means,

$$\sigma_{\bar{x}} = \frac{\sigma}{\sqrt{N}} \sqrt{\left(\frac{N_p - N}{N_p - 1}\right)} = \frac{0.5}{\sqrt{60}} \sqrt{\left(\frac{1500 - 60}{1500 - 1}\right)}$$

$$= 0.0633\,\text{kg}$$

Thus, the sample under consideration is part of a normal distribution of mean value 6.5 kg and a standard error of the means of 0.0633 kg.

(a) If the combined mass of 60 ingots is between 378 and 396 kg, then the mean mass of each of the 60 ingots lies between $\dfrac{378}{60}$ and $\dfrac{396}{60}$ kg, i.e. between 6.3 kg and 6.6 kg.

Since the masses are normally distributed, it is possible to use the techniques of the normal distribution to determine the probability of the mean mass lying between 6.3 and 6.6 kg. The normal standard variate value, z, is given by

$$z = \frac{x - \bar{x}}{\sigma}$$

hence for the sampling distribution of means, this becomes,

$$z = \frac{x - \mu_{\bar{x}}}{\sigma_{\bar{x}}}$$

Thus, 6.3 kg corresponds to a z-value of $\dfrac{6.3 - 6.5}{0.0633} = -3.16$ standard deviations.

Similarly, 6.6 kg corresponds to a z-value of $\dfrac{6.6 - 6.5}{0.0633} = 1.58$ standard deviations.

Using Table 89.1 (page 966), the areas corresponding to these values of standard deviations are 0.4992 and 0.4430 respectively. Hence **the probability of the mean mass lying between 6.3 kg and 6.6 kg is 0.4992 + 0.4430 = 0.9422.** (This means that if 10 000 samples are drawn, 9422 of these samples will have a combined mass of between 378 and 396 kg.)

(b) If the combined mass of 60 ingots is 399 kg, the mean mass of each ingot is $\dfrac{399}{60}$, that is, 6.65 kg.

The z-value for 6.65 kg is $\dfrac{6.65 - 6.5}{0.0633}$, i.e. 2.37 standard deviations. From Table 89.1 (page 966), the area corresponding to this z-value is 0.4911. But this is the area between the ordinate $z = 0$ and ordinate $z = 2.37$. The 'more than' value required is the total area to the right of the $z = 0$ ordinate, less the value between $z = 0$ and $z = 2.37$, i.e. $0.5000 - 0.4911$. Thus, since areas are proportional to probabilities for the standardised normal curve, **the probability of the mean mass being more than 6.65 kg is 0.5000 − 0.4911**, i.e. **0.0089**. (This means that only 89 samples in 10 000, for example, will have a combined mass exceeding 399 kg.)

Now try the following Practice Exercise

Practice Exercise 344   Further problems on the sampling distribution of means (answers on page 1191)

1. The lengths of 1500 bolts are normally distributed with a mean of 22.4 cm and a standard deviation of 0.0438 cm. If 30 samples are drawn at random from this population, each sample being 36 bolts, determine the mean of the sampling distribution and standard error of the means when sampling is done with replacement.

2. Determine the standard error of the means in Problem 1, if sampling is done without replacement, correct to four decimal places.

3. A power punch produces 1800 washers per hour. The mean inside diameter of the washers is 1.70 cm and the standard deviation is 0.013 cm. Random samples of 20 washers are drawn every 5 minutes. Determine the mean of the sampling distribution of means and the standard error of the means for the one hour's output from the punch, (a) with replacement and (b) without replacement, correct to three significant figures.

A large batch of electric light bulbs have a mean time to failure of 800 hours and the standard deviation of the batch is 60 hours. Use this data and also Table 89.1 on page 966 to solve Problems 4 to 6.

4. If a random sample of 64 light bulbs is drawn from the batch, determine the probability that the mean time to failure will be less than 785 hours, correct to three decimal places.

5. Determine the probability that the mean time to failure of a random sample of 16 light bulbs will be between 790 hours and 810 hours, correct to three decimal places.

6. For a random sample of 64 light bulbs, determine the probability that the mean time to failure will exceed 820 hours, correct to two significant figures.

7. The contents of a consignment of 1200 tins of a product have a mean mass of 0.504 kg and a standard deviation of 92 g. Determine the probability that a random sample of 40 tins drawn from the consignment will have a combined mass of (a) less than 20.13 kg, (b) between 20.13 kg and 20.17 kg, and (c) more than 20.17 kg, correct to three significant figures.

92.4 The estimation of population parameters based on a large sample size

When a population is large, it is not practical to determine its mean and standard deviation by using the basic formulae for these parameters. In fact, when a population is infinite, it is impossible to determine these values. For large and infinite populations the values of the mean and standard deviation may be estimated by using the data obtained from samples drawn from the population.

Point and interval estimates

An estimate of a population parameter, such as mean or standard deviation, based on a single number is called a **point estimate**. An estimate of a population parameter given by two numbers between which the parameter may be considered to lie is called an **interval estimate**. Thus if an estimate is made of the length of an object and the result is quoted as 150 cm, this is a point estimate. If the result is quoted as 150 ± 10 cm, this is an interval estimate and indicates that the length lies between 140 and 160 cm. Generally, a point estimate does not indicate how close the value is to the true value of the quantity and should be accompanied by additional information on which its merits may be judged. A statement of the error or the precision of an estimate is often called its **reliability**. In statistics, when estimates are made of population parameters based on samples, usually interval estimates are used. The word estimate does not suggest that we adopt the approach 'let's guess that the mean value is about…', but rather that a value is carefully selected and the degree of confidence which can be placed in the estimate is given in addition.

Confidence intervals

It is stated in Section 92.3 that when samples are taken from a population, the mean values of these samples are approximately normally distributed, that is, the mean values forming the sampling distribution of means is approximately normally distributed. It is also true that if the standard deviations of each of the samples is found, then the standard deviations of all the samples are approximately normally distributed, that is, the standard deviations of the sampling distribution of standard deviations are approximately normally distributed. Parameters such as the mean or the standard deviation of a sampling distribution are called **sampling statistics**, S. Let μ_s be the mean value of a sampling statistic of the sampling distribution, that is, the mean value of the means of the samples or the mean value of the standard deviations of the samples. Also let σ_s be the standard deviation of a sampling statistic of the sampling distribution, that is, the standard deviation of the means of the samples or the standard deviation of the standard deviations of the samples. Because the sampling distribution of the means and of the standard deviations are normally distributed, it is possible to predict the probability of the sampling statistic lying in the intervals:

mean ± 1 standard deviation,

mean ± 2 standard deviations,

or mean ± 3 standard deviations,

by using tables of the partial areas under the standardised normal curve given in Table 89.1 on page 966. From this table, the area corresponding to a z-value of $+1$ standard deviation is 0.3413, thus the area corresponding to ± 1 standard deviation is 2×0.3413, that is, 0.6826. Thus the percentage probability of a sampling statistic lying between the mean ± 1 standard deviation is 68.26%. Similarly, the probability of a sampling statistic lying between the mean ± 2 standard deviations is 95.44% and of lying between the mean ± 3 standard deviations is 99.74%

The values 68.26%, 95.44% and 99.74% are called the **confidence levels** for estimating a sampling statistic. A confidence level of 68.26% is associated with two distinct values, these being, $S - $ (1 standard deviation), i.e. $S - \sigma_s$ and $S + $ (1 standard deviation), i.e. $S + \sigma_s$. These two values are called the **confidence limits** of the estimate and the distance between the confidence limits

is called the **confidence interval**. A confidence interval indicates the expectation or confidence of finding an estimate of the population statistic in that interval, based on a sampling statistic. The list in Table 92.1 is based on values given in Table 89.1, and gives some of the confidence levels used in practice and their associated z-values (some of the values given are based on interpolation). When the table is used in this context, z-values are usually indicated by 'z_c' and are called the **confidence coefficients**.

Table 92.1

Confidence level, %	Confidence coefficient, z_c
99	2.58
98	2.33
96	2.05
95	1.96
90	1.645
80	1.28
50	0.6745

Any other values of confidence levels and their associated confidence coefficients can be obtained using Table 89.1.

Problem 4. Determine the confidence coefficient corresponding to a confidence level of 98.5%

98.5% is equivalent to a per unit value of 0.9850. This indicates that the area under the standardised normal curve between $-z_c$ and $+z_c$, i.e. corresponding to $2z_c$, is 0.9850 of the total area. Hence the area between the mean value and z_c is $\dfrac{0.9850}{2}$ i.e. 0.4925 of the total area. The z-value corresponding to a partial area of 0.4925 is 2.43 standard deviations from Table 89.1. Thus, **the confidence coefficient corresponding to a confidence limit of 98.5% is 2.43**

(a) Estimating the mean of a population when the standard deviation of the population is known

When a sample is drawn from a large population whose standard deviation is known, the mean value of the sample, $\overline{x}$, can be determined. This mean value can be used to make an estimate of the mean value of the population, μ. When this is done, the estimated mean value of the population is given as lying between two values, that is, lying in the confidence interval between the confidence limits.

If a high level of confidence is required in the estimated value of μ, then the range of the confidence interval will be large. For example, if the required confidence level is 96%, then from Table 92.1 the confidence interval is from $-z_c$ to $+z_c$, that is, $2 \times 2.05 = 4.10$ standard deviations wide. Conversely, a low level of confidence has a narrow confidence interval and a confidence level of, say, 50%, has a confidence interval of 2×0.6745, that is 1.3490 standard deviations. The 68.26% confidence level for an estimate of the population mean is given by estimating that the population mean, μ, is equal to the same mean, $\overline{x}$, and then stating the confidence interval of the estimate. Since the 68.26% confidence level is associated with '± 1 standard deviation of the means of the sampling distribution', then the 68.26% confidence level for the estimate of the population mean is given by:

$$\overline{x} \pm 1\sigma_{\overline{x}}$$

In general, any particular confidence level can be obtained in the estimate, by using $\overline{x} \pm z_c\sigma_{\overline{x}}$, where z_c is the confidence coefficient corresponding to the particular confidence level required. Thus for a 96% confidence level, the confidence limits of the population mean are given by $\overline{x} \pm 2.05\sigma_{\overline{x}}$. Since only one sample has been drawn, the standard error of the means, $\sigma_{\overline{x}}$, is not known. However, it is shown in Section 92.3 that

$$\sigma_{\overline{x}} = \frac{\sigma}{\sqrt{N}}\sqrt{\left(\frac{N_p - N}{N_p - 1}\right)}$$

Thus, **the confidence limits of the mean of the population are**:

$$\overline{x} \pm \frac{z_c\sigma}{\sqrt{N}}\sqrt{\left(\frac{N_p - N}{N_p - 1}\right)} \qquad (4)$$

for a **finite population of size** N_p.
The **confidence limits for the mean of the population are**:

$$\overline{x} \pm \frac{z_c\sigma}{\sqrt{N}} \qquad (5)$$

for an **infinite population**.
Thus for a sample of size N and mean $\overline{x}$, drawn from an infinite population having a standard deviation of σ, the mean value of the population is estimated to be, for example,

$$\overline{x} \pm \frac{2.33\sigma}{\sqrt{N}}$$

for a confidence level of 98%. This indicates that the mean value of the population lies between

$$\overline{x} - \frac{2.33\sigma}{\sqrt{N}} \text{ and } \overline{x} + \frac{2.33\sigma}{\sqrt{N}}$$

with 98% confidence in this prediction.

Problem 5. It is found that the standard deviation of the diameters of rivets produced by a certain machine over a long period of time is 0.018 cm. The diameters of a random sample of 100 rivets produced by this machine in a day have a mean value of 0.476 cm. If the machine produces 2500 rivets per day, determine (a) the 90% confidence limits, and (b) the 97% confidence limits for an estimate of the mean diameter of all the rivets produced by the machine in a day

For the population:

standard deviation, $\sigma = 0.018$ cm

number in the population, $N_p = 2500$

For the sample:

number in the sample, $N = 100$

mean, $\overline{x} = 0.476$ cm

There is a finite population and the standard deviation of the population is known, hence expression (4) is used for determining an estimate of the confidence limits of the population mean, i.e.

$$\overline{x} \pm \frac{z_c\sigma}{\sqrt{N}}\sqrt{\left(\frac{N_p - N}{N_p - 1}\right)}$$

(a) For a 90% confidence level, the value of z_c, the confidence coefficient, is 1.645 from Table 92.1. Hence, the estimate of the confidence limits of the population mean, μ, is

$$0.476 \pm \left(\frac{(1.645)(0.018)}{\sqrt{100}}\right)\sqrt{\left(\frac{2500 - 100}{2500 - 1}\right)}$$

i.e. $0.476 \pm (0.00296)(0.9800)$

$$= 0.476 \pm 0.0029 \text{ cm}$$

Thus, **the 90% confidence limits are 0.473 cm and 0.479 cm**.
This indicates that if the mean diameter of a sample of 100 rivets is 0.476 cm, then it is predicted that the mean diameter of all the rivets will be

between 0.473 cm and 0.479 cm and this prediction is made with confidence that it will be correct nine times out of ten.

(b) For a 97% confidence level, the value of z_c has to be determined from a table of partial areas under the standardised normal curve given in Table 89.1, as it is not one of the values given in Table 92.1. The total area between ordinates drawn at $-z_c$ and $+z_c$ has to be 0.9700. Because the standardised normal curve is symmetrical, the area between $z_c = 0$ and z_c is $\dfrac{0.9700}{2}$, i.e. 0.4850. From Table 89.1 an area of 0.4850 corresponds to a z_c value of 2.17. Hence, the estimated value of the confidence limits of the population mean is between

$$\bar{x} \pm \frac{z_c \sigma}{\sqrt{N}} \sqrt{\left(\frac{N_p - N}{N_p - 1} \right)}$$

$$= 0.476 \pm \left(\frac{(2.17)(0.018)}{\sqrt{100}} \right) \sqrt{\left(\frac{2500 - 100}{2500 - 1} \right)}$$

$$= 0.476 \pm (0.0039)(0.9800)$$

$$= 0.476 \pm 0.0038$$

Thus, **the 97% confidence limits are 0.472 cm and 0.480 cm**.

It can be seen that the higher value of confidence level required in part (b) results in a larger confidence interval.

> **Problem 6.** The mean diameter of a long length of wire is to be determined. The diameter of the wire is measured in 25 places selected at random throughout its length and the mean of these values is 0.425 mm. If the standard deviation of the diameter of the wire is given by the manufacturers as 0.030 mm, determine (a) the 80% confidence interval of the estimated mean diameter of the wire, and (b) with what degree of confidence it can be said that 'the mean diameter is 0.425 ± 0.012 mm'

For the population: $\sigma = 0.030$ mm
For the sample: $N = 25, \bar{x} = 0.425$ mm
Since an infinite number of measurements can be obtained for the diameter of the wire, the population is infinite and the estimated value of the confidence interval of the population mean is given by expression (5).

(a) For an 80% confidence level, the value of z_c is obtained from Table 92.1 and is 1.28

The 80% confidence level estimate of the confidence interval of

$$\mu = \bar{x} \pm \frac{z_c \sigma}{\sqrt{N}} = 0.425 \pm \frac{(1.28)(0.030)}{\sqrt{25}}$$

$$= 0.425 \pm 0.0077 \text{ mm}$$

i.e. **the 80% confidence interval is from 0.417 mm to 0.433 mm**.

This indicates that the estimated mean diameter of the wire is between 0.417 mm and 0.433 mm and that this prediction is likely to be correct 80 times out of 100

(b) To determine the confidence level, the given data is equated to expression (5), giving

$$0.425 \pm 0.012 = \bar{x} \pm z_c \frac{\sigma}{\sqrt{N}}$$

But $\bar{x} = 0.425$ therefore

$$\pm z_c \frac{\sigma}{\sqrt{N}} = \pm 0.012$$

i.e. $z_c = \dfrac{0.012\sqrt{N}}{\sigma} = \pm \dfrac{(0.012)(5)}{0.030} = \pm 2$

Using Table 89.1 of partial areas under the standardised normal curve, a z_c value of 2 standard deviations corresponds to an area of 0.4772 between the mean value ($z_c = 0$) and $+2$ standard deviations. Because the standardised normal curve is symmetrical, the area between the mean and ± 2 standard deviations is 0.4772 × 2, i.e. 0.9544

Thus the confidence level corresponding to 0.425 ± 0.012 mm is 95.44%

(b) Estimating the mean and standard deviation of a population from sample data

The standard deviation of a large population is not known and, in this case, several samples are drawn from the population. The mean of the sampling distribution of means, $\mu_{\bar{x}}$ and the standard deviation of the sampling distribution of means (i.e. the standard error of the means), $\sigma_{\bar{x}}$, may be determined. The confidence limits of the mean value of the population, μ, are given by

$$\mu_{\bar{x}} \pm z_c \sigma_{\bar{x}} \tag{6}$$

where z_c is the confidence coefficient corresponding to the confidence level required.

To make an estimate of the standard deviation, σ, of a normally distributed population:

(i)   a sampling distribution of the standard deviations of the samples is formed, and

(ii)   the standard deviation of the sampling distribution is determined by using the basic standard deviation formula.

This standard deviation is called the standard error of the standard deviations and is usually signified by σ_s. If s is the standard deviation of a sample, then the confidence limits of the standard deviation of the population are given by:

$$s \pm z_c \sigma_s \qquad (7)$$

where z_c is the confidence coefficient corresponding to the required confidence level.

> **Problem 7.**   Several samples of 50 fuses selected at random from a large batch are tested when operating at a 10% overload current and the mean time of the sampling distribution before the fuses failed is 16.50 minutes. The standard error of the means is 1.4 minutes. Determine the estimated mean time to failure of the batch of fuses for a confidence level of 90%

For the sampling distribution: the mean, $\mu_{\bar{x}} = 16.50$, the standard error of the means, $\sigma_{\bar{x}} = 1.4$

The estimated mean of the population is based on sampling distribution data only and so expression (6) is used, i.e. the confidence limits of the estimated mean of the population are $\mu_{\bar{x}} \pm z_c \sigma_{\bar{x}}$

For a 90% confidence level, $z_c = 1.645$ (from Table 92.1), thus,

$$\mu_{\bar{x}} \pm z_c \sigma_{\bar{x}} = 16.50 \pm (1.645)(1.4)$$

$$= 16.50 \pm 2.30 \, m$$

Thus, **the 90% confidence level of the mean time to failure is from 14.20 minutes to 18.80 minutes**.

> **Problem 8.**   The sampling distribution of random samples of capacitors drawn from a large batch is found to have a standard error of the standard deviations of 0.12 µF. Determine the 92% confidence interval for the estimate of the standard deviation of the whole batch, if in a particular sample, the standard deviation is 0.60 µF. It can be assumed that the values of capacitance of the batch are normally distributed

For the sample: the standard deviation, $s = 0.60 \, \mu F$. For the sampling distribution: the standard error of the standard deviations,

$$\sigma_s = 0.12 \, \mu F$$

When the confidence level is 92%, then by using Table 89.1 of partial areas under the standardised normal curve,

$$\text{area} = \frac{0.9200}{2} = 0.4600,$$

giving z_c as ± 1.751 standard deviations (by interpolation).

Since the population is normally distributed, the confidence limits of the standard deviation of the population may be estimated by using expression (7), i.e. $s \pm z_c \sigma_s = 0.60 \pm (1.751)(0.12) = 0.60 \pm 0.21 \, \mu F$. Thus, **the 92% confidence interval for the estimate of the standard deviation for the batch is from 0.39 µF to 0.81 µF**.

Now try the following Practice Exercise

> **Practice Exercise 345   Further problems on the estimation of population parameters based on a large sample size (answers on page 1191)**
>
> 1.   Measurements are made on a random sample of 100 components drawn from a population of size 1546 and having a standard deviation of 2.93 mm. The mean measurement of the components in the sample is 67.45 mm. Determine the 95% and 99% confidence limits for an estimate of the mean of the population.
>
> 2.   The standard deviation of the masses of 500 blocks is 150 kg. A random sample of 40 blocks has a mean mass of 2.40 Mg.
>
>   (a)   Determine the 95% and 99% confidence intervals for estimating the mean mass of the remaining 460 blocks.
>
>   (b)   With what degree of confidence can it be said that the mean mass of the remaining 460 blocks is 2.40 ± 0.035 Mg?

3. In order to estimate the thermal expansion of a metal, measurements of the change of length for a known change of temperature are taken by a group of students. The sampling distribution of the results has a mean of $12.81 \times 10^{-4} \, \text{m}\,^\circ\text{C}^{-1}$ and a standard error of the means of $0.04 \times 10^{-4} \, \text{m}\,^\circ\text{C}^{-1}$. Determine the 95% confidence interval for an estimate of the true value of the thermal expansion of the metal, correct to two decimal places.

4. The standard deviation of the time to failure of an electronic component is estimated as 100 hours. Determine how large a sample of these components must be, in order to be 90% confident that the error in the estimated time to failure will not exceed (a) 20 hours and (b) 10 hours.

5. A sample of 60 slings of a certain diameter, used for lifting purposes, are tested to destruction (that is, loaded until they snapped). The mean and standard deviation of the breaking loads are 11.09 tonnes and 0.73 tonnes, respectively. Find the 95% confidence interval for the mean of the snapping loads of all the slings of this diameter produced by this company.

6. The time taken to assemble a servo-mechanism is measured for 40 operatives and the mean time is 14.63 minutes with a standard deviation of 2.45 minutes. Determine the maximum error in estimating the true mean time to assemble the servo-mechanism for all operatives, based on a 95% confidence level.

92.5   Estimating the mean of a population based on a small sample size

The methods used in Section 92.4 to estimate the population mean and standard deviation rely on a relatively large sample size, usually taken as 30 or more. This is because when the sample size is large the sampling distribution of a parameter is approximately normally distributed. When the sample size is small, usually taken as less than 30, the techniques used for estimating the population parameters in Section 92.4 become more and more inaccurate as the sample size becomes smaller, since the sampling distribution no longer approximates

to a normal distribution. Investigations were carried out into the effect of small sample sizes on the estimation theory by W. S. Gosset in the early twentieth century and, as a result of his work, tables are available which enable a realistic estimate to be made, when sample sizes are small. In these tables, the t-value is determined from the relationship

$$t = \frac{(\bar{x} - \mu)}{s} \sqrt{(N-1)}$$

where $\bar{x}$ is the mean value of a sample, μ is the mean value of the population from which the sample is drawn, s is the standard deviation of the sample and N is the number of independent observations in the sample. He published his findings under the pen name of 'Student', and these tables are often referred to as the **'Student's t distribution'**.

The confidence limits of the mean value of a population based on a small sample drawn at random from the population are given by

$$\bar{x} \pm \frac{t_c s}{\sqrt{(N-1)}} \tag{8}$$

In this estimate, t_c is called the confidence coefficient for small samples, analogous to z_c for large samples, s is the standard deviation of the sample, $\bar{x}$ is the mean value of the sample and N is the number of members in the sample. Table 92.2 is called 'percentile values for Student's t distribution'. The columns are headed t_p where p is equal to $0.995, 0.99, 0.975, \ldots, 0.55$. For a confidence level of, say, 95%, the column headed $t_{0.95}$ is selected, and so on. The rows are headed with the Greek letter 'nu', ν, and are numbered from 1 to 30 in steps of 1, together with the numbers 40, 60, 120 and ∞. These numbers represent a quantity called the **degrees of freedom**, which is defined as follows:

'*The sample number, N, minus the number of population parameters which must be estimated for the sample.*'

When determining the t-value, given by

$$t = \frac{(\bar{x} - \mu)}{s} \sqrt{(N-1)}$$

it is necessary to know the sample parameters $\bar{x}$ and s and the population parameter μ. $\bar{x}$ and s can be calculated for the sample, but usually an estimate has to be made of the population mean μ, based on the sample mean value. The number of degrees of freedom, ν, is given by the number of independent observations in the sample, N, minus the number of population parameters

Table 92.2 Percentile values (t_p) for Student's t distribution with ν degrees of freedom (shaded area $= p$)

ν	$t_{0.995}$	$t_{0.99}$	$t_{0.975}$	$t_{0.95}$	$t_{0.90}$	$t_{0.80}$	$t_{0.75}$	$t_{0.70}$	$t_{0.60}$	$t_{0.55}$
1	63.66	31.82	12.71	6.31	3.08	1.376	1.000	0.727	0.325	0.158
2	9.92	6.96	4.30	2.92	1.89	1.061	0.816	0.617	0.289	0.142
3	5.84	4.54	3.18	2.35	1.64	0.978	0.765	0.584	0.277	0.137
4	4.60	3.75	2.78	2.13	1.53	0.941	0.741	0.569	0.271	0.134
5	4.03	3.36	2.57	2.02	1.48	0.920	0.727	0.559	0.267	0.132
6	3.71	3.14	2.45	1.94	1.44	0.906	0.718	0.553	0.265	0.131
7	3.50	3.00	2.36	1.90	1.42	0.896	0.711	0.549	0.263	0.130
8	3.36	2.90	2.31	1.86	1.40	0.889	0.706	0.546	0.262	0.130
9	3.25	2.82	2.26	1.83	1.38	0.883	0.703	0.543	0.261	0.129
10	3.17	2.76	2.23	1.81	1.37	0.879	0.700	0.542	0.260	0.129
11	3.11	2.72	2.20	1.80	1.36	0.876	0.697	0.540	0.260	0.129
12	3.06	2.68	2.18	1.78	1.36	0.873	0.695	0.539	0.259	0.128
13	3.01	2.65	2.16	1.77	1.35	0.870	0.694	0.538	0.259	0.128
14	2.98	2.62	2.14	1.76	1.34	0.868	0.692	0.537	0.258	0.128
15	2.95	2.60	2.13	1.75	1.34	0.866	0.691	0.536	0.258	0.128
16	2.92	2.58	2.12	1.75	1.34	0.865	0.690	0.535	0.258	0.128
17	2.90	2.57	2.11	1.74	1.33	0.863	0.689	0.534	0.257	0.128
18	2.88	2.55	2.10	1.73	1.33	0.862	0.688	0.534	0.257	0.127
19	2.86	2.54	2.09	1.73	1.33	0.861	0.688	0.533	0.257	0.127
20	2.84	2.53	2.09	1.72	1.32	0.860	0.687	0.533	0.257	0.127
21	2.83	2.52	2.08	1.72	1.32	0.859	0.686	0.532	0.257	0.127
22	2.82	2.51	2.07	1.72	1.32	0.858	0.686	0.532	0.256	0.127
23	2.81	2.50	2.07	1.71	1.32	0.858	0.685	0.532	0.256	0.127
24	2.80	2.49	2.06	1.71	1.32	0.857	0.685	0.531	0.256	0.127
25	2.79	2.48	2.06	1.71	1.32	0.856	0.684	0.531	0.256	0.127
26	2.78	2.48	2.06	1.71	1.32	0.856	0.684	0.531	0.256	0.127
27	2.77	2.47	2.05	1.70	1.31	0.855	0.684	0.531	0.256	0.127
28	2.76	2.47	2.05	1.70	1.31	0.855	0.683	0.530	0.256	0.127
29	2.76	2.46	2.04	1.70	1.31	0.854	0.683	0.530	0.256	0.127
30	2.75	2.46	2.04	1.70	1.31	0.854	0.683	0.530	0.256	0.127
40	2.70	2.42	2.02	1.68	1.30	0.851	0.681	0.529	0.255	0.126
60	2.66	2.39	2.00	1.67	1.30	0.848	0.679	0.527	0.254	0.126
120	2.62	2.36	1.98	1.66	1.29	0.845	0.677	0.526	0.254	0.126
∞	2.58	2.33	1.96	1.645	1.28	0.842	0.674	0.524	0.253	0.126

which have to be estimated, k, i.e. $\nu = N - k$. For the equation

$$t = \frac{(\bar{x} - \mu)}{s}\sqrt{(N - 1)}$$

only μ has to be estimated, hence $k = 1$, and $\nu = N - 1$. When determining the mean of a population based on a small sample size, only one population parameter is to be estimated, and hence ν can always be taken as $(N - 1)$. The method used to estimate the mean of a population based on a small sample is shown in Problems 9 to 11.

Problem 9. A sample of 12 measurements of the diameter of a bar is made and the mean of the sample is 1.850 cm. The standard deviation of the sample is 0.16 mm. Determine (a) the 90% confidence limits and (b) the 70% confidence limits for an estimate of the actual diameter of the bar

---

For the sample: the sample size, $N = 12$; mean, $\bar{x} = 1.850$ cm; standard deviation $s = 0.16$ mm $= 0.016$ cm.

Since the sample number is less than 30, the small sample estimate as given in expression (8) must be used. The number of degrees of freedom, i.e. sample size minus the number of estimations of population parameters to be made, is $12 - 1$, i.e. 11

(a) The percentile value corresponding to a confidence coefficient value of $t_{0.90}$ and a degree of freedom value of $\nu = 11$ can be found by using Table 92.2, and is 1.36, that is, $t_c = 1.36$. The estimated value of the mean of the population is given by

$$\bar{x} \pm \frac{t_c s}{\sqrt{(N - 1)}} = 1.850 \pm \frac{(1.36)(0.016)}{\sqrt{11}}$$

$$= 1.850 \pm 0.0066 \text{ cm}$$

Thus, **the 90% confidence limits are 1.843 cm and 1.857 cm.**

This indicates that the actual diameter is likely to lie between 1.843 cm and 1.857 cm and that this prediction stands a 90% chance of being correct.

(b) The percentile value corresponding to $t_{0.70}$ and to $\nu = 11$ is obtained from Table 92.2, and is 0.540, that is, $t_c = 0.540$

The estimated value of the 70% confidence limits is given by:

$$\bar{x} \pm \frac{t_c s}{\sqrt{(N - 1)}} = 1.850 \pm \frac{(0.540)(0.016)}{\sqrt{11}}$$

$$= 1.850 \pm 0.0026 \text{ cm}$$

Thus, **the 70% confidence limits are 1.847 cm and 1.853 cm**, i.e. the actual diameter of the bar is between 1.847 cm and 1.853 cm and this result has a 70% probability of being correct.

Problem 10. A sample of 9 electric lamps are selected randomly from a large batch and are tested until they fail. The mean and standard deviations of the time to failure are 1210 hours and 26 hours, respectively. Determine the confidence level based on an estimated failure time of 1210 ± 6.5 hours

---

For the sample: sample size, $N = 9$; standard deviation, $s = 26$ hours; mean, $\bar{x} = 1210$ hours. The confidence limits are given by:

$$\bar{x} \pm \frac{t_c s}{\sqrt{(N - 1)}}$$

and these are equal to 1210 ± 6.5
Since $\bar{x} = 1210$ hours,

then

$$\pm \frac{t_c s}{\sqrt{(N - 1)}} = \pm 6.5$$

i.e.

$$t_c = \pm \frac{6.5\sqrt{(N - 1)}}{s} = \pm \frac{(6.5)\sqrt{8}}{26}$$

$$= \pm 0.707$$

From Table 92.2, a t_c value of 0.707, having a ν value of $N - 1$, i.e. 8, gives a t_p value of $t_{0.75}$
Hence, **the confidence level of an estimated failure time of 1210 ± 6.5 hours is 75%**, i.e. it is likely that 75% of all of the lamps will fail between 1203.5 and 1216.5 hours.

Problem 11. The specific resistance of some copper wire of nominal diameter 1 mm is estimated by determining the resistance of 6 samples of the wire. The resistance values found (in ohms per metre) were:

2.16, 2.14, 2.17, 2.15, 2.16 and 2.18

Determine the 95% confidence interval for the true specific resistance of the wire

---

For the sample: sample size, $N = 6$, and mean,

$$\bar{x} = \frac{2.16 + 2.14 + 2.17 + 2.15 + 2.16 + 2.18}{6}$$

$$= 2.16 \, \Omega \, \text{m}^{-1}$$

standard deviation,

$$s = \sqrt{\left\{\frac{\begin{array}{l}(2.16-2.16)^2+(2.14-2.16)^2\\+(2.17-2.16)^2+(2.15-2.16)^2\\+(2.16-2.16)^2+(2.18-2.16)^2\end{array}}{6}\right\}}$$

$$= \sqrt{\frac{0.001}{6}} = 0.0129\,\Omega\,\mathrm{m}^{-1}$$

The percentile value corresponding to a confidence coefficient value of $t_{0.95}$ and a degree of freedom value of $N-1$, i.e. $6-1=5$ is 2.02 from Table 92.2. The estimated value of the 95% confidence limits is given by:

$$\bar{x} \pm \frac{t_c s}{\sqrt{(N-1)}} = 2.16 \pm \frac{(2.02)(0.0129)}{\sqrt{5}}$$

$$= 2.16 \pm 0.01165\,\Omega\,\mathrm{m}^{-1}$$

Thus, **the 95% confidence limits are 2.148 $\Omega\,\mathrm{m}^{-1}$ and 2.172 $\Omega\,\mathrm{m}^{-1}$**, which indicates that there is a 95% chance that the true specific resistance of the wire lies between $2.148\,\Omega\,\mathrm{m}^{-1}$ and $2.172\,\Omega\,\mathrm{m}^{-1}$

Now try the following Practice Exercise

Practice Exercise 346 Further problems on estimating the mean of a population based on a small sample size (answers on page 1191)

1. The value of the ultimate tensile strength of a material is determined by measurements on 10 samples of the materials. The mean and standard deviation of the results are found to be 5.17 MPa and 0.06 MPa, respectively. Determine the 95% confidence interval for the mean of the ultimate tensile strength of the material.

2. Use the data given in Problem 1 above to determine the 97.5% confidence interval for the mean of the ultimate tensile strength of the material.

3. The specific resistance of a reel of German silver wire of nominal diameter 0.5 mm is estimated by determining the resistance of 7 samples of the wire. These were found to have resistance values (in ohms per metre) of:

    1.12, 1.15, 1.10, 1.14, 1.15, 1.10 and 1.11

    Determine the 99% confidence interval for the true specific resistance of the reel of wire.

4. In determining the melting point of a metal, five determinations of the melting point are made. The mean and standard deviation of the five results are 132.27°C and 0.742°C. Calculate the confidence with which the prediction 'the melting point of the metal is between 131.48°C and 133.06°C' can be made.

For fully worked solutions to each of the problems in Practice Exercises 344 to 346 in this chapter, go to the website:
www.routledge.com/cw/bird

Section L

Chapter 93

Significance testing

Why it is important to understand: Significance testing

In statistical testing, a result is called statistically significant if it is unlikely to have occurred by chance, and hence provides enough evidence to reject the hypothesis of 'no effect'. The tests involve comparing the observed values with theoretical values. The tests establish whether there is a relationship between the variables, or whether pure chance could produce the observed results. For most scientific research, a statistical significance test eliminates the possibility that the results arose by chance, allowing a rejection of the null hypothesis. This chapter introduces the principles of significance testing.

At the end of this chapter, you should be able to:

- understand hypotheses
- appreciate type I and type II errors
- calculate type I and type II errors using binomial and Poisson approximations
- appreciate significance tests for population means
- determine hypotheses using significance testing
- compare two sample means given a level of significance

93.1 Hypotheses

Industrial applications of statistics are often concerned with making decisions about populations and population parameters. For example, decisions about which is the better of two processes or decisions about whether to discontinue production on a particular machine because it is producing an economically unacceptable number of defective components are often based on deciding the mean or standard deviation of a population, calculated using sample data drawn from the population. In reaching these decisions, certain assumptions are made, which may or may not be true. A **statistical hypothesis** is an assumption about the distribution of a random variable, and is usually concerned with statements about probability distributions of populations.

For example, in order to decide whether a dice is fair, that is, unbiased, a hypothesis can be made that a particular number, say 5, should occur with a probability of one in six, since there are six numbers on a dice. Such a hypothesis is called a **null hypothesis** and is an initial statement. The symbol H_0 is used to indicate a null hypothesis. Thus, if p is the probability of throwing a 5, then H_0: $p = \frac{1}{6}$ means, 'the null hypothesis that the probability of throwing a 5 is $\frac{1}{6}$'. Any hypothesis which differs from a given hypothesis is called an **alternative hypothesis**, and is indicated by the symbol H_1. Thus, if after many trials, it is found that the dice is biased and that a 5 only occurs, on average, one in every seven throws, then several alternative hypotheses may be formulated. For example: H_1: $p = \frac{1}{7}$ or H_1: $p < \frac{1}{6}$ or H_1: $p > \frac{1}{8}$ or H_1: $p \neq \frac{1}{6}$ are all possible alternative hypotheses to the null hypothesis that $p = \frac{1}{6}$

Hypotheses may also be used when comparisons are being made. If we wish to compare, say, the strength of two metals, a null hypothesis may be formulated that there is **no difference** between the strengths of the two metals. If the forces that the two metals can withstand are F_1 and F_2, then the null hypothesis is $H_0: F_1 = F_2$. If it is found that the null hypothesis has to be rejected, that is, that the strengths of the two metals are not the same, then the alternative hypotheses could be of several forms. For example, $H_1: F_1 > F_2$ or $H_1: F_2 > F_1$ or $H_1: F_1 \neq F_2$. These are all alternative hypotheses to the original null hypothesis.

93.2   Type I and type II errors

To illustrate what is meant by type I and type II errors, let us consider an automatic machine producing, say, small bolts. These are stamped out of a length of metal and various faults may occur. For example, the heads or the threads may be incorrectly formed, the length might be incorrect, and so on. Assume that, say, 3 bolts out of every 100 produced are defective in some way. If a sample of 200 bolts is drawn at random, then the manufacturer might be satisfied that his defect rate is still 3% provided there are 6 defective bolts in the sample. Also, the manufacturer might be satisfied that his defect rate is 3% or less, provided that there are 6 or fewer bolts defective in the sample. He might then formulate the following hypotheses:

$$H_0: p = 0.03 \text{ (the null hypothesis that the defect rate is 3\%)}$$

The null hypothesis is that the defect rate is 3%. Suppose that he also makes a decision that should the defect rate rise to 5% or more, he will take some action. Then the alternative hypothesis is:

$$H_1: p \geq 0.05 \text{ (the alternative hypothesis that the defect rate is equal to or greater than 5\%)}$$

The manufacturer's decisions, which are related to these hypotheses, might well be:

(i)   a null hypothesis that the defect rate is 3%, on the assumption that the associated number of defective bolts is insufficient to endanger his firm's good name;

(ii)   if the null hypothesis is rejected and the defect rate rises to 5% or over, stop the machine and adjust or renew parts as necessary; since the machine is not then producing bolts, this will reduce his profit.

These decisions may seem logical at first sight, but by applying the statistical concepts introduced in previous

chapters it can be shown that the manufacturer is not necessarily making very sound decisions. This is shown as follows.

When drawing a random sample of 200 bolts from the machine with a defect rate of 3%, by the laws of probability, some samples will contain no defective bolts, some samples will contain one defective bolt, and so on.

A **binomial distribution** can be used to determine the probabilities of getting 0, 1, 2, ..., 9 defective bolts in the sample. Thus the probability of getting 10 or more defective bolts in a sample, **even with a 3% defect rate**, is given by: $1 -$ (the sum of probabilities of getting 0, 1, 2, ..., 9 defective bolts). This is an extremely large calculation, given by:

$$1 - \left(0.97^{200} + 200 \times 0.97^{199} \times 0.03 \right.$$

$$\left. + \frac{200 \times 199}{2} \times 0.97^{198} \times 0.03^2 \text{ to 10 terms} \right)$$

An alternative way of calculating the required probability is to use the **normal approximation** to the binomial distribution. This may be stated as follows:

'If the probability of a defective item is p and a non-defective item is q, then if a sample of N items is drawn at random from a large population, provided both Np and Nq are greater than 5, the binomial distribution approximates to a normal distribution of mean Np and standard deviation $\sqrt{(Npq)}$'

The defect rate is 3%, thus $p = 0.03$. Since $q = 1 - p$, $q = 0.97$. Sample size $N = 200$. Since Np and Nq are greater than 5, a normal approximation to the binomial distribution can be used.

The mean of the normal distribution,

$$\bar{x} = Np = 200 \times 0.03 = 6$$

The standard deviation of the normal distribution

$$\sigma = \sqrt{(Npq)}$$

$$= \sqrt{[(200)(0.03)(0.97)]} = 2.41$$

The normal standard variate for 10 bolts is

$$z = \frac{\text{variate} - \text{mean}}{\text{standard deviation}}$$

$$= \frac{10 - 6}{2.41} = 1.66$$

Section L

Table 89.1 on page 966 is used to determine the area between the mean and a z-value of 1.66, and is 0.4515. The probability of having 10 or more defective bolts is the total area under the standardised normal curve minus the area to the left of the $z=1.66$ ordinate, i.e. $1-(0.5+0.4515)$, i.e. $1-0.9515=0.0485 \approx 5\%$. Thus the probability of getting 10 or more defective bolts in a sample of 200 bolts, **even though the defect rate is still 3%**, is 5%. It follows that as a result of the manufacturer's decisions, for 5 times in every 100 the number of defects in the sample will exceed 10, the alternative hypothesis will be adopted and the machine will be stopped (and profit lost) unnecessarily. In general terms:

'A hypothesis has been rejected when it should have been accepted.'

When this occurs, it is called a **type I error**, and, in this example, the probability of a type I error is 5%. Thus a type I error is rejecting the null hypothesis when it is in fact true.

Assume now that the defect rate has risen to 5%, i.e. the expectancy of a defective bolt is now 10. A second error resulting from this decisions occurs, due to the probability of getting fewer than 10 defective bolts in a random sample, even though the defect rate has risen to 5%. Using the normal approximation to a binomial distribution: $N=200$, $p=0.05$, $q=0.95$. Np and Nq are greater than 5, hence a normal approximation to a binomial distribution is a satisfactory method. The normal distribution has:

mean, $\bar{x}=Np=(200)(0.05)=10$

standard deviation,

$$\sigma = \sqrt{(Npq)}$$

$$= \sqrt{[(200)(0.05)(0.95)]} = 3.08$$

The normal standard variate for 9 defective bolts,

$$z = \frac{\text{variate} - \text{mean}}{\text{standard deviation}}$$

$$= \frac{9-10}{3.08} = -0.32$$

Using Table 89.1 of partial areas under the standardised normal curve given on page 966, a z-value of -0.32 corresponds to an area between the mean and the ordinate at $z=-0.32$ to 0.1255. Thus, the probability of there being 9 or fewer defective bolts in the sample is

given by the area to the left of the $z=0.32$ ordinate, i.e. $0.5000-0.1255$, that is, 0.3745. Thus, the probability of getting 9 or fewer defective bolts in a sample of 200 bolts, **even though the defect rate has risen to 5%**, is 37%. It follows that as a result of the manufacturer's decisions, for 37 samples in every 100, the machine will be left running even though the defect rate has risen to 5%. In general terms:

'A hypothesis has been accepted when it should have been rejected.'

When this occurs, it is called a **type II error**, and, in this example, the probability of a type II error is 37%. Thus a type II error is not rejecting the null hypothesis when in fact the alternative hypothesis is true.

Tests of hypotheses and rules of decisions should be designed to minimise the errors of decision. This is achieved largely by trial and error for a particular set of circumstances. Type I errors can be reduced by increasing the number of defective items allowable in a sample, but this is at the expense of allowing a larger percentage of defective items to leave the factory, increasing the criticism from customers. Type II errors can be reduced by increasing the percentage defect rate in the alternative hypothesis. If a higher percentage defect rate is given in the alternative hypothesis, the type II errors are reduced very effectively, as shown in the second of the two tables below, relating the decision rule to the magnitude of the type II errors. Some examples of the magnitude of type I errors are given below, for a sample of 1000 components being produced by a machine with a mean defect rate of 5%.

Decision rule Stop production if the number of defective components is equal to or greater than:	Type I error (%)
52	38.6
56	19.2
60	7.35
64	2.12
68	0.45

The magnitude of the type II errors for the output of the same machine, again based on a random sample of 1000 components and a mean defect rate of 5%, is given below.

Decision rule	Type II error
Stop production when the number of defective components is 60, when the defect rate is (%):	(%)
5.5	75.49
7	10.75
8.5	0.23
10	0.00

When testing a hypothesis, the largest value of probability which is acceptable for a type I error is called the **level of significance** of the test. The level of significance is indicated by the symbol α (alpha) and the levels commonly adopted are 0.1, 0.05, 0.01, 0.005 and 0.002. A level of significance of, say, 0.05 means that 5 times in 100 the hypothesis has been rejected when it should have been accepted.

In significance tests, the following terminology is frequently adopted:

(i) if the level of significance is 0.01 or less, i.e. the confidence level is 99% or more, the results are considered to be **highly significant**, i.e. the results are considered likely to be correct,

(ii) if the level of significance is 0.05 or between 0.05 and 0.01, i.e. the confidence level is 95% or between 95% and 99%, the results are considered to be **probably significant**, i.e. the results are probably correct,

(iii) if the level of significance is greater than 0.05, i.e. the confidence level is less than 95%, the results are considered to be **not significant**, that is, there are doubts about the correctness of the results obtained.

This terminology indicates that the use of a level of significance of 0.05 for 'probably significant' is, in effect, a rule of thumb. Situations can arise when the probability changes with the nature of the test being done and the use being made of the results.

The example of a machine producing bolts, used to illustrate type I and type II errors, is based on a single random sample being drawn from the output of the machine. In practice, sampling is a continuous process and using the data obtained from several samples, sampling distributions are formed. From the concepts introduced in Chapter 92, the means and standard deviations of samples are normally distributed, thus for a particular sample its mean and standard deviation are part of a

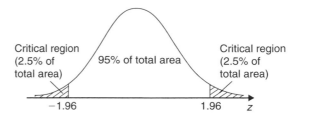

Figure 93.1

normal distribution. For a set of results to be probably significant a confidence level of 95% is required for a particular hypothesis being probably correct. This is equivalent to the hypothesis being rejected when the level of significance is greater than 0.05. For this to occur, the z-value of the mean of the samples will lie between -1.96 and $+1.96$ (since the area under the standardised normal distribution curve between these z-values is 95%). The shaded area in Figure 93.1 is based on results which are probably significant, i.e. having a level of significance of 0.05, and represents the probability of rejecting a hypothesis when it is correct. The z-values of less than -1.96 and more than 1.96 are called **critical values** and the shaded areas in Figure 93.1 are called the **critical regions** or regions for which the hypothesis is rejected. Having formulated hypotheses, the rules of decision and a level of significance, the magnitude of the type I error is given. Nothing can now be done about type II errors and in most cases they are accepted in the hope that they are not too large.

When critical regions occur on both sides of the mean of a normal distribution, as shown in Figure 93.1, they are as a result of **two-tailed** or **two-sided tests**. In such tests, consideration has to be given to values on both sides of the mean. For example, if it is required to show that the percentage of metal, p, in a particular alloy is $x\%$, then a two-tailed test is used, since the null hypothesis is incorrect if the percentage of metal is either less than x or more than x. The hypothesis is then of the form:

$$H_0: p = x\% \quad H_1: p \neq x\%$$

However, for the machine producing bolts, the manufacturer's decision is not affected by the fact that a sample contains say 1 or 2 defective bolts. He is only concerned with the sample containing, say, 10 or more defective bolts. Thus a 'tail' on the left of the mean is not required. In this case a **one-tailed test** or a **one-sided test** is really required. If the defect rate is, say, d and the per unit values economically acceptable to the manufacturer are u_1 and u_2, where u_1 is an acceptable defect rate and u_2 is the maximum acceptable defect rate, then the hypotheses in this case are of the form:

$$H_0: d = u_1 \quad H_1: d > u_2$$

and the critical region lies on the right-hand side of the mean, as shown in Figure 93.2(a). A one-tailed test can have its critical region either on the right-hand side or on the left-hand side of the mean. For example, if lamps are being tested and the manufacturer is only interested in those lamps whose life length does not meet a certain minimum requirement, then the hypotheses are of the form:

$$H_0: l = h \quad H_1: l < h$$

where l is the life length and h is the number of hours to failure. In this case the critical region lies on the left-hand side of the mean, as shown in Figure 93.2(b).

The z-values for various levels of confidence are given in Table 92.1 on page 988. The corresponding levels of significance (a confidence level of 95% is equivalent to a level of significance of 0.05 in a two-tailed test) and their z-values for both one-tailed and two-tailed tests are given in Table 93.1. It can be seen that two values of z are given for one-tailed tests, the negative value for critical regions lying to the left of the mean and a positive value for critical regions lying to the right of the mean.

The problem of the machine producing 3% defective bolts can now be reconsidered from a significance testing point of view. A random sample of 200 bolts is drawn, and the manufacturer is interested in a change in the defect rate in a specified direction (i.e. an increase), hence the hypotheses tests are designed accordingly. If the manufacturer is willing to accept a defect rate of 3%, but wants adjustments made to the machine if the defect rate exceeds 3%, then the hypotheses will be:

(i) a null hypothesis such that the defect rate, p, is equal to 3%,

   i.e. $H_0: p = 0.03$, and

(ii) an alternative hypothesis such that the defect rate is greater than 3%,

   i.e. $H_1: p > 0.03$

The first rule of decision is as follows: let the level of significance, α, be 0.05; this will limit the type I error, that is, the error due to rejecting the hypothesis when it should be accepted, to 5%, which means that the results are probably correct. The second rule of decision is to decide the number of defective bolts in a sample for which the machine is stopped and adjustments are made. For a one-tailed test, a level of significance of 0.05 and the critical region lying to the right of the mean of the standardised normal distribution, the z-value from Table 93.1 is 1.645. If the defect rate p is 0.03%, the mean of the normal distribution is given by $Np = 200 \times 0.03 = 6$ and the standard deviation is $\sqrt{(Npq)} = \sqrt{(200 \times 0.03 \times 0.97)} = 2.41$, using the normal approximation to a binomial distribution. Since the z-value is $\dfrac{\text{variate} - \text{mean}}{\text{standard deviation}}$, then $1.645 = \dfrac{\text{variate} - 6}{2.41}$ giving a variate value of 9.96. This variate is the number of defective bolts in a sample such that when this number is reached or exceeded the null hypothesis is rejected. For 95 times out of 100 this will

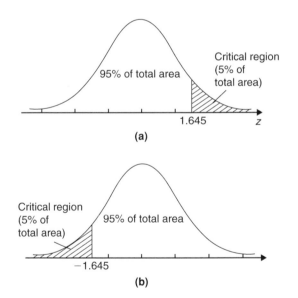

95% of total area

Critical region (5% of total area)

1.645 z

(a)

Critical region (5% of total area)

95% of total area

−1.645

(b)

Figure 93.2

Table 93.1

Level of significance, α	0.1	0.05	0.01	0.005	0.002
z-value, one-tailed test	$\begin{cases} -1.28 \\ \text{or } 1.28 \end{cases}$	$\begin{array}{c} -1.645 \\ \text{or } 1.645 \end{array}$	$\begin{array}{c} -2.33 \\ \text{or } 2.33 \end{array}$	$\begin{array}{c} -2.58 \\ \text{or } 2.58 \end{array}$	$\begin{array}{c} -2.88 \\ \text{or } 2.88 \end{array}$
z-value, two-tailed test	$\begin{cases} -1.645 \\ \text{and } 1.645 \end{cases}$	$\begin{array}{c} -1.96 \\ \text{and } 1.96 \end{array}$	$\begin{array}{c} -2.58 \\ \text{and } 2.58 \end{array}$	$\begin{array}{c} -2.81 \\ \text{and } 2.81 \end{array}$	$\begin{array}{c} -3.08 \\ \text{and } 3.08 \end{array}$

be the correct thing to do. The second rule of decision will thus be 'reject H_0 if the number of defective bolts in a random sample is equal to or exceeds 10, otherwise accept H_0'. That is, the machine is adjusted when the number of defective bolts in a random sample reaches 10 and this will be the correct decision for 95% of the time. The type II error can now be calculated, but there is little point, since having fixed the sample number and the level of significance, there is nothing that can be done about it.

A two-tailed test is used when it is required to test for changes in an **unspecified direction**. For example, if the manufacturer of bolts, used in the previous example, is inspecting the diameter of the bolts, he will want to know whether the diameters are too large or too small. Let the nominal diameter of the bolts be 2 mm. In this case the hypotheses will be:

$$H_0: d = 2.00 \text{ mm} \quad H_1: d \neq 2.00 \text{ mm}$$

where d is the mean diameter of the bolts. His first decision is to set the level of significance, to limit his type I error. A two-tailed test is used, since adjustments must be made to the machine if the diameter does not lie within specified limits. The method of using such a significance test is given in Section 93.3.

When determining the magnitude of type I and type II errors, it is often possible to reduce the amount of work involved by using a normal or a Poisson distribution rather than binomial distribution. A summary of the criteria for the use of these distributions and their form is given below, for a sample of size N, a probability of defective components p and a probability of non-defective components q.

Binomial distribution

From Chapter 88, the probability of having $0, 1, 2, 3, \ldots$ defective components in a random sample of N components is given by the successive terms of the expansion of $(q + p)^N$, taken from the left. Thus:

Number of defective components	Probability
0	q^N
1	$Nq^{N-1}p$
2	$\dfrac{N(N-1)}{2!}q^{N-2}p^2$
3	$\dfrac{N(N-1)(N-2)}{3!}q^{N-3}p^3 \ldots$

Poisson approximation to a binomial distribution

When $N \geq 50$ and $Np < 5$, the Poisson distribution is approximately the same as the binomial distribution. In the Poisson distribution, the expectation $\lambda = Np$ and from Chapter 88, the probability of $0, 1, 2, 3, \ldots$ defective components in a random sample of N components is given by the successive terms of

$$e^{-\lambda}\left(1 + \lambda + \frac{\lambda^2}{2!} + \frac{\lambda^3}{3!} + \cdots\right)$$

taken from the left. Thus,

Number of defective components	0	1	2	3
Probability	$e^{-\lambda}$	$\lambda e^{-\lambda}$	$\dfrac{\lambda^2 e^{-\lambda}}{2!}$	$\dfrac{\lambda^3 e^{-\lambda}}{3!}$

Normal approximation to a binomial distribution

When both Np and Nq are greater than 5, the normal distribution is approximately the same as the binomial distribution. The normal distribution has a mean of Np and a standard deviation of $\sqrt{(Npq)}$

Problem 1. Wood screws are produced by an automatic machine and it is found over a period of time that 7% of all the screws produced are defective. Random samples of 80 screws are drawn periodically from the output of the machine. If a decision is made that production continues until a sample contains more than 7 defective screws, determine the type I error based on this decision for a defect rate of 7%. Also determine the magnitude of the type II error when the defect rate has risen to 10%

$$N = 80, \quad p = 0.07, \quad q = 0.93$$

Since both Np and Nq are greater than 5, a normal approximation to the binomial distribution is used.

Mean of the normal distribution,

$$Np = 80 \times 0.07 = 5.6$$

Standard deviation of the normal distribution,

$$\sqrt{(Npq)} = \sqrt{(80 \times 0.07 \times 0.93)} = 2.28$$

A **type I error** is the probability of rejecting a hypothesis when it is correct, hence, the type I error in this

Problem is the probability of stopping the machine, that is, the probability of getting more than 7 defective screws in a sample, even though the defect rate is still 7%. The z-value corresponding to 7 defective screws is given by:

$$\frac{\text{variate} - \text{mean}}{\text{standard deviation}} = \frac{7 - 5.6}{2.28} = 0.61$$

Using Table 89.1 of partial areas under the standardised normal curve given on page 966, the area between the mean and a z-value of 0.61 is 0.2291. Thus, the probability of more than 7 defective screws is the area to the right of the z ordinate at 0.61, that is,

[total area − (area to the left of mean

+ area between mean and $z = 0.61$)]

i.e. $1 - (0.5 + 0.2291)$. This gives a probability of 0.2709. It is usual to express type I errors as a percentage, giving

$$\textbf{type I error} = \textbf{27.1\%}$$

A **type II error** is the probability of accepting a hypothesis when it should be rejected. The type II error in this Problem is the probability of a sample containing fewer than 7 defective screws, even though the defect rate has risen to 10%. The values are now:

$$N = 80, \quad p = 0.1, \quad q = 0.9$$

As Np and Nq are both greater than 5, a normal approximation to a binomial distribution is used, in which the mean Np is $80 \times 0.1 = 8$ and the standard deviation $\sqrt{(Npq)} = \sqrt{(80 \times 0.1 \times 0.9)} = 2.68$

The z-value for a variate of 7 defective screws is $\frac{7 - 8}{2.68} = -0.37$

Using Table 89.1 of partial areas given on page 966, the area between the mean and $z = -0.37$ is 0.1443. Hence, the probability of getting fewer than 7 defective screws, even though the defect rate is 10% is (area to the left of mean − area between mean and a z-value of −0.37), i.e. $0.5 - 0.1443 = 0.3557$. It is usual to express type II errors as a percentage, giving

$$\textbf{type II error} = \textbf{35.6\%}$$

Problem 2. The sample size in Problem 1 is reduced to 50. Determine the type I error if the defect rate remains at 7% and the type II error when the defect rate rises to 9%. The decision is now to stop the machine for adjustment if a sample contains 4 or more defective screws

$$N = 50, \quad p = 0.07$$

When $N \geq 50$ and $Np < 5$, the Poisson approximation to a binomial distribution is used. The expectation $\lambda = Np = 3.5$. The probabilities of $0, 1, 2, 3, \ldots$ defective screws are given by $e^{-\lambda}, \lambda e^{-\lambda}, \dfrac{\lambda^2 e^{-\lambda}}{2!}, \dfrac{\lambda^3 e^{-\lambda}}{3!}, \ldots$ Thus,

probability of a sample containing no defective screws,	$e^{-\lambda} = 0.0302$
probability of a sample containing 1 defective screw,	$\lambda e^{-\lambda} = 0.1057$
probability of a sample containing 2 defective screws,	$\dfrac{\lambda^2 e^{-\lambda}}{2!} = 0.1850$
probability of a sample containing 3 defective screws,	$\dfrac{\lambda^3 e^{-\lambda}}{3} = 0.2158$
probability of a sample containing 0, 1, 2, or 3 defective screws is	0.5367

Hence, the probability of a sample containing 4 or more defective screws is $1 - 0.5367 = 0.4633$. Thus the **type I error**, that is, rejecting the hypothesis when it should be accepted or stopping the machine for adjustment when it should continue running, is **46.3%**

When the defect rate has risen to 9%, $p = 0.09$ and $Np = \lambda = 4.5$. Since $N \geq 50$ and $Np < 5$, the Poisson approximation to a binomial distribution can still be used. Thus,

probability of a sample containing no defective screws,	$e^{-\lambda} = 0.0111$
probability of a sample containing 1 defective screw,	$\lambda e^{-\lambda} = 0.0500$
probability of a sample containing 2 defective screws,	$\dfrac{\lambda^2 e^{-\lambda}}{2!} = 0.1125$
probability of a sample containing 3 defective screws,	$\dfrac{\lambda^3 e^{-\lambda}}{3!} = 0.1687$
probability of a sample containing 0, 1, 2, or 3 defective screws is	0.3423

That is, the probability of a sample containing fewer than 4 defective screws is 0.3423. Thus, the **type II error**, that is, accepting the hypothesis when it should have been rejected or leaving the machine running when it should be stopped, is **34.2%**

> **Problem 3.**   The sample size in Problem 1 is now reduced to 25. Determine the type I error if the defect rate remains at 7%, and the type II error when the defect rate rises to 10%. The decision is now to stop the machine for adjustment if a sample contains 3 or more defective screws

$$N = 25, \quad p = 0.07, \quad q = 0.93$$

The criteria for a normal approximation to a binomial distribution and for a Poisson approximation to a binomial distribution are not met, hence the binomial distribution is applied.

Probability of no defective screws in a sample,

$$q^N = 0.93^{25} \qquad\qquad = 0.1630$$

Probability of 1 defective screw in a sample,

$$Nq^{N-1}p = 25 \times 0.93^{24} \times 0.07 \qquad = 0.3066$$

Probability of 2 defective screws in a sample,

$$\frac{N(N-1)}{2} q^{N-2} p^2$$

$$= \frac{25 \times 24}{2} \times 0.93^{23} \times 0.07^2 \qquad = 0.2770$$

Probability of 0, 1, or 2 defective screws in a sample $\qquad = 0.7466$

Thus, the probability of a type I error, i.e. stopping the machine even though the defect rate is still 7%, is $1 - 0.7466 = 0.2534$. Hence, the **type I error is 25.3%**

When the defect rate has risen to 10%:

$$N = 25, \quad p = 0.1, \quad q = 0.9$$

Probability of no defective screws in a sample,

$$q^N = 0.9^{25} \qquad\qquad = 0.0718$$

Probability of 1 defective screw in a sample,

$$Nq^{N-1}p = 25 \times 0.9^{24} \times 0.1 \qquad = 0.1994$$

Probability of 2 defective screws in a sample,

$$\frac{N(N-1)}{2} q^{N-2} p^2$$

$$= \frac{25 \times 24}{2} \times 0.9^{23} \times 0.1^2 \qquad = 0.2659$$

Probability of 0, 1, or 2 defective screws in a sample $\qquad = 0.5371$

That is, the probability of a **type II error**, i.e. leaving the machine running even though the defect rate has risen to 10%, **is 53.7%**

Now try the following Practice Exercise

> **Practice Exercise 347   Further problems on type I and type II errors (answers on page 1191)**
>
> Problems 1 and 2 refer to an automatic machine producing piston rings for car engines. Random samples of 1000 rings are drawn from the output of the machine periodically for inspection purposes. A defect rate of 5% is acceptable to the manufacturer, but if the defect rate is believed to have exceeded this value, the machine producing the rings is stopped and adjusted.
>
> In Problem 1, determine the type I errors which occur for the decision rules stated.
>
> 1.   Stop production and adjust the machine if a sample contains (a) 54, (b) 62 and (c) 70 or more defective rings.
>
> In Problem 2, determine the type II errors which are made if the decision rule is to stop production if there are more than 60 defective components in the sample.
>
> 2.   When the actual defect rate has risen to (a) 6% (b) 7.5% and (c) 9%
>
> 3.   A random sample of 100 components is drawn from the output of a machine whose defect rate is 3%. Determine the type I error if the decision rule is to stop production when the sample contains: (a) 4 or more defective components, (b) 5 or more defective components, and (c) 6 or more defective components.
>
> 4.   If there are 4 or more defective components in a sample drawn from the machine given in Problem 3 above, determine the type II error when the actual defect rate is: (a) 5% (b) 6% and (c) 7%

93.3   Significance tests for population means

When carrying out tests or measurements, it is often possible to form a hypothesis as a result of these tests.

For example, the boiling point of water is found to be: 101.7°C, 99.8°C, 100.4°C, 100.3°C, 99.5°C and 98.9°C, as a result of six tests. The mean of these six results is 100.1°C. Based on these results, how confidently can it be predicted, that at this particular height above sea level and at this particular barometric pressure, water boils at 100.1°C? In other words, are the results based on sampling **significantly different** from the true result? There are a variety of ways of testing significance, but only one or two of these in common use are introduced in this section. Usually, in significance tests, some predictions about population parameters, based on sample data, are required. In significance tests for population means, a random sample is drawn from the population and the mean value of the sample, $\bar{x}$, is determined. The testing procedure depends on whether or not the standard deviation of the population is known.

(a) When the standard deviation of the population is known

A null hypothesis is made that there is no difference between the value of a sample mean $\bar{x}$ and that of the population mean, μ, i.e. H_0: $x = \mu$. If many samples had been drawn from a population and a sampling distribution of means had been formed, then, provided N is large (usually taken as $N \geq 30$) the samples would form a normal distribution, having a mean value of $\mu_{\bar{x}}$ and a standard deviation or standard error of the means (see Section 92.3).

The particular value of $\bar{x}$ of a large sample drawn for a significance test is therefore part of a normal distribution and it is possible to determine by how much $\bar{x}$ is likely to differ from $\mu_{\bar{x}}$ in terms of the normal standard variate z.

The relationship is $z = \dfrac{\bar{x} - \mu_{\bar{x}}}{\sigma_{\bar{x}}}$

However, with reference to Chapter 92, page 989,

$$\sigma_{\bar{x}} = \frac{\sigma}{\sqrt{N}} \sqrt{\left(\frac{N_p - N}{N_p - 1} \right)} \text{ for finite populations,}$$

$$= \frac{\sigma}{\sqrt{N}} \text{ for infinite populations, and } \mu_{\bar{x}} = \mu$$

where N is the sample size, N_p is the size of the population, μ is the mean of the population and σ the standard deviation of the population.

Substituting for $\mu_{\bar{x}}$ and $\sigma_{\bar{x}}$ in the equation for z gives:

$$z = \frac{\bar{x} - \mu}{\dfrac{\sigma}{\sqrt{N}}} \text{ for infinite populations,} \qquad (1)$$

$$z = \frac{\bar{x} - \mu}{\dfrac{\sigma}{\sqrt{N}} \sqrt{\left(\dfrac{N_p - N}{N_p - 1} \right)}} \qquad (2)$$

for populations of size N_p

In Table 93.1 on page 1000, the relationship between z-values and levels of significance for both one-tailed and two-tailed tests are given. It can be seen from this table for a level of significance of, say, 0.05 and a two-tailed test, the z-value is $+1.96$, and z-values outside of this range are not significant. Thus, for a given level of significance (i.e. a known value of z), the mean of the population, μ, can be predicted by using equations (1) and (2) above, based on the mean of a sample $\bar{x}$. Alternatively, if the mean of the population is known, the significance of a particular value of z, based on sample data, can be established. If the z-value based on the mean of a random sample for a two-tailed test is found to be, say, 2.01, then at a level of significance of 0.05, that is, the results being probably significant, the mean of the sampling distribution is said to differ significantly from what would be expected as a result of the null hypothesis (i.e. that $\bar{x} = \mu$), due to the result of the test being classed as 'not significant' (see page 999). The hypothesis would then be rejected and an alternative hypothesis formed, i.e. H_1: $\bar{x} \neq \mu$. The rules of decision for such a test would be:

(i) reject the hypothesis at a 0.05 level of significance, i.e. if the z-value of the sample mean is outside of the range -1.96 to $+1.96$

(ii) accept the hypothesis otherwise.

For small sample sizes (usually taken as $N < 30$), the sampling distribution is not normally distributed, but approximates to Student's t-distributions (see Section 92.5). In this case, t-values rather than z-values are used and the equations analogous to equations (1) and (2) are:

$$|t| = \frac{\bar{x} - \mu}{\dfrac{\sigma}{\sqrt{N}}} \text{ for infinite populations} \qquad (3)$$

$$|t| = \frac{\bar{x} - \mu}{\dfrac{\sigma}{\sqrt{N}} \sqrt{\left(\dfrac{N_p - N}{N_p - 1} \right)}} \qquad (4)$$

for populations of size N_p

where $|t|$ means the modulus of t, i.e. the positive value of t.

(b) When the standard deviation of the population is not known

It is found, in practice, that if the standard deviation of a sample is determined, its value is less than the value of the standard deviation of the population from which it is drawn. This is as expected, since the range of a sample is likely to be less than the range of the population. The difference between the two standard deviations becomes more pronounced when the sample size is small. Investigations have shown that the variance, s^2, of a sample of N items is approximately related to the variance, σ^2, of the population from which it is drawn by:

$$s^2 = \left(\frac{N-1}{N}\right)\sigma^2$$

The factor $\left(\frac{N-1}{N}\right)$ is known as **Bessel's correction**. This relationship may be used to find the relationship between the standard deviation of a sample, s, and an estimate of the standard deviation of a population, $\hat{\sigma}$, and is:

$$\hat{\sigma}^2 = s^2\left(\frac{N}{N-1}\right) \text{ i.e. } \hat{\sigma} = s\sqrt{\left(\frac{N}{N-1}\right)}$$

For large samples, say, a minimum of N being 30, the factor $\sqrt{\left(\frac{N}{N-1}\right)}$ is $\sqrt{\frac{30}{29}}$ which is approximately equal to 1.017. Thus, for large samples s is very nearly equal to $\hat{\sigma}$ and the factor $\sqrt{\left(\frac{N}{N-1}\right)}$ can be omitted without introducing any appreciable error. In equations (1) and (2), s can be written for σ, giving:

$$z = \frac{\overline{x}-\mu}{\dfrac{s}{\sqrt{N}}} \text{ for infinite populations} \qquad (5)$$

and

$$z = \frac{\overline{x}-\mu}{\dfrac{s}{\sqrt{N}}\sqrt{\left(\dfrac{N_p-N}{N_p-1}\right)}} \qquad (6)$$

for populations of size N_p

For small samples, the factor $\sqrt{\left(\frac{N}{N-1}\right)}$ cannot be disregarded and substituting $\sigma = s\sqrt{\left(\frac{N}{N-1}\right)}$ in equations (3) and (4) gives:

$$|t| = \frac{\overline{x}-\mu}{\dfrac{s\sqrt{\left(\dfrac{N}{N-1}\right)}}{\sqrt{N}}} = \frac{(\overline{x}-\mu)\sqrt{(N-1)}}{s} \qquad (7)$$

for infinite populations, and

$$|t| = \frac{\overline{x}-\mu}{\dfrac{s\sqrt{\left(\dfrac{N}{N-1}\right)}}{\sqrt{N}}\sqrt{\left(\dfrac{N_p-N}{N_p-1}\right)}}$$

$$= \frac{(\overline{x}-\mu)\sqrt{(N-1)}}{s\sqrt{\left(\dfrac{N_p-N}{N_p-1}\right)}} \qquad (8)$$

for populations of size N_p.

The equations given in this section are parts of tests which are applied to determine population means. The way in which some of them are used is shown in the following Problems.

> **Problem 4.** Sugar is packed in bags by an automatic machine. The mean mass of the contents of a bag is 1.000 kg. Random samples of 36 bags are selected throughout the day and the mean mass of a particular sample is found to be 1.003 kg. If the manufacturer is willing to accept a standard deviation on all bags packed of 0.01 kg and a level of significance of 0.05, above which values the machine must be stopped and adjustments made, determine if, as a result of the sample under test, the machine should be adjusted

Population mean $\mu = 1.000$ kg, sample mean $\overline{x} = 1.003$ kg, population standard deviation $\sigma = 0.01$ kg and sample size, $N = 36$

A null hypothesis for this problem is that the sample mean and the mean of the population are equal, i.e. $H_0\colon \overline{x} = \mu$

Since the manufacturer is interested in deviations on both sides of the mean, the alternative hypothesis is that the sample mean is not equal to the population mean, i.e. $H_1\colon \overline{x} \neq \mu$

The decision rules associated with these hypotheses are:

(i) reject H_0 if the z-value of the sample mean is outside of the range of the z-values corresponding to a level of significance of 0.05 for a two-tailed test, i.e. stop machine and adjust, and

(ii) accept H_0 otherwise, i.e. keep the machine running.

The sample size is over 30 so this is a 'large sample' problem and the population can be considered to be infinite. Because values of $\bar{x}, \mu, \sigma$ and N are all known, equation (1) can be used to determine the z-value of the sample mean,

i.e. $z = \dfrac{\bar{x} - \mu}{\dfrac{\sigma}{\sqrt{N}}} = \dfrac{1.003 - 1.000}{\dfrac{0.01}{\sqrt{36}}} = \pm\dfrac{0.003}{0.0016}$

$$= \pm 1.8$$

The z-value corresponding to a level of significance of 0.05 for a two-tailed test is given in Table 93.1 on page 1000 and is ± 1.96. Since the z-value of the sample is within this range, **the null hypothesis is accepted and the machine should not be adjusted**.

> **Problem 5.** The mean lifetime of a random sample of 50 similar torch bulbs drawn from a batch of 500 bulbs is 72 hours. The standard deviation of the lifetime of the sample is 10.4 hours. The batch is classed as inferior if the mean lifetime of the batch is less than the population mean of 75 hours. Determine whether, as a result of the sample data, the batch is considered to be inferior at a level of significance of (a) 0.05 and (b) 0.01

Population size, $N_p = 500$, population mean, $\mu = 75$ hours, mean of sample, $\bar{x} = 72$ hours, standard deviation of sample, $s = 10.4$ hours, size of sample, $N = 50$.
The null hypothesis is that the mean of the sample is equal to the mean of the population, i.e. $H_0: \bar{x} = \mu$.
The alternative hypothesis is that the mean of the sample is less than the mean of the population, i.e. $H_1: \bar{x} < \mu$.
(The fact that $\bar{x} = 72$ should not lead to the conclusion that the batch is necessarily inferior. At a level of significance of 0.05, the result is 'probably significant', but since this corresponds to a confidence level of 95%, there are still 5 times in every 100 when the result can be significantly different, that is, be outside of the range of z-values for this data. This particular sample result may be one of these 5 times.)
The decision rules associated with the hypotheses are:

(i) reject H_0 if the z-value (or t-value) of the sample mean is less than the z-value (or t-value) corresponding to a level of significance of (a) 0.05 and (b) 0.01, i.e. the batch is inferior,

(ii) accept H_0 otherwise, i.e. the batch is not inferior.

The data given is $N, N_p, \bar{x}, s$ and μ. The alternative hypothesis indicates a one-tailed distribution and since $N > 30$ the 'large sample' theory applies.

From equation (6),

$$z = \dfrac{\bar{x} - \mu}{\dfrac{s}{\sqrt{N}}\sqrt{\left(\dfrac{N_p - N}{N_p - 1}\right)}} = \dfrac{72 - 75}{\dfrac{10.4}{\sqrt{50}}\sqrt{\left(\dfrac{500 - 50}{500 - 1}\right)}}$$

$$= \dfrac{-3}{(1.471)(0.9496)} = -2.15$$

(a) For a level of significance of 0.05 and a one-tailed test, all values to the left of the z-ordinate at -1.645 (see Table 93.1 on page 1000) indicate that the results are 'not significant', that is, they differ significantly from the null hypothesis. Since the z-value of the sample mean is -2.15, i.e. less than -1.645, **the batch is considered to be inferior at a level of significance of 0.05**

(b) The z-value for a level of significance of 0.01 for a one-tailed test is -2.33 and in this case, z-values of sample means lying to the left of the z-ordinate at -2.33 are 'not significant'. Since the z-value of the sample lies to the right of this ordinate, it does not differ significantly from the null hypothesis and **the batch is not considered to be inferior at a level of significance of 0.01**

(At first sight, for a mean value to be significant at a level of significance of 0.05, but not at 0.01, appears to be incorrect. However, it is stated earlier in the chapter that for a result to be probably significant, i.e. at a level of significance of between 0.01 and 0.05, the range of z-values is less than the range for the result to be highly significant, that is, having a level of significance of 0.01 or better. Hence the results of the problem are logical.)

> **Problem 6.** An analysis of the mass of carbon in six similar specimens of cast iron, each of mass 425.0 g, yielded the following results:
>
> 17.1 g, 17.3 g, 16.8 g, 16.9 g, 17.8 g and 17.4 g
>
> Test the hypothesis that the percentage of carbon is 4.00% assuming an arbitrary level of significance of (a) 0.2 and (b) 0.1

The sample mean,

$$\bar{x} = \frac{17.1 + 17.3 + 16.8 + 16.9 + 17.8 + 17.4}{6}$$

$$= 17.22$$

The sample standard deviation,

$$s = \sqrt{\left\{ \frac{\begin{array}{c}(17.1 - 17.22)^2 + (17.3 - 17.22)^2 \\ + (16.8 - 17.22)^2 + \cdots + (17.4 - 17.22)^2\end{array}}{6} \right\}}$$

$$= 0.334$$

The null hypothesis is that the sample and population means are equal, i.e. $H_0: \bar{x} = \mu$

The alternative hypothesis is that the sample and population means are not equal, i.e. $H_1: \bar{x} \neq \mu$

The decision rules are:

(i)   reject H_0 if the z- or t-value of the sample mean is outside of the range of the z- or t-value corresponding to a level of significance of (a) 0.2 and (b) 0.1, i.e. the mass of carbon is not 4.00%

(ii) accept H_0 otherwise, i.e. the mass of carbon is 4.00%

The number of tests taken, N, is 6 and an infinite number of tests could have been taken, hence the population is considered to be infinite. Because $N < 30$, a t-distribution is used.

If the mean mass of carbon in the bulk of the metal is 4.00%, the mean mass of carbon in a specimen is 4.00% of 425.0, i.e. 17.00 g, thus $\mu = 17.00$

From equation (7),

$$|t| = \frac{(\bar{x} - \mu)\sqrt{(N - 1)}}{s}$$

$$= \frac{(17.22 - 17.00)\sqrt{(6 - 1)}}{0.334}$$

$$= \mathbf{1.473}$$

In general, for any two-tailed distribution there is a critical region both to the left and to the right of the mean of the distribution. For a level of significance of 0.2, 0.1 of the percentile value of a t-distribution lies to the left of the mean and 0.1 of the percentile value lies to the right of the mean. Thus, for a level of significance of α, a value $t_{\left(1 - \frac{\alpha}{2}\right)}$, is required for a two-tailed distribution when using Table 92.2 on page 993. This conversion is necessary because the

t-distribution is given in terms of levels of confidence and for a one-tailed distribution. The row t-value for a value of α of 0.2 is $t_{\left(1 - \frac{0.2}{2}\right)}$, i.e. $t_{0.90}$. The degrees of freedom ν are $N - 1$, that is 5. From Table 92.2 on page 993, the percentile value corresponding to ($t_{0.90}$, $\nu = 5$) is 1.48, and for a two-tailed test, ± 1.48. Since the mean value of the sample is within this range, the hypothesis is accepted at a level of significance of 0.2. The t-value for $\alpha = 0.1$ is $t_{\left(1 - \frac{0.1}{2}\right)}$, i.e. $t_{0.95}$. The percentile value corresponding to $t_{0.95}$, $\nu = 5$ is 2.02 and since the mean value of the sample is within the range ± 2.02, the hypothesis is also accepted at this level of significance. **Thus, it is probable that the mass of metal contains 4% carbon at levels of significance of 0.2 and 0.1**

Now try the following Practice Exercise

Practice Exercise 348   Further problems on significance tests for population means (answers on page 1191)

1. A batch of cables produced by a manufacturer have a mean breaking strength of 2000 kN and a standard deviation of 100 kN. A sample of 50 cables is found to have a mean breaking strength of 2050 kN. Test the hypothesis that the breaking strength of the sample is greater than the breaking strength of the population from which it is drawn at a level of significance of 0.01

2. Nine estimations of the percentage of copper in a bronze alloy have a mean of 80.8% and standard deviation of 1.2%. Assuming that the percentage of copper in samples is normally distributed, test the null hypothesis that the true percentage of copper is 80% against an alternative hypothesis that it exceeds 80%, at a level of significance of 0.05

3. The internal diameter of a pipe has a mean diameter of 3.0000 cm with a standard deviation of 0.015 cm. A random sample of 30 measurements are taken and the mean of the samples is 3.0078 cm. Test the hypothesis that the mean diameter of the pipe is 3.0000 cm at a level of significance of 0.01

4. A fishing line has a mean breaking strength of 10.25 kN. Following a special treatment on the

line, the following results are obtained for 20 specimens taken from the line

Breaking strength (kN)	Frequency
9.8	1
10	1
10.1	4
10.2	5
10.5	3
10.7	2
10.8	2
10.9	1
11.0	1

Test the hypothesis that the special treatment has improved the breaking strength at a level of significance of 0.05

5. A machine produces ball bearings having a mean diameter of 0.50 cm. A sample of 10 ball bearings is drawn at random and the sample mean is 0.53 cm with a standard deviation of 0.03 cm. Test the hypothesis that the mean diameter is 0.50 cm at a level of significance of (a) 0.05 and (b) 0.01

6. Six similar switches are tested to destruction at an overload of 20% of their normal maximum current rating. The mean number of operations before failure is 8200 with a standard deviation of 145. The manufacturer of the switches claims that they can be operated at least 8000 times at a 20% overload current. Can the manufacturer's claim be supported at a level of significance of (a) 0.1 and (b) 0.02?

93.4 Comparing two sample means

The techniques introduced in Section 93.3 can be used for comparison purposes. For example, it may be necessary to compare the performance of, say, two similar lamps produced by different manufacturers or different operators carrying out a test or tests on the same items using different equipment. The null hypothesis adopted for tests involving two different populations is that there is **no difference** between the mean values of the populations.

The technique is based on the following theorem:

If x_1 and x_2 are the means of random samples of size N_1 and N_2 drawn from populations having means of μ_1 and μ_2 and standard deviations of σ_1 and σ_2, then the sampling distribution of the differences of the means, $(\bar{x}_1 - \bar{x}_2)$, is a close approximation to a normal distribution, having a mean of zero and a standard deviation of

$$\sqrt{\left(\frac{\sigma_1^2}{N_1} + \frac{\sigma_2^2}{N_2}\right)}$$

For large samples, when comparing the mean values of two samples, the variate is the difference in the means of the two samples, $\bar{x}_1 - \bar{x}_2$; the mean of sampling distribution (and hence the difference in population means) is zero and the standard error of the sampling distribution $\sigma_{\bar{x}}$ is

$$\sqrt{\left(\frac{\sigma_1^2}{N_1} + \frac{\sigma_2^2}{N_2}\right)}$$

Hence, the z-value is

$$\frac{(\bar{x}_1 - \bar{x}_2) - 0}{\sqrt{\left(\dfrac{\sigma_1^2}{N_1} + \dfrac{\sigma_2^2}{N_2}\right)}} = \frac{\bar{x}_1 - \bar{x}_2}{\sqrt{\left(\dfrac{\sigma_1^2}{N_1} + \dfrac{\sigma_2^2}{N_2}\right)}} \qquad (9)$$

For small samples, Student's t-distribution values are used and in this case:

$$|t| = \frac{\bar{x}_1 - \bar{x}_2}{\sqrt{\left(\dfrac{\sigma_1^2}{N_1} + \dfrac{\sigma_2^2}{N_2}\right)}} \qquad (10)$$

where $|t|$ means the modulus of t, i.e. the positive value of t.

When the standard deviation of the population is not known, then Bessel's correction is applied to estimate it from the sample standard deviation (i.e. the estimate of the population variance, $\sigma^2 = s^2\left(\dfrac{N}{N-1}\right)$ (see page 1005)). For large populations, the factor $\left(\dfrac{N}{N-1}\right)$ is small and may be neglected. However, when $N < 30$, this correction factor should be included. Also, since estimates of both σ_1 and σ_2 are being made, the k factor in the degrees of freedom in Student's t-distribution tables becomes 2 and ν is given by $(N_1 + N_2 - 2)$. With these factors taken into account, when testing the hypotheses that samples come from the same population, or that there is no difference between the mean values of two populations, the t-value is given by:

$$|t| = \frac{\overline{x}_1 - \overline{x}_2}{\sigma\sqrt{\left(\dfrac{1}{N_1} + \dfrac{1}{N_2}\right)}} \qquad (11)$$

An estimate of the standard deviation σ is based on a concept called 'pooling'. This states that if one estimate of the variance of a population is based on a sample, giving a result of $\sigma_1^2 = \dfrac{N_1 s_1^2}{N_1 - 1}$ and another estimate is based on a second sample, giving $\sigma_2^2 = \dfrac{N_2 s_2^2}{N_2 - 1}$, then a better estimate of the population variance, σ^2, is given by:

$$\sigma^2 = \frac{N_1 s_1^2 + N_2 s_2^2}{(N_1 - 1) + (N_2 - 1)}$$

i.e. $\sigma = \sqrt{\left(\dfrac{N_1 s_1^2 + N_2 s_2^2}{N_1 + N_2 - 2}\right)} \qquad (12)$

Problem 7. An automatic machine is producing components, and as a result of many tests the standard deviation of their size is 0.02 cm. Two samples of 40 components are taken, the mean size of the first sample being 1.51 cm and the second 1.52 cm. Determine whether the size has altered appreciably if a level of significance of 0.05 is adopted, i.e. that the results are probably significant

Since both samples are drawn from the same population, $\sigma_1 = \sigma_2 = \sigma = 0.02$ cm. Also $N_1 = N_2 = 40$ and $\overline{x}_1 = 1.51$ cm, $\overline{x}_2 = 1.52$ cm.
The level of significance, $\alpha = 0.05$
The null hypothesis is that the size of the component has not altered, i.e. $\overline{x}_1 = \overline{x}_2$, hence it is $H_0: \overline{x}_1 - \overline{x}_2 = 0$. The alternative hypothesis is that the size of the components has altered, i.e. that $\overline{x}_1 \neq \overline{x}_2$, hence it is $H_1: \overline{x}_1 - \overline{x}_2 \neq 0$
For a large sample having a known standard deviation of the population, the z-value of the difference of means of two samples is given by equation (9), i.e.

$$z = \frac{\overline{x}_1 - \overline{x}_2}{\sqrt{\left(\dfrac{\sigma_1^2}{N_1} + \dfrac{\sigma_2^2}{N_2}\right)}}$$

Since $N_1 = N_2 =$ say, N, and $\sigma_1 = \sigma_2 = \sigma$, this equation becomes

$$z = \frac{\overline{x}_1 - \overline{x}_2}{\sigma\sqrt{\left(\dfrac{2}{N}\right)}} = \frac{1.51 - 1.52}{0.02\sqrt{\left(\dfrac{2}{40}\right)}} = -2.236$$

Since the difference between $\overline{x}_1$ and $\overline{x}_2$ has no specified direction, a two-tailed test is indicated. The z-value corresponding to a level of significance of 0.05 and a two-tailed test is $+1.96$ (see Table 93.1, page 1000). The result for the z-value for the difference of means is outside of the range $+1.96$, that is, **it is probable that the size has altered appreciably at a level of significance of 0.05**

Problem 8. The electrical resistances of two products are being compared. The parameters of product 1 are:

sample size 40, mean value of sample 74 ohms, standard deviation of whole of product 1 batch is 8 ohms
Those of product 2 are:

sample size 50, mean value of sample 78 ohms, standard deviation of whole of product 2 batch is 7 ohms

Determine if there is any significant difference between the two products at a level of significance of (a) 0.05 and (b) 0.01

Let the mean of the batch of product 1 be μ_1, and that of product 2 be μ_2
The null hypothesis is that the means are the same, i.e. $H_0: \mu_1 - \mu_2 = 0$
The alternative hypothesis is that the means are not the same, i.e. $H_1: \mu_1 - \mu_2 \neq 0$
The population standard deviations are known, i.e. $\sigma_1 = 8$ ohms and $\sigma_2 = 7$ ohms, the sample means are known, i.e. $\overline{x}_1 = 74$ ohms and $\overline{x}_2 = 78$ ohms. Also the sample sizes are known, i.e. $N_1 = 40$ and $N_2 = 50$. Hence, equation (9) can be used to determine the z-value of the difference of the sample means. From equation (9),

$$z = \frac{\overline{x}_1 - \overline{x}_2}{\sqrt{\left(\dfrac{\sigma_1^2}{N_1} + \dfrac{\sigma_2^2}{N_2}\right)}} = \frac{74 - 78}{\sqrt{\left(\dfrac{8^2}{40} + \dfrac{7^2}{50}\right)}}$$

$$= \frac{-4}{1.606} = -2.49$$

(a) For a two-tailed test, the results are probably significant at a 0.05 level of significance when z lies between -1.96 and $+1.96$. Hence the z-value of the difference of means shows there is 'no significance', i.e. that **product 1 is significantly different from product 2 at a level of significance of 0.05**

(b) For a two-tailed test, the results are highly significant at a 0.01 level of significance when z lies between -2.58 and $+2.58$. Hence there is **no significant difference between product 1 and product 2 at a level of significance of 0.01**

> **Problem 9.** The reaction times in seconds of two people, A and B, are measured by electrodermal responses and the results of the tests are as shown below.
>
> | Person A (s) | 0.243 | 0.243 | 0.239 |
> | Person B (s) | 0.238 | 0.239 | 0.225 |
>
> | Person A (s) | 0.232 | 0.229 | 0.241 |
> | Person B (s) | 0.236 | 0.235 | 0.234 |
>
> Find if there is any significant difference between the reaction times of the two people at a level of significance of 0.1

The mean, $\bar{x}$, and standard deviation, s, of the response times of the two people are determined.

$$\bar{x}_A = \frac{\begin{array}{c}0.243 + 0.243 + 0.239 + 0.232 \\ + 0.229 + 0.241\end{array}}{6}$$

$$= 0.2378\,s$$

$$\bar{x}_B = \frac{\begin{array}{c}0.238 + 0.239 + 0.225 + 0.236 \\ + 0.235 + 0.234\end{array}}{6}$$

$$= 0.2345\,s$$

$$s_A = \sqrt{\left[\frac{\begin{array}{c}(0.243 - 0.2378)^2 + (0.243 - 0.2378)^2 \\ + \cdots + (0.241 - 0.2378)^2\end{array}}{6}\right]}$$

$$= 0.00543\,s$$

$$s_B = \sqrt{\left[\frac{\begin{array}{c}(0.238 - 0.2345)^2 + (0.239 - 0.2345)^2 \\ + \cdots + (0.234 - 0.2345)^2\end{array}}{6}\right]}$$

$$= 0.00457\,s$$

The null hypothesis is that there is no difference between the reaction times of the two people, i.e. $H_0: \bar{x}_A - \bar{x}_B = 0$ The alternative hypothesis is that the reaction times are different, i.e. $H_1: \bar{x}_A - \bar{x}_B \neq 0$, indicating a two-tailed test.

The sample numbers (combined) are less than 30 and a t-distribution is used. The standard deviation of all the reaction times of the two people is not known, so an estimate based on the standard deviations of the samples is used. Applying Bessel's correction, the estimate of the standard deviation of the population,

$$\sigma^2 = s^2 \left(\frac{N}{N-1}\right)$$

gives $\quad \sigma_A = (0.00543)\sqrt{\left(\frac{6}{5}\right)} = 0.00595$

and $\quad \sigma_B = (0.00457)\sqrt{\left(\frac{6}{5}\right)} = 0.00501$

From equation (10), the t-value of the difference of the means is given by:

$$|t| = \frac{\bar{x}_A - \bar{x}_B}{\sqrt{\left(\dfrac{\sigma_A^2}{N_A} + \dfrac{\sigma_B^2}{N_B}\right)}}$$

$$= \frac{0.2378 - 0.2345}{\sqrt{\left(\dfrac{0.00595^2}{6} + \dfrac{0.00501^2}{6}\right)}}$$

$$= \mathbf{1.039}$$

For a two-tailed test and a level of significance of 0.1, the column heading in the t-distribution of Table 92.2 (on page 993) is $t_{0.95}$ (refer to Problem 6). The degrees of freedom due to k being 2 is $\nu = N_1 + N_2 - 2$, i.e. $6 + 6 - 2 = 10$. The corresponding t-value from Table 92.2 is 1.81. Since the t-value of the difference of the means is within the range ± 1.81, there is **no significant difference between the reaction times at a level of significance of 0.1**

Problem 10. An analyst carries out 10 analyses on equal masses of a substance which is found to contain a mean of 49.20 g of a metal, with a standard deviation of 0.41 g. A trainee operator carries out 12 analyses on equal masses of the same substance which is found to contain a mean of 49.30 g, with a standard deviation of 0.32 g. Is there any significance between the results of the operators?

Let μ_1 and μ_2 be the mean values of the amounts of metal found by the two operators.

The null hypothesis is that there is no difference between the results obtained by the two operators, i.e. $H_0: \mu_1 = \mu_2$

The alternative hypothesis is that there is a difference between the results of the two operators, i.e. $H_1: \mu_1 \neq \mu_2$

Under the hypothesis H_0 the standard deviations of the amount of metal, σ, will be the same, and from equation (12)

$$\sigma = \sqrt{\left(\frac{N_1 s_1^2 + N_2 s_2^2}{N_1 + N_2 - 2}\right)}$$

$$= \sqrt{\left(\frac{(10)(0.41)^2 + (12)(0.32)^2}{10 + 12 - 2}\right)}$$

$$= \mathbf{0.3814}$$

The t-value of the results obtained is given by equation (11), i.e.

$$|t| = \frac{\overline{x}_1 - \overline{x}_2}{\sigma \sqrt{\left(\frac{1}{N_1} + \frac{1}{N_2}\right)}} = \frac{49.20 - 49.30}{(0.3814)\sqrt{\left(\frac{1}{10} + \frac{1}{12}\right)}}$$

$$= \mathbf{-0.612}$$

For the results to be probably significant, a two-tailed test and a level of significance of 0.05 is taken. H_0 is rejected outside of the range $t_{-0.975}$ and $t_{0.975}$. The number of degrees of freedom is $N_1 + N_2 - 2$. For $t_{0.975}$, $\nu = 20$, from Table 92.2 on page 993, the range is from -2.09 to $+2.09$. Since the t-value based on the sample data is within this range, **there is no significant difference between the results of the two operators at a level of significance of 0.05**

Now try the following Practice Exercise

Practice Exercise 349   Further problems on comparing two sample means (answers on page 1192)

1. A comparison is being made between batteries used in calculators. Batteries of type A have a mean lifetime of 24 hours with a standard deviation of 4 hours, this data being calculated from a sample of 100 of the batteries. A sample of 80 of the type B batteries has a mean lifetime of 40 hours with a standard deviation of 6 hours. Test the hypothesis that the type B batteries have a mean lifetime of at least 15 hours more than those of type A, at a level of significance of 0.05

2. Two randomly selected groups of 50 operatives in a factory are timed during an assembly operation. The first group take a mean time of 112 minutes with a standard deviation of 12 minutes. The second group take a mean time of 117 minutes with a standard deviation of 9 minutes. Test the hypothesis that the mean time for the assembly operation is the same for both groups of employees at a level of significance of 0.05

3. Capacitors having a nominal capacitance of $24\,\mu F$ but produced by two different companies are tested. The values of actual capacitance are:

Company 1	21.4	23.6	24.8	22.4	26.3
Company 2	22.4	27.7	23.5	29.1	25.8

Test the hypothesis that the mean capacitance of capacitors produced by company 2 is higher than that of those produced by company 1 at a level of significance of 0.01

$$\left(\text{Bessel's correction is } \hat{\sigma}^2 = \frac{s^2 N}{N - 1}\right)$$

4. A sample of 100 relays produced by manufacturer A operated on average 1190 times before failure occurred, with a standard deviation of 90.75. Relays produced by manufacturer B operated on average 1220 times before failure with a standard deviation of 120. Determine if the number of operations before failure is

significantly different for the two manufacturers at a level of significance of (a) 0.05 and (b) 0.01

5. A sample of 12 car engines produced by manufacturer A showed that the mean petrol consumption over a measured distance was 4.8 litres with a standard deviation of 0.40 litres. Twelve similar engines for manufacturer B were tested over the same distance and the mean petrol consumption was 5.1 litres with a standard deviation of 0.36 litres. Test the hypothesis that the engines produced by manufacturer A are more economical than those produced by manufacturer B at a level of significance of (a) 0.01 and (b) 0.05

6. Unleaded and premium petrol is tested in 5 similar cars under identical conditions. For unleaded petrol, the cars covered a mean distance of 21.4 kilometres with a standard deviation of 0.54 kilometres for a given mass of petrol. For the same mass of premium petrol, the mean distance covered was 22.6 kilometres with a standard deviation of 0.48 kilometres. Test the hypothesis that premium petrol gives more kilometres per litre than unleaded petrol at a level of significance of 0.05

For fully worked solutions to each of the problems in Practice Exercises 347 to 349 in this chapter, go to the website:
www.routledge.com/cw/bird

Chapter 94

Chi-square and distribution-free tests

Why it is important to understand: **Chi-square and distribution-free tests**

Chi-square and distribution-free tests are used in science and engineering. Chi-square is a statistical test commonly used to compare observed data with data we would expect to obtain according to a specific hypothesis. Distribution-free methods do not rely on assumptions that the data are drawn from a given probability distribution. Non-parametric methods are widely used for studying populations that take on a ranked order. These test are explained in this chapter.

At the end of this chapter, you should be able to:

- calculate a Chi-square value for a given distribution
- test hypotheses on fitting data to theoretical distribution using the Chi-square distribution
- recognise distribution-free test
- use the sign test for two samples
- use the Wilcoxon signed-rank test for two samples
- use the Mann–Whitney test for two samples

94.1 Chi-square values

The significance tests introduced in Chapter 93 rely very largely on the normal distribution. For large sample numbers where z-values are used, the mean of the samples and the standard error of the means of the samples are assumed to be normally distributed (central limit theorem). For small sample numbers where t-values are used, the population from which samples are taken should be approximately normally distributed for the t-values to be meaningful. **Chi-square tests** (pronounced *KY* and denoted by the Greek letter χ), which are introduced in this chapter, do not rely on the population or a sampling statistic such as the mean or standard

error of the means being normally distributed. Significance tests based on z- and t-values are concerned with the parameters of a distribution, such as the mean and the standard deviation, whereas Chi-square tests are concerned with the individual members of a set and are associated with **non-parametric tests**.

Observed and expected frequencies

The results obtained from trials are rarely exactly the same as the results predicted by statistical theories. For example, if a coin is tossed 100 times, it is unlikely that the result will be exactly 50 heads and 50 tails. Let us assume that, say, 5 people each toss a coin 100 times and note the number of, say, heads obtained. Let the results obtained be as shown below.

Person	A	B	C	D	E
Observed frequency	43	54	60	48	57
Expected frequency	50	50	50	50	50

A measure of the discrepancy existing between the observed frequencies shown in row 2 and the expected frequencies shown in row 3 can be determined by calculating the Chi-square value. The Chi-square value is defined as follows:

$$\chi^2 = \sum \left\{ \frac{(o-e)^2}{e} \right\}$$

where o and e are the observed and expected frequencies, respectively.

Problem 1. Determine the Chi-square value for the coin-tossing data given above

The χ^2-value for the given data may be calculated by using a tabular approach, as shown below.

Person	Observed frequency, o	Expected frequency, e
A	43	50
B	54	50
C	60	50
D	48	50
E	57	50

$o-e$	$(o-e)^2$	$\dfrac{(o-e)^2}{e}$
−7	49	0.98
4	16	0.32
10	100	2.00
−2	4	0.08
7	49	0.98

$$\chi^2 = \sum \left\{ \frac{(o-e)^2}{e} \right\} = \underline{\mathbf{4.36}}$$

Hence the Chi-square value $\chi^2 = \mathbf{4.36}$

If the value of χ^2 is zero, then the observed and expected frequencies agree exactly. The greater the difference between the χ^2-value and zero, the greater the discrepancy between the observed and expected frequencies.

Now try the following Practice Exercise

Practice Exercise 350   Problems on determining Chi-square values (answers on page 1192)

1. A dice is rolled 240 times and the observed and expected frequencies are as shown.

Face	Observed frequency	Expected frequency
1	49	40
2	35	40
3	32	40
4	46	40
5	49	40
6	29	40

Determine the χ^2-value for this distribution.

2. The numbers of telephone calls received by the switchboard of a company in 200 five-minute intervals are shown in the distribution below.

Number of calls	Observed frequency	Expected frequency
0	11	16
1	44	42
2	53	52
3	46	42
4	24	26
5	12	14
6	7	6
7	3	2

Calculate the χ^2-value for this data.

94.2 Fitting data to theoretical distributions

For theoretical distributions such as the binomial, Poisson and normal distributions, expected frequencies can be calculated. For example, from the theory of the binomial distribution, the probability of having $0, 1, 2, \ldots, n$ defective items in a sample of n items can be determined from the successive terms of $(q + p)^n$, where p is the defect rate and $q = 1 - p$. These probabilities can be used to determine the expected frequencies of having $0, 1, 2, \ldots, n$ defective items. As a result of counting the number of defective items when sampling, the observed frequencies are obtained. The expected and observed frequencies can be compared by means of a Chi-square test and predictions can be made as to whether the differences are due to random errors, due to some fault in the method of sampling, or due to the assumptions made.

As for normal and t distributions, a table is available for relating various calculated values of χ^2 to those likely because of random variations, at various levels of confidence. Such a table is shown in Table 94.1. In Table 94.1, the column on the left denotes the number of degrees of freedom, ν, and when the χ^2-values refer to fitting data to theoretical distributions, the number of degrees of freedom is usually $(N - 1)$, where N is the number of rows in the table from which χ^2 is calculated. However, when the population parameters such as the mean and standard deviation are based on sample data, the number of degrees of freedom is given by $\nu = N - 1 - M$, where M is the number of estimated population parameters. An application of this is shown in Problem 4.

The columns of the table headed $\chi^2_{0.995}$, $\chi^2_{0.99}$, $\ldots$ give the percentile of χ^2-values corresponding to levels of confidence of 99.5%, 99%, $\ldots$ (i.e. levels of significance of 0.005, 0.01, $\ldots$). On the far right of the table, the columns headed $\ldots$, $\chi^2_{0.01}$, $\chi^2_{0.005}$ also correspond to levels of confidence of $\ldots$ 99%, 99.5%, and are used to predict the 'too good to be true' type results, where the fit obtained is so good that the method of sampling must be suspect. The method in which χ^2-values are used to test the goodness of fit of data to probability distributions is shown in the following problems.

Problem 2.   As a result of a survey carried out of 200 families, each with five children, the distribution shown below was produced. Test the null hypothesis that the observed frequencies are consistent with male and female births being equally probable, assuming a binomial distribution, a level of significance of 0.05 and a 'too good to be true' fit at a confidence level of 95%

Number of boys (B) and girls (G)	Number of families
5B, 0G	11
4B, 1G	35
3B, 2G	69

Number of boys (B) and girls (G)	Number of families
2B, 3G	55
1B, 4G	25
OB, 5G	5

To determine the expected frequencies

Using the usual binomial distribution symbols, let p be the probability of a male birth and $q = 1 - p$ be the probability of a female birth. The probabilities of having 5 boys, 4 boys, $\ldots$, 0 boys are given by the successive terms of the expansion of $(q + p)^n$. Since there are 5 children in each family, $n = 5$, and

$$(q + p)^5 = q^5 + 5q^4 p + 10q^3 p^2 + 10q^2 p^3 + 5qp^4 + p^5$$

When $q = p = 0.5$, the probabilities of 5 boys, 4 boys, $\ldots$, 0 boys are

$$0.03125, 0.15625, 0.3125, 0.3125,$$

$$0.15625 \text{ and } 0.3125$$

For 200 families, the expected frequencies, rounded off to the nearest whole number, are: 6, 31, 63, 63, 31 and 6 respectively.

To determine the χ^2-value

Using a tabular approach, the χ^2-value is calculated using $\chi^2 = \sum \left\{ \dfrac{(o - e)^2}{e} \right\}$

Section L

Table 94.1 Chi-square distribution

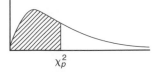

χ_p^2

Percentile values (χ_p^2) for the Chi-square distribution with ν degrees of freedom

ν	$\chi_{0.995}^2$	$\chi_{0.99}^2$	$\chi_{0.975}^2$	$\chi_{0.95}^2$	$\chi_{0.90}^2$	$\chi_{0.75}^2$	$\chi_{0.50}^2$	$\chi_{0.25}^2$	$\chi_{0.10}^2$	$\chi_{0.05}^2$	$\chi_{0.025}^2$	$\chi_{0.01}^2$	$\chi_{0.005}^2$
1	7.88	6.63	5.02	3.84	2.71	1.32	0.455	0.102	0.0158	0.0039	0.0010	0.0002	0.0000
2	10.6	9.21	7.38	5.99	4.61	2.77	1.39	0.575	0.211	0.103	0.0506	0.0201	0.0100
3	12.8	11.3	9.35	7.81	6.25	4.11	2.37	1.21	0.584	0.352	0.216	0.115	0.072
4	14.9	13.3	11.1	9.49	7.78	5.39	3.36	1.92	1.06	0.711	0.484	0.297	0.207
5	16.7	15.1	12.8	11.1	9.24	6.63	4.35	2.67	1.61	1.15	0.831	0.554	0.412
6	18.5	16.8	14.4	12.6	10.6	7.84	5.35	3.45	2.20	1.64	1.24	0.872	0.676
7	20.3	18.5	16.0	14.1	12.0	9.04	6.35	4.25	2.83	2.17	1.69	1.24	0.989
8	22.0	20.1	17.5	15.5	13.4	10.2	7.34	5.07	3.49	2.73	2.18	1.65	1.34
9	23.6	21.7	19.0	16.9	14.7	11.4	8.34	5.90	4.17	3.33	2.70	2.09	1.73
10	25.2	23.2	20.5	18.3	16.0	12.5	9.34	6.74	4.87	3.94	3.25	2.56	2.16
11	26.8	24.7	21.9	19.7	17.3	13.7	10.3	7.58	5.58	4.57	3.82	3.05	2.60
12	28.3	26.2	23.3	21.0	18.5	14.8	11.3	8.44	6.30	5.23	4.40	3.57	3.07
13	29.8	27.7	24.7	22.4	19.8	16.0	12.3	9.30	7.04	5.89	5.01	4.11	3.57
14	31.3	29.1	26.1	23.7	21.1	17.1	13.3	10.2	7.79	6.57	5.63	4.66	4.07
15	32.8	30.6	27.5	25.0	22.3	18.2	14.3	11.0	8.55	7.26	6.26	5.23	4.60
16	34.3	32.0	28.8	26.3	23.5	19.4	15.3	11.9	9.31	7.96	6.91	5.81	5.14
17	35.7	33.4	30.2	27.6	24.8	20.5	16.3	12.8	10.1	8.67	7.56	6.41	5.70
18	37.2	34.8	31.5	28.9	26.0	21.6	17.3	13.7	10.9	9.39	8.23	7.01	6.26
19	38.6	36.2	32.9	30.1	27.2	22.7	18.3	14.6	11.7	10.1	8.91	7.63	6.84
20	40.0	37.6	34.4	31.4	28.4	23.8	19.3	15.5	12.4	10.9	9.59	8.26	7.43
21	41.4	38.9	35.5	32.7	29.6	24.9	20.3	16.3	13.2	11.6	10.3	8.90	8.03
22	42.8	40.3	36.8	33.9	30.8	26.0	21.3	17.2	14.0	12.3	11.0	9.54	8.64
23	44.2	41.6	38.1	35.2	32.0	27.1	22.3	18.1	14.8	13.1	11.7	10.2	9.26
24	45.6	43.0	39.4	36.4	33.2	28.2	23.3	19.0	15.7	13.8	12.4	10.9	9.89
25	46.9	44.3	40.6	37.7	34.4	29.3	24.3	19.9	16.5	14.6	13.1	11.5	10.5
26	48.3	45.9	41.9	38.9	35.6	30.4	25.3	20.8	17.3	15.4	13.8	12.2	11.2
27	49.6	47.0	43.2	40.1	36.7	31.5	26.3	21.7	18.1	16.2	14.6	12.9	11.8

Table 94.1 (*Continued*)

v	$\chi^2_{0.995}$	$\chi^2_{0.99}$	$\chi^2_{0.975}$	$\chi^2_{0.95}$	$\chi^2_{0.90}$	$\chi^2_{0.75}$	$\chi^2_{0.50}$	$\chi^2_{0.25}$	$\chi^2_{0.10}$	$\chi^2_{0.05}$	$\chi^2_{0.025}$	$\chi^2_{0.01}$	$\chi^2_{0.005}$
28	51.0	48.3	44.5	41.3	37.9	32.6	27.3	22.7	18.9	16.9	15.3	13.6	12.5
29	52.3	49.6	45.7	42.6	39.1	33.7	28.3	23.6	19.8	17.7	16.0	14.3	13.1
30	53.7	50.9	47.7	43.8	40.3	34.8	29.3	24.5	20.6	18.5	16.8	15.0	13.8
40	66.8	63.7	59.3	55.8	51.8	45.6	39.3	33.7	29.1	26.5	24.4	22.2	20.7
50	79.5	76.2	71.4	67.5	63.2	56.3	49.3	42.9	37.7	34.8	32.4	29.7	28.0
60	92.0	88.4	83.3	79.1	74.4	67.0	59.3	52.3	46.5	43.2	40.5	37.5	35.5
70	104.2	100.4	95.0	90.5	85.5	77.6	69.3	61.7	55.3	51.7	48.8	45.4	43.3
80	116.3	112.3	106.6	101.9	96.6	88.1	79.3	71.1	64.3	60.4	57.2	53.5	51.2
90	128.3	124.1	118.1	113.1	107.6	98.6	89.3	80.6	73.3	69.1	65.6	61.8	59.2
100	140.2	135.8	129.6	124.3	118.5	109.1	99.3	90.1	82.4	77.9	74.2	70.1	67.3

Number of boys (B) and girls (G)	Observed frequency, o	Expected frequency, e
5B, 0G	11	6
4B, 1G	35	31
3B, 2G	69	63
2B, 3G	55	63
1B, 4G	25	31
0B, 5G	5	6

$o-e$	$(o-e)^2$	$\dfrac{(o-e)^2}{e}$
5	25	4.167
4	16	0.516
6	36	0.571
−8	64	1.016
−6	36	1.161
−1	1	0.167

$$\chi^2 = \sum \left\{ \frac{(o-e)^2}{e} \right\} = \mathbf{7.598}$$

To test the significance of the χ^2-value

The number of degrees of freedom is given by $v = N - 1$ where N is the number of rows in the table above, thus

$v = 6 - 1 = 5$. For a level of significance of 0.05, the confidence level is 95%, i.e. 0.95 per unit. From Table 94.1 for the $\chi^2_{0.95}$, $v = 5$ value, the percentile value χ^2_p is 11.1. Since the calculated value of χ^2 is less than χ^2_p **the null hypothesis that the observed frequencies are consistent with male and female births being equally probable is accepted**.

For a confidence level of 95%, the $\chi^2_{0.05}$, $v = 5$ value from Table 94.1 is 1.15 and because the calculated value of χ^2 (i.e. 7.598) is greater than this value, **the fit is not so good as to be unbelievable**.

Problem 3. The deposition of grit particles from the atmosphere is measured by counting the number of particles on 200 prepared cards in a specified time. The following distribution was obtained.

Number of particles	0	1	2	3	4	5	6
Number of cards	41	69	44	27	12	6	1

Test the null hypothesis that the deposition of grit particles is according to a Poisson distribution at a level of significance of 0.01 and determine if the data is 'too good to be true' at a confidence level of 99%

To determine the expected frequency

The expectation or average occurrence is given by:

$$\lambda = \frac{\text{total number of particles deposited}}{\text{total number of cards}}$$

$$= \frac{69 + 88 + 81 + 48 + 30 + 6}{200} = 1.61$$

The expected frequencies are calculated using a Poisson distribution, where the probabilities of there being $0, 1, 2, \ldots, 6$ particles deposited are given by the successive terms of $e^{-\lambda}\left(1 + \lambda + \dfrac{\lambda^2}{2!} + \dfrac{\lambda^3}{3!} + \cdots\right)$ taken from left to right,

i.e. $e^{-\lambda}, \lambda e^{-\lambda}, \dfrac{\lambda^2 e^{-\lambda}}{2!}, \dfrac{\lambda^3 e^{-\lambda}}{3!} \cdots$

Calculating these terms for $\lambda = 1.61$ gives:

Number of particles deposited	Probability	Expected frequency
0	0.1999	40
1	0.3218	64
2	0.2591	52
3	0.1390	28
4	0.0560	11
5	0.0180	4
6	0.0048	1

To determine the χ^2-valve

The χ^2-value is calculated using a tabular method as shown below.

Number of grit particles	Observed frequency, o	Expected frequency, e
0	41	40
1	69	64
2	44	52
3	27	28
4	12	11
5	6	4
6	1	1

$o - e$	$(o - e)^2$	$\dfrac{(o - e)^2}{e}$
1	1	0.0250
5	25	0.3906
−8	64	1.2308
−1	1	0.0357
1	1	0.0909
2	4	1.0000
0	0	0.0000

$$\chi^2 = \Sigma\left\{\frac{(o - e)^2}{e}\right\} = \overline{2.773}$$

To test the significance of the χ^2-value

The number of degrees of freedom is $v = N - 1$, where N is the number of rows in the table above, giving $v = 7 - 1 = 6$. The percentile value of χ^2 is determined from Table 94.1, for $(\chi^2_{0.99}, v = 6)$, and is 16.8. Since the calculated value of χ^2 (i.e. 2.773) is smaller than the percentile value, **the hypothesis that the grit deposition is according to a Poisson distribution is accepted**. For a confidence level of 99%, the $(\chi^2_{0.01}, v = 6)$ value is obtained from Table 94.1, and is 0.872. Since the calculated value of χ^2 is greater than this value, **the fit is not 'too good to be true'**.

Problem 4. The diameters of a sample of 500 rivets produced by an automatic process have the following size distribution.

Diameter (mm)	Frequency
4.011	12
4.015	47
4.019	86
4.023	123
4.027	107
4.031	97
4.035	28

Test the null hypothesis that the diameters of the rivets are normally distributed at a level of significance of 0.05 and also determine if the distribution gives a 'too good' fit at a level of confidence of 90%

To determine the expected frequencies

In order to determine the expected frequencies, the mean and standard deviation of the distribution are required. These population parameters, μ and σ, are based on sample data, $\bar{x}$ and s, and an allowance is made in the number of degrees of freedom used for estimating the population parameters from sample data.
The sample mean,

$$\bar{x} = \frac{\begin{array}{c}12(4.011) + 47(4.015) + 86(4.019) + 123(4.023)\\ +107(4.027) + 97(4.031) + 28(4.035)\end{array}}{500}$$

$$= \frac{2012.176}{500} = \mathbf{4.024}$$

The sample standard deviation s is given by:

$$s = \sqrt{\left[\frac{\begin{array}{c}12(4.011 - 4.024)^2 + 47(4.015 - 4.024)^2\\ + \cdots + 28(4.035 - 4.024)^2\end{array}}{500}\right]}$$

$$= \sqrt{\frac{0.017212}{500}} = \mathbf{0.00587}$$

The class boundaries for the diameters are 4.009 to 4.013, 4.013 to 4.017, and so on, and are shown in column 2 of Table 94.2. Using the theory of the normal probability distribution, the probability for each class and hence the expected frequency is calculated as shown in Table 94.2.
In column 3, the z-values corresponding to the class boundaries are determined using $z = \dfrac{x - \bar{x}}{s}$ which in this case is $z = \dfrac{x - 4.024}{0.00587}$. The area between a z-value in column 3 and the mean of the distribution at $z = 0$ is determined using the table of partial areas under the standardised normal distribution curve given in Table 89.1 on page 966, and is shown in column 4. By subtracting the area between the mean and the z-value of the lower class boundary from that of the upper class boundary, the area and hence the probability of a particular class is obtained, and is shown in column 5.

There is one exception in column 5, corresponding to class boundaries of 4.021 and 4.025, where the areas are added to give the probability of the 4.023 class. This is because these areas lie immediately to the left and right of the mean value. Column 6 is obtained by multiplying the probabilities in column 5 by the sample number, 500. The sum of column 6 is not equal to 500 because the areas under the standardised normal curve for z-values of less than -2.56 and more than 2.21 are neglected. The error introduced by doing this is 10 in 500, i.e. 2%, and is acceptable in most problems of this type. If it is not acceptable, each expected frequency can be increased by the percentage error.

To determine the χ^2-value

The χ^2-value is calculated using a tabular method as shown below.

Diameter of rivets	Observed frequency, o	Expected, frequency, e
4.011	12	13
4.015	47	43
4.019	86	94
4.023	123	131
4.027	107	117
4.031	97	67
4.035	28	25

$o - e$	$(o - e)^2$	$\dfrac{(o - e)^2}{e}$
-1	1	0.0769
4	16	0.3721
-8	64	0.6809
-8	64	0.4885
-10	100	0.8547
30	900	13.4328
3	9	0.3600

$$\chi^2 = \sum\left\{\frac{(o - e)^2}{e}\right\} = \overline{16.2659}$$

Table 94.2

1 Class mid-point	2 Class boundaries, x	3 z-value for class boundary	4 Area from 0 to z	5 Area for class	6 Expected frequency
	4.009	-2.56	0.4948		
4.011				0.0255	13
	4.013	-1.87	0.4693		
4.015				0.0863	43
	4.017	-1.19	0.3830		
4.019				0.1880	94
	4.021	-0.51	0.1950		
4.023				0.2628	131
	4.025	0.17	0.0678		
4.027				0.2345	117
	4.029	0.85	0.3023		
4.031				0.1347	67
	4.033	1.53	0.4370		
4.035				0.0494	25
	4.037	2.21	0.4864		
				Total	490

To test the significance of the χ^2-value

The number of degrees of freedom is given by $N - 1 - M$, where M is the number of estimated parameters in the population. Both the mean and the standard deviation of the population are based on the sample value, $M = 2$, hence $\nu = 7 - 1 - 2 = 4$. From Table 94.1, the χ_p^2-value corresponding to $\chi_{0.95}^2$ and ν_4 is 9.49. **Hence the null hypothesis that the diameters of the rivets are normally distributed is rejected**. For $\chi_{0.10}^2$, ν_4, the χ_p^2-value is 1.06, hence **the fit is not 'too good'**. Since the null hypothesis is rejected, the second significance test need not be carried out.

Now try the following Practice Exercise

Practice Exercise 351   Further problems on fitting data to theoretical distributions (answers on page 1192)

1. Test the null hypothesis that the observed data given below fits a binomial distribution of the form $250(0.6 + 0.4)^7$ at a level of significance of 0.05

| Observed frequency | 8 | 27 | 62 | 79 | 45 | 24 | 5 | 0 |

Is the fit of the data 'too good' at a level of confidence of 90%?

2. The data given below refers to the number of people injured in a city by accidents for weekly periods throughout a year. It is believed that the data fits a Poisson distribution. Test the goodness of fit at a level of significance of 0.05

Number of people injured in the week	Number of weeks
0	5
1	12
2	13
3	9
4	7
5	4
6	2

3. The resistances of a sample of carbon resistors are as shown below.

Resistance (MΩ)	Frequency
1.28	7
1.29	19
1.30	41
1.31	50
1.32	73
1.33	52
1.34	28
1.35	17
1.36	9

Test the null hypothesis that this data corresponds to a normal distribution at a level of significance of 0.05

4. The quality assurance department of a firm selects 250 capacitors at random from a large quantity of them and carries out various tests on them. The results obtained are as follows:

Number of tests failed	Number of capacitors
0	113
1	77
2	39
3	16
4	4
5	1
6 and over	0

Test the goodness of fit of this distribution to a Poisson distribution at a level of significance of 0.05

5. Test the hypothesis that the maximum load before breaking supported by certain cables produced by a company follows a normal distribution at a level of significance of 0.05, based on the experimental data given below. Also test to see if the data is 'too good' at a level of significance of 0.05

Maximum load (MN)	Number of cables
8.5	2
9.0	5
9.5	12
10.0	17
10.5	14
11.0	6
11.5	3
12.0	1

94.3 Introduction to distribution-free tests

Sometimes, sampling distributions arise from populations with unknown parameters. Tests that deal with such distributions are called **distribution-free tests**; since they do not involve the use of parameters, they are known as **non-parametric tests**. Three such tests are explained in this chapter – the **sign test** in Section 94.4 following, the **Wilcoxon signed-rank test** in Section 94.5 and the **Mann–Whitney test** in Section 94.6.

94.4 The sign test

The sign test is the simplest, quickest and oldest of all non-parametric tests.

Procedure

(i) State for the data the null and alternative hypotheses, H_0 and H_1

(ii) Know whether the stated significance level, α, is for a one-tailed or a two-tailed test. Let, for example, H_0: $x = \phi$, then if H_1: $x \neq \phi$ then a two-tailed test is suggested because x could be less than or more than ϕ (thus use α_2 in Table 94.3), but if say H_1: $x < \phi$ or H_1: $x > \phi$ then a one-tailed test is suggested (thus use α_1 in Table 94.3).

(iii) Assign plus or minus signs to each piece of data – compared with ϕ (see Problems 5 and 6) or assign plus and minus signs to the difference for paired observations (see Problem 7).

(iv) Sum either the number of plus signs or the number of minus signs. For the two-tailed test, whichever is the smallest is taken; for a one-tailed test, the one which would be expected to have the smaller value when H_1 is true is used. The sum decided upon is denoted by S

(v) Use Table 94.3 for given values of n, and α_1 or α_2 to read the critical region of S. For example, if, say, $n = 16$ and $\alpha_1 = 5\%$, then from Table 94.3, $S \leq 4$. Thus if S in part (iv) is greater than 4 we accept the null hypothesis H_0 and if S is less than or equal to 4 we accept the alternative hypothesis H_1

This procedure for the sign test is demonstrated in the following Problems.

Problem 5. A manager of a manufacturer is concerned about suspected slow progress in dealing with orders. He wants at least half of the orders received to be processed within a working day (i.e. 7 hours). A little later he decides to time 17 orders selected at random, to check if his request has been met. The times spent by the 17 orders being processed were as follows:

$$4\tfrac{3}{4}\,\text{h} \quad 9\tfrac{3}{4}\,\text{h} \quad 15\tfrac{1}{2}\,\text{h} \quad 11\,\text{h} \quad 8\tfrac{1}{4}\,\text{h} \quad 6\tfrac{1}{2}\,\text{h}$$

$$9\,\text{h} \quad 8\tfrac{3}{4}\,\text{h} \quad 10\tfrac{3}{4}\,\text{h} \quad 3\tfrac{1}{2}\,\text{h} \quad 8\tfrac{1}{2}\,\text{h} \quad 9\tfrac{1}{2}\,\text{h}$$

$$15\tfrac{1}{4}\,\text{h} \quad 13\,\text{h} \quad 8\,\text{h} \quad 7\tfrac{3}{4}\,\text{h} \quad 6\tfrac{3}{4}\,\text{h}$$

Use the sign test at a significance level of 5% to check if the manager's request for quicker processing is being met

Using the above procedure:

(i) The hypotheses are H_0: $t = 7\,\text{h}$ and H_1: $t > 7\,\text{h}$, where t is time.

(ii) Since H_1 is $t > 7\,\text{h}$, a one-tailed test is assumed, i.e. $\alpha_1 = 5\%$

(iii) In the sign test each value of data is assigned $a + \text{or} - \text{sign}$. For the above data let us assign $a +$ for times greater than 7 hours and $a -$ for less than 7 hours. This gives the following pattern:

$$- \ + \ + \ + \ + \ - \ + \ + \ +$$
$$- \ + \ + \ + \ + \ + \ -$$

(iv) The test statistic, S, in this case is the number of minus signs ($-$ if H_0 were true there would be an equal number of $+$ and $-$ signs). Table 94.3 gives critical values for the sign test and is given in terms of small values; hence in this case S is the number of $-$ signs, i.e. $S = 4$

(v) From Table 94.3, with a sample size $n = 17$, for a significance level of $\alpha_1 = 5\%$, $S \leq 4$

Since $S = 4$ in our data, the result **is significant** at $\alpha_1 = 5\%$, i.e. **the alternative hypothesis is accepted – it appears that the manager's request for quicker processing of orders is not being met**.

Table 94.3 Critical values for the sign test

n	$\alpha_1 = 5\%$ $\alpha_2 = 10\%$	$2\frac{1}{2}\%$ 5%	1% 2%	$\frac{1}{2}\%$ 1%	n	$\alpha_1 = 5\%$ $\alpha_2 = 10\%$	$2\frac{1}{2}\%$ 5%	1% 2%	$\frac{1}{2}\%$ 1%
1	—	—	—	—	26	8	7	6	6
2	—	—	—	—	27	8	7	7	6
3	—	—	—	—	28	9	8	7	6
4	—	—	—	—	29	9	8	7	7
5	0	—	—	—	30	10	9	8	7
6	0	0	—	—	31	10	9	8	7
7	0	0	0	—	32	10	9	8	8
8	1	0	0	0	33	11	10	9	8
9	1	1	0	0	34	11	10	9	9
10	1	1	0	0	35	12	11	10	9
11	2	1	1	0	36	12	11	10	9
12	2	2	1	1	37	13	12	10	10
13	3	2	1	1	38	13	12	11	10
14	3	2	2	1	39	13	12	11	11
15	3	3	2	2	40	14	13	12	11
16	4	3	2	2	41	14	13	12	11
17	4	4	3	2	42	15	14	13	12
18	5	4	3	3	43	15	14	13	12
19	5	4	4	3	44	16	15	13	13
20	5	5	4	3	45	16	15	14	13
21	6	5	4	4	46	16	15	14	13
22	6	5	5	4	47	17	16	15	14
23	7	6	5	4	48	17	16	15	14
24	7	6	5	5	49	18	17	15	15
25	7	7	6	5	50	18	17	16	15

Problem 6. The following data represents the number of hours that a portable car vacuum cleaner operates before recharging is required.

Operating time (h)	1.4	2.3	0.8	1.4	1.8	1.5
	1.9	1.4	2.1	1.1	1.6	

Use the sign test to test the hypothesis, at a 5% level of significance, that this particular vacuum cleaner operates, on average, 1.7 hours before needing a recharge

Using the procedure:

(i) Null hypothesis $H_0\!:\! t = 1.7\,\text{h}$
Alternative hypothesis $H_1\!:\! t \neq 1.7\,\text{h}$

(ii) Significance level, $\alpha_2 = 5\%$ (since this is a two-tailed test).

(iii) Assuming a $+$ sign for times > 1.7 and a $-$ sign for times < 1.7 gives:

$$- \quad + \quad - \quad - \quad + \quad - \quad + \quad - \quad + \quad - \quad -$$

(iv) There are 4 plus signs and 7 minus signs; taking the smallest number, $S = 4$

(v) From Table 94.3, where $n = 11$ and $\alpha_2 = 5\%$, $S \le 1$

Since $S = 4$ falls in the acceptance region (i.e. in this case is greater than 1), **the null hypothesis is accepted, i.e. the average operating time is not significantly different from 1.7 h**

Problem 7. An engineer is investigating two different types of metering devices, A and B, for an electronic fuel injection system to determine if they differ in their fuel mileage performance. The system is installed on 12 different cars, and a test is run with each metering system in turn on each car. The observed fuel mileage data (in miles/gallon) is shown below:

A	18.7	20.3	20.8	18.3	16.4	16.8
B	17.6	21.2	19.1	17.5	16.9	16.4

A	17.2	19.1	17.9	19.8	18.2	19.1
B	17.7	19.2	17.5	21.4	17.6	18.8

Use the sign test at a level of significance of 5% to determine whether there is any difference between the two systems

Using the procedure:

(i) $H_0: F_A = F_B$ and $H_1: F_A \neq F_B$ where F_A and F_B are the fuels in miles/gallon for systems A and B, respectively.

(ii) $\alpha_2 = 5\%$ (since it is a two-tailed test).

(iii) The difference between the observations is determined and a $+$ or a $-$ sign assigned to each as shown below:

$$(A - B) \quad +1.1 \quad -0.9 \quad +1.7 \quad +0.8$$
$$-0.5 \quad +0.4 \quad -0.5 \quad -0.1$$
$$+0.4 \quad -1.6 \quad +0.6 \quad +0.3$$

(iv) There are 7 '$+$ signs' and 5 '$-$ signs'. Taking the smallest number, $S = 5$

(v) From Table 94.3, with $n = 12$ and $\alpha_2 = 5\%$, $S \le 2$

Since from (iv), S is not equal to or less than 2, **the null hypothesis cannot be rejected, i.e. the two metering devices produce the same fuel mileage performance**.

Now try the following Practice Exercise

Practice Exercise 352   Further problems on the sign test (answers on page 1192)

1. The following data represent the number of hours of flight training received by 16 trainee pilots prior to their first solo flight:

11.5 h	20 h	9 h	12.5 h	15 h	19 h
11 h	10.5 h	13 h	22 h	14.5 h	16.5 h
17 h	18 h	14 h	12 h		

Use the sign test at a significance level of 2% to test the claim that, on average, the trainees solo after 15 hours of flight training.

2. In a laboratory experiment, 18 measurements of the coefficient of friction, μ, between metal and leather gave the following results:

0.60	0.57	0.51	0.55	0.66	0.56
0.52	0.59	0.58	0.48	0.59	0.63
0.61	0.69	0.57	0.51	0.58	0.54

Use the sign test at a level of significance of 5% to test the null hypothesis $\mu = 0.56$ against an alternative hypothesis $\mu \neq 0.56$

3. 18 random samples of two types of 9 V batteries are taken and the mean lifetimes (in hours) of each are:

Type A	8.2	7.0	11.3	13.9	9.0
	13.8	16.2	8.6	9.4	3.6
	7.5	6.5	18.0	11.5	13.4
	6.9	14.2	12.4		

Type B	15.3	15.4	11.2	16.1	18.1
	17.1	17.7	8.4	13.5	7.8
	9.8	10.6	16.4	12.7	16.8
	9.9	12.9	14.7		

Use the sign test, at a level of significance of 5%, to test the null hypothesis that the two samples come from the same population.

94.5 Wilcoxon signed-rank test

The sign test represents data by using only plus and minus signs, all other information being ignored. The Wilcoxon signed-rank test does make some use of the sizes of the differences between the observed values and the hypothesised median. However, the distribution needs to be continuous and reasonably symmetric.

Procedure

(i) State for the data the null and alternative hypotheses, H_0 and H_1

(ii) Know whether the stated significance level, α, is for a one-tailed or a two-tailed test (see (ii) in the procedure for the sign test on page 1022).

(iii) Find the difference of each piece of data compared with the null hypothesis (see Problems 8 and 9) or assign plus and minus signs to the difference for paired observations (see Problem 10).

(iv) Rank the differences, ignoring whether they are positive or negative.

(v) The Wilcoxon signed-rank statistic T is calculated as the sum of the ranks of either the positive differences or the negative differences – whichever is the smaller for a two-tailed test, and the one which would be expected to have the smaller value when H_1 is true for a one-tailed test.

(vi) Use Table 94.4 for given values of n, and α_1 or α_2 to read the critical region of T. For example, if, say, $n = 16$ and $\alpha_1 = 5\%$, then from Table 94.4, $T \leq 35$. Thus if T in part (v) is greater than 35 we accept the null hypothesis H_0 and if T is less than or equal to 35 we accept the alternative hypothesis H_1

This procedure for the Wilcoxon signed-rank test is demonstrated in the following Problems.

Problem 8. A manager of a manufacturer is concerned about suspected slow progress in dealing with orders. He wants at least half of the orders received to be processed within a working day (i.e. 7 hours). A little later he decides to time 17 orders selected at random, to check if his request has been

met. The times spent by the 17 orders being processed were as follows:

$4\frac{3}{4}$ h	$9\frac{3}{4}$ h	$15\frac{1}{2}$ h	11 h	$8\frac{1}{4}$ h	$6\frac{1}{2}$ h
9 h	$8\frac{3}{4}$ h	$10\frac{3}{4}$ h	$3\frac{1}{2}$ h	$8\frac{1}{2}$ h	$9\frac{1}{2}$ h
$15\frac{1}{4}$ h	13 h	8 h	$7\frac{3}{4}$ h	$6\frac{3}{4}$ h	

Use the Wilcoxon signed-rank test at a significance level of 5% to check if the manager's request for quicker processing is being met

(This is the same as Problem 5 where the sign test was used.)

Using the procedure:

(i) The hypotheses are H_0: $t = 7$ h and H_1: $t > 7$ h, where t is time.

(ii) Since H_1 is $t > 7$ h, a one-tailed test is assumed, i.e. $\alpha_1 = 5\%$

(iii) Taking the difference between the time taken for each order and 7 h gives:

$-2\frac{1}{4}$ h	$+2\frac{3}{4}$ h	$+8\frac{1}{2}$ h	$+4$ h	$+1\frac{1}{4}$ h
$-\frac{1}{2}$ h	$+2$ h	$+1\frac{3}{4}$ h	$+3\frac{3}{4}$ h	$-3\frac{1}{2}$ h
$+1\frac{1}{2}$ h	$+2\frac{1}{2}$ h	$+8\frac{1}{4}$ h	$+6$ h	$+1$ h
$+\frac{3}{4}$ h	$-\frac{1}{4}$ h			

(iv) These differences may now be ranked from 1 to 17, ignoring whether they are positive or negative:

Rank	1	2	3	4	5	6
Difference	$-\frac{1}{4}$	$-\frac{1}{2}$	$\frac{3}{4}$	1	$1\frac{1}{4}$	$1\frac{1}{2}$

Rank	7	8	9	10	11	12
Difference	$1\frac{3}{4}$	2	$-2\frac{1}{4}$	$2\frac{1}{2}$	$2\frac{3}{4}$	$-3\frac{1}{2}$

Rank	13	14	15	16	17
Difference	$3\frac{3}{4}$	4	6	$8\frac{1}{4}$	$8\frac{1}{2}$

(v) The Wilcoxon signed-rank statistic T is calculated as **the sum of the ranks** of the negative differences for a one-tailed test.

Table 94.4 Critical values for the Wilcoxon signed-rank test

n	$\alpha_1 = 5\%$ $\alpha_2 = 10\%$	$2\frac{1}{2}\%$ 5%	1% 2%	$\frac{1}{2}\%$ 1%	n	$\alpha_1 = 5\%$ $\alpha_2 = 10\%$	$2\frac{1}{2}\%$ 5%	1% 2%	$\frac{1}{2}\%$ 1%
1	—	—	—	—	26	110	98	84	75
2	—	—	—	—	27	119	107	92	83
3	—	—	—	—	28	130	116	101	91
4	—	—	—	—	29	140	126	110	100
5	0	—	—	—	30	151	137	120	109
6	2	0	—	—	31	163	147	130	118
7	3	2	0	—	32	175	159	140	128
8	5	3	1	0	33	187	170	151	138
9	8	5	3	1	34	200	182	162	148
10	10	8	5	3	35	213	195	173	159
11	13	10	7	5	36	227	208	185	171
12	17	13	9	7	37	241	221	198	182
13	21	17	12	9	38	256	235	211	194
14	25	21	15	12	39	271	249	224	207
15	30	25	19	15	40	286	264	238	220
16	35	29	23	19	41	302	279	252	233
17	41	34	27	23	42	319	294	266	247
18	47	40	32	27	43	336	310	281	261
19	53	46	37	32	44	353	327	296	276
20	60	52	43	37	45	371	343	312	291
21	67	58	49	42	46	389	361	328	307
22	75	65	55	48	47	407	378	345	322
23	83	73	62	54	48	426	396	362	339
24	91	81	69	61	49	446	415	379	355
25	100	89	76	68	50	466	434	397	373

The sum of the ranks for the negative values is:
$$T = 1 + 2 + 9 + 12 = \mathbf{24}$$

(vi) Table 94.4 gives the critical values of T for the Wilcoxon signed-rank test. For $n = 17$ and a significance level $\alpha_1 = 5\%$, $T \leq 41$

Hence the conclusion is that since $T = 24$ the result is within the 5% critical region. **There is therefore strong evidence to support H_1, the alternative hypothesis, that the median processing time is greater than 7 hours**.

Problem 9. The following data represents the number of hours that a portable car vacuum cleaner operates before recharging is required.

Operating time (h)	1.4	2.3	0.8	1.4	1.8	1.5
	1.9	1.4	2.1	1.1	1.6	

Use the Wilcoxon signed-rank test to test the hypothesis, at a 5% level of significance, that this particular vacuum cleaner operates, on average, 1.7 hours before needing a recharge

(This is the same as Problem 6 where the sign test was used.)

Using the procedure:

(i) $H_0: t = 1.7\,\text{h}$ and $H_1: t \neq 1.7\,\text{h}$

(ii) Significance level, $\alpha_2 = 5\%$ (since this is a two-tailed test).

(iii) Taking the difference between each operating time and 1.7h gives:

$$-0.3\text{h} \quad +0.6\text{h} \quad -0.9\text{h} \quad -0.3\text{h}$$

$$+0.1\text{h} \quad -0.2\text{h} \quad +0.2\text{h} \quad -0.3\text{h}$$

$$+0.4\text{h} \quad -0.6\text{h} \quad -0.1\text{h}$$

(iv) These differences may now be ranked from 1 to 11 (ignoring whether they are positive or negative).

Some of the differences are equal to each other. For example, there are two 0.1s (ignoring signs) that would occupy positions 1 and 2 when ordered. We average these as far as rankings are concerned, i.e. each is assigned a ranking of $\dfrac{1+2}{2}$ i.e. 1.5. Similarly, the two 0.2 values in positions 3 and 4 when ordered are each assigned rankings of $\dfrac{3+4}{2}$ i.e. 3.5, and the three 0.3 values in positions 5, 6, and 7 are each assigned a ranking of $\dfrac{5+6+7}{3}$ i.e. 6, and so on. The rankings are therefore:

Rank	1.5	1.5	3.5	3.5
Difference	+0.1	−0.1	−0.2	+0.2

Rank	6	6	6	8
Difference	−0.3	−0.3	−0.3	+0.4

Rank	9.5	9.5	11
Difference	+0.6	−0.6	−0.9

(v)   There are 4 positive terms and 7 negative terms. Taking the smaller number, the four positive terms have rankings of 1.5, 3.5, 8 and 9.5. Summing the positive ranks gives: $T = 1.5 + 3.5 + 8 + 9.5 = \mathbf{22.5}$

(vi) From Table 94.4, when $n = 11$ and $\alpha_2 = 5\%$, $\mathbf{T \leq 10}$

Since $T = 22.5$ falls in the acceptance region (i.e. in this case is greater than 10), **the null hypothesis is accepted, i.e. the average operating time is not significantly different from 1.7 h**

(Note that if, say, a piece of the given data was 1.7h, such that the difference was zero, that data is ignored and n would be 10 instead of 11 in this case.)

Problem 10.   An engineer is investigating two different types of metering devices, A and B, for an electronic fuel injection system to determine if they differ in their fuel mileage performance. The system is installed on 12 different cars, and a test is run with each metering system in turn on each car. The observed fuel mileage data (in miles/gallon) is shown below:

A	18.7	20.3	20.8	18.3	16.4	16.8
B	17.6	21.2	19.1	17.5	16.9	16.4

A	17.2	19.1	17.9	19.8	18.2	19.1
B	17.7	19.2	17.5	21.4	17.6	18.8

Use the Wilcoxon signed-rank test, at a level of significance of 5%, to determine whether there is any difference between the two systems

(This is the same as Problem 7 where the sign test was used.)

Using the procedure:

(i) $H_0: F_A = F_B$ and $H_1: F_A \neq F_B$ where F_A and F_B are the fuels in miles/gallon for systems A and B, respectively.

(ii) $\alpha_2 = 5\%$ (since it is a two-tailed test).

(iii) The difference between the observations is determined and a + or a − sign assigned to each as shown below:

$$(A - B) \quad \mathbf{+1.1} \quad \mathbf{-0.9} \quad \mathbf{+1.7} \quad \mathbf{+0.8}$$

$$\mathbf{-0.5} \quad \mathbf{+0.4} \quad \mathbf{-0.5} \quad \mathbf{-0.1}$$

$$\mathbf{+0.4} \quad \mathbf{-1.6} \quad \mathbf{+0.6} \quad \mathbf{+0.3}$$

(iv) The differences are now ranked from 1 to 12 (ignoring whether they are positive or negative).

When ordered, 0.4 occupies positions 3 and 4; their average is 3.5 and both are assigned this value when ranked. Similarly 0.5 occupies positions 5 and 6 and their average of 5.5 is assigned to each when ranked.

Rank	1	2	3.5	3.5
Difference	−0.1	+0.3	+0.4	+0.4

Rank	5.5	5.5	7	8
Difference	−0.5	−0.5	+0.6	+0.8

Rank	9	10	11	12
Difference	−0.9	+1.1	−1.6	+1.7

(v) There are 7 '+ signs' and 5 '− signs'. Taking the smaller number, the negative signs have rankings of 1, 5.5, 5.5, 9 and 11

Summing the negative ranks gives: $T = 1 + 5.5 + 5.5 + 9 + 11 = \mathbf{32}$

(vi) From Table 94.4, when $n = 12$ and $\alpha_2 = 5\%$, $T \leq \mathbf{13}$

Since from (v), T is not equal or less than 13, **the null hypothesis cannot be rejected, i.e. the two metering devices produce the same fuel mileage performance**.

Now try the following Practice Exercise

Practice Exercise 353   Further problems on the Wilcoxon signed-rank test (answers on page 1192)

1. The time to repair an electronic instrument is a random variable. The repair times (in hours) for 16 instruments are as follows:

218	275	264	210	161	374	178	265
150	360	185	171	215	100	474	248

   Use the Wilcoxon signed-rank test, at a 5% level of significance, to test the hypothesis that the mean repair time is 220 hours.

2. 18 samples of serum are analysed for their sodium content. The results, expressed as ppm, are as follows:

169	151	166	155	149	154
164	151	147	142	168	152
149	129	153	154	149	143

At a level of significance of 5%, use the Wilcoxon signed-rank test to test the null hypothesis that the average value for the method of analysis used is 150 ppm.

3. A paint supplier claims that a new additive will reduce the drying time of their acrylic paint. To test his claim, 12 pieces of wood are painted, one half of each piece with paint containing the regular additive and the other half with paint containing the new additive. The drying times (in hours) were measured as follows:

New additive	4.5	5.5	3.9	3.6	4.1	6.3
Regular additive	4.7	5.9	3.9	3.8	4.4	6.5

New additive	5.9	6.7	5.1	3.6	4.0	3.0
Regular additive	6.9	6.5	5.3	3.6	3.9	3.9

Use the Wilcoxon signed-rank test at a significance level of 5% to test the hypothesis that there is no difference, on average, in the drying times of the new and regular additive paints.

94.6   The Mann–Whitney test

As long as the sample sizes are not too large, for tests involving two samples the Mann–Whitney test is easy to apply, is powerful and is widely used.

Procedure

(i) State for the data the null and alternative hypotheses, H_0 and H_1

(ii) Know whether the stated significance level, α, is for a one-tailed or a two-tailed test (see (ii) in the procedure for the sign test on page 1022).

(iii) Arrange all the data in ascending order whilst retaining their separate identities.

(iv) If the data is now a mixture of, say, As and Bs, write under each letter A the number of Bs that precede it in the sequence (or vice versa).

(v) Add together the numbers obtained from (iv) and denote total by U. U is defined as whichever type of count would be expected to be smallest when H_1 is true.

(vi) Use Table 94.5 on pages 1030–1033 for given values of n_1 and n_2, and α_1 or α_2 to read the critical region of U. For example, if, say, $n_1 = 10$ and $n_2 = 16$ and $\alpha_2 = 5\%$, then from Table 94.5, $U \le 42$. If U in part (v) is greater than 42 we accept the null hypothesis H_0, and if U is equal to or less than 42, we accept the alternative hypothesis H_1

The procedure for the Mann–Whitney test is demonstrated in the following problems.

Problem 11. 10 British cars and 8 non-British cars are compared for faults during their first 10 000 miles of use. The percentage of cars of each type developing faults were as follows:

Non-British cars, P	5	8	14	10	15

British cars, Q	18	9	25	6	21

Non-British cars, P	7	12	4

British cars, Q	20	28	11	16	34

Use the Mann–Whitney test, at a level of significance of 1%, to test whether non-British cars have better average reliability than British models

Using the above procedure:

(i) The hypotheses are:

H_0: Equal proportions of British and non-British cars have breakdowns.

H_1: A higher proportion of British cars have breakdowns.

(ii) Level of significance $\alpha_1 = 1\%$

(iii) Let the sizes of the samples be n_P and n_Q, where $n_P = 8$ and $n_Q = 10$. The Mann–Whitney test compares every item in sample P in turn with every item in sample Q, a record being kept of the number of times, say, that the item from P is greater than Q, or vice versa. In this case there are $n_P n_Q$, i.e. $(8)(10) = 80$ comparisons to be made. All the data is arranged into ascending order whilst retaining their separate identities – an easy way is to arrange a linear scale as shown in Figure 94.1, on page 1033.

From Figure 94.1, a list of Ps and Qs can be ranked giving:

$P\ P\ Q\ P\ P\ Q\ P\ Q\ P\ P\ P\ Q\ Q\ Q$

$Q\ Q\ Q\ Q$

(iv) Write under each letter P the number of Qs that precede it in the sequence, giving:

P	P	Q	P	P	Q	P	Q	P	P	P	Q
0	0		1	1		2		3	3	3	

$Q\ Q\ Q\ Q\ Q\ Q$

(v) Add together these 8 numbers, denoting the sum by U, i.e.

$$U = 0 + 0 + 1 + 1 + 2 + 3 + 3 + 3 = \mathbf{13}$$

(vi) The critical regions are of the form $U \le$ critical region.

From Table 94.5, for a sample size 8 and 10 at significance level $\alpha_1 = 1\%$ the critical region is $U \le \mathbf{13}$

The value of U in our case, from (v), is 13 which is significant at 1% significance level.

The Mann–Whitney test has therefore confirmed that **there is evidence that the non-British cars have better reliability than the British cars in the first 10 000 miles, i.e. the alternative hypothesis applies**.

Problem 12. Two machines, A and B, are used to measure vibration in a particular rubber product. The data given below are the vibrational forces, in kilograms, of random samples from each machine (continued on page 1033):

Table 94.5 Critical values for the Mann–Whitney test

n_1	n_2	$\alpha_1 = 5\%$ $\alpha_2 = 10\%$	$2\frac{1}{2}\%$ 5%	1% 2%	$\frac{1}{2}\%$ 1%	n_1	n_2	$\alpha_1 = 5\%$ $\alpha_2 = 10\%$	$2\frac{1}{2}\%$ 5%	1% 2%	$\frac{1}{2}\%$ 1%
2	2	—	—	—	—	3	13	6	4	2	1
2	3	—	—	—	—	3	14	7	5	2	1
2	4	—	—	—	—	3	15	7	5	3	2
2	5	0	—	—	—	3	16	8	6	3	2
2	6	0	—	—	—	3	17	9	6	4	2
2	7	0	—	—	—	3	18	9	7	4	2
2	8	1	0	—	—	3	19	10	7	4	3
2	9	1	0	—	—	3	20	11	8	5	3
2	10	1	0	—	—						
2	11	1	0	—	—	4	4	1	0	—	—
2	12	2	1	—	—	4	5	2	1	0	—
2	13	2	1	0	—	4	6	3	2	1	0
2	14	3	1	0	—	4	7	4	3	1	0
2	15	3	1	0	—	4	8	5	4	2	1
2	16	3	1	0	—	4	9	6	4	3	1
2	17	3	2	0	—	4	10	7	5	3	2
2	18	4	2	0	—	4	11	8	6	4	2
2	19	4	2	1	0	4	12	9	7	5	3
2	20	4	2	1	0	4	13	10	8	5	3
						4	14	11	9	6	4
3	3	0	—	—	—	4	15	12	10	7	5
3	4	0	—	—	—	4	16	14	11	7	5
3	5	1	0	—	—	4	17	15	11	8	6
3	6	2	1	—	—	4	18	16	12	9	6
3	7	2	1	0	—	4	19	17	13	9	7
3	8	3	2	0	—	4	20	18	14	10	8
3	9	4	2	1	0						
3	10	4	3	1	0	5	5	4	2	1	0
3	11	5	3	1	0	5	6	5	3	2	1
3	12	5	4	2	1	5	7	6	5	3	1

Table 94.5 (*Continued*)

n_1	n_2	$\alpha_1 = 5\%$ $\alpha_2 = 10\%$	$2\frac{1}{2}\%$ 5%	1% 2%	$\frac{1}{2}\%$ 1%	n_1	n_2	$\alpha_1 = 5\%$ $\alpha_2 = 10\%$	$2\frac{1}{2}\%$ 5%	1% 2%	$\frac{1}{2}\%$ 1%
5	8	8	6	4	2	7	7	11	8	6	4
5	9	9	7	5	3	7	8	13	10	7	6
5	10	11	8	6	4	7	9	15	12	9	7
5	11	12	9	7	5	7	10	17	14	11	9
5	12	13	11	8	6	7	11	19	16	12	10
5	13	15	12	9	7	7	12	21	18	14	12
5	14	16	13	10	7	7	13	24	20	16	13
5	15	18	14	11	8	7	14	26	22	17	15
5	16	19	15	12	9	7	15	28	24	19	16
5	17	20	17	13	10	7	16	30	26	21	18
5	18	22	18	14	11	7	17	33	28	23	19
5	19	23	19	15	12	7	18	35	30	24	21
5	20	25	20	16	13	7	19	37	32	26	22
						7	20	39	34	28	24
6	6	7	5	3	2						
6	7	8	6	4	3	8	8	15	13	9	7
6	8	10	8	6	4	8	9	18	15	11	9
6	9	12	10	7	5	8	10	20	17	13	11
6	10	14	11	8	6	8	11	23	19	15	13
6	11	16	13	9	7	8	12	26	22	17	15
6	12	17	14	11	9	8	13	28	24	20	17
6	13	19	16	12	10	8	14	31	26	22	18
6	14	21	17	13	11	8	15	33	29	24	20
6	15	23	19	15	12	8	16	36	31	26	22
6	16	25	21	16	13	8	17	39	34	28	24
6	17	26	22	18	15	8	18	41	36	30	26
6	18	28	24	19	16	8	19	44	38	32	28
6	19	30	25	20	17	8	20	47	41	34	30
6	20	32	27	22	18						

Table 94.5 (*Continued*)

n_1	n_2	$\alpha_1 = 5\%$ $\alpha_2 = 10\%$	$2\frac{1}{2}\%$ 5%	1% 2%	$\frac{1}{2}\%$ 1%	n_1	n_2	$\alpha_1 = 5\%$ $\alpha_2 = 10\%$	$2\frac{1}{2}\%$ 5%	1% 2%	$\frac{1}{2}\%$ 1%
9	9	21	17	14	11	11	16	54	47	41	36
9	10	24	20	16	13	11	17	57	51	44	39
9	11	27	23	18	16	11	18	61	55	47	42
9	12	30	26	21	18	11	19	65	58	50	45
9	13	33	28	23	20	11	20	69	62	53	48
9	14	36	31	26	22						
9	15	39	34	28	24	12	12	42	37	31	27
9	16	42	37	31	27	12	13	47	41	35	31
9	17	45	39	33	29	12	14	51	45	38	34
9	18	48	42	36	31	12	15	55	49	42	37
9	19	51	45	38	33	12	16	60	53	46	41
9	20	54	48	40	36	12	17	64	57	49	44
						12	18	68	61	53	47
10	10	27	23	19	16	12	19	72	65	56	51
10	11	31	26	22	18	12	20	77	69	60	54
10	12	34	29	24	21						
10	13	37	33	27	24	13	13	51	45	39	34
10	14	41	36	30	26	13	14	56	50	43	38
10	15	44	39	33	29	13	15	61	54	47	42
10	16	48	42	36	31	13	16	65	59	51	45
10	17	51	45	38	34	13	17	70	63	55	49
10	18	55	48	41	37	13	18	75	67	59	53
10	19	58	52	44	39	13	19	80	72	63	57
10	20	62	55	47	42	13	20	84	76	67	60
11	11	34	30	25	21	14	14	61	55	47	42
11	12	38	33	28	24	14	15	66	59	51	46
11	13	42	37	31	27	14	16	71	64	56	50
11	14	46	40	34	30	14	17	77	69	60	54
11	15	50	44	37	33	14	18	82	74	65	58

Table 94.5 (*Continued*)

n_1	n_2	$\alpha_1 = 5\%$ $\alpha_2 = 10\%$	$2\frac{1}{2}\%$ 5%	1% 2%	$\frac{1}{2}\%$ 1%	n_1	n_2	$\alpha_1 = 5\%$ $\alpha_2 = 10\%$	$2\frac{1}{2}\%$ 5%	1% 2%	$\frac{1}{2}\%$ 1%
14	19	87	78	69	63	17	17	96	87	77	70
14	20	92	83	73	67	17	18	102	92	82	75
						17	19	109	99	88	81
15	15	72	64	56	51	17	20	115	105	93	86
15	16	77	70	61	55						
15	17	83	75	66	60	18	18	109	99	88	81
15	18	88	80	70	64	18	19	116	106	94	87
15	19	94	85	75	69	18	20	123	112	100	92
15	20	100	90	80	73						
						19	19	123	112	101	93
16	16	83	75	66	60	19	20	130	119	107	99
16	17	89	81	71	65						
16	18	95	86	76	70	20	20	138	127	114	105
16	19	101	92	82	74						
16	20	107	98	87	79						

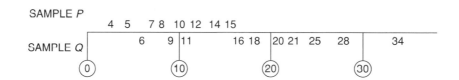

Figure 94.1

(continued from page 1029)

A   9.7   10.2   11.2   12.4   14.1   22.3
     29.6   31.7   33.0   33.2   33.4   46.2
     50.7   52.5   55.4

B   20.6   25.3   29.2   35.2   41.9   48.5
     54.1   57.1   59.8   63.2   68.5

Use the Mann–Whitney test at a significance level of 5% to determine if there is any evidence of the two machines producing different results

Using the procedure:

(i) H_0: There is no difference in results from the machines, on average.

      H_1: The results from the two machines are different, on average.

(ii) $\alpha_2 = 5\%$

(iii)   Arranging the data in order gives:

9.7	10.2	11.2	12.4	14.1	20.6	22.3
A	A	A	A	A	B	A

25.3	29.2	29.6	31.7	33.0	33.2	33.4
B	B	A	A	A	A	A

35.2	41.9	46.2	48.5	50.7	52.5	54.1
B	B	A	B	A	A	B

55.4	57.1	59.8	63.2	68.5
A	B	B	B	B

(iv)   The number of Bs preceding the As in the sequence is as follows:

A	A	A	A	A	B	A	B	B
0	0	0	0	0		1		

A	A	A	A	A	B	B	A	B
3	3	3	3	3			5	

A	A	B	A	B	B	B	B
6	6		7				

(v) Adding the numbers from (iv) gives:

$$U = 0+0+0+0+0+1+3+3+3+3$$
$$+3+5+6+6+7 = \mathbf{40}$$

(vi) From Table 94.5, for $n_1 = 11$ and $n_2 = 15$, and $\alpha_2 = 5\%$, $U \leq \mathbf{44}$

Since our value of U from (v) is less than 44, H_0 is rejected and H_1 accepted, i.e. **the results from the two machines are different**.

Now try the following Practice Exercise

Practice Exercise 354   Further problems on the Mann–Whitney test (answers on page 1192)

1. The tar content of two brands of cigarettes (in mg) was measured as follows:

Brand P	22.6	4.1	3.9	0.7	3.2
Brand Q	3.4	6.2	3.5	4.7	6.3

Brand P	6.1	1.7	2.3	5.6	2.0
Brand Q	5.5	3.8	2.1		

Use the Mann–Whitney test at a 0.05 level of significance to determine if the tar contents of the two brands are equal.

2. A component is manufactured by two processes. Some components from each process are selected at random and tested for breaking strength to determine if there is a difference between the processes. The results are:

Process A	9.7	10.5	10.1	11.6	9.8
Process B	11.3	8.6	9.6	10.2	10.9

Process A	8.9	11.2	12.0	9.2
Process B	9.4	10.8		

At a level of significance of 10%, use the Mann–Whitney test to determine if there is a difference between the mean breaking strengths of the components manufactured by the two processes.

3. An experiment, designed to compare two preventive methods against corrosion gave the following results for the maximum depths of pits (in mm) in metal strands:

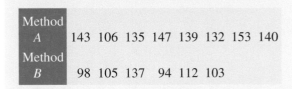

Method A	143	106	135	147	139	132	153	140
Method B	98	105	137	94	112	103		

Use the Mann–Whitney test, at a level of significance of 0.05, to determine whether the two tests are equally effective.

4. Repeat Problem 3 of Exercise 352, page 1024, using the Mann–Whitney test.

For fully worked solutions to each of the problems in Practice Exercises 350 to 354 in this chapter, go to the website:
www.routledge.com/cw/bird

This assignment covers the material contained in Chapters 92 to 94. *The marks available are shown in brackets at the end of each question.*

1. 1200 metal bolts have a mean mass of 7.2 g and a standard deviation of 0.3 g. Determine the standard error of the means. Calculate also the probability that a sample of 60 bolts chosen at random, without replacement, will have a mass of (a) between 7.1 g and 7.25 g, and (b) more than 7.3 g   (12)

2. A sample of 10 measurements of the length of a component are made and the mean of the sample is 3.650 cm. The standard deviation of the samples is 0.030 cm. Determine (a) the 99% confidence limits, and (b) the 90% confidence limits for an estimate of the actual length of the component.   (10)

3. An automated machine produces metal screws, and over a period of time it is found that 8% are defective. Random samples of 75 screws are drawn periodically.
   (a) If a decision is made that production continues until a sample contains more than 8 defective screws, determine the type I error based on this decision for a defect rate of 8%.
   (b) Determine the magnitude of the type II error when the defect rate has risen to 12%.

   The above sample size is now reduced to 55 screws. The decision now is to stop the machine for adjustment if a sample contains 4 or more defective screws.
   (c) Determine the type I error if the defect rate remains at 8%.
   (d) Determine the type II error when the defect rate rises to 9%   (22)

4. In a random sample of 40 similar light bulbs drawn from a batch of 400, the mean lifetime is found to be 252 hours. The standard deviation of the lifetime of the sample is 25 hours. The batch is classed as inferior if the mean lifetime of the batch is less than the population mean of 260 hours. As a result of the sample data, determine whether the batch is considered to be inferior at a level of significance of (a) 0.05, and (b) 0.01   (9)

5. The lengths of two products are being compared.
   Product 1: sample size = 50, mean value of sample = 6.5 cm, standard deviation of whole of batch = 0.40 cm.
   Product 2: sample size = 60, mean value of sample = 6.65 cm, standard deviation of whole of batch = 0.35 cm.

   Determine if there is any significant difference between the two products at a level of significance of (a) 0.05 and (b) 0.01   (7)

6. The resistances of a sample of 400 resistors produced by an automatic process have the following distribution.

Resistance (Ω)	Frequency
50.11	9
50.15	35
50.19	61
50.23	102
50.27	89
50.31	83
50.35	21

   Calculate for the sample: (a) the mean and (b) the standard deviation. (c) Test the null hypothesis that the resistances of the resistors are normally distributed at a level of significance of 0.05, and determine if the distribution gives a 'too good' fit at a level of confidence of 90%   (25)

7. A fishing line is manufactured by two processes, *A* and *B*. To determine if there is a difference in the mean breaking strengths of the lines, 8 lines by each process are selected and tested for breaking strength. The results are as follows:

| Process A | 8.6 | 7.1 | 6.9 | 6.5 | 7.9 | 6.3 | 7.8 | 8.1 |
| Process B | 6.8 | 7.6 | 8.2 | 6.2 | 7.5 | 8.9 | 8.0 | 8.7 |

Determine if there is a difference between the mean breaking strengths of the lines manufactured by the two processes, at a significance level of 0.10, using (a) the sign test, (b) the Wilcoxon signed-rank test, (c) the Mann–Whitney test. (15)

Multiple choice questions Test 9
Statistics
This test covers the material in Chapters 85 to 94

All questions have only one correct answer (answers on page 1200).

1. A pie diagram is shown in Figure M9.1 where P, Q, R and S represent the salaries of four employees of a firm. P earns £24 000 p.a. Employee S earns:

    (a) £40 000 (b) £36 000 (c) £20 000 (d) £24 000

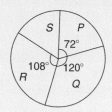

Figure M9.1

Questions 2 to 5 relate to the following information.

The capacitance (in pF) of 6 capacitors is as follows: {5, 6, 8, 5, 10, 2}

2. The median value is:

    (a) 36 pF   (b) 6 pF   (c) 5.5 pF   (d) 5 pF

3. The modal value is:

    (a) 36 pF   (b) 6 pF   (c) 5.5 pF   (d) 5 pF

4. The mean value is:

    (a) 36 pF   (b) 6 pF   (c) 5.5 pF   (d) 5 pF

5. The standard deviation is:

    (a) 2.66 pF   (b) 2.52 pF   (c) 2.45 pF   (d) 6.33 pF

6. The curve obtained by joining the co-ordinates of cumulative frequency against upper class boundary values is called;

    (a)   a histogram

    (b)   a frequency polygon

    (c)   a tally diagram

    (d)   an ogive

Questions 7 to 9 relate to the following information.

A box contains 35 brass washers, 40 steel washers and 25 aluminium washers. Three washers are drawn at random from the box without replacement.

7. The probability that all three are steel washers is:

    (a) 0.0611   (b) 1.200   (c) 0.0640   (d) 1.182

8. The probability that there are no aluminium washers is:

    (a) 2.250   (b) 0.418   (c) 0.014   (d) 0.422

9. The probability that there are two brass washers and either a steel or an aluminium washer is:

    (a) 0.071   (b) 0.687   (c) 0.239   (d) 0.343

10. The coefficient of correlation between two variables is -0.75. This indicates:

    (a)   a very good direct correlation

    (b)   a fairly good direct correlation

    (c)   a very good indirect correlation

    (d)   a fairly good indirect correlation

11. The number of faults occurring on a production line in a 9-week period are as shown:

    32 29 27 26 29 39 33 29 37

    The third quartile value is:

    (a) 29   (b) 35   (c) 31   (d) 28

12. 2% of the components produced by a manufacturer are defective. Using the Poisson distribution the percentage probability that more than two will be defective in a sample of 100 components is:

    (a) 13.5%   (b) 32.3%   (c) 27.1%   (d) 59.4%

Questions 13 and 14 relate to the following information.

The frequency distribution for the values of resistance in ohms of 40 transistors is as follows:

15.5–15.9 3   16.0–16.4 10   16.5–16.9 13

17.0–17.4 8   17.5–17.9 6

13. The mean value of the resistance is:

    (a) 16.75 Ω   (b) 1.0 Ω   (c) 15.85 Ω   (d) 16.95 Ω

14. The standard deviation is:

    (a) $0.335\,\Omega$ (b) $0.251\,\Omega$ (c) $0.682\,\Omega$ (d) $0.579\,\Omega$

15. A manufacturer estimates that 4% of components produced are defective. Using the binomial distribution, the percentage probability that less than two components will be defective in a sample of 10 components is:

    (a) 0.40% (b) 5.19% (c) 0.63% (d) 99.4%

Questions 16 to 18 relate to the following information. The probability of a component failing in one year due to excessive temperature is $\dfrac{1}{16}$, due to excessive vibration is $\dfrac{1}{20}$ and due to excessive humidity is $\dfrac{1}{40}$.

16. The probability that a component fails due to excessive temperature and excessive vibration is:

    (a) $\dfrac{285}{320}$ (b) $\dfrac{1}{320}$ (c) $\dfrac{9}{80}$ (d) $\dfrac{1}{800}$

17. The probability that a component fails due to excessive vibration or excessive humidity is:

    (a) 0.00125 (b) 0.00257 (c) 0.0750 (d) 0.1125

18. The probability that a component will not fail because of both excessive temperature and excessive humidity is:

    (a) 0.914 (b) 1.913 (c) 0.00156 (d) 0.0875

Questions 19 and 20 relate to the following information. A set of measurements (in mm) is as follows: {4, 5, 2, 11, 7, 6, 5, 1, 5, 8, 12, 6}

19. The median is:

    (a) 6 mm (b) 5 mm (c) 72 mm (d) 5.5 mm

20. The mean is:

    (a) 6 mm (b) 5 mm (c) 72 mm (d) 5.5 mm

21. The lengths of 2000 components are normally distributed with a mean of 80 mm and a standard deviation of 4 mm. If random samples are taken of 50 components without replacement, the standard error of the means is:

    (a) 0.559 mm (b) 8 mm
    (c) 0.566 mm (d) 0.088 mm

Questions 22 to 24 relate to the following information. The mean height of 400 people is 170 cm and the standard deviation is 8 cm. Assume a normal distribution.

22. The number of people likely to have heights of between 154 cm and 186 cm is:

    (a) 390 (b) 380 (c) 190 (d) 185

23. The number of people likely to have heights less than 162 cm is:

    (a) 133 (b) 380 (c) 67 (d) 185

24. The number of people likely to have a height of more than 186 cm is:

    (a) 10 (b) 67 (c) 137 (d) 20

25. Five numbers between 1 and 25 are chosen when playing a lottery. The probability of winning the top prize (i.e. 5 correct numbers) if 20 tickets were purchased and five different numbers were chosen on each ticket is:

    (a) $\dfrac{1}{6375600}$ (b) $\dfrac{1}{53130}$ (c) $\dfrac{1}{255024}$ (d) $\dfrac{2}{5313}$

Mean, median, mode and standard deviation

If x = variate and f = frequency then:

mean $\bar{x} = \dfrac{\sum fx}{\sum f}$

The **median** is the middle term of a ranked set of data.

The **mode** is the most commonly occurring value in a set of data

Standard deviation $\sigma = \sqrt{\left[\dfrac{\sum\left\{f(x-\bar{x})^2\right\}}{\sum f}\right]}$ for a population

Probability

The addition law
The probability of events A **or** B **or** C **or** ... occurring is given by:

$$p_A + p_B + p_C + \ldots$$

The multiplication law
The probability of events A **and** B **and** C **and** ... occurring is given by:

$$p_A \times p_B \times p_C \times \ldots$$

Binomial probability distribution

If n = number in sample, p = probability of the occurrence of an event and $q = 1 - p$, then the probability of 0, 1, 2, 3, ... occurrences is given by:

$$q^n, nq^{n-1}p, \frac{n(n-1)}{2!}q^{n-2}p^2, \frac{n(n-1)(n-2)}{3!}q^{n-3}p^3, \ldots$$

(i.e. successive terms of the $(q+p)^n$ expansion)

Normal approximation to a binomial distribution

Mean $= np$ Standard deviation $\sigma = \sqrt{(npq)}$

Poisson distribution

If λ is the expectation of the occurrence of an event then the probability of 0, 1, 2, 3, .. occurrences is given by:

$$e^{-\lambda}, \lambda e^{-\lambda}, \lambda^2 \frac{e^{-\lambda}}{2!}, \lambda^3 \frac{e^{-\lambda}}{3!}, \ldots$$

Product-moment formula for the linear correlation coefficient

Coefficient of correlation $r = \dfrac{\sum xy}{\sqrt{[(\sum x^2)(\sum y^2)]}}$

where $x = X - \bar{X}$ and $y = Y - \bar{Y}$ and (X_1, Y_1), (X_2, Y_2), ... denote a random sample from a bivariate normal distribution and $\bar{X}$ and $\bar{Y}$ are the means of the X and Y values, respectively.

Normal probability distribution: Partial areas under the standardised normal curve – see Table 89.1 on page 966.

Student's t distribution: Percentile values (t_p) for Student's t distribution with ν degrees of freedom – see Table 92.2 on page 993.

Chi-square distribution: Percentile values (χ_p^2) for the Chi-square distribution with ν degrees of freedom – see Table 94.1 on page 1016.

$\chi^2 = \sum \left\{ \dfrac{(o-e)^2}{e} \right\}$ where o and e are the observed and expected frequencies.

Symbols

Population: number of members N_p mean μ standard deviation σ
Sample: number of members N mean $\bar{x}$ standard deviation s
Sampling distributions: mean of sampling distribution of means $\mu_{\bar{x}}$ standard error of means $\sigma_{\bar{x}}$ standard error of the standard deviations σ_s

Standard error of the means

Standard error of the means of a sample distribution, i.e. the standard deviation of the means of samples, is:

$\sigma_{\bar{x}} = \dfrac{\sigma}{\sqrt{N}}\sqrt{\left(\dfrac{N_p - N}{N_p - 1}\right)}$ for a finite population and/or for sampling without replacement, and

$\sigma_{\bar{x}} = \dfrac{\sigma}{\sqrt{N}}$ for an infinite population and/or for sampling with replacement.

The relationship between sample mean and population mean

$\mu_{\bar{x}} = \mu$ for all possible samples of size N which are drawn from a population of size N_p

Estimating the mean of a population (σ known)

The confidence coefficient for a large sample size ($N \geq 30$) is z_c where:

Confidence level %	Confidence coefficient z_c
99	2.58
98	2.33
96	2.05
95	1.96
90	1.645
80	1.28
50	0.6745

The confidence limits of a population mean based on sample data are given by:

$$\bar{x} \pm \frac{z_c \sigma}{\sqrt{N}} \sqrt{\left(\frac{N_p - N}{N_p - 1} \right)} \text{ for a finite population of size } N_p,$$

and by $\bar{x} \pm \dfrac{z_c \sigma}{\sqrt{N}}$ for an infinite population.

Estimating the mean of a population (σ unknown)

The confidence limits of a population mean based on sample data are given by:

$$\mu_{\bar{x}} \pm z_c \sigma_{\bar{x}}$$

Estimating the standard deviation of a population

The confidence limits of the standard deviation of a population based on sample data are given by:

$$s \pm z_c \sigma_s$$

Estimating the mean of a population based on a small sample size

The confidence coefficient for a small sample size ($N < 30$) is t_c which can be determined using Table 92.1, page 988. The confidence limits of a population mean based on sample data given by:

$$\bar{x} \pm \frac{t_c s}{\sqrt{(N-1)}}$$

Section L

Section M

Laplace transforms

Chapter 95

Introduction to Laplace transforms

Why it is important to understand: Introduction to Laplace transforms

The Laplace transform is a very powerful mathematical tool applied in various areas of engineering and science. With the increasing complexity of engineering problems, Laplace transforms help in solving complex problems with a very simple approach; the transform is an integral transform method which is particularly useful in solving linear ordinary differential equations. It has very wide applications in various areas of physics, electrical engineering, control engineering, optics, mathematics and signal processing. This chapter just gets us started in understanding some standard Laplace transforms.

At the end of this chapter, you should be able to:

- define a Laplace transform
- recognise common notations used for the Laplace transform
- derive Laplace transforms of elementary functions
- use a standard list of Laplace transforms to determine the transform of common functions

95.1 Introduction

The solution of most electrical circuit problems can be reduced ultimately to the solution of differential equations. The use of **Laplace* transforms** provides an alternative method to those discussed in Chapters 77 to 82 for solving linear differential equations.

95.2 Definition of a Laplace transform

The Laplace transform of the function $f(t)$ is defined by the integral $\int_0^\infty e^{-st} f(t)\, dt$, where s is a parameter assumed to be a real number.

Common notations used for the Laplace transform

There are various commonly used notations for the Laplace transform of $f(t)$ and these include:

(i) $\mathcal{L}\{f(t)\}$ or $L\{f(t)\}$

(ii) $\mathcal{L}(f)$ or Lf

(iii) $\overline{f}(s)$ or $f(s)$

Also, the letter p is sometimes used instead of s as the parameter. The notation adopted in this book will be $f(t)$ for the original function and $\mathcal{L}\{f(t)\}$ for its Laplace transform.

---

* **Who was Laplace**? See page 909 for image and resume of Laplace. To find out more go to **www.routledge.com/cw/bird**

Hence, from above:

$$\mathcal{L}\{f(t)\} = \int_0^\infty e^{-st} f(t)\,dt \qquad (1)$$

95.3 Linearity property of the Laplace transform

From equation (1),

$$\mathcal{L}\{kf(t)\} = \int_0^\infty e^{-st} k\, f(t)\,dt$$

$$= k \int_0^\infty e^{-st} f(t)\,dt$$

i.e. $\quad \mathcal{L}\{k\,f(t)\} = k\mathcal{L}\{f(t)\} \qquad (2)$

where k is any constant.
Similarly,

$$\mathcal{L}\{a\,f(t) + bg(t)\} = \int_0^\infty e^{-st}(a\,f(t) + bg(t))\,dt$$

$$= a \int_0^\infty e^{-st} f(t)\,dt$$

$$+ b \int_0^\infty e^{-st} g(t)\,dt$$

i.e. $\quad \mathcal{L}\{af(t) + bg(t)\} = a\mathcal{L}\{f(t)\} + b\mathcal{L}\{g(t)\}, \qquad (3)$

where a and b are any real constants.
The Laplace transform is termed a **linear operator** because of the properties shown in equations (2) and (3).

95.4 Laplace transforms of elementary functions

Using the definition of the Laplace transform in equation (1), a number of elementary functions may be transformed. For example:

(a) $\quad f(t) = 1$. From equation (1),

$$\mathcal{L}\{1\} = \int_0^\infty e^{-st}(1)\,dt = \left[\frac{e^{-st}}{-s}\right]_0^\infty$$

$$= -\frac{1}{s}[e^{-s(\infty)} - e^0] = -\frac{1}{s}[0 - 1]$$

$$= \frac{1}{s} \text{ (provided } s > 0)$$

(b) $\quad f(t) = k$. From equation (2),

$$\mathcal{L}\{k\} = k\mathcal{L}\{1\}$$

Hence $\mathcal{L}\{k\} = k\left(\dfrac{1}{s}\right) = \dfrac{k}{s}$, from (a) above.

(c) $\quad f(t) = e^{at}$ (where a is a real constant $\neq 0$).
From equation (1),

$$\mathcal{L}\{e^{at}\} = \int_0^\infty e^{-st}(e^{at})\,dt = \int_0^\infty e^{-(s-a)t}\,dt,$$

from the laws of indices,

$$= \left[\frac{e^{-(s-a)t}}{-(s-a)}\right]_0^\infty = \frac{1}{-(s-a)}(0-1)$$

$$= \frac{1}{s-a}$$

(provided $(s - a) > 0$, i.e. $s > a$)

(d) $\quad f(t) = \cos at$ (where a is a real constant).
From equation (1),

$$\mathcal{L}\{\cos at\} = \int_0^\infty e^{-st} \cos at\,dt$$

$$= \left[\frac{e^{-st}}{s^2 + a^2}(a\sin at - s\cos at)\right]_0^\infty$$

by integration by parts twice (see page 753),

$$= \left[\frac{e^{-s(\infty)}}{s^2 + a^2}(a\sin a(\infty) - s\cos a(\infty))\right.$$

$$\left. - \frac{e^0}{s^2 + a^2}(a\sin 0 - s\cos 0)\right]$$

$$= \frac{s}{s^2 + a^2} \text{ (provided } s > 0)$$

(e) $\quad f(t) = t$. From equation (1),

$$\mathcal{L}\{t\} = \int_0^\infty e^{-st} t\,dt = \left[\frac{te^{-st}}{-s} - \int \frac{e^{-st}}{-s}\,dt\right]_0^\infty$$

$$= \left[\frac{te^{-st}}{-s} - \frac{e^{-st}}{s^2}\right]_0^\infty$$

by integration by parts,

$$= \left[\frac{\infty e^{-s(\infty)}}{-s} - \frac{e^{-s(\infty)}}{s^2}\right] - \left[0 - \frac{e^0}{s^2}\right]$$

$$= (0 - 0) - \left(0 - \frac{1}{s^2}\right)$$

since $(\infty \times 0) = 0$,

$$= \frac{1}{s^2} \text{ (provided } s > 0)$$

(f) $\quad f(t) = t^n$ (where $n = 0, 1, 2, 3, \ldots$).

By a similar method to (e) it may be shown that $\mathcal{L}\{t^2\} = \dfrac{2}{s^3}$ and $\mathcal{L}\{t^3\} = \dfrac{(3)(2)}{s^4} = \dfrac{3!}{s^4}$. These results can be extended to n being any positive integer.

Thus $\mathcal{L}\{t^n\} = \dfrac{n!}{s^{n+1}}$ (provided $s > 0$)

(g) $f(t) = \sinh at$. From Chapter 24, $\sinh at = \dfrac{1}{2}(e^{at} - e^{-at})$. Hence,

$$\mathcal{L}\{\sinh at\} = \mathcal{L}\left\{\dfrac{1}{2}e^{at} - \dfrac{1}{2}e^{-at}\right\}$$

$$= \dfrac{1}{2}\mathcal{L}\{e^{at}\} - \dfrac{1}{2}\mathcal{L}\{e^{-at}\}$$

from equations (2) and (3),

$$= \dfrac{1}{2}\left[\dfrac{1}{s-a}\right] - \dfrac{1}{2}\left[\dfrac{1}{s+a}\right]$$

from (c) above,

$$= \dfrac{1}{2}\left[\dfrac{1}{s-a} - \dfrac{1}{s+a}\right]$$

$$= \dfrac{a}{s^2 - a^2} \text{ (provided } s > a)$$

A list of elementary standard Laplace transforms are summarised in Table 95.1.

95.5 Worked problems on standard Laplace transforms

Problem 1. Using a standard list of Laplace transforms, determine the following:
(a) $\mathcal{L}\left\{1 + 2t - \dfrac{1}{3}t^4\right\}$ (b) $\mathcal{L}\{5e^{2t} - 3e^{-t}\}$

(a) $\mathcal{L}\left\{1 + 2t - \dfrac{1}{3}t^4\right\}$

$$= \mathcal{L}\{1\} + 2\mathcal{L}\{t\} - \dfrac{1}{3}\mathcal{L}\{t^4\}$$

from equations (2) and (3)

$$= \dfrac{1}{s} + 2\left(\dfrac{1}{s^2}\right) - \dfrac{1}{3}\left(\dfrac{4!}{s^{4+1}}\right)$$

from (i), (vi) and (viii) of Table 95.1

$$= \dfrac{1}{s} + \dfrac{2}{s^2} - \dfrac{1}{3}\left(\dfrac{4.3.2.1}{s^5}\right)$$

$$= \dfrac{1}{s} + \dfrac{2}{s^2} - \dfrac{8}{s^5}$$

Table 95.1 Elementary standard Laplace transforms

Function $f(t)$	Laplace transforms $\mathcal{L}\{f(t)\} = \int_0^\infty e^{-st} f(t)\, dt$
(i) 1	$\dfrac{1}{s}$
(ii) k	$\dfrac{k}{s}$
(iii) e^{at}	$\dfrac{1}{s-a}$
(iv) $\sin at$	$\dfrac{a}{s^2 + a^2}$
(v) $\cos at$	$\dfrac{s}{s^2 + a^2}$
(vi) t	$\dfrac{1}{s^2}$
(vii) t^2	$\dfrac{2!}{s^3}$
(viii) t^n $(n = 1, 2, 3, \ldots)$	$\dfrac{n!}{s^{n+1}}$
(ix) $\cosh at$	$\dfrac{s}{s^2 - a^2}$
(x) $\sinh at$	$\dfrac{a}{s^2 - a^2}$

(b) $\mathcal{L}\{5e^{2t} - 3e^{-t}\} = 5\mathcal{L}(e^{2t}) - 3\mathcal{L}\{e^{-t}\}$

from equations (2) and (3)

$$= 5\left(\dfrac{1}{s-2}\right) - 3\left(\dfrac{1}{s-(-1)}\right)$$

from (iii) of Table 95.1

$$= \dfrac{5}{s-2} - \dfrac{3}{s+1}$$

$$= \dfrac{5(s+1) - 3(s-2)}{(s-2)(s+1)}$$

$$= \dfrac{2s+11}{s^2 - s - 2}$$

Problem 2. Find the Laplace transforms of:
(a) $6\sin 3t - 4\cos 5t$ (b) $2\cosh 2\theta - \sinh 3\theta$

(a) $\mathcal{L}\{6\sin 3t - 4\cos 5t\}$

$$= 6\mathcal{L}\{\sin 3t\} - 4\mathcal{L}\{\cos 5t\}$$

$$= 6\left(\frac{3}{s^2 + 3^2}\right) - 4\left(\frac{s}{s^2 + 5^2}\right)$$

from (iv) and (v) of Table 95.1

$$= \frac{18}{s^2 + 9} - \frac{4s}{s^2 + 25}$$

(b) $\mathcal{L}\{2\cosh 2\theta - \sinh 3\theta\}$

$$= 2\mathcal{L}\{\cosh 2\theta\} - \mathcal{L}\{\sinh 3\theta\}$$

$$= 2\left(\frac{s}{s^2 - 2^2}\right) - \left(\frac{3}{s^2 - 3^2}\right)$$

from (ix) and (x) of Table 95.1

$$= \frac{2s}{s^2 - 4} - \frac{3}{s^2 - 9}$$

Problem 3. Prove that
(a) $\mathcal{L}\{\sin at\} = \dfrac{a}{s^2 + a^2}$ (b) $\mathcal{L}\{t^2\} = \dfrac{2}{s^3}$

(c) $\mathcal{L}\{\cosh at\} = \dfrac{s}{s^2 - a^2}$

(a) From equation (1),

$$\mathcal{L}\{\sin at\} = \int_0^\infty e^{-st} \sin at \, dt$$

$$= \left[\frac{e^{-st}}{s^2 + a^2}(-s\sin at - a\cos at)\right]_0^\infty$$

by integration by parts,

$$= \frac{1}{s^2 + a^2}[e^{-s(\infty)}(-s\sin a(\infty)$$
$$- a\cos a(\infty)) - e^0(-s\sin 0$$
$$- a\cos 0)]$$

$$= \frac{1}{s^2 + a^2}[(0) - 1(0 - a)]$$

$$= \frac{a}{s^2 + a^2} \text{ (provided } s > 0)$$

(b) From equation (1),

$$\mathcal{L}\{t^2\} = \int_0^\infty e^{-st} t^2 \, dt$$

$$= \left[\frac{t^2 e^{-st}}{-s} - \frac{2t e^{-st}}{s^2} - \frac{2e^{-st}}{s^3}\right]_0^\infty$$

by integration by parts twice,

$$= \left[(0 - 0 - 0) - \left(0 - 0 - \frac{2}{s^3}\right)\right]$$

$$= \frac{2}{s^3} \text{ (provided } s > 0)$$

(c) From equation (1),

$$\mathcal{L}\{\cosh at\} = \mathcal{L}\left\{\frac{1}{2}(e^{at} + e^{-at})\right\},$$

from Chapter 24

$$= \frac{1}{2}\mathcal{L}\{e^{at}\} + \frac{1}{2}\mathcal{L}\{e^{-at}\},$$

from equations (2) and (3)

$$= \frac{1}{2}\left(\frac{1}{s-a}\right) + \frac{1}{2}\left(\frac{1}{s-(-a)}\right)$$

from (iii) of Table 95.1

$$= \frac{1}{2}\left[\frac{1}{s-a} + \frac{1}{s+a}\right]$$

$$= \frac{1}{2}\left[\frac{(s+a) + (s-a)}{(s-a)(s+a)}\right]$$

$$= \frac{s}{s^2 - a^2} \text{ (provided } s > a)$$

Problem 4. Determine the Laplace transforms of:
(a) $\sin^2 t$ (b) $\cosh^2 3x$

(a) Since $\cos 2t = 1 - 2\sin^2 t$ then

$$\sin^2 t = \frac{1}{2}(1 - \cos 2t). \text{ Hence,}$$

$$\mathcal{L}\{\sin^2 t\} = \mathcal{L}\left\{\frac{1}{2}(1 - \cos 2t)\right\}$$

$$= \frac{1}{2}\mathcal{L}\{1\} - \frac{1}{2}\mathcal{L}\{\cos 2t\}$$

$$= \frac{1}{2}\left(\frac{1}{s}\right) - \frac{1}{2}\left(\frac{s}{s^2 + 2^2}\right)$$

from (i) and (v) of Table 95.1

$$= \frac{(s^2 + 4) - s^2}{2s(s^2 + 4)} = \frac{4}{2s(s^2 + 4)}$$

$$= \frac{2}{s(s^2 + 4)}$$

(b) Since $\cosh 2x = 2\cosh^2 x - 1$ then

$$\cosh^2 x = \frac{1}{2}(1 + \cosh 2x) \text{ from Chapter 24.}$$

Hence $\cosh^2 3x = \frac{1}{2}(1 + \cosh 6x)$

Thus $\mathcal{L}\{\cosh^2 3x\} = \mathcal{L}\left\{\dfrac{1}{2}(1+\cosh 6x)\right\}$

$$= \dfrac{1}{2}\mathcal{L}\{1\} + \dfrac{1}{2}\mathcal{L}\{\cosh 6x\}$$

$$= \dfrac{1}{2}\left(\dfrac{1}{s}\right) + \dfrac{1}{2}\left(\dfrac{s}{s^2-6^2}\right)$$

$$= \dfrac{2s^2-36}{2s(s^2-36)} = \dfrac{s^2-18}{s(s^2-36)}$$

Problem 5. Find the Laplace transform of $3\sin(\omega t + \alpha)$, where ω and α are constants

Using the compound angle formula for $\sin(A+B)$, from Chapter 44, $\sin(\omega t + \alpha)$ may be expanded to $(\sin\omega t\cos\alpha + \cos\omega t\sin\alpha)$. Hence,

$\mathcal{L}\{3\sin(\omega t + \alpha)\}$

$$= \mathcal{L}\{3(\sin\omega t\cos\alpha + \cos\omega t\sin\alpha)\}$$

$$- 3\cos\alpha\mathcal{L}\{\sin\omega t\} + 3\sin\alpha\mathcal{L}\{\cos\omega t\},$$

since α is a constant

$$= 3\cos\alpha\left(\dfrac{\omega}{s^2+\omega^2}\right) + 3\sin\alpha\left(\dfrac{s}{s^2+\omega^2}\right)$$

from (iv) and (v) of Table 95.1

$$= \dfrac{3}{(s^2+\omega^2)}(\omega\cos\alpha + s\sin\alpha)$$

Now try the following Practice Exercise

Practice Exercise 355   Further problems on an introduction to Laplace transforms (answers on page 1192)

Determine the Laplace transforms in Problems 1 to 9.

1. (a) $2t-3$ (b) $5t^2+4t-3$

2. (a) $\dfrac{t^3}{24} - 3t + 2$ (b) $\dfrac{t^5}{15} - 2t^4 + \dfrac{t^2}{2}$

3. (a) $5e^{3t}$ (b) $2e^{-2t}$

4. (a) $4\sin 3t$ (b) $3\cos 2t$

5. (a) $7\cosh 2x$ (b) $\dfrac{1}{3}\sinh 3t$

6. (a) $2\cos^2 t$ (b) $3\sin^2 2x$

7. (a) $\cosh^2 t$ (b) $2\sinh^2 2\theta$

8. $4\sin(at+b)$, where a and b are constants.

9. $3\cos(\omega t - \alpha)$, where ω and α are constants.

10. Show that $\mathcal{L}(\cos^2 3t - \sin^2 3t) = \dfrac{s}{s^2+36}$

For fully worked solutions to each of the problems in Practice Exercise 355 in this chapter, go to the website:
www.routledge.com/cw/bird

Chapter 96

Properties of Laplace transforms

Why it is important to understand: Properties of Laplace transforms

As stated in the preceding chapter, the Laplace transform is a widely used integral transform with many applications in engineering, where it is used for analysis of linear time-invariant systems such as electrical circuits, harmonic oscillators, optical devices and mechanical systems. The Laplace transform is also a valuable tool in solving differential equations, such as in electronic circuits, and in feedback control systems, such as in stability and control of aircraft systems. This chapter considers further transforms together with the Laplace transform of derivatives that are needed when solving differential equations.

At the end of this chapter, you should be able to:

- derive the Laplace transform of $e^{at} f(t)$
- use a standard list of Laplace transforms to determine transforms of the form $e^{at} f(t)$
- derive the Laplace transforms of derivatives
- state and use the initial and final value theorems

96.1   The Laplace transform of $e^{at} f(t)$

From Chapter 95, the definition of the Laplace transform of $f(t)$ is:

$$\mathcal{L}\{f(t)\} = \int_0^\infty e^{-st} f(t)\, dt \qquad (1)$$

Thus $\mathcal{L}\{e^{at} f(t)\} = \int_0^\infty e^{-st} (e^{at} f(t))\, dt$

$$= \int_0^\infty e^{-(s-a)} f(t)\, dt \qquad (2)$$

(where a is a real constant)

Hence the substitution of $(s-a)$ for s in the transform shown in equation (1) corresponds to the multiplication of the original function $f(t)$ by e^{at}. This is known as a shift theorem.

96.2   Laplace transforms of the form $e^{at} f(t)$

From equation (2), Laplace transforms of the form $e^{at} f(t)$ may be deduced. For example:

(i) $\mathcal{L}\{e^{at} t^n\}$

Since $\mathcal{L}\{t^n\} = \dfrac{n!}{s^{n+1}}$ from (viii) of Table 95.1, page 1045.

then $\mathcal{L}\{e^{at}\,t^n\}=\dfrac{n!}{(s-a)^{n+1}}$ from equation (2) above (provided $s>a$)

(ii) $\mathcal{L}\{e^{at}\sin\omega t\}$

Since $\mathcal{L}\{\sin\omega t\}=\dfrac{\omega}{s^2+\omega^2}$ from (iv) of Table 95.1, page 1045.

then $\mathcal{L}\{e^{at}\sin\omega t\}=\dfrac{\omega}{(s-a)^2+\omega^2}$ from equation (2) (provided $s>a$)

(iii) $\mathcal{L}\{e^{at}\cosh\omega t\}$

Since $\mathcal{L}\{\cosh\omega t\}=\dfrac{s}{s^2-\omega^2}$ from (ix) of Table 95.1, page 1045.

then $\mathcal{L}\{e^{at}\cosh\omega t\}=\dfrac{s-a}{(s-a)^2-\omega^2}$ from equation (2) (provided $s>a$)

A summary of Laplace transforms of the form $e^{at}f(t)$ is shown in Table 96.1

Table 96.1 Laplace transforms of the form $e^{at}f(t)$

Function $e^{at}f(t)$ (a is a real constant)	Laplace transform $\mathcal{L}\{e^{at}f(t)\}$
(i) $e^{at}t^n$	$\dfrac{n!}{(s-a)^{n+1}}$
(ii) $e^{at}\sin\omega t$	$\dfrac{\omega}{(s-a)^2+\omega^2}$
(iii) $e^{at}\cos\omega t$	$\dfrac{s-a}{(s-a)^2+\omega^2}$
(iv) $e^{at}\sinh\omega t$	$\dfrac{\omega}{(s-a)^2-\omega^2}$
(v) $e^{at}\cosh\omega t$	$\dfrac{s-a}{(s-a)^2-\omega^2}$

Problem 1. Determine (a) $\mathcal{L}\{2t^4e^{3t}\}$ (b) $\mathcal{L}\{4e^{3t}\cos 5t\}$

(a)   From (i) of Table 96.1,

$$\mathcal{L}\{2t^4e^{3t}\}=2\mathcal{L}\{t^4e^{3t}\}=2\left(\frac{4!}{(s-3)^{4+1}}\right)$$

$$=\frac{2(4)(3)(2)}{(s-3)^5}=\frac{48}{(s-3)^5}$$

(b)   From (iii) of Table 96.1,

$$\mathcal{L}\{4e^{3t}\cos 5t\}=4\mathcal{L}\{e^{3t}\cos 5t\}$$

$$=4\left(\frac{s-3}{(s-3)^2+5^2}\right)$$

$$=\frac{4(s-3)}{s^2-6s+9+25}$$

$$=\frac{4(s-3)}{s^2-6s+34}$$

Problem 2. Determine (a) $\mathcal{L}\{e^{-2t}\sin 3t\}$ (b) $\mathcal{L}\{3e^{\theta}\cosh 4\theta\}$

(a)   From (ii) of Table 96.1,

$$\mathcal{L}\{e^{-2t}\sin 3t\}=\frac{3}{(s-(-2))^2+3^2}=\frac{3}{(s+2)^2+9}$$

$$=\frac{3}{s^2+4s+4+9}=\frac{3}{s^2+4s+13}$$

(b)   From (v) of Table 96.1,

$$\mathcal{L}\{3e^{\theta}\cosh 4\theta\}=3\mathcal{L}\{e^{\theta}\cosh 4\theta\}=\frac{3(s-1)}{(s-1)^2-4^2}$$

$$=\frac{3(s-1)}{s^2-2s+1-16}=\frac{3(s-1)}{s^2-2s-15}$$

Problem 3. Determine the Laplace transforms of (a) $5e^{-3t}\sinh 2t$ (b) $2e^{3t}(4\cos 2t-5\sin 2t)$

(a)   From (iv) of Table 96.1,

$$\mathcal{L}\{5e^{-3t}\sinh 2t\}=5\mathcal{L}\{e^{-3t}\sinh 2t\}$$

$$=5\left(\frac{2}{(s-(-3))^2-2^2}\right)$$

$$=\frac{10}{(s+3)^2-2^2}=\frac{10}{s^2+6s+9-4}$$

$$=\frac{10}{s^2+6s+5}$$

(b) $\mathcal{L}\{2e^{3t}(4\cos 2t-5\sin 2t)\}$

$$=8\mathcal{L}\{e^{3t}\cos 2t\}-10\mathcal{L}\{e^{3t}\sin 2t\}$$

$$=\frac{8(s-3)}{(s-3)^2+2^2}-\frac{10(2)}{(s-3)^2+2^2}$$

from (iii) and (ii) of Table 96.1

$$=\frac{8(s-3)-10(2)}{(s-3)^2+2^2}=\frac{8s-44}{s^2-6s+13}$$

Problem 4. Show that

$$\mathcal{L}\left\{3e^{-\frac{1}{2}x}\sin^2 x\right\} = \frac{48}{(2s+1)(4s^2+4s+17)}$$

Since $\cos 2x = 1 - 2\sin^2 x$, $\sin^2 x = \frac{1}{2}(1 - \cos 2x)$.

Hence,

$$\mathcal{L}\left\{3e^{-\frac{1}{2}x}\sin^2 x\right\}$$

$$= \mathcal{L}\left\{3e^{-\frac{1}{2}x}\frac{1}{2}(1 - \cos 2x)\right\}$$

$$= \frac{3}{2}\mathcal{L}\left\{e^{-\frac{1}{2}x}\right\} - \frac{3}{2}\mathcal{L}\left\{e^{-\frac{1}{2}x}\cos 2x\right\}$$

$$= \frac{3}{2}\left(\frac{1}{s - \left(-\frac{1}{2}\right)}\right) - \frac{3}{2}\left(\frac{\left(s - \left(-\frac{1}{2}\right)\right)}{\left(s - \left(-\frac{1}{2}\right)\right)^2 + 2^2}\right)$$

from (iii) of Table 95.1 (page 1045) and (iii) of Table 96.1 above,

$$= \frac{3}{2\left(s + \frac{1}{2}\right)} - \frac{3\left(s + \frac{1}{2}\right)}{2\left[\left(s + \frac{1}{2}\right)^2 + 2^2\right]}$$

$$= \frac{3}{2s+1} - \frac{6s+3}{4\left(s^2 + s + \frac{1}{4} + 4\right)}$$

$$= \frac{3}{2s+1} - \frac{6s+3}{4s^2+4s+17}$$

$$= \frac{3(4s^2+4s+17) - (6s+3)(2s+1)}{(2s+1)(4s^2+4s+17)}$$

$$= \frac{12s^2 + 12s + 51 - 12s^2 - 6s - 6s - 3}{(2s+1)(4s^2+4s+17)}$$

$$= \frac{48}{(2s+1)(4s^2+4s+17)}$$

Now try the following Practice Exercise

Practice Exercise 356   Further problems on Laplace transforms of the form $e^{at}f(t)$ (answers on page 1193)

Determine the Laplace transforms of the following functions:

1. (a) $2te^{2t}$ (b) $t^2 e^t$

2. (a) $4t^3 e^{-2t}$ (b) $\frac{1}{2}t^4 e^{-3t}$

3. (a) $e^t \cos t$ (b) $3e^{2t}\sin 2t$

4. (a) $5e^{-2t}\cos 3t$ (b) $4e^{-5t}\sin t$

5. (a) $2e^t \sin^2 t$ (b) $\frac{1}{2}e^{3t}\cos^2 t$

6. (a) $e^t \sinh t$ (b) $3e^{2t}\cosh 4t$

7. (a) $2e^{-t}\sinh 3t$ (b) $\frac{1}{4}e^{-3t}\cosh 2t$

8. (a) $2e^t(\cos 3t - 3\sin 3t)$

   (b) $3e^{-2t}(\sinh 2t - 2\cosh 2t)$

96.3   The Laplace transforms of derivatives

First derivative

Let the first derivative of $f(t)$ be $f'(t)$ then, from equation (1),

$$\mathcal{L}\{f'(t)\} = \int_0^\infty e^{-st}f'(t)\,dt$$

From Chapter 68, when integrating by parts

$$\int u\frac{dv}{dt}\,dt = uv - \int v\frac{du}{dt}\,dt$$

When evaluating $\int_0^\infty e^{-st}f'(t)\,dt$

let $u = e^{-st}$ and $\frac{dv}{dt} = f'(t)$

from which,

$$\frac{du}{dt} = -se^{-st} \text{ and } v = \int f'(t)\,dt = f(t)$$

Hence $\int_0^\infty e^{-st}f'(t)\,dt$

$$= \left[e^{-st}f(t)\right]_0^\infty - \int_0^\infty f(t)(-se^{-st})\,dt$$

$$= [0 - f(0)] + s \int_0^\infty e^{-st} f(t)\, dt$$

$$= -f(0) + s\mathcal{L}\{f(t)\}$$

assuming $e^{-st} f(t) \to 0$ as $t \to \infty$, and $f(0)$ is the value of $f(t)$ at $t=0$. Hence,

$$\left. \begin{array}{l} \mathcal{L}\{f'(t)\} = s\mathcal{L}\{f(t)\} - f(0) \\ \text{or} \quad \mathcal{L}\left\{\dfrac{dy}{dx}\right\} = s\mathcal{L}\{y\} - y(0) \end{array} \right\} \quad (3)$$

where $y(0)$ is the value of y at $x=0$

Second derivative

Let the second derivative of $f(t)$ be $f''(t)$, then from equation (1),

$$\mathcal{L}\{f''(t)\} = \int_0^\infty e^{-st} f''(t)\, dt$$

Integrating by parts gives:

$$\int_0^\infty e^{-st} f''(t)\, dt = \left[e^{-st} f'(t) \right]_0^\infty + s \int_0^\infty e^{-st} f'(t)\, dt$$

$$= [0 - f'(0)] + s\mathcal{L}\{f'(t)\}$$

assuming $e^{-st} f'(t) \to 0$ as $t \to \infty$, and $f'(0)$ is the value of $f'(t)$ at $t=0$. Hence
$\{f''(t)\} = -f'(0) + s[s(f(t)) - f(0)]$, from equation (3),

$$\text{i.e.} \quad \left. \begin{array}{l} \mathcal{L}\{f''(t)\} \\ = s^2\mathcal{L}\{f(t)\} - sf(0) - f'(0) \\ \text{or} \quad \mathcal{L}\left\{\dfrac{d^2y}{dx^2}\right\} \\ = s^2\mathcal{L}\{y\} - sy(0) - y'(0) \end{array} \right\} \quad (4)$$

where $y'(0)$ is the value of $\dfrac{dy}{dx}$ at $x=0$

Equations (3) and (4) are important and are used in the solution of differential equations (see Chapter 99) and simultaneous differential equations (Chapter 100).

Problem 5. Use the Laplace transform of the first derivative to derive:

(a) $\mathcal{L}\{k\} = \dfrac{k}{s}$ (b) $\mathcal{L}\{2t\} = \dfrac{2}{s^2}$

(c) $\mathcal{L}\{e^{-at}\} = \dfrac{1}{s+a}$

From equation (3), $\mathcal{L}\{f'(t)\} = s\mathcal{L}\{f(t)\} - f(0)$

(a) Let $f(t)=k$, then $f'(t)=0$ and $f(0)=k$

Substituting into equation (3) gives:

$$\mathcal{L}\{0\} = s\mathcal{L}\{k\} - k$$

i.e. $\quad k = s\mathcal{L}\{k\}$

Hence $\quad \mathcal{L}\{k\} = \dfrac{k}{s}$

(b) Let $f(t)=2t$ then $f'(t)=2$ and $f(0)=0$

Substituting into equation (3) gives:

$$\mathcal{L}\{2\} = s\mathcal{L}\{2t\} - 0$$

i.e. $\quad \dfrac{2}{s} = s\mathcal{L}\{2t\}$

Hence $\quad \mathcal{L}\{2t\} = \dfrac{2}{s^2}$

(c) Let $f(t)=e^{-at}$ then $f'(t)=-ae^{-at}$ and $f(0)=1$

Substituting into equation (3) gives:

$$\mathcal{L}\{-ae^{-at}\} = s\mathcal{L}\{e^{-at}\} - 1$$

$$-a\mathcal{L}\{e^{-at}\} = s\mathcal{L}\{e^{-at}\} - 1$$

$$1 = s\mathcal{L}\{e^{-at}\} + a\mathcal{L}\{e^{-at}\}$$

$$1 = (s+a)\mathcal{L}\{e^{-at}\}$$

Hence $\mathcal{L}\{e^{-at}\} = \dfrac{1}{s+a}$

Problem 6. Use the Laplace transform of the second derivative to derive

$$\mathcal{L}\{\cos at\} = \dfrac{s}{s^2 + a^2}$$

From equation (4),

$$\mathcal{L}\{f''(t)\} = s^2\mathcal{L}\{f(t)\} - sf(0) - f'(0)$$

Let $f(t)=\cos at$, then $f'(t) = -a\sin at$ and $f''(t) = -a^2\cos at$, $f(0) = 1$ and $f'(0) = 0$

Substituting into equation (4) gives:

$$\mathcal{L}\{-a^2\cos at\} = s^2\{\cos at\} - s(1) - 0$$

i.e. $-a^2\mathcal{L}\{\cos at\}=s^2\mathcal{L}\{\cos at\}-s$

Hence $s=(s^2+a^2)\mathcal{L}\{\cos at\}$

from which, $\mathcal{L}\{\cos at\}=\dfrac{s}{s^2+a^2}$

Now try the following Practice Exercise

Practice Exercise 357   Further problems on the Laplace transforms of derivatives (answers on page 1193)

1. Derive the Laplace transform of the first derivative from the definition of a Laplace transform. Hence derive the transform

$$\mathcal{L}\{1\}=\frac{1}{s}$$

2. Use the Laplace transform of the first derivative to derive the transforms:

    (a) $\mathcal{L}\{e^{at}\}=\dfrac{1}{s-a}$ (b) $\mathcal{L}\{3t^2\}=\dfrac{6}{s^3}$

3. Derive the Laplace transform of the second derivative from the definition of a Laplace transform. Hence derive the transform

$$\mathcal{L}\{\sin at\}=\frac{a}{s^2+a^2}$$

4. Use the Laplace transform of the second derivative to derive the transforms:

    (a) $\mathcal{L}\{\sinh at\}=\dfrac{a}{s^2-a^2}$

    (b) $\mathcal{L}\{\cosh at\}=\dfrac{s}{s^2-a^2}$

96.4   The initial and final value theorems

There are several Laplace transform theorems used to simplify and interpret the solution of certain problems. Two such theorems are the initial value theorem and the final value theorem.

The initial value theorem

$$\lim_{t\to 0}\,[f(t)]=\lim_{s\to\infty}\,[s\mathcal{L}\{f(t)\}]$$

For example, if $f(t)=3e^{4t}$ then

$$\mathcal{L}\{3e^{4t}\}=\frac{3}{s-4}$$

from (iii) of Table 95.1, page 1045.

By the initial value theorem,

$$\lim_{t\to 0}[3e^{4t}]=\lim_{s\to\infty}\left[s\left(\frac{3}{s-4}\right)\right]$$

i.e. $3e^0=\infty\left(\dfrac{3}{\infty-4}\right)$

i.e.   **$3=3$**, which illustrates the theorem.

Problem 7.   Verify the initial value theorem for the voltage function $(5+2\cos 3t)$ volts, and state its initial value

Let $f(t)=5+2\cos 3t$

$$\mathcal{L}\{f(t)\}=\mathcal{L}\{5+2\cos 3t\}=\frac{5}{s}+\frac{2s}{s^2+9}$$

from (ii) and (v) of Table 95.1, page 1045.

By the initial value theorem,

$$\lim_{t\to 0}[f(t)]=\lim_{s\to\infty}[s\mathcal{L}\{f(t)\}]$$

i.e. $\lim_{t\to 0}[5+2\cos 3t]=\lim_{s\to\infty}\left[s\left(\frac{5}{s}+\frac{2s}{s^2+9}\right)\right]$

$$=\lim_{s\to\infty}\left[5+\frac{2s^2}{s^2+9}\right]$$

i.e. $5+2(1)=5+\dfrac{2\infty^2}{\infty^2+9}=5+2$

i.e. **$7=7$**, which verifies the theorem in this case.

The initial value of the voltage is thus **7 V**

Problem 8.   Verify the initial value theorem for the function $(2t-3)^2$ and state its initial value

Let $f(t)=(2t-3)^2=4t^2-12t+9$

Let $\mathcal{L}\{f(t)\}=\mathcal{L}(4t^2-12t+9)$

$$=4\left(\frac{2}{s^3}\right)-\frac{12}{s^2}+\frac{9}{s}$$

from (vii), (vi) and (ii) of Table 95.1, page 1045.

By the initial value theorem,

$$\lim_{t\to 0}[(2t-3)^2]=\lim_{s\to\infty}\left[s\left(\frac{8}{s^3}-\frac{12}{s^2}+\frac{9}{s}\right)\right]$$

$$=\lim_{s\to\infty}\left[\frac{8}{s^2}-\frac{12}{s}+9\right]$$

i.e. $(0-3)^2=\dfrac{8}{\infty^2}-\dfrac{12}{\infty}+9$

i.e. **9=9**, which verifies the theorem in this case.

The initial value of the given function is thus **9**

The final value theorem

$$\lim_{t\to\infty} [f(t)] = \lim_{s\to 0} [s\mathcal{L}\{f(t)\}]$$

For example, if $f(t) = 3e^{-4t}$ then:

$$\lim_{t\to\infty}[3e^{-4t}] = \lim_{s\to 0}\left[s\left(\frac{3}{s+4}\right)\right]$$

i.e. $3e^{-\infty} = (0)\left(\frac{3}{0+4}\right)$

i.e. **0=0**, which illustrates the theorem.

Problem 9. Verify the final value theorem for the function $(2 + 3e^{-2t}\sin 4t)$ cm, which represents the displacement of a particle. State its final steady value

Let $f(t) = 2 + 3e^{-2t}\sin 4t$

$\mathcal{L}\{f(t)\} = \mathcal{L}\{2 + 3e^{-2t}\sin 4t\}$

$\qquad = \dfrac{2}{s} + 3\left(\dfrac{4}{(s-(-2))^2 + 4^2}\right)$

$\qquad = \dfrac{2}{s} + \dfrac{12}{(s+2)^2 + 16}$

from (ii) of Table 95.1, page 1045 and (ii) of Table 96.1 on page 1049.

By the final value theorem,

$$\lim_{t\to\infty}[f(t)] = \lim_{s\to 0}[s\mathcal{L}\{f(t)\}]$$

i.e. $\displaystyle\lim_{t\to\infty}[2 + 3e^{-2t}\sin 4t]$

$\qquad = \displaystyle\lim_{s\to 0}\left[s\left(\frac{2}{s} + \frac{12}{(s+2)^2 + 16}\right)\right]$

$\qquad = \displaystyle\lim_{s\to 0}\left[2 + \frac{12s}{(s+2)^2 + 16}\right]$

i.e. $2+0=2+0$

i.e. **2=2**, which verifies the theorem in this case.

The final value of the displacement is thus 2 cm.

The initial and final value theorems are used in pulse circuit applications where the response of the circuit for small periods of time, or the behaviour immediately after the switch is closed, are of interest. The final value theorem is particularly useful in investigating the stability of systems (such as in automatic aircraft-landing systems) and is concerned with the steady-state response for large values of time t, i.e. after all transient effects have died away.

Now try the following Practice Exercise

Practice Exercise 358   Further problems on initial and final value theorems (answers on page 1193)

1.   State the initial value theorem. Verify the theorem for the functions (a) $3 - 4\sin t$, (b) $(t-4)^2$ and state their initial values.

2.   Verify the initial value theorem for the voltage functions (a) $4 + 2\cos t$, (b) $t - \cos 3t$, and state their initial values.

3.   State the final value theorem and state a practical application where it is of use. Verify the theorem for the function $4 + e^{-2t}(\sin t + \cos t)$ representing a displacement, and state its final value.

4.   Verify the final value theorem for the function $3t^2 e^{-4t}$ and determine its steady-state value.

For fully worked solutions to each of the problems in Practice Exercises 356 to 358 in this chapter, go to the website:
www.routledge.com/cw/bird

Chapter 97

Inverse Laplace transforms

Why it is important to understand: Inverse Laplace transforms

Laplace transforms and their inverses are a mathematical technique which allows us to solve differential equations by primarily using algebraic methods. This simplification in the solving of equations, coupled with the ability to directly implement electrical components in their transformed form, makes the use of Laplace transforms widespread in both electrical engineering and control systems engineering. Laplace transforms have many further applications in mathematics, physics, optics, signal processing and probability. This chapter specifically explains how the inverse Laplace transform is determined, which can also involve the use of partial fractions. In addition, poles and zeros of transfer functions are briefly explained; these are of importance in stability and control systems.

At the end of this chapter, you should be able to:

- define the inverse Laplace transform
- use a standard list to determine the inverse Laplace transforms of simple functions
- determine inverse Laplace transforms using partial fractions
- define a pole and a zero
- determine poles and zeros for transfer functions, showing them on a pole–zero diagram

---

97.1 Definition of the inverse Laplace transform

If the Laplace transform of a function $f(t)$ is $F(s)$, i.e. $\mathcal{L}\{f(t)\}=F(s)$, then $f(t)$ is called the **inverse Laplace transform** of $F(s)$ and is written as $f(t)=\mathcal{L}^{-1}\{F(s)\}$.

For example, since $\mathcal{L}\{1\}=\dfrac{1}{s}$ then $\boldsymbol{\mathcal{L}^{-1}\left\{\dfrac{1}{s}\right\}=1}$

Similarly, since $\mathcal{L}\{\sin at\}=\dfrac{a}{s^2+a^2}$ then

$$\mathcal{L}^{-1}\left\{\frac{a}{s^2+a^2}\right\}=\boldsymbol{\sin at},\text{ and so on.}$$

97.2 Inverse Laplace transforms of simple functions

Tables of Laplace transforms, such as the tables in Chapters 95 and 96 (see pages 1045 and 1049), may be used to find inverse Laplace transforms.

However, for convenience, a summary of inverse Laplace transforms is shown in Table 97.1.

Problem 1. Find the following inverse Laplace transforms:

(a) $\mathcal{L}^{-1}\left\{\dfrac{1}{s^2+9}\right\}$ (b) $\mathcal{L}^{-1}\left\{\dfrac{5}{3s-1}\right\}$

Table 97.1 Inverse Laplace transforms

$F(s) = \mathcal{L}\{f(t)\}$	$\mathcal{L}^{-1}\{F(s)\} = f(t)$
(i) $\dfrac{1}{s}$	1
(ii) $\dfrac{k}{s}$	k
(iii) $\dfrac{1}{s-a}$	e^{at}
(iv) $\dfrac{a}{s^2+a^2}$	$\sin at$
(v) $\dfrac{s}{s^2+a^2}$	$\cos at$
(vi) $\dfrac{1}{s^2}$	t
(vii) $\dfrac{2!}{s^3}$	t^2
(viii) $\dfrac{n!}{s^{n+1}}$	t^n
(ix) $\dfrac{a}{s^2-a^2}$	$\sinh at$
(x) $\dfrac{s}{s^2-a^2}$	$\cosh at$
(xi) $\dfrac{n!}{(s-a)^{n+1}}$	$e^{at}t^n$
(xii) $\dfrac{\omega}{(s-a)^2+\omega^2}$	$e^{at}\sin \omega t$
(xiii) $\dfrac{s-a}{(s-a)^2+\omega^2}$	$e^{at}\cos \omega t$
(xiv) $\dfrac{\omega}{(s-a)^2-\omega^2}$	$e^{at}\sinh \omega t$
(xv) $\dfrac{s-a}{(s-a)^2-\omega^2}$	$e^{at}\cosh \omega t$

(a) From (iv) of Table 97.1,

$$\mathcal{L}^{-1}\left\{\frac{a}{s^2+a^2}\right\} = \sin at$$

Hence $\mathcal{L}^{-1}\left\{\dfrac{1}{s^2+9}\right\} = \mathcal{L}^{-1}\left\{\dfrac{1}{s^2+3^2}\right\}$

$$= \frac{1}{3}\mathcal{L}^{-1}\left\{\frac{3}{s^2+3^2}\right\}$$

$$= \frac{1}{3}\sin 3t$$

(b) $\mathcal{L}^{-1}\left\{\dfrac{5}{3s-1}\right\} = \mathcal{L}^{-1}\left\{\dfrac{5}{3\left(s-\dfrac{1}{3}\right)}\right\}$

$$= \frac{5}{3}\mathcal{L}^{-1}\left\{\frac{1}{\left(s-\dfrac{1}{3}\right)}\right\} = \frac{5}{3}e^{\frac{1}{3}t}$$

from (iii) of Table 97.1

Problem 2. Find the following inverse Laplace transforms:

(a) $\mathcal{L}^{-1}\left\{\dfrac{6}{s^3}\right\}$ (b) $\mathcal{L}^{-1}\left\{\dfrac{3}{s^4}\right\}$

(a) From (vii) of Table 97.1, $\mathcal{L}^{-1}\left\{\dfrac{2}{s^3}\right\} = t^2$

Hence $\mathcal{L}^{-1}\left\{\dfrac{6}{s^3}\right\} = 3\mathcal{L}^{-1}\left\{\dfrac{2}{s^3}\right\} = 3t^2$

(b) From (viii) of Table 97.1, if s is to have a power of 4 then $n=3$

Thus $\mathcal{L}^{-1}\left\{\dfrac{3!}{s^4}\right\} = t^3$ i.e. $\mathcal{L}^{-1}\left\{\dfrac{6}{s^4}\right\} = t^3$

Hence $\mathcal{L}^{-1}\left\{\dfrac{3}{s^4}\right\} = \dfrac{1}{2}\mathcal{L}^{-1}\left\{\dfrac{6}{s^4}\right\} = \dfrac{1}{2}t^3$

Problem 3. Determine

(a) $\mathcal{L}^{-1}\left\{\dfrac{7s}{s^2+4}\right\}$ (b) $\mathcal{L}^{-1}\left\{\dfrac{4s}{s^2-16}\right\}$

(a) $\mathcal{L}^{-1}\left\{\dfrac{7s}{s^2+4}\right\} = 7\mathcal{L}^{-1}\left\{\dfrac{s}{s^2+2^2}\right\} = 7\cos 2t$

from (v) of Table 97.1

(b) $\mathcal{L}^{-1}\left\{\dfrac{4s}{s^2-16}\right\} = 4\mathcal{L}^{-1}\left\{\dfrac{s}{s^2-4^2}\right\}$

$$= 4\cosh 4t$$

from (x) of Table 97.1

Problem 4. Find

(a) $\mathcal{L}^{-1}\left\{\dfrac{3}{s^2-7}\right\}$ (b) $\mathcal{L}^{-1}\left\{\dfrac{2}{(s-3)^5}\right\}$

(a) From (ix) of Table 97.1,

$$\mathcal{L}^{-1}\left\{\frac{a}{s^2-a^2}\right\}=\sinh at$$

Thus

$$\mathcal{L}^{-1}\left\{\frac{3}{s^2-7}\right\}=3\mathcal{L}^{-1}\left\{\frac{1}{s^2-(\sqrt{7})^2}\right\}$$

$$=\frac{3}{\sqrt{7}}\mathcal{L}^{-1}\left\{\frac{\sqrt{7}}{s^2-(\sqrt{7})^2}\right\}$$

$$=\frac{\mathbf{3}}{\sqrt{\mathbf{7}}}\sinh\sqrt{\mathbf{7}}t$$

(b) From (xi) of Table 97.1,

$$\mathcal{L}^{-1}\left\{\frac{n!}{(s-a)^{n+1}}\right\}=e^{at}t^n$$

Thus $\mathcal{L}^{-1}\left\{\dfrac{1}{(s-a)^{n+1}}\right\}=\dfrac{1}{n!}e^{at}t^n$

and comparing with $\mathcal{L}^{-1}\left\{\dfrac{2}{(s-3)^5}\right\}$ shows that $n=4$ and $a=3$

Hence

$$\mathcal{L}^{-1}\left\{\frac{2}{(s-3)^5}\right\}=2\mathcal{L}^{-1}\left\{\frac{1}{(s-3)^5}\right\}$$

$$=2\left(\frac{1}{4!}e^{3t}t^4\right)=\frac{\mathbf{1}}{\mathbf{12}}e^{\mathbf{3}t}t^{\mathbf{4}}$$

Problem 5. Determine

(a) $\mathcal{L}^{-1}\left\{\dfrac{3}{s^2-4s+13}\right\}$

(b) $\mathcal{L}^{-1}\left\{\dfrac{2(s+1)}{s^2+2s+10}\right\}$

(a) $\mathcal{L}^{-1}\left\{\dfrac{3}{s^2-4s+13}\right\}=\mathcal{L}^{-1}\left\{\dfrac{3}{(s-2)^2+3^2}\right\}$

$$=e^{2t}\sin 3t$$

from (xii) of Table 97.1

(b) $\mathcal{L}^{-1}\left\{\dfrac{2(s+1)}{s^2+2s+10}\right\}=\mathcal{L}^{-1}\left\{\dfrac{2(s+1)}{(s+1)^2+3^2}\right\}$

$$=2e^{-t}\cos 3t$$

from (xiii) of Table 97.1

Problem 6. Determine

(a) $\mathcal{L}^{-1}\left\{\dfrac{5}{s^2+2s-3}\right\}$

(b) $\mathcal{L}^{-1}\left\{\dfrac{4s-3}{s^2-4s-5}\right\}$

(a) $\mathcal{L}^{-1}\left\{\dfrac{5}{s^2+2s-3}\right\}=\mathcal{L}^{-1}\left\{\dfrac{5}{(s+1)^2-2^2}\right\}$

$$=\mathcal{L}^{-1}\left\{\frac{\frac{5}{2}(2)}{(s+1)^2-2^2}\right\}$$

$$=\frac{\mathbf{5}}{\mathbf{2}}e^{-t}\sinh 2t$$

from (xiv) of Table 97.1

(b) $\mathcal{L}^{-1}\left\{\dfrac{4s-3}{s^2-4s-5}\right\}=\mathcal{L}^{-1}\left\{\dfrac{4s-3}{(s-2)^2-3^2}\right\}$

$$=\mathcal{L}^{-1}\left\{\frac{4(s-2)+5}{(s-2)^2-3^2}\right\}$$

$$=\mathcal{L}^{-1}\left\{\frac{4(s-2)}{(s-2)^2-3^2}\right\}$$

$$+\mathcal{L}^{-1}\left\{\frac{5}{(s-2)^2-3^2}\right\}$$

$$=4e^{2t}\cosh 3t+\mathcal{L}^{-1}\left\{\frac{\frac{5}{3}(3)}{(s-2)^2-3^2}\right\}$$

from (xv) of Table 97.1

$$=4e^{2t}\cosh 3t+\frac{\mathbf{5}}{\mathbf{3}}e^{2t}\sinh 3t$$

from (xiv) of Table 97.1

Now try the following Practice Exercise

Determine the inverse Laplace transforms of the following:

1. (a) $\dfrac{7}{s}$ (b) $\dfrac{2}{s-5}$

2. (a) $\dfrac{3}{2s+1}$ (b) $\dfrac{2s}{s^2+4}$

3. (a) $\dfrac{1}{s^2+25}$ (b) $\dfrac{4}{s^2+9}$

4. (a) $\dfrac{5s}{2s^2+18}$ (b) $\dfrac{6}{s^2}$

5. (a) $\dfrac{5}{s^3}$ (b) $\dfrac{8}{s^4}$

6. (a) $\dfrac{3s}{\frac{1}{2}s^2-8}$ (b) $\dfrac{7}{s^2-16}$

7. (a) $\dfrac{15}{3s^2-27}$ (b) $\dfrac{4}{(s-1)^3}$

8. (a) $\dfrac{1}{(s+2)^4}$ (b) $\dfrac{3}{(s-3)^5}$

9. (a) $\dfrac{s+1}{s^2+2s+10}$ (b) $\dfrac{3}{s^2+6s+13}$

10. (a) $\dfrac{2(s-3)}{s^2-6s+13}$ (b) $\dfrac{7}{s^2-8s+12}$

11. (a) $\dfrac{2s+5}{s^2+4s-5}$ (b) $\dfrac{3s+2}{s^2-8s+25}$

97.3   Inverse Laplace transforms using partial fractions

Sometimes the function whose inverse is required is not recognisable as a standard type, such as those listed in Table 97.1. In such cases it may be possible, by using partial fractions, to resolve the function into simpler fractions which may be inverted on sight. For example, the function,

$$F(s) = \frac{2s-3}{s(s-3)}$$

cannot be inverted on sight from Table 97.1. However, by using partial fractions, $\dfrac{2s-3}{s(s-3)} \equiv \dfrac{1}{s} + \dfrac{1}{s-3}$ which may be inverted as $1+e^{3t}$ from (i) and (iii) of Table 97.1. Partial fractions are discussed in Chapter 19, and a summary of the forms of partial fractions is given in Table 19.1 on page 179.

Problem 7.   Determine $\mathcal{L}^{-1}\left\{\dfrac{4s-5}{s^2-s-2}\right\}$

$$\frac{4s-5}{s^2-s-2} \equiv \frac{4s-5}{(s-2)(s+1)} \equiv \frac{A}{(s-2)} + \frac{B}{(s+1)}$$

$$\equiv \frac{A(s+1)+B(s-2)}{(s-2)(s+1)}$$

Hence $4s-5 \equiv A(s+1)+B(s-2)$

When $s=2$, $3=3A$, from which, $A=1$

When $s=-1$, $-9=-3B$, from which, $B=3$

Hence $\mathcal{L}^{-1}\left\{\dfrac{4s-5}{s^2-s-2}\right\}$

$$\equiv \mathcal{L}^{-1}\left\{\frac{1}{s-2} + \frac{3}{s+1}\right\}$$

$$= \mathcal{L}^{-1}\left\{\frac{1}{s-2}\right\} + \mathcal{L}^{-1}\left\{\frac{3}{s+1}\right\}$$

$$= \mathbf{e^{2t} + 3e^{-t}} \text{ from (iii) of Table 97.1}$$

Problem 8.   Find $\mathcal{L}^{-1}\left\{\dfrac{3s^3+s^2+12s+2}{(s-3)(s+1)^3}\right\}$

$$\frac{3s^3+s^2+12s+2}{(s-3)(s+1)^3}$$

$$\equiv \frac{A}{s-3} + \frac{B}{s+1} + \frac{C}{(s+1)^2} + \frac{D}{(s+1)^3}$$

$$\equiv \frac{\left(\begin{array}{c} A(s+1)^3 + B(s-3)(s+1)^2 \\ +C(s-3)(s+1) + D(s-3) \end{array}\right)}{(s-3)(s+1)^3}$$

Hence

$$3s^3+s^2+12s+2 \equiv A(s+1)^3 + B(s-3)(s+1)^2$$
$$+C(s-3)(s+1) + D(s-3)$$

When $s=3$, $128=64A$, from which, $A=2$

When $s=-1, -12=-4D$, from which, $D=3$

Equating s^3 terms gives: $3=A+B$, from which, $B=1$

Equating constant terms gives:

$$2 = A - 3B - 3C - 3D,$$

i.e. $\quad 2 = 2 - 3 - 3C - 9,$

from which, $3C = -12$ and $C = -4$

Hence

$$\mathcal{L}^{-1}\left\{\frac{3s^3 + s^2 + 12s + 2}{(s-3)(s+1)^3}\right\}$$

$$\equiv \mathcal{L}^{-1}\left\{\frac{2}{s-3} + \frac{1}{s+1} - \frac{4}{(s+1)^2} + \frac{3}{(s+1)^3}\right\}$$

$$= 2e^{3t} + e^{-t} - 4e^{-t}t + \frac{3}{2}e^{-t}t^2$$

from (iii) and (xi) of Table 97.1

Problem 9. Determine
$$\mathcal{L}^{-1}\left\{\frac{5s^2 + 8s - 1}{(s+3)(s^2+1)}\right\}$$

$$\frac{5s^2 + 8s - 1}{(s+3)(s^2+1)} \equiv \frac{A}{s+3} + \frac{Bs+C}{(s^2+1)}$$

$$\equiv \frac{A(s^2+1) + (Bs+C)(s+3)}{(s+3)(s^2+1)}$$

Hence $5s^2 + 8s - 1 \equiv A(s^2+1) + (Bs+C)(s+3)$.

When $s = -3, 20 = 10A$, from which, $A = 2$

Equating s^2 terms gives: $5 = A + B$, from which, $B = 3$, since $A = 2$

Equating s terms gives: $8 = 3B + C$, from which, $C = -1$, since $B = 3$

Hence $\mathcal{L}^{-1}\left\{\frac{5s^2 + 8s - 1}{(s+3)(s^2+1)}\right\}$

$$\equiv \mathcal{L}^{-1}\left\{\frac{2}{s+3} + \frac{3s-1}{s^2+1}\right\}$$

$$\equiv \mathcal{L}^{-1}\left\{\frac{2}{s+3}\right\} + \mathcal{L}^{-1}\left\{\frac{3s}{s^2+1}\right\}$$

$$- \mathcal{L}^{-1}\left\{\frac{1}{s^2+1}\right\}$$

$$= 2e^{-3t} + 3\cos t - \sin t$$

from (iii), (v) and (iv) of Table 97.1

Problem 10. Find $\mathcal{L}^{-1}\left\{\frac{7s+13}{s(s^2+4s+13)}\right\}$

$$\frac{7s+13}{s(s^2+4s+13)} \equiv \frac{A}{s} + \frac{Bs+C}{s^2+4s+13}$$

$$\equiv \frac{A(s^2+4s+13) + (Bs+C)(s)}{s(s^2+4s+13)}$$

Hence $7s + 13 \equiv A(s^2+4s+13) + (Bs+C)(s)$

When $s = 0, 13 = 13A$, from which, $A = 1$

Equating s^2 terms gives: $0 = A + B$, from which, $B = -1$

Equating s terms gives: $7 = 4A + C$, from which, $C = 3$

Hence $\mathcal{L}^{-1}\left\{\frac{7s+13}{s(s^2+4s+13)}\right\}$

$$\equiv \mathcal{L}^{-1}\left\{\frac{1}{s} + \frac{-s+3}{s^2+4s+13}\right\}$$

$$\equiv \mathcal{L}^{-1}\left\{\frac{1}{s}\right\} + \mathcal{L}^{-1}\left\{\frac{-s+3}{(s+2)^2+3^2}\right\}$$

$$\equiv \mathcal{L}^{-1}\left\{\frac{1}{s}\right\} + \mathcal{L}^{-1}\left\{\frac{-(s+2)+5}{(s+2)^2+3^2}\right\}$$

$$\equiv \mathcal{L}^{-1}\left\{\frac{1}{s}\right\} - \mathcal{L}^{-1}\left\{\frac{s+2}{(s+2)^2+3^2}\right\}$$

$$+ \mathcal{L}^{-1}\left\{\frac{5}{(s+2)^2+3^2}\right\}$$

$$\equiv 1 - e^{-2t}\cos 3t + \frac{5}{3}e^{-2t}\sin 3t$$

from (i), (xiii) and (xii) of Table 97.1

Now try the following Practice Exercise

Practice Exercise 360   Further problems on inverse Laplace transforms using partial fractions (answers on page 1193)

Use partial fractions to find the inverse Laplace transforms of the following functions:

1. $\dfrac{11 - 3s}{s^2 + 2s - 3}$

2. $\dfrac{2s^2 - 9s - 35}{(s+1)(s-2)(s+3)}$

3. $\dfrac{5s^2 - 2s - 19}{(s+3)(s-1)^2}$

4. $\dfrac{3s^2 + 16s + 15}{(s+3)^3}$

5. $\dfrac{7s^2+5s+13}{(s^2+2)(s+1)}$

6. $\dfrac{3+6s+4s^2-2s^3}{s^2(s^2+3)}$

7. $\dfrac{26-s^2}{s(s^2+4s+13)}$

97.4 Poles and zeros

It was seen in the previous section that Laplace transforms, in general, have the form $f(s)=\dfrac{\phi(s)}{\theta(s)}$. This is the same form as most transfer functions for engineering systems, a **transfer function** being one that relates the response at a given pair of terminals to a source or stimulus at another pair of terminals.

Let a function in the s domain be given by: $f(s)=\dfrac{\phi(s)}{(s-a)(s-b)(s-c)}$ where $\phi(s)$ is of less degree than the denominator.

Poles: The values $a, b, c, \ldots$ that make the denominator zero, and hence $f(s)$ infinite, are called the system poles of $f(s)$.

If there are no repeated factors, the poles are **simple poles**.

If there are repeated factors, the poles are **multiple poles**.

Zeros: Values of s that make the numerator $\phi(s)$ zero, and hence $f(s)$ zero, are called the system zeros of $f(s)$.

For example: $\dfrac{s-4}{(s+1)(s-2)}$ has simple poles at $s=-1$ and $s=+2$, and a zero at $s=4$ $\dfrac{s+3}{(s+1)^2(2s+5)}$ has a simple pole at $s=-\dfrac{5}{2}$ and double poles at $s=-1$, and a zero at $s=-3$ and $\dfrac{s+2}{s(s-1)(s+4)(2s+1)}$ has simple poles at $s=0, +1, -4,$ and $-\dfrac{1}{2}$ and a zero at $s=-2$

Pole–zero diagram

The poles and zeros of a function are values of complex frequency s and can therefore be plotted on the complex frequency or s-plane. The resulting plot is the **pole–zero diagram** or **pole–zero map**. On the rectangular axes, the real part is labelled the σ-**axis** and the imaginary part the $j\omega$-axis.

The location of a pole in the s-plane is denoted by a cross ($\times$) and the location of a zero by a small circle (o). This is demonstrated in the following examples. From the pole–zero diagram it may be determined that the magnitude of the transfer function will be larger when it is closer to the poles and smaller when it is closer to the zeros. This is important in understanding what the system does at various frequencies and is crucial in the study of **stability** and **control theory** in general.

Problem 11. Determine for the transfer function:

$$R(s)=\dfrac{400\,(s+10)}{s\,(s+25)(s^2+10s+125)}$$

(a) the zero and (b) the poles. Show the poles and zero on a pole-zero diagram

(a) For the numerator to be zero, $(s+10)=0$

Hence, $s=-10$ **is a zero** of $R(s)$

(b) For the denominator to be zero, $s=0$ or $s=-25$ or $s^2+10s+125=0$

Using the quadratic formula.

$$s=\dfrac{-10\pm\sqrt{10^2-4(1)(125)}}{2}=\dfrac{-10\pm\sqrt{-400}}{2}$$

$$=\dfrac{-10\pm j20}{2}$$

$$=(-5\pm j10)$$

Hence, **poles occur at $s=0$, $s=-25$, $(-5+j10)$ and $(-5-j10)$**

The pole–zero diagram is shown in Figure 97.1.

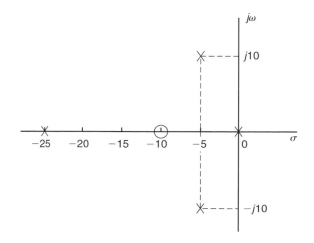

Figure 97.1

Problem 12. Determine the poles and zeros for the function: $F(s) = \dfrac{(s+3)(s-2)}{(s+4)(s^2+2s+2)}$ and plot them on a pole–zero map

For the numerator to be zero, $(s+3)=0$ and $(s-2)=0$, hence **zeros occur at $s=-3$** and at **$s=+2$**. Poles occur when the denominator is zero, i.e. when $(s+4)=0$, i.e. $s=-4$, and when $s^2+2s+2=0$,

i.e. $s = \dfrac{-2 \pm \sqrt{2^2 - 4(1)(2)}}{2} = \dfrac{-2 \pm \sqrt{-4}}{2}$

$= \dfrac{-2 \pm j2}{2} = (-1+j)$ or $(-1-j)$

The poles and zeros are shown on the pole–zero map of $F(s)$ in Figure 97.2.

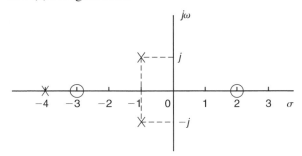

Figure 97.2

It is seen from these problems that poles and zeros are always real or complex conjugate.

Now try the following Practice Exercise

Practice Exercise 361   Further problems on poles and zeros (answers on page 1193)

1.   Determine for the transfer function:
$R(s) = \dfrac{50(s+4)}{s(s+2)(s^2-8s+25)}$
(a) the zero and (b) the poles. Show the poles and zeros on a pole–zero diagram.

2.   Determine the poles and zeros for the function: $F(s) = \dfrac{(s-1)(s+2)}{(s+3)(s^2-2s+5)}$ and plot them on a pole–zero map.

3.   For the function $G(s) = \dfrac{s-1}{(s+2)(s^2+2s+5)}$ determine the poles and zeros and show them on a pole–zero diagram.

4.   Find the poles and zeros for the transfer function: $H(s) = \dfrac{s^2-5s-6}{s(s^2+4)}$ and plot the results in the s-plane.

Chapter 98

The Laplace transform of the Heaviside function

Why it is important to understand: **The Laplace transform of the Heaviside function**

The Heaviside unit step function is used in the mathematics of control theory and signal processing to represent a signal that switches on at a specified time and stays switched on indefinitely. It is also used in structural mechanics to describe different types of structural loads. The Heaviside function has applications in engineering where periodic functions are represented. In many physical situations things change suddenly: brakes are applied, a switch is thrown, collisions occur; the Heaviside unit function is very useful for representing sudden change.

At the end of this chapter, you should be able to:

- define the Heaviside unit step function
- use a standard list to determine the Laplace transform of $H(t-c)$
- use a standard list to determine the Laplace transform of $H(t-c).f(t-c)$
- determine the inverse transforms of Heaviside functions

98.1 Heaviside unit step function

In engineering applications, functions are frequently encountered whose values change abruptly at specified values of time t. One common example is when a voltage is switched on or off in an electrical circuit at a specified value of time t.

The switching process can be described mathematically by the function called the **Unit Step Function** – otherwise known as the **Heaviside unit step function**. Figure 98.1 shows a function that maintains a zero value for all values of t up to $t = c$ and a value of 1 for all values of $t \geq c$. This is the Heaviside unit step function and is denoted by:

$$f(t) = H(t-c) \text{ or } u(t-c)$$

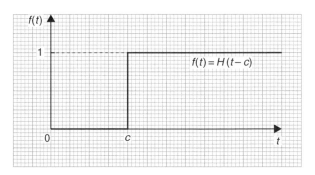

Figure 98.1

where the c indicates the value of t at which the function changes from a value of zero to a value of unity (i.e. 1).

It follows that $f(t) = H(t - 5)$ is as shown in Figure 98.2 and $f(t) = 3H(t - 4)$ is as shown in Figure 98.3.

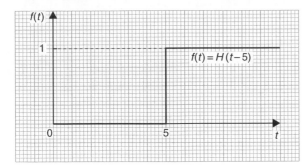

Figure 98.2

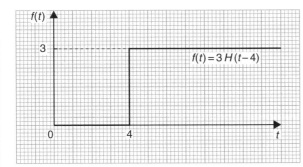

Figure 98.3

If the unit step occurs at the origin, then $c = 0$ and $f(t) = H(t - 0)$, i.e. $\mathbf{H(t)}$ as shown in Figure 98.4.

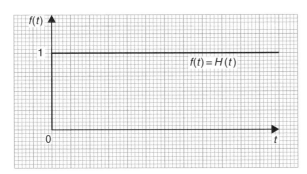

Figure 98.4

Figure 98.5(a) shows a graph of $f(t) = t^2$; the graph shown in Figure 98.5(b) is $f(t) = H(t - 2).t^2$
where for $t < 2$, $H(t - 2).t^2 = 0$ and when $t \geq 2$, $H(t - 2).t^2 = t^2$. The function $H(t - 2)$ suppresses the function t^2 for all values of t up to $t = 2$ and then 'switches on' the function t^2 at $t = 2$

A common situation in an electrical circuit is for a voltage V to be applied at a particular time, say, $t = a$, and

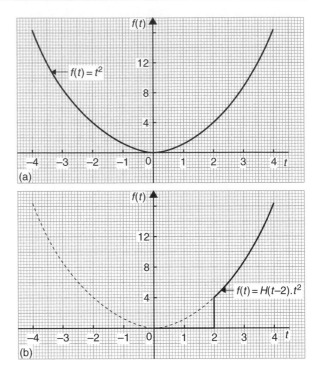

Figure 98.5

removed later, say at $t = b$. Such a situation is written using step functions as:

$$V(t) = H(t - a) - H(t - b)$$

For example, Figure 98.6 shows the function $f(t) = H(t - 2) - H(t - 5)$

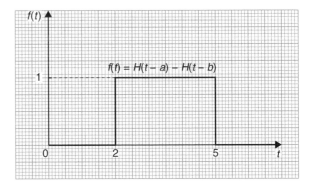

Figure 98.6

Representing the Heaviside unit step function is further explored in the following Problems.

Problem 1. A 12 V source is switched on at time $t = 3$ s. Sketch the waveform and write the function in terms of the Heaviside step function

The function is shown sketched in Figure 98.7.

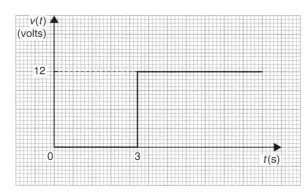

Figure 98.7

The Heaviside step function is: $V(t) = 12\,H(t-3)$

Problem 2. Write the function
$$V(t) = \begin{cases} 1 & \text{for } 0\langle t\langle a \\ 0 & \text{for } t\rangle a \end{cases} \text{ in terms of the Heaviside}$$
step function and sketch the waveform

The voltage has a value of 1 until time $t = a$; then it is turned off.
The function is shown sketched in Figure 98.8.

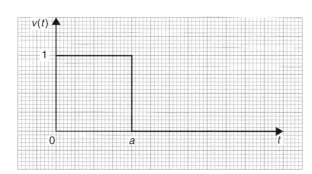

Figure 98.8

The Heaviside step function is: $V(t) = H(t) - H(t-a)$

Problem 3. Sketch the graph of $f(t) = 5H(t-2)$

A function $H(t-2)$ has a maximum value of 1 and starts when $t = 2$
A function $5H(t-2)$ has a maximum value of 5 and starts when $t = 2$, as shown in Figure 98.9.

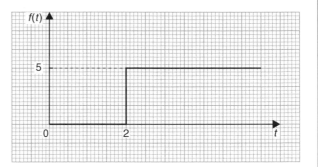

Figure 98.9

Problem 4. Sketch the graph of
$f(t) = H(t - \pi/3).\sin t$

Figure 98.10(a) shows a graph of $f(t) = \sin t$; the graph shown in Figure 98.10(b) is $f(t) = H(t - \pi/3).\sin t$ where the graph of $\sin t$ does not 'switch on' until $t = \pi/3$

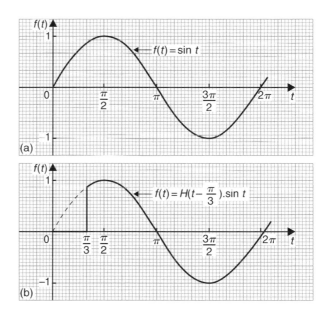

Figure 98.10

Problem 5. Sketch the graph of
$f(t) = 2H(t - 2\pi/3).\sin(t - \pi/6)$

Figure 98.11(a) shows a graph of $f(t) = 2\sin(t - \pi/6)$; the graph shown in Figure 98.11(b) is $f(t) = 2H(t - 2\pi/3).\sin(t - \pi/6)$, where the graph of $2\sin(t - \pi/6)$ does not 'switch on' until $t = 2\pi/3$

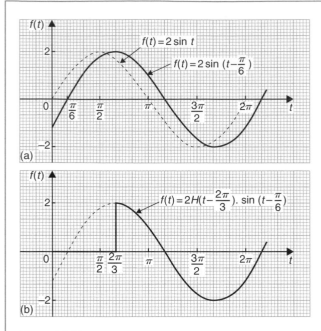

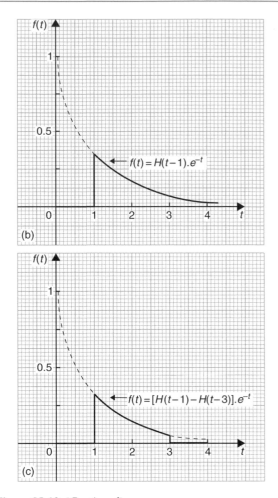

Figure 98.11

Figure 98.12 *(Continued)*

Problem 6. Sketch the graphs of
(a) $f(t) = H(t-1) \cdot e^{-t}$
(b) $f(t) = [H(t-1) - H(t-3)] \cdot e^{-t}$

Figure 98.12(a) shows a graph of $f(t) = e^{-t}$

(a) The graph shown in Figure 98.12(b) is $f(t) = H(t-1) \cdot e^{-t}$ where the graph of e^{-t} does not 'switch on' until $t = 1$

(b) Figure 98.12(c) shows the graph of $f(t) = [H(t-1) - H(t-3)] \cdot e^{-t}$ where the graph of e^{-t} does not 'switch on' until $t = 1$, but then 'switches off' at $t = 3$

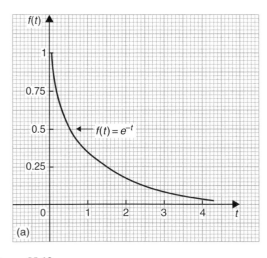

Figure 98.12

Now try the following Practice Exercise

Practice Exercise 362 Heaviside unit step function (answers on page 1194)

1. A 6 V source is switched on at time $t = 4$ s. Write the function in terms of the Heaviside step function and sketch the waveform.

2. Write the function $V(t) = \begin{cases} 2 & \text{for } 0 < t < 5 \\ 0 & \text{for } t > 5 \end{cases}$ in terms of the Heaviside step function and sketch the waveform.

In Problems 3 to 12, sketch graphs of the given functions.

3. $f(t) = H(t-2)$

4. $f(t) = H(t)$

5. $f(t) = 4H(t-1)$

6. $f(t) = 7H(t-5)$

7. $f(t) = H\left(t - \dfrac{\pi}{4}\right).\cos t$

8. $f(t) = 3H\left(t - \dfrac{\pi}{2}\right).\cos\left(t - \dfrac{\pi}{6}\right)$

9. $f(t) = H(t-1).t^2$

10. $f(t) = H(t-2).e^{-\frac{t}{2}}$

11. $f(t) = [H(t-2) - H(t-5)].e^{-\frac{t}{4}}$

12. $f(t) = 5H\left(t - \dfrac{\pi}{3}\right).\sin\left(t + \dfrac{\pi}{4}\right)$

98.2 Laplace transform of $H(t-c)$

From the definition of a Laplace transform,

$$\mathcal{L}\{H(t-c)\} = \int_0^\infty e^{-st} H(t-c)\,dt$$

However, $e^{-st}H(t-c) = \begin{cases} 0 & \text{for } 0 < t < c \\ e^{-st} & \text{for } t \geq c \end{cases}$

Hence, $\mathcal{L}\{H(t-c)\} = \int_0^\infty e^{-st} H(t-c)\,dt$

$$= \int_c^\infty e^{-st}\,dt = \left[\dfrac{e^{-st}}{-s}\right]_c^\infty = \left[\dfrac{e^{-s(\infty)}}{-s} - \dfrac{e^{-sc}}{-s}\right]$$

$$= \left[0 - \dfrac{e^{-sc}}{-s}\right] = \dfrac{e^{-sc}}{s}$$

When $c = 0$ (i.e. a unit step at the origin),

$$\mathcal{L}\{H(t)\} = \dfrac{e^{-s(0)}}{s} = \dfrac{1}{s}$$

Summarising, $\mathcal{L}\{H(t)\} = \dfrac{1}{s}$ and $\mathcal{L}\{H(t-c)\} = \dfrac{e^{-cs}}{s}$

From the definition of $H(t)$: $\mathcal{L}\{1\} = \{1 . H(t)\}$

$$\mathcal{L}\{t\} = \{t . H(t)\}$$

and $\qquad \mathcal{L}\{f(t)\} = \{f(t) . H(t)\}$

98.3 Laplace transform of $H(t-c) . f(t-c)$

It may be shown that:

$$\mathcal{L}\{H(t-c) . f(t-c)\} = e^{-cs}\mathcal{L}\{f(t)\} = e^{-cs}F(s)$$

$$\text{where } F(s) = \mathcal{L}\{f(t)\}$$

This is demonstrated in the following Problems.

Problem 7. Determine $\mathcal{L}\{4H(t-5)\}$

From above, $\mathcal{L}\{H(t-c) . f(t-c)\} = e^{-cs}F(s)$ where in this case, $F(s) = \mathcal{L}\{4\}$ and $c = 5$

Hence, $\quad \mathcal{L}\{4H(t-5)\} = e^{-5s}\left(\dfrac{4}{s}\right) \quad$ from (ii) of Table 95.1, page 1045

$$= \dfrac{4e^{-5s}}{s}$$

Problem 8. Determine $\mathcal{L}\{H(t-3).(t-3)^2\}$

From above, $\mathcal{L}\{H(t-c) . f(t-c)\} = e^{-cs}F(s)$ where in this case, $F(s) = \mathcal{L}\{t^2\}$ and $c = 3$

Note that $F(s)$ is the transform of t^2 and not of $(t-3)^2$

Hence, $\quad \mathcal{L}\{H(t-3) . f(t-3)^2\} = e^{-3s}\left(\dfrac{2!}{s^3}\right) \quad$ from (vii) of Table 95.1, page 1045

$$= \dfrac{2e^{-3s}}{s^3}$$

Problem 9. Determine $\mathcal{L}\{H(t-2).\sin(t-2)\}$

From above, $\mathcal{L}\{H(t-c) . f(t-c)\} = e^{-cs}F(s)$ where in this case, $F(s) = \mathcal{L}\{\sin t\}$ and $c = 2$

Hence, $\quad \mathcal{L}\{H(t-2) . \sin(t-2)\} = e^{-2s}\left(\dfrac{1}{s^2 + 1^2}\right)$ from (iv) of Table 95.1, page 1045

$$= \dfrac{e^{-2s}}{s^2 + 1}$$

Problem 10. Determine $\mathcal{L}\{H(t-1).\sin 4(t-1)\}$

From above, $\mathcal{L}\{H(t-c) . f(t-c)\} = e^{-cs}F(s)$ where in this case, $F(s) = \mathcal{L}\{\sin 4t\}$ and $c = 1$

Hence, $\quad \mathcal{L}\{H(t-1) . \sin 4(t-1)\} = e^{-s}\left(\dfrac{4}{s^2 + 4^2}\right)$ from (iv) of Table 95.1, page 1045

$$= \dfrac{4e^{-s}}{s^2 + 16}$$

Problem 11. Determine $\mathcal{L}\{H(t-3).e^{t-3}\}$

From above, $\mathcal{L}\{H(t-c) . f(t-c)\} = e^{-cs}F(s)$ where in this case, $F(s) = \mathcal{L}\{e^t\}$ and $c = 3$

Hence, $\mathcal{L}\{H(t-3).e^{t-3}\} = e^{-3s}\left(\dfrac{1}{s-1}\right)$ from (iii)

of Table 95.1, page 1045

$$= \frac{e^{-3s}}{s-1}$$

Problem 12. Determine

$$\mathcal{L}\left\{H\left(t-\frac{\pi}{2}\right).\cos 3\left(t-\frac{\pi}{2}\right)\right\}$$

From above, $\mathcal{L}\{H(t-c).f(t-c)\} = e^{-cs}F(s)$ where in this case, $F(s) = \mathcal{L}\{\cos 3t\}$ and $c = \dfrac{\pi}{2}$

Hence, $\mathcal{L}\left\{\left(t-\dfrac{\pi}{2}\right).\cos 3\left(1-\dfrac{\pi}{2}\right)\right\} =$

$e^{-\frac{\pi}{2}s}\left(\dfrac{s}{s^2+3^2}\right)$ from (v) of Table 95.1, page 1045

$$= \frac{s\,e^{-\frac{\pi}{2}s}}{s^2+9}$$

Now try the following Practice Exercise

Practice Exercise 363 Laplace transforms of $H(t-c).f(t-c)$ (answers on page 1196)

1. Determine $\mathcal{L}\{H(t-1)\}$

2. Determine $\mathcal{L}\{7H(t-3)\}$

3. Determine $\mathcal{L}\{H(t-2).(t-2)^2\}$

4. Determine $\mathcal{L}\{H(t-3).\sin(t-3)\}$

5. Determine $\mathcal{L}\{H(t-4).e^{t-4}\}$

6. Determine $\mathcal{L}\{H(t-5).\sin 3(t-5)\}$

7. Determine $\mathcal{L}\{H(t-1).(t-1)^3\}$

8. Determine $\mathcal{L}\{H(t-6).\cos 3(t-6)\}$

9. Determine $\mathcal{L}\{5H(t-5).\sinh 2(t-5)\}$

10. Determine $\mathcal{L}\left\{H\left(t-\dfrac{\pi}{3}\right).\cos 2\left(t-\dfrac{\pi}{3}\right)\right\}$

11. Determine $\mathcal{L}\{2H(t-3).e^{t-3}\}$

12. Determine $\mathcal{L}\{3H(t-2).\cosh(t-2)\}$

98.4 Inverse Laplace transforms of Heaviside functions

In the previous section it was stated that: $\mathcal{L}\{H(t-c).f(t-c)\} = e^{-cs}F(s)$ where $F(s) = \mathcal{L}\{f(t)\}$

Written in reverse, this becomes:

if $F(s) = \mathcal{L}\{f(t)\}$, then $e^{-cs}F(s) = \mathcal{L}\{H(t-c).f(t-c)\}$

This is known as the **second shift theorem** and is used when finding **inverse Laplace transforms**, as demonstrated in the following Problems.

Problem 13. Determine $\mathcal{L}^{-1}\left\{\dfrac{3e^{-2s}}{s}\right\}$

Part of the numerator corresponds to e^{-cs} where $c = 2$. This indicates $H(t-2)$

Then $\dfrac{3}{s} = F(s) = \mathcal{L}\{3\}$ from (ii) of Table 95.1, page 1045

Hence, $\mathcal{L}^{-1}\left\{\dfrac{3e^{-2s}}{s}\right\} = 3H(t-2)$

Problem 14. Determine the inverse of $\dfrac{e^{-3s}}{s^2}$

The numerator corresponds to e^{-cs} where $c = 3$. This indicates $H(t-3)$

$\dfrac{1}{s^2} = F(s) = \mathcal{L}\{t\}$ from (vi) of Table 95.1, page 1045

Then $\mathcal{L}^{-1}\left\{\dfrac{e^{-3s}}{s^2}\right\} = H(t-3).(t-3)$

Problem 15. Determine $\mathcal{L}^{-1}\left\{\dfrac{8e^{-4s}}{s^2+4}\right\}$

Part of the numerator corresponds to e^{-cs} where $c = 4$. This indicates $H(t-4)\dfrac{8}{s^2+4}$, which may be written as:

$$4\left(\frac{2}{s^2+2^2}\right)$$

Then $4\left(\dfrac{2}{s^2+2^2}\right) = F(s) = \mathcal{L}\{4\sin 2t\}$ from (iv) of Table 95.1, page 1045

Hence, $\mathcal{L}^{-1}\left\{\dfrac{8e^{-4s}}{s^2+4}\right\} = H(t-4).4\sin 2(t-4)$

$$= 4H(t-4).\sin 2(t-4)$$

Problem 16. Determine $\mathcal{L}^{-1}\left\{\dfrac{5se^{-2s}}{s^2+9}\right\}$

Part of the numerator corresponds to e^{-cs} where $c=2$. This indicates $H(t-2)$

$\dfrac{5s}{s^2+9}$, which may be written as: $5\left(\dfrac{s}{s^2+3^2}\right)$

Then $5\left(\dfrac{s}{s^2+3^2}\right)=F(s)=\mathcal{L}\{5\cos 3t\}$ from (v) of Table 95.1, page 1045

Hence, $\mathcal{L}^{-1}\left\{\dfrac{5se^{-2s}}{s^2+3^2}\right\}=H(t-2)\,.5\cos 3(t-2)$

$\qquad\qquad\qquad = 5H(t-2)\,.\cos 3(t-2)$

Problem 17. Determine $\mathcal{L}^{-1}\left\{\dfrac{7e^{-3s}}{s^2-1}\right\}$

Part of the numerator corresponds to e^{-cs} where $c=3$. This indicates $H(t-3)\dfrac{7}{s^2-1}$ which may be written as:

$7\left(\dfrac{1}{s^2-1^2}\right)$

Then $7\left(\dfrac{1}{s^2-1^2}\right)=F(s)=\mathcal{L}\{7\sinh t\}$ from (x) of Table 95.1, page 1045

Hence, $\mathcal{L}^{-1}\left\{\dfrac{7e^{-3s}}{s^2-1}\right\}=H(t-3)\,.7\sinh(t-3)$

$\qquad\qquad\qquad = 7H(t-3)\,.\sinh(t-3)$

Now try the following Practice Exercise

Practice Exercise 364   Inverse Laplace transforms of Heaviside functions (answers on page 1196)

1. Determine $\mathcal{L}^{-1}\left\{\dfrac{e^{-9s}}{s}\right\}$

2. Determine $\mathcal{L}^{-1}\left\{\dfrac{4e^{-3s}}{s}\right\}$

3. Determine $\mathcal{L}^{-1}\left\{\dfrac{2e^{-2s}}{s^2}\right\}$

4. Determine $\mathcal{L}^{-1}\left\{\dfrac{5e^{-2s}}{s^2+1}\right\}$

5. Determine $\mathcal{L}^{-1}\left\{\dfrac{3se^{-4s}}{s^2+16}\right\}$

6. Determine $\mathcal{L}^{-1}\left\{\dfrac{6e^{-2s}}{s^2-1}\right\}$

7. Determine $\mathcal{L}^{-1}\left\{\dfrac{3e^{-6s}}{s^3}\right\}$

8. Determine $\mathcal{L}^{-1}\left\{\dfrac{2se^{-4s}}{s^2-16}\right\}$

9. Determine $\mathcal{L}^{-1}\left\{\dfrac{2se^{-\frac{1}{2}s}}{s^2+5}\right\}$

10. Determine $\mathcal{L}^{-1}\left\{\dfrac{4e^{-7s}}{s-1}\right\}$

For fully worked solutions to each of the problems in Practice Exercises 362 to 364 in this chapter, go to the website:
www.routledge.com/cw/bird

Chapter 99

The solution of differential equations using Laplace transforms

---

Why it is important to understand: **The solution of differential equations using Laplace transforms**

Laplace transforms and their inverses are a mathematical technique which allows us to solve differential equations, by primarily using algebraic methods. This simplification in the solving of equations, coupled with the ability to directly implement electrical components in their transformed form, makes the use of Laplace transforms widespread in both electrical engineering and control systems engineering. The procedures explained in the previous chapters are all used in this chapter, which demonstrates how differential equations are solved using Laplace transforms.

At the end of this chapter, you should be able to:

- understand the procedure to solve differential equations using Laplace transforms
- solve differential equations using Laplace transforms

99.1 Introduction

An alternative method of solving differential equations to that used in Chapters 77 to 82 is possible by using Laplace transforms.

99.2 Procedure to solve differential equations by using Laplace transforms

(i) Take the Laplace transform of both sides of the differential equation by applying the formulae for the Laplace transforms of derivatives (i.e. equations (3) and (4) of Chapter 96) and, where necessary, using a list of standard Laplace transforms, such as Tables 95.1 and 96.1 on pages 1045 and 1049.

(ii) Put in the given initial conditions, i.e. $y(0)$ and $y'(0)$

(iii) Rearrange the equation to make $\mathcal{L}\{y\}$ the subject.

(iv) Determine y by using, where necessary, partial fractions, and taking the inverse of each term by using Table 97.1 on page 1055.

99.3 Worked problems on solving differential equations using Laplace transforms

Problem 1. Use Laplace transforms to solve the differential equation

$$2\frac{d^2 y}{dx^2} + 5\frac{dy}{dx} - 3y = 0, \text{ given that when } x = 0,$$

$$y = 4 \text{ and } \frac{dy}{dx} = 9$$

This is the same problem as Problem 1 of Chapter 81, page 865 and a comparison of methods can be made. Using the above procedure:

(i) $2\mathcal{L}\left\{\dfrac{d^2 y}{dx^2}\right\} + 5\mathcal{L}\left\{\dfrac{dy}{dx}\right\} - 3\mathcal{L}\{y\} = \mathcal{L}\{0\}$

$2[s^2 \mathcal{L}\{y\} - sy(0) - y'(0)] + 5[s\mathcal{L}\{y\}$

$$- y(0)] - 3\mathcal{L}\{y\} = 0$$

from equations (3) and (4) of Chapter 96.

(ii) $y(0) = 4$ and $y'(0) = 9$

Thus $2[s^2 \mathcal{L}\{y\} - 4s - 9] + 5[s\mathcal{L}\{y\} - 4]$

$$- 3\mathcal{L}\{y\} = 0$$

i.e. $2s^2 \mathcal{L}\{y\} - 8s - 18 + 5s\mathcal{L}\{y\} - 20$

$$- 3\mathcal{L}\{y\} = 0$$

(iii) Rearranging gives:

$$(2s^2 + 5s - 3)\mathcal{L}\{y\} = 8s + 38$$

i.e. $\mathcal{L}\{y\} = \dfrac{8s + 38}{2s^2 + 5s - 3}$

(iv) $y = \mathcal{L}^{-1}\left\{\dfrac{8s + 38}{2s^2 + 5s - 3}\right\}$

$\dfrac{8s + 38}{2s^2 + 5s - 3} \equiv \dfrac{8s + 38}{(2s - 1)(s + 3)}$

$$\equiv \dfrac{A}{2s - 1} + \dfrac{B}{s + 3}$$

$$\equiv \dfrac{A(s + 3) + B(2s - 1)}{(2s - 1)(s + 3)}$$

Hence $8s + 38 = A(s + 3) + B(2s - 1)$

When $s = \dfrac{1}{2}$, $42 = 3\dfrac{1}{2}A$, from which, $A = 12$

When $s = -3$, $14 = -7B$, from which, $B = -2$

Hence $y = \mathcal{L}^{-1}\left\{\dfrac{8s + 38}{2s^2 + 5s - 3}\right\}$

$$= \mathcal{L}^{-1}\left\{\dfrac{12}{2s - 1} - \dfrac{2}{s + 3}\right\}$$

$$= \mathcal{L}^{-1}\left\{\dfrac{12}{2\left(s - \frac{1}{2}\right)}\right\} - \mathcal{L}^{-1}\left\{\dfrac{2}{s + 3}\right\}$$

Hence $y = 6e^{\frac{1}{2}x} - 2e^{-3x}$, from (iii) of

Table 97.1

Problem 2. Use Laplace transforms to solve the differential equation:

$$\frac{d^2 y}{dx^2} + 6\frac{dy}{dx} + 13y = 0, \text{ given that when } x = 0, \ y = 3$$

and $\dfrac{dy}{dx} = 7$

This is the same as Problem 3 of Chapter 81, page 866. Using the above procedure:

(i) $\mathcal{L}\left\{\dfrac{d^2 x}{dy^2}\right\} + 6\mathcal{L}\left\{\dfrac{dy}{dx}\right\} + 13\mathcal{L}\{y\} = \mathcal{L}\{0\}$

Hence $[s^2 \mathcal{L}\{y\} - sy(0) - y'(0)]$

$$+ 6[s\mathcal{L}\{y\} - y(0)] + 13\mathcal{L}\{y\} = 0$$

from equations (3) and (4) of Chapter 96.

(ii) $y(0) = 3$ and $y'(0) = 7$

Thus $s^2 \mathcal{L}\{y\} - 3s - 7 + 6s\mathcal{L}\{y\}$

$$- 18 + 13\mathcal{L}\{y\} = 0$$

(iii) Rearranging gives:

$$(s^2 + 6s + 13)\mathcal{L}\{y\} = 3s + 25$$

i.e. $\mathcal{L}\{y\} = \dfrac{3s + 25}{s^2 + 6s + 13}$

(iv) $y = \mathcal{L}^{-1}\left\{\dfrac{3s + 25}{s^2 + 6s + 13}\right\}$

$$= \mathcal{L}^{-1}\left\{\dfrac{3s + 25}{(s + 3)^2 + 2^2}\right\}$$

$$= \mathcal{L}^{-1}\left\{\dfrac{3(s + 3) + 16}{(s + 3)^2 + 2^2}\right\}$$

$$= \mathcal{L}^{-1}\left\{\frac{3(s+3)}{(s+3)^2+2^2}\right\}$$

$$+ \mathcal{L}^{-1}\left\{\frac{8(2)}{(s+3)^2+2^2}\right\}$$

$$= 3e^{-3t}\cos 2t + 8e^{-3t}\sin 2t, \text{ from (xiii)}$$

and (xii) of Table 97.1

Hence $y = e^{-3t}(3\cos 2t + 8\sin 2t)$

Problem 3. Use Laplace transforms to solve the differential equation:

$\dfrac{d^2 y}{dx^2} - 3\dfrac{dy}{dx} = 9$, given that when $x=0$, $y=0$ and $\dfrac{dy}{dx}=0$

This is the same problem as Problem 2 of Chapter 82, page 872. Using the procedure:

(i) $\mathcal{L}\left\{\dfrac{d^2 y}{dx^2}\right\} - 3\mathcal{L}\left\{\dfrac{dy}{dx}\right\} = \mathcal{L}\{9\}$

Hence $[s^2\mathcal{L}\{y\} - sy(0) - y'(0)]$

$$- 3[s\mathcal{L}\{y\} - y(0)] = \frac{9}{s}$$

(ii) $y(0)=0$ and $y'(0)=0$

Hence $s^2\mathcal{L}\{y\} - 3s\mathcal{L}\{y\} = \dfrac{9}{s}$

(iii) Rearranging gives:

$$(s^2 - 3s)\mathcal{L}\{y\} = \frac{9}{s}$$

i.e. $\mathcal{L}\{y\} = \dfrac{9}{s(s^2 - 3s)} = \dfrac{9}{s^2(s-3)}$

(iv) $y = \mathcal{L}^{-1}\left\{\dfrac{9}{s^2(s-3)}\right\}$

$$\frac{9}{s^2(s-3)} \equiv \frac{A}{s} + \frac{B}{s^2} + \frac{C}{s-3}$$

$$\equiv \frac{A(s)(s-3) + B(s-3) + Cs^2}{s^2(s-3)}$$

Hence $9 \equiv A(s)(s-3) + B(s-3) + Cs^2$

When $s=0$, $9=-3B$, from which, $B=-3$

When $s=3$, $9=9C$, from which, $C=1$

Equating s^2 terms gives: $0 = A+C$, from which, $A=-1$, since $C=1$. Hence,

$$\mathcal{L}^{-1}\left\{\frac{9}{s^2(s-3)}\right\} = \mathcal{L}^{-1}\left\{-\frac{1}{s} - \frac{3}{s^2} + \frac{1}{s-3}\right\}$$

$$= -1 - 3x + e^{3x}, \text{ from (i),}$$

(vi) and (iii) of Table 97.1

i.e. $y = e^{3x} - 3x - 1$

Problem 4. Use Laplace transforms to solve the differential equation:

$\dfrac{d^2 y}{dx^2} - 7\dfrac{dy}{dx} + 10y = e^{2x} + 20$, given that when $x=0$, $y=0$ and $\dfrac{dy}{dx} = -\dfrac{1}{3}$

Using the procedure:

(i) $\mathcal{L}\left\{\dfrac{d^2 y}{dx^2}\right\} - 7\mathcal{L}\left\{\dfrac{dy}{dx}\right\} + 10\mathcal{L}\{y\} = \mathcal{L}\{e^{2x} + 20\}$

Hence $[s^2\mathcal{L}\{y\} - sy(0) - y'(0)] - 7[s\mathcal{L}\{y\}$

$$- y(0)] + 10\mathcal{L}\{y\} = \frac{1}{s-2} + \frac{20}{s}$$

(ii) $y(0)=0$ and $y'(0)=-\dfrac{1}{3}$

Hence $s^2\mathcal{L}\{y\} - 0 - \left(-\dfrac{1}{3}\right) - 7s\mathcal{L}\{y\} + 0$

$$+ 10\mathcal{L}\{y\} = \frac{21s - 40}{s(s-2)}$$

(iii) $(s^2 - 7s + 10)\mathcal{L}\{y\} = \dfrac{21s-40}{s(s-2)} - \dfrac{1}{3}$

$$= \frac{3(21s - 40) - s(s-2)}{3s(s-2)}$$

$$= \frac{-s^2 + 65s - 120}{3s(s-2)}$$

Hence $\mathcal{L}\{y\} = \dfrac{-s^2 + 65s - 120}{3s(s-2)(s^2 - 7s + 10)}$

$$= \frac{1}{3}\left[\frac{-s^2 + 65s - 120}{s(s-2)(s-2)(s-5)}\right]$$

$$= \frac{1}{3}\left[\frac{-s^2 + 65s - 120}{s(s-5)(s-2)^2}\right]$$

(iv) $y = \dfrac{1}{3} \mathcal{L}^{-1} \left\{ \dfrac{-s^2 + 65s - 120}{s(s-5)(s-2)^2} \right\}$

$\dfrac{-s^2 + 65s - 120}{s(s-5)(s-2)^2}$

$\equiv \dfrac{A}{s} + \dfrac{B}{s-5} + \dfrac{C}{s-2} + \dfrac{D}{(s-2)^2}$

$\equiv \dfrac{\left(\begin{array}{c} A(s-5)(s-2)^2 + B(s)(s-2)^2 \\ + C(s)(s-5)(s-2) + D(s)(s-5) \end{array} \right)}{s(s-5)(s-2)^2}$

Hence

$-s^2 + 65s - 120$

$\equiv A(s-5)(s-2)^2 + B(s)(s-2)^2$

$+ C(s)(s-5)(s-2) + D(s)(s-5)$

When $s=0$, $-120 = -20A$, from which, $A=6$

When $s=5$, $180 = 45B$, from which, $B=4$

When $s=2$, $6 = -6D$, from which, $D=-1$

Equating s^3 terms gives: $0 = A+B+C$, from which, $C=-10$

Hence $\dfrac{1}{3} \mathcal{L}^{-1} \left\{ \dfrac{-s^2 + 65s - 120}{s(s-5)(s-2)^2} \right\}$

$= \dfrac{1}{3} \mathcal{L}^{-1} \left\{ \dfrac{6}{s} + \dfrac{4}{s-5} - \dfrac{10}{s-2} - \dfrac{1}{(s-2)^2} \right\}$

$= \dfrac{1}{3} [6 + 4e^{5x} - 10e^{2x} - xe^{2x}]$

Thus $y = 2 + \dfrac{4}{3} e^{5x} - \dfrac{10}{3} e^{2x} - \dfrac{x}{3} e^{2x}$

Problem 5.   The current flowing in an electrical circuit is given by the differential equation $Ri + L(di/dt) = E$, where E, L and R are constants. Use Laplace transforms to solve the equation for current i given that when $t=0$, $i=0$

Using the procedure:

(i) $\mathcal{L}\{Ri\} + \mathcal{L}\left\{ L\dfrac{di}{dt} \right\} = \mathcal{L}\{E\}$

i.e. $R\mathcal{L}\{i\} + L[s\mathcal{L}\{i\} - i(0)] = \dfrac{E}{s}$

(ii) $i(0) = 0$, hence $R\mathcal{L}\{i\} + Ls\mathcal{L}\{i\} = \dfrac{E}{s}$

(iii)   Rearranging gives:

$(R + Ls)\mathcal{L}\{i\} = \dfrac{E}{s}$

i.e. $\mathcal{L}\{i\} = \dfrac{E}{s(R + Ls)}$

(iv) $i = \mathcal{L}^{-1} \left\{ \dfrac{E}{s(R + Ls)} \right\}$

$\dfrac{E}{s(R + Ls)} \equiv \dfrac{A}{s} + \dfrac{B}{R + Ls}$

$\equiv \dfrac{A(R + Ls) + Bs}{s(R + Ls)}$

Hence $E = A(R + Ls) + Bs$

When $s = 0$, $E = AR$

from which, $A = \dfrac{E}{R}$

When $s = -\dfrac{R}{L}$, $E = B\left(-\dfrac{R}{L} \right)$

from which, $B = -\dfrac{EL}{R}$

Hence $\mathcal{L}^{-1} \left\{ \dfrac{E}{s(R + Ls)} \right\}$

$= \mathcal{L}^{-1} \left\{ \dfrac{E/R}{s} + \dfrac{-EL/R}{R + Ls} \right\}$

$= \mathcal{L}^{-1} \left\{ \dfrac{E}{Rs} - \dfrac{EL}{R(R + Ls)} \right\}$

$= \mathcal{L}^{-1} \left\{ \dfrac{E}{R}\left(\dfrac{1}{s} \right) - \dfrac{E}{R}\left(\dfrac{1}{\dfrac{R}{L} + s} \right) \right\}$

$= \dfrac{E}{R} \mathcal{L}^{-1} \left\{ \dfrac{1}{s} - \dfrac{1}{\left(s + \dfrac{R}{L} \right)} \right\}$

Hence **current** $i = \dfrac{E}{R}\left(1 - e^{-\frac{Rt}{L}} \right)$

Now try the following Practice Exercise

> **Practice Exercise 365   Further problems on solving differential equations using Laplace transforms (answers on page 1196)**

1. A first-order differential equation involving current i in a series $R-L$ circuit is given by:

   $$\frac{di}{dt} + 5i = \frac{E}{2} \text{ and } i = 0 \text{ at time } t = 0$$

   Use Laplace transforms to solve for i when (a) $E = 20$, (b) $E = 40e^{-3t}$ and (c) $E = 50 \sin 5t$

In Problems 2 to 9, use Laplace transforms to solve the given differential equations.

2. $9\dfrac{d^2 y}{dt^2} - 24\dfrac{dy}{dt} + 16y = 0$, given $y(0) = 3$ and $y'(0) = 3$

3. $\dfrac{d^2 x}{dt^2} + 100x = 0$, given $x(0) = 2$ and $x'(0) = 0$

4. $\dfrac{d^2 i}{dt^2} + 1000\dfrac{di}{dt} + 250000i = 0$, given $i(0) = 0$ and $i'(0) = 100$

5. $\dfrac{d^2 x}{dt^2} + 6\dfrac{dx}{dt} + 8x = 0$, given $x(0) = 4$ and $x'(0) = 8$

6. $\dfrac{d^2 y}{dx^2} - 2\dfrac{dy}{dx} + y = 3e^{4x}$, given $y(0) = -\dfrac{2}{3}$ and $y'(0) = 4\dfrac{1}{3}$

7. $\dfrac{d^2 y}{dx^2} - 3\dfrac{dy}{dx} - 4y = 3 \sin x$, given $y(0) = 0$, $y'(0) = 0$

8. $\dfrac{d^2 y}{dx^2} + \dfrac{dy}{dx} - 2y = 3\cos 3x - 11 \sin 3x$, given $y(0) = 0$ and $y'(0) = 6$

9. $\dfrac{d^2 y}{dx^2} - 2\dfrac{dy}{dx} + 2y = 3e^x \cos 2x$, given $y(0) = 2$ and $y'(0) = 5$

10. The free oscillations of a lightly damped elastic system are given by the equation:

    $$\frac{d^2 y}{dt^2} + 2\frac{dy}{dt} + 5y = 0$$

    where y is the displacement from the equilibrium position. If when time $t = 0$, $y = 2$ and $\dfrac{dy}{dt} = 0$, determine an expression for the displacement.

11. Solve, using Laplace transforms, Problems 4 to 9 of Exercise 310, page 867 and Problems 1 to 5 of Exercise 311, page 869.

12. Solve, using Laplace transforms, Problems 3 to 6 of Exercise 312, page 874, Problems 5 and 6 of Exercise 313, page 876, Problems 4 and 7 of Exercise 314, page 878 and Problems 5 and 6 of Exercise 315, page 880.

For fully worked solutions to each of the problems in Practice Exercise 365 in this chapter, go to the website:
www.routledge.com/cw/bird

Chapter 100

The solution of simultaneous differential equations using Laplace transforms

Why it is important to understand: The solution of simultaneous differential equations using Laplace transforms

As stated in the previous chapters, Laplace transforms have many applications in mathematics, physics, optics, electrical engineering, control engineering, signal processing and probability, and that Laplace transforms and their inverses are a mathematical technique which allows us to solve differential equations, by primarily using algebraic methods. Specifically, this chapter explains the procedure for solving simultaneous differential equations; this requires all of the knowledge gained in the preceding chapters.

At the end of this chapter, you should be able to:

- understand the procedure to solve simultaneous differential equations using Laplace transforms
- solve simultaneous differential equations using Laplace transforms

100.1 Introduction

It is sometimes necessary to solve simultaneous differential equations. An example occurs when two electrical circuits are coupled magnetically where the equations relating the two currents i_1 and i_2 are typically:

$$L_1 \frac{di_1}{dt} + M \frac{di_2}{dt} + R_1 i_1 = E_1$$

$$L_2 \frac{di_2}{dt} + M \frac{di_1}{dt} + R_2 i_2 = 0$$

where L represents inductance, R resistance, M mutual inductance and E_1 the p.d. applied to one of the circuits.

100.2 Procedure to solve simultaneous differential equations using Laplace transforms

(i) Take the Laplace transform of both sides of each simultaneous equation by applying the formulae for the Laplace transforms of derivatives (i.e. equations (3) and (4) of Chapter 96, page 1051) and using a list of standard Laplace transforms, as in Table 95.1, page 1045 and Table 96.1, page 1049.

(ii) Put in the initial conditions, i.e. $x(0)$, $y(0)$, $x'(0)$, $y'(0)$

(iii) Solve the simultaneous equations for $\mathcal{L}\{y\}$ and $\mathcal{L}\{x\}$ by the normal algebraic method.

(iv) Determine y and x by using, where necessary, partial fractions, and taking the inverse of each term.

100.3 Worked problems on solving simultaneous differential equations by using Laplace transforms

Problem 1. Solve the following pair of simultaneous differential equations

$$\frac{dy}{dt} + x = 1$$

$$\frac{dx}{dt} - y + 4e^t = 0$$

given that at $t=0$, $x=0$ and $y=0$

Using the above procedure:

(i) $\mathcal{L}\left\{\dfrac{dy}{dt}\right\} + \mathcal{L}\{x\} = \mathcal{L}\{1\}$ \hfill (1)

$\mathcal{L}\left\{\dfrac{dx}{dt}\right\} - \mathcal{L}\{y\} + 4\mathcal{L}\{e^t\} = 0$ \hfill (2)

Equation (1) becomes:

$$[s\mathcal{L}\{y\} - y(0)] + \mathcal{L}\{x\} = \frac{1}{s} \qquad (1')$$

from equation (3), page 1051 and Table 95.1, page 1045.

Equation (2) becomes:

$$[s\mathcal{L}\{x\} - x(0)] - \mathcal{L}\{y\} = -\frac{4}{s-1} \qquad (2')$$

(ii) $x(0)=0$ and $y(0)=0$ hence

Equation $(1')$ becomes:

$$s\mathcal{L}\{y\} + \mathcal{L}\{x\} = \frac{1}{s} \qquad (1'')$$

and equation $(2')$ becomes:

$$s\mathcal{L}\{x\} - \mathcal{L}\{y\} = -\frac{4}{s-1}$$

or $-\mathcal{L}\{y\} + s\mathcal{L}\{x\} = -\dfrac{4}{s-1}$ \hfill $(2'')$

(iii) $1 \times$ equation $(1'')$ and $s \times$ equation $(2'')$ gives:

$$s\mathcal{L}\{y\} + \mathcal{L}\{x\} = \frac{1}{s} \qquad (3)$$

$$-s\mathcal{L}\{y\} + s^2\mathcal{L}\{x\} = -\frac{4s}{s-1} \qquad (4)$$

Adding equations (3) and (4) gives:

$$(s^2+1)\mathcal{L}\{x\} = \frac{1}{s} - \frac{4s}{s-1}$$

$$= \frac{(s-1) - s(4s)}{s(s-1)}$$

$$= \frac{-4s^2 + s - 1}{s(s-1)}$$

from which, $\mathcal{L}\{x\} = \dfrac{-4s^2 + s - 1}{s(s-1)(s^2+1)}$ \hfill (5)

Using partial fractions

$$\frac{-4s^2 + s - 1}{s(s-1)(s^2+1)} \equiv \frac{A}{s} + \frac{B}{(s-1)} + \frac{Cs+D}{(s^2+1)}$$

$$= \frac{\left(\begin{array}{c} A(s-1)(s^2+1) + Bs(s^2+1) \\ + (Cs+D)s(s-1) \end{array}\right)}{s(s-1)(s^2+1)}$$

Hence

$$-4s^2 + s - 1 = A(s-1)(s^2+1) + Bs(s^2+1)$$
$$+ (Cs+D)s(s-1)$$

When $s=0$, $-1 = -A$ hence $A = 1$

When $s=1$, $-4 = 2B$ hence $B = -2$

Equating s^3 coefficients:

$$0 = A + B + C \quad \text{hence} \quad C = 1$$
$$(\text{since } A = 1 \text{ and } B = -2)$$

Equating s^2 coefficients:

$$-4 = -A + D - C \quad \text{hence} \quad D = -2$$
$$(\text{since } A = 1 \text{ and } C = 1)$$

Thus $\mathcal{L}\{x\} = \dfrac{-4s^2 + s - 1}{s(s-1)(s^2+1)}$

$$= \frac{1}{s} - \frac{2}{(s-1)} + \frac{s-2}{(s^2+1)}$$

(iv)   Hence

$$x = \mathcal{L}^{-1}\left\{\frac{1}{s} - \frac{2}{(s-1)} + \frac{s-2}{(s^2+1)}\right\}$$

$$= \mathcal{L}^{-1}\left\{\frac{1}{s} - \frac{2}{(s-1)} + \frac{s}{(s^2+1)} - \frac{2}{(s^2+1)}\right\}$$

i.e. $x = 1 - 2e^t + \cos t - 2\sin t$,

from Table 97.1, page 1055.

From the second equation given in the question,

$$\frac{dx}{dt} - y + 4e^t = 0$$

from which,

$$y = \frac{dx}{dt} + 4e^t$$

$$= \frac{d}{dt}(1 - 2e^t + \cos t - 2\sin t) + 4e^t$$

$$= -2e^t - \sin t - 2\cos t + 4e^t$$

i.e. $y = 2e^t - \sin t - 2\cos t$

(Alternatively, to determine y, return to equations (1″) and (2″).)

Problem 2.   Solve the following pair of simultaneous differential equations

$$3\frac{dx}{dt} - 5\frac{dy}{dt} + 2x = 6$$

$$2\frac{dy}{dt} - \frac{dx}{dt} - y = -1$$

given that at $t = 0$, $x = 8$ and $y = 3$

Using the above procedure:

(i) $3\mathcal{L}\left\{\dfrac{dx}{dt}\right\} - 5\mathcal{L}\left\{\dfrac{dy}{dt}\right\} + 2\mathcal{L}\{x\} = \mathcal{L}\{6\}$ (1)

$2\mathcal{L}\left\{\dfrac{dy}{dt}\right\} - \mathcal{L}\left\{\dfrac{dx}{dt}\right\} - \mathcal{L}\{y\} = \mathcal{L}\{-1\}$ (2)

Equation (1) becomes:

$$3[s\mathcal{L}\{x\} - x(0)] - 5[s\mathcal{L}\{y\} - y(0)]$$
$$+ 2\mathcal{L}\{x\} = \frac{6}{s}$$

from equation (3), page 1051, and Table 95.1, page 1045.

i.e. $3s\mathcal{L}\{x\} - 3x(0) - 5s\mathcal{L}\{y\}$
$$+ 5y(0) + 2\mathcal{L}\{x\} = \frac{6}{s}$$

i.e. $(3s + 2)\mathcal{L}\{x\} - 3x(0) - 5s\mathcal{L}\{y\}$
$$+ 5y(0) = \frac{6}{s}$$ (1′)

Equation (2) becomes:

$$2[s\mathcal{L}\{y\} - y(0)] - [s\mathcal{L}\{x\} - x(0)]$$
$$- \mathcal{L}\{y\} = -\frac{1}{s}$$

from equation (3), page 1051, and Table 95.1, page 1045.

i.e. $2s\mathcal{L}\{y\} - 2y(0) - s\mathcal{L}\{x\}$
$$+ x(0) - \mathcal{L}\{y\} = -\frac{1}{s}$$

i.e. $(2s - 1)\mathcal{L}\{y\} - 2y(0) - s\mathcal{L}\{x\}$
$$+ x(0) = -\frac{1}{s}$$ (2′)

(ii) $x(0) = 8$ and $y(0) = 3$, hence equation (1′) becomes

$$(3s + 2)\mathcal{L}\{x\} - 3(8) - 5s\mathcal{L}\{y\}$$
$$+ 5(3) = \frac{6}{s}$$ (1″)

and equation (2′) becomes

$$(2s - 1)\mathcal{L}\{y\} - 2(3) - s\mathcal{L}\{x\}$$
$$+ 8 = -\frac{1}{s}$$ (2″)

i.e. $(3s + 2)\mathcal{L}\{x\} - 5s\mathcal{L}\{y\} = \dfrac{6}{s} + 9$ (1″)

$$\left.\begin{array}{l}(3s+2)\mathcal{L}\{x\} - 5s\mathcal{L}\{y\} \\ \quad = \dfrac{6}{s} + 9 \qquad\qquad (1''') \\[2mm] -s\mathcal{L}\{x\} + (2s-1)\mathcal{L}\{y\} \\ \quad = -\dfrac{1}{s} - 2 \qquad\quad (2''')\end{array}\right\} \quad \text{(A)}$$

Section M

(iii) $s \times$ equation $(1''')$ and $(3s+2) \times$ equation $(2''')$ gives:

$$s(3s+2)\mathcal{L}\{x\} - 5s^2\mathcal{L}\{y\} = s\left(\frac{6}{s}+9\right) \quad (3)$$

$$-s(3s+2)\mathcal{L}\{x\} + (3s+2)(2s-1)\mathcal{L}\{y\}$$

$$= (3s+2)\left(-\frac{1}{s}-2\right) \quad (4)$$

i.e. $s(3s+2)\mathcal{L}\{x\} - 5s^2\mathcal{L}\{y\} = 6+9s \quad (3')$

$$-s(3s+2)\mathcal{L}\{x\} + (6s^2+s-2)\mathcal{L}\{y\}$$

$$= -6s - \frac{2}{s} - 7 \quad (4')$$

Adding equations $(3')$ and $(4')$ gives:

$$(s^2+s-2)\mathcal{L}\{y\} = -1+3s-\frac{2}{s}$$

$$= \frac{-s+3s^2-2}{s}$$

from which, $\mathcal{L}\{y\} = \dfrac{3s^2-s-2}{s(s^2+s-2)}$

Using partial fractions

$$\frac{3s^2-s-2}{s(s^2+s-2)}$$

$$\equiv \frac{A}{s} + \frac{B}{(s+2)} + \frac{C}{(s-1)}$$

$$= \frac{A(s+2)(s-1) + Bs(s-1) + Cs(s+2)}{s(s+2)(s-1)}$$

i.e. $3s^2-s-2 = A(s+2)(s-1)$

$$+ Bs(s-1) + Cs(s+2)$$

When $s=0, -2=-2A,$ hence $A=1$

When $s=1, 0=3C,$ hence $C=0$

When $s=-2, 12=6B,$ hence $B=2$

Thus $\mathcal{L}\{y\} = \dfrac{3s^2-s-2}{s(s^2+s-2)} = \dfrac{1}{s} + \dfrac{2}{(s+2)}$

(iv) Hence $y = \mathcal{L}^{-1}\left\{\dfrac{1}{s} + \dfrac{2}{s+2}\right\} = \mathbf{1+2e^{-2t}}$

Returning to equations (A) to determine $\mathcal{L}\{x\}$ and hence x:

$(2s-1) \times$ equation $(1''')$ and $5s \times (2''')$ gives:

$$(2s-1)(3s+2)\mathcal{L}\{x\} - 5s(2s-1)\mathcal{L}\{y\}$$

$$= (2s-1)\left(\frac{6}{s}+9\right) \quad (5)$$

and $-s(5s)\mathcal{L}\{x\} + 5s(2s-1)\mathcal{L}\{y\}$

$$= 5s\left(-\frac{1}{s}-2\right) \quad (6)$$

i.e. $(6s^2+s-2)\mathcal{L}\{x\} - 5s(2s-1)\mathcal{L}\{y\}$

$$= 12+18s-\frac{6}{s}-9 \quad (5')$$

and $-5s^2\mathcal{L}\{x\} + 5s(2s-1)\mathcal{L}\{y\}$

$$= -5-10s \quad (6')$$

Adding equations $(5')$ and $(6')$ gives:

$$(s^2+s-2)\mathcal{L}\{x\} = -2+8s-\frac{6}{s}$$

$$= \frac{-2s+8s^2-6}{s}$$

from which, $\mathcal{L}\{x\} = \dfrac{8s^2-2s-6}{s(s^2+s-2)}$

$$= \frac{8s^2-2s-6}{s(s+2)(s-1)}$$

Using partial fractions

$$\frac{8s^2-2s-6}{s(s+2)(s-1)}$$

$$\equiv \frac{A}{s} + \frac{B}{(s+2)} + \frac{C}{(s-1)}$$

$$= \frac{A(s+2)(s-1) + Bs(s-1) + Cs(s+2)}{s(s+2)(s-1)}$$

i.e. $8s^2-2s-6 = A(s+2)(s-1)$

$$+ Bs(s-1) + Cs(s+2)$$

When $s=0$, $-6=-2A$, hence $A=3$

When $s=1$, $0=3C$, hence $C=0$

When $s=-2$, $30=6B$, hence $B=5$

Thus $\mathcal{L}\{x\} = \dfrac{8s^2-2s-6}{s(s+2)(s-1)} = \dfrac{3}{s} + \dfrac{5}{(s+2)}$

Hence $x = \mathcal{L}^{-1}\left\{\dfrac{3}{s} + \dfrac{5}{s+2}\right\} = 3 + 5e^{-2t}$

Therefore the solutions of the given simultaneous differential equations are

$$y = 1 + 2e^{-2t} \quad \text{and} \quad x = 3 + 5e^{-2t}$$

(These solutions may be checked by substituting the expressions for x and y into the original equations.)

> **Problem 3.** Solve the following pair of simultaneous differential equations
>
> $$\dfrac{d^2x}{dt^2} - x = y$$
>
> $$\dfrac{d^2y}{dt^2} + y = -x$$
>
> given that at $t=0$, $x=2$, $y=-1$, $\dfrac{dx}{dt}=0$
>
> and $\dfrac{dy}{dt}=0$

Using the procedure:

(i) $[s^2\mathcal{L}\{x\} - sx(0) - x'(0)] - \mathcal{L}\{x\} = \mathcal{L}\{y\}$ (1)

$[s^2\mathcal{L}\{y\} - sy(0) - y'(0)] + \mathcal{L}\{y\} = -\mathcal{L}\{x\}$ (2)

from equation (4), page 1051

(ii) $x(0)=2$, $y(0)=-1$, $x'(0)=0$ and $y'(0)=0$

hence $s^2\mathcal{L}\{x\} - 2s - \mathcal{L}\{x\} = \mathcal{L}\{y\}$ (1′)

$s^2\mathcal{L}\{y\} + s + \mathcal{L}\{y\} = -\mathcal{L}\{x\}$ (2′)

(iii) Rearranging gives:

$(s^2-1)\mathcal{L}\{x\} - \mathcal{L}\{y\} = 2s$ (3)

$\mathcal{L}\{x\} + (s^2+1)\mathcal{L}\{y\} = -s$ (4)

Equation $(3) \times (s^2+1)$ and equation $(4) \times 1$ gives:

$(s^2+1)(s^2-1)\mathcal{L}\{x\} - (s^2+1)\mathcal{L}\{y\}$
$$= (s^2+1)2s \quad (5)$$

$\mathcal{L}\{x\} + (s^2+1)\mathcal{L}\{y\} = -s$ (6)

Adding equations (5) and (6) gives:

$[(s^2+1)(s^2-1) + 1]\mathcal{L}\{x\} = (s^2+1)2s - s$

i.e. $s^4\mathcal{L}\{x\} = 2s^3 + s = s(2s^2+1)$

from which, $\mathcal{L}\{x\} = \dfrac{s(2s^2+1)}{s^4} = \dfrac{2s^2+1}{s^3}$

$$= \dfrac{2s^2}{s^3} + \dfrac{1}{s^3} = \dfrac{2}{s} + \dfrac{1}{s^3}$$

(iv) Hence $x = \mathcal{L}^{-1}\left\{\dfrac{2}{s} + \dfrac{1}{s^3}\right\}$

i.e. $x = 2 + \dfrac{1}{2}t^2$

Returning to equations (3) and (4) to determine y:

$1 \times$ equation (3) and $(s^2-1) \times$ equation (4) gives:

$(s^2-1)\mathcal{L}\{x\} - \mathcal{L}\{y\} = 2s$ (7)

$(s^2-1)\mathcal{L}\{x\} + (s^2-1)(s^2+1)\mathcal{L}\{y\}$
$$= -s(s^2-1) \quad (8)$$

Equation (7) − equation (8) gives:

$[-1 - (s^2-1)(s^2+1)]\mathcal{L}\{y\}$
$$= 2s + s(s^2-1)$$

i.e. $-s^4\mathcal{L}\{y\} = s^3 + s$

and $\mathcal{L}\{y\} = \dfrac{s^3+s}{-s^4} = -\dfrac{1}{s} - \dfrac{1}{s^3}$

from which, $y = \mathcal{L}^{-1}\left\{-\dfrac{1}{s} - \dfrac{1}{s^3}\right\}$

i.e. $y = -1 - \dfrac{1}{2}t^2$

Now try the following Practice Exercise

Practice Exercise 366   Further problems on solving simultaneous differential equations using Laplace transforms (answers on page 1196)

Solve the following pairs of simultaneous differential equations:

1. $2\dfrac{dx}{dt} + \dfrac{dy}{dt} = 5e^t$

    $\dfrac{dy}{dt} - 3\dfrac{dx}{dt} = 5$

    given that when $t=0$, $x=0$ and $y=0$

2. $2\dfrac{dy}{dt} - y + x + \dfrac{dx}{dt} - 5\sin t = 0$

    $3\dfrac{dy}{dt} + x - y + 2\dfrac{dx}{dt} - e^t = 0$

    given that at $t=0$, $x=0$ and $y=0$

3. $\dfrac{d^2x}{dt^2} + 2x = y$

    $\dfrac{d^2y}{dt^2} + 2y = x$

    given that at $t=0$, $x=4$, $y=2$, $\dfrac{dx}{dt}=0$

    and $\dfrac{dy}{dt}=0$

For fully worked solutions to each of the problems in Practice Exercise 366 in this chapter,
go to the website:
www.routledge.com/cw/bird

Revision Test 33   Laplace transforms

This assignment covers the material contained in Chapters 95 to 100. *The marks available are shown in brackets at the end of each question.*

1. Find the Laplace transforms of the following functions:

    (a) $2t^3 - 4t + 5$ (b) $3e^{-2t} - 4\sin 2t$

    (c) $3\cosh 2t$ (d) $2t^4 e^{-3t}$

    (e) $5e^{2t}\cos 3t$ (f) $2e^{3t}\sinh 4t$ (16)

2. Find the inverse Laplace transforms of the following functions:

    (a) $\dfrac{5}{2s+1}$ (b) $\dfrac{12}{s^5}$

    (c) $\dfrac{4s}{s^2+9}$ (d) $\dfrac{5}{s^2-9}$

    (e) $\dfrac{3}{(s+2)^4}$ (f) $\dfrac{s-4}{s^2-8s-20}$

    (g) $\dfrac{8}{s^2-4s+3}$ (17)

3. Use partial fractions to determine the following:

    (a) $\mathcal{L}^{-1}\left\{\dfrac{5s-1}{s^2-s-2}\right\}$

    (b) $\mathcal{L}^{-1}\left\{\dfrac{2s^2+11s-9}{s(s-1)(s+3)}\right\}$

    (c) $\mathcal{L}^{-1}\left\{\dfrac{13-s^2}{s(s^2+4s+13)}\right\}$ (24)

4. Determine the poles and zeros for the transfer function: $F(s)=\dfrac{(s+2)(s-3)}{(s+3)(s^2+2s+5)}$ and plot them on a pole–zero diagram.   (10)

5. Sketch the graphs of (a) $f(t)=4H(t-3)$

    (b) $f(t)=3[H(t-2)-H(t-5)]$ (5)

6. Determine (a) $\mathcal{L}\{H(t-2).e^{t-2}\}$

    (b) $\mathcal{L}\{5H(t-1).\sin(t-1)\}$ (6)

7. Determine (a) $\mathcal{L}^{-1}\left\{\dfrac{3e^{-2s}}{s^2}\right\}$ (b) $\mathcal{L}^{-1}\left\{\dfrac{3se^{-5s}}{s^2+4}\right\}$ (9)

8. In a galvanometer the deflection θ satisfies the differential equation:

    $$\frac{d^2\theta}{dt^2} + 2\frac{d\theta}{dt} + \theta = 4$$

    Use Laplace transforms to solve the equation for θ given that when $t=0$, $\theta=0$ and $\dfrac{d\theta}{dt}=0$ (13)

9. Solve the following pair of simultaneous differential equations:

    $$3\frac{dx}{dt} = 3x + 2y$$

    $$2\frac{dy}{dt} + 3x = 6y$$

    given that when $t=0$, $x=1$ and $y=3$ (20)

Formulae/revision hints for Section M
Laplace transforms

Laplace transforms

Function $f(t)$	Laplace transforms $\mathcal{L}\{f(t)\} = \int_0^\infty e^{-st}f(t)\,dt$
1	$\dfrac{1}{s}$
k	$\dfrac{k}{s}$
e^{at}	$\dfrac{1}{s-a}$
$\sin at$	$\dfrac{a}{s^2+a^2}$
$\cos at$	$\dfrac{s}{s^2+a^2}$
t	$\dfrac{1}{s^2}$
t^n (n = positive integer)	$\dfrac{n!}{s^{n+1}}$
$\cosh at$	$\dfrac{s}{s^2-a^2}$
$\sinh at$	$\dfrac{a}{s^2-a^2}$
$e^{-at}t^n$	$\dfrac{n!}{(s+a)^{n+1}}$
$e^{-at}\sin \omega t$	$\dfrac{\omega}{(s+a)^2+\omega^2}$
$e^{-at}\cos \omega t$	$\dfrac{s+a}{(s+a)^2+\omega^2}$
$e^{-at}\cosh \omega t$	$\dfrac{s+a}{(s+a)^2-\omega^2}$
$e^{-at}\sinh \omega t$	$\dfrac{\omega}{(s+a)^2-\omega^2}$

The Laplace transforms of derivatives

First derivative $\mathcal{L}\left\{\dfrac{dy}{dx}\right\} = s\{y\} - y(0)$ where $y(0)$ is the value of y at $x = 0$

Second derivative $\mathcal{L}\left\{\dfrac{dy}{dx}\right\} = s^2\{y\} - sy(0) - y'(0)$

where $y'(0)$ is the value of $\dfrac{dy}{dx}$ at $x = 0$

Heaviside unit step function

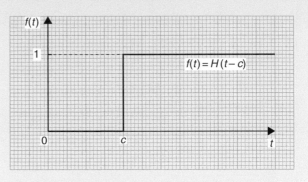

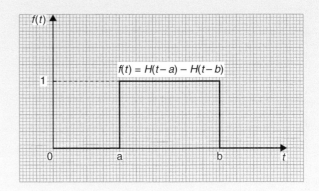

Laplace transform of $H(t-c)$

$$\mathcal{L}\{H(t-c)\} = \frac{e^{-cs}}{s} \quad \text{and} \quad \mathcal{L}\{H(t)\} = \frac{1}{s}$$

Laplace transform of $H(t-c).f(t-c)$

$$\mathcal{L}\{H(t-c).f(t-c)\} = e^{-cs}F(s) \quad \text{where } F(s) = \mathcal{L}\{f(t)\}$$

Inverse Laplace transform of Heaviside functions

if $F(s) = \mathcal{L}\{f(t)\}$, then $e^{-cs}F(s) = \mathcal{L}\{H(t-c).f(t-c)\}$

Section N

Fourier series

Chapter 101

Fourier series for periodic functions of period 2π

Why it is important to understand: Fourier series for periodic functions of period 2π

A Fourier series changes a periodic function into an infinite expansion of a function in terms of sines and cosines. In engineering and physics, expanding functions in terms of sines and cosines is useful because it makes it possible to more easily manipulate functions that are just too difficult to represent analytically. The fields of electronics, quantum mechanics and electrodynamics all make great use of Fourier series. The Fourier series has become one of the most widely used and useful mathematical tools available to any scientist. This chapter introduces and explains Fourier series.

At the end of this chapter, you should be able to:

- describe a Fourier series
- understand periodic functions
- state the formula for a Fourier series and Fourier coefficients
- obtain Fourier series for given functions

101.1 Introduction

Fourier* series provides a method of analysing periodic functions into their constituent components. Alternating currents and voltages, displacement, velocity and acceleration of slider-crank mechanisms and acoustic waves are typical practical examples in engineering and science where periodic functions are involved and often requiring analysis.

101.2 Periodic functions

A function $f(x)$ is said to be **periodic** if $f(x+T) = f(x)$ for all values of x, where T is some positive number. T is the interval between two successive repetitions and is called the **period** of the functions $f(x)$. For example, $y = \sin x$ is periodic in x with period 2π since $\sin x = \sin(x + 2\pi) = \sin(x + 4\pi)$, and so on. In general, if $y = \sin \omega t$ then the period of the waveform is $2\pi/\omega$. The function shown in Figure 101.1 is also periodic of period 2π and is defined by:

* **Who was Fourier? Jean Baptiste Joseph Fourier** (21 March 1768–16 May 1830) was a French mathematician and physicist best known for initiating the investigation of Fourier series and their applications to problems of heat transfer and vibrations. Fourier is also credited with the discovery of the greenhouse effect. To find out more go to **www.routledge.com/cw/bird**

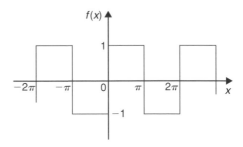

Figure 101.1

$$f(x) = \begin{cases} -1, & \text{when } -\pi < x < 0 \\ 1, & \text{when } \quad 0 < x < \pi \end{cases}$$

If a graph of a function has no sudden jumps or breaks it is called a **continuous function**, examples being the graphs of sine and cosine functions. However, other graphs make finite jumps at a point or points in the interval. The square wave shown in Figure 101.1 has finite discontinuities at $x = \pi$, 2π, 3π, and so on. A great advantage of Fourier series over other series is that it can be applied to functions which are discontinuous as well as those which are continuous.

101.3 Fourier series

(i) The basis of a Fourier series is that all functions of practical significance which are defined in the interval $-\pi \le x \le \pi$ can be expressed in terms of a convergent trigonometric series of the form:

$$f(x) = a_0 + a_1 \cos x + a_2 \cos 2x$$
$$+ a_3 \cos 3x + \cdots + b_1 \sin x$$
$$+ b_2 \sin 2x + b_3 \sin 3x + \cdots$$

when $a_0, a_1, a_2, \ldots b_1, b_2, \ldots$ are real constants, i.e.

$$f(x) = a_0 + \sum_{n=1}^{\infty} (a_n \cos nx + b_n \sin nx) \quad (1)$$

where for the range $-\pi$ to π:

$$a_0 = \frac{1}{2\pi} \int_{-\pi}^{\pi} f(x)\,\mathrm{d}x$$

$$a_n = \frac{1}{\pi} \int_{-\pi}^{\pi} f(x)\cos nx\,\mathrm{d}x$$

$$(n = 1, 2, 3, \ldots)$$

and $b_n = \dfrac{1}{\pi} \displaystyle\int_{-\pi}^{\pi} f(x) \sin nx \, \mathrm{d}x$

$$(n = 1, 2, 3, \ldots)$$

(ii) a_0, a_n and b_n are called the **Fourier coefficients** of the series and if these can be determined, the series of equation (1) is called the **Fourier series** corresponding to $f(x)$.

(iii) An alternative way of writing the series is by using the $a\cos x + b\sin x = c\sin(x + \alpha)$ relationship introduced in Chapter 44, i.e.

$$f(x) = a_0 + c_1 \sin(x + \alpha_1) + c_2 \sin(2x + \alpha_2)$$

$$+ \cdots + c_n \sin(nx + \alpha_n),$$

where a_0 is a constant,

$$c_1 = \sqrt{(a_1^2 + b_1^2)}, \ldots c_n = \sqrt{(a_n^2 + b_n^2)}$$

are the amplitudes of the various components, and phase angle

$$\alpha_n = \tan^{-1} \dfrac{a_n}{b_n}$$

(iv) For the series of equation (1): the term $(a_1 \cos x + b_1 \sin x)$ or $c_1 \sin(x + \alpha_1)$ is called the **first harmonic** or the **fundamental**, the term $(a_2 \cos 2x + b_2 \sin 2x)$ or $c_2 \sin(2x + \alpha_2)$ is called the **second harmonic**, and so on.

For an exact representation of a complex wave, an infinite number of terms are, in general, required. In many practical cases, however, it is sufficient to take the first few terms only (see Problem 2).

The sum of a Fourier series at a point of discontinuity is given by the arithmetic mean of the two limiting values of $f(x)$ as x approaches the point of discontinuity from the two sides. For example, for the waveform shown in

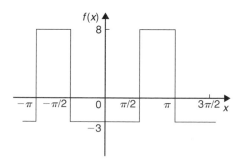

Figure 101.2

Figure 101.2, the sum of the Fourier series at the points of discontinuity (i.e. at $\dfrac{\pi}{2}, \pi, \ldots$ is given by:

$$\frac{8 + (-3)}{2} = \frac{5}{2} \text{ or } 2\frac{1}{2}$$

101.4 Worked problems on Fourier series of periodic functions of period 2π

Problem 1. Obtain a Fourier series for the periodic function $f(x)$ defined as:

$$f(x) = \begin{cases} -k, & \text{when } -\pi < x < 0 \\ +k, & \text{when } 0 < x < \pi \end{cases}$$

The function is periodic outside of this range with period 2π

The square wave function defined is shown in Figure 101.3. Since $f(x)$ is given by two different expressions in the two halves of the range the integration is performed in two parts, one from $-\pi$ to 0 and the other from 0 to π

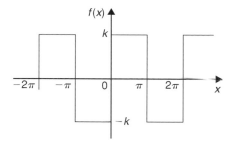

Figure 101.3

From Section 101.3(i):

$$a_0 = \frac{1}{2\pi} \int_{-\pi}^{\pi} f(x) \, \mathrm{d}x$$

$$= \frac{1}{2\pi} \left[\int_{-\pi}^{0} -k \, \mathrm{d}x + \int_{0}^{\pi} k \, \mathrm{d}x \right]$$

$$= \frac{1}{2\pi} \{ [-kx]_{-\pi}^{0} + [kx]_{0}^{\pi} \} = 0$$

(a_0 is in fact the **mean value** of the waveform over a complete period of 2π and this could have been deduced on sight from Figure 101.3)

Section N

From Section 101.3(i):

$$a_n = \frac{1}{\pi} \int_{-\pi}^{\pi} f(x) \cos nx \, dx$$

$$= \frac{1}{\pi} \left\{ \int_{-\pi}^{0} -k \cos nx \, dx + \int_{0}^{\pi} k \cos nx \, dx \right\}$$

$$= \frac{1}{\pi} \left\{ \left[\frac{-k \sin nx}{n} \right]_{-\pi}^{0} + \left[\frac{k \sin nx}{n} \right]_{0}^{\pi} \right\} = 0$$

Hence a_1, a_2, a_3, ... are all zero (since $\sin 0 = \sin(-n\pi) = \sin n\pi = 0$), and therefore no cosine terms will appear in the Fourier series.

From Section 101.3(i):

$$b_n = \frac{1}{\pi} \int_{-\pi}^{\pi} f(x) \sin nx \, dx$$

$$= \frac{1}{\pi} \left\{ \int_{-\pi}^{0} -k \sin nx \, dx + \int_{0}^{\pi} k \sin nx \, dx \right\}$$

$$= \frac{1}{\pi} \left\{ \left[\frac{k \cos nx}{n} \right]_{-\pi}^{0} + \left[\frac{-k \cos nx}{n} \right]_{0}^{\pi} \right\}$$

When n is odd:

$$b_n = \frac{k}{\pi} \left\{ \left[\left(\frac{1}{n} \right) - \left(-\frac{1}{n} \right) \right] \right.$$

$$\left. + \left[-\left(-\frac{1}{n} \right) - \left(-\frac{1}{n} \right) \right] \right\}$$

$$= \frac{k}{\pi} \left\{ \frac{2}{n} + \frac{2}{n} \right\} = \frac{4k}{n\pi}$$

Hence $b_1 = \frac{4k}{\pi}$, $b_3 = \frac{4k}{3\pi}$, $b_5 = \frac{4k}{5\pi}$, and so on.

When n is even:

$$b_n = \frac{k}{\pi} \left\{ \left[\frac{1}{n} - \frac{1}{n} \right] + \left[-\frac{1}{n} - \left(-\frac{1}{n} \right) \right] \right\} = 0$$

Hence, from equation (1), the Fourier series for the function shown in Figure 101.3 is given by:

$$f(x) = a_0 + \sum_{n=1}^{\infty} (a_n \cos nx + b_n \sin nx)$$

$$= 0 + \sum_{n=1}^{\infty} (0 + b_n \sin nx)$$

i.e. $$f(x) = \frac{4k}{\pi} \sin x + \frac{4k}{3\pi} \sin 3x + \frac{4k}{5\pi} \sin 5x + \cdots$$

i.e. $$f(x) = \frac{4k}{\pi} \left(\sin x + \frac{1}{3} \sin 3x + \frac{1}{5} \sin 5x + \cdots \right)$$

Problem 2. For the Fourier series of Problem 1, let $k = \pi$. Show by plotting the first three partial sums of this Fourier series that as the series is added together term by term the result approximates more and more closely to the function it represents.

If $k = \pi$ in the Fourier series of Problem 1 then:

$$f(x) = 4(\sin x + \tfrac{1}{3} \sin 3x + \tfrac{1}{5} \sin 5x + \cdots)$$

$4 \sin x$ is termed the first partial sum of the Fourier series of $f(x)$, $(4 \sin x + \frac{4}{3} \sin 3x)$ is termed the second partial sum of the Fourier series and $(4 \sin x + \frac{4}{3} \sin 3x + \frac{4}{5} \sin 5x)$ is termed the third partial sum, and so on.

Let $P_1 = 4 \sin x$

$$P_2 = \left(4 \sin x + \tfrac{4}{3} \sin 3x \right)$$

and $P_3 = \left(4 \sin x + \tfrac{4}{3} \sin 3x + \tfrac{4}{5} \sin 5x \right)$

Graphs of P_1, P_2 and P_3, obtained by drawing up tables of values and adding waveforms, are shown in Figure 101.4(a) to (c) and they show that the series is convergent, i.e. continually approximating towards a definite limit as more and more partial sums are taken, and in the limit will have the sum $f(x) = \pi$

Even with just three partial sums, the waveform is starting to approach the rectangular wave the Fourier series is representing.

Problem 3. If in the Fourier series of Problem 1, $k = 1$, deduce a series for $\frac{\pi}{4}$ at the point $x = \frac{\pi}{2}$

If $k = 1$ in the Fourier series of Problem 1:

$$f(x) = \frac{4}{\pi} \left(\sin x + \frac{1}{3} \sin 3x + \frac{1}{5} \sin 5x + \cdots \right)$$

When $x = \frac{\pi}{2}$, $f(x) = 1$,

$$\sin x = \sin \frac{\pi}{2} = 1,$$

$$\sin 3x = \sin \frac{3\pi}{2} = -1,$$

$$\sin 5x = \sin \frac{5\pi}{2} = 1, \text{ and so on.}$$

Hence $$1 = \frac{4}{\pi} \left[1 + \frac{1}{3}(-1) + \frac{1}{5}(1) + \frac{1}{7}(-1) + \cdots \right]$$

i.e. $$\frac{\pi}{4} = 1 - \frac{1}{3} + \frac{1}{5} - \frac{1}{7} + \cdots$$

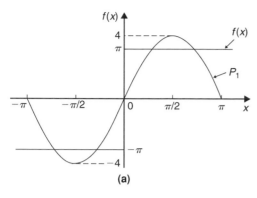

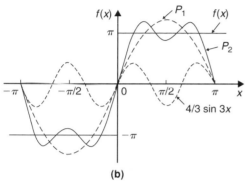

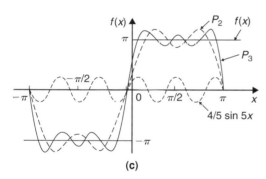

Figure 101.4

Problem 4. Determine the Fourier series for the full wave rectified sine wave $i = 5\sin\dfrac{\theta}{2}$ shown in Figure 101.5.

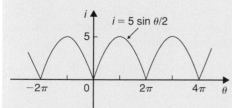

Figure 101.5

$i = 5\sin\dfrac{\theta}{2}$ is a periodic function of period 2π.
Thus

$$i = f(\theta) = a_0 + \sum_{n=1}^{\infty}(a_n\cos n\theta + b_n\sin n\theta)$$

In this case it is better to take the range 0 to 2π instead of $-\pi$ to $+\pi$ since the waveform is continuous between 0 and 2π

$$a_0 = \frac{1}{2\pi}\int_0^{2\pi} f(\theta)\,d\theta = \frac{1}{2\pi}\int_0^{2\pi} 5\sin\frac{\theta}{2}\,d\theta$$

$$= \frac{5}{2\pi}\left[-2\cos\frac{\theta}{2}\right]_0^{2\pi}$$

$$= \frac{5}{\pi}\left[\left(-\cos\frac{2\pi}{2}\right) - (-\cos 0)\right]$$

$$= \frac{5}{\pi}[(1) - (-1)] = \frac{10}{\pi}$$

$$a_n = \frac{1}{\pi}\int_0^{2\pi} 5\sin\frac{\theta}{2}\cos n\theta\,d\theta$$

$$= \frac{5}{\pi}\int_0^{2\pi}\frac{1}{2}\left\{\sin\left(\frac{\theta}{2}+n\theta\right)\right.$$

$$\left. + \sin\left(\frac{\theta}{2}-n\theta\right)\right\}d\theta$$
(see Chapter 65, page 735)

$$= \frac{5}{2\pi}\left[\frac{-\cos\left[\theta\left(\frac{1}{2}+n\right)\right]}{\left(\frac{1}{2}+n\right)}\right.$$

$$\left. - \frac{\cos\left[\theta\left(\frac{1}{2}-n\right)\right]}{\left(\frac{1}{2}-n\right)}\right]_0^{2\pi}$$

$$= \frac{5}{2\pi}\left\{\left[\frac{-\cos\left[2\pi\left(\frac{1}{2}+n\right)\right]}{\left(\frac{1}{2}+n\right)}\right.\right.$$

$$\left. - \frac{\cos\left[2\pi\left(\frac{1}{2}-n\right)\right]}{\left(\frac{1}{2}-n\right)}\right]$$

$$\left. - \left[\frac{-\cos 0}{\left(\frac{1}{2}+n\right)} - \frac{\cos 0}{\left(\frac{1}{2}-n\right)}\right]\right\}$$

When n is both odd and even,

$$a_n = \frac{5}{2\pi}\left\{\left[\frac{1}{\left(\frac{1}{2}+n\right)} + \frac{1}{\left(\frac{1}{2}-n\right)}\right]\right.$$

$$-\left[\frac{-1}{\left(\frac{1}{2}+n\right)}-\frac{1}{\left(\frac{1}{2}-n\right)}\right]\right\}$$

$$=\frac{5}{2\pi}\left\{\frac{2}{\left(\frac{1}{2}+n\right)}+\frac{2}{\left(\frac{1}{2}-n\right)}\right\}$$

$$=\frac{5}{\pi}\left\{\frac{1}{\left(\frac{1}{2}+n\right)}+\frac{1}{\left(\frac{1}{2}-n\right)}\right\}$$

Hence

$$a_1=\frac{5}{\pi}\left[\frac{1}{\frac{3}{2}}+\frac{1}{-\frac{1}{2}}\right]=\frac{5}{\pi}\left[\frac{2}{3}-\frac{2}{1}\right]=\frac{-20}{3\pi}$$

$$a_2=\frac{5}{\pi}\left[\frac{1}{\frac{5}{2}}+\frac{1}{-\frac{3}{2}}\right]=\frac{5}{\pi}\left[\frac{2}{5}-\frac{2}{3}\right]=\frac{-20}{(3)(5)\pi}$$

$$a_3=\frac{5}{\pi}\left[\frac{1}{\frac{7}{2}}+\frac{1}{-\frac{5}{2}}\right]=\frac{5}{\pi}\left[\frac{2}{7}-\frac{2}{5}\right]=\frac{-20}{(5)(7)\pi}$$

and so on.

$$b_n=\frac{1}{\pi}\int_0^{2\pi}5\sin\frac{\theta}{2}\sin n\theta\,d\theta$$

$$=\frac{5}{\pi}\int_0^{2\pi}-\frac{1}{2}\left\{\cos\left[\theta\left(\frac{1}{2}+n\right)\right]\right.$$

$$\left.-\cos\left[\theta\left(\frac{1}{2}-n\right)\right]\right\}d\theta$$

from Chapter 65

$$=\frac{5}{2\pi}\left[\frac{\sin\left[\theta\left(\frac{1}{2}-n\right)\right]}{\left(\frac{1}{2}-n\right)}-\frac{\sin\left[\theta\left(\frac{1}{2}+n\right)\right]}{\left(\frac{1}{2}+n\right)}\right]_0^{2\pi}$$

$$=\frac{5}{2\pi}\left\{\left[\frac{\sin 2\pi\left(\frac{1}{2}-n\right)}{\left(\frac{1}{2}-n\right)}-\frac{\sin 2\pi\left(\frac{1}{2}+n\right)}{\left(\frac{1}{2}+n\right)}\right]\right.$$

$$\left.-\left[\frac{\sin 0}{\left(\frac{1}{2}-n\right)}-\frac{\sin 0}{\left(\frac{1}{2}+n\right)}\right]\right\}$$

When n is both odd and even, $b_n=0$ since $\sin(-\pi)$, $\sin 0$, $\sin\pi$, $\sin 3\pi$, ... are all zero. Hence the Fourier series for the rectified sine wave,

$i=5\sin\dfrac{\theta}{2}$ is given by:

$$f(\theta)=a_0+\sum_{n=1}^{\infty}(a_n\cos n\theta+b_n\sin n\theta)$$

i.e. $i=f(\theta)=\dfrac{10}{\pi}-\dfrac{20}{3\pi}\cos\theta-\dfrac{20}{(3)(5)\pi}\cos 2\theta$

$$-\frac{20}{(5)(7)\pi}\cos 3\theta-\cdots$$

i.e. $\boldsymbol{i=\dfrac{20}{\pi}\left(\dfrac{1}{2}-\dfrac{\cos\theta}{(3)}-\dfrac{\cos 2\theta}{(3)(5)}-\dfrac{\cos 3\theta}{(5)(7)}-\cdots\right)}$

Now try the following Practice Exercise

Practice Exercises 367 Further problems on Fourier series of periodic functions of period 2π (answers on page 1196)

1. Determine the Fourier series for the periodic function:

$$f(x)=\begin{cases}-2,&\text{when }-\pi<x<0\\+2,&\text{when }\quad 0<x<\pi\end{cases}$$

   which is periodic outside this range of period 2π

2. For the Fourier series in Problem 1, deduce a series for $\dfrac{\pi}{4}$ at the point where $x=\dfrac{\pi}{2}$

3. For the waveform shown in Figure 101.6 determine (a) the Fourier series for the function and (b) the sum of the Fourier series at the points of discontinuity.

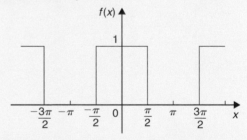

Figure 101.6

4. For Problem 3, draw graphs of the first three partial sums of the Fourier series and show that as the series is added together term by term the result approximates more and more closely to the function it represents.

5. Find the term representing the third harmonic for the periodic function of period 2π given by:

$$f(x)=\begin{cases}0,&\text{when }-\pi<x<0\\1,&\text{when }\quad 0<x<\pi\end{cases}$$

6. Determine the Fourier series for the periodic function of period 2π defined by:

$$f(t) = \begin{cases} 0, & \text{when} \quad -\pi < t < 0 \\ 1, & \text{when} \quad 0 < t < \dfrac{\pi}{2} \\ -1, & \text{when} \quad \dfrac{\pi}{2} < t < \pi \end{cases}$$

The function has a period of 2π

7. Show that the Fourier series for the periodic function of period 2π defined by

$$f(\theta) = \begin{cases} 0, & \text{when} \quad -\pi < \theta < 0 \\ \sin\theta, & \text{when} \quad 0 < \theta < \pi \end{cases}$$

is given by:

$$f(\theta) = \frac{2}{\pi}\left(\frac{1}{2} - \frac{\cos 2\theta}{(3)} - \frac{\cos 4\theta}{(3)(5)}\right.$$

$$\left. - \frac{\cos 6\theta}{(5)(7)} - \cdots\right)$$

For fully worked solutions to each of the problems in Practice Exercise 367 in this chapter, go to the website:
www.routledge.com/cw/bird

Chapter 102

Fourier series for a non-periodic function over range 2π

Why it is important to understand: **Fourier series for a non-periodic function over range 2π**

As was stated in the previous chapter, there are many practical uses of Fourier series in science and engineering. The technique has practical applications in the resolution of sound waves into their different frequencies; for example, in an MP3 player, in telecommunications and Wi-Fi, in computer graphics and image processing, in climate variation, in water waves, and much more. In this chapter, the Fourier series for non-periodic functions is explained.

At the end of this chapter, you should be able to:

- appreciate that Fourier expansions of non-periodic functions have limited range
- determine Fourier series for non-periodic functions over a range of 2π

102.1 Expansion of non-periodic functions

If a function $f(x)$ is not periodic then it cannot be expanded in a Fourier series for **all** values of x. However, it is possible to determine a Fourier series to represent the function over any range of width 2π.

Given a non-periodic function, a new function may be constructed by taking the values of $f(x)$ in the given range and then repeating them outside of the given range at intervals of 2π. Since this new function is, by construction, periodic with period 2π, it may then be expanded in a Fourier series for all values of x. For

example, the function $f(x) = x$ is not a periodic function. However, if a Fourier series for $f(x) = x$ is required then the function is constructed outside of this range so that it is periodic with period 2π as shown by the broken lines in Figure 102.1.

For non-periodic functions, such as $f(x) = x$, the sum of the Fourier series is equal to $f(x)$ at all points in the given range but it is not equal to $f(x)$ at points outside of the range.

For determining a Fourier series of a non-periodic function over a range 2π, exactly the same formulae for the Fourier coefficients are used as in Section 101.3(i).

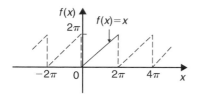

Figure 102.1

102.2 Worked problems on Fourier series of non-periodic functions over a range of 2π

Problem 1. Determine the Fourier series to represent the function $f(x)=2x$ in the range $-\pi$ to $+\pi$

The function $f(x)=2x$ is not periodic. The function is shown in the range $-\pi$ to π in Figure 102.2 and is then constructed outside of that range so that it is periodic of period 2π (see broken lines) with the resulting saw-tooth waveform.

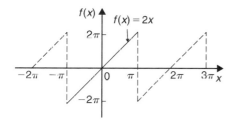

Figure 102.2

For a Fourier series:

$$f(x) = a_0 + \sum_{n=1}^{\infty}(a_n\cos nx + b_n\sin nx)$$

From Section 101.3(i),

$$a_0 = \frac{1}{2\pi}\int_{-\pi}^{\pi}f(x)\,dx$$

$$= \frac{1}{2\pi}\int_{-\pi}^{\pi}2x\,dx = \frac{2}{2\pi}\left[\frac{x^2}{2}\right]_{-\pi}^{\pi} = 0$$

$$a_n = \frac{1}{\pi}\int_{-\pi}^{\pi}f(x)\cos nx\,dx = \frac{1}{\pi}\int_{-\pi}^{\pi}2x\cos nx\,dx$$

$$= \frac{2}{\pi}\left[\frac{x\sin nx}{n} - \int\frac{\sin nx}{n}\,dx\right]_{-\pi}^{\pi}$$

by parts (see Chapter 68)

$$= \frac{2}{\pi}\left[\frac{x\sin nx}{n} + \frac{\cos nx}{n^2}\right]_{-\pi}^{\pi}$$

$$= \frac{2}{\pi}\left[\left(0 + \frac{\cos n\pi}{n^2}\right) - \left(0 + \frac{\cos n(-\pi)}{n^2}\right)\right] = 0$$

$$b_n = \frac{1}{\pi}\int_{-\pi}^{\pi}f(x)\sin nx\,dx = \frac{1}{\pi}\int_{-\pi}^{\pi}2x\sin nx\,dx$$

$$= \frac{2}{\pi}\left[\frac{-x\cos nx}{n} - \int\left(\frac{-\cos nx}{n}\right)dx\right]_{-\pi}^{\pi}$$

by parts

$$= \frac{2}{\pi}\left[\frac{-x\cos nx}{n} + \frac{\sin nx}{n^2}\right]_{-\pi}^{\pi}$$

$$= \frac{2}{\pi}\left[\left(\frac{-\pi\cos n\pi}{n} + \frac{\sin n\pi}{n^2}\right)\right.$$
$$\left. - \left(\frac{-(-\pi)\cos n(-\pi)}{n} + \frac{\sin n(-\pi)}{n^2}\right)\right]$$

$$= \frac{2}{\pi}\left[\frac{-\pi\cos n\pi}{n} - \frac{\pi\cos(-n\pi)}{n}\right] = \frac{-4}{n}\cos n\pi$$

When n is odd, $b_n = \dfrac{4}{n}$. Thus $b_1 = 4$, $b_3 = \dfrac{4}{3}$, $b_5 = \dfrac{4}{5}$, and so on.

When n is even, $b_n = \dfrac{-4}{n}$. Thus $b_2 = -\dfrac{4}{2}$, $b_4 = -\dfrac{4}{4}$, $b_6 = -\dfrac{4}{6}$, and so on.

Thus $f(x) = 2x = 4\sin x - \dfrac{4}{2}\sin 2x + \dfrac{4}{3}\sin 3x$
$$- \frac{4}{4}\sin 4x + \frac{4}{5}\sin 5x - \frac{4}{6}\sin 6x + \cdots$$

i.e. $\mathbf{2x = 4\left(\sin x - \dfrac{1}{2}\sin 2x + \dfrac{1}{3}\sin 3x - \dfrac{1}{4}\sin 4x\right.}$
$$\left.+ \frac{1}{5}\sin 5x - \frac{1}{6}\sin 6x + \cdots\right) \qquad (1)$$

for values of $f(x)$ between $-\pi$ and π. For values of $f(x)$ outside the range $-\pi$ to $+\pi$ the sum of the series is not equal to $f(x)$.

Problem 2. In the Fourier series of Problem 1, by letting $x = \pi/2$, deduce a series for $\pi/4$

When $x = \pi/2$, $f(x) = \pi$ from Figure 102.2.

Section N

Thus, from the Fourier series of equation (1):

$$2\left(\frac{\pi}{2}\right) = 4\left(\sin\frac{\pi}{2} - \frac{1}{2}\sin\frac{2\pi}{2} + \frac{1}{3}\sin\frac{3\pi}{2}\right.$$

$$-\frac{1}{4}\sin\frac{4\pi}{2} + \frac{1}{5}\sin\frac{5\pi}{2}$$

$$\left. -\frac{1}{6}\sin\frac{6\pi}{2} + \cdots\right)$$

$$\pi = 4\left(1 - 0 - \frac{1}{3} - 0 + \frac{1}{5} - 0 - \frac{1}{7} - \cdots\right)$$

i.e. $\quad \dfrac{\pi}{4} = 1 - \dfrac{1}{3} + \dfrac{1}{5} - \dfrac{1}{7} + \cdots$

> **Problem 3.** Obtain a Fourier series for the function defined by:
>
> $$f(x) = \begin{cases} x, & \text{when } 0 < x < \pi \\ 0, & \text{when } \pi < x < 2\pi \end{cases}$$

The defined function is shown in Figure 102.3 between 0 and 2π. The function is constructed outside of this range so that it is periodic of period 2π, as shown by the broken line in Figure 102.3.

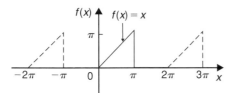

Figure 102.3

For a Fourier series:

$$f(x) = a_0 + \sum_{n=1}^{\infty}(a_n\cos nx + b_n\sin nx)$$

It is more convenient in this case to take the limits from 0 to 2π instead of from $-\pi$ to $+\pi$. The values of the Fourier coefficients are unaltered by this change of limits. Hence

$$a_0 = \frac{1}{2\pi}\int_0^{2\pi} f(x)\,dx = \frac{1}{2\pi}\left[\int_0^{\pi} x\,dx + \int_\pi^{2\pi} 0\,dx\right]$$

$$= \frac{1}{2\pi}\left[\frac{x^2}{2}\right]_0^{\pi} = \frac{1}{2\pi}\left(\frac{\pi^2}{2}\right) = \frac{\pi}{4}$$

$$a_n = \frac{1}{\pi}\int_0^{2\pi} f(x)\cos nx\,dx$$

$$= \frac{1}{\pi}\left[\int_0^{\pi} x\cos nx\,dx + \int_\pi^{2\pi} 0\,dx\right]$$

$$= \frac{1}{\pi}\left[\frac{x\sin nx}{n} + \frac{\cos nx}{n^2}\right]_0^{\pi}$$

(from Problem 1, by parts)

$$= \frac{1}{\pi}\left\{\left[\frac{\pi\sin n\pi}{n} + \frac{\cos n\pi}{n^2}\right] - \left[0 + \frac{\cos 0}{n^2}\right]\right\}$$

$$= \frac{1}{\pi n^2}(\cos n\pi - 1)$$

When n is even, $a_n = 0$

When n is odd, $a_n = \dfrac{-2}{\pi n^2}$

Hence $a_1 = \dfrac{-2}{\pi}, a_3 = \dfrac{-2}{3^2\pi}, a_5 = \dfrac{-2}{5^2\pi}$, and so on.

$$b_n = \frac{1}{\pi}\int_0^{2\pi} f(x)\sin nx\,dx$$

$$= \frac{1}{\pi}\left[\int_0^{\pi} x\sin nx\,dx - \int_\pi^{2\pi} 0\,dx\right]$$

$$= \frac{1}{\pi}\left[\frac{-x\cos nx}{n} + \frac{\sin nx}{n^2}\right]_0^{\pi}$$

(from Problem 1, by parts)

$$= \frac{1}{\pi}\left\{\left[\frac{-\pi\cos n\pi}{n} + \frac{\sin n\pi}{n^2}\right] - \left[0 + \frac{\sin 0}{n^2}\right]\right\}$$

$$= \frac{1}{\pi}\left[\frac{-\pi\cos n\pi}{n}\right] = \frac{-\cos n\pi}{n}$$

Hence $b_1 = -\cos\pi = 1, b_2 = -\dfrac{1}{2}, b_3 = \dfrac{1}{3}$, and so on.

Thus the Fourier series is:

$$f(x) = a_0 + \sum_{n=1}^{\infty}(a_n\cos nx + b_n\sin nx)$$

i.e. $\quad f(x) = \dfrac{\pi}{4} - \dfrac{2}{\pi}\cos x - \dfrac{2}{3^2\pi}\cos 3x$

$$-\frac{2}{5^2\pi}\cos 5x - \cdots + \sin x$$

$$-\frac{1}{2}\sin 2x + \frac{1}{3}\sin 3x - \cdots$$

i.e. $\boldsymbol{f(x)}$

$$= \frac{\pi}{4} - \frac{2}{\pi}\left(\cos x + \frac{\cos 3x}{3^2} + \frac{\cos 5x}{5^2} + \cdots\right)$$

$$+ \left(\sin x - \frac{1}{2}\sin 2x + \frac{1}{3}\sin 3x - \cdots\right)$$

Problem 4. For the Fourier series of Problem 3: (a) What is the sum of the series at the point of discontinuity (i.e. at $x = \pi$)? (b) What is the amplitude and phase angle of the third harmonic? (c) Let $x = 0$, and deduce a series for $\pi^2/8$

(a) The sum of the Fourier series at the point of discontinuity is given by the arithmetic mean of the two limiting values of $f(x)$ as x approaches the point of discontinuity from the two sides.

Hence, sum of the series at $x = \pi$ is

$$\frac{\pi - 0}{2} = \frac{\pi}{2}$$

(b) The third harmonic term of the Fourier series is

$$\left(-\frac{2}{3^2 \pi} \cos 3x + \frac{1}{3} \sin 3x \right)$$

This may also be written in the form $c \sin(3x + \alpha)$,

where amplitude, $c = \sqrt{\left[\left(\frac{-2}{3^2 \pi} \right)^2 + \left(\frac{1}{3} \right)^2 \right]}$

$$= 0.341$$

and phase angle,

$$\alpha = \tan^{-1}\left(\frac{\dfrac{-2}{3^2 \pi}}{\dfrac{1}{3}} \right)$$

$$= -11.98° \quad \text{or} \quad -0.209 \text{ radians}$$

Hence the third harmonic is given by

0.341 sin(3x − 0.209)

(c) When $x = 0$, $f(x) = 0$ (see Figure 102.3)

Hence, from the Fourier series:

$$0 = \frac{\pi}{4} - \frac{2}{\pi}\left(\cos 0 + \frac{1}{3^2}\cos 0 + \frac{1}{5^2}\cos 0 + \cdots \right) + (0)$$

i.e. $\quad -\dfrac{\pi}{4} = -\dfrac{2}{\pi}\left(1 + \dfrac{1}{3^2} + \dfrac{1}{5^2} + \dfrac{1}{7^2} + \cdots \right)$

Hence $\quad \dfrac{\pi^2}{8} = 1 + \dfrac{1}{3^2} + \dfrac{1}{5^2} + \dfrac{1}{7^2} + \cdots$

Problem 5. Deduce the Fourier series for the function $f(\theta) = \theta^2$ in the range 0 to 2π

$f(\theta) = \theta^2$ is shown in Figure 102.4 in the range 0 to 2π. The function is not periodic but is constructed outside of this range so that it is periodic of period 2π, as shown by the broken lines.

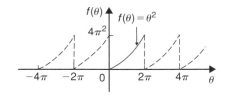

Figure 102.4

For a Fourier series:

$$f(x) = a_0 + \sum_{n=1}^{\infty}(a_n \cos nx + b_n \sin nx)$$

$$a_0 = \frac{1}{2\pi}\int_0^{2\pi} f(\theta)\,d\theta = \frac{1}{2\pi}\int_0^{2\pi}\theta^2\,d\theta$$

$$= \frac{1}{2\pi}\left[\frac{\theta^3}{3} \right]_0^{2\pi} = \frac{1}{2\pi}\left[\frac{8\pi^3}{3} - 0 \right] = \frac{4\pi^2}{3}$$

$$a_n = \frac{1}{\pi}\int_0^{2\pi} f(\theta)\cos n\theta\,d\theta$$

$$= \frac{1}{\pi}\int_0^{2\pi}\theta^2 \cos n\theta\,d\theta$$

$$= \frac{1}{\pi}\left[\frac{\theta^2 \sin n\theta}{n} + \frac{2\theta \cos n\theta}{n^2} - \frac{2 \sin n\theta}{n^3} \right]_0^{2\pi}$$

$$\text{by parts}$$

$$= \frac{1}{\pi}\left[\left(0 + \frac{4\pi \cos 2\pi n}{n^2} - 0 \right) - (0) \right]$$

$$= \frac{4}{n^2}\cos 2\pi n = \frac{4}{n^2} \quad \text{when } n = 1, 2, 3, \ldots$$

Hence $a_1 = \dfrac{4}{1^2}$, $a_2 = \dfrac{4}{2^2}$, $a_3 = \dfrac{4}{3^2}$ and so on

$$b_n = \frac{1}{\pi}\int_0^{2\pi} f(\theta)\sin n\theta\,d\theta = \frac{1}{\pi}\int_0^{2\pi}\theta^2 \sin n\theta\,d\theta$$

$$= \frac{1}{\pi}\left[\frac{-\theta^2 \cos n\theta}{n} + \frac{2\theta \sin n\theta}{n^2} + \frac{2 \cos n\theta}{n^3} \right]_0^{2\pi}$$

$$\text{by parts}$$

$$= \frac{1}{\pi}\left[\left(\frac{-4\pi^2\cos 2\pi n}{n}+0+\frac{2\cos 2\pi n}{n^3}\right)\right.$$

$$\left.-\left(0+0+\frac{2\cos 0}{n^3}\right)\right]$$

$$= \frac{1}{\pi}\left[\frac{-4\pi^2}{n}+\frac{2}{n^3}-\frac{2}{n^3}\right]=\frac{-4\pi}{n}$$

Hence $b_1 = \dfrac{-4\pi}{1}$, $b_2 = \dfrac{-4\pi}{2}$, $b_3 = \dfrac{-4\pi}{3}$, and so on.

Thus $f(\theta) = \theta^2$

$$= \frac{4\pi^2}{3}+\sum_{n=1}^{\infty}\left(\frac{4}{n^2}\cos n\theta - \frac{4\pi}{n}\sin n\theta\right)$$

i.e. $\boldsymbol{\theta^2 =}$

$$\frac{4\pi^2}{3}+4\left(\cos\theta + \frac{1}{2^2}\cos 2\theta + \frac{1}{3^2}\cos 3\theta + \cdots\right)$$

$$-4\pi\left(\sin\theta + \frac{1}{2}\sin 2\theta + \frac{1}{3}\sin 3\theta + \cdots\right)$$

for values of θ between 0 and 2π

Problem 6. In the Fourier series of Problem 5, let $\theta = \pi$ and determine a series for $\dfrac{\pi^2}{12}$

When $\theta = \pi$, $f(\theta) = \pi^2$

Hence $\pi^2 = \dfrac{4\pi^2}{3}+4\left(\cos\pi + \dfrac{1}{4}\cos 2\pi\right.$

$$+\frac{1}{9}\cos 3\pi + \frac{1}{16}\cos 4\pi + \cdots\Bigg)$$

$$-4\pi\left(\sin\pi + \frac{1}{2}\sin 2\pi\right.$$

$$\left.+\frac{1}{3}\sin 3\pi + \cdots\right)$$

i.e. $\pi^2 - \dfrac{4\pi^2}{3} = 4\left(-1+\dfrac{1}{4}-\dfrac{1}{9}\right.$

$$\left.+\frac{1}{16}-\cdots\right)-4\pi(0)$$

$$-\frac{\pi^2}{3}=4\left(-1+\frac{1}{4}-\frac{1}{9}+\frac{1}{16}-\cdots\right)$$

$$\frac{\pi^2}{3}=4\left(1-\frac{1}{4}+\frac{1}{9}-\frac{1}{16}+\cdots\right)$$

Hence $\quad \dfrac{\pi^2}{12}=1-\dfrac{1}{4}+\dfrac{1}{9}-\dfrac{1}{16}+\cdots$

or $\quad \dfrac{\pi^2}{12}=1-\dfrac{1}{2^2}+\dfrac{1}{3^2}-\dfrac{1}{4^2}+\cdots$

Now try the following Practice Exercise

Practice Exercise 368 **Further problems on Fourier series of non-periodic functions over a range of 2π (answers on page 1196)**

1. Show that the Fourier series for the function $f(x) = x$ over the range $x = 0$ to $x = 2\pi$ is given by:

$$f(x) = \pi - 2\left(\sin x + \tfrac{1}{2}\sin 2x\right.$$
$$\left.+ \tfrac{1}{3}\sin 3x + \tfrac{1}{4}\sin 4x + \cdots\right)$$

2. Determine the Fourier series for the function defined by:

$$f(t) = \begin{cases} 1-t, & \text{when } -\pi < t < 0 \\ 1+t, & \text{when } 0 < t < \pi \end{cases}$$

Draw a graph of the function within and outside of the given range.

3. Find the Fourier series for the function $f(x) = x + \pi$ within the range $-\pi < x < \pi$

4. Determine the Fourier series up to and including the third harmonic for the function defined by:

$$f(x) = \begin{cases} x, & \text{when } 0 < x < \pi \\ 2\pi - x, & \text{when } \pi < x < 2\pi \end{cases}$$

Sketch a graph of the function within and outside of the given range, assuming the period is 2π

5. Expand the function $f(\theta) = \theta^2$ in a Fourier series in the range $-\pi < \theta < \pi$

   Sketch the function within and outside of the given range.

6. For the Fourier series obtained in Problem 5, let $\theta = \pi$ and deduce the series for $\sum_{n=1}^{\infty} \frac{1}{n^2}$

7. Sketch the waveform defined by:

$$f(x) = \begin{cases} 1 + \dfrac{2x}{\pi}, & \text{when } -\pi < x < 0 \\ 1 - \dfrac{2x}{\pi}, & \text{when } 0 < x < \pi \end{cases}$$

   Determine the Fourier series in this range.

8. For the Fourier series of Problem 8, deduce a series for $\dfrac{\pi^2}{8}$

For fully worked solutions to each of the problems in Practice Exercise 368 in this chapter, go to the website: www.routledge.com/cw/bird

Chapter 103

Even and odd functions and half-range Fourier series

> **Why it is important to understand:** Even and odd functions and half-range Fourier series
>
> It has already been noted in previous chapters that Fourier series is a very useful tool. The Fourier series has many applications; in fact, any field of physical science that uses sinusoidal signals, such as engineering, applied mathematics and chemistry, will make use of Fourier series. Applications are found in electrical engineering, such as in determining the harmonic components in a.c. waveforms, in vibration analysis, acoustics, optics, signal processing, image processing and in quantum mechanics. If it can be found 'on sight' that a function is even or odd, then determining the Fourier series becomes an easier exercise. This is explained in this chapter, together with half-range Fourier series.

At the end of this chapter, you should be able to:

- define even and odd functions
- determine Fourier cosine series and Fourier sine series
- determine Fourier half-range cosine series and Fourier half-range sine series

103.1 Even and odd functions

Even functions

A function $y = f(x)$ is said to be **even** if $f(-x) = f(x)$ for all values of x. Graphs of even functions are always **symmetrical about the y-axis** (i.e. is a mirror image). Two examples of even functions are $y = x^2$ and $y = \cos x$ as shown in Figure 36.25, page 401.

Odd functions

A function $y = f(x)$ is said to be **odd** if $f(-x) = -f(x)$ for all values of x. Graphs of odd functions are always **symmetrical about the origin**. Two examples of odd functions are $y = x^3$ and $y = \sin x$ as shown in Figure 36.26, page 401.

Many functions are neither even nor odd, two such examples being shown in Figure 36.27, page 401. See also Problems 3 and 4, page 402.

103.2 Fourier cosine and Fourier sine series

(a) Fourier cosine series

The Fourier series of an **even periodic** function $f(x)$ having period 2π contains **cosine terms only** (i.e. contains no sine terms) and may contain a constant term.

Hence $f(x) = a_0 + \sum_{n=1}^{\infty} a_n \cos nx$

where　$a_0 = \dfrac{1}{2\pi} \displaystyle\int_{-\pi}^{\pi} f(x)\,dx$

$= \dfrac{1}{\pi} \displaystyle\int_{0}^{\pi} f(x)\,dx$

(due to symmetry)

and　$a_n = \dfrac{1}{\pi} \displaystyle\int_{-\pi}^{\pi} f(x) \cos nx\,dx$

$= \dfrac{2}{\pi} \displaystyle\int_{0}^{\pi} f(x) \cos nx\,dx$

(b) Fourier sine series

The Fourier series of an **odd** periodic function $f(x)$ having period 2π contains sine terms only (i.e. contains no constant term and no cosine terms).

Hence　$f(x) = \sum_{n=1}^{\infty} b_n \sin nx$

where　$b_n = \dfrac{1}{\pi} \displaystyle\int_{-\pi}^{\pi} f(x) \sin nx\,dx$

$= \dfrac{2}{\pi} \displaystyle\int_{0}^{\pi} f(x) \sin nx\,dx$

Problem 1.　Determine the Fourier series for the periodic function defined by:

$$f(x) = \begin{cases} -2, & \text{when } -\pi < x < -\dfrac{\pi}{2} \\[2mm] 2, & \text{when } -\dfrac{\pi}{2} < x < \dfrac{\pi}{2} \\[2mm] -2, & \text{when } \dfrac{\pi}{2} < x < \pi \end{cases}$$

and has a period of 2π

The square wave shown in Figure 103.1 is an even function since it is symmetrical about the $f(x)$ axis.

Hence from para. (a) the Fourier series is given by:

$$f(x) = a_0 + \sum_{n=1}^{\infty} a_n \cos nx$$

(i.e. the series contains no sine terms).

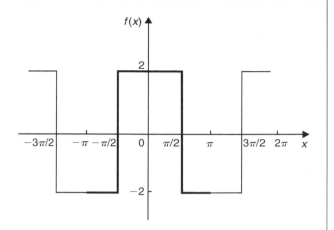

Figure 103.1

From para. (a),

$a_0 = \dfrac{1}{\pi} \displaystyle\int_{0}^{\pi} f(x)\,dx$

$= \dfrac{1}{\pi} \left\{ \displaystyle\int_{0}^{\pi/2} 2\,dx + \displaystyle\int_{\pi/2}^{\pi} -2\,dx \right\}$

$= \dfrac{1}{\pi} \left\{ [2x]_{0}^{\pi/2} + [-2x]_{\pi/2}^{\pi} \right\}$

$= \dfrac{1}{\pi} \{(\pi) + [(-2\pi) - (-\pi)]\} = 0$

$a_n = \dfrac{2}{\pi} \displaystyle\int_{0}^{\pi} f(x) \cos nx\,dx$

$= \dfrac{2}{\pi} \left\{ \displaystyle\int_{0}^{\pi/2} 2\cos nx\,dx + \displaystyle\int_{\pi/2}^{\pi} -2\cos nx\,dx \right\}$

$= \dfrac{4}{\pi} \left\{ \left[\dfrac{\sin nx}{n} \right]_{0}^{\pi/2} + \left[\dfrac{-\sin nx}{n} \right]_{\pi/2}^{\pi} \right\}$

$= \dfrac{4}{\pi} \left\{ \left(\dfrac{\sin(\pi/2)n}{n} - 0 \right) \right.$

$\left. \qquad\qquad + \left(0 - \dfrac{-\sin(\pi/2)n}{n} \right) \right\}$

$= \dfrac{4}{\pi} \left(\dfrac{2\sin(\pi/2)n}{n} \right) = \dfrac{8}{\pi n} \left(\sin \dfrac{n\pi}{2} \right)$

When n is even, $a_n = 0$

When n is odd, $a_n = \dfrac{8}{\pi n}$ for $n = 1, 5, 9, \ldots$

and　$a_n = \dfrac{-8}{\pi n}$ for $n = 3, 7, 11, \ldots$

Hence $a_1 = \dfrac{8}{\pi}$, $a_3 = \dfrac{-8}{3\pi}$, $a_5 = \dfrac{8}{5\pi}$, and so on.

Hence the Fourier series for the waveform of Figure 103.1 is given by:

$$f(x) = \frac{8}{\pi}\left(\cos x - \frac{1}{3}\cos 3x + \frac{1}{5}\cos 5x \right.$$
$$\left. - \frac{1}{7}\cos 7x + \cdots\right)$$

Problem 2. In the Fourier series of Problem 1, let $x=0$ and deduce a series for $\pi/4$

When $x=0$, $f(x)=2$ (from Figure 103.1)

Thus, from the Fourier series,

$$2 = \frac{8}{\pi}\left(\cos 0 - \frac{1}{3}\cos 0 + \frac{1}{5}\cos 0 \right.$$
$$\left. - \frac{1}{7}\cos 0 + \cdots\right)$$

Hence $\dfrac{2\pi}{8} = 1 - \dfrac{1}{3} + \dfrac{1}{5} - \dfrac{1}{7} + \cdots$

i.e. $\dfrac{\pi}{4} = 1 - \dfrac{1}{3} + \dfrac{1}{5} - \dfrac{1}{7} + \cdots$

Problem 3. Obtain the Fourier series for the square wave shown in Figure 103.2

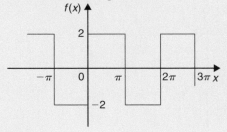

$f(x)$

Figure 103.2

The square wave shown in Figure 103.2 is an odd function since it is symmetrical about the origin.

Hence, from para. (b), the Fourier series is given by:

$$f(x) = \sum_{n=1}^{\infty} b_n \sin nx$$

The function is defined by:

$$f(x) = \begin{cases} -2, & \text{when } -\pi < x < 0 \\ 2, & \text{when } 0 < x < \pi \end{cases}$$

From para. (b), $b_n = \dfrac{2}{\pi}\displaystyle\int_0^{\pi} f(x)\sin nx\, dx$

$$= \frac{2}{\pi}\int_0^{\pi} 2\sin nx\, dx$$

$$= \frac{4}{\pi}\left[\frac{-\cos nx}{n}\right]_0^{\pi}$$

$$= \frac{4}{\pi}\left[\left(\frac{-\cos n\pi}{n}\right) - \left(-\frac{1}{n}\right)\right]$$

$$= \frac{4}{\pi n}(1 - \cos n\pi)$$

When n is even, $b_n = 0$

When n is odd, $b_n = \dfrac{4}{\pi n}(1 - (-1)) = \dfrac{8}{\pi n}$

Hence $\qquad b_1 = \dfrac{8}{\pi}$, $b_3 = \dfrac{8}{3\pi}$, $b_5 = \dfrac{8}{5\pi}$,

and so on

Hence the Fourier series is:

$$f(x) = \frac{8}{\pi}\left(\sin x + \frac{1}{3}\sin 3x + \frac{1}{5}\sin 5x \right.$$
$$\left. + \frac{1}{7}\sin 7x + \cdots\right)$$

Problem 4. Determine the Fourier series for the function $f(\theta) = \theta^2$ in the range $-\pi < \theta < \pi$. The function has a period of 2π

A graph of $f(\theta) = \theta^2$ is shown in Figure 103.3 in the range $-\pi$ to π with period 2π. The function is symmetrical about the $f(\theta)$ axis and is thus an even function. Thus a Fourier cosine series will result of the form:

$$f(\theta) = a_0 + \sum_{n=1}^{\infty} a_n \cos n\theta$$

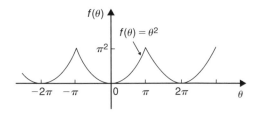

Figure 103.3

From para. (a),

$$a_0 = \frac{1}{\pi} \int_0^{\pi} f(\theta)\,d\theta = \frac{1}{\pi} \int_0^{\pi} \theta^2 \, d\theta$$

$$= \frac{1}{\pi}\left[\frac{\theta^3}{3}\right]_0^{\pi} = \frac{\pi^2}{3}$$

and $a_n = \dfrac{2}{\pi} \displaystyle\int_0^{\pi} f(\theta)\cos n\theta \, d\theta$

$$= \frac{2}{\pi} \int_0^{\pi} \theta^2 \cos n\theta \, d\theta$$

$$= \frac{2}{\pi}\left[\frac{\theta^2 \sin n\theta}{n} + \frac{2\theta \cos n\theta}{n^2} - \frac{2\sin n\theta}{n^3}\right]_0^{\pi}$$

by parts

$$= \frac{2}{\pi}\left[\left(0 + \frac{2\pi \cos n\pi}{n^2} - 0\right) - (0)\right]$$

$$= \frac{4}{n^2}\cos n\pi$$

When n is odd, $a_n = \dfrac{-4}{n^2}$. Hence $a_1 = \dfrac{-4}{1^2}$, $a_3 = \dfrac{-4}{3^2}, a_5 = \dfrac{-4}{5^2}$, and so on.

When n is even, $a_n = \dfrac{4}{n^2}$. Hence $a_2 = \dfrac{4}{2^2}, a_4 = \dfrac{4}{4^2}$, and so on.

Hence the Fourier series is:

$$f(\theta) = \theta^2 = \frac{\pi^2}{3} - 4\left(\cos\theta - \frac{1}{2^2}\cos 2\theta + \frac{1}{3^2}\cos 3\theta\right.$$

$$\left. - \frac{1}{4^2}\cos 4\theta + \frac{1}{5^2}\cos 5\theta - \cdots\right)$$

Problem 5. For the Fourier series of Problem 4, let $\theta = \pi$ and show that $\displaystyle\sum_{n=1}^{\infty} \frac{1}{n^2} = \frac{\pi^2}{6}$

When $\theta = \pi$, $f(\theta) = \pi^2$ (see Figure 103.3). Hence from the Fourier series:

$$\pi^2 = \frac{\pi^2}{3} - 4\left(\cos\pi - \frac{1}{2^2}\cos 2\pi + \frac{1}{3^2}\cos 3\pi\right.$$

$$\left. - \frac{1}{4^2}\cos 4\pi + \frac{1}{5^2}\cos 5\pi - \cdots\right)$$

i.e.

$$\pi^2 - \frac{\pi^2}{3} = -4\left(-1 - \frac{1}{2^2} - \frac{1}{3^2} - \frac{1}{4^2} - \frac{1}{5^2} - \cdots\right)$$

$$\frac{2\pi^2}{3} = 4\left(1 + \frac{1}{2^2} + \frac{1}{3^2} + \frac{1}{4^2} + \frac{1}{5^2} + \cdots\right)$$

i.e. $\dfrac{2\pi^2}{(3)(4)} = 1 + \dfrac{1}{2^2} + \dfrac{1}{3^2} + \dfrac{1}{4^2} + \dfrac{1}{5^2} + \cdots$

i.e. $\dfrac{\pi^2}{6} = \dfrac{1}{1^2} + \dfrac{1}{2^2} + \dfrac{1}{3^2} + \dfrac{1}{4^2} + \dfrac{1}{5^2} + \cdots$

Hence

$$\sum_{n=1}^{\infty} \frac{1}{n^2} = \frac{\pi^2}{6}$$

Now try the following Practice Exercise

Practice Exercise 369   Further problems on Fourier cosine and Fourier sine series (answers on page 1197)

1. Determine the Fourier series for the function defined by:

$$f(x) = \begin{cases} -1, & -\pi < x < -\dfrac{\pi}{2} \\ 1, & -\dfrac{\pi}{2} < x < \dfrac{\pi}{2} \\ -1, & \dfrac{\pi}{2} < x < \pi \end{cases}$$

which is periodic outside of this range of period 2π

2. Obtain the Fourier series of the function defined by:

$$f(t) = \begin{cases} t + \pi, & -\pi < t < 0 \\ t - \pi, & 0 < t < \pi \end{cases}$$

which is periodic of period 2π. Sketch the given function.

3. Determine the Fourier series defined by

$$f(x) = \begin{cases} 1 - x, & -\pi < x < 0 \\ 1 + x, & 0 < x < \pi \end{cases}$$

which is periodic of period 2π

4. In the Fourier series of Problem 3, let $x = 0$ and deduce a series for $\pi^2/8$

Section N

5. Show that the Fourier series for the triangular waveform shown in Figure 103.4 is given by:

$$y = \frac{8}{\pi^2}\left(\sin\theta - \frac{1}{3^2}\sin 3\theta + \frac{1}{5^2}\sin 5\theta\right.$$

$$\left. - \frac{1}{7^2}\sin 7\theta + \cdots\right)$$

The function is periodic of period 2π

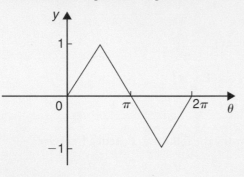

Figure 103.4

103.3 Half-range Fourier series

(a) When a function is defined over the range, say, 0 to π instead of from 0 to 2π it may be expanded in a series of sine terms only or of cosine terms only. The series produced is called a **half-range Fourier series**.

(b) If a **half-range cosine series** is required for the function $f(x) = x$ in the range 0 to π, then an **even** periodic function is required. In Figure 103.5, $f(x) = x$ is shown plotted from $x = 0$ to $x = \pi$. Since an even function is symmetrical about the $f(x)$ axis the line AB is constructed as shown. If the triangular waveform produced is assumed to be periodic of period 2π outside of this range then the waveform is as shown in Figure 103.5. When a half-range cosine series is required then the Fourier coefficients a_0 and a_n are calculated as in Section 103.2(a), i.e.

$$f(x) = a_0 + \sum_{n=1}^{\infty} a_n \cos nx$$

where $a_0 = \dfrac{1}{\pi}\displaystyle\int_0^\pi f(x)\,dx$

and $a_n = \dfrac{2}{\pi}\displaystyle\int_0^\pi f(x)\cos nx\,dx$

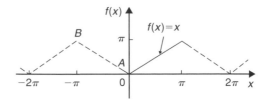

Figure 103.5

(c) If a **half-range sine series** is required for the function $f(x) = x$ in the range 0 to π, then an odd periodic function is required. In Figure 103.6, $f(x) = x$ is shown plotted from $x = 0$ to $x = \pi$. Since an odd function is symmetrical about the origin the line CD is constructed as shown. If the saw-tooth waveform produced is assumed to be periodic of period 2π outside of this range, then the waveform is as shown in Figure 103.6. When a half-range sine series is required then the Fourier coefficient b_n is calculated as in Section 103.2(b), i.e.

$$f(x) = \sum_{n=1}^{\infty} b_n \sin nx$$

where $b_n = \dfrac{2}{\pi}\displaystyle\int_0^\pi f(x)\sin nx\,dx$

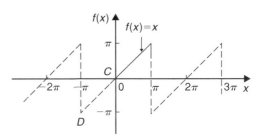

Figure 103.6

Problem 6. Determine the half-range Fourier cosine series to represent the function $f(x) = 3x$ in the range $0 \le x \le \pi$

From para. (b), for a half-range cosine series:

$$f(x) = a_0 + \sum_{n=1}^{\infty} a_n \cos nx$$

When $f(x) = 3x$,

$$a_0 = \frac{1}{\pi}\int_0^\pi f(x)\,dx = \frac{1}{\pi}\int_0^\pi 3x\,dx$$

$$= \frac{3}{\pi}\left[\frac{x^2}{2}\right]_0^\pi = \frac{3\pi}{2}$$

$$a_n = \frac{2}{\pi} \int_0^\pi f(x)\cos nx \, dx$$

$$= \frac{2}{\pi} \int_0^\pi 3x \cos nx \, dx$$

$$= \frac{6}{\pi} \left[\frac{x \sin nx}{n} + \frac{\cos nx}{n^2} \right]_0^\pi \quad \text{by parts}$$

$$= \frac{6}{\pi} \left[\left(\frac{\pi \sin n\pi}{n} + \frac{\cos n\pi}{n^2} \right) - \left(0 + \frac{\cos 0}{n^2} \right) \right]$$

$$= \frac{6}{\pi} \left(0 + \frac{\cos n\pi}{n^2} - \frac{\cos 0}{n^2} \right)$$

$$= \frac{6}{\pi n^2} (\cos n\pi - 1)$$

When n is even, $a_n = 0$

When n is odd, $a_n = \frac{6}{\pi n^2}(-1 - 1) = \frac{-12}{\pi n^2}$

Hence $a_1 = \frac{-12}{\pi}$, $a_3 = \frac{-12}{\pi 3^2}$, $a_5 = \frac{-12}{\pi 5^2}$, and so on.

Hence the half-range Fourier cosine series is given by:

$$f(x) = 3x = \frac{3\pi}{2} - \frac{12}{\pi} \left(\cos x + \frac{1}{3^2} \cos 3x \right.$$

$$\left. + \frac{1}{5^2} \cos 5x + \cdots \right)$$

Problem 7. Find the half-range Fourier sine series to represent the function $f(x) = 3x$ in the range $0 \le x \le \pi$

From para. (c), for a half-range sine series:

$$f(x) = \sum_{n=1}^\infty b_n \sin nx$$

When $f(x) = 3x$,

$$b_n = \frac{2}{\pi} \int_0^\pi f(x) \sin nx \, dx = \frac{2}{\pi} \int_0^\pi 3x \sin nx \, dx$$

$$= \frac{6}{\pi} \left[\frac{-x \cos nx}{n} + \frac{\sin nx}{n^2} \right]_0^\pi \quad \text{by parts}$$

$$= \frac{6}{\pi} \left[\left(\frac{-\pi \cos n\pi}{n} + \frac{\sin n\pi}{n^2} \right) - (0 + 0) \right]$$

$$= -\frac{6}{n} \cos n\pi$$

When n is odd, $b_n = \frac{6}{n}$

Hence $b_1 = \frac{6}{1}$, $b_3 = \frac{6}{3}$, $b_5 = \frac{6}{5}$ and so on.

When n is even, $b_n = -\frac{6}{n}$

Hence $b_2 = -\frac{6}{2}$, $b_4 = -\frac{6}{4}$, $b_6 = -\frac{6}{6}$ and so on.

Hence the half-range Fourier sine series is given by:

$$f(x) = 3x = 6 \left(\sin x - \frac{1}{2} \sin 2x + \frac{1}{3} \sin 3x \right.$$

$$\left. - \frac{1}{4} \sin 4x + \frac{1}{5} \sin 5x - \cdots \right)$$

Problem 8. Expand $f(x) = \cos x$ as a half-range Fourier sine series in the range $0 \le x \le \pi$, and sketch the function within and outside of the given range

When a half-range sine series is required then an odd function is implied, i.e. a function symmetrical about the origin. A graph of $y = \cos x$ is shown in Figure 103.7 in the range 0 to π. For $\cos x$ to be symmetrical about the origin the function is as shown by the broken lines in Figure 103.7 outside of the given range. From para. (c), for a half-range Fourier sine series:

$$f(x) = \sum_{n=1}^\infty b_n \sin nx \, dx$$

$$b_n = \frac{2}{\pi} \int_0^\pi f(x) \sin nx \, dx$$

$$= \frac{2}{\pi} \int_0^\pi \cos x \sin nx \, dx$$

$$= \frac{2}{\pi} \int_0^\pi \frac{1}{2} [\sin(x + nx) - \sin(x - nx)] \, dx$$

$$= \frac{1}{\pi} \left[\frac{-\cos[x(1 + n)]}{(1 + n)} + \frac{\cos[x(1 - n)]}{(1 - n)} \right]_0^\pi$$

$$= \frac{1}{\pi} \left[\left(\frac{-\cos[\pi(1 + n)]}{(1 + n)} + \frac{\cos[\pi(1 - n)]}{(1 - n)} \right) \right.$$

$$\left. - \left(\frac{-\cos 0}{(1 + n)} + \frac{\cos 0}{(1 - n)} \right) \right]$$

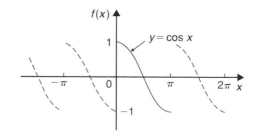

Figure 103.7

When n is odd,

$$b_n = \frac{1}{\pi}\left[\left(\frac{-1}{(1+n)} + \frac{1}{(1-n)}\right)\right.$$
$$\left.- \left(\frac{-1}{(1+n)} + \frac{1}{(1-n)}\right)\right] = 0$$

When n is even,

$$b_n = \frac{1}{\pi}\left[\left(\frac{1}{(1+n)} - \frac{1}{(1-n)}\right)\right.$$
$$\left.- \left(\frac{-1}{(1+n)} + \frac{1}{(1-n)}\right)\right]$$

$$= \frac{1}{\pi}\left(\frac{2}{(1+n)} - \frac{2}{(1-n)}\right)$$

$$= \frac{1}{\pi}\left(\frac{2(1-n) - 2(1+n)}{1-n^2}\right)$$

$$= \frac{1}{\pi}\left(\frac{-4n}{1-n^2}\right) = \frac{4n}{\pi(n^2-1)}$$

Hence $b_2 = \dfrac{8}{3\pi}$, $b_4 = \dfrac{16}{15\pi}$, $b_6 = \dfrac{24}{35\pi}$ and so on.

Hence the half-range Fourier sine series for $f(x)$ in the range 0 to π is given by:

$$f(x) = \frac{8}{3\pi}\sin 2x + \frac{16}{15\pi}\sin 4x$$
$$+ \frac{24}{35\pi}\sin 6x + \cdots$$

or $f(x) = \dfrac{8}{\pi}\left(\dfrac{1}{3}\sin 2x + \dfrac{2}{(3)(5)}\sin 4x\right.$

$$\left.+ \frac{3}{(5)(7)}\sin 6x + \cdots\right)$$

Now try the following Practice Exercise

Practice Exercise 370   Further problems on half-range Fourier series (answers on page 1197)

1. Determine the half-range sine series for the function defined by:

$$f(x) = \begin{cases} x, & 0 < x < \dfrac{\pi}{2} \\[2mm] 0, & \dfrac{\pi}{2} < x < \pi \end{cases}$$

2. Obtain (a) the half-range cosine series and (b) the half-range sine series for the function

$$f(t) = \begin{cases} 0, & 0 < t < \dfrac{\pi}{2} \\[2mm] 1, & \dfrac{\pi}{2} < t < \pi \end{cases}$$

3. Find the half-range Fourier sine series for the function $f(x) = \sin^2 x$ in the range $0 \le x \le \pi$. Sketch the function within and outside of the given range.

4. Determine the half-range Fourier cosine series in the range $x = 0$ to $x = \pi$ for the function defined by:

$$f(x) = \begin{cases} x, & 0 < x < \dfrac{\pi}{2} \\[2mm] (\pi - x), & \dfrac{\pi}{2} < x < \pi \end{cases}$$

For fully worked solutions to each of the problems in Practice Exercises 369 and 370 in this chapter, go to the website:
www.routledge.com/cw/bird

Chapter 104

Fourier series over any range

Why it is important to understand: Fourier series over any range

As has been mentioned in preceding chapters, the Fourier series has many applications; in fact, any field of physical science that uses sinusoidal signals, such as engineering, applied mathematics and chemistry, will make use of Fourier series. In communications, Fourier series is essential to understanding how a signal behaves when it passes through filters, amplifiers and communications channels. In astronomy, radar and digital signal processing Fourier analysis is used to map the planet. In geology, seismic research uses Fourier analysis, and in optics, Fourier analysis is used in light diffraction. This chapter explains how to determine the Fourier series of a periodic function over any range.

At the end of this chapter, you should be able to:

- understand the Fourier series of a periodic function of period L
- determine the Fourier series of a periodic function of period L
- determine the half-range Fourier series for functions of period L

104.1 Expansion of a periodic function of period L

(a) A periodic function $f(x)$ of period L repeats itself when x increases by L, i.e. $f(x+L)=f(x)$. The change from functions dealt with previously having period 2π to functions having period L is not difficult since it may be achieved by a change of variable.

(b) To find a Fourier series for a function $f(x)$ in the range $-\dfrac{L}{2} \leq x \leq \dfrac{L}{2}$ a new variable u is introduced such that $f(x)$, as a function of u, has period 2π. If $u=\dfrac{2\pi x}{L}$ then, when $x=-\dfrac{L}{2}$, $u=-\pi$ and when $x=\dfrac{L}{2}, u=+\pi$. Also, let

$f(x)=f\left(\dfrac{Lu}{2\pi}\right)=F(u)$. The Fourier series for $F(u)$ is given by:

$$F(u) = a_0 + \sum_{n=1}^{\infty}(a_n \cos nu + b_n \sin nu),$$

where $a_0 = \dfrac{1}{2\pi}\displaystyle\int_{-\pi}^{\pi} F(u)\,\mathrm{d}u,$

$$a_n = \dfrac{1}{\pi}\int_{-\pi}^{\pi} F(u)\cos nu\,\mathrm{d}u$$

and $b_n = \dfrac{1}{\pi}\displaystyle\int_{-\pi}^{\pi} F(u)\sin nu\,\mathrm{d}u$

(c) It is, however, more usual to change the formula of para. (b) to terms of x. Since $u=\dfrac{2\pi x}{L}$, then

$$du = \frac{2\pi}{L} dx$$

and the limits of integration are $-\frac{L}{2}$ to $+\frac{L}{2}$ instead of from $-\pi$ to $+\pi$. Hence the Fourier series expressed in terms of x is given by:

$$f(x) = a_0 + \sum_{n=1}^{\infty} \left[a_n \cos\left(\frac{2\pi n x}{L}\right) + b_n \sin\left(\frac{2\pi n x}{L}\right) \right]$$

where, in the range $-\frac{L}{2}$ to $+\frac{L}{2}$:

$$a_0 = \frac{1}{L} \int_{-\frac{L}{2}}^{\frac{L}{2}} f(x) \, dx$$

$$a_n = \frac{2}{L} \int_{-\frac{L}{2}}^{\frac{L}{2}} f(x) \cos\left(\frac{2\pi n x}{L}\right) dx$$

and

$$b_n = \frac{2}{L} \int_{-\frac{L}{2}}^{\frac{L}{2}} f(x) \sin\left(\frac{2\pi n x}{L}\right) dx$$

The limits of integration may be replaced by any interval of length L, such as from 0 to L.

Problem 1. The voltage from a square wave generator is of the form:

$$v(t) = \begin{cases} 0, & -4 < t < 0 \\ 10, & 0 < t < 4 \end{cases}$$
$$\text{and has a period of 8 ms}$$

Find the Fourier series for this periodic function

The square wave is shown in Figure 104.1. From para. (c), the Fourier series is of the form:

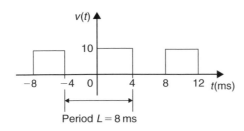

Figure 104.1

$$v(t) = a_0 + \sum_{n=1}^{\infty} \left[a_n \cos\left(\frac{2\pi n t}{L}\right) + b_n \sin\left(\frac{2\pi n t}{L}\right) \right]$$

$$a_0 = \frac{1}{L} \int_{-\frac{L}{2}}^{\frac{L}{2}} v(t) \, dt = \frac{1}{8} \int_{-4}^{4} v(t) \, dt$$

$$= \frac{1}{8} \left\{ \int_{-4}^{0} 0 \, dt + \int_{0}^{4} 10 \, dt \right\} = \frac{1}{8} [10t]_0^4 = 5$$

$$a_n = \frac{2}{L} \int_{-\frac{L}{2}}^{\frac{L}{2}} v(t) \cos\left(\frac{2\pi n t}{L}\right) dt$$

$$= \frac{2}{8} \int_{-4}^{4} v(t) \cos\left(\frac{2\pi n t}{8}\right) dt$$

$$= \frac{1}{4} \left\{ \int_{-4}^{0} 0 \cos\left(\frac{\pi n t}{4}\right) dt \right.$$

$$\left. + \int_{0}^{4} 10 \cos\left(\frac{\pi n t}{4}\right) dt \right\}$$

$$= \frac{1}{4} \left[\frac{10 \sin\left(\frac{\pi n t}{4}\right)}{\left(\frac{\pi n}{4}\right)} \right]_0^4 = \frac{10}{\pi n} [\sin \pi n - \sin 0]$$

$$= 0 \text{ for } n = 1, 2, 3, \dots$$

$$b_n = \frac{2}{L} \int_{-\frac{L}{2}}^{\frac{L}{2}} v(t) \sin\left(\frac{2\pi n t}{L}\right) dt$$

$$= \frac{2}{8} \int_{-4}^{4} v(t) \sin\left(\frac{2\pi n t}{8}\right) dt$$

$$= \frac{1}{4} \left\{ \int_{-4}^{0} 0 \sin\left(\frac{\pi n t}{4}\right) dt \right.$$

$$\left. + \int_{0}^{4} 10 \sin\left(\frac{\pi n t}{4}\right) dt \right\}$$

$$= \frac{1}{4} \left[\frac{-10 \cos\left(\frac{\pi n t}{4}\right)}{\left(\frac{\pi n}{4}\right)} \right]_0^4$$

$$= \frac{-10}{\pi n} [\cos \pi n - \cos 0]$$

When n is even, $b_n = 0$

When n is odd, $b_1 = \frac{-10}{\pi}(-1-1) = \frac{20}{\pi}$,

$$b_3 = \frac{-10}{3\pi}(-1-1) = \frac{20}{3\pi},$$

$$b_5 = \frac{20}{5\pi}, \text{ and so on.}$$

Thus the Fourier series for the function $v(t)$ is given by:

$$v(t) = 5 + \frac{20}{\pi}\left[\sin\left(\frac{\pi t}{4}\right) + \frac{1}{3}\sin\left(\frac{3\pi t}{4}\right)\right.$$
$$\left. + \frac{1}{5}\sin\left(\frac{5\pi t}{4}\right) + \cdots\right]$$

Problem 2. Obtain the Fourier series for the function defined by:

$$f(x) = \begin{cases} 0, & \text{when } -2 < x < -1 \\ 5, & \text{when } -1 < x < 1 \\ 0, & \text{when } 1 < x < 2 \end{cases}$$

The function is periodic outside of this range of period 4

The function $f(x)$ is shown in Figure 104.2 where period, $L = 4$. Since the function is symmetrical about the $f(x)$ axis it is an even function and the Fourier series contains no sine terms (i.e. $b_n = 0$)

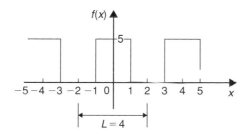

Figure 104.2

Thus, from para. (c),

$$f(x) = a_0 + \sum_{n=1}^{\infty} a_n \cos\left(\frac{2\pi nx}{L}\right)$$

$$a_0 = \frac{1}{L}\int_{-\frac{L}{2}}^{\frac{L}{2}} f(x)\,dx = \frac{1}{4}\int_{-2}^{2} f(x)\,dx$$

$$= \frac{1}{4}\left\{\int_{-2}^{-1} 0\,dx + \int_{-1}^{1} 5\,dx + \int_{1}^{2} 0\,dx\right\}$$

$$= \frac{1}{4}[5x]_{-1}^{1} = \frac{1}{4}[(5) - (-5)] = \frac{10}{4} = \frac{5}{2}$$

$$a_n = \frac{2}{L}\int_{-\frac{L}{2}}^{\frac{L}{2}} f(x)\cos\left(\frac{2\pi nx}{L}\right)dx$$

$$= \frac{2}{4}\int_{-2}^{2} f(x)\cos\left(\frac{2\pi nx}{4}\right)dx$$

$$= \frac{1}{2}\left\{\int_{-2}^{-1} 0\cos\left(\frac{\pi nx}{2}\right)dx\right.$$
$$+ \int_{-1}^{1} 5\cos\left(\frac{\pi nx}{2}\right)dx$$
$$\left. + \int_{1}^{2} 0\cos\left(\frac{\pi nx}{2}\right)dx\right\}$$

$$= \frac{5}{2}\left[\frac{\sin\frac{\pi nx}{2}}{\frac{\pi n}{2}}\right]_{-1}^{1}$$

$$= \frac{5}{\pi n}\left[\sin\left(\frac{\pi n}{2}\right) - \sin\left(\frac{-\pi n}{2}\right)\right]$$

When n is even, $a_n = 0$
When n is odd,

$$a_1 = \frac{5}{\pi}(1 - (-1)) = \frac{10}{\pi}$$

$$a_3 = \frac{5}{3\pi}(-1 - 1) = \frac{-10}{3\pi}$$

$$a_5 = \frac{5}{5\pi}(1 - (-1)) = \frac{10}{5\pi} \text{ and so on.}$$

Hence the Fourier series for the function $f(x)$ is given by:

$$f(x) = \frac{5}{2} + \frac{10}{\pi}\left[\cos\left(\frac{\pi x}{2}\right) - \frac{1}{3}\cos\left(\frac{3\pi x}{2}\right)\right.$$

$$\left. + \frac{1}{5}\cos\left(\frac{5\pi x}{2}\right) - \frac{1}{7}\cos\left(\frac{7\pi x}{2}\right) + \cdots\right]$$

Problem 3. Determine the Fourier series for the function $f(t) = t$ in the range $t = 0$ to $t = 3$

The function $f(t) = t$ in the interval 0 to 3 is shown in Figure 104.3. Although the function is not periodic it may be constructed outside of this range so that it is periodic of period 3, as shown by the broken lines in Figure 104.3.

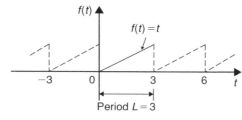

Figure 104.3

From para. (c), the Fourier series is given by:

$$f(t) = a_0 + \sum_{n=1}^{\infty}\left[a_n\cos\left(\frac{2\pi nt}{L}\right) + b_n\sin\left(\frac{2\pi nt}{L}\right)\right]$$

$$a_0 = \frac{1}{L}\int_{\frac{-L}{2}}^{\frac{L}{2}} f(t)\,dx = \frac{1}{L}\int_{0}^{L} f(t)\,dx$$

$$= \frac{1}{3}\int_{0}^{3} t\,dt = \frac{1}{3}\left[\frac{t^2}{2}\right]_{0}^{3} = \frac{3}{2}$$

$$a_n = \frac{2}{L}\int_{\frac{-L}{2}}^{\frac{L}{2}} f(t)\cos\left(\frac{2\pi nt}{L}\right)dt$$

$$= \frac{2}{L}\int_{0}^{L} t\cos\left(\frac{2\pi nt}{L}\right)dt$$

$$= \frac{2}{3}\int_{0}^{3} t\cos\left(\frac{2\pi nt}{3}\right)dt$$

$$= \frac{2}{3}\left[\frac{t\sin\left(\frac{2\pi nt}{3}\right)}{\left(\frac{2\pi n}{3}\right)} + \frac{\cos\left(\frac{2\pi nt}{3}\right)}{\left(\frac{2\pi n}{3}\right)^2}\right]_{0}^{3}$$

by parts

$$= \frac{2}{3}\left[\left\{\frac{3\sin 2\pi n}{\left(\frac{2\pi n}{3}\right)} + \frac{\cos 2\pi n}{\left(\frac{2\pi n}{3}\right)^2}\right\}\right.$$

$$\left. - \left\{0 + \frac{\cos 0}{\left(\frac{2\pi n}{3}\right)^2}\right\}\right] = 0$$

$$b_n = \frac{2}{L}\int_{\frac{-L}{2}}^{\frac{L}{2}} f(t)\sin\left(\frac{2\pi nt}{L}\right)dt$$

$$= \frac{2}{L}\int_{0}^{L} t\sin\left(\frac{2\pi nt}{L}\right)dt$$

$$= \frac{2}{3}\int_{0}^{3} t\sin\left(\frac{2\pi nt}{3}\right)dt$$

$$= \frac{2}{3}\left[\frac{-t\cos\left(\frac{2\pi nt}{3}\right)}{\left(\frac{2\pi n}{3}\right)} + \frac{\sin\left(\frac{2\pi nt}{3}\right)}{\left(\frac{2\pi n}{3}\right)^2}\right]_{0}^{3}$$

by parts

$$= \frac{2}{3}\left[\left\{\frac{-3\cos 2\pi n}{\left(\frac{2\pi n}{3}\right)} + \frac{\sin 2\pi n}{\left(\frac{2\pi n}{3}\right)^2}\right\}\right.$$

$$\left. - \left\{0 + \frac{\sin 0}{\left(\frac{2\pi n}{3}\right)^2}\right\}\right]$$

$$= \frac{2}{3}\left[\frac{-3\cos 2\pi n}{\left(\frac{2\pi n}{3}\right)}\right] = \frac{-3}{\pi n}\cos 2\pi n = \frac{-3}{\pi n}$$

Hence $b_1 = \dfrac{-3}{\pi}$, $b_2 = \dfrac{-3}{2\pi}$, $b_3 = \dfrac{-3}{3\pi}$ and so on.

Thus the Fourier series for the function $f(t)$ in the range 0 to 3 is given by:

$$f(t) = \frac{3}{2} - \frac{3}{\pi}\left[\sin\left(\frac{2\pi t}{3}\right) + \frac{1}{2}\sin\left(\frac{4\pi t}{3}\right)\right.$$

$$\left. + \frac{1}{3}\sin\left(\frac{6\pi t}{3}\right) + \cdots\right]$$

Now try the following Practice Exercise

Practice Exercise 371 Further problems on Fourier series over any range L (answers on page 1197)

1. The voltage from a square wave generator is of the form:

$$v(t) = \begin{cases} 0, & -10 < t < 0 \\ 5, & 0 < t < 10 \end{cases}$$

and is periodic of period 20. Show that the Fourier series for the function is given by:

$$v(t) = \frac{5}{2} + \frac{10}{\pi}\left[\sin\left(\frac{\pi t}{10}\right) + \frac{1}{3}\sin\left(\frac{3\pi t}{10}\right)\right.$$

$$\left. + \frac{1}{5}\sin\left(\frac{5\pi t}{10}\right) + \cdots\right]$$

2. Find the Fourier series for $f(x) = x$ in the range $x = 0$ to $x = 5$

3. A periodic function of period 4 is defined by:

$$f(x) = \begin{cases} -3, & -2 < x < 0 \\ +3, & 0 < x < 2 \end{cases}$$

Sketch the function and obtain the Fourier series for the function.

4. Determine the Fourier series for the half wave rectified sinusoidal voltage $V \sin t$ defined by:

$$f(t) = \begin{cases} V \sin t, & 0 < t < \pi \\ \\ 0, & \pi < t < 2\pi \end{cases}$$

which is periodic of period 2π

104.2 Half-range Fourier series for functions defined over range L

(a) By making the substitution $u = \dfrac{\pi x}{L}$ (see Section 104.1), the range $x = 0$ to $x = L$ corresponds to the range $u = 0$ to $u = \pi$. Hence a function may be expanded in a series of either cosine terms or sine terms only, i.e. a **half-range Fourier series**.

(b) A **half-range cosine series** in the range 0 to L can be expanded as:

$$f(x) = a_0 + \sum_{n=1}^{\infty} a_n \cos\left(\frac{n\pi x}{L}\right)$$

where

$$a_0 = \frac{1}{L} \int_0^L f(x)\, dx$$

and

$$a_n = \frac{2}{L} \int_0^L f(x) \cos\left(\frac{n\pi x}{L}\right) dx$$

(c) A **half-range sine series** in the range 0 to L can be expanded as:

$$f(x) = \sum_{n=1}^{\infty} b_n \sin\left(\frac{n\pi x}{L}\right)$$

where

$$b_n = \frac{2}{L} \int_0^L f(x) \sin\left(\frac{n\pi x}{L}\right) dx$$

Problem 4. Determine the half-range Fourier cosine series for the function $f(x) = x$ in the range $0 \le x \le 2$. Sketch the function within and outside of the given range

A half-range Fourier cosine series indicates an even function. Thus the graph of $f(x) = x$ in the range 0 to 2 is shown in Figure 104.4 and is extended outside of this range so as to be symmetrical about the $f(x)$ axis as shown by the broken lines.

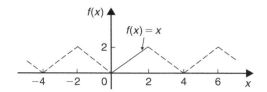

Figure 104.4

From para. (b), for a half-range cosine series:

$$f(x) = a_0 + \sum_{n=1}^{\infty} a_n \cos\left(\frac{n\pi x}{L}\right)$$

$$a_0 = \frac{1}{L} \int_0^L f(x)\, dx = \frac{1}{2} \int_0^2 x\, dx$$

$$= \frac{1}{2}\left[\frac{x^2}{2}\right]_0^2 = 1$$

$$a_n = \frac{2}{L} \int_0^L f(x) \cos\left(\frac{n\pi x}{L}\right) dx$$

$$= \frac{2}{2} \int_0^2 x \cos\left(\frac{n\pi x}{2}\right) dx$$

$$= \left[\frac{x \sin\left(\dfrac{n\pi x}{2}\right)}{\left(\dfrac{n\pi}{2}\right)} + \frac{\cos\left(\dfrac{n\pi x}{2}\right)}{\left(\dfrac{n\pi}{2}\right)^2}\right]_0^2$$

$$= \left[\left(\frac{2\sin n\pi}{\left(\dfrac{n\pi}{2}\right)} + \frac{\cos n\pi}{\left(\dfrac{n\pi}{2}\right)^2}\right) - \left(0 + \frac{\cos 0}{\left(\dfrac{n\pi}{2}\right)^2}\right)\right]$$

$$= \left[\frac{\cos n\pi}{\left(\dfrac{n\pi}{2}\right)^2} - \frac{1}{\left(\dfrac{n\pi}{2}\right)^2}\right]$$

$$= \left(\frac{2}{\pi n}\right)^2 (\cos n\pi - 1)$$

When n is even, $a_n = 0$

$$a_1 = \frac{-8}{\pi^2}, \quad a_3 = \frac{-8}{\pi^2 3^2}, \quad a_5 = \frac{-8}{\pi^2 5^2} \text{ and so on.}$$

Hence the half-range Fourier cosine series for $f(x)$ in the range 0 to 2 is given by:

$$f(x) = 1 - \frac{8}{\pi^2}\left[\cos\left(\frac{\pi x}{2}\right) + \frac{1}{3^2}\cos\left(\frac{3\pi x}{2}\right)\right.$$
$$\left. + \frac{1}{5^2}\cos\left(\frac{5\pi x}{2}\right) + \cdots\right]$$

Section N

> **Problem 5.** Find the half-range Fourier sine series for the function $f(x)=x$ in the range $0 \leq x \leq 2$. Sketch the function within and outside of the given range

A half-range Fourier sine series indicates an odd function. Thus the graph of $f(x)=x$ in the range 0 to 2 is shown in Figure 104.5 and is extended outside of this range so as to be symmetrical about the origin, as shown by the broken lines.

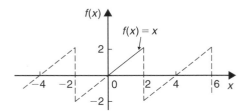

Figure 104.5

From para. (c), for a half-range sine series:

$$f(x) = \sum_{n=1}^{\infty} b_n \sin\left(\frac{n\pi x}{L}\right)$$

$$b_n = \frac{2}{L}\int_0^L f(x)\sin\left(\frac{n\pi x}{L}\right)dx$$

$$= \frac{2}{2}\int_0^2 x\sin\left(\frac{n\pi x}{L}\right)dx$$

$$= \left[\frac{-x\cos\left(\frac{n\pi x}{2}\right)}{\left(\frac{n\pi}{2}\right)} + \frac{\sin\left(\frac{n\pi x}{2}\right)}{\left(\frac{n\pi}{2}\right)^2}\right]_0^2$$

$$= \left[\left(\frac{-2\cos n\pi}{\left(\frac{n\pi}{2}\right)} + \frac{\sin n\pi}{\left(\frac{n\pi}{2}\right)^2}\right) - \left(0 + \frac{\sin 0}{\left(\frac{n\pi}{2}\right)^2}\right)\right]$$

$$= \frac{-2\cos n\pi}{\frac{n\pi}{2}} = \frac{-4}{n\pi}\cos n\pi$$

Hence $b_1 = \frac{-4}{\pi}(-1) = \frac{4}{\pi}$

$$b_2 = \frac{-4}{2\pi}(1) = \frac{-4}{2\pi}$$

$$b_3 = \frac{-4}{3\pi}(-1) = \frac{4}{3\pi} \quad \text{and so on.}$$

Thus the half-range Fourier sine series in the range 0 to 2 is given by:

$$f(x) = \frac{4}{\pi}\left[\sin\left(\frac{\pi x}{2}\right) - \frac{1}{2}\sin\left(\frac{2\pi x}{2}\right)\right.$$

$$\left. + \frac{1}{3}\sin\left(\frac{3\pi x}{2}\right) - \frac{1}{4}\sin\left(\frac{4\pi x}{2}\right) + \cdots\right]$$

Now try the following Practice Exercise

> **Practice Exercise 372   Further problems on half-range Fourier series over range L (answers on page 1197)**
>
> 1. Determine the half-range Fourier cosine series for the function $f(x)=x$ in the range $0 \leq x \leq 3$. Sketch the function within and outside of the given range.
>
> 2. Find the half-range Fourier sine series for the function $f(x)=x$ in the range $0 \leq x \leq 3$. Sketch the function within and outside of the given range.
>
> 3. Determine the half-range Fourier sine series for the function defined by:
>
> $$f(t) = \begin{cases} t, & 0 < t < 1 \\ (2-t), & 1 < t < 2 \end{cases}$$
>
> 4. Show that the half-range Fourier cosine series for the function $f(\theta) = \theta^2$ in the range 0 to 4 is given by:
>
> $$f(\theta) = \frac{16}{3} - \frac{64}{\pi^2}\left(\cos\left(\frac{\pi\theta}{4}\right)\right.$$
>
> $$- \frac{1}{2^2}\cos\left(\frac{2\pi\theta}{4}\right)$$
>
> $$\left. + \frac{1}{3^2}\cos\left(\frac{3\pi\theta}{4}\right) - \cdots\right)$$
>
> Sketch the function within and outside of the given range.

For fully worked solutions to each of the problems in Practice Exercises 371 and 372 in this chapter, go to the website:
www.routledge.com/cw/bird

Chapter 105

A numerical method of harmonic analysis

Why it is important to understand: A numerical method of harmonic analysis

In music, if a note has frequency f, integer multiples of that frequency, $2f$, $3f$, $4f$, and so on, are known as harmonics. As a result, the mathematical study of overlapping waves is called harmonic analysis; this analysis is a diverse field and may be used to produce a Fourier series. Signal processing, medical imaging, astronomy, optics and quantum mechanics are some of the fields that use harmonic analysis extensively. This chapter explains a simple method of harmonic analysis using the trapezoidal rule.

At the end of this chapter, you should be able to:

- define harmonic analysis
- perform a harmonic analysis on data in tabular or graphical form
- consider complex waveform considerations to reduce working of harmonic analysis

105.1 Introduction

Many practical waveforms can be represented by simple mathematical expressions, and, by using Fourier series, the magnitude of their harmonic components determined, as shown in Chapters 101 to 104. For waveforms not in this category, analysis may be achieved by numerical methods. **Harmonic analysis** is the process of resolving a periodic, non-sinusoidal quantity into a series of sinusoidal components of ascending order of frequency.

105.2 Harmonic analysis on data given in tabular or graphical form

The Fourier coefficients a_0, a_n and b_n used in Chapters 101 to 104 all require functions to be integrated, i.e.

$$a_0 = \frac{1}{2\pi} \int_{-\pi}^{\pi} f(x)\mathrm{d}x = \frac{1}{2\pi} \int_{0}^{2\pi} f(x)\,\mathrm{d}x$$

$= $ mean value of $f(x)$ in the range $-\pi$ to π or 0 to 2π

$$a_n = \frac{1}{\pi} \int_{-\pi}^{\pi} f(x) \cos nx \, dx$$

$$= \frac{1}{\pi} \int_{0}^{2\pi} f(x) \cos nx \, dx$$

= twice the mean value of $f(x) \cos nx$ in the range 0 to 2π

$$b_n = \frac{1}{\pi} \int_{-\pi}^{\pi} f(x) \sin nx \, dx$$

$$= \frac{1}{\pi} \int_{0}^{2\pi} f(x) \sin nx \, dx$$

= twice the mean value of $f(x) \sin nx$ in the range 0 to 2π

However, irregular waveforms are not usually defined by mathematical expressions and thus the Fourier coefficients cannot be determined by using calculus. In these cases, approximate methods, such as the **trapezoidal rule**, can be used to evaluate the Fourier coefficients. Most practical waveforms to be analysed are periodic. Let the period of a waveform be 2π and be divided into p equal parts, as shown in Figure 105.1. The width of each interval is thus $\frac{2\pi}{p}$. Let the ordinates be labelled $y_0, y_1, y_2, \ldots y_p$ (note that $y_0 = y_p$). The trapezoidal rule states:

$$\text{Area} = (\text{width of interval}) \left[\frac{1}{2}(\text{first} + \text{last ordinate}) \right.$$
$$\left. + \text{sum of remaining ordinates} \right]$$

$$\approx \frac{2\pi}{p} \left[\frac{1}{2}(y_0 + y_p) + y_1 + y_2 + y_3 + \cdots \right]$$

Since $y_0 = y_p$, then $\frac{1}{2}(y_0 + y_p) = y_0 = y_p$

Hence area $\approx \dfrac{2\pi}{p} \displaystyle\sum_{k=1}^{p} y_k$

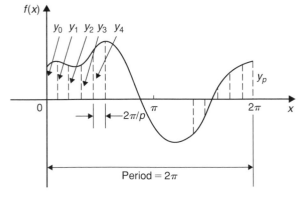

Figure 105.1

Mean value $= \dfrac{\text{area}}{\text{length of base}}$

$$\approx \frac{1}{2\pi} \left(\frac{2\pi}{p} \right) \sum_{k=1}^{p} y_k \approx \frac{1}{p} \sum_{k=1}^{p} y_k$$

However, $a_0 =$ mean value of $f(x)$ in the range 0 to 2π

Thus $\quad \boldsymbol{a_0 \approx \dfrac{1}{p} \displaystyle\sum_{k=1}^{p} y_k} \qquad\qquad (1)$

Similarly, $a_n =$ twice the mean value of $f(x) \cos nx$ in the range 0 to 2π,

thus $\quad \boldsymbol{a_n \approx \dfrac{2}{p} \displaystyle\sum_{k=1}^{p} y_k \cos nx_k} \qquad (2)$

and $b_n =$ twice the mean value of $f(x) \sin nx$ in the range 0 to 2π,

thus $\quad \boldsymbol{b_n \approx \dfrac{2}{p} \displaystyle\sum_{k=1}^{p} y_k \sin nx_k} \qquad (3)$

Problem 1. The values of the voltage v volts at different moments in a cycle are given by:

$\theta°$ (degrees)	V (volts)	$\theta°$ (degrees)	V (volts)
30	62	210	−28
60	35	240	24
90	−38	270	80
120	−64	300	96
150	−63	330	90
180	−52	360	70

Draw the graph of voltage V against angle θ and analyse the voltage into its first three constituent harmonics, each coefficient correct to 2 decimal places

The graph of voltage V against angle θ is shown in Figure 105.2. The range 0 to 2π is divided into 12 equal intervals giving an interval width of $\frac{2\pi}{12}$, i.e. $\frac{\pi}{6}$ rad or $30°$. The values of the ordinates $y_1, y_2, y_3, \ldots$ are 62, 35, −38, ... from the given table of values. If a larger number of intervals are used, results having a greater accuracy are achieved. The data is tabulated in the proforma shown in Table 105.1, on page 1112.

From equation (1), $a_0 \approx \dfrac{1}{p} \displaystyle\sum_{k=1}^{p} y_k = \dfrac{1}{12}(212)$

$$= 17.67 \text{ (since } p = 12)$$

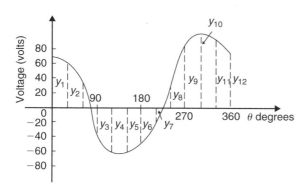

Figure 105.2

From equation (2), $a_n \approx \dfrac{2}{p} \sum\limits_{k=1}^{p} y_k \cos n x_k$

hence $\qquad a_1 \approx \dfrac{2}{12}(417.94) = 69.66$

$$a_2 \approx \dfrac{2}{12}(-39) = -6.50$$

and $\qquad a_3 \approx \dfrac{2}{12}(-49) = -8.17$

From equation (3), $b_n \approx \dfrac{2}{p} \sum\limits_{k=1}^{p} y_k \sin n x_k$

hence $\qquad b_1 \approx \dfrac{2}{12}(-278.53) = -46.42$

$$b_2 \approx \dfrac{2}{12}(29.43) = 4.91$$

and $\qquad b_3 \approx \dfrac{2}{12}(55) = 9.17$

Substituting these values into the Fourier series:

$$f(x) = a_0 + \sum_{n=1}^{\infty}(a_n \cos nx + b_n \sin nx)$$

gives: $\quad v = 17.67 + 69.66 \cos\theta - 6.50 \cos 2\theta$

$$- 8.17 \cos 3\theta + \cdots - 46.42 \sin\theta$$

$$+ 4.91 \sin 2\theta + 9.17 \sin 3\theta + \cdots \qquad (4)$$

Note that in equation (4), $(-46.42 \sin\theta + 69.66 \cos\theta)$ comprises the fundamental, $(4.91 \sin 2\theta - 6.50 \cos 2\theta)$ comprises the second harmonic and $(9.17 \sin 3\theta - 8.17 \cos 3\theta)$ comprises the third harmonic.

It is shown in Chapter 44 that:

$$a \sin\omega t + b \cos\omega t = R \sin(\omega t + \alpha)$$

where $\quad a = R\cos\alpha, \quad b = R\sin\alpha, \quad R = \sqrt{a^2 + b^2}$ and

$\alpha = \tan^{-1} \dfrac{b}{a}$

For the fundamental, $R = \sqrt{(-46.42)^2 + (69.66)^2}$

$$= 83.71$$

If $\quad a = R\cos\alpha$, then $\cos\alpha = \dfrac{a}{R} = \dfrac{-46.42}{83.71}$

which is negative,

and if $\quad b = R\sin\alpha$, then $\sin\alpha = \dfrac{b}{R} = \dfrac{69.66}{83.71}$

which is positive.

The only quadrant where $\cos\alpha$ is negative *and* $\sin\alpha$ is positive is the second quadrant.

Hence $\alpha = \tan^{-1} \dfrac{b}{a} = \tan^{-1} \dfrac{69.66}{-46.42}$

$$= 123.68° \text{ or } 2.16 \,\text{rad}$$

Thus $(-46.42 \sin\theta + 69.66 \cos\theta)$

$$= 83.71 \sin(\theta + 2.16)$$

By a similar method it may be shown that the second harmonic

$$(4.91 \sin 2\theta - 6.50 \cos 2\theta) = 8.15 \sin(2\theta - 0.92)$$

and the third harmonic

$$(9.17 \sin 3\theta - 8.17 \cos 3\theta) = 12.28 \sin(3\theta - 0.73)$$

Hence equation (4) may be re-written as:

$$v = 17.67 + 83.71 \sin(\theta + 2.16)$$

$$+ 8.15 \sin(2\theta - 0.92)$$

$$+ 12.28 \sin(3\theta - 0.73) \text{ volts}$$

which is the form used in Chapter 50 with complex waveforms.

Now try the following Practice Exercise

Practice Exercise 373   Further problems on numerical harmonic analysis (answers on page 1197)

Determine the Fourier series to represent the periodic functions given by the tables of values in Problems 1 to 3, up to and including the third harmonic and each coefficient correct to 2 decimal places. Use 12 ordinates in each case.

1.

Angle $\theta°$	30	60	90	120	150	180
Displacement y	40	43	38	30	23	17

Angle $\theta°$	210	240	270	300	330	360
Displacement y	11	9	10	13	21	32

Section N

Table 105.1

Ordinates	$\theta°$	V	$\cos\theta$	$V\cos\theta$	$\sin\theta$	$V\sin\theta$	$\cos2\theta$	$V\cos2\theta$	$\sin2\theta$	$V\sin2\theta$	$\cos3\theta$	$V\cos3\theta$	$\sin3\theta$	$V\sin3\theta$
y_1	30	62	0.866	53.69	0.5	31	0.5	31	0.866	53.69	0	0	1	62
y_2	60	35	0.5	17.5	0.866	30.31	−0.5	−17.5	0.866	30.31	−1	−35	0	0
y_3	90	−38	0	0	1	−38	−1	38	0	0	0	0	−1	38
y_4	120	−64	−0.5	32	0.866	−55.42	−0.5	32	−0.866	55.42	1	−64	0	0
y_5	150	−63	−0.866	54.56	0.5	−31.5	0.5	−31.5	−0.866	54.56	0	0	1	−63
y_6	180	−52	−1	52	0	0	1	−52	0	0	−1	52	0	0
y_7	210	−28	−0.866	24.25	−0.5	14	0.5	−14	0.866	−24.25	0	0	−1	28
y_8	240	24	−0.5	−12	−0.866	−20.78	−0.5	−12	0.866	20.78	1	24	0	0
y_9	270	80	0	0	−1	−80	−1	−80	0	0	0	0	1	80
y_{10}	300	96	0.5	48	−0.866	−83.14	−0.5	−48	−0.866	−83.14	−1	−96	0	0
y_{11}	330	90	0.866	77.94	−0.5	−45	0.5	45	−0.866	−77.94	0	0	−1	−90
y_{12}	360	70	1	70	0	0	1	70	0	0	1	70	0	0
$\sum_{k=1}^{12} y_k = (212)$				$\sum_{k=1}^{12} y_k\cos\theta_k$ $= 417.94$		$\sum_{k=1}^{12} y_k\sin\theta_k$ $= -278.53$		$\sum_{k=1}^{12} y_k\cos2\theta_k$ $= -39$		$\sum_{k=1}^{12} y_k\sin2\theta_k$ $= 29.43$		$\sum_{k=1}^{12} y_k\cos3\theta_k$ $= -49$		$\sum_{k=1}^{12} y_k\sin3\theta_k$ $= 55$

2.

Angle $\theta°$	0	30	60	90	120	150
Voltage v	−5.0	−1.5	6.0	12.5	16.0	16.5

Angle $\theta°$	180	210	240	270	300	330
Voltage v	15.0	12.5	6.5	−4.0	−7.0	−7.5

3.

Angle $\theta°$	30	60	90	120	150	180
Current i	0	−1.4	−1.8	−1.9	−1.8	−1.3

Angle $\theta°$	210	240	270	300	330	360
Current i	0	2.2	3.8	3.9	3.5	2.5

105.3 Complex waveform considerations

It is sometimes possible to predict the harmonic content of a waveform on inspection of particular waveform characteristics.

(i) If a periodic waveform is such that the area above the horizontal axis is equal to the area below then the mean value is zero. Hence $a_0 = 0$ (see Figure 105.3(a)).

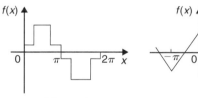

(a) $a_0 = 0$ **(b)** Contains no sine terms

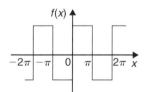 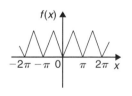

(c) Contains no cosine terms   **(d)** Contains only even harmonics

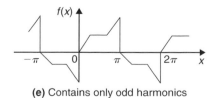

(e) Contains only odd harmonics

Figure 105.3

(ii) An **even function** is symmetrical about the vertical axis and contains **no sine terms** (see Figure 105.3(b)).

(iii) An **odd function** is symmetrical about the origin and contains **no cosine terms** (see Figure 105.3(c)).

(iv) $f(x) = f(x+\pi)$ represents a waveform which repeats after half a cycle and **only even harmonics** are present (see Figure 105.3(d)).

(v) $f(x) = -f(x+\pi)$ represents a waveform for which the positive and negative cycles are identical in shape and **only odd harmonics** are present (see Figure 105.3(e)).

Problem 2. Without calculating Fourier coefficients, state which harmonics will be present in the waveforms shown in Figure 105.4.

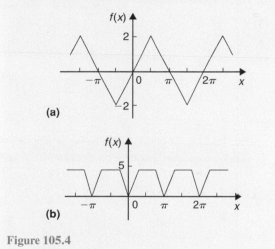

Figure 105.4

(a) The waveform shown in Figure 105.4(a) is symmetrical about the origin and is thus an odd function. An odd function contains no cosine terms. Also, the waveform has the characteristic $f(x) = -f(x+\pi)$, i.e. the positive and negative half cycles are identical in shape. Only odd harmonics can be present in such a waveform. Thus the waveform shown in Figure 105.4(a) contains **only odd sine terms**. Since the area above the x-axis is equal to the area below, $a_0 = 0$

(b) The waveform shown in Figure 105.4(b) is symmetrical about the $f(x)$ axis and is thus an even function. An even function contains no sine terms. Also, the waveform has the characteristic

$f(x) = f(x+\pi)$, i.e. the waveform repeats itself after half a cycle. Only even harmonics can be present in such a waveform. Thus the waveform shown in Figure 105.4(b) contains **only even cosine terms** (together with a constant term, a_0).

Problem 3. An alternating current i amperes is shown in Figure 105.5. Analyse the waveform into its constituent harmonics as far as and including the fifth harmonic, correct to 2 decimal places, by taking $30°$ intervals.

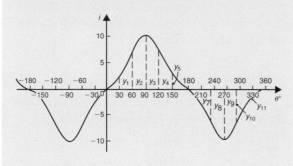

Figure 105.5

With reference to Figure 105.5, the following characteristics are noted:

(i) The mean value is zero since the area above the θ axis is equal to the area below it. Thus the constant term, or d.c. component, $a_0 = 0$

(ii) Since the waveform is symmetrical about the origin the function i is odd, which means that there are no cosine terms present in the Fourier series.

(iii) The waveform is of the form $f(\theta) = -f(\theta + \pi)$, which means that only odd harmonics are present.

Investigating waveform characteristics has thus saved unnecessary calculations and in this case the Fourier series has only odd sine terms present, i.e.

$$i = b_1 \sin\theta + b_3 \sin 3\theta + b_5 \sin 5\theta + \cdots$$

A proforma, similar to Table 105.1, but without the 'cosine terms' columns and without the 'even sine terms' columns is shown in Table 105.2, up to, and including, the fifth harmonic, from which the Fourier coefficients b_1, b_3 and b_5 can be determined. Twelve co-ordinates are chosen and labelled y_1, y_2, y_3, ... y_{12}, as shown in Figure 105.5.

Table 105.2

Ordinate	θ	i	$\sin\theta$	$i\sin\theta$	$\sin 3\theta$	$i\sin 3\theta$	$\sin 5\theta$	$i\sin 5\theta$
y_1	30	2	0.5	1	1	2	0.5	1
y_2	60	7	0.866	6.06	0	0	−0.866	−6.06
y_3	90	10	1	10	−1	−10	1	10
y_4	120	7	0.866	6.06	0	0	−0.866	−6.06
y_5	150	2	0.5	1	1	2	0.5	1
y_6	180	0	0	0	0	0	0	0
y_7	210	−2	−0.5	1	−1	2	−0.5	1
y_8	240	−7	−0.866	6.06	0	0	0.866	−6.06
y_9	270	−10	−1	10	1	−10	−1	10
y_{10}	300	−7	−0.866	6.06	0	0	0.866	−6.06
y_{11}	330	−2	−0.5	1	−1	2	−0.5	1
y_{12}	360	0	0	0	0	0	0	0
				$\sum_{k=1}^{12} y_k \sin\theta_k = 48.24$		$\sum_{k=1}^{12} y_k \sin 3\theta_k = -12$		$\sum_{k=1}^{12} y_k \sin 5\theta_k = -0.24$

From equation (3), Section 105.2,

$$b_n = \frac{2}{p} \sum_{k=1}^{p} i_k \sin n\theta_k, \text{ where } p = 12$$

Hence $b_1 \approx \dfrac{2}{12}(48.24) = 8.04,$

$$b_3 \approx \frac{2}{12}(-12) = -2.00,$$

and $b_5 \approx \dfrac{2}{12}(-0.24) = -0.04$

Thus the Fourier series for current i is given by:

$$i = 8.04 \sin \theta - 2.00 \sin 3\theta - 0.04 \sin 5\theta$$

Now try the following Practice Exercise

Practice Exercise 374   Further problems on a numerical method of harmonic analysis (answers on page 1198)

1. Without performing calculations, state which harmonics will be present in the waveforms shown in Figure 105.6.

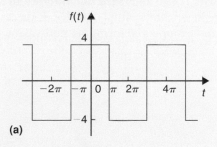

(a)

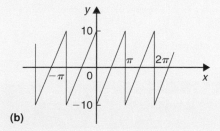

(b)

Figure 105.6

2. Analyse the periodic waveform of displacement y against angle θ in Figure 105.7(a) into its constituent harmonics as far as and including the third harmonic, by taking $30°$ intervals.

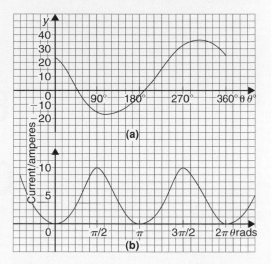

Figure 105.7

3. For the waveform of current shown in Figure 105.7(b), state why only a d.c. component and even cosine terms will appear in the Fourier series and determine the series, using $\pi/6$ rad intervals, up to and including the sixth harmonic.

4. Determine the Fourier series as far as the third harmonic to represent the periodic function y given by the waveform in Figure 105.8. Take 12 intervals when analysing the waveform.

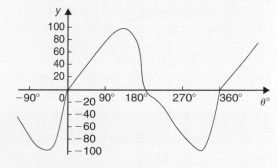

Figure 105.8

For fully worked solutions to each of the problems in Practice Exercises 373 and 374 in this chapter, go to the website:
www.routledge.com/cw/bird

Chapter 106

The complex or exponential form of a Fourier series

Why it is important to understand: The complex or exponential form of a Fourier series

A Fourier series may be represented not only as a sum of sines and cosines, as in previous chapters, but as a sum of complex exponentials. The complex exponentials provide a more convenient and compact way of expressing the Fourier series than the trigonometric form. It also allows the magnitude and phase spectra to be easily calculated. This form is widely used by engineers, for example, in circuit theory and control theory. This chapter explains how the trigonometric and exponential forms are equivalent.

At the end of this chapter, you should be able to:

- derive the exponential or complex form of a Fourier series
- derive the complex coefficients for a Fourier series
- determine the complex Fourier series for a given function
- deduce the complex coefficient symmetry relationships
- understand the frequency spectrum of a waveform
- determine phasors in exponential form for various sinusoidal voltages

106.1   Introduction

The form used for the Fourier series in Chapters 101 to 105 consisted of cosine and sine terms. However, there is another form that is commonly used – one that directly gives the amplitude terms in the frequency spectrum and relates to phasor notation. This form involves the use of complex numbers (see Chapters 45 and 46). It is called the **exponential** or **complex form** of a Fourier series.

106.2   Exponential or complex notation

It was shown on page 540, equations (4) and (5) that:

$$e^{j\theta} = \cos\theta + j\sin\theta \tag{1}$$

$$\text{and} \quad e^{-j\theta} = \cos\theta - j\sin\theta \tag{2}$$

Adding equations (1) and (2) gives:

$$e^{j\theta} + e^{-j\theta} = 2\cos\theta$$

---

from which, $\cos\theta = \dfrac{e^{j\theta} + e^{-j\theta}}{2}$ (3)

Similarly, equation (1) – equation (2) gives:

$$e^{j\theta} - e^{-j\theta} = 2j\sin\theta$$

from which, $\sin\theta = \dfrac{e^{j\theta} - e^{-j\theta}}{2j}$ (4)

Thus, from page 1104, the Fourier series $f(x)$ over any range L,

$$f(x) = a_0 + \sum_{n=1}^{\infty}\left[a_n\cos\left(\frac{2\pi nx}{L}\right) + b_n\sin\left(\frac{2\pi nx}{L}\right)\right]$$

may be written as:

$$f(x) = a_0 + \sum_{n=1}^{\infty}\left[a_n\left(\frac{e^{j\frac{2\pi nx}{L}} + e^{-j\frac{2\pi nx}{L}}}{2}\right)\right.$$

$$\left. + b_n\left(\frac{e^{j\frac{2\pi nx}{L}} - e^{-j\frac{2\pi nx}{L}}}{2j}\right)\right]$$

Multiplying top and bottom of the b_n term by $-j$ (and remembering that $j^2 = -1$) gives:

$$f(x) = a_0 + \sum_{n=1}^{\infty}\left[a_n\left(\frac{e^{j\frac{2\pi nx}{L}} + e^{-j\frac{2\pi nx}{L}}}{2}\right)\right.$$

$$\left. - jb_n\left(\frac{e^{j\frac{2\pi nx}{L}} - e^{-j\frac{2\pi nx}{L}}}{2}\right)\right]$$

Rearranging gives:

$$f(x) = a_0 + \sum_{n=1}^{\infty}\left[\left(\frac{a_n - jb_n}{2}\right)e^{j\frac{2\pi nx}{L}}\right.$$

$$\left. + \left(\frac{a_n + jb_n}{2}\right)e^{-j\frac{2\pi nx}{L}}\right]$$ (5)

The Fourier coefficients a_0, a_n and b_n may be replaced by complex coefficients c_0, c_n and c_{-n} such that

$$c_0 = a_0$$ (6)

$$c_n = \frac{a_n - jb_n}{2}$$ (7)

and $c_{-n} = \dfrac{a_n + jb_n}{2}$ (8)

where c_{-n} represents the complex conjugate of c_n (see page 526).

Thus, equation (5) may be rewritten as:

$$f(x) = c_0 + \sum_{n=1}^{\infty}c_n e^{j\frac{2\pi nx}{L}} + \sum_{n=1}^{\infty}c_{-n}e^{-j\frac{2\pi nx}{L}}$$ (9)

Since $e^0 = 1$, the c_0 term can be absorbed into the summation since it is just another term to be added to the summation of the c_n term when $n = 0$. Thus,

$$f(x) = \sum_{n=0}^{\infty}c_n e^{j\frac{2\pi nx}{L}} + \sum_{n=1}^{\infty}c_{-n}e^{-j\frac{2\pi nx}{L}}$$ (10)

The c_{-n} term may be rewritten by changing the limits $n = 1$ and $n = \infty$ to $n = -1$ and $n = -\infty$. Since n has been made negative, the exponential term becomes $e^{j\frac{2\pi nx}{L}}$ and c_{-n} becomes c_n. Thus,

$$f(x) = \sum_{n=0}^{\infty}c_n e^{j\frac{2\pi nx}{L}} + \sum_{n=-1}^{-\infty}c_n e^{j\frac{2\pi nx}{L}}$$

Since the summations now extend from $-\infty$ to -1 and from 0 to $+\infty$, equation (10) may be written as:

$$f(x) = \sum_{n=-\infty}^{\infty}c_n e^{j\frac{2\pi nx}{L}}$$ (11)

Equation (11) is the **complex** or **exponential form** of the Fourier series.

106.3   The complex coefficients

From equation (7), the complex coefficient c_n was defined as: $c_n = \dfrac{a_n - jb_n}{2}$

However, a_n and b_n are defined (from page 1104) by:

$$a_n = \frac{2}{L}\int_{-\frac{L}{2}}^{\frac{L}{2}}f(x)\cos\left(\frac{2\pi nx}{L}\right)dx \quad\text{and}$$

$$b_n = \frac{2}{L}\int_{-\frac{L}{2}}^{\frac{L}{2}}f(x)\sin\left(\frac{2\pi nx}{L}\right)dx$$

Thus, $c_n = \dfrac{\left(\begin{array}{c}\frac{2}{L}\int_{-\frac{L}{2}}^{\frac{L}{2}}f(x)\cos\left(\frac{2\pi nx}{L}\right)dx \\[4pt] -j\frac{2}{L}\int_{-\frac{L}{2}}^{\frac{L}{2}}f(x)\sin\left(\frac{2\pi nx}{L}\right)dx\end{array}\right)}{2}$

$$= \frac{1}{L}\int_{-\frac{L}{2}}^{\frac{L}{2}}f(x)\cos\left(\frac{2\pi nx}{L}\right)dx$$

$$- j\frac{1}{L}\int_{-\frac{L}{2}}^{\frac{L}{2}}f(x)\sin\left(\frac{2\pi nx}{L}\right)dx$$

Section N

From equations (3) and (4),

$$c_n = \frac{1}{L}\int_{-\frac{L}{2}}^{\frac{L}{2}} f(x)\left(\frac{e^{j\frac{2\pi nx}{L}}+e^{-j\frac{2\pi nx}{L}}}{2}\right)dx$$
$$-j\frac{1}{L}\int_{-\frac{L}{2}}^{\frac{L}{2}} f(x)\left(\frac{e^{j\frac{2\pi nx}{L}}-e^{-j\frac{2\pi nx}{L}}}{2j}\right)dx$$

from which,

$$c_n = \frac{1}{L}\int_{-\frac{L}{2}}^{\frac{L}{2}} f(x)\left(\frac{e^{j\frac{2\pi nx}{L}}+e^{-j\frac{2\pi nx}{L}}}{2}\right)dx$$
$$-\frac{1}{L}\int_{-\frac{L}{2}}^{\frac{L}{2}} f(x)\left(\frac{e^{j\frac{2\pi nx}{L}}-e^{-j\frac{2\pi nx}{L}}}{2}\right)dx$$

i.e. $\qquad c_n = \frac{1}{L}\int_{-\frac{L}{2}}^{\frac{L}{2}} f(x)e^{-j\frac{2\pi nx}{L}}dx \qquad (12)$

Care needs to be taken when determining c_0. If n appears in the denominator of an expression the expansion is invalid when $n=0$. In such circumstances it is usually simpler to evaluate c_0 by using the relationship:

$$c_0 = a_0 = \frac{1}{L}\int_{-\frac{L}{2}}^{\frac{L}{2}} f(x)dx \qquad (13)$$

This is merely equation (12) with $n=0$

Problem 1. Determine the complex Fourier series for the function defined by:

$$f(x) = \begin{cases} 0, & \text{when } -2 \le x \le -1 \\ 5, & \text{when } -1 \le x \le 1 \\ 0, & \text{when } 1 \le x \le 2 \end{cases}$$

The function is periodic outside this range of period 4

This is the same Problem as Problem 2 on page 1105 and we can use this to demonstrate that the two forms of Fourier series are equivalent.

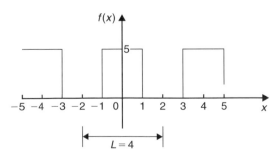

$f(x)$

Figure 106.1

The function $f(x)$ is shown in Figure 106.1, where the period $L=4$

From equation (11), the complex Fourier series is given by:

$$f(x) = \sum_{n=-\infty}^{\infty} c_n e^{j\frac{2\pi nx}{L}}$$

where c_n is given by:

$$c_n = \frac{1}{L}\int_{-\frac{L}{2}}^{\frac{L}{2}} f(x)\,e^{-j\frac{2\pi nx}{L}}dx \quad \text{(from equation (12))}.$$

With reference to Figure 106.1, when $L=4$,

$$c_n = \frac{1}{4}\left\{\int_{-2}^{-1} 0\,dx + \int_{-1}^{1} 5e^{-j\frac{2\pi nx}{4}}dx + \int_{1}^{2} 0\,dx\right\}$$

$$= \frac{1}{4}\int_{-1}^{1} 5e^{-j\frac{\pi nx}{2}}dx = \frac{5}{4}\left[\frac{e^{-\frac{j\pi nx}{2}}}{-\frac{j\pi n}{2}}\right]_{-1}^{1}$$

$$= \frac{-5}{j2\pi n}\left[e^{-\frac{j\pi nx}{2}}\right]_{-1}^{1} = \frac{-5}{j2\pi n}\left(e^{-\frac{j\pi n}{2}} - e^{\frac{j\pi n}{2}}\right)$$

$$= \frac{5}{\pi n}\left(\frac{e^{j\frac{\pi n}{2}} - e^{-j\frac{\pi n}{2}}}{2j}\right)$$

$$= \frac{5}{\pi n}\sin\frac{\pi n}{2} \quad \text{(from equation (4))}.$$

Hence, from equation (11), **the complex form of the Fourier series** is given by:

$$f(x) = \sum_{n=-\infty}^{\infty} c_n e^{j\frac{2\pi nx}{L}} = \sum_{n=-\infty}^{\infty} \frac{5}{\pi n}\sin\frac{\pi n}{2}\,e^{j\frac{\pi nx}{2}} \qquad (14)$$

Let us show how this result is equivalent to the result involving sine and cosine terms determined on page 1105.

From equation (13),

$$c_0 = a_0 = \frac{1}{L}\int_{-\frac{L}{2}}^{\frac{L}{2}} f(x)dx = \frac{1}{4}\int_{-1}^{1} 5\,dx$$

$$= \frac{5}{4}[x]_{-1}^{1} = \frac{5}{4}[1-(-1)] = \frac{5}{2}$$

Since $c_n = \frac{5}{\pi n}\sin\frac{\pi n}{2}$, then

$$c_1 = \frac{5}{\pi}\sin\frac{\pi}{2} = \frac{5}{\pi}$$

$$c_2 = \frac{5}{2\pi}\sin\pi = 0$$

(in fact, **all even terms will be zero** since $\sin n\pi = 0$)

$$c_3 = \frac{5}{\pi n} \sin \frac{\pi n}{2} = \frac{5}{3\pi} \sin \frac{3\pi}{2} = -\frac{5}{3\pi}$$

By similar substitution,

$$c_5 = \frac{5}{5\pi} \qquad c_7 = -\frac{5}{7\pi}, \text{ and so on.}$$

Similarly,

$$c_{-1} = \frac{5}{-\pi} \sin \frac{-\pi}{2} = \frac{5}{\pi}$$

$$c_{-2} = -\frac{5}{2\pi} \sin \frac{-2\pi}{2} = 0 = c_{-4} = c_{-6}, \text{ and so on.}$$

$$c_{-3} = -\frac{5}{3\pi} \sin \frac{-3\pi}{2} = -\frac{5}{3\pi}$$

$$c_{-5} = -\frac{5}{5\pi} \sin \frac{-5\pi}{2} = \frac{5}{5\pi}, \text{ and so on.}$$

Hence, the extended complex form of the Fourier series shown in equation (14) becomes:

$$f(x) = \frac{5}{2} + \frac{5}{\pi} e^{j\frac{\pi x}{2}} - \frac{5}{3\pi} e^{j\frac{3\pi x}{2}} + \frac{5}{5\pi} e^{j\frac{5\pi x}{2}}$$

$$- \frac{5}{7\pi} e^{j\frac{7\pi x}{2}} + \cdots + \frac{5}{\pi} e^{-j\frac{\pi x}{2}}$$

$$- \frac{5}{3\pi} e^{-j\frac{3\pi x}{2}} + \frac{5}{5\pi} e^{-j\frac{5\pi x}{2}}$$

$$- \frac{5}{7\pi} e^{-j\frac{7\pi x}{2}} + \cdots$$

$$= \frac{5}{2} + \frac{5}{\pi} \left(e^{j\frac{\pi x}{2}} + e^{-j\frac{\pi x}{2}} \right)$$

$$- \frac{5}{3\pi} \left(e^{j\frac{3\pi x}{2}} + e^{-j\frac{3\pi x}{2}} \right)$$

$$+ \frac{5}{5\pi} \left(e^{\frac{5\pi x}{2}} + e^{-j\frac{5\pi x}{2}} \right) - \cdots$$

$$= \frac{5}{2} + \frac{5}{\pi}(2)\left(\frac{e^{j\frac{\pi x}{2}} + e^{-j\frac{\pi x}{2}}}{2} \right)$$

$$- \frac{5}{3\pi}(2)\left(\frac{e^{j\frac{3\pi x}{2}} + e^{-j\frac{3\pi x}{2}}}{2} \right)$$

$$+ \frac{5}{5\pi}(2)\left(\frac{e^{j\frac{5\pi x}{2}} + e^{-j\frac{5\pi x}{2}}}{2} \right) - \cdots$$

$$= \frac{5}{2} + \frac{10}{\pi} \cos\left(\frac{\pi x}{2}\right) - \frac{10}{3\pi} \cos\left(\frac{3\pi x}{2}\right)$$

$$+ \frac{10}{5\pi} \cos\left(\frac{5\pi x}{2}\right) - \cdots$$

(from equation (3))

i.e. $f(x) = \dfrac{5}{2} + \dfrac{10}{\pi}\left[\cos\left(\dfrac{\pi x}{2}\right) - \dfrac{1}{3}\cos\left(\dfrac{3\pi x}{2}\right) \right.$

$$\left. + \frac{1}{5}\cos\left(\frac{5\pi x}{2}\right) - \cdots \right]$$

which is the same as obtained on page 1105.

Hence, $\displaystyle\sum_{n=-\infty}^{\infty} \frac{5}{\pi n} \sin \frac{n\pi}{2} \, e^{j\frac{\pi n x}{2}}$ is equivalent to

$$\frac{5}{2} + \frac{10}{\pi}\left[\cos\left(\frac{\pi x}{2}\right) - \frac{1}{3}\cos\left(\frac{3\pi x}{2}\right) \right.$$

$$\left. + \frac{1}{5}\cos\left(\frac{5\pi x}{2}\right) - \cdots \right]$$

Problem 2. Show that the complex Fourier series for the function $f(t) = t$ in the range $t = 0$ to $t = 1$, and of period 1, may be expressed as:

$$f(t) = \frac{1}{2} + \frac{j}{2\pi} \sum_{n=-\infty}^{\infty} \frac{e^{j2\pi n t}}{n}$$

The saw-tooth waveform is shown in Figure 106.2.

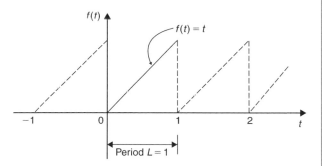

Figure 106.2

From equation (11), the complex Fourier series is given by:

$$f(t) = \sum_{n=-\infty}^{\infty} c_n e^{j\frac{2\pi n t}{L}}$$

and when the period $L = 1$, then:

$$f(t) = \sum_{n=-\infty}^{\infty} c_n e^{j2\pi n t}$$

where, from equation (12),

$$c_n = \frac{1}{L}\int_{-\frac{L}{2}}^{\frac{L}{2}} f(t)\,e^{-j\frac{2\pi nt}{L}}\,dt = \frac{1}{L}\int_0^L f(t)\,e^{-j\frac{2\pi nt}{L}}\,dt$$

and when $L=1$ and $f(t)=t$, then:

$$c_n = \frac{1}{1}\int_0^1 t\,e^{-j\frac{2\pi nt}{1}}\,dt = \int_0^1 t\,e^{-j2\pi nt}\,dt$$

Using integration by parts (see Chapter 68), let $u=t$, from which, $\dfrac{du}{dt}=1$, and $dt=du$, and

let $dv=e^{-j2\pi nt}$, from which,

$$v = \int e^{-j2\pi nt}\,dt = \frac{e^{-j2\pi nt}}{-j2\pi n}$$

Hence, $c_n = \displaystyle\int_0^1 t\,e^{-j2\pi nt} = uv - \int v\,du$

$$= \left[t\frac{e^{-j2\pi nt}}{-j2\pi n}\right]_0^1 - \int_0^1 \frac{e^{-j2\pi nt}}{-j2\pi n}\,dt$$

$$= \left[t\frac{e^{-j2\pi nt}}{-j2\pi n} - \frac{e^{-j2\pi nt}}{(-j2\pi n)^2}\right]_0^1$$

$$= \left(\frac{e^{-j2\pi n}}{-j2\pi n} - \frac{e^{-j2\pi n}}{(-j2\pi n)^2}\right)$$

$$- \left(0 - \frac{e^0}{(-j2\pi n)^2}\right)$$

From equation (2),

$$c_n = \left(\frac{\cos 2\pi n - j\sin 2\pi n}{-j2\pi n} - \frac{\cos 2\pi n - j\sin 2\pi n}{(-j2\pi n)^2}\right)$$
$$+ \frac{1}{(-j2\pi n)^2}$$

However, $\cos 2\pi n = 1$ and $\sin 2\pi n = 0$ for all positive and negative integer values of n.

Thus, $c_n = \dfrac{1}{-j2\pi n} - \dfrac{1}{(-j2\pi n)^2} + \dfrac{1}{(-j2\pi n)^2}$

$$= \frac{1}{-j2\pi n} = \frac{1(j)}{-j2\pi n(j)}$$

i.e. $\quad \boldsymbol{c_n = \dfrac{j}{2\pi n}}$

From equation (13),

$$c_0 = a_0 = \frac{1}{L}\int_{-\frac{L}{2}}^{\frac{L}{2}} f(t)\,dt$$

$$= \frac{1}{L}\int_0^L f(t)\,dt = \frac{1}{1}\int_0^1 t\,dt$$

$$= \left[\frac{t^2}{2}\right]_0^1 = \left[\frac{1}{2}-0\right] = \frac{1}{2}$$

Hence, the complex Fourier series is given by:

$$f(t) = \sum_{n=-\infty}^{\infty} c_n\,e^{j\frac{2\pi nt}{L}} \quad \text{from equation (11)}$$

i.e. $f(t) = \dfrac{1}{2} + \displaystyle\sum_{n=-\infty}^{\infty} \dfrac{j}{2\pi n}\,e^{j2\pi nt}$

$$= \frac{1}{2} + \frac{j}{2\pi}\sum_{n=-\infty}^{\infty} \frac{e^{j2\pi nt}}{n}$$

Problem 3. Show that the exponential form of the Fourier series for the waveform described by:

$$f(x) = \begin{cases} 0 \text{ when } -4 \le x \le 0 \\ 10 \text{ when } \;\; 0 \le x \le 4 \end{cases}$$

and which has a period of 8, is given by:

$$f(x) = \sum_{n=-\infty}^{\infty} \frac{5j}{n\pi}(\cos n\pi - 1)\,e^{j\frac{n\pi x}{4}}$$

From equation (12),

$$c_n = \frac{1}{L}\int_{-\frac{L}{2}}^{\frac{L}{2}} f(x)\,e^{-j\frac{2\pi nx}{L}}\,dx$$

$$= \frac{1}{8}\left[\int_{-4}^0 0\,e^{-j\frac{\pi nx}{4}}\,dx + \int_0^4 10\,e^{-j\frac{\pi nx}{4}}\,dx\right]$$

$$= \frac{10}{8}\left[\frac{e^{-j\frac{\pi nt}{4}}}{-j\frac{\pi n}{4}}\right]_0^4 = \frac{10}{8}\left(\frac{4}{-j\pi n}\right)\left[e^{-j\pi n}-1\right]$$

$$= \frac{5j}{-j^2\pi n}\left(e^{-j\pi n}-1\right) = \frac{5j}{\pi n}\left(e^{-j\pi n}-1\right)$$

From equation (2), $e^{-j\theta} = \cos\theta - j\sin\theta$, thus $e^{-j\pi n} = \cos\pi n - j\sin\pi n = \cos\pi n$ for all integer values of n. Hence,

$$c_n = \frac{5j}{\pi n}\left(e^{-j\pi n}-1\right) = \frac{5j}{\pi n}(\cos n\pi - 1)$$

From equation (11), the exponential Fourier series is given by:

$$f(x) = \sum_{n=-\infty}^{\infty} c_n e^{j\frac{2\pi nx}{L}}$$

$$= \sum_{n=-\infty}^{\infty} \frac{5j}{n\pi}(\cos n\pi - 1)e^{j\frac{n\pi x}{4}}$$

Now try the following Practice Exercise

Practice Exercise 375   Further problems on the complex form of a Fourier series (answers on page 1198)

1. Determine the complex Fourier series for the function defined by:

$$f(t) = \begin{cases} 0, & \text{when } -\pi \le t \le 0 \\ 2, & \text{when } \ \ 0 \le t \le \pi \end{cases}$$

The function is periodic outside of this range of period 2π

2. Show that the complex Fourier series for the waveform shown in Figure 106.3, that has period 2, may be represented by:

$$f(t) = 2 + \sum_{\substack{n=-\infty \\ (n \ne 0)}}^{\infty} \frac{j2}{\pi n}(\cos n\pi - 1)e^{j\pi nt}$$

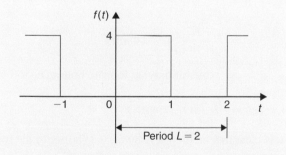

Figure 106.3

3. Show that the complex Fourier series of Problem 2 is equivalent to:

$$f(t) = 2 + \frac{8}{\pi}\left(\sin \pi t + \frac{1}{3}\sin 3\pi t \right.$$

$$\left. + \frac{1}{5}\sin 5\pi t + \cdots \right)$$

4. Determine the exponential form of the Fourier series for the function defined by: $f(t) = e^{2t}$ when $-1 < t < 1$ and has period 2

106.4   Symmetry relationships

If even or odd symmetry is noted in a function, then time can be saved in determining coefficients.

The Fourier coefficients present in the complex Fourier series form are affected by symmetry. Summarising from previous chapters:

An **even function** is symmetrical about the vertical axis and contains no sine terms, i.e. $b_n = 0$

For even symmetry,

$$a_0 = \frac{1}{L}\int_0^L f(x)\mathrm{d}x \quad \text{and}$$

$$a_n = \frac{2}{L}\int_0^L f(x)\cos\left(\frac{2\pi nx}{L}\right)\mathrm{d}x$$

$$= \frac{4}{L}\int_0^{\frac{L}{2}} f(x)\cos\left(\frac{2\pi nx}{L}\right)\mathrm{d}x$$

An **odd function** is symmetrical about the origin and contains no cosine terms, $a_0 = a_n = 0$

For odd symmetry,

$$b_n = \frac{2}{L}\int_0^L f(x)\sin\left(\frac{2\pi nx}{L}\right)\mathrm{d}x$$

$$= \frac{4}{L}\int_0^{\frac{L}{2}} f(x)\sin\left(\frac{2\pi nx}{L}\right)\mathrm{d}x$$

From equation (7), page 1117, $c_n = \dfrac{a_n - jb_n}{2}$

Thus, for **even symmetry**, $b_n = 0$ and

$$c_n = \frac{a_n}{2} = \frac{2}{L}\int_0^{\frac{L}{2}} f(x)\cos\left(\frac{2\pi nx}{L}\right)\mathrm{d}x \qquad (15)$$

For **odd symmetry**, $a_n = 0$ and

$$c_n = \frac{-jb_n}{2} = -j\frac{2}{L}\int_0^{\frac{L}{2}} f(x)\sin\left(\frac{2\pi nx}{L}\right)\mathrm{d}x \qquad (16)$$

For example, in Problem 1 on page 1118, the function $f(x)$ is even, since the waveform is symmetrical about the $f(x)$ axis. Thus equation (15) could have been used, giving:

$$c_n = \frac{2}{L} \int_0^{\frac{L}{2}} f(x) \cos\left(\frac{2\pi nx}{L}\right) dx$$

$$= \frac{2}{4} \int_0^2 f(x) \cos\left(\frac{2\pi nx}{4}\right) dx$$

$$= \frac{1}{2} \left\{ \int_0^1 5 \cos\left(\frac{\pi nx}{2}\right) dx + \int_1^2 0 \, dx \right\}$$

$$= \frac{5}{2} \left[\frac{\sin\left(\frac{\pi nx}{2}\right)}{\frac{\pi n}{2}} \right]_0^1 = \frac{5}{2} \left(\frac{2}{\pi n}\right) \left(\sin\frac{n\pi}{2} - 0\right)$$

$$= \frac{5}{\pi n} \sin\frac{n\pi}{2}$$

which is the same answer as in Problem 1; however, a knowledge of even functions has produced the coefficient more quickly.

Problem 4. Obtain the Fourier series, in complex form, for the square wave shown in Figure 106.4

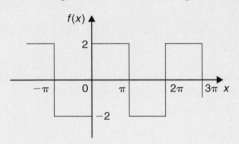

Figure 106.4

Method A

The square wave shown in Figure 106.4 is an **odd function** since it is symmetrical about the origin.
The period of the waveform $L = 2\pi$
Thus, using equation (16):

$$c_n = -j\frac{2}{L} \int_0^{\frac{L}{2}} f(x) \sin\left(\frac{2\pi nx}{L}\right) dx$$

$$= -j\frac{2}{2\pi} \int_0^\pi 2 \sin\left(\frac{2\pi nx}{2\pi}\right) dx$$

$$= -j\frac{2}{\pi} \int_0^\pi \sin nx \, dx = -j\frac{2}{\pi} \left[\frac{-\cos nx}{n}\right]_0^\pi$$

$$= -j\frac{2}{\pi n} \left((-\cos \pi n) - (-\cos 0)\right)$$

i.e. $c_n = -j\dfrac{2}{\pi n}[1 - \cos\pi n]$ \hfill (17)

Method B

If it had **not** been noted that the function was odd, equation (12) would have been used, i.e.

$$c_n = \frac{1}{L} \int_{-\frac{L}{2}}^{\frac{L}{2}} f(x) e^{-j\frac{2\pi nx}{L}} dx$$

$$= \frac{1}{2\pi} \int_{-\pi}^\pi f(x) e^{-j\frac{2\pi nx}{2\pi}} dx$$

$$= \frac{1}{2\pi} \left\{ \int_{-\pi}^0 -2 e^{-jnx} dx + \int_0^\pi 2 e^{-jnx} dx \right\}$$

$$= \frac{1}{2\pi} \left\{ \left[\frac{-2 e^{-jnx}}{-jn}\right]_{-\pi}^0 + \left[\frac{2 e^{-jnx}}{-jn}\right]_0^\pi \right\}$$

$$= \frac{1}{2\pi} \left(\frac{2}{jn}\right) \left\{ \left[e^{-jnx}\right]_{-\pi}^0 - \left[e^{-jnx}\right]_0^\pi \right\}$$

$$= \frac{1}{2\pi} \left(\frac{2}{jn}\right) \left\{ \left[e^0 - e^{+jn\pi}\right] - \left[e^{-jn\pi} - e^0\right] \right\}$$

$$= \frac{1}{j\pi n} \left\{ 1 \quad e^{jn\pi} - e^{-jn\pi} + 1 \right\}$$

$$= \frac{1}{jn\pi} \left\{ 2 - 2\left(\frac{e^{jn\pi} + e^{-jn\pi}}{2}\right) \right\}$$

by rearranging

$$= \frac{2}{jn\pi} \left\{ 1 - \left(\frac{e^{jn\pi} + e^{-jn\pi}}{2}\right) \right\}$$

$$= \frac{2}{jn\pi} \{1 - \cos n\pi\} \qquad \text{from equation (3)}$$

$$= \frac{-j2}{-j(jn\pi)} \{1 - \cos n\pi\}$$

by multiplying top and bottom by $-j$

i.e. $c_n = -j\dfrac{2}{n\pi}(1 - \cos n\pi)$ \hfill (17)

It is clear that method A is by far the shorter of the two methods.
From equation (11), the complex Fourier series is given by:

$$f(x) = \sum_{n=-\infty}^\infty c_n e^{j\frac{2\pi nx}{L}}$$

$$= \sum_{n=-\infty}^\infty -j\frac{2}{n\pi}(1 - \cos n\pi) e^{jnx} \hfill (18)$$

Problem 5. Show that the complex Fourier series obtained in Problem 4 above is equivalent to

$$f(x) = \frac{8}{\pi}\left(\sin x + \frac{1}{3}\sin 3x + \frac{1}{5}\sin 5x \right.$$
$$\left. + \frac{1}{7}\sin 7x + \cdots\right)$$

(which was the Fourier series obtained in terms of sines and cosines in Problem 3 on page 1098)

From equation (17) above, $c_n = -j\dfrac{2}{n\pi}(1-\cos n\pi)$

When $n=1$,

$$c_1 = -j\frac{2}{(1)\pi}(1-\cos\pi)$$

$$= -j\frac{2}{\pi}\left(1-(-1)\right) = -\frac{j4}{\pi}$$

When $n=2$,

$$c_2 = -j\frac{2}{2\pi}(1-\cos 2\pi) = 0;$$

in fact, all even values of c_n will be zero.

When $n=3$,

$$c_3 = -j\frac{2}{3\pi}(1-\cos 3\pi)$$

$$= -j\frac{2}{3\pi}(1-(-1)) = -\frac{j4}{3\pi}$$

By similar reasoning,

$$c_5 = -\frac{j4}{5\pi}, \quad c_7 = -\frac{j4}{7\pi}, \quad \text{and so on.}$$

When $n=-1$,

$$c_{-1} = -j\frac{2}{(-1)\pi}(1-\cos(-\pi))$$

$$= +j\frac{2}{\pi}(1-(-1)) = +\frac{j4}{\pi}$$

When $n=-3$,

$$c_{-3} = -j\frac{2}{(-3)\pi}(1-\cos(-3\pi))$$

$$= +j\frac{2}{3\pi}(1-(-1)) = +\frac{j4}{3\pi}$$

By similar reasoning,

$$c_{-5} = +\frac{j4}{5\pi}, \quad c_{-7} = +\frac{j4}{7\pi}, \quad \text{and so on.}$$

Since the waveform is odd, $c_0 = a_0 = 0$

From equation (18) above,

$$f(x) = \sum_{n=-\infty}^{\infty} -j\frac{2}{n\pi}(1-\cos n\pi)\,e^{jnx}$$

Hence,

$$f(x) = -\frac{j4}{\pi}e^{jx} - \frac{j4}{3\pi}e^{j3x} - \frac{j4}{5\pi}e^{j5x}$$
$$-\frac{j4}{7\pi}e^{j7x} - \cdots + \frac{j4}{\pi}e^{-jx} + \frac{j4}{3\pi}e^{-j3x}$$
$$+ \frac{j4}{5\pi}e^{-j5x} + \frac{j4}{7\pi}e^{-j7x} + \cdots$$

$$= \left(-\frac{j4}{\pi}e^{jx} + \frac{j4}{\pi}e^{-jx}\right)$$

$$+ \left(-\frac{j4}{3\pi}e^{3x} + \frac{j4}{3\pi}e^{-3x}\right)$$

$$+ \left(-\frac{j4}{5\pi}e^{5x} + \frac{j4}{5\pi}e^{-5x}\right) + \cdots$$

$$= -\frac{j4}{\pi}\left(e^{jx} - e^{jx}\right) - \frac{j4}{3\pi}\left(e^{3x} - e^{-3x}\right)$$

$$- \frac{j4}{5\pi}\left(e^{5x} - e^{-5x}\right) + \cdots$$

$$= \frac{4}{j\pi}\left(e^{jx} - e^{-jx}\right) + \frac{4}{j3\pi}\left(e^{3x} - e^{-3x}\right)$$

$$+ \frac{4}{j5\pi}\left(e^{5x} - e^{-5x}\right) + \cdots$$

by multiplying top and bottom by j

$$= \frac{8}{\pi}\left(\frac{e^{jx} - e^{-jx}}{2j}\right) + \frac{8}{3\pi}\left(\frac{e^{j3x} - e^{-j3}}{2j}\right)$$

$$+ \frac{8}{5\pi}\left(\frac{e^{j5x} - e^{-j5x}}{2j}\right) + \cdots$$

by rearranging

$$= \frac{8}{\pi}\sin x + \frac{8}{3\pi}\sin 3x + \frac{8}{3x}\sin 5x + \cdots$$

from equation (4), page 1117

i.e.

$$f(x) = \frac{8}{\pi}\left(\sin x + \frac{1}{3}\sin 3x + \frac{1}{5}\sin 5x\right.$$
$$\left. + \frac{1}{7}\sin 7x + \cdots\right)$$

Section N

Hence,

$$f(x) = \sum_{n=-\infty}^{\infty} -j\frac{2}{n\pi}(1-\cos n\pi)\,e^{jnx}$$

$$\equiv \frac{8}{\pi}\left(\sin x + \frac{1}{3}\sin 3x + \frac{1}{5}\sin 5x\right.$$

$$\left. + \frac{1}{7}\sin 7x + \cdots\right)$$

Now try the following Practice Exercise

Practice Exercise 376   Further problems on symmetry relationships (answers on page 1198)

1. Determine the exponential form of the Fourier series for the periodic function defined by:

$$f(x) = \begin{cases} -2, & \text{when } -\pi \leq x \leq -\dfrac{\pi}{2} \\[2mm] 2, & \text{when } -\dfrac{\pi}{2} \leq x \leq +\dfrac{\pi}{2} \\[2mm] -2, & \text{when } +\dfrac{\pi}{2} \leq x \leq +\pi \end{cases}$$

and which has a period of 2π

2. Show that the exponential form of the Fourier series in Problem 1 above is equivalent to:

$$f(x) = \frac{8}{\pi}\left(\cos x - \frac{1}{3}\cos 3x + \frac{1}{5}\cos 5x\right.$$

$$\left. - \frac{1}{7}\cos 7x + \cdots\right)$$

3. Determine the complex Fourier series to represent the function $f(t) = 2t$ in the range $-\pi$ to $+\pi$

4. Show that the complex Fourier series in Problem 3 above is equivalent to:

$$f(t) = 4\left(\sin t - \frac{1}{2}\sin 2t + \frac{1}{3}\sin 3t\right.$$

$$\left. - \frac{1}{4}\sin 4t + \cdots\right)$$

106.5   The frequency spectrum

In the Fourier analysis of periodic waveforms seen in previous chapters, although waveforms physically exist in the time domain, they can be regarded as comprising components with a variety of frequencies. The amplitude and phase of these components are obtained from the Fourier coefficients a_n and b_n; this is known as a **frequency domain**. A **spectrum** is a graph of the amplitude of the various frequencies plotted against their frequencies. A simple example is demonstrated in Problem 6 following.

Problem 6.   A pulse of height 20 and width 2 has a period of 10. Sketch the spectrum of the waveform

The pulse is shown in Figure 106.5.
The complex coefficient is given by equation (12):

$$c_n = \frac{1}{L}\int_{-\frac{L}{2}}^{\frac{L}{2}} f(t)e^{-j\frac{2\pi nt}{L}}\,dt$$

$$= \frac{1}{10}\int_{-1}^{1} 20e^{-j\frac{2\pi nt}{10}}\,dt = \frac{20}{10}\left[\frac{e^{-j\frac{\pi nt}{5}}}{-j\pi n / 5}\right]_{-1}^{1}$$

$$= \frac{20}{10}\left(\frac{5}{-j\pi n}\right)\left[e^{-j\frac{\pi n}{5}} - e^{j\frac{\pi n}{5}}\right]$$

$$= \frac{20}{\pi n}\left[\frac{e^{j\frac{\pi n}{5}} - e^{-j\frac{\pi n}{5}}}{2j}\right]$$

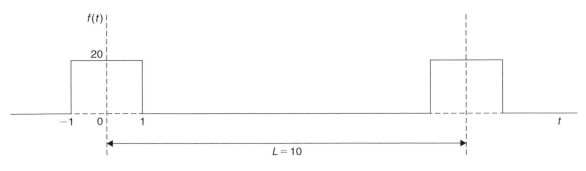

Figure 106.5

i.e. $c_n = \dfrac{20}{\pi n} \sin \dfrac{n\pi}{5}$

from equation (4), page 1117.

From equation (13),

$$c_0 = \frac{1}{L} \int_{-\frac{L}{2}}^{\frac{L}{2}} f(x)\,\mathrm{d}x = \frac{1}{10} \int_{-1}^{1} 20\,\mathrm{d}t$$

$$= \frac{1}{10}[20t]_{-1}^{1} = \frac{1}{10}[20 - (-20)] = 4$$

$$c_1 = \frac{20}{\pi} \sin \frac{\pi}{5} = 3.74 \text{ and}$$

$$c_{-1} = -\frac{20}{\pi} \sin\left(-\frac{\pi}{5}\right) = 3.74$$

Further values of c_n and c_{-n}, up to $n = 10$, are calculated and are shown in the following table.

n	c_n	c_{-n}
0	4	4
1	3.74	3.74
2	3.03	3.03
3	2.02	2.02
4	0.94	0.94
5	0	0
6	−0.62	−0.62
7	−0.86	−0.86
8	−0.76	−0.76
9	−0.42	−0.42
10	0	0

A graph of $|c_n|$ plotted against the number of the harmonic, n, is shown in Figure 106.6.

Figure 106.7 shows the corresponding plot of c_n against n.

Since c_n is real (i.e. no j terms) then the phase must be either $0°$ or $\pm 180°$, depending on the sign of the sine, as shown in Figure 106.8.

When c_n is positive, i.e. between $n = -4$ and $n = +4$, angle $\alpha_n = 0°$.

When c_n is negative, then $\alpha_n = \pm 180°$; between $n = +6$ and $n = +9$, α_n is taken as $+180°$, and between $n = -6$ and $n = -9$, α_n is taken as $-180°$.

Figures 106.6 to 106.8 together form the spectrum of the waveform shown in Figure 106.5.

106.6   Phasors

Electrical engineers in particular often need to analyse alternating current circuits, i.e. circuits containing a sinusoidal input and resulting sinusoidal currents and voltages within the circuit.

It was shown in Chapter 39, page 467, that a general sinusoidal voltage function can be represented by:

$$v = V_m \sin(\omega t + \alpha) \text{ volts} \tag{19}$$

where V_m is the maximum voltage or amplitude of the voltage v, ω is the angular velocity ($= 2\pi f$, where f is the frequency) and α is the phase angle compared with $v = V_m \sin \omega t$.

Similarly, a sinusoidal expression may also be expressed in terms of cosine as:

$$v = V_m \cos(\omega t + \alpha) \text{ volts} \tag{20}$$

It is quite complicated to add, subtract, multiply and divide quantities in the time domain form of equations (19) and (20). As an alternative method of analysis a waveform representation called a **phasor** is used. A phasor has two distinct parts – a magnitude and an angle; for example, the polar form of a complex number, say $5\angle\pi/6$, can represent a phasor, where 5 is the magnitude or modulus, and $\pi/6$ radians is the angle or argument. Also, it was shown on page 540 that $5\angle\pi/6$ may be written as $5\,\mathrm{e}^{j\pi/6}$ in exponential form.

In chapter 46, equation (4), page 540, it is shown that:

$$\mathrm{e}^{j\theta} = \cos\theta + j\sin\theta \tag{21}$$

which is known as **Euler's* formula**.

From equation (21),

$$\mathrm{e}^{j(\omega t + \alpha)} = \cos(\omega t + \alpha) + j\sin(\omega t + \alpha)$$

$$\text{and} \quad V_m \mathrm{e}^{j(\omega t + \alpha)} = V_m \cos(\omega t + \alpha)$$

$$+ j V_m \sin(\omega t + \alpha)$$

Thus a sinusoidal varying voltage such as in equation (19) or equation (20) can be considered to be either the real or the imaginary part of $V_m\,\mathrm{e}^{j(\omega t + \alpha)}$, depending on whether the cosine or sine function is being considered.

$V_m\,\mathrm{e}^{j(\omega t + \alpha)}$ may be rewritten as $V_m\,\mathrm{e}^{j\omega t}\mathrm{e}^{j\alpha}$ since $a^{m+n} = a^m \times a^n$ from the laws of indices, page 57. The $\mathrm{e}^{j\omega t}$ term can be considered to arise from the fact that a radius is rotated with an angular velocity ω, and

---

* **Who was Euler?** See page 847 for image and resume of Euler. To find out more go to **www.routledge.com/cw/bird**

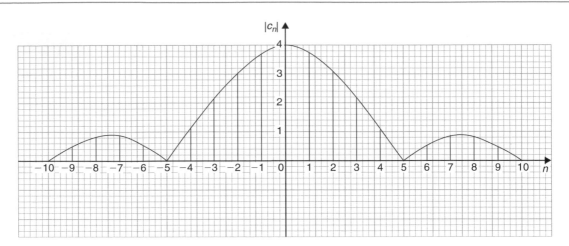

Figure 106.6

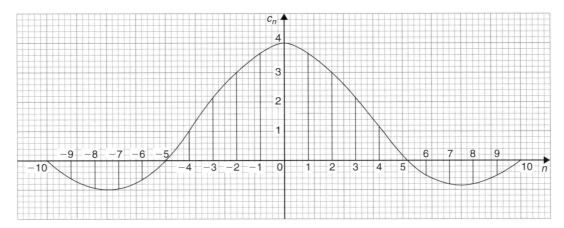

Figure 106.7

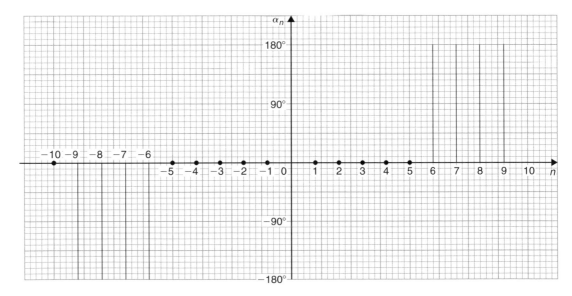

Figure 106.8

α is the angle at which the radius starts to rotate at time $t = 0$ (see Chapter 39, page 467).

Thus, $V_m e^{j\omega t} e^{j\alpha}$ defines a **phasor**. In a particular circuit the angular velocity ω is the same for all the elements thus the phasor can be adequately described by $V_m \angle \alpha$, as suggested above.

Alternatively, if

$$v = V_m \cos(\omega t + \alpha) \text{ volts}$$

and

$$\cos\theta = \frac{1}{2}\left(e^{j\vartheta} + e^{-j\theta}\right)$$

from equation (3), page 1117,

then

$$v = V_m\left[\frac{1}{2}\left(e^{j(\omega t + \alpha)} + e^{-j(\omega t + \alpha)}\right)\right]$$

i.e.

$$v = \frac{1}{2}V_m e^{j\omega t} e^{j\alpha} + \frac{1}{2}V_m e^{-j\omega t} e^{-j\alpha}$$

Thus, v is the sum of two phasors, each with half the amplitude, with one having a positive value of angular velocity (i.e. rotating anticlockwise) and a positive value of α, and the other having a negative value of angular velocity (i.e. rotating clockwise) and a negative value of α, as shown in Figure 106.9.

The two phasors are $\frac{1}{2}V_m \angle \alpha$ and $\frac{1}{2}V_m \angle -\alpha$

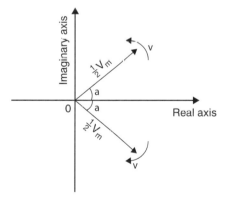

Figure 106.9

From equation (11), page 1117, the Fourier representation of a waveform in complex form is:

$$c_n e^{j\frac{2\pi nt}{L}} = c_n e^{j\omega nt} \quad \text{for positive values of } n$$

$$\left(\text{since } \omega = \frac{2\pi}{L}\right)$$

and $\quad c_n e^{-j\omega nt} \quad$ for negative values of n

It can thus be considered that these terms represent phasors, those with positives powers being phasors rotating with a positive angular velocity (i.e. anticlockwise), and those with negative powers being phasors rotating with a negative angular velocity (i.e. clockwise).

In the above equations,

$n = 0$ represents a non-rotating component, since $e^0 = 1$,

$n = 1$ represents a rotating component with angular velocity of 1ω,

$n = 2$ represents a rotating component with angular velocity of 2ω, and so on.

Thus we have a set of phasors, the algebraic sum of which at some instant of time gives the magnitude of the waveform at that time.

Problem 7. Determine the pair of phasors that can be used to represent the following voltages:

(a) $v = 8 \cos 2t$ (b) $v = 8 \cos(2t - 1.5)$

(a)   From equation (3), page 1117,

$$\cos\theta = \frac{1}{2}(e^{j\theta} + e^{-j\theta})$$

Hence,

$$v = 8\cos 2t = 8\left[\frac{1}{2}\left(e^{j2t} + e^{-j2t}\right)\right]$$

$$= 4e^{j2t} + 4e^{-j2t}$$

This represents a phasor of length 4 rotating anticlockwise (i.e. in the positive direction) with an angular velocity of 2 rad/s, and another phasor of length 4 and rotating clockwise (i.e. in the negative direction) with an angular velocity of 2 rad/s. Both phasors have zero phase angle. Figure 106.10 shows the two phasors.

(b)   From equation (3), page 1117,

$$\cos\theta = \frac{1}{2}\left(e^{j\theta} + e^{-j\theta}\right)$$

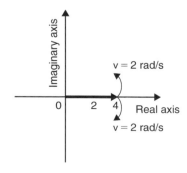

Figure 106.10

Hence, $v = 8\cos(2t - 1.5)$

$$= 8\left[\frac{1}{2}\left(e^{j(2t-1.5)} + e^{-j(2t-1.5)}\right)\right]$$

$$= 4e^{j(2t-1.5)} + 4e^{-j(2t-1.5)}$$

i.e. $\boldsymbol{v = 4e^{2t}\,e^{-j\,1.5} + 4e^{-j2t}\,e^{j1.5}}$

This represents a phasor of length 4 and phase angle -1.5 radians rotating anticlockwise (i.e. in the positive direction) with an angular velocity of 2 rad/s, and another phasor of length 4 and phase angle $+1.5$ radians rotating clockwise (i.e. in the negative direction) with an angular velocity of 2 rad/s. Figure 106.11 shows the two phasors.

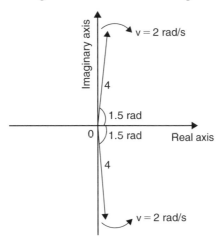

Figure 106.11

> **Problem 8.**   Determine the pair of phasors that can be used to represent the third harmonic
>
> $$v = 8\cos 3t - 20\sin 3t$$

Using $\cos t = \dfrac{1}{2}\left(e^{jt} + e^{-jt}\right)$

and $\sin t = \dfrac{1}{2j}\left(e^{jt} - e^{-jt}\right)$ from page 1117

gives: $v = 8\cos 3t - 20\sin 3t$

$$= 8\left[\frac{1}{2}\left(e^{j3t} + e^{-j3t}\right)\right]$$

$$- 20\left[\frac{1}{2j}\left(e^{j3t} - e^{-j3t}\right)\right]$$

$$= 4e^{j3t} + 4e^{-j3t} - \frac{10}{j}e^{j3t} + \frac{10}{j}e^{-j3t}$$

$$= 4e^{j3t} + 4e^{-j3t} - \frac{10(j)}{j(j)}e^{j3t} + \frac{10(j)}{j(j)}e^{-j3t}$$

$$= 4e^{j3t} + 4e^{-j3t} + 10je^{j3t} - 10je^{-j3t}$$

since $j^2 = -1$

$$= (4 + j10)e^{j3t} + (4 - j10)e^{-j3t}$$

$$(4 + j10) = \sqrt{4^2 + 10^2}\angle\tan^{-1}\left(\frac{10}{4}\right)$$

$$= 10.77\angle 1.19$$

and $(4 - j10)$

$$= 10.77\angle -1.19$$

Hence, $\boldsymbol{v = 10.77\,\angle 1.19 + 10.77\,\angle -1.19}$

Thus v comprises a phasor $10.77\angle 1.19$ rotating anti-clockwise with an angular velocity of 3 rad/s, and a phasor $10.77\angle -1.19$ rotating clockwise with an angular velocity of 3 rad/s.

Now try the following Practice Exercise

> **Practice Exercise 377   Further problems on phasors (answers on page 1198)**
>
> 1.   Determine the pair of phasors that can be used to represent the following voltages:
>
>   (a) $v = 4\cos 4t$ (b) $v = 4\cos(4t + \pi/2)$
>
> 2.   Determine the pair of phasors that can represent the harmonic given by:
> $v = 10\cos 2t - 12\sin 2t$
>
> 3.   Find the pair of phasors that can represent the fundamental current: $i = 6\sin t + 4\cos t$

For fully worked solutions to each of the problems in Practice Exercises 375 to 377 in this chapter, go to the website:
www.routledge.com/cw/bird

This assignment covers the material contained in Chapters 101 to 106. *The marks available are shown in brackets at the end of each question.*

1. Obtain a Fourier series for the periodic function $f(x)$ defined as follows:

$$f(x) = \begin{cases} -1, & \text{when } -\pi \le x \le 0 \\ 1, & \text{when } \quad 0 \le x \le \pi \end{cases}$$

   The function is periodic outside of this range with period 2π (13)

2. Obtain a Fourier series to represent $f(t) = t$ in the range $-\pi$ to $+\pi$ (13)

3. Expand the function $f(\theta) = \theta$ in the range $0 \le \theta \le \pi$ into (a) a half range cosine series, and (b) a half range sine series. (18)

4. (a) Sketch the waveform defined by:

$$f(x) = \begin{cases} 0, & \text{when } -4 \le x \le -2 \\ 3, & \text{when } -2 \le x \le 2 \\ 0, & \text{when } \quad 2 \le x \le 4 \end{cases}$$

   and which is periodic outside of this range of period 8

   (b) State whether the waveform in (a) is odd, even or neither odd nor even.
   (c) Deduce the Fourier series for the function defined in (a). (15)

5. Displacement y on a point on a pulley when turned through an angle of θ degrees is given by:

θ	y
30	3.99
60	4.01
90	3.60
120	2.84
150	1.84
180	0.88
210	0.27
240	0.13
270	0.45
300	1.25
330	2.37
360	3.41

   Sketch the waveform and construct a Fourier series for the first three harmonics. (23)

6. A rectangular waveform is shown in Figure RT34.1.

   (a) State whether the waveform is an odd or even function.

   (b) Obtain the Fourier series for the waveform in complex form.

   (c) Show that the complex Fourier series in (b) is equivalent to:

$$f(x) = \frac{20}{\pi}\left(\sin x + \frac{1}{3}\sin 3x + \frac{1}{5}\sin 5x + \frac{1}{7}\sin 7x + \cdots \right)$$ (18)

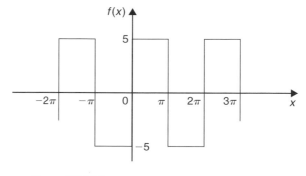

Figure RT34.1

Section N

Fourier series

If $f(x)$ is a **periodic function of period 2π** then its Fourier series is given by:

$$f(x) = a_0 + \sum_{n=1}^{\infty} (a_n \cos nx + b_n \sin nx)$$

where, for the range $-\pi$ to $+\pi$:

$$a_0 = \frac{1}{2\pi} \int_{-\pi}^{\pi} f(x)\, dx$$

$$a_n = \frac{1}{\pi} \int_{-\pi}^{\pi} f(x) \cos nx\, dx \quad (n = 1, 2, 3, ..)$$

$$b_n = \frac{1}{\pi} \int_{-\pi}^{\pi} f(x) \sin nx\, dx \quad (n = 1, 2, 3, ..)$$

Even functions

If $f(x)$ is **even** in the range $-\pi < x < \pi$ then $f(-x) = f(x)$ and the Fourier series has **no sine terms**,

i.e. $\quad f(x) = a_0 + \sum_{n=1}^{\infty} a_n \cos nx$

where

$$a_0 = \frac{1}{\pi} \int_0^{\pi} f(x)\, dx \qquad a_n = \frac{2}{\pi} \int_0^{\pi} f(x) \cos nx\, dx$$

Odd functions

If $f(x)$ is **odd** in the range $-\pi < x < \pi$ then $f(-x) = -f(x)$ and the Fourier series has **no constant term** and **no cosine terms**, i.e.

$$f(x) = \sum_{n=1}^{\infty} b_n \sin nx$$

where

$$b_n = \frac{2}{\pi} \int_0^{\pi} f(x) \sin nx\, dx$$

If $f(x)$ is a **periodic function of period L** then its Fourier series is given by:

$$f(x) = a_0 + \sum_{n=1}^{\infty} \left\{ a_n \cos\left(\frac{2\pi nx}{L}\right) + b_n \sin\left(\frac{2\pi nx}{L}\right) \right\}$$

where for the range $-\frac{L}{2}$ to $+\frac{L}{2}$:

$$a_0 = \frac{1}{L} \int_{-L/2}^{L/2} f(x)\, dx$$

$$a_n = \frac{2}{L} \int_{-L/2}^{L/2} f(x) \cos\left(\frac{2\pi nx}{L}\right) dx \quad (n = 1, 2, 3, ..)$$

$$b_n = \frac{2}{L} \int_{-L/2}^{L/2} f(x) \sin\left(\frac{2\pi nx}{L}\right) dx \quad (n = 1, 2, 3, ..)$$

For an even function,

$$a_0 = \frac{2}{L} \int_0^{L/2} f(x)\, dx$$

$$a_n = \frac{4}{L} \int_0^{L/2} f(x) \cos\left(\frac{2\pi nx}{L}\right) dx$$

and $\qquad b_n = 0$

For an odd function, $\quad a_0 = 0$

$$a_n = 0$$

and $\qquad b_n = \frac{4}{L} \int_0^{L/2} f(x) \sin\left(\frac{2\pi nx}{L}\right) dx$

Complex or exponential Fourier series

$$f(x) = \sum_{n=-\infty}^{\infty} c_n e^{j\frac{2\pi nx}{L}} \quad \text{where} \quad c_n = \frac{1}{L} \int_{-L/2}^{L/2} f(x)\, e^{-j\frac{2\pi nx}{L}}\, dx$$

For even symmetry, $c_n = \dfrac{2}{L} \displaystyle\int_0^{\frac{L}{2}} f(x) \cos\left(\frac{2\pi nx}{L}\right) dx$

For odd symmetry, $c_n = -j\dfrac{2}{L} \displaystyle\int_0^{\frac{L}{2}} f(x) \sin\left(\frac{2\pi nx}{L}\right) dx$

Section O

Z-transforms

Chapter 107

An introduction to z-transforms

Why it is important to understand: An introduction to z-transforms

In mathematics and signal processing, the z-transform converts a discrete-time signal, which is a sequence of real or complex numbers, into a complex frequency domain representation. It can be considered as a discrete-time equivalent of the Laplace transform.

Laplace transform methods are widely used for analysis in linear systems and are used when a system is described by a linear differential equation, with constant coefficients. However, there are numerous systems that are described by difference equations – not differential equations – and these systems are common and different from those described by differential equations.

Systems that satisfy difference equations include computer controlled systems – systems that take measurements with digital input/output boards or GPIB instruments (digital 8-bit parallel communications interface with data transfer rates up to 1 Mbyte/s), calculate an output voltage and output that voltage digitally. Frequently these systems run a program loop that executes in a fixed interval of time. Other systems that satisfy difference equations are those systems with digital filters – which are found anywhere digital signal processing/digital filtering is undertaken – that includes digital signal transmission systems like the telephone system or systems that process audio signals. A CD contains digital signal information, and when it is read off the CD, it is initially a digital signal that can be processed with a digital filter. There are an incredible number of systems used every day that have digital components which satisfy difference equations. In continuous systems Laplace transforms play a unique role. They allow system and circuit designers to analyse systems and predict performance, and to think in different terms – like frequency responses – to help understand linear continuous systems. They are a very powerful tool that shapes how engineers think about those systems. Z-transforms play the role in sampled systems that Laplace transforms play in continuous systems. In sampled systems, inputs and outputs are related by difference equations and z-transform techniques are used to solve those difference equations. In continuous systems, Laplace transforms are used to represent systems with transfer functions, while in sampled systems, z-transforms are used to represent systems with transfer functions.

This chapter merely provides an introduction to z-transforms – how to read z-transforms from a table, to appreciate various properties of z-transforms, how to determine inverse z-transforms and how to solve difference equations using z-transforms.

At the end of this chapter, you should be able to:

- define the z-transform
- prove z-transforms for simple sequences
- use a table of z-transforms to determine simple transforms
- understand the properties of z-transforms – linearity, first and second shift theorems, translations, final and initial value theorems, and derivatives of transforms
- use a table of z-transforms to determine inverse transforms
- use z-transforms to solve difference equations

107.1   Sequences

The sequence $\ldots, 2^{-3}, 2^{-2}, 2^{-1}, 2^0, 2^1, 2^2, 2^3, \ldots$ has a general term of the form 2^k and this series can be written as $\{2^k\}_{-\infty}^{\infty}$ indicating that the power or index of the number 2 has a range from $-\infty$ to ∞.

The sum $\displaystyle\sum_{k=-\infty}^{\infty} \left(\frac{2}{z}\right)^k = \ldots + \left(\frac{2}{z}\right)^{-3} + \left(\frac{2}{z}\right)^{-2} +$

$\left(\frac{2}{z}\right)^{-1} + \left(\frac{2}{z}\right)^{0} + \left(\frac{2}{z}\right)^{1} + \left(\frac{2}{z}\right)^{2} + \ldots$

is called the **z-transform** of the sequence $Z\{2^k\}_{-\infty}^{\infty}$ and is denoted by $F(z)$, where the complex number z is chosen to ensure that the sum is finite.

Thus, $\{2^k\}_{-\infty}^{\infty}$ and $Z\{2^k\}_{-\infty}^{\infty} = F(z) = \displaystyle\sum_{k=-\infty}^{\infty} \left(\frac{2}{z}\right)^k$

forming what is referred to as a '**z-transform pair**'. For simplicity, the range will be restricted to $z\{x_k\}_0^{\infty}$ where $x_k = 0$ for k < 0 and denote it by $\{x_k\}$

i.e. $Z\{x_k\} = F(z) = \displaystyle\sum_{k=0}^{\infty} \frac{x_k}{z^k}$

This is the **definition of the z-transform** of the sequence $\{x_k\}$ and is used in the following worked problems.

Problem 1. Determine the z-transform for the unit impulse $\{\delta_k\} = \{1, 0, 0, 0, \ldots\}$

The z-transform of $\{\delta_k\}$ is given by:

$Z\{\delta_k\} = F(z) = \displaystyle\sum_{k=0}^{\infty} \frac{\delta_k}{z^k} = \frac{1}{z^0} + \frac{0}{z^1} + \frac{0}{z^2} + \ldots = 1$

i.e. $Z\{\delta_k\} = 1$ valid for all values of z

Problem 2. Determine the z-transform for the unit step sequence $\{u_k\} = \{1, 1, 1, 1, \ldots\} = \{1\}$

The z-transform of $\{u_k\}$ is given by:

$Z\{u_k\} = F(z) = \displaystyle\sum_{k=0}^{\infty} \frac{u_k}{z^k} = \displaystyle\sum_{k=0}^{\infty} \frac{1}{z^k}$

$= \frac{1}{z^0} + \frac{1}{z^1} + \frac{1}{z^2} + \frac{1}{z^3} + \frac{1}{z^4} + \ldots$

i.e. $Z\{u_k\} = 1 + \dfrac{1}{z} + \dfrac{1}{z^2} + \dfrac{1}{z^3} + \dfrac{1}{z^4} + \ldots$ (1)

Using the binomial theorem for $(1+x)^n$, the series expansion of $\dfrac{1}{1-x}$ may be determined:

$\dfrac{1}{1-x} = (1-x)^{-1}$

$= 1 + (-1)(-x) + \dfrac{(-1)(-2)}{2!}(-x)^2$

$+ \dfrac{(-1)(-2)(-3)}{3!}(-x)^3 + \ldots$

$= 1 + x + x^2 + x^3 + \ldots$ valid for $|x| < 1$ (2)

Comparing equations (1) and (2) gives: $F(z) = \dfrac{1}{1 - \dfrac{1}{z}}$

provided $\left|\dfrac{1}{z}\right| < 1$

$\dfrac{1}{1 - \dfrac{1}{z}} = \dfrac{1}{\dfrac{z-1}{z}} = \dfrac{z}{z-1}$ hence, $z\{u_k\} = \dfrac{z}{z-1}$ provided $|z| > 1$

Problem 3. Show that the z-transform for the unit step sequence
$\{x_k\} = \{1, a, a^2, a^3, a^4, \ldots\} = \{a^k\}$ is given by
$\dfrac{z}{z-a}$

The z-transform of $\{x_k\}$ is given by:

$z\{x_k\} = z\{a^k\} = \displaystyle\sum_{k=0}^{\infty} \frac{a^k}{z^k} = \displaystyle\sum_{k=0}^{\infty} \left(\frac{a}{z}\right)^k$

$= \left(\frac{a}{z}\right)^0 + \left(\frac{a}{z}\right)^1 + \left(\frac{a}{z}\right)^2 + \left(\frac{a}{z}\right)^3 + \left(\frac{a}{z}\right)^4 + \ldots$

$= 1 + \frac{a}{z} + \left(\frac{a}{z}\right)^2 + \left(\frac{a}{z}\right)^3 + \left(\frac{a}{z}\right)^4 + \ldots$

Comparing this with the series expansion of $\dfrac{1}{1-x} = 1 + x + x^2 + x^3 + \ldots$ which is valid for $|x| < 1$ i.e. equation (2) above, shows that:

$F(z) = 1 + \frac{a}{z} + \left(\frac{a}{z}\right)^2 + \left(\frac{a}{z}\right)^3 + \left(\frac{a}{z}\right)^4 + \ldots$

$= \dfrac{1}{1 - \dfrac{a}{z}}$ provided $\left|\dfrac{a}{z}\right| < 1$

Hence, $\dfrac{1}{1 - \dfrac{a}{z}} = \dfrac{1}{\dfrac{z-a}{z}} = \dfrac{z}{z-a}$ and $F(z) = \dfrac{z}{z-a}$

provided $|z| > |a|$

Problem 4. Determine the z-transform for the unit step sequence $\{x_k\} = \{0,\ 1,\ 2,\ 3,\ 4,\ \ldots\} = \{k\}$ and show that it is equivalent to $\dfrac{z}{(z-1)^2}$

The z-transform of $\{u_k\}$ is given by:

$$Z\{x_k\} = F(z) = \sum_{k=0}^{\infty} \frac{x_k}{z^k} = \sum_{k=0}^{\infty} \frac{k}{z^k}$$

$$= \frac{0}{z^0} + \frac{1}{z^1} + \frac{2}{z^2} + \frac{3}{z^3} + \frac{4}{z^4} + \ldots$$

$$= 0 + \frac{1}{z} + \frac{2}{z^2} + \frac{3}{z^3} + \frac{4}{z^4} + \ldots \qquad (3)$$

It was shown earlier, equation (2), that

$$\frac{1}{1-x} = (1-x)^{-1} = 1 + x + x^2 + x^3 + \ldots$$

Now $\dfrac{d}{dx}\left(1 + x + x^2 + x^3 + x^4 + \ldots\right)$

$$= 1 + 2x + 3x^2 + 4x^3 + \ldots \qquad (4)$$

and $\dfrac{d}{dx}\left[(1-x)^{-1}\right] = -(1-x)^{-2}(-1) = \dfrac{1}{(1-x)^2}$

using the function of a function rule,

i.e. $\qquad 1 + 2x + 3x^2 + 4x^3 + \ldots = \dfrac{1}{(1-x)^2}$

Comparing equations (3) and (4) shows that by multiplying $F(z)$ by z,

then $zF(z) = z\left(0 + \dfrac{1}{z} + \dfrac{2}{z^2} + \dfrac{3}{z^3} + \dfrac{4}{z^4} + \ldots\right)$

$$= 1 + \frac{2}{z} + \frac{3}{z^2} + \frac{4}{z^3} + \ldots = \frac{1}{\left(1 - \dfrac{1}{z}\right)^2}$$

Dividing both sides by z gives:

$$F(z) = \frac{1}{z\left(1 - \dfrac{1}{z}\right)^2} = \frac{1}{z\left(\dfrac{z-1}{z}\right)^2} = \frac{1}{z\dfrac{(z-1)^2}{z^2}}$$

$$= \frac{1}{\dfrac{(z-1)^2}{z}} = \frac{z}{(z-1)^2}$$

Hence, $Z\{x_k\} = F(z) = 0 + \dfrac{1}{z} + \dfrac{2}{z^2} + \dfrac{3}{z^3} + \dfrac{4}{z^4} + \ldots$

$$= \frac{z}{(z-1)^2}$$

From the results obtained in Problems 1 to 4, together with some additional results, a summary of some z-transforms is shown in Table 107.1, on page 1136, which may now be accepted and used.
Here are some further problems using Table 107.1.

Problem 5. Determine the z-transform of $5k^2$

From 4 in Table 107.1, $Z\{k^2\} = \dfrac{z(z+1)}{(z-1)^3}$

Hence, $\qquad Z\{5k^2\} = 5Z\{k^2\} = \dfrac{5z(z+1)}{(z-1)^3}$

Problem 6. Determine the z-transform of (a) 3^k (b) $(-3)^k$

(a) From 6 in Table 107.1, $Z\{a^k\} = \dfrac{z}{z-a}$

If $a = 3$, then $\qquad Z\{3^k\} = \dfrac{z}{z-3}$

(b) From 6 in Table 107.1, $Z\{a^k\} = \dfrac{z}{z-a}$

If $a = -3$, then $Z\{(-3)^k\} = \dfrac{z}{z--3} = \dfrac{z}{z+3}$

Problem 7. Determine the z-transform of $2e^{-3k}$

From 9 in Table 107.1, $Z\{e^{-ak}\} = \dfrac{z}{z-e^{-a}}$

Hence, $Z\{2e^{-3k}\} = 2Z\{e^{-3k}\} = \dfrac{2z}{z-e^{-3}}$

Problem 8. Determine $Z\{\cos 3k\}$

From 11 in Table 107.1, $Z\{\cos ak\} = \dfrac{z(z-\cos a)}{z^2 - 2z\cos a + 1}$

Hence, since $a = 3$, $Z\{\cos 3k\} = \dfrac{z(z-\cos 3)}{z^2 - 2z\cos 3 + 1}$

Problem 9. Determine the z-transform of $3\sin 2k$

From 10 in Table 107.1, $Z\{\sin ak\} = \dfrac{z\sin a}{z^2 - 2z\cos a + 1}$

Hence, since $a = 2$, $Z\{3\sin 2k\} = 3Z\{\sin 2k\}$

$$= 3\left(\frac{z\sin 2}{z^2 - 2z\cos 2 + 1}\right)$$

$$= \frac{3z\sin 2}{z^2 - 2z\cos 2 + 1}$$

Problem 10. Determine $Z\{e^{-2k}\cos 4k\}$

From 13 in Table 107.1,

$$Z\{e^{-ak}\cos bk\} = \frac{z^2 - ze^{-a}\cos b}{z^2 - 2ze^{-a}\cos b + e^{-2a}}$$

Table 107.1

Sequence	Transform $F(z)$	
1. $\{\delta_k\} = \{1,\ 0,\ 0,\ \ldots\}$	1	for all values of z
2. $\{u_k\} = \{1,\ 1,\ 1,\ \ldots\}$	$\dfrac{z}{z-1}$	for $\lvert z \rvert > 1$
3. $\{k\} = \{0,\ 1,\ 2,\ 3,\ \ldots\}$	$\dfrac{z}{(z-1)^2}$	for $\lvert z \rvert > 1$
4. $\left\{k^2\right\} = \{0,\ 1,\ 4,\ 9,\ \ldots\}$	$\dfrac{z(z+1)}{(z-1)^3}$	for $\lvert z \rvert > 1$
5. $\left\{k^3\right\} = \{0,\ 1,\ 8,\ 27,\ \ldots\}$	$\dfrac{z(z^2+4z+1)}{(z-1)^4}$	for $\lvert z \rvert > 1$
6. $\left\{a^k\right\} = \{1,\ a,\ a^2,\ a^3,\ \ldots\}$	$\dfrac{z}{z-a}$	for $\lvert z \rvert > \lvert a \rvert$
7. $\left\{ka^k\right\} = \{0,\ a,\ 2a^2,\ 3a^3,\ \ldots\}$	$\dfrac{az}{(z-a)^2}$	for $\lvert z \rvert > \lvert a \rvert$
8. $\left\{k^2a^k\right\} = \{0,\ a,\ 4a^2,\ 9a^3,\ \ldots\}$	$\dfrac{az(z+a)}{(z-a)^3}$	for $\lvert z \rvert > \lvert a \rvert$
9. $\left\{e^{-ak}\right\} = \left\{e^{-a},\ e^{-2a},\ e^{-3a},\ \ldots\right\}$	$\dfrac{z}{z-e^{-a}}$	
10. $\sin ak = \{\sin a,\ \sin 2a,\ \ldots\}$	$\dfrac{z\sin a}{z^2-2z\cos a+1}$	
11. $\cos ak = \{\cos a,\ \cos 2a,\ \ldots\}$	$\dfrac{z(z-\cos a)}{z^2-2z\cos a+1}$	
12. $e^{-ak}\sin bk = \left\{e^{-a}\sin b,\ e^{-2a}\sin 2b,\ \ldots\right\}$	$\dfrac{ze^{-a}\sin b}{z^2-2ze^{-a}\cos b+e^{-2a}}$	
13. $e^{-ak}\cos bk = \left\{e^{-a}\cos b,\ e^{-2a}\cos 2b,\ \ldots\right\}$	$\dfrac{z^2-ze^{-a}\cos b}{z^2-2ze^{-a}\cos b+e^{-2a}}$	

Hence, since $a = 2$ and $b = 4$,

$$Z\left\{e^{-2k}\cos 4k\right\} = \frac{z^2 - ze^{-2}\cos 4}{z^2 - 2ze^{-2}\cos 4 + e^{-4}}$$

Now try the following Practice Exercise

Practice Exercise 378   Determining z-transforms (answers on page 1198)

Use Table 107.1 to find the z-transforms of the following:

1. $2k$	2. $3k^2$	3. $4k^3$
4. $\sin 3k$	5. 2^k	6. $k(2)^k$
7. $5\cos 2k$	8. $3e^{-2k}$	9. $\sin\frac{1}{2}k$
10. $4e^{3k}\cos 2k$	11. $(-4)^k$	12. $3k^2(2)^k$
13. $3(-5)^k$	14. $k(-3)^k$	15. $3e^{5k}$
16. $2e^{-4k}\sin 2k$		

107.2 Some properties of z-transforms

(a) Linearity property

The z-transform is a **linear transform**,

i.e. $\mathbf{Z}(a\{x_k\} + b\{y_k\}) = a\mathbf{Z}\{x_k\} + b\mathbf{Z}\{y_k\}$ (4)

where a and b are constants

Problem 11. Determine the z-transform of

$$2\{k\} - 3\left\{e^{-2k}\right\}$$

Now, $2Z\{k\} = 2\left\{\dfrac{z}{(z-1)^2}\right\}$ from 4 in Table 107.1

and since $Z\{e^{-ak}\} = \dfrac{z}{z - e^{-a}}$ from 9 in Table 107.1,

then $Z\{e^{-2k}\} = \dfrac{z}{z - e^{-2}}$

Hence, $Z\left(2\{k\} - 3\left\{e^{-2k}\right\}\right) = 2Z\{k\} - 3Z\left\{e^{-2k}\right\}$

$$= 2\left\{\frac{z}{(z-1)^2}\right\} - 3\left\{\frac{z}{z - e^{-2}}\right\} \quad \text{from equation (5)}$$

$$= \left\{\frac{2z}{(z-1)^2}\right\} - \left\{\frac{3z}{z - e^{-2}}\right\}$$

$$= \frac{2z\left(z - e^{-2}\right) - 3z\left(z-1\right)^2}{(z-1)^2\left(z - e^{-2}\right)}$$

$$= \frac{2z^2 - 2ze^{-2} - 3z\left(z^2 - 2z + 1\right)}{(z-1)^2\left(z - e^{-2}\right)}$$

$$= \frac{2z^2 - 2ze^{-2} - 3z^3 + 6z^2 - 3z}{(z-1)^2\left(z - e^{-2}\right)}$$

i.e. $2Z\{k\} - 3Z\left\{e^{-2k}\right\}$

$$= \frac{-3z^3 + 8z^2 - z\left(2e^{-2} + 3\right)}{(z-1)^2\left(z - e^{-2}\right)}$$

Now try the following Practice Exercise

Practice Exercise 379   Linearity property of z-transforms (answers on page 1198)

Use Table 107.1 to find the z-transforms of the following:

1. $5\{k\} - 4\left\{e^{-3k}\right\}$

2. $4\{k^2\} + 3\{k\}$

3. $3\{\sin 2k - \cos 2k\}$

4. $2\left\{(3)^k\right\} + 4\left\{(-3)^k\right\}$

5. $\left\{k + e^{-k}\right\}$

(b) First shift theorem (shifting to the left)

It may be shown by the **first shift theorem (shifting to the left)**, that

if $\quad Z\{x_k\} = F(z)$

then $Z\{x_{k+m}\}$

$$= z^m F(z) - \left[z^m x_0 + z^{m-1}x_1 + \ldots + zx_{m-1}\right]$$

(6)

is the z-transform of the sequence that has been shifted by m places to the left.

This theorem is often needed when solving difference equations (see Section 107.4).

Problem 12. Determine $Z\left\{3^{k+2}\right\}$

Since from equation (6),

$Z\{x_{k+m}\} = z^m F(z) - \left[z^m x_0 + z^{m-1}x_1 + \ldots + zx_{m-1}\right]$

then $\quad Z\left\{3^{k+2}\right\} = z^2 Z\left\{3^k\right\} - \left[z^2 3^0 + z3^1\right]$ (7)

From 6 of Table 107.1, $Z\left\{a^k\right\} = \dfrac{z}{z - a}$ thus

$$Z\left\{3^k\right\} = \frac{z}{z - 3}$$

Hence, substituting in equation (7),

$$Z\left\{3^{k+2}\right\} = z^2\left(\frac{z}{z-3}\right) - \left[z^2 + 3z\right]$$

$$= \frac{z^3}{z-3} - \left[z^2 + 3z\right]$$

$$= \frac{z^3 - (z-3)\left[z^2 + 3z\right]}{z - 3}$$

$$= \frac{z^3 - \left[z^3 + 3z^2 - 3z^2 - 9z\right]}{z - 3}$$

$$= \frac{z^3 - \left[z^3 - 9z\right]}{z - 3}$$

i.e. $\quad Z\left\{3^{k+2}\right\} = \dfrac{9z}{z - 3}$

This is the z-transform of the sequence $\{9, 27, 81, \ldots\}$ by shifting the sequence $\{1, 3, 9, 27, \ldots\}$ two places to the left and losing the first two terms.

Section 0

Problem 13. Determine $Z\left\{5^{k+3}\right\}$

Since from equation (6),

$$Z\left\{x_{k+m}\right\} = z^m F(z) - \left[z^m x_0 + z^{m-1} x_1 + \ldots + z x_{m-1}\right]$$

then $\quad Z\left\{5^{k+3}\right\} = z^3 Z\left\{5^k\right\} - \left[z^3 5^0 + z^2 5^1 + z 5^2\right]$

$$\tag{8}$$

From 6 of Table 107.1, $Z\left\{a^k\right\} = \dfrac{z}{z-a}$

thus $\quad Z\left\{5^k\right\} = \dfrac{z}{z-5}$

Hence, substituting in equation (8),

$$Z\left\{5^{k+3}\right\} = z^3\left(\frac{z}{z-5}\right) - \left[z^3 + 5z^2 + 25z\right]$$

$$= \frac{z^4}{z-5} - \left[z^3 + 5z^2 + 25z\right]$$

$$= \frac{z^4 - (z-5)\left[z^3 + 5z^2 + 25z\right]}{z-5}$$

$$= \frac{z^4 - \left[z^4 + 5z^3 + 25z^2 - 5z^3 - 25z^2 - 125z\right]}{z-5}$$

$$= \frac{z^4 - \left[z^4 - 125z\right]}{z-5}$$

i.e. $\quad \boldsymbol{Z\left\{5^{k+3}\right\} = \dfrac{125z}{z-5}}$

This is the z-transform of the sequence $\{125, 625, 3125, \ldots\}$ by shifting the sequence $\{1, 5, 25, 125, \ldots\}$ three places to the left and losing the first three terms.

Problem 14. Determine $Z\{k+1\}$

From 3 of Table 107.1, $Z\{k\} = \dfrac{z}{(z-1)^2}$

Since from equation (6),

$$Z\{x_{k+m}\} = z^m F(z) - \left[z^m x_0 + z^{m-1} x_1 + \ldots + z x_{m-1}\right]$$

then $Z\{k+1\} = z^1 Z(k) - \left[z^1 x_0\right]$

$$= z^1\left(\frac{z}{(z-1)^2}\right) - \left[z^1 \times 0\right] \text{ from 3 of Table 107.1}$$

i.e. $\quad \boldsymbol{Z\{k+1\} = \dfrac{z^2}{(z-1)^2}}$

Now try the following Practice Exercise

Practice Exercise 380   First shift theorem of z-transforms (answers on page 1198)

Use Table 107.1 to find the z-transforms of the following:

1. $\left\{2^{k+3}\right\}$ 2. $\left\{4^{k+1}\right\}$

3. $6\{k+1\}$ 4. $\left\{3^{k+2} - 2^{k+1}\right\}$

(c) Second shift theorem (shifting to the right)

It may be shown by the **second shift theorem (shifting to the right)**, that

if $Z\{x_k\} = F(z)$

then $\quad \boldsymbol{Z\{x_{k-m}\} = z^{-m} F(z)}$ $\tag{9}$

is the z-transform of the sequence that has been shifted by m places to the right.

Problem 15. Determine $Z\{x_{k-2}\}$

Since from equation (9), $Z\{x_{k-m}\} = z^{-m} F(z)$

then $\quad Z\{x_{k-2}\} = z^{-2} F(z)$

$$= z^{-2}\left(\frac{z}{z-1}\right)$$

since $Z\{x_k\} = \dfrac{z}{z-1}$ from 2 of Table 107.1

i.e. $\quad \boldsymbol{Z\{x_{k-2}\} = \left(\dfrac{z^{-1}}{z-1}\right) = \dfrac{1}{z(z-1)}}$

This is the z-transform of the sequence $\{0, 0, 1, 1, 1, \ldots\}$ by shifting the sequence $\{1, 1, 1, 1, \ldots\}$ two places to the right and defining the first two terms as zeros.

Problem 16. Determine $Z\left\{a^{k-1}\right\}$

Since from equation (9), $Z\{x_{k-m}\} = z^{-m} F(z)$

then $Z\left\{a^{k-1}\right\} = z^{-1} F(z) = z^{-1}\left(\dfrac{z}{z-a}\right)$

since $Z\left\{a^k\right\} = \dfrac{z}{z-a}$ from 6 of Table 107.1

i.e. $\quad \boldsymbol{Z\left\{a^{k-1}\right\} = \left(\dfrac{z^{-1} \times z}{z-a}\right) = \dfrac{1}{(z-a)}}$

which is the z-transform of $\left\{a^k\right\}$ shifted one place to the right.

Now try the following Practice Exercise

Practice Exercise 381   Second shift theorem of z-transforms (answers on page 1198)

Use Table 107.1 to find the z-transforms of the following:

1. $\{x_{k-1}\}$

2. $\{x_{k-3}\}$

3. $\{a^{k-2}\}$

4. $\{a^{k-3}\}$

5. $\{3^{k-4}\}$

(d) Translation

If the sequence $\{x_k\}$ has the z-transform

$$Z\left\{a^k x_k\right\} = F(z),$$

then the sequence $\left\{a^k x_k\right\}$ has the z-transform

$$Z\left\{a^k x_k\right\} = F\left(a^{-1} z\right)$$

Problem 17. Determine $Z\left\{3^k k\right\}$

Since $Z(k) = \dfrac{z}{(z-1)^2}$ from 3 of Table 107.1

then by the translation property,

$$Z\left\{3^k k\right\} = F\left(3^{-1} z\right) = \frac{3^{-1} z}{(3^{-1} z - 1)^2}$$

$$= \frac{3^{-1} z}{\left[3^{-1}\left(z - \frac{1}{3^{-1}}\right)\right]^2}$$

$$= \frac{3^{-1} z}{3^{-2}(z - 3)^2} = \frac{3^2 z}{3^1 (z - 3)^2}$$

i.e. $\qquad Z\left\{3^k k\right\} = \dfrac{3z}{(z-3)^2}$

(e) Final value theorem

For the sequence $\{x_k\}$ with the z-transform $F(z)$,

$$\underset{k \to \infty}{Lim}\ x_k = \underset{z \to 1}{Lim}\left\{\left(\frac{z-1}{z}\right) F(z)\right\}\ \text{provided that } \underset{k \to \infty}{Lim}\ x_k$$
$$\text{exists.}$$

Problem 18. Determine $\underset{k \to \infty}{Lim}\left\{\left(\frac{1}{3}\right)^k\right\}$

$$F\left\{x^k\right\} = \frac{z}{z-1}\ \text{and}\ F\left\{\left(\frac{1}{3}\right)^k\right\} = \frac{z}{z - \frac{1}{3}} = \frac{3z}{3z - 1}$$

Now $\underset{k \to \infty}{Lim}\left\{\left(\frac{1}{3}\right)^k\right\} = \underset{z \to 1}{Lim}\left\{\left(\frac{z-1}{z}\right) F(z)\right\}$

$$= \underset{z \to 1}{Lim}\left\{\left(\frac{z-1}{z}\right)\frac{3z}{3z-1}\right\}$$

$$= \underset{z \to 1}{Lim}\left\{\frac{3(z-1)}{3z-1}\right\} = 0$$

i.e. $\qquad \underset{k \to \infty}{Lim}\left\{\left(\frac{1}{3}\right)^k\right\} = \mathbf{0}$

Problem 19. Determine $\underset{k \to \infty}{Lim}\left\{\dfrac{33z^2 - 25z}{(z-1)(3z-1)^2}\right\}$

By the final value theorem,

$$\underset{k \to \infty}{Lim}\left\{\frac{33z^2 - 25z}{(z-1)(3z-1)^2}\right\} = \underset{z \to 1}{Lim}\left\{\left(\frac{z-1}{z}\right) F(z)\right\}$$

$$= \underset{z \to 1}{Lim}\left\{\left(\frac{z-1}{z}\right)\frac{33z^2 - 25z}{(z-1)(3z-1)^2}\right\}$$

$$= \underset{z \to 1}{Lim}\left\{\frac{33z - 25}{(3z-1)^2}\right\}$$

$$= \frac{33 - 25}{2^2} = \frac{8}{4} = \mathbf{2}$$

(f) The initial value theorem

For the sequence $\{x_k\}$ with the z-transform $F(z)$,

initial value, $x_0 = \underset{z \to \infty}{Lim}\ \{F(z)\}$

Problem 20. Determine $\underset{z \to \infty}{Lim}\ \left\{a^k\right\}$

$F(z) = F\left\{a^k\right\} = \dfrac{z}{z-a}$ from 6 of Table 107.1

and $\quad \underset{z \to \infty}{Lim}\ \{F(z)\} = \underset{z \to \infty}{Lim}\ \left\{\dfrac{z}{z-a}\right\}$

$$= \underset{z \to \infty}{Lim}\left\{\frac{\frac{d}{dz}(z)\,dz}{\frac{d}{dz}(z-a)\,dz}\right\} = \underset{z \to \infty}{Lim}\left\{\frac{1}{1}\right\}$$

by L'Hôpital's rule (see Chapter 22)

i.e. $\qquad \underset{z \to \infty}{Lim}\ \left\{a^k\right\} = \mathbf{1}$

(g) The derivative of the transform

If $\qquad Z\{x_k\} = F(z)$

then $\qquad -z\,F'(z) = Z\{k x_k\}$

Problem 21. Determine the derivative of $Z\{ka^k\}$

$F(z) = F\{a^k\} = \dfrac{z}{z-a}$ from 6 of Table 107.1

Now $Z\{kx_k\} = -zF'(z)$ from above

i.e. $\quad Z\{kx_k\} = -zF'\left(\dfrac{z}{z-a}\right)$

$= -z\left[\dfrac{(z-a)(1) - z(1)}{(z-a)^2}\right]$ using the quotient rule

$= -z\left[\dfrac{z-a-z}{(z-a)^2}\right] = -z\left[\dfrac{-a}{(z-a)^2}\right] = \dfrac{az}{(z-a)^2}$

i.e. **the derivative of $Z\{ka^k\} = -zF'(z) = \dfrac{az}{(z-a)^2}$**

which confirms result 7 in Table 107.1.

Now try the following Practice Exercise

Practice Exercise 382   Properties of z-transforms (answers on page 1199)

1. Determine $Z\{2^k k\}$

2. Determine $Z\{4^k k\}$

3. Determine $\underset{k\to\infty}{Lim}\left\{\left(\tfrac{1}{2}\right)^k\right\}$

4. Determine $\underset{k\to\infty}{Lim}\left\{\dfrac{15z^2 + 5z}{(z-1)(4z-1)^2}\right\}$

5. Determine $\underset{z\to\infty}{Lim}\{3^k\}$

6. Determine the derivative of $Z\{k(2)^k\}$

107.3   Inverse z-transforms

If the sequence $\{x_k\}$ has a Z transform $Z\{x_k\} = F(z)$, then the **inverse z-transform** is defined as:

$$Z^{-1}F(z) = \{x_k\}$$

In the following problems, some inverses may be determined directly from Table 107.1, albeit with a little manipulation; others sometimes require the use of partial fractions – just as with inverse Laplace transforms.

Problem 22. Determine the inverse z-transform of $F(z) = \dfrac{z}{z+5}$

From 6 in Table 107.1, $Z\{a^k\} = \dfrac{z}{z-a}$

hence $\qquad Z^{-1}\left\{\dfrac{z}{z-a}\right\} = a^k$

Comparing $\dfrac{z}{z+5}$ with $\dfrac{z}{z-a}$ shows that $a = -5$

Thus, $\quad \mathbf{Z^{-1}\left\{\dfrac{z}{z+5}\right\} = (-5)^k}$

Problem 23. Determine the inverse z-transform of

$$F(z) = \dfrac{2z}{2z+1}$$

$$\dfrac{2z}{2z+1} = \dfrac{2z}{2\left(z+\tfrac{1}{2}\right)} = \dfrac{z}{z+\tfrac{1}{2}}$$

From 6 in Table 107.1, $Z\{a^k\} = \dfrac{z}{z-a}$

hence $\qquad Z^{-1}\left\{\dfrac{z}{z-a}\right\} = a^k$

Comparing $\dfrac{z}{z+\tfrac{1}{2}}$ with $\dfrac{z}{z-a}$ shows that $a = -\dfrac{1}{2}$

Thus, $\quad \mathbf{Z^{-1}\left\{\dfrac{2z}{2z+1}\right\} = Z^{-1}\left\{\dfrac{z}{z+\tfrac{1}{2}}\right\} = \left(-\dfrac{1}{2}\right)^k}$

Problem 24. If $F(z) = \dfrac{3z}{3z-1}$ determine $Z^{-1}F(z)$

$$\dfrac{3z}{3z-1} = \dfrac{3z}{3\left(z-\tfrac{1}{3}\right)} = \dfrac{z}{z-\tfrac{1}{3}}$$

From 6 in Table 107.1, $Z\{a^k\} = \dfrac{z}{z-a}$

hence $\qquad Z^{-1}\left\{\dfrac{z}{z-a}\right\} = a^k$

Comparing $\dfrac{z}{z-\tfrac{1}{3}}$ with $\dfrac{z}{z-a}$ shows that $a = \dfrac{1}{3}$

Thus, $\quad \mathbf{Z^{-1}\left\{\dfrac{3z}{3z-1}\right\} = Z^{-1}\left\{\dfrac{z}{z-\tfrac{1}{3}}\right\} = \left(\dfrac{1}{3}\right)^k}$

Problem 25. Determine the inverse z-transform of $F(z) = \dfrac{z}{z - e^2}$

From 9 in Table 107.1, $Z\{e^{-ak}\} = \dfrac{z}{z - e^{-a}}$

hence $\qquad Z^{-1}\left\{\dfrac{z}{z - e^{-a}}\right\} = e^{-ak}$

Comparing $\dfrac{z}{z-e^2}$ with $\dfrac{z}{z-e^{-a}}$ shows that $a = -2$

Thus, $\mathbf{Z^{-1}\left\{\dfrac{z}{z-e^2}\right\} = e^{--2k} = e^{2k}}$

Problem 26.   If $F(z) = \dfrac{z}{z^2+1}$ determine $Z^{-1}F(z)$

In Table 107.1, results 10 and 11 have z^2 in their denominators.

If the numerator is to be z in result 10, then $\sin a$ has to equal 1, i.e. $a = \frac{\pi}{2}$

From 10 in Table 107.1, $Z\{\sin ak\} = \dfrac{z\sin a}{z^2 - 2z\cos a + 1}$

hence $Z^{-1}\left\{\dfrac{z\sin a}{z^2 - 2z\cos a + 1}\right\} = \sin ak$

When $a = \dfrac{\pi}{2}, \dfrac{z\sin a}{z^2 - 2z\cos a + 1} = \dfrac{z\sin\frac{\pi}{2}}{z^2 - 2z\cos\frac{\pi}{2} + 1}$

$$= \dfrac{z}{z^2 + 1}$$

Thus, $\mathbf{Z^{-1}\left\{\dfrac{z}{z^2+1}\right\} = \sin\dfrac{\pi}{2}k}$

Problem 27.   Determine the inverse z-transform of $F(z) = \dfrac{z}{z^2 - 7z + 12}$

Using partial fractions, let

$$\dfrac{z}{z^2 - 7z + 12} = \dfrac{z}{(z-4)(z-3)} = \dfrac{A}{(z-4)} + \dfrac{B}{(z-3)}$$

$$= \dfrac{A(z-3) + B(z-4)}{(z-4)(z-3)}$$

from which, $z = A(z-3) + B(z-4)$

Letting $z = 4$ gives: $4 = A$

Letting $z = 3$ gives: $3 = -B$ i.e. $B = -3$

Hence, $F(z) = \dfrac{z}{z^2 - 7z + 12} = \dfrac{4}{(z-4)} - \dfrac{3}{(z-3)}$

The nearest transform in Table 107.1 to either of these partial fractions is $Z\{a^k\} = \dfrac{z}{z-a}$

Rearranging gives: $F(z) = \dfrac{4}{(z-4)} - \dfrac{3}{(z-3)}$

$$= \dfrac{4}{z} \times \dfrac{z}{(z-4)} - \dfrac{3}{z} \times \dfrac{z}{(z-3)}$$

$$= 4 \times z^{-1}Z\left\{4^k\right\} - 3 \times z^{-1}Z\left\{3^k\right\}$$

Hence, $Z^{-1}F(z) = 4 \times \left\{4^{k-1}\right\} - 3 \times \left\{3^{k-1}\right\}$ by the second shift theorem

$$= \left\{4^k\right\} - \left\{3^k\right\} = \left\{4^k - 3^k\right\}$$

i.e.   **the sequence is $x_k = 4^k - 3^k$**

With the denominator of $F(z) = \dfrac{z}{z^2 - 7z + 12}$ being z, there is an alternative, and more straightforward method of determining the inverse transform,

i.e. by initially rearranging as: $\dfrac{F(z)}{z} = \dfrac{1}{z^2 - 7z + 12}$

Using partial fractions,

$$\dfrac{1}{z^2 - 7z + 12} = \dfrac{1}{(z-4)(z-3)}$$

$$= \dfrac{A}{(z-4)} + \dfrac{B}{(z-3)} = \dfrac{A(z-3) + B(z-4)}{(z-4)(z-3)}$$

from which, $1 = A(z-3) + B(z-4)$

Letting $z = 4$ gives: $1 = A$

Letting $z = 3$ gives: $1 = -B$ i.e. $B = -1$

Hence, $\dfrac{F(z)}{z} = \dfrac{1}{z^2 - 7z + 12} = \dfrac{1}{(z-4)} - \dfrac{1}{(z-3)}$

and $F(z) = \dfrac{z}{(z-4)} - \dfrac{z}{(z-3)}$

and $z^{-1}F(z) = \left\{4^k\right\} - \left\{3^k\right\} = \left\{4^k - 3^k\right\}$ from 6 in Table 107.1

Problem 28.   Determine the inverse z-transform of $F(z) = \dfrac{z}{z^2 - 3z + 2}$

Since $F(z) = \dfrac{z}{z^2 - 3z + 2}$ then $\dfrac{F(z)}{z} = \dfrac{1}{z^2 - 3z + 2}$

Using partial fractions, let

$$\dfrac{1}{z^2 - 3z + 2} = \dfrac{1}{(z-1)(z-2)} = \dfrac{A}{(z-1)} + \dfrac{B}{(z-2)}$$

$$= \dfrac{A(z-2) + B(z-1)}{(z-1)(z-2)}$$

from which, $1 = A(z-2) + B(z-1)$

Letting $z = 1$ gives: $1 = -A$ i.e. $A = -1$

Letting $z = 2$ gives: $1 = B$

Hence,

$$\dfrac{F(z)}{z} = \dfrac{-1}{(z-1)} + \dfrac{1}{(z-2)} = \dfrac{1}{(z-2)} - \dfrac{1}{(z-1)}$$

and $\quad F(z) = \dfrac{z}{(z-2)} - \dfrac{z}{(z-1)}$

Thus, $\quad Z^{-1}F(z) = Z^{-1}\left\{\dfrac{z}{(z-2)} - \dfrac{z}{(z-1)}\right\}$

$$= Z^{-1}\left\{\dfrac{z}{(z-2)}\right\} - Z^{-1}\left\{\dfrac{z}{(z-1)}\right\}$$

From 6 in Table 107.1, $\mathbf{Z^{-1}F(z) = (2)^k - (1)^k = (2)^k - 1}$

> **Problem 29.** Determine the sequence $\{x_k\}$ which has the z-transform $F(z) = \dfrac{4z}{(z+3)\left(z^2 - 4z + 4\right)}$

$\dfrac{F(z)}{z} = \dfrac{1}{z} \times \dfrac{4z}{(z+3)\left(z^2-4z+4\right)} = \dfrac{4}{(z+3)\left(z^2-4z+4\right)}$

Using partial fractions,

$$\dfrac{4}{(z+3)\left(z^2-4z+4\right)} = \dfrac{4}{(z+3)(z-2)^2}$$

$$= \dfrac{A}{(z+3)} + \dfrac{B}{(z-2)} + \dfrac{C}{(z-2)^2}$$

$$= \dfrac{A(z-2)^2 + B(z+3)(z-2) + C(z+3)}{(z+3)(z-2)^2}$$

from which, $4 = A(z-2)^2 + B(z+3)(z-2) + C(z+3)$

Letting $z = -3$ gives: $4 = 25A$ i.e. $A = 4/25$

Letting $z = 2$ gives: $4 = 5C$ i.e. $C = 4/5$

Equating z^2 coefficients gives: $0 = A + B$ from which,
$$B = -4/25$$

Hence, $\quad \dfrac{F(z)}{z} = \dfrac{4}{(z+3)(z-2)^2}$

$$= \dfrac{4/25}{(z+3)} - \dfrac{4/25}{(z-2)} + \dfrac{4/5}{(z-2)^2}$$

and $\quad F(z) = \dfrac{(4/25)z}{(z+3)} - \dfrac{(4/25)z}{(z-2)} + \dfrac{(4/5)z}{(z-2)^2}$

$$= \dfrac{4}{25} \times \dfrac{z}{(z+3)} - \dfrac{4}{25} \times \dfrac{z}{(z-2)} + \dfrac{4}{5} \times \dfrac{z}{(z-2)^2}$$

$$= \dfrac{4}{25} \times \dfrac{z}{(z+3)} - \dfrac{4}{25} \times \dfrac{z}{(z-2)} + \dfrac{2}{5} \times \dfrac{2z}{(z-2)^2}$$

i.e. $Z^{-1}F(z) = \dfrac{4}{25}\left\{(-3)^k\right\} - \dfrac{4}{25}\left\{(2)^k\right\} + \dfrac{2}{5}\left\{k(2)^k\right\}$

from 6 and 7 of Table 107.1

i.e. $\quad \{\mathbf{x_k}\} = \dfrac{4}{25}\left\{\mathbf{(-3)^k}\right\} - \dfrac{4}{25}\left\{\mathbf{(2)^k}\right\} + \dfrac{2}{5}\left\{\mathbf{k(2)^k}\right\}$

Now try the following Practice Exercise

Practice Exercise 383   Inverse z-transforms (answers on page 1199)

In Problems 1 to 12, use Table 107.1 to find the sequences which have the following z-transforms

1. $\dfrac{z}{z-1}$

2. $\dfrac{z}{z-2}$

3. $\dfrac{z}{z+1}$

4. $\dfrac{z}{z+4}$

5. $\dfrac{z}{z-\dfrac{1}{3}}$

6. $\dfrac{4z}{4z+1}$

7. $\dfrac{5z}{5z-1}$

8. $\dfrac{z}{z-e^{-5}}$

9. $\dfrac{z}{z-e^3}$

10. $\dfrac{3z}{(z-1)^2}$

11. $\dfrac{5z(z+1)}{(z-1)^3}$

12. $\dfrac{z}{2(z-2)^2}$

In Problems 13 to 20, use partial fractions to determine the inverse z-transforms of the following.

13. $\dfrac{z}{(z+1)(z-2)}$

14. $\dfrac{3z}{(z-1)(z+3)}$

15. $\dfrac{z}{(z-1)(z+2)}$

16. $\dfrac{z}{(z-3)(2z+1)}$

17. $\dfrac{z}{2-3z+z^2}$

18. $\dfrac{z(3z+1)}{z^2-5z+6}$

19. $\dfrac{2z}{2z^2+z-1}$

20. $\dfrac{3z}{z^2+3z-10}$

107.4 Using z-transforms to solve difference equations

In Chapter 99, Laplace transforms were used to solve differential equations; in this section, the solution of difference equations using z-transforms is demonstrated. Difference equations arise in a number of different ways – sometimes from the direct modelling of systems in discrete time, or as an approximation to a differential equation describing the behaviour of a system modelled as a continuous-time system. The z-transform method is based on the first shift theorem (see earlier, page 1137), and the method of solution is explained through the following problems.

Problem 30. Solve the difference equation $x_{k+1} - 2x_k = 0$ given the initial condition that $x_0 = 3$

Taking the z-transform of each term gives:

$$Z\{x_{k+1}\} - 2Z\{x_k\} = Z\{0\}$$

Since from equation (6),

$$Z\{x_{k+m}\} = z^m F(z) - \left[z^m x_0 + z^{m-1} x_1 + \ldots + zx_{m-1}\right]$$

then $\left(z^1 Z\{k\} - \left[z^1(3)\right]\right) - 2Z\{x_k\} = 0$

i.e. $zZ\{x_k\} - 3z - 2Z\{x_k\} = 0$

i.e. $(z-2)Z\{x_k\} = 3z$

and $Z\{x_k\} = \dfrac{3z}{z-2}$

Taking the inverse z-transform gives:

$$\{x_k\} = Z^{-1}\left(\frac{3z}{z-2}\right) = 3Z^{-1}\left(\frac{z}{z-2}\right)$$

i.e. $\{x_k\} = 3(2^k)$ from 6 of Table 107.1

Problem 31. Solve the difference equation: $x_{k+2} - 3x_{k+1} + 2x_k = 1$ given that $x_0 = 0$ and $x_1 = 2$

Taking the z-transform of each term gives:

$$Z\{x_{k+2}\} - 3Z\{x_{k+1}\} + 2Z\{x_k\} = Z\{1\}$$

Since from equation (6),

$$Z\{x_{k+m}\} = z^m F(z) - \left[z^m x_0 + z^{m-1} x_1 + \ldots + zx_{m-1}\right]$$

$$\left(z^2 Z\{x_k\} - \left[z^2(0) + z^1(2)\right]\right)$$
$$-3\left(z^1 Z\{x_k\} - \left[z^1(0)\right]\right) + 2Z\{x_k\} = \frac{z}{z-1}$$

i.e. $z^2 Z\{x_k\} - 2z - 3zZ\{x_k\} + 2Z\{x_k\} = \dfrac{z}{z-1}$

and $(z^2 - 3z + 2)Z\{x_k\} = \dfrac{z}{z-1} + 2z$

$$= \frac{z + 2z(z-1)}{z-1} = \frac{2z^2 - z}{z-1} = \frac{z(2z-1)}{z-1}$$

from which, $Z\{x_k\} = \dfrac{z(2z-1)}{(z-1)(z^2 - 3z + 2)}$

$$= \frac{z(2z-1)}{(z-1)(z-2)(z-1)}$$

or $\dfrac{Z\{x_k\}}{z} = \dfrac{(2z-1)}{(z-1)(z-2)(z-1)}$

$$= \frac{(2z-1)}{(z-1)^2(z-2)}$$

Using partial fractions, let

$$\frac{(2z-1)}{(z-1)^2(z-2)} = \frac{A}{(z-1)} + \frac{B}{(z-1)^2} + \frac{C}{(z-2)}$$

$$= \frac{A(z-1)(z-2) + B(z-2) + C(z-1)^2}{(z-1)^2(z-2)}$$

and $2z - 1 = A(z-1)(z-2) + B(z-2) + C(z-1)^2$

Letting $z = 1$ gives: $1 = -B$ i.e. $B = -1$

Letting $z = 2$ gives: $3 = C$

Equating z^2 coefficients gives: $0 = A + C$ i.e. $A = -3$

Hence, $\dfrac{Z\{x_k\}}{z} = \dfrac{(2z-1)}{(z-1)^2(z-2)} = \dfrac{-3}{(z-1)} + \dfrac{-1}{(z-1)^2}$

$$+ \frac{3}{(z-2)}$$

Therefore, $Z\{x_k\} = 3\left(\dfrac{z}{(z-2)}\right) - 3\left(\dfrac{z}{(z-1)}\right)$

$$- \frac{z}{(z-1)^2}$$

Taking the inverse z-transform gives:

$$\{x_k\} = 3Z^{-1}\left(\frac{z}{(z-2)}\right) - 3Z^{-1}\left(\frac{z}{(z-1)}\right)$$

$$- Z^{-1}\left(\frac{z}{(z-1)^2}\right)$$

$$= 3(2)^k - 3(1)^k - k \text{ from 6 and 3 of Table 107.1}$$

i.e. $\{x_k\} = 3(2^k) - 3 - k$

Problem 32. Solve the difference equation: $x_{k+2} - x_k = 1$ given that $x_0 = 0$ and $x_1 = -1$

Taking the z-transform of each term gives:

$$Z\{x_{k+2}\} - Z\{x_k\} = Z\{1\}$$

Since from equation (6),

$$Z\{x_{k+m}\} = z^m F(z) - \left[z^m x_0 + z^{m-1} x_1 + \ldots + zx_{m-1}\right]$$

$$z^2 Z\{x_k\} - \left[z^2(0) + z^1(-1)\right] - Z\{x_k\} = \frac{z}{z-1}$$

i.e. $z^2 Z\{x_k\} + z - Z\{x_k\} = \dfrac{z}{z-1}$

and $\left(z^2 - 1\right) Z\{x_k\} = \dfrac{z}{z-1} - z = \dfrac{z - z(z-1)}{z-1}$

$$= \dfrac{2z - z^2}{z-1}$$

from which, $Z\{x_k\} = \dfrac{2z - z^2}{(z-1)\left(z^2 - 1\right)}$

$$= \dfrac{2z - z^2}{(z-1)(z-1)(z+1)} = \dfrac{2z - z^2}{(z-1)^2(z+1)}$$

and $\dfrac{Z\{x_k\}}{z} = \dfrac{2-z}{(z-1)^2(z+1)}$

Using partial fractions, let

$$\dfrac{2-z}{(z-1)^2(z+1)} = \dfrac{A}{(z-1)} + \dfrac{B}{(z-1)^2} + \dfrac{C}{(z+1)}$$

$$= \dfrac{A(z-1)(z+1) + B(z+1) + C(z-1)^2}{(z-1)^2(z+1)}$$

and $2 - z = A(z-1)(z+1) + B(z+1) + C(z-1)^2$

Letting $z = 1$ gives: $1 = 2B$ i.e. $B = 1/2$

Letting $z = -1$ gives: $3 = 4C$ i.e. $C = 3/4$

Equating z^2 coefficients gives: $0 = A + C$
i.e. $\qquad\qquad\qquad\qquad A = -3/4$

Hence, $\dfrac{Z\{x_k\}}{z} = \dfrac{2-z}{(z-1)^2(z+1)}$

$$= \dfrac{-3/4}{(z-1)} + \dfrac{1/2}{(z-1)^2} + \dfrac{3/4}{(z+1)}$$

Therefore, $Z\{x_k\} = -\dfrac{3}{4}\left(\dfrac{z}{z-1}\right) + \dfrac{1}{2}\left(\dfrac{z}{(z-1)^2}\right)$

$$-\dfrac{3}{4}\left(\dfrac{z}{z+1}\right)$$

Taking the inverse z-transform gives:

$$\{x_k\} = -\dfrac{3}{4}Z^{-1}\left(\dfrac{z}{(z-2)}\right) + \dfrac{1}{2}Z^{-1}\left(\dfrac{z}{(z-1)^2}\right)$$

$$-\dfrac{3}{4}Z^{-1}\left(\dfrac{z}{z+1}\right)$$

i.e. $\{x_k\} = -\dfrac{3}{4}\left(2^k\right) + \dfrac{1}{2}k - \dfrac{3}{4}(-1)^k$ from 6 and 3 of

Table 107.1

Problem 33. Solve the difference equation:
$x_{k+2} - 3x_{k+1} + 2x_k = 1$ given that $x_0 = 0$ and $x_1 = 1$

Taking the z-transform of both sides of the equation
gives:

$$Z\{x_{k+2} - 3x_{k+1} + 2x_k\} = Z\{1\}$$

i.e. $\qquad Z\{x_{k+2}\} - 3Z\{x_{k+1}\} + 2Z\{x_k\} = Z\{1\}$

Using the first shift theorem and $Z\{x_k\} = F(z)$ gives:

$$\left(z^2 F(z) - z^2 x_0 - zx_1\right) - 3\left(zF(z) - zx_0\right) + 2F(z)$$

$$= \dfrac{z}{z-1}$$

$x_0 = 0$ and $x_1 = 1$, hence $\left(z^2 F(z) - z^2(0) - z(1)\right) -$

$$3\left(zF(z) - z(0)\right) + 2F(z) = \dfrac{z}{z-1}$$

i.e. $\qquad z^2 F(z) - z - 3zF(z) + 2F(z) = \dfrac{z}{z-1}$

and $\left(z^2 - 3z + 2\right) F(z) = \dfrac{z}{z-1} + z = \dfrac{z}{z-1} + \dfrac{z}{1}$

$$= \dfrac{z + z(z-1)}{z-1} = \dfrac{z + z^2 - z}{z-1} = \dfrac{z^2}{z-1}$$

Hence, $F(z) = \dfrac{z^2}{\left(z^2 - 3z + 2\right)(z-1)}$

$$= \dfrac{z^2}{(z-2)(z-1)(z-1)} = \dfrac{z^2}{(z-2)(z-1)^2}$$

and $\qquad \dfrac{F(z)}{z} = \dfrac{z}{(z-2)(z-1)^2}$

Using partial fractions, let

$$\dfrac{z}{(z-2)(z-1)^2} = \dfrac{A}{(z-2)} + \dfrac{B}{(z-1)} + \dfrac{C}{(z-1)^2}$$

$$= \dfrac{A(z-1)^2 + B(z-2)(z-1) + C(z-2)}{(z-2)(z-1)^2}$$

from which, $z = A(z-1)^2 + B(z-2)(z-1) + C(z-2)$

Letting $z = 2$ gives: $2 = A(1)^2$ i.e. $A = 2$

Letting $z = 1$ gives: $1 = C(-1)$ i.e. $C = -1$

Equating z^2 coefficients gives: $0 = A + B$ i.e. $B = -2$

Therefore, $\dfrac{F(z)}{z} = \dfrac{2}{(z-2)} - \dfrac{2}{(z-1)} - \dfrac{1}{(z-1)^2}$

or $\qquad F(z) = \dfrac{2z}{(z-2)} - \dfrac{2z}{(z-1)} - \dfrac{z}{(z-1)^2}$

Taking the inverse z-transform of $F(z)$ gives:

$$Z^{-1}F(z) = 2Z^{-1}\left(\frac{z}{(z-2)}\right) - 2Z^{-1}\left(\frac{z}{(z-1)}\right)$$

$$-Z^{-1}\left(\frac{z}{(z-1)^2}\right)$$

$$= 2\left(2^k\right) - 2(1) - k$$

from 2, 6 and 7 of Table 107.1

i.e. $$\{x_k\} = 2^{k+1} - 2 - k$$

Now try the following Practice Exercise

Practice Exercise 384   Solving difference equations (answers on page 1199)

1. Solve the difference equation:
   $x_{k+1} - 3x_k = 0$ given $x_0 = 4$

2. Solve the difference equation:
   $x_{k+1} - 3x_k = -6$ given $x_0 = 1$

3. Solve the difference equation:
   $2x_{k+2} - 5x_{k+1} + 4x_k = 0$ given $x_0 = 3$ and $x_1 = 2$

4. Solve the difference equation:
   $2x_{k+1} - x_k = 2^k$ given $x_0 = 2$

5. Solve the difference equation:
   $x_{k+2} + 3x_{k+1} + 2x_k = 0$ given $x_0 = 0$ and $x_1 = 1$

6. Solve the difference equation:
   $x_{k+2} - 5x_{k+1} + 6x_k = 5$ given $x_0 = 0$ and $x_1 = 1$

For fully worked solutions to each of the problems in Practice Exercises 378 to 384 in this chapter, go to the website:
www.routledge.com/cw/bird

Section O

Revision Test 35 Z-transforms

This assignment covers the material in Chapter 107. *The marks available are shown in brackets at the end of each question.*

1. Determine the z-transform of
   (a) 2^k (b) $5e^{-2k}$ (c) $2e^{-3k}\cos 5k$ (d) 4^{k+2} (10)

2. Determine the inverse z-transform of
   (a) $\dfrac{2z}{4z+1}$ (b) $\dfrac{z}{z-e^4}$ (c) $\dfrac{z}{z^2-z-2}$ (10)

3. Solve the difference equation:
   $$x_{k+2} - 2x_{k+1} - 3x_k = 1 \text{ given } x_0 = 0 \text{ and } x_1 = 1$$
   (20)

Sequence	Transform $F(z)$
1. $\{\delta_k\} = \{1,\ 0,\ 0,\ ...\}$	1 for all values of z
2. $\{u_k\} = \{1,\ 1,\ 1,\ ...\}$	$\dfrac{z}{z-1}$ for $\|z\| > 1$
3. $\{k\} = \{0,\ 1,\ 2,\ 3,\ ...\}$	$\dfrac{z}{(z-1)^2}$ for $\|z\| > 1$
4. $\{k^2\} = \{0,\ 1,\ 4,\ 9,\ ...\}$	$\dfrac{z(z+1)}{(z-1)^3}$ for $\|z\| > 1$
5. $\{k^3\} = \{0,\ 1,\ 8,\ 27,\ ...\}$	$\dfrac{z(z^2+4z+1)}{(z-1)^4}$ for $\|z\| > 1$
6. $\{a^k\} = \{1,\ a,\ a^2,\ a^3,\ ...\}$	$\dfrac{z}{z-a}$ for $\|z\| > \|a\|$
7. $\{ka^k\} = \{0,\ a,\ 2a^2,\ 3a^3,\ ...\}$	$\dfrac{az}{(z-a)^2}$ for $\|z\| > \|a\|$
8. $\{k^2a^k\} = \{0,\ a,\ 4a^2,\ 9a^3,\ ...\}$	$\dfrac{az(z+a)}{(z-a)^3}$ for $\|z\| > \|a\|$
9. $\{e^{-ak}\} = \{e^{-a},\ e^{-2a},\ e^{-3a},\ ...\}$	$\dfrac{z}{z-e^{-a}}$
10. $\sin ak = \{\sin a,\ \sin 2a,\ ...\}$	$\dfrac{z\sin a}{z^2 - 2z\cos a + 1}$
11. $\cos ak = \{\cos a,\ \cos 2a,\ ...\}$	$\dfrac{z(z-\cos a)}{z^2 - 2z\cos a + 1}$
12. $e^{-ak}\sin bk = \{e^{-a}\sin b,\ e^{-2a}\sin 2b,\ ...\}$	$\dfrac{ze^{-a}\sin b}{z^2 - 2ze^{-a}\cos b + e^{-2a}}$
13. $e^{-ak}\cos bk = \{e^{-a}\cos b,\ e^{-2a}\cos 2b,\ ...\}$	$\dfrac{z^2 - ze^{-a}\cos b}{z^2 - 2ze^{-a}\cos b + e^{-2a}}$

For a copy of these formulae/revision hints, go to:
www.routledge.com/cw/bird

Answers to practice exercises

Chapter 1

Exercise 1 (page 5)

1. $19\,\text{kg}$ 2. $16\,\text{m}$ 3. $479\,\text{mm}$
4. -66 5. £565   6. -225
7. -2136 8. $-36\,121$ 9. £107 701
10. -4 11. 1487 12. 5914
13. $189\,\text{g}$ 14. $-70\,872$ 15. $15\,333$
16. $d = 64\,\text{mm}$, $A = 136\,\text{mm}$, $B = 10\,\text{mm}$

Exercise 2 (page 7)

1. (a) 468   (b) 868   2. (a) £1827   (b) £4158
3. (a) $8613\,\text{kg}$ (b) $584\,\text{kg}$
4. (a) $351\,\text{mm}$ (b) $924\,\text{mm}$
5. (a) $10\,304$ (b) -4433 6. (a) $48\,\text{m}$ (b) $89\,\text{m}$
7. (a) 259   (b) 56   8. (a) 1648   (b) -1060
9. (a) 8067   (b) 3347   10. $18\,\text{kg}$
11. $89.25\,\text{cm}$ 12. 29

Exercise 3 (page 8)

1. (a) 4   (b) 24   2. (a) 12   (b) 360
3. (a) 10   (b) 350   4. (a) 90   (b) 2700
5. (a) 2   (b) 210   6. (a) 3   (b) 180
7. (a) 5   (b) 210   8. (a) 15   (b) 6300
9. (a) 14   (b) 420 420   10. (a) 14   (b) 53 900

Exercise 4 (page 10)

1. 59   2. 14   3. 88   4. 5   5. 33
6. 22   7. 68   8. 5   9. 2   10. 5
11. -1

Chapter 2

Exercise 5 (page 13)

1. $2\frac{1}{7}$ 2. $7\frac{2}{5}$ 3. $\frac{22}{9}$ 4. $\frac{71}{8}$
5. $\frac{4}{11}$ 6. $\frac{3}{7}, \frac{4}{9}, \frac{1}{2}, \frac{3}{5}, \frac{5}{8}$ 7. $\frac{8}{25}$ 8. $\frac{11}{15}$
9. $\frac{17}{30}$ 10. $\frac{9}{10}$ 11. $\frac{3}{16}$ 12. $\frac{43}{77}$
13. $\frac{47}{63}$ 14. $1\frac{1}{15}$ 15. $\frac{4}{27}$ 16. $8\frac{51}{52}$
17. $1\frac{9}{40}$ 18. $1\frac{16}{21}$ 19. $\frac{17}{60}$ 20. $\frac{17}{20}$

Exercise 6 (page 15)

1. $\frac{8}{35}$ 2. $2\frac{2}{9}$ 3. $\frac{6}{11}$ 4. $\frac{5}{12}$ 5. $\frac{3}{28}$
6. $\frac{3}{5}$ 7. 11   8. $\frac{1}{13}$ 9. $1\frac{1}{2}$ 10. $\frac{8}{15}$
11. $2\frac{2}{5}$ 12. $\frac{5}{12}$ 13. $3\frac{3}{4}$ 14. $\frac{12}{23}$ 15. 4
16. $\frac{3}{4}$ 17. $\frac{1}{9}$ 18. 13   19. 15   20. 400 litres
21. (a) £60,   (b) P£36, Q£16   22. 2880 litres

Exercise 7 (page 17)

1. $2\frac{1}{18}$ 2. $-\frac{1}{9}$ 3. $1\frac{1}{6}$ 4. $4\frac{3}{4}$ 5. $\frac{13}{20}$
6. $\frac{7}{15}$ 7. $4\frac{19}{20}$ 8. 2   9. $7\frac{1}{3}$ 10. $\frac{1}{15}$
11. 4   12. $2\frac{17}{20}$

Chapter 3

Exercise 8 (page 21)

1. $\dfrac{13}{20}$ 2. $\dfrac{9}{250}$ 3. $\dfrac{7}{40}$ 4. $\dfrac{6}{125}$

5. (a) $\dfrac{33}{50}$ (b) $\dfrac{21}{25}$ (c) $\dfrac{1}{80}$ (d) $\dfrac{141}{500}$ (e) $\dfrac{3}{125}$

6. $4\dfrac{21}{40}$ 7. $23\dfrac{11}{25}$ 8. $10\dfrac{3}{200}$ 9. $6\dfrac{7}{16}$

10. (a) $1\dfrac{41}{50}$ (b) $4\dfrac{11}{40}$ (c) $14\dfrac{1}{8}$ (d) $15\dfrac{7}{20}$ (e) $16\dfrac{17}{80}$

11. 0.625   12. 6.6875   13. 0.21875   14. 11.1875

15. 0.28125

Exercise 9 (page 22)

1. 14.18   2. 2.785   3. 65.38   4. 43.27
5. 1.297   6. 0.000528

Exercise 10 (page 23)

1. 80.3   2. 329.3   3. 72.54   4. -124.83
5. 295.3   6. 18.3   7. 12.52 mm

Exercise 11 (page 24)

1. 4.998   2. 47.544   3. 385.02   4. 582.42
5. 456.9   6. 434.82   7. 626.1   8. 1591.6
9. 0.444   10. 0.62963   11. 1.563   12. 53.455
13. 13.84   14. 8.69   15. (a) 24.81 (b) 24.812
16. (a) 0.00639 (b) 0.0064   17. (a) $8.\dot{4}$ (b) $62.\dot{6}$
18. 2400

Chapter 4

Exercise 12 (page 26)

1. 30.797   2. 11927   3. 13.62   4. 53.832
5. 84.42   6. 1.0944   7. 50.330   8. 36.45
9. 10.59   10. 12.325

Exercise 13 (page 27)

1. 12.25   2. 0.0361   3. 46.923   4. 1.296×10^{-3}
5. 2.4430   6. 2.197   7. 30.96   8. 0.0549
9. 219.26   10. 5.832×10^{-6}

Exercise 14 (page 28)

1. 0.571   2. 40   3. 0.13459   4. 137.9
5. 14.96   6. 19.4481   7. 515.36×10^{-6}
8. 1.0871   9. 15.625×10^{-9} 10. 52.70

Exercise 15 (page 29)

1. 2.182   2. 11.122   3. 185.82
4. 0.8307   5. 0.1581   6. 2.571
7. 5.273   8. 1.2555   9. 0.30366
10. 1.068   11. 3.5×10^6 12. 37.5×10^3
13. 4.2×10^{-6} 14. 202.767×10^{-3} 15. 18.32×10^6

Exercise 16 (page 30)

1. 0.4667   2. $\dfrac{13}{14}$ 3. 4.458   4. 2.732

5. $\dfrac{1}{21}$ 6. 0.7083   7. $-\dfrac{9}{10}$ 8. $3\dfrac{1}{3}$

9. 2.567   10. 0.0776

Exercise 17 (page 31)

1. 0.9205   2. 0.7314   3. 2.9042   4. 0.2719
5. 0.4424   6. 0.0321   7. 0.4232   8. 0.1502
9. -0.6992 10. 5.8452

Exercise 18 (page 31)

1. 4.995   2. 5.782   3. 25.72   4. 69.42
5. 0.6977   6. 52.92   7. 591.0   8. 17.90
9. 3.520   10. 0.3770

Exercise 19 (page 33)

1. Order of magnitude error – should be 2.1
2. Rounding-off error – should add 'correct to 4 significant figures' or 'correct to 1 decimal place'
3. Blunder
4. Measured values, hence $c = 55800\,\text{Pa}\,\text{m}^3$
5. Order of magnitude error and rounding-off error – should be 0.0225, correct to 3 significant figures or 0.0225, correct to 4 decimal places
6. ≈ 30 (29.61 by calculator)
7. ≈ 2 (1.988, correct to 4 s.f., by calculator)
8. ≈ 10 (9.481, correct to 4 s.f., by calculator)

Exercise 20 (page 33)

1. R   2. I   3. R   4. R   5. I
6. R   7. I   8. R   9. R   10. R

Exercise 21 (page 34)

1. $A = 66.59\,\text{cm}^2$ **2.** $C = 52.78\,\text{mm}$ **3.** $R = 37.5$
4. $159\,\text{m/s}$ **5.** $0.407\,\text{A}$ **6.** $5.02\,\text{mm}$
7. $0.144\,\text{J}$ **8.** $628.8\,\text{m}^2$ **9.** 224.5
10. $14230\,\text{kg/m}^3$ **11.** $281.1\,\text{m/s}$ **12.** $2.526\,\Omega$

Exercise 22 (page 36)

1. £589.27 **2.** $508.1\,\text{W}$ **3.** $V = 2.61\,\text{V}$
4. $F = 854.5$ **5.** $I = 3.81\,\text{A}$ **6.** $t = 14.79\,\text{s}$
7. $E = 3.96\,\text{J}$ **8.** $I = 12.77\,\text{A}$ **9.** $s = 17.25\,\text{m}$
10. $A = 7.184\,\text{cm}^2$ **11.** $v = 7.327$
12. (a) $12.53\,\text{h}$ (b) $1\,\text{h}\ 40\,\text{min}$, 33 m.p.h.
(c) $13.02\,\text{h}$ (d) $13.15\,\text{h}$

Chapter 5

Exercise 23 (page 40)

1. 0.32% **2.** 173.4% **3.** 5.7% **4.** 37.4%
5. 128.5% **6.** 0.20 **7.** 0.0125 **8.** 68.75%
9. 38.462% **10.** (a) 21.2% (b) 79.2% (c) 169%
11. (b), (d), (c), (a) **12.** $\dfrac{13}{20}$ **13.** $\dfrac{5}{16}$ **14.** $\dfrac{9}{16}$
15. $A = \dfrac{1}{2}$, $B = 50\%$, $C = 0.25$, $D = 25\%$, $E = 0.30$,
$F = \dfrac{3}{10}$, $G = 0.60$, $H = 60\%$, $I = 0.85$, $J = \dfrac{17}{20}$
16. $779\,\Omega$ to $861\,\Omega$

Exercise 24 (page 41)

1. $21.8\,\text{kg}$ **2.** $9.72\,\text{m}$
3. (a) $496.4\,\text{t}$ (b) $8.657\,\text{g}$ (c) $20.73\,\text{s}$ **4.** 2.25%
5. (a) 14% (b) 15.67% (c) 5.36% **6.** $37.8\,\text{g}$
7. 14 minutes 57 seconds **8.** $76\,\text{g}$ **9.** £624
10. 37.49% **11.** 39.2% **12.** 17% **13.** 38.7%
14. 2.7% **15.** $5.60\,\text{m}$ **16.** 3.5%
17. (a) (i) $656\,\Omega$ (ii) $984\,\Omega$
(b) (i) $44.65\,\text{k}\Omega$ (ii) $49.35\,\text{k}\Omega$
18. $2592\,\text{rev/min}$

Exercise 25 (page 43)

1. 2.5% **2.** 18% **3.** £310 **4.** £175000
5. £260 **6.** £20000 **7.** £8446.05 **8.** £50.25
9. £39.60 **10.** £917.70 **11.** £185000 **12.** 7.2%
13. $A\ 0.6\,\text{kg}$, $B\ 0.9\,\text{kg}$, $C\ 0.5\,\text{kg}$

14. $54\%, 31\%, 15\%, 0.3\,\text{t}$
15. $20000\,\text{kg}$ (or 20 tonnes)
16. $13.5\,\text{mm}, 11.5\,\text{mm}$ **17.** $600\,\text{kW}$

Chapter 6

Exercise 26 (page 48)

1. $36:1$ **2.** $3.5:1$ or $7:2$ **3.** $47:3$
4. $96\,\text{cm}, 240\,\text{cm}$ **5.** $5\dfrac{1}{4}$ hours or 5 hours 15 minutes
6. £3680, £1840, £920 **7.** $12\,\text{cm}$ **8.** £2172

Exercise 27 (page 49)

1. $1:15$ **2.** $76\,\text{ml}$ **3.** 25% **4.** $12.6\,\text{kg}$
5. $14.3\,\text{kg}$ **6.** $25000\,\text{kg}$

Exercise 28 (page 51)

1. £556 **2.** £66 **3.** $264\,\text{kg}$ **4.** $450\,\text{N}$ **5.** $14.56\,\text{kg}$
6. (a) 0.00025 (b) $48\,\text{MPa}$ **7.** (a) $440\,\text{K}$ (b) $5.76\,\text{litres}$
8. 8960

Exercise 29 (page 52)

1. (a) $2\,\text{mA}$ (b) $25\,\text{V}$ **2.** $434\,\text{fr}$ **3.** $685.8\,\text{mm}$
4. $83\,\text{lb}\ 10\,\text{oz}$ **5.** (a) $159.1\,\text{litres}$ (b) $16.5\,\text{gallons}$
6. $29.4\,\text{MPa}$ **7.** $584.2\,\text{mm}$ **8.** $\$891$

Exercise 30 (page 54)

1. 3.5 weeks **2.** 20 days
3. (a) 9.18 (b) 6.12 (c) 0.3375 **4.** 50 minutes
5. (a) 300×10^3 (b) $0.375\,\text{m}^2$ (c) $24 \times 10^3\,\text{Pa}$
6. (a) $32\,\text{J}$ (b) $0.5\,\text{m}$

Chapter 7

Exercise 31 (page 57)

1. 27 **2.** 128 **3.** 100000 **4.** 96 **5.** 2^4
6. ± 5 **7.** ± 8 **8.** 100 **9.** 1 **10.** 64

Exercise 32 (page 59)

1. 128 **2.** 3^9 **3.** 16 **4.** $\dfrac{1}{9}$ **5.** 1 **6.** 8
7. 100 **8.** 1000 **9.** $\dfrac{1}{100}$ or 0.01 **10.** 5 **11.** 7^6

12. 3^6 **13.** 3^6 **14.** 3^4 **15.** 1 **16.** 25

17. $\dfrac{1}{3^5}$ or $\dfrac{1}{243}$ **18.** 49 **19.** $\dfrac{1}{2}$ or 0.5 **20.** 1

Exercise 33 (page 60)

1. $\dfrac{1}{3 \times 5^2}$ **2.** $\dfrac{1}{7^3 \times 3^7}$ **3.** $\dfrac{3^2}{2^5}$ **4.** $\dfrac{1}{2^{10} \times 5^2}$

5. 9 **6.** 3 **7.** $\dfrac{1}{2}$ **8.** $\pm\dfrac{2}{3}$

9. $\dfrac{147}{148}$ **10.** $-1\dfrac{19}{56}$ **11.** $-3\dfrac{13}{45}$ **12.** $\dfrac{1}{9}$

13. $-\dfrac{17}{18}$ **14.** 64 **15.** $4\dfrac{1}{2}$

Chapter 8

Exercise 34 (page 66)

1. cubic metre, m^3 **2.** farad, F
3. square metre, m^2 **4.** metre per second, m/s
5. kilogram per cubic metre, kg/m^3
6. joule, J **7.** coulomb, C **8.** watt, W
9. radian or degree **10.** volt, V **11.** mass
12. electrical resistance **13.** frequency
14. acceleration **15.** electric current
16. inductance **17.** length
18. temperature **19.** pressure
20. angular velocity **21.** $\times 10^9$ **22.** m, $\times 10^{-3}$
23. $\times 10^{-12}$ **24.** M, $\times 10^6$

Exercise 35 (page 67)

1. (a) 7.39×10 (b) 2.84×10 (c) 1.9762×10^2
2. (a) 2.748×10^3 (b) 3.317×10^4 (c) 2.74218×10^5
3. (a) 2.401×10^{-1} (b) 1.74×10^{-2} (c) 9.23×10^{-3}
4. (a) 1.7023×10^3 (b) 1.004×10 (c) 1.09×10^{-2}
5. (a) 5×10^{-1} (b) 1.1875×10
   (c) 1.306×10^2 (d) 3.125×10^{-2}
6. (a) 1010 (b) 932.7 (c) 54100 (d) 7
7. (a) 0.0389 (b) 0.6741 (c) 0.008
8. (a) 1.35×10^2 (b) 1.1×10^5
9. (a) 2×10^2 (b) 1.5×10^{-3}
10. (a) $2.71 \times 10^3\,kg\,m^{-3}$ (b) 4.4×10^{-1}
   (c) $3.7673 \times 10^2\,\Omega$ (d) $5.11 \times 10^{-1}\,MeV$
   (e) $9.57897 \times 10^7\,C\,kg^{-1}$
   (f) $2.241 \times 10^{-2}\,m^3\,mol^{-1}$

Exercise 36 (page 69)

1. 60 kPa **2.** 150 µW or 0.15 mW
3. 50 MV **4.** 55 nF
5. 100 kW **6.** 0.54 mA or 540 µA
7. 1.5 MΩ **8.** 22.5 mV
9. 35 GHz **10.** 15 pF
11. 17 µA **12.** 46.2 kΩ
13. 3 µA **14.** 2.025 MHz
15. 50 kN **16.** 0.3 nF
17. 62.50 m **18.** 0.0346 kg
19. 13.5×10^{-3} **20.** 4×10^3
21. 3.8×10^5 km **22.** 0.053 nm
23. 5.6×10^6 Pa = 5.6 MPa
24. 4.3×10^{-3} m = 4.3 mm

Exercise 37 (page 70)

1. 2450 mm **2.** 167.5 cm **3.** 658 mm
4. 25.4 m **5.** 56.32 m **6.** 4356 mm
7. 87.5 cm **8.** (a) 4650 mm (b) 4.65 m
9. (a) 504 cm (b) 5.04 m
10. 148.5 mm to 151.5 mm

Exercise 38 (page 71)

1. 8×10^4 cm^2 **2.** 240×10^{-4} m^2 or 0.024 m^2
3. 3.6×10^6 mm^2 **4.** 350×10^{-6} m^2 or 0.35×10^{-3} m^2
5. 5000 mm^2 **6.** 2.5 cm^2
7. (a) 288000 mm^2 (b) 2880 cm^2 (c) 0.288 m^2

Exercise 39 (page 73)

1. 2.5×10^6 cm^3 **2.** 400×10^{-6} m^3
3. 0.87×10^9 mm^3 or 870×10^6 mm^3 **4.** 2.4 m^3
5. 1500×10^{-9} m^3 or 1.5×10^{-6} m^3
6. 400×10^{-3} cm^3 or 0.4 cm^3 **7.** 6.4×10^3 mm^3
8. 7500×10^{-3} cm^3 or 7.5 cm^3
9. (a) 0.18 m^3 (b) 180000 cm^3 (c) 180 litres

Exercise 40 (page 76)

1. 18.74 in **2.** 82.28 in
3. 37.95 yd **4.** 18.36 mile
5. 41.66 cm **6.** 1.98 m
7. 14.36 m **8.** 5.91 km
9. (a) 22,745 yd (b) 20.81 km **10.** 9.688 in^2
11. 18.18 yd^2 **12.** 30.89 acres

13. 21.89 mile^2
14. 41 cm^2
15. 28.67 m^2
16. 10117 m^2
17. 55.17 km^2
18. 12.24 in^3
19. 7.572 ft^3
20. 17.59 yd^3
21. 31.7 fluid pints
22. 35.23 cm^3
23. 4.956 m^3
24. 3.667 litres
25. 47.32 litres
26. 34.59 oz
27. 121.3 lb
28. 4.409 short tons
29. 220.6 g
30. 1.630 kg
31. (a) 280 lb (b) 127.2 kg
32. $131°\text{F}$
33. $75° \text{ C}$

Chapter 9

Exercise 41 (page 81)

1. $-3a$
2. $a + 2b + 4c$
3. $3x - 3x^2 - 3y - 2y^2$
4. $6ab - 3a$
5. $6x - 5y + 8z$
6. $1 + 2x$
7. $4x + 2y + 2$
8. $3a + 5b$
9. $-2a - b + 2c$
10. $3x^2 - y^2$

Exercise 42 (page 83)

1. $p^2 q^3 r$
2. $8a^2$
3. $6q^2$
4. 46
5. $5\frac{1}{3}$
6. $-\frac{1}{2}$
7. 6
8. $\frac{1}{7y}$
9. $5xz^2$
10. $3a^2 + 2ab - b^2$
11. $6a^2 - 13ab + 3ac - 5b^2 + bc$
12. $\frac{1}{3b}$
13. $2ab$
14. $3x$
15. $2x - y$
16. $3p + 2q$
17. $2a^2 + 2b^2$

Exercise 43 (page 85)

1. z^8
2. a^8
3. n^3
4. b^{11}
5. b^{-3}
6. c^4
7. m^4
8. x^{-3} or $\frac{1}{x^3}$
9. x^{12}
10. y^{-6} or $\frac{1}{y^6}$
11. t^8
12. c^{14}
13. a^{-9} or $\frac{1}{a^9}$
14. b^{-12} or $\frac{1}{b^{12}}$
15. b^{10}
16. s^{-9}
17. $p^6 q^7 r^5$
18. $x^{-2} y z^{-2}$ or $\frac{y}{x^2 z^2}$
19. $x^5 y^4 z^3, 13\frac{1}{2}$
20. $a^3 b^{-2} c$ or $\frac{a^3 c}{b^2}, 9$

Exercise 44 (page 86)

1. $a^2 b^{1/2} c^{-2}, \pm 4\frac{1}{2}$
2. $\frac{1+a}{b}$
3. $a b^6 c^{3/2}$
4. $a^{-4} b^5 c^{11}$
5. $\frac{p^2 q}{q - p}$
6. $x y^3 \sqrt[6]{z^{13}}$
7. $\frac{1}{ef^2}$
8. $a^{11/6} b^{1/3} c^{-3/2}$ or $\frac{\sqrt[6]{a^{11}} \sqrt[3]{b}}{\sqrt{c^3}}$

Chapter 10

Exercise 45 (page 88)

1. $x^2 + 5x + 6$
2. $2x^2 + 9x + 4$
3. $4x^2 + 12x + 9$
4. $2j^2 + 2j - 12$
5. $4x^2 + 22x + 30$
6. $2pqr + p^2 q^2 + r^2$
7. $a^2 + 2ab + b^2$
8. $x^2 + 12x + 36$
9. $a^2 - 2ac + c^2$
10. $25x^2 + 30x + 9$
11. $4x^2 - 24x + 36$
12. $4x^2 - 9$
13. $64x^2 + 64x + 16$
14. $r^2 s^2 + 2rst + t^2$
15. $3ab - 6a^2$
16. $2x^2 - 2xy$
17. $2a^2 - 3ab - 5b^2$
18. $13p - 7q$
19. $7x - y - 4z$
20. $4a^2 - 25b^2$
21. $x^2 - 4xy + 4y^2$
22. $9a^2 - 6ab + b^2$
23. 0
24. $4 - a$
25. $4ab - 8a^2$
26. $3xy + 9x^2 y - 15x^2$
27. $2 + 5b^2$
28. $11q - 2p$

Exercise 46 (page 90)

1. $2(x + 2)$
2. $2x(y - 4z)$
3. $p(b + 2c)$
4. $2x(1 + 2y)$
5. $4d(d - 3f^5)$
6. $4x(1 + 2x)$
7. $2q(q + 4n)$
8. $r(s + p + t)$
9. $x(1 + 3x + 5x^2)$
10. $bc(a + b^2)$
11. $3xy(xy^3 - 5y + 6)$
12. $2pq^2(2p^2 - 5q)$
13. $7ab(3ab - 4)$
14. $2xy(y + 3x + 4x^2)$
15. $2xy(x - 2y^2 + 4x^2 y^3)$
16. $7y(4 + y + 2x)$
17. $\frac{3x}{y}$
18. 0
19. $\frac{2r}{t}$
20. $(a + b)(y + 1)$
21. $(p + q)(x + y)$
22. $(x - y)(a + b)$
23. $(a - 2b)(2x + 3y)$
24. $\frac{A^2}{pg}\left(\frac{A}{pg^2} - \frac{1}{g} + A^3\right)$

Exercise 47 (page 92)

1. $2x + 8x^2$ or $2x(1+4x)$ **2.** $12y^2 - 3y$ or $3y(4y-1)$

3. $4b - 15b^2$ or $b(4-15b)$ **4.** $4 + 3a$

5. $\dfrac{3}{2} - 4x$ **6.** 1

7. $10y^2 - 3y + \dfrac{1}{4}$ **8.** $9x^2 + \dfrac{1}{3} - 4x$

9. $6a^2 + 5a - \dfrac{1}{7}$ **10.** $-15t$

11. $\dfrac{1}{5} - x - x^2$ **12.** $10a^2 - 3a + 2$

Chapter 11

Exercise 48 (page 97)

1. 1   **2.** 2   **3.** 6   **4.** -4 **5.** 2

6. 1   **7.** 2   **8.** $\dfrac{1}{2}$ **9.** 0   **10.** 3

11. 2   **12.** -10 **13.** 6   **14.** -2 **15.** 2.5

16. 2   **17.** 6   **18.** -3

Exercise 49 (page 99)

1. 5   **2.** -2 **3.** $-4\dfrac{1}{2}$ **4.** 2   **5.** 12

6. 15   **7.** -4 **8.** $5\dfrac{1}{3}$ **9.** 2   **10.** 13

11. -10 **12.** 2   **13.** 3   **14.** -11 **15.** -6

16. 9   **17.** $6\dfrac{1}{4}$ **18.** 3   **19.** 4   **20.** 10

21. ± 12 **22.** $-3\dfrac{1}{3}$ **23.** ± 3 **24.** ± 4

Exercise 50 (page 101)

1. 10^{-7} **2.** $8\,\text{m/s}^2$ **3.** 3.472

4. (a) $1.8\,\Omega$ (b) $30\,\Omega$

5. digital camera battery £9, camcorder battery £14

6. $800\,\Omega$ **7.** $30\,\text{m/s}^2$ **8.** $176\,\text{MPa}$

Exercise 51 (page 103)

1. $12\,\text{cm}, 240\,\text{cm}^2$ **2.** 0.004   **3.** 30

4. $45°\text{C}$ **5.** 50   **6.** £312, £240

7. $30\,\text{kg}$ **8.** $12\,\text{m}, 8\,\text{m}$ **9.** $3.5\,\text{N}$

Chapter 12

Exercise 52 (page 107)

1. $d = c - e - a - b$ **2.** $x = \dfrac{y}{7}$

3. $v = \dfrac{c}{p}$ **4.** $a = \dfrac{v - u}{t}$

5. $R = \dfrac{V}{I}$ **6.** $y = \dfrac{1}{3}(t - x)$

7. $r = \dfrac{c}{2\pi}$ **8.** $x = \dfrac{y - c}{m}$

9. $T = \dfrac{I}{PR}$ **10.** $c = \dfrac{Q}{m\,\Delta T}$

11. $L = \dfrac{X_L}{2\pi f}$ **12.** $R = \dfrac{E}{I}$

13. $x = a(y - 3)$ **14.** $C = \dfrac{5}{9}(F - 32)$

15. $f = \dfrac{1}{2\pi\, CX_C}$ **16.** $R = \dfrac{pV}{mT}$

Exercise 53 (page 109)

1. $r = \dfrac{S - a}{S}$ or $1 - \dfrac{a}{S}$

2. $x = \dfrac{d}{\lambda}(y + \lambda)$ or $d + \dfrac{yd}{\lambda}$

3. $f = \dfrac{3F - AL}{3}$ or $f = F - \dfrac{AL}{3}$

4. $D = \dfrac{AB^2}{5Cy}$ **5.** $t = \dfrac{R - R_0}{R_0\alpha}$

6. $R_2 = \dfrac{RR_1}{R_1 - R}$

7. $R = \dfrac{E - e - Ir}{I}$ or $R = \dfrac{E - e}{I} - r$

8. $V_2 = \dfrac{P_1 V_1 T_2}{P_2 T_1}$ **9.** $b = \sqrt{\left(\dfrac{y}{4ac^2}\right)}$

10. $x = \dfrac{ay}{\sqrt{(y^2 - b^2)}}$ **11.** $L = \dfrac{gt^2}{4\pi^2}$

12. $u = \sqrt{v^2 - 2as}$ **13.** $R = \sqrt{\left(\dfrac{360A}{\pi\theta}\right)}$

14. $T_2 = \dfrac{P_2 V_2 T_1}{P_1 V_1}$ **15.** $a = N^2 y - x$

16. $L = \dfrac{\sqrt{Z^2 - R^2}}{2\pi f}, 0.080$ **17.** $v = \sqrt{\left(\dfrac{2L}{\rho\,a\,c}\right)}$

18. $V = \dfrac{k^2 H^2 L^2}{\theta^2}$

Exercise 54 (page 111)

1. $a = \sqrt{\left(\dfrac{xy}{m-n}\right)}$　**2.** $R = \sqrt[4]{\left(\dfrac{M}{\pi} + r^4\right)}$

3. $r = \dfrac{3(x+y)}{(1-x-y)}$　**4.** $L = \dfrac{mrCR}{\mu - m}$

5. $b = \dfrac{c}{\sqrt{1-a^2}}$　**6.** $r = \sqrt{\left(\dfrac{x-y}{x+y}\right)}$

7. $b = \dfrac{a(p^2 - q^2)}{2(p^2 + q^2)}$　**8.** $v = \dfrac{uf}{u-f}, 30$

9. $t_2 = t_1 + \dfrac{Q}{mc}, 55$　**10.** $v = \sqrt{\left(\dfrac{2dgh}{0.03L}\right)}, 0.965$

11. $l = \dfrac{8S^2}{3d} + d, 2.725$

12. $C = \dfrac{1}{\omega\left\{\omega L - \sqrt{Z^2 - R^2}\right\}}, 63.1 \times 10^{-6}$

13. $64\,\text{mm}$　**14.** $\lambda = \sqrt[5]{\left(\dfrac{a\mu}{\rho CZ^4 n}\right)^2}$

15. $w = \dfrac{2R-F}{L}, 3\,\text{kN/m}$　**16.** $t_2 = t_1 - \dfrac{Qd}{kA}$

17. $r = \dfrac{v}{\omega}\left(1 - \dfrac{s}{100}\right)$　**18.** $F = EI\left(\dfrac{n\pi}{L}\right)^2, 13.61\,\text{MN}$

19. $r = \sqrt[4]{\dfrac{8\eta \ell V}{\pi p}}$　**20.** $\ell = \sqrt[3]{\left(\dfrac{20gH^2}{I\rho^4 D^2}\right)^2}$

Chapter 13

Exercise 55 (page 115)

1. $x = 4, y = 2$　**2.** $x = 3, y = 4$
3. $x = 2, y = 1.5$　**4.** $x = 4, y = 1$
5. $p = 2, q = -1$　**6.** $x = 1, y = 2$
7. $x = 3, y = 2$　**8.** $a = 2, b = 3$
9. $a = 5, b = 2$　**10.** $x = 1, y = 1$
11. $s = 2, t = 3$　**12.** $x = 3, y = -2$
13. $m = 2.5, n = 0.5$　**14.** $a = 6, b = -1$
15. $x = 2, y = 5$　**16.** $c = 2, d = -3$

Exercise 56 (page 117)

1. $p = -1, q = -2$　**2.** $x = 4, y = 6$
3. $a = 2, b = 3$　**4.** $s = 4, t = -1$
5. $x = 3, y = 4$　**6.** $u = 12, v = 2$
7. $x = 10, y = 15$　**8.** $a = 0.30, b = 0.40$

Exercise 57 (page 119)

1. $x = \dfrac{1}{2}, y = \dfrac{1}{4}$　**2.** $a = \dfrac{1}{3}, b = -\dfrac{1}{2}$

3. $p = \dfrac{1}{4}, q = \dfrac{1}{5}$　**4.** $x = 10, y = 5$

5. $c = 3, d = 4$　**6.** $r = 3, s = \dfrac{1}{2}$

7. $x = 5, y = 1\dfrac{3}{4}$　**8.** 1

Exercise 58 (page 122)

1. $a = 0.2, b = 4$　**2.** $I_1 = 6.47, I_2 = 4.62$
3. $u = 12, a = 4, v = 26$　**4.** £15 500, £12 800
5. $m = -0.5, c = 3$
6. $\alpha = 0.00426, R_0 = 22.56\,\Omega$　**7.** $a = 15, b = 0.50$
8. $a = 4, b = 10$　**9.** $F_1 = 1.5, F_2 = -4.5$
10. $R_1 = 5.7\,\text{kN}, R_2 = 6.3\,\text{kN}$

Exercise 59 (page 123)

1. $x = 2, y = 1, z = 3$　**2.** $x = 2, y = -2, z = 2$
3. $x = 5, y = -1, z = -2$　**4.** $x = 4, y = 0, z = 3$
5. $x = 2, y = 4, z = 5$　**6.** $x = 1, y = 6, z = 7$
7. $x = 5, y = 4, z = 2$　**8.** $x = -4, y = 3, z = 2$
9. $x = 1.5, y = 2.5, z = 4.5$
10. $i_1 = -5, i_2 = -4, i_3 = 2$
11. $F_1 = 2, F_2 = -3, F_3 = 4$

Chapter 14

Exercise 60 (page 129)

1. 4 or -4　**2.** 4 or -8　**3.** 2 or -6

4. -1.5 or 1.5　**5.** 0 or $-\dfrac{4}{3}$　**6.** 2 or -2

7. 4　**8.** -5　**9.** 1

10. -2 or -3　**11.** -3 or -7　**12.** 2 or -1
13. 4 or -3　**14.** 2 or 7　**15.** -4
16. 2　**17.** -3　**18.** 3 or -3

19. -2 or $-\dfrac{2}{3}$　**20.** -1.5　**21.** $\dfrac{1}{8}$ or $-\dfrac{1}{8}$

22. 4 or -7　**23.** -1 or 1.5　**24.** $\dfrac{1}{2}$ or $\dfrac{1}{3}$

25. $\dfrac{1}{2}$ or $-\dfrac{4}{5}$　**26.** $1\dfrac{1}{3}$ or $-\dfrac{1}{7}$　**27.** $\dfrac{3}{8}$ or -2

28. $\dfrac{2}{5}$ or -3　**29.** $\dfrac{4}{3}$ or $-\dfrac{1}{2}$　**30.** $\dfrac{5}{4}$ or $-\dfrac{3}{2}$

31. $x^2 - 4x + 3 = 0$ **32.** $x^2 + 3x - 10 = 0$
33. $x^2 + 5x + 4 = 0$ **34.** $4x^2 - 8x - 5 = 0$
35. $x^2 - 36 = 0$ **36.** $x^2 - 1.7x - 1.68 = 0$

Exercise 61 (page 131)

1. -3.732 or -0.268 **2.** -3.137 or 0.637
3. 1.468 or -1.135 **4.** 1.290 or 0.310
5. 2.443 or 0.307 **6.** -2.851 or 0.351

Exercise 62 (page 132)

1. 0.637 or -3.137 **2.** 0.296 or -0.792
3. 2.781 or 0.719 **4.** 0.443 or -1.693
5. 3.608 or -1.108 **6.** 1.434 or 0.232
7. 0.851 or -2.351 **8.** 2.086 or -0.086
9. 1.481 or -1.081 **10.** 4.176 or -1.676
11. 4 or 2.167 **12.** 7.141 or -3.641
13. 4.562 or 0.438

Exercise 63 (page 134)

1. $1.191\,\text{s}$ **2.** $0.345\,\text{A}$ or $0.905\,\text{A}$
3. $7.84\,\text{cm}$ **4.** $0.619\,\text{m}$ or $19.38\,\text{m}$
5. 0.0133 **6.** $1.066\,\text{m}$
7. $86.78\,\text{cm}$ **8.** $1.835\,\text{m}$ or $18.165\,\text{m}$
9. $7\,\text{m}$ **10.** $12\,\text{ohms}, 28\,\text{ohms}$
11. $5.73\,\text{ms}, 0.52\,\text{s}$ **12.** $400\,\text{rad/s}$

Exercise 64 (page 135)

1. $x = 1, y = 3$ and $x = -3, y = 7$

2. $x = \dfrac{2}{5}, y = -\dfrac{1}{5}$ and $x = -1\dfrac{2}{3}, y = -4\dfrac{1}{3}$

3. $x = 0, y = 4$ and $x = 3, y = 1$

Chapter 15

Exercise 65 (page 140)

1. 4 **2.** 4 **3.** 3 **4.** -3 **5.** $\dfrac{1}{3}$

6. 3 **7.** 2 **8.** -2 **9.** $1\dfrac{1}{2}$ **10.** $\dfrac{1}{3}$

11. 2 **12.** 10000 **13.** 100000 **14.** 9 **15.** $\dfrac{1}{32}$

16. 0.01 **17.** $\dfrac{1}{16}$ **18.** e^3

Exercise 66 (page 142)

1. $\log 6$ **2.** $\log 15$ **3.** $\log 2$ **4.** $\log 3$
5. $\log 12$ **6.** $\log 500$ **7.** $\log 100$ **8.** $\log 6$
9. $\log 10$ **10.** $\log 1 = 0$ **11.** $\log 2$
12. $\log 243$ or $\log 3^5$ or $5\log 3$
13. $\log 16$ or $\log 2^4$ or $4\log 2$
14. $\log 64$ or $\log 2^6$ or $6\log 2$
15. 0.5 **16.** 1.5 **17.** $x = 2.5$ **18.** $t = 8$
19. $b = 2$ **20.** $x = 2$ **21.** $a = 6$ **22.** $x = 5$

Exercise 67 (page 144)

1. 1.690 **2.** 3.170 **3.** 0.2696 **4.** 6.058 **5.** 2.251
6. 3.959 **7.** 2.542 **8.** -0.3272 **9.** 316.2 **10.** $0.057\,\text{m}^3$

Chapter 16

Exercise 68 (page 147)

1. (a) 0.1653 (b) 0.4584 (c) 22030
2. (a) 5.0988 (b) 0.064037 (c) 40.446
3. (a) 4.55848 (b) 2.40444 (c) 8.05124
4. (a) 48.04106 (b) 4.07482 (c) -0.08286
5. 2.739 **6.** $120.7\,\text{m}$

Exercise 69 (page 149)

1. 2.0601 **2.** (a) 7.389 (b) 0.7408

3. $1 - 2x^2 - \dfrac{8}{3}x^3 - 2x^4$

4. $2x^{1/2} + 2x^{5/2} + x^{9/2} + \dfrac{1}{3}x^{13/2} + \dfrac{1}{12}x^{17/2} + \dfrac{1}{60}x^{21/2}$

Exercise 70 (page 150)

1. $3.95, 2.05$ **2.** $1.65, -1.30$
3. (a) $28\,\text{cm}^3$ (b) $116\,\text{min}$ **4.** (a) $70°\text{C}$ (b) 5 minutes

Exercise 71 (page 153)

1. (a) 0.55547 (b) 0.91374 (c) 8.8941
2. (a) 2.2293 (b) -0.33154 (c) 0.13087
3. -0.4904 **4.** -0.5822 **5.** 2.197 **6.** 816.2
7. 0.8274 **8.** 11.02 **9.** 1.522 **10.** 1.485
11. 1.962 **12.** 3 **13.** 4
14. 147.9 **15.** 4.901 **16.** 3.095
17. $t = e^{b + a\ln D} = e^b e^{a\ln D} = e^b e^{\ln D^a}$ i.e. $t = e^b D^a$

18. 500   **19.** $W = PV \ln\left(\dfrac{U_2}{U_1}\right)$

20. 992 m/s   **21.** 348.5 Pa

Exercise 72 (page 156)

1. (a) 150°C (b) 100.5°C   **2.** 99.21 kPa
3. (a) 29.32 volts (b) 71.31×10^{-6} s
4. (a) 1.993 m (b) 2.293 m   **5.** (a) 50°C (b) 55.45 s
6. 30.37 N   **7.** (a) 3.04 A (b) 1.46 s
8. 2.45 mol/cm^3 **9.** (a) 7.07 A (b) 0.966 s
10. £2424
11. (a) 100% (b) 67.03% (c) 1.83%
12. 2.45 mA   **13.** 142 ms
14. 99.752%   **15.** 20 min 38 s

Chapter 17

Exercise 73 (page 159)

1. (a) $t > 2$ (b) $x < 5$ **2.** (a) $x > 3$ (b) $x \geq 3$
3. (a) $t \leq 1$ (b) $x \leq 6$ **4.** (a) $k \geq \dfrac{3}{2}$ (b) $z > \dfrac{1}{2}$
5. (a) $y \geq -\dfrac{4}{3}$ (b) $x \geq -\dfrac{1}{2}$

Exercise 74 (page 160)

1. $-5 < t < 3$ **2.** $-5 \leq y \leq -1$
3. $-\dfrac{3}{2} < x < \dfrac{5}{2}$ **4.** $t > 3$ and $t < \dfrac{1}{3}$
5. $k \geq 4$ and $k \leq -2$

Exercise 75 (page 161)

1. $-4 \leq x \leq 3$ **2.** $t > 5$ or $t < -9$
3. $-5 < z \leq 14$ **4.** $-3 < x \leq -2$

Exercise 76 (page 162)

1. $z > 4$ or $z < -4$ **2.** $-4 < z < 4$
3. $x \geq \sqrt{3}$ or $x \leq -\sqrt{3}$ **4.** $-2 \leq k \leq 2$
5. $-5 \leq t \leq 7$ **6.** $t \geq 7$ or $t \leq -5$
7. $y \geq 2$ or $y \leq -2$ **8.** $k > -\dfrac{1}{2}$ or $k < -2$

Exercise 77 (page 163)

1. $x > 3$ or $x < -2$ **2.** $-4 \leq t \leq 2$
3. $-2 < x < \dfrac{1}{2}$ **4.** $y \geq 5$ or $y \leq -4$

5. $-4 \leq z \leq 0$
6. $(-\sqrt{3} - 3) \leq x \leq (\sqrt{3} - 3)$
7. $t \geq (2 + \sqrt{11})$ or $t \leq (2 - \sqrt{11})$
8. $k \geq \left(\sqrt{\dfrac{13}{4}} + \dfrac{1}{2}\right)$ or $k \leq \left(-\sqrt{\dfrac{13}{4}} + \dfrac{1}{2}\right)$
Alternatively, $k \geq \dfrac{1}{2}(1 + \sqrt{13})$ or $k \leq \dfrac{1}{2}(1 - \sqrt{13})$

Chapter 18

Exercise 78 (page 173)

1. $2x - y$ **2.** $3x - 1$
3. $5x - 2$ **4.** $7x + 1$
5. $x^2 + 2xy + y^2$ **6.** $5x + 4 + \dfrac{8}{x - 1}$
7. $3x^2 - 4x + 3 - \dfrac{2}{x + 2}$
8. $5x^3 + 18x^2 + 54x + 160 + \dfrac{481}{x - 3}$

Exercise 79 (page 175)

1. $(x - 1)(x + 3)$ **2.** $(x + 1)(x + 2)(x - 2)$
3. $(x + 1)(2x^2 + 3x - 7)$ **4.** $(x - 1)(x + 3)(2x - 5)$
5. $x^3 + 4x^2 + x - 6 = (x - 1)(x + 3)(x + 2)$
  $x = 1, x = -3$ and $x = -2$
6. $x = 1, x = 2$ and $x = -1$

Exercise 80 (page 177)

1. (a) 6 (b) 9   **2.** (a) -39 (b) -29
3. $(x - 1)(x - 2)(x - 3)$
4. $x = -1, x = -2$ and $x = -4$ **5.** $a = -3$
6. $x = 1, x = -2$ and $x = 1.5$

Chapter 19

Exercise 81 (page 181)

1. $\dfrac{2}{(x - 3)} - \dfrac{2}{(x + 3)}$ **2.** $\dfrac{5}{(x + 1)} - \dfrac{1}{(x - 3)}$
3. $\dfrac{3}{x} + \dfrac{2}{(x - 2)} - \dfrac{4}{(x - 1)}$
4. $\dfrac{7}{(x + 4)} - \dfrac{3}{(x + 1)} - \dfrac{2}{(2x - 1)}$
5. $1 + \dfrac{2}{(x + 3)} + \dfrac{6}{(x - 2)}$ **6.** $1 - \dfrac{2}{(x - 3)} + \dfrac{3}{(x + 1)}$
7. $3x - 2 + \dfrac{1}{(x - 2)} - \dfrac{5}{(x + 2)}$

Exercise 82 (page 182)

1. $\dfrac{4}{(x+1)} - \dfrac{7}{(x+1)^2}$ **2.** $\dfrac{1}{x^2} + \dfrac{2}{x} - \dfrac{1}{(x+3)}$

3. $\dfrac{5}{(x-2)} - \dfrac{10}{(x-2)^2} + \dfrac{4}{(x-2)^3}$

4. $\dfrac{2}{(x-5)} - \dfrac{3}{(x+2)} + \dfrac{4}{(x+2)^2}$

Exercise 83 (page 183)

1. $\dfrac{2x+3}{(x^2+7)} - \dfrac{1}{(x-2)}$ **2.** $\dfrac{1}{(x-4)} + \dfrac{2-x}{(x^2+3)}$

3. $\dfrac{1}{x} + \dfrac{3}{x^2} + \dfrac{2-5x}{(x^2+5)}$

4. $\dfrac{3}{(x-1)} + \dfrac{2}{(x-1)^2} + \dfrac{1-2x}{(x^2+8)}$ **5.** Proof

Chapter 20

Exercise 84 (page 186)

1. $21, 25$ **2.** $48, 96$ **3.** $14, 7$ **4.** $-3, -8$

5. $50, 65$ **6.** $0.001, 0.0001$ **7.** $54, 79$

Exercise 85 (page 187)

1. $1, 3, 5, 7, \ldots$ **2.** $7, 10, 13, 16, 19, \ldots$

3. $6, 11, 16, 21, \ldots$ **4.** $5n$ **5.** $6n - 2$

6. $2n + 1$ **7.** $4n - 2$ **8.** $3n + 6$

9. $6^3 (= 216), 7^3 (= 343)$

Exercise 86 (page 188)

1. 68 **2.** 6.2 **3.** 85.25 **4.** 23.5 **5.** 11

6. 209 **7.** 346.5

Exercise 87 (page 189)

1. -0.5 **2.** $1.5, 3, 4.5$ **3.** 7808 **4.** 25

5. $8.5, 12, 15.5, 19$ **6.** (a) 120 (b) $26\,070$ (c) 250.5

7. £19840, £223680   **8.** £8720

Exercise 88 (page 191)

1. 2560 **2.** 273.25 **3.** $512, 4096$

4. 812.5 **5.** 8 **6.** $1\dfrac{2}{3}$

Exercise 89 (page 193)

1. (a) 3 (b) 2 (c) 59022   **2.** 10th

3. £1566, 11 years   **4.** 56.68 million

5. $71.53\,\text{g}$ **6.** (a) £599.14   (b) 19 years

7. $100, 139, 193, 268, 373, 518, 720, 1000\,\text{rev/min}$

Exercise 90 (page 194)

1. (a) 84 (b) 3   **2.** (a) 15 (b) 56

3. (a) 12 (b) 840   **4.** (a) 720 (b) 6720

Chapter 21

Exercise 91 (page 198)

1. $x^7 - 7x^6 y + 21x^5 y^2 - 35x^4 y^3$
$\quad + 35x^3 y^4 - 21x^2 y^5 + 7xy^6 - y^7$

2. $32a^5 + 240a^4 b + 720a^3 b^2$
$\quad + 1080a^2 b^3 + 810ab^4 + 243b^5$

Exercise 92 (page 200)

1. $a^4 + 8a^3 x + 24a^2 x^2 + 32ax^3 + 16x^4$

2. $64 - 192x + 240x^2 - 160x^3 + 60x^4 - 12x^5 + x^6$

3. $16x^4 - 96x^3 y + 216x^2 y^2 - 216xy^3 + 81y^4$

4. $32x^5 + 160x^3 + 320x + \dfrac{320}{x} + \dfrac{160}{x^3} + \dfrac{32}{x^5}$

5. $p^{11} + 22p^{10} q + 220p^9 q^2 + 1320p^8 q^3 + 5280p^7 q^4$

6. $34749\,p^8 q^5$ **7.** $700\,000\,a^4 b^4$

8. (a) 1.0243 (b) 0.8681 **9.** 4373.88

Exercise 93 (page 202)

1. $1 + x + x^2 + x^3 + \cdots, |x| < 1$

2. $1 - 2x + 3x^2 - 4x^3 + \cdots, |x| < 1$

3. $\dfrac{1}{8}\left(1 - \dfrac{3x}{2} + \dfrac{3x^2}{2} - \dfrac{5x^3}{4} + \cdots\right) |x| < 2$

4. $\sqrt{2}\left(1 + \dfrac{x}{4} - \dfrac{x^2}{32} + \dfrac{x^3}{128} - \cdots\right) |x| < 2$

5. $\left(1 - \dfrac{3}{2}x + \dfrac{27}{8}x^2 - \dfrac{135}{16}x^3 + \cdots\right) |x| < \dfrac{1}{3}$

6. $\dfrac{1}{64}\left(1 - 9x + \dfrac{189}{4}x^2\right) |x| < \dfrac{2}{3}$ **7.** Proof

8. $4 - \dfrac{31}{15}x$ **9.** (a) $1 - x + \dfrac{1}{2}x^2, \ |x| < 1$

   (b) $1 - x - \dfrac{7}{2}x^2, \ |x| < \dfrac{1}{3}$

Exercise 94 (page 203)

1. 0.6% decrease   **2.** 3.5% decrease

3. (a) 4.5% increase   (b) 3.0% increase

4. 2.2% increase   **5.** 4.5% increase

6. Proof   **7.** 7.5% decrease

8. 2.5% increase   **9.** 0.9% too small

10. +7%   **11.** Proof

12. 5.5%

Chapter 22

Exercise 95 (page 210)

1. $2x - \dfrac{4}{3}x^3 + \dfrac{4}{15}x^5 - \dfrac{8}{315}x^7 + \cdots$

2. $1 + \dfrac{9}{2}x^2 + \dfrac{27}{8}x^4 + \dfrac{81}{80}x^6$ **3.** $\ln 2 + \dfrac{x}{2} + \dfrac{x^2}{8}$

4. $1 - 8t^2 + \dfrac{32}{3}t^4 - \dfrac{256}{45}t^6$

5. $1 + \dfrac{3}{2}x + \dfrac{9}{8}x^2 + \dfrac{9}{16}x^3$ **6.** $1 + 2x^2 + \dfrac{10}{3}x^4$

7. $1 + 2\theta - \dfrac{5}{2}\theta^2$ **8.** $x^2 - \dfrac{1}{3}x^4 + \dfrac{2}{45}x^6 \cdots$

9. $81 + 216t + 216t^2 + 96t^3 + 16t^4$

Exercise 96 (page 212)

1. 1.784 **2.** 0.88 **3.** 0.53 **4.** 0.061

Exercise 97 (page 214)

1. $\dfrac{1}{9}$ **2.** 1   **3.** 1   **4.** -1 **5.** $\dfrac{1}{3}$ **6.** $\dfrac{1}{2}$

7. $\dfrac{1}{3}$ **8.** 1   **9.** $\dfrac{1}{2}$

Chapter 23

Exercise 98 (page 220)

1. 1.19   **2.** 1.146   **3.** 1.20   **4.** 3.146   **5.** 1.849

Exercise 99 (page 223)

1. $-3.36, 1.69$ **2.** -2.686

3. $-1.53, 1.68$ **4.** $-12.01, 1.000$

Exercise 100 (page 225)

1. $-2.742, 4.742$ **2.** 2.313   **3.** $-1.721, 2.648$

4. $-1.386, 1.491$ **5.** 1.147

6. $-1.693, -0.846, 0.744$ **7.** 2.05

8. See earlier answers **9.** 0.0399 **10.** 4.19

11. 2.9143

Chapter 24

Exercise 101 (page 229)

1. (a) 0.6846 (b) 4.376   **2.** (a) 1.271 (b) 5.910

3. (a) 0.5717 (b) 0.9478   **4.** (a) 1.754 (b) 0.08849

5. (a) 0.9285 (b) 0.1859   **6.** (a) 2.398 (b) 1.051

7. 56.38   **8.** 30.71

9. 5.042

Exercise 102 (page 233)

1. Proof   **2.** Proof   **3.** Proof   **4.** Proof
5. $P = 2, Q = -4$ **6.** $A = 9, B = 1$

Exercise 103 (page 235)

1. (a) 0.8814 (b) −1.6209

2. (a) ±1.2384 (b) ±0.9624

3. (a) −0.9962 (b) 1.0986

4. (a) ±2.1272 (b) ±0.4947

5. (a) 0.6442 (b) 0.9832   **6.** (a) 0.4162 (b) −0.6176

7. −0.8959   **8.** 0.6389 or −2.2484

9. 0.2554   **10.** (a) 67.30 (b) ±26.42

Exercise 104 (page 236)

1. (a) 2.3524 (b) 1.3374   **2.** (a) 0.5211 (b) 3.6269

3. (a) $3x + \dfrac{9}{2}x^3 + \dfrac{81}{40}x^5$ (b) $1 + 2x^2 + \dfrac{2}{3}x^4$

4. Proof   **5.** Proof

Chapter 25

Exercise 105 (page 240)

1. (a) 6_{10} (b) 11_{10} (c) 14_{10} (d) 9_{10}

2. (a) 21_{10} (b) 25_{10} (c) 45_{10} (d) 51_{10}

3. (a) 42_{10} (b) 56_{10} (c) 65_{10} (d) 184_{10}

4. (a) 0.8125_{10} (b) 0.78125_{10} (c) 0.21875_{10}
(d) 0.34375_{10}

5. (a) 26.75_{10} (b) 23.375_{10} (c) 53.4375_{10}
(d) 213.71875_{10}

Exercise 106 (page 242)

1. (a) 101_2 (b) 1111_2 (c) 10011_2 (d) 11101_2

2. (a) 11111_2 (b) 101010_2 (c) 111001_2
(d) 111111_2

3. (a) 101111_2 (b) 111100_2 (c) 1001001_2
(d) 1010100_2

4. (a) 0.01_2 (b) 0.00111_2 (c) 0.01001_2
(d) 0.10011_2

5. (a) 101111.01101_2 (b) 11110.1101_2
(c) 110101.11101_2 (d) 111101.10101_2

Exercise 107 (page 243)

1. 101   **2.** 1011   **3.** 10100

4. 101100   **5.** 1001000   **6.** 100001010

7. 1010110111   **8.** 1001111101 **9.** 111100

10. 110111   **11.** 110011

12. 1101110   **13.** 100100

Exercise 108 (page 245)

1. (a) $1010101 11_2$ (b) 1000111100_2
(c) 10011110001_2

2. (a) 0.01111_2 (b) 0.1011_2 (c) 0.10111_2

3. (a) 11110111.00011_2 (b) 1000000010.0111_2
(c) 11010110100.11001_2

4. (a) 7.4375_{10} (b) 41.25_{10} (c) 7386.1875_{10}

Exercise 109 (page 247)

1. 231_{10} **2.** 44_{10} **3.** 152_{10} **4.** 753_{10}

5. 36_{16} **6.** $C8_{16}$ **7.** $5B_{16}$ **8.** EE_{16}

Exercise 110 (page 248)

1. $D7_{16}$ **2.** EA_{16} **3.** $8B_{16}$

4. $A5_{16}$ **5.** 110111_2 **6.** 11101101_2

7. 10011111_2 **8.** 101000100001_2

Chapter 26

Exercise 111 (page 253)

1. $Z = C \cdot \left(A \cdot B + \overline{A} \cdot B \right)$

A	B	C	$A \cdot B$	$\overline{A}$	$\overline{A} \cdot B$	$A \cdot B + \overline{A} \cdot B$	$Z = C \cdot (A \cdot B + \overline{A} \cdot B)$
0	0	0	0	1	0	0	0
0	0	1	0	1	0	0	0
0	1	0	0	1	1	1	0
0	1	1	0	1	1	1	1
1	0	0	0	0	0	0	0
1	0	1	0	0	0	0	0
1	1	0	1	0	0	1	0
1	1	1	1	0	0	1	1

2. $Z = C \cdot \left(A \cdot \overline{B} + \overline{A} \right)$

A	B	C	$\overline{A}$	$\overline{B}$	$A \cdot \overline{B}$	$A \cdot \overline{B} + \overline{A}$	$Z = C \cdot (A \cdot \overline{B} + \overline{A})$
0	0	0	1	1	0	1	0
0	0	1	1	1	0	1	1
0	1	0	1	0	0	1	0
0	1	1	1	0	0	1	1
1	0	0	0	1	1	1	0
1	0	1	0	1	1	1	1
1	1	0	0	0	0	0	0
1	1	1	0	0	0	0	0

3. $Z = A \cdot B \cdot \left(B \cdot \overline{C} + \overline{B} \cdot C + \overline{A} \cdot B \right)$

A	B	C	$\overline{A}$	$\overline{B}$	$\overline{C}$	$B \cdot \overline{C}$	$\overline{B} \cdot C$	$\overline{A} \cdot B$	$(B \cdot \overline{C} + \overline{B} \cdot C + \overline{A} \cdot B)$	$Z = A \cdot B \cdot (B \cdot \overline{C} + \overline{B} \cdot C + \overline{A} \cdot B)$
0	0	0	1	1	1	0	0	0	0	0
0	0	1	1	1	0	0	1	0	1	0
0	1	0	1	0	1	1	0	1	1	0
0	1	1	1	0	0	0	0	1	1	0
1	0	0	0	1	1	0	0	0	0	0
1	0	1	0	1	0	0	1	0	1	0
1	1	0	0	0	1	1	0	0	1	1
1	1	1	0	0	0	0	0	0	0	0

4. $Z = C \cdot \left(B \cdot C \cdot \overline{A} + A \cdot (B + \overline{C}) \right)$

A	B	C	$\overline{A}$	$\overline{C}$	$B \cdot C \cdot \overline{A}$	$B + \overline{C}$	$A \cdot (B + \overline{C})$	$B \cdot C \cdot \overline{A} + A \cdot (B + \overline{C})$	Z
0	0	0	1	1	0	1	0	0	0
0	0	1	1	0	0	0	0	0	0
0	1	0	1	1	0	1	0	0	0
0	1	1	1	0	1	1	0	1	1
1	0	0	0	1	0	1	1	1	0
1	0	1	0	0	0	0	0	0	0
1	1	0	0	1	0	1	1	1	0
1	1	1	0	0	0	1	1	1	1

5.

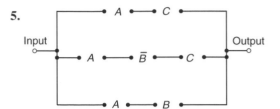

6.

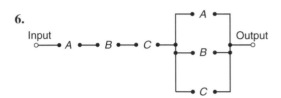

7.

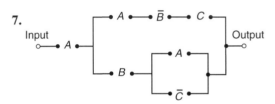

8. $\overline{A} \cdot \overline{B} \cdot C + A \cdot B \cdot \overline{C}$

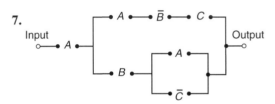

9. $\overline{A} \cdot \overline{B} \cdot \overline{C} + \overline{A} \cdot B \cdot C + A \cdot \overline{B} \cdot C$

10. $\overline{A} \cdot \overline{B} \cdot \overline{C} + \overline{A} \cdot B \cdot \overline{C} + A \cdot \overline{B} \cdot \overline{C} + A \cdot \overline{B} \cdot C$

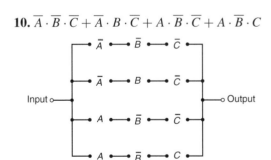

Exercise 112 (page 256)

1. $\overline{P}$ **2.** $\overline{P} + P \cdot Q$ **3.** $\overline{G}$

4. F **5.** $P \cdot Q$ **6.** $H \cdot (\overline{F} + F \cdot \overline{G})$

7. $F \cdot \overline{G} \cdot \overline{H} + G \cdot H$ **8.** $\overline{Q} \cdot \overline{R} + \overline{P} \cdot Q \cdot R$

9. $\overline{G}$ **10.** $F \cdot H + G \cdot \overline{H}$

11. $P \cdot R + \overline{P} \cdot \overline{R}$ **12.** $P + \overline{Q} \cdot \overline{R}$

Exercise 113 (page 257)

1. $\overline{A} \cdot \overline{B}$ **2.** $\overline{A} + \overline{B} + C$

3. $\overline{A} \cdot \overline{B} + A \cdot B \cdot C$ **4.** 1   **5.** $\overline{P} \cdot (Q + \overline{R})$

Exercise 114 (page 261)

1. Y **2.** $\overline{X} + Y$

3. $\overline{P} \cdot \overline{Q}$ **4.** $B + A \cdot \overline{C} + \overline{A} \cdot C$

5. $\overline{R} \cdot (\overline{P} + Q)$ **6.** $P \cdot (Q + R) + \overline{P} \cdot \overline{Q} \cdot \overline{R}$

7. $\overline{A} \cdot \overline{C} \cdot (B + D)$ **8.** $\overline{B} \cdot C \cdot (\overline{A} + D)$

9. $D \cdot (A + B \cdot \overline{C})$ **10.** $A \cdot \overline{D} + \overline{A} \cdot \overline{B} \cdot \overline{C} \cdot D$

11. $\overline{A} \cdot C + A \cdot \overline{C} \cdot \overline{D} + B \cdot \overline{D} \cdot (\overline{A} + C)$

Exercise 115 (page 265)

1.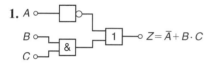

2.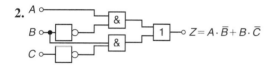

3.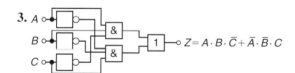

4.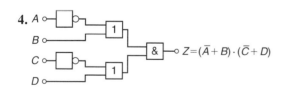

5. $Z_1 = A \cdot B + C$

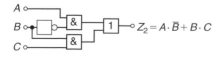

6. $Z_2 = A \cdot \overline{B} + B \cdot C$

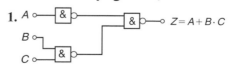

7. $Z_3 = A \cdot C + B$

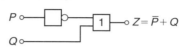

8. $Z = \overline{P} + Q$

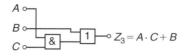

9. $\overline{R} \cdot (P + \overline{Q})$

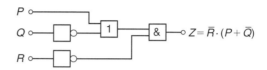

10. $\overline{Q} \cdot (P + \overline{R})$

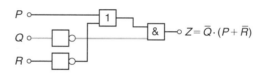

11. $\overline{D} \cdot (\overline{A} \cdot C + \overline{B})$

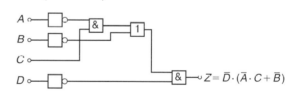

12. $\overline{P} \cdot (\overline{Q} + \overline{R})$

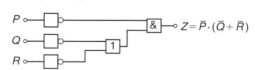

Exercise 116 (page 267)

1.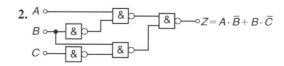

2. $Z = A \cdot \overline{B} + B \cdot \overline{C}$

3. 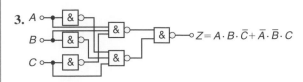 $Z = A \cdot B \cdot \bar{C} + \bar{A} \cdot \bar{B} \cdot C$

4. $Z = (\bar{A} + B) \cdot (\bar{C} + D)$

5. $Z = A \cdot \bar{B} + B \cdot \bar{C} + C \cdot \bar{D}$

6. $Z = \bar{P} \cdot Q + P \cdot (Q + R)$

7. $\bar{P} \cdot (Q + \bar{R})$

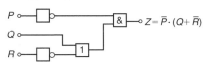 $Z = \bar{P} \cdot (Q + \bar{R})$

8. $Z = A \cdot (B + C + D)$

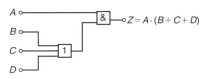 $Z = A \cdot (B + C + D)$

9. $Z = A \cdot (B + C) + B \cdot C$

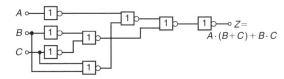 $Z = A \cdot (B + C) + B \cdot C$

10. $Z = A \cdot (B + C)$

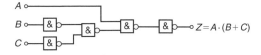 $Z = A \cdot (B + C)$

Chapter 27

Exercise 117 (page 279)

1. $p = 105°, q = 35°$ **2.** $r = 142°, s = 95°$
3. $t = 146°$

Exercise 118 (page 283)

1. (i) rhombus (a) $14\,\text{cm}^2$ (b) $16\,\text{cm}$ (ii) parallelogram
(a) $180\,\text{mm}^2$ (b) $80\,\text{mm}$ (iii) rectangle (a) $3600\,\text{mm}^2$
(b) $300\,\text{mm}$ (iv) trapezium (a) $190\,\text{cm}^2$ (b) $62.91\,\text{cm}$

2. $35.7\,\text{cm}^2$ **3.** (a) $80\,\text{m}$ (b) $170\,\text{m}$ **4.** $27.2\,\text{cm}^2$
5. $18\,\text{cm}$ **6.** $1200\,\text{mm}$
7. (a) $29\,\text{cm}^2$ (b) $650\,\text{mm}^2$ **8.** $560\,\text{m}^2$
9. $3.4\,\text{cm}$ **10.** $6750\,\text{mm}^2$ **11.** $43.30\,\text{cm}^2$
12. 32 **13.** $230,400$

Exercise 119 (page 285)

1. $482\,\text{m}^2$
2. (a) $50.27\,\text{cm}^2$ (b) $706.9\,\text{mm}^2$ (c) $3183\,\text{mm}^2$
3. $2513\,\text{mm}^2$ **4.** (a) $20.19\,\text{mm}$ (b) $63.41\,\text{mm}$
5. (a) $53.01\,\text{cm}^2$ (b) $129.9\,\text{mm}^2$ **6.** $5773\,\text{mm}^2$
7. $1.89\,\text{m}^2$ **8.** $15\,710\,\text{mm}^2$, $471\,\text{mm}$
9. £4712 **10.** $6597\,\text{m}^2$

Exercise 120 (page 287)

1. $1932\,\text{mm}^2$ **2.** $1624\,\text{mm}^2$ **3.** (a) $0.918\,\text{ha}$ (b) $456\,\text{m}$

Exercise 121 (page 288)

1. $80\,\text{ha}$ **2.** $80\,\text{m}^2$ **3.** $3.14\,\text{ha}$

Chapter 28

Exercise 122 (page 291)

1. $45.24\,\text{cm}$ **2.** $259.5\,\text{mm}$ **3.** $2.629\,\text{cm}$ **4.** $47.68\,\text{cm}$
5. $38.73\,\text{cm}$ **6.** $12730\,\text{km}$ **7.** $97.13\,\text{mm}$

Exercise 123 (page 292)

1. (a) $\dfrac{\pi}{6}$ (b) $\dfrac{5\pi}{12}$ (c) $\dfrac{5\pi}{4}$
2. (a) 0.838 (b) 1.481 (c) 4.054
3. (a) $210°$ (b) $80°$ (c) $105°$
4. (a) $0°43'$ (b) $154°8'$ (c) $414°53'$ **5.** $104.7\,\text{rad/s}$

Exercise 124 (page 294)

1. $113\,\text{cm}^2$ **2.** $2376\,\text{mm}^2$ **3.** $1790\,\text{mm}^2$
4. $802\,\text{mm}^2$ **5.** $1709\,\text{mm}^2$ **6.** $1269\,\text{m}^2$
7. $1548\,\text{m}^2$ **8.** (a) $106.0\,\text{cm}$ (b) $783.9\,\text{cm}^2$
9. $21.46\,\text{m}^2$ **10.** $17.80\,\text{cm}$, $74.07\,\text{cm}^2$
11. (a) $59.86\,\text{mm}$ (b) $197.8\,\text{mm}$ **12.** $26.2\,\text{cm}$
13. $202\,\text{mm}^2$ **14.** $8.67\,\text{cm}$, $54.48\,\text{cm}$ **15.** $82°30'$
16. 748 **17.** (a) $0.698\,\text{rad}$ (b) $804.2\,\text{m}^2$

18. $10.47\,\text{m}^2$ **19.** (a) $396\,\text{mm}^2$ (b) 42.24%
20. $701.8\,\text{mm}$ **21.** $7.74\,\text{mm}$

Exercise 125 (page 297)

1. (a) 2 (b) $(3, -4)$ **2.** Centre at $(3, -2)$, radius 4
3. Circle, centre $(0, 1)$, radius 5
4. Circle, centre $(0, 0)$, radius 6

Exercise 126 (page 298)

1. $\omega = 90\,\text{rad/s}$, $v = 13.5\,\text{m/s}$
2. $v = 10\,\text{m/s}$, $\omega = 40\,\text{rad/s}$
3. (a) $75\,\text{rad/s}$, $716.2\,\text{rev/min}$ (b) $1074\,\text{revs}$

Exercise 127 (page 300)

1. $2\,\text{N}$ **2.** $988\,\text{N}$, $5.14\,\text{km/h}$ **3.** $1.49\,\text{m/s}^2$

Chapter 29

Exercise 128 (page 307)

1. $1.2\,\text{m}^3$ **2.** $5\,\text{cm}^3$ **3.** $8\,\text{cm}^3$
4. (a) $3840\,\text{mm}^3$ (b) $1792\,\text{mm}^2$
5. $972\,\text{litres}$ **6.** $15\,\text{cm}^3$, $135\,\text{g}$ **7.** $500\,\text{litres}$
8. $1.44\,\text{m}^3$ **9.** (a) $35.3\,\text{cm}^3$ (b) $61.3\,\text{cm}^2$
10. (a) $2400\,\text{cm}^3$ (b) $2460\,\text{cm}^2$ **11.** $37.04\,\text{m}$
12. $1.63\,\text{cm}$ **13.** $8796\,\text{cm}^3$
14. $4.709\,\text{cm}$, $153.9\,\text{cm}^2$
15. $2.99\,\text{cm}$ **16.** $28060\,\text{cm}^3$, $1.099\,\text{m}^2$
17. $8.22\,\text{m}$ by $8.22\,\text{m}$ **18.** $62.5\,\text{min}$
19. $4\,\text{cm}$ **20.** $4.08\,\text{m}^3$

Exercise 129 (page 310)

1. $201.1\,\text{cm}^3$, $159.0\,\text{cm}^2$ **2.** $7.68\,\text{cm}^3$, $25.81\,\text{cm}^2$
3. $113.1\,\text{cm}^3$, $113.1\,\text{cm}^2$ **4.** $5.131\,\text{cm}$ **5.** $3\,\text{cm}$
6. $2681\,\text{mm}^3$ **7.** (a) $268083\,\text{mm}^3$ or $268.083\,\text{cm}^3$
    (b) $20106\,\text{mm}^2$ or $201.06\,\text{cm}^2$
8. $8.53\,\text{cm}$ **9.** (a) $512 \times 10^6\,\text{km}^2$ (b) $1.09 \times 10^{12}\,\text{km}^3$
10. 664   **11.** $92\ \text{m}^3$, $92000\,\text{litres}$

Exercise 130 (page 314)

1. $5890\,\text{mm}^2$ or $58.90\,\text{cm}^2$
2. (a) $56.55\,\text{cm}^3$ (b) $84.82\,\text{cm}^2$ **3.** $13.57\,\text{kg}$
4. $29.32\,\text{cm}^3$ **5.** $393.4\,\text{m}^2$
6. (i) (a) $670\,\text{cm}^3$ (b) $523\,\text{cm}^2$ (ii) (a) $180\,\text{cm}^3$
    (b) $154\,\text{cm}^2$ (iii) (a) $56.5\,\text{cm}^3$ (b) $84.8\,\text{cm}^2$
    (iv) (a) $10.4\,\text{cm}^3$ (b) $32.0\,\text{cm}^2$ (v) (a) $96.0\,\text{cm}^3$
    (b) $146\,\text{cm}^2$ (vi) (a) $86.5\,\text{cm}^3$ (b) $142\,\text{cm}^2$
    (vii) (a) $805\,\text{cm}^3$ (b) $539\,\text{cm}^2$

7. (a) $17.9\,\text{cm}$ (b) $38.0\,\text{cm}$ **8.** $125\,\text{cm}^3$
9. $10.3\,\text{m}^3$, $25.5\,\text{m}^2$ **10.** $6560\,\text{litres}$
11. $657.1\,\text{cm}^3$, $1027\,\text{cm}^2$ **12.** $220.7\,\text{cm}^3$
13. (a) $1458\,\text{litres}$ (b) $9.77\,\text{m}^2$ (c) £140.45

Exercise 131 (page 319)

1. $147\,\text{cm}^3$, $164\,\text{cm}^2$ **2.** $403\,\text{cm}^3$, $337\,\text{cm}^2$
3. $10480\,\text{m}^3$, $1852\,\text{m}^2$ **4.** $1707\,\text{cm}^2$
5. $10.69\,\text{cm}$ **6.** $55910\,\text{cm}^3$, $6051\,\text{cm}^2$
7. $5.14\,\text{m}$

Exercise 132 (page 321)

1. $11210\,\text{cm}^3$, $1503\,\text{cm}^2$ **2.** $259.2\,\text{cm}^3$, $118.3\,\text{cm}^2$
3. $1150\,\text{cm}^3$, $531\,\text{cm}^2$, $2.60\,\text{cm}$, $326.7\,\text{cm}^3$
4. $14.84\,\text{cm}^3$ **5.** $35.34\,\text{litres}$

Exercise 133 (page 323)

1. $1500\,\text{cm}^3$ **2.** $418.9\,\text{cm}^3$
3. $31.20\,\text{litres}$ **4.** $1.267 \times 10^6\,\text{litre}$

Exercise 134 (page 324)

1. $8 : 125$ **2.** $137.2\,\text{g}$

Chapter 30

Exercise 135 (page 327)

1. $4.5\,\text{square units}$ **2.** $54.7\,\text{square units}$ **3.** $63.33\,\text{m}$
4. $4.70\,\text{ha}$ **5.** $143\,\text{m}^2$

Exercise 136 (page 328)

1. $42.59\,\text{m}^3$ **2.** $147\,\text{m}^3$ **3.** $20.42\,\text{m}^3$

Exercise 137 (page 332)

1. (a) $2\,\text{A}$ (b) $50\,\text{V}$ (c) $2.5\,\text{A}$ **2.** (a) $2.5\,\text{mV}$ (b) $3\,\text{A}$
3. $0.093\,\text{As}$, $3.1\,\text{A}$ **4.** (a) $31.83\,\text{V}$ (b) 0
5. $49.13\,\text{cm}^2$, $368.5\,\text{kPa}$

Chapter 31

Exercise 138 (page 346)

1. (a) Horizontal axis: $1\,\text{cm} = 4\,\text{V}$ (or $1\,\text{cm} = 5\,\text{V}$),
    vertical axis: $1\,\text{cm} = 10\,\Omega$
    (b) Horizontal axis: $1\,\text{cm} = 5\,\text{m}$, vertical axis:
    $1\,\text{cm} = 0.1\,\text{V}$

(c) Horizontal axis: $1\,\text{cm} = 10\,\text{N}$, vertical axis:
$1\,\text{cm} = 0.2\,\text{mm}$

2. (a) -1 (b) -8 (c) -1.5 (d) 4   3. 14.5
4. (a) -1.1 (b) -1.4
5. The 1010 rev/min reading should be 1070 rev/min;
(a) 1000 rev/min (b) 167 V

Exercise 139 (page 352)

1. Missing values: $-0.75, 0.25, 0.75, 1.75, 2.25, 2.75$;
Gradient $= \dfrac{1}{2}$
2. (a) $4, -2$ (b) $-1, 0$ (c) $-3, -4$ (d) $0, 4$
3. (a) $2, \dfrac{1}{2}$ (b) $3, -2\dfrac{1}{2}$ (c) $\dfrac{1}{24}, \dfrac{1}{2}$
4. (a) $6, -3$ (b) $-2, 4$ (c) $3, 0$ (d) $0, 7$
5. (a) $2, -\dfrac{1}{2}$ (b) $-\dfrac{2}{3}, -1\dfrac{2}{3}$ (c) $\dfrac{1}{18}, 2$ (d) $10, -4\dfrac{2}{3}$
6. (a) $\dfrac{3}{5}$ (b) -4 (c) $-1\dfrac{5}{6}$
7. (a) and (c), (b) and (e)
8. $(2, 1)$ 9. $(1.5, 6)$ 10. $(1, 2)$
11. (a) 89 cm (b) 11 N (c) 2.4 (d) $l = 2.4W + 48$
12. $P = 0.15W + 3.5$ 13. $a = -20, b = 412$

Exercise 140 (page 357)

1. (a) $40°\text{C}$ (b) $128\,\Omega$
2. (a) 850 rev/min (b) 77.5 V
3. (a) 0.25 (b) 12 (c) $F = 0.25L + 12$
(d) 89.5 N (e) 592 N (f) 212 N
4. $-0.003, 8.73\,\text{N/cm}^2$
5. (a) 22.5 m/s (b) 6.5 s (c) $v - 0.7t + 15.5$
6. $m = 26.8L$
7. (a) $1.25t$ (b) 21.6% (c) $F = -0.095w + 2.2$
8. (a) $96 \times 10^9\,\text{Pa}$ (b) 0.00022 (c) $29 \times 10^6\,\text{Pa}$
9. (a) $\dfrac{1}{5}$ (b) 6 (c) $E = \dfrac{1}{5}L + 6$ (d) 12 N (e) 65 N
10. $a = 0.85, b = 12,\ 254.3\,\text{kPa}, 275.5\,\text{kPa}, 280\,\text{K}$

Chapter 32

Exercise 141 (page 363)

1. (a) y (b) x^2 (c) c (d) d 2. (a) y (b) $\sqrt{x}$ (c) b (d) a
3. (a) y (b) $\dfrac{1}{x}$ (c) f (d) e 4. (a) $\dfrac{y}{x}$ (b) x (c) b (d) c

5. (a) $\dfrac{y}{x}$ (b) $\dfrac{1}{x^2}$ (c) a (d) b
6. $a = 1.5, b = 0.4, 11.78\,\text{mm}^2$
7. $y = 2x^2 + 7,\ 5.15$
8. (a) 950 (b) 317 kN
9. $a = 0.4, b = 8.6$ (a) 94.4 (b) 11.2

Exercise 142 (page 367)

1. (a) $\lg y$ (b) x (c) $\lg a$ (d) $\lg b$
2. (a) $\lg y$ (b) $\lg x$ (c) L (d) $\lg k$
3. (a) $\ln y$ (b) x (c) n (d) $\ln m$
4. $I = 0.0012\,\text{V}^2, 6.75$ candelas
5. $a = 3.0, b = 0.5$
6. $a = 5.6, b = 2.6, 37.86, 3.0$
7. $R_0 = 25.1, c = 1.42$
8. $y = 0.08e^{0.24x}$
9. $T_0 = 35.3\,\text{N}, \mu = 0.27, 64.8\,\text{N}, 1.29$ radians

Chapter 33

Exercise 143 (page 372)

1. $a = 12, n = 1.8, 451, 28.5$ 2. $k = 1.5, n = -1$
3. $m = 3, n = 2.5$

Exercise 144 (page 374)

1. (i) $a = -8, b = 5.3, p = -8(5.3)^q$ (ii) -224.7
(iii) 3.31

Exercise 145 (page 376)

1. $a = 76, k = -7 \times 10^{-5}, p = 76e^{-7 \times 10^{-5}h}, 37.74\,\text{cm}$
2. $\theta_0 = 152, k = -0.05$

Chapter 34

Exercise 146 (page 382)

1. (a) Similar to Figure 34.2, where $a = 3$
(b) Similar to Figure 34.3, where $a = 4$
2. (a) Similar to Figure 34.5, where $a = 2$
(b) Similar to Figure 34.6, where $a = 7$
3. (a) Similar to Figure 34.8, where $a = 5$
(b) Similar to Figure 34.9(a), where $a = 4$
4. (a) Similar to Figure 34.9(c), where $a = 3$
(b) Similar to Figure 34.9(b), where $a = 9$
5. (a) Similar to Figure 34.10
(b) Similar to Figure 34.12, where $a = 2$

6. (a) Similar to Figure 34.13(a), where $a = 1$
and $b = 2$
(b) Similar to Figure 34.13(b), where $a = 3$
and $b = 1$

Chapter 35

Exercise 147 (page 384)

1. $x = 2, y = 4$ **2.** $x = 1, y = 1$
3. $x = 3.5, y = 1.5$ **4.** $x = -1, y = 2$
5. $x = 2.3, y = -1.2$ **6.** $x = -2, y = -3$
7. $a = 0.4, b = 1.6$

Exercise 148 (page 388)

1. (a) Minimum $(0, 0)$ (b) Minimum $(0, -1)$
(c) Maximum $(0, 3)$ (d) Maximum $(0, -1)$
2. -0.4 or 0.6 **3.** -3.9 or 6.9
4. -1.1 or 4.1 **5.** -1.8 or 2.2
6. $x = -1.5$ or -2, Minimum at $(-1.75, -0.1)$
7. $x = -0.7$ or 1.6
8. (a) ± 1.63 (b) 1 or -0.3

9. $(-2.6, 13.2), (0.6, 0.8); x = -2.6$ or 0.6
10. $x = -1.2$ or 2.5 (a) -30 (b) 2.75 and -1.50
(c) 2.3 or -0.8

Exercise 149 (page 389)

1. $x = 4, y = 8$ and $x = -0.5, y = -5.5$
2. (a) $x = -1.5$ or 3.5 (b) $x = -1.24$ or 3.24
(c) $x = -1.5$ or 3.0

Exercise 150 (page 390)

1. $x = -2.0, -0.5$ or 1.5
2. $x = -2, 1$ or 3, Minimum at $(2.1, -4.1)$,
Maximum at $(-0.8, 8.2)$
3. $x = 1$ **4.** $x = -2.0, 0.4$ or 2.6
5. $x = -1.2$ or 0.70 or 2.5
6. $x = -2.3, 1.0$ or 1.8 **7.** $x = -1.5$

Chapter 36

Exercise 151 (page 400)

See Figure 36.39

1.

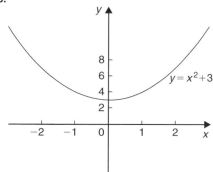

2.

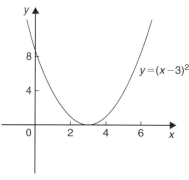

3.

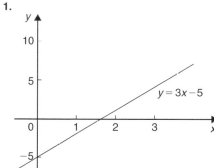

4.

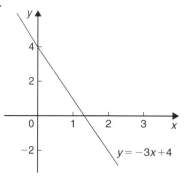

Figure 36.39 Graphical solutions to Exercise 151, page 400.

5.

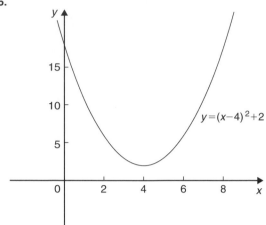

$y = (x-4)^2 + 2$

6.

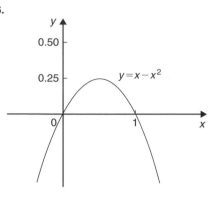

$y = x - x^2$

7.

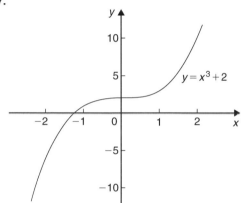

$y = x^3 + 2$

8.

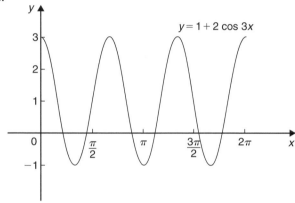

$y = 1 + 2 \cos 3x$

9.

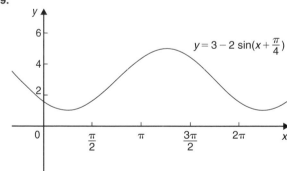

$y = 3 - 2\sin\left(x + \frac{\pi}{4}\right)$

10.

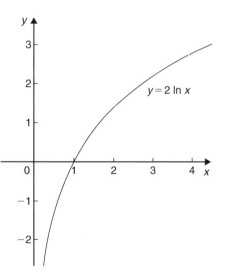

$y = 2 \ln x$

Figure 36.39 (*Continued*)

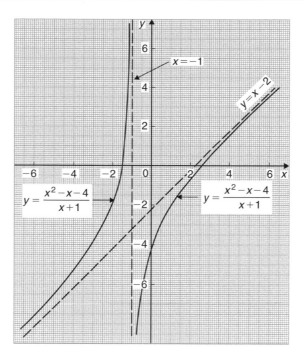

Figure 36.40

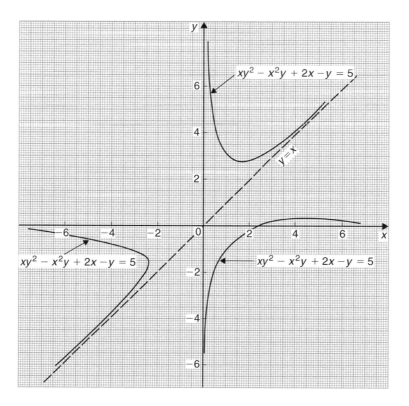

Figure 36.41

Exercise 152 (page 402)

1. (a) even   (b) odd   **2.** (a) odd   (b) even
   (c) neither   (d) even   (c) odd   (d) neither
3. (a) even   (b) odd

Exercise 153 (page 404)

1. $f^{-1}(x) = x - 1$ **2.** $f^{-1}(x) = \frac{1}{5}(x+1)$

3. $f^{-1}(x) = \sqrt[3]{x-1}$ **4.** $f^{-1}(x) = \frac{1}{x-2}$

5. $-\frac{\pi}{2}$ or -1.5708 rad   **6.** $\frac{\pi}{3}$ or 1.0472 rad

7. $\frac{\pi}{4}$ or 0.7854 rad   **8.** 0.4636 rad

9. 0.4115 rad   **10.** 0.8411 rad

11. $\frac{\pi}{4}$ or 0.7854 rad   **12.** 0.257 **13.** 1.533

Exercise 154 (page 410)

1. $y = 1, x = -1$ **2.** $x = 3, y = 1$ and $y = -1$
3. $x = -1, x = -2$ and $y = 1$ **4.** $x = 0, y = x$ and $y = -x$
5. $y = 4, y = -4$ and $x = 0$
6. $x = -1, y = x - 2$, see   **7.** $x = 0, y = 0, y = x$, see
   Figure 36.40, page 1167   Figure 36.41, page 1167

Exercise 155 (page 414)

1. (a) Parabola with minimum value at $(-1.5, -5)$
        and passing through $(0, 1.75)$
   (b) Parabola with maximum value at $(2, 70)$
        and passing through $(0, 50)$
2. circle, centre $(0, 0)$, radius 4 units
3. parabola, symmetrical about x-axis, vertex at $(0, 0)$
4. hyperbola, symmetrical about x- and y-axes, distance between vertices 8 units along x-axis
5. ellipse, centre $(0, 0)$, major axis 10 units along x-axis, minor axis $2\sqrt{10}$ units along y-axis
6. hyperbola, symmetrical about x- and y-axes, distance between vertices 6 units along x-axis
7. rectangular hyperbola, lying in first and third quadrants only
8. ellipse, centre $(0, 0)$, major axis 4 units along x-axis, minor axis $2\sqrt{2}$ units along y-axis
9. Centre at $(2, -5)$, radius 2
10. ellipse, centre $(0, 0)$, major axis $2\sqrt{3}$ units along y-axis, minor axis 2 units along x-axis
11. hyperbola, symmetrical about x- and y-axes, vertices 2 units apart along x-axis

12. circle, centre $(0, 0)$, radius 3 units
13. rectangular hyperbola, lying in first and third quadrants, symmetrical about x- and y-axes
14. parabola, vertex at $(0, 0)$, symmetrical about the x-axis
15. ellipse, centre $(0, 0)$, major axis $2\sqrt{8}$ units along the y-axis, minor axis 4 units along the x-axis

Chapter 37

Exercise 156 (page 426)

1. $82°27'$ **2.** $27°54'$ **3.** $51°11'$
4. $100°6'52''$ **5.** $15°44'17''$ **6.** $86°49'1''$
7. $72.55°$ **8.** $27.754°$ **9.** $37°57'$
10. $58°22'52''$

Exercise 157 (page 428)

1. reflex **2.** obtuse **3.** acute   **4.** right angle
5. (a) $21°$ (b) $62°23'$ (c) $48°56'17''$
6. (a) $102°$ (b) $165°$ (c) $10°18'49''$
7. (a) $60°$ (b) $110°$ (c) $75°$ (d) $143°$ (e) $140°$
   (f) $20°$ (g) $129.3°$ (h) $79°$ (i) $54°$
8. Transversal (a) 1 & 3, 2 & 4, 5 & 7, 6 & 8
   (b) 1 & 2, 2 & 3, 3 & 4, 4 & 1, 5 & 6, 6 & 7, 7 & 8, 8 & 5, 3 & 8, 1 & 6, 4 & 7 or 2 & 5
   (c) 1 & 5, 2 & 6, 4 & 8, 3 & 7 (d) 3 & 5 or 2 & 8
9. $59°20'$ **10.** $a = 69°, b = 21°, c = 82°$ **11.** $51°$
12. 1.326 rad **13.** 0.605 rad **14.** $40°55'$

Exercise 158 (page 432)

1. (a) acute-angled scalene triangle
   (b) isosceles triangle   (c) right-angled triangle
   (d) obtuse-angled scalene triangle
   (e) equilateral triangle (f) right-angled triangle
2. $a = 40°, b = 82°, c = 66°,$
   $d = 75°, e = 30°, f = 75°$
3. DF, DE **4.** $52°$ **5.** $122.5°$
6. $\phi = 51°, x = 161°$
7. $40°, 70°, 70°, 125°$, isosceles
8. $a = 18°50', b = 71°10', c = 68°, d = 90°,$
   $e = 22°, f = 49°, g = 41°$
9. $a = 103°, b = 55°, c = 77°, d = 125°,$
   $e = 55°, f = 22°, g = 103°, h = 77°,$
   $i = 103°, j = 77°, k = 81°$
10. $17°$ **11.** $A = 37°, B = 60°, E = 83°$

Exercise 159 (page 434)

1. (a) congruent *BAC*, *DAC* (SAS)
   (b) congruent *FGE*, *JHI* (SSS)
   (c) not necessarily congruent
   (d) congruent *QRT*, *SRT* (RHS)
   (e) congruent *UVW*, *XZY* (ASA)
2. Proof

Exercise 160 (page 436)

1. $x = 16.54\,\text{mm}$, $y = 4.18\,\text{mm}$ 2. $9\,\text{cm}$, $7.79\,\text{cm}$
3. (a) $2.25\,\text{cm}$ (b) $4\,\text{cm}$ 4. $3\,\text{m}$

Exercise 161 (page 438)

1–5. Constructions – see similar constructions in
   Problems 30 to 33 on pages 437–438.

Chapter 38

Exercise 162 (page 441)

1. $9\,\text{cm}$ 2. $24\,\text{m}$ 3. $9.54\,\text{mm}$
4. $20.81\,\text{cm}$ 5. $7.21\,\text{m}$ 6. $11.18\,\text{cm}$
7. $24.11\,\text{mm}$ 8. $8^2 + 15^2 = 17^2$
9. (a) $27.20\,\text{cm}$ each   (b) $45°$
10. $20.81\,\text{km}$ 11. $3.35\,\text{m}$, $10\,\text{cm}$
12. 132.7 nautical miles   13. $2.94\,\text{mm}$
14. $24\,\text{mm}$

Exercise 163 (page 444)

1. $\sin Z = \dfrac{9}{41}$, $\cos Z = \dfrac{40}{41}$, $\tan X = \dfrac{40}{9}$, $\cos X = \dfrac{9}{41}$

2. $\sin A = \dfrac{3}{5}$, $\cos A = \dfrac{4}{5}$, $\tan A = \dfrac{3}{4}$, $\sin B = \dfrac{4}{5}$,

   $\cos B = \dfrac{3}{5}$, $\tan B = \dfrac{4}{3}$

3. $\sin A = \dfrac{8}{17}$, $\tan A = \dfrac{8}{15}$

4. $\sin X = \dfrac{15}{113}$, $\cos X = \dfrac{112}{113}$

5. (a) $\dfrac{15}{17}$ (b) $\dfrac{15}{17}$ (c) $\dfrac{8}{15}$

6. (a) $\sin\theta = \dfrac{7}{25}$ (b) $\cos\theta = \dfrac{24}{25}$

7. (a) 9.434 (b) -0.625

Exercise 164 (page 446)

1. 2.7550 2. 4.846 3. 36.52
4. (a) 0.8660 (b) -0.1010 (c) 0.5865
5. $42.33°$ 6. $15.25°$ 7. $73.78°$ 8. $7°56'$
9. $31°22'$ 10. $41°54'$ 11. $29.05°$ 12. $20°21'$
13. 0.3586 14. 1.803 15. (a) $40°$ (b) $6.79\,\text{m}$

Exercise 165 (page 448)

1. $\sin X = \dfrac{24}{25}$, $\tan X = \dfrac{24}{7}$, $\sec X = \dfrac{25}{7}$,

   $\operatorname{cosec} X = \dfrac{25}{24}$, $\cot X = \dfrac{7}{24}$

2. $\operatorname{cosec}\theta = 2.50$, $\sec\theta = 2.00$, $\tan\theta = 0.80$,
   $\cot\theta = 1.25$
3. (a) 3.4203 (b) 3.5313 (c) -1.0974
4. (a) -1.8361 (b) 3.7139 (c) -1.3421
5. (a) 0.3443 (b) -1.8510 (c) -1.2519
6. (a) 1.0824 (b) 5.5675 (c) -1.7083
7. $51.92°$, $51°55'$, 0.906 rad
8. $23.69°$, $23°41'$, 0.413 rad
9. $27.01°$, $27°1'$, 0.471 rad   10. 0.7199
11. 1.8258, 1.1952, 0.6546 12. 12.85
13. -1.710

Exercise 166 (page 450)

1. $\dfrac{1}{2}$ 2. $\dfrac{7}{2}\sqrt{3}$ 3. 1 4. $2-\sqrt{3}$ 5. $\dfrac{1}{\sqrt{3}}$

Exercise 167 (page 451)

1. (a) 12.22 (b) 5.619 (c) 14.87 (d) 8.349
   (e) 5.595 (f) 5.275
2. (a) $AC = 5.831\,\text{cm}$, $\angle A = 59.04°$, $\angle C = 30.96°$
   (b) $DE = 6.928\,\text{cm}$, $\angle D = 30°$, $\angle F = 60°$
   (c) $\angle J = 62°$, $HJ = 5.634\,\text{cm}$, $GH = 10.60\,\text{cm}$
   (d) $\angle L = 63°$, $LM = 6.810\,\text{cm}$, $KM = 13.37\,\text{cm}$
   (e) $\angle N = 26°$, $ON = 9.124\,\text{cm}$, $NP = 8.201\,\text{cm}$
   (f) $\angle S = 49°$, $RS = 4.346\,\text{cm}$, $QS = 6.625\,\text{cm}$
3. $6.54\,\text{m}$ 4. $9.40\,\text{mm}$ 5. $5.63\,\text{m}$

Exercise 168 (page 454)

1. $36.15\,\text{m}$ 2. $48\,\text{m}$ 3. $249.5\,\text{m}$ 4. $110.1\,\text{m}$
5. $53.0\,\text{m}$ 6. $9.50\,\text{m}$ 7. $107.8\,\text{m}$
8. $9.43\,\text{m}$, $10.56\,\text{m}$ 9. $60\,\text{m}$

Chapter 39

Exercise 169 (page 462)

1. (a) $42.78°$ and $137.22°$ (b) $188.53°$ and $351.47°$
2. (a) $29.08°$ and $330.92°$ (b) $123.86°$ and $236.14°$
3. (a) $44.21°$ and $224.21°$ (b) $113.12°$ and $293.12°$
4. $t = 122°7'$ and $237°53'$
5. $\alpha = 218°41'$ and $321°19'$
6. $\theta = 39°44'$ and $219°44'$

Exercise 170 (page 466)

1. 5   2. $180°$ 3. 30   4. $120°$
5. $1, 120°$ 6. $2, 144°$ 7. $3, 90°$ 8. $5, 720°$
9. $3.5, 960°$ 10. $6, 360°$ 11. $4, 180°$ 12. 5 ms
13. 40 Hz   14. $100\,\mu s$ or 0.1 ms
15. 625 Hz   16. leading   17. leading

Exercise 171 (page 468)

1. (a) 40 mA (b) 25 Hz (c) 0.04 s or 40 ms
   (d) 0.29 rad (or $16.62°$) leading $40\sin 50\pi t$
2. (a) 75 cm (b) 6.37 Hz (c) 0.157 s
   (d) 0.54 rad (or $30.94°$) lagging $75\sin 40t$
3. (a) 300 V (b) 100 Hz (c) 0.01 s or 10 ms
   (d) 0.412 rad (or $23.61°$) lagging $300\sin 200\pi t$
4. (a) $v = 120\sin 100\pi t$ volts
   (b) $v = 120\sin(100\pi t + 0.43)$ volts
5. $i = 20\sin\left(80\pi t - \dfrac{\pi}{6}\right)$ A or
   $i = 20\sin(80\pi t - 0.524)$ A
6. $3.2\sin(100\pi t + 0.488)$ m
7. (a) 5A, 50 Hz, 20 ms, $24.75°$ lagging
   (b) -2.093 A (c) 4.363 A (d) 6.375 ms (e) 3.423 ms

Exercise 172 (page 473)

1. (a) $i = (70.71\sin 628.3t + 16.97\sin 1885t)$ A
2. (a) $v = 300\sin 314.2t + 90\sin(628.3t - \pi/2)$
   $\qquad\qquad + 30\sin(1256.6t + \pi/3)$ V
3. Sketch
4. $i = 16\sin 2\pi 10^3 t + 3.2\sin 6\pi 10^3 t$
   $\qquad\qquad + 1.6\sin\pi 10^4 t$ A
5. (a) 60 Hz, 180 Hz, 300 Hz (b) 40% (c) 10%

Chapter 40

Exercise 173 (page 476)

1. $(5.83, 59.04°)$ or $(5.83, 1.03\,\text{rad})$
2. $(6.61, 20.82°)$ or $(6.61, 0.36\,\text{rad})$
3. $(4.47, 116.57°)$ or $(4.47, 2.03\,\text{rad})$
4. $(6.55, 145.58°)$ or $(6.55, 2.54\,\text{rad})$
5. $(7.62, 203.20°)$ or $(7.62, 3.55\,\text{rad})$
6. $(4.33, 236.31°)$ or $(4.33, 4.12\,\text{rad})$
7. $(5.83, 329.04°)$ or $(5.83, 5.74\,\text{rad})$
8. $(15.68, 307.75°)$ or $(15.68, 5.37\,\text{rad})$

Exercise 174 (page 477)

1. $(1.294, 4.830)$ 2. $(1.917, 3.960)$
3. $(-5.362, 4.500)$ 4. $(-2.884, 2.154)$
5. $(-9.353, -5.400)$ 6. $(-2.615, -3.027)$
7. $(0.750, -1.299)$ 8. $(4.252, -4.233)$
9. (a) $40\angle 18°$, $40\angle 90°$, $40\angle 162°$, $40\angle 234°$, $40\angle 306°$
   (b) $(38.04, 12.36)$, $(0, 40)$, $(-38.04, 12.36)$,
   $(-23.51, -32.36)$, $(23.51, -32.36)$
10. 47.0 mm

Chapter 41

Exercise 175 (page 481)

1. $C = 83°, a = 14.1\,\text{mm}, c = 28.9\,\text{mm}$,
   area $= 189\,\text{mm}^2$
2. $A = 52°2', c = 7.568\,\text{cm}, a = 7.152\,\text{cm}$,
   area $= 25.65\,\text{cm}^2$
3. $D = 19°48', E = 134°12', e = 36.0\,\text{cm}$,
   area $= 134\,\text{cm}^2$
4. $E = 49°0', F = 26°38', f = 15.09\,\text{mm}$,
   area $= 185.6\,\text{mm}^2$
5. $J = 44°29', L = 99°31', l = 5.420\,\text{cm}$,
   area $= 6.133\,\text{cm}^2$, or $J = 135°31', L = 8°29'$,
   $l = 0.811\,\text{cm}$, area $= 0.917\,\text{cm}^2$
6. $K = 47°8', J = 97°52', j = 62.2\,\text{mm}$,
   area $= 820.2\,\text{mm}^2$, or $K = 132°52', J = 12°8'$,
   $j = 13.19\,\text{mm}$, area $= 174.0\,\text{mm}^2$

Exercise 176 (page 483)

1. $p = 13.2\,\text{cm}, Q = 47.34°, R = 78.66°$,
   area $= 77.7\,\text{cm}^2$
2. $p = 6.127\,\text{m}, Q = 30.83°, R = 44.17°$,
   area $= 6.938\,\text{m}^2$

3. $X = 83.33°, Y = 52.62°, Z = 44.05°,$
area $= 27.8\,\text{cm}^2$

4. $X = 29.77°, Y = 53.50°, Z = 96.73°,$
area $= 355\,\text{mm}^2$

Exercise 177 (page 485)

1. 193 km   **2.** (a) 122.6 m (b) 94.80°, 40.66°, 44.54°
3. (a) 11.4 m (b) 17.55°   **4.** 163.4 m
5. $BF = 3.9$ m, $EB = 4.0$ m   **6.** 6.35 m, 5.37 m
7. 32.48 A, 14.31°

Exercise 178 (page 487)

1. 80.42°, 59.38°, 40.20°   **2.** (a) 15.23 m (b) 50.07°
3. 40.25 cm, 126.05°   **4.** 19.8 cm   **5.** 36.2 m
6. $x = 69.3$ mm, $y = 142$ mm   **7.** 130°   **8.** 13.66 mm

Chapter 42

Exercise 179 (page 491)

1. to **6.** Proofs

Exercise 180 (page 493)

1. $\theta = 34.85°$ or $145.15°$ **2.** $A = 213.06°$ or $326.94°$
3. $t = 66.75°$ or $246.75°$ **4.** 60°, 300°
5. 59°, 239°   **6.** 41.81°, 138.19°
7. $\pm131.81°$ **8.** 39.81°, $-140.19°$
9. $-30°, -150°$ **10.** 33.69°, 213.69°
11. 101.31°, 281.31°

Exercise 181 (page 493)

1. $y = 50.77°, 129.23°, 230.77°$ or $309.23°$
2. $\theta = 60°, 120°, 240°$ or $300°$
3. $\theta = 60°, 120°, 240°$ or $300°$ **4.** $D = 90°$ or $270°$
5. $\theta = 32.31°, 147.69°, 212.31°$ or $327.69°$

Exercise 182 (page 494)

1. $A = 19.47°, 160.53°, 203.58°$ or $336.42°$
2. $\theta = 51.34°, 123.69°, 231.34°$ or $303.69°$
3. $t = 14.48°, 165.52°, 221.81°$ or $318.19°$
4. $\theta = 60°$ or $300°$

Exercise 183 (page 495)

1. $\theta = 90°, 210°, 330°$ **2.** $t = 190.1°, 349.9°$
3. $\theta = 38.67°, 321.33°$ **4.** $\theta = 0°, 60°, 300°, 360°$
5. $\theta = 48.19°, 138.59°, 221.41°$ or $311.81°$
6. $x = 52.94°$ or $307.06°$ **7.** $A = 90°$

8. $t = 107.83°$ or $252.17°$ **9.** $a = 27.83°$ or $152.17°$
10. $\beta = 60.17°, 161.02°, 240.17°$ or $341.02°$
11. 51.83°, 308.17°   **12.** 30°, 150°

Chapter 43

Exercise 184 (page 497)

1. to **5.** Proofs

Exercise 185 (page 499)

1. $1 - \tanh^2\theta = \text{sech}^2\theta$
2. $\cosh(\theta + \phi) = \cosh\theta\cosh\phi + \sinh\theta\sinh\phi$
3. $\sinh(\theta - \phi) = \sinh\theta\cosh\phi - \cosh\theta\sinh\phi$
4. $\tanh 2\theta = \dfrac{2\tanh\theta}{1 + \tanh^2\theta}$
5. $\cosh\theta\sinh\phi = \dfrac{1}{2}[\sinh(\theta + \phi) - \sinh(\theta - \phi)]$
6. $\sinh^3\theta = \dfrac{1}{4}\sinh 3\theta - \dfrac{3}{4}\sinh\theta$
7. $\coth^2\theta(1 - \text{sech}^2\theta) = 1$

Chapter 44

Exercise 186 (page 502)

1. (a) $\sin 58°$ (b) $\sin 4t$
2. (a) $\cos 104°$ (b) $\cos\dfrac{\pi}{12}$ **3.** Proof   **4.** Proof
5. (a) 0.3136 (b) 0.9495 (c) -2.4678
6. 64.72° or 244.72°   **7.** 67.52° or 247.52°

Exercise 187 (page 506)

1. $9.434\sin(\omega t + 1.012)$ **2.** $5\sin(\omega t - 0.644)$
3. $8.062\sin(\omega t + 2.622)$ **4.** $6.708\sin(\omega t - 2.034)$
5. (a) 74.44° or 338.70° (b) 64.69° or 189.05°
6. (a) 72.74° or 354.64° (b) 11.15° or 311.98°
7. (a) 90° or 343.74° (b) 0°, 53.14°
8. (a) 82.92°, 296°
(b) 32.36°, 97°, 152.36°, 217°, 272.36° and 337°
9. $8.13\sin(3\theta + 2.584)$
10. $x = 4.0\sin(\omega t + 0.927)$ m
11. $9.434\sin(\omega t + 2.583)$ V   **12.** $7.07\sin\left(2t + \dfrac{\pi}{4}\right)$ cm

Exercise 188 (page 507)

1. $\dfrac{V^2}{2R}(1 + \cos 2t)$ **2.** Proof   **3.** $\cos 3\theta = 4\cos^3\theta - 3\cos\theta$
4. $-90°, 30°, 150°$ **5.** $-160.53°, -90°, -19.47°, 90°$
6. $-150°, -90°, -30°, 90°$
7. $-90°$ **8.** 45°, $-135°$

Exercise 189 (page 509)

1. $\frac{1}{2}(\sin 9t + \sin 5t)$ 2. $\frac{1}{2}(\sin 10x - \sin 6x)$

3. $\cos 4t - \cos 10t$ 4. $2(\cos 4\theta + \cos 2\theta)$

5. $\frac{3}{2}\left(\sin\frac{\pi}{2} + \sin\frac{\pi}{6}\right)$ 6. $-\dfrac{\cos 4t}{4} - \dfrac{\cos 2t}{2} + c$

7. $-\dfrac{20}{21}$ 8. $30°, 90°$ or $150°$

Exercise 190 (page 510)

1. $2\sin 2x \cos x$ 2. $\cos 8\theta \sin\theta$

3. $2\cos 4t \cos t$ 4. $-\dfrac{1}{4}\sin 3t \sin 2t$

5. $\cos\dfrac{7\pi}{24}\cos\dfrac{\pi}{24}$ 6. Proof

7. $22.5°, 45°, 67.5°, 112.5°, 135°, 157.5°$

8. $0°, 45°, 135°, 180°$ 9. $21.47°$ or $158.53°$

10. $0°, 60°, 90°, 120°, 180°, 240°, 270°, 300°, 360°$

Chapter 45

Exercise 191 (page 524)

1. $\pm j5$ 2. $x = 1 \pm j$

3. $x = 2 \pm j$ 4. $x = 3 \pm j$

5. $x = 0.5 \pm j0.5$ 6. $x = 2 \pm j2$

7. $x = 0.2 \pm j0.2$

8. $-\dfrac{3}{4} \pm j\dfrac{\sqrt{23}}{4}$ or $-0.750 \pm j1.199$

9. $\dfrac{5}{8} \pm j\dfrac{\sqrt{87}}{8}$ or $0.625 \pm j1.166$

10. (a) 1 (b) $-j$ (c) $-j2$

Exercise 192 (page 527)

1. (a) $8 + j$ (b) $-5 + j8$ 2. (a) $3 - j4$ (b) $2 + j$

3. (a) 5 (b) $1 - j2$ (c) $j5$ (d) $16 + j3$ (e) 5 (f) $3 + j4$

4. (a) $7 - j4$ (b) $-2 - j6$

5. (a) $10 + j5$ (b) $13 - j13$

6. (a) $-13 - j2$ (b) $-35 + j20$

7. (a) $\dfrac{-2}{25} + j\dfrac{11}{25}$ (b) $\dfrac{-19}{85} + j\dfrac{43}{85}$

8. (a) $\dfrac{3}{26} + j\dfrac{41}{26}$ (b) $\dfrac{45}{26} - j\dfrac{9}{26}$

9. (a) $-j$ (b) $\dfrac{1}{2} - j\dfrac{1}{2}$ 10. Proof

Exercise 193 (page 528)

1. $a = 8, b = -1$ 2. $x = \dfrac{3}{2},\ y = -\dfrac{1}{2}$

3. $a = -5, b = -12$ 4. $x = 3, y = 1$

5. $Z = 10 + j13.75$

Exercise 194 (page 531)

1. (a) $4.472, 63.43°$ (b) $5.385, -158.20°$
   (c) $2.236, 63.43°$

2. (a) $\sqrt{13}\angle 56.31°$ (b) $4\angle 180°$ (c) $\sqrt{37}\angle 170.54°$

3. (a) $3\angle -90°$ (b) $\sqrt{125}\angle 100.30°$ (c) $\sqrt{2}\angle -135°$

4. (a) $4.330 + j2.500$ (b) $1.500 + j2.598$
   (c) $4.950 + j4.950$

5. (a) $-3.441 + j4.915$ (b) $-4.000 + j0$
   (c) $-1.750 - j3.031$

6. (a) $45\angle 65°$ (b) $10.56\angle 44°$

7. (a) $3.2\angle 42°$ (b) $2\angle 150°$

8. (a) $6.986\angle 26.79°$ (b) $7.190\angle 85.77°$

Exercise 195 (page 533)

1. (a) $R = 3\,\Omega, L = 25.5\,\text{mH}$
   (b) $R = 2\,\Omega, C = 1061\,\mu\text{F}$
   (c) $R = 0, L = 44.56\,\text{mH}$
   (d) $R = 4\,\Omega, C = 459.4\,\mu\text{F}$

2. $15.76\,\text{A}, 23.20°$ lagging

3. $27.25\,\text{A}, 3.37°$ lagging

4. $14.42\,\text{A}, 43.83°$ lagging, 0.721

5. $14.6\,\text{A}, 2.51°$ leading

6. $8.394\,\text{N}, 208.68°$ from force A

7. $(10 + j20)\,\Omega,\ 22.36\angle 63.43°\,\Omega$ 8. $\pm\dfrac{mh}{2\pi}$

9. (a) $922\,\text{km/h}$ at $77.47°$ (b) $922\,\text{km/h}$ at $-102.53°$

10. (a) $3.770\angle 8.17°$ (b) $1.488\angle 100.37°$

11. Proof   12. gain $= 353.6$, phase $= -45°$

13. $275\angle -36.87°\,\text{mA}$

Chapter 46

Exercise 196 (page 538)

1. (a) $7.594\angle 75°$ (b) $125\angle 20.61°$

2. (a) $81\angle 164°, -77.86 + j22.33$
   (b) $55.90\angle -47.18°, 38 - j41$

3. $\sqrt{10}\angle -18.43°, 3162\angle -129°$

4. $476.4\angle 119.42°, -234 + j415$

5. $45530\angle 12.78°, 44400 + j10070$

6. $2809\angle 63.78°, 1241 + j2520$

7. $(38.27 \times 10^6)\angle 176.15°,$
   $10^6(-38.18 + j2.570)$

Exercise 197 (page 539)

1. (a) $\pm(1.099 + j0.455)$ (b) $\pm(0.707 + j0.707)$
2. (a) $\pm(2 - j)$ (b) $\pm(0.786 - j1.272)$
3. (a) $\pm(2.291 + j1.323)$ (b) $\pm(-2.449 + j2.449)$
4. Moduli 1.710, arguments $17.71°$, $137.71°$ and $257.71°$
5. Modulus 1.223, arguments $38.36°$, $128.36°$, $218.36°$ and $308.36°$
6. Modulus 2.795, arguments $109.90°$, $289.90°$
7. Modulus 0.3420, arguments $24.58°$, $144.58°$ and $264.58°$
8. $Z_0 = 390.2\angle -10.43°\,\Omega$, $\gamma = 0.1029\angle 61.92°$

Exercise 198 (page 541)

1. $5.83e^{j0.54}$ 2. $4.89e^{j2.11}$
3. $-1.50 + j3.27$ 4. $34.79 + j20.09$
5. $-4.52 - j3.38$
6. (a) $\ln 7 + j2.1$ (b) $2.86\angle 47.18°$ or $2.86\angle 0.82$
7. $3.51\angle -34.72°$ or $3.51\angle -0.61$
8. (a) $2.06\angle 35.26°$ or $2.06\angle 0.615$
   (b) $4.11\angle 66.96°$ or $4.11\angle 1.17$

9. $Ae^{-\frac{ht}{2m}} \cos\left(\frac{\sqrt{(4mf - h^2)}}{2m - a}\right) t$

Exercise 199 (page 544)

1. (a) $x^2 + y^2 = 4$ (b) a circle, centre $(0, 0)$ and radius 2
2. (a) $x^2 + y^2 = 25$ (b) a circle, centre $(0, 0)$ and radius 5
3. (a) $y = \sqrt{3}(x - 2)$ (b) a straight line
4. (a) $y = \dfrac{1}{\sqrt{3}}(x + 1)$ (b) a straight line
5. (a) $x^2 - 4x - 12 + y^2 = 0$ or $(x - 2)^2 + y^2 = 4^2$
   (b) a circle, centre $(2, 0)$ and radius 4
6. (a) $x^2 + 6x - 16 + y^2 = 0$ or $(x + 3)^2 + y^2 = 5^2$
   (b) a circle, centre $(-3, 0)$ and radius 5
7. (a) $2x^2 - 5x + 2 + 2y^2 = 0$ or
   $\left(x - \dfrac{5}{4}\right)^2 + y^2 = \left(\dfrac{3}{4}\right)^2$
   (b) a circle, centre $\left(\dfrac{5}{4}, 0\right)$ and radius $\dfrac{3}{4}$
8. (a) $x^2 + 2x - 1 + y^2 = 0$ or $(x + 1)^2 + y^2 = 2$
   (b) a circle, centre $(-1, 0)$ and radius $\sqrt{2}$

9. (a) $x^2 - x - y + y^2 = 0$ or
   $\left(x - \dfrac{1}{2}\right)^2 + \left(y - \dfrac{1}{2}\right)^2 = \dfrac{1}{2}$
   (b) a circle, centre $\left(\dfrac{1}{2}, \dfrac{1}{2}\right)$ and radius $\dfrac{1}{\sqrt{2}}$
10. (a) $x^2 + 2x + 2y + y^2 = 0$ or $(x + 1)^2 + (y + 1)^2 = 2$
    (b) a circle, centre $(-1, -1)$ and radius $\sqrt{2}$
11. (a) $y = 2x + 1.5$ (b) a straight line
12. (a) $y = 2x - 3$ (b) a straight line
13. (a) $x = \dfrac{1}{2}$ (b) a straight line

Chapter 47

Exercise 200 (page 552)

1. $\begin{pmatrix} 8 & 1 \\ -5 & 13 \end{pmatrix}$ 2. $\begin{pmatrix} 7 & -1 & 8 \\ 3 & 1 & 7 \\ 4 & 7 & -2 \end{pmatrix}$
3. $\begin{pmatrix} -2 & -3 \\ -3 & 1 \end{pmatrix}$ 4. $\begin{pmatrix} 9.3 & -6.4 \\ -7.5 & 16.9 \end{pmatrix}$
5. $\begin{pmatrix} 45 & 7 \\ -26 & 71 \end{pmatrix}$ 6. $\begin{pmatrix} 4.6 & -5.6 & -7.6 \\ 17.4 & -16.2 & 28.6 \\ -14.2 & 0.4 & 17.2 \end{pmatrix}$
7. $\begin{pmatrix} -11 \\ 43 \end{pmatrix}$ 8. $\begin{pmatrix} 16 & 0 \\ -27 & 34 \end{pmatrix}$
9. $\begin{pmatrix} -6.4 & 26.1 \\ 22.7 & -56.9 \end{pmatrix}$ 10. $\begin{pmatrix} 135 \\ -52 \\ -85 \end{pmatrix}$
11. $\begin{pmatrix} 5 & 6 \\ 12 & -3 \\ 1 & 0 \end{pmatrix}$ 12. $\begin{pmatrix} 55.4 & 3.4 & 10.1 \\ -12.6 & 10.4 & -20.4 \\ -16.9 & 25.0 & 37.9 \end{pmatrix}$
13. $A \times C = \begin{pmatrix} -6.4 & 26.1 \\ 22.7 & -56.9 \end{pmatrix}$
    $C \times A = \begin{pmatrix} -33.5 & 53.1 \\ 23.1 & -29.8 \end{pmatrix}$
    Hence they are not equal

Exercise 201 (page 554)

1. 17   2. -3 3. -13.43 4. $-5 + j3$
5. $(-19.75 + j19.79)$ or $27.96\angle 134.94°$
6. $x = 6, x = -1$

Exercise 202 (page 555)

1. $\begin{pmatrix} \dfrac{7}{17} & \dfrac{1}{17} \\[2mm] \dfrac{4}{17} & \dfrac{3}{17} \end{pmatrix}$ 2. $\begin{pmatrix} 7\dfrac{5}{7} & 8\dfrac{4}{7} \\[2mm] -4\dfrac{2}{7} & -6\dfrac{3}{7} \end{pmatrix}$

3. $\begin{pmatrix} 0.290 & 0.551 \\ 0.186 & 0.097 \end{pmatrix}$ correct to 3 decimal places

Exercise 203 (page 556)

1. $\begin{pmatrix} -16 & 8 & -34 \\ -14 & -46 & 63 \\ -24 & 12 & 2 \end{pmatrix}$ 2. $\begin{pmatrix} -16 & -8 & -34 \\ 14 & -46 & -63 \\ -24 & -12 & 2 \end{pmatrix}$

3. -212 4. -328 5. -242.83 6. $-2-j$
7. $26.94\angle -139.52°$ or $(-20.49 - j17.49)$
8. (a) $\lambda = 3$ or 4 (b) $\lambda = 1$ or 2 or 3

Exercise 204 (page 557)

1. $\begin{pmatrix} 4 & -2 & 5 \\ -7 & 4 & 7 \\ 6 & 0 & -4 \end{pmatrix}$ 2. $\begin{pmatrix} 3 & 5 & -1 \\ 6 & -\dfrac{2}{3} & 0 \\ \dfrac{1}{2} & 7 & \dfrac{3}{5} \end{pmatrix}$

3. $\begin{pmatrix} -16 & 14 & -24 \\ -8 & -46 & -12 \\ -34 & -63 & 2 \end{pmatrix}$ 4. $\begin{pmatrix} -\dfrac{2}{5} & -3\dfrac{3}{5} & 42\dfrac{1}{3} \\ -10 & 2\dfrac{3}{10} & -18\dfrac{1}{2} \\ -\dfrac{2}{3} & -6 & -32 \end{pmatrix}$

5. $-\dfrac{1}{212}\begin{pmatrix} -16 & 14 & -24 \\ -8 & -46 & -12 \\ -34 & -63 & 2 \end{pmatrix}$

6. $-\dfrac{15}{923}\begin{pmatrix} -\dfrac{2}{5} & -3\dfrac{3}{5} & 42\dfrac{1}{3} \\ -10 & 2\dfrac{3}{10} & -18\dfrac{1}{2} \\ -\dfrac{2}{3} & -6 & -32 \end{pmatrix}$

Chapter 48

Exercise 205 (page 561)

1. $x = 4, y = -3$ 2. $p = 1.2, q = -3.4$
3. $x = 1, y = -1, z = 2$ 4. $a = 2.5, b = 3.5, c = 6.5$
5. $p = 4.1, q = -1.9, r = -2.7$
6. $I_1 = 2, I_2 = -3$ 7. $s = 2, v = -3, a = 4$
8. $\ddot{x} = 0.5, \dot{x} = 0.77, x = 1.4$

Exercise 206 (page 564)

1. $x = -1.2, y = 2.8$ 2. $m = -6.4, n = -4.9$
3. $x = 1, y = 2, z = -1$
4. $p = 1.5, q = 4.5, r = 0.5$
5. $x = \dfrac{7}{20}, y = \dfrac{17}{40}, z = -\dfrac{5}{24}$
6. $F_1 = 1.5, F_2 = -4.5$
7. $I_1 = 10.77\angle 19.23°\,A, I_2 = 10.45\angle -56.73°\,A$
8. $i_1 = -5, i_2 = -4, i_3 = 2$
9. $F_1 = 2, F_2 = -3, F_3 = 4$
10. $I_1 = 3.317\angle 22.57°\,A$
    $I_2 = 1.963\angle 40.97°\,A$
    $I_3 = 1.010\angle -148.32°\,A$

Exercise 207 (page 566)

1. See earlier answers   2. See earlier answers

Exercise 208 (page 568)

1. $\ddot{x} = -0.30, \dot{x} = 0.60, x = 1.20$
2. $T_1 = 0.8, T_2 = 0.4, T_3 = 0.2$
3. See earlier answers   4. See earlier answers

Exercise 209 (page 572)

1. (a) $\lambda_1 = 3, \lambda_2 = -2$ (b) $\begin{pmatrix} -4 \\ 1 \end{pmatrix}, \begin{pmatrix} 1 \\ 1 \end{pmatrix}$

2. (a) $\lambda_1 = 1, \lambda_2 = 6$ (b) $\begin{pmatrix} -3 \\ 1 \end{pmatrix}, \begin{pmatrix} 2 \\ 1 \end{pmatrix}$

3. (a) $\lambda_1 = 1, \lambda_2 = 2$ (b) $\begin{pmatrix} 1 \\ -2 \end{pmatrix}, \begin{pmatrix} 1 \\ -1 \end{pmatrix}$

4. (a) $\lambda_1 = 2, \lambda_2 = 6, \lambda_3 = -2$
   (b) $\begin{pmatrix} 1 \\ -2 \\ 1 \end{pmatrix}, \begin{pmatrix} 0 \\ 1 \\ 1 \end{pmatrix}, \begin{pmatrix} 1 \\ 1 \\ 0 \end{pmatrix}$

5. (a) $\lambda_1 = 0, \lambda_2 = 1, \lambda_3 = 3$
   (b) $\begin{pmatrix} 1 \\ 1 \\ 1 \end{pmatrix}, \begin{pmatrix} 1 \\ 0 \\ -1 \end{pmatrix}, \begin{pmatrix} 1 \\ -2 \\ 1 \end{pmatrix}$

6. (a) $\lambda_1 = 1, \lambda_2 = 2, \lambda_3 = 4$
   (b) $\begin{pmatrix} -2 \\ 1 \\ 0 \end{pmatrix}, \begin{pmatrix} -2 \\ 1 \\ 1 \end{pmatrix}, \begin{pmatrix} 0 \\ 1 \\ 1 \end{pmatrix}$

7. (a) $\lambda_1 = 1, \lambda_2 = 2, \lambda_3 = 3$
   (b) $\begin{pmatrix} 0 \\ 2 \\ -1 \end{pmatrix}, \begin{pmatrix} 1 \\ 1 \\ 0 \end{pmatrix}, \begin{pmatrix} 2 \\ 2 \\ 1 \end{pmatrix}$

Chapter 49

Exercise 210 (page 578)

1. See page 577 **2.** S **3.** S **4.** V **5.** S
6. S   **7.** V **8.** S **9.** V

Exercise 211 (page 585)

1. 17.35 N at 18.00° to the 12 N force
2. 13 m/s at 22.62° to the 12 m/s velocity
3. 16.40 N at 37.57° to the 13 N force
4. 28.43 N at 129.30° to the horizontal
5. 32.31 m at 21.80° to the 30 m displacement
6. 14.72 N at −14.72° to the 5 N force
7. 29.15 m/s at 29.04° to the horizontal
8. 9.28 N at 16.70°   **9.** 6.89 m/s at 159.56°
10. 15.62 N at 26.33° to the 10 N force
11. 21.07 knots, E 9.22° S

Exercise 212 (page 588)

1. (a) 54.0 N at 78.16° (b) 45.64 N at 4.66°
2. (a) 31.71 m/s at 121.81° (b) 19.55 m/s at 8.63°

Exercise 213 (page 589)

1. 83.5 km/h at 71.6° to the vertical
2. 4 minutes 55 seconds, 60°
3. 22.79 km/h, E 9.78° N

Exercise 214 (page 590)

1. $i - j - 4k$ **2.** $4i + j - 6k$
3. $-i + 7j - k$ **4.** $5i - 10k$
5. $-3i + 27j - 8k$ **6.** $-5i + 10k$
7. $i + 7.5j - 4k$ **8.** $20.5j - 10k$
9. $3.6i + 4.4j - 6.9k$ **10.** $2i + 40j - 43k$

Chapter 50

Exercise 215 (page 593)

1. $4.5\sin(A + 63.5°)$
2. (a) $20.9\sin(\omega t + 0.63)$ volts
   (b) $12.5\sin(\omega t - 1.36)$ volts
3. $13\sin(\omega t + 0.393)$

Exercise 216 (page 595)

1. $4.5\sin(\theta + 63.5°)$
2. (a) $20.9\sin(\omega t + 0.62)$ volts
   (b) $12.5\sin(\omega t - 1.33)$ volts
3. $13\sin(\omega t + 0.40)$

Exercise 217 (page 596)

1. $4.472\sin(A + 63.44°)$
2. (a) $20.88\sin(\omega t + 0.62)$ volts
   (b) $12.50\sin(\omega t - 1.33)$ volts
3. $13\sin(\omega t + 0.395)$ **4.** $11.11\sin(\omega t + 0.324)$
5. $8.73\sin(\omega t - 0.173)$ **6.** $1.01\sin(\omega t - 0.698)$ A

Exercise 218 (page 598)

1. $11.11\sin(\omega t + 0.324)$ **2.** $8.73\sin(\omega t - 0.173)$
3. $i = 21.79\sin(\omega t - 0.639)$
4. $x = 14.38\sin(\omega t + 1.444)$
5. (a) $305.3\sin(314.2t - 0.233)$V (b) 50 Hz
6. (a) $10.21\sin(628.3t + 0.818)$V
   (b) 100 Hz (c) 10 ms
7. (a) $79.83\sin(300\pi t + 0.352)$V
   (b) 150 Hz (c) 6.667 ms
8. $150.6\sin(\omega t - 0.247)$V

Exercise 219 (page 600)

1. $12.07\sin(\omega t + 0.297)$ **2.** $14.51\sin(\omega t - 0.315)$
3. $9.173\sin(\omega t + 0.396)$ **4.** $16.168\sin(\omega t + 1.451)$
5. (a) $371.95\sin(314.2t - 0.239)$V (b) 50 Hz
6. (a) $11.44\sin(200\pi t + 0.715)$V (b) 100 Hz
   (c) 10 ms
7. (a) $79.73\sin(300\pi - 0.536)$ V (b) 150 Hz
   (c) 6.667 ms (d) 56.37 V
8. $354.6\angle32.41°$A

Chapter 51

Exercise 220 (page 607)

1. (i) 7 (ii) 0   **2.** (a) −12 (b) −4
3. (a) 11 (b) 11   **4.** (a) $\sqrt{13}$ (b) $\sqrt{14}$
5. (a) −16 (b) 38   **6.** (a) $\sqrt{19}$ (b) 7.347
7. (a) 143.82° (b) 44.52°
8. (a) 0.555, −0.832, 0 (b) 0, 0.970, −0.243
   (c) 0.267, 0.535, −0.802
9. 11.54°   **10.** 66.40°   **11.** 53 N m

Exercise 221 (page 610)

1. (a) $4i - 7j - 6k$ (b) $-4i + 7j + 6k$
2. (a) 11.92 (b) 13.96
3. (a) $-36i - 30j + 54k$ (b) $11i + 4j - k$
4. (a) $-22i - j + 33k$ (b) $18i + 162j + 102k$
5. (i) −15 (ii) $-4i + 4j + 10k$ (iii) 11.49
   (iv) $4i - 4j - 10k$ (v) 142.55°

6. (i) $-62\frac{1}{2}$ (ii) $-1\frac{1}{2}i - 4j + 11k$ (iii) 11.80

   (iv) $1\frac{1}{2}i + 4j - 11k$ (v) $169.31°$

7. $10\,\text{Nm}$
8. $M = (5i + 8j - 2k)\,\text{Nm}$, $|M| = 9.64\,\text{Nm}$
9. $v = -14i + 7j + 12k$, $|v| = 19.72\,\text{m/s}$
10. $6i - 10j - 14k$, $18.22\,\text{m/s}$

Exercise 222 (page 612)

1. $r = (5 + 2\lambda)i + (7\lambda - 2)j + (3 - 4\lambda)k$;
   $r = 9i + 12j - 5k$

2. $\dfrac{x-5}{2} = \dfrac{y+2}{7} = \dfrac{3-z}{4} = \lambda$

3. $r = \dfrac{1}{3}(1 + 4\lambda)i + \dfrac{1}{5}(2\lambda - 1)j + (4 - 3\lambda)k$

4. $r = \dfrac{1}{2}(\lambda - 1)i + \dfrac{1}{4}(1 - 5\lambda)j + \dfrac{1}{3}(1 + 4\lambda)k$

Chapter 52

Exercise 223 (page 622)

1. $1, 5, 21, 9, 61$ 2. $0, 11, -10, 21$ 3. Proof
4. $8, -a^2 - a + 8, -a^2 - a, -a - 1$

Exercise 224 (page 623)

1. $16, 8$

Exercise 225 (page 626)

1. 1 2. 7 3. $8x$ 4. $15x^2$
5. $-4x + 3$ 6. 0 7. 9 8. $\dfrac{2}{3}$
9. $18x$ 10. $-21x^2$ 11. $2x + 15$ 12. 0
13. $12x^2$ 14. $6x$

Exercise 226 (page 627)

1. $28x^3$ 2. $\dfrac{1}{2\sqrt{x}}$ 3. $\dfrac{3}{2}\sqrt{t}$ 4. $-\dfrac{3}{x^4}$
5. $3 + \dfrac{1}{2\sqrt{x^3}} - \dfrac{1}{x^2}$ 6. $-\dfrac{10}{x^3} + \dfrac{7}{2\sqrt{x^9}}$
7. $6t - 12$ 8. $3x^2 + 6x + 3$
9. See earlier answers   10. $12x - 3$ (a) -15 (b) 21
11. $6x^2 + 6x - 4, 32$ 12. $-6x^2 + 4, -9.5$

Exercise 227 (page 629)

1. (a) $12\cos 3x$ (b) $-12\sin 6x$
2. $6\cos 3\theta + 10\sin 2\theta$
3. $270.2\,\text{A/s}$ 4. $1393.4\,\text{V/s}$
5. $12\cos(4t + 0.12) + 6\sin(3t - 0.72)$

Exercise 228 (page 630)

1. (a) $15e^{3x}$ (b) $-\dfrac{4}{7e^{2x}}$ 2. $\dfrac{1}{\theta}$
3. 16 4. 664

Chapter 53

Exercise 229 (page 633)

1. (a) $25x^4$ (b) $8.4x^{2.5}$ (c) $-\dfrac{1}{x^2}$

2. (a) $\dfrac{8}{x^3}$ (b) 0 (c) 2

3. (a) $\dfrac{1}{\sqrt{x}}$ (b) $5\sqrt[3]{x^2}$ (c) $-\dfrac{2}{\sqrt{x^3}}$

4. (a) $\dfrac{1}{\sqrt[3]{x^4}}$ (b) $2(x - 1)$ (c) $6\cos 3x$

5. (a) $8\sin 2x$ (b) $12e^{6x}$ (c) $\dfrac{-15}{e^{5x}}$

6. (a) $\dfrac{4}{x}$ (b) $\dfrac{e^x + e^{-x}}{2}$ (c) $\dfrac{-1}{x^2} + \dfrac{1}{2\sqrt{x^3}}$

7. $-1, 16$ 8. $\left(\dfrac{1}{2}, \dfrac{3}{4}\right)$

9. (a) $\dfrac{-4}{\theta^3} + \dfrac{2}{\theta} + 10\sin 5\theta - 12\cos 2\theta + \dfrac{6}{e^{3\theta}}$
   (b) 22.30

10. 3.29 11. $x = \dfrac{mg}{k}$ 12. $27.0\,\text{V}$

Exercise 230 (page 634)

1. $x\cos x + \sin x$ 2. $2xe^{2x}(x + 1)$
3. $x(1 + 2\ln x)$ 4. $6x^2(\cos 3x - x\sin 3x)$
5. $\sqrt{x}\left(1 + \dfrac{3}{2}\ln 3x\right)$ 6. $e^{3t}(4\cos 4t + 3\sin 4t)$
7. $e^{4\theta}\left(\dfrac{1}{\theta} + 4\ln 3\theta\right)$
8. $e^t\left\{\left(\dfrac{1}{t} + \ln t\right)\cos t - \ln t\sin t\right\}$
9. 8.732 10. 32.31

Exercise 231 (page 636)

1. $\dfrac{x\cos x - \sin x}{x^2}$ 2. $\dfrac{-6}{x^4}(x\sin 3x + \cos 3x)$

3. $\dfrac{2(1-x^2)}{(x^2+1)^2}$ 4. $\dfrac{\dfrac{\cos x}{2\sqrt{x}} + \sqrt{x}\sin x}{\cos^2 x}$

5. $\dfrac{3\sqrt{\theta}(3\sin 2\theta - 4\theta\cos 2\theta)}{4\sin^2 2\theta}$

6. $\dfrac{1 - \dfrac{1}{2}\ln 2t}{\sqrt{t^3}}$ 7. $\dfrac{2e^{4x}}{\sin^2 x}\{(1+4x)\sin x - x\cos x\}$

8. -18 9. 3.82

Exercise 232 (page 637)

1. $12(2x-1)^5$ 2. $5(6x^2-5)(2x^3-5x)^4$
3. $6\cos(3\theta - 2)$ 4. $-10\cos^4\alpha\sin\alpha$

5. $\dfrac{5(2-3x^2)}{(x^3-2x+1)^6}$ 6. $10e^{2t+1}$

7. $-20t\,\mathrm{cosec}^2(5t^2+3)$ 8. $18\sec^2(3y+1)$
9. $2\sec^2\theta\,e^{\tan\theta}$ 10. 1.86
11. (a) 24.21 mm/s (b) -70.46 mm/s

Exercise 233 (page 638)

1. (a) $36x^2 + 12x$ (b) $72x + 12$
2. (a) $\dfrac{4}{5} - \dfrac{12}{t^5} + \dfrac{6}{t^3} + \dfrac{1}{4\sqrt{t^3}}$ (b) -4.95

3. (a) $\dfrac{dq}{dt} = -\dfrac{V}{R}e^{-\frac{t}{CR}}$ (b) $\dfrac{d^2q}{dt^2} = \dfrac{V}{CR^2}e^{-\frac{t}{CR}}$

4. (a) $-(12\sin 2t + \cos t)$ (b) $\dfrac{-2}{\theta^2}$

5. (a) $4(\sin^2 x - \cos^2 x)$ (b) $48(2x-3)^2$
6. 18 7. Proof   8. Proof   9. Proof

Chapter 54

Exercise 234 (page 641)

1. 3000π A/s   2. (a) 0.24 cd/V (b) 250 V
3. (a) -625 V/s (b) -220.5 V/s   4. -1.635 Pa/m
5. -390 m^3/min

Exercise 235 (page 644)

1. (a) 100 m/s (b) 4 s (c) 200 m (d) -100 m/s
2. (a) 90 km/h (b) 62.5 m   3. (a) 4 s (b) 3 rads
4. (a) 3 m/s; -1 m/s^2 (b) 6 m/s; -4 m/s^2 (c) $\dfrac{3}{4}$ s

5. (a) $\omega = 1.40$ rad/s (b) $\alpha = -0.37$ rad/s^2
   (c) $t = 6.28$ s
6. (a) 6 m/s, -23 m/s^2 (b) 117 m/s, 97 m/s^2
   (c) $\dfrac{3}{4}$ s or $\dfrac{2}{5}$ s (d) $1\dfrac{1}{2}$ s (e) $75\dfrac{1}{6}$ m
7. $t = 3$ s

Exercise 236 (page 648)

1. $(3, -9)$ Minimum   2. $(1, 9)$ Maximum
3. $(2, -1)$ Minimum
4. $(0, 3)$ Minimum, $(2, 7)$ Maximum
5. Minimum at $\left(\dfrac{2}{3}, \dfrac{2}{3}\right)$
6. Maximum at $(3, 9)$ 7. Minimum $(2, -88)$
                          Maximum $(-2.5, 94.25)$
8. Minimum at $(0.4000, 3.8326)$
9. Maximum at $(0.6931, -0.6137)$
10. Minimum at $(1, 2.5)$ Maximum at $\left(-\dfrac{2}{3}, 4\dfrac{22}{27}\right)$
11. Minimum at $(0.5, 6)$
12. Maximum of 13 at $337.38°$,   13. Proof
    Minimum of -13 at $157.38°$

Exercise 237 (page 651)

1. 54 km/h   2. $90\,000$ m^2 3. 48 m   4. 11.42 m^2
5. radius -4.607 cm, height -9.212 cm   6. 6.67 cm
7. Proof   8. height $= 5.42$ cm, radius $= 2.71$ cm
9. 44.72 10. 42.72 volts
11. 50.0 miles/gallon, 52.6 miles/hour
12. $45°$ 13. $r = 0.607$

Exercise 238 (page 653)

1. $\left(\dfrac{1}{2}, -1\right)$ 2. $\left(-\dfrac{1}{4}, 4\right)$ 3. (π, π)
4. $(3, -100)$ 5. $(2, 0.541)$
6. Max at $(0, 10)$, Min at $(2, -2)$, point of inflexion at $(1, 4)$

Exercise 239 (page 654)

1. (a) $y = 4x - 2$ (b) $4y + x = 9$
2. (a) $y = 10x - 12$ (b) $10y + x = 82$
3. (a) $y = \dfrac{3}{2}x + 1$ (b) $6y + 4x + 7 = 0$
4. (a) $y = 5x + 5$ (b) $5y + x + 27 = 0$
5. (a) $9\theta + t = 6$
   (b) $\theta = 9t - 26\dfrac{2}{3}$ or $3\theta = 27t - 80$

Exercise 240 (page 655)

1. (a) -0.03 (b) -0.008 2. $-0.032, -1.6\%$
3. (a) $60\,\text{cm}^3$ (b) $12\,\text{cm}^2$
4. (a) $-6.03\,\text{cm}^2$ (b) $-18.10\,\text{cm}^3$ 5. 12.5%

Chapter 55

Exercise 241 (page 659)

1. $\dfrac{1}{3}(2t-1)$ 2. 2

3. (a) $-\dfrac{1}{4}\cot\theta$ (b) $-\dfrac{1}{16}\text{cosec}^3\theta$

4. 4 5. -6.25 6. $y=-1.155x+4$

7. $y=-\dfrac{1}{4}x+5$

Exercise 242 (page 661)

1. (a) 3.122 (b) -14.43 2. $y=-2x+3$
3. $y=-x+\pi$ 4. 0.02975
5. (a) 13.14 (b) 5.196

Chapter 56

Exercise 243 (page 664)

1. (a) $15y^4\dfrac{dy}{dx}$ (b) $-8\sin4\theta\dfrac{d\theta}{dx}$ (c) $\dfrac{1}{2\sqrt{k}}\dfrac{dk}{dx}$

2. (a) $\dfrac{5}{2t}\dfrac{dt}{dx}$ (b) $\dfrac{3}{2}e^{2y+1}\dfrac{dy}{dx}$ (c) $6\sec^2 3y\dfrac{dy}{dx}$

3. (a) $6\cos2\theta\dfrac{d\theta}{dy}$ (b) $6\sqrt{x}\dfrac{dx}{dy}$ (c) $\dfrac{-2}{e^t}\dfrac{dt}{dy}$

4. (a) $\dfrac{-6}{(3x+1)^2}\dfrac{dx}{du}$ (b) $6\sec2\theta\tan2\theta\dfrac{d\theta}{du}$

   (c) $\dfrac{-1}{\sqrt{y^3}}\dfrac{dy}{du}$

Exercise 244 (page 665)

1. $3xy^2\left(3x\dfrac{dy}{dx}+2y\right)$ 2. $\dfrac{2}{5x^2}\left(x\dfrac{dy}{dx}-y\right)$

3. $\dfrac{3}{4v^2}\left(v-u\dfrac{dv}{du}\right)$

4. $3\left(\dfrac{\cos3x}{2\sqrt{y}}\right)\dfrac{dy}{dx}-9\sqrt{y}\sin3x$

5. $2x^2\left(\dfrac{x}{y}+3\ln y\dfrac{dx}{dy}\right)$

Exercise 245 (page 667)

1. $\dfrac{2x+4}{3-2y}$ 2. $\dfrac{3}{1-6y^2}$ 3. $-\dfrac{\sqrt{5}}{2}$

4. $\dfrac{-(x+\sin4y)}{4x\cos4y}$ 5. $\dfrac{4x-y}{3y+x}$ 6. $\dfrac{x(4y+9x)}{\cos y-2x^2}$

7. $\dfrac{1-2\ln y}{3+(2x/y)-4y^3}$ 8. 5 9. ±0.5774

10. ±1.5 11. -6

Chapter 57

Exercise 246 (page 669)

1. $\dfrac{2}{2x-5}$ 2. $-3\tan3x$ 3. $\dfrac{9x^2+1}{3x^3+x}$

4. $\dfrac{10x+10}{5x^2+10x-7}$ 5. $\dfrac{1}{x}$ 6. $\dfrac{2x}{x^2-1}$

7. $\dfrac{3}{x}$ 8. $2\cot x$ 9. $\dfrac{12x^2-12x+3}{4x^3-6x^2+3x}$

Exercise 247 (page 671)

1. $\dfrac{(x-2)(x+1)}{(x-1)(x+3)}\left\{\dfrac{1}{(x-2)}+\dfrac{1}{(x+1)}\right.$
   $\left.-\dfrac{1}{(x-1)}-\dfrac{1}{(x+3)}\right\}$

2. $\dfrac{(x+1)(2x+1)^3}{(x-3)^2(x+2)^4}\left\{\dfrac{1}{(x+1)}+\dfrac{6}{(2x+1)}\right.$
   $\left.-\dfrac{2}{(x-3)}-\dfrac{4}{(x+2)}\right\}$

3. $\dfrac{(2x-1)\sqrt{(x+2)}}{(x-3)\sqrt{(x+1)^3}}\left\{\dfrac{2}{(2x-1)}+\dfrac{1}{2(x+2)}\right.$
   $\left.-\dfrac{1}{(x-3)}-\dfrac{3}{2(x+1)}\right\}$

4. $\dfrac{e^{2x}\cos3x}{\sqrt{(x-4)}}\left\{2-3\tan3x-\dfrac{1}{2(x-4)}\right\}$

5. $3\theta\sin\theta\cos\theta\left\{\dfrac{1}{\theta}+\cot\theta-\tan\theta\right\}$

6. $\dfrac{2x^4\tan x}{e^{2x}\ln2x}\left\{\dfrac{4}{x}+\dfrac{1}{\sin x\cos x}-2-\dfrac{1}{x\ln2x}\right\}$

7. $\dfrac{13}{16}$ 8. -6.71

Exercise 248 (page 672)

1. $2x^{2x}(1+\ln x)$ **2.** $(2x-1)^x\left\{\dfrac{2x}{2x-1}+\ln(2x-1)\right\}$

3. $\sqrt[x]{(x+3)}\left\{\dfrac{1}{x(x+3)}-\dfrac{\ln(x+3)}{x^2}\right\}$

4. $3x^{4x+1}\left\{4+\dfrac{1}{x}+4\ln x\right\}$ **5.** Proof **6.** $\dfrac{1}{3}$ **7.** Proof

Chapter 58

Exercise 249 (page 676)

1. (a) $6\,\mathrm{ch}\,2x$ (b) $10\,\mathrm{sh}\,5\theta$ (c) $36\,\mathrm{sech}^2\,9t$

2. (a) $-\dfrac{10}{3}\,\mathrm{sech}\,5x\,\mathrm{th}\,5x$ (b) $-\dfrac{5}{16}\,\mathrm{cosech}\,\dfrac{t}{2}\,\mathrm{coth}\,\dfrac{t}{2}$

(c) $-14\,\mathrm{cosech}^2\,7\theta$

3. (a) $2\,\mathrm{coth}\,x$ (b) $\dfrac{3}{8}\,\mathrm{sech}\,\dfrac{\theta}{2}\,\mathrm{cosech}\,\dfrac{\theta}{2}$

4. (a) $2(\mathrm{sh}^2\,2x+\mathrm{ch}^2\,2x)$ (b) $6e^{2x}(\mathrm{sech}^2\,2x+\mathrm{th}\,2x)$

5. (a) $\dfrac{12x\,\mathrm{ch}\,4x-9\,\mathrm{sh}\,4x}{2x^4}$

(b) $\dfrac{2(\cos 2t\,\mathrm{sh}\,2t+\mathrm{ch}\,2t\,\sin 2t)}{\cos^2 2t}$

Chapter 59

Exercise 250 (page 682)

1. (a) $\dfrac{4}{\sqrt{1-16x^2}}$ (b) $\dfrac{1}{\sqrt{4-x^2}}$

2. (a) $\dfrac{-3}{\sqrt{1-9x^2}}$ (b) $\dfrac{-2}{3\sqrt{9-x^2}}$

3. (a) $\dfrac{6}{1+4x^2}$ (b) $\dfrac{1}{4\sqrt{x}(1+x)}$

4. (a) $\dfrac{2}{t\sqrt{4t^2-1}}$ (b) $\dfrac{4}{x\sqrt{9x^2-16}}$

5. (a) $\dfrac{-5}{\theta\sqrt{\theta^2-4}}$ (b) $\dfrac{-2}{x\sqrt{x^4-1}}$

6. (a) $\dfrac{-6}{1+4t^2}$ (b) $\dfrac{-1}{\theta\sqrt{\theta^2-1}}$ **7.** Proof

8. (a) $\dfrac{6x}{\sqrt{1-9x^2}}+2\sin^{-1}3x$

(b) $\dfrac{t}{\sqrt{4t^2-1}}+2t\sec^{-1}2t$

9. (a) $2\theta\cos^{-1}(\theta^2-1)-\dfrac{2\theta^2}{\sqrt{2-\theta^2}}$

(b) $\left(\dfrac{1-x^2}{1+x^2}\right)-2x\tan^{-1}x$

10. (a) $\dfrac{-2\sqrt{t}}{1+t^2}+\dfrac{1}{\sqrt{t}}\cot^{-1}t$

(b) $\mathrm{cosec}^{-1}\sqrt{x}-\dfrac{1}{2\sqrt{(x-1)}}$

11. (a) $\dfrac{1}{x^3}\left\{\dfrac{3x}{\sqrt{1-9x^2}}-2\sin^{-1}3x\right\}$

(b) $\dfrac{-1+\dfrac{x}{\sqrt{1-x^2}}\cos^{-1}x}{(1-x^2)}$

Exercise 251 (page 683)

1. (a) 0.4812 (b) 2.0947 (c) 0.8089
2. (a) 0.6931 (b) 1.7627 (c) 2.1380
3. (a) 0.2554 (b) 0.7332 (c) 0.8673

Exercise 252 (page 686)

1. (a) $\dfrac{1}{\sqrt{(x^2+9)}}$ (b) $\dfrac{4}{\sqrt{(16x^2+1)}}$

2. (a) $\dfrac{2}{\sqrt{(t^2-9)}}$ (b) $\dfrac{1}{\sqrt{(4\theta^2-1)}}$

3. (a) $\dfrac{10}{25-4x^2}$ (b) $\dfrac{9}{(1-9x^2)}$

4. (a) $\dfrac{-4}{x\sqrt{(16-9x^2)}}$ (b) $\dfrac{1}{2x\sqrt{(1-4x^2)}}$

5. (a) $\dfrac{-4}{x\sqrt{(x^2+16)}}$ (b) $\dfrac{-1}{2x\sqrt{(16x^2+1)}}$

6. (a) $\dfrac{14}{49-4x^2}$ (b) $\dfrac{3}{4(1-9t^2)}$

7. (a) $\dfrac{2}{\sqrt{(x^2-1)}}$ (b) $\dfrac{1}{2\sqrt{(x^2+1)}}$

8. (a) $\dfrac{-1}{(x-1)\sqrt{[x(2-x)]}}$ (b) 1

9. (a) $\dfrac{-1}{(t-1)\sqrt{(2t-1)}}$ (b) $-\mathrm{cosec}\,x$

10. (a) $\dfrac{\theta}{\sqrt{(\theta^2+1)}}+\sinh^{-1}\theta$

(b) $\dfrac{\sqrt{x}}{\sqrt{(x^2-1)}}+\dfrac{\cosh^{-1}x}{2\sqrt{x}}$

11. (a) $\dfrac{-1}{t^3}\left\{\dfrac{1}{\sqrt{(1-t)}}+4\,\mathrm{sech}^{-1}\sqrt{t}\right\}$

(b) $\dfrac{1+2x\tanh^{-1}x}{(1-x^2)^2}$

12. Proof

13. (a) $\sinh^{-1}\dfrac{x}{3}+c$ (b) $\dfrac{3}{2}\sinh^{-1}\dfrac{2x}{5}+c$

14. (a) $\cosh^{-1}\dfrac{x}{4}+c$ (b) $\cosh^{-1}\dfrac{t}{\sqrt{5}}+c$

15. (a) $\dfrac{1}{6}\tan^{-1}\dfrac{\theta}{6}+c$ (b) $\dfrac{3}{2\sqrt{8}}\tanh^{-1}\dfrac{x}{\sqrt{8}}+c$

Chapter 60

Exercise 253 (page 690)

1. $\dfrac{\partial z}{\partial x}=2y \quad \dfrac{\partial z}{\partial y}=2x$ **2.** $\dfrac{\partial z}{\partial x}=3x^2-2y$

$\qquad\qquad\qquad\qquad\qquad\qquad \dfrac{\partial z}{\partial y}=-2x+2y$

3. $\dfrac{\partial z}{\partial x}=\dfrac{1}{y}$ **4.** $\dfrac{\partial z}{\partial x}=4\cos(4x+3y)$

$\quad\dfrac{\partial z}{\partial y}=\dfrac{-x}{y^2}$ $\dfrac{\partial z}{\partial y}=3\cos(4x+3y)$

5. $\dfrac{\partial z}{\partial x}=3x^2y^2+\dfrac{2y}{x^3}$

$\quad\dfrac{\partial z}{\partial y}=2x^3y-\dfrac{1}{x^2}-\dfrac{1}{y^2}$

6. $\dfrac{\partial z}{\partial x}=-3\sin 3x\sin 4y$

$\quad\dfrac{\partial z}{\partial y}=4\cos 3x\cos 4y$

7. $\dfrac{\partial V}{\partial h}=\dfrac{1}{3}\pi r^2 \quad \dfrac{\partial V}{\partial r}=\dfrac{2}{3}\pi rh$ **8.** Proof

9. $\dfrac{\partial y}{\partial t}=\dfrac{n\pi b}{L}\sin\left(\dfrac{n\pi}{L}\right)x\left\{c\cos\left(\dfrac{n\pi b}{L}\right)t\right.$

$\qquad\qquad\qquad\qquad\qquad\qquad \left. -k\sin\left(\dfrac{n\pi b}{L}\right)t\right\}$

$\quad\dfrac{\partial y}{\partial x}=\dfrac{n\pi}{L}\cos\left(\dfrac{n\pi}{L}\right)x\left\{k\cos\left(\dfrac{n\pi b}{L}\right)t\right.$

$\qquad\qquad\qquad\qquad\qquad\qquad \left. +c\sin\left(\dfrac{n\pi b}{L}\right)t\right\}$

10. (a) $\dfrac{\partial k}{\partial T}=\dfrac{A\,\Delta H}{RT^2}e^{\frac{T\Delta S-\Delta H}{RT}}$

    (b) $\dfrac{\partial A}{\partial T}=-\dfrac{k\,\Delta H}{RT^2}e^{\frac{\Delta H-T\Delta S}{RT}}$

    (c) $\dfrac{\partial(\Delta S)}{\partial T}=-\dfrac{\Delta H}{T^2}$

    (d) $\dfrac{\partial(\Delta H)}{\partial T}=\Delta S-R\ln\left(\dfrac{k}{A}\right)$

Exercise 254 (page 692)

1. (a) 8   (b) 18   (c) -12 (d) -12

2. (a) $\dfrac{-2}{x^2}$ (b) $\dfrac{-2}{y^2}$ (c) 0   (d) 0

3. (a) $\dfrac{-4y}{(x+y)^3}$ (b) $\dfrac{4x}{(x+y)^3}$ (c) $\dfrac{2(x-y)}{(x+y)^3}$

    (d) $\dfrac{2(x-y)}{(x+y)^3}$

4. (a) $\sinh x\cosh 2y$ (b) $4\sinh x\cosh 2y$

    (c) $2\cosh x\sinh 2y$ (d) $2\cosh x\sinh 2y$

5. (a) $(2-x^2)\sin(x-2y)+4x\cos(x-2y)$

    (b) $-4x^2\sin(x-2y)$

6. (a) $\dfrac{\partial^2 z}{\partial x^2}=\dfrac{-x}{\sqrt{(y^2-x^2)^3}}$

    (b) $\dfrac{\partial^2 z}{\partial y^2}=\dfrac{-x}{\sqrt{(y^2-x^2)}}\left\{\dfrac{1}{y^2}+\dfrac{1}{(y^2-x^2)}\right\}$

    (c) $\dfrac{\partial^2 z}{\partial x\partial y}=\dfrac{\partial^2 z}{\partial y\partial x}=\dfrac{y}{\sqrt{(y^2-x^2)^3}}$

7. $-\dfrac{1}{\sqrt{2}}$ **8.** Proof

Chapter 61

Exercise 255 (page 695)

1. $3x^2\,dx+2y\,dy$ **2.** $(2y+\sin x)\,dx+2x\,dy$

3. $\dfrac{2y}{(x+y)^2}\,dx-\dfrac{2x}{(x+y)^2}\,dy$ **4.** $\ln y\,dx+\dfrac{x}{y}\,dy$

5. $\left(y+\dfrac{1}{2y\sqrt{x}}\right)dx+\left(x-\dfrac{\sqrt{x}}{y^2}\right)dy$

6. $b(2+c)\,da+(2a-6bc+ac)\,db+b(a-3b)\,dc$

7. Proof

Exercise 256 (page 697)

1. $+226.2\,\text{cm}^3/\text{s}$ **2.** 2520 units/s

3. $515.5\,\text{cm/s}$ **4.** $1.35\,\text{cm}^3/\text{s}$ **5.** $17.4\,\text{cm}^2/\text{s}$

Exercise 257 (page 699)

1. $+21$ watts   **2.** $+2\%$ **3.** -1% **4.** $+1.35\,\text{cm}^4$

5. $-0.179\,\text{cm}$ **6.** $+6\%$ **7.** $+2.2\%$

Chapter 62

Exercise 258 (page 704)

1. Minimum at $(0, 0)$

2. (a) Minimum at $(1, -2)$

    (b) Saddle point at $(1, 2)$ (c) Maximum at $(0, 1)$

3. Maximum point at $(0,0)$, saddle point at $(4,0)$
4. Minimum at $(0,0)$
5. saddle point at $(0,0)$, minimum at $\left(\frac{1}{3},\frac{1}{3}\right)$

Exercise 259 (page 708)

1. Minimum at $(-4, 4)$
2. 4 m by 4 m by 2 m, surface area $= 48\text{m}^2$
3. Minimum at $(1, 0)$, minimum at $(-1, 0)$, saddle point at $(0, 0)$
4. Maximum at $(0, 0)$, saddle point at $(4, 0)$
5. Minimum at $(1, 2)$, maximum at $(-1, -2)$, saddle points at $(1, -2)$ and $(-1, 2)$
6. 150m^2

Chapter 63

Exercise 260 (page 723)

1. (a) $4x + c$ (b) $\dfrac{7x^2}{2} + c$

2. (a) $\dfrac{2}{15}x^3 + c$ (b) $\dfrac{5}{24}x^4 + c$

3. (a) $\dfrac{3x^2}{2} - 5x + c$ (b) $4\theta + 2\theta^2 + \dfrac{\theta^3}{3} + c$

4. (a) $\dfrac{-4}{3x} + c$ (b) $\dfrac{-1}{4x^3} + c$

5. (a) $\dfrac{4}{5}\sqrt{x^5} + c$ (b) $\dfrac{1}{9}\sqrt[4]{x^9} + c$

6. (a) $\dfrac{10}{\sqrt{t}} + c$ (b) $\dfrac{15}{7}\sqrt[5]{x} + c$

7. (a) $\dfrac{3}{2}\sin 2x + c$ (b) $-\dfrac{7}{3}\cos 3\theta + c$

8. (a) $\dfrac{1}{4}\tan 3x + c$ (b) $-\dfrac{1}{2}\cot 4\theta + c$

9. (a) $-\dfrac{5}{2}\operatorname{cosec} 2t + c$ (b) $\dfrac{1}{3}\sec 4t + c$

10. (a) $\dfrac{3}{8}e^{2x} + c$ (b) $\dfrac{-2}{15e^{5x}} + c$

11. (a) $\dfrac{2}{3}\ln x + c$ (b) $\dfrac{u^2}{2} - \ln u + c$

12. (a) $8\sqrt{x} + 8\sqrt{x^3} + \dfrac{18}{5}\sqrt{x^5} + c$

  (b) $-\dfrac{1}{t} + 4t + \dfrac{4t^3}{3} + c$

Exercise 261 (page 724)

1. (a) 105 (b) $-\dfrac{1}{2}$ **2.** (a) 6 (b) $-1\dfrac{1}{3}$
3. (a) 0 (b) 4 **4.** (a) 1 (b) 4.248

5. (a) 0.2352 (b) 2.598 **6.** (a) 0.2527 (b) 2.638
7. (a) 19.09 (b) 2.457 **8.** (a) 0.2703 (b) 9.099
9. 55.65 J/K **10.** Proof
11. 7.26 **12.** $77.7\,\text{m}^3$

Chapter 64

Exercise 262 (page 728)

1. $-\dfrac{1}{2}\cos(4x+9) + c$ **2.** $\dfrac{3}{2}\sin(2\theta - 5) + c$

3. $\dfrac{4}{3}\tan(3t+1) + c$ **4.** $\dfrac{1}{70}(5x-3)^7 + c$

5. $-\dfrac{3}{2}\ln(2x-1) + c$ **6.** $e^{3\theta+5} + c$ **7.** 227.5

8. 4.333 **9.** 0.9428 **10.** 0.7369

11. 1.6 years

Exercise 263 (page 730)

1. $\dfrac{1}{12}(2x^2 - 3)^6 + c$ **2.** $-\dfrac{5}{6}\cos^6 t + c$

3. $\dfrac{1}{2}\sec^2 3x + c$ or $\dfrac{1}{2}\tan^2 3x + c$

4. $\dfrac{2}{9}\sqrt{(3t^2 - 1)^3} + c$ **5.** $\dfrac{1}{2}(\ln\theta)^2 + c$

6. $\dfrac{3}{2}\ln(\sec 2t) + c$ **7.** $4\sqrt{(e^t + 4)} + c$

8. 1.763 **9.** 0.6000 **10.** 0.09259

11. $V = 2\pi\sigma\left\{\sqrt{(9^2 + r^2)} - r\right\}$ **12.** $\dfrac{8\pi^2 IkT}{h^2}$

13. Proof **14.** 11 min 50 s

Chapter 65

Exercise 264 (page 733)

1. $\dfrac{1}{2}\left(x - \dfrac{\sin 4x}{4}\right) + c$ **2.** $\dfrac{3}{2}\left(t + \dfrac{\sin 2t}{2}\right) + c$

3. $5\left(\dfrac{1}{3}\tan 3\theta - \theta\right) + c$ **4.** $-(\cot 2t + 2t) + c$

5. $\dfrac{\pi}{2}$ or 1.571 **6.** $\dfrac{\pi}{8}$ or 0.3927

7. -4.185 **8.** 0.6311

Exercise 265 (page 734)

1. $-\cos\theta + \dfrac{\cos^3\theta}{3} + c$ **2.** $\sin 2x - \dfrac{\sin^3 2x}{3} + c$

3. $\dfrac{-2}{3}\cos^3 t + \dfrac{2}{5}\cos^5 t + c$ **4.** $\dfrac{-\cos^5 x}{5} + \dfrac{\cos^7 x}{7} + c$

5. $\dfrac{3\theta}{4} - \dfrac{1}{4}\sin 4\theta + \dfrac{1}{32}\sin 8\theta + c$

6. $\dfrac{t}{8} - \dfrac{1}{32}\sin 4t + c$

Exercise 266 (page 735)

1. $-\dfrac{1}{2}\left(\dfrac{\cos 7t}{7} + \dfrac{\cos 3t}{3}\right) + c$

2. $\dfrac{\sin 2x}{2} - \dfrac{\sin 4x}{4} + c$ **3.** $\dfrac{3}{2}\left(\dfrac{\sin 7x}{7} + \dfrac{\sin 5x}{5}\right) + c$

4. $\dfrac{1}{4}\left(\dfrac{\cos 2\theta}{2} - \dfrac{\cos 6\theta}{6}\right) + c$ **5.** $\dfrac{3}{7}$ or 0.4286

6. 0.5973   **7.** 0.2474   **8.** −0.1999

Exercise 267 (page 737)

1. $5\sin^{-1}\dfrac{t}{2} + c$ **2.** $3\sin^{-1}\dfrac{x}{3} + c$

3. $2\sin^{-1}\dfrac{x}{2} + \dfrac{x}{2}\sqrt{(4-x^2)} + c$

4. $\dfrac{8}{3}\sin^{-1}\dfrac{3t}{4} + \dfrac{t}{2}\sqrt{(16-9t^2)} + c$

5. $\dfrac{\pi}{2}$ or 1.571   **6.** 2.760

Exercise 268 (page 738)

1. $\dfrac{3}{2}\tan^{-1}\dfrac{t}{2} + c$ **2.** $\dfrac{5}{12}\tan^{-1}\dfrac{3\theta}{4} + c$

3. 2.356   **4.** 2.457

Exercise 269 (page 739)

1. $2\sinh^{-1}\dfrac{x}{4} + c$ **2.** $\dfrac{3}{\sqrt{5}}\sinh^{-1}\dfrac{\sqrt{5}}{3}x + c$

3. $\dfrac{9}{2}\sinh^{-1}\dfrac{x}{3} + \dfrac{x}{2}\sqrt{(x^2+9)} + c$

4. $\dfrac{25}{4}\sinh^{-1}\dfrac{2t}{5} + \dfrac{t}{2}\sqrt{(4t^2+25)} + c$

5. 3.525   **6.** 4.348

Exercise 270 (page 741)

1. $\cosh^{-1}\dfrac{t}{4} + c$ **2.** $\dfrac{3}{2}\cosh^{-1}\dfrac{2x}{3} + c$

3. $\dfrac{\theta}{2}\sqrt{(\theta^2-9)} - \dfrac{9}{2}\cosh^{-1}\dfrac{\theta}{3} + c$

4. $\theta\sqrt{\left(\theta^2 - \dfrac{25}{4}\right)} - \dfrac{25}{4}\cosh^{-1}\dfrac{2\theta}{5} + c$

5. 2.634   **6.** 1.429

Chapter 66

Exercise 271 (page 744)

1. $2\ln(x-3) - 2\ln(x+3) + c$ or $\ln\left\{\dfrac{x-3}{x+3}\right\}^2 + c$

2. $5\ln(x+1) - \ln(x-3) + c$ or $\ln\left\{\dfrac{(x+1)^5}{(x-3)}\right\} + c$

3. $7\ln(x+4) - 3\ln(x+1) - \ln(2x-1) + c$ or

$\ln\left\{\dfrac{(x+4)^7}{(x+1)^3(2x-1)}\right\} + c$

4. $x + 2\ln(x+3) + 6\ln(x-2) + c$ or
$x + \ln\{(x+3)^2(x-2)^6\} + c$

5. $\dfrac{3x^2}{2} - 2x + \ln(x-2) - 5\ln(x+2) + c$

6. 0.6275   **7.** 0.8122   **8.** $\dfrac{1}{3}$ **9.** Proof

10. 19.05 ms

Exercise 272 (page 746)

1. $4\ln(x+1) + \dfrac{7}{(x+1)} + c$

2. $5\ln(x-2) + \dfrac{10}{(x-2)} - \dfrac{2}{(x-2)^2} + c$

3. 1.663   **4.** 1.089   **5.** Proof

Exercise 273 (page 747)

1. $\ln(x^2+7) + \dfrac{3}{\sqrt{7}}\tan^{-1}\dfrac{x}{\sqrt{7}} - \ln(x-2) + c$

2. 0.5880   **3.** 0.2939   **4.** 0.1865   **5.** Proof

Chapter 67

Exercise 274 (page 750)

1. $\dfrac{-2}{1+\tan\dfrac{\theta}{2}} + c$ **2.** $\ln\left\{\dfrac{\tan\dfrac{x}{2}}{1+\tan\dfrac{x}{2}}\right\} + c$

3. $\dfrac{2}{\sqrt{5}}\tan^{-1}\left(\dfrac{1}{\sqrt{5}}\tan\dfrac{\alpha}{2}\right) + c$

4. $\dfrac{1}{5}\ln\left\{\dfrac{2\tan\dfrac{x}{2} - 1}{\tan\dfrac{x}{2} + 2}\right\} + c$

Exercise 275 (page 752)

1. $\dfrac{2}{3}\tan^{-1}\left(\dfrac{5\tan\dfrac{\theta}{2}+4}{3}\right)+c$

2. $\dfrac{1}{\sqrt{3}}\ln\left\{\dfrac{\tan\dfrac{x}{2}+2-\sqrt{3}}{\tan\dfrac{x}{2}+2+\sqrt{3}}\right\}+c$

3. $\dfrac{1}{\sqrt{11}}\ln\left\{\dfrac{\tan\dfrac{p}{2}-4-\sqrt{11}}{\tan\dfrac{p}{2}-4+\sqrt{11}}\right\}+c$

4. $\dfrac{1}{\sqrt{7}}\ln\left\{\dfrac{3\tan\dfrac{\theta}{2}-4-\sqrt{7}}{3\tan\dfrac{\theta}{2}-4+\sqrt{7}}\right\}+c$

5. Proof   **6.** Proof   **7.** Proof

Chapter 68

Exercise 276 (page 755)

1. $\left[\dfrac{e^{2x}}{2}\left(x-\dfrac{1}{2}\right)\right]+c$ **2.** $-\dfrac{4}{3}e^{-3x}\left(x+\dfrac{1}{3}\right)+c$

3. $-x\cos x+\sin x+c$ **4.** $\dfrac{5}{2}\left(\theta\sin 2\theta+\dfrac{1}{2}\cos 2\theta\right)+c$

5. $\dfrac{3}{2}e^{2t}\left(t^{2}-t+\dfrac{1}{2}\right)+c$ **6.** 16.78

7. 0.2500   **8.** 0.4674   **9.** 15.78

Exercise 277 (page 757)

1. $\dfrac{2}{3}x^{3}\left(\ln x-\dfrac{1}{3}\right)+c$ **2.** $2x(\ln 3x-1)+c$

3. $\dfrac{\cos 3x}{27}(2-9x^{2})+\dfrac{2}{9}x\sin 3x+c$

4. $\dfrac{2}{29}e^{5x}(2\sin 2x+5\cos 2x)+c$

5. $2[\theta\tan\theta-\ln(\sec\theta)]+c$ **6.** 0.6363

7. 11.31   **8.** -1.543 **9.** 12.78

10. Proof   **11.** $C=0.66, S=0.41$

Chapter 69

Exercise 278 (page 761)

1. $e^{x}(x^{4}-4x^{3}+12x^{2}-24x+24)+c$

2. $e^{2t}\left(\dfrac{1}{2}t^{3}-\dfrac{3}{4}t^{2}+\dfrac{3}{4}t-\dfrac{3}{8}\right)+c$ **3.** 6.493

Exercise 279 (page 764)

1. $x^{5}\sin x+5x^{4}\cos x-20x^{3}\sin x-60x^{2}\cos x$
$+120x\sin x+120\cos x+c$

2. -134.87

3. $-x^{5}\cos x+5x^{4}\sin x+20x^{3}\cos x-60x^{2}\sin x$
$-120x\cos x+120\sin x+c$

4. 62.89

Exercise 280 (page 767)

1. $-\dfrac{1}{7}\sin^{6}x\cos x-\dfrac{6}{35}\sin^{4}x\cos x$
$\qquad\qquad-\dfrac{8}{35}\sin^{2}x\cos x-\dfrac{16}{35}\cos x+c$

2. 4   **3.** $\dfrac{8}{15}$

4. $\dfrac{1}{6}\cos^{5}x\sin x+\dfrac{5}{24}\cos^{3}x\sin x+\dfrac{5}{16}\cos x\sin x$
$\quad+\dfrac{5}{16}x+c$

5. $\dfrac{16}{35}$

Exercise 281 (page 768)

1. $\dfrac{8}{105}$ **2.** $\dfrac{13}{15}-\dfrac{\pi}{4}$ **3.** $\dfrac{8}{315}$

4. $x(\ln x)^{4}-4x(\ln x)^{3}+12x(\ln x)^{2}-24x\ln x$
$+24x+c$

5. Proof

Chapter 70

Exercise 282 (page 771)

1. 12   **2.** 3.5   **3.** 0.5   **4.** -174
5. 405   **6.** -157.5 **7.** 12π or 37.70   **8.** 112
9. 5   **10.** $170\,\text{cm}^{4}$

Exercise 283 (page 772)

1. 15   **2.** 15   **3.** 60   **4.** -8

5. $\dfrac{\pi^{4}}{2}$ or 48.70   **6.** -9 **7.** 8   **8.** 18

Chapter 71

Exercise 284 (page 775)

1. 1.569   **2.** 6.979   **3.** 0.672   **4.** 0.843

Exercise 285 (page 777)

1. 3.323 **2.** 0.997 **3.** 0.605 **4.** 0.799

Exercise 286 (page 780)

1. 1.187 **2.** 1.034 **3.** 0.747
4. 0.571 **5.** 1.260
6. (a) 1.875 (b) 2.107 (c) 1.765 (d) 1.916
7. (a) 1.585 (b) 1.588 (c) 1.583 (d) 1.585
8. (a) 10.194 (b) 10.007 (c) 10.070
9. (a) 0.677 (b) 0.674 (c) 0.675
10. 28.8 m **11.** 0.485 m

Chapter 72

Exercise 287 (page 787)

1. Proof **2.** 32 **3.** 37.5 **4.** 7.5
5. 1 **6.** 8.389 **7.** 1.67 **8.** 2.67

Exercise 288 (page 789)

1. 16 square units **2.** 5.545 square units
3. 29.33 N m **4.** 10.67 square units
5. 51 sq. units, 90 sq. units **6.** 73.83 sq. units
7. 140 m **8.** 693.1 kJ

Exercise 289 (page 791)

1. (0, 0) and (3, 3), 3 sq. units **2.** 20.83 square units
3. 0.4142 square units **4.** 2.5 sq. units

Chapter 73

Exercise 290 (page 794)

1. (a) 4 (b) $\dfrac{2}{\pi}$ or 0.637 (c) 69.17 **2.** 19

3. 9 m/s **4.** 4.17 **5.** 2.67 km **6.** $\dfrac{2c}{\pi}$

Exercise 291 (page 795)

1. (a) 6.928 (b) 4.919 (c) $\dfrac{25}{\sqrt{2}}$ or 17.68

2. (a) $\dfrac{1}{\sqrt{2}}$ or 0.707 (b) 1.225 (c) 2.121

3. 2.58 **4.** 19.10 A, 21.21 A **5.** 216 V, 240 V

6. 1.11 **7.** $\sqrt{\dfrac{E_1^2 + E_3^2}{2}}$

Chapter 74

Exercise 292 (page 799)

1. 525π **2.** 55π **3.** 75.6π **4.** 48π
5. 1.5π **6.** 4π **7.** 2.67π **8.** 2.67π
9. (a) 329.4π (b) 81π

Exercise 293 (page 801)

1. 428.8π **2.** π **3.** 42.67π **4.** 1.622π
5. 53.33π **6.** 8.1π **7.** 57.07π **8.** 113.33π
9. (a) $(-2, 6)$ and $(5, 27)$ (b) 57.17 square units
(c) 1326π cubic units

Chapter 75

Exercise 294 (page 805)

1. (2, 2) **2.** (2.50, 4.75) **3.** (3.036, 24.36)
4. (1.60, 4.57) **5.** $(-0.833, 0.633)$

Exercise 295 (page 807)

1. (2, 1.6) **2.** $(1, -0.4)$ **3.** (0, 3.6)
4. (0.375, 2.40) **5.** (3.0, 1.875) **6.** (2.4, 0)
7. (0, 0) and (4, 4), (1.80, 1.80)
8. $(-1, 7)$ and (3, 23), (1, 10.20)

Exercise 296 (page 810)

1. 189.6 cm^3 **2.** 1131 cm^3
3. On the centre line, distance 2.40 cm from the centre, i.e. at co-ordinates (1.70, 1.70)
4. (a) 45 square units
(b) (i) 1215π cubic units (ii) 202.5π cubic units
(c) (2.25, 13.5)
5. 64.90 cm^3, 16.86%, 506.2 g

Chapter 76

Exercise 297 (page 816)

1. (a) 72 cm^4, 1.73 cm (b) 128 cm^4, 2.31 cm
(c) 512 cm^4, 4.62 cm
2. (a) 729 cm^4, 3.67 cm (b) 2187 cm^4, 6.36 cm
(c) 243 cm^4, 2.12 cm
3. (a) 201 cm^4, 2.0 cm (b) 1005 cm^4, 4.47 cm
4. 3927 mm^4, 5.0 mm

5. (a) $335\,\text{cm}^4$, $4.73\,\text{cm}$ (b) $22\,030\,\text{cm}^4$, $14.3\,\text{cm}$
(c) $628\,\text{cm}^4$, $7.07\,\text{cm}$
6. $0.866\,\text{m}$ **7.** $0.245\,\text{m}^4$, $0.599\,\text{m}$
8. $14\,280\,\text{cm}^4$, $5.96\,\text{cm}$

Exercise 298 (page 818)

1. (a) $12\,190\,\text{mm}^4$, $10.9\,\text{mm}$ (b) $549.5\,\text{cm}^4$, $4.18\,\text{cm}$
2. $I_{AA} = 4224\,\text{cm}^4$, $I_{BB} = 6718\,\text{cm}^4$,
$I_{CC} = 37\,300\,\text{cm}^4$
3. $1350\,\text{cm}^4$, $5.67\,\text{cm}$

Chapter 77

Exercise 299 (page 830)

1. sketches   **2.** $y = x^2 + 3x - 1$

Exercise 300 (page 832)

1. $y = \dfrac{\sin 4x}{4} - x^2 + c$ **2.** $y = \dfrac{3}{2}\ln x - \dfrac{x^3}{6} + c$

3. $y = 3x - \dfrac{x^2}{2} - \dfrac{1}{2}$ **4.** $y = \dfrac{1}{3}\cos\theta + \dfrac{1}{2}$

5. $y = \dfrac{1}{6}\left(x^2 - 4x + \dfrac{2}{e^x} + 4\right)$ **6.** $y = \dfrac{3}{2}x^2 - \dfrac{x^3}{6} - 1$

7. $v = u + at$ **8.** $15.9\,\text{m}$

Exercise 301 (page 833)

1. $x = \dfrac{1}{3}\ln(2 + 3y) + c$ **2.** $\tan y = 2x + c$

3. $\dfrac{y^2}{2} + 2\ln y = 5x - 2$ **4.** Proof

5. $x = a(1 - e^{-kt})$

6. (a) $Q = Q_0 e^{\frac{-t}{CR}}$ (b) $9.30\,\text{C}$, $5.81\,\text{C}$

7. $273.3\,\text{N}$, $2.31\,\text{rads}$ **8.** $8\,\text{m}\,40\,\text{s}$

Exercise 302 (page 836)

1. $\ln y = 2\sin x + c$ **2.** $y^2 - y = x^3 + x$
3. $e^y = \dfrac{1}{2}e^{2x} + \dfrac{1}{2}$ **4.** $\ln(x^2 y) = 2x - y - 1$
5. Proof   **6.** $y = \dfrac{1}{\sqrt{(1 - x^2)}}$
7. $y^2 = x^2 - 2\ln x + 3$

8. (a) $V = E\left(1 - e^{\frac{-t}{CR}}\right)$ (b) $13.2\,\text{V}$ **9.** 3

Chapter 78

Exercise 303 (page 839)

1. $-\dfrac{1}{3}\ln\left(\dfrac{x^3 - y^3}{x^3}\right) = \ln x + c$

2. $y = x(c - \ln x)$ **3.** $x^2 = 2y^2\left(\ln y + \dfrac{1}{2}\right)$

4. $-\dfrac{1}{2}\ln\left(1 + \dfrac{2y}{x} - \dfrac{y^2}{x^2}\right) = \ln x + c$
or $x^2 + 2xy - y^2 = k$

5. $x^2 + xy - y^2 = 1$

Exercise 304 (page 840)

1. $y^4 = 4x^4(\ln x + c)$

2. $\dfrac{1}{5}\left\{\dfrac{3}{13}\ln\left(\dfrac{13y - 3x}{x}\right) - \ln\left(\dfrac{y - x}{x}\right)\right\} = \ln x + c$

3. $(x + y)^2 = 4x^3$ **4.** Proof

5. $y^3 = x^3(3\ln x + 64)$ **6.** Proof

Chapter 79

Exercise 305 (page 843)

1. $y = 3 + \dfrac{c}{x}$ **2.** $y = \dfrac{1}{2} + ce^{-x^2}$

3. $y = \dfrac{5t}{2} + \dfrac{c}{t}$ **4.** $y = \dfrac{x^3}{5} - \dfrac{x}{3} + \dfrac{47}{15x^2}$

5. $y = 1 + ce^{-x^2/2}$ **6.** $y = \dfrac{x}{2} + \dfrac{1}{4} + ce^{2x}$

Exercise 306 (page 844)

1. $y = \dfrac{1}{2} + \cos^2 x$ **2.** $\theta = \dfrac{1}{t}(\sin t - \pi\cos t)$

3. Proof   **4.** Proof   **5.** Proof

6. $v = \dfrac{b}{a^2} - \dfrac{bt}{a} + \left(u - \dfrac{b}{a^2}\right)e^{-at}$

7. Proof   **8.** Proof

9. $v = k\left\{\dfrac{1}{c} - \dfrac{e^{-t}}{c - 1} + \dfrac{e^{-ct}}{c(c - 1)}\right\}$

Chapter 80

Exercise 307 (page 851)

1.

x	y
1.0	2
1.1	2.1
1.2	2.209091
1.3	2.325000
1.4	2.446154
1.5	2.571429

2.

x	y	$(y')_0$
0	1	0
0.2	1	-0.2
0.4	0.96	-0.368
0.6	0.8864	-0.46368
0.8	0.793664	-0.469824
1.0	0.699692	

3. (a)

x	y
2.0	1
2.1	0.85
2.2	0.709524
2.3	0.577273
2.4	0.452174
2.5	0.333334

(b) 1.206%

4. (a)

x	y
2.0	1
2.2	1.2
2.4	1.421818
2.6	1.664849
2.8	1.928718
3.0	2.213187

(b) 1.596%

Exercise 308 (page 856)

1.

x	y	y'
1.0	2	1
1.1	2.10454546	1.08677686
1.2	2.216666672	1.152777773
1.3	2.33461539	1.204142008
1.4	2.457142859	1.244897958
1.5	2.5883333335	

2. (a) 0.412% (b) 0.000000214%

3. (a)

x	y	y'
0	1	1
0.1	1.11	1.21
0.2	1.24205	1.44205
0.3	1.398465	1.698465
0.4	1.581804	1.981804
0.5	1.794893	

(b) 0.117%

4.

x	y	$(y')_0$
0	1	0
0.2	0.98	-0.192
0.4	0.925472	-0.3403776
0.6	0.84854666	-0.41825599
0.8	0.76433779	-0.42294046
1.0	0.68609380	

Exercise 309 (page 861)

1.

n	x_n	y_n
0	1.0	2.0
1	1.1	2.104545
2	1.2	2.216667
3	1.3	2.334615
4	1.4	2.457143
5	1.5	2.533333

2.

n	x_n	y_n
0	0	1.0
1	0.2	0.980395
2	0.4	0.926072
3	0.6	0.848838
4	0.8	0.763649
5	1.0	0.683952

3. (a)

n	x_n	y_n
0	2.0	1.0
1	2.1	0.854762
2	2.2	0.718182
3	2.3	0.589130
4	2.4	0.466667
5	2.5	0.340000

(b) no error

Chapter 81

Exercise 310 (page 867)

1. $y = Ae^{\frac{2}{3}t} + Be^{-\frac{1}{2}t}$ **2.** $\theta = (At + B)e^{-\frac{1}{2}t}$

3. $y = e^{-x}(A\cos 2x + B\sin 2x)$

4. $y = 3e^{\frac{2}{3}x} + 2e^{-\frac{3}{2}x}$ **5.** $y = 4e^{\frac{1}{4}t} - 3e^{t}$

6. $y = 2xe^{-\frac{5}{3}x}$ **7.** $x = 2(1 - 3t)e^{3t}$

8. $y = 2e^{-3x}(2\cos 2x + 3\sin 2x)$

9. $\theta = e^{-2.5t}(3\cos 5t + 2\sin 5t)$

Exercise 311 (page 869)

1. Proof **2.** $s = c\cos at$

3. $\theta = e^{-2t}(0.3\cos 6t + 0.1\sin 6t)$

4. $x = \{s + (u + ns)t\}e^{-nt}$

5. $i = \dfrac{1}{20}\left(e^{-160t} - e^{-840t}\right)$ **6.** $s = 4te^{-\frac{3}{2}t}$

Chapter 82

Exercise 312 (page 874)

1. $y = Ae^{\frac{1}{2}x} + Be^{-3x} - 2$

2. $y = Ae^{\frac{1}{3}x} + Be^{-x} - 2 - \frac{3}{2}x$

3. $y = \frac{2}{7}(3e^{-\frac{4}{3}x} + 4e^{x}) - 2$

4. $y = -\left(2 + \frac{3}{4}x\right)e^{\frac{2}{3}x} + 2 + \frac{3}{4}x$

5. (a) $q = \dfrac{1}{20} - \left(\dfrac{5}{2}t + \dfrac{1}{20}\right)e^{-50t}$

(b) $q = \dfrac{1}{20}(1 - \cos 50t)$

6. $\theta = 2(te^{-2t} + 1)$

Exercise 313 (page 876)

1. $y = Ae^{3x} + Be^{-2x} - \frac{1}{3}e^{x}$

2. $y = Ae^{4x} + Be^{-x} - \frac{3}{5}xe^{-x}$

3. $y = A\cos 3x + B\sin 3x + 2c^{2x}$

4. $y = (At + B)e^{\frac{1}{3}t} + \frac{2}{3}t^{2}e^{\frac{1}{3}t}$

5. $y = \dfrac{5}{44}\left(e^{-2x} - e^{\frac{1}{5}x}\right) + \dfrac{1}{4}e^{x}$

6. $y = 2e^{3t}(1 - 3t + t^{2})$

Exercise 314 (page 878)

1. $y = Ae^{\frac{3}{2}x} + Be^{-x} - \frac{1}{5}(11\sin 2x - 2\cos 2x)$

2. $y = (Ax + B)e^{2x} - \frac{4}{5}\sin x + \frac{3}{5}\cos x$

3. $y = A\cos x + B\sin x + 2x\sin x$

4. $y = \dfrac{1}{170}(6e^{4x} - 51e^{-x}) - \dfrac{1}{34}(15\sin x - 9\cos x)$

5. $y = \dfrac{k}{n^{4} - p^{4}}\left(p^{2}\left(\sin pt - \dfrac{p}{n}\sin nt\right)\right.$
$\left. + n^{2}(\cos pt - \cos nt)\right)$

6. Proof **7.** $q = (10t + 0.01)e^{-1000t}$
$+ 0.024\sin 200t - 0.010\cos 200t$

Exercise 315 (page 880)

1. $y = Ae^{\frac{x}{4}} + Be^{\frac{x}{2}} + 2x + 12 + \frac{8}{17}(6\cos x - 7\sin x)$

2. $y = Ae^{2\theta} + Be^{\theta} + \frac{1}{2}(\sin 2\theta + \cos 2\theta)$

3. $y = Ae^x + Be^{-2x} - \frac{3}{4} - \frac{1}{2}x - \frac{1}{2}x^2 + \frac{1}{4}e^{2x}$

4. $y = e^t(A\cos t + B\sin t) - \frac{t}{2}e^t\cos t$

5. $y = \frac{4}{3}e^{5x} - \frac{10}{3}e^{2x} - \frac{1}{3}xe^{2x} + 2$

6. $y = 2e^{-\frac{3}{2}x} - 2e^{2x} + \frac{3e^x}{29}(3\sin x - 7\cos x)$

Chapter 83

Exercise 316 (page 883)

1. (a) $16e^{2x}$ (b) $\frac{1}{4}e^{\frac{t}{2}}$

2. (a) $81\sin 3t$ (b) $-1562.5\cos 5\theta$

3. (a) $256\cos 2x$ (b) $-\frac{2^9}{3^8}\sin\frac{2}{3}t$

4. (a) $(9!)x^2$ (b) $630t$

5. (a) $32\cosh 2x$ (b) $1458\sinh 3x$

6. (a) $128\sinh 2x$ (b) $729\cosh 3x$

7. (a) $-\frac{12}{\theta^4}$ (b) $\frac{240}{t^7}$

Exercise 317 (page 885)

1. $x^2 y^{(n)} + 2nxy^{(n-1)} + n(n-1)y^{(n-2)}$
2. $y^{(n)} = e^{2x}2^{n-3}\{8x^3 + 12nx^2 + n(n-1)(6x) + n(n-1)(n-2)\}$

   $y^{(3)} = e^{2x}(8x^3 + 36x^2 + 36x + 6)$
3. $y^{(4)} = 2e^{-x}(x^3 - 12x^2 + 36x - 24)$
4. $y^{(5)} = (60x - x^3)\sin x + (15x^2 - 60)\cos x$
5. $y^{(4)} = -4e^{-t}\sin t$ 6. $y^{(3)} = x^2(47 + 60\ln 2x)$
7. Proof 8. $y^{(5)} = e^{2x}2^4(2x^3 + 19x^2 + 50x + 35)$

Exercise 318 (page 888)

1. $y = \left(1 - \frac{x^2}{2!} + \frac{5x^4}{4!} - \frac{5 \times 9x^6}{6!}\right.$

   $\left. + \frac{5 \times 9 \times 13x^8}{8!} - \cdots\right)$

   $+ 2\left(x - \frac{3x^3}{3!} + \frac{3 \times 7x^5}{5!} - \frac{3 \times 7 \times 11x^7}{7!} + \cdots\right)$

2. Proof

3. $y = 1 + x + 2x^2 + \frac{x^3}{2} - \frac{x^5}{8} + \frac{x^7}{16} + \cdots$

4. $y = \left\{1 - \frac{x^2}{2^2} + \frac{x^4}{2^2 \times 4^2} - \frac{x^6}{2^2 \times 4^2 \times 6^2} + \cdots\right\}$

   $+ 2\left\{x - \frac{x^3}{3^2} + \frac{x^5}{3^2 \times 5^2} - \frac{x^7}{3^2 \times 5^2 \times 7^2} + \cdots\right\}$

Exercise 319 (page 895)

1. $y = A\left\{1 + x + \frac{x^2}{(2 \times 3)} + \frac{x^3}{(2 \times 3)(3 \times 5)} + \cdots\right\}$

   $+ Bx^{\frac{1}{2}}\left\{1 + \frac{x}{(1 \times 3)} + \frac{x^2}{(1 \times 2)(3 \times 5)}\right.$

   $\left. + \frac{x^3}{(1 \times 2 \times 3)(3 \times 5 \times 7)} + \cdots\right\}$

2. $y = A\left(1 - \frac{x^2}{2!} + \frac{x^4}{4!} - \cdots\right) + B\left(x - \frac{x^3}{3!} + \frac{x^5}{5!} - \cdots\right)$

   $= P\cos x + Q\sin x$

3. $y = A\left\{1 + \frac{x}{(1 \times 4)} + \frac{x^2}{(1 \times 2)(4 \times 7)}\right.$

   $\left. + \frac{x^3}{(1 \times 2 \times 3)(4 \times 7 \times 10)} + \cdots\right\}$

   $+ Bx^{-\frac{1}{3}}\left\{1 + \frac{x}{(1 \times 2)} + \frac{x^2}{(1 \times 2)(2 \times 5)}\right.$

   $\left. + \frac{x^3}{(1 \times 2 \times 3)(2 \times 5 \times 8)} + \cdots\right\}$

4. Proof

Exercise 320 (page 899)

1. $y = Ax^2\left\{1 - \frac{x^2}{12} + \frac{x^4}{384} - \cdots\right\}$

2. $J_3(x) = \left(\frac{x}{2}\right)^3\left\{\frac{1}{\Gamma 4} - \frac{x^2}{2^2\Gamma 5} + \frac{x^4}{2^5\Gamma 6} - \cdots\right\}$

3. $J_0(x) = 0.765$, $J_1(x) = 0.440$

Exercise 321 (page 904)

1. (a) $y = a_0 + a_1\left(x + \frac{x^3}{3} + \frac{x^5}{5} + \cdots\right)$

   (b) $y = a_0\{1 - 3x^2\} + a_1\left\{x - \frac{2}{3}x^3 - \frac{1}{5}x^5\right\}$

2. (a) x (b) $\dfrac{1}{8}\left(35x^4 - 30x^2 + 3\right)$

   (c) $\dfrac{1}{8}\left(63x^5 - 70x^3 + 15x\right)$

Chapter 84

Exercise 322 (page 908)

1. $u = 2ty^2 + f(t)$ **2.** $u = t^2(\cos\theta - 1) + 2t$

3. Proof   **4.** Proof

5. $u = -4e^y \cos 2x - \cos x + 4\cos 2x + 2y^2 - 4e^y + 4$

6. $u = y\left(\dfrac{x^4}{3} - \dfrac{x^2}{2}\right) + x\cos 2y + \sin y$

7. $u = -\sin(x+t) + x + \sin x + 2t + \sin t$

8. Proof

9. $u = \sin x \sin y + \dfrac{x^2}{2} + 2\cos y - \dfrac{\pi^2}{2}$

10. Proof

Exercise 323 (page 910)

1. $T = A e^{3t} + B e^{-3t}$ **2.** $T = A\cos 3t + B\sin 3t$

3. $X = A e^x + B e^{-x}$ **4.** $X = A\cos x + B\sin x$

Exercise 324 (page 913)

1. $u(x,t) = \dfrac{12}{\pi^2} \displaystyle\sum_{n=1}^{\infty} \dfrac{1}{n^2} \sin\dfrac{n\pi}{2} \sin\dfrac{n\pi x}{40} \cos\dfrac{3n\pi t}{40}$

2. $u(x,t) = \dfrac{8}{\pi^2} \displaystyle\sum_{n=1}^{\infty} \dfrac{1}{n^2} \sin\dfrac{n\pi}{2} \sin\dfrac{n\pi x}{80} \cos\dfrac{n\pi t}{10}$

Exercise 325 (page 915)

1. $u(x,t) = \dfrac{40}{\pi} \displaystyle\sum_{n(\text{odd})=1}^{\infty} \dfrac{1}{n} e^{-\frac{n^2\pi^2 c^2 t}{16}} \sin\dfrac{n\pi x}{4}$

2. $u(x,t) = \left(\dfrac{8}{\pi}\right)^3 \displaystyle\sum_{n(\text{odd})=1}^{\infty} \dfrac{1}{n^3} e^{-\frac{n^2\pi^2 t}{64}} \sin\dfrac{n\pi x}{8}$

3. $u(x,t) = \dfrac{320}{\pi^2} \displaystyle\sum_{n(\text{odd})=1}^{\infty} \dfrac{1}{n^2} \sin\dfrac{n\pi}{2} \sin\dfrac{n\pi x}{20} e^{-\left(\frac{n^2\pi^2 t}{400}\right)}$

Exercise 326 (page 918)

1. $u(x,y) = \dfrac{20}{\pi} \displaystyle\sum_{n(\text{odd})=1}^{\infty} \dfrac{1}{n} \operatorname{cosech} n\pi \sin n\pi x \sinh n\pi (y-2)$

2. $u(x,y) = \dfrac{216}{\pi^3} \displaystyle\sum_{n(\text{odd})=1}^{\infty} \dfrac{1}{n^3} \operatorname{cosech}\dfrac{2n\pi}{3} \sin\dfrac{n\pi x}{3} \sinh\dfrac{n\pi}{3}(2-y)$

Chapter 85

Exercise 327 (page 926)

1. (a) continuous (b) continuous (c) discrete
   (d) continuous

2. (a) discrete (b) continuous (c) discrete (d) discrete

Exercise 328 (page 929)

1. If one symbol is used to represent 10 vehicles, working correct to the nearest 5 vehicles gives: 3.5, 4.5, 6, 7, 5 and 4 symbols, respectively.

2. If one symbol represents 200 components, working correct to the nearest 100 components gives: Mon 8, Tues 11, Wed 9, Thurs 12 and Fri 6.5.

3. 6 equally spaced horizontal rectangles, whose lengths are proportional to 35, 44, 62, 68, 49 and 41, respectively.

4. 5 equally spaced horizontal rectangles, whose lengths are proportional to 1580, 2190, 1840, 2385 and 1280 units, respectively.

5. 6 equally spaced vertical rectangles, whose heights are proportional to 35, 44, 62, 68, 49 and 41 units, respectively.

6. 5 equally spaced vertical rectangles, whose heights are proportional to 1580, 2190, 1840, 2385 and 1280 units, respectively.

7. Three rectangles of equal height, subdivided in the percentages shown in the columns of the question. P increases by 20% at the expense of Q and R.

8. Four rectangles of equal height, subdivided as follows: week 1: 18%, 7%, 35%, 12%, 28%; week 2: 20%, 8%, 32%, 13%, 27%; week 3: 22%, 10%, 29%, 14%, 25%; week 4: 20%, 9%, 27%, 19%, 25%. Little change in centres A and B, a reduction of about 8% in C, an increase of about 7% in D and a reduction of about 3% in E.

9. A circle of any radius, subdivided into sectors having angles of $7.5°, 22.5°, 52.5°, 167.5°$ and $110°$, respectively.

10. A circle of any radius, subdivided into sectors having angles of $107°, 156°, 29°$ and $68°$, respectively.

11. (a) £495 (b) 88   **12.** (a) £16 450 (b) 138

Exercise 329 (page 935)

1. There is no unique solution, but one solution is:

  39.3–39.4   1;   39.5–39.6   5;   39.7–39.8   9;
  39.9–40.0 17; 40.1–40.2 15; 40.3–40.4   7;
  40.5–40.6   4; 40.7–40.8   2

2. Rectangles, touching one another, having midpoints of $39.35, 39.55, 39.75, 39.95, \ldots$ and heights of $1, 5, 9, 17, \ldots$

3. There is no unique solution, but one solution is:
   20.5–20.9 3; 21.0–21.4 10; 21.5–21.9 11;
   22.0–22.4 13; 22.5–22.9 9; 23.0–23.4 2

4. There is no unique solution, but one solution is:
   1–10 3; 11–19 7; 20–22 12; 23–25 11;
   26–28 10; 29–38 5; 39–48 2

5. 20.95 3; 21.45 13; 21.95 24; 22.45 37; 22.95 46;
   23.45 48

6. Rectangles, touching one another, having midpoints of 5.5, 15, 21, 24, 27, 33.5 and 43.5. The heights of the rectangles (frequency per unit class range) are 0.3, 0.78, 4, 3.67, 3.33, 0.5 and 0.2

7. (10.95 2), (11.45 9), (11.95 19), (12.45 31), (12.95 42), (13.45, 50)

8. A graph of cumulative frequency against upper class boundary having co-ordinates given in the answer to Problem 7.

9. (a) There is no unique solution, but one solution is:
      2.05–2.09 3; 2.10–2.14 10; 2.15–2.19 11;
      2.20–2.24 13; 2.25–2.29 9; 2.30–2.34 2

   (b) Rectangles, touching one another, having midpoints of $2.07, 2.12, \ldots$ and heights of $3, 10, \ldots$

   (c) Using the frequency distribution given in the solution to part (a) gives 2.095 3; 2.145 13; 2.195 24; 2.245 37; 2.295 46; 2.345 48

   (d) A graph of cumulative frequency against upper class boundary having the co-ordinates given in part (c).

Chapter 86

Exercise 330 (page 939)

1. Mean 7.33, median 8, mode 8
2. Mean 27.25, median 27, mode 26
3. Mean 4.7225, median 4.72, mode 4.72
4. Mean 115.2, median 126.4, no mode

Exercise 331 (page 940)

1. 23.85 kg   2. 171.7 cm
3. Mean 89.5, median 89, mode 88.2
4. Mean 2.02158 cm, median 2.02152 cm, mode 2.02167 cm

Exercise 332 (page 942)

1. 4.60   2. 2.83 μF
3. Mean 34.53 MPa, standard deviation 0.07474 MPa
4. 0.296 kg   5. 9.394 cm   6. 0.00544 cm

Exercise 333 (page 943)

1. 30, 27.5, 33.5 days   2. 27, 26, 33 faults
3. $Q_1 = 164.5$ cm, $Q_2 = 172.5$ cm, $Q_3 = 179$ cm, 7.25 cm
4. 37 and 38; 40 and 41   5. 40, 40, 41; 50, 51, 51

Chapter 87

Exercise 334 (page 947)

1. (a) $\dfrac{2}{9}$ or 0.2222 (b) $\dfrac{7}{9}$ or 0.7778

2. (a) $\dfrac{23}{139}$ or 0.1655 (b) $\dfrac{47}{139}$ or 0.3381
   (c) $\dfrac{69}{139}$ or 0.4964

3. (a) $\dfrac{1}{6}$ (b) $\dfrac{1}{6}$ (c) $\dfrac{1}{36}$ 4. $\dfrac{5}{36}$

5. (a) $\dfrac{2}{5}$ (b) $\dfrac{1}{5}$ (c) $\dfrac{4}{15}$ (d) $\dfrac{13}{15}$

6. (a) $\dfrac{1}{250}$ (b) $\dfrac{1}{200}$ (c) $\dfrac{9}{1000}$ (d) $\dfrac{1}{50000}$

Exercise 335 (page 949)

1. (a) 0.6 (b) 0.2 (c) 0.15   2. (a) 0.64 (b) 0.32
3. 0.0768   4. (a) 0.4912 (b) 0.4211
5. (a) 89.38% (b) 10.25%
6. (a) 0.0227 (b) 0.0234 (c) 0.0169

Exercise 336 (page 951)

1. (a) 210   (b) 3024   2. (a) 792   (b) 15
3. 210   4. 3003
5. $\dfrac{10}{^{49}C_6} = \dfrac{10}{13983816} = \dfrac{1}{1398382}$ or 715×10^{-9}

Exercise 337 (page 952)

1. 53.26%   2. 70.32%   3. 7.48%

Chapter 88

Exercise 338 (page 959)

1. (a) 0.0186 (b) 0.9976   2. (a) 0.2316 (b) 0.1408
3. (a) 0.7514 (b) 0.0019   4. (a) 0.9655 (b) 0.0028

Answers to practice exercises **1191**

5. Vertical adjacent rectangles, whose heights are proportional to 0.0313, 0.1563, 0.3125, 0.3125, 0.1563 and 0.0313
6. Vertical adjacent rectangles, whose heights are proportional to 0.0280, 0.1306, 0.2613, 0.2903, 0.1935, 0.0774, 0.0172 and 0.0016
7. 0.0574

Exercise 339 (page 962)

1. 0.0613 2. 0.5768
3. (a) 0.1771 (b) 0.5153 4. 0.9856
5. The probabilities of the demand for 0, 1, 2, … tools are 0.0067, 0.0337, 0.0842, 0.1404, 0.1755, 0.1755, 0.1462, 0.1044, 0.0653, … This shows that the probability of wanting a tool 8 times a day is 0.0653, i.e. less than 10%. Hence 7 should be kept in the store
6. Vertical adjacent rectangles having heights proportional to 0.4966, 0.3476, 0.1217, 0.0284, 0.0050 and 0.0007

Chapter 89

Exercise 340 (page 967)

1. 6 2. 22 3. (a) 479 (b) 63 (c) 21
4. 4 5. (a) 131 (b) 553
6. (a) 15 (b) 4 7. (a) 65 (b) 209 (c) 89
8. (a) 1 (b) 85 (c) 13

Exercise 341 (page 970)

1. Graphically, $\bar{x} = 27.1, \sigma = 0.3$; by calculation, $\bar{x} = 27.079,$ $\sigma = 0.3001$
2. (a) $\bar{x} = 23.5\,\text{kN},$ $\sigma = 2.9\,\text{kN}$ (b) $\bar{x} = 23.364\,\text{kN},$ $\sigma = 2.917\,\text{kN}$

Chapter 90

Exercise 342 (page 974)

1. 0.999 2. -0.916 3. 0.422 4. 0.999
5. -0.962 6. 0.632 7. 0.937

Chapter 91

Exercise 343 (page 980)

1. $Y = -256 + 80.6X$ 2. $Y = 0.0477 + 0.216X$
3. $X = 3.20 + 0.0124Y$ 4. $X = -0.056 + 4.56Y$
5. $Y = 1.142 + 2.268X$ 6. $X = -0.483 + 0.440Y$
7. (a) 7.95 V (b) 17.1 mA 8. $Y = 0.881 - 0.0290X$
9. $X = 30.194 - 34.039Y$ 10. (a) 0.417 s (b) 21.7 N

Chapter 92

Exercise 344 (page 987)

1. $\mu_{\bar{x}} = 22.4\,\text{cm}, \sigma_{\bar{x}} = 0.0080\,\text{cm}$
2. $\sigma_{\bar{x}} = 0.0079\,\text{cm}$
3. (a) $\mu_{\bar{x}} = 1.70\,\text{cm}, \sigma_{\bar{x}} = 2.91 \times 10^{-3}\,\text{cm}$
   (b) $\mu_{\bar{x}} = 1.70\,\text{cm}, \sigma_{\bar{x}} = 2.89 \times 10^{-3}\,\text{cm}$
4. 0.023 5. 0.497 6. 0.0038
7. (a) 0.0179 (b) 0.740 (c) 0.242

Exercise 345 (page 991)

1. 66.89 and 68.01 mm, 66.72 and 68.18 mm
2. (a) 2.355 Mg to 2.445 Mg; 2.341 Mg to 2.459 Mg
   (b) 86%
3. $12.73 \times 10^{-4}\,\text{m}^\circ\text{C}^{-1}$ to $12.89 \times 10^{-4}\,\text{m}^\circ\text{C}^{-1}$
4. (a) at least 68 (b) at least 271
5. 10.91 t to 11.27 t 6. 45.6 seconds

Exercise 346 (page 995)

1. 5.133 MPa to 5.207 MPa
2. 5.125 MPa to 5.215 MPa
3. $1.10\,\Omega\,\text{m}^{-1}$ to $1.15\,\Omega\,\text{m}^{-1}$ 4. 95%

Chapter 93

Exercise 347 (page 1003)

1. (a) 28.1% (b) 4.09% (c) 0.19%
2. (a) 55.2% (b) 4.65% (c) 0.07%
3. (a) 35.3% (b) 18.5% (c) 8.4%
4. (a) 32.3% (b) 20.1% (c) 11.9%

Exercise 348 (page 1007)

1. z (sample) $= 3.54, z_\alpha = 2.58$, hence hypothesis is rejected, where z_α is the z-value corresponding to a level of significance of α
2. $t_{0.95}, \nu_8 = 1.86, |t| = 1.89,$ hence null hypothesis rejected
3. z (sample) $= 2.85, z_\alpha = \pm 2.58$, hence hypothesis is rejected
4. $\bar{x} = 10.38, s = 0.33, t_{0.95}\nu_{19} = 1.73, |t| = 1.72,$ hence hypothesis is accepted
5. $|t| = 3.00,$
   (a) $t_{0.975}\nu_9 = 2.26,$ hence hypothesis rejected,
   (b) $t_{0.995}\nu_9 = 3.25,$ hence hypothesis is accepted
6. $|t| = 3.08,$
   (a) $t_{0.95}\nu_5 = 2.02,$ hence claim supported,
   (b) $t_{0.99}\nu_5 = 3.36,$ hence claim not supported

Exercise 349 (page 1011)

1. Take $\bar{x}$ as $24 + 15$, i.e. 39 hours, $z = 1.28$, $z_{0.05}$, one-tailed test $= 1.645$, hence hypothesis is accepted
2. $z = 2.357$, $z_{0.05}$, two-tailed test $= \pm 1.96$, hence hypothesis is rejected
3. $\bar{x}_1 = 23.7$, $s_1 = 1.73$, $\sigma_1 = 1.93$, $\bar{x}_2 = 25.7$, $s_2 = 2.50$, $\sigma_2 = 2.80$, $|t| = 1.32$, $t_{0.995}\nu_8 = 3.36$, hence hypothesis is accepted
4. z (sample) $= 1.99$,
   (a) $z_{0.05}$, two-tailed test $= \pm 1.96$, no significance,
   (b) $z_{0.01}$, two-tailed test $= \pm 2.58$, significant difference
5. Assuming null hypothesis of no difference, $\sigma = 0.397$, $|t| = 1.85$,
   (a) $t_{0.995}$, $\nu_{22} = 2.82$, hypothesis rejected,
   (b) $t_{0.95}$, $\nu_{22} = 1.72$, hypothesis accepted
6. $\sigma = 0.571$, $|t| = 3.32$, $t_{0.95}$, $\nu_8 = 1.86$, hence hypothesis is rejected

Chapter 94

Exercise 350 (page 1014)

1. 10.2 2. 3.16

Exercise 351 (page 1020)

1. Expected frequencies: $7, 33, 65, 73, 48, 19, 4, 0$; χ^2-value $= 3.62$, $\chi^2_{0.95}$, $\nu_7 = 14.1$, hence hypothesis accepted. $\chi^2_{0.10}$, $\nu_7 = 2.83$, hence data is not 'too good'
2. $\lambda = 2.404$; expected frequencies: $11, 27, 33, 26, 16, 8, 3i\,\chi^2$-value $= 42.24$; $\chi^2_{0.95}$, $\nu_6 = 12.6$, hence the data does not fit a Poisson distribution at a level of significance of 0.05
3. $\bar{x} = 1.32$, $s = 0.0180$; expected frequencies, $6, 17, 36, 55, 65, 55, 36, 17, 6$; χ^2-value $= 5.98$; $\chi^2_{0.95}$, $\nu_6 = 12.6$, hence the null hypothesis is accepted, i.e. the data does correspond to a normal distribution
4. $\lambda = 0.896$; expected frequencies are $102, 91, 41, 12, 3, 0, 0$; χ^2-value $= 5.10$. $\chi^2_{0.95}$, $\nu_6 = 12.6$, hence this data fits a Poisson distribution at a level of significance of 0.05
5. $\bar{x} = 10.09\,\text{MN}$; $\sigma = 0.733\,\text{MN}$; expected frequencies, $2, 5, 12, 16, 14, 8, 3, 1$; χ^2-value $= 0.563$; $\chi^2_{0.95}$, $\nu_5 = 11.1$. Hence hypothesis accepted. $\chi^2_{0.05}$, $\nu_5 = 1.15$, hence the results are 'too good to be true'

Exercise 352 (page 1024)

1. H_0: $t = 15\,\text{h}$, H_1: $t \neq 15\,\text{h}$
   $S = 7$. From Table 94.3,
   $S \leq 2$, hence accept H_0
2. $S = 6$. From Table 94.3, $S \leq 4$, hence null hypothesis accepted
3. H_0: $\text{mean}_A = \text{mean}_B$, H_1: $\text{mean}_A \neq \text{mean}_B$, $S = 4$
   From Table 94.3, $S \leq 4$, hence H_1 is accepted

Exercise 353 (page 1028)

1. H_0: $t = 220\,\text{h}$, H_1: $t \neq 220\,\text{h}$, $T = 74$.
   From Table 94.4, $T \leq 29$, hence H_0 is accepted
2. H_0: $s = 150$, H_1: $s \neq 150$, $T = 59.5$
   From Table 94.4, $T \leq 40$, hence alternative hypothesis H_0 is accepted
3. H_0: $N = R$, H_0: $N \neq R$, $T = 5$ From Table 94.4, with $n = 10$ (since two differences are zero), $T \leq 8$, hence there is a significant difference in the drying times

Exercise 354 (page 1034)

1. H_0: $T_A = T_B$, H_1: $T_A \neq T_B$, $U = 30$.
   From Table 94.5, $U \leq 17$, hence accept H_0, i.e. there is no difference between the brands
2. H_0: $B.S._A = B.S._B$, H_1: $B.S._A \neq B.S._B$, $\alpha_2 = 10\%$, $U = 28$. From Table 94.5, $U \leq 15$, hence accept H_0, i.e. there is no difference between the processes
3. H_0: $A = B$, H_1: $A \neq B$, $\alpha_2 = 5\%$, $U = 4$. From Table 94.5, $U \leq 8$, hence null hypothesis is rejected, i.e. the two methods are not equally effective
4. H_0: $\text{mean}_A = \text{mean}_B$, H_1: $\text{mean}_A \neq \text{mean}_B$, $\alpha_2 = 5\%$, $U = 90$ From Table 94.5, $U \leq 99$, hence H_0 is rejected and H_1 accepted

Chapter 95

Exercise 355 (page 1047)

1. (a) $\dfrac{2}{s^2} - \dfrac{3}{s}$ (b) $\dfrac{10}{s^3} + \dfrac{4}{s^2} - \dfrac{3}{s}$

2. (a) $\dfrac{1}{4s^4} - \dfrac{3}{s^2} + \dfrac{2}{s}$ (b) $\dfrac{8}{s^6} - \dfrac{48}{s^5} + \dfrac{1}{s^3}$

3. (a) $\dfrac{5}{s-3}$ (b) $\dfrac{2}{s+2}$

4. (a) $\dfrac{12}{s^2+9}$ (b) $\dfrac{3s}{s^2+4}$

5. (a) $\dfrac{7s}{s^2-4}$ (b) $\dfrac{1}{s^2-9}$

6. (a) $\dfrac{2(s^2+2)}{s(s^2+4)}$ (b) $\dfrac{24}{s(s^2+16)}$

7. (a) $\dfrac{s^2-2}{s(s^2-4)}$ (b) $\dfrac{16}{s(s^2-16)}$

8. $\dfrac{4}{s^2+a^2}(a\cos b+s\sin b)$

9. $\dfrac{3}{s^2+\omega^2}(s\cos\alpha+\omega\sin\alpha)$ **10.** Proof

Chapter 96

Exercise 356 (page 1050)

1. (a) $\dfrac{2}{(s-2)^2}$ (b) $\dfrac{2}{(s-1)^3}$

2. (a) $\dfrac{24}{(s+2)^4}$ (b) $\dfrac{12}{(s+3)^5}$

3. (a) $\dfrac{s-1}{s^2-2s+2}$ (b) $\dfrac{6}{s^2-4s+8}$

4. (a) $\dfrac{5(s+2)}{s^2+4s+13}$ (b) $\dfrac{4}{s^2+10s+26}$

5. (a) $\dfrac{1}{s-1}-\dfrac{s-1}{s^2-2s+5}$

(b) $\dfrac{1}{4}\left(\dfrac{1}{s-3}+\dfrac{s-3}{s^2-6s+13}\right)$

6. (a) $\dfrac{1}{s(s-2)}$ (b) $\dfrac{3(s-2)}{s^2-4s-12}$

7. (a) $\dfrac{6}{s^2+2s-8}$ (b) $\dfrac{s+3}{4(s^2+6s+5)}$

8. (a) $\dfrac{2(s-10)}{s^2-2s+10}$ (b) $\dfrac{-6(s+1)}{s(s+4)}$

Exercise 357 (page 1052)

1. Proof   **2.** Proof   **3.** Proof   **4.** Proof

Exercise 358 (page 1053)

1. (a) 3 (b) 16   **2.** (a) 6 (b) -1 **3.** 4   **4.** 0

Chapter 97

Exercise 359 (page 1057)

1. (a) 7 (b) $2\,e^{5t}$ **2.** (a) $\dfrac{3}{2}\,e^{-\frac{1}{2}t}$ (b) $2\cos 2t$

3. (a) $\dfrac{1}{5}\sin 5t$ (b) $\dfrac{4}{3}\sin 3t$

4. (a) $\dfrac{5}{2}\cos 3t$ (b) $6t$ **5.** (a) $\dfrac{5}{2}t^2$ (b) $\dfrac{4}{3}t^3$

6. (a) $6\cosh 4t$ (b) $\dfrac{7}{4}\sinh 4t$

7. (a) $\dfrac{5}{3}\sinh 3t$ (b) $2\,e^t t^2$

8. (a) $\dfrac{1}{6}\,e^{-2t}t^3$ (b) $\dfrac{1}{8}\,e^{3t}t^4$

9. (a) $e^{-t}\cos 3t$ (b) $\dfrac{3}{2}\,e^{-3t}\sin 2t$

10. (a) $2\,e^{3t}\cos 2t$ (b) $\dfrac{7}{2}\,e^{4t}\sinh 2t$

11. (a) $2\,e^{-2t}\cosh 3t+\dfrac{1}{3}\,e^{-2t}\sinh 3t$

(b) $3\,e^{4t}\cos 3t+\dfrac{14}{3}\,e^{4t}\sin 3t$

Exercise 360 (page 1058)

1. $2e^t-5e^{-3t}$ **2.** $4e^{-t}-3e^{2t}+e^{-3t}$

3. $2e^{-3t}+3e^t-4e^t t$ **4.** $e^{-3t}(3-2t-3t^2)$

5. $2\cos\sqrt{2}t+\dfrac{3}{\sqrt{2}}\sin\sqrt{2}t+5e^{-t}$

6. $2+t+\sqrt{3}\sin\sqrt{3}t-4\cos\sqrt{3}t$

7. $2-3e^{-2t}\cos 3t-\dfrac{2}{3}e^{-2t}\sin 3t$

Exercise 361 (page 1060)

1. (a) $s=-4$ (b) $s=0,s=-2,s=4+j3,s=4-j3$

2. poles at $s=-3,s=1+j2,s=1-j2$,
zeros at $s=+1,s=-2$

3. poles at $s=-2,s=-1+j2,s=-1-j2$,
zero at $s=1$

4. poles at $s=0,s=+j2,s=-j2$,
zeros at $s=-1,s=6$

Chapter 98

Exercise 362 (page 1064)

1. $V(t) = 6(H - 4)$

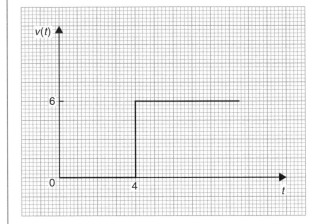

2. $2H(t) - H(t - 5)$

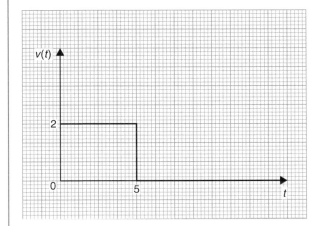

3.

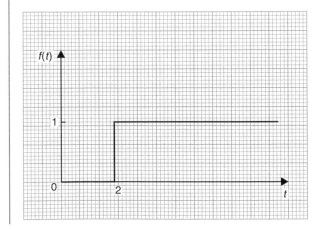

4.

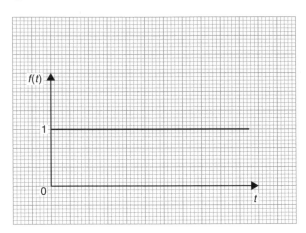

5.

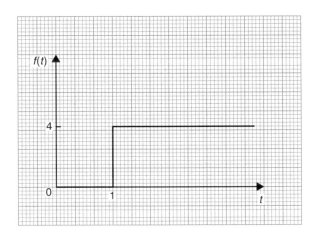

6.

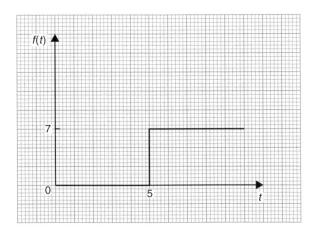

7.

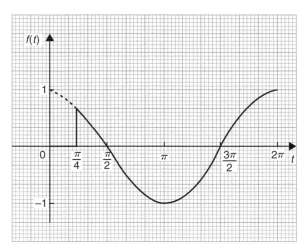

8.

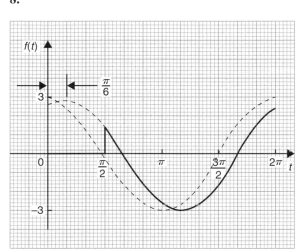

9.

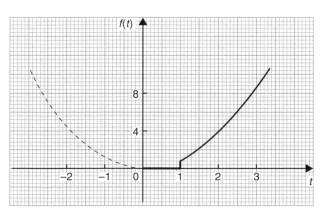

10.

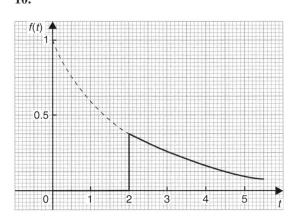

11.

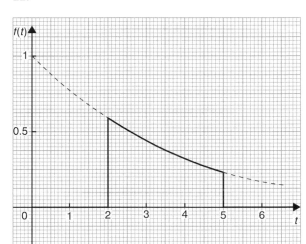

12.

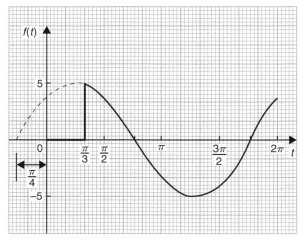

Exercise 363 (page 1066)

1. $\dfrac{e^{-s}}{s}$ 2. $\dfrac{7e^{-3s}}{s}$ 3. $\dfrac{2e^{-2s}}{s^3}$

4. $\dfrac{e^{-3s}}{s^2+1}$ 5. $\dfrac{e^{-4s}}{s-1}$ 6. $\dfrac{3e^{-5s}}{s^2+9}$

7. $\dfrac{6e^{-s}}{s^4}$ 8. $\dfrac{se^{-6s}}{s^2+9}$ 9. $\dfrac{10e^{-5s}}{s^2-4}$

10. $\dfrac{se^{-\frac{\pi}{3}s}}{s^2+4}$ 11. $\dfrac{2e^{-3s}}{s-1}$ 12. $\dfrac{3se^{-2s}}{s^2-1}$

Exercise 364 (page 1067)

1. $H(t-9)$ 2. $4H(t-3)$
3. $2H(t-2).(t-2)$ 4. $5H(t-2).\sin(t-2)$
5. $3H(t-4).\cos 4(t-4)$ 6. $6H(t-2).\sinh(t-2)$
7. $1.5H(t-6).(t-6)^2$
8. $2H(t-4).\cosh 4(t-4)$
9. $2H\left(t-\dfrac{1}{2}\right).\cos\sqrt{5}\left(t-\dfrac{1}{2}\right)$
10. $4H(t-7).e^{t-7}$

Chapter 99

Exercise 365 (page 1072)

1. (a) $i=2(1-e^{-5t})$ (b) $i=10(e^{-3t}-e^{-5t})$
   (c) $i=\dfrac{5}{2}(e^{-5t}-\cos 5t+\sin 5t)$

2. $y=(3-t)e^{\frac{4}{3}t}$ 3. $x=2\cos 10t$

4. $i=100te^{-500t}$ 5. $x=4(3e^{-2t}-2e^{-4t})$

6. $y=(4x-1)e^x+\dfrac{1}{3}e^{4x}$

7. $y=\dfrac{3}{85}e^{4x}-\dfrac{3}{10}e^{-x}+\dfrac{9}{34}\cos x-\dfrac{15}{34}\sin x$
8. $y=e^x-e^{-2x}+\sin 3x$

9. $y=3e^x(\cos x+\sin x)-e^x\cos 2x$

10. $y=e^{-t}(2\cos 2t+\sin 2t)$ 11. See earlier answers

12. See earlier answers

Chapter 100

Exercise 366 (page 1078)

1. $x=e^t-t-1$ and $y=2t-3+3e^t$
2. $x=5\cos t+5\sin t-e^{2t}-e^t-3$ and
   $y=e^{2t}+2e^t-3-5\sin t$
3. $y=3\cos t+\cos(\sqrt{3}t)$ and $z=3\cos t-\cos(\sqrt{3}t)$

Chapter 101

Exercise 367 (page 1088)

1. $f(x)=\dfrac{8}{\pi}\left(\sin x+\dfrac{1}{3}\sin 3x+\dfrac{1}{5}\sin 5x+\cdots\right)$

2. $\dfrac{\pi}{4}=1-\dfrac{1}{3}+\dfrac{1}{5}-\dfrac{1}{7}+\cdots$

3. (a) $f(x)=\dfrac{1}{2}+\dfrac{2}{\pi}\left(\cos x-\dfrac{1}{3}\cos 3x+\dfrac{1}{5}\cos 5x-\cdots\right)$
   (b) $\dfrac{1}{2}$

4. Graph sketching 5. $\dfrac{2}{3\pi}\sin 3x$

6. $f(t)=\dfrac{2}{\pi}\left(\begin{array}{l}\cos t-\dfrac{1}{3}\cos 3t+\dfrac{1}{5}\cos 5t-\cdots\\[2mm]+\sin 2t+\dfrac{1}{3}\sin 6t+\dfrac{1}{5}\sin 10t+\cdots\end{array}\right)$

7. Proof

Chapter 102

Exercise 368 (page 1094)

1. Proof

2. $f(t)=\dfrac{\pi}{2}+1-\dfrac{4}{\pi}\left(\cos t+\dfrac{\cos 3t}{3^2}+\dfrac{\cos 5t}{5^2}+\cdots\right)$

3. $f(x)=\pi+2\left(\sin x-\dfrac{1}{2}\sin 2x+\dfrac{1}{3}\sin 3x-\cdots\right)$

4. $f(x)=\dfrac{\pi}{2}-\dfrac{4}{\pi}\left(\cos x+\dfrac{\cos 3x}{3^2}+\dfrac{\cos 5x}{5^2}+\cdots\right)$

5. $f(\theta)=\dfrac{\pi^2}{3}-4\left(\cos\theta-\dfrac{1}{2^2}\cos 2\theta+\dfrac{1}{3^2}\cos 3\theta-\cdots\right)$

6. $1+\dfrac{1}{2^2}+\dfrac{1}{3^2}+\dfrac{1}{4^2}+\dfrac{1}{5^2}+\cdots=\dfrac{\pi^2}{6}$

7. $f(x)=\dfrac{8}{\pi^2}\left(\cos x+\dfrac{1}{3^2}\cos 3x+\dfrac{1}{5^2}\cos 5x\right.$
   $\left.+\dfrac{1}{7^2}\cos 7x+\cdots\right)$

8. $\dfrac{\pi^2}{8}=1+\dfrac{1}{3^2}+\dfrac{1}{5^2}+\dfrac{1}{7^2}+\dfrac{1}{9^2}+\cdots$

Chapter 103

Exercise 369 (page 1099)

1. $f(x) = \dfrac{4}{\pi}\left(\cos x - \dfrac{1}{3}\cos 3x + \dfrac{1}{5}\cos 5x\right.$

$$\left. - \dfrac{1}{7}\cos 7x + \cdots\right)$$

2. $f(t) = -2\left(\sin t + \dfrac{1}{2}\sin 2t + \dfrac{1}{3}\sin 3t\right.$

$$\left. + \dfrac{1}{4}\sin 4t + \cdots\right)$$

3. $f(x) = \dfrac{\pi}{2} + 1$

$$- \dfrac{4}{\pi}\left(\cos x + \dfrac{1}{3^2}\cos 3x + \dfrac{1}{5^2}\cos 5x + \cdots\right)$$

4. $\dfrac{\pi^2}{8} = 1 + \dfrac{1}{3^2} + \dfrac{1}{5^2} + \dfrac{1}{7^2} + \cdots$ **5.** Proof

Exercise 370 (page 1102)

1. $f(x) = \dfrac{2}{\pi}\left(\sin x + \dfrac{\pi}{4}\sin 2x - \dfrac{1}{9}\sin 3x\right.$

$$\left. - \dfrac{\pi}{8}\sin 4x + \cdots\right)$$

2. (a) $f(t) = \dfrac{1}{2} - \dfrac{2}{\pi}\left(\cos t - \dfrac{1}{3}\cos 3t\right.$

$$\left. + \dfrac{1}{5}\cos 5t - \cdots\right)$$

(b) $f(t) = \dfrac{2}{\pi}\left(\sin t - \sin 2t + \dfrac{1}{3}\sin 3t + \dfrac{1}{5}\sin 5t\right.$

$$\left. - \dfrac{1}{3}\sin 6t + \cdots\right)$$

3. $f(x) = \dfrac{8}{\pi}\left(\dfrac{\sin x}{(1)(3)} - \dfrac{\sin 3x}{(1)(3)(5)} - \dfrac{\sin 5x}{(3)(5)(7)}\right.$

$$\left. - \dfrac{\sin 7x}{(5)(7)(9)} - \cdots\right)$$

4. $f(x) = \dfrac{\pi}{4} - \dfrac{2}{\pi}\left(\cos 2x + \dfrac{\cos 6x}{3^2} + \dfrac{\cos 10x}{5^2} + \cdots\right)$

Chapter 104

Exercise 371 (page 1106)

1. Proof

2. $f(x) = \dfrac{5}{2} - \dfrac{5}{\pi}\left[\sin\left(\dfrac{2\pi x}{5}\right) + \dfrac{1}{2}\sin\left(\dfrac{4\pi x}{5}\right)\right.$

$$\left. + \dfrac{1}{3}\sin\left(\dfrac{6\pi x}{5}\right) + \cdots\right]$$

3. $f(x) = \dfrac{12}{\pi}\left(\sin\left(\dfrac{\pi x}{2}\right) + \dfrac{1}{3}\sin\left(\dfrac{3\pi x}{2}\right)\right.$

$$\left. + \dfrac{1}{5}\sin\left(\dfrac{5\pi x}{2}\right) + \cdots\right)$$

4. $f(t) = \dfrac{V}{\pi} + \dfrac{V}{2}\sin t - \dfrac{2V}{\pi}\left(\dfrac{\cos 2t}{(1)(3)} + \dfrac{\cos 4t}{(3)(5)}\right.$

$$\left. + \dfrac{\cos 6t}{(5)(7)} + \cdots\right)$$

Exercise 372 (page 1108)

1. $f(x) = \dfrac{3}{2} - \dfrac{12}{\pi^2}\left\{\cos\left(\dfrac{\pi x}{3}\right) + \dfrac{1}{3^2}\cos\left(\dfrac{3\pi x}{3}\right)\right.$

$$\left. + \dfrac{1}{5^2}\cos\left(\dfrac{5\pi x}{3}\right) + \cdots\right\}$$

2. $f(x) = \dfrac{6}{\pi}\left(\sin\left(\dfrac{\pi x}{3}\right) - \dfrac{1}{2}\sin\left(\dfrac{2\pi x}{3}\right)\right.$

$$\left. + \dfrac{1}{3}\sin\left(\dfrac{3\pi x}{3}\right) - \dfrac{1}{4}\sin\left(\dfrac{4\pi x}{3}\right) + \cdots\right)$$

3. $f(t) = \dfrac{8}{\pi^2}\left(\sin\left(\dfrac{\pi t}{2}\right) - \dfrac{1}{3^2}\sin\left(\dfrac{3\pi t}{2}\right)\right.$

$$\left. + \dfrac{1}{5^2}\sin\left(\dfrac{5\pi t}{2}\right) - \cdots\right)$$

4. Proof

Chapter 105

Exercise 373 (page 1111)

1. $y = 23.92 + 7.81\cos\theta + 14.61\sin\theta + 0.17\cos 2\theta$
$\quad + 2.31\sin 2\theta - 0.33\cos 3\theta + 0.50\sin 3\theta$

2. $v = 5.00 - 10.78\cos\theta + 6.83\sin\theta + 0.13\cos 2\theta$
$\quad + 0.79\sin 2\theta + 0.58\cos 3\theta - 1.08\sin 3\theta$

3. $i = 0.64 + 1.58\cos\theta - 2.73\sin\theta - 0.23\cos 2\theta$
$\quad - 0.42\sin 2\theta + 0.27\cos 3\theta + 0.05\sin 3\theta$

Exercise 374 (page 1115)

1. (a) only odd cosine terms present
   (b) only even sine terms present
2. $y = 9.4 + 13.2\cos\theta - 24.1\sin\theta + 0.92\cos 2\theta$
   $- 0.14\sin 2\theta + 0.83\cos 3\theta + 0.67\sin 3\theta$
3. $I = 4.00 - 4.67\cos 2\theta + 1.00\cos 4\theta - 0.66\cos 6\theta$
4. $y = 1.83 - 25.67\cos\theta + 83.89\sin\theta + 1.0\cos 2\theta$
   $- 0.29\sin 2\theta + 15.83\cos 3\theta + 10.5\sin 3\theta$

Chapter 106

Exercise 375 (page 1121)

1. $f(t) = \displaystyle\sum_{n=-\infty}^{\infty} \frac{j}{n\pi}(\cos n\pi - 1)\,e^{jnt}$

   $= 1 - j\dfrac{2}{\pi}\left(e^{jt} + \dfrac{1}{3}e^{j3t} + \dfrac{1}{5}e^{j5t} + \cdots\right)$

   $+ j\dfrac{2}{\pi}\left(e^{-jt} + \dfrac{1}{3}e^{-j3t} + \dfrac{1}{5}e^{-j5t} + \cdots\right)$

2. Proof   3. Proof

4. $f(t) = \dfrac{1}{2}\displaystyle\sum_{n=-\infty}^{\infty}\left(\dfrac{e^{(2-j\pi n)} - e^{-(2-j\pi n)}}{2 - j\pi n}\right)e^{j\pi nt}$

Exercise 376 (page 1124)

1. $f(x) = \displaystyle\sum_{n=-\infty}^{\infty}\left(\dfrac{4}{n\pi}\sin\dfrac{n\pi}{2}\right)e^{jnx}$ 2. Proof

3. $f(t) = \displaystyle\sum_{n=-\infty}^{\infty}\left(\dfrac{j2}{n}\cos n\pi\right)e^{jnt}$ 4. Proof

Exercise 377 (page 1128)

1. (a) $2e^{j4t} + 2e^{-j4t}$, $2\angle 0°$ anticlockwise, $2\angle 0°$
   clockwise, each with $\omega = 4$ rad/s
   (b) $2e^{j4t}e^{j\pi/2} + 2e^{-j4t}e^{-j\pi/2}$, $2\angle \pi/2$
   anticlockwise, $2\angle -\pi/2$ clockwise, each with
   $\omega = 4$ rad/s
2. $(5 + j6)e^{j2t} + (5 - j6)e^{-j2t}$, $7.81\angle 0.88$ rotating
   anticlockwise, $7.81\angle -0.88$ rotating clockwise, each
   with $\omega = 2$ rad/s
3. $(2 - j3)e^{jt} + (2 + j3)e^{-jt}$, $3.61\angle -0.98$ rotating
   anticlockwise, $3.61\angle 0.98$ rotating clockwise, each
   with $\omega = 1$ rad/s

Chapter 107

Exercise 378 (page 1136)

1. $\dfrac{2z}{(z-1)^2}$ 2. $\dfrac{3z(z+1)}{(z-1)^3}$

3. $\dfrac{4z(z^2 + 4z + 1)}{(z-1)^4}$ 4. $\dfrac{z\sin 3}{z^2 - 2z\cos 3 + 1}$

5. $\dfrac{z}{z-2}$ 6. $\dfrac{2z}{(z-2)^2}$

7. $\dfrac{5z(z - \cos 2)}{z^2 - 2z\cos 2 + 1}$ 8. $\dfrac{3z}{z - e^{-2}}$

9. $\dfrac{z\sin\dfrac{1}{2}}{z^2 - 2z\cos\dfrac{1}{2} + 1}$ 10. $4\left(\dfrac{z^2 - ze^3\cos 2}{z^2 - 2ze^3\cos 2 + e^6}\right)$

11. $\dfrac{z}{z+4}$ 12. $\dfrac{6z(z+2)}{(z-2)^3}$

13. $\dfrac{3z}{z+5}$ 14. $\dfrac{-3z}{(z+3)^2}$

15. $\dfrac{3z}{z - e^5}$ 16. $\dfrac{2ze^{-4}\sin 2}{z^2 - 2ze^{-4}\cos 2 + e^{-8}}$

Exercise 379 (page 1137)

1. $\left\{\dfrac{5z}{(z-1)^2}\right\} - \left\{\dfrac{4z}{z - e^{-3}}\right\}$ or $\dfrac{-4z^3 + 13z^2 - z\left(5e^{-3} + 4\right)}{(z-1)^2\left(z - e^{-3}\right)}$

2. $\dfrac{z(7z+1)}{(z-1)^3}$

3. $\dfrac{3z(\sin 2 - z + \cos 2)}{z^2 - 2z\cos 2 + 1}$

4. $\dfrac{6z(z-1)}{z^2 - 9}$

5. $\dfrac{z}{(z-1)^2} + \dfrac{z}{\left(z - e^{-1}\right)}$ or $\dfrac{z^3 - z^2 + z\left(1 - e^{-1}\right)}{(z-1)^2\left(z - e^{-1}\right)}$

Exercise 380 (page 1138)

1. $\dfrac{8z}{z-2}$ 2. $\dfrac{4z}{z-4}$ 3. $\dfrac{6z^2}{(z-1)^2}$

4. $\dfrac{z(7z - 12)}{z^2 - 5z + 6}$

Exercise 381 (page 1139)

1. $\dfrac{1}{(z-1)}$ 2. $\dfrac{1}{z^2(z-1)}$ 3. $\dfrac{1}{z(z-a)}$

4. $\dfrac{1}{z^2(z-a)}$ 5. $\dfrac{1}{z^3(z-3)}$

Exercise 382 (page 1140)

1. $\dfrac{2z}{(z-2)^2}$ **2.** $\dfrac{4z}{(z-4)^2}$ **3.** 0

4. $20/9$ or $2\dfrac{2}{9}$ **5.** 1 **6.** $\dfrac{2z}{(z-2)^2}$

Exercise 383 (page 1142)

1. 1 **2.** $(2)^k$

3. $(-1)^k$ **4.** $(-4)^k$

5. $\left(\dfrac{1}{3}\right)^k$ **6.** $\left(-\dfrac{1}{4}\right)^k$

7. $\left(\dfrac{1}{5}\right)^k$ **8.** e^{-5k}

9. e^{3k} **10.** $3k$

11. $5k^2$ **12.** $\dfrac{1}{2}k$

13. $\dfrac{1}{3}\{(2)^k-(-1)^k\}$ **14.** $\dfrac{3}{4}\{1-(-3)^k\}$

15. $\dfrac{1}{3}\{1-(-2)^k\}$ **16.** $\dfrac{1}{7}\left\{(3)^k-\left(-\dfrac{1}{2}\right)^k\right\}$

17. $(2)^k-1$ **18.** $10(3)^k-7(2)^k$

19. $\dfrac{2}{3}\left\{\left(\dfrac{1}{2}\right)^k-(-1)^k\right\}$ **20.** $\dfrac{3}{7}\{(2)^k-(-5)^k\}$

Exercise 384 (page 1145)

1. $4\left(3^k\right)$ **2.** $3-2(3)^k$

3. $\dfrac{1}{3}\{10-(4)^k\}$ **4.** $\dfrac{1}{3}\left\{(2)^k-5\left(\dfrac{1}{2}\right)^k\right\}$

5. $(-1)^k-(-2)^k$ **6.** $\dfrac{5}{2}-6\,(2)^k+\dfrac{7}{2}(3)^k$

Answers to multiple choice questions

Test 1 (page 93)

1. (a)	**6.** (a)	**11.** (b)	**16.** (b)	**21.** (a)
2. (c)	**7.** (d)	**12.** (b)	**17.** (d)	**22.** (c)
3. (c)	**8.** (d)	**13.** (c)	**18.** (d)	**23.** (c)
4. (c)	**9.** (a)	**14.** (d)	**19.** (a)	**24.** (a)
5. (b)	**10.** (b)	**15.** (a)	**20.** (c)	**25.** (d)

Test 2 (page 136)

1. (b)	**6.** (a)	**11.** (b)	**16.** (d)	**21.** (c)
2. (b)	**7.** (a)	**12.** (c)	**17.** (c)	**22.** (c)
3. (b)	**8.** (d)	**13.** (a)	**18.** (c)	**23.** (a)
4. (a)	**9.** (d)	**14.** (a)	**19.** (c)	**24.** (d)
5. (c)	**10.** (c)	**15.** (d)	**20.** (b)	**25.** (d)

Test 3 (page 270)

1. (c)	**6.** (b)	**11.** (a)	**16.** (b)	**21.** (d)
2. (d)	**7.** (d)	**12.** (d)	**17.** (a)	**22.** (c)
3. (b)	**8.** (b)	**13.** (a)	**18.** (c)	**23.** (b)
4. (d)	**9.** (c)	**14.** (b)	**19.** (c)	**24.** (a)
5. (a)	**10.** (d)	**15.** (c)	**20.** (d)	**25.** (b)

Test 4 (page 416)

1. (c)	**9.** (b)	**17.** (d)	**25.** (b)	**33.** (d)
2. (c)	**10.** (a)	**18.** (b)	**26.** (b)	**34.** (c)
3. (d)	**11.** (d)	**19.** (b)	**27.** (b)	**35.** (c)
4. (d)	**12.** (a)	**20.** (c)	**28.** (a)	**36.** (b)
5. (d)	**13.** (a)	**21.** (b)	**29.** (a)	**37.** (d)
6. (c)	**14.** (c)	**22.** (c)	**30.** (a)	**38.** (b)
7. (a)	**15.** (c)	**23.** (a)	**31.** (c)	**39.** (a)
8. (d)	**16.** (b)	**24.** (d)	**32.** (a)	**40.** (b)

Test 5 (page 515)

1. (d)	**9.** (a)	**17.** (b)	**25.** (d)	**33.** (c)
2. (a)	**10.** (b)	**18.** (a)	**26.** (d)	**34.** (c)
3. (b)	**11.** (d)	**19.** (c)	**27.** (b)	**35.** (c)
4. (a)	**12.** (c)	**20.** (a)	**28.** (c)	**36.** (d)
5. (b)	**13.** (d)	**21.** (b)	**29.** (b)	**37.** (a)
6. (a)	**14.** (d)	**22.** (a)	**30.** (d)	**38.** (d)
7. (c)	**15.** (b)	**23.** (d)	**31.** (a)	**39.** (b)
8. (c)	**16.** (d)	**24.** (b)	**32.** (b)	**40.** (a)

Test 6 (page 614)

1. (b)	**6.** (b)	**11.** (c)	**16.** (c)	**21.** (c)
2. (a)	**7.** (d)	**12.** (d)	**17.** (c)	**22.** (b)
3. (d)	**8.** (b)	**13.** (a)	**18.** (b)	**23.** (a)
4. (c)	**9.** (a)	**14.** (c)	**19.** (d)	**24.** (d)
5. (a)	**10.** (c)	**15.** (d)	**20.** (a)	**25.** (b)

Test 7 (page 711)

1. (b)	**6.** (c)	**11.** (c)	**16.** (a)	**21.** (d)
2. (a)	**7.** (d)	**12.** (b)	**17.** (d)	**22.** (b)
3. (c)	**8.** (b)	**13.** (b)	**18.** (c)	**23.** (a)
4. (a)	**9.** (d)	**14.** (a)	**19.** (d)	**24.** (d)
5. (b)	**10.** (c)	**15.** (c)	**20.** (d)	**25.** (a)

Test 8 (page 821)

1. (d)	**6.** (c)	**11.** (d)	**16.** (a)	**21.** (d)
2. (a)	**7.** (a)	**12.** (c)	**17.** (c)	**22.** (b)
3. (b)	**8.** (b)	**13.** (d)	**18.** (b)	**23.** (a)
4. (c)	**9.** (d)	**14.** (b)	**19.** (b)	**24.** (c)
5. (d)	**10.** (a)	**15.** (d)	**20.** (c)	**25.** (a)

Test 9 (page 1037)

1. (c)	**6.** (d)	**11.** (b)	**16.** (b)	**21.** (a)
2. (c)	**7.** (a)	**12.** (b)	**17.** (c)	**22.** (b)
3. (d)	**8.** (b)	**13.** (a)	**18.** (a)	**23.** (c)
4. (b)	**9.** (c)	**14.** (d)	**19.** (d)	**24.** (a)
5. (b)	**10.** (d)	**15.** (c)	**20.** (a)	**25.** (d)

Index